idi-, *idios*, one's own: idiopathic
ile-, *ileum*: ileocolic sphincter
ili-, ilio-, *ilium*: iliac
in-, in, within, or denoting negative effect: inactivate
infra-, *infra*, beneath: infraorbital
inter-, *inter*, between: interventricular
intra-, *intra*, within: intracapsular
ipsi-, *ipse*, itself: ipsilateral
iso-, *isos*, equal: isotonic
-itis, *-itis*, inflammation: dermatitis
karyo-, *karyon*, body: megakaryocyte
kerato-, *keros*, horn: keratin
kino-, -kinin, *kinein*, to move: bradykinin
lact-, lacto-, -lactin, *lac*, milk: prolactin
lapar-, *lapara*, flank or loins: laparoscopy
-lemma, *lemma*, husk: plasmalemma
leuko-, *leukos*, white: leukocyte
liga-, *ligare*, to bind together: ligase
lip-, lipo-, *lipos*, fat: lipoid
lith-, *lithos*, stone: cholelithiasis
lyso-, -lysis, -lyze, *lysis*, dissolution: hydrolysis
macr-, *mukros*, large: macrophage
mal-, *mal*, abnormal: malabsorption
mamilla-, *mamilla*, little breast: mamillary
mast-, masto-, *mastos*, breast: mastoid
mega-, *megas*, big: megakaryocyte
melan-, *melas*, black: melanocyte
men-, *men*, month: menstrual
mero-, *meros*, part: merocrine
meso-, *mesos*, middle: mesoderm
meta-, *meta*, after, beyond: metaphase
micr-, *mikros*, small: microscope
mono-, *monos*, single: monocyte
morpho-, *morphe*, form: morphology
multi-, *multus*, much, many: multicellular
-mural, *murus*, wall: intramural
myelo-, *myelos*, marrow: myeloblast
myo-, *mys*, muscle: myofilament
narc-, *narkoun*, to numb or deaden: narcotics
nas-, *nasus*, nose: nasolacrimal duct
natri-, *natrium*, sodium: natriuretic
necr-, *nekros*, corpse: necrosis
nephr-, *nephros*, kidney: pronephros
neur-, neuro-, *neuron*, nerve: neuromuscular
oculo-, *oculus*, eye: oculomotor
odont-, *odontos*, tooth: odontoid process
-oid, *eidos*, form, resemblance: odontoid process
oligo-, *oligos*, little, few: oligopeptide
-ology, *logos*, the study of: physiology
-oma, *-oma*, swelling: carcinoma
onco-, *onkos*, mass, tumor: oncology
oo-, *oon*, egg: oocyte
ophthalm-, *ophthalmos*, eye: ophthalmic nerve
-opia, *ops*, eye: optic
orb-, *orbita*, a circle: orbicularis oris
orchi-, *orchis*, testis: orchiectomy
orth-, *orthos*, correct, straight: orthopedist
-osis, *-osis*, state, condition: neurosis
osteon, osteo-, *os*, bone: osteocyte
oto-, *otikos*, ear: otoconia
para-, *para*, beyond: paraplegia
patho-, -path, -pathy, *pathos*, disease: pathology
pedia-, *paidos*, child: pediatrician
per-, *per*, through, throughout: percutaneous
peri-, *peri*, around: perineurium
phag-, *phagein*, to eat: phagocyte

-phasia, *phasis*, speech: aphasia
-phil, -philia, *philus*, love: hydrophilic
phleb-, *phleps*, a vein: phlebitis
-phobe, -phobia, *phobos*, fear: hydrophobic
phot-, *phos*, light: photoreceptor
-phylaxis, *phylax*, a guard: prophylaxis
physio-, *physis*, nature: physiology
-plasia, *plasis*, formation: dysplasia
platy-, *platys*, flat: platysma
-plegia, *plege*, a blow, paralysis: paraplegia
-plexy, *plessein*, to strike: apoplexy
pneum-, *pneuma*, air: pneumotaxic center
podo-, *podon*, foot: podocyte
-poiesis, *poiesis*, making: hemopoiesis
poly-, *polys*, many: polysaccharide
post-, *post*, after: postanal
pre-, *prae*, before: precapillary sphincter
presby-, *presbys*, old: presbyopia
pro-, *pro*, before: prophase
proct-, *proktos*, anus: proctology
pterygo-, *pteryx*, wing: pterygoid
pulmo-, *pulmo*, lung: pulmonary
pulp-, *pulpa*, flesh: pulpitis
pyel-, *pyelos*, trough or pelvis: pyelitis
quadr-, *quadrans*, one quarter: quadriplegia
re-, back, again: reinfection
retro-, *retro*, backward: retroperitoneal
rhin-, *rhis*, nose: rhinitis
-rrhage, *rhegnymi*, to burst forth: hemorrhage
-rrhea, *rhein*, flow, discharge: amenorrhea
sarco-, *sarkos*, flesh: sarcomere
scler-, sclero-, *skleros*, hard: sclera
-scope, *skopeo*, to view: colonoscope
-sect, *sectio*, to cut: transect
semi-, *semis*, half: semitendinosus
-septic, *septikos*, putrid: antiseptic
-sis, state or condition: metastasis
som-, -some, *soma*, body: somatic
spino-, *spina*, spine, vertebral column: spinodeltoid
-stalsis, *staltikos*, contractile: peristalsis
sten-, *stenos*, a narrowing: stenosis
-stomy, *stoma*, mouth, opening: colostomy
stylo-, *stylus*, stake, pole: styloid
sub-, *sub*, below: subcutaneous
super-, *super*, above or beyond: superficial
supra-, *supra*, on the upper side: supraspinous fossa
syn-, *syn*, together: synthesis
tachy-, *tachys*, swift: tachycardia
telo-, *telos*, end: telophase
tetra-, *tettares*, four: tetralogy of Fallot
therm-, thermo-, *therme*, heat: thermoregulation
thorac-, *thorax*, chest: thoracentesis
thromb-, *thrombos*, clot: thrombocyte
-tomy, *temnein*, to cut: appendectomy
tox-, *toxikon*, poison: toxemia
trans-, *trans*, through: transudate
tri-, *tres*, three: trimester
-tropic, *trope*, turning: adrenocorticotropic
tropho-, *trophe*, nutrition: trophoblast
-trophy, *trophikos*, nourishing: atrophy
tropo-, *tropikos*, turning: troponin
uni-, *unus*, one: unicellular
uro-, -uria, *ouron*, urine: glycosuria
vas-, *vas*, vessel: vascular
zyg-, *zygotos*, yoked: zygote

FOURTH EDITION

Fundamentals of Anatomy and Physiology

FREDERIC H. MARTINI, Ph.D.

with

WILLIAM C. OBER, M.D.
■ *Art coordinator and illustrator*

CLAIRE W. GARRISON, R.N.
■ *Illustrator*

KATHLEEN WELCH, M.D.
■ *Clinical consultant*

RALPH T. HUTCHINGS
■ *Biomedical photographer*

 PRENTICE HALL, *Upper Saddle River, New Jersey 07458*

Library of Congress Cataloging-in-Publication Data

Martini, Frederic.
    Fundamentals of anatomy and physiology / Frederic H. Martini;
with William C. Ober, art coordinator and illustrator; Claire W.
Garrison, illustrator; Kathleen Welch, clinical consultant; Ralph T.
Hutchings, biomedical photographer–4th ed.
        p.   cm.
    ISBN 0-13-736265-X
    1. Human physiology.   2. Human anatomy.   I. Title.
    [DNLM: 1. Anatomy.   2. Physiology.   QS 4 M3855f 1997]
    QP34.5.M4615   1998
    612–dc21
    DNLM/DLC   97-11122
    for Library of Congress   CIP

Executive Editor: *David Kendric Brake*
Development Editor: *Karen Karlin*
Production Editors: *Karen Malley, Kim Dellas*
Assistant Vice President of Production & Manufacturing:
    David W. Riccardi
Executive Managing Editor: *Kathleen Schiaparelli*
Acquisitions Editor: *Linda Schreiber*
Assistant Managing Editor: *Shari Toron*
Formatting Manager: *John J. Jordan*
Page Layout: *Karen Noferi, Jeff Henn, Erik Unhjem,*
    Matthew S. Garrison, Terrance Cummings
Additional Formatting: *David Tay*
Template: *Richard Foster*
Editorial Director: *Tim Bozik*
Editor in Chief: *Paul F. Corey*
Editor in Chief, Development: *Ray Mullaney*
Associate Editor in Chief, Development: *Carol Trueheart*
Director of Marketing: *Kelly McDonald*
Marketing Manager: *Jennifer Welchans*
Manufacturing Manager: *Trudy Pisciotti*
Copy Editor: *Margo Quinto*
Editorial Assistance: *Byron Smith, Adam Velthaus, David Stack*

Creative Director: *Paula Maylahn*
Art Director: *Heather Scott*
Interior Design: *Sheree Goodman, Lisa A. Jones*
Cover Designer: *Tom Nery*
Art Manager: *Gus Vibal*
Art Editor: *Warren Ramezzana*
Photo Research: *Stuart Kenter Associates*
Photo Editor: *Carolyn Gauntt*
Illustrators: *William C. Ober, M.D.; Claire Garrison, R.N.*
Clinical Consultant: *Kathleen Welch, M.D.*
Biomedical Photographer: *Ralph T. Hutchings*
Production Assistance: *Erik R. Trinidad, Steven Graydonus,*
    Wanda España, Rebecca Wald, Ariel Colón, Diane
    Koromhas, Eric Hulsizer
Art Coordinators: *Ray Caramanna, Pat Gutierrez, Charles*
    Pelletreau, Liz Nemeth, Shea Oakley
Special Projects Manager: *Barbara A. Murray*
Supplements Production Editor: *James Buckley*
Media Product Development Editor: *Laura Edwards*
Cover Photograph: *Andrea Weber and Andrew Pacho Photo,*
    ©1993 Lois Greenfield

ISBN 0-13-736265-X

Prentice-Hall International (UK) Limited, *London*
Prentice-Hall of Australia Pty. Limited, *Sydney*
Prentice-Hall Canada Inc., *Toronto*
Prentice-Hall Hispanoamericana, S.A., *Mexico*
Prentice-Hall of India Private Limited, *New Delhi*
Prentice-Hall of Japan, Inc., *Tokyo*
Simon & Schuster Asia Pte. Ltd., *Singapore*
Editora Prentice-Hall do Brasil, Ltda., *Rio de Janeiro*

Welcome to the Fascinating Study of
Anatomy & Physiology!

Yu have just made a valuable investment in your education by purchasing Martini's *Fundamentals of Anatomy and Physiology*. You will find that this textbook, designed with **you** in mind, is a very useful learning tool that will help you master this subject. The accompanying free *Applications Manual* provides additional real-world/clinical examples as well as self-test material at the end of sections.

Let us suggest 3 ways to IMPROVE YOUR GRADE:

1 Turn to the STUDENT WALK-THROUGH (pages xxix–xli) to learn how to use this textbook to your advantage.

2 Take advantage of our WEB SITE (http://www.prenhall.com/martini/fap) for additional hints, suggestions, quizzes, and other materials designed to help you succeed.

3 Order any of the following TOOLS from the Martini "learning system"—products that have been developed to fit your style of studying.

Interactive CD-ROM Study Guide
The next best thing to having a personal tutor–

Prentice Hall's **Interactive CD-ROM Study Guide for Anatomy and Physiology** provides a truly interactive learning experience, almost like having a personal tutor. Each chapter provides questions and labeling exercises with **instant feedback:** You get a "green light" for a correct response or, if your answer is incorrect, **a helpful hint to guide** you to the answer. Plus, the program **tracks your improvement** from one session to the next.

Material on the CD-ROM is tied to Martini's *Fundamentals of Anatomy and Physiology*, Fourth Edition. The CD-ROM also offers:

- Over **1,000** multiple-choice questions, true/false questions, and fill-in-the-blank, sequencing, and labeling exercises.

- **Color graphics** with **"drag-and-drop" labeling** exercises that will help you learn anatomical terms and physiological processes.

- A **built-in glossary**.

- **Print-out capability**, allowing you to generate a hard copy of all the material on your monitor.

(0-13-751827-7)

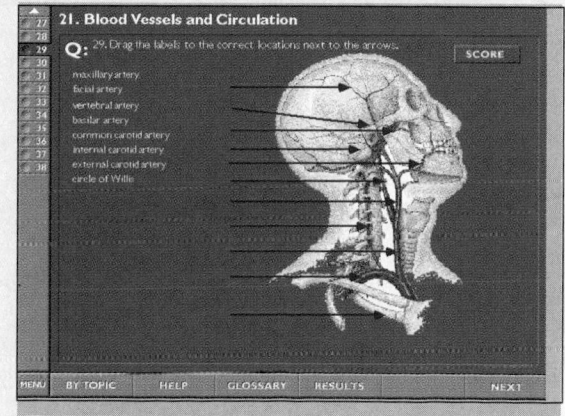

The Anatomy & Physiology Video Tutor

You may not have access to a CD-ROM drive, but you probably have access to a VCR–

The **Anatomy and Physiology Video Tutor** is a unique collection of professionally narrated, computer-generated **animations on videotape** that guides you through the **most-difficult concepts** in anatomy and physiology. Self-check questions have been incorporated into each segment to help test your knowledge.

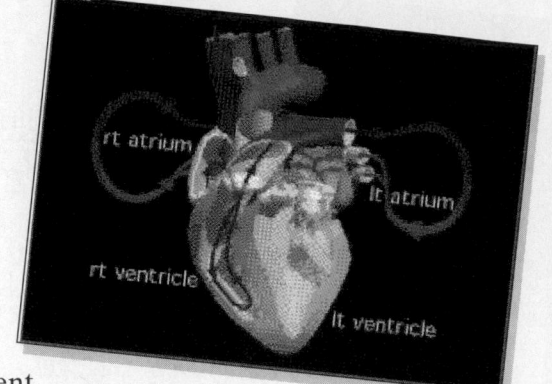

Eleven segments include:

membrane transport	nervous system	vision
protein synthesis	heart function	hearing
muscle function	urinary system	human development
respiratory system	immune system	

(0-13-751843-9)

Audio Cassette Study Tapes

Perfect for the student who has too much to do in too little time–

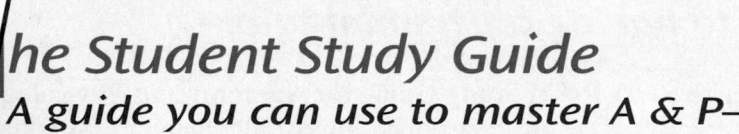

Designed to be used in conjunction with Martini's *Fundamentals*, the eight 90-minute **Audio Cassette Study Tapes** provide:

- A review of **chapter objectives.**
- **Definitions** of important vocabulary terms.
- **Pronunciation exercises.**

(0-13-751835-8)

The Student Study Guide

A guide you can use to master A & P–

Study Guide
Charles Seiger

FUNDAMENTALS OF
Anatomy &
Physiology
MARTINI

The **Student Study Guide** parallels the text's unique **three-level learning system:**

- Level 1 questions **test your understanding of the key terminology and chapter objectives** with multiple-choice questions and matching, fill-in-the-blank, and labeling exercises keyed to the chapter objectives.
- Level 2 questions require you to **synthesize the material from the chapter;** they incorporate concept mapping and short-answer essays in addition to multiple-choice questions and fill-in-the-blank exercises.
- Level 3 problems **encourage you to think critically** and apply the information you have learned to real-life situations.

(0-13-751819-6)

Check first with your bookstore to see if these items are in stock. If not, ask if they can be special-ordered for you. Otherwise, call Prentice Hall at 1-800-947-7700.

Text and Illustration Team

Frederic H. Martini received his Ph.D. from Cornell University in comparative and functional anatomy. He has broad interests in vertebrate biology, with special expertise in anatomy, physiology, histology, and embryology. Dr. Martini's publications include journal articles, technical reports, contributed chapters, magazine articles, and a book on the biology and geology of tropical islands. He has coauthored *Human Anatomy* (2e, 1997) with Prof. Michael J. Timmons and *Essentials of Anatomy and Physiology* (1e, 1997) with Prof. Edwin F. Bartholomew. *Human Anatomy* (1e) and *Fundamentals of Anatomy and Physiology* (3e) received Textbook Excellence Awards for 1995 and 1996, respectively, from the Text and Academic Author's Association. Dr. Martini has been involved in teaching undergraduate courses in anatomy and physiology (comparative and/or human) since 1970. He is currently on the faculty of University of Hawaii and remains affiliated with the Shoals Marine Laboratory (SML), a joint venture between Cornell University and the University of New Hampshire. Dr. Martini is involved with all aspects of the A & P learning system that accompanies each of his textbooks. He also maintains an active research program on vertebrate physiology, anatomy, and ecology. Dr. Martini is a member of the Human Anatomy and Physiology Society, for which he serves on the Curriculum Committee. He is also a member of the National Association of Biology Teachers, the Society for College Science Teachers, the Society for Integrative and Comparative Biology, the Western Society of Naturalists, the International Society of Vertebrate Morphologists, the Author's Guild, the Text and Academic Author's Association, and the National Association of Underwater Instructors. For more information on his teaching and research interests, visit the A & P home page at http://www.prenhall.com/martini/fap.

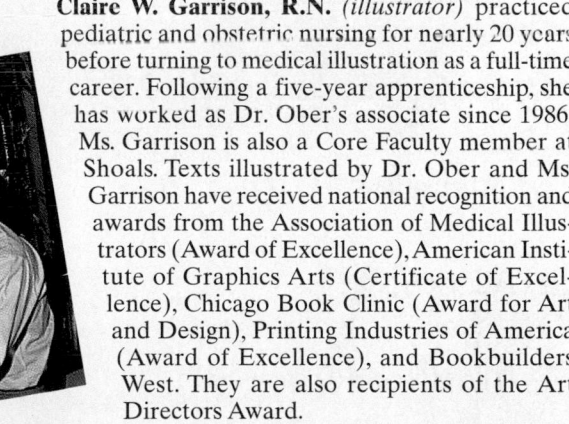

Dr. William C. Ober *(art coordinator and illustrator)* received his undergraduate degree from Washington and Lee University and his M.D. from the University of Virginia in Charlottesville. While in medical school, he also studied in the Department of Art as Applied to Medicine at Johns Hopkins University. After graduation, Dr. Ober completed a residency in family practice, and is currently on the faculty of the University of Virginia as a Clinical Assistant Professor in the Department of Family Medicine. He is also part of the Core Faculty at Shoals Marine Laboratory, where he teaches biological illustration in the summer program. Dr. Ober now devotes his full attention to medical and scientific illustration.

Claire W. Garrison, R.N. *(illustrator)* practiced pediatric and obstetric nursing for nearly 20 years before turning to medical illustration as a full-time career. Following a five-year apprenticeship, she has worked as Dr. Ober's associate since 1986. Ms. Garrison is also a Core Faculty member at Shoals. Texts illustrated by Dr. Ober and Ms. Garrison have received national recognition and awards from the Association of Medical Illustrators (Award of Excellence), American Institute of Graphics Arts (Certificate of Excellence), Chicago Book Clinic (Award for Art and Design), Printing Industries of America (Award of Excellence), and Bookbuilders West. They are also recipients of the Art Directors Award.

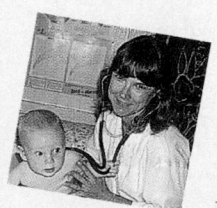

Dr. Kathleen Welch, clinical consultant for all Martini projects, received her M.D. from the University of Seattle and did her residency at the University of North Carolina in Chapel Hill. For two years, she served as Director of Maternal and Child Health at the LBJ Tropical Medical Center in American Samoa. She then joined the Department of Family Practice at the Kaiser Permanente Clinic in Lahaina, Hawaii. She has been in private practice since 1987. Dr. Welch is a Fellow of the American Academy of Family Practice. She is a member of the Hawaii Medical Association and the Human Anatomy and Physiology Society. Drs. Martini and Welch were married in 1979; they have one child, "P. K.," born in January 1995.

Ralph T. Hutchings is a biomedical photographer who was associated with the Royal College of Surgeons for 20 years. An engineer by training, Mr. Hutchings has focused his attention for years now on photographing the structure of the human body. The result has been a series of color atlases, including the *Color Atlas of Human Anatomy*, *The Color Atlas of Surface Anatomy,* and *The Human Skeleton* (all published by Mosby-Yearbook Publishing, St. Louis, Missouri, USA). Mr. Hutchings makes his home in North London, where he tries to balance the demands of his photographic assignments with his hobbies of early motor cars and airplanes.

Contents in Brief

Contents

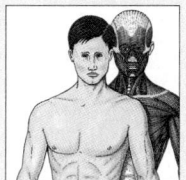
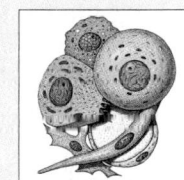

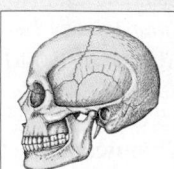

7

The Axial Skeleton 197

8

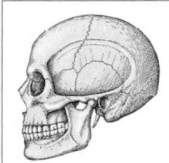

The Appendicular Skeleton 233

9

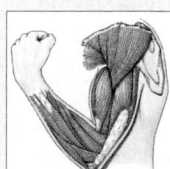

Articulations 253

10

Muscle Tissue 276

11
The Muscular System 315

12
Neural Tissue 368

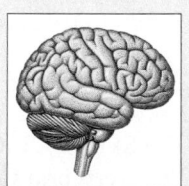

13
The Spinal Cord and Spinal Nerves   416

14
The Brain and Cranial Nerves   444

15
Integrative Functions   490

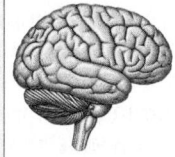

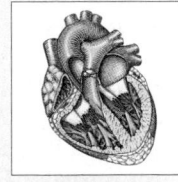

23

The Respiratory System *814*

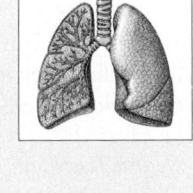

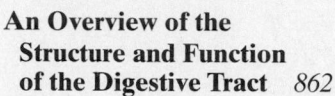
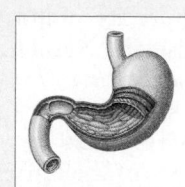
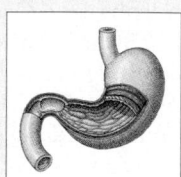

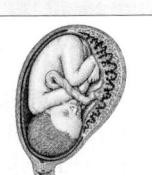

26

27

28

29

Development and Inheritance   *1086*

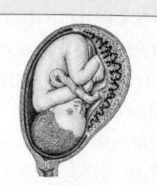

LIST OF BOXES AND CLINICAL DISCUSSIONS

 Boxes

Clinical Discussions

Preface

WELCOME TO A NEW WORLD

The world has changed a great deal since the first edition of *Fundamentals of Anatomy and Physiology* was published in 1989. Our understanding of the intricate inner workings of the human body has increased dramatically. For example, we now know much more about the biochemical and molecular foundations of many normal physiological processes than we did in 1988, and as a result the molecular origins of many disease processes are now understood. This enhanced understanding has led to the development of effective new treatments, protocols, and interventions to combat disease, reduce suffering, and promote good health. Many of these advances could not have been made if they were not preceded by improvements in the speed and reliability of computers. Modern technology, especially computers and the communication network known as the Internet, has changed our daily lives and is reshaping our societies in many ways.

The emerging medical technologies allow us to perform complex surgical procedures on individual organs or to target specific cells within an organ. We can manipulate the DNA within cells, and we can even clone genetically identical animals, although we are still unable to repair even the simplest genetic defect in humans. Many people are horrified by what we can *now* do with technology; others are intent on pushing ahead without a clear idea of the ultimate destination. Nevertheless, all of us are affected by modern technology. Ironically, some of the most-heated arguments about whether new technologies are a blessing or a curse occur on the Internet, a recent technological innovation.

Like it or not, the technology is here to stay, and it will probably continue to advance. Every day, medical professionals around the world trade messages by electronic mail, and they upload and download X-rays, photomicrographs, and laparoscopic video clips. Technology has become a means for acquiring and distributing information, and medical professionals have learned to use technology to increase their effectiveness as well as to improve the quality of medical care.

By enhancing the availability of *specific* information, modern technology has affected how each of us deals with information *in general*. In many cases, knowing where to look for information is preferable to trying to memorize specific data. That is certainly the trend in the training of medical and allied health professionals today. Students begin by mastering the terminology and memorizing a substantial core of basic conceptual information. In the process, they are also given (1) a "mental framework" for organizing new information; (2) an ability to access additional information when needed, by consulting relevant print or electronic data sources; and (3) an understanding of how to apply their knowledge to solve specific problems. The same skills are equally important to people in other career tracks. To be effective in almost any job today, you must know how to access and absorb new information, to use (or learn to use) available technology, and to solve problems.

TODAY'S A & P PROCESS

A great many instructors and students are aware of this relationship among information, technology, and problem solving. Throughout North America, Australia, and New Zealand, I have visited instructors and students involved in anatomy and physiology courses. I am encouraged by the dedication so many of these instructors exhibit and by the potential of so many of their students, but I am also concerned about the amount of information that must be covered in their courses. For many students, this course is their first college-level science course. As such, they must learn not only a new vocabulary but also new ways of studying and organizing information. For most instructors, teaching anatomy and physiology has never been more challenging; in addition to teaching terminology, facts, and concepts, instructors must help their students be good problem solvers. Mastering terminology, facts, and concepts is of little value if that mastery cannot be used to solve problems. It would be analogous to a person's knowing all the parts of an automobile and the concepts behind the internal combustion engine without being able to back the car out of the garage and drive it away.

The ultimate goal of any anatomy and physiology course should be to empower students to use their conceptual understanding to solve problems. Preparation for this goal involves a substantial amount of memorization, but the burden of memorization is reduced if the relevance of the information is evident. *Fundamentals of Anatomy and Physiology* has been designed to place the information in a meaningful context and to help students develop their problem-solving skills. (We shall discuss specific examples later, when we introduce the pedagogical framework of the text.) At the companion World Wide Web site, both students and instructors can begin learning how to use technology to access and manage information.

The focus of this text and its support materials has been to simplify the processes of teaching and learning anatomy and physiology. Much as I would like to, I cannot create materials that will give any of us more time. But a carefully designed text and supplements package can help you make better use of the time you *do* have, whether you are a student or an instructor. All A & P books are about the same size and length, and they all have a similar organization, colorful illustrations, and an assortment of supplements. However, that does not mean that all A & P books are exactly alike, any more than all cars are alike simply because they all have engines and tires and can transport you from one place to another. This text has been designed to meet specific needs. How were those needs determined? In part, through personal experience, but no one person has a complete view of any situation. So, wherever and whenever possible, I have met with instructors and students to learn more about shared problems, solutions, and perspectives. My publisher has sponsored student focus groups and instructor focus groups, giving me the opportunity to get direct feedback and to "test-drive" new features. I have received hundreds of letters, phone calls, and e-mail messages from instructors and students with comments and suggestions for improving this book and its supplements. The result is a book that is more than just "reliable transportation"—it will help you negotiate the curves and avoid the potholes. The book's important and distinctive features are explained in "To the Student" (page xxix). I urge you to read it before your course gets under way.

The right supplement can also help students and instructors make better use of their time. Which is the right supplement? That depends on the student. For some students, it might be a traditional study guide, which reinforces material in the text, or it might be the multimedia study guide, which takes advantage of CD-ROM technology. For the busy student who works part-time and commutes to work and school, it might be the audio study tapes, which let the student turn "downtime" on the road into productive study time. For the visually oriented student, the *Anatomy and Physiology Video Tutor* may be the answer. For a student with Internet access, the best supplement may be the World Wide Web site, which has options and features that none of us could have imagined in 1989. Every course, every instructor, and every student is different. These varied, integrated supplements were developed and reviewed by instructors who have used this text in their classroom. Together, we have tried to create learning tools that will suit unique combinations of teaching styles, learning styles, hardware, and budgets. (See the supplements section of this preface for a more detailed overview of each supplement available with this text.)

SO, WHAT'S DIFFERENT ABOUT THE FOURTH EDITION?

Each new edition requires some revising and updating. The basic format and organization have not changed significantly, but you will find numerous improvements to the art program and narrative that were based on feedback from the classroom and from other instructors, reviewers, and students.

The Art Program

Both anatomical structures and physiological processes must be visualized if students are to understand them. The art program in previous editions was so well received and praised by users of the text that it may seem odd to find it listed under the heading of "new and improved." Yet, there is always room for improvement in even the finest art program. A detailed examination of any chapter in this edition will reveal subtle improvements and additions to the illustrations. Larger-scale changes in the art program for this edition fall into the following categories:

- Chapters 7–9: Many of the paintings of bones in the skeletal system have been supplemented with or replaced by bone photographs supplied by Ralph Hutchings, a well-known biomedical photographer.
- Chapter 10: Several new illustrations were added to help students better visualize the key aspects of muscle physiology.
- Chapter 11: New anatomical paintings depicting the muscles of the forearm, hand, and hip have been added.
- Chapters 14, 20, 26, and 28 (among others): There are many new cadaver photographs paired with the anatomical art. These photos were also supplied by Ralph Hutchings.
- Chapters 20 and 21: Several illustrations have been replaced or expanded to clarify important concepts of cardiovascular physiology and details of cardiovascular anatomy.
- Virtually every chapter that includes complex physiology processes contains new graphs and flowcharts.

- The *Applications Manual*: This free supplement includes many new illustrations, including new labeled images in the "Surface Anatomy and Cadaver Atlas," as well as a new section that details the origins and insertions of major muscles.

Throughout this project, I have been fortunate to work with William Ober, M.D., a physician who is also an award-winning medical illustrator. Dr. Ober and his associate, Claire Garrison, R.N., play a key role in coordinating the art program for this text. Having a single illustration team responsible for the visual presentation of information helps ensure that structures and processes are depicted in a consistent manner from figure to figure and from chapter to chapter. *Fundamentals of Anatomy and Physiology* is the only text on the market with an art program created and managed by a medical illustrator. (To learn more about Dr. Ober and Ms. Garrison, please see the brief biographical sketch opposite the title page.)

Content and Organizational Changes

In response to instructor and student feedback, I have made the following content changes and enhancements:

- Chapter 10 ("Muscle Tissue"): expanded coverage of muscle physiology, especially with regard to tension production, internal versus external tension, and eccentric, concentric, and isotonic contractions
- Chapter 12 ("Neural Tissue"): slightly reorganized and rewritten to clarify the presentation of action potential propagation
- Chapter 17 ("Sensory Function"): two additional classes of taste receptors introduced
- Chapter 18 ("The Endocrine System"): revised discussion of hormone modes of action, including pulsatile release, and addition of the probable roles of leptin, a newly described hormone
- Chapter 21 ("Blood Vessels and Circulation"): revised discussion of key aspects of cardiovascular regulation and addition of details about circulation to the thoracic and abdominopelvic viscera
- Chapter 22 ("The Lymphatic System and Immunity"): revision of some of the sections dealing with cytokine regulation and update of the AIDS narrative
- Chapter 23 ("The Respiratory System"): revision of the discussion of hemoglobin and the effects of O_2 bonding
- Chapter 26 ("The Urinary System"): clarified discussions of countercurrent multiplication and the function of the vasa recta
- Chapter 28 ("The Reproductive System"): updated coverage of female reproductive physiology to include the role of GnRH pulses in the establishment and maintenance of the ovarian cycle

This text and the *Applications Manual* provide current statistics on the incidence of cancer and various infectious diseases and on the frequency of major surgical procedures. The data were obtained from sources, such as the American Cancer Society, that project the annual incidence based on trends observed in previous years.

The Format and Design of the Text

Introductory A & P students often need a lot of help with organizing and integrating the material, so this text offers a variety of pedagogical aids. The red dots, red checks, and various icons that distinguish this text from all others are important and effective learning aids that were, in many cases, suggested by students themselves. Focus group participants and reviewers who used the third edition—as students or as instructors—were virtually unanimous in telling me to retain these features in the fourth edition. These pedagogical features were retained, but the design has been changed to soften their visual impact. As a bonus, this new design has enabled me to coordinate the narrative and illustrations more closely.

 The Martini World Wide Web Site
http://www.prenhall.com/martini/fap

Can technology simplify your life? The answer depends on how the technology is used. The Martini Web site has been designed to make it easier to teach and learn anatomy and physiology. Notice that I did *not* say that the Web site makes anatomy and physiology *easy*; rather, it makes the process of learning anatomy and physiology less stressful and more interesting.

The World Wide Web is changing the way we view and utilize information, and this Web site takes full advantage of the range of available options. You can use the on-line study guide to test your understanding of basic concepts. If you would like to assess your problem-solving skills, you can visit the Web site's "Problem Center." If you are having a problem with a particular concept, you can visit the "Frequently Asked Questions" section of the appropriate chapter to see if the same problem has been posted and discussed before. If it has not, you can submit the question and get a reply from me or from another A & P instructor who monitors the site. You can also follow hot links from this Web site to other sites, where you will discover interesting new medical procedures and other real-life applications. It is truly a new world, and you may find yourself taking advantage of the new technologies in ways you never imagined.

I hope that as the year passes, you will use the Web site to contact me with suggestions for future improvements or new illustrations or with interesting research developments. In the past year, I have held e-mail discussions on various aspects of the fourth edition with A & P instructors from around the globe—from Sydney, Australia, to Umea, Sweden. Distance is no longer a limiting factor in our "global A & P village," and I hope that even more instructors and students will participate in the development of future editions.

THE SUPPLEMENTS PACKAGE

The supplements team consists of talented A & P instructors, most of whom have been using this text since its first or second edition. Their efforts are coordinated by supplements editors and multimedia editors at Prentice Hall. During the development of the fourth edition, they have worked with me as an extra team of reviewers. In return, I have provided them with additional suggestions and comments.

For the Instructor

INSTRUCTOR'S PRESENTATION BOX
Prentice Hall and I have tried to simplify the task of preparing lectures for this course. The Instructor's Presentation Box contains three folders for each chapter. The first folder contains traditional instructor's manual materials, including detailed lecture notes and a list of visual resources for that chapter. This folder also contains suggested demonstrations, analogies, and mnemonic activities designed to engage students' interest and provide some additional flavor for your lectures. (These demonstrations, analogies, and mnemonic exercises have been collected from A & P instructors around the world and then categorized by chapter for your use.) The second folder contains items each instructor may want to make available as handouts—items such as crossword puzzles, career profiles, line art without labels, and answers to end-of-chapter questions. The third folder contains all the transparency acetates for that chapter. (0-13-902851-X)

Lecture Presentation Notebook   To help you better organize your classroom lectures, we are including a special three-ring binder with an easel-back spine. This binder can stand upright on any flat surface. The notebook contains brief lecture outlines for each chapter on laminated stock, with plenty of room for notes. Included with the notebook is a dry-erase marker (for making notes and observations on the laminated lecture outline) and a special pocket to hold the *Image Bank and Presentation Manager*. The little things in life can mean a lot, and we think the addition of this presentation notebook is indeed a little thing that can make your lectures a little easier to prepare and present.

Transparencies   We have included more than 450 acetates with art and labels enlarged for easy lecture hall viewing. The selection of acetates and each image have been reviewed by transparency users to ensure that we have the most important images and the best presentation available. (0-13-751769-6)

COLLEGE NEWSLINK
The ultimate news resource for the academic use of the World Wide Web, College Newslink delivers information compiled from more than 40 major national and international news services daily. It has Web links to relevant sites that offer more detailed coverage. Visit http://www.ssnewslink.com to learn more about College Newslink.

TEST ITEM FILE AND COMPUTER TEST MANAGER
The Test Bank has been thoroughly reviewed and revised. A bank of more than 3000 questions organized around the three-level learning system (as seen in the text and study guide) will help you design a variety of tests and quizzes. (0-13-751736-X)

IMAGE BANK AND PRESENTATION MANAGER
Available in both Mac and Windows formats, the Image Bank contains more than 1000 illustrations, including line art from the text, cadaver and cat dissection photos, MRI and CT scans, X-rays, and laboratory model photos. In addition, we have included dynamic animations of the most difficult concepts in physiology as well as video clips of laboratory dissections of the sheep brain, kidney, heart, and eye. Organized with the new *PresMan 3.0* software (embedded on the CD-ROM), the Image Bank enables you to search by image title, caption, or key term. The PresMan 3.0 software also allows you to preview selected images and add custom overlays to any image. IBM (0-13-751793-9); Macintosh (0-13-779273-5)

LASER DISK AND BAR CODE MANUAL
For those of you who do not have CD-ROM access in your classroom, most of the Image Bank material is also available on a laser disk (packaged with a bar code manual). (0-13-751777-7)

For the Student

APPLICATIONS MANUAL
Each new copy of the text comes packaged with an *Applications Manual*. This unique supplement provides you with access to interesting and relevant clinical and diagnostic information. Critical-thinking questions, clinical problems, and case studies offer you the opportunity to develop sound problem-solving skills. Discussions of the scientific method, diagnostic procedures, and current research topics give you a frame of reference for

your study of anatomy and physiology. The *Applications Manual* includes a full-color surface anatomy and cadaver atlas and atlas of sectional scans. I wrote the *Applications Manual* in collaboration with my wife, Kathleen Welch, M.D., who also serves as the clinical consultant for the text. (0-13-751868-4)

MARTINI WORLD WIDE WEB SITE
http://www.prenhall.com/martini/fap
See our earlier discussion on page xxiii.

BIOLOGY ON THE INTERNET: A STUDENT'S GUIDE
This guide by Andrew Stull is a handy reference to get you up to speed on the World Wide Web's vast resources, including the Martini *Fundamentals* Web site. The guide, a unique resource, gives clear steps to help you access regularly updated biology resources as well as an overview of general navigation strategies. (0-13-890120-1)

VIDEO TUTOR
The *Prentice Hall Video Tutor for Anatomy and Physiology* generated a lot of excitement when it debuted with the third edition of *Fundamentals*. For the fourth edition, we have made it an even more valuable study aid. Each segment includes a professionally narrated animation that walks you through the most difficult physiological concepts and offers self-check questions that allow you to test your understanding of the material. Segments of the video tutor include membrane transport, protein synthesis, muscle contraction, action potential propagation, vision, auditory function, heart function, urine formation, and the immune response. (0-13-751843-9)

INTERACTIVE STUDY GUIDE
The revision of this popular CD-ROM features more graphically based questions, including coloring and sequencing exercises. Each chapter of the text is thoroughly covered to provide you with an excellent "interactive review." Immediate feedback accompanies all answers, and incorrect answers generate additional reading assignments from the text. Mac and Windows versions, as well as a networkable version, are available. (0-13-751827-7)

AUDIO CASSETTE STUDY TAPES
Eight 90-minute tapes teach you the pronunciation of difficult terms while reinforcing chapter objectives. These tapes have been designed to teach you the language and concepts of anatomy and physiology in an efficient, interactive fashion. (0-13-751835-8)

STUDY GUIDE
This very popular study guide, written by Charles Seiger of Atlantic Community College, is an excellent way to review terminology and concepts from the book. Organized around the three-level learning system used in the text, the study guide promotes an understanding of basic facts and concepts and helps you develop problem-solving skills. (0-13-751819-6)

THE NEW YORK TIMES CONTEMPORARY VIEW
This program is sponsored jointly by Prentice Hall and *The New York Times* to enhance student access to current, relevant information. Articles are selected by the text author and compiled into a free supplement that helps you make connections between your classroom and the outside world.

For the Laboratory
Prentice Hall publishes a variety of laboratory manuals to meet the diverse needs of anatomy and physiology labs. Please see your Prentice Hall sales representative for more details. Here is a list of those manuals:

Laboratory Text for Anatomy and Physiology, 1st edition, by Michael T. Wood. This full-color, 700-page lab manual has been designed to complement the Martini text by using compatible line art and terminology. It features cat dissections. (0-13-255670-7)

Laboratory Manual for Fundamentals of Anatomy and Physiology, 4th edition, by Roberta Meehan. This lab manual features both cat and fetal pig dissections. (0-13-751850-1)

Laboratory Exercises in Anatomy and Physiology with Cat Dissections, 5th edition, by Gerard J. Tortora and Robert Tallitsch. (0-13-237579-6)

Anatomy and Physiology Laboratory Manual, 5th edition, by Gerard J. Tortora. (0-13-576240-5)

ACKNOWLEDGMENTS

This textbook is not the product of any single individual; rather, it represents a group effort, and the members of the group deserve to be acknowledged. Foremost on my thank-you list are the faculty and reviewers throughout the world who offered suggestions that helped guide me through the revision process. Their interest in the subject, concern for the accuracy and method of presentation, and experience with students of widely varying abilities and backgrounds made the revision process an educational experience for me. To them, I express my sincere thanks and best wishes:

Maxine A'Hearn, *Prince George's Community College*
Ahmed Naguy Ali, *Alexandria, Egypt*
Timothy Alan Ballard, *University of North Carolina–Wilmington*
Steven Bassett, *Southeast Community College*
Alan Bretag, *University of South Australia School of Pharmacy*
Lucia Cepriano, *SUNY Farmingdale*
Beng Cheah, *University of Newcastle*
Chin Moi Chow, *Cumberland College of Health Sciences, University of Sydney*

Kim Cooper, *Arizona State University*
William F. Crowley, *Harvard Medical School*
Brent DeMars, *Lakeland Community College*
Gerald R. Dotson, *Front Range Community College*
Denise Friedman, *Hudson Valley Community College*
Freda Glaser, *University of Maryland–Baltimore County*
Mac E. Hadley, *University of Arizona*
Cecil Hampton, *Jefferson College*
Ruth Lanier Hays, *Clemson University*
Jean Helgeson, *Collin County Community College*
Jacqueline A. Homan, *South Plains College*
Aaron James, *Gateway Community College*
Eileen Kalmar, *St. John Fisher College*
George Karleskint, *St. Louis Community College*
Frank Kitakis, *Wayne County Community College*
Michael Kokkinn, *University of South Australia School of Pharmacy*
Bob Kucera, *University of Newcastle*
Alice Gerke McAfee, *University of Toledo Technical College*
Paul McGrath, *University of Newcastle*
Mike McKinley, *Glendale Community College*
Roberta Meehan, *University of Northern Colorado*
Richard F. Meginniss, *College of Lake County*
Erik Nyholm, *Umea University, Umea, Sweden*
John M. Olson, *Villanova University*
Brian K. Paulsen, *California University of Pennsylvania*
Ingrid Persson, *Umea University, Umea, Sweden*
Joel Reicherter, *SUNY Farmingdale*
Peta Reid, *University of Sydney, Faculty of Nursing*
Frank Schwartz, *Cuyahoga Community College*
Charles Seiger, *Atlantic Community College*
P. George Simone, *Eastern Michigan University*
Tom Smeaton, *University of South Australia School of Pharmacy*
Jeffery Smith, *Delgado Community College*
Michael Soules, Dept. of OBGYN, *University of Washington*
Jenna Sullivan, *University of Arizona*
Kathleen A. Tatum, *Iowa State University*
Kathy Taylor, *University of Arizona*
Caryl Tickner, *Stark Technical College*
Marge Torode, *Cumberland College of Health Sciences, University of Sydney*
Sheila Van Holst, *University of Sydney, Faculty of Nursing*
Michael Vennig, *University of South Australia School of Pharmacy*
Debra A. Wellner, *Wichita State University*
J. Wilkinson, University of Sydney, *School of Biological Sciences*
Stephen Williams, *Glendale Community College*

After the initial drafts were completed, two dedicated A & P instructors functioned as technical editors:

Mike McKinley, *Glendale Community College*
Eva Weinreb, *Community College of Philadelphia*

I extend special thanks to them for providing additional reviews and for assisting with the proofreading of text, art, and supplements. They have been immensely helpful to me during the revision of this edition as well as in previous editions.

Focus groups and casual meetings with students around the world helped me concentrate on the needs of individual students. I first undertook this project in 1983 to address the needs of my own students, and their feedback continues to be very important to me. I also thank the many instructors and students who took the time to contact me by phone or e-mail with specific suggestions, kudos, questions, or problems.

Over time, a textbook evolves. As with organisms, each evolutionary step builds on a preexisting framework. Thus, I also thank the individuals who helped with the development of previous editions. That list includes the following:

Maxine A'Hearn, *Prince George's Community College*
Paul Anderson, *Massachusetts Bay Community College*
Debra Joan Barnes, *Contra Cost College*
Edwin Bartholomew, *Maui Community College*
CeCe Barto, *Tomball College*
Steven Bassett, *Southeast Community College*
Robert Bauman, Jr., *Amarillo College*
Dean Beckwith, *Illinois Central College*
Doris Benfer, *Montgomery Community College*
Latsy Best, *Palm Beach Community College–North*
Charles Biggers, *University of Memphis*
Cynthia Bottrell, *Scott Community College*
Spencer R. Bowers, *Oakton Community College*
Mimi Bres, *Prince George's Community College*
C. David Bridges, *Purdue University*
Sandra Bruner, *Polk Community College*
Gene Carella, *Niagara County Community College*
Robert M. Carey, *University of Arizona*
Wayne Carley, *Lamar University*
William M. Chamberlain, *Indiana State University*
William D. Chapple, *University of Connecticut*
Anthony Chee, *Houston Community College*
Suzette Chopin, *Texas A&M University at Corpus Christi*
O. D. Cockrum, *Texas State Technical College*
Kim Cooper, *Arizona State University*
Richard Coppings, *Chattanooga State College*
Grant Dahmer, *University of Arizona*
Charles Daniels, *Kapiolani Community College*
Gerald R. Dotson, *Front Range Community College*
Ellen Dupre, *Indian River Community College*
John Dziak, *Community College of Allegheny*
Lee Famiano, *Cuyahoga Community College*

Lee Farello, *Niagara County Community College*
Marion Fintel, *Jefferson State Community College*
Ruby Fogg, *New Hampshire Tech*
Mildred Fowler, *Tidewater Community College*
Sharon Fowler, *Dutchess Community College*
Ann Funkhouser, *University of the Pacific*
Mildred Galliher, *Cochise College*
Jeff Gerst, *North Dakota State University*
Louis Giacinti, *Milwaukee Area Technical College*
Delaine Gilcrease, *Mesa Community College*
Kathleen M. Gorczyca, *North Shore Community College*
Bonnie Gordon, *Memphis State University*
Mac E. Hadley, *University of Arizona*
William Hairston, *H.A.A.C.*
Cecil Hampton, *Jefferson College*
Ernest Harber, *San Antonio College*
John P. Harley, *Eastern Kentucky University*
Ann Harmer, *Orange Coast Community College*
Linden Haynes, *Hinds Community College*
Mary Healey, *Springfield College*
Jean Helgeson, *Collin County Community College*
Vickie S. Hennessy, *Sinclair Community College*
Donna Hoel, *Stark Community College*
Elvis J. Holt, *Purdue University*
James Horwitz, *Palm Beach Community College*
Beth Howard, *Rutgers University*
Yvette Huet-Hudson, *University of North Carolina–Charlotte*
Angie Huxley, *Pima Community College*
Aaron James, *Gateway Community College*
Desiree Jett, *Essex County College*
Drusilla Jolly, *Forsyth Technical Community College*
David Kalichstein, *Ocean County College*
George Karleskint, *St. Louis Community College*
C. Ward Kischer, *University of Arizona*
William Kleinelp, *Middlesex County College*
Jerry K. Lindsey, *Tarrant County Junior College*
Mary Lockwood, *University of New Hampshire*
Susan Lustick, *San Jacinto College North*
Greg Maravellas, *Bristol Community College*
Dan Mark, *Penn Valley Community College*
Jane Marks, *Scottsdale Community College*
Thomas W. McCort, *Cuyahoga Community College*
Mike McCusker, *Eastern College*
Bob McDonough, *DeKalb College*
Ruth McFarland, *Mt. Hood Community College*
Roberta Meehan, *University of Northern Colorado*
Ann Miller, *Middlesex Community College*
Alice Mills, *Middle Tennessee State University*
Melvin Mills, *Scottsdale Community College*
Ron Mobley, *Wake Technical Community College*

Rose Morgan, *Minot State University*
Aubrey Morris, *Pensacola Junior College*
Robert L. Moskowitz, *Community College of Philadelphia*
Richard Mostardi, *Akron University*
Elizabeth Naugle, *Lane Community College*
Martha Newsome, *Tomball College*
Bill Nicholson, *University of Arizona*
Richard Northrup, *Delta College*
Betty Orr, *Sinclair Community College*
Mary Theresa Ortiz, *Kinsborough Community College*
David L. Parker, *Northern Virginia Community College*
Mark Paulissen, *Slippery Rock University*
Lois Peck, *Philadelphia College of Pharmacy*
Philip Penner, *Borough of Manhattan Community College*
Clifford Pohl, *Duquesne University*
Robert Pollack, *Nassau Community College*
Robert L. Preston, *Illinois State University*
Gary Quick, *Paradise Valley Community College*
Anil Rao, *Metro State College of Denver*
Jean Rigden, *Scottsdale Community College*
Carolyn C. Robertson, *Tarrant County Junior College*
Kevin Ryan, *Stark Technical College*
Kristi Sather-Smith, *Hinds Junior College*
Charles Seiger, *Atlantic Community College*
Mark Shoop, *Macon College*
Robert A. Sinclair, *Sun Antonio College*
David S. Smith, *San Antonio College*
Jeffery Smith, *Delgado Community College*
Philip Sokolove, *University of Maryland–Baltimore County*
Thomas S. Spurgeon, *Colorado State University*
Eric Sun, *Macon College*
Richard Symmons, *California State University at Hayward*
Dennis Taylor, *Hiram College*
Kathy Taylor, *University of Arizona*
Jay Templin, *Widener University*
Caryl Tickner, *Stark Technical College*
Lucia Tranel, *St. Louis Community College* and *St. Louis College*
Steve Trautwein, *Southeast Missouri State University*
Pat Turner, *Howard Community College*
Kent M. Van De Graaff, *Brigham Young University*
Jane Wallace, *Chattanooga State Technical Community College*
Eva Weinreb, *Community College of Philadelphia*
Rosamund Wendt, *Community College of Philadelphia*
Stephen Williams, *Glendale Community College*
Eric Wise, *University of California at Santa Barbara*
Michael Wood, *Del Mar College*
Jamie Young, *Chattanooga State Technical Community College*

Karen Karlin, Senior Development Editor at Prentice Hall, also played a vital role in the shaping of the fourth edition. Throughout the process, she helped me keep the general tone and level of presentation consistent. The accuracy and currency of the clinical material in this edition and in the *Applications Manual* in large part reflect the detailed clinical reviews performed by my wife, Kathleen Welch, M.D. Virtually without exception, reviewers stressed the importance of accurate, integrated, and visually attractive illustrations in aiding the students to understand essential material. The revision of the art program was primarily directed by Bill Ober, M.D., and Claire Garrison, R.N. Their suggestions about topics of clinical importance, presentation sequence, and revisions to the proposed art were of incalculable value to me and to the project.

Many of the text's illustrations include color photographs or micrographs collected from a variety of sources. Much of the work in tracking down these materials was performed by Stuart Kenter, whose efforts are greatly appreciated. I gratefully acknowledge Ralph Hutchings for the use of his cadaver photographs in the text and *Applications Manual.* Dr. Eugene C. Wasson, III, and the staff of Maui Radiology Consultants, Inc., provided valuable assistance in the selection and printing of the CT and MRI scans found in the *Applications Manual.*

I also express my appreciation to the editors and support staff at Prentice Hall who made the entire project possible. I owe special thanks to Tim Bozik, Editorial Director, and Paul Corey, Editor in Chief, for their support of the project; David K. Brake, Executive Editor for Biology, for being the driving force and project coordinator; Ray Mullaney, Editor in Chief of Engineering, Science and Mathematics Development, for his calm and patient supervision; Linda Schreiber, Acquisitions Editor, and Laura Edwards, Media Product Development Editor, for their outstanding job in initiating and overseeing a highly complex supplements program; David Riccardi, Assistant Vice President of Production, Kathleen Schiaparelli, Executive Managing Editor for Science and Mathematics Production, and Shari Toron, Assistant Managing Editor, for providing the extraordinary resources (human as well as technological) required to produce a book of this size and complexity; Karen Malley and Kim Dellas, Production Editors, for somehow managing to keep people, text, and illustrations moving in the proper direction at the appropriate times; John J. Jordan, Formatting Manager, and his intrepid crew—Karen Noferi, Jeff Henn, Erik Unhjem, Scott Garrison, Terrance Cummings, Richard Foster, and David Tay—for their extraordinary dedication and craftsmanship in shaping these pages; Paula Maylahn, Creative Director, Heather Scott, Art Director, and Gus Vibal, Art Manager, for their supervision of the design, the art program, and much else; Warren Ramezzana, Art Editor, for his help with the organization and trafficking of the art; Ray Caramanna, Patti Gutierrez, Charles Pelletreau, Liz Nemeth, and Shea Oakley, Art Coordinators, for their assistance in drafting and revising technical art; Jennifer Welchans, Marketing Manager, and Kelly McDonald, Director of Marketing, for their enthusiasm and tireless efforts in promoting this book and its supplements; Editorial Assistants Byron Smith, Adam Velthaus, and David Stack for their help with numerous tasks; and Barbara Murray, Special Projects Manager, and James Buckley, Supplements Production Editor, for their work on the transparencies and on the *Applications Manual,* which is a text unto itself.

No one person could expect to produce a flawless textbook of this scope and complexity. Any errors or oversights are strictly my own rather than those of the reviewers, artists, or editors. To help improve future editions, I encourage you to send any pertinent information, suggestions, or comments about the organization or content of this textbook to me directly, using the e-mail address below. I will deeply appreciate any and all comments and suggestions and will carefully consider them in the preparation of the fifth edition.

Frederic H. Martini
Haiku, Hawaii
martini@maui.net

To the Student

HOW TO GET THE MOST OUT OF THIS BOOK

Dear Student,

This text was designed to help you master the terminology and basic concepts of human anatomy and physiology. It should also allow you to begin applying what you have learned to everyday problems and situations. These goals have shaped every aspect of the book and its various supplements.

I have no doubt that you will find the study of anatomy and physiology to be one of the most interesting, challenging, and satisfying of all your educational experiences. I also know that as you start the course, you are going to be intimidated by the volume of material you will be expected to learn and understand. Although there is a lot of information between the covers of this text, there are themes and patterns that appear again and again. The material will therefore be much easier to deal with if you can organize the information in a way that highlights those patterns for you. With this fact in mind, I have taken extra care to point out important patterns and help you organize new information.

Many learning aids are built into the text to simplify your study of this material and to make that process rewarding. These learning aids were developed during brainstorming sessions with students in campuses across the United States. Many students, in person or by phone or e-mail, have told me that this system really does work—but it can help you only if you learn to use it (preferably, well before your first exam). This book can help you get the most from the time and energy you invest if you examine this overview carefully and consult your instructor if you have any questions.

Good luck and best wishes,

Ric

CHAPTER-OPENING PHOTOGRAPH AND CAPTION

Each chapter begins with a **photograph** that dramatizes some aspect of the material covered in the chapter.

The accompanying **caption** relates the photograph to themes that will be developed in the course of the chapter, providing you with a preview of the chapter's main topics.

CHAPTER OUTLINE AND OBJECTIVES

Before you begin a chapter, it is important to know what it will cover and what you need to learn from it. For this reason, every chapter opens with an **outline** that gives you an overview of its contents. Each entry in this outline corresponds to a major heading in the text, and a page number is provided for easy reference.

The basic themes of the chapter are set forth in a series of **learning objectives**. These objectives will help you structure your reading of the chapter and keep you focused on the key points. Each objective is placed under the heading where the relevant text material is located, creating a framework for learning that can guide your study right from the start.

CHAPTER 28

The Reproductive System

Of all our organ systems, only one is not needed for an individual's survival. Indeed, it does not seem to confer any direct benefit—at least physically. In a sense, our reproductive systems do not exist for us at all; they exist for the human race. But how can we measure the value to us—to the part of us that is not just cells and tissues and organs—of the ability to have children? Of contributing something to the next generation of human beings and in the process making a little part of ourselves live on? In this chapter, we shall study the system that has enabled humans to exist on Earth for a very long time—and, with luck, to stick around for at least a while longer.

CHAPTER OUTLINE AND OBJECTIVES

KEY TERMS
Throughout the text, the most important new terms are presented in **boldface** type.

The Blood–Brain Barrier

Neural tissue in the CNS is isolated from the general circulation by the **blood–brain barrier.** This barrier exists because the endothelial cells that line the capillaries of the CNS are extensively interconnected by tight junctions. These junctions prevent the diffusion of materials between adjacent endothelial cells. In general, only lipid-soluble compounds (including carbon dioxide, oxygen, ammonia, lipids, such as steroids or prostaglandins, and small alcohols) can diffuse across the lipid bilayer membranes of endothelial cells into the interstitial fluid of the brain and spinal cord. Water and ions must pass through channels in the inner and outer cell membranes. Larger water-soluble compounds can cross the capillary walls only through active or passive transport. The restricted permeability characteristics of the endothelial lining of brain capillaries are in some way dependent on chemicals secreted by astrocytes. We described these cells, which are in close contact with CNS capillaries, in Chapter 12. ↪ *[p. 376]* The outer surfaces of the endothelial cells are covered by the processes of astrocytes. Because the astrocytes release chemicals that control the permeabilities of the endothelium, these cells play a key supporting role in the blood–brain barrier. If the astrocytes are damaged or stop stimulating the endothelial cells, the blood–brain barrier disappears.

The choroid plexus lies outside the neural tissue of the brain, and there are no astrocytes in contact with the endothelial cells. As a result, capillaries there are highly permeable. Substances do not have free access to the CNS, however, because a **blood–CSF barrier** is created by specialized ependymal cells. These cells, interconnected by tight junctions, surround the capillaries of the choroid plexus.

Transport across the blood–brain and blood–CSF barriers is selective and directional. Even the passage of small ions, such as sodium, hydrogen, potassium, or chloride, is controlled. As a result, the pH and concentrations of sodium, potassium, calcium, and magnesium ions in the blood and CSF are different. Some organic compounds are readily transported, and others cross only in minute amounts. Neurons have a constant need for glucose. This need must be met regardless of the relative concentrations in the blood and interstitial fluid. Even when circulating glucose levels are low, endothelial cells continue to transport glucose from the blood to the interstitial fluid of the brain. In contrast, only trace amounts of circulating norepinephrine, epinephrine, dopamine, or serotonin pass into the interstitial fluid or CSF of the brain. This limitation is important because these compounds are neurotransmitters, and their entry from the circulation (where concentrations can be relatively high) could result in the uncontrolled stimulation of neurons throughout the brain.

The blood–brain barrier remains intact throughout the CNS, with four noteworthy exceptions:

1. In portions of the hypothalamus, the capillary endothelium is extremely permeable. This permeability exposes hypothalamic nuclei to circulating hormones and permits the diffusion of hypothalamic hormones into the circulation.

2. Capillaries in the posterior pituitary gland are highly permeable. At this site, the hormones *ADH* and *oxytocin*, produced by hypothalamic neurons, are released into the circulation.

3. Capillaries in the *pineal gland* are also very permeable. The pineal gland, an endocrine structure, is located in the posterior, superior surface of the diencephalon. The capillary permeability allows pineal secretions into the general circulation.

4. Capillaries at the choroid plexus are extremely permeable. Although the capillary characteristics of the blood–brain barrier are lost there, the transport activities of specialized ependymal cells within the choroid plexus maintain the blood–CSF barrier.

PENETRATING THE BLOOD–BRAIN AND BLOOD–CSF BARRIERS Physicians must sometimes get specific compounds into the interstitial fluid of the brain to fight CNS infections or to treat other neural disorders. To do this, they must understand the limitations of the blood–brain and blood–CSF barriers. For example, *Parkinson's disease*, discussed on page 471, results from inadequate release of the neurotransmitter dopamine in a nucleus in the cerebrum. Although dopamine will not cross the blood–brain or blood–CSF barrier, a related compound, *L dopa*, passes readily and is converted to dopamine inside the brain. In treatments for meningitis or other CNS infections, the antibiotics *penicillin* and *tetracycline* are excluded from the brain, but *erythromycin*, *sulfisoxazole*, and *sulfadiazine* enter the CNS very rapidly.

When the CNS is damaged, the blood–brain barrier breaks down. As a result, the locations of injury sites, infections, and tumors can be determined by injecting into the bloodstream tracers that cannot enter the CNS tissues. One common method is to use *albumin*, a plasma protein, labeled with radioactive iodine. Any radioactivity within the brain will be easily detected in a brain scan, using procedures described in Section 2 of the *Applications Manual*. **AM** *Medical Importance of Radioisotopes*

☑ What would happen if an interventricular foramen became blocked?

☑ How would decreased diffusion across the arachnoid granulations affect the volume of cerebrospinal fluid in the ventricles?

☑ Many water-soluble molecules found in the blood in relatively large amounts occur in small amounts or not at all in the extracellular fluid of the brain. Why?

CLINICAL DISCUSSIONS
Important **clinical topics** are presented in context, set off by a caduceus icon, title, and vertical red bar alongside. These topics have been selected not only for their medical importance but also for their value in helping you learn essential principles of anatomy and physiology. Many of the clinical discussions deal with diseases or therapies; they are designed to show how an understanding of abnormal conditions can shed light on normal function and vice versa. Others deal with **personal health and fitness**.

APPLICATIONS MANUAL REFERENCES
Many additional topics of clinical significance are treated in the *Applications Manual*, which accompanies each copy of this text. The *Applications Manual* also contains extensive and detailed coverage of clinical topics that are introduced in the text and some nonclinical topics as well. Topics in the *Applications Manual* are identified in the text by the **AM icon**, followed by the title of the relevant discussion (in blue).

CONCEPT LINKS
To understand anatomy and physiology, you must remember many terms and be able to relate concepts that are introduced in different chapters. Concept links provide a quick visual signal that new material being presented is related to or builds on previous discussions. Each **"links" icon** is followed by a page reference to help you find the relevant material from an earlier part of the book.

CONCEPT CHECK QUESTIONS
These questions, located at the ends of major sections in each chapter, encourage you to stop and assess your understanding of the basic concepts presented in the preceding pages. Because concepts build on one another, you need to have mastered important material before you go on to the next section. These questions are designed to be answered quickly and easily *if* you have grasped the preceding discussion. If the questions seem difficult, you probably should review that material before you move on. **Answers** to the Concept Check Questions are provided in Appendix I.

BOXES

Clinical or health-related topics of particular importance are presented in Boxes set off from the main text by means of a background tint and a caduceus icon. These essays cover major diseases, such as breast cancer, atherosclerosis, heart disease, and other subjects of special interest, such as steroid abuse.

 Spinal Taps and Myelography

Tissue samples, or **biopsies**, are taken from many organs to assist in diagnosis. For example, when a liver or skin disorder is suspected, small plugs of tissue are removed and examined for signs of infec-tion or cell damage or are used to identify the bacteria that cause an infection. Unlike many other tissues, however, neural tissue consists largely of cells rather than extracellular fluids or fibers. Neural tissue samples are seldom removed for analysis, because any extracted or damaged neurons will not be replaced. Instead, small volumes of cerebrospinal fluid are collected and analyzed. Cerebrospinal fluid is intimately associated with the neural tissue of the CNS, and pathogens, cell debris, or metabolic wastes in the CNS will therefore be detectable in the CSF.

The withdrawal of cerebrospinal fluid, a procedure called a *spinal tap*, must be done with care to avoid injuring the spinal cord. The adult spinal cord extends inferiorly only as far as vertebra L_1 or L_2. Between vertebra L_2 and the sacrum, the meningeal layers remain intact, but they enclose only the relatively sturdy components of the cauda equina and a significant quantity of CSF. When the vertebral column is flexed, a needle can be inserted between the lower lumbar vertebrae and into the subarachnoid space with minimal risk to the cauda equina. In this procedure, known as a **lumbar puncture**, 3–9 ml of fluid is taken from the subarachnoid space between vertebrae L_3 and L_4 (Figure 13-3a●). Spinal taps are performed when CNS infection is suspected or when severe back pain, headaches, disc problems, and some types of strokes are diagnosed.

Myelography is the introduction of radiopaque dyes into the CSF of the subarachnoid space. Because the dyes are opaque to X-rays, the CSF appears white on an X-ray image (Figure 13-3b●). Any tumors, inflammation, or adhesions that distort or divert CSF circulation will be shown in silhouette. In the event of severe infection, inflammation, or leukemia (cancer of the white blood cells), antibiotics, steroids, or anti-cancer drugs can be injected into the subarachnoid space.

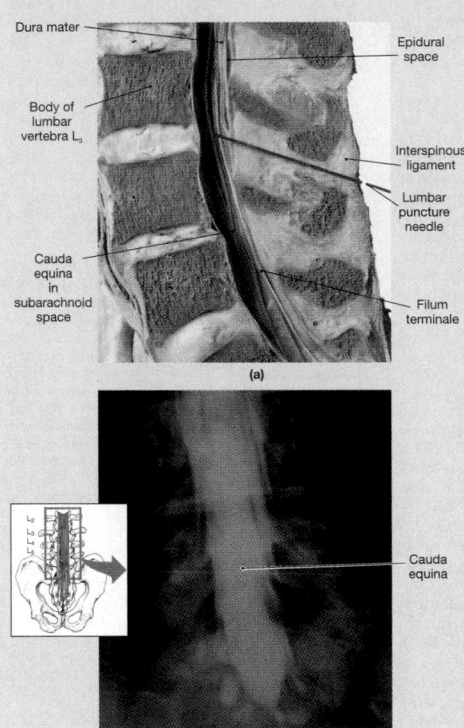

Dura mater

Body of lumbar vertebra L_3

Cauda equina in subarachnoid space

Epidural space

Interspinous ligament

Lumbar puncture needle

Filum terminale

(a)

Cauda equina

(b)

●**FIGURE 13-3 Spinal Taps and Myelography. (a)** The lumbar puncture needle is in the subarachnoid space, between the nerves of the cauda equina. The needle has been inserted in the midline between the third and fourth lumbar vertebral spines, pointing at a superior angle toward the umbilicus. Once the needle punctures the dura mater and enters the subarachnoid space, a sample of CSF may be obtained. **(b)** A myelogram—an X-ray image of the spinal cord after a radiopaque dye has been introduced into the CSF—showing the cauda equina in the lower lumbar region.

The FOCUS format is designed to present related text, art, and tabular material as an **integrated package**. These features, identified by a vertical blue rule alongside, can help you assimilate complex information as efficiently as possible.

204 CHAPTER 7 / The Axial Skeleton

FOCUS The Individual Bones of the Skull

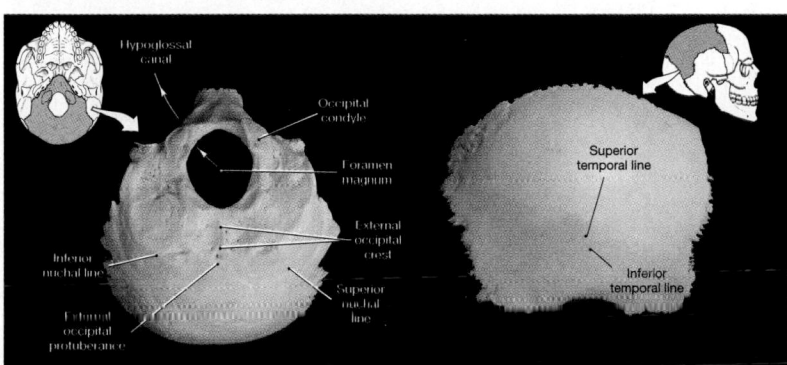

(a) Occipital bone, inferior (external) view

(b) Right parietal bone, lateral view

•**FIGURE 7-5** **The Occipital and Parietal Bones.** **(a)** Inferior view of the occipital bone. **(b)** Lateral view of the right parietal bone.

Bones of the Cranium

The Occipital Bone (Figure 7-5a•)

GENERAL FUNCTIONS: Forms much of the posterior and inferior surfaces of the cranium.

ARTICULATIONS: The occipital bone articulates with the parietal bones, the temporal bones, the sphenoid bone, and the first cervical vertebra (the atlas) (Figures 7-3a–c,e and 7-4•).

REGIONS/LANDMARKS: The **external occipital protuberance** is a small bump at the midline on the inferior surface.

The **occipital crest,** which begins at the external occipital protuberance, marks the attachment of a ligament that helps stabilize the neck

The **inferior** and **superior nuchal** (NOO-kul) **lines** are ridges that intersect the crest. They mark the attachment sites of muscles and ligaments that stabilize the articulation at the occipital condyles and balance the weight of the head over the vertebrae of the neck.

The **occipital condyles** are the site of articulation between the skull and the first cervical vertebra.

The concave internal surface of the occipital bone (Figure 7-4a•) closely follows the contours of the brain. The grooves follow the paths of major vessels, and the ridges mark the attachment sites of membranes that stabilize the position of the brain.

FORAMINA: The **foramen magnum** connects the cranial cavity with the spinal cavity, which is enclosed by the vertebral column (Figure 7-4b•). The foramen magnum surrounds the connection between the brain and spinal cord.

The **jugular foramen** lies between the occipital bone and temporal bone. The *internal jugular vein* passes through this foramen, carrying venous blood from the brain.

The **hypoglossal canals** (Figure 7-3e•) begin at the lateral base of each occipital condyle and end on the inner surface of the occipital bone near the foramen magnum. The *hypoglossal nerves,* cranial nerves that control the tongue muscles, pass through these canals.

The Parietal Bones (Figure 7-5b•)

GENERAL FUNCTIONS: Form part of the superior and lateral surfaces of the cranium.

ARTICULATIONS: The parietal bones articulate with one another and with the occipital, temporal, frontal, and sphenoid bones (Figures 7-3a–d and 7-4•).

REGIONS/LANDMARKS: The **superior** and **inferior temporal lines** are low ridges that mark the attachment sites of the *temporalis muscle,* a large muscle that closes the mouth.

Grooves on the inner surface of the parietal bones mark the paths of cranial blood vessels (Figure 7-4a•).

FIGURE REFERENCE LOCATORS

In studying a highly visual subject such as anatomy and physiology, you must frequently look back and forth between the text and the numerous figures. **Red dots** next to figure references in the text serve as place markers, making it easy to return to your spot in the narrative after you have consulted an illustration. A red dot also precedes each figure caption, tying the art and text together.

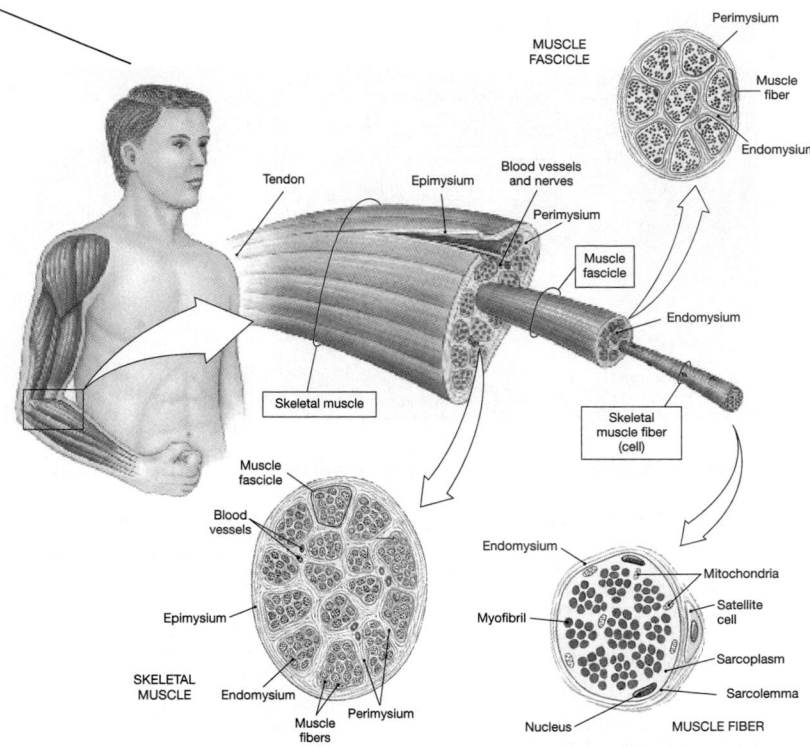

•*FIGURE 10-1* **Organization of Skeletal Muscles.** A skeletal muscle consists of fascicles (bundles of muscle fibers) enclosed by the epimysium. The bundles are separated by connective tissue fibers of the perimysium, and within each bundle the muscle fibers are surrounded by the endomysium. Each muscle fiber has many superficial nuclei as well as mitochondria and other organelles detailed in Figure 10-2.

Connective Tissue Organization

Three layers of connective tissue are part of each muscle: an outer *epimysium*, a central *perimysium*, and an inner *endomysium*. These layers and their relationships are diagrammed in Figure 10-1•.

The entire muscle is surrounded by the **epimysium** (ep-i-MĪZ-ē-um; *epi-*, oñ + *mys*, muscle), a dense layer of collagen fibers. The epimysium separates the muscle from surrounding tissues and organs. It is one component of the *deep fascia*, the dense connective tissue layer described in Chapter 4 (see Figure 4-16•). ∞ [p.134]

The connective tissue fibers of the **perimysium** (per-i-MĪZ-ē-um; *peri-*, around) divide the skeletal muscle into a series of compartments, each containing a bundle of muscle fibers called a **fascicle** (FA-sik-ul; *fasciculus*, a bundle). In addition to collagen and elastic fibers, the perimysium contains blood vessels and nerves that maintain blood flow and innervate the fascicles. Each fascicle receives branches of the blood vessels and nerves in the perimysium.

Within the fascicle, the delicate connective tissue of the **endomysium** (en-dō-MĪZ-ē-um; *endo-*, inside) surrounds the individual skeletal muscle fibers and ties together adjacent muscle fibers. Scattered **satellite cells**

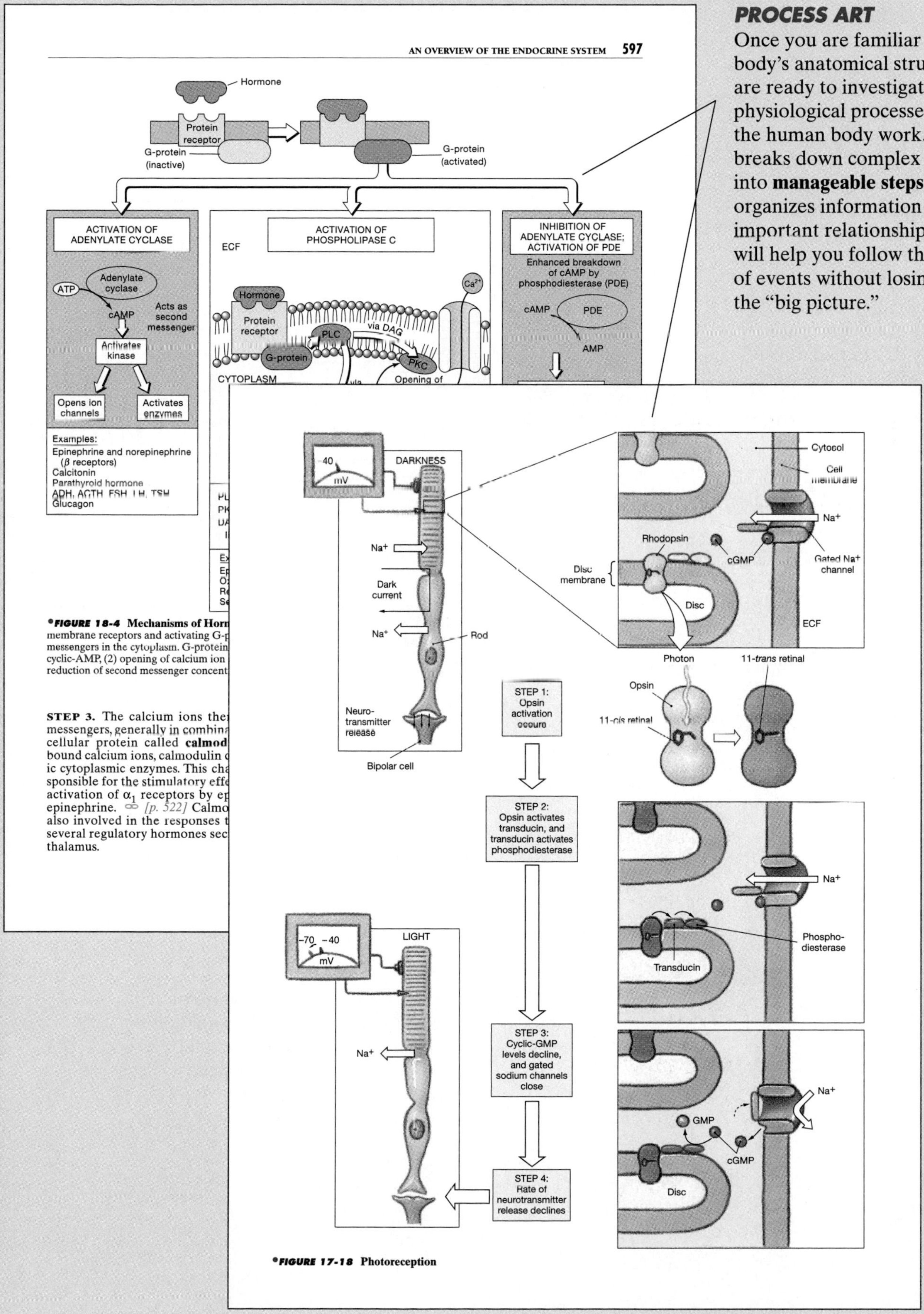

Hormone

Protein
receptor

G-protein
(inactive)

G-protein
(activated)

ACTIVATION OF ADENYLATE CYCLASE	ACTIVATION OF PHOSPHOLIPASE C	INHIBITION OF ADENYLATE CYCLASE; ACTIVATION OF PDE

ACTIVATION OF ADENYLATE CYCLASE

ATP

Adenylate
cyclase

cAMP

Acts as
second
messenger

Activates
kinase

Opens ion
channels

Activates
enzymes

Examples:
Epinephrine and norepinephrine
  (β receptors)
Calcitonin
Parathyroid hormone
ADH, ACTH, FSH, LH, TSH
Glucagon

ACTIVATION OF PHOSPHOLIPASE C

ECF

Hormone

Protein
receptor

G-protein

via DAG

PLC

PKC

CYTOPLASM

Opening of

Ca^{2+}

INHIBITION OF ADENYLATE CYCLASE; ACTIVATION OF PDE

Enhanced breakdown
of cAMP by
phosphodiesterase (PDE)

cAMP

PDE

AMP

FIGURE 18-4 Mechanisms of Horm
membrane receptors and activating G-p
messengers in the cytoplasm. G-protein
cyclic-AMP, (2) opening of calcium ion
reduction of second messenger concent

STEP 3. The calcium ions the
messengers, generally in combina
cellular protein called **calmod**
bound calcium ions, calmodulin of
ic cytoplasmic enzymes. This cha
sponsible for the stimulatory effe
activation of α_1 receptors by ep
epinephrine. [p. 522] Calmo
also involved in the responses t
several regulatory hormones sec
thalamus.

DARKNESS

−40
mV

Na^+

Dark
current

Na^+

Rod

Neuro-
transmitter
release

Bipolar cell

STEP 1:
Opsin
activation
occurs

STEP 2:
Opsin activates
transducin, and
transducin activates
phosphodiesterase

STEP 3:
Cyclic-GMP
levels decline,
and gated
sodium channels
close

STEP 4:
Rate of
neurotransmitter
release declines

Cytosol

Cell
membrane

Rhodopsin

Na^+

Disc
membrane

cGMP

Gated Na^+
channel

Disc

ECF

Photon

11-*trans* retinal

Opsin

11-*cis* retinal

Na^+

Phospho-
diesterase

Transducin

GMP

Na^+

cGMP

Disc

−70 −40
mV

LIGHT

Na^+

FIGURE 17-18 Photoreception

PROCESS ART
Once you are familiar with the body's anatomical structures, you are ready to investigate the physiological processes that make the human body work. Process art breaks down complex processes into **manageable steps** and organizes information to highlight important relationships. Such art will help you follow the sequence of events without losing track of the "big picture."

EMBRYOLOGY SUMMARIES

Located throughout the book, these features use a combination of art and text to present the formation of significant organs, structures, and systems during **embryological and fetal development**. The Embryology Summaries are self-contained so that they can be integrated with the rest of the chapter, studied separately, or omitted as you and your instructor see fit.

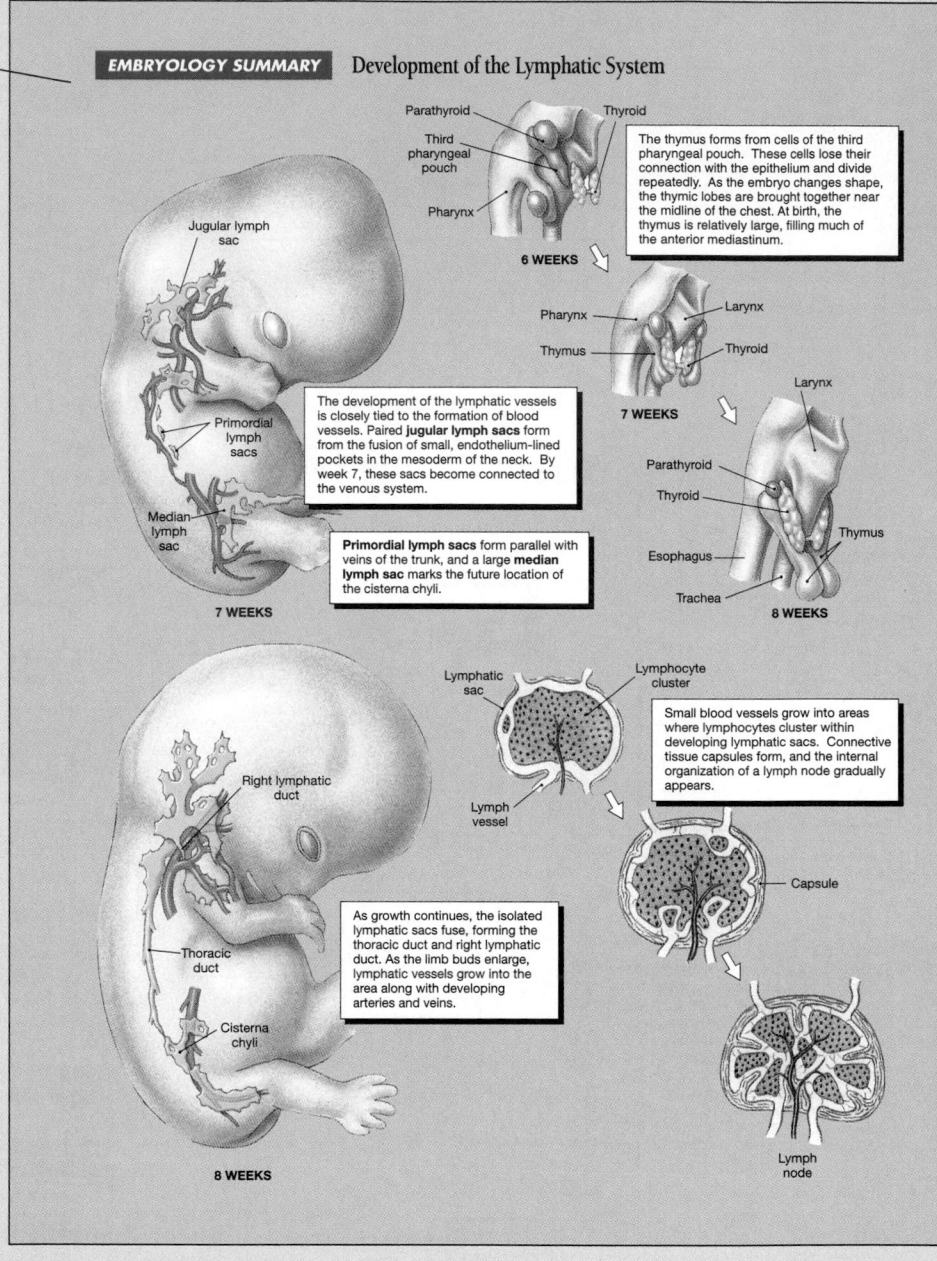

EMBRYOLOGY SUMMARY Development of the Lymphatic System

The thymus forms from cells of the third pharyngeal pouch. These cells lose their connection with the epithelium and divide repeatedly. As the embryo changes shape, the thymic lobes are brought together near the midline of the chest. At birth, the thymus is relatively large, filling much of the anterior mediastinum.

6 WEEKS

7 WEEKS

8 WEEKS

The development of the lymphatic vessels is closely tied to the formation of blood vessels. Paired **jugular lymph sacs** form from the fusion of small, endothelium-lined pockets in the mesoderm of the neck. By week 7, these sacs become connected to the venous system.

Primordial lymph sacs form parallel with veins of the trunk, and a large **median lymph sac** marks the future location of the cisterna chyli.

7 WEEKS

Small blood vessels grow into areas where lymphocytes cluster within developing lymphatic sacs. Connective tissue capsules form, and the internal organization of a lymph node gradually appears.

As growth continues, the isolated lymphatic sacs fuse, forming the thoracic duct and right lymphatic duct. As the limb buds enlarge, lymphatic vessels grow into the area along with developing arteries and veins.

8 WEEKS

xxxvi

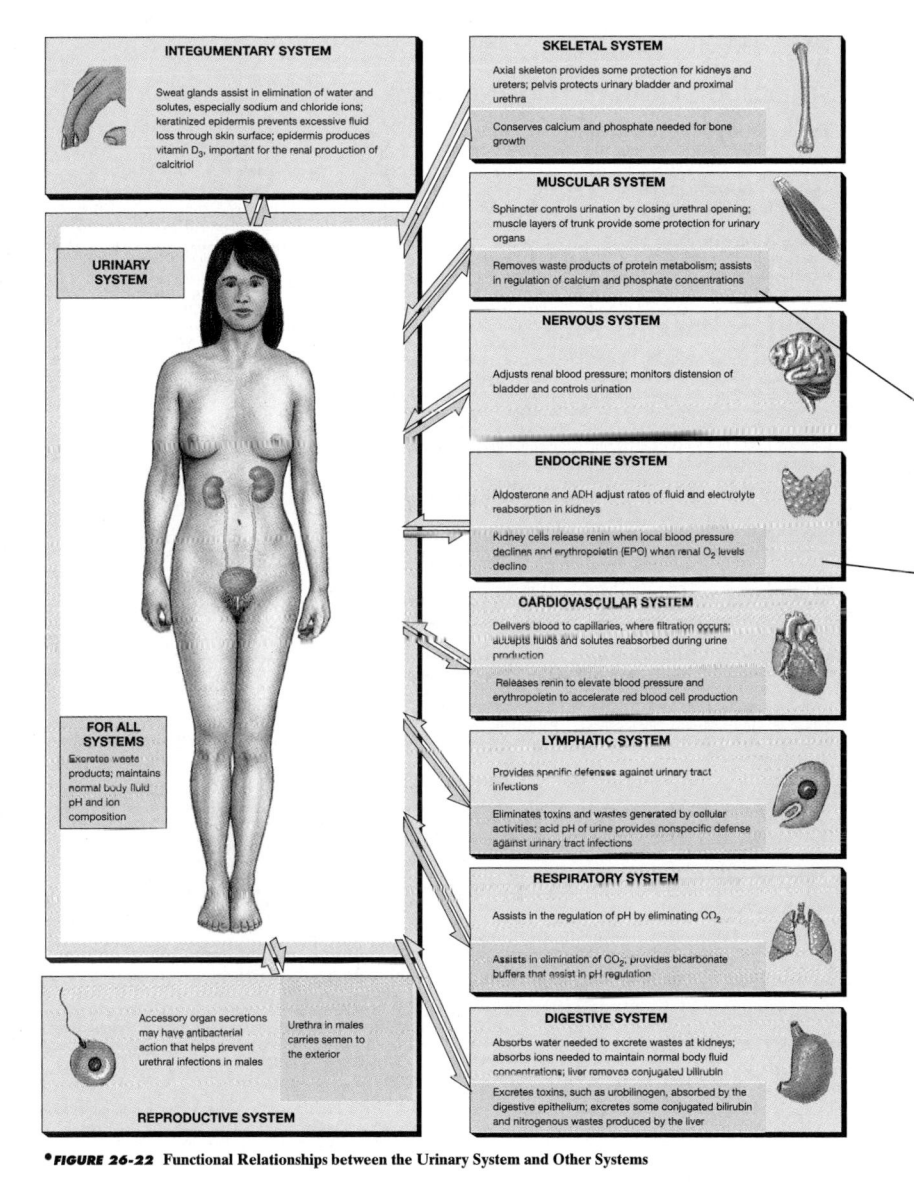

FIGURE 26-22 Functional Relationships between the Urinary System and Other Systems

SYSTEM INTEGRATORS

These figures, located at the end of each systems chapter, show how the system under consideration interacts with each of the body's other systems. They reinforce the concept that the body functions as an integrated unit.

The **interactions among systems** are color-coded:

- Beige is used to highlight the contributions of the system in question to each of the body's other systems.

- Blue is used to indicate the contributions other systems make to the system being studied.

SELECTED CLINICAL TERMINOLOGY

The first part of this section reviews the definitions and (where necessary) the pronunciations of the most important medical terms discussed **in the chapter**. Page references are provided for easy review.

The second part of this section gives definitions of medical terms that appear in related discussions **in the *Applications Manual***.

STUDY OUTLINE

The Study Outline provides you with a concise but detailed **summary** of the chapter's most important facts and concepts.

The short numbered statements are organized under the **main chapter headings**, which appear with their **page references**. You can easily review the organization of the chapter as well as its contents.

Key terms that are boldfaced in the text appear in **boldface** in the summary.

The **figures and tables** relevant to each section are also listed.

The Web icons remind you to refer to our text-specific **Web site**, which includes an on-line Student Study Guide with hot links to key A & P-related Web sites around the world.

SELECTED CLINICAL TERMINOLOGY

Terms Discussed in This Chapter

Alzheimer's disease: A progressive disorder marked by the loss of higher cerebral functions. *(p. 509 and AM)*

amnesia: A temporary or permanent loss of memory. *(p. 506 and AM)*

amyotrophic lateral sclerosis (ALS): A progressive, degenerative disorder affecting motor neurons of the spinal cord, brain stem, and cerebral hemispheres. *(p. 502 and AM)*

anencephaly (an-en-SEF-a-lē): A rare condition in which the brain fails to develop at levels above the mesencephalon or lower diencephalon. *(p. 503)*

cerebral palsy: A disorder that affects voluntary motor performance and arises in infancy or childhood as a result of pre-natal trauma, drug exposure, or a congenital defect. *(p. 499)*

electroencephalogram (EEG): A printed record of the brain's electrical activity over time. *(p. 503)*

epilepsy: One of more than 40 different conditions characterized by a recurring pattern of seizures over extended periods. *(p. 503 and AM)*

Huntington's disease: An inherited disease marked by a progressive deterioration of mental abilities and by motor disturbances. *(p. 508 and AM)*

schizophrenia: A psychological disorder marked by pronounced disturbances of mood, thought patterns, and behavior. *(p. 508)*

seizure: A temporary disorder of cerebral function, accompanied by abnormal movements, unusual sensations, and/or inappropriate behavior; includes *focal seizures, temporal lobe seizures, convulsive seizures, generalized seizures, grand mal seizures,* and *petit mal seizures.* *(p. 503 and AM)*

senile dementia: A progressive loss of memory, spatial orientation, language, and personality as a consequence of aging. *(p. 509)*

AM   **Additional Terms Discussed in the *Applications Manual***

anterograde amnesia: Loss of the ability to store new memories after an injury, although earlier memories remain intact and accessible.

aphasia: A disorder affecting the ability to speak or read.

delirium: A state characterized by wild oscillations in the level of wakefulness, little or no grasp of reality, and possible hallucinations. Also called an *acute confusional state* (ACS).

dementia: A stable, chronic state of consciousness characterized by deficits in memory, spatial orientation, language, or personality.

hypersomnia: Unusually long periods of sleep that is otherwise normal.

insomnia: Shortened sleeping periods and difficulty in getting to sleep.

parasomnias: Abnormal behaviors performed during sleep.

post-traumatic amnesia (PTA): Amnesia caused by head injury, in which the individual can neither remember the past nor consolidate memories of the present.

retrograde amnesia (*retro-,* behind): The loss of memories of events that occurred prior to an injury.

sleep apnea: A condition in which a sleeping individual stops breathing for short intervals throughout the night.

CHAPTER REVIEW

 On-line resources for this chapter are on our World Wide Web site at: http://www.prenhall.com/martini/fap

STUDY OUTLINE

INTRODUCTION, p. 491

1. There is continuous communication among the brain, spinal cord, and peripheral nerves.

SENSORY AND MOTOR PATHWAYS, p. 491

Sensory Pathways, p. 491

1. A **sensation** arrives in the form of action potentials in an afferent fiber. The **posterior column pathway** carries fine touch, pressure, and proprioceptive sensations. The axons ascend within the **fasciculus gracilis** and **fasciculus cuneatus** and relay information to the thalamus via the **medial lemniscus.** Before the axons enter the medial lemniscus, they *decussate* (cross over) to the opposite side of the brain stem. *(Figures 15-1, 15-2; Table 15-1)*

2. The **spinothalamic pathway** carries poorly localized sensations of touch, pressure, pain, and temperature. The axons involved decussate in the spinal cord and ascend within the **anterior** and **lateral spinothalamic tracts** to the **ventral posterolateral** and **ventral posteromedial nuclei** of the thalamus. *(Figures 15-2, 15-3; Table 15-1)*

3. The **spinocerebellar pathway,** including the **posterior** and **anterior spinocerebellar tracts,** carries sensations to the cerebellum concerning the position of muscles, tendons, and joints. *(Figure 15-4; Table 15-1)*

Motor Pathways, p. 497

4. Somatic motor pathways always involve an **upper motor neuron** (whose soma lies in a CNS processing center) and a **lower motor neuron** (whose soma is located in a motor nucleus of the brain stem or spinal cord). *(Figure 15-5)*

Vertebral Anatomy, p. 219

3. A typical vertebra has a **body** (*centrum*) and a **vertebral arch** (*neural arch*) and articulates with adjacent vertebrae at the **superior** and **inferior articular processes.** *(Figure 7-18)*
4. Adjacent vertebrae are separated by **intervertebral discs.** Spaces between successive **pedicles** form the **intervertebral foramina.** *(Figure 7-18)*

Vertebral Regions, p. 220

5. Cervical vertebrae are distinguished by the shape of the body, the relative size of the vertebral foramen, the presence of **costal processes** with **transverse foramina,** and notched **spinous processes.** These vertebrae include the **atlas, axis,** and **vertebra prominens.** *(Figure 7-19; Table 7-2)*
6. Thoracic vertebrae have distinctive heart-shaped bodies, long, slender spinous processes, and articulations for the ribs. *(Figures 7-20, 7-23; Table 7-2)*
7. The lumbar vertebrae are the most massive and least mobile; they are subjected to the greatest strains. *(Figure 7-21; Table 7-2)*

8. The sacrum protects reproductive, digestive, and urinary organs and articulates with the pelvic girdle and with the fused elements of the coccyx. *(Figure 7-22)*

THE THORACIC CAGE, p. 225

1. The skeleton of the thoracic cage consists of the thoracic vertebrae, the ribs, and the sternum. The **ribs** and **sternum** form the *rib cage. (Figure 7-23)*

The Ribs, p. 225

2. Ribs 1–7 are **true,** or **vertebrosternal, ribs.** Ribs 8–12 are called **false ribs;** they include the **vertebrochondral ribs** and two pairs of **floating (vertebral) ribs.** A typical rib has a **head,** or **capitulum;** a **neck;** a **tubercle,** or **tuberculum;** an **angle;** and a **body,** or **shaft.** A **costal groove** marks the path of nerves and blood vessels. *(Figures 7-23, 7-24)*

The Sternum, p. 227

3. The sternum consists of a **manubrium,** a **body,** and a **xiphoid process.** *(Figure 7-23)*

REVIEW QUESTIONS

Level 1   Reviewing Facts and Terms

Match each numbered item with the most closely related lettered item. Use letters for answers in the spaces provided.

___ 1. foramina
___ 2. sinuses
___ 3. sutures
___ 4. calvaria
___ 5. auditory ossicles
___ 6. hypophyseal fossa
___ 7. lacrimal bones
___ 8. accommodation curve
___ 9. compensation curve
___ 10. costae
___ 11. C_1
___ 12. C_2

a. skullcap
b. tear ducts
c. atlas
d. lumbar and cervical
e. air-filled chambers
f. axis
g. ear bones
h. ribs
i. passageways
j. sella turcica
k. thoracic and sacral
l. immovable joints

13. The axial skeleton consists of the bones of the
   (a) pectoral and pelvic girdles
   (b) skull, thorax, and vertebral column
   (c) arms, legs, hands, and feet
   (d) limbs, pectoral girdle, and pelvic girdle
14. The appendicular skeleton consists of the bones of the
   (a) pectoral and pelvic girdles
   (b) skull, thorax, and vertebral column
   (c) arms, legs, hands, and feet
   (d) limbs, pectoral girdle, and pelvic girdle
15. Which of the following lists contains *only* bones of the cranium?
   (a) frontal, parietal, occipital, sphenoid
   (b) frontal, occipital, zygomatic, parietal
   (c) occipital, sphenoid, temporal, lacrimal
   (d) mandible, maxillary, nasal, zygomatic

23. The hyoid bone
   (a) forms the inferior portion of the skull
   (b) is attached to the zygomatic bone
   (c) does not articulate with any other bone
   (d) is a cartilaginous structure
24. The membranous areas between the cranial bones of the fetal skull are
   (a) fontanels   (b) sutures
   (c) Wormian bones   (d) foramina
25. The part of the vertebra that transfers weight along the axis of the vertebral column is (are) the
   (a) neural arch   (b) lamina
   (c) pedicles   (d) body
26. Progressing inferiorly, the components of the sternum include the
   (a) xiphoid process, body, manubrium
   (b) body, manubrium, xiphoid process
   (c) manubrium, body, xiphoid process
   (d) manubrium, xiphoid process, body

27. Which five bones make up the facial complex?
28. Which four major sutures make up the boundaries between the skull bones?
29. What is the primary function of the vomer?
30. Which bones contain the paranasal sinuses?
31. What characteristic of the hyoid bone makes that bone unique?
32. What purpose do the fontanels serve during delivery?
33. Beginning at the skull, cite the regions of the vertebral column and list the number of vertebrae in each region of the adult vertebral column.
34. List the four spinal curves that are apparent in the adult spinal column when viewed from the side.
35. What two primary functions are performed by the thoracic cage?
36. Why are the upper seven pairs of ribs called *true* ribs and the lower five pairs called *false* ribs?

Level 2   Reviewing Concepts

37. The atlas (C_1) can be distinguished from the other vertebrae by
   (a) the presence of anterior and posterior vertebral arches
   (b) the lack of a body
   (c) the presence of superior facets and inferior articular facets
   (d) a, b, and c are correct
38. What is the relationship between the temporal bone and the ear?
39. What is the relationship between the sphenoid bone and the pituitary gland?
40. Unlike that of other animals, the skull of humans is balanced above the vertebral column, allowing an upright posture. What anatomical adaptations are necessary in humans to maintain the balance of the skull?

41. The structural features and skeletal components of the sternum make it a part of the axial skeleton that is important in a variety of clinical situations. If you were teaching this information to prospective nursing students, what clinical applications would you cite?
42. Why are ruptured intervertebral discs more common in lumbar vertebrae and dislocations and fractures more common in cervical vertebrae?
43. We generally associate a "runny" nose with a cold. Does this nasal condition serve a purpose? What is the cause of this type of reaction?
44. Why is it important to keep your back straight when you lift a heavy object?

Level 3   Critical Thinking and Clinical Applications

45. Tess is diagnosed with a disease that affects the membranes surrounding the brain. The physician tells her family that the disease is caused by an airborne virus. Explain how this virus could have entered the cranium.
46. Joe is 40 years old and 30 pounds overweight. Like many middle-aged men, Joe carries most of this extra weight in his abdomen and jokes with his friends about his "beer gut." During an annual physical, Joe's physician advises Joe that his spine is developing an abnormal curvature. Why is the curvature of Joe's spine changing, and what is this condition called?
47. A babysitter is on trial for the death of a 10-month-old infant. The prosecutor contends that the child died as the result of being violently shaken. The defense claims that the child's head became stuck in the slats of his crib and, in trying to twist

free, the child broke his neck. The medical examiner testifies that the child died as the result of a compression of the spinal cord between the sixth and seventh cervical vertebrae. The superior articular processes of the seventh cervical vertebra and the inferior articular processes of the sixth cervical vertebra were fractured, and the processes on the right side were laterally displaced, causing the sixth vertebra to slide laterally across the seventh, damaging the spinal cord. On the basis of this evidence, whom do you believe? Why?
48. While working at an excavation, an archaeologist finds several small skull bones. She examines the frontal, parietal, and occipital bones and concludes that the skulls are those of children not yet 1 year old. How can she tell their ages from examining the bones?

APPLICATIONS MANUAL

The *Applications Manual,* which accompanies each copy of this text, provides a wealth of supplemental material to enrich your learning experience. Although the text covers all essential principles of anatomy and physiology, the *Applications Manual* allows you to explore many clinical, diagnostic, and other topics in greater depth.

- The introductory sections of the *Applications Manual*—**An Introduction to Diagnostics** and **Applied Research Topics**—include an overview of basic diagnostic principles and procedures, an explanation of the scientific method, and discussions of the application of chemistry and cellular biology and molecular biology to the diagnosis and treatment of disease.

- **The Body Systems** section parallels the text chapter by chapter and system by system. Included are more-detailed discussions of many topics introduced in the text as well as discussions of additional diseases, syndromes, and diagnostic techniques not covered in the text.

- An atlas of **Origins and Insertions** shows the attachment sites of major muscles and muscle groups. Use this section to get help in integrating information about the skeletal and muscular systems.

- A **Scanning Atlas** of photographs produced by various modern imaging techniques lets you view the interior of the human body section by section. These images will help you develop an understanding of three-dimensional relationships within the body. The Scanning Atlas also includes a number of unlabeled images. By labeling them yourself, you can test your knowledge of anatomical structure while developing your powers of visualization in three dimensions.

- A full-color **Surface Anatomy and Cadaver Atlas** contains dissection photographs that allow you to visualize the internal structure of all major body regions and organs. Surface anatomy photographs of live models are included for comparison with the dissection views.

- Extensive **cross-referencing** integrates the text and *Applications Manual*, making it easy to move back and forth between related material in the two books.

- More than a collection of information on clinically related topics, the *Applications Manual* is designed to supplement and extend the text's **three-level learning system**. It includes many features that can strengthen your grasp of the subject, improve your powers of analysis, and develop the ability to apply what you have learned.

- The **Critical-Thinking Questions** at the end of each system help sharpen your ability to think analytically. These questions are comparable to the Level 3 Review Questions in the text.

- The **Clinical Problems**, located after groups of related systems, assist you in integrating information about several body systems and enable you to practice making reasonable inferences in realistic situations.

- The **Case Studies** that follow the Body Systems section provide further opportunities for you to develop your powers of analysis, integration, and problem solving. The studies give you the chance to learn how to work through complex, multisystem problems in a series of small steps. Based on actual case histories, the studies presented draw on material from the entire text—as they would in real life. The accompanying questions, keyed to crucial points in each presentation, help you identify the relevant facts and form plausible hypotheses.

xl

The Nervous System 77

assisted programs such as FES can improve the quality of life for thousands of paralyzed individuals. The 1995 horseback-riding injury of actor Christopher Reeve has brought publicity and research funds to this area.

Shingles and Hansen's Disease
FAP p. 425

In **shingles**, or *herpes zoster*, the *herpes varicella-zoster* virus attacks neurons within the dorsal roots of spinal nerves and sensory ganglia of cranial nerves. This disorder produces a painful rash whose distribution corresponds to that of the affected sensory nerves (Figure A-26a). Shingles develops in adults who were first exposed to the virus as children. The initial infection produces symptoms known as chickenpox. After this encounter, the virus remains dormant within neurons of the anterior gray horns of the spinal cord. It is not known what triggers reactivation of this pathogen. Fortunately for those affected, attacks of shingles generally heal and leave behind only unpleasant memories.

Most people suffer only a single episode of shingles in their adult lives. However, the problem may recur in people with weakened immune systems, including those with AIDS or some forms of cancer. Treatment for shingles typically involves large doses of the antiviral drug *acyclovir (Zovirax®).*

The condition traditionally called **leprosy,** now more commonly known as **Hansen's disease,** is an infectious disease caused by a bacterium, *Mycobacterium leprae.* It is a disease that progresses slowly, and symptoms may not appear for up to 30 years after infection. The bacterium invades peripheral nerves, especially those in the skin, producing initial sensory losses. Over time, motor paralysis develops, and the combination of sensory and

motor loss can lead to recurring injuries and infections. The eyes, nose, hands, and feet may develop deformities as a result of neglected injuries (Figure A-26b). There are several forms of this disease; peripheral nerves are always affected, but some forms also involve extensive lesions of the skin and mucous membranes.

Only about 5 percent of those exposed to *Mycobacterium leprae* develop symptoms; people living in the tropics are at greatest risk. There are about 2000 cases in the United States, and an estimated 12–20 million cases worldwide. If detected before deformities occur, the disease can generally be treated successfully with drugs such as rifampin and dapsone. Treated individuals are not infectious, and the practice of confining "lepers" in isolated compounds has been discontinued.

Peripheral Neuropathies FAP p. 426

Peripheral nerve palsies, or peripheral neuropathies, are characterized by regional losses of sensory and motor function as the result of nerve trauma or compression. **Brachial palsies** result from injuries to the brachial plexus or its branches. **Crural palsies** involve the nerves of the lumbosacral plexus.

Palsies may appear for several reasons. The *pressure palsies* are especially interesting; a familiar but mild example is the experience of having an arm or leg "fall asleep." The limb becomes numb, and afterward an uncomfortable "pins-and-needles" sensation, or **paresthesia,** accompanies the return to normal function.

These incidents are seldom of clinical significance, but they provide graphic examples of the effects of more serious palsies that can last for days to months. In **radial nerve palsy,** pressure on the back of the arm interrupts the function of the radial nerve, so the extensors of the wrist and fingers

13

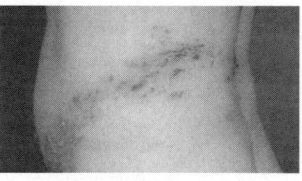

(a)

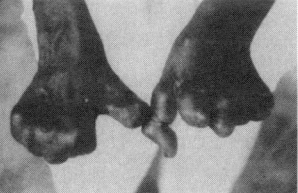

(b)

Figure A-26 Shingles and Hansen's Disease.
(a) The left side of a person with shingles. The skin eruptions follow the distribution of dermatomal innervation. (b) The distal extremities are gradually deformed as untreated Hansen's disease progresses.

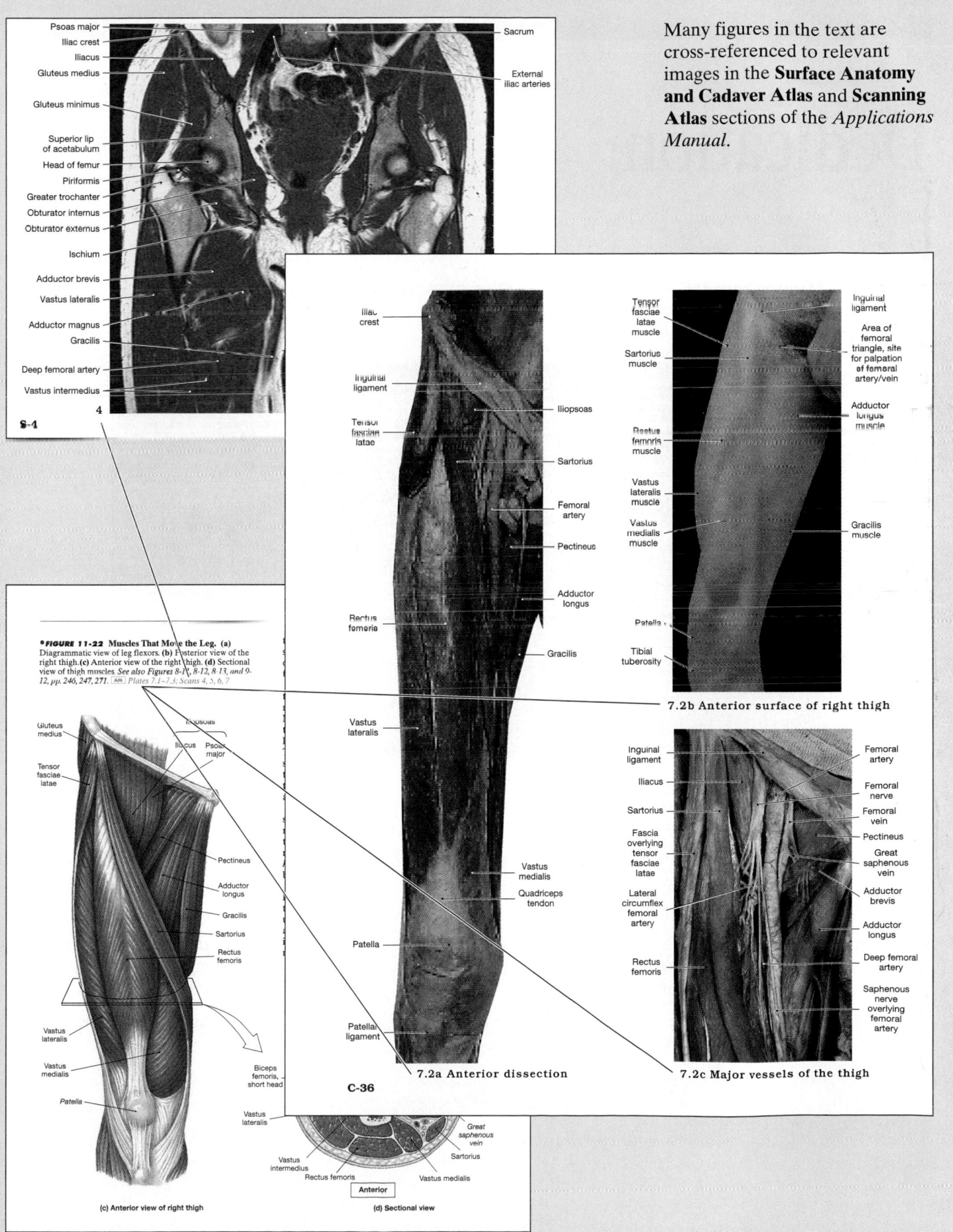

Many figures in the text are cross-referenced to relevant images in the **Surface Anatomy and Cadaver Atlas** and **Scanning Atlas** sections of the *Applications Manual.*

S-1

4

Psoas major
Iliac crest
Iliacus
Gluteus medius
Gluteus minimus
Superior lip of acetabulum
Head of femur
Piriformis
Greater trochanter
Obturator internus
Obturator externus
Ischium
Adductor brevis
Vastus lateralis
Adductor magnus
Gracilis
Deep femoral artery
Vastus intermedius

Sacrum
External iliac arteries

Iliac crest
Inguinal ligament
Tensor fasciae latae
Rectus femoris
Vastus lateralis
Vastus medialis
Quadriceps tendon
Patella
Patellar ligament

Iliopsoas
Sartorius
Femoral artery
Pectineus
Adductor longus
Gracilis

7.2a Anterior dissection

Tensor fasciae latae muscle
Sartorius muscle
Rectus femoris muscle
Vastus lateralis muscle
Vastus medialis muscle
Patella
Tibial tuberosity

Inguinal ligament
Area of femoral triangle, site for palpation of femoral artery/vein
Adductor longus muscle
Gracilis muscle

7.2b Anterior surface of right thigh

Inguinal ligament
Iliacus
Sartorius
Fascia overlying tensor fasciae latae
Lateral circumflex femoral artery
Rectus femoris

Femoral artery
Femoral nerve
Femoral vein
Pectineus
Great saphenous vein
Adductor brevis
Adductor longus
Deep femoral artery
Saphenous nerve overlying femoral artery

7.2c Major vessels of the thigh

●*FIGURE 11-22* **Muscles That Move the Leg. (a)** Diagrammatic view of leg flexors. **(b)** Posterior view of the right thigh. **(c)** Anterior view of the right thigh. **(d)** Sectional view of thigh muscles. *See also Figures 8-11, 8-12, 8-13, and 9-12, pp. 246, 247, 271.* Plates 7.1–7.5; Scans 4, 5, 6, 7

Gluteus medius
Tensor fasciae latae
Vastus lateralis
Vastus medialis
Patella

Iliacus
Psoas major
Pectineus
Adductor longus
Gracilis
Sartorius
Rectus femoris
Biceps femoris, short head

(c) Anterior view of right thigh

C-36

Vastus lateralis
Vastus intermedius
Rectus femoris

Great saphenous vein
Sartorius
Vastus medialis

Anterior

(d) Sectional view

Dedication

To my son, P.K., and to my dad, who missed this edition of the book but loved this addition to the family.

Frederic C. Martini
(1918–1995)

Frederic P. K. Martini
(1995+)

An Introduction to Anatomy and Physiology

Every day we perform a balancing act. It is an amazingly complex performance, and it goes on continuously, without a break or intermission. Every part of our physical being is involved in this balance, from individual atoms to the largest organs and organ systems. Our bodies face constant challenges: stresses (both physical and mental), injury, and disease. Every change, whether it occurs inside us or outside, threatens this delicate balance. Adjusting to these changes is crucial, for if the body loses its balance it may plunge out of control, into dysfunction, illness, and even death.

CHAPTER OUTLINE AND OBJECTIVES

The world around us contains an enormous diversity of living organisms that vary widely in appearance and lifestyle. One aim of **biology**, the study of life, is to discover the unity and patterns that underlie this diversity. Biologists have found that all living things share certain basic characteristics, including the following:

- **Responsiveness**: Organisms respond to changes in their immediate environment; this property is also called *irritability*. You move your hand away from a hot stove; your dog barks at approaching strangers; fish are scared by loud noises; and amoebas glide toward potential prey. Organisms also make longer-term changes as they adjust to their environments. For example, an animal may either grow a heavier coat as winter approaches or migrate to a warmer climate. The potential to make such adjustments is termed *adaptability*.

- **Growth** and **differentiation**: Over a lifetime, organisms grow larger, increasing in size through an increase in the size or number of their cells. In multicellular organisms, the individual cells become specialized to perform particular functions. This specialization is called *differentiation*. Growth and differentiation often produce changes in form and function. For example, the body proportions and functional capabilities of an adult human are quite different from those of an infant.

- **Reproduction**: Organisms reproduce, creating subsequent generations of similar organisms.

- **Movement**: Organisms are capable of producing movement, which may be internal (transporting food, blood, or other materials inside the body) or external (moving through the environment).

- **Metabolism** and **excretion**: Organisms rely on complex chemical reactions to provide the energy for responsiveness, growth, reproduction, and movement. They must also synthesize complex chemicals, such as proteins. The term *metabolism* refers to all the chemical operations under way in the body. Many normal metabolic operations require the absorption of certain materials from the environment. These materials, called *nutrients*, are used for growth and maintenance; many nutrients are also important as energy sources. To generate energy efficiently, most cells require oxygen, an atmospheric gas. The term *respiration* refers to the absorption, transport, and use of oxygen by cells. Metabolic operations often generate unneeded or potentially harmful waste products that must be eliminated through the process of *excretion*.

These are basic characteristics of living things. Several additional functions can be distinguished when you consider animals as complex as fish, cats, or human beings. Whereas very small organisms can absorb materials across exposed surfaces, creatures larger than a few millimeters seldom absorb nutrients directly from their environment. For example, human beings cannot absorb steaks, apples, or ice cream without pro-

cessing them first. That processing, called *digestion*, occurs in specialized areas where complex foods are broken down into simpler components that can be absorbed easily. Respiration and excretion are also more complicated for large organisms than for small ones. Humans have specialized structures responsible for gas exchange (lungs) and waste elimination (kidneys). Finally, because absorption, respiration, and excretion are performed in different portions of the body, humans require an internal transportation system, or *cardiovascular system*.

Biology includes subspecialties, each with a slightly different perspective. This text considers two biological subjects, *anatomy* (ah-NAT-o-mē) and *physiology* (fiz-ē-OL-o-jē). In your journey through these 29 chapters, you will become familiar with the basic anatomy and physiology of the human body.

THE SCIENCES OF ANATOMY AND PHYSIOLOGY

The word *anatomy* has Greek origins, as do many other anatomical terms and phrases. A literal translation would be "a cutting open." **Anatomy** is the study of internal and external structures and the physical relationships among body parts. For example, someone studying anatomy might examine how a particular muscle attaches to the skeleton. **Physiology**, another word adopted from Greek, is the study of how living organisms perform their vital functions. Someone studying physiology might consider how a muscle contracts or what forces a contracting muscle exerts on the skeleton.

Anatomy and physiology are closely integrated both theoretically and practically. Anatomical information provides clues about probable functions, and physiological mechanisms can be explained only in terms of the underlying anatomy. This observation leads to a very important concept: *All specific functions are performed by specific structures.*

Anatomists and physiologists approach the relationship between structure and function from different perspectives. To understand the difference, consider a simple nonbiological analogy. Suppose that an anatomist and a physiologist were asked to examine and report on a pickup truck. The anatomist might begin by measuring and photographing the various parts of the truck and, if possible, taking it apart and putting it back together. The anatomist would then be able to explain its key structural relationships—for example, how the movement of the pistons in the engine cylinders caused the drive shaft to rotate and how this motion was then conveyed by the transmission to the wheels. The physiologist also would note the relationships among the components, but the primary focus would be on functional characteristics, such as the amount of power that the engine could generate, the amount of force transmitted to the wheels in different gears, and so on.

The link between structure and function is always present but not always understood. For example, the superficial anatomy of the heart was clearly described in the fifteenth century, but almost 200 years passed before the pumping action of the heart was demonstrated.

This text will introduce anatomical structures and the physiological processes that make human life possible. The information will enable you to understand important mechanisms of disease and will help you make intelligent decisions about personal health.

Anatomy

Anatomy can be divided into *microscopic anatomy* and *gross (macroscopic) anatomy* on the basis of the degree of structural detail under consideration. Other anatomical specialties focus on specific processes or medical applications.

Microscopic Anatomy

Microscopic anatomy deals with structures that cannot be seen without magnification. The boundaries of microscopic anatomy are established by the limits of the equipment used. With a light microscope, you can see basic details of cell structure; with an electron microscope, you can see individual molecules that are only a few nanometers across. As we proceed through the text, we will be considering details at all levels, from macroscopic to microscopic. (Readers unfamiliar with the terms used to describe measurements and weights should consult the reference tables in Appendix II.)

Microscopic anatomy includes *cytology* and *histology*. **Cytology** (sī-TOL-o-jē) is the analysis of the internal structure of individual cells, the simplest units of life. Living cells are composed of chemical substances in various combinations, and our lives depend on the chemical processes occurring in the trillions of cells that form our body. For this reason we will consider basic chemistry (Chapter 2) before we examine cell structure (Chapter 3).

Histology (his-TOL-o-jē) involves the examination of **tissues**, groups of specialized cells and cell products that work together to perform specific functions (Chapter 4). Tissues combine to form **organs**, such as the heart, kidney, liver, and brain. Many organs are easily examined without a microscope, and at the organ level we cross the boundary into gross anatomy.

Gross Anatomy

Gross anatomy, or *macroscopic anatomy*, involves the examination of relatively large structures and features usually visible with the unaided eye. There are many ways to approach gross anatomy:

- *Surface anatomy* is the study of general form and superficial markings.
- *Regional anatomy* focuses on the anatomical organization of specific areas of the body, such as the head, neck, or trunk. Advanced courses in anatomy often stress a regional approach because it emphasizes the spatial relationships between structures already familiar to students.
- *Systemic anatomy* is the study of the structure of *organ systems*, such as the *skeletal system* or the *muscular system*. **Organ systems** are groups of organs that function together in a coordinated manner. For example, the heart, blood, and blood vessels form the *cardiovascular system*, which distributes oxygen and nutrients throughout the body. Introductory texts present systemic anatomy because that approach clarifies functional relationships among the component organs. There are 11 organ systems in the human body, and we will introduce them later in the chapter.
- *Developmental anatomy* deals with the changes in form that occur during the period between conception and physical maturity. Because it considers anatomical structures over such a broad range of sizes (from a single cell to an adult human), techniques used in developmental anatomy are similar to those used in both microscopic and gross anatomy. The most extensive structural changes occur during the first 2 months of development. The study of these early developmental processes is called *embryology* (em-brē-OL-o-jē).

There are several anatomical specialties that are important in a clinical setting. Examples include *medical anatomy* (anatomical features that change during illness), *radiographic anatomy* (anatomical structures as seen by using specialized imaging techniques, discussed later in this chapter), and *surgical anatomy* (anatomical landmarks important in surgery). Physicians normally use a combination of anatomical, physiological, and behavioral information when they evaluate patients. The evaluation process follows a logical sequence; this sequence is an example of what is known as the *scientific method*. AM *The Scientific Method*

Physiology

Physiology is the study of the function of anatomical structures, and **human physiology** is the study of the functions of the human body. These functions are complex and much more difficult to examine than most anatomical structures. As a result, there are even more specialties in physiology than in anatomy.

- *Cell physiology*, the study of the functions of living cells, is the cornerstone of human physiology. Cell physiology deals with events at the chemical and molecular levels: both chemical processes within cells and chemical interactions between cells. Chapters 2–4 focus on the chemical structure, internal organization, and control mechanisms of living cells and tissues.
- *Special physiology* is the study of the physiology of specific organs. For example, *cardiac physiology* is the study of heart function.
- *Systemic physiology* includes all aspects of the function of specific organ systems. Cardiovascular physiology, respiratory physiology, and reproductive physiology are examples of systemic physiology.
- *Pathological physiology* is the study of the effects of diseases on organ or system functions. (*Pathos* is the Greek word for "disease.") Modern medicine depends on an understanding of both normal and pathological physiology. You will find extensive information on clinically important topics in subsequent chapters and in the *Applications Manual* that accompanies this text.

LEVELS OF ORGANIZATION

Our study of the human body will begin with an overview of microscopic anatomy and then proceed to the gross and microscopic anatomy of each organ system. When considering events from the microscopic to the macroscopic scale, we are examining several interdependent levels of organization. Figure 1-1● presents an example of the relationships among these various levels of organization:

- *Atoms*, the smallest stable units of matter, can combine to form *molecules* with complex shapes. This is the *chemical*, or *molecular*, *level of organization*.
- Molecules can interact to form *organelles*, such as the protein filaments found in muscle cells. Each type of organelle has specific functions. For example, interactions among protein filaments produce the contractions of muscle cells in the heart. Cells are the smallest living units in the body; organelles are their structural and functional components. This is the *cellular level of organization*.
- Cells can combine to form *tissues*. Heart muscle cells, or cardiac muscle cells (*cardium,* heart), interact with other cell types and extracellular materials to form a distinctive type of *muscle tissue*. Cardiac muscle tissue is an example of the *tissue level of organization*.
- Tissues in combination create *organs*. Layers of cardiac muscle tissue in combination with *connective tissue,* another tissue type, form the bulk of the wall of the *heart*, a hollow, three-dimensional organ. This is an example of the *organ level of organization*.
- Organs interact in *organ systems*. Each time it contracts, the heart pushes blood into a network of blood vessels. Together the heart, blood, and blood vessels form the cardiovascular system, an example of the *organ system level of organization*.
- All the organ systems of the body work together to maintain life and health. This brings us to the highest level of organization, that of the *organism*—in this case, a human being.

The organization at each level determines the characteristics and functions of higher levels. For example, the arrangement of atoms and molecules at the molecular level creates the protein filaments that, at the cellular level, give cardiac muscle cells the ability to contract powerfully. At the tissue level these cells are linked together, forming cardiac muscle tissue. The structure of the tissue ensures that the contractions are coordinated, producing a heartbeat. When that beat occurs, the internal anatomy of the heart, an organ, enables it to function as a pump. The heart is filled with blood and connected to the blood vessels, and the pumping action circulates blood through the cardiovascular system. Through interactions with the respiratory, digestive, urinary, and other systems, the cardiovascular system performs a variety of functions essential to the survival of the organism.

Something that affects a system will ultimately affect all its components. For example, the heart cannot pump blood effectively after a massive blood loss. If the heart cannot pump and blood cannot flow, oxygen and nutrients cannot be distributed. In a very short time, the cardiac muscle tissue begins to break down as individual muscle cells die from oxygen and nutrient starvation. These changes will not be restricted to the cardiovascular system; all the cells, tissues, and organs in the body will be damaged.

☑ What characteristics of life are white blood cells exhibiting when they migrate to an injury site in response to chemicals released from damaged cells?

☑ A histologist investigates structures at what level of organization?

☑ A researcher studies the factors that cause heart failure. What might this specialty be called?

AN INTRODUCTION TO ORGAN SYSTEMS

Figure 1-2● introduces the 11 organ systems in the human body: the **integumentary system**, the **skeletal system**, the **muscular system**, the **nervous system**, the **endocrine system**, the **cardiovascular system**, the **lymphatic system**, the **respiratory system**, the **digestive system**, the **urinary system**, and the **reproductive system**.

•FIGURE 1-1   **Levels of Organization.**
Interacting atoms form molecules that
combine in the protein fibers of heart
muscle cells. These cells interlock, creating
heart muscle tissue that constitutes most
of the walls of a three-dimensional organ,
the heart. The heart is one component of
the cardiovascular system, which also
includes the blood and blood vessels. The
combined organ systems form an
organism, a living human being.

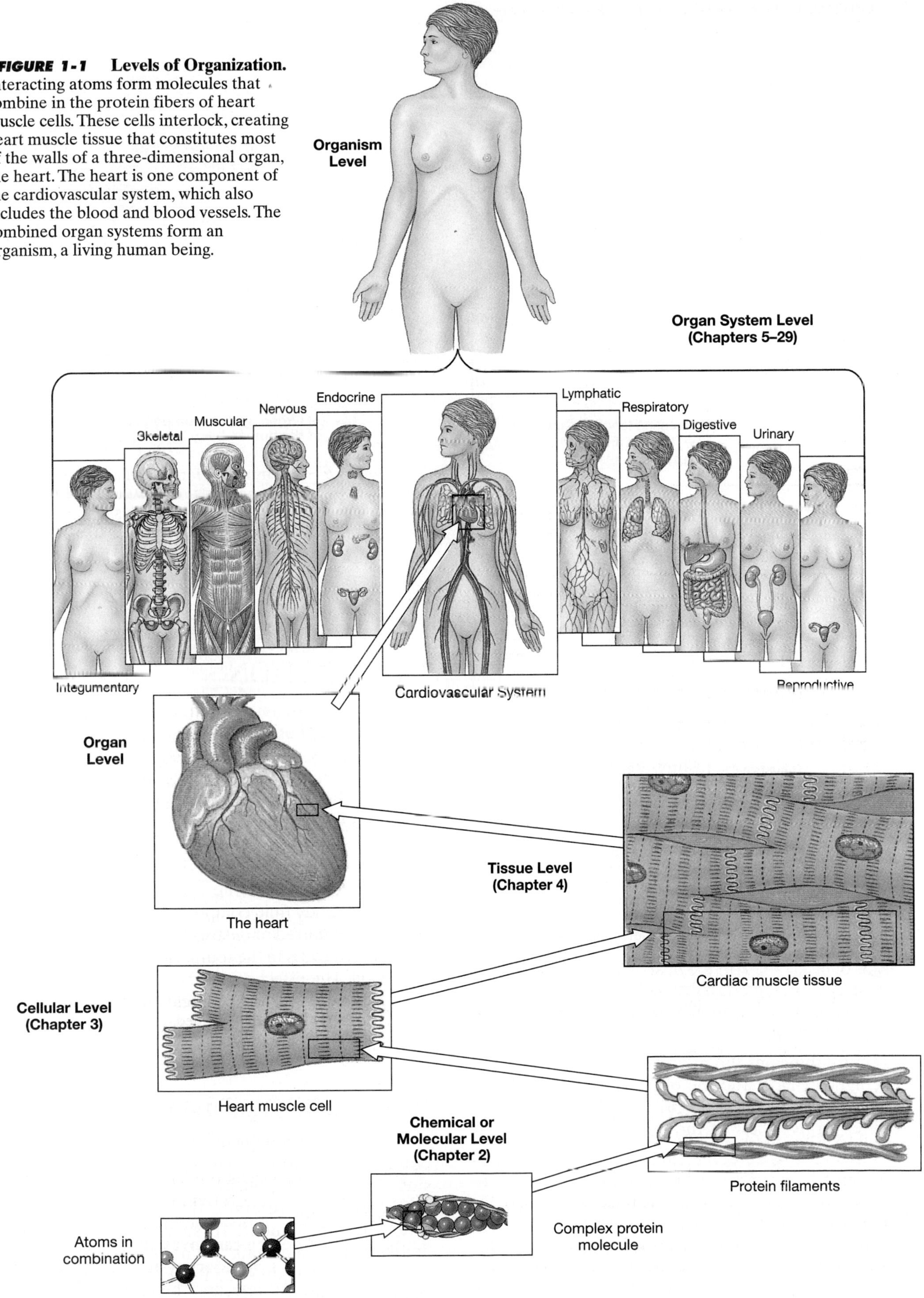

**Organism
Level**

**Organ System Level
(Chapters 5–29)**

Skeletal

Muscular

Nervous

Endocrine

Lymphatic

Respiratory

Digestive

Urinary

Integumentary

Cardiovascular System

Reproductive

**Organ
Level**

The heart

**Tissue Level
(Chapter 4)**

Cardiac muscle tissue

**Cellular Level
(Chapter 3)**

Heart muscle cell

**Chemical or
Molecular Level
(Chapter 2)**

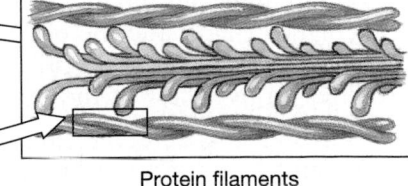

Protein filaments

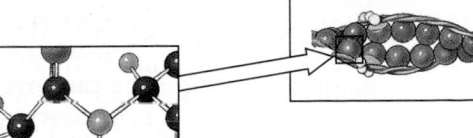

Atoms in
combination

Complex protein
molecule

Organ System		Major Functions
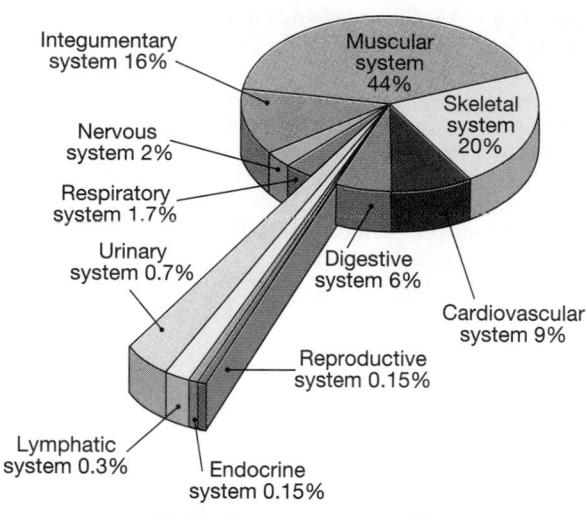 Integumentary system		Protection from environmental hazards; temperature control
Skeletal system		Support; protection of soft tissues; mineral storage; blood formation
Muscular system		Locomotion; support; heat production
Nervous system		Directing immediate responses to stimuli, generally by coordinating the activities of other organ systems
Endocrine system		Directing long-term changes in the activities of other organ systems
Cardiovascular system		Internal transport of cells and dissolved materials, including nutrients, wastes, and gases
Lymphatic system		Defense against infection and disease
Respiratory system		Delivery of air to sites where gas exchange can occur between the air and circulating blood
Digestive system		Processing of food and absorption of nutrients, minerals, vitamins, and water
Urinary system		Elimination of excess water, salts, and waste products
Reproductive system		Production of sex cells and hormones

(a)

(b) Organ system composition of the human body by weight

Integumentary system 16%
Muscular system 44%
Skeletal system 20%
Nervous system 2%
Respiratory system 1.7%
Urinary system 0.7%
Digestive system 6%
Cardiovascular system 9%
Reproductive system 0.15%
Lymphatic system 0.3%
Endocrine system 0.15%

●*FIGURE 1-2* **An Introduction to Organ Systems.** Here you see **(a)** the major functions of each system and **(b)** the relative percentage each system contributes to total body weight.

This figure indicates the major functions of each system (Figure 1-2a●) and the percentage of total body weight each represents (Figure 1-2b●). Figure 1-3● provides an overview of the individual systems and their major components.

☑ The ability to move and produce heat are major functions of which body system?

☑ Individuals suffering from AIDS exhibit an impaired ability to defend themselves against infection. Which organ system is affected?

HOMEOSTASIS AND SYSTEM INTEGRATION

Organ systems are interdependent, interconnected, and packaged together in a relatively small space. The cells, tissues, organs, and systems of the body live together in a shared environment, like the inhabitants of a large city. City dwellers breathe the city air and drink the water provided by the local water company; cells in the human body absorb oxygen and nutrients from fluids that surround them. If a city is blanketed in smog or the water is contaminated, the inhabitants will become ill. Similarly, if body fluid composition becomes abnormal, cells will be injured or destroyed. For example, suppose there are changes in the temperature or salt content of the blood. The effect on the heart could range from a minor adjustment (heart muscle tissue contracts more often, and the heart rate goes up) to a total disaster (the heart stops beating, and the individual dies).

Homeostatic Regulation

A variety of physiological mechanisms act to prevent potentially disruptive changes in the environment inside the body. **Homeostasis** (*homeo*, unchanging + *stasis*, standing) refers to the existence of a stable internal environment. To survive, every living organism must maintain homeostasis. *Homeostatic regulation* is the adjustment of physiological systems to preserve homeostasis.

•*FIGURE 1-3*
The Organ Systems of the Human Body

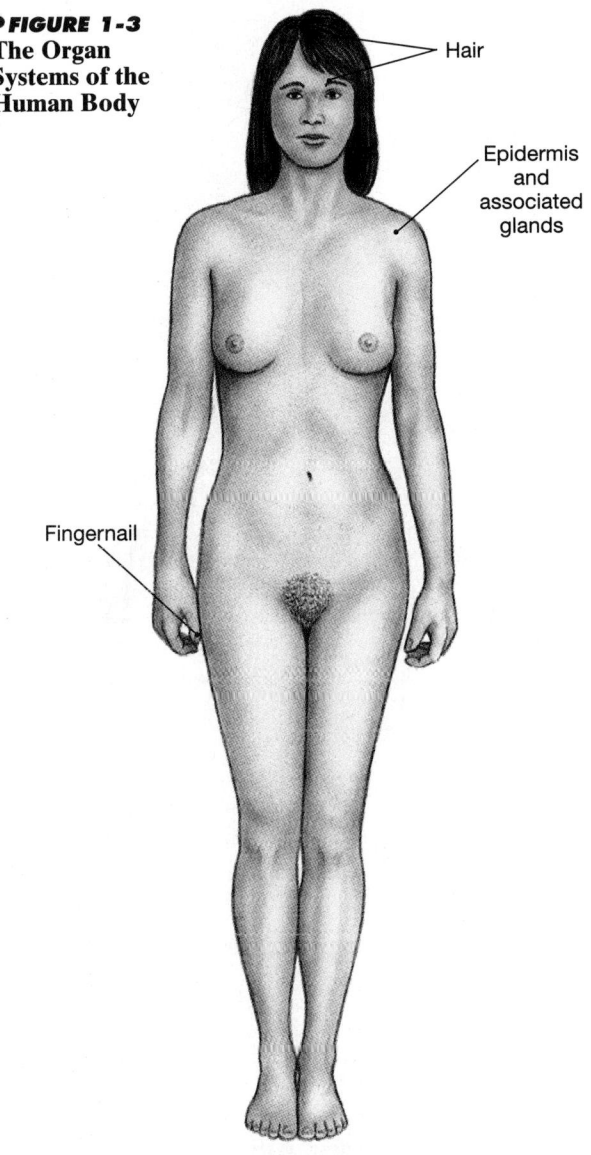

(a) **The Integumentary System**

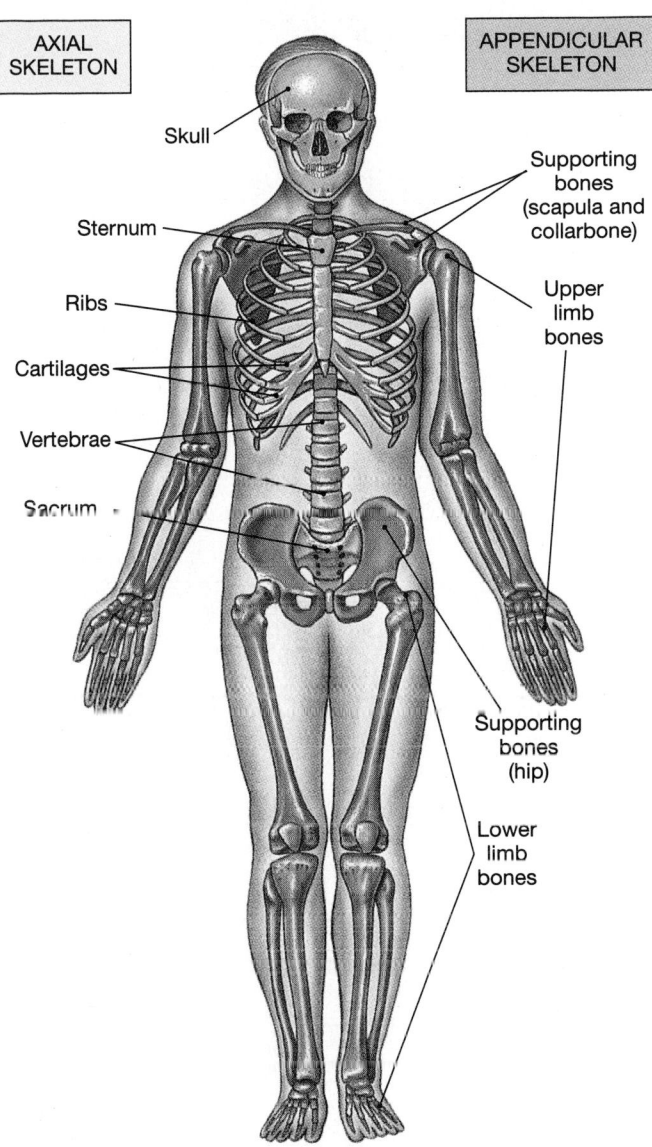

(b) **The Skeletal System**

Organ	Primary Functions
CUTANEOUS MEMBRANE	
Epidermis	Covers surface; protects underlying tissues
Dermis	Nourishes epidermis; provides strength; contains glands
HAIR FOLLICLES	Produce hair
Hairs	Provide sensation; provide some protection for head
Sebaceous glands	Secrete lipid coating that lubricates hair shaft
SWEAT GLANDS	Produce perspiration for evaporative cooling
NAILS	Protect and stiffen distal tips of digits
SENSORY RECEPTORS	Provide sensations of touch, pressure, temperature, pain
SUBCUTANEOUS LAYER	Stores lipids; attaches skin to deeper structures

Organ	Primary Functions
BONES (206), CARTILAGES, AND LIGAMENTS	Support, protect soft tissues; store minerals
Axial skeleton (skull, vertebrae, sacrum, ribs, sternum)	Protects brain, spinal cord, sense organs, and soft tissues of thoracic cavity; supports the body weight over the lower limbs
Appendicular skeleton (limbs and supporting bones)	Provides internal support and positioning of the limbs; supports and moves axial skeleton
BONE MARROW	Primary site of blood cell production

•*FIGURE 1-3 (continued)*

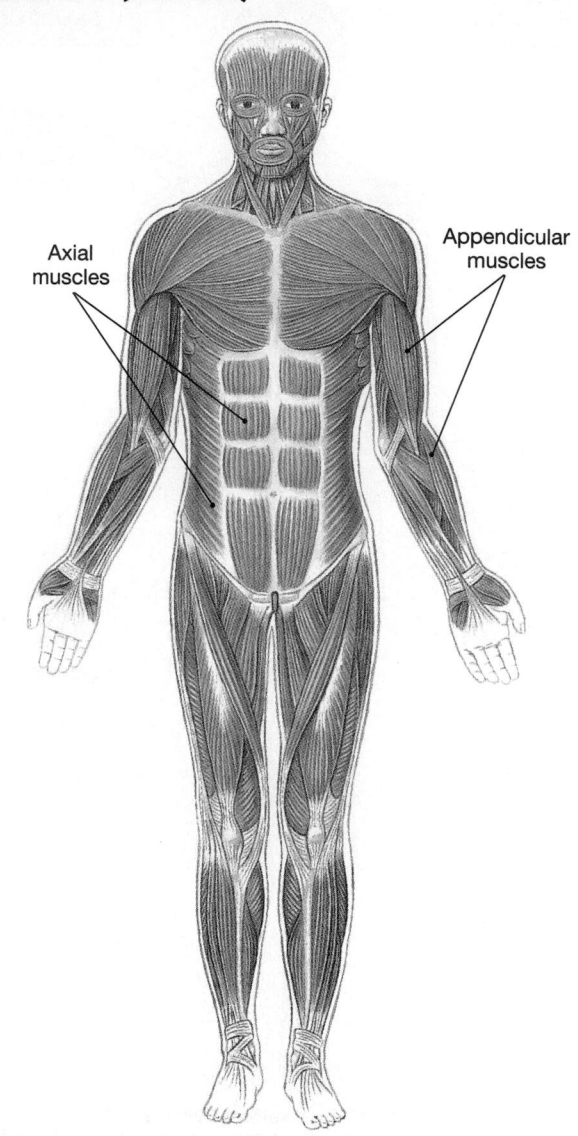

Axial muscles

Appendicular muscles

(c)   The Muscular System

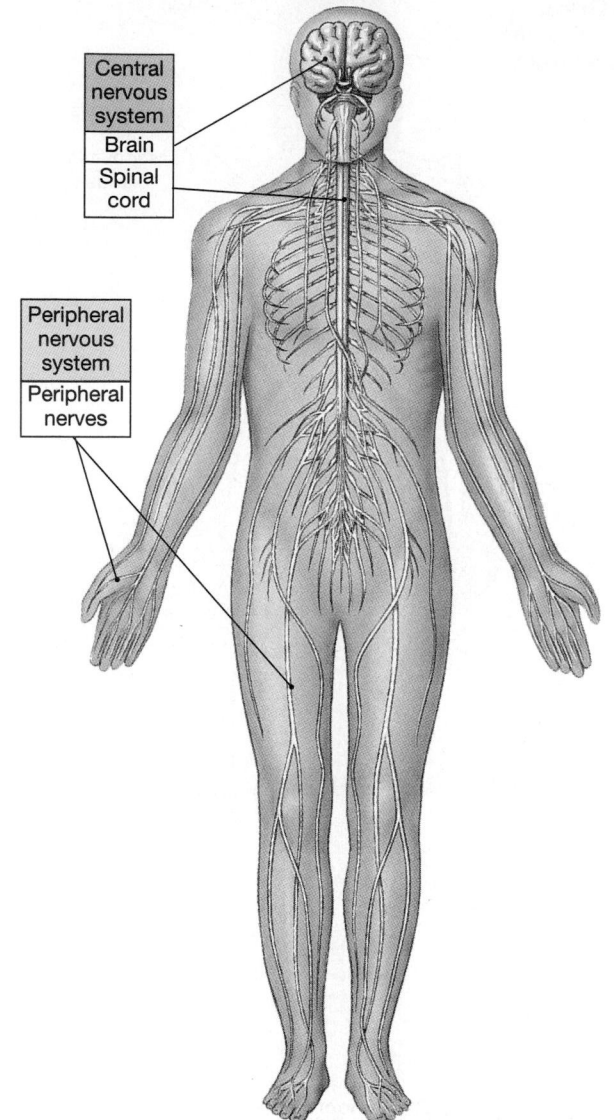

Central nervous system

Brain

Spinal cord

Peripheral nervous system

Peripheral nerves

(d)   The Nervous System

Organ	Primary Functions
SKELETAL MUSCLES (700)	Provide skeletal movement; control entrances and exits of digestive tract; produce heat; support skeletal position; protect soft tissues
Axial muscles	Support and position axial skeleton
Appendicular muscles	Support, move, and brace limbs
TENDONS, APONEUROSES	Harness forces of contraction to perform specific tasks

Organ	Primary Functions
CENTRAL NERVOUS SYSTEM (CNS)	Control center for nervous system: processes information; provides short-term control over activities of other systems
Brain	Performs complex integrative functions; controls voluntary activities
Spinal cord	Relays information to and from the brain; performs less-complex integrative functions; directs many simple involuntary activities
PERIPHERAL NERVOUS SYSTEM (PNS)	Links CNS with other systems and with sense organs

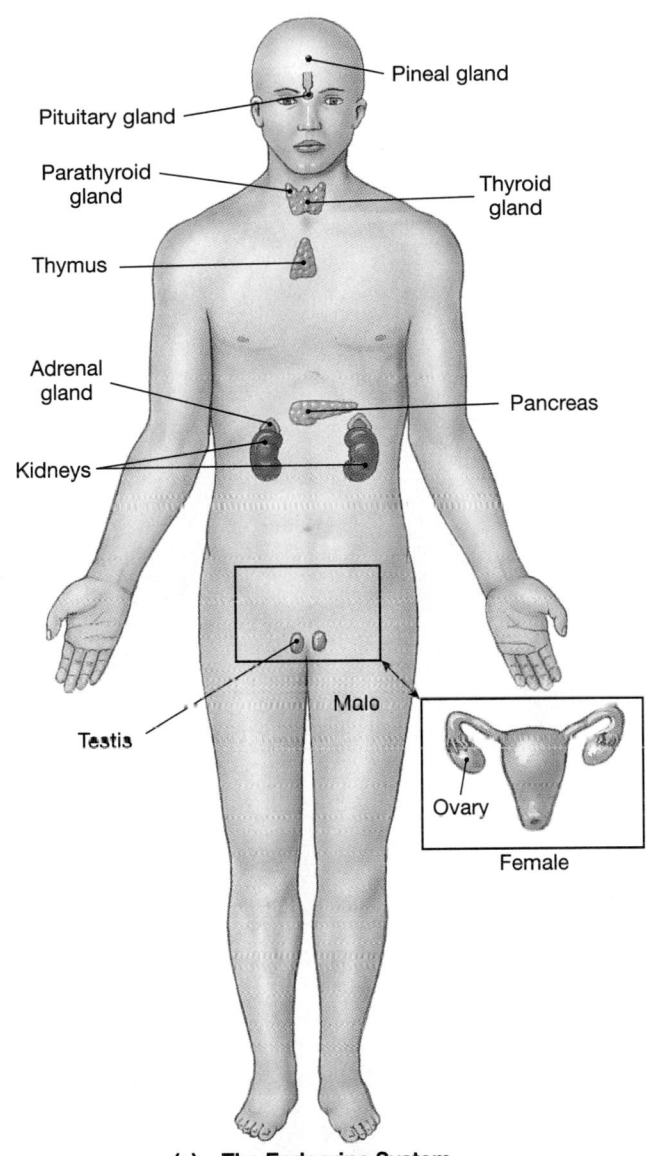

(e) The Endocrine System

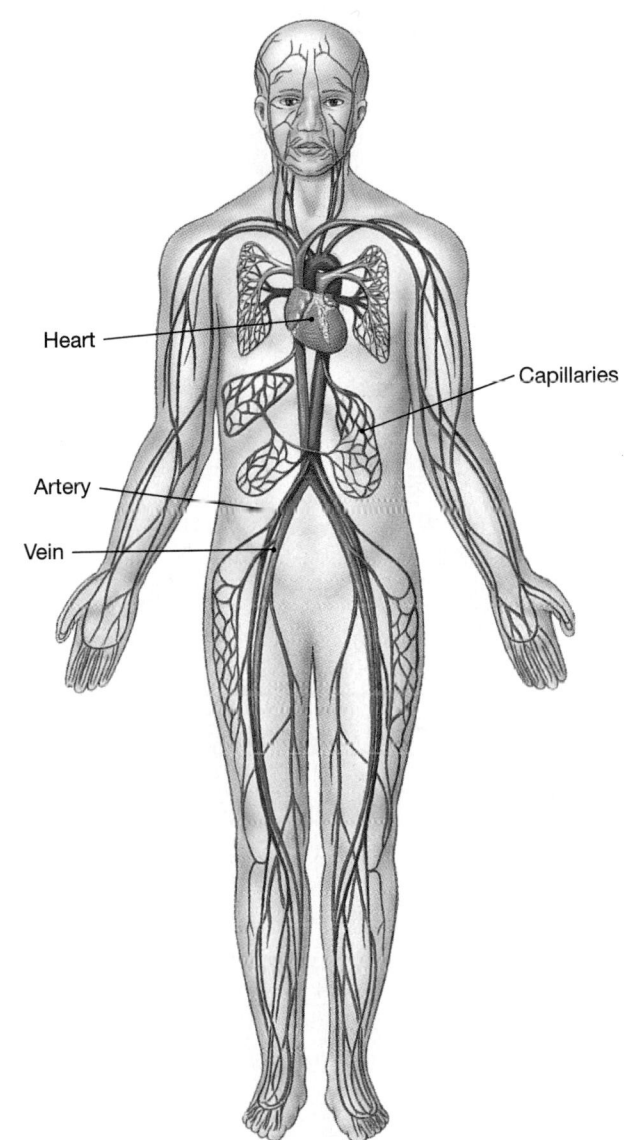

(f) The Cardiovascular System

Organ	Primary Functions
PINEAL GLAND	May control timing of reproduction and set day/night rhythms
PITUITARY GLAND	Controls other endocrine glands; regulates growth and fluid balance
THYROID GLAND	Controls tissue metabolic rate; regulates calcium levels
PARATHYROID GLAND	Regulates calcium levels (with thyroid)
THYMUS	Controls maturation of lymphocytes
ADRENAL GLANDS	Adjust water balance, tissue metabolism, and cardiovascular and respiratory activity
KIDNEYS	Control red blood cell production and elevate blood pressure
PANCREAS	Regulates blood glucose levels
GONADS	
Testes	Support male sexual characteristics and reproductive functions *(see Figure 1-3k)*
Ovaries	Support female sexual characteristics and reproductive functions *(see Figure 1-3k)*

Organ	Primary Functions
HEART	Propels blood; maintains blood pressure
BLOOD VESSELS	Distribute blood around the body
Arteries	Carry blood from heart to capillaries
Capillaries	Site of diffusion between blood and interstitial fluids
Veins	Return blood from capillaries to the heart
BLOOD	Transports oxygen, carbon dioxide, and blood cells; delivers nutrients and hormones; removes waste products; assists in temperature regulation and defense against disease

●*FIGURE 1-3 (continued)*

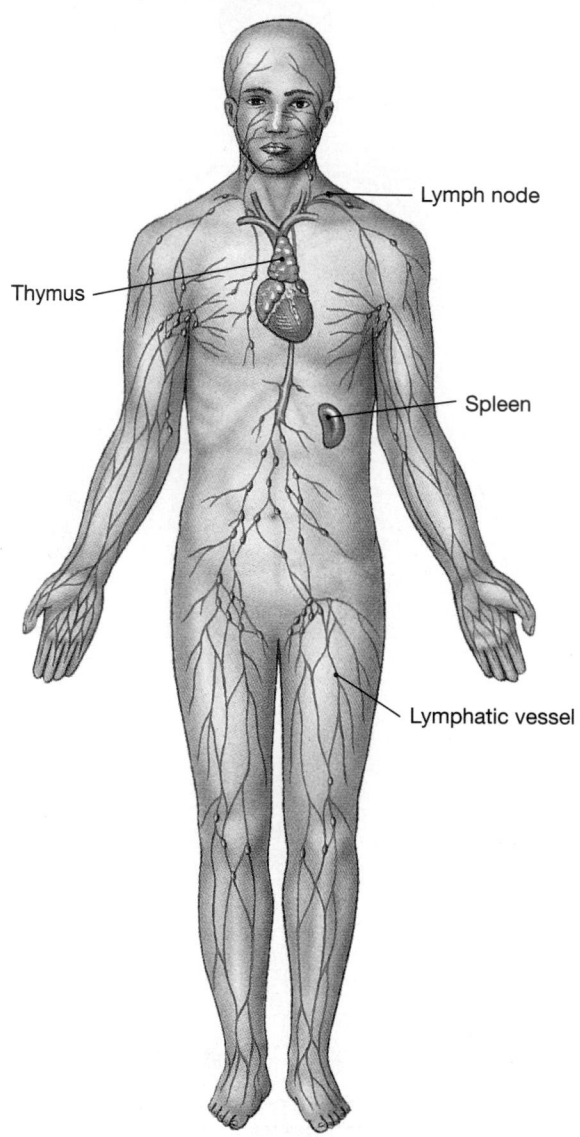

(g) **The Lymphatic System**

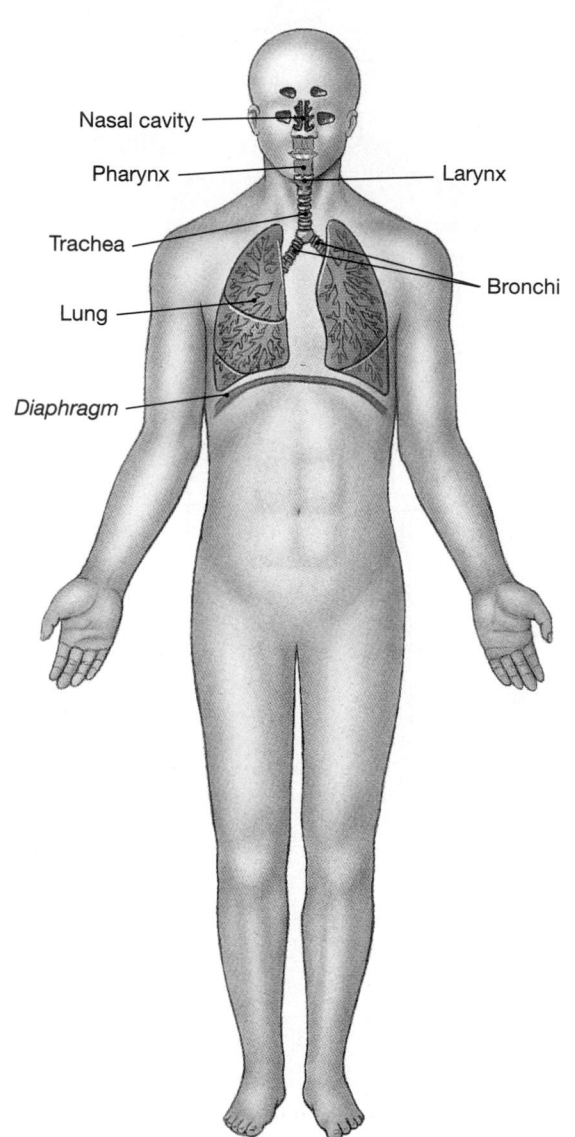

(h) **The Respiratory System**

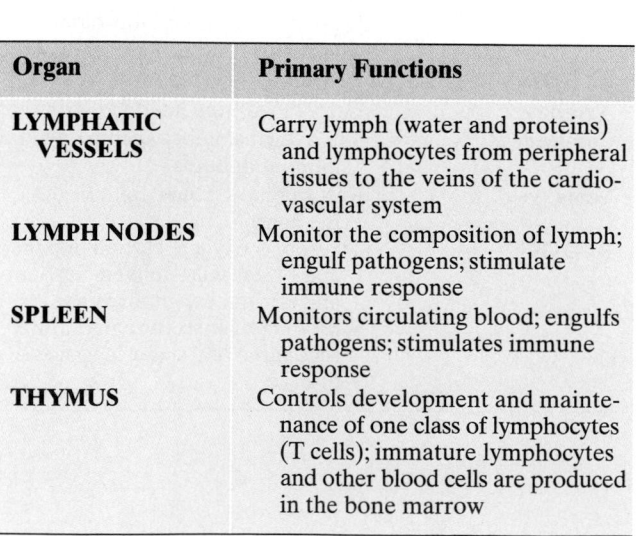

Organ	Primary Functions
LYMPHATIC VESSELS	Carry lymph (water and proteins) and lymphocytes from peripheral tissues to the veins of the cardio-vascular system
LYMPH NODES	Monitor the composition of lymph; engulf pathogens; stimulate immune response
SPLEEN	Monitors circulating blood; engulfs pathogens; stimulates immune response
THYMUS	Controls development and mainte-nance of one class of lymphocytes (T cells); immature lymphocytes and other blood cells are produced in the bone marrow

Organ	Primary Functions
NASAL CAVITIES, PARANASAL SINUSES	Filter, warm, humidify air; detect smells
PHARYNX	Chamber shared with digestive tract; conducts air to larynx
LARYNX	Protects opening to trachea and contains vocal cords
TRACHEA	Filters air, traps particles in mucus; cartilages keep airway open
BRONCHI	Same functions as trachea
LUNGS	Include airways and alveoli; volume changes due to move-ments of ribs and diaphragm are responsible for air movement
ALVEOLI	Sites of gas exchange between air and blood

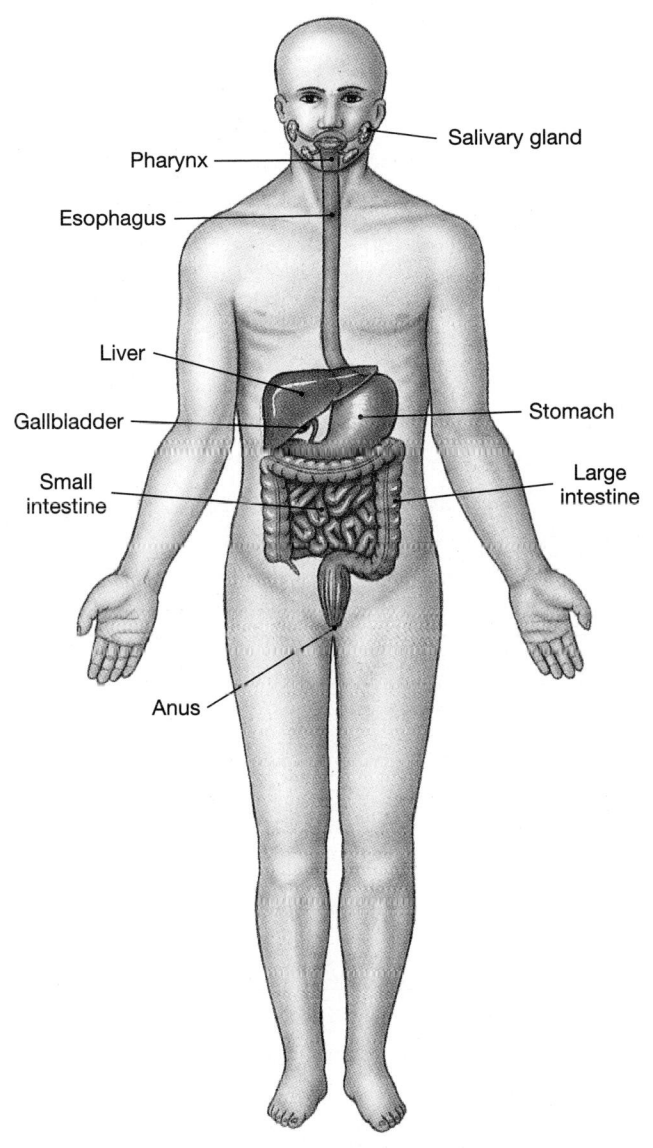

(i) The Digestive System

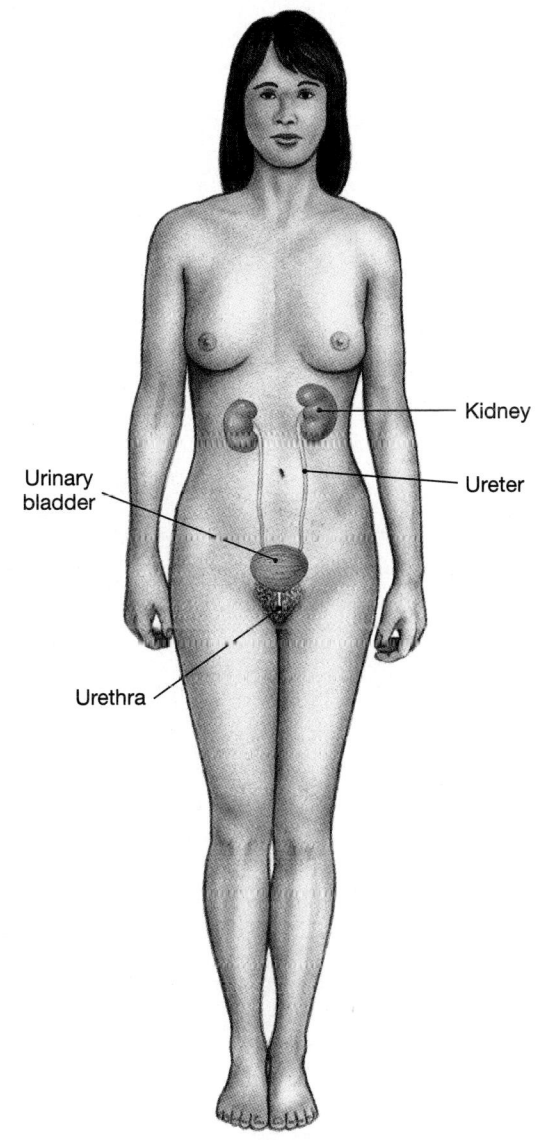

(j) The Urinary System

Organ	Primary Functions
SALIVARY GLANDS	Provide buffers and lubrication; produce enzymes that begin digestion
PHARYNX	Passageway connected to esophagus and trachea
ESOPHAGUS	Delivers food to stomach
STOMACH	Secretes acids and enzymes
SMALL INTESTINE	Secretes digestive enzymes, buffers, and hormones; absorbs nutrients
LIVER	Secretes bile; regulates blood composition of nutrients
GALLBLADDER	Stores bile for release into small intestine
PANCREAS	Secretes digestive enzymes and buffers; contains endocrine cells (*see Figure 1-3e*)
LARGE INTESTINE	Removes water from fecal material; stores wastes

Organ	Primary Functions
KIDNEYS	Form and concentrate urine; regulate blood pH and ion concentrations; endocrine functions noted in *Figure 1-3e*
URETERS	Conduct urine from kidneys to urinary bladder
URINARY BLADDER	Stores urine for eventual elimination
URETHRA	Conducts urine to exterior

● *FIGURE 1-3* **(continued)**

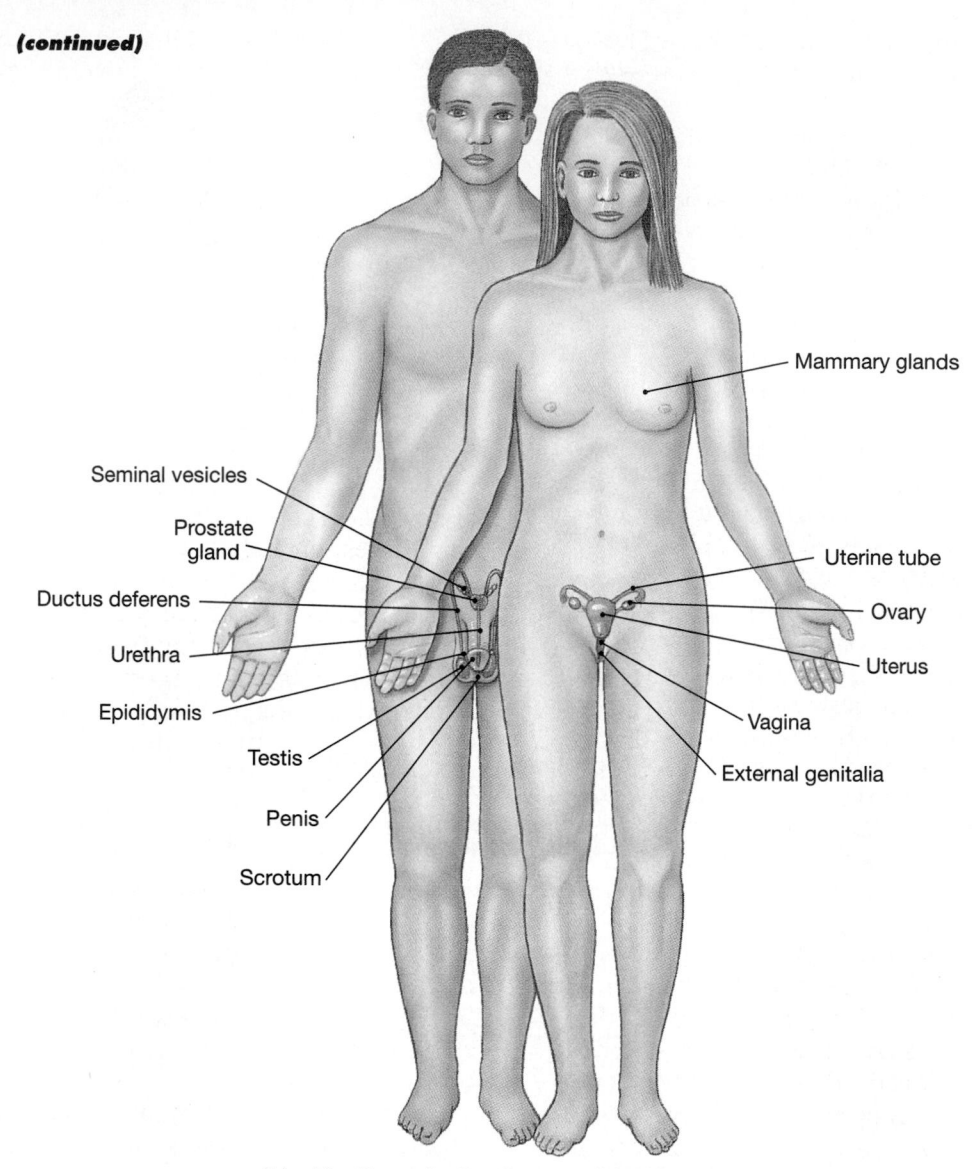

Mammary glands

Seminal vesicles

Prostate gland

Ductus deferens

Urethra

Epididymis

Testis

Penis

Scrotum

Uterine tube

Ovary

Uterus

Vagina

External genitalia

(k)   The Reproductive System of the Male and Female

Organ	Primary Functions
TESTES	Produce sperm and hormones (*see Figure 1-3e*)
ACCESSORY ORGANS	
Epididymis	Site of sperm maturation
Ductus deferens (sperm duct)	Conducts sperm between epididymis and prostate
Seminal vesicles	Secrete fluid that makes up much of the volume of semen
Prostate gland	Secretes fluid and enzymes
Urethra	Conducts semen to exterior
EXTERNAL GENITALIA	
Penis	Erectile organ used to deposit sperm in the vagina of a female; produces pleasurable sensations during sexual act
Scrotum	Surrounds the testes and controls their temperature

Organ	Primary Functions
OVARIES	Produce oocytes and hormones (*see Figure 1-3e*)
UTERINE TUBES	Deliver ova or embryo to uterus; normal site of fertilization
UTERUS	Site of embryonic development and diffusion between maternal and embryonic bloodstreams
VAGINA	Site of sperm deposition; birth canal at delivery; provides passage of fluids during menstruation
EXTERNAL GENITALIA	
Clitoris	Erectile organ; produces pleasurable sensations during sexual act
Labia	Contain glands that lubricate entrance to vagina
MAMMARY GLANDS	Produce milk that nourishes newborn infant

Homeostasis is absolutely vital; a failure of homeostatic regulation leads to illness or even death within a relatively short period. The principle of homeostasis is the central theme of this text and the foundation for all modern physiology. An understanding of homeostatic regulation will help you make accurate predictions about the body's responses to both normal and abnormal conditions.

Two general mechanisms are involved in homeostatic regulation: *autoregulation* and *extrinsic regulation*:

1. **Autoregulation** occurs when the activities of a cell, tissue, organ, or organ system change automatically when faced with some environmental variation. For example, when the cells within a tissue need more oxygen, they release chemicals that dilate blood vessels in the area. This dilation increases the rate of blood flow and provides more oxygen to the region.

2. **Extrinsic regulation** results from the activities of the nervous system or endocrine system, organ systems that can control or adjust the activities of many different systems simultaneously. For example, when you are exercising, your nervous system issues commands that increase your heart rate so that blood will circulate faster. Your nervous system also reduces blood flow to organs, such as the digestive tract, that are relatively inactive. The oxygen in the circulating blood is thus saved for the active muscles.

In general, the nervous system performs crisis management by directing rapid, short-term, and very specific responses. When you accidentally set your hand on a hot stove, the rising temperature produces a painful, localized disturbance of homeostasis. Your nervous system responds by ordering the contraction of specific muscles that will pull your hand away from the stove. The effects last only as long as the neural activity continues, usually a matter of seconds.

In contrast, the endocrine system releases chemical messengers, called *hormones*, that affect tissues and organs throughout the body. The responses may not be immediately apparent, but when the effects appear, they often persist for days or weeks. Examples of endocrine function include the long-term regulation of blood volume and composition and the adjustment of organ system function during development or during starvation.

Regardless of the system involved, homeostatic regulation always attempts to keep the characteristics of the internal environment within desirable limits. The regulatory mechanism consists of three parts: (1) a **receptor**, a sensor that is sensitive to a particular environmental change, or *stimulus*; (2) a **control center**, or *integration center*, which receives and processes the information supplied by the receptor;

and (3) an **effector**, a cell or organ that responds to the commands of the control center and whose activity opposes or enhances the stimulus. You are probably already familiar with comparable regulatory mechanisms. As an example, consider the operation of the thermostat in your home or apartment (Figure 1-4●).

A thermostat is a control center that receives information about room temperature from an internal or remote thermometer (a receptor). The dial on the thermostat establishes the *set point*, or desired value, which in this case is the temperature you prefer. In our example, the set point is 22°C (about 72°F). The function of the thermostat is to keep room temperature within acceptable limits, usually within a degree or so of the set point. In summer the thermostat accomplishes this function by controlling an air conditioner (an effector). When the temperature at the thermometer rises above the acceptable range, the thermostat turns on the air conditioner. The air conditioner then cools the room (Figure 1-4a●). When the temperature at the thermometer approaches the set point, the thermostat turns off the air conditioner (Figure 1-4b●).

Negative Feedback

The essential feature of thermostatic temperature control can be summarized very simply: A variation outside the desired range triggers an automatic response that corrects the situation. This method of homeostatic regulation is called **negative feedback** because the effector(s) activated by the control center either oppose or eliminate the stimulus.

Most homeostatic regulatory mechanisms involve negative feedback. For example, consider the control of body temperature, a process called *thermoregulation*. Thermoregulation involves altering the relationship between heat loss, which occurs primarily at the body surface, and heat production, which occurs in all active tissues.

The thermoregulatory control center is located in the brain (Figure 1-5a●, p. 15) in a region known as the *hypothalamus*. This control center receives information from two sets of temperature receptors, one in the skin and the other inside the thermoregulatory center. At the normal set point, the general body temperature will be approximately 37°C (98.6°F). If body temperature rises above 37.2°C, activity in the control center targets two different effectors: (1) muscle tissue in the walls of blood vessels supplying the skin and (2) sweat glands. The muscle tissue relaxes and the blood vessels dilate, increasing blood flow through vessels near the body surface, and the sweat glands accelerate their secretion. The skin then acts like a radiator, losing heat to the en-

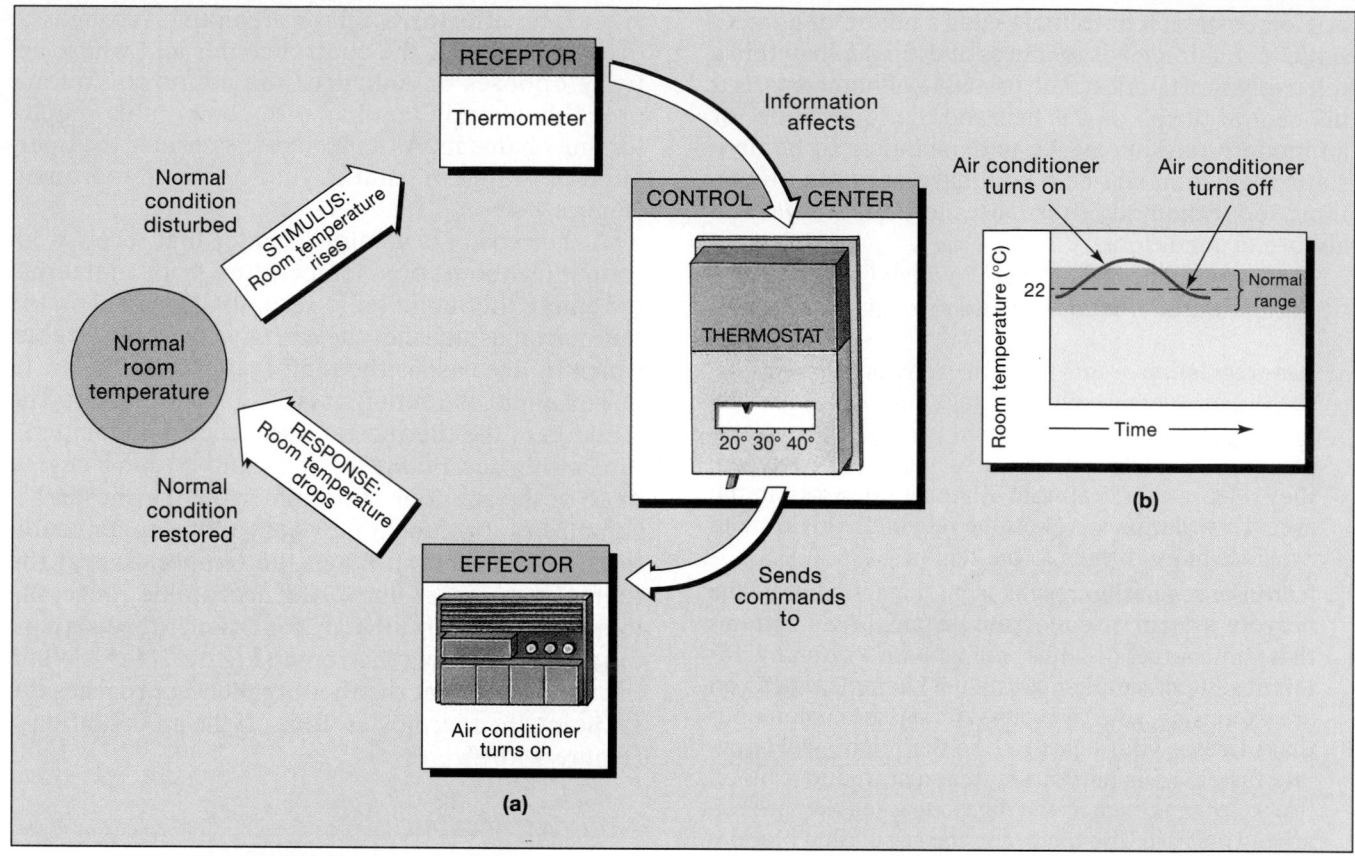

•FIGURE 1-4 **The Control of Room Temperature. (a)** A thermostat stabilizes room temperature by turning on an air conditioner (or heater) as needed to keep room temperature near the desired set point. Whether the room temperature rises or falls, the thermostat (a control center) triggers an effector response that restores normal temperature. When room temperature rises, the thermostat turns on the air conditioner, and the room temperature returns to normal levels. **(b)** With this regulatory system, room temperature oscillates around the set point.

vironment, and the evaporation of sweat speeds the process. When body temperature returns to normal, the thermoregulatory control center becomes less active, and superficial blood flow and sweat gland activity decrease to previous levels.

Negative feedback is the primary mechanism of homeostatic regulation, and it provides long-term control over internal conditions and systems. Homeostatic mechanisms using negative feedback normally ignore minor variations, and they maintain a normal *range* rather than a fixed value. In the previous example, body temperature oscillated around the set-point temperature (Figure 1-5b•). The regulatory process itself is dynamic, because the set point may vary with changing environments or differing activity levels. For example, when you are asleep, your thermoregulatory set point drifts lower, whereas when you work outside on a hot day (or when you have a fever), it climbs. Thus body temperature can vary from moment to moment or day to day for any

individual due to (1) small oscillations around the set point or (2) changes in the set point. Comparable variations occur in all other physiological characteristics.

The variability among individuals is even greater, for each person has homeostatic set points that may differ due to genetic factors, age, gender, general health, or environmental conditions. It is therefore impractical to define "normal" homeostatic conditions very precisely. By convention, physiological values are reported either as averages, the average value obtained by sampling a large number of individuals, or as a range that includes 95 percent or more of the sample population. For instance, 5 percent of healthy adults have resting body temperatures that are below 36.7°C or above 37.2°C. But these temperatures are perfectly normal for them, and the variations have no clinical significance. Physicians must keep this variability in mind when they review lab reports or clinical discussions, be-

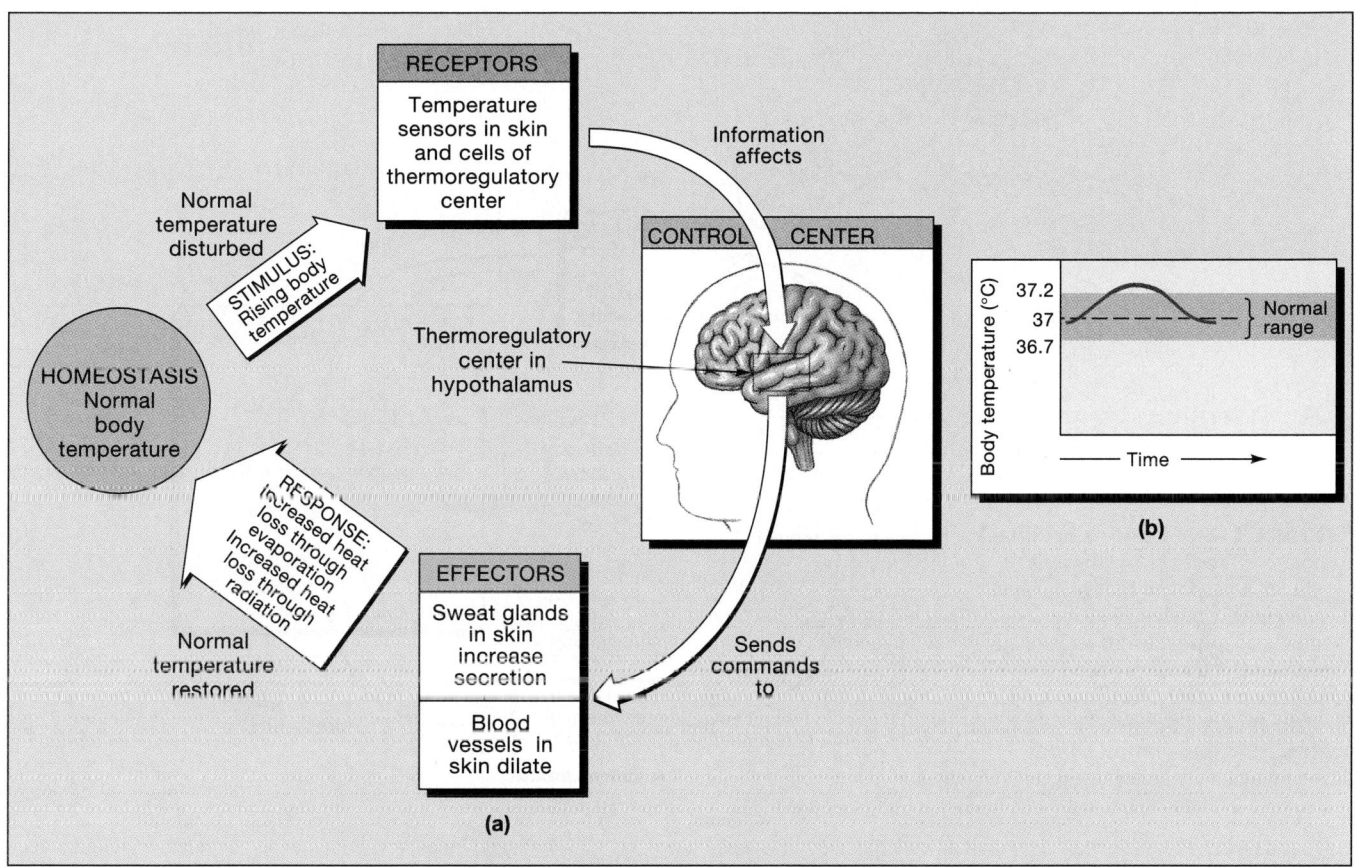

●**FIGURE 1-5** Negative Feedback in Body Temperature Control. **(a)** Events comparable to those shown in Figure 1-4 occur in the regulation of body temperature. A control center in the brain functions as a thermostat with a set point of 37°C. If body temperature climbs above 37.2°C, heat loss is increased through enhanced blood flow to the skin and increased sweating. **(b)** The thermoregulatory center keeps body temperature oscillating within an acceptable range, usually between 36.7 and 37.2°C.

cause unusual values—even those outside the normal range—may represent individual variation rather than disease.

Positive Feedback

In **positive feedback**, the initial stimulus produces a response that exaggerates or enhances its effects. For example, suppose that the thermostat in Figure 1-4a● is rewired so that it is connected to a heater rather than to an air conditioner. Now when room temperature rises, the thermostat turns on the heater, causing a further rise in room temperature. Room temperature will continue to rise until someone switches off the thermostat, turns off the heater, or intervenes in some other way.

Control mechanisms relying on positive feedback are not nearly as common as those involving negative feedback. However, positive feedback is important in controlling physiological processes that must

be completed quickly. For example, consider the process of labor that results in the delivery of a newborn infant. Once labor begins, it is important that it be completed as soon as possible to limit stress that can be dangerous to both mother and infant. The primary stimulus initiating labor and delivery is distortion of the uterus by the growing fetus. The trigger for the initiation of labor contractions is a rise in the uterine level of oxytocin (oks-ē-TŌ-sin), a hormone that stimulates the contractions of uterine muscles. One source of oxytocin, and the regulatory mechanism that controls its secretion, has been diagrammed in Figure 1-6●. Stretch receptors in the uterine wall are monitored by an endocrine control center in the brain. When sufficient uterine distortion occurs, the control center releases oxytocin. As uterine contractions occur, the fetus is pushed toward the vagina. This movement causes extreme distortion of the lower portion of the uterus, triggering the release of additional oxytocin. The uterine contractions then

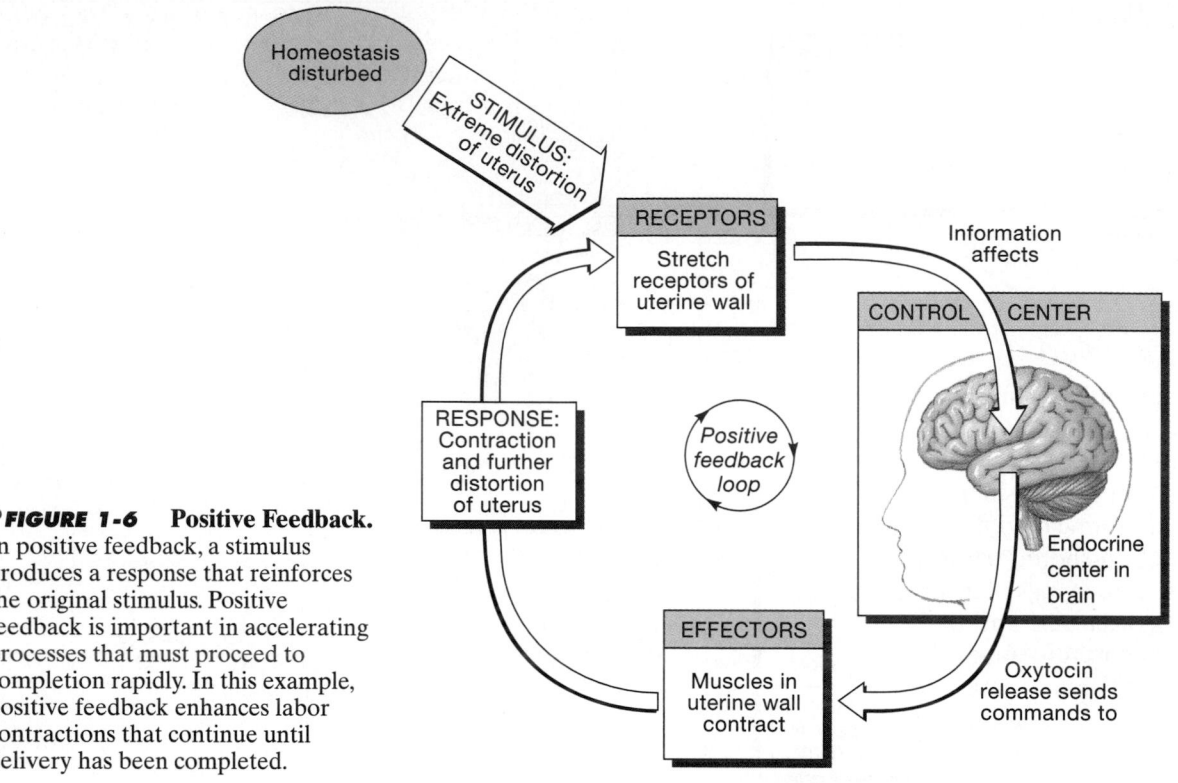

•FIGURE 1-6   Positive Feedback.
In positive feedback, a stimulus
produces a response that reinforces
the original stimulus. Positive
feedback is important in accelerating
processes that must proceed to
completion rapidly. In this example,
positive feedback enhances labor
contractions that continue until
delivery has been completed.

become more forceful, leading to greater movement
and distortion. Each time the control center responds,
the action of the effectors causes an increase in receptor stimulation. This kind of cycle, a positive feedback loop, can be broken only by some external force
or process—in this instance, the delivery of the newborn infant, which eliminates the uterine distortion.
We shall consider labor and delivery in detail in
Chapter 29 and blood clotting, another important example of positive feedback, in Chapter 19.

HOMEOSTASIS AND DISEASE When homeostat-
ic regulation fails, whether as a result of infection, injury, or genetic abnormality, organ systems begin to
malfunction. The result is a state known as illness, or **disease**.
The chapters that follow devote considerable attention to the
mechanisms responsible for a variety of human diseases. An
understanding of normal homeostatic mechanisms can usually enable you to draw conclusions about what might be responsible for the *signs* and *symptoms* characteristic of many
diseases. Major organizational and functional patterns relevant to clinical practice are discussed in the *Applications
Manual*. ☒ *Homeostasis and Disease*

☑ Why is homeostatic regulation important to human beings?

☑ What happens to the body when homeostasis breaks down?

☑ Why is positive feedback helpful in blood clotting but unsuitable for regulation of body temperature?

A FRAME OF REFERENCE FOR ANATOMICAL STUDIES

Early anatomists faced serious problems in communication. Stating that a bump is "on the back" does
not give very precise information about its location.
So anatomists created maps of the human body. The
landmarks are prominent anatomical structures; distances are measured in centimeters or inches; and specialized directional terms are used. In effect, anatomy
uses a special language that must be learned almost at
the start.

A familiarity with Latin roots and patterns makes
anatomical terms more understandable. As new terms
are introduced, notes on pronunciation and relevant
word roots will be provided. Additional information
on word roots, prefixes, suffixes, and combining forms
can be found on the front endpapers.

Latin and Greek terms are not the only ones that
have been imported into the anatomical vocabulary
over the centuries, and the vocabulary continues to expand. Many anatomical structures and clinical conditions were initially named after either the discoverer
or, in the case of diseases, the most famous victim.
Over the last 100 years, most of these commemorative
names, or *eponyms*, have been replaced by more precise terms. Appendix V lists the most important
eponyms and related historical details.

Superficial Anatomy

A familiarity with anatomical landmarks and directional references will make subsequent chapters more understandable. As you encounter new terms, you must create your own mental maps with the information provided in the accompanying anatomical illustrations.

Anatomical Landmarks

Important anatomical landmarks are presented in Figure 1-7•. The anatomical terms are given in boldface, the common names in plain type, and the anatomical adjectives in parentheses. You should become familiar with all three types of terms. Understanding the terms and their origins will help you remember the location of a particular structure as well as its name. For example, the term *brachium* refers to the arm; in later chapters, we will consider the *brachialis muscle* and branches of the *brachial artery*.

Standard anatomical illustrations show the human form in the **anatomical position**. In this position the hands are at the sides with the palms facing forward. Figure 1-7 shows an individual in the anatomical position as seen from the front (Figure 1-7a•) and back (Figure 1-7b•). Unless otherwise noted, all the descriptions given in this text refer to the body in the anatomical position. A person lying down in the anatomical position is said to be **supine** (SOO-pīn) when lying face up and **prone** when lying face down.

Anatomical Regions

Major regions of the body are listed in Table 1-1• and shown in Figure 1-7•. In addition to specific landmarks, anatomists and clinicians

often need to use regional terms to describe a general area of interest or injury. Two methods are used, both concerned with mapping the surface of the abdominopelvic region. Clinicians refer to the

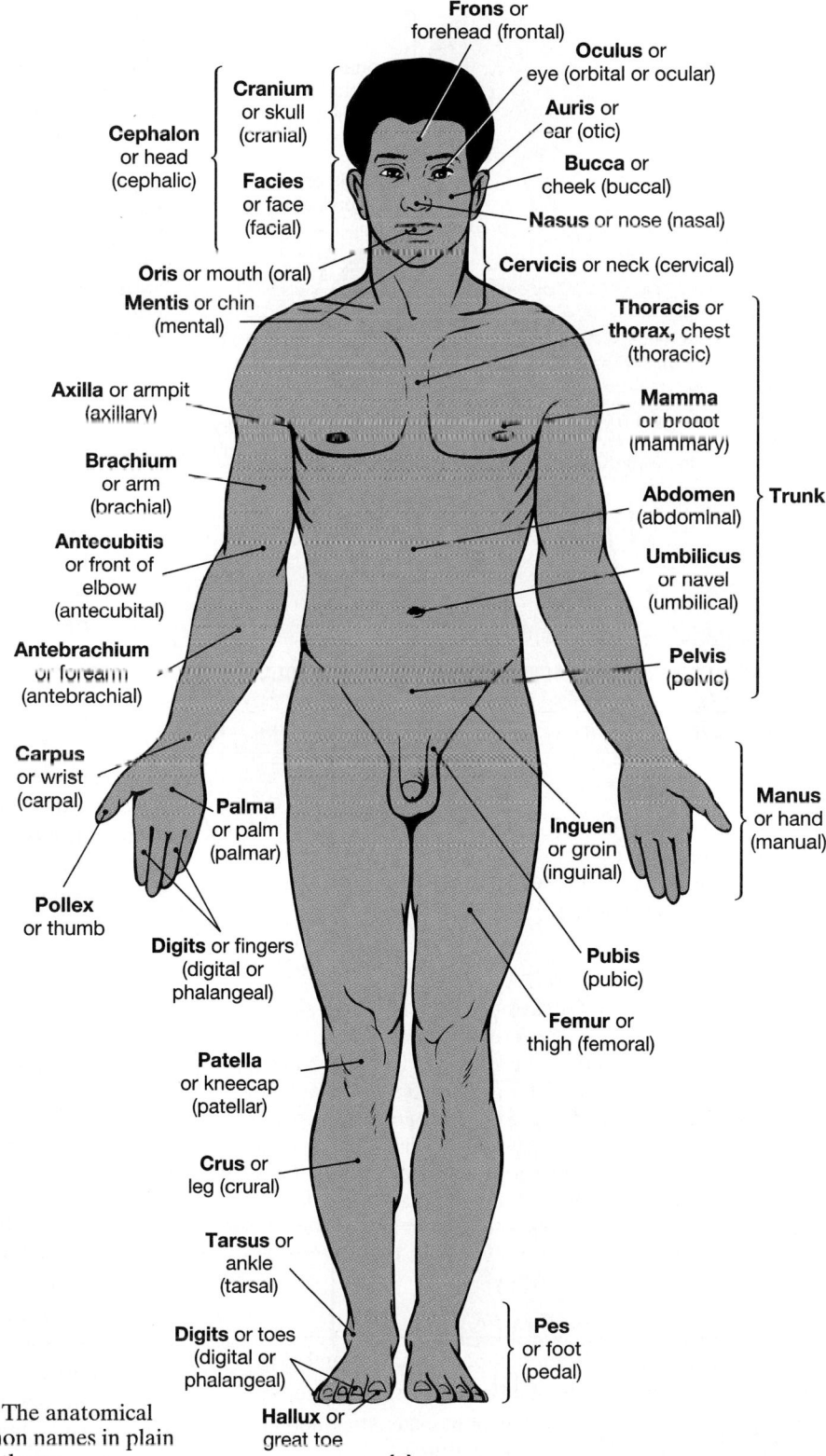

•**FIGURE 1-7** **Anatomical Landmarks.** The anatomical terms are shown in boldface type, the common names in plain type, and the anatomical adjectives in parentheses.

(a)

abdominopelvic quadrants. The region is divided into four segments by using a pair of imaginary lines that intersect at the umbilicus (navel). This simple method,

TABLE 1-1 Regions of the Human Body (*see Figure 1-7*)

Structure	Region
Cephalon (head)	Cephalic region
Cervicis (neck)	Cervical region
Thoracis (thorax or chest)	Thoracic region
Brachium (arm)	Brachial region
Antebrachium (forearm)	Antebrachial region
Manus (hand)	Manual region
Abdomen	Abdominal region
Lumbus (loin)	Lumbar region
Gluteus (buttock)	Gluteal region
Pelvis	Pelvic region
Pubis (anterior pelvis)	Pubic region
Inguen (groin)	Inguinal region
Femur (thigh)	Femoral region
Crus (anterior leg)	Crural region
Sura (calf)	Sural region
Pes (foot)	Pedal region
Planta (sole)	Plantar region

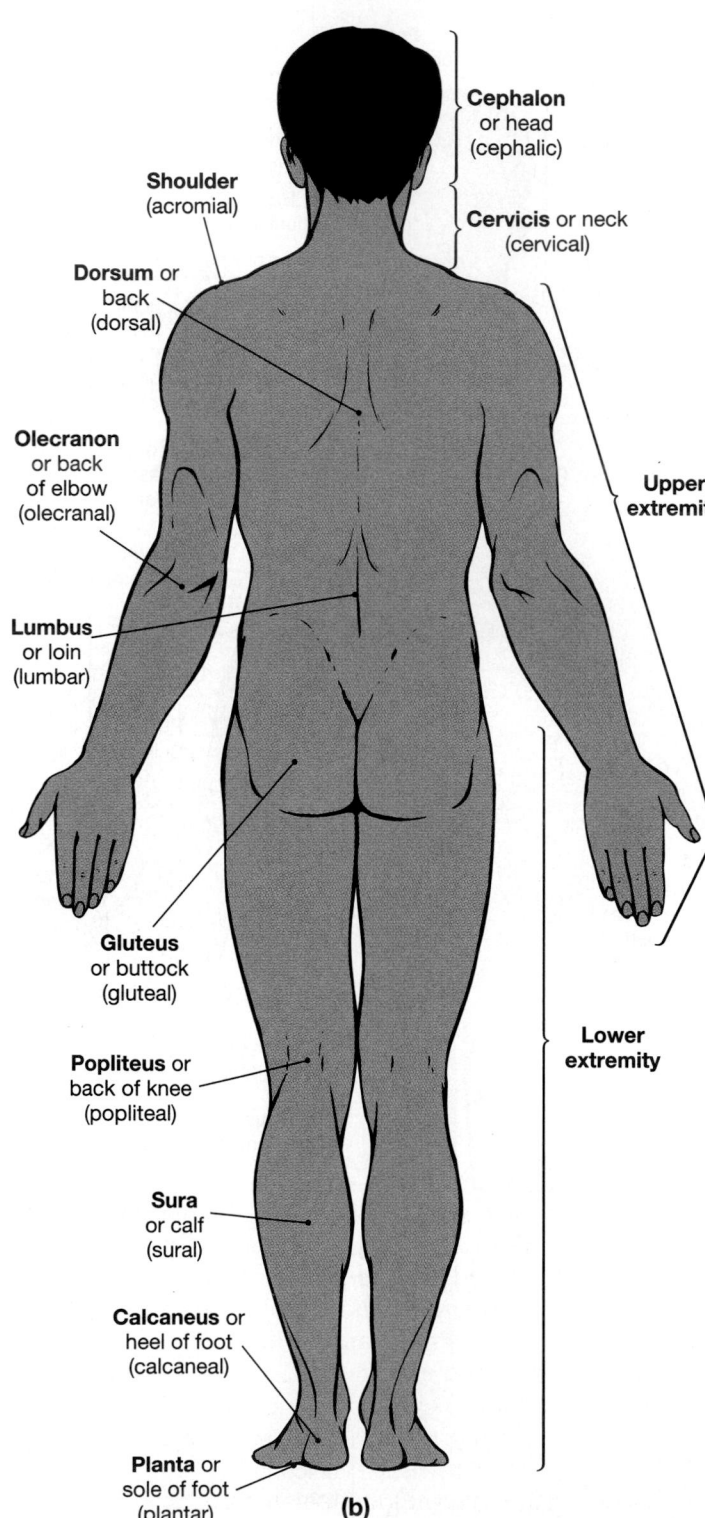

Cephalon or head (cephalic)

Shoulder (acromial)

Cervicis or neck (cervical)

Dorsum or back (dorsal)

Olecranon or back of elbow (olecranal)

Upper extremity

Lumbus or loin (lumbar)

Gluteus or buttock (gluteal)

Lower extremity

Popliteus or back of knee (popliteal)

Sura or calf (sural)

Calcaneus or heel of foot (calcaneal)

Planta or sole of foot (plantar)

(b)

•*FIGURE 1-7 (continued)*

shown in Figure 1-8a•, provides useful references for the description of aches, pains, and injuries. The location can help the physician decide the possible cause; for example, tenderness in the right lower quadrant (RLQ) is a symptom of appendicitis, whereas tenderness in the right upper quadrant (RUQ) may indicate gallbladder or liver problems.

Anatomists like to use more-precise regional distinctions to describe the location and orientation of internal organs. They recognize nine **abdominopelvic regions** (Figure 1-8b•). Figure 1-8c• shows the relationships among quadrants, regions, and internal organs.

Anatomical Directions

Figure 1-9• and Table 1-2 (p. 20) show the principal directional terms and examples of their use. There are many different terms, and some can be used interchangeably. For example, *anterior* refers to the front of the body viewed in the anatomical position; in human beings, this term is equivalent to *ventral*, which actually refers to the belly. Although your instructor may have additional recommendations, the terms that appear frequently in later chapters have been emphasized. Before you continue, review Table 1-2 in detail, and practice using these terms at every opportunity. If you are familiar with the basic vocabulary, the descriptions in future chapters will be easier to follow.

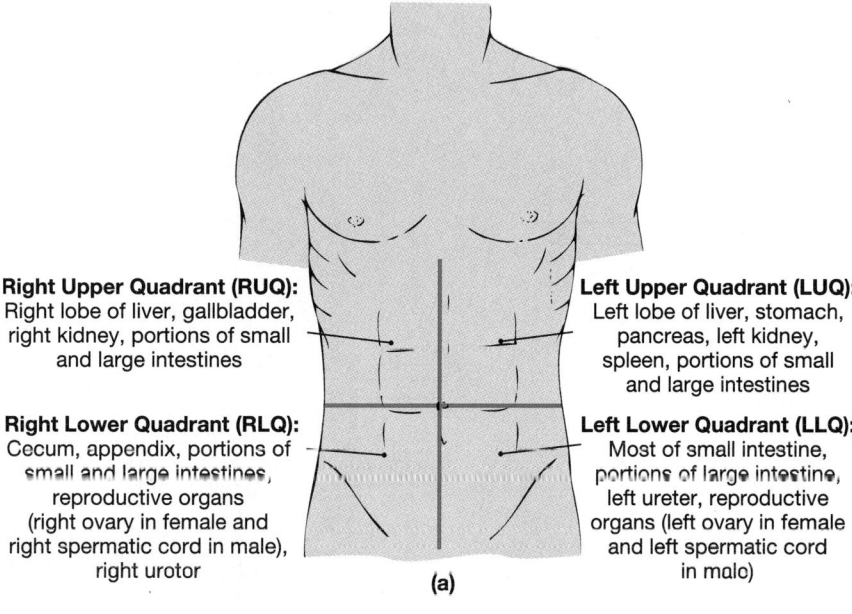

Right Upper Quadrant (RUQ):
Right lobe of liver, gallbladder, right kidney, portions of small and large intestines

Left Upper Quadrant (LUQ):
Left lobe of liver, stomach, pancreas, left kidney, spleen, portions of small and large intestines

Right Lower Quadrant (RLQ):
Cecum, appendix, portions of small and large intestines, reproductive organs (right ovary in female and right spermatic cord in male), right ureter

Left Lower Quadrant (LLQ):
Most of small intestine, portions of large intestine, left ureter, reproductive organs (left ovary in female and left spermatic cord in male)

(a)

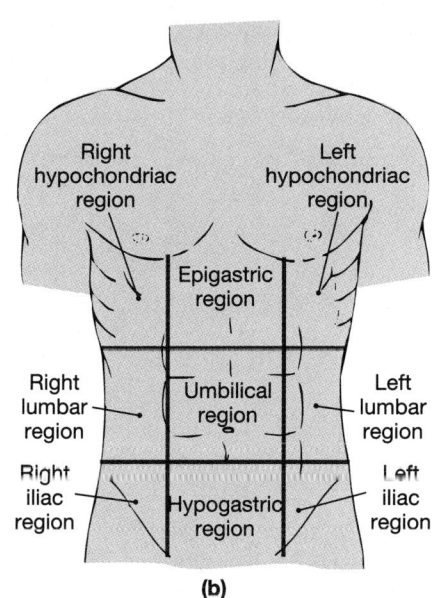

Right hypochondriac region
Left hypochondriac region
Epigastric region
Right lumbar region
Umbilical region
Left lumbar region
Right iliac region
Hypogastric region
Left iliac region

(b)

•FIGURE 1-8   Abdominopelvic Quadrants and Regions. (a) Abdominopelvic quadrants divide the area into four sections. These terms, or their abbreviations, are most often used in clinical discussions. **(b)** More-precise regional descriptions are provided by reference to the appropriate abdominopelvic region. **(c)** Quadrants or regions are useful because there is a known relationship between superficial anatomical landmarks and underlying organs.

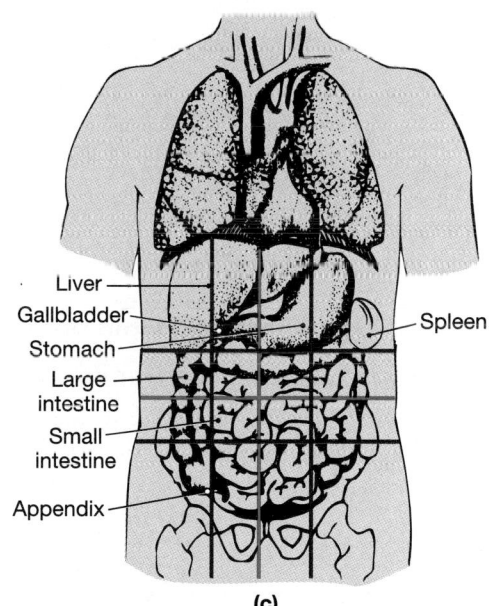

Liver
Gallbladder
Stomach
Large intestine
Small intestine
Appendix
Spleen

(c)

When reading anatomical descriptions, you will find it useful to remember that the terms *left* and *right* always refer to the left and right sides of the subject, not of the observer.

Sectional Anatomy

A presentation in sectional view is sometimes the only way to illustrate the relationships among the parts of a three-dimensional object. An understanding of sectional views has become increasingly important since the development of electronic imaging techniques that enable us to see inside the living body without resorting to surgery.

Planes and Sections

Any slice through a three-dimensional object can be described with reference to three **sectional planes**, indicated in Figure 1-10• and Table 1-3 (p. 21). The **transverse plane** lies at right angles to the long axis of the body, dividing it into *superior* and *inferior* sections. A cut in this plane is called a **transverse section**, or *cross section*. The **frontal plane,** or *coronal*

plane, and the **sagittal plane** parallel the long axis of the body. The frontal plane extends from side to side, dividing the body into *anterior* and *posterior* sections. The sagittal plane extends from front to back, dividing the body into left and right sections. A cut that passes along the midline and divides the body into left and right halves is a *midsagittal section*, or *median section*; a cut parallel to the midsagittal line is a *parasagittal section*.

Sometimes it is helpful to compare the information provided by sections made along different planes. You can experiment with this procedure by mentally sec-

•FIGURE 1-9
Directional References.
Important directional terms
used in this text are
indicated by arrows;
definitions and descriptions
are included in Table 1-2.

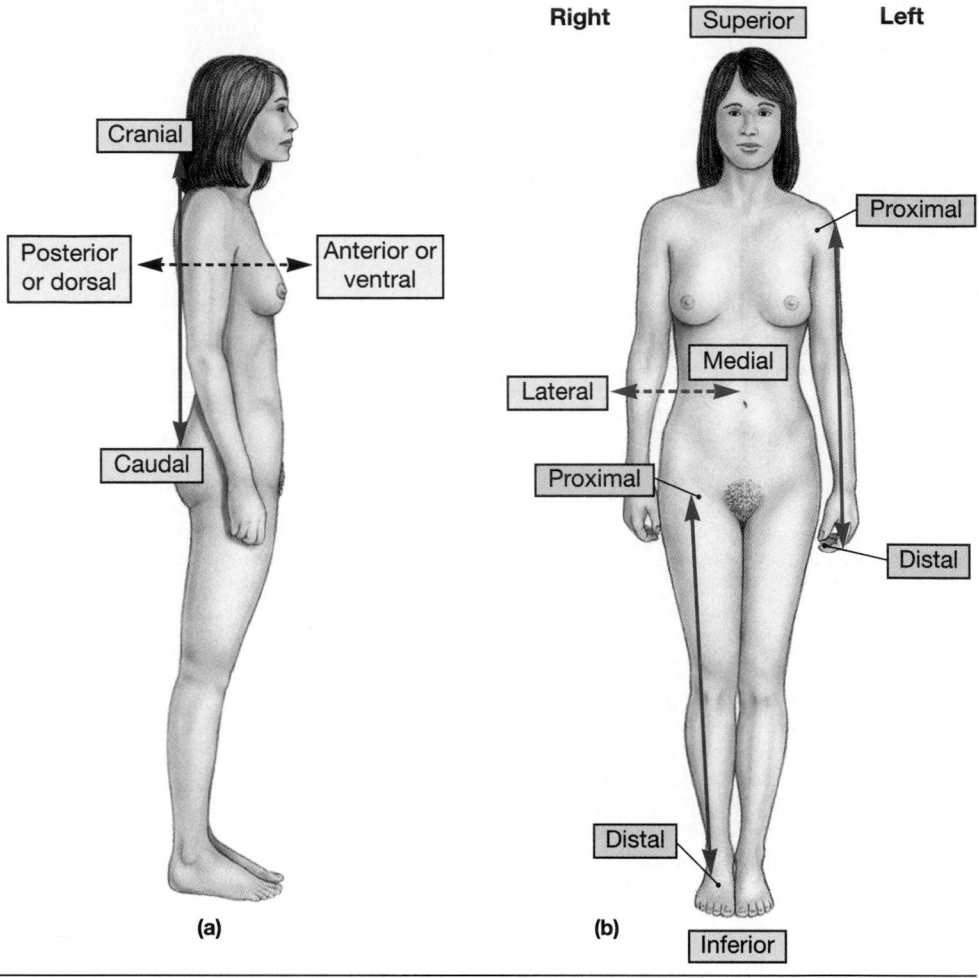

(a)

(b)

TABLE 1-2 Directional Terms (*see Figure 1-9*)

Term	Region or Reference	Example
Anterior	The front; before	The navel is on the *anterior* surface of the trunk.
Ventral	The belly side (equivalent to anterior when referring to human body)	In humans, the navel is on the *ventral* surface.
Posterior	The back; behind	The shoulder blade is located *posterior* to the rib cage.
Dorsal	The back (equivalent to posterior when referring to human body)	The *dorsal* body cavity encloses the brain and spinal cord.
Cranial or cephalic	The head	The *cranial*, or *cephalic*, border of the pelvis is on the side toward the head rather than toward the thigh.
Superior	Above; at a higher level (in human body, toward the head)	In humans, the cranial border of the pelvis is *superior* to the thigh.
Caudal	The tail (coccyx in humans)	The hips are *caudal* to the waist.
Inferior	Below; at a lower level	The knees are *inferior* to the hips.
Medial	Toward the body's longitudinal axis	The *medial* surfaces of the thighs may be in contact; moving medially from the arm across the chest surface brings you to the sternum.
Lateral	Away from the body's longitudinal axis	The thigh articulates with the *lateral* surface of the pelvis; moving laterally from the nose brings you to the eyes.
Proximal	Toward an attached base	The thigh is *proximal* to the foot; moving proximally from the wrist brings you to the elbow.
Distal	Away from an attached base	The fingers are *distal* to the wrist; moving distally from the elbow brings you to the wrist.
Superficial	At, near, or relatively close to the body surface	The skin is *superficial* to underlying structures.
Deep	Farther from the body surface	The bone of the thigh is *deep* to the surrounding skeletal muscles.

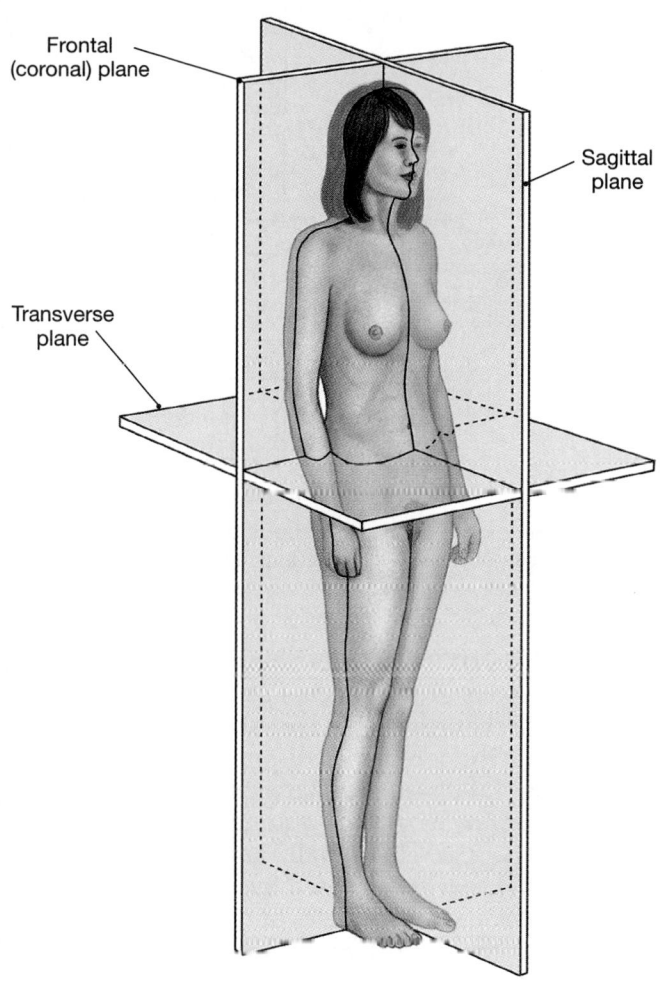

Frontal (coronal) plane

Sagittal plane

Transverse plane

°FIGURE 1-10 **Planes of Section.** The three primary planes of section are indicated here. Table 1-3 defines and describes them.

tioning this book, as in Figure 1-11a●. (Performing this experiment is not recommended, unless you are dropping the course.) Each sectional plane provides a different perspective on the book's structure. When combined with observations of the external anatomy, they create a reasonably complete picture.

Obtaining a more accurate and detailed picture would entail choosing one sectional plane and making a series of sections at small intervals. This process, called *serial reconstruction*, permits the analysis of relatively complex structures. Figure 1-11b● shows the serial reconstruction of a simple bent tube. The same procedure could be used to visualize the path of a small blood vessel or to follow a loop of the intestine. Serial reconstruction is an important method for studying histological structure and for analyzing the images produced by sophisticated clinical procedures.

Body Cavities

Many vital organs are suspended in internal chambers called *body cavities.* These cavities have two essential functions: (1) They protect delicate organs, such as the brain and spinal cord, from accidental shocks and cushion them from the thumps and bumps that occur during walking, jumping, and running; and (2) they permit significant changes in the size and shape of internal organs. For example, because they are situated within body cavities, the lungs, heart, stomach, intestines, urinary bladder, and many other organs can expand and contract without distorting surrounding tissues and disrupting the activities of nearby organs.

The *dorsal body cavity* contains the brain and spinal cord. The much larger *ventral body cavity* contains or-

TABLE 1-3	**Terms That Indicate Planes of Section** (*see Figure 1-10*)		
Orientation of Plane	**Plane**	**Directional Reference**	**Description**
Perpendicular to long axis	Transverse or horizontal	Transversely or horizontally	A *transverse*, or *horizontal*, *section* separates superior and inferior portions of the body.
Parallel to long axis	Sagittal	Sagittally	A *sagittal section* separates right and left portions. You examine a sagittal section, but you section sagittally.
	Midsagittal		In a *midsagittal section* the plane passes through the midline, dividing the body in half and separating right and left sides.
	Parasagittal		A *parasagittal section* misses the midline, separating right and left portions of unequal size.
	Frontal or coronal	Frontally or coronally	A *frontal*, or *coronal*, *section* separates anterior and posterior portions of the body; *coronal* usually refers to sections passing through the skull.

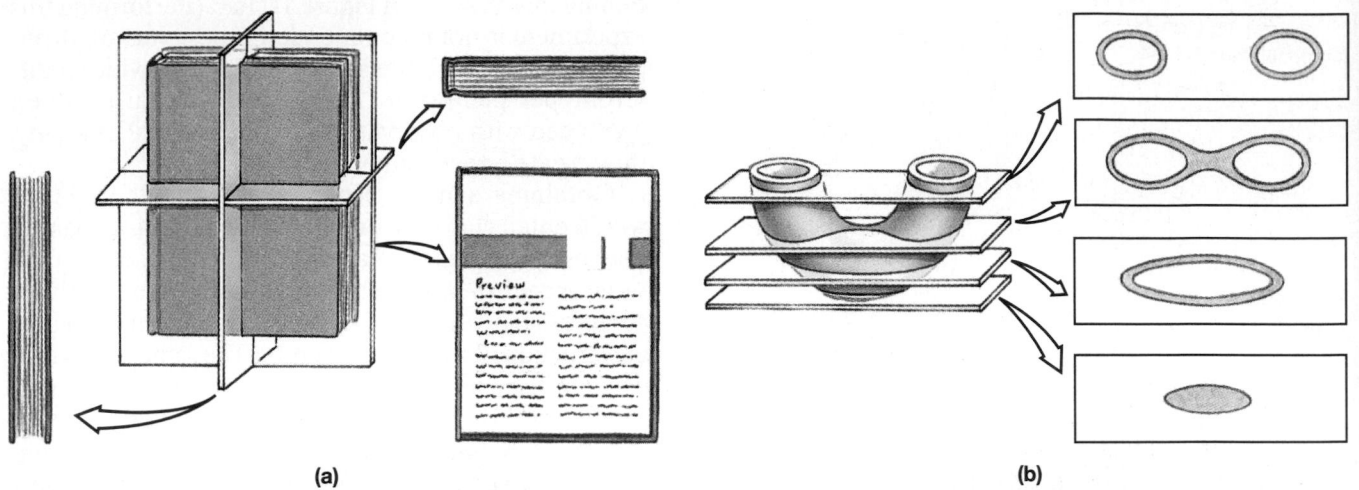

•FIGURE 1-11   Sectional Planes and Visualization. (a) Taking three different sections through a book provides detailed information about its three-dimensional structure. **(b)** More complete pictures can be assembled by taking a series of sections at small intervals. This process is called serial reconstruction.

gans of the respiratory, cardiovascular, digestive, urinary, and reproductive systems. Relationships among the dorsal and ventral body cavities and their various subdivisions are indicated in Figure 1-12• and shown in Figure 1-13•.

DORSAL BODY CAVITY   The **dorsal body cavity** (Figure 1-13a•) is a fluid-filled space whose limits are established by the cranium, the bones of the skull that surround the brain, and by the **vertebrae** of the spinal column. The dorsal body cavity is subdivided into the **cranial cavity**, which contains the brain, and the **spinal cavity**, which contains the spinal cord.

VENTRAL BODY CAVITY   As development proceeds, internal organs grow and change their relative positions. These changes lead to the subdivision of the **ventral body cavity**, or *coelom* (SĒ-lōm; *koila*, cavity). The **diaphragm** (DĪ-ah-fram), a flat muscular sheet, divides the ventral body cavity into a superior **thoracic cavity**, bounded by the chest wall, and an inferior **abdominopelvic cavity**, enclosed by the abdominal wall and by the bones and muscles of the pelvis.

Many of the organs within these cavities change size and shape as they perform their functions. For example, the stomach swells at each meal and shrinks between meals. These organs project into moist in-

•FIGURE 1-12   Relationships of the Various Body Cavities

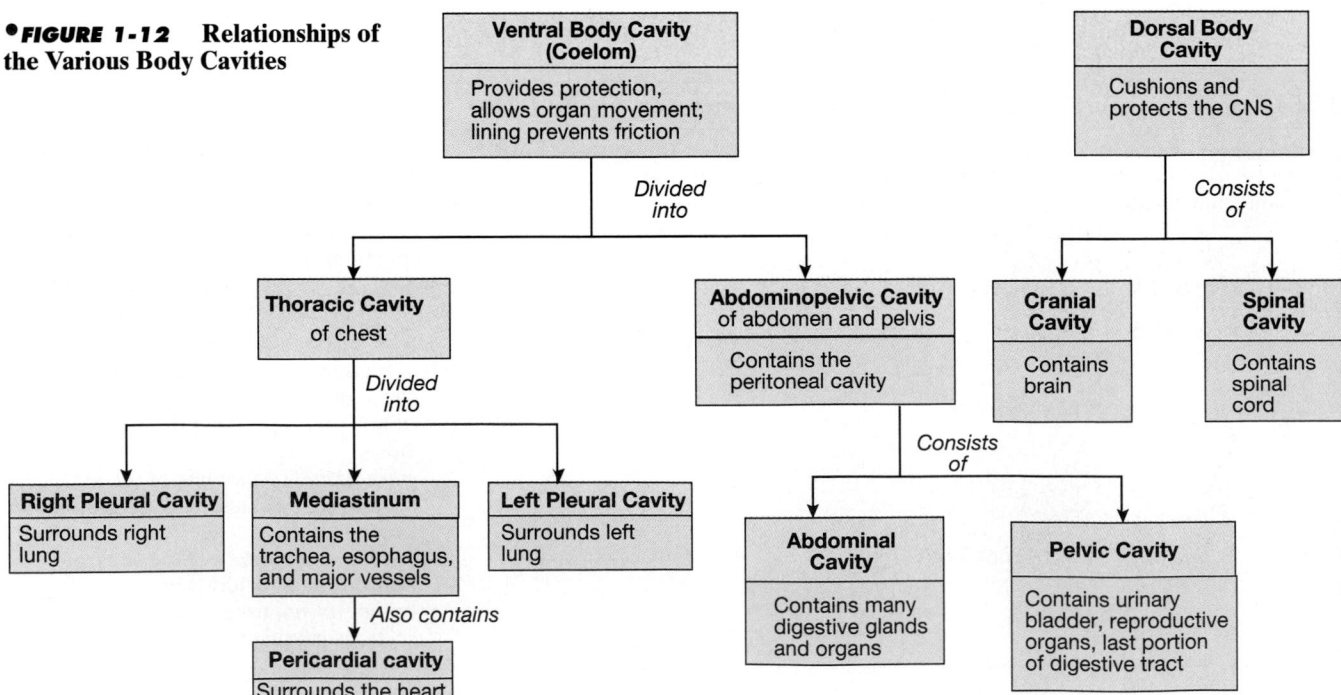

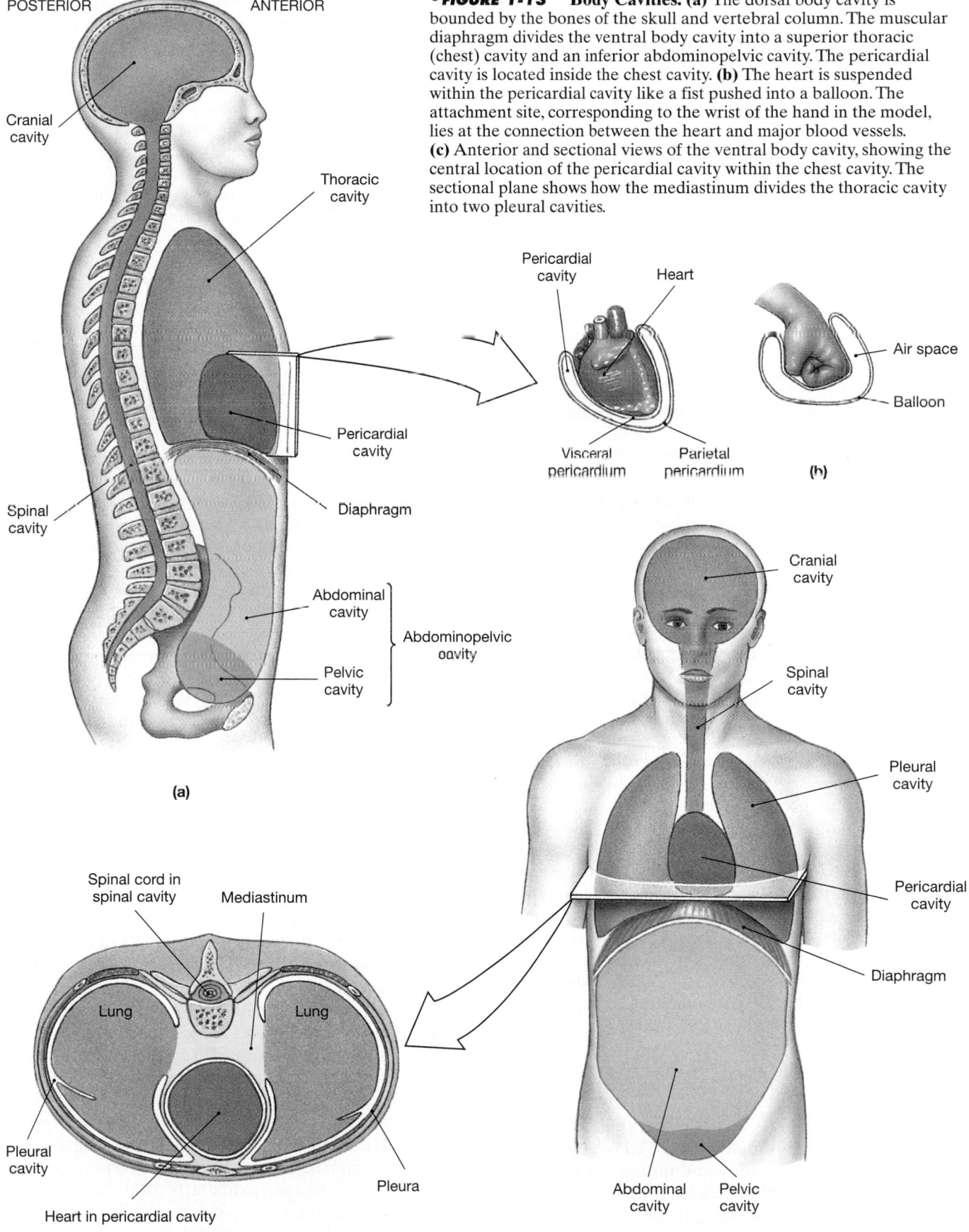

•FIGURE 1-13 **Body Cavities. (a)** The dorsal body cavity is bounded by the bones of the skull and vertebral column. The muscular diaphragm divides the ventral body cavity into a superior thoracic (chest) cavity and an inferior abdominopelvic cavity. The pericardial cavity is located inside the chest cavity. **(b)** The heart is suspended within the pericardial cavity like a fist pushed into a balloon. The attachment site, corresponding to the wrist of the hand in the model, lies at the connection between the heart and major blood vessels. **(c)** Anterior and sectional views of the ventral body cavity, showing the central location of the pericardial cavity within the chest cavity. The sectional plane shows how the mediastinum divides the thoracic cavity into two pleural cavities.

POSTERIOR   ANTERIOR

Cranial cavity

Thoracic cavity

Pericardial cavity

Diaphragm

Spinal cavity

Abdominal cavity

Abdominopelvic cavity

Pelvic cavity

(a)

Pericardial cavity   Heart

Air space

Balloon

Visceral pericardium   Parietal pericardium

(b)

Cranial cavity

Spinal cavity

Pleural cavity

Pericardial cavity

Diaphragm

Abdominal cavity   Pelvic cavity

(c)

Spinal cord in spinal cavity   Mediastinum

Lung   Lung

Pleural cavity

Pleura

Heart in pericardial cavity

(d)

ternal spaces that permit expansion and limited movement but prevent friction. There are three such chambers in the thoracic cavity and one in the abdominopelvic cavity. The internal organs that project into these cavities are called **viscera** (VIS-e-ruh). Delicate *serous membranes* line the walls of those cavities and cover the surfaces of the enclosed viscera. Those smooth, moist surfaces secrete a watery fluid that reduces friction within the internal cavities.

The Thoracic Cavity. The walls of the thoracic cavity surround the lungs and heart, associated organs of the respiratory, cardiovascular, and lymphatic systems, the inferior portions of the esophagus, and the thymus (Figure 1-13a●). The thoracic cavity contains three internal chambers: a single *pericardial cavity* and a pair of *pleural cavities*. These cavities are lined by shiny, slippery, and delicate serous membranes.

- The heart projects into the **pericardial cavity**. The relationship between the heart and the cavity resembles that of a fist pushing into a balloon (Figure 1-13b●). The wrist corresponds to the *base* (attached portion) of the heart, and the balloon corresponds to the serous membrane lining the pericardial cavity. The serous membrane covering the heart is called the *pericardium* (*peri-*, around + *kardia*, heart). The layer covering the heart is the *visceral pericardium*, and the opposing surface is the *parietal pericardium*. The space between the visceral pericardium and the parietal pericardium is the pericardial cavity. During each beat, the heart changes in size and shape. The pericardial cavity permits these changes, and the slippery pericardial lining prevents friction between the heart and adjacent structures in the thoracic cavity.

  The pericardium lies within the **mediastinum** (mē-dē-as-TĪ-num or mē-dē-AS-ti-num). The mediastinum is the portion of the thoracic cavity that lies between the left and right pleural cavities (Figure 1-13d●). The connective tissue of the mediastinum surrounds the pericardial cavity and heart, the large arteries and veins attached to the heart, and the thymus, trachea, and esophagus.

- There is one **pleural cavity** on each side of the mediastinum. Each pleural cavity encloses a lung. The relationship between a lung and a pleural cavity is comparable to that between the heart and the pericardial cavity. The serous membrane lining a pleural cavity is called a *pleura* (PLOO-rah). The outer surfaces of the lungs are covered by the *visceral pleura*, and the *parietal pleura* covers the opposing mediastinal surfaces and the inner body wall.

The Abdominopelvic Cavity. The abdominopelvic cavity extends from the diaphragm to the pelvis. It is subdivided into a superior **abdominal cavity** and an inferior **pelvic cavity** (Figure 1-13a,c●). The abdominopelvic cavity contains the *peritoneal* (per-i-tō-NĒ-al) *cavity*, a chamber lined by a serous membrane

known as the *peritoneum* (per-i-tō-NĒ-um). The parietal peritoneum lines the inner surface of the body wall. A narrow, fluid-filled space separates the *parietal peritoneum* from the *visceral peritoneum* that covers the enclosed organs.

- *The abdominal cavity.* The abdominal cavity extends from the inferior surface of the diaphragm to an imaginary plane extending from the inferior surface of the lowest vertebra to the anterior and superior margin of the pelvis. The abdominal cavity contains the liver, stomach, spleen, small intestine, and most of the large intestine. (You can see the positions of most of these organs in Figure 1-8c●, p. 19). These organs project partially or completely into the peritoneal cavity, much as the heart or lungs project into the pericardial or pleural cavities. A few organs, such as the kidneys and pancreas, lie between the peritoneal lining and the muscular wall of the abdominal cavity. Those organs are said to be *retroperitoneal* (*retro,* behind).

- *The pelvic cavity.* The portion of the ventral body cavity inferior to the abdominal cavity is the pelvic cavity. The bones of the pelvis form the walls of that cavity, and a layer of muscle forms its floor. The pelvic cavity contains the last segments of the large intestine, the urinary bladder, and various reproductive organs. For example, the pelvic cavity of a female contains the ovaries, uterine tubes, and uterus; in a male, it contains the prostate gland and seminal vesicles. The inferior portion of the peritoneal cavity extends partway into the pelvic cavity. The superior portion of the urinary bladder in both sexes, and the uterine tubes, the ovaries, and the superior portion of the uterus in females are covered by peritoneum.

☑ Which type of section would separate the two eyes?

☑ If a surgeon makes an incision just inferior to the diaphragm, which body cavity will be opened?

This chapter provided an overview of the locations and functions of the major components of each organ system. It also introduced the anatomical vocabulary needed for you to follow more-detailed anatomical descriptions in later chapters. Many of the figures in later chapters contain images produced by the procedures outlined in Figures 1-14● and 1-15● (p. 26), which summarize the most common methods of visualizing anatomical structures in living individuals.

The next three chapters will take you on a tour of the principal levels of organization, from individual atoms to individual human beings. As we proceed through the text, we will emphasize major structural and functional patterns. To sharpen your analytical skills, we have included critical-thinking questions at the end of each chapter and in the *Applications Manual.*

FOCUS   Sectional Anatomy and Clinical Technology

The term *radiological procedures* includes (1) scanning techniques that involve the use of beams of radiation, such as X-rays, to create a photographic or computer-generated image of internal structures and (2) methods that involve the administration of radioactive materials. Physicians who specialize in the performance of these procedures and the analysis of the resulting images are called *radiologists*. Radiological procedures can provide detailed information about internal systems. Figures 1-14● and 1-15● compare the views provided by several techniques used by radiologists and by other clinicians. These figures include examples of *X-rays, CT scans, MRI scans,* and *ultrasound* images. You will find other examples of clinical technology in later chapters and in the *Applications Manual.* As you encounter them, it will be helpful to remember that when anatomical diagrams or clinical procedures present cross-sectional views of the body, the sections are oriented as though the observer were standing at the feet and looking toward the head of the subject.

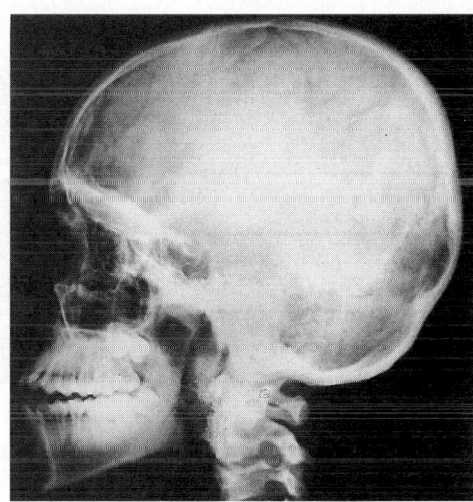 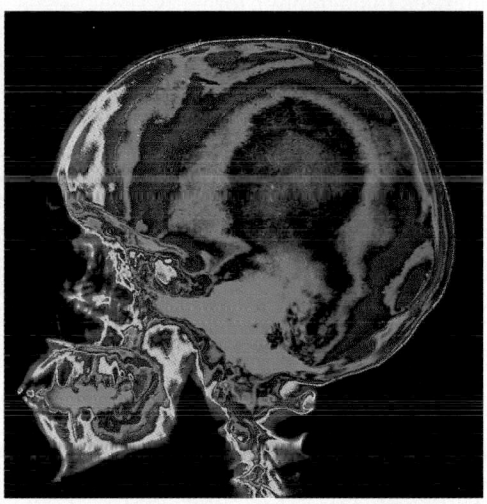

(a)

●**FIGURE 1-14**   **X-rays. (a)** X-rays of the skull, taken from the left side. **X-rays** are a form of high-energy radiation that can penetrate living tissues. In the most familiar procedure, a beam of X-rays travels through the body and strikes a photographic plate. All the projected X-rays do not arrive at the film; some are absorbed or deflected as they pass through the body. The resistance to X-ray penetration is called **radiodensity**. In the human body, the order of increasing radiodensity is as follows: air, fat, liver, blood, muscle, bone. The result is an image with radiodense tissues, such as bone, appearing in white, and less-dense tissues in shades of gray to black. (The image on the right has been color-enhanced.) These are two-dimensional images of a three-dimensional object. In such an image it is difficult to decide whether a particular feature is on the left side (toward the viewer) or on the right side (away from the viewer).

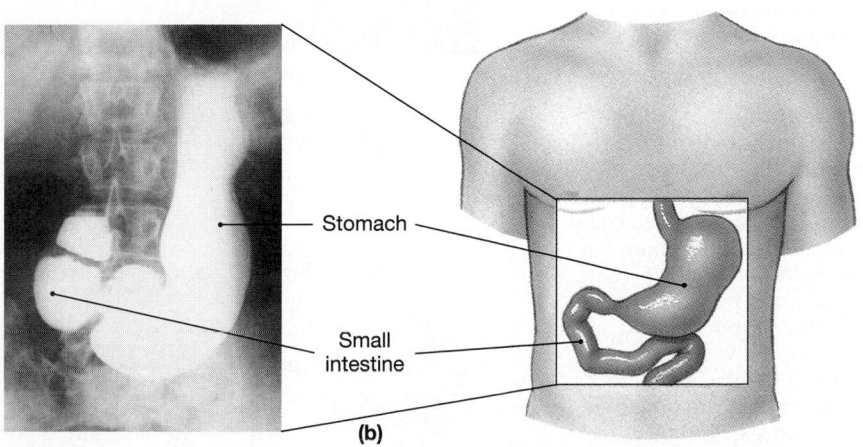

(b)

(b) A barium-contrast X-ray of the upper digestive tract. Barium is very dense, and the contours of the gastric and intestinal lining can be seen outlined against the white of the barium solution.

•FIGURE 1-15   Scanning Techniques

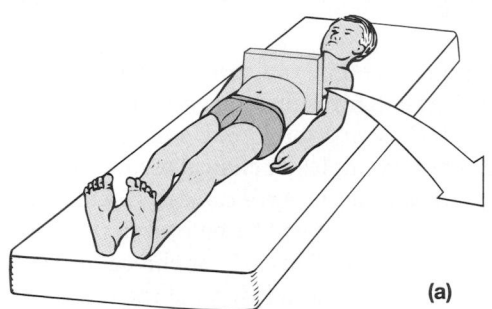

(a)

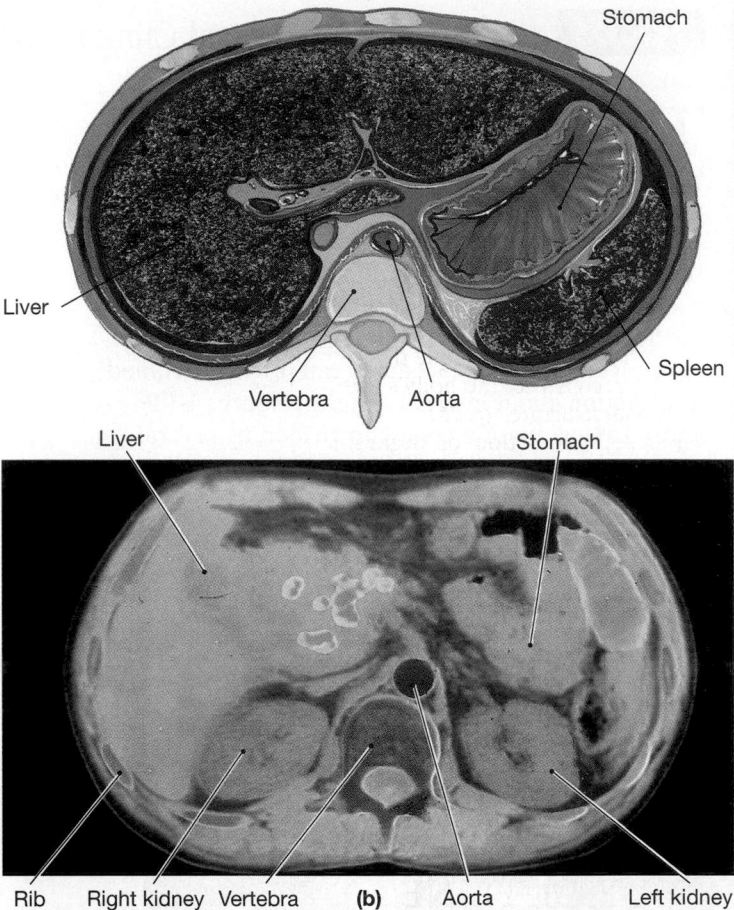

(a) A diagrammatic view showing the relative position and orientation of the scans shown here.

(b) A color-enhanced CT scan of the abdomen. **CT** (**c**omputerized **t**omography), formerly called **CAT** (**c**omputerized **a**xial **t**omography), uses computers to reconstruct sectional views. A single X-ray source rotates around the body, and the X-ray beam strikes a sensor monitored by the computer. The source completes one revolution around the body every few seconds; it then moves a short distance and repeats the process. The result is usually displayed as a sectional view in black and white, but it can be colorized for visual effect. CT scans show three-dimensional relationships and soft-tissue structure more clearly than do standard X-rays.

(c) A color-enhanced MRI scan of the same region. **M**agnetic **r**esonance **i**maging (**MRI**) surrounds part or all of the body with a magnetic field about 3000 times as strong as that of Earth. This field affects protons within atomic nuclei throughout the body. The protons line up along the magnetic lines of force like compass needles in Earth's magnetic field. When struck by a radio wave of the proper frequency, a proton will absorb energy. When the wave pulse ends, that energy is released, and the source of the radiation is detected. Each element differs in terms of the radio frequency required to affect its protons.

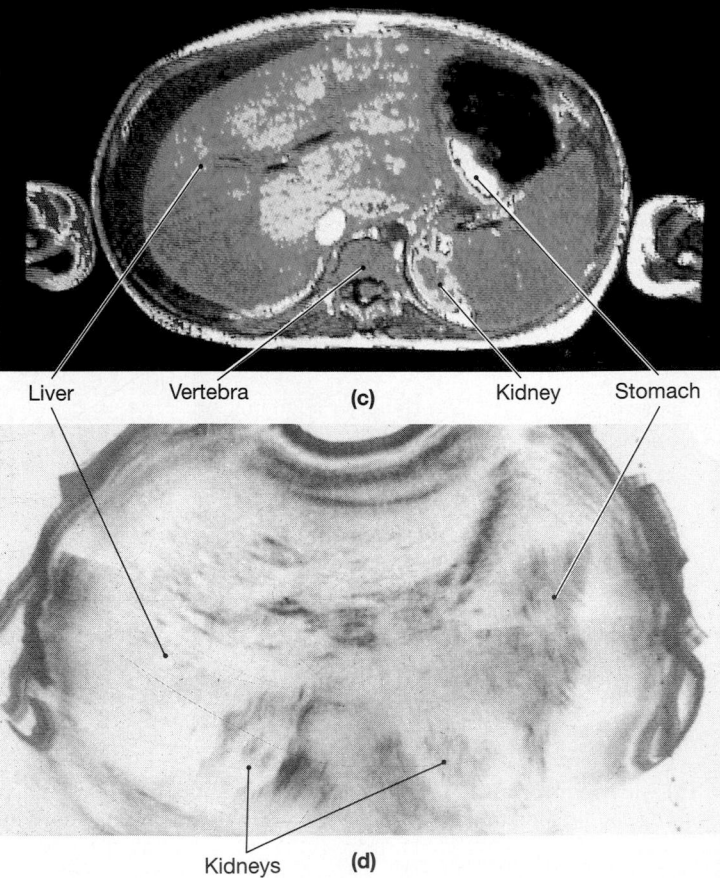

(d) An ultrasound scan of the abdomen. In **ultrasound** procedures, a small transmitter contacting the skin broadcasts a brief, narrow burst of high-frequency sound and then picks up the echoes. The sound waves are reflected by internal structures. A picture, or **echogram,** can be assembled from the pattern of echoes. These images lack the clarity of other procedures, but no adverse effects have been reported, and fetal development can be monitored without a significant risk of birth defects. Special methods of transmission and processing permit analysis of the beating heart, without the complications that can accompany dye injections. Note the differences in detail between this image, the CT scan, and the MRI image.

Selected Clinical Terminology

Terms Discussed in This Chapter

abdominopelvic quadrant: One of four divisions of the anterior abdominal surface. *(p. 18)*

abdominopelvic region: One of nine divisions of the anterior abdominal surface. *(p. 18)*

CT, CAT (computerized [axial] tomography): An imaging technique that uses X-rays to reconstruct the body's three-dimensional structure. *(p. 26)*

disease: A malfunction of organs or organ systems resulting from failure of homeostatic regulation. *(p. 16)*

echogram: An image created by ultrasound. *(p. 26)*

embryology: The study of structural changes during the first two months of development. *(p. 3)*

histology (his-TOL-o-jē): The study of tissues. *(p. 3)*

MRI (magnetic resonance imaging): An imaging technique that employs a magnetic field and radio waves to portray subtle structural differences. *(p. 26)*

radiologist: A physician who specializes in performing and analyzing radiological procedures. *(p. 25)*

ultrasound: An imaging technique that uses brief bursts of high-frequency sound reflected by internal structures. *(p. 26)*

X-rays: High-energy radiation that can penetrate living tissues. *(p. 25)*

AM Additional Terms Discussed in the *Applications Manual*

auscultation (aws-kul-TĀ-shun): Listening to a patient's body sounds using a stethoscope.

inspection: A careful observation of a patient's appearance and actions.

palpation: Using hands and fingers to feel the patient's body as part of a physical exam.

percussion: Tapping with the fingers or hand to obtain information about the densities of a patient's underlying tissues.

CHAPTER REVIEW

 On-line resources for this chapter are on our World Wide Web site at:
http://www.prenhall.com/martini/fap

STUDY OUTLINE

INTRODUCTION, p. 2

1. **Biology** is the study of life; one of its goals is to discover the unity and patterns that underlie the diversity of living organisms.
2. All living things have certain common characteristics, including **responsiveness, growth** and **differentiation, reproduction, movement,** and **metabolism** and **excretion.**

THE SCIENCES OF ANATOMY AND PHYSIOLOGY, p. 2

1. **Anatomy** is the study of internal and external structure and the physical relationships among body parts. **Physiology** is the study of how living organisms perform vital functions. All specific functions are performed by specific structures.

Anatomy, p. 3

2. The boundaries of *microscopic anatomy* are established by the equipment used. In **cytology,** we analyze the internal structure of individual cells. In **histology,** we examine **tissues,** groups of cells that have specific functional roles. Tissues combine to form **organs,** anatomical units with multiple functions. Organs combine to form **organ systems,** groups of organs that function together.
3. In *gross (macroscopic) anatomy,* we consider features visible without a microscope. It includes *surface anatomy* (general form and superficial markings); *regional anatomy* (superficial and internal features in a specific area of the body); and *systemic anatomy* (structure of major organ systems).
4. In *developmental anatomy,* we examine the changes in form that occur between conception and physical maturity. In *embryology,* we study processes that occur during the first two months of development.

Physiology, p. 3

5. **Human physiology** is the study of the functions of the human body. It is based on *cell physiology,* the study of the functions of living cells. In *special physiology,* we study the physiology of specific organs. In *systemic physiology,* we consider all aspects of the function of specific organ systems. In *pathological physiology,* we study the effects of diseases on organ or system functions.

LEVELS OF ORGANIZATION, p. 4

1. Anatomical structures and physiological mechanisms are arranged in a series of interacting levels of organization. *(Figure 1-1)*

AN INTRODUCTION TO ORGAN SYSTEMS, p. 4

1. The 11 organ systems of the body are the **integumentary, skeletal, muscular, nervous, endocrine, cardiovascular, lymphatic, respiratory, digestive, urinary,** and **reproductive systems.** *(Figures 1-2, 1-3)*

HOMEOSTASIS AND SYSTEM INTEGRATION, p. 6

Homeostatic Regulation, p. 6

1. **Homeostasis** is the presence of a stable environment within the body. Through *homeostatic regulation,* physiological systems preserve homeostasis.
2. **Autoregulation** occurs when the activities of a cell, tissue, organ, or system change automatically in response to an environmental change. **Extrinsic regulation** results from the activities of the nervous or endocrine system.
3. Homeostatic regulation usually involves a **receptor** sensitive to a particular stimulus, a **control center** that receives

and processes the information from the receptor, and an **effector** whose activities are regulated by the control center and whose actions have a direct or indirect effect on the same stimulus.

4. **Negative feedback** is a corrective mechanism involving an action that directly opposes a variation from normal limits. *(Figures 1-4, 1-5)*

5. In **positive feedback** the initial stimulus produces a response that exaggerates the stimulus. *(Figure 1-6)*

A FRAME OF REFERENCE FOR ANATOMICAL STUDIES, p. 16

Superficial Anatomy, p. 17

1. Standard anatomical illustrations show the body in the **anatomical position**. If the figure is shown lying down, it can be either *supine* (face up) or *prone* (face down). *(Figure 1-7)*

2. **Abdominopelvic quadrants** and **abdominopelvic regions** represent two different approaches to describing anatomical regions of the body. *(Figure 1-8, Table 1-1)*

3. The use of special directional terms provides clarity for the description of anatomical structures. *(Figure 1-9; Table 1-2)*

Sectional Anatomy, p. 19

4. The three **sectional planes** (**frontal**, or *coronal*, **plane**; **sagittal plane**; and **transverse plane**) describe relationships among the parts of the three-dimensional human body. *(Figure 1-10; Table 1-3)*

5. *Serial reconstruction* is an important technique for studying histological structure and analyzing images produced by radiological procedures. *(Figure 1-11)*

6. *Body cavities* protect delicate organs and permit changes in the size and shape of visceral organs. The **dorsal body cavity** contains the **cranial cavity** (enclosing the brain) and **spinal cavity** (surrounding the spinal cord). The **ventral body cavity**, or *coelom*, surrounds developing respiratory, cardiovascular, digestive, urinary, and reproductive organs. *(Figure 1-12)*

7. During development, the **diaphragm** divides the ventral body cavity into the superior thoracic and inferior peritoneal cavities. By birth, the thoracic cavity contains two **pleural cavities** (each containing a lung) and a **pericardial cavity** (which surrounds the heart). The **abdominopelvic cavity** consists of the **abdominal cavity** and the **pelvic cavity**. It contains the *peritoneal cavity*, an internal chamber lined by a serous membrane, the *peritoneum*. *(Figure 1-13)*

FOCUS: Sectional Anatomy and Clinical Technology, p. 25

8. Important *radiological procedures* (which can provide detailed information about internal systems) include **X-rays**, **CT**, **MRI**, and **ultrasound**. Each technique has advantages and disadvantages. *(Figures 1-14, 1-15)*

REVIEW QUESTIONS

Level 1   Reviewing Facts and Terms

Match each numbered item with the most closely related lettered item. Use letters for answers in the spaces provided.

___	1. cytology	**a.**	study of tissues
___	2. physiology	**b.**	constant internal environment
___	3. histology	**c.**	face up
___	4. metabolism	**d.**	study of functions
___	5. homeostasis	**e.**	positive feedback
___	6. muscle	**f.**	system
___	7. heart	**g.**	study of cells
___	8. endocrine	**h.**	negative feedback
___	9. temperature regulation	**i.**	brain and spinal cord
___	10. labor and delivery	**j.**	all chemical activity in body
___	11. supine	**k.**	thoracic and abdominopelvic
___	12. prone	**l.**	tissue
___	13. ventral body cavity	**m.**	serous membrane
___	14. dorsal body cavity	**n.**	organ
___	15. pericardium	**o.**	face down

16. The process by which an organism increases the size and/or number of cells is
(a) reproduction   (b) adaptation
(c) growth   (d) metabolism

17. The following is a list of several levels of organization that make up the human body:
1. tissue   2. cell   3. organ
4. molecule   5. organism   6. organ system

The correct order from the smallest to the largest level is
(a) 2, 4, 1, 3, 6, 5   (b) 4, 2, 1, 3, 6, 5
(c) 4, 2, 1, 6, 3, 5   (d) 4, 2, 3, 1, 6, 5
(e) 2, 1, 4, 3, 5, 6

18. The study of internal and external structure and the physical relationships among body parts is
(a) histology   (b) anatomy
(c) physiology   (d) embryology

19. The specialist who attempts to determine the physical and chemical processes responsible for vital functions is a
(a) physiologist   (b) physician
(c) pathologist   (d) cytologist

20. The relative stability of an organism's internal environment is
(a) homeopathy   (b) uniformity
(c) equilibrium   (d) homeostasis

21. The automatic change that occurs when the activities of a cell, tissue, organ, or system are faced with some environmental variation is
(a) autoregulation
(b) extrinsic regulation
(c) deregulation
(d) disease

22. Crisis management by directing rapid, short-term, specific responses is a function of the
(a) endocrine system
(b) nervous system
(c) hormones
(d) autoregulatory system

23. Growth and sexual maturation are regulated primarily by the
(a) nervous system   (b) circulatory system
(c) endocrine system   (d) digestive system

24. When a variation outside normal limits triggers a response that restores the normal condition, the regulatory mechanism involves
   (a) negative feedback **(b)** positive feedback
   (c) compensation **(d)** adaptation
25. In positive feedback, the initial stimulus produces a response that
   (a) suppresses the stimulus
   (b) has no effect on the stimulus
   (c) interferes with completion of the process
   (d) exaggerates the stimulus
26. Failure of homeostatic regulation in the body results in
   (a) autoregulation
   (b) extrinsic regulation
   (c) disease
   (d) positive feedback
27. The terms that apply to the front of the body when in anatomical position are
   (a) posterior, dorsal
   (b) back, front
   (c) medial, lateral
   (d) anterior, ventral

28. A plane through the body that passes perpendicular to the long axis of the body and divides the body into a superior and an inferior section is a
   (a) sagittal section **(b)** transverse section
   (c) coronal section **(d)** frontal section
29. The cranial and spinal cavity are found in the
   (a) ventral body cavity **(b)** thoracic cavity
   (c) dorsal body cavity **(d)** abdominopelvic cavity
30. The diaphragm divides the ventral body cavity into a superior _____ cavity and an inferior _____ cavity:
   (a) pleural, pericardial
   (b) abdominal, pelvic
   (c) thoracic, abdominopelvic
   (d) cranial, thoracic
31. The mediastinum is the region between the
   (a) lungs and heart
   (b) two pleural cavities
   (c) chest and abdomen
   (d) heart and pericardium

Level 2   Reviewing Concepts

32. What basic functions are performed by all living things?
33. (a) Define the term *anatomy*.
   (b) Define the term *physiology*.
34. Why is it important to understand the different levels of organization in the human body?
35. Beginning with the molecular level, list in correct sequence the levels of organization from the simplest level to the most complex level.
36. What is homeostatic regulation, and what is its physiological importance?
37. What distinguishes autoregulation from extrinsic regulation?
38. What necessary components are involved in homeostatic regulation? What is the function of each?
39. How does negative feedback differ from positive feedback?
40. Describe the position of the body when it is in the anatomical position.

41. What is the role of serous membranes in the body?
42. As a surgeon, you perform an invasive (surgical) procedure that necessitates cutting through the peritoneum. Are you more likely to be operating on the heart or on the stomach?
43. In which body cavity would each of the following organs or systems be found?
   (a) brain and spinal cord
   (b) cardiovascular, digestive, and urinary systems
   (c) heart, lungs
   (d) stomach, intestines
44. What are the names of the serous membranes that line each of the body cavities containing the following organs?
   (a) heart
   (b) lungs
   (c) intestines

Level 3   Critical Thinking and Clinical Applications

45. A hormone called *calcitonin* from the thyroid gland is released in response to increased levels of calcium ions in the blood. If the control of this hormone is by negative feedback, what effect would calcitonin have on blood calcium levels?
46. During exercise, blood flow to skeletal muscles increases. The initial response that increases blood flow is automatic and independent of the nervous and endocrine systems. What type of homeostatic regulation is this? Why?
47. A chemical imbalance in a heart muscle cell can cause the heart to cease pumping blood, which in turn will cause other

tissues and organs to cease functioning. This observation supports the view that
   (a) all organisms are composed of cells
   (b) all levels of organization within an organism are interdependent
   (c) chemical molecules make up cells
   (d) all cells are independent of each other
   (e) congenital defects can be life threatening

The Chemical Level of Organization

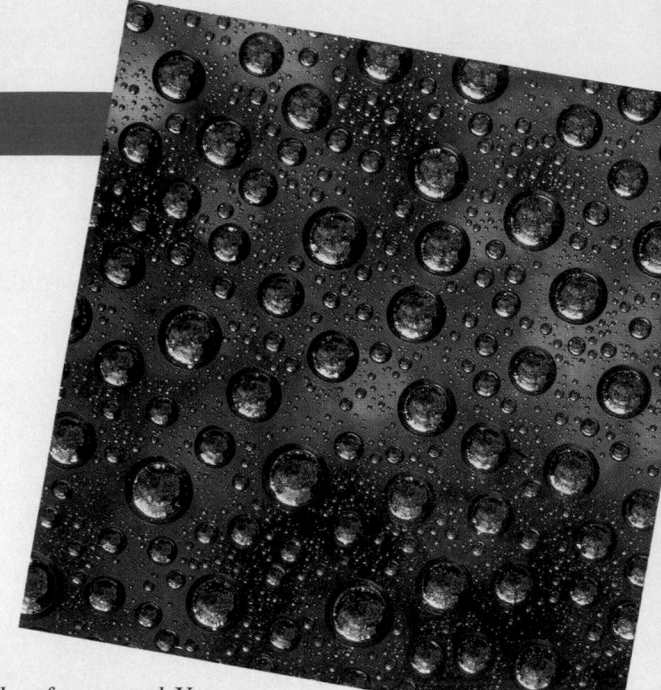

Water is so common, and seems so ordinary, that it is often taken for granted. Yet all living things, including human beings, are totally dependent on water. You can survive for weeks without food but for only a few days without water. The basic reason may surprise you: In large part, you are water. Water makes up 50–60 percent of your body weight.

Chemically, water is a simple substance: two atoms of hydrogen joined to one atom of oxygen. Yet, when these three atoms are linked by chemical bonds, they produce a substance with many unusual properties. In this chapter, you will learn how atoms are bound together to form molecules, the building blocks of living cells. You will also meet the larger molecules that form the structural framework of the body and that enable our cells to grow, divide, and perform other essential functions.

CHAPTER OUTLINE AND OBJECTIVES

Our study of the human body begins at the chemical level of organization. *Chemistry* is the science that deals with the structure of *matter*, defined as anything that takes up space and has *mass*. Mass is a physical property that determines the weight of an object in Earth's magnetic field. For our purposes the mass of an object is the same as its weight. If we were to ride on the space shuttle, however, we would find that the two are not always equivalent. In orbit we would be weightless, but our mass would remain unchanged.

Atoms are the smallest stable units of matter. Air, elephants, oranges, oceans, rocks, and people are all composed of atoms in varying combinations. The unique characteristics of each object, living or non-living, result from the types of atoms involved and the ways those atoms combine and interact. You will need to become familiar with the chemical principles governing those interactions before you can understand the anatomy and physiology of the cells, tissues, organs, and organ systems of the human body.

ATOMS AND MOLECULES

Atoms are composed of **subatomic particles**. Although there are dozens of different subatomic particles, only three are important for understanding chemical properties: *protons*, *neutrons*, and *electrons*. Protons and neutrons are similar in size and mass, but **protons** (p^+) bear a positive electrical charge, whereas **neutrons** (n or n^0) are electrically *neutral*, or uncharged. **Electrons** (e^-) are much lighter than protons—only 1/1836th as massive—and bear a negative electrical charge. The mass of an atom is therefore determined primarily by the number of protons and neutrons in the nucleus. The mass of a large object, such as your body, is the sum of the masses of all the component atoms.

Atomic Structure

Atoms normally contain equal numbers of protons and electrons. The number of protons in an atom is known as its **atomic number**. *Hydrogen* (H) is the simplest atom, with an atomic number of 1. Thus an atom of hydrogen contains one proton, and it contains one electron as well. The proton is located in the center of the atom and forms the nucleus. Hydrogen atoms seldom contain neutrons, but when neutrons are present, they are also found in the nucleus. The electron whirls around the nucleus at high speed, forming a three-dimensional **electron cloud** (Figure 2-1a•). The electron orbits the nucleus because the proton (+) and the electron (−) are oppositely charged, and opposite electrical charges are attracted to one another. This attraction is an ex-

ample of an *electrical force*. For convenience, we often illustrate atomic structure in the simplified form shown in Figure 2-1b•. In this two-dimensional representation, the electrons occupy a circular **electron shell**.

The dimensions of the electron cloud determine the overall size of the atom. To get an idea of the scale involved, consider that if the nucleus were the size of a tennis ball, the electron cloud would have a radius of 10 km (roughly 6 miles). In reality, atoms are so small that atomic measurements are most conveniently reported in terms of *nanometers* (NA-nō-mē-terz) (nm). One nanometer is 10^{-9} meters (0.000000001 m). The very largest atoms approach 0.5 nm in diameter (0.00000005 cm, or 0.00000002 in.).

All atoms other than the simple hydrogen atom diagrammed above have at least one neutron in the nucleus. The number of neutrons present does not affect the properties of an atom, other than its mass.

Elements and Isotopes

Atoms can be assigned to groups called **elements**. Each element includes all the atoms that have the

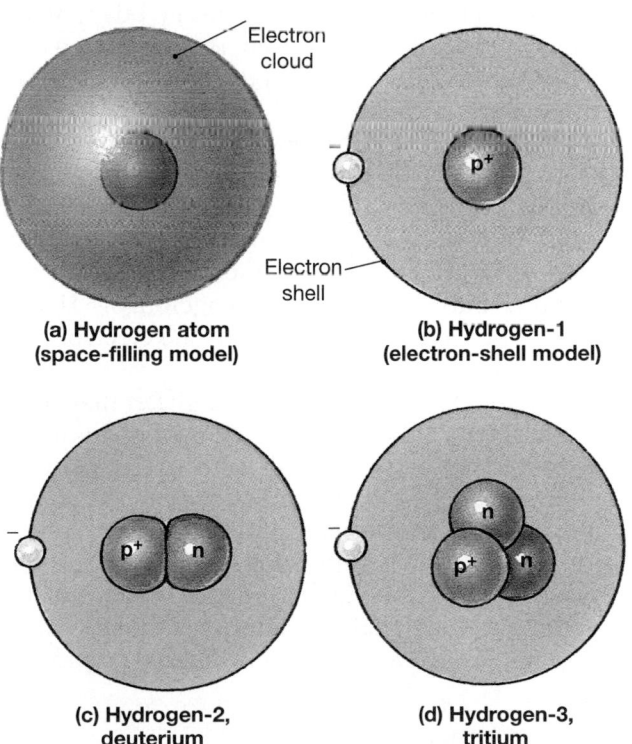

(a) Hydrogen atom (space-filling model)

(b) Hydrogen-1 (electron-shell model)

(c) Hydrogen-2, deuterium

(d) Hydrogen-3, tritium

•**FIGURE 2-1** **Hydrogen Atoms.** **(a)** In a three-dimensional model, the electron cloud of a hydrogen atom is formed by the orbiting of an electron around the nucleus. **(b)** A two-dimensional model depicting the electron in an electron shell makes it easier to visualize the components of the atom. A typical hydrogen nucleus contains a single proton and no neutrons. **(c)** A deuterium (^{2}H) nucleus contains a proton and a neutron. **(d)** A tritium (^{3}H) nucleus contains a pair of neutrons in addition to the proton.

same number of protons and thus the same atomic number. Only 92 different elements occur in nature, although 20 more have been made by means of nuclear reactions in the laboratory. Each element has a chemical symbol, an abbreviation recognized by scientists everywhere. Most of the symbols are easily connected with the English names of the elements, but a few are abbreviations of their Latin names. For example, the symbol for sodium, Na, comes from the Latin word *natrium*. (Appendix III, the *periodic table*, gives the chemical symbols and atomic numbers for each element.) The relative contributions of each of the 13 most abundant elements in the human body to the total body weight are shown in Table 2-1. The list is incomplete, for the human body contains atoms of another 13 elements—called *trace elements*—that are found in such small amounts that percentage values are meaningless.

The atoms of a single element can differ in terms of the number of neutrons in the nucleus. For example, although most hydrogen nuclei consist of a single proton, 0.015 percent also contain one neutron, and a very small percentage contain two neutrons. The *isotopes* of an element are atoms whose nuclei contain different numbers of neutrons. Because the presence or absence of neutrons has no effect on the chemical properties of an atom, isotopes are usually indistinguishable except on the basis of mass. The **mass number**—the total number of protons plus neutrons in the nucleus—is used to designate a particular isotope. Thus, the three isotopes of hydrogen are hydrogen-1, or ^1H, with one proton and one electron (Figure 2-1b•); hydrogen-2, or ^2H, also known as *deuterium*, with one proton, one electron, and one neutron (Figure 2-1c•); and hydrogen-3, or ^3H, also known as *tritium*, with one proton, one electron, and two neutrons (Figure 2-1d•).

Radioisotopes are isotopes with nuclei that spontaneously emit subatomic particles or radiation in measurable amounts. Strongly radioactive isotopes are dangerous, because the emissions can break molecules apart, destroy cells, and otherwise damage living tissues. Weakly radioactive isotopes are sometimes used in diagnostic procedures to monitor the structural or functional characteristics of internal organs. [AM] *Medical Importance of Radioisotopes*

Atomic Weights

A typical atom of *oxygen*, which has an atomic number of 8, contains eight protons, but it also contains eight neutrons. The mass number of this isotope is therefore 16. The mass numbers of other isotopes of oxygen will differ, depending on the number of neutrons present. Mass numbers are useful because they tell us the number of subatomic particles in the nuclei

TABLE 2-1   Principal Elements in the Human Body

Element (% of total body weight)	Significance
Oxygen (65)	A component of water and other compounds; gaseous form is essential for respiration
Carbon (18.6)	Found in all organic molecules
Hydrogen (9.7)	A component of water and most other compounds in the body
Nitrogen (3.2)	Found in proteins, nucleic acids, and other organic compounds
Calcium (1.8)	Found in bones and teeth; important for membrane function, nerve impulses, muscle contraction, and blood clotting
Phosphorus (1.0)	Found in bones and teeth, nucleic acids, and high-energy compounds
Potassium (0.4)	Important for proper membrane function, nerve impulses, and muscle contraction
Sodium (0.2)	Important for membrane function, nerve impulses, and muscle contraction
Chlorine (0.2)	Important for membrane function and water absorption
Magnesium (0.06)	A cofactor for several enzymes
Sulfur (0.04)	Found in many proteins
Iron (0.007)	Essential for oxygen transport and energy capture
Iodine (0.0002)	A component of hormones of the thyroid gland
Trace elements: silicon (Si), fluorine (F), copper (Cu), manganese (Mn), zinc (Zn), selenium (Se), cobalt (Co), molybdenum (Mo), cadmium (Cd), chromium (Cr), tin (Sn), aluminum (Al), and boron (B)	Some function as cofactors; the functions of many trace elements are poorly understood

of different atoms. However, they do not tell us the *actual* mass of the atoms. For example, they do not take into account the masses of the electrons or the slight difference between the mass of a proton and that of a neutron. The actual mass of an atom is known as its **atomic weight**.

The unit used to express the atomic weight is the **dalton** (also known as the *atomic mass unit*, or *amu*). One dalton is very close to the weight of a single proton. Thus the atomic weight of the most common isotope of hydrogen is very close to 1 and that of the most common isotope of oxygen is very close to 16.

The atomic weight of an element is an average mass number that reflects the proportions of different isotopes. For example, even though the atomic number of hydrogen is 1, the atomic weight of hydrogen is 1.0079, primarily because some hydrogen atoms (0.015 percent) have a mass number of 2, and an even smaller percentage have a mass number of 3. The atomic weight of an element is usually very close to the mass number of the most common isotope of that element. The atomic weights of the elements are included in Appendix III.

A quantity of any element that has a weight in grams equal to the atomic weight of that element is called a **mole** (abbreviated *mol*). One mole of a given element contains the same number of atoms as a mole of any other element. That number (called *Avogadro's number*) is 6.023×10^{23}, or about 600 billion trillion. Expressing relationships in moles rather than in grams makes it much easier to keep track of the relative numbers of atoms in chemical samples and processes. For example, if a report stated that a sample contains 0.5 mol of hydrogen atoms and 0.5 mol of oxygen atoms, you would know immediately that they were present in equal numbers. That would not be so evident if the report stated that there were 0.505 g of hydrogen atoms and 8 g of oxygen atoms. Most chemical analyses and laboratory reports today record concentrations in terms of moles (mol), millimoles (mmol), or micromoles (μmol) in a specific volume (generally ml, dl, or l). $\boxed{\text{AM}}$ *Solutions and Concentrations*

Electrons and Energy Levels

Atoms are electrically neutral because every positively charged proton is balanced by a negatively charged electron. Thus, each increase in the atomic number is accompanied by a comparable increase in the number of electrons orbiting the nucleus. These electrons occupy an orderly series of **energy levels** within the electron cloud. Although the electrons within an energy level may travel in complex orbits around the nucleus, for our purposes they can be diagrammed as simple electron shells.

The number and arrangement of electrons in an atom's outer energy level determine the chemical properties of that element. Each energy level can accommodate a specific number of electrons. For example, the innermost level of any atom can hold up to two electrons. As indicated in Figure 2-2•, a hydrogen atom has one electron in the innermost energy level, and the level is thus unfilled. Atoms with unfilled energy levels will react with other atoms. A *helium atom* has two electrons in its first energy level. Because its outer energy level is filled, a helium atom is very stable, and it will not ordinarily react with other atoms. A *lithium atom* has three electrons, so its first level is filled, and the third electron occupies a second energy level. The second level of any atom can hold up to eight electrons. This level is filled in a *neon atom*, with an atomic num-

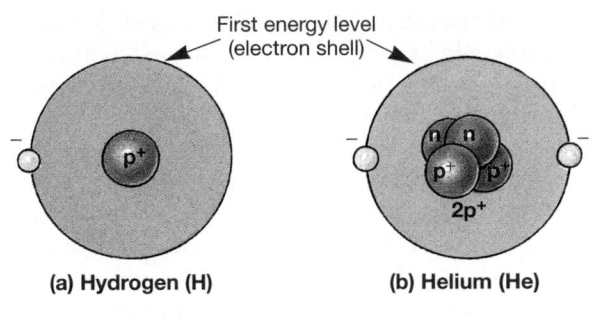

(a) Hydrogen (H)   **(b) Helium (He)**

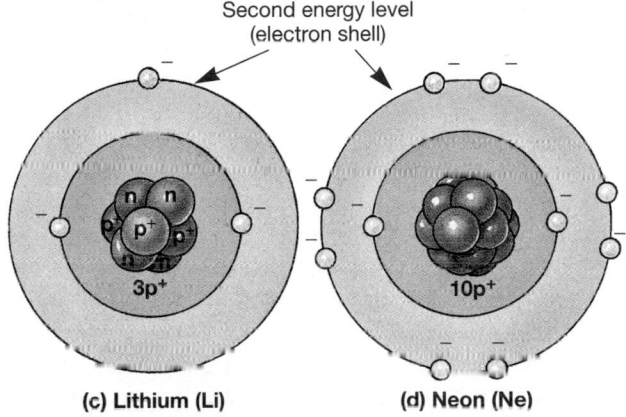

(c) Lithium (Li)   **(d) Neon (Ne)**

•**FIGURE 2-2** **Atoms and Energy Levels.** (a) A typical hydrogen atom has one proton and one electron. The electron orbiting the nucleus occupies the first energy level, diagrammed as an electron shell. (b) An atom of helium has two protons, two neutrons, and two electrons. The two electrons orbit in the same energy level. (c) The first energy level can hold only two electrons. In a lithium atom, with three protons, four neutrons, and three electrons, the third electron occupies a second energy level. (d) The second level can hold up to eight electrons. A neon atom has 10 protons, 10 neutrons, and 10 electrons; thus both the first and second energy levels are filled.

ber of 10. Neon atoms, like helium atoms, are thus very stable. (To see how other elements are related to those discussed here, consult Appendix III.)

Chemical Bonds

Elements that do not readily participate in chemical processes are said to be *inert*. Helium, neon, and argon are called *inert gases* because their atoms, having full outer energy levels, neither react with one another nor combine with atoms of other elements. Atoms with unfilled outer energy levels are called *reactive* because they will readily interact or combine with other atoms. In doing so, these atoms achieve stability by gaining, losing, or sharing electrons to fill their outer energy levels. The interactions often involve the formation of **chemical bonds,** which hold the participating atoms together once the reaction has ended. We will consider three basic types of chemical bonds:

ionic bonds, covalent bonds, and *hydrogen bonds*. The term **molecule** refers to any chemical structure consisting of atoms held together by covalent bonds. A **compound** is a chemical structure that contains two or more elements. (Whether the compound is a molecule or not depends on the type of bonding involved.) A compound has properties that can be quite different from those of its component elements. For example, a mixture of hydrogen and oxygen is highly flammable, but the chemical combination of hydrogen and oxygen atoms produces water, a compound used to put out fires.

Ionic Bonds

An atom or molecule that has a positive or negative charge is called an **ion**. Ions with a positive charge are called **cations** (KAT-ī-onz). Ions with a negative charge are called **anions** (AN-ī-onz). **Ionic bonds** are chemical bonds created by the electrical attraction between anions and cations.

Atoms become ions by losing or gaining electrons. If we assign a value of +1 to the charge on a proton, the charge on an electron is –1. As long as the number of protons is equal to the number of electrons, an atom will be electrically neutral. An atom that loses an electron will become a cation with a charge of +1, because it will have one proton that lacks a corresponding electron. Losing a second electron would give the cation a charge of +2. Adding an extra electron to a neutral atom produces an anion with a charge of –1; adding a second electron gives the anion a charge of –2.

In the formation of an ionic bond:

- One atom—the *electron donor*—loses one or more electrons and becomes a cation, with a positive (+) charge.
- Another atom—the *electron acceptor*—gains those same electrons and becomes an anion, with a negative (–) charge.
- Attraction between the opposite charges then draws the two ions together.

The formation of an ionic bond is illustrated in Figure 2-3a●. The sodium atom diagrammed in Step 1 has an atomic number of 11, so this atom normally contains 11 protons and 11 electrons. (Because neutrons are electrically neutral, their presence has no effect on the formation of ions or ionic bonds.) Electrons fill the first and second energy levels, and a single electron occupies the outermost level. Losing that one electron would give the sodium atom a full outer energy level—the second level—and would produce a **sodium ion,** with a +1 charge. (The chemical shorthand for sodium abbreviates this ion as Na^+.) But a sodium atom cannot simply throw away the electron; the elec-

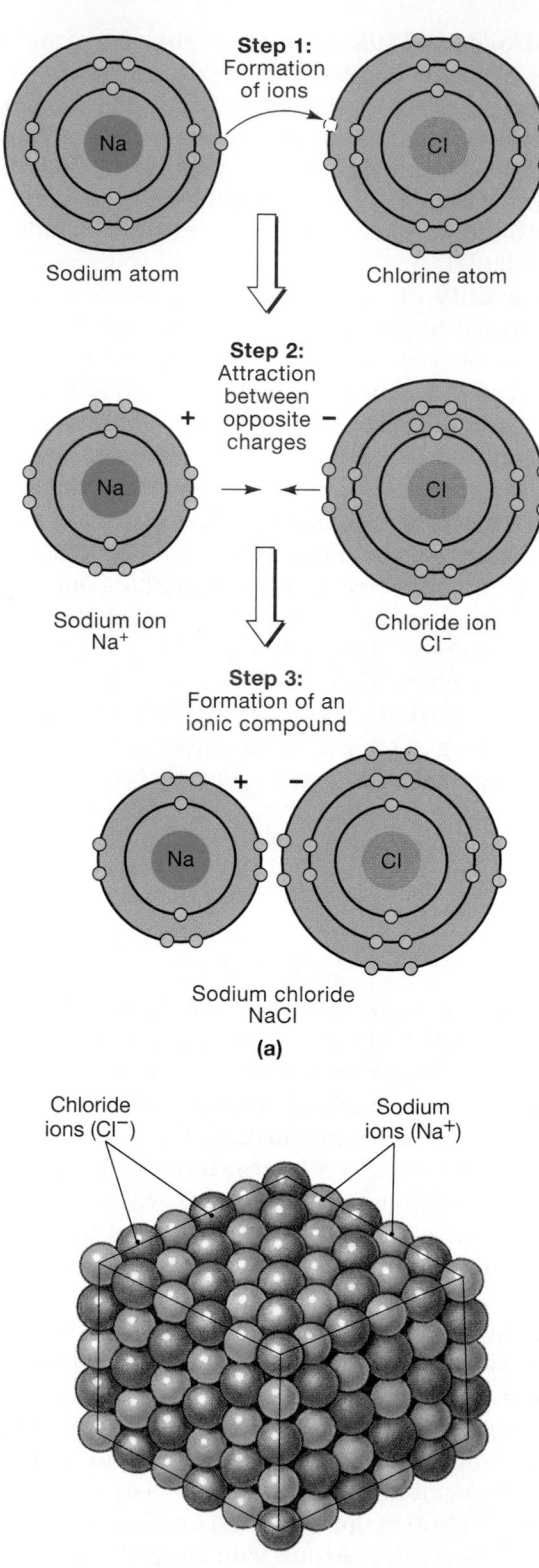

●**FIGURE 2-3   Ionic Bonding.   (a)** Step 1: A sodium (Na) atom loses an electron, which is accepted by a chlorine (Cl) atom. Step 2: Because the sodium (Na^+) and chloride (Cl^-) ions have opposite charges, they are attracted to one another. Step 3: The association of sodium and chloride ions forms the ionic compound sodium chloride. **(b)** Large numbers of sodium and chloride ions form a crystal of sodium chloride.

tron must be donated to an electron acceptor. A chlorine atom has seven electrons in its outermost energy level, so it needs only one electron to achieve stability. A sodium atom can provide the extra electron. In the process (Step 2), the chlorine atom becomes a **chloride ion** (Cl^-), with a –1 charge.

Both atoms have now become stable ions with filled outer energy levels. But the two ions do not move apart after the electron transfer, because the positively charged sodium ion is attracted to the negatively charged chloride ion (Step 3). The combination of oppositely charged ions forms the *ionic compound* **sodium chloride**, otherwise known as table salt. Large numbers of sodium and chloride ions interact to form highly structured crystals (Figure 2-3b•), held together only by the electrical attraction of oppositely charged ions. Although sodium chloride and other ionic compounds are common in body fluids, they are not present as intact crystals. When placed in water, ionic compounds dissolve, and the component anions and cations separate. This property explains why eating a heavily salted meal causes a rise in the sodium ion and chloride ion concentrations of your body fluids.

Covalent Bonds

Atoms can complete their outer electron shells not only by gaining or losing electrons but also by sharing electrons with other atoms. The result is a molecule—two or more atoms held together by **covalent** (KŌ vā lent) **bonds**.

Individual hydrogen atoms, as diagrammed in Figure 2-2a, are not found in nature. Instead, we find hydrogen molecules (Figure 2-4a•). Molecular hydrogen consists of a pair of hydrogen atoms. In chemical shorthand, molecular hydrogen is indicated by H_2, where H is the chemical symbol for hydrogen and the subscript 2 indicates the number of atoms involved. Molecular hydrogen is a gas present in the atmosphere in very small quantities. The two hydrogen atoms share their electrons, and each electron whirls around both nuclei. The sharing of one pair of electrons creates a **single covalent bond**.

Notice in Figure 2-4a• that three different methods can be used to show the structure of a hydrogen molecule: (1) an *electron shell model*, which diagrams the positions of the electrons in concentric shells that represent the energy levels; (2) a *structural formula*, which uses a single connecting line to indicate a shared electron pair; and (3) a *space-filling model*, which shows the molecule in three dimensions.

Oxygen, with an atomic number of 8, has two electrons in its first energy level and six in the second. The oxygen atoms diagrammed in Figure 2-4b• attain a stable electron configuration by sharing two pairs of electrons and thereby forming a **double covalent bond**. In a structural formula, a double covalent bond is represented by two connecting lines. Molecular oxygen (O_2) is an atmospheric gas that is very important to most living organisms. Our cells would die without a relatively constant supply of oxygen.

In our bodies, chemical processes that consume oxygen also produce **carbon dioxide** (CO_2) as a waste product. The oxygen atoms in a carbon dioxide molecule form double covalent bonds with the carbon atom, as indicated in Figure 2-4c•.

A triple covalent bond, such as the one joining two nitrogen molecules (N_2), is indicated by three connecting lines ($N\equiv N$). Molecular nitrogen accounts for roughly 79 percent of our planet's atmosphere, but our cells ignore it completely. In fact, deep-sea divers live for long periods breathing artificial air that does not contain nitrogen. (We shall discuss the reasons for eliminating nitrogen under these conditions in Chapter 23.)

Covalent bonds are very strong, because the shared electrons hold the atoms together. In typical covalent bonds the atoms remain electrically neutral, because each shared electron spends just as much time "at home" as away. (If you and a friend were tossing a pair of baseballs back and forth as fast as you could, on average, each of you would have just one baseball.) Many covalent bonds involve an equal sharing of electrons. Such bonds, which occur, for instance, between two atoms of the same type, are called **nonpolar covalent bonds**. Nonpolar covalent bonds, especially those between carbon atoms, create the stable framework of the large molecules that make up most of the structural components of the human body.

Covalent bonds usually form molecules that complete the outer energy levels of the atoms involved. An ion or molecule that contains unpaired electrons in its outermost energy level is called a *free radical.* Free radicals are highly reactive. Almost as fast as it forms, a free radical enters additional reactions that are typically destructive. For example, free radicals can damage or destroy vital compounds, such as proteins. Free radicals sometimes form in the course of normal metabolism, but living cells have several methods of removing or inactivating them. However, *nitric oxide* (NO) (Figure 2-4d•) is a free radical that has important functions in the body. It is involved in chemical communication in the nervous system, in the control of blood vessel diameter, in blood clotting, and in the defense against bacteria and other pathogens.

POLAR COVALENT BONDS Covalent bonds involving different types of atoms may instead involve a very unequal sharing of electrons, because the elements

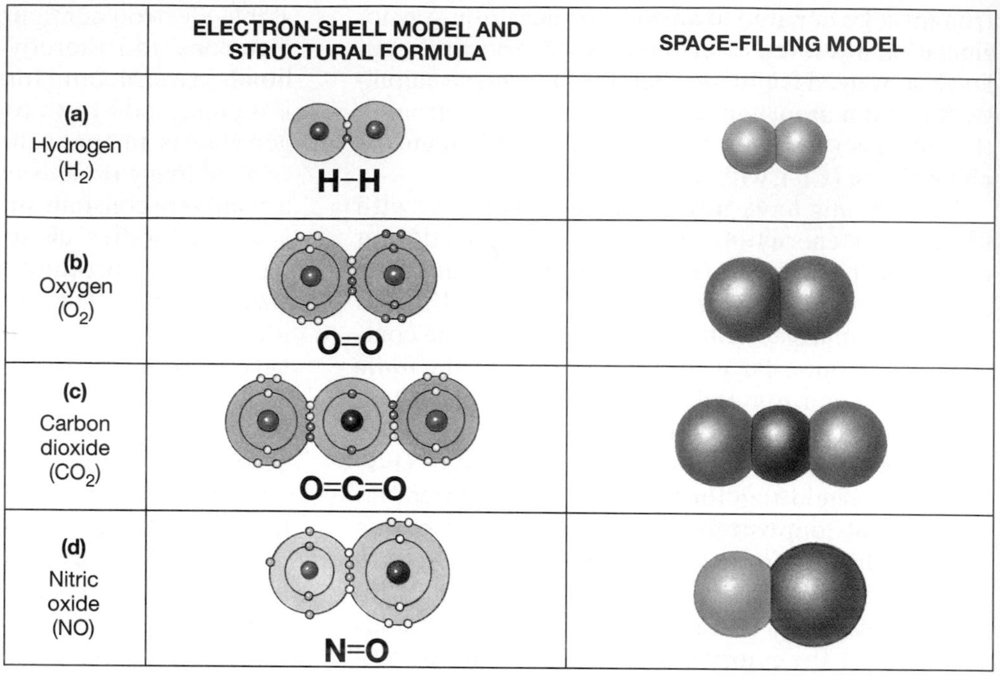

•*FIGURE 2-4* **Covalent Bonds.** **(a)** In a molecule of hydrogen, two hydrogen atoms share their electrons such that each has a filled outer electron shell. This sharing creates a single covalent bond. **(b)** A molecule of oxygen consists of two oxygen atoms that share two pairs of electrons. The result is a double covalent bond. **(c)** In a molecule of carbon dioxide, a central carbon atom forms double covalent bonds with a pair of oxygen atoms. **(d)** A molecule of nitric oxide is held together by a double covalent bond, but the outer electron shell of the nitrogen atom requires an additional electron to be complete. Thus, nitric oxide is a free radical, which will readily react with another atom or molecule.

	ELECTRON-SHELL MODEL AND STRUCTURAL FORMULA	SPACE-FILLING MODEL
(a) Hydrogen (H_2)	H–H	
(b) Oxygen (O_2)	O=O	
(c) Carbon dioxide (CO_2)	O=C=O	
(d) Nitric oxide (NO)	N=O	

differ in how strongly they attract electrons. An unequal sharing of electrons creates a **polar covalent bond**. For example, in a molecule of water (Figure 2-5•), an oxygen atom forms covalent bonds with two hydrogen atoms. The oxygen nucleus has a much stronger attraction for the shared electrons than the hydrogen atoms do. As a result, the electrons spend more time orbiting the oxygen nucleus than orbiting the hydrogen nuclei. Because it has two extra electrons most of the time, the oxygen atom develops a slight (partial) negative charge, indicated by δ^-. At the same time, each hydrogen atom develops a slight (partial) positive charge, δ^+, because its electron is away much of the time. (Let us return to our baseball analogy. Suppose you and a friend were tossing a pair of baseballs back and forth, but one of you returned them back as fast as possible, and the other held onto them for a while before throwing them back. One of you would now, on average, have more than one baseball, and the other would have fewer than one.) The unequal sharing of electrons makes polar covalent bonds somewhat weaker than other covalent bonds.

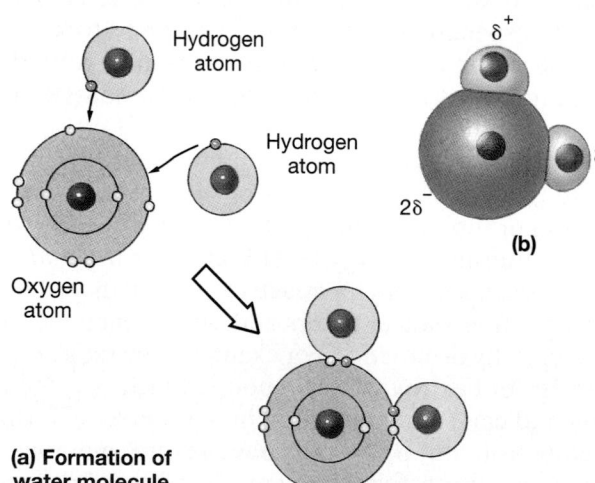

Hydrogen atom

Hydrogen atom

Oxygen atom

δ^+

$2\delta^-$

δ^+

(b)

(a) Formation of water molecule

•*FIGURE 2-5* **Polar Covalent Bonds and the Structure of Water.** **(a)** In forming a water molecule, an oxygen atom completes its outer energy level by sharing electrons with a pair of hydrogen atoms. The sharing is unequal because the oxygen atom holds the electrons more tightly than do the hydrogen atoms. **(b)** Because the oxygen atom has two extra electrons much of the time, it develops a slight negative charge, and the hydrogen atoms become weakly positive. The bonds in a water molecule are polar covalent bonds.

Hydrogen Bonds

Covalent and ionic bonds tie atoms together to form molecules and compounds. There are also comparatively weak forces that act between adjacent molecules and even between atoms within a large molecule. **Hydrogen bonds** are the most important of these weak attractive forces. A hydrogen bond is the attraction between a δ^+ on the hydrogen atom involved in a polar covalent bond and a δ^- on an oxygen or nitrogen atom in another polar covalent bond. The polar covalent bond containing the oxygen or nitrogen atom can be in a different molecule or in the same molecule as the hydrogen atom.

Hydrogen bonds are too weak to create molecules, but they can change molecular shapes or pull molecules together. For example, hydrogen bonding occurs between water molecules (Figure 2-6a●). At the water surface, the attraction between molecules slows the rate of evaporation and creates the phenomenon known as surface tension (Figure 2-6b●). **Surface tension** acts as a barrier that keeps small objects from entering the water; it is the reason that insects can walk across the surface of a pond or puddle. Surface tension in a layer of tears keeps small objects such as dust particles from touching the surface of the eye. At the cellular level, hydrogen bonds affect the shapes and properties of complex molecules, such as proteins and nucleic acids (including DNA), and they may also determine the three-dimensional relationships between molecules.

States of Matter

Matter in our environment can be in one of three states: solid, liquid, or gaseous. *Solids* maintain their volume and shape at ordinary temperatures and pressures. A lump of granite, a brick, and a textbook are examples of solid objects. *Liquids* have a constant volume but no fixed shape. The shape of a liquid is determined by the shape of its container. Water, coffee, and soda are examples of liquids. A *gas* has neither a constant volume nor a fixed shape. Gases can be compressed or expanded, and they will fill a container of any shape. The air in our atmosphere is the gas with which we are most accustomed.

Whether a particular substance exists as a solid, a liquid, or a gas depends on the degree of interaction among the atoms or molecules involved. For

example, hydrogen molecules have little attraction for one another; in our environment, molecular hydrogen exists as a gas. Water is the only substance that occurs as a solid (ice), a liquid (water), and a gas (water vapor) at temperatures compatible with life. Water exists in the liquid state over a broad range of temperatures primarily due to hydrogen bonding among the water molecules. We will talk more about the unusual properties of water in a later section.

Molecular Weights

The **molecular weight** of a molecule is equal to the sum of the atomic weights of its component atoms. In other words, the molecular weight in grams is equal to the weight of one mole of molecules. Molecular weights are important because you cannot handle individual molecules; nor could you easily count the billions of molecules involved in chemical reactions in the body. Using molecular weights, you can calculate the quantities of reactants needed to perform a specific reaction and determine the amount of product generated. For example, suppose you wanted to create water from hydrogen and oxygen, according to the balanced equation

$$2\,H_2 + O_2 \longrightarrow 2\,H_2O$$

The first step in performing such an experiment would be to calculate the molecular weights involved. The atomic weight of hydrogen is close to 1.0, so one hydrogen molecule (H_2) has a molecular weight near 2.0. Oxygen has an atomic weight of about 16, so the molecular weight of an oxygen mol-

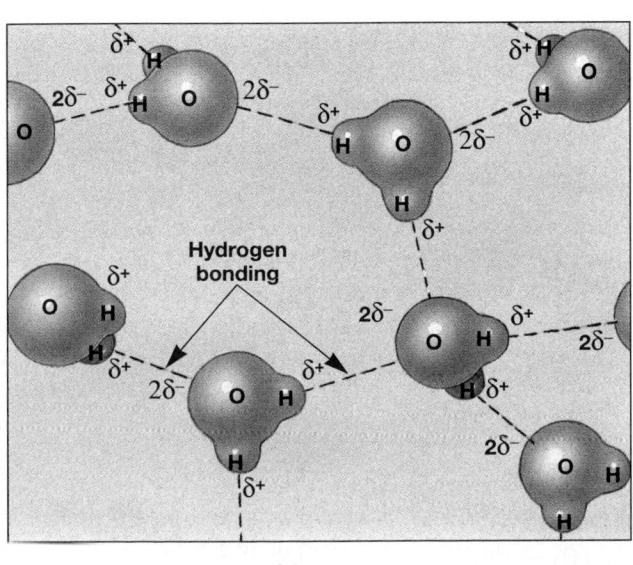

(a)

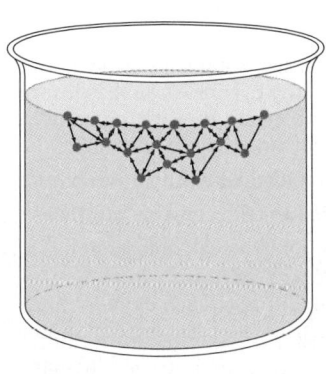

(b)

●**FIGURE 2-6** **Hydrogen Bonds.** **(a)** The hydrogen atoms of a water molecule have a slight positive charge, and the oxygen atom has a slight negative charge (see Figure 2-5). Attraction between a hydrogen atom of one water molecule and the oxygen atom of another is a hydrogen bond (indicated by dashed lines). **(b)** Hydrogen bonding between water molecules at a free surface restricts evaporation and creates surface tension.

ecule (O_2) is roughly 32. In practical terms, if you wanted to perform the experiment, you would combine 4 g of hydrogen with 32 g of oxygen to produce 36 g of water. You could also work with ounces, pounds, or tons, as long as the proportions remained the same.

☑ Both oxygen and neon are gases at room temperature. Oxygen combines readily with other elements, but neon does not. Why?

☑ How is it possible for two samples of hydrogen to contain the same number of atoms but have different weights?

☑ Which kind of bond holds atoms in a water molecule together? What attracts water molecules to one another?

CHEMICAL REACTIONS

Living cells remain alive and functional by controlling chemical reactions. In a **chemical reaction**, new chemical bonds form between atoms, or existing bonds between atoms are broken. These changes occur as atoms in the reacting substances, or **reactants**, are rearranged to form different substances, or **products**. In effect, each cell is a chemical factory. For example, growth, maintenance and repair, secretion, and contraction all involve complex chemical reactions. The term **metabolism** (me-TAB-ō-lizm) refers to all the chemical reactions that occur in an organism. Living cells use chemical reactions to provide the energy needed to maintain homeostasis and to perform essential functions.

FOCUS   Chemical Notation

Before we can consider the specific compounds that occur in the human body, we must be able to describe chemical compounds and reactions effectively. The use of sentences to describe chemical structures and events often leads to confusion. A simple form of "chemical shorthand" makes communication much more efficient. The chemical shorthand we will use is known as **chemical notation**. Chemical notation enables us to describe complex events briefly and precisely. The rules of chemical notation are summarized in Table 2-2.

TABLE 2-2 **Rules of Chemical Notation**

1. The symbol of an element indicates one atom of that element:

    H = one atom of hydrogen; O = one atom of oxygen

2. A number preceding the symbol of an element indicates more than one atom:

    2 H = two individual atoms of hydrogen
    2 O = two individual atoms of oxygen

3. A subscript following the symbol of an element indicates a molecule with that number of atoms of that element:

    H_2 = a hydrogen molecule, composed of two hydrogen atoms
    O_2 = an oxygen molecule, composed of two oxygen atoms
    H_2O = a water molecule, composed of two hydrogen atoms and one oxygen atom

4. In a description of a chemical reaction, the interacting participants are called reactants, and the reaction generates one or more products. An arrow indicates the direction of the reaction, from reactants (usually on the left) to products (usually on the right). In the following reaction, two atoms of hydrogen combine with one atom of oxygen to produce a single molecule of water:

    $$2\,H + O \rightarrow H_2O$$

5. A superscript plus or minus sign following the symbol of an element indicates an ion. A single plus sign indicates a cation with a charge of +1 (the original atom has lost one electron). A single minus sign indicates an anion with a charge of –1 (gain of one electron). If more than one electron has been lost or gained, the charge on the ion is indicated by a number preceding the plus or minus:

    Na^+ = one sodium ion (the sodium atom has lost 1 electron)
    Cl^- = one chloride ion (the chlorine atom has gained 1 electron)
    Ca^{2+} = one calcium ion (the calcium atom has lost 2 electrons)

6. Chemical reactions neither create nor destroy atoms—they merely rearrange atoms into new combinations. Therefore, the numbers of atoms of each element must always be the same on both sides of the equation for a chemical reaction. When this is the case, the equation is balanced:

    Unbalanced: $H_2 + O_2 \rightarrow H_2O$
    Balanced: $2\,H_2 + O_2 \rightarrow 2\,H_2O$

Basic Energy Concepts

Before we can consider various types of reactions, we must understand some basic principles concerning the relationships between matter and energy. **Work** is movement or a change in the physical structure of matter. **Energy** is the capacity to perform work; movement or physical change cannot occur unless energy is provided. There are two major types of energy: kinetic energy and potential energy:

1. **Kinetic energy** is the energy of motion—energy that is doing work. When a car hits a tree, it is kinetic energy that does the damage.

2. **Potential energy** is stored energy—energy that has the potential to do work. It may result from the position of an object (a book on a high shelf) or from an object's physical or chemical structure (a stretched spring or a charged battery).

Kinetic energy must be used in lifting the book to the shelf, in stretching the spring, or in charging the battery. The potential energy is converted back into kinetic energy when the book falls, the spring recoils, or the battery discharges. The kinetic energy can then be used to perform work.

Energy cannot be lost; it can only be converted from one form to another. A conversion between potential energy and kinetic energy is never 100 percent efficient. Each time an energy exchange occurs, some of the energy is released in the form of heat. *Heat* is an increase in random molecular motion. The temperature of an object is proportional to its heat content.

Living cells perform work in many forms. For example, a skeletal muscle at rest contains potential energy in the form of the positions of protein filaments and the covalent bonds between molecules inside the cell. When a muscle contracts, it performs work; potential energy is converted into kinetic energy, and heat is released. The amount of heat is proportional to the amount of work done. As a result, when you exercise, your body temperature rises.

Types of Reactions

Three types of chemical reactions are important to the study of physiology: (1) decomposition reactions, (2) synthesis reactions, and (3) exchange reactions.

Decomposition

Decomposition is a reaction that breaks a molecule into smaller fragments. You could diagram a typical *decomposition reaction* as

$$AB \longrightarrow A + B$$

Decomposition reactions occur outside cells as well as inside them. For example, a typical meal contains molecules of fats, sugars, and proteins that are too large and too complex to be absorbed and used by your body. Decomposition reactions in the digestive tract break these down into smaller fragments before absorption begins.

Catabolism (ka-TAB-ō-lizm; *katabole*, a throwing down) is the decomposition of molecules within cells. When a covalent bond—a form of potential energy—is broken, it releases kinetic energy that can perform work. By harnessing the energy released in this way, living cells perform vital functions such as growth, movement, and reproduction.

Synthesis

Synthesis (SIN-the-sis) is the opposite of decomposition. A *synthesis reaction* assembles larger molecules from smaller components. Simple synthetic reactions could be diagrammed as

$$A + B \longrightarrow AB$$

Synthesis reactions may involve individual atoms or the combination of molecules to form even larger products. The formation of water from hydrogen and oxygen molecules is a synthesis reaction. Synthesis always involves the formation of new chemical bonds, whether the reactants are atoms or molecules.

Anabolism (a-NAB-ō-lizm; *anabole*, a throwing upward) is the synthesis of new compounds within the body. Because it takes energy to create a chemical bond, anabolism is usually an "uphill" process. Living cells must balance their energy budgets, with catabolism providing the energy to support anabolism as well as other vital functions.

Exchange

In an **exchange reaction**, parts of the reacting molecules are shuffled around. The following equation represents a simple exchange reaction:

$$AB + CD \longrightarrow AD + CB$$

In this example, there are two reactants and two products. Although the reactants and products contain the same components (A, B, C, and D), those components are present in different combinations. In an exchange reaction, the reactant molecules AB and CD break apart (a decomposition) before they interact with one another to form AD and CB (a synthesis). Specific chemical reactions may absorb or release energy, generally in the form of heat. If more energy is released by the breaking of the old bonds than it takes to create the new ones, the exchange reaction will release energy. Reactions that release energy are said to be **exergonic** (*exo-,* outside). If more energy is required for the synthesis part of the reaction than is released by the associated decomposition, additional energy must be provided. Reactions that absorb energy are termed **endergonic** (*endo-,* inside).

Reversible Reactions

Chemical reactions are (at least theoretically) reversible, so if $A + B \longrightarrow AB$, then $AB \longrightarrow A + B$. Many important biological reactions are freely reversible. Such reactions can be represented as an equation:

$$A + B \rightleftharpoons AB$$

This equation reminds you that there are really two reactions occurring simultaneously—one, a synthesis $(A + B \longrightarrow AB)$, and the other, a decomposition $(AB \longrightarrow A + B)$. At **equilibrium** (ē-kwi-LIB-rē-um), the rates at which the two equations proceed are in balance. As fast as one molecule of AB forms, another degrades into $A + B$.

It is possible to predict the result of a disturbance in the equilibrium condition. In our example, the synthesis reaction rate is directly proportional to the frequency of encounters between A and B. The frequency of encounters depends on the degree of crowding: In a crowded room, you are much more likely to run into another person than in a room that is almost empty. The **concentration** of a substance is the number of atoms or molecules (or moles) in a specified volume, typically a liter. *The concentrations of the reactants and products in a chemical reaction have a direct effect on the reaction rate.* When the concentration of a reactant increases, the rate of reaction will increase. In the preceding example, the addition of more AB molecules will increase the rate of conversion of AB to A and B. The concentrations of A and B will then increase, and this increase will lead to an increase in the rate of the reverse reaction—the formation of AB from A and B. Eventually, an equilibrium is again established.

Not all chemical reactions, however, are easily reversed. The requirements for two paired reactions may differ, so at any given time and place the overall reaction will proceed chiefly in one direction only. For example, the synthesis reaction may occur only when the A and B molecules are heated; the decomposition reaction may occur only when AB molecules are placed in water. In this case, the reaction would be written as

$$A + B \xrightleftharpoons[\text{H}_2\text{O}]{\text{heat}} AB$$

Decomposition reactions involving water are important in the breakdown of complex molecules in the body. We will encounter several examples of this process, called *hydrolysis,* later in this chapter. In **hydrolysis** (hī-DROL-i-sis; *hydro-,* water + *lysis,* dissolution), one of the bonds in a complex molecule is broken, and the components of a water molecule (H and OH) are added to the resulting fragments:

$$\text{A–B–C–D–E} + \text{H}_2\text{O} \longrightarrow \text{A–B–C–H} + \text{HO–D–E}$$

Enzymes and Chemical Reactions

Most chemical reactions do not occur spontaneously, or they occur so slowly that they would be of little value to living cells. Before a reaction can proceed, enough energy must be provided to activate the reactants. **Activation energy** is the amount of energy required to start a reaction. Although many reactions can be activated by changes in temperature or acidity, such changes are deadly to cells. For example, every day your cells break down complex sugars as part of your normal metabolism. Yet to break down a complex sugar in a laboratory, you must boil it in an acid solution. Your cells don't have that option; temperatures that high and solutions that corrosive would immediately destroy living tissues. Instead, your cells use special proteins called *enzymes* to perform most of the complex synthesis and decomposition reactions in your body.

Enzymes promote chemical reactions by lowering the activation energy requirements, making it possible for chemical reactions, such as the breakdown of sugars, to proceed under conditions compatible with life. Enzymes belong to a class of substances called *catalysts* (KAT-uh-lists; *katalysis,* dissolution), compounds that accelerate chemical reactions without themselves being permanently changed or consumed in the process. An enzyme molecule is made by a living cell to promote a specific reaction. Enzymatic reactions, which are reversible, can be written as

$$A + B \xrightleftharpoons{\text{enzyme}} AB$$

Although the presence of an appropriate enzyme can accelerate a reaction, an enzyme affects only the rate of the reaction, not the direction of the reaction or the products that will be formed. An enzyme cannot bring about a reaction that would otherwise be impossible.

The complex reactions that support life proceed in a series of interlocking steps, and each step is controlled by a specific enzyme. Such a reaction sequence is called a *pathway.* A synthetic pathway could be diagrammed as

$$A \underset{\text{Step 1}}{\xrightleftharpoons{\text{enzyme 1}}} B \underset{\text{Step 2}}{\xrightleftharpoons{\text{enzyme 2}}} C \underset{\text{Step 3}}{\xrightleftharpoons{\text{enzyme 3}}} \text{and so on}$$

In some cases, the steps in the synthetic pathway differ from those of the decomposition pathway, and separate enzymes may be involved.

☑ In living cells, glucose, a six-carbon molecule, is converted into two three-carbon molecules by a reaction that releases energy. How would you classify this reaction?

☑ If the product of a reversible reaction is continuously removed, what effect do you think its removal will have on the equilibrium?

INORGANIC COMPOUNDS

Although the human body is very complex, it contains relatively few elements, as indicated in Table 2-1, p. 32. But knowing the identity and quantity of each element will not help you understand a human being any more than studying the alphabet will help you understand this textbook. The rest of this chapter focuses on *nutrients* and *metabolites*. **Nutrients** are the essential elements and molecules normally obtained from the diet. **Metabolites** (me-TAB-o-līts; *metabole*, change) include all the molecules synthesized or broken down by chemical reactions inside our bodies. Nutrients and metabolites can be broadly categorized as *inorganic* or *organic*. **Inorganic compounds** generally do not contain carbon and hydrogen atoms as the primary structural ingredients, whereas carbon and hydrogen *always* form the basis for **organic compounds**.

The most important inorganic substances in the body are (1) *carbon dioxide*, a byproduct of cell metabolism; (2) *oxygen*, an atmospheric gas required in important metabolic reactions; (3) *water*, which accounts for most of our body weight; and (4) *inorganic acids, bases*, and *salts*, compounds held together partially or completely by ionic bonds. In this section, we will be concerned primarily with water, its properties, and how those properties establish the conditions necessary for life. Most of the other inorganic molecules and compounds in the body are found in association with water, the primary component of our body fluids. For example, carbon dioxide and oxygen are gas molecules that occur both in the atmosphere and in body fluids; all the inorganic acids, bases, and salts we will discuss are dissolved in body fluids.

Water and Its Properties

Water, H_2O, is the single most important constituent of the body, accounting for almost two-thirds of total body weight. A change in body water content can have fatal consequences because virtually all physiological systems will be affected.

Although familiar to everyone, water really has some very unusual properties. These properties are a direct result of the hydrogen bonding that occurs between adjacent water molecules:

1. A remarkable number of inorganic and organic molecules will *dissolve* (break up) in water, creating a solution. Every **solution** is a uniform mixture of two or more substances—whether in solid, liquid, or gaseous form. A solution consists of a medium, or **solvent**, in which atoms, ions, or molecules of another substance, or **solute**, are dispersed. In **aqueous solutions**, water is the solvent. As they dissolve, solute molecules become uniformly dispersed throughout the solution. The solvent properties of water are so important that we will consider them in detail in the next section.

2. In our bodies, chemical reactions occur in water and water molecules are also participants in some reactions. For example, water molecules participate in some catabolic reactions that break large molecules into smaller fragments. Such a reaction occurs during the process of *hydrolysis*.

3. Water has a very high heat capacity, and its freezing and boiling points are far apart. *Heat capacity* is the ability to absorb and retain heat. Water has an unusually high heat capacity because water molecules in the liquid state are attracted to one another through hydrogen bonding. Important consequences of this attraction include the following:

   - The temperature must be quite high before individual molecules have enough energy to break free and become water vapor. Consequently, water stays in the liquid state over a broad range of environmental temperatures.
   - Water carries a great deal of heat away with it when it finally does change from a liquid to a vapor. This feature accounts for the cooling effect of perspiration on the skin.
   - It takes an unusually large amount of heat energy to change the temperature of 1 g of water by 1°C. Thus, once a quantity of water has reached a particular temperature, it will change temperature only slowly. This property is called *thermal inertia*. Because water accounts for 50–60 percent of the weight of the human body, thermal inertia helps stabilize body temperature. Similarly, water's high heat capacity allows the blood plasma to transport and redistribute large amounts of heat as it circulates within the body. For example, heat absorbed as the blood flows through active muscles will be released when the blood reaches vessels in the relatively cool body surface.

4. Water is an effective lubricant. There is little friction between water molecules. Thus, if two opposing sur-

faces are separated by even a thin layer of water, friction between those surfaces will be greatly reduced. (That is why driving on wet roads can be tricky; your tires may start sliding on a layer of water rather than maintaining contact with the road.) Within joints such as the knee, an aqueous solution prevents friction between the opposing surfaces, and a small amount of fluid in the ventral body cavities prevents friction between internal organs, such as the heart or lungs, and the body wall. ∞ *[p. 24]*

Aqueous Solutions

Its chemical structure makes water an unusually effective solvent. Because the hydrogen atoms are attached to the oxygen atom by polar covalent bonds, the hydrogen atoms have a slight positive charge and the oxygen atom has a slight negative charge (Figure 2-7a●). The bonds in a water molecule are oriented so as to place the hydrogen atoms relatively close together. As a result, the water molecule has positive and negative poles. A water molecule is therefore called a *polar molecule*, or a *dipole*.

Many inorganic compounds are held together partially or completely by ionic bonds. In water, these compounds undergo *ionization* (ī-on-i-ZĀ-shun), or *dissociation* (di-sō-sē-Ā-shun). In this process (Figure 2-7b●), ionic bonds are broken apart as the individual ions interact with the positive or negative ends of polar water molecules. The result is a mixture of cations and anions that are surrounded by water molecules. The water molecules around each ion form a *hydration sphere*.

An aqueous solution containing anions and cations will conduct an electrical current. Cations (+) move toward the negative side, or negative *terminal*, and anions (−) move toward the positive terminal. Electrical forces across cell membranes affect the functioning of all cells, and small electrical currents carried by ions are essential to muscle contraction and nerve function. Chapters 10 and 12 will discuss these processes in more detail.

Soluble inorganic molecules whose ions will conduct an electrical current in solution are called **electrolytes** (e-LEK-trō-līts). Sodium ions (Na^+), potassium ions (K^+), calcium ions (Ca^{2+}), and chloride ions (Cl^-) are released by the dissociation of electrolytes in blood and other body fluids. Table 2-3 lists important electrolytes and the ions released by their dissociation.

Alterations in the body fluid concentrations of electrolytes will disturb almost every vital function. For example, declining potassium levels will lead to a general muscular paralysis, and rising concentrations will cause weak and irregular heartbeats. The concentrations of ions in body fluids are carefully regulated, primarily by the coordination of activities at the kidneys (ion excretion), the digestive tract (ion absorption), and the skeletal system (ion storage or release).

Organic molecules typically contain polar covalent bonds, which also attract water molecules. The hydration spheres that form then carry these molecules into solution, as shown in Figure 2-7c●. The molecule in this figure is an important soluble sugar, *glucose*. Molecules that readily interact with water molecules in this way are called *hydrophilic* (hī-drō-FI-lik; *hydro-*, water + *philos*, loving).

Molecules that do not readily interact with water are called *hydrophobic* (hī-drō-FŌ-bik; *hydro-*, water + *phobos*, fear). Hydrophobic molecules have few if

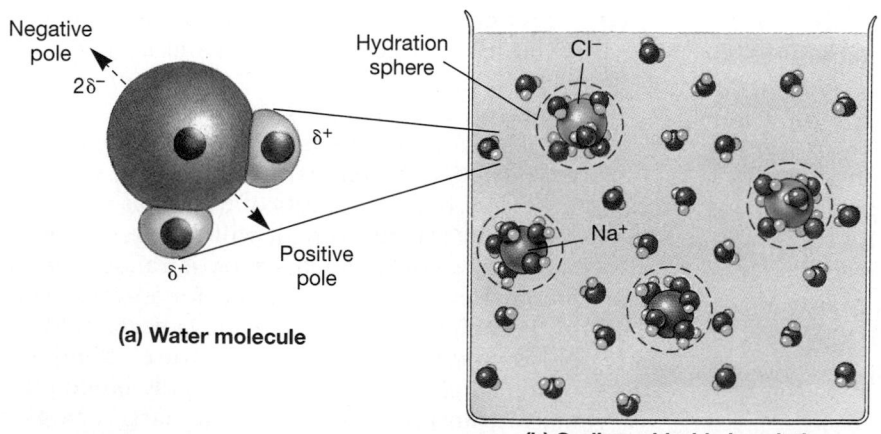

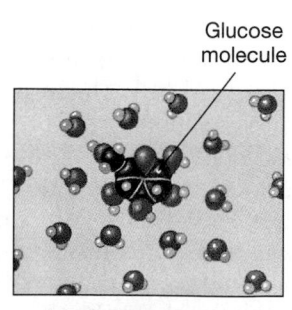

(a) Water molecule

(b) Sodium chloride in solution

(c) Glucose in solution

●*FIGURE 2-7*   **Water Molecules and Solutions.**   **(a)** In a water molecule, oxygen forms polar covalent bonds with two hydrogen atoms. Because the hydrogen atoms are positioned toward one end of the molecule, the molecule has an uneven distribution of charges that creates positive and negative poles. **(b)** Ionic compounds dissociate in water as the polar water molecules disrupt the ionic bonds. Each ion in solution is surrounded by water molecules, creating hydration spheres. **(c)** Hydration spheres also form around an organic molecule containing polar covalent bonds. If the molecule is small, it will be carried into solution, as shown here with glucose.

TABLE 2-3 Important Electrolytes That Dissociate in Body Fluids

NaCl (sodium chloride)	$\rightarrow$	Na^+	$+$	Cl^-
KCl (potassium chloride)	$\rightarrow$	K^+	$+$	Cl^-
CaCl$_2$ (calcium chloride)	$\rightarrow$	Ca^{2+}	$+$	$2\,Cl^-$
NaHCO$_3$ (sodium bicarbonate)	$\rightarrow$	Na^+	$+$	HCO_3^-
MgCl$_2$ (magnesium chloride)	$\rightarrow$	Mg^{2+}	$+$	$2\,Cl^-$
Na$_2$HPO$_4$ (disodium phosphate)	$\rightarrow$	$2\,Na^+$	$+$	HPO_4^{2-}
Na$_2$SO$_4$ (sodium sulfate)	$\rightarrow$	$2\,Na^+$	$+$	SO_4^{2-}

any polar covalent bonds; they are said to be *nonpolar*. When nonpolar molecules are exposed to water, hydration spheres do not form, and the molecules do not dissolve. Body fat deposits consist of large, insoluble droplets of hydrophobic molecules trapped in the watery interior of cells. Gasoline, heating oil, and diesel fuel are examples of hydrophobic molecules not found in the body.

SOLUTE CONCENTRATIONS Physiologists and clinicians often monitor inorganic and organic **solute concentrations** in body fluids such as blood or urine. Solute concentrations may be reported in several different ways, and we will introduce two methods here. In the first method, we report the number of solute atoms, molecules, or ions in a specific volume of solution. Values are reported in terms of moles (mol) per liter or millimoles (mmol) per liter. As you will recall from earlier in the chapter, a mole is the number of atoms or molecules present in a sample that has a weight in grams equal to the atomic weight or molecular weight, respectively. Physiological concentrations today are most often reported in terms of millimoles per liter (mmol/l).

You can report concentrations in terms of millimoles only when you are dealing with an ion or molecule of known molecular weight. When the chemical structure is unknown or you are dealing with a complex mixture of materials, concentration is reported in terms of the weight of material dissolved in a unit volume of solution. Values are most often expressed in terms of milligrams (mg) or grams (g) per deciliter (dl, or 100 ml). This is the method used, for example, when reporting the concentration of plasma proteins in a blood sample. The *Applications Manual* contains a detailed analysis of the compositions of blood and other body fluids. **AM** *Solutions and Concentrations*

Colloids and Suspensions

Body fluids may contain large and complex organic molecules, such as proteins and protein complexes, that are held in solution by their association with water molecules, as in Figure 2-7c•. A solution containing dispersed proteins or other large molecules is called a *colloid*. Liquid Jell-O™ is a familiar viscous (thick) colloid.

The particles or molecules in a colloid will remain in solution indefinitely. A *suspension* contains even larger particles that will, if undisturbed, settle out of solution due to the force of gravity. Stirring beach sand into a bucket of water creates a temporary suspension that will last only until the sand settles to the bottom. Whole blood is another temporary suspension, because the blood cells are suspended in the blood plasma. If clotting is prevented, the cells in a blood sample will gradually settle to the bottom of the container.

Hydrogen Ions in Body Fluids

The concentration of hydrogen ions in body fluids is precisely regulated. Hydrogen ions are extremely reactive in solution. In excessive numbers, they will break chemical bonds, change the shapes of complex molecules, and generally disrupt cell and tissue functions.

A few hydrogen ions are normally present, even in a sample of pure water, because some of the water molecules dissociate, releasing cations and anions. The dissociation of water is a reversible reaction that can be represented as

$$H_2O \rightleftharpoons H^+ + OH^-$$

The dissociation of one water molecule yields one *hydrogen ion*, H^+, and one *hydroxide* (hī-DROK-sīd) *ion*, OH^-. Very few water molecules ionize in pure water, and the number of hydrogen and hydroxide ions is very small. The quantities are usually reported in moles, making it easy to keep track of the relative numbers of hydrogen and hydroxide ions. One liter of pure water contains about 0.0000001 mol of hydrogen ions and an equal number of hydroxide ions. In other words, the *concentration* of hydrogen ions in a solution of pure water is 0.0000001 mol per liter. This can be written as

$$[H^+] = 1 \times 10^{-7} \text{ mol/l}$$

The brackets around H^+ indicate "the concentration of," another example of chemical notation.

pH The hydrogen ion concentration in body fluids is so important to physiological processes that a special shorthand is used to express it. The **pH** of a solution is defined as the negative logarithm of the hydrogen ion concentration in moles per liter. Thus, instead of using the previous equation for $[H^+]$, we say that the pH of pure water is $-(-7)$, or 7. Using pH values saves space, but you must always remember that the pH number is an *exponent* and that the pH scale is logarithmic. For instance, a pH of 6 ($[H^+] = 1 \times 10^{-6}$, or 0.000001) means

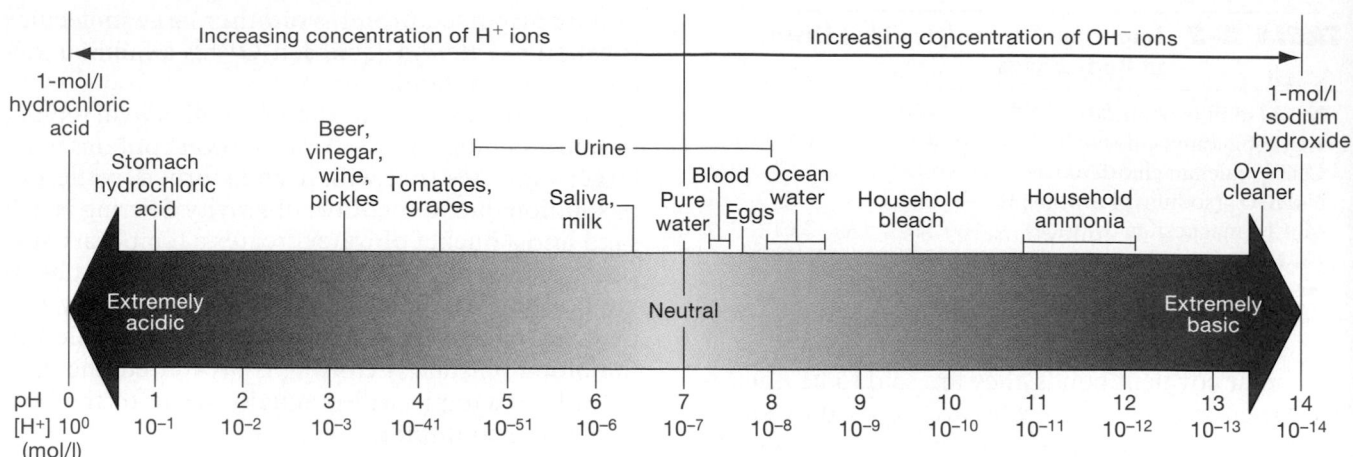

•FIGURE 2-8   **pH and Hydrogen Ion Concentration.**   Note that the scale is logarithmic; an increase or decrease of one unit corresponds to a 10-fold change in H^+ concentration.

that the concentration of hydrogen ions is *10 times as great* as it is at a pH of 7 ($[H^+] = 1 \times 10^{-7}$, or 0.0000001). For common liquids, the pH scale, included in Figure 2-8•, ranges from 0 to 14.

Although pure water has a pH of 7, solutions display a wide range of pH values, depending on the nature of the solutes involved. A solution with a pH of 7 is said to be **neutral**, because it contains equal numbers of hydrogen and hydroxide ions. A solution with a pH below 7 is **acidic** (a-SI-dik), meaning that hydrogen ions predominate. A pH above 7 is **basic**, or *alkaline* (AL-kuh-lin), with hydroxide ions in the majority.

The pH of the blood normally ranges from 7.35 to 7.45. Abnormal fluctuations in pH can damage cells and tissues by breaking chemical bonds, changing the shapes of proteins, and altering cellular functions. The human body generates significant quantities of acids that threaten homeostasis and must be neutralized. Under unusual circumstances, the loss of acids in body fluids may promote an equally disruptive increase in pH. *Acidosis* refers to an abnormal physiological state caused by low blood pH (below 7.35); a pH below 7 can produce coma. *Alkalosis* results from an abnormally high pH (above 7.45); a blood pH above 7.8 generally causes uncontrollable, sustained skeletal muscle contractions.

Inorganic Acids and Bases

The body contains both inorganic and organic *acids* and *bases* that can potentially cause acidosis or alkalosis, respectively. An **acid** is any solute that dissociates in solution and releases hydrogen ions, thereby lowering the pH. Because a hydrogen atom that loses its electron consists solely of a proton, hydrogen ions are often referred to simply as protons, and acids as *proton donors*.

A *strong acid* dissociates completely in solution, and the reaction occurs essentially in one direction only. *Hydrochloric acid* (HCl) is a representative strong acid; in water it ionizes as

$$HCl \longrightarrow H^+ + Cl^-$$

The stomach produces this powerful acid to assist in the breakdown of food. Hardware stores sell HCl under the name muriatic acid, for cleaning sidewalks and swimming pools.

A **base** is a solute that removes hydrogen ions from a solution and thereby acts as a *proton acceptor*. In solution, many bases release a hydroxide ion (OH^-). Hydroxide ions have a strong affinity for hydrogen ions and quickly react with them to form water molecules. A *strong base* dissociates completely in solution. *Sodium hydroxide*, NaOH, is a strong base, and the reaction that releases sodium ions and hydroxide ions can be written as

$$NaOH \longrightarrow Na^+ + OH^-$$

Strong bases have a variety of industrial and household uses. Drain openers (Drano™) and lye are two familiar examples.

Weak acids and *weak bases* fail to dissociate completely, and at equilibrium a significant number of molecules remain intact in the solution. For the same number of molecules in solution, weak acids and weak bases therefore have less of an impact on pH than do strong acids and strong bases. *Carbonic acid* (H_2CO_3) is an example of a weak acid found in body fluids. In solution, carbonic acid reversibly dissociates into a hydrogen ion and a *bicarbonate ion*, HCO_3^-:

$$H_2CO_3 \rightleftharpoons H^+ + HCO_3^-$$

Salts

A **salt** is an ionic compound consisting of any cation except a hydrogen ion and any anion except a hydroxide ion. Because they are held together by ionic bonds, many salts dissociate completely in water, releasing cations and anions. For example, sodium chloride, or table salt, immediately dissociates in water, releasing Na^+ and Cl^- ions. Sodium and chloride are the most abundant ions in body fluids. However, many other ions are present in lesser amounts as a result of the dissociation of other inorganic compounds. Ionic concentrations in the body are regulated by mechanisms we shall describe in Chapters 26 and 27.

The ionization of sodium chloride does not affect the local concentrations of hydrogen ions or hydroxide ions, so NaCl, like many salts, is a "neutral" solute. Through their interactions with water molecules, however, other salts may indirectly affect the concentrations of H^+ and OH^- ions. Thus, the dissociation of some salts may make a solution slightly acidic or slightly basic.

Buffers and pH Control

Buffers are compounds that stabilize the pH of a solution by removing or replacing hydrogen ions. *Buffer systems* typically involve a weak acid and its related salt, which functions as a weak base. For example, the carbonic acid–bicarbonate buffer system, detailed in Chapter 27, consists of carbonic acid and sodium bicarbonate, $NaHCO_3$, otherwise known as baking soda. Buffers and buffer systems in body fluids maintain pH within normal limits. Figure 2-8● includes the pH of several body fluids. Antacids such as Alka-Seltzer® and Rolaids® use sodium bicarbonate to neutralize excess hydrochloric acid in the stomach. The effects of neutralization are most evident when you add a strong acid to a strong base. For example, by adding hydrochloric acid to sodium hydroxide, you neutralize both the strong acid and the strong base:

$$HCl + NaOH \longrightarrow H_2O + NaCl$$

This reaction produces water and a salt—in this case, the neutral salt sodium chloride.

☑ Why does a solution of table salt conduct electricity but a sugar solution does not?

☑ How does an antacid help decrease stomach discomfort?

ORGANIC COMPOUNDS

Organic compounds always contain the elements carbon and hydrogen and generally oxygen as well. Many organic molecules contain long chains of carbon atoms linked by covalent bonds. These carbon atoms typically form additional covalent bonds with hydrogen or oxygen atoms and, less commonly, with nitrogen, phosphorus, sulfur, iron, or other elements.

Many organic molecules are soluble in water. Although the previous discussion focused on inorganic acids and bases, there are also important organic acids and bases. For example, *lactic acid* is an organic acid, generated by active muscle tissues, that must be neutralized by the buffers in body fluids.

In this discussion, we will consider four major classes of organic compounds: *carbohydrates, lipids, proteins,* and *nucleic acids.* We will also introduce *high-energy compounds,* which are insignificant in terms of their abundance in the body but absolutely vital to the survival of our cells. In addition, the human body contains small quantities of many other organic compounds whose structures and functions we will consider in later chapters.

Carbohydrates

A **carbohydrate** (kar-bō-HĪ-drāt) is an organic molecule that contains carbon, hydrogen, and oxygen in a ratio near 1:2:1. Familiar carbohydrates include the *sugars* and *starches* that make up roughly half of the typical U.S. diet. Our tissues can break down most carbohydrates. Although they may have other functions, carbohydrates are most important as sources of energy. We will focus on *monosaccharides, disaccharides,* and *polysaccharides.*

Monosaccharides

A *simple sugar,* or **monosaccharide** (mon-ō-SAK-uh-rīd; *mono-,* single + *sakcharon,* sugar), is a carbohydrate containing from three to seven carbon atoms. A simple sugar may be called a *triose* (three-carbon), *tetrose* (four-carbon), *pentose* (five-carbon), *hexose* (six-carbon), or *heptose* (seven-carbon). The hexose **glucose** (GLOO-kōs), $C_6H_{12}O_6$, is the most important metabolic "fuel" in the body. The atoms in a glucose molecule may form a straight chain (Figure 2-9a●) or a ring (Figure 2-9b●). In the body, the ring form is more common. A space-filling model shows the three-dimensional arrangement of atoms in the ring most clearly (Figure 2-9c●).

The three-dimensional shape of a molecule is an important characteristic, because the shape of an organic molecule commonly determines its fate or its function. **Isomers** are molecules that have the same molecular formula but different shapes. In other words, they contain the same types and numbers of atoms, but they have different three-dimensional structures. The body usually treats different isomers as distinct molecules. For example, the simple sugars

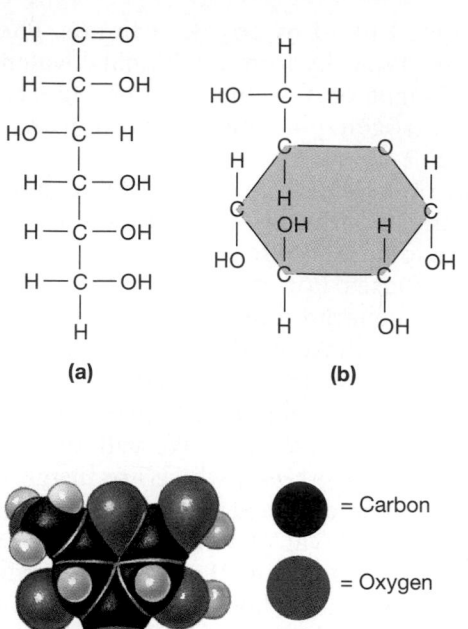

(a)   (b)

= Carbon

= Oxygen

= Hydrogen

(c)

•FIGURE 2-9 **Glucose.** (a) The straight-chain structural formula. (b) The ring form that is most common in nature. (c) A space-filling model that shows the actual three-dimensional organization of the atoms in the ring.

glucose and fructose are isomers. *Fructose* is a hexose found in many fruits and in secretions of the male reproductive tract. Although its chemical formula—$C_6H_{12}O_6$—is the same as that of glucose, separate enzymes and reaction sequences control its breakdown and synthesis. Simple sugars such as glucose and fructose dissolve readily in water, and they are rapidly distributed throughout the body by the blood and other body fluids. AM *The Pharmaceutical Use of Isomers*

Disaccharides and Polysaccharides

Carbohydrates other than simple sugars are complex molecules composed of monosaccharide building blocks. Two simple sugars joined together form a **disaccharide** (dī-SAK-uh-rīd; *di-*, two). Disaccharides such as *sucrose* (table sugar) have a sweet taste and, like monosaccharides, are quite water-soluble. The formation of sucrose (Figure 2-10a•) is an example of a process called **dehydration synthesis**, or a *condensation reaction*. This process links molecules together by the removal of a water molecule. The breakdown of sucrose into simple sugars (Figure 2-10b•) is an example of *hydrolysis*, a process introduced on page 40.

Many foods contain disaccharides, but all carbohydrates except simple sugars must be disassembled through hydrolysis before they can provide useful energy. Most popular junk foods, such as candies and sodas, abound in simple sugars (commonly fructose) and disaccharides (generally sucrose). Some people cannot tolerate sugar for medical reasons; others avoid it in an effort to control their weight. (Excess sugars are converted to fat for long-term storage.) Many such people use *artificial sweeteners* in their foods and beverages. These compounds have a very sweet taste, but they either cannot be broken down in the body or are used in insignificant amounts. AM *Artificial Sweeteners*

Dehydration synthesis can continue adding monosaccharides or combining disaccharides to create very complex carbohydrates. These large molecules are called **polysaccharides** (pol-ē-SAK-uh-rīdz; *poly-*, many). Polysaccharide chains can be straight or highly branched. *Starches* are glucose-based polysaccharides. Most starches are manufactured by plants. The digestive tract of humans can break these molecules into simple sugars. Starches such as those in potatoes and grains represent a major dietary energy source. *Cellulose*, a structural component of many plants, is a polysaccharide that our bodies cannot digest. Foods such as celery—which contains cellulose, water, and little else—contribute to the bulk of feces but are useless as an energy source. (In fact, you expend more energy chewing a stalk of celery than you obtain by digesting it.)

Glycogen (GLĪ-kō-jen), or *animal starch*, is a branched polysaccharide composed of interconnected glucose molecules (Figure 2-11•). Like most other polysaccharides, glycogen does not dissolve in water or other body fluids. Liver and muscle tissues manufacture and store glycogen. When these tissues have a high demand for glucose, glycogen molecules are broken down; when the demand is low, these tissues absorb glucose from the bloodstream and rebuild glycogen reserves.

Despite their metabolic importance, carbohydrates account for less than 3 percent of our total body weight. Table 2-4 (p. 48) summarizes information concerning the carbohydrates.

Lipids

Like carbohydrates, **lipids** (*lipos,* fat) contain carbon, hydrogen, and oxygen, but the ratios in lipids do not approximate 1:2:1. In general, a lipid molecule contains much less oxygen than a carbohydrate with the same number of carbon atoms. Lipids may also contain small quantities of other elements, such as phosphorus, nitrogen, or sulfur. Familiar lipids include *fats, oils*, and *waxes*. Most lipids are insoluble in water, but

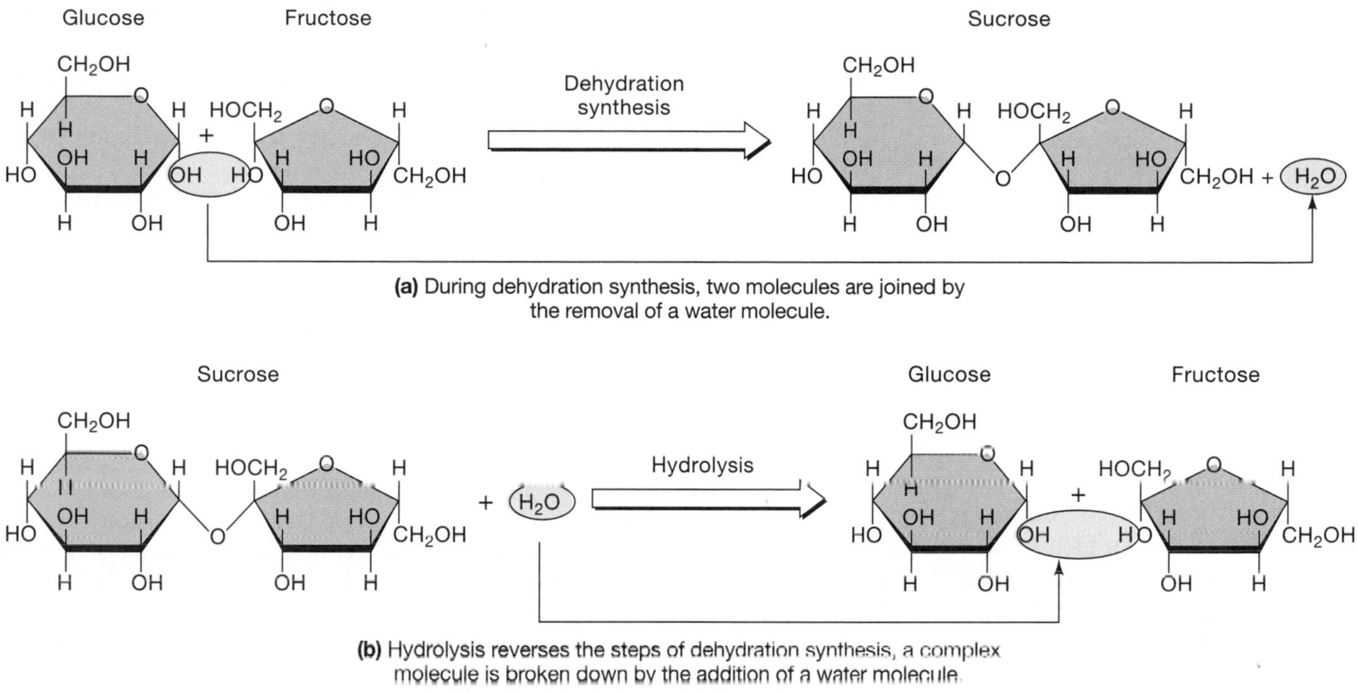

Glucose   Fructose   Sucrose

CH_2OH ... CH_2OH

(a) During dehydration synthesis, two molecules are joined by the removal of a water molecule.

Sucrose   Glucose   Fructose

(b) Hydrolysis reverses the steps of dehydration synthesis, a complex molecule is broken down by the addition of a water molecule.

•FIGURE 2-10   The Formation and Breakdown of Complex Sugars

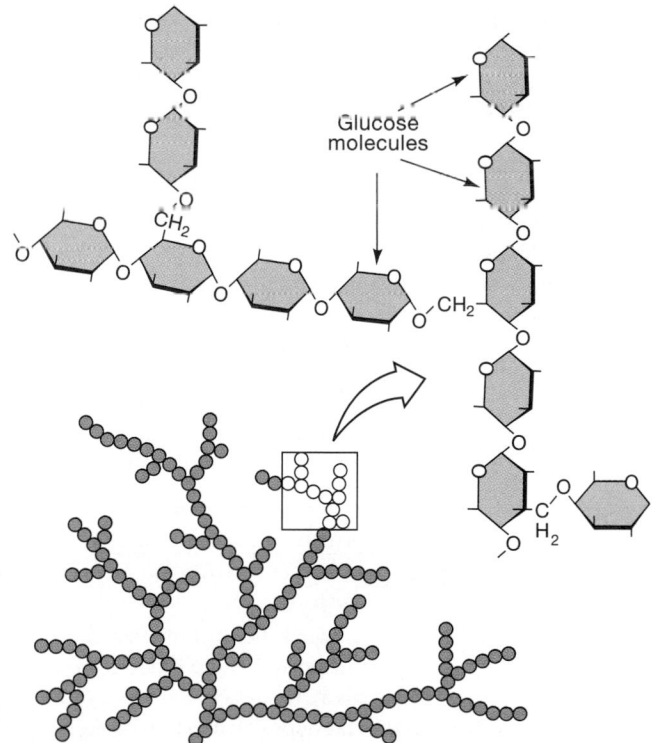

Glucose molecules

•FIGURE 2-11   Structure of a Polysaccharide.   Liver and muscle cells store glucose as the polysaccharide glycogen, a long, branching chain of glucose molecules. This figure introduces a different method of representing a carbon ring structure; at each corner of the solid hexagon is a carbon atom. The position of an oxygen atom in each glucose ring is shown.

special transport mechanisms carry them in the circulating blood.

Lipids form essential structural components of all cells. In addition, lipid deposits are important as energy reserves. On average, lipids provide roughly twice as much energy as carbohydrates do, gram for gram, when broken down in the body. When the supply of lipids exceeds the demand for energy, the excess is stored in fat deposits. For this reason, there has been great interest in developing *fat substitutes* that provide less energy but have the same taste and texture as lipids. **AM** *Fat Substitutes*

Overall, lipids normally account for 10–12 percent of our total body weight. There are many different kinds of lipids in the body. Major lipid types are presented in Table 2-5. We will consider five classes of lipid molecules: *fatty acids, eicosanoids, glycerides, steroids,* and *phospholipids and glycolipids.*

Fatty Acids

Fatty acids are long carbon chains with hydrogen atoms attached. One end of the carbon chain always bears a *carboxylic* (kar-bok-SIL-ik) *acid group*, a structure that can be represented as

$$R \ldots -\underset{\displaystyle}{\overset{\displaystyle OH}{C}} = O$$

TABLE 2-4   Carbohydrates in the Body

Structure	Example(s)	Primary Function(s)	Remarks
Monosaccharides (simple sugars)	Glucose, fructose	Energy source	Manufactured in the body and obtained from food; found in body fluids
Disaccharides	Sucrose, lactose, maltose	Energy source	Sucrose is table sugar, lactose is present in milk; all must be broken down to monosaccharides before absorption
Polysaccharides	Glycogen	Storage of glucose molecules	Glycogen is found in animal cells; other starches and cellulose occur in plant cells

The carbon chain is abbreviated by R, and the carboxylic acid group by COOH. The name *carboxyl* should help you remember that a carbon and a hydroxyl (—OH) group are the important structural features of fatty acids. The end opposite the carboxylic acid group is known as a hydrocarbon "tail." Figure 2-12a● shows a representative fatty acid, *lauric acid*.

In a fatty acid, the polar covalent bond between the oxygen and hydrogen atoms of the carboxylic acid group breaks down in solution, releasing a hydrogen ion. This reaction can be summarized as

$$\ldots - \overset{\overset{\displaystyle OH}{\diagup}}{C} = O \longrightarrow \ldots - \overset{\overset{\displaystyle O^-}{\diagup}}{C} = O + H^+$$

When a fatty acid is placed in solution, only the carboxyl end associates with water molecules, for this is the only hydrophilic portion of the molecule. The rest of the carbon chain is hydrophobic, so fatty acids have a very limited solubility in water.

In a *saturated* fatty acid, each carbon atom in the hydrocarbon tail has four single covalent bonds. If some of the carbon-to-carbon bonds are double covalent bonds, the fatty acid is said to be *unsaturated*. These terms refer to the number of hydrogen atoms in the fatty acid. Replacing a double bond between carbon atoms with a single bond allows the molecule to accept two more hydrogen atoms; hence, a double-bonded chain is unsaturated (Figure 2-12b●). A *monounsaturated* fatty acid has a single unsaturated bond in the hydrocarbon tail. A *polyunsaturated* fatty acid contains multiple unsaturated bonds.

FATTY ACIDS AND HEALTH Both saturated and unsaturated fatty acids can be broken down for energy, but a diet containing large amounts of saturated fatty acids increases the risk of heart disease and other circulatory problems. Butter, fatty meat, and ice cream are popular dietary sources of saturated fatty acids. Vegetable oils, such as olive oil or corn oil, contain a mixture of mono-unsaturated and polyunsaturated fatty acids. Current research indicates that the monounsaturated fats may be more effective than polyunsaturated fats in lowering the risk of heart disease. In fact, the manufacture of margarine and vegetable shortening from polyunsaturated fats produces compounds called *trans* fatty acids that may actually increase the risk of heart disease. Increasing the proportion of oleic acid, an 18-carbon monounsaturated fatty acid, in cooking oils and other products could therefore yield health benefits.

The carbons in a fatty acid molecule are numbered, beginning at the carboxylic acid end; the last carbon in the chain is called the *omega* carbon. Eskimos have lower rates of heart disease than other populations do, despite the fact that the Eskimo diet is high in fats and cholesterol. The fatty acids in the Eskimo diet have an unsaturated bond three carbons before the omega carbon, a position known as "omega minus 3," or *omega-3* (Figure 2-12c●). Omega-3 fatty acids are found in fish flesh and fish oils. For unknown reasons, the presence of omega-3 fatty acids (or some unidentified component of fish flesh) in the diet reduces the risks of heart disease, rheumatoid arthritis, and other inflammatory diseases.

TABLE 2-5   Representative Lipids and Their Functions in the Body

Lipid Type	Example(s)	Primary Function(s)	Remarks
Fatty acids	Lauric acid	Energy source	Absorbed from food or synthesized in cells; transported in the blood for use in many tissues
Glycerides	Monoglycerides, diglycerides, triglycerides	Energy source, energy storage, insulation, and physical protection	Stored in fat deposits; must be broken down to fatty acids and glycerol before they can be used as an energy source
Eicosanoids	Prostaglandins, leukotrienes	Chemical messengers coordinating local cellular activities	Prostaglandins are produced in most body tissues
Steroids	Cholesterol	Structural component of cell membranes, hormones, digestive secretions in bile	All have the same carbon ring framework
Phospholipids, glycolipids		Structural components of cell membranes	Derived from fatty acids and nonlipid components

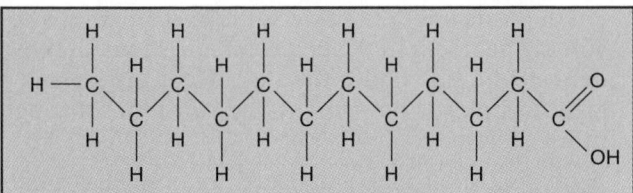

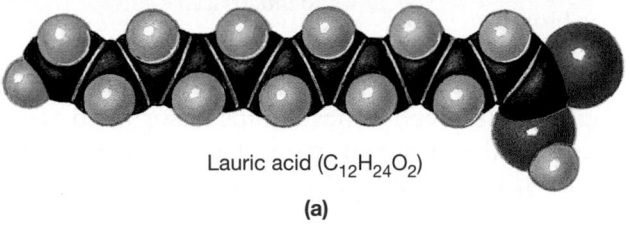

Lauric acid (C₁₂H₂₄O₂)

(a)

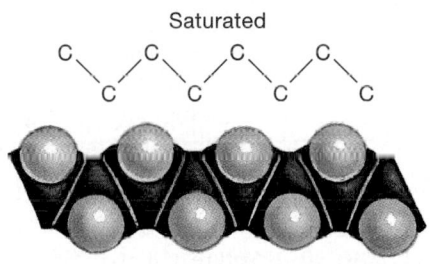

Saturated

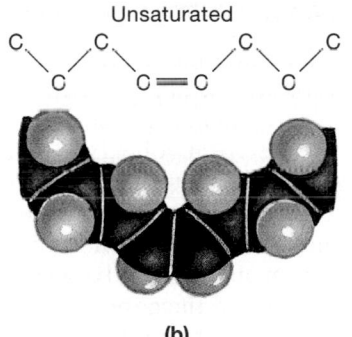

Unsaturated

(b)

Eicosanoids

Eicosanoids (ī-KŌ-sa-noyds) are lipids derived from *arachidonic* (ah-rak-i-DON-ik) *acid,* a fatty acid that must be absorbed in the diet because it cannot be synthesized by the body. There are two major classes of eicosanoids: (1) *prostaglandins* and (2) *leukotrienes.* We will consider only prostaglandins at this time, because virtually all tissues synthesize and respond to prostaglandins. We shall consider leukotrienes in Chapters 18 and 22.

PROSTAGLANDINS Prostaglandins (pros-tuh-GLAN-dinz) are short-chain fatty acids that have five of their carbon atoms joined in a ring (Figure 2-13•). Cells release prostaglandins to coordinate or direct local cellular activities. Prostaglandins are extremely powerful and effective in minute quantities. Almost every tissue in the body releases and responds to prostaglandins. Their effects vary depending on the nature of the prostaglandin and the site of release. Consider two examples:

1. Prostaglandins released by damaged tissues stimulate nerve endings and produce the sensation of pain (Chapter 17).
2. Prostaglandins released in the uterus help trigger the start of labor contractions (Chapter 29).

Prostaglandins are chemical messengers released to coordinate local activities; they usually do not travel through the circulatory system to reach their target cells. As a result, prostaglandins are often called *local hormones; hormones* are chemical messengers produced at one site that journey through the bloodstream to regulate activities under way in a different portion of the body.

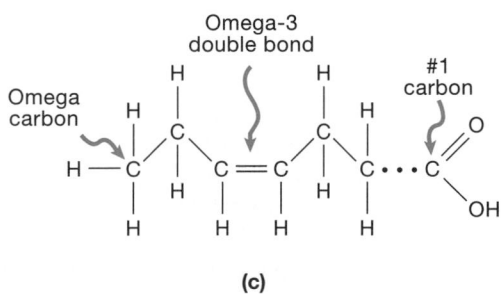

(c)

•*FIGURE 2-12* **Fatty Acids.** **(a)** Lauric acid demonstrates two structural characteristics common to all fatty acids: a long backbone of carbon atoms and a carboxylic acid group (—COOH) at one end. **(b)** A fatty acid may be saturated or unsaturated. Unsaturated fatty acids have double covalent bonds, and their presence causes a sharp bend in the molecule. **(c)** An omega-3 fatty acid has an unsaturated bond three carbons before the end of the carbon chain.

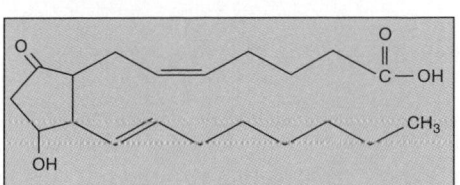

•*FIGURE 2-13* **Prostaglandins.** Prostaglandins are unusual short-chain fatty acids. Compare the complete structural formula of a typical prostaglandin with the abbreviated formula.

Glycerides

Individual fatty acids cannot be strung together in a chain by dehydration synthesis, as simple sugars can. But they can be attached to another compound, **glycerol** (GLI-se-rol), through a similar reaction. The result is a lipid known as a **glyceride** (GLI-se-rīd). Dehydration synthesis can produce a **monoglyceride** (mo-nō-GLI-se-rīd) consisting of glycerol plus one fatty acid. Subsequent reactions can yield a **diglyceride** (glycerol + two fatty acids) and then a **triglyceride** (glycerol + three fatty acids) Figure 2-14•. Hydrolysis breaks the glycerides into fatty acids and glycerol. Compare Figure 2-14 with Figure 2-10• to convince yourself that dehydration synthesis and hydrolysis operate in the same way, whether the molecules involved are carbohydrates or lipids.

Triglycerides, also known as neutral fats, have three important functions:

1. Fatty deposits in specialized sites of the body represent a significant energy reserve. In times of need, the triglycerides are disassembled to yield fatty acids that can be broken down to provide energy.

2. Fat deposits under the skin serve as insulation, preventing heat loss to the environment. Many of these deposits lie just under the skin. Heat loss across a layer of lipids is only about one-third that through other tissues.

3. A fat deposit around a delicate organ such as a kidney provides a cushion that protects against shocks or blows.

Triglycerides are stored in the body as lipid droplets within cells. These lipids absorb and accumulate lipid-soluble vitamins, drugs, or toxins that appear in body fluids. This accumulation has both positive and negative effects. For example, the body's lipid reserves retain both valuable lipid-soluble vitamins (A, D, E, K) and potentially dangerous lipid-soluble pesticides, such as DDT.

Steroids

Steroids are large lipid molecules that share a distinctive carbon framework (Figure 2-15a•). They differ in the carbon chains attached to this basic structure. Figure 2-15b• shows the structural formula of the steroid **cholesterol** (ko-LES-ter-ol; *chole-*, bile + *stereos*, solid). Cholesterol and related steroids are important for the following reasons:

- All animal cell membranes contain cholesterol. Cells need cholesterol to maintain their cell membranes as well as for cell growth and division.
- Steroid hormones are involved in the regulation of sexual function. Examples include the sex hormones, such as *testosterone* and the *estrogens*.
- Steroid hormones are important in the regulation of tissue metabolism and mineral balance. Examples include the hormones of the adrenal cortex, called *corticosteroids*, and *calcitriol*, a hormone important in calcium ion regulation.
- Steroid derivatives called *bile salts* are required for the normal processing of dietary fats. Bile salts, which are produced in the liver and secreted in bile, interact with lipids in the intestinal tract and facilitate the digestion and absorption of lipids.

Cholesterol is obtained from two sources: (1) by absorption from animal products in the diet and (2) by synthesis within the body. Meat, cream, and egg yolks are especially rich dietary sources of cholesterol. A diet high in cholesterol can be harmful, because a strong link exists between high blood cholesterol levels and heart disease. Current nutritional advice suggests that you reduce cholesterol intake to under 300 mg per day. This amount represents a 40 percent reduction for the average adult in the United States. Unfortunately, because the body can synthesize cholesterol as well, it is sometimes difficult to control blood cholesterol levels by dietary re-

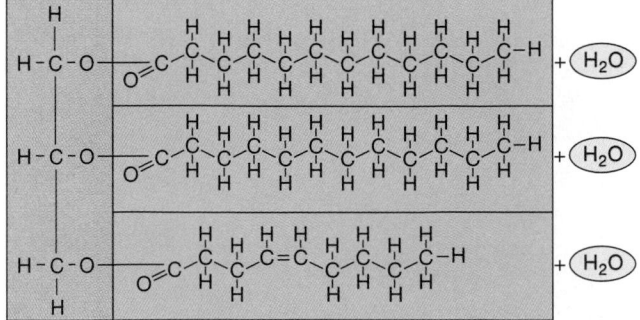

•*FIGURE 2-14*   **Triglyceride Formation.**   The formation of a triglyceride involves the attachment of fatty acids to the carbons of a glycerol molecule. This example shows the formation of a triglyceride by the attachment of one unsaturated and two saturated fatty acids to a glycerol molecule.

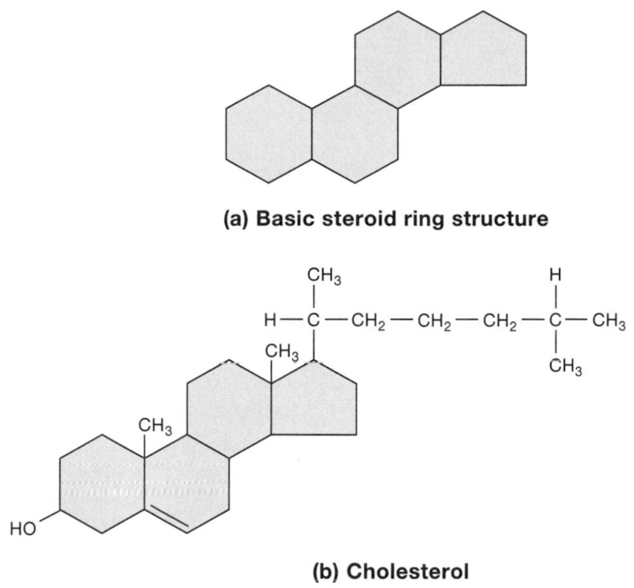

(a) Basic steroid ring structure

(b) Cholesterol

•*FIGURE 2-15* **Steroids.** **(a)** All steroids share this complex four-ring structure. **(b)** Individual steroids differ in the side chains attached to the carbon rings. This is a molecule of cholesterol.

striction alone. We shall examine the connection between blood cholesterol levels and heart disease more closely in later chapters.

Phospholipids and Glycolipids

Phospholipids (FOS-fō-lip-idz) and **glycolipids** (GLĪ-cō-lip-idz) are structurally related, and our cells can synthesize both types of lipids, primarily from fatty acids. In a *phospho*lipid, a *phosphate group* (PO_4^{3-}) links a diglyceride to a nonlipid group (Figure 2-16a•). In a *glyco*lipid, a carbohydrate is attached to a diglyceride (Figure 2-16b•). Note that placing -*lipid* last in these names indicates that the molecule consists primarily of lipid.

The long fatty acid "tails" of phospholipids and glycolipids are hydrophobic, but the nonlipid "heads" are hydrophilic. In water, large numbers of these molecules tend to form droplets, or *micelles* (mī-SELZ), with the hydrophilic portions on the outside (Figure 2-16c•). Most meals that you eat contain a mixture of lipids and other organic molecules, and micelles form as the food breaks down in your digestive tract. In addition to phospholipids and glycolipids, micelles may contain other insoluble lipids, such as steroids, glycerides, and long-chain fatty acids.

Cholesterol, phospholipids, and glycolipids are called *structural lipids* because they form and maintain intracellular structures called *membranes*. Membranes are sheets or layers composed primarily of hydrophobic lipids. A membrane surrounds each cell

and separates the aqueous solution inside the cell from the aqueous solution in the extracellular environment. Because the two solutions are separated by a membrane, their compositions can be very different. A variety of internal membranes subdivide the interior of the cell into specialized compartments, each with a distinctive chemical nature and, as a result, a different function.

☑ A food contains organic molecules with the elements C, H, and O in a ratio of 1:2:1. Which type of compound is this, and what is its major function in the body?

☑ When two monosaccharides undergo a dehydration synthesis reaction, which type of molecule is formed?

☑ Which kind of lipid would be found in a sample of fatty tissue taken from beneath the skin?

☑ Which lipids would you expect to find in an analysis of human cell membranes?

Proteins

Proteins are the most abundant organic components of the human body and in many ways the most important. There are roughly 100,000 different kinds of proteins, and they account for about 20 percent of the total body weight. All proteins contain carbon, hydrogen, oxygen, and nitrogen; smaller quantities of sulfur may also be present. Table 2-6 (p. 53) identifies the major types of proteins in the body.

Proteins perform a variety of essential functions. These include seven major functional categories:

1. **Support.** *Structural proteins* create a three-dimensional framework for the body, providing strength, organization, and support for cells, tissues, and organs.

2. **Movement.** *Contractile proteins* are responsible for muscular contraction; related proteins are responsible for the movement of individual cells.

3. **Transport.** Insoluble lipids, respiratory gases, special minerals, such as iron, and several hormones must be bound to special *transport proteins* before they can be transported in the blood. Other specialized proteins transport materials from one part of a cell to another.

4. **Buffering.** Proteins provide a considerable buffering action and thereby help prevent potentially dangerous changes in pH in our cells and tissues.

5. **Metabolic regulation.** *Enzymes* accelerate chemical reactions in living cells. The sensitivity of enzymes to environmental factors is extremely important in controlling the pace and direction of metabolic operations.

6. **Coordination** and **control.** Protein hormones can influence the metabolic activities of every cell in the

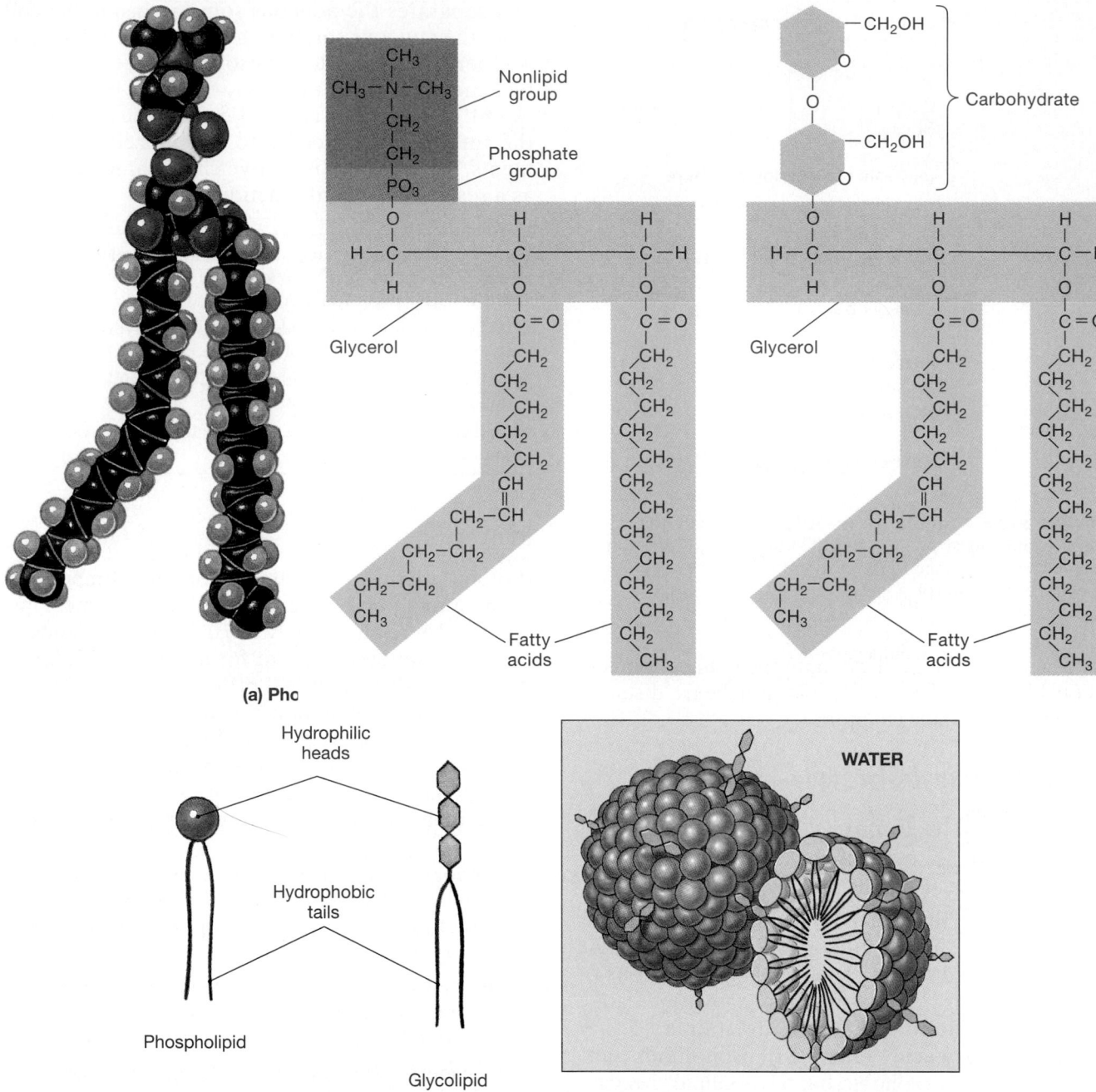

(a) Pho [...]

Phospholipid

Glycolipid

(c) Micelle structure

•**FIGURE 2-16** **Phospholipids and Glycolipids.** **(a)** In a phospholipid, a phosphate group links a nonlipid molecule to a diglyceride. This phospholipid is a molecule of *lecithin*. **(b)** In a glycolipid, a carbohydrate is attached to a diglyceride. **(c)** When present in large numbers, these molecules will form micelles, with the hydrophilic heads facing the water molecules and the hydrophobic tails on the inside of each droplet.

body or affect the function of specific organs or organ systems.

7. **Defense.** The tough, waterproof proteins of the skin, hair, and nails protect the body from environmental hazards. Proteins called *antibodies*, components of the *immune response*, help protect us from disease. Special *clotting proteins* restrict bleeding after an injury to the cardiovascular system.

Structure of Proteins

Proteins consist of long chains of organic molecules called **amino acids**. There are 20 amino acids that occur in significant quantities in the human body. A typical protein contains 1000 amino acids, but the largest protein complexes may have 100,000 or more. Each amino acid consists of a central carbon atom to which four groups are attached: (1) a hy-

TABLE 2-6 Representative Proteins and Their Functions

Protein Type	Example(s)	Representative Location(s)	Function(s)
Structural proteins	Keratin	Skin surface, hair, nails	Provides strength and waterproofing
	Collagen	Dermis of skin, tendons	Provides tensile strength
Contractile proteins	Actin, myosin	Muscle cells	Perform contraction and movement
Transport proteins	Albumin	Circulating blood	Transports fatty acids and steroid and thyroid hormones
	Transferrin	Circulating blood	Transports iron
	Apolipoproteins	Circulating blood	Transport glycerides
	Hemoglobin	Circulating blood	Transports oxygen in blood (within red blood cells)
Buffers	Intracellular and extracellular proteins	Within cells and body fluids	Stabilize pH
Enzymes	Hydrolases	All cells	Catalyze hydrolysis of organic molecules
	Kinases	All cells	Attach phosphate groups to organic substrates
	Proteases	All cells; digestive secretions of stomach, pancreas	Break down proteins
	Carbohydrases	All cells; digestive secretions of salivary glands, pancreas	Break down carbohydrates
	Lipases	All cells; digestive secretions of pancreas	Break down lipids
Antibodies (immuno-globulins)	Gamma globulins	Circulating blood	Attack foreign proteins and pathogens
Hormones	Insulin, glucagon	Circulating blood	Coordinate and/or control metabolic activities

drogen atom; (2) an *amino group* (—NH₂); (3) a *carboxylic acid group* (—COOH); and (4) a variable group, known as an *R group*, or *side chain* (Figure 2-17a•). Different R groups distinguish one amino acid from another, giving each its individual chemical properties. The name *amino acid* refers to the presence of the *amino* group and the carboxylic *acid* group. Amino acids are relatively small, water-soluble molecules; the 20 amino acids commonly found in proteins are diagrammed in Appendix IV. In the

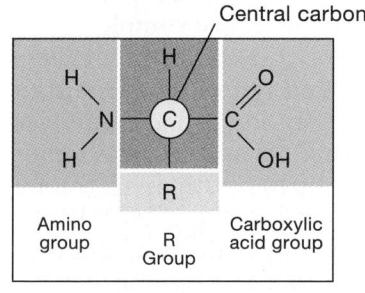

(a) Structure of an amino acid

•**FIGURE 2-17** **Amino Acids and Peptide Bonds.** **(a)** Each amino acid consists of a central carbon atom to which four different groups are attached: a hydrogen atom, an amino group (—NH₂), a carboxylic acid group (—COOH), and a variable group generally designated R. **(b)** Peptides form as dehydration synthesis creates a peptide bond between the carboxylic acid group of one amino acid and the amino group of another. In this example, the amino acids glycine and alanine are linked to form a dipeptide.

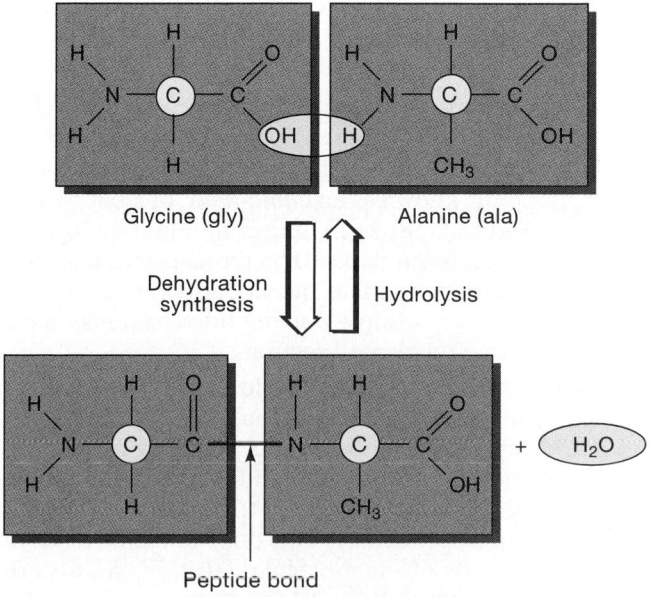

(b) Peptide bond formation

normal pH range of body fluids, the carboxylic acid groups on many amino acids give up their hydrogens. When the carboxylic acid group changes from —COOH to —COO⁻, these amino acids become negatively charged.

Figure 2-17b• shows two representative amino acids, *glycine* and *alanine*. As the figure indicates, dehydration synthesis can link two amino acids. This process creates a covalent bond between the carboxylic acid group of one amino acid and the amino group of another. Such a bond is known as a **peptide bond**; the molecule created in this example is called a *dipeptide* because it contains two amino acids.

The chain can be lengthened by the addition of more amino acids. Attaching a third amino acid produces a *tripeptide*, and there are *tetrapeptides, pentapeptides*, and so forth. Tripeptides and larger peptide chains are called **polypeptides**. Polypeptides containing more than 100 amino acids are usually called proteins. You are probably already familiar with the names of several important proteins, including *hemoglobin* in red blood cells and *keratin* in fingernails and hair. Because each protein contains amino acids that are negatively charged, the entire protein has a net negative charge. For this reason, proteins are often indicated by the abbreviation *Pr⁻*.

Protein Shape

The characteristics of a particular protein are determined in part by the R groups on its component amino acids. But the properties of a protein are more than just the sum of the properties of its parts, for polypeptides can have very complex shapes. Proteins have four levels of structural complexity:

1. **Primary structure** is the sequence of amino acids along the length of a single polypeptide. The primary structure of a short peptide chain is diagrammed in Figure 2-18a•.

2. **Secondary structure** results from bonds that develop between atoms at different parts of the polypeptide chain. Hydrogen bonding, for example, may create a simple spiral, known as an *alpha-helix*, or a flat *pleated sheet* (Figure 2-18b•). Whether an alpha-helix or a pleated sheet forms depends on the sequence of amino acids in the peptide chain; the alpha-helix is most common. However, a single polypeptide chain may have both helical and pleated sections.

3. **Tertiary structure** is the complex coiling and folding that gives the protein its final three-dimensional shape. Tertiary structure results primarily from interactions between the polypeptide chain and the surrounding water molecules and to a lesser extent from interactions between the R groups of amino acids in different parts of the molecule. *Disulfide bonds* are covalent bonds that may form between two molecules of the amino acid *cysteine*, each molecule located at a different site along the chain. Disulfide bonds create permanent loops or coils in a polypeptide chain. Figure 2-18c• shows the tertiary structure of *myoglobin*, a protein found in muscle cells. Myoglobin is an example of a **globular protein**. Globular proteins are compact, generally rounded, and readily enter solution.

4. **Quaternary structure** is the interaction between individual polypeptide chains to form a protein complex (Figure 2-18d•). Each of the polypeptide subunits has its own secondary and tertiary structures. The protein *hemoglobin* contains four globular subunits, each structurally similar to myoglobin. Hemoglobin is found within red blood cells, where it binds and transports oxygen. In *keratin* and *collagen* three alpha-helical polypeptides are wound together like the strands of a rope. Keratin is the tough, water-resistant protein found at the surface of the skin and in nails and hair. Collagen is the most abundant structural protein, as collagen fibers form the framework that supports cells in most tissues. Keratin and collagen are examples of **fibrous proteins**. Fibrous proteins are tough, durable, and generally insoluble.

SHAPE AND FUNCTION Proteins are very versatile, and they have a variety of functions. The shape of a protein determines its functional properties, and the ultimate determinant of shape is the sequence of amino acids. The 20 common amino acids can be linked in an astonishing number of combinations, creating proteins of enormously varied shape and function. Changing the identity of a single amino acid out of the 10,000 or more amino acids in a protein can significantly alter the protein's functional properties. For example, several cancers and *sickle cell anemia*, a blood disorder, result from single changes in the amino acid sequences of complex proteins.

The tertiary and quaternary shapes of complex proteins depend not only on their amino acid sequence, but also on the local environmental characteristics. Small changes in the ionic composition, temperature, or pH of their surroundings can thus affect the function of proteins. Protein shape can also be affected by hydrogen bonding to other molecules in solution. The significance of these factors is most striking when we consider the function of enzymes, for these proteins are essential to the metabolic operations under way in every cell in the body. We will now take a closer look at enzyme function, a topic introduced on p. 40.

Enzyme Function

The reactants in enzymatic reactions are usually called **substrates**. As in other types of chemical reactions,

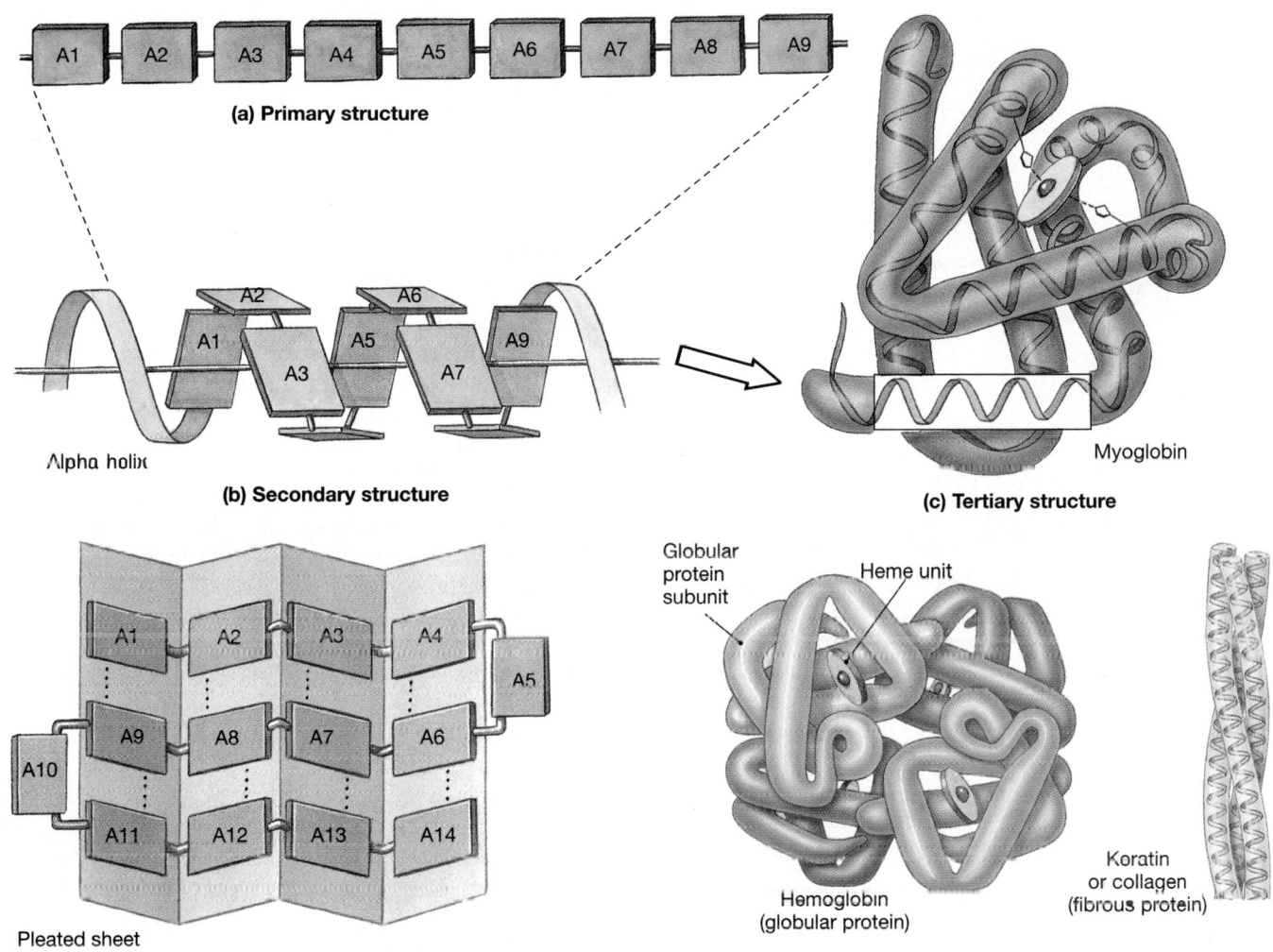

(a) Primary structure

Alpha holix

(b) Secondary structure

Pleated sheet

Myoglobin

(c) Tertiary structure

Globular protein subunit

Heme unit

Hemoglobin (globular protein)

Koratin or collagen (fibrous protein)

(d) Quaternary structure

•FIGURE 2-18   Protein Structure.   (a) The primary structure of a polypeptide is the sequence of amino acids along its length. **(b)** The secondary structure is primarily the result of hydrogen bonding along the length of the polypeptide chain. Such bonding often produces a simple spiral (an alpha-helix) or a flattened arrangement known as a pleated sheet. **(c)** The tertiary structure is the coiling and folding of a polypeptide. This is the structure of myoglobin, a globular protein involved in the storage of oxygen in muscle tissue. Within the cylindrical segments, the polypeptide chain is arranged in an alpha-helix. **(d)** The quaternary structure develops when separate polypeptide subunits interact to form a larger molecule. A single hemoglobin molecule contains four globular subunits, each structurally similar to myoglobin. Hemoglobin transports oxygen in the blood; the oxygen binds reversibly to the heme units. In keratin and collagen, three fibrous subunits intertwine. Keratin is a tough, water-resistant protein found in skin, hair, and nails. Collagen is the principal extracellular protein in most organs.

the interactions among substrates yield specific products. Before an enzyme can function as a catalyst in a reaction, the substrates must bind to a special region of the enzyme called the **active site**. The tertiary or quaternary structure of the enzyme molecule determines the shape of the active site, typically a groove or pocket into which one or more substrates nestle. This physical fit is reinforced by weak electrical attractive forces, such as hydrogen bonding.

Figure 2-19• diagrams enzyme function in a synthesis reaction. Substrates bind to the enzyme at its active site (Step 1). The enzyme then promotes prod-

uct formation (Step 2). It appears likely that substrate binding results in a change in the shape of the protein and that this shape change promotes the reaction by placing physical stresses on the substrate molecules. The completed product then detaches from the active site (Step 3), and the enzyme is free to repeat the process. Enzymes work quickly, cycling rapidly between substrates and products. For example, an enzyme providing energy during a muscular contraction performs its reaction sequence 100 times per second.

Our example in Figure 2-19• was an enzyme that catalyzed a synthesis reaction. Other enzymes may cat-

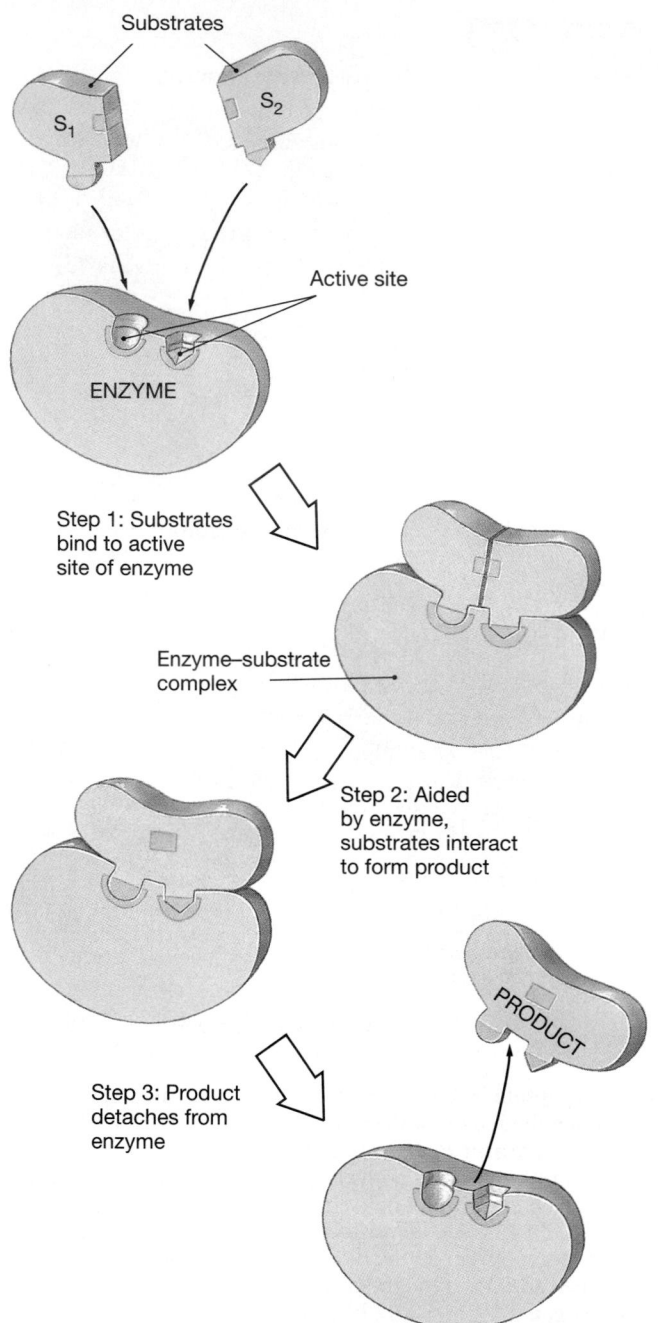

•FIGURE 2-19 **A Simplified View of Enzyme Structure and Function.** Each enzyme contains a specific active site somewhere on its exposed surface. *Step 1:* A pair of substrate molecules (S_1 and S_2) bind to the active site. *Step 2:* The substrates interact, forming a product. *Step 3:* That product detaches from the active site. Because the structure of the enzyme has not been affected, the entire process can be repeated.

alyze decomposition reactions or exchange reactions. All enzymes share three basic characteristics:

1. *Specificity.* Each enzyme catalyzes only one type of reaction and can accommodate only one type of substrate molecule. This property is called speci-

ficity. The specificity of an enzyme is determined by the ability of a particular substrate to bind to the active site, a relatively small portion of the entire protein. Differences in the quaternary structure of the enzyme that do not affect the active site or change the response of the enzyme to substrate binding have no effect on enzyme function. *Isozymes* are enzymes that differ in structure but catalyze the same reaction. Different tissues typically contain different isozymes.

2. *Saturation limits.* The rate of an enzymatic reaction is directly proportional to the concentration of substrate molecules and enzymes. An enzyme molecule must encounter appropriate substrates before it can catalyze a reaction; the higher the substrate concentration, the more frequent encounters will be. When substrate concentrations are high enough that every enzyme molecule is cycling through its reaction sequence at top speed, further increases in substrate concentration will have no effect on the rate of reaction unless additional enzyme molecules are provided. The substrate concentration required to have the maximum rate of reaction is called the *saturation limit.* An enzyme that has reached the saturation limit is said to be **saturated**.

3. *Regulation.* A variety of factors can turn enzymes "on" or "off" and thereby control reaction rates inside the cell. We will consider only one example here, the presence or absence of *cofactors* (considered next). In fact, virtually anything that changes the tertiary or quaternary shape of an enzyme may turn enzyme function on or off. Each cell contains an assortment of enzymes, and any particular enzyme may be active under one set of conditions and inactive under another. Because the change is immediate, enzyme activation or inactivation is an important method of short-term control over reaction rates and pathways. If you are interested in a more detailed discussion of enzyme function and its regulation, refer to the *Applications Manual.* AM *A Closer Look: The Control of Enzyme Activity*

COFACTORS AND ENZYME FUNCTION **Cofactors** are ions or molecules that must bind to the enzyme before substrate binding can occur. Without a cofactor, the enzyme is intact but nonfunctional; with the cofactor, the enzyme can catalyze a specific reaction. Examples of cofactors include ions such as calcium (Ca^{2+}) and magnesium (Mg^{2+}), which bind at the enzyme's active site. Cofactors may also bind at other sites, as long as they produce a change in the shape of the active site that makes substrate binding possible.

Coenzymes are nonprotein organic molecules that function as cofactors. Our bodies convert many vitamins into essential coenzymes. *Vitamins*, detailed n Chapter 25, are structurally related to lipids or carbohydrates, but they have unique functional roles. The human body cannot synthesize most of he vitamins it needs, so you must obtain them from your diet.

TEMPERATURE AND pH Each enzyme works best at an optimal combination of temperature and pH values. As temperatures rise, proteins change shape and enzyme function deteriorates. Very high body temperatures (above 43°C, or 110°F) cause death, because at these temperatures proteins undergo **denaturation**, a temporary or permanent change in their tertiary or quaternary structure. Denatured proteins are nonfunctional, and the loss of structural proteins and enzymes causes irreparable damage to organs and organ systems. You see denaturation in progress each time you fry an egg, for the clear egg white contains abundant dissolved proteins. As the temperature rises, the protein structure changes. Eventually, the egg proteins become completely and irreversibly denatured, forming an insoluble white mass.

Enzymes are equally sensitive to pH changes. *Pepsin*, an enzyme that breaks down proteins in the contents of your stomach, works best at a pH of 2.0 (strongly acidic). Your small intestine contains *trypsin*, another enzyme that attacks proteins. Trypsin works only in an alkaline environment, with an optimum pH of 7.7 (somewhat basic).

Glycoproteins and Proteoglycans

Glycoproteins (GLĪ-kō-prō-tēnz) and **proteoglycans** (prō-tē-ō-GLĪ-kanz) are combinations of protein and carbohydrate molecules. Glyco*proteins* are large proteins with small carbohydrate groups attached. Glycoproteins may function as enzymes, antibodies, hormones, or protein components of cell membranes. Glycoproteins in cell membranes play a major role in the identification of normal versus abnormal cells as well as in the initiation and coordination of the immune response. (We shall detail these mechanisms in Chapter 22.) Proteo*glycans* are large polysaccharide molecules linked by polypeptide chains. Proteoglycan secretions coat the surfaces of the respiratory and digestive tracts, providing lubrication, and the proteoglycans in tissue fluids give them a syrupy consistency.

☑ Proteins are chains of what small organic molecules?

☑ Which level of protein structure would be affected by an agent that breaks hydrogen bonds?

☑ Why does boiling a protein affect its structural and functional properties?

☑ How might a change in an enzyme's active site affect its function?

Nucleic Acids

Nucleic (noo-KLĀ-ik) **acids** are large organic molecules composed of carbon, hydrogen, oxygen, nitrogen, and phosphorus. Nucleic acids store and process information at the molecular level, inside living cells. There are two classes of nucleic acid molecules: (1) **deoxyribonucleic** (dē-ok-sē-rī-bō-noo-KLĀ-ik) **acid**, or **DNA**, and (2) **ribonucleic** (rī-bō-noo-KLĀ-ik) **acid**, or **RNA**. As we shall see, these two classes of nucleic acids differ in composition, structure, and function.

The DNA in our cells determines our inherited characteristics, such as eye color, hair color, blood type, and so on. DNA affects all aspects of body structure and function, because DNA molecules encode the information needed to build proteins. By directing the synthesis of structural proteins, DNA controls the shape and physical characteristics of our bodies. By controlling the manufacture of enzymes, DNA regulates not only protein synthesis but all aspects of cellular metabolism, including the creation and destruction of lipids, carbohydrates, and other vital molecules.

Several forms of RNA cooperate to manufacture specific proteins by using the information provided by DNA. We shall detail the functional relationships between DNA and RNA in Chapter 3.

Structure of Nucleic Acids

A nucleic acid consists of a series of nucleotides linked through the process of dehydration synthesis. A single **nucleotide** has three basic components: (1) a sugar, (2) a phosphate group, and (3) a **nitrogenous** (nitrogen-containing) **base** (Figure 2-20a•). The sugar is always a *pentose* (five-carbon sugar), either *ribose* (in RNA) or *deoxyribose* (in DNA). Each pentose is attached to a phosphate group (PO_4^{3-}) and to a nitrogenous base. There are five nitrogenous bases: **adenine (A), guanine (G), cytosine (C), thymine (T),** and **uracil (U).** Adenine and guanine are double-ringed structures called *purines*; the others are single-ringed molecules called *pyrimidines*. Both RNA and DNA contain adenine, guanine, and cytosine. Uracil is found only in RNA, and thymine occurs only in DNA.

In the formation of a nucleic acid, a nitrogenous base is attached to a pentose sugar molecule, creating a **nucleoside** that is named after the nitrogenous base. For example, the attachment of a guanine molecule to ribose or deoxyribose produces a guanine nucleoside. A nucleotide forms when a phosphate group binds to the sugar of a nucleoside. In the creation of a nucleic acid, dehydration synthesis attaches the phosphate group of one nucleotide to the carbohydrate of another. The "backbone" of a nucleic acid molecule is thus a linear sugar-to-phosphate-to-sugar sequence, with the nitrogenous bases projecting to one side.

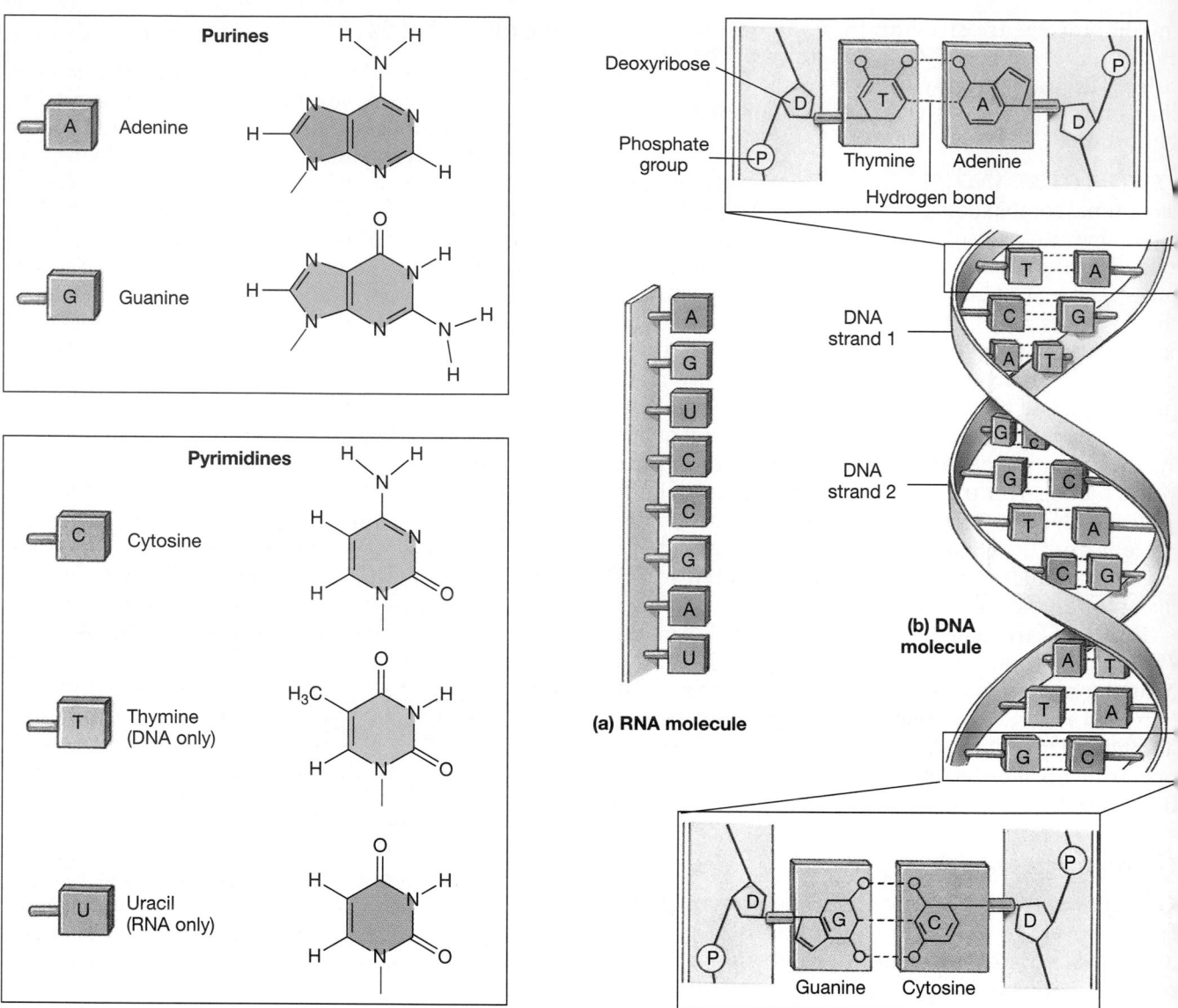

•FIGURE 2-20   Nucleic Acids: RNA and DNA.   Nucleic acids are long chains of nucleotides. Each molecule starts at the sugar–nitrogenous base of the first nucleotide and ends at the phosphate group of the last member of the chain. **(a)** An RNA molecule consists of a single nucleotide chain. Its shape is determined by the sequence of nucleotides and by the interactions between them. **(b)** A DNA molecule consists of a pair of nucleotide chains linked by hydrogen bonding between complementary base pairs.

RNA and DNA

There are important structural differences between RNA and DNA. A molecule of RNA consists of a single chain of nucleotides (Figure 2-20a•). Its shape depends on the order of the nucleotides and the interactions among them. There are three types of RNA in our cells: *messenger RNA (mRNA), transfer RNA (tRNA),* and *ribosomal RNA (rRNA).* These types have different shapes and functions, but all three are required for the synthesis of proteins. (We shall discuss this process further in Chapter 3).

A DNA molecule consists of a *pair* of nucleotide chains (Figure 2-20b•). Hydrogen bonding between opposing nitrogenous bases holds the two strands together. Because of the shapes of the nitrogenous bases,

adenine can bond only with thymine, and cytosine only with guanine. As a result, adenine–thymine and cytosine–guanine are known as **complementary base pairs**.

The two strands of DNA twist around one another in a double helix that resembles a spiral staircase. Each step of the staircase corresponds to one complementary base pair. Table 2-7 summarizes the differences between RNA and DNA.

High-Energy Compounds

Living cells must use energy to perform their vital functions. The energy is obtained through the enzymatic catabolism of organic substrates. To be useful, that energy must be captured and transferred from molecule to molecule or from one part of the cell to

TABLE 2-7 A Comparison of RNA and DNA

Characteristic	RNA	DNA
Sugar	Ribose	Deoxyribose
Nitrogenous bases	Adenine Guanine Cytosine Uracil	Adenine Guanine Cytosine Thymine
Number of nucleotides in typical molecule	Varies from fewer than 100 to about 50,000	Always more than 45 million
Shape of molecule	Varies, depending on hydrogen bonding along the length of the strand; 3 main types (mRNA, rRNA, tRNA)	Paired strands coiled in a double helix
Function	Performs protein synthesis as directed by DNA	Stores genetic information that controls protein synthesis by RNA

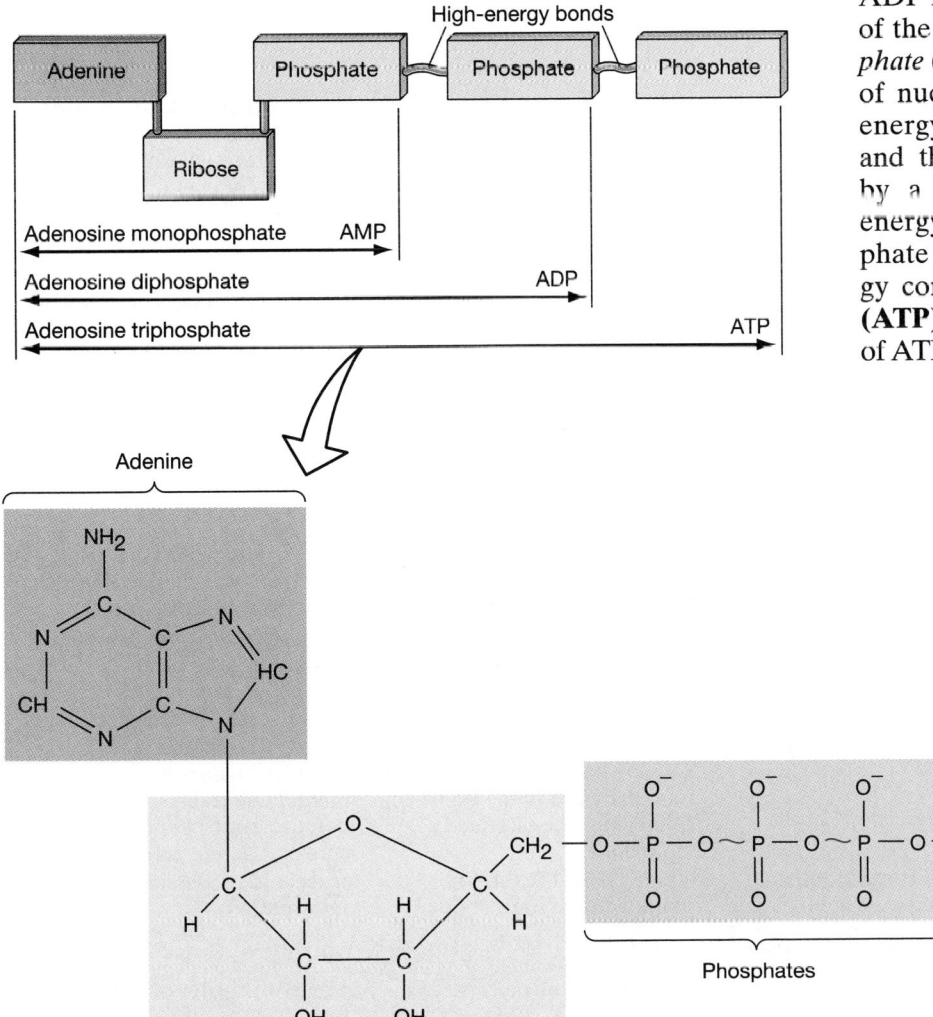

another. The usual method of energy transfer involves the creation of *high-energy bonds*. A high-energy bond is a covalent bond whose breakdown releases energy that can be harnessed by the cell. In your cells, a high-energy bond generally connects a phosphate group (PO_4^{3-}) to an organic molecule. The resulting complex is called a **high-energy compound**.

Phosphorylation (fos-for-i-LĀ-shun) is the attachment of a phosphate group to another molecule. This process does not necessarily produce high-energy bonds. The creation of a high-energy compound requires (1) a phosphate group, (2) appropriate enzymes, and (3) suitable organic substrates. The most important substrate is the adenine nucleoside *adenosine,* with two phosphate groups attached. This combination is called **adenosine diphosphate (ADP)**. ADP is created by the phosphorylation of the nucleotide *adenosine monophosphate (AMP),* one of the building blocks of nucleic acids. It takes a significant energy input to convert AMP to ADP, and the second phosphate is attached by a high-energy bond. Even more energy is required to add a third phosphate and thereby create the high-energy compound **adenosine triphosphate (ATP)**. Figure 2-21• details the structure of ATP.

•**FIGURE 2-21**
The Structure of ATP.
A molecule of ATP consists of an adenine nucleoside to which three phosphate groups have been joined. Cells most often transfer energy by attaching a third phosphate group to ADP with a high-energy bond and then removing that phosphate group at another site, where the associated release of energy performs cellular work.

The conversion of ADP to ATP and the reversion of ATP to ADP are the most common methods of energy transfer in our cells. The arrangement can be summarized as

$$ADP + \text{phosphate group} + \text{energy} \rightleftharpoons ATP + H_2O$$

Throughout life, our cells are continuously generating ATP from ADP and using the energy provided by the ATP to perform vital functions, such as the synthesis of proteins or the contraction of muscles.

Although ATP is the most abundant high-energy compound, there are others—typically other nucleotides that have undergone phosphorylation. For example, *guanosine triphosphate* (*GTP*) and *uridine triphosphate* (*UTP*) are nucleotide-based high-energy compounds that transfer energy in specific enzymatic reactions.

Table 2-8 summarizes information about the inorganic and organic compounds that we covered in this chapter.

CHEMICALS AND LIVING CELLS

The human body is more than a collection of chemicals. Biochemical building blocks form functional units called **cells**. Each cell behaves like a miniature organism, responding to internal and external stimuli. A phospholipid membrane separates the cell from its environment, and internal membranes create compartments with specific functions. Proteins form an internal supporting framework and, as enzymes, accelerate and control the chemical reactions that maintain homeostasis. Nucleic acids direct the syn-

TABLE 2-8 Structure and Function of Biologically Important Compounds

Class	Building Blocks	Sources	Functions
Inorganic			
Water (pp. 41–44)	Hydrogen and oxygen atoms	Absorbed as liquid water or generated by metabolism	Solvent; transport medium for dissolved materials and heat; cooling through evaporation; medium for chemical reactions; reactant in hydrolysis
Acids, bases, salts (pp. 44–45)	H^+, OH^-, various anions and cations	Obtained from the diet or generated by metabolism	Structural components; buffers; sources of ions
Dissolved gases (p. 41)	O, C, N, and other atoms	Atmosphere	O_2: required for normal cellular metabolism CO_2: generated by cells as a waste product NO: chemical messenger involved in cardiovascular, nervous, and lymphatic systems
Organic			
Carbohydrates (pp. 45–46)	C, H, O, in some cases N; CHO in a 1:2:1 ratio	Obtained from the diet or manufactured in the body	Energy source; some structural role when attached to lipids or proteins; energy storage
Lipids (pp. 46–51)	C, H, O, in some cases N or P; CHO not in 1:2:1 ratio	Obtained from the diet or manufactured in the body	Energy source; energy storage; insulation; structural components; chemical messengers; physical protection
Proteins (pp. 51–57)	C, H, O, N, commonly S	20 common amino acids; roughly half can be manufactured in the body, others must be obtained from the diet	Catalysts for metabolic reactions; structural components; movement; transport; buffers; defense; control and coordination of activities
Nucleic acids (pp. 57–58)	C, H, O, N, and P; nucleotides composed of phosphates, sugars, and nitrogenous bases	Obtained from the diet or manufactured in the body	Storage and processing of genetic information
High-energy compounds (pp. 58–60)	Nucleotides joined to phosphates by high-energy bonds	Synthesized by all cells	Storage or transfer of energy

hesis of all cellular proteins, including the enzymes that enable the cell to synthesize a wide variety of other substances. Carbohydrates provide energy (transferred by high-energy compounds) to support vital activities, and they form part of specialized compounds such as proteoglycans and glycolipids.

Cells are dynamic structures that adapt to changes in their environment. That adaptation may involve changes in the chemical organization of the cell. Such changes are easily made because organic molecules other than DNA are temporary rather than permanent components of the cell. Their continuous removal and replacement are part of the process of **metabolic turnover**.

Most of the organic molecules in the cell are replaced at intervals ranging from hours to months. The average time between synthesis and recycling is known as the *turnover rate*. Table 2-9 indicates the turnover rates of the organic components of representative cells.. AM *Metabolic Anomalies*

TABLE 2-9 Turnover Rates

Cell Type	Component	Average Recycling Time*
Liver	Total protein	5–6 days
	Enzymes	1 hour to several days, depending on the enzyme
	Glycogen	1–2 days
	Cholesterol	5–7 days
Muscle cell	Total protein	30 days
	Glycogen	12–24 hours
Neuron	Phospholipids	200 days
	Cholesterol	100+ days
Fat cell	Triglycerides	15–20 days

*Most values were obtained from studies on mammals other than humans.

☑ Analysis of a large organic molecule indicates that it is composed of the sugar ribose, nitrogenous bases, and phosphate groups. Which nucleic acid is this?

☑ What molecule is produced by the phosphorylation of ADP?

SELECTED CLINICAL TERMINOLOGY

Terms Discussed in This Chapter

cholesterol: A steroid important in the structure of cellular membranes that in high concentrations increases the risks of heart disease. *(p. 50)*

mole (mol), **millimole** (mmol): The number of atoms or molecules that has a weight in grams equal to the atomic weight or the molecular weight, respectively. 1 mmol = 0.001 mol. *(p. 33)*

omega 3 fatty acids: Fatty acids, abundant in fish flesh and fish oils, that have a double bond three carbons away from the end of the hydrocarbon chain. Their presence in the diet has been linked to reduced risks of heart disease and other conditions. *(p. 48)*

radioisotopes: Isotopes with unstable nuclei, which spontaneously emit subatomic particles or radiation in measurable amounts. *(p. 32)*

AM Additional Terms Discussed in the *Applications Manual*

nuclear imaging: A procedure in which an image is created on a photographic plate or video screen by the radiation emitted by injected radioisotopes.

PET (positron emission tomography) scan: A nuclear imaging technique in which the emitted radiation is analyzed and the image is created by a computer.

tracer: A compound labeled with a radioisotope that can be tracked within the body by the radiation it releases.

CHAPTER REVIEW

 On-line resources for this chapter are on our World Wide Web site at: *http://www.prenhall.com/martini/fap*

STUDY OUTLINE

INTRODUCTION, p. 31

ATOMS AND MOLECULES, p. 31

1. **Atoms** are the smallest units of matter. They consist of **protons, neutrons,** and **electrons**. *(Figure 2-1)*

Atomic Structure, p. 31

2. An **element** consists of atoms in which the number of protons (the **atomic number**) generally equals the number of electrons. Within an atom, an **electron cloud** surrounds the nucleus. *(Figure 2-1; Table 2-1)*

3. The **mass number** of an atom is the total number of protons and neutrons in its nucleus. **Isotopes** are atoms of the same element whose nuclei contain different numbers of neutrons.

4. Electrons occupy a series of **energy levels** that are often illustrated as **electron shells**. The electrons in the outermost energy level determine an atom's chemical properties. *(Figure 2-2)*

Chemical Bonds, p. 33

5. Atoms can combine through chemical reactions that create **chemical bonds**. The term **molecule** refers to any chemical structure consisting of atoms held together by covalent bonds. A **compound** is a chemical structure containing atoms of two or more elements.

6. An **ionic bond** results from the attraction between **ions**, atoms that have gained or lost electrons. **Cations** are positively charged; **anions** are negatively charged. (*Figure 2-3*)

7. Atoms that share electrons to form a molecule are held together by **covalent bonds**. Sharing one pair of electrons equally creates a **single covalent bond**; sharing two pairs equally forms a **double covalent bond**. A bond with equal sharing of electrons is a **nonpolar covalent bond**; unequal sharing of electrons creates a **polar covalent bond**. (*Figures 2-4, 2-5*)

8. A **hydrogen bond** is a weak but important force that can affect the shapes and properties of molecules. (*Figure 2-6*)

9. Matter can exist as a *solid*, a *liquid*, or a *gas*, depending on the nature of the interactions among the component atoms or molecules.

CHEMICAL REACTIONS, p. 38

1. **Metabolism** refers to all the **chemical reactions** in the body. Through metabolism, cells capture, store, and use energy to maintain homeostasis and to support essential functions.

FOCUS: Chemical Notation, p. 38

2. **Chemical notation** allows us to describe reactions between **reactants** that generate one or more **products**. (*Table 2-2*)

Basic Energy Concepts, p. 39

3. **Work** is movement of an object or a change in its physical structure. **Energy** is the capacity to perform work. There are two major types of energy: *kinetic* and *potential*.

4. **Kinetic energy** is the energy of motion. **Potential energy** is stored energy that results from the position or structure of an object. Conversions from potential to kinetic energy are not 100 percent efficient; every such energy exchange releases *heat*.

Types of Reactions, p. 39

5. Cells gain energy to power their functions by **catabolism**, the breakdown of complex molecules. Much of this energy supports **anabolism**, the synthesis of new molecules.

6. A chemical reaction may be classified as a **decomposition, synthesis,** or **exchange reaction**. **Exergonic reactions** release energy; **endergonic reactions** require energy.

Reversible Reactions, p. 40

7. At **equilibrium**, the rates of two opposing reactions are in balance.

Enzymes and Chemical Reactions, p. 40

8. **Activation energy** is the amount of energy required to start a reaction. **Enzymes** control many chemical reactions within our bodies. Enzymes are organic *catalysts*—substances that accelerate chemical reactions without themselves being permanently changed or used up.

9. The complex reactions that support life proceed in a series of interlocking steps called *pathways*. Each step in a pathway is controlled by a different enzyme.

INORGANIC COMPOUNDS, p. 41

1. **Nutrients** and **metabolites** can be broadly classified as **inorganic** or **organic**.

Water and Its Properties, p. 41

2. Water is the most important component of the body.

3. A **solution**, a uniform mixture of two or more substances, consists of a medium, or **solvent**, in which atoms, ions, or molecules of another substance, or **solute**, are dispersed. In **aqueous solutions**, water is the solvent.

4. Many inorganic compounds, called **electrolytes**, undergo *ionization*, or *dissociation*, in water to form ions. (*Figure 2-7; Table 2-3*)

5. The **pH** of a solution indicates the concentration of hydrogen ions it contains. Solutions can be classified as **neutral, acidic,** or **basic** (*alkaline*) on the basis of pH. (*Figure 2-8*)

Inorganic Acids and Bases, p. 44

6. An **acid** releases hydrogen ions, and a **base** removes hydrogen ions from a solution. *Strong acids* and *strong bases* ionize completely, whereas *weak acids* and *weak bases* do not.

Salts, p. 45

7. A **salt** is an electrolyte whose cation is not hydrogen (H^+) and whose anion is not hydroxide (OH^-).

Buffers and pH Control, p. 45

8. **Buffers** remove or replace hydrogen ions in solution. Buffers and *buffer systems* maintain the pH of body fluids within normal limits.

ORGANIC COMPOUNDS, p. 45

1. Organic compounds always contain carbon and hydrogen and generally oxygen as well. Major classes of organic compounds include *carbohydrates, lipids, proteins,* and *nucleic acids. High-energy compounds* are small in terms of their abundance but absolutely vital to the survival of our cells.

Carbohydrates, p. 45

2. **Carbohydrates** are most important as an energy source for metabolic processes. The three major types are **monosaccharides** (*simple sugars*), **disaccharides,** and **polysaccharides**. Disaccharides and polysaccharides form by **dehydration synthesis**. (*Figures 2-9–2-11; Table 2-4*)

Lipids, p. 46

3. **Lipids** are water-insoluble molecules that include *fats, oils,* and *waxes*. There are six important classes of lipids: **fatty acids, eicosanoids, glycerides, steroids, phospholipids,** and **glycolipids**. (*Figures 2-12–2-16; Table 2-5*)

4. **Triglycerides** (neutral fats) consist of three fatty acid molecules attached by dehydration synthesis to a molecule of **glycerol; diglycerides** consist of two fatty acids and glycerol; and **monoglycerides** consist of one fatty acid plus glycerol.

5. **Steroids** (1) are involved in the structure of cell membranes, (2) include sex hormones and hormones regulating metabolic activities, and (3) are important in lipid digestion.

Proteins, p. 51

6. **Proteins** perform a great variety of functions in the body. Important types of proteins include structural proteins, *contractile proteins, transport proteins, enzymes,* and *antibodies.* (*Table 2-6*)

7. Proteins are chains of **amino acids**. Each amino acid consists of an *amino group,* a *carboxylic acid group,* and an *R group (side chain)*. A protein, or **polypeptide,** is a linear sequence of amino acids held together by **peptide bonds**. (*Figures 2-17, 2-18*)

8. There are four levels of protein structure: **primary structure** (amino acid sequence), **secondary structure** (amino

acid interactions, such as hydrogen bonds), **tertiary structure** (complex folding, *disulfide bonds*, and interaction with water molecules), and **quaternary structure** (formation of protein complexes from individual subunits). **Globular proteins**, such as *myoglobin*, are generally rounded and water-soluble. **Fibrous proteins**, such as *keratin* and *collagen*, are tough, durable, and generally insoluble.

9. The reactants in an enzymatic reaction, called **substrates**, interact to yield a product by binding to the enzyme at the **active site**. **Cofactors** are ions or molecules that must bind to the enzyme before substrate binding can occur; **coenzymes** are organic cofactors commonly derived from *vitamins*. *(Figure 2-19)*

10. The shape of a protein determines its functional characteristics. Each protein works best at an optimal combination of temperature and pH and will undergo reversible or irreversible **denaturation** at temperatures or pH values outside the normal range.

Nucleic Acids, p. 57

11. **Nucleic acids** store and process information at the molecular level. There are two kinds of nucleic acids: **deoxyribonucleic acid (DNA)** and **ribonucleic acid (RNA)**. *(Figure 2-20; Table 2-7)*

12. Nucleic acids are chains of **nucleotides**. Each nucleotide contains a sugar, a phosphate group, and a **nitrogenous base**. The sugar is *ribose* in RNA and *deoxyribose* in DNA. The nitrogenous bases found in DNA, which is a two-stranded double helix, are **adenine, guanine, cytosine,** and **thymine**. RNA, which consists of a single strand, contains **uracil** instead of thymine.

High-Energy Compounds, p. 58

13. Cells store energy in the *high-energy bonds* of **high-energy compounds** for later use. The most important high-energy compound is **ATP (adenosine triphosphate)**. Cells make ATP by adding a phosphate group to **ADP (adenosine diphosphate)** through the process of **phosphorylation**. When ATP is broken down to ADP and phosphate, energy is released, and this energy may be used by the cell to power essential activities. *(Figure 2-21; Table 2-8)*

CHEMICALS AND LIVING CELLS, p. 60

1. Biochemical building blocks form functional units called **cells**.

2. The continuous removal and replacement of cellular organic molecules (other than DNA), a process called **metabolic turnover**, allows cells to change and adapt to changes in their environment. *(Table 2-9)*

REVIEW QUESTIONS

Level 1 Reviewing Facts and Terms

Match each numbered item with the most closely related lettered item. Use letters for answers in the spaces provided.

___ 1. atomic number
___ 2. mass number
___ 3. covalent bond
___ 4. ionic bond
___ 5. molecular weight
___ 6. catabolism
___ 7. anabolism
___ 8. exchange reaction
___ 9. reversible reaction
___ 10. hydrophobic

___ 11. acid
___ 12. enzyme
___ 13. buffer
___ 14. organic compounds
___ 15. inorganic compounds

a. sum of atomic weights
b. water-insoluble
c. synthesis
d. catalyst
e. sharing of electrons
f. $A + B \rightleftharpoons AB$
g. stabilize pH
h. number of protons
i. decomposition
j. carbohydrates, lipids, proteins
k. loss or gain of electrons
l. water, salts
m. H^+ donor
n. number of protons plus neutrons
o. $AB + CD \longrightarrow AD + CB$

16. In atoms, protons and neutrons are found
   (a) only in the nucleus
   (b) outside the nucleus
   (c) inside and outside the nucleus
   (d) in the electron shells

17. Isotopes of an element differ from each other in the number of
   (a) protons in the nucleus
   (b) neutrons in the nucleus
   (c) electrons in the outer shells
   (d) protons and electrons in the nucleus

18. The number and arrangement of electrons in an atom's outer energy level determines the atom's
   (a) atomic weight   (b) atomic number
   (c) electrical properties (d) chemical properties

19. The bond between sodium and chlorine in the compound sodium chloride (NaCl) is
   (a) an ionic bond
   (b) a single covalent bond
   (c) a nonpolar covalent bond
   (d) a double covalent bond

20. The oxygen atoms in a molecule of oxygen are held together by
   (a) a single covalent bond
   (b) a double covalent bond
   (c) a polar covalent bond
   (d) an ionic bond

21. All the chemical reactions that occur in the human body are collectively referred to as
   (a) anabolism   (b) catabolism
   (c) metabolism   (d) homeostasis

22. Which of the following equations illustrates a typical decomposition reaction?
   (a) $A + B \longrightarrow AB$ (b) $AB + CD \longrightarrow AD + CB$
   (c) $2 A_2 + B_2 \longrightarrow 2 A_2 B$ (d) $AB \longrightarrow A + B$

23. Of the following choices, the pH of the least acidic solution is
   (a) 6.0   (b) 4.5
   (c) 2.3   (d) 1.0

24. A pH of 7.8 in the human body typifies a condition referred to as
   (a) acidosis   (b) alkalosis
   (c) metabolic acidosis (d) homeostasis

25. A(n) _____ is a solute that dissociates to release hydrogen ions, and a(n) _____ is a solute that removes hydrogen ions from solution:

    (a) base, acid   **(b)** salt, base

    (c) acid, salt   **(d)** acid, base

26. Carbohydrates, lipids, and proteins are formed from their basic building blocks by the

    (a) removal of a water molecule between the building blocks

    (b) addition of a water molecule between the building blocks

    (c) addition of carbon to each molecule

    (d) addition of oxygen to each molecule

27. Complementary base pairing in DNA includes

    (a) adenine–uracil; cytosine–guanine

    (b) adenine–thymine; cytosine–guanine

    (c) adenine–guanine; cytosine–thymine

    (d) guanine–uracil; cytosine–thymine

28. What are the three stable fundamental particles in atoms?

29. What is the difference between potential and kinetic energy?

30. What is the role of enzymes in chemical reactions?

31. List six inorganic molecules found in abundance in the human body.

32. What four major classes of organic compounds are found in the body?

33. List three important functions of triglycerides (neutral fats) in the body.

34. List seven major functions performed by proteins.

35. **(a)** What three basic components make up a nucleotide of DNA?

    (b) What three basic components make up a nucleotide of RNA?

36. What three components are required to create the high-energy compound ATP?

Level 2   Reviewing Concepts

37. If an isotope of oxygen has 8 protons, 10 neutrons, and 8 electrons, its mass number is

    (a) 26   **(b)** 16

    (c) 18   **(d)** 8

38. Hydrophilic molecules readily associate with

    (a) lipid molecules

    (b) insoluble droplets of hydrophobic molecules

    (c) water molecules

    (d) a, b, and c are all correct

39. The speed, or rate, of a chemical reaction is influenced by

    (a) the presence of catalysts

    (b) the temperature

    (c) the concentration of the reactants

    (d) a, b, and c are all correct

40. Using the periodic table of the elements as a reference, determine the following information about calcium:

    (a) number of protons   **(b)** number of electrons

    (c) number of neutrons   **(d)** atomic number

    (e) atomic weight   **(f)** number of electrons in each shell

41. Explain the differences among nonpolar covalent bonds, polar covalent bonds, and ionic bonds.

42. How does thermal inertia help stabilize body temperature?

43. Why does pure water have a neutral pH?

44. What role do buffer systems play in the human body?

45. Analyzing a sample that contains an organic molecule, a biologist finds the following constituents: carbon, hydrogen, oxygen, nitrogen, and phosphorus. On the basis of this information, is the molecule a carbohydrate, a lipid, a protein, or a nucleic acid?

Level 3   Critical Thinking and Clinical Applications

46. The element sulfur has an atomic number of 16 and an atomic weight of 32. How many neutrons are in the nucleus of a sulfur atom? Assuming that sulfur forms covalent bonds with hydrogen, how many hydrogen atoms could bond to one sulfur atom?

47. An important buffer system in the human body involves carbon dioxide (CO_2) and bicarbonate ion (HCO_3^-) as shown:

$$CO_2 + H_2O \rightleftharpoons H_2CO_3 \rightleftharpoons H^+ + HCO_3^-$$

If a person becomes excited and exhales large amounts of CO_2, how will the pH of the person's body be affected?

48. A student needs to prepare 1 liter of a solution of NaCl that contains 1.2 moles of ions per liter of solution. How many moles of NaCl should the student use to prepare the solution? What will that quantity weigh, in grams?

CHAPTER 3

The Cellular Level of Organization

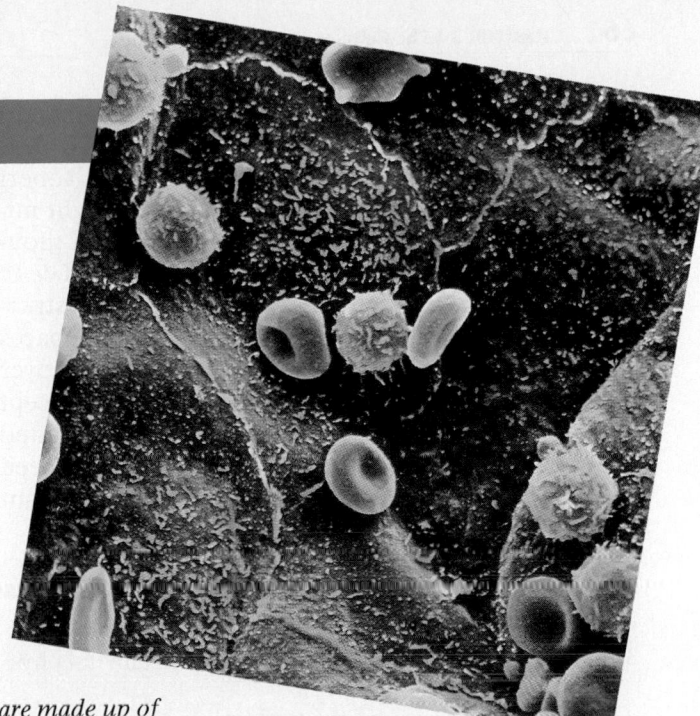

*M*any familiar structures, from pyramids to patchwork quilts, are made up of numerous small, similar components. The human body is built on the same principle, for it is made up of several trillion tiny units called cells. Cells, however, differ from many other building blocks in a number of ways. For one thing, they vary widely in size and appearance. Here, several relatively small blood cells lie on a layer of large, flattened cells that line a blood vessel. An individual brick can perform few (if any) of the functions of a building. But a cell can perform many of the functions of the body of which it is a part. Indeed, cells are the smallest entities that can perform all the basic life functions discussed in Chapter 1. In addition, each cell in the body performs specialized functions and plays a role in the maintenance of homeostasis. As you can imagine, cells are complex structures with many specialized components. In this chapter, we will explore the functional organization of a representative cell.

CHAPTER OUTLINE AND OBJECTIVES

Atoms are the building blocks of molecules; cells are the building blocks of the human body. Cells were first described by the English scientist Robert Hooke around 1665. Hooke used an early light microscope to examine dried cork. He observed thousands of tiny, empty chambers, which he named *cells.* Later that decade, other scientists, observing the structure of living plants, realized that in life these spaces were filled with a gelatinous material. Research over the next 175 years led to the *cell theory,* the concept that cells are the fundamental units of all plant and animal tissues. Since that time, the cell theory has been expanded to incorporate five basic concepts relevant to our discussion of the human body:

1. Cells are the building blocks of all plants and animals.
2. Cells are produced by the division of preexisting cells.
3. Cells are the smallest units that perform all vital physiological functions.
4. Each cell maintains homeostasis at the cellular level.
5. Homeostasis at the tissue, organ, organ system, and organism levels reflects the combined and coordinated actions of many cells.

Cells have a variety of forms and functions. Figure 3-1• gives examples of the range of cell sizes and shapes found in the human body. The relative proportions of the cells in this figure are correct, but all have been magnified roughly 500 times. Together, these and other types of cells create and maintain all anatomical structures and perform all vital physiological functions.

The human body contains trillions of cells, and all our activities—from running to thinking—result from the combined and coordinated responses of millions or even billions of cells. Yet each cell also functions as an individual entity, responding to a variety of environmental cues. As a result, anyone interested in understanding how the human body functions must first become familiar with basic concepts of cell biology. AM *A Closer Look: The Nature of Pathogens*

STUDYING CELLS

Cytology (sī-TOL-o-jē; *cyto-*, cell + *-ology*, the study of) is the study of the structure and function of cells. What we have learned since the 1950s has given us new insights into cellular physiology and the mechanisms of homeostatic control. Today, cytology is part of the broader discipline **cell biology,** which incorporates aspects of biology, chemistry, and physics. The two most common methods used to study cell and tissue structure are *light microscopy* and *electron microscopy.* For information on these techniques and their limitations, see the *Applications Manual.* AM *Methods of Microanatomy*

In this chapter, we describe the structure of a typical animal cell, some of the ways in which cells interact with their environment, and how cells reproduce. Later chapters deal with the coordination of cellular activities in physiological systems.

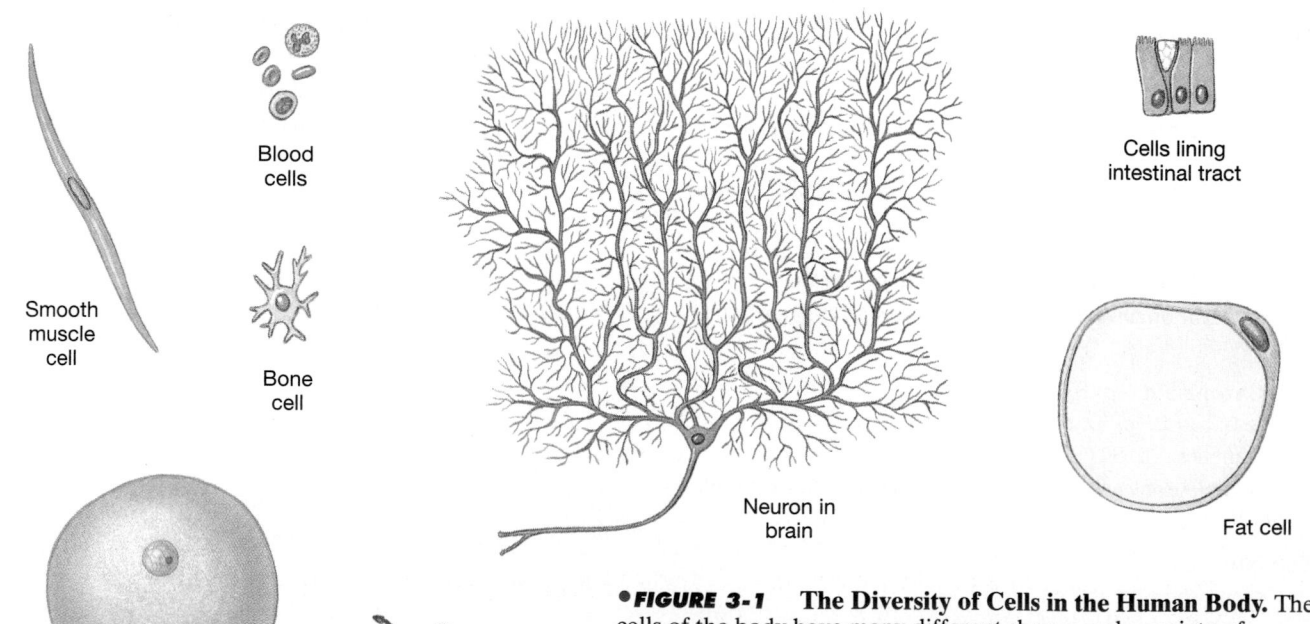

Smooth muscle cell

Blood cells

Bone cell

Oocyte

Sperm

Neuron in brain

Cells lining intestinal tract

Fat cell

•**FIGURE 3-1**   **The Diversity of Cells in the Human Body.** The cells of the body have many different shapes and a variety of special functions. These examples give an indication of the range of forms. All the cells are shown with the dimensions they would have if magnified approximately 500 times.

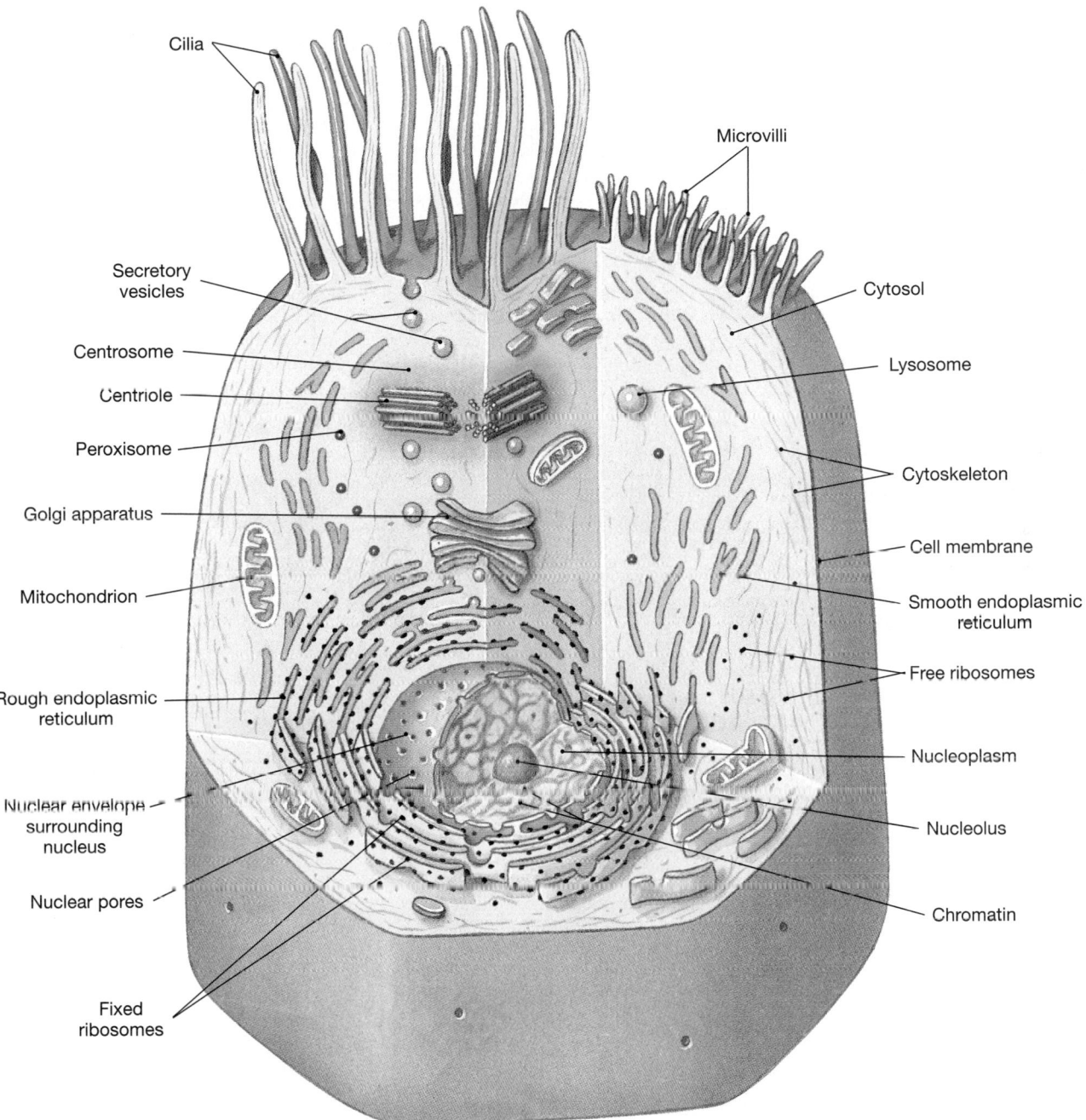

Cilia

Microvilli

Secretory
vesicles

Cytosol

Centrosome

Lysosome

Centriole

Peroxisome

Cytoskeleton

Golgi apparatus

Cell membrane

Mitochondrion

Smooth endoplasmic
reticulum

Free ribosomes

Rough endoplasmic
reticulum

Nucleoplasm

Nuclear envelope
surrounding
nucleus

Nucleolus

Nuclear pores

Chromatin

Fixed
ribosomes

●*FIGURE 3-2*   **The Anatomy of a Representative Cell.** See Table 3-1 for a summary of the functions associated with
the various cell structures.

An Overview of Cellular Anatomy

The human body contains two classes of cells: *somatic cells* and *sex cells*. There are only two types of **sex
cells** (also called *germ cells* or *reproductive cells*):
(1) the sperm of a male and (2) the oocytes of a female. **Somatic cells** (*soma,* body) include all the other
cells in the human body. In this chapter, we will examine somatic cells; we shall discuss sex cells in chapters dealing with the reproductive system.

The "typical" somatic cell is like the "average" person. Any description masks enormous individual variations. Our model cell will share features with most
cells of the body without being identical to any one.
Figure 3-2● shows such a cell, and Table 3-1 summarizes the structures and functions of its parts.

Our cell is surrounded by a watery medium known
as the **extracellular fluid.** The extracellular fluid found
in most tissues is called **interstitial** (in-ter-STISH-ul)
fluid (*interstitium,* something standing between). A *cell*

TABLE 3-1 Organelles of a Representative Cell

Appearance	Structure	Composition	Function
	CELL MEMBRANE	Lipid bilayer, containing phospholipids, steroids, and proteins	Isolation; protection; sensitivity; support; controls entrance/exit of materials
	CYTOSOL	Fluid component of cytoplasm	Distributes materials by diffusion
	NONMEMBRANOUS ORGANELLES		
	Cytoskeleton Microtubule Microfilament	Proteins organized in fine filaments or slender tubes	Strength and support; movement of cellular structures and materials
	Microvilli	Membrane extensions containing microfilaments	Increase surface area to facilitate absorption of extracellular materials
	Cilia	Membrane extensions containing microtubule doublets in a 9 + 2 array	Movement of materials over cell surface
	Centrosome Centriole	Cytoplasm containing two centrioles, at right angles; each centriole is composed of 9 microtubule triplets	Essential for movement of chromosomes during cell division; organization of microtubules in cytoskeleton
	Ribosomes	RNA + proteins; fixed ribosomes bound to endoplasmic reticulum, free ribosomes scattered in cytoplasm	Protein synthesis
	MEMBRANOUS ORGANELLES		
	Mitochondria	Double membrane, with inner membrane folds (cristae) enclosing important metabolic enzymes	Produce 95% of the ATP required by the cell
	Endoplasmic reticulum (ER)	Network of membranous channels extending throughout the cytoplasm	Synthesis of secretory products; intracellular storage and transport
	Rough ER	Has ribosomes bound to membranes	Modification and packaging of newly synthesized proteins
	Smooth ER	Lacks attached ribosomes	Lipid and carbohydrate synthesis
	Golgi apparatus	Stacks of flattened membranes (saccules) containing chambers (cisternae)	Storage, alteration, and packaging of secretory products and lysosomal enzymes
	Lysosomes	Vesicles containing powerful digestive enzymes	Intracellular removal of damaged organelles or of pathogens
	Peroxisomes	Vesicles containing degradative enzymes	Neutralization of toxic compounds
	NUCLEUS	Nucleoplasm containing nucleotides, enzymes, nucleoproteins, and chromatin; surrounded by double membrane (nuclear envelope)	Control of metabolism; storage and processing of genetic information; control of protein synthesis
	Nucleolus	Dense region in nucleoplasm containing DNA and RNA	Site of rRNA synthesis and assembly of ribosomal subunits

membrane separates the cell contents, or *cytoplasm,* from the interstitial fluid. The cytoplasm, which surrounds the membranous *nucleus,* can be further subdivided into a liquid, the *cytosol,* and intracellular structures collectively known as *organelles* (or-gan-ELZ; "little organs").

THE CELL MEMBRANE

The **cell membrane,** also called the **plasma membrane,** or *plasmalemma* (*lemma,* husk), forms the outer boundary of the cell. The general functions of the cell membrane include:

- *Physical isolation.* The cell membrane is a physical barrier that separates the inside of the cell from the surrounding extracellular fluid. Conditions inside and outside the cell are very different, and those differences must be maintained to preserve homeostasis. For example, the cell membrane is a barrier that keeps enzymes and structural proteins inside the cell.
- *Regulation of exchange with the environment.* The cell membrane controls the entry of ions and nutrients, such as glucose; the elimination of wastes; and the release of secretions.
- *Sensitivity.* The cell membrane is the first part of the cell affected by changes in the composition, concentration, or pH of the extracellular fluid. It also contains a variety of receptors that allow the cell to recognize and respond to specific molecules in its environment. For example, the cell membrane may bind chemical signals from other cells. A single alteration in the cell membrane may trigger the activation or deactivation of enzymes that affect many cellular activities.
- *Structural support.* Specialized connections between cell membranes or between membranes and extracellular materials give tissues a stable structure. For example, the cells at the surface of the skin are bound together, and those in the deepest layers are attached to extracellular protein fibers in underlying tissues.

Membrane Structure

The cell membrane is extremely thin and delicate, ranging from 6 to 10 nm in thickness. The cell membrane contains lipids, proteins, and carbohydrates. In terms of relative abundance, phospholipids are the largest contributors to membrane structure, followed by proteins, glycolipids, and cholesterol.

Membrane Lipids

Figure 3-3 presents a diagrammatic view of cell membrane structure. Lipids form most of the membrane surface but account for only about 42 percent of its weight. Proteins, which are relatively dense, account for more than half the weight of the membrane. The cell membrane is called a **phospholipid bilayer** because the phospholipids form two distinct layers. In each layer, the phospholipid molecules lie with the hydrophilic heads at the membrane surface and the hydrophobic tails on the inside, in association with cholesterol and small quantities of other lipids.

Note the similarities in lipid organization between a micelle, described in Chapter 2 (Figure 2-16c, p. 52), and the cell membrane. Ions and water-soluble compounds cannot enter the interior of a micelle because

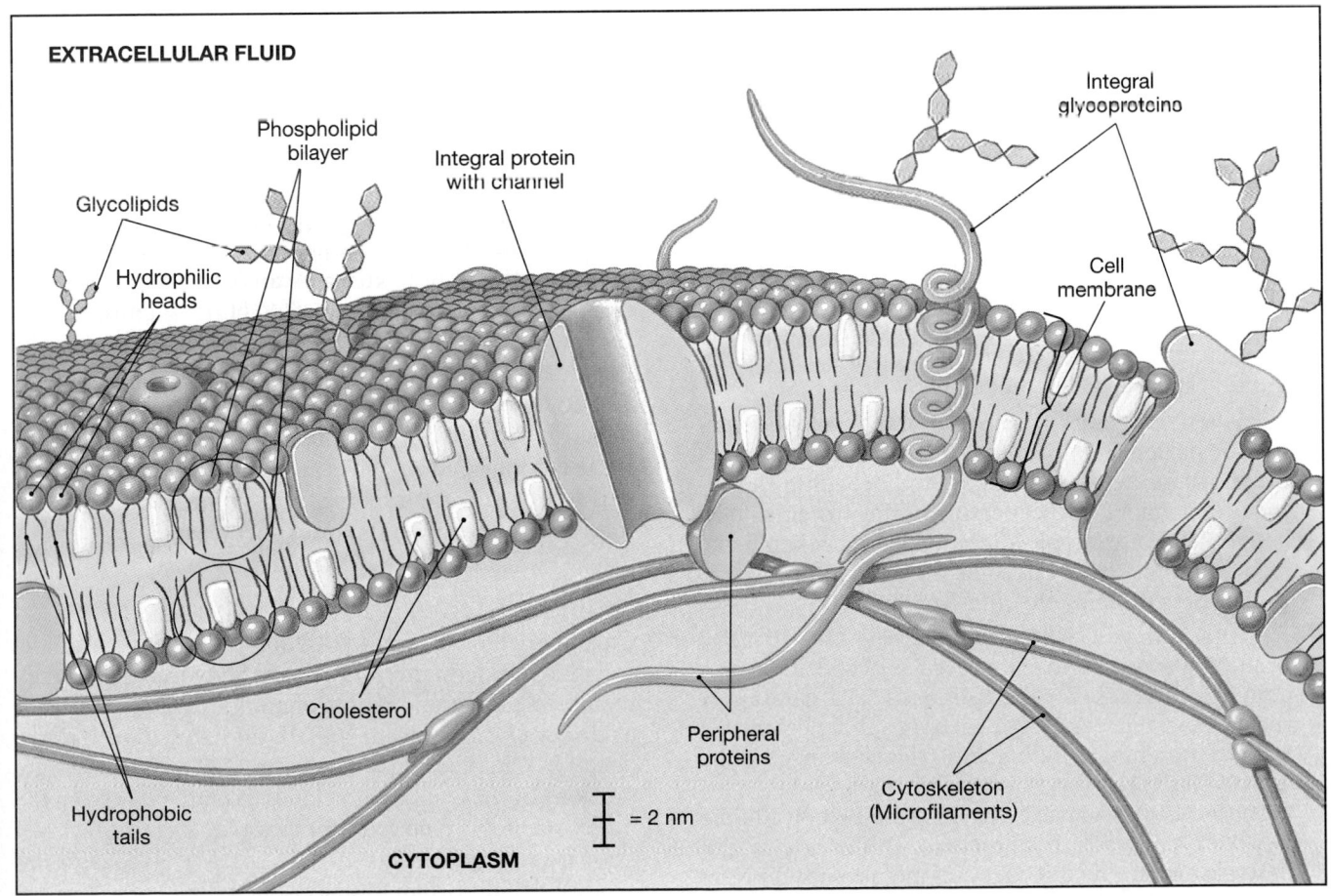

•FIGURE 3-3 **The Cell Membrane**

the lipid tails of the phospholipid molecules are hydrophobic and will not associate with water molecules. For the same reason, water and solutes cannot cross the lipid portion of a cell membrane. This feature makes the membrane very effective in isolating the cytoplasm from the surrounding fluid environment. Such isolation is important because the composition of the cytoplasm is very different from that of the extracellular fluid, and the cell cannot survive if those differences are eliminated.

Membrane Proteins

Proteins account for roughly 55 percent of the weight of a cell membrane. Several types of proteins are involved. **Integral proteins** are part of the membrane structure. Most integral proteins span the entire width of the membrane one or more times and are therefore known as *transmembrane proteins*. **Peripheral proteins** are bound to the inner or outer surfaces of the membrane and are easily separated from it. Integral proteins greatly outnumber peripheral proteins. Membrane proteins (Figure 3-4●) may function specifically as *anchoring proteins, identifiers, enzymes, receptors, carriers,* or *channels.*

1. *Anchoring proteins.* Membrane proteins may attach the cell membrane to other structures and stabilize its position. Inside the cell, membrane proteins are bound to the *cytoskeleton,* a network of supporting filaments within the cytoplasm. Outside the cell, other membrane proteins may attach the cell to extracellular protein fibers or to another cell.

2. *Recognition proteins (identifiers).* The cells of the immune system recognize other cells as normal or abnormal on the basis of the presence or absence of characteristic **recognition proteins**. Many important recognition proteins are glycoproteins. (We shall discuss one group, the *MHC proteins* involved in the immune response, in Chapter 22.)

3. *Enzymes.* Enzymes in cell membranes may be integral or peripheral proteins. These enzymes catalyze reactions in the extracellular fluid or within the cytosol, depending on the location of the protein and its active site. For example, dipeptides are broken down into amino acids by enzymes on the exposed membranes of cells lining the intestinal tract.

4. *Receptor proteins.* **Receptor proteins** in the cell membrane are sensitive to the presence of specific extracellular molecules called **ligands** (LĪ-gandz). A receptor protein exposed to an appropriate ligand will bind to it, and that binding may trigger changes in the activity of the cell. For example, the binding of the hormone *insulin* to a specific membrane receptor is the key step that leads to an increase in the rate of glucose absorption by the cell. Cell membranes differ in the type and number of receptor proteins they contain; these differences account for their differing sensitivities to hormones and other solutes.

5. *Carrier proteins.* **Carrier proteins** bind solutes and transport them across the cell membrane. The transport process involves a change in the shape of the carrier protein—the shape changes when solute binding occurs and returns to its original form when the solute is released. Carrier proteins may or may not require ATP as an energy source. For example, virtually all cells have carrier proteins that can bring glucose into the cytoplasm without expending ATP, but these cells must expend ATP to transport ions such as sodium and calcium across the cell membrane and out of the cytoplasm.

6. *Channels.* Some integral proteins contain a central pore, or **channel,** that forms a passageway that permits the movement of water and small solutes across the cell membrane. Ions do not dissolve in lipids, and they cannot cross the phospholipid bilayer. Thus, ions and other small water-soluble materials can cross the membrane only by passing through channels. Ion movements through channels are involved in a variety of physiological mechanisms. Physiologists speak of sodium channels, calcium channels, potassium channels, and so forth when referring to channels that will permit the movement of only specific ions. There are two major kinds of channels: (1) **leak channels,** which permit water and ion movement at all times (although the rate may vary from moment to moment), and (2) **gated channels,** which can open or close to regulate ion passage. Channels account for about 0.2 percent of the total membrane surface area.

Membrane structure is not rigid. The embedded proteins drift from place to place across the surface of the membrane like the ice cubes in a bowl of punch. In addition, the composition of the cell membrane can change over time as components of the membrane are removed and recycled in the process of metabolic turnover. ∞ *[p. 61]*

The inner and outer surfaces of the cell membrane differ in protein and lipid composition. For example, some cytoplasmic enzymes are found only on the inner surface of the membrane, and some receptors are found exclusively on its outer surface.

Membrane Carbohydrates

Carbohydrates account for roughly 3 percent of the weight of a cell membrane. The carbohydrates in the cell membrane are components of complex molecules such as *proteoglycans, glycoproteins,* and *glycolipids,* introduced in Chapter 2. ∞ *[pp. 51, 57]* The carbohydrate portions of these large molecules extend away from the outer surface of the cell membrane, forming a layer known as the **glycocalyx** (*calyx,* cup). The glycocalyx has a variety of important functions, including the following:

■ The glycoproteins and glycolipids form a viscous layer that lubricates and protects the cell membrane.

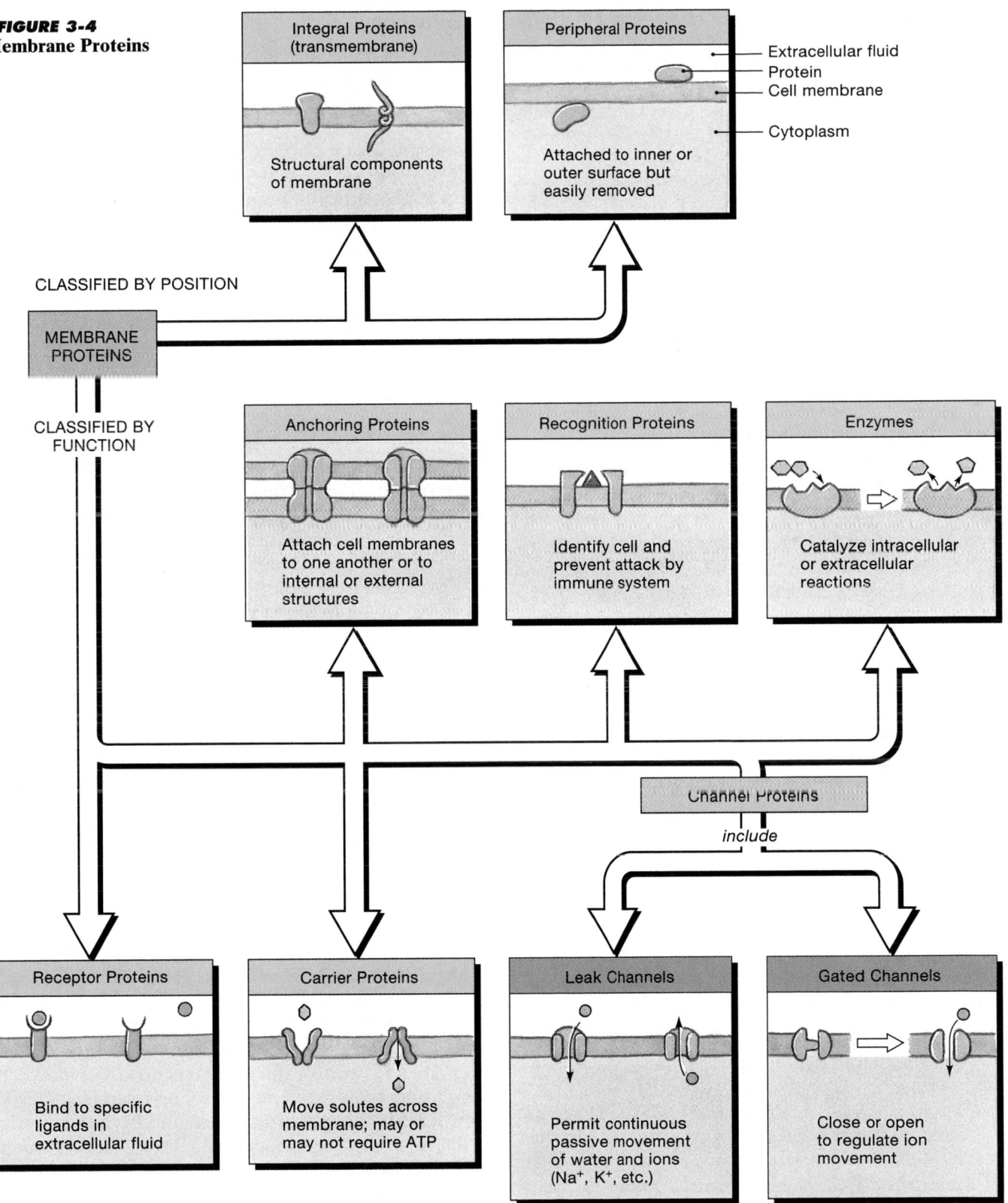

•*FIGURE 3-4*
Membrane Proteins

- Because the components are sticky, the glycocalyx can help anchor the cell in place. It also participates in the locomotion of specialized cells.
- Glycoproteins and glycolipids can function as receptors, binding specific extracellular compounds. Such

binding can alter the properties of the cell surface and indirectly affect the behavior of the cell.
- Glycoproteins and glycolipids are recognized as normal or abnormal by cells involved with the immune response. The characteristics of the glycocalyx are

genetically determined. For example, your blood type (A, B, AB, or O) is determined by the presence or absence of membrane glycolipids on circulating red blood cells. The body's immune system can recognize normal membrane glycoproteins and glycolipids as "self" rather than as "foreign." This recognition system keeps your immune system from attacking your blood cells but enables it to recognize and destroy foreign blood cells, should they appear in the circulation.

☑ What component of the cell membrane is primarily responsible for the membrane's ability to form a physical barrier between the cell's internal and external environments?

☑ What kind of integral proteins allow water and small ions to pass through the cell membrane?

Membrane Permeability

The **permeability** of the cell membrane determines precisely which substances can enter or leave the cytoplasm. If nothing could cross the cell membrane, the membrane would be described as **impermeable**. If any substance could cross without difficulty, the membrane would be **freely permeable**. The permeability of cell membranes lies somewhere between those extremes, and our cell membranes are best described as **selectively permeable**. A selectively permeable membrane permits the free passage of some materials and restricts the passage of others. The distinction may be on the basis of size, electrical charge, molecular shape, lipid solubility, or a combination of factors. Cells differ in their permeabilities according to variations in the organization and identity of membrane lipids and proteins.

Passage across the membrane is either passive or active. *Passive processes* move ions or molecules across the cell membrane without any expenditure of energy by the cell. *Active processes* require that the cell expend energy, generally in the form of ATP.

Transport processes are also categorized by the nature of the mechanism involved. The four major categories are:

1. *Diffusion,* which results from the random motion and collisions of ions and molecules. Diffusion is a passive process.
2. *Filtration,* which occurs when hydrostatic pressure forces fluid and solutes across a membrane barrier. Filtration is also a passive process.
3. *Carrier-mediated transport,* which requires the presence of specialized integral membrane proteins. Carrier-mediated transport may be passive or active, depending on the substance transported and the nature of the transport mechanism.
4. *Vesicular transport,* which involves the movement of materials within small membranous sacs, or *vesicles*. Vesicular transport is an active process.

Diffusion

Ions and molecules are constantly in motion, colliding and bouncing off one another and off any obstacles in their paths. The movement is random: A molecule can bounce in any direction. One result of this continuous random motion is that, over time, the molecules in any given space will tend to become evenly distributed. This distribution process is known as **diffusion**. As the molecules move around, there will be a net movement of material from areas of relatively high concentration to areas of relatively low concentration. The difference between the high and low concentrations represents a **concentration gradient**. Diffusion tends to eliminate that gradient. After the gradient has been eliminated, the molecular motion continues, but there is no longer net movement in any particular direction. For convenience, we restrict use of the term *diffusion* to the directional movement that eliminates concentration gradients; this process is called *net diffusion*. Because diffusion tends to spread materials from a region of high concentration to one of lower concentration, it is often described as proceeding "down a concentration gradient" or "downhill."

All of us have experienced the effects of diffusion, which occurs in air as well as in water. The scent of fresh flowers in a vase sweetens the air in a large room, just as a drop of ink spreads to color an entire glass of water. In each case you begin with an extremely high concentration of molecules in a very localized area. Consider ink dropped in a water glass (Figure 3-5●). Placing that drop in a large volume of clear water (Step 1) establishes a steep concentration gradient for the dye: The dye concentration is high at the drop and negligible everywhere else. As diffusion proceeds, the dye molecules spread through the solution (Step 2) until they are distributed evenly (Step 3).

Diffusion is important in body fluids because it tends to eliminate local concentration gradients. For example, every cell in the body generates carbon dioxide, and the intracellular concentration is relatively high. Carbon dioxide concentrations are lower in the surrounding interstitial fluid and lower still in the circulating blood. Because cell membranes are freely permeable to carbon dioxide, it can diffuse down its concentration gradient—traveling from the cell's interior into the interstitial fluid, and from the interstitial fluid into the bloodstream, for eventual delivery to the lungs.

To be effective, the diffusion of nutrients, waste products, and dissolved gases must be able to keep pace with the demands of active cells. Important factors that influence diffusion rates include:

■ *Distance.* Concentration gradients are eliminated quickly over short distances. The greater the distance, the longer the time required. In the human body, dif-

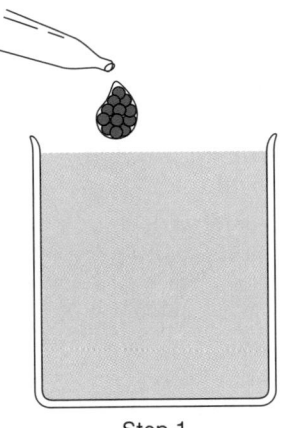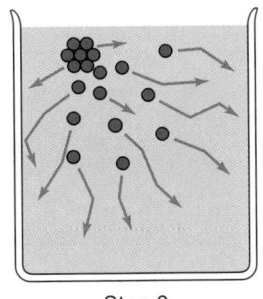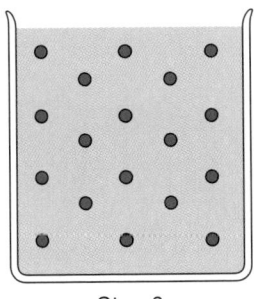

Step 1   Step 2   Step 3

•FIGURE 3-5 **Diffusion.** *Step 1:* Placing an ink drop in a glass of water establishes a strong concentration gradient because there are many ink molecules in one location and none elsewhere. *Step 2:* As diffusion occurs, the ink molecules spread through the solution. *Step 3:* Eventually, diffusion eliminates the concentration gradient, and the ink molecules are distributed evenly. Molecular motion continues, but there is no directional movement.

fusion distances are generally small. For example, few living cells are farther than 125 μm from a blood vessel.

- *Size of the gradient.* The larger the concentration gradient, the faster diffusion proceeds. When cells become more active, the intracellular concentration of oxygen declines. This change increases the concentration gradient for oxygen between the inside of the cell (relatively low) and the interstitial fluid outside (relatively high). The rate of oxygen diffusion into the cell then increases.
- *Molecule size.* Ions and small organic molecules such as glucose diffuse more rapidly than large proteins.
- *Temperature.* The higher the temperature, the faster the diffusion rate. The human body maintains a temperature of about 37°C, and diffusion proceeds more rapidly at this temperature than at normal environmental temperatures.
- *Electrical forces.* The interior of the cell membrane has a net negative charge relative to the exterior surface, in part due to the high concentration of proteins (Pr^-) in the cell. ∞ *[p. 54]* Opposite charges (+ and –) are attracted to one another; similar charges (+ and + or – and –) repel one another. The negative charge on the inside of the cell membrane tends to pull positive ions from the extracellular fluid into the cell and to oppose the entry of negative ions. For example, compared with the cytosol, interstitial fluid contains relatively high concentrations of sodium ions (Na^+) and chloride ions (Cl^-). Both the concentration gradient, or *chemical gradient,* and the electrical gradient favor the diffusion of the positively charged sodium ions into the cell. In contrast, diffusion of the negatively charged chloride ions is favored by the chemical gradient but opposed by the electrical gradient. For any given ion, the net result of the chemical and electrical forces acting on it is called the **electrochemical gradient.**

DIFFUSION ACROSS CELL MEMBRANES In the extracellular fluids of the body, water and dissolved solutes diffuse freely. A cell membrane, however, acts as a barrier that selectively restricts diffusion: Some substances can pass through easily, whereas others cannot penetrate the membrane. There are only two ways

for an ion or molecule to diffuse across a cell membrane: (1) by passing through a membrane channel or (2) by moving across the lipid portion of the membrane (Figure 3-6•). If there is an electrochemical gradient for a particular ion or molecule, whether that substance moves across the cell membrane will depend on two major factors: its lipid solubility and its size relative to the sizes of the membrane channels.

Lipid Solubility Alcohol, fatty acids, and steroids can enter cells easily because they can diffuse through the lipid portions of the membrane. Dissolved gases, such as oxygen and carbon dioxide, and lipid-soluble drugs also enter and leave our cells by diffusing through the lipid bilayer. |AM| *Drugs and the Cell Membrane*

Size To diffuse into the cytoplasm, a water-soluble compound must pass through a membrane channel. These channels are very small—on average, about 0.8 nm in diameter. Water molecules can enter or exit freely, as can ions such as sodium, potassium, calcium, hydrogen, and chloride, but even a small organic molecule, such as glucose, is too big to fit through the channels.

OSMOSIS: A SPECIAL CASE OF DIFFUSION The net diffusion of water across a membrane is so important that it is given a special name, **osmosis** (oz-MŌ-sis; *osmos,* thrust). For convenience, we will always use the term *osmosis* when we consider water movement and will restrict use of the term *diffusion* to the movement of solutes.

Intracellular and extracellular fluids are solutions that contain a variety of dissolved materials. Each solute tends to diffuse as if it were the only material in solution. For example, the diffusion of sodium ions occurs only in response to the existence of a concentration gradient for sodium. Thus, a concentration gradient for another ion will have no effect on the rate or direction of sodium ion diffusion. Some solutes diffuse into the cytoplasm, others diffuse out, and a few, such as proteins, are unable to diffuse

•FIGURE 3-6 Diffusion across the Cell Membrane

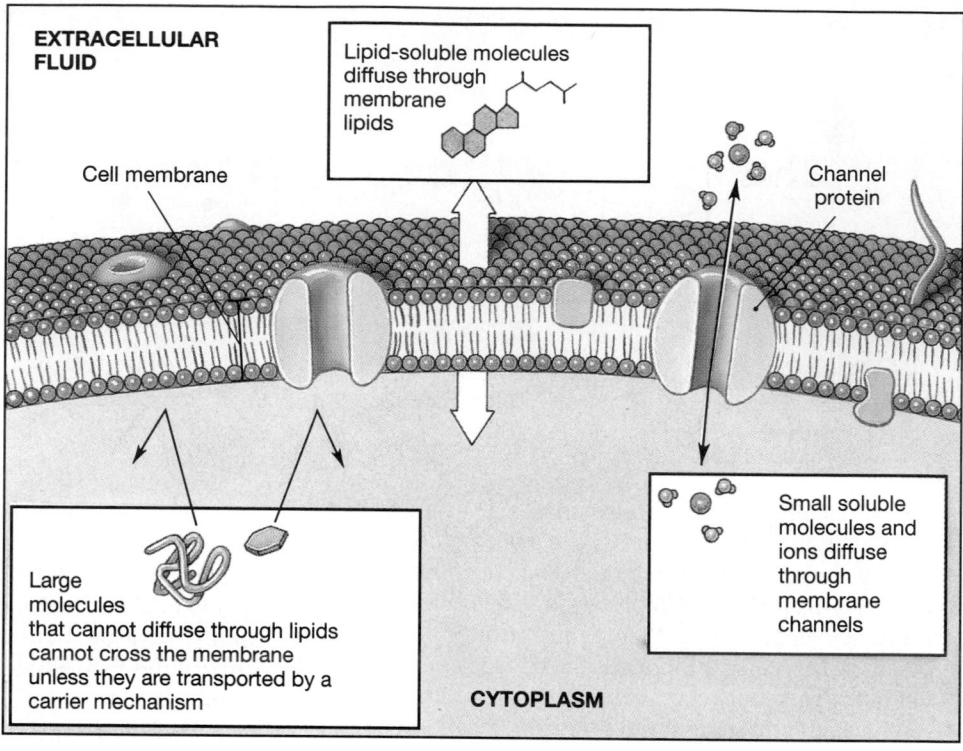

EXTRACELLULAR FLUID

Lipid-soluble molecules diffuse through membrane lipids

Cell membrane

Channel protein

Large molecules that cannot diffuse through lipids cannot cross the membrane unless they are transported by a carrier mechanism

Small soluble molecules and ions diffuse through membrane channels

CYTOPLASM

across the cell membrane at all. But if we ignore the individual identities and simply count ions and molecules, we find that the *total* concentration of dissolved ions and molecules on either side of the cell membrane stays the same.

This state of equilibrium persists because the entire cell membrane is freely permeable to water. Whenever a solute concentration gradient exists, there is also a concentration gradient for water. Because dissolved solute molecules occupy space that would otherwise be taken up by water molecules, the higher the solute concentration, the lower the water concentration. As a result, *water molecules tend to diffuse across a membrane toward the solution containing the higher solute concentration,* because this movement is down the concentration gradient for water.

Remember the following three characteristics of osmosis:

1. Osmosis is the diffusion of water molecules across a membrane.
2. Osmosis occurs across a selectively permeable membrane that is freely permeable to water but not freely permeable to solutes.
3. In osmosis, *water will flow across a membrane toward the solution that has the higher concentration of solutes,* because that is where the concentration of water is lowest.

OSMOSIS AND OSMOTIC PRESSURE Figure 3-7• diagrams the process of osmosis. Step 1 shows two so-

lutions (A and B) with differing solute concentrations separated by a selectively permeable membrane. As osmosis occurs, water molecules cross the membrane until the solute concentrations in the two solutions are identical (Step 2a). Thus the volume of solution B increases while that of solution A decreases. The greater the initial difference in solute concentrations, the stronger the osmotic flow. The **osmotic pressure** of a solution is an indication of the force of water movement *into that solution* as a result of its solute concentration. We can measure a solution's osmotic pressure in several ways. For example, a strong enough opposing pressure can prevent the osmotic flow of water into the solution. Pushing against a fluid generates **hydrostatic pressure**. In Step 2b, hydrostatic pressure opposes the osmotic pressure of solution B, and no net osmotic flow occurs.

Osmosis eliminates solute concentration differences much more quickly than we might predict on the basis of diffusion rates for other molecules. When water molecules cross a membrane, they move in groups held together by hydrogen bonding. So, whereas solute molecules usually diffuse through membrane channels one at a time, water molecules move together in large numbers. This phenomenon is called *bulk flow.*

OSMOLARITY AND TONICITY The total solute concentration in an aqueous solution is the solution's **osmotic concentration**, or **osmolarity**. A solution that contains solutes in the same concentration as the cytosol

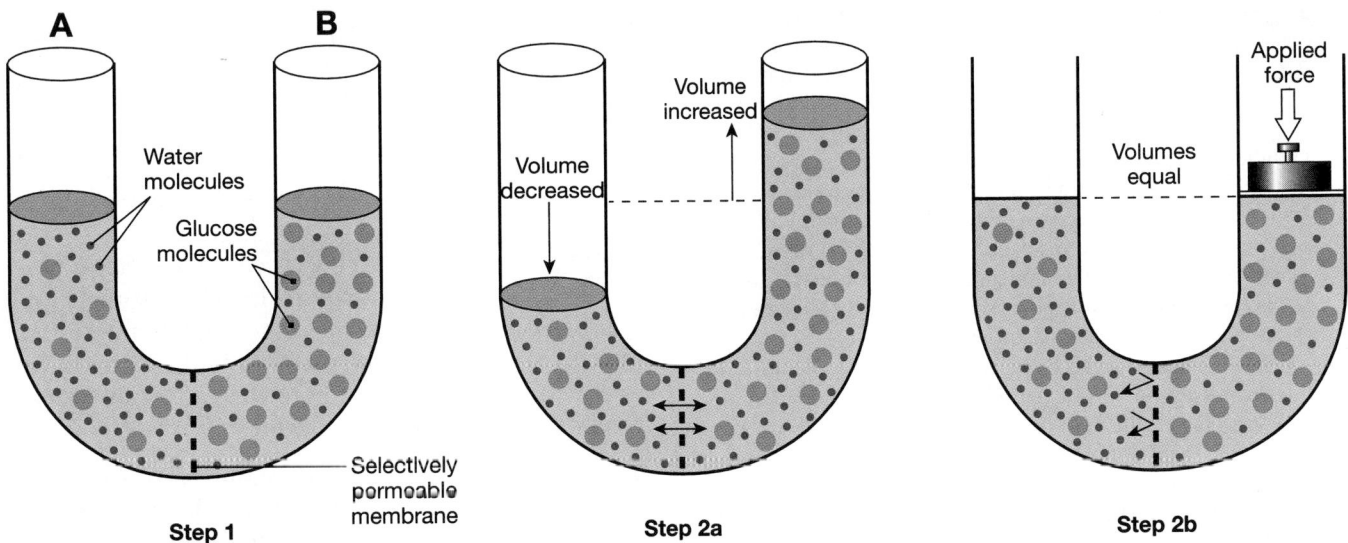

• **FIGURE 3-7** Osmosis. **Step 1:** Two solutions containing different solute concentrations are separated by a selectively permeable membrane. Water molecules (small blue dots) begin to cross the membrane toward solution B, the solution with the higher concentration of solutes (larger red circles). **Step 2a:** At equilibrium the solute concentrations on the two sides of the membrane are equal. The volume of solution B has increased at the expense of that of solution A. **Step 2b:** Osmosis can be prevented by resisting the volume change. The osmotic pressure of solution B is equal to the amount of hydrostatic pressure required to stop the osmotic flow.

is said to be *isosmotic*. A solution with a higher solute concentration than that of the cytosol is *hyperosmotic* (*hyper*, above); one with a lower solute concentration is *hyposmotic* (*hypo*, below). The effect of such solutions on the cell may vary, depending on the nature of the solutes (see the clinical discussion "Tonicity and Normal Saline"). Therefore, when we describe the effects of various osmotic solutions on living cells, we use the term **tonicity** instead of osmolarity. If a solution does not cause an osmotic flow of water into or out of a cell, the solution is called **isotonic** (*iso*, same + *tonos*, tension).

Figure 3-8a● shows the appearance of a red blood cell immersed in an isotonic solution. If a red blood cell is in a **hypotonic** solution, water will flow into the cell, causing it to swell up like a balloon (Figure 3-8b●). Ultimately, the cell may rupture, releasing its contents. This event is known as **hemolysis** (*hemo-*, blood + *lysis*, dissolution). A cell placed in a **hypertonic** solution will lose water through osmosis. As it does, the cell shrivels and dehydrates. The shrinking of red blood cells is called **crenation** (Figure 3-8c●).

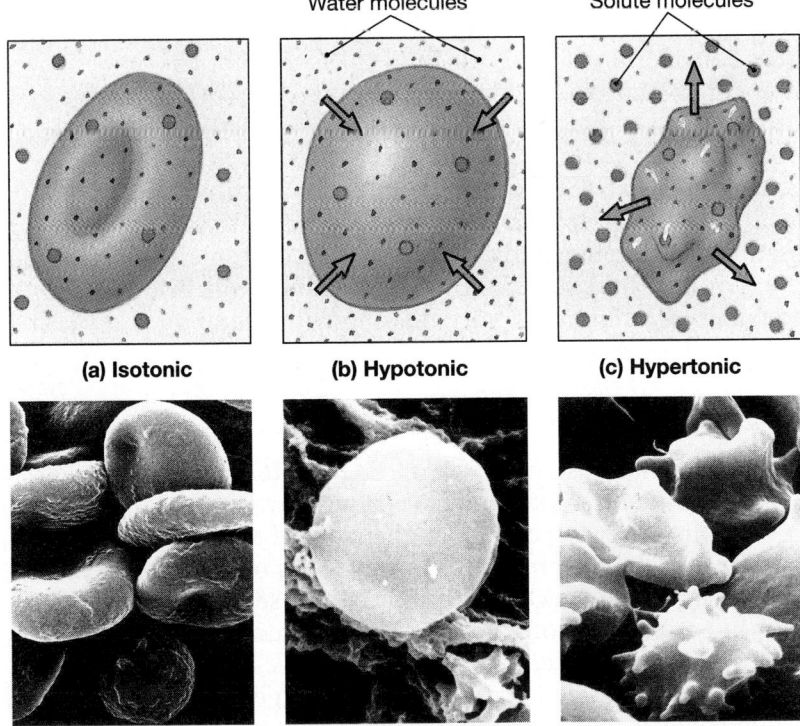

(a) Isotonic   (b) Hypotonic   (c) Hypertonic

• **FIGURE 3-8** **Osmotic Flow across a Cell Membrane.** Blue arrows indicate the direction of osmotic water movement. **(a)** Because these red blood cells are immersed in an isotonic saline solution, no osmotic flow occurs, and the cells are normal in appearance. **(b)** Immersion in a hypotonic saline solution results in the osmotic flow of water into the cells. The swelling may continue until the cell membrane ruptures, or lyses. **(c)** Exposure to a hypertonic solution results in the movement of water out of the cells. The red blood cells shrivel and become crenated. (SEM × 833)

TONICITY AND NORMAL SALINE Although often used interchangeably, the terms *osmolarity* and *tonicity* do not always mean the same thing. For example, consider a solution that has the same osmolarity as the intracellular fluid but a higher concentration of one or more individual ions. If any of those ions can diffuse into the cell, the osmolarity of the intracellular fluid will increase and that of the extracellular solution will decrease. Osmosis will then occur, moving water into the cell. If the process continues, the cell will eventually burst. In this case, the extracellular solution and the intracellular fluid were isosmotic (equal in osmolarity), but the solution was not isotonic.

It is often necessary to give patients large volumes of fluid after severe blood loss or dehydration. One fluid often administered is a 0.9 percent (0.9 g/dl) solution of sodium chloride (NaCl). This solution, which approximates the normal osmotic concentration of the extracellular fluids, is called *normal saline*. It is used because sodium and chloride are the most abundant ions in the extracellular fluid. There is little net movement of either ion across cell membranes; thus, normal saline is essentially isotonic with respect to body cells. An alternative treatment involves the use of an isotonic solution containing *dextran,* a carbohydrate that cannot cross cell membranes.

FLUID SHIFTS Minor fluctuations in intracellular and extracellular solute concentrations are eliminated in a matter of seconds by *fluid shifts,* the osmotic movement of water into or out of cells. It can take considerably longer for fluid shifts to compensate for system-wide changes in solute concentrations. For example, after you drink a large glass of pure water, it may take a half-hour for your intracellular and extracellular fluids to become isotonic again. Severe alterations in body water content, such as those occurring in dehydration, are extremely dangerous. Fluid shifts and water balance will be considered in later chapters dealing with metabolism and kidney function.

Filtration

In **filtration,** hydrostatic pressure forces water across a membrane, and solute molecules may be transported with the water, depending on their size. If they are smaller than the membrane pores, molecules of solute will be carried along with the water. We can see filtration in action in a coffeemaker. Gravity forces hot water through the filter, and the water carries with it a variety of dissolved compounds. The large coffee grounds never reach the pot because they cannot fit through the fine pores in the filter.

In the body, the heart pushes blood through the cardiovascular system while generating hydrostatic pressure. Filtration occurs across the walls of small blood vessels, pushing water and dissolved solutes into the tissues of the body. Filtration across specialized blood vessels in the kidneys is an essential step in the production of urine.

☑ How would a decrease in the concentration of oxygen in the lungs affect the diffusion of oxygen into the blood?

☑ Some pediatricians recommend the use of a 10 percent salt solution to relieve congestion for infants with stuffy noses. What effect would such a solution have on the cells lining the nasal cavity, and why?

Carrier-Mediated Transport

In **carrier-mediated transport,** integral proteins bind specific ions or organic substrates and facilitate their movement across the cell membrane. All forms of carrier-mediated transport share the following characteristics in common with enzymes:

1. *Specificity.* Carrier proteins show *specificity.* That is, each carrier protein in the cell membrane is selective about what substances it will bind and transport. For example, the carrier protein that transports glucose will not transport other simple sugars.

2. *Saturation.* The rate of transport into or out of the cell is limited by the availability of substrate molecules and carrier proteins, just as enzymatic reaction rates are limited by substrate and enzyme concentrations. When all the available carrier molecules are operating at maximum speed, the carrier system is said to be *saturated.* Because no more carrier proteins are available, and the existing ones cannot work any faster, the rate of transport cannot increase, regardless of the size of the concentration gradient.

3. *Regulation.* Just as enzyme activity often depends on the presence of cofactors, carrier protein activity can be altered by the binding of other molecules, such as hormones. Hormones thus provide an important means of coordinating carrier protein activity throughout the body. The interplay between hormones and cell membranes will be examined further in chapters dealing with the endocrine system (Chapter 18) and metabolism (Chapter 25).

Many examples of carrier-mediated transport involve the movement of a single substrate molecule across the cell membrane. A few carrier mechanisms transport more than one substrate at a time. In **cotransport**, or *symport,* the carrier transports two different substances in the same direction simultaneously, either into or out of the cell. In **countertransport**, or *antiport*, one substance moves into the cell, and the other moves out.

We will consider two major examples of carrier-mediated transport at this time: *facilitated diffusion* and *active transport.*

FACILITATED DIFFUSION Many essential nutrients, such as glucose and amino acids, are insoluble in lipids but too large to fit through membrane channels. These substances can be passively transported across the membrane by carrier proteins in a process called **facilitated diffusion** (Figure 3-9a●). The molecule to be

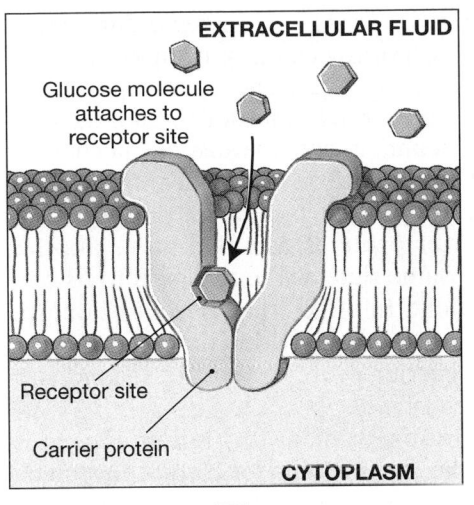

EXTRACELLULAR FLUID

Glucose molecule attaches to receptor site

Receptor site

Carrier protein

CYTOPLASM

Change in shape of carrier protein

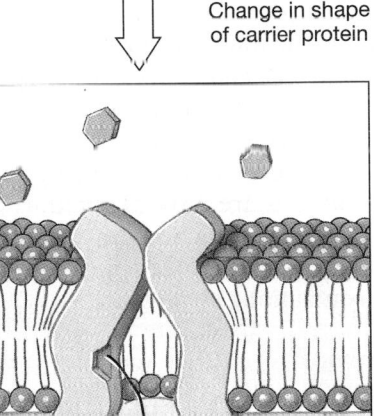

Glucose released into cytoplasm

(a)

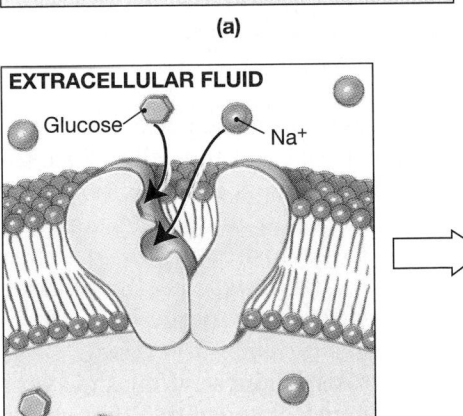

EXTRACELLULAR FLUID

Glucose

Na⁺

CYTOPLASM

(c)

•FIGURE 3-9 **Carrier-Mediated Transport. (a)** Facilitated diffusion. In this process an extracellular molecule, such as glucose, binds to a receptor site on a carrier protein. The binding alters the shape of the protein, which then releases the molecule to diffuse into the cytoplasm. **(b)** The sodium–potassium exchange pump. The operation of the sodium–potassium exchange pump is an example of active transport. For each ATP molecule converted to ADP, this protein, called sodium–potassium ATPase, carries three Na^+ ions out of the cell and two K^+ ions into the cell. **(c)** Secondary active transport. Glucose transport by this carrier will occur only after the carrier has bound a sodium ion. In three cycles, three glucose molecules and three sodium ions are transported into the cytoplasm. The cell now pumps the sodium ions across the cell membrane via the sodium–potassium exchange pump, at a cost of one ATP.

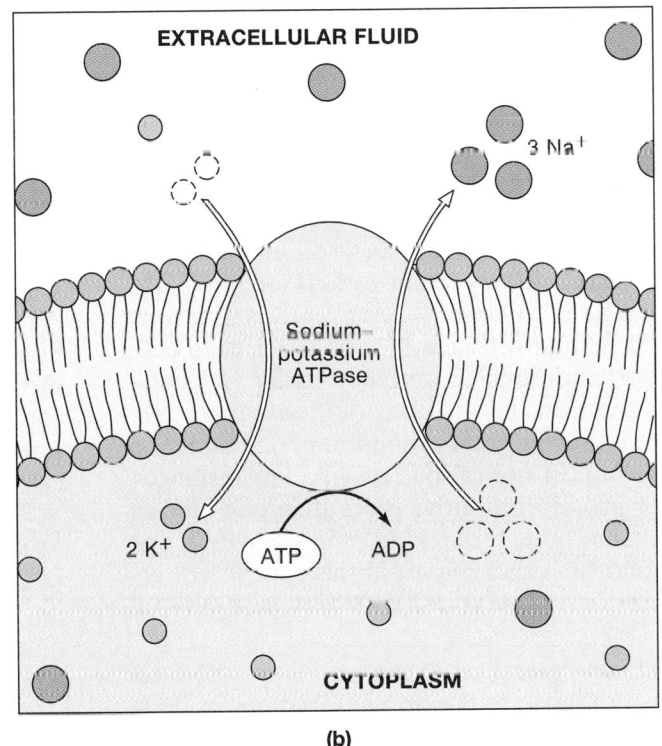

EXTRACELLULAR FLUID

3 Na⁺

Sodium–potassium ATPase

2 K⁺   ATP   ADP

CYTOPLASM

(b)

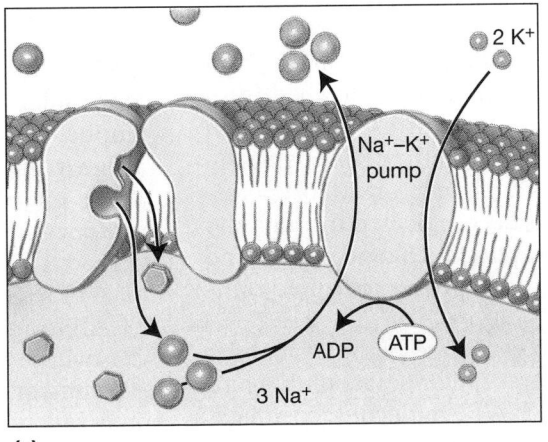

2 K⁺

Na⁺–K⁺ pump

ADP   ATP

3 Na⁺

transported must first bind to a **receptor site** on the protein. The shape of the protein then changes, moving the molecule to the inside of the cell membrane, where it is released into the cytoplasm.

As in the case of simple diffusion, no ATP is expended in facilitated diffusion, and the molecules move from an area of higher concentration to one of lower concentration. However, facilitated diffusion differs from simple diffusion because once the carrier proteins are saturated, the rate of transport cannot increase, regardless of further changes in the concentration gradient.

All cells move glucose across their membranes through facilitated diffusion. However, there are several different forms of carrier proteins involved. In muscle cells, fat cells, and many other cell types, the glucose transporter functions only when stimulated by the hormone *insulin*. Inadequate production of this hormone is one cause of *diabetes mellitus*, a metabolic disorder that we shall detail in Chapter 18.

ACTIVE TRANSPORT In **active transport,** the high-energy bond in ATP (or another high-energy compound) provides the energy needed to move ions or molecules across the membrane. Although it has an energy cost, active transport offers one great advantage: It is not dependent on a concentration gradient. As a result, the cell can import or export specific substrates *regardless of their intracellular or extracellular concentrations.*

All living cells contain carrier proteins called **ion pumps** that actively transport the cations sodium (Na$^+$), potassium (K$^+$), calcium (Ca^{2+}), and magnesium (Mg^{2+}) across their cell membranes. Specialized cells can transport additional ions such as iodide (I$^-$), chloride (Cl$^-$), and iron (Fe^{2+}). Many of these carrier proteins move a specific cation or anion in one direction only, either into or out of the cell. In a few instances, one carrier protein will move more than one kind of ion at the same time. If one kind of ion moves in one direction and the other moves in the opposite direction, the carrier protein is called an **exchange pump.**

The Sodium–Potassium Exchange Pump Sodium and potassium ions are the principal cations in body fluids. Sodium ion concentrations are high in the extracellular fluids but relatively low in the cytoplasm. The distribution of potassium in the body is just the opposite—low in the extracellular fluids and high in the cytoplasm. As a result, sodium ions slowly diffuse into the cell, and potassium ions diffuse out through leak channels. Homeostasis within the cell depends on the ejection of sodium ions and recapture of lost potassium ions. This exchange is accomplished through the activity of a **sodium–potassium exchange pump**. The carrier protein involved in this process is called *sodium–potassium ATPase.*

The sodium–potassium exchange pump exchanges intracellular sodium for extracellular potassium (Figure 3-9b●). On average, for each ATP molecule consumed, three sodium ions are ejected and two potassium ions are reclaimed by the cell. If ATP is readily available, the rate of transport depends on the concentration of sodium ions in the cytoplasm. When that concentration rises, the pump becomes more active. The energy demands are impressive; sodium–potassium ATPase may consume up to 40 percent of the ATP produced by a resting cell.

Secondary Active Transport In **secondary active transport**, the transport mechanism itself does not require energy, but the cell often needs to expend ATP at a later time to preserve homeostasis. Like facilitated transport, a secondary active transport mechanism moves a specific substrate down a concentration gradient. Unlike the proteins in facilitated transport, however, these carrier proteins can also move another substrate at the same time, without regard to its concentration gradient. In effect, the concentration gradient for one substance provides the driving force needed by the carrier protein, and the second substance gets a free ride.

The concentration gradient for sodium ions most often provides the driving force for cotransport mechanisms that move materials into the cell. For example, sodium-linked cotransport is important in the absorption of glucose and amino acids along the intestinal tract. Although the transport activity proceeds without a direct energy expense, the cell must expend ATP to pump the arriving sodium ions out of the cell by using the sodium–potassium exchange pump (Figure 3-9c●). Sodium ions are also involved with many countertransport mechanisms. Sodium–calcium countertransport is responsible for keeping intracellular calcium ion concentrations very low.

Vesicular Transport

In **vesicular transport,** materials move into or out of the cell by means of **vesicles,** small membranous sacs that form at or fuse with the cell membrane. Because large volumes of fluid and solutes are transported in this way, this process is also known as *bulk transport.* There are two major categories of vesicular transport: *endocytosis* and *exocytosis.*

ENDOCYTOSIS Endocytosis is the packaging of extracellular materials in a vesicle at the cell surface for import into the cell. This process involves relatively large volumes of extracellular material. There are three major types of endocytosis: *receptor-mediated endocytosis, pinocytosis,* and *phagocytosis.* All three are active processes that require energy in the form of ATP.

All forms of endocytosis produce cytoplasmic vesicles whose contents remain isolated within the vesicle. Movement of materials into the surrounding cytoplasm may involve active transport, simple or facilitated diffusion, or the destruction of the vesicle membrane.

Receptor-Mediated Endocytosis **Receptor-mediated endocytosis** is a selective process that involves the formation of small vesicles at the membrane surface. This process produces vesicles that contain a specific target molecule in high concentrations. Receptor-mediated endocytosis begins when materials in the extracellular fluid bind to receptors on the membrane surface (Figure 3-10●). Most of the receptor molecules are glycoproteins, and each binds to a specific target, or *ligand,* such as a transport protein or hormone. Some receptors are distributed widely over the surface of the cell membrane. Others may be restricted to pits on the cell surface. There is evidence that these pits are always present, forming a special class of membranous structures called *caveolae* (ka-vē-Ō-lē; little hole). The structure and function of caveolae are now topics of intense research activity.

Receptors bound to ligands cluster together. Once an area of the cell membrane has become covered with ligands, it forms grooves or pockets that pinch off to form vesicles. These **coated vesicles** are surrounded by a protein–fiber network essential to their formation and movement. Inside the cell, the coated vesicles fuse with *primary lysosomes,* vesicles containing digestive enzymes. The fusion of one or more primary lysosomes with another vesicle creates a *secondary lysosome.* Lysosomal enzymes then free the ligands from their receptors, and the ligands enter the cytosol by diffusion or active transport. After the vesicle membrane pinches off from the secondary lysosome and returns to the cell surface, its receptors are ready to bind more ligands.

Many important substances, including cholesterol and iron ions (Fe^{2+}), are distributed through the body attached to special transport proteins. The transport proteins are too large to pass through membrane pores, but they can enter the cell through receptor-mediated endocytosis.

Pinocytosis   **Pinocytosis** (pi-nō-sī-TŌ-sis), or *cell drinking,* is the formation of small vesicles filled with extracellular fluid. This process is not as selective as receptor-mediated endocytosis because there are no receptor proteins involved. The target appears to be the fluid contents in general rather than specific bound ligands. In pinocytosis, a deep groove or pocket forms in the cell membrane and then pinches off (Figure 3-11a●). The steps involved in the formation and fate of a pinocytotic vesicle are the same as those of a vesicle formed through receptor-mediated endocytosis, except that ligand binding is not involved in initiating the process.

Phagocytosis   **Phagocytosis** (fa-gō-sī-TŌ-sis), or *cell eating,* produces vesicles containing *solid objects* that may be as large as the cell itself. In this process, cytoplasmic extensions called **pseudopodia** (soo-dō-PŌ-dē-ah; *pseudo-,* false + *podon,* foot) surround the object, and their membranes fuse to form a vesicle (Figure 3-11b●). The vesicle, called a *phagosome,* then fuses with many lysosomes, whereupon its contents are digested by lysosomal enzymes. Most cells display pinocytosis, but phagocytosis is performed only by spe-

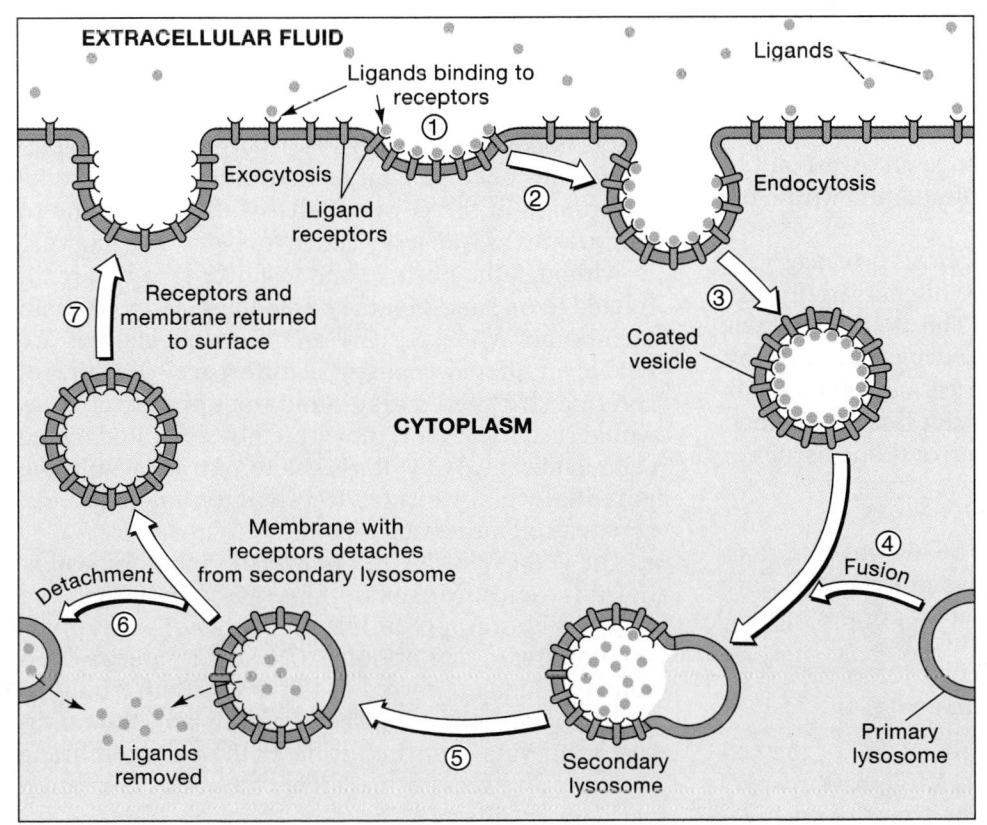

●**FIGURE 3-10**   **Receptor-Mediated Endocytosis.** In this process, ① specific target molecules called ligands bind to receptors, generally glycoproteins, in the membrane surface. ② Membrane areas coated with ligands pinch off to form ③ vesicles that ④ fuse with primary lysosomes. ⑤ The ligands are freed from the receptors and, if necessary, broken down by enzymes before diffusing or being transported into the surrounding cytoplasm. ⑥ The membrane containing the receptor molecules detaches from the membrane of the lysosome and ⑦ returns to the cell surface to bind additional ligands.

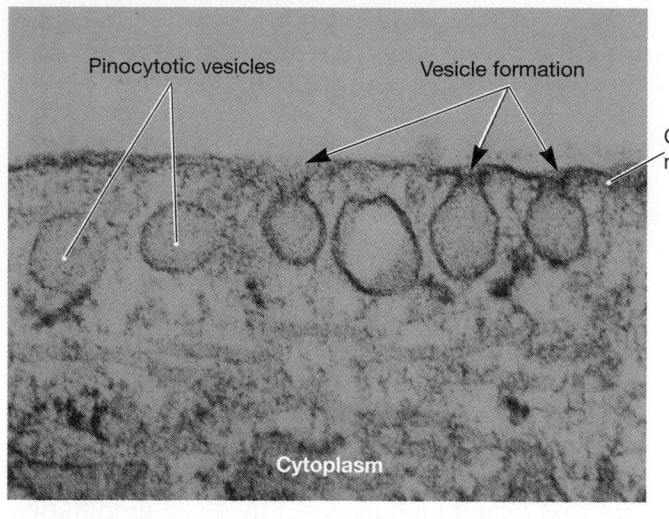

(a)

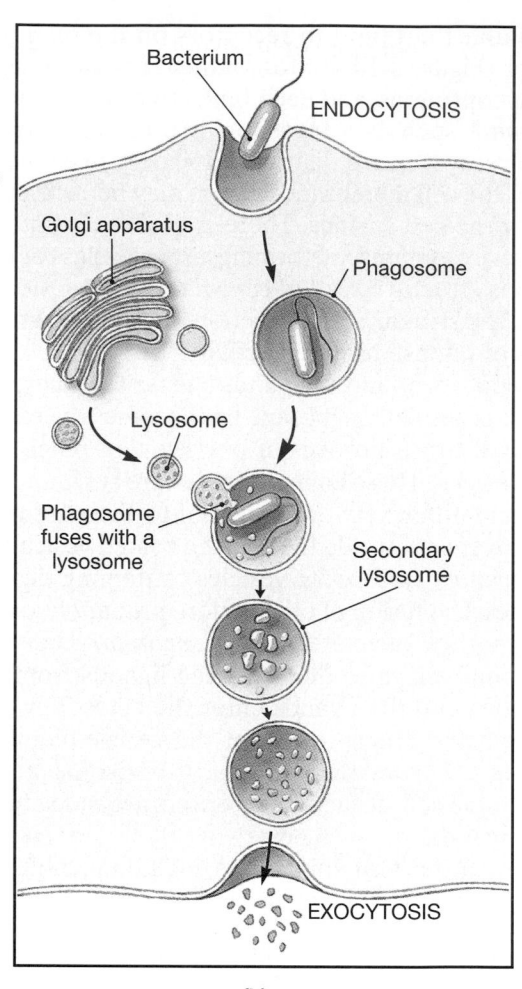

(b)

•FIGURE 3-11 **Pinocytosis and Phagocytosis. (a)** An electron micrograph showing pinocytosis at the bases of microvilli in a cell lining the intestinal tract. **(b)** Material brought into the cell through phagocytosis is enclosed in a phagosome and subsequently exposed to lysosomal enzymes. After absorption of nutrients from the vesicle, the residue is discharged through exocytosis.

cialized cells, such as the *macrophages* that protect tissues by engulfing bacteria, cell debris, and other abnormal materials.

EXOCYTOSIS Exocytosis (ek-sō-sī-TŌ-sis) is the functional reverse of endocytosis. In exocytosis, seen in Figure 3-11b•, a vesicle created inside the cell fuses with the cell membrane and discharges its contents into the extracellular environment. The ejected material may be a secretory product, such as mucus, or waste products such as those accumulating within endocytic vesicles.

In a few specialized cells, endocytosis produces vesicles on one side of the cell that are discharged through exocytosis on the opposite side. This method of bulk transport is common in cells lining *capillaries,* the most delicate blood vessels. These cells use a combination of pinocytosis and exocytosis to transfer fluid and solutes from the bloodstream into the surrounding tissues.

A Review of Membrane Permeability

Many different mechanisms can be moving materials into and out of the cell at any given moment. Before proceeding further, review and compare the mechanisms summarized in Table 3-2.

The Transmembrane Potential

As we noted previously (p. 73), the inside of the cell membrane has a slight negative charge with respect to the outside. The cause is a slight excess of positively charged ions outside the cell membrane and a slight excess of negatively charged ions inside the cell membrane. This unequal charge distribution is created by differences in the permeability of the membrane to various ions as well as by active transport mechanisms.

Although the positive and negative charges are attracted to one another, they are separated by the lipid membrane. When positive and negative charges are held apart, there is a **potential difference**—like the recoil of a stretched spring—and the opposite charges would rush together if they were not separated by the cell membrane. We will use the term **transmembrane potential** when we refer to the potential difference across a cell membrane.

The *volt* (V) is the unit of measurement for potential difference. Most cars have 12-V batteries. The transmembrane potentials of living cells are much smaller, averaging about 0.07 V for a neuron. This value is usually reported as –70 mV, with mV indicating *millivolts* (thousandths of a volt) and the minus sign signifying that the inside of the cell membrane contains an excess of negative charges as compared with the outside.

TABLE 3-2 **Summary of Mechanisms Involved in Movement across Cell Membranes**

Mechanism	Process	Factors Affecting Rate	Substances Involved (location)
Diffusion	Molecular movement of solutes; direction determined by relative concentrations	Size of gradient; molecular size; charge; lipid solubility, temperature	Small inorganic ions; lipid-soluble materials (all cells)
Osmosis	Movement of water molecules toward solution containing relatively higher solute concentration; requires selectively permeable membrane	Concentration gradient; opposing osmotic or hydrostatic pressure	Water only (all cells)
Filtration	Movement of water, usually with solute, by hydrostatic pressure; requires filtration membrane	Amount of pressure; size of pores in filter	Water and small ions (blood vessels)
Carrier-Mediated Transport			
Facilitated diffusion	Carrier proteins passively transport solutes across a membrane down a concentration gradient	Size of gradient, temperature, and availability of carrier protein	Glucose and amino acids (all cells, but several different regulatory mechanisms exist)
Active transport	Carrier proteins actively transport solutes across a membrane regardless of any concentration gradients	Availability of carrier, substrate, and ATP	Na^+, K^+, Ca^{2+}, Mg^{2+} (all cells); other solutes by specialized cells
Secondary active transport	Carrier proteins passively transport two solutes, with one (normally Na^+) moving down its concentration gradient; the cell must later expend ATP to reject the Na^+	Availability of carrier, substrates, and ATP	Glucose and amino acids (specialized cells)
Vesicular Transport			
Endocytosis	Creation of membranous vesicles containing fluid or solid material	Stimulus and mechanics incompletely understood; requires ATP	Fluids, nutrients (all cells), debris, pathogens (specialized cells)
Exocytosis	Fusion of vesicles containing fluids and/or solids within the cell membrane	Stimulus and mechanics incompletely understood; requires ATP	Fluids, debris (all cells)

ROLE OF THE TRANSMEMBRANE POTENTIAL If the lipid barrier were removed, the positive and negative charges would move together. The cell membrane thus acts like a dam across a stream. A dam resists the water pressure that builds up on the upstream side; a cell membrane resists electrochemical forces that would otherwise drive ions into or out of the cell. The water retained behind a dam and the ions held on either side of the cell membrane have *potential energy.* People have designed many ways to make use of the potential energy stored behind a dam—for example, turning a mill wheel or a turbine. Similarly, cells have ways of utilizing the potential energy stored in the transmembrane potential. For example, it is the transmembrane potential that makes possible the transmission of information in the nervous system, and thus our perceptions and thoughts. As we shall see in later chapters, changes in the transmembrane potential also trigger the contraction of muscles and the secretion of glands.

THE RESTING POTENTIAL The **resting potential** is the transmembrane potential in an undisturbed cell. Each cell type has a characteristic resting potential between –10 mV and –100 mV. Examples include fat cells (–hr 40 mV), thyroid cells (–50 mV), neurons (–70 mV), skeletal muscle cells (–85 mV), and cardiac muscle cells (–90 mV). The normal resting potential is negative primarily because the interior of the cell contains an abundance of negatively charged proteins, which cannot cross the cell membrane.

A cell must expend energy to maintain its resting potential because leak channels permit the entry of sodium ions and the departure of potassium ions. The sodium–potassium exchange pump stabilizes the resting potential by ejecting sodium ions from the cytosol and reclaiming potassium ions from the extracellular fluid. Figure 3-12● provides a diagrammatic view of a cell membrane at its resting potential.

•FIGURE 3-12 Ion Distribution across the Cell Membrane.
Potassium ions are continually leaking out of the cell through potassium leak channels, and sodium ions are entering through sodium leak channels. At the resting potential, three sodium ions enter the cell for every two potassium ions that leave the cell. These movements are counteracted by the activities of the sodium–potassium exchange pump, which removes three sodium ions from the cytoplasm in exchange for two potassium ions from the extracellular fluid (Figure 3-9b). The net result is that the exchange pump maintains a stable ion distribution across the cell membrane

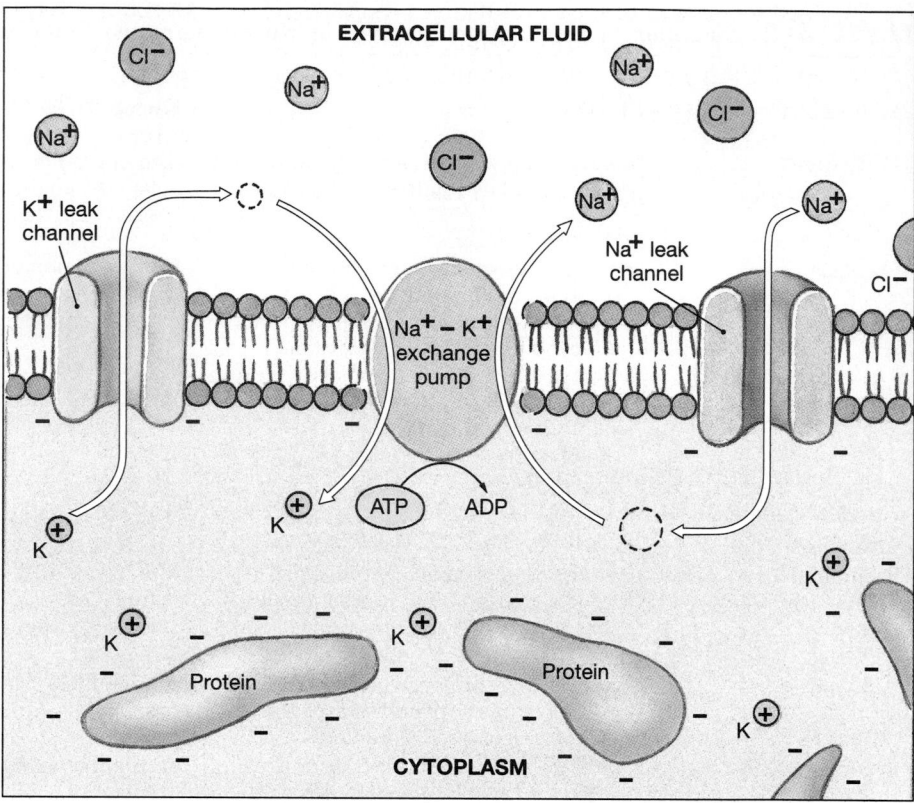

☑ During digestion in the stomach, the concentration of hydrogen (H^+) ions rises to many times that found in cells of the stomach. Which transport process could be responsible?

☑ If the cell membrane were freely permeable to sodium ions (Na^+), how would the transmembrane potential be affected?

☑ When they encounter bacteria, certain types of white blood cells are able to engulf the bacteria and bring them into the cell. What is this process called?

THE CYTOPLASM

Cytoplasm is a general term for the material located inside the cell membrane and outside the membrane surrounding the nucleus. Cytoplasm contains many more proteins than does the extracellular fluid, and proteins account for 15–30 percent of the weight of the cell. The cytoplasm includes two major subdivisions:

1. **Cytosol,** or **intracellular fluid,** which contains dissolved nutrients, ions, soluble and insoluble proteins, and waste products. The cell membrane separates the cytosol from the surrounding extracellular fluid.

2. **Organelles,** which are structures that perform specific functions within the cell.

The Cytosol

The three most important differences between cytosol and extracellular fluid are:

1. The cytosol contains a relatively high concentration of potassium ions and a low sodium concentration, whereas extracellular fluid contains a high concentration of sodium ions and a low potassium concentration.

2. The cytosol contains a relatively high concentration of suspended proteins. Many of the proteins are enzymes that regulate metabolic operations; others are associated with the various organelles. The cytosol is a colloid with a consistency that varies between that of thin maple syrup and almost-set gelatin. ∞ *[p. 43]*

3. The cytosol usually contains relatively small quantities of carbohydrates and large reserves of amino acids and lipids. The carbohydrates are broken down to provide energy, and the amino acids are used to manufacture proteins. Lipids are used primarily as an energy source when carbohydrates are unavailable.

The cytosol may also contain masses of insoluble materials known as **inclusions.** Among the most common inclusions are stored nutrients. For example, glycogen granules in liver or in skeletal muscle cells, and lipid droplets in fat cells are inclusions.

Organelles

Each organelle performs specific functions that are essential to normal cell structure, maintenance, and metabolism. Cellular organelles can be divided into two broad categories. **Nonmembranous organelles** are always in direct contact with the cytosol. **Membranous organelles** are surrounded by lipid membranes that isolate them from the cytosol, just as the cell membrane isolates the cytosol from the extracellular fluid.

The cell's nonmembranous organelles include the *cytoskeleton, microvilli, centrioles, cilia, flagella,* and *ribosomes.* Membranous organelles include the *mitochondria, the endoplasmic reticulum, the Golgi apparatus, lysosomes,* and *peroxisomes.* (The *nucleus,* which is also surrounded by a membranous envelope and therefore sometimes included with the membranous organelles, will be the focus of a separate section.)

The Cytoskeleton

The **cytoskeleton** (Figure 3-13●) is an internal protein framework that gives the cytoplasm strength and flexibility. It has four major components: *microfilaments, intermediate filaments, thick filaments,* and *microtubules.* Thick filaments appear only in muscle cells. The other filaments are found in varying numbers in all cells. In a typical cell, the microfilaments, intermediate filaments, and microtubules form a dynamic network. The organizational details are as yet poorly understood, because the network is extremely delicate and thus hard to study in an intact state. Figure 3-13a● is based on our current knowledge of cytoskeletal structure.

MICROFILAMENTS **Microfilaments** are slender protein strands, generally less than 6 nm in diameter. The most abundant microfilaments are composed of the protein **actin**. In most cells, these *actin filaments* are common in the periphery of the cell but relatively rare in the region immediately surrounding the nucleus. In cells that form a layer or lining, such as the lining of the intestinal tract, actin filaments also form a layer, the *terminal web,* just inside the membrane at the exposed surface of the cell.

These microfilaments have three major functions:

1. Microfilaments anchor the cytoskeleton to integral proteins of the cell membrane. They provide additional mechanical strength to the cell and attach the cell membrane to the enclosed cytoplasm.

2. Actin filaments interacting with other proteins determine the consistency of the cytoplasm. The consistency can vary from one region to another, from a viscous gel to a fluid, depending on the nature of these interactions. For example, actin filaments in a dense, flexible network give the cytoplasm a gelatinous consistency, whereas fluid cytoplasm contains dispersed microfilaments.

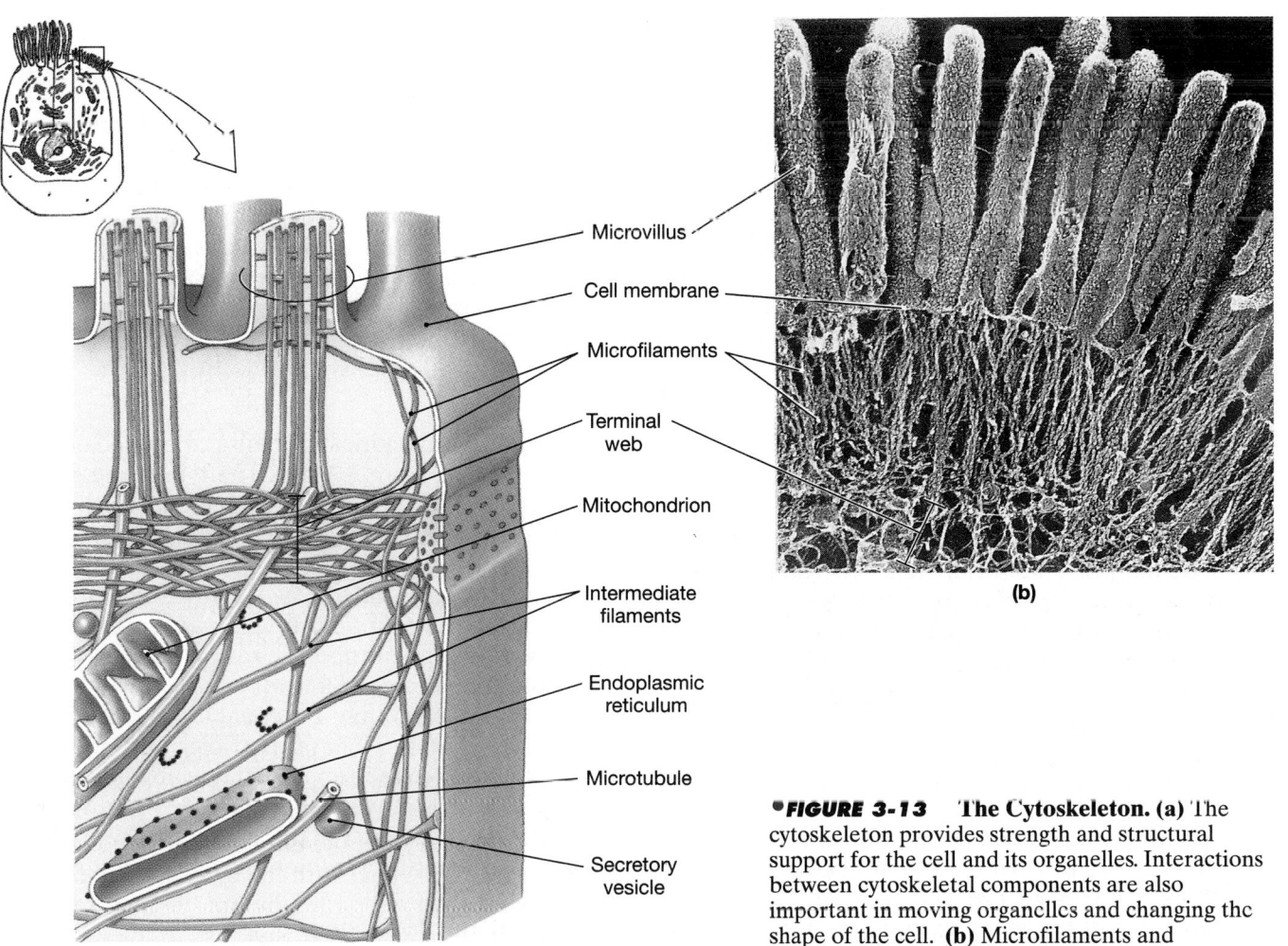

●*FIGURE 3-13* The Cytoskeleton. **(a)** The cytoskeleton provides strength and structural support for the cell and its organelles. Interactions between cytoskeletal components are also important in moving organelles and changing the shape of the cell. **(b)** Microfilaments and microvilli of an intestinal cell.

3. Actin can interact with the protein **myosin** to produce active movement of a portion of a cell or to change the shape of the entire cell.

INTERMEDIATE FILAMENTS   **Intermediate filaments** are intermediate in size between microfilaments and thick filaments. The protein composition of intermediate filaments, which range from 7 to 11 nm in diameter, varies from one cell type to another. Intermediate filaments (1) provide strength and stability to cell shape, (2) stabilize the positions of organelles, and (3) stabilize the position of the cell with respect to surrounding cells through specialized attachment to the cell membrane. Intermediate filaments are insoluble fibers, and these are the most durable of the cytoskeletal elements. Many cells contain specialized intermediate filaments with unique functions. For example, neurons contain *neurofilaments*, which provide structural support for cell processes called *axons*, that may approach a meter in length.

THICK FILAMENTS   **Thick filaments** are relatively massive bundles of myosin protein subunits. Thick filaments may reach 15 nm in diameter. They appear only in muscle cells, where they interact with actin filaments to produce powerful contractions.

MICROTUBULES   **Microtubules**, found in all our cells, are hollow tubes built from the globular protein **tubulin**. Microtubules are the largest components of the cytoskeleton, with diameters of about 25 nm. The number and distribution of microtubules within a cell change over time. A microtubule forms through the aggregation of tubulin molecules. It is continuously undergoing remodeling, with tubulin molecules dispersing at one end of the microtubule and attaching at the other. The entire structure persists for a time and then disassembles into individual tubulin molecules once again. The microtubular array within a cell is centered near the nucleus, in a region known as the *centrosome*. From the centrosome, microtubules extend outward into the periphery of the cell.

Microtubules have the following functions:

1. Microtubules form the primary components of the cytoskeleton, giving the cell strength and rigidity and anchoring the position of major organelles.
2. Disassembly of microtubules provides a mechanism for changing the shape of the cell, perhaps assisting in cell movement.
3. Microtubules can attach to organelles and other intracellular materials that then move along the axis of the microtubule. The movement is performed by proteins called *molecular motors*. These proteins, which attach to the microtubule and to the vesicle or other organelle, "walk" their way along the microtubule. Whether the transport occurs toward one end of the microtubule or the other depends on the identity of the molecular motor. For example, if *kinesin* moves in one direction, *dynein* will move in the opposite direction. Regardless of the direction of transport or the nature of the motor, the process requires ATP, and it is essential to normal cellular function.
4. During cell division, microtubules form the *spindle apparatus* that distributes the duplicated chromosomes to opposite ends of the dividing cell. We shall consider this process in more detail in a later section.
5. Microtubules form structural components of organelles such as *centrioles, cilia,* and *flagella.*

In addition to these functions, the cytoskeleton has other functions that are only now being investigated. For example, many intracellular enzymes, especially those involved with metabolism and energy production, and the ribosomes and messenger RNAs responsible for protein synthesis are attached to the microfilaments and microtubules of the cytoskeleton. Thus, the cytoskeleton plays a role in the metabolic organization of the cell by determining where in the cytoplasm key enzymatic reactions occur and where specific proteins are synthesized.

Microvilli

Microvilli are small, finger-shaped projections of the cell membrane (Figure 3-13b●) that greatly increase the surface area of the cell exposed to the extracellular environment. Microvilli cover the surfaces of cells that are actively absorbing materials from the extracellular fluid, such as the cells lining the digestive tract. Microvilli differ from other membranous processes, such as pseudopodia, in their degree of integration with the cytoskeleton. A core of microfilaments stiffens each microvillus and anchors it to the cytoskeleton at the terminal web.

Centrioles

A **centriole** is a cylindrical structure composed of short microtubules (Figure 3-14a●). There are nine groups of microtubules, with three in each group. Each of the nine triplets is connected to its nearest neighbors on either side. Because there are no central microtubules, this organization is called a *9 + 0 array*. Some preparations show an axial structure, with radial spokes leading toward the microtubular groups. The function of this axial complex is not known.

All animal cells that are capable of reproducing themselves contain a pair of centrioles. During cell division, the centrioles form the spindle apparatus associated with the movement of DNA strands. (Cell division is detailed later in this chapter.) There are no centrioles in mature red blood cells, skeletal muscle or cardiac muscle cells, or typical neurons; these cells are not capable of dividing.

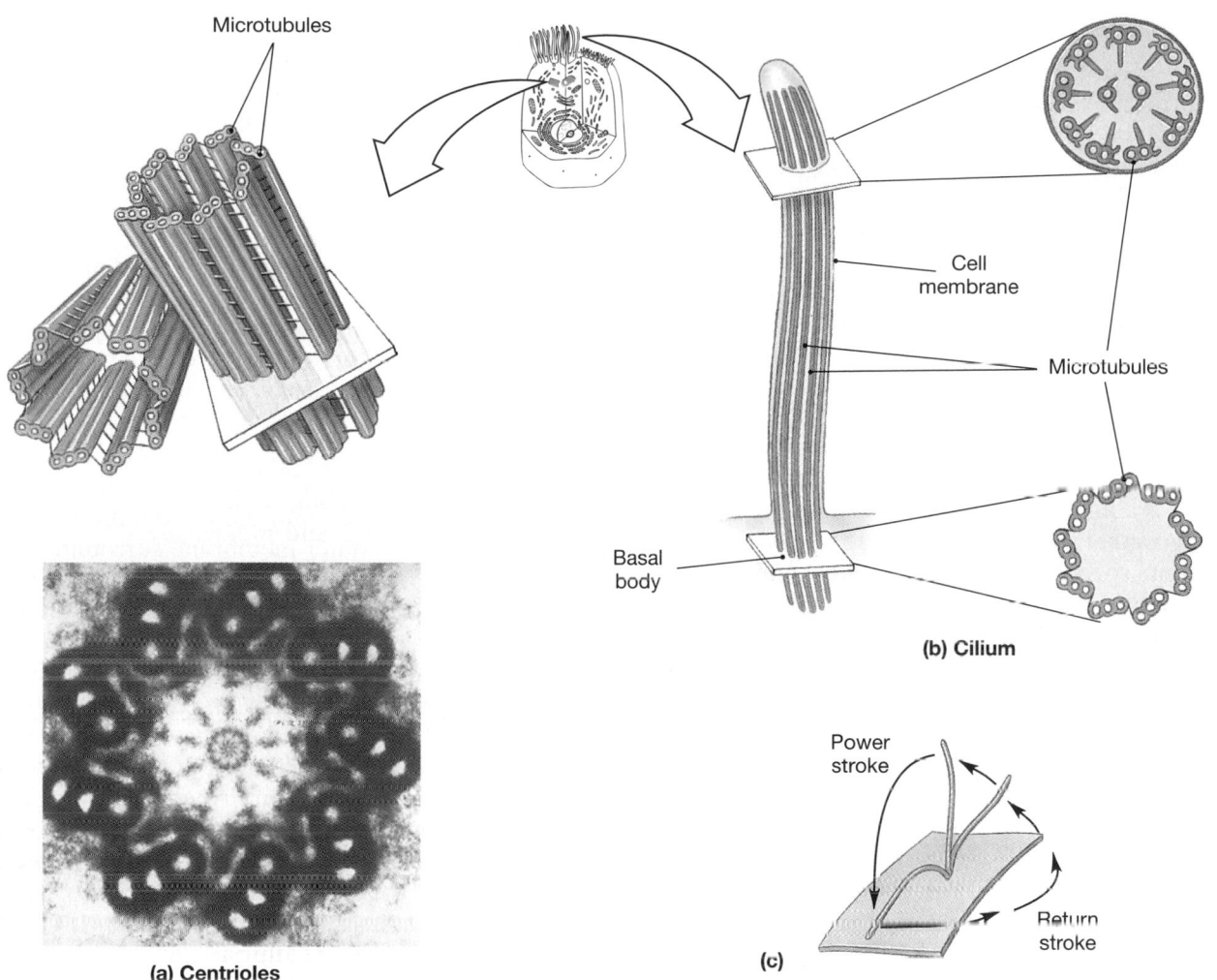

Microtubules

Cell
membrane

Microtubules

Basal
body

(b) Cilium

Power
stroke

Return
stroke

(c)

(a) Centrioles

•**FIGURE 3-14** **Centrioles and Cilia. (a)** A centriole consists of nine microtubule triplets (known as a 9 + 0 array). The centrosome contains a pair of centrioles oriented at right angles to one another. **(b)** A cilium contains nine pairs of microtubules surrounding a central pair (9 + 2 array). The basal body to which the cilium is anchored has a structure similar to that of a centriole. **(c)** A single cilium swings forward and then returns to its original position. During the power stroke, the cilium is relatively stiff, but during the return stroke, it bends and moves parallel to the cell surface.

The **centrosome,** the cytoplasm surrounding the centrioles, is the heart of the cytoskeletal system. Microtubules of the cytoskeleton generally begin at the centrosome and radiate through the cytoplasm.

Cilia

Cilia (singular, *cilium*) contain nine pairs of microtubules surrounding a central pair (Figure 3-14b•). This organization is known as a *9 + 2 array*. Cilia are anchored to a compact **basal body** situated just beneath the cell surface. In section, the organization of microtubules in the basal body resembles that of a centriole.

The exposed portion of the cilium is completely covered by the cell membrane. Cilia "beat" rhythmically (Figure 3-14c•). The cilium is relatively stiff during the effective *power stroke* and flexible during the *return stroke*. The coordinated action of cilia moves fluids or secretions across the cell surface. For example, cilia lining the respiratory tract beat in a synchronized manner to move sticky mucus and trapped dust particles toward the throat and away from delicate respiratory surfaces. If the cilia are damaged or immobilized by heavy smoking or some metabolic problem, the cleansing action is lost, and the irritants will no longer be removed. As a result, chronic respiratory infections develop.

Flagella

Flagella (fla-JEL-uh; singular, *flagellum,* whip) resemble cilia, but they are much longer. Flagella move a cell through the surrounding fluid rather than moving the fluid past a stationary cell. The sperm cell is the only human cell that has a flagellum, which is used to move the cell along the female reproductive tract. If sperm flagella are paralyzed or otherwise abnormal, the individual will be sterile because immobile sperm cannot perform fertilization.

Ribosomes

Ribosomes are organelles that manufacture proteins, using information provided by the DNA of the nucleus. The number of ribosomes within a particular cell varies depending on the type of cell. For example, liver cells, which manufacture blood proteins, contain far more ribosomes than do fat cells, which synthesize triglycerides.

Ribosomes cannot be seen clearly with the light microscope. In an electron micrograph, ribosomes appear as dense granules roughly 25 nm in diameter. Each ribosome consists of roughly 60 percent RNA and 40 percent protein.

A functional ribosome consists of two subunits that are normally separate and distinct. The subunits differ in size; one is a **light ribosomal subunit** and the other a **heavy ribosomal subunit** (Figure 3-15•). These subunits contain special proteins and **ribosomal RNA (rRNA)**, one of the RNA types introduced in Chapter 2. ∞ *[p. 58]* Before protein synthesis can begin, a light and a heavy ribosomal subunit must join together with a strand of *messenger RNA* to create a functional ribosome.

There are two major types of functional ribosomes: *free ribosomes* and *fixed ribosomes* (Figure 3-2•, p. 67). **Free ribosomes** are scattered throughout the cytoplasm. The proteins they manufacture enter the cytosol. **Fixed ribosomes** are attached to the *endoplasmic reticulum (ER),* a membranous organelle. Proteins manufactured by fixed ribosomes enter the ER, where they are modified and packaged for secretion. We shall examine ribosomal structure and functions in later sections when we deal with the endoplasmic reticulum and protein synthesis.

☑ Cells lining the small intestine have numerous fingerlike projections on their free surface. What are these structures, and what is their function?

☑ How would the absence of a flagellum affect a sperm cell?

Mitochondria

Mitochondria (mī-tō-KON-drē-uh; singular, *mitochondrion; mitos,* thread + *chondros,* cartilage) are small organelles that can have a variety of shapes, from long and slender to short and fat. The number of mitochondria in a particular cell varies depending on the cell's energy demands. No mitochondria are found in red blood cells, but these organelles may account for 20 percent of the volume of an active liver cell.

Mitochondria have an unusual double membrane (Figure 3-16a•). The outer membrane surrounds the entire organelle. The inner membrane contains numerous folds, called **cristae,** which increase the surface area exposed to the fluid contents, or **matrix,** of the mitochondrion. Metabolic enzymes in the matrix perform the reactions that provide energy for cellular functions.

Most of the chemical reactions that release energy occur in the mitochondria, but most of the cellular activities that require energy occur in the surrounding cytoplasm. Cells must therefore store energy in a form that can be moved from place to place. Energy is stored and transferred in the form of *high-energy bonds.* Such a bond generally attaches a phosphate group (PO_4^{3-}) to a suitable molecule, creating a *high-energy compound.* ∞ *[p. 58]* The most important high-energy compound is *adenosine triphosphate (ATP).* Living cells break the high-energy bond under controlled conditions, reconverting ATP to ADP and phosphate and thereby releasing energy for the cell's use.

MITOCHONDRIAL ENERGY PRODUCTION Most cells generate ATP and other high-energy compounds through the breakdown of carbohydrates, especially glucose. We shall examine the entire process in Chapter 25, but you will need to understand a few basic concepts before you can follow discussions of muscle contraction, neuron function, and endocrine function in Chapters 10–18.

Although most of the ATP production occurs inside mitochondria, the first steps take place in the cytosol. In this reaction sequence, called **glycolysis,** each six-carbon glucose molecule is broken down into two three-carbon molecules of *pyruvic acid.* The pyruvic acid molecules are then absorbed by mitochondria (Figure 3-16b•).

Inside the mitochondrial matrix, a CO_2 molecule is removed from each absorbed pyruvic acid molecule, leaving a two-carbon fragment for entry into the **tricarboxylic acid cycle,** or **TCA cycle.** The TCA cycle is an en-

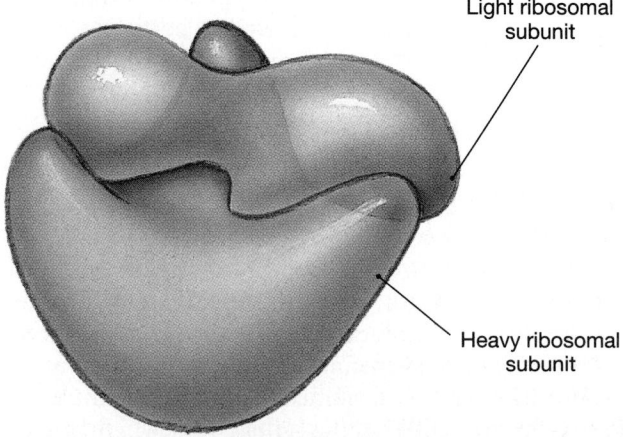

Light ribosomal subunit

Heavy ribosomal subunit

•*FIGURE 3-15* **Ribosomes.** A diagrammatic view of the structure of an intact ribosome. The subunits are separate unless the ribosome is engaged in protein synthesis.

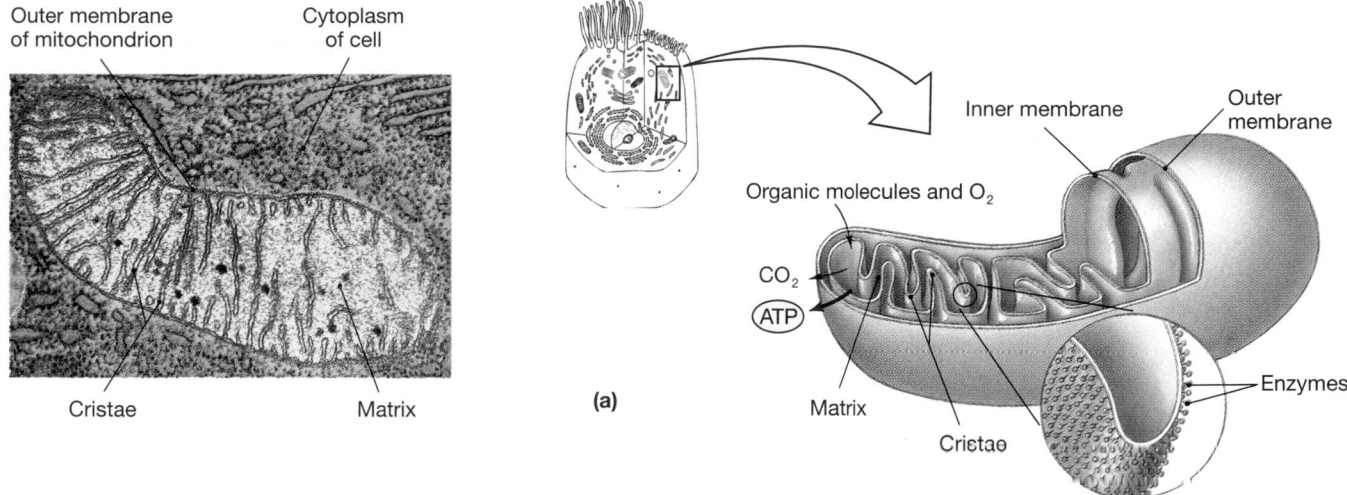

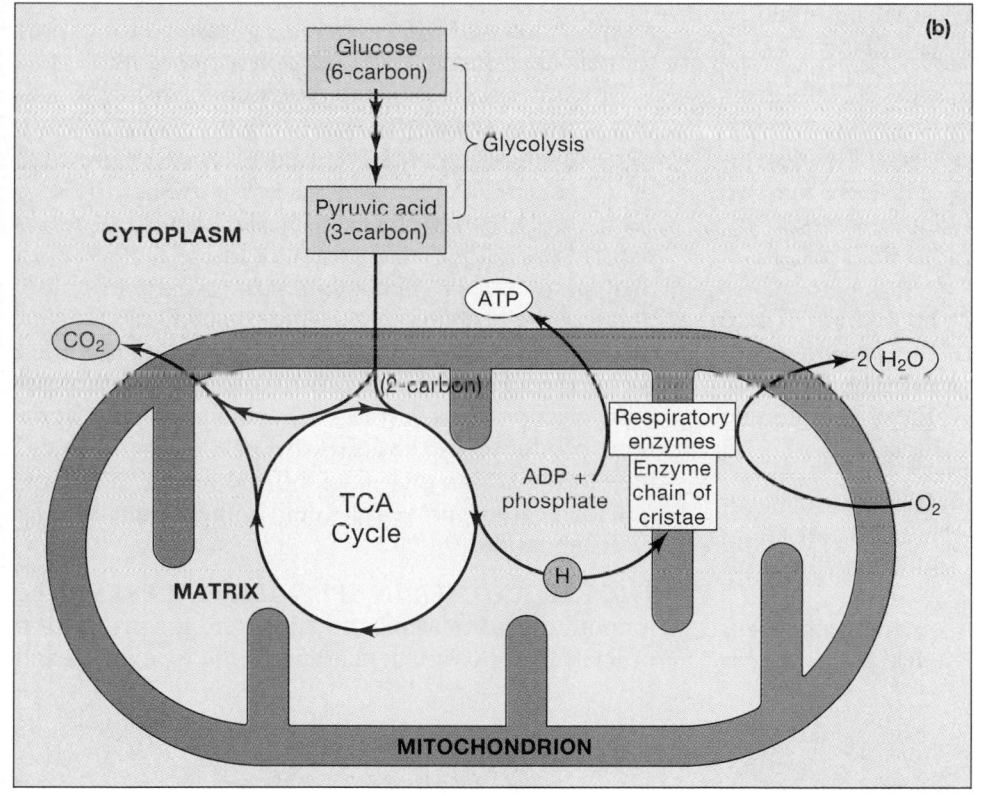

●FIGURE 3-16
Mitochondria. (a) This electron micrograph shows a typical mitochondrion in section, and the sketch details its three-dimensional organization. (TEM × 46,332) **(b)** Overview of the role of mitochondria in energy production. Mitochondria absorb short carbon chains (such as pyruvic acid) and oxygen and generate carbon dioxide and ATP.

zymatic pathway that breaks down organic molecules. The remnants of pyruvic acid molecules contain carbon, oxygen, and hydrogen atoms. The carbon and oxygen atoms are released as carbon dioxide, which diffuses out of the cell. The hydrogen atoms, which contain much of the potential energy of the original glucose molecule, are transferred to a chain of enzymes and coenzymes located in the cristae. The electrons are removed and passed from coenzyme to coenzyme. The energy released during these steps performs the conversion of ADP to ATP. The electrons are ultimately donated to oxygen atoms, which combine with hydrogen ions to form water. (The entire

reaction sequence will be detailed in Chapter 25.) Because mitochondrial activity requires oxygen, this method of ATP production is known as **aerobic metabolism** *(aero-,* air + *bios,* life), or *cellular respiration.* Aerobic metabolism in mitochondria produces about 95 percent of the ATP needed to keep a cell alive.

Several inheritable disorders result from abnormal mitochondrial activity. The mitochondria involved have defective enzymes that reduce their ability to generate ATP. Cells throughout the body may be affected, but symptoms involving muscle cells, neurons, and the receptor cells in the eye are most common because these cells have especially high energy demands. AM *Mitochondrial DNA, Disease, and Evolution*

The Endoplasmic Reticulum

The **endoplasmic reticulum** (en-dō-PLAZ-mik re-TIK-ū-lum), or **ER**, is a network of intracellular membranes that is connected to the *nuclear envelope* surrounding the nucleus. This network has four major functions:

1. *Synthesis.* Specialized regions of the ER synthesize proteins, carbohydrates, and lipids.
2. *Storage.* The ER can contain synthesized molecules or materials absorbed from the cytosol without affecting other cellular operations.
3. *Transport.* Materials can travel from place to place in the ER.
4. *Detoxification.* Drugs or toxins can be absorbed by the ER and neutralized by enzymes within it.

The endoplasmic reticulum (Figure 3-17•) forms hollow tubes, flattened sheets, and round chambers. The chambers are called **cisternae** (sis-TUR-nē; singular, *cisterna,* a reservoir for water). There are two distinct types of endoplasmic reticulum—*smooth endoplasmic reticulum* and *rough endoplasmic reticulum.*

THE SMOOTH ENDOPLASMIC RETICULUM   There are no ribosomes associated with the **smooth endoplasmic reticulum (SER)**. The SER has a variety of functions that center around the synthesis of lipids and carbohydrates. Those functions include:

- Synthesis of the phospholipids and cholesterol needed for maintenance and growth of the cell membrane, ER, nuclear membrane, and Golgi apparatus in all cells.
- Synthesis of steroid hormones, such as *androgens* (the dominant sex hormones in males) and *estrogens* (the dominant sex hormones in females) in the reproductive organs and the steroid hormones of the adrenal glands.
- Synthesis and storage of glycerides, especially triglycerides, in liver and fat cells.
- Detoxification or inactivation of drugs in the SER of liver and kidney cells.
- Synthesis and storage of glycogen in skeletal muscle and liver cells.
- Removal and storage of calcium ions (Ca^{2+}) or larger molecules from the cytosol. Calcium ions are stored in the SER of skeletal muscle cells, neurons, and many other cell types.

THE ROUGH ENDOPLASMIC RETICULUM   The **rough endoplasmic reticulum (RER)** functions as a combination workshop and shipping depot. It is where many newly synthesized proteins undergo chemical modification and where they are packaged for export to their next destination, the *Golgi apparatus.*

The outer surface of the rough endoplasmic reticulum contains fixed ribosomes (Figure 3-17b•). Ribosomes synthesize proteins by using instructions provided by a type of RNA known as *messenger RNA* (mRNA). As the polypeptide chains grow, they enter the cisternae of the endoplasmic reticulum. Inside the ER, each protein assumes its secondary and tertiary structure. Some of the proteins are enzymes that will function inside the ER. Other proteins are chemically modified within the ER by the attachment of carbohydrates, creating glycoproteins. Most of the proteins and glycoproteins produced by the RER are packaged into small membranous sacs that pinch off the tips of the cisternae. These **transport vesicles** deliver the products to the Golgi apparatus.

THE RER AND SER IN SPECIALIZED CELLS   The amount of endoplasmic reticulum and the proportion of RER to SER vary depending on the type of cell and

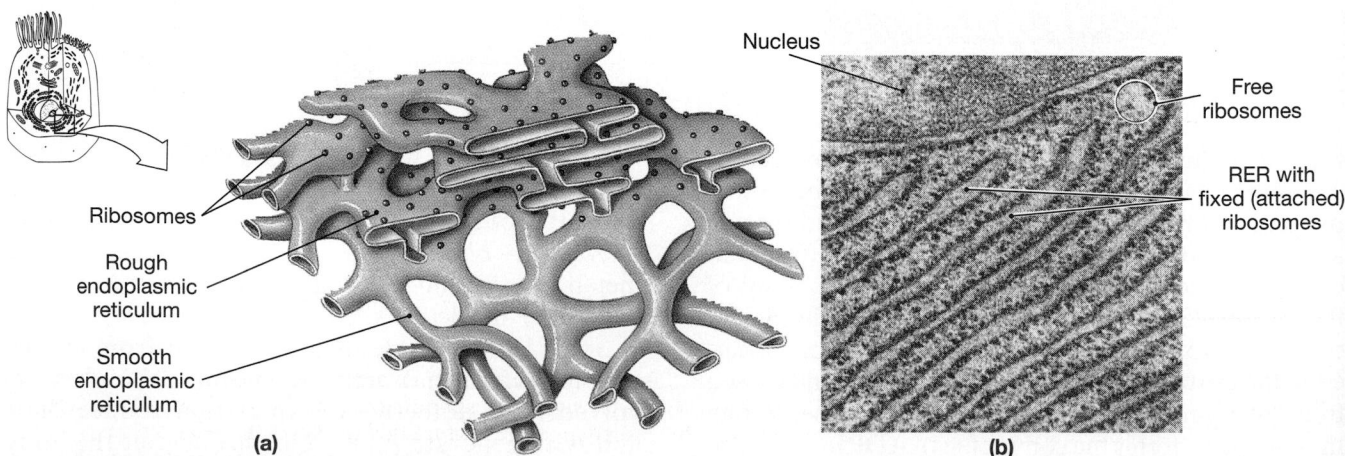

Ribosomes
Rough endoplasmic reticulum
Smooth endoplasmic reticulum
(a)

Nucleus
Free ribosomes
RER with fixed (attached) ribosomes
(b)

•**FIGURE 3-17**   **The Endoplasmic Reticulum. (a)** This diagrammatic sketch indicates the three-dimensional relationships between the rough and smooth endoplasmic reticula. **(b)** Rough endoplasmic reticulum and free ribosomes in the cytoplasm of a cell. (TEM × 73,600)

its ongoing activities. For example, pancreatic cells that manufacture digestive enzymes contain an extensive RER, but the SER is relatively small. The situation is just the reverse in the cells that synthesize steroid hormones in the reproductive system.

The Golgi Apparatus

The **Golgi** (GŌL-jē) **apparatus** (Figure 3-18a,b●) consists of flattened membrane discs called *saccules*. A typical Golgi apparatus consists of five to six saccules. A single cell may contain several sets, each resembling a stack of dinner plates. Most often these stacks lie near the nucleus of the cell. The Golgi saccules communicate with the ER and with the cell

surface through the formation, movement, and fusion of vesicles.

The three major functions of the Golgi apparatus are:

1. Synthesis and packaging of secretions, such as hormones or enzymes, for release through exocytosis.
2. Renewal or modification of the cell membrane.
3. Packaging of special enzymes within vesicles for use in the cytosol.

SECRETORY VESICLES AND EXOCYTOSIS Figure 3-18c● diagrams the role of the Golgi apparatus in packaging secretions. Protein and glycoprotein synthesis occur in the RER, and transport vesicles then move these products to the Golgi apparatus. The vesi-

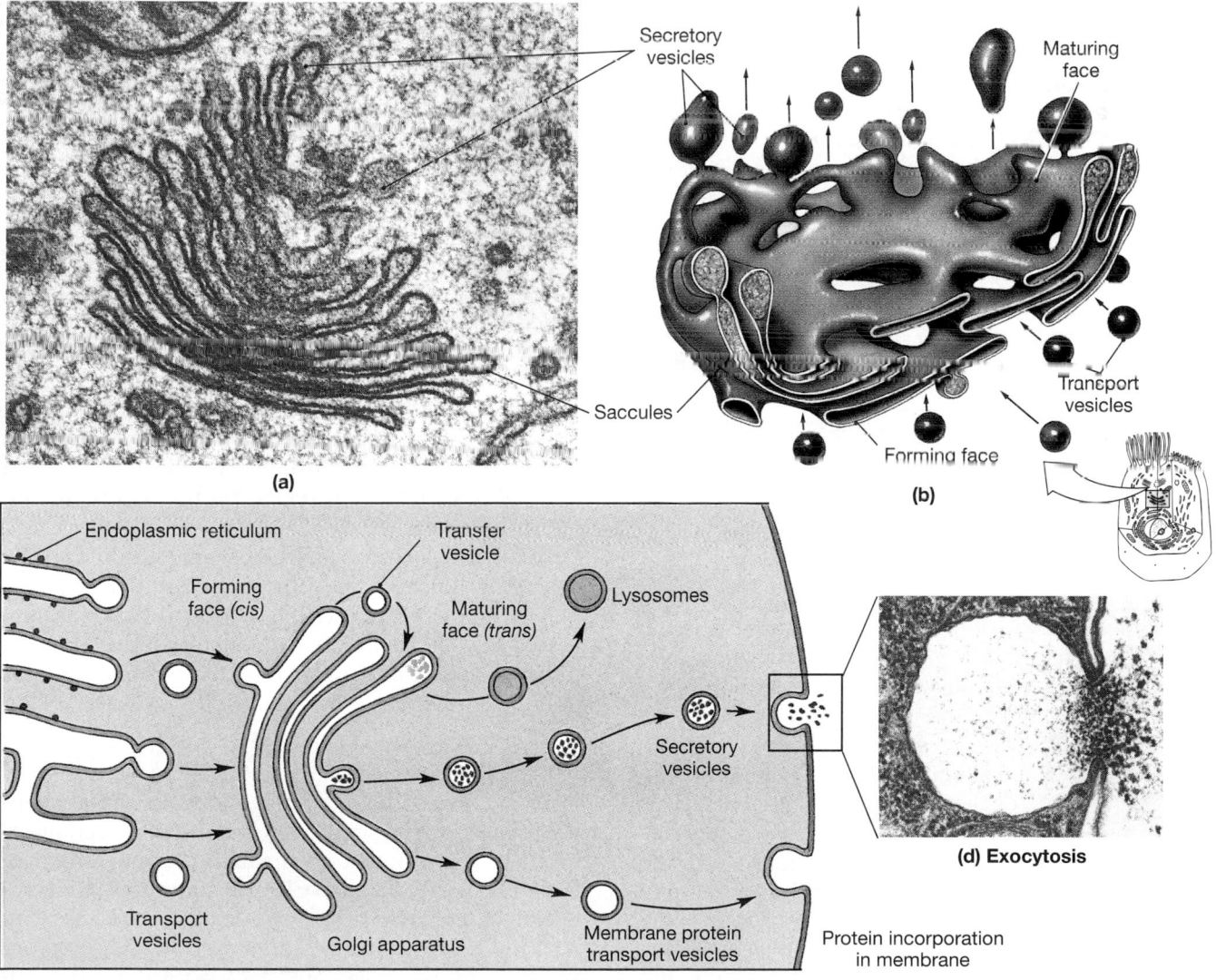

●FIGURE 3-18 **The Golgi Apparatus.** **(a)** A sectional view of the Golgi apparatus of an active secretory cell. (TEM ×57,660) **(b)** A three-dimensional view of the Golgi apparatus with a cut edge corresponding to part (a). **(c)** Transport vesicles carry the secretory product from the endoplasmic reticulum to the Golgi apparatus (simplified to clarify the relationships between the membranes). Transfer vesicles move membrane and materials between the Golgi saccules. At the maturing face, three functional categories of vesicles develop. Secretory vesicles carry the secretion from the Golgi to the cell surface, where exocytosis releases the contents into the extracellular fluid. Other vesicles add surface area and integral proteins to the cell membrane. Lysosomes, which remain in the cytoplasm, are vesicles filled with enzymes. **(d)** Exocytosis at the surface of a cell.

cles generally arrive at a convex saccule known as the *forming face,* which is usually directed toward the nucleus. The transport vesicles then fuse with the Golgi membrane, emptying their contents into the cisterna. Inside the Golgi apparatus, enzymes modify the arriving proteins and glycoproteins. For example, the enzymes may change the carbohydrate structure of a glycoprotein, or they may attach a phosphate group, sugar, or fatty acid to a protein.

Material moves from saccule to saccule by means of small **transfer vesicles.** Ultimately, the product arrives at the *maturing face,* which is usually oriented toward the cell surface. At the maturing face, vesicles form that carry materials away from the Golgi apparatus. Three types of vesicles form at the Golgi apparatus:

1. **Secretory vesicles.** Secretory vesicles are vesicles containing secretions that will be discharged from the cell through exocytosis (Figure 3-18d●).

2. **New cell membrane components.** As vesicles produced at the Golgi apparatus fuse with the surface of the cell, they are adding new lipids and proteins to the cell membrane. At the same time, other areas of the cell membrane are being removed during endocytosis. The Golgi apparatus can thus change the properties of the cell membrane over time. For example, new glycoprotein receptors can be added, making the cell more sensitive to a particular stimulus; alternatively, receptors can be removed, making the cell less sensitive. Such changes can profoundly alter the sensitivity and functions of the cell.

3. **Packaging of intracellular enzymes.** A third class of vesicles produced at the Golgi apparatus never leaves the cytoplasm. These vesicles, called *lysosomes,* contain digestive enzymes. Their varied functions will be detailed in the next section.

Lysosomes

Lysosomes (LĪ-sō-sōmz; *lyso-,* dissolution + *soma,* body) are vesicles filled with digestive enzymes. They are produced at the Golgi apparatus (Figure 3-18c●). *Primary lysosomes* contain inactive enzymes. Enzyme activation occurs when the lysosome fuses with the membranes of damaged organelles, such as mitochondria or fragments of the endoplasmic reticulum. This fusion creates a *secondary lysosome,* which contains active enzymes. These enzymes then break down the lysosomal contents. Nutrients reenter the cytosol, and the remaining material is eliminated by exocytosis.

Lysosomes also function in the defense against disease. Cells may remove bacteria as well as liquids and organic debris through endocytosis. (The role of lysosomes in receptor-mediated endocytosis was detailed in Figures 3-10 and 3-11●, pp. 79, 80.) Lysosomes fuse with the vesicles created through endocytosis, and the digestive enzymes then break down the contents and release usable substances, such as sugars or amino acids.

In this way, the cell both protects itself against pathogenic organisms and obtains valuable nutrients.

Lysosomes perform essential cleanup and recycling functions inside the cell. For example, when muscle cells are inactive, lysosomes gradually break down their contractile proteins; if the cells become active once again, this destruction ceases. This regulatory mechanism fails in a damaged or dead cell. Lysosomes then disintegrate, releasing active enzymes into the cytosol. These enzymes rapidly destroy the proteins and organelles of the cell, a process called **autolysis** (aw-TAH-li-sis; *auto-,* self). We do not know how to control lysosomal activities or why the enclosed enzymes do not digest the lysosomal walls unless the cell is damaged.

Problems with lysosomal enzyme production cause more than 30 serious diseases affecting children. In these conditions, called *lysosomal storage diseases,* the lack of a specific lysosomal enzyme results in the buildup of waste products and debris normally removed and recycled by lysosomes. Affected individuals may die when vital cells, such as those of the heart, can no longer function. **AM** *Lysosomal Storage Diseases*

Peroxisomes

Peroxisomes are smaller than lysosomes and carry a different group of enzymes. In contrast to lysosomes, which are produced at the Golgi apparatus, peroxisomes probably originate at the RER. Peroxisomes absorb and neutralize toxins, such as alcohol or hydrogen peroxide (H_2O_2), that are absorbed from the interstitial fluid or generated by chemical reactions in the cytoplasm. Peroxisomes are most abundant in liver cells, which are responsible for removing and neutralizing toxins absorbed by the digestive tract. Peroxisomes protect all cells from the potentially damaging effects of free radicals produced during normal metabolic reactions. ∞ *[p. 35]* It has been suggested that the cumulative damage produced by free radicals inside and outside our cells is a major factor in the aging process.

Membrane Flow

With the exception of the mitochondria, all the membranous organelles in the cell are either interconnected or in communication through the movement of vesicles. The RER and SER are continuous and connected to the nuclear envelope. Transport vesicles connect the ER with the Golgi apparatus, and secretory vesicles link the Golgi apparatus with the cell membrane. Finally, vesicles forming at the exposed surface of the cell remove and recycle segments of the cell membrane. This continuous movement and exchange is called **membrane flow.** In an actively secreting cell, an area equal to the entire membrane surface may be replaced each hour.

Membrane flow is another example of the dynamic nature of cells. It provides a mechanism by which cells change the characteristics of their cell membranes—lipids, receptors, channels, anchors, and enzymes—as they grow, mature, or respond to a specific environmental stimulus.

☑ Microscopic examination of a cell reveals that it contains many mitochondria. What does this observation imply about the cell's energy requirements?

☑ Certain cells in the ovaries and testes contain large amounts of smooth endoplasmic reticulum (SER). Why?

THE NUCLEUS

The **nucleus** is the control center of cellular operations. A single nucleus stores all the information needed to control the synthesis of the approximately 100,000 different proteins in the human body. The nucleus determines the structural and functional characteristics of the cell by controlling which proteins are synthesized, and in what amounts. A cell without a nucleus could be compared to a car without a driver. However, a car can sit idle for years, but a cell without a nucleus will disintegrate within 3–4 months.

Most cells contain a single nucleus, but there are exceptions. For example, skeletal muscle cells have many nuclei, and mature red blood cells have none. Figure 3-19● details the structure of a typical nucleus. A **nuclear envelope** surrounds the nucleus and separates it from the cytosol. The nuclear envelope is a double membrane with its two layers separated by a narrow **perinuclear space** (*peri-*, around). At several locations, the nuclear envelope is connected to the rough endoplasmic reticulum, as shown in Figure 3-2●, p. 67.

The nucleus directs processes that take place in the cytosol and must in turn receive information about conditions and activities in the cytosol. Chemical communication between the nucleus and cytosol occurs through **nuclear pores.** These pores, which cover about 10 percent of the surface of the nucleus, are large enough to permit the movement of ions and small molecules but too small for the free passage of proteins or DNA.

The term *nucleoplasm* refers to the fluid contents of the nucleus. The nucleoplasm contains a network of fine filaments, the **nuclear matrix,** that provides structural support and may be involved in the regulation of genetic activity. The nucleoplasm also contains ions, enzymes, RNA and DNA nucleotides, small amounts of RNA, and DNA. The DNA strands are located in

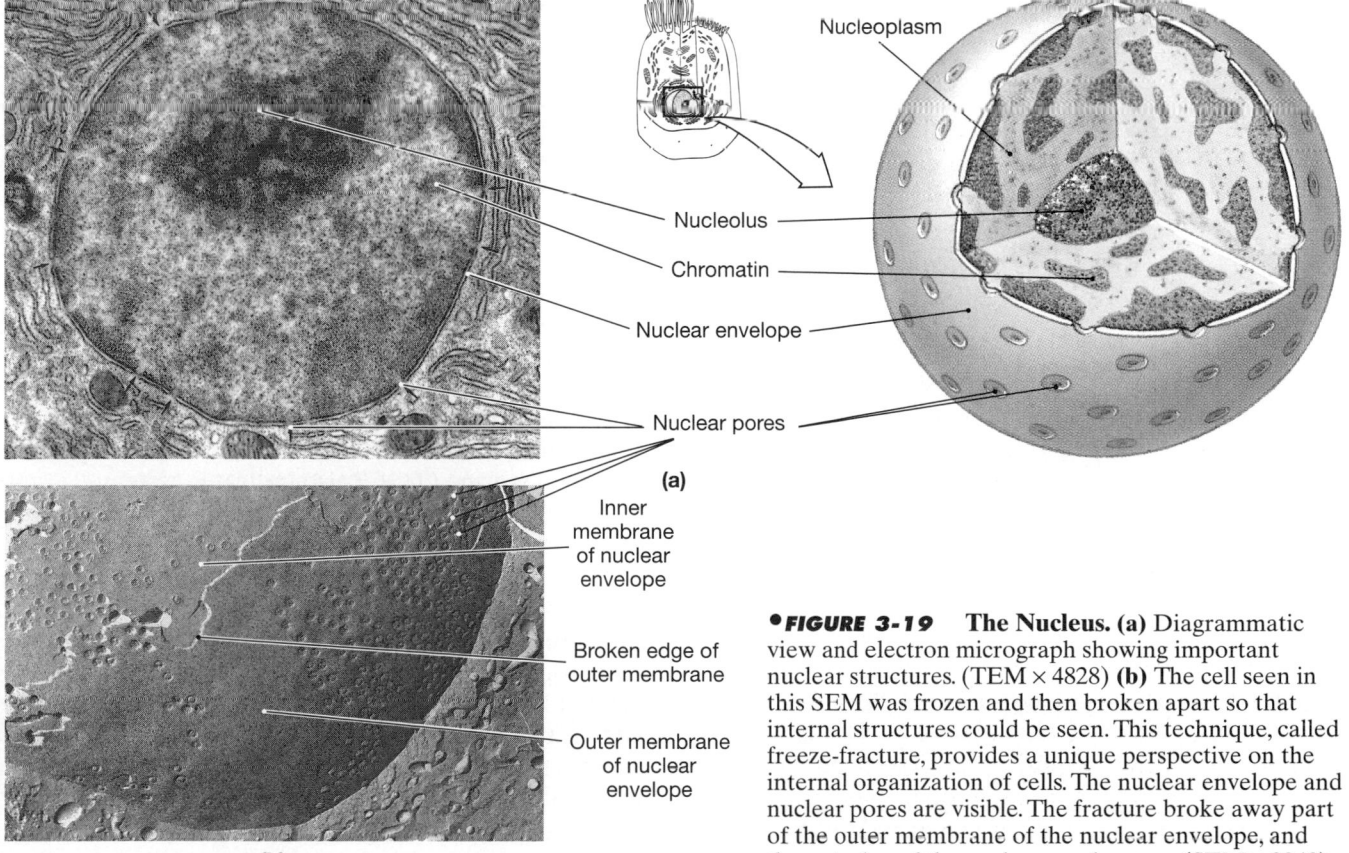

Nucleoplasm

Nucleolus

Chromatin

Nuclear envelope

Nuclear pores

(a)

Inner membrane of nuclear envelope

Broken edge of outer membrane

Outer membrane of nuclear envelope

(b)

●**FIGURE 3-19** **The Nucleus. (a)** Diagrammatic view and electron micrograph showing important nuclear structures. (TEM × 4828) **(b)** The cell seen in this SEM was frozen and then broken apart so that internal structures could be seen. This technique, called freeze-fracture, provides a unique perspective on the internal organization of cells. The nuclear envelope and nuclear pores are visible. The fracture broke away part of the outer membrane of the nuclear envelope, and the cut edge of the nucleus can be seen. (SEM × 9240)

complex structures known as *chromosomes* (*chroma,* color).

Chromosome Structure

It is the DNA in the nucleus that stores the vital information, and this DNA is found in **chromosomes.** Each chromosome contains DNA strands bound to special proteins called **histones.** The DNA strands coil around the histones, allowing a great deal of DNA to be packaged in a small space. Interactions between the DNA and the histones help determine what information is available to the cell at any given moment.

The nuclei of somatic cells (all cells other than sex cells) in humans contain 23 pairs of chromosomes. One member of each pair is derived from the mother, and one from the father. The structure of a typical chromosome is diagrammed in Figure 3-20●.

At intervals, the DNA strands wind around the histones, forming a complex known as a **nucleosome.** The entire chain of nucleosomes may coil around other proteins. The degree of coiling determines whether the chromosome is long and thin or short and fat. Chromosomes in a dividing cell are very tightly coiled; they can be seen clearly as separate structures

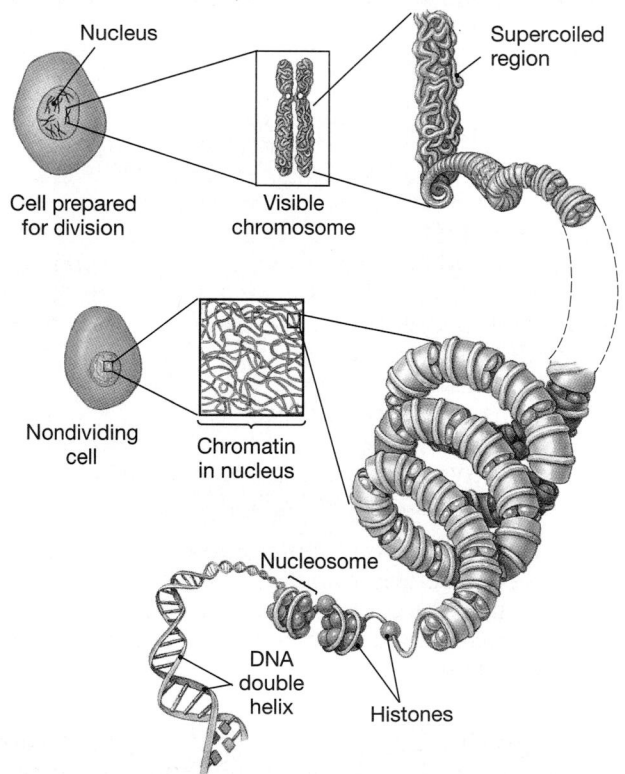

●*FIGURE 3-20* **Chromosome Structure.** DNA strands are coiled around histones to form nucleosomes. Nucleosomes form coils that may be very tight or rather loose. In cells that are not dividing, the DNA is loosely coiled, forming a tangled network known as chromatin. When the coiling becomes tighter, as it does in preparation for cell division, the DNA becomes visible as distinct structures called chromosomes.

in light or electron micrographs. In cells that are not dividing, the chromosomal material is loosely coiled, forming a tangle of fine filaments known as **chromatin** that gives the nucleus a clumped, grainy appearance.

Most nuclei contain one to four dark-staining areas called **nucleoli** (noo-KLĒ-ō-lī); singular, *nucleolus*). Nucleoli are nuclear organelles that synthesize *ribosomal RNA* (rRNA). They also assemble the ribosomal subunits, which reach the cytoplasm by carrier-mediated transport at the nuclear pores. A nucleolus contains histones and enzymes as well as RNA. Nucleoli form around a chromosomal region that includes the DNA that bears the instructions for producing ribosomal proteins and RNA. Nucleoli are most prominent in cells that manufacture large amounts of proteins, such as liver cells and muscle cells, because these cells need large numbers of ribosomes.

The Genetic Code

The **genetic code** is the method of information storage within the DNA strands of the nucleus. An understanding of the genetic code has enabled us to determine how cells build proteins and how various structural and functional characteristics are inherited from generation to generation. Researchers have begun to experiment with the manipulation of the genetic information in human cells; the techniques may revolutionize the treatment of some serious diseases.

We described the basic structure of nucleic acids in Chapter 2. ∞ *[pp. 57–58]* A single DNA molecule consists of a pair of DNA strands held together by hydrogen bonding between complementary nitrogenous bases. *Information is stored in the sequence of nitrogenous bases along the length of the DNA strands.* Those nitrogenous bases are *adenine,* A; *thymine,* T; *cytosine,* C; and *guanine,* G.

The genetic code is called a *triplet code* because a sequence of three nitrogenous bases can specify the identity of a single amino acid. For example, the DNA triplet thymine-guanine-thymine (TGT) indicates the amino acid cysteine. More than one triplet may represent the same amino acid, however. The DNA triplet thymine-guanine-cytosine (TGC) also indicates cysteine.

Each **gene** contains all the triplets needed to produce a specific protein. The number of triplets varies from gene to gene, depending on the size of the peptide represented. A relatively short peptide chain might require fewer than 100 triplets, whereas the instructions for building a large protein might involve 1000 or more. The entire DNA molecule is not devoted to carrying instructions for proteins; some segments contain instructions for the synthesis of transfer or ribosomal RNA, and others have no apparent function.

Every gene contains segments responsible for regulating its own activity. In effect, these are triplets that say "do (or do not) read this message," "message starts here," and "message ends here." The "read me," "don't read me," and "start" signals form a special *control segment* at the start of each gene. Each gene ends with a "stop" signal.

DNA FINGERPRINTING Every nucleated somatic cell in the body carries an identical set of 46 chromosomes. Not all the DNA of these chromosomes codes for proteins, however, and a significant percentage of DNA segments have no known function. Some of the "useless" segments contain the same nucleotide sequence repeated over and over. The number of segments and the number of repetitions vary from individual to individual. The chances of any two individuals, other than identical twins, having the same pattern of repeating segments is less than one in 9 billion. In other words, it is extremely unlikely that you will ever encounter someone else who has the same pattern of repeating nucleotide sequences as that present in your DNA.

Individual identification can therefore be made on the basis of a DNA pattern analysis, just as it can on the basis of a fingerprint. Skin scrapings, blood, semen, hair, or other tissues can be used as a sample source. Information from **DNA fingerprinting** has already been used to convict (and to acquit) persons accused of committing violent crimes, such as rape or murder.

Gene Activation and Protein Synthesis

Each molecule of DNA contains thousands of individual genes and therefore holds the information needed to synthesize thousands of proteins. These genes are normally tightly coiled, and histones bound to the control segments prevent their activation. Before a specific gene can be activated to begin **protein synthesis,** enzymes must temporarily disrupt the weak bonds between the nitrogenous bases and detach the histone that guards the control segment. The regulation of this process is only partially understood. Another enzyme, **RNA polymerase,** then binds to the initial segment of the gene.

Protein synthesis can be divided into two stages:

1. **Transcription,** the production of RNA from a DNA template. In protein synthesis, polymerase uses the genetic information to assemble a strand of mRNA.
2. **Translation,** the assembling of a functional protein by ribosomes, which use the information carried by an mRNA strand.

Transcription

Activated genes do not leave the nucleus, nor do they lose their connections with other genes along the length of the DNA strand. Instead, a messenger carries the instructions from the nucleus to the cytoplasm, where the amino acids that will make up the new proteins are located. The carrier is a single strand of **messenger RNA (mRNA).** The term *transcription* is appropriate to the process of mRNA formation because the mRNA is transcribing, or "taking notes," from the gene.

The two DNA strands are mirror images. One of the two strands in the gene contains the triplets that specify the sequence of amino acids in the polypeptide. This is known as the **coding strand.** The other strand, known as the **template strand,** contains complementary triplets that will be used as a template for mRNA production. The resulting mRNA will have a nucleotide sequence identical to that of the coding strand but with uracil substituted for thymine. Figure 3-21● details the process of transcription.

STEP 1. Once the DNA strands have separated and the control segment has been exposed, transcription can begin. The key event is the attachment of RNA polymerase to the template strand.

STEP 2. RNA polymerase promotes hydrogen bonding between the nitrogenous bases of the template strand and complementary nucleotides in the nucleoplasm. It then strings nucleotides together by covalent bonding. The RNA polymerase interacts with only a small portion of the template strand at any one time as it travels along the DNA strand. Moving from triplet to triplet, the enzyme collects additional nucleotides and attaches them to the growing chain. The nucleotides involved are those characteristic of RNA, not of DNA; RNA polymerase may attach adenine, guanine, cytosine, or uracil but never thymine. Thus, wherever an A occurs in the DNA strand, the polymerase will attach a U rather than a T to the growing mRNA strand. In this way, RNA polymerase assembles a complete strand of mRNA. The nucleotide sequence of the template strand determines the nucleotide sequence of the mRNA strand. Thus, each DNA triplet will correspond to a sequence of three nucleotide bases in the mRNA strand. Such a three-base mRNA sequence is called a **codon** (KŌ-don). Codons contain nitrogenous bases that are complementary to those of the triplets in the template strand. For example, if the DNA triplet is TCG, the corresponding codon on the mRNA strand will be AGC.

STEP 3. At the "stop" signal, the enzyme and the mRNA strand detach from the DNA strand, and transcription ends. The complementary DNA strands now reassociate as hydrogen bonding occurs between complementary base pairs.

Each gene includes a number of triplets that do not contain information needed to build a functional protein. As a result, the mRNA strand assembled during transcription, sometimes called immature mRNA or *pre-mRNA,* must be "edited" before it leaves the nucleus to direct protein synthesis. In this **RNA processing,** nonsense regions, called **introns,** are snipped out, and the remaining segments, or **exons,** are spliced

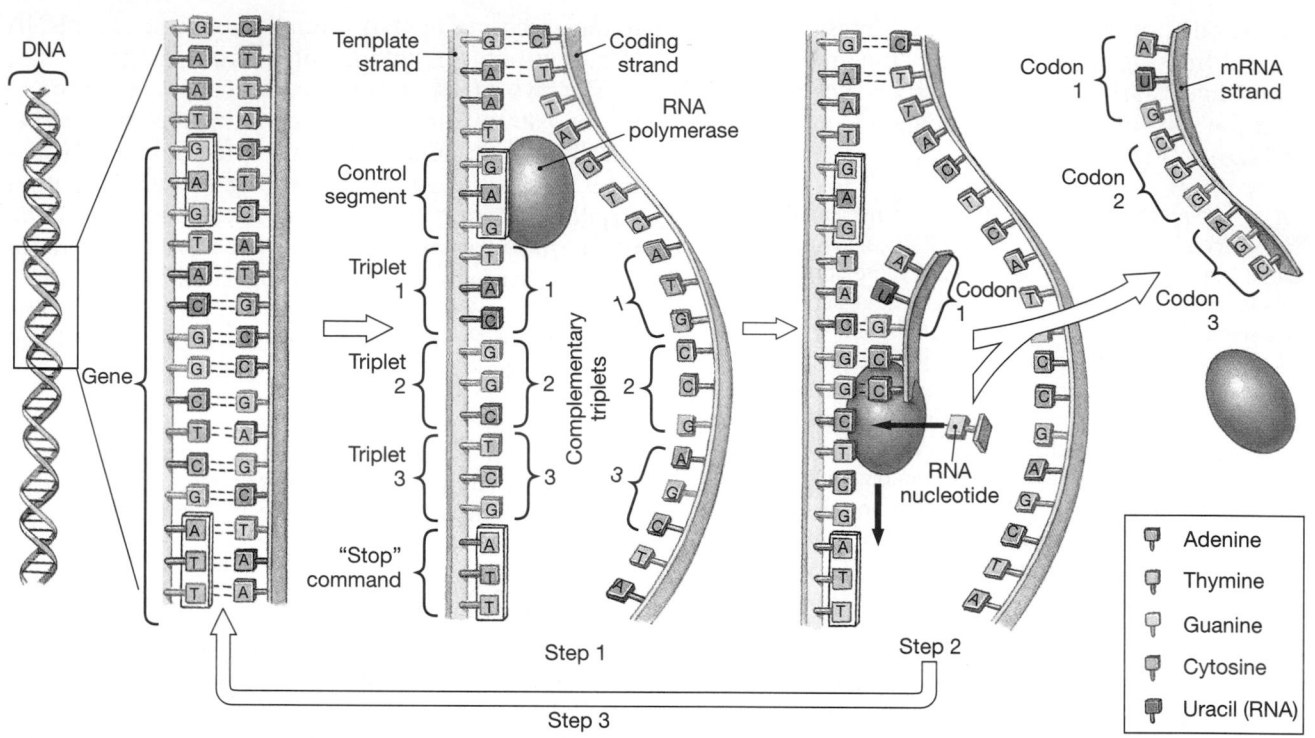

•**FIGURE 3-21**   **mRNA Transcription.** This diagram shows a small portion of a single DNA molecule, containing a single gene available for transcription. *Step 1:* The two DNA strands separate, and RNA polymerase binds to the control segment of the gene. *Step 2:* The RNA polymerase moves from one nucleotide to another along the length of the template strand. At each site, complementary RNA nucleotides form hydrogen bonds with the DNA nucleotides of the template strand. The RNA polymerase then strings the arriving nucleotides together into a strand of mRNA. *Step 3:* Upon reaching the stop signal at the end of the gene, the RNA polymerase and the mRNA strand detach, and the two DNA strands reassociate.

together. This process creates a much shorter, functional strand of mRNA that then enters the cytoplasm through one of the nuclear pores.

Translation

Translation is the construction of a functional polypeptide by using the information in an mRNA strand. The codons of an mRNA strand carry the message spelled out in the triplets of the coding strand of the gene. Each mRNA codon designates a particular amino acid to be incorporated into the polypeptide chain. During translation, the sequence of codons will determine the sequence of amino acids in the polypeptide. The amino acids are provided by a relatively small and mobile type of RNA, **transfer RNA (tRNA)** (Figure 3-22•). Each tRNA molecule binds and delivers an amino acid of a specific type. There are more than 20 different kinds of transfer RNA, at least one for each of the amino acids used in protein synthesis.

Each tRNA molecule has a tail that binds an amino acid. Roughly midway along its length, the nucleotide chain of the tRNA forms a tight loop. It is this loop that can interact with an mRNA strand. The loop con-

tains three nitrogenous bases that form an **anticodon.** The anticodon can bond complementarily with an appropriate mRNA codon during translation. The base sequence of the anticodon indicates the type of amino acid carried by the tRNA. For example, a tRNA with the anticodon GGG always carries the amino acid *proline,* whereas a tRNA with the anticodon CGG carries *alanine.* Figure 3-22• includes examples of other codons and anticodons that specify individual amino acids and summarizes the relationships among DNA, codons, and anticodons.

The tRNA molecules thus provide the physical link between codons and amino acids. During translation, each codon along the mRNA strand will bind a complementary anticodon on a tRNA molecule. Thus, if the mRNA has the codons (CCC)-(GCC)-(UUA), it will bind to tRNAs with anticodons (GGG)-(CGG)-(AAU). The amino acid sequence of the peptide chain created is determined by the sequence of delivery by tRNAs, and that sequence depends on the arrangement of codons along the mRNA strand. In this case, the amino acid sequence in the peptide would be proline-alanine-leucine.

Translation has three phases: *initiation,* which begins the process; *elongation,* which produces the pep-

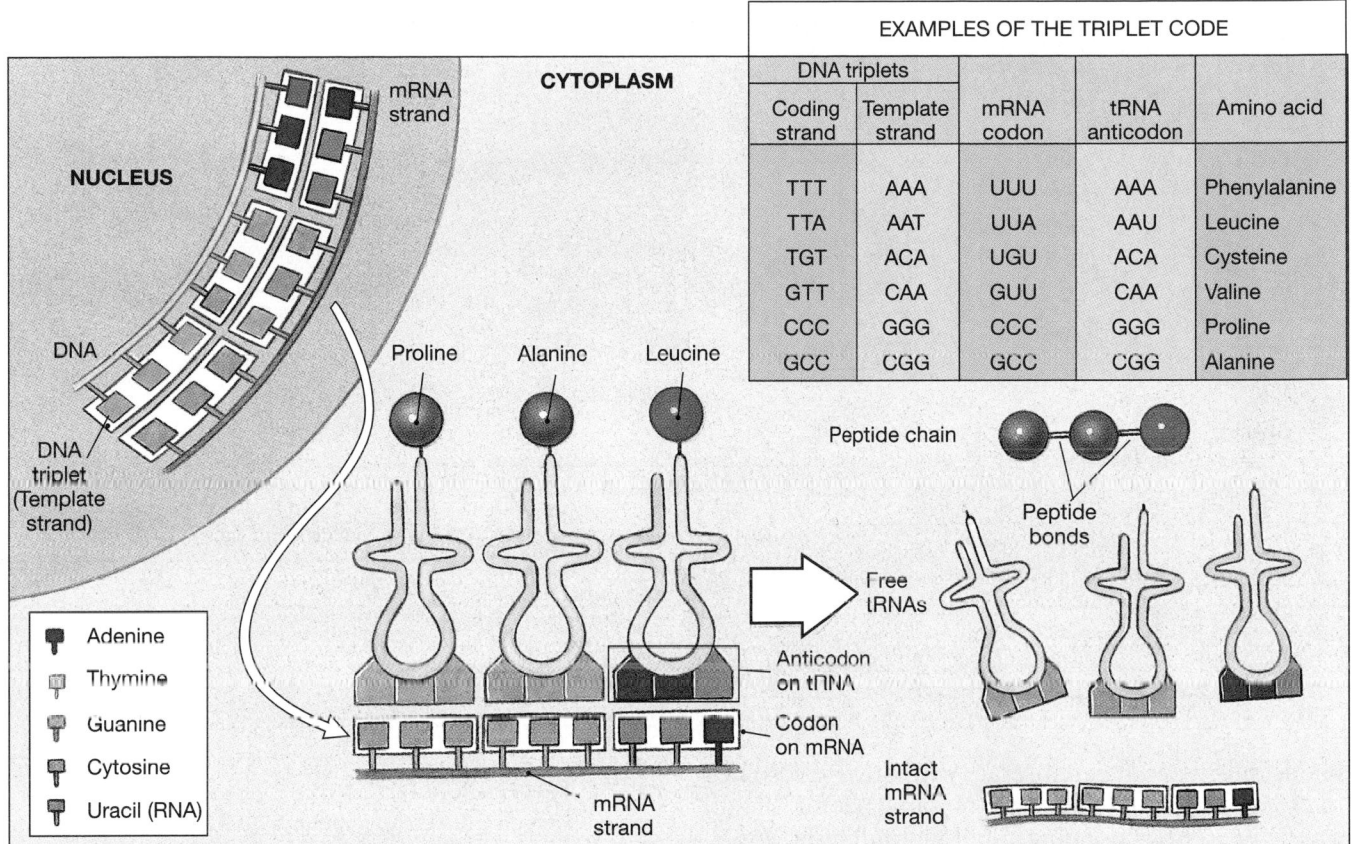

EXAMPLES OF THE TRIPLET CODE				
DNA triplets				
Coding strand	Template strand	mRNA codon	tRNA anticodon	Amino acid
TTT	AAA	UUU	AAA	Phenylalanine
TTA	AAT	UUA	AAU	Leucine
TGT	ACA	UGU	ACA	Cysteine
GTT	CAA	GUU	CAA	Valine
CCC	GGG	CCC	GGG	Proline
GCC	CGG	GCC	CGG	Alanine

•FIGURE 3-22 An Overview of Protein Synthesis. A mRNA strand assembled during transcription contains codons that are complementary to triplets on one of the DNA strands (the template strand). The mRNA then detaches from the DNA strand and enters the cytoplasm. Molecules of tRNA contain anticodons that are complementary to the mRNA codons. During translation, different amino acids are delivered to the mRNA strand by tRNAs with appropriate anticodons; examples are indicated. The sequence of amino acids in the completed peptide will reflect the sequence of tRNA arrival. The tRNAs arrive one after the other, and the peptide chain grows one amino acid at a time. Figure 3-23 shows the sequence of events.

tide chain; and *termination,* which ends the process and releases the completed peptide (Figure 3-23•).

Initiation

STEP 1. Initiation begins as the mRNA strand binds to a light ribosomal subunit. The first codon, or *start codon,* of the mRNA strand always has the base sequence AUG. It therefore binds a tRNA with the complementary anticodon sequence UAC. This tRNA, which carries the amino acid *methionine,* attaches to the first of two tRNA binding sites on the light ribosomal subunit. (The initial methionine will ultimately be removed from the finished protein.)

STEP 2. When this tRNA binding occurs, a heavy ribosomal subunit joins the complex to create a complete ribosome. The mRNA strand nestles in the gap between the light and heavy ribosomal subunits.

Elongation

STEP 3. A second tRNA now arrives at the second tRNA binding site of the ribosome, and its anticodon binds to the next codon of the mRNA strand.

STEP 4. Enzymes of the heavy ribosomal subunit then break the linkage between the tRNA molecule and its amino acid. At the same time, the enzymes attach the amino acid to its neighbor by means of a peptide bond. The ribosome then moves one codon down the mRNA strand.

STEP 5. The cycle is then repeated with the arrival of another molecule of tRNA. The tRNA already stripped of its amino acid drifts away. It will soon bind to another amino acid and be ready to participate in protein synthesis at a later time.

Termination

STEP 6. Elongation continues, adding amino acids to the growing polypeptide chain, until the ribosome reaches a "stop" signal, or *stop codon,* at the end of the mRNA strand. The ribosomal subunits now detach, leaving an intact strand of mRNA and a completed polypeptide.

Translation proceeds swiftly, producing a typical protein in about 20 seconds. The mRNA strand re-

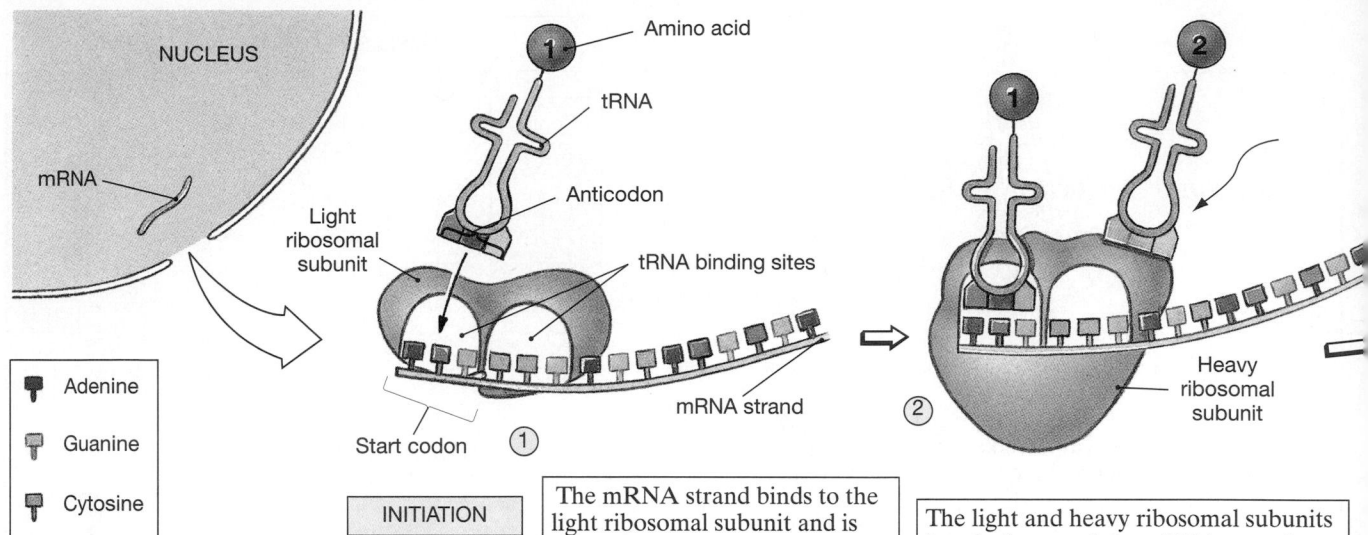

Adenine

Guanine

Cytosine

Uracil

INITIATION

The mRNA strand binds to the light ribosomal subunit and is joined at the start codon by the first tRNA, which carries the amino acid methionine. Binding occurs between complementary base pairs of the codon and anticodon.

The light and heavy ribosomal subunits interlock around the mRNA strand.

•*FIGURE 3-23* **The Process of Translation**

mains intact, and it can interact with other ribosomes to create additional copies of the same peptide chain. The process does not continue indefinitely, however, because after a few minutes to a few hours, mRNA strands are broken down and the nucleotides recycled.

Polyribosomes

During translation, only two mRNA codons are "read" by a ribosome at any one time, although the entire strand may contain thousands of codons. As a result, there is ample room for many ribosomes to bind to a single mRNA strand. At any given moment, each will be reading a different part of the same message, but each will end up constructing a copy of the same protein. The arrangement is similar to a line of people at a buffet lunch—each person will assemble the same meal, but always a step behind the person ahead. A series of ribosomes attached to the same mRNA strand is called a **polyribosome,** or *polysome* (Figure 3-24•). Polyribosome formation greatly increases the rate of protein synthesis.

DNA and the Control of Cell Structure and Function

The DNA of the nucleus controls the cell by directing the synthesis of specific proteins. Through control of protein synthesis, virtually every aspect of cell structure and function can be regulated. There are two levels of control involved:

1. The DNA of the nucleus has *direct* control over the synthesis of structural proteins, such as cytoskeletal

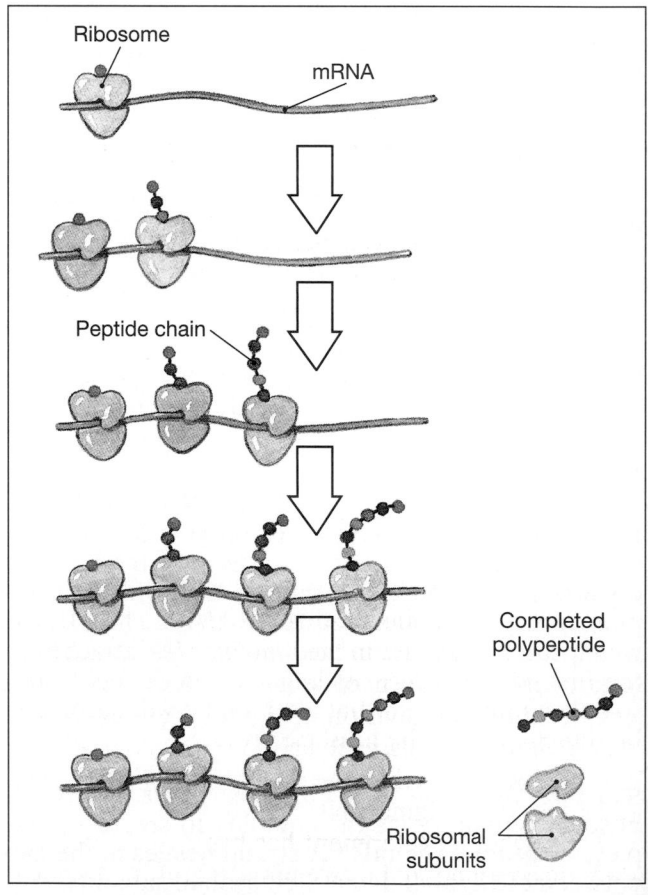

•*FIGURE 3-24* **Polyribosomes.** Each polyribosome consists of one strand of mRNA that is being read simultaneously by a number of ribosomes. In this way, a single strand of mRNA can produce many polypeptide molecules in a short time.

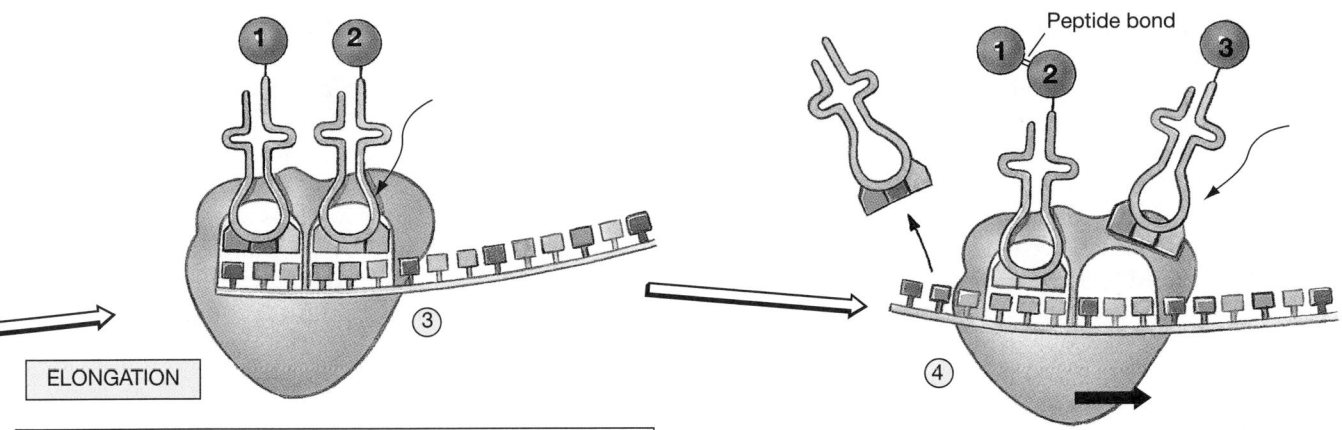

ELONGATION

A second tRNA arrives at the adjacent site of the ribosome. The anticodon of the second tRNA binds to the next mRNA codon.

The first amino acid is detached from its tRNA and is joined to the second amino acid by a peptide bond. The ribosome moves one codon farther along the mRNA strand; the first tRNA detaches as another tRNA arrives.

components, membrane proteins (including receptors), and secretory products. By issuing appropriate orders, the nucleus can alter the internal structure of the cell, its sensitivity to substances in its environment, or its secretory functions to meet changing circumstances and needs.

2. The DNA of the nucleus has *indirect* control over all other aspects of cellular metabolism because it regulates the synthesis of enzymes. By ordering the production or stopping the production of appropriate enzymes, the nucleus can regulate all the metabolic activities and functions of the cell. For example, the nucleus can accelerate the rate of glycolysis by increasing the number of needed enzymes in the cytoplasm.

☑ How does the nucleus control the activities of a cell?
☑ What process would be affected by the lack of the enzyme RNA polymerase?

THE CELL LIFE CYCLE

Between fertilization and physical maturity, a human being goes from a single cell to roughly 75 trillion cells. This amazing increase in the individual's size and complexity involves a form of cellular reproduction called **cell division.** The division of a single cell produces a pair of **daughter cells,** each half the size of the original. Before dividing, each of the daughter cells will grow to the size of the original cell.

Even when development has been completed, cell division continues to be essential to survival. Although cells are highly adaptable, they can be damaged by physical wear and tear, toxic chemicals, temperature changes, or other environmental stresses. In addition, cells, like individuals, are subject to aging. The life span of a cell varies from hours to decades, depending on

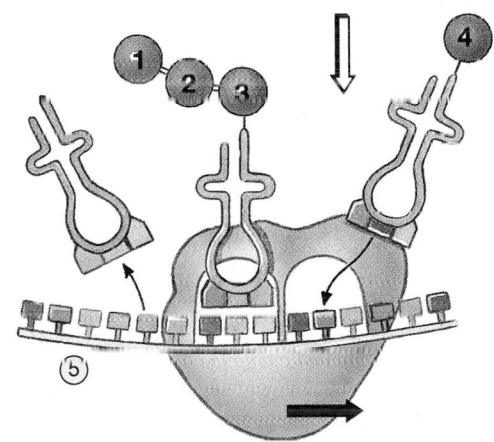

This cycle is repeated as the ribosome moves along the length of the mRNA strand, binds new tRNAs, and incorporates their amino acids into the polypeptide chain.

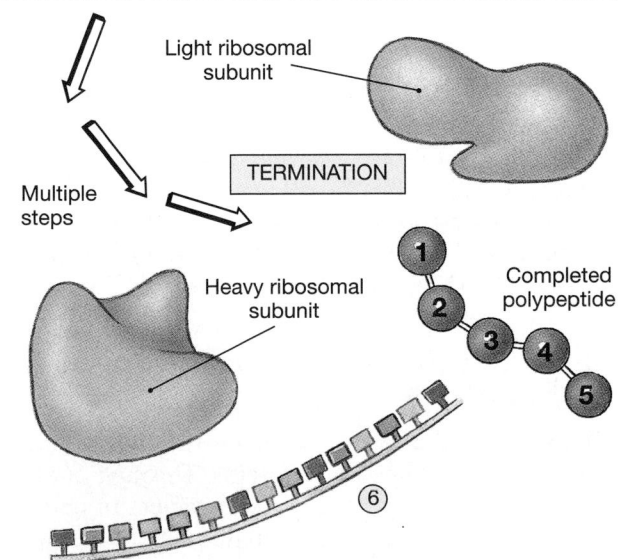

Elongation continues until the stop codon is reached; the components then separate.

the type of cell and the environmental stresses involved. Many cells appear to be programmed to self-destruct after a certain period of time. Their destruction results from the activation of specific "suicide genes" in the nucleus. The genetically controlled death of cells is called **apoptosis** (ap-op-TŌ-sis; *ptosis,* a falling away). A gene involved in the regulation of this process has been identified. This gene, called *bcl-2,* appears to prevent apoptosis and to keep a cell alive and functional. If something interferes with the function of this gene, the cell self-destructs.

Because a typical cell does not live nearly as long as a typical person, cell populations must be maintained over time by cell division. For cell reproduction to be successful, the cell's genetic material must be duplicated accurately, and that duplicated material must be distributed to each of the daughter cells formed by division. This process of nuclear division is called **mitosis** (mī-TŌ-sis). Mitosis occurs during the division of somatic cells. The production of sperm and ova involves a different process, **meiosis** (mī-Ō-sis), which we shall describe in Chapter 28.

Figure 3-25● presents the life cycle of a typical cell in greater detail. That life cycle includes a relatively brief period of mitosis alternating with an *interphase* period of variable duration.

Interphase

Most cells spend only a small part of their time actively engaged in cell division. Somatic cells spend the majority of their functional lives in a state known as **interphase.** During interphase, a cell performs all its normal functions, and, if necessary, it makes preparations for cell division. In a cell preparing to divide, interphase can be divided into the G_1, S, and G_2 *phases.* An interphase cell in the G_0 *phase* is not preparing for division but is performing all other normal cell functions. Some mature cells, such as skeletal muscle cells and most neurons, remain in G_0 indefinitely and never divide. In contrast, stem cells, which divide repeatedly with very brief interphase periods, never enter G_0.

The G_1 Phase

A cell that is going to divide first enters the **G_1 phase.** In this phase, the cell manufactures enough mitochondria, cytoskeletal elements, endoplasmic reticulum, ribosomes, Golgi membranes, and cytosol to make two functional cells. Centriole replication begins in G_1 and commonly continues until G_2. In cells dividing at top speed, G_1 may last as little as 8–12 hours. Such cells pour all their energy into mitosis, and all other activities cease. If G_1 lasts for days, weeks, or months, preparation for mitosis occurs as the cells perform their normal functions.

The S Phase

When the activities of G_1 have been completed, the cell enters the **S phase.** Over the next 6–8 hours the cell duplicates its chromosomes. In this process, it copies its DNA and combines it with histones and other proteins in the nucleus.

Throughout the life of a cell, the DNA strands in the nucleus remain intact. DNA synthesis, or **DNA replication,** occurs during the S phase in cells preparing to undergo mitosis or meiosis. The goal of this replication is to copy the genetic information in the nucleus so that one set of chromosomes can be given to each of the two daughter cells produced.

DNA REPLICATION   A DNA molecule consists of a pair of DNA strands held together by hydrogen bonding between complementary nitrogenous bases. Figure 3-26● diagrams the process of DNA replication. It begins when various enzymes unwind the strands and disrupt the weak bonds between the nitrogenous bases. As the strands unwind, molecules of **DNA polymerase** bind to the exposed nitrogenous bases. This enzyme (1) promotes bonding between the nitrogenous bases of the DNA strand and complementary DNA nucleotides dissolved in the nucleoplasm and (2) links the nucleotides by covalent bonds.

Many molecules of DNA polymerase are working simultaneously along the DNA strands. On one strand, DNA polymerase moves forward as the original strands unwind, producing an intact complementary DNA strand (Figure 3-26●). On the other strand, DNA poly-

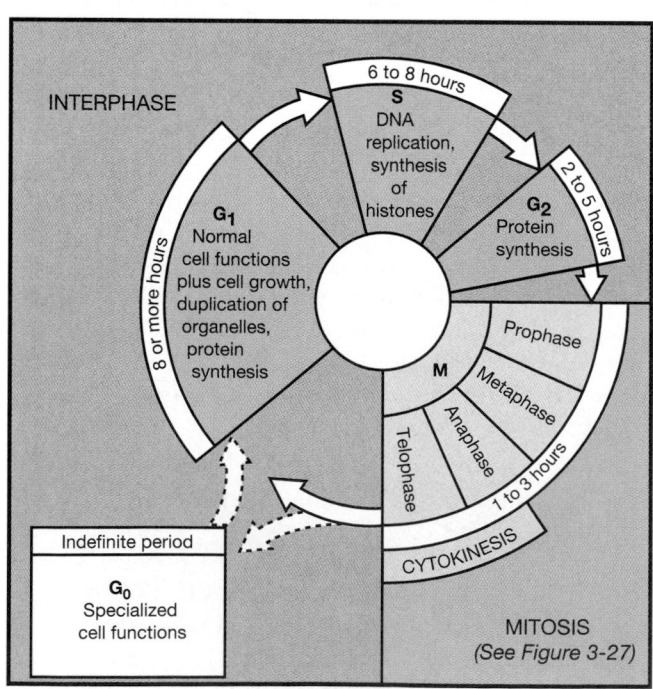

●*FIGURE 3-25*   **The Cell Life Cycle**

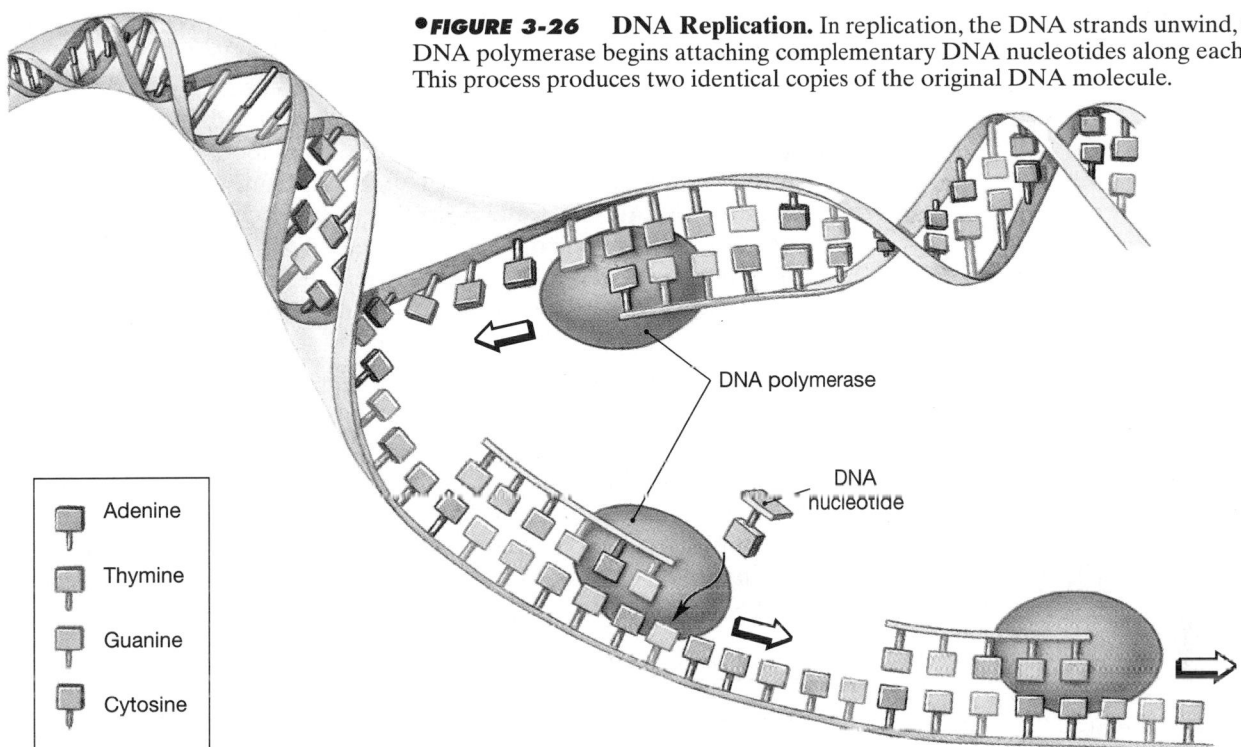

●FIGURE 3-26 **DNA Replication.** In replication, the DNA strands unwind, and DNA polymerase begins attaching complementary DNA nucleotides along each strand. This process produces two identical copies of the original DNA molecule.

DNA polymerase

DNA nucleotide

Adenine

Thymine

Guanine

Cytosine

merase works *away* from the point of unwinding, creating short nucleotide chains that must be linked together by enzymes called **ligases** (LĪ-gās-ez; *liga,* to tie). The final result is a pair of identical DNA molecules.

The G₂ Phase

Once DNA replication has ended, there is a brief (2–5 hour) **G₂ phase** devoted to last-minute protein synthesis and to the completion of centriole replication. The cell then enters the **M phase,** and mitosis begins.

MUTATIONS **Mutations** are permanent alterations in a cell's DNA that affect the nucleotide sequence of one or more genes. The simplest is a **point mutation,** a change in a single nucleotide that affects one codon. The triplet code has some flexibility because several different codons can specify the same amino acid. But a point mutation that produces a codon that specifies a different amino acid will usually change the structure of the completed protein. A single change in the amino acid sequence of a structural protein or enzyme can prove fatal. Several cancers and two potentially lethal blood disorders, *thalassemia* and *sickle cell anemia,* result from variations in a single nucleotide. ∞ *[p. 54]*

More than 100 inherited disorders have been traced to abnormalities in enzyme or protein structure that reflect single alterations in nucleotide sequence. More elaborate mutations, such as additions or deletions of nucleotides, can affect multiple codons within a gene, several adjacent genes, or the structure of one or more chromosomes.

Because most mutations occur during DNA replication, they are most likely to involve cells that are undergoing cell

division. A single cell, a group of cells, or an entire individual may be affected. This last prospect will occur if the changes are made early in development. For example, a mutation affecting the DNA of an individual's gametes (sperm or oocytes) will be inherited by that individual's children. Our understanding of genetic structure is opening the possibility of diagnosing and correcting some of these problems. For a discussion of the principles and technologies involved, see the *Applications Manual.* **AM** *Genetic Engineering and Gene Therapy*

Mitosis

Mitosis is a process that separates the duplicated chromosomes of the original cell into two identical nuclei. Mitosis specifically refers to the division and duplication of the nucleus of the cell. Division of the cytoplasm to form two distinct new cells involves a separate but related process known as **cytokinesis** (sī-tō-ki-NĒ-sis; *cyto-,* cell + *kinesis,* motion). Figure 3-27a● depicts interphase, and Figures 3-27b–f● summarize the four stages of mitosis: *prophase* (early and late), *metaphase, anaphase,* and *telophase.*

Stage 1: Prophase

Prophase (PRŌ-fāz; *pro,* before) begins when the chromosomes coil so tightly that they become visible as individual structures. As a result of DNA replication during the S phase, there are now two copies of each chromosome. Each copy, called a **chromatid** (KRŌ-ma-tid), is connected to its duplicate at a single point, the **centromere** (SEN-trō-mēr). The region of

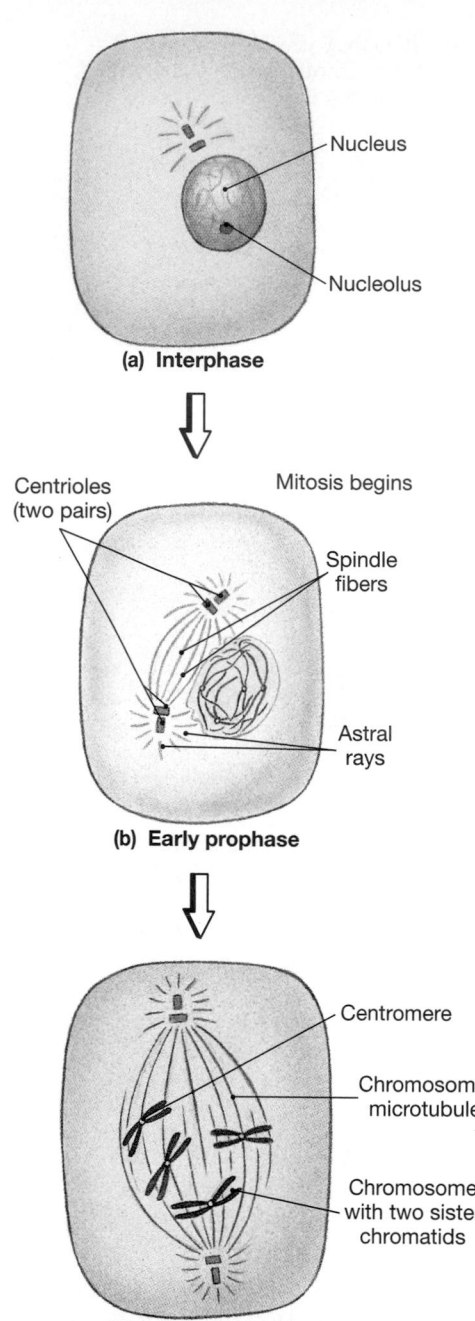

(a) Interphase

- Nucleus
- Nucleolus

Mitosis begins

Centrioles (two pairs)

Spindle fibers

Astral rays

(b) Early prophase

Centromere

Chromosomal microtubule

Chromosome with two sister chromatids

(c) Late prophase

the centromere is surrounded by a protein complex called a **kineto-chore** (ki-NE-tō-kor).

As the chromosomes appear, the nucleoli disappear. At around this time, the two pairs of centrioles manufactured during the G_1–G_2 period move toward opposite poles of the nucleus. An array of microtubules called **spindle fibers** extends between the centriole pairs. Smaller microtubules called *astral rays* radiate into the surrounding cytoplasm. Late in prophase, the nuclear envelope disappears. The spindle fibers now enter the nuclear region, and the chromatids begin attaching to spindle fibers called *chromosomal microtubules*. The attachment occurs at the kinetochore on opposite sides of the centromere linking each pair of chromatids.

Stage 2: Metaphase

Metaphase (MET-a-fāz; *meta,* after) begins after the disintegration of the nuclear envelope and attachment of chromatids to chromosomal microtubules. The chromatids move to a narrow central zone called the **metaphase plate.** Metaphase ends when all the chromatids are aligned in the plane of the metaphase plate.

Stage 3: Anaphase

Anaphase (AN-a-fāz; *ana,* apart) begins when the kinetochore of each chromatid pair splits and the chromatids separate. The two **daughter chromosomes** are now pulled toward opposite ends of the cell along the chromosomal microtubules. This movement involves an interaction between the kinetochore and the microtubule. Anaphase ends when the daughter chromosomes arrive near the centrioles at opposite ends of the cell.

Stage 4: Telophase

During **telophase** (TĒL-ō-fāz; *telo,* end) each cell prepares to return to the interphase state. The nuclear membranes form, the nuclei enlarge, and the chromosomes gradually uncoil. Once the chromosomes have relaxed and the fine filaments of chromatin become visible, nu-

•**FIGURE 3-27**   **Interphase and Mitosis.** Diagrammatic view.

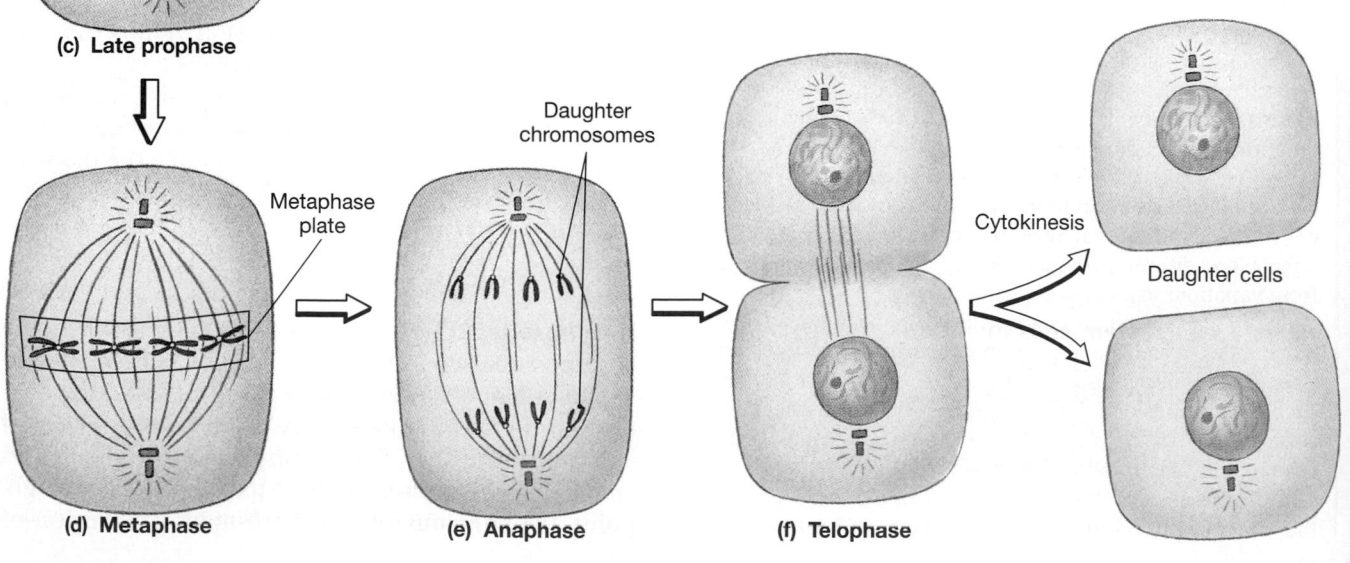

Metaphase plate

(d) Metaphase

Daughter chromosomes

(e) Anaphase

(f) Telophase

Cytokinesis

Daughter cells

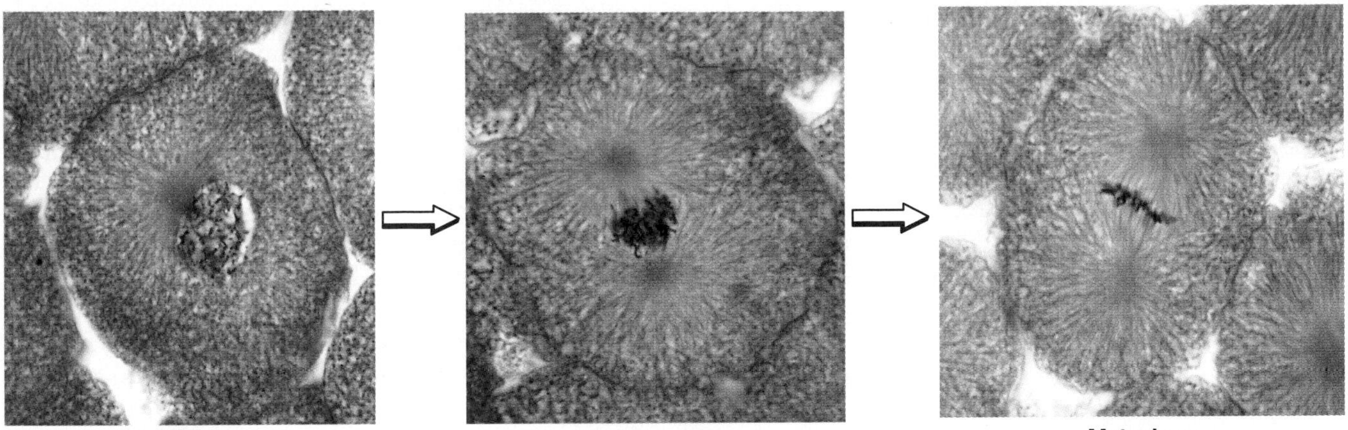

| Interphase | Late prophase | Metaphase |

●FIGURE 3-27 (*continued*). (LM × 581)

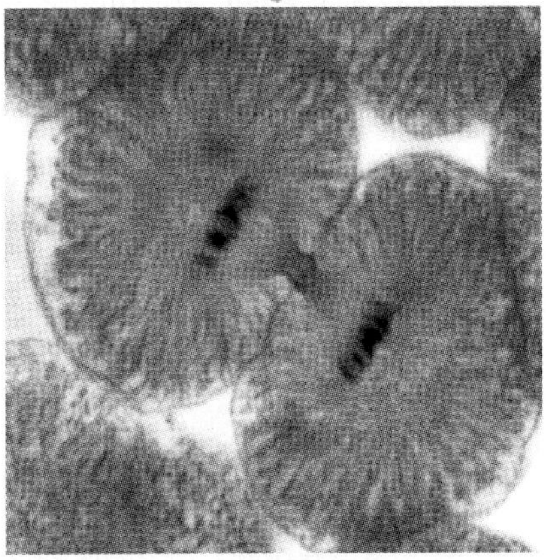

cleoli reappear and the nuclei resemble those of interphase cells. This stage marks the end of mitosis.

Cytokinesis

Cytokinesis, the cytoplasmic division of the daughter cells, usually begins in late anaphase. As the daughter chromosomes approach the ends of the spindle apparatus, the cytoplasm constricts along the plane of the metaphase plate. This process continues throughout telophase and is usually completed sometime after a nuclear membrane has reformed around each daughter nucleus. The completion of cytokinesis marks the end of the process of cell division.

The Mitotic Rate and Energetics

The preparations for cell division that occur between G_1 and the end of the S phase are difficult to recognize in a light micrograph. However, the start of mitosis is easy to recognize, because the chromosomes become condensed and highly visible. The frequency of cell division can thus be estimated by the number of cells in mitosis at any given time. As a result, we often use the term **mitotic rate** when we discuss rates of cell division. In general, the longer the life expectancy of a cell type, the slower the mitotic rate. Relatively long-lived cells, such as muscle cells and neurons, either never divide or do so only under special circumstances. Other cells, such as those covering the surface of the skin or the lining of the digestive tract, are subject to attack by chemicals, pathogens, and abrasion. They survive for only days or even hours. Special cells called **stem cells** maintain these cell populations through repeated cycles of cell division.

Stem cells are relatively unspecialized, and their only function is the production of daughter cells. Each time a stem cell divides, one of its daughter cells develops functional specializations while the other prepares for further stem cell divisions. The rate of stem cell division can vary, depending on the

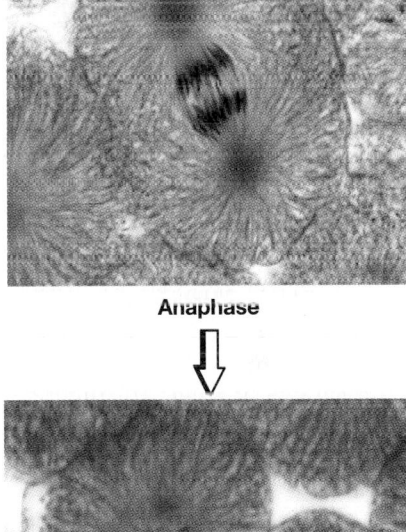

Anaphase

Telophase

tissue and the demand for new cells. In heavily abraded skin, stem cells may divide more than once a day, but stem cells in adult connective tissues may remain inactive for years.

Dividing cells use an unusually large amount of energy. For example, they must synthesize new organic materials and move organelles and chromosomes within the cell. All these processes require

ATP in substantial amounts. Cells that do not have adequate energy sources cannot divide. In a person who is starving, normal cell growth and maintenance grind to a halt. For this reason, prolonged starvation slows wound healing, lowers resistance to disease, thins the skin, and changes the lining of the digestive tract. The same changes are seen in the late stages of many cancers, because the cancer cells are "stealing" the nutrients that would otherwise be used to support normal cell growth and maintenance.

Regulation of the Cell Life Cycle

In normal tissues, the rates of cell division balance cell loss or destruction. Mitotic rates are genetically controlled, and many different stimuli may be responsible for activating genes that promote cell division. The most important of these stimuli appear to be extracellular compounds, generally peptides, that stimulate the division of specific cell types. These compounds include several hormones and a variety of **growth factors.** Table 3-3 lists several of these stimulatory compounds and their target tissues; we shall discuss these hormones and factors in later chapters. Each of these compounds appears to exert its effects by binding to receptors on the cell membrane. This binding initiates a series of biochemical events that ultimately trigger cell division. The effects of these stimulatory factors may be opposed by a poorly understood class of peptides called *chalones* (KĀ-lōnz) (Table 3-3).

Many of the peptide growth factors bind to membrane receptors at the cell surface. Binding activates enzymes inside the cell, starting a chain reaction leading to activation of genes that prepare the cell to undergo cell division. The presence or absence of the appropriate binding protein determines whether a particular cell will respond to a particular growth factor. The mechanism inside the cell appears to be very similar, however, regardless of the type of membrane receptor. Binding at the membrane surface triggers the activation and release of intermediaries known as *Ras proteins.* These proteins in turn activate intracellular enzymes and promote gene activation. Growth factors that do not use Ras proteins may use other intermediaries or may enter the cell and exert their effects directly on the nucleus.

Genetic mechanisms for inhibiting cell division have recently been identified. The genes involved are known as *repressor genes.* One gene, called *p53,* controls a protein that resides in the nucleus and activates genes that direct the production of growth-inhibiting factors inside the cell. Roughly half of all cancers are associated with abnormal forms of the *p53* gene.

Cell Division and Cancer

When the rate of cell division and growth exceeds the rate of cell death, a tissue begins to enlarge. A **tumor,** or **neoplasm,** is a mass or swelling produced by abnormal cell growth and division. In a **benign tumor,** the cells usually remain within a connective tissue capsule. Such a tumor seldom threatens an individual's life. The tumor can commonly be surgically removed if its size or position disturbs tissue function.

Cells in a **malignant tumor** no longer respond to normal control mechanisms. These cells spread into surrounding tissues from the *primary tumor* (or *primary neoplasm*). This process is called **invasion.** Cancer cells may also travel to distant tissues and organs and establish *secondary tumors.* This dispersion is called **metastasis** (me-TAS-ta-sis; *meta,* after + *stasis,* standing still). Metastasis is dangerous and difficult to control.

The term **cancer** refers to an illness characterized by malignant cells. Cancer develops in the series of steps diagrammed in Figure 3-28●. Initially, the cancer cells are restricted to the primary tumor. In most

TABLE 3-3 Representative Chemical Factors Affecting Cell Division

Factor	Source	Effect	Target
Growth hormone	Anterior pituitary gland	Stimulation of growth, cell division, differentiation	All cells, especially in epithelia and connective tissues
Prolactin	Anterior pituitary gland	Stimulation of growth, cell division, development	Gland and duct cells of mammary glands
Nerve growth factor (NGF)	Salivary glands; other sources suspected	Stimulation of nerve cell repair and development	Neurons and glial cells
Epidermal growth factor (EGF)	Duodenal glands; other sources suspected	Stimulation of stem cell divisions and epithelial repairs	Epidermis of skin
Fibroblast growth factor (FGF)	Unknown	Division and differentiation of fibroblasts and related cells	Connective tissues
Erythropoietin	Kidneys (primary source)	Stimulation of stem cell divisions and maturation of red blood cells	Bone marrow
Thymosins and related compouds	Thymus	Stimulation of division and differentiation of lymphocytes (especially T cells)	Thymus and other lymphoid tissues and organs
Chalones	Many tissues	Inhibition of cell division	Cells in the immediate area

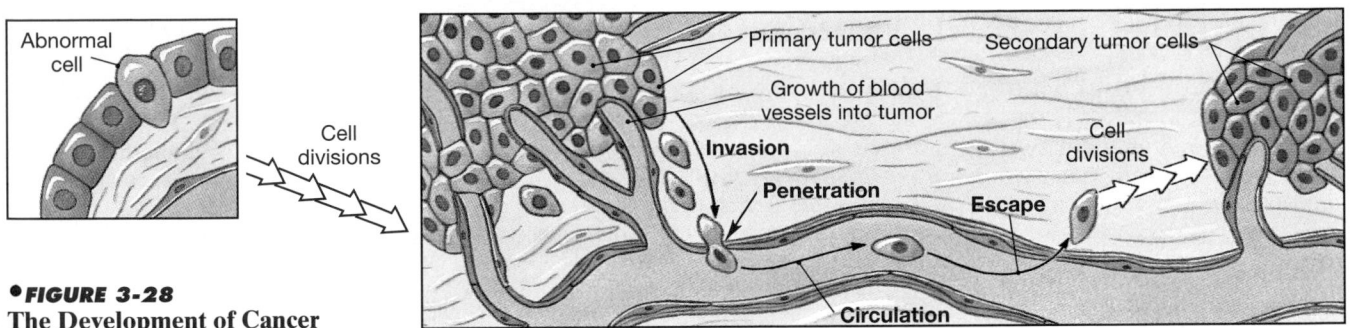

●FIGURE 3-28
The Development of Cancer

cases, all the cells in the tumor are the daughter cells of a single malignant cell. Cancer cells gradually lose their resemblance to normal cells. They change size and shape, typically becoming abnormally large or small. At first, the growth of the primary tumor distorts the tissue, but the basic tissue organization remains intact. Metastasis begins with invasion as tumor cells "break out" of the primary tumor and invade the surrounding tissue. They may then enter the lymphatic system and accumulate in nearby lymph nodes. When metastasis involves the penetration of blood vessels, the cancer cells circulate throughout the body.

Responding to cues that are as yet unknown, cancer cells within the circulatory system ultimately escape out of the blood vessels to establish secondary tumors at other sites. These tumors are extremely active metabolically, and their presence stimulates the growth of blood vessels into the area. The increased circulatory supply provides additional nutrients and further accelerates tumor growth and metastasis.

Organ function begins to deteriorate as the number of cancer cells increases. The cancer cells may not perform their original functions at all, or they may perform normal functions in an unusual way. For example, endocrine cancer cells may produce normal hormones but in excessively large amounts. Cancer cells do not use energy very efficiently. They grow and multiply at the expense of healthy tissues, competing for space and nutrients with normal cells. This competition accounts for the starved appearance of many patients in the late stages of cancer. Death may occur as a result of the compression of vital organs when nonfunctional cancer cells have killed or replaced the healthy cells in those organs or when the cancer cells have starved normal tissues of essential nutrients.

The growth of blood vessels into the tumor is a vital step in the development and spread of the cancer. Without those vessels, the growth and metastasis of the cancer cells will be limited by the availability of oxygen and nutrients. A peptide called *antiangiogenesis factor* can prevent the growth of blood vessels and can slow the growth of cancers. This peptide, produced in normal human cartilage, can be extracted in large quantities from sharks, whose skeletons are entirely cartilaginous.

Sharks are now being collected to obtain antiangiogenesis factor for use in experimental cancer therapies. We will return to the subject of cancer in later chapters that deal with specific systems, and the *Applications Manual* contains detailed information on cancer formation, development, and treatment. [AM] *A Closer Look: Cancer*

CELL DIVERSITY AND DIFFERENTIATION

The liver cells, fat cells, and neurons of an individual contain the same chromosomes and genes, but in each case a different set of genes has been turned *off*. In other words, liver cells differ from other cells because different genes are accessible for transcription.

When a gene is functionally eliminated, the cell loses the ability to create a particular protein—and thus to perform any functions involving that protein. Each time another gene switches off, the cell's functional abilities become more restricted. This specialization process is called **differentiation**.

Fertilization produces a single cell with all its genetic potential intact. A period of repeated cell divisions follows, and differentiation begins as the number of cells increases. Differentiation produces specialized cells with limited capabilities. These cells form organized collections known as *tissues,* each with discrete functional roles. In Chapter 4, we examine the structure and function of tissues and consider the role of tissue interactions in the maintenance of homeostasis.

☑ A living cell is actively manufacturing enough organelles to serve two functional cells. This cell is probably in what phase of its life cycle?

☑ During the process of DNA replication, a nucleotide was deleted from a sequence that coded for a polypeptide. What effect would this deletion have on the amino acid sequence of the polypeptide?

☑ What would happen if spindle fibers failed to form in a cell during mitosis?

SELECTED CLINICAL TERMINOLOGY

Terms Discussed in This Chapter

benign tumor: A mass or swelling in which the cells usually remain within a connective tissue capsule; rarely life-threatening. *(p. 102)*

cancer: An illness characterized by the metastasis of malignant tumor cells. *(p. 102)*

dextran: A carbohydrate that cannot cross cell membranes; often administered in solution to patients after blood loss or dehydration. *(p. 76)*

DNA fingerprinting: Identifying an individual on the basis of repeating nu-cleotide sequences in his or her DNA. *(p. 93)*

invasion: The spread of cancer cells from a primary tumor into surrounding tissues. *(p. 102)*

malignant tumor: A mass or swelling in which the cells no longer respond to normal control mechanisms but divide rapidly *(p. 102)*

metastasis (me-TAS-ta-sis): The spread of malignant cells into distant tissues and organs, where secondary tumors subsequently develop. *(p. 102)*

normal saline: A solution that approximates the normal osmotic concentration of extracellular fluids. *(p. 76)*

primary tumor *(primary neoplasm)*: The site at which a cancer cell initially develops. *(p. 102)*

secondary tumor: A colony of cancerous cells formed by metastasis. *(p. 102)*

tumor (neoplasm): A mass or swelling produced by abnormal cell growth and division. *(p. 102)*

AM Additional Terms Discussed in the *Applications Manual*

carcinogen (kar-SIN-ō-jen): An environmental factor that stimulates the conversion of a normal cell to a cancer cell.

genetic engineering: The popular term for research and experiments related to changing the genetic makeup of an organism.

karyotyping (KAR-ē-ō-tī-ping): The determination of the shape, size, and cen-tromere position of an individual's chromosomes.

mutagen (MŪ-ta-jen): A factor that can damage DNA strands and sometimes cause chromosomal breakage, stimulating the development of cancer cells.

oncogene (ON-kō-jēn): A cancer-causing gene created by a somatic mutation in a normal gene involved with growth, differentiation, or cell division.

recombinant DNA: DNA created by splicing a specific gene from one organism into the DNA strand of another organism.

CHAPTER REVIEW

On-line resources for this chapter are on our World Wide Web site at:
http://www.prenhall.com/martini/fap

STUDY OUTLINE

INTRODUCTION, p. 66

1. Contemporary *cell theory* incorporates several basic concepts: (1) Cells are the building blocks of all plants and animals; (2) cells are produced by the division of preexisting cells; (3) cells are the smallest units that perform all vital physiological functions; (4) each cell maintains homeostasis at the cellular level; and (5) homeostasis at the tissue, organ, organ system, and organism levels reflects the combined and coordinated actions of many cells. *(Figure 3-1)*

STUDYING CELLS, p. 66

1. **Cytology** is the study of the structure and function of cells; it is part of **cell biology**.

An Overview of Cellular Anatomy, p. 67

2. The human body contains two types of cells: **sex cells** (sperm or oocytes) and **somatic cells** (all other cells).

THE CELL MEMBRANE, p. 68

1. A typical cell is surrounded by **extracellular fluid,** specifically the **interstitial fluid** of a tissue. The cell's outer boundary is the **cell membrane,** or **plasma membrane.** The membrane's functions include physical isolation, regulation of exchange with the environment, sensitivity, and structural support. *(Figure 3-2; Table 3-1)*

Membrane Structure, p. 69

2. The cell membrane, which is a **phospholipid bilayer,** contains lipids, proteins, and carbohydrates. Phospholipids are the largest contributors to membrane structure. *(Figure 3-3)*

3. **Integral proteins** are part of the membrane itself; **peripheral proteins** are attached but can separate from it.

4. Membrane proteins can act as anchors, identifiers (**recognition proteins**), enzymes, receptors (**receptor proteins**), carriers (**carrier proteins**), or **channels.** Channels allow water and ions to move across the membrane. Some are called **gated channels** because they can open or close to regulate ion passage; other channels are **leak channels**. *(Figure 3-4)*

Membrane Permeability, p. 72

5. **Permeability** is the ease with which substances can cross the cell membrane. Nothing can pass through an **impermeable** barrier; anything can pass through a **freely permeable** barrier. Cell membranes are **selectively permeable.**

6. **Diffusion** is the net movement of material from an area where its concentration is relatively high to an area where its concentration is lower. Diffusion occurs until the **concentration gradient** is eliminated. *(Figures 3-5, 3-6)*

7. The diffusion of water across a membrane in response to differences in solute concentration is **osmosis. Osmotic**

pressure is the force of water movement into a solution as the result of osmotic forces. **Hydrostatic pressure** can oppose osmotic pressure. Water molecules undergo *bulk flow*, movement in groups across a membrane. *(Figures 3-7, 3-8)*

8. **Tonicity** describes the effects of osmotic solutions on living cells. A solution that does not cause an osmotic flow is **isotonic**. A solution that causes water to flow into a cell is **hypotonic** and can lead to **hemolysis**. A solution that causes water to flow out of a cell is **hypertonic** and can lead to **crenation**.

9. In **filtration,** hydrostatic pressure forces water across a membrane. If membrane pores are large enough, molecules of solute will be carried along.

10. In **facilitated diffusion,** a type of **carrier-mediated transport,** compounds are transported across a membrane after binding to a **receptor site** of a carrier protein. *(Figure 3-9)*

11. **Active transport** mechanisms consume ATP but are independent of concentration gradients. Some **ion pumps** are **exchange pumps. Secondary active transport** may involve **cotransport** or **countertransport.** *(Figure 3-9)*

12. In **vesicular transport,** material moves into or out of the cell in membranous sacs called **vesicles.** Movement into the cell is accomplished through **endocytosis,** an active process that can take three forms: **receptor-mediated endocytosis** (by means of **coated vesicles**), **pinocytosis,** and **phagocytosis** (using **pseudopodia**). The ejection of materials from the cytoplasm is accomplished through **exocytosis.** *(Figures 3-10, 3-11, 3-18; Table 3-2)*

13. The **potential difference** between the two sides of a cell membrane is a **transmembrane potential.** The transmembrane potential in an undisturbed cell is its **resting potential.** *(Figure 3-12)*

THE CYTOPLASM, p. 82

The Cytosol, p. 82

1. The **cytoplasm** contains a fluid **cytosol** that surrounds **organelles.**

Organelles, p. 82

2. **Nonmembranous organelles** are always in contact with the cytosol. They include the cytoskeleton, microvilli, centrioles, cilia, flagella, and ribosomes. *(Table 3-1)*

3. **Membranous organelles** are surrounded by lipid membranes that isolate them from the cytosol. They include the mitochondria, endoplasmic reticulum, Golgi apparatus, lysosomes, and peroxisomes. *(Table 3-1)*

4. The **cytoskeleton** gives the cytoplasm strength and flexibility. It has four components: **microfilaments** (typically made of **actin), intermediate filaments, thick filaments** (made of **myosin**), and **microtubules** (made of **tubulin**). *(Figure 3-13)*

5. **Microvilli** are small projections of the cell membrane that increase the surface area exposed to the extracellular environment. *(Figure 3-13)*

6. **Centrioles** direct the movement of chromosomes during cell division and organize the cytoskeleton. The **centrosome** is the cytoplasm surrounding the centrioles. *(Figure 3-14)*

7. **Cilia,** anchored by a **basal body,** beat rhythmically to move fluids or secretions across the cell surface. *(Figure 3-14)*

8. **Flagella** move a cell through surrounding fluid rather than moving fluid past a stationary cell.

9. A **ribosome,** composed of **light** and **heavy ribosomal subunits** that contain **ribosomal RNA (rRNA)**, is an intracellular factory that manufactures proteins. There are **free**

ribosomes in the cytoplasm and **fixed ribosomes** attached to the endoplasmic reticulum. *(Figure 3-15)*

10. **Mitochondria** are responsible for 95 percent of the ATP production within a typical cell. The fluid contents, or **matrix,** of a mitochondrion lie inside the folds, or **cristae,** of an inner membrane. *(Figure 3-16)*

11. The **endoplasmic reticulum (ER)** is a network of intracellular membranes that function in synthesis, storage, transport, and detoxification. The ER forms hollow tubes, flattened sheets, and round chambers called **cisternae.** There are two types: rough and smooth. **Rough endoplasmic reticulum (RER)** contains ribosomes on its outer surface and forms **transport vesicles; smooth endoplasmic reticulum (SER)** does not. *(Figure 3-17)*

12. The **Golgi apparatus** moves materials via **transfer vesicles,** forms **secretory vesicles** and new membrane components, and packages lysosomes. Secretions are discharged from the cell through exocytosis. *(Figure 3-18)*

13. **Lysosomes** are vesicles filled with digestive enzymes.

14. **Peroxisomes** carry enzymes that absorb and neutralize toxins.

THE NUCLEUS, p. 91

1. The **nucleus** is the control center of cellular operations. It is surrounded by a **nuclear envelope** (a double membrane with a **perinuclear space**), through which it communicates with the cytosol by way of **nuclear pores.** The nucleus contains a supportive **nuclear matrix.** *(Figure 3-19)*

Chromosome Structure, p. 92

2. The nucleus controls the cell by directing the synthesis of specific proteins using information stored in **chromosomes,** which consist of DNA bound to **histones.** In nondividing cells, chromosomes form a tangle of filaments called **chromatin.** The nucleus may also contain **nucleoli.** *(Figures 3-19, 3-20)*

The Genetic Code, p. 92

3. The cell's information storage system, the **genetic code,** is called a *triplet code* because a sequence of three nitrogenous bases identifies a single amino acid. Each **gene** contains all the triplets needed to produce a specific polypeptide chain.

Gene Activation and Protein Synthesis, p. 93

4. Before a gene can be activated to begin **protein synthesis, RNA polymerase** must bind to the gene. Protein synthesis consists of two processes: transcription and translation.

Transcription, p. 93

5. **Transcription** is the process of forming RNA from a DNA template. After mRNA transcription, a strand of **messenger RNA (mRNA)** carries instructions from the nucleus to the cytoplasm. *(Figures 3-21, 3-22)*

Translation, p. 94

6. During **translation,** a functional polypeptide is constructed by using the information contained in the sequence of **codons** along an mRNA strand. The sequence of codons determines the sequence of amino acids in the polypeptide. Translation proceeds in three steps: *initiation, elongation,* and *termination.* *(Figures 3-23, 3-24)*

7. By complementary base pairing of **anticodons** to mRNA codons, molecules of **transfer RNA (tRNA)** bring amino acids to the ribosomal complex. *(Figures 3-23, 3-24)*

DNA and the Control of Cell Structure and Function, p. 96

8. The DNA of the nucleus has both direct and indirect control over protein synthesis.

THE CELL LIFE CYCLE, p. 97

1. **Cell division** is the reproduction of cells; **apoptosis** is the genetically controlled death of cells. **Mitosis** refers to the nuclear division of somatic cells. Sex cells (sperm and oocytes) are produced by **meiosis.**

Interphase, p. 98

2. Most somatic cells spend most of their time in **interphase**, which includes **DNA replication.** *(Figures 3-25, 3-26)*

Mitosis, p. 99

3. Mitosis proceeds in four stages: **prophase, metaphase, anaphase,** and **telophase.** *(Figure 3-27)*

Cytokinesis, p. 101

4. During **cytokinesis,** the cytoplasm is divided, generally producing two identical **daughter cells.** *(Figure 3-27)*

The Mitotic Rate and Energetics, p. 101

5. In general, the longer the life expectancy of a cell type, the slower the **mitotic rate. Stem cells** undergo frequent mitoses to replace other, more-specialized cells.

6. Cell division is genetically controlled.

Regulation of the Cell Life Cycle, p. 102

7. A variety of **growth factors** can stimulate cell division and growth. *(Table 3-3)*

8. Produced by abnormal cell growth and division, a **tumor,** or **neoplasm,** can be **benign** or **malignant.** Malignant cells undergo **invasion** and possibly **metastasis. Cancer** is the illness caused by malignant cells. *(Figure 3-28)*

CELL DIVERSITY AND DIFFERENTIATION, p. 103

1. **Differentiation** is the process of specialization that is due to gene regulation and produces cells with limited capabilities. These specialized cells form organized collections called *tissues,* each of which has certain functional roles.

REVIEW QUESTIONS

Level 1   Reviewing Facts and Terms

1. All the following membrane transport mechanisms are passive processes except
   (a) diffusion
   (b) facilitated diffusion
   (c) vesicular transport
   (d) filtration

2. The principal cations in body fluids are
   (a) calcium and magnesium
   (b) chloride and bicarbonate
   (c) sodium and potassium
   (d) sodium and chloride

3. _____ ion concentrations are high in the extracellular fluids, and _____ ion concentrations are high in the cytoplasm:
   (a) calcium, magnesium
   (b) chloride, sodium
   (c) potassium, sodium
   (d) sodium, potassium

4. In a resting transmembrane potential, the inside of the cell is _____ _____, and the cell exterior is _____:
   (a) slightly negative, slightly positive
   (b) slightly positive, slightly negative
   (c) slightly positive, neutral
   (d) slightly negative, neutral

5. The process that generates carbon dioxide while removing hydrogen atoms from organic molecules is
   (a) the TCA cycle
   (b) active transport
   (c) glycolysis
   (d) phosphorylation

6. The reaction sequence in which glucose is broken down into pyruvic acid is
   (a) aerobic metabolism
   (b) the TCA cycle
   (c) mitochondrial energy production
   (d) glycolysis

7. The construction of a functional polypeptide by using the information provided by an mRNA strand is
   (a) translation
   (b) transcription
   (c) replication
   (d) gene activation

8. Our somatic cell nuclei contain _____ pairs of chromosomes:
   (a) 8   **(b)** 16
   (c) 23   **(d)** 46

9. Termination is the final stage in the production of a(n)
   (a) DNA molecule
   (b) mRNA molecule
   (c) tRNA molecule
   (d) protein

10. The term *differentiation* refers to
   (a) the loss of genes from cells
   (b) the acquisition of new functional capabilities by cells
   (c) the production of functionally specialized cells
   (d) the division of genes among different types of cells

11. The interphase of the cell life cycle is divided into the following phases:
   (a) prophase, metaphase, anaphase, and telophase
   (b) G_0, G_1, S, and G_2
   (c) mitosis and cytokinesis
   (d) a, b, and c are correct

12. List the five basic concepts that make up the modern-day cell theory.

13. What are four general functions of the cell membrane?

14. What are the primary functions of membrane proteins?

15. By what four major transport mechanisms do substances get into and out of cells?

16. List four important factors that influence diffusion rates.

17. State three general characteristics of osmosis that define its role in the body.

18. List **(a)** the nonmembranous organelles and **(b)** the membranous organelles of a typical cell.
19. What are the four major functions of the endoplasmic reticulum?

20. List the four stages of mitosis in their correct order of occurrence.

Level 2   Reviewing Concepts

21. Diffusion is important in body fluids because it tends to
  (a) increase local concentration gradients
  (b) eliminate local concentration gradients
  (c) move substances against concentration gradients
  (d) create concentration gradients
22. Osmotic pressure differs from hydrostatic pressure because the osmotic pressure of a solution is an indication of the force of water movement resulting from
  (a) its solute concentration
  (b) the volume of water
  (c) the permeability of the membrane
  (d) a, b, and c are correct
23. When a cell is placed in a _____ solution, the cell will lose water through osmosis. The process results in the _____ of red blood cells:
  (a) hypotonic, crenation   **(b)** hypertonic, crenation
  (c) isotonic, hemolysis   **(d)** hypotonic, hemolysis
24. Suppose that a DNA segment has the following nucleotide sequence: CTC ATA CGA TTC AAG TTA. Which nucleo-tide sequences would be found in a complementary mRNA strand?
  (a) GAG UAU GAU AAC UUG AAU
  (b) GAG TAT GCT AAG TTC AAT
  (c) GAG UAU GCU AAG UUC AAU
  (d) GUG UAU GGA UUG AAC GGU

25. How many amino acids are coded in the DNA segment in Question 24?
  (a) 18   **(b)** 9
  (c) 6   **(d)** 3
26. How does hydrostatic pressure differ from osmotic pressure?
27. What general characteristics are important in carrier-mediated transport mechanisms?
28. **(a)** What are the similarities between facilitated diffusion and active transport?
  (b) What are the differences?
29. What role does the sodium–potassium exchange pump play in stabilizing the resting membrane potential?
30. How does the cytosol differ in composition from the interstitial fluid?
31. Differentiate between transcription and translation.
32. List in sequence the phases of the interphase state of the cell life cycle, and briefly describe what happens in each.
33. List the stages of mitosis, and briefly describe the events that occur in each.
34. **(a)** What is cytokinesis?
  (b) What is its role in the cell cycle?

Level 3   Critical Thinking and Clinical Applications

35. Experimental evidence shows that the transport of a certain molecule exhibits the following characteristics: (1) The molecule moves along its concentration gradient; (2) at concentrations above a given level, there is no increase in the rate of transport; and (3) cellular energy is not required for transport to occur. What type of transport process is at work?
36. Solutions A and B are separated by a semipermeable barrier. Over time, the level of fluid on side A increases. Which solution initially had the higher concentration of solute?

37. In kidney dialysis, a person's blood is passed through a bath that contains a mixture of several ions and molecules. The blood is separated from the dialysis fluid by a membrane that allows water, small ions, and small molecules to pass but does not allow large proteins or blood cells to pass. What should the composition of the dialysis fluid be for it to remove urea (a small molecule) without changing blood volume (removing water from the blood)?

CHAPTER 4

The Tissue Level of Organization

*E*xotic creatures on the deep-sea floor? Well, no. The "creatures" that you see here are much smaller than the strange life forms photographed in the ocean depths, and they are much closer to home. These are cells that line the trachea, the airway leading to your lungs. The "tentacles" are cilia that help remove dirt and harmful microbes from air you inhale. Notice that the tracheal lining contains more than one type of cell; some cells have long cilia, and others stumpy microvilli. Groups of cells specialized to perform a particular set of functions are called tissues, and these cells form an epithelium, one of the four primary tissue types. Each of your body's tissues has its own distinctive structure and its own role to play in maintaining homeostasis. We will meet all of them in this chapter.

CHAPTER OUTLINE AND OBJECTIVES

No single cell contains the metabolic machinery and organelles needed to perform all the many functions of the human body. Instead, through the process of differentiation, each cell develops a characteristic set of structural features and a limited number of functions. ∞ *[p. 103]* These structures and functions can be quite distinct from those of nearby cells. Nevertheless, all the cells in a given location work together.

A detailed examination of the body reveals a number of patterns at the cellular level. Although there are trillions of cells in the human body, there are only about 200 types of cells. These cell types combine to form **tissues,** collections of specialized cells and cell products that perform a relatively limited number of functions. **Histology** is the study of tissues.

There are four basic *tissue types: epithelial tissue, connective tissue, muscle tissue,* and *neural tissue:*

1. *Epithelial tissue* covers exposed surfaces, lines internal passageways and chambers, and forms glands.
2. *Connective tissue* fills internal spaces, provides structural support for other tissues, transports materials within the body, and stores energy reserves.
3. *Muscle tissue* contracts to perform specific movements and in the process generates heat that warms the body.
4. *Neural tissue* carries information from one part of the body to another in the form of electrical impulses.

In this chapter, we will discuss the characteristics of each tissue type, focusing on the relationship between cellular organization and tissue function. In later chapters, our primary interest will be the patterns of tissue interaction in various organs and organ systems.

EPITHELIAL TISSUE

Epithelial tissue includes *epithelia* and *glands,* secretory structures derived from epithelia. An **epithelium** (e-pi-THĒ-lē-um) is a layer of cells that forms a barrier with specific properties. Epithelia cover every exposed body surface. The surface of the skin is a good example, but epithelia also line the digestive, respiratory, reproductive, and urinary tracts—passageways that communicate with the outside world. Epithelia also line internal cavities and passageways, such as the chest cavity, fluid-filled chambers in the brain, eye, and inner ear, and the inner surfaces of blood vessels and the heart.

Important characteristics of epithelia include:

- *Cellularity.* Epithelia are composed almost entirely of cells bound closely together by one or more types of cell junctions. In other tissue types, the cells are often widely separated by extracellular materials.
- *Polarity.* An epithelium always has an exposed surface, or *apical surface*, that faces the exterior of the body or some internal space. It also has an attached *basal surface* where the epithelium is attached to underlying tissues. Organelles and other cytoplasmic components are usually distributed unevenly between the apical and basal surfaces of the cell. This property is called **polarity.**
- *Attachment.* The basal surface of an epithelium is bound to a thin **basement membrane**. The basement membrane is a complex structure produced by the basal surface of the epithelium and the underlying connective tissue.
- *Avascularity.* Epithelia do not contain blood vessels. Because of this **avascular** (ā-VAS-kū-lar; *a-,* without + *vas,* vessel) condition, epithelial cells must obtain nutrients by diffusion or absorption across the apical or basal surface.
- *Regeneration.* Epithelial cells damaged or lost at the apical surface are continuously being replaced through the divisions of stem cells within the epithelium.

Functions of Epithelial Tissue

Epithelia perform essential functions that can be summarized as follows:

1. *Provide physical protection.* Epithelia protect exposed and internal surfaces from abrasion, dehydration, and destruction by chemical or biological agents.
2. *Control permeability.* Any substance that enters or leaves your body has to cross an epithelium. Some epithelia are relatively impermeable, whereas others are easily crossed by compounds as large as proteins. Many epithelia contain the molecular "machinery" needed for selective absorption or secretion. The epithelial barrier can be regulated and modified in response to various stimuli. For example, hormones can affect the transport of ions and nutrients through epithelial cells. Even physical stress can alter the structure and properties of epithelia—think of the calluses that form on your hands when you do rough physical work for a period of time.
3. *Provide sensation.* Most epithelia are extensively innervated by sensory nerves—that is, they have a large sensory nerve supply. Specialized epithelial cells can detect changes in their environment and can convey information about such changes to the nervous system. For example, touch receptors in the deepest epithelial layers of the skin respond to pressure by stimulating adjacent sensory nerves. A *neuroepithelium* is an epithelium containing sensory cells that provide sensations of smell, taste, sight, equilibrium, and hearing.
4. *Produce specialized secretions.* Epithelial cells that produce secretions are called *gland cells*. Individual gland cells are typically scattered among other cell types in an epithelium. In a **glandular epithelium**, most or all of the epithelial cells produce secretions. We can classify these secretions according to their discharge location:

a. *Exocrine* (*exo-*, outside + *krinein*, to secrete) *secretions* are discharged onto the surface of the skin or other epithelial surface. Enzymes entering the digestive tract, perspiration on the skin, and milk produced by mammary glands are examples.

b. *Endocrine* (*endo-*, inside) *secretions* are released into the surrounding interstitial fluid and blood. These secretions, called *hormones*, regulate or coordinate the activities of other tissues, organs, and organ systems. Endocrine secretions are produced in organs such as the pancreas, thyroid, and anterior pituitary gland.

Specializations of Epithelial Cells

Epithelial cells have several specializations that distinguish them from other body cells. Many epithelial cells are specialized for (1) the production of secretions, (2) the movement of fluids over the epithelial surface, or (3) the movement of fluids through the epithelium. These specialized epithelial cells generally show a strong polarity. The actual arrangement of organelles varies with the functions of the epithelium. The cells in Figure 4-1a● show a common type of polarity.

Many epithelial cells that line internal passageways have microvilli on their exposed surfaces (Figure 4-1b●). There may be just a few, or microvilli may carpet the entire surface. Microvilli are especially abundant on epithelial surfaces where absorption and secretion take place, such as along portions of the digestive and urinary tracts. The epithelial cells in these locations are transport specialists. A cell with microvilli has at least 20 times more surface area available for material transport than a cell without them. **Stereocilia** are structurally similar to microvilli, but they are very long (up to 250 μm) and incapable of active movement. These structures occur in two places only: (1) along portions of the male reproductive tract and (2) on receptor cells of the inner ear.

Figure 4-1b● shows the surface of a **ciliated epithelium.** A typical ciliated cell contains about 250 cilia that beat in a coordinated fashion. Substances are moved over the epithelial surface by the synchronized beating of cilia, like a continuously moving escalator. For example, the ciliated epithelium that lines the respiratory tract moves mucus up from the lungs and toward the throat. The mucus traps foreign particles such as dust, pollen, and pathogens and carries them away from more delicate surfaces deeper in the lungs. Injury to the cilia or to the epithelial cells in general can stop ciliary movement and block the protective

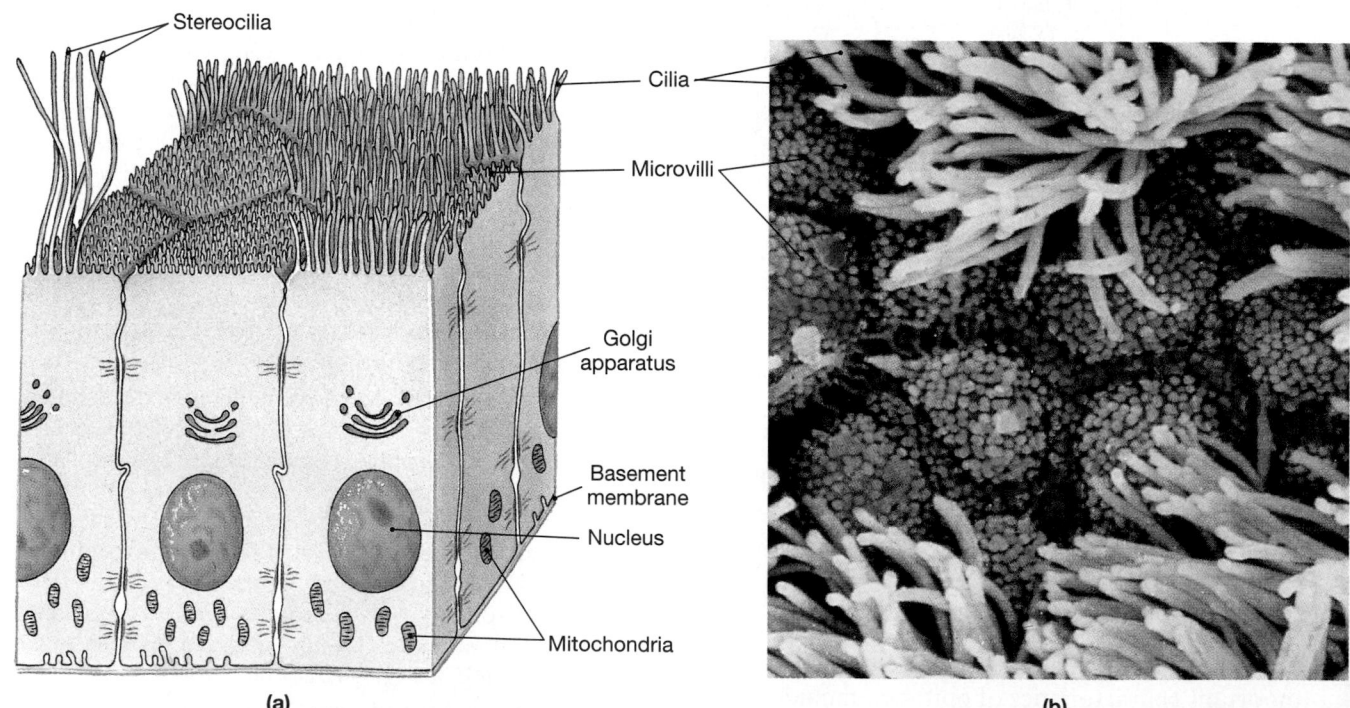

(a)   (b)

●*FIGURE 4-1* **Polarity of Epithelial Cells.** **(a)** Many epithelial cells differ in internal organization along an axis between the free surface (here, the top) and the basement membrane. In many cases, the free surface bears microvilli; in some cases, this surface may have cilia or (very rarely) stereocilia. (All three would not normally be found on the same group of cells but are depicted here for purposes of illustration.) In some epithelia, such as the lining of the kidney tubules, mitochondria are concentrated near the base of the cell, probably to provide energy for the cell's transport activities. **(b)** Micrograph showing the surface of a ciliated epithelium that lines most of the respiratory tract. The small, bristly areas are microvilli on the exposed surfaces of mucus-producing cells that are scattered among the ciliated epithelial cells. (SEM × 15,200)

flow of mucus. This is one effect of smoking (we shall examine other unpleasant effects in later sections and chapters).

Maintaining the Integrity of Epithelia

Three factors are involved in maintaining the physical integrity of an epithelium: (1) intercellular connections, (2) attachment to the basement membrane, and (3) epithelial maintenance and repair.

Intercellular Connections

Cells in an epithelium are firmly attached to one another, and the epithelium as a unit is attached to extracellular fibers of the basement membrane. Many other cells in your body form permanent or tempo-

rary bonds to other cells or extracellular materials. However, epithelial cells are specialists in intercellular connection (Figure 4-2a●).

Intercellular connections may involve extensive areas of opposing cell membranes, or they may be concentrated at specialized attachment sites. Large areas of opposing cell membranes may be interconnected by transmembrane proteins called **cell adhesion molecules (CAMs),** which bind to each other and to other extracellular materials. For example, CAMs on the attached base of an epithelium help bind the basal surface to the underlying basement membrane. The membranes of adjacent cells may also be bonded by **intercellular cement,** a thin layer of proteoglycans. These proteoglycans contain polysaccharide derivatives known as *glycosaminoglycans,* most notably **hyaluronan** (*hyaluronic acid*).

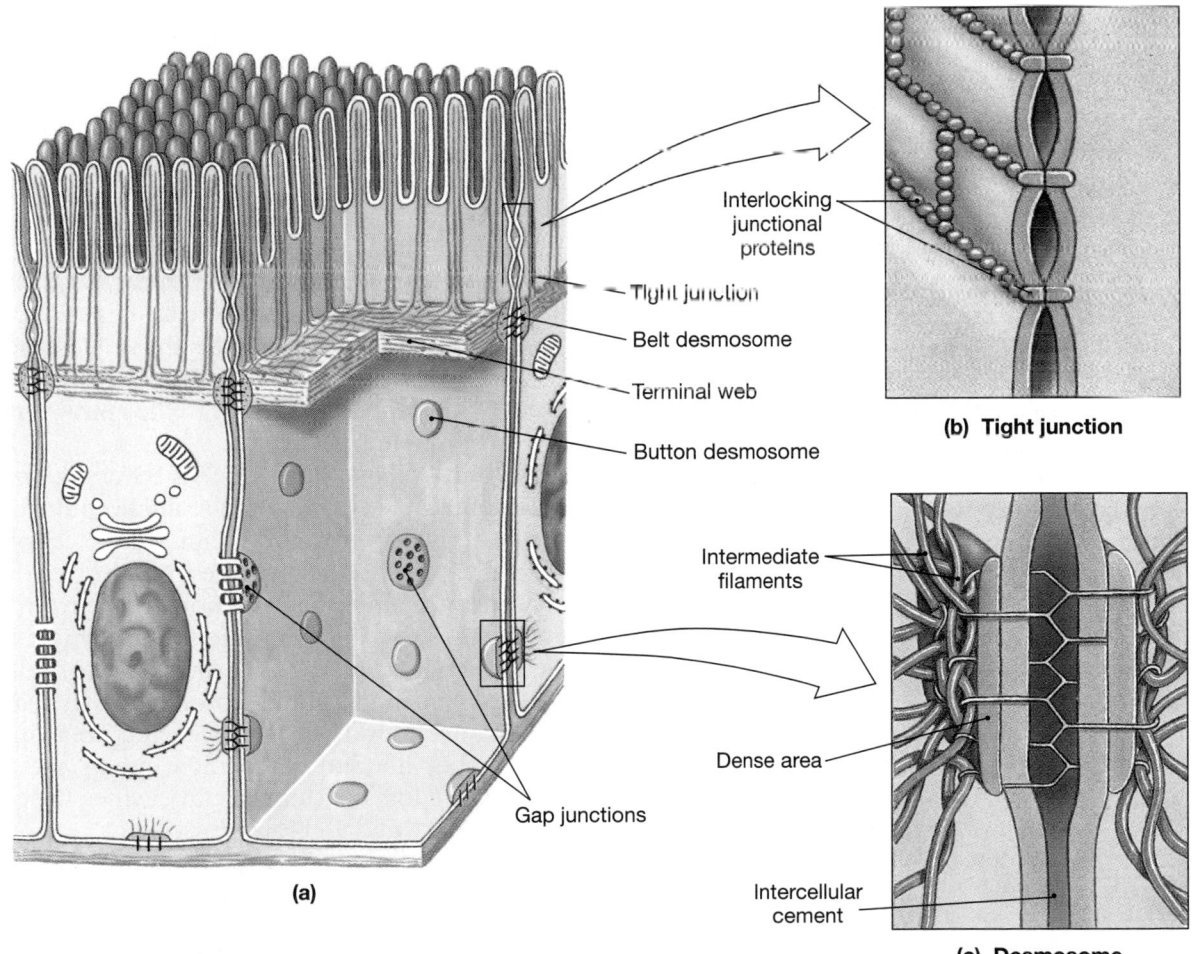

●FIGURE 4-2 Cell Attachments. (a) Diagrammatic view of an epithelial cell, showing the major types of intercellular connections. **(b)** A tight junction is formed by the fusion of the outer layers of two cell membranes. Bands of tight junctions encircle the apical portion of cuboidal and columnar epithelial cells, preventing the diffusion of fluids and solutes between the cells. **(c)** A desmosome has a more organized network of intermediate filaments. Desmosomes attach one cell to another or attach a cell to extracellular structures, such as the protein fibers in connective tissues. A continuous belt of desmosomes lies deep to the tight junctions. This belt is tied to the microfilaments of the terminal web.

There are three major types of **cell junctions:** (1) *tight junctions,* (2) *desmosomes,* and (3) *gap junctions.* At a **tight junction,** there is a partial fusion of the lipid portions of the two cell membranes (Figure 4-2b ●). Because the membranes are fused together, tight junctions block the passage of water and solutes between the cells. For example, tight junctions are found near the apical surfaces of cells that line the digestive tract, thereby keeping enzymes, acids, and wastes from damaging delicate underlying tissues.

At a **desmosome** (DEZ-mō-sōm; *desmos,* ligament + *soma,* body) there are CAMs and proteoglycans linking the opposing cell membranes. Within the cell is a complex known as a *dense area* that is connected to the cytoskeleton (Figure 4-2c●). A *belt desmosome* forms a band that encircles the cell. This band is attached to the microfilaments of the *terminal web.* ∞ *[p. 83] Button desmosomes* are small discs connected to bands of intermediate fibers. The intermediate fibers function like cross-braces to stabilize the shape of the cell. Desmosomes are very strong, and they can resist stretching and twisting. These connections are most abundant between cells in the superficial layers of the skin, where desmosomes create links so strong that dead skin cells are shed in thick sheets rather than individually. Desmosomes are also common in cardiac muscle tissue, interconnecting the muscle cells.

At a **gap junction** (Figure 4-2a●), two cells are held together by an interlocking of membrane proteins called *connexons.* Because these are channel proteins, the result is a narrow passageway that lets small molecules and ions pass from cell to cell. Gap junctions are common among epithelial cells, where they help coordinate functions such as the beating of cilia. Gap junctions are most abundant in cardiac muscle and smooth muscle tissue, in which they are essential to the coordination of muscle cell contractions.

The Basement Membrane

Not only do epithelial cells hold onto one another, but they also remain firmly connected to the rest of the body. The inner surface of each epithelium is attached to a special two-part *basement membrane.* The layer closer to the epithelium, called the **basal lamina** (LA-mi-nah; *lamina,* thin layer), contains glycoproteins and a network of fine protein filaments (Figure 4-2a●). The basal lamina, secreted by the adjacent layer of epithelial cells, provides a barrier that restricts the movement of proteins and other large molecules from the underlying connective tissue into the epithelium. The deeper portion of the basement membrane, the **reticular lamina,** contains bundles of coarse protein fibers produced by connective tissue cells. The reticular lamina gives the basement membrane its strength. Attachments between the fibers of the basal lamina and those of the reticular lamina hold the two layers together.

Epithelial Maintenance and Repair

An epithelium must continuously repair and renew itself. Epithelial cells lead hard lives, for they may be exposed to disruptive enzymes, toxic chemicals, pathogenic bacteria, or mechanical abrasion. Consider the lining of the small intestine, where epithelial cells are exposed to a variety of enzymes and abrasion from partially digested food. In this extreme environment, an epithelial cell may last just a day or two before it is shed or destroyed. The only way the epithelium can maintain its structure over time is by the continual division of *stem cells.* ∞ *[p. 101]* Most stem cells, also called **germinative cells,** are located in the deepest layers of the epithelium, near the basement membrane.

☑ Microscopic examination of an epithelial surface reveals the presence of many microvilli. What is the probable function of this epithelium?

☑ What is the functional significance of gap junctions?

☑ Why do you find the same epithelial organization in the pharynx, esophagus, anus, and vagina?

Classification of Epithelia

Epithelia are classified according to the number of cell layers and the shape of the exposed cells. The classification scheme recognizes two types of layering—*simple* and *stratified*—and three cell shapes—*squamous, cuboidal,* and *columnar.* The shape classification is based on the appearance of the cell when seen in a section that is perpendicular to the apical surface and to the basement membrane.

If only a single layer of cells covers the basement membrane, the epithelium is a **simple epithelium.** Simple epithelia are relatively thin. All the cells have the same polarity, so the distance from the nucleus to the basement membrane does not change from one cell to the next. Because they are so thin, simple epithelia are also relatively fragile. A single layer of cells cannot provide much mechanical protection, and simple epithelia are found only in protected areas inside the body. They line internal compartments and passageways, including the ventral body cavities, the chambers of the heart, and all blood vessels.

Simple epithelia are also characteristic of regions where secretion or absorption occurs, such as the lining of the intestines and the gas-exchange surfaces of the lungs. In these places, the thinness of simple epithelia is an advantage, for it reduces the time required for materials to cross the epithelial barrier.

A **stratified epithelium** has several layers of cells covering the basement membrane. Stratified epithelia are generally found in areas subject to mechanical or chemical stresses, such as the surface of the skin and the lining of the mouth.

By combining the two basic epithelial layouts (simple and stratified) and the three possible cell shapes (squamous, cuboidal, and columnar), we can describe almost every epithelium in the body. In most stratified epithelia, the shapes of the cells change as they approach the exposed surface. As a result, the classification scheme for stratified epithelia is based only on the appearance of the cells in the superficial layer.

Squamous Epithelia

In a **squamous epithelium** (SKWĀ-mus; *squama,* plate or scale) the cells are thin, flat, and somewhat irregular in shape, like puzzle pieces (Figure 4-3●). In a sec-

tional view, the nucleus occupies the thickest portion of each cell; from the surface, the cells look like fried eggs laid side by side.

A **simple squamous epithelium** is the most delicate type of epithelium in the body (Figure 4-3a●). This type of epithelium is found in protected regions where absorption takes place or where a slick, slippery surface reduces friction. Examples include the respiratory exchange surfaces *(alveoli)* of the lungs, the lining of the ventral body cavities, and the lining of the heart and blood vessels.

Special names have been given to simple squamous epithelia that line chambers and passageways that do not communicate with the outside world. The simple squa-

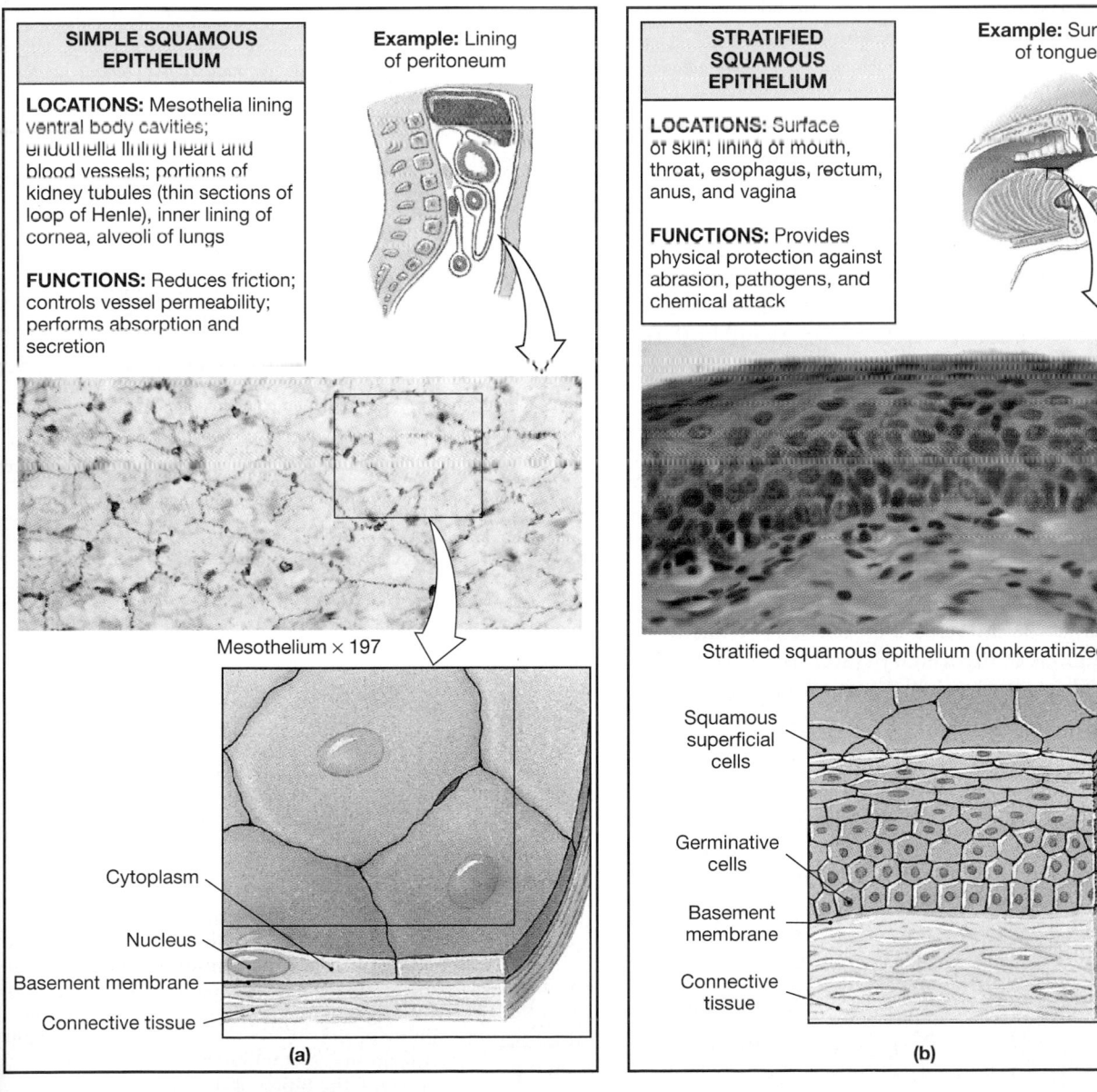

●**FIGURE 4-3 Squamous Epithelia. (a)** A superficial view of the simple squamous epithelium (mesothelium) that lines the peritoneal cavity. The three-dimensional drawing shows the epithelium in superficial and sectional view. **(b)** Sectional and diagrammatic views of the stratified squamous epithelium that covers the tongue.

mous epithelium that lines the ventral body cavities is known as a **mesothelium** (mez-ō-THĒ-lē-um; *mesos,* middle). The pleura, peritoneum, and pericardium each contain a superficial layer of mesothelium. The simple squamous epithelium lining the heart and all blood vessels is called an **endothelium** (en-dō-THĒ-lē-um).

A **stratified squamous epithelium** (Figure 4-3b●) is generally located where mechanical stresses are severe. Note how the cells form a series of layers, like a stack of plywood sheets. The surface of the skin and the lining of the mouth, esophagus, and anus are areas where this epithelial type provides protection from physical and chemical attack. On exposed body surfaces, where mechanical stress and dehydration are potential problems, the apical layers of epithelial cells are packed with filaments of the protein *keratin.* As a result, the superficial layers are both tough and water-resistant, and the epithelium is said to be *keratinized.* A *nonkeratinized* stratified squamous epithelium provides resistance to abrasion but will dry out and deteriorate unless kept moist. Nonkeratinized stratified squamous epithelia are found in the oral cavity, pharynx, esophagus, anus, and vagina.

Cuboidal Epithelia

The cells of a **cuboidal epithelium** resemble little hexagonal boxes; they appear square in typical sectional views. The nuclei are near the center of each cell. The distance between adjacent nuclei is roughly equal to the height of the epithelium. A **simple cuboidal epithelium** provides limited protection and occurs in regions where secretion or absorption takes place. Such an epithelium lines portions of the kidney tubules (Figure 4-4a●). In the pancreas and salivary glands, simple cuboidal epithelia secrete enzymes and buffers and line portions of the ducts that discharge those secretions. The thyroid gland contains chambers called *thyroid follicles* that are lined by a cuboidal secretory epithelium. Thyroid hormones accumulate within the follicles before being released into the bloodstream.

Stratified cuboidal epithelia are relatively rare; they are located along the ducts of sweat glands (Figure 4-4b●) and in the larger ducts of the mammary glands. A **transitional epithelium** (Figure 4-4c●) is unusual because, unlike most epithelia, it tolerates considerable stretching. It is called transitional because the appearance of the epithelium changes as the stretching occurs. A transitional epithelium is found in regions of the urinary system, such as the urinary bladder, where large changes in volume occur. In an empty urinary bladder, the epithelium seems to have many layers, and the outermost cells are typically plump cuboidal cells. The layered appearance results from overcrowding; the actual structure of the epithelium can be seen in the full urinary bladder, when the pressure of the urine has stretched the lining to its natural thickness.

Columnar Epithelia

Columnar epithelial cells are also hexagonal in cross section, but they are taller and more slender than cuboidal epithelial cells. The nuclei are crowded into a narrow band close to the basement membrane. The height of the epithelium is several times the distance between two nuclei (Figure 4-5●, p. 116). A **simple columnar epithelium** typically occurs where absorption or secretion is underway, such as inside the small intestine (Figure 4-5a●). Inside the stomach and large intestine, the secretions of simple columnar epithelia provide protection from chemical stresses.

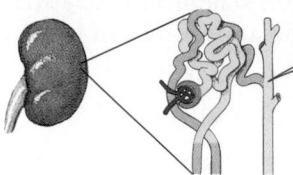

SIMPLE CUBOIDAL EPITHELIUM

LOCATIONS: Glands, ducts, portions of kidney tubules, thyroid gland
FUNCTIONS: Limited protection, secretion and/or absorption

Example: Convoluted tubule of kidney

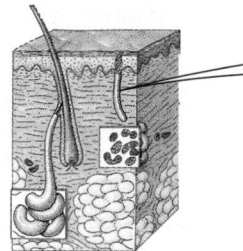

STRATIFIED CUBOIDAL EPITHELIUM

LOCATIONS: Lining of some ducts (rare)
FUNCTIONS: Protection, secretion, absorption

Example: Duct of sweat gland

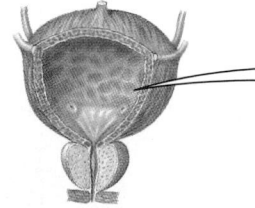

TRANSITIONAL EPITHELIUM

LOCATIONS: Urinary bladder, renal pelvis, ureters
FUNCTIONS: Permit expansion and recoil after stretching

Example: Lining of urinary bladder

●**FIGURE 4-4** Cuboidal Epithelia. **(a)** A section through the cuboidal epithelial cells of a kidney tubule. **(b)** Sectional view of the stratified cuboidal epithelium lining a sweat gland duct in the skin. **(c)** At left, the lining of the empty urinary bladder, showing transitional epithelium in the relaxed state. At right, the lining of the full bladder, showing the effects of stretching on the appearance of cells in the epithelium. (LM × 408)

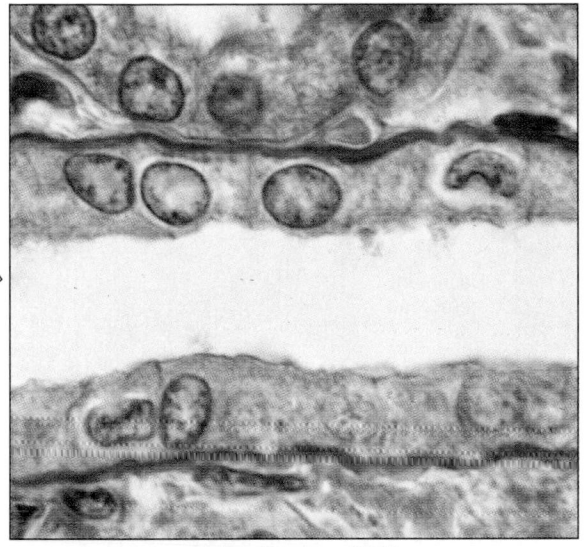
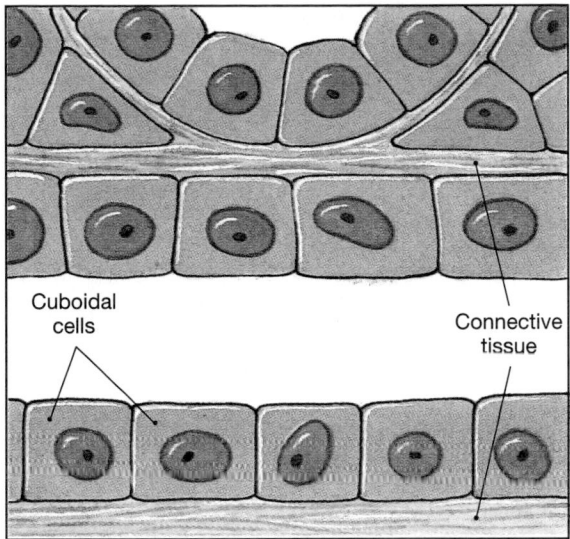

Kidney tubule × 2000

Cuboidal cells

Connective tissue

(a)

Basement membrane   Connective tissue

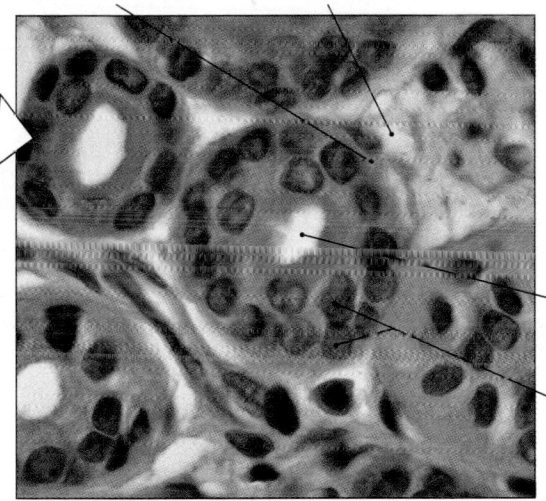
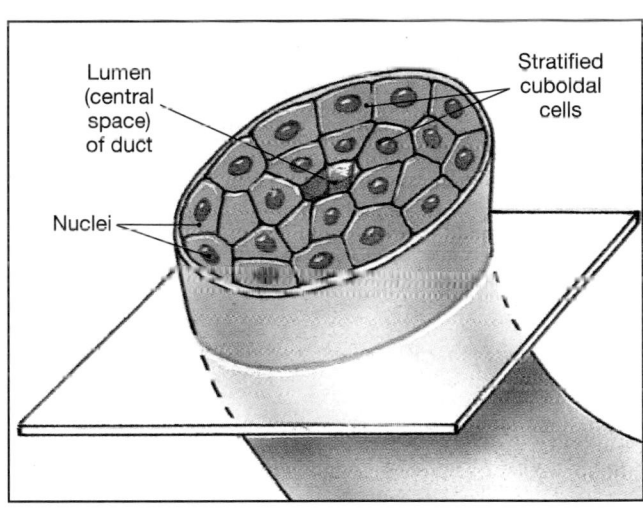

Lumen of duct

Nuclei of stratified cuboidal cells

Sweat gland duct × 1413

Lumen (central space) of duct

Stratified cuboidal cells

Nuclei

(b)

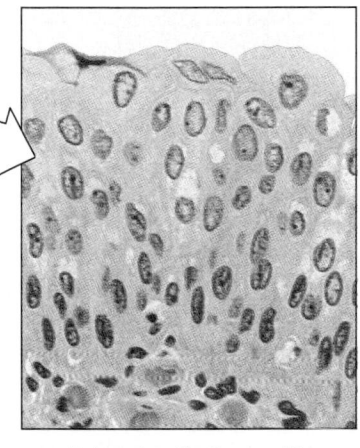
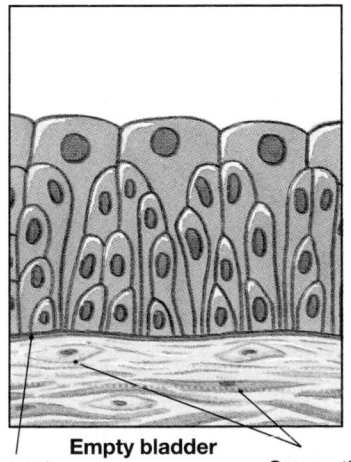
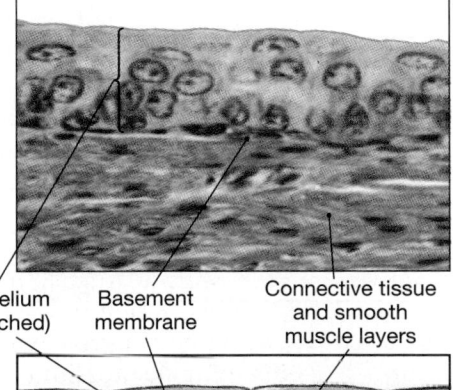

Relaxed epithelium × 454

Basement membrane

Empty bladder

Connective tissue and smooth muscle layers

Epithelium (stretched)

Basement membrane

Connective tissue and smooth muscle layers

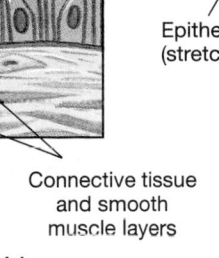

Full bladder

(c)

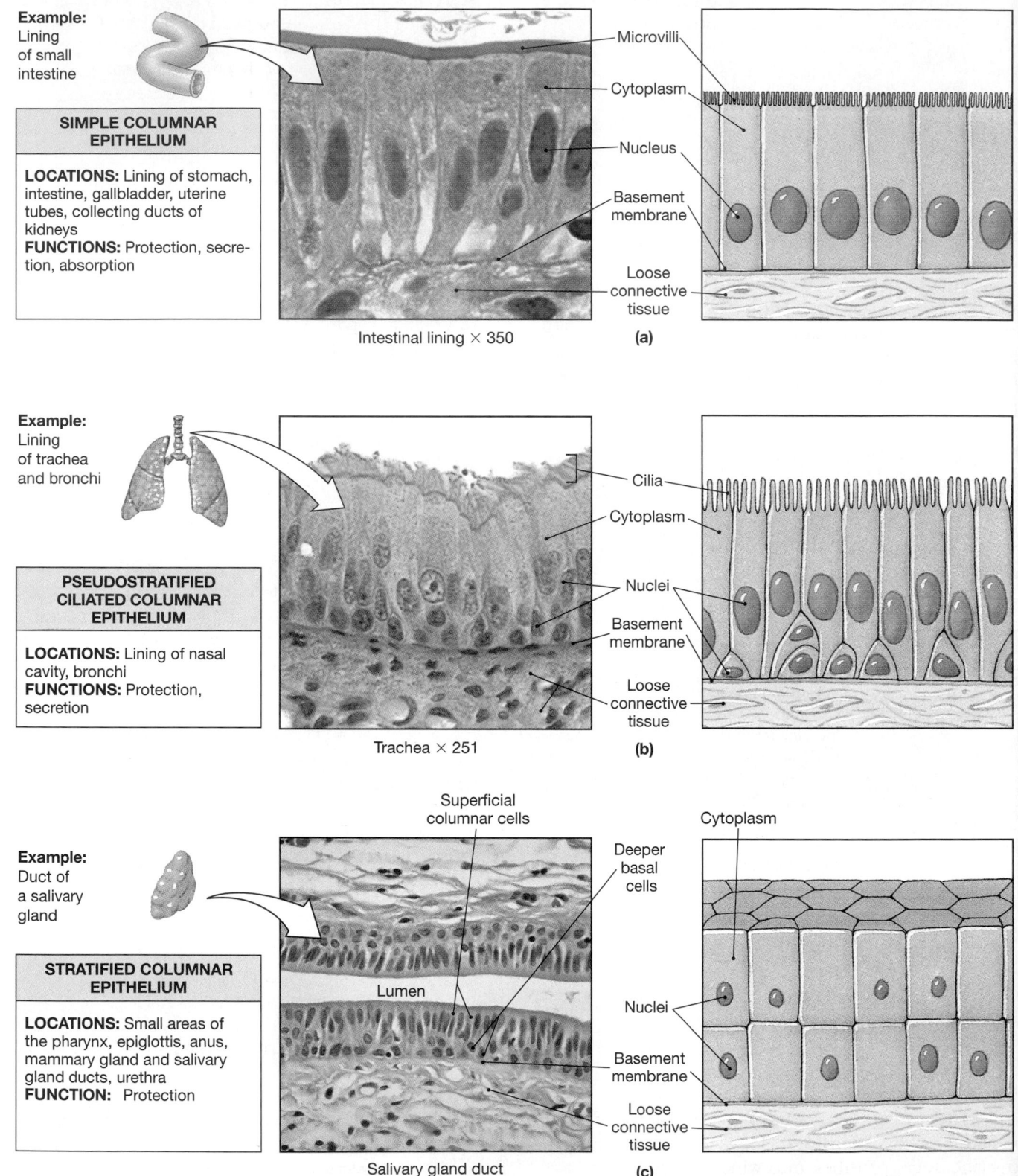

Example:
Lining
of small
intestine

**SIMPLE COLUMNAR
EPITHELIUM**

LOCATIONS: Lining of stomach,
intestine, gallbladder, uterine
tubes, collecting ducts of
kidneys
FUNCTIONS: Protection, secretion, absorption

Microvilli

Cytoplasm

Nucleus

Basement
membrane

Loose
connective
tissue

Intestinal lining × 350

(a)

Example:
Lining
of trachea
and bronchi

**PSEUDOSTRATIFIED
CILIATED COLUMNAR
EPITHELIUM**

LOCATIONS: Lining of nasal
cavity, bronchi
FUNCTIONS: Protection,
secretion

Cilia

Cytoplasm

Nuclei

Basement
membrane

Loose
connective
tissue

Trachea × 251

(b)

Example:
Duct of
a salivary
gland

**STRATIFIED COLUMNAR
EPITHELIUM**

LOCATIONS: Small areas of
the pharynx, epiglottis, anus,
mammary gland and salivary
gland ducts, urethra
FUNCTION: Protection

Superficial
columnar cells

Deeper
basal
cells

Lumen

Cytoplasm

Nuclei

Basement
membrane

Loose
connective
tissue

Salivary gland duct

(c)

●**FIGURE 4-5 Columnar Epithelia. (a)** Micrograph showing the characteristics of simple columnar epithelium. **(b)** The pseudostratified, ciliated, columnar epithelium of the respiratory tract. Note the uneven layering of the nuclei. **(c)** A stratified columnar epithelium occurs along some large ducts, such as this salivary gland duct. Note the thickness of the epithelium and the location and orientation of the nuclei.

Portions of the respiratory tract contain a columnar epithelium that includes several different cell types with varying shapes and functions. Because the distances between the cell nuclei and the surface vary, the epithelium appears to be layered, or stratified. But it is not truly stratified, because all the epithelial cells contact the basement membrane. Thus it is known as a **pseudostratified columnar epithelium** (Figure 4-5b●). Pseudostratified columnar epithelial cells typically possess cilia. This type of epithelium lines most of the nasal cavity, the trachea (windpipe), bronchi, and portions of the male reproductive tract.

Stratified columnar epithelia are relatively rare, providing protection along portions of the pharynx, epiglottis, urethra, and anus as well as along a few large excretory ducts. The epithelium may have either two layers (Figure 4-5c●) or multiple layers. When multiple layers exist, only the superficial cells are columnar.

EXFOLIATIVE CYTOLOGY Exfoliative cytology (eks-FŌ-lē-a-tiv; *ex*, from + *folium*, leaf) is the study of cells shed or collected from epithelial surfaces. The cells may be examined for a variety of reasons—for example, to check for cellular changes that indicate cancer formation or to identify the pathogens involved in an infection. The cells are collected either by sampling the fluids that cover the epithelia lining the respiratory, digestive, urinary, or reproductive tract or by removing fluid from one of the ventral body cavities. The sampling procedure is often called a *Pap test*, named after Dr. George Papanicolaou. Probably the most familiar Pap test is the test for cervical cancer, which involves the scraping of a small number of cells from the tip of the *cervix*, the portion of the uterus that projects into the vagina.

Amniocentesis is another important test based on exfoliative cytology. In this procedure, exfoliated epithelial cells are collected from a sample of *amniotic fluid*, the fluid that surrounds and protects a developing fetus. Examination of these cells can determine if the fetus has a genetic abnormality, such as *Down syndrome*, that affects the number or structure of chromosomes.

Glandular Epithelia

Many epithelia contain gland cells that produce secretions. In general, glands and gland cells are classified as *endocrine* or *exocrine* according to the final distribution of their secretions.

Endocrine Glands

Endocrine glands release their secretions into the surrounding interstitial fluid. Because there are no endocrine ducts, or tubes, into which the secretions are released, endocrine glands are often called *ductless glands*. From the interstitial fluids, these secretions, called *hormones*, enter the circulation for distribution throughout the body. Hormones regulate or coordinate the activities of other cells that may reside in the same tissue or in other tissues and organs.

Endocrine cells may be part of an epithelial surface, such as the lining of the digestive tract, or they may be separate, as in the pancreas, thyroid gland, thymus, and pituitary gland. We shall consider endocrine cells, organs, and hormones further in Chapter 18.

Exocrine Glands

Exocrine glands produce secretions that are discharged onto epithelial surfaces. Most exocrine glands release their secretions into tubular passageways, called **ducts**, that empty onto the surface of the skin or onto an epithelial surface lining one of the internal passageways that communicates with the exterior. There are many kinds of exocrine secretions; sweat, saliva, and tears are familiar examples.

Only a few complex glands produce both exocrine and endocrine secretions. For example, the pancreas contains endocrine cells that secrete hormones as well as exocrine cells that produce digestive enzymes and buffers.

Exocrine glands may be classified by their mode of secretion, the type of secretion, and the organization of the gland cells and the associated ducts.

MODES OF SECRETION A glandular epithelial cell may use one of three methods to release its secretions: merocrine secretion, apocrine secretion, or holocrine secretion. In **merocrine secretion** (MER-ō-krin; *meros*, part + *krinein*, to separate), the product is released from secretory vesicles through exocytosis (Figure 4-6a●). This is the most common mode of secretion. One type of merocrine secretion mixes with water to form **mucus**, an effective lubricant, a protective barrier, and a sticky trap for foreign particles and microorganisms. The mucous secretions of merocrine glands coat the passageways of the digestive and respiratory tracts. In the skin, merocrine sweat glands produce the watery perspiration that helps cool your body on a hot day.

Apocrine secretion (AP-ō-krin; *apo-*, off) involves the loss of cytoplasm as well as the secretory product (Figure 4-6b●). The apical portion of the cytoplasm becomes packed with secretory vesicles before it is shed. The thick, sticky underarm perspiration targeted by the deodorant industry results from apocrine secretion. Milk production in the mammary glands involves a combination of merocrine and apocrine secretion.

Merocrine and apocrine secretions leave a cell relatively intact and able to continue secreting. **Holocrine secretion** (HOL-ō-krin; *holos*, entire) destroys the gland cell. During holocrine secretion, the entire cell becomes packed with secretory products and then bursts (Figure 4-6c●), releasing the secretion but killing the cell. Further secretion depends on the replacement of destroyed gland cells by the division of stem cells. Sebaceous glands, associated with hair follicles, produce an oily hair coating by means of holocrine secretion.

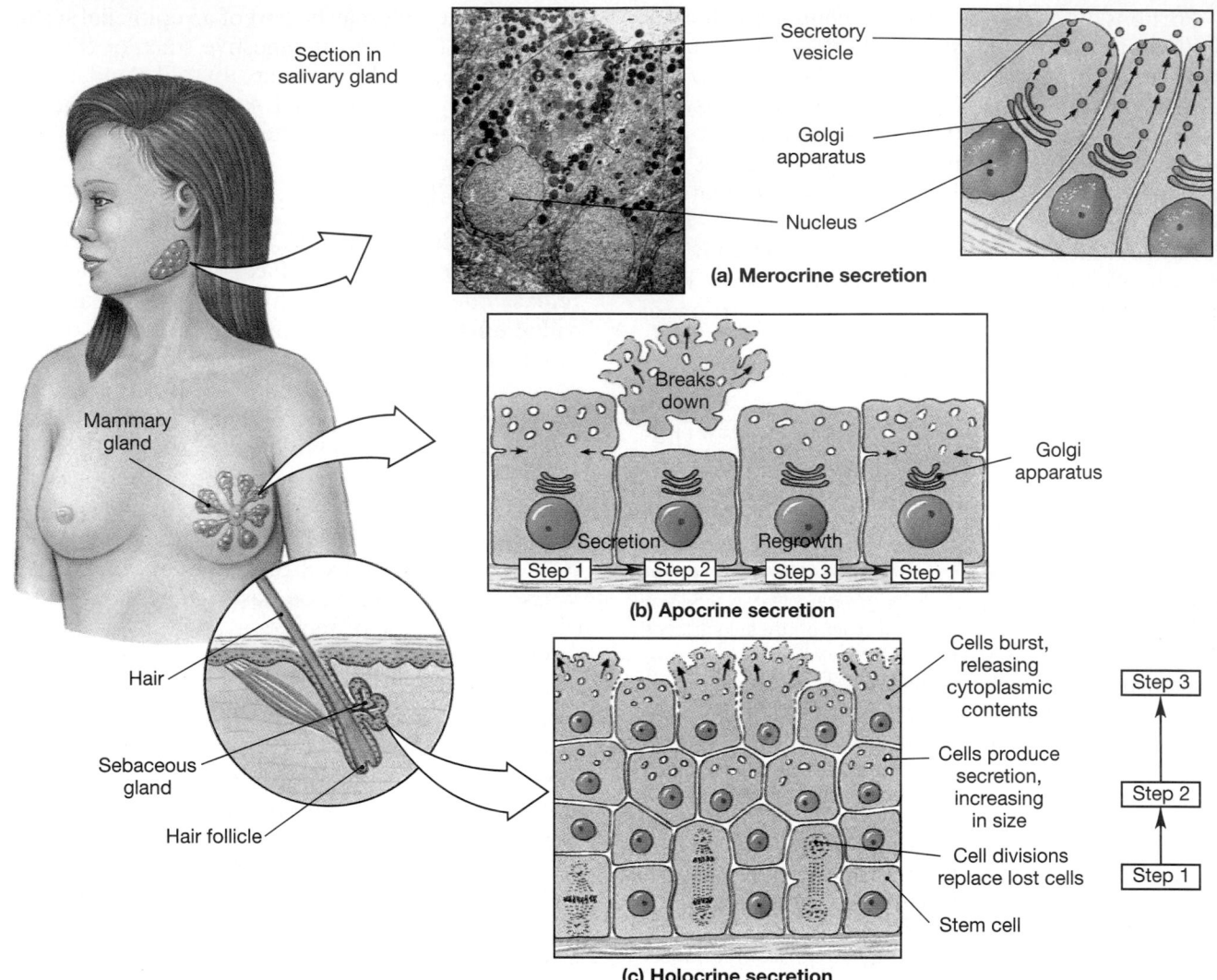

• FIGURE 4-6 **Mechanisms of Glandular Secretion.** **(a)** In merocrine secretion, secretory vesicles are discharged at the surface of the gland cell through exocytosis. **(b)** Apocrine secretion involves the loss of apical cytoplasm. Inclusions, secretory vesicles, and other cytoplasmic components are shed in the process. The gland cell then undergoes a period of growth and repair before it releases additional secretions. **(c)** Holocrine secretion occurs as superficial gland cells burst. Continued secretion involves the replacement of these cells through the mitotic divisions of underlying stem cells.

TYPES OF SECRETIONS Exocrine glands may be categorized as one of three types according to the nature of the secretion produced:

1. **Serous glands** secrete a watery solution that contains enzymes. The parotid salivary glands are examples of serous glands.

2. **Mucous glands** secrete *mucins*, glycoproteins that upon hydration form mucus, a slippery lubricant. The sublingual salivary glands and the submucosal glands of the small intestine are examples of mucous glands.

3. **Mixed exocrine glands** contain more than one type of gland cell and may produce two different exocrine secretions, one serous and the other mucous. The submandibular salivary glands are examples of mixed exocrine glands.

GLAND STRUCTURE In epithelia that contain independent, scattered gland cells, the individual secretory cells are called **unicellular glands. Multicellular glands** include both glandular epithelia and aggregations of gland cells that produce exocrine or endocrine secretions.

The only **unicellular exocrine glands** in the body are **goblet cells.** Goblet cells, which secrete mucins, are scattered among other epithelial cells. For example, both the pseudostratified ciliated columnar epithelium that lines the trachea and the columnar epithelium of the small and large intestines contain an abundance of goblet cells.

The simplest **multicellular exocrine gland** is called a **secretory sheet.** In a secretory sheet, glandular cells dominate the epithelium and release their secretions into an inner compartment. The mucin-secreting cells that line the stomach are an example. Their continuous secretion protects the stomach from its own acids and enzymes. Most other multicellular exocrine glands are found in pockets set back from the epithelial surface; their secretory products travel through one or more ducts to reach the epithelial surface. Examples include the salivary glands, which produce mucins and digestive enzymes.

Multicellular exocrine glands can be complex in structure. Three characteristics are used to describe the organization of such a gland:

1. *Shape of the secretory portion of the gland.* Glands whose glandular cells form tubes are called *tubular.* Those that form blind pockets are called *alveolar* (al-VĒ-ō-lar; *alveolus,* sac) or **acinar** (A-si-nar; *acinus,* chamber). Glands whose secretory cells form both tubes and pockets are called *tubuloalveolar* or *tubuloacinar.*

2. *Branching pattern of the duct.* A gland is called *simple* if there is a single duct that does not divide on its way to the gland cells. The gland is called *compound* if the duct divides repeatedly.

3. *Relationship between the ducts and the glandular areas.* A gland is called *branched* if several secretory areas (tubular or acinar) share a common duct. Note that the term *branched* always refers to the glandular areas and not to the duct.

Figure 4-7● diagrams this method of classification on the basis of gland structure. We shall discuss specific examples of each gland type in later chapters.

☑ With a light microscope, you examine a tissue and see a simple squamous epithelium on the outer surface. Can this be a sample of the skin surface?

☑ The secretory cells of sebaceous glands fill with secretions and then rupture, releasing their contents. What type of secretion is this?

☑ A gland has no ducts to carry the glandular secretions. Chemical tests reveal that the gland's secretions are released directly into the extracellular fluid. What type of gland is this?

CONNECTIVE TISSUES

Connective tissues include tissues such as bone, fat, and blood, which are quite different in appearance and function. Nevertheless, all connective tissues

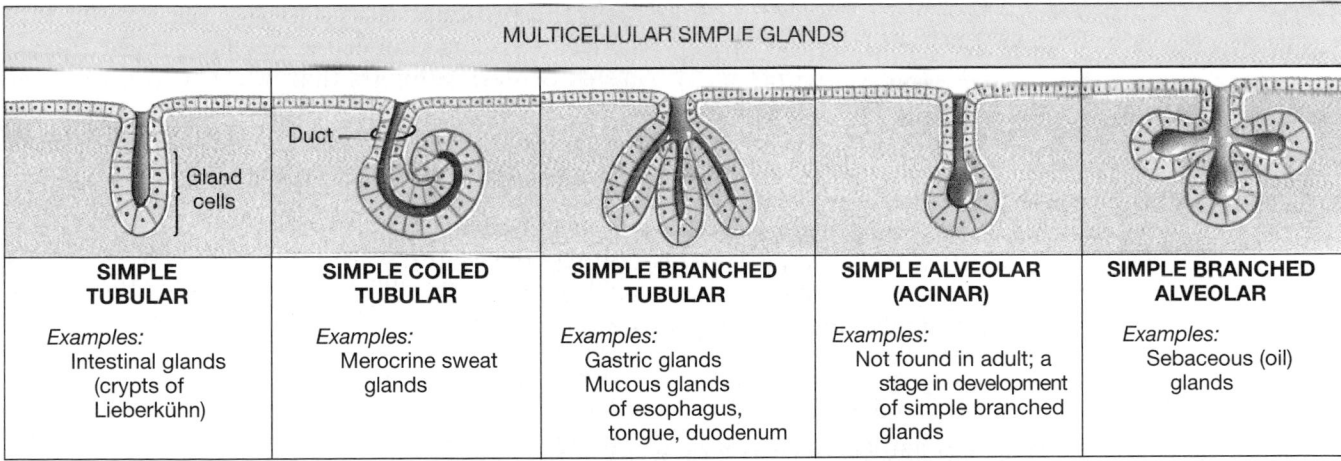

MULTICELLULAR SIMPLE GLANDS				
SIMPLE TUBULAR	**SIMPLE COILED TUBULAR**	**SIMPLE BRANCHED TUBULAR**	**SIMPLE ALVEOLAR (ACINAR)**	**SIMPLE BRANCHED ALVEOLAR**
Examples: Intestinal glands (crypts of Lieberkühn)	*Examples:* Merocrine sweat glands	*Examples:* Gastric glands Mucous glands of esophagus, tongue, duodenum	*Examples:* Not found in adult; a stage in development of simple branched glands	*Examples:* Sebaceous (oil) glands

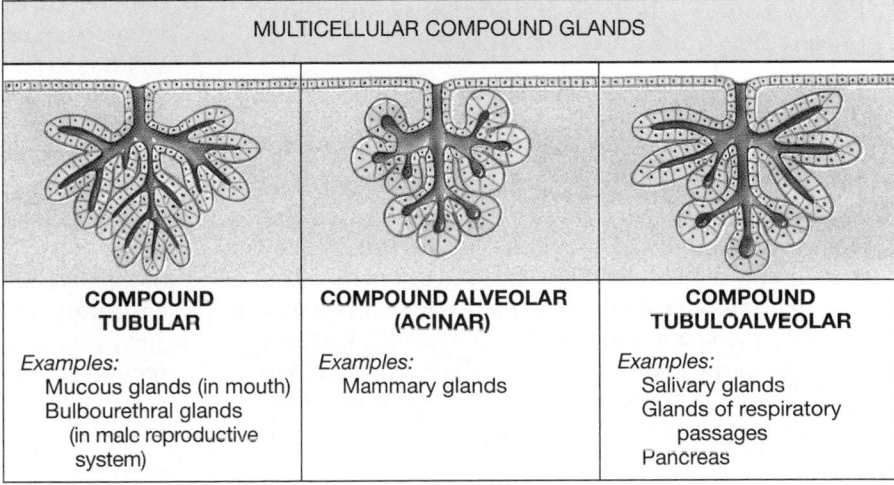

MULTICELLULAR COMPOUND GLANDS		
COMPOUND TUBULAR	**COMPOUND ALVEOLAR (ACINAR)**	**COMPOUND TUBULOALVEOLAR**
Examples: Mucous glands (in mouth) Bulbourethral glands (in male reproductive system)	*Examples:* Mammary glands	*Examples:* Salivary glands Glands of respiratory passages Pancreas

●**FIGURE 4-7**
A Structural Classification of Exocrine Glands

have three basic components: (1) specialized cells, (2) extracellular protein fibers, and (3) a fluid known as the **ground substance.** The extracellular fibers and ground substance constitute the **matrix** that surrounds the cells. Whereas cells make up the bulk of epithelial tissue, the extracellular matrix typically accounts for most of the volume of connective tissues.

Connective tissues are found throughout the body but are never exposed to the environment outside the body. Many connective tissues are highly vascular (that is, they have many blood vessels) and contain sensory receptors that provide pain, pressure, temperature, and other sensations.

The functions of connective tissues include:

- Establishing a structural framework for the body.
- Transporting fluids and dissolved materials from one region of the body to another.
- Protecting delicate organs.
- Supporting, surrounding, and interconnecting other tissue types.
- Storing energy reserves, especially in the form of lipids.
- Defending the body from invading microorganisms.

Classification of Connective Tissues

Connective tissue can be classified into three categories: *connective tissue proper, fluid connective tissues,* and *supporting connective tissues.*

1. **Connective tissue proper** refers to connective tissues with many types of cells and extracellular fibers in a syrupy ground substance. These connective tissues differ in the number of cell types they contain and the relative properties and proportions of fibers and ground substance. For example, both *adipose tissue* (fat) and *tendons* are connective tissue proper, but they have very different structural and functional characteristics. Connective tissue proper is divided into (1) *loose connective tissues* and (2) *dense connective tissues* on the basis of the relative proportions of cells, fibers, and ground substance.

2. **Fluid connective tissues** have a distinctive population of cells suspended in a watery matrix that contains dissolved proteins. There are two fluid connective tissues: *blood* and *lymph.*

3. **Supporting connective tissues** are of two types, *cartilage* and *bone.* These tissues have a less diverse cell population than connective tissue proper and a matrix that contains closely packed fibers. The matrix of cartilage is a gel whose characteristics vary depending on the predominant fiber type. The matrix of bone is said to be **calcified** because it contains mineral deposits, primarily calcium salts. These minerals give bone its rigidity.

Connective Tissue Proper

Connective tissue proper contains extracellular fibers, a viscous ground substance, and a varied cell population. Some of these cells are involved with local maintenance, repair, and energy storage. These cells include *fibroblasts, adipocytes,* and *mesenchymal cells.* Other cells are responsible for the defense and repair of damaged tissues. These cells are not permanent residents of the tissue but migrate through healthy connective tissues and aggregate at sites of tissue injury. Examples include *macrophages, mast cells, lymphocytes, plasma cells,* and *microphages.*

The number of cells and cell types within a tissue at any given moment varies depending on local conditions. Refer to Figure 4-8a● as we describe the cells and fibers of connective tissue proper.

The Cell Population

Fibroblasts (FĪ-brō-blasts) are the most abundant permanent residents in connective tissue proper. They are also the *only* cells that are *always* present in every connective tissue proper. Fibroblasts secrete hyaluronan (a polysaccharide) and proteins. These two components interact in the extracellular fluid to form the proteoglycans that give the ground substance its viscous properties. Each fibroblast also secretes protein subunits that interact to form large extracellular fibers.

In addition to fibroblasts, connective tissues proper may contain the following cell types:

- **Macrophages** (MAC-rō-fā-jez; *phagein,* to eat) are large, amoeboid cells that are scattered throughout the matrix. These cells engulf pathogens or damaged cells that enter the tissue. Although they are not abundant, they play an important role in mobilizing the body's defenses. When stimulated, they release chemicals that activate the immune system and attract large numbers of additional macrophages and other cells involved in tissue defense. Macrophages are often divided into *fixed macrophages,* which spend long periods within a particular tissue, and *free macrophages,* which migrate rapidly through the tissue. In effect, the fixed macrophages in a tissue provide a "front-line" defense that can be reinforced by the arrival of free macrophages and other specialized cells.

- **Adipocytes** (AD-i-pō-sīts) are also known as fat cells, or *adipose cells.* A typical adipocyte contains a single, enormous lipid droplet. The nucleus and other organelles are squeezed to one side, making the cell in section resemble a class ring. The number of fat cells varies from one type of connective tissue to another, from one region of the body to another, and from individual to individual.

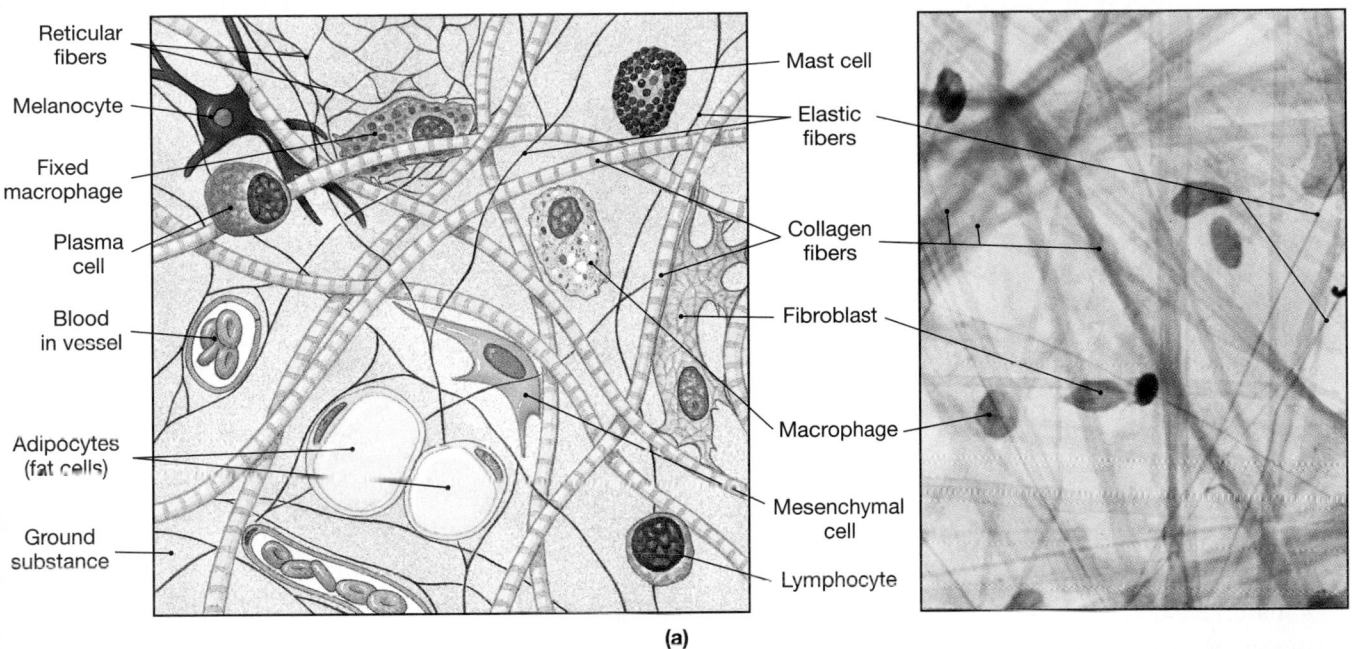

Reticular fibers
Melanocyte
Fixed macrophage
Plasma cell
Blood in vessel
Adipocytes (fat cells)
Ground substance

Mast cell
Elastic fibers
Collagen fibers
Fibroblast
Macrophage
Mesenchymal cell
Lymphocyte

(a)

•*FIGURE 4-8* **The Cells and Fibers of Connective Tissue Proper.**
(a) A summary of the cell types and fibers of connective tissue proper.
(LM × 384) **(b)** Mesenchyme, the first connective tissue to appear in
the embryo. (LM × 136) **(c)** Mucous connective tissue. This sample was
taken from the umbilical cord of a fetus. Mucous connective tissue in
this location is also known as *Wharton's jelly.* (LM × 136)

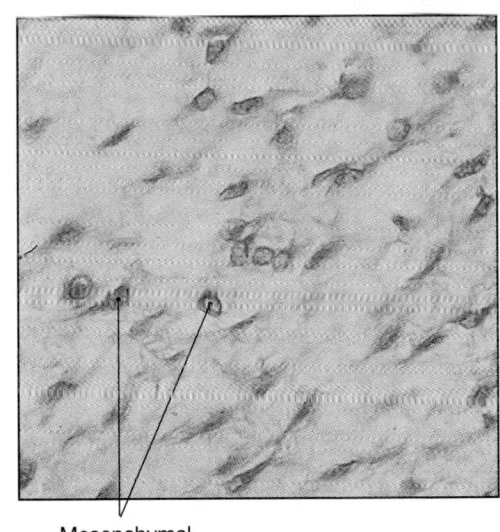

Mesenchymal cells   (b)

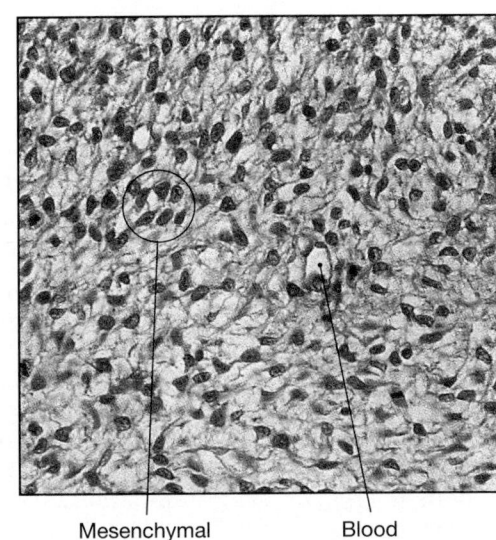

Mesenchymal cells   Blood vessel   (c)

- **Mesenchymal cells** are stem cells that are present in many connective tissues. These cells respond to local injury or infection
  by dividing to produce daughter cells that differentiate into fibroblasts, macrophages, or other connective tissue cells.
- **Melanocytes** (me-LAN-ō-sīts) synthesize and store the brown
  pigment **melanin** (ME-la-nin), which gives the tissue a dark
  color. Melanocytes are common in the epithelium of the skin,
  where they play a major role in determining skin color. However, melanocytes are also abundant in connective tissues of
  the eye, and they are present in the dermis of the skin, although there are regional and individual differences in the
  number present.
- **Mast cells** are small, mobile connective tissue cells commonly
  found near blood vessels. The cytoplasm of a mast cell is filled
  with granules of **histamine** (HIS-tuh-mēn) and **heparin** (HEP-
  uh-rin). These chemicals, released after injury or infection,
  stimulate local inflammation. Histamine and heparin granules
  are also found in *basophils,* blood cells that enter damaged
  tissues and enhance the inflammation process.
- **Lymphocytes** (LIM-fō-sīts) migrate throughout the body.
  Their numbers increase markedly wherever tissue damage occurs, and some of the lymphocytes may then develop into **plasma cells.** Plasma cells are responsible for the production of
  antibodies, proteins involved in defending the body against
  disease.

Development of Tissues

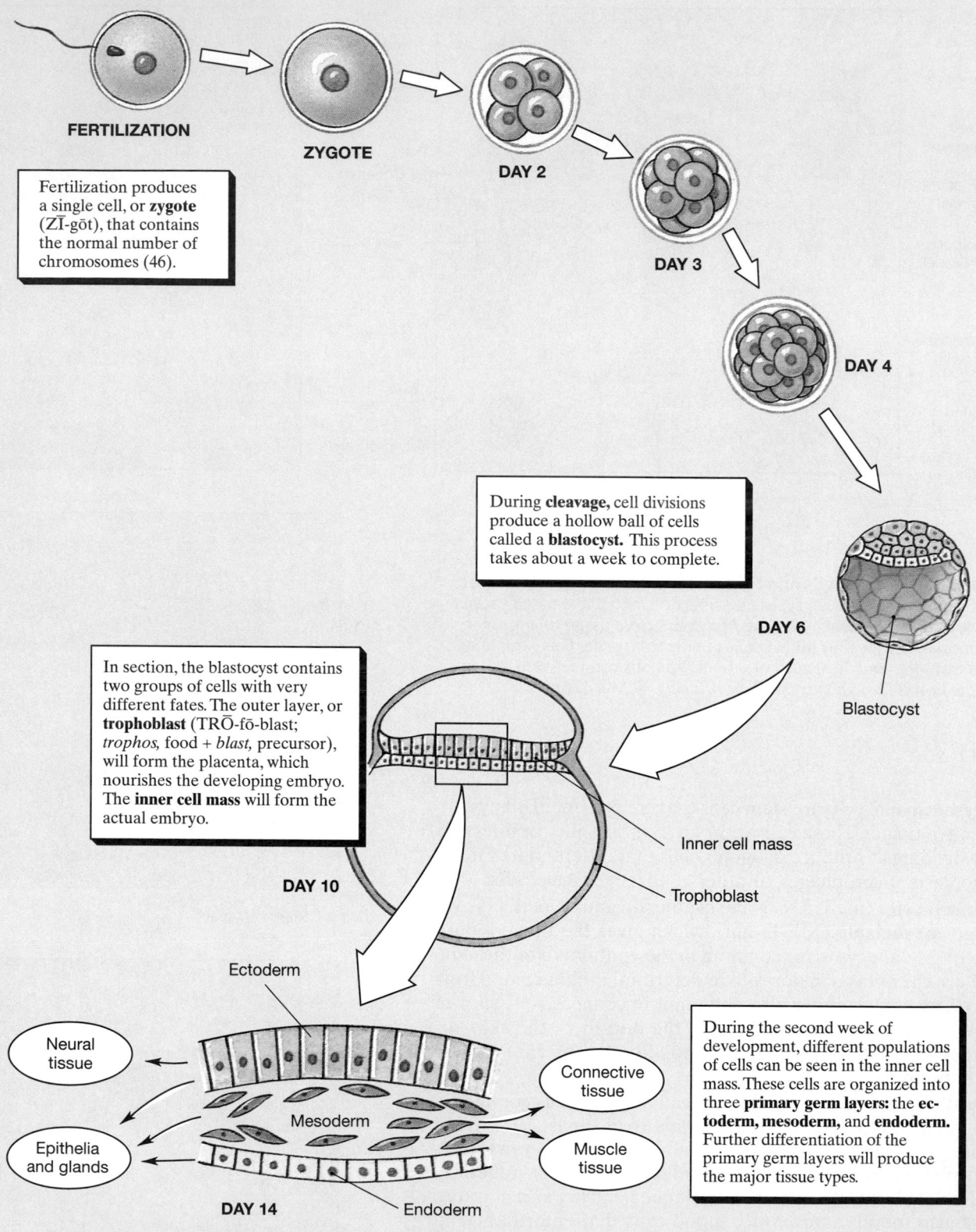

FERTILIZATION

ZYGOTE

DAY 2

DAY 3

DAY 4

DAY 6

Blastocyst

Fertilization produces a single cell, or **zygote** (ZĪ-gōt), that contains the normal number of chromosomes (46).

During **cleavage,** cell divisions produce a hollow ball of cells called a **blastocyst.** This process takes about a week to complete.

In section, the blastocyst contains two groups of cells with very different fates. The outer layer, or **trophoblast** (TRŌ-fō-blast; *trophos,* food + *blast,* precursor), will form the placenta, which nourishes the developing embryo. The **inner cell mass** will form the actual embryo.

Inner cell mass

Trophoblast

DAY 10

Ectoderm

Neural tissue

Mesoderm

Connective tissue

Muscle tissue

Epithelia and glands

DAY 14

Endoderm

During the second week of development, different populations of cells can be seen in the inner cell mass. These cells are organized into three **primary germ layers:** the **ectoderm, mesoderm,** and **endoderm.** Further differentiation of the primary germ layers will produce the major tissue types.

All three germ layers participate in the formation of functional organs and organ systems. Their interactions will be detailed in later Embryology Summaries dealing with specific systems.

■ **Microphages** (*neutrophils* and *eosinophils*) are phagocytic blood cells that move through connective tissues in small numbers. When an infection or injury occurs, chemicals released by macrophages and mast cells attract microphages in large numbers.

Connective Tissue Fibers

Three basic types of fibers are found in connective tissue: *collagen fibers, reticular fibers,* and *elastic fibers.* Fibroblasts form all three types of fibers through the secretion of protein subunits that interact within the matrix.

1. **Collagen fibers** are long, straight, and unbranched. These are the most common fibers in connective tissue proper. Each collagen fiber consists of a bundle of fibrous protein subunits wound together like the strands of a rope. Like a rope, a collagen fiber is flexible, but it is stronger than steel when pulled from either end. *Tendons,* which connect skeletal muscles to bones, consist almost entirely of collagen fibers. Typical *ligaments* are similar to tendons, but they connect one bone to another. Tendons and ligaments can withstand tremendous forces. Uncontrolled muscle contractions or skeletal movements are more likely to break a bone than to snap a tendon or ligament.

2. **Reticular fibers** (*reticulum,* network) contain the same protein subunits as collagen fibers, but they are arranged differently. These fibers are thinner than collagen fibers, and they form a branching, interwoven framework that is tough but flexible. Because they form a network rather than share a common alignment, reticular fibers can resist forces applied from many different directions. This interwoven network, called a *stroma,* stabilizes the relative positions of the functional cells, or **parenchyma** (pa-RENG-ki-ma). The reticular fibers also stabilize the relative positions of an organ's blood vessels, nerves, and other structures despite changing positions and the pull of gravity.

3. **Elastic fibers** contain the protein *elastin.* Elastic fibers are branched and wavy. After stretching, they will return to their original length. **Elastic ligaments** are dominated by elastic fibers. They are relatively rare but have important functions, such as interconnecting the vertebrae.

Ground Substance

Ground substance fills the spaces between cells and surrounds the connective tissue fibers (Figure 4-8a●). Ground substance in normal connective tissue proper is clear, viscous, and colorless. It typically has a consistency similar to that of maple syrup, due to the presence of proteoglycans and glycoproteins. ∞ *[p. 57]* The ground substance is dense enough that bacteria have trouble moving through it—imagine swimming in molasses. This density slows the spread of pathogens through the tissue and makes them easier for phagocytes to catch.

MARFAN'S SYNDROME **Marfan's syndrome** is an inherited condition caused by the production of an abnormal form of *fibrillin*, a glycoprotein important to normal connective tissue strength and elasticity. Because connective tissues are found in most organs, the effects of this defect are widespread. The most visible sign of Marfan's syndrome involves the skeleton; most individuals with this condition are tall, with abnormally long arms, legs, and fingers. But the most serious consequences involve the cardiovascular system. Roughly 90 percent of people with Marfan's syndrome have structural abnormalities in their cardiovascular system. The most dangerous potential result is that the weakened elastic connective tissues in the walls of major arteries, such as the aorta, may burst, causing a sudden, fatal loss of blood.

Embryonic Connective Tissues

Mesenchyme, or *embryonic connective tissue,* is the first connective tissue to appear in the developing embryo. Mesenchyme contains star-shaped stem cells (mesenchymal cells) separated by a matrix that contains very fine protein filaments. This connective tissue (Figure 4-8b●) gives rise to all other forms of connective tissue, including fluid connective tissues, cartilage, and bone. **Mucous connective tissue** (Figure 4-8c●) is a loose connective tissue found in many portions of the embryo, including the umbilical cord.

Neither of these embryonic forms of connective tissue is found in the adult. However, many adult connective tissues contain scattered mesenchymal stem cells that can assist in tissue repair after an injury.

Loose Connective Tissues

Loose connective tissue, or **areolar tissue** (*areola,* little space), is the "packing material" of the body. It fills spaces between organs, provides cushioning, and supports epithelia. Loose connective tissue also surrounds and supports blood vessels and nerves, stores lipids, and provides a route for the diffusion of materials.

Typical loose connective tissue (Figure 4-8a●) is the least specialized connective tissue in the adult body. This tissue may contain all the cells and fibers found in any connective tissue proper. Loose connective tissue has an open framework, and ground substance accounts for most of its volume. This syrupy fluid cushions shocks. Because the fibers are loosely organized, loose connective tissue can distort without damage. The presence of elastic fibers makes it fairly resilient, so this tissue returns to its original shape after external pressure is relieved.

Loose connective tissue forms a layer that separates the skin from deeper structures. In addition to providing padding, the elastic properties of this layer allow a considerable amount of independent movement. Thus, if you pinch the skin of your arm, you will not affect the underlying muscle. Conversely, contractions of the underlying muscles do not pull against your skin—as the muscle bulges, the loose connective tissue stretches. Because this tissue has an extensive circulatory supply, the loose connective tissue layer under the skin is a common injection site for drugs.

In addition to delivering oxygen and nutrients and removing carbon dioxide and waste products, the capillaries in loose connective tissue carry wandering cells to and from the tissue. Epithelia generally cover a layer of loose connective tissue, and fibroblasts are responsible for maintaining the reticular lamina of the basement membrane. The epithelial cells rely on the diffusion of oxygen and nutrients across that membrane from capillaries in the underlying connective tissue.

There are two specialized varieties of loose connective tissue that are relatively common: *adipose tissue* and *reticular tissue*.

ADIPOSE TISSUE The distinction between other loose connective tissues and fat, or **adipose tissue**, can be somewhat arbitrary. Adipocytes account for most of the volume of adipose tissue (Figure 4-9a●) but only a fraction of the volume of loose connective tissue. Adipose tissue provides padding, cushions shocks, acts as an insulator to slow heat loss through the skin, and serves as packing or filler around structures. Adipose tissue is common under the skin of the groin, sides, buttocks, and breasts. It fills the bony sockets behind the eyes, surrounds the kidneys, and dominates extensive areas of loose connective tissue in the pericardial and abdominal cavities.

Brown Fat In infants and young children, the adipose tissue between the shoulder blades, around the neck, and possibly elsewhere in the upper body is different from most of the adipose tissue in the adult. It is highly vascularized, and the individual adipocytes contain numerous mitochondria. Together, these characteristics give the tissue a deep, rich color responsible for the name **brown fat**. When these cells are stimulated by the nervous system, lipid breakdown accelerates. The cells do not capture the energy released, and it is absorbed by the surrounding tissues as heat. This heat warms the circulating blood, which distributes the heat throughout the body. In this way, an infant can increase metabolic heat generation by 100 percent very quickly. There is little if any brown fat in the adult. With increased body size, skeletal muscle mass, and insulation, shivering is significantly more effective in elevating the body temperature of adults.

ADIPOSE TISSUE AND WEIGHT CONTROL Adipocytes are metabolically active cells—their lipids are constantly being broken down and replaced. When nutrients are scarce, adipocytes deflate like collapsing balloons. This deflation is what occurs during a weight-loss program. Because the cells are not killed but merely reduced in size, the lost weight can easily be regained in the same areas of the body. In adults, adipocytes are incapable of dividing. The number of fat cells in peripheral tissues is established in the first few weeks of a newborn's life, perhaps in response to the amount of fats in the diet. However, this is not the end of the story, because loose connective tissues also contain mesenchymal cells. If circulating lipid levels are chronically elevated, the mesenchymal cells will divide, giving rise to cells that differentiate into fat cells. As a result, areas of loose connective tissue can become adipose tissue in times of nutritional plenty, even in the adult. In the procedure known as *liposuction*, unwanted adipose tissue is surgically removed. Because adipose tissue can regenerate through differentiation of mesenchymal cells, liposuction provides only a temporary solution to the problem of excess weight.

RETICULAR TISSUE **Reticular tissue** is found in organs such as the spleen and liver, where reticular fibers create a complex three-dimensional network, or stroma, that supports the functional cells of these organs. This fibrous framework is also found in the lymph nodes and bone marrow (Figure 4-9b●). Fixed macrophages and fibroblasts are associated with the reticular fibers, but these cells are seldom visible because the organs are dominated by specialized cells with other functions.

Dense Connective Tissues

Most of the volume of **dense connective tissues** is occupied by fibers. Dense connective tissues are often called **collagenous** (ko-LA-jin-us) **tissues** because collagen fibers are the dominant fiber type. Two types of dense connective tissues occur in the body: (1) *dense regular connective tissue* and (2) *dense irregular connective tissue*.

In **dense regular connective tissue,** the collagen fibers are arranged parallel to each other, packed tightly, and aligned with the forces applied to the tissue. **Tendons** (Figure 4-10a● , p. 126) are cords of dense regular connective tissue that attach skeletal muscles to bones. The collagen fibers run along the longitudinal axis of the tendon and transfer the pull of the contracting muscle to the bone. An **aponeurosis** is a tendinous sheet. **Ligaments** resemble tendons but connect one bone to another. Large numbers of fibroblasts are found between the collagen fibers. **Elastic tissue** contains large numbers of elastic fibers. Because elastic fibers outnumber collagen fibers, the tissue has a springy, resilient

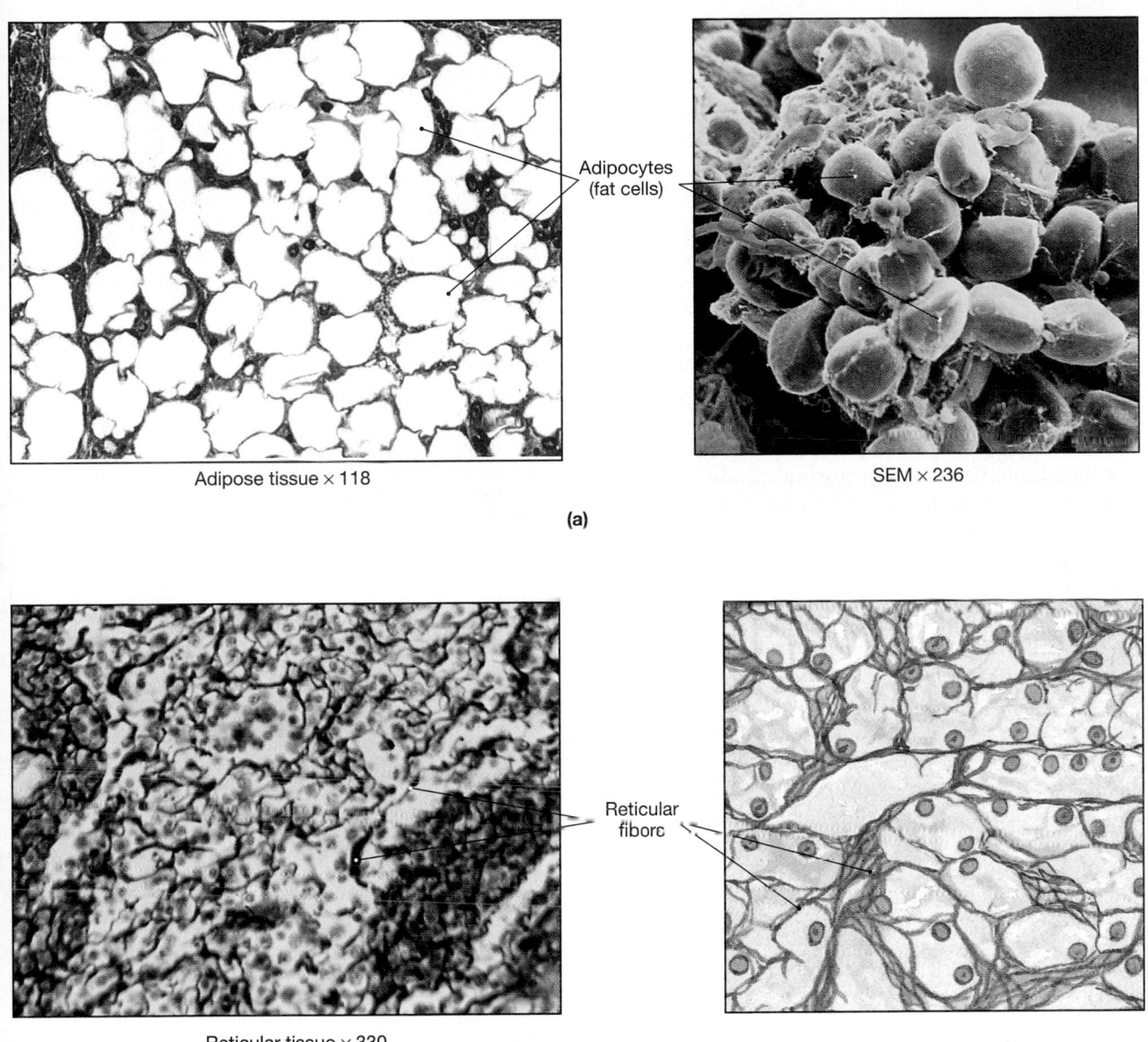

(a)

Adipocytes
(fat cells)

Adipose tissue × 118

SEM × 236

Reticular
fibers

Reticular tissue × 330

(b)

●**FIGURE 4-9 Adipose and Reticular Tissues. (a)** Adipose tissue is a loose connective tissue that is dominated by adipocytes. In standard histological preparations, the tissue looks empty because the lipids in the fat cells dissolve in the alcohol used in tissue processing. **(b)** Reticular tissue has an open framework of reticular fibers. These fibers are usually very difficult to see because of the large numbers of cells around them.

nature. This ability to stretch and rebound allows it to tolerate cycles of expansion and contraction. Elastic tissue typically underlies transitional epithelia. It is also located in the walls of blood vessels, and it surrounds the respiratory passageways. Ribbons of elastic fibers form **elastic ligaments** that help stabilize the positions of the vertebrae of the spinal column (Figure 4-10b●).

The fibers in **dense irregular connective tissue** form an interwoven meshwork and do not show any consistent pattern (Figure 4-10c●). These tissues provide

strength and support to areas subjected to stresses from many directions. A layer of dense irregular connective tissue gives skin its strength. A piece of cured leather (animal skin) provides an excellent illustration of the interwoven nature of this tissue. Except at joints, dense irregular connective tissue forms a sheath around cartilages (the *perichondrium*) and bones (the *periosteum*). Dense irregular connective tissue also forms a thick fibrous layer called a **capsule**, which surrounds internal organs, such as liver, kidneys, and spleen, and encloses the cavities of joints.

☑Lack of vitamin C in the diet interferes with the ability of fibroblasts to produce collagen. What effect might this interference have on connective tissue?

☑Many allergy sufferers take antihistamines to relieve their allergy symptoms. What type of cell produces the molecule that this medication blocks?

☑What type of connective tissue contains primarily triglycerides?

Fluid Connective Tissues

Blood and *lymph* are connective tissues that contain distinctive collections of cells in a fluid matrix. The watery matrix of blood and lymph contains cells and many different types of suspended proteins that do not form insoluble fibers under normal conditions.

Blood contains blood cells and fragments of cells, collectively known as *formed elements* (Figure 4-11•, p. 128). A single cell type, the **red blood cell,** or **erythrocyte** (e-RITH-rō-sīt; *erythros,* red), accounts for almost half the volume of blood. Red blood cells are responsible for the transport of oxygen and, to a lesser degree, of carbon dioxide in the blood. The watery ground substance of blood, called **plasma,** also contains small numbers of **white blood cells,** or **leukocytes** (LOO-kō-sīts; *leuko,* white). White blood cells include the phagocytic microphages (*neutrophils* and *eosinophils*), *basophils*, *lymphocytes,* and macrophages called *monocytes.* The white blood cells are important components of the immune system, which protects the body from infection and disease. Tiny membrane-covered packets of cytoplasm called **platelets** contain enzymes and special proteins. Platelets function in the clotting response that seals breaks in the endothelial lining.

The extracellular fluid of the body includes three major subdivisions: *plasma, interstitial fluid,* and *lymph.* Plasma is normally confined to the vessels of the circulatory system, and the contractions of the heart keep it in constant motion. **Arteries** are vessels that carry blood away from the heart toward fine, thin-walled vessels called **capillaries. Veins** are vessels that drain the capillaries and return blood to the heart, completing the circuit. In the tissues, filtration moves water and small solutes out of the capillaries and into the interstitial fluid that bathes the cells of the body. The major difference between plasma and interstitial fluid is that plasma contains large numbers of suspended proteins.

Lymph forms as interstitial fluid enters small passageways, or **lymphatics,** that return it to the cardiovascular system. Along the way, cells of the immune system monitor the composition of the lymph and respond to signs of injury or infection. The number of cells in lymph may vary, but ordinarily 99 percent of them are lymphocytes. (The rest are primarily macrophages or microphages.)

DENSE REGULAR CONNECTIVE TISSUE	
LOCATIONS: Between skeletal muscles and skeleton (tendons and aponeuroses); between bones (ligaments); covering skeletal muscles (deep fasciae) **FUNCTIONS:** Provides firm attachment; conducts pull of muscles; reduces friction between muscles; stabilizes relative positions of bones	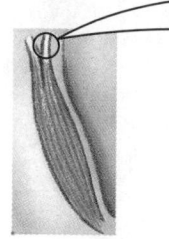 **Example:** Tendon

ELASTIC TISSUE	
LOCATION: Between vertebrae of the spinal column (ligamentum flava and ligamentum nuchae) **FUNCTIONS:** Stabilizes positions of vertebrae; cushions shocks	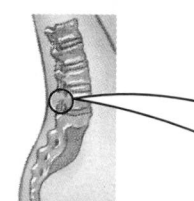 **Example:** Elastic ligament

DENSE IRREGULAR CONNECTIVE TISSUE	
LOCATIONS: Capsules of visceral organs; periostea and perichondria; nerve and muscle sheaths; dermis **FUNCTIONS:** Provides strength to resist forces applied from many directions; helps prevent over expansion of organs such as the urinary bladder	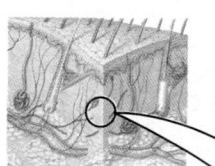 **Example:** Deep dermis of skin

•*FIGURE 4-10* **Dense Connective Tissues.** **(a)** The dense regular connective tissue in a tendon. Notice the densely packed, parallel bundles of collagen fibers. The fibroblast nuclei can be seen flattened between the bundles. **(b)** An elastic ligament from between the vertebrae of the spinal column. The bundles are fatter than those of a tendon or ligament composed of collagen. **(c)** The deep dermis of the skin contains a thick layer of dense irregular connective tissue.

Supporting Connective Tissues

Cartilage and bone are called supporting connective tissues because they provide a strong framework that supports the rest of the body. In these connective tissues, the matrix contains numerous fibers and, in some cases, deposits of insoluble calcium salts.

Cartilage

The matrix of **cartilage** is a firm gel that contains polysaccharide derivatives called **chondroitin sulfates** (kon-

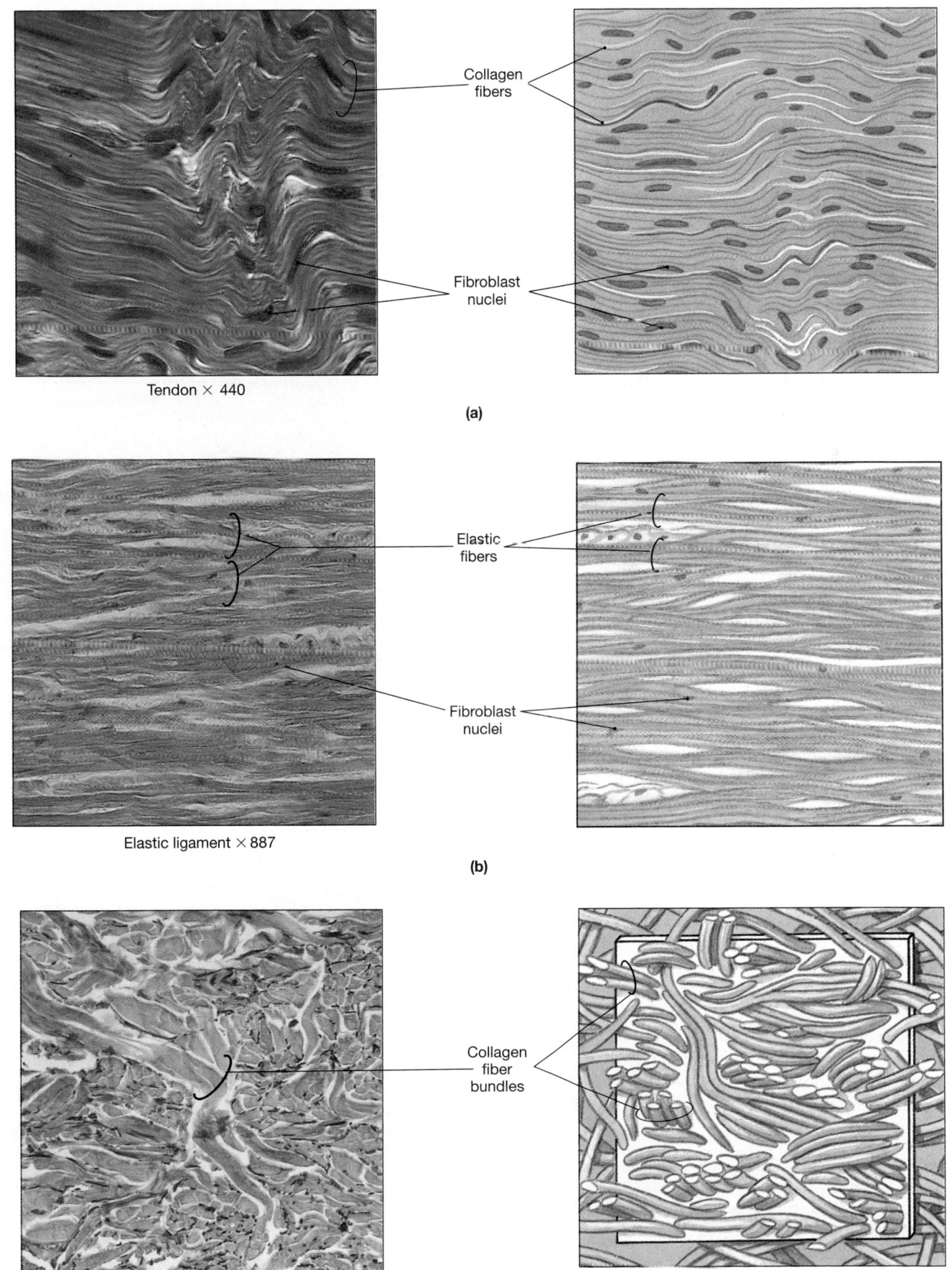

Tendon × 440

Collagen fibers

Fibroblast nuclei

(a)

Elastic ligament × 887

Elastic fibers

Fibroblast nuclei

(b)

Deep dermis × 111

Collagen fiber bundles

(c)

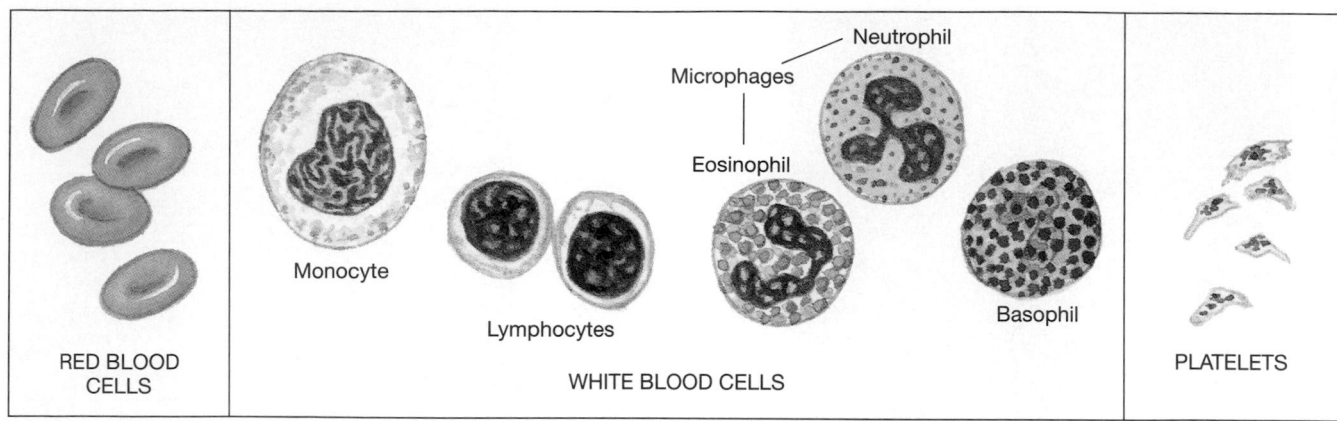

●*FIGURE 4-11* **Formed Elements of the Blood**

DROY-tin; *chondros,* cartilage). The chondroitin sulfates form complexes with proteins in the ground substance, forming proteoglycans. Cartilage cells, or **chondrocytes** (KON-drō-sīts), are the only cells found within the matrix. These cells live in small pockets known as **lacunae** (la-KOO-nē; *lacus,* pool). The physical properties of cartilage depend on the type and abundance of extracellular fibers as well as the proteoglycan components.

Cartilage is avascular, and all nutrient and waste product exchange must occur by diffusion through the matrix. Blood vessels do not grow into cartilage because chondrocytes produce a chemical that discourages their formation. This chemical has been named **antiangiogenesis factor** (*anti-,* against + *angeion,* vessel + *genno,* to produce). There is now interest in this compound as a potential anticancer agent. Tumors enlarge rapidly because blood vessels usually grow into areas where cells are crowded and active, improving nutrient and oxygen delivery. This growth could theoretically be prevented by antiangiogenesis factor, but the quantities produced in normal human cartilage are extremely small.

A cartilage is generally set apart from surrounding tissues by a fibrous **perichondrium** (pe-rē-KON-drē-um; *peri-,* around) (Figure 4-12a ●). The perichondrium contains two distinct layers—an outer, fibrous region of dense irregular connective tissue and an inner, cellular layer. The fibrous layer provides mechanical support and protection and attaches the cartilage to other structures. The cellular layer is important to the growth and maintenance of the cartilage.

TYPES OF CARTILAGE Three major types of cartilage are found in the body: *hyaline cartilage, elastic cartilage,* and *fibrocartilage:*

1. **Hyaline cartilage** (HĪ-uh-lin; *hyalos,* glass) is the most common type of cartilage. The matrix of hyaline cartilage contains closely packed collagen fibers, making

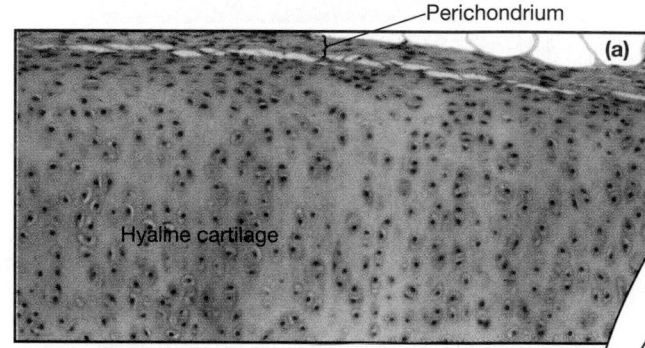

HYALINE CARTILAGE	**Example:** shoulder joint
LOCATIONS: Between tips of ribs and bones of sternum; covering bone surfaces at synovial joints; supporting larynx (voicebox), trachea, and bronchi; forming part of nasal septum **FUNCTIONS:** Provides stiff but somewhat flexible support; reduces friction between bony surfaces	

ELASTIC CARTILAGE	**Example:** pinna of external ear
LOCATIONS: Tip of nose; pinna; epiglottis **FUNCTIONS:** Provides support but tolerates distortion without damage and returns to original shape	

FIBROCARTILAGE	**Example:** discs separating vertebrae
LOCATIONS: Pads within knee joint; between pubic bones of pelvis; intervertebral discs **FUNCTIONS:** Resists compression; prevents bone-to-bone contact; limits relative movement	

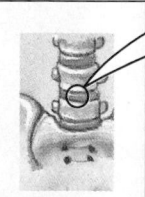

●*FIGURE 4-12* **The Perichondrium and Types of Cartilage.** (a) A perichondrium separates cartilage from other tissues. (b) *Hyaline cartilage.* Note the translucent matrix and the absence of prominent fibers. (c) *Elastic cartilage.* The closely packed elastic fibers are visible between the chondrocytes. (d) *Fibrocartilage.* The collagen fibers are extremely dense, and the chondrocytes are relatively far apart.

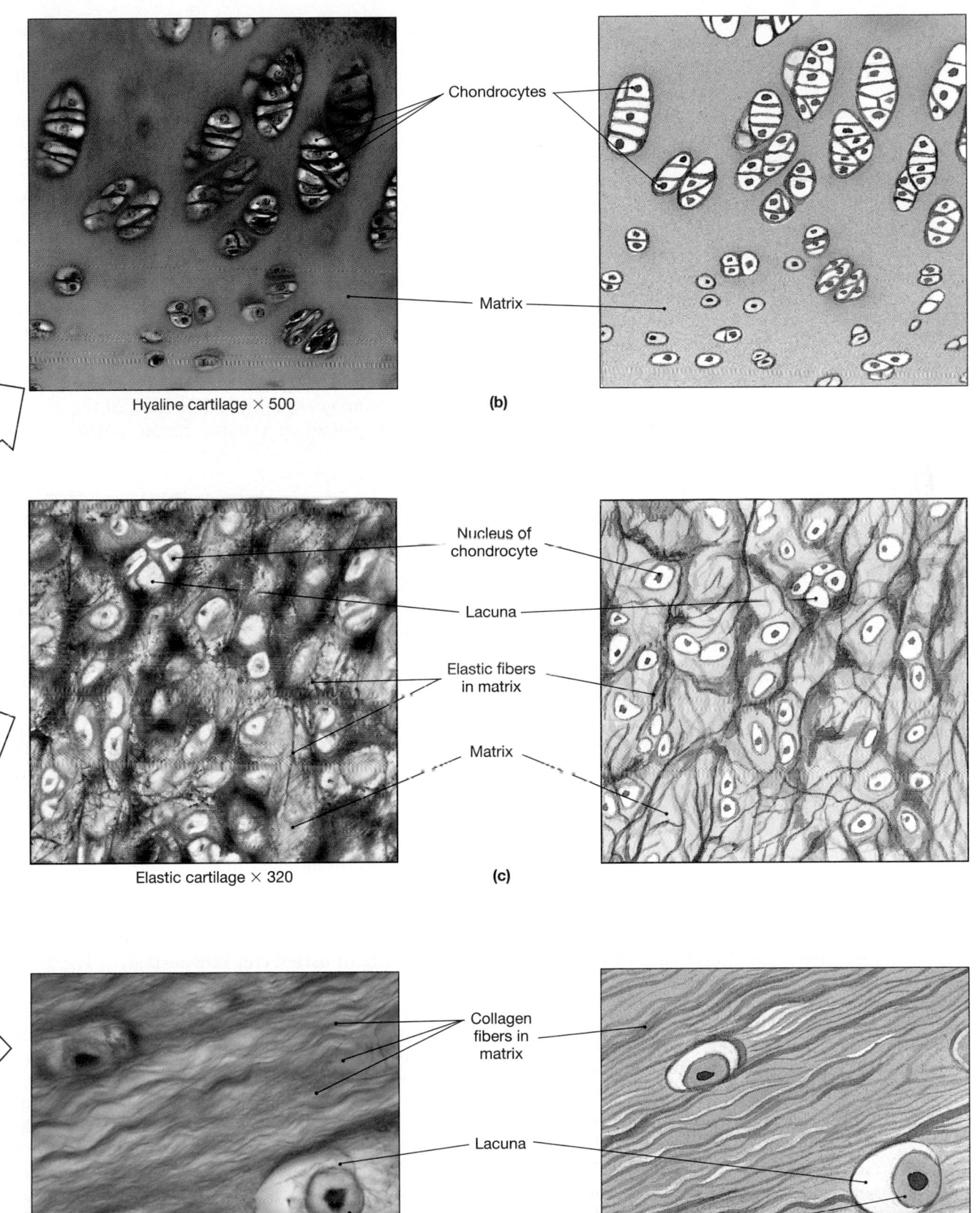

Chondrocytes

Matrix

Hyaline cartilage × 500

(b)

Nucleus of
chondrocyte

Lacuna

Elastic fibers
in matrix

Matrix

Elastic cartilage × 320

(c)

Collagen
fibers in
matrix

Lacuna

Chondrocyte

Fibrocartilage × 750

(d)

it tough but somewhat flexible. Because the fibers do not stain well, they are not always apparent in light microscopy (Figure 4-12b●). Examples of this type of cartilage in the adult body include (1) the connections between the ribs and the sternum, (2) the nasal cartilages and the supporting cartilages along the conducting passageways of the respiratory tract, and (3) the *articular cartilages,* which cover opposing bone surfaces within many joints, such as the elbow and knee.

2. **Elastic cartilage** (Figure 4-12c●) contains numerous elastic fibers that make it extremely resilient and flexible. Elastic cartilage forms the external flap (*pinna*) of the outer ear, the epiglottis, an airway to the middle ear cavity (the *auditory tube*), and small cartilages in the larynx (the *cuneiform cartilages*).

3. **Fibrocartilage** has little ground substance, and its matrix is dominated by collagen fibers (Figure 4-12d●). The collagen fibers are densely interwoven, making this tissue extremely durable and tough. Fibrocartilaginous pads lie between the spinal vertebrae, between the pubic bones of the pelvis, and around or within a few joints and tendons. In these positions, fibrocartilage resists compression, absorbs shocks, and prevents damaging bone-to-bone contact. Cartilages in general heal poorly, and damaged fibrocartilages in joints such as the knee can interfere with normal movements.

CARTILAGES AND KNEE INJURIES The knee is an extremely complex joint that contains both hyaline and fibrocartilage. The hyaline cartilage covers bony surfaces, and fibrocartilage pads within the joint prevent bone contact when movements are under way. Many sports injuries result in the tearing of the fibrocartilage pads. This tearing places more strain on the articular cartilages and leads to further joint damage. Because cartilages are avascular, they heal poorly, and joint cartilages heal even more slowly than other cartilages do. Surgery generally produces only a temporary or incomplete repair.

Recent advances in tissue culture have enabled researchers to grow fibrocartilages in the laboratory. Chondrocytes removed from the knees of injured dogs are cultured in an artificial framework of collagen fibers. They eventually produce masses of fibrocartilage that can be inserted into the damaged joints. Over time, the pads change shape and grow, restoring normal joint function. In the future, this technique may be used to treat severe knee injuries in humans.

CARTILAGE GROWTH Cartilages grow by two mechanisms: *interstitial growth* and *appositional growth*. In **interstitial growth** (Figure 4-13a●), chondrocytes within the cartilage matrix undergo cell division, and the daughter cells produce additional matrix. This process enlarges the cartilage from within, rather like inflating a balloon by forcing more air into it. In **appositional growth,** new layers of cartilage are added to the surface (Figure 4-13b●). In this process, cells of the inner layer of the perichondrium undergo repeated cycles of division. The innermost cells then begin producing cartilage matrix and differentiate into chondrocytes. This process gradually increases the size of the cartilage by adding to its outer surface. Interstitial growth is most important during embryonic development. Appositional growth begins early in development and continues through adolescence. Neither interstitial nor appositional growth occurs in the cartilages of normal adults. However, appositional growth may occur in unusual circumstances, such as after cartilage damage or under excessive stimulation with *growth hormone* from the pituitary gland. Minor damage to cartilage can be repaired by appositional growth at the damaged surface. After more-severe damage, the injured portion of the cartilage will be replaced by a dense fibrous patch.

Bone

Because we shall examine the detailed histology of **bone,** or **osseous** (OS-ē-us) **tissue** (*os,* bone), in Chapter 6, here we will focus on significant differences between cartilage and bone. The volume of ground substance in bone is very small. Roughly one-third of the matrix of bone consists of collagen fibers. The balance is a mixture of calcium salts, primarily calcium phosphate with lesser amounts of calcium carbonate. This combination gives bone truly remarkable properties. By themselves, calcium salts are hard but rather brittle. Collagen fibers are stronger but relatively flexible. In bone, the minerals are organized around the collagen fibers. The result is a strong, somewhat flexible combination that is very resistant to shattering. In its overall properties, bone can compete with the best steel-reinforced concrete.

Figure 4-14● shows the general organization of osseous tissue. Lacunae within the matrix contain bone cells, or **osteocytes** (OS-tē-ō-sīts). The lacunae are typically organized around blood vessels that branch through the bony matrix. Although diffusion cannot occur through the calcium salts, osteocytes communicate with the blood vessels and with one another by means of slender cytoplasmic extensions. These extensions run through long, slender passages in the matrix. These passageways, called **canaliculi** (kan-a-LIK-ū-lē; little canals), form a branching network for the exchange of materials between the blood vessels and the osteocytes.

Except within joint cavities, where they are covered by a layer of hyaline cartilage, bone surfaces are sheathed by a **periosteum** (pe-rē-OS-tē-um) composed of fibrous (outer) and cellular (inner) layers. The periosteum assists in the attachment of a

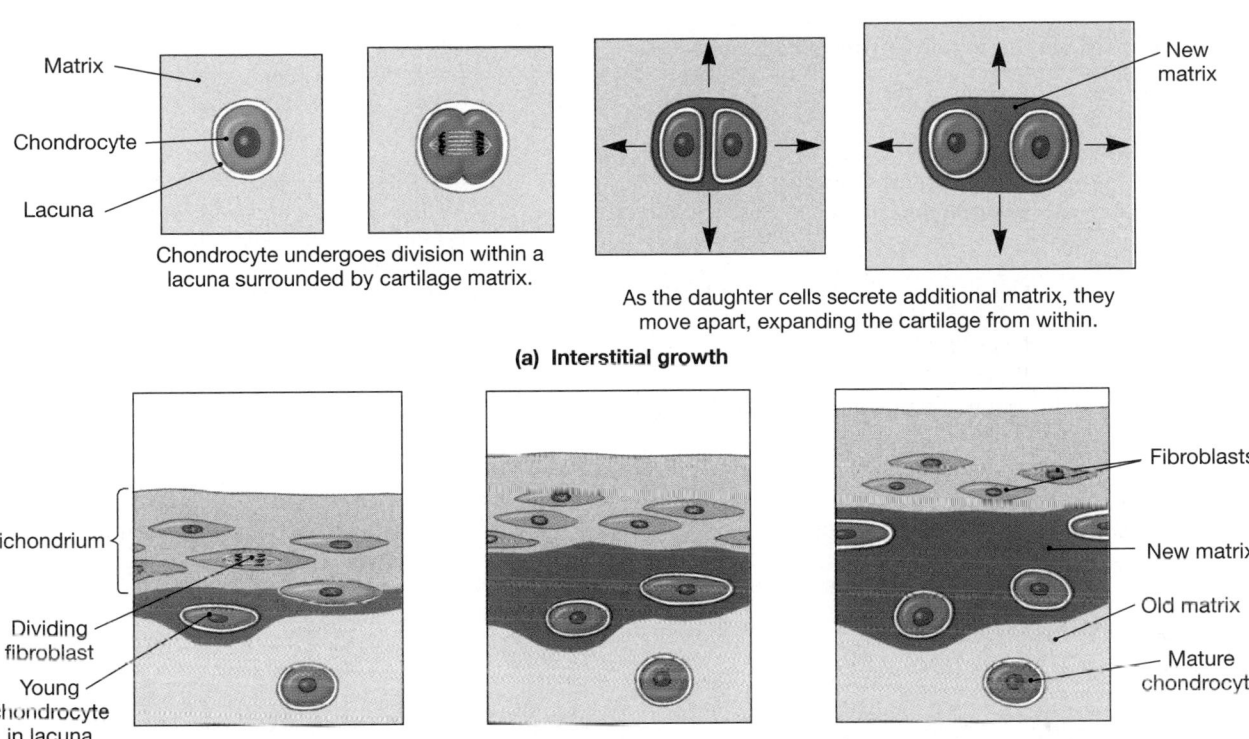

● FIGURE 4-13 Formation and Growth of Cartilage. (a) Interstitial growth. The cartilage expands from within as chondrocytes in the matrix divide, grow, and produce new matrix. **(b)** Appositional growth. The cartilage grows at its external surface through the differentiation of fibroblasts into chondrocytes within the cellular layer of the perichondrium.

bone to surrounding tissues and to associated tendons and ligaments. The cellular layer functions in appositional bone growth and participates in repairs after an injury. Unlike cartilage, bone undergoes extensive remodeling throughout life, and complete repairs can be made even after severe damage has occurred. Bones also respond to the stresses placed on them, growing thicker and stronger with exercise and thin and brittle with inactivity.

Table 4-1 summarizes the similarities and differences between cartilage and bone.

☑ Why does cartilage heal so slowly?

☑ If a person has a herniated intervertebral disc, what type of cartilage has been damaged?

☑ Which two types of connective tissue have a matrix that is fluid?

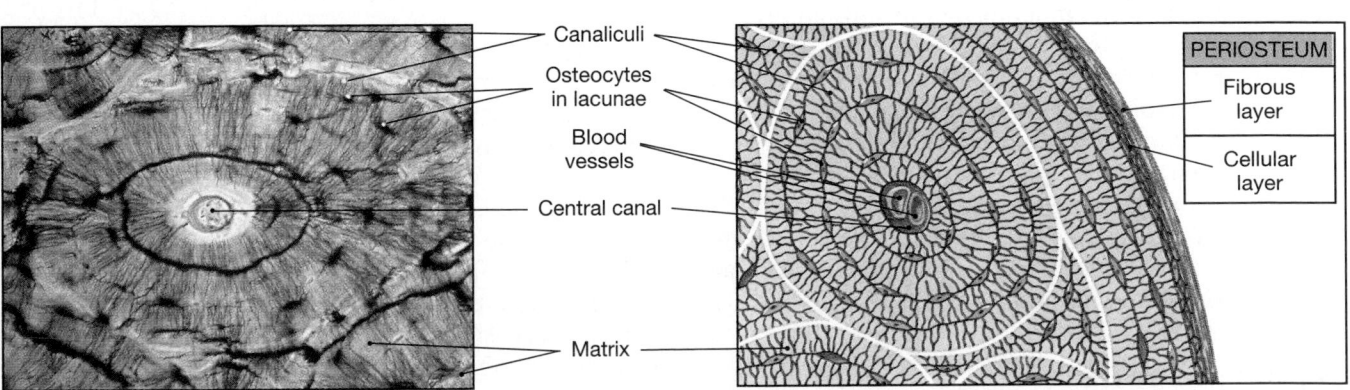

● FIGURE 4-14 Bone. The osteocytes in bone are generally organized in groups around a central space that contains blood vessels. Bone dust produced during grinding fills the lacunae and the central canal, making them appear dark. (LM × 362)

TABLE 4-1 A Comparison of Cartilage and Bone

Characteristic	Cartilage	Bone
Structural Features		
Cells	Chondrocytes in lacunae	Osteocytes in lacunae
Ground substance	Chondroitin sulfate (in proteoglycan) and water	A small volume of liquid surrounding insoluble crystals of calcium phosphate and calcium carbonate
Fibers	Collagen, elastic, and reticular fibers (proportions vary)	Collagen fibers predominate
Vascularity	None	Extensive
Covering	Perichondrium, two-part	Periosteum, two-part
Strength	Limited: bends easily but hard to break	Strong: resists distortion until breaking point
Metabolic Features		
Oxygen demands	Relatively low	Relatively high
Nutrient delivery	By diffusion through matrix	By diffusion through cytoplasm and fluid in canaliculi
Growth	Interstitial and appositional	Appositional only
Repair capabilities	Limited ability	Extensive ability

MEMBRANES

Epithelia and connective tissues combine to form membranes that cover and protect other structures and tissues in the body. Four such membranes occur in the body: (1) *mucous membranes,* (2) *serous membranes,* (3) the *cutaneous membrane,* and (4) *synovial membranes.*

Mucous Membranes

Mucous membranes line cavities that communicate with the exterior, including the digestive, respiratory, reproductive, and urinary tracts (Figure 4-15a●). The epithelial surfaces are kept moist at all times. They are lubricated either by mucus produced by goblet cells or multicellular glands or by exposure to fluids such as urine or semen. The loose connective tissue component of a mucous membrane is called the **lamina propria** (PRŌ-pre-uh). We will consider the organization of specific mucous membranes in greater detail in later chapters.

Many mucous membranes are lined by simple epithelia that perform absorptive or secretory functions, such as the simple columnar epithelium of the digestive tract. Other types of epithelia may be involved, however. For example, a stratified squamous epithelium is part of the mucous membrane of the mouth, and the mucous membrane along most of the urinary tract has a transitional epithelium.

Serous Membranes

Serous membranes line the sealed, internal divisions of the ventral body cavity. These membranes consist of a *mesothelium* supported by loose connective tissue (Figure 4-15b●). You may recall from Chapter 1 (∞ [p. 24]) that there are three types of serous membranes: (1) The *pleura* lines the pleural cavities and covers the lungs; (2) the *peritoneum* lines the peritoneal cavity and covers the surfaces of the enclosed organs; and (3) the *pericardium* lines the pericardial cavity and covers the heart. Serous membranes are very thin, but they are firmly attached to the body wall and to the organs they cover. When you are looking at an organ such as the heart or stomach, you are really seeing the tissues of the organ through a transparent serous membrane.

The opposing (*parietal* and *visceral*) surfaces of a serous membrane are in close contact at all times. Minimizing friction between these opposing surfaces is the primary function of serous membranes. Because the mesothelia are very thin, serous membranes are relatively permeable, and tissue fluids diffuse onto the exposed surface, keeping it moist and slippery.

The fluid formed on the surfaces of a serous membrane is called a **transudate** (TRAN-sū-dāt; *trans,* across). Specific transudates are called **pleural fluid, peritoneal fluid,** or **pericardial fluid,** depending on their source. In healthy individuals, the total volume of transudate is extremely small, just enough to prevent friction between the walls of the cavities and the surfaces of internal organs. But after an injury or in certain disease states, the volume of transudate may increase dramatically, complicating existing medical problems or producing new ones. Examples of conditions characterized by accelerated transudate production include *pleural effusions, pericardial effusions,* and *ascites.*

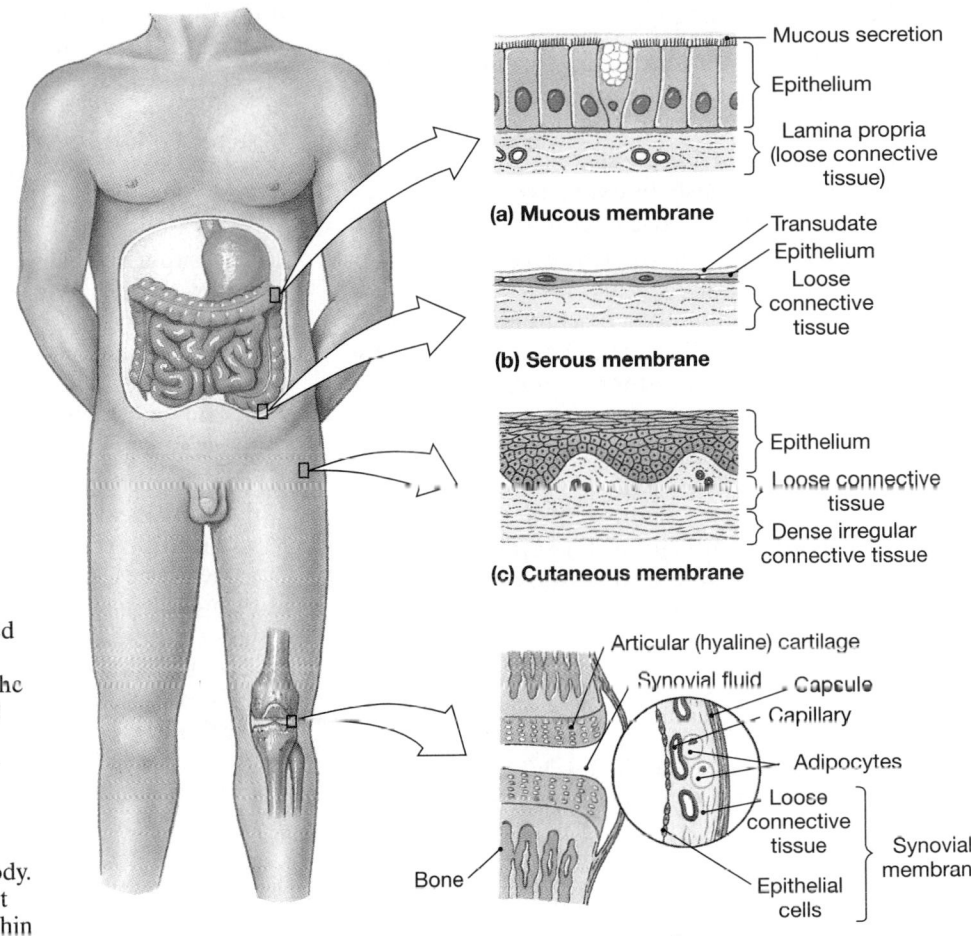

•FIGURE 4-15 Membranes.
(a) Mucous membranes are coated with the secretions of mucous glands. Mucous membranes line the digestive, respiratory, urinary, and reproductive tracts. **(b)** Serous membranes line the ventral body cavities (the peritoneal, pleural, and pericardial cavities). **(c)** The cutaneous membrane, or skin, covers the outer surface of the body. **(d)** Synovial membranes line joint cavities and produce the fluid within the joint.

PROBLEMS WITH SEROUS MEMBRANES Several clinical conditions, including infection and chronic irritation, can cause the abnormal buildup of fluid within one of the ventral body cavities. *Pleuritis,* or *pleurisy,* is an inflammation of the pleural cavities. At first the membranes become dry, and the opposing membranes may scratch against one another, producing a sound known as a *pleural rub.* Friction between opposing layers of serous membrane may promote the formation of **adhesions**, fibrous connections that lock the membranes together and eliminate the friction. But they also severely restrict the movement of the affected organ(s), and they may compress blood vessels or nerves. Adhesions seldom form between the serous membranes of the pleural cavities. More commonly, continued inflammation and rubbing lead to a gradual increase in the production of fluid to levels well above normal. Fluid then accumulates in the pleural cavities, producing a condition known as *pleural effusion.* Pleural effusion can also be caused by heart conditions that elevate the pressure in blood vessels of the lungs. As fluids build up in the pleural cavities, the lungs are compressed, making breathing difficult. The combination of severe pleural effusion and heart disease can be lethal.

Pericarditis is an inflammation of the pericardium. This condition typically leads to a *pericardial effusion,* an ab-normal accumulation of the fluid in the pericardial cavity. When sudden or severe, the fluid buildup can seriously reduce the efficiency of the heart and restrict blood flow through major vessels.

Peritonitis, an inflammation of the peritoneum, can follow an infection of or injury to the peritoneal lining. Peritonitis is a potential complication of any surgical procedure in which the peritoneal cavity is opened. Liver disease, kidney disease, or heart failure can cause an increase in the rate of fluid movement through the peritoneal lining. *Ascites* (a-SĪ-tēz), the accumulation of peritoneal fluid, creates a characteristic abdominal swelling. The distortion of internal organs by the contained fluid can result in a variety of symptoms; heartburn, indigestion, and low-back pain are common complaints.

The Cutaneous Membrane

The **cutaneous membrane,** or skin, covers the surface of the body. It consists of a stratified squamous epithelium and a layer of loose connective tissue reinforced by underlying dense connective tissue (Figure 4-15c•). In contrast to serous and mucous membranes, the cutaneous membrane is thick, relatively waterproof, and usually dry.

Synovial Membranes

Bones contact one another at joints, or *articulations*. The joints that permit significant movement are surrounded by a fibrous capsule, and they contain a joint cavity lined by a **synovial membrane** (sin-Ō-vē-ul). Such a membrane consists of extensive areas of specialized loose connective tissue containing a matrix of interwoven collagen fibers, proteoglycans, and glycoproteins. The lining of the joint cavity consists of an incomplete layer of squamous or cuboidal cells (Figure 4-15d●). Although often called an epithelium by histologists, it differs from other epithelia in four respects: (1) The lining cells are macrophages and specialized fibroblasts, which regulate the composition of the synovial fluid within the cavity; (2) there is no basement membrane; (3) there may be gaps of up to 1 μm between adjacent cells; and (4) there is a continuous exchange of fluid and solutes between the **synovial fluid** and capillaries in the underlying connective tissue.

THE CONNECTIVE TISSUE FRAMEWORK OF THE BODY

Connective tissues create the internal framework of the body. Layers of connective tissue connect the organs within the dorsal and ventral body cavities with the rest of the body. These layers (1) provide strength and stability; (2) maintain the relative positions of internal organs; and (3) provide a route for the distrib-

ution of blood vessels, lymphatics, and nerves. **Fasciae** (FASH-ē-ē); singular, *fascia*) are connective tissue layers and wrappings that support and surround organs. We can divide the fasciae of the body into three types of layers—the *superficial fascia*, the *deep fascia*, and the *subserous fascia;* the functional anatomy of these layers is illustrated in Figure 4-16●:

1. The **superficial fascia,** or **subcutaneous layer** (*sub,* below + *cutis,* skin) is also termed the **hypodermis** (*hypo,* below + *dermis,* skin). This layer of loose connective tissue and fat separates the skin from underlying tissues and organs. It provides insulation and padding and lets the skin and underlying structures move independently.

2. The **deep fascia** consists of dense connective tissue. The organization of the fibers resembles that of plywood: All the fibers in an individual layer run in the same direction, but the orientation of the fibers changes from one layer to another. This arrangement helps the tissue resist forces applied from many different directions. The tough capsules that surround most organs, including the kidneys and the organs in the thoracic and peritoneal cavities, are components of the deep fascia. The perichondrium around cartilages, the periosteum around bones and the ligaments that interconnect them, and the connective tissues of muscle, including tendons, are all part of the deep fascia. The dense connective tissue components are interwoven. For example, the deep fascia around a muscle blends into the tendon, whose fibers intermingle with those of the periosteum. This arrangement creates a strong, fibrous network for the body and ties structural elements together.

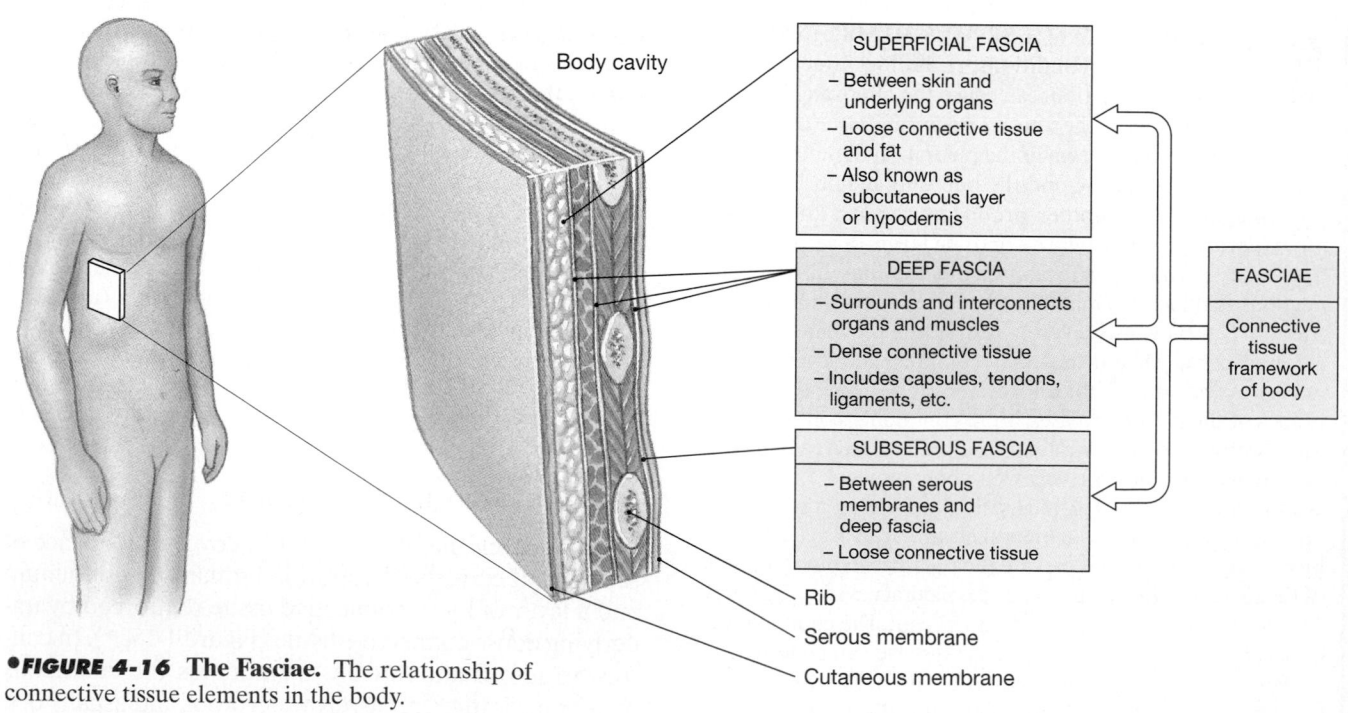

●**FIGURE 4-16** **The Fasciae.** The relationship of connective tissue elements in the body.

3. The **subserous fascia** is a layer of loose connective tissue that lies between the deep fascia and the serous membranes that line body cavities. Because this layer separates the serous membranes from the deep fascia, movements of muscles or muscular organs do not severely distort the delicate lining.

☑ Which cavities in the body are lined by serous membranes?

☑ The lining of the nasal cavity is normally moist, contains numerous goblet cells, and rests on a layer of connective tissue called the lamina propria. What type of membrane is this?

☑ A sheet of tissue has many layers of collagen fibers that run in different directions in successive layers. What type of tissue is this?

MUSCLE TISSUE

Muscle tissue is specialized for contraction. Muscle cells possess organelles and properties distinct from those of other cells. They are capable of powerful contractions that shorten the cell along its longitudinal axis. Because they are different from "typical" cells, the term **sarcoplasm** is used to refer to the cytoplasm of a muscle cell, and we refer to the **sarcolemma** rather than to the cell membrane.

Three types of muscle tissue are found in the body: (1) *skeletal muscle,* (2) *cardiac muscle,* and (3) *smooth muscle.* The contraction mechanism is similar in all three, but they differ in their internal organization. We will describe each muscle type in greater detail in Chapter 10. Here we will focus on general characteristics rather than on specific details.

Skeletal Muscle Tissue

Skeletal muscle tissue contains very large muscle cells. Skeletal muscle cells are very unusual because they may be 0.3 meter (1 ft) or more in length. Because the individual muscle cells are relatively long and slender, they are usually called **muscle fibers.** Each fiber is **multinucleated**—that is, it contains many nuclei. There are generally several hundred nuclei just under the surface of the sarcolemma (Figure 4-17a●). Skeletal muscle fibers are incapable of dividing, but new muscle fibers can be produced through the divisions of **satellite cells,** stem cells that persist in adult skeletal muscle tissue. As a result, skeletal muscle tissue can at least partially repair itself after an injury.

Because the actin and myosin filaments are arranged in organized groups, skeletal muscle fibers have a banded, or *striated,* appearance. Skeletal muscle fibers do not usually contract unless stimulated by nerves, and the nervous system provides voluntary control over their activities. Thus, skeletal muscle is called **striated voluntary muscle.**

Skeletal muscle fibers are tied together by loose connective tissue. The collagen and elastic fibers surrounding each cell and group of cells may blend into those of a tendon that conducts the force of contraction, generally to a bone. When the muscle tissue contracts, it pulls on the attached bone.

Cardiac Muscle Tissue

Cardiac muscle tissue is located only in the heart. A typical cardiac muscle cell, also known as a **cardiocyte,** or *cardiac myocyte,* is smaller than a skeletal muscle cell. A typical cardiac muscle cell has one centrally placed nucleus, although cardiocytes may occasionally have as many as five. The prominent striations visible in Figure 4-17b● resemble those of skeletal muscle because the actin and myosin filaments are arranged the same way in both cell types.

Cardiac muscle cells form extensive connections with one another. The connections occur at specialized regions known as **intercalated discs.** At an intercalated disc, the membranes are locked together by desmosomes, intercellular cement, and gap junctions. As a result, cardiac muscle tissue consists of a branching network of interconnected muscle cells. The desmosomes and intercellular cement keep the cells locked together during a contraction, and ion movement through gap junctions helps coordinate the contractions of the cardiac muscle cells. Like skeletal muscle fibers, cardiac muscle cells are incapable of dividing. Because this tissue lacks satellite cells, cardiac muscle tissue damaged by injury or disease cannot regenerate.

Cardiac muscle cells do not rely on nerve activity to start a contraction. Instead, specialized cardiac muscle cells, called **pacemaker cells,** establish a regular rate of contraction. Although the nervous system can alter the rate of pacemaker cell activity, it does not provide voluntary control over individual cardiac muscle cells. Therefore, cardiac muscle is called **striated involuntary muscle.**

Smooth Muscle Tissue

Smooth muscle tissue is located (1) in the walls of blood vessels; (2) around hollow organs such as the urinary bladder; and (3) in layers around the respiratory, circulatory, digestive, and reproductive tracts. A smooth muscle cell is a small, spindle-shaped cell with tapering ends and a single oval nucleus (Figure 4-17c●). Smooth muscle cells can divide, and smooth muscle tissue can regenerate after an injury. The actin and myosin filaments in smooth muscle cells are organized differently from those of skeletal and cardiac muscle, and there are no striations. Smooth muscle cells may contract on

SKELETAL MUSCLE TISSUE

LOCATIONS: Combined with connective tissues and nervous tissue in skeletal muscles, organs such as the skeletal muscles of the limbs

FUNCTIONS: Moves or stabilizes the position of the skeleton; guards entrances and exits to the digestive, respiratory, and urinary tracts; generates heat; protects internal organs

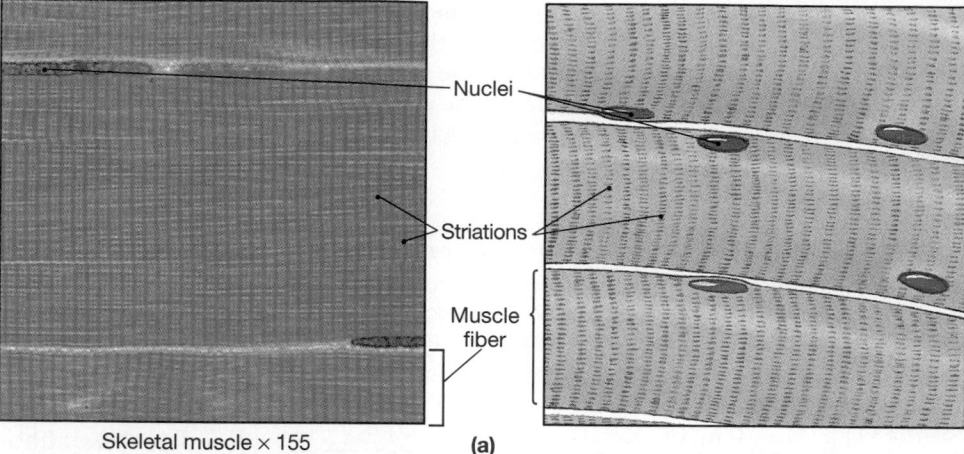

Skeletal muscle × 155

Nuclei

Striations

Muscle fiber

(a)

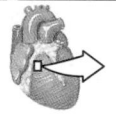

CARDIAC MUSCLE TISSUE

LOCATION: Heart

FUNCTIONS: Circulates blood; maintains blood (hydrostatic) pressure

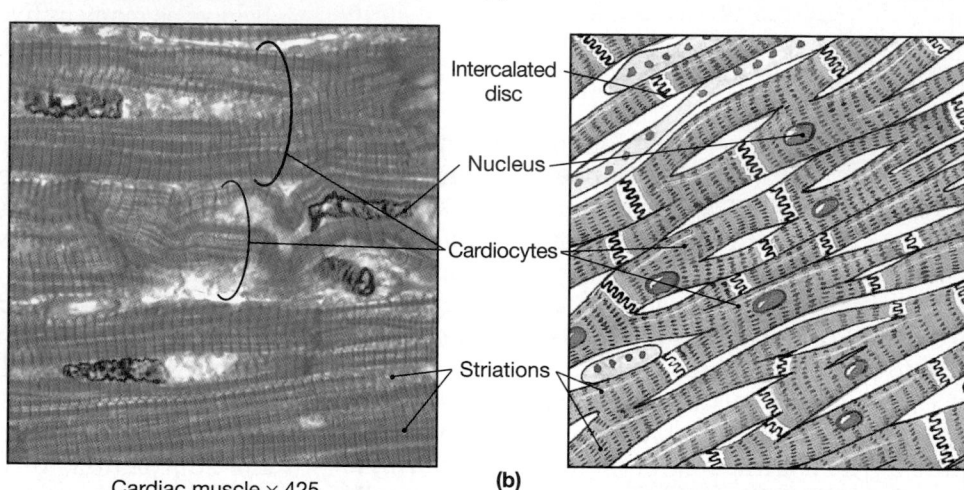

Cardiac muscle × 425

Intercalated disc

Nucleus

Cardiocytes

Striations

(b)

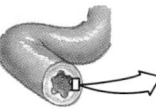

SMOOTH MUSCLE TISSUE

LOCATIONS: Encircles blood vessels; in the walls of digestive, respiratory, urinary, and reproductive organs

FUNCTIONS: Moves food, urine, and reproductive tract secretions; controls diameters of respiratory passageways; regulates diameter of blood vessels and thereby contributes to regulation of tissue blood flow

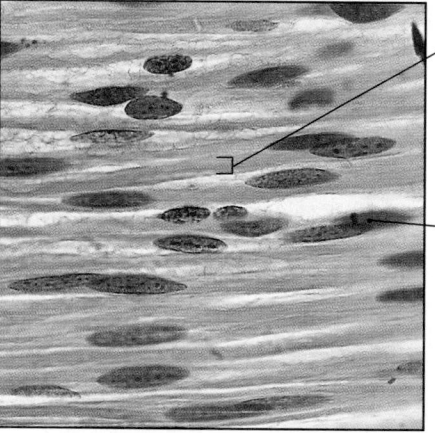

Smooth muscle × 330

Smooth muscle cell

Nucleus

(c)

•*FIGURE 4-17* **Muscle Tissue.** **(a)** Skeletal muscle fibers are large, have multiple, peripherally located nuclei, and exhibit a prominent banding pattern and an unbranched arrangement. **(b)** Cardiac muscle cells differ from skeletal muscle fibers in three major ways: size (cardiac muscle cells are smaller), organization (cardiac muscle cells branch), and number and location of nuclei (a typical cardiac muscle cell has one centrally placed nucleus). Both contain actin and myosin filaments in an organized array that produces striations. **(c)** Smooth muscle cells are small and spindle-shaped, with a central nucleus. They do not branch, and there are no striations.

their own, with gap junctions between adjacent cells coordinating the contractions of individual cells. The contraction of some smooth muscle tissue can be controlled by the nervous system, but the contractile activity is not under your voluntary control. Because your nervous system usually does not provide voluntary control over smooth muscle contractions, smooth muscle is known as **nonstriated involuntary muscle.**

NEURAL TISSUE

Neural tissue, which is also known as *nervous tissue* or *nerve tissue,* is specialized for the conduction of electrical impulses from one region of the body to another region. Ninety-eight percent of the neural tissue in the body is concentrated in the brain and spinal cord, which are the control centers of the nervous system.

Neural tissue contains two basic types of cells: (1) **neurons** (NOO-ronz; *neuro,* nerve) and (2) several different kinds of supporting cells, collectively called **neuroglia** (noo-RŌG-lē-uh), or *glial cells* (*glia,* glue). Our conscious and unconscious thought processes reflect the communication among neurons in the brain. Such communication involves the propagation of electrical impulses in the form of changes in the transmembrane potential. Information is conveyed by the frequency and pattern of impulse generation. Neuroglia have different functions, such as providing a supporting framework for neural tissue and helping supply nutrients to neurons.

The longest cells in your body are neurons; many neurons reach a meter (39 in.) in length. Most neurons are incapable of dividing under normal circumstances, and they have a very limited ability to repair themselves after injury. A typical neuron has a cell body, or **soma,** that contains a large nucleus with a prominent nucleolus (Figure 4-18●). Extending from the soma are various branching processes (projections or outgrowths) termed **dendrites** (DEN-drīts; *dendron,* a tree) and a single **axon.** Dendrites receive information, typically from other neurons, and axons carry information to other cells. Because axons tend to be very long and slender, they are also called **nerve fibers.** In Chapter 12, we shall discuss the properties of neural tissue in detail.

☑ Which type of muscle tissue has small, tapering cells with single nuclei and no obvious striations?

☑ A tissue contains irregularly shaped cells with many fibrous projections, some several centimeters long. These are probably which type of cell?

☑ If skeletal muscle cells in adults are incapable of dividing, how is new skeletal muscle formed?

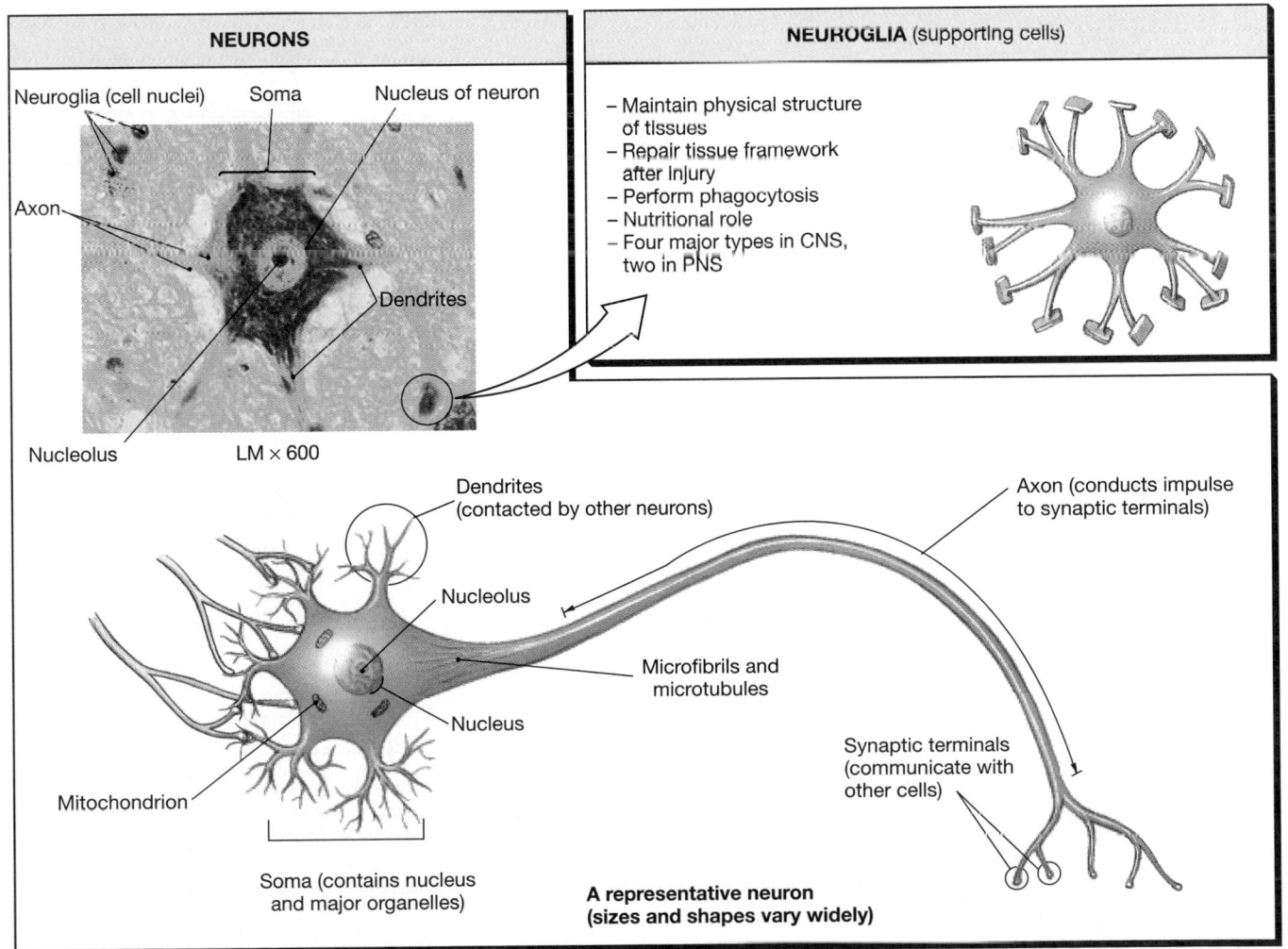

NEURONS	NEUROGLIA (supporting cells)
Neuroglia (cell nuclei) Soma Nucleus of neuron Axon Dendrites Nucleolus LM × 600	– Maintain physical structure of tissues – Repair tissue framework after injury – Perform phagocytosis – Nutritional role – Four major types in CNS, two in PNS

Dendrites (contacted by other neurons)

Nucleolus

Nucleus

Microfibrils and microtubules

Mitochondrion

Soma (contains nucleus and major organelles)

Axon (conducts impulse to synaptic terminals)

Synaptic terminals (communicate with other cells)

A representative neuron (sizes and shapes vary widely)

●**FIGURE 4-18** Neural Tissue

Development of Organ Systems

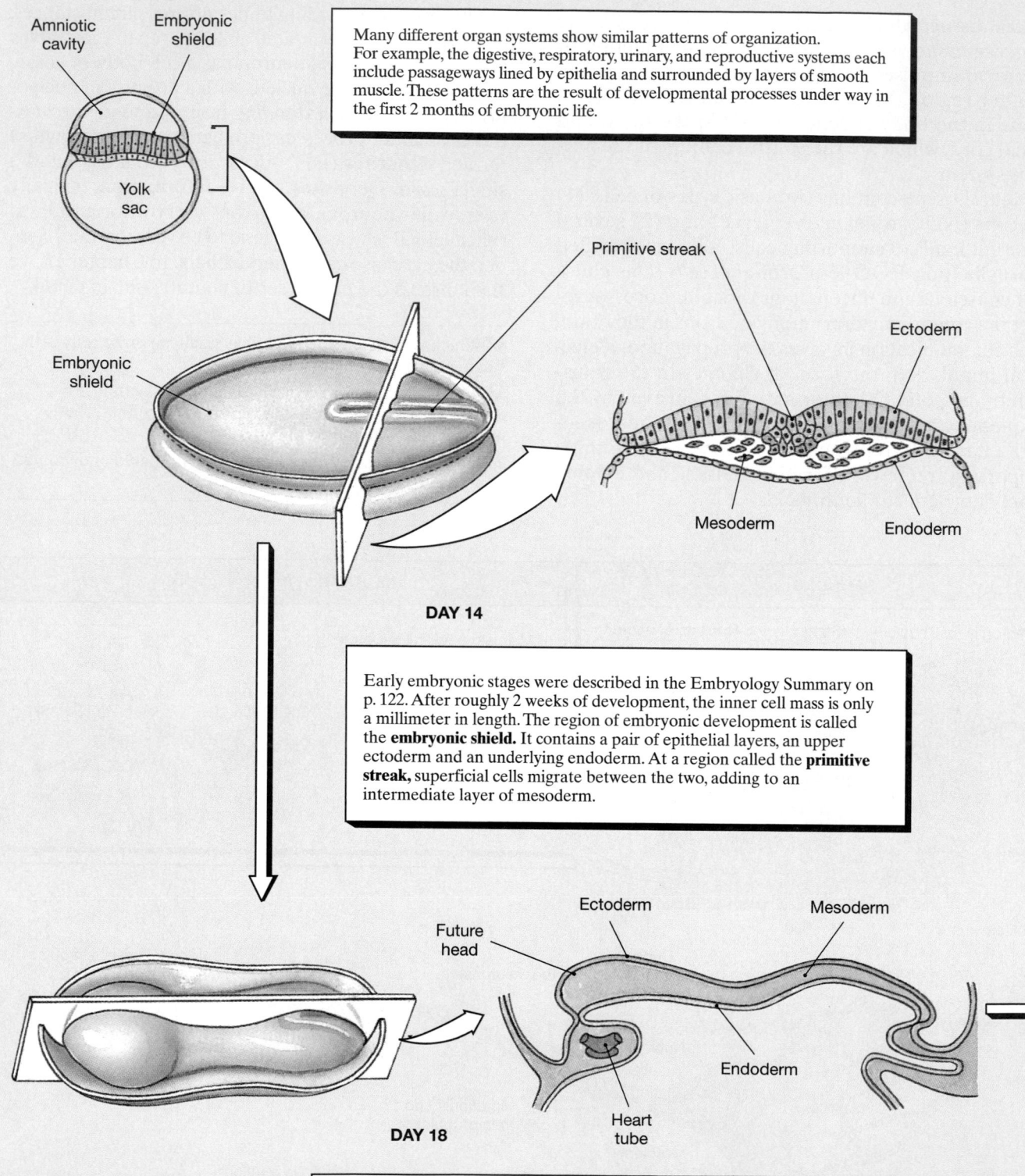

Amniotic cavity

Embryonic shield

Yolk sac

Many different organ systems show similar patterns of organization. For example, the digestive, respiratory, urinary, and reproductive systems each include passageways lined by epithelia and surrounded by layers of smooth muscle. These patterns are the result of developmental processes under way in the first 2 months of embryonic life.

Primitive streak

Ectoderm

Embryonic shield

Mesoderm

Endoderm

DAY 14

Early embryonic stages were described in the Embryology Summary on p. 122. After roughly 2 weeks of development, the inner cell mass is only a millimeter in length. The region of embryonic development is called the **embryonic shield.** It contains a pair of epithelial layers, an upper ectoderm and an underlying endoderm. At a region called the **primitive streak,** superficial cells migrate between the two, adding to an intermediate layer of mesoderm.

Ectoderm

Mesoderm

Future head

Endoderm

Heart tube

DAY 18

By day 18, the embryo has begun to lift off the surface of the embryonic shield. The heart and blood vessels have already formed, well ahead of the other organ systems. Unless otherwise noted, discussions of organ system development in later chapters will begin at this stage.

DERIVATIVES OF PRIMARY GERM LAYERS

Ectoderm forms:	Epidermis of integumentary system and the hair follicles, nails, and glands (sweat, milk, and sebum) communicating with the skin surface Lining of the mouth, salivary glands, nasal passageways, and anus Nervous system, including brain and spinal cord Portions of endocrine system (pituitary and parts of adrenal glands) Pharyngeal cartilages
Mesoderm forms:	Lining of the body cavities (pleural, pericardial, peritoneal) Muscular, skeletal, cardiovascular, and lymphatic systems Kidneys and part of the urinary tract Gonads and most of the reproductive tract Connective tissues supporting all organ systems Portions of endocrine system (parts of adrenal glands and endocrine tissues of reproductive tract)
Endoderm forms:	Most of the digestive system: epithelium (except mouth and anus), exocrine glands (except salivary glands), the liver and pancreas Most of the respiratory system: epithelium (except nasal passageways) and mucous glands Portions of urinary and reproductive systems (ducts and sex cells) Portions of endocrine system (thymus, thyroid, parathyroids, pancreas)

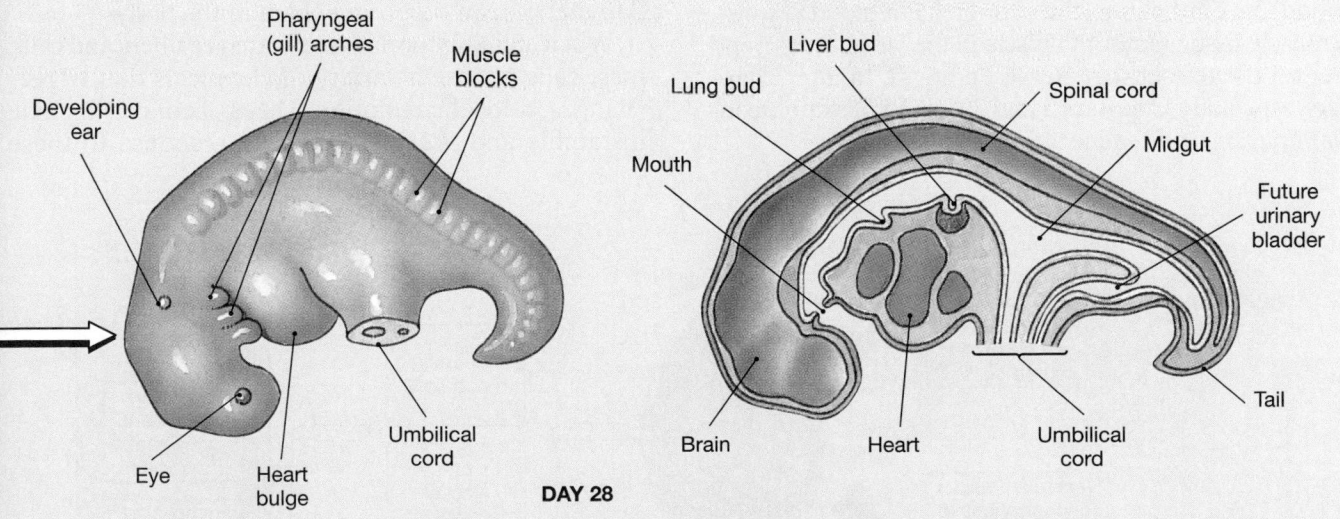

DAY 28

After 1 month, major organ systems have begun to form. The role of each of the primary germ layers in the formation of organs is summarized in the accompanying table; details will be found in later chapters.

TISSUE INJURIES AND AGING

Tissues in the body are not isolated; they combine to form organs with diverse functions. Any injury to the body affects several different tissue types simultaneously. These tissues must respond in a coordinated manner to preserve homeostasis.

Inflammation and Regeneration

The restoration of homeostasis after a tissue has been injured involves two related processes. First, the area is isolated while damaged cells, tissue components, and any dangerous microorganisms are cleaned up. This phase, which coordinates the activities of several different types of tissues, is called **inflammation,** or the **inflammatory response.** Inflammation begins immediately after an injury occurs. It produces several familiar sensations, including swelling, redness, warmth, and pain. An **infection** is an inflammation resulting from the presence of pathogens, such as harmful bacteria.

Second, the damaged tissues are replaced or repaired to restore normal function. This repair process is called **regeneration.** Inflammation and regeneration are controlled at the tissue level. The two phases overlap; isolation establishes a framework that guides the cells responsible for reconstruction, and repairs are under way well before cleanup operations have ended. We will now consider the basics of the repair process after an injury. At this time, we will focus on the interaction among different tissues. Our example includes two connective tissues (loose connective tissue and blood), an epithelium (the endothelia of blood vessels), a muscle tissue (smooth muscle in the vessel walls), and neural tissue (sensory nerve endings). In later chapters, especially Chapters 5 and 22, we shall examine inflammation and regeneration in more detail.

First Phase: Inflammation

Many stimuli—including impact, abrasion, distortion, chemical irritation, infection by pathogenic organisms (such as bacteria or viruses), and extreme temperatures (hot or cold)—can produce inflammation. Each of these stimuli either kills cells, damages fibers, or injures the tissue in some other way. These changes alter the chemical composition of the interstitial fluid: Damaged cells release prostaglandins, proteins, and potassium ions, and the injury itself may have introduced foreign proteins or pathogens.

Immediately after the injury, tissue conditions become even more abnormal. **Necrosis** (ne-KRŌ-sis) refers to the tissue degeneration that occurs after cells have been injured or destroyed. The process begins several hours after the original injury, and the damage is caused by lysosomal enzymes. Through widespread autolysis, lysosomes release enzymes that first destroy the injured cells and then attack surrounding tissues. ∞ *[p. 90]* The result may be an accumulation of debris, fluid, dead and dying cells, and necrotic tissue components collectively known as *pus.* An accumulation of pus in an enclosed tissue space is called an **abscess**.

These tissue changes trigger the inflammatory response by stimulating mast cells, connective tissue cells introduced earlier in this chapter. Figure 4-19● follows the events set in motion by the activation of mast cells. Although in this example the injury has occurred in a loose connective tissue, the process would be basically the same after an injury to any connective tissue proper. Because all organs have connective tissues, inflammation can occur anywhere in the body.

When an injury occurs that damages fibers and cells, mast cells release a variety of chemicals that trigger changes in local circulation. These chemicals include histamine and prostaglandins. In response to these

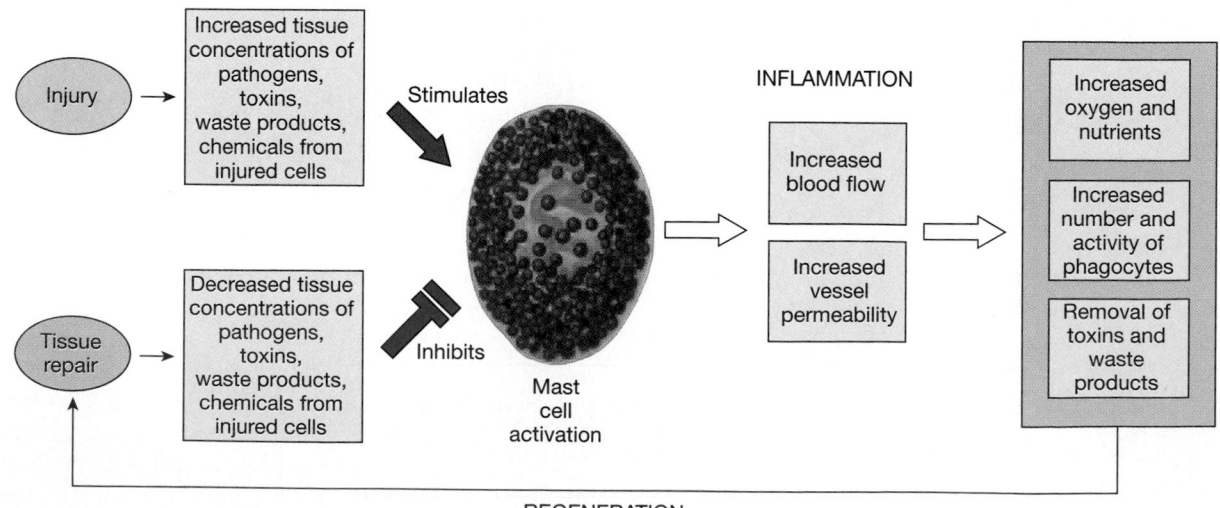

●*FIGURE 4-19* **An Introduction to Inflammation**

chemicals, the smooth muscle tissue that surrounds local blood vessels relaxes, and the vessels enlarge in diameter, or *dilate*. This dilation increases blood flow through the tissue, giving the region a reddish color and making it warm to the touch. The combination of abnormal tissue conditions and chemicals released by mast cells stimulates sensory nerve endings that produce sensations of pain. At the same time, the chemicals released by mast cells make the endothelial cells of local capillaries more permeable. Plasma, including blood proteins, now diffuses into the injured tissue, and the area becomes swollen.

The increased blood flow accelerates the delivery of nutrients and oxygen and the removal of dissolved waste products and toxic chemicals. It also brings white blood cells to the region. These phagocytic cells migrate into the injury site and assist in the defense and cleanup operations. Stimulated macrophages and microphages protect the tissue from infection and perform cleanup by engulfing both debris and bacteria. Over a period of hours to days, this cleanup process generally succeeds in eliminating the inflammatory stimulus.

Second Phase: Regeneration

Although the situation is under control and no further injury will occur, many cells in the area have died, as a result of either the initial injury or subsequent regional changes. As tissue conditions return to normal, fibroblasts move into the necrotic area, laying down a network of collagen fibers that stabilizes the injury site. This process produces a dense, collagenous framework known as *scar tissue*. Over time, scar tissue is remodeled and assumes a more normal appearance. The cell population in the area gradually increases; some cells migrate to the site, and others are produced through the division of mesenchymal stem cells.

Each organ has a different ability to regenerate after injury. That ability can be directly linked to the pattern of tissue organization in the injured organ. Epithelia and connective tissues regenerate well, smooth and skeletal muscle tissues do so relatively poorly, and cardiac muscle tissue and neural tissue cannot regenerate at all. Your skin, which is dominated by epithelia and connective tissues, regenerates rapidly and completely after injury. (We shall detail this process in Chapter 5.) In contrast, damage to the heart is much more serious, because, although the connective tissues of the heart can be repaired, lost cardiac muscle cells are replaced only by scar tissue.

Aging and Tissue Repair

Tissues change with age, and the speed and effectiveness of tissue repairs decrease. In general, repair and maintenance activities throughout the body slow down, and a combination of hormonal changes and alterations in lifestyle affect the structure and chemical composition of many tissues. Epithelia get thinner and connective tissues more fragile. Individuals bruise easily and bones become brittle; joint pain and broken bones are common complaints among the elderly. Because cardiac muscle cells and neurons cannot be replaced, cumulative damage can eventually cause major health problems such as cardiovascular disease or deterioration in mental function.

In future chapters, we will consider the effects of aging on specific organs and systems. Some of these changes are genetically programmed. For example, the chondrocytes of older individuals produce a slightly different form of proteoglycan than those of younger people do. The difference probably accounts for the thinner and less resilient cartilage of older people. In some cases, the tissue degeneration can be temporarily slowed or even reversed. The age-related reduction in bone strength in women, a condition called *osteoporosis*, typically results from a combination of inactivity, low dietary calcium levels, and a reduction in circulating *estrogens* (sex hormones). A program of exercise, calcium supplements, and hormone replacement therapies can generally maintain normal bone structure for many years.

Aging and Cancer Incidence

Cancer rates increase with age, and roughly 25 percent of all people in the United States develop cancer at some point in their lives. It has been estimated that 70–80 percent of cancer cases result from chemical exposure, environmental factors, or some combination of the two, and 40 percent of those cancers are caused by cigarette smoke. Each year in the United States more than 500,000 individuals die of cancer, making it Public Health Enemy Number 2, second only to heart disease. We covered cancer development and growth in Chapter 3. ∞ *[p. 103]* The *Applications Manual* contains a detailed discussion of cancer causes, incidence, and treatment. [AM] *A Closer Look: Cancer*

This chapter introduced the four basic tissue types that occur in the body. In terms of their contribution to total body weight, muscle tissue is most important (about 50 percent), followed by connective tissues (45 percent), epithelium and glands (3 percent), and neural tissue (2 percent). In combination, these tissues form all the organs and systems we shall discuss in subsequent chapters. In Chapter 5 we shall consider the *integument,* an easily accessible structure with structural characteristics intermediate between those of organs and organ systems.

Tissue Structure and Disease

Pathologists (pa-THOL-o-jists) are physicians who specialize in the study of disease processes. Diagnosis, rather than treatment, is usually the main focus of their activities. In their analyses, pathologists integrate anatomical and histological observations to determine the nature and severity of the disease. Because disease processes affect the histological organization of tissues and organs, tissue samples, or **biopsies**, often play a key role in their diagnoses.

Figure 4-20● diagrams the histological changes induced by one relatively common irritating stimulus, cigarette smoke. The first abnormality to be observed is **dysplasia** (dis-PLĀ-zē-uh), a change in the normal shape, size, and organization of tissue cells. Dysplasia is generally a response to chronic irritation or inflammation, and the changes are reversible. The normal trachea (windpipe) and its branches are lined by a pseudostratified, ciliated, columnar epithelium. The cilia move a mucous layer that traps foreign particles and moistens incoming air. The drying and chemical effects of smoking first paralyze the cilia, halting the movement of mucus (Figure 4-20a●). As mucus builds up, the individual coughs to dislodge it (the well-known "smoker's cough").

Epithelia and connective tissues may undergo more-radical changes in structure, caused by the division and differentiation of stem cells. **Metaplasia** (me-tuh-PLĀ-zē-uh) is a structural change that dramatically alters the character of the tissue. In our example, heavy smoking first paralyzes the cilia, and over time the epithelial cells lose their cilia altogether.

As metaplasia occurs, the epithelial cells produced by stem cell divisions no longer differentiate into ciliated columnar cells. Instead, they form a stratified squamous epithelium that provides a greater resistance to drying and chemical irritation (Figure 4-20b●). This epithelium protects the underlying tissues more effectively, but it completely eliminates the moisturization and cleaning properties of the epithelium. The cigarette smoke will now have an even greater effect on more-delicate portions of the respiratory tract. Fortunately, metaplasia is reversible, and the epithelium gradually returns to normal if the individual quits smoking.

During **anaplasia** (a-nuh-PLĀ-zē-uh), tissue organization breaks down. Tissue cells change size and shape, typically becoming unusually large or small (Figure 4-20c●). In anaplasia, which occurs in smokers who develop one form of lung cancer, the cells divide more frequently, but not all divisions proceed in the normal way. Many of the tumor cells have abnormal chromosomes. Unlike dysplasia and metaplasia, anaplasia is irreversible.

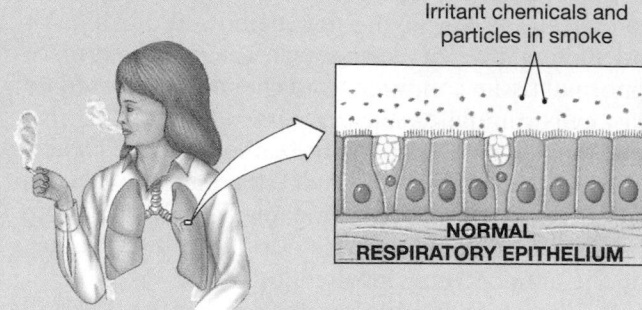

Irritant chemicals and particles in smoke

NORMAL RESPIRATORY EPITHELIUM

Reversible

(a) The cilia of respiratory epithelial cells are damaged and paralyzed by exposure to cigarette smoke. These changes cause the local buildup of mucus and reduce the effectiveness of the epithelium in protecting deeper, more delicate portions of the respiratory tract.

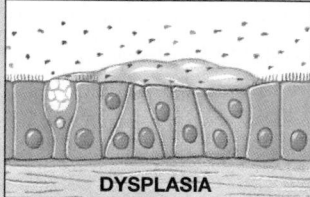

DYSPLASIA

Reversible

(b) In metaplasia, a tissue changes its structure. In this case the stressed respiratory surface converts to a stratified epithelium that protects underlying connective tissues but does nothing for other areas of the respiratory tract.

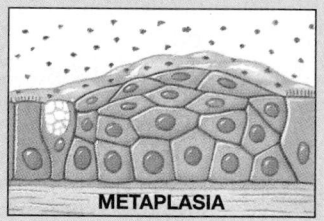

METAPLASIA

Irreversible

(c) In anaplasia, the tissue cells become tumor cells; anaplasia produces a cancerous tumor.

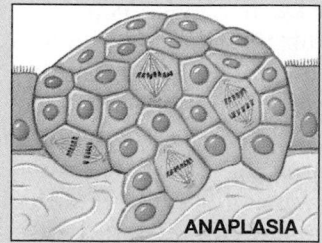

ANAPLASIA

●**FIGURE 4-20** **Changes in a Tissue under Stress.** **(a)** The cilia of respiratory epithelial cells are damaged and paralyzed by exposure to cigarette smoke. These changes cause the local buildup of mucus and reduce the effectiveness of the epithelium in protecting deeper, more-delicate portions of the respiratory tract. **(b)** In metaplasia, a tissue changes in structure. In this case, the stressed respiratory surface converts to a stratified epithelium that protects underlying connective tissues but does nothing for other areas of the respiratory tract. **(c)** In anaplasia, the tissue cells become tumor cells; anaplasia produces cancers.

SELECTED CLINICAL TERMINOLOGY

Terms Discussed in This Chapter

abscess: Accumulation of pus within an enclosed tissue space. *(p. 140)*

adhesions: Restrictive fibrous connections that can result from surgery, infection, or other injuries to serous membranes. *(p. 133)*

anaplasia (a-nuh-PLĀ-zē-uh): An irreversible change in the size and shape of tissue cells. *(p. 142)*

antiangiogenesis factor: A secretion produced by chondrocytes that inhibits the growth of blood vessels. *(p. 128)*

ascites (a-SĪ-tēz): Accumulation of fluid in the peritoneal cavity, usually caused by liver or kidney disease or heart failure. *(p. 133)*

dysplasia (dis-PLĀ-zē-uh): A change in the normal shape, size, and organization of tissue cells. *(p. 142)*

exfoliative cytology: The study of cells shed or collected from epithelial surfaces. *(p. 117)*

liposuction: A surgical procedure to remove unwanted adipose tissue by sucking it out through a tube. *(p. 124)*

metaplasia (me-tuh-PLĀ-zē-uh): A structural change that alters the character of a tissue. *(p. 142)*

necrosis: Tissue degeneration that occurs after cells have been injured or destroyed; due to the release of lysosomal enzymes through autolysis. *(p. 140)*

pathologists (pa-THOL-o-jists): Physicians who specialize in the study of disease processes. *(p. 142)*

pericarditis: An inflammation of the pericardial lining that may lead to the accumulation of pericardial fluid (a *pericardial effusion*). *(p. 133)*

peritonitis: Inflammation of the peritoneum after infection or injury. *(p. 133)*

pleural effusion: Accumulation of fluid within the pleural cavities as the result of chronic infection or inflammation of the pleura. *(p. 133)*

pleuritis *(pleurisy):* An inflammation of the pleural cavities. This condition may cause the production of a sound known as the pleural rub. *(p. 133)*

regeneration: The repairing of injured tissues that follows inflammation. *(p. 140)*

AM Additional Terms Discussed in the *Applications Manual*

chemotherapy: Administration of drugs that either kill cancerous tissues or prevent mitotic divisions.

immunotherapy: Administration of drugs that help the immune system recognize and attack cancer cells.

oncologists (on-KOL-o-jists): Physicians who specialize in identifying and treating cancers.

remission: A stage in which a tumor stops growing or becomes smaller; the goal of cancer treatment.

CHAPTER REVIEW

 On-line resources for this chapter are on our World Wide Web site at:
http://www.prenhall.com/martini/fap

STUDY OUTLINE

INTRODUCTION, p. 109

1. **Tissues** are collections of specialized cells and cell products that are organized to perform a relatively limited number of functions. There are four *tissue types: epithelial tissue, connective tissue, muscle tissue,* and *neural tissue.*
2. **Histology** is the study of tissues.

EPITHELIAL TISSUE, p. 109

1. **Epithelial tissue** includes epithelia and glands. An **epithelium** is an **avascular** layer of cells that forms a barrier with certain properties.

Functions of Epithelial Tissue, p. 109

2. Epithelia provide physical protection, control permeability, provide sensation, and produce specialized secretions. Gland cells are epithelial cells that produce secretions. In **glandular epithelia**, most cells produce secretions.
3. A **basement membrane** attaches epithelia to underlying connective tissues.

Specializations of Epithelial Cells, p. 110

4. Epithelial cells are specialized to allow them to maintain the physical integrity of the epithelium and to perform secretory or transport functions. *(Figure 4-1)*
5. Epithelial cells may show **polarity.** Many have microvilli, and some have **stereocilia.**

6. The coordinated beating of the cilia on a **ciliated epithelium** moves materials across the epithelial surface.

Maintaining the Integrity of Epithelia, p. 111

7. Cells can attach to other cells or to extracellular protein fibers by means of **cell adhesion molecules (CAMs)** or at specialized attachment sites called **cell junctions.** There are three major types of cell junctions: **tight junctions, desmosomes,** and **gap junctions.** *(Figure 4-2)*
8. The inner surface of each epithelium is connected to a two-part basement membrane consisting of a **basal lamina** and a **reticular lamina.** Divisions by **germinative cells** continually replace the short-lived epithelial cells.

Classification of Epithelia, p. 112

9. Epithelia are classified on the basis of the number of cell layers and the shape of the exposed cells.
10. A **simple epithelium** has a single layer of cells covering the basement membrane; a **stratified epithelium** has several layers. The cells in a **squamous epithelium** are thin and flat. Cells in a **cuboidal epithelium** resemble little hexagonal boxes; those in a **columnar epithelium** are taller and more slender. *(Figures 4-3, 4-4, 4-5)*

Glandular Epithelia, p. 117

11. **Exocrine glands** discharge secretions onto the body surface or into **ducts** that communicate with the exterior. *Hor-*

mones, the secretions of **endocrine glands**, are released by gland cells into the surrounding interstitial fluid.

12. A glandular epithelial cell may release its secretions through merocrine, apocrine, or holocrine mechanisms. In **merocrine secretion,** the most common secretion method, the product is released through exocytosis. **Apocrine secretion** involves the loss of both secretory product and cytoplasm. Unlike the other two methods, **holocrine secretion** destroys the cell, which becomes packed with secretions and bursts. *(Figure 4-6)*

13. In epithelia that contain scattered gland cells, individual secretory cells are called **unicellular glands. Multicellular glands** are organs that contain glandular epithelia that produce exocrine or endocrine secretions.

14. Exocrine glands can be classified as **unicellular exocrine glands (goblet cells)** or as **multicellular exocrine glands** on the basis of structure. Multicellular exocrine glands can be further classified according to structure. *(Figure 4-7)*

CONNECTIVE TISSUES, p. 119

1. **Connective tissues** are internal tissues with many important functions: establishing a structural framework; transporting fluids and dissolved materials; protecting delicate organs; supporting, surrounding, and interconnecting tissues; storing energy reserves; and defending the body from microorganisms.

2. All connective tissues have specialized cells and a **matrix** composed of extracellular protein fibers and a **ground substance.**

Classification of Connective Tissues, p. 120

3. **Connective tissue proper** refers to connective tissues that contain varied cell populations and fiber types surrounded by a syrupy ground substance.

4. **Fluid connective tissues** have a distinctive population of cells suspended in a watery ground substance that contains dissolved proteins. The two types are blood and lymph.

5. **Supporting connective tissues** have a less diverse cell population than connective tissue proper does and a dense ground substance with closely packed fibers. The two types of supporting connective tissues are cartilage and bone.

Connective Tissue Proper, p. 120

6. Connective tissue proper contains fibers, a viscous ground substance, and a varied population of cells, including **fibroblasts, macrophages, adipocytes, mesenchymal cells, melanocytes, mast cells, lymphocytes,** and **microphages.**

7. There are three types of fibers in connective tissue: **collagen fibers, reticular fibers,** and **elastic fibers.**

8. The first connective tissue to appear in an embryo is **mesenchyme,** or *embryonic connective tissue.*

9. Connective tissue proper is classified as **loose connective tissue (areolar tissue)** or **dense connective tissue.** There are two specialized types of loose connective tissues: **adipose tissue,** including **brown fat,** and **reticular tissue.** Most of the volume in dense connective tissue consists of fibers. There are two types of dense connective tissue: **dense regular connective tissue** and **dense irregular connective tissue.** *(Figures 4-8, 4-9, 4-10)*

Fluid Connective Tissues, p. 126

10. **Blood** and **lymph** are connective tissues that contain distinctive collections of cells in a fluid matrix.

11. Blood contains *formed elements:* **red blood cells (erythrocytes), white blood cells (leukocytes),** and **platelets.** Its watery ground substance is called **plasma.** *(Figure 4-11)*

12. **Arteries** carry blood from the heart and toward **capillaries,** where water and small solutes move into the interstitial fluid of surrounding tissues. **Veins** return blood to the heart.

13. Lymph forms as interstitial fluid enters the **lymphatics,** which return lymph to the cardiovascular system.

Supporting Connective Tissues, p. 126

14. Cartilage and bone are called supporting connective tissues because they support the rest of the body.

15. The matrix of **cartilage** is a firm gel that contains **chondroitin sulfates** (used to form proteoglycans) and cells called **chondrocytes,** cells that live in pockets called **lacunae.** A fibrous **perichondrium** separates cartilage from surrounding tissues. There are three types of cartilage: **hyaline cartilage, elastic cartilage,** and **fibrocartilage.** *(Figure 4-12)*

16. Cartilage grows by two different mechanisms—**interstitial growth** and **appositional growth.** *(Figure 4-13)*

17. Chondrocytes rely on diffusion through the avascular matrix to obtain nutrients.

18. **Bone,** or **osseous tissue,** has **osteocytes,** which also live in lacunae, and a matrix consisting of collagen fibers and calcium salts, giving it unique properties. *(Figure 4-14; Table 4-1)*

19. Osteocytes depend on diffusion through **canaliculi** for nutrient intake.

20. Each bone is surrounded by a **periosteum** with fibrous and cellular layers.

MEMBRANES, p. 132

1. Membranes form a barrier or interface. Epithelia and connective tissues combine to form membranes that cover and protect other structures and tissues. There are four types of membranes: *mucous, serous, cutaneous,* and *synovial.* *(Figure 4-15)*

Mucous Membranes, p. 132

2. **Mucous membranes** line cavities that communicate with the exterior. They contain loose connective tissue called the **lamina propria.**

Serous Membranes, p. 132

3. **Serous membranes** line the body's sealed internal cavities. They form a fluid called a **transudate.**

The Cutaneous Membrane, p. 133

4. The **cutaneous membrane,** or skin, covers the body surface.

Synovial Membranes, p. 134

5. **Synovial membranes** form an incomplete lining within the cavities of synovial joints.

THE CONNECTIVE TISSUE FRAMEWORK OF THE BODY, p. 134

1. Internal organs and systems are tied together by a network of connective tissue proper that includes the **superficial fascia** (the **subcutaneous layer,** or **hypodermis,** separating the skin from underlying tissues and organs), the **deep fascia** (dense connective tissue), and the **subserous fascia** (the layer between the deep fascia and the serous membranes that line body cavities). *(Figure 4-16)*

MUSCLE TISSUE, p. 135

1. **Muscle tissue** is specialized for contraction. There are three types of muscle tissue: *skeletal muscle, cardiac muscle,* and *smooth muscle.* *(Figure 4-17)*

Skeletal Muscle Tissue, p. 135

2. The cells of **skeletal muscle tissue** are **multinucleated**. Skeletal muscle, or **striated voluntary muscle**, produces new fibers by the division of **satellite cells**.

Cardiac Muscle Tissue, p. 135

3. **Cardiocytes**, the cells of **cardiac muscle tissue**, occur only in the heart. Cardiac muscle, or **striated involuntary muscle**, relies on **pacemaker cells** for regular contraction.

Smooth Muscle Tissue, p. 135

4. **Smooth muscle tissue** is not striated. Smooth muscle cells, or **nonstriated involuntary muscle**, can divide and therefore regenerate after injury has occurred.

NEURAL TISSUE, p. 137

1. **Neural tissue** conducts electrical impulses that convey information from one area of the body to another.

2. Cells in neural tissue are either neurons or neuroglia. **Neurons,** whose axons are often called **nerve fibers,** transmit information as electrical impulses. There are several kinds of **neuroglia,** but their basic functions include supporting neural tissue and helping supply nutrients to neurons. *(Figure 4-18)*

3. A typical neuron has a **soma, dendrites,** and an **axon.** The axon carries information to other cells.

TISSUE INJURIES AND AGING , p. 140

Inflammation and Regeneration, p. 140

1. Any injury affects several tissue types simultaneously, and they respond in a coordinated manner. Homeostasis is restored by two processes: inflammation and regeneration.

2. **Inflammation,** or the **inflammatory response,** isolates the injured area while damaged cells, tissue components, and any dangerous microorganisms (which could cause **infection**) are cleaned up. **Regeneration** is the repair process that restores normal function. *(Figure 4-19)*

Aging and Tissue Repair, p. 141

3. Tissues change with age. Repair and maintenance grow less efficient, and the structure and chemical composition of many tissues are altered.

Aging and Cancer Incidence, p. 141

4. Cancer incidence increases with age, with roughly three-quarters of all cases caused by exposure to chemicals or by other environmental factors.

REVIEW QUESTIONS

| Level 1 | Reviewing Facts and Terms |

Match each numbered item with the most closely related lettered item. Use letters for answers in the spaces provided.

___ **1.** histology
___ **2.** microvilli

___ **3.** gap junction
___ **4.** tight junction
___ **5.** ground substance
___ **6.** basal and reticular laminae
___ **7.** germinative cells
___ **8.** mesothelium
___ **9.** endothelium
___ **10.** mucus
___ **11.** destroys gland cells
___ **12.** hormones
___ **13.** goblet cells
___ **14.** adipocytes
___ **15.** macrophages
___ **16.** mast cells

___ **17.** bone-to-bone attachment
___ **18.** muscle-to-bone attachment
___ **19.** skeletal muscle
___ **20.** cardiac muscle

a. hyaluronan
b. lines heart and blood vessels
c. repair and renewal
d. ligament
e. endocrine secretion
f. merocrine secretion

g. absorption and secretion
h. fat cells
i. holocrine secretion
j. study of tissues
k. unicellular exocrine glands
l. tendon
m. intercellular connection

n. histamine and heparin
o. interlocking of membrane proteins
p. lines ventral body cavities
q. intercalated discs

r. striated, voluntary
s. defense and repair
t. basement membrane

21. The four basic tissue types found in the body are
  (a) epithelial, connective, muscle, and neural
  (b) simple, cuboidal, squamous, and stratified
  (c) fibroblasts, adipocytes, melanocytes, and mesenchymal
  (d) lymphocytes, macrophages, microphages, and adipocytes

22. Long microvilli incapable of movement are
  (a) cilia   (b) flagella
  (c) stereocilia   (d) a, b, and c are correct

23. A type of junction common in cardiac and smooth muscle tissue is the
  (a) desmosome   (b) basement membrane
  (c) tight junction (d) gap junction

24. The most abundant connections between cells in the superficial layers of the skin are
  (a) connexons   (b) gap junctions
  (c) desmosomes   (d) tight junctions

25. The bulk movement of fluids across the epithelial lining of a capillary is
  (a) active transport (b) epithelial transport
  (c) facilitated diffusion (d) simple diffusion

26. Mucous secretions that coat the passageways of the digestive and respiratory tracts result from _____ secretion:
  (a) apocrine   (b) merocrine
  (c) holocrine   (d) endocrine

27. The type of tissue that contains a fluid known as the ground substance is
  (a) epithelial   (b) neural
  (c) muscle   (d) connective

28. The three basic types of fibers found in connective tissue are
  (a) tendons, ligaments, and elastic ligaments
  (b) loose, dense, and irregular
  (c) cartilage, bone, and collagen
  (d) collagen, reticular, and elastic

29. The three types of loose connective tissue are
    (a) collagen, reticular, and elastic
    (b) areolar, adipose, and reticular
    (c) collagen, bone, and cartilage
    (d) fluid, supporting, and connective tissue proper
30. Two major examples of dense regular connective tissue are
    (a) cartilage and bone
    (b) elastic tissue and bone
    (c) tendons and elastic tissue
    (d) collagen and tendons
31. The three major types of cartilage in the body are
    (a) collagen, reticular, and elastic
    (b) areolar, adipose, and reticular
    (c) hyaline, elastic, and fibrocartilage
    (d) tendons, reticular, and elastic
32. The primary function of serous membranes in the body is
    (a) to minimize friction between opposing surfaces
    (b) to line cavities that communicate with the exterior
    (c) to perform absorptive and secretory functions
    (d) to cover the surface of the body
33. The layer of loose connective tissue that separates the skin from underlying tissues and organs is the
    (a) superficial fascia   (b) subcutaneous layer
    (c) hypodermis   (d) a, b, and c are correct
34. Intercalated discs and pacemaker cells are characteristic of
    (a) smooth muscle tissue
    (b) cardiac muscle tissue
    (c) skeletal muscle tissue
    (d) a, b, and c are correct
35. Axons, dendrites, and a soma are characteristics of cells located in
    (a) neural tissue   (b) muscle tissue
    (c) connective tissue   (d) epithelial tissue

36. The repair process necessary to restore normal function in damaged tissues is
    (a) isolation   (b) regeneration
    (c) reconstruction   (d) a, b, and c are correct
37. What four major characteristics are typical of epithelial tissue?
38. What are the four essential functions of epithelial tissue?
39. What four types of cell junctions serve as cellular attachments in the body?
40. What three types of layering make epithelial tissue recognizable?
41. What three cell shapes describe almost every epithelium in the body?
42. What three methods do various glandular epithelial cells use to release their secretions?
43. List three basic components of connective tissues.
44. What six basic functions characterize the role of connective tissue in the body?
45. What fluid connective tissues and supporting connective tissues are found in the human body?
46. What three types of fibers are found in connective tissues?
47. Give four major examples of dense regular connective tissue.
48. What three major types of cartilage are found in the body?
49. What four kinds of membranes composed of epithelial and connective tissue cover and protect other structures and tissues in the body?
50. What are the names of the transudates formed on the surfaces of the serous membranes of the (a) lungs, (b) heart, and (c) organs in the abdomen?
51. What three types of muscle tissue are found in the body?
52. What two cell populations make up neural tissue? What is the function of each?

Level 2   Reviewing Concepts

53. In surfaces of the body where mechanical stresses are severe, the dominant epithelium is
    (a) stratified squamous epithelium
    (b) simple cuboidal epithelium
    (c) simple columnar epithelium
    (d) stratified cuboidal epithelium
54. Why is the presence of cilia on the respiratory surfaces important?
55. Why does holocrine secretion require continuous cell division?
56. What is the difference between an exocrine and an endocrine secretion?
57. A significant structural feature in the digestive system is the presence of tight junctions near the exposed surfaces of cells lining the digestive tract. Why are these junctions so important?
58. Why are dead skin cells generally shed in thick sheets rather than individually?

59. Why are infections always a serious threat after a severe burn or abrasion?
60. Why can't adults generate metabolic heat as readily as infants and young children can?
61. After a weight-loss program, why is the lost weight often regained quickly in the same areas of the body?
62. Cartilage heals poorly and in many instances does not heal or recover at all after a severe injury. Why not?
63. What characteristics make the cutaneous membrane different from the serous and mucous membranes?
64. Why is cardiac muscle tissue that has been damaged by injury or disease incapable of regeneration?
65. Where are you most likely to find elastic ligaments in the body, and why?

Level 3   Critical Thinking and Clinical Applications

66. Assuming that you had the necessary materials to perform a detailed chemical analysis of body secretions, how could you determine whether a given secretion was merocrine or apocrine?
67. After many years of chronic smoking, Mr. Butts finds himself plagued by a hacking cough. Explain the causes of this cough.

68. A biology student accidentally loses the labels of two prepared slides she is studying. One is a slide of animal intestine, and the other of animal esophagus. You volunteer to help her sort them out. How would you decide which slide is which?
69. You are asked to develop a two-step scheme that can be used to identify the three different types of muscle tissue. What would the two steps be?

5

The Integumentary System

*H*ow many of the sun-lovers who crowd the beaches on a warm day ever stop to consider the contributions and sacrifices made by their skin? This remarkable structure absorbs ultraviolet radiation, prevents dehydration, preserves normal body temperature, and tolerates the chafing and abrasion of the sand. Although few of us think of it in these terms, the skin is an organ— the largest organ of the body. In this chapter, we will examine its varied functions.

CHAPTER OUTLINE AND OBJECTIVES

The **integument** (in-TEG-ū-ment) can be considered to be either a large, highly complex organ or a structurally integrated organ system. Most people recognize both possibilities and consider the terms *integument* and **integumentary system** as interchangeable. The components of the integumentary system include the **cutaneous membrane**, or *skin*, and the associated hairs, nails, and exocrine glands.

Few organs are as accessible, large, varied in function, and underappreciated. Your integument accounts for about 16 percent of your total body weight. Its 1.5–2-m² surface is continuously abused, abraded, attacked by microorganisms, irradiated by sunlight, and exposed to environmental chemicals. Of all the body systems, the integument is the only one you see every day.

Because others see this system as well, you devote a lot of time to improving the appearance of the skin and associated structures. Washing your face, brushing and trimming your hair, taking showers, and applying deodorant are activities that modify the appearance or properties of the integument. And when something goes wrong with your integument, the effects are immediately apparent. You will notice even a minor skin condition or blemish at once, whereas you may ignore more serious problems in other systems. In addition to watching for abnormal skin signs, physicians note the general appearance of the skin, because changes in the color, flexibility, elasticity, or sensitivity of the integument may indicate a dysfunction in another system. **AM** *Examination of the Skin*

The integumentary system has two major components—the cutaneous membrane and the accessory structures.

1. The cutaneous membrane has two components—the superficial epithelium, or **epidermis** (*epi-*, above), and the underlying connective tissues of the **dermis**.
2. The accessory structures include hair, nails, and multicellular exocrine glands. These structures are located in the dermis and protrude through the epidermis to the skin surface.

The integument does not function in isolation. An extensive network of blood vessels branches through the dermis, and sensory receptors that monitor touch, pressure, temperature, and pain provide valuable information to the central nervous system about the state of the body. The general structure of the integument is shown in Figure 5-1●. Beneath the dermis, the loose connective tissue of the **subcutaneous layer**, also known as the superficial fascia, or *hypodermis*, separates the integument from the deep fascia around other organs, such as muscles and bones. ∞ [p. 134] Although it is often not considered a part of the integument, we will consider the subcutaneous layer in this chapter because of its extensive interconnections with the dermis.

The general functions of the skin include:

- **Protection** of underlying tissues and organs.
- **Excretion** of salts, water, and organic wastes.

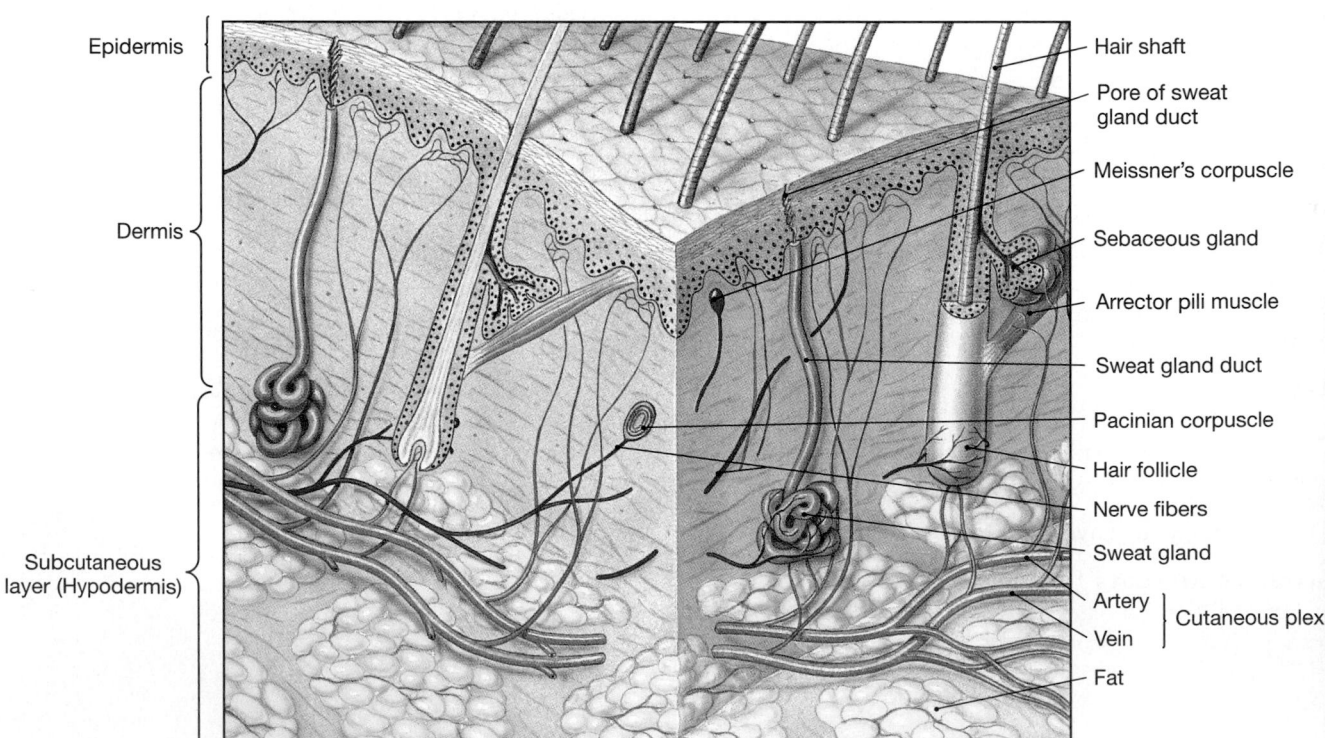

●**FIGURE 5-1** **Components of the Integumentary System.** Relationships among the major components of the integumentary system (with the exception of nails, shown in Figure 5-10).

- **Maintenance** of normal body temperature.
- **Synthesis** of a steroid, *vitamin D$_3$*, that is subsequently converted to the hormone calcitriol, important to normal calcium metabolism.
- **Storage** of nutrients.
- **Detection** of touch, pressure, pain, and temperature stimuli and the relay of that information to the nervous system.

We shall explore each of these functions more fully as we discuss each component of the integument.

THE EPIDERMIS

The epidermis provides mechanical protection and helps keep microorganisms outside the body. This layer consists of a stratified squamous epithelium (Figures 5-1 and 5-2●).

The most abundant epithelial cells, called **keratinocytes** (ker-A-tin-ō-sīts), form several different layers. The precise boundaries between these layers can be difficult to see in a light micrograph. In **thick skin,** found on the palms of the hands and soles of the feet, five layers can be distinguished. Only four layers can be distinguished in the **thin skin** that covers the rest of the body. The terms *thick* and *thin* refer to the relative thickness of the epidermis, not to the integument as a whole.

Most of the body is covered by thin skin. In a sample of thin skin (Figure 5-2a,b●), the epidermis is a mere 0.08 mm thick. The epidermis in a sample of thick skin (Figure 5-2c●) may be as much as six times as thick.

Layers of the Epidermis

Figure 5-3● shows the layers in a section of thick skin. Beginning at the basement membrane and traveling toward the free surface, we find the *stratum germina-*

tivum, the *stratum spinosum,* the *stratum granulosum,* the *stratum lucidum,* and the *stratum corneum.*

Stratum Germinativum

The innermost epidermal layer is the **stratum germinativum** (STRA-tum jer-mi-na-TĒ-vum), or *stratum basale* (Figure 5-3●). This layer is firmly attached to the basement membrane that separates the epidermis from the loose connective tissue of the adjacent dermis. The stratum germinativum forms **epidermal ridges** that extend into the dermis, increasing the area of contact between the two regions. Dermal projections called **dermal papillae** (singular, *papilla,* a nipple-shaped mound) extend between adjacent epidermal ridges (Figure 5-2a●).

The contours of the skin surface follow the ridge patterns, which vary from small conical pegs (in thin skin) to the complex whorls seen on the thick skin of the palms and soles. Ridges on the palms and soles increase the surface area of the skin and increase friction, ensuring a secure grip. Ridge shapes are genetically determined: Your epidermal ridges are unique to you and do not change during your lifetime. Fingerprints are ridge patterns on the tips of the fingers that can be used to identify individuals. These patterns have been used in criminal investigations for more than a century.

Large germinative cells, or **basal cells,** dominate the stratum germinativum. These are stem cells whose divisions replace the more superficial keratinocytes that are lost or shed at the epithelial surface. Skin surfaces that lack hair contain specialized epithelial cells known as **Merkel cells.** These cells are located among the deepest cells of the stratum germinativum. They are sensitive to touch; when compressed or disturbed, Merkel cells release chemicals that stimulate sensory nerve endings. (There are many other kinds of sensory receptors in the dermis, and we will describe those in later sections.) The brown tones of the skin result from the synthetic activ-

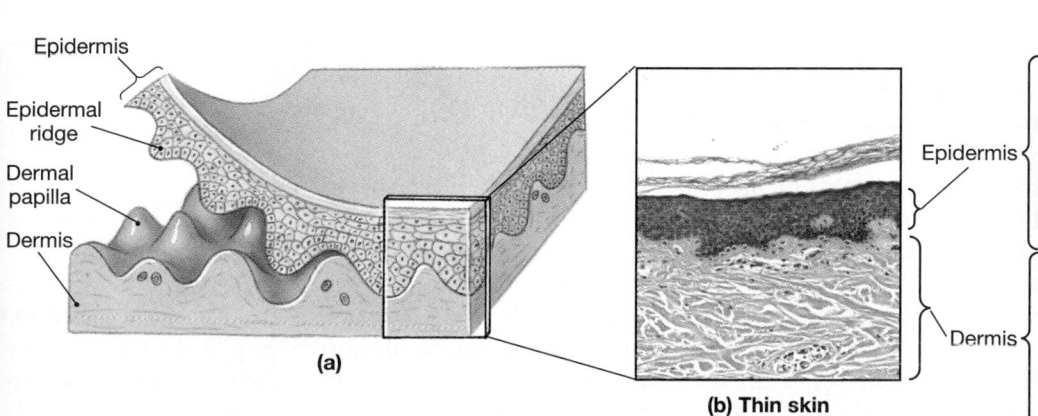

Epidermis

Epidermal ridge

Dermal papilla

Dermis

(a)

Epidermis

Dermis

(b) Thin skin

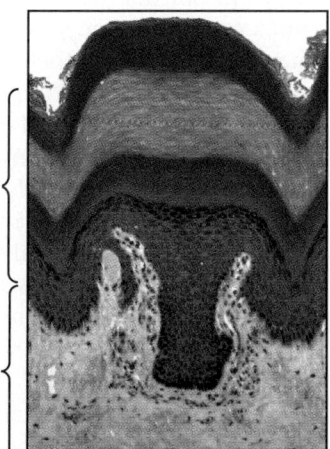

(c) Thick skin

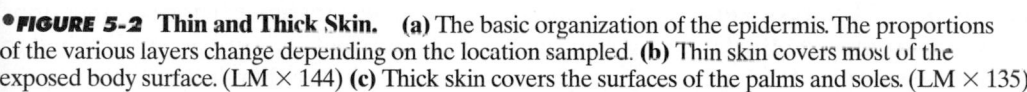

●**FIGURE 5-2** **Thin and Thick Skin.** **(a)** The basic organization of the epidermis. The proportions of the various layers change depending on the location sampled. **(b)** Thin skin covers most of the exposed body surface. (LM × 144) **(c)** Thick skin covers the surfaces of the palms and soles. (LM × 135)

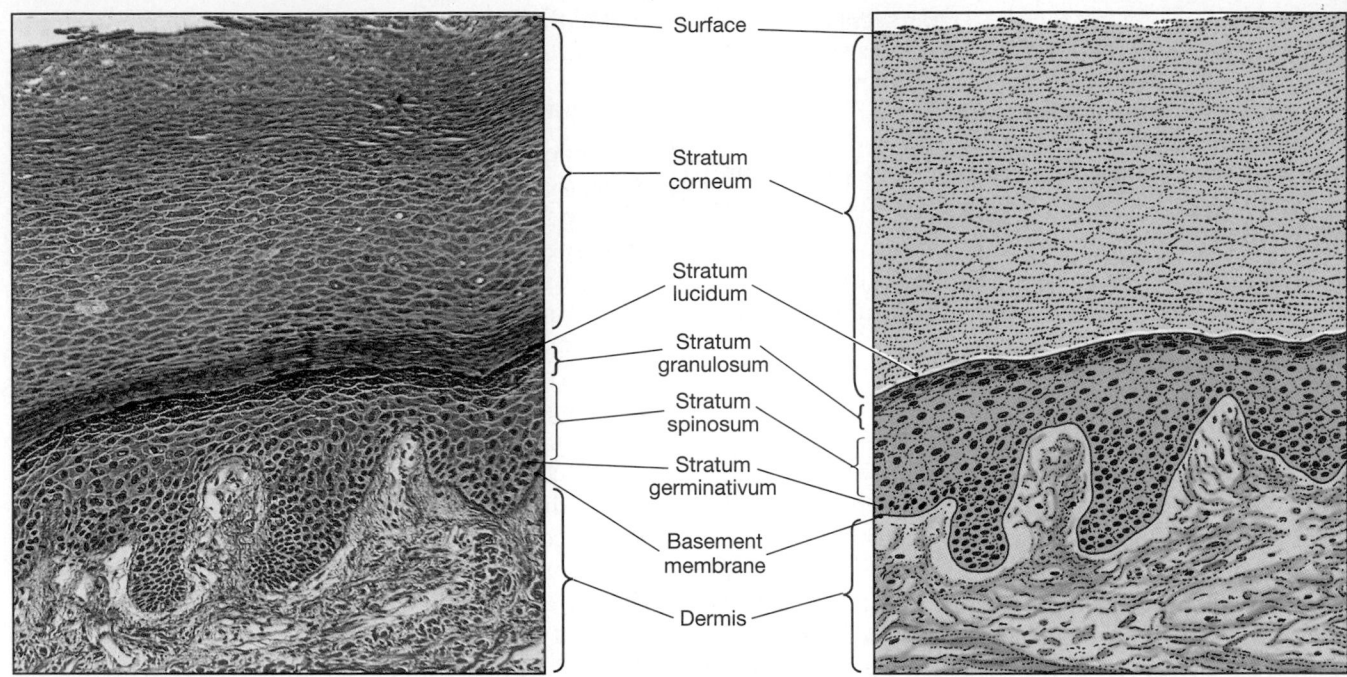

•*FIGURE 5-3* **The Structure of the Epidermis.** A portion of the epidermis in thick skin, showing the major stratified layers of epidermal cells. (LM × 210)

ities of *melanocytes*, pigment cells introduced in Chapter 4. ∞ *[p. 121]* These cells are scattered throughout the stratum germinativum, with cell processes extending into more-superficial layers.

Stratum Spinosum

Each time a stem cell divides, one of the daughter cells is pushed above the germinativum into the next layer, the **stratum spinosum,** which means spiny layer (Figure 5-3•). The stratum spinosum consists of 8–10 layers of cells, and the keratinocytes are bound together by desmosomes. Standard histological procedures shrink the cytoplasm, but the cytoskeletal elements and desmosomes remain intact, so the cells look like miniature pincushions. Some of the cells entering this layer from the stratum germinativum continue to divide, further increasing the thickness of the epithelium. The stratum spinosum also contains **Langerhans cells**, part of the immune response. These cells are responsible for stimulating a defense against (1) microorganisms that manage to penetrate the superficial layers of the epidermis and (2) superficial skin cancers. Langerhans cells cannot be distinguished in standard histological preparations.

Stratum Granulosum

The layer of cells superficial to the stratum spinosum is the **stratum granulosum,** or grainy layer (Figure 5-3•). The stratum granulosum consists of 3–5 layers of keratinocytes displaced from the stratum spinosum. By the time cells reach this layer, most have stopped dividing.

They begin manufacturing large quantities of the proteins **keratohyalin** (ker-a-tō-HĪ-a-lin) and **keratin** (KER-a-tin; *keros*, horn). ∞ *[p. 54]* In humans, keratin, a fibrous protein, also is the basic structural component of hair and nails. As keratin fibers are developing, the cells become thinner and flatter. The cell membranes thicken and become less permeable. The nuclei and other organelles disintegrate, the cells die, and their subsequent dehydration creates a tightly interlocked layer of keratin fibers surrounded by keratohyalin and sandwiched between phospholipid membranes.

Stratum Lucidum

In the thick skin of the palms and soles, a glassy **stratum lucidum** (clear layer) covers the stratum granulosum (Figure 5-3•). The cells in this layer are flattened, densely packed, and filled with keratin.

Stratum Corneum

The **stratum corneum** (KOR-nē-um; *cornu,* horn) is found at the surface of both thick and thin skin (Figure 5-3•). There are normally 15–30 layers of keratinized cells in the stratum corneum. *Keratinization,* or *cornification*, occurs on all exposed skin surfaces except the anterior surfaces of the eyes. The dead cells within each layer of the stratum corneum remain tightly interconnected by desmosomes. Those connections are so secure that cornified cells are generally shed in large groups or sheets rather than individually.

It takes 15–30 days for a cell to move from the stratum germinativum to the stratum corneum. The dead cells generally remain in the exposed stratum corneum layer for an additional 2 weeks before they are shed or washed away. This arrangement places the deeper portions of the epithelium and underlying tissues beneath a protective barrier composed of dead, durable, and expendable cells. Normally, the surface of the stratum corneum is relatively dry, so it is unsuitable for the growth of many microorganisms. Maintenance of this barrier involves coating the surface with lipid secretions from sebaceous and sweat glands.

Although the stratum corneum is water-resistant, it is not waterproof. Water from the interstitial fluids slowly penetrates the surface, to be evaporated into the surrounding air. You lose roughly 500 ml (about 1 pt) of water in this way each day. This process is called **insensible perspiration** to distinguish it from the **sensible perspiration** produced by active sweat glands. Damage to the epidermis can increase the rate of fluid movement. If the damage breaks connections between superficial and deeper layers of the epidermis, this fluid will accumulate in pockets, or *blisters*, within the epidermis. (Blisters can also form between the epidermis and dermis if that attachment is disrupted.) If the damage to the epidermis causes the stratum corneum to lose its effectiveness as a water barrier, the rate of insensible perspiration skyrockets, and a potentially dangerous fluid loss occurs. This is a serious consequence of severe burns and a complication in the condition known as *xerosis* (excessively dry skin). [AM] *Disorders of Keratin Production*

When the skin is immersed in water, osmotic forces may move water into or out of the epithelium. ∞ *[p. 73]* Sitting in a freshwater bath causes water to move into the epidermis because fresh water is hypotonic (has fewer dissolved materials) compared with body fluids. The epithelial cells may swell to four times their normal volumes, a phenomenon particularly noticeable in the thickly keratinized areas of the palms and soles. Swimming in the ocean reverses the direction of osmotic flow, because the ocean is a hypertonic solution. As a result, water leaves the body, crossing the epidermis from the underlying tissues. The process is slow, but long-term exposure to seawater endangers survivors of a shipwreck by accelerating dehydration.

✔ Excessive shedding of cells from the outer layer of skin in the scalp causes dandruff. What is the name of this skin layer?

✔ A splinter pierces the palm of your hand and lodges in the third layer of the epidermis. Identify this layer.

✔ Why does swimming in fresh water for an extended period cause epidermal swelling?

✔ Some criminals sand the tips of their fingers so as not to leave recognizable fingerprints. Would this practice permanently remove fingerprints? Why or why not?

Skin Color

The color of your skin is due to an interaction between (1) pigment composition and concentration and (2) the dermal blood supply.

Skin Pigmentation

The epidermis contains variable quantities of two pigments, *carotene* and *melanin:*

1. **Carotene** (KAR-ō-tēn) is an orange-yellow pigment that normally accumulates inside epidermal cells. It is most apparent in cells of the stratum corneum of light-skinned individuals. This pigment is also distributed in fatty tissues in the dermis. Carotene is found in a variety of orange-colored vegetables, such as carrots and squashes. The skin of a Caucasian vegetarian with a special fondness for carrots can actually turn orange from an overabundance of carotenes. The skin of people of other races will pick up the orange tone, but the color change is less striking. Carotene can be converted to vitamin A, which is required for (1) the normal maintenance of epithelia and (2) the synthesis of photoreceptor pigments in the eye.

2. **Melanin** is a brown, yellow-brown, or black pigment produced by melanocytes.

MELANOCYTES **Melanocytes** are located in the stratum germinativum, squeezed between or deep to the epithelial cells (Figure 5-4●). Melanocytes manufacture melanin pigments from molecules of the amino acid *tyrosine*. The melanin is packaged in intracellular vesicles called *melanosomes*. These vesicles travel within the melanocyte processes and are transferred intact to keratinocytes. This transfer of pigmentation colors the keratinocyte temporarily, until the melanosomes are destroyed by fusion with lysosomes. In Caucasians, this transfer occurs in the stratum germinativum and stratum spinosum, and the cells of more-superfical layers lose their pigmentation. In black people, the melanosomes are larger, and the transfer may occur in the stratum granulosum as well; the pigmentation is thus darker and more persistent.

The melanin in keratinocytes protects your epidermis and dermis from the harmful effects of sunlight, which contains significant amounts of **ultraviolet (UV) radiation.** A small amount of UV radiation is beneficial, for it stimulates synthetic activity in the epidermis (a process discussed in a later section). However, UV radiation can damage DNA, causing mutations and promoting cancer development. Within the keratinocytes, melanosomes become concentrated in the region around the nucleus, where the melanin pigments provide some UV protection for the DNA in these cells. In addition, UV radiation can produce the

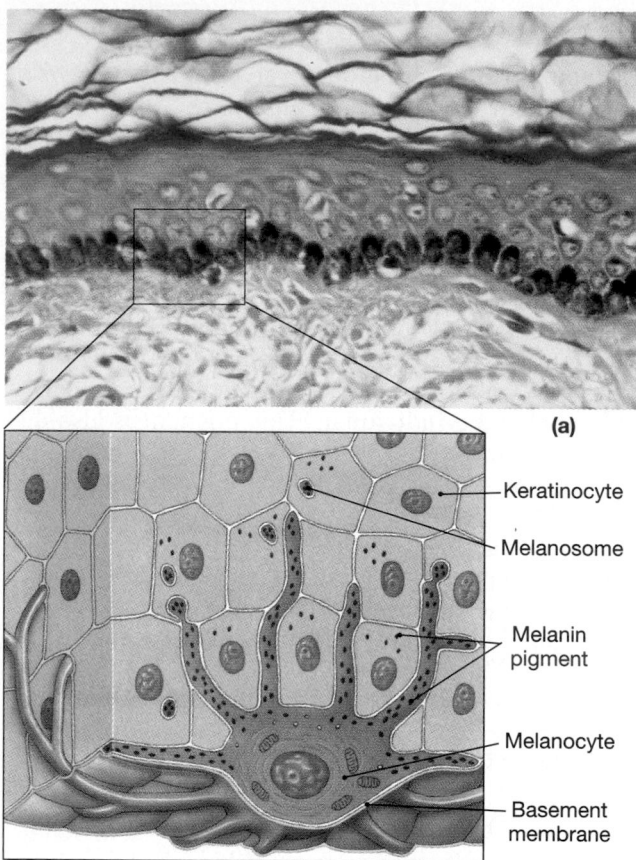

Keratinocyte
Melanosome
Melanin pigment
Melanocyte
Basement membrane

(a)

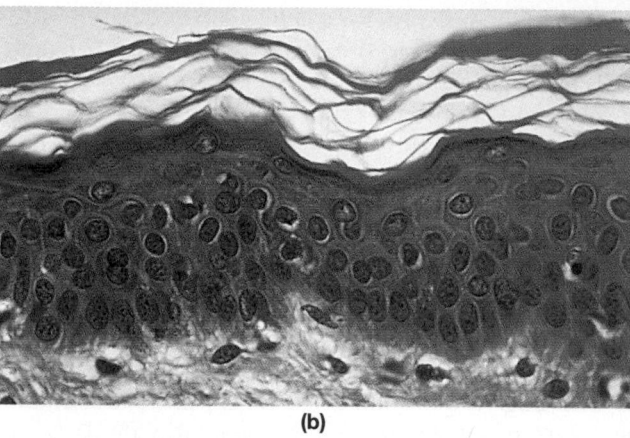

(b)

•**FIGURE 5-4** **Melanocytes.** **(a)** The location and orientation of melanocytes in the stratum germinativum of a black person. **(b)** Comparable micrograph of the skin of a Caucasian person.

immediate effects of mild or even serious burns, damaging both the epidermis and dermis. Thus, the presence of a pigment layer in the deep layers of the epidermis helps protect both epidermal and dermal tissues. However, although melanocytes respond to UV exposure by increasing their activity, the response is not rapid enough to prevent a sunburn the first day you spend at the beach. Melanin synthesis accelerates slowly, peaking about 10 days after the initial exposure. Individuals of any skin color can suffer sun damage to the integument, but dark-skinned individuals have greater initial protection against the effects of UV radiation.

Over time, cumulative damage to the integument by UV exposure can harm fibroblasts, leading to impaired maintenance of the dermis. The resulting structural alterations lead to premature wrinkling. In addition, skin cancers can result from chromosomal damage in germinative cells or melanocytes. One of the major consequences of the global depletion of the ozone layer in the upper atmosphere will likely be a sharp increase in the rate of skin cancers, such as *malignant melanoma*. Such an increase has been reported in Australia, which has already experienced a significant loss of ozone. AM

Skin Cancers

The ratio between melanocytes and germinative cells ranges between 1:4 and 1:20, depending on

the region of the body surveyed. For example, there are about 1000 melanocytes per square millimeter in the skin covering most areas of the body. Higher concentrations (about 2000/mm^2) are found in the cheeks and forehead, in the nipples, and in the genital region (the scrotum of males and the labia majora of females). Observed differences in skin color among individuals and even among races do not reflect different numbers of melanocytes but merely different levels of synthetic activity. Even the melanocytes of *albino* individuals are distributed normally, although these cells are incapable of producing melanin.

Dermal Circulation

Blood contains red blood cells filled with the pigment *hemoglobin*. Hemoglobin binds and transports oxygen in the bloodstream. When bound to oxygen, hemoglobin has a bright red color, giving blood vessels in the dermis a reddish tint that is most apparent in lightly pigmented individuals. When those vessels are dilated, as during inflammation, the red tones become much more pronounced.

When the circulatory supply is temporarily reduced, the skin becomes relatively pale; when frightened, a Caucasian may "turn white" because of a sudden drop in blood supply to the skin. During a sustained reduction in circulatory supply, the tissue oxygen levels decline, and the hemoglobin in these tissues releases oxygen and changes color to a much darker red tone. Seen from the surface, the skin takes on a bluish coloration called **cyanosis** (sī-uh-NŌ-sis; *kyanos,* blue). In individuals of any skin color, cyanosis is most apparent in areas of thin skin, such as the lips or beneath the nails. It can be a response to extreme cold or a result of circulatory or respiratory disorders, such as heart failure or severe asthma.

 # Drug Administration Through the Skin

Drugs in oils or other lipid-soluble carriers can penetrate the epidermis. The movement is slow, particularly through the layers of cell membranes in the stratum corneum. But once a drug reaches the underlying tissues it will be absorbed into the circulation. Placing a drug in a solvent that is lipid-soluble can assist its movement through the lipid barriers. Drugs can also be administered by packaging them in *liposomes*, artificially produced lipid droplets. Liposomes containing DNA fragments have been used experimentally to introduce normal genes into abnormal human cells. For example, genes carried by liposomes have been used to alter the cell membranes of skin cancer cells so that the tumor cells will be attacked by the immune system. Another experimental procedure involves the creation of a transient change in skin permeability by administering a brief pulse of electricity. This electrical pulse temporarily changes the positions of the cells in the stratum corneum, creating channels that allow the drug to penetrate.

A useful technique for long-term drug administration involves the placement of a sticky patch containing a drug over an area of thin skin. To overcome the relatively slow rate of diffusion, the patch must contain an extremely high concentration of the drug. This procedure, called *transdermal administration*, has the advantage that a single patch may work for several days, making daily pills unnecessary. Transdermal administration of small amounts of nicotine, the addictive compound in tobacco, is being used to help people quit smoking. The nicotine from the patch depresses the craving for a cigarette, and the transdermal dosage can be gradually reduced in small, controlled steps.

Several drugs are now routinely administered transdermally:

- Transdermal scopolamine, a drug that affects the nervous system, can be used to control the nausea associated with motion sickness.
- Transdermal nitroglycerin is used to improve blood flow within heart muscle and prevent a heart attack.
- Transdermal estrogens are administered to women to reduce symptoms of menopause.
- Transdermal nicotine is used to suppress the urge to smoke cigarettes.

In addition, pain medications and drugs to control high blood pressue may be administered via transdermal patches.

DMSO (dimethyl sulfoxide) is a transdermal drug intended for the treatment of injuries to the muscles and joints of domesticated animals, such as horses or cows. It is a solvent that rapidly crosses the skin; drugs dissolved in DMSO are carried into the body along with it. DMSO has not been tested and approved for the treatment of humans in the United States, either for joint or muscle injuries or as a transdermal solvent. However, it can be prescribed in Canada and Europe. The long-term risks associated with it are unknown; reported short-term side effects include nausea, vomiting, cramps, and chills.

VARIATIONS IN SKIN PIGMENTATION *Freckles* are small pigmented spots that appear on the skin of pale-skinned individuals. These spots, which typically have an irregular border, represent the area serviced by melanocytes that are producing larger-than-average amounts of melanin. Freckles tend to be most abundant on surfaces such as the face, probably due to sun exposure. *Lentigos* are similar to freckles, but they have regular borders and contain abnormal melanocytes. *Senile lentigos*, or *liver spots*, are variably pigmented areas that develop on exposed skin in older Caucasians.

Several diseases that have primary impacts on other systems may have secondary effects on skin color and pigmentation. Because the skin is easily observed, these color changes can be useful in diagnosis. For example:

- In *jaundice* (JAWN-dis), the liver is unable to excrete bile, and a yellowish pigment accumulates in body fluids. In advanced stages, the skin and whites of the eyes turn yellow.

- Some tumors affecting the pituitary gland result in the secretion of large amounts of *melanocyte-stimulating hormone* (MSH). This hormone causes a darkening of the skin, as if the individual has an extremely deep bronze tan.
- In *Addison's disease,* the pituitary gland secretes large quantities of *adrenocorticotropic hormone (ACTH),* a hormone that is structurally similar to MSH. The effect of ACTH on the skin coloration is also similar to that of MSH.
- In *vitiligo* (vi-ti-LĪ-gō), individuals lose their melanocytes. The condition develops in about 1 percent of the population, and the incidence increases among individuals with thyroid gland disorders, Addison's disease, and several other disorders. It is suspected that this disorder develops when the immune defenses malfunction and antibodies attack normal melanocytes. The primary problem with vitiligo is cosmetic, especially for individuals with darkly pigmented skin. Michael Jackson is said to suffer from vitiligo.

The Epidermis and Vitamin D$_3$

Although strong sunlight can damage epithelial cells and deeper tissues, limited exposure to sunlight is very beneficial. When exposed to ultraviolet radiation, epidermal cells in the stratum spinosum and stratum germinativum convert a cholesterol-related steroid into **vitamin D$_3$**, or **cholecalciferol** (kō-le-kal-SIF-e-rol). The liver converts cholecalciferol into an intermediary product used by the kidneys to synthesize the hormone **calcitriol** (kal-si-TRĪ-ol). Calcitriol is essential for normal calcium and phosphorus absorption by the small intestine. An inadequate supply of calcitriol leads to impaired bone maintenance and growth.

If present in the diet, cholecalciferol can also be absorbed by the digestive tract. This fact accounts for the use of the term *vitamin,* even though the body can synthesize its own cholecalciferol. (*Vitamin* usually refers to an essential organic nutrient that the body either cannot make or makes in insufficient amounts.) Children who live in areas with overcast skies and whose diets lack cholecalciferol can have abnormal bone development. This condition has largely been eliminated in the United States because dairy companies add cholecalciferol, usually identified on the label as "vitamin D," to the milk sold in grocery stores. In Chapter 6, we shall consider the hormonal control of bone growth in greater detail.

Epidermal Growth Factor

Epidermal growth factor (**EGF**) is one of the peptide growth factors introduced in Chapter 3. ∞ *[p. 102]* Produced by the salivary glands and glands of the duodenum, EGF has widespread effects on epithelia, especially the epidermis. Its effects include:

- Promoting the divisions of germinative cells in the stratum germinativum and stratum spinosum.
- Accelerating the production of keratin in differentiating epidermal cells.
- Stimulating epidermal development and epidermal repair after injury.
- Stimulating synthetic activity and secretion by epithelial glands.

Epidermal growth factor has such a pronounced effect that it can be used to stimulate the growth and division of epidermal cells outside the body, a practice known as *tissue culture.* Sheets of epidermal cells produced in this fashion have been used in the treatment of extensive burns. Burned areas can be covered by epidermal sheets "grown" from a small sample of intact skin from another part of the burn victim's body. (We will consider this treatment on page 168 in the discussion of burns.)

☑ Why does exposure to sunlight or sunlamps darken skin?

☑ Why does the skin appear red when it is warm?

☑ In some cultures, women are required to be completely covered except for their eyes when they go outside. These women exhibit a high incidence of problems with their bones. Why?

THE DERMIS

The dermis lies beneath the epidermis (Figure 5-1•, p. 148). It has two major components—a superficial *papillary layer* and a deeper *reticular layer.*

Dermal Organization

The **papillary layer** consists of loose connective tissue. This region contains the capillaries and the sensory neurons that supply the surface of the skin. The papillary layer derives its name from the dermal papillae that project between the epidermal ridges (Figure 5-2a•, p. 149).

The **reticular layer** deep to the papillary layer consists of an interwoven meshwork of dense irregular connective tissue. ∞ *[p. 125]* Bundles of collagen fibers leave the reticular layer to blend into those of the papillary layer above, so the boundary between these layers is indistinct. Collagen fibers of the reticular layer also extend into the underlying subcutaneous layer.

In addition to extracellular protein fibers, the dermis contains all the cells of connective tissue proper. ∞ *[p. 120]* Accessory organs of epidermal origin, such as hair follicles and sweat glands, extend into the dermis. In addition, the reticular and papillary layers of the dermis contain networks of blood vessels, lymph vessels, and nerve fibers (Figure 5-1•, p. 148).

Wrinkles and Stretch Marks

Individual collagen fibers are very strong and resist stretching, although they are easily bent or twisted. Elastic fibers permit stretching and will then recoil to their original length. These fiber types are interwoven within the dermis. As a result, the dermis tolerates limited stretching. Your skin remains flexible and somewhat elastic, but the collagen fibers stop the distortion before tissue damage occurs. The water content of the skin also helps maintain its flexibility and resilience, properties known as *skin turgor.* As a result, dehydration causes a temporary decline in skin flexibility. Aging, hormones, and the destructive effects of ultraviolet radiation permanently reduce the amount of elastin in the dermis, producing wrinkles and sagging skin. The extensive distortion of the dermis that occurs over the abdomen during pregnancy or after a substantial weight gain can exceed the elastic capabilities of the skin. The resulting damage to the dermis prevents it from recoiling to its original size after delivery or a rigorous diet. The skin then wrinkles and creases, creating a network of **stretch marks.**

Tretinoin (Retin-A™) is a derivative of vitamin A that can be applied to the skin as a cream or gel. This drug was originally developed to treat acne, but it also increases blood flow to the dermis and stimulates dermal repair. As a result, the rate of wrinkle formation decreases, and existing wrinkles become smaller. The degree of improvement varies among individuals.

Lines of Cleavage

Most of the collagen and elastic fibers at any one location are arranged in parallel bundles. The orientation of these bundles depends on the stress placed on the skin during normal movement: The bundles are aligned to resist the applied forces. The resulting pattern of fiber bundles establishes the **lines of cleavage** of the skin. Lines of cleavage, shown in Figure 5-5●, are clinically significant because a cut parallel to a cleavage line will usually remain closed, whereas a cut at right angles to a cleavage line will be pulled open as cut elastic fibers recoil. Surgeons choose their incision patterns accordingly, for a neatly closed incision made parallel to the lines of cleavage will heal faster and with less scarring than will an incision that crosses lines of cleavage.

Dermal Circulation and Innervation

The Dermal Blood Supply

Arteries supplying the skin form a network in the subcutaneous layer along its border with the reticular layer of the dermis. This network is called the *cutaneous plexus.* Tributaries of these arteries supply the adipose tissues of the subcutaneous layer and the tissues of the integument. As small arteries travel toward the epidermis, branches supply the hair follicles, sweat glands, and other structures in the dermis. On reaching the papillary layer, these small arteries form another branching network, the *papillary plexus,* which provides arterial blood to capillary loops that follow the contours of the epidermis-dermis boundary. These capillaries empty into a network of small veins connected to larger veins in the subcutaneous layer.

An **ulcer** is a localized shedding of an epithelium. **Decubitis ulcers,** or bedsores, may affect bedridden or mobile patients with circulatory restrictions, especially when splints, casts, or bedding continuously presses against superficial blood vessels. Such sores most commonly affect the skin near joints or projecting bones, where the dermal blood vessels are pressed against deeper structures. The chronic lack of circulation kills epidermal cells, removing a barrier to bacterial infection. Eventually, the dermal tissues deteriorate as well. (This type of tissue degeneration, or *necrosis,* will occur in any tissue deprived of ade-

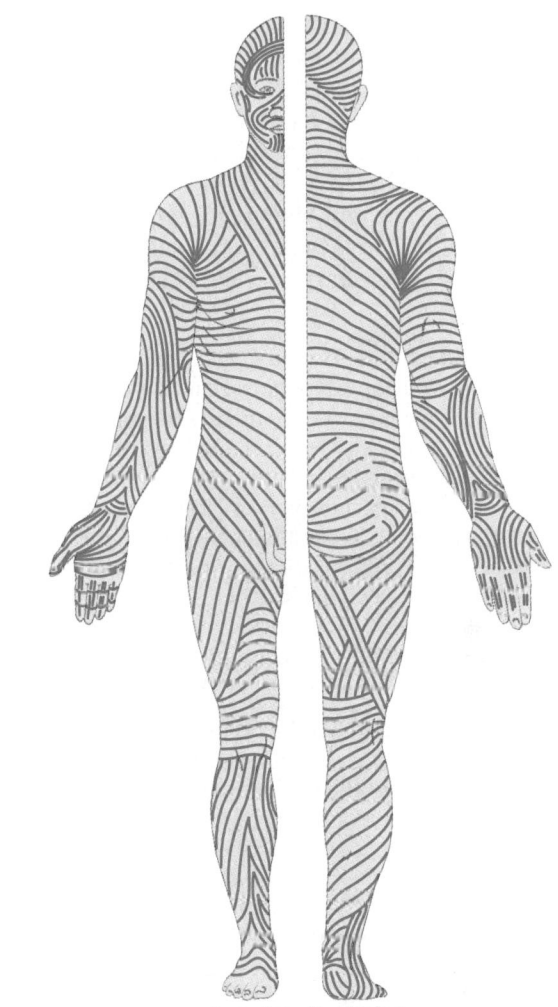

Anterior Posterior

●**FIGURE 5-5** Lines of Cleavage of the Skin. Lines of cleavage follow lines of tension in the skin. They reflect the orientation of collagen fiber bundles in the dermis.

quate circulation.) Bedsores can be prevented or treated by frequently changing body position to vary the pressures applied to specific blood vessels.

Tumors sometimes form in the dermal blood vessels during development. The result is a temporary or permanent *birthmark.* A *capillary hemangioma* involves capillaries of the papillary layer of the dermis. It generally enlarges after birth but subsequently fades and disappears. *Cavernous hemangiomas,* or "portwine stains," affect larger vessels in the dermis. Most such birthmarks last a lifetime, but some can be removed by laser surgery.

The Innervation of the Skin

Nerve fibers in the skin control blood flow, adjust gland secretion rates, and monitor sensory receptors in the dermis and the deeper layers of the epidermis. We have already noted the presence of Merkel cells in the deeper layers of the epidermis. These cells are monitored by sensory terminals known as *Merkel's discs.* The epidermis also contains the extensions of sensory neurons that provide sensations of pain and temperature. The dermis contains similar receptors as well as other, more specialized receptors. Examples discussed in Chapter 17 and shown in Figure 5-1●, p. 148, include receptors sensitive to light touch (*Meissner's corpuscles,* located in dermal papillae) and to deep pressure and vibration (*Pacinian corpuscles,* in the reticular layer).

DERMATITIS Due to the abundance of sensory receptors in the skin, regional infection or inflammation can be very painful. **Dermatitis** (der-muh-TĪ-tis) is an inflammation of the skin that primarily involves the papillary layer. In typical dermatitis, inflammation begins in a portion of the skin exposed to infection or irritated by chemicals, radiation, or mechanical stimuli. Dermatitis may cause no physical discomfort, or it may produce an annoying itch, as in poison ivy. Other forms of this condition can be quite painful, and the inflammation can spread rapidly across the entire integument.

There are many forms of dermatitis, some of them quite common:

- **Contact dermatitis** generally occurs in response to strong chemical irritants. It produces an itchy rash that may spread to other areas; poison ivy is an example.
- **Eczema** (EK-se-muh) is a dermatitis that can be triggered by temperature changes, fungi, chemical irritants, greases, detergents, or stress. Hereditary, environmental factors, or both can promote the development of eczema.
- **Diaper rash** is a localized dermatitis caused by a combination of moisture, irritating chemicals from fecal or urinary wastes, and flourishing microorganisms.
- **Urticaria** (ur-ti-KAR-ē-uh), or *hives,* is an extensive allergic response to a food, drugs, an insect bite, infection, stress, or other stimulus.

☑ Where would you find the capillaries and sensory neurons that supply the epidermis?

☑ What accounts for the ability of the dermis to undergo repeated stretching?

THE SUBCUTANEOUS LAYER

The connective tissue fibers of the reticular layer are extensively interwoven with those of the subcutaneous layer, or **hypodermis,** and the boundary between the two is generally indistinct (Figure 5-1●, p. 148). Although the subcutaneous layer is not a part of the integument, it is important in stabilizing the position of the skin in relation to underlying tissues, such as skeletal muscles or other organs, while permitting independent movement.

The subcutaneous layer consists of loose connective tissue with abundant fat cells. Most infants and small children have extensive "baby fat," which helps

them reduce heat loss. Subcutaneous fat also serves as a substantial energy reserve and a shock absorber for the rough-and-tumble activities of our early years.

As we grow, the distribution of our subcutaneous fat changes. Men accumulate subcutaneous fat at the neck, arms, along the lower back, and over the buttocks. In women, the breasts, buttocks, hips, and thighs are the primary sites of subcutaneous fat storage. In adults of either gender, the subcutaneous layer of the backs of the hands and the upper surfaces of the feet contain few fat cells, whereas distressing amounts of adipose tissue can accumulate in the abdominal region, producing a prominent "potbelly."

The subcutaneous layer is quite elastic. Only the superficial region contains large arteries and veins. The rest contains a limited number of capillaries and no vital organs. This last characteristic makes **subcutaneous injection**—by means of a **hypodermic needle**—a useful method of administering drugs.

ACCESSORY STRUCTURES

The accessory structures of the integument include hair follicles, sebaceous glands, sweat glands, and nails. During embryological development, these structures originate from the epidermis, so they are also known as *epidermal derivatives.* Although located in the dermis, they project through the epidermis to the integumentary surface.

Hair Follicles and Hair

Hairs project above the surface of the skin almost everywhere except over the sides and soles of the feet, the palms of the hands, the sides of the fingers and toes, the lips, and portions of the external genitalia. There are about 5 million hairs on the human body, and 98 percent of them are on the general body surface, not on the head. Hairs originate in complex organs called **hair follicles.** Hair production is a complex process involving cooperation between the dermis and epidermis.

Hair Production

Hair follicles extend deep into the dermis, typically projecting into the underlying subcutaneous layer (Figure 5-6●). At the base of a hair follicle is a small **hair papilla,** a peg of connective tissue containing capillaries and nerves. The **hair bulb** consists of epithelial cells that surround the hair papilla.

Hair production involves a specialization of the cornification process. The epithelial layer involved is called the **hair matrix.** Basal cells near the center of the hair matrix divide, producing daughter cells that are gradually pushed toward the surface. Those cells pro-

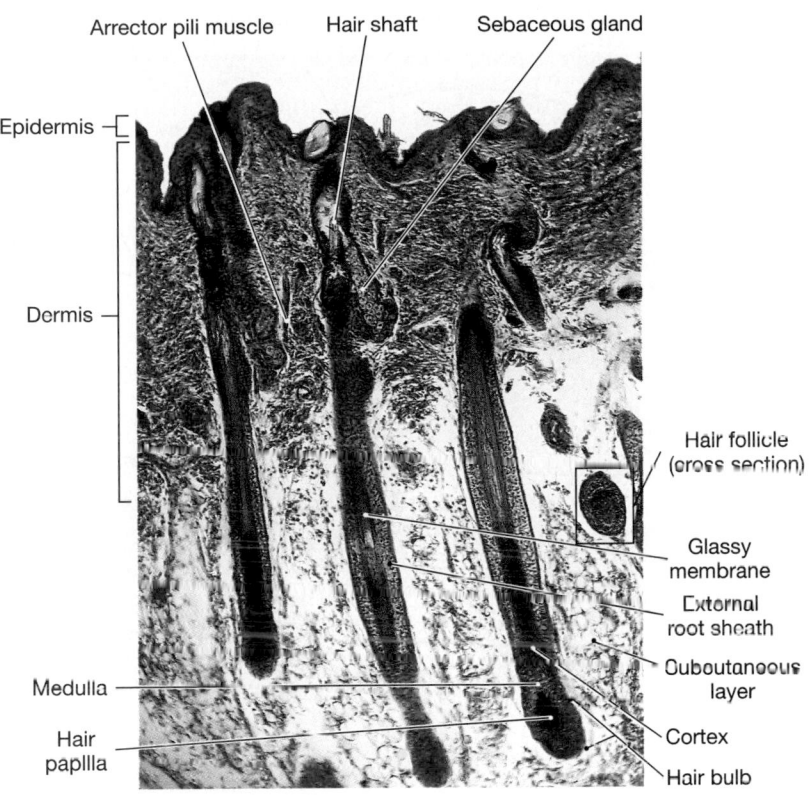

Arrector pili muscle   Hair shaft   Sebaceous gland

Epidermis

Dermis

Hair follicle
(cross section)

Glassy
membrane

External
root sheath

Subcutaneous
layer

Medulla

Hair
papilla

Cortex

Hair bulb

•FIGURE 5-6   Accessory Structures of the Skin. A sectional view of the skin of the scalp. (LM × 73)

duced closest to the center of the matrix form the **medulla,** or core, of the hair, whereas cells closer to the edge of the developing hair form the relatively hard **cortex.** The medulla contains flexible **soft keratin. Hard keratin** in the cortex gives the hair its stiffness. Dead cells at the surface of the hair form the **cuticle,** a layer of hard keratin that coats the hair (Figure 5-7●).

The **root** of the hair extends from the hair bulb to the point where the internal organization of the hair is complete, generally about halfway to the skin surface. The **hair shaft** extends from this point to the exposed tip of the hair. The size, shape, and color of the hair shaft are highly variable.

Follicle Structure

The cells of the follicle walls are organized into several concentric layers (Figure 5-7a●). Moving outward from the hair cuticle, these layers include:

- The **internal root sheath,** which surrounds the hair root and the deeper portion of the shaft. It is produced by the cells at the periphery of the hair matrix. The cells of the internal root sheath disintegrate relatively quickly. This layer does not extend the entire length of the follicle.
- The **external root sheath,** which extends from the skin surface to the hair matrix. Over most of that distance,

it has all the cell layers found in the superficial epidermis. However, where the external root sheath joins the hair matrix, all the cells resemble those of the stratum germinativum.

- The **glassy membrane,** a thickened basement membrane wrapped in a dense connective tissue sheath.

Functions of Hair

The 5 million hairs on your body have important functions. The roughly 100,000 hairs on your head protect your scalp from ultraviolet light, help cushion a blow to the head, and insulate the skull. The hairs guarding the entrances to your nostrils and external ear canals help prevent the entry of foreign particles and insects, and your eyelashes perform a similar function for the surface of the eye. A **root hair plexus** of sensory nerves surrounds the base of each hair follicle. As a result, you can feel the movement of the shaft of even a single hair. This sensitivity provides an early-warning system that may help prevent injury. For example, you may be able to swat a mosquito before it reaches your skin surface.

Ribbons of smooth muscle, called **arrector pili** (a-REK-tor PI-lē) muscles (Figures 5-6, 5-7a●), extend from the papillary layer of the dermis to the connective tissue sheath surrounding the hair follicle. When stimulated, the arrector pili pull on the follicles and force the hairs to stand erect. Contraction may be the result of emotional states, such as fear or rage, or a response to cold, producing the characteristic "goose bumps." In a furry mammal, this action increases the thickness of the insulating coat, rather like putting on an extra sweater. Although humans do not receive any comparable insulating benefits, the reflex persists.

Types of Hairs

There are two major types of hairs in the adult integument—*vellus hairs* and *terminal hairs.*

- **Vellus hairs** are the fine "peach fuzz" hairs located over much of the body surface.
- **Terminal hairs** are heavy, more deeply pigmented, and sometimes curly. The hair on your head, including your eyebrows and eyelashes, are terminal hairs.

Hair follicles may alter the structure of the hairs in response to circulating hormones. This fact accounts for many of the changes in the distribution of terminal hairs that begin at puberty.

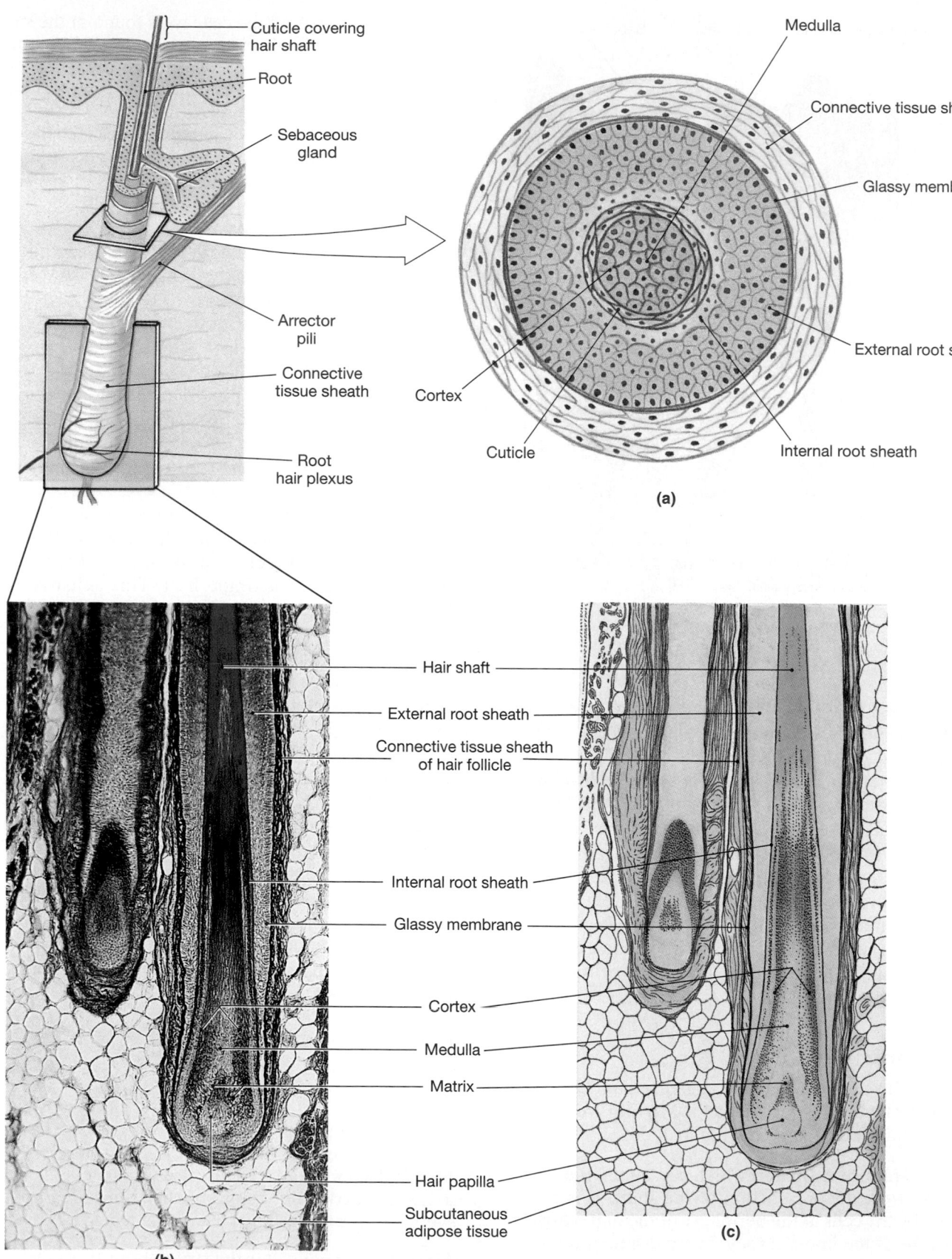

Cuticle covering hair shaft

Root

Sebaceous gland

Arrector pili

Connective tissue sheath

Root hair plexus

Medulla

Connective tissue sheath

Glassy membrane

External root sheath

Internal root sheath

Cortex

Cuticle

(a)

Hair shaft

External root sheath

Connective tissue sheath of hair follicle

Internal root sheath

Glassy membrane

Cortex

Medulla

Matrix

Hair papilla

Subcutaneous adipose tissue

(b)

(c)

•*FIGURE 5-7* **Hair Follicles. (a)** A longitudinal section and a cross section through a hair follicle. **(b)** A closer view of the base of the follicle and hair shaft and the matrix and papilla at the hair root. (LM ×46) **(c)** A section along the longitudinal axis of a hair follicle.

Hair Color

Variations in hair color reflect differences in structure and variations in the pigment produced by melanocytes at the hair papilla. These characteristics are genetically determined, but hormonal and environmental factors can influence the condition of your hair. As pigment production decreases with age, the hair color lightens toward gray. White hair results from the combination of a lack of pigment and the presence of air bubbles within the medulla of the hair shaft. Because the hair itself is dead and inert, changes in coloration are gradual.

Growth and Replacement of Hair

Our hairs grow and are shed according to a **hair growth cycle.** A hair in the scalp grows for 2–5 years, at a rate of about 0.33 mm per day. Variations in the growth rate of hair and in the duration of the hair growth cycle account for individual differences in the length of uncut hair.

While hair growth is under way, the cells of the hair root absorb nutrients and incorporate them into the hair structure. Clipping or collecting hair for analysis can be helpful in diagnosing several disorders. For example, the hairs of individuals suffering from lead or other heavy-metal poisoning contain high quantities of those metal ions. Hair samples can also be used for identification purposes through the process of DNA fingerprinting, discussed in Chapter 3. ∞ *[p. 93]*

As it grows, the root of the hair is firmly attached to the matrix of the follicle. At the end of the growth cycle, the follicle becomes inactive. The hair is now termed a **club hair.** The follicle gets smaller, and over time the connections between the hair matrix and the root of the club hair break down. When another growth cycle begins, the follicle produces a new hair, and the old club hair gets pushed toward the surface and is eventually shed.

HAIR LOSS If you are a healthy adult, you lose about 50 hairs from your head each day. Several factors affect this rate. Sustained losses of more than 100 hairs per day generally indicate that something is wrong. Temporary increases in hair loss can result from drugs, dietary factors, radiation, vitamin A excess, high fever, stress, or hormonal factors related to pregnancy. In males, changes in the level of the sex hormones circulating in the blood can affect the scalp, causing a shift from terminal hair to vellus hair production, beginning at the temples and the crown of the head. This alteration is called **male pattern baldness.** Some cases of male pattern baldness respond to drug therapies, such as topical application of *minoxidil (Rogaine™)*. For additional information, see the *Applications Manual*. ⃞AM *Baldness and Hirsutism*

☑ What condition is produced by the contraction of the arrector pili muscles?

☑ A person suffers a burn on the forearm that destroys the epidermis and extensive areas of the deep dermis. When the injury heals, would you expect to find hair growing again in the area of the injury?

Glands in the Skin

The skin contains two types of exocrine glands: *sebaceous glands* and *sweat glands*. Sebaceous glands produce an oily lipid that coats hair shafts and the epidermis. Sweat glands produce a watery solution and perform other special functions.

Sebaceous (Oil) Glands

Sebaceous (se-BĀ-shus) **glands,** or *oil glands,* are holocrine glands that discharge a waxy, oily secretion into hair follicles (Figure 5-8●). Sebaceous glands that communicate with a single follicle share a duct and thus are classified as *simple branched alveolar glands.*

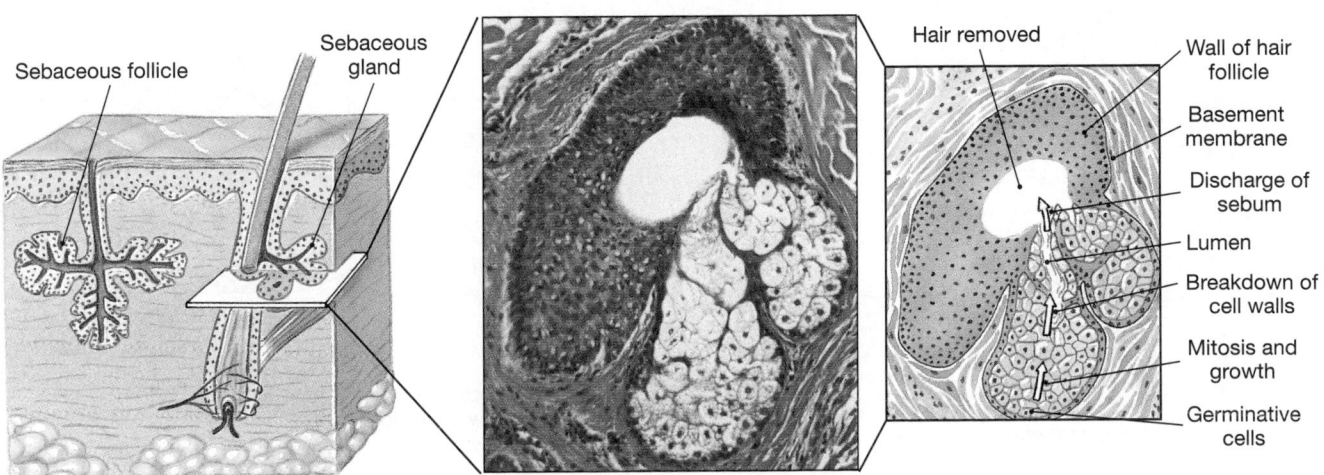

Sebaceous gland (LM × 150)

●**FIGURE 5-8** **Sebaceous Glands and Follicles.** The structure of sebaceous glands and sebaceous follicles in the skin.

∞ [p. 119] The gland cells manufacture large quantities of lipids as they mature, and the lipid product is released through *holocrine secretion*, a process that involves the rupture of the secretory cells. ∞ [p. 117]

The lipids released from these cells enter the open passageway, or **lumen,** of the gland. Contraction of the arrector pili muscle that erects the hair squeezes the sebaceous gland, forcing an oily secretion into the follicle and onto the surface of the skin. This secretion, called **sebum** (SĒ-bum), is a mixture of triglycerides, cholesterol, proteins, and electrolytes. Sebum provides lubrication and inhibits the growth of bacteria. Keratin is a tough protein, but dead, cornified cells become dry and brittle once exposed to the environment. Sebum lubricates and protects the keratin of the hair shaft and conditions the surrounding skin. Shampooing removes the natural oily coating; excessive washing can make hairs stiff and brittle.

Sebaceous follicles are large sebaceous glands that communicate directly with the epidermis. These follicles, which never produce hairs, are located on the integument covering the face, back, chest, nipples, and male sex organs.

Although sebum has bactericidal (bacteria-killing) properties, under some conditions bacteria can invade sebaceous glands or follicles. The presence of bacteria in sebaceous glands or follicles can produce a local inflammation known as *folliculitis* (fo-lik-ū-LĪ-tis). If the duct of a sebaceous gland becomes blocked, a distinctive abscess called a *furuncle* (FUR-ung-kl), or boil,

develops. The usual treatment for a furuncle is to cut it open, or "lance" it, so that normal drainage and healing can occur.

Sebaceous glands and sebaceous follicles are very sensitive to changes in the concentrations of sex hormones, and their secretory activities accelerate at puberty. For this reason, an individual with large sebaceous glands may be especially prone to develop **acne** during adolescence. In acne, sebaceous ducts become blocked and secretions accumulate, causing inflammation and a raised "pimple." The trapped secretions provide a fertile environment for bacterial infection. AM *Acne*

Seborrheic dermatitis is an inflammation around abnormally active sebaceous glands. The affected area becomes red, and there is usually some epidermal scaling. Sebaceous glands of the scalp are most often involved. In infants, mild cases are called *cradle cap*. This condition is one cause of dandruff in adults. Anxiety, stress, and food allergies can exaggerate the problem.

Sweat Glands

The skin contains two different types of sweat glands, or **sudoriferous glands**—*apocrine sweat glands* and *merocrine sweat glands* (Figure 5-9●). These names refer to the mechanism of secretion, introduced in Chapter 4. ∞ [p. 117]

APOCRINE SWEAT GLANDS In the armpits (axillae), around the nipples, and in the groin, **apocrine**

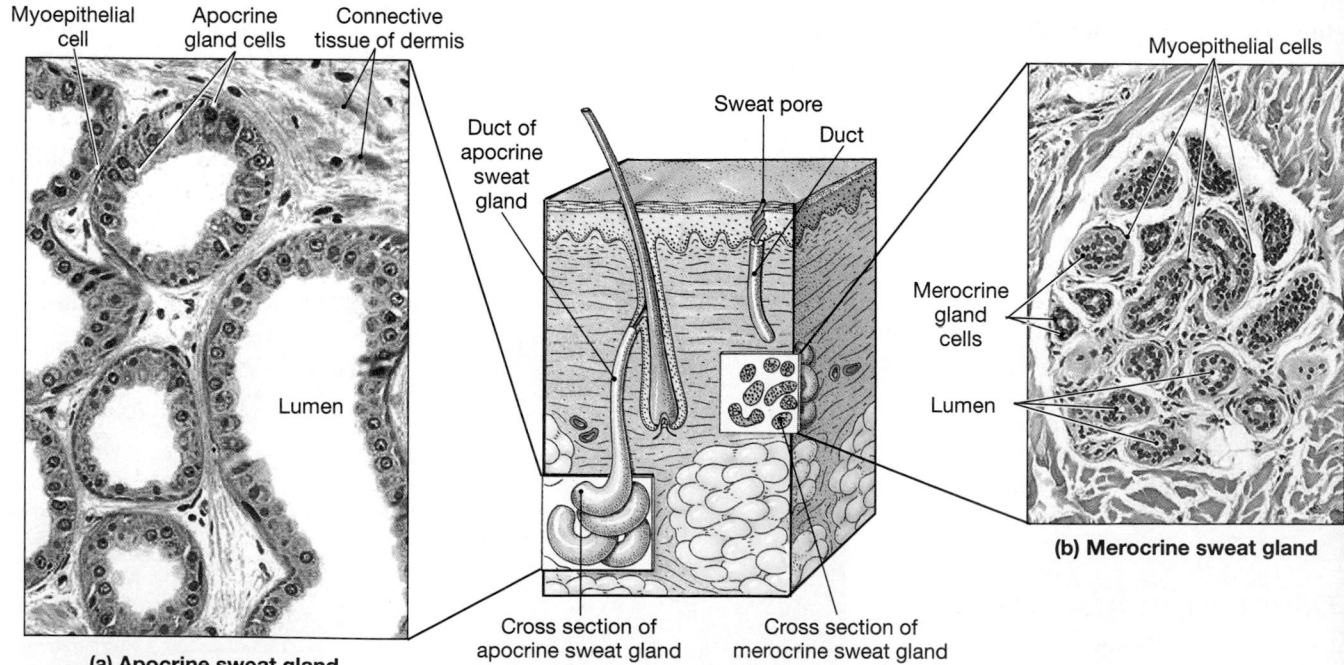

●*FIGURE 5-9* Sweat Glands. **(a)** Apocrine sweat glands are located in the axillae, groin, and nipples. These glands produce a thick, odorous fluid by apocrine secretion. (LM × 369) **(b)** Merocrine sweat glands produce a watery fluid by merocrine secretion. (LM × 194)

sweat glands communicate with hair follicles (Figure 5-9a ●). These are coiled tubular glands that produce a sticky, cloudy, and potentially odorous secretion. Apocrine sweat glands begin secreting at puberty. The sweat produced is a nutrient source for bacteria, which intensify its odor. Special **myoepithelial cells** (*myo-*, muscle) surround the secretory cells. Myoepithelial cells contract, squeezing the gland and thereby discharging the accumulated secretion into the hair follicles. The secretory activities of the gland cells and the contractions of myoepithelial cells are controlled by the nervous system and by circulating hormones.

MEROCRINE (ECCRINE) SWEAT GLANDS **Merocrine sweat glands**, also known as **eccrine** (EK-rin) **sweat glands**, are far more numerous and widely distributed than apocrine glands (Figure 5-9b●). The adult integument contains 2–5 million merocrine sweat glands. They are smaller than apocrine sweat glands, and they do not extend as far into the dermis. Palms and soles have the highest numbers; the palm of the hand has an estimated 500 merocrine sweat glands per square centimeter (3000 per in.^{2}). These are coiled tubular glands that discharge their secretions directly onto the surface of the skin.

The sweat produced by merocrine sweat glands is called sensible perspiration. Sweat is 99 percent water, but it also contains some electrolytes (chiefly sodium chloride), organic nutrients, and waste products. It has a pH of 4–6.8, and the presence of sodium chloride gives sweat a salty taste. See Appendix VI for a complete analysis of the composition of normal sweat.

The functions of merocrine sweat gland activity include:

- *Cooling the surface of the skin to reduce body temperature.* This is the primary function of sensible perspiration, and the degree of secretory activity is regulated by neural and hormonal mechanisms. When all the merocrine sweat glands are working at their maximum, the rate of perspiration may exceed a gallon per hour, and dangerous fluid and electrolyte losses can occur. For this reason, athletes in endurance sports must pause at regular intervals to drink fluids.
- *Excretion of water and electrolytes.* A number of ingested drugs are excreted as well.
- *Protection from environmental hazards.* Sweat dilutes harmful chemicals and discourages the growth of microorganisms.

Other Integumentary Glands

Merocrine sweat glands are widely distributed across the body surface, and sebaceous glands are located wherever there are hair follicles. Apocrine sweat glands are located in relatively restricted areas. The skin also contains a variety of specialized glands that are restricted to specific locations. We shall encounter many of them in later chapters but will cite two important examples here:

1. The **mammary glands** of the breasts are anatomically related to apocrine sweat glands. A complex interaction between sex hormones and pituitary hormones controls their development and secretion. We shall discuss mammary gland structure and function in Chapter 28.
2. **Ceruminous** (se-ROO-mi-nus) **glands** are modified sweat glands located in the external auditory canal. Their secretions combine with those of nearby sebaceous glands, forming a mixture called **cerumen,** or ear wax. Ear wax, together with tiny hairs along the ear canal, helps trap foreign particles or small insects and keeps them from reaching the eardrum.

Control of Glandular Secretions

Sebaceous glands and apocrine sweat glands can be collectively turned on or off by the autonomic nervous system, but no regional control is possible. When one sebaceous or apocrine gland is activated, so are all the other glands of that type in the body. Merocrine sweat glands are much more precisely controlled, and the amount of secretion and the area of the body involved can be varied independently. For example, when you are nervously awaiting an anatomy and physiology exam, your palms may begin to sweat.

As we have seen, the primary function of sensible perspiration is to cool the surface of the skin and to reduce body temperature. When you sweat in the hot sun, all your merocrine glands are working together. The blood vessels beneath your epidermis are flushed with blood, and your skin reddens. Your skin surface is warm and wet. As the moisture evaporates, your skin cools. If your body temperature then falls below normal, sensible perspiration ceases, blood flow to the skin declines, and the cool, dry surface releases little heat into the environment. We introduced the negative feedback mechanisms of thermoregulation (temperature control) in Chapter 1 and shall provide additional details in Chapter 25. ∞ *[pp. 13–15]*

☑ What are the functions of sebaceous secretions?

☑ Deodorants are used to mask the effects of secretions from which type of skin gland?

☑ Which type of skin gland is most affected by the hormonal changes that occur during puberty?

Nails

Nails form on the dorsal surfaces of the tips of the fingers and toes. The nails protect the exposed tips of the fingers and toes and help limit their distortion when they are subjected to mechanical stress—for example,

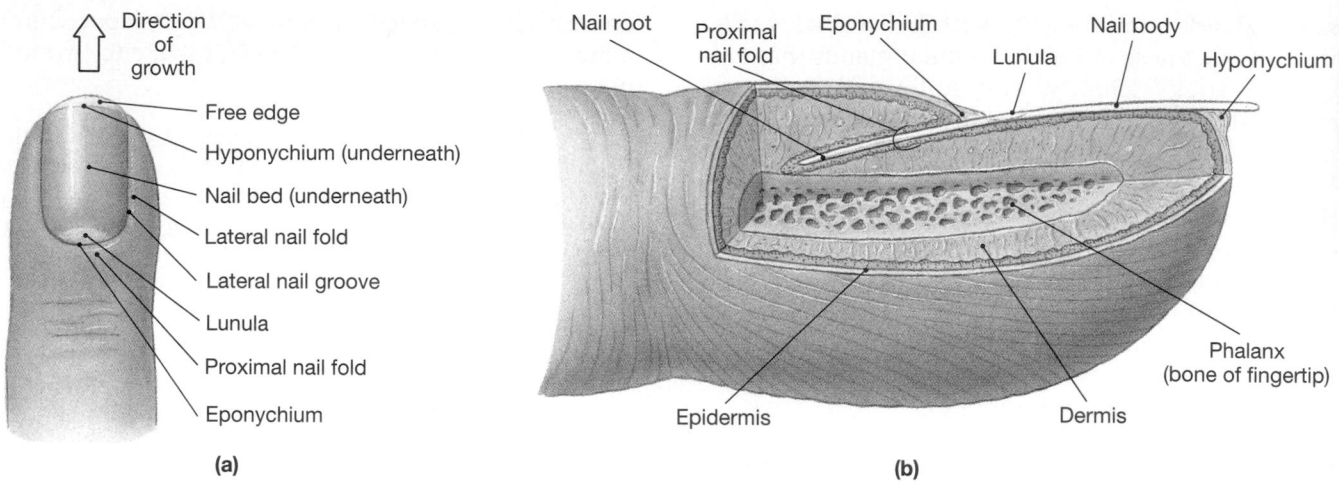

•FIGURE 5-10 Structure of a Nail. The prominent features of a typical fingernail as viewed **(a)** from the surface and **(b)** in section.

when you run or grasp objects. The **nail body** covers the **nail bed** (Figure 5-10a•). Nail production occurs at the **nail root,** an epithelial fold not visible from the surface (Figure 5-10b•). The deepest portion of the nail root lies very close to the periosteum of the bone of the fingertip.

A portion of the stratum corneum of the nail root extends over the exposed nail, forming the **cuticle,** or **eponychium** (ep-ō-NIK-ē-um; *epi-,* over + *onyx,* nail). Underlying blood vessels give the nail its characteristic pink color, but near the root these vessels may be obscured, leaving a pale crescent known as the **lunula** (LOO-nū-la; *luna,* moon). The nail body is recessed beneath the level of the surrounding epithelium, and it is bounded by **nail grooves** and **nail folds.** The **free edge** of the nail extends over a thickened stratum corneum, the **hyponychium** (hī-pō-NIK-ē-um).

The body of the nail consists of dead, tightly compressed cells packed with keratin. The cells producing the nails can be affected by conditions that alter body metabolism, so changes in the shape, structure, or appearance of the nails can assist in diagnosis. For example, the nails may turn yellow in patients who have chronic respiratory disorders, thyroid gland disorders, or AIDS. Nails may become pitted and distorted in *psoriasis* (a condition marked by rapid stem cell division in the stratum germinativum) and concave in some blood disorders.

LOCAL CONTROL OF INTEGUMENTARY FUNCTION

The integumentary system displays a significant degree of functional independence. It often responds di-

rectly and automatically to local influences without the involvement of the nervous or endocrine system. For example, when the skin is subjected to mechanical stresses, stem cells in the stratum germinativum divide more rapidly, and the depth of the epithelium increases. That is why calluses form on your palms when you perform manual labor. A more dramatic display of local regulation can be seen after an injury to the skin.

Injury and Repair

The skin can regenerate effectively even after considerable damage has occurred, because stem cells persist in both the epithelial and connective tissue components. Germinative cell divisions replace lost epidermal cells, and mesenchymal cell divisions replace lost dermal cells. This process can be slow. When large surface areas are involved, problems of infection and fluid loss complicate the situation. The relative speed and effectiveness of skin repair vary according to the type of wound involved. A slender, straight cut, or *incision,* may heal relatively quickly compared with a deep scrape, or *abrasion,* which involves a much greater surface area of skin to be repaired. ⏻AM⏻ *A Classification of Wounds*

Figure 5-11• illustrates the four stages in the regeneration of the skin after an injury. When damage extends through the epidermis and into the dermis, bleeding generally occurs (Step 1). The blood clot, or **scab,** that forms at the surface temporarily restores the integrity of the epidermis and restricts the entry of additional microorganisms into the area (Step 2). The bulk of the clot consists of an insoluble network

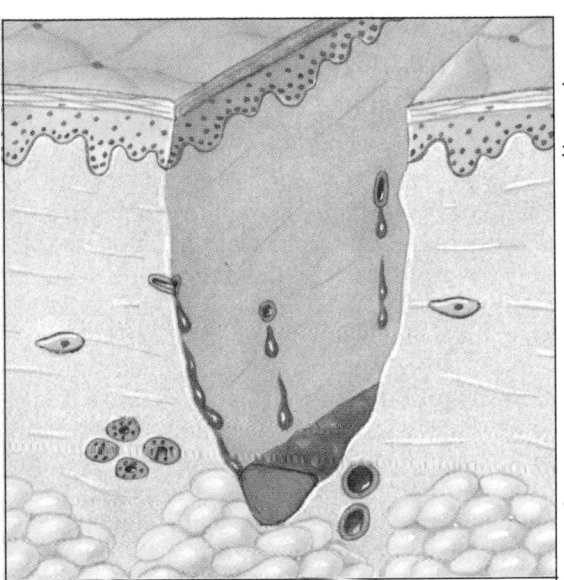

Epidermis

Dermis

Step 1: Bleeding occurs at the injury site immediately after the injury, and mast cells in the region trigger an inflammatory response. ∞ *[p. 140]*

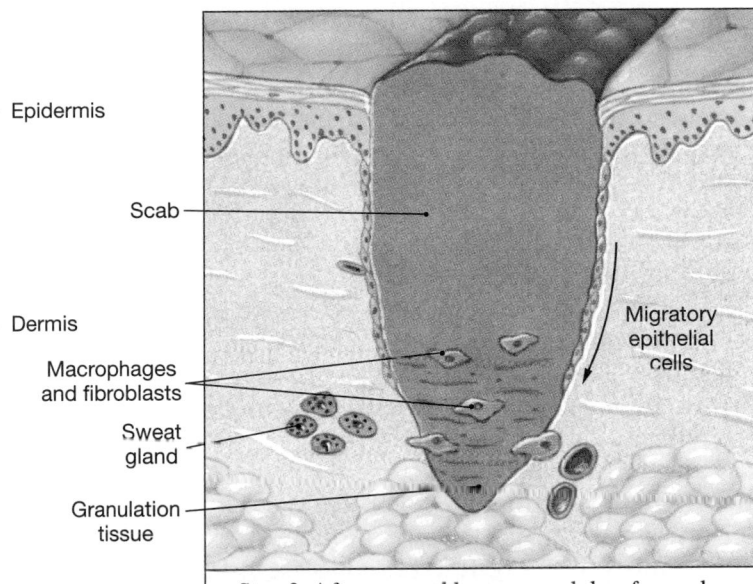

Scab

Migratory epithelial cells

Macrophages and fibroblasts

Sweat gland

Granulation tissue

Step 2: After several hours, a scab has formed and cells of the stratum germinativum are migrating along the edges of the wound. Phagocytic cells are removing debris, and more of these cells are arriving via the enhanced circulation. Clotting around the edges of the affected area partially isolates the region.

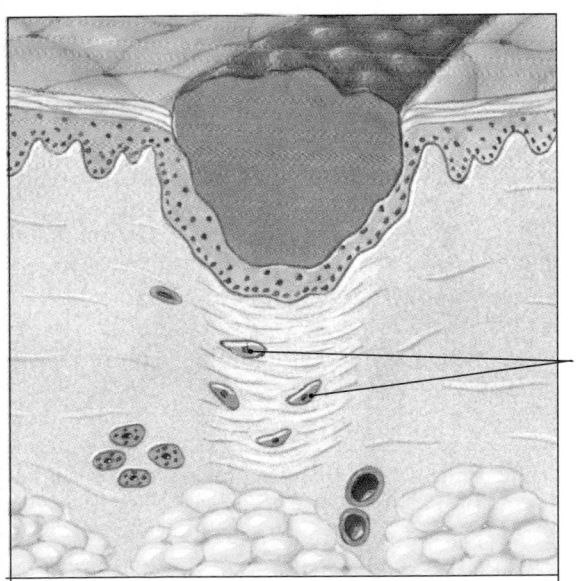

Fibroblasts

Step 3: One week after the injury, the scab has been undermined by epidermal cells migrating over the meshwork produced by fibroblast activity. Phagocytic activity around the site has almost ended, and the fibrin clot is disintegrating.

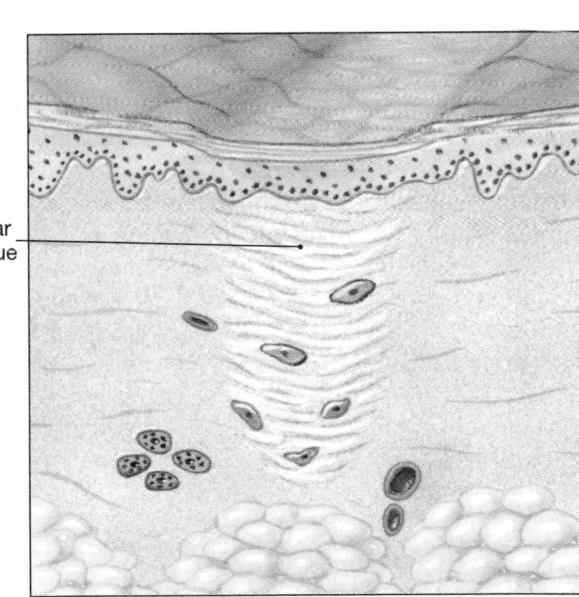

Scar tissue

Step 4: After several weeks, the scab has been shed, and the epidermis is complete. A shallow depression marks the injury site, but fibroblasts in the dermis continue to create scar tissue that will gradually elevate the overlying epidermis.

•*FIGURE 5-11* **Integumentary Repair**

of *fibrin*, a fibrous protein that forms from blood proteins during the clotting response. The clot's color reflects the presence of trapped red blood cells. Cells of the stratum germinativum undergo rapid divisions and begin to migrate along the sides of the wound in an attempt to replace the missing epidermal cells. Meanwhile, macrophages patrol the damaged area of the dermis, phagocytizing any debris and pathogens.

If the wound covers an extensive area or involves a region covered by thin skin, dermal repairs must be under way before epithelial cells can cover the surface. Fibroblast and mesenchymal cell divisions produce mobile cells that invade the deeper areas of injury (Step 3). Endothelial cells of damaged blood vessels also begin to divide, and capillaries follow the fibroblasts, providing a circulatory supply. The combination of blood clot, fibroblasts, and an extensive capillary network is called **granulation tissue.** Over time, the clot dissolves and the number of capillaries declines (Step 4). Fibroblast activity leads to the appearance of collagen fibers and typical ground substance.

These repairs do not restore the integument to its original condition, however, for the dermis will contain an abnormally large number of collagen fibers and relatively few blood vessels. Severely damaged hair follicles, sebaceous or sweat glands, muscle cells, and nerves are seldom repaired, and they too are replaced by fibrous tissue. The formation of this rather inflexible, fibrous, noncellular **scar tissue** can be considered a practical limit to the healing process.

We do not know what regulates the extent of scar tissue formation, and the process is highly variable. For example, surgical procedures performed on a fetus do not leave scars, perhaps because damaged fetal tissues do not produce the same types of growth factors that adult tissues do. In some adults, most often those with dark skin, scar tissue formation may continue beyond the requirements of tissue repair. The result is a flattened mass of scar tissue that begins at the injury site and grows into the surrounding dermis. This thickened area of scar tissue, called a **keloid** (KĒ-loyd) is covered by a shiny, smooth epidermal surface. Keloids most commonly develop on the upper back, shoulders, anterior chest, and earlobes. They are harmless, and some aboriginal cultures intentionally produce keloids as a form of body decoration.

⬚AGING AND THE INTEGUMENTARY SYSTEM

Aging affects all the components of the integumentary system. These changes include:

- The epidermis thins as germinative cell activity declines, making older people more prone to injury and skin infections.
- The number of Langerhans cells decreases to about 50 percent of levels seen at maturity (roughly, age 21). This decrease may reduce the sensitivity of the immune system and further encourage skin damage and infection.
- Vitamin D_3 production declines by about 75 percent. The result can be reduced calcium and phosphate ab-

sorption, eventually leading to muscle weakness and a reduction in bone strength.
- Melanocyte activity declines, and in Caucasians the skin becomes very pale. With less melanin in the skin, people become more sensitive to sun exposure and more likely to experience sunburn.
- Glandular activity declines. The skin becomes dry and often scaly because sebum production is reduced. Merocrine sweat glands are also less active; with impaired perspiration, older people cannot lose heat as fast as younger people can. Thus, the elderly are at greater risk of overheating in warm environments.
- The blood supply to the dermis is reduced at the same time that sweat glands become less active. Because the circulation decreases, the skin becomes cool; this cooling can stimulate thermoreceptors, making the person "feel" cold, even in a warm room. However, with reduced circulation and sweat gland function, the elderly are less able than younger people to lose body heat. As a result, overexertion or exposure to high temperatures (such as a sauna or hot tub) can cause dangerously high body temperatures.
- Hair follicles stop functioning or produce thinner, finer hairs. With decreased melanocyte activity, these hairs are gray or white.
- The dermis thins, and the elastic fiber network decreases in size. The integument therefore becomes weaker and less resilient; sagging and wrinkling occur. These effects are most pronounced in areas of the body that have been exposed to the sun.
- With changes in levels of sex hormones, secondary sexual characteristics in hair and body-fat distribution begin to fade. In consequence, people age 90–100 of both sexes and all races tend to look alike.
- Skin repairs proceed relatively slowly, and recurring infections may result. Skin repairs proceed most rapidly in young, healthy individuals. For example, barring infection, it takes 3–4 weeks to complete the repairs to a blister site in a young adult. The same repairs at age 65–75 take 6–8 weeks.

✔ What is the name given to the combination of fibrin clot, fibroblasts, and the extensive network of capillaries in healing tissue?

✔ Why can skin regenerate effectively even after considerable damage has occurred?

✔ Older individuals do not tolerate the summer heat as well as they did when they were young, and they are more prone to heat-related illness. What accounts for this change?

INTEGRATION WITH OTHER SYSTEMS

Although it can function independently, many activities of the integumentary system are integrated with those of other systems. Figure 5-12● diagrams the major functional relationships.

SKELETAL SYSTEM

Provides structural support.

Synthesizes vitamin D₃, essential for calcium and phosphorus absorption (bone maintenance and growth).

MUSCULAR SYSTEM

Contractions of skeletal muscles pull against skin of face, producing facial expressions important in communication.

Synthesizes vitamin D_3, essential for normal calcium absorption (calcium ions play an essential role in muscle contraction).

NERVOUS SYSTEM

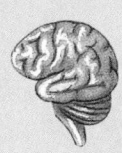

Controls blood flow and sweat gland activity for thermoregulation; stimulates contraction of arrector pili muscles to elevate hairs.

Receptors in dermis and deep epidermis provide sensations of touch, pressure, vibration, temperature, and pain.

ENDOCRINE SYSTEM

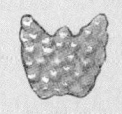

Sex hormones stimulate sebaceous gland activity; male and female sex hormones influence hair growth, distribution of subcutaneous fat, and apocrine sweat gland activity; adrenal hormones alter dermal blood flow and help mobilize lipids from

Synthesizes vitamin D_3, precursor of calcitriol

CARDIOVASCULAR SYSTEM

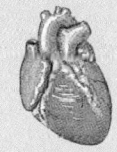

Provides O₂ and nutrients; delivers hormones and cells of immune system; carries away CO₂, waste products, and toxins; provides heat to maintain normal skin temperature.

Stimulation by mast cells produces localized changes in blood flow and capillary permeability,

LYMPHATIC SYSTEM

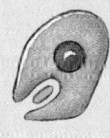

Assists in defending the integument by providing additional macrophages and mobilizing lymphocytes.

Provides physical barriers that prevent pathogen entry; Langerhans cells and macrophages resist infection; mast cells trigger inflammation and initiate the immune response.

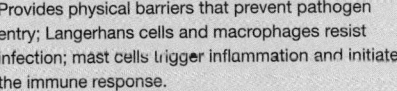

RESPIRATORY SYSTEM

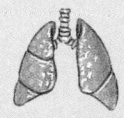

Provides oxygen and eliminates carbon dioxide.

Hairs guard entrance to nasal cavity.

DIGESTIVE SYSTEM

Provides nutrients for all cells and lipids for storage by adipocytes.

Synthesizes vitamin D_3, needed for absorption of calcium and phosphorus.

URINARY SYSTEM

Excretes waste products, maintains normal body fluid pH and ion composition.

Assists in elimination of water and solutes; keratinized epidermis limits fluid loss through skin.

THE INTEGUMENTARY SYSTEM

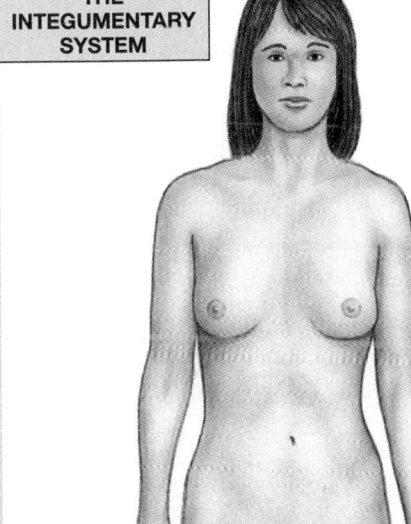

FOR ALL SYSTEMS

Provides mechanical protection against environmental hazards.

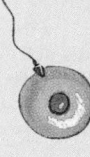

Covers external genitalia; provides sensations that stimulate sexual behaviors; mammary gland secretions provide nourishment for newborn infant.

Sex hormones affect hair distribution, adipose tissue distribution in subcutaneous layer, and mammary gland development.

REPRODUCTIVE SYSTEM

• **FIGURE 5-12** **Functional Relationships between the Integumentary System and Other Systems**

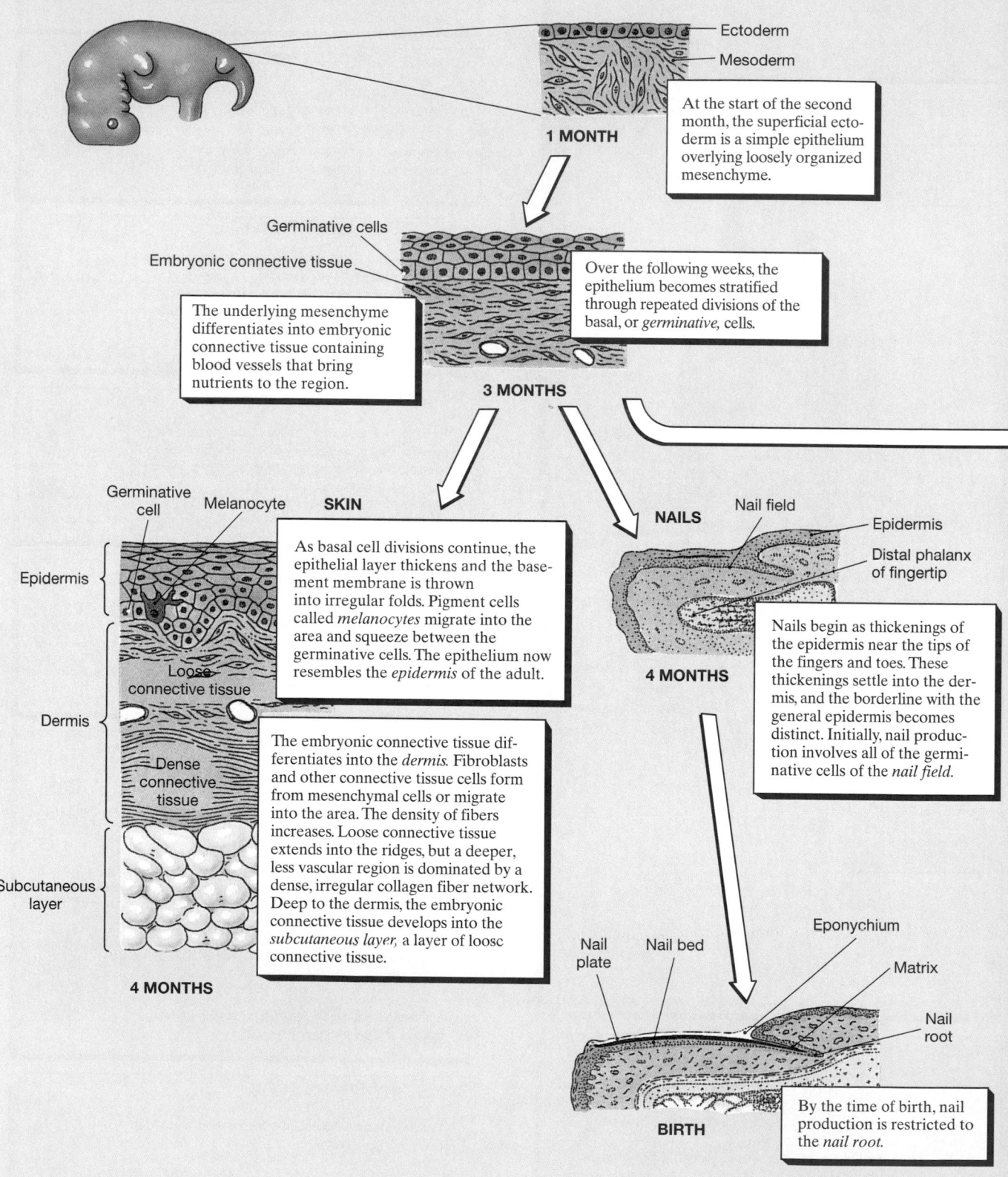

1 MONTH

Ectoderm

Mesoderm

At the start of the second month, the superficial ectoderm is a simple epithelium overlying loosely organized mesenchyme.

Germinative cells

Embryonic connective tissue

Over the following weeks, the epithelium becomes stratified through repeated divisions of the basal, or *germinative*, cells.

The underlying mesenchyme differentiates into embryonic connective tissue containing blood vessels that bring nutrients to the region.

3 MONTHS

Germinative cell   Melanocyte   **SKIN**

Epidermis

Dermis

Subcutaneous layer

Loose connective tissue

Dense connective tissue

As basal cell divisions continue, the epithelial layer thickens and the basement membrane is thrown into irregular folds. Pigment cells called *melanocytes* migrate into the area and squeeze between the germinative cells. The epithelium now resembles the *epidermis* of the adult.

The embryonic connective tissue differentiates into the *dermis*. Fibroblasts and other connective tissue cells form from mesenchymal cells or migrate into the area. The density of fibers increases. Loose connective tissue extends into the ridges, but a deeper, less vascular region is dominated by a dense, irregular collagen fiber network. Deep to the dermis, the embryonic connective tissue develops into the *subcutaneous layer,* a layer of loose connective tissue.

4 MONTHS

NAILS   Nail field

Epidermis

Distal phalanx of fingertip

4 MONTHS

Nails begin as thickenings of the epidermis near the tips of the fingers and toes. These thickenings settle into the dermis, and the borderline with the general epidermis becomes distinct. Initially, nail production involves all of the germinative cells of the *nail field*.

Nail plate   Nail bed   Eponychium

Matrix

Nail root

BIRTH

By the time of birth, nail production is restricted to the *nail root*.

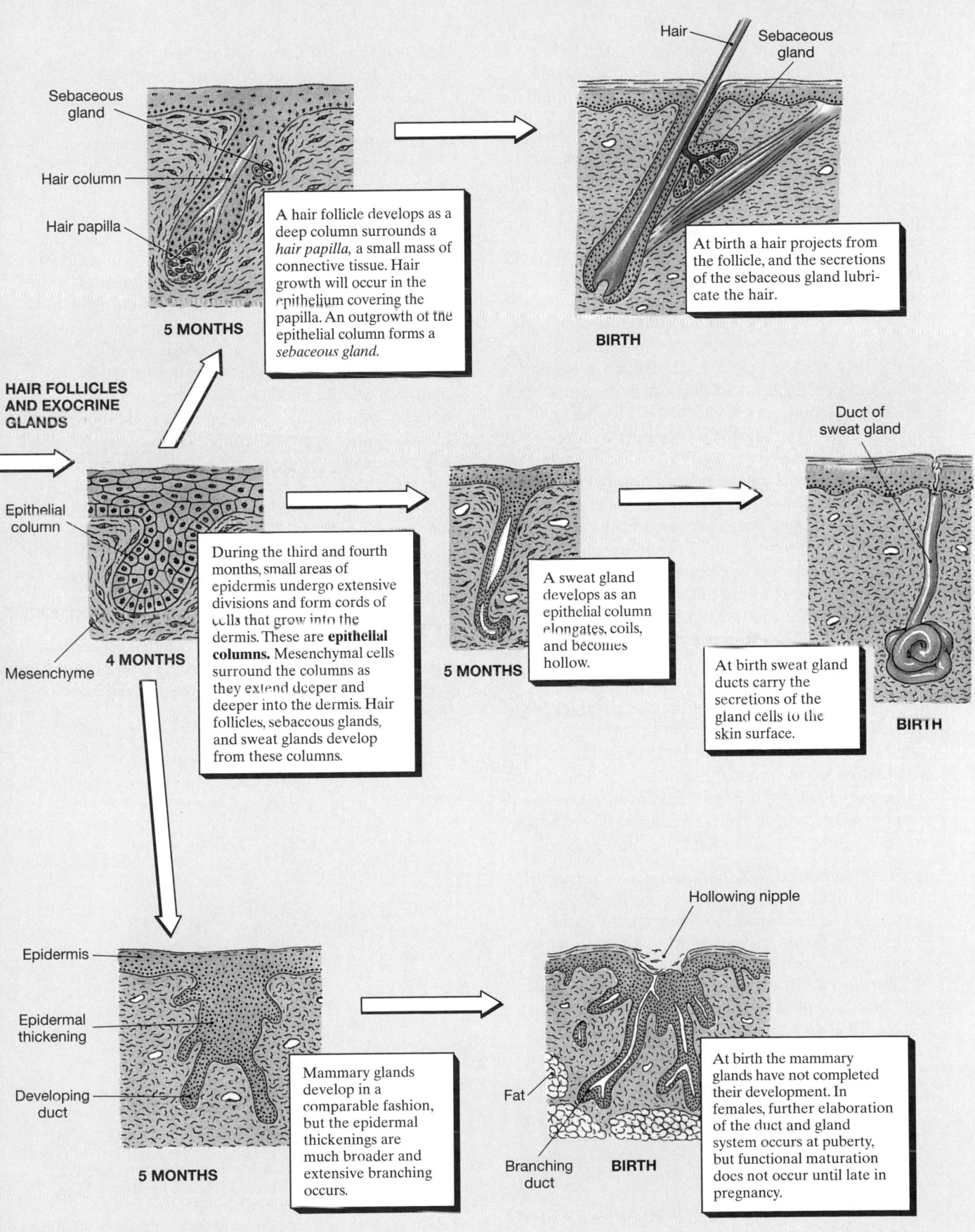

Sebaceous gland

Hair column

Hair papilla

5 MONTHS

A hair follicle develops as a deep column surrounds a *hair papilla*, a small mass of connective tissue. Hair growth will occur in the epithelium covering the papilla. An outgrowth of the epithelial column forms a *sebaceous gland.*

HAIR FOLLICLES AND EXOCRINE GLANDS

Epithelial column

Mesenchyme

4 MONTHS

Hair

Sebaceous gland

At birth a hair projects from the follicle, and the secretions of the sebaceous gland lubricate the hair.

BIRTH

During the third and fourth months, small areas of epidermis undergo extensive divisions and form cords of cells that grow into the dermis. These are **epithelial columns.** Mesenchymal cells surround the columns as they extend deeper and deeper into the dermis. Hair follicles, sebaceous glands, and sweat glands develop from these columns.

A sweat gland develops as an epithelial column elongates, coils, and becomes hollow.

5 MONTHS

Duct of sweat gland

At birth sweat gland ducts carry the secretions of the gland cells to the skin surface.

BIRTH

Epidermis

Epidermal thickening

Developing duct

5 MONTHS

Mammary glands develop in a comparable fashion, but the epidermal thickenings are much broader and extensive branching occurs.

Hollowing nipple

Fat

Branching duct

BIRTH

At birth the mammary glands have not completed their development. In females, further elaboration of the duct and gland system occurs at puberty, but functional maturation does not occur until late in pregnancy.

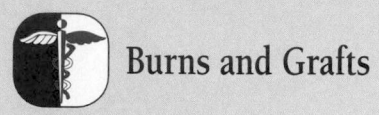

Burns and Grafts

Burns result from exposure of the skin to heat, radiation, electrical shock, or strong chemical agents. The severity of the burn reflects the depth of penetration and the total area affected.

First- and second-degree burns are also called *partial-thickness burns* because damage is restricted to the superficial layers of the skin. Only the surface of the epidermis is affected by a *first-degree burn*. In this type of burn, including most sunburns, the skin reddens and can be painful. The redness, a sign called **erythema** (er-i-THĒ-ma), results from inflammation of the sun-damaged tissues. In a *second-degree burn*, the entire epidermis and perhaps some of the dermis are damaged. Accessory structures such as hair follicles and glands are generally not affected, but blistering, pain, and swelling occur. If the blisters rupture at the surface, infection can easily develop. Healing typically takes 1–2 weeks, and some scar tissue may form. *Full-thickness burns,* or *third-degree burns,* destroy the epidermis and dermis, extending into subcutaneous tissues. Despite swelling, these burns are less painful than second-degree burns because sensory nerves are destroyed along with accessory structures, blood vessels, and other dermal components. Extensive third-degree burns cannot repair themselves, because granulation tissue cannot form and epithelial cells are unable to cover the injury site. As a result, the affected area remains open to potential infection.

Each year in the United States, roughly 10,000 people die from the effects of burns. The larger the area burned, the more significant the effects on integumentary function. Figure 5-13● presents a standard reference for calculating the percentage of total surface area damaged. Burns that cover more than 20 percent of the skin surface represent serious threats to life because they affect the following functions:

- *Fluid and electrolyte balance.* Even areas with partial-thickness burns lose their effectiveness as barriers to fluid and electrolyte losses. In full-thickness burns, the rate of fluid loss through the skin may reach five times the normal level.
- *Thermoregulation.* Increased fluid loss means increased evaporative cooling. More energy must be expended to keep body temperature within acceptable limits.
- *Protection from attack.* The epidermal surface, damp from uncontrolled fluid losses, encourages bacterial growth. If the skin is broken at a blister or the site of a third-degree burn, infection is likely. Widespread bacterial infection, or **sepsis** (*septikos*, rotting), is the leading cause of death in burn victims.

Effective treatment of full-thickness burns focuses on the following four procedures:

1. Replacing lost fluids and electrolytes.
2. Providing sufficient nutrients to meet increased metabolic demands for thermoregulation and healing.
3. Preventing infection by cleaning and covering the burn while administering antibiotic drugs.
4. Assisting tissue repairs.

Because full-thickness burns cannot heal unaided, surgical procedures are necessary to encourage healing. In a **skin graft,** areas of intact skin are transplanted to cover the burn site. A *split-thickness graft* takes a shaving of the epidermis and superficial portions of the dermis. A *full-thickness graft* involves the epidermis and both layers of the dermis.

With fluid-replacement therapies, infection control methods, and grafting techniques, young patients with burns over 80 percent of the body now have about a 50 percent chance of recovery. Recent advances in cell culture techniques may improve survival rates further. A small section of undamaged epidermis can be removed and grown under controlled laboratory conditions. Over time, the germinative cell divisions produce large sheets of epidermal cells that can then be used to cover the burn area. From initial samples the size of a postage stamp, square meters of epidermis have been grown and transplanted onto body surfaces. Although questions remain concerning the strength and flexibility of the repairs, skin cultivation represents a substantial advance in the treatment of serious burns. AM *Synthetic Skin*

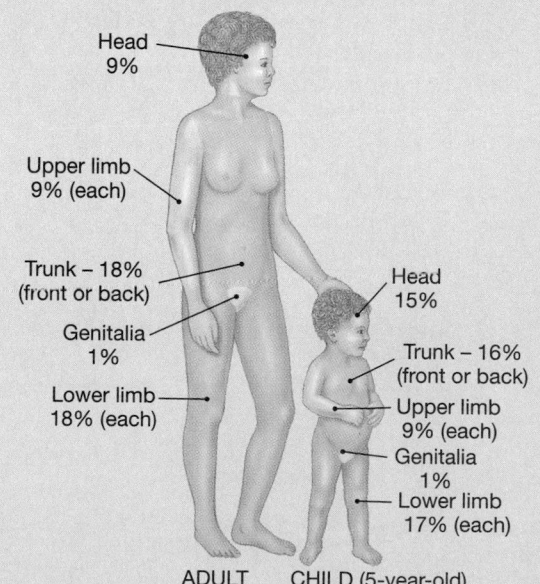

●**FIGURE 5-13 A Quick Method of Estimating the Percentage of Surface Area Affected by Burns.** The method is called the *rule of nines* because the surface area in adults is in multiples of nine. This rule must be modified for children because their proportions are quite different.

SELECTED CLINICAL TERMINOLOGY

Terms Discussed in This Chapter

acne: A sebaceous gland inflammation caused by an accumulation of secretions. *(p. 160 and AM)*

capillary hemangioma: A birthmark caused by a tumor in the capillaries of the papillary layer of the dermis. It generally enlarges after birth but subsequently fades and disappears. *(p. 155)*

cavernous hemangioma ("port-wine stain"): A birthmark caused by a tumor affecting large vessels in the dermis; generally lasts a lifetime. *(p. 155)*

contact dermatitis: A dermatitis generally caused by strong chemical irritants. It produces an itchy rash that may spread to other areas as scratching distributes the chemical agent; includes poison ivy. *(p. 156)*

cyanosis (sī-uh-NŌ-sis): Bluish skin color as a result of reduced oxygenation of the blood in superficial vessels. *(p. 152)*

decubitis ulcers (bedsores): Ulcers that occur in areas subject to restricted circulation, especially common in hospitalized or bedridden persons. *(p. 155)*

dermatitis: An inflammation of the skin that primarily involves the papillary region of the dermis. *(p. 156)*

diaper rash: A localized dermatitis caused by a combination of moisture, irritating chemicals from fecal or urinary wastes, and flourishing microorganisms. *(p. 156)*

eczema (EK-se-muh): A dermatitis that can be triggered by temperature changes, fungi, chemical irritants, greases, detergents, or stress and that can be related to hereditary and/or environmental factors. *(p. 156)*

folliculitis (fo-lik-ū-LĪ-tis): A local inflammation caused by bacterial infection of a sebaceous gland or sebaceous follicle. *(p. 160)*

furuncle (FUR-ung-kl): A boil, an abscess that develops when the duct of a sebaceous gland is blocked. *(p. 160)*

granulation tissue: A combination of fibrin, fibroblasts, and capillaries that forms during tissue repair after inflammation. *(p. 164)*

hypodermic needle: A needle used to administer drugs via subcutaneous injection. *(p. 156)*

keloid (KĒ-loyd): A thickened area of scar tissue covered by a shiny, smooth epidermal surface. *(p. 164)*

male pattern baldness: Hair loss in an adult male due to changes in levels of circulating sex hormones. *(p. 159)*

malignant melanoma (mel-uh-NŌ-muh): A skin cancer originating in malignant melanocytes *(p. 152 and AM)*

psoriasis (sō-RĪ-uh-sis): A painless condition characterized by rapid stem cell divisions in the stratum germinativum of the scalp, elbows, palms, soles, groin, and nails. Affected areas appear dry and scaly. *(p. 162 and AM)*

scab: A blood clot that forms at the surface of a wound to the skin. *(p. 162)*

seborrheic dermatitis: An inflammation around abnormally active sebaceous glands. *(p. 160)*

sepsis: A dangerous, widespread bacterial infection; the leading cause of death in burn patients. *(p. 168)*

skin graft: Transplantation of a section of skin (partial thickness or full thickness) to cover an extensive injury site, such as a third-degree burn. *(p. 168)*

ulcer: A localized shedding of an epithelium. *(p. 155)*

urticaria (ur-ti-KAR ē-uh), or *hives*: An extensive dermatitis resulting from an allergic reaction to food, drugs, an insect bite, infection, stress, or other stimulus. *(p. 156)*

AM Additional Terms Discussed in the *Applications Manual*

basal cell carcinoma: A malignant cancer that originates in the stratum germinativum; the most common skin cancer. Roughly two-thirds of these cancers appear in areas subjected to chronic UV exposure. Metastasis seldom occurs.

erysipelas (er-i-SIP-e-lus): A widespread inflammation of the dermis caused by bacterial infection.

pruritis (proo-RĪ-tis): An irritating itching sensation, common in skin conditions.

skin signs: Characteristic abnormalities in the skin surface that can assist in diagnosing skin conditions. They include flat macules, wheals, papules, nodules, vesicles (blisters), pustules, erosions (ulcers), crusts, scales, fissures.

squamous cell carcinoma: A form of skin cancer less common than basal cell carcinoma, almost totally restricted to areas of sun-exposed skin. Metastasis seldom occurs.

xerosis (ze-RŌ-sis): "Dry skin," a common complaint of older persons and almost anyone living in an arid climate.

CHAPTER REVIEW

 On-line resources for this chapter are on our World Wide Web site at: http://www.prenhall.com/martini/fap

STUDY OUTLINE

INTRODUCTION, p. 148

1. The **integumentary system** consists of the **cutaneous membrane,** or *skin*, which includes the **epidermis** and **dermis,** and the **accessory structures.** Underneath the dermis lies the **subcutaneous layer.** *(Figure 5-1)*

THE EPIDERMIS, p. 149

1. **Thin skin,** formed by four layers of **keratinocytes,** covers most of the body. Heavily abraded body surfaces may be covered by **thick skin,** formed by five layers. *(Figure 5-2)*

2. The epidermis provides mechanical protection, prevents fluid loss, and helps keep microorganisms out of the body.

Layers of the Epidermis, p. 149

3. Cell divisions in the **stratum germinativum,** the innermost epidermal layer, replace more-superficial cells. *(Figure 5-3)*

4. As epidermal cells age, they pass through the **stratum spinosum,** the **stratum granulosum,** the **stratum lucidum** (if thick skin), and the **stratum corneum.** In the process, they accumulate large amounts of **keratin.** Ultimately, the cells are shed or lost. *(Figure 5-3)*

5. **Epidermal ridges,** such as those on the palms and soles, improve our gripping ability and increase the skin's sensitivity. They interlock with **dermal papillae** of the underlying dermis.

6. **Langherhans cells** are part of the immune system. **Merkel cells** provide information about objects that touch the skin.

Skin Color, p. 151

7. The color of the epidermis depends on two factors: blood supply and pigment composition and concentration.

8. The epidermis contains the pigments **carotene** and **melanin. Melanocytes,** which produce melanin, protect us from **ultraviolet (UV) radiation.** *(Figure 5-4)*

9. Interruptions of the dermal blood supply can lead to such conditions as **cyanosis** or **decubitis ulcers.**

The Epidermis and Vitamin D₃, p. 154

10. Epidermal cells synthesize **vitamin D₃,** or **cholecalciferol,** when exposed to the UV radiation in sunlight.

Epidermal Growth Factor, p. 154

11. **Epidermal growth factor** (**EGF**) stimulates the maintenance and repair of the epidermis and the secretion of epithelial glands.

THE DERMIS, p. 154

Dermal Organization, p. 154

1. The dermis consists of the **papillary layer** and the deeper **reticular layer.**

2. The papillary layer of the dermis contains blood vessels, lymphatics, and sensory nerves. This layer supports and nourishes the overlying epidermis. The reticular layer consists of a meshwork of collagen and elastic fibers oriented to resist tension in the skin.

3. Extensive distortion of the dermis can cause **stretch marks.**

4. The pattern of collagen and elastic fiber bundles forms **lines of cleavage.** *(Figure 5-5)*

Dermal Circulation and Innervation, p. 155

5. **Dermatitis** is an inflammation that generally affects the papillary layer.

THE SUBCUTANEOUS LAYER, p. 156

1. The subcutaneous layer, or **hypodermis,** stabilizes the skin's position against underlying organs and tissues. *(Figure 5-1)*

ACCESSORY STRUCTURES, p. 156

Hair Follicles and Hair, p. 156

1. **Hairs** originate in complex organs called **hair follicles.** Each hair has a **root** and a **shaft.** At the base of the root are a **hair papilla,** surrounded by a **hair bulb,** and a **root hair plexus** of sensory nerves. Hair production involves cell specialization of the **hair matrix** to form a soft core, or **medulla,** surrounded by a **cortex.** The **cuticle** is a hard layer that coats the hair. *(Figures 5-6, 5-7)*

2. The **arrector pili** muscles can erect the hairs. *(Figures 5-6, 5-7)*

3. Our bodies have both **vellus hairs** ("peach fuzz") and heavy **terminal hairs.** A hair that has stopped growing is called a **club hair.**

4. Our hairs grow and are shed according to the **hair growth cycle.** A single hair grows for 2–5 years and is subsequently shed.

Glands in the Skin, p. 159

5. A typical **sebaceous gland** discharges the waxy **sebum** into a **lumen** and ultimately into a hair follicle. **Sebaceous follicles** are large sebaceous glands that empty directly onto the surface of the epidermis. Inflammation of sebaceous glands can lead to **acne** or **seborrheic dermatitis.** *(Figure 5-8)*

6. There are two classes of sweat glands, or **sudoriferous glands. Apocrine sweat glands** produce an odorous secretion. The more numerous **merocrine,** or **eccrine, sweat glands** produce a watery secretion known as **sensible perspiration.** *(Figure 5-9)*

7. **Mammary glands** of the breasts are related to apocrine sweat glands. **Ceruminous glands** in the ear produce a waxy **cerumen.**

Nails, p. 161

8. The **nail body** of a **nail** covers the **nail bed.** Nail production occurs at the **nail root,** which is overlain by the **cuticle,** or **eponychium.** The **free edge** of the nail extends over the **hyponychium.** *(Figure 5-10)*

LOCAL CONTROL OF INTEGUMENTARY FUNCTION, p. 162

Injury and Repair, p. 162

1. The skin can regenerate effectively even after considerable damage. The process includes formation of a **scab** and **granulation tissue.** *(Figure 5-11)*

AGING AND THE INTEGUMENTARY SYSTEM, p. 164

1. Aging affects all the components of the integumentary system.

INTEGRATION WITH OTHER SYSTEMS, p. 164

1. Many activities of the integumentary system are integrated with those of other systems. *(Figure 5-12)*

REVIEW QUESTIONS

Level 1   Reviewing Facts and Terms

1. The two major components of the integumentary system are
   (a) the cutaneous membrane and the accessory structures
   (b) the epidermis and the hypodermis
   (c) the hair and the nails
   (d) the dermis and the subcutaneous layer

2. Beginning at the basement membrane and traveling toward the free surface, the layers of the epidermis include the following strata:

   (a) corneum, lucidum, granulosum, spinosum, germinativum
   (b) granulosum, lucidum, spinosum, germinativum, corneum
   (c) germinativum, spinosum, granulosum, lucidum, corneum
   (d) lucidum, granulosum, spinosum, germinativum, corneum

3. The fibrous protein that forms the basic structural component of hair and nails is
  - **(a)** collagen
  - **(b)** melanin
  - **(c)** elastin
  - **(d)** keratin

4. The primary pigments contained in the epidermis are
  - **(a)** carotene and xanthophyll
  - **(b)** carotene and melanin
  - **(c)** melanin and chlorophyll
  - **(d)** xanthophyll and melanin

5. The two major components of the dermis are the
  - **(a)** superficial fascia and cutaneous membrane
  - **(b)** epidermis and hypodermis
  - **(c)** papillary layer and reticular layer
  - **(d)** stratum germinativum and stratum corneum

6. The cutaneous plexus and papillary plexus consist of
  - **(a)** a network of arteries providing the dermal blood supply
  - **(b)** a network of nerves providing dermal sensations
  - **(c)** specialized cells for cutaneous sensations
  - **(d)** gland cells that release cutaneous secretions

7. The accessory structures of the integument include the
  - **(a)** blood vessels, glands, muscles, and nerves
  - **(b)** Merkel cells, Pacinian corpuscles, and Meissner's corpuscles
  - **(c)** hair, skin, and nails
  - **(d)** hair follicles, sebaceous glands, and sweat glands

8. The two major types of hairs in the integument are
  - **(a)** thick, thin
  - **(b)** vellus, terminal
  - **(c)** primary, secondary
  - **(d)** short, long

9. The two types of exocrine glands in the skin are
  - **(a)** merocrine and sweat glands
  - **(b)** sebaceous and sweat glands
  - **(c)** apocrine and sweat glands
  - **(d)** eccrine and sweat glands

10. Sweat glands that communicate with hair follicles in the armpits and produce an odorous secretion are
  - **(a)** apocrine glands
  - **(b)** merocrine glands
  - **(c)** sebaceous glands
  - **(d)** a, b, and c are correct

11. The primary function of sensible perspiration is to
  - **(a)** get rid of wastes
  - **(b)** protect the skin from dryness
  - **(c)** maintain electrolyte balance
  - **(d)** reduce body temperature

12. The thickened stratum corneum over which the free edge of the nail extends is called the
  - **(a)** eponychium
  - **(b)** hyponychium
  - **(c)** cuticle
  - **(d)** cerumen

13. Muscle weakness and a reduction in bone strength in the elderly results from decreased
  - **(a)** vitamin D_3 production
  - **(b)** melanin production
  - **(c)** sebum production
  - **(d)** dermal blood supply

14. The reason older persons are more sensitive to sun exposure and more likely to experience sunburn than are younger persons is that with age,
  - **(a)** melanocyte activity declines
  - **(b)** vitamin D_3 production declines
  - **(c)** glandular activity declines
  - **(d)** skin thickness decreases

15. In which layer(s) of the epidermis does cell division occur?

16. What are the protein precursors of keratin, and in what layer(s) of the epidermis are these proteins produced?

17. What two skin pigments are found in the epidermis?

18. What widespread effects does epidermal growth factor (EGF) have on the integument?

19. What two major layers constitute the dermis, and what components are found in each layer?

20. Beginning at the hair cuticle, what three cell layers make up the walls of the hair follicle?

21. Which two groups of sweat glands are contained in the skin?

Level 2   Reviewing Concepts

22. How do insensible and sensible perspiration differ?

23. During transdermal administration of drugs, why are fat-soluble drugs more desirable than those that are water-soluble?

24. In our society, a tan body is associated with good health. However, medical research constantly warns about the dangers of excessive exposure to the sun. What are the benefits of a tan?

25. Why is it important for a surgeon to choose an incision pattern according to the lines of cleavage of the skin?

26. Why is regional infection or inflammation of the skin usually very painful?

27. Why is a subcutaneous injection with a hypodermic needle a useful method of administering drugs?

28. How are changes in the shape, structure, or appearance of the nails clinically significant?

29. Why is scab formation important in skin repair?

30. Why does skin sag and wrinkle as a person ages?

Level 3   Critical Thinking and Clinical Applications

31. A new mother notices that her 6-month-old child has a yellow-orange complexion. Fearful that the child may have jaundice (a condition caused by a toxic yellow-orange pigment in the blood), she takes him to her pediatrician. After examining the child, the pediatrician declares him perfectly healthy and advises the mother to watch the child's diet. Why?

32. Vanessa's 80-year-old grandmother sets her thermostat at 80°F and wears a sweater on balmy spring days. When asked why, the grandmother says she is cold. Can you explain why?

33. Exposure to optimum amounts of sunlight is necessary for proper bone maintenance and growth in children. **(a)** What does sunlight do to promote bone maintenance and growth? **(b)** If a child lives in an area where exposure to sunlight is rare because of pollution or overcast skies, what can be done to minimize impaired maintenance and growth of bone?

34. One of the factors to which lie detectors respond is an increase in skin conductivity due to the presence of moisture. Explain the physiological basis for the use of this indicator.

Osseous Tissue and Skeletal Structure

*T*he Tin Man wore his skeleton on the outside. The armor plating was useful, but there were drawbacks, such as having to oil the joints. We avoid such problems by having internal skeletons with self-lubricating joints. Our skeletons are made of bone, a remarkable tissue that is as strong as steel-reinforced concrete, immune to rust (unlike the Tin Man's), and capable of repairing itself even after serious injuries. This chapter focuses on the nature of bone.

CHAPTER OUTLINE AND OBJECTIVES

Thïs chapter begins our examination of the skeletal system. This system includes the bones of the skeleton and the cartilages, ligaments, and other connective tissues that stabilize or connect them. Skeletal elements are more than just racks that muscles hang from; they have a great variety of vital functions. In addition to supporting the weight of the body, bones work together with muscles to maintain body position and to produce controlled, precise movements. Without the skeleton to pull against, contracting muscle fibers would be unable to make us sit, stand, walk, or run. Without something to hold onto, contracting muscles merely get shorter and fatter.

We begin our study of the skeletal system by identifying its primary functions:

1. *Support.* The skeletal system provides structural support for the entire body. Individual bones or groups of bones provide a framework for the attachment of soft tissues and organs.

2. *Storage of minerals and lipids.* As we shall learn in Chapter 25, *minerals* are inorganic ions that contribute to the osmolarity of body fluids. They also participate in various physiological processes, and several minerals are important as enzyme cofactors. The calcium salts of bone represent a valuable mineral reserve that maintains normal concentrations of calcium and phosphate ions in body fluids. Calcium is the most abundant mineral in the human body. In addition to acting as a mineral reserve, the bones of the skeleton store energy reserves as lipids in areas of *yellow marrow.*

3. *Blood cell production.* Red blood cells, white blood cells, and other blood elements are produced within the *red marrow* that fills the internal cavities of many bones. We shall describe the role of the bone marrow in blood cell formation when we deal with the cardiovascular and lymphatic systems (Chapters 19 and 22).

4. *Protection.* Many delicate tissues and organs are surrounded by skeletal elements. The ribs protect the heart and lungs, the skull encloses the brain, the vertebrae shield the spinal cord, and the pelvis cradles delicate digestive and reproductive organs.

5. *Leverage.* Many bones function as *levers* that can change the magnitude and direction of the forces generated by skeletal muscles. The movements produced range from the delicate motion of a fingertip to powerful changes in the position of the entire body.

A CLASSIFICATION OF BONES

Bone Shapes

Every adult skeleton contains 206 major bones. We can divide these bones into six broad categories according to their individual shapes (Figure 6-1●):

1. **Long bones** (Figure 6-1a●) are relatively long and slender. Long bones are found in the arm and forearm, thigh and leg, palms, soles, fingers, and toes. The femur, the long bone of the thigh, is the largest and heaviest bone in the body.

2. **Short bones** are boxlike in appearance (Figure 6-1b●). Examples of short bones include the *carpal bones* (wrists) and *tarsal bones* (ankles).

3. **Flat bones** have thin, roughly parallel surfaces. Flat bones form the roof of the skull (Figure 6-1c●), the sternum, the ribs, and the scapula. They provide protection for underlying soft tissues and offer an extensive surface area for the attachment of skeletal muscles.

4. **Irregular bones** have complex shapes with short, flat, notched, or ridged surfaces (Figure 6-1d●). The spinal vertebrae and several skull bones are irregular bones.

5. **Sesamoid bones** are generally small, flat, and shaped somewhat like a sesame seed (Figure 6-1e●). They develop inside tendons and are most commonly located near joints at the knees, the hands, and the feet. Everyone has sesamoid **patellae** (pa-TEL-ē), or kneecaps, but individuals vary in terms of the location and abundance of other sesamoid bones. (Sesamoid bones may form in at least 26 locations.)

6. **Sutural bones,** or *Wormian bones,* are small, flat, irregularly shaped bones between the flat bones of the skull (Figure 6-1f●). There are individual variations in the number, shape, and position of the sutural bones. Their borders are like puzzle pieces, and they range in size from a grain of sand to the size of a quarter.

Each bone in the skeleton contains two different forms of **osseous tissue,** or bone tissue: (1) *compact bone* and (2) *spongy bone.* **Compact bone,** or *dense bone,* is relatively solid, whereas **spongy bone** forms an open network of struts and plates. Compact bone is always located on the surface of a bone, where it forms a sturdy protective layer. Spongy bone is in the interior of a bone. The relationship between compact and spongy bone and their relative proportions vary with the shape of the bone. We will consider two extremes at this time: long bones and flat bones.

Figure 6-2● (p. 175) introduces the anatomy of the femur, a representative long bone. It has a long tubular *shaft,* or **diaphysis** (dī-AF-i-sis). At each end, there is an expanded area known as the **epiphysis** (ē-PIF-i-sis). The diaphysis is connected to each epiphysis at a narrow zone known as the **metaphysis** (me-TAF-i-sis). The wall of the diaphysis consists of a layer of compact bone that surrounds the **marrow cavity,** also known as the *medullary cavity* (*medulla,* innermost part) (Figure 6-2a●). The epiphyses consist largely of spongy bone with a thin covering, or **cortex,** of compact bone. The epiphyses of the femur form complex joints at the hip and knee. These joints have large joint cavities lined by *synovial membranes* and filled with *synovial fluid.* ∞ *[p. 134]* The portion of an epiphysis within such a joint cavity is covered by a thin layer of hyaline cartilage called the *articular cartilage.*

The marrow cavity of the diaphysis and the spaces between the struts and plates of the epiphyses contain

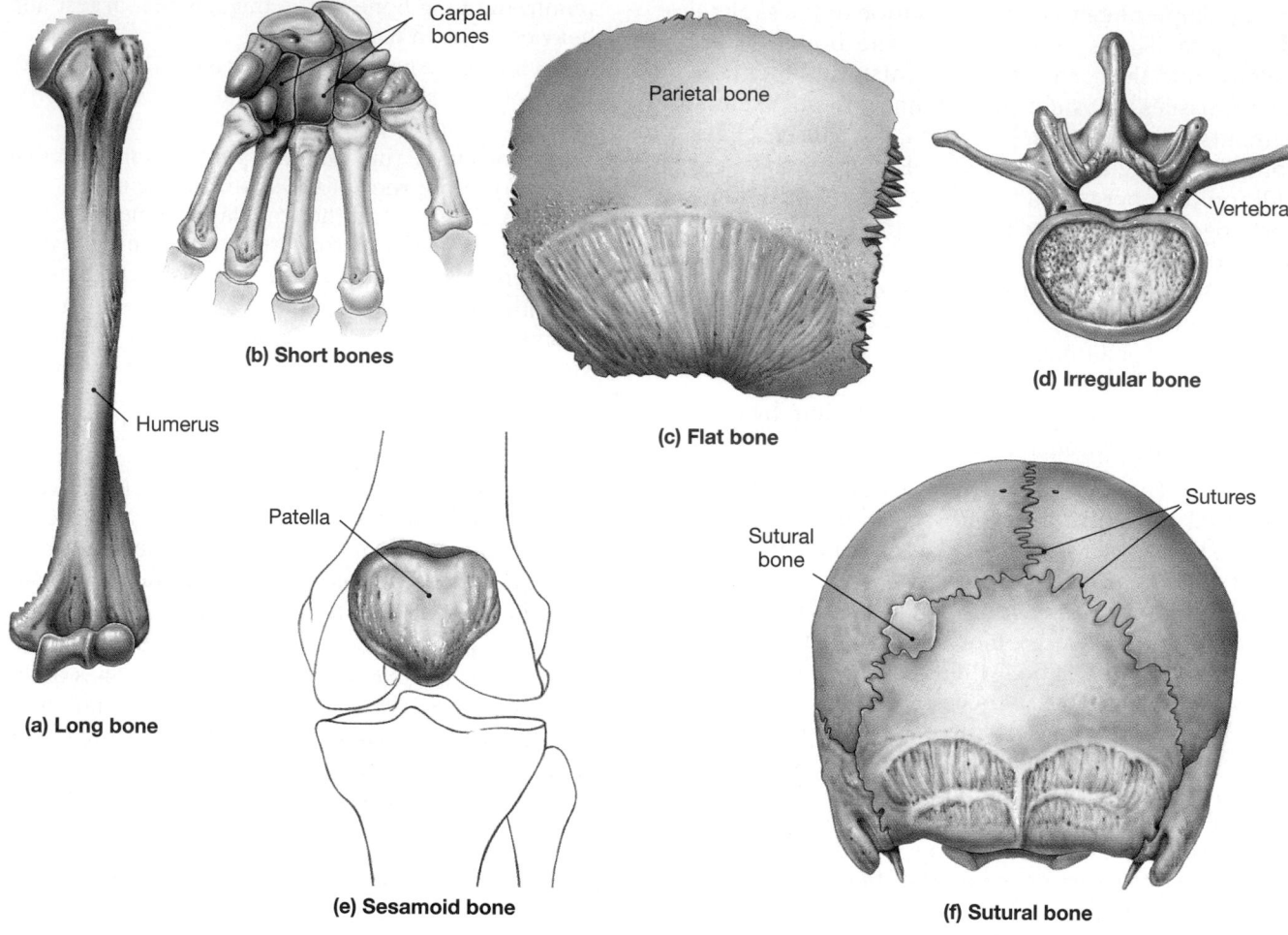

Carpal bones

(b) Short bones

Humerus

(a) Long bone

Patella

(e) Sesamoid bone

Parietal bone

(c) Flat bone

Vertebra

(d) Irregular bone

Sutures

Sutural bone

(f) Sutural bone

•*FIGURE 6-1* Classification of Bones by Shape

bone marrow, a loose connective tissue. Bone marrow may be dominated by fat cells (**yellow bone marrow**) or by a mixture of mature and immature red blood cells, white blood cells, and the stem cells that produce them (**red bone marrow**). Yellow bone marrow is an important energy reserve; areas of red bone marrow are important sites of blood cell formation.

Figure 6-2b• details the structure of a flat bone from the skull, such as one of the *parietal bones*. A flat bone resembles a spongy bone sandwich, with layers of compact bone covering a core of spongy bone. Although bone marrow is present within the area of spongy bone, there is no marrow cavity. We often use special terms when we describe the flat bones of the skull. Their relatively thick layers of compact bone are called the *internal* and *external tables*. The layer of spongy bone between the tables is called the *diploë* (DIP-lō-ē).

Many people think of the skeleton as a rather dull collection of bony props. This is far from the truth. Our bones are actually complex, dynamic organs that are constantly changing to adapt to the demands we place on them. We will now consider the internal organization of a typical bone and see how that seemingly inert structure can be remodeled and repaired.

BONE HISTOLOGY

Osseous tissue is one of the supporting connective tissues. (You may wish to review the sections on dense connective tissues, cartilage, and bone in Chapter 4 at this time.) ∞ *[pp. 124–132]* Like other connective tissues, osseous tissue contains specialized cells and a matrix consisting of extracellular protein fibers and a ground substance. The matrix of bone tissue is solid and sturdy because of the deposition of calcium salts around the protein fibers.

In Chapter 4, which introduced the organization of bone tissue, we discussed four basic features of a representative sample of bone:

1. The matrix of bone is very dense and contains deposits of calcium salts.
2. The matrix contains bone cells, or *osteocytes,* within pockets, or **lacunae.** The lacunae are typically organized around blood vessels that branch through the bony matrix.
3. Narrow passageways through the matrix, called **canaliculi,** extend between the lacunae and nearby blood vessels, forming a branching network for the exchange of nutrients, waste products, and gases.

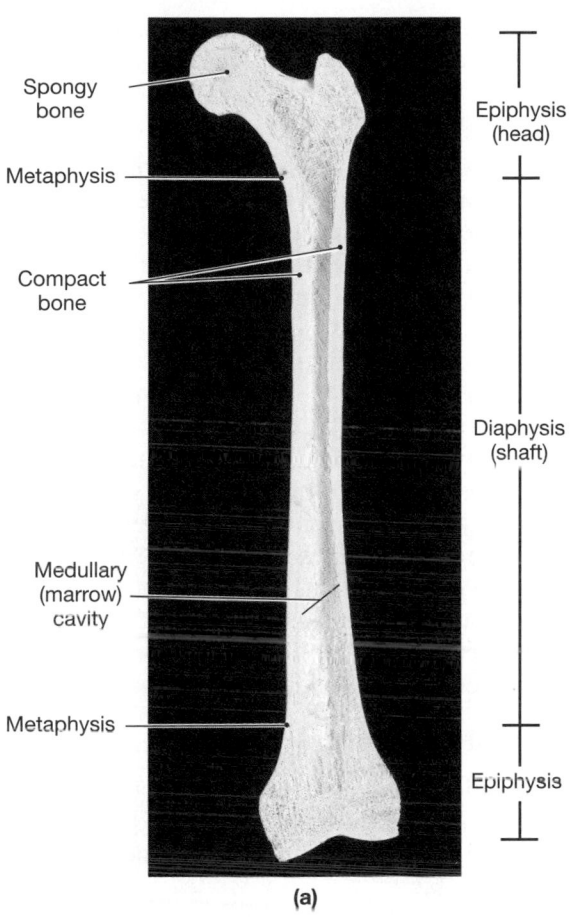

(a)

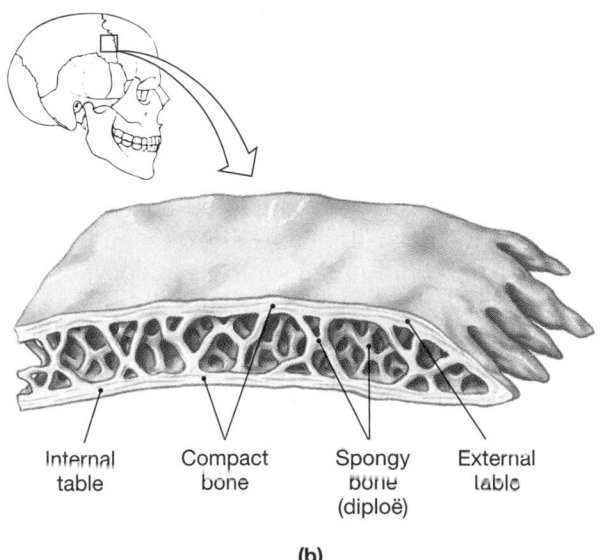

(b)

•**FIGURE 6-2** **Bone Structure.** **(a)** Structure of a representative long bone. **(b)** Structure of a flat bone.

4. Except at joints, the outer surfaces of bones are covered by a *periosteum* that has outer fibrous and inner cellular layers.

We will now take a closer look at the organization of the matrix and cells of bone.

The Matrix of Bone

Calcium phosphate, $Ca_3(PO_4)_2$, accounts for almost two-thirds of the weight of bone. The calcium phosphate interacts with calcium hydroxide $[Ca(OH)_2]$ to form crystals of **hydroxyapatite,** $Ca_{10}(PO_4)_6(OH)_2$. As they form, these crystals also incorporate other calcium salts, such as calcium carbonate ($CaCO_3$), and ions such as sodium, magnesium, and fluoride. Roughly one-third of the weight of bone is contributed by collagen fibers. Osteocytes and other cell types account for only 2 percent of the mass of a typical bone.

Calcium phosphate crystals are very hard but relatively inflexible and quite brittle. They can withstand compression, but the crystals are likely to shatter when exposed to bending, twisting, or sudden impacts. Collagen fibers are remarkably strong; when subjected to tension (pull), they are stronger than steel. Flexible as well as tough, they can easily tolerate twisting and bending but offer little resistance to

compression. When compressed, they simply bend out of the way.

In bone, the collagen fibers provide an organic framework for the formation of hydroxyapatite crystals. These crystals form small plates and rods that are locked into the collagen fibers. The result is a protein-crystal combination with properties intermediate between those of collagen and those of pure mineral crystals. Bone is strong, somewhat flexible, and very resistant to shattering. In its overall properties, bone is on a par with the best steel-reinforced concrete. In reality, bone is far superior to concrete because it can be remodeled—that is, replaced with new bone—on a regular basis and can repair itself after injury.

Cells in Bone

Although osteocytes are most abundant, bone contains four different cell types: (1) *osteoprogenitor cells*, (2) *osteoblasts*, (3) *osteocytes*, and (4) *osteoclasts*.

Bone contains small numbers of mesenchymal cells called **osteoprogenitor** (os-tē-ō-prō-JEN-i-tor) **cells** (*progenitor,* ancestor). These stem cells can divide to produce daughter cells that differentiate into osteoblasts. Osteoprogenitor cells maintain populations of osteoblasts and play an important role in the repair of a *fracture* (a break or crack in a bone). They are found (1) in the inner, cellular layer of the periosteum, (2) in an inner layer, or *endosteum*, that lines the marrow cavities of bone, and (3) in the lining of vascular passageways within the matrix.

Osteoblasts (OS-tē-ō-blasts; *blast,* precursor) are responsible for the production of new bone matrix, a process called **osteogenesis** (os-tē-ō-JEN-e-sis; *gennan,* to produce). These cells synthesize and release the proteins and other organic components of the bone matrix. The matrix prior to calcification is called **osteoid** (OS-tē-oyd). The osteoblasts also assist in elevating local concentrations of calcium phosphate and promoting the deposition of calcium salts within the organic matrix. This process converts osteoid to bone. When an osteoblast becomes completely surrounded by bone matrix, that cell differentiates into an osteocyte.

Osteocytes (*osteon,* bone) are mature bone cells that account for most of the cell population. Osteocytes are found in lacunae that are sandwiched between layers of calcified matrix. These layers of matrix are called **lamellae** (lah-MEL-lē; singular, *lamella,* a thin plate). Osteocytes cannot divide, and there is never more than one osteocyte within each lacuna. Canaliculi penetrate the lamellae, radiating through the matrix and connecting lacunae with one another and with nutrient sources. The canaliculi contain cytoplasmic extensions of the osteocytes. Neighboring osteocytes are linked by gap junctions that permit the exchange of ions and small molecules, including nutrients and hormones, between the cells. The interstitial fluid that surrounds the osteocytes and their extensions provides an additional route for the diffusion of nutrients and waste products.

Osteocytes have two major functions:

1. *They recycle the calcium salts in the matrix around them.* Osteocytes release chemicals that dissolve the minerals in the adjacent matrix. These minerals enter the circulation, ultimately to be replaced through the deposition of new hydroxyapatite crystals.

2. *They can participate in the repair of damaged bone.* If released from their lacunae, these cells can convert to a less specialized type of cell, such as an osteoblast or an osteoprogenitor cell.

Osteoclasts (OS-tē-ō-klasts; *clast,* to break), giant cells with 50 or more nuclei, remove bone matrix. Osteoclasts are not related to osteoprogenitor cells or their descendants. Instead, they are derived from *monocytes,* the circulating form of the macrophages found in loose and dense connective tissues. Acids and proteolytic (protein-digesting) enzymes secreted by osteoclasts dissolve the matrix and release the stored minerals. This process, called **osteolysis** (os-tē-OL-i-sis), or *resorption,* is important in the regulation of calcium and phosphate concentrations in body fluids. Osteoclasts are always removing matrix, and osteoblasts are always adding to it. The balance between the activities of osteoblasts and osteoclasts is very important. When osteoclasts remove calcium salts faster than osteoblasts deposit them, bones become weaker. When osteoblast activity predominates, bones become stronger and more massive.

Compact and Spongy Bone

The composition of the matrix in compact bone is the same as that in spongy bone, but they differ in the three-dimensional arrangement of osteocytes, canaliculi, and lamellae.

Compact Bone

The basic functional unit of mature compact bone is the **osteon** (OS-tē-on), or *Haversian system* (Figures 6-3 and 6-4a●). Within an osteon, the osteocytes are

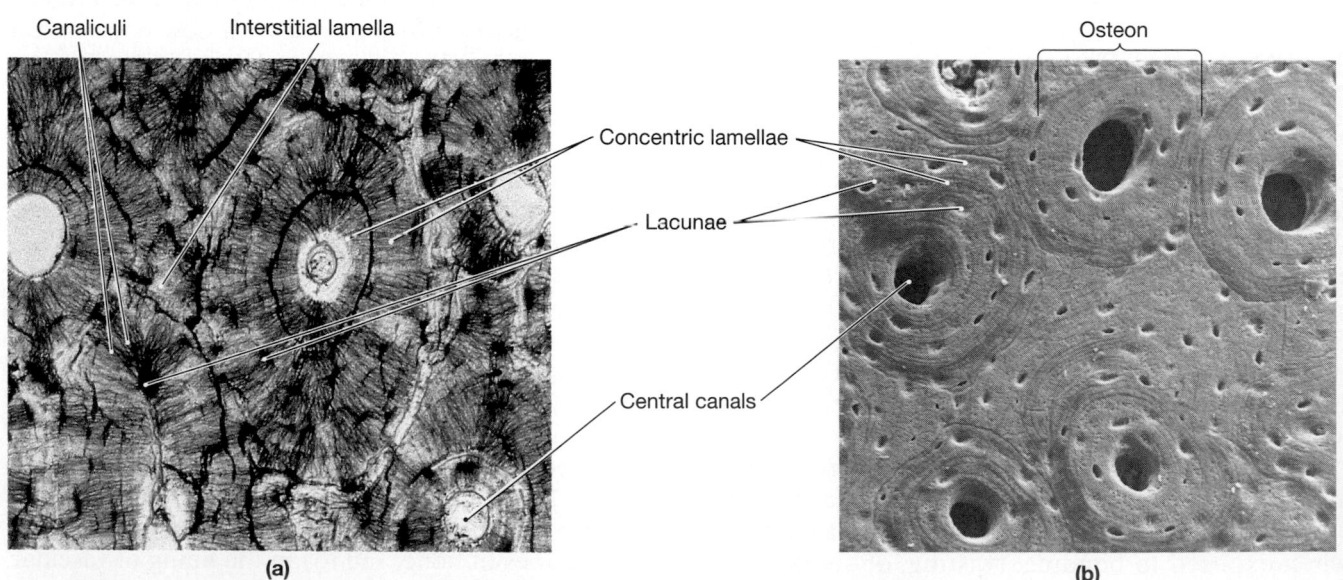

●*FIGURE 6-3* **Histology of Compact Bone.** **(a)** A thin section through compact bone. By this procedure, the intact matrix and central canals appear white, and the lacunae and canaliculi are shown in black. (LM × 343) **(b)** A micrograph of several osteons in compact bone. (SEM × 182) Reproduced from R. G. Kessel and R. H. Kardon, "Tissues and Organs: A Text-Atlas of Scanning Electron Microscopy," W. H. Freeman & Co., 1979.

arranged in concentric layers around a **central canal**, or *Haversian canal*. This canal contains one or more blood vessels (normally a capillary and a *venule*, a very small vein) that carry blood to and from the osteon. Central canals generally run parallel to the surface of the bone. Other passageways, known as **perforating canals**, or the *canals of Volkmann*, extend roughly perpendicular to the surface. Blood vessels in the perforating canals supply blood to osteons deeper in the bone and to tissues of the marrow cavity.

The lamellae of each osteon are cylindrical and aligned parallel to the long axis of the central canal. These are known as *concentric lamellae*. Collectively, the concentric lamellae resemble a bull's-eye around the central canal. Canaliculi radiating through the lamellae interconnect the lacunae of the osteon with one another and with the central canal. *Interstitial lamellae* fill in the spaces between the osteons in compact bone. These lamellae are remnants of osteons whose matrix components have been almost completely recycled by osteoclasts. *Circumferential* (ser-kum-fer-EN-shul) *lamellae* (*circum-,* around + *ferre,* to bear) are covered by the periosteum or endosteum (Figure 6-4a,b●). These lamellae are produced during the growth of the bone.

FUNCTION OF COMPACT BONE Compact bone is thickest where stresses arrive from a limited range of directions. All osteons in compact bone are aligned the same way, making such bones very strong when

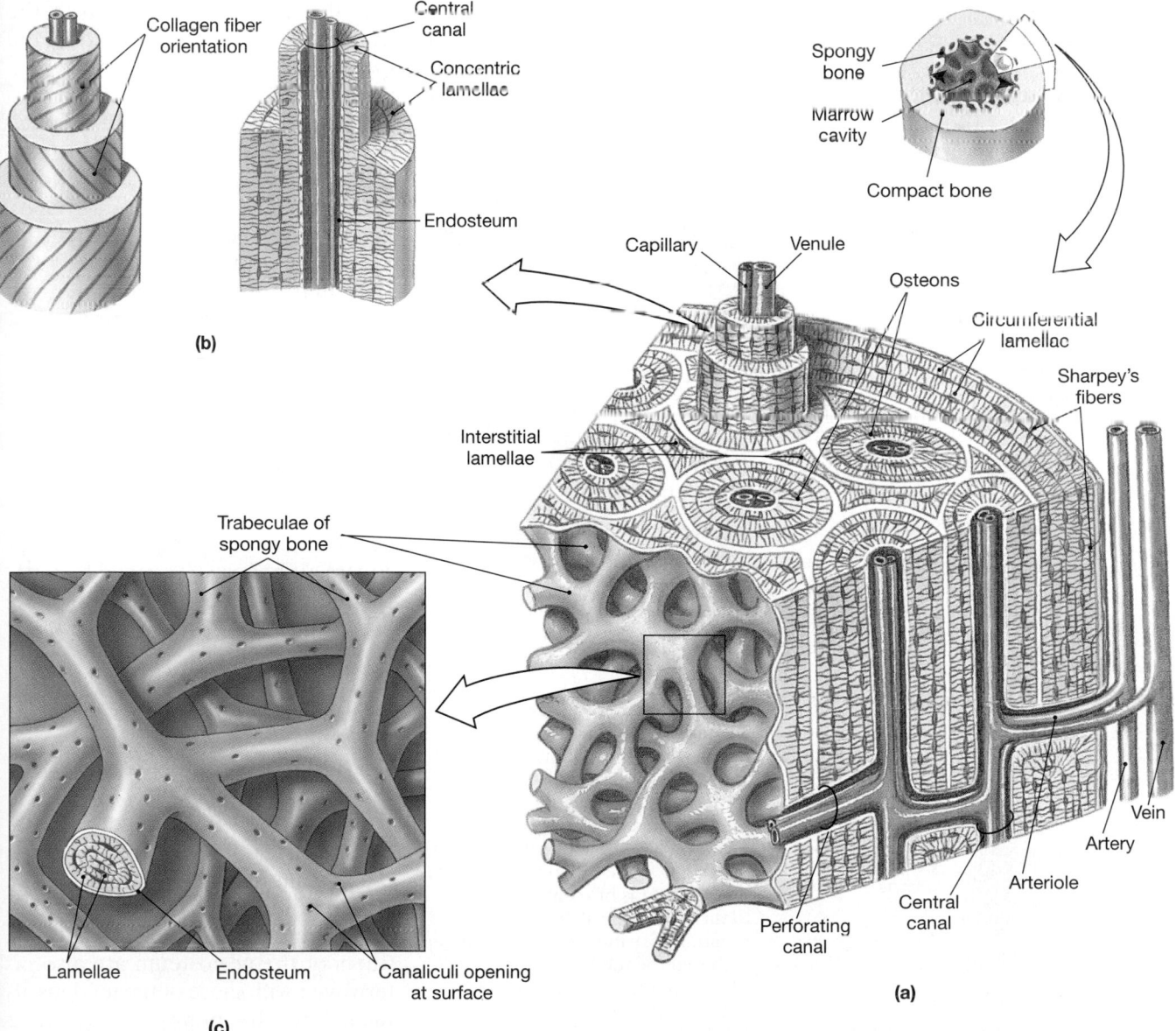

●FIGURE 6-4 **Structure of Osseous Tissue.** **(a)** The relationships among spongy bone, compact bone, and the marrow cavity. The insets show **(b)** the orientation of collagen fibers in adjacent lamellae and **(c)** details of the organization of spongy bone.

stressed along that axis. You might envision a single osteon as a drinking straw with very thick walls. When you attempt to push the ends of a straw together or to pull them apart, the straw is quite strong. But if you hold the ends and push from the side, it will break easily.

The osteons in the diaphysis are parallel to the long axis of the shaft. As a result, the shaft does not bend even when extreme forces are applied to either end. (The femur can withstand many times the weight of the body without bending or breaking.) Yet a relatively small impact to the side of the shaft can break the femur.

Figure 6-5a● shows the distribution of forces applied to the compact and spongy bone of the femur. The hip joint consists of the head of the femur and a corresponding socket on the lateral surface of the hip bone. The femoral head projects medially, and body weight compresses the medial side of the diaphysis. Because the force is applied off-center, the bone must resist the tendency to bend into a lateral bow. The other side of the shaft, which resists this bending, is placed under a stretching load, or *tension*.

Spongy Bone

There are neither osteons nor blood vessels in spongy bone, or *cancellous* (KAN-sel-us) *bone*. The matrix in spongy bone forms struts and plates called **trabeculae** (tra-BEK-ū-lē) (Figure 6-4a,c●). The thin trabeculae typically branch, creating an open network. Nutrients reach the osteocytes by diffusion along canaliculi that open onto the surfaces of the trabeculae.

FUNCTION OF SPONGY BONE Spongy bone is located where bones are not heavily stressed or where stresses arrive from many directions. Figure 6-5● diagrams the alignment of trabeculae in the proximal epiphysis of the femur (the end nearest the hip). The trabeculae are oriented along the stress lines but with extensive cross-bracing. At the proximal epiphysis, the trabeculae transfer forces from the hip to the compact bone of the femoral shaft; at the distal epiphysis, the trabeculae transfer the weight from the shaft to the leg, across the knee joint.

In addition to being able to withstand stresses applied from many directions, spongy bone is much lighter than compact bone. Spongy bone reduces the weight of the skeleton and makes it easier for muscles to move the bones. Finally, the trabecular framework supports and protects the cells of the bone marrow. The red bone marrow within the spongy bone of the femoral epiphyses is a key site of blood cell formation.

The Periosteum and Endosteum

Except within joint cavities, the superficial layer of compact bone that covers all bones is wrapped by a **periosteum.** The periosteum has a fibrous outer layer and a cellular inner layer (Figure 6-6●). The periosteum (1) isolates the bone from surrounding tissues, (2) provides a route for circulatory and nervous supply, and (3) actively participates in bone growth and repair.

Near joints, the periosteum becomes continuous with the connective tissues that lock the bones together. At a synovial joint, the periosteum is continuous with the joint capsule. The fibers of the periosteum are also interwoven with those of the tendons attached to the bone. As the bone grows, these tendon fibers are cemented into the superficial lamellae by osteoblasts from the cellular layer

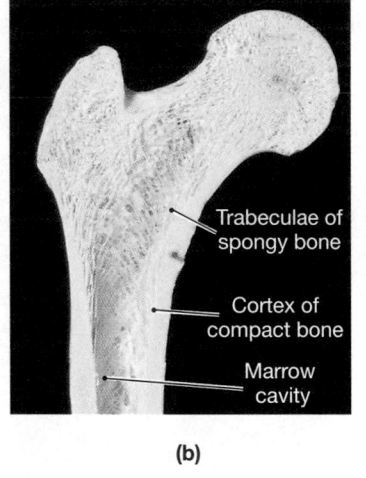

(b)

●***FIGURE 6-5*** **Lamellar Organization in a Long Bone.** **(a)** The femur, or thigh bone, has a diaphysis (shaft) with walls of compact bone and epiphyses (heads) filled with spongy bone. The body weight is transferred to the femur at the hip joint. Because the hip joint is off-center relative to the axis of the shaft, the body weight is distributed along the bone such that the medial (inner) portion of the shaft is compressed and the lateral (outer) portion is stretched. **(b)** A photograph showing the epiphysis after sectioning. Compare the orientation of trabeculae with the stress lines indicated in (a).

Body weight (applied force)

Trabeculae of spongy bone

Epiphysis (head)

Point of zero stress

Diaphysis (shaft)

Compression lines (medial portion of shaft)

Tension lines (lateral portion of shaft)

Tension

Compression

Marrow cavity

Compact bone

Epiphysis

Weight distributed across knee joint to leg

(a)

Trabeculae of spongy bone

Cortex of compact bone

Marrow cavity

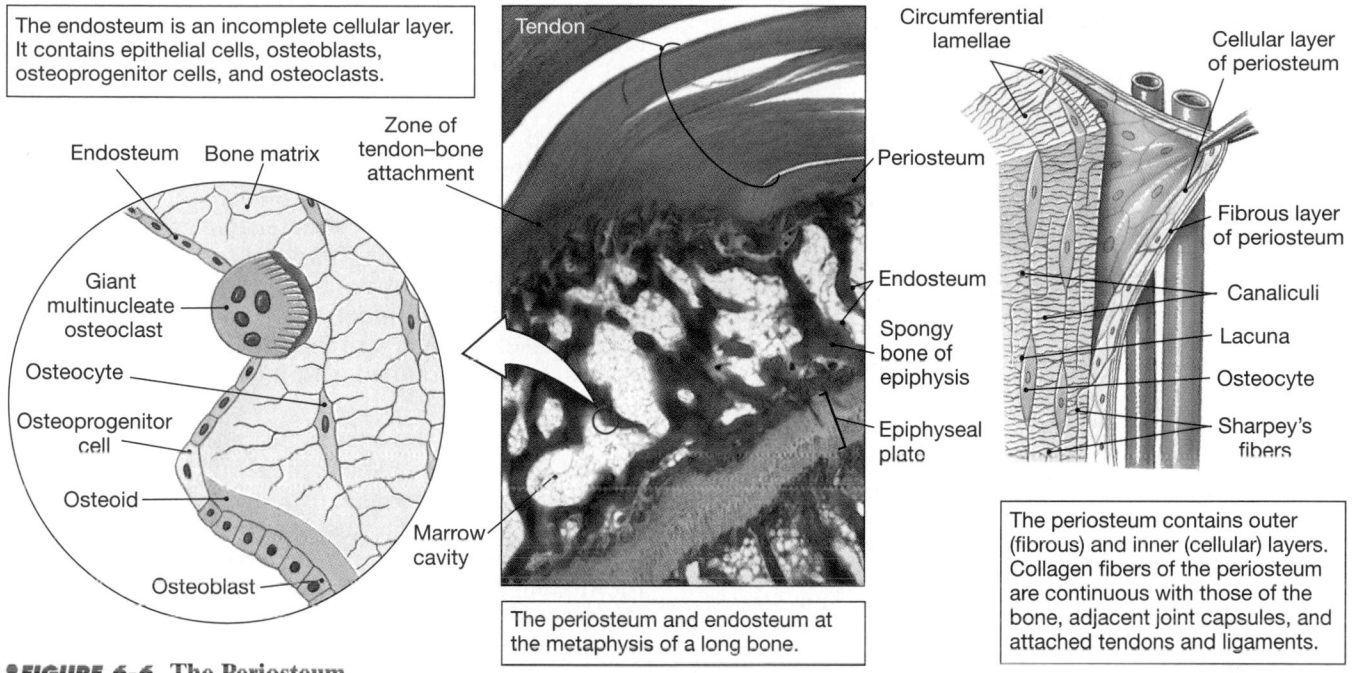

The endosteum is an incomplete cellular layer. It contains epithelial cells, osteoblasts, osteoprogenitor cells, and osteoclasts.

Endosteum | Bone matrix
Zone of tendon–bone attachment

Giant multinucleate osteoclast
Osteocyte
Osteoprogenitor cell
Osteoid
Osteoblast
Marrow cavity

Tendon
Periosteum
Endosteum
Spongy bone of epiphysis
Epiphyseal plate

The periosteum and endosteum at the metaphysis of a long bone.

Circumferential lamellae
Cellular layer of periosteum
Fibrous layer of periosteum
Canaliculi
Lacuna
Osteocyte
Sharpey's fibers

The periosteum contains outer (fibrous) and inner (cellular) layers. Collagen fibers of the periosteum are continuous with those of the bone, adjacent joint capsules, and attached tendons and ligaments.

•**FIGURE 6-6** The Periosteum and Endosteum

of the periosteum. Collagen fibers incorporated into bone tissue from tendons and ligaments, as well as from the superficial periosteum, are called *perforating fibers* (*Sharpey's fibers*). This method of attachment bonds the tendons and ligaments into the general structure of the bone, providing a much stronger attachment than would otherwise be possible. An extremely powerful pull on a tendon or ligament will usually break a bone rather than snap the collagen fibers at the bone surface.

A cellular layer, the **endosteum,** lines the marrow cavity. This layer covers the trabeculae of spongy bone and lines the inner surfaces of the central canals. The endosteum is active during bone growth, repair, and remodeling. The endosteum consists of a simple flattened layer of osteoprogenitor cells that covers the bone matrix, generally without any intervening connective tissue fibers. The cellular layer is not complete, and the matrix is occasionally exposed. At these exposed sites, osteoclasts and osteoblasts can remove or deposit matrix components. The osteoclasts generally occur in shallow depressions (*Howship's lacunae*) that they have eroded into the matrix.

☑ How would the strength of a bone be affected if the ratio of collagen to hydroxyapatite increased?

☑ A sample of bone shows concentric lamellae surrounding a central canal. Is the sample from the cortex or the marrow cavity of a long bone?

☑ If the activity of osteoclasts exceeds the activity of osteoblasts in a bone, how will the mass of the bone be affected?

BONE DEVELOPMENT AND GROWTH

The growth of the skeleton determines the size and proportions of our body. The bony skeleton begins to form about 6 weeks after fertilization, when the embryo is approximately 12 mm (0.5 in.) long. (At this stage the existing skeletal elements are cartilaginous.) During subsequent development, the bones undergo a tremendous increase in size. Bone growth continues through adolescence, and portions of the skeleton generally do not stop growing until age 25. The entire process is carefully regulated, and a breakdown in regulation will ultimately affect all body systems. In this section, we will consider the physical process of osteogenesis (bone formation) and bone growth. In the next section, we will examine the maintenance and replacement of mineral reserves in the adult skeleton.

AM *Examination of the Skeletal System*

Bone may develop directly from mesenchyme or by the replacement of cartilage. The process of replacing other tissues with bone is called **ossification.** Ossification refers specifically to the formation of bone. The process of **calcification,** the deposition of calcium salts, occurs during ossification, but it can also occur in other tissues. (The result is a calcified tissue, such as calcified cartilage, that does not resemble bone.) There are two major forms of ossification: (1) intramembranous and (2) endochondral ossifications. In *intramembranous ossification,* bone develops from mesenchyme or fibrous connective tissue. In *endochondral ossification,* bone replaces existing cartilage.

Intramembranous Ossification

Intramembranous (in-tra-MEM-bra-nus) **ossification,** or *dermal ossification,* begins when osteoblasts differentiate within a mesenchymal or fibrous connective tissue. This type of ossification normally occurs in the deeper layers of the dermis. The bones that result are often called **dermal bones.** Examples of dermal bones include the roofing bones of the skull, the *mandible* (lower jaw), and the *clavicle* (collarbone). In response to abnormal stresses, bone may form in other dermal areas or in connective tissues within tendons, around joints, in the kidneys, or in skeletal muscles. Intramembranous bones forming in abnormal locations are called *heterotopic bones* (*hetero-,* different + *topos,* place), or *ectopic bones* (*ektos,* outside). ▢AM *Heterotopic Bone Formation*

The steps in the process of intramembranous ossification (Figure 6-7●) may be summarized as follows:

STEP 1. Mesenchymal cells first cluster together and start to secrete the organic components of the matrix. The resulting osteoid then becomes mineralized through the crystallization of calcium salts. (The enzyme *alkaline phosphatase* plays a role in this process.) As calcification occurs, the mesenchymal cells differentiate into osteoblasts. The location in a tissue where ossification begins is called an **ossification center.** As

ossification proceeds, it traps some osteoblasts inside bony pockets; these cells differentiate into osteocytes.

STEP 2. The developing bone grows outward from the ossification center in small struts called **spicules.** Although osteoblasts are still being trapped in the expanding bone, mesenchymal cell divisions continue to produce additional osteoblasts. Bone growth is an active process, and osteoblasts require oxygen and a reliable supply of nutrients. Blood vessels trapped among the spicules meet these demands, and additional vessels branch into the area. These vessels will, in turn, become trapped within the growing bone.

STEP 3. Initially, the intramembranous bone consists only of spongy bone. Subsequent remodeling around trapped blood vessels can produce osteons typical of compact bone. As the rate of growth slows, the connective tissue around the bone becomes organized into the fibrous layer of the periosteum. The osteoblasts closest to the bone surface become less active but remain as the inner, cellular layer of the periosteum.

Figure 6-8● shows two stages of intramembranous ossification in the skull bones of a fetus.

Endochondral Ossification

Endochondral (en-dō-KON-drul) **ossification** (*endo-,* inside + *chondros,* cartilage) begins with the formation of a hyaline cartilage model. Most bones in the

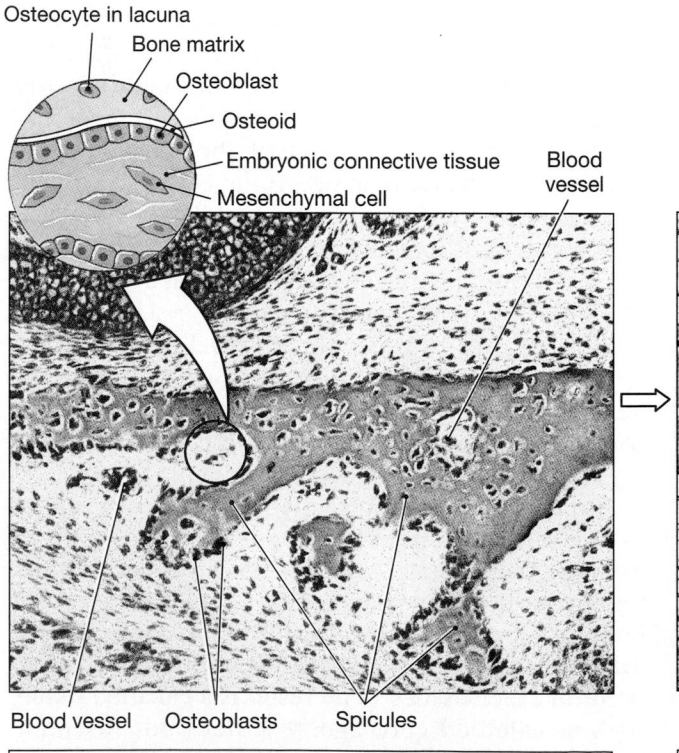

Osteocyte in lacuna
Bone matrix
Osteoblast
Osteoid
Embryonic connective tissue
Mesenchymal cell
Blood vessel
Blood vessel   Osteoblasts   Spicules

Step 1: Mesenchymal cells aggregate, differentiate, and begin the ossification process. The bone expands as a series of spicules that spread into surrounding tissues. (LM× 22)

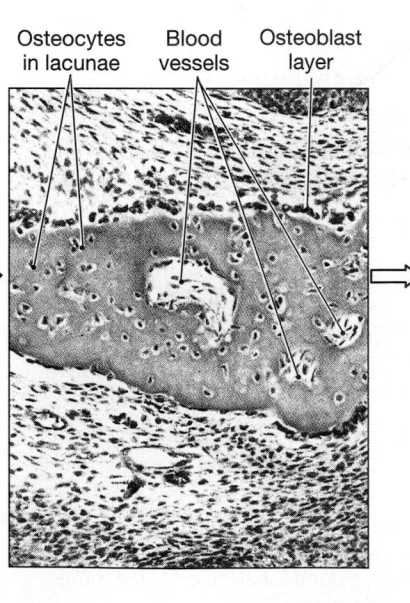

Osteocytes in lacunae   Blood vessels   Osteoblast layer

Step 2: As the spicules interconnect, they trap blood vessels within the bone. (LM× 23)

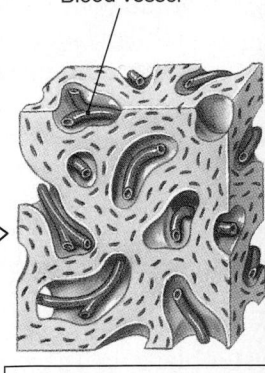

Blood vessel

Step 3: Over time, the bone assumes the structure of spongy bone. Areas of spongy bone may later be removed, creating marrow cavities. Through remodeling, spongy bone formed in this way can be converted to compact bone.

●**FIGURE 6-7** Intramembranous Ossification

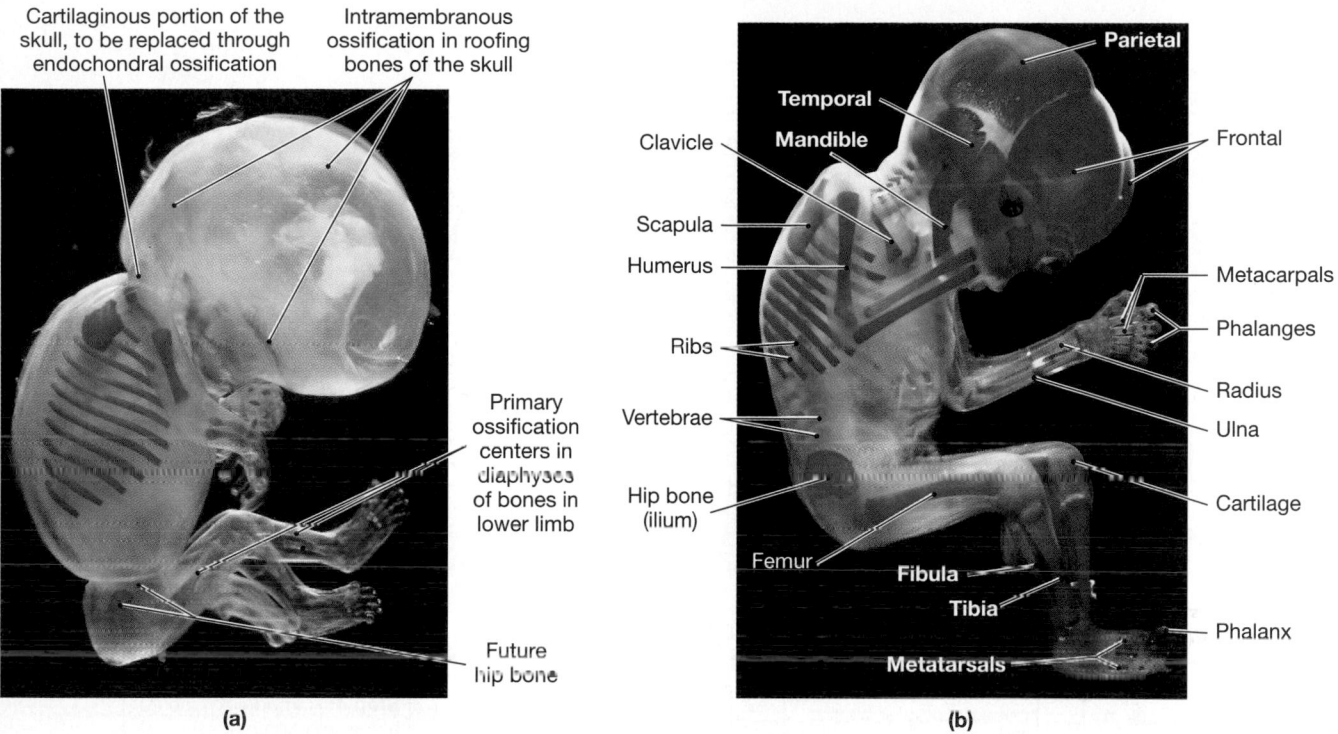

Cartilaginous portion of the skull, to be replaced through endochondral ossification

Intramembranous ossification in roofing bones of the skull

Primary ossification centers in diaphyses of bones in lower limb

Future hip bone

(a)

Parietal

Temporal

Mandible

Clavicle

Scapula

Humerus

Ribs

Vertebrae

Hip bone (ilium)

Femur

Fibula

Tibia

Metatarsals

Frontal

Metacarpals

Phalanges

Radius

Ulna

Cartilage

Phalanx

(b)

●*FIGURE 6-8* **Intramembranous and Endochondral Ossification in a Fetus. (a)** This 10-week-old human fetus has been specially stained to show developing skeletal elements. **(b)** At 16 weeks, intramembranous bones are completing the roof of the skull. Most of the appendicular skeleton at this stage consists of cartilage that will be replaced through endochondral ossification.

body form in this way. As an example, we will follow the steps in limb bone development. By the time an embryo is 6 weeks old, the proximal bone of the limb, either the humerus (arm) or femur (thigh), is present, but it is composed entirely of cartilage. This cartilage model continues to grow by expansion of the cartilage matrix (*interstitial growth*) and the production of new cartilage at the outer surface (*appositional growth*). (We introduced these growth mechanisms in Chapter 4. ∞ *[p. 130]*) Steps in the growth and ossification of a limb bone are diagrammed in Figure 6-9a●:

STEP 1. As the cartilage enlarges, chondrocytes near the center of the shaft begin to increase greatly in size. As these cells enlarge, their lacunae expand, and the matrix is reduced to a series of thin struts. These struts soon begin to calcify. The enlarged chondrocytes are now deprived of nutrients (diffusion cannot occur through calcified cartilage), and they soon die and disintegrate.

STEP 2. Blood vessels grow into the perichondrium surrounding the shaft of the cartilage. (We introduced the structure of the perichondrium and its role in cartilage formation in Chapter 4.) ∞ *[p. 128]* The cells of the inner layer of the perichondrium in this region then differentiate into osteoblasts. The perichondrium is now a periosteum, and the inner **osteogenic layer** soon produces a thin layer of bone around the shaft of the cartilage.

STEP 3. While these changes are under way, the blood supply to the periosteum increases, and capillaries and fibroblasts migrate into the heart of the cartilage, invading the spaces left by the disintegrating chondrocytes. The calcified cartilaginous matrix breaks down; the fibroblasts differentiate into osteoblasts that replace it with spongy bone. Bone development begins at this **primary center of ossification** and spreads toward both ends of the cartilaginous model. While the diameter is small, the entire diaphysis is filled with spongy bone.

STEP 4. As the bone enlarges, osteoclasts appear and begin eroding the trabeculae in the central portion, creating a marrow cavity. Further growth involves two distinct processes: an increase in *length* and an enlargement in *diameter*.

Increasing the Length of a Developing Bone

During the initial stages of osteogenesis, osteoblasts move away from the center of ossification toward the epiphyses. But they do not complete the ossification of the model immediately, because the epiphyseal cartilages continue to enlarge. The region where the cartilage is being replaced by bone lies at the metaphysis. On the shaft side of the metaphysis, osteoblasts are continuously invading the cartilage and replacing it with bone. But on the epiphyseal side, new cartilage is continuously being added. The osteoblasts are therefore moving toward the epiphysis, which is being pushed ahead

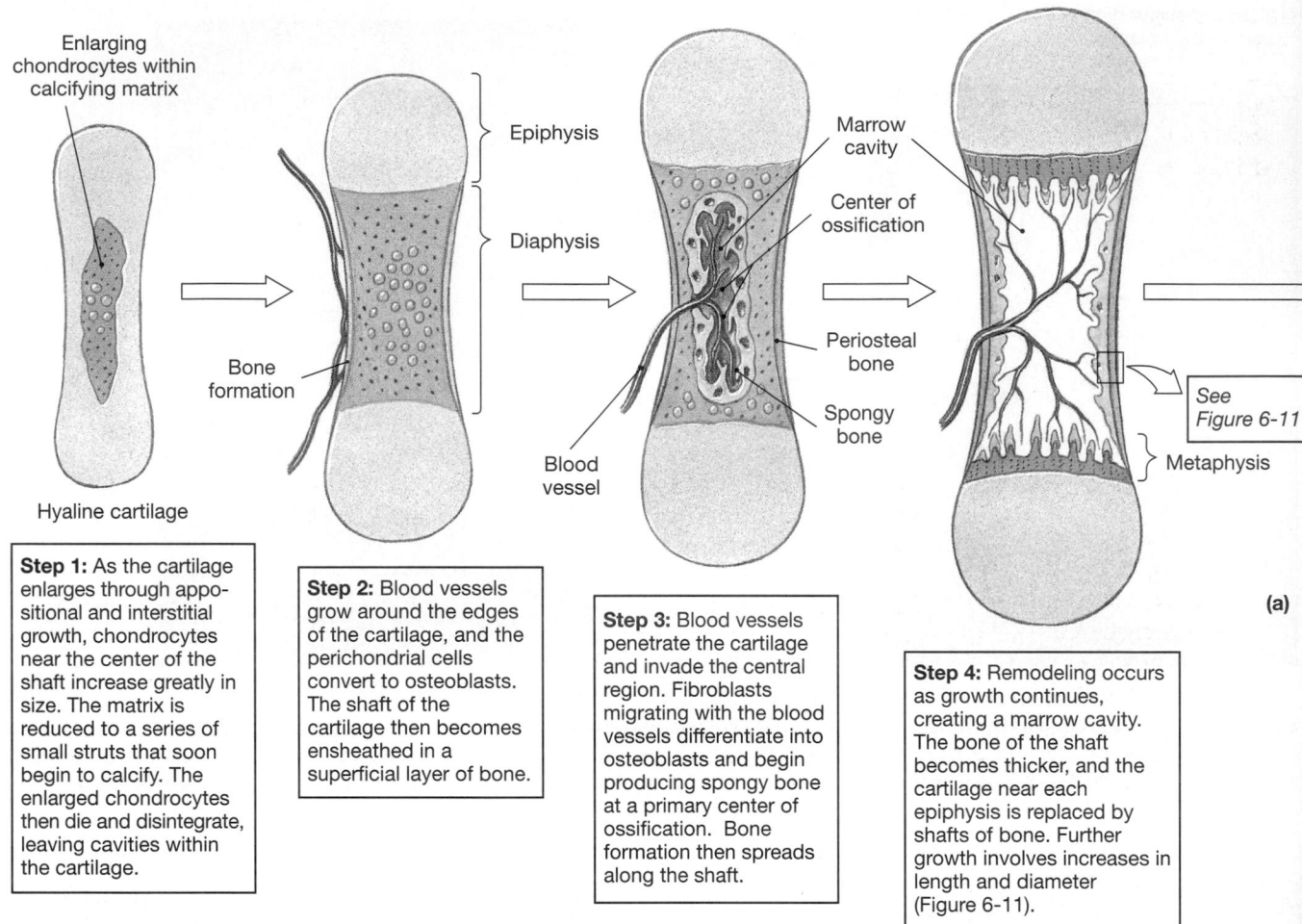

Enlarging chondrocytes within calcifying matrix

Hyaline cartilage

Epiphysis

Diaphysis

Bone formation

Blood vessel

Marrow cavity

Center of ossification

Periosteal bone

Spongy bone

See Figure 6-11

Metaphysis

(a)

Step 1: As the cartilage enlarges through appositional and interstitial growth, chondrocytes near the center of the shaft increase greatly in size. The matrix is reduced to a series of small struts that soon begin to calcify. The enlarged chondrocytes then die and disintegrate, leaving cavities within the cartilage.

Step 2: Blood vessels grow around the edges of the cartilage, and the perichondrial cells convert to osteoblasts. The shaft of the cartilage then becomes ensheathed in a superficial layer of bone.

Step 3: Blood vessels penetrate the cartilage and invade the central region. Fibroblasts migrating with the blood vessels differentiate into osteoblasts and begin producing spongy bone at a primary center of ossification. Bone formation then spreads along the shaft.

Step 4: Remodeling occurs as growth continues, creating a marrow cavity. The bone of the shaft becomes thicker, and the cartilage near each epiphysis is replaced by shafts of bone. Further growth involves increases in length and diameter (Figure 6-11).

•**FIGURE 6-9** **Endochondral Ossification.** **(a)** Steps in endochondral ossification. **(b)** Light micrograph showing the interface between the degenerating cartilage and the advancing osteoblasts.

by the growth of the cartilage. The situation is like a pair of joggers, one in front of the other. As long as they are running at the same speed, they can run for miles without colliding. The osteoblasts don't catch up to the epiphysis, but both the osteoblasts and the epiphysis "run away" from the primary ossification center. As they do so, the bone grows longer and longer.

STEP 5. The next major change occurs when the centers of the epiphyses begin to calcify. Capillaries and osteoblasts then migrate into these areas, creating **secondary ossification centers**. The time of appearance of secondary ossification centers varies from one bone to another and from individual to individual. Secondary ossification centers may be found at birth in both ends of the *humerus* (arm), *femur* (thigh), and *tibia* (leg), but the ends of some other bones remain cartilaginous until early adulthood.

STEP 6. The epiphyses eventually become filled with spongy bone. A thin cap of the original cartilage model will remain exposed to the joint cavity as the **articular cartilage**. This cartilage prevents damaging bone-to-bone contact within the joint. At the metaphysis, a relatively narrow cartilaginous **epiphyseal plate** now

separates the epiphysis from the diaphysis. The light micrograph in Figure 6-9b• shows the interface between the degenerating cartilage and the advancing osteoblasts. As long as the rate of cartilage growth keeps pace with the rate of osteoblast invasion, the shaft grows longer but the epiphyseal plate survives.

At puberty, the combination of rising levels of sex hormones, growth hormone, and thyroid hormones stimulates bone growth dramatically. Osteoblasts now begin producing bone faster than the rate of epiphyseal cartilage expansion. As a result, the epiphyseal plate gets narrower and narrower until it ultimately disappears. The former location of the plate can still be detected in X-rays as a distinct **epiphyseal line** that remains after epiphyseal growth has ended (Figure 6-10•).

Increasing the Diameter of a Developing Bone

The diameter of a bone enlarges through appositional growth at the outer surface. In this process, periosteal cells differentiate into osteoblasts and contribute to the growth of the bone matrix. Eventually they become surrounded by matrix and differentiate into osteocytes. As this process continues to add layers of bone tissue, blood vessels and

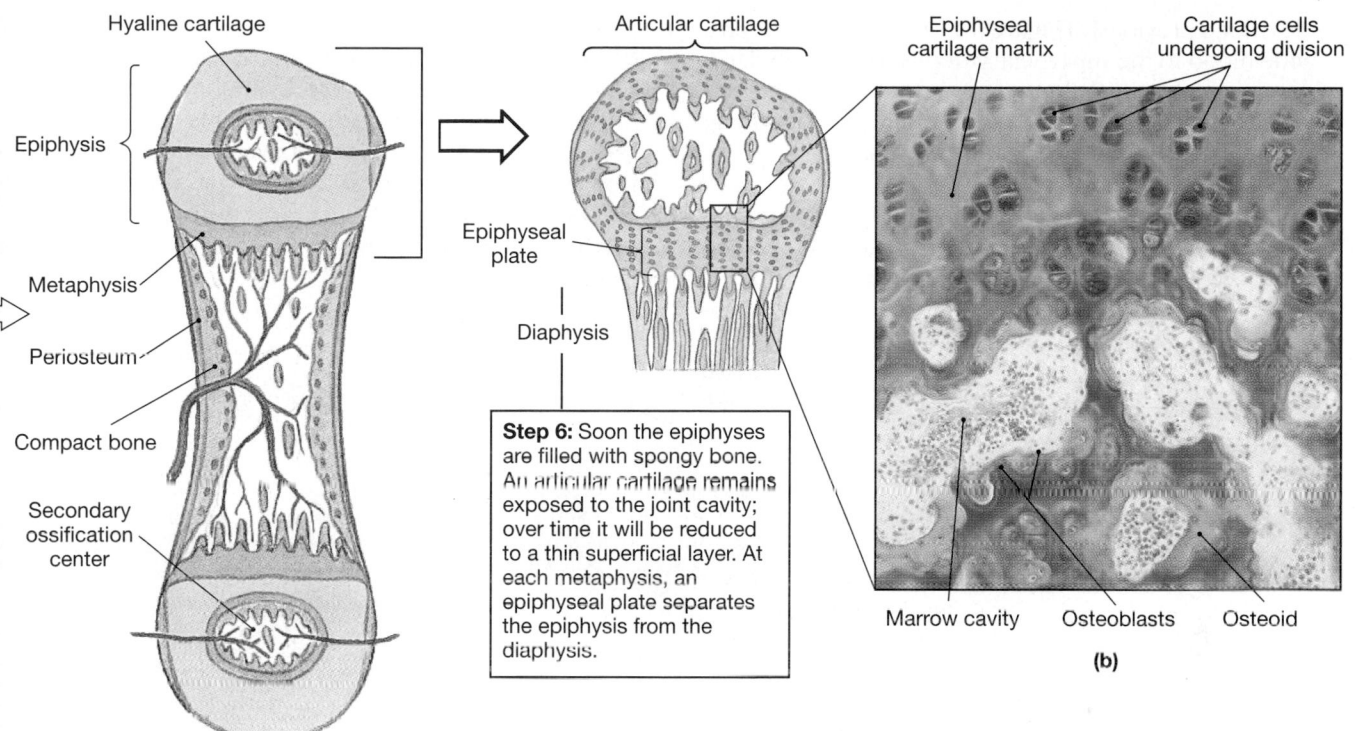

Hyaline cartilage

Epiphysis

Metaphysis

Periosteum

Compact bone

Secondary ossification center

Articular cartilage

Epiphyseal plate

Diaphysis

Epiphyseal cartilage matrix

Cartilage cells undergoing division

Step 6: Soon the epiphyses are filled with spongy bone. An articular cartilage remains exposed to the joint cavity; over time it will be reduced to a thin superficial layer. At each metaphysis, an epiphyseal plate separates the epiphysis from the diaphysis.

Marrow cavity   Osteoblasts   Osteoid

(b)

Step 5: Capillaries and osteoblasts migrate into the epiphyses, creating secondary ossification centers.

collagen fibers of the periosteum become incorporated into the bony structure. Steps in appositional bone growth are described and illustrated in Figure 6-11•.

While bone matrix is being added to the outer surface of the growing bone, osteoclasts are removing bone matrix at the inner surface. As a result, the marrow cavity gradually enlarges as the bone gets larger in diameter.

☑ During the process of intramembranous ossification, what type(s) of tissue is (are) replaced by bone?

☑ In endochondral ossification, what is the original source of osteoblasts?

☑ How could X-rays of the femur be used to determine whether a person has reached full height?

The Blood and Nerve Supply

Osseous tissue is very vascular, and the bones of the skeleton have an extensive blood supply. In a typical bone such as the humerus, three major sets of blood vessels develop (Figure 6-12•, p. 185):

1. *The nutrient artery and vein.* These vessels, which supply the diaphysis, form as blood vessels invade the cartilage model as endochondral ossification begins. There is generally only one *nutrient artery* and one *nutrient vein*, but a few bones, including the femur, have more than one of each. The vessels enter the bone through one or more *nutrient foramina* in the diaphysis. Branches of these vessels extend along the length of the shaft and into the osteons of the surrounding cortex.

2. *Metaphyseal vessels.* *Metaphyseal vessels* supply blood to the inner (diaphyseal) surface of each epiphyseal plate, where bone is replacing the cartilage.

3. *Periosteal vessels.* Blood vessels from the periosteum are incorporated into the developing bone surface as

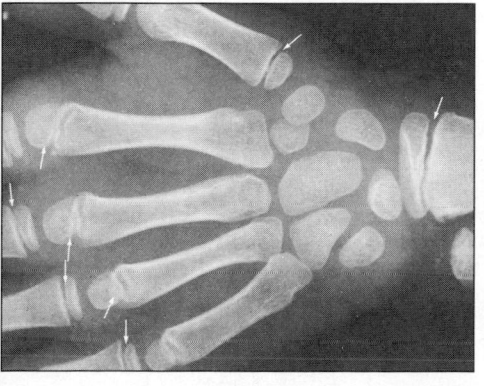

(a)

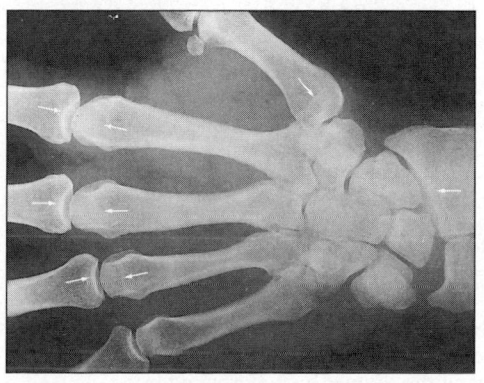

(b)

•*FIGURE 6-10* **Bone Growth at the Epiphyseal Plate. (a)** An X-ray of growing epiphyseal plates (arrows). **(b)** Epiphyseal lines in an adult (arrows).

described previously (Figure 6-11●). These vessels provide blood to the superficial osteons of the shaft. During endochondral bone formation, branches of periosteal vessels also enter the epiphyses, providing blood to the secondary ossification centers.

Following the complete ossification, or *closure*, of the epiphyses, all three sets of vessels become extensively interconnected.

The periosteum contains an extensive network of lymphatic vessels and sensory nerves. The lymphatics collect lymph from branches that enter the bone and reach individual osteons via the perforating canals. The sensory nerves penetrate the cortex with the nutrient artery to innervate the endosteum, marrow cavity, and epiphyses. Because of the rich sensory innervation, injuries to bones are usually very painful.

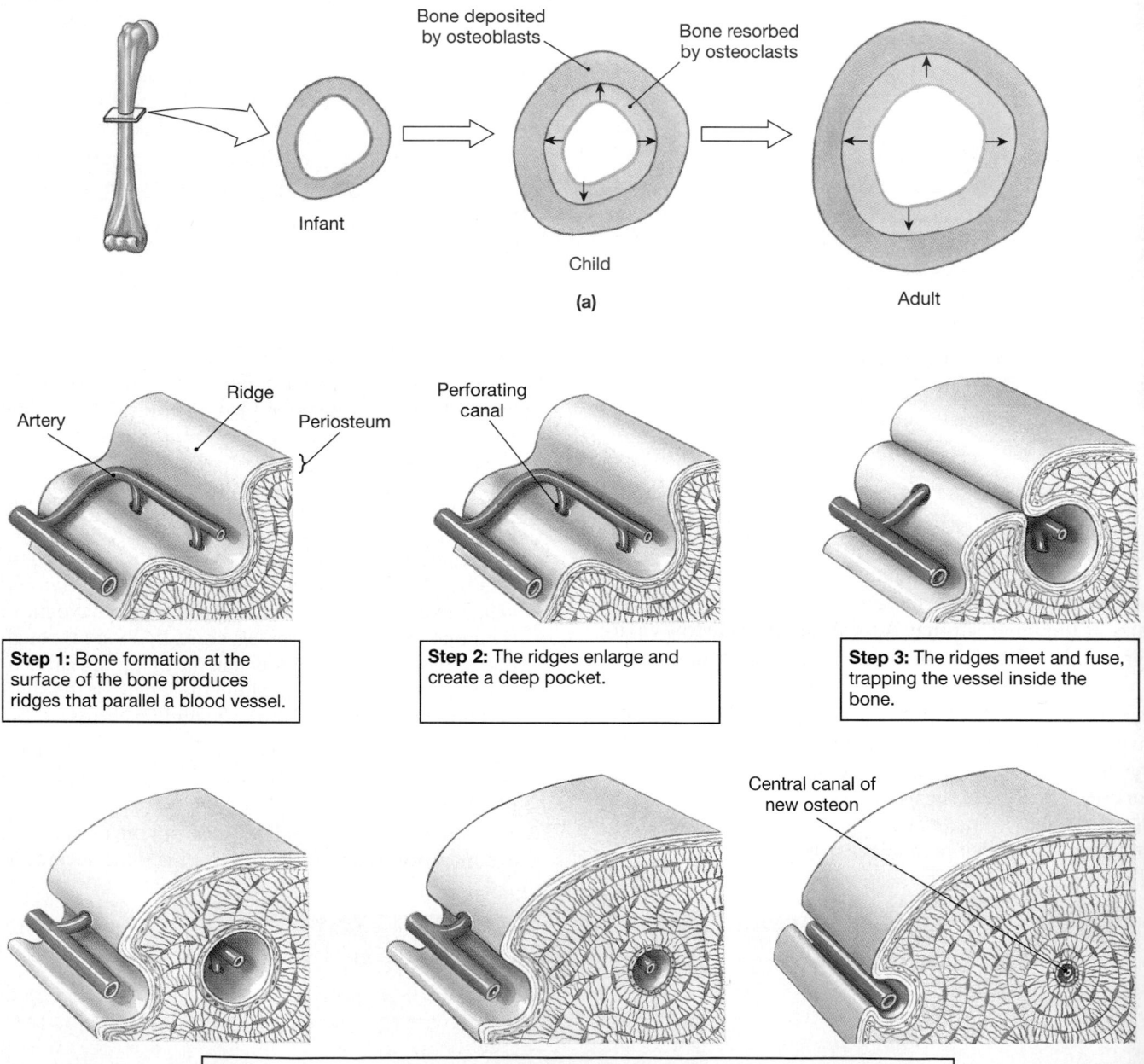

Step 1: Bone formation at the surface of the bone produces ridges that parallel a blood vessel.

Step 2: The ridges enlarge and create a deep pocket.

Step 3: The ridges meet and fuse, trapping the vessel inside the bone.

Steps 4–6: Bone deposition then proceeds inward toward the vessel, creating a typical osteon. Meanwhile, additional circumferential lamellae are deposited and the bone continues to increase in diameter. As it does so, additional blood vessels will be enclosed.

●**FIGURE 6-11** **Appositional Bone Growth.** **(a)** A bone grows in diameter as new bone is added to the outer surface. At the same time, osteoclasts resorb bone on the inside, enlarging the marrow cavity. **(b)** Three-dimensional diagrams illustrating the mechanism responsible for increasing the diameter of a growing bone.

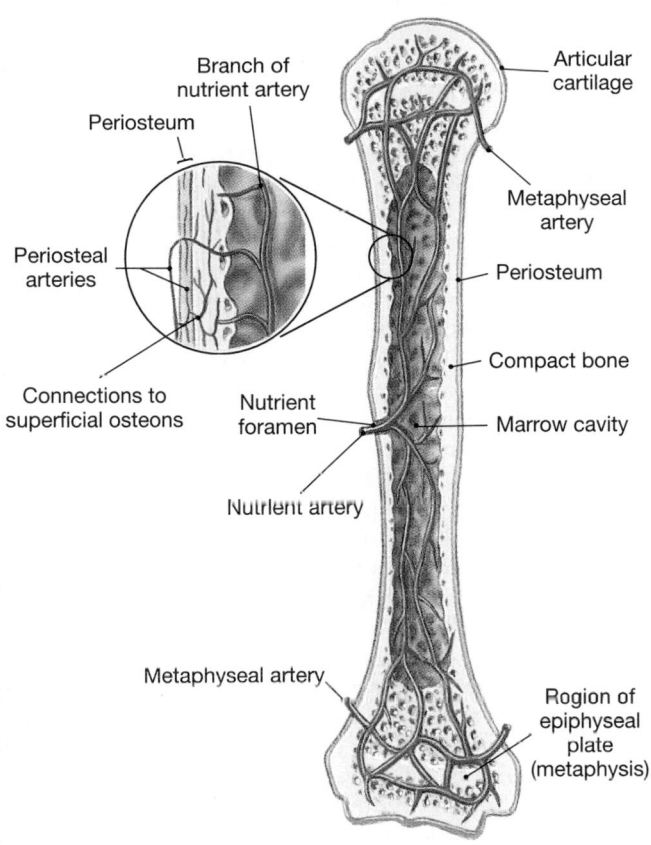

Branch of
nutrient artery

Periosteum

Articular
cartilage

Metaphyseal
artery

Periosteal
arteries

Periosteum

Connections to
superficial osteons

Nutrient
foramen

Compact bone

Nutrient artery

Marrow cavity

Metaphyseal artery

Region of
epiphyseal
plate
(metaphysis)

•FIGURE 6-12 **Circulatory Supply to a Mature Bone**

THE DYNAMIC NATURE OF BONE

The organic and mineral components of the bone matrix are continuousy being recycled and renewed through the process of **remodeling.** Bone remodeling is under way throughout life, as part of normal bone maintenance. Remodeling can replace the matrix but leave the bone as a whole unchanged, or there may be alterations in the shape, internal architecture, or mineral content of the bone.

Bone remodeling involves an interplay between the activities of osteocytes, osteoblasts, and osteoclasts. In the adult, osteocytes are continuously removing and replacing the surrounding calcium salts. But osteoclasts and osteoblasts also remain active, even after the epiphyseal plates have closed. Normally, their activities are balanced: As one osteon forms through the activity of osteoblasts, another is destroyed by osteoclasts. The turnover rate of bone is quite high. In a young adult, each year almost one-fifth of the adult skeleton is demolished and then rebuilt or replaced. Every part of every bone may not be affected, as there are regional and even local differences in the rate of turnover. For example, the spongy bone in the head of the femur may be replaced two or three times each

year, whereas the compact bone along the shaft remains largely unchanged.

Exercise Effects on Bone

The turnover and recycling of minerals give each bone the ability to adapt to new stresses. Osteoblast sensitivity to electrical events has been theorized as the mechanism that controls the internal organization and structure of bone. Whenever a bone is stressed, the mineral crystals generate minute electrical fields. Osteoblasts are apparently attracted to these electrical fields, and once in the area, they begin to produce bone. (Electrical fields are also used to stimulate the repair of severe fractures.)

Because bones are adaptable, their shapes reflect the forces applied to them. For example, bumps and ridges on the surface of a bone mark the sites where tendons attach to the bone. If muscles become more powerful, the corresponding bumps and ridges enlarge to withstand the increased forces. Heavily stressed bones become thicker and stronger, whereas bones not subjected to ordinary stresses will become thin and brittle. Regular exercise is therefore important as a stimulus that maintains normal bone structure. Champion weight lifters have massive bones with thick, prominent ridges where muscles attach. In nonathletes (especially couch potatoes), moderate amounts of physical activity and weight bearing are essential to stimulate normal bone maintenance and maintain adequate bone strength.

Degenerative changes in the skeleton occur after relatively brief periods of inactivity. For example, you may use a crutch to take weight off an injured leg while you wear a cast. After a few weeks, your unstressed bones will lose up to a third of their mass. The bones rebuild just as quickly when you resume normal weight loading. However, the removal of calcium salts can be a potentially serious health hazard for astronauts remaining in a weightless environment and for bedridden or paralyzed patients who spend months or years without stressing their skeleton.

Hormonal and Nutritional Effects on Bone

Normal bone growth and maintenance depend on a combination of nutritional and hormonal factors:

- Normal bone growth and maintenance cannot occur without a constant dietary source of calcium and phosphate salts. Lesser amounts of other minerals, such as magnesium, fluoride, iron, and manganese, are also required.
- The hormone *calcitriol,* synthesized in the kidneys, is essential for normal calcium and phosphate ion absorption in the digestive tract. Calcitriol synthesis is dependent on the availability of a related steroid, *cholecalciferol* (vitamin D_3), which may be synthesized in the skin or absorbed from the diet. ∞ [p. 154]

- Adequate levels of vitamin C must be present in the diet. This vitamin is required by certain key enzymatic reactions involved with collagen synthesis. It also stimulates osteoblast differentiation. One of the signs of vitamin C deficiency, a condition called *scurvy,* is a loss of bone mass and strength.
- Three other vitamins have significant effects on bone structure. Vitamin A, which stimulates osteoblast activity, is particularly important for normal bone growth in children. Vitamins K and B$_{12}$ are required for the synthesis of proteins in normal bone.
- *Growth hormone,* produced by the pituitary gland, and *thyroxine,* from the thyroid gland, stimulate bone growth. Growth hormone stimulates protein synthesis and cell growth throughout the body. Thyroxine stimulates cell metabolism and increases the rate of osteoblast activity. In proper balance, these hormones maintain normal activity at the epiphyseal plates until roughly the time of puberty.
- At puberty, rising levels of sex hormones *(estrogens* in the female and *androgens* in the male) stimulate osteoblasts to produce bone faster than the rate of epiphyseal cartilage expansion. Over time, the epiphyseal plates narrow and eventually close. There are differences from bone to bone and individual to individual as to the timing of epiphyseal closure. The toes may complete their ossification by age 11, whereas portions of the pelvis or the wrist may continue to enlarge until age 25. Differences in the male and female sex hormones account for the variation between the genders and for related variations in body size and proportions. (For example, because estrogens cause faster epiphyseal closure than do androgens, women are generally shorter than men at maturity.)

Two other hormones are important in the homeostatic control of calcium and phosphate levels in body fluids. We will consider the interactions between these hormones—*calcitonin* (kal-si-TŌ-nin) from the thyroid gland and *parathyroid hormone* from the parathyroid gland—in the next section. The major hormones affecting the growth and maintenance of the skeletal system are summarized in Table 6-1.

 ABNORMAL BONE GROWTH AND DEVELOPMENT A variety of endocrine or metabolic problems can result in characteristic skeletal changes. **Gigantism** results from an overproduction of growth hormone before puberty. (The world record for height is 272 cm, or 8 ft 11 in., with a weight of 216 kg, or 475 lb.) The opposite extreme is **pituitary growth failure** (*pituitary dwarfism*), in which inadequate growth hormone production leads to reduced epiphyseal activity and abnormally short bones. This form of growth failure is becoming increasingly rare in the United States because children can be treated with human growth hormone.

Several inherited metabolic problems that affect many systems influence the growth and development of the skeletal system. These conditions are often recognizable because of characteristic variations in body proportions. For example, many individuals with *Marfan's syndrome* are very tall, with long, slender limbs, due to excessive cartilage formation at the epiphyseal plates. Although this is an obvious physical distinction, the characteristic body proportions are not in themselves dangerous. However, the underlying mutation, which affects connective tissue structure, commonly causes life-threatening problems with the cardiovascular system. ∞ *[p. 123]*

If growth hormone levels rise abnormally after epiphyseal plate closure, the skeleton does not grow larger, but cartilage growth and alterations in soft tissue structure lead to changes in bone density and in physical features, such as the contours of the face. These physical changes occur in the disorder called *acromegaly.* AM *Abnormalities of Skeletal Development*

✔ Why would you expect the arm bones of a weight lifter to be thicker and heavier than those of a jogger?

✔ A child who enters puberty several years later than the average age is generally taller than average as an adult. Why?

✔ A 7-year-old child has a pituitary tumor involving the cells that secrete growth hormone (GH), resulting in increased levels of GH. How will this condition affect the child's growth?

TABLE 6-1 Hormones Involved in the Regulation of Bone Growth

Hormone	Primary Source	Effects on Skeletal System
Calcitriol	Kidneys	Promotes calcium and phosphate ion absorption along the digestive tract
Growth hormone	Pituitary gland	Stimulates osteoblast activity and the synthesis of bone matrix
Thyroxine	Thyroid gland (follicle cells)	With growth hormone, stimulates osteoblast activity and synthesis of bone matrix
Sex hormones	Ovaries (estrogens) Testes (androgens)	Stimulate osteoblast activity and synthesis of bone matrix
Parathyroid hormone	Parathyroid glands	Stimulates osteoclast activity; elevates calcium ion concentrations in body fluids
Calcitonin	Thyroid gland (C cells)	Inhibits osteoclast activity; reduces calcium ion concentrations in body fluids

The Skeleton as a Calcium Reserve

The bones of the skeleton are important mineral reservoirs. For the moment, we will focus on the homeostatic regulation of calcium ion concentrations in body fluids; we shall consider other minerals in later chapters. Calcium is the most abundant mineral in the human body (Figure 6-13●). A typical human body contains 1–2 kg (2.2–4.4 lb) of calcium, with roughly 99 percent of it deposited in the skeleton.

Calcium ion concentrations must be closely controlled to prevent damage to essential physiological systems. Even small variations from normal concentrations will have some effect on cellular operations. Larger changes can cause a clinical crisis. Calcium ions are particularly important to the membranes and intracellular activities of neurons and muscle cells, especially cardiac muscle cells. If the calcium concentration in body fluids increases by 30 percent, neurons and muscle cells become relatively unresponsive. If calcium levels decrease by 35 percent, neurons become so excitable that convulsions may occur. A 50 percent reduction in calcium concentrations generally causes death. Such gross disturbances in calcium metabolism are relatively rare, for the calcium ion concentrations are so closely regulated that daily fluctuations of more than 10 percent are very unusual.

Hormones and Calcium Balance

The calcium ion concentration in body fluids depends on activities under way in three organs: (1) the bones, (2) the intestinal tract, and (3) the kidneys.

Calcium ion homeostasis is maintained by a negative feedback system involving a pair of hormones with opposing effects. These hormones, parathyroid hormone and calcitonin, coordinate the storage, absorption, and excretion of calcium ions. The target organs are the bones (storage), digestive tract (absorption), and kidneys (excretion). Figure 6-14 shows interactions among these components. Figure 6-14a● indicates factors that elevate calcium levels; Figure 6-14b● indicates factors that depress calcium ion concentrations.

When calcium ion concentrations in the blood fall below normal, cells of the **parathyroid gland** release **parathyroid hormone** (**PTH**) into the bloodstream. Parathyroid hormone has three major effects:

1. *Stimulating osteoclast activity* and enhancing the recycling of minerals by osteocytes.
2. *Increasing the rate of intestinal absorption of calcium ions* by enhancing the action of calcitriol. Under normal circumstances, calcitriol is always present and parathyroid hormone controls its impact on the intestinal epithelium.
3. *Decreasing the rate of excretion of calcium ions at the kidneys.*

Under these conditions, more calcium ions enter body fluids, losses are restricted, and calcium ion concentrations increase to normal levels.

If the calcium ion concentration of the blood rises above normal, special cells (*parafollicular cells*, or *C cells*) within the thyroid gland secrete **calcitonin.** Calcitonin has two major functions:

1. *Inhibiting osteoclast activity.*
2. *Increasing the rate of excretion of calcium ions at the kidneys.*

Less calcium *enters* body fluids because osteoclasts leave the mineral matrix alone. More calcium *leaves* body fluids because osteoblasts continue to produce new bone matrix while calcium ion excretion at the kidneys accelerates. The net result is a decline in the calcium ion concentration in body fluids, restoring homeostasis.

By providing a calcium reserve, the skeleton plays the primary role in the homeostatic maintenance of normal calcium ion concentrations in body fluids. This function can have a direct effect on the shape and strength of the bones in the skeleton. When large numbers of calcium ions are mobilized in fluids, the bones become weaker; when calcium salts are deposited, bones become denser and stronger. Because the bone matrix contains protein fibers as well as mineral deposits, changes in mineral content do not

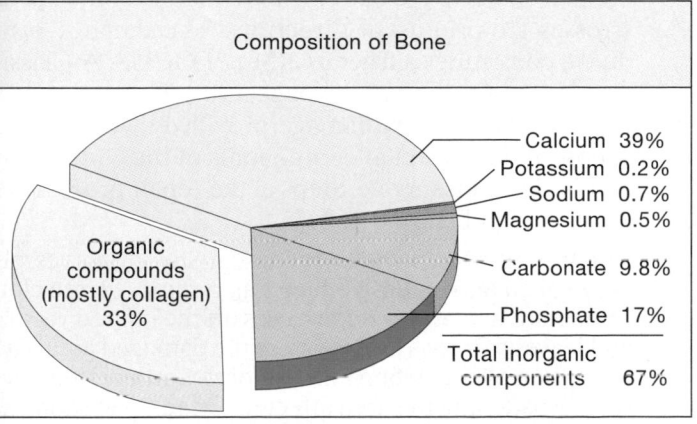

Composition of Bone

- Calcium 39%
- Potassium 0.2%
- Sodium 0.7%
- Magnesium 0.5%
- Carbonate 9.8%
- Phosphate 17%

Organic compounds (mostly collagen) 33%

Total inorganic components 67%

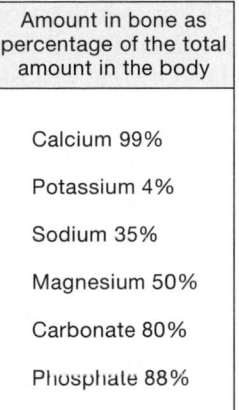

Amount in bone as percentage of the total amount in the body
Calcium 99%
Potassium 4%
Sodium 35%
Magnesium 50%
Carbonate 80%
Phosphate 88%

●*FIGURE 6-13* A Chemical Analysis of Bone

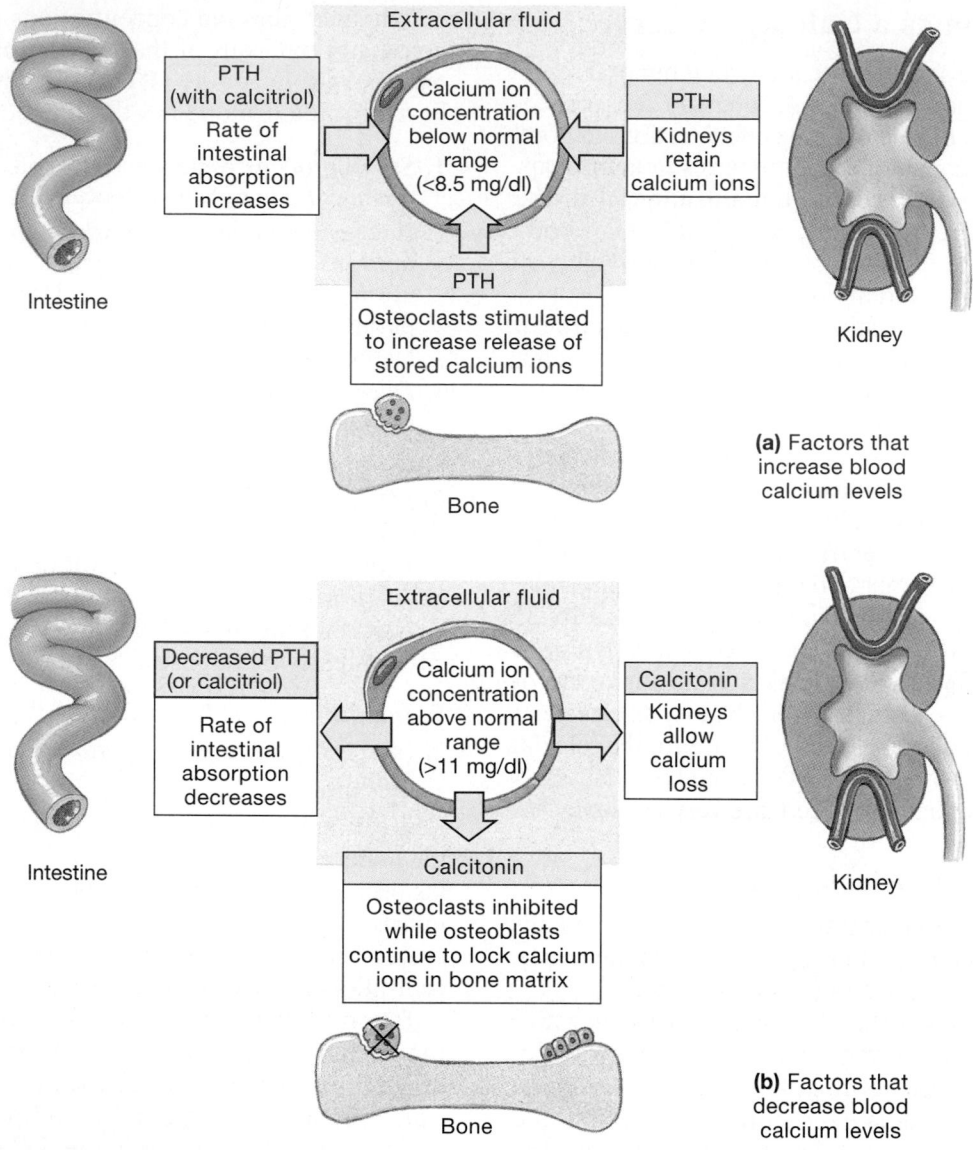

Extracellular fluid

PTH (with calcitriol) | Rate of intestinal absorption increases

Calcium ion concentration below normal range (<8.5 mg/dl)

PTH | Kidneys retain calcium ions

Intestine

PTH | Osteoclasts stimulated to increase release of stored calcium ions

Bone

Kidney

(a) Factors that increase blood calcium levels

Extracellular fluid

Decreased PTH (or calcitriol) | Rate of intestinal absorption decreases

Calcium ion concentration above normal range (>11 mg/dl)

Calcitonin | Kidneys allow calcium loss

Intestine

Calcitonin | Osteoclasts inhibited while osteoblasts continue to lock calcium ions in bone matrix

Bone

Kidney

(b) Factors that decrease blood calcium levels

• *FIGURE 6-14* **Factors That Alter the Concentration of Calcium Ions in Body Fluids.** **(a)** Factors that increase blood calcium ion concentrations. **(b)** Factors that decrease blood calcium ion concentrations.

necessarily affect the shape of the bone. In *osteomalacia* (os-tē-ō-ma-LĀ-shē-uh; *malakia,* softness) the bones appear normal although they are relatively weak and flexible because of poor mineralization. *Rickets,* a form of osteomalacia affecting children, generally results from a vitamin D_3 deficiency caused by inadequate exposure to sunlight and inadequate dietary supply. The bones of children with rickets are 'so poorly mineralized that they become very flexible. Affected individuals develop a bowlegged appearance as the thigh and leg bones bend under the weight of the body.

☑ Why does a child with rickets have difficulty walking?

☑ What effect would increased PTH secretion have on blood calcium levels?

☑ How does calcitonin help lower blood calcium levels?

Fracture Repair

Despite its mineral strength, bone may crack or even break if subjected to extreme loads, sudden impacts, or stresses from unusual directions. The damage produced constitutes a **fracture.** (See "FOCUS: A Classification of Fractures" on page 190.) Most fractures heal even after severe damage, provided that the blood supply and the cellular components of the endosteum and periosteum survive. Steps in the repair process are illustrated in Figure 6-15•:

STEP 1. In even a small fracture, many blood vessels are broken and extensive bleeding occurs. A large clot, or **fracture hematoma**, soon closes off the injured vessels and leaves a fibrous meshwork in the damaged area. The disruption of circulation kills osteocytes around the fracture, broadening the area affected. Dead bone soon extends along the shaft in either direction from the break.

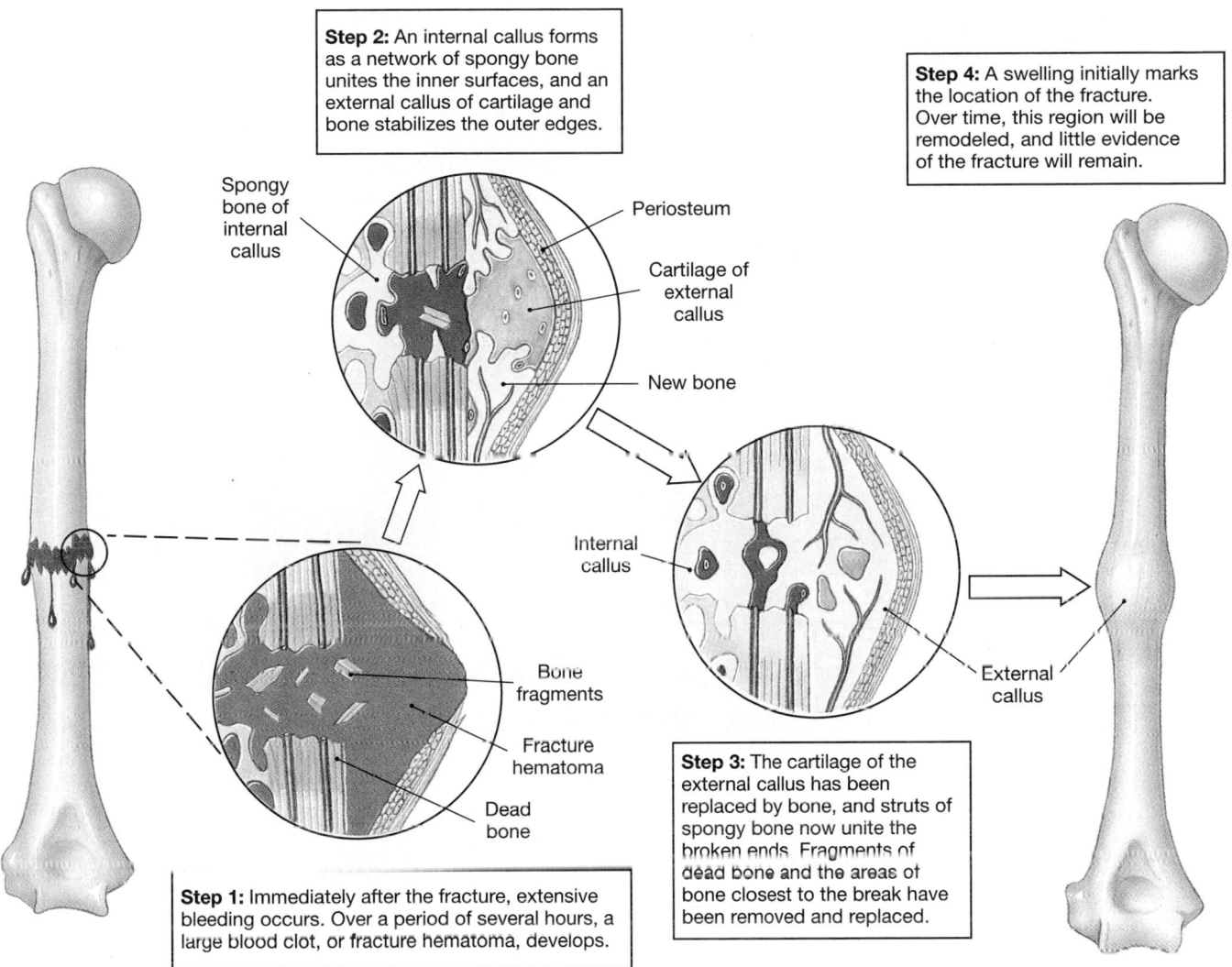

Step 2: An internal callus forms as a network of spongy bone unites the inner surfaces, and an external callus of cartilage and bone stabilizes the outer edges.

Step 4: A swelling initially marks the location of the fracture. Over time, this region will be remodeled, and little evidence of the fracture will remain.

Spongy bone of internal callus

Periosteum

Cartilage of external callus

New bone

Internal callus

Bone fragments

Fracture hematoma

Dead bone

External callus

Step 1: Immediately after the fracture, extensive bleeding occurs. Over a period of several hours, a large blood clot, or fracture hematoma, develops.

Step 3: The cartilage of the external callus has been replaced by bone, and struts of spongy bone now unite the broken ends. Fragments of dead bone and the areas of bone closest to the break have been removed and replaced.

•**FIGURE 6-15** Steps in the Repair of a Fracture

STEP 2. In an adult, the cells of the periosteum and endosteum are relatively inactive. When a fracture occurs, the cells of the intact endosteum and periosteum undergo rapid mitoses, and the daughter cells migrate into the fracture zone. An **external callus** (*callum,* hard skin), or enlarged collar of cartilage and bone, forms and encircles the bone at the level of the fracture. An extensive **internal callus** organizes within the marrow cavity and between the broken ends of the shaft. At the center of the external callus, cells differentiate into chondrocytes and produce blocks of cartilage. At the edges of each callus, the cells differentiate into osteoblasts and begin creating a bridge between the bone fragments on either side of the fracture.

STEP 3. As the repair continues, osteoblasts replace the central cartilage of the external callus with spongy bone. When this conversion is complete, the external and internal calluses form an extensive and continuous brace at the fracture site. Struts of spongy bone now unite the broken ends.

STEP 4. Osteoclasts and osteoblasts then remodel the region for a period ranging from 4 months to well over

a year. When the remodeling is complete, the trabecular bone of the calluses will be gone and only living compact bone will remain. The repair may be "good as new," with no indications that a fracture ever occurred, or the bone may be slightly thicker and stronger than normal at the fracture site. Under comparable stresses, a second fracture will generally occur at a different site.

Comparable events occur after more complex breaks or even after transplantation of a bone fragment. Each year in the United States, roughly 200,000 people receive *bone grafts* to stimulate bone repair. The bone fragments are commonly taken from another part of the body, such as the hip. However, because the inserted bone is ultimately destroyed and replaced, dead and sterilized bone fragments from other donors or even other species of animals can be used to establish a framework for the repair process. In situations in which the fracture is so severe that normal repair processes cannot occur, even with bone grafts, surgeons may rely on synthetic bone or the use of strong electrical fields to stimulate osteoblast activity. AM *Stimulation of Bone Growth and Repair*

FOCUS   A Classification of Fractures

Fractures are classified according to their external appearance, the site of the fracture, and the nature of the crack or break in the bone. Important fracture types are indicated here with representative X-rays. **Closed**, or *simple*, fractures are completely internal. They do not involve a break in the skin. **Open**, or *compound*, fractures project through the skin. They are more dangerous than closed fractures due to the possibility of infection or uncontrolled bleeding. Many fractures fall into more than one category. For example, a *Colles' fracture* is a transverse fracture that, depending on the injury, may also be an open or closed comminuted fracture.

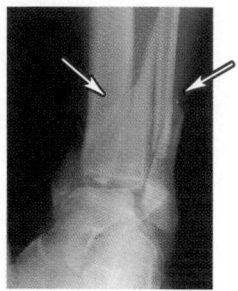

Pott's fracture—dislocation of the tibia and fibula

A **Pott's fracture** occurs at the ankle and affects both bones of the leg.

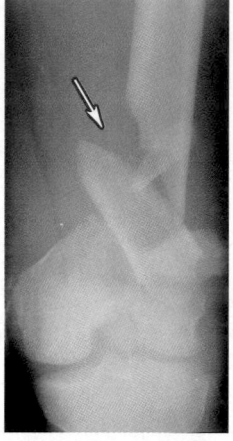

Comminuted fracture of distal femur

Comminuted fractures shatter the affected area into a multitude of bony fragments.

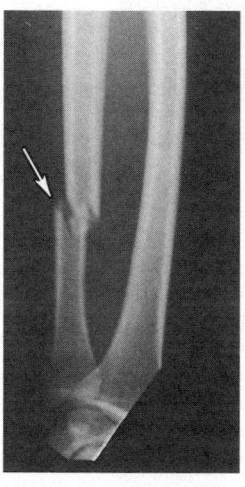

Transverse fractures break a shaft bone across its long axis.

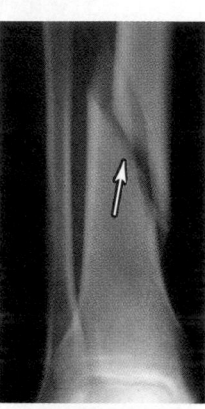

Spiral fracture of tibia

Spiral fractures, produced by twisting stresses, spread along the length of the bone.

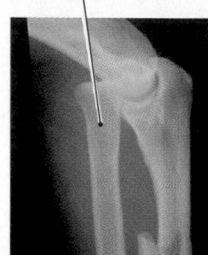

Dislocated radius

Displaced ulnar fracture

Displaced fractures produce new and abnormal arrangements of bony elements.

Nondisplaced fractures retain the normal alignment of the bone elements or fragments.

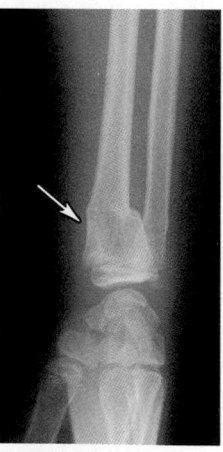

A **Colles' fracture**, a break in the distal portion of the radius (the slender bone of the forearm), is typically the result of reaching out to cushion a fall.

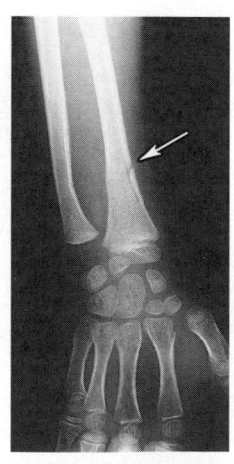

In a **greenstick fracture**, only one side of the shaft is broken, and the other is bent. This type generally occurs in children, whose long bones have yet to ossify fully.

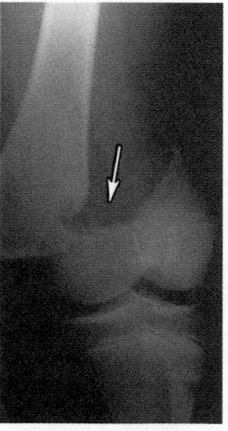

Epiphyseal fracture of femur

Epiphyseal fractures tend to occur where the bone matrix is undergoing calcification and chondrocytes are dying. A clean transverse fracture along this line generally heals well. Unless carefully treated, fractures between the epiphysis and the epiphyseal plate can permanently halt further longitudinal growth; surgery is typically required.

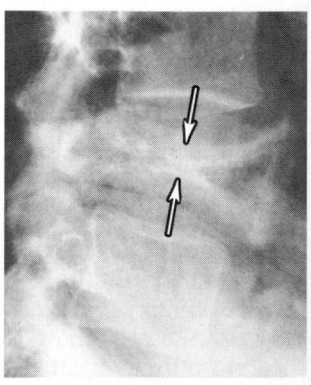

Compression fracture of vertebra

Compression fractures occur in vertebrae subjected to extreme stresses, as when you land on your seat in a fall.

⊠Aging and the Skeletal System

The bones of the skeleton become thinner and relatively weaker as a normal part of the aging process. Inadequate ossification is called **osteopenia** (os-tē-ō-PĒ-nē-uh; *penia,* lacking), and all of us become slightly osteopenic as we age. This reduction in bone mass begins between the ages of 30 and 40. Over this period, osteoblast activity begins to decline while osteoclast activity continues at previous levels. Once the reduction begins, women lose roughly 8 percent of their skeletal mass every decade, whereas the skeletons of men deteriorate at a slower rate, about 3 percent per decade. All parts of the skeleton are not equally affected. Epiphyses, vertebrae, and the jaws lose more than their share, resulting in fragile limbs, a reduction in height, and the loss of teeth.

When reduction in bone mass is sufficient to compromise normal function, the condition is known as **osteoporosis** (os-tē-ō-por-Ō-sis; *porosus,* porous). The extent of the loss of spongy bone mass due to osteoporosis is shown in Figure 6-16•; the reduction in compact bone mass is equally severe. The fragile bones that result are likely to break when exposed to stresses that would be easily tolerated by younger individuals. For example, a hip fracture may occur when a woman in her nineties tries to stand up. Any fractures that do occur lead to a loss of independence and an immobility that further weakens the skeleton.

Sex hormones are important in maintaining normal rates of bone deposition. A significant percentage of women over age 45 suffer from osteoporosis. The condition accelerates after menopause due to a decline in circulating estrogens. Because men continue to produce androgens until relatively late in life, severe osteoporosis is less common in males below age 60.

Osteoporosis can also develop as a secondary effect of many cancers. Cancers of the bone marrow, breast, or other tissues release a chemical known as **osteoclast-activating factor.** This compound increases both the number and activity of osteoclasts and produces a severe osteoporosis. See the *Applications Manual* for more coverage of age-related conditions that affect bones. [AM]
Osteoporosis and Age-Related Skeletal Abnormalities

BONE MARKINGS (SURFACE FEATURES)

Each bone in the body has characteristic external and internal features. Elevations or projections form where tendons and ligaments attach and where adjacent bones articulate. Depressions, grooves, and tunnels in bone indicate sites where blood vessels and nerves lie alongside or penetrate the bone. Detailed examination of these **bone markings,** or *surface features,* can yield an abundance of anatomical information. For example, anthropologists, criminologists, and pathologists can often determine the size, weight, sex, and general appearance of an individual on the basis of incomplete skeletal remains. (We shall discuss this topic further in Chapter 8.)

We will ignore minor variations of individual bones and focus on prominent features that identify the bone (Table 6-2). These markings are useful because they provide fixed landmarks that can help us determine the position of the soft tissue components of other systems. Specific anatomical terms are used to describe the various elevations and depressions.

INTEGRATION WITH OTHER SYSTEMS

Although the bones may seem inert, you should now realize that they are dynamic structures. The entire skeletal system is intimately associated with other systems. For instance, the bones of the skeleton are attached to the muscular system, extensively connected to the cardiovascular and lymphatic systems, and largely under the physiological control of the endocrine system. The digestive and

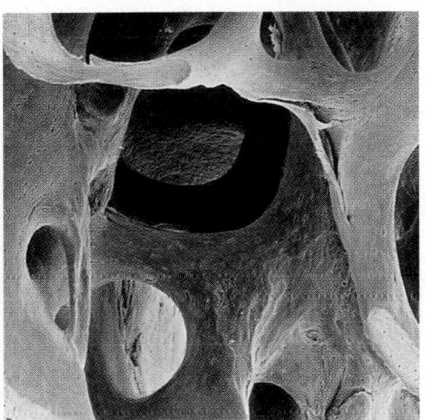

(a)

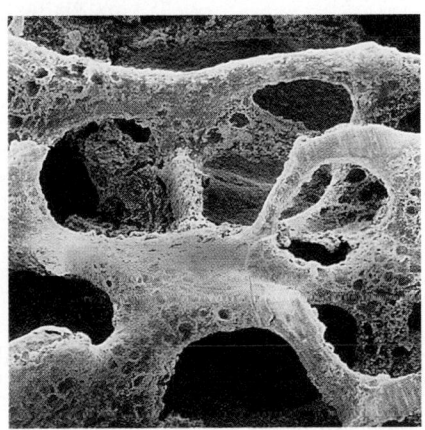

(b)

•*FIGURE 6-16* **The Effects of Osteoporosis.** **(a)** Normal spongy bone from the epiphysis of a young adult. (SEM × 23) **(b)** Spongy bone from a person with osteoporosis. (SEM × 19)

TABLE 6-2 An Introduction to Skeletal Terminology

General Description	Anatomical Term	Definition
Elevations and projections (general)	**Process** **Ramus**	Any projection or bump An extension of a bone making an angle to the rest of the structure
Processes formed where tendons or ligaments attach	**Trochanter** **Tuberosity** **Tubercle** **Crest** **Line**	A large, rough projection A smaller, rough projection A small, rounded projection A prominent ridge A low ridge
Processes formed for articulation with adjacent bones	**Head** **Neck** **Condyle** **Trochlea** **Facet** **Spine**	The expanded articular end of an epiphysis, separated from the shaft by the neck A narrow connection between the epiphysis and diaphysis A smooth, rounded articular process A smooth, grooved articular process shaped like a pulley A small, flat articular surface A pointed process
Depressions	**Fossa** **Sulcus**	A shallow depression A narrow groove
Openings	**Foramen** **Fissure** **Sinus or antrum**	A rounded passageway for blood vessels and/or nerves An elongate cleft A chamber within a bone, normally filled with air

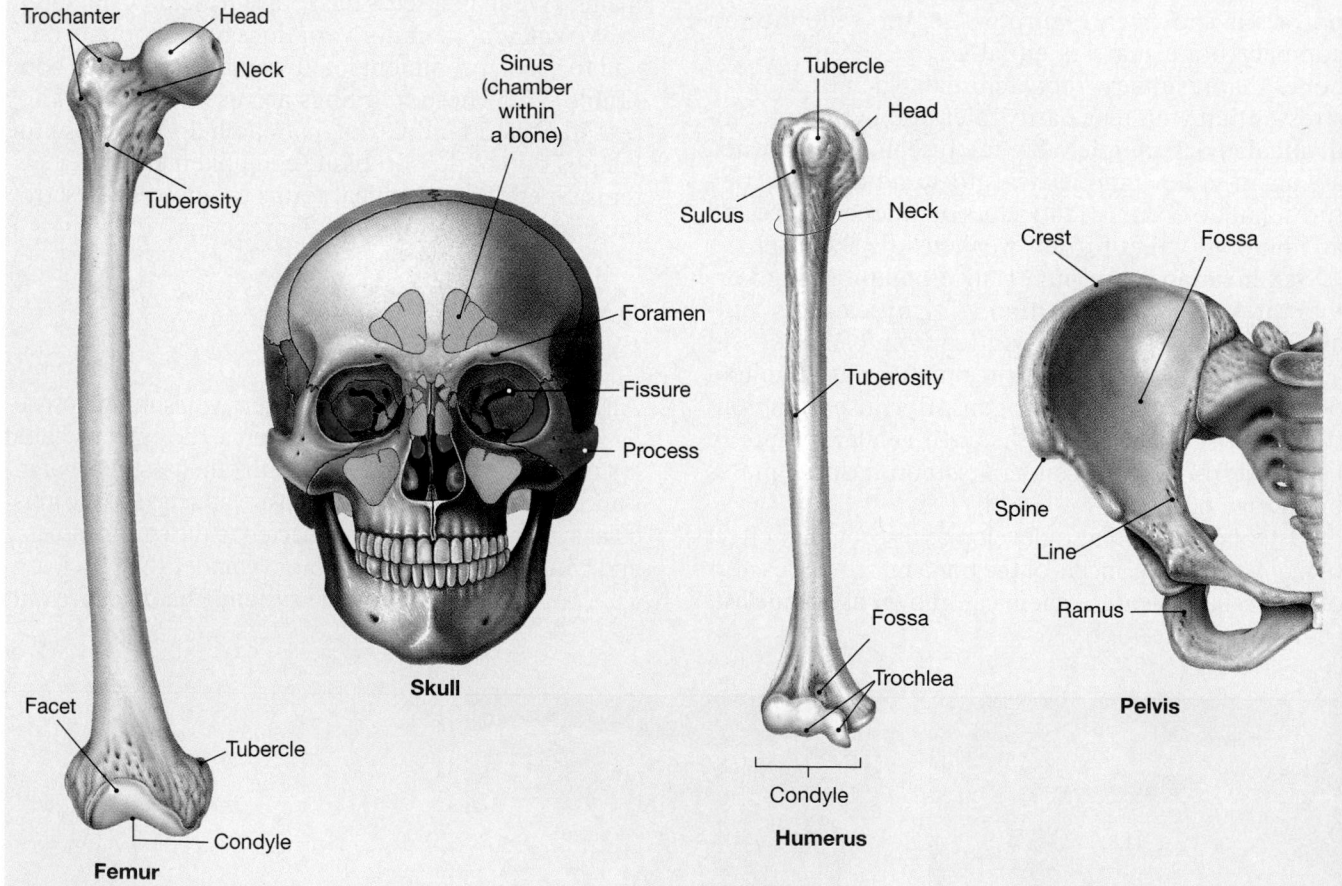

urinary systems also play important roles in providing the calcium and phosphate minerals needed for bone growth. In return, the skeleton represents a reserve of calcium, phosphate, and other minerals that can compensate for changes in the dietary supply of those ions. The functional relationships between the skeletal system and other systems are diagrammed in Figure 6-17●.

☑ At what point in fracture repair would you find an external callus?

☑ Why is the condition known as osteoporosis more common in older women than in older men?

☑ X-rays of a child's skeleton indicate bones forming in the tendons of several joints. What kind of bones are these?

INTEGUMENTARY SYSTEM

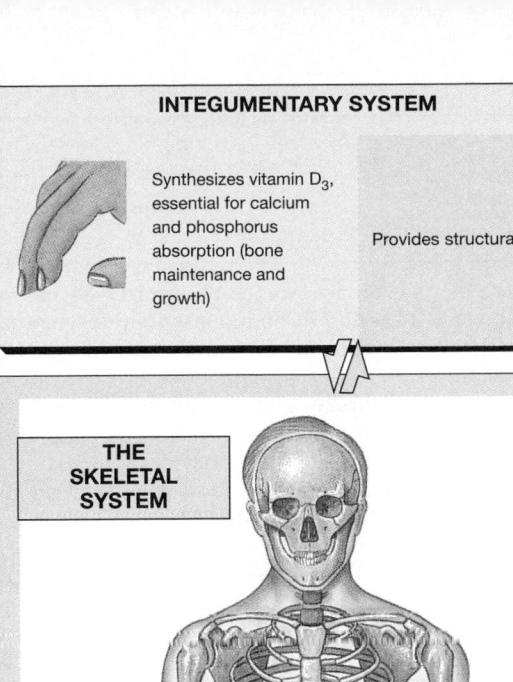

Synthesizes vitamin D₃, essential for calcium and phosphorus absorption (bone maintenance and growth)

Provides structural support

MUSCULAR SYSTEM

Stabilizes bone positions; tension in tendons stimulates bone growth and maintenance

Provides calcium needed for normal muscle contraction; bones act as levers to produce body movements

NERVOUS SYSTEM

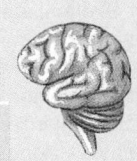

Regulates bone position by controlling muscle contractions

Provides calcium for neural function; protects brain, spinal cord; receptors at joints provide information about body position

ENDOCRINE SYSTEM

Skeletal growth regulated by growth hormone, thyroid hormones, and sex hormones; calcium mobilization regulated by parathyroid hormone and calcitonin

Protects endocrine organs, especially in brain, chest, and pelvic cavity

CARDIOVASCULAR SYSTEM

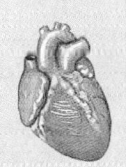

Provides oxygen, nutrients, hormones, blood cells; removes waste products and carbon dioxide

Provides calcium needed for cardiac muscle contraction; blood cells produced in bone marrow

LYMPHATIC SYSTEM

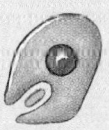

Lymphocytes assist in the defense and repair of bone following injuries

Lymphocytes and other cells of the immune response are produced and stored in bone marrow

RESPIRATORY SYSTEM

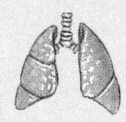

Provides oxygen and eliminates carbon dioxide

Movements of ribs important in breathing; axial skeleton surrounds and protects lungs

DIGESTIVE SYSTEM

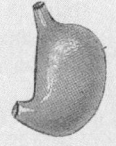

Provides nutrients, calcium, and phosphate

Ribs protect portions of liver, stomach, and intestines

URINARY SYSTEM

Conserves calcium and phosphate needed for bone growth; disposes of waste products

Axial skeleton provides some protection for kidneys and ureters; pelvis protects urinary bladder and proximal urethra

---

THE SKELETAL SYSTEM

FOR ALL SYSTEMS

Provides mechanical support
Stores energy reserves
Stores calcium and phosphate reserves

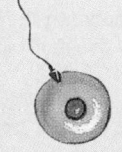

Sex hormones stimulate growth and maintenance of bones; surge of sex hormones at puberty causes acceleration of growth and closure of epiphyseal plates

Pelvis protects reproductive organs of female; protects portion of ductus deferens and accessory glands in male

REPRODUCTIVE SYSTEM

● **FIGURE 6-17** **Functional Relationships between the Skeletal System and Other Systems**

SELECTED CLINICAL TERMINOLOGY

Terms Discussed in This Chapter

acromegaly: A condition caused by excess secretion of growth hormone after puberty. Skeletal abnormalities develop, affecting the cartilages and various small bones. *(p. 186 and AM)*

external callus: A toughened layer of connective tissue that encircles and stabilizes a bone at a fracture site. *(p. 189)*

fracture: A crack or break in a bone. *(p. 188)*

fracture hematoma: A large blood clot that closes off the injured vessels around a fracture and leaves a fibrous meshwork in the damaged area of bone; the first step in fracture repair. *(p. 188)*

gigantism: A condition resulting from an overproduction of growth hormone before puberty. *(p. 186)*

internal callus: A bridgework of trabecular bone that unites the broken ends of a bone on the marrow side of a fracture. *(p. 189)*

Marfan's syndrome: An inherited condition linked to defective production of a connective tissue glycoprotein. Extreme height and long, slender limbs are the most obvious physical indications. *(p. 186 and AM)*

osteoclast-activating factor: A compound released by cancers of the bone marrow, breast, or other tissues. It produces a severe osteoporosis. *(p. 191)*

osteomalacia (os-tē-ō-ma-LĀ-shē-uh): A softening of bone due to a decrease in the mineral content. *(p. 188)*

osteopenia (os-tē-ō-PĒ-nē-uh): Inadequate ossification, leading to thinner, weaker bones. *(p. 191)*

osteoporosis (os-tē-ō-por-Ō-sis): A reduction in bone mass to a degree that compromises normal function. *(p. 191 and AM)*

pituitary growth failure: A type of dwarfism caused by inadequate growth hormone production. *(p. 186)*

rickets: A childhood disorder that reduces the amount of calcium salts in the skeleton; typically characterized by a bowlegged appearance because the leg bones bend. *(p. 188)*

scurvy: A condition involving weak, brittle bones as a result of a vitamin C deficiency. *(p. 186)*

AM **Additional Terms Discussed in the *Applications Manual***

achondroplasia (ā-kon-drō-PLĀ-sē-uh): A condition resulting from abnormal epiphyseal activity; the epiphyseal plates grow unusually slowly, and the individual develops short, stocky limbs. The trunk is normal in size, and sexual and mental development remain unaffected.

hyperostosis (hī-per-os-TŌ-sis): The excessive formation of bone tissue.

osteitis deformans (os-tē-Ī-tis de-FOR-manz), or **Paget's disease**: A condition characterized by gradual deformation of the skeleton.

osteogenesis imperfecta (im-per-FEK-tuh): An inherited condition affecting the organization of collagen fibers. Osteoblast function is impaired, growth is abnormal, and the bones are very fragile,

leading to progressive skeletal deformation and repeated fractures.

osteomyelitis (os-tē-ō-mī-e-LĪ-tis): A painful infection in a bone, generally caused by bacteria.

osteopetrosis (os-tē-ō-pe-TRŌ-sis): A condition caused by a decrease in osteoclast activity, causing increased bone mass and various skeletal deformities.

CHAPTER REVIEW

On-line resources for this chapter are on our World Wide Web site at:
http://www.prenhall.com/martini/fap

STUDY OUTLINE

INTRODUCTION, p. 173

1. The skeletal system includes the bones of the skeleton and the cartilages, ligaments, and other connective tissues that stabilize or interconnect bones. Its functions include structural support, storage of minerals and lipids, blood cell production, protection, and leverage.

A CLASSIFICATION OF BONES, p. 173

Bone Shapes, p. 173

1. Categories of bones include: **long bones, short bones, flat bones, irregular bones, sesamoid bones,** and **sutural bones** (*Wormian bones*). *(Figure 6-1)*
2. A representative long bone has a **diaphysis, epiphyses, metaphyses, articular cartilages,** and a **marrow cavity.**
3. There are two types of bone: **compact** *(dense)* **bone** and **spongy** *(cancellous)* **bone.** The spaces within spongy bone and the marrow cavity contain **bone marrow,** either **yellow bone marrow** (for lipid storage) or **red bone marrow** (for blood cell formation). *(Figure 6-2)*

BONE HISTOLOGY, p. 174

1. **Osseous tissue** is a supporting connective tissue with a solid matrix.

The Matrix of Bone, p. 175

2. Bone matrix consists largely of crystals of **hydroxyapatite**.

Cells in Bone, p. 175

3. **Osteoprogenitor cells** differentiate into osteoblasts. **Osteoblasts** synthesize the matrix in the process of **osteogenesis. Osteocytes,** located in lacunae, are the most abundant bone cells. **Osteoclasts** dissolve the bony matrix through **osteolysis.**

Compact and Spongy Bone, p. 176

4. The basic functional unit of compact bone is the **osteon,** containing osteocytes arranged around a **central canal. Lamellae** are layers of calcified matrix. **Canaliculi** within and between the lamellae interconnect the **lacunae.** *(Figures 6-3, 6-4)*
5. Spongy bone contains **trabeculae,** typically in an open network. *(Figure 6-4)*

6. Compact bone occurs where stresses come from a limited range of directions, such as along the diaphysis of some bones. Spongy bone is located where stresses are few or come from many different directions, such as at the epiphyses of some bones. *(Figure 6-5)*

The Periosteum and Endosteum, p. 178

7. A bone is covered by a **periosteum** and lined with an **endosteum.** *(Figure 6-6)*

BONE DEVELOPMENT AND GROWTH, p. 179

1. **Ossification** is the process of converting other tissues to bone. **Calcification** is the process of depositing calcium salts within a tissue.

Intramembranous Ossification, p. 180

2. **Intramembranous ossification** begins when osteoblasts differentiate within connective tissue and can produce spongy or compact bone. Such ossification begins at an **ossification center**. *(Figures 6-7, 6-8)*

Endochondral Ossification, p. 180

3. **Endochondral ossification** begins by forming a cartilage model that is gradually replaced by bone at the metaphysis. In this way, bone length increases. *(Figures 6-8, 6-9)*
4. There are differences between bones and between individuals regarding the timing of *closure* of the **epiphyseal plate.** *(Figure 6-10)*
5. Bone diameter increases through *appositional growth*. *(Figure 6-11)*

The Blood and Nerve Supply, p. 183

6. An extensive supply of blood is supplied to bone through three major sets of blood vessels. *(Figure 6-12)*

THE DYNAMIC NATURE OF BONE, p. 185

1. The organic and mineral components of bone are continuously recycled and renewed through the process of **remodeling**.

Exercise Effects on Bone, p. 185

2. The shapes and thicknesses of bones reflect the stresses applied to them.

Hormonal and Nutritional Effects on Bone, p. 185

3. Normal osteogenesis requires a reliable source of minerals, vitamins, and hormones.
4. *Growth hormone* and *thyroxine* stimulate bone growth. *(Table 6-1)*

The Skeleton as a Calcium Reserve, p. 187

5. Calcium is the most common mineral in the human body, with roughly 99 percent of it located in the skeleton. *(Figure 6-13)*
6. Interactions among the bones, intestinal tract, and kidneys affect calcium ion concentrations. *(Figure 6-14)*
7. Two hormones, **calcitonin** and **parathyroid hormone (PTH),** regulate calcium ion homeostasis. Calcitonin leads to a decline in calcium concentrations, whereas parathyroid hormone increases calcium concentrations. *(Figure 6-14)*

Fracture Repair, p. 188

8. A **fracture** is a crack or break in a bone. Repair of a fracture involves the formation of a **fracture hematoma,** an **external callus,** and an **internal callus.** *(Figure 6-15)*

FOCUS: A Classification of Fractures, p. 190

9. Various criteria are used to classify fractures. *(Focus: A Classification of Fractures)*

Aging and the Skeletal System, p. 191

10. Effects of aging on the skeleton can include **osteopenia** and **osteoporosis.** *(Figure 6-16)*

BONE MARKINGS (SURFACE FEATURES), p. 191

1. **Bone markings** can be used to identify specific bones within each category. *(Table 6-2)*

INTEGRATION WITH OTHER SYSTEMS, p. 194

1. The skeletal system is dynamically associated with all the other systems of the body. *(Figure 6-17)*

REVIEW QUESTIONS

Level 1   Reviewing Facts and Terms

1. The bones of the skeleton store energy reserves as lipids in called areas of
   - (a) red marrow
   - (b) yellow marrow
   - (c) the matrix of bone tissue
   - (d) the ground substance
2. Two-thirds of the weight of bone is accounted for by
   - (a) crystals of calcium phosphate
   - (b) collagen fibers
   - (c) osteocytes
   - (d) calcium carbonate
3. The two types of osseous tissue are
   - (a) compact bone and spongy bone
   - (b) dense bone and compact bone
   - (c) spongy bone and cancellous bone
   - (d) a, b, and c are correct

4. The basic functional units of mature compact bone are called
   - (a) lacunae
   - (b) osteocytes
   - (c) osteons
   - (d) canaliculi
5. Unlike compact bone, spongy bone contains concentric lamellae that form struts or plates called
   - (a) canaliculi   (b) canals of Volkmann
   - (c) osteons   (d) trabeculae
6. The vitamins essential for normal adult bone maintenance and repair are
   - (a) A and E
   - (b) C and D
   - (c) B and E
   - (d) B complex and K

7. The hormones that coordinate the storage, absorption, and excretion of calcium ions are
    (a) growth hormone and thyroxine
    (b) calcitonin and parathyroid hormone
    (c) calcitriol and cholecalciferol
    (d) estrogens and androgens

8. Inadequate ossification due to the aging process is
    (a) osteomalacia   (b) osteolysis
    (c) osteopenia   (d) osteogenesis

9. The primary reason that osteoporosis accelerates after menopause in women is
    (a) reduced levels of circulating estrogens
    (b) reduced levels of vitamin C
    (c) diminished osteoclast activity
    (d) increased osteoblast activity

10. A child suffering from rickets would have
    (a) oversized facial bones
    (b) long arms and legs
    (c) weak, brittle bones
    (d) bowlegs

11. What are the five primary functions of the skeletal system?

12. List the four distinctive cell populations that occur in osseous tissue.

13. What is the functional difference between an osteoblast and an osteoclast?

14. What are the three primary functions of the periosteum?

15. What is the primary difference between intramembranous ossification and endochondral ossification?

16. What three major sets of blood vessels develop in typical bones in the skeleton?

17. (a) What nutritional factors are essential for normal bone growth and maintenance?
    (b) What hormonal factors are necessary for normal bone growth and maintenance?

18. What three organs and/or tissues interact to assist in the regulation of calcium ion concentration in body fluids?

19. What major effects of parathyroid hormone oppose those of calcitonin?

20. What are the major functions of the hormone calcitonin?

Level 2   Reviewing Concepts

21. If there are no osteons in spongy bone, how do nutrients reach the osteocytes?

22. Why are stresses or impacts to the side of the shaft in a long bone more dangerous than stress applied to the long axis of the shaft?

23. Why do extended periods of inactivity cause degenerative changes in the skeleton?

24. What are the functional relationships between the skeleton and both the digestive and the urinary systems?

25. During the growth of a long bone, how is the epiphysis forced farther from the shaft?

26. What role do sex hormones in males and females play in bone growth and development?

27. How might damage to the thyroid gland influence calcium regulation in the body?

28. Why does a second fracture in the same bone tend to occur at a site different from that of the first fracture?

29. Why are individuals in their eighties shorter than they were in their forties?

30. How might bone markings be useful in identifying the remains of a criminal who has been shot and killed?

31. What purpose do elevations or projections serve on bones?

Level 3   Critical Thinking and Clinical Applications

32. While playing on her swing set, 10-year-old Sally falls and breaks her right leg. At the emergency room, the doctor tells her parents that the proximal end of the tibia where the epiphysis meets the diaphysis is fractured. The fracture is properly set and eventually heals. During a routine physical when she is 18, Sally learns that her right leg is 2 cm shorter than her left, probably because of her accident. What might account for this difference?

33. Sherry is a pregnant teenager. Her diet before she was pregnant consisted mostly of junk food, and that hasn't changed since she became pregnant. Approximately 8–10 weeks into her pregnancy, she falls and breaks her arm. She thinks this is unusual, because it wasn't a hard fall. Test results determine that a significant amount of bone demineralization is occurring. Explain what is happening to Sherry.

34. As a result of vitamin D_3 deficiency, would you expect to see changes in blood levels of the hormones calcitonin and PTH? Explain.

35. Why might a person suffering from kidney failure exhibit symptoms similar to those of osteoporosis?

7

The Axial Skeleton

*M*any societies throughout the world have chosen this sensible method of carrying heavy loads. Rather than bearing the load in their arms and possibly straining their backs, people place bundles on their heads. The weight is thus distributed along the longitudinal axis of the body via the axial skeleton. One or both hands are free to perform other tasks. Weight bearing is a major function of the axial skeleton, but, as you will see, these bones have many other important roles.

C H A P T E R   O U T L I N E   A N D   O B J E C T I V E S

Bone first appeared more than 500 million years ago in primitive fish. It became very common in many different groups within 100 million years. This was dermal bone (∞ *[p. 180]*) that provided a mineral reserve and a protective armor plating for defense against predators. The internal skeletons of these fishes were composed of cartilage. In the evolutionary branch that led toward humans, most of those cartilages were replaced by bones.

As we saw in Chapter 6, bones are remarkable structures. They are as strong as or stronger than reinforced concrete but considerably lighter. ∞ *[p. 175]* Better yet, bones can be remodeled and reshaped to meet metabolic demands or to adapt to changing activity patterns. The basic features of the human skeleton have been shaped by evolution, but the detailed characteristics of each bone reflect the stresses placed on it. As a result, the skeleton changes in the course of a lifetime. Examples discussed in Chapter 6 included the proportional changes at puberty and the gradual osteopenia of aging. In this chapter, we will consider additional examples, such as the changes that occurred in the shape of your vertebral column as you made the transition from crawling to walking.

The skeletal system is divided into *axial* and *appendicular divisions* (Figure 7-1●). The skeletal system includes at least 206 separate bones and a number of associated cartilages and ligaments. The **axial skeleton** (shown in Figure 7-1a● and highlighted in Figure 7-1b●) consists of the bones of the skull, the vertebral column, the ribs, and the sternum (breastbone). These elements form the longitudinal axis of the body. There are 80 bones in the axial skeleton, roughly 40 percent of the bones in the human body. The axial components include:

- The *skull* (8 *cranial bones* and 14 *facial bones*).
- Bones associated with the skull (6 *auditory ossicles* and the *hyoid bone*).
- The *thoracic cage* (the *sternum* and 24 *ribs*).
- The *vertebral column* (24 *vertebrae*, the *sacrum*, and the *coccyx*).

The function of the axial skeleton is to create a framework that supports and protects organs in the dorsal and ventral body cavities. In addition, it provides an extensive surface area for the attachment of muscles that (1) adjust the positions of the head, neck, and trunk; (2) perform respiratory movements; and (3) stabilize or position structures of the appendicular skeleton. The joints of the axial skeleton permit limited movement, but they are very strong and heavily reinforced with ligaments.

The **appendicular skeleton** consists of 126 bones. This division includes the bones of the limbs and the pectoral and pelvic girdles that attach the limbs to the trunk. We shall examine the appendicular skeleton in Chapter 8. This chapter describes the functional anatomy of the axial skeleton, and we will begin with the skull. Before proceeding, you may find it helpful to review the directional references included in Tables 1-1 and 1-2 and the terms introduced in Table 6-2. ∞ *[pp. 18, 20, 192]*

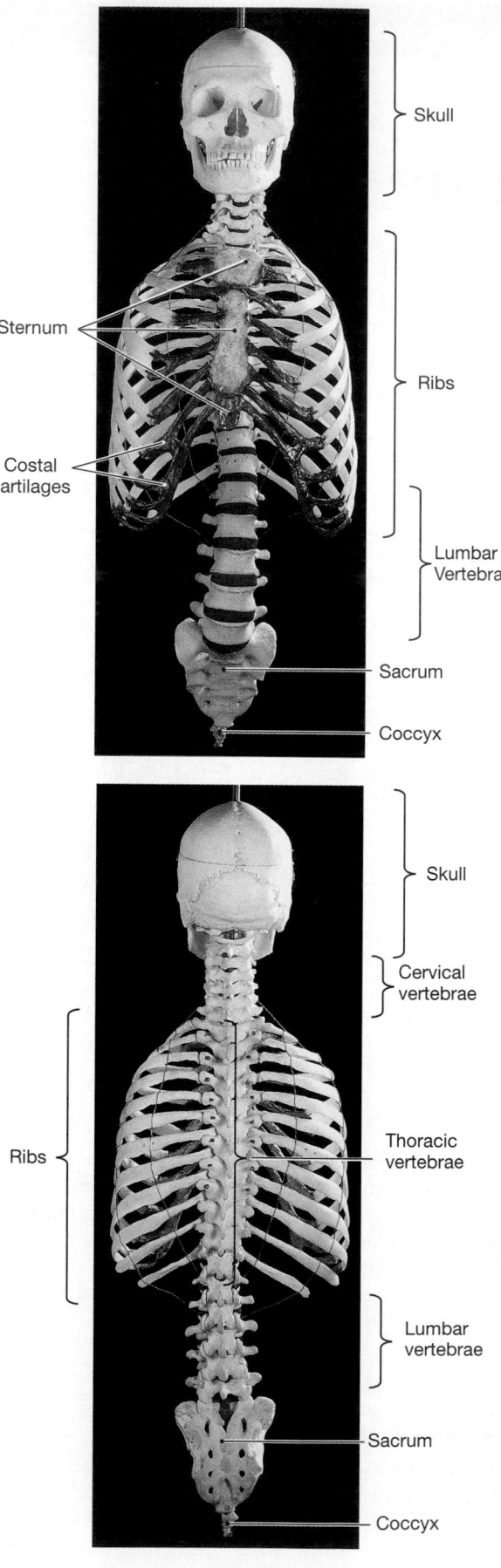

(a)

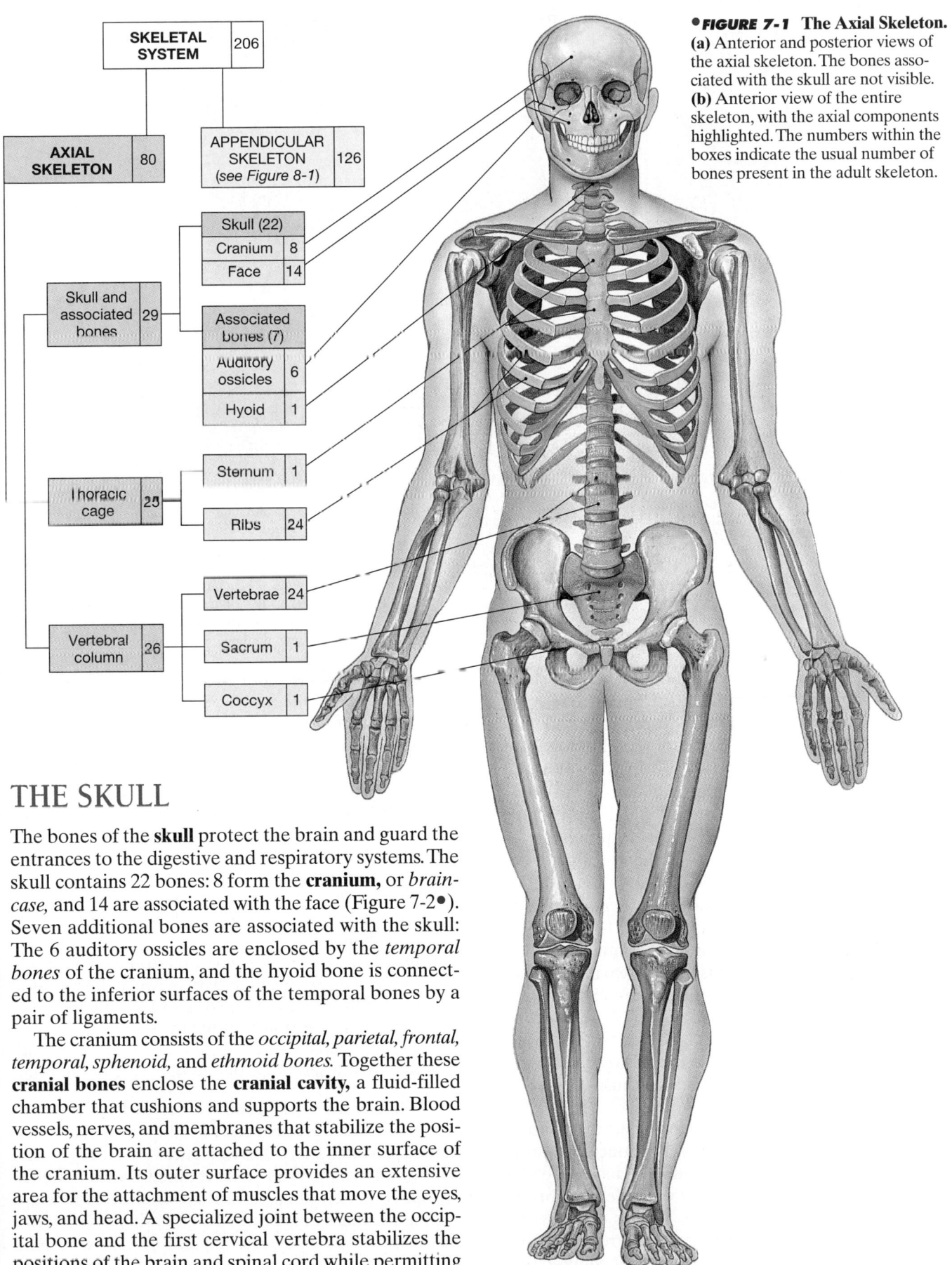

•**FIGURE 7-1** The Axial Skeleton.
(a) Anterior and posterior views of the axial skeleton. The bones associated with the skull are not visible.
(b) Anterior view of the entire skeleton, with the axial components highlighted. The numbers within the boxes indicate the usual number of bones present in the adult skeleton.

(b)

THE SKULL

The bones of the **skull** protect the brain and guard the entrances to the digestive and respiratory systems. The skull contains 22 bones: 8 form the **cranium,** or *brain-case,* and 14 are associated with the face (Figure 7-2•). Seven additional bones are associated with the skull: The 6 auditory ossicles are enclosed by the *temporal bones* of the cranium, and the hyoid bone is connected to the inferior surfaces of the temporal bones by a pair of ligaments.

The cranium consists of the *occipital, parietal, frontal, temporal, sphenoid,* and *ethmoid bones.* Together these **cranial bones** enclose the **cranial cavity,** a fluid-filled chamber that cushions and supports the brain. Blood vessels, nerves, and membranes that stabilize the position of the brain are attached to the inner surface of the cranium. Its outer surface provides an extensive area for the attachment of muscles that move the eyes, jaws, and head. A specialized joint between the occipital bone and the first cervical vertebra stabilizes the positions of the brain and spinal cord while permitting a considerable range of head movements.

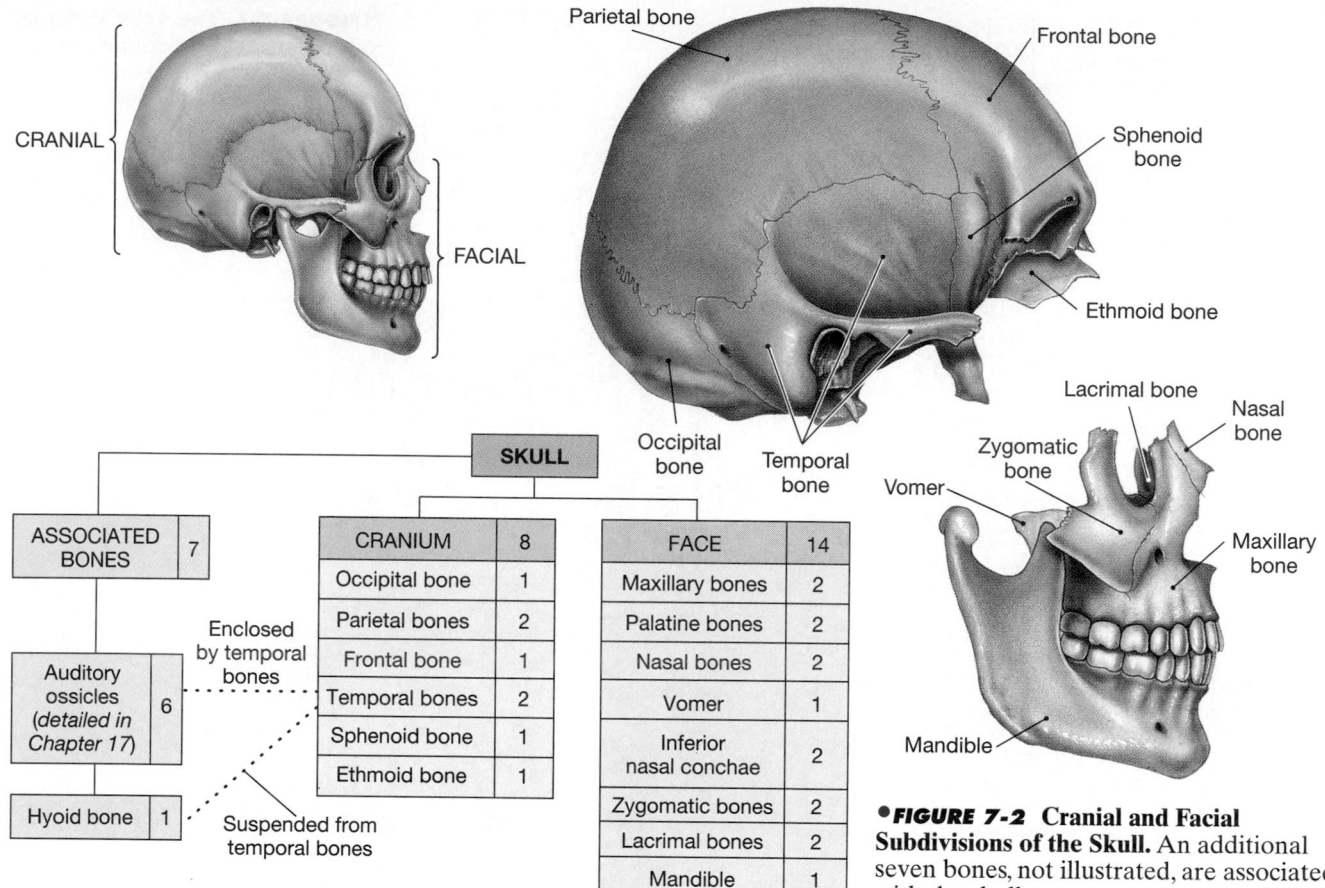

CRANIUM	8
Occipital bone	1
Parietal bones	2
Frontal bone	1
Temporal bones	2
Sphenoid bone	1
Ethmoid bone	1

FACE	14
Maxillary bones	2
Palatine bones	2
Nasal bones	2
Vomer	1
Inferior nasal conchae	2
Zygomatic bones	2
Lacrimal bones	2
Mandible	1

ASSOCIATED BONES	7
Auditory ossicles (*detailed in Chapter 17*)	6
Hyoid bone	1

•**FIGURE 7-2** **Cranial and Facial Subdivisions of the Skull.** An additional seven bones, not illustrated, are associated with the skull.

If the cranium is the house where the brain resides, the *facial complex* is the front porch. **Facial bones** protect and support the entrances to the digestive and respiratory tracts. The superficial facial bones (the *maxillary, lacrimal, nasal,* and *zygomatic bones* and the *mandible*) (Figure 7-2•) provide areas for the attachment of muscles that control facial expressions and assist in the manipulation of food. The deeper facial bones (the *palatine bone* and the *vomer,* respectively) either help separate the oral and nasal cavities or contribute to the **nasal septum** (*septum,* wall), which subdivides the nasal cavity.

Several of the bones of the skull contain air-filled chambers called **sinuses.** Sinuses have two major functions: (1) The presence of a sinus makes a bone much lighter than it would be otherwise, and (2) the mucous membrane lining the sinuses produces mucus that helps moisten and clean the air in and adjacent to the sinus. We will consider the individual sinuses as we discuss specific bones.

Except where the mandible contacts the cranium, the connections between the skull bones of adults are immovable joints called **sutures.** At a suture, the bones are tied firmly together with dense fibrous connective tissue. Each of the sutures of the skull has a name, but now you need to know only four major sutures—the *lambdoidal, coronal, sagittal,* and *squamosal* sutures:

- *Lambdoidal suture.* The **lambdoidal** (lam-DOYD-ul) **suture** (Greek *lambda,* Λ + *eidos,* shape) arches across the posterior surface of the skull (Figure 7-3a•). This suture separates the *occipital bone* from the two *parietal bones.* One or more **sutural bones** (*Wormian bones*) may be present along this suture.
- *Coronal suture.* The **coronal suture** attaches the frontal bone to the parietal bones of either side (Figure 7-3b•). The occipital, parietal, and frontal bones form the **calvaria** (kal-VAR-ē-uh), or skullcap.
- *Sagittal suture.* The **sagittal suture** extends from the lambdoidal suture to the coronal suture, between the parietal bones (Figure 7-3b•).
- *Squamosal sutures.* A **squamosal suture** on each side of the skull marks the boundary between the *temporal bone* and the parietal bone of that side. Figure 7-3a• shows the intersection between the squamosal sutures and the lambdoidal suture. Figure 7-3c• shows the path of the squamosal suture on the right side of the skull.

"FOCUS: The Individual Bones of the Skull" on pages 204–212 considers each skull bone. Refer to the figures provided there and to Figures 7-3 and 7-4• (the adult skull in superficial and sectional views) to develop a three-dimensional perspective on the relationships among the individual bones. Ridges and *foramina* (fo-RAM-in-uh; passageways; singular, *foramen,* fo-RĀ-men) detailed here mark the attachment sites of muscles or the passage of nerves and blood vessels.

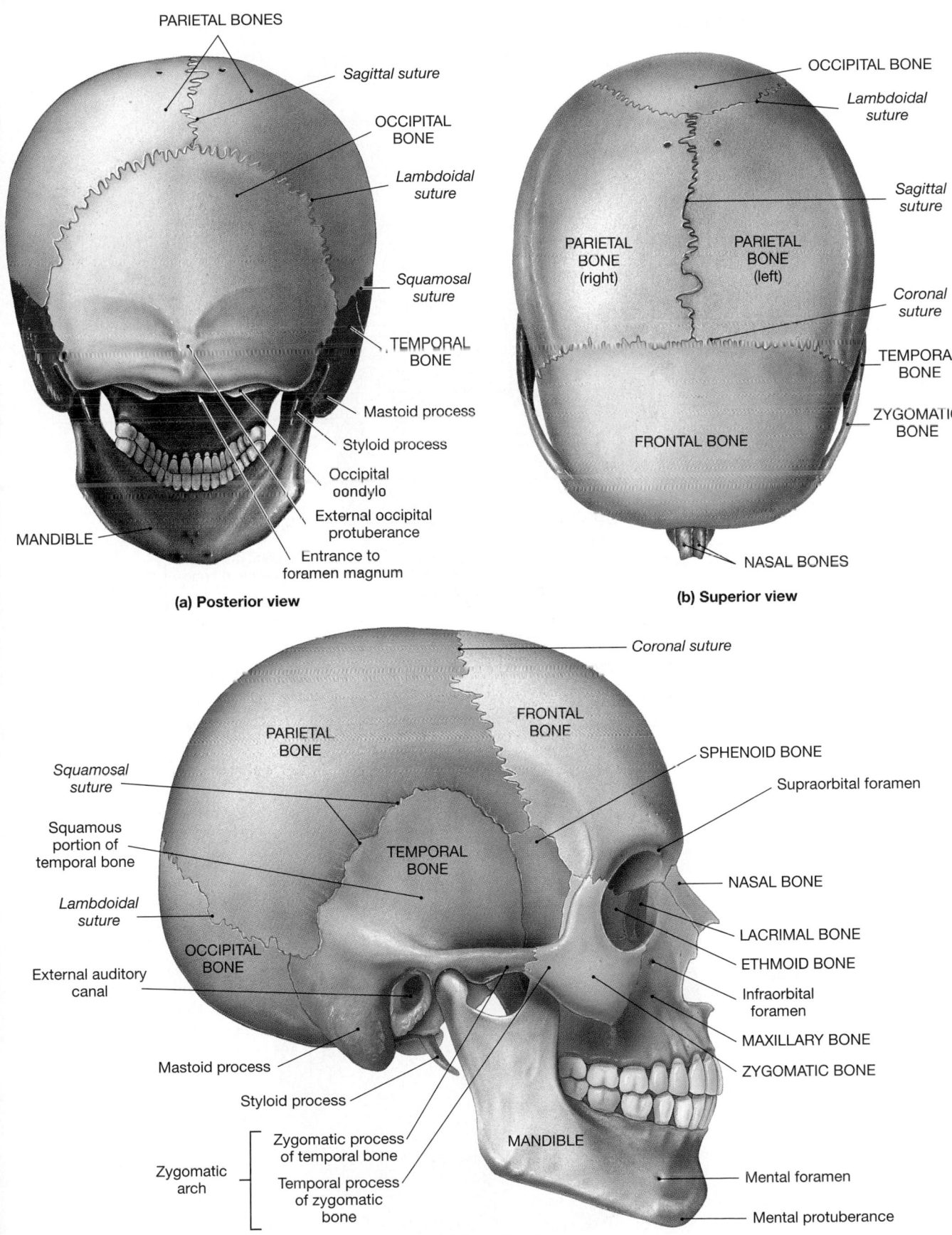

PARIETAL BONES

Sagittal suture

OCCIPITAL BONE

Lambdoidal suture

Squamosal suture

TEMPORAL BONE

Mastoid process

Styloid process

Occipital oondylo

External occipital protuberance

MANDIBLE

Entrance to foramen magnum

(a) Posterior view

OCCIPITAL BONE

Lambdoidal suture

Sagittal suture

PARIETAL BONE (right)

PARIETAL BONE (left)

Coronal suture

TEMPORAL BONE

ZYGOMATIC BONE

FRONTAL BONE

NASAL BONES

(b) Superior view

Coronal suture

PARIETAL BONE

FRONTAL BONE

SPHENOID BONE

Supraorbital foramen

Squamosal suture

Squamous portion of temporal bone

TEMPORAL BONE

Lambdoidal suture

OCCIPITAL BONE

External auditory canal

NASAL BONE

LACRIMAL BONE

ETHMOID BONE

Infraorbital foramen

MAXILLARY BONE

ZYGOMATIC BONE

Mastoid process

Styloid process

Zygomatic arch

Zygomatic process of temporal bone

Temporal process of zygomatic bone

MANDIBLE

Mental foramen

Mental protuberance

(c) Lateral view

● **FIGURE 7-3** **The Adult Skull.** The adult skull in **(a)** posterior view, **(b)** superior view, and **(c)** lateral view.

AM *Plates 2.1, 2.2*

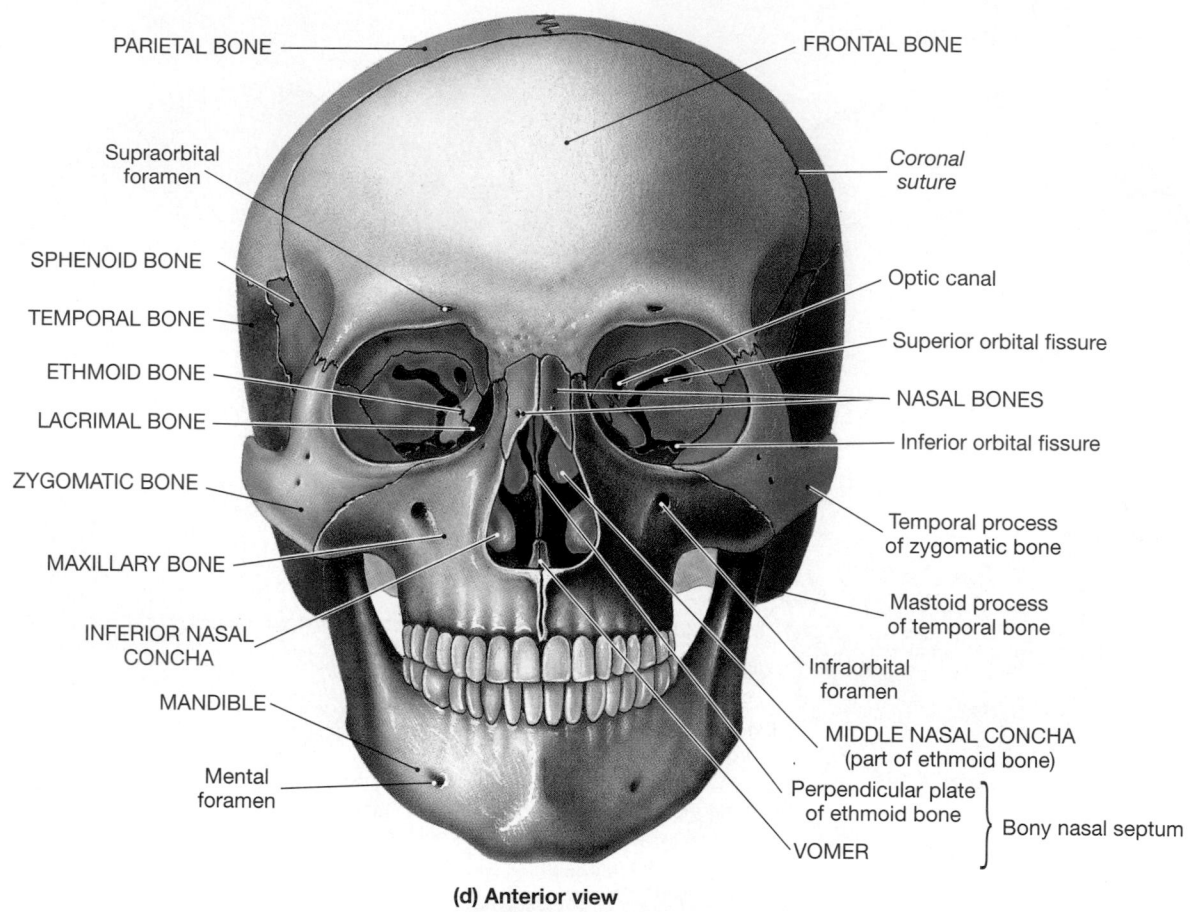

PARIETAL BONE

FRONTAL BONE

Supraorbital foramen

Coronal suture

SPHENOID BONE

Optic canal

TEMPORAL BONE

Superior orbital fissure

ETHMOID BONE

NASAL BONES

LACRIMAL BONE

Inferior orbital fissure

ZYGOMATIC BONE

Temporal process of zygomatic bone

MAXILLARY BONE

Mastoid process of temporal bone

INFERIOR NASAL CONCHA

Infraorbital foramen

MANDIBLE

MIDDLE NASAL CONCHA (part of ethmoid bone)

Mental foramen

Perpendicular plate of ethmoid bone

Bony nasal septum

VOMER

(d) Anterior view

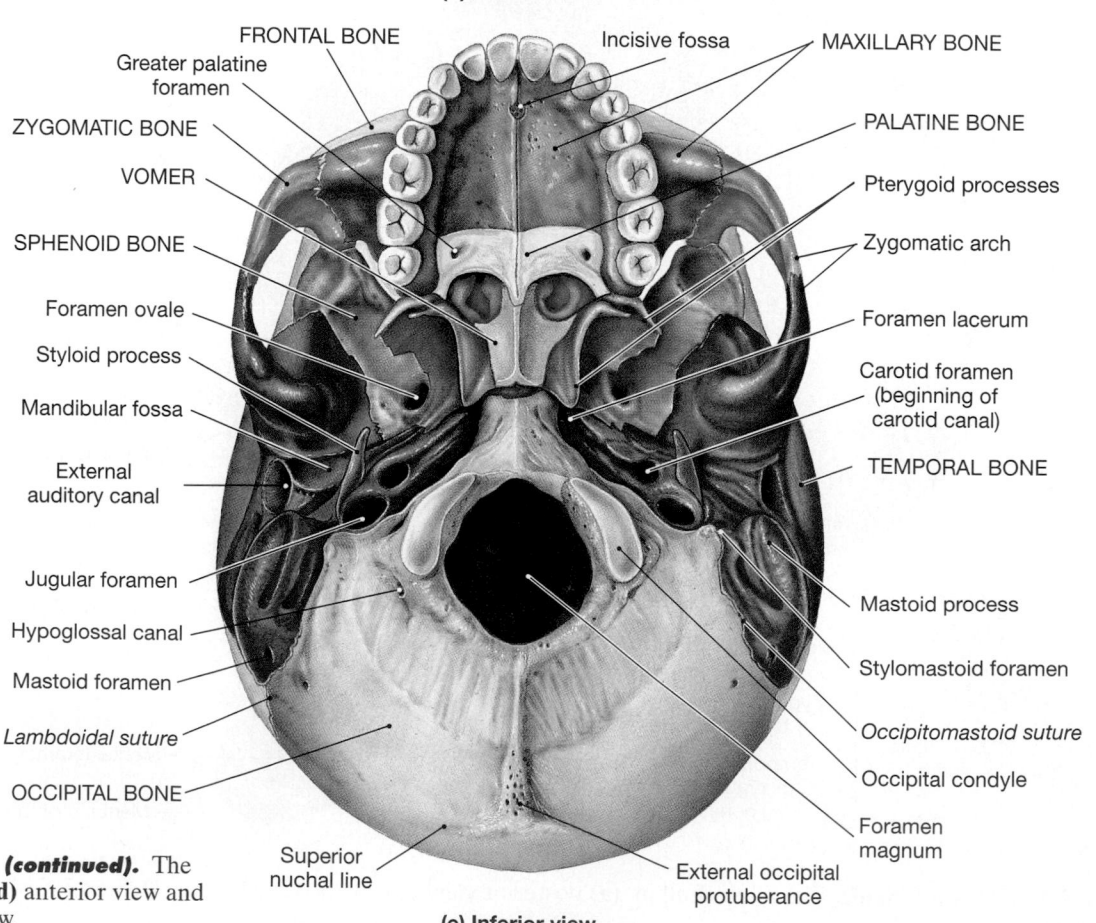

FRONTAL BONE

Incisive fossa

MAXILLARY BONE

Greater palatine foramen

ZYGOMATIC BONE

PALATINE BONE

VOMER

Pterygoid processes

SPHENOID BONE

Zygomatic arch

Foramen ovale

Foramen lacerum

Styloid process

Carotid foramen (beginning of carotid canal)

Mandibular fossa

External auditory canal

TEMPORAL BONE

Jugular foramen

Mastoid process

Hypoglossal canal

Stylomastoid foramen

Mastoid foramen

Occipitomastoid suture

Lambdoidal suture

Occipital condyle

OCCIPITAL BONE

Foramen magnum

•**FIGURE 7-3 (continued).** The adult skull in **(d)** anterior view and **(e)** inferior view.

Superior nuchal line

External occipital protuberance

(e) Inferior view

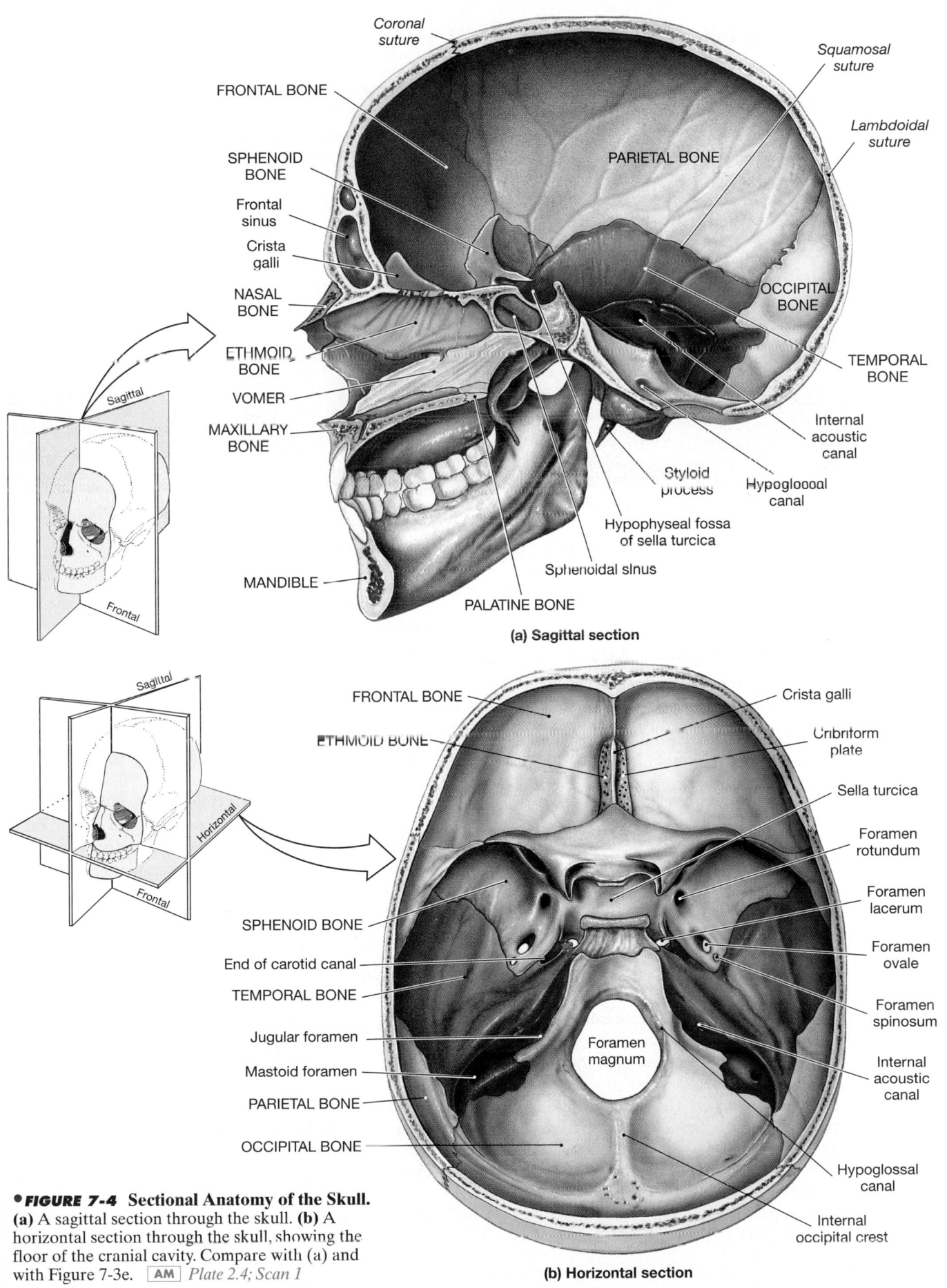

(a) Sagittal section

Coronal suture

FRONTAL BONE

SPHENOID BONE

Frontal sinus

Crista galli

NASAL BONE

ETHMOID BONE

VOMER

MAXILLARY BONE

MANDIBLE

PALATINE BONE

Sphenoidal sinus

Hypophyseal fossa of sella turcica

Styloid process

Hypoglossal canal

Internal acoustic canal

TEMPORAL BONE

OCCIPITAL BONE

Squamosal suture

Lambdoidal suture

PARIETAL BONE

Sagittal

Frontal

(b) Horizontal section

FRONTAL BONE

ETHMOID BONE

SPHENOID BONE

End of carotid canal

TEMPORAL BONE

Jugular foramen

Mastoid foramen

PARIETAL BONE

OCCIPITAL BONE

Crista galli

Cribriform plate

Sella turcica

Foramen rotundum

Foramen lacerum

Foramen ovale

Foramen spinosum

Internal acoustic canal

Hypoglossal canal

Internal occipital crest

Foramen magnum

Sagittal

Horizontal

Frontal

● **FIGURE 7-4 Sectional Anatomy of the Skull.**
(a) A sagittal section through the skull. **(b)** A
horizontal section through the skull, showing the
floor of the cranial cavity. Compare with (a) and
with Figure 7-3e. AM *Plate 2.4; Scan 1*

FOCUS   The Individual Bones of the Skull

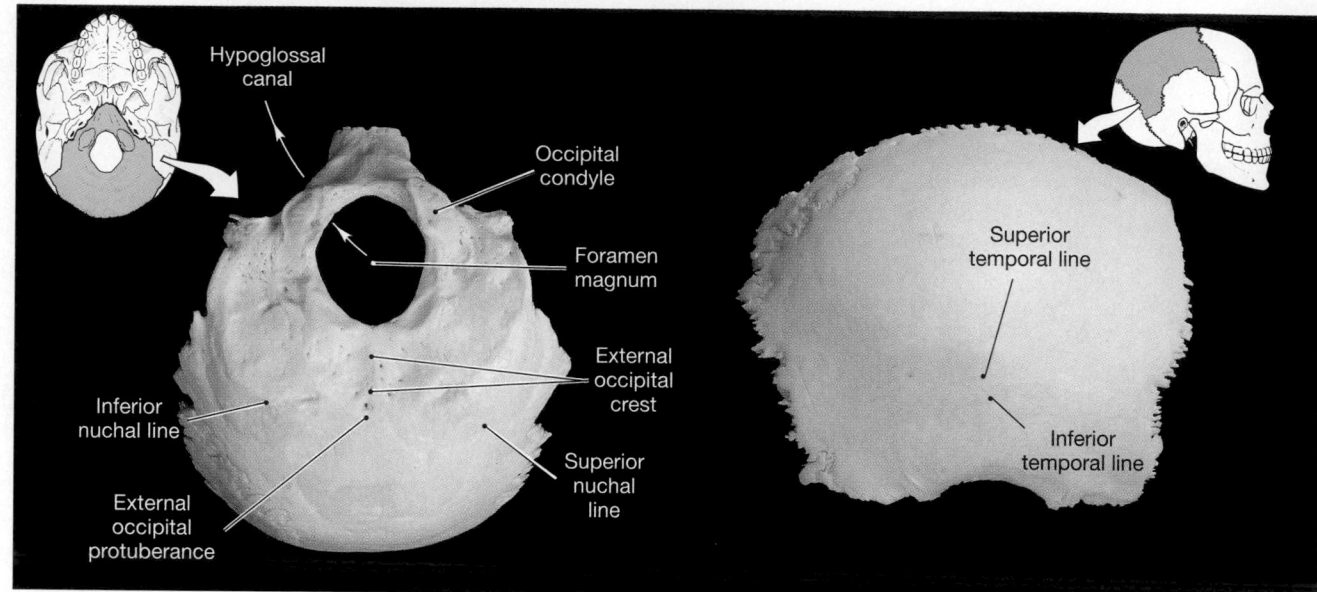

(a) Occipital bone, inferior (external) view   **(b) Right parietal bone, lateral view**

●**FIGURE 7-5** **The Occipital and Parietal Bones.** **(a)** Inferior view of the occipital bone. **(b)** Lateral view of the right parietal bone.

Bones of the Cranium

The Occipital Bone (Figure 7-5a●)

GENERAL FUNCTIONS: Forms much of the posterior and inferior surfaces of the cranium.

ARTICULATIONS: The occipital bone articulates with the parietal bones, the temporal bones, the sphenoid bone, and the first cervical vertebra (the atlas) (Figures 7-3a–c,e and 7-4●).

REGIONS/LANDMARKS: The **external occipital protuberance** is a small bump at the midline on the inferior surface.

The **occipital crest,** which begins at the external occipital protuberance, marks the attachment of a ligament that helps stabilize the neck.

The **inferior** and **superior nuchal** (NOO-kul) **lines** are ridges that intersect the crest. They mark the attachment sites of muscles and ligaments that stabilize the articulation at the occipital condyles and balance the weight of the head over the vertebrae of the neck.

The **occipital condyles** are the site of articulation between the skull and the first cervical vertebra.

The concave internal surface of the occipital bone (Figure 7-4a●) closely follows the contours of the brain. The grooves follow the paths of major vessels, and the ridges mark the attachment sites of membranes that stabilize the position of the brain.

FORAMINA: The **foramen magnum** connects the cranial cavity with the spinal cavity, which is enclosed by the vertebral column (Figure 7-4b●). The foramen magnum surrounds the connection between the brain and spinal cord.

The **jugular foramen** lies between the occipital bone and temporal bone. The *internal jugular vein* passes through this foramen, carrying venous blood from the brain.

The **hypoglossal canals** (Figure 7-3e●) begin at the lateral base of each occipital condyle and end on the inner surface of the occipital bone near the foramen magnum. The *hypoglossal nerves,* cranial nerves that control the tongue muscles, pass through these canals.

The Parietal Bones (Figure 7-5b●)

GENERAL FUNCTIONS: Form part of the superior and lateral surfaces of the cranium.

ARTICULATIONS: The parietal bones articulate with one another and with the occipital, temporal, frontal, and sphenoid bones (Figures 7-3a–d and 7-4●).

REGIONS/LANDMARKS: The **superior** and **inferior temporal lines** are low ridges that mark the attachment sites of the *temporalis muscle,* a large muscle that closes the mouth.

Grooves on the inner surface of the parietal bones mark the paths of cranial blood vessels (Figure 7-4a●).

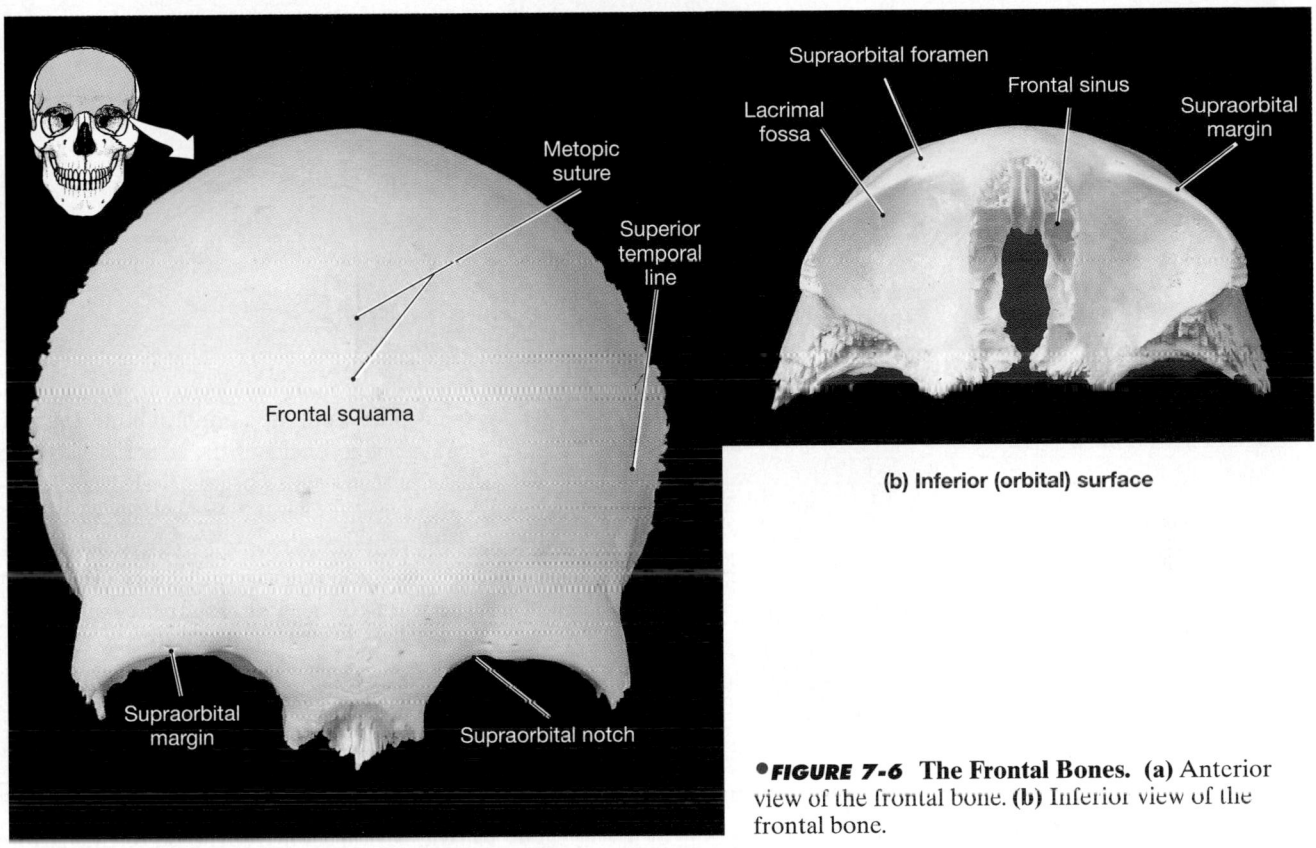

(a) Anterior surface

(b) Inferior (orbital) surface

•*FIGURE 7-6* **The Frontal Bones. (a)** Anterior view of the frontal bone. **(b)** Inferior view of the frontal bone.

The Frontal Bone (Figure 7-6a,b•)

GENERAL FUNCTIONS: Forms the anterior portion of the cranium and the roof of the orbits; mucous secretions of frontal sinuses help flush the surfaces of the nasal cavities.

ARTICULATIONS: The frontal bone articulates with the parietal, sphenoid, ethmoid, nasal, lacrimal, maxillary, and zygomatic bones (Figures 7-3b–e and 7-4•).

REGIONS/LANDMARKS: The **frontal squama**, or forehead, forms the anterior, superior portion of cranium and provides surface area for the attachment of facial muscles.

The superior temporal line is a continuation of that feature on the parietal bone.

The **supraorbital margin** helps protect the eye.

The **lacrimal fossa** on the inner surface of the orbit is a shallow depression that marks the location of the *lacrimal* (tear) *gland,* which lubricates the surface of the eye.

The **frontal sinuses** are extremely variable in size and time of appearance. They generally appear after age 6, but some people never develop them. We will describe the frontal sinuses and other sinuses of the cranium and face in a later section.

FORAMINA: The **supraorbital foramen** provides passage for blood vessels that supply the eyebrow, eyelids, and frontal sinuses. These structures are not always completely enclosed by bone; when not, there is a **supraorbital notch,** rather than a foramen, on the orbital rim.

REMARKS: During development, the bones of the cranium form by the fusion of separate centers of ossification. At birth, the fusions have not been completed—there are two frontal bones that articulate along the **metopic suture.** Although the suture generally disappears by age 8 as the bones fuse, the adult skull commonly retains traces of the suture line. This suture, or what remains of it, runs down the center of the frontal squama.

The center of each supraorbital margin contains either a single supraorbital foramen or a supraorbital notch, marking the path of blood vessels that supply the eyebrow, eyelids, and frontal sinuses.

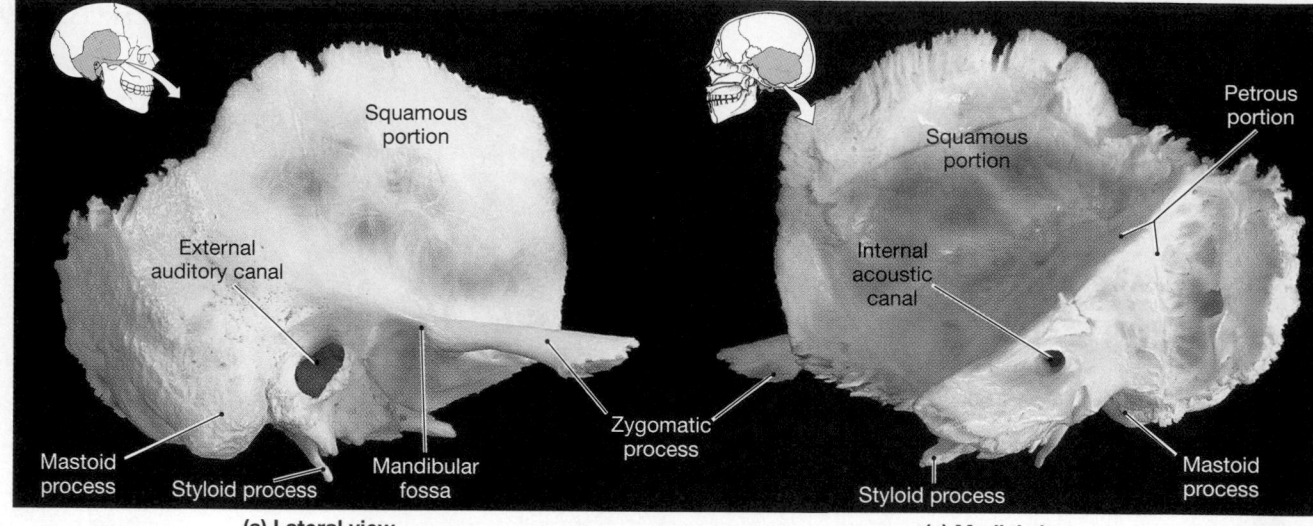

Squamous portion

External auditory canal

Mastoid process   Styloid process   Mandibular fossa   Zygomatic process

(a) Lateral view

Squamous portion

Internal acoustic canal

Petrous portion

Styloid process   Mastoid process

(c) Medial view

(b) Mastoid air cells

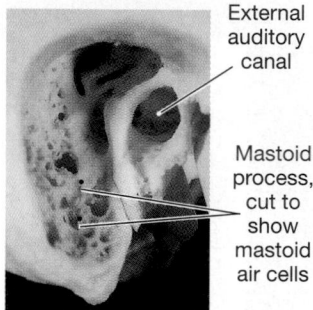

External auditory canal

Mastoid process, cut to show mastoid air cells

• **FIGURE 7-7** The Temporal Bone.
(a) Lateral view of the right temporal bone.
(b) A cutaway view of the mastoid air cells.
(c) Medial view of the right temporal bone.

The Temporal Bones (Figure 7-7•)

GENERAL FUNCTIONS: Forms part of both the lateral walls of the cranium and the *zygomatic arch;* forms the only articulation between the mandible and other facial bones; surrounds and protects the sense organs of the inner ear; acts as an attachment site for muscles that close the jaws and move the head.

ARTICULATIONS: The temporal bones articulate with the zygomatic, sphenoid, parietal, and occipital bones and with the mandible (Figures 7-3 and 7-4•).

REGIONS/LANDMARKS: The **squamous portion,** or *squama,* of the **temporal bone** is the convex, irregular surface that borders the squamosal suture.

The **zygomatic process,** inferior to the squamous portion, is attached to the **temporal process** of the zygomatic bone. Together these processes form the **zygomatic arch,** or *cheekbone.*

The **mandibular fossa** on the inferior surface marks the site of articulation with the mandible.

The **mastoid process** is an attachment site for muscles that rotate or extend the head.

Mastoid air cells within the mastoid process are connected to the middle ear cavity.

The **styloid** (STĪ-loyd; *stylos,* pillar) **process,** near the base of the mastoid process, is attached to ligaments that support the hyoid bone and to the tendons of several muscles associated with the hyoid bone, the tongue, and the pharynx.

The **petrous portion** of the temporal bone, located on its internal surface, encloses the structures associated with the *inner ear*—sense organs that provide information about hearing and balance. The **auditory ossicles** are located in the *tympanic cavity* or *middle ear,* a cavity within the petrous portion of the temporal bone. These tiny bones—three on each side—transfer sound vibrations from the delicate *tympanic membrane,* or *eardrum,* to the inner ear. (We shall discuss each bone and its role in hearing in Chapter 17.)

FORAMINA (Figures 7-4e and 7-7•): The jugular foramen, between the temporal and occipital bones, provides passage for the internal jugular vein.

The **carotid foramen,** the entrance to the *carotid canal,* provides passage for the internal carotid artery, a major artery to the brain.

The **foramen lacerum** (LA-se-rum; *lacerare,* to tear) is a jagged slit extending between the occipital and temporal bones. It contains hyaline cartilage and small arteries that supply the inner surface of the cranium.

The **external auditory canal** on the lateral surface ends at the eardrum, which disintegrates during the preparation of a dried skull.

The **mastoid foramen** penetrates the temporal bone and begins near the base of the mastoid process. Blood vessels travel through this passageway to reach the membranes surrounding the brain.

The **stylomastoid foramen** lies posterior to the base of the styloid process. The *facial nerve* passes through this foramen to control the facial muscles.

The *auditory tube,* or *pharyngotympanic tube,* begins at the articulation between the temporal bone and the sphenoid bone. Also known as the *Eustachian* (ū-STĀ-kē-an) *tube,* it ends at the tympanic cavity.

The **internal acoustic canal** begins on the medial surface of the petrous portion of the temporal bone. It carries blood vessels and nerves to the inner ear, and the facial nerve to the stylomastoid foramen.

REMARKS: If pathogens invade the mastoid air cells, *mastoiditis* develops. Symptoms include severe earaches, fever, and swelling behind the ear.

The Sphenoid Bone (Figure 7-8a,b●)

GENERAL FUNCTIONS: Forms part of the floor of the cranium; unites the cranial and facial bones; acts as a cross-brace that strengthens the sides of the skull; mucous secretions of *sphenoidal sinuses* help flush surfaces of nasal cavities.

ARTICULATIONS: The sphenoid bone articulates with the frontal, occipital, parietal, ethmoid, and temporal bones of the cranium and the palatine bones, zygomatic bones, maxillary bones, and vomer of the face (Figures 7-3c–e and 7-4●).

REGIONS/LANDMARKS: The shape of the sphenoid bone has been compared to a bat with its wings extended.

The **body** forms the central axis of the sphenoid bone.

The **greater wings** extend laterally from the body as part of the cranial floor. Vertical extensions of each wing contribute to the posterior wall of each orbit.

The **lesser wings** extend horizontally anterior to the sella turcica.

The **sella turcica** (TUR-si-kuh), which supposedly resembles a "Turkish saddle," is a bony enclosure on the superior surface of the body.

The **hypophyseal** (hī-pō-FIZ-ē-ul) **fossa** is the depression within the sella turcica. The *pituitary gland* sits within this fossa.

The **sphenoidal sinuses** are on either side of the body, inferior to the sella turcica.

The **pterygoid** (TER-i-goyd; *pterygion,* wing) **processes** are vertical projections that begin at the boundary between the greater wings and the body. Each process forms a pair of *pterygoid plates,* which are important as attachment sites for muscles that move the lower jaw and soft palate.

FORAMINA: The *optic canals* permit passage of the optic nerves from the eyes to the brain.

The **superior orbital fissure, foramen rotundum, foramen ovale** (ō-VAH-lē), and **foramen spinosum** penetrate the greater wings. These passages carry blood vessels and nerves to the orbit, face, and jaws.

REMARKS: Although this bone is relatively large, much of it is hidden by more superficial bones.

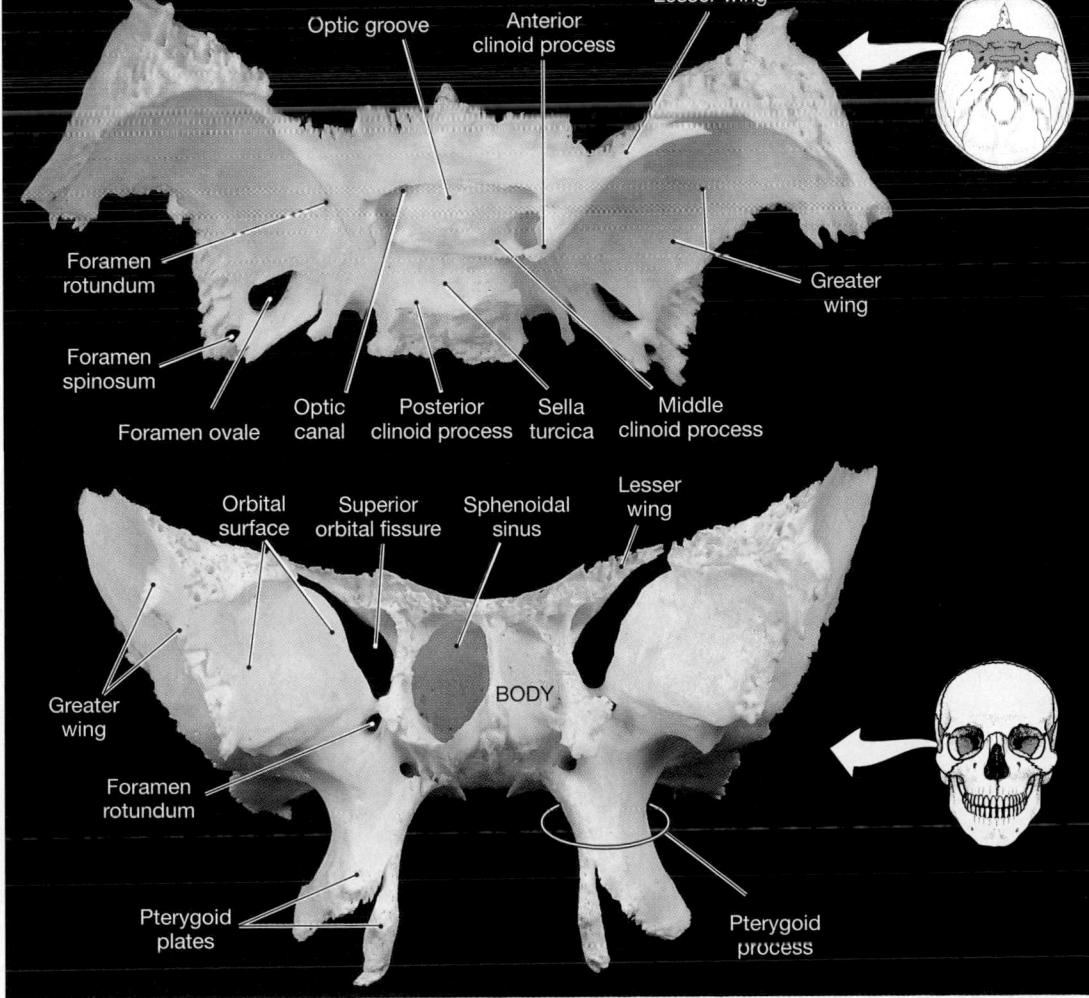

●**FIGURE 7-8**
The Sphenoid Bone. (a) The superior and (b) anterior surfaces of the sphenoid bone.

Optic groove

Anterior clinoid process

Lesser wing

Foramen rotundum

Foramen spinosum

Foramen ovale

Optic canal

Posterior clinoid process

Sella turcica

Middle clinoid process

Greater wing

(a) Superior surface

Orbital surface

Superior orbital fissure

Sphenoidal sinus

Lesser wing

Greater wing

Foramen rotundum

BODY

Pterygoid plates

Pterygoid process

(b) Anterior surface

The Ethmoid Bone (Figure 7-9a,b•)

GENERAL FUNCTIONS: Forms the anteromedial floor of the cranium, the roof of the nasal cavity, and part of the nasal septum; contributes to the medial orbital wall; mucous secretions from the ethmoidal sinuses flush the surfaces of the nasal cavities.

ARTICULATIONS: The ethmoid bone articulates with the frontal and sphenoid bones of the cranium and with the nasal, lacrimal, palatine, and maxillary bones and the inferior nasal conchae and vomer of the face (Figures 7-3c and 7-4•).

REGIONS/LANDMARKS: The ethmoid has three parts: the *cribriform plate*, the paired *lateral masses*, and the *perpendicular plate*.

1. The **cribriform plate** (*cribrum*, sieve) forms the anteromedial floor of the cranium and the roof of the nasal cavity.

The **crista galli** (*crista*, crest + *gallus*, chicken; cock's comb) is a bony ridge that projects superior to the cribriform plate. The *falx cerebri*, a membrane that stabilizes the position of the brain, attaches to this ridge.

2. The **lateral masses** contain the **ethmoidal sinuses**, or *ethmoidal air cells*, which open into the nasal cavity on each side.

The **superior nasal conchae** (KONG-kē; singular, *concha*, a snail shell) and the **middle nasal conchae** are delicate projections of the lateral masses.

3. The **perpendicular plate** forms part of the nasal septum, along with the vomer and a piece of hyaline cartilage.

FORAMINA: The **olfactory foramina** in the cribriform plate permit passage of the olfactory nerves, which provide your sense of smell.

REMARKS: *Olfactory* (smell) *receptors* are located in the epithelium that covers the inferior surfaces of the cribriform plate, the medial surfaces of the superior nasal conchae, and the superior portion of the perpendicular plate.

The nasal conchae break up the airflow in the nasal cavity, creating swirls, turbulence, and eddies that have three major functions: (1) The swirling throws any particles in the air against the sticky mucus that covers the walls of the nasal cavity; (2) the turbulence slows air movement, providing time for warming, humidification, and dust removal before the air reaches more delicate portions of the respiratory tract; and (3) the eddies direct air toward the superior portion of the nasal cavity, adjacent to the cribriform plate, where the olfactory receptors are located.

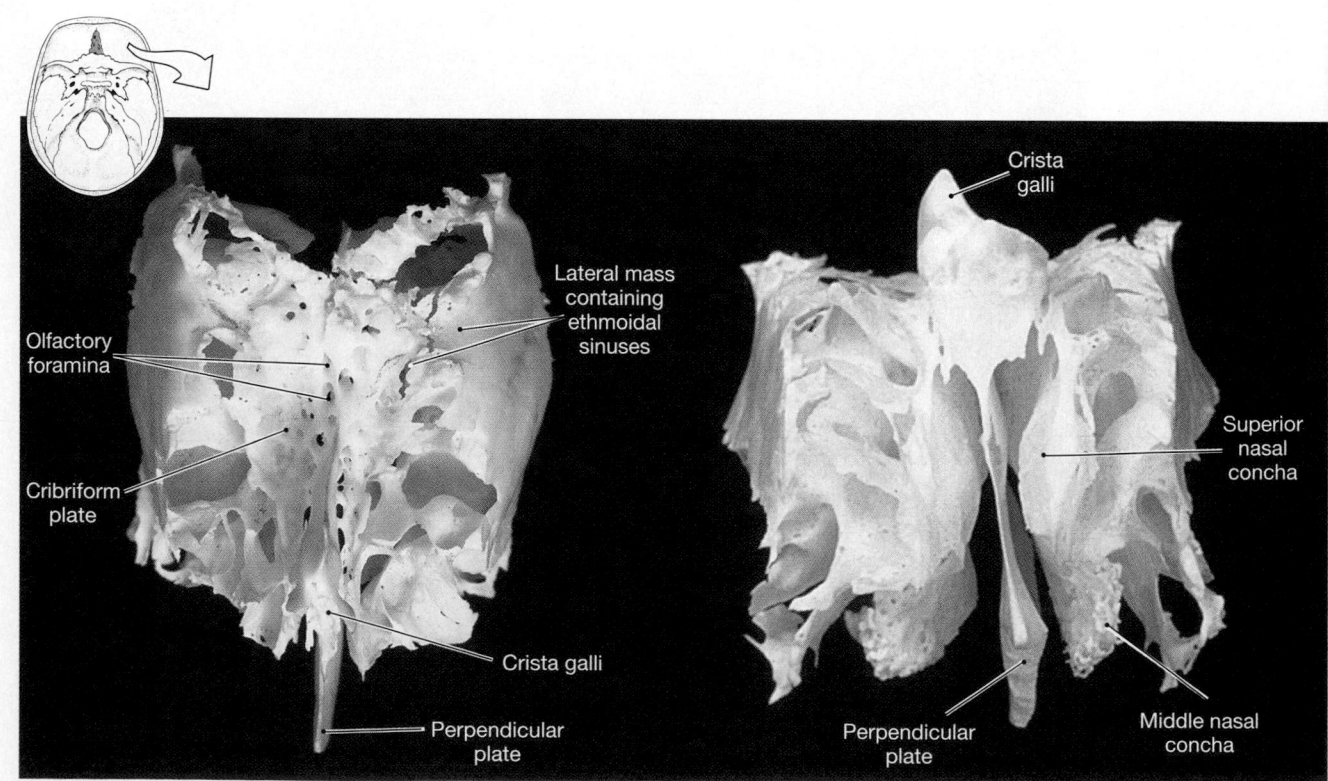

(a) Superior surface

Olfactory foramina
Cribriform plate
Lateral mass containing ethmoidal sinuses
Crista galli
Perpendicular plate

(b) Posterior surface

Crista galli
Superior nasal concha
Perpendicular plate
Middle nasal concha

•**FIGURE 7-9** **The Ethmoid Bone. (a)** The superior and **(b)** posterior surfaces of the ethmoid bone.

Bones of the Face

The Maxillary Bones (Figure 7-10a,b•)

GENERAL FUNCTIONS: Form the inferior orbital rim, the lateral margins of the external nares, the upper jaw, and most of the hard palate; support the teeth; maxillary sinuses produce mucous secretions that flush the inferior surfaces of the nasal cavities.

ARTICULATIONS: The maxillary bones, or **maxillae**, articulate with the frontal and ethmoid bones, with one another, and with all the other facial bones except the mandible (Figures 7-3c–e and 7-4a•).

REGIONS/LANDMARKS: The **orbital rim** protects the eye and other structures in the orbit.

The **alveolar processes** that border the mouth support the upper teeth.

The **palatine processes** form most of the **hard palate,** or bony roof of the mouth.

The **maxillary sinuses** lighten the portion of the maxillary bone superior to the teeth.

The **lacrimal fossa,** formed by the maxillary and lacrimal bones, protects the *lacrimal sac* and the *nasolacrimal duct,* which carries tears from the orbit to the nasal cavity.

The **incisive fossa** is an indentation in the anterior, inferior portion of the hard palate.

FORAMINA: The **infraorbital foramen** marks the entrance of a major sensory nerve from the face; that nerve exits the skull through the foramen rotundum of the sphenoid bone.

The **inferior orbital fissure,** which lies between the maxillary bone and the sphenoid bone, permits passage of cranial nerves and blood vessels.

REMARKS: These are the largest facial bones; the maxillary sinuses are the largest sinuses. One type of *cleft palate,* a developmental disorder, results when the maxillary bones fail to meet along the midline of the hard palate. (See the Embryology Summary on p. 217.)

The Palatine Bones (Figure 7-10b,c•)

GENERAL FUNCTIONS: Form the posterior portion of the hard palate and contribute to the floor of each orbit.

ARTICULATIONS: The palatine bones articulate with one another, with the maxillary, sphenoid, and ethmoid bones, with the inferior nasal conchae, and with the vomer (Figures 7-3e and 7-4a•).

REGIONS/LANDMARKS: These bones are L-shaped.

The **horizontal plate** forms the posterior portion of the hard palate, and the **vertical plate** extends from the horizontal plate to the *orbital process.*

The **orbital process** forms part of the floor of the orbit. This process contains a small sinus that normally opens into the sphenoidal sinus.

FORAMINA: The **greater palatine foramen** permits passage of the *greater palatine nerve,* a sensory nerve supplying the roof of the mouth.

The **lesser palatine foramina** permit passage of the *lesser palatine nerves,* smaller branches of the greater palatine nerve.

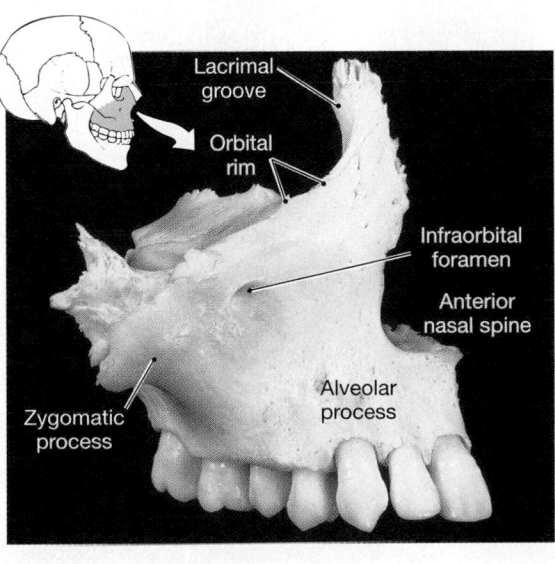

(a) Lateral surface

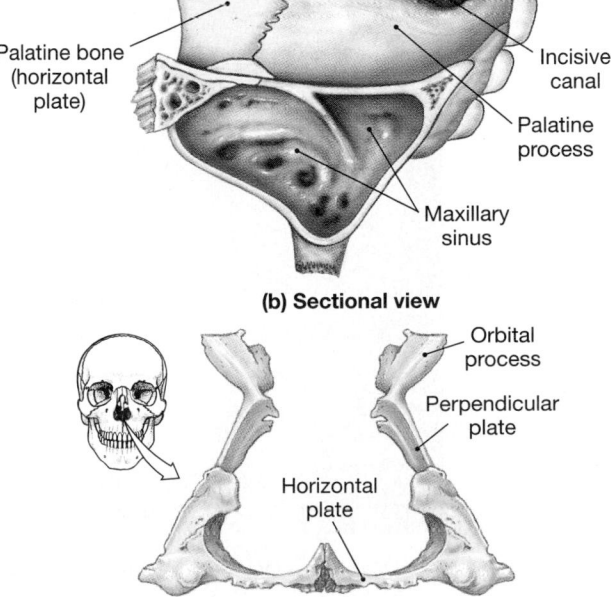

(b) Sectional view

(c) Anterior view

•**FIGURE 7-10** **The Maxillary and Palatine Bones.** The right maxillary bone. **(a)** Anterolateral view. **(b)** Horizontal section; note the size and orientation of the maxillary sinus. **(c)** Anterior view of the two palatine bones.

The Nasal Bones (Figure 7-11•)

GENERAL FUNCTIONS: Support the superior portion of the bridge of the nose and form the superior border of the **external nares** (NA-rēz), the entrance to the nasal cavity.

ARTICULATIONS: The paired nasal bones articulate with one another and with the frontal, ethmoid, and maxillary bones (Figures 7-3c,d and 7-4a•).

The Vomer (Figure 7-11•)

GENERAL FUNCTIONS: Forms the inferior portion of the bony nasal septum.

ARTICULATIONS: The vomer articulates with the sphenoid, ethmoid, palatine, and maxillary bones and with the cartilaginous portion of the nasal septum, which extends into the fleshy portion of the nose (Figures 7-3d and 7-4a•).

The Inferior Nasal Conchae (Figure 7-11•)

GENERAL FUNCTIONS: Create turbulence in air passing through the nasal cavity; increase epithelial surface area to promote warming and humidification of inhaled air.

ARTICULATIONS: The inferior nasal conchae articulate with the ethmoid, maxillary, palatine, and lacrimal bones (Figure 7-3d•).

The Zygomatic Bones (Figure 7-11•)

GENERAL FUNCTIONS: Contribute to the rim and lateral wall of the orbit and form part of the zygomatic arch.

ARTICULATIONS: The zygomatic bone articulates with the frontal, temporal, sphenoid, and maxillary bones (Figure 7-3c•).

REGIONS/LANDMARKS: The temporal process curves posteriorly to meet the zygomatic process of the temporal bone.

FORAMINA: The **zygomaticofacial foramen** on the anterior surface of each zygomatic bone carries a sensory nerve that innervates the cheek.

The Lacrimal Bones (Figure 7-11•)

GENERAL FUNCTIONS: Form part of medial wall of orbit.

ARTICULATIONS: The lacrimal bones articulate with the frontal, maxillary, and ethmoid bones (Figure 7-3c,d•).

REGIONS/LANDMARKS: The **lacrimal groove** is a depression along the edge of the lacrimal bone where it articulates with the maxillary bone to form the nasolacrimal canal, which opens into the nasal cavity.

REMARKS: These are the smallest of the facial bones.

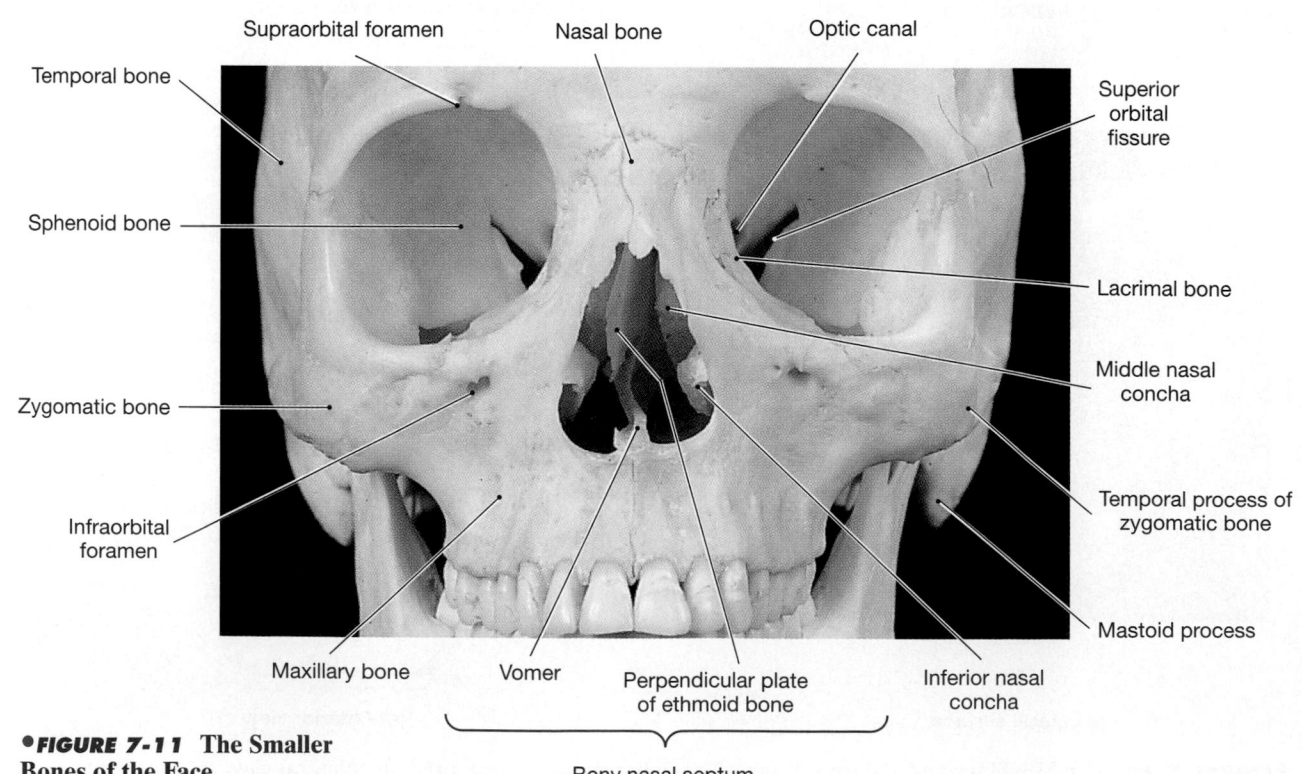

•**FIGURE 7-11** The Smaller Bones of the Face

The Mandible (Figure 7-12a,b•)

GENERAL FUNCTIONS: The mandible forms the lower jaw.

ARTICULATIONS: The mandible articulates with the mandibular fossae of the temporal bones (Figures 7-3c and 7-3e•).

REGIONS/LANDMARKS: The **mandibular body** is the horizontal portion of the mandible.

The **alveolar processes** support the teeth.

The **mental protuberance**, or *chin*, is the attachment site for several facial muscles.

A prominent depression on the medial surface marks the position of the *submandibular salivary gland*.

The **mylohyoid line** marks the insertion of the *mylohyoid muscle*, which supports the floor of the mouth.

The **mandibular ramus** is the ascending portion of the mandible that begins at the *mandibular angle* on either side. On each ramus:

The **condylar process** articulates with the temporal bone at the *temporomandibular joint*.

The **coronoid** (ko-RŌ-noyd) **process** is the insertion point for the *temporalis muscle*, a powerful muscle that closes the jaws.

The **mandibular notch** is the depression that separates the condylar and coronoid processes.

FORAMINA: The **mental foramina** (*mentalis*, chin) are openings for nerves that carry sensory information from the lips and chin to the brain.

The **mandibular foramen** is the entrance to the *mandibular canal*, a passageway for blood vessels and nerves that service the lower teeth. Before they work on the lower teeth, dentists typically anesthetize the sensory nerve that enters this canal.

The Hyoid Bone (Figure 7-12c•)

GENERAL FUNCTIONS: Support of larynx; attachment site for muscles of the larynx, pharynx, and tongue.

ARTICULATIONS: *Stylohyoid ligaments* connect the *lesser cornua* to the styloid processes of the temporal bones.

REGIONS/PROCESSES: The **body** is an attachment site for muscles of larynx, tongue, and pharynx.

The **greater cornua** (*cornua*, horns; singular, *cornu*) help support the larynx and are attached to muscles that move the tongue.

The **lesser cornua** are attached to the stylohyoid ligaments; from these ligaments, the hyoid and larynx hang beneath the skull like a child's swing from the limb of a tree.

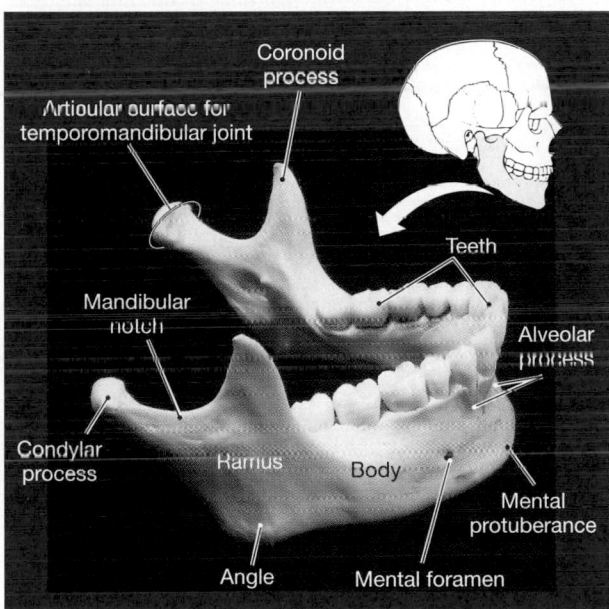

(a) Lateral view

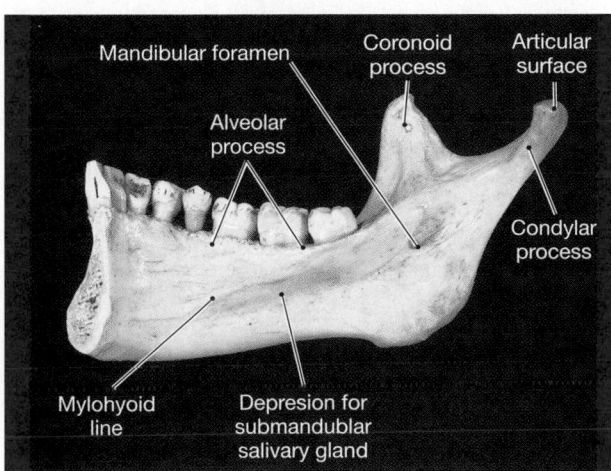

(b) Medial view

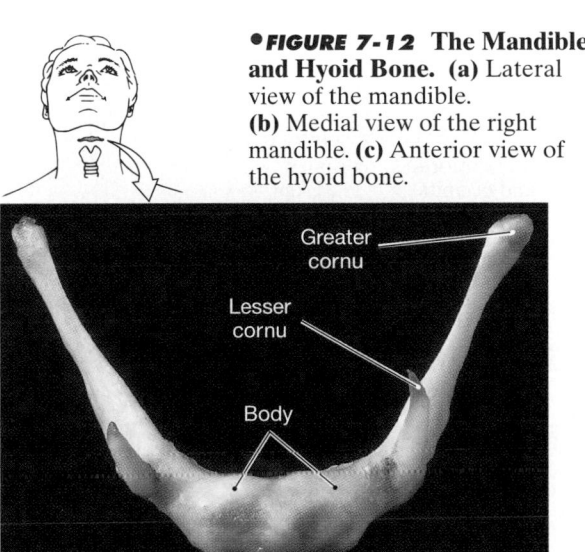

•**FIGURE 7-12** The Mandible and Hyoid Bone. **(a)** Lateral view of the mandible. **(b)** Medial view of the right mandible. **(c)** Anterior view of the hyoid bone.

(c) Anterior view

A Summary of the Foramina and Fissures of the Skull

Table 7-1 summarizes information about the foramina and fissures introduced thus far. This reference will be especially important to you in later chapters when you deal with the nervous and cardiovascular systems.

TABLE 7-1 A Key to the Foramina and Fissures of the Skull

Bone	Foramen/Fissure	Major Structures Using Passageway	
		Neural Tissue*	Vessels and Other Structures
Occipital bone	Foramen magnum	Medulla oblongata (most caudal portion of brain) and accessory nerve (XI)	Vertebral arteries; supporting membranes around central nervous system
	Hypoglossal canal	Hypoglossal nerve (XII) provides motor control to muscles of the tongue	
With temporal bone	Jugular foramen	Glossopharyngeal nerve (IX), vagus nerve (X), accessory nerve (XI). Nerve IX provides taste sensation; X is important for visceral functions; XI innervates important muscles of the back and neck	Internal jugular vein; important vein returning blood from brain to heart
Frontal bone	Supraorbital foramen (or notch)	Supraorbital nerve, sensory branch of ophthalmic nerve, innervating the eyebrow, eyelid, and frontal sinus	Supraorbital artery delivers blood to same region
Lacrimal bone	Lacrimal groove, nasolacrimal canal (with maxillary bone)		Tear duct; drains into the nasal cavity
Temporal bone	Mastoid foramen		Vessels to membranes around central nervous system
	Stylomastoid foramen	Facial nerve (VII) provides motor control of facial muscles	
	Carotid foramen		Internal carotid artery; supplying blood to the brain
	External auditory canal		Air in canal conducts sound to eardrum
	Internal acoustic canal	Vestibulocochlear nerve (VIII) from sense organs for hearing and balance. Facial nerve (VII) enters here, exits at stylomastoid foramen	Internal acoustic artery to inner ear
Sphenoid bone	Optic canal	Optic nerve (II) brings information from the eye to the brain	Ophthalmic artery brings blood into orbit
	Superior orbital fissure	Oculomotor nerve (III), trochlear nerve (IV), ophthalmic branch of trigeminal nerve (V), abducens nerve (VI). Ophthalmic nerve provides sensory information about eye and orbit; other nerves control muscles that move the eye	Ophthalmic vein returns blood from orbit
	Foramen rotundum	Maxillary branch of trigeminal nerve (V) provides sensation from the face	
	Foramen ovale	Mandibular branch of trigeminal nerve (V) controls the muscles that move the lower jaw and provides sensory information from that area	
	Foramen spinosum		Vessels to membranes around central nervous system
With temporal and occipital bones	Foramen lacerum		Internal carotid artery leaves carotid canal, enters cranium via foramen lacerum
With maxillary bone	Inferior orbital fissure	Maxillary branch of trigeminal nerve (V). *See Foramen rotundum.*	
Ethmoid bone	Cribriform plate	Olfactory nerve (I) provides sense of smell	
Maxillary bone	Infraorbital foramen	Infraorbital nerve, maxillary branch of trigeminal nerve (V) from the inferior orbital fissure to face	Infraorbital artery with same distribution
Mandible	Mental foramen	Mental nerve, sensory branch of the mandibular nerve, provides sensation from the chin and lips	Mental vessels to chin and lips
	Mandibular foramen	Inferior alveolar nerve, sensory branch of mandibular nerve, provides sensation from the gums, teeth	Inferior alveolar vessels supply same region
Zygomatic bone	Zygomaticofacial foramen	Zygomaticofacial nerve, sensory branch of mandibular nerve to cheek	

* There are 12 pairs of cranial nerves, numbered I–XII. Their functions and distribution are detailed in Chapter 14, pp. 474–484.

TMJ SYNDROME The *temporomandibular joint* (*TMJ*) is quite mobile; your jaw moves while you chew or talk. The disadvantage of such mobility is that your jaw can easily be dislocated by forceful forward or lateral displacement. The connective tissue sheath, or *capsule*, that surrounds the joint is relatively loose, and the opposing bone surfaces are separated by a fibrocartilage pad. In **TMJ syndrome**, or *myofacial pain syndrome*, the mandible is pulled slightly out of alignment, generally by spasms in one of the jaw muscles. The individual experiences (1) facial pain that radiates around the ear on the affected side and (2) an inability to open the mouth fully.

TMJ syndrome is a repeating cycle of muscle spasm → misalignment → pain → muscle spasm. It has been linked to involuntary behaviors, such as grinding of the teeth during sleep, and to emotional stress. Treatment focuses on breaking the cycle of muscle spasm and pain and, when necessary, providing emotional support. The application of heat to the affected joint, coupled with the use of anti-inflammatory drugs, local anesthetics, or both, may help. If teeth grinding is suspected, special mouth guards may be worn at night.

☑ In which bone is the foramen magnum located?

☑ Tomás suffers a blow to the skull that fractures the right superior lateral surface of his cranium. Which bone is fractured?

☑ The internal jugular veins are important blood vessels of the head. Between which bones do these blood vessels pass?

☑ Which bone contains the depression called the sella turcica? What is located in this depression?

☑ Which of the five senses would be affected if the cribriform plate of the ethmoid bone failed to form?

The Orbital and Nasal Complexes

Together, the cranial bones and facial bones form the *orbital complex,* which surrounds each eye, and the *nasal complex,* which surrounds the nasal cavities.

The **orbits** are the bony recesses that contain the eyes. Each orbit is made up of the seven bones of the **orbital complex** (Figure 7-13●). The frontal bone forms the roof, and the maxillary bone provides most of the orbital floor. The orbital rim and the first portion of the medial wall are formed by the maxillary bone, the lacrimal bone, and the lateral mass of the ethmoid bone. The lateral mass articulates with the sphenoid bone and a small process of the palatine bone. Several prominent foramina and fissures penetrate the sphenoid or lie between it and the maxillary bone. Laterally, the sphenoid bone and maxillary bone articulate with the zygomatic bone, which forms the lateral wall and rim of the orbit.

The **nasal complex** includes the bones that enclose the nasal cavities and the **paranasal sinuses**, air-filled chambers connected to the nasal cavities. The frontal, sphenoid, and ethmoid bones form the superior wall of the nasal cavities. The lateral walls are formed by the maxillary and lacrimal bones, the ethmoid bone (the superior and middle nasal conchae), and the inferior nasal conchae (Figure 7-14●). Much of the anterior margin of the nasal cavity is formed by the soft tissues of the nose, but the bridge of the nose is supported by the maxillary and nasal bones.

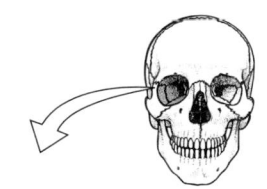

●**FIGURE 7-13** **The Orbital Complex.** The right orbital region. AM *Plate 2.3*

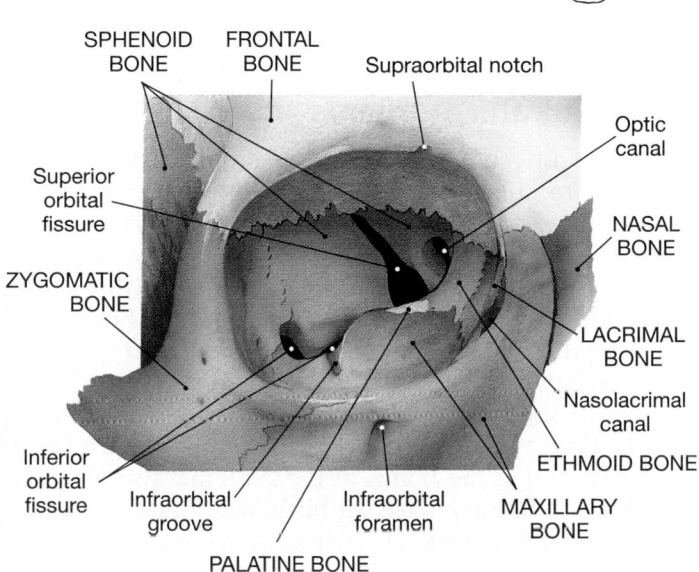

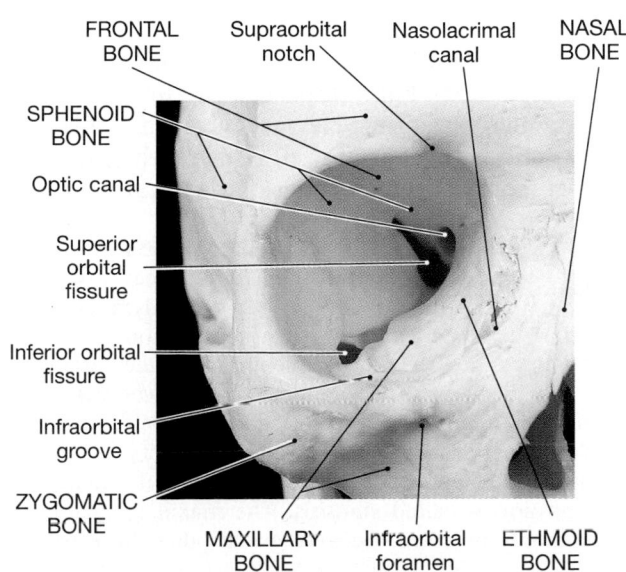

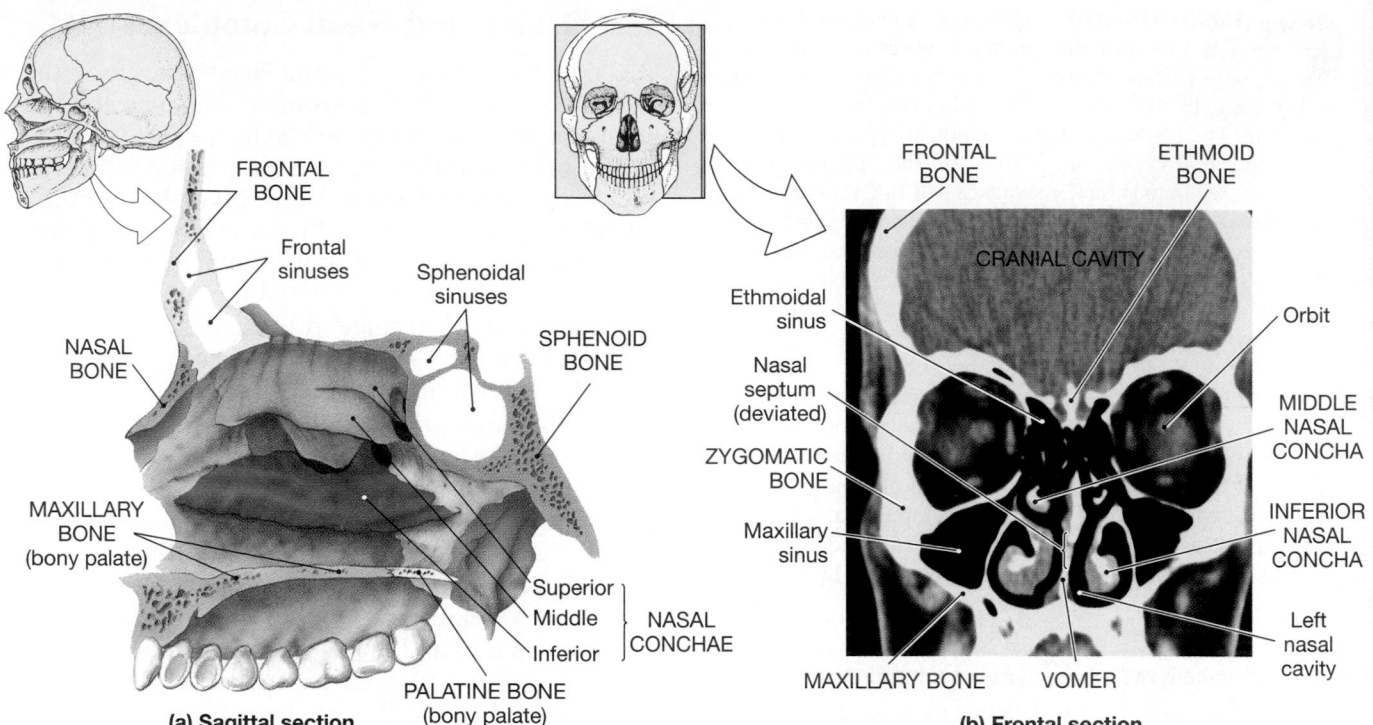

(a) Sagittal section

(b) Frontal section

•*FIGURE 7-14* **The Nasal Complex.** **(a)** Sagittal section through the skull, with the nasal septum removed to show major features of the wall of the right nasal cavity. **(b)** MRI scan showing a frontal section through the ethmoidal and maxillary sinuses, part of the paranasal sinuses. (The sphenoidal sinus is visible in (a).) [AM] *Scans 1, 2*

Paranasal Sinuses

The frontal, sphenoid, ethmoid, palatine, and maxillary bones contain the paranasal sinuses. Figure 7-14a• shows the location of the frontal and sphenoidal sinuses. The ethmoidal and maxillary sinuses are shown in Figure 7-14b•. (The tiny palatine sinuses, not shown, generally open into the sphenoidal sinus.) The sinuses lighten the various skull bones and provide an extensive area of mucous epithelium. The mucous secretions are released into the nasal cavities. The ciliated epithelium passes the mucus back toward the throat, where it is eventually swallowed. Incoming air is humidified and warmed as it flows across this thick carpet of mucus. Foreign particulate matter, such as dust or microorganisms, becomes trapped in this sticky mucus and is then swallowed. This mechanism helps protect more delicate portions of the respiratory tract.

SINUS PROBLEMS AND SEPTAL DEFECTS Flushing the nasal epithelium with mucus produced in the paranasal sinuses often succeeds in removing a mild irritant. But a viral or bacterial infection produces an inflammation of the mucous membrane of the nasal cavity. As swelling occurs, the communicating passageways narrow. Drainage of mucus slows, congestion increases, and the individual experiences headaches and a feeling of pressure within the facial bones. This condition of sinus inflammation and congestion is called **sinusitis.** The maxillary sinuses are commonly involved. Because gravity does little to assist mucus drainage from these sinuses, the effectiveness of the flushing action is reduced, and pressure on the sinus walls typically increases.

Temporary sinus problems may accompany allergies or the exposure of the mucous epithelium to chemical irritants or invading microorganisms. Chronic sinusitis may occur as the result of a **deviated** (nasal) **septum.** In this condition, the nasal septum has a bend in it, generally at the junction between the bony and cartilaginous regions. Septal deviation often blocks drainage of one or more sinuses, producing chronic cycles of infection and inflammation. A deviated septum can result from developmental abnormalities or from injuries to the nose. The condition can usually be corrected or improved by surgery.

The Skulls of Infants and Children

Many different centers of ossification are involved in the formation of the skull. As development proceeds, the centers fuse, producing a smaller number of composite bones. For example, the sphenoid bone begins as 14 separate ossification centers. At birth, fusion has not been completed: There are two frontal bones, four occipital bones, and several sphenoid and temporal elements.

The skull organizes around the developing brain. As the time of birth approaches, the brain enlarges rapidly. Although the bones of the skull are also growing, they fail to keep pace. At birth, the cranial bones are connected by areas of fibrous connective tissue. These

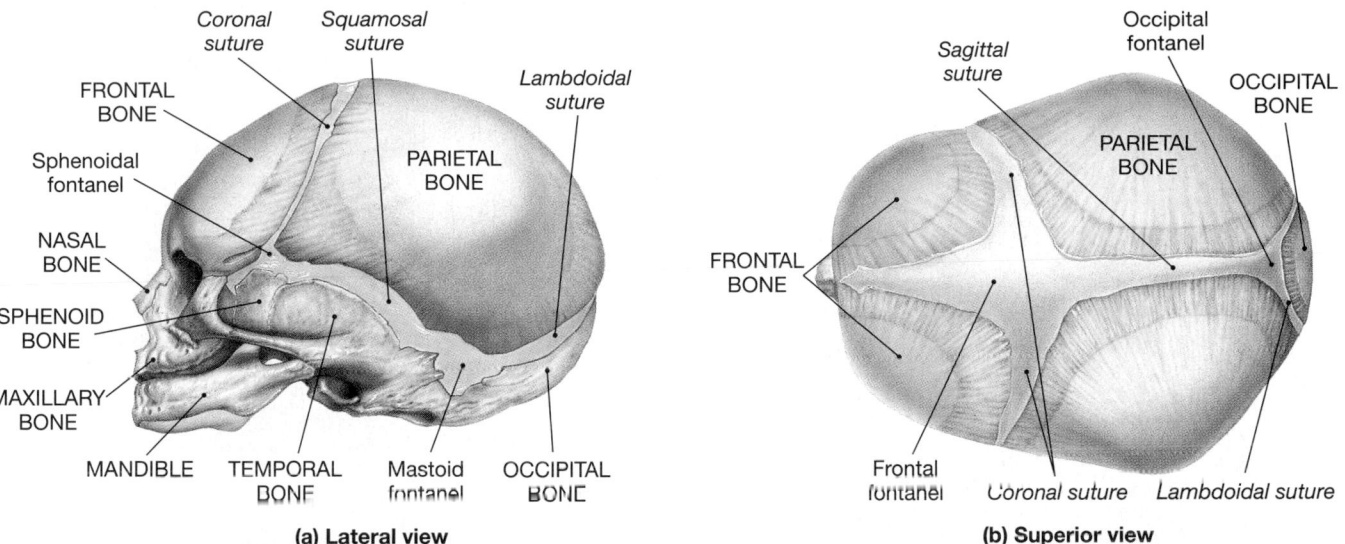

(a) Lateral view

(b) Superior view

•*FIGURE 7-15* **The Skull of an Infant. (a)** Lateral view. The skull of an infant contains more individual bones than that of an adult. Many of the bones will eventually fuse; thus the adult skull will have fewer bones. The flat bones of the skull are separated by areas of fibrous connective tissue, allowing for cranial expansion and the distortion of the skull during birth. The large fibrous areas are called fontanels. By about age 4, these areas will disappear. **(b)** Superior view. AM *Plates 2.5*

connections are quite flexible, and the skull can be distorted without damage. Such distortion normally occurs during delivery and eases the passage of the infant through the birth canal. The fibrous areas between the cranial bones are known as **fontanels** (fon-tuh-NELZ; sometimes spelled *fontanelles*) (Figure 7-15•):

- The *frontal fontanel* is the largest fontanel. It lies at the intersection of the frontal, sagittal, and coronal sutures.
- The *occipital fontanel* is at the junction between the lambdoidal and sagittal sutures.
- The *sphenoidal fontanels* are at the junctions between the squamosal sutures and the coronal suture.
- The *mastoid fontanels* are at the junctions between the squamosal sutures and the lambdoidal suture.

The occipital, sphenoidal, and mastoid fontanels disappear within a month or two after birth. The frontal fontanel generally persists until a child is nearly 2 years old. Even after the fontanels disappear, the bones of the skull remain separated by fibrous connections.

The skulls of infants and adults differ in terms of the shape and structure of cranial elements. This difference accounts for variations in proportions as well as in size. The most significant growth in the skull occurs before age 5, for at that time the brain stops growing and the cranial sutures develop. As a result, the cranium of a young child, when compared with the skull as a whole, is relatively larger than that of an adult. The growth of the cranium is generally coordinated with the expansion of the brain. If one or more sutures form before the brain stops growing, the skull will be abnormal in shape, size, or both.

CRANIOSTENOSIS AND MICROCEPHALY Unusual distortions of the skull result from the premature closure of one or more fontanels, a condition called *craniostenosis* (krā-nē-ō-sten-Ō-sis; *stenosis*, narrowing). As the brain continues to enlarge, the rest of the skull distorts to accommodate it. A long and narrow head will be produced by early closure of the sagittal suture, whereas a very broad skull results if the coronal suture forms prematurely. Closure of all cranial sutures restricts the development of the brain, and surgery must be performed to prevent brain damage. If the brain enlargement stops because of genetic or developmental abnormalities, however, skull growth ceases as well. This condition, which results in a very undersized head, is called *microcephaly* (mī-krō-SEF-uh-lē). AM *Phrenology Then and Now*

THE VERTEBRAL COLUMN

The rest of the axial skeleton is subdivided on the basis of vertebral structure. The adult **vertebral column** consists of 26 bones: the **vertebrae** (24), the **sacrum,** and the **coccyx** (KOK-siks). The vertebrae provide a column of support, bearing the weight of the head, neck, and trunk and ultimately transferring the weight to the appendicular skeleton of the lower limbs. The vertebrae also protect the spinal cord and help maintain an upright body position, as in sitting or standing.

The vertebral column is divided into *cervical, thoracic, lumbar, sacral,* and *coccygeal* regions (Figure 7-16•). Seven **cervical vertebrae** constitute the neck and extend inferiorly to the trunk. Twelve **thoracic vertebrae** form the upper back; each articulates with one or more pairs of ribs. Five **lumbar vertebrae** form

Development of the Skull

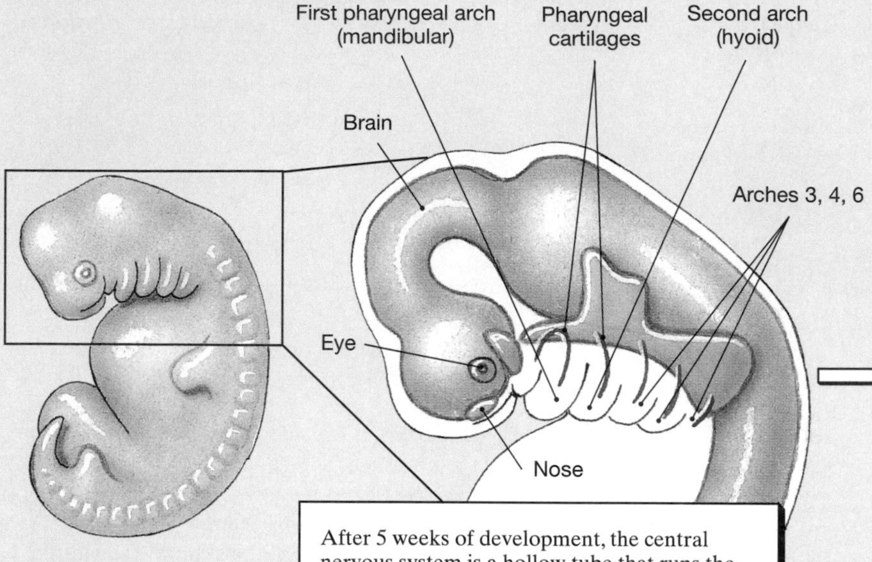

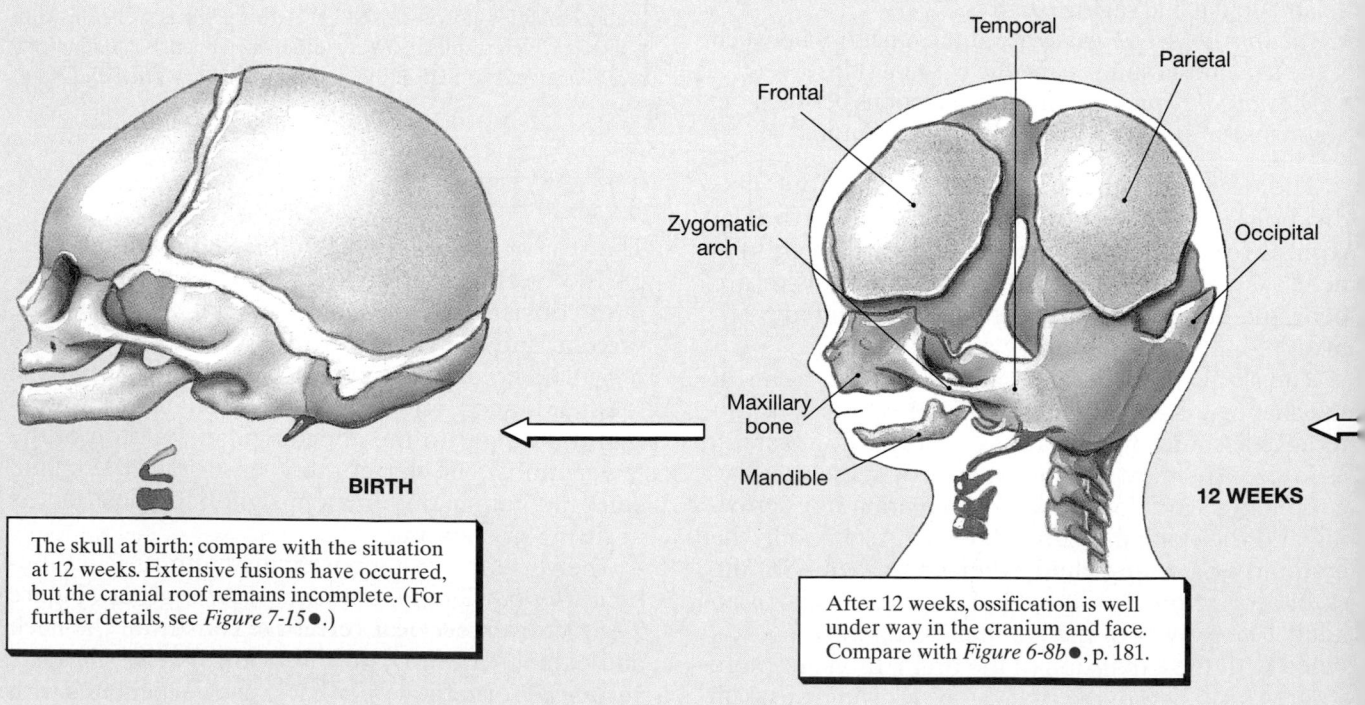

First pharyngeal arch (mandibular)

Pharyngeal cartilages

Second arch (hyoid)

Brain

Arches 3, 4, 6

Eye

Nose

5 WEEKS

After 5 weeks of development, the central nervous system is a hollow tube that runs the length of the body. A series of cartilages appears in the mesenchyme of the head beneath and alongside the expanding brain and around the developing nose, eyes, and ears. These cartilages are shown in light blue. Five additional pairs of cartilages develop in the walls of the pharynx. These cartilages, shown in dark blue, are located within the **pharyngeal,** or **branchial, arches.** (*Branchial* refers to gills—in fish the caudal arches develop into skeletal supports for the gills.) The first arch, or **mandibular arch,** is the largest.

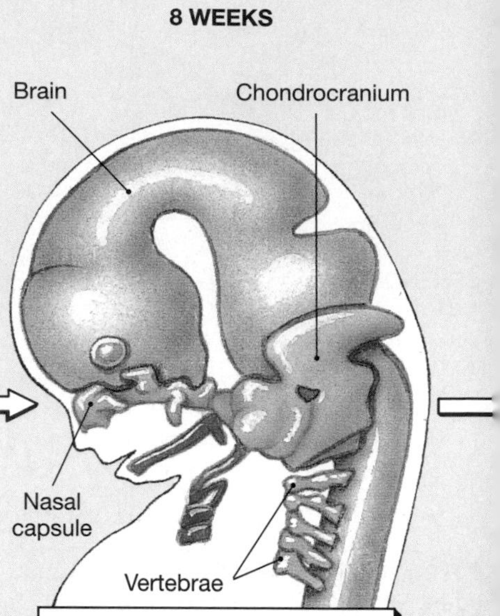

8 WEEKS

Brain

Chondrocranium

Nasal capsule

Vertebrae

The cartilages associated with the brain enlarge and fuse, forming a cartilaginous **chondrocranium** (kon-drō-KRĀ-nē-yum; *chondros,* cartilage + *cranium,* skull), which cradles the brain and sense organs. At 8 weeks, its walls and floor are incomplete, and there is no roof.

Temporal

Parietal

Frontal

Zygomatic arch

Occipital

Maxillary bone

Mandible

BIRTH

The skull at birth; compare with the situation at 12 weeks. Extensive fusions have occurred, but the cranial roof remains incomplete. (For further details, see *Figure 7-15●*.)

12 WEEKS

After 12 weeks, ossification is well under way in the cranium and face. Compare with *Figure 6-8b●*, p. 181.

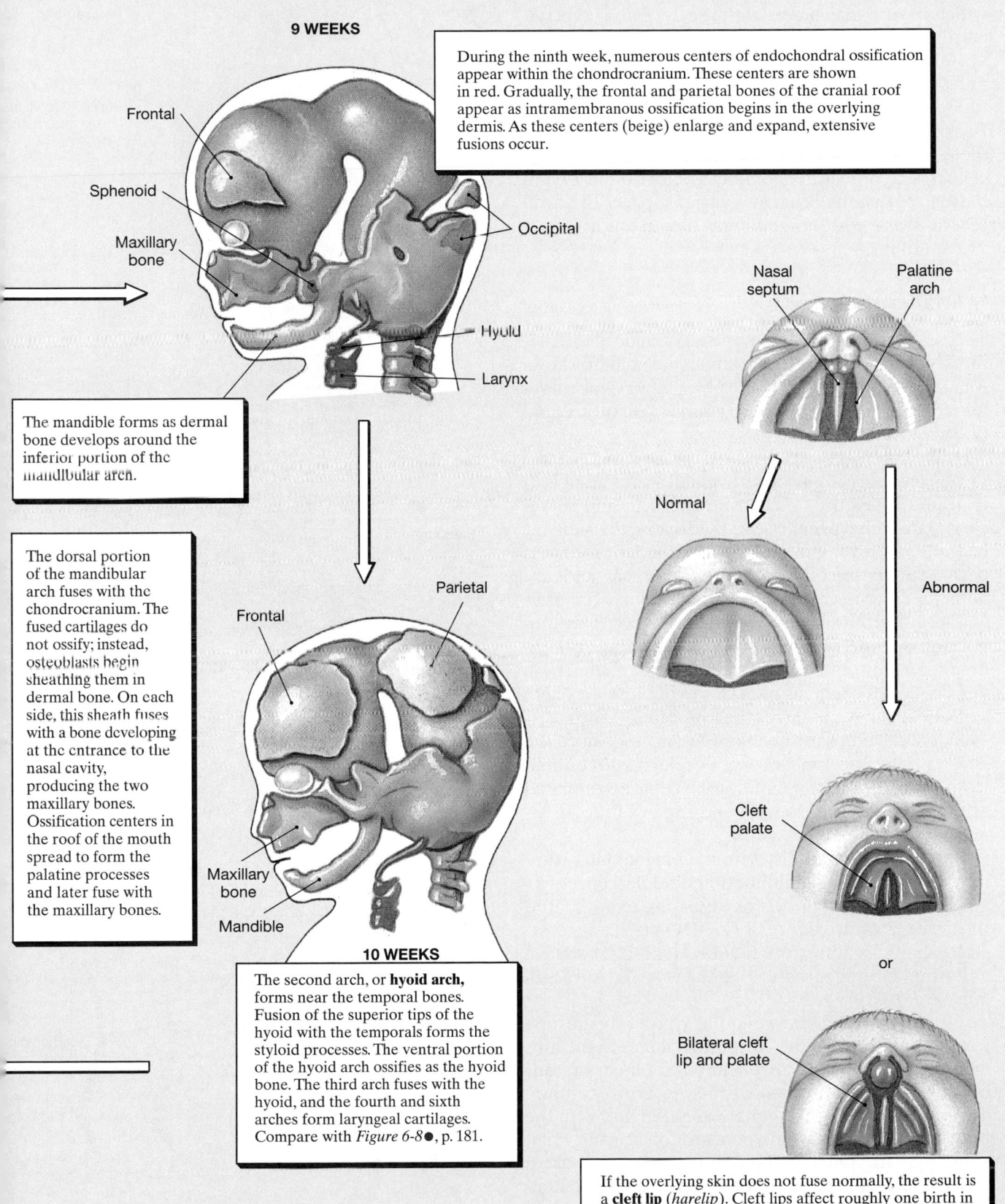

9 WEEKS

Frontal

Sphenoid

Maxillary bone

Occipital

Hyoid

Larynx

During the ninth week, numerous centers of endochondral ossification appear within the chondrocranium. These centers are shown in red. Gradually, the frontal and parietal bones of the cranial roof appear as intramembranous ossification begins in the overlying dermis. As these centers (beige) enlarge and expand, extensive fusions occur.

Nasal septum

Palatine arch

Normal

Abnormal

The mandible forms as dermal bone develops around the inferior portion of the mandibular arch.

The dorsal portion of the mandibular arch fuses with the chondrocranium. The fused cartilages do not ossify; instead, osteoblasts begin sheathing them in dermal bone. On each side, this sheath fuses with a bone developing at the entrance to the nasal cavity, producing the two maxillary bones. Ossification centers in the roof of the mouth spread to form the palatine processes and later fuse with the maxillary bones.

Frontal

Parietal

Maxillary bone

Mandible

10 WEEKS

Cleft palate

or

Bilateral cleft lip and palate

The second arch, or **hyoid arch,** forms near the temporal bones. Fusion of the superior tips of the hyoid with the temporals forms the styloid processes. The ventral portion of the hyoid arch ossifies as the hyoid bone. The third arch fuses with the hyoid, and the fourth and sixth arches form laryngeal cartilages. Compare with *Figure 6-8●*, p. 181.

If the overlying skin does not fuse normally, the result is a **cleft lip** (*harelip*). Cleft lips affect roughly one birth in a thousand. A split extending into the orbit and palate is called a **cleft palate.** Cleft palates are half as common as cleft lips. Both conditions can be corrected surgically.

the lower back; the fifth articulates with the sacrum, which in turn articulates with the coccyx. The cervical, thoracic, and lumbar regions consist of individual vertebrae. During development, the sacrum originates as a group of five vertebrae, and the coccyx, or tailbone, begins as three to five very small vertebrae. In general, the vertebrae of the sacrum are completely fused by age 25. Ossification of the distal coccygeal vertebrae is not complete before puberty, and thereafter fusion occurs at a variable pace. The total length of the vertebral column of an adult averages 71 cm (28 in).

Spinal Curvature

The vertebrae do not form a straight and rigid structure. A lateral view of the adult spinal column shows four **spinal curves** (Figure 7-16●): (1) **cervical curvature,** (2) **thoracic curvature,** (3) **lumbar curvature,** and (4) **sacral curvature.**

The sequence of appearance of the spinal curvatures from fetus to newborn, to child, and to adult is illustrated in Figure 7-17●. The thoracic and sacral curves are called **primary curves,** because they appear late in fetal development, or **accommodation curves,** because they accommodate the thoracic and abdominopelvic viscera. Only the primary curves are present in the vertebral column of the newborn. The lumbar and cervical curves, known as **secondary curves,** do not appear until several months after birth. These are also called **compensation curves** because they help shift the trunk weight over the lower limbs. The cervical curve develops as the infant learns to balance the head upright. The lumbar curve develops with the ability to stand. Both compensations become accentuated as the toddler learns to walk and run. All four curves are fully developed by age 10.

Several abnormal distortions of spinal curvature may appear during childhood and adolescence. Examples include *kyphosis* (kī-FŌ-sis), exaggerated thoracic curvature; *lordosis* (lor-DŌ-sis), exaggerated lumbar curvature; and *scoliosis* (skō-lē-Ō-sis), an abnormal lateral curvature. AM *Kyphosis, Lordosis, and Scoliosis*

When you stand, the weight of your body must be transmitted through the vertebral column to the hips and ultimately to the lower limbs. Yet most of your body weight lies in front of the vertebral column. The various curves bring that weight in line with the body axis. Consider what you do automatically when standing with a heavy object hugged to your chest. You avoid toppling forward by exaggerating the lumbar curvature and moving the weight back toward the body axis. This posture can lead to discomfort at the base of the spinal column. For example, women in the last 3 months of pregnancy often develop chronic back pain from the changes in lumbar cur-

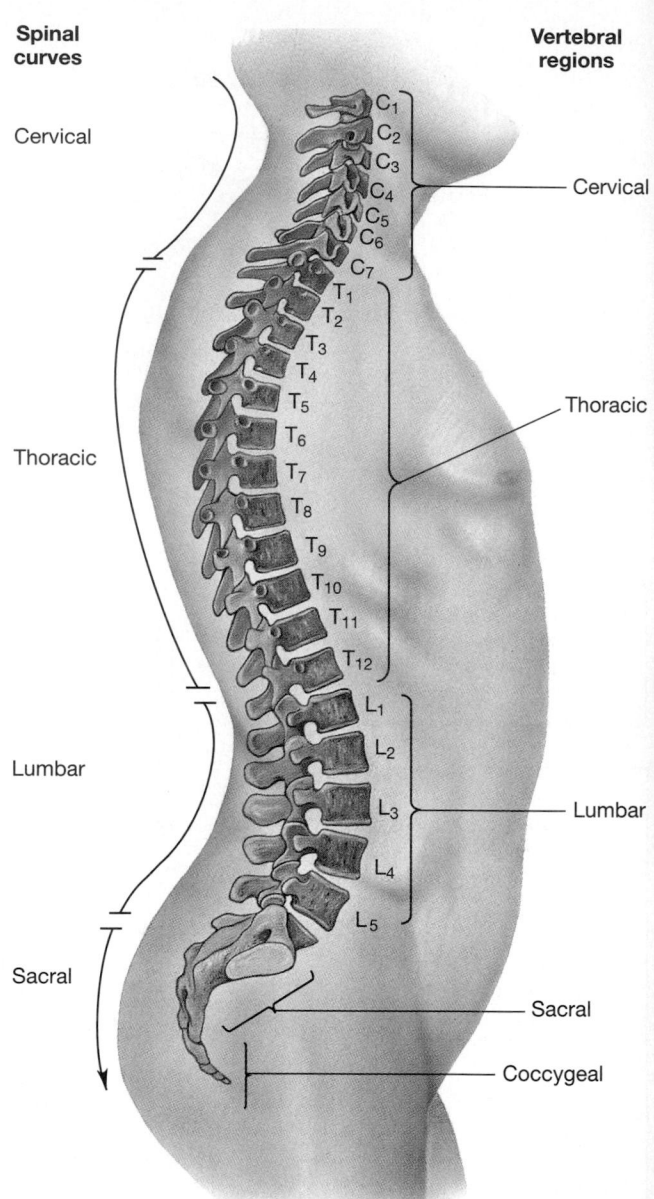

●**FIGURE 7-16 The Vertebral Column.** The major divisions of the vertebral column; note the four spinal curves. AM *Scan 3*

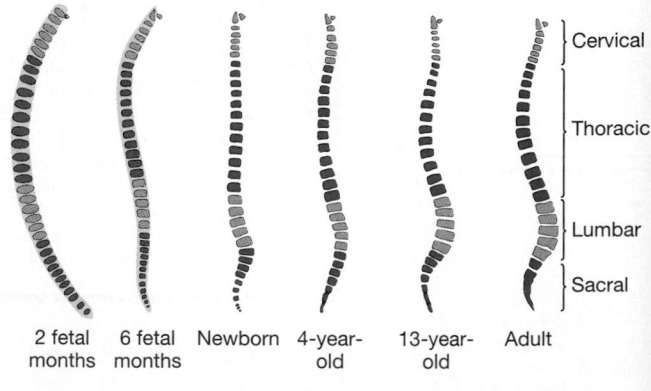

●**FIGURE 7-17 The Development of Spinal Curvature.** Primary curves are shown in blue. Secondary curves (pink) balance the body weight over the legs. These curves do not develop fully until the individual begins walking.

vature that must adjust for the increasing weight of the fetus. African and South American natives often balance heavy objects on their heads, as seen in the chapter-opening photograph. This practice increases the load on the vertebral column, but the spinal curves are not affected because the weight is aligned with the axis of the spine.

Vertebral Anatomy

Each vertebra consists of three basic parts: (1) a *body*, (2) a *vertebral arch*, and (3) *articular processes* (Figure 7-18●).

The Vertebral Body

The **body,** or *centrum* (plural, *centra*), is the part of a vertebra that transfers weight along the axis of the ver-

tebral column. The bodies of adjacent vertebrae are interconnected by ligaments but are separated by pads of fibrocartilage, the **intervertebral discs.**

The Vertebral Arch

The **vertebral arch,** also called the *neural arch*, forms the posterior margin of each **vertebral foramen.** Together, the vertebral foramina enclose the *spinal canal*, which contains the spinal cord. The vertebral arch has walls, called **pedicles** (PE-di-kulz), and a roof, formed by the **laminae** (LA-mi-nē; singular, *lamina,* a thin plate). The pedicles arise along the posterior and lateral margins of the body. The laminae on either side extend dorsally and medially to complete the roof.

The vertebral foramina of successive vertebrae collectively form the **vertebral canal,** which encloses the entire spinal cord. In the condition called *spina bifida* (SPĪ-nuh BI-fi-duh), the vertebral laminae fail to unite during development. The neural arch is incomplete, and the membranes, or *meninges,* that line the dorsal body cavity bulge outward. Mild cases involving the sacral and lumbar regions may pass unnoticed; in severe cases, the entire spinal column and skull are affected.

A **spinous process,** also known as a *spinal process,* projects posteriorly from the point where the vertebral laminae fuse to complete the vertebral arch. You can see—and feel—the spinous processes through the skin of the back when the spine is flexed. **Transverse processes** project laterally or dorsolaterally on both sides from the point where the laminae join the pedicles. These processes are sites of muscle attachment, and they may also articulate with the ribs.

The Articular Processes

Like the transverse processes, the **articular processes** arise at the junction between the pedicles and laminae. A **superior** and an **inferior articular process** lie on each side of the vertebra.

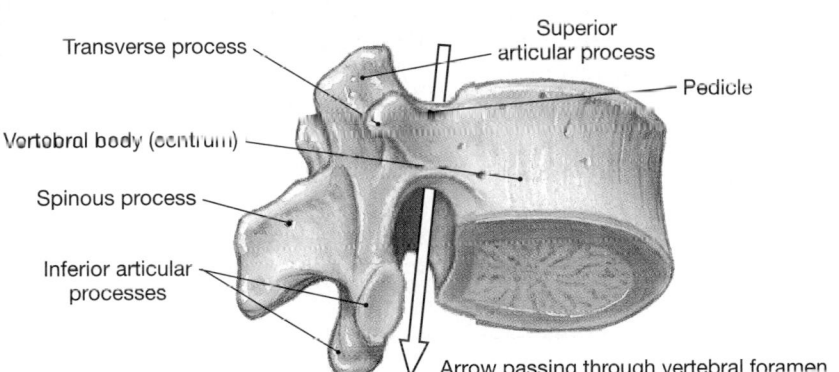

Lateral and inferior view

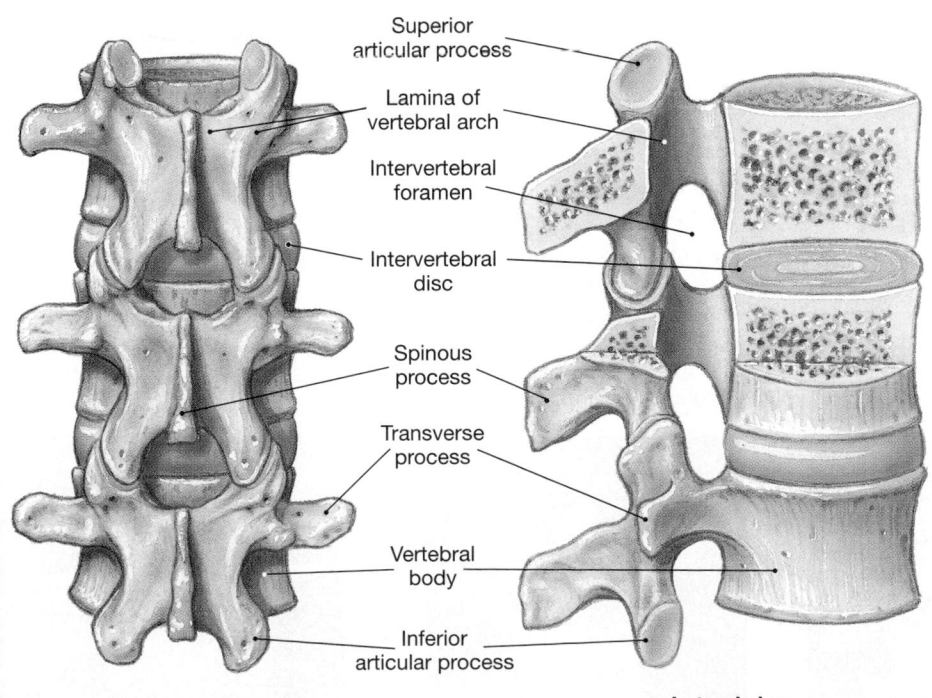

Posterior view | **Lateral view**

●*FIGURE 7-18* **Vertebral Anatomy.** The anatomy of a typical vertebra and the arrangement of articulations between vertebrae.

Vertebral Articulation

The inferior articular processes of one vertebra articulate with the superior articular processes of the next vertebra. Each articular process has a polished concave surface called an **articular facet.** The superior processes have articular facets on their dorsal surfaces, whereas the inferior processes articulate along their ventral surfaces.

Adjacent vertebral bodies are separated by intervertebral discs, and there are gaps between the pedicles of successive vertebrae. These gaps, called **intervertebral foramina,** permit the passage of nerves running to or from the enclosed spinal cord.

☑ Why are there fewer vertebrae in the vertebral column of an adult than in the vertebral column of a newborn?

☑ What is the importance of the secondary curves of the spine?

☑ When you run your finger along a person's spine, what part of the vertebrae are you feeling just beneath the skin?

Vertebral Regions

When we refer to the vertebrae, we use a capital letter to indicate each vertebral region and a number to indicate each vertebra, starting with the cervical vertebra closest to the skull. For example, C_3 is the third cervical vertebra, with C_1 in contact with the skull; L_4 is the fourth lumbar vertebra, with L_1 in contact with the last thoracic vertebra (Figure 7-16●). We shall use this shorthand throughout the text.

Although each vertebra bears characteristic markings and articulations, we will focus on the general characteristics of each region and on how the regional variations determine the vertebral group's function. Table 7-2 compares typical vertebrae from the cervical, thoracic, and lumbar regions of the vertebral column.

Cervical Vertebrae

The seven cervical vertebrae (Figure 7-19a●) are the smallest of the vertebrae. They extend from the occipital bone of the skull to the thorax. The body of a cervical vertebra is small compared with the size of the vertebral foramen (Figure 7-19b●). At this level, the spinal cord still contains most of the axons that connect the brain to the rest of the body. As you proceed caudally along the vertebral canal, the diameter of the spinal cord decreases, and so does the diameter of the vertebral arch. However, cervical vertebrae support only the weight of the head, so the vertebral body can be relatively small and light. As you continue toward the sacrum, the loading increases and the vertebral bodies gradually enlarge.

TABLE 7-2 **Regional Differences in Vertebral Structure and Function**

Feature	Type (Number)		
	Cervical Vertebrae (7)	**Thoracic Vertebrae (12)**	**Lumbar Vertebrae (5)**
Location	Neck	Chest	Lower back
Body	Small, oval, curved faces	Medium, heart-shaped, flat faces; facets for rib articulations	Massive, oval, flat faces
Vertebral foramen	Large	Smaller	Smallest
Spinous process	Long; split tip; points inferiorly	Long, slender; not split; points inferiorly	Blunt, broad, points posteriorly
Transverse process	Has transverse foramen	All but two (T_{11}, T_{12}) have facets for rib articulations	Short; no articular facets or transverse foramen
Functions	Support skull, stabilize relative positions of brain and spinal cord, and allow controlled head movement	Support weight of head, neck, upper limbs, chest; articulate with ribs to allow changes in volume of thoracic cage	Support weight of head, neck, upper limbs, and trunk

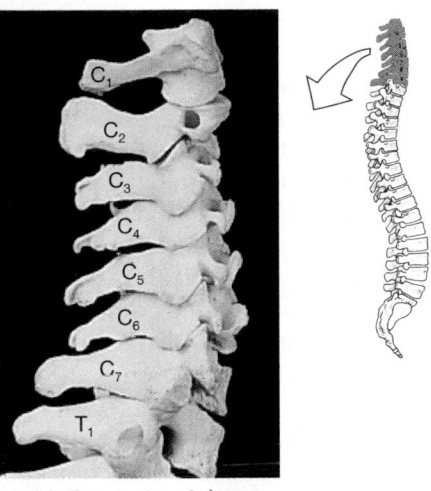

(a) Cervical vertebra

•FIGURE 7-19 **The Cervical Vertebrae.** **(a)** Lateral view of the cervical vertebrae, C_3–C_7. **(b)** Lateral view of a typical cervical vertebra (C_3–C_6). **(c)** Superior view of the same vertebra. Note the characteristic features listed in Table 7-2. **(d)** The atlas (C_1) and axis (C_2). AM *Scan 3*

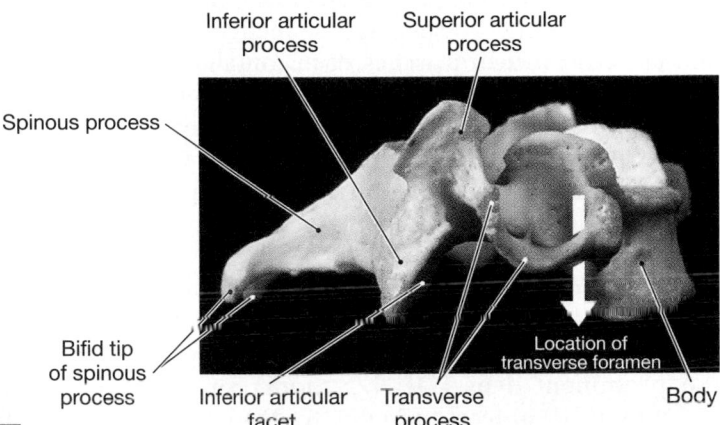

(b) Typical cervical vertebra (lateral view)

(c) Typical cervical vertebra (superior view)

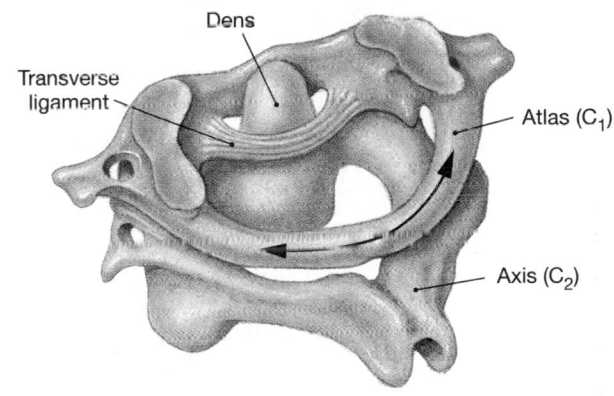

(d) The atlas/axis complex

Your head is relatively massive. It sits atop the cervical vertebrae like a soup bowl on the tip of a finger. With this arrangement, small muscles can produce significant effects by tipping the balance one way or another. But if you suddenly change position, as in a fall or during rapid acceleration (a jet takeoff) or deceleration (a car crash), the balancing muscles are not strong enough to stabilize the head. A dangerous partial or complete dislocation of the cervical vertebrae can result, with injury to muscles and ligaments and potential injury to the spinal cord. The term **whiplash** is used to describe such an injury, because the movement of the head resembles the cracking of a whip.

In a cervical vertebra, the superior surface of the body is concave from side to side, and it slopes, with the anterior edge inferior to the posterior edge (Figure 7-19c•). The spinous process is relatively stumpy, generally shorter than the diameter of the vertebral foramen. The tip of each process other than that of C_7 bears

a prominent notch. A notched spinous process is described as **bifid** (BĪ-fid; *bifidus*, cut into two parts).

Laterally, the transverse processes are fused to the **costal processes,** which originate near the ventrolateral portion of the body. The costal and transverse processes encircle prominent, round, **transverse foramina.** These passageways protect the *vertebral arteries* and *vertebral veins,* important blood vessels that service the brain.

This description would be adequate to identify all but the first two cervical vertebrae. When cervical vertebrae C_3–C_7 articulate, their interlocking bodies permit more flexibility than do those of other regions. The first two cervical vertebrae are unique, and the seventh is modified. Table 7-2 summarizes the features of cervical vertebrae.

THE ATLAS (C_1) The **atlas,** cervical vertebra C_1, holds up the head, articulating with the occipital condyles of the skull (Figure 7-19d•). It is named after Atlas, who, according to Greek myth, holds the world on his shoulders. The articulation between the occipital condyles and the atlas is a joint that permits you to nod (such as when you indicate "yes"). The atlas can be distinguished from the other vertebrae by the (1) lack of a body or spinous process; (2) possession of semicircular **anterior** and **posterior vertebral arches,** each containing **anterior** and **posterior tubercles;** and (3) presence of oval **superior facets** and round **inferior articular facets.**

The atlas articulates with the second cervical vertebra, the *axis*. This articulation permits rotation (such as when you shake your head to indicate "no").

THE AXIS (C_2) During development, the body of the atlas fuses to the body of the second cervical vertebra, called the **axis** (C_2) (Figure 7-19d●). This fusion creates the prominent **dens** (DENZ; *tooth*), or **odontoid** (ō-DON-toyd) **process** (*odontos,* tooth), of the axis. A transverse ligament binds the dens to the inner surface of the atlas, forming a pivot for rotation of the atlas and skull. Important muscles controlling the position of the head and neck attach to the especially robust spinous process of the axis.

In children, the fusion between the dens and axis is incomplete. Impacts or even severe shaking can cause dislocation of the dens and severe damage to the spinal cord. In adults, a blow to the base of the skull can be equally dangerous because a dislocation of the axis-atlas joint can force the dens into the base of the brain, with fatal results.

THE VERTEBRA PROMINENS (C_7) The transition from one vertebral region to another is not abrupt, and the last vertebra of one region generally resembles the first vertebra of the next. The **vertebra prominens,** or seventh cervical vertebra (C_7), has a long, slender spinous process that ends in a broad tubercle that you can feel beneath the skin at the base of the neck. This vertebra is the interface between the cervical curve, which arches forward, and the thoracic curve, which arches backward (Figure 7-20a●). The transverse processes are large, providing additional surface area for muscle attachment. A large elastic ligament, the **ligamentum nuchae** (li-guh-MEN-tum NOO-kē; *nucha,* nape) begins at the vertebra prominens and extends to an insertion along the external occipital crest. Along the way, it attaches to the spinous processes of the other cervical vertebrae. When your head is upright, this ligament acts like the string on a bow, maintaining the cervical curvature without muscular effort. If you have bent your neck forward, the elasticity in this ligament helps return your head to an upright position.

Thoracic Vertebrae

There are 12 thoracic vertebrae. A typical thoracic vertebra (Figure 7-20b●) has a distinctive heart-shaped body that is more massive than that of a cervical vertebra; the vertebral foramen is relatively smaller, and the long, slender spinous process projects posteriorly and inferiorly. The spinous processes of T_{10}, T_{11}, and T_{12} increasingly resemble those of the lumbar series, as the transition between the thoracic and lumbar curvatures approaches. Because the lower thoracic and lumbar vertebrae carry so much weight, the transition between the thoracic and lumbar curves is difficult to stabilize. As a result, compression fractures or compression-dislocation fractures after a hard fall tend to involve the last thoracic and first two lumbar vertebrae.

Each thoracic vertebra articulates with ribs along the dorsolateral surfaces of the body. The location and structure of the articulations vary somewhat from vertebra to vertebra (Figure 7-20a●). Rib pairs 2 through 8 originate between adjacent vertebrae, so vertebrae T_2–T_8 have **superior** and **inferior demifacets** (*partial facets*) on each side. The first rib originates at the body of T_1, so that vertebra has a *whole facet* and an inferior demifacet on each side. Vertebra T_9 has only a superior demifacet on each side, whereas T_{10}, T_{11}, and T_{12} have a single whole facet on each side.

The transverse processes of vertebrae T_1–T_{10} are relatively thick, and they contain **transverse costal facets** for rib articulation. Thus, rib pairs 1 through 10 contact their vertebrae at two points—at a whole facet or demifacet and at a transverse costal facet. Table 7-2 summarizes the features of the thoracic vertebrae.

Lumbar Vertebrae

The five lumbar vertebrae are the largest of the vertebrae. The body of a typical lumbar vertebra (Figure 7-21●, p. 224) is thicker than that of a thoracic vertebra, and the superior and inferior surfaces are oval rather than heart-shaped. Other noteworthy features include the following: (1) Lumbar vertebrae have neither whole facets nor demifacets on the body; (2) the slender transverse processes, which lack costal facets, project dorsolaterally; (3) the vertebral foramen is triangular; (4) the stumpy spinous processes project dorsally; (5) the superior articular processes face medially ("up and in"); and (6) the inferior articular processes face laterally ("down and out").

The lumbar vertebrae bear the most weight. Their massive spinous processes provide surface area for the attachment of lower back muscles that reinforce or adjust the lumbar curvature. Table 7-2 summarizes the characteristics of lumbar vertebrae.

The Sacrum

The sacrum consists of the fused components of five sacral vertebrae. These vertebrae begin fusing shortly after puberty and in general are completely fused between ages 25 and 30. The sacrum provides protection for reproductive, digestive, and urinary organs and, via

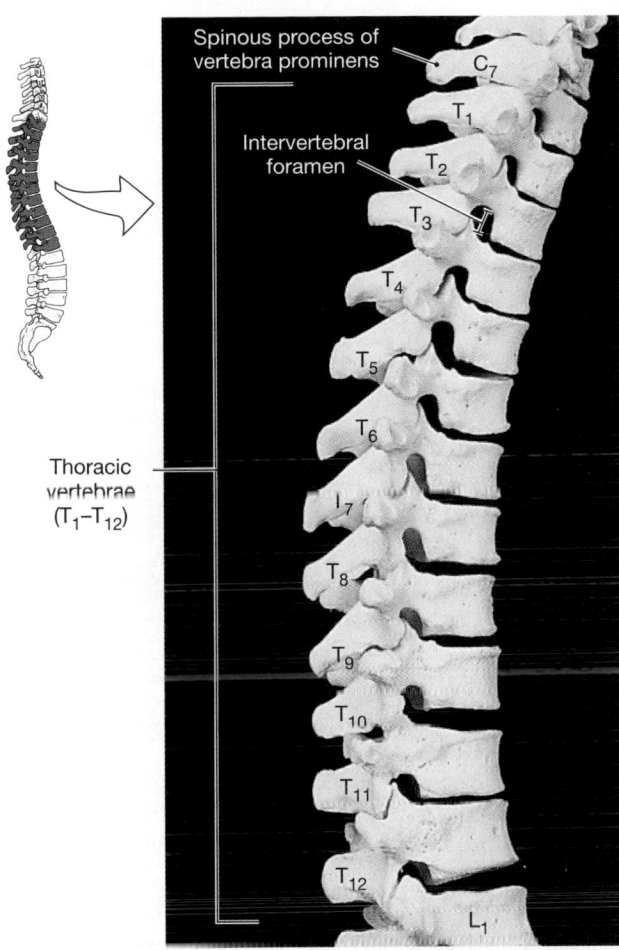

(a) Thoracic vertebrae, lateral view

Labels on image (a):
Spinous process of vertebra prominens
C₇
T₁
Intervertebral foramen
T₂
T₃
T₄
T₅
T₆
I₇
T₈
T₉
T₁₀
T₁₁
T₁₂
L₁
Thoracic vertebrae (T₁–T₁₂)

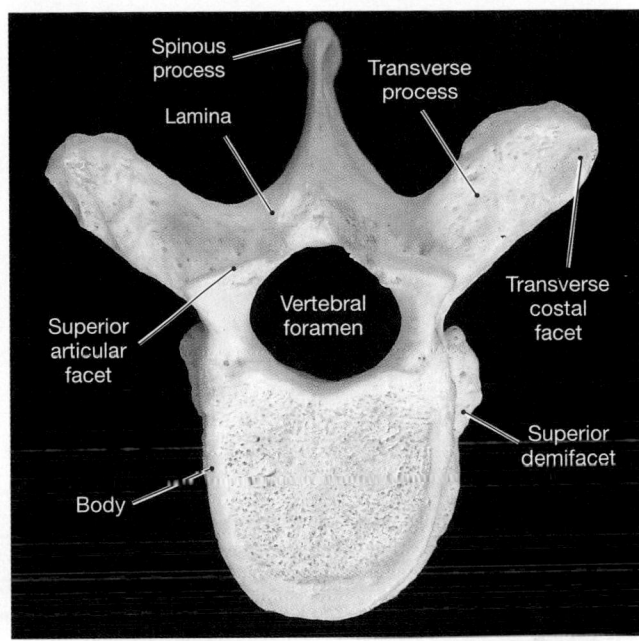

(b) Thoracic vertebra, superior view

Labels on image (b):
Spinous process
Transverse process
Lamina
Vertebral foramen
Transverse costal facet
Superior articular facet
Superior demifacet
Body

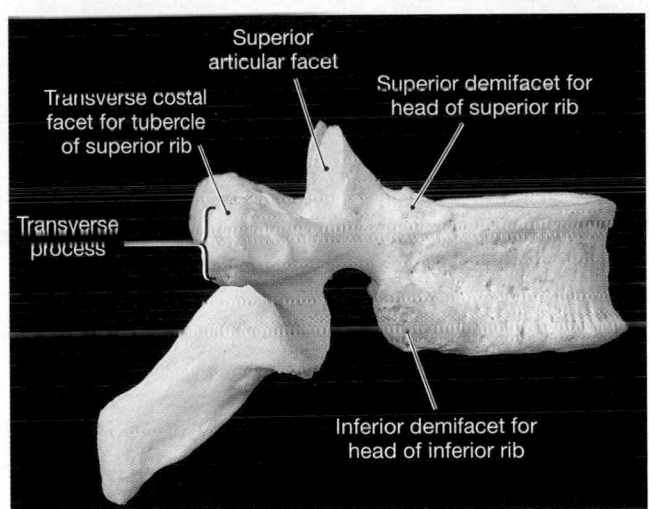

(c) Thoracic vertebra, lateral view

Labels on image (c):
Superior articular facet
Transverse costal facet for tubercle of superior rib
Superior demifacet for head of superior rib
Transverse process
Inferior demifacet for head of inferior rib

●**FIGURE 7-20 The Thoracic Vertebrae.** (a) Lateral view of the thoracic region of the vertebral column. The vertebra prominens (C₇) resembles T₁ but lacks facets for rib articulation. Vertebra T₁₂ resembles the first lumbar vertebra (L₁) but has a facet for rib articulation. (b) Thoracic vertebra, superior view. (c) Thoracic vertebra, lateral view. Note the characteristic features listed in Table 7-2.

paired articulations, attaches the axial skeleton to the pelvic girdle of the appendicular skeleton (Figure 7-1b●, p. 199). The broad surface area of the sacrum (Figure 7-22●) provides an extensive area for the attachment of muscles, especially those responsible for movement of the thigh.

The sacrum is curved, with a convex posterior surface (Figure 7-22a●). The narrow, inferior portion is the sacral **apex,** whereas the broad superior surface forms the **base.** The **articular processes** form synovial articulations with the last lumbar vertebra. The **sacral canal** begins between those processes and extends the length of the sacrum. Nerves and membranes that line the vertebral canal in the spinal cord continue into the sacral canal. The **sacral promontory,** a prominent bulge at the anterior tip of the base, is an important land-

mark in females during pelvic examinations and during labor and delivery.

The spinous processes of the five fused sacral vertebrae form a series of elevations along the **median sacral crest.** The laminae of the fifth sacral vertebra fail to contact one another at the midline, and they form the **sacral cornua** (KORN-ū-uh; singular, *cornu*). These ridges establish the margins of the **sacral hiatus** (hī-Ā-tus), the end of the sacral canal. This opening is covered by connective tissues. On either side of the median sacral crest, the **sacral foramina** represent the intervertebral foramina, now enclosed by the fused sacral bones. A broad sacral **ala,** or **wing,** extends laterally from each **lateral sacral crest.** The median and lateral sacral crests provide surface area for the attachment of muscles of the lower back and hip.

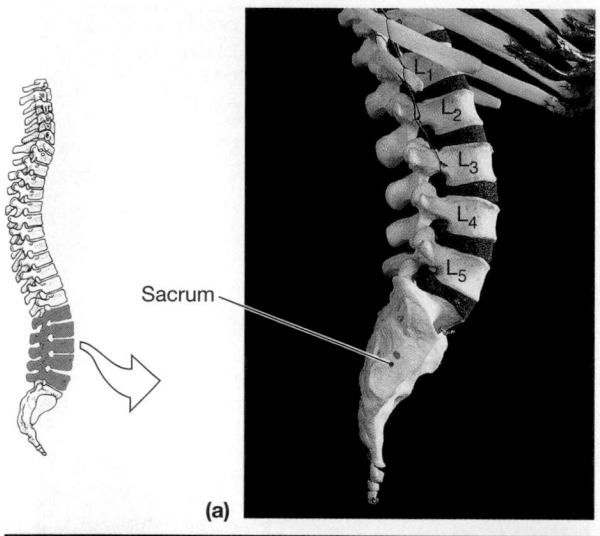

•FIGURE 7-21 **The Lumbar Vertebrae.** **(a)** Right lateral view. **(b)** Lateral view of a typical lumbar vertebra. **(c)** Superior view.

Sacrum

(a)

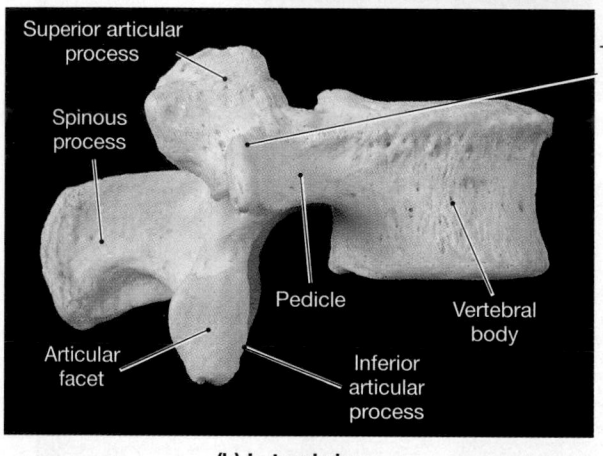

Superior articular process

Spinous process

Articular facet

Pedicle

Inferior articular process

Vertebral body

Transverse process

(b) Lateral view

Spinous process

Lamina

Vertebral foramen

Superior articular facet and process

Transverse process

Pedicle

Vertebral body

(c) Superior view

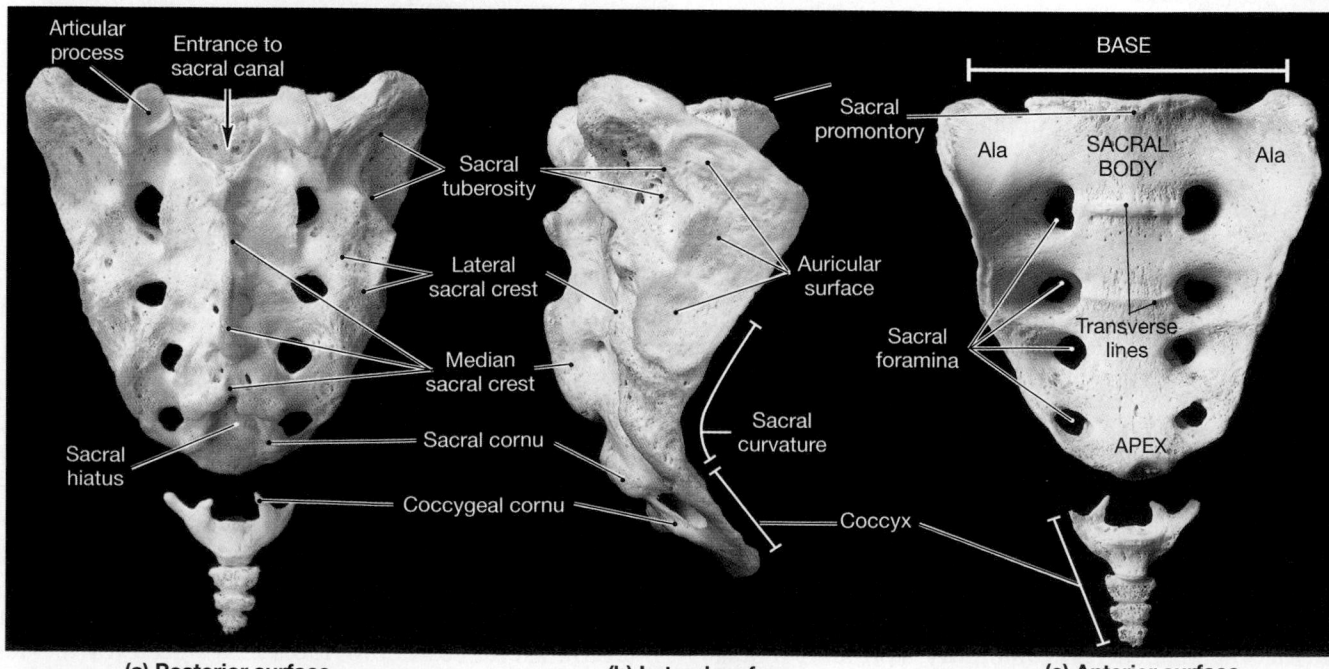

Articular process

Entrance to sacral canal

Sacral tuberosity

Lateral sacral crest

Median sacral crest

Sacral hiatus

Sacral cornu

Coccygeal cornu

Auricular surface

Sacral curvature

Coccyx

BASE

Sacral promontory

Ala

SACRAL BODY

Ala

Sacral foramina

Transverse lines

APEX

(a) Posterior surface   **(b) Lateral surface**   **(c) Anterior surface**

•FIGURE 7-22 **The Sacrum and Coccyx.** **(a)** Posterior view. **(b)** Lateral view from the right side. **(c)** Anterior view.

The sacral curvature is most apparent when viewed from the side (Figure 7-22b●) and is more pronounced in males than in females. A thickened, flattened area, the **auricular surface,** marks the site of articulation with the pelvic girdle, the *sacroiliac joint.* Dorsal to the auricular surface is the **sacral tuberosity,** a roughened area marking the attachment site of a ligament that stabilizes this articulation. The anterior surface of the sacrum is concave (Figure 7-22c●). After fusion is completed, prominent *transverse lines* mark the former boundaries of individual vertebrae. At the apex, a flattened area marks the site of articulation with the coccyx.

The Coccyx

The small coccyx consists of 3–5 (typically 4) coccygeal vertebrae that have generally begun fusing by age 26 (Figure 7-22●). The coccyx provides an attachment site for a number of ligaments and for a muscle that constricts the anal opening. The first two coccygeal vertebrae have transverse processes and unfused neural arches. The prominent laminae of the first coccygeal vertebrae are known as the **coccygeal cornua.** They curve to meet the cornua of the sacrum. The coccygeal vertebrae do not fuse completely until late in adulthood. In very old people, the coccyx may fuse with the sacrum.

☑ Joe suffered a hairline fracture at the base of the odontoid process. What bone is fractured, and where would you find it?

☑ Examining a human vertebra, you notice that in addition to the large foramen for the spinal cord, there are two smaller foramina on either side of the bone in the region of the transverse processes. From which region of the spinal column did this vertebra come?

☑ Why are the bodies of the lumbar vertebrae so large?

THE THORACIC CAGE

The skeleton of the chest, or **thoracic cage,** consists of the thoracic vertebrae, the ribs, and the sternum (breastbone) (Figure 7-23●). The ribs and the sternum form the *rib cage* and support the walls of the thoracic cavity. The thoracic cage serves two functions:

1. It protects the heart, lungs, thymus, and other structures in the thoracic cavity.

2. It serves as an attachment point for muscles involved in (1) respiration, (2) the position of the vertebral column, and (3) movements of the pectoral girdle and upper extremity.

The Ribs

Ribs, or **costae,** are elongate, curved, flattened bones that (1) originate on or between the thoracic vertebrae and (2) end in the wall of the thoracic cavity. There are

12 pairs of ribs (Figure 7-23●). The first seven pairs are called **true ribs,** or **vertebrosternal ribs.** They reach the anterior body wall and are connected to the sternum by separate cartilaginous extensions, the **costal cartilages.** Beginning with the first rib, the vertebrosternal ribs gradually increase in length and in the radius of curvature.

Ribs 8–12 are called **false ribs** because they do not attach directly to the sternum. The costal cartilages of ribs 8–10, the **vertebrochondral ribs,** fuse together and merge with the cartilages of rib pair 7 before they reach the sternum (Figure 7-23a●). The last two pairs of ribs (11 and 12) are called **floating ribs,** because they have no connection with the sternum, or **vertebral ribs,** because they are attached only to the vertebrae (Figure 7-23a,b●).

Figure 7-24a● (p. 227) shows the superior surface of a typical rib. The *vertebral end* of the rib articulates with the vertebral column at the **head,** or **capitulum** (ka-PIT-Ū-lum) (Figure 7-24a●). The **interarticular crest** divides the articular surface of the head into **superior** and **inferior articular facets** (Figure 7-24b●). After a short **neck,** the **tubercle,** or **tuberculum** (too-BER-kū-lum), projects dorsally. The inferior portion of the tubercle contains an articular facet that contacts the transverse process of the thoracic vertebra.

Ribs 1 and 10 originate at whole facets on vertebrae T_1 and T_{10}, respectively, and their tubercular facets articulate with those vertebrae. Ribs 2–9 originate at demifacets, and their tubercular facets articulate with the inferior member of the vertebral pair. Ribs 11–12 originate at whole facets on T_{11} and T_{12}. These ribs do not have tubercular facets and do not contact the transverse processes of T_{11} or T_{12}. The difference in rib orientation can be seen by comparing Figure 7-20a, p. 223, with Figure 7-23b●.

The bend, or **angle,** of the rib is the site where the tubular **body,** or **shaft,** begins curving toward the sternum. The internal rib surface is concave, and a prominent **costal groove** along its inferior border marks the path of nerves and blood vessels. The superficial surface is convex and provides an attachment site for muscles of the pectoral girdle and trunk. The *intercostal muscles,* which move the ribs, are attached to the superior and inferior surfaces.

With their complex musculature, dual articulations at the vertebrae, and flexible connection to the sternum, the ribs are quite mobile. Note how the ribs curve away from the vertebral column to angle inferiorly (Figure 7-23b●). A typical rib acts as if it were the handle on a bucket, lying just below the horizontal plane. Pushing the handle down forces it inward; pulling it up swings it outward (Figure 7-24c●). In addition, because of the curvature of the ribs, the same movements change the position of the sternum. Depressing the ribs pulls the sternum inward, whereas elevation moves it outward. As a result, movements of the ribs affect both the width and the depth of the thoracic cage, increasing or decreasing its volume accordingly.

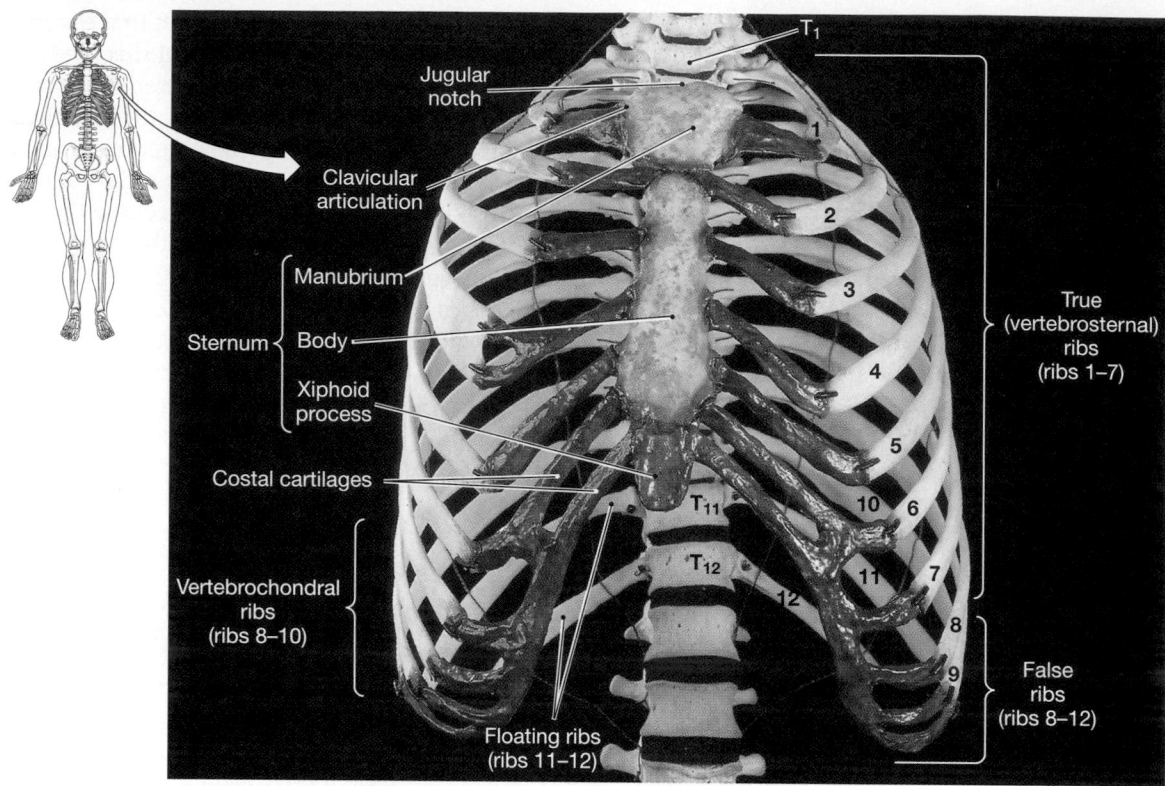

(a) Anterior view

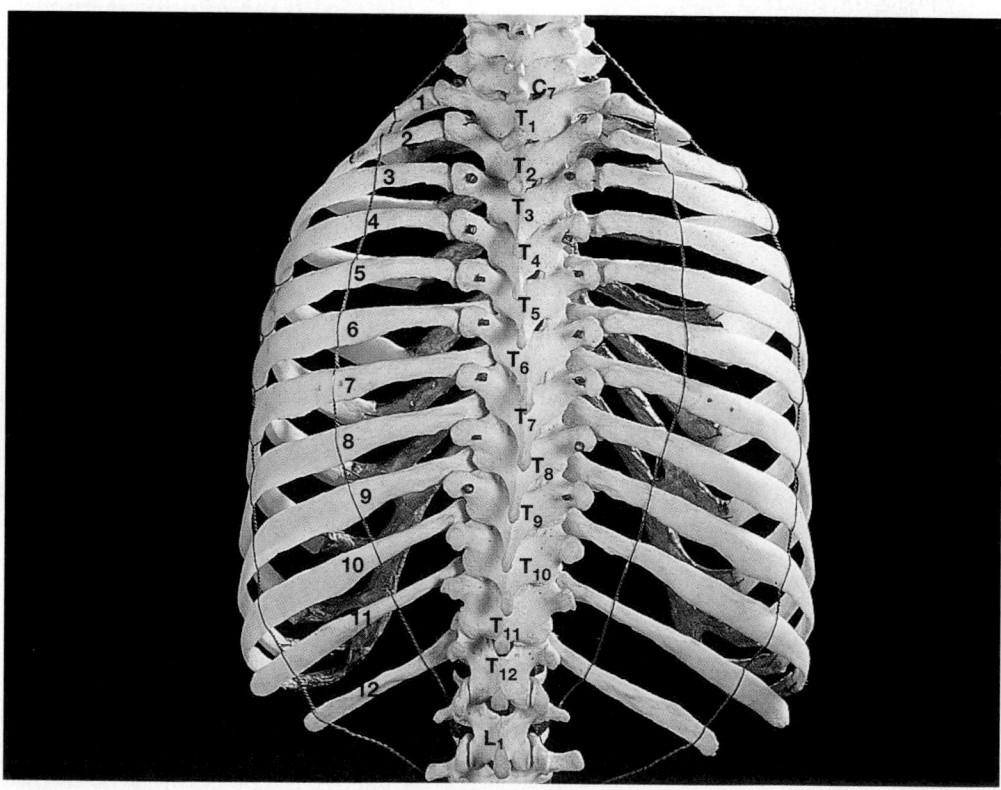

(b) Posterior view

●*FIGURE 7-23* **The Thoracic Cage.** **(a)** Anterior view of the thoracic cage and sternum. **(b)** Posterior view of the thoracic cage.

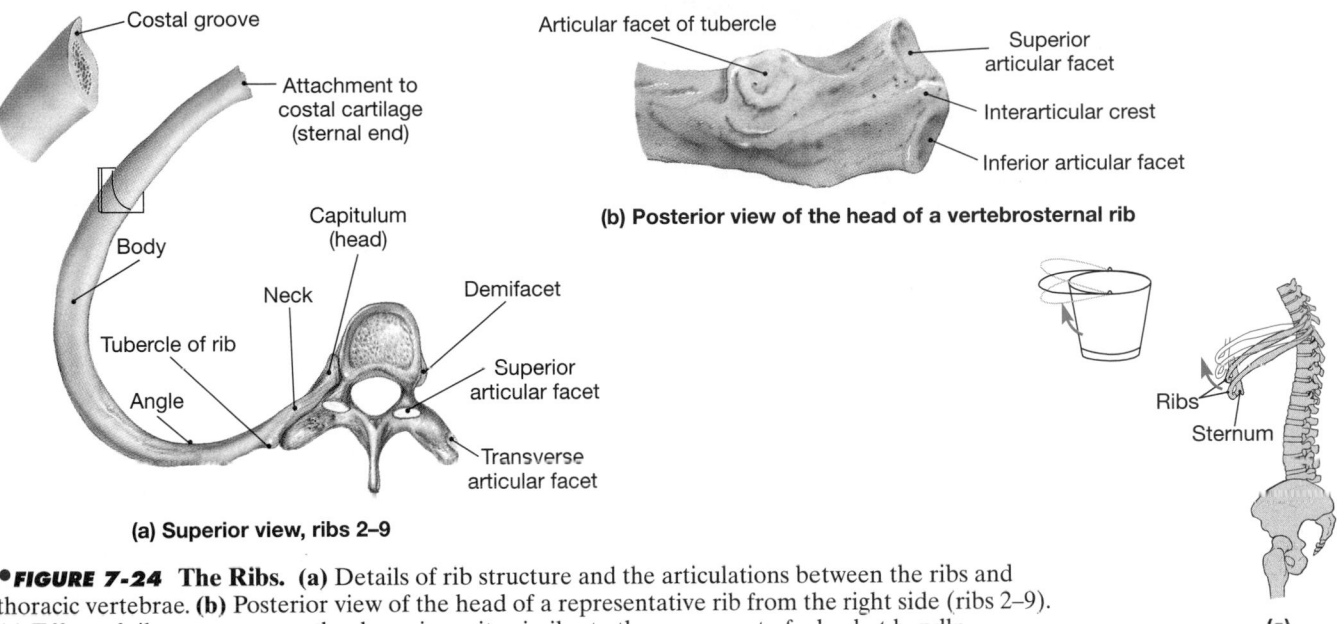

(b) Posterior view of the head of a vertebrosternal rib

(a) Superior view, ribs 2–9

(c)

●**FIGURE 7-24** **The Ribs. (a)** Details of rib structure and the articulations between the ribs and thoracic vertebrae. **(b)** Posterior view of the head of a representative rib from the right side (ribs 2–9). **(c)** Effect of rib movement on the thoracic cavity, similar to the movement of a bucket handle.

The ribs can bend and move to cushion shocks and absorb blows, but severe and sudden impacts can cause painful rib fractures. Because the ribs are tightly bound in connective tissues, a cracked rib can heal without a cast or splint. But compound fractures of the ribs can send bone splinters or fragments into the thoracic cavity, with potential damage to internal organs.

Surgery on the heart, lungs, or other organs in the thorax typically involves entering the thoracic cavity. The mobility of the ribs and the cartilaginous connections with the sternum allow the ribs to be temporarily moved out of the way. "Rib-spreaders" are used to push the ribs apart in much the same way that a jack lifts a car off the ground for a tire change. If more extensive access is required, the sternal cartilages can be cut and the entire sternum can be folded out of the way. Once replaced, scar tissue reunites the cartilages, and the ribs heal fairly rapidly.

THORACENTESIS After thoracic surgery, *chest tubes* may be inserted through the wall of the thoracic cavity to permit drainage of fluids. To install a chest tube or obtain a sample of pleural fluid, a physician must penetrate the wall of the thorax. This process, called *thoracentesis* (tho-ra-sen-TĒ-sis) or *thoracocentesis* (tho-ra-kō-sen-TĒ-sis; *kentesis,* perforating), involves the penetration of the thoracic wall along the superior border of one of the ribs. Penetration at this location avoids damage to vessels and nerves within the costal groove.

The Sternum

The adult **sternum,** or breastbone, is a flat bone that forms in the anterior midline of the thoracic wall (Figure 7-23a●). The sternum has three components:

1. The broad, triangular **manubrium** (ma-NOO-bre-um) articulates with the *clavicles* (collarbones) and the cartilages of the first pair of ribs. This is the widest and most superior portion of the sternum. Only the first pair of ribs is attached by cartilage to this portion of the sternum. The **jugular notch,** located between the clavicular articulations, is a shallow indentation on the superior surface of the manubrium.

2. The tongue-shaped **body** attaches to the inferior surface of the manubrium and extends inferiorly along the midline. Individual costal cartilages from rib pairs 2–7 are attached to this portion of the sternum.

3. The **xiphoid** (ZĪ-foyd) **process,** the smallest part of the sternum, is attached to the inferior surface of the body. The muscular *diaphragm* and *rectus abdominis muscles* attach to the xiphoid process.

Ossification of the sternum begins at 6 to 10 different centers, and fusion is not completed until at least age 25. Before age 25, the sternal body consists of four separate bones. In adults, their boundaries can be detected as a series of transverse lines crossing the sternum. The xiphoid process is generally the last of the sternal components to ossify and fuse. Its connection to the body of the sternum can be broken by an impact or strong pressure, creating a spear of bone that can severely damage the liver. Cardiopulmonary resuscitation (CPR) training strongly emphasizes the proper positioning of the hand to reduce the chances of breaking ribs or the xiphoid process.

✔ How could you distinguish between true ribs and false ribs?

✔ Improper administration of CPR (cardiopulmonary resuscitation) could result in a fracture of which bone(s)?

Development of the Spinal Column

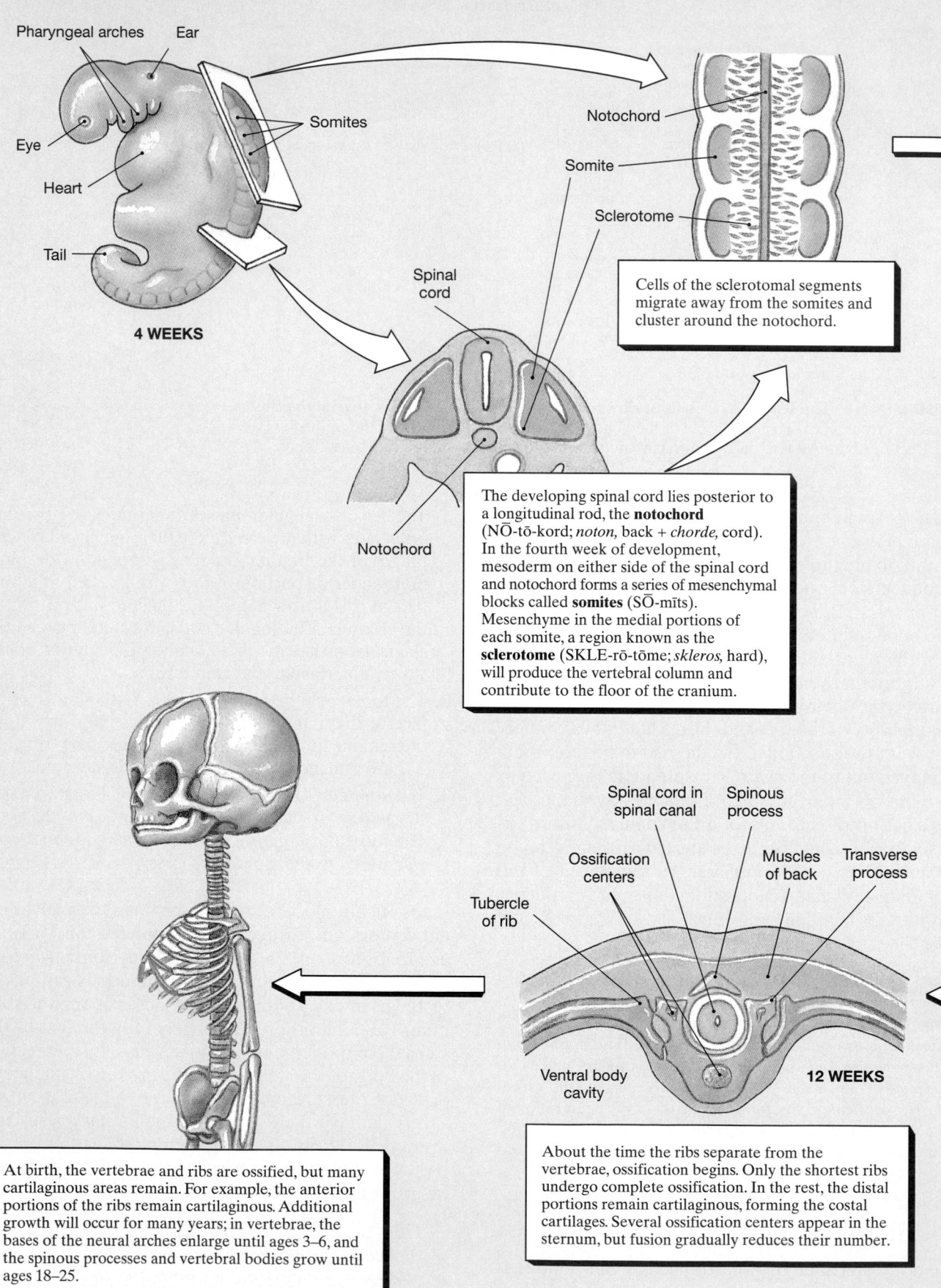

4 WEEKS

Pharyngeal arches

Ear

Eye

Heart

Tail

Somites

Notochord

Somite

Sclerotome

Cells of the sclerotomal segments migrate away from the somites and cluster around the notochord.

Spinal cord

Notochord

The developing spinal cord lies posterior to a longitudinal rod, the **notochord** (NŌ-tō-kord; *noton,* back + *chorde,* cord). In the fourth week of development, mesoderm on either side of the spinal cord and notochord forms a series of mesenchymal blocks called **somites** (SŌ-mīts). Mesenchyme in the medial portions of each somite, a region known as the **sclerotome** (SKLE-rō-tōme; *skleros,* hard), will produce the vertebral column and contribute to the floor of the cranium.

Spinal cord in spinal canal

Spinous process

Ossification centers

Muscles of back

Transverse process

Tubercle of rib

Ventral body cavity

12 WEEKS

About the time the ribs separate from the vertebrae, ossification begins. Only the shortest ribs undergo complete ossification. In the rest, the distal portions remain cartilaginous, forming the costal cartilages. Several ossification centers appear in the sternum, but fusion gradually reduces their number.

At birth, the vertebrae and ribs are ossified, but many cartilaginous areas remain. For example, the anterior portions of the ribs remain cartilaginous. Additional growth will occur for many years; in vertebrae, the bases of the neural arches enlarge until ages 3–6, and the spinous processes and vertebral bodies grow until ages 18–25.

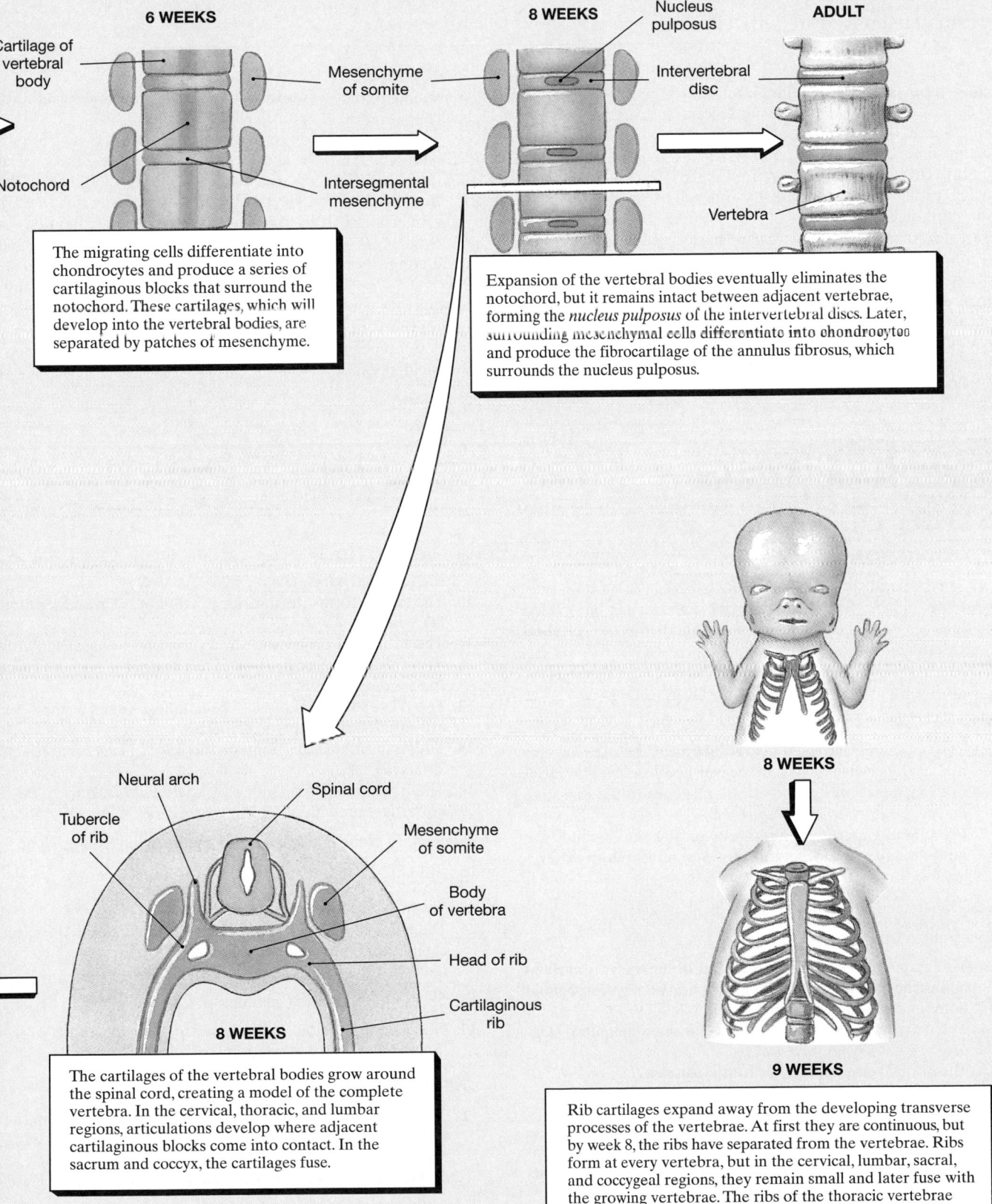

6 WEEKS

Cartilage of vertebral body

Notochord

The migrating cells differentiate into chondrocytes and produce a series of cartilaginous blocks that surround the notochord. These cartilages, which will develop into the vertebral bodies, are separated by patches of mesenchyme.

Mesenchyme of somite

Intersegmental mesenchyme

8 WEEKS

Nucleus pulposus

Intervertebral disc

ADULT

Vertebra

Expansion of the vertebral bodies eventually eliminates the notochord, but it remains intact between adjacent vertebrae, forming the *nucleus pulposus* of the intervertebral discs. Later, surrounding mesenchymal cells differentiate into chondrocytes and produce the fibrocartilage of the annulus fibrosus, which surrounds the nucleus pulposus.

8 WEEKS

Neural arch

Tubercle of rib

Spinal cord

Mesenchyme of somite

Body of vertebra

Head of rib

Cartilaginous rib

8 WEEKS

The cartilages of the vertebral bodies grow around the spinal cord, creating a model of the complete vertebra. In the cervical, thoracic, and lumbar regions, articulations develop where adjacent cartilaginous blocks come into contact. In the sacrum and coccyx, the cartilages fuse.

9 WEEKS

Rib cartilages expand away from the developing transverse processes of the vertebrae. At first they are continuous, but by week 8, the ribs have separated from the vertebrae. Ribs form at every vertebra, but in the cervical, lumbar, sacral, and coccygeal regions, they remain small and later fuse with the growing vertebrae. The ribs of the thoracic vertebrae continue to enlarge, following the curvature of the body wall. When they reach the ventral midline, they fuse with the cartilages of the sternum.

SELECTED CLINICAL TERMINOLOGY

Terms Discussed in This Chapter

chest tube: A drain installed after thoracic surgery to permit removal of blood and pleural fluid. *(p. 227)*

craniostenosis (krā-nē-ō-sten-Ō-sis): Premature closure of one or more fontanels, which can lead to unusual distortions of the skull. *(p. 215)*

deviated (nasal) septum: A bent nasal septum that slows or prevents sinus drainage. *(p. 214)*

kyphosis (kī-FŌ-sis): Abnormal exaggeration of the thoracic curvature that produces a humpback appearance. *(p. 218 and AM)*

lordosis (lor-DŌ-sis): Abnormal lumbar curvature giving a swayback appearance. *(p. 218 and AM)*

microcephaly (mī-krō-SEF-uh-lē): An undersized head resulting from genetic or developmental abnormalities. *(p. 215)*

scoliosis (skō-lē-Ō-sis): Abnormal lateral curvature of the spine. *(p. 218 and AM)*

sinusitis: Inflammation and congestion of the sinuses. *(p. 214)*

spina bifida (SPĪ-nuh BI-fi-duh): A condition resulting from failure of the vertebral laminae to unite during development; commonly associated with developmental

abnormalities of the brain and spinal cord. *(p. 219 and AM)*

thoracentesis (tho-ra-sen-TĒ -sis), or **thoracocentesis** (tho-ra-kō-sen-TĒ-sis): The penetration of the thoracic wall along the superior border of one of the ribs. *(p. 227)*

TMJ syndrome: A painful condition resulting from misalignment of the mandible at the temporomandibular joint. *(p. 213)*

whiplash: An injury resulting from a sudden change in the body position that can injure the cervical vertebrae. *(p. 221)*

CHAPTER REVIEW

On-line resources for this chapter are on our World Wide Web site at:
http://www.prenhall.com/martini/fap

STUDY OUTLINE

INTRODUCTION, p. 198

1. The skeletal system consists of the axial skeleton and the appendicular skeleton. The **axial skeleton** can be divided into the **skull,** the **auditory ossicles** and **hyoid,** the **vertebral column,** and the **thoracic cage.** *(Figure 7-1)*

2. The appendicular skeleton includes the pectoral and pelvic girdles that support the upper and lower limbs.

THE SKULL, p. 199

1. The **cranium,** composed of **cranial bones,** encloses the **cranial cavity,** a division of the dorsal body cavity. The **facial bones** protect and support the entrances to the digestive and respiratory tracts. *(Figure 7-2)*

2. Prominent superficial landmarks on the skull include the **lambdoidal, coronal, sagittal,** and **squamosal sutures.** *(Figure 7-3)*

FOCUS: The Individual Bones of the Skull, p. 204

Bones of the Cranium, p. 204

3. The cranial bones are the **occipital bone;** the two **parietal bones;** the **frontal bone;** the two **temporal bones;** the **sphenoid bone;** and the **ethmoid bone.** *(Figures 7-3–7-9)*

4. The occipital bone surrounds the **foramen magnum.** *(Figures 7-3–7-5)*

5. The frontal bone contains the **frontal sinuses.** *(Figures 7-4, 7-6)*

Bones of the Face, p. 209

6. The bones of the face are the **maxillary bones,** the **palatine bones,** the **nasal bones,** the **vomer,** the **inferior nasal conchae,** the **zygomatic bones,** the **lacrimal bones,** the **mandible,** and the **hyoid bone.** *(Figures 7-3, 7-4, 7-10–7-12)*

7. The left and right maxillary bones, or **maxillae,** are the largest facial bones; they form the upper jaw. *(Figures 7-3, 7-4, 7-10)*

8. The palatine bones are small, L-shaped bones that form the posterior portions of the hard palate and contribute to the floor of the orbital cavities. *(Figures 7-3, 7-4, 7-10)*

9. The paired nasal bones extend to the superior border of the **external nares.** *(Figures 7-3, 7-4, 7-11)*

10. The vomer forms the inferior portion of the **nasal septum.** *(Figures 7-3, 7-4, 7-11)*

11. The **temporal process** of the zygomatic bone articulates with the **zygomatic process** of the temporal bone to form the **zygomatic arch.** *(Figures 7-3, 7-11)*

12. The paired lacrimal bones, the smallest bones of the face, are situated medially in each **orbit.** *(Figures 7-3, 7-11)*

13. The mandible is the bone of the lower jaw. *(Figures 7-3, 7-4, 7-12)*

14. The hyoid bone, suspended by *stylohyoid ligaments,* does not articulate with other bones. *(Figure 7-12)*

A Summary of the Foramina and Fissures of the Skull, p. 212

The Orbital and Nasal Complexes, p. 213

15. Seven bones form each **orbital complex.** *(Figure 7-13)*

16. The **nasal complex** includes the bones that enclose the nasal cavities and the **paranasal sinuses,** hollow airways that connect with the nasal passages. *(Figure 7-14)*

The Skulls of Infants and Children, p. 214

17. Fibrous connections at **fontanels** permit the skulls of infants and children to continue growing. *(Figure 7-15)*

THE VERTEBRAL COLUMN, p. 215

1. There are 7 **cervical vertebrae** (the first articulates with the skull), 12 **thoracic vertebrae** (which articulate with ribs), and 5 **lumbar vertebrae** (the last articulates with the sacrum). The **sacrum** and **coccyx** consist of fused vertebrae. *(Figure 7-16)*

Spinal Curvature, p. 218

2. The spinal column has four **spinal curves.** The **thoracic** and **sacral curvatures** are called **primary,** or **accommodation, curves;** the **lumbar** and **cervical curvatures** are known as **secondary,** or **compensation, curves.** *(Figures 7-16, 7-17)*

Vertebral Anatomy, p. 219

3. A typical vertebra has a **body** (*centrum*) and a **vertebral arch** (*neural arch*) and articulates with adjacent vertebrae at the **superior** and **inferior articular processes.** *(Figure 7-18)*
4. Adjacent vertebrae are separated by **intervertebral discs.** Spaces between successive **pedicles** form the **intervertebral foramina.** *(Figure 7-18)*

Vertebral Regions, p. 220

5. Cervical vertebrae are distinguished by the shape of the body, the relative size of the vertebral foramen, the presence of **costal processes** with **transverse foramina,** and notched **spinous processes.** These vertebrae include the **atlas, axis,** and **vertebra prominens.** *(Figure 7-19; Table 7-2)*
6. Thoracic vertebrae have distinctive heart-shaped bodies, long, slender spinous processes, and articulations for the ribs. *(Figures 7-20, 7-23; Table 7-2)*
7. The lumbar vertebrae are the most massive and least mobile; they are subjected to the greatest strains. *(Figure 7-21; Table 7-2)*

8. The sacrum protects reproductive, digestive, and urinary organs and articulates with the pelvic girdle and with the fused elements of the coccyx. *(Figure 7-22)*

THE THORACIC CAGE, p. 225

1. The skeleton of the thoracic cage consists of the thoracic vertebrae, the ribs, and the sternum. The **ribs** and **sternum** form the *rib cage. (Figure 7-23)*

The Ribs, p. 225

2. Ribs 1–7 are **true,** or **vertebrosternal, ribs.** Ribs 8–12 are called **false ribs;** they include the **vertebrochondral ribs** and two pairs of **floating (vertebral) ribs.** A typical rib has a **head,** or **capitulum;** a **neck;** a **tubercle,** or **tuberculum;** an **angle;** and a **body,** or **shaft.** A **costal groove** marks the path of nerves and blood vessels. *(Figures 7-23, 7-24)*

The Sternum, p. 227

3. The sternum consists of a **manubrium,** a **body,** and a **xiphoid process.** *(Figure 7-23)*

REVIEW QUESTIONS

Level 1   Reviewing Facts and Terms

Match each numbered item with the most closely related lettered item. Use letters for answers in the spaces provided.

___ 1. foramina
___ 2. sinuses
___ 3. sutures
___ 4. calvaria
___ 5. auditory ossicles
___ 6. hypophyseal fossa
___ 7. lacrimal bones
___ 8. accommodation curve
___ 9. compensation curve
___ 10. costae
___ 11. C_1
___ 12. C_2

a. skullcap
b. tear ducts
c. atlas
d. lumbar and cervical
e. air-filled chambers
f. axis
g. ear bones
h. ribs
i. passageways
j. sella turcica
k. thoracic and sacral
l. immovable joints

13. The axial skeleton consists of the bones of the
 (a) pectoral and pelvic girdles
 (b) skull, thorax, and vertebral column
 (c) arms, legs, hands, and feet
 (d) limbs, pectoral girdle, and pelvic girdle
14. The appendicular skeleton consists of the bones of the
 (a) pectoral and pelvic girdles
 (b) skull, thorax, and vertebral column
 (c) arms, legs, hands, and feet
 (d) limbs, pectoral girdle, and pelvic girdle
15. Which of the following lists contains *only* bones of the cranium?
 (a) frontal, parietal, occipital, sphenoid
 (b) frontal, occipital, zygomatic, parietal
 (c) occipital, sphenoid, temporal, lacrimal
 (d) mandible, maxilla, nasal, zygomatic

16. The facial bones include the
 (a) lacrimal, nasal, maxillary, mandible
 (b) frontal, lacrimal, zygomatic, sphenoid
 (c) maxillary, mandible, ethmoid, sphenoid
 (d) ethmoid, sphenoid, temporal, parietal
17. The boundaries between skull bones are immovable joints called
 (a) foramina   (b) fontanels
 (c) lacunae   (d) sutures
18. The major sutures of the skull are the
 (a) frontal, parietal, occipital, sphenoid
 (b) frontal, lambdoidal, occipital, coronal
 (c) lambdoidal, coronal, sagittal, squamosal
 (d) coronal, sagittal, frontal, parietal
19. Of the following bones, which is unpaired?
 (a) vomer   (b) maxillary
 (c) palatine   (d) nasal
20. The cribriform plate, crista galli, and superior conchae are parts of the
 (a) parietal bone   (b) occipital bone
 (c) sphenoid bone   (d) ethmoid bone
21. The bone that houses the inner ear structures associated with hearing and balance is the
 (a) temporal   (b) occipital
 (c) parietal   (d) sphenoid
22. The bony enclosure that forms a prominent depression on the superior surface of the sphenoid bone and houses the pituitary gland is the
 (a) clinoid process
 (b) sella turcica
 (c) external auditory canal
 (d) auditory tube

23. The hyoid bone
    (a) forms the inferior portion of the skull
    (b) is attached to the zygomatic bone
    (c) does not articulate with any other bone
    (d) is a cartilaginous structure
24. The membranous areas between the cranial bones of the fetal skull are
    (a) fontanels   (b) sutures
    (c) Wormian bones   (d) foramina
25. The part of the vertebra that transfers weight along the axis of the vertebral column is (are) the
    (a) neural arch   (b) lamina
    (c) pedicles   (d) body
26. Progressing inferiorly, the components of the sternum include the
    (a) xiphoid process, body, manubrium
    (b) body, manubrium, xiphoid process
    (c) manubrium, body, xiphoid process
    (d) manubrium, xiphoid process, body

27. Which five bones make up the facial complex?
28. Which four major sutures make up the boundaries between the skull bones?
29. What is the primary function of the vomer?
30. Which bones contain the paranasal sinuses?
31. What characteristic of the hyoid bone makes that bone unique?
32. What purpose do the fontanels serve during delivery?
33. Beginning at the skull, cite the regions of the vertebral column and list the number of vertebrae in each region of the adult vertebral column.
34. List the four spinal curves that are apparent in the adult spinal column when viewed from the side.
35. What two primary functions are performed by the thoracic cage?
36. Why are the upper seven pairs of ribs called *true* ribs and the lower five pairs called *false* ribs?

Level 2   Reviewing Concepts

37. The atlas (C_1) can be distinguished from the other vertebrae by
    (a) the presence of anterior and posterior vertebral arches
    (b) the lack of a body
    (c) the presence of superior facets and inferior articular facets
    (d) a, b, and c are correct
38. What is the relationship between the temporal bone and the ear?
39. What is the relationship between the sphenoid bone and the pituitary gland?
40. Unlike that of other animals, the skull of humans is balanced above the vertebral column, allowing an upright posture. What anatomical adaptations are necessary in humans to maintain the balance of the skull?

41. The structural features and skeletal components of the sternum make it a part of the axial skeleton that is important in a variety of clinical situations. If you were teaching this information to prospective nursing students, what clinical applications would you cite?
42. Why are ruptured intervertebral discs more common in lumbar vertebrae and dislocations and fractures more common in cervical vertebrae?
43. We generally associate a "runny" nose with a cold. Does this nasal condition serve a purpose? What is the cause of this type of reaction?
44. Why is it important to keep your back straight when you lift a heavy object?

Level 3   Critical Thinking and Clinical Applications

45. Tess is diagnosed with a disease that affects the membranes surrounding the brain. The physician tells her family that the disease is caused by an airborne virus. Explain how this virus could have entered the cranium.
46. Joe is 40 years old and 30 pounds overweight. Like many middle-aged men, Joe carries most of this extra weight in his abdomen and jokes with his friends about his "beer gut." During an annual physical, Joe's physician advises Joe that his spine is developing an abnormal curvature. Why is the curvature of Joe's spine changing, and what is this condition called?
47. A babysitter is on trial for the death of a 10-month-old infant. The prosecutor contends that the child died as the result of being violently shaken. The defense claims that the child's head became stuck in the slats of his crib and, in trying to twist

free, the child broke his neck. The medical examiner testifies that the child died as the result of a compression of the spinal cord between the sixth and seventh cervical vertebrae. The superior articular processes of the seventh cervical vertebra and the inferior articular processes of the sixth cervical vertebra were fractured, and the processes on the right side were laterally displaced, causing the sixth vertebra to slide laterally across the seventh, damaging the spinal cord. On the basis of this evidence, whom do you believe? Why?
48. While working at an excavation, an archaeologist finds several small skull bones. She examines the frontal, parietal, and occipital bones and concludes that the skulls are those of children not yet 1 year old. How can she tell their ages from examining the bones?

CHAPTER 8

The Appendicular Skeleton

*C*linging to a cliff face by your fingers and toes places unusually severe stresses on the appendicular skeleton and on the associated muscles and supporting connective tissues. Even if you do not climb mountains, your appendicular skeleton plays a major role in your daily life. Make a mental list of all the things you have done with your arms or legs in the past day. Standing, walking, writing, turning pages, eating, dressing, shaking hands, waving—the list quickly becomes unwieldy. Your axial skeleton protects and supports internal organs, and it participates in vital functions, such as respiration. But it is your appendicular skeleton that gives you control over your environment, changes your position in space, and provides mobility. In this chapter, we examine the components of the appendicular skeleton; in Chapter 9, we will consider the ways those elements interact during normal movements.

CHAPTER OUTLINE AND OBJECTIVES

The **appendicular skeleton** includes the bones of the limbs and the supporting elements, or *girdles,* that connect them to the trunk (Figure 8-1•). The descriptions in this chapter emphasize surface features that have functional importance, such as the attachment sites for skeletal muscles or the paths of major nerves and blood vessels.

THE PECTORAL GIRDLE AND UPPER LIMBS

Each arm articulates (that is, forms a joint) with the trunk at the **shoulder girdle,** or **pectoral girdle.** The shoulder girdle consists of two S-shaped *clavicles*

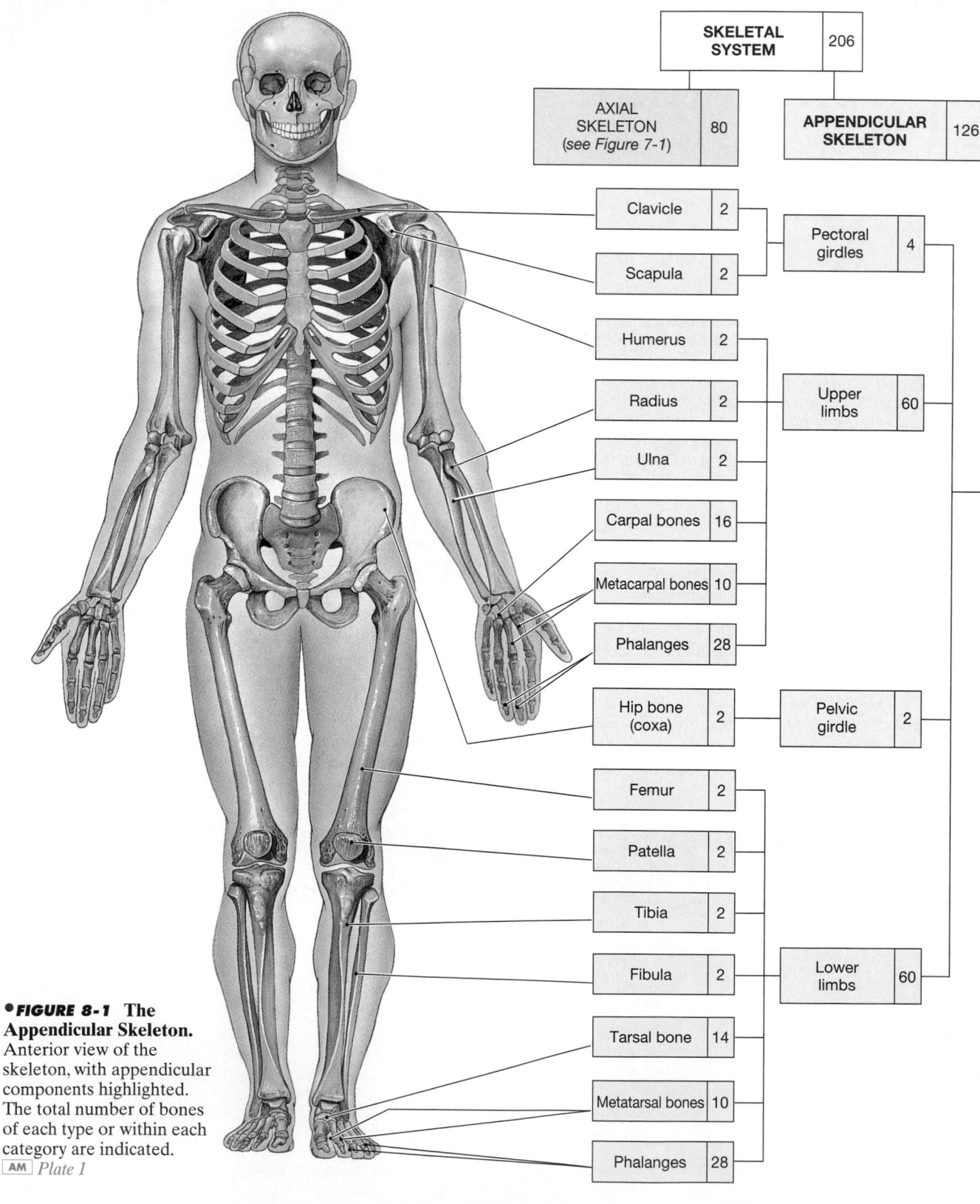

•*FIGURE 8-1* **The Appendicular Skeleton.** Anterior view of the skeleton, with appendicular components highlighted. The total number of bones of each type or within each category are indicated.

AM *Plate 1*

(KLAV-i-kulz; collarbones) and two broad, flat *scapulae* (SKAP-ū-lē; singular, *scapula,* SKAP-ū-luh; shoulder blade). Each clavicle articulates with the manubrium of the sternum. These articulations are the *only* direct connections between the pectoral girdle and the axial skeleton. Skeletal muscles support and position the scapulae, which have no direct bony or ligamentous connections to the thoracic cage. As a result, the shoulders are extremely mobile but not very strong.

The Pectoral Girdle

Movements of the clavicles and scapulae position the shoulder joints and provide a base for arm movement. Once the shoulder joints are in position, muscles that originate on the pectoral girdle help move the upper limbs. The surfaces of the scapulae and clavicles are extremely important as sites for muscle attachment. Where they attach, major muscles leave marks in the form of bony ridges and flanges. Other bone markings, such as sulci or foramina, indicate the position of nerves or the passage of blood vessels that control the muscles and nourish the muscles and bones, respectively.

The Clavicles

The **clavicles** are S-shaped bones that originate at the superior and lateral border of the manubrium of the sternum, lateral to the jugular notch (Figure 8-2•). From the roughly pyramidal **sternal end,** each clavicle curves laterally and dorsally until it articulates with a process of the scapula, the *acromion* (a-KRŌ-mē-on). The broad, flat, **acromial end** of the clavicle is broader than the sternal end.

The smooth, superior surface of the clavicle lies just beneath the skin. The rough, inferior surface of the acromial end is marked by prominent lines and tubercles. These surface features are attachment sites for muscles and ligaments of the shoulder.

You can explore the interaction between scapulae and clavicles. With your fingers in your jugular notch, locate the clavicle on either side. When you move your shoulders you can feel the clavicles change their positions. Because the clavicles are so close to the skin, you can trace one laterally until it articulates with the scapula.

The clavicles are relatively small and fragile, and fractures of the clavicle are fairly common. For example, you can fracture a clavicle as the result of a simple fall if you land on your hand with an outstretched arm. Because it breaks rather easily and causes intense pain, instructors teaching self-defense often advise their students to strike an opponent's clavicle in an emergency. Fortunately, in view of the clavicle's vulnerability, most clavicular fractures heal rapidly without a cast.

The Scapulae

The anterior surface of the **body** of each **scapula** forms a broad triangle (Figure 8-3a•). The three sides of that triangle are the **superior border;** the **medial border,** or *vertebral border;* and the **lateral border,** or *axillary border* (*axilla,* armpit). Muscles that position the scapula attach along these edges. The corners of the triangle are called the **superior angle,** the **inferior angle,** and the **lateral angle.** The lateral angle, or *head* of the scapula, forms a broad process that supports the cup-shaped **glenoid cavity.** At the glenoid cavity, the scapula articulates with the *humerus,* the proximal bone of the upper limb. This articulation is the shoulder joint, also known as the *glenohumeral,* or *scapulohumeral, joint.* The anterior surface of the body of the scapula is relatively smooth and concave. The depression in the anterior surface is called the **subscapular fossa.**

Two large scapular processes extend over the margin of the glenoid cavity (Figure 8-3b•) superior to the head of the humerus. The smaller, anterior projection is the **coracoid** (KOR-uh-koyd) **process.** The **acromion process** is the larger, posterior process. If you run your fingers along the superior surface of the shoulder joint, you will feel this process. The acromion articulates with the clavicle at the *acromioclavicular joint.* Both the acromion and coracoid processes are attached to ligaments and tendons associated with the shoulder joint, which we will consider further in Chapter 9.

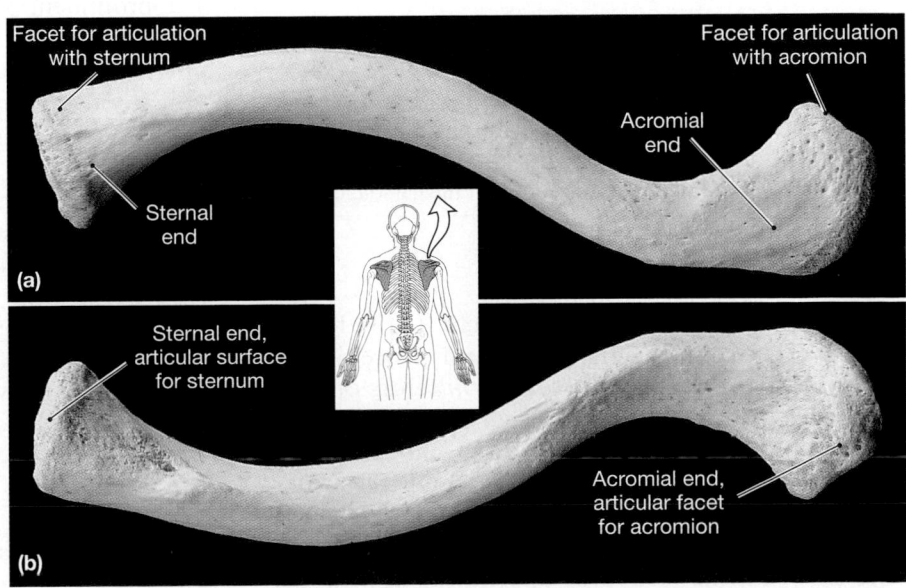

Facet for articulation with sternum

Facet for articulation with acromion

Acromial end

Sternal end

(a)

Sternal end, articular surface for sternum

Acromial end, articular facet for acromion

(b)

•*FIGURE 8-2* **The Clavicle. (a)** Superior and **(b)** inferior views of the right clavicle.

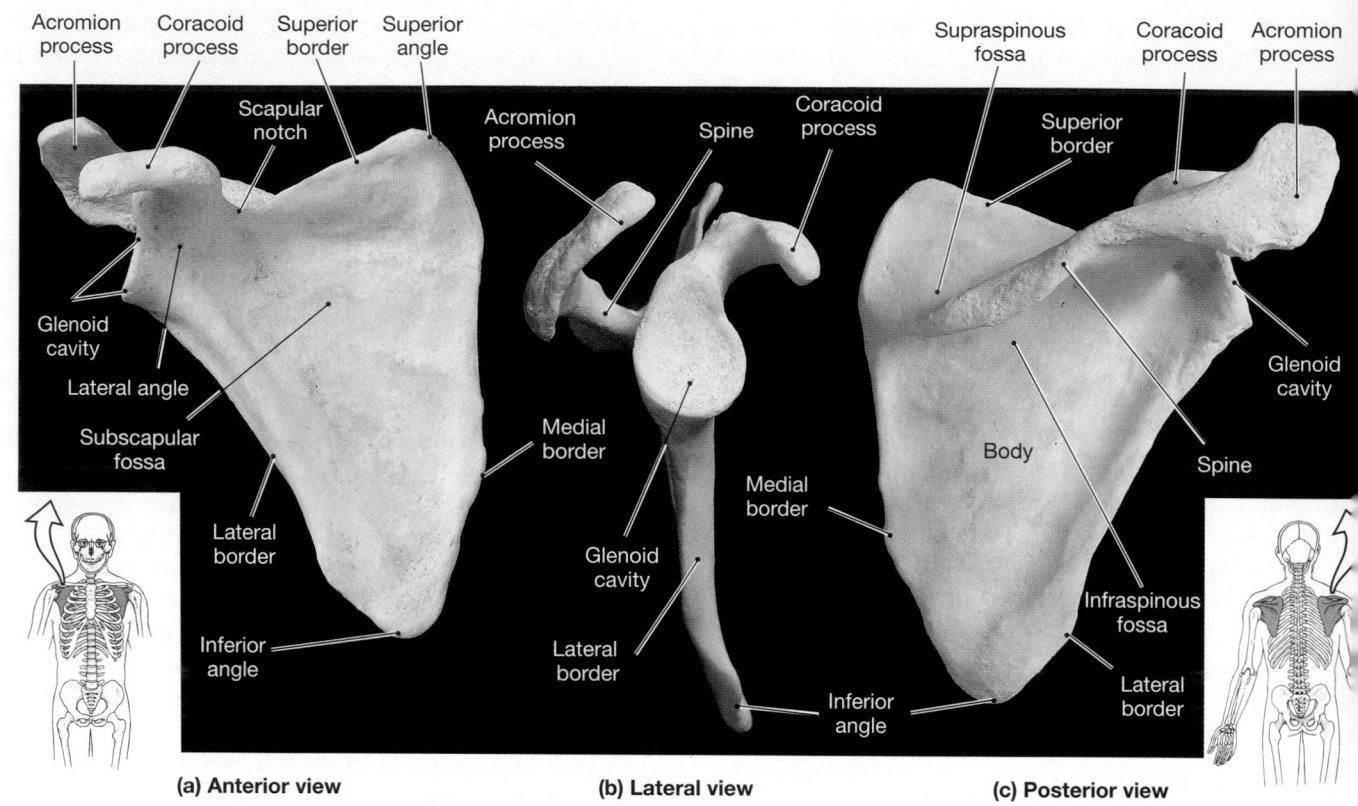

Acromion process · **Coracoid process** · **Superior border** · **Superior angle** · Scapular notch · Glenoid cavity · Lateral angle · Subscapular fossa · Lateral border · Inferior angle

Acromion process · Spine · **Coracoid process** · Medial border · Glenoid cavity · Lateral border

Supraspinous fossa · **Coracoid process** · **Acromion process** · Superior border · Glenoid cavity · Body · Spine · Medial border · Infraspinous fossa · Lateral border · Inferior angle

(a) Anterior view   **(b) Lateral view**   **(c) Posterior view**

•*FIGURE 8-3* **The Scapula.** **(a)** Anterior, **(b)** lateral, and **(c)** posterior views of the right scapula.

The acromion process is continuous with the **scapular spine** (Figure 8-3c●). This ridge crosses the scapular body before ending at the medial border. The scapular spine divides the convex dorsal surface of the body into two regions. The area superior to the spine constitutes the **supraspinous fossa** (*supra,* above); the region inferior to the spine is the **infraspinous fossa** (*infra,* beneath). The spine is an attachment site for two large muscles: the *supraspinatus* and the *infraspinatus.* The entire posterior surface is marked by small ridges and lines where smaller muscles attach to the scapula.

☑ Why would a broken clavicle affect the mobility of the scapula?

☑ What bone articulates with the scapula at the glenoid cavity?

The Upper Limbs

The skeleton of the upper limbs includes the bones of the arms, forearms, wrists, and hands. Note that in anatomical descriptions the term *arm* refers only to the proximal portion of the upper limb, not to the entire limb. The anatomical usage and common usage are thus quite different; we shall use *upper limb* and *arm* the way many people use *arm* (for entire limb) and *upper arm* (for the proximal portion). Note also that the term *digit* applies to both fingers and toes.

We will examine the bones of the right upper limb. The arm, or *brachium,* contains one bone, the **humerus.** The humerus extends from the scapula to the elbow. At the proximal end of the humerus, the round **head** articulates with the scapula (Figure 8-4●). At its distal end, the humerus articulates with the bones of the forearm (or *antebrachium*), the *radius* and the *ulna.*

The Humerus

The prominent **greater tubercle** of the humerus is located near the head, on the lateral surface of the epiphysis (Figure 8-4●). The greater tubercle establishes the lateral contour of the shoulder. You can verify its position by feeling for a bump situated a few centimeters from the tip of the acromion process. The **lesser tubercle** lies on the anterior and medial surface of the epiphysis, separated from the greater tubercle by the **intertubercular groove,** or *intertubercular sulcus.* Both tubercles are important sites for muscle attachment; a large tendon runs along the groove. The **anatomical neck** marks the extent of the joint capsule. The anatomical neck lies between the tubercles and the articular surface of the head. The narrower **surgical neck** corresponds to the metaphysis of the growing bone. This name reflects the fact that fractures typically occur at this site.

The proximal **shaft** of the humerus is round in section. The **deltoid tuberosity** runs along the lateral surface of the shaft, extending more than halfway down its length.

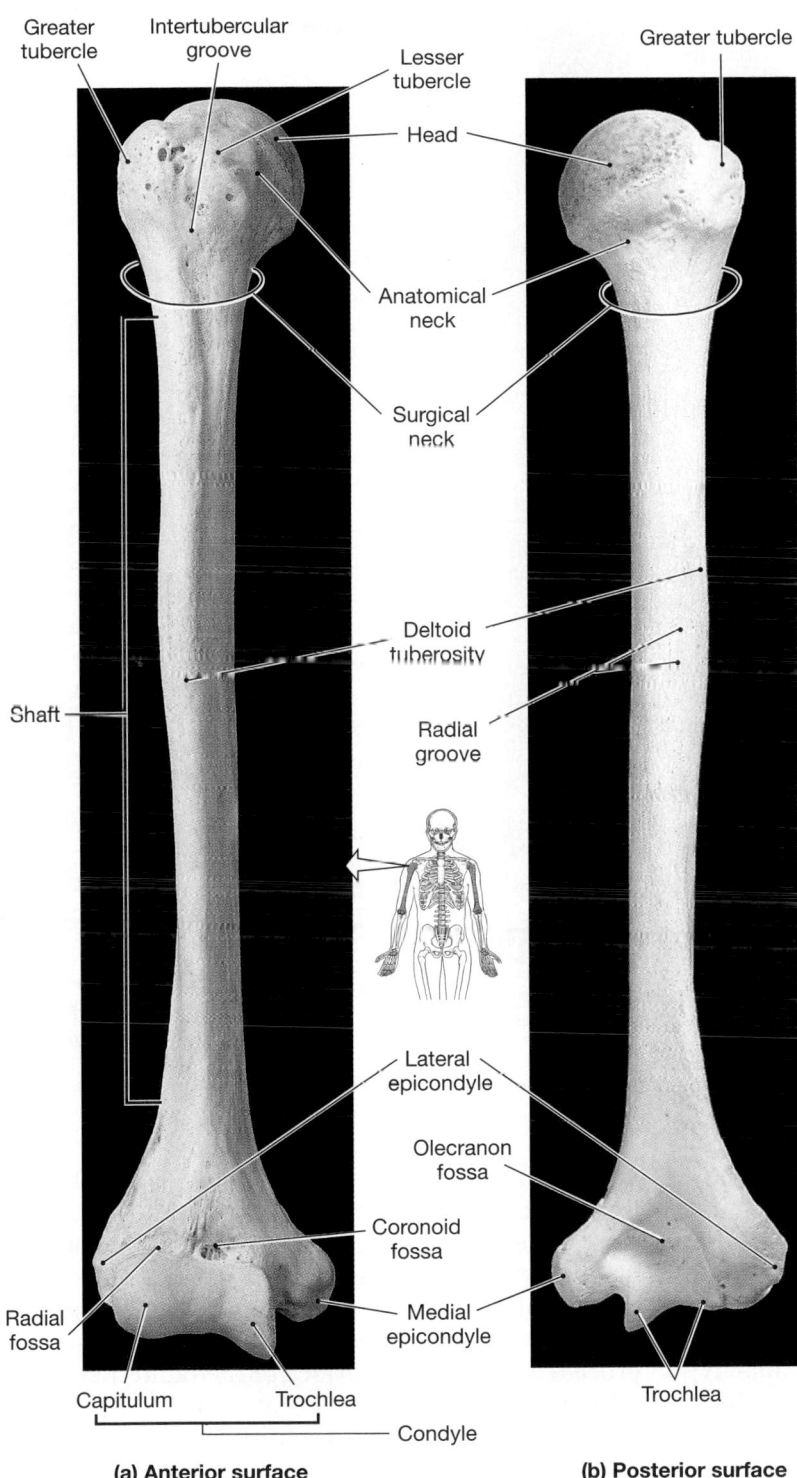

Greater tubercle
Intertubercular groove
Lesser tubercle
Head
Anatomical neck
Surgical neck
Shaft
Deltoid tuberosity
Radial groove
Lateral epicondyle
Olecranon fossa
Coronoid fossa
Radial fossa
Medial epicondyle
Capitulum
Trochlea
Condyle

(a) Anterior surface

Greater tubercle
Trochlea

(b) Posterior surface

•FIGURE 8-4 **The Humerus.** **(a)** The anterior and **(b)** the posterior surfaces of the right humerus.

It is named after the *deltoid muscle,* which attaches to it. On the anterior surface of the shaft, the intertubercular groove continues alongside the deltoid tuberosity.

On the posterior surface (Figure 8-4b•), the deltoid tuberosity ends at the **radial groove.** This depression marks the path of the *radial nerve,* a large nerve that provides sensory information from the back of the hand and motor control over large muscles that straighten the elbow. Distal to the radial groove, the posterior surface of the humerus is relatively flat. Near the distal end, the shaft expands to either side, forming a broad triangle. The **medial** and **lateral epicondyles** project to either side and provide additional surface area for muscle attachment. The *ulnar nerve* crosses the posterior surface of the medial epicondyle. A blow at the humeral side of the elbow joint can strike this nerve and produce a temporary numbness and paralysis of muscles on the anterior surface of the forearm. For this reason, the area is sometimes called the *funny bone.*

At the articular **condyle,** the humerus articulates with the bones of the forearm— the radius and the ulna. A low ridge crosses the articular condyle, dividing it into two articular regions, the trochlea and the capitulum (Figure 8-4a,b•). The **trochlea** (*trochlea,* a pulley) is the spool-shaped medial portion that extends from the base of the **coronoid** (*corona,* crown) **fossa** on the anterior surface to the **olecranon** (ō-LEK-ruh-non) **fossa** on the posterior surface (Figure 8-4b•). These depressions accept projections from the ulnar surface as the elbow approaches the limits of its range of motion. The rounded **capitulum** forms the lateral surface of the condyle. A shallow **radial fossa** next to the capitulum accommodates a portion of the radial head as the forearm approaches the humerus.

The Ulna

The **ulna** and **radius** are parallel bones that support the forearm. In the anatomical position, the ulna lies medial to the radius.

The **olecranon,** or *olecranon process,* is the point of the elbow (Figure 8-5a•). On the anterior surface of the proximal epiphysis (Figure 8-5b•), the **trochlear notch** (or *semilunar notch*) articulates with the trochlea of the humerus at the elbow joint, also known as the *olecranal joint,* or *humeroulnar joint.* The olecranon forms the superior lip of the trochlear notch, and the **coronoid process** forms its inferior lip. At the limit of *extension,* with the forearm and arm forming a straight line, the olecranon swings into the olecranon fossa on the posterior surface of the humerus. At the limit of *flexion,* with the arm and forearm forming a V, the coro-

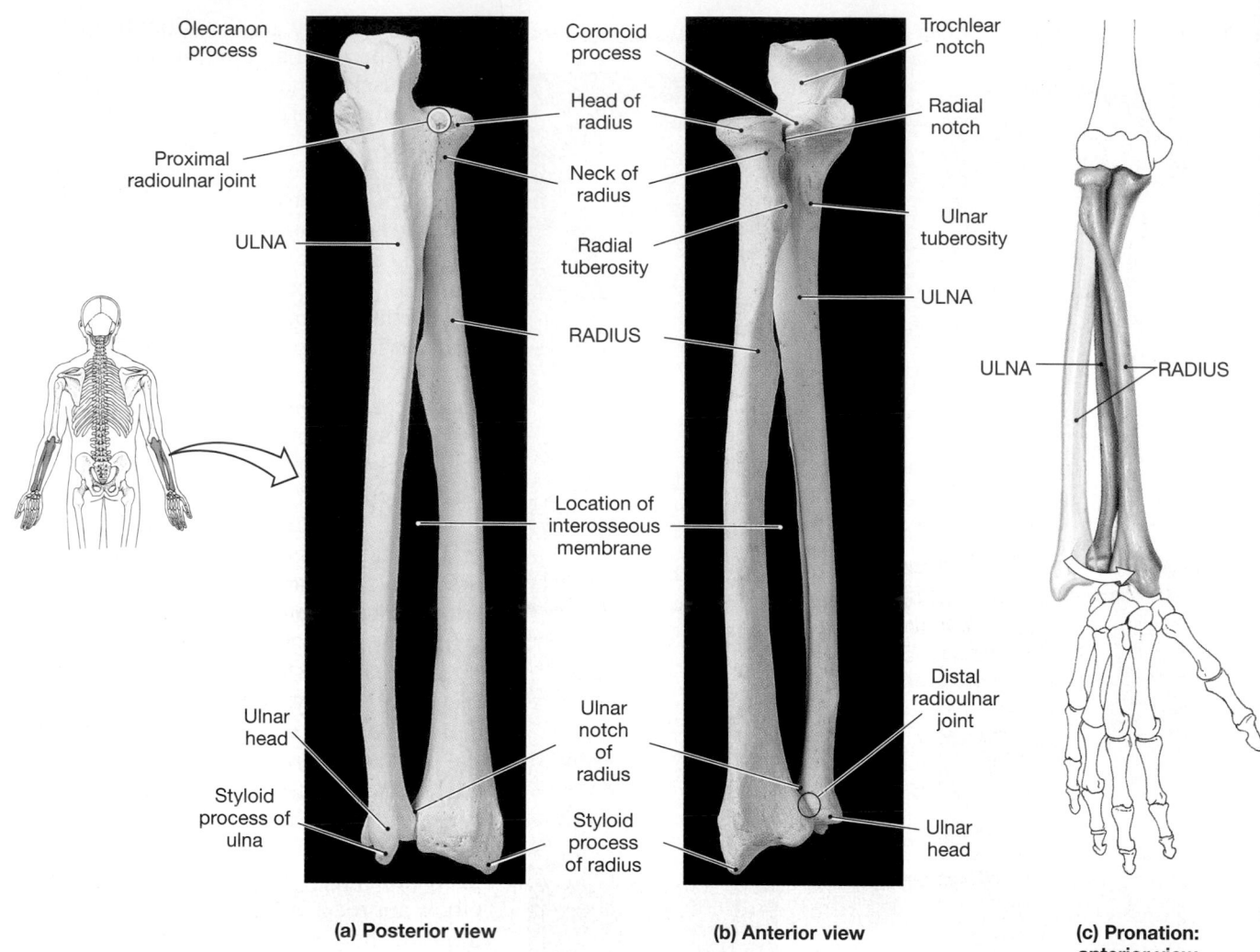

•FIGURE 8-5 **The Radius and Ulna.** The right radius and ulna in **(a)** posterior view, **(b)** anterior view, and **(c)** during pronation.

noid process projects into the coronoid fossa on the anterior humeral surface. Lateral to the coronoid process, a smooth **radial notch** accommodates the head of the radius at the *proximal radioulnar joint.*

When viewed in cross section, the shaft of the ulna is roughly triangular. A fibrous sheet, the **interosseous membrane,** connects the lateral margin of the ulna to the radius. Near the wrist, the ulnar shaft narrows before ending at a disc-shaped **head** that bears a short **styloid process** (*styloid,* long and pointed). A triangular **articular cartilage** attaches to the styloid process, and this cartilage separates the ulnar head from the bones of the wrist. As a result, the head of the ulna is not part of the wrist joint. The *distal radioulnar joint* lies near the lateral edge of the ulnar head.

The Radius

The radius is the lateral bone of the forearm. The disc-shaped radial **head** articulates with the capitulum of the humerus. A narrow **neck** extends from the radial head to the **radial tuberosity,** which marks the

attachment site of the *biceps brachii*, a muscle that flexes the forearm. The shaft of the radius curves along its length. It also enlarges, and the distal portion of the radius is considerably larger than the distal portion of the ulna. The distal end of the radius articulates with the bones of the wrist. The **styloid process** on the lateral surface of the radius helps stabilize this joint.

The medial surface of the distal end of the radius articulates with the ulnar head at the **ulnar notch.** The proximal radioulnar articulation permits *rotation*, letting the ulnar notch of the radius roll across the rounded surface of the ulnar head (Figure 8-5b•). This movement is called **pronation** (Figure 8-5c•). The reverse movement, which returns the radius and ulna to their anatomical position, is called **supination.**

The Carpal Bones

The eight **carpal bones** of the wrist, or **carpus,** form two rows. There are four **proximal carpal** and four **distal carpal bones.**

THE PROXIMAL CARPAL BONES The proximal carpal bones are the *scaphoid, lunate, triangular,* and *pisiform* (PIS-i-form) *bones* (Figure 8-6a●):

- The **scaphoid bone** is the proximal carpal bone on the lateral border of the wrist, and the carpal bone closest to the styloid process of the radius.
- The comma-shaped **lunate** (*luna,* moon) **bone** lies medial to the scaphoid bone. Like the scaphoid bone, the lunate bone articulates with the radius.
- The **triangular bone,** or *triquetral bone,* is medial to the lunate bone. It has the shape of a small pyramid. The triangular bone articulates with the cartilage that separates the ulnar head from the wrist.
- The small, pea-shaped **pisiform bone** sits anterior to the triangular bone.

THE DISTAL CARPAL BONES The distal carpal bones are the *trapezium, trapezoid, capitate,* and *hamate bones* (Figure 8-6a●):

- The **trapezium** is the lateral bone of the distal row; its proximal surface articulates with the scaphoid bone.
- The wedge-shaped **trapezoid bone** lies medial to the trapezium. Like the trapezium, it has a proximal articulation with the scaphoid bone.
- The **capitate bone,** the largest carpal bone, sits between the trapezoid bone and the hamate bone.
- The **hamate** (*hamatum,* hooked) **bone** is the medial distal carpal bone.

The carpal bones articulate with one another at joints that permit limited sliding and twisting. Ligaments interconnect the carpal bones and help stabilize the wrist joint. The tendons of muscles that flex the fingers pass across the anterior surface of the wrist, sandwiched between the intercarpal ligaments and a broad, superficial transverse ligament called the *flexor retinaculum.* Inflammation of the connective tissues between the flexor retinaculum and the carpal bones can compress the tendons and adjacent sensory nerves, producing pain and a loss of wrist mobility. This condition is called *carpal tunnel syndrome.*

The Hand

Five **metacarpal** (met-uh-KAR-pul) **bones** articulate with the distal carpal bones and support the hand (Figure 8-6●). Roman numerals I–V are used to identify the metacarpal bones, beginning with the lateral metacarpal bone that articulates with the trapezium. Distally, the metacarpal bones articulate with the proximal finger bones. There are 14 finger bones, or **phalanges** (fa-LAN-jēz; singular, *phalanx*) in each hand. The thumb, or **pollex** (POL-eks), has two phalanges (*proximal* and *distal*). Each of the other fingers has three phalanges (*proximal, middle,* and *distal*).

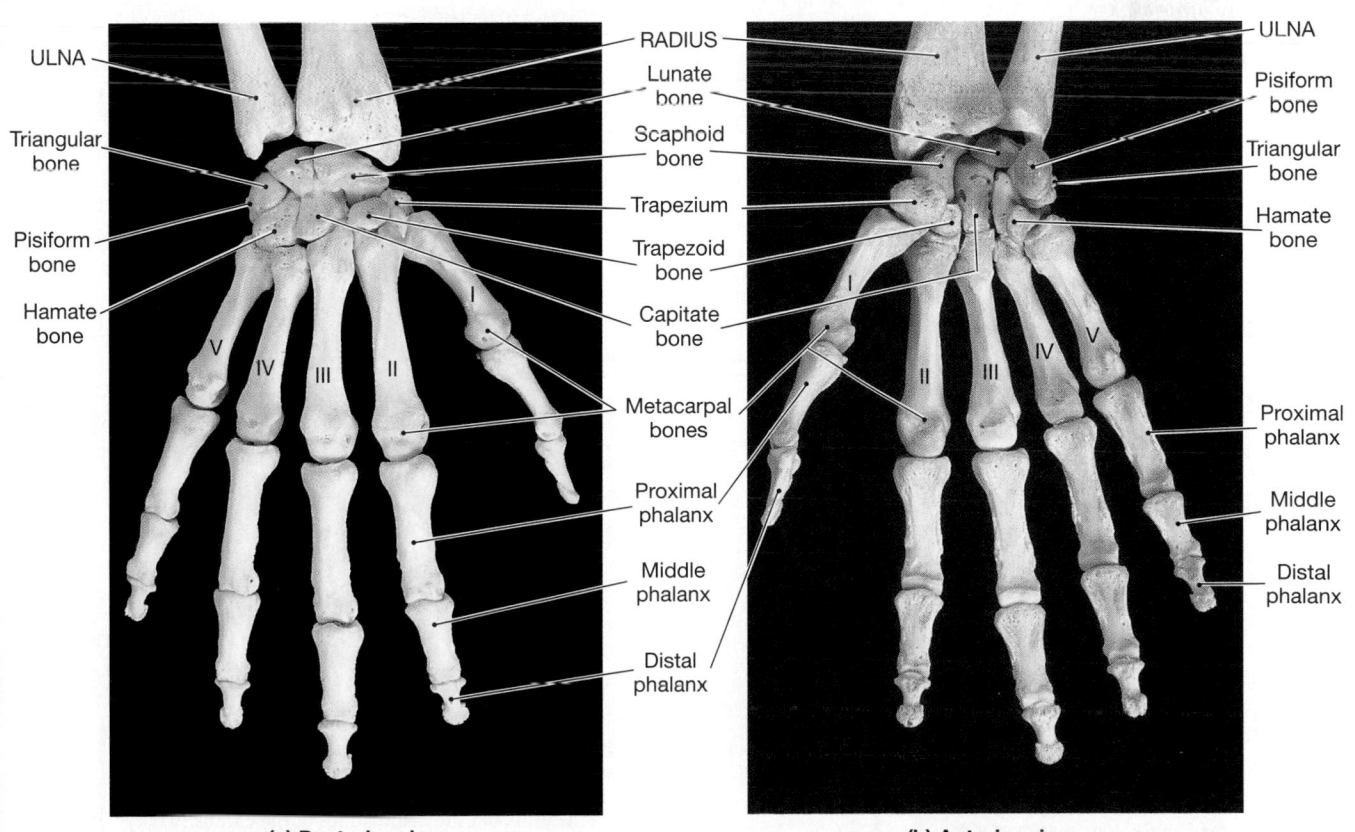

(a) Posterior view

(b) Anterior view

●*FIGURE 8-6* **Bones of the Wrist and Hand. (a)** Posterior view of the right hand. **(b)** Anterior view of the right hand.

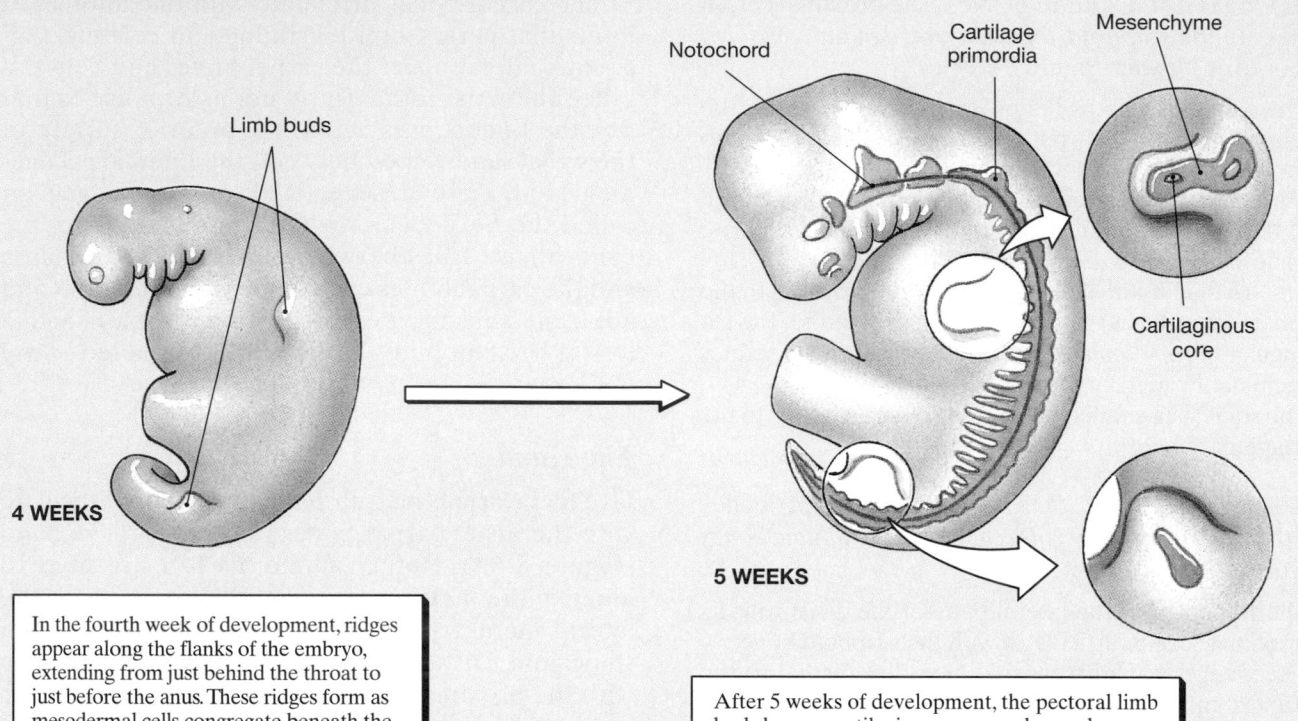

Limb buds

4 WEEKS

Notochord

Cartilage primordia

Mesenchyme

Cartilaginous core

5 WEEKS

In the fourth week of development, ridges appear along the flanks of the embryo, extending from just behind the throat to just before the anus. These ridges form as mesodermal cells congregate beneath the ectoderm of the flank. Mesoderm gradually accumulates at the end of each ridge, forming two pairs of limb buds.

After 5 weeks of development, the pectoral limb buds have a cartilaginous core and scapular cartilages are developing in the mesenchyme of the trunk.

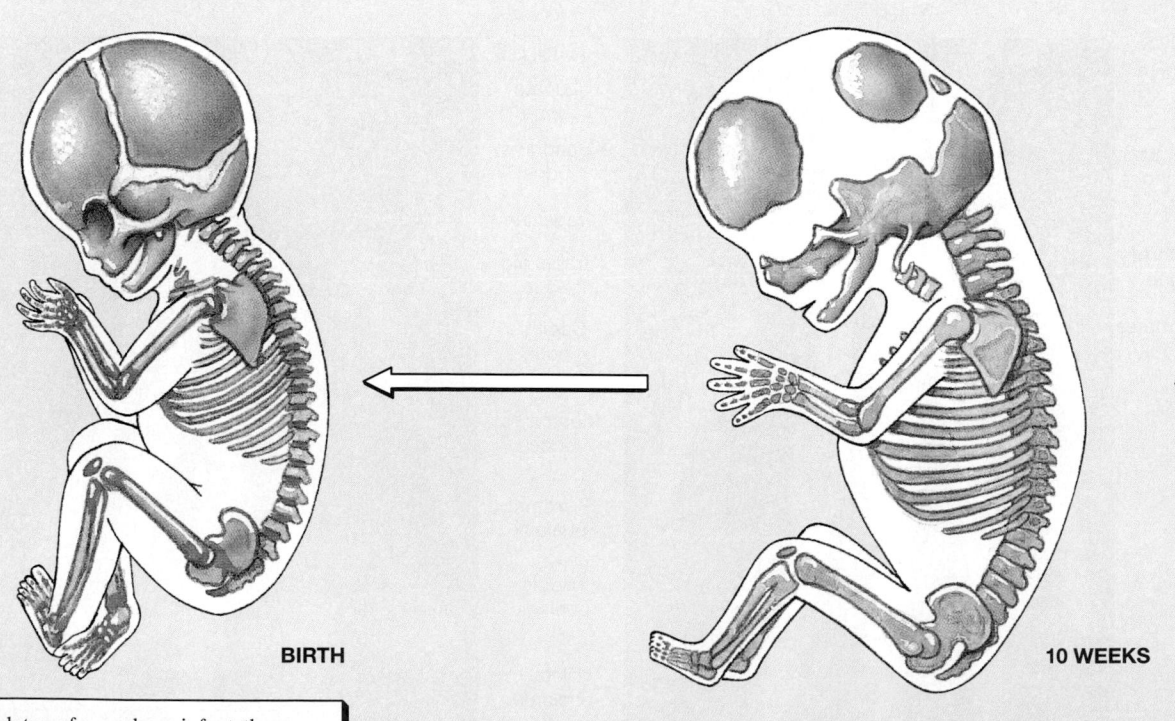

BIRTH

10 WEEKS

In the skeleton of a newborn infant, there are extensive areas of cartilage (blue) in the humeral head, in the wrist, between the bones of the palm and fingers, and in the coxae. Notice the appearance of the axial skeleton (refer to the Embryology Summaries in Chapter 7, pp. 216–217, 228–229).

Beige areas indicate the extent of ossification in the skeleton after approximately 10 weeks of development. The shafts of the limb bones undergo rapid ossification, but the distal bones of the carpus (wrist) and tarsus (ankle) remain cartilaginous.

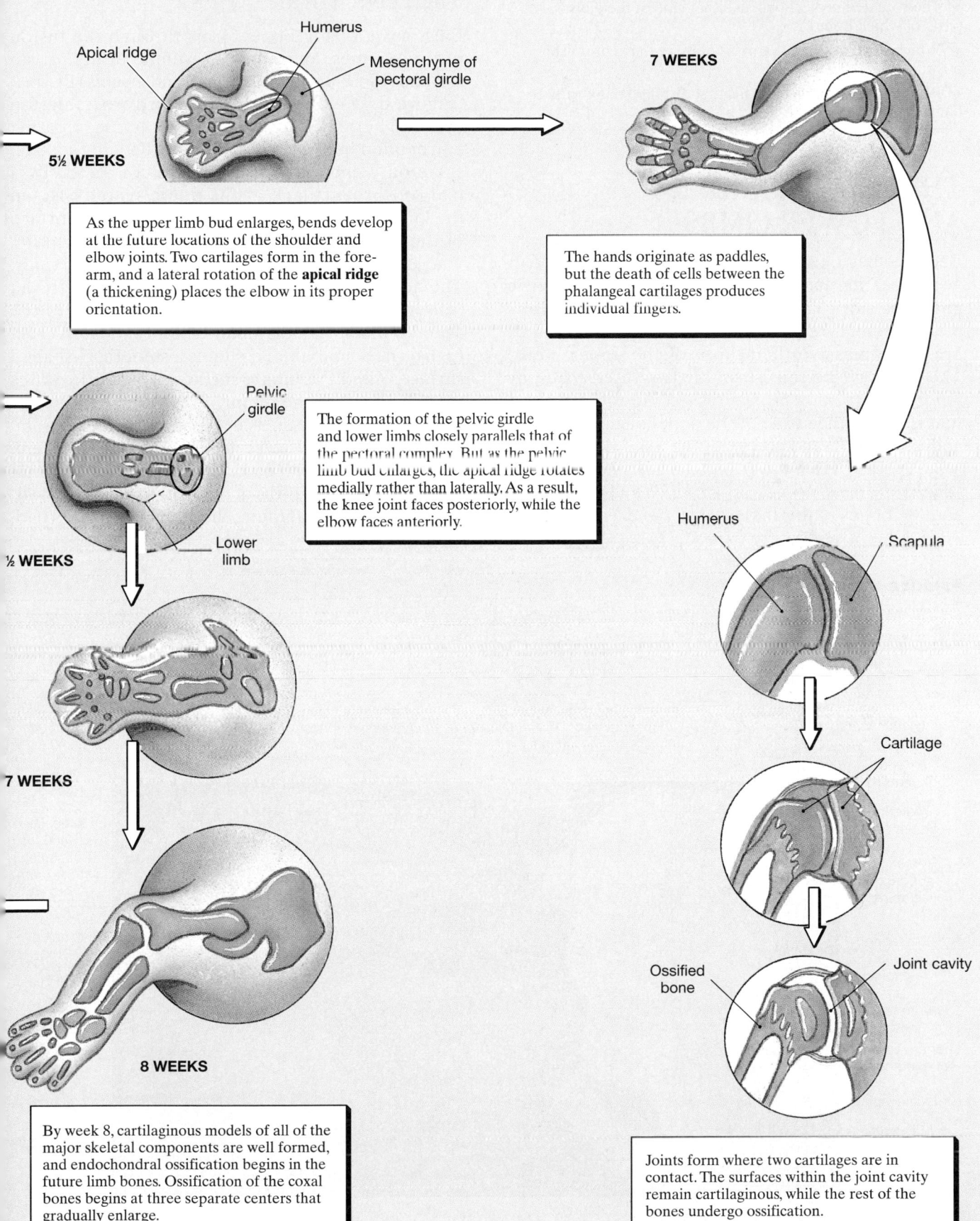

Apical ridge

Humerus

Mesenchyme of
pectoral girdle

5½ WEEKS

7 WEEKS

As the upper limb bud enlarges, bends develop at the future locations of the shoulder and elbow joints. Two cartilages form in the forearm, and a lateral rotation of the **apical ridge** (a thickening) places the elbow in its proper orientation.

The hands originate as paddles, but the death of cells between the phalangeal cartilages produces individual fingers.

Pelvic
girdle

The formation of the pelvic girdle and lower limbs closely parallels that of the pectoral complex. But as the pelvic limb bud enlarges, the apical ridge rotates medially rather than laterally. As a result, the knee joint faces posteriorly, while the elbow faces anteriorly.

Lower
limb

½ WEEKS

Humerus

Scapula

7 WEEKS

Cartilage

8 WEEKS

Ossified
bone

Joint cavity

By week 8, cartilaginous models of all of the major skeletal components are well formed, and endochondral ossification begins in the future limb bones. Ossification of the coxal bones begins at three separate centers that gradually enlarge.

Joints form where two cartilages are in contact. The surfaces within the joint cavity remain cartilaginous, while the rest of the bones undergo ossification.

☑ The rounded projections on either side of the elbow are parts of which bone?

☑ Which bone of the forearm is lateral in the anatomical position?

☑ Bill accidentally fractures his first distal phalanx with a hammer. Which finger is broken?

THE PELVIC GIRDLE AND LOWER LIMBS

Because they must withstand the stresses involved in weight bearing and locomotion, the bones of the **pelvic girdle** are more massive than those of the pectoral girdle. For similar reasons, the bones of the lower limbs are more massive than those of the upper limbs. The pelvic girdle consists of the two fused *coxae,* or *innominate bones.* The *pelvis* is a composite structure that includes the coxae of the appendicular skeleton and the sacrum and coccyx of the axial skeleton. ∞ *[pp. 222–225]* Each lower limb consists of the *femur* (thigh), the *patella* (kneecap), the *tibia* and *fibula* (leg), and the bones of the ankle (*tarsal bones*) and foot.

The Pelvic Girdle

Each **coxa,** or hip bone, forms through the fusion of three bones: an **ilium** (IL-ē-um; plural, *ilia*), an **ischium** (IS-kē-um; plural, *ischia*), and a **pubis** (PŪ-bis) (Figure 8-7●). The ilia have a sturdy articulation with the auricular surfaces of the sacrum, creating a firm bond between the pelvic girdle and the axial skeleton. Ventrally, the coxae are connected by a pad of fibrocartilage at the **pubic symphysis**. On the lateral surface of each coxa, the curved surface of the **acetabulum** (a-se-TAB-ū-lum; *acetabulum,* vinegar cup) articulates with the head of the femur. The acetabulum lies inferior and anterior to the center of the coxa (Figure 8-7a●). The walls of the acetabulum have a diameter of approximately 5 cm (2 in). The acetabulum contains a smooth, C-shaped surface called the **lunate surface,** which articulates with the head of the femur.

The three coxal bones meet inside the acetabulum, as if it were a pie sliced into three pieces. The ilium, the largest coxal bone, makes up the superior slice that includes about two-fifths of the acetabular surface. Superior to the acetabulum, the ilium forms a broad,

●FIGURE 8-7 The Coxa

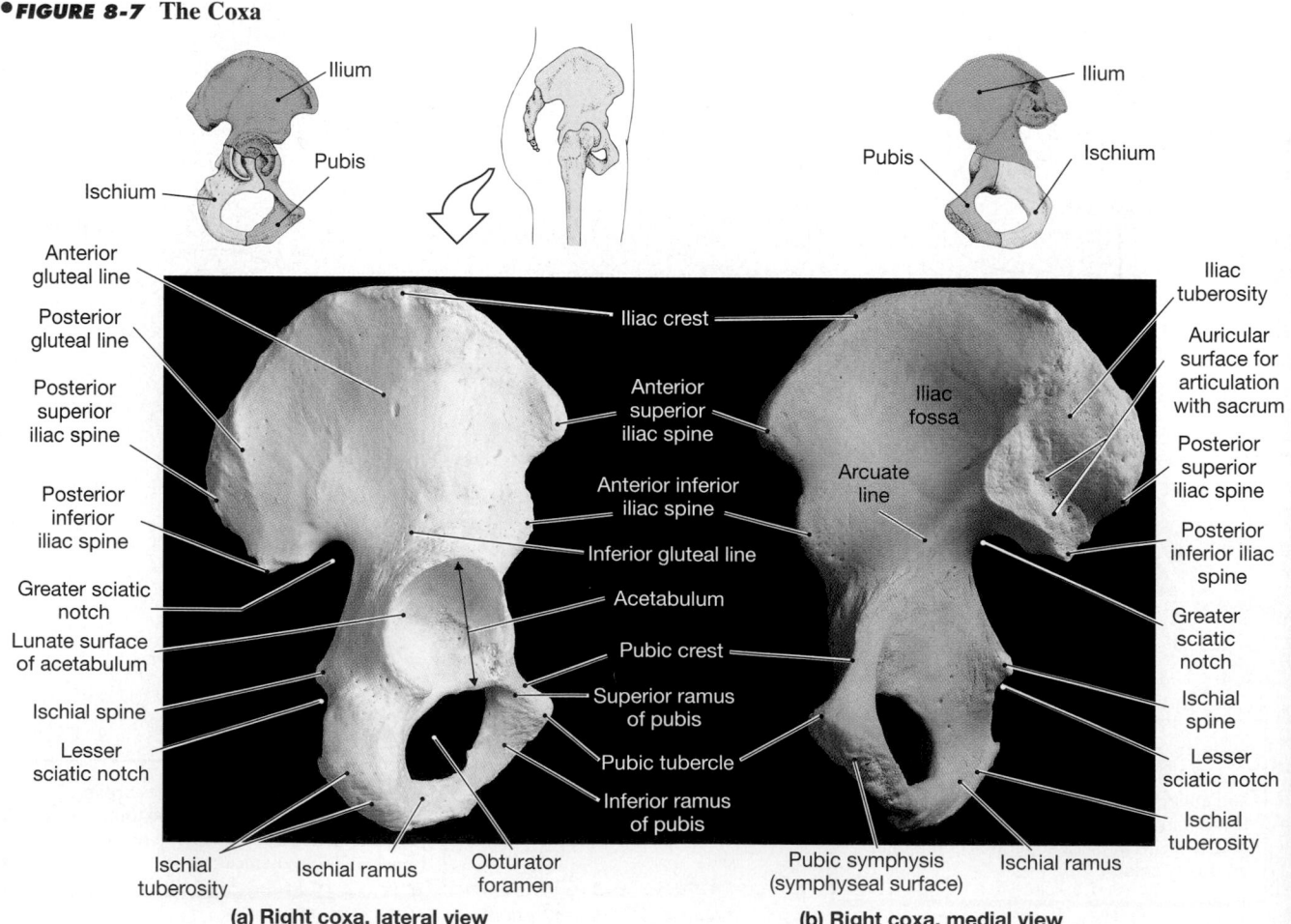

(a) Right coxa, lateral view

(b) Right coxa, medial view

curved surface that provides an extensive area for the attachment of muscles, tendons, and ligaments (Figure 8-7a●). The broadest part of the ilium extends between the **arcuate line** and the **iliac crest,** a ridge marking the attachments of both ligaments and muscles (Figure 8-7b●). Other important landmarks along the margin of the ilium include the **anterior superior** and **anterior inferior iliac spines** and the **posterior superior** and **posterior inferior iliac spines.** These spines are the attachment sites of important muscles and ligaments. The posterior inferior iliac spine sits above the **greater sciatic** (sī-A-tik) **notch,** through which a major nerve (the *sciatic nerve*) reaches the lower limb.

Near the edge of the acetabulum, the ilium fuses with the ischium. The ischium accounts for the posterior two-fifths of the acetabular surface. Posterior to the acetabulum, the prominent **ischial spine** projects above the **lesser sciatic notch,** through which blood vessels, nerves, and a small muscle pass. The **ischial tuberosity,** a roughened projection, is located at the posterior and lateral edge of the ischium. When you are seated, the ischial tuberosities bear your body's weight.

The narrow **ramus** (branch) of the ischium continues until it meets the **inferior ramus** of the pubis. The inferior ramus extends between the ramus of the ischium and the **pubic tubercle.** There it meets the **superior ramus** of the pubis, which originates near the acetabular margin. The anterior, superior surface of the superior ramus bears the **pubic crest,** a ridge that ends at the pubic tubercle. The pubic and ischial rami encircle the **obturator** (OB-tū-rā-tor) **foramen.** This space is closed by a sheet of collagen fibers whose inner and outer surfaces provide a firm base for the attachment of muscles and visceral structures.

The **iliac fossa** on the inner (medial) surface of the ilium extends between the iliac crest and the arcuate line. The concave surface of the iliac fossa helps support the abdominal organs and provides additional surface area for muscle attachment (Figure 8-7b●). The anterior and medial surface of the pubis contains a roughened area that marks the site of articulation with the pubis of the opposite side. At this articulation, the pubic symphysis, the two pubic bones are attached to a median fibrocartilage pad. The arcuate line extends diagonally across the pubis and continues toward the **auricular surface** of the ilium, site of the articu-

lation with the sacrum. Ligaments arising at the **iliac tuberosity** stabilize this joint, called the *sacroiliac joint.*

The Pelvis

Figure 8-8● shows anterior and posterior views of the **pelvis,** which consists of the coxae, the sacrum,

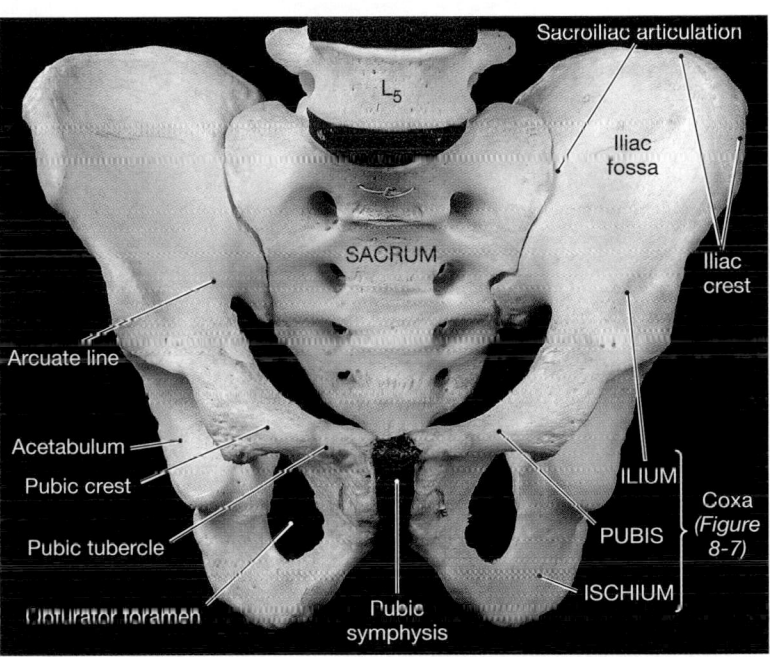

(a) Anterior view

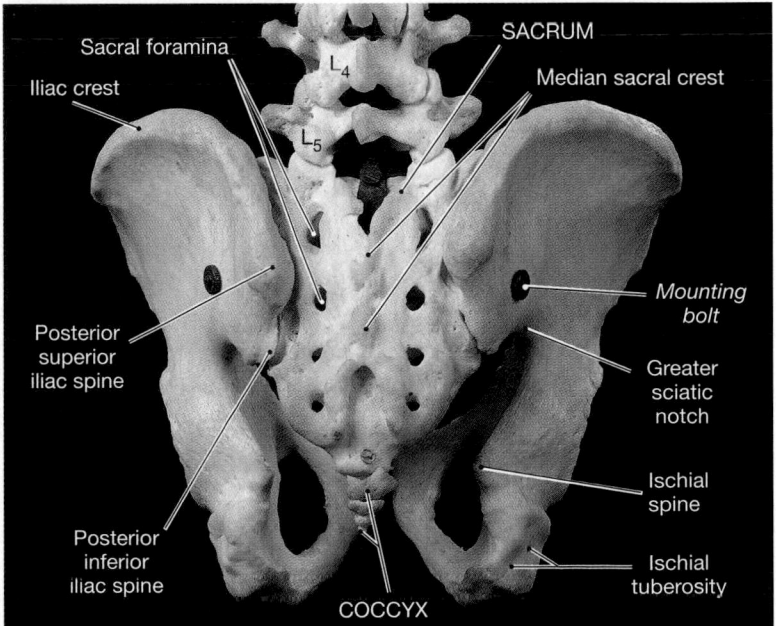

(b) Posterior view

●**FIGURE 8-8 The Pelvis.** The pelvis of an adult male is shown. (*See Figure 7-22 for a detailed view of the sacrum.*)

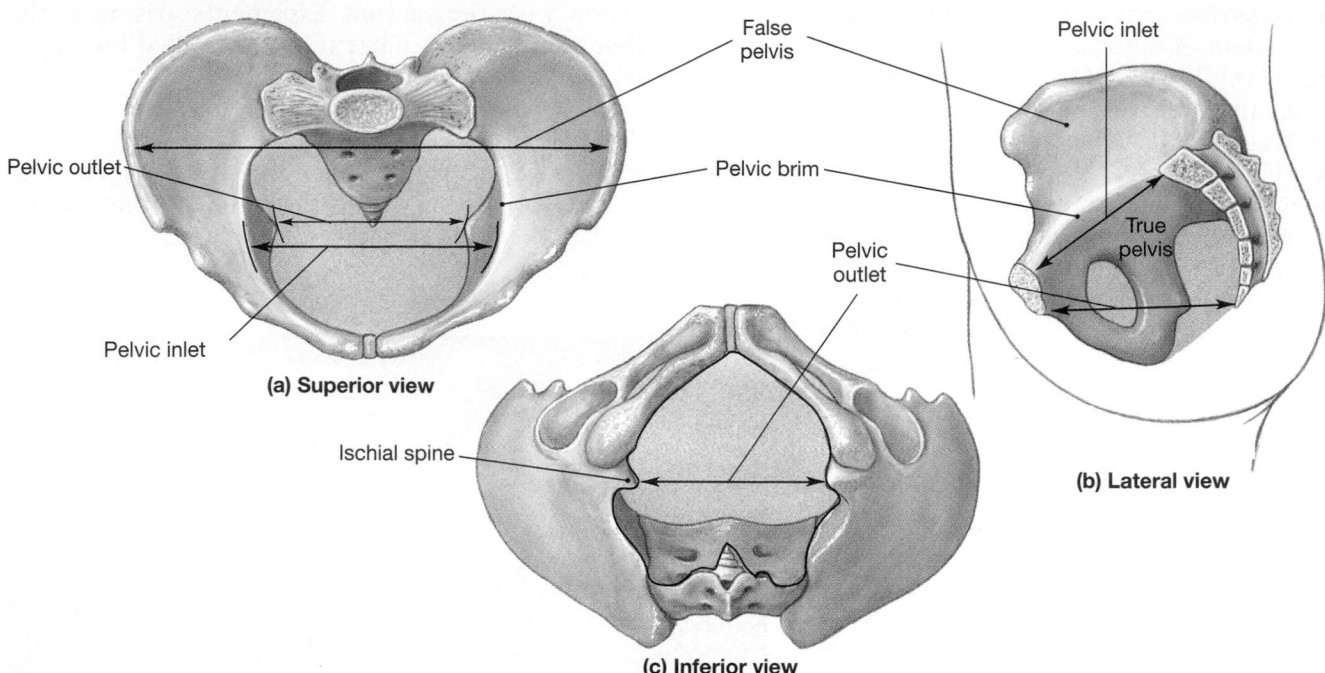

●FIGURE 8-9 **Divisions of the Pelvis.** **(a)** Superior view, showing the pelvic brim, pelvic inlet, and pelvic outlet. **(b)** Lateral view, showing the boundaries of the true (lesser) and false (greater) pelvis. **(c)** Inferior view, showing the limits of the pelvic outlet.

and the coccyx. An extensive network of ligaments connects the lateral borders of the sacrum with the iliac crest, the ischial tuberosity, the ischial spine, and the arcuate line. Other ligaments tie the ilia to the posterior lumbar vertebrae. These interconnections increase the stability of the pelvis.

The pelvis may be divided into the **false** (*greater*) **pelvis** and the **true** (*lesser*) **pelvis** (Figure 8-9a,b●). The false pelvis consists of the expanded, bladelike portions of each ilium superior to the arcuate line. Structures inferior to that line, including the inferior portions of each ilium, both pubic bones, the ischia, the sacrum, and the coccyx, form the true pelvis.

The true pelvis encloses the *pelvic cavity,* a subdivision of the abdominopelvic cavity. ∞ *[p. 24]* In lateral view (Figure 8-9b●), the superior limit of the true pelvis is a line that extends from either side of the base of the sacrum, along the arcuate line, to the superior margin of the pubic symphysis. The bony edge of the true pelvis is called the **pelvic brim,** and the enclosed space is the **pelvic inlet.**

The **pelvic outlet** is the opening bounded by the inferior edges of the pelvis (Figure 8-9b,c●). The region bounded by the coccyx, the ischial tuberosities, and the inferior border of the pubic symphysis is called the **perineum** (per-i-NĒ-um). Perineal muscles form the floor of the pelvic cavity and support the enclosed organs.

The shape of the pelvis of a female is somewhat different from that of a male (Figure 8-10●). Some of these differences are the result of variations in body size and in muscle mass. For example, in the female, the pelvis is generally smoother, lighter, and has less prominent markings. Females have other adaptations for childbearing, including:

- An enlarged pelvic outlet.
- Less curvature on the sacrum and coccyx, which in the male arc into the pelvic outlet.
- A wider, more circular pelvic inlet.
- A relatively broad, low pelvis.
- Ilia that project farther laterally but do not extend as far above the sacrum.
- A broader pubic angle (the inferior angle between the pubic bones); in females, it is greater than 100°.

These adaptations are related to (1) the support of the weight of the developing fetus and uterus and (2) passage of the newborn through the pelvic outlet at the time of delivery. In addition, the hormone *relaxin,* produced during pregnancy, loosens the pubic symphysis, allowing relative movement between the coxae that can further increase the size of the pelvic inlet and outlet.

☑ What three bones make up the coxa?

☑ How is the pelvis of the female adapted for childbearing?

☑ When you are seated, what part of the pelvis bears your body's weight?

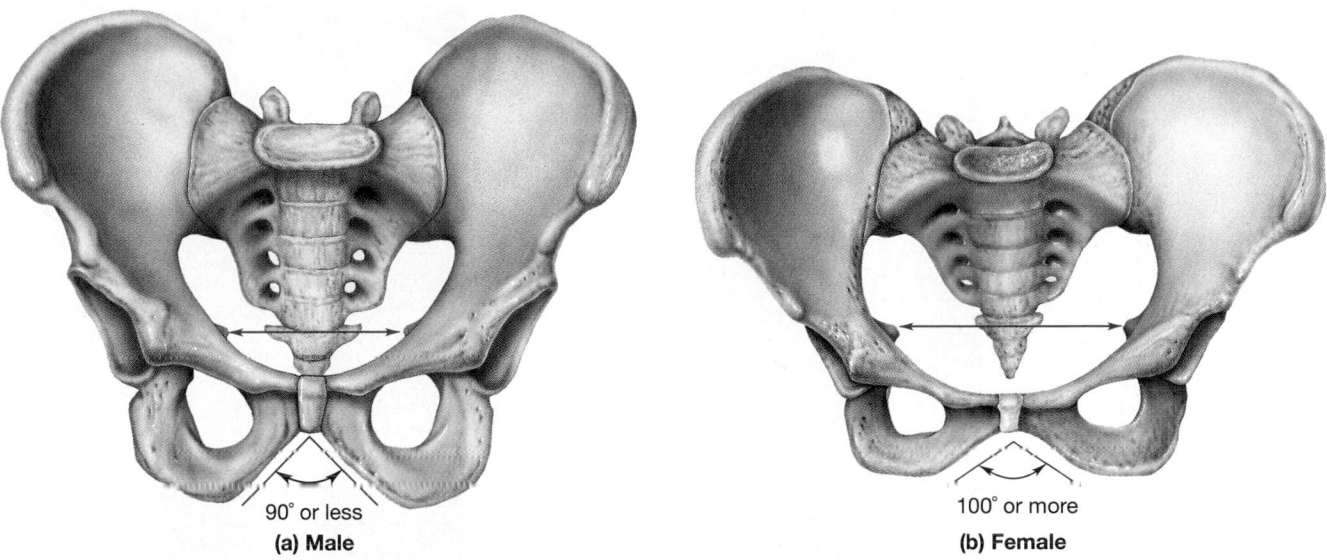

90° or less
(a) Male

100° or more
(b) Female

●**FIGURE 8-10** **Anatomical Differences in the Pelvis of a Male and a Female.** Note the much sharper pubic angle in **(a)** the pelvis of a male than in **(b)** that of a female.

The Lower Limbs

The skeleton of each lower limb consists of a femur, a patella, a tibia and fibula, and the bones of the ankle and foot. Once again, anatomical terminology differs from common usage. In anatomical terms, *leg* refers only to the distal portion of the limb, not to the entire lower limb. Thus we will use *thigh* and *leg* where people unfamiliar with proper terminology might use *upper leg* and *lower leg*, respectively.

The functional anatomy of the lower limbs is very different from that of the upper limbs, primarily because the lower limbs must transfer the body weight to the ground. We will now examine the bones of the right lower limb.

The Femur

The **femur** is the longest and heaviest bone in the body. It articulates with the coxa at the hip joint and with the tibia of the leg at the knee joint. The rounded epiphysis, or **head,** of the femur articulates with the pelvis at the acetabulum. A ligament attaches the acetabulum to the femur at the **fovea capitis**, a small pit in the center of the head. The **neck** of the femur joins the **shaft** at an angle of about 125° (Figure 8-11●). The **greater** and **lesser trochanters** project laterally from the junction of the neck and shaft. Both trochanters develop where large tendons attach to the femur. On the anterior surface of the femur, the raised **intertrochanteric** (in-ter-trō-kan-TER-ik) **line** marks the edge of the articular capsule. This line continues around to the posterior surface as the **intertrochanteric crest.**

The **linea aspera** (*aspera,* rough), a prominent elevation, runs along the center of the posterior surface, marking the attachment site of powerful muscles that move the femur (Figure 8-11b●). As it approaches the knee joint, the linea aspera divides into a pair of ridges that continue to the **medial** and **lateral epicondyles.** These smoothly rounded projections participate in the knee joint. On the posterior surface, the two condyles are separated by a deep **intercondylar fossa.**

The condyles continue across the inferior surface of the femur, but the intercondylar fossa does not extend completely around to the anterior surface. As a result, the smooth articular faces merge, producing an articular surface with elevated lateral borders (Figure 8-11a●). This is the **patellar surface,** over which the patella glides.

The Patella

The **patella** is a large sesamoid bone that forms within the tendon of the *quadriceps femoris,* a group of muscles that extend the leg. The patella has a rough, convex anterior surface and broad *base* (Figure 8-12a●). The roughened surface reflects the attachment of the quadriceps tendon (anterior and superior surfaces) and the *patellar ligament* (anterior and inferior surfaces). The patellar ligament connects the *apex* of the patella to the tibia. The posterior surface (Figure 8-12b●) presents two concave **facets (medial** and **lateral)** for articulation with the medial and lateral condyles of the femur.

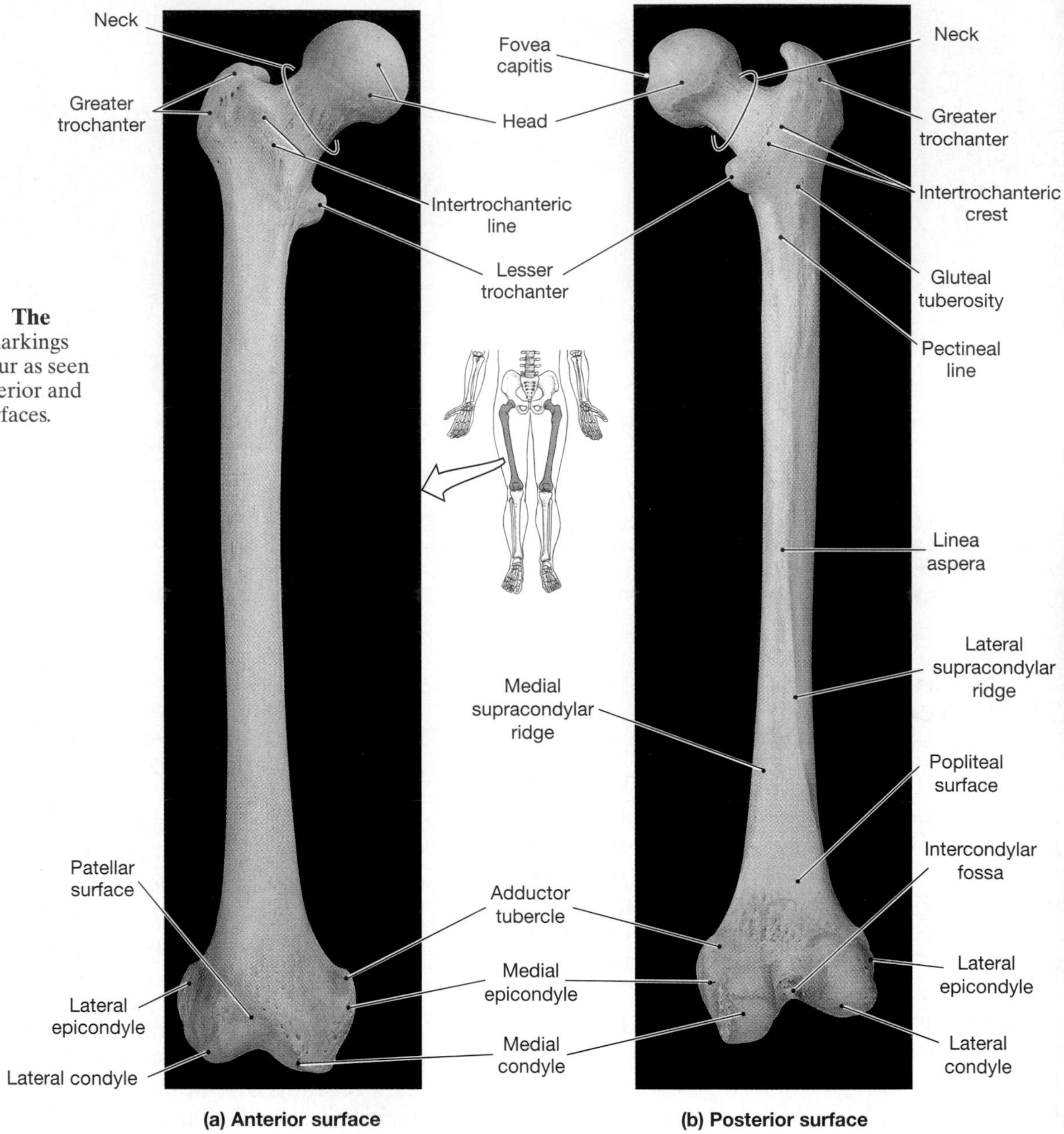

●**FIGURE 8-11** The **Femur.** Bone markings on the right femur as seen from the **(a)** anterior and **(b)** posterior surfaces.

Neck

Greater trochanter

Fovea capitis

Head

Intertrochanteric line

Lesser trochanter

Neck

Greater trochanter

Intertrochanteric crest

Gluteal tuberosity

Pectineal line

Linea aspera

Lateral supracondylar ridge

Medial supracondylar ridge

Popliteal surface

Intercondylar fossa

Patellar surface

Adductor tubercle

Medial epicondyle

Medial condyle

Lateral epicondyle

Lateral condyle

Lateral epicondyle

Lateral condyle

(a) Anterior surface

(b) Posterior surface

Base of patella

Attachment area for quadriceps tendon and patellar ligament

Medial facet, for medial condyle of femur

Lateral facet, for lateral condyle of femur

Base of femur

Apex

(a) Apex of patella

(b) Articular surface of patella

●**FIGURE 8-12** The Right Patella. **(a)** Anterior surface. **(b)** Posterior surface.

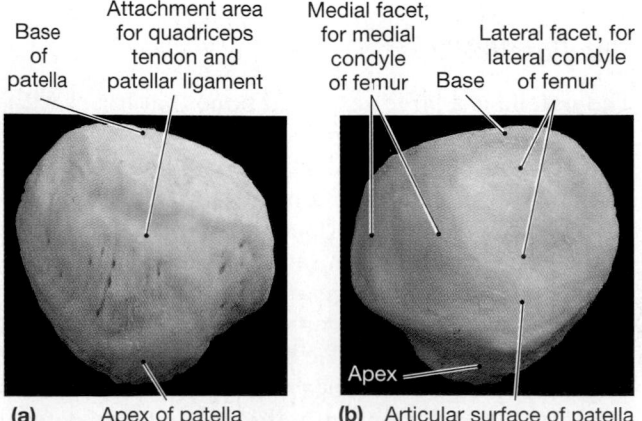

The Tibia

The **tibia** (TI-bē-uh) is the large medial bone of the leg (Figure 8-13a●). The medial and lateral condyles of the femur articulate with the **medial** and **lateral tibial condyles** at the proximal end of the tibia. The **intercondylar eminence,** a ridge, separates the two condyles (Figure 8-13b●). The anterior surface of the tibia near the condyles bears a prominent, rough **tibial tuberosity,** which you can easily feel beneath the skin of the leg. This tuberosity marks the attachment of the patellar ligament.

The **anterior crest** is a ridge that begins at the tibial tuberosity and extends distally along the anterior tibial surface. You can also easily feel the anterior crest of the tibia through the skin. As it approaches

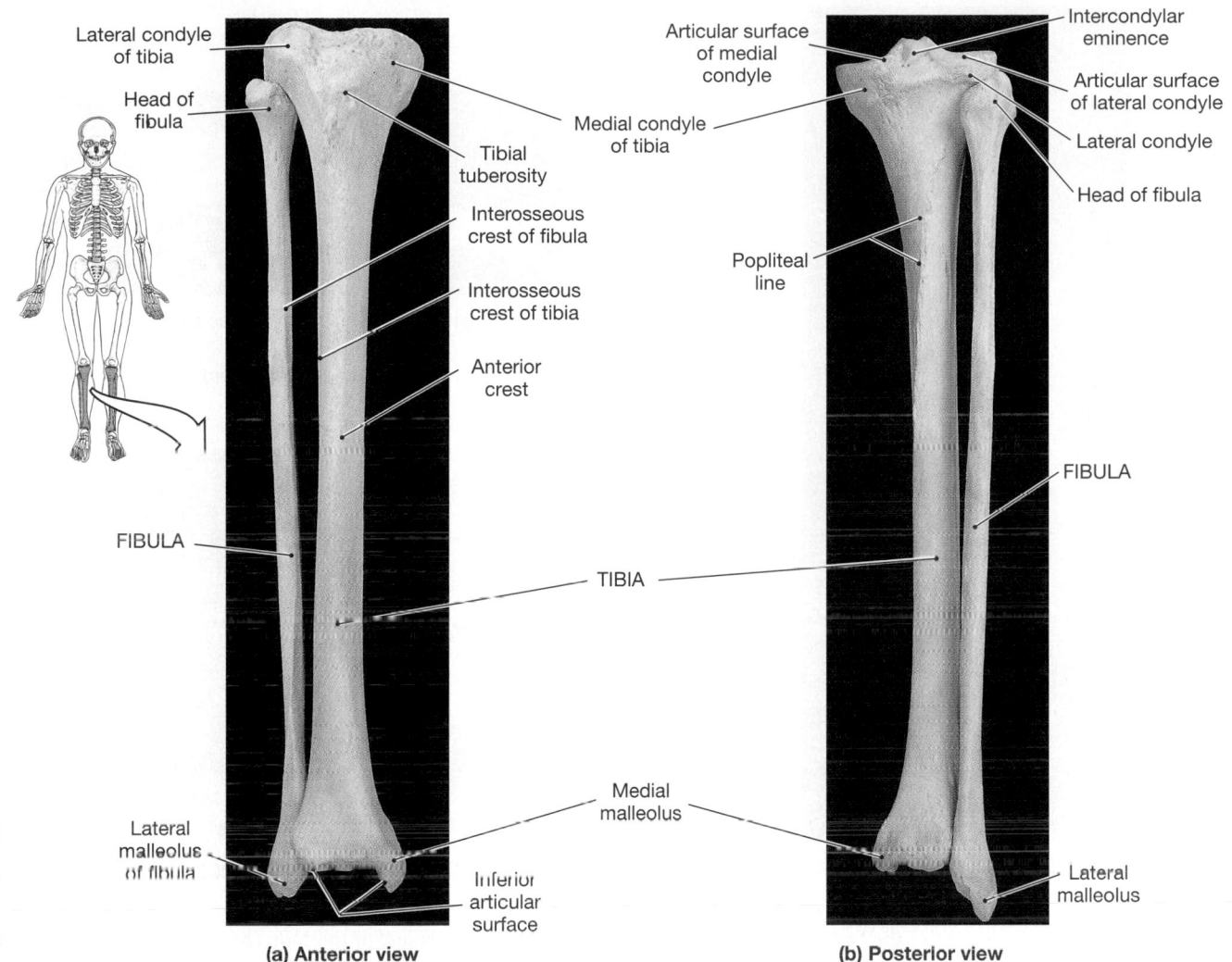

●FIGURE 8-13   The Tibia and Fibula. (a) Anterior view of the right tibia and fibula. **(b)** Posterior view.

the ankle joint, the tibia broadens, and the medial border ends in the **medial malleolus** (ma-LĒ-ō-lus; *malleolus,* hammer), a large process. The inferior surface of the tibia articulates with the proximal bone of the ankle; the medial malleolus provides medial support for this joint.

The Fibula

The slender **fibula** (FIB-ū-luh) parallels the lateral border of the tibia (Figure 8-13a,b●). The fibular **head** articulates with the tibia at an articular facet located near the lateral tibial condyle. The medial border of the thin **shaft** is bound to the tibia by the **interosseous membrane,** which extends from the **interosseous crest** of the fibula to that of the tibia. This membrane helps stabilize the positions of the two bones and provides additional surface area for muscle attachment.

The fibula does not articulate with the femur or play a role in the transfer of weight to the ankle and foot. However, it is important as a site for muscle attachment. In addition, the distal tip of the fibula extends lateral to the ankle joint. This fibular process, the lat-

eral malleolus, provides lateral stability to the ankle. The stability is somewhat limited, and forceful movement of the foot outward and backward can dislocate the ankle, breaking both the lateral malleolus of the fibula and the medial malleolus of the tibia. This injury is called a *Pott's fracture,* one of the fracture types introduced in Chapter 6. ∞ *[p. 190]*

The Ankle

The ankle, or **tarsus**, consists of seven **tarsal bones:** *talus, calcaneus, cuboid, navicular,* and three *cuneiform bones* (Figure 8-14●):

1. The **talus** transmits the weight of the body from the tibia toward the toes. The talus is the second largest bone of the foot. The primary tibial articulation occurs across the smooth superior surface of the **trochlea.** The trochlea has lateral and medial extensions that articulate with the lateral malleolus of the fibula and with the medial malleolus of the tibia. The lateral surfaces of the talus are roughened where ligaments connect it to the tibia and fibula, further stabilizing the ankle joint.

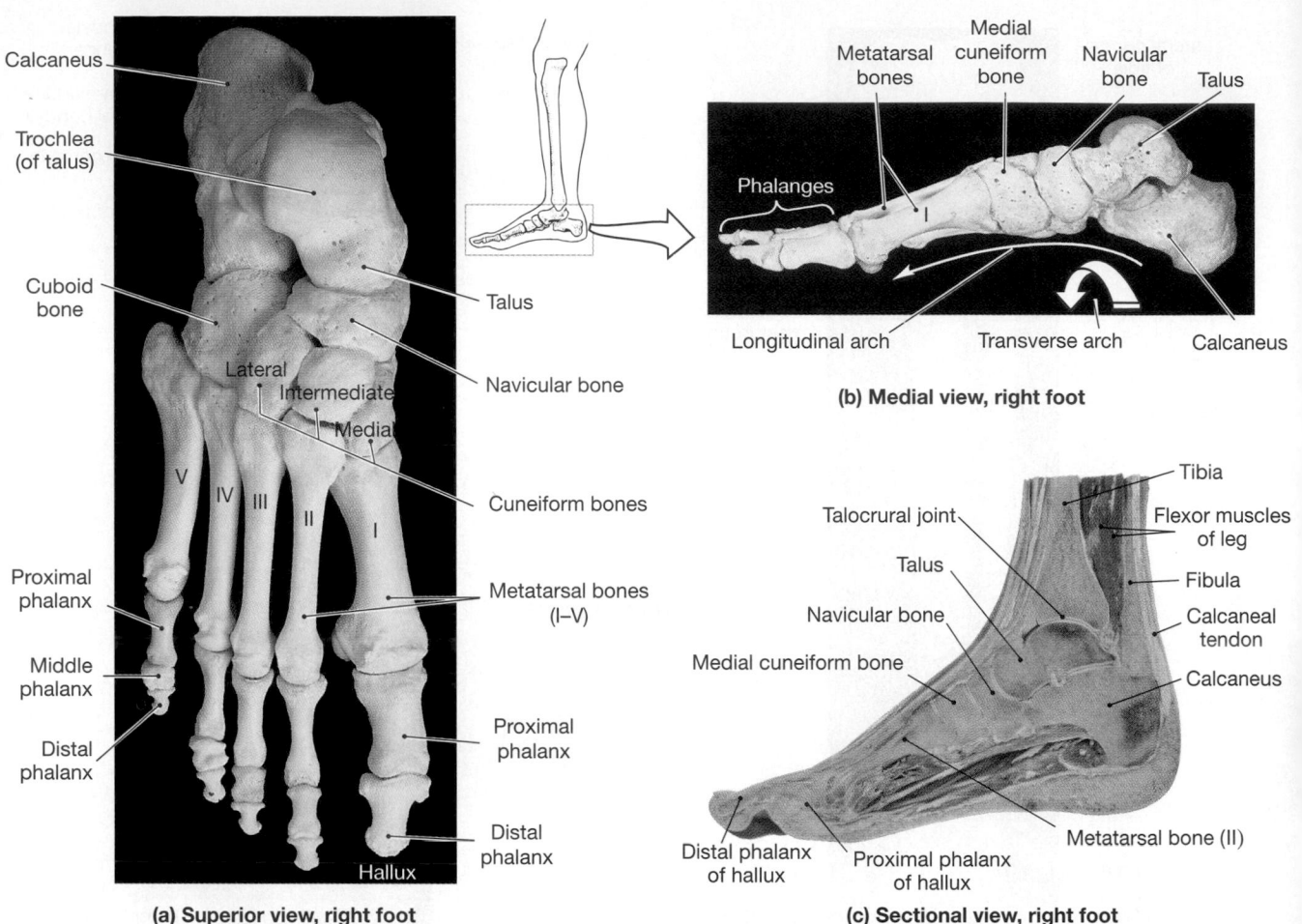

•FIGURE 8-14 **Bones of the Ankle and Foot.** **(a)** Bones of the right foot as viewed from above. Note the orientation of the tarsals, which convey the weight of the body to the heel and to the plantar surfaces (soles) of the foot. **(b)** Medial view of bones of the right foot. **(c)** Sectional view of the right foot. Note the muscles and tendons within the longitudinal arch. [AM] *Scans 8a, 8b*

2. The **calcaneus** (kal-KĀ-nē-us), or heel bone, is the largest of the tarsal bones. When you stand normally, most of your weight is transmitted from the tibia to the talus to the calcaneus, and then to the ground. The posterior portion of the calcaneus is a rough, knob-shaped projection. This is the attachment site for the *calcaneal tendon* (*Achilles tendon*), which arises at strong calf muscles. These muscles raise the heel and depress the sole, such as when you stand on tiptoes. The superior and anterior surfaces of the calcaneus bear smooth facets for articulation with other tarsal bones.

3. The **cuboid bone** articulates with the anterior, lateral surface of the calcaneus.

4. The **navicular bone** is located anterior to the talus, on the medial side of the ankle. It articulates with the talus and with the cuneiform bones.

5–7. The three **cuneiform bones** are wedge-shaped bones arranged in a row, with articulations between them. They are named according to their position: **medial cuneiform, intermediate cuneiform,** and **lateral cuneiform.** Proximally, the cuneiform bones articulate

with the anterior surface of the navicular bone. The lateral cuneiform bone also articulates with the medial surface of the cuboid bone.

The distal surfaces of the cuboid bone and the cuneiform bones articulate with the metatarsal bones of the foot.

The *ankle joint,* or *tibiotalar joint,* involves (1) the distal articular surface of the tibia, including the medial malleolus; (2) the lateral malleolus of the fibula; and (3) the trochlea and lateral articular facets of the talus. This joint works like a hinge that permits only superior/inferior movement of the foot. Articulations between the talus and other tarsal bones permit sliding and twisting of the foot.

The Foot

The **metatarsal bones** are five long bones that form the sole (*plantar surface*) of the foot (Figure 8-14•). The metatarsals are identified by Roman numerals I–V, proceeding from medial to lateral across the sole. Proximally, the first three metatarsal bones articulate with the three cuneiform bones, and the last two articulate with the cuboid bone. Distally, each metatarsal bone articulates with a proximal phalanx.

The **phalanges,** or toe bones (Figure 8-14●), have the same anatomical organization as the fingers. The toes contain 14 phalanges. The great toe, or **hallux,** has two phalanges (*proximal* and *distal*), and the other four toes have three phalanges apiece (*proximal, middle,* and *distal*).

ARCHES OF THE FOOT Weight transfer occurs along the **longitudinal arch** of the foot (Figure 8-14b●). Ligaments and tendons maintain this arch by tying the calcaneus to the distal portions of the metatarsals. The lateral, calcaneal side of the foot carries most of your body weight while you stand normally. This portion of the arch has much less curvature than the medial, talar portion. The talar arch also has considerably more elasticity than the calcaneal arch. As a result, the medial, plantar surface remains elevated, and the muscles, nerves, and blood vessels that supply the inferior surface of the foot are not squeezed between the metatarsals and the ground. In the condition known as *flatfeet,* the normal arches do not form. Individuals with this condition cannot walk long distances without discomfort; for this reason they cannot enlist in the U.S. Army.

The elasticity of the talar arch absorbs the shocks that accompany sudden changes in weight loading. For example, the stresses that running or ballet dancing place on the toes are cushioned by the elasticity of this portion of the arch. Because the degree of curvature changes from the medial to the lateral borders of the foot, a **transverse arch** also exists.

The amount of weight transferred forward depends on the position of the foot and the placement of body weight. During *dorsiflexion* of the foot, such as when you "dig in your heels," all your body weight rests on the calcaneus. During *plantar flexion,* when you "stand on tiptoe," the talus and calcaneus transfer your weight to the metatarsals and phalanges through more anterior tarsal bones.

CONGENITAL TALIPES EQUINOVARUS *Congenital talipes equinovarus (clubfoot)* results from an inherited developmental abnormality that affects 2 in 1000 births. Boys are affected roughly twice as often as girls. One or both feet may be involved, and the condition may be mild, moderate, or severe. The underlying problem is abnormal muscle development that distorts growing bones and joints. In most cases, the tibia, ankle, and foot are affected, and the feet are turned medially and inverted. The longitudinal arch is exaggerated. If both feet are involved, the soles face one another. Prompt treatment with casts or other supports in infancy helps alleviate the problem, and fewer than half of the cases require surgery. Kristi Yamaguchi, an Olympic gold medalist in figure skating, was born with this condition. AM *Problems with the Ankle and Foot*

☑ The fibula does not participate in the knee joint or bear weight. When it is fractured, however, walking becomes difficult. Why?

☑ While jumping off the back steps at his house, 10-year-old Joey lands on his right heel and breaks his foot. What foot bone is most likely broken?

☑ Which foot bone transmits the weight of the body from the tibia toward the toes?

☑ Ballet dancers strengthen their thigh muscles by placing one leg laterally on a waist-high bar and contracting against the resistance. How would the increased strength of these muscles affect the shape of the femur?

INDIVIDUAL VARIATION IN THE SKELETAL SYSTEM

A comprehensive study of a human skeleton can reveal important information about the individual. For example, there are characteristic racial differences in portions of the skeleton, especially the skull. We can estimate a person's muscular development and body weight from the development of various ridges and general bone mass. Details such as the condition of the teeth or the presence of healed fractures can provide information about the individual's medical history. Two important details, gender and age, can be determined or closely estimated on the basis of measurements indicated in Tables 8-1 and 8-2.

Table 8-1 identifies characteristic differences between the skeletons of males and females, but not every skeleton shows every feature in classic detail. Many differences, including markings on the skull, cranial capacity, and general skeletal features, reflect differences in average body size, muscle mass, and muscular strength. The general changes in the skeletal system that take place with age are summarized in Table 8-2. Note how these changes begin at age 1 and continue throughout life. For example, fusion of the epiphyseal plates begins at about age 3, and degenerative changes in the normal skeletal system, such as a reduction in mineral content in the bony matrix, typically do not begin until age 30–45.

An understanding of individual variation and of the normal timing of skeletal development is important in clinical diagnosis and treatment. Several professions focus on specific aspects of skeletal form and function. Each specialty has a different perspective with its own techniques, traditions, and biases. For example, a person with back pain may consult an *orthopedist,* an *osteopath,* or a *chiropractor.* Information concerning these clinical specialties is included in the *Applications Manual.* AM *A Matter of Perspective*

TABLE 8-1 Gender Differences in the Human Skeleton

Region/Feature	Male (as compared with female)	Female (as compared with male)
Skull		
General appearance	Heavier, rougher	Lighter, smoother
Forehead	Sloping	More vertical
Sinuses	Larger	Smaller
Cranium	About 10% larger (average)	About 10% smaller
Mandible	Larger, robust	Lighter, smaller
Teeth	Larger	Smaller
Pelvis		
General appearance	Narrow, robust, heavy, rough	Broader, lighter, smoother
Pelvic inlet	Heart-shaped	Oval to round
Iliac fossa	Relatively deep	Relatively shallow
Ilium	Extends farther above sacral articulation	More vertical; less extension above sacroiliac joint
Angle inferior to pubic symphysis	Under 90°	100° or more *(see Figure 8-10, p. 245)*
Acetabulum	Directed laterally	Faces slightly anteriorly as well as laterally
Obturator foramen	Oval	Triangular
Ischial spine	Points medially	Points posteriorly
Sacrum	Long, narrow triangle with pronounced sacral curvature	Broad, short triangle with less curvature
Coccyx	Points anteriorly	Points inferiorly
Other skeletal elements		
Bone weight	Heavier	Lighter
Bone markings	More prominent	Less prominent

TABLE 8-2 Age-Related Changes in the Skeleton

Region/Feature	Event(s)	Age (Years)
General skeleton		
Bony matrix	Reduction in mineral content	Values differ for males versus females between ages 45 and 65; similar reductions occur in both genders after age 65
Markings	Reduction in size, roughness	Gradual reduction with increasing age and decreasing muscular strength and mass
Skull		
Fontanels	Closure	Completed by age 2
Metopic suture	Fusion	2–8
Occipital bone	Fusion of ossification centers	1–4
Styloid process	Fusion with temporal bone	12–16
Hyoid bone	Complete ossification and fusion	25–30
Teeth	Loss of "baby teeth"; appearance of secondary dentition; eruption of posterior molars	Detailed in Chapter 24 (digestive system)
Mandible	Loss of teeth; reduction in bone mass; change in angle at mandibular notch	Accelerates in later years (60+)
Vertebrae		
Curvature	Development of major curves	Described in Figure 7-16, p. 218
Intervertebral discs	Reduction in size, percentage contribution to height	Accelerates in later years (60+)
Long bones		
Epiphyseal plates	Fusion	Ranges vary, but general analysis permits determination of approximate age
Pectoral and pelvic girdles		
Epiphyses	Fusion	Overlapping ranges somewhat narrower than the above, including 14–16, 16–18, 22–25

SELECTED CLINICAL TERMINOLOGY

Terms Discussed in This Chapter

carpal tunnel syndrome: Inflammation of the tissues beneath the flexor retinaculum, causing compression of the flexor tendons and adjacent sensory nerves. Symptoms are pain and a loss of wrist mobility. *(p. 239)*

congenital talipes equinovarus: A congenital deformity affecting one or both feet, commonly known as clubfoot. It develops secondary to abnormalities in muscular development. *(p. 249)*

flatfeet: The loss or absence of a longitudinal arch. *(p. 249)*

AM Additional Terms Discussed in the *Applications Manual*

dancer's fracture: A fracture of the fifth metatarsal, generally near its proximal articulation.

CHAPTER REVIEW   *On-line resources for this chapter are on our World Wide Web site at:* *http://www.prenhall.com/martini/fap*

STUDY OUTLINE

INTRODUCTION, p. 234

1. The **appendicular skeleton** includes the bones of the upper and lower extremities and the pectoral and pelvic girdles that connect the limbs to the trunk. *(Figure 8-1)*

THE PECTORAL GIRDLE AND UPPER LIMBS, p. 234

1. Each upper limb articulates with the trunk via the **shoulder girdle,** or **pectoral girdle,** which consists of the **scapulae** and **clavicles.**

The Pectoral Girdle, p. 235

2. On each side, a clavicle and scapula position the shoulder joint, help move the upper limb, and provide a base for muscle attachment. *(Figures 8-2, 8-3)*

3. Both the **coracoid** and **acromion processes** are attached to ligaments and tendons. The **scapular spine** crosses the scapular **body.** *(Figure 8-3)*

The Upper Limbs, p. 236

4. The scapula articulates with the **humerus** at the shoulder (*glenohumeral,* or *scapulohumeral*) joint. The **greater** and **lesser tubercles** are important sites of muscle attachment. *(Figure 8-4)*

5. The humerus articulates with the **radius** and **ulna,** the bones of the forearm, at the elbow (*olecranal or humeroulnar*) joint. *(Figure 8-5)*

6. The **carpal bones** of the wrist, or **carpus,** form two rows. The distal row articulates with the five **metacarpal bones.** Four of the fingers contain three **phalanges;** the **pollex** (thumb) has only two. *(Figure 8-6)*

THE PELVIC GIRDLE AND LOWER LIMBS, p. 242

1. The bones of the **pelvic girdle** are more massive than those of the pectoral girdle.

The Pelvic Girdle, p. 242

2. The pelvic girdle consists of two **coxae;** each coxa forms through the fusion of an **ilium,** an **ischium,** and a **pubis.** *(Figure 8-7)*

3. The ilium is the largest coxal bone. Inside the **acetabulum,** the ilium is fused to the ischium (posteriorly) and the pubis (anteriorly). The **pubic symphysis** limits movement between the pubic bones of the left and right coxae. *(Figures 8-7, 8-8)*

4. The **pelvis** consists of the coxae, sacrum, and coccyx. It may be subdivided into the **false** (*greater*) **pelvis** and the **true** (*lesser*) **pelvis.** *(Figures 8-8, 8-9, 8-10)*

The Lower Limbs, p. 245

5. The **femur** is the longest bone in the body. It articulates with the **tibia** at the knee joint. *(Figures 8-11, 8-13)*

6. The **patella,** or kneecap, is a large sesamoid bone. *(Figure 8-12)*

7. The **fibula** parallels the tibia laterally. *(Figure 8-13)*

8. The **tarsus,** or ankle, includes seven **tarsal bones.** *(Figure 8-14)*

9. The basic organizational pattern of the **metatarsal bones** and **phalanges** of the foot resembles that of the hand. All the toes have three phalanges except for the **hallux,** which has two. *(Figure 8-14)*

10. When a person stands normally, most of the body weight is transferred to the **calcaneus,** and the rest is passed on to the five metatarsal bones. Weight transfer occurs along the **longitudinal arch;** there is also a **transverse arch.** *(Figure 8-14)*

INDIVIDUAL VARIATION IN THE SKELETAL SYSTEM, p. 249

1. Studying a human skeleton can reveal important information such as race, medical history, weight, gender, body size, muscle mass, and age. *(Tables 8-1, 8-2)*

2. Age-related changes and events take place in the skeletal system. These changes begin at about age 1 and continue throughout life. *(Table 8-2)*

REVIEW QUESTIONS

Level 1 — Reviewing Facts and Terms

1. The only direct connection between the pectoral girdle and the axial skeleton is where the
(a) clavicle articulates with the humerus
(b) clavicle articulates with the manubrium of the sternum
(c) coxa articulates with the femur
(d) vertebral column articulates with the skull

2. The presence of foramina in bones indicates the position of
(a) tendons and ligaments (b) nerves and blood vessels
(c) ridges and flanges (d) a, b, and c are correct

3. At the glenoid cavity, the scapula articulates with the proximal end of the
(a) humerus (b) radius
(c) ulna (d) femur

4. In anatomical position, the ulna lies
(a) medial to the radius (b) lateral to the radius
(c) inferior to the radius (d) superior to the radius

5. The proximal carpal bones include the
(a) trapezium, trapezoid, capitate, hamate
(b) scaphoid, capitate, lunate, hamate
(c) trapezium, triangular, trapezoid, pisiform
(d) scaphoid, lunate, triangular, pisiform

6. The distal carpal bones are the
(a) scaphoid, pisiform, capitate, hamate
(b) trapezium, trapezoid, capitate, hamate
(c) scaphoid, lunate, triangular, pisiform
(d) trapezium, lunate, triangular, hamate

7. The bones of the hand articulate distally with the
(a) carpal bones (b) ulna and radius
(c) metacarpal bones (d) phalanges

8. Each coxa of the pelvic girdle consists of three fused bones:
(a) ulna, radius, humerus
(b) ilium, ischium, pubis
(c) femur, tibia, fibula
(d) hamate, capitate, trapezium

9. The large foramen between the pubic and ischial rami is the
(a) foramen magnum (b) suborbital foramen
(c) acetabulum (d) obturator foramen

10. Which of the following is an adaptation for childbearing?
(a) inferior angle between the pubic bones 100° or more
(b) a relatively broad, low pelvis
(c) less curvature of the sacrum and coccyx
(d) a, b, and c are correct

11. The epiphysis of the femur articulates with the pelvis at the
(a) pubic symphysis (b) acetabulum
(c) sciatic notch (d) obturator foramen

12. The large medial bone of the leg is the
(a) tibia (b) femur
(c) fibula (d) humerus

13. The selection that includes *only* tarsal bones is
(a) scaphoid, pisiform, capitate, hamate, talus
(b) calcaneus, navicular, lunate, capitate, talus
(c) talus, calcaneus, cuboid, navicular, cuneiforms
(d) navicular, scaphoid, capitate, talus, calcaneus

14. The calcaneus is the attachment site for the
(a) Achilles tendon (b) muscles of the calf
(c) metatarsal bones (d) a, b, and c are correct

15. Name the skeletal components of the pectoral girdle.

16. Which two large scapular processes are associated with the shoulder joint?

17. Which bones make up the arm and forearm?

18. Which two movements are associated with the proximal radioulnar articulation?

19. List the eight carpal bones of the wrist.

20. Which bones constitute the lower limb?

21. Name the components of each coxa of the pelvic girdle.

22. How does the anatomist distinguish between the false (greater) pelvis and the true (lesser) pelvis?

23. What seven bones make up the ankle (tarsus)?

24. Distinguish between the pollex and the hallux.

Level 2 — Reviewing Concepts

25. Why are clavicular injuries common?

26. What is the skeletal structural difference between the pelvic girdle and the pelvis?

27. Why is the tibia but not the fibula involved in the transfer of weight to the ankle and foot?

28. What are the direct anatomical connections between the skeletal and muscular systems?

29. Why would an instructor teaching self-defense advise a student to strike an assailant's clavicle in an attack?

30. Why is it necessary for the bones of the pelvic girdle to be more massive than the bones of the pectoral girdle?

Level 3 — Critical Thinking and Clinical Applications

31. In order to settle a bet, you need to measure the length of your lower limb (femur and tibia). What landmarks would you use to make the measurement?

32. While Fred, a fireman, is fighting a fire in a building, part of the ceiling collapses, and a beam strikes him on his left shoulder. He is rescued but has a great deal of pain in his shoulder. He cannot move his arm properly, especially in the anterior direction. His clavicle is not broken, and his humerus is intact. What is the probable nature of Fred's injury?

33. Cindy is anxiously awaiting the birth of her first child. As she gets closer to term, she undergoes an ultrasound. After seeing the results, her physician makes some calculations and informs Cindy that she will probably have to have a cesarean section. What clues might tell the physician that Cindy can't deliver the baby by natural childbirth?

9

Articulations

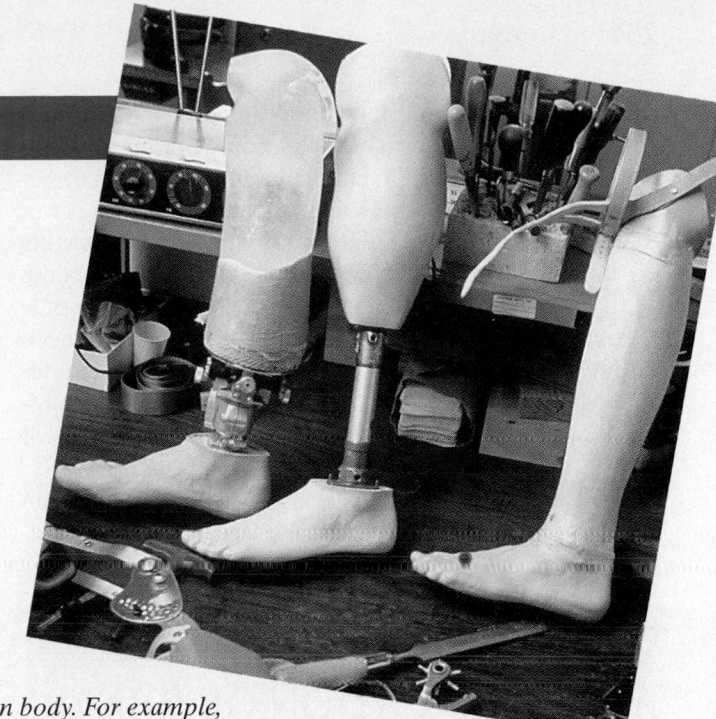

*O*ur movements must conform to the limitations of the human body. For example, it is not possible to bend the shaft of the humerus or femur; movements are restricted to joints, where bones articulate. The same is true of artificial limbs, which use mechanical equivalents of human joints. Very early in life, we learn to work within those limitations. Each joint—natural or artificial—will tolerate a specific range of motion. A variety of bony surfaces, cartilages, ligaments, tendons, and muscles— or metal and plastic balls, shafts, and screws—work together to keep movement within the normal range. The stronger the joint, the more restrictive the limitations are likely to be. In this chapter, we will examine representative joints and explore the relationship between structural stability and the range of motion.

CHAPTER OUTLINE AND OBJECTIVES

In the last two chapters, you have become familiar with the individual bones of the skeleton. Your understanding of the skeletal system is still incomplete, however, because in life the skeleton is always in motion. Most of our daily activities—from breathing or speaking, to writing or running, involve movements of the skeleton. The bones of the skeleton are rigid, and movements can occur only at joints, or **articulations,** where two bones interconnect.

The characteristic structure of a joint determines the type of movement that may occur. Each joint reflects a workable compromise between the need for strength and the need for mobility.

This chapter compares the relationships between articular form and function. We will use several different examples that range from relatively immobile but very strong (the intervertebral articulations) to highly mobile and relatively weak (the shoulder).

A CLASSIFICATION OF JOINTS

Two different classification methods are used to categorize joints. The first, and the one we will use in this chapter, considers the range of motion permitted (Table 9-1). These functional categories are further subdivided on the basis of the anatomical structure of the joint or the range of motion permitted:

1. An *immovable joint* is a **synarthrosis** (sin-ar-THRŌ-sis; *syn,* together + *arthros,* joint). A synarthrosis may be *fibrous* or *cartilaginous,* depending on the nature of the connection, or the two bones may fuse.

2. A *slightly movable joint* is an **amphiarthrosis** (am-fē-ar-THRŌ-sis; *amphi,* on both sides). An amphiarthrosis may be *fibrous* or *cartilaginous,* depending on the nature of the connection between the opposing bones.

3. A *freely movable joint* is a **diarthrosis** (dī-ar-THRŌ-sis; *dia,* through). Diarthroses are also called *synovial joints.* We introduced their basic structure in Chapter 4 in our discussion of synovial membranes. ∞ *[p. 134]* Diarthroses are subdivided according to the degree of movement permitted.

The second classification scheme relies solely on the anatomical organization of the joint, without regard to the degree of permitted movement. In this framework, joints are classified as *bony, fibrous, cartilaginous,* or *synovial* (Table 9-2). We will use the functional classification rather than the anatomical one, because we are primarily interested in how joints work.

Immovable Joints (Synarthroses)

At a synarthrosis, the bony edges are quite close together and may even interlock. These extremely strong joints are found where movement between the bones must be prevented. There are four major types of synarthrotic joints: *sutures, gomphoses, synchondroses,* and *synostoses.*

1. *Sutures.* A **suture** (*sutura,* a sewing together) is a synarthrotic joint located only between the bones of the skull. The edges of the bones are interlocked and bound together at the suture by dense connective tissue.

TABLE 9-1 A Functional Classification of Articulations

Functional Category	Structural Category	Description	Example
Synarthrosis (no movement)	**Fibrous**		
	Suture	Fibrous connections plus interdigitation	Between the bones of the skull
	Gomphosis	Fibrous connections plus insertion in alveolar process	Between the teeth and jaws
	Cartilaginous		
	Synchondrosis	Interposition of cartilage plate	Epiphyseal plates
	Bony fusion		
	Synostosis	Conversion of other articular form to solid mass of bone	Portions of the skull, epiphyseal lines
Amphiarthrosis (little movement)	**Fibrous**		
	Syndesmosis	Ligamentous connection	Between the tibia and fibula
	Cartilaginous		
	Symphysis	Connection by a fibrocartilage pad	Between right and left halves of pelvis between adjacent vertebral bodies along vertebral column
Diarthrosis (free movement)	**Synovial**	Complex joint bounded by joint capsule and containing synovial fluid	Numerous; subdivided by range of movement *(see Figure 9-6)*
	Monaxial	Permits movement in one plane	Elbow, ankle
	Biaxial	Permits movement in two planes	Ribs, wrist
	Triaxial	Permits movement in all three planes	Shoulder, hip

TABLE 9-2 A Structural Classification of Articulations

Structure	Type	Functional Category	Example
Bony fusion	Synostosis (illustrated)	Synarthrosis	Frontal bone — Metopic suture (fusion)
Fibrous joint	Suture (illustrated) Gomphosis Syndesmosis	Synarthrosis Synarthrosis Amphiarthrosis	Skull — Lambdoidal suture
Cartilaginous joint	Synchondrosis Symphysis (illustrated)	Synarthrosis Amphiarthrosis	Pelvis — Pubic symphysis
Synovial joint	Monaxial Biaxial Triaxial (illustrated)	Diarthroses	Synovial joint

2. *Gomphoses.* A **gomphosis** (gom-FŌ-sis; *gomphosis,* a bolting together) is a synarthrosis that binds the teeth to bony sockets in the maxillary bone and mandible. The fibrous connection between a tooth and its socket is a **periodontal** (pe-rē-ō-DON-tal) **ligament** (*peri,* around + *odontos,* tooth).

3. *Synchondroses.* A **synchondrosis** (sin-kon-DRŌ-sis; *syn,* together + *chondros,* cartilage) is a rigid, cartilaginous bridge between two articulating bones. The hyaline cartilage of an *epiphyseal plate* is a synchondrosis that connects the diaphysis with the epiphysis, even though the bones involved are part of the same skeletal element. ∞ *[p. 182]* Another example is the cartilaginous connection between the ends of the vertebrosternal ribs (ribs 1–7) and the sternum.

4. *Synostoses.* A **synostosis** (sin-os-TŌ-sis) is a totally rigid, immovable joint created when two separate bones fuse to the point at which the boundary between them disappears. The *metopic suture* of the frontal bone and the epiphyseal lines of mature bones are synostoses. ∞ *[pp. 205, 182]*

Slightly Movable Joints (Amphiarthroses)

An amphiarthrosis is another compromise between mobility and strength. It permits more movement than a synarthrosis but is much stronger than a freely movable joint. The articulating bones may be connected by collagen fibers or cartilage. There are two major types of amphiarthrotic joints:

1. At a **syndesmosis** (sin-dez-MŌ-sis; *desmos,* a band or ligament), bones are connected by a ligament. One example is the distal articulation between the tibia and fibula.

2. At a **symphysis**, the articulating bones are separated by a wedge or pad of fibrocartilage. The articulation between the bodies of adjacent vertebrae (at the *intervertebral disc*) and the anterior connection between the two pubic bones (the *pubic symphysis*) are examples of this type of joint.

Freely Movable Joints (Diarthroses)

Diarthroses, or **synovial** (si-NŌ-vē-ul) **joints,** permit a wide range of motion. A synovial joint is surrounded by a fibrous **articular capsule,** and a *synovial membrane* lines the articular cavity. We introduced the basic structure of a synovial joint during the discussion of synovial membranes in Chapter 4. ∞ *[p. 134]* Figure 9-1● provides additional information about the structure of synovial joints. These joints are typically found at the ends of long bones, such as those of the upper and lower limbs.

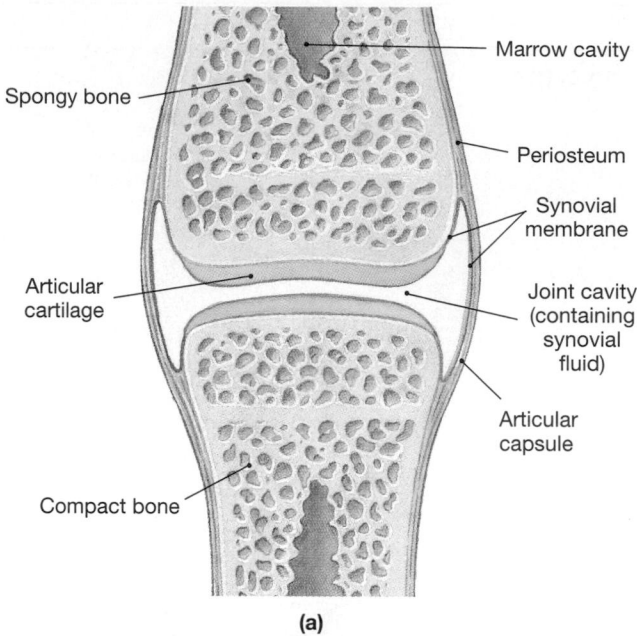

(a)

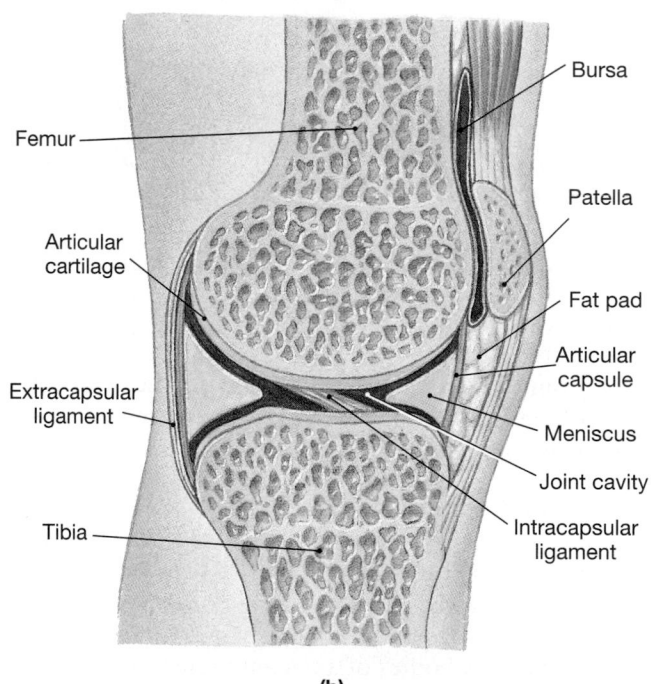

(b)

•FIGURE 9-1 **The Structure of a Synovial Joint.**
(a) Diagrammatic view of a simple articulation.
(b) A simplified sectional view of the knee joint.

Articular Cartilages

Under normal conditions, the bony surfaces at a synovial joint cannot contact one another, because the articulating surfaces are covered by special **articular cartilages.** Articular cartilages resemble hyaline cartilages elsewhere in the body. However, articular cartilages have no perichondrium (the fibrous sheath described in Chapter 4) ∞ *[p. 128]*, and the matrix contains more water than do other cartilages.

The surfaces of the articular cartilages are slick and smooth. This feature alone can reduce friction during movement at the joint. However, even when pressure is applied across a joint, the smooth articular cartilages do not actually touch one another, because they are separated by a thin film of *synovial fluid* in the joint cavity. This fluid acts as a lubricant, keeping friction at a minimum.

Proper synovial function can continue only if the articular cartilages retain their normal structure. If an articular cartilage is damaged, the matrix may begin to break down. The exposed surface will then change from a slick, smooth, gliding surface to a rough feltwork of bristly collagen fibers. This feltwork drastically increases friction at the joint.

RHEUMATISM AND ARTHRITIS **Rheumatism** (ROO-muh-tizm) is a general term that indicates pain and stiffness affecting the skeletal system, the muscular system, or both. There are several major forms of rheumatism. **Arthritis** (ar-THRĪ-tis) encompasses all the rheumatic diseases that affect synovial joints. Arthritis always involves damage to the articular cartilages, but the specific cause may vary. For example, arthritis can result from bacterial or viral infection, injury to the joint, metabolic problems, or severe physical stresses.

Osteoarthritis (os-tē-ō-ar-THRĪ-tis), also known as *degenerative arthritis* or *degenerative joint disease (DJD),* generally affects individuals age 60 or older. DJD may result from cumulative wear and tear at the joint surfaces or from genetic factors affecting collagen formation. In the U.S. population, 25 percent of women and 15 percent of men over age 60 show signs of this disease. *Rheumatoid arthritis* is an inflammatory condition that affects roughly 2.5 percent of the adult population. At least some cases occur when the immune response mistakenly attacks the joint tissues. Allergies, bacteria, viruses, and genetic factors have all been proposed as contributing to or triggering the destructive inflammation.

Regular exercise, physical therapy, and drugs that reduce inflammation, such as aspirin, can slow the progress of the disease. Surgical procedures can realign or redesign the affected joint. In extreme cases involving the hip, knee, elbow, or shoulder, the defective joint can be replaced by an artificial one. Additional information concerning the various forms of arthritis is in the *Applications Manual*. AM *Rheumatism, Arthritis, and Synovial Function*

Degenerative changes comparable to those seen in arthritis may result from joint immobilization. When motion ceases, so does the circulation of synovial fluid, and the cartilages begin to suffer. **Continuous passive motion (CPM)** of any injured joint appears to encourage the repair process by improving the circulation of synovial fluid. The movement is often performed by a physical therapist during the recovery process.

Synovial Fluid

Synovial fluid resembles interstitial fluid but contains a high concentration of proteoglycans secreted by fibroblasts of the synovial membrane. It is a thick, viscous solution with the consistency of heavy molasses. The synovial fluid within a joint has three primary functions:

1. *Lubrication.* The articular cartilages are like sponges filled with synovial fluid. When part of an articular cartilage is compressed, some of the synovial fluid is squeezed out of the cartilage and into the space between the opposing surfaces. This thin layer of fluid markedly reduces friction between moving surfaces, just as a thin film of water reduces friction between a car's tires and a highway. When the compression stops, synovial fluid is sucked back into the articular cartilages.

2. *Nutrient distribution.* The total quantity of synovial fluid in a joint is normally less than 3 ml, even in a large joint such as the knee. This small volume of fluid must circulate continuously to provide nutrients and a waste-disposal route for the chondrocytes of the articular cartilages. The synovial fluid circulates whenever the joint moves, and the compression and reexpansion of the articular cartilages pumps synovial fluid into and out of the cartilage matrix.

3. *Shock absorption.* Synovial fluid cushions shocks in joints that are subjected to compression. For example, your hip, knee, and ankle joints are compressed as you walk and are more severely compressed when you jog or run. When the pressure across a joint suddenly increases, the synovial fluid lessens the shock by distributing it evenly across the articular surfaces.

Accessory Structures

Synovial joints may have a variety of accessory structures, including pads of cartilage or fat, ligaments, tendons, and bursae (Figure 9-1b●).

CARTILAGES AND FAT PADS In several joints, including the knee (Figure 9-1b●), accessory structures may lie between the opposing articular surfaces. These include *menisci* and *fat pads.* A **meniscus** (men-IS-kus; a crescent; plural, *menisci*) is a pad of fibrocartilage situated between opposing bones within a synovial joint. Menisci, or *articular discs,* may subdivide a synovial cavity, channel the flow of synovial fluid, or allow for variations in the shapes of the articular surfaces.

Fat pads are localized masses of adipose tissue covered by a layer of synovial membrane. They are commonly superficial to a joint capsule. Fat pads protect the articular cartilages and act as packing material for the joint. When the bones move, the fat pads fill in the spaces created as the joint cavity changes shape.

LIGAMENTS The joint capsule that surrounds the entire joint is continuous with the periostea of the articulating bones. **Accessory ligaments** are localized thickenings of the capsule. These ligaments reinforce and strengthen the capsule, and they may also limit rotation at the joint. **Extracapsular ligaments** interconnect the articulating bones and pass across the out-

side of the capsule. These ligaments provide additional support to the wall of the joint. **Intracapsular ligaments**, found inside the capsule, help prevent extreme movements that might otherwise damage the joint. In a **sprain**, a ligament is stretched to the point at which some of the collagen fibers are torn, but the ligament as a whole survives, and the joint is not damaged. Ligaments are very strong, and one of the attached bones typically breaks before the ligament tears. In general, a broken bone heals much more quickly and effectively than does a torn ligament.

TENDONS While not part of the articulation itself, tendons passing across or around a joint may limit the range of motion and provide mechanical support. For example, tendons associated with the muscles of the arm provide much of the bracing for the shoulder joint.

BURSAE Bursae are small, fluid-filled pockets in connective tissue. They contain synovial fluid and are lined by a synovial membrane. Bursae may be connected to the joint cavity or may be completely separate from it. They form where a tendon or ligament rubs against other tissues. Their function is to reduce friction and act as a shock absorber. Bursae are found around most synovial joints, such as the shoulder joint. **Synovial tendon sheaths** are tubular bursae that surround tendons where they pass across bony surfaces. Bursae may also appear beneath the skin covering a bone or within other connective tissues exposed to friction or pressure. Bursae that develop in abnormal locations, or because of abnormal stresses, are called *adventitious bursae.*

BURSITIS When bursae become inflamed, causing pain in the affected area whenever the tendon or ligament moves, a condition of **bursitis** exists. Inflammation can result from the friction associated with repetitive motion, pressure over the joint, irritation by chemical stimuli, infection, or trauma. Bursitis associated with repetitive motion typically occurs at the shoulder; for example, musicians, golfers, pitchers, and tennis players may develop bursitis at this location. The most common pressure-related bursitis is a **bunion**. Bunions form over the base of the great toe as a result of the friction and distortion of the first metatarsophalangeal joint by tight shoes, especially those with pointed toes.

We have special names for bursitis at other locations, indicating the occupations most often associated with them. In "housemaid's knee," which accompanies prolonged kneeling, the affected bursa lies between the patella and the skin. The condition of "student's elbow" is a form of bursitis that can result from propping your head up with your arm on a desk while you read your A & P textbook.

Factors That Stabilize Joints

A joint cannot be both highly mobile and very strong. The greater the range of motion at a joint, the weaker it becomes. A synarthrosis, the strongest type of joint, permits no movement, whereas a diarthrosis such as the shoulder permits a broad range of movements. Any mobile diarthrosis will be damaged by movement beyond its normal range of motion. Several factors are responsible for limiting the range of motion, stabilizing the joint, and reducing the chance of injury:

- The collagen fibers of the joint capsule and any accessory, extracapsular, or intracapsular ligaments.
- The shapes of the articulating surfaces, which may prevent movement in specific directions.
- The presence of other bones, skeletal muscles, or fat pads around the joint.
- Tension in tendons attached to the articulating bones. When a skeletal muscle contracts and pulls on a tendon, movement in a specific direction may be either encouraged or opposed.

The pattern of stabilizing structures varies from one joint to another. For example, the hip joint is stabilized by the shapes of the bones (the head of the femur projects into the acetabulum), a heavy capsule, intracapsular and extracapsular ligaments, tendons, and massive muscles. It is therefore very strong and very stable. In contrast, the elbow, another very stable joint, gains its stability primarily from the interlocking of the articulating bones, with additional support of the capsule and associated ligaments. In general, the more stable the joint, the more restricted its range of motion. The shoulder joint, the most mobile synovial joint, relies only on the surrounding ligaments, muscles, and tendons to provide stability. It is therefore a relatively weak joint.

When reinforcing structures cannot protect a joint from extreme stresses, a **dislocation,** or **luxation** (luks-Ā-shun), results. In a dislocation, the articulating surfaces are forced out of position. This displacement can damage the articular cartilages, tear ligaments, or distort the joint capsule. Although the *inside* of a joint has no pain receptors, nerves that monitor the capsule, ligaments, and tendons are quite sensitive, and dislocations are very painful. The damage accompanying a partial dislocation, or **subluxation** (sub-luks-Ā-shun), is less severe. People who are "double-jointed" have joints that are weakly stabilized. Although their joints permit a greater range of motion than do those of other individuals, they are more likely to suffer partial or complete dislocations.

☑ What common characteristics are found in typical synarthrotic and amphiarthrotic joints?

☑ In a newborn infant, the large bones of the skull are joined by fibrous connective tissue. What type of joints are these? These bones later grow, interlock, and form immovable joints. What type of joints are these?

☑ Why would improper circulation of synovial fluid lead to degeneration of articular cartilages in the affected joint?

ARTICULAR FORM AND FUNCTION

To *understand* human movement you must become aware of the relationship between structure and function at each articulation. To *describe* human movement you need a frame of reference that permits accurate and precise communication. We can classify the synovial joints according to their anatomical and functional properties. To demonstrate the basis for that classification, we will use a simple model to describe the movements that can occur at a typical synovial joint.

Describing Dynamic Motion

Take a pencil (or pen) as your model and stand it upright on the surface of a desk or table (Figure 9-2a●).

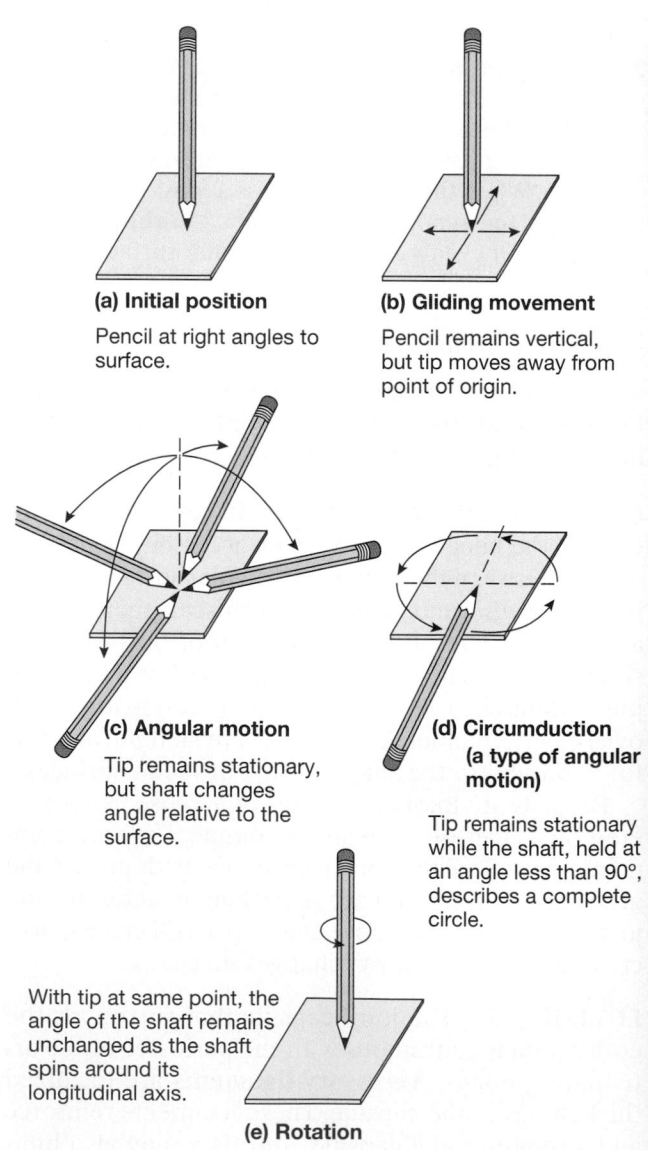

(a) Initial position
Pencil at right angles to surface.

(b) Gliding movement
Pencil remains vertical, but tip moves away from point of origin.

(c) Angular motion
Tip remains stationary, but shaft changes angle relative to the surface.

(d) Circumduction (a type of angular motion)
Tip remains stationary while the shaft, held at an angle less than 90°, describes a complete circle.

(e) Rotation
With tip at same point, the angle of the shaft remains unchanged as the shaft spins around its longitudinal axis.

●**FIGURE 9-2** A Simple Model of Articular Motion

The pencil represents a bone, and the desk is an articular surface. A little imagination and a lot of twisting, pushing, and pulling will demonstrate that there are only three ways to move the model. Considering them one at a time will provide a frame of reference for us to analyze complex movements:

Possible movement 1: The point can move. If you hold the pencil upright but do not secure the point, you can push the pencil across the surface. This kind of motion is called *gliding* (Figure 9-2b●), and it is an example of **linear motion.** You could slide the point forward or backward, from one side to the other, or diagonally. However you move the pencil, the motion can be described by using two lines of reference. One line represents forward/backward motion, and the other left/right movement. For example, a simple movement along one axis could be described as "forward 1 cm" or "left 2 cm." A diagonal movement could be described by using both axes, as in "backward 1 cm and to the right 2.5 cm."

Possible movement 2: The shaft can change its angle with the surface. With the tip held in position, you can move the free (eraser) end forward and backward or from side to side. These movements, which change the angle between the shaft and the articular surface, are examples of **angular motion** (Figure 9-2c●). We can describe such motion by the angle the shaft makes with the surface.

Any angular movement can be described with reference to the same two axes (forward/backward, left/right) and the angular change (in degrees). In one instance, however, a special term is used to describe a complex angular movement. Grasp the free end of the pencil and move it in any direction until the shaft is no longer vertical. Now swing that end through a complete circle (Figure 9-2d●). This movement, which corresponds to the path of your arm when you draw a large circle on a chalkboard, is very difficult to describe. Anatomists avoid the problem by using a special term, **circumduction** (sir-kum-DUK-shun; *circum,* around), for this type of angular motion.

Possible movement 3: The shaft can rotate. If you prevent movement of the base and keep the shaft vertical, you can spin the shaft around its longitudinal axis. This movement is called **rotation** (Figure 9-2e●). Several articulations will permit partial rotation, but none can rotate freely. Such a movement would hopelessly tangle the blood vessels, nerves, and muscles that cross the joint.

An articulation that permits movement along only one axis is called **monaxial** (mon-AKS-ē-ul). In the above model, if an articulation permits only angular movement in the forward/backward plane or prevents any movement other than rotation around its longitudinal axis, it is monaxial. If movement can occur along two axes, the articulation is **biaxial** (bī-AKS-ē-ul). If the pencil could undergo angular motion in the forward/backward *and* left/right planes, but not rota-

tion, it would be biaxial. The most mobile joints permit a combination of angular movement and rotation. These are said to be **triaxial** (tri-AKS-ē-ul). Joints that permit gliding allow only small amounts of movement. These joints may be called *nonaxial,* because they permit only small sliding movements, or *multiaxial,* because that sliding may occur in any direction.

Types of Movements

In descriptions of motion at synovial joints, phrases such as "bend the leg" or "raise the arm" are not sufficiently precise. Anatomists use descriptive terms that have specific meanings. We will consider these motions with reference to the basic categories of movement we discussed previously.

Gliding

In **gliding,** two opposing surfaces slide past one another, as in possible movement 1. Gliding occurs between the surfaces of articulating carpal bones, between tarsal bones, and between the clavicles and the sternum. The movement can occur in almost any direction, but the amount of movement is slight, and rotation is generally prevented by the capsule and associated ligaments.

Angular Motion

Examples of angular motion include *flexion, extension, abduction, adduction,* and *circumduction.* The descriptions of the movements are based on reference to an individual in the anatomical position.

FLEXION AND EXTENSION Flexion (FLEK-shun) is defined as movement in the anterior/posterior plane that *reduces the angle between the articulating elements.* **Extension** occurs in the same plane, but it *increases the angle between articulating elements* (Figure 9-3a●). When you bring your head toward your chest, you flex your neck. When you bend down to touch your toes, you flex the spine. Extension reverses these movements, returning you to the anatomical position.

Flexion at the shoulder or hip joint moves the limbs anteriorly, whereas extension moves them posteriorly. Flexion of the wrist joint moves the palm anteriorly, and extension moves it posteriorly. In each of these examples, extension can be continued past the anatomical position. Extension past the anatomical position is called **hyperextension** (Figure 9-3a●). You can hyperextend your neck, a movement that allows you to gaze at the ceiling. Hyperextension of many joints is prevented by ligaments, bony processes, or soft tissues.

ABDUCTION AND ADDUCTION Abduction (*ab,* from) is movement *away from the longitudinal axis of the body* in the frontal plane (Figure 9-3b●). For example, swinging the upper limb to the side is abduction of the limb. Moving it back to the anatomical position con-

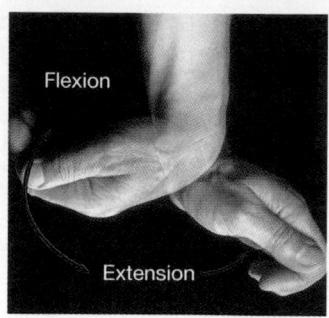

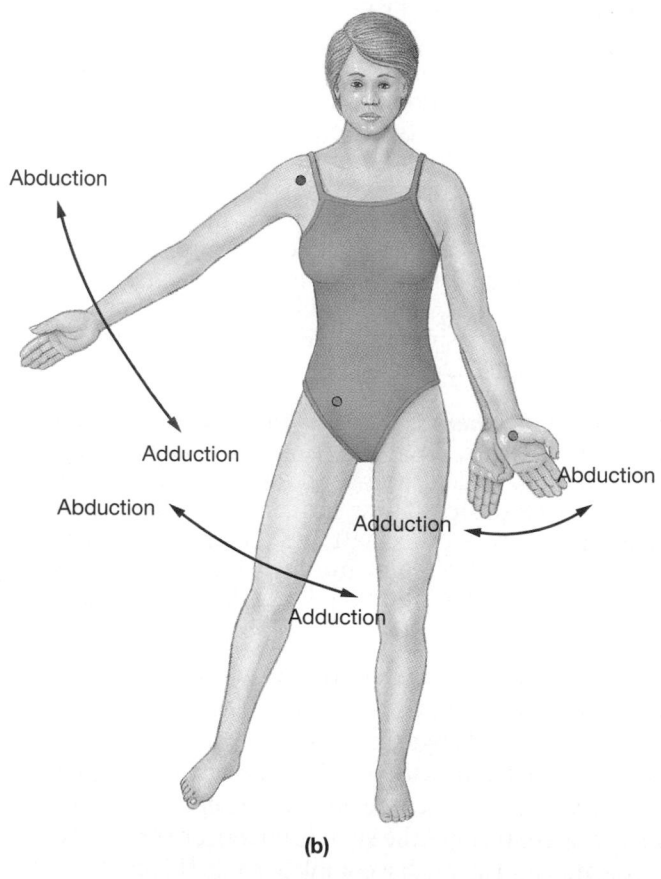

•**FIGURE 9-3** **Angular Movements.** The red dots indicate the locations of the joints involved in the illustrated movement.

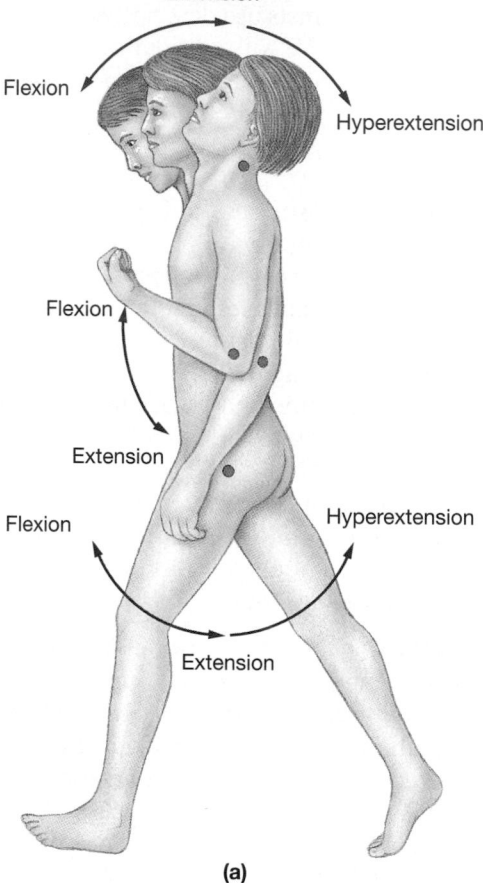

(a)

(b)

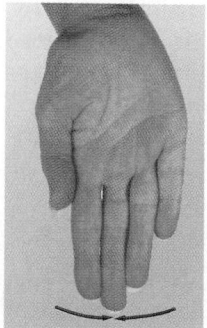

Adduction

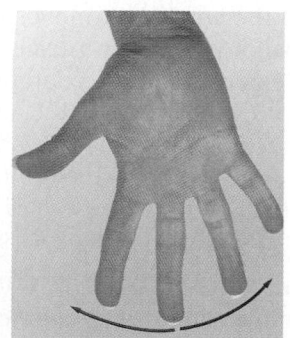

Abduction

(c)

Circumduction

(d)

stitutes **adduction** (*ad*, to). Adduction of the wrist moves the heel of the hand *toward* the body, whereas abduction moves it farther away. Spreading the fingers or toes apart abducts them, because they move *away* from a central digit (Figure 9-3c●). Bringing them together constitutes adduction. Abduction and adduction always refer to movements of the appendicular skeleton, not to those of the axial skeleton.

CIRCUMDUCTION We introduced a special type of angular motion, *circumduction*, in our model. A familiar example of circumduction is moving your arm in a loop (Figure 9-3d●), as when you draw a large circle on a chalkboard. Although your hand moves in a circle, your arm does not rotate.

Rotation

Rotational movements are also described with reference to a figure in the anatomical position (Figure 9-4●). Rotation of the head may involve **left rotation** or **right rotation.** Limb rotation may be described by reference to the longitudinal axis of the trunk. During **medial rotation**, also known as *internal rotation* or *inward rotation*, the anterior surface of a limb turns toward the long axis of the trunk. The reverse movement is called **lateral rotation,** *external rotation*, or *inward rotation.*

The articulations between the radius and ulna permit the rotation of the distal end of the radius across the anterior surface of the ulna. This rotation moves the wrist and hand from palm-facing-front to palm-facing-back. This motion is **pronation** (prō-NĀ-shun). The opposing movement, in which the palm is turned anteriorly, is **supination** (soo-pi-NĀ-shun).

Special Movements

Several special terms apply to specific articulations or unusual types of movement (Figure 9-5●):

- **Inversion** (*in*, into + *vertere*, to turn) is a twisting motion of the foot that turns the sole inward. The opposite movement is called **eversion** (ē-VER-zhun; *e*, out).
- **Dorsiflexion** is flexion of the ankle and elevation of the sole, as when you dig in your heel. **Plantar flexion** (*planta*, sole), the opposite movement, extends the ankle and elevates the heel, as when you stand on tiptoe.
- **Opposition** is a special movement of the thumb that

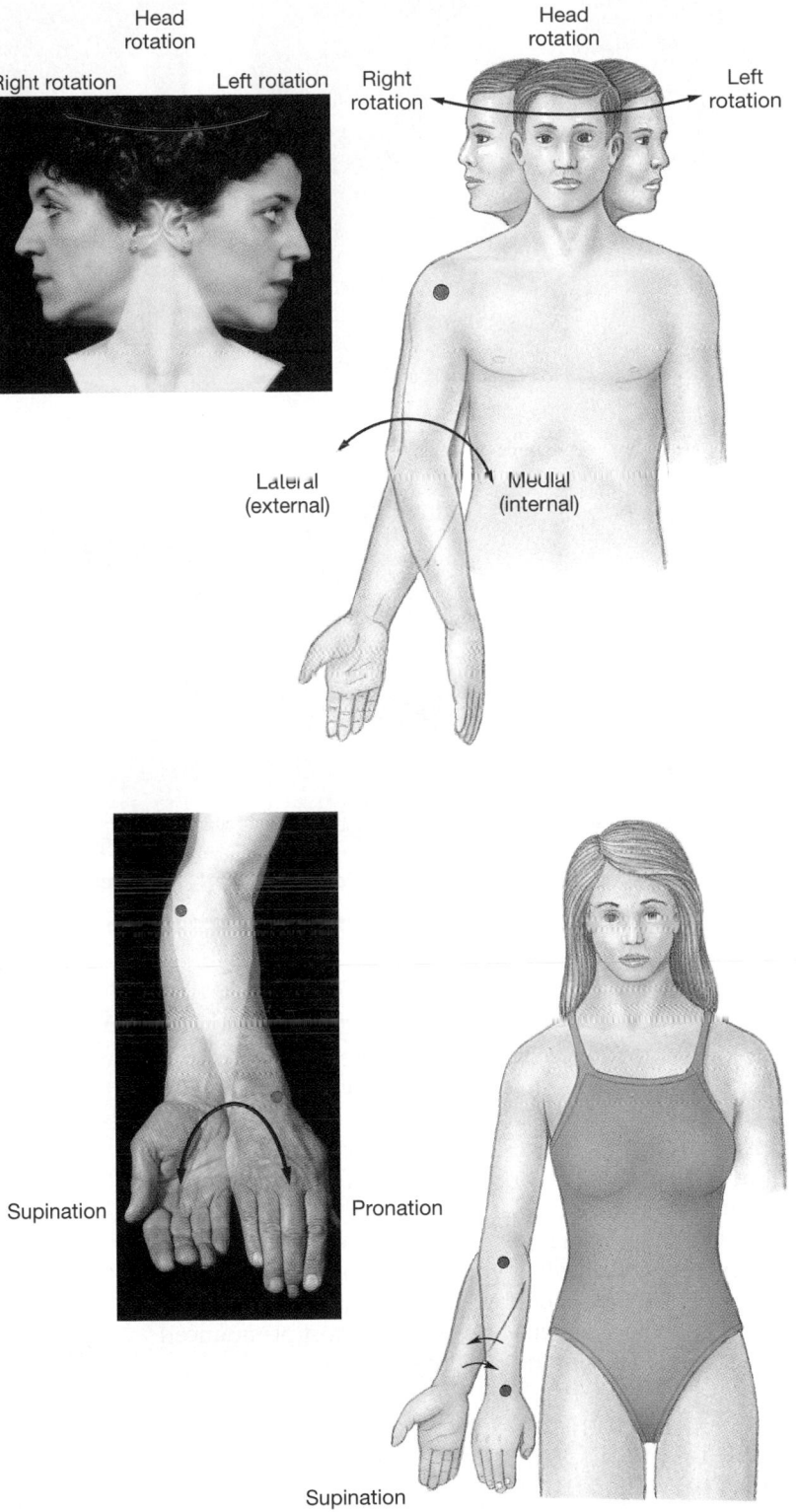

●**FIGURE 9-4** Rotational Movements

involves the carpometacarpal and metacarpophalangeal joints. Opposition permits you to grasp and hold an object with your thumb and palm.

- **Protraction** entails moving a part of the body anteriorly in the horizontal plane. **Retraction** is the reverse

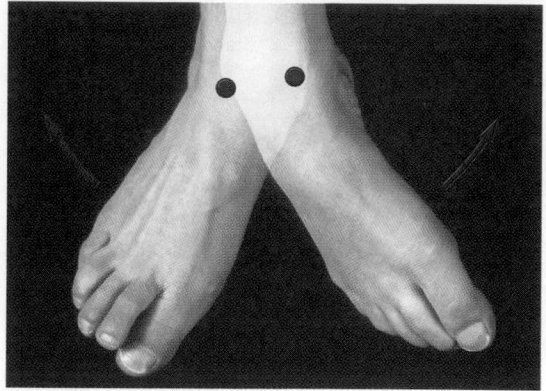
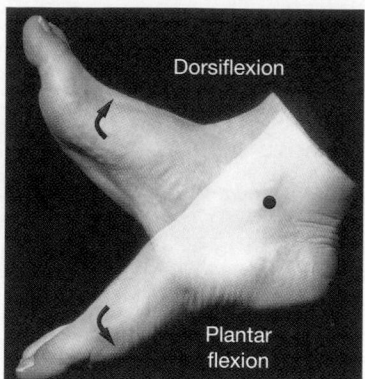

Eversion   Inversion

Dorsiflexion

Plantar flexion

Opposition

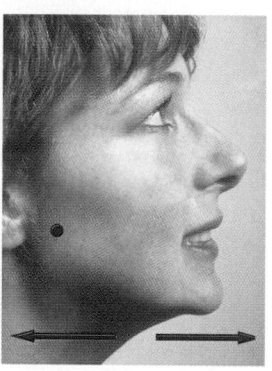
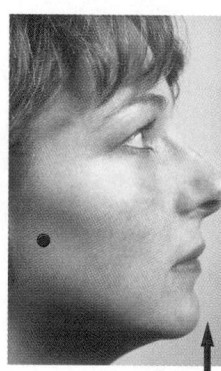
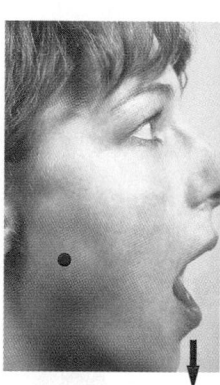
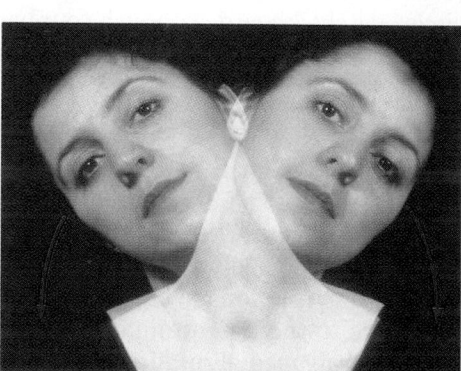

Retraction   Protraction   Elevation   Depression   Lateral flexion

•**FIGURE 9-5** Special Movements

movement. You protract your jaw when you grasp your upper lip with your lower teeth, and you protract your clavicles when you cross your arms.

- **Elevation** and **depression** occur when a structure moves in a superior or inferior direction, respectively. You depress your mandible when you open your mouth; you elevate it as you close it. Another familiar elevation occurs when you shrug your shoulders.
- **Lateral flexion** occurs when your vertebral column bends to the side. This movement is most pronounced in the cervical and thoracic regions.

A Structural Classification of Synovial Joints

Synovial joints can be described as *gliding, hinge, pivot, ellipsoidal, saddle,* or *ball-and-socket joints* on the basis of the shapes of the articulating surfaces. Each type of joint permits a different type and range of motion.

Gliding Joints

Gliding joints (Figure 9-6a•), also called *planar joints,* have flattened or slightly curved faces. The relatively flat articular surfaces slide across one another, but the amount of movement is very slight. Although rotation is theoretically possible at such a joint, ligaments usually prevent or restrict such movement. Gliding joints are found at the ends of the clavicles, between the carpal bones, between the tarsal bones, and between the articular facets of adjacent spinal vertebrae.

Hinge Joints

Hinge joints (Figure 9-6b•) permit angular movement in a single plane, like the opening and closing of a door. A hinge joint is a monaxial joint. Examples include the joint between the occipital bone and atlas of the axial skeleton, and the elbow, knee, ankle, and interphalangeal joints of the appendicular skeleton.

Pivot Joints

Pivot joints (Figure 9-6c•) are also monaxial, but they permit only rotation. A pivot joint between the atlas and axis allows you to rotate your head to either side, and another between the head of the radius and the proximal shaft of the ulna permits pronation and supination of the palm.

•FIGURE 9-6 A Functional Classification of Synovial Joints

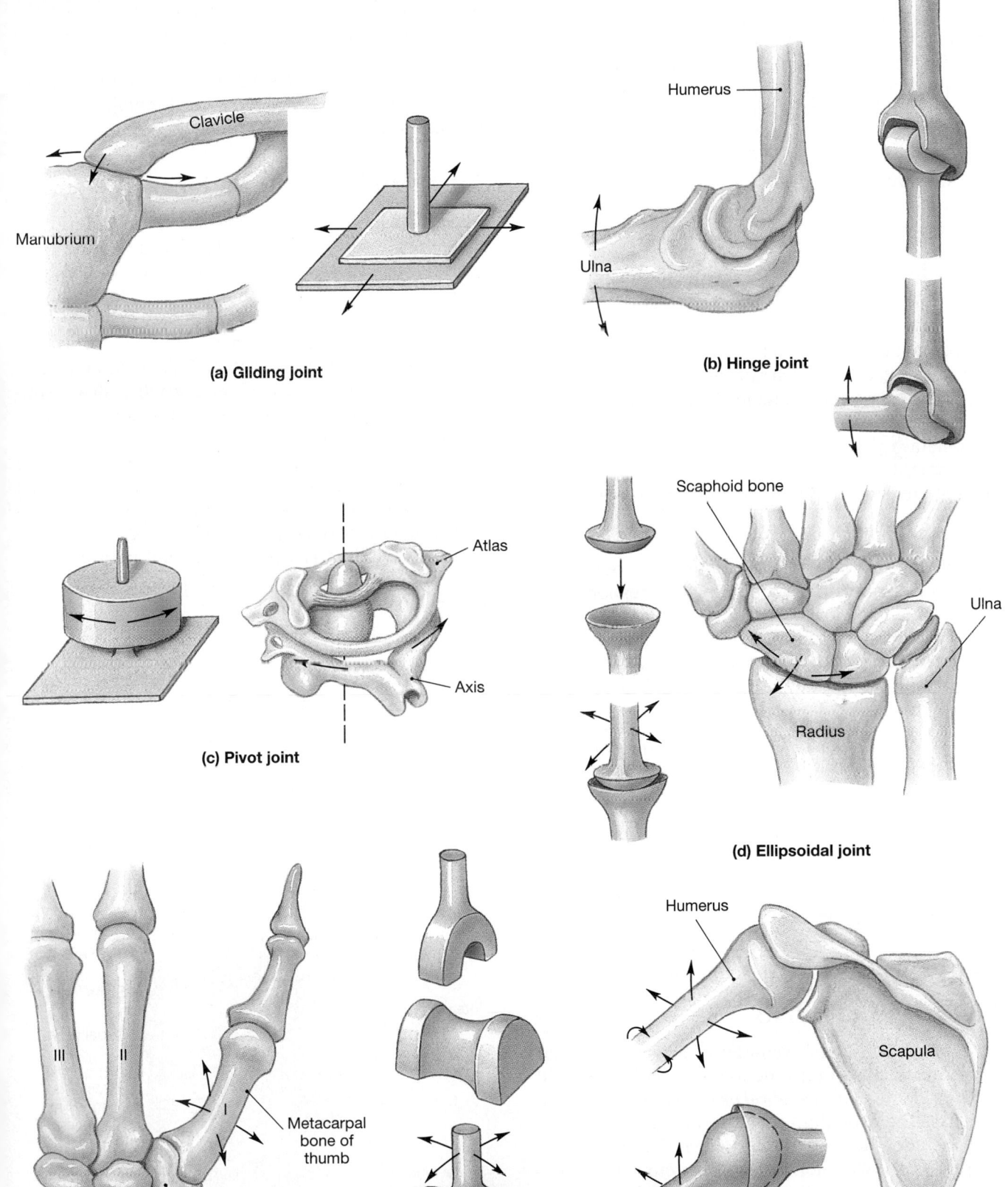

(a) Gliding joint

(b) Hinge joint

(c) Pivot joint

(d) Ellipsoidal joint

(e) Saddle joint

(f) Ball-and-socket joint

Ellipsoidal Joints

In an **ellipsoidal joint** (Figure 9-6d●), or *condyloid joint*, an oval articular face nestles within a depression in the opposing surface. With such an arrangement, angular motion occurs in two planes, along or across the length of the oval. It is thus a biaxial joint. Any form of angular movement, including circumduction, is permitted, but rotation cannot occur. Ellipsoidal joints connect the radius with the proximal carpal bones and the phalanges of the fingers and toes with the metacarpal bones and metatarsal bones, respectively.

Saddle Joints

Saddle joints (Figure 9-6e●), or *sellaris joints*, have articular faces that resemble saddles. Each face is concave on one axis and convex on the other, and the opposing faces nest together. This arrangement permits angular motion, including circumduction, but prevents rotation. Saddle joints are usually considered to be biaxial. The carpometacarpal joint at the base of the thumb is the best example of a saddle joint, and twiddling your thumbs will demonstrate the possible movements.

Ball-and-Socket Joints

In a **ball-and-socket joint** (Figure 9-6f●), the round head of one bone rests within a cup-shaped depression in another. All combinations of angular and rotational movements, including circumduction and rotation, can be performed at ball-and-socket joints. These are triaxial joints. Examples include the shoulder and hip joints.

☑ Give the proper term for each of the following types of motion: (a) moving the humerus away from the midline of the body; (b) turning the palms so that they face forward; (c) bending the elbow.

☑ When you do jumping jacks, what lower limb movements are necessary?

☑ What types of movement are usually associated with hinge joints?

REPRESENTATIVE ARTICULATIONS

This section considers examples of articulations that demonstrate important functional principles. We will first consider the *intervertebral articulations* of the axial skeleton. The articulations between adjacent vertebrae include gliding articulations and cartilaginous symphyses. We will then discuss the synovial articulations of the appendicular skeleton. The shoulder has great mobility, the elbow has great strength, and the wrist makes fine adjustments in the orientation of the palm and fingers. The functional requirements of the joints in the lower limb are very different from those of the upper limb. Articulations at the hip, knee, and ankle

must transfer the body weight to the ground; during movements such as running, jumping, or twisting, the applied forces are considerably greater than the weight of the body. Although we consider only representative articulations here, Tables 9-3, 9-4, and 9-5 (pp. 265, 269, 272) summarize data concerning the majority of articulations in the body.

Intervertebral Articulations

The articulations between the superior and inferior articular processes of adjacent vertebrae are gliding joints that permit small movements associated with flexion and rotation of the vertebral column. Little gliding occurs between adjacent vertebral bodies. Figure 9-7● illustrates the structure of these joints. From axis to sacrum, the vertebrae are separated and cushioned by pads of fibrocartilage called **intervertebral discs.** Thus the bodies of vertebrae form symphyseal joints. Intervertebral discs and symphyseal joints are not found in the sacrum or coccyx, where vertebrae have fused, or between the first and second cervical vertebrae. The first cervical vertebra has no vertebral body and no intervertebral disc; the only articulation between the first two cervical vertebrae is a pivot joint that permits much more rotation than the symphyseal joints between other cervical vertebrae.

The Intervertebral Discs

Each intervertebral disc has a tough outer layer of fibrocartilage, the **annulus fibrosus** (AN-ū-lus fī-BRŌ-sus). The collagen fibers of the annulus fibrosis attach the disc to the bodies of adjacent vertebrae. The an-

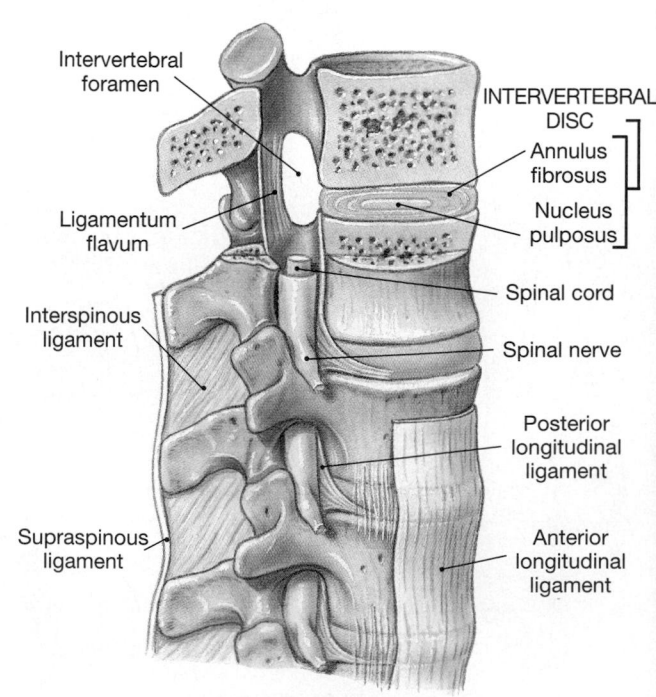

Intervertebral foramen

INTERVERTEBRAL DISC

Annulus fibrosus

Nucleus pulposus

Ligamentum flavum

Interspinous ligament

Spinal cord

Spinal nerve

Posterior longitudinal ligament

Supraspinous ligament

Anterior longitudinal ligament

●*FIGURE 9-7* **Intervertebral Articulations**

nulus fibrosus surrounds the **nucleus pulposus** (pul-PŌ-sus), a soft, elastic, and gelatinous core. The nucleus pulposus gives the disc resiliency and enables it to act as a shock absorber. About 75 percent of the core is water; the rest is scattered reticular and elastic fibers.

Movement of the vertebral column compresses the nucleus pulposus and displaces it in the opposite direction. This displacement permits smooth gliding movements by each vertebra while still maintaining their alignment. The discs make a significant contribution to an individual's height: They account for roughly one-quarter the length of the vertebral column above the sacrum. As we grow older, the water content of the nucleus pulposus within each disc decreases. The discs gradually become less effective as a cushion, and the chances for vertebral injury increase. Loss of water by the discs also causes shortening of the vertebral column; this loss accounts for the characteristic decrease in height with advancing age.

Intervertebral Ligaments

Numerous ligaments are attached to the bodies and processes of all vertebrae to bind them together and stabilize the vertebral column. Ligaments interconnecting adjacent vertebrae include:

- The **anterior longitudinal ligament**, which connects the anterior surfaces of adjacent vertebral bodies.
- The **posterior longitudinal ligament**, which parallels the anterior longitudinal ligament and connects the posterior surfaces of adjacent vertebral bodies.
- The **ligamentum flavum** (plural, *ligamenta flava*), which connects the laminae of adjacent vertebrae.
- The **interspinous ligament**, which connects the spinous processes of adjacent vertebrae.
- The **supraspinous ligament**, which interconnects the tips of the spinous processes from C_7 to the sacrum. The *ligamentum nuchae*, discussed in Chapter 7, extends from vertebra C_7 to the base of the skull. It is continuous with supraspinous ligament. ∽ [p. 222]

Vertebral Movements

The following movements of the vertebral column are possible: (1) *flexion*, bending forward; (2) *extension*, bending backward; (3) *lateral flexion*, bending to the side; and (4) *rotation*. Table 9-3 summarizes information concerning other articulations of the axial skeleton.

TABLE 9-3 Articulations of the Axial Skeleton

Element	Joint	Type of Articulation	Movements
Skull			
Cranial and facial bones of skull	Various	Synarthroses (suture or synostosis)	None
Maxillary bone/teeth and mandible/teeth		Synarthrosis (gomphosis)	None
Temporal bone/mandible	Temporomandibular	Combined gliding joint and hinge diarthrosis	Elevation, depression, and lateral gliding
Vertebral column			
Occipital bone/atlas	Atlanto-occipital	Ellipsoidal diarthrosis	Flexion/extension
Atlas/axis	Atlanto-axial	Pivot diarthrosis	Rotation
Other vertebral elements	Intervertebral (between vertebral bodies)	Amphiarthrosis (symphysis)	Slight movement
	Intervertebral (between articular processes)	Gliding diarthrosis	Slight rotation and flexion/extension
L_5/sacrum	Between L_5 body and	Amphiarthrosis (symphysis) sacral body	Slight movement
	Between inferior articular processes of L_5 and articular processes of sacrum	Gliding diarthrosis	Slight flexion/extension
Sacrum/coxae	Sacroiliac	Gliding diarthrosis	Slight movement
Sacrum/coccyx	Sacrococcygeal	Gliding diarthrosis (may	Slight movement become fused)
Coccygeal bones		Synarthrosis (synostosis)	No movement
Thoracic cage			
Thoracic vertebrae and ribs	Vertebrocostal	Gliding diarthrosis	Elevation/depression
Ribs and sternum	Sternocostal	Synarthrosis (synchondrosis)	No movement
Sternum and clavicle	Sternoclavicular	Gliding diarthrosis	Protraction/retraction, depression/elevation

Problems with the Intervertebral Discs

An intervertebral disc compressed beyond its normal limits may become temporarily or permanently damaged. If the posterior longitudinal ligaments are weakened, as often occurs with advancing age, the compressed nucleus pulposus may distort the annulus fibrosus, forcing it partway into the vertebral canal. This condition is often called a **slipped disc** (Figure 9-8a●), although the disc does not actually slip. The most common sites for disc problems are at C_5–C_6, L_4–L_5, and between L_5 and S_1. A disc problem can occur at any age as the result of an accidental injury, such as a hard fall or a "whiplash" injury to the neck. But, with advanced age, the supporting ligaments may become so weak that the problem could occur without warning or apparent cause.

If the nucleus pulposus breaks through the annulus fibrosis, it too may protrude into the vertebral canal. This condition is called a **herniated disc** (Figure 9-8b●). When a disc herniates, sensory nerves are distorted, and the protruding mass can also compress the nerve roots passing through the adjacent intervertebral foramen. The result is severe backache, abnormal posture (abnormal vertebral flexion), abnormal sensory function, typically a burning or tingling sensation from the lower back and lower limbs, and in some cases a partial loss of control over skeletal muscles innervated by the compressed nerve fibers. The location of the injured disc can generally be determined by noting the distribution of abnormal sensations. For example, someone with a herniated disc at L_4–L_5 will experience pain in the hip, groin, the posterior and lateral surfaces of the thigh, the lateral surface of the calf, and the superior surface of the foot. A herniation at L_5–S_1 produces pain in the buttocks, the posterior thigh, the posterior calf, and the sole of the foot.

Most lumbar disc problems can be treated successfully with some combination of rest, back braces, analgesic (pain-killing) drugs, and physical therapy. Surgery to relieve the symptoms is required in only about 10 percent of cases involving lumbar disc herniation. The primary method of treatment involves removing the offending disc, and, if necessary, fusing the vertebral bodies together to prevent relative movement. Accessing the disc requires removal of the nearest vertebral arch by shaving away the laminae. For this reason the procedure is known as a *laminectomy* (la-mi-NEK-to-mē).

In cases in which the herniated portion of the disc does not extend well into the vertebral foramen, portions of the disc may be removed with a suction cutter guided to the site by radiological imaging. Although this procedure is faster and easier than a laminectomy, relatively few herniated discs fall within this category.

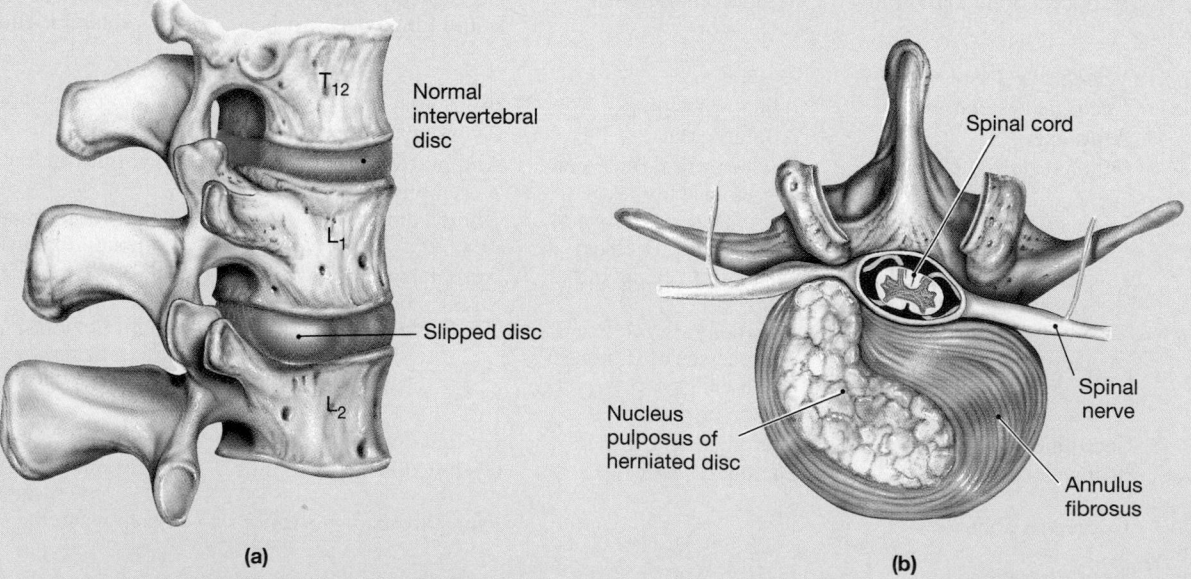

(a) (b)

●**FIGURE 9-8** **Damage to the Intervertebral Discs.** **(a)** Lateral view of the lumbar region of the spinal column, showing a distorted intervertebral disc (a "slipped" disc). **(b)** Sectional view through a herniated disc, showing release of the nucleus pulposus and its effect on the spinal cord and adjacent nerves.

☑ What regions of the vertebral column do not contain intervertebral discs? Why is the lack of intervertebral discs significant?

☑ What vertebral movements are involved in (a) bending forward, (b) bending to the side, and (c) moving the head to signify "no"?

The Shoulder Joint

The shoulder joint, or *glenohumeral (or scapulahumeral) joint,* permits the greatest range of motion of any joint in the body. Because it is also the most frequently dislocated joint, it provides an excellent demonstration of the principle that strength and stability must be sacrificed to obtain mobility.

This joint is a ball-and-socket diarthrosis formed by the articulation of the head of the humerus with the glenoid cavity of the scapula (Figure 9-9●). The surface of the glenoid cavity is covered by a fibrocartilaginous **glenoid labrum** (*labrum,* lip or edge). The relatively loose articular capsule extends from the scapular neck to the humerus. It is a somewhat oversized capsule that permits an extensive range of motion. The bones of the pectoral girdle provide some stability to the superior surface, because the acromion and coracoid processes project laterally superior to the humeral head. Howev-

er, most of the stability at this joint is provided by ligaments and surrounding skeletal muscles and their associated tendons. Bursae reduce friction between the tendons and other tissues at the joint.

The major ligaments involved with stabilizing the shoulder joint include the **glenohumeral, coracohumeral, coracoacromial, coracoclavicular,** and **acromioclavicular ligaments.** The acromioclavicular ligament reinforces the capsule of the acromioclavicular joint and provides support for the superior surface of the shoulder. A **shoulder separation** is a relatively common injury involving partial or complete dislocation of the acromioclavicular joint. This injury can result from a blow to the upper surface of the shoulder. The acromion is forcibly depressed, but the clavicle is held back by strong muscles.

The muscles that move the humerus do more to stabilize the shoulder joint than do all the ligaments and capsular fibers combined. Muscles originating on the trunk, shoulder girdle, and humerus cover the anterior, superior, and posterior surfaces of the capsule. The tendons of the *supraspinatus, infraspinatus, subscapularis,* and *teres minor* reinforce the joint capsule and limit the range of movement. These muscles, known as the muscles of the *rotator cuff,* are the primary mechanism for supporting the shoulder joint and limiting the range of movement. Damage to the rotator cuff typically occurs when individuals are engaged in

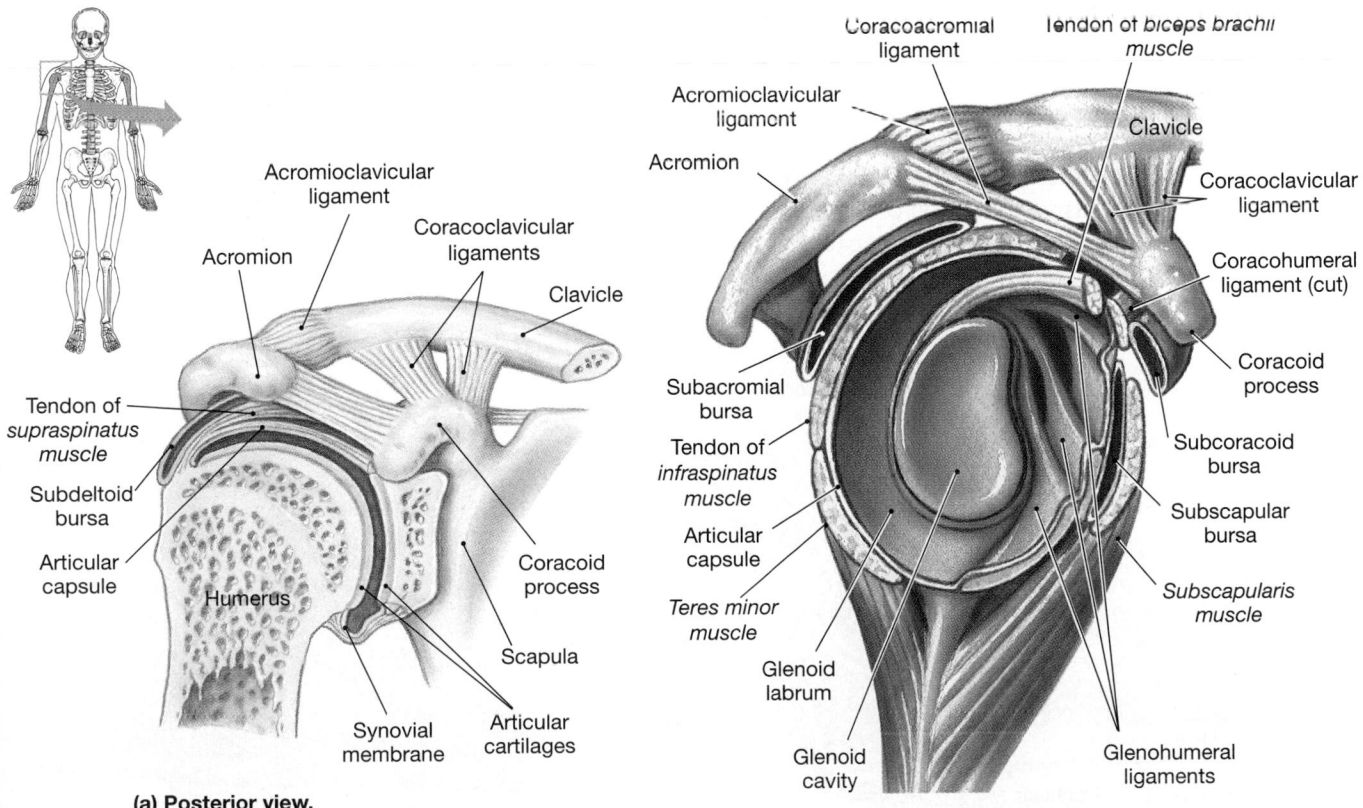

(a) Posterior view, right shoulder joint

(b) Lateral view of shoulder joint, humerus removed

●**FIGURE 9-9** **The Shoulder Joint.** **(a)** Sectional view showing major structural features. **(b)** Lateral view of the shoulder joint with the humerus removed.

sports that place severe strains on the shoulder. White-water kayakers, baseball pitchers, and quarterbacks are all at high risk for rotator cuff injuries.

The anterior, superior, and posterior surfaces of the shoulder joint are reinforced by ligaments, muscles, and tendons, but the inferior capsule is poorly reinforced. As a result, a dislocation caused by an impact or violent muscle contraction is most likely to occur at this site. Such a dislocation can tear the inferior capsular wall and the glenoid labrum. The healing process typically leaves a weakness that increases the chances for future dislocations.

As at other joints, bursae at the shoulder reduce friction where large muscles and tendons pass across the joint capsule. The shoulder has a relatively large number of important bursae, such as the *subacromial bursa*, the *subcoracoid bursa*, the *subdeltoid bursa*, and the *subscapular bursa* (Figure 9-9●). Inflammation of one or more of these bursae can restrict motion and produce the painful symptoms of bursitis (p. 257).

The Elbow Joint

The elbow joint, or *olecranal joint*, is a hinge diarthrosis that permits only flexion and extension. The trochlea of the humerus articulates with the trochlear notch of the ulna, and the capitulum of the humerus articulates with the head of the radius (Figure 9-10a●).

Muscles that extend the elbow attach to the rough surface of the olecranon process. These muscles are primarily under the control of the *radial nerve*, which passes along the *radial groove* of the humerus. ∞ *[p. 237]* The large *biceps brachii* muscle covers the anterior surface of the arm. Its tendon is attached to the radius at the *radial tuberosity*. Contraction of this muscle produces supination of the forearm and flexion of the elbow.

The elbow joint is extremely stable because (1) the bony surfaces of the humerus and ulna interlock;

(2) the articular capsule is very thick; and (3) the capsule is reinforced by strong ligaments. The **radial collateral ligament** stabilizes the lateral surface of the joint. It extends between the lateral epicondyle of the humerus and the **annular ligament**, which binds the proximal radial head to the ulna. The medial surface of the joint is stabilized by the **ulnar collateral ligament**. This ligament extends from the medial epicondyle of the humerus anteriorly to the coronoid processes of the ulna and posteriorly to the olecranon (Figure 9-10b●).

Despite the strength of the capsule and ligaments, the elbow joint can be damaged by severe impacts or unusual stresses. For example, if you fall on a hand with a partially flexed elbow, contractions of muscles that extend the elbow may break the ulna at the center of the trochlear notch. Less violent stresses can produce dislocations or other injuries to the elbow, especially if epiphyseal growth has not been completed. For example, parents in a hurry may drag a toddler along behind them, exerting an upward, twisting pull on the elbow joint that can result in a partial dislocation known as *nursemaid's elbow*.

Table 9-4 summarizes the characteristics of the joints of the upper limb.

☑ Which tissues or structures provide most of the stability for the shoulder joint?

☑ Would a tennis player or a jogger be more likely to develop inflammation of the subscapular bursa? Why?

☑ A football player received a blow to the upper surface of his shoulder, causing a shoulder separation. What does this mean?

☑ Terry suffers an injury to his forearm and elbow. After the injury, he notices an unusually large degree of motion between the radius and the ulna at the elbow. What ligament did Terry most likely damage?

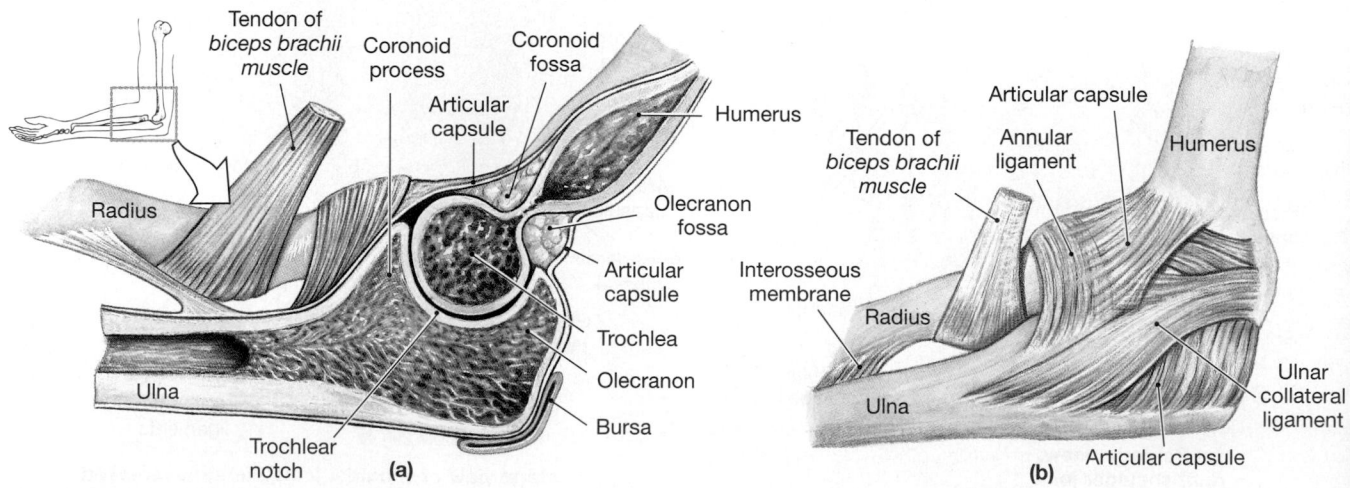

●**FIGURE 9-10** **The Elbow Joint.** **(a)** Longitudinal section through the right elbow joint. **(b)** Medial view of the right elbow joint, showing ligaments that stabilize the joint.

TABLE 9-4 Articulations of the Pectoral Girdle and Upper Limb

Element	Joint	Type of Articulation	Movement
Clavicle/sternum	Sternoclavicular	Gliding diarthrosis	Gliding
Scapula/clavicle	Acromioclavicular	Gliding diarthrosis	Gliding
Scapula/humerus	Shoulder, or glenohumeral (scapulohumeral)	Ball-and-socket diarthrosis	Flexion/extension, adduction/abduction, circumduction, rotation
Humerus/ulna and radius	Elbow, or olecranal	Hinge diarthrosis	Flexion/extension
Radius/ulna	Proximal radioulnar	Pivot diarthrosis	Pronation/supination
	Distal radioulnar	Pivot diarthrosis	Pronation/supination
Radius/carpal bones	Wrist, or radiocarpal	Ellipsoidal diarthrosis	Flexion/extension, adduction/abduction, circumduction
Carpal bone to carpal bone	Intercarpal	Gliding diarthrosis	Gliding
Carpal bone to metacarpal bone (I)	Carpometacarpal of thumb	Saddle diarthrosis	Flexion/extension, adduction/abduction, circumduction, opposition
Carpal bone to metacarpal bone (II–V)	Carpometacarpal	Gliding diarthrosis	Slight flexion/extension, adduction/abduction
Metacarpal bone to phalanx	Metacarpophalangeal	Ellipsoidal diarthrosis	Flexion/extension, adduction/abduction, circumduction
Phalanx/phalanx	Interphalangeal	Hinge diarthrosis	Flexion/extension

The Hip Joint

The hip joint is a sturdy ball-and-socket diarthrosis that permits flexion and extension, adduction and abduction, circumduction, and rotation. Figure 9-11 introduces the structure of the hip joint. A fibrocartilage pad covers the articular surface of the acetabulum and extends like a horseshoe along the sides of the **acetabular notch** (Figure 9-11a). A fat pad covered by synovial membrane covers the central portion of the acetabulum. This pad acts as a shock absorber, and the adipose tissue stretches and distorts without damage.

The articular capsule of the hip joint is extremely dense and strong. It extends from the lateral and inferior surfaces of the pelvic girdle to the intertrochanteric line and intertrochanteric crest of the femur, enclosing both the femoral head and neck. This arrangement helps keep the head from moving away from the acetabulum.

Four broad ligaments reinforce the articular capsule (Figure 9-11a,b,c). Three of them—the **iliofemoral, pubofemoral,** and **ischiofemoral ligaments**—are regional thickenings of the capsule. The **transverse acetabular ligament** crosses the acetabular notch and completes the inferior border of the acetabular fossa. A fifth ligament, the **ligament of the femoral head**, or *ligamentum teres* (*teres,* long and round), originates along the transverse acetabular ligament (Figure 9-11a) and attaches to the center of the femoral head at the *fovea capitis.* ∞ [*p. 245*] This ligament tenses only when the hip is flexed and the thigh is undergoing lateral rotation. Much more important stabilization is provided by the bulk of the surrounding muscles, aided by ligaments and capsular fibers.

The combination of an almost complete bony socket, a strong articular capsule, supporting ligaments, and muscular padding makes the hip joint an extremely stable joint. Fractures of the femoral neck or between the trochanters are more common than hip dislocations. As we noted in Chapter 6, hip fractures are relatively common in elderly individuals with severe osteoporosis. ∞ [*p. 191*]

HIP FRACTURES Hip fractures are most often suffered by individuals over age 60, when osteoporosis has weakened the thigh bones. These injuries may be accompanied by dislocation of the hip or by pelvic fractures. For individuals with osteoporosis, healing proceeds very slowly. In addition, the powerful muscles that surround the joint can easily prevent proper alignment of the bone fragments. Trochanteric fractures generally heal well if the joint can be stabilized; steel frames, pins, screws, or some combination of these devices may be needed to preserve alignment and permit healing to proceed normally.

Severe hip fractures are most common among those over age 60, but in recent years the frequency of hip fractures has increased dramatically among young, healthy professional athletes. Probably the best-known example is the case of Bo Jackson, discussed in the *Applications Manual.* [AM] *Hip Fractures, Aging, and Professional Athletes*

The Knee Joint

The hip joint passes weight to the femur, and the knee joint transfers the weight from the femur to the tibia. The shoulder is mobile; the hip, stable; and the knee, . . .?

•**FIGURE 9-11** **The Hip Joint.** **(a)** Lateral view of the right hip joint with the femur removed. **(b)** Anterior view of the right hip joint. **(c)** Posterior view of the right hip joint, showing additional ligaments that add strength to the capsule. AM *Plate 7.3a; Scan 4*

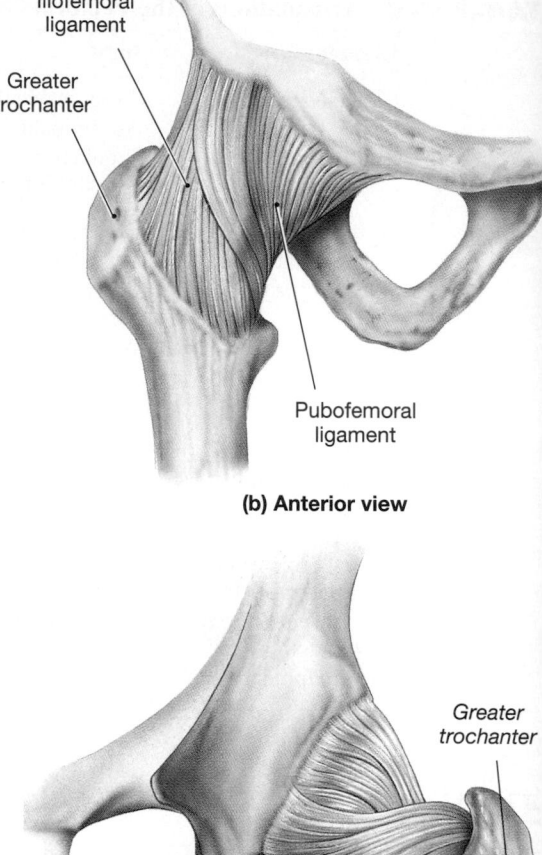

(b) Anterior view

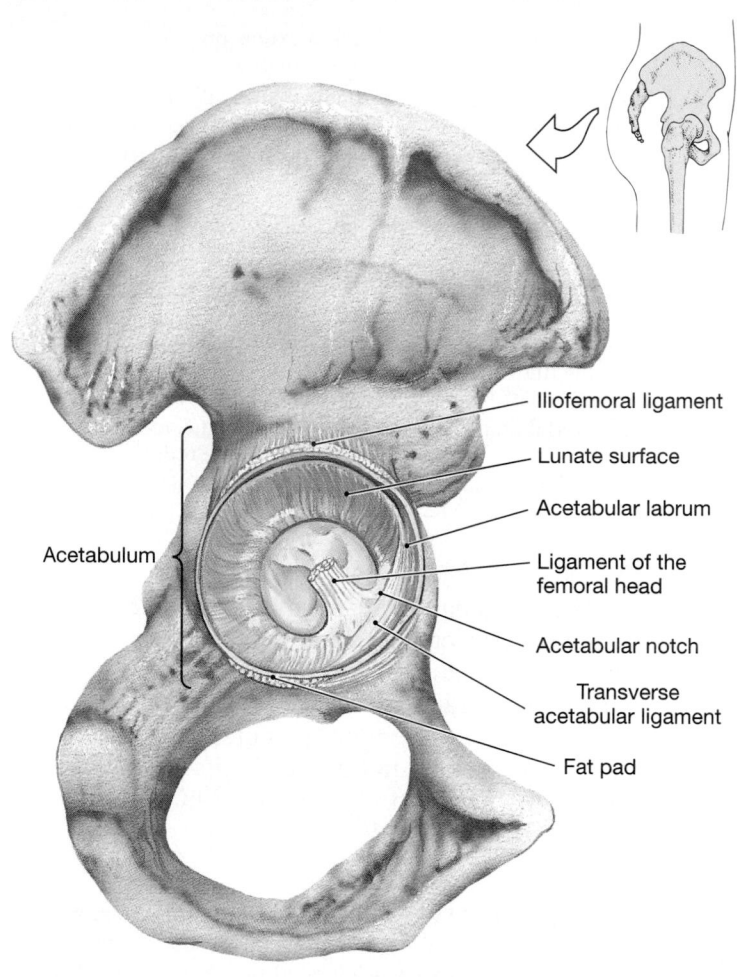

(a) Lateral view

(c) Posterior view

If you had to choose one word, it would probably be "complicated." Although the knee functions as a hinge joint, the articulation is far more complex than that of the elbow or even the ankle. The rounded femoral condyles roll across the top of the tibia, so the points of contact are constantly changing. The joint permits flexion and extension and very limited rotation.

Structurally, the knee resembles three separate joints—two between the femur and tibia (medial condyle to medial condyle and lateral condyle to lateral condyle) and one between the patella and the patellar surface of the femur.

The Articular Capsule and Joint Cavity

There is no single, unified capsule at the knee joint; nor is there a common synovial cavity (Figure 9-12a•). A pair of fibrocartilage pads, the **medial** and **lateral**

menisci, lie between the femoral and tibial surfaces (Figure 9-12a,b,c•). The menisci (1) act as cushions, (2) conform to the shape of the articulating surfaces as the femur changes position, and (3) provide lateral stability to the joint. Prominent fat pads cushion the margins of the joint and assist the many bursae in reducing friction between the patella and other tissues.

Supporting Ligaments

A complete dislocation of the knee is very rare, largely because seven major ligaments stabilize the knee joint:

1. The tendon from the muscles responsible for extending the knee passes over the anterior surface of the joint. The patella is embedded within this tendon, and the **patellar ligament** continues to its attachment on the anterior surface of the tibia. The patellar ligament

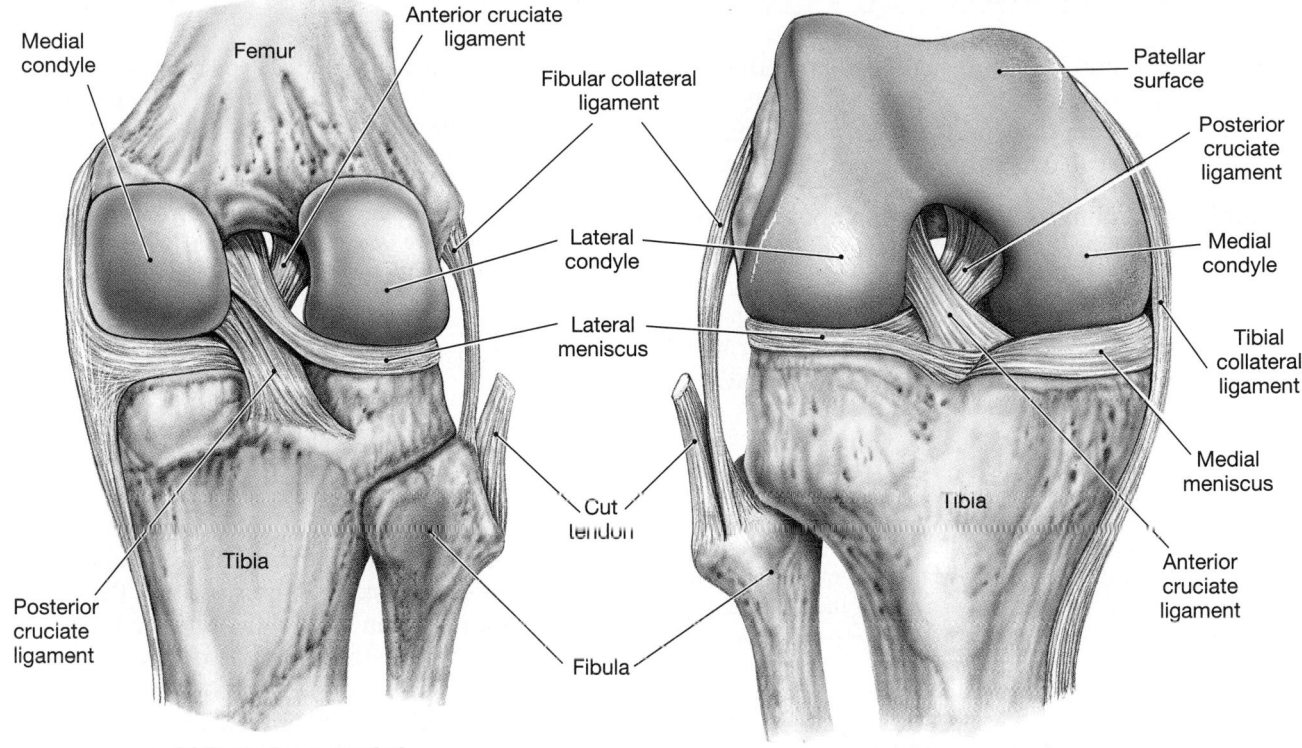

(a) Posterior, extended

(b) Anterior, flexed

provides support to the anterior surface of the knee joint (Figure 9-12c●).

2, 3. Two **popliteal ligaments** (not shown) extend between the femur and the heads of the tibia and fibula. These ligaments reinforce the knee joint's posterior surface.

4, 5. Inside the joint capsule, the **anterior cruciate** and **posterior cruciate ligaments** attach the intercondylar area of the tibia to the condyles of the femur. *Anterior* and *posterior* refer to their sites of origin on the tibia. They cross one another as they proceed to their destinations on the femur (Figure 9-12a,b●). (The term *cruciate* is derived from the Latin word *crucialis*, meaning a cross.) These ligaments limit the anterior and posterior movement of the femur and maintain the alignment of the femoral and tibial condyles.

6, 7. The **tibial collateral ligament** reinforces the medial surface of the knee joint, and the **fibular collateral ligament** reinforces the lateral surface (Figure 9-12a,b●). These ligaments tighten only at full extension. In this position, they stabilize the joint.

The knee is structurally complex and subjected to severe stresses in the course of normal activities. Painful knee injuries are all too familiar to both amateur and professional athletes. Treatment is often costly and prolonged, and repairs seldom make the joint "good as new."

Locking of the Knee

Your knee joint can "lock" in the extended position. At full extension, a slight lateral rotation of the tibia tightens the anterior cruciate ligament and jams the meniscus between the tibia and femur. This mechanism lets you stand for prolonged periods without using (and tiring) the muscles that extend your leg. Unlocking the

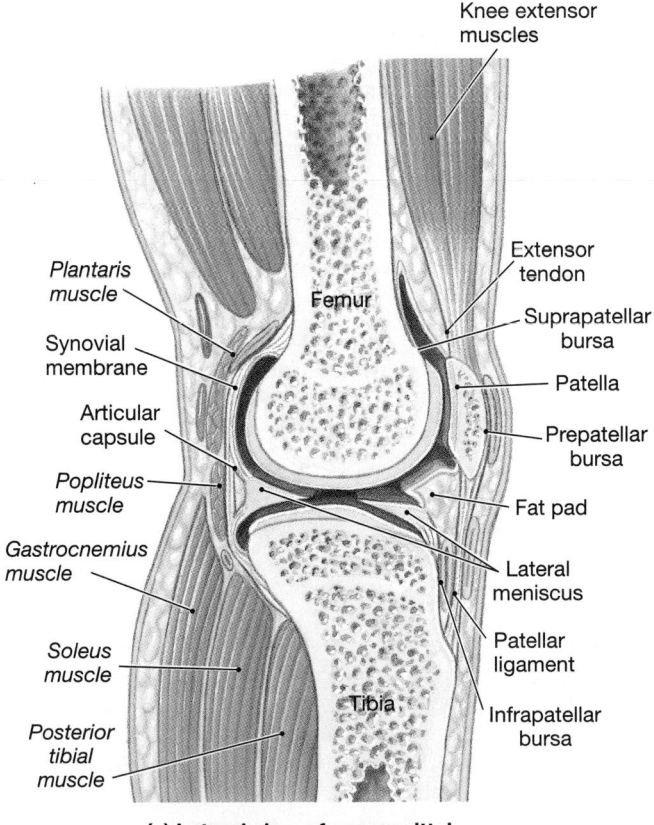

(c) Lateral view of parasagittal section through right knee

●FIGURE 9-12 **The Knee Joint.** **(a)** Posterior view of the right knee at full extension. **(b)** Anterior view of the right knee at full flexion. **(c)** Lateral view of the extended right knee in parasagittal section, showing major anatomical features. [AM] *Plate 7.3c; Scans 5, 6, 7*

joint requires muscular contractions that medially rotate the tibia or laterally rotate the femur. If the locked knee is struck from the side, the menisci can tear, and serious damage can occur to the supporting ligaments.

Table 9-5 summarizes information about the articulations of the pelvic girdle and lower limb.

☑ Where would you find the following ligaments: iliofemoral ligament, pubofemoral ligament, and ischiofemoral ligament?

☑ What symptoms would you expect to see in an individual who has damaged the menisci of the knee joint?

☑ Why is the condition known as "clergyman's knee" (a type of bursitis) common in carpet layers and roofers?

KNEE INJURIES Athletes place tremendous stresses on their knees. Ordinarily, the medial and lateral menisci move as the position of the femur changes. Placing a lot of weight on the knee while it is partially flexed can trap a meniscus between the tibia and femur, resulting in a break or tear in the cartilage. In the most common injury, the lateral surface of the leg is driven medially, tearing the medial meniscus. In addition to being quite painful, the torn cartilage may restrict movement at the joint. It can also lead to chronic problems and the development of a "trick knee," a knee that feels unstable. Sometimes the meniscus can be heard and felt popping in and out of position when the knee is extended.

Less common knee injuries involve tearing one or more stabilizing ligaments or damaging the patella. Torn ligaments tend to be difficult to correct surgically, and healing is slow. The patella can be injured in a number of ways. For example, if the leg is immobilized (as it might be in a football pileup) while you try to extend the knee, the muscles are powerful enough to pull the patella apart. Impacts to the anterior surface of the knee may also shatter the kneecap. (This was the intended outcome of the assault on figure skater Nancy Kerrigan prior to the 1994 U.S. National Championship.) Treatment of a fractured patella is difficult and time-consuming. The fragments must be surgically removed, and the tendons and ligaments repaired. The joint must then be immobilized. Total knee replacements are rarely performed on young people, but they are becoming increasingly common among elderly patients with severe arthritis. **AM** *Arthroscopic Surgery*

BONES AND MUSCLES

The skeletal and muscular systems are structurally and functionally interdependent. Their interactions are so extensive that they are often considered to be parts of a single *musculoskeletal system*. There are direct physical connections between the two systems. Many of the anatomical landmarks identified in this chapter are attachment sites of skeletal muscles, and the connective tissues that surround individual muscle fibers are continuous with those that form the organic framework of an attached bone. Muscles and bones are also physiologically linked through mutual dependence on stable calcium ion concentrations in body fluids. With most of the body's calcium tied up in the skeleton, bone abnormalities can have a direct effect on the muscle.

The bones, in turn, are directly affected by muscular activity. When a muscle enlarges due to regular exercise, the increased forces exerted on the skeleton will make bones become stronger and more massive. The reverse holds true as well, for both muscle and bones decrease in size and strength after periods of inactivity.

In the next two chapters, we will examine the structure and function of the muscular system and discuss how muscular contractions perform specific movements.

TABLE 9-5 **Articulations of the Pelvic Girdle and Lower Limb**

	Element	Joint	Type of Articulation	Movements
	Sacrum/ilium of coxa	Sacroiliac	Gliding diarthrosis	Gliding
	Coxa/coxa	Pubic symphysis	Amphiarthrotic symphysis	Slight
	Coxa/femur	Hip	Ball-and-socket diarthrosis	Flexion/extension, adduction/abduction, circumduction, rotation
	Femur/tibia	Knee	Complex, functions as hinge	Flexion/extension, limited rotation
	Tibia/fibula	Tibiofibular (proximal)	Gliding diarthrosis	Gliding
		Tibiofibular (distal)	Gliding diarthrosis and amphiarthrotic syndesmosis	Slight gliding
	Tibia and fibula with talus	Ankle, or tibiotalar	Hinge diarthrosis	Dorsiflexion/plantar flexion
	Tarsal bone to tarsal bone	Intertarsal	Gliding diarthrosis	Gliding
	Tarsal bone to metatarsal bone	Tarsometatarsal	Gliding diarthrosis	Gliding
	Metatarsal bone to phalanx	Metatarsophalangeal	Ellipsoidal diarthrosis	Flexion/extension, adduction/abduction
	Phalanx/phalanx	Interphalangeal	Hinge diarthrosis	Flexion/extension

SELECTED CLINICAL TERMINOLOGY

Terms Discussed in This Chapter

arthritis: (ar-THRĪ-tis): Rheumatic diseases that affect synovial joints. Arthritis always involves damage to the articular cartilages, but the specific cause may vary. The diseases of arthritis are usually classified as either *degenerative* or *inflammatory*. *(p. 256 and AM)*

bunion: The most common pressure-related bursitis, involving a tender nodule formed around bursae over the base of the great toe. *(p. 257)*

bursitis: Inflammation of a bursa, causing pain whenever the associated tendon or ligament moves. *(p. 257)*

continuous passive motion (CPM): A therapeutic procedure involving passive movement of an injured joint to stimulate circulation of synovial fluid. The goal is to prevent degeneration of the articular cartilages. *(p. 256 and AM)*

herniated disc: A condition caused by intervertebral compression severe enough to rupture the annulus fibrosus and release the nucleus pulposus, which may protrude beyond the intervertebral space. *(p. 266)*

laminectomy: (la-mi-NEK-to-mē): Removal of vertebral laminae; may be performed to access the vertebral canal and relieve symptoms of a herniated disc. *(p. 266)*

luxation: (luks-Ā-shun): A dislocation; a condition in which the articulating surfaces are forced out of position. *(p. 258)*

osteoarthritis (os-tē-ō-ar-THRĪ-tis) (*degenerative arthritis* or *degenerative joint disease, DJD*): An arthritic condition resulting from (1) cumulative wear and tear on joint surfaces or (2) genetic predisposition. In the U.S. population, 25 percent of women and 15 percent of men over age 60 show signs of this disease. *(p. 256 and AM)*

rheumatism: (ROO-muh-tizm): A general term that indicates pain and stiffness affecting the skeletal system, the muscular system, or both. *(p. 256 and AM)*

rheumatoid arthritis: An inflammatory arthritis that affects roughly 2.5 percent of the adult U.S. population. The cause is uncertain, although allergies, bacteria, viruses, and genetic factors have all been proposed. The primary symptom is *synovitis*, swelling and inflammation of the synovial membrane. *(p. 256 and AM)*

shoulder separation: The partial or complete dislocation of the acromioclavicular joint. *(p. 267)*

slipped disc: A common name for a condition caused by distortion of an intervertebral disc. The distortion applies pressure to spinal nerves, causing pain and limiting range of motion. *(p. 266)*

sprain: A condition caused when a ligament is stretched to the point at which some of the collagen fibers are torn. The ligament remains functional, and the structure of the joint is not affected. *(p. 257)*

subluxation (sub-luks-Ā-shun): A partial dislocation; displacement of articulating surfaces sufficient to cause discomfort but resulting in less physical damage to the joint than occurs during a dislocation. *(p. 258)*

 Additional Terms Discussed in the *Applications Manual*

---

ankylosis (ang-ki-LŌ-sis): An abnormal fusion between articulating bones in response to trauma and friction within a joint.

arthroscope: A fiber-optic instrument used to view the inside of joint cavities.

arthroscopic surgery: The surgical modification of a joint by using an arthroscope.

bony crepitus: A crackling or popping sound produced during movement of an abnormal joint.

meniscectomy: The surgical removal of a damaged meniscus.

monoarthritic: An arthritic condition affecting a single articulation.

polyarthritic: An arthritic condition affecting multiple articulations.

CHAPTER REVIEW

On-line resources for this chapter are on our World Wide Web site at:
http://www.prenhall.com/martini/fap

STUDY OUTLINE

INTRODUCTION, p. 254

1. **Articulations** (joints) exist wherever two bones interact.

A CLASSIFICATION OF JOINTS, p. 254

1. *Immovable joints* are **synarthroses,** *slightly movable joints* are **amphiarthroses,** and joints that are *freely movable* are called **diarthroses,** or **synovial joints.** *(Table 9-1)*

2. Alternatively, joints are classified as *bony, fibrous, cartilaginous,* or *synovial.* *(Table 9-2)*

Immovable Joints (Synarthroses), p. 254

3. The four major types of synarthroses are a **suture** (skull bones bound together by dense connective tissue), a **gomphosis** (teeth bound to bony sockets by **periodontal ligaments**), a

synchondrosis (two bones joined by a rigid cartilaginous bridge), and a **synostosis** (two bones completely fused).

Slightly Movable Joints (Amphiarthroses), p. 255

4. The two major types of amphiarthroses are a **syndesmosis** (bones connected by a ligament) and a **symphysis** (bones separated by fibrocartilage).

Freely Movable Joints (Diarthroses), p. 255

5. The bony surfaces at diarthroses are enclosed within an **articular capsule,** covered by **articular cartilages,** and lubricated by **synovial fluid.**

6. Other synovial structures can include **menisci,** or *articular discs;* **fat pads; accessory ligaments;** and **bursae.** *(Figure 9-1)*

7. A **dislocation** occurs when articulating surfaces are forced out of position.

ARTICULAR FORM AND FUNCTION, p. 258

Describing Dynamic Motion, p. 258

1. Possible movements can be classified as **linear motion, angular motion,** or **rotation.** *(Figure 9-2)*
2. Joints are called **monaxial, biaxial,** or **triaxial,** depending on the degree of movement they allow.

Types of Movements, p. 259

3. In **gliding,** two surfaces slide past one another.
4. Important terms that describe angular motion are **flexion, extension, hyperextension, circumduction, abduction,** and **adduction.** *(Figure 9-3)*
5. Rotational movement can be **left** or **right rotation, medial** *(internal)* or **lateral** *(external)* **rotation,** or, in the bones of the forearm, **pronation** or **supination.** *(Figure 9-4)*
6. Movements of the foot include **inversion** and **eversion.** The ankle undergoes **dorsiflexion** and **plantar flexion. Opposition** is the thumb movement that enables us to grasp objects. *(Figure 9-5)*
7. **Protraction** involves moving something anteriorly; **retraction** involves moving it posteriorly. **Depression** and **elevation** occur when we move a structure inferiorly or superiorly. **Lateral flexion** occurs when the vertebral column bends to one side. *(Figure 9-5)*

A Structural Classification of Synovial Joints, p. 262

8. **Gliding joints** permit limited movement, generally in a single plane. *(Figure 9-6)*
9. **Hinge joints** are monaxial joints that permit only angular movement in one plane. *(Figure 9-6)*
10. **Pivot joints** are monaxial joints that permit only rotation. *(Figure 9-6)*
11. **Ellipsoidal joints** are biaxial joints with an oval articular face that nestles within a depression in the opposing surface. *(Figure 9-6)*
12. **Saddle joints** are biaxial joints with articular faces that are concave on one axis and convex on the other. *(Figure 9-6)*
13. **Ball-and-socket joints** are triaxial joints that permit rotation as well as other movements. *(Figure 9-6)*

REPRESENTATIVE ARTICULATIONS, p. 264

Intervertebral Articulations, p. 264

1. The articular processes of vertebrae form gliding joints with those of adjacent vertebrae. The bodies form symphyseal joints. They are separated and cushioned by **intervertebral discs,** which contain an inner **nucleus pulposus** and an outer **annulus fibrosus.** Several ligaments stabilize the vertebral column. *(Figures 9-7, 9-8; Table 9-3)*

The Shoulder Joint, p. 267

2. The shoulder joint, or *glenohumeral (scapulohumeral) joint,* is formed by the glenoid cavity and the head of the humerus. This articulation permits the greatest range of motion of any joint in the body. It is a ball-and-socket diarthrosis with various stabilizing ligaments. Strength and stability are sacrificed to obtain mobility. *(Figure 9-9; Table 9-4)*

The Elbow Joint, p. 268

3. The elbow joint, or *olecranal joint,* permits only flexion/extension. It is a hinge diarthrosis whose capsule is reinforced by strong ligaments. *(Figure 9-10; Table 9-4)*

The Hip Joint, p. 269

4. The hip joint is a ball-and-socket diarthrosis formed by the union of the acetabulum with the head of the femur. The joint permits flexion/extension, adduction/abduction, circumduction, and rotation; it is stabilized by numerous ligaments. *(Figure 9-11; Table 9-5)*

The Knee Joint, p. 269

5. The knee joint is a hinge joint formed by the union of the condyles of the femur with the superior condylar surfaces of the tibia. The joint permits flexion/extension and limited rotation, and it has various supporting ligaments. *(Figure 9-12; Table 9-5)*

Bones and Muscles, p. 272

6. The interactions of the skeletal and muscular systems are so extensive that we often refer to a single *musculoskeletal system.*

REVIEW QUESTIONS

Level 1   Reviewing Facts and Terms

1. A synarthrotic joint located only between the bones of the skull is a
- (a) symphysis
- (b) syndesmosis
- (c) synchondrosis
- (d) suture

2. The distal articulation between the tibia and fibula is a
- (a) syndesmosis
- (b) symphysis
- (c) synchondrosis
- (d) synostosis

3. The anterior articulation between the two pubic bones is a
- (a) synchondrosis
- (b) synostosis
- (c) symphysis
- (d) synarthrosis

4. Joints typically found at the end of long bones are
- (a) synarthroses
- (b) amphiarthroses
- (c) diarthroses
- (d) symphyses

5. The function of the synovial fluid is
- (a) to nourish chondrocytes
- (b) to provide lubrication
- (c) to absorb shock
- (d) a, b, and c are correct

6. The structures responsible for channeling the flow of synovial fluid are
- (a) menisci
- (b) bursae
- (c) carpal tunnels
- (d) articular cartilages

7. The structures that limit the range of motion and provide mechanical support across or around a joint are
- (a) bursae
- (b) tendons
- (c) menisci
- (d) a, b, and c are correct

8. A partial dislocation of an articulating surface is a
- (a) circumduction
- (b) hyperextension
- (c) subluxation
- (d) supination

9. Abduction and adduction always refer to movements of the
- (a) axial skeleton
- (b) appendicular skeleton
- (c) skull
- (d) vertebral column

10. Rotation of the forearm that makes the palm face posteriorly is
- (a) supination
- (b) pronation
- (c) proliferation
- (d) projection

11. Spreading the fingers or toes apart is
  (a) dorsiflexion   (b) circumduction
  (c) adduction   (d) abduction
12. Standing on tiptoe is an example of
  (a) elevation   (b) dorsiflexion
  (c) plantar flexion (d) retraction
13. Examples of monaxial joints, which permit angular movement in a single plane, are the
  (a) carpals and tarsals   (b) shoulder and hip
  (c) elbow and knee   (d) a, b, and c are correct
14. Joints that connect the fingers and toes with the metacarpals and metatarsals are
  (a) ellipsoidal joints   (b) saddle joints
  (c) pivot joints   (d) hinge joints
15. Movements that occur at the shoulder and the hip represent the action that occurs at a _____ joint:
  (a) hinge   (b) ball-and-socket
  (c) pivot   (d) gliding

16. The annulus fibrosus and nucleus pulposus are structures associated with the
  (a) intervertebral discs
  (b) knee and elbow
  (c) shoulder and hip
  (d) carpals and tarsals
17. Subacromial, subcoracoid, and subscapular bursae reduce friction in the _____ joint:
  (a) hip   (b) knee
  (c) elbow   (d) shoulder
18. The functional anatomy of the lower limbs is very different from that of the upper limbs primarily because
  (a) the upper limbs are used for grasping
  (b) the upper limbs contain more bones than the lower limbs do
  (c) the lower limbs must transfer the body weight to the ground
  (d) the lower limbs are used for bipedal mobility

Level 2   Reviewing Concepts

19. The hip is an extremely stable joint because it has
  (a) a complete bony socket
  (b) a strong articular capsule
  (c) supporting ligaments
  (d) a, b, and c are correct
20. Complete dislocation of the knee is an extremely rare event because
  (a) it is protected by the patella
  (b) the femur articulates with the tibia at the knee
  (c) it contains seven major ligaments
  (d) it contains fat pads to absorb shocks
21. How does a meniscus (articular disc) function in a joint?

22. The greater the range of motion at a joint, the weaker the joint becomes. Why?
23. How do articular cartilages differ from other cartilages in the body?
24. What is the significance of the fact that the pubic symphysis is a slightly movable joint?
25. How would you explain to your grandmother the characteristic decrease in height with advancing age?
26. When the biceps brachii muscle contracts, what movements does it produce?
27. When you stand for a long period of time, why should you "lock" your knee in extended position? How does the knee lock?

Level 3   Critical Thinking and Clinical Applications

28. While playing tennis, Dave "overturns" his ankle. He experiences swelling and pain. After being examined, he is told that there are no torn ligaments and that the structure of the ankle is not affected. On the basis of the symptoms and the examination results, what do you think happened to Dave's ankle?
29. During a basketball game, Bob injured his right knee when he jumped to retrieve the ball and then landed off-balance on his right leg. Since then he has pain and limited mobility of his right knee joint. What type of injury do you think Bob sustained?
30. A high-school student comes to the emergency room complaining of persistent pain beneath her right shoulder blade. In talking with her, you discover that she has been spending many hours trying to improve her pitching skills for her school's softball team. What do you think is causing the pain?

Muscle Tissue

*A*stronaut Mae C. Jemison seems to be enjoying the experience of zero gravity on
board the space shuttle Endeavour, *but the situation is less than ideal for her bones and muscles.
Our bones and muscles work together. Under weightless conditions, bones are not stressed by the weight
of the body, and muscles no longer work to oppose the force of gravity. "What you don't use, you lose,"
and prolonged weightlessness results in a loss of bone and muscle mass. Because this loss can be
potentially dangerous on return to normal gravity, astronauts in orbit have exercise periods with
workouts to stress their muscles and prevent or reduce the loss of muscle mass. This chapter focuses on
how muscles work and how exercise—in space or on Earth—affects strength and endurance.*

CHAPTER OUTLINE AND OBJECTIVES

Think for a moment what life would be like without muscle tissue. Imagine being unable to sit, stand, walk, speak, or grasp objects in your environment. Imagine, too, how your internal functions would be affected. Blood would not circulate, because you would have no heartbeat to propel it through the vessels. Your lungs could not rhythmically empty and fill, nor could food move along your digestive tract. In fact, there would be practically no movement through any of your internal passageways.

This is not to say that all life depends on muscle tissue. There are large organisms that get by very nicely without it; we call them plants. But life as *we* live it would be impossible, for many of our physiological processes, and virtually all of our dynamic interactions with the environment, involve muscle tissue. Muscle cells are specialized to contract. In muscle tissue, the individual cells are tied together, primarily by collagen fibers. When the muscle cells contract, they pull on those fibers the way a group of people might pull on a rope. The pull, called **tension**, is an active force—energy must be expended to produce it. Tension is always applied *to* some object, whether a rope, a rubber band, or a book on a tabletop.

Tension tends to produce movement, but before movement can occur, the applied tension must overcome the **resistance** of the object. Resistance is a passive force that opposes movement. The amount of resistance can depend on the weight of the object, its shape, friction, and other factors. When the applied tension exceeds the resistance, the object is pulled *toward* the source of the tension. Conversely, *compression*, another active force, tends to push a resistance *away* from the source. Muscle cells can only actively shorten and generate tension; they cannot actively lengthen to generate compression.

As we noted in Chapter 6, collagen fibers cannot withstand compression; when compressed, they simply bend out of the way. ∞ *[p. 175]* But collagen fibers can resist considerable tension, so when a muscle pulls on collagen fibers, those fibers transmit the force and pull on something else. What happens next depends largely on what the collagen fibers are attached to and on how the muscle cells are arranged. We will encounter several different examples in this chapter as well as in Chapters 11, 20, and 24.

SKELETAL MUSCLE TISSUE AND THE MUSCULAR SYSTEM

Muscle tissue, one of the four primary tissue types, consists chiefly of muscle cells that are highly specialized for contraction. There are three types of muscle tissue: (1) *skeletal muscle*, (2) *cardiac muscle*, and (3) *smooth muscle*. Without these muscle tissues, introduced in Chapter 4, nothing in the body would move, and no body movement could occur. ∞ *[p. 135]* Skeletal muscles

move the body by pulling on bones of the skeleton, making it possible for us to walk, dance, bite into an apple, or play the ukulele. Cardiac muscle tissue pushes blood through the circulatory system. Smooth muscle tissue pushes fluids and solids along the digestive tract, regulates the diameters of small arteries, and performs a variety of other functions.

This chapter deals primarily with the structure and function of skeletal muscle tissue, in preparation for our discussion of the muscular system (Chapter 11). This chapter will also provide an overview of the differences among skeletal, cardiac, and smooth muscle tissue.

Skeletal muscle tissue is found within **skeletal muscles**, organs that also contain connective tissues, nerves, and blood vessels. As the name implies, skeletal muscles are directly or indirectly attached to the bones of the skeleton. The skeletal muscle tissue in our skeletal muscles performs the following five functions:

1. *Produce skeletal movement.* Skeletal muscle contractions pull on tendons and move the bones of the skeleton. The effects range from simple motions such as extending the arm to the highly coordinated movements of swimming, skiing, or typing.

2. *Maintain posture and body position.* Tension in our skeletal muscles maintains body posture—for example, holding your head in position when you read a book or balancing your body weight above your feet when you walk. Without constant muscular activity, we could not sit upright without collapsing into a heap or stand without toppling over.

3. *Support soft tissues.* The abdominal wall and the floor of the pelvic cavity consist of layers of skeletal muscle. These muscles support the weight of visceral organs and shield internal tissues from injury.

4. *Guard entrances and exits.* The openings of the digestive and urinary tracts are encircled by skeletal muscles. These muscles provide voluntary control over swallowing, defecation, and urination.

5. *Maintain body temperature.* Muscle contractions require energy, and whenever energy is used in the body, some of it is converted to heat. The heat released by working muscles keeps our body temperature in the range required for normal functioning.

We will begin our discussion with the gross anatomy of a typical skeletal muscle. We will then consider at the microscopic level the structural features that make contractions possible.

ANATOMY OF SKELETAL MUSCLE

Figure 10-1• illustrates the appearance and organization of a representative skeletal muscle. A skeletal muscle contains connective tissues, blood vessels, nerves, and skeletal muscle tissue.

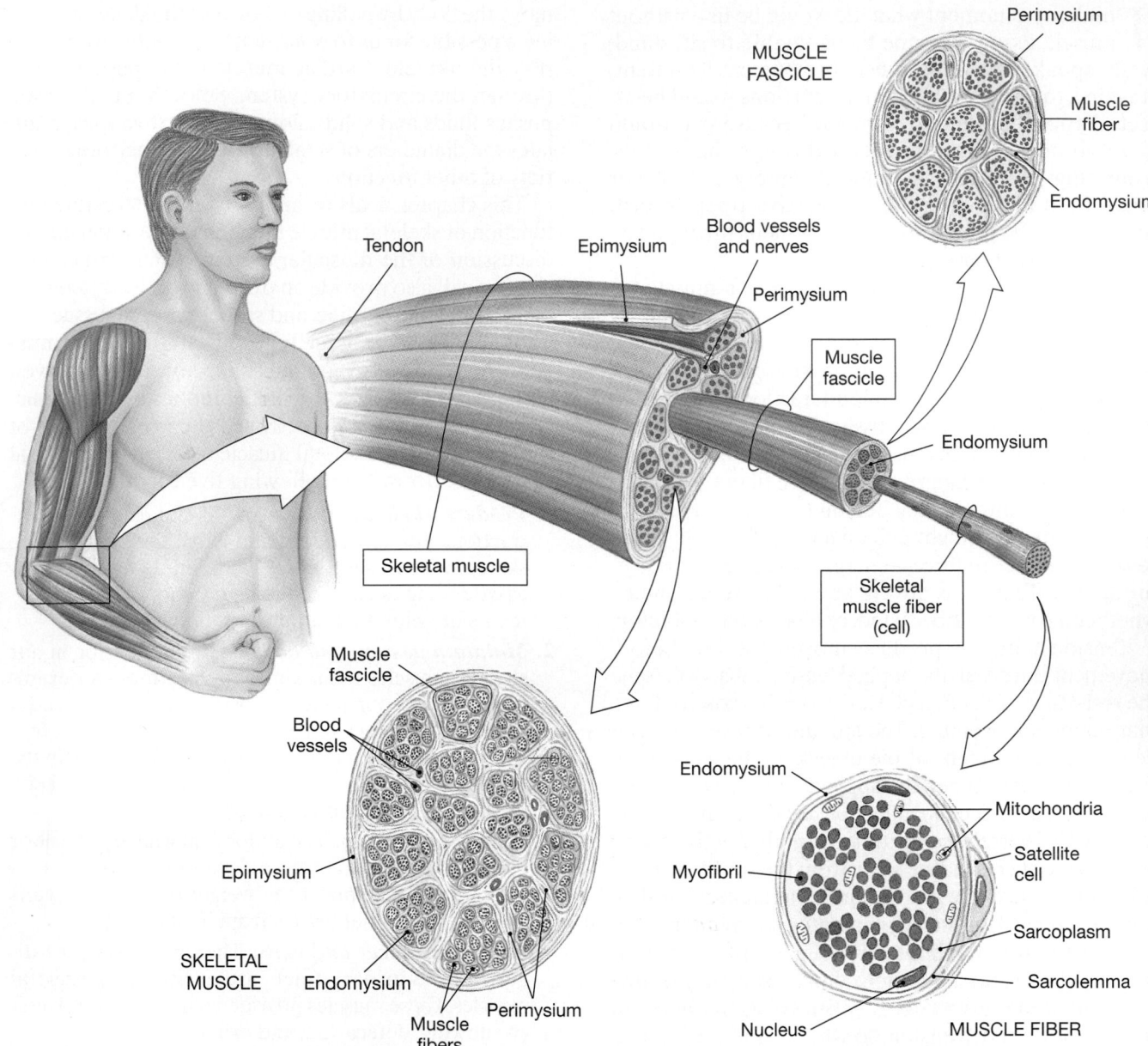

● **FIGURE 10-1** **Organization of Skeletal Muscles.** A skeletal muscle consists of fascicles (bundles of muscle fibers) enclosed by the epimysium. The bundles are separated by connective tissue fibers of the perimysium, and within each bundle the muscle fibers are surrounded by the endomysium. Each muscle fiber has many superficial nuclei as well as mitochondria and other organelles detailed in Figure 10-2.

Connective Tissue Organization

Three layers of connective tissue are part of each muscle: an outer *epimysium*, a central *perimysium*, and an inner *endomysium*. These layers and their relationships are diagrammed in Figure 10-1●.

The entire muscle is surrounded by the **epimysium** (ep-i-MĪZ-ē-um; *epi-*, on + *mys*, muscle), a dense layer of collagen fibers. The epimysium separates the muscle from surrounding tissues and organs. It is one component of the *deep fascia*, the dense connective tissue layer described in Chapter 4 (see Figure 4-16●). ∞ *[p.134]*

The connective tissue fibers of the **perimysium** (per-i-MĪZ-ē-um; *peri-*, around) divide the skeletal muscle into a series of compartments, each containing a bundle of muscle fibers called a **fascicle** (FA-sik-ul; *fasciculus*, a bundle). In addition to collagen and elastic fibers, the perimysium contains blood vessels and nerves that maintain blood flow and innervate the fascicles. Each fascicle receives branches of the blood vessels and nerves in the perimysium.

Within the fascicle, the delicate connective tissue of the **endomysium** (en-dō-MĪZ-ē-um; *endo-*, inside) surrounds the individual skeletal muscle fibers and ties together adjacent muscle fibers. Scattered **satellite cells** lie between the endomysium and the muscle fibers. These embryonic stem cells, introduced in Chapter 4, function in the repair of damaged muscle tissue. ∞ *[p. 135]* AM *Disruption of Normal Muscle Organization*

The collagen fibers of the endomysium and perimysium are interwoven and blend into one another. At each end of the muscle, the collagen fibers of the epimysium, perimysium, and endomysium come together to form a bundle known as a **tendon** or a broad tendinous sheet called an **aponeurosis**. Tendons and aponeuroses, examples of dense regular connective tissue, attach skeletal muscles to bones. ∞ *[p. 124]* The tendon fibers are interwoven with the periosteum and extend into the bone matrix, providing a firm attachment. As a result, any contraction of the muscle will exert a pull on its tendon and thereby on the attached bone(s).

Blood Vessels and Nerves

The connective tissues of the epimysium and perimysium contain the blood vessels and nerves that supply the muscle fibers. Muscle contraction requires tremendous quantities of energy. An extensive vascular network delivers the necessary oxygen and nutrients and carries away the metabolic wastes generated by active skeletal muscles. The blood vessels and the nerve supply generally enter the muscle together; the vessels and nerves follow the same branching pattern through the perimysium. Within the endomysium, arterioles supply blood to a capillary network that surrounds each individual muscle fiber.

Skeletal muscles contract only under stimulation from the central nervous system. Axons, or *nerve fibers*, penetrate the epimysium, branch through the perimysium, and enter the endomysium to innervate individual muscle fibers. Skeletal muscles are often called *voluntary muscles* because we have voluntary control over their contractions. Many of these skeletal muscles may also be controlled at a subconscious level. For example, skeletal muscles involved with breathing, such as the diaphragm, usually work outside our conscious awareness.

Next we will examine the microscopic structure of a typical skeletal muscle fiber and relate that microstructure to the physiology of the contraction process.

Microanatomy of Skeletal Muscle Fibers

Skeletal muscle fibers are quite different from the "typical" cells we described in Chapter 3. One obvious difference is size, for skeletal muscle fibers are enormous. A muscle fiber from a thigh muscle could have a diameter of 100 μm and a length equal to that of the entire muscle (up to 30 cm, or 12 in.). A second obvious difference is that skeletal muscle fibers are *multinucleate:* Each skeletal muscle fiber contains hundreds of nuclei just beneath the cell membrane. The genes contained in these nuclei direct the production of enzymes and structural proteins required for normal muscle contraction, and the presence of multiple copies of these genes speeds up the process. This feature is particularly important because metabolic turnover tends to be very rapid in skeletal muscle fibers. ∞ *[p. 61]*

The distinctive features of size and multiple nuclei are related. During development, groups of embryonic cells called **myoblasts** fuse, creating individual skeletal muscle fibers (Figure 10-2a●). Each nucleus in a skeletal muscle fiber reflects the contribution of a single myoblast. Some myoblasts, however, do not fuse with developing muscle fibers. These unfused cells remain in adult skeletal muscle tissue as the satellite cells seen in Figure 10-1●. After an injury, satellite cells may enlarge, divide, and fuse with damaged muscle fibers, thereby assisting in the regeneration of the tissue.

The Sarcolemma and Transverse Tubules

The **sarcolemma** (sar-cō-LEM-uh; *sarkos*, flesh + *lemma*, husk), or cell membrane of a muscle fiber, surrounds the **sarcoplasm** (SAR-kō-plazm), or cytoplasm of the muscle fiber (Figure 10-2b●). Like other cell membranes, the sarcolemma has a characteristic transmembrane potential due to the unequal distribution of positive and negative charges across the membrane. ∞ *[p. 80]* In a skeletal muscle fiber, a sudden change in the transmembrane potential is the first step that leads to a contraction. A skeletal muscle fiber is very large, but all regions of the cell must contract simultaneously. Thus, the signal to contract must be distributed quickly throughout the interior of the cell. This signal conduction occurs at the transverse tubules. **Transverse tubules,** or **T tubules,** are narrow tubes that are continuous with the sarcolemma and extend into the sarcoplasm at right angles to the membrane surface (Figure 10-2c●). Filled with extracellular fluid, they form passageways through the muscle fiber, like a series of tunnels through a mountain. The T tubules open onto the sarcolemma, and the membrane that lines each tubule has the same general properties as the sarcolemma. Electrical impulses conducted by the sarcolemma travel along these T tubules. These impulses, or *action potentials*, are the trigger for muscle fiber contraction.

Myofibrils

Inside the muscle fiber, branches of the transverse tubules encircle cylindrical structures called *myofibrils*. A **myofibril** (Figure 10-2b,c●) is 1–2 μm in diameter and as long as the entire cell. Each skeletal muscle fiber contains hundreds to thousands of myofibrils.

Myofibrils consist of bundles of **myofilaments**, protein filaments composed primarily of *actin* and *myosin*. The actin is found in **thin filaments**, and the myosin forms **thick filaments**. We introduced both types of filaments in Chapter 3. ∞ *[pp. 83–84]* Myofibrils can actively shorten; they are responsible for skeletal muscle

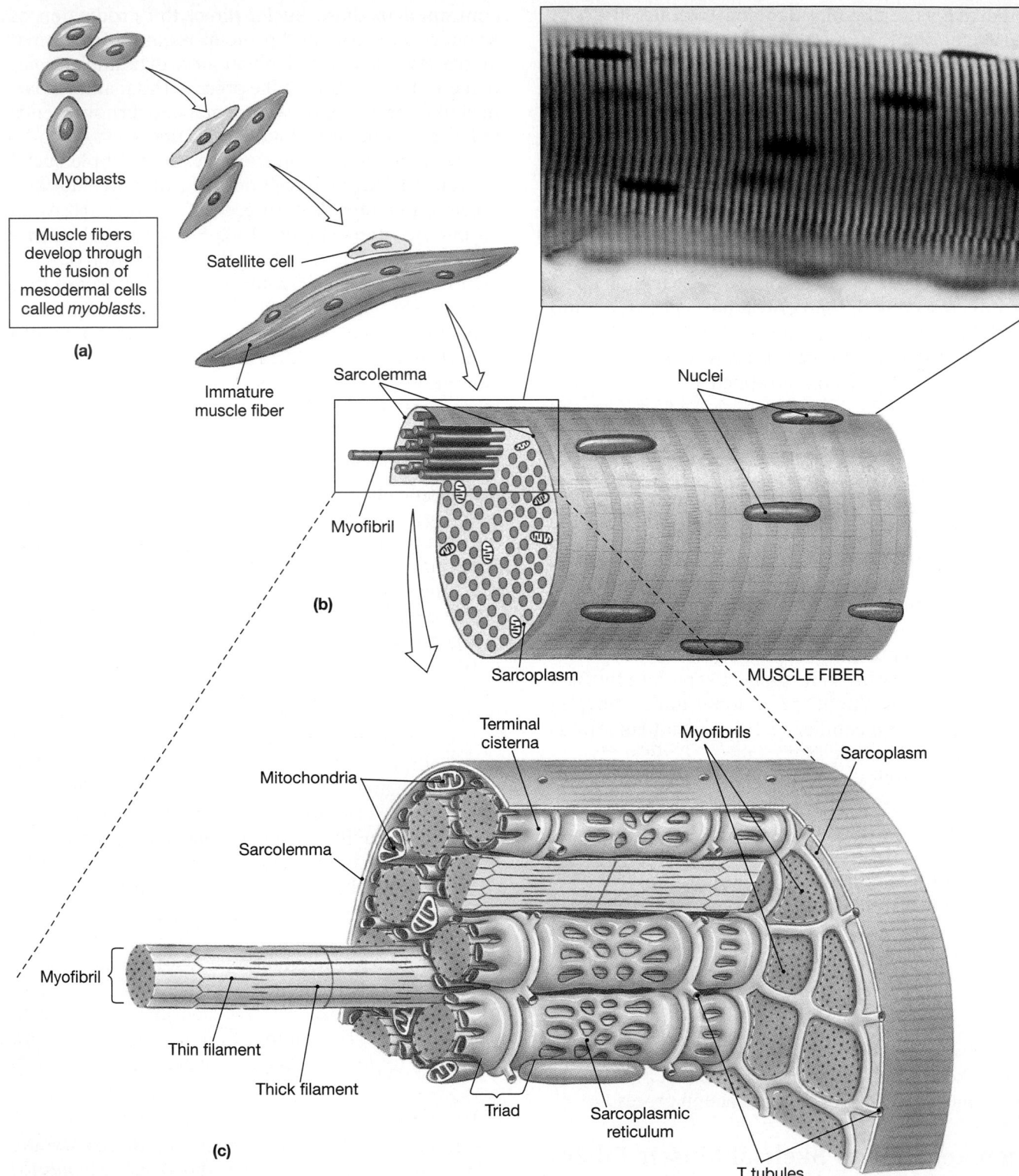

Myoblasts

Muscle fibers develop through the fusion of mesodermal cells called *myoblasts*.

(a)

Satellite cell

Immature muscle fiber

Sarcolemma

Nuclei

Myofibril

Sarcoplasm

MUSCLE FIBER

(b)

Terminal cisterna

Myofibrils

Mitochondria

Sarcolemma

Sarcoplasm

Myofibril

Thin filament

Thick filament

Triad

Sarcoplasmic reticulum

T tubules

(c)

● **FIGURE 10-2** **Formation and Structure of a Skeletal Muscle Fiber.** **(a)** The formation of muscle fibers through the fusion of myoblasts. **(b)** Micrograph and diagrammatic view of one muscle fiber. (LM × 612) **(c)** The internal organization of a muscle fiber.

fiber contraction. In Chapter 3, we noted that proteins, including those of the cytoskeleton, can be bound to the inner surface of the cell membrane. ∞ *[p. 70]* At each end of the skeletal muscle fiber, the myofibrils are

attached to the sarcolemma. In Chapter 4, we noted that proteins in cell membranes can attach the cells to extracellular structures, such as connective tissue fibers. The sarcolemma at each end of a muscle fiber is firmly bound to collagen fibers that extend into the tendon or aponeurosis of the skeletal muscle. As a result, when the myofibrils contract, the entire cell shortens, and in doing so, it pulls on the tendon. Scattered among and around the myofibrils are mitochondria and granules of glycogen, the storage form of glucose. Glucose breakdown through *glycolysis* and mitochondrial activity provide the ATP needed to power muscular contractions.

The Sarcoplasmic Reticulum

Wherever a transverse tubule encircles a myofibril, the tubule makes close contact with the membranes of the sarcoplasmic reticulum. The **sarcoplasmic reticulum (SR)** is a membrane complex similar to the smooth endoplasmic reticulum of other cells. In skeletal muscle fibers, the SR forms a tubular network around each individual myofibril (Figure 10-2c●). On either side of a transverse tubule, the tubules of the SR enlarge, fuse, and form expanded chambers called **terminal cisternae**. The combination of a pair of terminal cisternae plus a transverse tubule is known as a **triad**. Although the membranes of the triad are tightly bound together, their fluid contents are separate and distinct.

In Chapter 3, we noted the existence of special ion pumps that keep the intracellular concentration of calcium ions (Ca^{2+}) very low. ∞ *[p 78]* Most cells pump the calcium ions across their cell membranes and into the extracellular fluid. Although skeletal muscle fibers do pump Ca^{2+} out of the cell in this way, they also remove calcium ions from the cytosol by actively transporting them into the cisternae of the sarcoplasmic reticulum. The sarcoplasm of a resting skeletal muscle fiber contains very low concentrations of Ca^{2+}, around 10^{-7} mmol/l. The free Ca^{2+} concentration inside the terminal cisternae may be as much as 1000 times higher. In addition, the cisternae contain a protein, *calsequestrin*, that reversibly binds Ca^{2+}. Including both the free calcium and the bound calcium, the total concentration of Ca^{2+} inside the cisternae may be 40,000 times that of the surrounding sarcoplasm.

A muscle contraction begins when stored calcium ions are released into the sarcoplasm. These ions then diffuse into individual contractile units called *sarcomeres*.

Sarcomeres

We have seen that myofibrils are bundles of thin and thick filaments. These myofilaments are organized in repeating functional units called **sarcomeres** (SAR-kō-mērz; *sarkos*, flesh + *meros*, part) (Figure 10-3●).

SARCOMERE ORGANIZATION A myofibril consists of approximately 10,000 sarcomeres placed end to end. Each sarcomere has a resting length of 1.6–2.6 μm. *Sar-*

comeres are the smallest functional units of the muscle fiber. Interactions between the thick and thin filaments of sarcomeres are responsible for muscle contraction. A sarcomere contains (1) thick filaments, (2) thin filaments, (3) proteins that stabilize the positions of thick and thin filaments, and (4) proteins that regulate the interactions between thick and thin filaments.

Figure 10-3a,b● indicates the structure of an individual sarcomere. Differences in the size, density, and distribution of thick filaments and thin filaments account for the banded appearance of the sarcomere. There are dark bands (**A bands**) and light bands (**I bands**) (Figure 10-3b,c,d●). The terms *A band* and *I band* are derived from *anisotropic* and *isotropic*, which refer to the appearance of these bands when they are viewed under polarized light. You may find it helpful to remember that A bands are d*A*rk and that I bands are l*I*ght.

A Band The thick filaments are located at the center of a sarcomere, within the A band. The length of the A band is equal to the length of a typical thick filament. The A band, which also includes portions of thin filaments, includes several subdivisions:

1. *M line.* The central portion of each thick filament is connected to its neighbors by proteins of the **M line**. These densely staining proteins help stabilize the positions of the thick filaments.

2. *H zone.* In a resting sarcomere, the **H zone**, or *H band*, is a lighter region on either side of the M line. The H zone contains thick filaments but no thin filaments.

3. *Zone of overlap.* In the **zone of overlap**, thin filaments are found between the thick filaments. In this region, each thin filament sits in a triangle formed by three thick filaments, and each thick filament is surrounded by six thin filaments.

I Band Each I band, which contains thin filaments but not thick filaments, extends from the A band of one sarcomere to the A band of the next sarcomere. **Z lines** mark the boundary lines between adjacent sarcomeres. The Z lines consist of proteins called *connectins*, which interconnect thin filaments associated with adjacent sarcomeres. From the Z lines at either end of the sarcomere, thin filaments extend toward the M line and into the zone of overlap. Strands of the protein **titin** extend from the tips of the thick filaments to attachment sites at the Z line. Titin helps keep the thick and thin filaments in proper alignment, and it also helps the muscle fiber resist extreme stretching that would otherwise disrupt the contraction mechanism.

Two transverse tubules encircle each sarcomere, and there are triads on each side of the M line in the region of the zone of overlap. As a result, calcium ions released by the SR enter the regions where thick and thin filaments can interact.

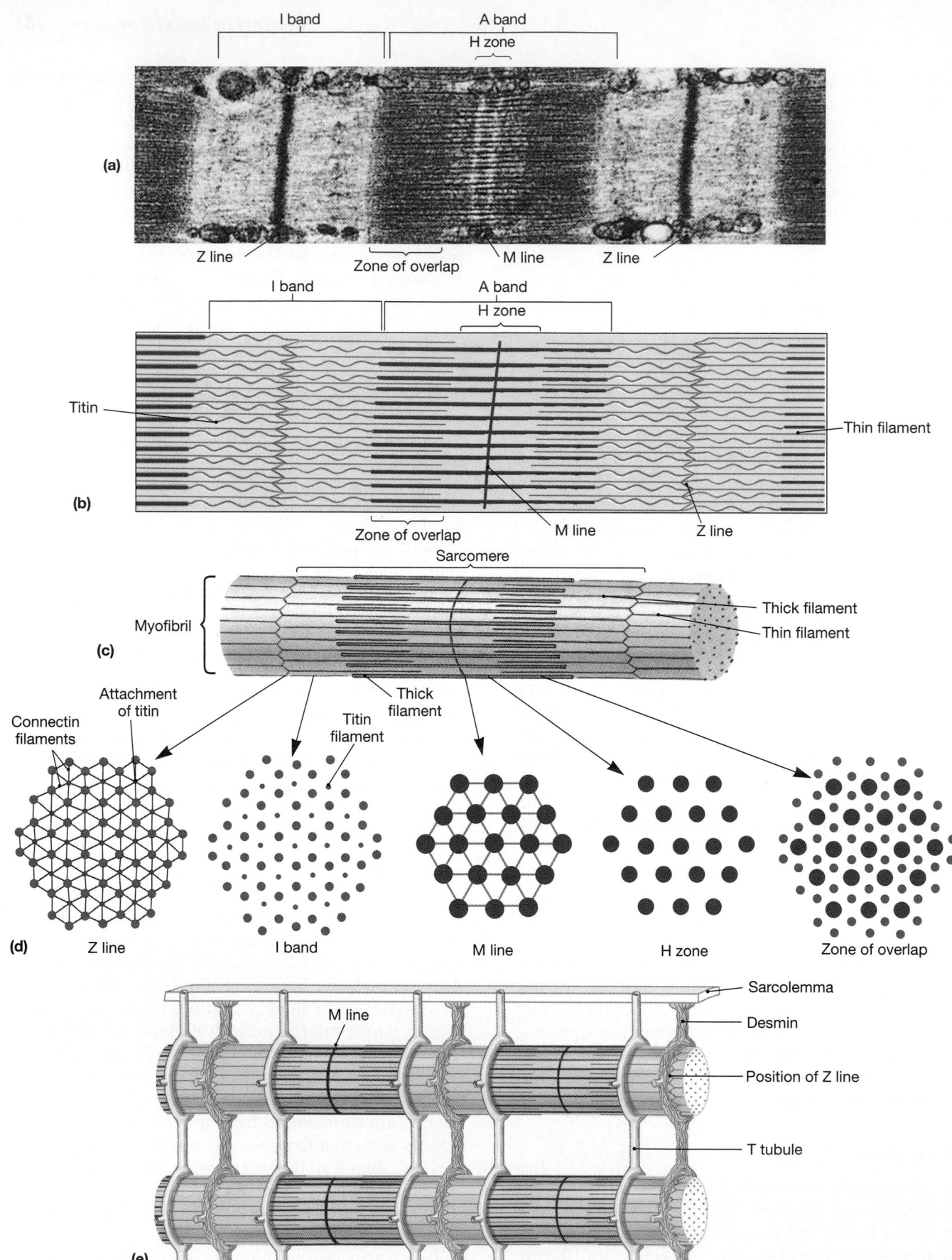

•*FIGURE* 10-3 Sarcomere Structure. (a) A sarcomere in a myofibril from a muscle fiber in the *gastrocnemius* muscle of the calf. (TEM × 64,000) **(b)** Simplified diagrammatic view of a sarcomere. **(c)** Organization of thick and thin filaments in a sarcomere. **(d)** Cross-sectional views of different portions of the sarcomere. **(e)** The organization of the cytoskeleton, which maintains the alignment of sarcomeres and myofibrils within the muscle fiber. The interconnections extend to the myofibrils on both sides.

Each Z line is surrounded by a meshwork of intermediate filaments composed of the protein **desmin**. Additional desmin filaments interconnect adjacent myofibrils. The myofibrils closest to the sarcolemma are bound to attachment sites on the inside of the membrane (Figure 10-3e●). Because the Z lines of all the myofibrils are aligned in this way, the muscle fiber as a whole has a banded appearance. Skeletal muscle tissue is also known as *striated muscle* because these bands, or *striations*, are visible with the light microscope (Figure 10-2b●). ∞ *[p. 135]*

Figure 10-4● reviews the levels of organization we have discussed so far. We will now proceed to the molecular level of organization and consider the structure of the myofilaments responsible for muscle contraction.

THE MUSCULAR DYSTROPHIES Abnormalities in the genes that code for structural and functional proteins in muscle fibers are responsible for a number of inherited diseases collectively known as the **muscular dystrophies** (DIS-trō-fēz) These conditions, which cause a progressive muscular weakness and deterioration, are the result of abnormalities in the sarcolemma or in the structure of internal proteins. The best-known example is *Duchenne's muscular dystrophy,* which typically develops in males from 3 to 7 years old. [AM] *The Muscular Dystrophies*

THIN FILAMENTS A typical thin filament is 5–6 nm in diameter and 1 μm in length (Figure 10-5a●). A single thin filament contains three proteins: *F actin, tropomyosin,* and *troponin:*

1. **F actin** is a twisted strand composed of 300–400 individual globular molecules of **G actin** (Figure 10-5b●). Each molecule of G actin contains an **active site** that can bind to a thick filament in much the same way that a substrate molecule binds to the active site of an enzyme. Under resting conditions, myosin binding is prevented by the *troponin–tropomyosin complex.*

2. Strands of **tropomyosin** (trō-pō-MĪ-ō-sin) cover the active sites and prevent actin–myosin interaction (Figure 10-5b●). A single tropomyosin molecule is a protein strand that covers seven active sites. It is bound to one molecule of troponin midway along its length.

3. A **troponin** (TRŌ-pō-nin; *trope,* turning) molecule consists of three globular subunits: One subunit binds to tropomyosin, locking them together; a second subunit binds to G actin, holding the troponin–tropomyosin complex in position; and the third subunit has a receptor that binds a calcium ion. In a resting muscle, intracellular Ca^{2+} concentrations are very low, and that binding site is empty.

A contraction cannot occur unless there is a change in the position of the troponin–tropomyosin complex that exposes the active sites on F actin. The necessary change in position occurs when calcium ions bind to receptors on the troponin molecules.

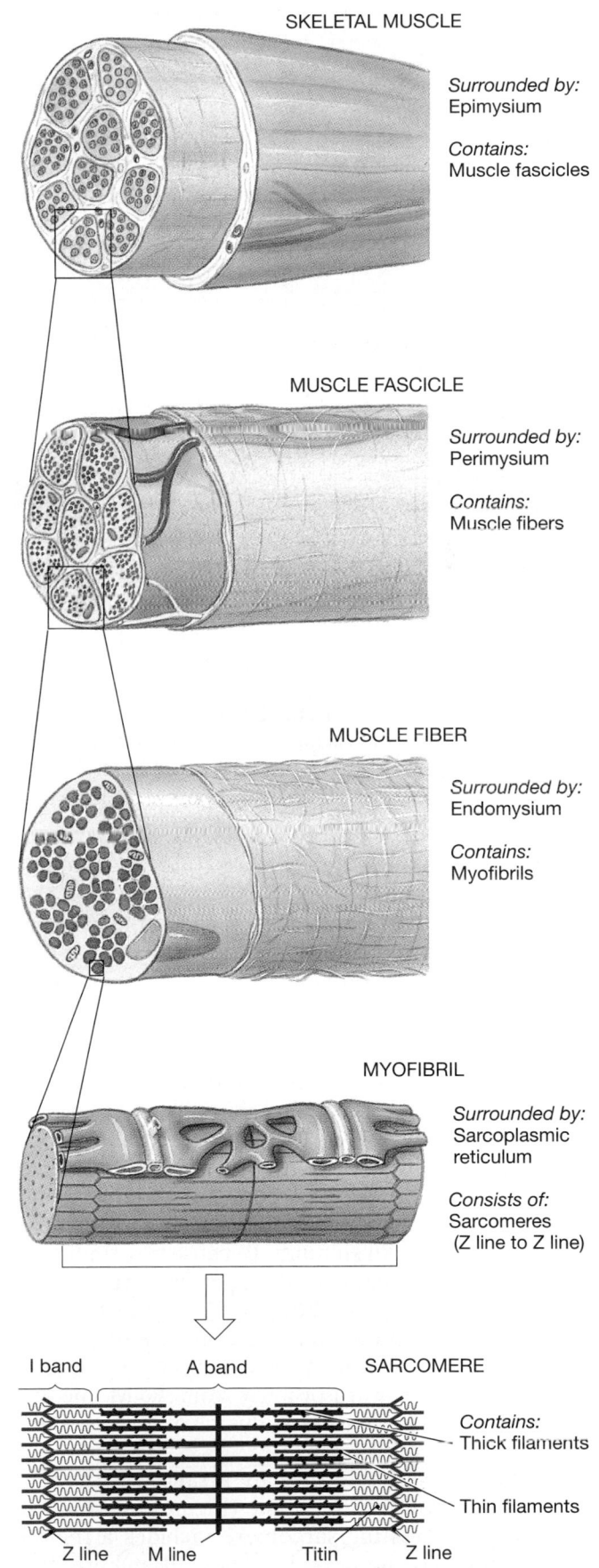

SKELETAL MUSCLE

Surrounded by:
Epimysium

Contains:
Muscle fascicles

MUSCLE FASCICLE

Surrounded by:
Perimysium

Contains:
Muscle fibers

MUSCLE FIBER

Surrounded by:
Endomysium

Contains:
Myofibrils

MYOFIBRIL

Surrounded by:
Sarcoplasmic reticulum

Consists of:
Sarcomeres
(Z line to Z line)

I band   A band   SARCOMERE

Contains:
Thick filaments

Thin filaments

Z line   M line   Titin   Z line

●**FIGURE 10-4** **Levels of Functional Organization in a Skeletal Muscle Fiber**

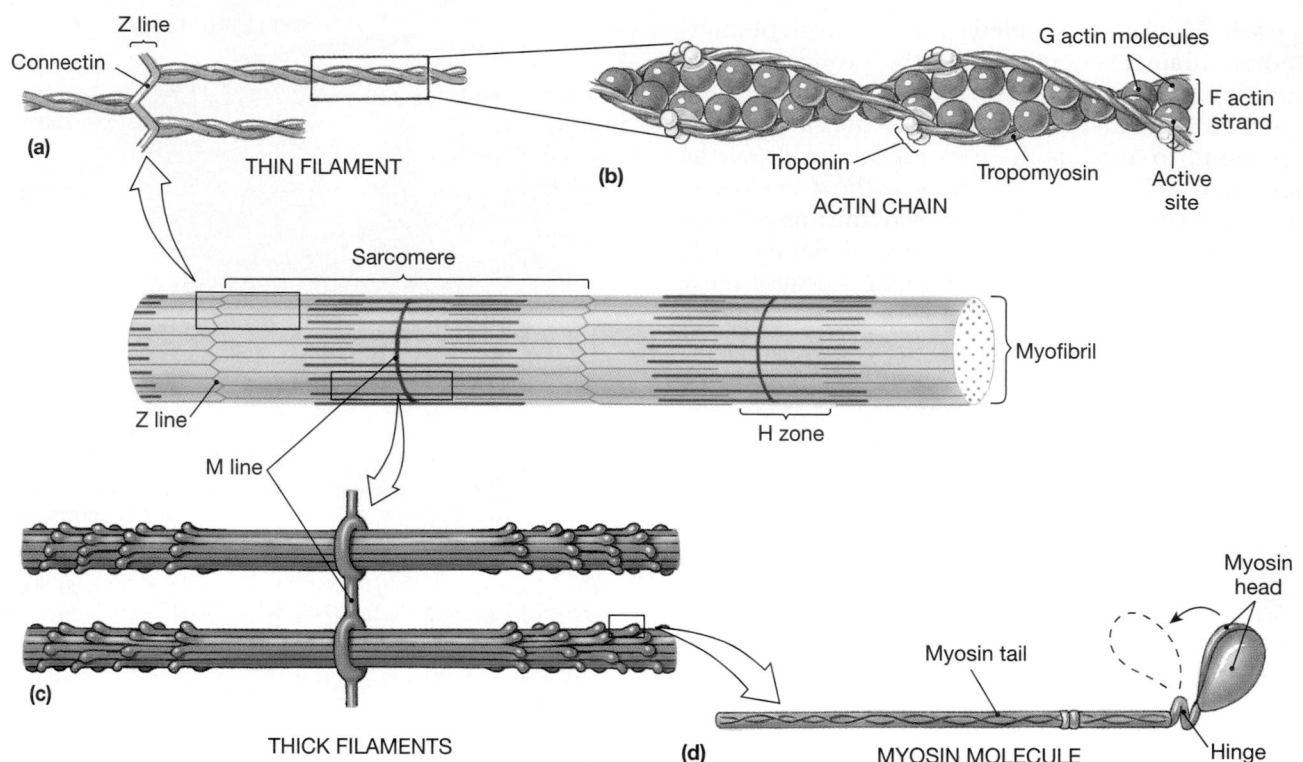

●**FIGURE 10-5** **Thick and Thin Filaments.** **(a)** Gross structure of a thin filament, showing the attachment at the Z line. **(b)** The organization of G actin subunits in an F actin strand and the position of the troponin–tropomyosin complex. **(c)** Structure of a thick filament, showing the orientation of the myosin molecules along the thick filaments. **(d)** Structure of a myosin molecule.

At either end of the sarcomere, the thin filaments are attached to the Z line (Figure 10-5a●). Although called a *line* because it looks like a dark line on the surface of the myofibril, in sectional view the Z line is more like an open meshwork (Figure 10-3d●). For this reason, the Z line is often called the *Z disc.*

THICK FILAMENTS Thick filaments are 10–12 nm in diameter and 1.6 μm long (Figure 10-5c●). Each thick filament consists of roughly 500 myosin molecules. Each myosin molecule consists of a pair of myosin subunits twisted around one another (Figure 10-5d●). The long, attached **tail** is bound to other myosin molecules in the thick filament. The free **head**, which projects outward toward the nearest thin filament, consists of two globular protein subunits. Because myosin heads interact with thin filaments during a contraction, they are also known as **cross-bridges**. The connection between the head and the tail functions as a hinge that lets the head pivot at its base. When pivoting occurs, the head swings toward or away from the M line. As we will see in a later section, this pivoting is the key step in muscle contraction.

All the myosin molecules are arranged with their tails pointing toward the M line (Figure 10-5c●). The H zone in the resting sarcomere includes a region where there are no myosin heads. Elsewhere, the myosin molecules are arranged in a spiral, with the myosin heads facing the surrounding thin filaments.

Each thick filament has a core of titin. On either side of the M line, a strand of titin extends the length of the thick filament and then continues across the I band to the Z line on that side. The portion of the titin strand exposed within the I band is highly elastic and will recoil after stretching. In the normal resting sarcomere, the titin strands are completely relaxed; they become tense only when some external force stretches the sarcomere.

The Sliding Filament Theory

When a skeletal muscle fiber contracts, (1) the H bands and I bands get smaller, (2) the zones of overlap get larger, (3) the Z lines move closer together, and (4) the width of the A bands remains constant throughout the contraction. The contraction ends once the fiber has shortened by about 30 percent, with the elimination of the I bands. These observations make sense only if *the thin filaments are sliding toward the center of the sarcomere, alongside the thick filaments* (Figure 10-6●). This explanation for the physical changes that occur during contraction is the **sliding filament theory**.

☑ How would severing the tendon attached to a muscle affect the muscle's ability to move a body part?

☑ Why does skeletal muscle appear striated when viewed through a microscope?

☑ Where would you expect to find the greatest concentration of Ca^{2+} in resting skeletal muscle?

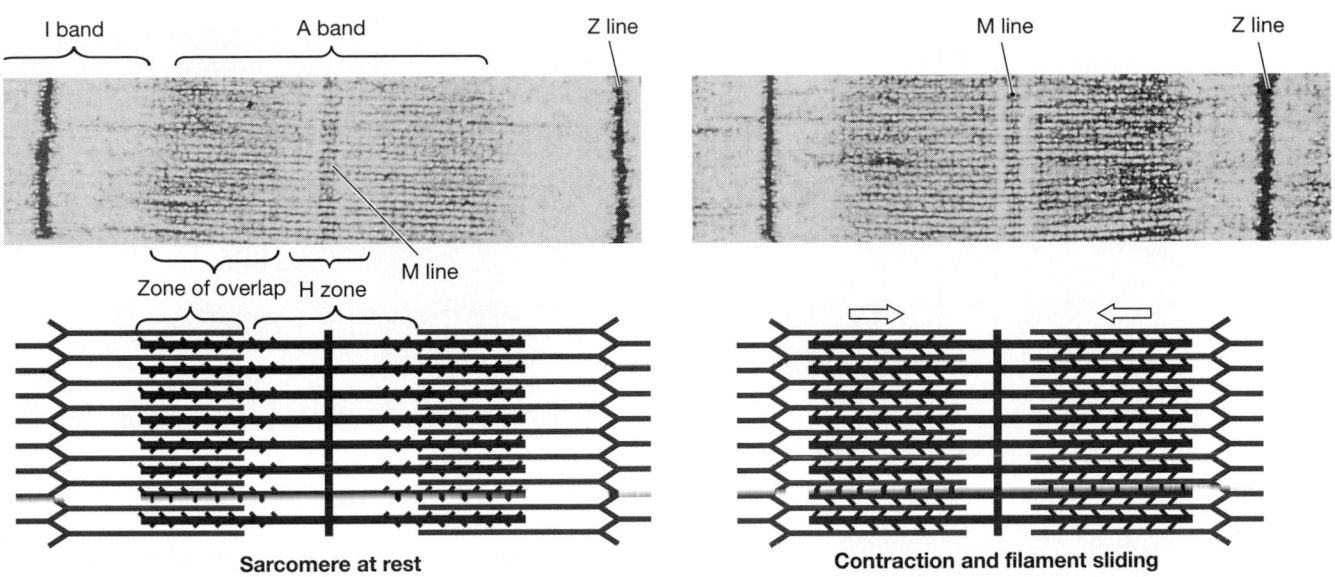

***FIGURE 10-6** Changes in the Appearance of a Sarcomere during Contraction of a Skeletal Muscle Fiber.** During a contraction, the A band stays the same width, but the Z lines move closer together and the I band gets smaller. For clarity, titin fibers are not shown.

CONTRACTION OF SKELETAL MUSCLE

The sliding filament theory explains what happens to the sarcomere during a contraction. With that in mind, we will now follow the contraction sequence in greater detail.

The Control of Skeletal Muscle Activity

Skeletal muscle fibers contract only under the control of the nervous system. Communication between the nervous system and a skeletal muscle fiber occurs at a specialized intercellular connection known as a **neuromuscular junction (NMJ)**, or *myoneural junction*. One such junction is shown in Figure 10-7a*.

The Neuromuscular Junction

Each skeletal muscle fiber is controlled by a neuron at a single neuromuscular junction midway along the fiber's length. Figure 10-7b* (Step 1) summarizes key features of this structure. A single axon branches within the perimysium to form a number of fine branches. Each of these branches ends at an expanded **synaptic terminal**. The cytoplasm of the synaptic terminal contains mitochondria and vesicles filled with molecules of **acetylcholine** (as-ē-til-KŌ-lēn), usually abbreviated **ACh**. ACh is a *neurotransmitter*, a chemical released by neurons to change the membrane properties of other cells. The release of ACh from the synaptic terminal can result in changes in the sarcolemma that trigger the contraction of the muscle fiber.

A narrow space, the **synaptic cleft**, separates the synaptic terminal of the neuron from the opposing sarcolemmal surface. This surface, which contains membrane receptors that bind ACh, is known as the **motor end plate**. The motor end plate has deep creases called *junctional folds*, which increase the membrane surface area and thus the number of available ACh receptors. The synaptic cleft and sarcolemma also contain the enzyme **acetylcholinesterase** (**AChE**, or *cholinesterase*), which breaks down ACh.

When a neuron stimulates a muscle fiber, the process occurs in a series of steps (Figure 10-7b*):

STEP 1. *The arrival of an action potential.* The stimulus for ACh release is the arrival of an electrical impulse, or **action potential**, at the synaptic terminal. An action potential is a sudden change in the transmembrane potential propagated along the length of the axon.

STEP 2. *The release of ACh.* When that impulse reaches the synaptic terminal, permeability changes in the membrane trigger the exocytosis of ACh into the synaptic cleft. This exocytosis is accomplished when vesicles in the synaptic terminal fuse with the membrane of the neuron.

STEP 3. *ACh binding at the motor end plate.* ACh molecules diffuse across the synaptic cleft and bind to ACh receptors on the motor end plate's surface. When ACh binding occurs, it changes the permeability of the sarcolemma to sodium ions. As you will recall from Chapter 3, the extracellular fluid contains a high concentration of sodium ions, whereas sodium ion concentrations inside the cell are very low. ∞ *[pp. 78, 81]* When the membrane permeability to sodium increases, sodium ions rush into the sarcoplasm. This influx continues until AChE removes the ACh from the receptors.

STEP 4. *Appearance of an action potential in the sarcolemma.* The sudden inrush of sodium ions results in the generation of an action potential in the sarcolemma

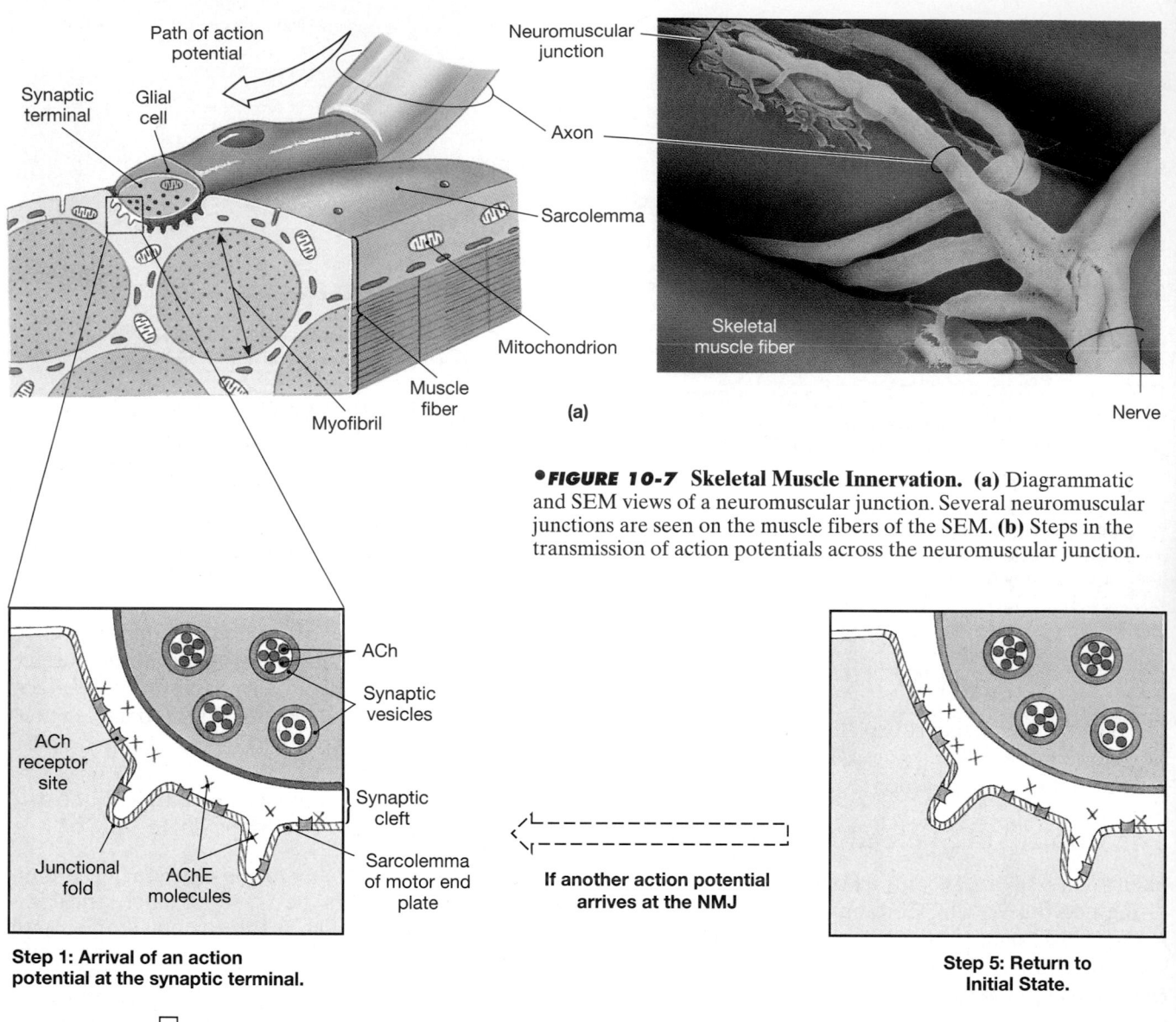

Path of action
potential

Synaptic
terminal

Glial
cell

Neuromuscular
junction

Axon

Sarcolemma

Mitochondrion

Muscle
fiber

Myofibril

Skeletal
muscle fiber

Nerve

(a)

●*FIGURE 10-7* **Skeletal Muscle Innervation. (a)** Diagrammatic and SEM views of a neuromuscular junction. Several neuromuscular junctions are seen on the muscle fibers of the SEM. **(b)** Steps in the transmission of action potentials across the neuromuscular junction.

ACh

Synaptic
vesicles

ACh
receptor
site

Synaptic
cleft

Junctional
fold

AChE
molecules

Sarcolemma
of motor end
plate

If another action potential
arrives at the NMJ

**Step 1: Arrival of an action
potential at the synaptic terminal.**

**Step 5: Return to
Initial State.**

(b)

Excitable membrane

Na⁺

Action
potential
propagation

Na⁺

Na⁺

Step 2: Release of Acetylcholine.
Vesicles in the synaptic terminal fuse with the neuronal membrane and dump their contents into the synaptic cleft.

Step 3: ACh Binding at the Motor End Plate. The binding of ACh to the receptors increases the membrane permeability to sodium ions. Sodium ions then enter the cell at an increased rate.

Step 4: Appearance of an Action Potential in the Sarcolemma. The action potential spreads across the membrane surface and down each T tubule, triggering the release of Ca^{2+} at the terminal cisternae. While this occurs, AChE removes the ACh.

at the edges of the motor end plate. This electrical impulse sweeps across the entire membrane surface and travels along each of the transverse tubules. The arrival of an action potential at the synaptic terminal thus leads to the appearance of an action potential in the sarcolemma.

STEP 5. *Return to the initial state.* Even before the action potential has spread across the entire membrane, the ACh has been broken down by AChE. Some of the breakdown products will be absorbed by the synaptic terminal and used to resynthesize ACh for subsequent release. This sequence of events can now be repeated, should another action potential arrive at the synaptic terminal.

Excitation–Contraction Coupling

The link between the generation of an action potential in the sarcolemma and the start of a muscle contraction is called **excitation–contraction coupling**. This coupling occurs at the triads. When an action potential reaches a triad, it triggers the release of Ca^{2+} from the cisternae of the sarcoplasmic reticulum. The change in the permeability of the SR to Ca^{2+} is temporary, lasting only about 0.03 second. Yet within a millisecond the Ca^{2+} concentration in and around the sarcomere reaches 100 times resting levels. Because the cisternae are situated at the zones of overlap, where the thick and thin filaments interact, the effect of calcium release on the sarcomere is almost instantaneous. The binding of Ca^{2+} to troponin exposes the active sites along the thin filaments, initiating the contraction. The *contraction cycle* now begins.

During the contraction cycle, cross-bridge binding occurs, and the subsequent pivoting of the myosin heads shortens the sarcomeres. Each myosin head continues to attach, pivot, and detach as long as Ca^{2+} and ATP remain available. Details of the contraction process are presented in Figure 10-8●, in "FOCUS: The Contraction Cycle" (pp. 288–289).

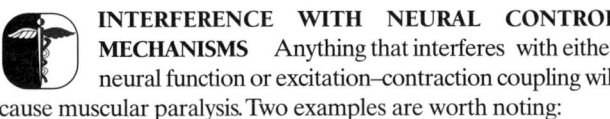 **INTERFERENCE WITH NEURAL CONTROL MECHANISMS** Anything that interferes with either neural function or excitation–contraction coupling will cause muscular paralysis. Two examples are worth noting:

1. Cases of *botulism* result from the consumption of contaminated canned or smoked foods that contain a toxin. The toxin, produced by bacteria, prevents the release of ACh at the synaptic terminals, leading to a potentially fatal muscular paralysis.
2. The progressive muscular paralysis seen in *myasthenia gravis* results from the loss of ACh receptors at the junctional folds. The primary cause is a misguided attack on the ACh receptors by the immune system. Genetic factors play a role in predisposing individuals to develop this condition. **AM** *Problems with the Control of Muscle Activity*

Relaxation

The duration of a contraction depends on the duration of stimulation at the neuromuscular junction. The ACh

released after the arrival of a single action potential at the synaptic terminal does not remain intact for long. ACh that is bound to the sarcolemma or free in the synaptic cleft is rapidly broken down and inactivated by AChE. Inside the muscle fiber, the permeability changes in the SR are also very brief. A contraction will therefore continue only if multiple additional action potentials arrive at the synaptic terminal in rapid succession. When they do, the continual release of ACh into the synaptic cleft produces a series of action potentials in the sarcolemma that keeps a high concentration of Ca^{2+} in the sarcoplasm. Under these conditions, the contraction cycle will be repeated over and over.

If just one action potential arrives at the NMJ, Ca^{2+} concentrations in the sarcoplasm will quickly return to normal resting levels. Two mechanisms are involved in this process: (1) active Ca^{2+} transport across the cell membrane into the extracellular fluid and (2) active Ca^{2+} transport into the SR. Of the two, transport into the SR is the more important. Virtually as soon as the calcium ions have been released, the SR returns to its normal permeability and begins actively absorbing Ca^{2+} from the surrounding sarcoplasm. As Ca^{2+} concentrations in the sarcoplasm fall, (1) calcium ions detach from troponin, (2) troponin returns to its original position, and (3) the active sites are covered by tropomyosin. The contraction is now at an end.

Once the contraction has ended, the sarcomere does not automatically return to its original length. Sarcomeres shorten actively, but there is no active mechanism for reversing the process. External forces must act on the contracted muscle fiber to stretch the myofibrils and sarcomeres to their original dimensions. We will describe those forces in a later section.

Before you proceed, read the FOCUS section on pages 288–289. Then review the entire sequence of events from neural activation through excitation–contraction coupling to the completion of a contraction. Table 10-1 (p. 290) provides a summary of the contraction process, from ACh release to the end of the contraction.

Length–Tension Relationships

When many people are pulling on a rope, the amount of tension produced is proportional to the number of people involved. In a muscle fiber, the amount of tension generated during a contraction depends on the number of cross-bridge interactions that occur in all the sarcomeres along all the myofibrils. The number of cross-bridge interactions is determined by the degree of overlap between thick and thin filaments. *When the muscle fiber is stimulated to contract, only myosin heads within the zone of overlap can bind to active sites and produce tension.* The tension produced by the intact muscle fiber can thus be related to the structure of an individual sarcomere.

●FIGURE 10-8 Molecular Events of the Contraction Process

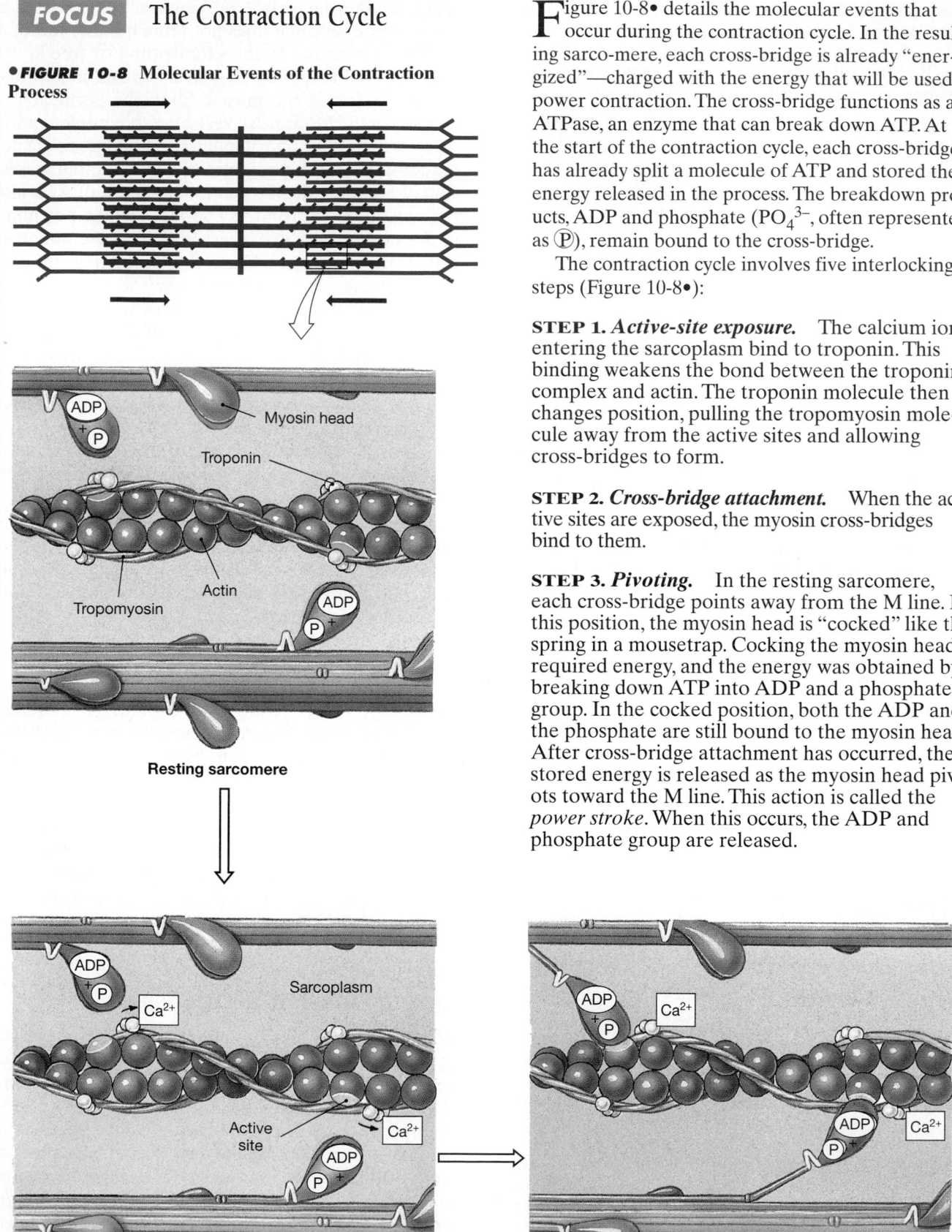

Resting sarcomere

Step 1: Active-site exposure

Step 2: Cross-bridge attachment

Figure 10-8● details the molecular events that occur during the contraction cycle. In the resulting sarco-mere, each cross-bridge is already "energized"—charged with the energy that will be used to power contraction. The cross-bridge functions as an ATPase, an enzyme that can break down ATP. At the start of the contraction cycle, each cross-bridge has already split a molecule of ATP and stored the energy released in the process. The breakdown products, ADP and phosphate (PO_4^{3-}, often represented as Ⓟ), remain bound to the cross-bridge.

The contraction cycle involves five interlocking steps (Figure 10-8●):

STEP 1. *Active-site exposure.* The calcium ions entering the sarcoplasm bind to troponin. This binding weakens the bond between the troponin complex and actin. The troponin molecule then changes position, pulling the tropomyosin molecule away from the active sites and allowing cross-bridges to form.

STEP 2. *Cross-bridge attachment.* When the active sites are exposed, the myosin cross-bridges bind to them.

STEP 3. *Pivoting.* In the resting sarcomere, each cross-bridge points away from the M line. In this position, the myosin head is "cocked" like the spring in a mousetrap. Cocking the myosin head required energy, and the energy was obtained by breaking down ATP into ADP and a phosphate group. In the cocked position, both the ADP and the phosphate are still bound to the myosin head. After cross-bridge attachment has occurred, the stored energy is released as the myosin head pivots toward the M line. This action is called the *power stroke*. When this occurs, the ADP and phosphate group are released.

STEP 4. *Cross-bridge detachment.* When an ATP binds to the myosin head, the link between the active site on the actin molecule and the myosin head is broken. The active site is now exposed and able to interact with another cross-bridge.

STEP 5. *Myosin reactivation.* Myosin reactivation occurs when the free myosin head splits the ATP into ADP and a phosphate group. The energy released in this process is used to recock the myosin head. The entire cycle can now be repeated. If calcium ion concentrations remain elevated and ATP reserves are sufficient, each myosin head will repeat this cycle about five times per second. Each power stroke shortens the sarcomere by about 1 percent, so each second the sarcomere can shorten by roughly 5 percent. Because all the sarcomeres contract together, the entire muscle shortens at the same rate.

To appreciate the overall effect, imagine that you are pulling on a large rope. You are the myosin head, and the rope is a thin filament. You reach forward, grab the rope with both hands, and pull it toward you. This action corresponds to cross-bridge attachment and pivoting. You now release the rope, reach forward, and grab it once again. By repeating the cycle over and over, you can gradually pull in the rope.

Now consider the situation when several people are lined up, all pulling on the same rope, as in a tug-of-war team. Each person reaches forward, grabs the rope, pulls it, releases it, and then grabs it again to repeat the cycle. The individual actions are not coordinated: At any one moment, some people are grabbing, some are pulling, and others are letting go. The amount of tension produced is a function of how many people are pulling at any given instant. A comparable situation applies to tension in a muscle fiber, where the myosin heads along a thick filament work together to pull a thin filament toward the center of the sarcomere.

RIGOR MORTIS When death occurs, circulation ceases and the skeletal muscles are deprived of nutrients and oxygen. Within a few hours, the skeletal muscle fibers have run out of ATP, and the sarcoplasmic reticulum becomes unable to pump Ca^{2+} out of the sarcoplasm. Calcium ions diffusing into the sarcoplasm from the extracellular fluid or leaking out of the sarcoplasmic reticulum then trigger a sustained contraction. Without ATP, the cross-bridges cannot detach from the active sites, and skeletal muscles throughout the body become locked in the contracted position. Because all the body's skeletal muscles are involved, the individual becomes "stiff as a board." This physical state, called **rigor mortis**, lasts until the lysosomal enzymes released by autolysis break down the myofilaments 15–25 hours later. The timing is dependent on environmental factors, such as temperature. Forensic pathologists can estimate the time of death on the basis of the degree of rigor mortis and the characteristics of the local environment.

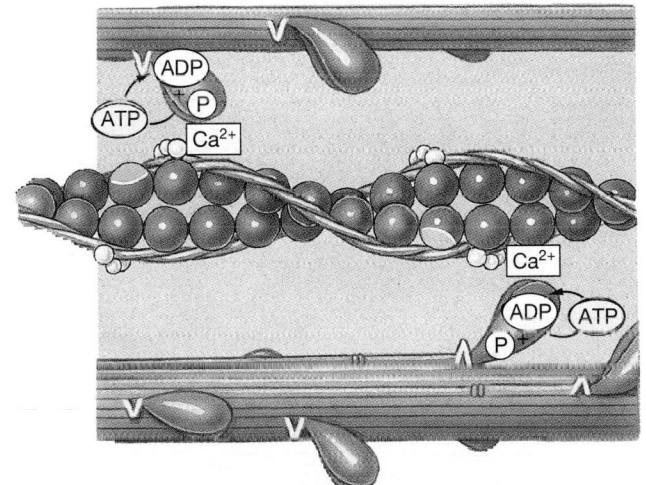

Step 5: Myosin reactivation

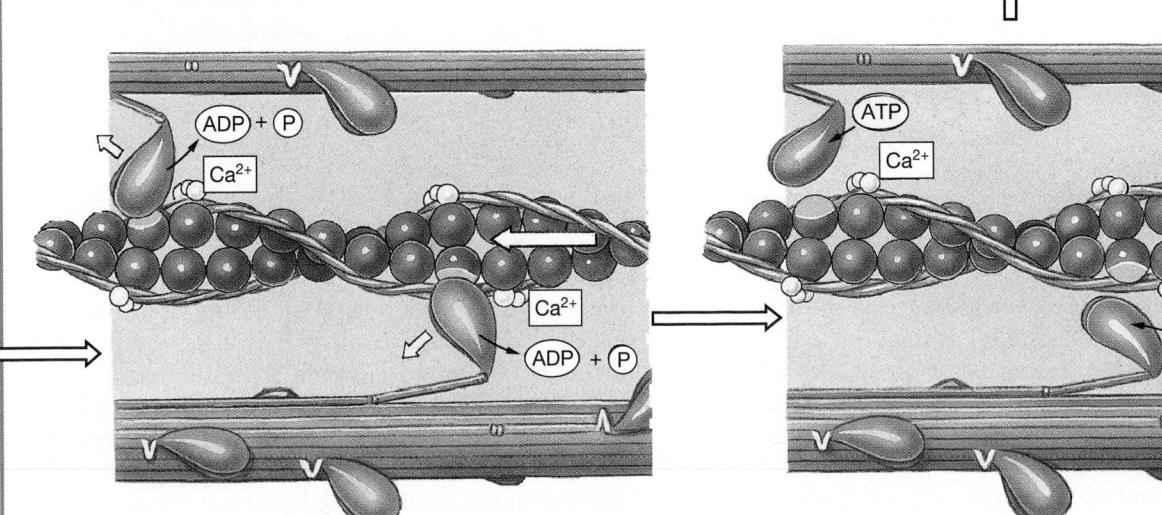

Step 3: Pivoting of myosin head　　　　**Step 4: Cross-bridge detachment**

TABLE 10-1 A Summary of the Steps Involved in Skeletal Muscle Contraction

Key steps in the initiation of a contraction include the following:

1. At the neuromuscular junction (NMJ), ACh released by the synaptic terminal binds to receptors on the sarcolemma.
2. The resulting change in the transmembrane potential of the muscle fiber leads to the production of an action potential that spreads across its entire surface and along the T tubules.
3. The sarcoplasmic reticulum (SR) releases stored calcium ions, increasing the calcium concentration of the sarcoplasm in and around the sarcomeres.
4. Calcium ions bind to troponin, producing a change in the orientation of the troponin–tropomyosin complex that exposes active sites on the thin (actin) filaments. Myosin cross-bridges form when myosin heads bind to active sites.
5. Repeated cycles of cross-bridge binding, pivoting, and detachment occur, powered by the hydrolysis of ATP. These events produce filament sliding, and the muscle fiber shortens.

This process continues for a brief period, until:

6. Action potential generation ceases as ACh is broken down by acetylcholinesterase (AChE).
7. The SR reabsorbs calcium ions, and the concentration of calcium ions in the sarcoplasm declines.
8. When calcium ion concentrations approach normal resting levels, the troponin–tropomyosin complex returns to its normal position. This change covers the active sites and prevents further cross-bridge interaction.
9. Without cross-bridge interactions, further sliding will not take place, and the contraction will end.
10. Muscle relaxation occurs, and the muscle returns passively to resting length.

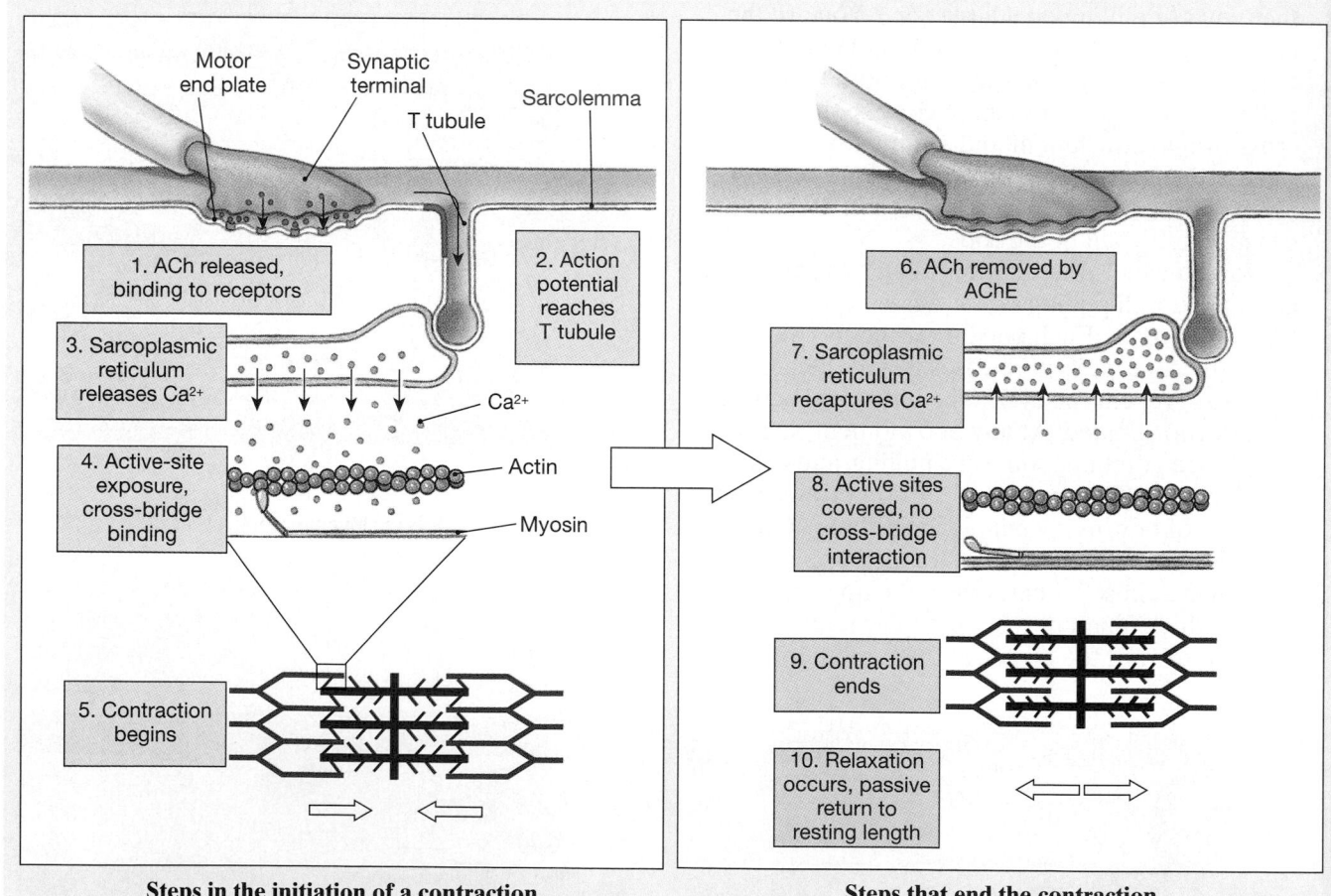

Steps in the initiation of a contraction

Steps that end the contraction

When the sarcomeres are as short as they can be, the thick filaments are jammed against the Z lines. Although cross-bridge binding can occur, the myosin heads cannot pivot, and no tension is produced (Figure 10-9a●). Even if the sarcomeres are somewhat longer, the thin filaments that extend across the center of the sarcomere collide with or overlap the thin filaments of the opposite side (Figure 10-9b●). This disruption of the normal arrangement of thick and thin filaments interferes with the binding of cross-bridges to active sites. The result is that little tension is produced when the muscle fiber is stimulated.

Within the optimal range of sarcomere lengths (Figure 10-9c●), the maximum number of cross-bridges can form and the tension produced is highest. Any further increase in sarcomere length reduces the tension produced by reducing the size of the zone of overlap and the number of potential cross-bridge interactions (Figure

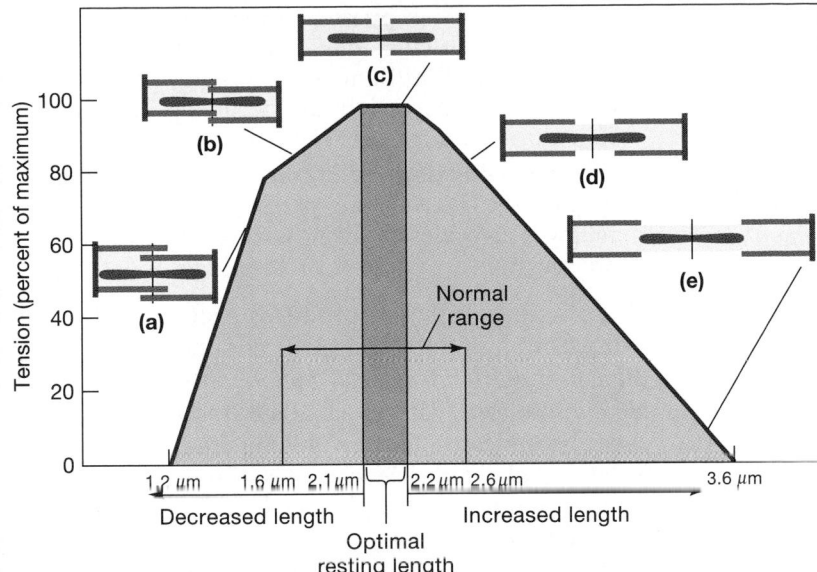

●**FIGURE 10-9** **The Effect of Sarcomere Length on Tension.** **(a,b)** At short resting lengths, thin filaments extending across the center of the sarcomere interfere with the normal orientation of thick and thin filaments, reducing tension production. In addition, very little contraction can occur before the thick filaments crash into the Z lines. **(c)** The tension produced reaches a maximum when the zone of overlap is large but the thin filaments do not extend across the center of the sarcomere. If the sarcomeres are stretched too far, the zone of overlap is **(d)** reduced or **(e)** disappears, and cross-bridge interactions are reduced or cannot occur. The light purple area represents the normal range of sarcomere lengths in the body.

10-9d●). When the zone of overlap disappears altogether, thin and thick filaments cannot interact (Figure 10-9e●). The muscle fiber cannot actively produce any tension and a contraction cannot occur. Such extreme stretching of a muscle fiber is normally prevented by the titin filaments within the muscle fiber, which tie the thick filaments to the Z lines, and by surrounding connective tissues, which limit the degree of muscle distortion.

In summary, muscle fibers can contract most forcefully when stimulated over a relatively narrow range of resting lengths. The normal range of sarcomere lengths in the body, indicated in Figure 10-9●, is from 75 to 130 percent of the optimal length. The arrangement of skeletal muscles, connective tissues, and bones normally prevents extreme compression or excessive stretching that would correspond to Figure 10-9a or 10-9e●. For example, straightening your elbow stretches your *biceps brachii* muscle, but stabilizing ligaments and the bony structure of the elbow stop this movement before the muscle fibers stretch too far.

During normal movements, your muscle fibers perform over a broad range of intermediate lengths; the tension produced varies, depending on the initial length of the muscle fibers. During an activity such as walking, in which muscles contract and relax cyclically, muscle fibers are stretched to a length very close to "ideal" before they are stimulated to contract. Some muscles must contract over a large range of resting lengths. When these muscles are contracting at inefficient resting lengths, they are generally assisted by the contractions of other muscles whose lengths are closer to ideal. (We will discuss the mechanical principles involved in Chapter 11.)

☑ How would a drug that interferes with cross-bridge formation affect muscle contraction?

☑ What would you expect to happen to a resting skeletal muscle if the sarcolemma suddenly became very permeable to Ca^{2+}?

☑ Predict what would happen to a muscle if the motor end plate failed to produce any acetylcholinesterase.

MUSCLE MECHANICS

Next, we will consider the coordination, metabolism, and performance of skeletal muscles as functional organs.

Tension Production

Now that you are familiar with the basic mechanisms of muscle contraction at the level of the individual muscle fiber, we can begin to examine the performance of *skeletal muscles*, organs of the muscular system. In this section, we will consider the coordinated contractions of an entire population of skeletal muscle fibers.

The amount of tension produced by an individual muscle fiber depends solely on the number of cross-bridge interactions. If a muscle fiber at a given resting length is stimulated to contract, it will always produce the same amount of tension. There is no mechanism to regulate the amount of tension produced in that contraction by having only a few sarcomeres contract. When calcium ions are released, they are released at all the triads in the muscle fiber. As a result, all the myosin heads within zones of overlap will interact with thin filaments. Thus a muscle fiber is either "ON" (producing as much tension as possible at that resting length) or "OFF" (relaxed). This feature of muscle mechanics is known as the **all-or-none principle**. The amount of tension produced in the skeletal muscle *as a whole* is therefore determined by (1) the frequency of stimulation and (2) the number of muscle fibers stimulated.

The Frequency of Muscle Stimulation

A **twitch** is a single stimulus–contraction–relaxation sequence in a muscle fiber. Twitches vary in duration.

Twitches in one eye muscle fiber can be as brief as 7.5 msec, but a twitch in a muscle fiber from the *soleus*, a small calf muscle, lasts about 100 msec. Figure 10-10a● is a **myogram**, or graph of tension development in various muscles during a twitch contraction.

Figure 10-10b● details the phases of a 40-msec twitch in the *gastrocnemius* muscle, a prominent calf muscle. A single twitch can be divided into a *latent period*, a *contraction phase*, and a *relaxation phase*:

1. The **latent period** begins at stimulation and typically lasts about 2 msec. Over this period, the action potential sweeps across the sarcolemma, and calcium ions are released by the sarcoplasmic reticulum. During the latent period, the muscle fiber does not produce tension because the contraction cycle has yet to begin.

2. In the **contraction phase,** tension rises to a peak. As tension rises, calcium ions are binding to troponin, active sites on thin filaments are being exposed, and cross-bridge interactions are occurring. The contraction phase ends roughly 15 msec after stimulation.

3. The **relaxation phase** then continues for about another 25 msec. During this period, calcium levels are falling, active sites are being covered by tropomyosin, and the number of active cross-bridges is declining.

A single stimulation produces a single twitch, but twitches in a skeletal muscle do not accomplish anything useful. *All normal activities involve sustained muscle contractions.* The mechanism involved can most easily be understood by examining the responses of an isolated skeletal muscle when it is stimulated under laboratory conditions. Under these conditions, the myogram of tension production changes dramatically as the rate of stimulation increases (Figure 10-11●).

WAVE SUMMATION AND INCOMPLETE TETANUS
If a second stimulus arrives before the relaxation phase has ended, a second, more powerful contraction occurs. The addition of one twitch to another in this way constitutes the **summation of twitches,** or **wave summation** (Figure 10-11a●). The frequency of stimulation necessary to produce wave summation depends on the duration of a single twitch in that particular muscle fiber. For example, if a twitch lasts 20 msec (1/50 sec), stimulation at a frequency of less than 50 per second will produce individual twitches. Stimulation at a higher frequency will result in wave summation.

If the stimulation continues and the muscle is never allowed to relax completely, tension will rise to a peak (Figure 10-11b●). A muscle producing peak tension during rapid cycles of contraction and relaxation is said to be in **incomplete tetanus** (*tetanos*, convulsive tension).

COMPLETE TETANUS   **Complete tetanus** can be obtained by increasing the rate of stimulation until the relaxation phase is eliminated (Figure 10-11c●). During complete tetanus, the action potentials are arriving so rapidly that the sarcoplasmic reticulum does not have time to reclaim the calcium ions. The high Ca^{2+} concentration in the cytoplasm prolongs the state of contraction, making it continuous. *Virtually all normal muscular contractions involve the complete tetanus of the participating muscle fibers.*

TREPPE   If a skeletal muscle is stimulated a second time immediately after the relaxation phase has ended, the contraction that occurs will develop a slightly higher maximum tension than did the contraction after the first stimulation. The increase in peak tension indicated in Figure 10-11d● will continue over the first 30–50 stimulations. Thereafter, the amount of tension produced will remain constant at roughly 25 percent of the maximal tension that would be produced in a complete tetanic contraction. Because the tension rises in stages, like the steps in

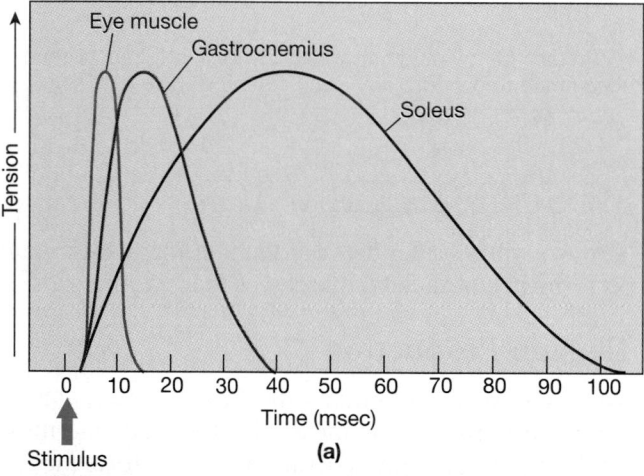

(a)

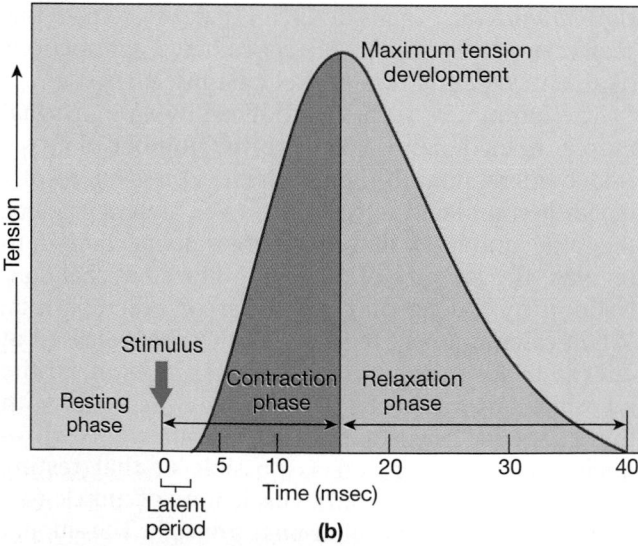

(b)

●**FIGURE 10-10** **The Twitch and Development of Tension.** (a) Myogram showing differences in the time course of a twitch contraction in different skeletal muscles in the body. (b) Details of the time course of a single twitch contraction in the gastrocnemius muscle. Note the presence of a latent period, which corresponds to the time needed for the conduction of action potential and the subsequent release of calcium ions by the sarcoplasmic reticulum.

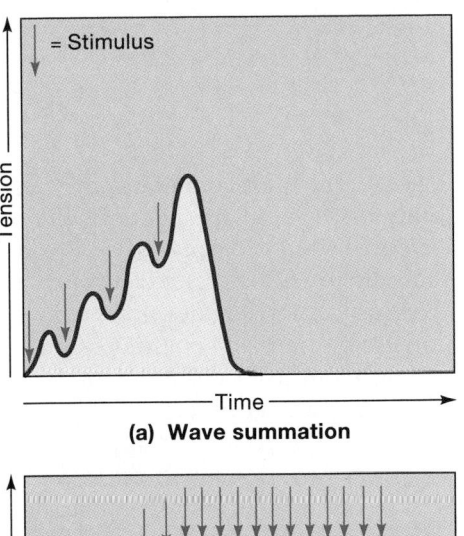

(a) **Wave summation**

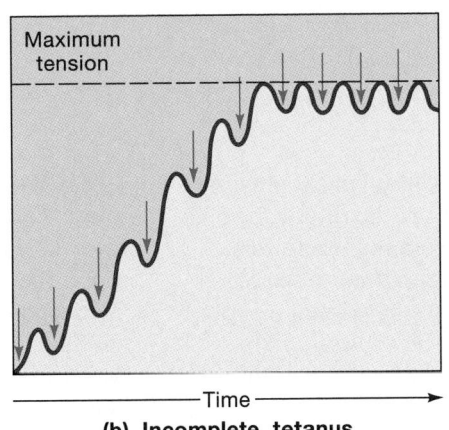

(b) **Incomplete tetanus**

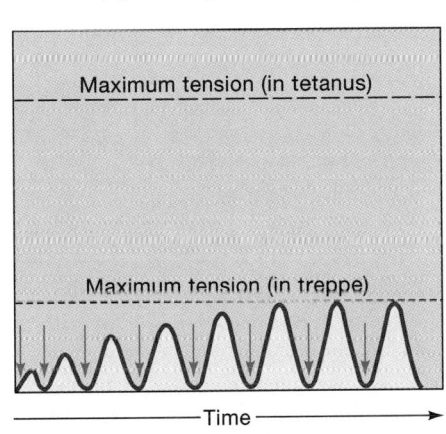

(c) **Complete tetanus**

(d) **Treppe**

•FIGURE 10-11 Effects of Repeated Stimulations.
(a) Wave summation occurs when successive stimuli arrive before relaxation (the downturn of the curve) has been completed.
(b) Incomplete tetanus occurs if the rate of stimulation increases further. Tension production will rise to a peak, and the periods of relaxation will be very brief.
(c) During complete tetanus, the frequency of stimulation is so high that the relaxation phase has been entirely eliminated and tension plateaus at maximal levels. **(d)** Treppe is an increase in peak tension following repeated stimuli delivered shortly after the completion of the relaxation phase of each twitch.

a staircase, this phenomenon is called **treppe** (TREP-e), a German word meaning "stairs." The rise is thought to result from a gradual increase in the concentration of calcium ions in the cytoplasm, in part because the ion pumps in the sarcoplasmic reticulum are unable to recapture them in the time between stimulations.

Internal Tension and External Tension

We have already discussed the relationship between resting length and tension production for individual muscle fibers. When we consider an entire skeletal muscle, the situation is complicated by the fact that the muscle fibers are not directly connected to the structures they pull against. The extracellular fibers of the endomysium, perimysium, epimysium, and tendons are flexible and somewhat elastic. When a skeletal muscle contracts, the myofibrils inside the muscle fibers generate *internal tension*. This internal tension is applied to the extracellular fibers. The tension in the extracellular fibers is called *external tension*.

As the external tension rises, the extracellular fibers stretch; for this reason, they are called *series elastic elements*. The series elastic elements behave like fat rubber bands. They stretch easily at first, but as they elongate they become stiffer and more effective at transferring tension to the resistance. As a result, external tension does not climb as quickly as internal tension when a contraction occurs.

To understand this relationship, attach a rubber band to one of the rings in a three-ring notebook. Put a finger through the rubber band and use it to pull the notebook across a table. Your finger represents the muscle fibers; the rubber band, the attached tendon; and the notebook, a bone of the skeleton. When you first apply tension, the rubber band stretches and becomes stiffer. Over this period, external tension is rising. The notebook starts to move when the rubber band becomes sufficiently taut—that is, when the tension in the rubber band (the external tension) overcomes friction and the weight of the notebook (the resistance).

If you now relax your hand, the rubber band will pull your finger toward the notebook. The same thing happens in a muscle: When the contraction ends, the series elastic elements recoil and pull on the muscle. This recoil helps return the muscle to its original resting length.

A myogram performed in the laboratory generally measures the tension in a tendon and so is reporting external tension rather than internal tension. Figure 10-12a• compares the internal and external tensions during a twitch contraction. A single twitch is so short in duration that there isn't enough time for the external tension to rise as high as the internal tension. Twitches are therefore ineffective in terms of performing useful work. If a second twitch occurs before the relaxation phase has ended, the peak tension increases because the second contraction begins while the external tension is still elevated from the previous contrac-

Tetanus

Children are often told to be careful of rusty nails. Parents are not worrying about the rust or the nail but about infection with a very common bacterium, *Clostridium tetani*. This bacterium can cause **tetanus**, a disease that has no relationship to the normal muscle response to neural stimulation. The *Clostridium* bacterium, although found virtually everywhere in the environment, can thrive only in tissues that contain abnormally low amounts of oxygen. For this reason, a deep puncture wound, such as that from a nail, carries a much greater risk of producing tetanus than does a shallow, open cut that bleeds freely.

When active in body tissues, these bacteria release a powerful toxin that affects the central nervous system. Motor neurons, which carry information from the central to the peripheral nervous system, are particularly sensitive to it. The toxin suppresses the mechanism responsible for regulating motor neuron activity. The result is a sustained, powerful contraction of skeletal muscles throughout the body.

After exposure to the bacteria, the incubation period (the time before symptoms develop) is gen-erally less than 2 weeks. The most common complaints are headache, muscle stiffness, and difficulty in swallowing. Because it soon becomes difficult to open the mouth, this disease is also called *lockjaw*. Widespread muscle spasms typically develop within 2–3 days of the initial symptoms and continue for a week before subsiding. After 2–4 weeks, patients who survive recover with no aftereffects.

Severe tetanus has a 40–60 percent mortality rate—that is, for every 100 people who develop severe tetanus, 40 to 60 will die. Fortunately, immunization is effective in preventing the disease. There are approximately 500,000 cases of tetanus worldwide each year, but only about 100 of them occur in the United States, thanks to an effective immunization program. ("Tetanus shots" are recommended, with booster shots every 10 years.) Severe symptoms in unimmunized patients can be prevented by early administration of an antitoxin, in most cases *human tetanus immune globulin*. Such treatment does not reduce symptoms that have already appeared, however, and there is no generally effective treatment for this disease.

tion. Thus, a greater proportion of the internal tension will be conveyed to the series elastic elements. This mechanism is now thought to be the primary basis for wave summation. During a tetanic contraction, there is sufficient time for internal and external tensions to equalize (Figure 10-12b●).

Motor Units and Tension Production

You have a remarkable ability to control the amount of tension exerted by your skeletal muscles. During a normal contraction, tension rises smoothly, not jerkily, because activated muscle fibers are stimulated to complete tetanus. The *total force* exerted by the skeletal muscle depends on how many muscle fibers are activated.

A typical skeletal muscle contains thousands of muscle fibers. Although some motor neurons control a few muscle fibers, most control hundreds of them. All the muscle fibers controlled by a single motor neu-

●FIGURE 10-12 Internal and External Tension. Internal tension rises as the muscle fiber contracts. External tension rises more slowly as the series elastic elements are stretched. **(a)** During a single twitch contraction, external tension cannot rise as high as internal tension before the relaxation phase begins. **(b)** During a tetanic contraction, external tension soon plateaus at a level that is roughly equivalent to internal tension. External tension remains elevated for the duration of the contraction.

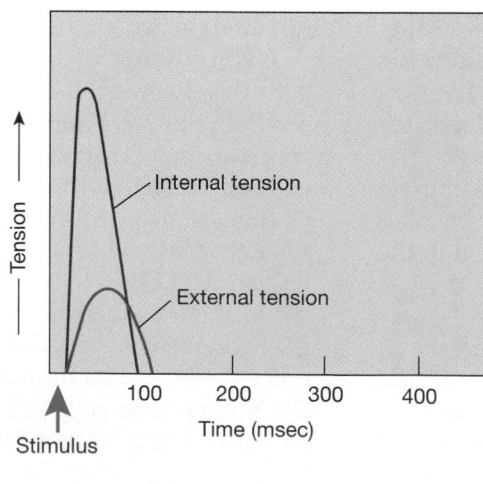

(a) Twitch

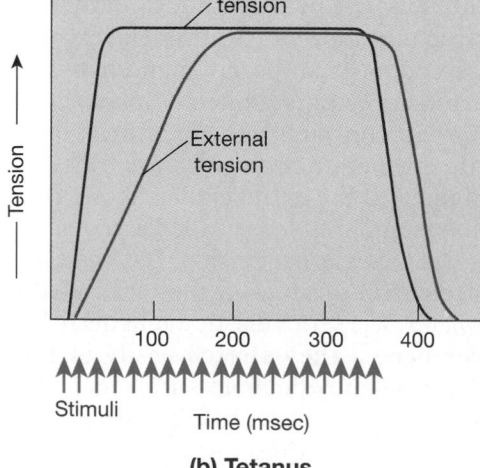

(b) Tetanus

ron constitute a **motor unit**. The size of a motor unit is an indication of how fine the control of movement can be. In the muscles of the eye, where precise control is extremely important, a motor neuron may control 4–6 muscle fibers. We have much less precise control over our leg muscles, where a single motor neuron may control 1000–2000 muscle fibers (depending on the specific muscle and the reference consulted). As Figure 10-13• indicates, the muscle fibers of each motor unit are intermingled with those of other motor units. Because of this intermingling, the direction of pull exerted on the tendon does not change despite variations in the numbers of activated motor units.

When you decide to perform a specific movement, specific groups of motor neurons within your central nervous system are stimulated. The contraction begins with the activation of the smallest motor units in the stimulated muscle. These motor units generally contain muscle fibers that contract relatively slowly. Over time, larger motor units containing faster and more powerful muscle fibers are activated, and tension production rises steeply. The smooth but steady increase in muscular tension produced by increasing the number of active motor units is called **recruitment**, or **multiple motor unit summation**.

Peak tension production occurs when all the motor units in the muscle are contracting in complete tetanus. Such powerful contractions do not last long, however, because the individual muscle fibers soon use up their available energy reserves. During a sustained tetanic con-

traction, motor units are activated on a rotating basis, so some of them are resting and recovering while others are actively contracting. This "relay team" approach, called *asynchronous motor unit summation*, lets each motor unit recover somewhat before it is stimulated again. As a result, when your muscles contract for sustained periods, they produce slightly less than maximal tension.

MUSCLE TONE Some of the motor units within any particular muscle are always active, even when the entire muscle is not contracting. Their contractions do not produce enough tension to cause movement, but they do tense and firm the muscle. This resting tension in a skeletal muscle is called **muscle tone**. A muscle with little muscle tone appears limp and flaccid, whereas one with moderate muscle tone is quite firm and solid. The identity of the stimulated motor units changes constantly, so a constant tension in the attached tendon is maintained, but individual muscle fibers can relax.

Resting muscle tone stabilizes the position of bones and joints. For example, in muscles involved with balance and posture, enough motor units are stimulated to produce the tension needed to maintain body position. Muscle tone also helps prevent sudden, uncontrolled changes in the position of bones and joints. In addition to bracing the skeleton, the elastic nature of muscles and tendons lets skeletal muscles act as shock absorbers that cushion the impact of a sudden bump or shock.

Scattered within each skeletal muscle are clusters of specialized muscle fibers called *intrafusal fibers*. They are located in specialized sensory structures called *muscle spindles*. The intrafusal fibers are monitored by sensory neurons that deliver information about fiber length to the central nervous system. A *reflex* is an automatic motor response to sensory stimulation. Reflexes triggered by the distortion of muscle spindles adjust the tone of skeletal muscles. Such reflexes play an important role in the automatic adjustment of body position and posture, a topic we shall explore in more detail in Chapter 13.

The activity and sensitivity of muscle spindles increase with exercise training, producing an increase in background muscle tone. Heightened muscle tone accelerates the recruitment process during a voluntary contraction, because some of the motor units are already stimulated. Strong muscle tone also makes skeletal muscles appear firm and well defined, even at rest.

Isotonic and Isometric Contractions

We can classify muscle contractions as *isotonic* or *isometric* on the basis of the pattern of tension production.

ISOTONIC CONTRACTIONS In an **isotonic contraction** (*iso-*, equal + *tonos*, tension), tension reaches a plateau and the muscle shortens. Consider the experiment summarized in Figure 10-14•. A skeletal muscle

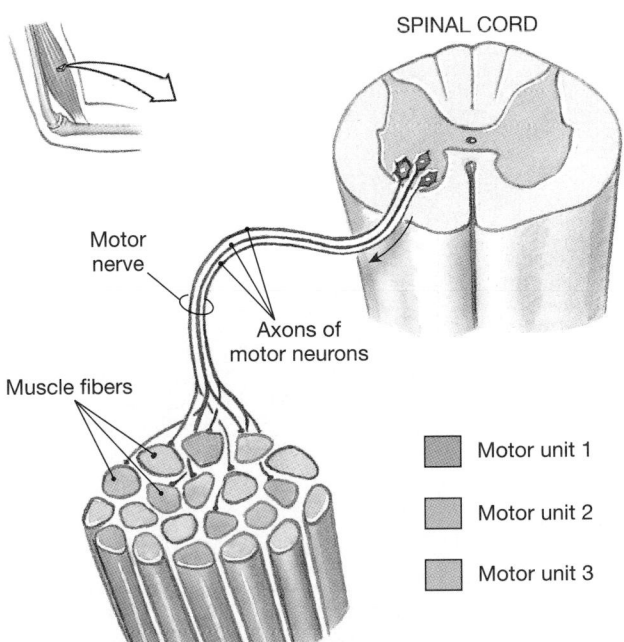

SPINAL CORD

Motor nerve

Axons of motor neurons

Muscle fibers

☐ Motor unit 1
☐ Motor unit 2
☐ Motor unit 3

•**FIGURE 10-13 Arrangement of Motor Units Within a Skeletal Muscle.** Muscle fibers of different motor units are intermingled, so the net distribution of force applied to the tendon remains constant even when individual muscle groups cycle between contraction and relaxation.

1 cm² in cross-sectional area can develop roughly 4 kg of force in complete tetanus. If we hang a 2-kg weight from that muscle and stimulate it, the muscle will shorten (Figure 10-14a●). Before the muscle can shorten, the cross-bridges must produce enough tension to overcome the resistance—in this case, the 2-kg weight. Over this period, internal tension in the muscle fibers rises until the external tension in the tendon exceeds the amount of resistance. As the muscle shortens, the internal and external tensions in the skeletal muscle remain constant at a value that just exceeds the resistance (Figure 10-14b●). Lifting an object off a desk, walking, and running involve isotonic contractions of this kind.

There are two types of isotonic contractions: (1) *concentric* and (2) *eccentric*. In a **concentric contraction**, the muscle tension exceeds the resistance and the muscle shortens, as in Figure 10-14a,b●. The speed of shortening varies with the difference between the amount of tension produced and the amount of resistance. If all the muscle units are stimulated and the resistance is relatively small, the muscle will shorten very quickly. In contrast, if the muscle barely produces enough tension to overcome the resistance, it will shorten very slowly.

In an **eccentric contraction**, the peak tension developed is less than the resistance, and the muscle elongates due to the contraction of another muscle or the pull of

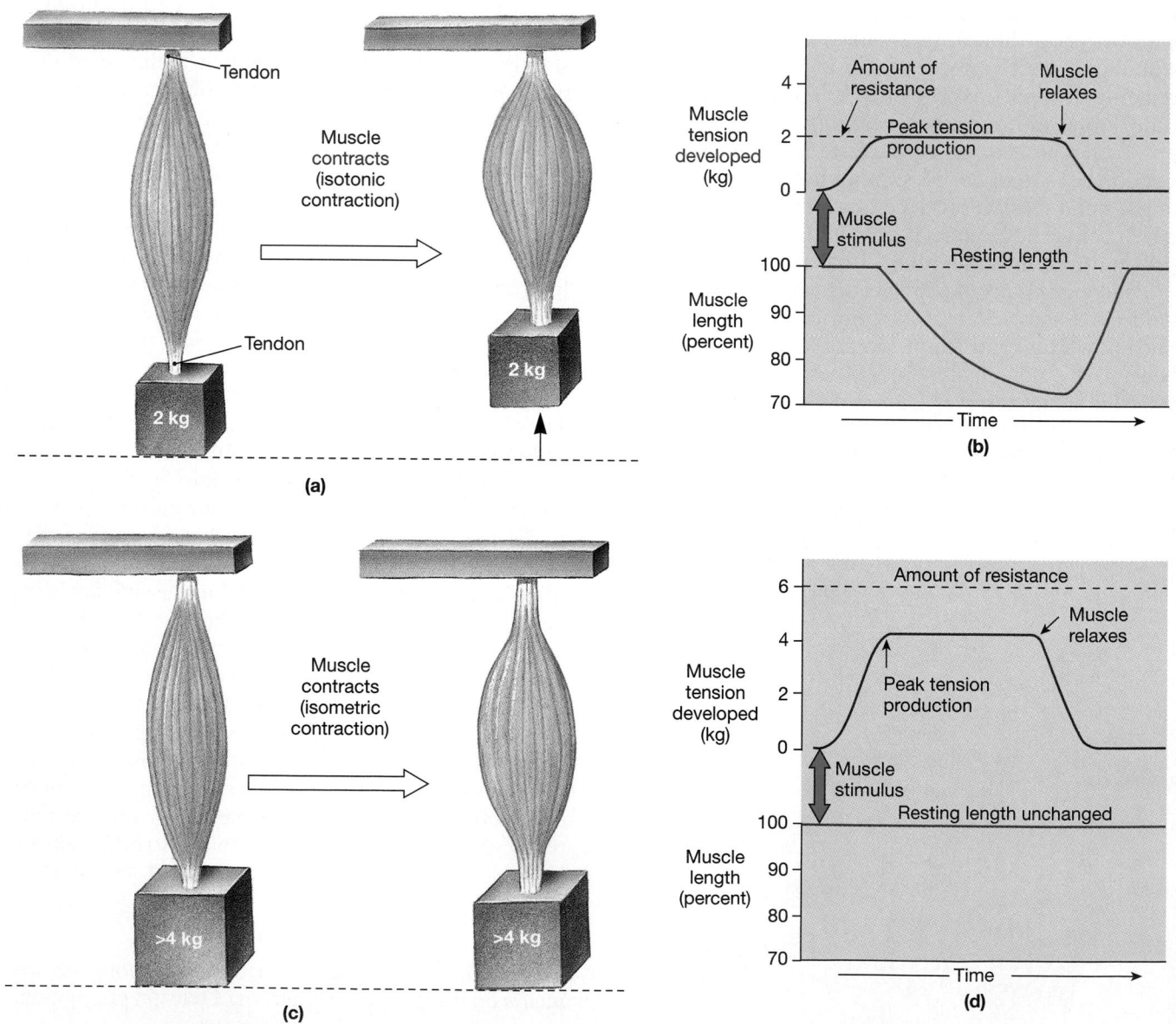

● **FIGURE 10-14** **Isotonic and Isometric Contractions.** **(a,b)** This muscle is attached to a weight less than its peak tension capabilities. On stimulation, it develops enough tension to lift the weight. Tension remains constant for the duration of the contraction, although the length of the muscle changes. This is an example of isotonic contraction. **(c,d)** The same muscle is attached to a weight that exceeds its peak tension capabilities. On stimulation, tension will rise to a peak, but the muscle as a whole cannot shorten. This is an isometric contraction.

gravity. Think of a tug-of-war team trying to stop a moving car. Although everyone pulls as hard as they can, the rope slips through their fingers. The speed of elongation depends on the difference between the amount of tension developed by the active muscle fibers and the amount of resistance. In our analogy, the team might slow down a small car but would have little effect on a large truck.

Eccentric contractions are very common, and they are an important part of a variety of movements. In these movements, you exert precise control over the amount of tension produced. By varying the tension in an eccentric contraction, you can control the rate of elongation, just as you can vary the tension in a concentric contraction. For example, for you to lower a book slowly onto a desk, the tension generated by your muscles must be slightly less than the weight of the book, whether it is a 200-g paperback or a 4-kg dictionary. Comparable eccentric contractions occur each time you walk down stairs or settle into a chair. During physical training, people commonly perform cycles of concentric and eccentric contractions, as when you hold a weight in your hand and flex and extend your elbow.

Resistance and Speed of Contraction

The reason you can lift a light object more rapidly than you can lift a heavy one is that there is an inverse relationship between the amount of resistance and the speed of contraction. If the resistance is less than the tension produced, an isotonic, concentric contraction will occur; the muscle will shorten. The heavier the resistance, the longer it takes for the movement to begin, because muscle tension, which increases gradually, must exceed the resistance before shortening can occur (Figure 10-15 •). The contraction itself proceeds more slowly; at the cellular level, the speed of cross-bridge pivoting is reduced as the load increases.

For each muscle, there is an optimal combination of tension and speed for any given resistance. If you have ever ridden a 10-speed bicycle, you are probably already aware of this fact. When you are cruising along comfortably, your thigh and leg muscles are working at an optimal combination of speed and tension. When you come to a hill, the resistance increases. Your muscles must now develop more tension, and they move more slowly; they are no longer working at optimal efficiency. You then shift to a lower gear. The load on your muscles decreases, the speed increases, and the muscles are once again working efficiently.

ISOMETRIC CONTRACTIONS In an **isometric contraction** (*metric*, measure), the muscle as a whole does not change length, and the tension produced never exceeds the resistance. Figure 10-14c• shows what happens if we attach a weight heavier than 4 kg to the experimental muscle and then stimulate the muscle. Although cross-bridges form and tension rises to peak values, the muscle cannot overcome the resistance of the weight, and so it cannot shorten (Figure 10-14d•). Examples of isometric contractions include holding a heavy weight above the ground, pushing against a locked door, or trying to pick up a car. These are rather unusual movements. However, many of the reflexive muscle contractions that keep your body upright when you stand or sit involve the isometric contractions of muscles that oppose the force of gravity.

You may have noticed that when you perform an isometric contraction, the contracting muscle bulges (although not as much as it does during an isotonic contraction). In an isometric contraction, although the muscle *as a whole* does not shorten, the individual muscle fibers shorten until the tendons are taut and the external tension equals the internal tension generated by the muscle fibers. The muscle fibers cannot shorten further, because the external tension does not exceed the resistance.

Normal daily activities therefore involve a combination of isotonic and isometric muscular contractions. As you sit and read this text, isometric contractions of postural muscles stabilize your vertebrae and maintain your upright position. When you next turn a page, the movements of your arm, forearm, hand, and fingers are produced by a combination of concentric and eccentric isotonic contractions.

Muscle Relaxation and the Return to Resting Length

As we noted earlier, there is no active mechanism for muscle fiber elongation. The sarcomeres in a muscle fiber can shorten and develop tension, but the power stroke cannot be reversed to push the Z lines farther apart. After a contraction, a muscle fiber returns to its original length through a combination of *elastic forces, opposing muscle contractions*, and *gravity*.

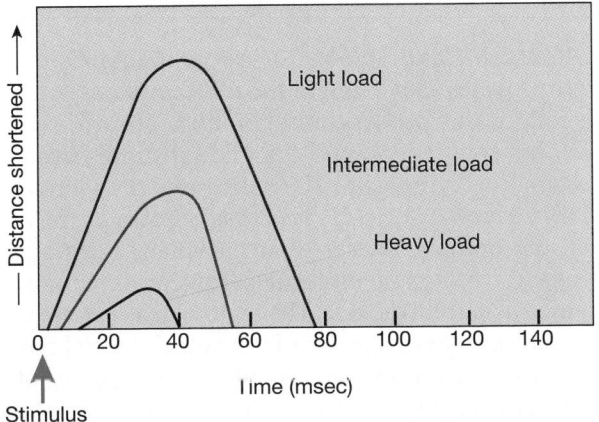

•*FIGURE 10-15* **Resistance and Speed of Contraction.** The heavier the resistance on a muscle, the longer it will take for the muscle to begin to shorten, and the less the muscle will shorten.

ELASTIC FORCES When the contraction ends, some of the energy initially "spent" in stretching the series elastic elements is recovered as they recoil. The recoil of the series elastic elements helps return the muscle fiber toward its original resting length.

OPPOSING MUSCLE CONTRACTIONS Elastic forces gradually stretch relaxed muscle fibers toward their normal resting lengths. More rapid returns to resting length result from the contraction of opposing muscles, sometimes aided by the pull of gravity. Consider the muscles of the arm that flex or extend the elbow. Contraction of the *biceps brachii* muscle on the anterior part of the arm flexes the elbow; contraction of the *triceps brachii* muscle on the posterior part of the arm extends the elbow. When the biceps brachii contracts, the triceps brachii is stretched. When the biceps brachii relaxes, contraction of the triceps brachii extends the elbow and stretches the muscle fibers of the biceps brachii to their original length.

GRAVITY Gravity may assist opposing muscle groups in returning a muscle to its resting length after a contraction. For example, imagine the biceps brachii fully contracted with the elbow pointed at the ground. When the muscle relaxes, gravity will pull the forearm down and stretch the muscle. Although gravity can provide assistance in stretching muscles, some active muscle tension is needed to control the rate of movement and prevent damage to the joint. In the previous example, eccentric contraction of the biceps brachii could control the movement.

☑ Why is it difficult to contract a muscle that has been overstretched?

☑ During treppe, why does tension in a muscle gradually increase even though the stimulus strength and frequency are constant?

☑ Is it possible for a skeletal muscle to contract without shortening? Explain.

Energetics of Muscular Activity

A single muscle fiber may contain 15 billion thick filaments. When actively contracting, each thick filament breaks down roughly 2500 ATP molecules per second. Because even a small skeletal muscle contains thousands of muscle fibers, the ATP demands of a contracting skeletal muscle are enormous. In practical terms, the demand for ATP in a contracting muscle fiber is so high that it would be impossible to have all the necessary energy available as ATP before the contraction begins. Instead, a resting muscle fiber contains only enough ATP and other high-energy compounds to sustain a contraction until additional ATP can be generated. Throughout

the rest of the contraction, the muscle fiber will generate ATP at roughly the same rate as it is used.

ATP and CP Reserves

The primary function of ATP is the transfer of energy from one location to another, rather than the long-term storage of energy. At rest, a skeletal muscle fiber produces more ATP than it needs. Under these conditions, ATP transfers energy to creatine. *Creatine* is a small molecule that muscle cells and neurons assemble from fragments of amino acids. The energy transfer creates another high-energy compound, **creatine phosphate (CP)**, or *phosphorylcreatine*:

$$ATP + creatine \rightarrow ADP + creatine\ phosphate$$

During a contraction, each myosin cross-bridge breaks down ATP, producing ADP and a phosphate group. The energy stored in creatine phosphate is then used to "recharge" ADP, converting it back to ATP through the reverse reaction:

$$ADP + creatine\ phosphate \rightarrow ATP + creatine$$

The enzyme that facilitates this reaction is called **creatine phosphokinase (CPK or CK)**. When muscle cells are damaged, CPK leaks across the cell membranes and into the circulation. Thus, a high blood concentration of CPK usually indicates serious muscle damage.

The energy reserves of a representative muscle fiber are indicated in Table 10-2. A resting skeletal muscle fiber contains about six times as much creatine phosphate as ATP. But when a muscle fiber is undergoing a sustained contraction, these energy reserves are exhausted in only about 17 seconds, and the muscle fiber must then rely on other mechanisms to convert ADP to ATP.

ATP Generation

We learned in Chapter 3 that most cells in the body generate ATP through *aerobic metabolism* in mitochondria and through *glycolysis* in the cytoplasm. ∞ *[p. 87]*

AEROBIC METABOLISM **Aerobic metabolism** normally provides 95 percent of the ATP demands of a resting cell. In this process, mitochondria absorb oxygen, ADP, phosphate ions, and organic substrates from the surrounding cytoplasm. The substrates then enter the *TCA (tricarboxylic acid) cycle* (also known as the *citric acid cycle* or the *Krebs cycle*), an enzymatic pathway that breaks down organic molecules. The carbon atoms are released as carbon dioxide. The hydrogen atoms are shuttled to respiratory enzymes of the mitochondrial cristae, where their electrons are removed. After a series of intermediate steps, the protons and electrons are combined with oxygen to form water. Along the way, large amounts of energy are released and used to make ATP. The entire process is very efficient; for each molecule "fed" to the TCA cycle, the cell will gain 17 ATP.

TABLE 10-2 **Sources of Energy Stored in a Typical Muscle Fiber**

Energy Stored as	Utilized through	Initial Quantity	Number of Twitches Supported by Each Energy Source Alone	Duration of Isometric Tetanic Contraction Supported by Each Energy Source Alone
ATP	ATP → ADP + P	3 mmol	10	2 sec
CP	ADP + CP → ATP + C	20 mmol	70	15 sec
Glycogen	Glycolysis (anaerobic)	100 mmol	670	130 sec
	Aerobic metabolism		12,000	2400 sec (40 min)

Resting skeletal muscle fibers rely almost exclusively on the aerobic metabolism of fatty acids to generate ATP. When the muscle starts contracting, the mitochondria begin breaking down molecules of pyruvic acid instead of fatty acids. The pyruvic acid is provided by the enzymatic pathway of glycolysis.

GLYCOLYSIS Glycolysis is the breakdown of glucose to pyruvic acid in the cytoplasm of a cell. It is called an **anaerobic** process, because it does not require oxygen. The reaction sequence of glycolysis provides a net gain of 2 ATP and generates 2 pyruvic acid molecules per glucose molecule. The ATP produced by glycolysis is thus only a small fraction of that produced by aerobic metabolism, in which the breakdown of the 2 pyruvic acid molecules in mitochondria would generate 34 ATP. *Yet, because it can proceed in the absence of oxygen, glycolysis can be very important when the availability of oxygen limits the rate of mitochondrial ATP production.*

In most skeletal muscles, glycolysis is the primary source of ATP during peak periods of activity. The glucose broken down under these conditions is obtained primarily by mobilizing the reserves of glycogen present in the sarcoplasm. Glycogen, diagrammed in Figure 2-12●, page 49, is a polysaccharide chain of glucose molecules. Typical skeletal muscle fibers contain large glycogen reserves that may account for 1.5 percent of the total muscle weight. When the muscle fiber begins to run short of ATP and CP, enzymes split the glycogen molecules apart, releasing glucose, which can be used to generate more ATP. As the level of muscular activity increases and these reserves are mobilized, the pattern of energy production and use changes.

Energy Use and the Level of Muscle Activity

In a resting skeletal muscle (Figure 10-16a●), the demand for ATP is low. More than enough oxygen is available for the mitochondria to meet that demand, and they produce a surplus of ATP. The extra ATP is used to build up reserves of CP and glycogen. Resting muscle fibers absorb fatty acids and glucose that are delivered by the bloodstream. The fatty acids are broken down in the mitochondria, and the ATP generated is used to convert creatine to creatine phosphate and glucose to glycogen.

At moderate levels of activity (Figure 10-16b●), the demand for ATP increases. This demand is met by the mitochondria. As the rate of mitochondrial ATP production rises, so does the rate of oxygen consumption. Oxygen availability is not a limiting factor, for oxygen can diffuse into the muscle fiber fast enough to meet mitochondrial needs. But all the ATP produced is needed by the muscle fiber, and no surplus is available. The skeletal muscle now relies primarily on the aerobic metabolism of pyruvic acid to generate ATP. The pyruvic acid is provided by glycolysis, which breaks down glucose molecules obtained from glycogen inside the muscle fiber. If glycogen reserves are low, the muscle fiber can also break down other substrates, such as lipids or amino acids. As long as the demand for ATP can be met by mitochondrial activity, the ATP provided by glycolysis makes a relatively minor contribution to the total energy budget of the muscle fiber.

At peak levels of activity, the ATP demands are enormous, and mitochondrial ATP production rises to a maximum. This maximum rate is determined by the availability of oxygen, and oxygen cannot diffuse into the muscle fiber fast enough to enable the mitochondria to produce the required ATP. At peak levels of exertion, mitochondrial activity can provide only about one-third of the ATP needed. The remainder is produced through glycolysis (Figure 10-16c●).

When glycolysis produces pyruvic acid faster than it can be utilized by the mitochondria, pyruvic acid levels rise in the sarcoplasm. Under these conditions, the pyruvic acid is converted to **lactic acid**, a related three-carbon molecule. The enzyme responsible is called **lactate dehydrogenase (LDH)**.

The anaerobic process of glycolysis enables the cell to generate additional ATP when the mitochondria are unable to meet the current energy demands. However, anaerobic energy production has its drawbacks:

- Lactic acid is an organic acid that in body fluids dissociates into a hydrogen ion and a negatively charged *lactate ion*. The production of lactic acid can therefore lower the intracellular pH. Buffers in the sarcoplasm can resist pH shifts, but these defenses are limited. Eventually, changes in pH will alter the functional characteristics of key enzymes. The muscle fiber will then become unable to continue contracting.

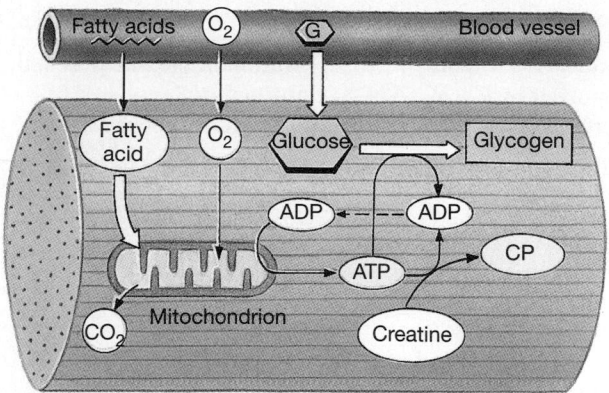

(a) Resting muscle: Fatty acids are catabolized; the ATP produced is used to build energy reserves of ATP, CP, and glycogen.

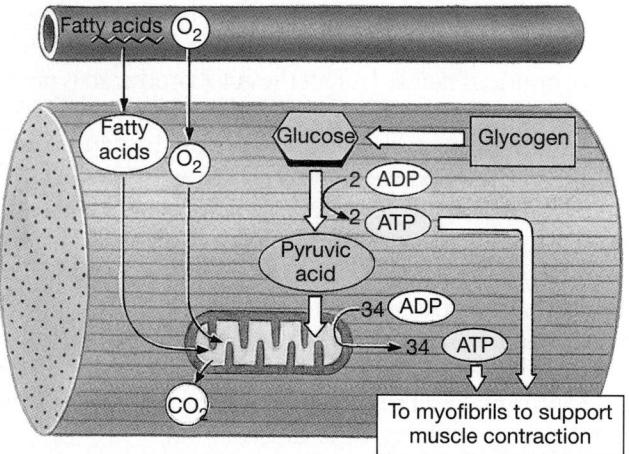

(b) Moderate activity: Glucose and fatty acids are catabolized; the ATP produced is used to power contraction.

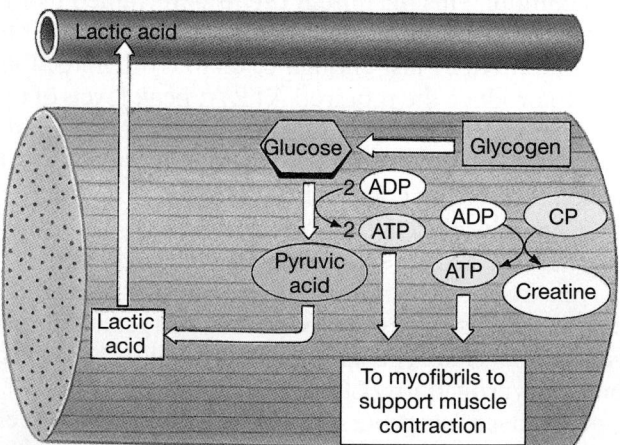

(c) Peak activity: Most ATP is produced through glycolysis, with lactic acid as a byproduct.

●*FIGURE 10-16* **Muscle Metabolism. (a)** A resting muscle breaks down fatty acids via aerobic respiration to make ATP. Surplus ATP is used to build reserves of creatine phosphate (CP) and glycogen. **(b)** At modest-rate activity levels, mitochondria can meet the ATP demands through aerobic metabolism of fatty acids and glucose. **(c)** At peak levels of activity, the mitochondria cannot get enough oxygen to meet ATP demands. Most of the ATP is provided by glycolysis, leading to the production of lactic acid.

- Glycolysis is a relatively inefficient way to generate ATP. Under anaerobic conditions, each glucose molecule generates 2 pyruvic acid molecules that are converted to lactic acid. In return, the cell gains 2 ATP through glycolysis. Had those 2 pyruvic acid molecules been catabolized aerobically in a mitochondrion, the cell would have gained *34 additional ATP.*

Muscle Fatigue

A skeletal muscle fiber is said to be **fatigued** when it can no longer contract despite continued neural stimulation. The cause of muscle fatigue varies with the level of muscle activity. After short peak levels of activity, such as running a 100-meter dash, fatigue may result from the exhaustion of ATP and CP reserves or from the drop in pH that accompanies the buildup of lactic acid. After prolonged exertion, such as running a marathon, fatigue may involve physical damage to the sarcoplasmic reticulum that interferes with the regulation of intracellular Ca^{2+} concentrations. Muscle fatigue is cumulative, and the effects become more pronounced as more muscle fibers are affected. The result is a gradual reduction in the capabilities of the entire skeletal muscle.

If the muscle fiber is contracting at moderate levels and ATP demands can be met through aerobic metabolism, fatigue will not occur until glycogen, lipid, and amino acid reserves are depleted. This type of fatigue affects the muscles of long-distance athletes, such as marathon runners, after hours of exertion.

When a muscle produces a sudden, intense burst of activity at peak levels, most of the ATP is provided by glycolysis. After a relatively short time (seconds to minutes), the rising lactic acid levels lower the tissue pH, and the muscle can no longer function normally. Athletes who run sprints, such as the 100-meter dash, suffer from this type of muscle fatigue. We will return to the topics of fatigue, athletic training, and metabolic activity later in the chapter.

Normal muscle function requires (1) substantial intracellular energy reserves, (2) a normal circulatory supply, and (3) a normal blood oxygen concentration. Anything that interferes with one or more of these factors will promote premature muscle fatigue. For example, reduced blood flow from tight clothing, a circulatory disorder, or loss of blood slows the delivery of oxygen and nutrients, accelerates the buildup of lactic acid, and promotes fatigue.

The Recovery Period

When a muscle fiber contracts, the conditions in the sarcoplasm are changed. Energy reserves are consumed, heat is released, and, if the contraction was at peak levels, lactic acid is generated. In the **recovery period**, conditions inside the muscle fibers are returned to normal, preexertion levels. It may take several hours for muscle fibers to recover from a period of moderate activity. After sustained activity at higher levels, complete recovery may take a week.

During the recovery period, the muscle fibers must rebuild their energy reserves. Even when contracting at moderate levels, active skeletal muscles gradually deplete their glycogen reserves. When muscles are contracting at maximal levels, the depletion occurs much more rapidly, because glycolysis is relatively inefficient. Under these conditions, the energy reserves will be restored through the recycling of lactic acid generated during the period of contraction. Rebuilding glycogen reserves is a synthetic activity that requires ATP. Oxygen is readily available during the recovery period, so the ATP is provided through aerobic metabolism in the mitochondria.

LACTIC ACID REMOVAL AND RECYCLING

Glycolysis enables a skeletal muscle to continue contracting even when mitochondrial activity is limited by the availability of oxygen. As we have seen, however, lactic acid production is not an ideal way to generate ATP. It is wasteful because it squanders the glucose reserves of the muscle fibers, and it is potentially dangerous because lactic acid can alter the pH of the blood and tissues.

During the recovery period, when oxygen is available in abundance, that lactic acid can be recycled by conversion back to pyruvic acid. The pyruvic acid can then be used either by mitochondria to generate ATP or as a substrate for enzyme pathways that synthesize glucose and rebuild glycogen reserves.

During a period of exertion, lactic acid diffuses out of the muscle fibers and into the bloodstream. This process continues after the exertion has ended, because intracellular lactic acid concentrations are still relatively high. The liver absorbs the lactic acid and converts it to pyruvic acid. Roughly 30 percent of these molecules is broken down in the TCA cycle, providing the ATP needed to convert the other pyruvic acid molecules to glucose. (We shall cover these processes more fully in Chapter 25.) The glucose molecules are then released into the circulation, where they are absorbed by skeletal muscle fibers and used to rebuild their glycogen reserves. This shuffling of lactate to the liver and glucose back to muscle cells is called the **Cori cycle**.

THE OXYGEN DEBT

During the recovery period, the body's oxygen demand goes up considerably. The recovery period is powered by the ATP generated through aerobic metabolism. The more ATP required, the more oxygen will be needed; on average, for each molecule of oxygen absorbed by a mitochondrion, the cell gains 4 ATP. The **oxygen debt** created during a period of exercise is the amount of oxygen needed to restore normal, preexertion conditions. The major tissues involved in the additional oxygen consumption are skeletal muscle fibers, which must restore ATP, creatine phosphate, and glycogen concentrations to their former

levels, and liver cells, which generate the ATP needed to convert excess lactic acid to glucose. Additional ATP is needed to restore normal conditions in other tissues. For example, sweat glands must increase their secretory activity to restore normal body temperature. While the oxygen debt is being repaid, the breathing rate and depth are increased. As a result, you continue to breathe heavily long after you stop exercising.

HEAT LOSS

Muscular activity generates substantial amounts of heat. When a catabolic reaction occurs, such as the breakdown of glycogen or the reactions of glycolysis, the muscle fiber captures only a portion of the released energy. ∞ *[p. 39]* The rest is released as heat. A resting muscle fiber relying on aerobic metabolism captures about 42 percent of the energy released in catabolism. The other 58 percent warms the sarcoplasm, interstitial fluid, and circulating blood. Active skeletal muscles release roughly 85 percent of the heat needed to maintain normal body temperature.

When muscles become active, their consumption of energy skyrockets. As anaerobic energy production becomes the primary method of ATP generation, muscle fibers become less efficient at capturing energy. At peak levels of exertion, only about 30 percent of the released energy is captured as ATP, and the remaining 70 percent warms the muscle and surrounding tissues. Body temperature soon begins to climb, and heat loss at the skin accelerates through mechanisms introduced in Chapters 1 and 5. ∞ *[pp. 13 14, 161]*

Even after the exercise period ends and **initial heat** production stops, energy use remains high as the oxygen debt is repaid. The **recovery heat** produced keeps us perspiring for some time thereafter.

Hormones and Muscle Metabolism

Metabolic activities in skeletal muscle fibers are adjusted by hormones of the endocrine system. *Growth hormone* from the pituitary gland and *testosterone* (the primary sex hormone in males) stimulate the synthesis of contractile proteins and the enlargement of skeletal muscles. *Thyroid hormones* elevate the rate of energy consumption by resting and active skeletal muscles. During a sudden crisis, hormones of the adrenal gland, notably *epinephrine* (adrenaline), stimulate muscle metabolism and increase both the duration of stimulation and the force of contraction. We shall further examine the effects of hormones on muscle and other tissues in Chapter 18.

Muscle Performance

Muscle performance can be considered in terms of sheer **power**, the maximum amount of tension produced by a particular muscle or muscle group, and **endurance**, the amount of time for which the individual can perform a particular activity. Two major factors determine the performance capabilities of a particular skeletal muscle: (1) the types of muscle fibers within the muscle and (2) physical conditioning or training.

Types of Skeletal Muscle Fibers

There are three major types of skeletal muscle fibers in the human body: *fast fibers, slow fibers*, and *intermediate fibers*.

FAST FIBERS   Most of the skeletal muscle fibers in the body are called **fast fibers** because they can contract in 0.01 second or less after stimulation. Fast fibers are large in diameter; they contain densely packed myofibrils, large glycogen reserves, and relatively few mitochondria. The tension produced by a muscle fiber is directly proportional to the number of sarcomeres, so muscles dominated by fast fibers produce powerful contractions. However, because these contractions use ATP in massive amounts, prolonged activity is supported primarily by anaerobic metabolism, and fast fibers fatigue rapidly.

SLOW FIBERS   **Slow fibers** are only about half the diameter of fast fibers, and they take three times as long to contract after stimulation. Slow fibers are specialized to enable them to continue contracting for extended periods, long after a fast muscle would have become fatigued. The most important specializations improve mitochondrial performance. Slow muscle tissue contains a more extensive network of capillaries than is typical of fast muscle tissue and so has a dramatically increased oxygen supply. In addition, slow muscle fibers contain the red pigment **myoglobin** (MĪ-ō-glō-bin). This globular protein (Figure 2-18●, p. 55) is structurally related to hemoglobin, the oxygen-carrying pigment found in the blood. Both myoglobin and hemoglobin are red pigments that reversibly bind oxygen molecules. Although small amounts of myoglobin are found in other muscle fiber types, it is most abundant in slow muscle fibers. As a result, resting slow muscle fibers contain substantial oxygen reserves that can be mobilized during a contraction. Because slow muscle fibers have (1) an extensive capillary supply and (2) a high concentration of myoglobin, skeletal muscles dominated by slow muscle fibers are dark red.

With oxygen reserves and a more efficient blood supply, the mitochondria of slow muscle fibers are able to contribute a greater amount of ATP while contractions are under way. Thus, slow muscle fibers are less dependent on anaerobic metabolism than are fast muscle fibers. Some of the mitochondrial energy production involves the breakdown of stored lipids rather than glycogen, so glycogen reserves are smaller than those of fast muscle cells. Slow muscles also contain more mitochondria than do fast muscle fibers. Figure 10-17● compares the appearance of fast and slow skeletal muscle fibers.

INTERMEDIATE FIBERS   **Intermediate fibers** have properties intermediate between those of fast fibers

and slow fibers. In appearance, intermediate fibers most closely resemble fast fibers, for they contain little myoglobin and are relatively pale. They have a more extensive capillary network around them, however, and are more resistant to fatigue than are fast fibers. The three types of muscle fibers are compared in Table 10-3. In muscles that contain a mixture of fast and intermediate muscle fibers, the proportion can change with physical conditioning. For example, if a muscle is used repeatedly for endurance events, some of the fast fibers will develop the appearance and functional capabilities of intermediate fibers. The muscle as a whole will thus become more resistant to fatigue.

Distribution of Muscle Fibers and Muscle Performance

The percentage of fast, intermediate, and slow muscle fibers in a particular skeletal muscle can be quite variable. Muscles dominated by fast fibers appear pale, and they are often called **white muscles**. Chicken breasts contain "white meat" because chickens use their wings only for brief intervals, as when fleeing from a predator, and the power for flight comes from fast fibers in their breast muscles. As we learned earlier, the extensive blood vessels and myoglobin in slow muscle fibers give these fibers a reddish color; muscles dominated by slow fibers are therefore known as **red muscles**. Chickens walk around all day, and the movements are performed by the slow muscle fibers in the "dark meat" of their legs.

Most human muscles contain a mixture of fiber types and so appear pink. However, there are no slow fibers in muscles of the eye and hand, where swift but brief contractions are required. Many back and calf muscles are dominated by slow fibers; these muscles contract almost continuously to maintain an upright posture. The percentage of fast versus slow fibers in each muscle is genetically determined. As we noted earlier, the proportion of intermediate fibers to fast fibers can increase as the result of athletic training.

Muscle Hypertrophy

As a result of repeated, exhaustive stimulation, muscle fibers develop more mitochondria, a higher concentration of glycolytic enzymes, and larger glycogen reserves. These muscle fibers have more myofibrils, and each myofibril contains more thick and thin filaments. The net effect is an enlargement, or **hypertrophy**, of the stimulated muscle. The number of muscle fibers does not change, but the muscle as a whole enlarges because each muscle fiber increases in diameter. Hypertrophy occurs in muscles that have been repeatedly stimulated to produce near-maximal tension. The intracellular changes that occur increase the amount of tension produced when these muscles contract. A champion weight lifter or bodybuilder is an excellent example of hypertrophied muscular development.

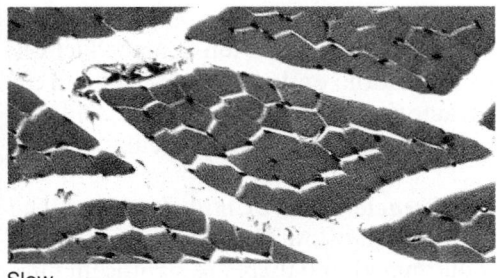

Slow

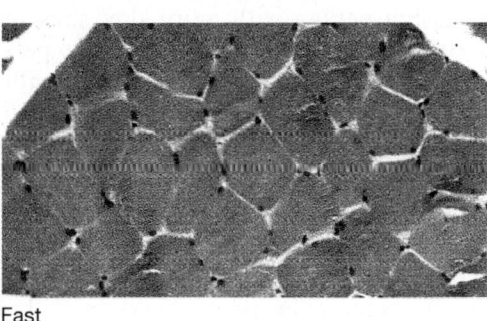

Fast

(a)

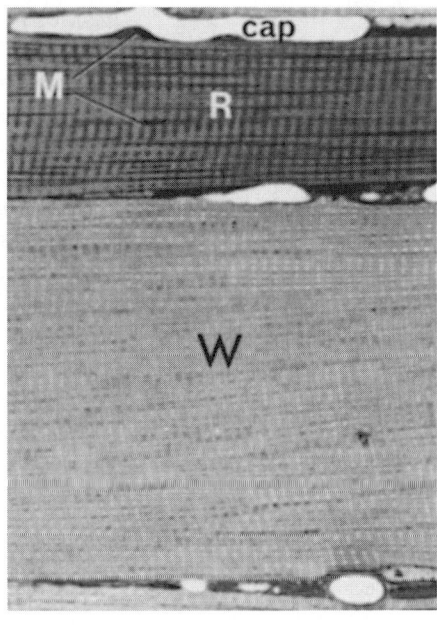

(b)

•**FIGURE 10-17** Fast and Slow Muscle Fibers. **(a)** Note the difference in the size of slow muscle fibers, above, and fast muscle fibers, below. (LM × 171) **(b)** The slender, slow muscle fiber (red, R) has more mitochondria (M) and a more extensive capillary supply (cap) than the fast muscle fiber (white, W). (LM × 783)

In general, the muscle mass of men exceeds that of women, because muscle fibers are stimulated by testosterone. In recent years, amateur and professional athletes have used this steroid hormone, or related drugs, to stimulate the development of hypertrophied muscles. The steroids appear to have the desired effect only when taken by individuals who are already engaged in an intensive weight-training program and are on a high-protein, high-calorie diet. It should be noted that this combination of training and diet would *by itself* produce muscular hypertrophy, even without the use of steroids. The benefits of steroids thus do not outweigh the known risks, which include liver failure and infertility. (We shall discuss the use of steroids and other hormones by athletes further in Chapter 18.)

Muscle Atrophy

A skeletal muscle that is not stimulated by a motor neuron on a regular basis loses muscle tone and mass. The muscle becomes flaccid, and the muscle fibers become smaller and weaker. This reduction in muscle size, tone, and power is called **atrophy**. Individuals paralyzed by spinal injuries or other damage to the nervous system will gradually lose muscle tone and size in the areas affected. Even a temporary reduction in muscle use can lead to muscular atrophy; you can easily observe this effect by comparing "before and after" limb muscles in someone who has worn a cast. Muscle atrophy is initially reversible, but dying muscle fibers are not replaced. In extreme atrophy, the functional losses are permanent. That is why physical therapy is crucial in cases in which people are temporarily unable to move normally.

Because skeletal muscles depend on their motor neurons for stimulation, disorders that affect the nervous system can indirectly affect the muscular system. In *polio*, a virus attacks motor neurons in the spinal cord and brain, producing muscular paralysis and atrophy. ⬛AM *Problems with the Control of Muscle Activity*

TABLE 10-3 Properties of Skeletal Muscle Fiber Types

Property	Slow	Intermediate	Fast
Cross-sectional diameter	Small	Intermediate	Large
Tension	Low	Intermediate	High
Contraction speed	Slow	Fast	Fast
Fatigue resistance	High	Intermediate	Low
Color	Red	White	White
Myoglobin content	High	Low	Low
Capillary supply	Dense	Intermediate	Scarce
Mitochondria	Many	Intermediate	Few
Glycolytic enzyme concentration in sarcoplasm	Low	High	High
Substrates used for ATP generation during contraction	Lipids, carbohydrates, amino acids (aerobic)	Primarily carbohydrates (anaerobic)	Carbohydrates (anaerobic)
Alternative names	Type I, S (slow), SO (slow oxidizing)	Type II, FR (fast resistant)	Type II, FF (fast fatigue)

Physical Conditioning

Physical conditioning and training schedules enable athletes to improve both power and endurance. In practice, the training schedule varies depending on whether the activity is supported primarily by aerobic or anaerobic energy production.

ANAEROBIC ENDURANCE **Anaerobic endurance** is the length of time muscular contraction can continue to be supported by glycolysis and by the existing energy reserves of ATP and CP. Anaerobic endurance is limited by (1) the amount of ATP and CP already at hand, (2) the amount of glycogen available for breakdown, and (3) the ability of the muscle to tolerate the lactic acid generated during the anaerobic period. Typically, the onset of muscle fatigue occurs within 2 minutes of the start of maximal activity.

Examples of activities that require anaerobic endurance include a 50-meter dash or swim, a pole vault, or a weight-lifting competition. Such activities are performed by fast muscle fibers. When these muscles begin contracting at maximal levels, the energy for the first 10–20 seconds comes from the ATP and CP reserves of the cytoplasm. As these reserves dwindle, glycogen breakdown and glycolysis provide additional energy. Athletes training to develop anaerobic endurance perform frequent, brief, intensive workouts that stimulate muscle hypertrophy.

AEROBIC ENDURANCE **Aerobic endurance** is the length of time that a muscle can continue to contract while supported by mitochondrial activities. Aerobic endurance is determined primarily by the availability of substrates for aerobic respiration, which the muscle fibers can obtain by breaking down carbohydrates, lipids, or amino acids. Initially, many of the nutrients catabolized by the muscle fiber are obtained from reserves within the muscle fibers themselves. Prolonged aerobic activity, however, must be supported by nutrients provided by the circulating blood. During exercise, blood vessels in the skeletal muscles dilate, and the increased blood flow brings oxygen and nutrients to the active muscle tissue. Warm-up periods are therefore important not only to take advantage of treppe, the increase in tension production noted on page 292, but also because they stimulate circulation in the muscles before the serious workout begins. Because glucose is a preferred energy source, aerobic athletes such as marathon runners, typically "load" or "bulk up," on carbohydrates for the last three days before an event. They may also consume glucose-rich "sports drinks" during a competition. We shall consider the risks and benefits of these practices in Chapter 25.

Training to improve aerobic endurance generally involves sustained low levels of muscular activity. Examples include jogging, distance swimming, and other exercises that do not require peak tension production. Improvements in aerobic endurance result from altering the characteristics of muscle fibers and improving the performance of the cardiovascular system:

1. ***Altering the characteristics of muscle fibers.*** The composition of fast and slow fibers in each muscle is genetically determined, and there are significant individual differences. These variations affect aerobic endurance, because a person with more slow muscle fibers in a particular muscle will be better able to perform under aerobic conditions than will a person with fewer. However, skeletal muscle cells respond to changes in the pattern of neural stimulation. Fast muscle fibers trained for aerobic competition develop the characteristics of intermediate fibers, and this change improves aerobic endurance.

2. ***Improving cardiovascular performance.*** Cardiovascular activity affects muscular performance by delivering oxygen and nutrients to active muscles. Physical training alters cardiovascular function by accelerating blood flow, thus improving oxygen and nutrient availability. We shall examine factors involved in improving cardiovascular performance in Chapter 21.

Aerobic activities do not promote hypertrophy. Many athletes train using a combination of aerobic and anaerobic exercises so that their muscles will enlarge and both anaerobic and aerobic endurance will improve. These athletes alternate an aerobic activity, such as swimming, with sprinting or weight lifting. The combination is known as *interval training*. Interval training is particularly useful for those engaged in racquet sports, such as tennis or squash. These sports are dominated by aerobic activities but are punctuated by brief periods of anaerobic effort.

POWER, ENDURANCE, AND ENERGY RESERVES Figure 10-18● compares the power/endurance curves for anaerobic and aerobic activities. The first half-minute of peak activity is totally supported by the mobilization of ATP and CP reserves (Figure 10-18a●). Thereafter, roughly two-thirds of the energy requirements of skeletal muscles operating at peak activity levels are met through glycolysis, with associated lactic acid generation.

At modest activity levels, a skeletal muscle can rely on aerobic respiration to provide ATP. At peak levels of activity, the muscle relies primarily on anaerobic metabolism. The level of activity at which the muscle must begin relying on anaerobic metabolism to meet its energy demands is called the **anaerobic threshold**. If energy demands are kept below the anaerobic threshold, muscular activity can be continued until nutrient sources are exhausted. In a trained athlete, muscle fatigue may not occur for several hours (Figure 10-18b●).

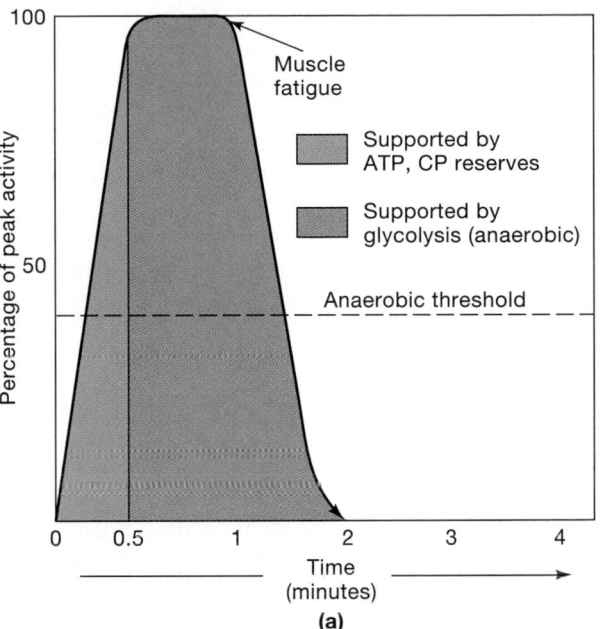

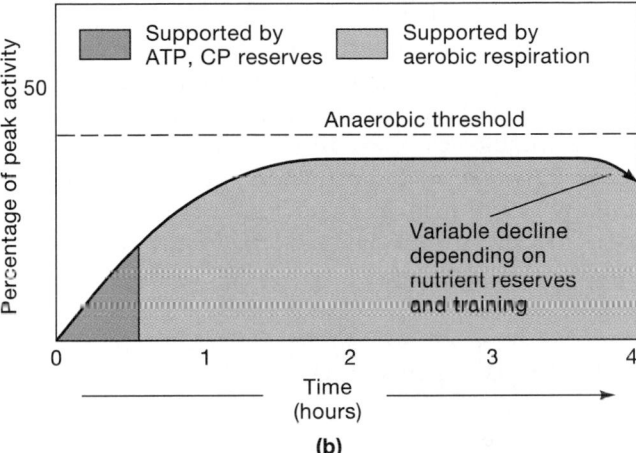

●*FIGURE 10-18* **Muscular Performance and Endurance.** **(a)** At peak levels of activity, skeletal muscles rely primarily on glycolysis for ATP production, with associated lactic acid production. Initial burst activity is supported by ATP and CP reserves. Muscles operating at peak levels fatigue rapidly. **(b)** Muscular activity can continue for extended periods when ATP demands are kept below the anaerobic threshold.

DELAYED-ONSET MUSCLE SORENESS You have probably experienced muscle soreness the day after a period of physical exertion. Considerable controversy exists over the source and significance of this pain, which is known as *delayed-onset muscle soreness* (DOMS). There are several interesting characteristics of DOMS:

- The soreness of DOMS is distinct from soreness you experience immediately after the exercise period ends. The short-term soreness is probably related to the biochemical events associated with muscle fatigue.
- Delayed-onset muscle soreness generally begins several hours after the exercise period ends and may last 3–4 days.
- The amount of soreness increases when the activity involves eccentric contractions. Activities dominated by concentric or isometric contractions produce less soreness.
- CPK and myoglobin levels are elevated in the blood, indicating damage to muscle cell membranes. The nature of the activity (eccentric versus concentric versus isometric) has no effect on these levels, nor can the levels be used to predict the degree of soreness experienced.

Three mechanisms have been proposed to explain DOMS:

1. There may be small tears in the muscle tissue, leaving muscle fibers with damaged membranes. The sarcolemma of each damaged muscle fiber permits the loss of enzymes, myoglobin, and other chemicals that may stimulate pain receptors in the region.
2. The pain may result from muscle spasms in the affected skeletal muscles. In some studies, stretching of the muscle involved can reduce the degree of soreness.
3. The pain may result from tears in the connective tissue framework and tendons of the skeletal muscle.

There is supporting evidence for each of those mechanisms, but none may tell the entire story. For example, muscle fiber damage is certainly supported by the biochemical findings, but if that were the only factor, the type of activity would correlate with the level of pain experienced.

⊠AGING AND THE MUSCULAR SYSTEM

As the body ages, there is a general reduction in the size and power of all muscle tissues. The effects on the muscular system can be summarized as follows:

- **Skeletal muscle fibers become smaller in diameter.** This reduction in size reflects primarily a decrease in the number of myofibrils. In addition, the muscle fibers contain smaller ATP, CP, and glycogen reserves and less myoglobin. The overall effect is a reduction in skeletal muscle size, strength, and endurance, combined with a tendency to fatigue rapidly. Because cardiovascular performance also decreases with age, blood flow to active muscles does not increase with exercise as rapidly as it does in younger people. These factors interact to produce decreases in anaerobic and aerobic performance of 30–50 percent by age 65.
- **Skeletal muscles become less elastic.** Aging skeletal muscles develop increasing amounts of fibrous connective tissue, a process called **fibrosis**. Fibrosis makes the muscle less flexible, and the collagen fibers can restrict movement and circulation.
  Tolerance for exercise decreases. A lower tolerance for exercise results in part from the tendency for rapid fatigue and in part from the reduction in thermoregulatory ability described in Chapter 5. ∞ *[p. 164]* Individuals over age 65 cannot eliminate the heat their muscles generate during contraction as effectively as younger people can and thus are subject to overheating.
- **Ability to recover from muscular injuries decreases.** The number of satellite cells steadily decreases with age, and the amount of fibrous tissue increases. As a

result, when an injury occurs, repair capabilities are limited, and scar tissue formation is the usual result.

The *rate* of decline in muscular performance is the same in all individuals, regardless of their exercise patterns or lifestyle. Therefore, to be in good shape late in life, you must be in *very* good shape early in life. Regular exercise helps control body weight, strengthens bones, and generally improves the quality of life at all ages. Extremely demanding exercise is not as important as regular exercise. In fact, extreme exercise in the elderly may lead to problems with tendons, bones, and joints. Although it has obvious effects on the quality of life, there is no clear evidence that exercise prolongs life expectancy.

INTEGRATION WITH OTHER SYSTEMS

To operate at maximum efficiency, the muscular system must be supported by many other systems. The changes that occur during exercise provide a good example of such interaction. As we noted in earlier sections, active muscles consume oxygen and generate carbon dioxide and heat. Responses of other systems include:

- *Cardiovascular system.* Dilation of blood vessels in the active muscles and the skin and an increase in the heart rate. These adjustments accelerate oxygen delivery and carbon dioxide removal at the muscle and bring heat to the skin for radiation into the environment.
- *Respiratory system.* Increased respiratory rate and depth of respiration. Air moves into and out of the lungs more quickly, keeping pace with the increased rate of blood flow through the lungs.
- *Integumentary system.* Dilation of blood vessels and increased sweat gland secretion. This combination helps promote evaporation at the skin surface and removes the excess heat generated by muscular activity.
- *Nervous and endocrine systems.* Direct the responses of other systems by controlling heart rate, respiratory rate, and sweat gland activity.

The muscular system has extensive interactions with other systems even at rest. Figure 10-19● summarizes the range of interactions between the muscular system and other vital systems.

☑ Why would a sprinter experience muscle fatigue before a marathon runner would?

☑ Which activity would be more likely to create an oxygen debt, swimming laps or lifting weights?

☑ What type of muscle fibers would you expect to predominate in the large leg muscles of someone who excels at endurance activities, such as cycling or long-distance running?

CARDIAC MUSCLE TISSUE

We introduced **cardiac muscle tissue** in Chapter 4 and briefly compared its properties with those of other muscle types. Cardiac muscle cells, also called **cardiocytes** or *cardiac myocytes*, are relatively small, averaging 10–20 µm in diameter and 50–100 µm in length. A typical cardiocyte (Figure 10-20●, p. 308) has a single, centrally placed nucleus, although a few may have two or more. As the name implies, cardiac muscle tissue is found only in the heart. Table 10-4 (p. 309) compares skeletal, cardiac, and smooth muscle tissues in greater detail.

Differences between Cardiac and Skeletal Muscle Tissues

As in skeletal muscle fibers, each cardiac muscle cell contains organized myofibrils, and the presence of many aligned sarcomeres gives it a striated appearance. Cardiac muscle cells are much smaller than skeletal muscle fibers, and there are significant structural and functional differences between the two.

Structural Differences

Important structural differences between skeletal muscle fibers and cardiac muscle cells include the following:

- The T tubules in a cardiac muscle cell are short and broad, and there are no triads. The T tubules encircle the sarcomeres at the Z lines rather than at the zone of overlap.
- The SR lacks terminal cisternae, and its tubules contact the cell membrane as well as the T tubules. As in skeletal muscle fibers, the appearance of an action potential triggers calcium release from the SR and the contraction of sarcomeres; it also increases the permeability of the sarcolemma to extracellular calcium ions.
- Cardiac muscle cells are almost totally dependent on aerobic metabolism to obtain the energy needed to continue contracting. The sarcoplasm of a cardiac muscle cell thus contains large numbers of mitochondria and abundant reserves of myoglobin (to store oxygen). Energy reserves are maintained in the form of glycogen and lipid inclusions.
- Each cardiac muscle cell contacts several others at specialized sites known as **intercalated** (in-TER-ka-lā-ted) **discs.** ∞ *[p. 135]* Because intercalated discs play a vital role in the function of cardiac muscle, we will discuss them in greater detail next.

INTERCALATED DISCS At an intercalated disc (Figure 10-20a,b●), the cell membranes of two cardiac muscle cells are extensively intertwined and bound together by gap junctions and desmosomes. These connections help stabilize the relative positions of adjacent cells and maintain the three-dimensional structure of the tissue. The gap junctions allow ions and small molecules to move from one cell to another. This arrangement creates a di-

INTEGUMENTARY SYSTEM

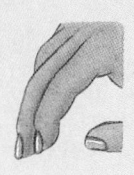

Removes excess body heat; synthesizes vitamin D$_3$ for Ca^{2+} and PO$_4^{3-}$ absorption; protects underlying muscles

Skeletal muscles pulling on skin of face produce facial expressions

SKELETAL SYSTEM

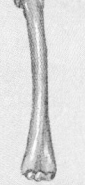

Maintains normal calcium and phosphate levels in body fluids; supports skeletal muscles; provides sites of attachment

Provides movement and support; stresses exerted by tendons maintain bone mass; stabilizes bones and joints

NERVOUS SYSTEM

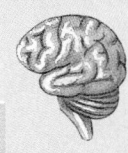

Controls skeletal muscle contractions; adjusts activities of respiratory and cardiovascular systems during periods of muscular activity

Muscle spindles monitor body position; facial muscles express emotions; intrinsic laryngeal muscles permit speech

ENDOCRINE SYSTEM

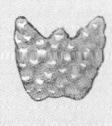

Hormones adjust muscle metabolism and growth; parathyroid hormone and calcitonin regulate calcium and phosphate ion concentrations

Skeletal muscles provide protection for some endocrine organs

CARDIOVASCULAR SYSTEM

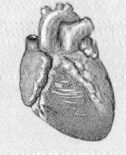

Delivers oxygen and nutrients, removes carbon dioxide, lactic acid, and heat

Skeletal muscle contractions assist in moving blood through veins; protects deep blood vessels

LYMPHATIC SYSTEM

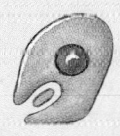

Defends skeletal muscles against infection and assists in tissue repairs after injury

Protects superficial lymph nodes and the lymphatic vessels in the abdominopelvic cavity

RESPIRATORY SYSTEM

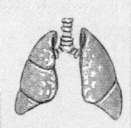

Provides oxygen and eliminates carbon dioxide

Muscles generate CO$_2$; control entrances to respiratory tract, fill and empty lungs, control airflow through larynx, and produce sounds

DIGESTIVE SYSTEM

Provides nutrients; liver regulates blood glucose and fatty acid levels and removes lactic acid from circulation

Protects and supports soft tissues in abdominal cavity; controls entrances to and exits from digestive tract

URINARY SYSTEM

Removes waste products of protein metabolism; assists in regulation of calcium and phosphate concentrations

External sphincter controls urination by constricting urethra

---

THE MUSCULAR SYSTEM

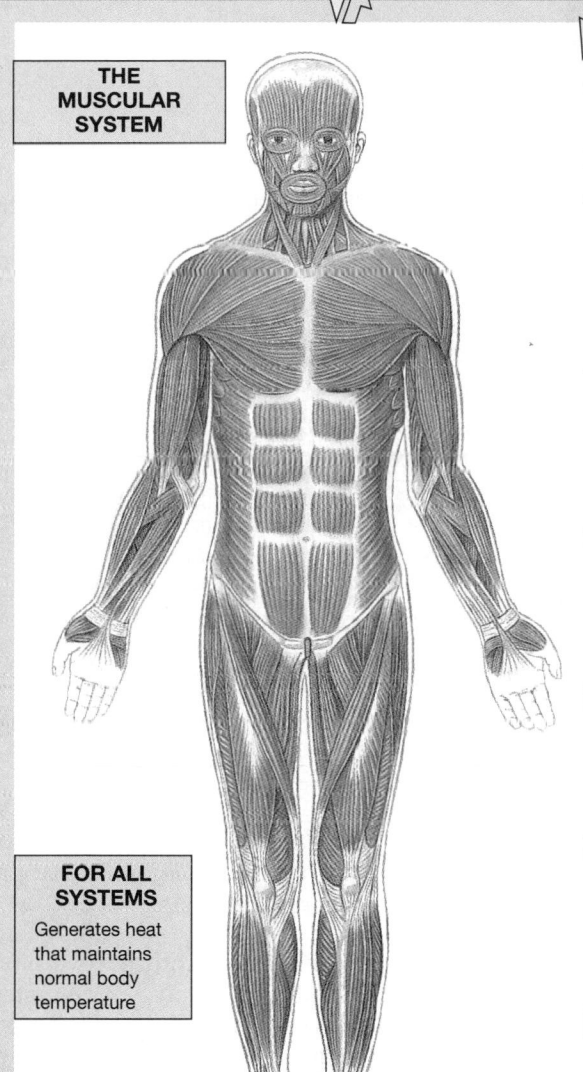

FOR ALL SYSTEMS

Generates heat that maintains normal body temperature

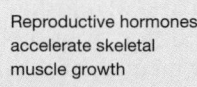

Reproductive hormones accelerate skeletal muscle growth

Contractions of skeletal muscles eject semen from male reproductive tract; muscle contractions during sex act produce pleasurable sensations

REPRODUCTIVE SYSTEM

● **FIGURE 10-19** **Functional Relationships between the Muscular System and Other Systems**

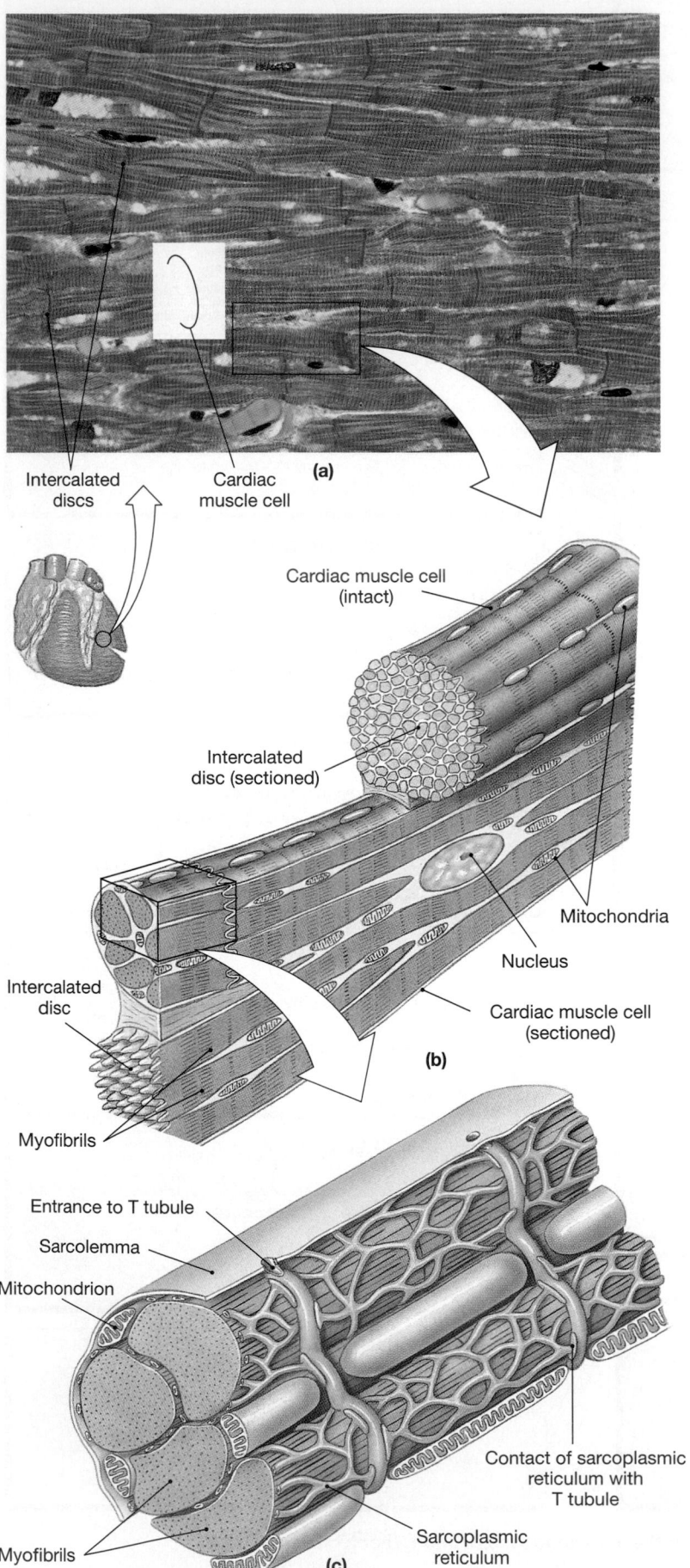

•FIGURE 10-20 Cardiac Muscle Tissue.
(a) Cardiac muscle tissue as seen with the light microscope. Note the striations and the intercalated discs. **(b,c)** Structure of a cardiac muscle cell; compare with Figure 10-2.

rect electrical connection between the two muscle cells. An action potential can travel across an intercalated disc, moving quickly from one cardiac muscle cell to another.

Myofibrils in the two interlocking muscle cells are firmly anchored to the membrane at the intercalated disc. Because their myofibrils are essentially locked together, the two muscle cells can "pull together" with maximum efficiency. Because the cardiac muscle cells are mechanically, chemically, and electrically connected to one another, the entire tissue resembles a single, enormous muscle cell. For this reason, cardiac muscle has been called a *functional syncytium* (sin-SISH-ē-um; a fused mass of cells).

Functional Differences

In Chapter 20, we will examine cardiac muscle physiology in detail; here we will briefly summarize four major functional specialties of cardiac muscle:

1. Cardiac muscle tissue contracts without neural stimulation. This property is called **automaticity**. The timing of contractions is normally determined by specialized cardiac muscle cells called **pacemaker cells**.

2. Innervation by the nervous system can alter the pace established by the pacemaker cells and adjust the amount of tension produced during a contraction.

3. Cardiac muscle cell contractions last roughly 10 times longer than do those of skeletal muscle fibers.

4. The properties of cardiac muscle cell membranes differ from those of skeletal muscle fiber membranes. As a result, individual cardiocyte twitches cannot undergo wave summation, and cardiac muscle tissue cannot produce tetanic contractions. This difference is important because a heart in a sustained tetanic contraction could not pump blood.

TABLE 10-4 A Comparison of Skeletal, Cardiac, and Smooth Muscle Tissues

Property	Skeletal Muscle Fiber	Cardiac Muscle Cell	Smooth Muscle Cell
Fiber dimensions (diameter × length)	100 μm × up to 30 cm	10–20 μm × 50–100 μm	5–10 μm × 30–200 μm
Nuclei	Multiple, near sarcolemma	Generally single, centrally located	Single, centrally located
Filament organization	In sarcomeres along myofibrils	In sarcomeres along myofibrils	Scattered throughout sarcoplasm
SR	Terminal cisternae in triads at zones of overlap	SR tubules contact T tubules at Z lines	Dispersed throughout sarcoplasm, no T tubules
Control mechanism	Neural, at single neuromuscular junction	Automaticity (pacemaker cells)	Automaticity (pacesetter cells), neural or hormonal control
Ca²⁺ source	Release from SR	From extracellular fluid and release from SR	Extracellular fluid
Contraction	Rapid onset, may be tetanized, rapid fatigue	Slower onset, cannot be tetanized, resistant to fatigue	Slow onset, may be tetanized, resistant to fatigue
Energy source	Aerobic metabolism at moderate levels of activity; glycolysis (anaerobic) during peak activity	Aerobic metabolism, usually lipid or carbohydrate substrates	Primarily aerobic metabolism

SMOOTH MUSCLE TISSUE

Smooth muscle cells range from 5 to 10 μm in diameter and from 30 to 200 μm in length. Each cell is spindle-shaped and has a single nucleus, centrally located. Figure 10-21a● shows typical smooth muscle fibers as seen by light microscopy, and Figure 10-21b● shows diagrammatic views of relaxed and contracted smooth muscle cells.

Smooth muscle tissue occurs within almost every organ, forming sheets, bundles, or sheaths around other tissues. In the skeletal, muscular, nervous, and endocrine systems, smooth muscles around blood ves-sels regulate blood flow through vital organs. In the digestive and urinary systems, rings of smooth muscle, called *sphincters*, regulate movement along internal passageways. Smooth muscles in bundles, layers, or sheets play a variety of other roles:

- ***Integumentary system.*** Smooth muscles around blood vessels regulate the flow of blood to the superficial dermis, smooth muscles of the arrector pili elevate the hairs. ∞ *[pp. 161, 157]*

- ***Cardiovascular system.*** Smooth muscles encircling vessels of the circulatory system provide control over the distribution of blood and help regulate blood pressure.

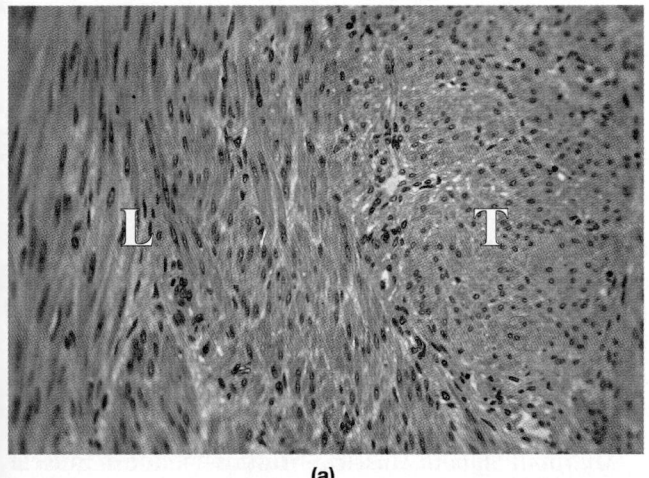

(a)

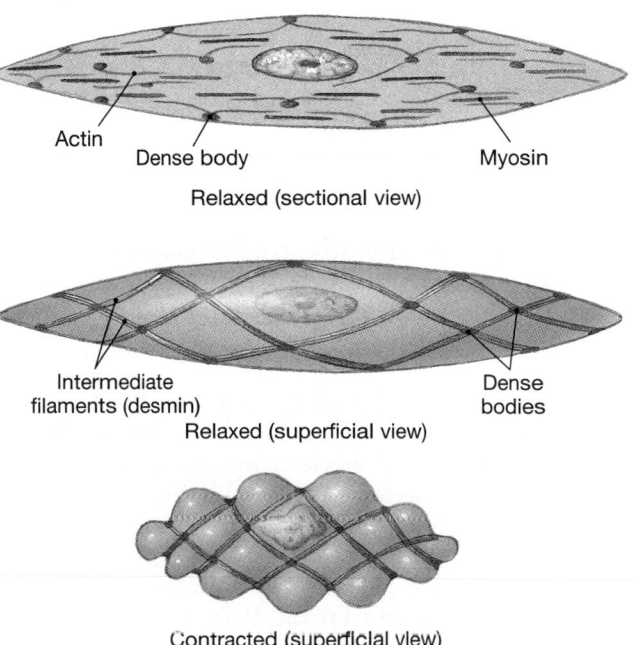

Actin   Dense body   Myosin

Relaxed (sectional view)

Intermediate filaments (desmin)   Dense bodies

Relaxed (superficial view)

Contracted (superficial view)

(b)

●**FIGURE 10-21** Smooth Muscle Tissue. **(a)** Many visceral organs contain several layers of smooth muscle tissue oriented in different ways. As a result, a single sectional view shows a smooth muscle cell in longitudinal (L) and transverse (T) sections. **(b)** A single relaxed smooth muscle cell is spindle-shaped and has no striations. Note the changes in cell shape as contraction occurs.

- *Respiratory system.* Smooth muscle contraction or relaxation alters the diameters of the respiratory passageways and changes the resistance to airflow.
- *Digestive system.* Extensive layers of smooth muscle in the walls of the digestive tract play an essential role in moving materials along the tract. Smooth muscle in the walls of the gallbladder contract to eject bile into the digestive tract.
- *Urinary system.* Smooth muscle tissue in the walls of small blood vessels alters the rate of filtration at the kidneys. Layers of smooth muscle in the walls of the ureters transport urine to the urinary bladder; contraction of the smooth muscle in the wall of the urinary bladder forces urine out of the body.
- *Reproductive system.* Layers of smooth muscle are important in the male for the movement of sperm along the reproductive tract and for the ejection of glandular secretions from the accessory glands into the reproductive tract. In the female, layers of smooth muscle are important in the movement of oocytes (and perhaps sperm) along the reproductive tract, and contraction of the smooth muscle in the walls of the uterus expels the fetus at delivery.

Differences between Smooth Muscle Tissue and Other Muscle Tissues

The structure and organization of smooth muscle tissue differs from that of skeletal or cardiac muscle tissue (Table 10-4).

Structural Differences

Actin and myosin are present in all three muscle types. In skeletal and cardiac muscle cells, these proteins are organized in sarcomeres, with thin and thick filaments. The internal organization of a smooth muscle cell is very different:

- There are no T tubules in a smooth muscle fiber, and the sarcoplasmic reticulum forms a loose network throughout the sarcoplasm. There are no myofibrils or sarcomeres in smooth muscle tissue. As a result, this tissue also has no striations, and it is therefore called **non-striated** muscle.
- Thick filaments are scattered throughout the sarcoplasm of a smooth muscle cell. The myosin proteins are organized differently than in skeletal or cardiac muscle cells, and there are more cross-bridges per thick filament.
- The thin filaments in a smooth muscle cell are attached to **dense bodies**, which are distributed throughout the sarcoplasm in a network of intermediate filaments composed of the protein desmin (Figure 10-21b•). Some of the dense bodies are firmly attached to the sarcolemma. The dense bodies and intermediate filaments anchor the thin filaments such that, when sliding occurs between thin and thick filaments, the cell shortens. Dense bodies are not arranged in straight

lines, so when a contraction occurs, the muscle cell twists like a corkscrew.
- Adjacent smooth muscle cells are bound together at dense bodies, transmitting the contractile forces from cell to cell throughout the tissue.
- Although smooth muscle cells are surrounded by connective tissue, the collagen fibers never unite to form tendons or aponeuroses as they do in skeletal muscles.

Functional Differences

Smooth muscle tissue differs from other muscle types in terms of (1) excitation–contraction coupling, (2) length–tension relationships, and (3) control mechanisms.

EXCITATION–CONTRACTION COUPLING The trigger for smooth muscle contraction is the appearance of free calcium ions in the cytoplasm. Most of these calcium ions enter the cell from the extracellular fluid. Once in the sarcoplasm, the calcium ions interact with *calmodulin*, a calcium-binding protein. Calmodulin then activates the enzyme **myosin light chain kinase**, which breaks down ATP and initiates the contraction. This situation is quite different from the one in skeletal and cardiac muscles, in which the trigger for contraction is the binding of calcium ions to troponin.

LENGTH–TENSION RELATIONSHIPS Because the thick and thin filaments are not rigidly organized, there is no direct relationship between tension development and resting length in smooth muscle. A stretched smooth muscle soon adapts to its new length and retains the ability to contract on demand. This ability to function over a wide range of lengths is called **plasticity**. Smooth muscle can contract over a range of lengths four times greater than that of skeletal muscle. This ability is especially important for digestive organs that undergo great changes in volume, such as the stomach. Despite the lack of sarcomere organization, smooth muscle contractions can be just as powerful as those of skeletal muscles. Like skeletal muscle fibers, smooth muscle cells often undergo sustained tetanic contractions.

CONTROL OF CONTRACTIONS Many smooth muscle fibers are not innervated by motor neurons, and the neurons that do innervate smooth muscles are not under voluntary control. The nature of the connection with the nervous system provides a means of categorizing smooth muscle cells as *multiunit* or *visceral*.

Multiunit Smooth Muscle **Multiunit smooth muscle cells** are innervated in motor units comparable to those of skeletal muscles, but each smooth muscle cell may be connected to several motor neurons rather than to just one. As in skeletal and cardiac muscles, an action potential is generated and propagated over the sarcolemma, but the contractions are more leisurely.

Multiunit smooth muscle cells are not typical of the digestive tract. They are found in the iris of the eye, where they regulate the diameter of the pupil; along portions of the male reproductive tract; within the walls of large arteries; and in the arrector pili muscles of the skin.

Visceral Smooth Muscle   In contrast, many **visceral smooth muscle cells** lack a direct contact with any motor neuron. Visceral smooth muscle cells are found in the walls of the digestive tract, the gallbladder, the urinary bladder, and many other internal organs.

Visceral smooth muscle cells are arranged in sheets or layers. Within each layer, adjacent muscle fibers are connected by gap junctions. Because they are connected in this way, whenever one muscle cell contracts, the electrical impulse that triggered the contraction can travel to adjacent smooth muscle cells. The contraction therefore spreads in a wave that soon involves every smooth muscle cell in the layer.

The initial stimulus may be the activation of a motor neuron that contacts one of the muscle cells in the region. But smooth muscle cells will contract or relax in response to chemicals, hormones, local concentrations of oxygen or carbon dioxide, or physical factors such as extreme stretching or irritation.

Many visceral smooth muscle networks show rhythmic cycles of activity in the absence of neural stimulation. These cycles are especially characteristic of the smooth muscle cells in the wall of the digestive tract, where **pacesetter cells** undergo spontaneous depolarization and trigger contraction of entire muscular sheets.

SMOOTH MUSCLE TONE   Both multiunit and visceral smooth muscle tissues show a normal background level of activity, or smooth muscle tone. The regulatory mechanisms detailed above stimulate contraction and increase muscle tone. Neural, hormonal, or chemical factors can also stimulate smooth muscle relaxation, producing a decrease in muscle tone. For example, smooth muscle cells at the entrances to capillaries regulate the amount of blood flow into each vessel. If the tissue becomes oxygen-starved, the smooth muscle cells relax, and the blood flow increases. This increase delivers additional oxygen. As conditions return to normal, the smooth muscle regains its normal muscle tone.

☑ What feature of cardiac muscle tissue allows the heart to act as a functional syncytium?

☑ Why are cardiac and smooth muscle contractions more affected by changes in extracellular Ca^{2+} than are skeletal muscle contractions?

☑ Smooth muscle can contract over a wider range of resting lengths than skeletal muscle can. Why?

SELECTED CLINICAL TERMINOLOGY

Terms Discussed in This Chapter

botulism: A severe, potentially fatal paralysis of skeletal muscles, resulting from consumption of a bacterial toxin. *(p. 287 and AM)*

Duchenne's muscular dystrophy (DMD): One of the most common and best understood of the muscular dystrophies. *(p. 283 and AM)*

fibrosis: A process in which increasing amounts of fibrous connective tissue develop, making muscles less flexible. *(p. 305)*

muscular dystrophies (DIS-trō-fēz): A varied collection of congenital diseases that produce progressive muscle weakness and deterioration. *(p. 283 and AM)*

myasthenia gravis (mī-as-THĒ-ne-uh GRA-vis): A general muscular weakness resulting from a reduction in the number of ACh receptors on the motor end plate. *(p. 287 and AM)*

polio: A disease resulting from viral destruction of motor neurons and characterized by the paralysis and atrophy of motor units. *(p. 303 and AM)*

rigor mortis: A state following death during which muscles are locked in the contracted position, making the body extremely stiff. *(p. 289)*

tetanus: A disease caused by a bacterial toxin that causes sustained, powerful contractions of skeletal muscles. *(p. 294)*

AM   ## Additional Terms Discussed in the *Applications Manual*

cholinesterase inhibitor: A compound that "ties up" the active sites at which acetylcholinesterase normally binds ACh.

chronic fatigue syndrome: A condition resembling fibromyalgia that may develop suddenly soon after a viral infection.

fibromyalgia syndrome: An inflammatory disorder characterized by a distinctive pattern of symptoms, including tender points on the body surface.

trichinosis (trik-i-NŌ-sis): A condition resulting from a parasitic infection by a nematode worm.

CHAPTER REVIEW

On-line resources for this chapter are on our World Wide Web site at:
http://www.prenhall.com/martini/fap

STUDY OUTLINE

INTRODUCTION, p. 277

1. Muscle cells are specialized to contract by pulling on collagen fibers. In muscle tissue, muscle cells exert a pull, or **tension**, on those fibers.

SKELETAL MUSCLE TISSUE AND THE MUSCULAR SYSTEM, p. 277

1. There are three types of muscle tissue: skeletal muscle, cardiac muscle, and smooth muscle.
2. **Skeletal muscles** attach to bones directly or indirectly. Their functions are (1) producing skeletal movement, (2) maintaining posture and body position, (3) supporting soft tissues, (4) guarding entrances and exits, and (5) maintaining body temperature.

ANATOMY OF SKELETAL MUSCLE, p. 277

Connective Tissue Organization, p. 278

1. Each muscle fiber is surrounded by an **endomysium**. Bundles of muscle fibers are sheathed by a **perimysium**, and the entire muscle is covered by an **epimysium**. At the ends of the muscle are **tendons** or **aponeuroses** that attach the muscle to other structures. *(Figure 10-1)*

Blood Vessels and Nerves, p. 279

2. The epimysium and perimysium contain the blood vessels and nerves that supply the muscle fibers.

Microanatomy of Skeletal Muscle Fibers, p. 279

3. A skeletal muscle fiber has a cell membrane, or **sarcolemma, sarcoplasm** (cytoplasm), and **sarcoplasmic reticulum (SR)**, similar to the smooth endoplasmic reticulum of other cells. **Transverse (T) tubules** and **myofibrils** aid in contraction. Filaments in a myofibril are organized into repeating functional units called **sarcomeres**. *(Figures 10-2–10-4)*
4. **Myofilaments** form myofibrils and consist of **thin filaments** and **thick filaments**. *(Figures 10-2–10-4)*
5. Thin filaments consist of **F actin, tropomyosin**, and **troponin**. Tropomyosin molecules cover **active sites** on the **G actin** subunits that form the F actin strand. Troponin binds to G actin and tropomyosin and holds the tropomyosin in position. *(Figure 10-5)*
6. Thick filaments consist of a bundle of myosin molecules, each consisting of an elongate **tail** and a globular **head**, or **cross-bridge**. In a resting muscle cell, the interactions between the active sites on G actin and myosin cross-bridges are prevented by tropomyosin. *(Figures 10-3, 10-5)*
7. The **sliding filament theory** explains how the relationship between thick and thin filaments changes as the muscle contracts and shortens. *(Figure 10-6)*

CONTRACTION OF SKELETAL MUSCLE, p. 285

The Control of Skeletal Muscle Activity, p. 285

1. Neural control of muscle function involves a link between electrical activity in the sarcolemma and the initiation of a contraction.
2. The activity of a muscle fiber is controlled by a neuron at a **neuromuscular** *(myoneural)* **junction (NMJ)**. *(Figure 10-7)*

3. When an **action potential** arrives at the **synaptic terminal, acetylcholine (ACh)** is released into the **synaptic cleft**. ACh binding to receptors on the opposing *junctional folds* leads to the generation of an action potential in the sarcolemma. *(Figure 10-7)*
4. **Excitation–contraction coupling** occurs as the passage of an action potential along a T tubule triggers the release of Ca^{2+} from the cisternae of the SR.

FOCUS: The Contraction Cycle, p. 288

5. A contraction involves a cycle of attachment, pivoting, detachment, and return. It begins when the SR releases calcium ions. The calcium ions bind to troponin, which changes position and moves tropomyosin away from the active sites of actin. Cross-bridges of myosin heads then bind to actin. Next, each myosin head pivots at its base, pulling the actin filament toward the center of the sarcomere. *(Figure 10-8)*

Relaxation, p. 287

6. **Acetylcholinesterase (AChE)** breaks down ACh and limits the duration of stimulation. *(Table 10-1)*
7. The contraction continues until (1) Ca^{2+} concentrations return to resting levels or (2) the muscle fiber runs out of ATP.

Length–Tension Relationships, p. 287

8. The number of cross-bridge interactions in a muscle fiber determines the amount of tension produced during a contraction. The fibers can contract most forcefully when stimulated over a narrow range of resting lengths. *(Figure 10-9)*

MUSCLE MECHANICS, p. 291

Tension Production, p. 291

1. A **twitch** is a cycle of contraction and relaxation produced by a single stimulus. *(Figure 10-10)*
2. Repeated stimulation before the relaxation phase ends may produce **summation of twitches (wave summation)**, in which one twitch is added to another; **incomplete tetanus**, in which tension peaks because the muscle is never allowed to relax completely; or **complete tetanus**, in which the relaxation phase is eliminated. *(Figure 10-11)*
3. Repeated stimulation after the relaxation phase has been completed produces **treppe**. *(Figure 10-11)*
4. *Internal tension* is generated inside contracting skeletal muscle fibers; *external tension* is generated in the extracellular fibers. *(Figure 10-12)*
5. The number and size of a muscle's **motor units** indicate how precisely controlled its movements are. *(Figure 10-13)*
6. Resting **muscle tone** stabilizes bones and joints.
7. Normal activities generally include both **isometric contractions** (in which tension rises but the length of the muscle remains constant) and **isotonic contractions** (in which the tension in a muscle remains constant but the length of the muscle changes). *(Figure 10-14)*
8. The return to resting length after a contraction may involve *series elastic elements*, the contraction of opposing muscle groups, and gravity. *(Figure 10-15)*

Energetics of Muscular Activity, p. 298

9. Muscle contractions require large amounts of energy. A resting muscle generates ATP through **aerobic metabolism** in mitochondria. *(Table 10-2)*

10. **Creatine phosphate (CP)** can release stored energy to convert ADP to ATP.

11. When a muscle fiber runs short of ATP and CP, enzymes can break down glycogen molecules to release glucose, which can in turn be broken down via glycolysis.

12. At rest or at moderate levels of activity, aerobic metabolism can provide most of the necessary ATP to support muscle contractions. *(Figure 10-16)*

13. At peak levels of activity, the cell relies heavily on the **anaerobic** process of **glycolysis** to generate ATP, because the mitochondria cannot obtain enough oxygen to meet the existing ATP demands. *(Figure 10-16)*

14. A **fatigued** muscle can no longer contract, because of changes in pH due to buildup of **lactic acid**, a lack of energy, or other problems.

15. The **recovery period** begins immediately after a period of muscle activity and continues until conditions inside the muscle have returned to preexertion levels. The **oxygen debt** created during exercise is the amount of oxygen used in the recovery period to restore normal conditions.

16. Circulating hormones may alter metabolic activities in skeletal muscle fibers.

Muscle Performance, p. 301

17. The three types of skeletal muscle fibers are **fast fibers, slow fibers**, and **intermediate fibers**. *(Table 10-3)*

18. Fast fibers, which are large in diameter, contain densely packed myofibrils, large glycogen reserves, and relatively few mitochondria. They produce rapid and powerful contractions of relatively brief duration. *(Figure 10-17)*

19. Slow fibers are about half the diameter of fast fibers and take three times as long to contract after stimulation. Specializations such as abundant mitochondria, an extensive capillary supply, and high concentrations of **myoglobin** enable them to continue contracting for extended periods. *(Figure 10-17)*

20. Intermediate fibers are very similar to fast fibers but have a greater resistance to fatigue.

21. **Anaerobic endurance** is the time over which a muscle can support sustained, powerful muscle contractions through anaerobic mechanisms. Training to develop anaerobic endurance can lead to **hypertrophy** (enlargement) of the stimulated muscles.

22. **Aerobic endurance** is the time over which a muscle can continue to contract while supported by mitochondrial activities. For maximum endurance, energy demands should remain at or below the **anaerobic threshold**. *(Figure 10-18)*

AGING AND THE MUSCULAR SYSTEM, p. 305

1. The aging process reduces the size, elasticity, and power of all muscle tissues. Both exercise tolerance and the ability to recover from muscular injuries decrease.

INTEGRATION WITH OTHER SYSTEMS, p. 306

1. To operate at maximum efficiency, the muscular system must be supported by many other systems. *(Figure 10-19)*

CARDIAC MUSCLE TISSUE, p. 306

Differences between Cardiac and Skeletal Muscle Tissues, p. 306

1. Cardiac muscle cells and skeletal muscle fibers differ structurally in size, the number and location of nuclei, the size and location of the T tubules at the sarcomeres, their dependence on aerobic metabolism when contracting at peak levels, the nature of the metabolic reserves, and the presence or absence of **intercalated discs**. *(Figure 10-20; Table 10-4)*

2. Cardiac muscle cells and skeletal muscle fibers differ functionally in terms of (1) the source of the stimulus for contraction, (2) the duration of contractions, (3) the ability to undergo tetanic contractions, and (4) the relationship with the autonomic nervous system.

SMOOTH MUSCLE TISSUE, p. 309

Differences Between Smooth Muscle Tissue and Other Muscle Tissues, p. 310

1. Smooth muscle is nonstriated, involuntary muscle tissue that exhibits **plasticity**.

2. Smooth muscle lacks sarcomeres and the resulting striations. The thin filaments are anchored to **dense bodies**. *(Figure 10-21; Table 10-4)*

3. **Visceral smooth muscle cells** are not always innervated by motor neurons. **Multiunit smooth muscle cells** are innervated by more than one motor neuron. Neurons that innervate smooth muscle cells are not under voluntary control.

REVIEW QUESTIONS

Level 1   Reviewing Facts and Terms

1. A skeletal muscle contains
   (a) connective tissues
   (b) blood vessels and nerves
   (c) skeletal muscle tissue
   (d) a, b, and c are correct

2. The contraction of a muscle exerts a pull on a bone because
   (a) muscles are attached to bones by ligaments
   (b) muscles are directly attached to bones
   (c) muscles are attached to bones by tendons
   (d) a, b, and c are correct

3. The detachment of the myosin cross-bridges is directly triggered by
   (a) repolarization of T tubules
   (b) attachment of ATP to myosin heads
   (c) hydrolysis of ATP
   (d) calcium ions

4. A muscle producing peak tension during rapid cycles of contraction and relaxation is said to be in
   (a) incomplete tetanus   (b) treppe
   (c) complete tetanus   (d) twitch

5. The smooth, steady increase in muscular tension produced by increasing the number of active motor units is
   (a) complete tetanus   (b) treppe
   (c) wave summation   (d) recruitment

6. The type of contraction in which the tension rises but the resistance does not move is
   (a) a wave summation
   (b) a twitch
   (c) an isotonic contraction
   (d) an isometric contraction

7. Large-diameter, densely packed myofibrils, large glycogen reserves, and few mitochondria are characteristics of
   (a) slow fibers   (b) intermediate fibers
   (c) fast fibers   (d) red muscles

8. An action potential can travel quickly from one cardiac muscle cell to another because of the presence of
   (a) gap junctions   (b) tight junctions
   (c) intercalated discs   (d) a and b are correct

9. List the three types of muscle tissue found in the body.

10. What are the five functions of skeletal muscle?

11. What three layers of connective tissue are part of each muscle? What functional role does each layer play?

12. What is a motor unit?

13. What structural feature of a skeletal muscle fiber is responsible for conducting action potentials into the interior of the cell?

14. Under resting conditions, what two proteins prevent interactions between cross-bridges and active sites?

15. What five interlocking steps are involved in the contraction process?

16. What two factors affect the amount of tension produced when a skeletal muscle contracts?

17. What specialized type of muscle fiber plays an important role in the reflex control of position and posture?

18. What forms of energy reserves are found in resting skeletal muscle fibers?

19. What two mechanisms are used to generate ATP from glucose in muscle cells?

20. What three factors are necessary to prevent premature muscle fatigue?

21. What is the calcium-binding protein in smooth muscle tissue?

Level 2   Reviewing Concepts

22. An activity that would require anaerobic endurance is
    (a) a 50-meter dash
    (b) a pole vault
    (c) a weight-lifting competition
    (d) a, b, and c are correct

23. Areas of the body where you would not expect to find slow fibers include the
    (a) back and calf muscles   (b) eye and hand
    (c) chest and abdomen   (d) a, b, and c are correct

24. When contraction occurs,
    (a) the H and I bands get smaller
    (b) the Z lines move closer together
    (c) the width of the A band remains constant
    (d) a, b, and c are correct

25. The amount of tension produced during a contraction is *not* affected by
    (a) the amount of ATP available to the muscle
    (b) the degree of overlap between thick and thin filaments
    (c) the interaction of troponin and tropomyosin
    (d) the number of cross-bridge interactions within a muscle fiber

26. Describe the basic sequence of events that occurs at a neuromuscular junction.

27. Why is the multinucleate condition important in skeletal muscle fibers?

28. Describe the graphic events seen on a myogram as tension is developed in a stimulated calf muscle fiber during a twitch.

29. What three processes are involved in repaying the oxygen debt during the recovery period?

30. What are the effects of aging on the muscular system?

31. Why is cardiac muscle an example of a functional syncytium?

32. How does cardiac muscle tissue contract without neural stimulation?

33. Atracurium is a drug that blocks the binding of ACh. Give an example of a site where such binding normally occurs, and predict the physical effect of this drug.

34. Describe what happens to muscles that are not "used" on a regular basis. What can be done to offset this effect?

35. The time of a murder victim's death is commonly estimated by the flexibility of the body. Explain why this is possible.

36. Many visceral smooth muscle cells lack motor neuron innervation. How are their contractions coordinated and controlled?

37. A motor unit from a skeletal muscle contains 1500 muscle fibers. Would this muscle be involved in fine, delicate movements or powerful, gross movements? Explain.

Level 3   Critical Thinking and Clinical Applications

38. Many potent insecticides contain toxins called organophosphates, which interfere with the action of the enzyme acetylcholinesterase. Ivan is using an insecticide containing organophosphates and is very careless. He does not use gloves or a dust mask and absorbs some of the chemical through his skin. He inhales a large amount as well. What symptoms would you expect to observe in Ivan as a result of the organophosphate poisoning?

39. A rare (hypothetical) genetic disease causes the body to produce antibodies that compete with acetylcholine for receptors on the motor end plate. Patients suffering from this disease exhibit varying degrees of muscular weakness and flaccid paralysis in the affected muscles. If you were able to administer a drug that inhibits acetylcholinesterase or a drug that inhibits acetylcholine, which would you use to alleviate the symptoms of the disease described?

40. Mary has just completed a 10-km race. Thirty minutes later she begins to notice soreness and stiffness in her leg muscles. She wonders whether she may have damaged the muscles during the race. She visits her physician, who orders a blood test. How could the physician tell from a blood test whether muscle damage had occurred?

11

The Muscular System

Skeletal muscles have a remarkable ability to adapt to the demands placed on them. Intensive weight training and physical conditioning will lead to an increase in the size and strength of the muscles being exercised. However, even the most dedicated bodybuilders cannot work on developing all of the 700 or so skeletal muscles in the human body. Instead, they concentrate on the largest, most prominent muscles responsible for powerful movements of the axial and appendicular skeletons. This chapter will introduce those skeletal muscles and the movements they perform.

CHAPTER OUTLINE AND OBJECTIVES

The **muscular system** includes all the skeletal muscles that can be controlled voluntarily. Most of the muscle tissue in the body is part of this system, and approximately 700 skeletal muscles have been identified. Some are attached to bony processes and others to broad sheets of connective tissue, but all are directly or indirectly associated with the skeletal system. Rather than attempt to survey all 700 skeletal muscles, we will focus on a relatively small but representative number of muscles, about 20 percent of the total. To ease the strain of memorization, these muscles are organized into anatomical and functional groups.

The shape or appearance of each muscle provides clues to its primary function. Muscles involved with locomotion and posture work across joints, producing skeletal movement. Those that support soft tissue form slings or sheets between relatively stable bony elements, whereas those that guard an entrance or exit completely encircle the opening.

At the level of the individual skeletal muscle, two factors interact to determine the effects of its contraction: (1) the anatomical arrangement of the muscle fibers and (2) the way the muscle attaches to the bones of the skeletal system. We can understand the performance of muscles in the body in terms of basic mechanical laws. The analysis of biological systems in mechanical terms is the study of *biomechanics*. In this chapter, we examine the biomechanics and gross anatomy of the muscular system.

BIOMECHANICS AND MUSCLE ANATOMY

Although most skeletal muscle fibers contract at comparable rates and shorten to the same degree, variations in microscopic and macroscopic organization can dramatically affect the power, range, and speed of movement produced when a muscle contracts.

Organization of Skeletal Muscle Fibers

Muscle fibers within a skeletal muscle form bundles called *fascicles.* ∞ *[p. 278]* The muscle fibers within a single fascicle are arranged in parallel, but the organization of the fascicles in the skeletal muscle can vary, as can the relationship between the fascicles and the associated tendon. Four patterns of fascicle organization produce *parallel muscles, convergent muscles, pennate muscles,* and *circular muscles.*

Parallel Muscles

In a **parallel muscle,** the fascicles are parallel to the long axis of the muscle. Most of the skeletal muscles in the body are parallel muscles. Some are flat bands with broad attachments (*aponeuroses*) at each end; others are plump and cylindrical with tendons at one or both ends. In the latter case, the muscle is spindle-shaped (Figure 11-1a●), with a central **body,** also known as the *belly,* or *gaster* (GAS-ter; stomach). The *biceps brachii* muscle of the arm is a parallel muscle with a central body. When a parallel muscle contracts, it gets shorter and larger in diameter. You can see the bulge of the contracting biceps brachii on the anterior surface of your arm when you flex your elbow.

A skeletal muscle cell can contract effectively until it has shortened by roughly 30 percent. Because the muscle fibers are parallel to the long axis of the muscle, when they contract together, the entire muscle shortens by the same amount. For example, if the skeletal muscle is 10 cm long, the end of the tendon will move 3 cm when the muscle contracts. The tension developed by the muscle during this contraction depends on the total number of myofibrils it contains. Because the myofibrils are distributed evenly through the sarcoplasm of each cell, we can estimate the tension on the basis of the cross-sectional area of the resting muscle. A parallel skeletal muscle 6.25 cm^2 (1 in.^{2}) in cross-sectional area can develop approximately 23 kg (50 lb) of tension.

Convergent Muscles

In a **convergent muscle,** the muscle fibers are based over a broad area, but all the fibers come together at a common attachment site. They may pull on a tendon, a tendinous sheet, or a slender band of collagen fibers known as a **raphe** (RĀ-fē; seam). The muscle fibers typically spread out, like a fan or a broad triangle, with a tendon at the apex (Figure 11-1b●). The prominent chest muscles of the *pectoralis group* have this shape. Such a muscle has versatility, for the direction of pull can be changed by stimulating only one portion of the entire muscle. However, when the entire muscle contracts, the muscle fibers do not pull as hard on the tendon as would a parallel muscle of the same size. The reason is that the convergent muscle fibers are pulling in different directions rather than working together. (Those on opposite sides of the tendon, for example, are pulling against one another.)

Pennate Muscles

In a **pennate muscle** (*penna,* feather), the fascicles form a common angle with the tendon. Because the muscle cells pull at an angle, contracting pennate muscles do not move their tendons as far as parallel muscles do. But a pennate muscle contains more muscle fibers—and, as a result, produces more tension—than does a parallel muscle of the same size. (The more muscle fibers, the more myofibrils and sarcomeres, and tension production is proportional to the number of contracting sarcomeres.)

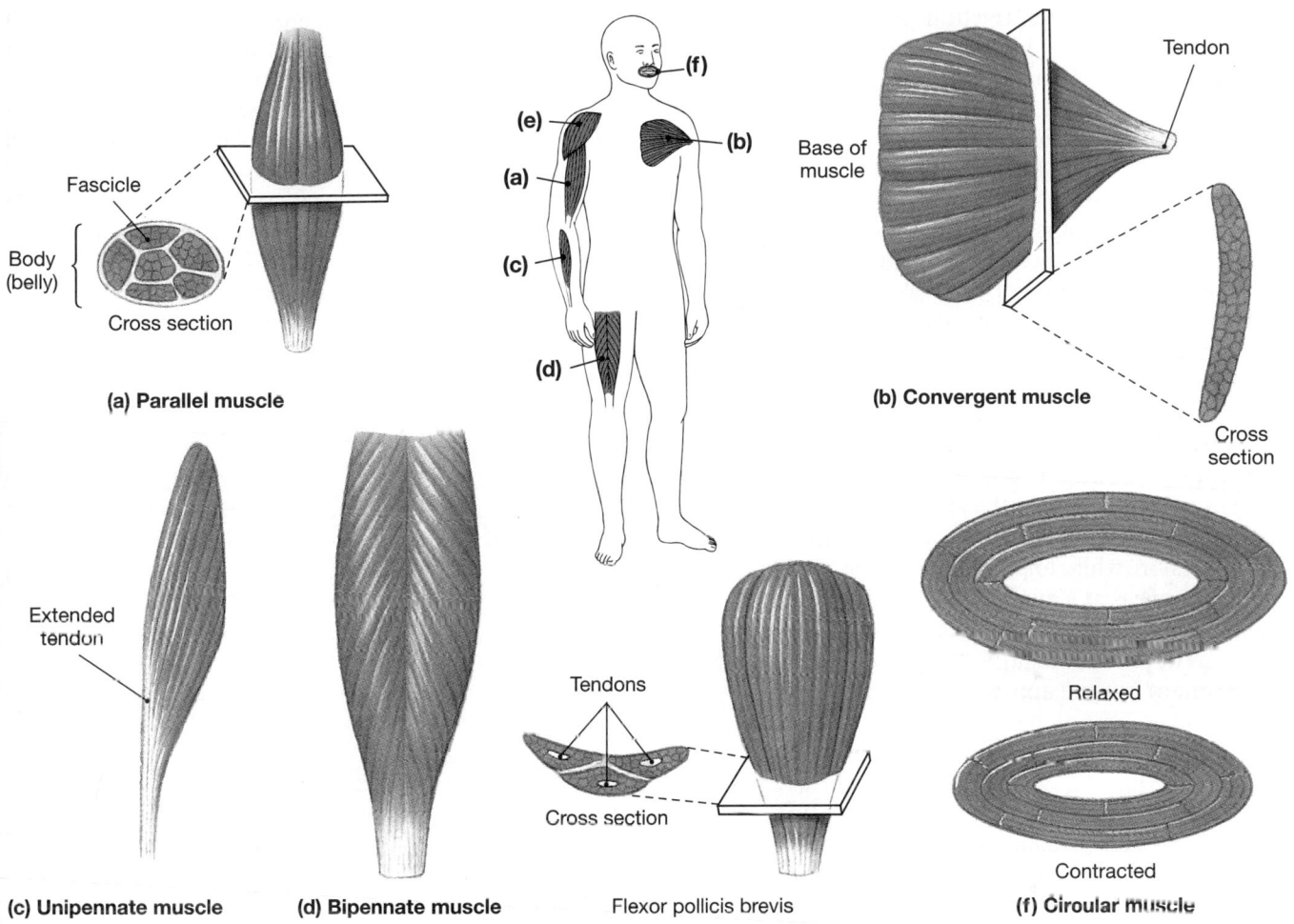

Body (belly)
Fascicle
Cross section
(a) Parallel muscle

(f)
(e)
(a)
(b)
(c)
(d)

Tendon
Base of muscle
(b) Convergent muscle
Cross section

Extended tendon
(c) Unipennate muscle

(d) Bipennate muscle

Tendons
Cross section
Flexor pollicis brevis

Relaxed
Contracted
(f) Circular muscle

•FIGURE 11-1 Different Arrangements of Skeletal Muscle Fibers

If all the muscle fibers are on the same side of the tendon, the pennate muscle is *unipennate* (Figure 11-1c•). The *extensor digitorum* (Figure 11-19c•, p. 349), a forearm muscle that extends the finger joints, is a unipennate muscle. More commonly, a pennate muscle has fibers on both sides of the tendon. Such a muscle is called *bipennate*. The *rectus femoris* (Figure 11-22c,d•, p. 355), a prominent muscle that extends the knee, is a bipennate muscle (Figure 11-1d•). If the tendon branches within a pennate muscle, the muscle is said to be *multipennate* (Figure 11-1e•). The triangular *deltoid* muscle of the shoulder (Figure 11-14•, pp. 340–341) is a multipennate muscle.

Circular Muscles

In a **circular muscle,** or **sphincter** (SFINK-ter), the fibers are concentrically arranged around an opening or recess. When the muscle contracts, the diameter of the opening decreases. Circular muscles guard entrances and exits of internal passageways such as the digestive and urinary tracts. An example is the *orbicularis oris* of the mouth (Figure 11-1f•).

Skeletal Muscle Length–Tension Relationships

Skeletal muscles produce maximum tension over a relatively narrow range of sarcomere lengths. ∞ *[p. 291]* Stretched too far, the muscle loses power because the overlap between thick and thin filaments is reduced. Excessive overlap also reduces tension production, by disrupting the normal relationship between thick and thin filaments.

Normally, the bulk of other skeletal muscles and associated tendons, bones, and ligaments limit the degree of stretching or compression of muscle fibers within a skeletal muscle. However, a skeletal muscle may still be stimulated over a range of resting lengths. The amount of tension produced in the resulting contraction varies with the resting sarcomere length. For example, consider what happens when you try to lift a heavy object by flexing your elbow. If you begin with the elbow fully extended, you find it very difficult. But if you start with the elbow partially flexed, the lifting seems easier, and the weight feels lighter. Over a full range of motion, such as when you go from full extension to full flexion, tension production rises and falls. In the exercises known as *curls,* you move your elbow from full extension to full flexion with a heavy weight in your hand. Although it is difficult to begin, flexion becomes easier at midrange before becoming more difficult as the movement nears completion.

When complex movements occur, muscles commonly work in groups rather than as individuals. Their cooperation helps improve the efficiency of a particular movement. For example, large limb muscles produce flexion or extension over an extended range of motion. Although these muscles cannot develop much tension at full extension, they are generally paired with one or more smaller muscles that provide assistance until the larger muscle can perform at maximum efficiency. At the start of the movement, the smaller muscle is producing maximum tension, while tension production by the larger muscle is at a minimum. The importance of this smaller "assistant" decreases as the movement proceeds and as the sarcomeres in the primary muscle approach optimal length.

Levers

Skeletal muscles do not work in isolation. When a muscle is attached to the skeleton, the nature and site of the connection will determine the force, speed, and range of the movement produced. These characteristics are interdependent, and the relationships can explain a great deal about the general organization of the muscular and skeletal systems.

The force, speed, or direction of movement produced by contraction of a muscle can be modified by attaching the muscle to a lever. A **lever** is a rigid structure—such as a board, a crowbar, or a bone—that moves on a fixed point called the **fulcrum.** In the body, each bone is a lever, and each joint a fulcrum. A child's teeter-totter, or seesaw, provides a more familiar example of lever action. Levers can change (1) the direction of an applied force, (2) the distance and speed of movement produced by an applied force, and (3) the effective strength of an applied force.

Classes of Levers

Three classes of levers are found in the human body. The seesaw is a **first-class lever.** In such a lever, the fulcrum lies between the applied force (AF) and the resistance (R). There are not many examples of first-class levers in the body. One example, the arrangement of muscles that extend the neck, is shown in Figure 11-2a•.

In a **second-class lever** (Figure 11-2b•), the resistance is located between the applied force and the fulcrum. A familiar example of such a lever is a loaded wheelbarrow. The weight of the load is the resistance, and the upward lift on the handle is the applied force. Because in this arrangement the force is always farther from the fulcrum than the resistance is, a small force can balance a larger weight. In other words, the effective force is in-

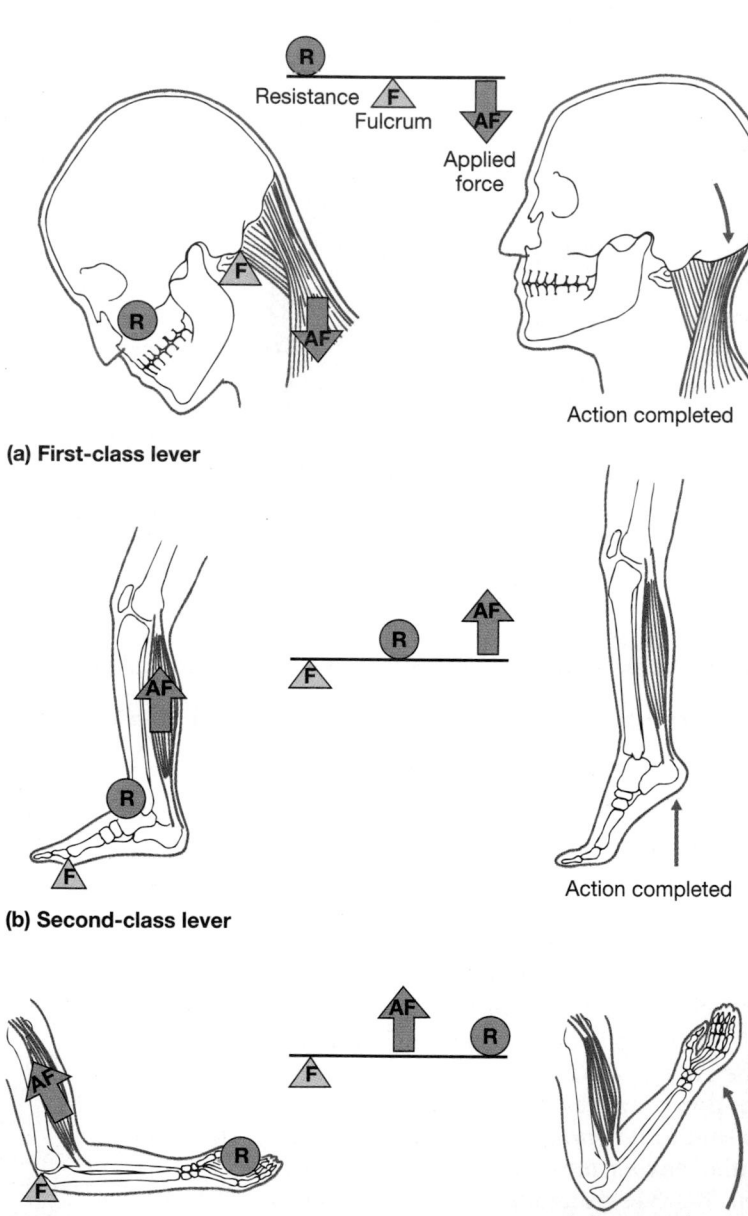

(a) First-class lever

(b) Second-class lever

(c) Third-class lever

•FIGURE 11-2 The Three Classes of Levers. (a) In a first-class lever, the applied force and the resistance are on opposite sides of the fulcrum. First-class levers can change the amount of force transmitted to the resistance and alter the direction and speed of movement. **(b)** In a second-class lever, the resistance lies between the applied force and the fulcrum. This arrangement magnifies force at the expense of distance and speed; the direction of movement remains unchanged. **(c)** In a third-class lever, the force is applied between the resistance and the fulcrum. This arrangement increases speed and distance moved but requires a larger applied force.

creased. Notice, however, that when a force moves the handle, the load moves more slowly and covers a shorter distance. There are few examples of second-class levers in the body. In performing plantar flexion, the calf muscles use a second-class lever (Figure 11-2b•).

Third-class levers are the most common levers in the body. In this lever system, a force is applied between the resistance and the fulcrum (Figure 11-2c•). The effect of this arrangement is just the reverse of that produced by a second-class lever: Speed and distance traveled are increased at the expense of effective force. In the example illustrated (the biceps brachii, which flexes the elbow), the resistance is six times farther from the fulcrum than is the applied force. The effective force is reduced accordingly. The muscle must generate 180 kg of tension at its attachment to the forearm to support 30 kg held in the hand. However, the distance traveled and the speed of movement are *increased* by the same 6:1 ratio: The load will travel 45 cm when the insertion point moves 7.5 cm.

Although every muscle does not operate as part of a lever system, the presence of levers provides speed and versatility far in excess of what we would predict on the basis of muscle physiology alone. Skeletal muscle fibers resemble one another closely, and their abilities to contract and generate tension are quite similar. Consider a skeletal muscle that can contract in 500 msec and shorten 1 cm while it exerts a 10-kg pull. Without using a lever, this muscle would be performing efficiently only when moving a 10-kg weight a distance of 1 cm. But, by using a lever, the same muscle operating at the same efficiency could move 20 kg a distance of 0.5 cm, 5 kg a distance of 2 cm, or 1 kg a distance of 10 cm.

✔ Why does a pennate muscle generate more tension than does a parallel muscle of the same size?

✔ What type of muscle would you expect to be guarding the opening between the stomach and the small intestine?

✔ The joint between the occipital bone of the skull and the first cervical vertebra (atlas) is part of what type of lever system?

MUSCLE TERMINOLOGY

This chapter focuses on the functional anatomy of skeletal muscles and of muscle groups. Once again you will be faced with a number of new terms, and this section attempts to give you some assistance in understanding them.

Origins and Insertions

Each muscle begins at an **origin,** ends at an **insertion,** and contracts to produce a specific **action.** In general, the origin remains stationary, whereas the insertion moves, or the origin is proximal to the insertion.

For example, the *triceps brachii* inserts on the olecranon of the ulna and originates closer to the shoulder. The decision as to which end is the origin and which is the insertion is usually based on movement from the anatomical position. Part of the fun of studying the muscular system is that you can actually do the movements and think about the muscles involved. As a result, laboratory discussions of the muscular system tend to resemble disorganized aerobics classes.

When the origins and insertions cannot be determined easily on the basis of common movement or position, other rules are used. If a muscle extends between a broad aponeurosis and a narrow tendon, the aponeurosis is the origin, and the tendon is the insertion. If there are several tendons at one end and just one at the other, there are multiple origins and a single insertion. These simple rules cannot cover every situation. Knowing which end is the origin and which is the insertion is ultimately less important than knowing where the two ends attach and what the muscle accomplishes when it contracts.

Actions

Almost all skeletal muscles either originate or insert on the skeleton. When a muscle moves a portion of the skeleton, that movement may involve *flexion, extension, adduction, abduction, protraction, retraction, elevation, depression, rotation, circumduction, pronation, supination, inversion, eversion, lateral flexion, or opposition.* Before proceeding, you may want to review the discussions of planes of motion and Figures 9-2 to 9-5. ∞ *[pp. 258–262]*

There are two methods of describing actions. The first method, used by most undergraduate textbooks and references such as *Gray's Anatomy,* describes actions in terms of the bone affected. Thus a muscle such as the biceps brachii is said to perform "flexion of the forearm." The second method, of increasing use among specialists such as kinesiologists, identifies the joint involved. With this method, the action of the biceps brachii would be "flexion of the elbow." Both methods are valid, and each has its advantages. We shall use whichever method is more efficient at describing the movement underway.

We can group muscles according to their **primary actions:**

- A **prime mover,** or **agonist,** is a muscle whose contraction is chiefly responsible for producing a particular movement. The biceps brachii is a prime mover that produces flexion of the elbow.
- **Antagonists** are prime movers whose actions oppose that of the agonist under consideration. The triceps brachii is a prime mover that extends the elbow. It is therefore an antagonist of the biceps brachii, and the biceps brachii is an antagonist of the triceps brachii.

Agonists and antagonists are functional opposites; if one produces flexion, the other will have extension as its primary action. When an agonist contracts to produce a particular movement, the corresponding antagonist will be stretched, but it will usually not relax completely. Instead, it will contract *eccentrically*, with the tension adjusted to control the speed of the movement and ensure its smoothness. ∞ *[p. 296]* Students sometimes learn about muscles in antagonist pairs on the basis of primary actions (flexors/extensors, abductors/adductors) at a specific joint. This method highlights the functions of the muscles involved, and it can help organize the information in a logical framework. The tables in this chapter are arranged to facilitate this approach.

■ When a **synergist** (*syn-*, together + *ergon*, work) contracts, it assists the prime mover in performing that action. Synergists may provide additional pull near the insertion or may stabilize the point of origin. Their importance in assisting a particular movement may change as the movement progresses: In many cases they are most useful at the start, when the prime mover is stretched and unable to develop maximum tension. For example, the *latissimus dorsi* is a large trunk muscle that extends, adducts, and medially rotates the arm at the shoulder joint. A much smaller muscle, the *teres* (TER-ēz) *major,* assists in starting such movements when the joint is at full flexion. Synergists may also assist an agonist by *preventing* movement at a joint and thereby stabilizing the origin of the agonist. These muscles are called **fixators**.

Names of Skeletal Muscles

You need not learn every one of the approximately 700 muscles in the human body, but you will have to become familiar with the most important ones. Fortunately, anatomists assigned names to the muscles that provided clues to their identification. If you can learn to recognize the clues, you will find it easier to remember the names and identify the muscles. The name of a muscle may include information concerning its *fascicle organization, location, relative position, structure, size, shape, origin and insertion,* or *action.*

Fascicle Organization

A muscle name may refer to the orientation of the muscle fibers within a particular skeletal muscle. **Rectus** means "straight," and rectus muscles are parallel muscles whose fibers generally run along the long axis of the body. Because there are several rectus muscles, the name typically includes a second term that refers to a precise region of the body. For example, the *rectus abdominis* is located on the abdomen, and the *rectus femoris* on the thigh. Other directional indicators include **transversus** and **obliquus** (oblique) for muscles whose fibers run across or at an oblique angle to the longitudinal axis of the body, respectively.

Location

Table 11-1 includes a useful summary of terms that designate specific regions of the body. They are common as modifiers that help identify individual muscles, as in the case of the rectus muscles noted previously. In a few cases, the muscle is such a prominent feature of the region that the regional name alone will identify it. Examples include the *temporalis* of the head and the *brachialis* (brā-kē-A-lis) of the arm.

Relative Position

Muscles visible at the body surface are often called **externus** or **superficialis,** whereas those beneath the surface are termed **internus** or **profundus.** Superficial muscles that position or stabilize an organ are called **extrinsic** muscles; muscles located entirely within the organ are called **intrinsic** muscles.

Structure, Size, and Shape

Some muscles are named after distinctive structural features. The *biceps brachii*, for example, has two tendons of origin (*bi-*, two + *caput*, head), the *triceps brachii* has three, and the *quadriceps group* four. Shape is sometimes an important clue to the name of a muscle. For example, *trapezius* (tra-PĒ-zē-us), *deltoid, rhomboideus* (rom-BOY-dē-us), and *orbicularis* (or-bik-ū-LAR-is) refer to prominent muscles that look like a trapezoid, a triangle, a rhomboid, and a circle, respectively. Long muscles are called **longus** (long) or **longissimus** (longest), and **teres** muscles are both long and round. Short muscles are called **brevis.** Large ones are called **magnus** (big), **major** (bigger), or **maximus** (biggest); and small ones are called **minor** (smaller) or **minimus** (smallest).

Origin and Insertion

Many names tell you the specific origin and insertion of each muscle. In such cases, the first part of the name indicates the origin, and the second part the insertion. The *genioglossus,* for example, originates at the chin (*geneion*) and inserts in the tongue (*glossus*). Although the names may be long and difficult to pronounce, Table 11-1 and the anatomical terms introduced in Chapter 1 can help you identify and remember them. ∞ *[pp. 17–24]*

Action

Many muscles are named *flexor, extensor, retractor, abductor,* and so on. These are such common actions that the names almost always include other clues as to the appearance or location of the muscle. For example, the *extensor carpi radialis longus* is a long muscle along the radial (lateral) border of the forearm. When it contracts, its primary function is extension of the carpus (wrist).

A few muscles are named after the specific movements associated with special occupations or habits.

TABLE 11-1 Muscle Terminology

Terms Indicating Direction Relative to Axes of the Body	Terms Indicating Specific Regions of the Body*	Terms Indicating Structural Characteristics of the Muscle	Terms Indicating Actions
Anterior (front)	Abdominis (abdomen)	**Origin**	**General**
Externus (superficial)	Anconeus (elbow)	Biceps (two heads)	Abductor
Extrinsic (outside)	Auricularis (auricle of ear)	Triceps (three heads)	Adductor
Inferioris (inferior)	Brachialis (brachium)	Quadriceps (four heads)	Depressor
Internus (deep, internal)	Capitis (head)		Extensor
Intrinsic (inside)	Carpi (wrist)	**Shape**	Flexor
Lateralis (lateral)	Cervicis (neck)	Deltoid (triangle)	Levator
Medialis/medius (medial, middle)	Cleido/clavius (clavicle)	Orbicularis (circle)	Pronator
Obliquus (oblique)	Coccygeus (coccyx)	Pectinate (comblike)	Rotator
Posterior (back)	Costalis (ribs)	Piriformis (pear-shaped)	Supinator
Profundus (deep)	Cutaneous (skin)	Platys- (flat)	Tensor
Rectus (straight, parallel)	Femoris (femur)	Pyramidal (pyramid)	
Superficialis (superficial)	Genio- (chin)	Rhomboideus (rhomboid)	**Specific**
Superioris (superior)	Glosso/glossal (tongue)	Serratus (serrated)	Buccinator (trumpeter)
Transversus (transverse)	Hallucis (great toe)	Splenius (bandage)	Risorius (laugher)
	Ilio- (ilium)	Teres (long and round)	Sartorius (like a tailor)
	Inguinal (groin)	Trapezius (trapezoid)	
	Lumborum (lumbar region)		
	Nasalis (nose)	**Other Striking Features**	
	Nuchal (back of neck)	Alba (white)	
	Oculo- (eye)	Brevis (short)	
	Oris (mouth)	Gracilis (slender)	
	Palpebrae (eyelid)	Lata (wide)	
	Pollicis (thumb)	Latissimus (widest)	
	Popliteus (behind knee)	Longissimus (longest)	
	Psoas (loin)	Longus (long)	
	Radialis (radius)	Magnus (large)	
	Scapularis (scapula)	Major (larger)	
	Temporalis (temples)	Maximus (largest)	
	Thoracis (thoracic region)	Minimus (smallest)	
	Tibialis (tibia)	Minor (smaller)	
	Ulnaris (ulna)	-tendinosus (tendinous)	
	Uro- (urinary)	Vastus (great)	

*For other regional terms, refer to Figure 1-7, pp. 17–18, which deals with anatomical landmarks.

The *sartorius* (sar-TO-rē-us) muscle is active when you cross your legs. Before sewing machines were invented, a tailor would sit on the floor cross-legged, and the name of this muscle was derived from *sartor,* the Latin word for "tailor." On the face, the *buccinator* (BUK-si-nā-tor) muscle compresses the cheeks, as when you purse your lips and blow forcefully. *Buccinator* translates as "trumpet player." Another facial muscle, the *risorius* (ri-SO-rē-us), was supposedly named after the mood expressed. The Latin term *risor,* however, means "laugher"; a more appropriate description for the effect would be "grimace."

Axial and Appendicular Muscles

The separation of the skeletal system into axial and appendicular divisions provides a useful guideline for subdividing the muscular system as well:

1. The **axial musculature** arises on the axial skeleton. It positions the head and spinal column and also moves the rib cage, assisting in the movements that make breathing possible. It does not play a role in movement or support of the pectoral or pelvic girdle or limbs. This category encompasses roughly 60 percent of the skeletal muscles in the body.

2. The **appendicular musculature** stabilizes or moves components of the appendicular skeleton and includes the remaining 40 percent of all skeletal muscles.

Figure 11-3• provides an overview of the axial and appendicular muscles of the human body. These are the superficial muscles, which tend to be relatively large. The superficial muscles cover deeper, smaller muscles that cannot be seen unless the overlying muscles are either removed or *reflected*—that is, cut and pulled out of the way. Later figures showing the deep muscles in

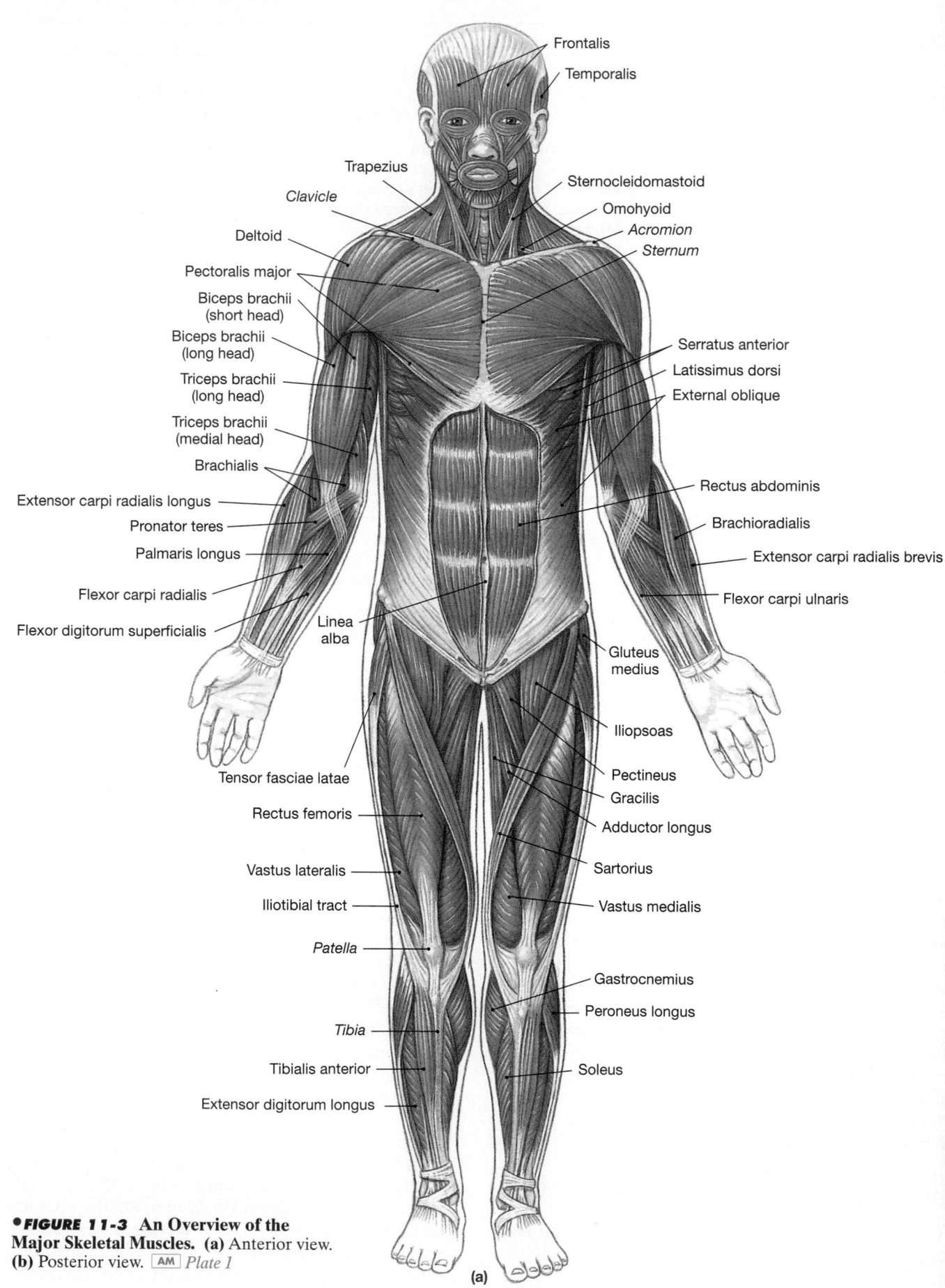

●*FIGURE 11-3* An Overview of the
Major Skeletal Muscles. (a) Anterior view.
(b) Posterior view. AM *Plate 1*

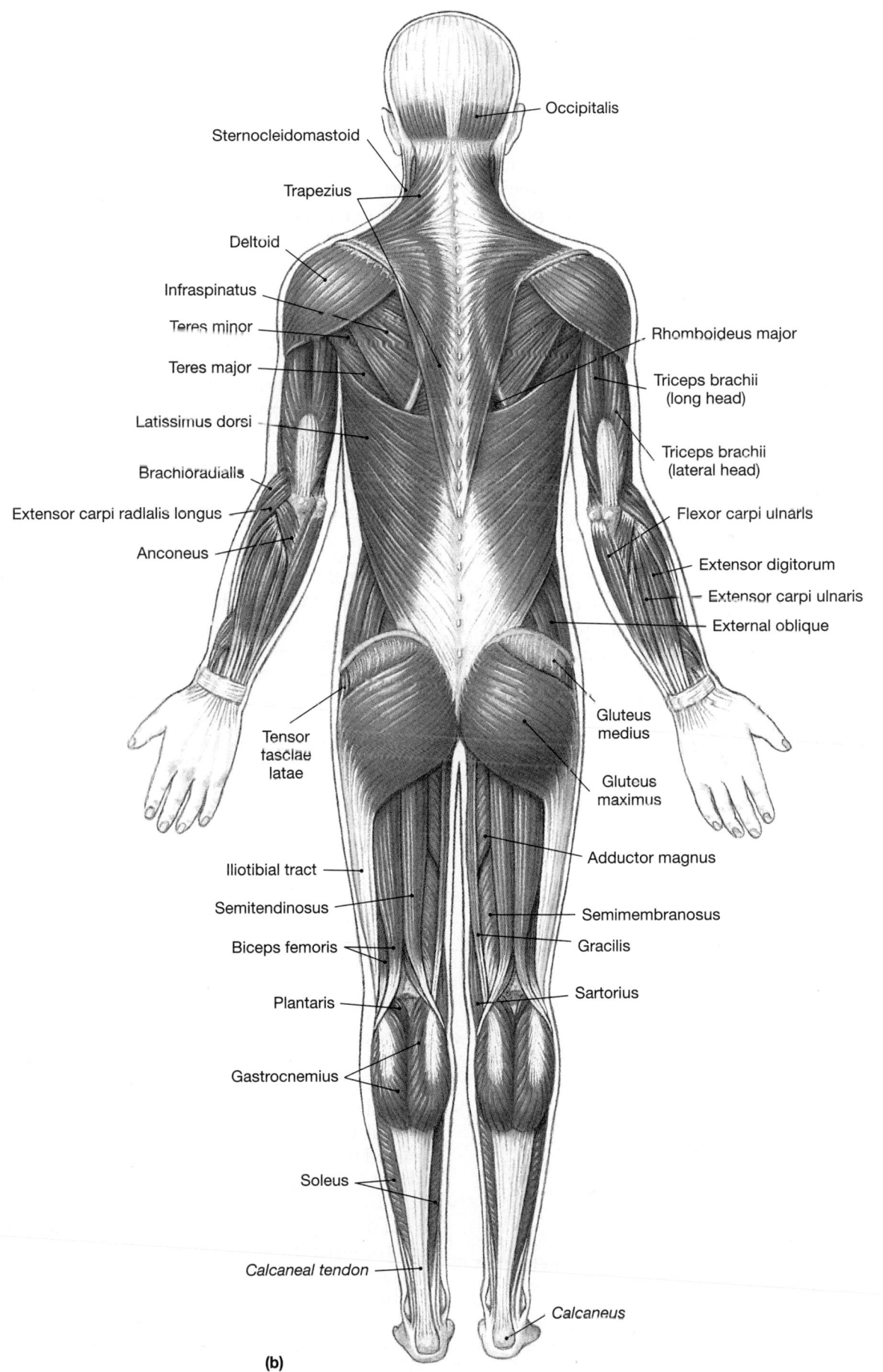

Occipitalis

Sternocleidomastoid

Trapezius

Deltoid

Infraspinatus

Teres minor

Teres major

Latissimus dorsi

Brachioradialis

Extensor carpi radialis longus

Anconeus

Rhomboideus major

Triceps brachii (long head)

Triceps brachii (lateral head)

Flexor carpi ulnaris

Extensor digitorum

Extensor carpi ulnaris

External oblique

Tensor fasciae latae

Gluteus medius

Gluteus maximus

Adductor magnus

Iliotibial tract

Semitendinosus

Biceps femoris

Plantaris

Gastrocnemius

Semimembranosus

Gracilis

Sartorius

Soleus

Calcaneal tendon

Calcaneus

(b)

specific regions will indicate whether superficial muscles have been removed or reflected for the sake of clarity.

Paying attention to patterns of origin, insertion, and action, we will now examine representatives of both muscular divisions. This discussion assumes that you already understand skeletal anatomy, and you may find it helpful to review appropriate figures in Chapters 7, 8, and 9 as you proceed. For convenience, the figure captions in this chapter indicate the relevant figures in those chapters. In addition, you should make frequent reference to the section titled "Origins and Insertions" in your *Applications Manual.*

☑ The *gracilis* muscle is attached to the anterior surface of the tibia at one end and to the pubis and ischium of the pelvis at the other. When the muscle contracts, hip flexion occurs. Which attachment point is the muscle's origin?

☑ Muscle A abducts the humerus, and muscle B adducts the humerus. What is the relationship between these two muscles?

☑ What does the name *flexor carpi radialis longus* tell you about this muscle?

THE AXIAL MUSCLES

The axial muscles fall into logical groups on the basis of location, function, or both. The groups do not always have distinct anatomical boundaries. For example, a function such as the extension of the spine involves muscles along the entire length of the spinal column. We will recognize four axial muscle groups:

1. The first group includes the *muscles of the head and neck* that are not associated with the spinal column. These muscles include those that move the face, tongue, and larynx. They are therefore responsible for verbal and nonverbal communication—laughing, talking, frowning, smiling, whistling, and so on. You also use this group of muscles when you eat—especially in sucking and chewing—and even when you look for something to eat, by controlling your eye movements.

2. The second group, the *muscles of the spine,* includes numerous flexors and extensors of the vertebral column.

3. The third group, the *oblique and rectus muscles,* forms the muscular walls of the thoracic and abdominopelvic cavities between the first thoracic vertebra and the pelvis. In the thoracic area these muscles are partitioned by the ribs, but over the abdominal surface they form broad muscular sheets. There are also oblique and rectus muscles in the neck. Although they do not form a complete muscular wall, they share a common developmental origin with the oblique and rectus muscles of the trunk.

4. The fourth group, the *muscles of the pelvic floor,* extend between the sacrum and pelvic girdle, forming the muscular *perineum,* which closes the pelvic outlet.

The tables that follow summarize information concerning the origin, insertion, and action of each mus-

cle. These tables also contain information concerning the innervation of the individual muscles. **Innervation** is the distribution of nerves to a region or organ; the tables indicate the nerves that control each muscle. Many of the muscles of the head and neck are innervated by cranial nerves, such as the *facial nerve,* or seventh cranial nerve (N VII), which innervates the facial musculature. Cranial nerves originate at the brain and pass through the foramina of the skull. Spinal nerves are connected to the spinal cord and pass through the intervertebral foramina. For example, spinal nerve L_1 passes between vertebrae L_1 and L_2. Spinal nerves may form a complex network, or *plexus*; one branch, such as the *sciatic nerve* of the thigh, may contain axons from several spinal nerves. Thus, many tables include the spinal nerves as well as the names of the peripheral nerves.

Muscles of the Head and Neck

We can divide the muscles of the head and neck into several groups. The *muscles of facial expression,* the *muscles of mastication* (chewing), the *muscles of the tongue,* and the *muscles of the pharynx* originate on the skull or hyoid bone. Muscles involved with sight and hearing are also based on the skull. We will also consider here the *extrinsic eye muscles*—those associated with movements of the eye. We shall discuss the intrinsic eye muscles, which control the diameter of the pupil and the shape of the lens, in Chapter 17 along with the muscles associated with the auditory ossicles. In the neck, the *extrinsic muscles of the larynx* adjust the position of the hyoid bone and larynx. We shall examine the intrinsic laryngeal muscles, including those of the vocal cords, in Chapter 23 (the respiratory system).

Muscles of Facial Expression

The muscles of facial expression originate on the surface of the skull (Figure 11-4● and Table 11-2, p. 326). At their insertions, the epimysial fibers are woven into those of the superficial fascia and the dermis of the skin: When they contract, the skin moves. These muscles are innervated by the facial nerve.

The largest group of facial muscles is associated with the mouth. The **orbicularis oris** constricts the opening, and other muscles move the lips or the corners of the mouth. The **buccinator** has two functions related to eating (in addition to its importance to musicians). During chewing, it cooperates with the masticatory muscles by moving food back across the teeth from the space inside the cheeks. In infants, the buccinator provides suction for suckling at the breast.

Smaller groups of muscles control movements of the eyebrows and eyelids, the scalp, the nose, and the external ear. The **epicranius** (ep-i-KRĀ-nē-us; *epi-,* on + *kranion,* skull), or scalp, contains two muscles, the **frontalis**

•*FIGURE 11-4* **Muscles of Facial Expression.** *See also Figure 7-3, pp. 201–202.* **(a)** Anterolateral view. **(b)** Anterior view. AM *Plates 4.2, 4.3*

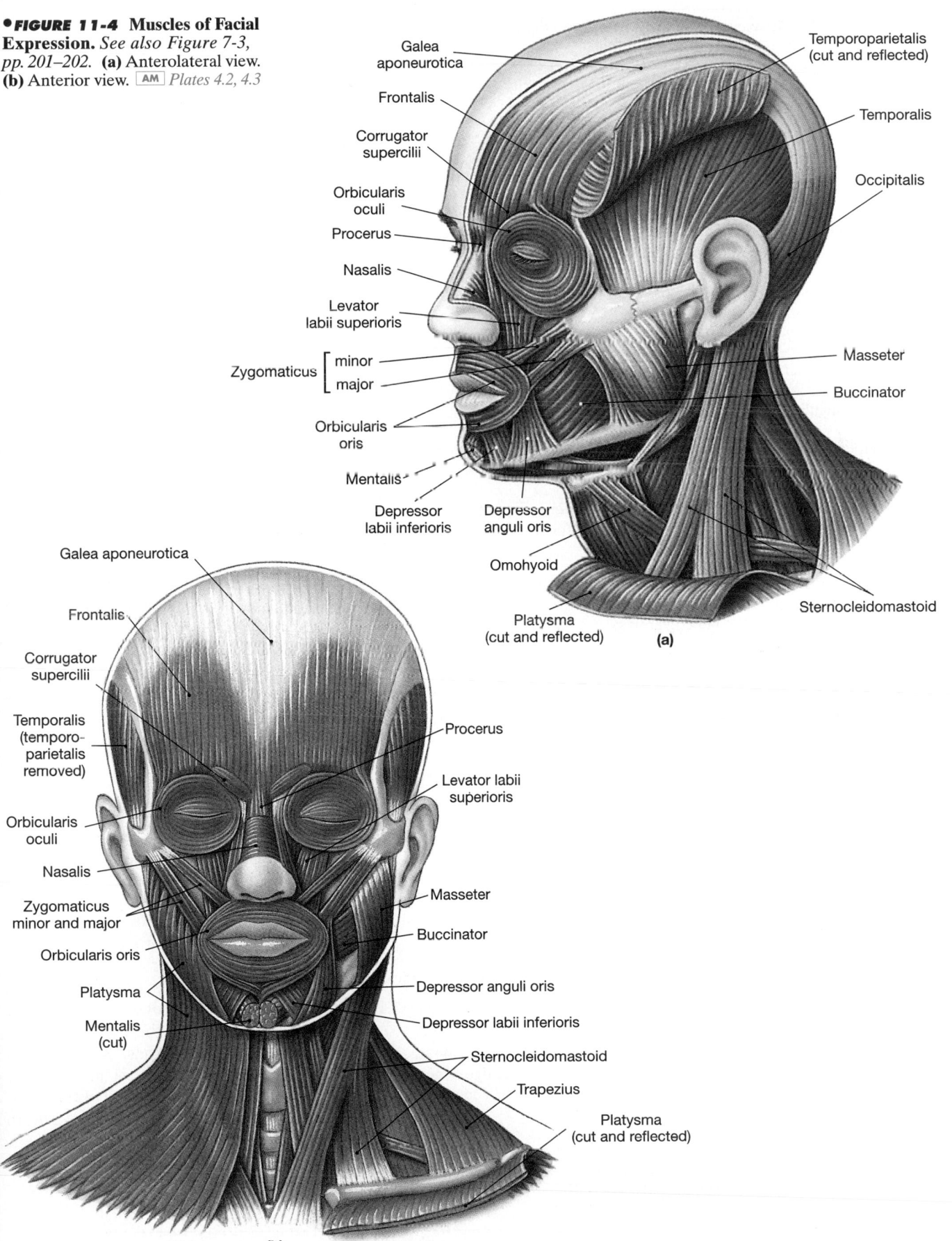

TABLE 11-2 Muscles of Facial Expression *(Figure 11-4)*

Region/Muscle	Origin	Insertion	Action	Innervation
Mouth				
Buccinator	Alveolar processes of maxillary bone and mandible	Blends into fibers of orbicularis oris	Compresses cheeks	Facial nerve (N VII)
Depressor labii inferioris	Mandible between the anterior midline and the mental foramen	Skin of lower lip	Depresses lip	As above
Levator labii superioris	Lower margin of orbit, superior to the infraorbital foramen	Orbicularis oris	Elevates upper lip	As above
Mentalis	Incisive fossa of mandible	Skin of chin	Elevates and protrudes lower lip	As above
Orbicularis oris	Maxillary bone and mandible	Lips	Compresses, purses lips	As above
Risorius	Fascia surrounding parotid salivary gland	Angle of mouth	Draws corner of mouth to the side	As above
Depressor anguli oris	Anterolateral surface of mandibular body	Skin at angle of mouth	Depresses corner of mouth	As above
Zygomaticus major	Zygomatic bone near the zygomatico-maxillary suture	Angle of mouth	Retracts and elevates corner of mouth	As above
Zygomaticus minor	Zygomatic bone posterior to zygomaticotemporal suture	Upper lip	Retracts and elevates corner of mouth	As above
Eye				
Corrugator supercilii	Orbital rim of frontal bone near nasal suture	Eyebrow	Pulls skin inferiorly and anteriorly; wrinkles brow	As above
Levator palpebrae superioris (not shown)	Tendinous band around optic foramen	Upper eyelid	Elevates upper eyelid	Oculomotor nerve (N III)[*]
Orbicularis oculi	Medial margin of orbit	Skin around eyelids	Closes eye	Facial nerve (N VII)
Nose				
Procerus	Nasal bones and lateral nasal cartilages	Aponeurosis at bridge of nose and skin of forehead	Moves nose, changes position and shape of nostrils	As above
Nasalis	Maxillary bone and alar cartilage of nose	Bridge of nose	Compresses bridge, depresses tip of nose; elevates corners of nostrils	As above
Ear (extrinsic)				
Temporoparietalis	Fascia around external ear	Galea aponeurotica	Tenses scalp, moves pinna of ear	As above
Scalp (Epicranius)				
Frontalis	Galea aponeurotica	Skin of eyebrow and bridge of nose	Raises eyebrows, wrinkles forehead	As above
Occipitalis	Superior nuchal line	Galea aponeurotica	Tenses and retracts scalp	As above
Neck				
Platysma	Upper thorax between cartilage of second rib and acromion of scapula	Mandible and skin of cheek	Tenses skin of neck, depresses mandible	As above

[*]This muscle originates in association with the extrinsic oculomotor muscles, so its innervation is unusual. *(See Figure 14-22, p. 477.)*

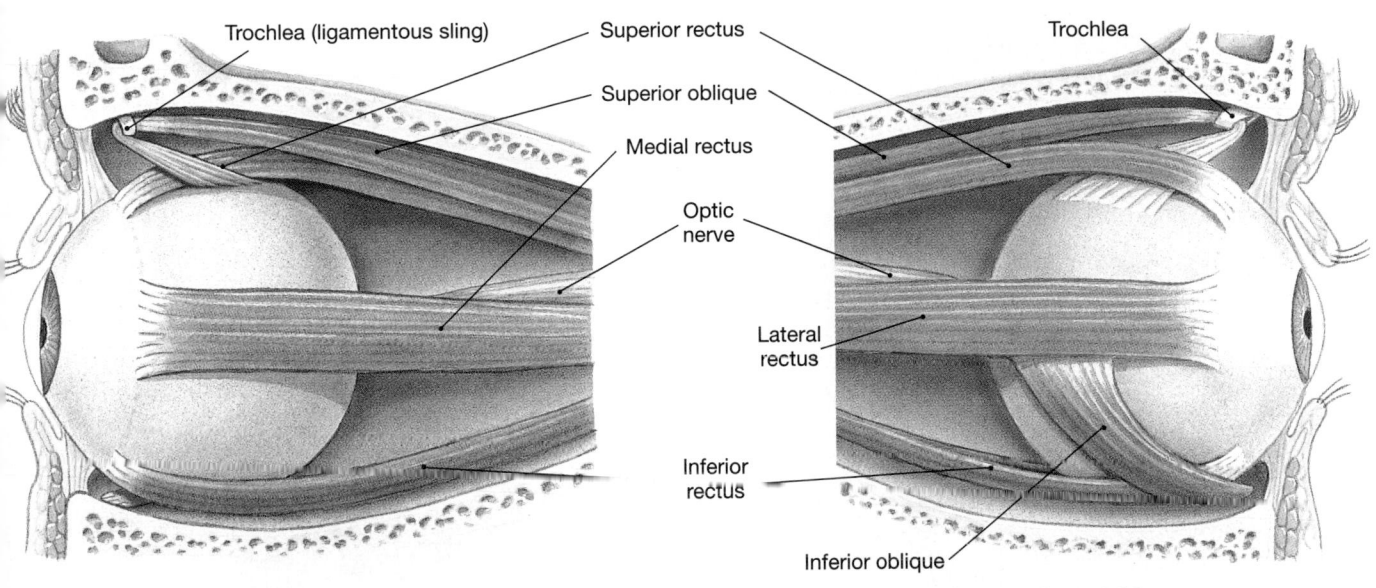

Trochlea (ligamentous sling) Superior rectus Trochlea

Superior oblique

Medial rectus

Optic nerve

Lateral rectus

Inferior rectus

Inferior oblique

(a) Medial surface, right eye **(b) Lateral surface, right eye**

•*FIGURE 11-5* Oculomotor Muscles. *See also Figure 7-13, p. 213.* AM *Plate 3.3*

and the **occipitalis,** separated by a collagenous sheet, the **galea aponeurotica** (GĀ-lē-uh ap-ō-nū-RO-ti-kuh; *galea,* a helmet + aponeurosis). The **platysma** (pla-TIZ-muh; *platys,* flat) covers the ventral surface of the neck, extending from the base of the neck to the periosteum of the mandible and the fascia at the corner of the mouth.

Extrinsic Eye Muscles

Six **extrinsic eye muscles,** or **oculomotor** (ok-ū-lō-MŌ-ter) muscles, originating on the surface of the orbit control the position of each eye. These muscles, shown in Figure 11-5• and detailed in Table 11-3, are the **inferior rectus, lateral rectus, medial rectus, superior rectus, inferior oblique,** and **superior oblique.** The oculomotor muscles are innervated by the third (*oculomotor*), fourth (*trochlear*), and sixth (*abducens*) cranial nerves.

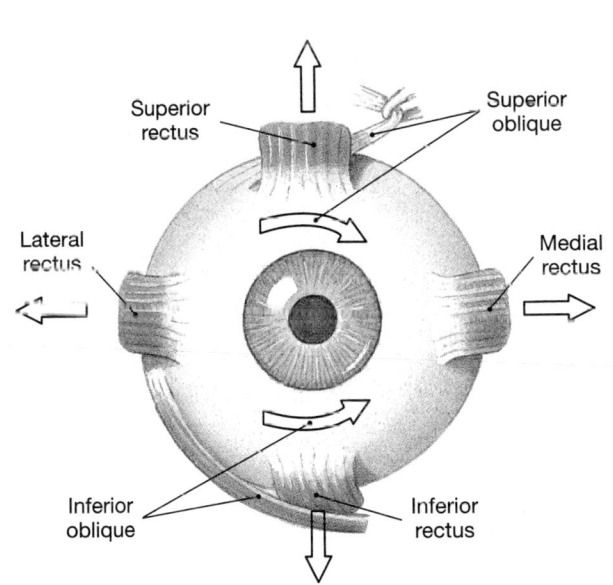

Superior rectus Superior oblique

Lateral rectus Medial rectus

Inferior oblique Inferior rectus

(c) Anterior view, right eye

TABLE 11-3 Extrinsic Oculomotor Muscles *(Figure 11-5)*

Muscle	Origin	Insertion	Action	Innervation
Inferior rectus	Sphenoid bone around optic canal	Inferior, medial surface of eyeball	Eye looks down	Oculomotor nerve (N III)
Medial rectus	As above	Medial surface of eyeball	Eye rotates medially	As above
Superior rectus	As above	Superior surface of eyeball	Eye looks up	As above
Inferior oblique	Maxillary bone at anterior portion of orbit	Inferior, lateral surface of eyeball	Eye rolls, looks up and to the side	As above
Superior oblique	Sphenoid bone around optic canal	Superior, lateral surface of eyeball	Eye rolls, looks down and to the side	Trochlear nerve (N IV)
Lateral rectus	As above	Lateral surface of eyeball	Eye rotates laterally	Abducens nerve (N VI)

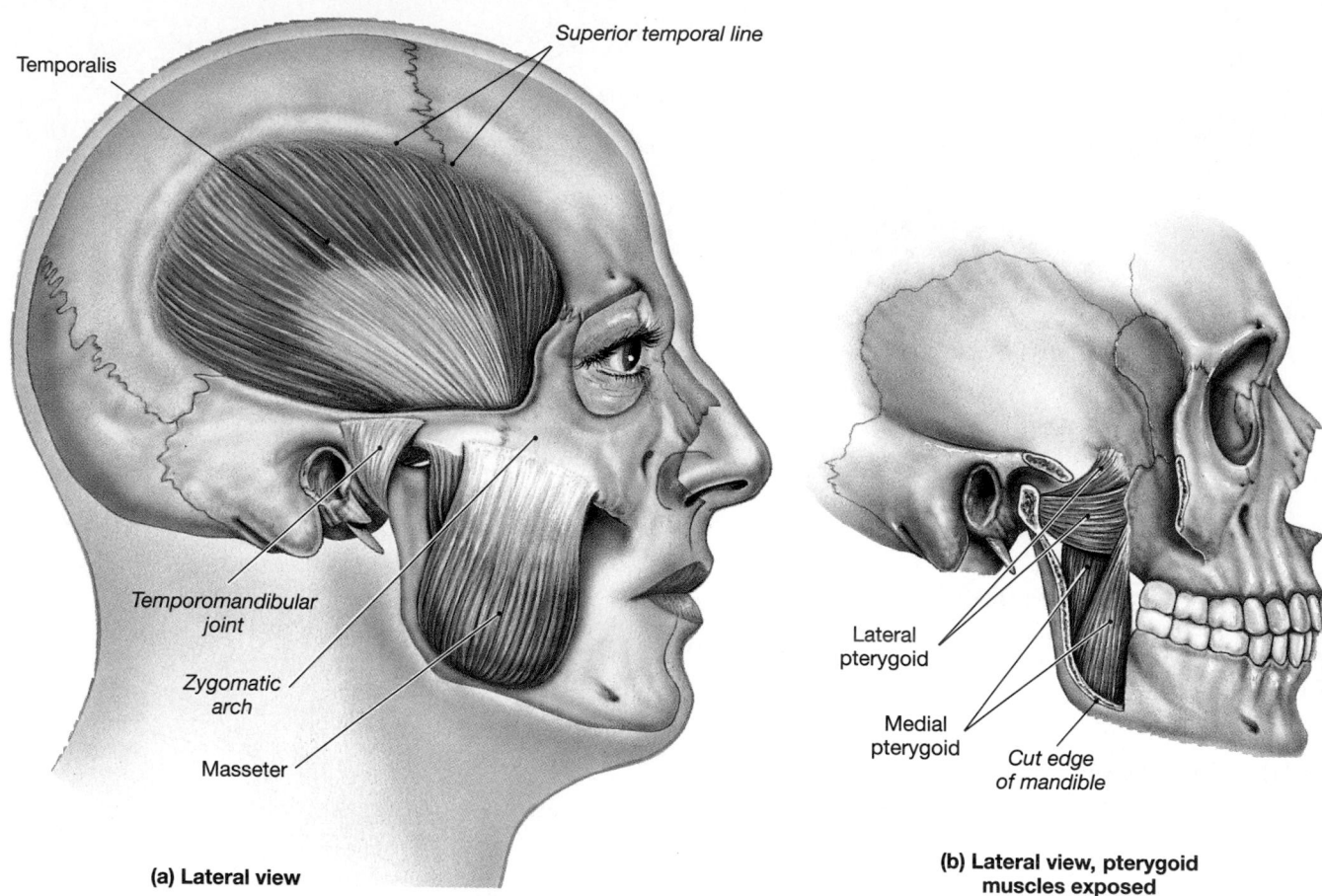

Temporalis

Superior temporal line

Temporomandibular joint

Zygomatic arch

Masseter

(a) Lateral view

Lateral pterygoid

Medial pterygoid

Cut edge of mandible

(b) Lateral view, pterygoid muscles exposed

•FIGURE 11-6 **Muscles of Mastication.** **(a)** The temporalis passes medial to the zygomatic arch to insert on the coronoid process of the mandible. The masseter inserts on the angle and lateral surface of the mandible. **(b)** The location and orientation of the pterygoid muscles can be seen after the overlying muscles, along with a portion of the mandible, are removed. *See also Figures 7-3 and 7-12, pp. 201–202, 211.* [AM] *Plate 4.3*

Muscles of Mastication

The muscles of mastication (Figure 11-6• and Table 11-4) move the mandible at the temporomandibular joint. The large **masseter** is the strongest jaw muscle. The **temporalis** assists in elevation of the mandible. The **pterygoid** muscles, used in various combinations, can elevate, depress, or protract the mandible or slide it from side to side, a movement called *lateral excursion*. These movements are important in making efficient use of your teeth while you chew foods of various consistencies. The muscles of mastication are innervated by the fifth cranial nerve, the *trigeminal nerve*.

TABLE 11-4 **Muscles of Mastication** *(Figure 11-6)*

Muscle	Origin	Insertion	Action	Innervation
Masseter	Zygomatic arch	Lateral surface of mandibular ramus	Elevates mandible	Trigeminal nerve (N V), mandibular branch
Temporalis	Along temporal lines of skull	Coronoid process of mandible	Elevates mandible	As above
Pterygoids (medial and lateral)	Lateral pterygoid plate	Medial surface of mandibular ramus	*Medial*: Elevates the mandible and closes the jaws, or moves mandible side to side	As above
			Lateral: Opens jaws, protrudes mandible, or moves mandible side to side	As above

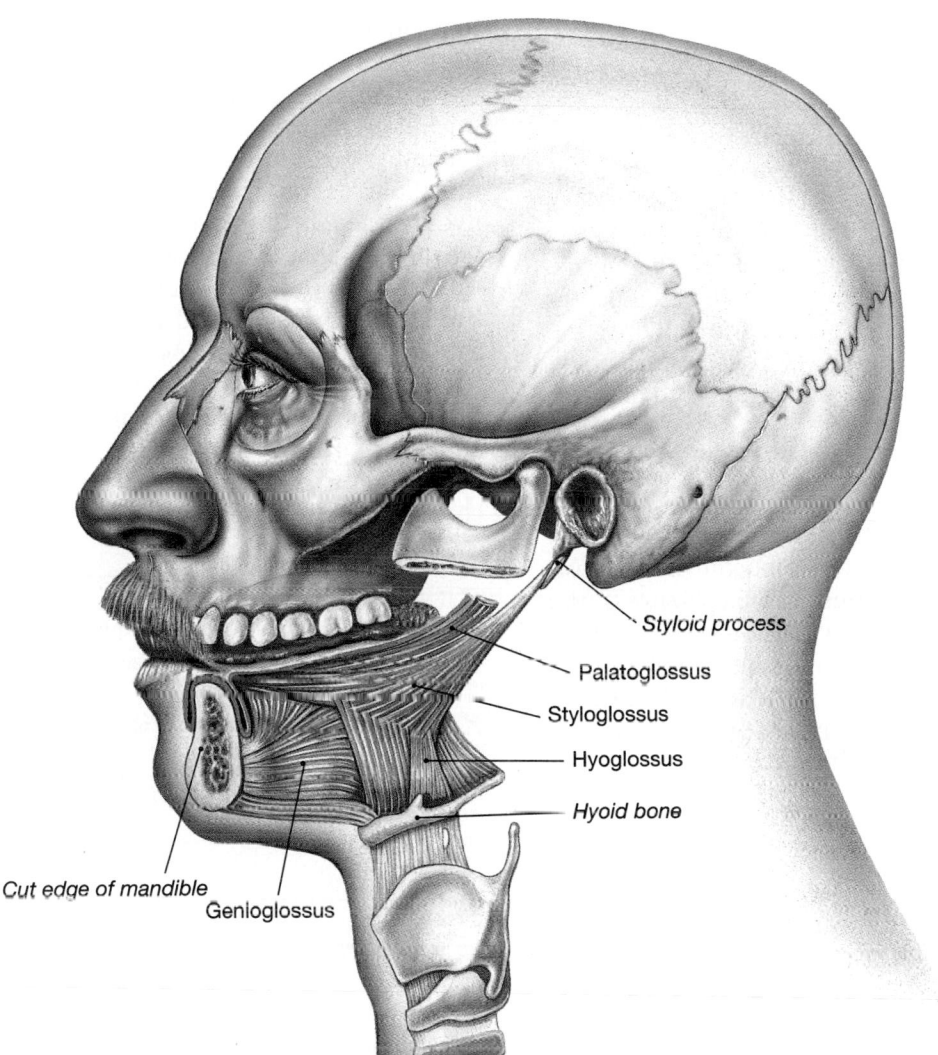

●**FIGURE 11-7** Muscles of the Tongue. *See also Figure 7-3, pp. 201–202.*

Muscles of the Tongue

The muscles of the tongue have names ending in *glossus,* the Greek word for "tongue." Once you can recall the structures referred to by *palato-, stylo-, genio-,* and *hyo-,* you shouldn't have much trouble with this group. The **palatoglossus** originates at the palate, the **styloglossus** at the styloid process, the **genioglossus** at the chin, and the **hyoglossus** at the hyoid bone (Figure 11-7●). These muscles, used in various combinations, move the tongue in the delicate and complex patterns necessary for speech, and they manipulate food within the mouth in preparation for swallowing. As indicated in Table 11-5, they are innervated by the *hypoglossal nerve* (N XII), a cranial nerve whose name indicates its function as well as its location.

TABLE 11-5 Muscles of the Tongue *(Figure 11-7)*

Muscle	Origin	Insertion	Action	Innervation
Genioglossus	Medial surface of mandible around chin	Body of tongue, hyoid bone	Depresses and protracts tongue	Hypoglossal nerve (N XII)
Hyoglossus	Body and greater cornu of hyoid bone	Side of tongue	Depresses and retracts tongue	As above
Palatoglossus	Anterior surface of soft palate	As above	Elevates tongue, depresses soft palate	As above
Styloglossus	Styloid process of temporal bone	Via side to the tip and base of tongue	Retracts tongue, elevates side	As above

Muscles of the Pharynx

The muscles of the pharynx (Figure 11-8● and Table 11-6) are responsible for initiating the swallowing process. The **pharyngeal constrictors** move materials into the esophagus. The **palatopharyngeus** (pal-a-tō-far-IN-jē-us), **salpingopharyngeus** (sal-pin-gō-far-IN-jē-us), and **stylopharyngeus** (stī-lō-far-IN-jē-us), collectively known as the **laryngeal elevators**, elevate the larynx. The **palatal muscles** raise the soft palate and adjacent portions of the pharyngeal wall and also pull open the entrance to the *auditory tube.* ∞ *[p. 206]* As a result, swallowing repeatedly can help you adjust to pressure changes when you fly or dive.

Anterior Muscles of the Neck

The anterior muscles of the neck (Figure 11-9● and Table 11-7) include (1) muscles that control the position of the larynx, (2) muscles that depress the mandible and tense the floor of the mouth, and (3) muscles that provide a stable foundation for muscles of the tongue and pharynx. The **digastric** (dī-GAS-trik) has two bellies, as the name implies (*di-*, two + *gaster*, stomach). One belly extends from the chin to the hyoid bone, and the other continues from the hyoid bone to the mastoid portion of the temporal bone. Depending on which of these bellies contracts and whether fixator muscles are stabilizing the position of the hyoid, this muscle can open the mouth by depressing the mandible, or it can elevate the larynx by raising the hyoid bone. The digastric overlies the broad, flat **mylohyoid,** which provides a muscular floor to the mouth. The **stylohyoid** forms a muscular connection between the hyoid apparatus and the styloid process of the skull. The **sternocleidomastoid** (ster-nō-klī-dō-MAS-toyd) extends from the clavicles and the sternum to the mastoid region of the skull (Figures 11-4, p. 325, and 11-9●). The other members of this group are straplike muscles that run between the sternum and the chin.

☑ If you were contracting and relaxing your masseter muscle, what would you probably be doing?

☑ What facial muscle would you expect to be well developed in a trumpet player?

☑ Why can swallowing help alleviate the pressure sensations at the eardrum when an airplane is changing altitude?

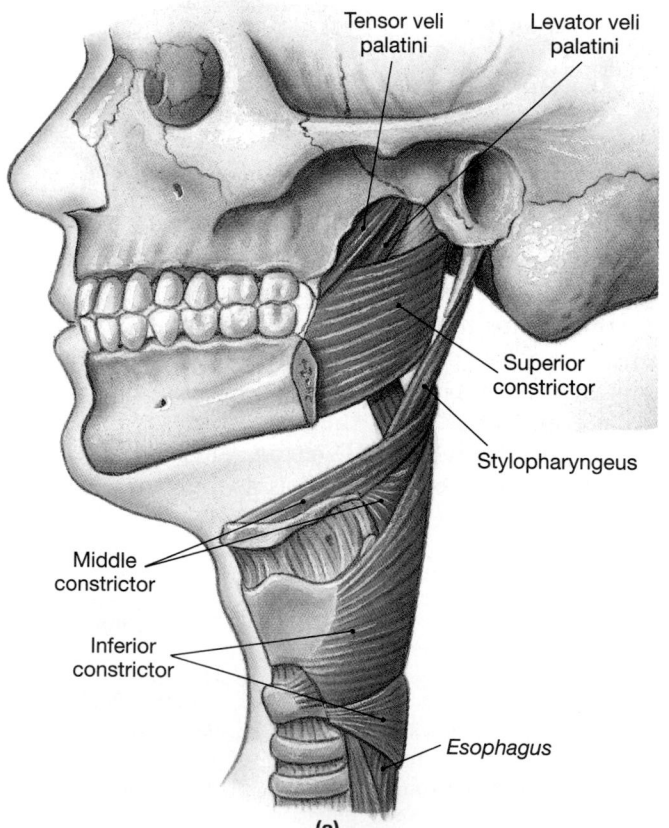

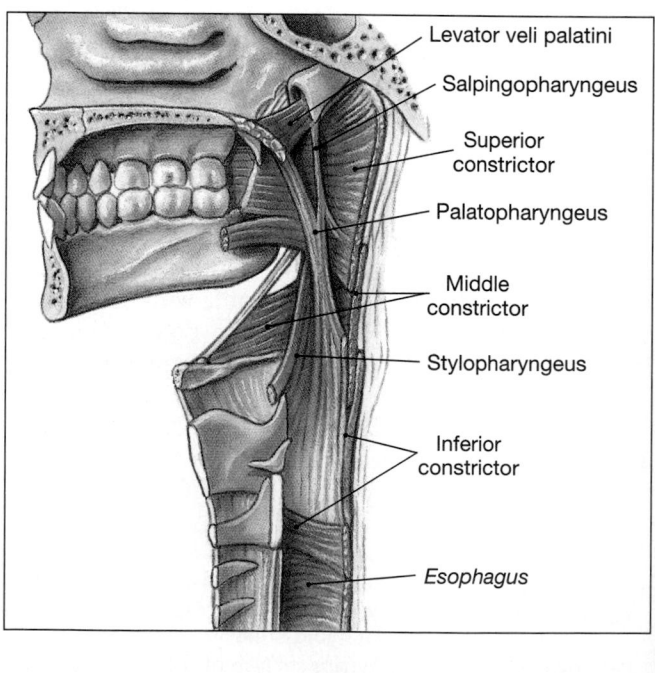

(a) **(b)**

●**FIGURE 11-8** **Muscles of the Pharynx.** **(a)** Lateral view. **(b)** Midsagittal section. *See also Figure 7-4, p. 203.*

TABLE 11-6 Muscles of the Pharynx *(Figure 11-8)*

Muscle	Origin	Insertion	Action	Innervation
Pharyngeal constrictors			Constrict pharynx to propel bolus into esophagus	Branches of pharyngeal plexus (N X)
Superior constrictor	Pterygoid process of sphenoid bone, medial surfaces of mandible	Median raphe attached to occipital bone		N X
Middle constrictor	Cornu of hyoid bone	Median raphe		N X
Inferior constrictor	Cricoid and thyroid cartilages of larynx	Median raphe		N X
Laryngeal elevators*			Elevate larynx	Branches of pharyngeal plexus (N IX & X)
Palatopharyngeus	Soft palate	Thyroid cartilage		N X
Salpingopharyngeus	Cartilage around the inferior portion of the auditory tube	Thyroid cartilage		N X
Stylopharyngeus	Styloid process of temporal bone	Thyroid cartilage		N X
Palatal muscles				
Levator veli palatini	Petrous portion of temporal bone, tissues around the auditory tube	Soft palate	Elevates soft palate	N XI
Tensor veli palatini	Spine of sphenoid bone, tissues around the auditory tube	Soft palate	As above	N V

*Assisted by the thyrohyoid, geniohyoid, stylohyoid, and hyoglossus muscles, discussed in Tables 11-5 and 11-7.

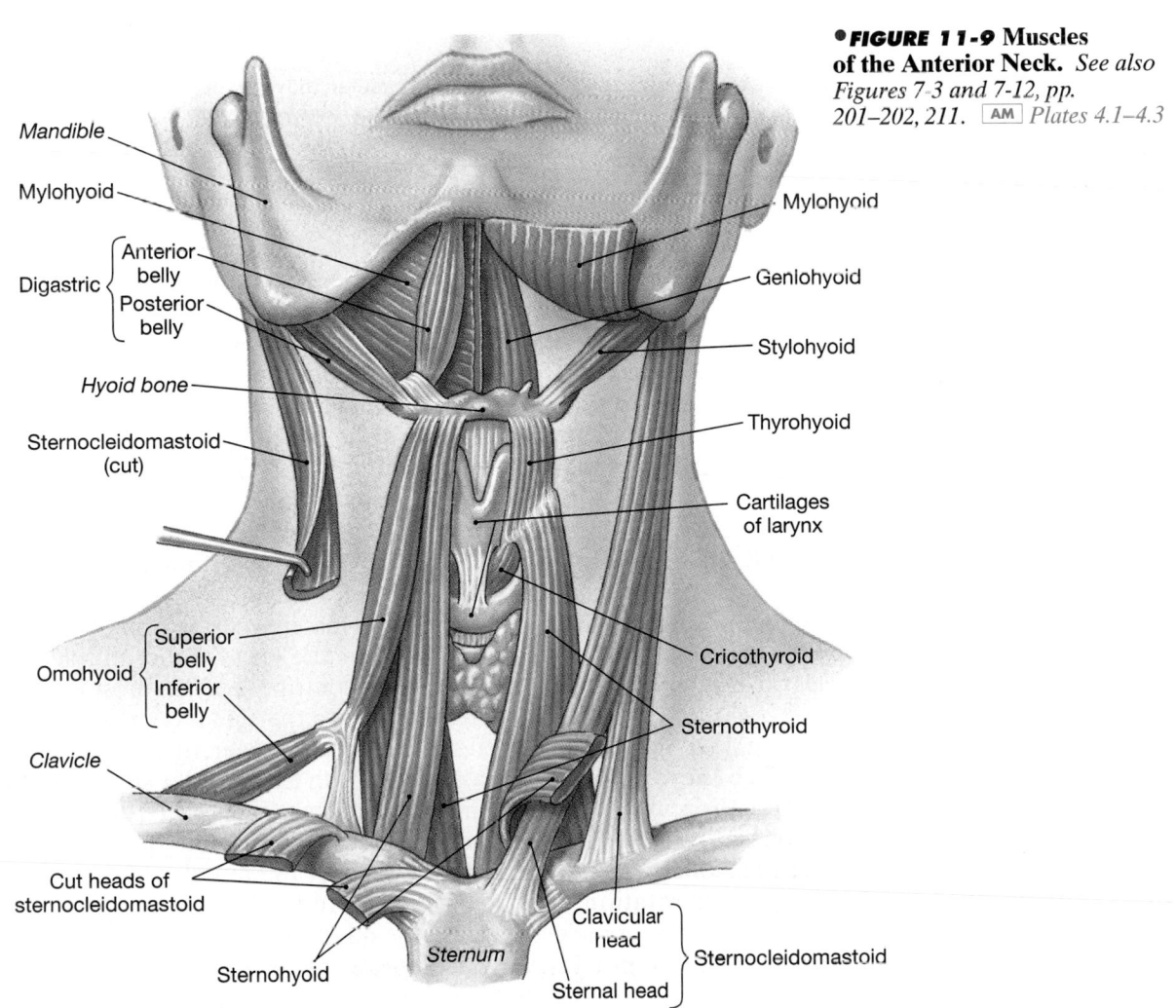

•FIGURE 11-9 Muscles of the Anterior Neck. *See also Figures 7-3 and 7-12, pp. 201–202, 211.* [AM] *Plates 4.1–4.3*

TABLE 11-7 Anterior Muscles of the Neck *(Figure 11-9)*

Muscle	Origin	Insertion	Action	Innervation
Digastric	Two bellies: *posterior* from mastoid region of temporal; *anterior* from inferior surface of mandible at chin	Hyoid bone	Depresses mandible and/or elevates larynx	Trigeminal nerve (N V), mandibular branch, to anterior belly Facial nerve (N VII) to posterior belly
Geniohyoid	Medial surface of mandible at chin	Hyoid bone	As above and pulls hyoid bone anteriorly	Cervical nerve C_1 via hypoglossal nerve (N XII)
Mylohyoid	Mylohyoid line of mandible	Median connective tissue band (raphe) that runs to hyoid bone	Elevates floor of mouth, elevates hyoid bone, and/or depresses mandible	Trigeminal nerve (N V), mandibular branch
Omohyoid	Central tendon attaches to clavicle and 1st rib	Two bellies: *superior* attaches to hyoid bone; *inferior* to superior margin of scapula	Depresses hyoid bone and larynx	Cervical spinal nerves C_2–C_3
Sternohyoid	Clavicle and manubrium	Hyoid bone	As above	Cervical spinal nerves C_1–C_3
Sternothyroid	Dorsal surface of manubrium and 1st costal cartilage	Thyroid cartilage of larynx	As above	As above
Stylohyoid	Styloid process of temporal bone	Hyoid bone	Elevates larynx	Facial nerve (N VII)
Thyrohyoid	Thyroid cartilage of larynx	Hyoid bone	Elevates thyroid, depresses hyoid bone	Cervical spinal nerves C_1–C_2 via hypoglossal nerve (N XII)
Sternocleido-mastoid	Two bellies: *clavicular head* attaches to sternal end of clavicle; *sternal head* attaches to manubrium	Mastoid region of skull	Together, they flex the neck; alone, one side bends head toward shoulder and turns face to opposite side	Accessory nerve (N XI) and cervical spinal nerves (C_2–C_3) of cervical plexus

Muscles of the Spine

The muscles of the spine are covered by more-superficial back muscles, such as the trapezius and latissimus dorsi (Figure 11-3●, pp. 322–323). The *spinal extensors*, or **erector spinae**, include superficial and deep layers. The relatively superficial layer can be divided into **spinalis, longissimus,** and **iliocostalis** divisions (Figure 11-10● and Table 11-8, p. 334). In the lower lumbar and sacral regions, the boundary between the longissimus and iliocostalis muscles becomes indistinct, and they are sometimes known as the **sacrospinalis** muscles. When contracting together, the erector spinae extend the spinal column. When the muscles on only one side contract, the spine is bent laterally.

Deep to the spinalis muscles, the deep muscles of the spine interconnect and stabilize the vertebrae. These muscles include the **semispinalis muscles** and the **multifidus, interspinales, intertransversarii,** and **rotatores** (Figure 11-10a,b●). In various combinations, they produce slight extension or rotation of the spinal column. They are also important in making delicate adjustments in the positions of individual vertebrae, and they stabilize adjacent vertebrae. If injured, these muscles can start a cycle of pain → muscle stimulation → contraction → pain. This cycle can lead to pressure on adjacent spinal nerves, leading to sensory losses as well as limiting mobility. Many of the warm-up and stretching exercises recommended before athletic events are intended to prepare these small but very important muscles for their supporting role.

The muscles of the spine include many dorsal extensors but few ventral flexors. The spinal column does not need a massive series of flexor muscles because (1) many of the large trunk muscles flex the spine when they contract, and (2) most of the body weight lies anterior to the spinal column, so gravity tends to flex the spine. However, there are a few spinal flexors associated with the anterior surface of the spinal column. In the neck (Figure 11-10c●), the **longus capitis** and the **longus colli** rotate or flex the neck, depending on whether the muscles of one or both sides are contracting. In the lumbar region (Figure 11-10a●), the large **quadratus lumborum** muscles flex the spine and depress the ribs.

•*FIGURE 11-10* **Muscles of the Spine.** *See also Figures 7-1 and 7-23, pp. 198–199, 226.* [AM] *Plate 1*

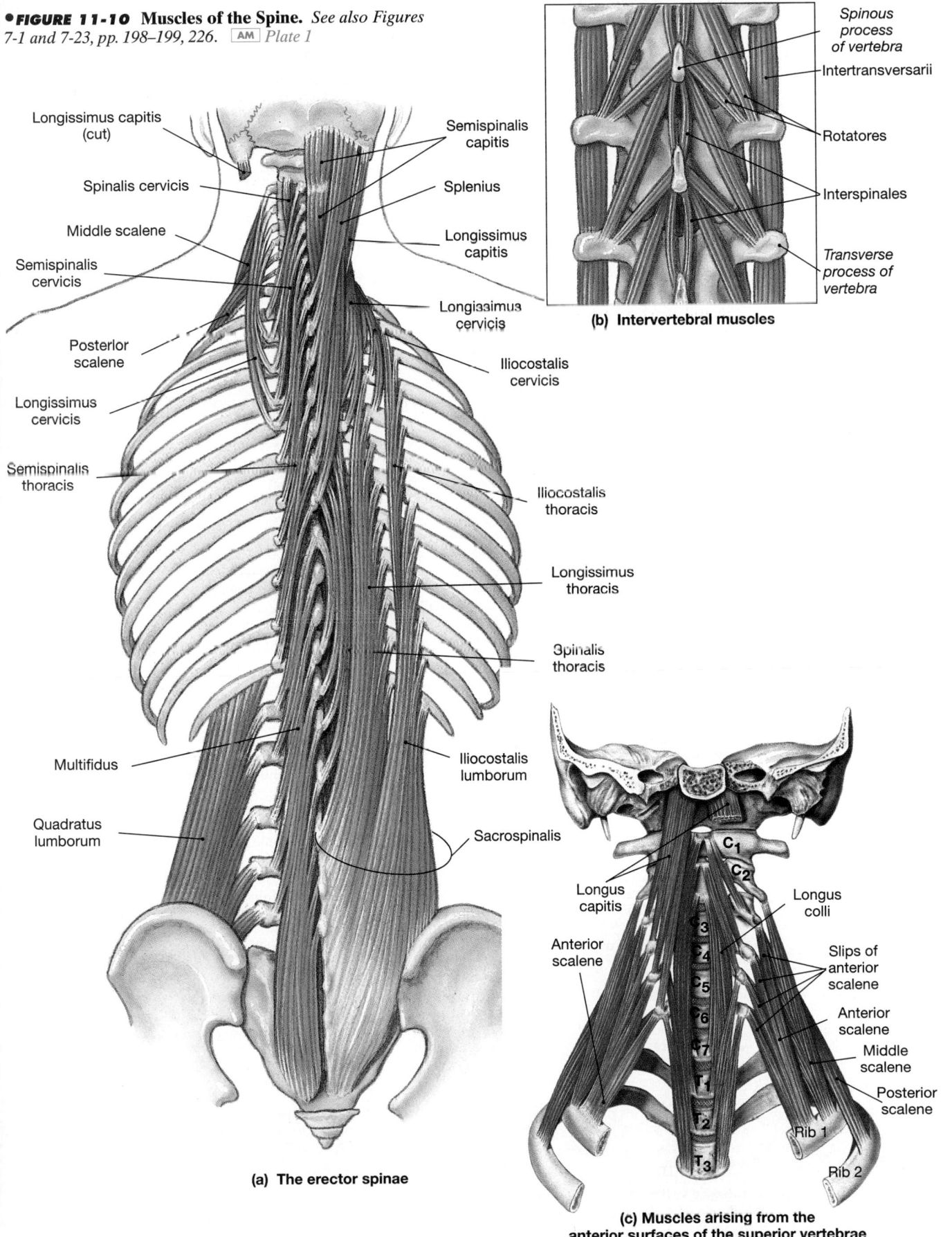

Longissimus capitis (cut)

Spinalis cervicis

Middle scalene

Semispinalis cervicis

Posterior scalene

Longissimus cervicis

Semispinalis thoracis

Multifidus

Quadratus lumborum

Semispinalis capitis

Splenius

Longissimus capitis

Longissimus cervicis

Iliocostalis cervicis

Iliocostalis thoracis

Longissimus thoracis

Spinalis thoracis

Iliocostalis lumborum

Sacrospinalis

(a) The erector spinae

Spinous process of vertebra

Intertransversarii

Rotatores

Interspinales

Transverse process of vertebra

(b) Intervertebral muscles

Longus capitis

Anterior scalene

Longus colli

Slips of anterior scalene

Anterior scalene

Middle scalene

Posterior scalene

C₁
C₂
C3
C4
C5
C6
C7
T1
T2
T3

Rib 1
Rib 2

(c) Muscles arising from the anterior surfaces of the superior vertebrae

TABLE 11-8 Muscles of the Spine *(Figure 11-10)*

Group/Muscle	Origin	Insertion	Action	Innervation
SUPERFICIAL LAYER				
Splenius (Splenius capitis, splenius cervicis)	Spinous processes and ligaments connecting lower cervical and upper thoracic vertebrae	Mastoid process, occipital bone of skull, and upper cervical vertebra	Together, the two sides extend head; alone, each rotates and tilts head to that side	Cervical spinal nerves
Erector spinae				
Spinalis group Spinalis cervicis	Inferior portion of ligamentum nuchae and spinous process of C_7	Spinous process of axis	Extends neck	As above
Spinalis thoracis	Spinous processes of lower thoracic and upper lumbar vertebrae	Spinous processes of upper thoracic vertebrae	Extends spinal column	Thoracic and lumbar spinal nerves
Longissimus group Longissimus capitis	Processes of lower cervical and upper thoracic vertebrae	Mastoid process of temporal bone	Together, the two sides extend head; alone, each rotates and tilts head to that side	Cervical and thoracic spinal nerves
Longissimus cervicis	Transverse processes of upper thoracic vertebrae	Transverse processes of middle and upper cervical vertebrae	As above	As above
Longissimus thoracis	Broad aponeurosis and at transverse processes of lower thoracic and upper lumbar vertebrae; joins iliocostalis to form "sacrospinalis"	Transverse processes of higher vertebrae and inferior surfaces of ribs	Extends and/or bends spine to the side	Thoracic and lumbar spinal nerves
Iliocostalis group Iliocostalis cervicis	Superior borders of vertebrosternal ribs near the angles	Transverse processes of middle and lower cervical vertebrae	Extends or bends neck, elevates ribs	Cervical and upper thoracic spinal nerves
Iliocostalis thoracis	Superior borders of lower 7 ribs medial to the angles	Upper ribs and transverse process of last cervical vertebra	Stabilizes thoracic vertebrae in extension	Thoracic spinal nerves
Iliocostalis lumborum	Sacrospinal aponeurosis and iliac crest	Inferior surfaces of lower 7 ribs near their angles	Extends spine, depresses ribs	Lower thoracic and lumbar spinal nerves
DEEP LAYER (TRANSVERSOSPINALIS)				
Semispinalis group Semispinalis capitis	Processes of lower cervical and upper thoracic vertebrae	Occipital bone, between nuchal lines	Together, the two sides extend head; alone, each extends and tilts head to that side	Cervical spinal nerves
Semispinalis cervicis	Transverse processes of T_1–T_5/T_6	Spinous processes of C_2–C_5	Extends vertebral column and rotates toward opposite side	As above
Semispinalis thoracis	Transverse processes of T_6–T_{10}	Spinous processes of C_5–T_4	As above	Thoracic spinal nerves
Multifidus	Sacrum and transverse processes of each vertebra	Spinous processes of the third or fourth more-superior vertebra	Extends vertebral column and rotates toward opposite side	Cervical, thoracic, and lumbar spinal nerves
Rotatores	Transverse processes of each vertebra	Spinous processes of adjacent, more-superior vertebra	Extends vertebral column and rotates toward opposite side	As above
Interspinales	Spinous processes of each vertebra	Spinous processes of preceding vertebra	Extends vertebral column	As above
Intertransversarii	Transverse processes of each vertebra	Transverse process of preceding vertebra	Bends the vertebral column laterally	As above
SPINAL FLEXORS				
Longus capitis	Transverse processes of each vertebrae	Base of the occipital bone	Together, the two sides bend head forward; alone, each rotates head to that side	Cervical spinal nerves
Longus colli	Anterior surfaces of cervical and upper thoracic vertebrae	Transverse processes of upper cervical vertebrae	Flexes and/or rotates neck; limits hyperextension	As above
Quadratus lumborum	Iliac crest and iliolumbar ligament	Last rib and transverse processes of lumbar vertebrae	Together, they depress ribs; alone, each side flexes spine laterally	Thoracic and lumbar spinal nerves

Oblique and Rectus Muscles

The muscles of the oblique and rectus groups (Table 11-9) lie within the body wall, between the spinous processes of vertebrae and the ventral midline (Figures 11-3, pp. 322–323, and 11-11•). The oblique muscles can compress underlying structures or rotate the spinal column, depending on whether one or both sides are contracting. The rectus muscles are important flexors of the spinal column, acting in opposition to the erector spinae. The oblique and rectus muscles are united by their common embryological origins. We can divide them into cervical, thoracic, and abdominal groups.

The oblique series includes the **scalenes** of the neck (Figure 11-10c•) and the **intercostal** and **transversus** muscles of the thoracic region (Figure 11-11a,b•). The scalenes (*anterior*, *middle*, and *posterior*) elevate the first two ribs and assist in flexion of the neck. In the thorax, the oblique muscles lie between the ribs, and the **external intercostals** cover the **internal intercostals.** Both groups of intercostal muscles are important in res-piratory movements of the ribs. A small **transversus thoracis** crosses the inner surface of the rib cage and is covered by the *pleura*, the serous membrane that lines the pleural cavities. ∞ *[p. 24]* The sternum occupies the place where we might otherwise expect to find thoracic rectus muscles.

The same basic pattern of musculature extends unbroken across the abdominopelvic surface (Figure 11-11a,c•). Here the muscles are called the **external obliques,** the **internal obliques,** the **transversus abdominis,** and the **rectus abdominis.** The rectus abdominis inserts at the xiphoid process and originates near the pubic symphysis. This muscle is longitudinally divided by a median collagenous partition, the **linea alba** (white line) (Figure 11-3a•, p. 322). Transverse **tendinous inscriptions** divide this muscle into segments.

When your abdominal muscles contract forcefully, pressures in your abdominopelvic cavity can skyrocket, and those pressures are applied to internal organs. If you exhale at the same time, the pressure is relieved, because your diaphragm can move upward as your

---

TABLE 11-9 Oblique and Rectus Muscles* *(Figures 11-11, 11-12)*

Group/Muscle	Origin	Insertion	Action	Innervation
OBLIQUE GROUP				
Cervical region				
Scalenes	Transverse and costal processes of cervical vertebrae	Superior surfaces of first two ribs	Elevates ribs and/or flexes neck	Cervical spinal nerves
Thoracic region				
External intercostals	Inferior border of each rib	Superior border of the next rib	Elevate ribs	Intercostal nerves (branches of thoracic spinal nerves)
Internal intercostals	Superior border of each rib	Inferior border of the previous rib	Depress ribs	As above
Transversus thoracis	Medial surface of sternum	Cartilages of ribs	As above	As above
Abdominal region				
External oblique	Lower eight ribs	Linea alba and iliac crest	Compresses abdomen, depresses ribs, flexes or bends spine	Intercostal, iliohypogastric, and ilioinguinal nerves
Internal oblique	Lumbodorsal fascia and iliac crest	Lower ribs, xiphoid process, and linea alba	As above	As above
Transversus abdominis	Cartilages of lower ribs, iliac crest, and lumbodorsal fascia	Linea alba and pubis	Compresses abdomen	As above
RECTUS GROUP				
Cervical region	*See muscles in Table 11-6*			
Thoracic region				
Diaphragm	Xiphoid process, cartilages of ribs 4–10, and anterior surfaces of lumbar vertebrae	Central tendinous sheet	Contraction expands thoracic cavity, compresses abdominopelvic cavity	Phrenic nerves $(C_3–C_5)$
Abdominal region				
Rectus abdominis	Superior surface of pubis around symphysis	Inferior surfaces of costal cartilages (ribs 5–7) and xiphoid process	Depresses ribs, flexes vertebral column	Intercostal nerves $(T_7–T_{12})$

*Where appropriate, spinal nerves involved are given in parentheses.

•**FIGURE 11-11** **Oblique and Rectus Muscles.** *See also Figures 7-1 and 7-23, pp. 198–199, 226.* [AM] *Plate 6.4a,b*

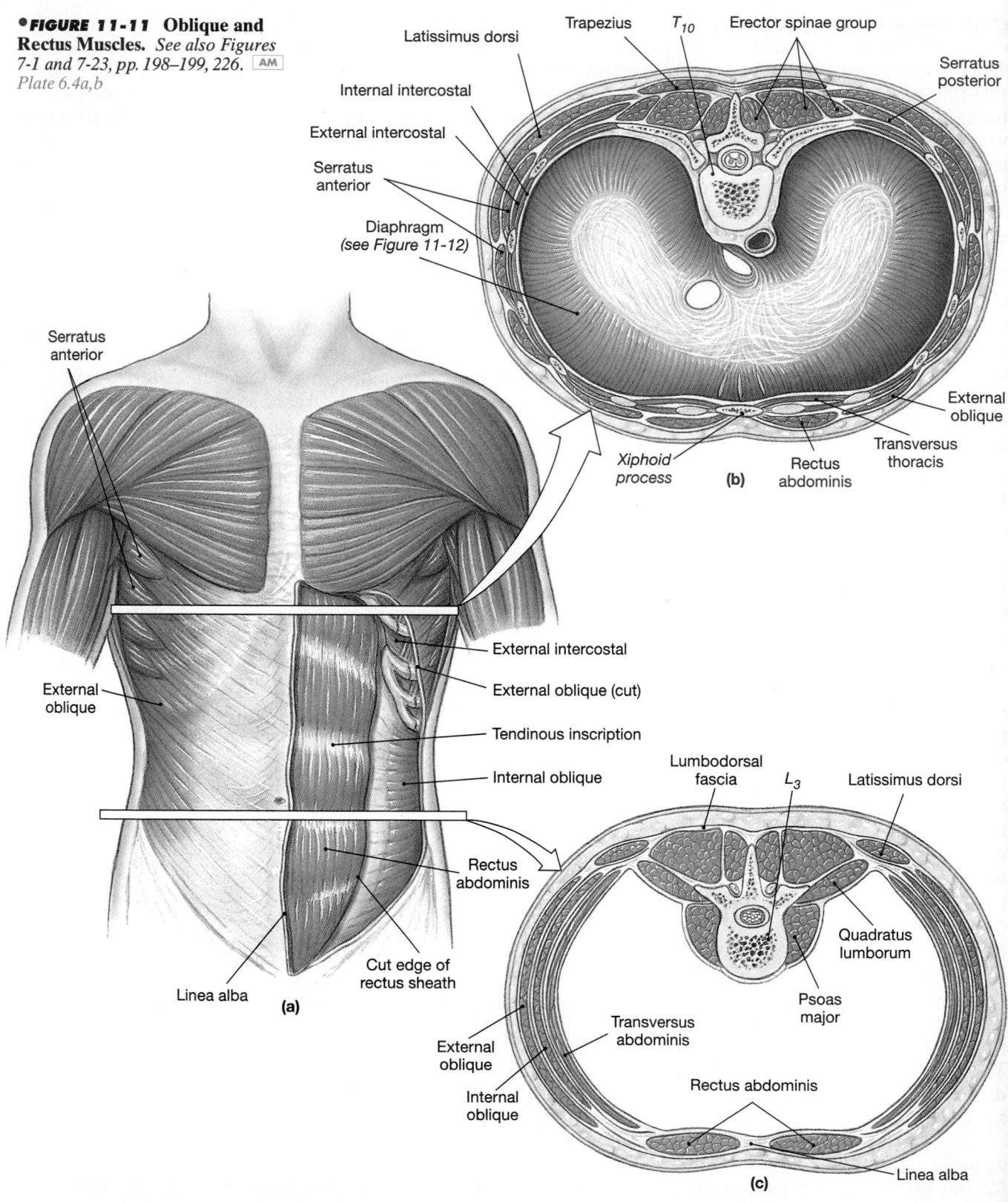

Serratus anterior

Latissimus dorsi

Internal intercostal

External intercostal

Serratus anterior

Diaphragm (see Figure 11-12)

Trapezius   T₁₀   Erector spinae group

Serratus posterior

External oblique

Transversus thoracis

Xiphoid process   (b)   Rectus abdominis

External intercostal

External oblique (cut)

Tendinous inscription

Internal oblique

Rectus abdominis

Cut edge of rectus sheath

External oblique

Linea alba   (a)

Lumbodorsal fascia   L₃   Latissimus dorsi

Quadratus lumborum

Psoas major

Transversus abdominis

External oblique

Internal oblique

Rectus abdominis

Linea alba   (c)

lungs collapse. But during vigorous isometric exercises or when you lift a weight while holding your breath, pressure in the abdominopelvic cavity can rise high enough to cause a variety of problems, among them the development of a hernia. A **hernia** develops when an organ protrudes through an abnormal opening. The most common hernias are inguinal and diaphragmatic hernias. *Inguinal hernias* typically occur in males, at the *inguinal canal,* the site where blood vessels, nerves, and reproductive ducts pass through the abdominal wall to reach the testes. Elevated abdominal pressure can force open the inguinal canal and push a portion of the intestine into the pocket created. *Diaphragmatic hernias* develop when visceral organs, such as a portion of the stomach, are forced into the left pleural cavity. If herniated structures become trapped or twisted, surgery may be required to prevent serious complications, such as intestinal blockage or tissue degeneration due to the interruption of blood flow. AM *Hernias*

The Diaphragm

The term *diaphragm* refers to any muscular sheet that forms a wall. When used without a modifier, however, **diaphragm,** or *diaphragmatic muscle* (Figure 11-12●), specifies the muscular partition that separates the abdominopelvic and thoracic cavities. We include this muscle here because it develops in association with the other muscles of the chest wall. The diaphragm is a major muscle of respiration.

Muscles of the Pelvic Floor

The muscles of the pelvic floor extend from the sacrum and coccyx to the ischium and pubis. These muscles (Figure 11-13● and Table 11-10, p. 339) (1) support the organs of the pelvic cavity, (2) flex the sacrum and coccyx, and (3) control the movement of materials through the urethra and anus.

The boundaries of the **perineum** are established by the inferior margins of the pelvis. If you draw a line between the ischial tuberosities, you will divide the perineum into two triangles—an anterior, **urogenital triangle** and a posterior, **anal triangle.** The superficial muscles of the anterior triangle are the muscles of the external genitalia. They cover deeper muscles that strengthen the pelvic floor and encircle the urethra. These muscles constitute the **urogenital diaphragm,** a muscular layer that extends between the pubic bones.

An even more extensive muscular sheet, the **pelvic diaphragm,** forms the muscular foundation of the anal triangle. This layer, covered by the the urogenital diaphragm, extends as far as the pubic symphysis.

The urogenital and pelvic diaphragms do not completely close the pelvic outlet, for the urethra, vagina, and anus pass through them to open on the external surface. Muscular sphincters surround their openings and permit voluntary control of urination and defecation. Muscles, nerves, and blood vessels also pass through the pelvic outlet as they travel to or from the lower limbs.

☑ Damage to the external intercostal muscles would interfere with what important process?

☑ If someone hit you in your rectus abdominis muscle, how would your body position change?

☑ After spending an afternoon carrying heavy boxes from his basement to his attic, Joe complains that the muscles in his back hurt. What muscle(s) is (are) most likely sore?

THE APPENDICULAR MUSCLES

The appendicular musculature positions and stabilizes the pectoral and pelvic girdles and moves the upper and lower limbs. There are two major groups of appendicular muscles: (1) the muscles of the shoulders and upper limbs and (2) the muscles of the pelvic girdle and lower limbs. The functions and required ranges of motion are very different from one group to another. In addition to increasing the mobility of the arms, the muscular connections between the pectoral girdle and the axial skeleton must act as shock absorbers. For example, while you jog, you can still perform delicate hand movements, because the muscular connections between the axial and appendicular skeletons smooth out the bounces in your stride. In contrast, the pelvic girdle has evolved to transfer weight from the axial to the appendicular skeleton. A mus-

●**FIGURE 11-12** **The Diaphragm.** *See also Figures 7-1 and 7-23, pp. 198–199, 226.*

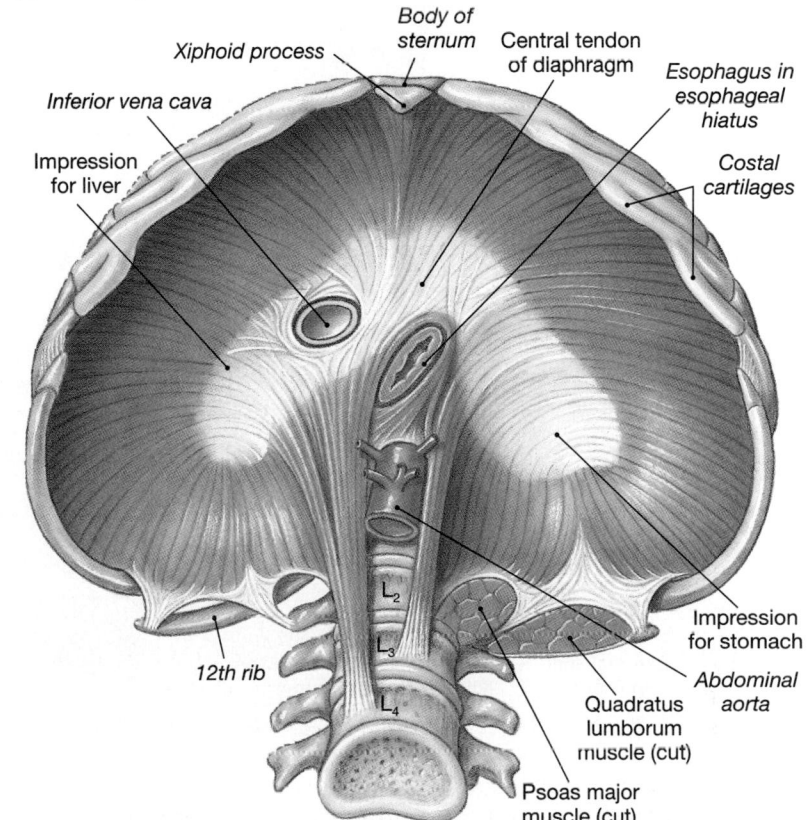

Xiphoid process

Body of sternum

Central tendon of diaphragm

Esophagus in esophageal hiatus

Inferior vena cava

Impression for liver

Costal cartilages

L₂

L₃

L₄

12th rib

Impression for stomach

Abdominal aorta

Quadratus lumborum muscle (cut)

Psoas major muscle (cut)

●FIGURE 11-13 **Muscles of the Pelvic Floor.** *See also Figures 7-1 and 8-9, pp. 198–199, 244.*

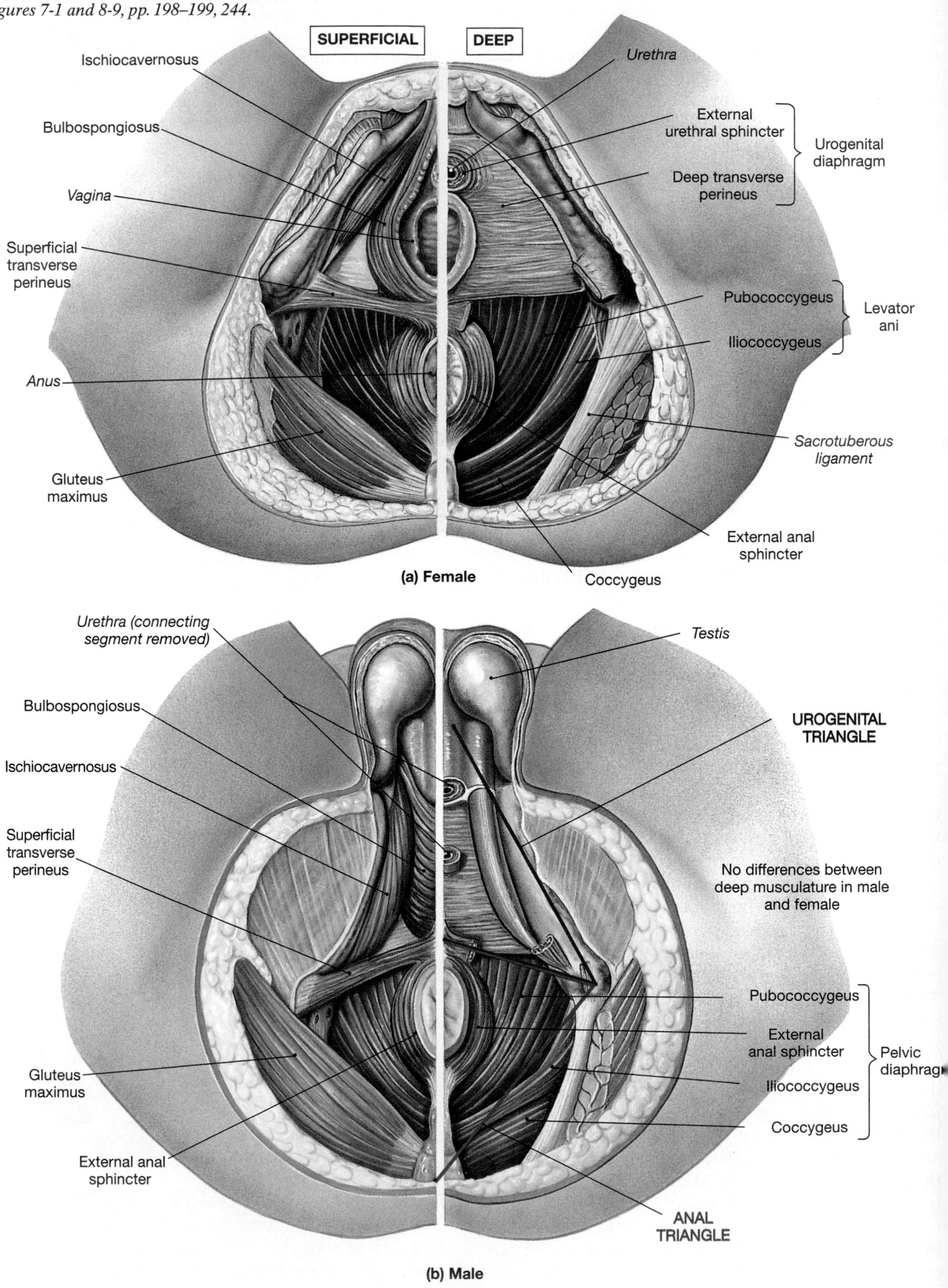

| SUPERFICIAL | DEEP |

Ischiocavernosus

Bulbospongiosus

Vagina

Superficial transverse perineus

Anus

Gluteus maximus

Urethra

External urethral sphincter

Deep transverse perineus

} Urogenital diaphragm

Pubococcygeus

Iliococcygeus

} Levator ani

Sacrotuberous ligament

External anal sphincter

Coccygeus

(a) Female

Urethra (connecting segment removed)

Bulbospongiosus

Ischiocavernosus

Superficial transverse perineus

Gluteus maximus

External anal sphincter

Testis

UROGENITAL TRIANGLE

No differences between deep musculature in male and female

Pubococcygeus

External anal sphincter

Iliococcygeus

Coccygeus

} Pelvic diaphragm

ANAL TRIANGLE

(b) Male

TABLE 11-10 Muscles of the Pelvic Floor *(Figure 11-13)*

Group/Muscle	Origin	Insertion	Action	Innervation
UROGENITAL TRIANGLE **Superficial muscles** Bulbospongiosus:				
Males	Collagen sheath at base of penis; fibers cross over urethra	Median raphe and central tendon of perineum	Compresses base and stiffens penis, ejects urine or semen	Pudendal nerve, perineal branch (S_2–S_4)
Females	Collagen sheath at base of clitoris; fibers run on either side of urethral and vaginal openings	Central tendon of perineum	Compresses and stiffens clitoris, narrows vaginal opening	As above
Ischiocavernosus	Ischial ramus and tuberosity	Pubic symphysis anterior to base of penis or clitoris	Compresses and stiffens penis or clitoris	As above
Superficial transverse perineus	Ischial ramus	Central tendon of perineum	Stabilizes central tendon of perineum	As above
Deep Muscles: The urogenital diaphragm Deep transverse perineus	Ischial ramus	Median raphe of urogenital diaphragm	Stabilizes central tendon of perineum	Pudendal nerve, perineal branch (S_2–S_4)
External urethral sphincter:				
Males	Ischial and pubic rami	To median raphe at base of penis; inner fibers encircle urethra	Closes urethra, compresses prostate and bulbourethral glands	As above
Females	Ischial and pubic rami	To median raphe; inner fibers encircle urethra	Closes urethra, compresses vagina and greater vestibular glands	As above
ANAL TRIANGLE **Pelvic diaphragm** Coccygeus	Ischial spine	Lateral, inferior borders of the sacrum	Flexes coccyx and coccygeal vertebrae	Lower sacral nerves (S_4–S_5)
External anal sphincter	Via tendon from coccyx	Encircles anal opening	Closes anal opening	Pudendal nerve, hemorrhoidal branch (S_2–S_4)
Levator ani: Iliococcygeus	Ischial spine, pubis	Coccyx	Tenses floor of pelvis, flexes coccyx, elevates and retracts anus	Pudendal nerve (S_2–S_4)
Pubococcygeus	Inner margins of pubis	Coccyx	As above	As above

cular connection would reduce the efficiency of the transfer, and the emphasis is on strength rather than versatility. Figure 11-14● provides an introduction to the organization of the appendicular muscles.

Muscles of the Shoulders and Upper Limbs

Muscles associated with the shoulders and upper limbs can be divided into four groups: (1) muscles that po-

sition the pectoral girdle, (2) muscles that move the arm, (3) muscles that move the forearm and hand, and (4) muscles that move the fingers.

Muscles That Position the Pectoral Girdle

The large, superficial **trapezius** muscles cover the back and portions of the neck, reaching to the base of the skull. These muscles originate along the midline of the neck and back and insert on the clavicles and the scapu-

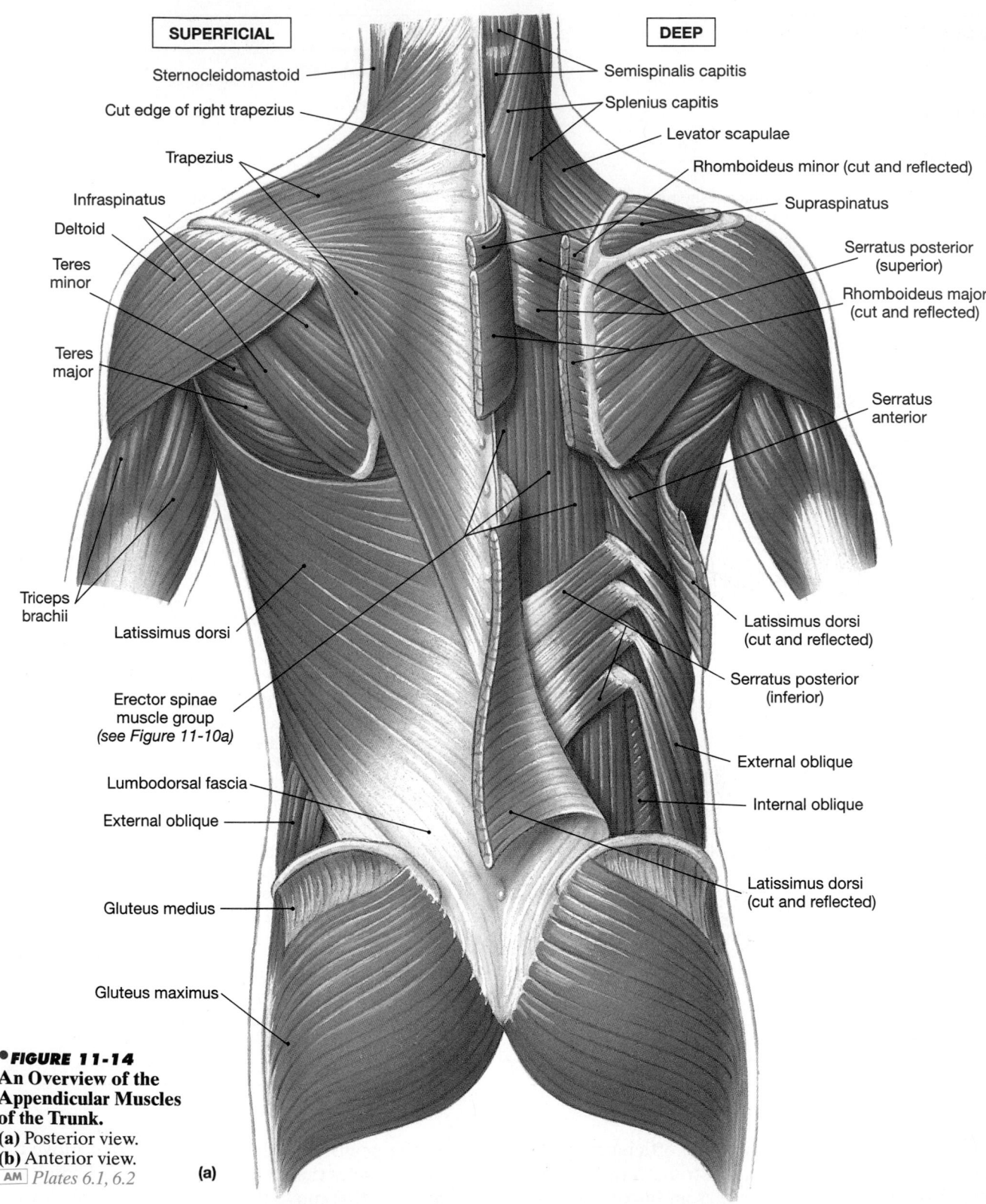

SUPERFICIAL

DEEP

Sternocleidomastoid

Cut edge of right trapezius

Trapezius

Infraspinatus

Deltoid

Teres minor

Teres major

Triceps brachii

Latissimus dorsi

Erector spinae muscle group *(see Figure 11-10a)*

Lumbodorsal fascia

External oblique

Gluteus medius

Gluteus maximus

Semispinalis capitis

Splenius capitis

Levator scapulae

Rhomboideus minor (cut and reflected)

Supraspinatus

Serratus posterior (superior)

Rhomboideus major (cut and reflected)

Serratus anterior

Latissimus dorsi (cut and reflected)

Serratus posterior (inferior)

External oblique

Internal oblique

Latissimus dorsi (cut and reflected)

●**FIGURE 11-14**
An Overview of the
Appendicular Muscles
of the Trunk.
(a) Posterior view.
(b) Anterior view.
AM *Plates 6.1, 6.2*

(a)

lar spines (Figures 11-14 and 11-15a●, p. 342). The trapezius muscles are innervated by more than one nerve (Table 11-11), and specific regions can be made to contract independently. As a result, their actions are quite varied.

Removing the trapezius reveals the **rhomboideus** muscles and the **levator scapulae** (Figure 11-15a●). These muscles are attached to the dorsal surfaces of the cervical and thoracic vertebrae. They insert along the vertebral border of each scapula, between the superior

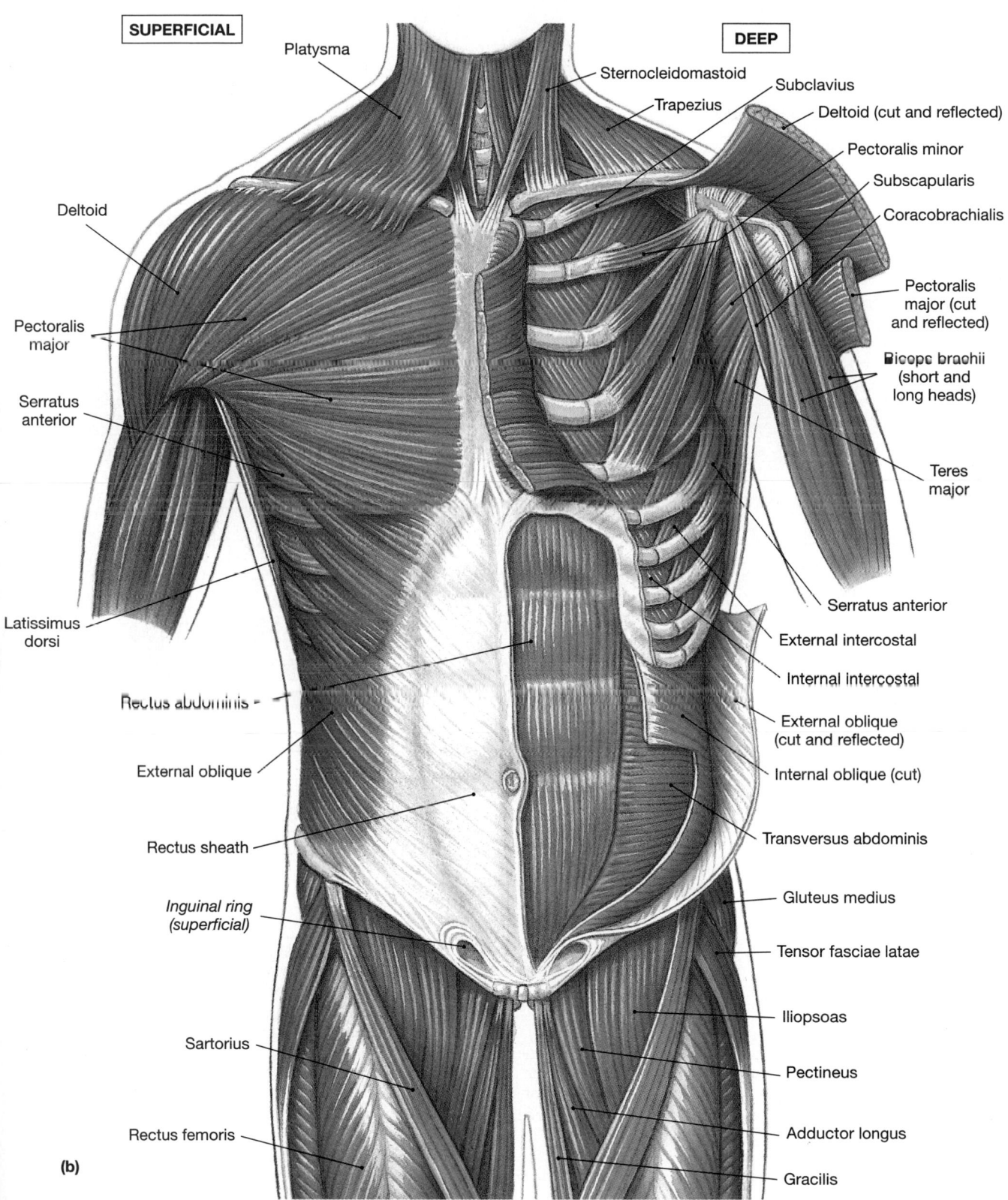

SUPERFICIAL

Platysma

Deltoid

Pectoralis
major

Serratus
anterior

Latissimus
dorsi

Rectus abdominis

External oblique

Rectus sheath

Inguinal ring
(superficial)

Sartorius

Rectus femoris

(b)

DEEP

Sternocleidomastoid

Trapezius

Subclavius

Deltoid (cut and reflected)

Pectoralis minor

Subscapularis

Coracobrachialis

Pectoralis
major (cut
and reflected)

Biceps brachii
(short and
long heads)

Teres
major

Serratus anterior

External intercostal

Internal intercostal

External oblique
(cut and reflected)

Internal oblique (cut)

Transversus abdominis

Gluteus medius

Tensor fasciae latae

Iliopsoas

Pectineus

Adductor longus

Gracilis

and inferior angles. Contraction of a rhomboideus muscle adducts (retracts) the scapula on that side. The levator scapulae, as its name implies, elevates the scapula.

On the chest, the **serratus anterior** originates along
the anterior surfaces of several ribs (Figures 11-3, pp.
322–323, and 11-15b●). This fan-shaped muscle inserts
along the anterior margin of the vertebral border of
the scapula. When the serratus anterior contracts, it
abducts (protracts) the scapula and swings the shoulder anteriorly.

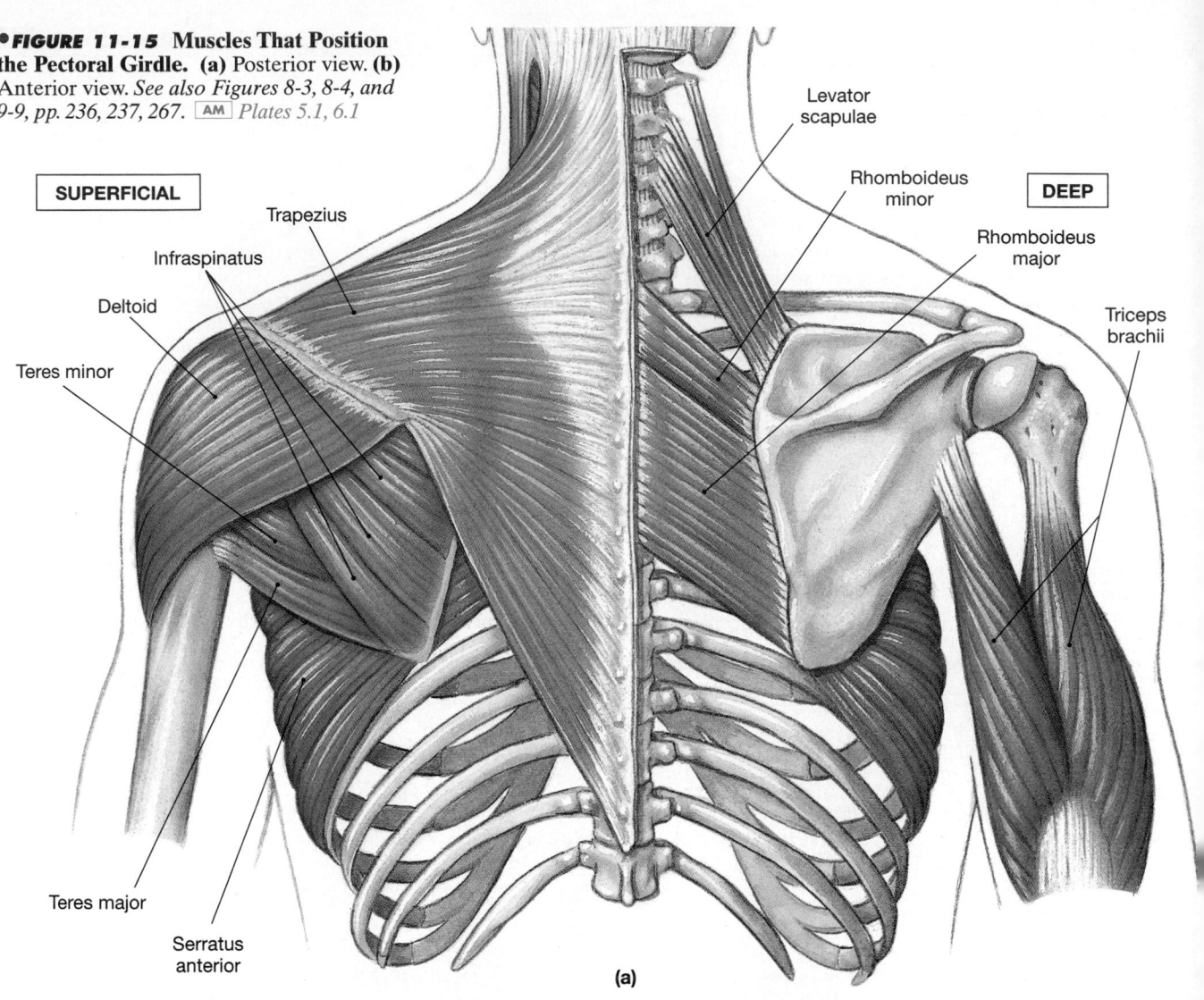

•*FIGURE 11-15* **Muscles That Position the Pectoral Girdle.** **(a)** Posterior view. **(b)** Anterior view. *See also Figures 8-3, 8-4, and 9-9, pp. 236, 237, 267.* [AM] *Plates 5.1, 6.1*

SUPERFICIAL

Trapezius

Infraspinatus

Deltoid

Teres minor

Teres major

Serratus anterior

Levator scapulae

Rhomboideus minor

Rhomboideus major

DEEP

Triceps brachii

(a)

TABLE 11-11 **Muscles That Move the Pectoral Girdlem** *(Figures 11-14, 11-15)*

Muscle	Origin	Insertion	Action	Innervation
Levator scapulae	Transverse processes of first 4 cervical vertebrae	Vertebral border of scapula near superior angle	Elevates scapula	Cervical nerves C_3–C_4 and dorsal scapular nerve (C_5)
Pectoralis minor	Ventral surfaces of ribs 3–5	Coracoid process of scapula	Depresses and protracts shoulder; rotates scapula so glenoid cavity moves inferiorly (downward rotation); elevates ribs if scapula is stationary	Medial pectoral nerve (C_8, T_1)
Rhomboideus major	Spinous processes of upper thoracic vertebrae	Vertebral border of scapula from spine to inferior angle	Adducts and performs downward rotation	Dorsal scapular nerve (C_5)
Rhomboideus minor	Spinous processes of vertebrae C_7–T_1	Vertebral border of scapula near spine	As above	As above
Serratus anterior	Anterior and superior margins of ribs 1–9	Anterior surface of vertebral border of scapula	Protracts shoulder, rotates scapula so glenoid cavity moves superiorly (upward rotation)	Long thoracic nerve (C_5–C_7)
Subclavius	First rib	Clavicle (inferior border)	Depresses and protracts shoulder	Subclavian nerve
Trapezius	Occipital bone, ligamentum nuchae, and spinous processes of thoracic vertebrae	Clavicle and scapula (acromion and scapular spine)	Depends on active region and state of other muscles; may elevate, retract, depress, or rotate scapula upward and/or elevate clavicle; can also extend head and neck	Accessory nerve (N XI) and cervical spinal nerves (C_3–C_4)

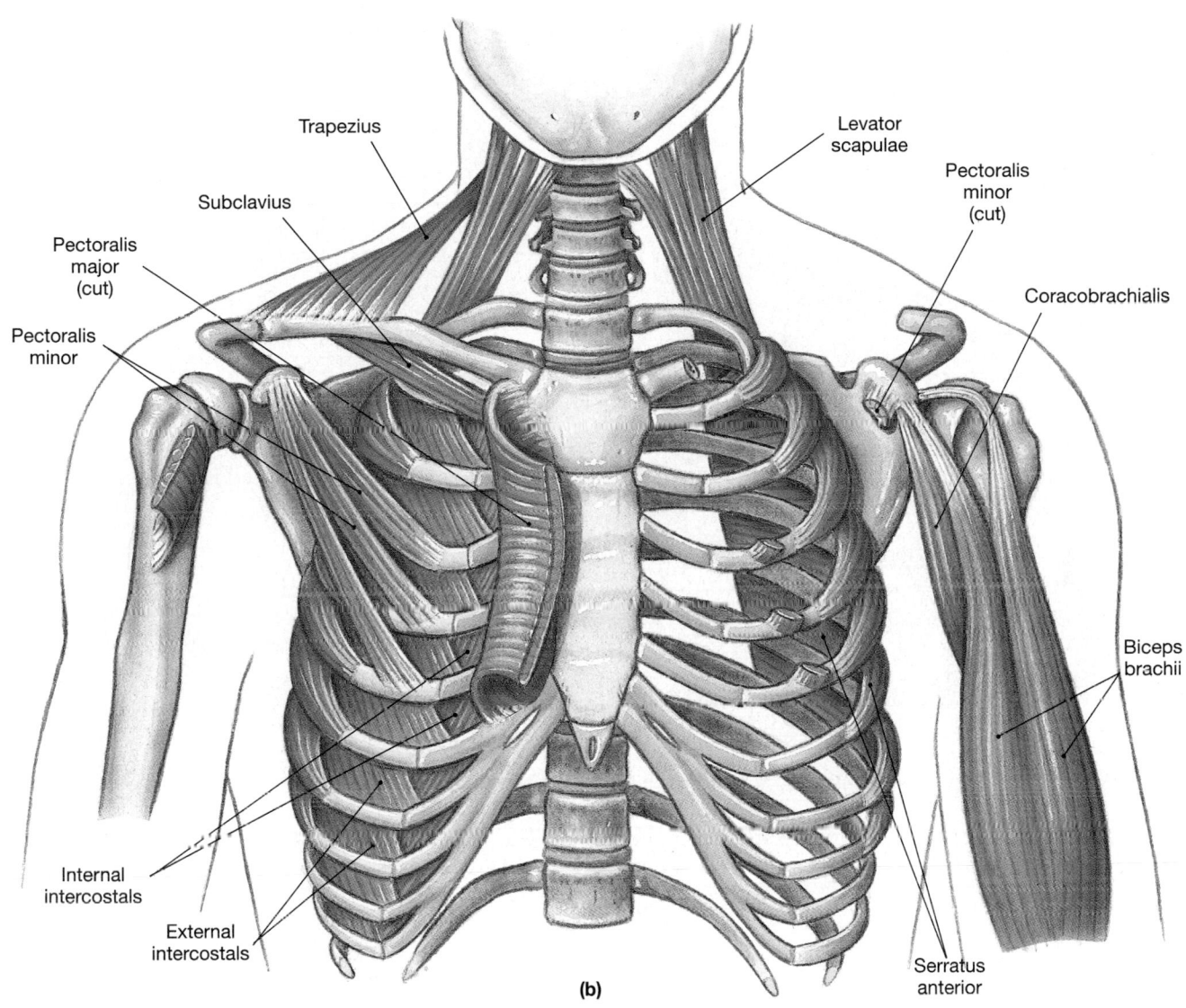

Trapezius

Subclavius

Pectoralis
major
(cut)

Pectoralis
minor

Internal
intercostals

External
intercostals

Levator
scapulae

Pectoralis
minor
(cut)

Coracobrachialis

Biceps
brachii

Serratus
anterior

(b)

Two other deep chest muscles arise along the ventral surfaces of the ribs on either side. The **subclavius** (sub-KLĀ-vē-us; *sub-,* below + *clavius,* clavicle) inserts on the inferior border of the clavicle (Figure 11-15b●). When it contracts, it depresses and protracts the scapular end of the clavicle. Because ligaments connect this end to the shoulder joint and scapula, those structures move as well. The **pectoralis** (pek-to-RA-lis) **minor** attaches to the coracoid process of the scapula. The contraction of this muscle generally complements that of the subclavius.

Muscles That Move the Arm

The muscles that move the arm (Figures 11-14 and 11-16●) are easiest to remember when grouped by primary actions (Table 11-12). The **deltoid** is the major abductor of the arm, but the **supraspinatus** (soo-pra-spī-NA-tus) assists at the start of this movement. The

subscapularis and **teres major** rotate the arm medially, whereas the **infraspinatus** and the **teres minor** perform lateral rotation. All these muscles originate on the scapula. The small **coracobrachialis** (KOR-uh-kō-brā-kē-A-lis) is the only muscle attached to the scapula that flexes and adducts the humerus. These muscles are shown in Figure 11-16a●.

The **pectoralis major** extends between the chest and the greater tubercle of the humerus. The **latissimus dorsi** (la-TIS-i-mus DOR-sē) extends between the thoracic vertebrae and the lesser tubercle of the humerus (Figure 11-16b●). The pectoralis major flexes the arm at the shoulder joint, and the latissimus dorsi extends it. These two muscles can also work together to produce adduction and medial rotation of the arm.

The supraspinatus, infraspinatus, subscapularis, and teres minor and associated tendons form the **rotator cuff,** a common site of sports injuries.

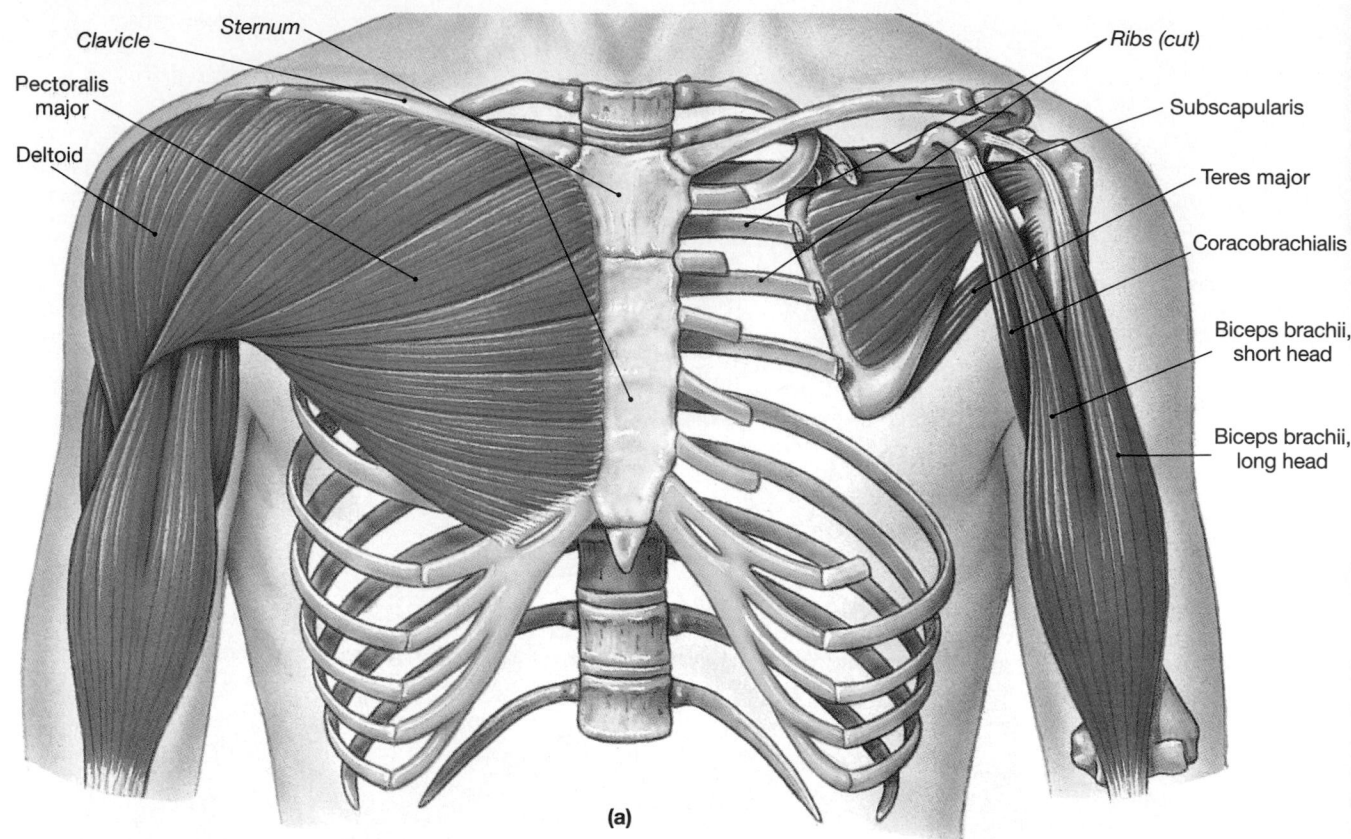

Clavicle

Sternum

Pectoralis major

Deltoid

Ribs (cut)

Subscapularis

Teres major

Coracobrachialis

Biceps brachii, short head

Biceps brachii, long head

(a)

●FIGURE 11-16 **Muscles That Move the Arm.** **(a)** Anterior view. **(b)** Posterior view. *See also Figures 7-23, 8-3, and 9-9, pp. 226, 236, 267.* AM *Plates 5.1, 6.1*

TABLE 11-12 Muscles That Move the Arm (*Figure 11-14 to 11-16*)

Muscle	Origin	Insertion	Action	Innervation
Coracobrachialis	Coracoid process	Medial margin of shaft of humerus	Adducts and flexes humerus	Musculocutaneous nerve (C_5–C_7)
Deltoid	Clavicle and scapula (acromion and adjacent scapular spine)	Deltoid tuberosity of humerus	Abducts humerus	Axillary nerve (C_5–C_6)
Supraspinatus	Supraspinous fossa of scapula	Greater tubercle of humerus	Abducts humerus	Suprascapular nerve (C_5)
Infraspinatus	Infraspinous fossa of scapula	Greater tubercle of humerus	Lateral rotation of humerus	Suprascapular nerve (C_5–C_6)
Subscapularis	Subscapular fossa of scapula	Lesser tubercle of humerus	Medial rotation of humerus	Subscapular nerves (C_5–C_6)
Teres major	Inferior angle of scapula	Medial lip of the intertubercular groove of the humerus	Extends, adducts, and medially rotates humerus	Lower subscapular nerve (C_5–C_6)
Teres minor	Lateral border of scapula	Greater tubercle of humerus	Lateral rotation of humerus	Axillary nerve (C_5)
Triceps brachii (long head)	*See Table 11-13*			
Latissimus dorsi	Spinous processes of lower thoracic and all lumbar vertebrae, ribs 8–12, and the lumbodorsal fascia	Floor of the intertubercular groove of the humerus	Extends, adducts, and medially rotates humerus	Thoracodorsal nerve (C_6–C_8)
Pectoralis major	Cartilages of ribs 2–6, body of sternum, and inferior, medial portion of clavicle	Crest of greater tubercle and lateral lip of intertubercular groove of humerus	Flexes, adducts, and medially rotates humerus	Pectoral nerves (C_5–T_1)

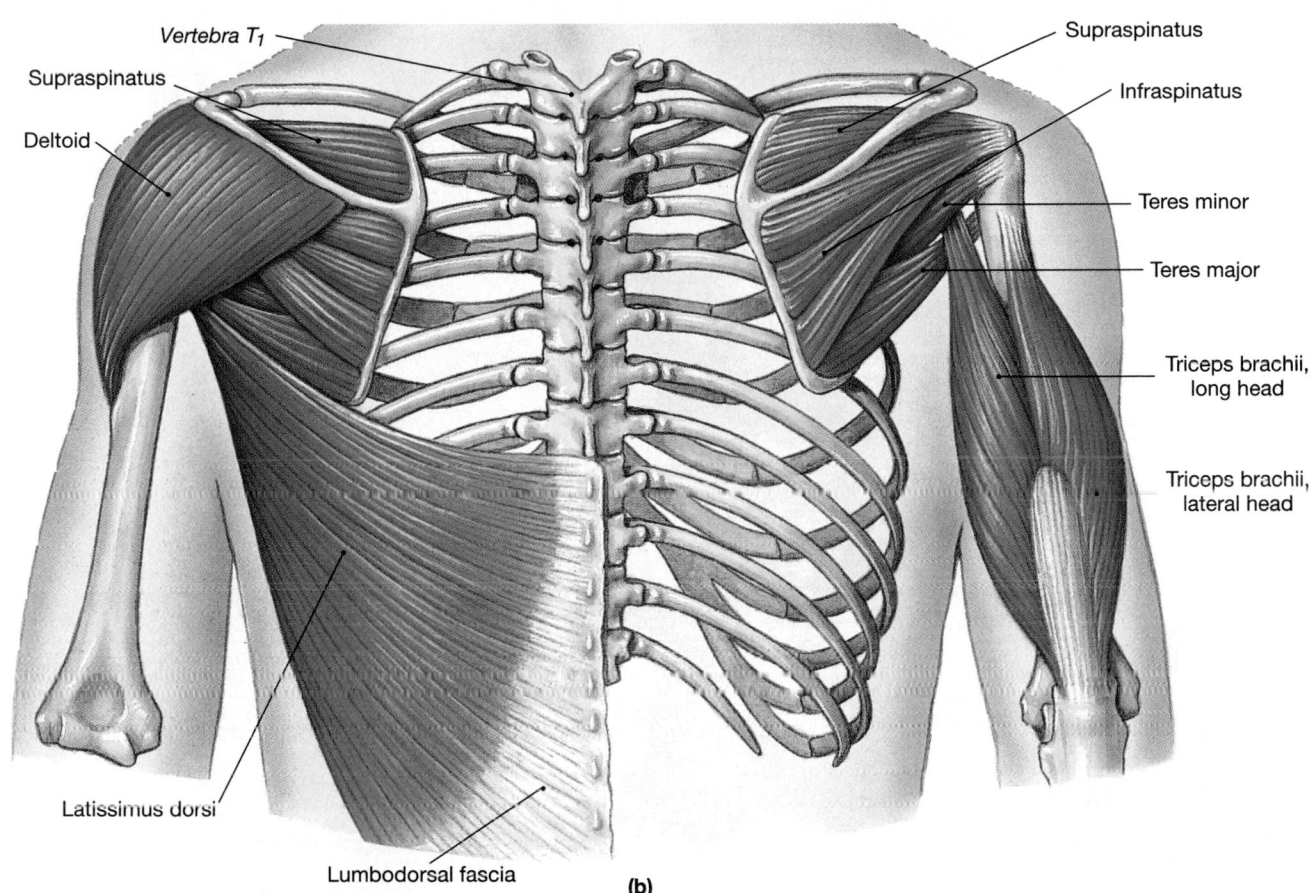

Vertebra T₁

Supraspinatus

Deltoid

Latissimus dorsi

Lumbodorsal fascia

(b)

Supraspinatus

Infraspinatus

Teres minor

Teres major

Triceps brachii, long head

Triceps brachii, lateral head

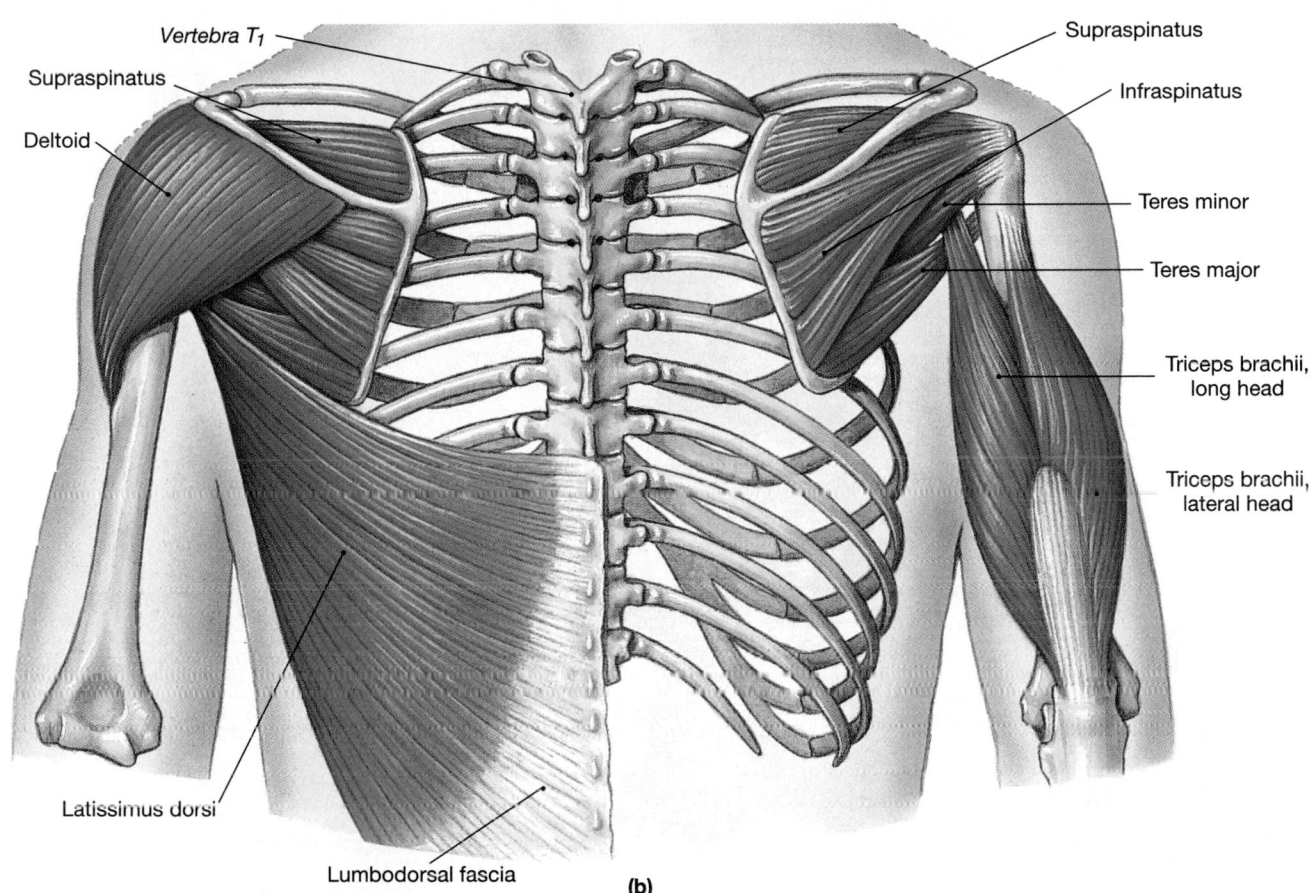

SPORTS INJURIES Exercise carries risks due to the stresses placed on muscles, joints, and connective tissues. Many of us participate in exercise programs and sports on a regular basis. More than 30 million people jog in the United States, and millions more participate in various amateur and professional sports. As a result, *sports injuries* are very common, and sports medicine has become an active area of professional and academic research interest. For information on the incidence and classification of sports injuries, see the *Applications Manual.* [AM] *Sports Injuries*

Muscles That Move the Forearm and Hand

Although most of the muscles that insert on the forearm and hand originate on the humerus, there are two noteworthy exceptions, the *biceps brachii* and the *triceps brachii*. These muscles are shown in their entirety in Figure 11-17•. The **biceps brachii** and the *long head* of the **triceps brachii** originate on the scapula and insert on the bones of the forearm. The triceps brachii inserts on the olecranon process. Contraction of the triceps brachii extends the elbow, as when you do push-ups. The biceps brachii inserts on the radial tuberosity, a roughened area on the anterior surface of the radius. ∞ *[p. 238]* Contraction of the biceps flexes the elbow and supinates the forearm. With the fore-

arm pronated (palm facing back), the biceps brachii cannot function effectively. As a result, you are strongest when you flex your elbow with a supinated forearm; the biceps brachii then makes a prominent bulge.

More muscles are shown in Figure 11-17• and detailed in Table 11-13 (p. 347). The **brachialis** and **brachioradialis** (brā-kē-ō-rā-dē-A-lis) flex the elbow and are opposed by the **anconeus** and the triceps brachii, respectively.

The **flexor carpi ulnaris,** the **flexor carpi radialis,** and the **palmaris longus** are superficial muscles that work together to produce flexion of the wrist. The flexor carpi radialis flexes and *ab*ducts, and the flexor carpi ulnaris flexes and *ad*ducts. *Pitcher's arm* is an inflammation at the origins of the flexor carpi muscles at the medial epicondyle. This condition results from forcibly flexing the wrist just before releasing a baseball.

The **extensor carpi radialis** muscles and the **extensor carpi ulnaris** have a similar relationship to that between the flexor carpi muscles. The extensor carpi radialis produces extension and *ab*duction, whereas the extensor carpi ulnaris produces extension and *ad*duction.

The **pronator teres** and the **supinator** arise on both the humerus and forearm. These muscles rotate the radius without either flexing or extending the elbow. The **pronator quadratus** arises on the ulna and assists the pronator teres in opposing the actions of the

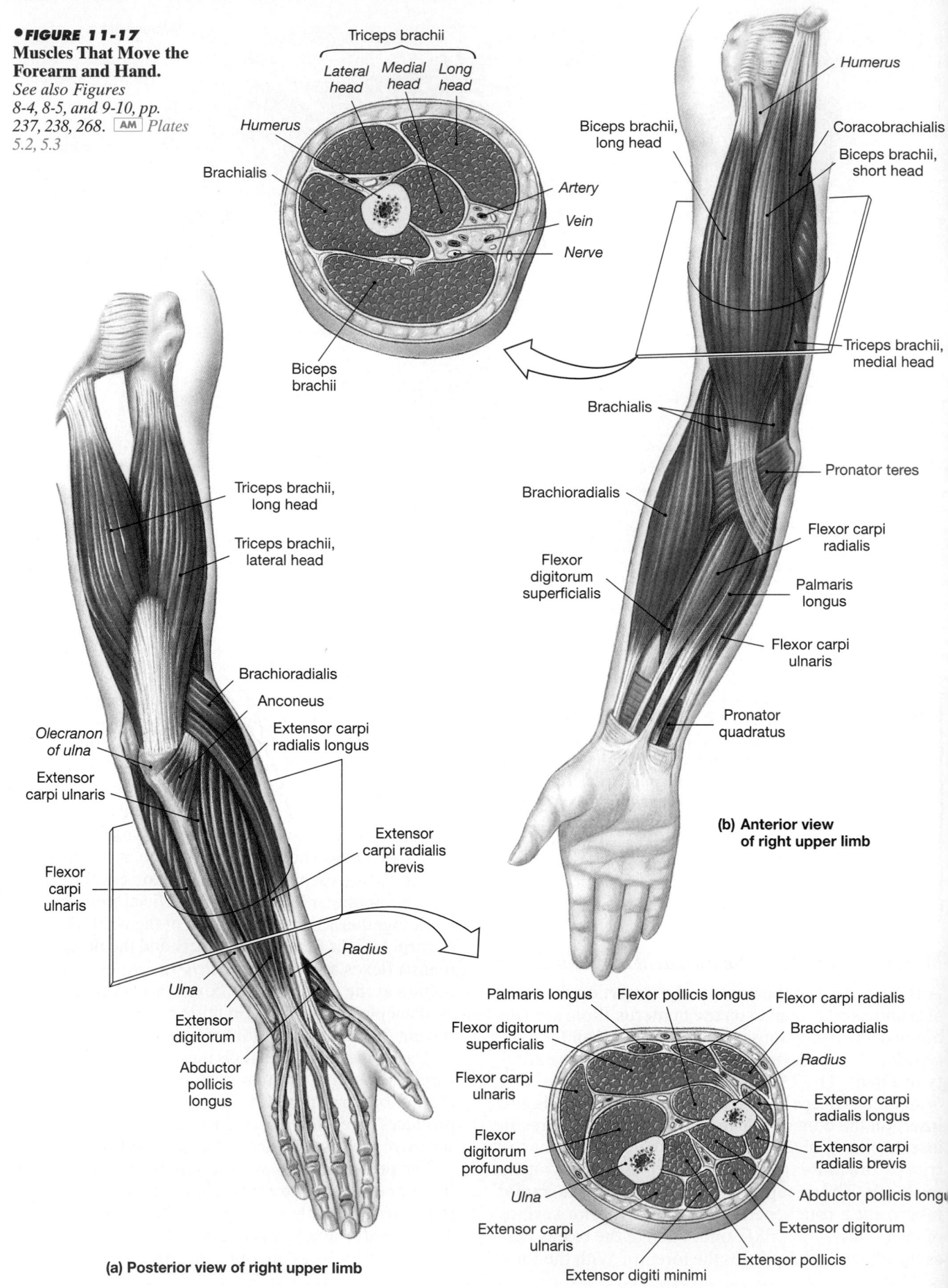

●*FIGURE 11-17*
Muscles That Move the Forearm and Hand.
See also Figures 8-4, 8-5, and 9-10, pp. 237, 238, 268. AM *Plates 5.2, 5.3*

Triceps brachii

Lateral head   Medial head   Long head

Humerus

Brachialis

Artery

Vein

Nerve

Biceps brachii

Humerus

Biceps brachii, long head

Coracobrachialis

Biceps brachii, short head

Triceps brachii, medial head

Brachialis

Pronator teres

Brachioradialis

Flexor carpi radialis

Flexor digitorum superficialis

Palmaris longus

Flexor carpi ulnaris

Pronator quadratus

(b) Anterior view of right upper limb

Triceps brachii, long head

Triceps brachii, lateral head

Brachioradialis

Anconeus

Extensor carpi radialis longus

Olecranon of ulna

Extensor carpi ulnaris

Extensor carpi radialis brevis

Flexor carpi ulnaris

Radius

Ulna

Extensor digitorum

Abductor pollicis longus

Palmaris longus   Flexor pollicis longus   Flexor carpi radialis

Flexor digitorum superficialis

Brachioradialis

Flexor carpi ulnaris

Radius

Flexor digitorum profundus

Extensor carpi radialis longus

Extensor carpi radialis brevis

Ulna

Abductor pollicis longus

Extensor carpi ulnaris

Extensor digitorum

Extensor digiti minimi

Extensor pollicis

(a) Posterior view of right upper limb

supinator or biceps brachii. The muscles involved in pronation and supination are shown in Figure 11-18●. Note the changes in orientation that occur as the pronator teres and pronator quadratus contract. During pronation, the tendon of the biceps brachii rotates with the radius. As a result, the biceps muscle cannot assist in flexion of the elbow when the forearm is pronated.

TABLE 11-13 Muscles That Move the Forearm and Hand (*Figure 11-17*)

Muscle	Origin	Insertion	Action	Innervation
PRIMARY ACTION AT THE ELBOW				
Flexors				
Biceps brachii	*Short head* from the coracoid process; *long head* from the supraglenoid tubercle (both on the scapula)	Tuberosity of radius	Flexes elbow and supinates forearm; flexes arm	Musculo-cutaneous nerve (C_5–C_6)
Brachialis	Anterior, distal surface of	Tuberosity of ulna	Flexes elbow	As above and radial nerve (C_7–C_8)
Brachioradialis	Lateral epicondyle of humerus	Lateral aspect of styloid process of radius	As above	Radial nerve (C_5 C_6)
Extensors				
Anconeus	Posterior surface of lateral epicondyle of humerus	Lateral margin of olecranon on ulna	Extends elbow; moves ulna laterally during pronation	Radial nerve (C_7–C_8)
Triceps brachii				
lateral head	Superior, lateral margin of humerus	Olecranon process of ulna	Extends elbow	As above
long head	Infraglenoid tubercle of scapula	As above	As above	As above
medial head	Posterior surface of humerus inferior to radial groove	As above	As above	As above
PRONATORS/SUPINATORS				
Pronator quadratus	Medial surface of distal portion of ulna	Anterolateral surface of distal portion of radius	Pronates forearm	Median nerve (C_8–T_1)
Pronator teres	Medial epicondyle of humerus and coronoid process of ulna	Distal lateral surface of radius	As above	Median nerve (C_6–C_7)
Supinator	Lateral epicondyle of humerus and ridge of proximal ulna	Anterolateral surface of radius distal to the radial tuberosity	Supinates forearm	Deep radial nerve (C_6)
PRIMARY ACTION AT THE HAND				
Flexors				
Flexor carpi radialis	Medial epicondyle of humerus	Bases of 2nd and 3rd metacarpal bones	Flexes and abducts wrist	Median nerve (C_6–C_7)
Flexor carpi ulnaris	Medial epicondyle of humerus; adjacent medial surface of olecranon and anteromedial portion of ulna	Pisiform bone, hamate bone, and base of 5th metacarpal bone	Flexes and adducts wrist	Ulnar nerve (C_8–T_1)
Palmaris longus	Medial epicondyle of humerus	Palmar aponeurosis and flexor retinaculum	Flexes wrist	Median nerve (C_6–C_7)
Extensors				
Extensor carpi radialis longus	Lateral supracondylar ridge of humerus	Base of 2nd metacarpal bone	Extends and abducts wrist	Radial nerve (C_6–C_7)
Extensor carpi radialis brevis	Lateral epicondyle of humerus	Base of 3rd metacarpal bone	As above	As above
Extensor carpi ulnaris	Lateral epicondyle of humerus; adjacent dorsal surface of ulna	Base of 5th metacarpal bone	Extends and adducts wrist	Deep radial nerve (C_6–C_8)

As you study the muscles included in Table 11-13, note that, in general, the extensor muscles lie along the posterior and lateral surfaces of the arm, whereas the flexors are on the anterior and medial surfaces.

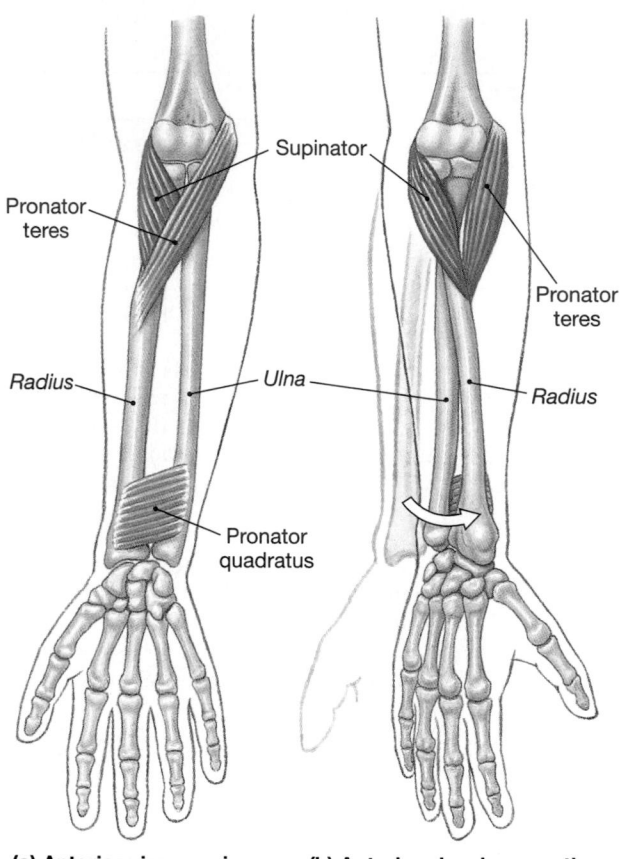

(a) Anterior view, supine

(b) Anterior view in pronation

•*FIGURE 11-18* Muscles of Pronation and Supination. *See also Figure 8-5, p. 238*

Muscles That Move the Fingers

Several superficial and deep muscles of the forearm (Table 11-14) flex and extend the finger joints. These muscles stop before reaching the wrist, and only their tendons cross the articulation. The muscles are relatively large (Figure 11-19•), and keeping them clear of the joints ensures maximum mobility at both the wrist and hand. The tendons that cross the dorsal and ventral surfaces of the wrist pass through **tendon sheaths,** elongate bursae that reduce friction. ∞ *[p. 257]* Figure 11-19• shows these muscles and their tendons in anterior and posterior views.

The muscles of the forearm provide strength and crude control of the hand and fingers. These muscles are known as the *extrinsic muscles of the hand.* Fine control of the hand involves small *intrinsic muscles,* which originate on the carpal and metacarpal bones. No muscles originate on the phalanges, and only tendons extend across the distal joints of the fingers. The intrinsic muscles of the hand are detailed in Figure 11-20• and Table 11-15 (p. 351).

The fascia of the forearm thickens on the posterior surface of the wrist, forming the **extensor retinaculum** (ret-i-NAK-ū-lum), a wide band of connective tissue (Figure 11-20•, p. 350). The extensor retinaculum holds the tendons of the extensor muscles in place. On the anterior surface, the fascia also thickens to form another wide band of connective tissue, the **flexor retinaculum,** which stabilizes the tendons of the flexor muscles. Inflammation of the retinacula and tendon sheaths can restrict movement and irritate the *median nerve,* a mixed (sensory and motor) nerve that innervates the hand. This condition, known as *carpal tunnel syndrome* (p. 350), causes chronic pain.

TABLE 11-14 Muscles That Move the Fingers and Hand (*Figure 11-19*)

Muscle	Origin	Insertion	Action	Innervation
Abductor pollicis longus	Proximal dorsal surfaces of ulna and radius	Lateral margin of 1st metacarpal bone	Abducts thumb	Deep radial nerve (C_6–C_7)
Extensor digitorum	Lateral epicondyle of humerus	Posterior surfaces of the phalanges, fingers 2–5	Extends fingers and hand	Deep radial nerve (C_6–C_8)
Extensor pollicis brevis	Shaft of radius distal to origin of adductor pollicis longus	Base of proximal phalanx of thumb	Extends thumb, abducts hand	Deep radial nerve (C_6–C_7)
Extensor pollicis longus	Posterior and lateral surfaces of ulna and interosseous membrane	Base of distal phalanx of thumb	Extends thumb, abducts hand	Deep radial nerve (C_6–C_8)
Extensor indicis	Posterior surface of ulna and interosseous membrane	Posterior surface of phalanges of little (5th) finger, with tendon of extensor digitorum	Extends and adducts little finger	As above
Extensor digiti minimi	Via extensor tendon to lateral epicondyle of humerus, and from intermuscular septa	Posterior surface of proximal phalanx of little finger	Extends little finger	As above

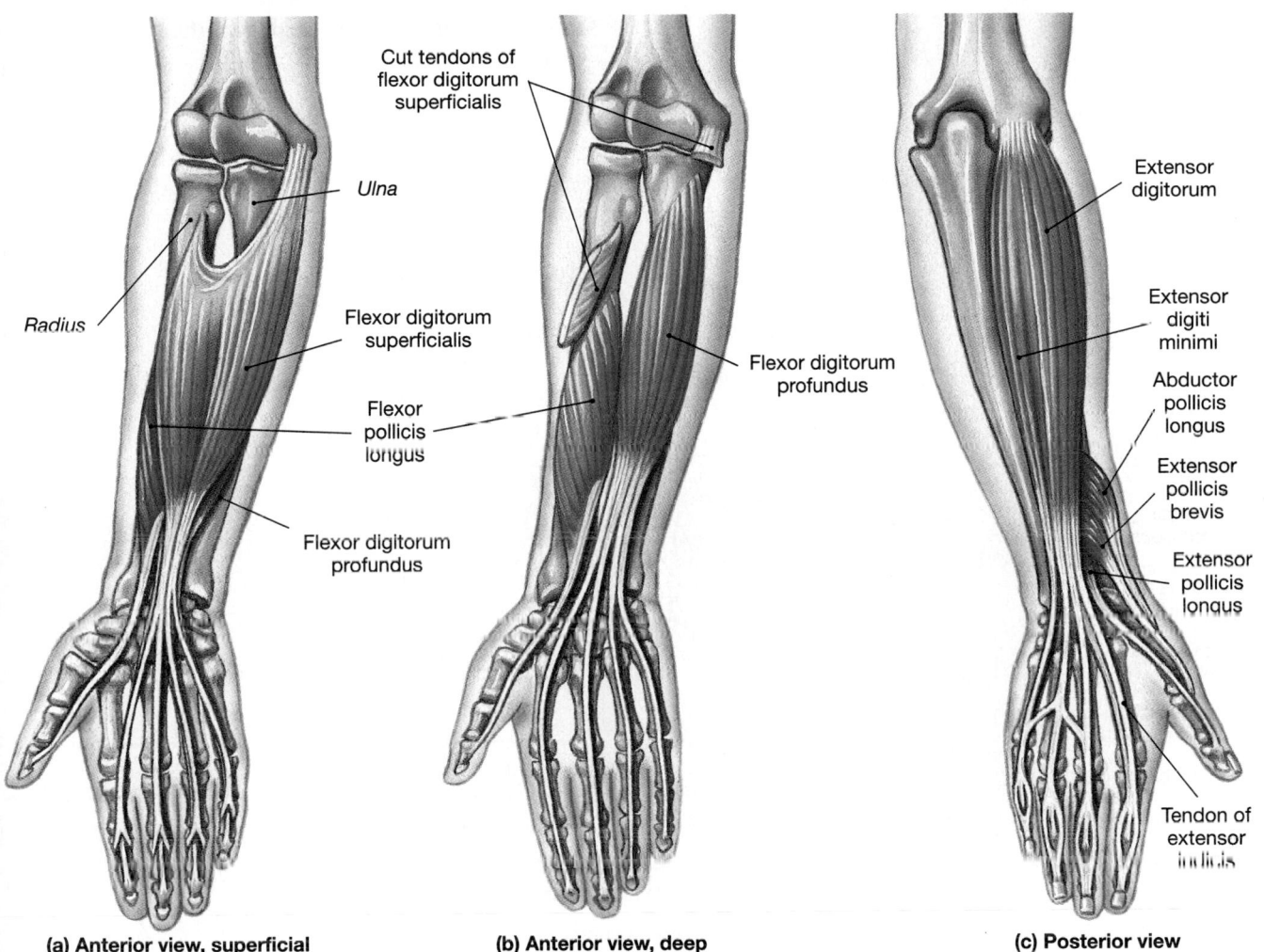

(a) Anterior view, superficial   (b) Anterior view, deep   (c) Posterior view

•**FIGURE 11-19** **Muscles That Move the Fingers.** The right forearm and hand. **(a)** Anterior view, showing superficial muscles. **(b)** Anterior view, showing deep digital flexors, the flexor digitorum profundus, and flexor pollicis longus. **(c)** Posterior view, showing the major digital extensors. *See also Figure 8-5, p. 238.* AM *Plates 5.2, 5.3*

TABLE 11-14 (continued) **Muscles That Move the Fingers and Hand**

Muscle	Origin	Insertion	Action	Innervation
Flexor digitorum superficialis	Medial epicondyle of humerus; adjacent anterior surfaces of ulna and radius	Midlateral surfaces of middle phalanges of fingers 2–5	Flexes fingers, specifically middle phalanx or proximal phalanx; flexes wrist	Median nerve $(C_7–T_1)$
Flexor digitorum profundus	Medial and posterior surfaces of ulna, medial surface of coronoid process, and interosseus membrane	Bases of distal phalanges of fingers 2–5	Flexes distal phalanges and, to a lesser degree, the other phalanges and hand	Palmar inter-osseous nerve, from median nerve and ulnar nerve $(C_8–T_1)$
Flexor pollicis longus	Anterior shaft of radius and interosseous membrane	Base of distal phalanx of thumb	Flexes thumb	Median nerve $(C_8–T_1)$

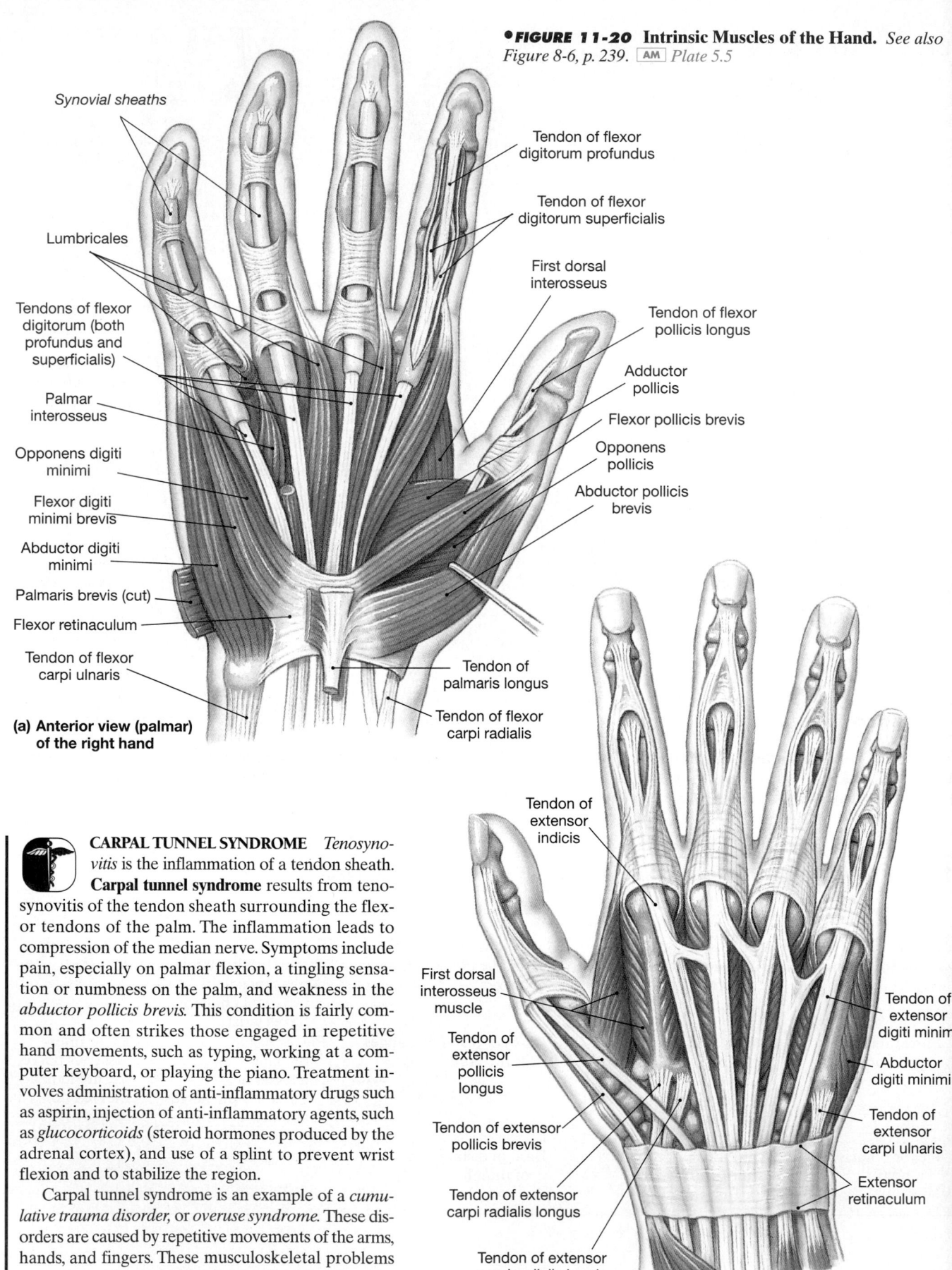

FIGURE 11-20 Intrinsic Muscles of the Hand. *See also* Figure 8-6, p. 239. [AM] *Plate 5.5*

Synovial sheaths

Lumbricales

Tendons of flexor digitorum (both profundus and superficialis)

Palmar interosseus

Opponens digiti minimi

Flexor digiti minimi brevis

Abductor digiti minimi

Palmaris brevis (cut)

Flexor retinaculum

Tendon of flexor carpi ulnaris

Tendon of flexor digitorum profundus

Tendon of flexor digitorum superficialis

First dorsal interosseus

Tendon of flexor pollicis longus

Adductor pollicis

Flexor pollicis brevis

Opponens pollicis

Abductor pollicis brevis

Tendon of palmaris longus

Tendon of flexor carpi radialis

(a) Anterior view (palmar) of the right hand

CARPAL TUNNEL SYNDROME *Tenosynovitis* is the inflammation of a tendon sheath. **Carpal tunnel syndrome** results from tenosynovitis of the tendon sheath surrounding the flexor tendons of the palm. The inflammation leads to compression of the median nerve. Symptoms include pain, especially on palmar flexion, a tingling sensation or numbness on the palm, and weakness in the *abductor pollicis brevis*. This condition is fairly common and often strikes those engaged in repetitive hand movements, such as typing, working at a computer keyboard, or playing the piano. Treatment involves administration of anti-inflammatory drugs such as aspirin, injection of anti-inflammatory agents, such as *glucocorticoids* (steroid hormones produced by the adrenal cortex), and use of a splint to prevent wrist flexion and to stabilize the region.

Carpal tunnel syndrome is an example of a *cumulative trauma disorder,* or *overuse syndrome.* These disorders are caused by repetitive movements of the arms, hands, and fingers. These musculoskeletal problems now account for over 50 percent of all work-related injuries in the United States.

Tendon of extensor indicis

First dorsal interosseus muscle

Tendon of extensor pollicis longus

Tendon of extensor pollicis brevis

Tendon of extensor carpi radialis longus

Tendon of extensor carpi radialis brevis

Tendon of extensor digiti minim

Abductor digiti minimi

Tendon of extensor carpi ulnaris

Extensor retinaculum

(b) Posterior view of the right hand

TABLE 11-15 Intrinsic Muscles of the Hand *(Figure 11-20)*

Muscle	Origin	Insertion	Action	Innervation
Adductor pollicis	Metacarpal and carpal bones	Proximal phalanx of thumb	Adducts thumb	Ulnar nerve, deep branch (C_8–T_1)
Opponens pollicis	Trapezium	First metacarpal bone	Opposition of thumb	Median nerve (C_6–C_7)
Palmaris brevis	Palmar aponeurosis	Skin of medial border of hand	Moves skin on medial border toward midline of palm	Ulnar nerve, superficial branch (C_8)
Abductor digiti minimi	Pisiform bone	Proximal phalanx of little finger	Abducts little finger and flexes its proximal phalanx	Ulnar nerve, deep branch (C_8–T_1)
Abductor pollicis brevis	Transverse carpal ligament, scaphoid bone, and trapezium	Radial side of the base of the proximal phalanx of the thumb	Abducts the thumb	Median nerve (C_6–C_7)
Flexor pollicis brevis*	Flexor retinaculum, trapezium, capitate bone, and ulnar side of first metacarpal bone	Ulnar side of the proximal phalanx of the thumb	Flexes and adducts the thumb	Branches of median and ulnar nerves
Flexor digiti minimi brevis	Hamate bone	Proximal phalanx of little finger	Flexes little finger	Ulnar nerve, deep branch (C_8–T_1)
Opponens digiti minimi	Hamate bone	Fifth metacarpal bone	Opposition of fifth metacarpal bone	As above
Lumbricales (4)	Tendons of flexor digitorum profundus	Tendons of extensor digitorum	Flexes metacarpophalangeal joints, extends middle and distal phalanges	No. 1 and no. 2 by median nerve, no. 3 and no. 4 by ulnar nerve, deep branch
Dorsal interossei (4)	Each originates from opposing faces of two metacarpal bones (I and II, II and III, III and IV, IV and V)	Bases of proximal phalanges of fingers 2–4	Abduct fingers 2–4 away from the midline axis of the middle finger (3), flex metacarpophalangeal joints, extend fingertips	Ulnar nerve, deep branch (C_8–T_1)
Palmar interossei (3–4)*	Sides of metacarpal bones II, IV, and V	Bases of proximal phalanges of fingers 2, 4, and 5	Adduct fingers 2, 4, and 5 toward the midline axis of the middle finger (3), flex metacarpophalangeal joints, extend fingertips	As above

*The portion of the flexor pollicis brevis originating on the first metacarpal bone is sometimes called the *first palmar interosseus muscle.*

☑ Which muscle are you using when you shrug your shoulders?

☑ Baseball pitchers sometimes suffer from rotator cuff injuries. What muscles are involved in this type of injury?

☑ Injury to the flexor carpi ulnaris would impair what two movements?

Muscles of the Lower Limbs

The pelvic girdle is tightly bound to the axial skeleton, and little relative movement is permitted. In our discussion of the axial musculature, we therefore encountered few muscles that can influence the position of the pelvis. The muscles of the lower limbs can be divided into three functional groups: (1) *muscles that move the thigh,* (2) *muscles that move the leg,* and (3) *muscles that move the foot and toes.*

Muscles That Move the Thigh

The muscles that move the thigh are detailed in Table 11-16. **Gluteal muscles** cover the lateral surfaces of the ilia (Figures 11-14, pp. 340–341, and 11-21a,b,d●, p. 353). The **gluteus maximus** is the largest and most posterior of the gluteal muscles. It originates along the edge of the posterior superior iliac spine and the ligaments that bind the sacrum to the ilium. Acting alone, this massive muscle extends and laterally rotates the thigh. The gluteus maximus shares an insertion with the **tensor fasciae latae** (FASH-ē-ē LA-tā), a muscle that originates on the anterior superior iliac spine. Together these muscles pull on the **iliotibial** (il-ē-ō-TIB-ē-ul) **tract,** a band of collagen fibers that extends along the lateral surface of the thigh and inserts upon the tibia. This tract provides a lateral brace for the knee that becomes particularly important when a person balances on one foot.

The **gluteus medius** and **gluteus minimus** (Figure 11-21b,d●) originate anterior to the origin of the gluteus maximus and insert on the greater trochanter of the femur. The anterior gluteal line on the lateral surface of the ilium marks the boundary between these muscles.

The **lateral rotators** arise at or inferior to the horizontal axis of the acetabulum. There are six muscles in all, of which the **piriformis** (pir-i-FOR-mis) and the **obturator** muscles are dominant (Figure 11-21c●).

The **adductors** (Figure 11-21d,e● and Table 11-16) are inferior to the acetabular surface. This muscle group includes the **adductor magnus**, the **adductor brevis**, the **adductor longus**, the **pectineus** (pek-TIN-ē-us), and the **gracilis** (GRAS-i-lis). All but the magnus

TABLE 11-16 **Muscles That Move the Thigh** *(Figure 11-21)*

Group/Muscle	Origin	Insertion	Action	Innervation
Gluteal group				
Gluteus maximus	Iliac crest of ilium, sacrum, coccyx, and lumbodorsal fascia	Iliotibial tract and gluteal tuberosity of femur	Extends and laterally rotates thigh	Inferior gluteal nerve (L_5–S_2)
Gluteus medius	Anterior iliac crest of ilium, lateral surface between posterior and anterior gluteal lines	Greater trochanter of femur	Abducts and medially rotates thigh	Superior gluteal nerve (L_4–S_1)
Gluteus minimus	Lateral surface of ilium between inferior and anterior gluteal lines	Greater trochanter of femur	Abducts and medially rotates thigh	As above
Tensor fasciae latae	Iliac crest and lateral surface of anterior superior iliac spine	Iliotibial tract	Flexes, abducts, and medially rotates thigh; tenses fascia lata, which laterally supports the knee	As above
Lateral rotator group				
Obturators (externus and internus)	Lateral and medial margins of obturator foramen	Trochanteric fossa of femur (externus); medial surface of greater trochanter (internus)	Laterally rotate thigh	Obturator nerve (externus: L_3–L_4) and special nerve from sacral plexus (internus: L_5–S_2)
Piriformis	Anterolateral surface of sacrum	Greater trochanter of femur	Laterally rotates and abducts thigh	Branches of sacral nerves (S_1–S_2)
Gemelli (superior and inferior)	Ischial spine and tuberosity	Medial surface of greater trochanter	Laterally rotates thigh	Nerves to obturator internus and quadratus femoris
Quadratus femoris	Lateral border of ischial tuberosity	Intertrochanteric crest of femur	As above	Special nerve from sacral plexus (L_4–S_1)
Adductor group				
Adductor brevis	Inferior ramus of pubis	Linea aspera of femur	Adducts, medially rotates, and flexes thigh	Obturator nerve (L_3–L_4)
Adductor longus	Inferior ramus of pubis anterior to adductor brevis	As above	Adducts, flexes, and medially rotates thigh	As above
Adductor magnus	Inferior ramus of pubis posterior to adductor brevis and ischial tuberosity	Linear aspera and adductor tubercle of femur	Adducts thigh; superior portion flexes and medially rotates thigh; inferior portion extends and laterally rotates thigh	Obturator and sciatic nerves
Pectineus	Superior ramus of pubis	Pectineal line inferior to lesser trochanter of femur	Flexes, medially rotates, and adducts thigh	Femoral nerve (L_2–L_4)
Gracilis	Inferior ramus of pubis	Medial surface of tibia inferior to medial condyle	Flexes leg, adducts, and medially rotates thigh	Obturator nerve (L_3–L_4)
Iliopsoas group				
Iliacus	Iliac fossa of ilium	Femur distal to lesser trochanter; tendon fused with that of psoas	Flexes hip and/or lumbar spine	Femoral nerve (L_2–L_3)
Psoas major	Anterior surfaces and transverse processes of vertebrae T_{12}–L_5	Lesser trochanter in company with iliacus	As above	Branches of the lumbar plexus (L_2–L_3)

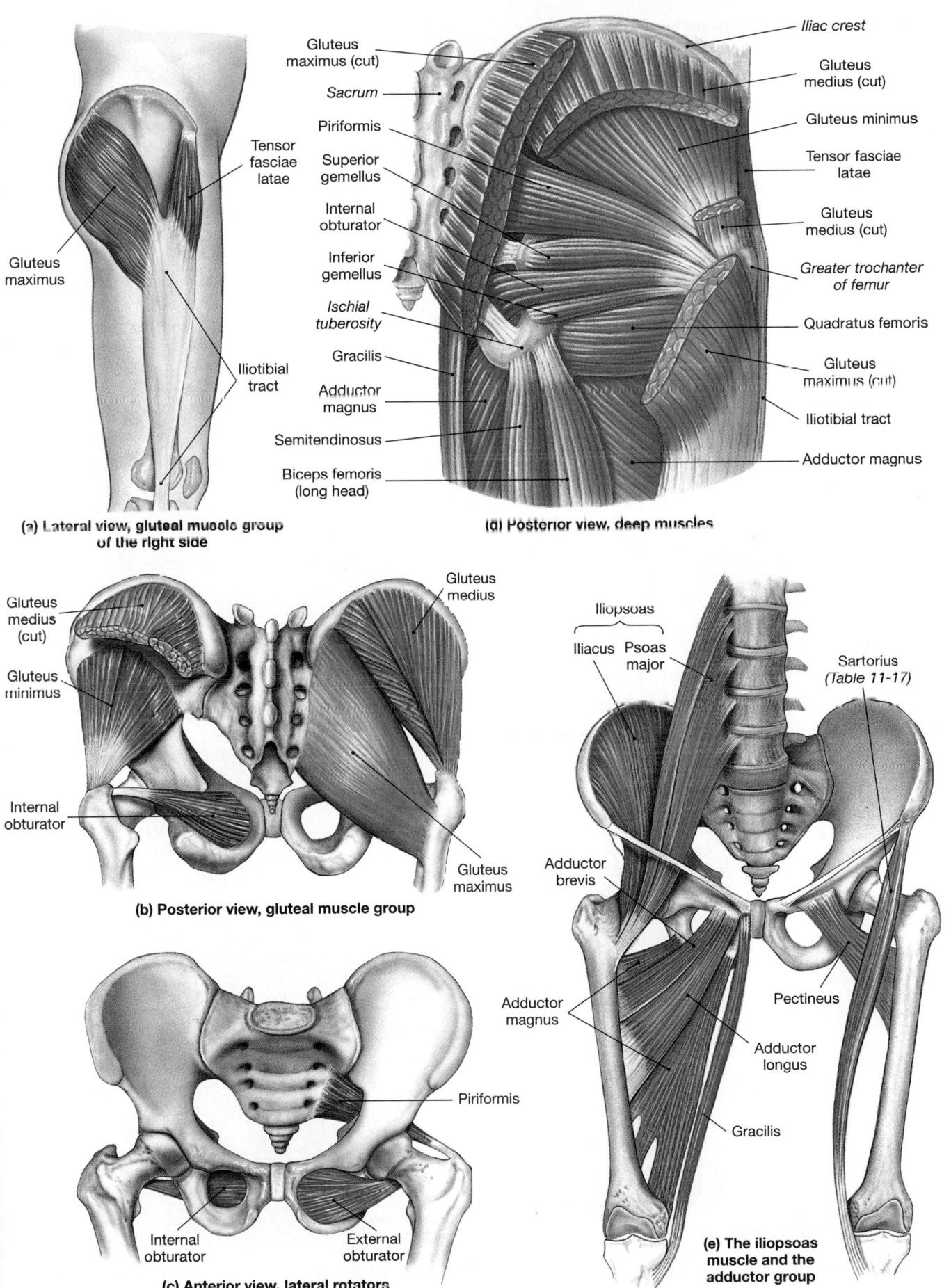

(a) Lateral view, gluteal muscle group of the right side

Tensor fasciae latae
Gluteus maximus
Iliotibial tract

(d) Posterior view, deep muscles

Gluteus maximus (cut)
Sacrum
Piriformis
Superior gemellus
Internal obturator
Inferior gemellus
Ischial tuberosity
Gracilis
Adductor magnus
Semitendinosus
Biceps femoris (long head)

Iliac crest
Gluteus medius (cut)
Gluteus minimus
Tensor fasciae latae
Gluteus medius (cut)
Greater trochanter of femur
Quadratus femoris
Gluteus maximus (cut)
Iliotibial tract
Adductor magnus

(b) Posterior view, gluteal muscle group

Gluteus medius (cut)
Gluteus minimus
Internal obturator
Gluteus medius
Gluteus maximus

(c) Anterior view, lateral rotators

Piriformis
Internal obturator
External obturator

(e) The iliopsoas muscle and the adductor group

Iliopsoas
Iliacus   Psoas major
Sartorius (Table 11-17)
Adductor brevis
Adductor magnus
Pectineus
Adductor longus
Gracilis

•**FIGURE 11-21** **Muscles That Move the Thigh.** *See also Figures 8-7, 8-8, 8-11, and 9-11, pp. 242, 243, 246, 270.* AM *Plates 7.1–7.3* are found both anterior and inferior to the joint, so they perform hip flexion as well as adduction. The adductor magnus can produce either adduction and flexion or adduction and extension, depending on the region stimulated. The adductor magnus may also produce medial or lateral rotation. The other muscles produce medial rotation. These muscles insert on low ridges along the posterior surface of the femur. When an athlete suffers a *pulled groin,* the problem is a strain in one of these adductor muscles.

The medial surface of the pelvis is dominated by a single pair of muscles. The large **psoas** (SŌ-us) **major**

arises alongside the lower thoracic and lumbar vertebrae, and its insertion lies on the lesser trochanter of the femur. Before reaching this insertion, its tendon merges with that of the **iliacus** (il-Ē-uh-kus), which lies nestled within the iliac fossa. These two muscles are powerful flexors of the thigh and are often referred to collectively as the **iliopsoas** (il-ē-ō-SŌ-us) muscle.

Muscles That Move the Leg

As in the upper limb, there is a pattern to muscle distribution in the lower limb (Figure 11-22●). Extensor muscles are located along the anterior and lateral surfaces of

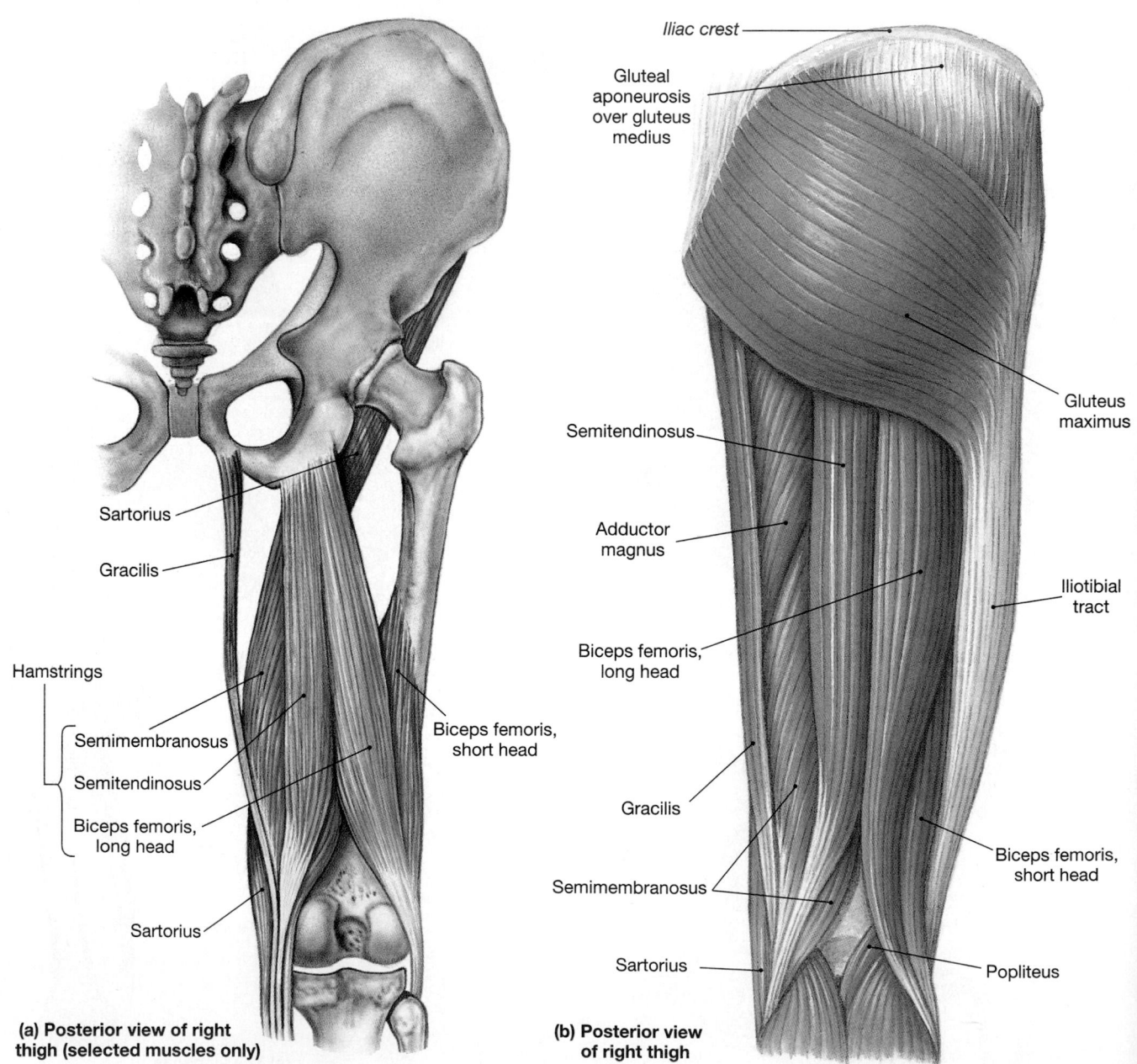

(a) Posterior view of right thigh (selected muscles only)

(b) Posterior view of right thigh

●**FIGURE 11-22** **Muscles That Move the Leg.** **(a)** Diagrammatic view of leg flexors. **(b)** Posterior view of the right thigh. **(c)** Anterior view of the right thigh. **(d)** Sectional view of thigh muscles. *See also Figures 8-11, 8-12, 8-13, and 9-12, pp. 246, 247, 271.*
AM *Plates 7.1–7.3; Scans 4, 5, 6, 7*

the leg, and flexors lie along the posterior and medial surfaces. Although the flexors and adductors originate on the pelvic girdle, most of the extensors originate on the femoral surface.

The *flexors of the knee* (Figure 11-22a,b,c●) include the **biceps femoris,** the **semimembranosus** (sem-ē-mem-bra-NŌ-sus), the **semitendinosus** (sem-ē-ten-di-NŌ-sus), and the **sartorius**. These muscles arise along the edges of the pelvis and insert on the tibia and fibula, and their contractions produce flexion of the knee. The sartorius is the only knee flexor that originates superior to the acetabulum, and its insertion lies along the medial aspect of the tibia. When the sartorius contracts, it flexes the knee and laterally rotates the thigh, as when you cross your legs.

Because the biceps femoris, semimembranosus, and semitendinosus originate on the pelvic surface inferior and posterior to the acetabulum, their contractions also produce extension of the hip. These three muscles are often called the **hamstrings.** A *pulled hamstring* is a relatively common sports injury caused by a strain affecting one of the hamstring muscles.

In Chapter 9, we noted that the knee joint can be locked at full extension by a slight lateral rotation of the tibia. ∞ *[p. 271]* The small **popliteus** (pop-LI-tē-us) muscle arises on the femur near the lateral condyle and inserts on the posterior tibial shaft. When flexion is initiated, this muscle contracts to produce a slight medial rotation of the tibia that unlocks the knee joint.

Collectively, the *knee extensors* are known as the **quadriceps femoris**. The three **vastus** muscles originate along the body of the femur and, with the assistance of the **rectus femoris** (Figure 11-22d●), cradle the femur the way a bun surrounds a hot dog. All four muscles insert on the patella and reach the tibial tuberosity by way of the patellar ligament. The rectus originates on the anterior inferior iliac spine, so in addition to extending the knee, it can assist in flexion of the hip. The flexors and extensors of the knee are detailed in Table 11-17.

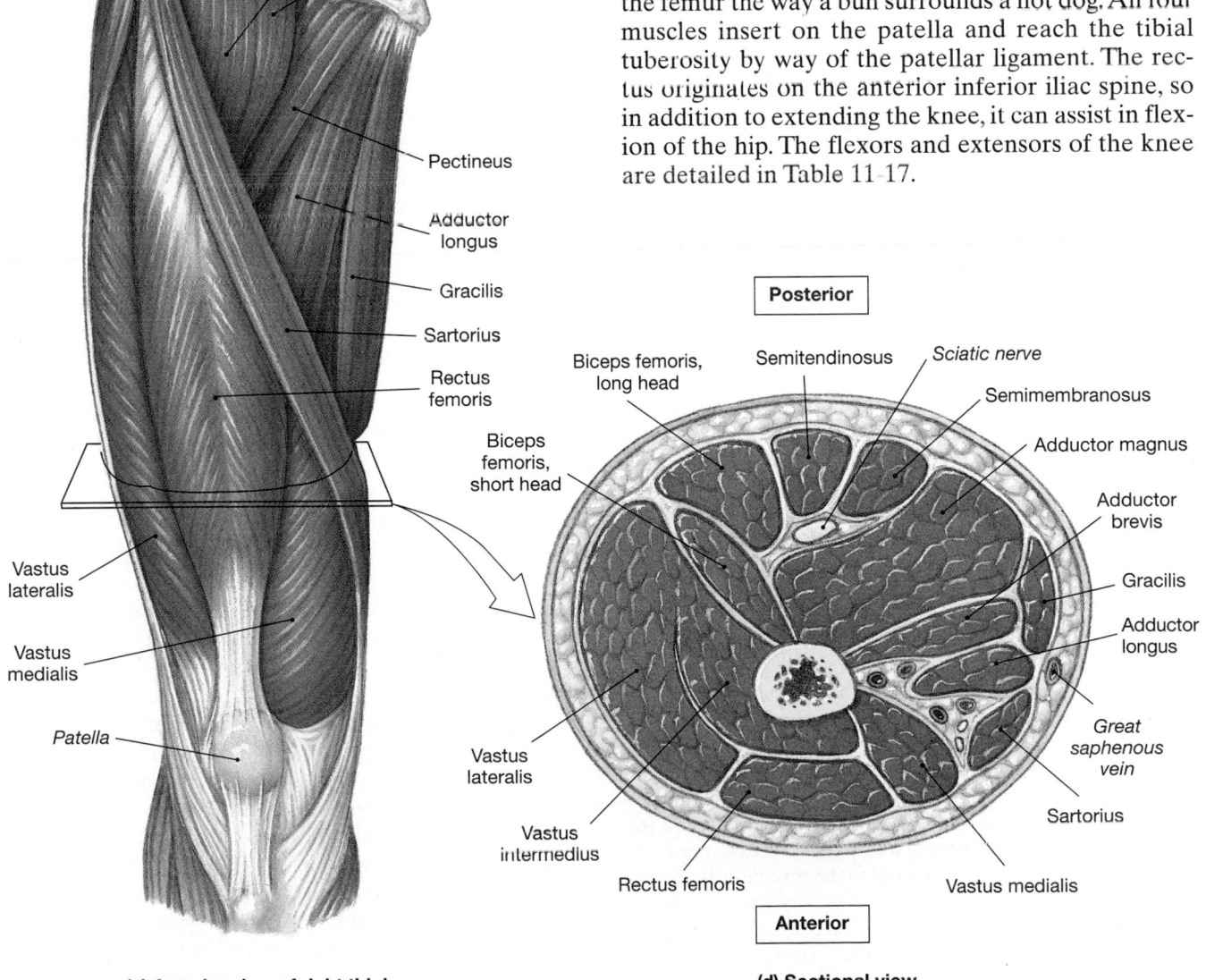

Gluteus medius

Tensor fasciae latae

Iliopsoas
Iliacus   Psoas major

Pectineus

Adductor longus

Gracilis

Sartorius

Rectus femoris

Vastus lateralis

Vastus medialis

Patella

(c) Anterior view of right thigh

Posterior

Biceps femoris, long head

Biceps femoris, short head

Semitendinosus

Sciatic nerve

Semimembranosus

Adductor magnus

Adductor brevis

Gracilis

Adductor longus

Great saphenous vein

Sartorius

Vastus medialis

Rectus femoris

Vastus intermedius

Vastus lateralis

Anterior

(d) Sectional view

TABLE 11-17 Muscles That Move the Leg *(Figure 11-22)*

Muscle	Origin	Insertion	Action	Innervation
Flexors of the Knee				
Biceps femoris	Ischial tuberosity and linea aspera of femur	Head of fibula, lateral condyle of tibia	Flexes knee, extends and laterally rotates thigh	Sciatic nerve; tibial portion (S_1–S_3; to long head) and common peroneal branch (L_5–S_2; to short head)
Semimembranosus	Ischial tuberosity	Posterior surface of medial condyle of tibia	Flexes knee, medially rotates leg, and extends hip	Sciatic nerve (tibial portion; L_5–S_2)
Semitendinosus	Ischial tuberosity	Proximal, medial surface of tibia near insertion of gracilis	As above	As above
Sartorius	Anterior superior iliac spine	Medial surface of tibia near tibial tuberosity	Flexes knee, flexes hip, and laterally rotates thigh	Femoral nerve (L_2–L_3)
Popliteus	Lateral condyle of femur	Posterior surface of proximal tibial shaft	Medially rotates tibia (or laterally rotates femur)	Tibial nerve (L_4–S_1)
Extensors of the Knee				
Rectus femoris	Anterior inferior iliac spine and superior acetabular rim of ilium	Tibial tuberosity via patellar ligament	Extends knee, flexes thigh	Femoral nerve (L_2–L_4)
Vastus intermedius	Anterolateral surface of femur and linea aspera (distal half)	As above	Extends knee	As above
Vastus lateralis	Anterior and inferior to greater trochanter of femur and along linea aspera (proximal half)	As above	As above	As above
Vastus medialis	Entire length of linea aspera of femur	As above	As above	As above

INTRAMUSCULAR INJECTIONS Drugs are commonly injected into tissues rather than directly into the circulation. This method enables the physician to introduce a large amount of a drug at one time yet have it enter the circulation gradually. In an **intramuscular (IM) injection**, the drug is introduced into the mass of a large skeletal muscle. Uptake is generally faster and accompanied by less tissue irritation than when drugs are administered *intradermally* or *subcutaneously* (injected into the dermis or subcutaneous layer, respectively). ∞ *[p. 156]* Up to 5 ml of fluid may be injected at one time, and multiple injections are possible.

The most common complications involve accidental injection into a blood vessel or piercing of a nerve. The sudden entry of massive quantities of a drug into the bloodstream can have unpleasant or even fatal consequences, and damage to a nerve can cause motor paralysis or sensory loss. As a result, the site of injection must be selected with care. Bulky muscles that contain few large vessels or nerves make ideal targets, and the gluteus medius or the posterior, lateral, superior portion of the gluteus maximus is commonly selected. The deltoid muscle of the arm, about 2.5 cm (1 in.) distal to the acromion, is another popular site. Probably the most satisfactory from a technical point of view is the vastus lateralis of the thigh, for an injection into this thick muscle will not encounter vessels or nerves. This is the preferred injection site in infants and young children, whose gluteal and deltoid muscles are relatively small.

Muscles That Move the Foot and Toes

The extrinsic muscles that move the foot and toes (Figure 11-23•) are detailed in Table 11-18 (pp. 358–359). Most of the muscles that move the ankle produce the plantar flexion involved with walking and running movements. The **gastrocnemius** (gas-trok-NĒ-mē-us; *gaster,* stomach + *kneme,* knee) of the calf is an important plantar flexor, but the slow muscle fibers of the underlying **soleus** (SŌ-lē-us) are more powerful. These muscles are best seen in posterior and lateral views (Figure11-23a,c•). The gastrocnemius arises from two heads located on the medial and lateral epicondyles of the femur just proximal to the knee. A sesamoid bone, the **fabella,** is generally present within the lateral head of the gastrocnemius. The gastrocnemius and soleus muscles share a common tendon, the **calcaneal tendon,** or *Achilles tendon.*

Deep to the gastrocnemius and soleus lie a pair of **peroneus** muscles (Figure 11-23b,c•). These muscles produce eversion of the foot as well as plantar flexion at the ankle. Inversion is caused by contraction of the **tibialis** (tib-ē-A-lis) muscles. The large **tibialis anterior** (Figure 11-23c,d•) opposes the gastrocnemius and dorsiflexes the ankle.

Important digital muscles originate on the surface of the tibia, the fibula, or both (Figure 11-23c,d● and Table 11-18). Large tendon sheaths surround the tendons of the tibialis anterior, **extensor digitorum longus,** and **extensor hallucis longus,** where they cross the ankle joint. The positions of these sheaths are stabilized by superior and inferior **extensor retinacula** (Figure 11-23d●).

Intrinsic muscles of the foot originate on the tarsal and metatarsal bones. Their contractions move the toes and contribute to the maintenance of the longitudinal arch of the foot. ∞ *[p. 248]* As in the hand, small **in**terosseus muscles (plural, *interossei*) originate on the lateral and medial surfaces of the metatarsal bones. Intrinsic muscles of the foot are detailed in Table 11-19 (pp. 360–361) and Figure 11-24● (p. 361).

☑ Which leg movement would be impaired by injury to the obturator muscle?

☑ You often hear of athletes who suffer a pulled hamstring. To what does this phrase refer?

☑ How would you expect a torn calcaneal tendon to affect movement of the foot?

●*FIGURE 11-23* **Extrinsic Muscles That Move the Foot and Toes.** *See also Figures 8-13 and 8-14, pp. 247, 248.* [AM] *Plate 7.4; Scans 5, 6, 7*

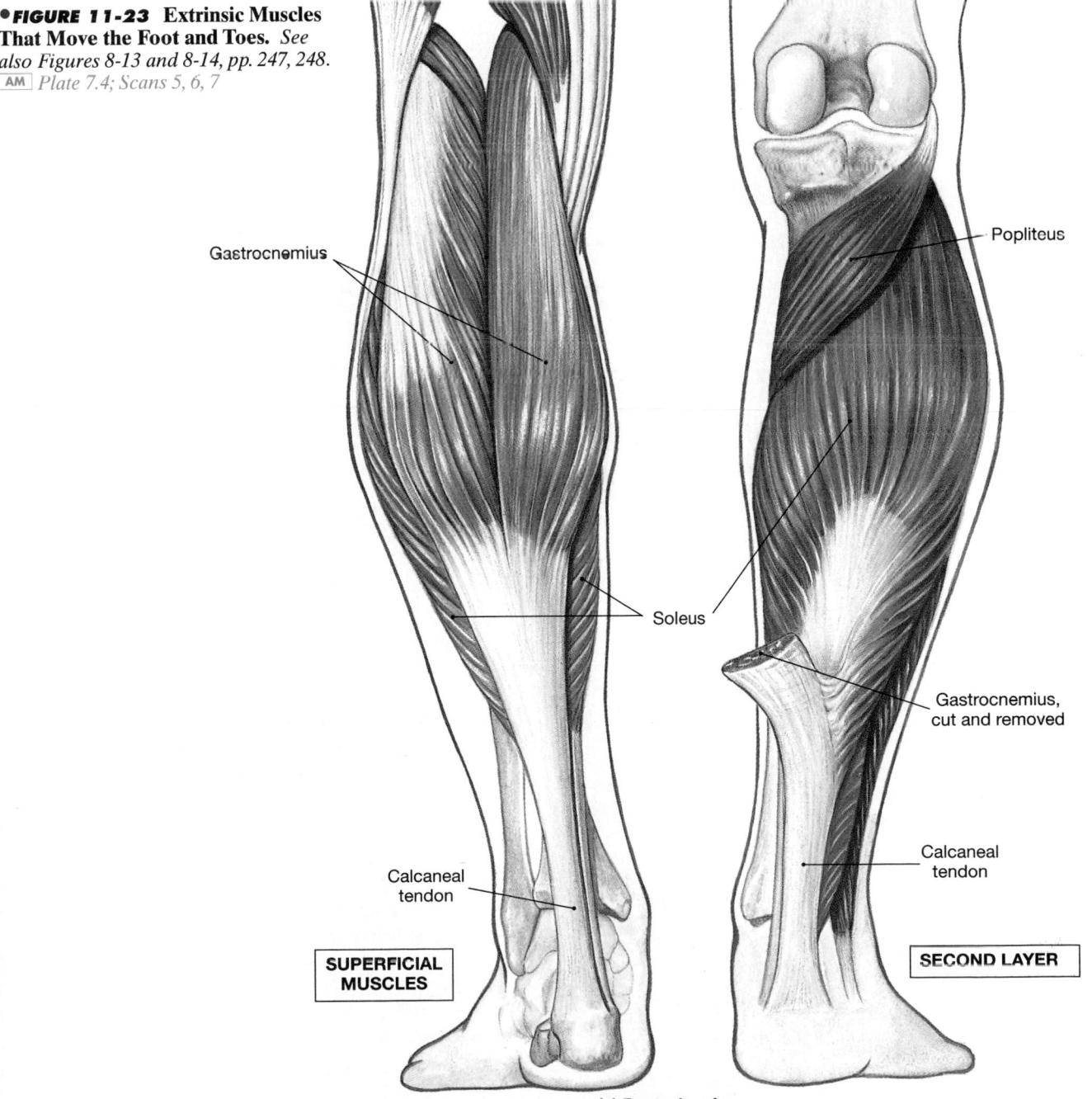

Gastrocnemius

Popliteus

Soleus

Gastrocnemius, cut and removed

Calcaneal tendon

Calcaneal tendon

SUPERFICIAL MUSCLES

SECOND LAYER

(a) Posterior view

TABLE 11-18 Extrinsic Muscles That Move the Foot and Toes *(Figure 11-23)*

Muscle	Origin	Insertion	Action	Innervation
PRIMARY ACTION AT THE ANKLE				
Dorsiflexors				
Tibialis anterior	Lateral condyle and proximal shaft of tibia	Base of 1st metatarsal bone and medial cuneiform bone	Dorsiflexes and inverts foot	Deep peroneal nerve (L_4–S_1)
Plantar flexors				
Gastrocnemius	Femoral condyles	Calcaneus via calcaneal tendon	Plantar flexes, inverts, and adducts foot; flexes knee	Tibial nerve (S_1–S_2)
Peroneus brevis	Midlateral margin of fibula	Base of 5th metatarsal bone	Everts and plantar flexes foot	Superficial peroneal nerve (L_4–S_1)
Peroneus longus	Lateral condyle of tibia, head and proximal shaft of fibula	Base of 1st metatarsal bone and medial cuneiform bone	Everts and plantar flexes foot; supports longitudinal arch	As above
Plantaris	Lateral supracondylar ridge	Posterior portion of calcaneus	Plantar flexes foot; flexes knee	Tibial nerve (L_4–S_1)

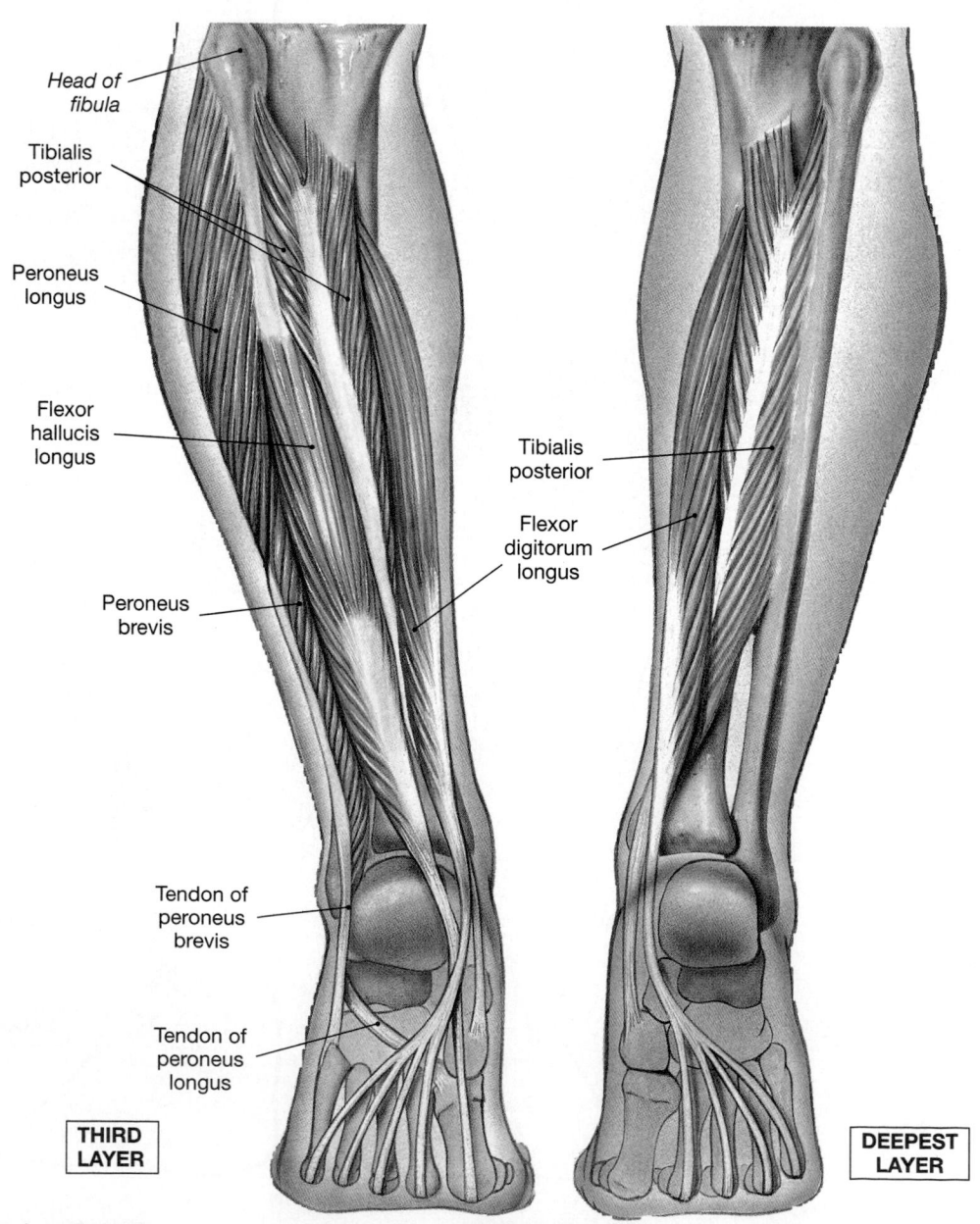

Head of fibula

Tibialis posterior

Peroneus longus

Flexor hallucis longus

Peroneus brevis

Tibialis posterior

Flexor digitorum longus

Tendon of peroneus brevis

Tendon of peroneus longus

THIRD LAYER

DEEPEST LAYER

•FIGURE 11-23 (continued)

(b) Posterior view

TABLE 11-18 (continued) Extrinsic Muscles That Move the Foot and Toes

Muscle	Origin	Insertion	Action	Innervation
Soleus	Head and proximal shaft of fibula, and adjacent posteromedial shaft of tibia	Calcaneus via calcaneal tendon (with gastrocnemius)	Plantar flexes, inverts, and adducts foot	Sciatic nerve, tibial branch (S_1–S_2)
Tibialis posterior	Interosseous membrane and adjacent shafts of tibia and fibula	Tarsal and metatarsal bones	Adducts, inverts, and plantar flexes foot	As above

PRIMARY ACTION AT THE TOES
Digital Flexors

Muscle	Origin	Insertion	Action	Innervation
Flexor digitorum longus	Posteromedial surface of tibia	Inferior surfaces of distal phalanges, toes 2–5	Plantar flexes toes 2–5	Sciatic nerve, tibial branch (L_5–S_1)
Flexor hallucis longus	Posterior surface of fibula	Inferior surface, distal phalanx of great toe	Plantar flexes great toe	Sciatic nerve, tibial branch (L_5–S_1)

Digital Extensors

Muscle	Origin	Insertion	Action	Innervation
Extensor digitorum longus	Lateral condyle of tibia, anterior surface of fibula	Superior surfaces of phalanges, toes 2–5	Extends toes 2–5	Deep peroneal nerve (L_4–S_1)
Extensor hallucis longus	Anterior surface of fibula	Superior surface, distal phalanx of great toe	Extends great toe	As above

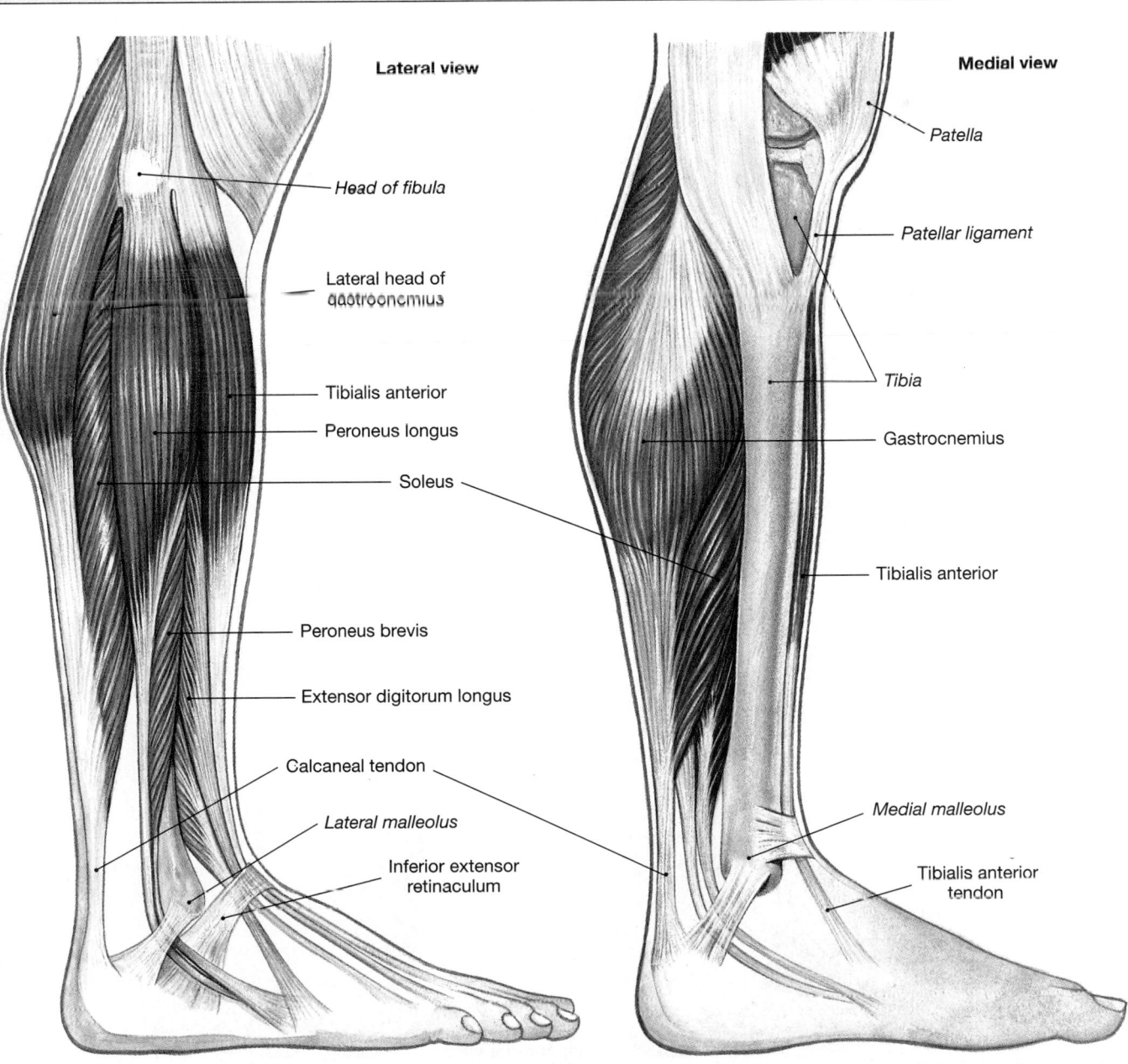

Lateral view

- Head of fibula
- Lateral head of gastrocnemius
- Tibialis anterior
- Peroneus longus
- Soleus
- Peroneus brevis
- Extensor digitorum longus
- Calcaneal tendon
- *Lateral malleolus*
- Inferior extensor retinaculum

Medial view

- *Patella*
- *Patellar ligament*
- *Tibia*
- Gastrocnemius
- Tibialis anterior
- *Medial malleolus*
- Tibialis anterior tendon

(c)

TABLE 11-19 Intrinsic Muscles of the Foot *(Figure 11-24)*

Muscle	Origin	Insertion
Extensor digitorum brevis	Calcaneus (superior and lateral surfaces)	Dorsal surfaces of toes 1–5
Abductor hallucis	Calcaneus (tuberosity on inferior surface)	Medial side of proximal phalanx of great toe
Flexor digitorum brevis	As above	Sides of middle phalanges, toes 2–5
Abductor digiti minimi	As above	Lateral side of proximal phalanx, toe 5
Quadratus plantae	Calcaneus (medial surface and lateral border of inferior surface)	Tendon of flexor digitorum longus
Lumbricales (4)	Tendons of flexor digitorum longus	Insertions of extensor digitorum longus
Flexor hallucis brevis	Cuboid and lateral cuneiform bones	Proximal phalanx of great toe
Adductor hallucis	Bases of metatarsal bones II–IV and plantar ligaments	Proximal phalanx of great toe
Flexor digiti minimi brevis	Base of metatarsal bone V	Lateral side of proximal phalanx of toe 5
Interossei dorsal (4)	Sides of metatarsal bones	Sides of toes 2–4
Interossei plantar (3)	Bases and medial sides of metatarsal bones	Sides of toes 3–5

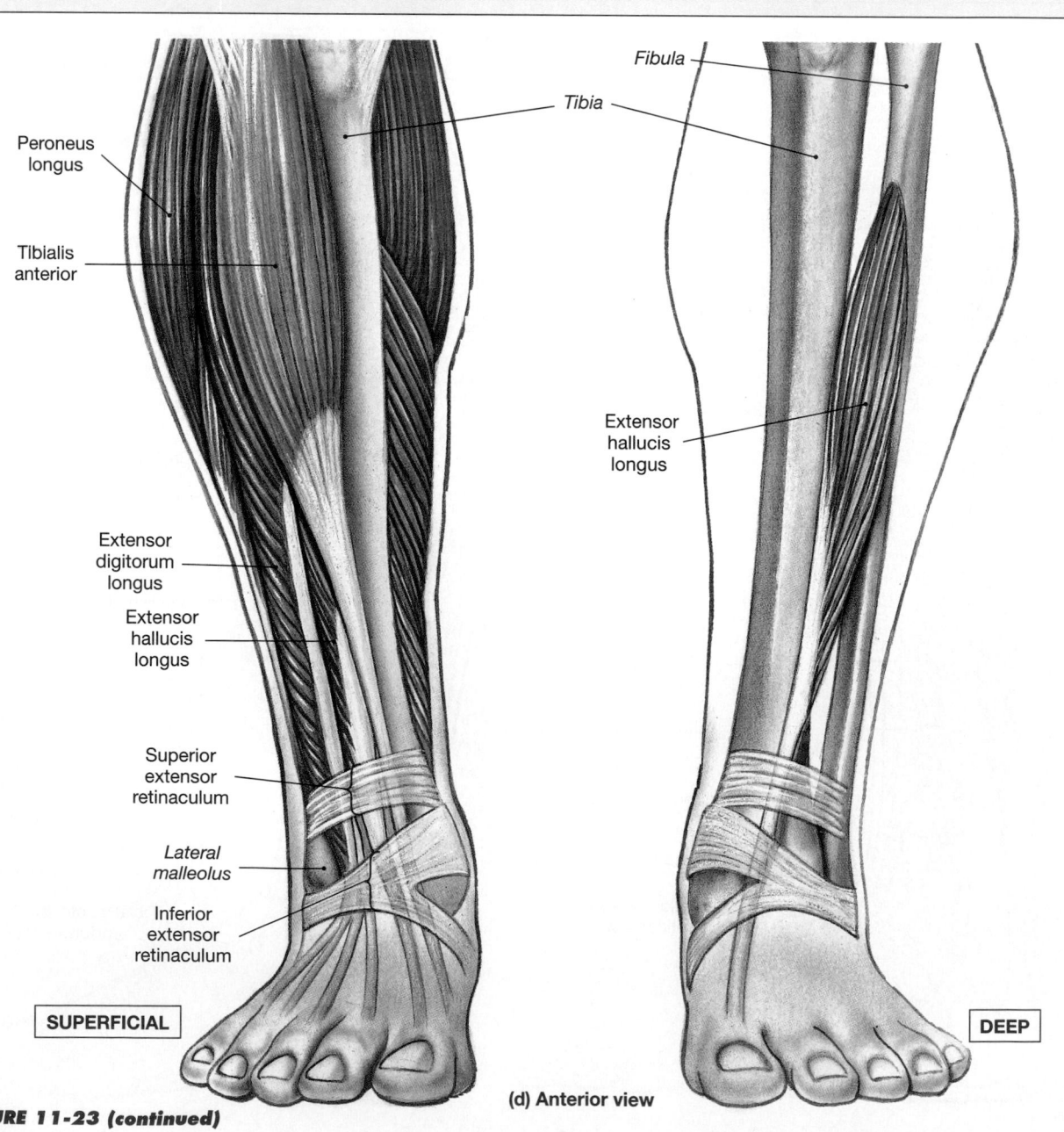

(d) Anterior view

● *FIGURE 11-23 (continued)*

Action	Innervation
Extends proximal phalanges of toes 1–4	Deep peroneal nerve (L_5-S_1)
Abducts great toe	Medial plantar nerve (L_4-L_5)
Flexes middle phalanx of toes 2–5	As above
Abducts toe 5	Lateral plantar nerve (L_4-L_5)
Flexes toes 2–5	As above
Flexes proximal phalanges; extends middle phalanges, toes 2–5	Medial plantar nerve (1), lateral plantar nerve (2–4)
Flexes proximal phalanx of great toe	Medial plantar nerve (L_4-L_5)
Adducts great toe	Lateral plantar nerve (S_1-S_2)
Flexes proximal phalanx of toe 5	As above
Abducts toes	As above
Adducts toes	As above

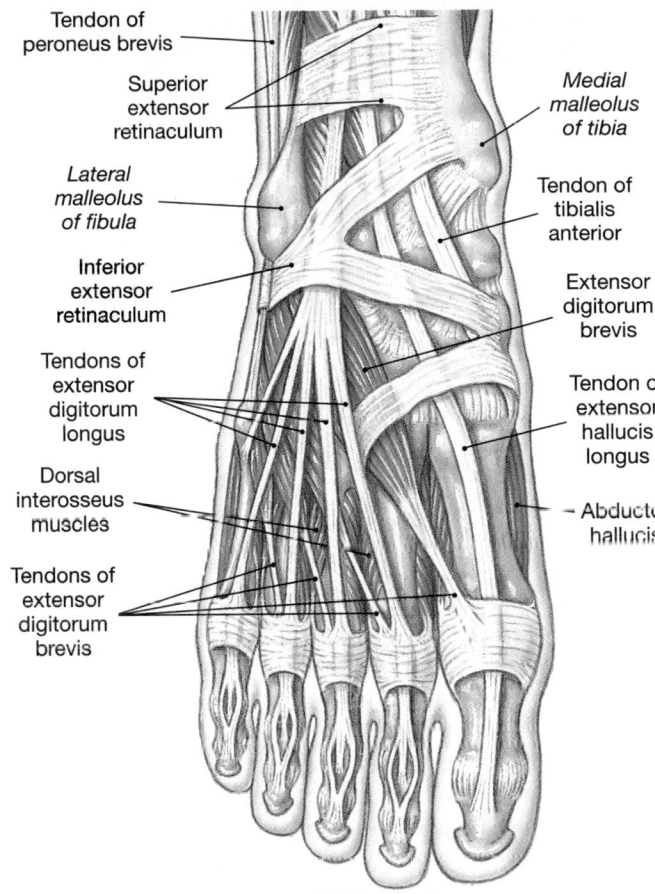

(a) Dorsal view

•*FIGURE* **11-24** **Intrinsic Muscles of the Foot.** *See also Figure 8-14, p. 248.* [AM] *Plate 7.5, Scan 8*

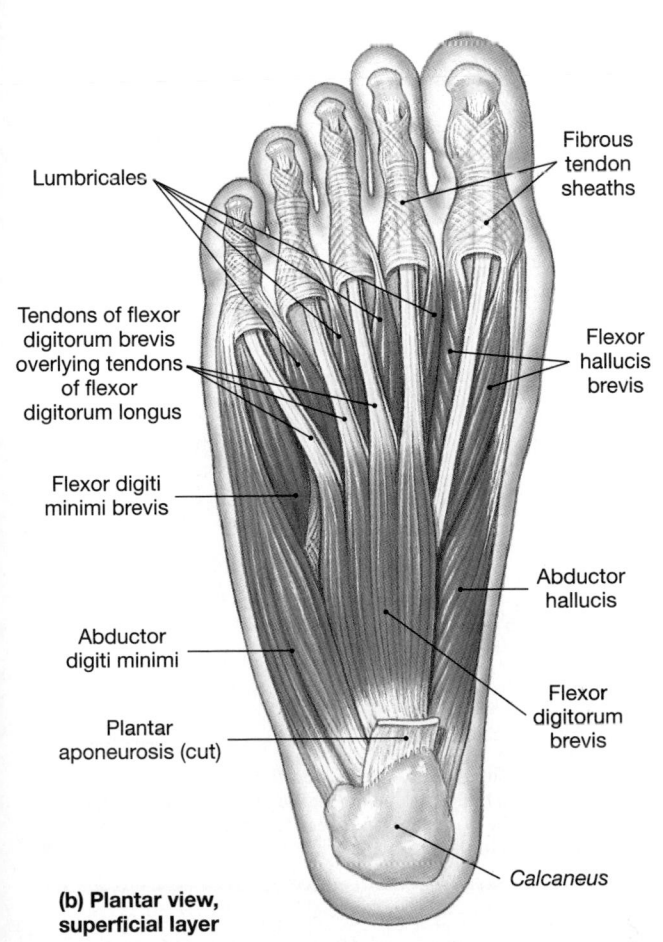

(b) Plantar view, superficial layer

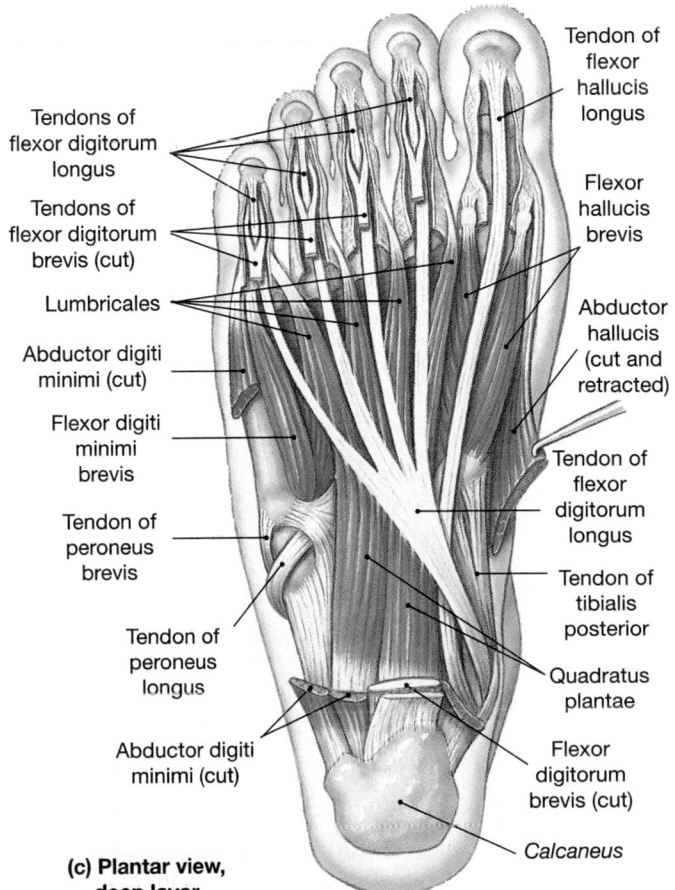

(c) Plantar view, deep layer

Development of the Muscular System

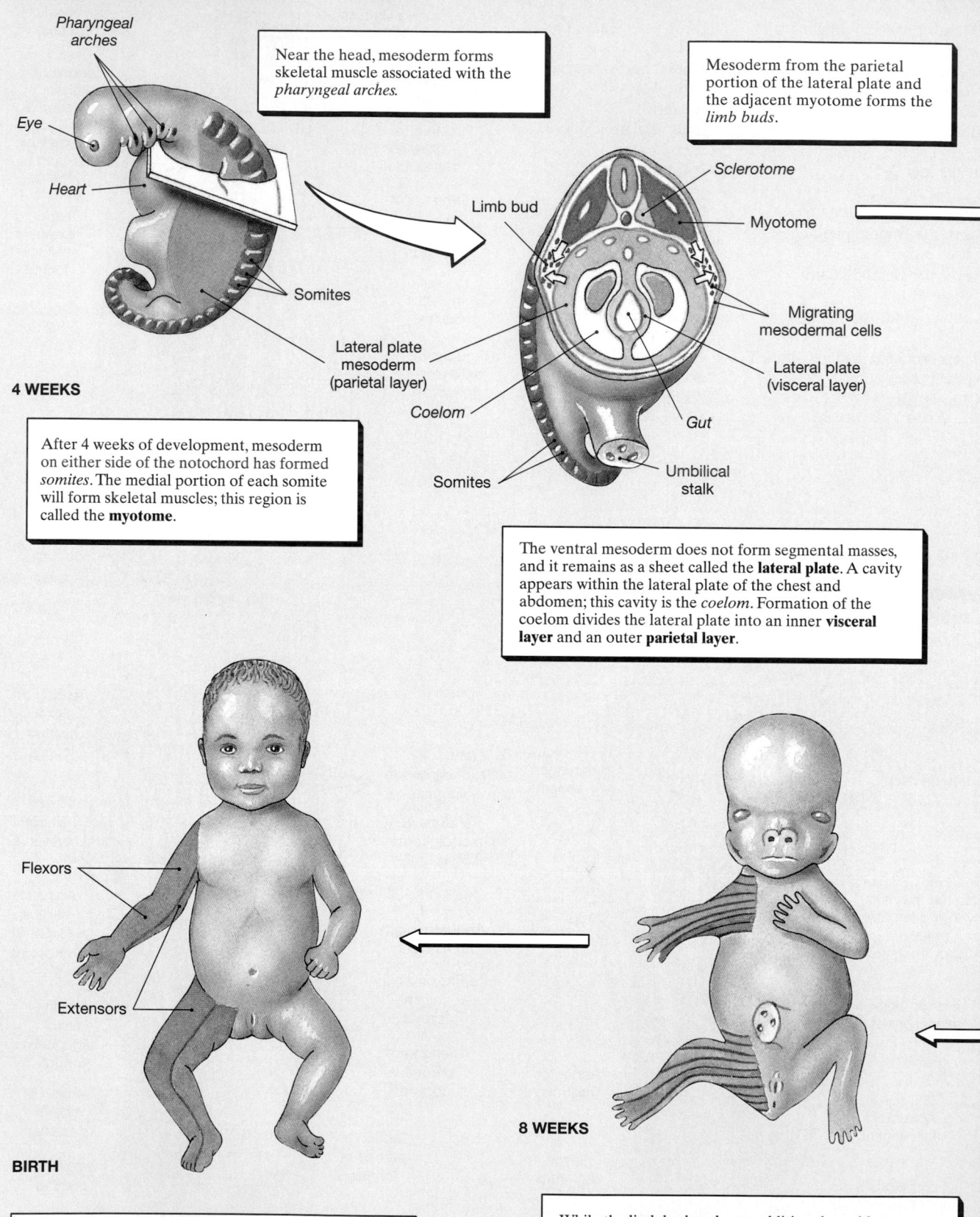

Pharyngeal arches

Eye

Heart

Somites

4 WEEKS

Near the head, mesoderm forms skeletal muscle associated with the *pharyngeal arches.*

Mesoderm from the parietal portion of the lateral plate and the adjacent myotome forms the *limb buds.*

Limb bud

Sclerotome

Myotome

Migrating mesodermal cells

Lateral plate (visceral layer)

Lateral plate mesoderm (parietal layer)

Coelom

Gut

Somites

Umbilical stalk

After 4 weeks of development, mesoderm on either side of the notochord has formed *somites.* The medial portion of each somite will form skeletal muscles; this region is called the **myotome.**

The ventral mesoderm does not form segmental masses, and it remains as a sheet called the **lateral plate.** A cavity appears within the lateral plate of the chest and abdomen; this cavity is the *coelom.* Formation of the coelom divides the lateral plate into an inner **visceral layer** and an outer **parietal layer.**

Flexors

Extensors

BIRTH

8 WEEKS

Rotation of the arm and leg buds produces a change in the position of these masses relative to the body axis.

While the limb buds enlarge, additional myoblasts invade the limb from myotomal segments nearby. Lines indicate the boundaries between myotomes providing myoblasts to the limb.

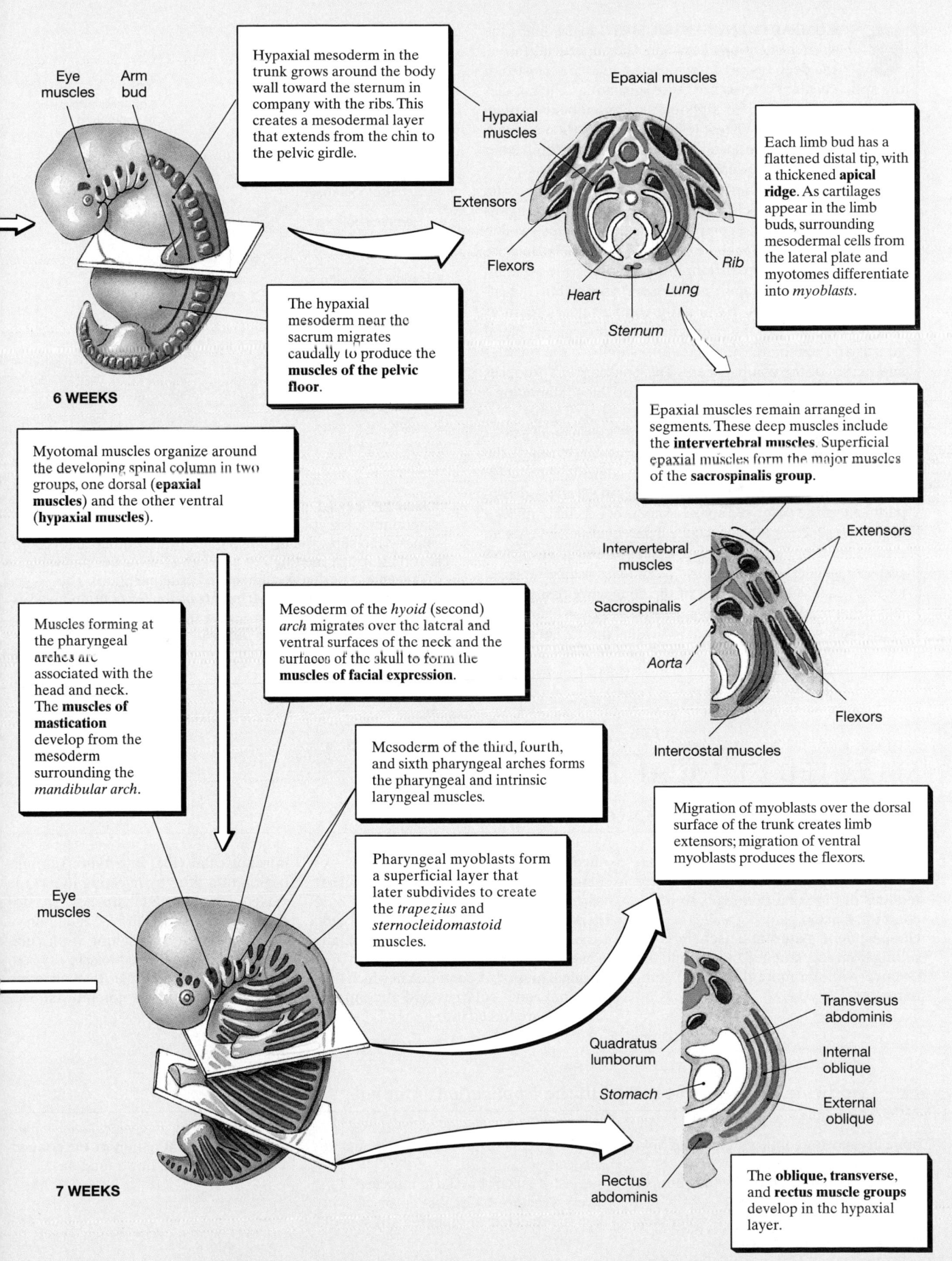

Eye muscles

Arm bud

Hypaxial mesoderm in the trunk grows around the body wall toward the sternum in company with the ribs. This creates a mesodermal layer that extends from the chin to the pelvic girdle.

The hypaxial mesoderm near the sacrum migrates caudally to produce the **muscles of the pelvic floor**.

6 WEEKS

Hypaxial muscles

Extensors

Flexors

Heart

Sternum

Lung

Rib

Epaxial muscles

Each limb bud has a flattened distal tip, with a thickened **apical ridge**. As cartilages appear in the limb buds, surrounding mesodermal cells from the lateral plate and myotomes differentiate into *myoblasts*.

Myotomal muscles organize around the developing spinal column in two groups, one dorsal (**epaxial muscles**) and the other ventral (**hypaxial muscles**).

Epaxial muscles remain arranged in segments. These deep muscles include the **intervertebral muscles**. Superficial epaxial muscles form the major muscles of the **sacrospinalis group**.

Intervertebral muscles

Sacrospinalis

Aorta

Extensors

Flexors

Intercostal muscles

Muscles forming at the pharyngeal arches are associated with the head and neck. The **muscles of mastication** develop from the mesoderm surrounding the *mandibular arch*.

Mesoderm of the *hyoid* (second) *arch* migrates over the lateral and ventral surfaces of the neck and the surfaces of the skull to form the **muscles of facial expression**.

Mesoderm of the third, fourth, and sixth pharyngeal arches forms the pharyngeal and intrinsic laryngeal muscles.

Pharyngeal myoblasts form a superficial layer that later subdivides to create the *trapezius* and *sternocleidomastoid* muscles.

Migration of myoblasts over the dorsal surface of the trunk creates limb extensors; migration of ventral myoblasts produces the flexors.

Eye muscles

Quadratus lumborum

Stomach

Rectus abdominis

Transversus abdominis

Internal oblique

External oblique

The **oblique, transverse,** and **rectus muscle groups** develop in the hypaxial layer.

7 WEEKS

COMPARTMENT SYNDROMES In the limbs, the interconnections between the superficial fascia, the deep fascia of the muscles, and the periostea of the appendicular skeleton are quite substantial. The muscles within a limb are in effect isolated in **compartments** formed by dense collagenous sheets (Figure 11-25●). Blood vessels and nerves traveling to specific muscles within the limb enter and branch within the appropriate compartments.

When a crushing injury, severe contusion, or muscle strain occurs, the blood vessels within one or more compartments may be damaged. These compartments then become swollen with blood and fluid leaked from damaged vessels. Because the connective tissue partitions are very strong, the accumulated fluid cannot escape, and pressure rises within the affected compartments. Eventually, compartment pressures may become so high that they compress the regional blood vessels and eliminate the circulatory supply to the muscles and nerves of the compartment. This compression produces a condition of **ischemia** (is-KĒ-mē-uh), or "blood starvation," known as the **compartment syndrome.**

Slicing into the compartment along its longitudinal axis or implanting a drain are emergency measures used to relieve the pressure. If such steps are not taken, the contents of the compartment will suffer severe damage. Nerves in the affected compartment will be destroyed after 2–4 hours of ischemia, although they can regenerate to some degree if the circulation is restored. After 6 hours or more, the muscle tissue will also be destroyed, and no regeneration will occur. The muscles will be replaced by scar tissue, and shortening of the connective tissue fibers may result in *contracture*, a permanent contraction of an entire muscle following the atrophy of individual muscle fibers.

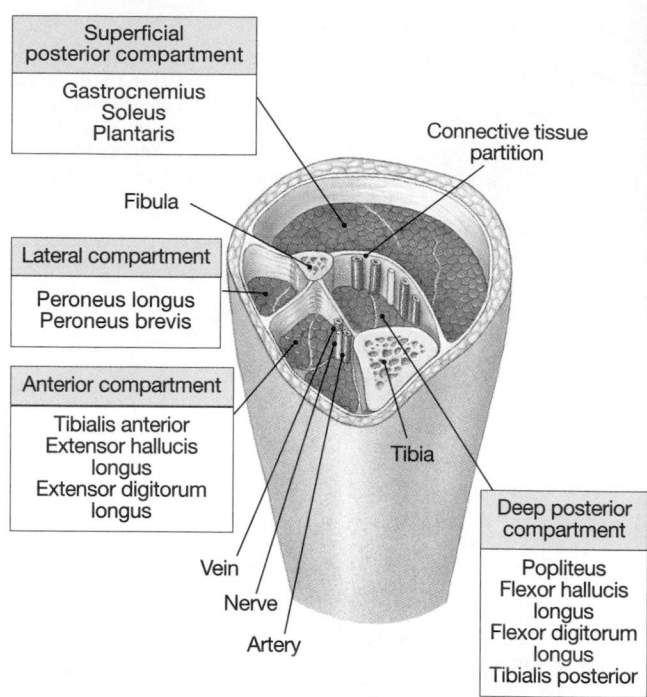

Superficial posterior compartment
Gastrocnemius
Soleus
Plantaris

Connective tissue partition

Fibula

Lateral compartment
Peroneus longus
Peroneus brevis

Anterior compartment
Tibialis anterior
Extensor hallucis longus
Extensor digitorum longus

Tibia

Deep posterior compartment
Popliteus
Flexor hallucis longus
Flexor digitorum longus
Tibialis posterior

Vein
Nerve
Artery

●**FIGURE 11-25 Musculoskeletal Compartments.** A diagrammatic section through the leg, with the muscles removed to show the arrangement of the compartments. A section through the thigh or arm would show a comparable arrangement of dense connective tissue partitions. The anterior and lateral compartments of the leg contain muscles of the extensor/dorsiflexor series, and the posterior compartments contain the flexor/plantar flexor muscles.

SELECTED CLINICAL TERMINOLOGY

Terms Discussed in This Chapter

carpal tunnel syndrome: An inflammation of the sheath surrounding the flexor tendons of the palm that leads to nerve compression and pain. *(p. 350)*
compartment syndrome: Ischemia resulting from accumulated blood and fluid trapped within a musculoskeletal compartment. *(p. 364)*

diaphragmatic hernia (*hiatal hernia*): A hernia that occurs when abdominal organs slide into the thoracic cavity. *(p. 337)*
hernia: A condition involving an organ or body part that protrudes through an abnormal opening. *(p. 336)*
inguinal hernia: A condition in which the inguinal canal enlarges and abdominal contents are forced into it. *(p. 336)*

intramuscular (IM) injection: Administration of a drug by injecting it into the mass of a large skeletal muscle. *(p. 356)*
ischemia (is-KĒ-mē-uh): A condition of "blood starvation" resulting from compression of regional blood vessels. *(p. 364)*
rotator cuff: The muscles that surround the shoulder joint; a common site of sports injuries. *(p. 343)*

AM ## Additional Terms Discussed in the *Applications Manual*

bone bruise: Bleeding within the periosteum of a bone.
bursitis: Inflammation of the bursae around one or more joints.
muscle cramps: Prolonged, involuntary, painful muscular contractions.

sprains: Tears or breaks in ligaments or tendons.
strains: Tears or breaks in muscles.
stress fractures: Cracks or breaks in bones subjected to repeated stresses or trauma.

tendinitis: Inflammation of the connective tissue surrounding a tendon.

CHAPTER REVIEW

On-line resources for this chapter are on our World Wide Web site at:
http://www.prenhall.com/martini/fap

STUDY OUTLINE

INTRODUCTION, p. 316

1. The **muscular system** includes all the skeletal muscle tissue that can be controlled voluntarily.

BIOMECHANICS AND MUSCLE ANATOMY, p. 316

Organization of Skeletal Muscle Fibers, p. 316

1. A muscle can be classified according to the arrangement of fibers and fascicles as a **parallel muscle, convergent muscle, pennate muscle,** or **circular muscle (sphincter).** A pennate muscle may be *unipennate, bipennate,* or *multipennate.* *(Figure 11-1)*

Skeletal Muscle Length–Tension Relationships, p. 317

2. Muscles commonly work in groups to optimize muscle length–tension relationships.

Levers, p. 318

3. A **lever** is a rigid structure that moves on a fixed point called the **fulcrum.** Levers can change the direction, speed, or distance of muscle movements and modify the force applied to muscles.

4. Levers may be classified as **first-class, second-class,** or **third-class levers;** the last are the most common type of lever in the body. *(Figure 11-2)*

MUSCLE TERMINOLOGY, p. 319

Origins and Insertions, p. 319

1. Each muscle may be identified by its **origin, insertion,** and **action.**

Actions, p. 319

2. According to its **primary action,** a muscle may be classified as a **prime mover** or **agonist,** an **antagonist,** a **synergist,** or a **fixator.**

Names of Skeletal Muscles, p. 320

3. The names of muscles commonly provide clues to their fascicle organization, location, relative position, structure, size, shape, origin and insertion, or action. *(Table 11-1)*

Axial and Appendicular Muscles, p. 321

4. The **axial musculature** arises on the axial skeleton; it positions the head and spinal column and moves the rib cage. The **appendicular musculature** stabilizes or moves components of the appendicular skeleton. *(Figure 11-3)*

THE AXIAL MUSCLES, p. 324

1. The axial muscles fall into logical groups on the basis of location, function, or both.

2. **Innervation** refers to the identity of the nerve that controls a given muscle.

Muscles of the Head and Neck, p. 324

3. The muscles of facial expression are the **orbicularis oris, buccinator, epicranius (frontalis** and **occipitalis),** and **platysma.** *(Figure 11-4; Table 11-2)*

4. Six **extrinsic eye muscles (oculomotor muscles)** control external eye movements: the **inferior** and **superior rectus,** lateral and **medial rectus,** and **inferior** and **superior obliques.** *(Figure 11-5; Table 11-3)*

5. The muscles of mastication (chewing) are the **masseter, temporalis,** and **pterygoid muscles.** *(Figure 11-6; Table 11-4)*

6. The muscles of the tongue are necessary for speech and swallowing and assist in mastication. They are the **genioglossus, hyoglossus, palatoglossus,** and **styloglossus.** *(Figure 11-7; Table 11-5)*

7. The muscles of the pharynx constrict the pharyngeal walls **(pharyngeal constrictors),** elevate the larynx **(laryngeal elevators),** and raise the soft palate **(palatal muscles).** *(Figure 11-8; Table 11-6)*

8. The muscles of the neck control the position of the larynx, depress the mandible, and provide a foundation for the muscles of the tongue and pharynx. *(Figure 11-9; Table 11-7)*

Muscles of the Spine, p. 332

9. The superficial muscles of the spine can be classified into the **spinalis, longissimus,** and **iliocostalis** divisions. In the lower lumbar and sacral regions, the longissimus and iliocostalis are sometimes called the **sacrospinalis** muscles. *(Figure 11-10; Table 11-8)*

10. Other muscles of the spine include the **longus capitis** and **longus colli** of the neck and the **quadratus lumborum** of the lumbar region. *(Figure 11-10; Table 11-8)*

Oblique and Rectus Muscles, p. 335

11. The oblique muscles include the **scalenes** and the **intercostal** and **transversus** muscles. The **external** and **internal intercostals** are important in respiratory movements of the ribs. Also important to respiration is the **diaphragm.** *(Figures 11-11, 11-12; Table 11-9)*

Muscles of the Pelvic Floor, p. 337

12. The **perineum** can be divided into an anterior **urogenital triangle** and a posterior **anal triangle.** The pelvic floor consists of the **urogenital diaphragm** and the **pelvic diaphragm.** *(Figure 11-13; Table 11-10)*

THE APPENDICULAR MUSCLES, p. 337

Muscles of the Shoulders and Upper Limbs, p. 339

1. The **trapezius** affects the positions of the shoulder girdle, head, and neck. Other muscles inserting on the scapula include the **rhomboideus,** the **levator scapulae,** the **serratus anterior,** the **subclavius,** and the **pectoralis minor.** *(Figures 11-14, 11-15; Table 11-11)*

2. The **deltoid** and the **supraspinatus** are important abductors. The **subscapularis** and the **teres major** rotate the arm medially; the **infraspinatus** and **teres minor** perform lateral rotation; and the **coracobrachialis** flexes and adducts the humerus at the shoulder joint. *(Figures 11-14, 11-16; Table 11-12)*

3. The **pectoralis major** flexes the humerus at the shoulder joint, and the **latissimus dorsi** extends it. *(Figures 11-14, 11-16; Table 11-12)*

4. The primary actions of the **biceps brachii** and the **triceps brachii** (long head) affect the elbow joint. The **brachialis** and **brachioradialis** flex the elbow, opposed by the

anconeus. The **flexor carpi ulnaris**, the **flexor carpi radialis,** and the **palmaris longus** cooperate to flex the wrist. They are opposed by the **extensor carpi radialis** and the **extensor carpi ulnaris.** The **pronator teres** and **pronator quadratus** pronate the forearm and are opposed by the **supinator.** *(Figures 11-16–11-20; Tables 11-13–11-15)*

Muscles of the Lower Limbs, p. 351

5. **Gluteal muscles** cover the lateral surfaces of the ilia. The largest is the **gluteus maximus,** which shares an insertion with the **tensor fasciae latae.** Together these muscles pull on the **iliotibial tract.** *(Figures 11-14, 11-21; Table 11-16)*

6. The **piriformis** and the **obturator** muscles are the most important **lateral rotators.** The **adductors** can produce a variety of movements. *(Figure 11-21; Table 11-16)*

7. The **psoas major** and the **iliacus** merge to form the **iliopsoas** muscle, a powerful flexor of the hip. *(Figure 11-21; Table 11-16)*

8. The flexors of the knee include the **biceps femoris, semimembranosus,** and **semitendinosus** (the three **hamstrings**), and the **sartorius.** The **popliteus** unlocks the knee joint. *(Figure 11-22; Table 11-17)*

9. Collectively, the knee extensors are known as the **quadriceps femoris.** This group includes the three **vastus** muscles and the **rectus femoris.** *(Figure 11-22; Table 11-17)*

10. The **gastrocnemius** and **soleus** muscles produce plantar flexion. A pair of **peroneus** muscles produces eversion as well as plantar flexion. *(Figure 11-23; Table 11-18)*

11. Smaller muscles of the calf and shin position the foot and move the toes. Precise control of the phalanges is provided by muscles originating at the tarsal and metatarsal bones. *(Figure 11-24; Table 11-19)*

REVIEW QUESTIONS

Level 1   Reviewing Facts and Terms

1. Muscle fibers within a skeletal muscle form bundles called
   - (a) aponeuroses
   - (b) fascicles
   - (c) sarcomeres
   - (d) myofibrils
2. Most muscles in the body are _____ muscles:
   - (a) convergent
   - (b) pennate
   - (c) circular
   - (d) parallel
3. Skeletal muscle fibers produce maximum tension over a relatively narrow range of
   - (a) complex movements
   - (b) extensions
   - (c) flexions
   - (d) sarcomere lengths
4. The bones, which serve as levers in the body, change
   - (a) the direction of an applied force
   - (b) the distance and speed of movement produced by a force
   - (c) the effective strength of a force
   - (d) a, b, and c are correct
5. A general rule to follow to understand the difference between an origin and insertion is
   - (a) the origin moves while the insertion remains stationary
   - (b) each muscle begins and ends at the origin
   - (c) the origin remains stationary while the insertion moves
   - (d) each muscle begins and ends at the insertion
6. If an agonist produces flexion, the primary action of the antagonist will be
   - (a) pronation
   - (b) adduction
   - (c) abduction
   - (d) extension
7. The muscles of facial expression are innervated by
   - (a) cranial nerve VII
   - (b) cranial nerve V
   - (c) cranial nerve IV
   - (d) cranial nerve VI
8. The strongest masticatory muscle is the
   - (a) pterygoid
   - (b) masseter
   - (c) temporalis
   - (d) mandible
9. The muscles of the tongue are innervated by the
   - (a) hypoglossal nerve (N XII)
   - (b) trochlear nerve (N IV)
   - (c) abducens nerve (N VI)
   - (d) a, b, and c are correct

10. The muscle that opens the mouth by depressing the mandible is the
    - (a) stylopharyngeus
    - (b) digastric
    - (c) sternocleidomastoid
    - (d) hypoglossus
11. Important flexors of the spinal column that act in opposition to the erector spinae are the
    - (a) rectus muscles
    - (b) longus capitis
    - (c) longus colli
    - (d) scalenes
12. The linea alba is a median collagenous partition that longitudinally divides the
    - (a) external obliques
    - (b) rectus abdominis
    - (c) external intercostals
    - (d) rectus femoris
13. The muscular partition that separates the abdominopelvic and thoracic cavities is the
    - (a) masseter
    - (b) perineum
    - (c) diaphragm
    - (d) transversus abdominis
14. The major abductor of the arm is the
    - (a) deltoid
    - (b) biceps brachii
    - (c) triceps brachii
    - (d) subscapularis
15. The muscles that rotate the radius without producing either flexion or extension of the elbow are
    - (a) the brachialis and the brachioradialis
    - (b) the pronator teres and the supinator
    - (c) the biceps and triceps brachii
    - (d) a, b, and c are correct
16. The powerful flexors of the hip are the
    - (a) piriformis
    - (b) obturators
    - (c) pectineus
    - (d) iliopsoas
17. Knee extensors known as the quadriceps include
    - (a) three vastus muscles and rectus femoris
    - (b) biceps femoris, gracilis, sartorius
    - (c) popliteus, iliopsoas, gracilis
    - (d) gastrocnemius, tibialis, peroneus
18. What two factors interact to determine the effects of a muscle contraction?
19. List the four fascicle organizations that produce the different patterns of skeletal muscles.
20. What are the distinguishing characteristics of unipennate, bipennate, and multipennate muscles?

21. How do first-, second-, and third-class levers differ?
22. What three primary actions are used to identify muscle groups? Give a brief description of each action.
23. What is the functional difference between the axial musculature and the appendicular musculature?
24. What four muscle groups make up the axial musculature?
25. Which axial muscles are used in the process of mastication?
26. What three functions are accomplished by the muscles of the pelvic floor?

27. What four groups of muscles are associated with the shoulders and upper extremities?
28. What three functional groups make up the muscles of the lower limbs?
29. Which three muscular sites are most desirable for intramuscular injections? Why?

Level 2 Reviewing Concepts

30. Of the following examples, the one that illustrates the action of a second-class lever is
    (a) leg extension
    (b) plantar flexion
    (c) flexion at the elbow
    (d) a, b, and c are correct
31. An example of a prime mover that produces flexion at the elbow is the
    (a) brachioradialis   (b) biceps brachii
    (c) brachialis   (d) biceps femoris
32. Removal of the trapezius muscle exposes the
    (a) serratus anterior and subclavian muscles
    (b) infraspinatus and teres minor muscles
    (c) deltoid and supraspinatus muscles
    (d) rhomboideus and levator scapulae muscles

33. The muscles of the spine include many dorsal extensors but few ventral flexors. Why?
34. What specific structural characteristic makes voluntary control of urination and defecation possible?
35. Why does a convergent muscle exhibit more versatility when contracting than does a parallel or pennate muscle?
36. Why can a pennate muscle generate more tension than can a parallel muscle of the same size?
37. Why is it difficult to lift a heavy object when the elbow is at full extension?
38. If the fifth cranial nerve (trigeminal) is damaged or severed, what function in the body will be affected?
39. What types of movements are affected when the hamstrings are injured?

Level 3 Critical Thinking and Clinical Applications

40. Mary sees Jill coming toward her and immediately contracts her frontalis and procerus muscles. She also contracts her levator labii on the right side. Is Mary glad to see Jill? How can you tell?
41. Jeff is interested in building up his leg muscles, specifically the quadriceps group. What exercises would you recommend to help Jeff accomplish his goal?

42. Shelly gives her son an ice-cream cone. The boy grasps the cone with his right hand, opens his mouth, and begins to lick at the ice cream. What muscles are used to perform these actions?

Neural Tissue

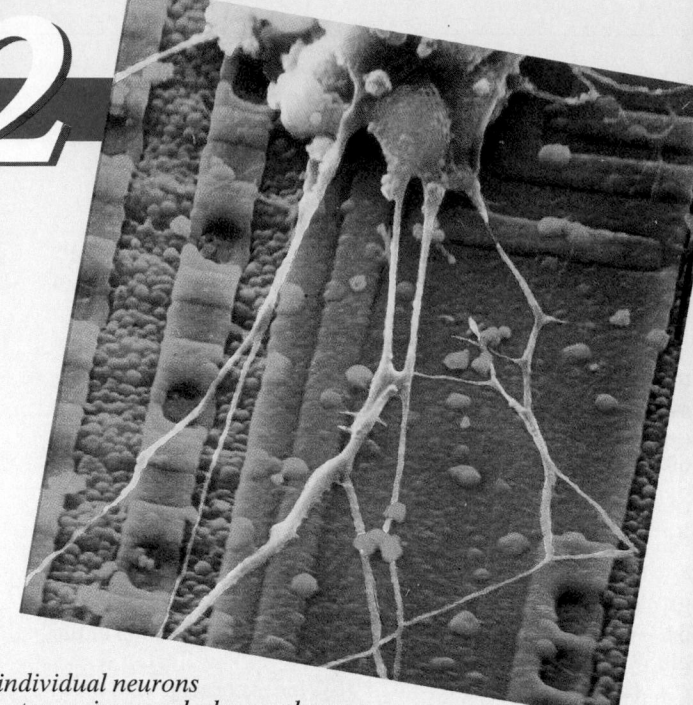

*I*f you think of the nervous system as an organic computer, individual neurons are the computer chips that make it work. This scanning electron micrograph shows a human neuron growing on the surface of a silicon chip. The circuits visible on the chip are quite small, but the branches of the neuron—the organic circuits—are considerably smaller. Any comparison between the nervous system and a computer is misleading, because even the most sophisticated computer lacks the versatility and adaptability of a single neuron. A neuron may process information from 100,000 different sources, and there are more than 20 billion neurons in the nervous system. In this chapter, we begin our examination of the nervous system by considering how neurons function and how groups of neurons interact.

CHAPTER OUTLINE AND OBJECTIVES

In the next seven chapters, our attention will shift to mechanisms that coordinate the activities of the body's organ systems. These activities are adjusted to meet changing situations and environmental conditions. You sit, stand, or walk by controlling muscular activities; your body temperature remains stable on a cold winter's day or in a warm kitchen because your rates of heat generation and heat loss are closely regulated.

Two organ systems, the *nervous system* and the *endocrine system,* provide the necessary regulation. These systems share several structural and functional characteristics, and they usually act in a complementary fashion. The endocrine system adjusts the metabolic operations of other systems in response to changes in the availability of nutrients and the demand for energy. It also directs activities that continue for extended periods, such as growth and maturation, sexual development, pregnancy, and responses to chronic environmental stresses. Endocrine responses tend to develop slowly but last much longer than those of the nervous system. The nervous system provides relatively swift but generally brief responses to stimuli by temporarily modifying the activities of other organ systems. The modifications may appear in a matter of milliseconds, but the effects disappear soon after neural activity ceases.

The nervous system, which accounts for a mere 3 percent of the total body weight, is the most complex organ system. It is vital not only to life but to our experience and appreciation of life as well. This chapter details the structure and function of neural tissue and introduces principles of neurophysiology that are vital to an understanding of the nervous system's capabilities and limitations. In subsequent chapters, we will examine increasing levels of structural and functional complexity. Table 12-1 provides an overview of the most important concepts and terms introduced in this chapter. [AM] *The Neurological Examination*

AN OVERVIEW OF THE NERVOUS SYSTEM

The nervous system includes all the *neural tissue* in the body. We introduced neural tissue in Chapter 4. ∞ *[p. 137]* The basic functional units of the nervous system are individual **neurons.** Supporting cells, or **neuroglia** (noo-RŌ-glē-ah; *glia,* glue), separate and protect the neurons, provide a supportive framework for neural tissue, act as phagocytes, and help regulate the composition of the interstitial fluid. Neuroglia, also called *glial cells,* far outnumber the neurons and account for roughly half the volume of the nervous system.

Neural tissue, with supporting blood vessels and connective tissues, forms the organs of the nervous system: the brain, the spinal cord, the receptors in complex sense organs, such as the eye and ear, and the *nerves* that inter-connect those organs and link the nervous system with other systems. In Chapter 1, we cited two major *anatomical* divisions of the nervous system: the *central nervous system* and the *peripheral nervous system.* ∞ *[p. 8]*

The **central nervous system (CNS)** consists of the brain and spinal cord. These are complex organs that include not only neural tissue but also blood vessels and the various connective tissues that provide physical protection and support. The CNS is responsible for integrating, processing, and coordinating sensory data and motor commands. Sensory data convey information about conditions inside or outside the body. Motor commands control or adjust the activities of peripheral organs, such as skeletal muscles. When you stumble, the CNS integrates information concerning balance and limb position and then coordinates your recovery by sending motor commands to appropriate skeletal muscles—all in a split second and without conscious effort. The CNS—specifically, the brain—is also the seat of higher functions, such as intelligence, memory, learning, and emotion.

The **peripheral nervous system (PNS)** includes all the neural tissue outside the CNS. The PNS delivers sensory information to the CNS and carries motor commands to peripheral tissues and systems. Bundles of **nerve fibers** (*axons*) carry sensory information and motor commands in the PNS. Such bundles, with associated blood vessels and connective tissues, are called *peripheral nerves,* or simply **nerves.** Nerves connected to the brain are called **cranial nerves;** those attached to the spinal cord are called **spinal nerves.**

Figure 12-1● (p. 371) diagrams the *functional* divisions of the nervous system. The **afferent division** (*ad,* to + *ferre,* to carry) of the PNS brings sensory information to the CNS from **receptors** in peripheral tissues and organs. Receptors are sensory structures, ranging from the processes of single cells to complex organs, that either detect changes in the internal environment or respond to the presence of specific stimuli. Receptors may be neurons or specialized cells of other tissues, such as the *Merkel cells* of the epidermis. ∞ *[p. 149]*

The **efferent division** (*ex,* from) of the PNS carries motor commands from the CNS to muscles and glands. These target organs, which respond by *doing* something, are called **effectors.** The efferent division has *somatic* and *autonomic* components. The **somatic nervous system (SNS)** controls skeletal muscle contractions. The contractions may be voluntary or involuntary. You are **conscious** when you are awake and alert. *Voluntary* contractions are under conscious control; you exert voluntary control over your arm as you raise a full glass of water to your lips. *Involuntary* contractions may be simple, automatic responses or complex movements directed at the **subconscious** level. If you accidentally place your hand on a hot stove, you will withdraw it immediately, usually before you even notice the pain. This type of automatic involuntary response is a **reflex.** The **auto-**

TABLE 12-1 An Introductory Glossary of the Nervous System

MAJOR ANATOMICAL AND FUNCTIONAL DIVISIONS

Central nervous system (CNS)	The brain and spinal cord, which contain control centers responsible for processing and integrating sensory information, planning and coordinating responses to stimuli, and providing short-term control over the activities of other systems.
Peripheral nervous system (PNS)	Neural tissue outside the CNS that links the CNS with sense organs and other systems.
Autonomic nervous system (ANS)	Components of the CNS and PNS that control visceral functions at the subconscious level.

GROSS ANATOMY

Center	A group of neuron cell bodies with a common function (p. 409).
Column	A group of tracts within a specific region of the spinal cord (p. 410).
Ganglia	An anatomically distinct collection of sensory or motor neuron cell bodies within the PNS (p. 378).
Nerve	A bundle of axons in the PNS (p. 369).
Nucleus	A CNS center with discrete anatomical boundaries (p. 409).
Pathways	Centers and tracts that connect the brain with other organs and systems in the body (p. 410).
Tract	A bundle of axons within the CNS that share a common origin, destination, and function (p. 410).

HISTOLOGY

Axon	A long, slender cytoplasmic process of a neuron; axons are capable of propagating nerve impulses (action potentials) (p. 371).
Dendrites	Neuronal processes that respond to specific stimuli in the extracellular environment (p. 371).
Gray matter	Neural tissue dominated by neuron cell bodies (p. 378).
Motor neuron	A neuron whose axon carries motor commands from the CNS toward effectors in the PNS (p. 375).
Myelin	A multilayered membranous wrapping, produced by glial cells, that ensheaths axons and speeds up action potential propagation; axons wrapped with myelin are said to be *myelinated* (p. 378).
Neural cortex	A layer of gray matter on the surface of the brain (p. 409).
Neuron	The basic functional unit of the nervous system; a highly specialized cell in neural tissue that performs intercellular communication; has a cell body (soma) and cytoplasmic processes that vary in number and length (p. 369).
Neuroglia, or glial cells	Supporting cells that interact with neurons and regulate the extracellular environment, defend against pathogens, and perform repairs within neural tissue (p. 369).
Sensory neuron	A neuron whose axon carries sensory information from the PNS toward the CNS (p. 374).
Soma	The cell body of a neuron (p. 371).
White matter	Neural tissue dominated by myelinated axons (p. 378).

FUNCTIONAL CATEGORIES

Action potentials	Sudden, transient changes in the membrane potential that are propagated along the surface of an axon or sarcolemma or by the membranes of other specialized cells (p. 388).
Conscious	Awake and alert; a motor response directed and controlled by an awake, alert person (p. 369).
Effector	A muscle, gland, or other specialized cell or organ that responds to neural stimulation by altering its activity and producing a specific effect (p. 369).
Involuntary	Reflexive, automatic; not under direct conscious control (p. 369).
Receptor	A specialized cell, dendrite, or organ that responds to specific stimuli in the extracellular environment and whose stimulation alters the level of activity in a sensory neuron (p. 369).
Reflex	A rapid, stereotyped response to a specific stimulus (p. 369).
Somatic	Pertaining to the control of skeletal muscle activity (*somatic motor*) or to sensory information from skeletal muscles, tendons, and joints (*somatic sensory*) (p. 375).
Subconscious	Pertaining to centers in the brain that operate outside a person's conscious awareness (p. 369).
Visceral	Pertaining to the control of visceral functions, such as digestion and circulation (*visceral motor*), or to sensory information from visceral organs (*visceral sensory*) (p. 375).
Voluntary	A motor response under direct conscious control (p. 369).

nomic nervous system (ANS), or *visceral motor system,* provides automatic, involuntary regulation of smooth muscle, cardiac muscle, and glandular activity or secretions. The ANS includes a *sympathetic division* and a *parasympathetic division.* These ANS divisions commonly have antagonistic effects. For example, activity of the sympathetic division accelerates the heart rate, whereas parasympathetic activity slows the heart rate.

NEURONS

Before we consider how neurons relay information, we need to take a closer look at the structure of an individual neuron. In this section, we will consider (1) the structure of a "model" neuron and (2) the structural and functional classification of neurons.

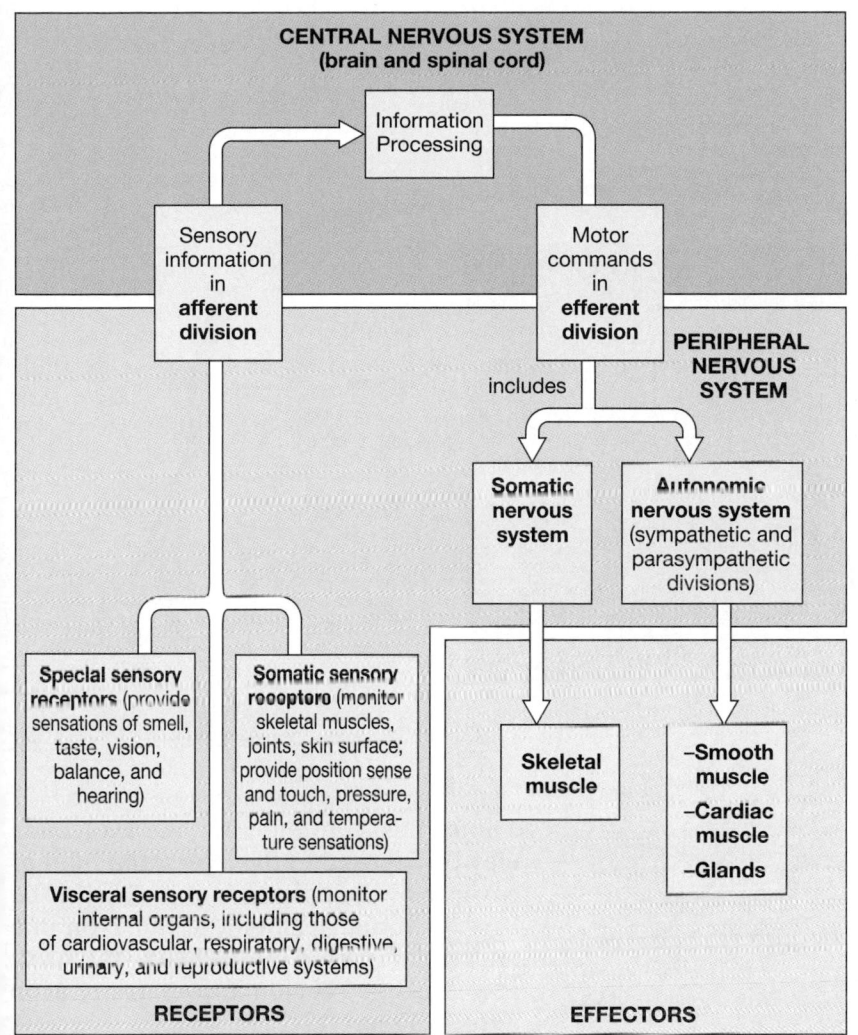

•*FIGURE 12-1* **Functional Overview of the Nervous System**

rough endoplasmic reticulum (RER) give the perikaryon a coarse, grainy appearance. Mitochondria generate ATP to meet the high energy demands of an active neuron. The ribosomes and RER synthesize peptides and proteins. Some areas of the perikaryon contain clusters of RER and free ribosomes. These regions, which stain darkly, are called *Nissl bodies,* because they were first described by the German microscopist Franz Nissl. Nissl bodies account for the gray color of areas containing neuron cell bodies—the *gray matter* seen in dissection.

Most neurons lack *centrioles,* important organelles involved in the organization of the cytoskeleton and the movement of chromosomes during mitosis. ▭ *[p. 84]* In some tissues, notably the olfactory (smell) epithelium of the nose, neural stem cells persist throughout life. Their divisions produce daughter cells that differentiate into highly specialized neurons, the olfactory receptors. However, stem cells are very rare inside the CNS, and CNS neurons generally lose their centrioles during differentiation. As a result, typical CNS neurons cannot divide, and they will not be replaced if lost to injury or disease.

Neuron Structure

Figure 12-2• shows the structure of a representative neuron in detail. Neurons can have a variety of shapes; the one shown here is a *multipolar neuron,* the most common type of neuron in the CNS. Such a neuron has only one axon but multiple dendrites.

The Soma

The **soma** (plural, *somata;* cell body) (Figure 12-2a•) contains a relatively large, round nucleus with a prominent nucleolus. The surrounding cytoplasm constitutes the **perikaryon** (per-i-KAR-ē-on; *karyon,* nucleus). The cytoskeleton of the perikaryon contains **neurofilaments** and **neurotubules.** Bundles of neurofilaments, called **neurofibrils,** extend into the dendrites and axon, providing internal support for these relatively slender processes.

The perikaryon contains organelles that provide energy and synthesize organic materials, especially neurotransmitters. The numerous mitochondria, free and fixed ribosomes, and membranes of the

The Dendrites

A variable number of **dendrites**, or sensitive processes, extend out from the soma (Figure 12-2b•). Typical dendrites are highly branched, and each branch bears fine processes called *dendritic spines.* In the CNS, a neuron receives information from other neurons via synaptic connections at the dendritic spines. Although some synaptic connections occur on the soma, most synapses occur on the dendrites, which represent 80–90 percent of the total surface area of the neuron. Chemicals released at these synapses cause localized changes in the transmembrane potential of the dendrites and soma.

The Axon

An **axon** is a long cytoplasmic process capable of propagating an action potential. The **axoplasm** (AK-sō-plazm), or cytoplasm of the axon, contains neurofibrils, neurotubules, small vesicles, lysosomes, mitochondria, and various enzymes. The axoplasm is surrounded by the *axolemma* (cell membrane;

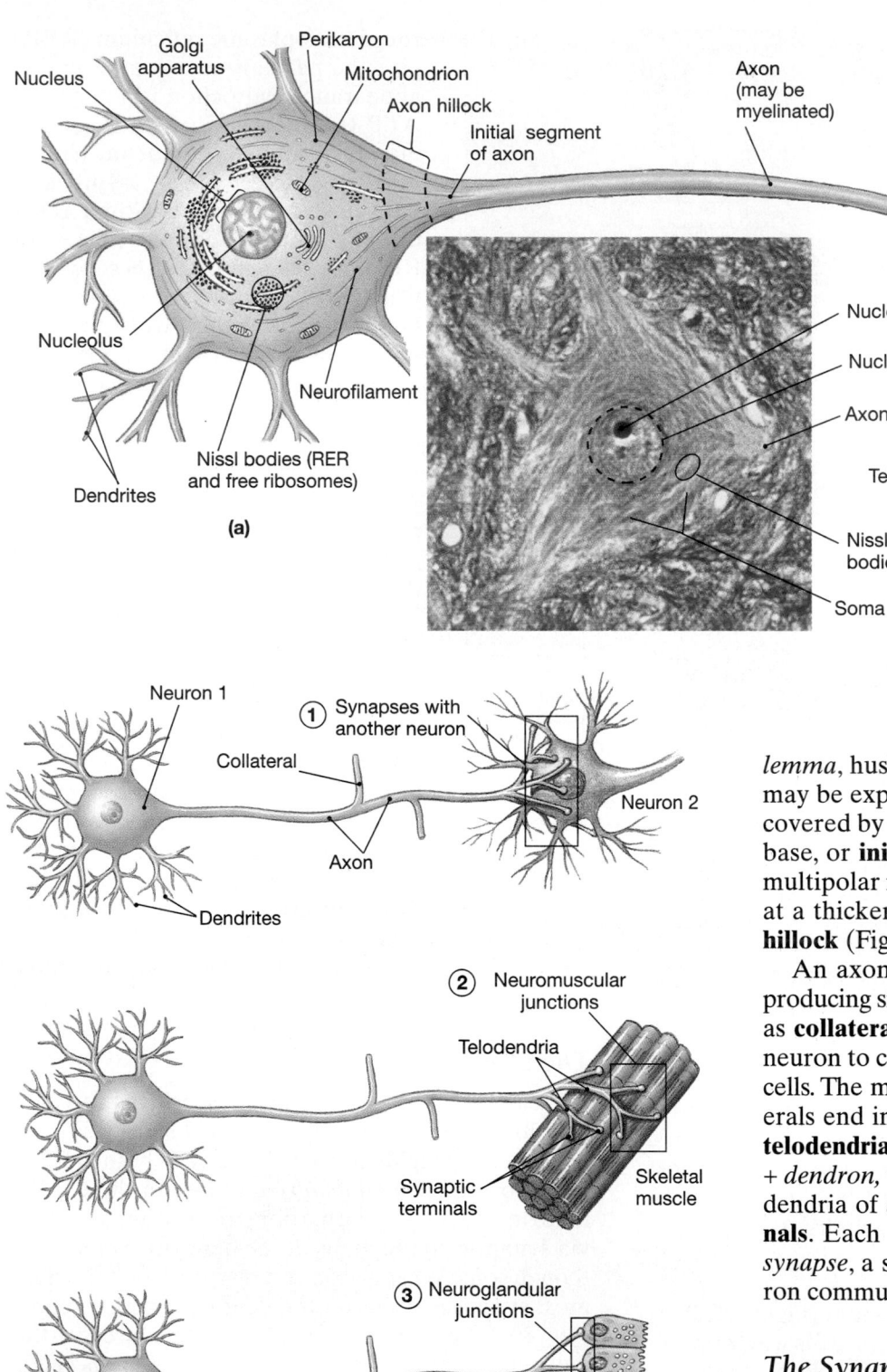

●*FIGURE 12-2* Anatomy of a Multipolar Neuron.
(a) Diagrammatic view of a neuron and micrograph of the soma, showing major organelles. (LM × 1250) **(b)** Distribution of the axon, showing collateral branches and three possible synaptic connections to ① another neuron, ② a skeletal muscle, or ③ gland cells.

lemma, husk). In the CNS, the axolemma may be exposed to the interstitial fluid or covered by the processes of glial cells. The base, or **initial segment,** of the axon in a multipolar neuron is attached to the soma at a thickened region known as the **axon hillock** (Figure 12-2a●).

An axon may branch along its length, producing side branches collectively known as **collaterals.** Collaterals enable a single neuron to communicate with several other cells. The main axon trunk and any collaterals end in a series of fine extensions, or **telodendria** (te-lō-DEN-drē-uh; *telo-,* end + *dendron,* tree) (Figure 12-2b●). The telodendria of an axon end at **synaptic terminals.** Each synaptic terminal is part of a *synapse,* a specialized site where the neuron communicates with another cell.

The Synapse

A **synapse** is a specialized site of intercellular communication. There are two cells at every synapse: (1) the *presynaptic* cell, which has the synaptic terminal and sends a message, and (2) the *postsynaptic* cell, which receives the message (Figure 12-3●). The communication between cells at a synapse most commonly involves the release of chemicals called **neurotransmitters** by the synaptic terminal. These

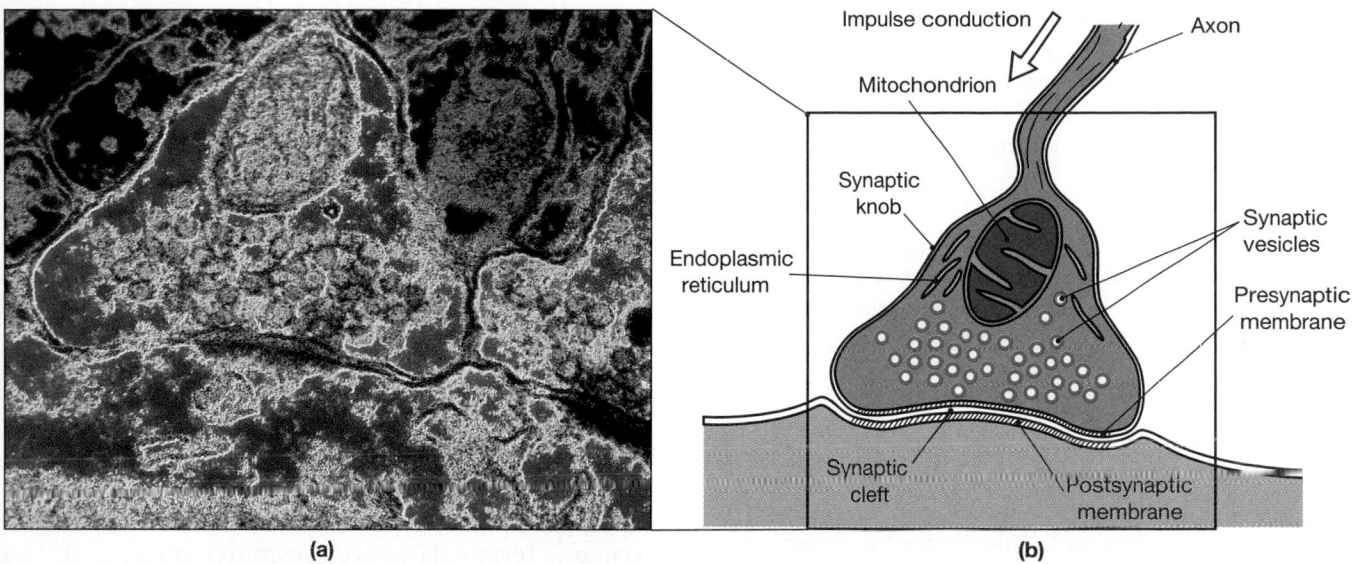

•FIGURE 12-3 **Structure of a Chemical Synapse.** **(a)** Micrograph of a chemical synapse between two neurons. (TEM, color-enhanced, × ???,000) **(b)** Diagrammatic view of the synapse.

chemicals, released by one neuron, affect the activity of another neuron or an effector. As we saw in Chapter 10, this release is triggered by electrical events, such as the arrival of an action potential. ∞ *[p. 285]* The neurotransmitters are typically packaged in *synaptic vesicles*.

When one neuron communicates with another, the synapse may occur on a dendrite, on the soma, or along the length of the axon of the receiving cell. A synapse between a neuron and another cell type is called a **neuroeffector junction.** There are two major classes of neuroeffector junctions: *neuromuscular junctions* and *neuroglandular junctions* (Figure 12-2b•). At a neuromuscular junction, the neuron communicates with a muscle cell. At a **neuroglandular junction,** a neuron controls or regulates the activity of a secretory cell. Neurons also innervate a variety of other cell types, such as adipocytes (fat cells), and we shall consider the nature of the innervation in later chapters.

The structure of the synaptic terminal varies with the type of postsynaptic cell. A relatively simple, round **synaptic knob** occurs where the postsynaptic cell is another neuron.[1] The synaptic terminal at a neuromuscular junction is much more complex. ∞ *[p. 285]* We will deal primarily with the structure of synaptic knobs at this time; we shall detail other types of synaptic terminals in later chapters.

AXOPLASMIC TRANSPORT Each synaptic knob contains mitochondria, portions of the endoplasmic reticulum, and thousands of vesicles filled with neurotransmitter molecules. Breakdown products of neurotransmitter released at the synapse are reabsorbed and reassembled at the synaptic knob. The synaptic knob also receives a continuous supply of neurotransmitter synthesized in the soma, along with enzymes and lysosomes. These products are exported to the synaptic knobs along the length of the axon. This movement occurs along neurotubules and involves *kinesins,* molecular motors that require ATP. ∞ *[p. 84]* Some materials travel slowly, at rates of a few millimeters per day. This transport mechanism is known as the "slow stream." Vesicles containing neurotransmitter move much more rapidly, traveling in the "fast stream" at 5–10 mm per hour.

The movement of materials between the soma and synaptic terminals is called **axoplasmic transport**. The flow of some materials from the soma to the periphery of the neuron is *anterograde flow*. At the same time, other substances are being transported toward the soma in *retrograde* (RET-rō-grād) *flow* (*retro,* backward). If debris or unusual chemicals appear in the synaptic knob, retrograde flow soon delivers them to the soma. The arriving materials may then alter the activity of the cell by turning appropriate genes on or off. For example, in Chapter 3 we noted the role of various *growth factors* on cell development and differentiation. ∞ *[p. 102] Nerve growth factor (NGF)* targets neurons in the CNS and PNS. NGF absorbed at a synaptic knob in peripheral tissues is transported to the soma by retrograde flow. Once within the soma, NGF activates genes that stimulate neuron growth and maintenance.

Rabies is perhaps the most dramatic example of a clinical condition directly related to retrograde flow. A bite on the hand from a rabid animal injects the rabies virus into peripheral tissues, where virus particles quickly enter synaptic knobs and peripheral axons. Retro-

---

[1]The term *synaptic knob* is widely recognized and will be used throughout this text. However, the same structures may be called terminal buttons, terminal boutons, end bulbs, or neuropodia.

grade flow then carries the virus into the CNS, with potentially fatal results. Many toxins, including heavy metals, some pathogenic bacteria, and other viruses also rely on some form of axoplasmic transport to bypass CNS defenses. [AM] *Axoplasmic Transport and Disease*

Neuron Classification

The billions of neurons in the nervous system are quite variable in form. Neurons are classified two ways: on the basis of (1) structure and (2) function.

Structural Classification

The anatomical classification of neurons is based on the relationship of the dendrites to the soma and the axon:

1. **Anaxonic** (an-ak-SON-ik) **neurons** are small and have no anatomical clues to distinguish dendrites from axons; all the cell processes look alike. Anaxonic neurons (Figure 12-4a●) are located in the brain and in special sense organs. Their functions are poorly understood.

2. **Bipolar neurons** have two distinct processes—one dendritic process that branches extensively at its distal tip, and one axon, with the cell body between them (Figure 12-4b●). Their processes are short, and these neurons measure less than 30 mm from tip to tip. Bipolar neurons are rare, but they occur in special sense organs, where they relay information about sight, smell, and hearing from the receptor cells to other neurons.

3. In a **unipolar neuron,** or *pseudounipolar neuron,* the dendritic and axonal processes are continuous, and the cell body lies off to one side (Figure 12-4c●). In such

a neuron, the initial segment lies at the base of the dendritic branches; the rest of the process, which carries action potentials, is usually considered to be an axon. Most sensory neurons of the peripheral nervous system are unipolar. Their axons may extend a meter or more, ending at synapses inside the CNS.

4. **Multipolar neurons** have several dendrites and a single axon with one or more branches. Multipolar neurons (Figure 12-4d●) are the most common type of neuron in the CNS. For example, all the motor neurons that control skeletal muscles are multipolar neurons. Their axons can be as long as those of unipolar neurons.

Bipolar neurons in the nose and ear, unipolar neurons, and multipolar neurons have very long axons. Information travels along those axons in the form of action potentials. When the action potentials reach the synaptic terminals, neurotransmitter is released. The processes of anaxonic neurons and of bipolar neurons in the eye are much shorter, and neurotransmitter release is triggered by more subtle changes in the transmembrane potential.

Functional Classification

Neurons are also categorized into functional groups as (1) *sensory neurons,* (2) *motor neurons,* and (3) *interneurons,* or *association neurons.* Their relationships are diagrammed in Figure 12-5●.

SENSORY NEURONS Sensory neurons form the *afferent division* of the PNS. They deliver information from sensory receptors to the CNS. The cell bodies of

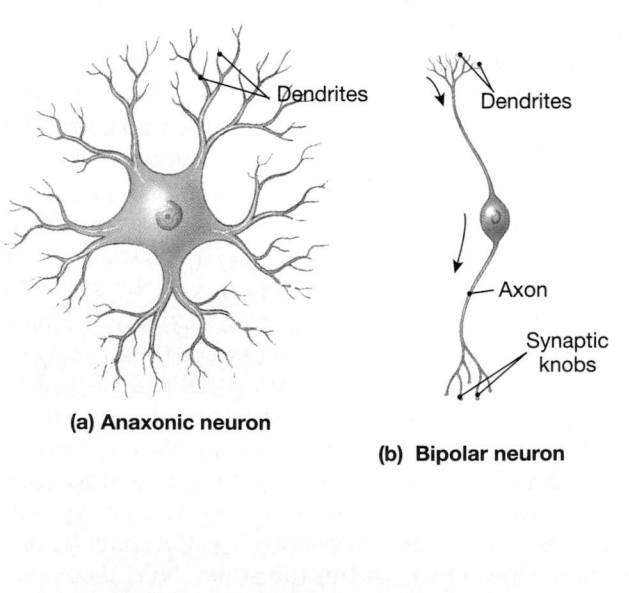

(a) Anaxonic neuron

(b) Bipolar neuron

●*FIGURE 12-4* **An Anatomical Classification of Neurons.** The arrows indicate the direction of action potential propagation, where applicable. The neurons are not drawn to scale; typical anaxonic neurons and bipolar neurons are many times smaller than typical unipolar or multipolar neurons.

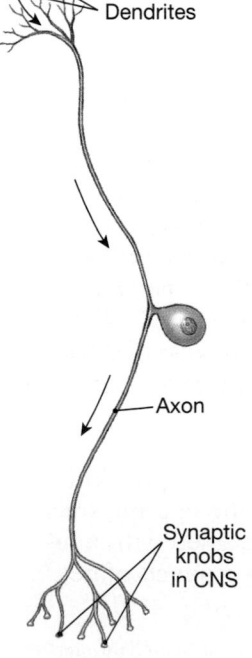

(c) Unipolar neuron

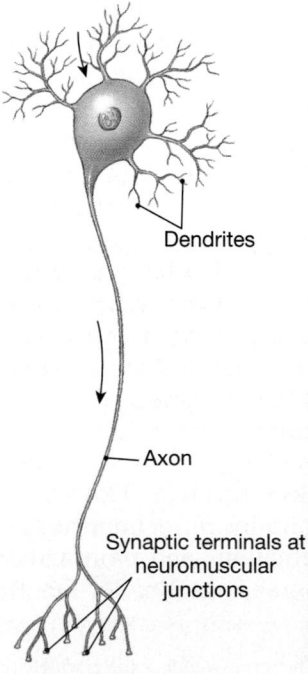

(d) Multipolar neuron (somatic motor neuron)

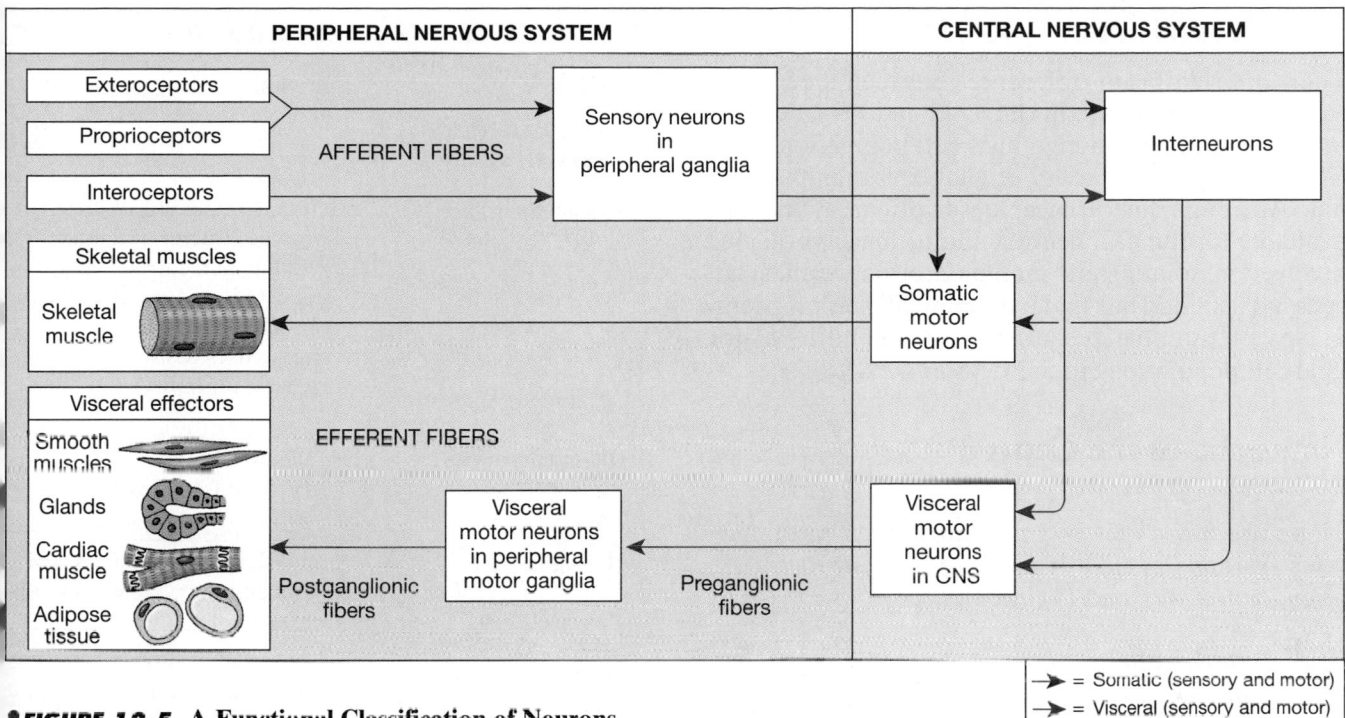

•*FIGURE 12-5* **A Functional Classification of Neurons**

sensory neurons are found in peripheral *sensory ganglia*. (A *ganglion* is a collection of neuron cell bodies in the PNS.) Sensory neurons are unipolar neurons with processes, known as **afferent fibers,** that extend between a sensory receptor and the spinal cord or brain. Sensory neurons collect information concerning the external or internal environment. The human body has about 10 million sensory neurons. **Somatic sensory neurons** monitor the outside world and our position within it. **Visceral sensory neurons** monitor internal conditions and the status of other organ systems.

Sensory receptors may be the processes of specialized sensory neurons or cells monitored by sensory neurons. Receptors are broadly categorized as follows:

1. **Exteroceptors** (*extero-,* outside) provide information about the external environment in the form of touch, temperature, and pressure sensations and the more complex senses of sight, smell, hearing, and taste.

2. **Proprioceptors** (prō-prē-ō-SEP-torz) monitor the position and movement of skeletal muscles and joints.

3. **Interoceptors** (*intero-,* inside) monitor the digestive, respiratory, cardiovascular, urinary, and reproductive systems and provide sensations of taste, deep pressure, and pain.

MOTOR NEURONS **Motor neurons,** or *efferent neurons*, form the *efferent division,* which carries instructions from the CNS to peripheral effectors. A motor neuron stimulates or modifies the activity of a peripheral tissue, organ, or organ system. Your body has about half a million motor neurons. Axons traveling away from the CNS are called **efferent fibers.** As we

learned earlier, there are two major efferent systems: the *somatic nervous system (SNS)* and the *autonomic (visceral) nervous system (ANS).* The somatic nervous system includes all the **somatic motor neurons** that innervate skeletal muscles. You have voluntary control over the activity of the SNS. The cell body of a somatic motor neuron lies inside the CNS, and its axon extends into the periphery to innervate one or more neuromuscular junctions.

The activities of the ANS are primarily controlled outside your conscious awareness. **Visceral motor neurons** innervate all peripheral effectors other than skeletal muscles. The axons of visceral motor neurons inside the CNS innervate neurons in peripheral *autonomic (motor) ganglia,* and the ganglionic neurons control peripheral effectors. Axons extending from the CNS to a ganglion are called **preganglionic fibers.** Axons connecting the ganglion cells with the peripheral effectors are known as **postganglionic fibers.** This arrangement, which does not occur in the somatic nervous system, clearly distinguishes the ANS from the SNS.

INTERNEURONS **Interneurons**, or *association neurons,* may be situated between sensory and motor neurons. Interneurons are located entirely within the brain and spinal cord. The human body's 20 billion or so interneurons outnumber all other types of neurons combined. Interneurons are responsible for the distribution of sensory information and the coordination of motor activity. The more complex the response to a given stimulus, the greater the number of interneurons involved.

NEUROGLIA

There are significant differences between the organization of neural tissue in the CNS and PNS, due primarily to their distinctive glial cell populations. The CNS has a greater variety of glial cells than the PNS has. Although histological descriptions have been available for the past century, the technical problems involved in isolating and manipulating individual glial cells have limited our understanding of their functions. Table 12-2 summarizes information about the major glial cell populations in the CNS and PNS.

Neuroglia of the Central Nervous System

There are four types of glial cells in the central nervous system: (1) *ependymal cells,* (2) *astrocytes,* (3) *oligodendrocytes,* and (4) *microglia.*

Ependymal Cells

The brain and spinal cord contain a fluid-filled central passageway. The thickness of the walls and the diameter of the passageway vary from one region to another. The narrow passageway within the spinal cord is called the *central canal;* the expanded chambers located in portions of the brain are *ventricles.* The ventricles and central canal are lined by a cellular layer of epithelial cells called the **ependyma** (e-PEN-di-muh) and filled with **cerebrospinal fluid (CSF).** This fluid, which also surrounds the brain and spinal cord, provides a protective cushion and transports dissolved gases, nutrients, wastes, and other materials.

Ependymal cells are cuboidal to columnar in shape (Figure 12-6a●). During embryonic development and early childhood, the free surfaces of ependymal cells are covered with cilia. The cilia persist in adults only within the ventricles of the brain, where they assist in the circulation of the CSF. In other areas, the ependymal cells typically have scattered microvilli. In a few parts of the brain, specialized ependymal cells participate in the secretion of the CSF. Other regions of the ependyma may have sensory functions. Unlike typical epithelial cells, ependymal cells have slender processes that branch extensively and make direct contact with glial cells in the surrounding neural tissue. The functions of these connections are not known.

Astrocytes

Astrocytes (AS-trō-sīts; *astro-,* star + *cyte,* cell) are the largest and most numerous glial cells. Astrocytes (Figure 12-6b●) have a variety of functions, many of them poorly understood. These functions include the following:

- *Maintaining the blood–brain barrier.* Compounds dissolved in the circulating blood do not have free access

TABLE 12-2 Glial Cells in the CNS and PNS

Cell Type	Functions
CENTRAL NERVOUS SYSTEM	
Astrocytes	Maintain blood–brain barrier; provide structural support; regulate ion, nutrient, and dissolved gas concentrations; absorb and recycle neurotransmitters; assist in tissue repair after injury
Oligodendrocytes	Myelinate CNS axons; provide structural framework
Microglia	Remove cell debris, wastes, and pathogens by phagocytosis
Ependymal cells	Line ventricles (brain) and central canal (spinal cavity); assist in production, circulation, and monitoring of cerebrospinal fluid
PERIPHERAL NERVOUS SYSTEM	
Satellite cells	Surround neuron cell bodies in ganglia
Schwann cells	Cover all axons in PNS; responsible for myelination of some peripheral axons; participate in repair process after injury

to the interstitial fluid of the CNS. Neural tissue must be isolated from the general circulation because hormones or other chemicals in the blood could alter neuron function. The endothelial cells lining CNS capillaries control the chemical exchange between the blood and interstitial fluid. These cells create a **blood–brain barrier** that isolates the CNS from the general circulation. The slender cytoplasmic extensions of astrocytes end in expanded "feet" that wrap around capillaries. Astrocytic processes form a complete blanket around the capillaries, interrupted only where other glial cells contact the capillary walls. Chemicals secreted by astrocytes are somehow responsible for maintaining the special permeability characteristics of the endothelial cells. (We shall discuss the blood–brain barrier further in Chapter 14.)

- *Creating a three-dimensional framework for the CNS.* Astrocytes are packed with microfilaments that extend

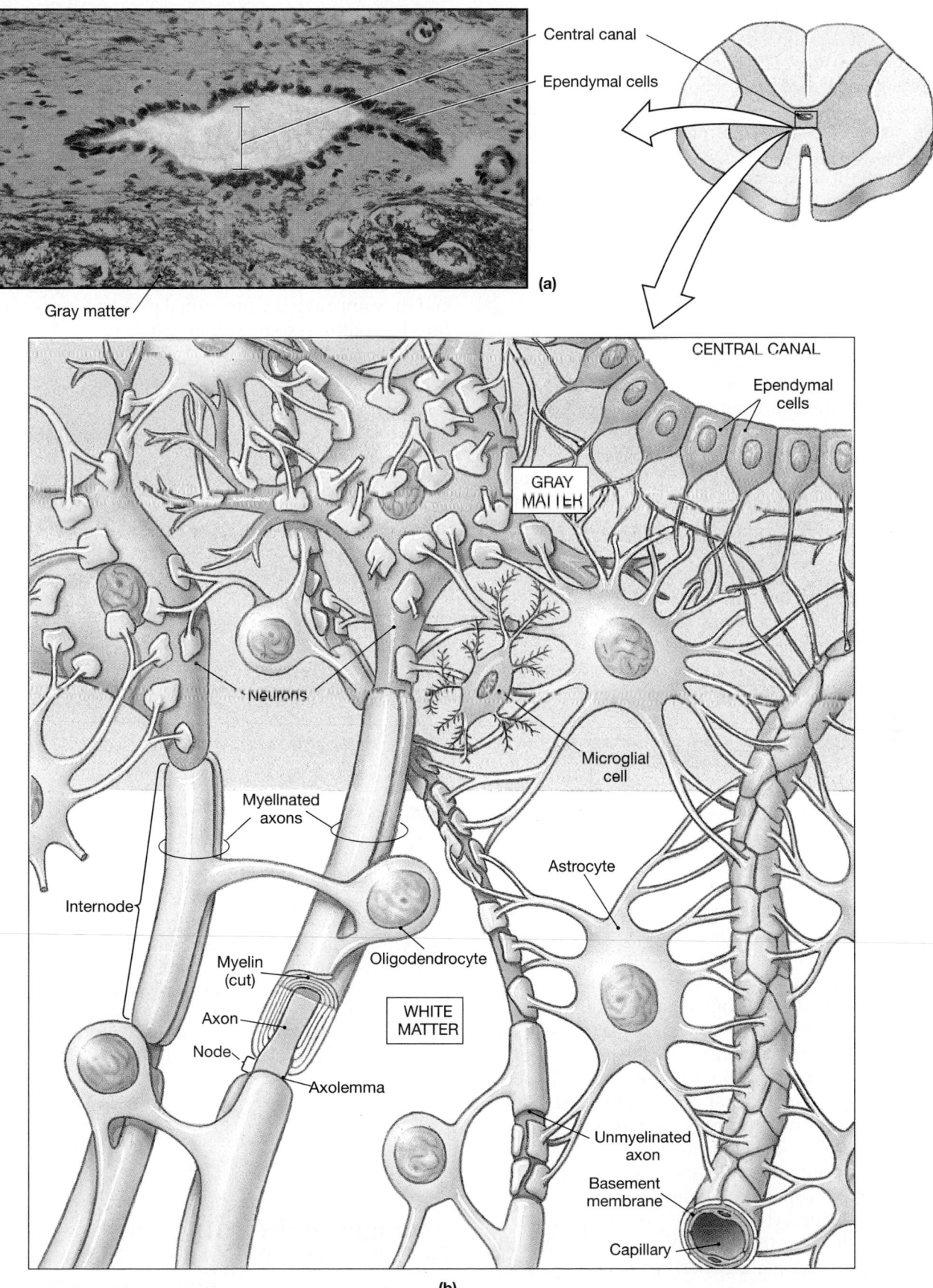

Central canal

Ependymal cells

Gray matter

(a)

CENTRAL CANAL

Ependymal cells

GRAY MATTER

Neurons

Microglial cell

Myelinated axons

Astrocyte

Internode

Myelin (cut)

Oligodendrocyte

Axon

WHITE MATTER

Node

Axolemma

Unmyelinated axon

Basement membrane

Capillary

(b)

•FIGURE 12-6 **Neuroglia in the CNS.** **(a)** Light micrograph, showing the ependymal lining of the central canal of the spinal cord. (LM × 236) **(b)** A diagrammatic view of neural tissue in the CNS, showing relationships between neuroglia and neurons.

from foot to foot across the breadth of the cell. This extensive cytoskeletal reinforcement assists them in providing a structural framework for the neurons of the brain and spinal cord.

- *Repairing damaged neural tissue.* In the CNS, damaged neural tissue seldom regains normal function. However, astrocytes moving into the injury site can make structural repairs that stabilize the tissue and prevent further injury. We will detail neural damage and subsequent repair in a later section (p. 407).
- *Guiding neuron development.* Astrocytes in the embryonic brain appear to be involved in directing the growth and interconnection of developing neurons.
- *Controlling the interstitial environment.* Although much remains to be learned about astrocyte physiology, there is evidence that astrocytes adjust the interstitial fluid composition by (1) regulating the concentration of sodium ions, potassium ions, and carbon dioxide; (2) providing a rapid-transit system for the transport of nutrients, ions, and dissolved gases between capillaries and neurons; (3) controlling the volume of blood flow through the capillaries; and (4) absorbing and recycling some neurotransmitters.

Oligodendrocytes

Like astrocytes, **oligodendrocytes** (o-li-gō-DEN-drō-sīts; *oligo-,* few) possess slender cytoplasmic extensions, but their cell bodies are smaller, and they have fewer processes (Figure 12-6b●). The processes generally contact the exposed surfaces of neurons. The functions of processes ending at the soma, or cell body, have yet to be determined; much more is known about the processes that end on the surfaces of axons. Many axons in the CNS are completely sheathed in the processes of oligodendrocytes.

The cell membrane of the axon is the **axolemma,** but the axolemma may be insulated from contact with the extracellular fluid by the processes of glial cells, especially oligodendrocytes. Near the tip of each oligodendritic process, the cytoplasm becomes very thin, but the cell membrane expands to form an enormous membranous pad. This flattened pancake somehow gets wound around the axon (Figure 12-6b●), creating a sheath composed of concentric layers of cell membrane. The composition of the sheath is roughly 80 percent lipid (primarily phospholipids) and 20 percent protein, the same as that of any other cell membrane. This multilayered membranous wrapping is called **myelin** (MĪ-e-lin), and the surrounded axon is said to be **myelinated.** Myelin increases the speed of action potential propagation along an axon.

Many oligodendrocytes cooperate in the formation of a myelin sheath along the length of an axon, with each oligodendrocyte myelinating short segments of several axons. Small gaps occur between adjacent wrappings. These gaps are called **nodes,** or the *nodes of Ranvier* (RAHN-vē-ā), and the relatively large areas wrapped in

myelin are called **internodes** (*inter,* between). In dissection, myelinated axons appear a glossy white, primarily because of the lipids present. Regions dominated by myelinated axons constitute the **white matter** of the CNS. Not all axons in the CNS are myelinated; **unmyelinated** axons may not be completely covered by glial cell processes. Unmyelinated axons are common in areas of **gray matter**, where relatively short axons and collaterals interconnect the densely packed neuron cell bodies. There areas have a dusky gray color.

In summary, oligodendrocytes play a role in structural organization by tying clusters of axons together, and they improve the functional performance of neurons by coating axons with myelin.

Microglia

Roughly 5 percent of the CNS glial cells are **microglia** (mī-KRŌG-lē-uh). Microglial cells (Figure 12-6b) are smaller than the other glial elements, and their slender cytoplasmic processes have many fine branches. They are capable of migrating through neural tissue. Microglia appear early in embryonic development. The mesodermal stem cells that produce microglia are related to those that produce monocytes and macrophages. ∞ *[pp. 126, 120]* The microglia migrate into the CNS as that system forms; thereafter they remain isolated within the neural tissue. They act as a wandering police force and janitorial service, engulfing cellular debris, waste products, and pathogens by phagocytosis. Under ordinary circumstances, the few microglia present are able to perform the necessary cleanup operations.

CNS TUMORS AND NEUROGLIA Tumors of the brain, spinal cord, and associated membranes result in approximately 90,000 deaths in the United States each year. Tumors that originate in the CNS are called *primary CNS tumors*, to distinguish them from *secondary CNS tumors,* which arise from the metastasis of cancer cells that originate elsewhere. Roughly 75 percent of CNS tumors are primary tumors. In adults, primary CNS tumors result from the divisions of abnormal glial cells rather than from the divisions of abnormal neurons, because typical neurons in the adult have neither centrosomes nor centrioles and therefore cannot divide. However, through the divisions of stem cells, neurons increase in number until children reach age 4. As a result, primary CNS tumors involving abnormal neurons can occur in young children. Symptoms of CNS tumors vary with the location affected. Treatment may involve surgery, radiation, or chemotherapy, alone or in combination.

Neuroglia of the Peripheral Nervous System

Neuron cell bodies in the PNS are clustered together in masses called **ganglia** (singular, *ganglion*). Neuron cell bodies and axons in the PNS are completely insu-

Introduction to the Development of the Nervous System

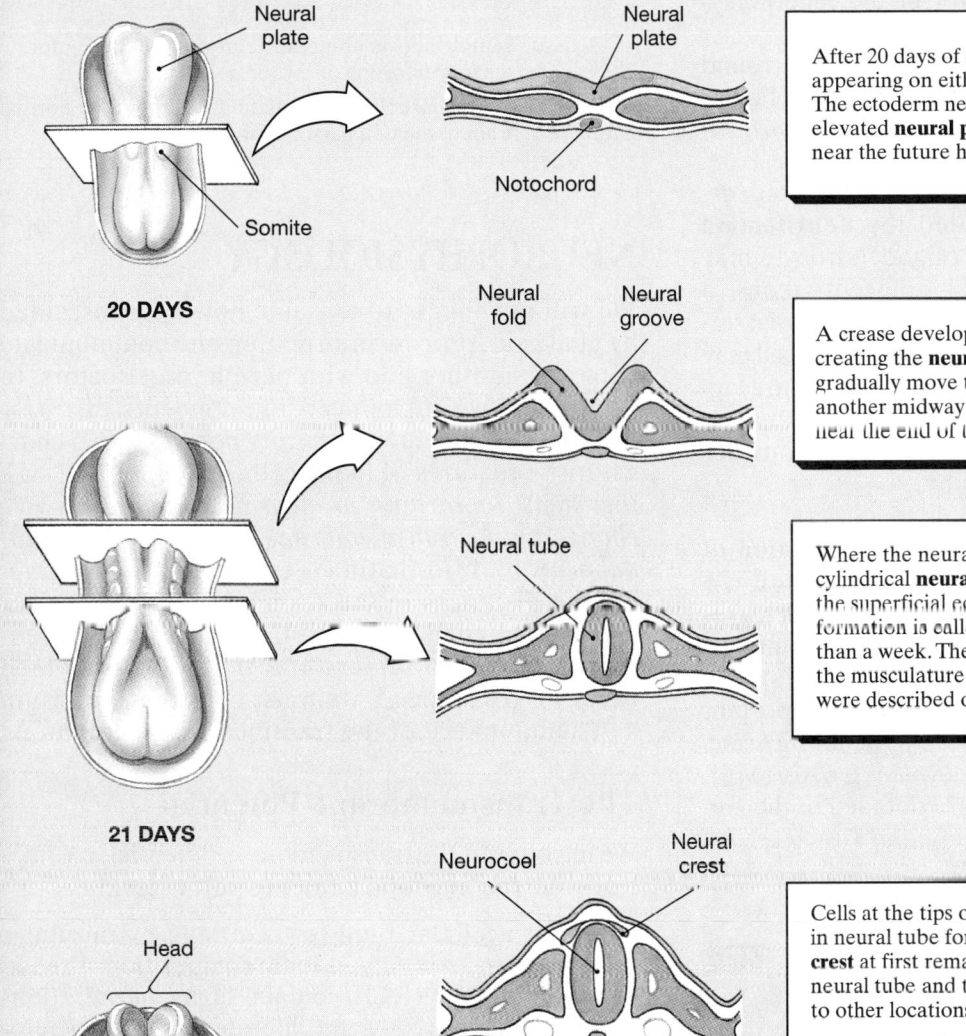

After 20 days of development, *somites* begin appearing on either side of the *notochord* (p. 362). The ectoderm near the midline thickens, forming an elevated **neural plate**. The neural plate is broadest near the future head of the developing embryo.

A crease develops along the axis of the neural plate, creating the **neural groove**. The edges, or **neural folds**, gradually move together. They first contact one another midway along the axis of the neural plate, near the end of the third week.

Where the neural folds meet, they fuse to form a cylindrical **neural tube** that loses its connection with the superficial ectoderm. The process of neural tube formation is called **neurulation**; it is completed in less than a week. The formation of the axial skeleton and the musculature around the developing neural tube were described on pages 362 and 363.

Cells at the tips of the neural folds do not participate in neural tube formation. These cells of the **neural crest** at first remain between the dorsal surface of the neural tube and the ectoderm, but they later migrate to other locations.

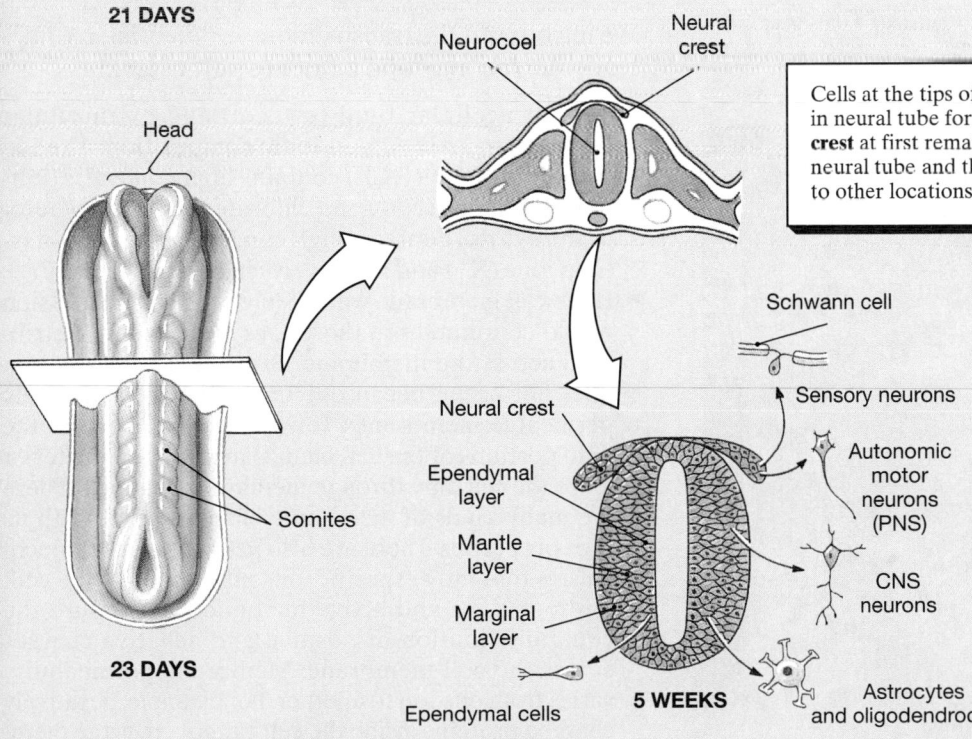

The first cells to appear in the mantle differentiate into neurons, and the last cells to arrive become astrocytes or oligodendrocytes. Further development of the CNS and PNS will be found in the Embryology Summaries on pages 436–437 and 454–455.

The neural tube increases in thickness as its epithelial lining undergoes repeated mitoses. By the middle of the fifth developmental week, there are three distinct layers. The **ependymal layer** lines the enclosed cavity, or **neurocoel**. The ependymal cells continue their mitotic activities, and daughter cells create the surrounding **mantle layer**. Axons from developing neurons form a superficial **marginal layer**.

lated from their surroundings by the processes of glial cells. The two glial cell types involved are called *satellite cells* and *Schwann cells*.

Satellite cells, or *amphicytes* (AM-fi-sīts), surround the neuron cell bodies in peripheral ganglia (Figure 12-7•). **Schwann cells,** or *neurilemmal cells (neurilemmocytes)*, form a sheath around every peripheral axon, whether unmyelinated or myelinated. The outer surface of these glial cells is called the **neurilemma** (noo-ri-LEM-uh). Whereas an oligodendrocyte may myelinate portions of several adjacent axons, a Schwann cell can myelinate only one segment of a single axon (Figure 12-8a•). However, a Schwann cell may *enclose* segments of several unmyelinated axons (Figure 12-8b•). The presumed mode of myelin formation in the PNS is summarized in Figure 12-8•; compare that illustration with Figure 12-6b•.

Demyelination is the progressive destruction of myelin sheaths in the CNS and PNS. The result is a gradual loss of sensation and motor control that leaves affected regions numb and paralyzed. Many unrelated conditions that result in myelin destruction can cause symptoms of demyelination. Several important examples of demyelinating disorders, including *heavy metal poisoning, diphtheria, multiple sclerosis* (MS), and *Guillain–Barré syndrome*, are detailed in the *Applications Manual*. **AM** *Demyelination Disorders*

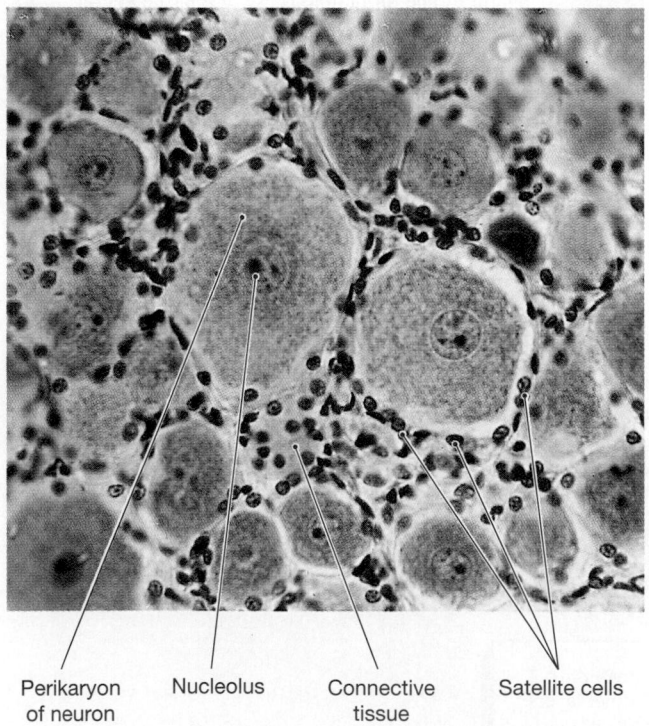

•*FIGURE 12-7* **Satellite Cells and Peripheral Neurons.**
Satellite cells surround neuron cell bodies in peripheral ganglia. (LM × 295)

Perikaryon of neuron   Nucleolus   Connective tissue   Satellite cells

☑ What would damage to the afferent division of the nervous system affect?

☑ A tissue sample shows unipolar neurons. Are these more likely to be sensory neurons or motor neurons?

☑ Which type of glial cell would you find in larger than normal numbers in brain tissue of a person with a CNS infection?

NEUROPHYSIOLOGY

We will now begin to examine how neurons, aided by glial cells, process information and communicate with one another and with peripheral effectors. In Chapter 3, we introduced the concepts of *transmembrane potential* and *resting potential,* two characteristic features of living cells. ∞ *[pp. 80–81] All the steps important to neural function involve changes in the transmembrane potentials of individual neurons.* Information is conveyed over relatively long distances in the form of *action potentials,* propagated changes in the transmembrane potential of axons. To understand these events, which occur at great speed, we must consider the origin and maintenance of the transmembrane potential.

The Transmembrane Potential

We introduced the transmembrane potential in Chapter 3, but you may find a brief review useful:

- The intracellular fluid (cytosol) and extracellular fluid differ markedly in ionic composition. The extracellular fluid (ECF) contains high concentrations of sodium ions (Na^+) and chloride ions (Cl^-), whereas the cytosol contains high concentrations of potassium ions (K^+) and negatively charged proteins (Pr^-).
- If the cell membrane were freely permeable, diffusion would continue until those ions were evenly distributed across the membrane. But an even distribution does not occur, because living cells have selectively permeable membranes. Ions cannot freely cross the lipid portions of the cell membrane. They can enter or leave the cell only through membrane channels. There are many kinds of membrane channels, each with its own properties. There are also active transport mechanisms that move specific ions into or out of the cell.
- These passive and active factors do not ensure the equal distribution of positive and negative charges across the cell membrane. Membrane permeability varies from one ion to another. For example, negatively charged proteins inside the cell cannot cross the membrane, and it is easier for K^+ to diffuse out of the cell through a potassium channel than it is for Na^+ to enter the cell through a sodium channel. As a result, there are differences in the distribution of positive and negative charges along the inner and outer surfaces of the cell membrane. *The inner surface contains an excess of negative charges with respect to the outer surface.*

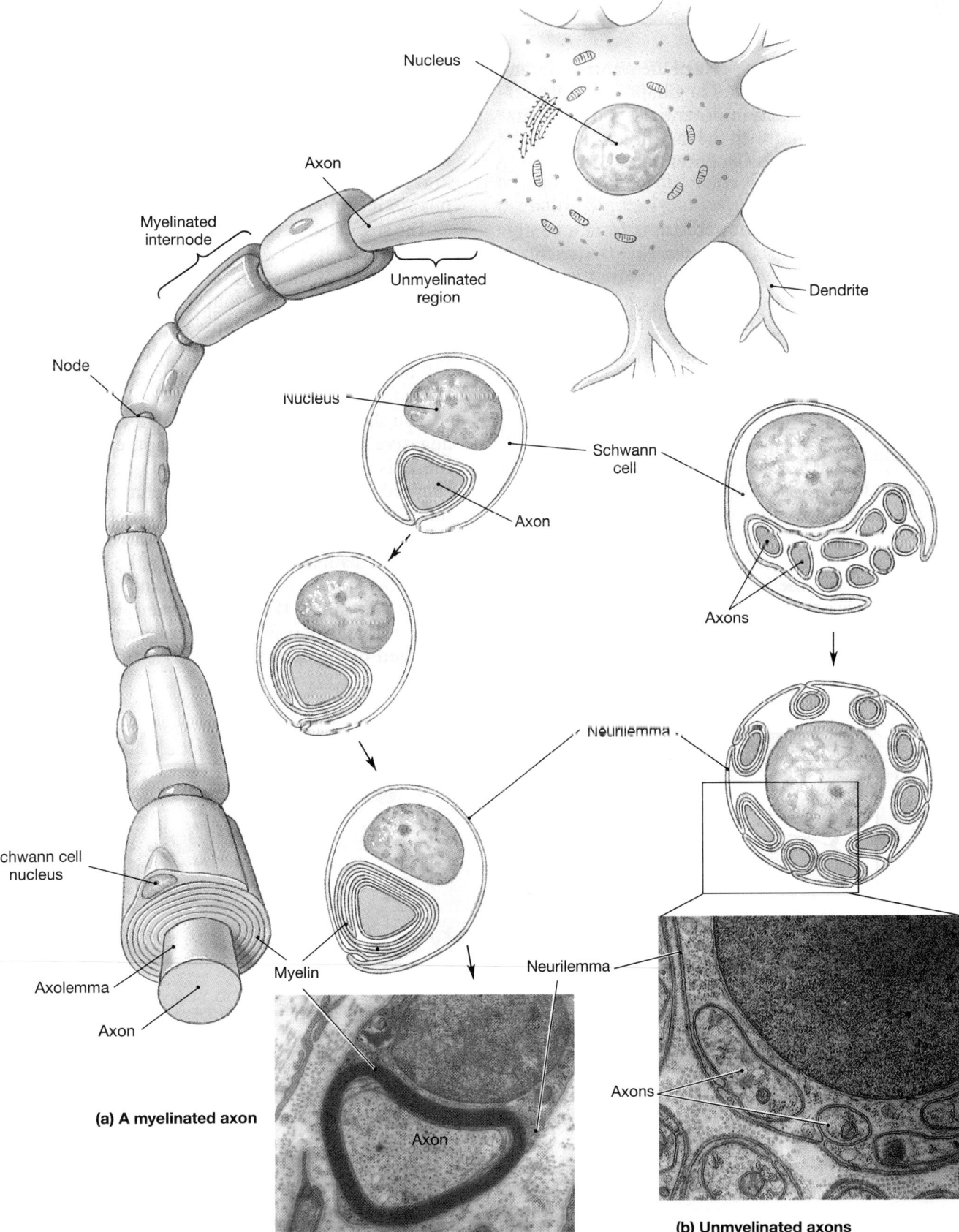

Nucleus

Axon

Myelinated
internode

Node

Unmyelinated
region

Dendrite

Nucleus

Schwann
cell

Axon

Axons

Schwann cell
nucleus

Neurilemma

Axolemma

Axon

Myelin

Neurilemma

Axons

(a) A myelinated axon

Axon

(b) Unmyelinated axons

•**FIGURE 12-8** **Schwann Cells and Peripheral Axons.** **(a)** A myelinated axon. (Left) A diagrammatic view of a myelinated axon, showing the organization of Schwann cells along the length of the axon. (Right) Stages in the formation of a myelin sheath by a single Schwann cell, which myelinates a portion of a single axon. This situation differs from the way myelin forms in the CNS; compare with Figure 12-6b. **(b)** A single Schwann cell can enfold a group of unmyelinated axons. A series of Schwann cells is required to cover the axons along their entire length. Every axon in the PNS is shielded from contact with the interstitial fluid by Schwann cells, whether the axon is myelinated or unmyelinated.

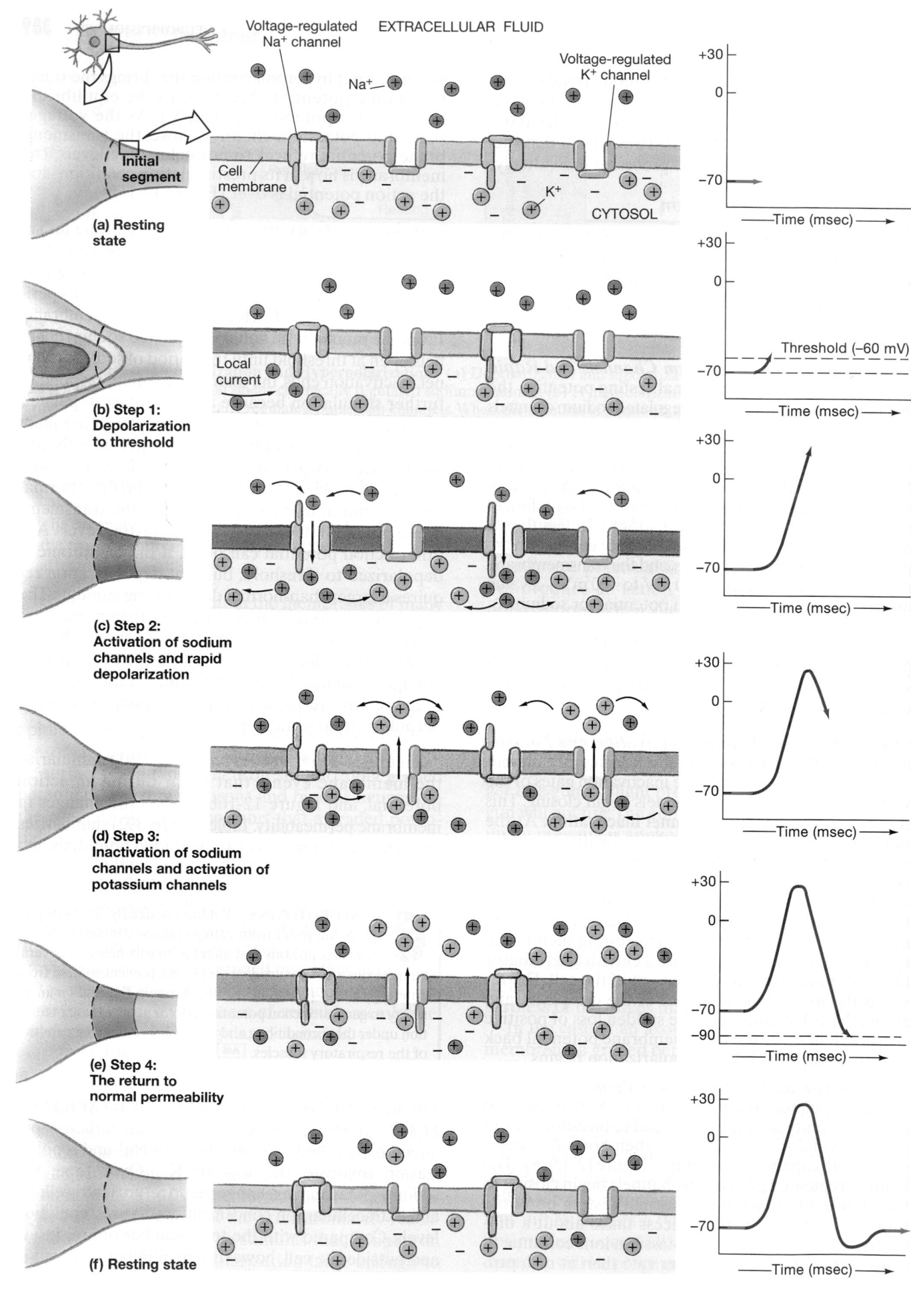

Voltage-regulated Na⁺ channel

EXTRACELLULAR FLUID

Voltage-regulated K⁺ channel

Na⁺

Cell membrane

K⁺

CYTOSOL

(a) Resting state

(b) Step 1: Depolarization to threshold

Local current

Threshold (−60 mV)

(c) Step 2: Activation of sodium channels and rapid depolarization

(d) Step 3: Inactivation of sodium channels and activation of potassium channels

(e) Step 4: The return to normal permeability

(f) Resting state

Time (msec)

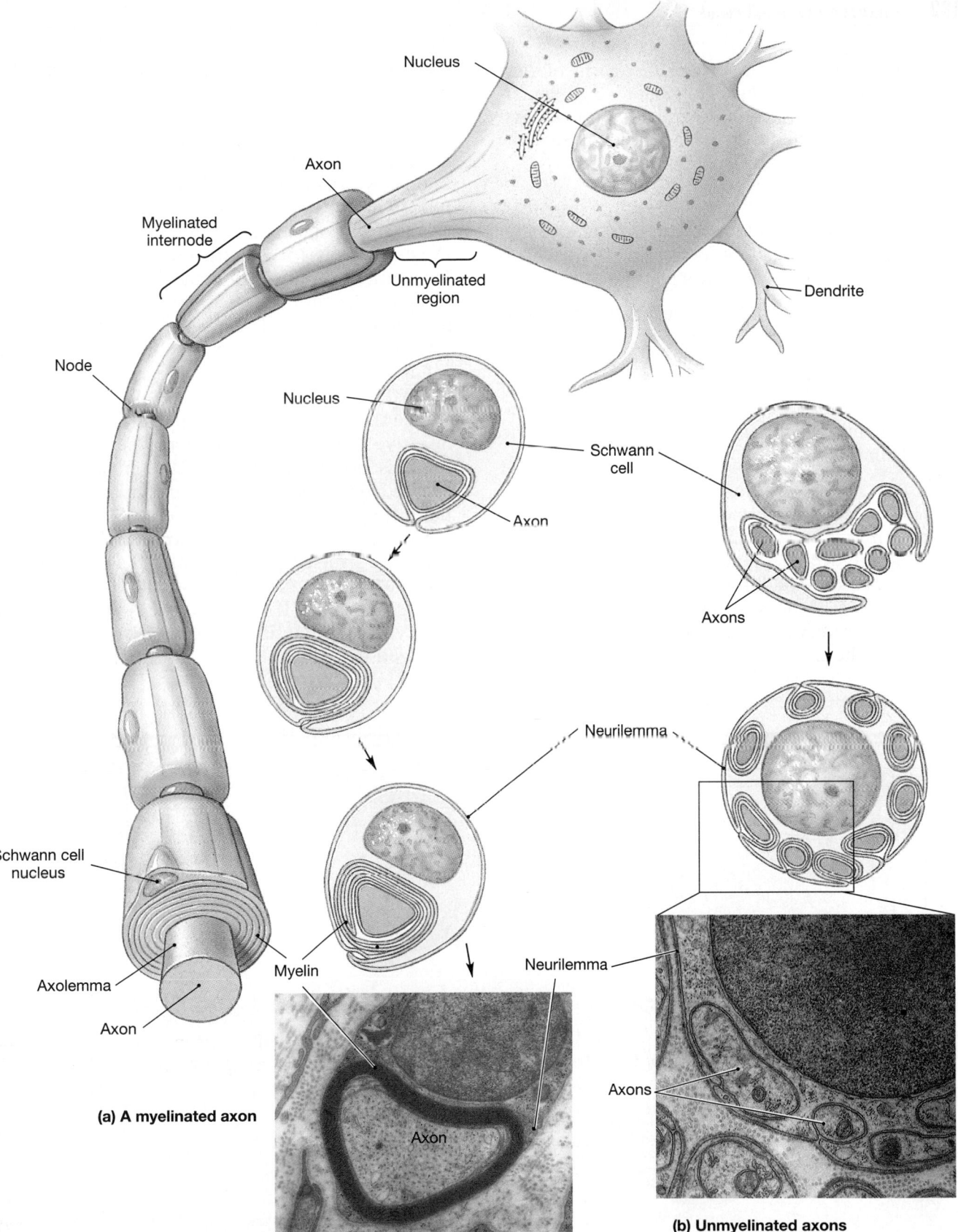

Nucleus

Axon

Myelinated
internode

Unmyelinated
region

Dendrite

Node

Nucleus

Schwann
cell

Schwann
cell

Axon

Axons

Neurilemma

Schwann cell
nucleus

Neurilemma

Myelin

Axolemma

Axons

Axon

Axon

(a) A myelinated axon

(b) Unmyelinated axons

• **FIGURE 12-8 Schwann Cells and Peripheral Axons. (a)** A myelinated axon. (Left) A diagrammatic view of a myelinated axon, showing the organization of Schwann cells along the length of the axon. (Right) Stages in the formation of a myelin sheath by a single Schwann cell, which myelinates a portion of a single axon. This situation differs from the way myelin forms in the CNS; compare with Figure 12-6b. **(b)** A single Schwann cell can enfold a group of unmyelinated axons. A series of Schwann cells is required to cover the axons along their entire length. Every axon in the PNS is shielded from contact with the interstitial fluid by Schwann cells, whether the axon is myelinated or unmyelinated.

Figure 12-9• reviews aspects of the transmembrane potential that we introduced in Chapter 3. To understand the nature of action potentials, you must first become more familiar with the factors involved in the establishment of the resting potential.

Passive Forces

The transmembrane potential results from a combination of passive and active forces. The passive forces are both chemical and electrical in nature.

CHEMICAL GRADIENTS   Because the intracellular concentration of potassium ions is relatively high, potassium ions tend to move out of the cell through open potassium channels. This movement is driven by a concentration gradient, or *chemical gradient*. Similarly, a chemical gradient for sodium ions tends to drive them into the cell.

ELECTRICAL GRADIENTS   Because the membrane is much more permeable to potassium than it is to sodium, potassium ions leave the cytoplasm more rapidly than sodium ions enter. As a result, the cytosol along the interior of the cell membrane experiences a net loss of positive charges, leaving an excess of negatively charged proteins. At the same time, the ECF near the exterior surface of the cell membrane experiences a net gain of positive charges. The positive and negative charges are separated by the cell membrane, which constitutes a **resistance,** because it resists the free movement of ions. Whenever positive and negative ions are separated by a resistance, such as a cell membrane, a *potential difference* exists.

The size of a potential difference is measured in volts (V) or millivolts (mV; thousandths of a volt). The *resting potential,* or transmembrane potential of an undisturbed cell, averages about 0.07 V for a neuron cell membrane. We usually express this value as –70 mV. The minus sign signifies that the inner surface of the cell membrane is negatively charged with respect to the exterior.

You should recall from Chapter 2 that positive and negative charges attract one another. ∞ *[p. 34]* If there is no resistance between them, oppositely charged ions move together and eliminate the potential difference. Such a movement of charges in response to voltage is called a **current.** The amount of current is inversely proportional to the resistance that separates the charges. If the resistance is high, the current is very small. If the resistance is low, the current is very large. The resistance of cell membranes can be changed by opening or closing ion channels. These

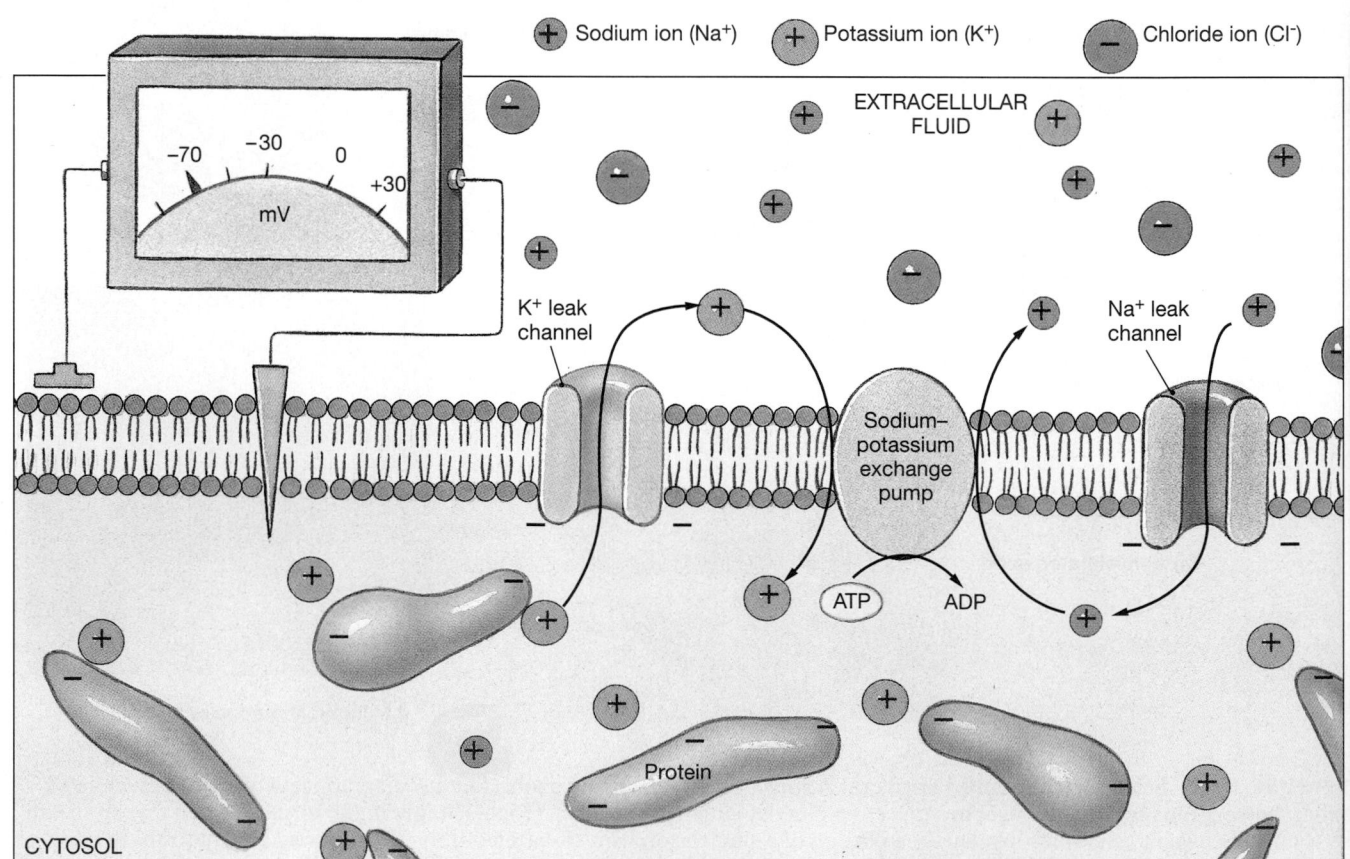

•*FIGURE 12-9* **The Cell Membrane at the Resting Potential**

changes result in currents that bring ions into or out of the cytoplasm.

THE ELECTROCHEMICAL GRADIENT Electrical gradients can reinforce or oppose the chemical gradient for each ion. The **electrochemical gradient** for a specific ion is the sum of the chemical and electrical forces acting across the cell membrane:

- Sodium ions in the ECF are attracted by the excess of negative charges inside the cell membrane, so both the chemical and the electrical forces push Na^+ into the cell (Figure 12-10a●).
- The chemical gradient for potassium ions tends to drive them out of the cell, but the movement is opposed by (1) attraction between K^+ and the negative charges on the inside of the cell membrane and (2) repulsion between K^+ and the positive charges on the outside of the membrane (Figure 12-10b●).

An electrochemical gradient is a form of *potential energy*. ∞ *[p. 39]* Potential energy is stored energy—the energy of position, as in a stretched spring, a charged battery, or water behind a dam.

Without a cell membrane, diffusion would eliminate electrochemical gradients. The cell membrane is a barrier that resists the electrochemical gradients that would otherwise tend to drive ions into or out of the cell. In effect, the cell membrane acts like a dam across a river. Without the dam, water would simply respond to gravity and flow downstream, gradually losing energy in the process. With a dam in place, even a small opening will release water under tremendous pressure. Similarly, any stimulus that increases the membrane permeability to potassium, sodium, or other ions for which an electrochemical gradient exists will produce a sudden and dramatic ion movement. For example, an increase in sodium ion permeability will cause an immediate rush of Na^+ into the cell. The size of this current has nothing to do with the size or nature of the stimulus. The stimulus needs only to open the door, and existing electrochemical gradients do the rest.

POTASSIUM IONS AND THE RESTING POTENTIAL
If the cell membrane were freely permeable to K^+ but impermeable to other positively charged ions, potassium ions would continue to leave the cell until the electrical gradient (holding K^+ inside the cell) was as large as the chemical gradient (pushing K^+ out of the cell). This equilibrium would occur at a transmembrane potential of about −90 mV, a value known as the *equilibrium potential* for potassium. Neuron mem-

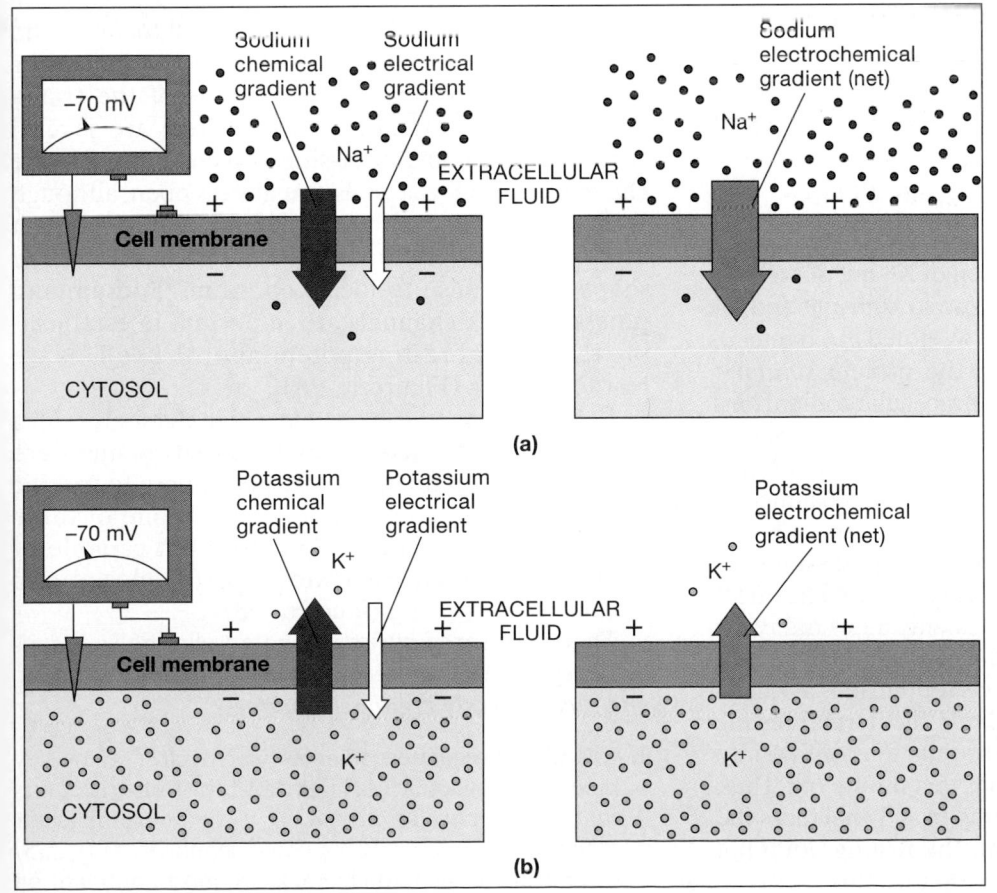

●FIGURE 12-10
Electrochemical Gradients.
(a) At the normal resting potential, chemical and electrical gradients tend to drive sodium ions into the cell. **(b)** At the normal resting potential, an electrical gradient opposes the chemical gradient for potassium ions. The net electrochemical gradient tends to force potassium ions out of the cell.

branes are very permeable to potassium, and this high permeability is the primary determinant of the resting potential. If other ion movements are prevented, the transmembrane potential of a neuron will reach –90 mV. Normally, however, the resting potential of a neuron is approximately –70 mV. The difference between the potassium equilibrium potential and the resting potential reflects primarily the small but significant membrane permeability to sodium ions.

SODIUM IONS AND THE RESTING POTENTIAL If the cell membrane were freely permeable to Na^+, these ions would continue to enter the cell until the interior of the cell membrane contained enough excess positive charges that the electrical gradient would reverse. The interior would now be positively charged, and the repulsion between similar charges (+ and +) would prevent any further net movement of sodium ions into the cell. This event would occur at a transmembrane potential of about +66 mV. The resting potential is nowhere near that value, because the membrane permeability to Na^+ is very low. In a typical neuron, sodium permeability has only a modest effect on the resting potential, shifting it from -90 mV, the potassium equilibrium potential, to –70 mV.

The electrochemical gradients for potassium and sodium are the primary factors affecting the resting potential of a neuron. We can ignore the contributions of the major negatively charged ions (Pr^- and Cl^-), because (1) proteins are too large to pass through the cell membrane, (2) the normal resting potential is near the equilibrium potential for chloride ions, and (3) the negative charges carried by the interior proteins repel the extracellular chloride ions.

Active Forces: Sodium–Potassium ATPase

At the normal resting potential, there is a slow leakage of Na^+ into the cell and a diffusion of K^+ out of the cell. The cell must use ATP to eject those sodium ions and reclaim the lost potassium ions. As we noted in Chapter 3, an ion pump that depends on the protein *sodium–potassium ATPase* exchanges 3 intracellular sodium ions for 2 extracellular potassium ions. ∞ *[p. 78]* Because the sodium–potassium exchange pump removes 3 positive ions from the cytoplasm for every 2 reclaimed, it contributes to the negativity of the transmembrane potential. For this reason, the sodium–potassium exchange pump is called an *electrogenic pump*. The effect is slight, however, and contributes only a few millivolts to the normal resting potential. The primary impact of this exchange pump is that it ejects sodium ions as quickly as they enter the cell. At –70 mV, the normal resting potential of a typical neuron, 3 sodium ions diffuse into the cell for every 2 potassium ions that diffuse out. Thus, the activity of the exchange pump exactly balances the passive forces of diffusion, and the resting potential remains stable.

Summary

To summarize,

- The resting potential reflects primarily the relatively high membrane permeability to K^+. It is therefore fairly close to –90 mV, the equilibrium potential for K^+.
- Although the electrochemical gradient for Na^+ is very large, sodium permeability is very small. Consequently, sodium ions have only a small effect on the normal resting potential, making it somewhat less negative than it would otherwise be.
- The sodium–potassium exchange pump stabilizes the resting potential when the ratio of passive Na^+ entry to passive K^+ loss is 3:2.
- The result of these passive and active factors is that a typical neuron has a resting potential of approximately –70 mV.

The transfer of information from neuron to neuron involves two integrated steps: (1) the propagation of one or more action potentials along an axon and (2) the chemical transmission of those signals across one or more synapses. Changes in membrane permeability, and resulting changes in the transmembrane potential, play a vital role in both steps.

Ion Channels

Ions cross the cell membrane through *membrane channels*. ∞ *[p. 73]* Our discussion will focus on the membrane permeability to sodium and potassium, as these are the primary determinants of the transmembrane potential of neurons. There are several types of sodium and potassium ion channels. **Passive channels,** or **leak channels,** are always open, although their permeability can vary from moment to moment as the proteins that make up the channel change shape in response to local conditions. Sodium and potassium leak channels are important in establishing the normal resting potential of the cell, as we learned earlier (Figure 12-9●).

Cell membranes also contain **active channels,** which are often called **gated channels.** Gated channels are ion channels that open or close in response to specific stimuli. Each gated channel can be in one of three states (Figure 12-11a●): (1) closed but capable of opening, (2) open **(activated),** or (3) closed and incapable of opening **(inactivated).**

There are three classes of gated channels: *chemically regulated channels, voltage-regulated channels,* and *mechanically regulated channels:*

1. **Chemically regulated channels** open or close when they bind specific chemicals (Figure 12-11b●). The receptors that bind *acetylcholine (ACh)* at the neuromuscular junction are chemically regulated channels. ∞ *[p. 285]* Chemically regulated channels are most abundant on

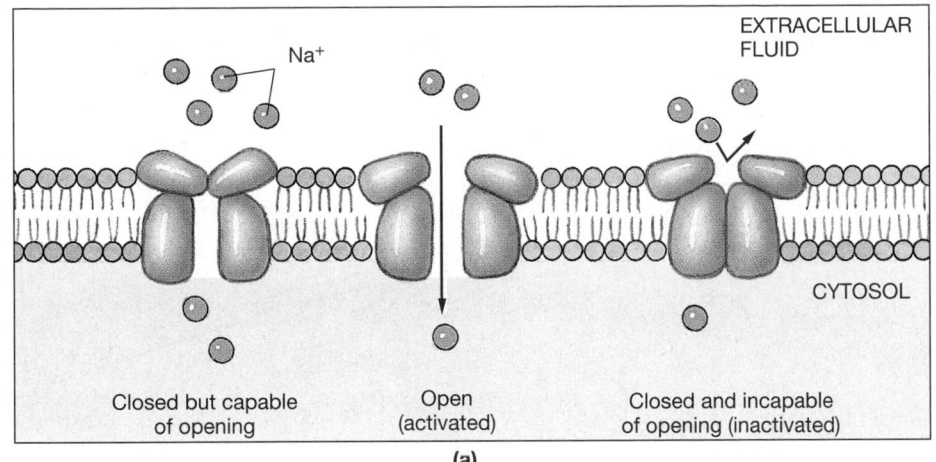

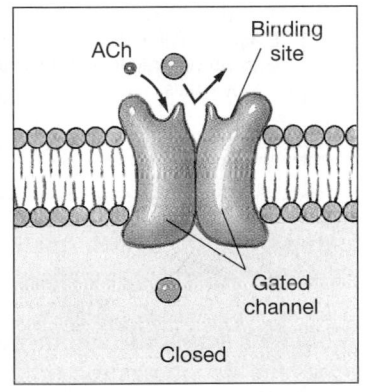

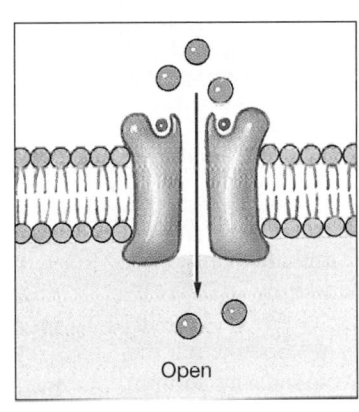

(b)

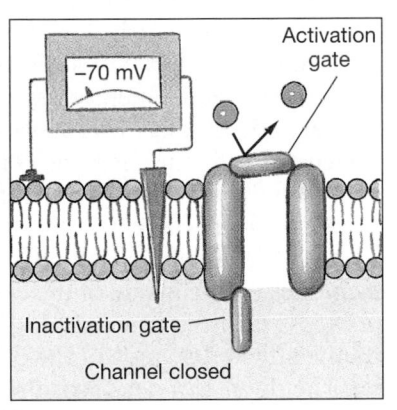

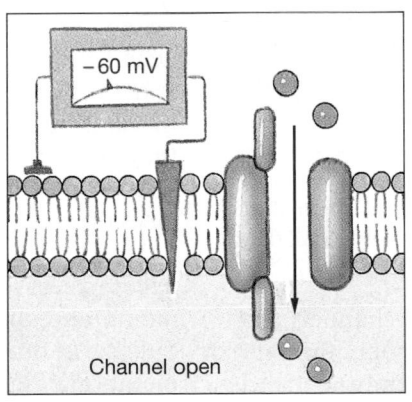

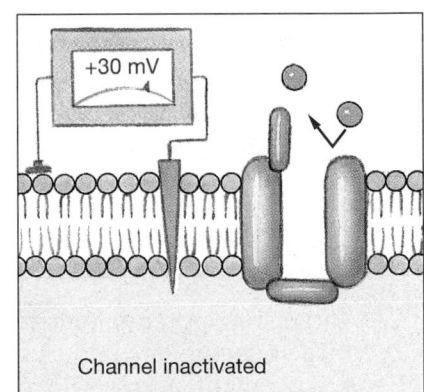

(c)

•FIGURE 12-11 Gated Channels. We use Na⁺ channels here, but comparable gated channels regulate the movements of other cations and axions. **(a)** Possible configurations of a gated ion channel. **(b)** A chemically regulated channel for Na⁺ that opens in response to the presence of ACh at a binding site. **(c)** A voltage-regulated sodium channel that responds to changes in the transmembrane potential. At the normal resting potential, the channel is closed. At a membrane potential of –60 mV, the channel will open. At a membrane potential of +30 mV, the channel is inactivated. We will consider factors that alter the membrane potential in a later section.

the dendrites and soma of a neuron, the areas where most synaptic communication occurs (Figure 12-12●). Chemically regulated channels are also important in specialized receptor cells of the nose and eye. Olfactory receptors are stimulated by the opening of ion channels in response to the presence of specific extracellular chemicals. When light strikes a visual receptor, chemicals are released that trigger the closing of chemically regulated ion channels. In Chapter 17, we will see how the resulting changes in permeability are the basis for our senses of smell and vision.

2. **Voltage-regulated channels** are characteristic of areas of **excitable membrane,** a membrane capable of generating and conducting an action potential. Examples of excitable membranes include the axons of unipolar and multipolar neurons (Figure 12-4●, p. 374) and the sarcolemma (including T tubules) of skeletal muscle fibers and cardiac muscle cells. Voltage-regulated channels open or close in response to changes in the transmembrane potential (Figure 12-11c●). The opening of voltage-regulated sodium channels is the key step in the generation of an action potential. These channels

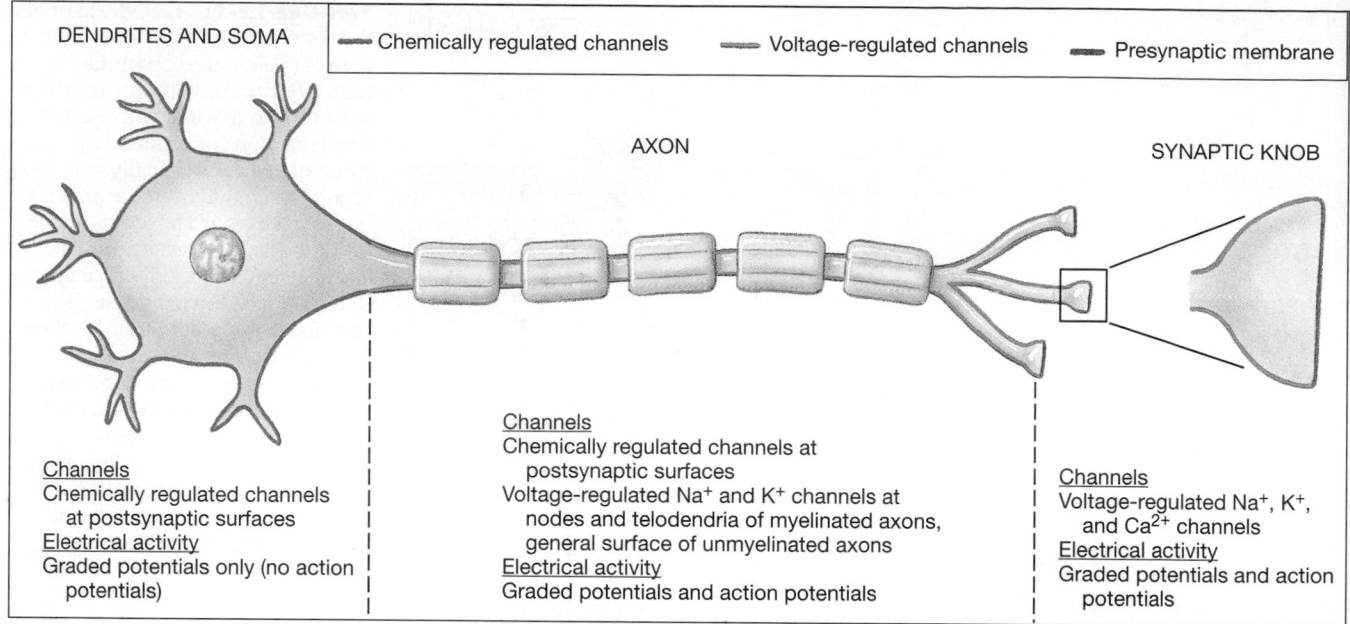

DENDRITES AND SOMA   ▬ Chemically regulated channels   ▬ Voltage-regulated channels   ▬ Presynaptic membrane

AXON

SYNAPTIC KNOB

Channels
Chemically regulated channels
    at postsynaptic surfaces
Electrical activity
Graded potentials only (no action
    potentials)

Channels
Chemically regulated channels at
    postsynaptic surfaces
Voltage-regulated Na$^+$ and K$^+$ channels at
    nodes and telodendria of myelinated axons,
    general surface of unmyelinated axons
Electrical activity
Graded potentials and action potentials

Channels
Voltage-regulated Na$^+$, K$^+$,
    and Ca^{2+} channels
Electrical activity
Graded potentials and action
    potentials

●*FIGURE 12-12* **The General Distribution of Gated Channels.** The properties of the cell membrane change from one region of the neuron to another.

have two gates—an *activation gate* and an *inactivation gate*. The two gates function independently. In the resting membrane, the activation gate is closed and the inactivation gate is open. Both gates are open when the proper chemical stimulation has occurred, and sodium ions can then enter. Sodium ion entry can be stopped by closing the inactivation gate. During the brief recovery period that follows, the activation gate closes and the inactivation gate reopens. Until the recovery period has passed, the ion channel will be unable to respond to voltage changes in the surrounding membrane.

3. **Mechanically regulated channels** open or close in response to physical distortion of the membrane surface. Such channels are important in sensory receptors that respond to touch, pressure, and vibration. We shall discuss these receptors in more detail in Chapter 17.

At the resting potential, most gated channels are closed. The opening of gated channels alters the rate of ion movement across the cell membrane and changes the transmembrane potential. The result may be a *graded potential* or an *action potential.*

Graded Potentials

Graded potentials, or *local potentials,* are changes in the transmembrane potential that cannot spread far from the area surrounding the site of stimulation. Figure 12-13a,b● considers what happens when a membrane is exposed to a chemical that opens chemically regulated sodium channels. Sodium ions enter the cell, and the arrival of additional positive charges shifts the transmembrane potential toward 0 mV. Any shift in the transmembrane potential away from resting levels toward 0 mV or above is called a **depolarization.** Thus, it

includes changes in potential from –70 mV to smaller negative values (–65 mV, –45 mV, –10 mV) as well as to membrane potentials past 0 mV (+10 mV, +30 mV).

The movement of sodium ions across the cell membrane at one location causes the depolarization of the surrounding membrane. Sodium ions entering the cell spread along the inner surface, attracted by the excess of negative charges (Figure 12-13c●). The arrival of these positive ions reduces the imbalance between negative and positive charges on the inside of the membrane. At the same time, extracellular sodium ions move toward the open channels, replacing those that enter the cell. The ion movement parallel to the inner and outer surfaces of a membrane is called a **local current.**

In the area affected by the local current, there are fewer excess negative charges on the inside of the cell membrane and fewer excess positive charges on the outside. In other words, the affected portion of the cell membrane depolarizes. The degree of depolarization decreases with distance, because the sodium ions entering at one point spread out in all directions along the inner surface of the membrane (Figure 12-13c●). *The change in the transmembrane potential and the area affected by the local currents are directly related to the number of sodium channels opened by the chemical stimulus.* The more open channels, the more sodium ions enter and spread along the inner surface of the membrane and the greater the depolarization.

Opening a potassium channel would have the opposite effect: The rate of potassium outflow would increase, and the interior of the cell would lose positive ions. The loss of positive ions produces **hyperpolarization,** a shift in the resting potential away from 0 mV, perhaps to –80 mV or more. Again, a local current

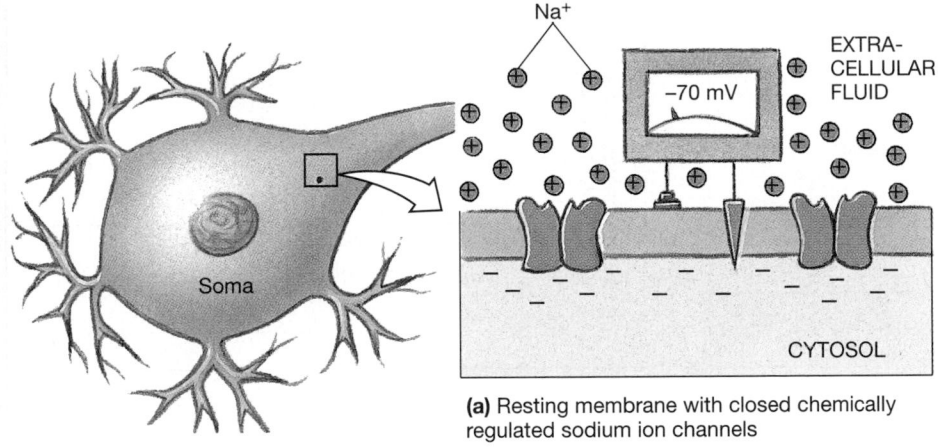

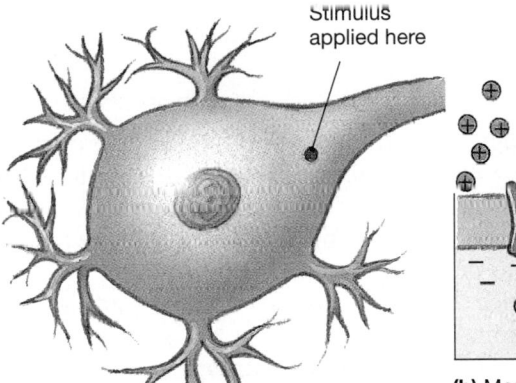

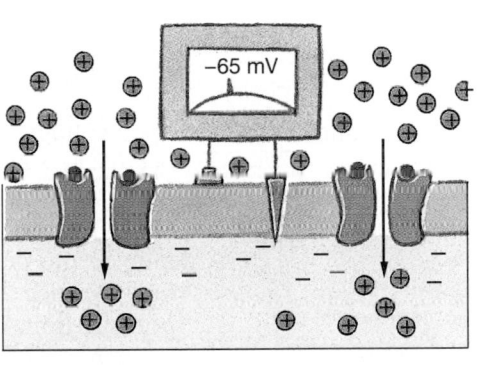

(a) Resting membrane with closed chemically regulated sodium ion channels

Stimulus applied here

(b) Membrane exposed to chemical that triggers the opening of these sodium ion channels

•*FIGURE 12-13* **Graded Potentials.** **(a)** Chemically regulated sodium channels are closed at the normal resting potential. **(b)** Opening these channels leads to a sudden influx of sodium ions, which depolarizes the membrane. **(c)** Local currents develop as sodium ions on the outside of the membrane diffuse into the area, replacing those that enter the cell, and sodium ions inside the cell spread along the inner surface of the membrane. The depolarizing effect radiates in all directions away from the source of stimulation. (For clarity, only gated channels are shown; leak channels as shown in Figure 12-9 are present but are not responsible for the production of graded potentials.)

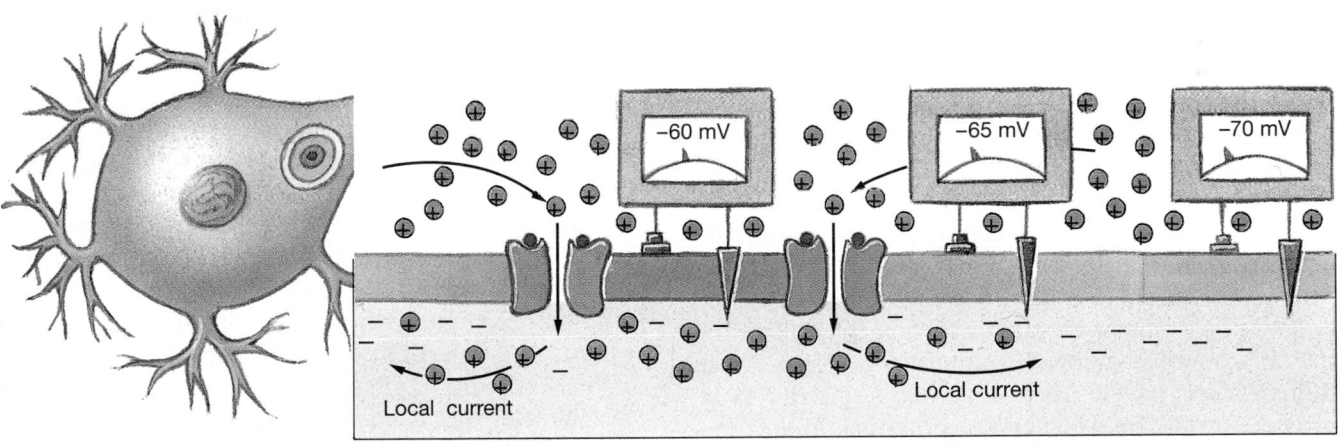

(c) Spread of sodium ions inside cell membrane produces a local current that depolarizes adjacent portions of the cell membrane

would distribute the effect to adjacent portions of the cell membrane.

Both depolarization and hyperpolarization move the transmembrane potential away from resting levels. When a chemical stimulus is removed and normal membrane permeability is restored, the transmembrane potential soon returns to resting levels (Figure 12-14•). The process of restoring the normal resting potential after depolarization is called **repolarization.** The return to the normal resting potential after hyperpolarization involves a slight depolarization.

Graded potentials, whether depolarizing or hyperpolarizing, share four basic characteristics:

1. The transmembrane potential is most affected at the site of stimulation, and the effect decreases with distance.

2. The effect spreads passively due to local currents.

3. The graded potential change may involve either depolarization or hyperpolarization. The nature of the change is determined by the properties of the membrane channels involved. For example, in a resting membrane, opening sodium channels will cause de-

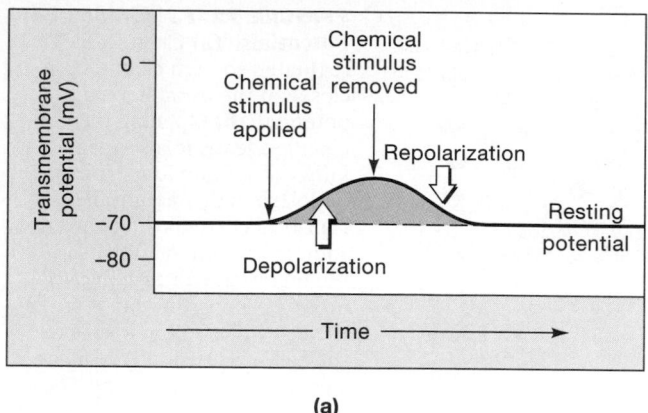

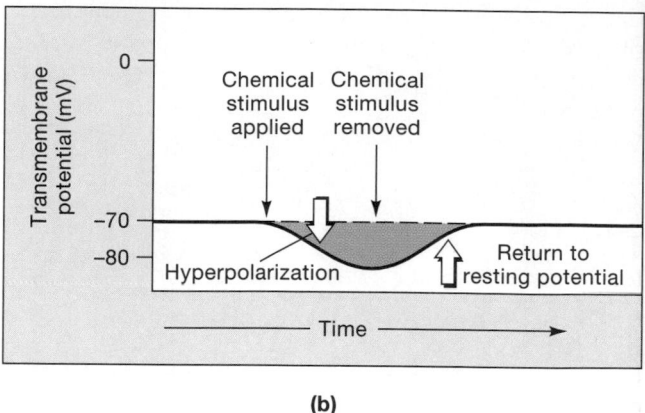

(a)

(b)

•FIGURE 12-14 Depolarization and Hyperpolarization. (a) Depolarization and repolarization in response to the addition and removal of a stimulus that opens chemically regulated sodium channels. **(b)** Hyperpolarization in response to the addition of a stimulus that opens chemically regulated potassium channels. When the stimulus is removed, the membrane potential returns to the resting level.

polarization, whereas opening potassium channels will cause hyperpolarization.

4. The stronger the stimulus, the greater the change in the transmembrane potential and the larger the area affected.

Graded potentials occur in the membranes of many different cell types. For example, the depolarization of the junctional folds at a neuromuscular junction by ACh is a graded potential that may result in the appearance of an action potential in adjacent portions of the sarcolemma. Graded potentials also occur in epithelial cells, gland cells, adipocytes, and a variety of sensory receptors. The dendrites, soma, and presynaptic surfaces of a typical neuron support only graded potentials. Anaxonic neurons and bipolar neurons in the eye have processes so short that a graded potential on a dendrite or soma can affect the transmembrane potential at the synaptic knobs and change the rate of neurotransmitter release. In other neurons, graded potentials on the dendrites and cell bodies are too far from the synaptic terminals to have any effect there. Action potentials propagated along the axon provide the link between graded potentials at the dendrites or soma and the synaptic terminals. Action potentials can occur only in areas of excitable membrane, which contain voltage-regulated channels.

Action Potentials

Action potentials are propagated changes in the transmembrane potential that, once initiated, spread across an entire excitable membrane. In a representative neuron, an action potential generally begins at the initial segment of the axon. The action potential is then propagated along the length of the axon, ultimately reaching the synaptic knobs. The stimulus that initiates an action potential is a depolarization large enough to

open voltage-regulated sodium channels. That opening occurs at a transmembrane potential known as the **threshold.** Threshold for an axon is typically between –60 mV and –55 mV. A stimulus that shifts the resting membrane potential from –70 mV to –62 mV will not produce an action potential, only a graded depolarization. When such a stimulus is removed, the transmembrane potential returns to resting levels. Depolarization of the initial segment is caused by local currents resulting from the graded depolarization of the soma, especially the axon hillock.

The All-or-None Principle

The initial depolarization acts like pressure on the trigger of a gun. A slight pressure can be applied, and the gun will not fire. The gun fires only when a certain minimum pressure is applied to the trigger. Once the trigger pressure reaches this preset level, however, the firing pin drops and the gun discharges. At that point, it no longer matters whether the pressure was applied gradually or suddenly, or whether the shooter is a 4-year-old who can barely squeeze the trigger or a weight lifter with a crushing grip. The speed and range of the bullet that leaves the gun do not change, regardless of the forces that were applied to the trigger.

In the case of an axon or another area of excitable membrane, a graded depolarization is the pressure on the trigger, and the action potential is the firing of the gun. All stimuli that bring the membrane to threshold will generate identical action potentials. In other words, *the properties of the action potential are independent of the relative strength of the depolarizing stimulus* as long as that stimulus exceeds threshold. This concept is called the **all-or-none principle,** because a given stimulus either triggers a typical action potential or does not produce one at all. If an action potential *is* produced, it will be propagated over the entire surface of the excitable membrane. The all-or-none principle applies to all excitable membranes. An action potential is *generated* at one site as the result of a localized depolarization. It is then *propagated* away from that site. We will consider each of these processes individually.

Action Potential Generation

The steps involved in the generation of an action potential from the resting state are shown in Figure 12-15●:

STEP 1: *Depolarization to Threshold.* Before an action potential can begin, an area of excitable membrane must be depolarized to threshold by local currents (Figure 12-15b●).

STEP 2: *Activation of Sodium Channels and Rapid Depolarization.* At the normal resting potential, the activation gates of the voltage-regulated sodium channels are closed. The closed gates are all that prevent sodium ion entry. At threshold, these activation gates open and the cell membrane becomes much more permeable to Na$^+$. Driven by the large electrochemical gradient, sodium ions rush into the cytoplasm, and rapid depolarization occurs at this site (Figure 12-15c●). In less than a millisecond, the inner membrane surface contains more positive ions than negative ones, and the transmembrane potential has changed from –60 mV to +30 mV—a value much closer to the equilibrium potential for sodium.

Notice that the first two steps in action potential generation are an example of positive feedback: A small depolarization acts as the trigger to produce a larger depolarization.

STEP 3: *Sodium Channel Inactivation and Potassium Channel Activation.* As the transmembrane potential approaches +30 mV, the inactivation gates of the voltage-regulated sodium channels begin closing. This step is known as **sodium channel inactivation.** At the same time that sodium channel inactivation is under way, voltage-regulated potassium channels are opening (Figure 12-15d●). At a transmembrane potential of +30 mV, the cytosol along the interior of the membrane contains an excess of positive charges. Thus, in contrast to the situation in the resting membrane (p. 383), both the electrical *and* chemical gradients favor potassium ion movement out of the cell. The result is the movement of K$^+$ out of the cytoplasm and across the cell membrane. The sudden loss of positive charges then shifts the transmembrane potential back toward resting levels, and repolarization begins.

STEP 4: *The Return to Normal Permeability.* The voltage-regulated sodium channels remain inactivated until the membrane has repolarized to threshold, about –60 mV. At this time, they regain their normal status—closed but capable of opening (Figure 12-15e●). The voltage-regulated potassium channels begin closing as the membrane reaches the normal resting potential (about –70 mV), but the process takes about a millisecond. Over that period, potassium ions continue to move out of the cell at a faster rate than at rest, pro-

ducing a brief hyperpolarization that brings the transmembrane potential very close to the equilibrium potential for potassium (–90 mV). As the voltage-regulated potassium channels close, the transmembrane potential returns to normal resting levels. The membrane is now in its prestimulation condition, and the action potential is over (Figure 12-15f●).

THE REFRACTORY PERIOD From the time an action potential begins until the normal resting potential has stabilized, the membrane will not respond normally to additional depolarizing stimuli. This period is known as the **refractory period** of the membrane. From the moment the voltage-regulated sodium channels open at threshold until the period of sodium channel inactivation ends, the membrane cannot respond to further stimulation because all the voltage-regulated sodium channels are either already open or are inactivated. This portion of the refractory period is the **absolute refractory period.** The **relative refractory period** begins when the sodium channels regain their normal resting condition and continues until the transmembrane potential stabilizes at normal resting levels. Another action potential can begin if the membrane is depolarized to threshold, but that depolarization requires a larger-than-normal depolarizing stimulus. The depolarizing stimulus must be larger than normal because (1) the local current must deliver enough Na$^+$ to counteract the loss of positively charged K$^+$ through voltage-regulated K$^+$ channels, and (2) through most of the relative refractory period, the membrane is hyperpolarized to some degree.

Figure 12-16a● and Table 12-3 (p. 391) summarize the membrane events that occur during an action potential, and Figure 12-16b● shows the changes in membrane permeability. Table 12-4 (p. 392) summarizes important differences between graded potentials and action potentials.

NEUROTOXINS Potentially deadly forms of poisoning result from eating seafood containing *neurotoxins,* poisons that affect primarily neurons. Several neurotoxins, such as *tetrodotoxin* (TTX), prevent sodium from entering voltage-regulated sodium channels. The result is an inability to generate action potentials. Motor neurons cannot function under these conditions, and death may result from paralysis of the respiratory muscles. [AM] *Neurotoxins in Seafood*

THE ROLE OF THE SODIUM–POTASSIUM EXCHANGE PUMP In the action potential detailed earlier, depolarization resulted from the influx of Na$^+$ and repolarization involved the loss of K$^+$. Over time, the sodium–potassium exchange pump returns intracellular and extracellular ion concentrations to prestimulation levels. Compared with the total number of ions inside and outside the cell, however, the number involved in

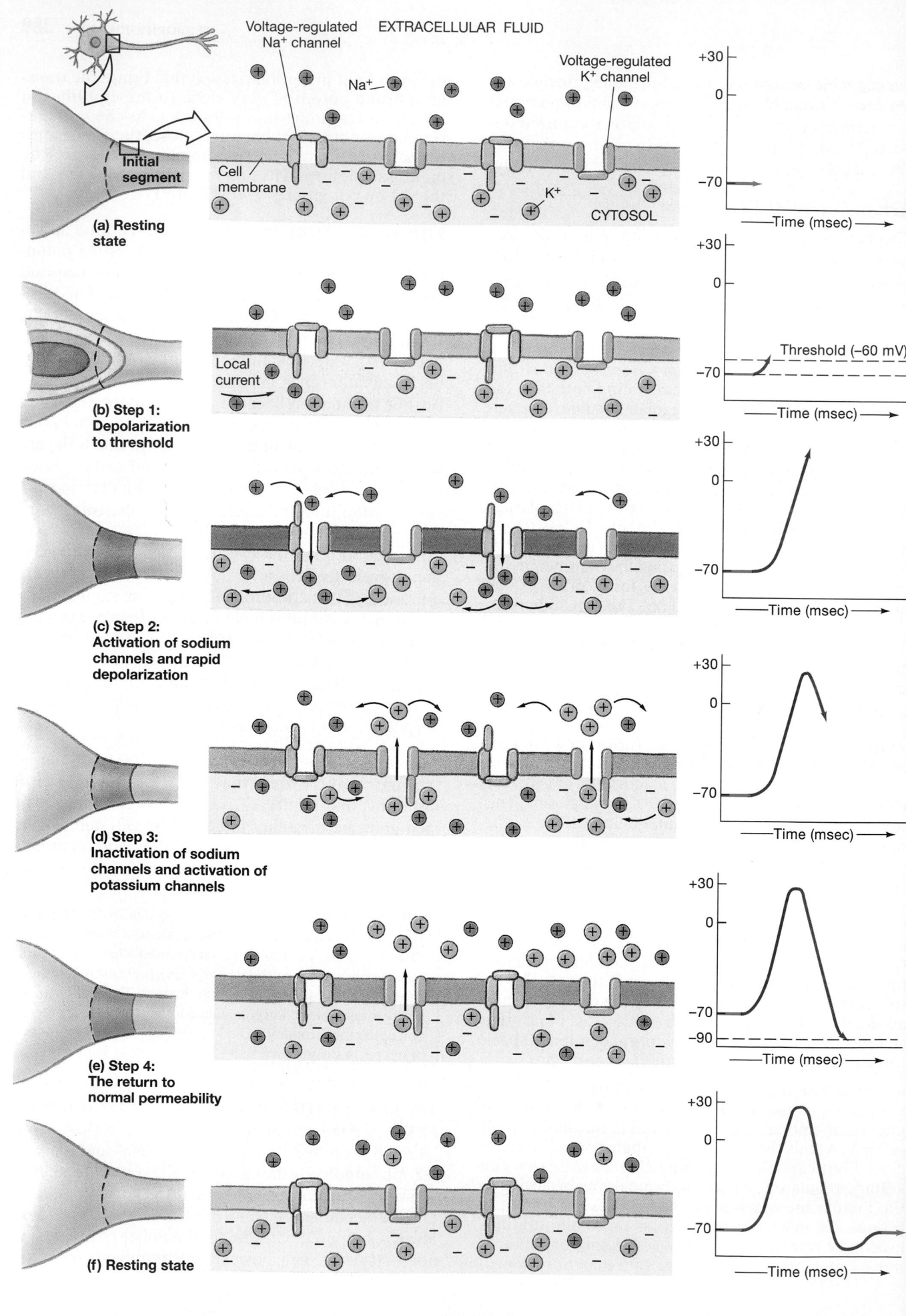

Voltage-regulated Na⁺ channel

EXTRACELLULAR FLUID

Voltage-regulated K⁺ channel

Initial segment

Cell membrane

Na⁺

K⁺

CYTOSOL

(a) Resting state

Time (msec)

(b) Step 1: Depolarization to threshold

Local current

Threshold (−60 mV)

Time (msec)

(c) Step 2: Activation of sodium channels and rapid depolarization

Time (msec)

(d) Step 3: Inactivation of sodium channels and activation of potassium channels

Time (msec)

(e) Step 4: The return to normal permeability

Time (msec)

(f) Resting state

Time (msec)

• FIGURE 12-15 (at left) **Generation of an Action Potential.**
(a) Resting condition. **(b)** *Step 1:* Depolarization to threshold.
Local currents associated with a graded depolarization of the
soma depolarize the proximal portion of the initial segment to
–60 mV. **(c)** *Step 2:* Activation of Na⁺ channels and rapid
depolarization. The rapid entry of Na⁺ drives the membrane
potential to +30 mV. **(d)** *Step 3:* Sodium channel inactivation
and potassium channel activation. Potassium ions leave the
cell, and repolarization begins. **(e)** *Step 4:* Return to normal
permeability. At threshold, the Na⁺ channels regain their
resting condition, and the K⁺ channels begin closing. The
transmembrane potential continues to drop until the voltage-
regulated K⁺ channels have closed. **(f)** The transmembrane
potential then returns to resting levels. For clarity, only gated
channels are shown.

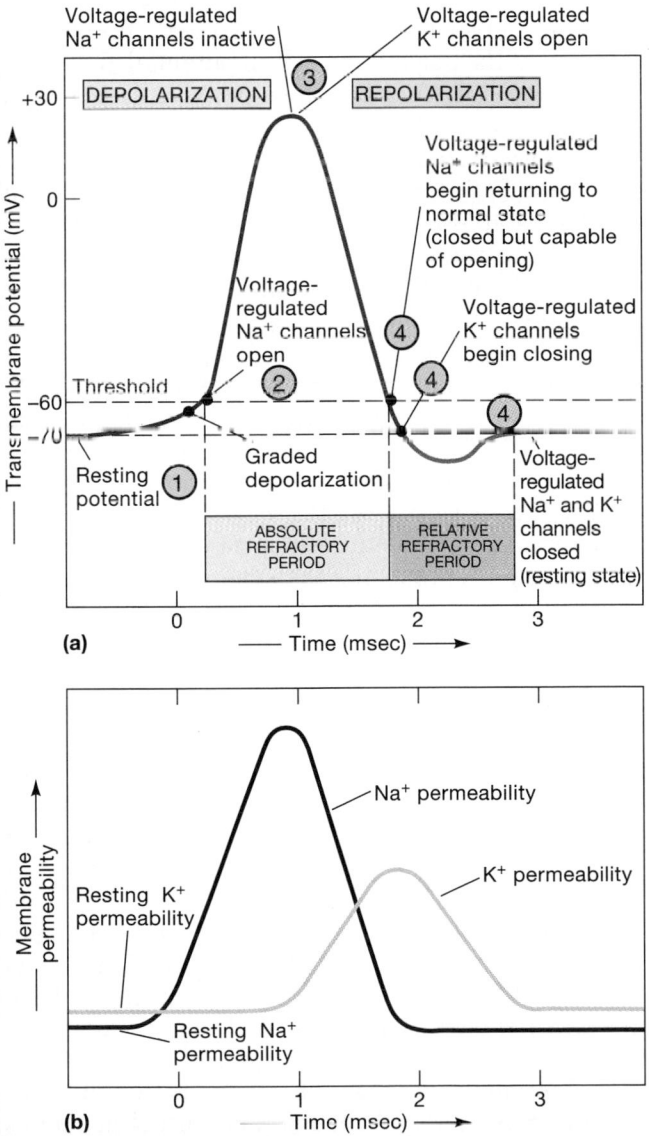

(a)

(b)

**• FIGURE 12-16 Changes in the Transmembrane Potential
and Membrane Permeabilities during an Action Potential.**
(a) Changes in the transmembrane potential during an action
potential and the factors involved. Circled numbers refer to
the steps in Figure 12-15 and in Table 12-3. **(b)** Permeability
changes in the membrane during an action potential.

TABLE 12-3 **Steps in the Generation of an Action
Potential**

Step 1: Depolarization to threshold

- A graded depolarization brings an area of excitable
  membrane to threshold (–60 mV).

Step 2: Activation of sodim channels and rapid depolarization

- The voltage-regulated sodium channels open (sodium
  channel activation).
- Sodium ions, driven by charge attraction and the
  concentration gradient, flood into the cell.
- The transmembrane potential goes from –60 mV, the
  threshold level, to +30 mV.

Step 3: Inactivation of sodium channels and activation of potassium channels

- The voltage-regulated sodium channels close (sodium
  channel inactivation).
- Voltage-regulated potassium channels are now open,
  and potassium ions move out of the cell.
- Repolarization begins.

Step 4: Return to normal permeability

- Voltage-regulated sodium channels regain their normal
  properties when threshold is reached (–60 mV).
- The voltage-regulated potassium channels begin
  closing at –70 mV. Because they close slowly, a
  temporary hyperpolarization occurs.
- At the end of the relative refractory period, all voltage-
  regulated channels have closed and the membrane is
  back to its resting state.

a single action potential is insignificant. *Tens of thousands of action potentials can occur before the intracellular ion concentrations change enough to disrupt the
entire mechanism.* Thus the exchange pump is not essential to any single action potential.

However, a maximally stimulated neuron may
generate action potentials at a rate of 1000 per second. Under these circumstances, the exchange pump
is needed if ion concentrations are to remain within acceptable limits over a prolonged period. The
sodium–potassium exchange pump requires energy
in the form of ATP. Each time the pump exchanges
two extracellular potassium ions for three intracellular sodium ions, one molecule of ATP is converted to ADP. The transmembrane protein of the
exchange pump is called sodium–potassium ATPase
because it splits ATP to ADP. If sodium–potassium
ATPase is inactivated by a metabolic poison or if
the cell runs out of ATP, a neuron will soon lose its
ability to function.

TABLE 12-4 Comparison of Graded Potentials and Action Potentials

Graded Potentials	Action Potentials
May be depolarizing or hyperpolarizing	Always depolarizing
No threshold value	Must depolarize to threshold before action potential begins
Amount of depolarization or hyperpolarization depends on intensity of stimulus	All-or-none phenomenon; any stimulus that exceeds threshold will produce an identical action potential
Passive spread from site of stimulation	Action potential at one site depolarizes adjacent sites to threshold
Effect on membrane potential decreases with distance from stimulation site	Propagated across entire membrane surface without decrease in strength
No refractory period	Refractory period
Occur in most cell membranes	Occur only in excitable membranes of specialized cells such as neurons and muscle cells

Action Potential Propagation

The sequence of events described earlier occurs in a relatively small portion of the total membrane surface. But we have already noted that, unlike a graded potential, which diminishes rapidly with distance, an action potential affects the entire excitable membrane. To understand the basic principle involved, imagine that you are standing by the doors of a movie theater at the start of a long line. Everyone is waiting for the doors to open. The manager steps outside and says to you, "Let everyone know that we're opening in 15 minutes." How would you spread the news? If you treated the line as an inexcitable membrane, you would shout, "The doors open in 15 minutes!" as loudly as you could. The closest people in the line would hear the news very clearly, but those farther away might not hear the entire message, and those at the end of the line might not hear you at all. If you treated the crowd as an excitable membrane, you would give the message to another person in line, with instructions to pass it on. In this way, the message would travel along the line until everyone had heard the news. Such a message "moves" as each person repeats it to someone else. Distance is not a factor—the line can contain 50 people or 5000.

This situation is comparable to the way an action potential spreads across an excitable membrane. An action potential (message) is relayed from one location to another in a series of steps. At each step, the message is repeated. Because the same events take place over and over, the term **propagation** is preferable to the term *conduction*.[2]

CONTINUOUS PROPAGATION

The basic mechanism of action potential propagation along an unmyelinated axon is shown in Figure 12-17●. For convenience, we will consider the membrane as a se-

ries of adjacent segments. The action potential begins at the initial segment (Figure 12-17a●). For a brief moment at the peak of the action potential, the transmembrane potential becomes positive rather than negative. A local current then develops (Figure 12-17b●), and sodium ions begin moving in the cytoplasm and in the extracellular fluid. The local current spreads in all directions, depolarizing adjacent portions of the membrane. The axon hillock cannot respond with an action potential, but when the initial segment of the axon is depolarized to threshold, an action potential develops there. The process then continues in a chain reaction (Figure 12-17c,d,e●). Eventually, the most distant portions of the cell membrane have been affected. As in our "crowd" model, the message is being relayed from one location to another. At each step along the way, the message is retold, so distance has no effect on the process. The action potential reaching the synaptic knob is identical to the one generated at the initial segment, and the net effect is the same as if a single action potential traveled across the membrane surface. This form of action potential propagation is known as **continuous propagation**, or *continuous conduction*.

Each time a local current develops, the action potential moves forward, not backward, because the previous segment of the axon is still in the absolute refractory period. As a result, *an action potential always proceeds away from the site of generation and cannot reverse direction.* By the time the refractory period in a segment of the membrane has ended, the action potential is too far away for local currents to have any effect. If there is to be a second action potential at the same site, a second stimulus must be applied.

In continuous propagation, an action potential appears to move across the membrane surface in a series of tiny steps. Even though the events at any one location take only about a millisecond, the sequence must be repeated at each step along the way. Continuous propagation along unmyelinated axons occurs at speeds of about 1 meter per second (approximately 2 mph).

[2]In electrical conduction, there is a flow of electrons along a *conductor,* such as a copper wire. Axons are relatively poor conductors of electricity.

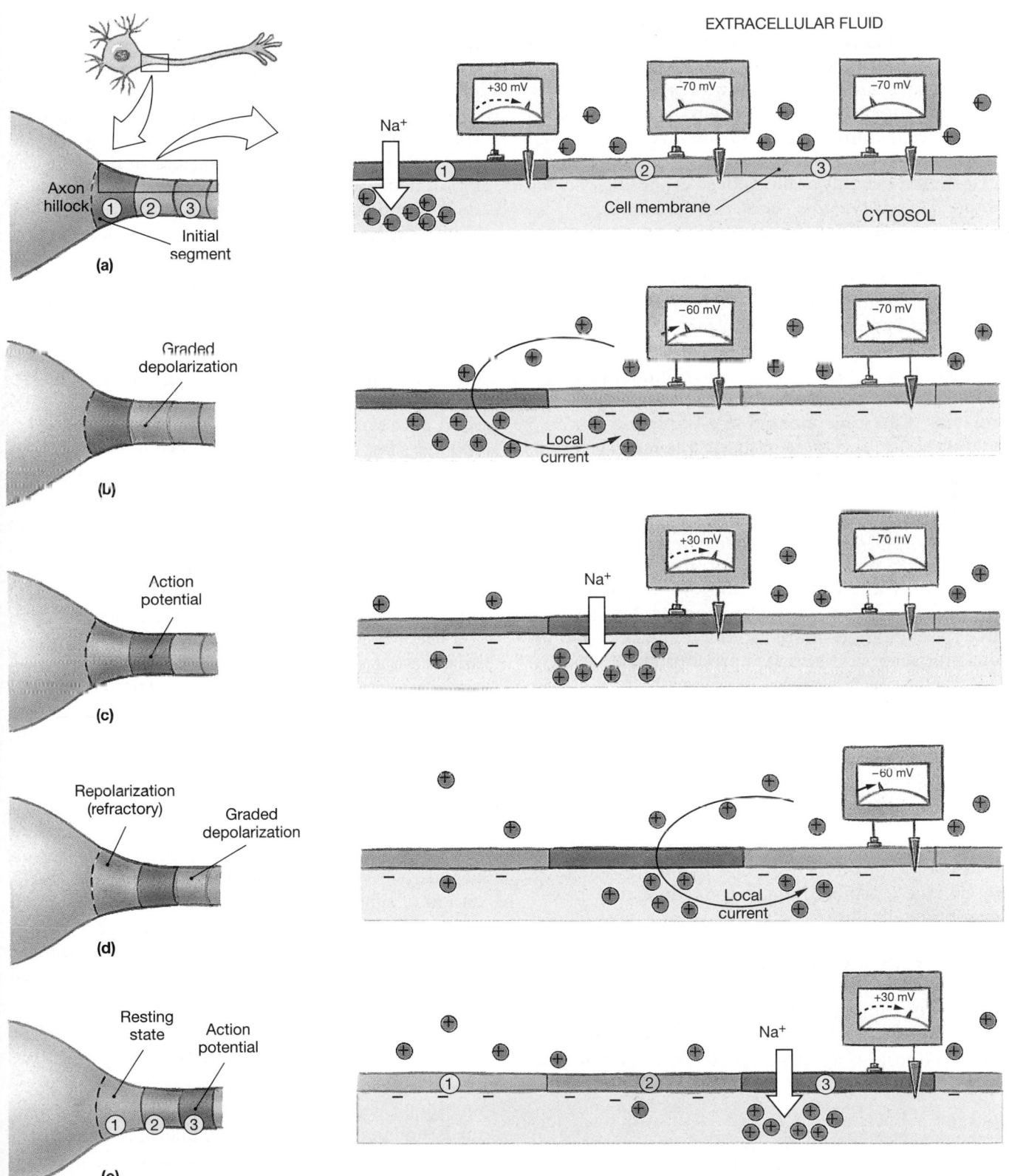

● **FIGURE 12-17** **Action Potential Propagation along an Unmyelinated Axon.** The axon can be viewed as a series of adjacent segments. **(a)** As an action potential develops in the initial segment, the transmembrane potential depolarizes to +30 mV. **(b)** A local current depolarizes the adjacent portion of the membrane to threshold. **(c)** An action potential then develops in the affected membrane. **(d)** A local current then depolarizes the next segment of the axon, and an action potential develops there. **(e)** The local current always drives the action potential forward because the previous segment is still in the absolute refractory period.

SALTATORY PROPAGATION In a myelinated axon, the axolemma is wrapped in layers of myelin. This wrapping is complete except at the nodes, where adjacent glial cells are in contact with one another. Continuous propagation cannot occur, because between the nodes the lipid content of the myelin blocks the flow of ions across the membrane. Ions can cross the cell membrane only at the nodes. As a result, *only the nodes can respond to a depolarizing stimulus.*

When an action potential appears at the initial segment, the local current skips the internodes and depolarizes the closest node to threshold (Figure 12-18●). Because the nodes may be 1–2 mm apart in a large myelinated axon, the action potential "jumps" from node to node rather than moving along the axon in a series of tiny steps. This process is **saltatory propagation,** or *saltatory conduction* (*saltare,* leaping). Imagine relaying a message along a line of people spaced 5 meters apart. Each person shouts the message to the next person in line; by the time the message has been repeated four times, it has moved 20 meters. By comparison, in our model of continuous propagation, in which people were closely packed, by the time the message had been repeated four times it would have moved only a few meters. Saltatory propagation in the CNS and PNS carries nerve impulses along an axon five to seven times more rapidly than does continuous propagation. It also uses proportionately less energy, because less surface area is involved and fewer sodium ions need to be pumped out of the cytoplasm.

Axon Diameter and Propagation Speed

Axon diameter also has an effect on propagation velocity. To depolarize adjacent portions of the cell membrane, ions must move through the cytoplasm. Cytoplasm also offers resistance to ion movement, although the resistance is much less than that of the cell membrane. In this instance, an axon behaves like an electrical cable: The larger the diameter, the lower the resistance. That is why motors with large current demands, such as the starter on a car, an electric stove, or a big air conditioner, use such thick wires.

Axons are classified into three groups according to the relationships among diameter, myelination, and propagation speed:

1. **Type A fibers** are the largest axons, with diameters ranging from 4 to 20 μm. These are myelinated axons that carry action potentials at speeds of up to 140 meters per second, the equivalent of over 300 mph.

2. **Type B fibers** are smaller myelinated axons, with diameters of 2–4 μm. Their propagation speeds average around 18 meters per second, or roughly 40 mph.

3. **Type C fibers** are unmyelinated and less than 2 μm in diameter. These axons propagate action potentials at the leisurely pace of 1 meter per second, a mere 2 mph.

We can understand the relative importance of myelin by noting that in going from Type C to Type A fibers, there is a tenfold increase in diameter, but the propagation speed increases by 140 times!

Type A fibers carry sensory information to the CNS concerning position, balance, and delicate touch and pressure sensations from the surface of the skin. The motor neurons that control skeletal muscles also send their commands over large, myelinated Type A axons. Type B fibers and Type C fibers carry less-urgent information concerning temperature, pain, and general touch and pressure sensations to the CNS and carry instructions to smooth muscle, cardiac muscle, glands, and other peripheral effectors.

When we need to tell a friend urgent news or receive an immediate response, we usually pick up the telephone. For general correspondence, we usually send a first-class letter or an e-mail message, which produces fast but not immediate responses. If we have to distribute an enormous volume of information to a huge mailing list and are in no particular rush, bulk mail offers efficiency at a considerable savings. Instead of representing a compromise between time and money, information transfer within the nervous system reflects a compromise between time and space. Axons carry the information, and the larger the axon, the faster the rate of transmission. But if all sensory information were carried by Type A fibers, your peripheral nerves would be the size of garden hoses, and your spinal cord would have the diameter of a garbage can. Instead, only around one-third of all axons carrying sensory information are myelinated, and most sensory information arrives over slender Type C fibers.

Myelination improves coordination and control by decreasing the time between reception of a sensation and initiation of an appropriate response. Myelination begins relatively late in development, and the myelination of sensory and motor fibers is not completed until early adolescence. In growing children, the pace of myelination and the pathways involved can be quite variable. This variability contributes to the observed range of physical capabilities within a given age group.

The Direction of Action Potential Propagation

Most synapses occur on the dendrites and soma of a multipolar neuron. Action potentials usually begin at the boundary between the axon hillock and the initial segment, traveling down the axon toward the synaptic terminals. Propagation in this case occurs in one direction only. However, synapses may also occur at a node or at a synaptic knob. An action potential originating at a node will be propagated along the axon in both directions; it will travel toward the soma as well as toward the telodendria. The action potential headed for the telodendria will depolarize the synaptic knob and cause the release of neurotransmitter. The action potential

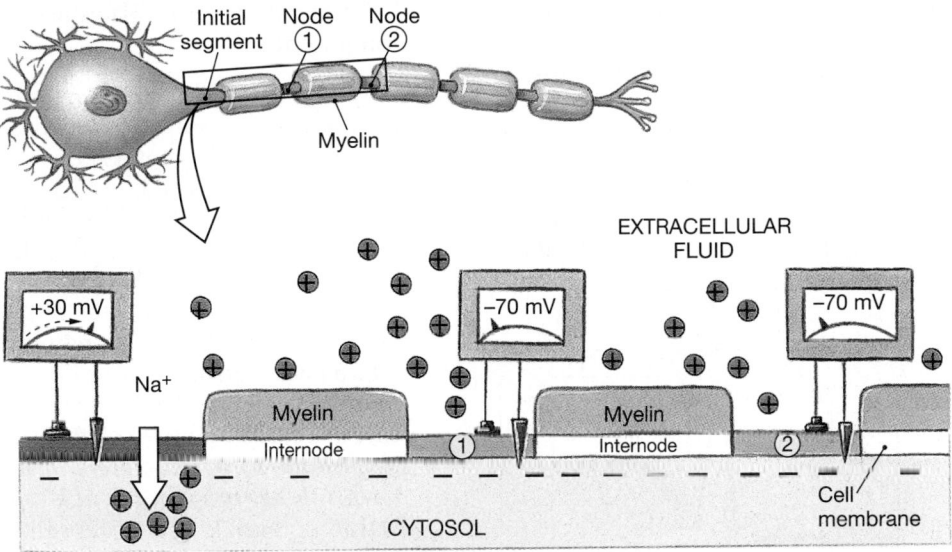

(a) Action potential at initial segment

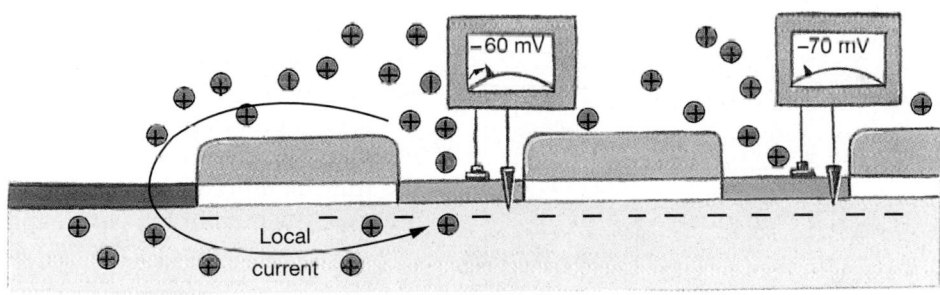

(b) Depolarization to threshold at node 1

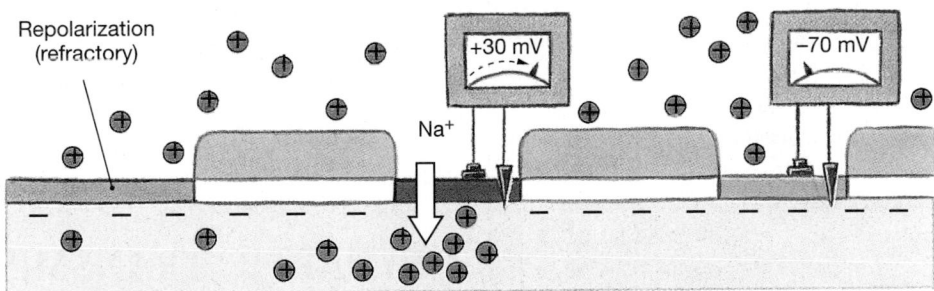

(c) Action potential at node 1

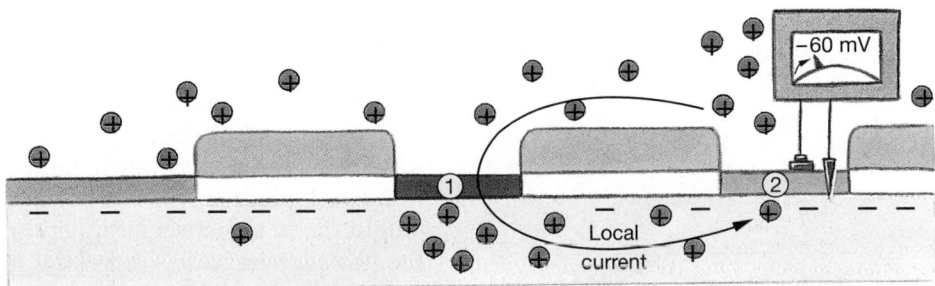

(d) Depolarization to threshold at node 2

•*FIGURE 12-18* **Saltatory Propagation along a Myelinated Axon.** **(a)** An action potential develops in the initial segment. **(b)** Local current depolarizes the next node to threshold. **(c)** An action potential develops at the node. **(d)** A local current depolarizes the membrane at the next node. This process will continue along the length of the axon.

headed toward the soma will have no effect, as it will not be propagated past the initial segment of the axon. As a result, the soma and dendrites will remain sensitive to depolarizing or hyperpolarizing stimuli.

Action Potentials in Muscle Tissues

Figure 12-19● compares action potentials in the excitable membranes of a neuron, a skeletal muscle fiber, a multiunit smooth muscle cell, and a cardiac muscle

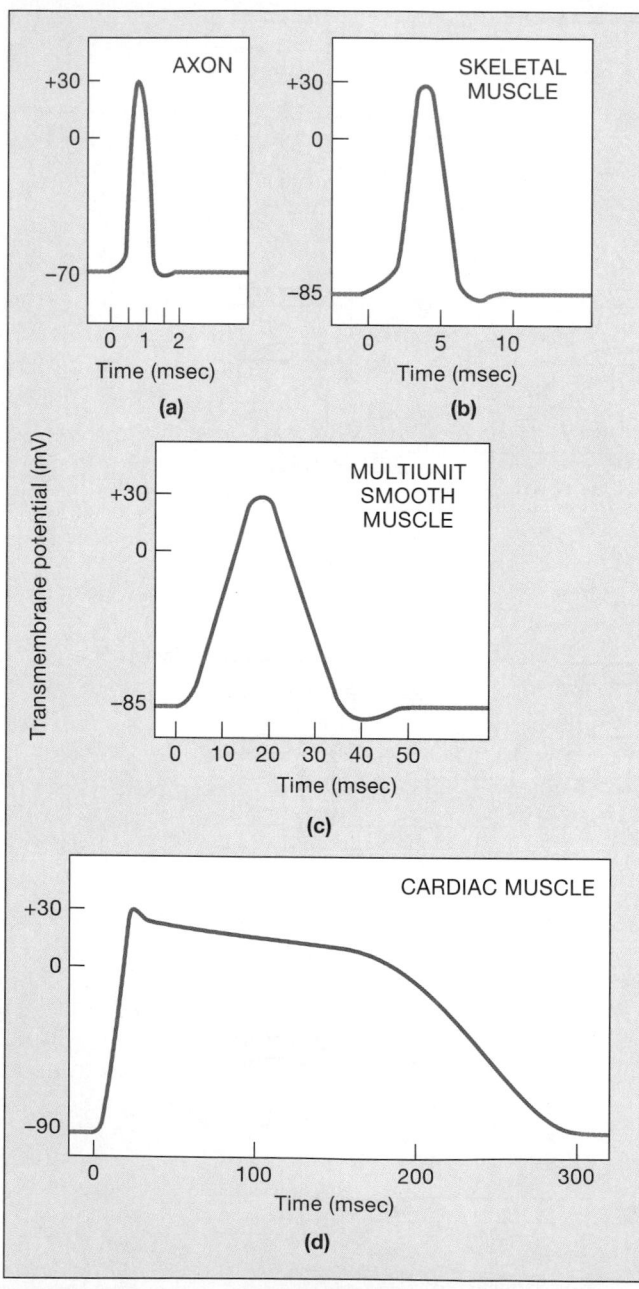

●*FIGURE 12-19* **Action Potentials in Nerve and Muscle.** A comparison of action potentials recorded in **(a)** an axon, **(b)** the sarcolemma of a skeletal muscle fiber, **(c)** a multiunit smooth muscle cell, and **(d)** a cardiac muscle cell from one of the ventricles. The vertical axes give the transmembrane potential in millivolts.

cell. Several important differences exist among these action potentials:

- ***The potential difference is larger in muscle cells than in axons.*** (Compare the blue lines in Figure 12-19●.) The resting potential of a skeletal, cardiac, or multiunit smooth muscle cell is −85 mV to −90 mV, close to the potassium equilibrium potential. Threshold values in these cells are comparable to those of axons.

- ***Action potentials last longer in muscle cells than in axons.*** (Compare the widths of the orange curves in Figure 12-19●.) An action potential in an axon may last 0.5 msec, versus 7.5 msec in a skeletal muscle fiber, 50 msec in a multiunit smooth muscle cell, and 250–300 msec in a cardiac muscle cell from one of the ventricles.

- ***Despite the large diameters of muscle cells, action potentials are propagated at a relatively slow speed.*** Action potentials of muscle cells travel by continuous propagation at speeds of 3–5 meters per second (6–10 mph).

Visceral smooth muscle cells do not have typical excitable membranes, and most have neither stable resting potentials nor action potentials comparable to those of other muscle types. When visceral smooth muscle cells are stimulated to contract, changes in the transmembrane potential reflect primarily alterations in the membrane permeability to calcium ions.

☑ How would a chemical that blocks the sodium channels in neuron cell membranes affect a neuron's ability to depolarize?

☑ What effect would decreasing the concentration of extracellular potassium ions have on the transmembrane potential of a neuron?

☑ One axon propagates action potentials at 50 meters per second; another carries them at 1 meter per second. Which axon is myelinated?

SYNAPTIC TRANSMISSION

In the nervous system, messages move from one location to another in the form of action potentials along axons. These electrical events are also known as **nerve impulses.** To be effective, a message must not only be propagated along an axon, it must also be transferred in some way to another cell. At a synapse involving two neurons, the impulse passes from the **presynaptic neuron** to the **postsynaptic neuron.** A synapse may also involve other types of postsynaptic cells. For example, the neuromuscular junction is a synapse where the postsynaptic cell is a skeletal muscle fiber.

At a synapse, a change in the transmembrane potential of the synaptic knob affects the activity of another cell. A synapse may be *electrical,* with direct physical contact between the cells, or *chemical,* involving a neurotransmitter.

Electrical Synapses

Electrical synapses are located in the CNS and PNS, but they are extremely rare. They are present in some centers of the brain, including the *vestibular nuclei*, and are found in the eye and in at least one pair of PNS ganglia (the *ciliary ganglia*). At an electrical synapse, the presynaptic and postsynaptic membranes are locked together at *gap junctions* (Figure 4-2●, p. 111). The lipid portions of the opposing membranes, separated by only 2 nm, are held in position by binding between integral membrane proteins called *connexons*. These proteins contain pores that permit the passage of ions between the cells. Because the two cells are linked in this way, changes in the transmembrane potential of one cell will produce local currents that affect the other cell as if the two shared a common cell membrane. As a result, an electrical synapse quickly and efficiently conducts action potentials from one cell to another.

Chemical Synapses

The situation at a **chemical synapse** is far more dynamic than that at an electrical synapse, because the cells are not directly coupled. For example, an action potential that reaches an electrical synapse will *always* be propagated to the next cell. But at a chemical synapse, an arriving action potential *may or may not* release enough neurotransmitter to bring the postsynaptic neuron to threshold. In addition, other factors may intervene and make the postsynaptic cell more or less sensitive to arriving stimuli. In essence, the postsynaptic cell at a chemical synapse is not a slave to the presynaptic neuron; its activity can be adjusted or "tuned" by a variety of factors.

Chemical synapses are by far the most abundant type of synapse. Most synapses between neurons and all communications between neurons and other cell types involve chemical synapses. Communication across a chemical synapse can normally occur in only one direction: from the presynaptic membrane to the postsynaptic membrane.

Although *acetylcholine* is the neurotransmitter that has received the most attention, there are many other important chemical transmitters. Neurotransmitters are often classified as **excitatory** or **inhibitory** on the basis of their effects on postsynaptic membranes. *Excitatory neurotransmitters* cause depolarization and promote action potential generation, whereas *inhibitory neurotransmitters* cause hyperpolarization and depress action potential generation.

This classification is useful but not always precise. For example, acetylcholine typically produces a depolarization in the postsynaptic membrane. But the acetylcholine released at neuromuscular junctions in the heart has an inhibitory effect, producing a transient **hyperpolarization** of the postsynaptic membrane. This situation highlights an important aspect of neurotransmitter function: *The effect of a neurotransmitter on the postsynaptic membrane depends on the properties of the receptor, not on the nature of the neurotransmitter.*

Cholinergic Synapses

We shall focus on chemical synapses that release the neurotransmitter **acetylcholine (ACh)**. These are known as **cholinergic synapses.** The neuromuscular junction described in Chapter 10 is one example of a cholinergic synapse. ∞ *[p. 285]* ACh, the most widespread (and best-studied) neurotransmitter, is released (1) at all neuromuscular junctions involving skeletal muscle fibers, (2) at many synapses inside the CNS, (3) at all neuron-to-neuron synapses in the PNS, and (4) at all neuroeffector junctions of the parasympathetic division of the ANS.

At a cholinergic synapse (Figure 12-20a●), the presynaptic membrane and the postsynaptic membrane are separated by a synaptic cleft that averages 20 nm (0.02 μm) in width. Most of the ACh in the synaptic knob is packaged within synaptic vesicles, each containing several thousand molecules of neurotransmitter. There may be a million such vesicles in a single synaptic knob.

Figure 12-20● diagrams the events that occur at a cholinergic synapse following the arrival of an action potential at a synaptic knob. The synapse under consideration is located at the boundary between the axon hillock and the initial segment of the axon.

STEP 1: *Arrival of an Action Potential and Depolarization of the Synaptic Knob* (Figure 12-20a●). The normal stimulus for neurotransmitter release is the depolarization of the presynaptic membrane following the arrival of an action potential.

STEP 2: *Extracellular Calcium Ions Enter the Synaptic Knob and Trigger Exocytosis of ACh* (Figure 12-20b●). The depolarization of the synaptic knob causes the opening of voltage-regulated calcium channels. During the brief period that these channels remain open, calcium ions flood into the axoplasm. Their arrival triggers exocytosis and the release of ACh into the synaptic cleft. The ACh is released in packets of roughly 3000 molecules, the average number of ACh molecules within a single vesicle. The release of ACh stops very soon because the calcium ions that triggered exocytosis are rapidly removed from the cytoplasm by active transport mechanisms. These mechanisms either pump the calcium ions out of the cell or transfer them into mitochondria, vesicles, or the endoplasmic reticulum.

STEP 3: *ACh Binding and Depolarization of the Postsynaptic Membrane* (Figure 12-20c●). The ACh

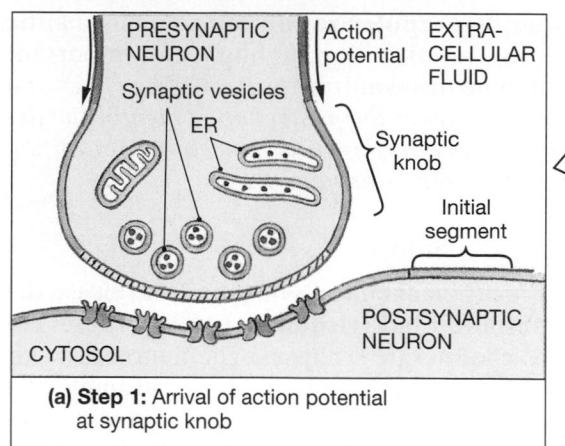

(a) **Step 1:** Arrival of action potential at synaptic knob

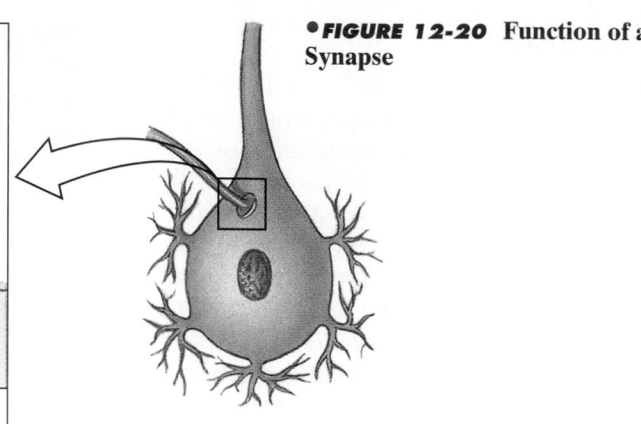

•**FIGURE 12-20** Function of a Cholinergic Synapse

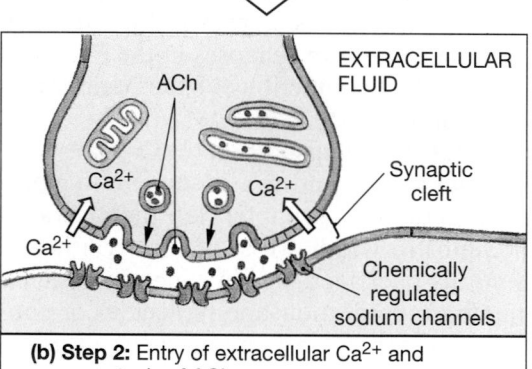

(b) **Step 2:** Entry of extracellular Ca^{2+} and exocytosis of ACh

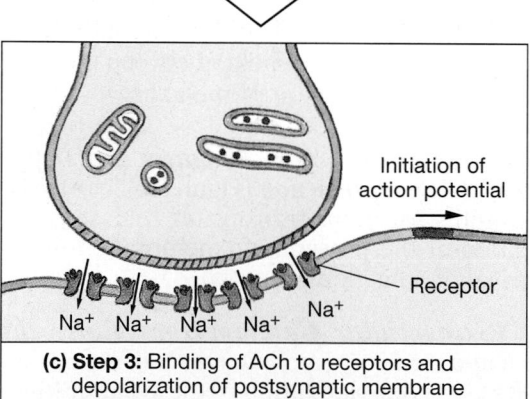

(c) **Step 3:** Binding of ACh to receptors and depolarization of postsynaptic membrane brings initial segment to threshold

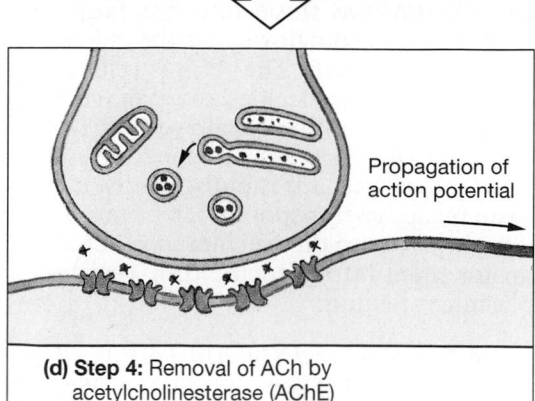

(d) **Step 4:** Removal of ACh by acetylcholinesterase (AChE)

released through exocytosis diffuses across the synaptic cleft toward receptors on the postsynaptic membrane. These receptors are chemically regulated ion channels. The primary response to ACh binding is an increased permeability to Na^+, producing a depolarization that lasts about 20 msec.[3] This depolarization is a graded potential; the more ACh released at the presynaptic membrane, the larger the depolarization. If the depolarization brings an adjacent area of excitable membrane to threshold, an action potential will appear in the postsynaptic neuron.

STEP 4: *Removal of ACh by AChE* (Figure 12-20d•). The effects on the postsynaptic membrane are temporary because the synaptic cleft and the postsynaptic membrane contain *acetylcholinesterase (AChE,* or *cholinesterase).* Roughly half of the ACh released at the presynaptic membrane is broken down before it reaches receptors on the postsynaptic membrane. ACh molecules that do succeed in binding to receptor sites are generally broken down within 20 msec of their arrival.

AChE breaks down molecules of ACh into **acetate** and **choline.** The choline is actively absorbed by the synaptic knob and used to synthesize more ACh, using acetate provided by *coenzyme A (CoA).* (As you may recall from Chapter 2, coenzymes derived from vitamins are required participants in many enzymatic reactions. ∞ *[p. 56])* Acetate diffusing away from the synapse can be absorbed and metabolized by the postsynaptic cell or by other cells and tissues.

Figure 12-21• shows the general pattern of ACh release and choline recycling, and Table 12-5 summarizes the events that occur at a cholinergic synapse.

SYNAPTIC DELAY There is a 0.2–0.5 msec **synaptic delay** between the arrival of the action potential at the synaptic knob and the effect on the postsynaptic membrane. Most of that delay reflects the time involved in calcium influx and neurotransmitter release, not in neurotransmitter diffusion—the synaptic cleft is relative-

---

[3]These channels also let potassium ions out of the cell. But, because sodium ions are driven by a much stronger electrochemical gradient, the net effect is a slight depolarization of the postsynaptic membrane.

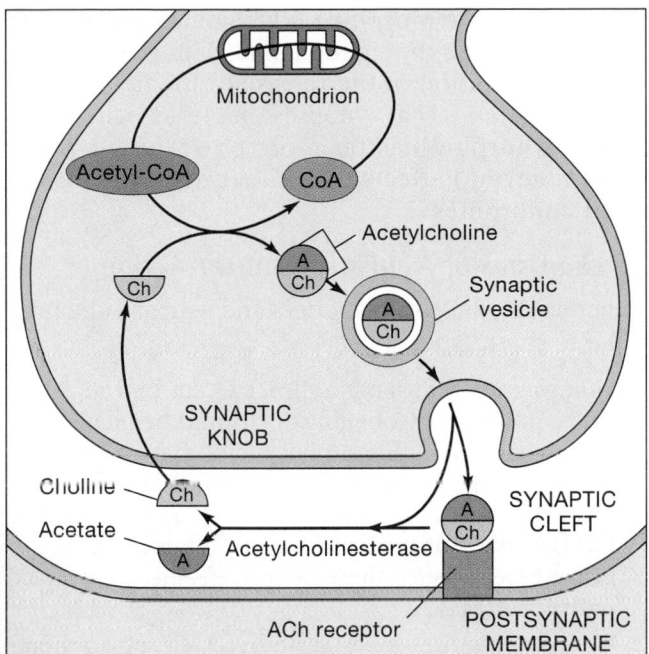

•FIGURE 12-21 ACh Release and Recycling at a Cholinergic Synapse. ACh at the synaptic knob is stored in synaptic vesicles. After it is released into the synaptic cleft, ACh is broken down by hydrolysis into acetate and choline by the enzyme acetylcholinesterase (AChE). The choline is reabsorbed by the synaptic knob and used to synthesize more ACh, which is packaged in vesicles at the ER.

TABLE 12-5 Sequence of Events at a Typical Cholinergic Synapse

Step 1:
- An arriving action potential depolarizes the synaptic knob and the presynaptic membrane.

Step 2:
- Calcium ions enter the cytoplasm of the synaptic knob.
- ACh release occurs primarily through exocytosis of neurotransmitter vesicles.

Step 3:
- ACh diffuses across the synaptic cleft and binds to receptors on the postsynaptic membrane.
- Chemically regulated sodium channels on the postsynaptic surface are activated, producing a graded depolarization.
- ACh release ceases because calcium ions are removed from the cytoplasm of the synaptic knob.

Step 4:
- The depolarization ends as ACh is broken down into acetate and choline by AChE.
- The synaptic knob reabsorbs choline from the synaptic cleft and uses it to resynthesize ACh.

ly narrow, and diffusion of neurotransmitters across it takes very little time. Although a delay of 0.5 msec is not very long, in that time an action potential may travel more than 7 cm (about 3 in.) along a myelinated axon. When information is being passed along a chain of interneurons in the CNS, the cumulative synaptic delay may exceed the propagation time. Reflexes, which provide rapid and automatic responses to stimuli, involve a small number of synapses. The fewer the number of synapses involved, the shorter the total synaptic delay and the faster the response. The fastest reflexes have just one synapse, with a sensory neuron directly controlling a motor neuron. The muscle spindle reflexes, introduced in Chapter 10 and detailed in Chapter 13, are an important example. ∞ *[p. 295]*

SYNAPTIC FATIGUE   Because ACh molecules are recycled, the synaptic knob is not totally dependent on the ACh synthesized in the soma and delivered by axoplasmic transport. But under intensive stimulation, the resynthesis and transport mechanisms may be unable to keep pace with the demand for neurotransmitter. **Synaptic fatigue** then occurs, and the synapse remains inactive until stores of ACh have been replenished.

Other Neurotransmitters

Other important neurotransmitters include *biogenic amines, amino acids, neuropeptides,* and *dissolved gases*

as well as a variety of other compounds. We will consider only a few of the most important neurotransmitters here; we shall encounter additional examples in later chapters:

- **Norepinephrine** (nor-ep-i-NEF-rin), or **NE,** is an important neurotransmitter in the brain and in portions of the ANS. Norepinephrine is also called **noradrenaline,** and synapses that release NE are described as **adrenergic.** Norepinephrine typically has an excitatory, depolarizing effect on the postsynaptic membrane, but the mechanism is quite distinct from that of ACh. We will consider the mechanism in detail in Chapter 16.

- **Dopamine** (DŌ-puh-mēn) is a CNS neurotransmitter released in a variety of centers in the brain. It may have inhibitory or excitatory effects. For example, inadequate dopamine production and release in one portion of the brain can lead to overstimulation of neurons that control skeletal muscle tone, due to a lack of inhibitory control. The result can be the characteristic rigidity and stiffness of *Parkinson's disease,* a condition we shall describe in Chapter 14. At other sites, dopamine can have excitatory effects. One of the effects of cocaine use is the inhibition of dopamine removal from synapses in specific areas of the brain. The resulting rise in dopamine concentrations at these synapses is responsible for the "high" experienced.

- **Serotonin** (ser-ō-TŌ-nin) is another important CNS neurotransmitter. Inadequate serotonin production can have widespread effects on attention and emotional states and may be responsible for many cases of severe chronic depression. *Fluoxetine (Prozac™), paroxetine, sertraline*, and related antidepressant drugs inhibit the reabsorption of serotonin by synaptic knobs. The inhibition leads to increased serotonin concentrations at synapses; over time, this increase may relieve the symptoms. Interactions among serotonin, norepinephrine, and other neurotransmitters are thought to be involved in the regulation of asleep/awake cycles.

- **Gamma aminobutyric** (a-MĒ-nō-bū-TIR-ik) **acid,** or **GABA,** generally has an inhibitory effect. GABA release appears to reduce anxiety, and some antianxiety drugs work by enhancing these effects.

There are many neurotransmitters whose functions are less well understood. In a clear demonstration of the principle "the more you look, the more you see," at least 50 neurotransmitters have now been identified, including certain amino acids, peptides, polypeptides, prostaglandins, and ATP. In addition, two gases have recently been identified as important neurotransmitters. Nitric oxide (NO) is generated by synaptic knobs that innervate smooth muscle in the walls of blood vessels in the PNS and at synapses in several regions of the brain. Carbon monoxide (CO), best known for its presence in automobile exhaust, can also be generated by specialized synaptic knobs in the brain, where it functions as a neurotransmitter. Our knowledge of the significance of these compounds and the mechanisms involved in their synthesis and release remains incomplete. Table 12-6 lists major neurotransmitters of the brain and spinal cord and their primary effects (if known).

NEUROMODULATORS   Chemical synapses are always active to some degree, and small numbers of neurotransmitter molecules continuously leak through the presynaptic membranes. Other chemicals may also be released by the synaptic knob, either through gradual diffusion or through exocytosis, in the company of neurotransmitter molecules. **Neuromodulators** (noo-rō-MOD-ū-lā-torz) are compounds that influence either the release of neurotransmitter by the presynaptic neuron or the postsynaptic cell's response to the neurotransmitter. Neuromodulators are typically **neuropeptides,** small peptide chains synthesized and released by the synaptic knob. Many neuromodulators act by binding to receptors in the presynaptic or postsynaptic membranes and activating cytoplasmic enzymes.

Endorphins.   Neuromodulators called **endorphins** (en-DOR-finz) are produced in the brain and spinal cord. The endorphins include several different **enkephalins** (en-KEF-a-linz), **beta-endorphin,** and **dynorphin** (dī-NOR-fin). Although their exact function is uncertain, their primary role is probably the relief of pain. These neuropeptides are structurally similar to morphine, which binds to the same receptor sites as the enkephalins. Pain relief occurs through inhibition of the release of the neurotransmitter *Substance P* at synapses that relay pain sensations. Dynorphin has far more powerful *analgesic* (pain-relieving) effects than either morphine or the other endorphins.

Mechanisms of Neurotransmitter Action

Functionally, neurotransmitters and neuromodulators fall into three groups:

1. Compounds that have a direct effect on the membrane potential by opening or closing chemically regulated channels. Examples include ACh and the amino acids *glutamate* and *aspartate.* In addition, some chemically regulated channels are sensitive to GABA and norepinephrine, although other membrane receptors for these neurotransmitters produce indirect effects.

2. Compounds that have an indirect effect on membrane potential through the activities of **second messengers.** Second messengers are intracellular intermediaries produced or released when neurotransmitters or other extracellular chemicals bond to specific membrane receptors. When epinephrine, norepinephrine, dopamine, serotonin, histamine, GABA, and any of the endorphins bind to membrane receptors, the receptors activate the intracellular enzyme **adenylate cyclase,** also known as *adenylyl* (or *adenyl*) *cyclase.* This enzyme catalyzes the formation of cyclic-AMP from ATP at the inner surface of the cell membrane. **Cyclic-AMP (cAMP)** is a second messenger that may open membrane channels and/or activate intracellular enzymes, depending on the nature of the postsynaptic cell. We shall encounter other second messengers, including *cyclic-GMP (cGMP)*, calcium ions, and *inositol trisphosphate* in later chapters.

3. Gases such as NO and CO are lipid-soluble, so they can diffuse through the membrane lipids and bind to enzymes on the inner surface of the membrane. These enzymes then promote the appearance of second messengers that can have various effects on cellular activity.

It can be very difficult to distinguish between neuro-transmitters and neuromodulators on either biochemical or functional grounds. In general, neuromodulators (1) have long-term effects that are relatively slow to appear; (2) trigger responses that involve second messengers; (3) may affect the presynaptic membrane, the postsynaptic membrane, or both; and (4) can be released alone or in the company of a neurotransmitter. However, the same compound may function in one site as a neurotransmitter and in another as a neuromodulator. For this reason, Table 12-6 does not distinguish between neurotransmitters and neuromodulators.

TABLE 12-6 **Representative Neurotransmitters and Their Effects**

Neurotransmitter	Mechanism of Action	Distribution (Examples)
Acetylcholine	Primarily direct, through binding to chemically regulated channels	*Widespread in CNS and PNS* *CNS:* Synapses throughout brain and spinal cord *PNS:* Neuromuscular junctions; preganglionic synapses of ANS; neuroeffector junctions of parasympathetic division and (rarely) sympathetic division of ANS
Biogenic Amines Norepinephrine (NE)	Indirect, using second messengers	*CNS and PNS* *CNS:* Cerebral cortex, hypothalamus, brain stem, cerebellum, spinal cord *PNS:* Most neuroeffector junctions of sympathetic division of ANS
Epinephrine	Indirect, using second messengers	*CNS:* Thalamus, hypothalamus, midbrain, spinal cord
Dopamine	Direct or indirect, depending on receptor type	*CNS:* Hypothalamus, limbic system
Serotonin	Direct or indirect, depending on receptor type	*CNS:* Limbic system, hypothalamus, cerebellum, spinal cord
Histamine	Indirect, using second messengers	*CNS:* Hypothalamus
Excitatory Amino Acids Glutamate	Direct or indirect, depending on receptor type	*CNS:* Cerebral cortex, brain stem
Aspartate	Direct or indirect, depending on receptor type	*CNS:* Spinal cord
Inhibitory Amino Acids Gamma aminobutyric acid (GABA)	Direct or indirect, depending on receptor type	*CNS:* Cerebellum, cerebral cortex, inhibitory interneurons throughout brain and spinal cord
Glycine	Direct	*CNS:* Inhibitory interneurons throughout brain and spinal cord
Neuropeptides Substance P	Indirect	*CNS:* Synapses of sensory axons within spinal cord, hypothalamus, other areas of brain
Enkephalins, endorphins	Indirect, using second messengers	*CNS:* Hypothalamus, thalamus, brain stem
Other Synaptic Chemicals High-energy compounds (ATP, GTP)	Indirect, using second messengers	*CNS:* Spinal cord
Hormones (ADH, oxytocin, insulin, glucagon, secretin, CCK, and many others)	Generally indirect	*CNS:* Brain; widespread
Prostaglandins	Indirect, using second messengers	*CNS:* Brain; widespread
Dissolved Gases Carbon monoxide (CO)	Indirect, using second messengers formation	*CNS:* Brain; some neuroeffector junctions
Nitric oxide (NO)	Indirect, using second messengers	*CNS:* Brain; some neuroeffector junctions *PNS:* Some sympathetic neuroeffector junctions (rare)

☑ What effect would blocking voltage-regulated calcium channels at a cholinergic synapse have on synaptic communication?

☑ While studying pathways in the central nervous system, you discover one pathway that consists of three neurons and another that consists of five neurons. If the neurons in both pathways are identical, which pathway will transmit impulses more rapidly?

☑ Norepinephrine causes hyperpolarization of smooth muscle cells in the blood vessels that serve skeletal muscles, but it causes depolarization of smooth muscle cells in blood vessels that supply the intestines. How is this possible?

 ## Drugs and Synaptic Function

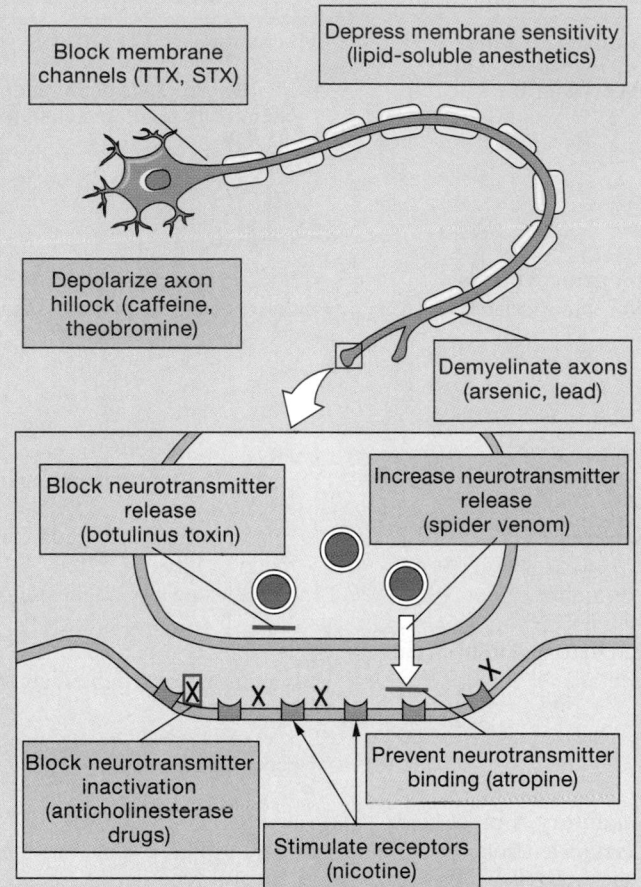

Many drugs interfere with key steps in the synaptic transmission process. These drugs may (1) interfere with transmitter synthesis, (2) alter the rate of transmitter release, (3) prevent transmitter inactivation, or (4) prevent transmitter binding to receptors. Our discussion here is limited to clinically important compounds that exert their effects at cholinergic synapses. Figure 12-22● indicates their sites of activity.

Some drugs affect the synaptic terminals. *Botulinus toxin* is responsible for the primary symptom of *botulism,* a widespread paralysis of skeletal muscles. ∞ *[p. 287]* Botulinus toxin blocks the release of ACh at the presynaptic membrane of cholinergic neurons. The venom of the black widow spider has the opposite effect. It causes a massive release of ACh that produces intense muscular cramps and spasms.

Other drugs primarily affect the postsynaptic membrane. **Anticholinesterase drugs,** sometimes called *cholinesterase inhibitors,* block the breakdown of ACh by acetylcholinesterase. The result is an exaggerated and prolonged stimulation of the postsynaptic membrane. At the neuromuscular junctions, this abnormal stimulation produces an extended and extreme state of contraction. Military nerve gases block cholinesterase activity for weeks, although few persons exposed are likely to live long enough to regain normal synaptic function. Most animals utilize ACh as a neurotransmitter, and anticholinesterase drugs, such as *malathion,* are in widespread use in pest-control projects.

Drugs such as **atropine** or **d-tubocurarine** prevent ACh from binding to the postsynaptic receptors. The latter compound is a derivative of *curare,* a plant extract used by certain South American tribes to paralyze their prey. Curare and related compounds induce paralysis by preventing stimulation of the neuromuscular junction by ACh. Atropine can also be administered intentionally to counteract the effects of anticholinesterase poisoning. Other compounds, including **nicotine,** an active ingredient in cigarette smoke, bind to the receptor sites and stimulate the

●**FIGURE 12-22** **Mechanism of Drug Action at a Cholinergic Synapse.** Factors that facilitate neural function and make neurons more excitable are shown in violet. Factors that inhibit or depress neural function are shown in blue.

postsynaptic membrane. There are no enzymes for removing these compounds, and the effects are relatively prolonged. AM *Neuroactive Drugs*

Figure 12-22 also indicates the sites of action for other chemicals mentioned in this chapter. Several toxins found in seafood (p. 389 and the *Applications Manual*), including *tetrodotoxin* (TTX), *saxitoxin* (STX), and *ciguatoxin* (CTX), block sodium ion channels. At low doses, these toxins can produce abnormal sensations and interfere with muscle control. Higher doses cause death, generally through paralysis of the respiratory muscles.

CELLULAR INFORMATION PROCESSING

At each synapse, an action potential arriving at a synaptic knob triggers chemical or electrical events that affect another cell. Some of the neurotransmitters arriving at the postsynaptic cell at any given moment may be excitatory, while others, from different presynaptic neurons, may be inhibitory. The net result may be a depolarization, a hyperpolarization, or no appreciable change in the transmembrane potential at the initial segment. The transmembrane potential at the initial segment therefore represents an integration of all the excitatory and inhibitory stimuli affecting the neuron at that moment. Excitatory and inhibitory stimuli are integrated through interactions between *postsynaptic potentials*. This interaction is the simplest level of **information processing** in the nervous system.

Postsynaptic Potentials

Postsynaptic potentials are graded potentials that develop in the postsynaptic membrane in response to a neurotransmitter. Two major types of postsynaptic potentials develop at neuron-to-neuron synapses: *excitatory postsynaptic potentials* and *inhibitory postsynaptic potentials.*

Excitatory Postsynaptic Potentials

An **excitatory postsynaptic potential,** or **EPSP,** is a graded depolarization caused by the arrival of a neurotransmitter at the postsynaptic membrane. An EPSP results from the opening of chemically regulated ion channels. The depolarization produced by the binding of ACh is an EPSP. Because it is a graded potential, an EPSP affects only the area immediately surrounding the synapse, as in Figure 12-13c●, p. 387.

Inhibitory Postsynaptic Potentials

We have already learned that not all neurotransmitters have an excitatory (depolarizing) effect. An **inhibitory postsynaptic potential,** or **IPSP,** is a transient hyperpolarization of the postsynaptic membrane. For example, an IPSP may result from the opening of chemically regulated potassium channels. While the hyperpolarization continues, the neuron is **inhibited** because a larger-than-usual depolarizing stimulus must be provided to bring the membrane potential to threshold. A stimulus sufficient to shift the transmembrane potential by 10 mV—from –70 mV to –60 mV—would normally produce an action potential. But if the transmembrane potential were reset at –85 mV by an IPSP, the same stimulus would depolarize it only to –75 mV—still well below threshold.

Summation

Summation is the mechanism responsible for the integration of EPSPs, IPSPs, or some combination of the two in the postsynaptic neuron. We will use the summation of EPSPs to demonstrate this process. An individual EPSP has a small effect on the transmembrane potential; a typical EPSP produces a depolarization of about 0.5 mV at the postsynaptic membrane. Before an action potential will appear in the initial segment, local currents must depolarize that region by at least 10 mV. A single EPSP will therefore not result in an action potential, even if the synapse is on the axon hillock. But individual EPSPs can combine through the process of summation. There are two forms of summation: *temporal summation* and *spatial summation.*

TEMPORAL SUMMATION **Temporal summation** (*tempus,* time) is the addition of stimuli occurring in rapid succession. Temporal summation occurs at a *single synapse* that is active *repeatedly.* Consider this simple analogy: You cannot fill a bathtub with a single bucket of water, but you can if you keep using the bucket over and over. The water in each bucket corresponds to the sodium ions that enter the cytoplasm during an EPSP. Consider what happens when a second EPSP arrives before the effects of the first have disappeared. A typical EPSP lasts about 20 msec, but under maximum stimulation one action potential may reach the synaptic knob each millisecond. Every time an action potential arrives, another group of vesicles discharges ACh into the synaptic cleft. Every time another batch of ACh molecules arrives at the postsynaptic membrane, more chemically regulated channels open, and the degree of depolarization increases. In this way, a series of small steps can eventually bring the initial segment to threshold (Figure 12-23a●).

SPATIAL SUMMATION **Spatial summation** occurs when simultaneous stimuli at different locations have a cumulative effect on the transmembrane potential. Spatial summation involves *multiple synapses* that are active *simultaneously.* In terms of our bucket analogy, you could fill a bathtub immediately if 50 friends standing by the tub emptied their buckets at the same time.

The activity of a single synapse produces a graded potential with localized effects. If more than one synapse is active at the same time, all "pour" sodium ions across the postsynaptic membrane. At each active synapse, the sodium ions that produce the EPSP spread out along the inner surface of the membrane *and mingle with those entering at other synapses.* As a result, the effects on the initial segment are cumulative (Figure 12-23b●). The degree of depolarization will depend on how many synapses are active at any given moment and on their distance from the initial segment. As in the case of temporal summation, an action potential will appear when the transmembrane potential at the initial segment reaches threshold.

FACILITATION Spatial or temporal summation of EPSPs will not necessarily depolarize the initial segment to threshold, but every step closer to threshold makes it easier for the *next* stimulus to trigger an action potential. A neuron that has been brought closer to threshold is said to be **facilitated.** The larger the degree of facilitation, the smaller the additional stimulus needed to trigger an action potential. In a highly facilitated neuron, even a small depolarizing stimulus will produce an action potential.

Facilitation can also result from neuron exposure to certain drugs. For example, nicotine stimulates postsynaptic ACh receptors, producing prolonged EPSPs

●*FIGURE 12-23* **Temporal and Spatial Summation.**
(a) Temporal summation occurs on a membrane that receives two depolarizing stimuli separated in time. The effects of the second stimulus are added to those of the first. **(b)** Spatial summation occurs when sources of stimulation arrive simultaneously but at different locations. Local currents spread the depolarizing effects, and areas of overlap experience the combined effects.

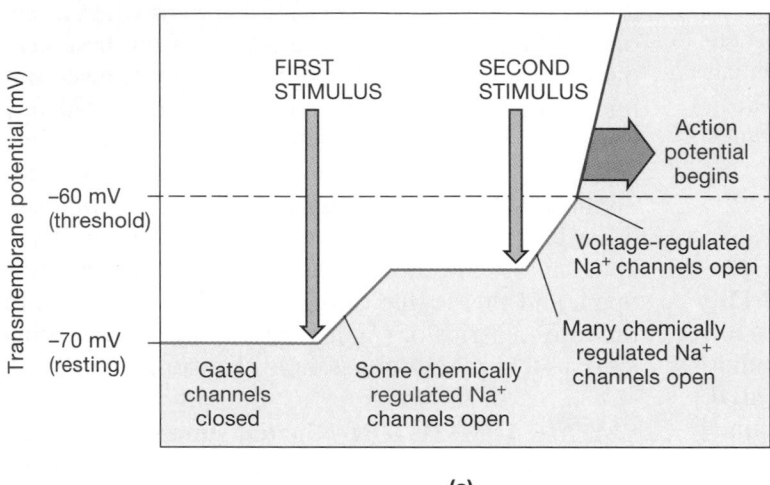

(a)

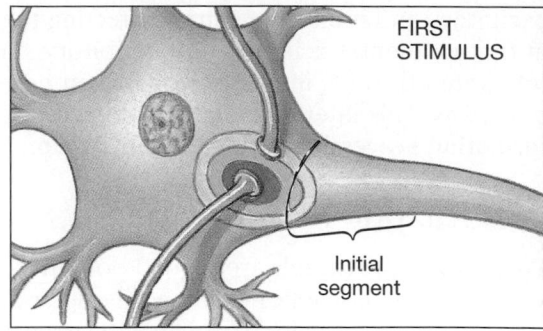

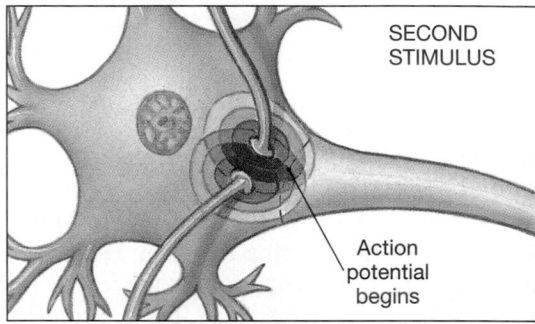

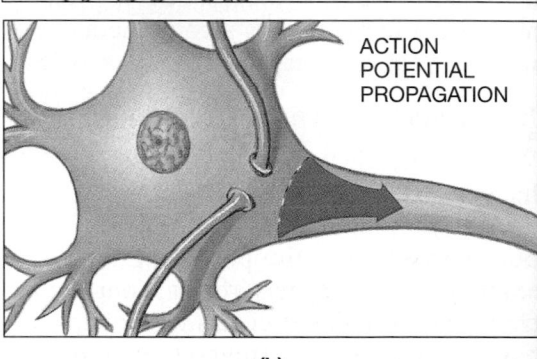

(b)

that facilitate CNS neurons. The active ingredients of coffee, colas, cocoa, and tea (caffeine, theobromine, and theophylline) also cause facilitation, but they act in a different way. These compounds lower the threshold at the initial segment, so a smaller-than-usual depolarization will cause an action potential. Once an action potential reaches the synaptic knob, these compounds also increase the amount of ACh released. The entire nervous system becomes more excitable, and you feel more alert. The phrase "coffee makes you jumpy" is not an exaggeration; after several cups, you may jump at a sudden noise because your reflexes are primed to respond to the slightest stimulus. Because nicotine and caffeine work in different ways, the combination of cigarettes and coffee has a very dramatic stimulatory effect on the CNS.

SUMMATION OF EPSPS AND IPSPS Like EPSPs, IPSPs can summate spatially and temporally. EPSPs and IPSPs reflect the activation of different types of chemically regulated channels, and they have opposing effects on the transmembrane potential. The antagonism between IPSPs and EPSPs is an important mechanism for cellular information processing. In terms of our bathtub analogy, EPSPs put water into the tub, and IPSPs take water out. If there are more buckets adding water than there are buckets removing water, the water level in the tub will rise. If there are more buckets removing water, the water level will fall. If a bucket of water is removed every time another

bucket is dumped into the tub, the water level will remain stable. Comparable interactions between EPSPs and IPSPs (Figure 12-24●) determine the transmembrane potential at the boundary between the axon hillock and the initial segment.

Neuromodulators, hormones, or both can change the sensitivity of the postsynaptic membrane to excitatory or inhibitory neurotransmitters. By shifting the balance between EPSPs and IPSPs, these compounds can promote facilitation or inhibition of neurons in the CNS and PNS.

Chloride Channels and EPSP/IPSP Summation. Many chemicals that inhibit EPSP formation open chemically regulated chloride channels rather than potassium channels. The opening of chemically regulated chloride channels in a resting membrane is difficult to detect, because the equilibrium potential for chloride ions is −70 mV, the same as the typical resting potential. However, with the chloride channels open, Cl^- is free to cross the membrane if the transmembrane potential changes. Consider what happens if the cell receives a depolarizing stimulus that would, under other circumstances,

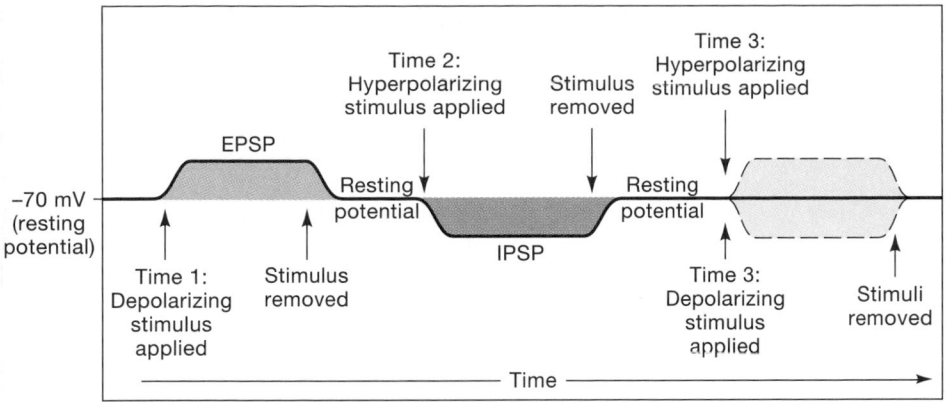

•FIGURE 12-24 EPSP/IPSP Interactions. At time 1, a small depolarizing stimulus produces an EPSP. At time 2, a small hyperpolarizing stimulus produces an IPSP of comparable magnitude. If the two stimuli are applied simultaneously, as at time 3, summation occurs. Because the two are equal in size, the membrane potential remains at the resting level. If the EPSP were larger, a net depolarization would result; if the IPSP were larger, a net hyperpolarization would result instead.

shift the transmembrane potential to 65 mV. As soon as the potential climbs above –70 mV, chloride ions start rushing into the cell. The situation now is like a bathtub with a large hole just above the normal water level; to raise the water level, you must add much more water—enough to overcome the water loss through the hole. In a similar way, once the chloride ion channels are open, you need a much larger-than-normal stimulus to depolarize the membrane to threshold.

The Rate of Action Potential Generation

In a computer, complex information is translated into a simple binary code of ones and zeros. In the nervous system, all sensory information and motor commands must be translated into action potentials that can be propagated along axons. On arrival, the message is often interpreted solely on the basis of the frequency of action potentials. For example, action potentials arriving at a neuromuscular junction at the rate of 1 per second may produce a series of isolated twitches in the associated skeletal muscle fiber, whereas at the rate of 100 per second they will cause a sustained tetanic contraction. A few action potentials per second along a sensory fiber may be perceived as a feather-light touch, whereas hundreds of action potentials per second along that same axon may be perceived as unbearable pressure. In this section, we will examine factors that vary the rate of action potential generation; we shall consider the functional significance of these changes in later chapters.

If a graded potential briefly depolarizes the axon hillock such that the initial segment reaches threshold, an action potential will be propagated along the axon. Now consider what happens if the axon hillock *remains* depolarized past threshold levels for an extended period. As the membrane repolarizes after the first action potential, it can generate a second one as soon as the absolute refractory period ends. Once the relative refractory period has begun, a larger-than-normal stimulus is required to produce

an action potential. With the axon hillock already depolarized, a stimulus is ready and waiting as soon as the relative refractory period begins. The greater the degree of depolarization at the axon hillock, the sooner the next action potential will occur. Thus, the greater the degree of sustained depolarization at the axon hillock, the higher the frequency of action potential generation. The rate of action potential generation reaches a maximum when the relative refractory period has been completely eliminated. The maximum theoretical frequency is therefore established by the duration of the absolute refractory period. The absolute refractory period is shortest in large-diameter axons, in which the theoretical maximum frequency is 2500 per second. However, the highest frequencies recorded from axons in the body range between 500 and 1000 per second.

Presynaptic Inhibition and Facilitation

Inhibitory or excitatory responses may occur at an *axoaxonal synapse,* a synapse between the axon of one neuron and the axon of another. An axoaxonal synapse that occurs on the synaptic knob (Figure 12-25•) can modify the rate of neurotransmitter release at the presynaptic membrane. In one form of **presynaptic inhibition** (Figure 12-25a•), GABA release inhibits the opening of voltage-regulated calcium channels in the synaptic knob. This inhibition reduces the amount of neurotransmitter released when an action potential arrives at the synaptic knob and thus limits the effects on the postsynaptic membrane. In **presynaptic facilitation** (Figure 12-25b•), activity at an axoaxonal synapse increases the amount of neurotransmitter released when an action potential arrives at the synaptic knob. This increase enhances and prolongs the effects of the neurotransmitter on the postsynaptic membrane. The neurotransmitter serotonin is involved in presynaptic facilitation. In the presence of serotonin released at an axoaxonal synapse, voltage-regulated calcium channels remain open for an extended period.

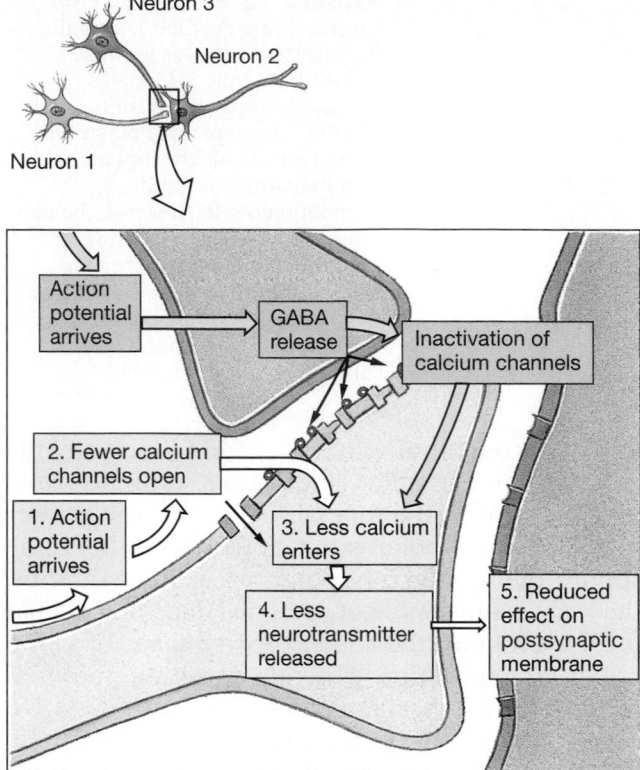

(a) Presynaptic inhibition

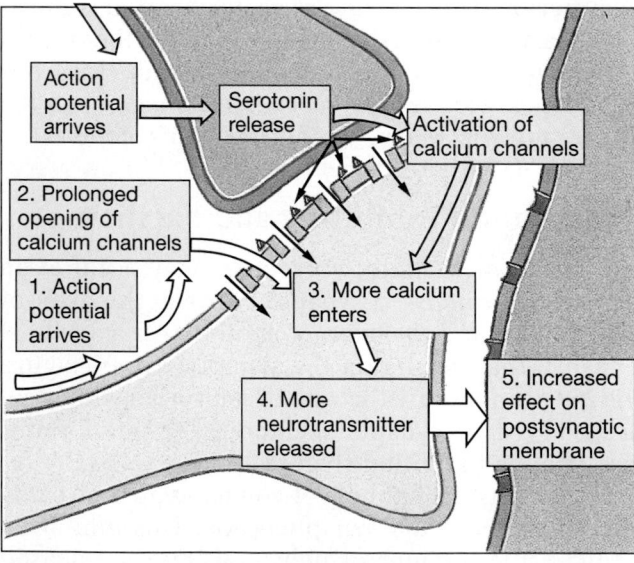

(b) Presynaptic facilitation

•*FIGURE 12-25* **Presynaptic Inhibition and Facilitation.**
(a) Steps in presynaptic inhibition. **(b)** Steps in presynaptic facilitation.

General Factors That Affect Neural Function

We will consider briefly two factors that can alter the function of neurons in the CNS and PNS: (1) changes in the extracellular environment and (2) the metabolic demands of active neurons.

Environmental Factors

Environmental factors, such as changes in pH, ionic composition, or temperature, can alter the resting membrane potential or disrupt the metabolic operations that support action potential generation:

- Changes in hydrogen ion concentration (pH) can have dramatic effects on neural activity. The normal extracellular pH averages 7.35–7.45. If the pH rises, neurons are facilitated: At a pH near 7.8, they begin to generate action potentials spontaneously, producing severe convulsions. If the pH drops, neurons are inhibited: At a pH around 7.0, the nervous system shuts down, and the individual becomes completely unresponsive. A variety of mechanisms exist to control the pH of the cerebrospinal fluid and other body fluids; we shall discuss those mechanisms in later chapters.

- Changes in the ionic composition of the extracellular fluids have a direct effect on neural function. Fluctuations in Na^+ or K^+ concentrations, such as those caused by dehydration or kidney disease, may facilitate or inhibit neural activity by depolarizing or hyperpolarizing the cell membrane. For example, an elevated extracellular K^+ concentration, a condition called **hyperkalemia** (hī-per-ka-LĒ-mē-uh; *kalium,* potassium + *haima,* blood) reduces the chemical gradient for K^+ across the cell membrane. The rate of potassium loss decreases as a result, and the retention of positive charges gradually depolarizes the membrane. Hyperkalemia has damaging effects on all excitable membranes; it can lead to a general paralysis of skeletal muscles and death by cardiac arrest. Abnormally high or low extracellular Ca^{2+} concentrations affect synaptic function by reducing or enhancing calcium entry into the synaptic knob, thereby modifying the amount of neurotransmitter released as well as changing the excitability of axonal membranes.

- Changes in body temperature have a direct effect on the metabolism of neurons, and this is one reason why body temperature regulation is so important. If your body temperature rises, your neurons become more excitable; an individual with a high fever may experience hallucinations or convulsions. If your body temperature falls, your neurons become inhibited; an individual whose body temperature has fallen outside normal limits will be lethargic and confused and may become unconscious.

Metabolic Processes

The brain accounts for just 2 percent of your body weight, but it accounts for 18 percent of your resting energy consumption. Active neurons need ATP to support (1) the synthesis, release, and recycling of neurotransmitter molecules; (2) the movement of materials to and from the soma by axoplasmic flow;

FOCUS Neural Responses to Injuries

•FIGURE 12-26 Peripheral
Nerve Regeneration after Injury

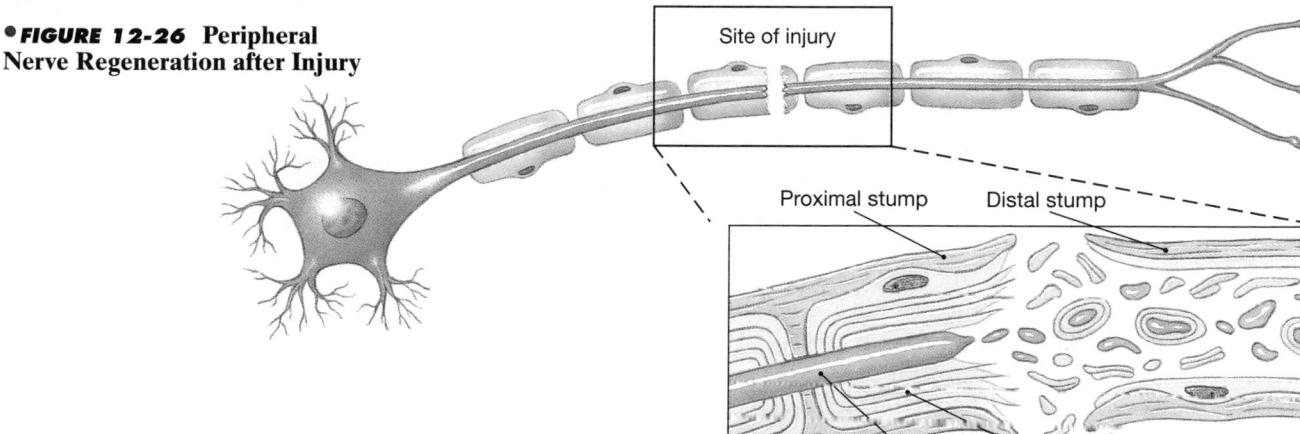

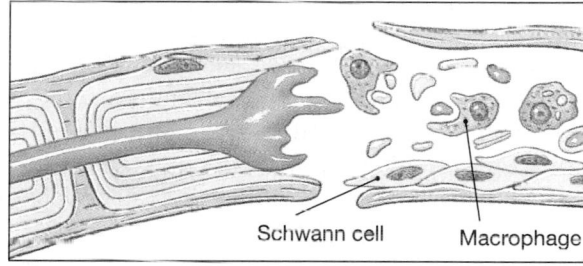

STEP 1:
Fragmentation of axon and myelin occurs in distal stump.

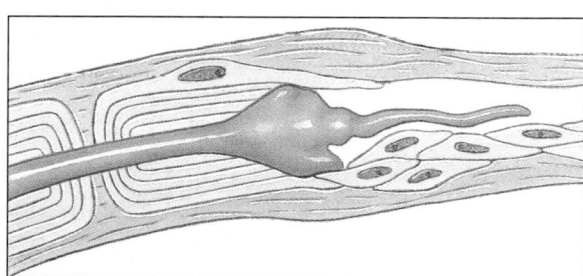

STEP 2:
Schwann cells form cord, grow into cut, and unite stumps.
Macrophages engulf degenerated axon and myelin.

A neuron responds to injury in a very limited, stereotyped fashion. Within the soma, the Nissl bodies disperse and the nucleus moves away from its centralized location as the cell increases its rate of protein synthesis. If the neuron recovers its functional abilities, it will return to a normal appearance. The key to recovery seems to be the events underway in the axon. Consider the response of a neuron to mechanical stresses, such as the pressure applied during a crushing injury. The pressure produces a local decrease in blood flow and oxygen availability, and the affected membrane becomes inexcitable. If the pressure is released after an hour or two, the neuron will recover within a few weeks. More severe or prolonged pressure will produce effects similar to those caused by cutting off the distal portion of the axon.

In the PNS, the Schwann cells participate in the repair of damaged nerves. In the process known as **Wallerian degeneration** (Figure 12-26•), the axon distal to the injury site degenerates, and macrophages migrate into the area to phagocytize the debris. The Schwann cells do not degenerate. Instead, they proliferate and form a solid cellular cord that follows the path of the original axon. As the neuron recovers, its axon grows into the injury site, and the Schwann cells wrap around it.

If the axon continues to grow into the periphery alongside the appropriate cord of Schwann cells, it may eventually reestablish its normal synaptic contacts. If it stops growing or wanders off in some new direction, normal function will not return. The growing axon is most likely to arrive at its appropriate destination if the cut edges of the original nerve bundle remain in contact.

Limited regeneration can occur inside the CNS, but the situation is more complicated because (1) many more axons are likely to be involved, (2) astrocytes produce scar tissue that can prevent axon growth across the damaged area, and (3) astrocytes release chemicals that block the regrowth of axons.

STEP 3:
Axon sends buds into network of Schwann cells and then starts growing along cord of Schwann cells.

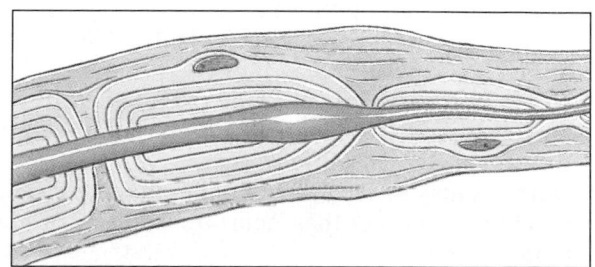

STEP 4:
Axon continues to grow into distal stump
and is enfolded by Schwann cells.

(3) the maintenance of the normal resting potential; and (4) the recovery from action potentials through the activity of the sodium–potassium exchange pump. Each time an action potential occurs, sodium ions enter and potassium ions leave the cell; over time, ATP must be expended to maintain normal cytoplasmic ion concentrations. When impulses are generated at high frequencies, the energy demands are enormous.

Neurons normally derive ATP solely through aerobic mechanisms. Because their cytoplasm does not contain glycogen reserves, these cells are totally dependent on a continuous and reliable supply of both oxygen and glucose from the blood. In cases of severe malnutrition, neural function deteriorates as the body's energy reserves are exhausted. Neural function is also impaired if the local circulation is restricted or, worse yet, shut off. If the circulation to a region is interrupted for just a few seconds, neurons in that region will be injured. The longer the interruption, the more severe the injury will be. In a **stroke,** the blood supply to the brain is interrupted by a circulatory blockage or vascular rupture. The degree of functional impairment after a stroke is determined by (1) the location and size of the region deprived of circulation and (2) the duration of the circulatory interruption. In subsequent chapters, we will consider the origins and treatment of strokes in greater detail.

Several inherited abnormalities in neural function are caused by metabolic problems within neurons. For example, *Tay-Sachs disease* is a genetic abnormality involving the metabolism of *gangliosides,* glycolipids that are important components of neuron cell membranes. AM *Tay-Sachs Disease*

Summary

Let's briefly review the basic principles of information processing introduced in earlier discussions:

- The neurotransmitters released at a synapse may have excitatory or inhibitory effects. The effect on the axon's initial segment reflects an integration of the stimuli that arrive at any given moment. The frequency of action potential generation is an indication of the degree of depolarization at the axon hillock.
- Neuromodulators can alter the rate of neurotransmitter release or the response of a postsynaptic neuron to specific neurotransmitters.
- Neurons may be facilitated or inhibited by extracellular chemicals other than neurotransmitters or neuromodulators.
- The effect of a presynaptic neuron's activation on a postsynaptic neuron at a synapse may be altered by other neurons.

HIGHER LEVELS OF ORGANIZATION AND PROCESSING

Functional Organization

The human body has about 10 million sensory neurons, 20 billion interneurons, and one-half million motor neurons. The interneurons are organized into a smaller number of neuronal pools. A **neuronal pool** is a group of interconnected neurons with specific functions. Neuronal pools are defined on the basis of function rather than on anatomical grounds. A pool may be diffuse, involving neurons in several different regions of the brain, or localized, with neurons restricted to one specific location in the brain or spinal cord. Estimates of the actual number of pools range between a few hundred and a few thousand. Each pool has a limited number of input sources and output destinations, and each may contain both excitatory and inhibitory neurons. The output of the entire pool may stimulate or depress the activity of other pools, or it may exert direct control over motor neurons or peripheral effectors.

The pattern of interaction among neurons provides clues to the functional characteristics of a neuronal pool. We can distinguish five patterns:

1. **Divergence** is the spread of information from one neuron to several neurons (Figure 12-27a•) or from one pool to multiple pools. Divergence permits the broad distribution of a specific input. Considerable divergence occurs when sensory neurons bring information into the CNS, for the information is distributed to neuronal pools throughout the spinal cord and brain.

2. In **convergence,** several neurons synapse on the same postsynaptic neuron (Figure 12-27b•). Several different patterns of activity in the presynaptic neurons can therefore have the same effect on the postsynaptic neuron. Through convergence, the same motor neurons can be subject to both conscious and subconscious control. For example, the movements of your diaphragm and ribs are now being subconsciously controlled by respiratory centers in your brain. But the same motor neurons can also be controlled consciously, as when you take a deep breath and hold it. Two different neuronal pools are involved, both synapsing on the same motor neurons.

3. Information may be relayed in a stepwise fashion, from one neuron to another or from one neuronal pool to the next. This pattern is called **serial processing** (Figure 12-27c•). Serial processing occurs as sensory information is relayed from one processing center to another in the brain.

4. **Parallel processing** occurs when several neurons or neuronal pools are processing the same information at one time (Figure 12-27d•). Divergence must take place before parallel processing can occur. Thanks to

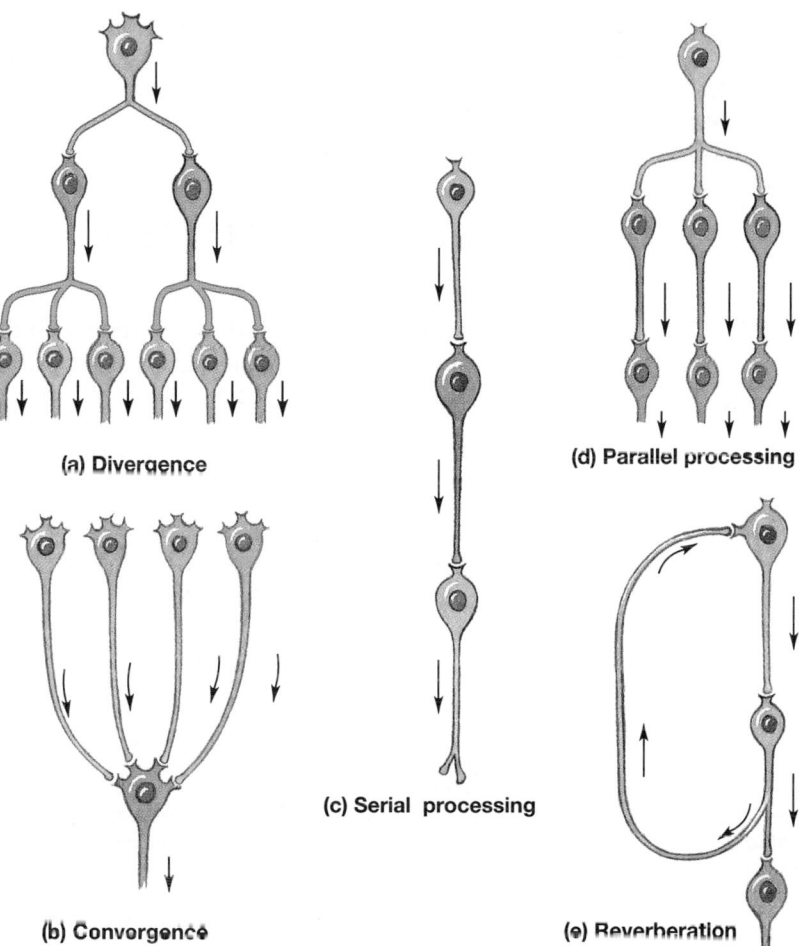

•FIGURE 12-27 Organization of Neuronal Pools. **(a)** Divergence, a mechanism for spreading stimulation to multiple neurons or neuronal pools in the CNS. **(b)** Convergence, a mechanism providing input to a single neuron from multiple sources. **(c)** Serial processing, in which neurons or pools work in a sequential manner. **(d)** Parallel processing, in which neurons or pools process information simultaneously. **(e)** Reverberation, a positive feedback mechanism.

(a) Divergence

(d) Parallel processing

(c) Serial processing

(b) Convergence

(e) Reverberation

parallel processing, many different responses can occur simultaneously. For example, when you step on a sharp object, sensory neurons that distribute the information to a number of neuronal pools are stimulated. As a result of parallel processing, you might withdraw your foot, shift your weight, move your arms, feel the pain, and shout "Ouch!" all at the same time.

5. Some neural circuits utilize positive feedback to produce **reverberation.** In this arrangement, collateral branches of axons somewhere along the sequence extend back toward the source of an impulse and further stimulate the presynaptic neurons. Once a reverberating circuit has been activated, it will continue to function until synaptic fatigue or inhibitory stimuli break the cycle. As with convergence or divergence, reverberation can occur within a single neuronal pool, or it may involve a series of interconnected pools. A simple example of reverberation is shown in Figure 12-27e•. Much more complicated examples of reverberation among neuronal pools in the brain may help maintain consciousness, muscular coordination, and normal breathing.

Anatomical Organization

The functions of the nervous system depend on the interactions among neurons in neuronal pools; the most

complex neural processing steps occur in the spinal cord and brain. Neurons and their axons are not randomly scattered in the CNS and PNS. Instead, they form masses or bundles with distinct anatomical boundaries, and they are identified by specific terms. We shall use these terms in all later chapters, so we offer a brief overview at this time:

In the PNS:

- Neuron cell bodies are found in *ganglia.*
- Axons are bundled together in *nerves,* with *spinal nerves* connected to the spinal cord and *cranial nerves* connected to the brain.

In the CNS:

- A collection of neuron cell bodies with a common function is called a **center.** A center with a discrete anatomical boundary is called a **nucleus.** Portions of the brain surface are covered by a thick layer of gray matter called the **neural cortex.** The term *higher centers* refers to the most complex integration centers, nuclei, and cortical areas of the brain.
- The white matter of the CNS contains bundles of axons that share common origins, destinations, and functions.

These bundles are called **tracts.** Tracts in the spinal cord form larger groups called **columns.**

- The centers and tracts that link the brain with the rest of the body are called **pathways.** For example, **sensory pathways** distribute information from peripheral receptors to processing centers in the brain, and **motor pathways** begin at CNS centers concerned with motor control and end at the effectors they control.

☑ A neuron from a cutaneous receptor synapses with a motor neuron located in the spinal cord. A motor neuron from the brain also synapses with the motor neuron in the spinal cord. What type of neuronal pool does this arrangement represent?

☑ In an injury involving a peripheral nerve, why is it important for the two ends of the damaged nerve to be closely aligned?

☑ How would damage to interneurons in the spinal cord affect nervous system function?

INTEGRATION WITH OTHER SYSTEMS

Figure 12-28● diagrams the relationships between the nervous system and other physiological systems. We shall explore many of these relationships in greater detail in subsequent chapters.

SELECTED CLINICAL TERMINOLOGY

Terms Discussed in This Chapter

anticholinesterase drug: A drug that blocks the breakdown of ACh by AChE. *(p. 402)*

atropine: A drug that prevents ACh from binding to the postsynaptic membrane. *(p. 402)*

demyelination: Destruction of the myelin sheaths around axons in the CNS and PNS. *(p. 380 and AM)*

d-tubocurarine: A drug, derived from curare, that prevents ACh from binding to the postsynaptic membrane. *(p. 402)*

endorphins (en-DOR-finz): Neuropeptides produced in the brain and spinal cord that appear to relieve pain and affect mood. *(p. 400)*

hyperkalemia (hī-per-ka-LĒ-mē-uh): An abnormal physiological state resulting from a high extracellular concentration of potassium. *(p. 406)*

neurotoxin: A compound that disrupts normal nervous system function by interfering with the generation or propagation of action potentials. Examples include *tetrodotoxin (TTX), saxitoxin (STX), paralytic shellfish poisoning (PSP),* and *ciguatoxin (CTX). (p. 389 and AM)*

nicotine: A compound, found in tobacco, that binds to specific ACh receptor sites and stimulates the postsynaptic membrane. *(p. 402)*

rabies: An acute disease caused by a virus that reaches the CNS by retrograde flow along peripheral axons. *(p. 373 and AM)*

Tay-Sachs disease: A genetic abnormality involving the metabolism of gangliosides, important components of neuron cell membranes. The result is a gradual deterioration of neurons due to the build-up of metabolic byproducts and release of lysosomal enzymes. *(p. 408 and AM)*

AM Additional Terms Discussed in the *Applications Manual*

diphtheria (dif-THĒ-rē-uh): A disease that results from a bacterial infection of the respiratory tract. Among other effects, the bacterial toxins damage Schwann cells and cause PNS demyelination.

Guillain–Barré syndrome: A progressive demyelination that appears to be linked with the immune response to a viral infection.

multiple sclerosis (skler-Ō-sis) **(MS):** A disease marked by recurrent incidents of demyelination affecting axons in the optic nerve, brain, and/or spinal cord.

CHAPTER REVIEW

 On-line resources for this chapter are on our World Wide Web site at: http://www.prenhall.com/martini/fap

STUDY OUTLINE

INTRODUCTION, p. 369

1. The nervous and endocrine systems coordinate the activities of other organ systems. The nervous system provides swift but brief responses to stimuli; the endocrine system adjusts metabolic operations and directs long-term changes.

AN OVERVIEW OF THE NERVOUS SYSTEM, p. 369

1. The nervous system includes all the *neural tissue* in the body. The basic functional unit is the **neuron.** The anatomical divisions of the nervous system are the **central nervous system (CNS)** (the brain and spinal cord) and the **peripheral nervous**

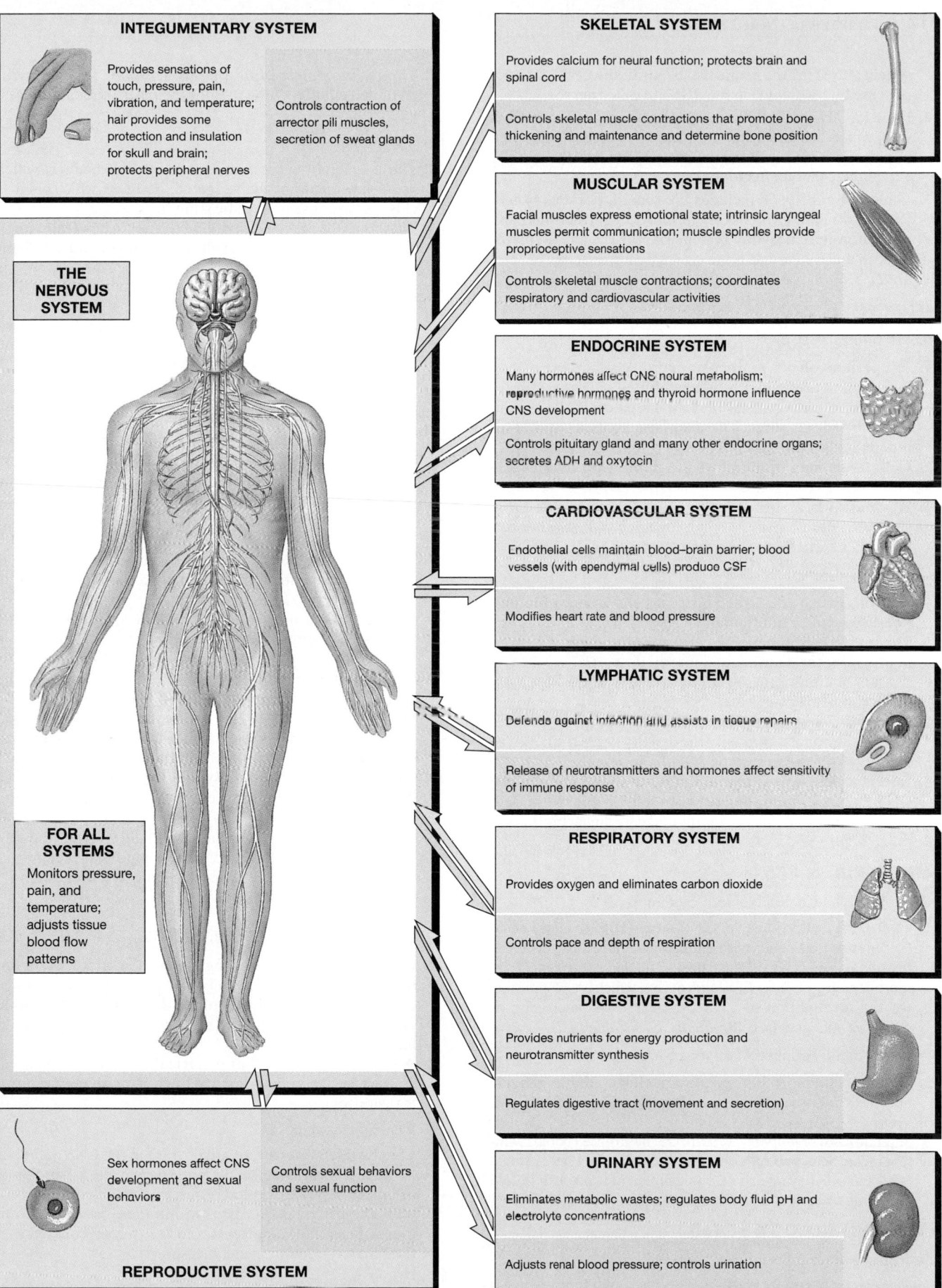

INTEGUMENTARY SYSTEM

Provides sensations of touch, pressure, pain, vibration, and temperature; hair provides some protection and insulation for skull and brain; protects peripheral nerves

Controls contraction of arrector pili muscles, secretion of sweat glands

SKELETAL SYSTEM

Provides calcium for neural function; protects brain and spinal cord

Controls skeletal muscle contractions that promote bone thickening and maintenance and determine bone position

MUSCULAR SYSTEM

Facial muscles express emotional state; intrinsic laryngeal muscles permit communication; muscle spindles provide proprioceptive sensations

Controls skeletal muscle contractions; coordinates respiratory and cardiovascular activities

ENDOCRINE SYSTEM

Many hormones affect CNS neural metabolism; reproductive hormones and thyroid hormone influence CNS development

Controls pituitary gland and many other endocrine organs; secretes ADH and oxytocin

CARDIOVASCULAR SYSTEM

Endothelial cells maintain blood–brain barrier; blood vessels (with ependymal cells) produce CSF

Modifies heart rate and blood pressure

LYMPHATIC SYSTEM

Defends against infection and assists in tissue repairs

Release of neurotransmitters and hormones affect sensitivity of immune response

RESPIRATORY SYSTEM

Provides oxygen and eliminates carbon dioxide

Controls pace and depth of respiration

DIGESTIVE SYSTEM

Provides nutrients for energy production and neurotransmitter synthesis

Regulates digestive tract (movement and secretion)

URINARY SYSTEM

Eliminates metabolic wastes; regulates body fluid pH and electrolyte concentrations

Adjusts renal blood pressure; controls urination

THE NERVOUS SYSTEM

FOR ALL SYSTEMS

Monitors pressure, pain, and temperature; adjusts tissue blood flow patterns

Sex hormones affect CNS development and sexual behaviors

Controls sexual behaviors and sexual function

REPRODUCTIVE SYSTEM

●*FIGURE 12-28* **Functional Relationships between the Nervous System and Other Systems**

system (PNS) (all the neural tissue outside the CNS). Bundles of **nerve fibers** (*axons*) in the PNS are called **nerves**.

2. Functionally, the PNS can be divided into an **afferent division,** which brings sensory information to the CNS, and an **efferent division,** which carries motor commands to muscles and glands, called **effectors**.

3. The efferent division includes the **somatic nervous system (SNS)** (which controls skeletal muscle contractions) and the **autonomic nervous system (ANS)** (which regulates smooth muscle, cardiac muscle, and glandular activity). *(Figure 12-1; Table 12-1)*

NEURONS, p. 370

Neuron Structure, p. 371

1. The **perikaryon** of a multipolar neuron contains organelles, including **neurofilaments, neurotubules,** and **neurofibrils.** The **axon hillock** connects the **initial segment** of the **axon** to the cell body, or **soma**. The **axoplasm** contains numerous organelles. *(Figure 12-2)*

2. **Collaterals** may branch from an axon, with **telodendria** branching from the axon's tip. *(Figure 12-2)*

3. A **synapse** is a site of intercellular communication. A synapse where a neuron communicates with another cell type is a **neuroeffector junction.** *(Figures 12-2, 12-3)*

Neuron Classification, p. 374

4. Neurons may be described histologically as **anaxonic, unipolar, bipolar,** or **multipolar.** *(Figure 12-4)*

5. There are three functional categories of neurons: sensory neurons, motor neurons, and interneurons. *(Figure 12-5)*

6. **Sensory neurons** form the afferent division of the PNS. They deliver information to the CNS. *(Figure 12-5)*

7. **Motor neurons** form the efferent division of the PNS. These neurons stimulate or modify the activity of a peripheral tissue, organ, or organ system. *(Figure 12-5)*

8. **Interneurons** (*association neurons*) are always found in the CNS. They may be located between sensory and motor neurons; they analyze sensory inputs and coordinate motor outputs. *(Figure 12-5)*

NEUROGLIA, p. 376

Neuroglia of the Central Nervous System, p. 376

1. There are four types of CNS **neuroglia**, or *glial cells*: (1) **ependymal cells,** with functions related to the **cerebrospinal fluid (CSF);** (2) **astrocytes,** the largest and most numerous neuroglia; (3) **oligodendrocytes,** which are responsible for the **myelination** of CNS axons; and (4) **microglia,** or migratory phagocytic cells. *(Figure 12-6; Table 12-2)*

Neuroglia of the Peripheral Nervous System, p. 378

2. Neuron cell bodies in the PNS are clustered into **ganglia**, and their axons form *peripheral nerves*. *(Figures 12-7, 12-8; Table 12-2)*

3. **Satellite cells,** or *amphicytes,* surround neuron cell bodies within ganglia. **Schwann cells** ensheath axons in the PNS. A single Schwann cell may myelinate one segment of an axon or enfold segments of several unmyelinated axons. *(Figures 12-7, 12-8)*

NEUROPHYSIOLOGY, p. 380

The Transmembrane Potential, p. 380

1. The **electrochemical gradient** is the sum of all the chemical and electrical forces acting across the membrane. *(Figures 12-9, 12-10)*

2. The *resting potential* of a neuron, about -70 mV, is determined chiefly by the membrane permeability to potassium ions.

Ion Channels, p. 384

3. The cell membrane contains **passive (leak) channels**, which are always open, and **active (gated) channels**, which open or close in response to specific stimuli.

4. There are three types of gated channels: **chemically regulated channels, voltage-regulated channels,** and **mechanically regulated channels**. *(Figures 12-11, 12-12)*

Graded Potentials, p. 386

5. **Depolarization** or **hyperpolarization** can lead to a **graded potential** (a change in potential that results when the degree of depolarization decreases with distance). *(Figures 12-13, 12-14; Table 12-4)*

Action Potentials, p. 388

6. An **action potential** appears when a region of excitable membrane depolarizes to **threshold.** The steps involved include membrane depolarization and activation of sodium channels; sodium channel inactivation; potassium channel activation; and the return to normal permeability. *(Figures 12-15, 12-16; Tables 12-3, 12-4)*

7. The activity of the sodium–potassium exchange pump is necessary to maintain ion concentrations within acceptable limits over time.

8. In **continuous propagation** (*continuous conduction*), an action potential spreads across the entire excitable membrane surface in a series of small steps. *(Figure 12-17)*

9. During **saltatory propagation** (*saltatory conduction*), the action potential appears to leap from node to node, skipping the intervening membrane surface. Saltatory propagation carries nerve impulses five to seven times more rapidly than does continuous propagation. *(Figure 12-18)*

10. Axons can be classified as **Type A fibers, Type B fibers,** or **Type C fibers** on the basis of diameter, myelination, and propagation speed.

11. Compared with action potentials in neural tissue, those in muscle tissue have (1) higher resting potentials, (2) longer-lasting action potentials, and (3) slower action potential propagation. *(Figure 12-19)*

SYNAPTIC TRANSMISSION, p. 396

1. An action potential traveling along an axon is a **nerve impulse.** At a synapse between two neurons, information passes from the **presynaptic neuron** to the **postsynaptic neuron.** A synapse may also involve other types of postsynaptic effector cells.

2. A synapse may be either electrical (with direct physical contact between cells) or chemical (involving a neurotransmitter).

Electrical Synapses, p. 397

3. **Electrical synapses** are relatively rare in the CNS and PNS. At an electrical synapse, the presynaptic and postsynaptic cell membranes are bound by interlocking membrane proteins at a gap junction. Pores within these proteins permit the passage of local currents, and the two neurons act as if they shared a common cell membrane.

Chemical Synapses, p. 397

4. **Chemical synapses** are more common than electrical synapses. **Excitatory** neurotransmitters cause depolarization

and promote action potential generation, whereas **inhibitory** neurotransmitters cause hyperpolarization and depress action potential generation.

5. The effect of a neurotransmitter on the postsynaptic membrane depends on the properties of the receptor, not on the nature of the neurotransmitter.

6. **Cholinergic synapses** release the neurotransmitter **acetylcholine (ACh).** Communication moves from the presynaptic neuron to the postsynaptic neuron. The separation of the presynaptic membrane and postsynaptic membrane by a synaptic cleft causes a **synaptic delay.** *(Figures 12-20–12-21; Table 12-5)*

7. The choline released during the breakdown of ACh in the synaptic cleft is reabsorbed and recycled by the synaptic knob. If stores of ACh are exhausted, **synaptic fatigue** can occur. *(Figure 12-21)*

8. **Adrenergic synapses** release **norepinephrine (NE),** also called **noradrenaline.** Other important neurotransmitters include **dopamine, serotonin,** and **gamma aminobutyric acid (GABA).** *(Table 12-6)*

9. **Neuromodulators** influence the postsynaptic cell's response to neurotransmitters.

CELLULAR INFORMATION PROCESSING, p. 402

1. Excitatory and inhibitory stimuli are integrated through interactions between **postsynaptic potentials.** This interaction is the simplest level of **information processing** in the nervous system.

Postsynaptic Potentials, p. 403

2. A depolarization caused by a neurotransmitter is an **excitatory postsynaptic potential (EPSP).** Individual EPSPs can combine through **summation.** Summation can be **temporal** (occuring at a single synapse when a second EPSP arrives before the effects of the first have disappeared) or **spatial** (resulting from the cumulative effects of multiple synapses at various locations). *(Figures 12-13, 12-23)*

3. Hyperpolarization of the postsynaptic membrane is an **inhibitory postsynaptic potential (IPSP).**

4. EPSP/IPSP interactions are the most important determinants of neural activity. *(Figure 12-24)*

The Rate of Action Potential Generation, p. 405

5. The greater the degree of sustained depolarization at the axon hillock, the higher the frequency of action potential generation.

Presynaptic Inhibition and Facilitation, p. 405

6. In **presynaptic inhibition,** GABA release at an axoaxonal synapse inhibits the opening of voltage-regulated calcium channels in the synaptic knob. This inhibition reduces the amount of neurotransmitter released when an action potential arrives at the synaptic knob. *(Figure 12-25)*

7. In **presynaptic facilitation,** activity at an axoaxonal synapse increases the amount of neurotransmitter released when an action potential arrives at the synaptic knob. This increase enhances and prolongs the effects of the neurotransmitter on the postsynaptic membrane. *(Figure 12-25)*

General Factors That Affect Neural Function, p. 406

8. Environmental factors, such as pH, ion, or temperature changes, can alter the resting membrane potential or disrupt the metabolic operations that support action potential generation.

9. Active neurons consume a great deal of ATP. Nutritional factors may therefore play a role in neural activity. Neurons are dependent on a reliable supply of glucose and also of oxygen.

FOCUS: Neural Responses to Injuries, p. 407

10. In the PNS, functional repair may follow **Wallerian degeneration.** In the CNS, many factors complicate the repair process and reduce the chances of functional recovery. *(Figure 12-26)*

Summary, p. 408

11. The neurotransmitters released at a synapse may have excitatory or inhibitory effects. The effect on the initial segment reflects an integration of the stimuli arriving at any given moment. The frequency of action potential generation is an indication of the degree of depolarization at the axon hillock.

12. Neuromodulators can alter the rate of neurotransmitter release or the response of a postsynaptic neuron to specific neurotransmitters. Neurons may be facilitated or inhibited by extracellular chemicals other than neurotransmitters or neuromodulators.

13. The effect of a presynaptic neuron's activation on a postsynaptic neuron at a synapse may be altered by other neurons.

HIGHER LEVELS OF ORGANIZATION AND PROCESSING, p. 408

Functional Organization, p. 408

1. The roughly 20 billion interneurons can be classified into **neuronal pools,** groups of interconnected neurons with specific functions.

2. **Divergence** is the spread of information from one neuron or neuronal pool to several neurons or pools. In **convergence,** several neurons synapse on the same postsynaptic neuron. Neurons or pools may also function in sequence **(serial processing),** or they may process the same information at one time **(parallel processing).** In **reverberation,** collateral axons establish a circuit that further stimulates presynaptic neurons. *(Figure 12-27)*

Anatomical Organization, p. 409

3. The functions of the nervous system as a whole depend on interactions among neurons in neuronal pools. In the PNS, **spinal nerves** communicate with the spinal cord, and **cranial nerves** are connected to the brain.

4. In the CNS, a collection of neuron cell bodies that share a particular function is a **center.** A center with a discrete anatomical boundary is called a **nucleus.** Portions of the brain surface are covered by a thick layer of gray matter called the **neural cortex.** The term *higher centers* refers to the most complex integration centers, nuclei, and cortical areas of the brain.

5. The white matter of the CNS contains bundles of axons called **tracts,** which share common origins, destinations, and functions. Tracts in the spinal cord form larger groups called **columns. Sensory pathways** carry information from peripheral receptors to the brain; **motor pathways** extend from CNS centers concerned with motor control to the associated effectors.

INTEGRATION WITH OTHER SYSTEMS, p. 410

1. The nervous system interacts with every other body system. *(Figure 12-28)*

REVIEW QUESTIONS

Level 1   Reviewing Facts and Terms

1. Regulation by the nervous system provides
(a) relatively slow but long-lasting responses to stimuli
(b) swift, long-lasting responses to stimuli
(c) swift but brief responses to stimuli
(d) relatively slow, short-lived responses to stimuli

2. The efferent division of the PNS
(a) brings sensory information to the CNS
(b) carries motor commands to muscles and glands
(c) processes and integrates sensory data
(d) is the seat of higher functions in the body

3. The part of the nervous system that provides voluntary control over skeletal muscle contractions is the
(a) somatic nervous system
(b) autonomic nervous system
(c) visceral motor system
(d) sympathetic division of the ANS

4. Smooth muscle, cardiac muscle, and glands are among the targets of the
(a) somatic nervous system
(b) sensory neurons
(c) afferent division of the PNS
(d) autonomic nervous system

5. The cellular layer that lines the ventricles of the brain and the central canal of the spinal cord is the
(a) microglia   (b) ganglia
(c) ependyma   (d) oligodendrocyte

6. Glial cells responsible for maintaining the blood–brain barrier are the
(a) microglia
(b) ependymal cells
(c) astrocytes
(d) oligodendrocytes

7. Phagocytic cells found in neural tissue of the CNS are
(a) astrocytes   (b) ependymal cells
(c) oligodendrocytes (d) microglia

8. Substances being transported from an axon terminal to the soma of the same neuron are delivered by
(a) axoplasmic transport
(b) synaptic conduction
(c) retrograde flow
(d) active transport

9. All the motor neurons that control skeletal muscles are
(a) multipolar neurons
(b) myelinated bipolar neurons
(c) unipolar, unmyelinated sensory neurons
(d) anaxonic neurons

10. The type of neural cells responsible for the analysis of sensory inputs and coordination of motor outputs are
(a) neuroglia   (b) interneurons
(c) sensory neurons (d) motor neurons

11. Depolarization of a neuron cell membrane will shift the membrane potential toward
(a) 0 mV   (b) –70 mV
(c) –90 mV   (d) a, b, and c are correct

12. The primary determinant of the resting membrane potential is the
(a) membrane permeability to sodium
(b) membrane permeability to potassium
(c) intracellular negatively charged proteins
(d) negatively charged chloride ions in the ECF

13. Receptors that bind acetylcholine at the postsynaptic membrane are
(a) chemically regulated channels
(b) voltage-regulated channels
(c) passive channels
(d) mechanically regulated channels

14. Gated channels that open or close in response to a change in the transmembrane potential are
(a) mechanically regulated channels
(b) voltage-regulated channels
(c) chemically regulated channels
(d) a, b, and c are correct

15. Changes in the transmembrane potential that are restricted to the area surrounding the site of stimulation are
(a) action potentials
(b) graded potentials
(c) inhibitory potentials
(d) hyperpolarizing potentials

16. Neuromodulators are compounds that influence the
(a) postsynaptic cell's response to a neurotransmitter
(b) synaptic vesicles in the synaptic knob
(c) release of calcium ions into the axoplasm
(d) a, b, and c are correct

17. A transient hyperpolarization of the postsynaptic membrane is
(a) a refractory period   (b) an EPSP
(c) an IPSP   (d) threshold

18. When simultaneous stimuli have a cumulative effect on the transmembrane potential, the process is
(a) reverberation
(b) spatial summation
(c) divergence
(d) temporal summation

19. The synapsing of several neurons on the same postsynaptic neuron is
(a) serial processing   (b) reverberation
(c) divergence   (d) convergence

20. Positive feedback in neural circuits produces
(a) reverberation
(b) divergence
(c) convergence
(d) temporal summation

21. (a) What are the major components of the central nervous system?
(b) What are the major components of the peripheral nervous system?

22. What two major cell populations are found in the nervous system? What is the primary function of each cell type?

23. (a) Which glial cells maintain the blood–brain barrier?
(b) Which glial cells serve a phagocytic function in the CNS?

24. Which two types of glial cells insulate neuron cell bodies and axons in the PNS from their surroundings?

25. Classifying neurons on the basis of *structure*, what four types are found in the nervous system?

26. What three *functional* groups of neurons are found in the nervous system? What is the function of each type of neuron?

27. What two integrated steps are necessary for transfer of nerve impulses from neuron to neuron?

28. What is the *functional* difference between voltage-regulated, chemically regulated, and mechanically regulated channels?

29. What four basic characteristics are associated with graded potentials?
30. State the all-or-none principle regarding action potentials.
31. Describe the steps involved in the generation of an action potential.

32. Describe the steps that take place at a typical cholinergic synapse.
33. What environmental factors play a role in altering the function of neurons in the CNS and PNS?

Level 2   Reviewing Concepts

34. If the resting membrane potential is –70 mV and the threshold is –55 mV, a membrane potential of –60 mV will
  (a) produce an action potential
  (b) make it easier to produce an action potential
  (c) make it harder to produce an action potential
  (d) hyperpolarize the membrane
35. A graded potential
  (a) decreases with distance from the point of stimulation
  (b) spreads passively because of local currents
  (c) may involve either depolarization or hyperpolarization
  (d) a, b, and c are correct
36. For an action potential to begin, an area of excitable membrane must
  (a) have its voltage-regulated gates inactivated
  (b) be hyperpolarized
  (c) be depolarized to threshold level
  (d) not be in a relative refractory period
37. During an absolute refractory period, the membrane
  (a) continues to hyperpolarize
  (b) cannot respond to further stimulation
  (c) can respond to a larger-than-normal depolarizing stimulus
  (d) will respond to summated stimulation
38. Action potentials are restricted to areas of excitable membranes that contain voltage-regulated channels:
  (a) true   **(b)** false
39. The all-or-none principle states that
  (a) the properties of an action potential are independent of the strength of the depolarizing stimulus
  (b) all stimuli will produce action potentials
  (c) all graded potentials will generate action potentials
  (d) any cell membrane can generate and propagate an action potential if stimulated to threshold

40. The loss of positive ions from the interior of a neuron produces
  (a) depolarization   **(b)** threshold
  (c) hyperpolarization   **(d)** an action potential
41. Continuous propagation of an action potential cannot occur in
  (a) myelinated axons
  (b) unmyelinated axons
  (c) Type A fibers
  (d) Type B fibers
42. Why can't most neurons in the CNS be replaced when they are lost to injury or disease?
43. What purpose do collaterals serve in the nervous system?
44. What is the difference between axoplasmic transport and retrograde flow?
45. What is the difference between continuous propagation and saltatory propagation?
46. How does an action potential in a skeletal muscle fiber differ from that in a neuron?
47. How does the action of a neurotransmitter differ from that of a neuromodulator?
48. What is the difference between temporal summation and spatial summation?
49. How does the neuronal activity of divergence differ from that of convergence?
50. Which functions of neurons necessitate the support of energy from ATP?
51. Why is an electrical synapse a more efficient carrier of nerve impulses from cell to cell than is a chemical synapse?
52. When a runner experiences the "runner's high," why is the suppression of pain common?
53. Multiple sclerosis (MS) is a demyelination disorder. How does this condition produce muscular paralysis and sensory losses?

Level 3   Critical Thinking and Clinical Applications

54. If neurons in the central nervous system lack centrioles and are unable to divide, how can a person develop brain cancer?
55. Harry suffers from a kidney condition that causes changes in his body's electrolyte levels (concentration of ions in the extracellular fluid). As a result of this problem, he is exhibiting tachycardia, an abnormally fast heart rate. What ion is involved, and how does a change in its concentration cause Harry's symptoms?
56. Twenty neurons synapse with a single receptor neuron. Fifteen of these neurons release neurotransmitters that produce EPSPs at the postsynaptic membrane, and the other five release neurotransmitters that produce IPSPs. Each time one of the neurons is stimulated, it releases enough neurotransmitter to

produce a 2-mV change in potential at the postsynaptic membrane. If the threshold of the postsynaptic neuron is 10 mV, how many of the excitatory neurons must be stimulated to produce an action potential in the receptor neuron if all five inhibitory neurons are stimulated? (Assume that spatial summation occurs.)
57. Myelination of peripheral neurons occurs rapidly throughout the first year of life. How can this process explain the increased abilities of infants during their first year?
58. A drug that blocks ATPase is introduced into an experimental neuron preparation. The neuron is then repeatedly stimulated, and recordings are made of the response. What effect would you expect to observe?

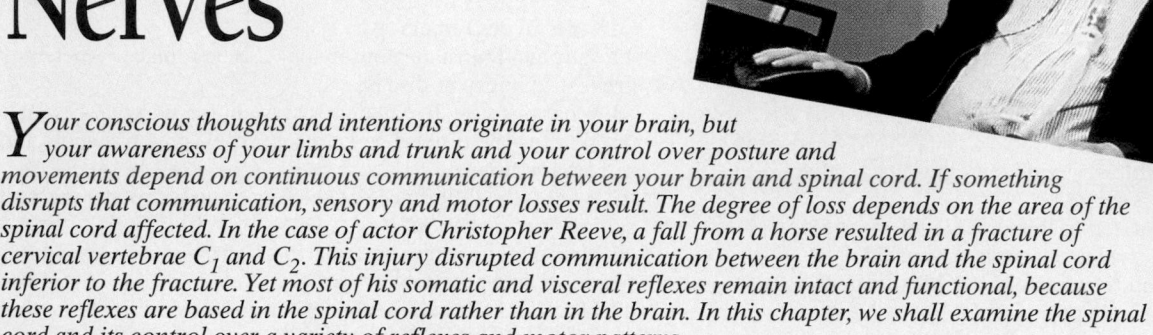

CHAPTER

13

The Spinal Cord and Spinal Nerves

Your conscious thoughts and intentions originate in your brain, but your awareness of your limbs and trunk and your control over posture and movements depend on continuous communication between your brain and spinal cord. If something disrupts that communication, sensory and motor losses result. The degree of loss depends on the area of the spinal cord affected. In the case of actor Christopher Reeve, a fall from a horse resulted in a fracture of cervical vertebrae C_1 and C_2. This injury disrupted communication between the brain and the spinal cord inferior to the fracture. Yet most of his somatic and visceral reflexes remain intact and functional, because these reflexes are based in the spinal cord rather than in the brain. In this chapter, we shall examine the spinal cord and its control over a variety of reflexes and motor patterns.

CHAPTER OUTLINE AND OBJECTIVES

As we saw in Chapter 12, the integration of thousands of EPSPs and IPSPs by a single neuron will determine (1) whether action potentials will appear and, if they do, (2) at what frequency they will be generated. ∞ *[p. 404]* When a neuron is active, that activity will affect all its synapses in the same way. For example, when a somatic motor neuron is active, an entire motor unit contracts; the neuron cannot stimulate just one of the muscle fibers and not the rest.

Within a single neuronal pool, there may be thousands of neurons communicating with one another across *millions* of active synapses. The integration process within a pool is significantly more complex and time-consuming than that for a single neuron, but the output is more flexible and varied. For example, a neuronal pool may control hundreds of somatic motor neurons. The output of the pool may stimulate some neurons, facilitate others, and inhibit a third group.

Over the next four chapters we will examine the functional organization of the nervous system at increasing levels of complexity. We will begin with the spinal cord and spinal nerves and consider the interactions between neuronal pools that direct relatively simple spinal reflexes. *Spinal reflexes* are stereotyped motor responses that are triggered by specific stimuli and controlled in the spinal cord. For example, a reflex controlled in the spinal cord, not in the brain, makes you drop a scalding-hot frying pan after you have accidentally grabbed it.

Your spinal cord is structurally and functionally integrated with your brain. Chapter 14 provides an overview of the major components and functions of the brain and cranial nerves. It also discusses the *cranial reflexes* comparable to those of the spinal cord. We then consider the nervous system as an integrated unit. Chapter 15 covers complex *higher-order functions,* such as memory and learning, that involve many different regions of the brain and affect all the activities of the nervous system. That chapter also deals with the interplay between centers in the brain and spinal cord that occurs in the processing of sensory information and in both the reflex control and the voluntary control of skeletal muscle activity. Chapter 16 completes the sequence with an examination of the autonomic nervous system. This system, which has processing centers in the brain, spinal cord, and PNS, is responsible for the control of visceral effectors, such as peripheral smooth muscles, cardiac muscle, and glands.

GROSS ANATOMY OF THE SPINAL CORD

The adult spinal cord (Figure 13-1a●) measures approximately 45 cm (18 in.) in length and has a maximum width of roughly 14 mm (0.55 in.). The posterior (dorsal) surface of the spinal cord bears a shallow longitudinal groove, the **posterior median sulcus.** The **anterior median fissure** is a deeper groove along the anterior (ventral) surface.

The amount of gray matter is greatest in segments of the spinal cord that deal with the sensory and motor control of the limbs. These areas are expanded, forming the **enlargements** of the spinal cord. The **cervical enlargement** supplies nerves to the shoulder girdles and upper limbs; the **lumbar enlargement** provides innervation to structures of the pelvis and lower limbs. Inferior to the lumbar enlargement, the spinal cord becomes tapered and conical; this region is the **conus medullaris.** The **filum terminale** ("terminal thread"), a slender strand of fibrous tissue, extends from the inferior tip of the conus medullaris. It continues along the length of the vertebral canal as far as the second sacral vertebra. There it provides longitudinal support to the spinal cord as a component of the *coccygeal ligament.*

Figure 13-1b● provides a series of sectional views that demonstrate the variations in the relative mass of gray and white matter in the cervical, thoracic, lumbar, and sacral regions of the spinal cord. The entire spinal cord can be divided into 31 segments on the basis of the origins of the spinal nerves. Each segment is identified by a letter and number designation, as used in the identification of individual vertebrae. For example, C_3, the segment in the uppermost section of Figure 13-1●, is the third cervical segment.

Every spinal segment is associated with a pair of **dorsal root ganglia** (Figure 13-1b●) that contains the cell bodies of sensory neurons. The **dorsal roots,** which contain the axons of these neurons, bring sensory information into the spinal cord. A pair of **ventral roots** contains the axons of motor neurons that extend into the periphery to control somatic and visceral effectors. On either side, the dorsal and ventral roots of each segment pass between the vertebral canal and the periphery at the *intervertebral foramen* between successive vertebrae. The dorsal root ganglion lies between the pedicles of the adjacent vertebrae. (You may wish to review the description of vertebral anatomy in Chapter 7, Figure 7-18●, p. 219.)

Distal to each dorsal root ganglion, the sensory and motor roots are bound together into a single **spinal nerve.** Spinal nerves are classified as **mixed nerves** because they contain both afferent (sensory) and efferent (motor) fibers. There are 31 pairs of spinal nerves, each identified by its association with adjacent vertebrae. For example, we may speak of "cervical spinal nerves" or even "cervical nerves" when we make a general reference to spinal nerves of the neck. When we indicate specific spinal nerves, it is customary to give them a regional number, as indicated in Figure 13-1●. Each spinal nerve caudal to the first thoracic vertebra takes its name from the vertebra immediately preceding it. Thus, spinal nerve T_1 emerges immediately caudal to

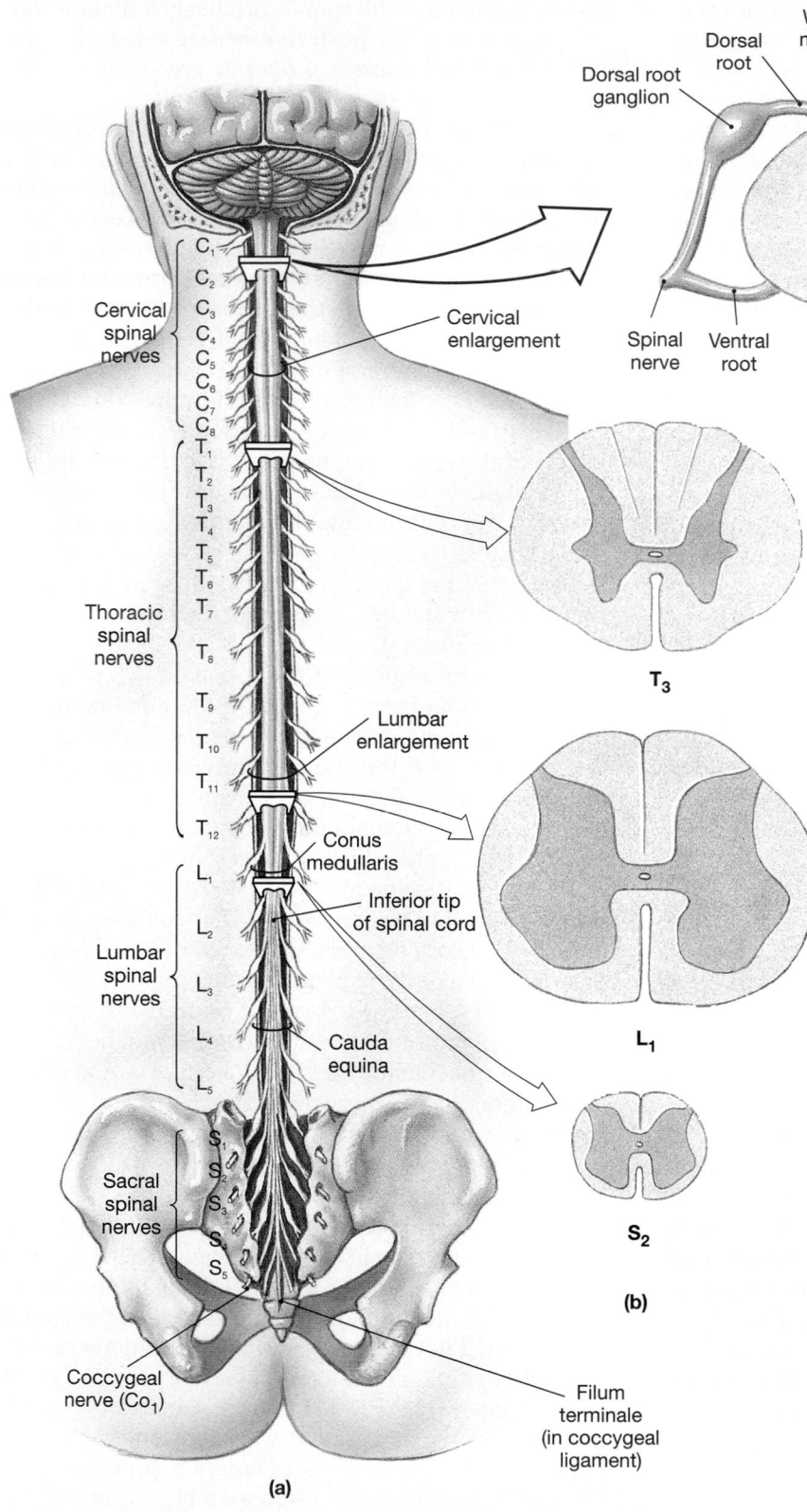

•FIGURE 13-1 Gross Anatomy of the Adult Spinal Cord. (a) Superficial anatomy and orientation of the adult spinal cord. The numbers to the left identify the spinal nerves and indicate where the nerve roots leave the vertebral canal. Note, however, that the adult spinal cord extends from the brain only to the level of vertebrae L_1–L_2. **(b)** Inferior views of cross sections through representative segments of the spinal cord, showing the arrangement of gray matter and white matter.

vertebra T_1, spinal nerve T_2 follows vertebra T_2, and so forth.

The arrangement differs in the cervical region because the first pair of spinal nerves, C_1, passes between the skull and the first cervical vertebra. For this reason, each cervical nerve takes its name from the vertebra immediately *following* it. In other words, cervical nerve C_2 *precedes* vertebra C_2, and the same system is used for the rest of the cervical series. The transition from one numbering system to another occurs between the last cervical and first thoracic vertebrae. Because the spinal nerve lying between them has been designated C_8, there are only seven cervical vertebrae but *eight* cervical nerves.

The spinal cord continues to enlarge and elongate until an individual is approximately 4 years old. Up to that time, enlargement of the spinal cord keeps pace with the growth of the vertebral column. Throughout this period, the ventral and dorsal roots are very short, and they enter the intervertebral foramina immediately adjacent to their spinal segment. After age 4, the vertebral column continues to elongate, but the spinal cord does not. This vertebral growth carries the dorsal roots and spinal nerves farther and farther from their original positions relative to the spinal cord. As a result, the dorsal and ventral roots gradually elongate, and the correspondence between spinal segment and the vertebral segment is lost. For example, in the adult the sacral segments of the spinal cord are at the level of vertebrae L_1–L_2 (Figure 13-1b•).

Because the adult spinal cord extends only to the level of the first or second lumbar vertebra, the dorsal and ventral roots of spinal segments L$_2$ to S$_5$ extend caudally, past the inferior tip of the conus medullaris. When seen in gross dissection, the filum terminale and the long ventral and dorsal roots resemble a horse's tail. As a result, early anatomists called the complex the **cauda equina** (KAW-duh ek-WĪ-nuh; *cauda*, tail + *equus*, horse).

Spinal Meninges

The vertebral column and its surrounding ligaments, tendons, and muscles isolate the spinal cord from the external environment. The delicate neural tissues must also be defended against damaging contacts with the surrounding bony walls of the vertebral canal. A series of specialized membranes, the **spinal meninges** (men-IN-jēz; *meninx*, membrane), provide the necessary physical stability and shock absorption. Blood vessels branching within these layers also deliver oxygen and nutrients to the spinal cord.

The relationships among the spinal meninges are shown in Figure 13-2a ●. There are three meningeal layers: (1) the *dura mater,* (2) the *arachnoid,* and (3) the *pia mater.* At the foramen magnum of the skull, the spinal meninges are continuous with the **cranial meninges** that surround the brain. (We shall discuss the cranial meninges, which have the same three layers, in Chapter 14.)

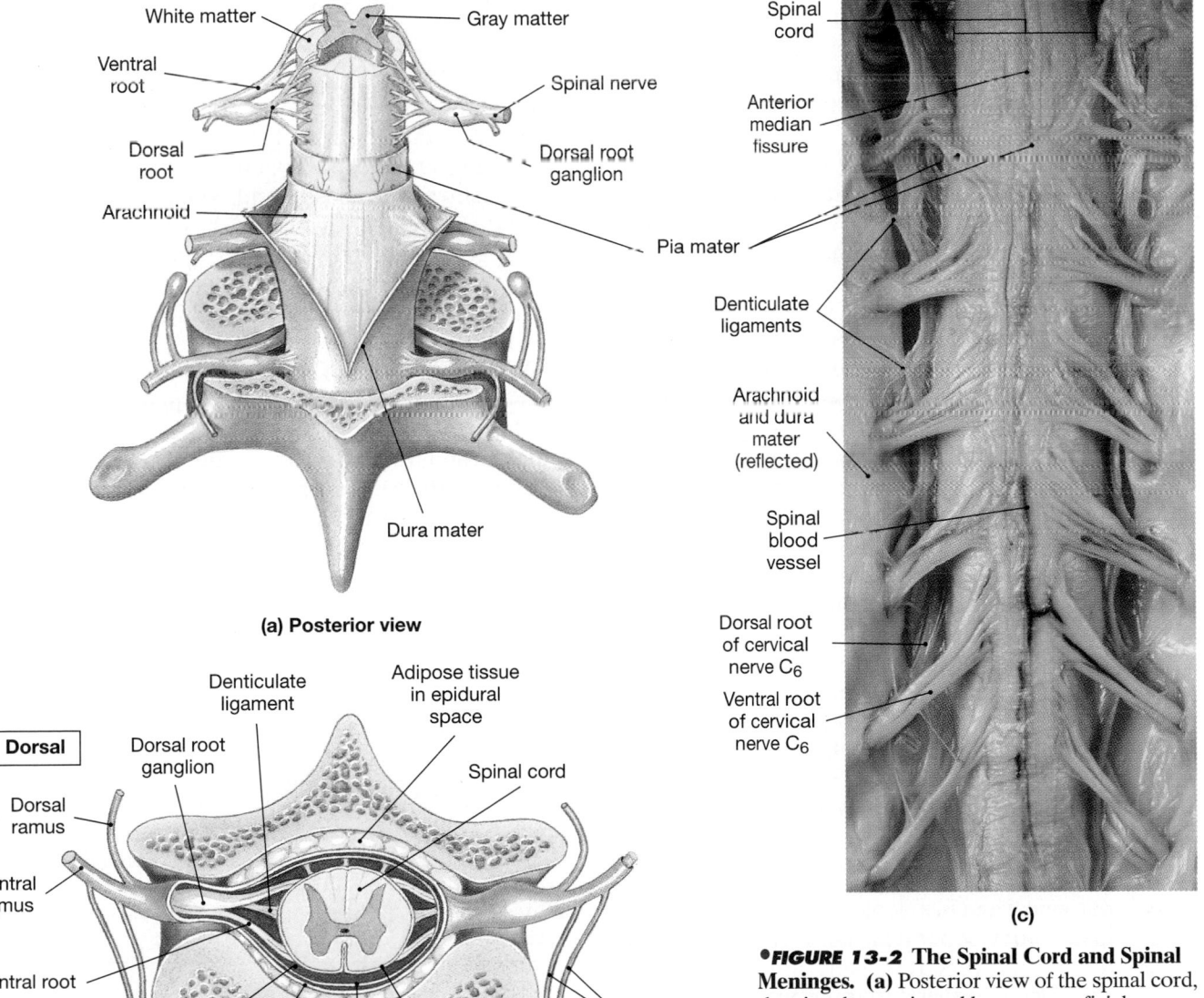

(a) Posterior view

(b)

(c)

●**FIGURE 13-2** **The Spinal Cord and Spinal Meninges.** **(a)** Posterior view of the spinal cord, showing the meningeal layers, superficial landmarks, and the distribution of gray and white matter. **(b)** Sectional view through the spinal cord and meninges, showing the peripheral distribution of the spinal nerves. **(c)** Anterior view of the spinal cord and spinal nerve roots within the vertebral canal. The dura mater and arachnoid membrane have been cut and reflected; note the blood vessels in the delicate pia mater.

Bacterial or viral infection can cause **meningitis**, or inflammation of the meningeal membranes. Meningitis is dangerous because it can disrupt the normal circulatory and cerebrospinal fluid supplies, damaging or killing neurons and glial cells in the affected areas. Although an initial diagnosis may specify the meninges of the spinal cord (*spinal meningitis*) or brain (*cerebral meningitis*), in later stages the entire meningeal system is usually affected. [AM] *Meningitis*

The Dura Mater

The tough, fibrous **dura mater** (DOO-ruh MĀ-ter; *dura,* hard + *mater,* mother) forms the outermost covering of the spinal cord (Figure 13-2a●). The dense collagen fibers of the dura mater are oriented along the longitudinal axis of the cord. Between the dura mater and the walls of the vertebral canal lies the **epidural space,** which contains loose connective tissue, blood vessels, and a protective padding of adipose tissue (Figure 13-2b●).

The dura mater does not have extensive, firm connections to the surrounding vertebrae. Longitudinal stability is provided by localized attachment sites at either end of the vertebral canal. Cranially, the outer layer of the dura mater fuses with the periosteum of the occipital bone around the margins of the foramen magnum. Within the sacral canal, the dura mater tapers from a sheath to a dense cord of collagen fibers that blend with components of the filum terminale to form the **coccygeal ligament.**

The coccygeal ligament continues along the sacral canal, ultimately blending into the periosteum of the coccyx. Lateral support for the dura mater is provided by loose connective tissue and adipose tissue within the epidural space. In addition, the dura mater extends between adjacent vertebrae at each intervertebral foramen, fusing with the connective tissues that surround the spinal nerves.

Injecting an anesthetic into the epidural space will affect only the spinal nerves in the immediate area of the injection. The result is a temporary sensory and motor paralysis known as an **epidural block.** Epidural blocks in the lower lumbar or sacral regions may be used to control pain during childbirth. [AM] *Spinal Anesthesia*

The Arachnoid

In most anatomical and histological preparations, a narrow **subdural space** separates the dura mater from deeper meningeal layers. It is likely, however, that in life no such space exists and that the inner surface of the dura mater is in contact with the outer surface of the **arachnoid** (a-RAK-noyd; *arachne,* spider) (Figure 13-2b●). The inner surface of the dura mater and the outer surface of the arachnoid are covered by simple squamous epithelia. The arachnoid includes the epithelium and the **subarachnoid space,** which contains the *arachnoid trabeculae,* a delicate network of colla-

gen and elastic fibers maintained by modified fibroblasts. The subarachnoid space is filled with **cerebrospinal fluid**, which acts as a shock absorber as well as a diffusion medium for dissolved gases, nutrients, chemical messengers, and waste products.

The arachnoid membrane extends caudally as far as the filum terminale, and the dorsal and ventral roots of the cauda equina travel within the fluid-filled subarachnoid space. In adults, the withdrawal of cerebrospinal fluid, a procedure known as a **spinal tap**, involves the insertion of a needle into the subarachnoid space in the lower lumbar region (Figure 13-3a●). This placement avoids the possibility of damage to the spinal cord. Spinal taps are performed when CNS infection is suspected or to diagnose severe back pain, headaches, disc problems, and some types of strokes.

The Pia Mater

The subarachnoid space bridges the gap between the arachnoid epithelium and the innermost meningeal layer, the **pia mater** (*pia,* delicate + *mater,* mother). The meshwork of elastic and collagen fibers of the pia mater are interwoven with those of the subarachnoid space. The blood vessels servicing the spinal cord are found here. Unlike more superficial meninges, the pia mater is firmly bound to the underlying neural tissue (Figure 13-2b●).

Along the length of the spinal cord, paired **denticulate ligaments** extend from the pia mater through the arachnoid to the dura mater. Denticulate ligaments, which originate along either side of the spinal cord (Figure 13-2b,c●), prevent lateral (side-to-side) movement. The dural connections at the foramen magnum and the coccygeal ligament prevent longitudinal (superior/inferior) movement.

The spinal meninges accompany the dorsal and ventral roots as they pass through the intervertebral foramina. As indicated in the sectional view of Figure 13-2b●, the meningeal membranes are continuous with the connective tissues that surround the spinal nerves and their peripheral branches.

☑ Damage to which root of a spinal nerve would interfere with motor function?

☑ Where is the cerebrospinal fluid that surrounds the spinal cord located?

Sectional Anatomy of the Spinal Cord

To understand the functional organization of the spinal cord, you must become familiar with its sectional organization. The *anterior median fissure* and the *posterior median sulcus* mark the division between left and right sides of the spinal cord. The su-

Spinal Taps and Myelography

Tissue samples, or **biopsies**, are taken from many organs to assist in diagnosis. For example, when a liver or skin disorder is suspected, small plugs of tissue are removed and examined for signs of infec-

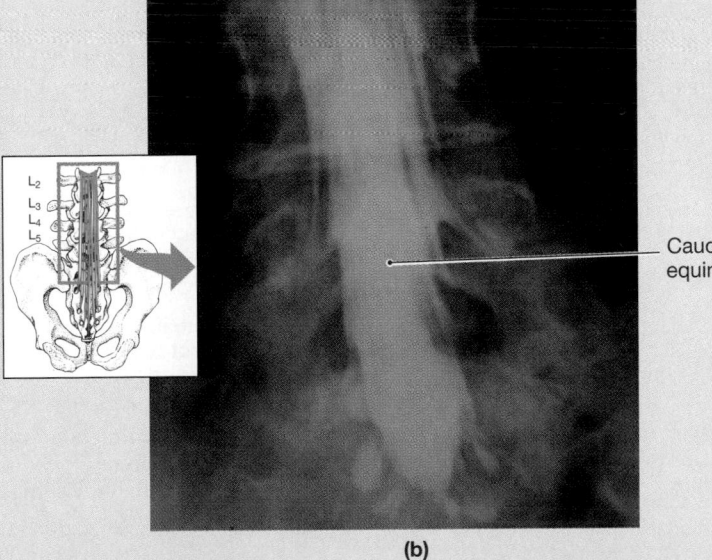

Dura mater

Epidural space

Body of lumbar vertebra L₃

Interspinous ligament

Lumbar puncture needle

Cauda equina in subarachnoid space

Filum terminale

(a)

L₂
L₃
L₄
L₅

Cauda equina

(b)

•**FIGURE 13-3** **Spinal Taps and Myelography.** (a) The lumbar puncture needle is in the subarachnoid space, between the nerves of the cauda equina. The needle has been inserted in the midline between the third and fourth lumbar vertebral spines, pointing at a superior angle toward the umbilicus. Once the needle punctures the dura mater and enters the subarachnoid space, a sample of CSF may be obtained. (b) A myelogram—an X-ray image of the spinal cord after a radiopaque dye has been introduced into the CSF—showing the cauda equina in the lower lumbar region.

tion or cell damage or are used to identify the bacteria that cause an infection. Unlike many other tissues, however, neural tissue consists largely of cells rather than extracellular fluids or fibers. Neural tissue samples are seldom removed for analysis, because any extracted or damaged neurons will not be replaced. Instead, small volumes of cerebrospinal fluid are collected and analyzed. Cerebrospinal fluid is intimately associated with the neural tissue of the CNS, and pathogens, cell debris, or metabolic wastes in the CNS will therefore be detectable in the CSF.

The withdrawal of cerebrospinal fluid, a procedure called a *spinal tap*, must be done with care to avoid injuring the spinal cord. The adult spinal cord extends inferiorly only as far as vertebra L_1 or L_2. Between vertebra L_2 and the sacrum, the meningeal layers remain intact, but they enclose only the relatively sturdy components of the cauda equina and a significant quantity of CSF. When the vertebral column is flexed, a needle can be inserted between the lower lumbar vertebrae and into the subarachnoid space with minimal risk to the cauda equina. In this procedure, known as a **lumbar puncture**, 3–9 ml of fluid is taken from the subarachnoid space between vertebrae L_3 and L_4 (Figure 13-3a•). Spinal taps are performed when CNS infection is suspected or when severe back pain, headaches, disc problems, and some types of strokes are diagnosed.

Myelography is the introduction of radiopaque dyes into the CSF of the subarachnoid space. Because the dyes are opaque to X-rays, the CSF appears white on an X-ray image (Figure 13-3b•). Any tumors, inflammation, or adhesions that distort or divert CSF circulation will be shown in silhouette. In the event of severe infection, inflammation, or leukemia (cancer of the white blood cells), antibiotics, steroids, or anti-cancer drugs can be injected into the subarachnoid space.

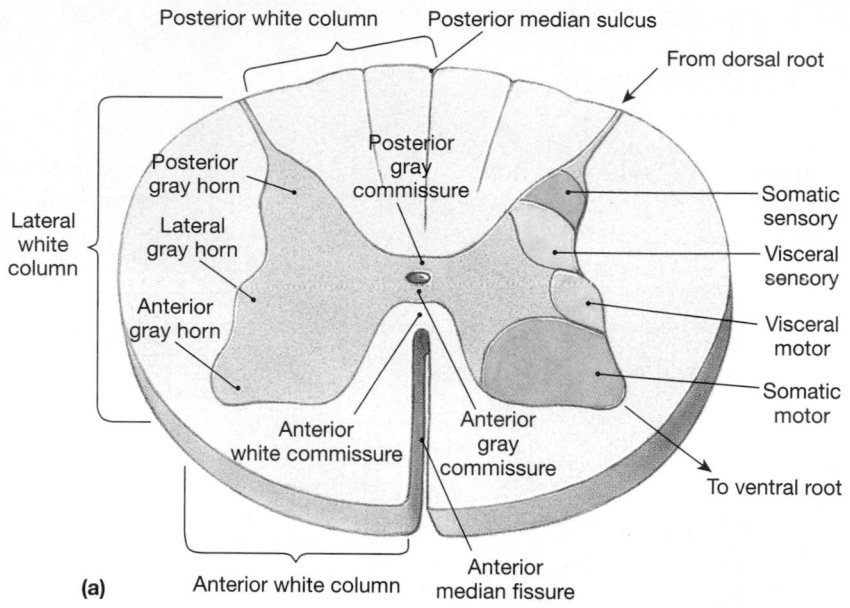

(a)

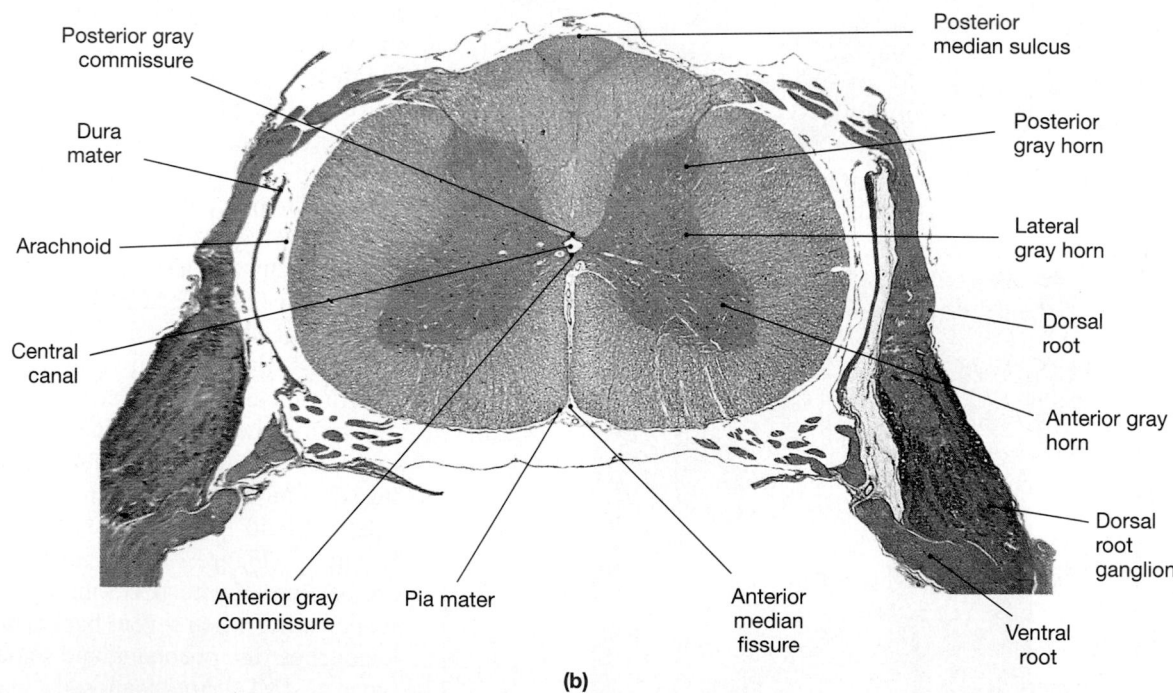

(b)

•FIGURE 13-4 Sectional Organization of the Spinal Cord. (a) The left half of this sectional view shows important anatomical landmarks and the major regions of white matter in the posterior white column. The right half indicates the functional organization of the gray matter in the anterior, lateral, and posterior gray horns. **(b)** Micrograph of a section through the spinal cord, showing major landmarks; compare with (a).

perficial *white matter* contains large numbers of myelinated and unmyelinated axons. The *gray matter,* dominated by the cell bodies of neurons, glial cells, and unmyelinated axons, surrounds the narrow **central canal** and forms an H or butterfly shape. The projections of gray matter toward the outer surface of the spinal cord are called **horns**. Figure 13-4• presents a typical section through the spinal cord.

Organization of Gray Matter

The cell bodies of neurons in the gray matter of the spinal cord are organized into functional groups called *nuclei.* **Sensory nuclei** receive and relay sensory information from peripheral receptors. **Motor nuclei** issue motor commands to peripheral effectors. Although sensory and motor nuclei appear rather small in transverse section, they may extend for a considerable distance along the length of the spinal cord.

A frontal section along the length of the central canal of the spinal cord will separate the sensory (posterior, or dorsal) nuclei from the motor (anterior, or ventral) nuclei. The **posterior gray horns** contain somatic and visceral sensory nuclei, whereas the **anterior gray horns** contain somatic motor nuclei. The **lateral gray horns,** located only in the thoracic and lumbar segments, contain visceral motor nuclei. The **gray commissures** (*commissura,* a joining together) posterior to and anterior to the central canal contain axons that cross from one side of the cord to the other before they reach a destination within the gray matter.

Figure 13-4a• shows the relationship between the function of a particular nucleus (sensory or motor) and its relative position within the gray matter of the spinal cord. The nuclei within each gray horn are also organized. For example, the anterior gray horns of the cervical enlargement contain nuclei whose motor neurons control the muscles of the upper limbs. On each side of the spinal cord, in medial to lateral sequence, are motor nuclei that control (1) muscles that position the shoulder girdle, (2) muscles that position the arm, (3) muscles that move the forearm and hand, and (4) muscles that move the hand and fingers. Within each of these regions, the motor neurons that control flexor muscles are grouped separately from those that control extensor muscles. Because the spinal cord is so highly organized, we can predict the muscles that will be affected by damage to a specific area of gray matter.

Organization of White Matter

The white matter on each side of the spinal cord can be divided into three regions called **columns,** or *funiculi* (Figure 13-4a•). The **posterior white columns** lie between the posterior gray horns and the posterior median sulcus. The **anterior white columns** lie between the anterior gray horns and the anterior median fissure. The anterior white columns are interconnected by the **anterior white commissure.** The white matter between the anterior and posterior columns on each side makes up the **lateral white column.**

Each column contains *tracts* whose axons share functional and structural characteristics. A **tract,** or *fasciculus (fa-SIK-ū-lus;* bundle), is a bundle of axons in the CNS that are relatively uniform with respect to diameter, myelination, and conduction speed. All the axons within a tract relay the same type of information (sensory or motor) in the same direction. Short tracts carry sensory or motor signals between segments of the spinal cord, and longer tracts connect the spinal cord with the brain. **Ascending tracts** carry sensory information toward the brain, and **descending tracts** convey motor commands into the spinal cord.

DAMAGE TO SPINAL TRACTS All the ascending sensory information and descending somatic motor commands associated with the trunk and limbs involve nuclei and tracts in the spinal cord. *Multiple sclerosis* (MS), a disorder that we mentioned in Chapter 12, produces muscular paralysis and sensory losses through demyelination. ∞ *[p. 380]* The initial symptoms appear as the result of myelin degeneration within the white matter of the lateral and posterior columns of the spinal cord or along tracts within the brain. During subsequent attacks, the effects may become more widespread, and the cumulative sensory and motor losses can eventually lead to a generalized muscular paralysis. **AM** *Multiple Sclerosis*

The viral disease called *polio* causes paralysis due to the destruction of somatic motor neurons. This disorder, introduced in Chapter 10, has been almost eliminated within the United States through an aggressive immunization program. Immunization continues because polio remains relatively common in other areas of the world. The disease could be brought into the United States at any time, leading to an epidemic among unimmunized children.

Paralysis can also result from physical damage to the spinal cord due to a severe auto crash or other accident; the damaged tracts seldom undergo even partial repairs. Extensive damage to the spinal cord at or above the fifth cervical vertebra will eliminate sensation and motor control of the upper and lower limbs. The extensive paralysis produced is called *quadriplegia. Paraplegia,* the loss of motor control of the lower limbs, may follow damage to the thoracic spinal cord. **AM** *Spinal Cord Injuries and Experimental Treatments*

Less severe injuries affecting the spinal cord or cauda equina produce symptoms of sensory loss or motor paralysis that reflect the specific nuclei, tracts, or spinal nerves involved. We will detail one example, the loss of peripheral sensation along the distribution of a spinal nerve, in a later section.

✔ A person with polio has lost the use of his leg muscles. In what area of the spinal cord would you expect the virally infected motor neurons to be in this individual?

✔ What portion of the spinal cord would be affected by a disease that damages myelin sheaths?

SPINAL NERVES

A series of connective tissue layers surrounds each spinal nerve and continues along all its peripheral branches. These layers, best seen in sectional view (Figure 13-5•), are comparable to those associated with skeletal muscles (Figure 10-1•, p. 278). The outermost layer, or **epineurium,** consists of a dense network of collagen fibers. The fibers of the **perineurium** extend inward from the epineurium, dividing the nerve into a series of compartments that contain bundles of axons, or *fascicles.* Delicate connective tissue fibers of the **endoneurium** extend from the perineurium and surround individual axons.

Arteries and veins penetrate the epineurium and branch within the perineurium. Capillaries leaving the perineurium branch in the endoneurium and supply the axons and Schwann cells of the nerve and the fibroblasts of the connective tissues.

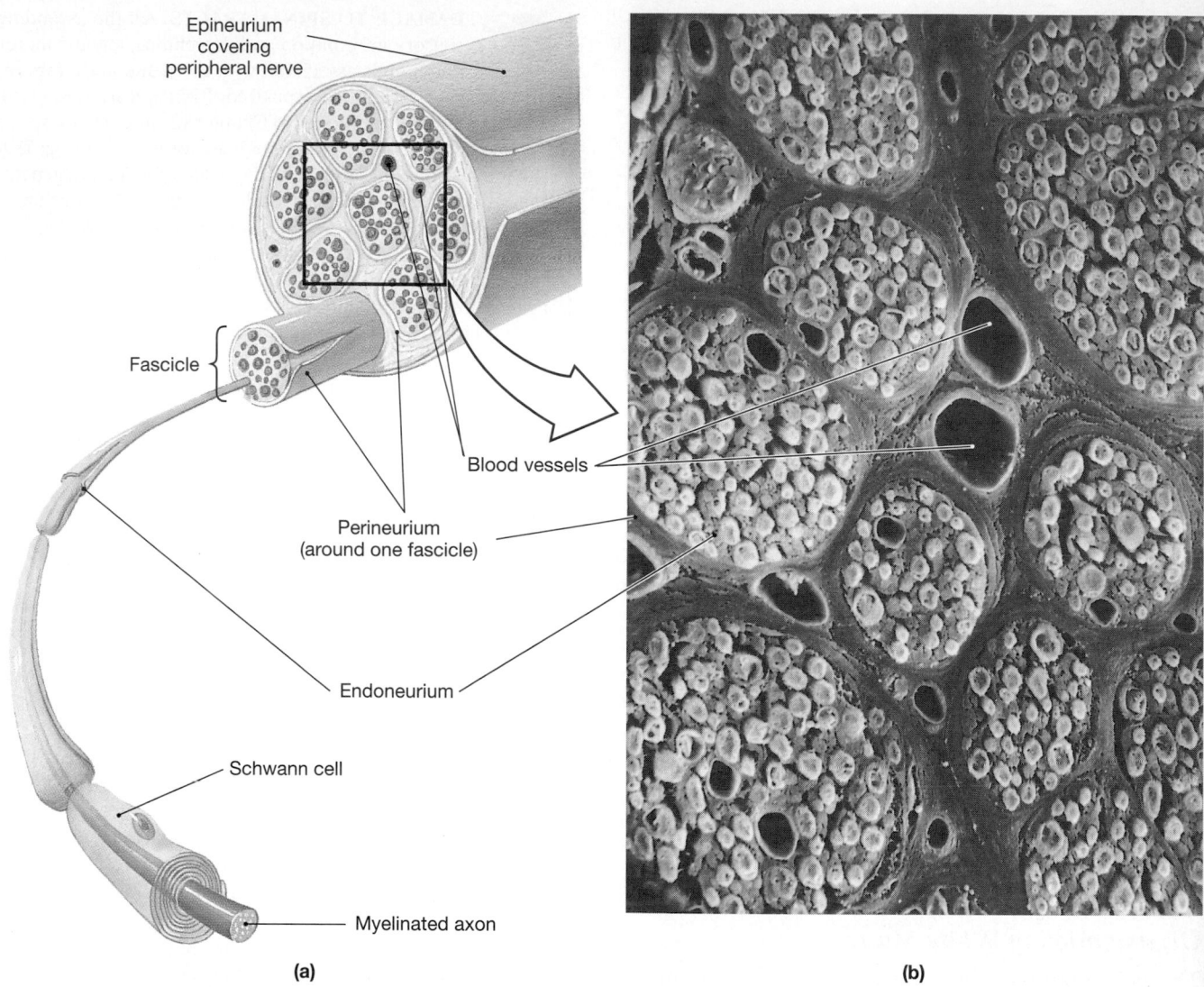

Epineurium
covering
peripheral nerve

Fascicle

Perineurium
(around one fascicle)

Blood vessels

Endoneurium

Schwann cell

Myelinated axon

(a)

(b)

●*FIGURE 13-5* **A Peripheral Nerve. (a)** A typical peripheral nerve and its connective tissue wrappings. **(b)** A micrograph showing the perineurium and endoneurium in great detail. (SEM × 340) Reproduced from R. G. Kessel and R. H. Kardon, "Tissues and Organs: A Text-Atlas of Scanning Electron Microscopy," W. H. Freeman & Co., 1979.

Peripheral Distribution of Spinal Nerves

Figure 13-6● shows the distribution, or pathway, of a typical spinal nerve that originates from the thoracic or upper lumbar segments of the spinal cord. The spinal nerve forms just lateral to the intervertebral foramen, where the dorsal and ventral roots unite. Let us follow the distribution of that nerve in the periphery.

In the thoracic and upper lumbar regions (segments T_1–L_2), the first branch from the spinal nerve carries visceral motor fibers to a nearby **autonomic ganglion** of the *sympathetic division* of the ANS. (The sympathetic division is, along with other functions, responsible for elevating the metabolic rate and for increasing alertness.) Because preganglionic axons are myelinated, this branch has a light color and is known as the **white ramus** ("branch"). Postganglionic fibers innervating glands and smooth muscles in the body

wall or limbs rejoin the spinal nerve. These fibers are unmyelinated and have a darker color; they form the **gray ramus**. The gray ramus is typically proximal to the white ramus. The gray and white rami are collectively termed the **rami communicantes,** or "communicating branches." Postganglionic fibers innervating visceral organs in the thoracic cavity form a series of separate **sympathetic nerves.** In the abdominal region, most of the sympathetic neurons innervating visceral organs in the abdominopelvic cavity are located in ganglia anterior to the spinal column rather than close to the bases of the spinal nerves. The preganglionic fibers traveling to these ganglia form the sympathetic nerves known as *splanchnic nerves.*

The **dorsal ramus** of each spinal nerve provides sensory and motor innervation to the skin and muscles of the back. The relatively large **ventral ramus** supplies the ventrolateral body surface, structures in

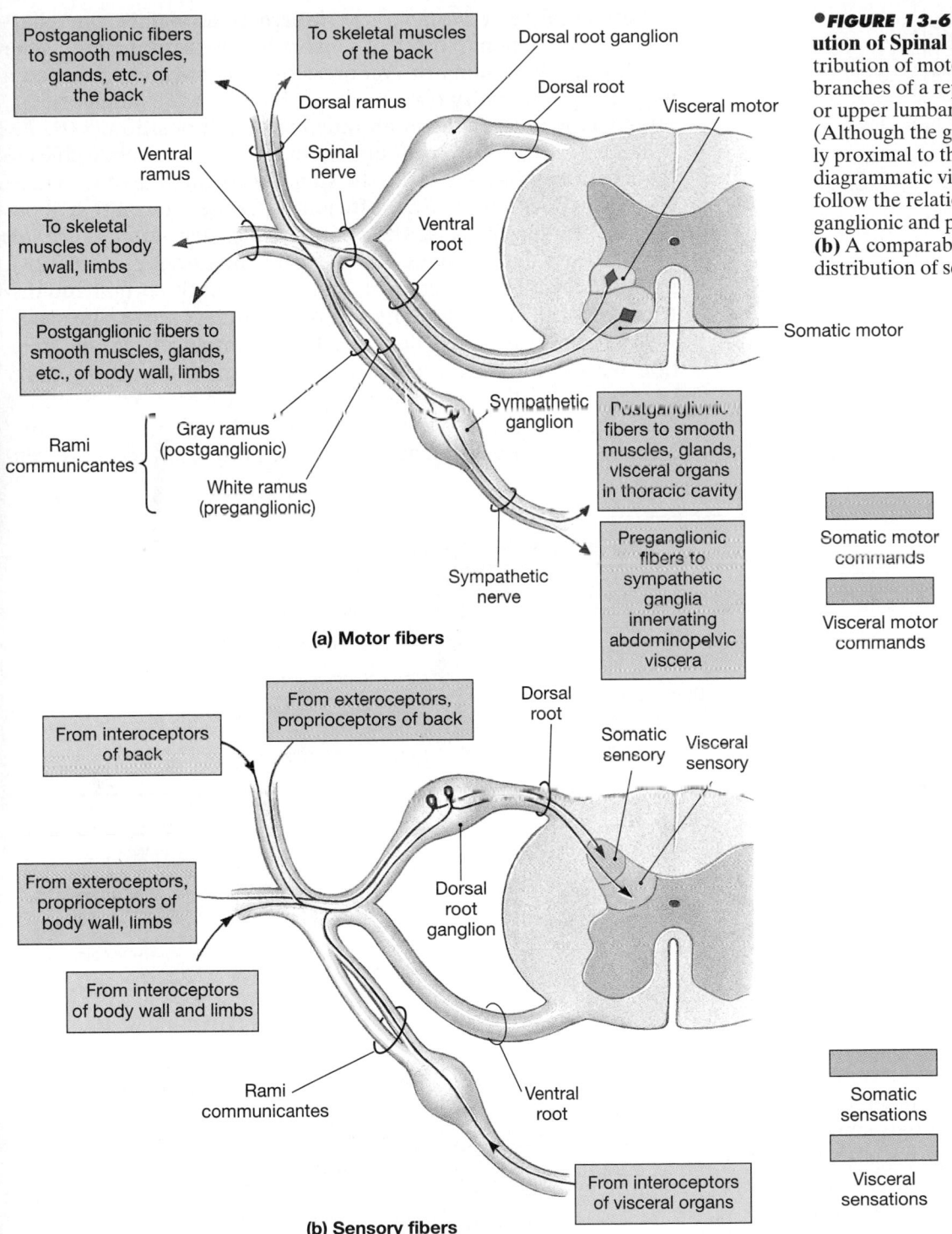

Postganglionic fibers to smooth muscles, glands, etc., of the back

To skeletal muscles of the back

Dorsal root ganglion

Dorsal root

Visceral motor

Dorsal ramus

Ventral ramus

Spinal nerve

Ventral root

To skeletal muscles of body wall, limbs

Postganglionic fibers to smooth muscles, glands, etc., of body wall, limbs

Somatic motor

Rami communicantes {

Gray ramus (postganglionic)

White ramus (preganglionic)

Sympathetic ganglion

Postganglionic fibers to smooth muscles, glands, visceral organs in thoracic cavity

Preganglionic fibers to sympathetic ganglia innervating abdominopelvic viscera

Sympathetic nerve

(a) Motor fibers

Somatic motor commands

Visceral motor commands

From interoceptors of back

From exteroceptors, proprioceptors of back

Dorsal root

Somatic sensory

Visceral sensory

From exteroceptors, proprioceptors of body wall, limbs

Dorsal root ganglion

From interoceptors of body wall and limbs

Rami communicantes

Ventral root

From interoceptors of visceral organs

(b) Sensory fibers

Somatic sensations

Visceral sensations

•*FIGURE 13-6* **Peripheral Distribution of Spinal Nerves. (a)** The distribution of motor fibers in the major branches of a representative thoracic or upper lumbar spinal nerve. (Although the gray ramus is normally proximal to the white ramus, this diagrammatic view makes it easier to follow the relationships between preganglionic and postganglionic fibers.) **(b)** A comparable view detailing the distribution of sensory fibers.

the body wall, and the limbs. Each pair of spinal nerves monitors a specific region of the body surface, an area known as a **dermatome.** Dermatomes (Figure 13-7•) are clinically important because damage or infection of a spinal nerve or dorsal root ganglion will produce a characteristic loss of sensation in the skin. For example, *shingles,* a virus that infects dorsal root ganglia, causes a painful rash whose distribution corresponds to that of the affected sensory nerves. [AM] *Shingles and Hansen's Disease*

Nerve Plexuses

The simple distribution pattern of dorsal and ventral rami illustrated in Figure 13-6• applies to spinal nerves T_2–T_{12}. But in segments controlling the skeletal musculature of the neck, upper limbs, or lower limbs, the situation is more complicated. During development, small skeletal muscles innervated by different ventral rami typically fuse to form larger muscles with compound origins. Although the anatomical distinctions between

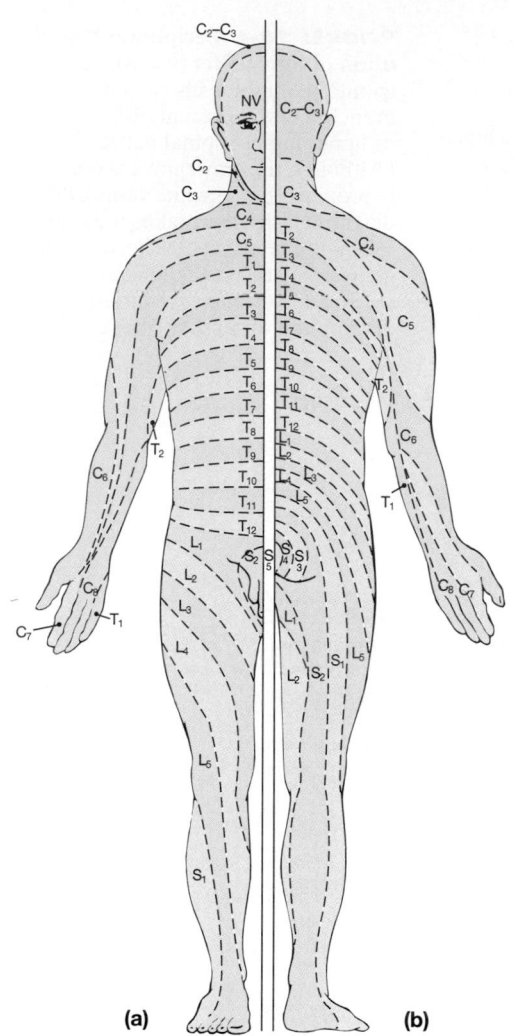

(a)   **(b)**

•**FIGURE 13-7** Dermatomes. (a) Anterior and (b) posterior distribution of dermatomes on the surface of the skin.

Peripheral *nerve palsies,* or **peripheral neuropathies,** are characterized by regional losses of sensory and motor function as the result of nerve trauma or compression. You have experienced a mild, temporary palsy if your arm or leg has ever "fallen asleep" after you leaned or sat in an uncomfortable position. $\boxed{AM}$ *Peripheral Neuropathies* Although dermatomes can provide clues to the location of injuries along the spinal cord, the loss of sensation at the skin does not provide sufficiently precise information about the site of injury because the boundaries of dermatomes are not exact, clearly defined lines. More exact conclusions can be drawn from the loss of motor control, on the basis of the origin and distribution of the peripheral nerves originating at nerve plexuses.

If a peripheral axon is damaged but not displaced, normal function may eventually return as the cut stump grows across the injury site away from the soma and along its former path. We described the mechanics of this process at the close of Chapter 12. ∞ *[p. 407]* Repairs made after damage to an entire peripheral *nerve* are gen-

the component muscles may disappear, separate ventral rami continue to provide sensory innervation and motor control to each portion of the compound muscle. As they converge, the ventral rami of adjacent spinal nerves blend their fibers, producing a series of compound nerve trunks. Such a complex interwoven network of nerves is a **nerve plexus** (PLEK-sus; *plexus,* braid). There are three major plexuses: the *cervical plexus,* the *brachial plexus,* and the *lumbosacral plexus* (Figure 13-8•).

In Chapter 11, we introduced the peripheral nerves that control the major axial and appendicular muscles. Look at the tables in that chapter as we proceed to review the innervation of the skeletal muscle groups. ∞ *[pp. 332–361]*

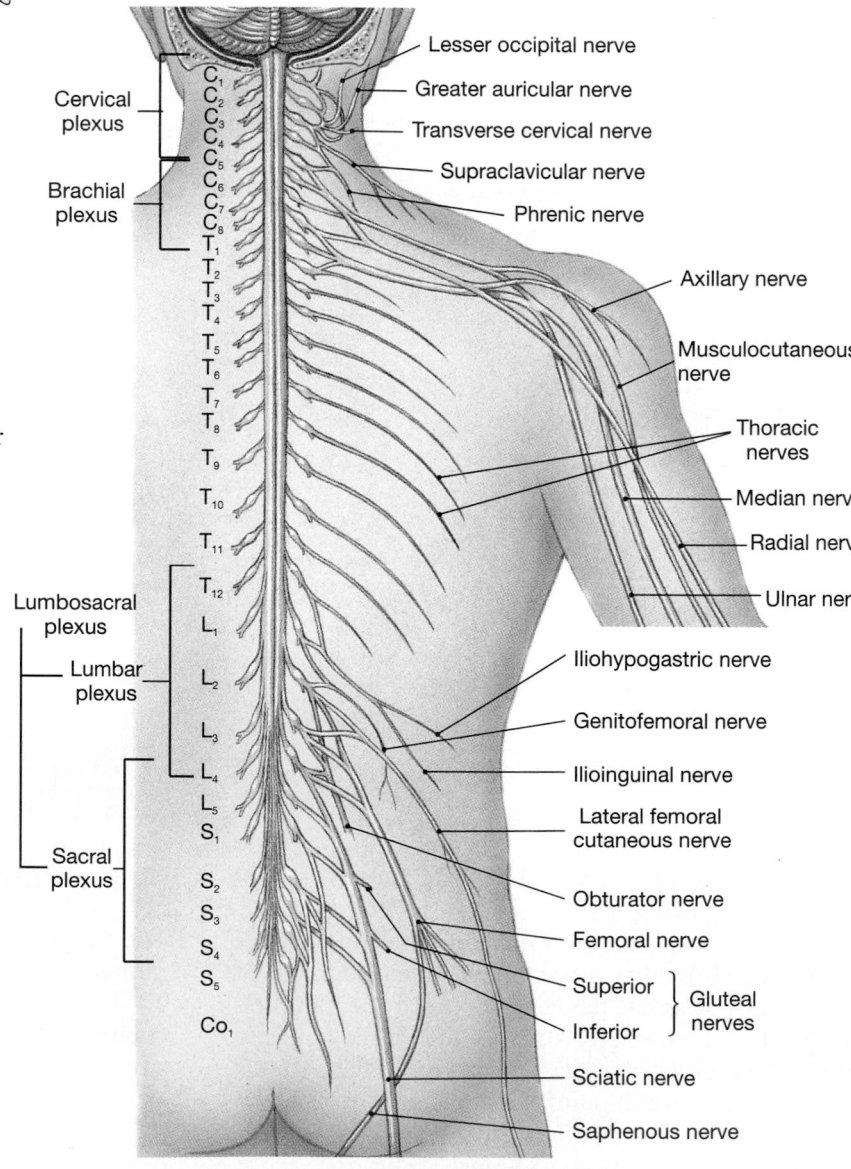

•**FIGURE 13-8** Peripheral Nerves and Plexuses

erally incomplete, primarily because of problems with axon alignment and regrowth. A variety of technologically sophisticated procedures designed to improve nerve regeneration and repair are currently under evaluation. An entire family of *nerve growth factors* has been discovered in recent years. Their use alone or in combination with other therapies may ultimately revolutionize the treatment of damaged neural tissue inside and outside the CNS. [AM] *Damage and Repair of Peripheral Nerves*

The Cervical Plexus

The **cervical plexus** (Figures 13-8, 13-9●; Table 13-1) consists of the ventral rami of spinal nerves C_1–C_5. Its branches innervate the neck's muscles and extend into the thoracic cavity, where they control the diaphragmatic muscles. The **phrenic nerve,** the major nerve of this plexus, provides the entire nerve supply to the *diaphragm,* a key respiratory muscle. Other branches are distributed to the skin of the neck and the upper part of the chest.

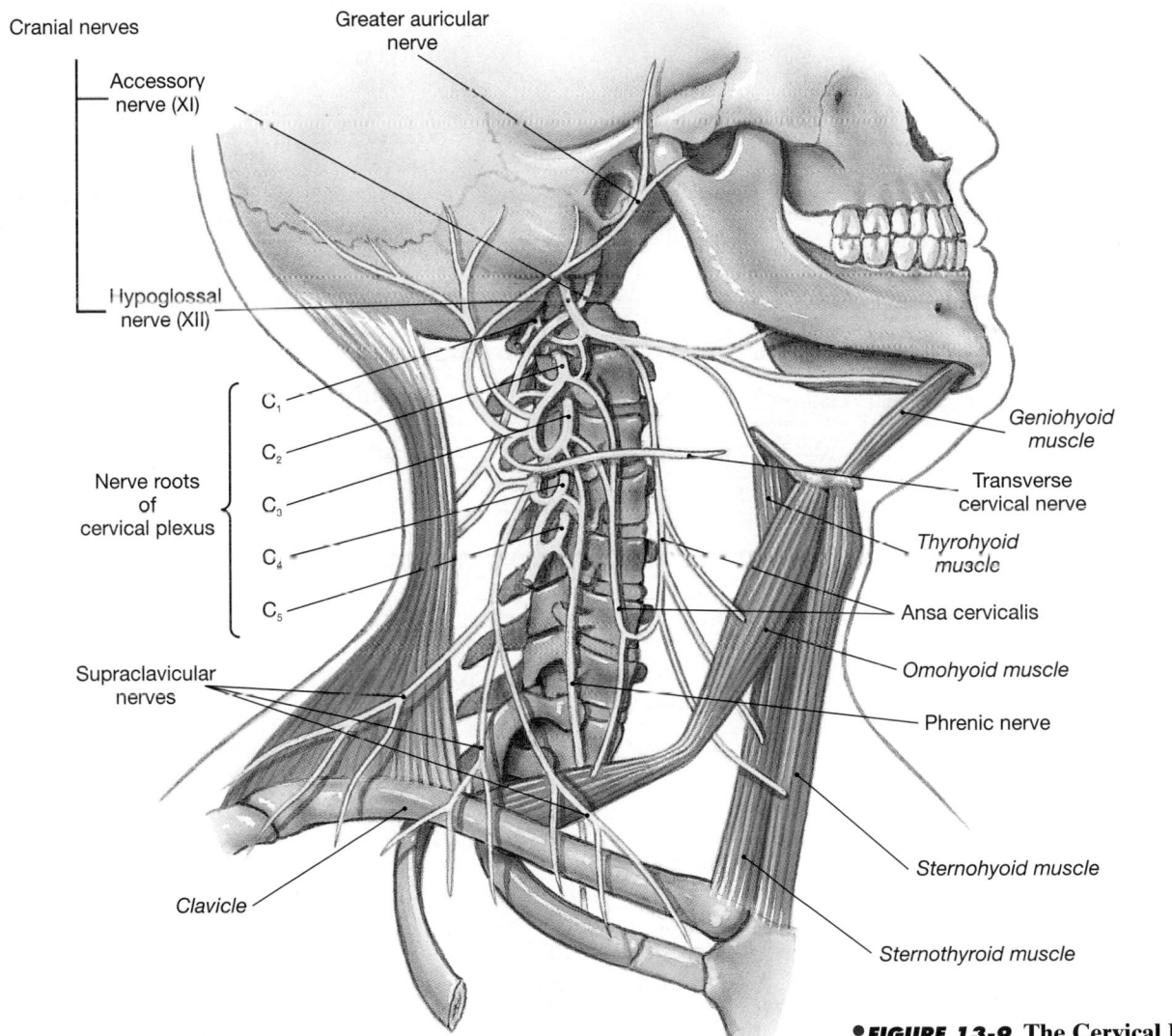

●**FIGURE 13-9** The Cervical Plexus

TABLE 13-1 The Cervical Plexus

Spinal Segment	Nerve(s)	Distribution
C_1–C_4	Ansa cervicalis (superior and inferior branches)	Five of the extrinsic laryngeal muscles: sternothyroid, sternohyoid, omohyoid, geniohyoid, and thyrohyoid (via N XII)
C_2–C_3	Lesser occipital, transverse cervical, supraclavicular, and greater auricular nerves	Skin of upper chest, shoulder, neck, and ear
C_3–C_5	Phrenic nerve	Diaphragm
C_1–C_5	Cervical nerves	Levator scapulae, scalenes, sternocleidomastoid, and trapezius (with N XI)

The Brachial Plexus

The **brachial plexus** (Table 13-2) innervates the shoulder girdle and upper limb, with contributions from the ventral rami of spinal nerves C_5–T_1 (Figures 13-8 and 13-10•). The nerves that form this plexus originate from trunks and cords named according to their location. Axons from the spinal nerves pass through the *superior, middle,* and *inferior trunks* to reach the *lateral, medial,* and *posterior cords,* respectively. The major nerves of the lateral cord are the **musculocutaneous nerve** and the **median nerve.** The **ulnar nerve** is the major nerve of the medial cord. The **axillary (circumflex) nerve** and **radial nerve** are the major nerves of the posterior cord.

TABLE 13-2 The Brachial Plexus

Spinal Segment	Nerve	Distribution
C_5, C_6	Axillary nerve	Deltoid and teres minor muscles
		Skin of shoulder
C_5–T_1	Radial nerve	Extensor muscles on the arm and forearm (triceps brachii, brachioradialis, extensor carpi radialis, and extensor carpi ulnaris)
		Digital extensors and abductor pollicis
		Skin over the posterolateral surface of the arm
C_5–C_7	Musculocutaneous nerve	Flexor muscles on the arm (biceps brachii, brachialis, and coracobrachialis)
		Skin over lateral surface of the forearm
C_6–T_1	Median nerve	Flexor muscles on the forearm (flexor carpi radialis and palmaris longus)
		Pronators (p. quadratus and p. teres)
		Digital flexors
		Skin over anterolateral surface of the hand
C_8, T_1	Ulnar nerve	Flexor muscle on the forearm (flexor carpi ulnaris)
		Adductor pollicis and small digital muscles
		Skin over medial surface of the hand

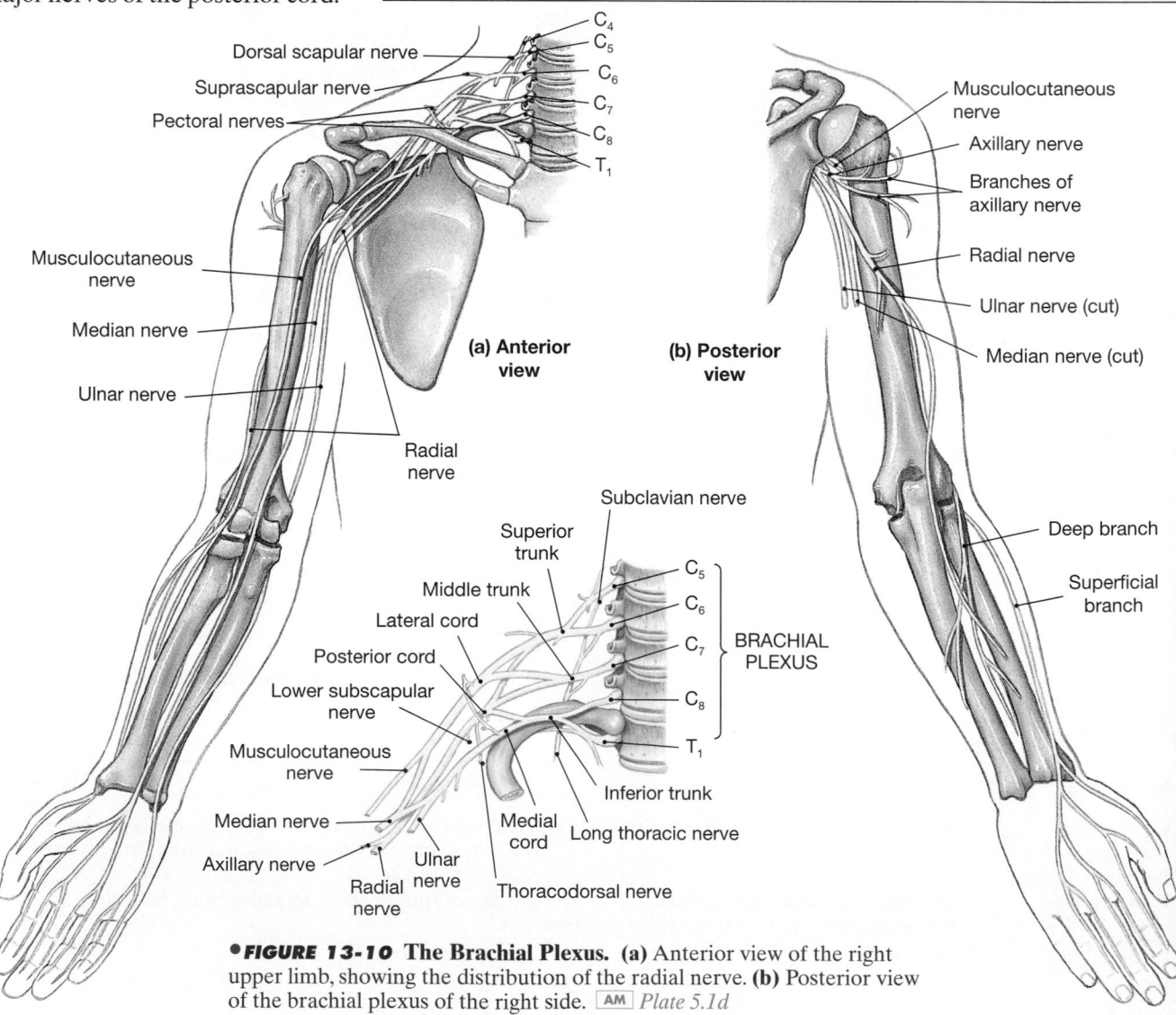

•**FIGURE 13-10** **The Brachial Plexus.** **(a)** Anterior view of the right upper limb, showing the distribution of the radial nerve. **(b)** Posterior view of the brachial plexus of the right side. AM *Plate 5.1d*

The Lumbosacral Plexus

The **lumbosacral plexus** (Figures 13-8 and 13-11●) arises from the lumbar and sacral segments of the spinal cord, and the ventral rami of these nerves supply the pelvic girdle and lower limbs. This plexus can be subdivided into a **lumbar plexus** (T_{12}–L_4) and a **sacral plexus** (L_4–S_4). The individual nerves that form the lumbosacral plexus and their distributions are detailed in Table 13-3.

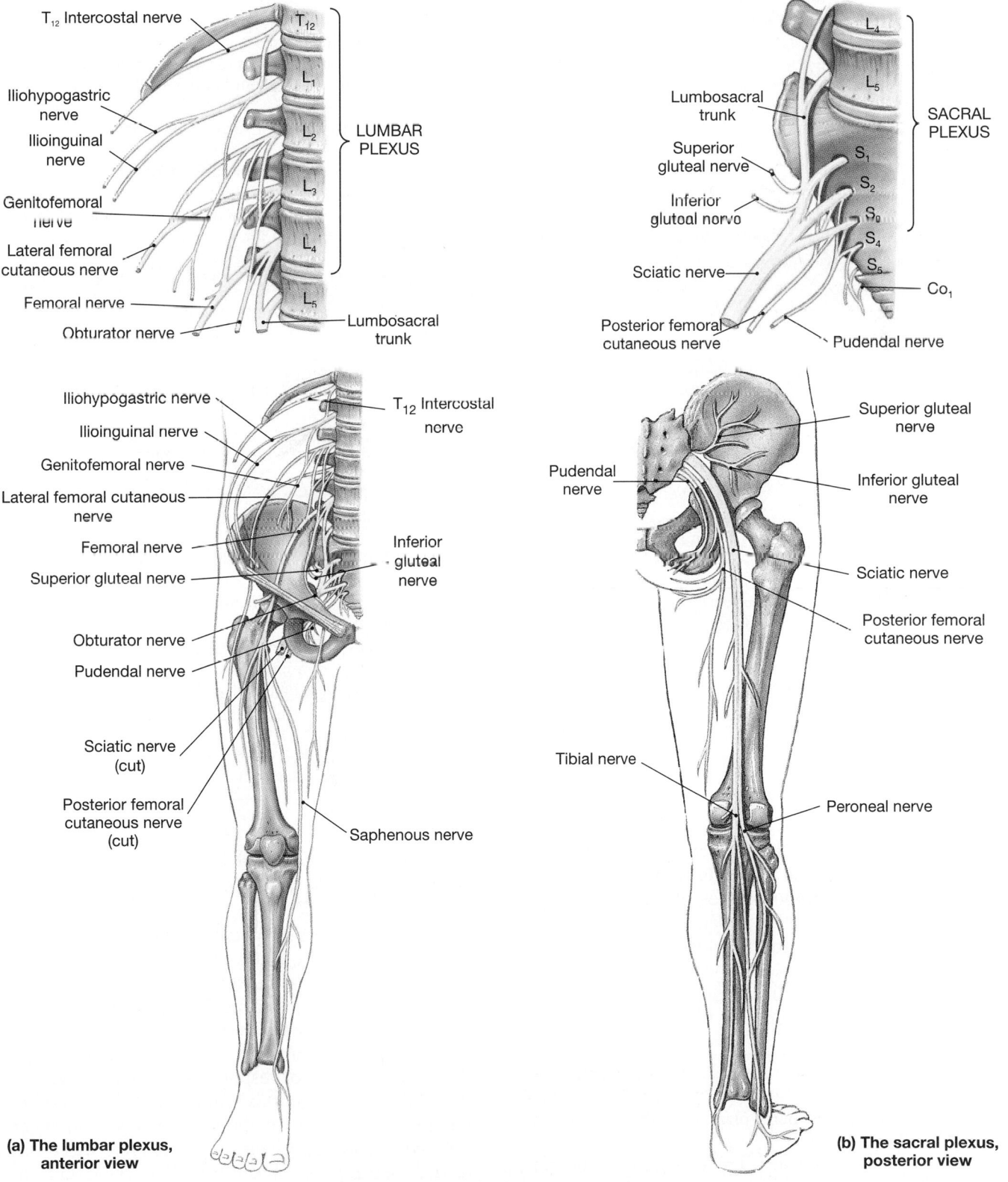

(a) The lumbar plexus, anterior view

(b) The sacral plexus, posterior view

●**FIGURE 13-11** The Lumbosacral Plexus

TABLE 13-3 The Lumbosacral Plexus

Spinal Segment	Nerve(s)	Distribution
Lumbar plexus		
T_{12}, L_1	Iliohypogastric nerve	Abdominal muscles (external and internal obliques, transversus abdominis)
		Skin over lower abdomen and buttocks
L_1	Ilioinguinal nerve	Abdominal muscles (with iliohypogastric)
		Skin over medial upper thigh and portions of external genitalia
L_1, L_2	Genitofemoral nerve	Skin over anteromedial surface of thigh and portions of external genitalia
L_2, L_3	Lateral femoral cutaneous nerve	Skin over anterior, lateral, and posterior surfaces of thigh
L_2–L_4	Femoral nerve	Anterior muscles of thigh (sartorius and quadriceps)
		Adductors of thigh (pectineus and iliopsoas)
		Skin over anteromedial surface of thigh, medial surface of leg and foot
L_2–L_4	Obturator nerve	Adductors of thigh (adductor magnus, brevis, longus)
		Gracilis muscle
		Skin over medial surface of thigh
L_2–L_4	Saphenous nerve	Skin over medial surface of leg
Sacral plexus		
L_4–S_2	Gluteal nerves:	
	Superior	Abductors of thigh (gluteus minimus, gluteus medius, and tensor fasciae latae)
	Inferior	Extensor of thigh (gluteus maximus)
	Posterior femoral cutaneous nerve	Skin of perineum and posterior surface of thigh and leg
L_4–S_3	Sciatic nerve:	Two of the hamstrings (semimembranosus and semitendinosus)
		Adductor magnus (with obturator nerve)
	Tibial nerve	Flexors of leg and plantar flexors of foot (popliteus, gastrocnemius, soleus, tibialis posterior, long head of biceps femoris)
		Flexors of toes
		Skin over posterior surface of leg, plantar surface of foot
	Peroneal nerve	Biceps femoris (short head)
		Peroneus (brevis and longus) and tibialis anterior
		Extensors of toes
		Skin over anterior surface of leg and dorsal surface of foot
S_2–S_4	Pudendal nerve	Muscles of perineum, including urogenital diaphragm and external anal and urethral sphincters
		Skin of external genitalia and related skeletal muscles (bulbospongiosus and ischiocavernosus)

The major nerves of the lumbar plexus are the **genitofemoral nerve, lateral femoral cutaneous nerve**, and **femoral nerve.** The major nerves of the sacral plexus are the **sciatic nerve** and the **pudendal nerve.** The sciatic nerve passes posterior to the femur, deep to the long head of the *biceps femoris* muscle. As it approaches the popliteal fossa, the sciatic nerve divides into two branches—the **peroneal nerve** and the **tibial nerve.**

☑ An anesthetic blocks the function of the dorsal rami of the cervical spinal nerves. What areas of the body will be affected?

☑ Injury to which of the nerve plexuses would interfere with the ability to breathe?

☑ Compression of which nerve produces the sensation that your leg has "fallen asleep"?

AN INTRODUCTION TO REFLEXES

Conditions inside or outside the body can change rapidly and unexpectedly. **Reflexes** are rapid, automatic responses to specific stimuli. Reflexes preserve homeostasis by making rapid adjustments in the function of organs or organ systems. The response shows little variability; each time a particular reflex is activated, it usually produces the same motor response. In Chapter 1, we introduced the basic functional components involved in all types of homeostatic regulation: a *receptor,* an *integration center,* and an *effector.* ∞ *[p. 13]* Here we consider *neural reflexes,* in which sensory fibers deliver information to the CNS and motor fibers carry motor commands

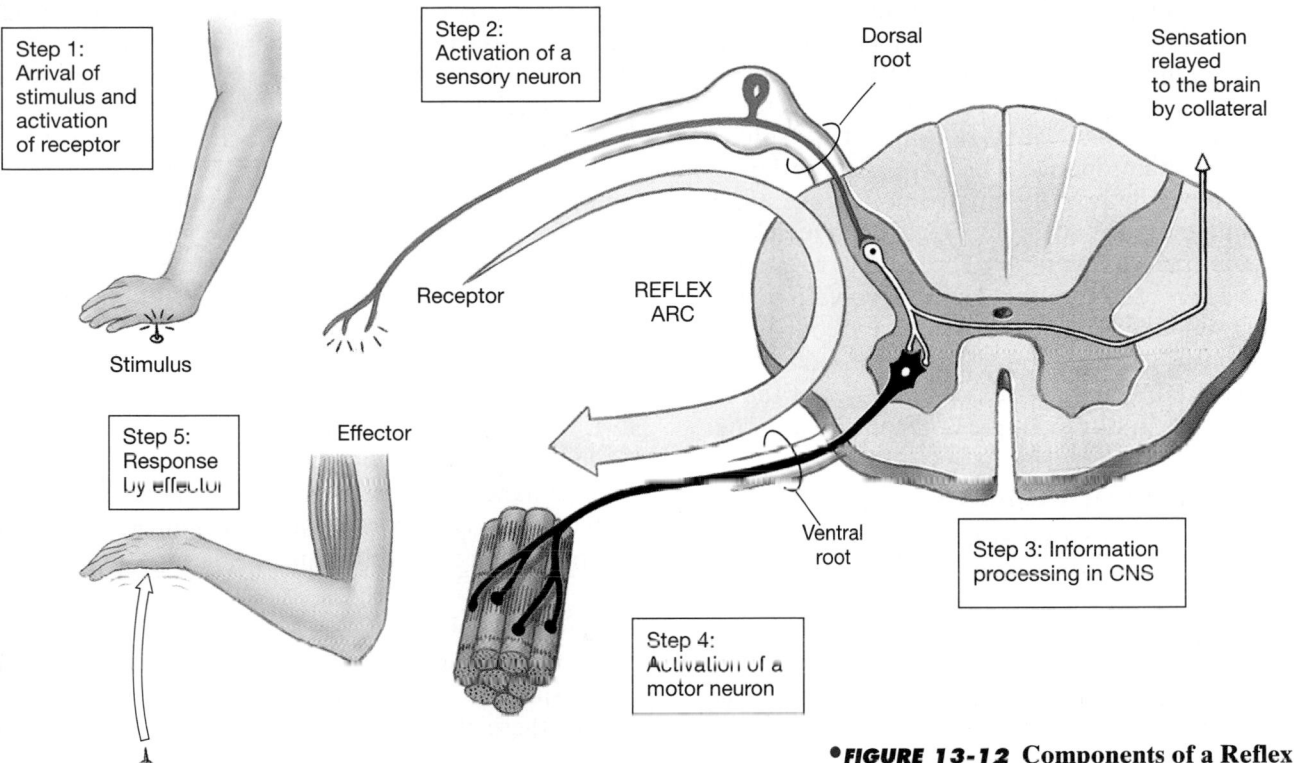

Step 1:
Arrival of
stimulus and
activation
of receptor

Step 2:
Activation of a
sensory neuron

Dorsal
root

Sensation
relayed
to the brain
by collateral

Receptor

REFLEX
ARC

Stimulus

Step 5:
Response
by effector

Effector

Ventral
root

Step 3: Information
processing in CNS

Step 4:
Activation of a
motor neuron

•*FIGURE 13-12* Components of a Reflex Arc

to peripheral effectors. In Chapter 18, we will examine *endocrine reflexes,* in which the commands to peripheral tissues and organs are delivered by hormones in the bloodstream.

The Reflex Arc

The "wiring" of a single reflex is called a **reflex arc.** A reflex arc begins at a receptor and ends at a peripheral effector, such as a muscle fiber or gland cell. Figure 13-12• diagrams the five steps involved in a neural reflex:

STEP 1: *Arrival of a Stimulus and Activation of a Receptor.* A *receptor* is a specialized cell or the dendrites of a sensory neuron. Receptors are sensitive to physical or chemical changes in the body or to the external environment. There are many types of sensory receptors; we introduced general categories in Chapter 12. ∞ *[p. 375]* When you lean on a tack, pain receptors in the palm of your hand are activated (Figure 13-12•). Pain receptors are the dendrites of sensory neurons; they respond to stimuli that cause or accompany tissue damage. (We shall discuss the linkage between receptor stimulation and sensory neuron activation further in Chapter 17.)

STEP 2: *Activation of a Sensory Neuron.* Stimulation of the pain receptors leads to the formation and propagation of action potentials along the axons of sensory neurons. This information reaches your spinal cord within one of the dorsal roots. In our example, Steps 1 and 2 involve the same cell. However, the two steps may in-

volve different cells. For example, reflexes triggered by loud sounds begin when receptor cells in the inner ear release neurotransmitters that stimulate sensory neurons.

STEP 3: *Information Processing.* Information processing begins when a neurotransmitter released by synaptic knobs of each sensory neuron affects the postsynaptic membrane of an interneuron, producing an EPSP. ∞ *[p. 403]* This stimulus will be integrated with other stimuli arriving at the postsynaptic neuron at that moment. The next step may involve a neuronal pool that coordinates an appropriate motor response by stimulating some motor neurons and inhibiting others. Interneurons are not always involved in a reflex arc; the sensory neuron sometimes innervates a motor neuron directly. In that case, the motor neuron performs the information processing as summation occurs at the axon hillock ∞ *[p. 404]*

STEP 4: *Activation of a Motor Neuron.* The axons of the stimulated motor neurons carry action potentials into the periphery—in this example, over the ventral root of a spinal nerve.

STEP 5: *Response of a Peripheral Effector.* The release of neurotransmitter by the motor neurons at synaptic knobs then leads to a response by a peripheral effector—in this case, a skeletal muscle whose contraction pulls your hand away from the tack.

A reflex response generally removes or opposes the original stimulus; here, the contracting muscle pulls your hand away from the painful stimulus. This reflex arc is therefore an example of a *negative feedback control mechanism.* ∞ *[p. 13]* By opposing potentially harmful changes in the internal or external environment, reflexes play an important role in homeostatic maintenance. The immediate reflex response is typically not the only response to a stimulus. In this example, you might wince, say "Ouch," and shake your hand a few times. These other responses, which are directed by your brain, involve multiple synapses and take longer to organize and coordinate. The responses are triggered by ascending action potentials carried by collaterals of the axons of the interneurons (as in our example) or sensory neurons.

Classification of Reflexes

Reflexes can be classified according to (1) their development, (2) the site where information processing occurs, (3) the nature of the resulting motor response, or (4) the complexity of the neural circuit involved. These categories are not mutually exclusive; they represent different ways of describing a single reflex (Figure 13-13●).

Development of Reflexes

Innate reflexes result from the connections that form between neurons during development. Such reflexes generally appear in a predictable sequence, from the simplest reflex responses (withdrawal from pain) to more complex motor patterns (chewing, suckling, or tracking objects with the eyes). The neural connections responsible for basic motor patterns of an innate reflex are genetically or developmentally programmed. Examples include the reflexive removal of your hand from a hot stovetop and blinking when your eyelashes are touched.

More complex, learned motor patterns are sometimes called **acquired reflexes.** An experienced driver steps on the brake when trouble appears on the road ahead; a professional skier makes rapid, automatic adjustments in body position while racing. These motor responses are rapid and automatic, but they were learned rather than preestablished. Such reflexes are enhanced by repetition. The distinction between innate and acquired reflexes is not absolute. Some people can learn motor patterns more quickly than others can, and the differences probably have a genetic basis.

Most reflexes, whether innate or acquired, can be modified over time or suppressed through conscious effort. For example, while walking a tightrope over the Grand Canyon, you would probably ignore a bee sting on your hand, although under other circumstances you would probably withdraw your hand immediately, with lots of shouting and thrashing.

Processing Sites

In a **spinal reflex,** the important interconnections and processing events occur inside the spinal cord. We will discuss these reflexes in detail in the next section. Reflexes processed in the brain are called **cranial reflexes.** For example, the reflex movements in response to a sudden loud noise or bright light are cranial reflexes directed by nuclei in the brain. We shall encounter many other examples of cranial reflexes in Chapters 15 and 16.

Nature of the Response

Somatic reflexes provide a mechanism for the involuntary control of the muscular system. **Visceral reflexes,** or *autonomic reflexes,* control the activities of other systems. We will consider somatic reflexes in detail in this chapter; we shall examine visceral reflexes in Chapter 16.

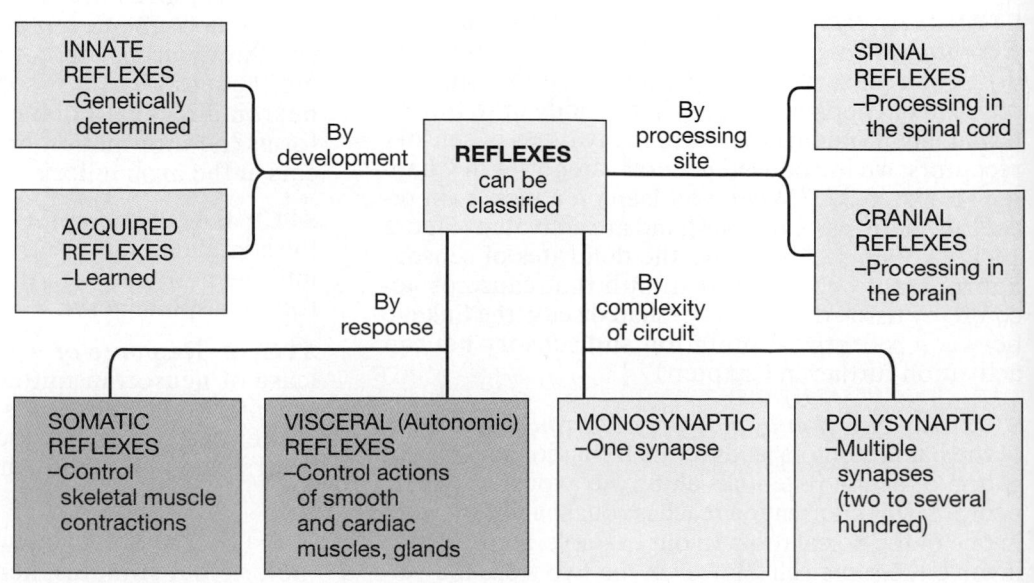

●**FIGURE 13-13** Methods of Classifying Reflexes

The movements directed by somatic reflexes are not delicate; nor are they precise. You might therefore wonder why they exist at all, because we have voluntary control over the same muscles. Somatic reflexes are absolutely vital, primarily because they are *immediate*. Making decisions and coordinating voluntary responses take time, and in an emergency situation—you slip while descending a flight of stairs, or you lean your hand against a knife edge—any delay increases the likelihood of a severe injury. Thus, somatic reflexes provide a rapid response that, if necessary, can later be modified by voluntary motor commands.

Complexity of the Circuit

In the simplest reflex arc, a sensory neuron synapses directly on a motor neuron, which itself serves as the processing center. Such a reflex is termed a **monosynaptic reflex.** Transmission across a chemical synapse always involves a synaptic delay—but, with only one synapse, the delay between the stimulus and the response is minimized. Circuit 1 in Figure 13-14a● is a typical monosynaptic reflex.

Most other reflexes have at least one interneuron between the sensory neuron and the motor neuron, as does Circuit 2 of Figure 13-14b●. These **polysynaptic reflexes** have a longer delay between stimulus and response; the length of the delay is proportional to the number of synapses involved. Polysynaptic reflexes can produce far more complicated responses than monosynaptic reflexes can, because the interneurons can control several different muscle groups.

SPINAL REFLEXES

Spinal reflexes range in complexity from simple monosynaptic reflexes involving a single segment of the spinal cord to polysynaptic reflexes that involve many different segments. In the most complicated spinal reflexes, called **intersegmental reflex arcs,** many segments interact to produce a coordinated motor response. These polysynaptic reflexes include excitatory and inhibitory synapses and produce highly variable motor patterns.

Monosynaptic Reflexes

In a monosynaptic reflex, there is little delay between sensory input and motor output. These reflexes control the most rapid, stereotyped motor responses of the nervous system to specific stimuli. The best-known example of a monosynaptic reflex is the *stretch reflex.*

The Stretch Reflex

The **stretch reflex** (Figure 13-15a●) provides automatic regulation of skeletal muscle length. The stimulus (increasing muscle length) activates a sensory neuron, which triggers an immediate motor response (contraction of the stretched muscle) that counteracts the stimulus. Action potentials traveling toward and away from the spinal cord are conducted along large, myelinated, Type A fibers (∞ *[p. 394]*), and the entire reflex is completed within 20–40 msec.

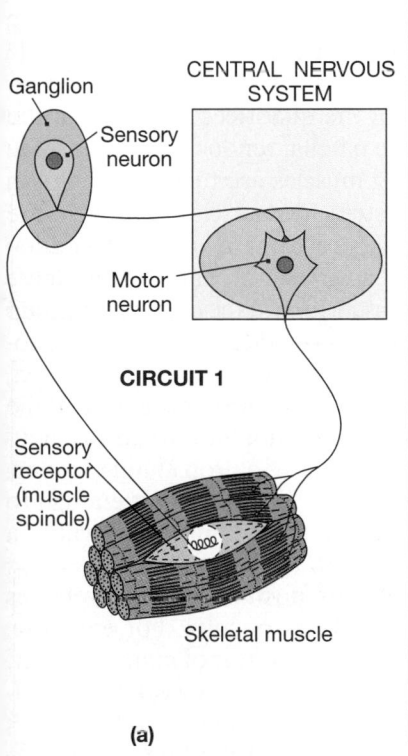

CIRCUIT 1

(a)

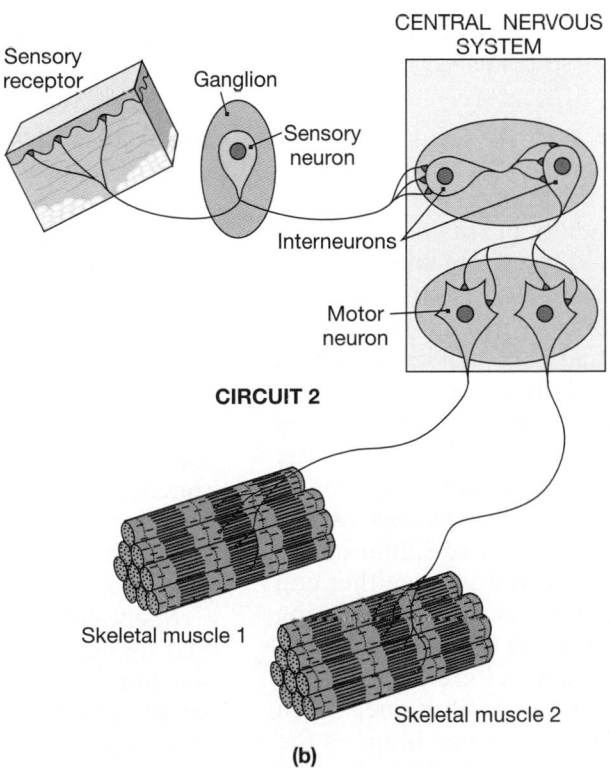

CIRCUIT 2

(b)

●**FIGURE 13-14**
Neural Organization and Simple Reflexes.
(a) A monosynaptic reflex involves a peripheral sensory neuron and a central motor neuron. In this example, stimulation of the receptor will lead to a reflexive contraction in a skeletal muscle.
(b) A polysynaptic reflex involves a sensory neuron, interneurons, and motor neurons. In this example, the stimulation of the receptor leads to the coordinated contractions of two different skeletal muscles.

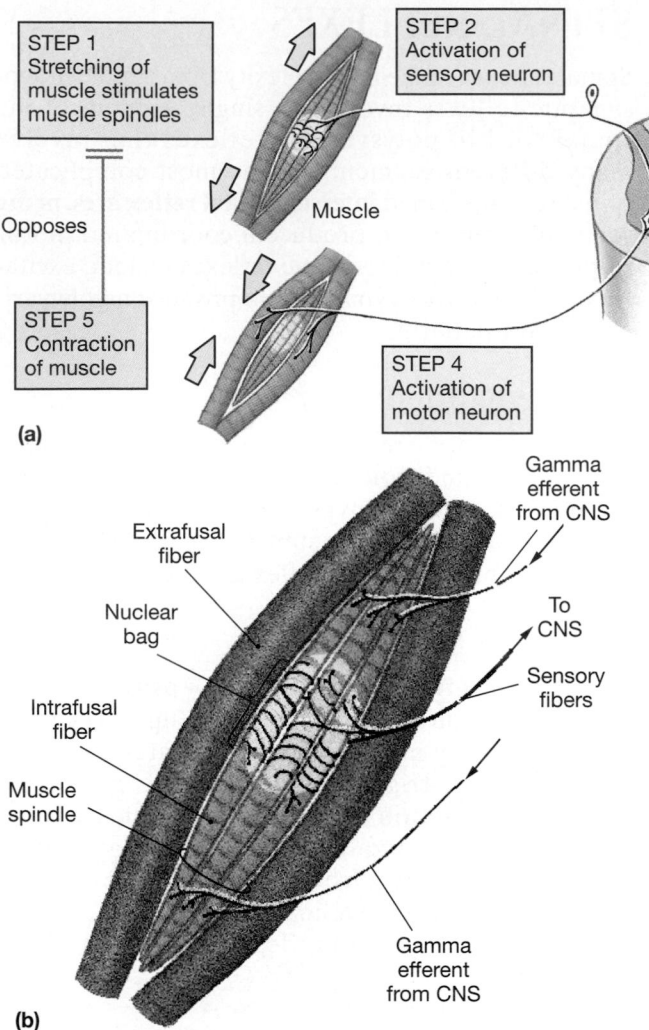

(a)

(b)

●*FIGURE 13-15* **Components of the Stretch Reflex.**
(a) Diagram of the activities in a stretch reflex. **(b)** Structure of a muscle spindle.

MUSCLE SPINDLES The sensory receptors involved in the stretch reflex are the **muscle spindles.** Each muscle spindle consists of a bundle of small, specialized skeletal muscle fibers called **intrafusal muscle fibers** (Figure 13-15b●). The muscle spindle is surrounded by larger **extrafusal muscle fibers**, which are responsible for the resting muscle tone and, at greater levels of stimulation, for the contraction of the entire muscle.

Each intrafusal fiber in the spindle is innervated by both sensory and motor neurons. The dendrites of the sensory neuron spiral around the central portion of the intrafusal fiber, a region known as the **nuclear bag.** Axons from a separate population of spinal motor neurons form neuromuscular junctions on either end of the intrafusal fiber. The motor neurons innervating intrafusal fibers are called **gamma (γ) motor neurons**, and their axons are called **gamma efferents.** There are two sets of myofibrils in an intrafusal fiber, one at each end, and each set is attached to the membrane of the nuclear bag.

The sensory neuron is always active, conducting impulses to the CNS. The axon enters the CNS in a dorsal root and synapses on motor neurons in the anterior gray horn of the spinal cord. Collaterals distribute the information to the brain, providing proprioceptive information about the skeletal muscle.

Stretching the nuclear bag distorts the dendrites and stimulates the sensory neuron, increasing the frequency of action potentials. Compressing the nuclear bag inhibits the sensory neuron, decreasing the frequency of action potential generation.

The axon of the sensory neuron synapses on CNS motor neurons that control the extrafusal muscle fibers *of the same muscle.* An increase in stimulation of the sensory neuron, caused by stretching of the nuclear bag, will lead to an increase in muscle tone. A decrease in stimulation of the sensory neuron, due to compression of the nuclear bag, will lead to a decrease in muscle tone.

STRETCH REFLEXES AND PASSIVE MUSCLE MOVE-MENTS When a skeletal muscle is stretched, muscle spindles elongate and muscle tone increases (Figures 13-15a and 13-16a●). This increase provides automatic resistance that reduces the chance of muscle damage due to overextension. When a skeletal muscle is compressed, muscle spindles are also compressed, and muscle tone decreases. This decrease helps reduce resistance to the movement under way.

The **knee jerk reflex,** or **patellar reflex** (Figure 13-16b●), is a stretch reflex triggered by passive muscle movement. We can demonstrate this reflex, which responds to stretching of the quadriceps muscles of the thigh, by tapping on the patellar tendon of a flexed knee when the thigh and leg muscles are relaxed. A tap on the patellar ligament stretches muscle spindles in the quadriceps group. This stretching produces a sudden burst of activity in the sensory neurons that monitor these muscle spindles, and muscle tone in the quadriceps group increases rapidly. The rise is so sudden that it usually produces a noticeable kick. As the extrafusal fibers contract, the muscle spindles return to their resting length, and the rate of action potential generation in the sensory neurons declines. This decline causes a drop in muscle tone, and the leg then swings back. If it swings far enough to stretch the muscle spindles a second time, there will be a small secondary kick (Figure 13-16c●).

Many stretch reflexes are **postural reflexes,** reflexes that maintain normal upright posture. For example, standing involves a cooperative effort of many different muscle groups. Some of these muscles work in opposition to one another, exerting forces that keep the body's weight balanced over the feet. If the body leans for-

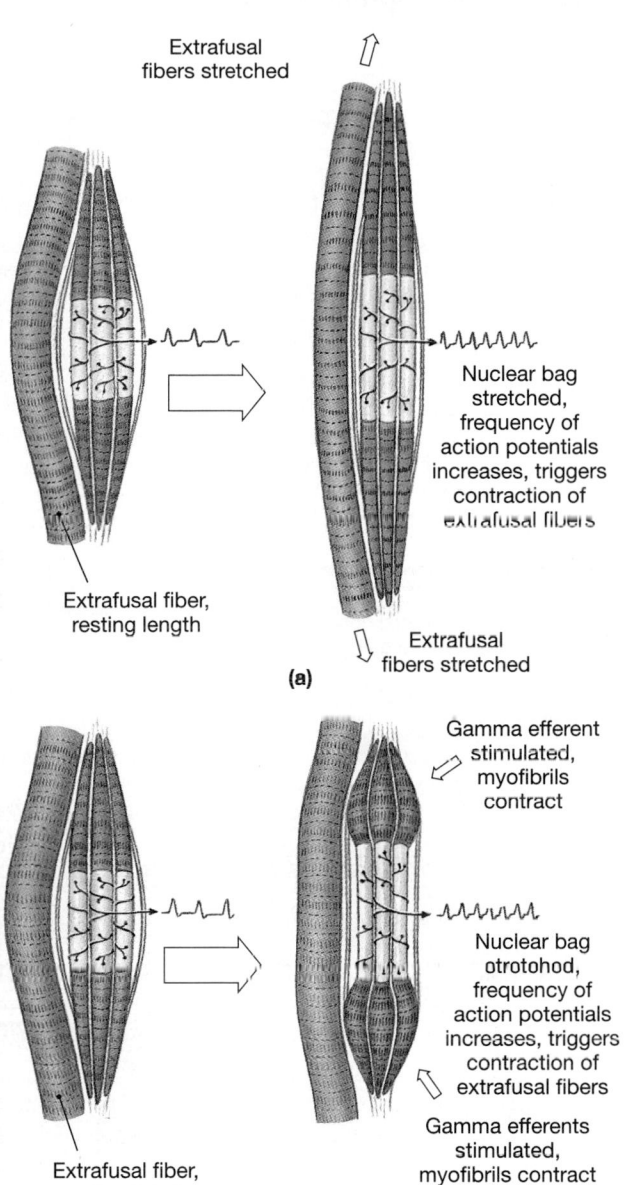

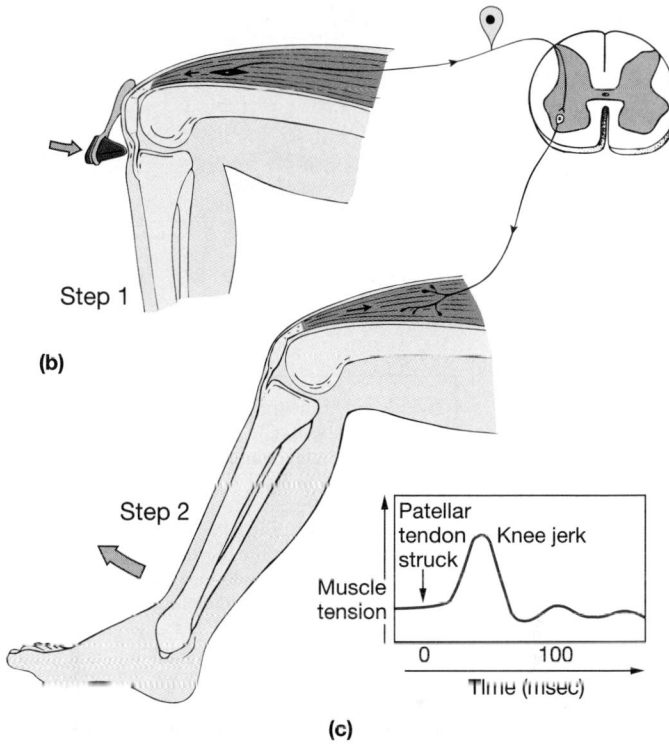

●**FIGURE 13-16** Intrafusal Fibers. **(a)** The nuclear bag will elongate when the entire skeletal muscle is stretched by an external force. This stretching increases the resistance to further stretching. **(b)** The patellar reflex is controlled by muscle spindles in the muscles that straighten the knee. Step 1: A reflex hammer strikes the muscle tendon, stretching the spindle fibers. This stretching results in a sudden increase in the activity of the sensory neurons. These neurons synapse on spinal motor neurons. Step 2: The activation of extrafusal motor units produces an immediate increase in muscle tone and a reflexive kick. **(c)** The changes in muscle tension can be graphically recorded. Note the small secondary kick that may occur if the leg rebounds past its resting position. **(d)** The nuclear bag will elongate when myofibrils contract within the intrafusal fiber. This contraction occurs under the command of gamma motor neurons. As a result, the muscle spindle remains sensitive to externally imposed changes in muscle length.

ward, stretch receptors in the calf muscles are stimulated, and those muscles respond by contracting. This contraction returns the body to an upright position. If the muscles overcompensate and the body begins to lean back, the calf muscles relax. But then stretch receptors in muscles of the shins and thighs are stimulated, and the problem is corrected immediately.

Postural muscles generally have a firm muscle tone and extremely sensitive stretch receptors. As a result, very fine adjustments are continually being made, and you are not aware of the cycles of contraction and relaxation that occur. Stretch reflexes are only one type of postural reflex; there are many complex polysynaptic postural reflexes.

STRETCH REFLEXES AND ACTIVE MUSCLE MOVEMENTS
The CNS adjusts the sensitivity of muscle spindles during a voluntary contraction. Throughout

the contraction, impulses arriving over gamma efferents cause the contraction of myofibrils within the intrafusal fibers of that muscle. The myofibrils pull on the nuclear bag (Figure 13-16d●). This tension keeps the muscle spindles sensitive to externally imposed changes in muscle length throughout the contraction. For example, if you hold your hand out palm up and someone drops a ball into it, the muscle spindles will automatically adjust muscle tone to compensate for the increased load.

☑ What is the minimum number of neurons needed for a reflex arc?

☑ One of the first somatic reflexes to develop is the suckling reflex. What type of reflex is this?

☑ How would stimulation of the muscle spindles involved in the patellar (knee jerk) reflex by gamma motor neurons affect the speed of the reflex?

Development of the Spinal Cord and Spinal Nerves

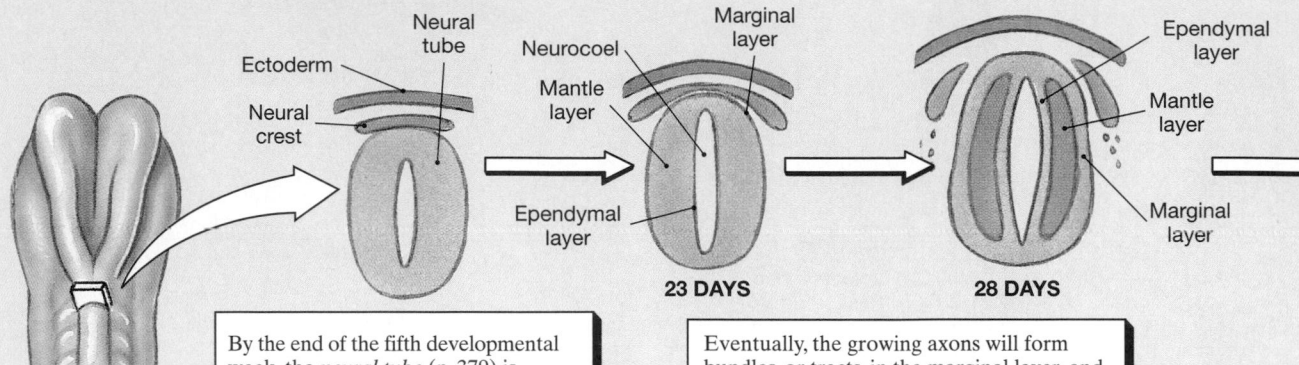

Ectoderm
Neural tube
Neural crest

Neurocoel
Mantle layer
Marginal layer
Ependymal layer

23 DAYS

Ependymal layer
Mantle layer
Marginal layer

28 DAYS

22 DAYS

By the end of the fifth developmental week, the *neural tube* (p. 379) is almost completely closed. In the spinal cord, the *mantle layer* that contains developing neurons and neuroglial cells will produce the gray matter that surrounds the *neurocoel*. As neurons develop in the mantle layer, their axons grow toward central or peripheral destinations. The axons leave the mantle layer and travel toward synaptic targets within a peripheral *marginal layer*.

Eventually, the growing axons will form bundles, or tracts, in the marginal layer, and these tracts will crowd together in the columns that form the white matter of the spinal cord.

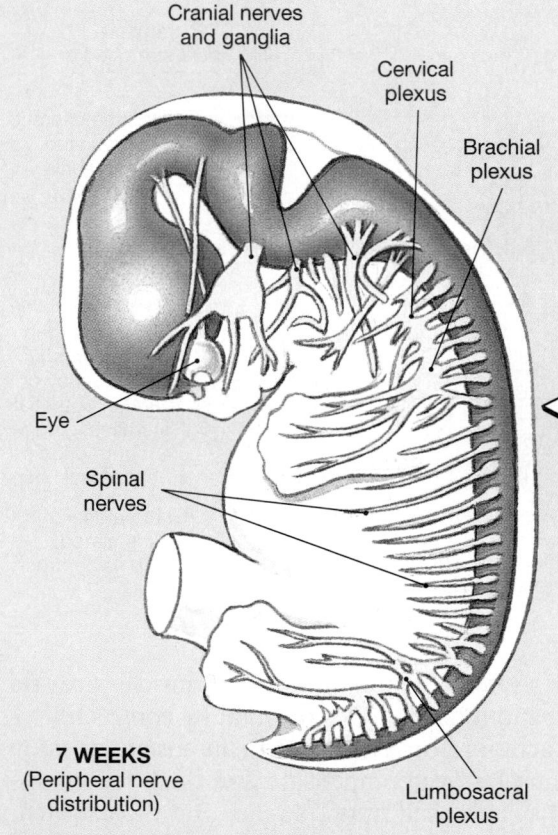

Cranial nerves and ganglia
Cervical plexus
Brachial plexus
Eye
Spinal nerves
Lumbosacral plexus

7 WEEKS
(Peripheral nerve distribution)

DEVELOPMENTAL ABNORMALITIES

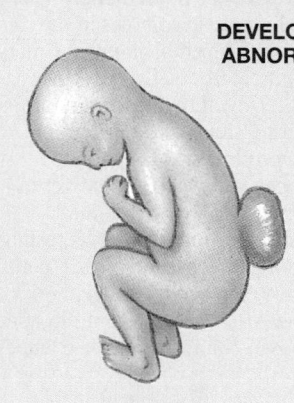

Spina bifida

Neural tube defect

Spina bifida (BI-fi-da) results when the developing vertebral laminae fail to unite. The neural arch is incomplete, and the meninges bulge outward beneath the skin of the back. The extent of the abnormality determines the severity of the defects. In mild cases, the condition may pass unnoticed; extreme cases involve much of the length of the spinal column and are usually associated with abnormal nerve function.

A **neural tube defect (NTD)** is a condition that is secondary to a developmental error in the formation of the spinal cord. Instead of forming a hollow tube, a portion of the spinal cord develops as a broad plate. This is often associated with spina bifida. Neural tube defects affect roughly one individual in 1000; prenatal testing can detect the existence of these defects with an 80–85% success rate.

Several spinal nerves innervate each developing limb. When embryonic muscle cells migrate away from the myotome, the nerves grow right along with them. If a large muscle in the adult is derived from several myotomal blocks, connective tissue partitions will often mark the original boundaries, and the innervation will always involve more than one spinal nerve.

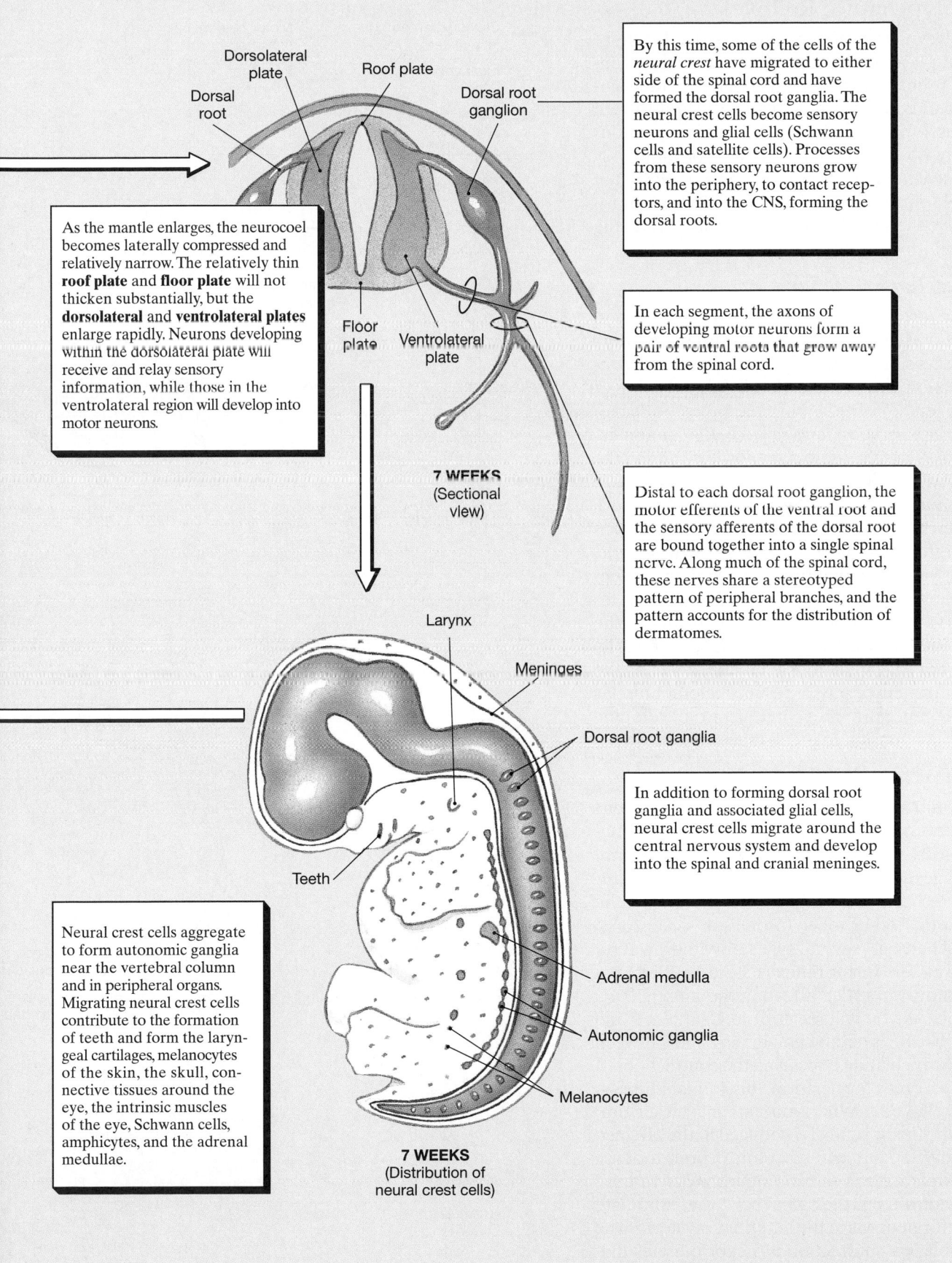

Dorsolateral plate

Dorsal root

Roof plate

Dorsal root ganglion

By this time, some of the cells of the *neural crest* have migrated to either side of the spinal cord and have formed the dorsal root ganglia. The neural crest cells become sensory neurons and glial cells (Schwann cells and satellite cells). Processes from these sensory neurons grow into the periphery, to contact receptors, and into the CNS, forming the dorsal roots.

As the mantle enlarges, the neurocoel becomes laterally compressed and relatively narrow. The relatively thin **roof plate** and **floor plate** will not thicken substantially, but the **dorsolateral** and **ventrolateral plates** enlarge rapidly. Neurons developing within the dorsolateral plate will receive and relay sensory information, while those in the ventrolateral region will develop into motor neurons.

Floor plate

Ventrolateral plate

In each segment, the axons of developing motor neurons form a pair of ventral roots that grow away from the spinal cord.

7 WEEKS
(Sectional view)

Distal to each dorsal root ganglion, the motor efferents of the ventral root and the sensory afferents of the dorsal root are bound together into a single spinal nerve. Along much of the spinal cord, these nerves share a stereotyped pattern of peripheral branches, and the pattern accounts for the distribution of dermatomes.

Larynx

Meninges

Dorsal root ganglia

In addition to forming dorsal root ganglia and associated glial cells, neural crest cells migrate around the central nervous system and develop into the spinal and cranial meninges.

Teeth

Neural crest cells aggregate to form autonomic ganglia near the vertebral column and in peripheral organs. Migrating neural crest cells contribute to the formation of teeth and form the laryngeal cartilages, melanocytes of the skin, the skull, connective tissues around the eye, the intrinsic muscles of the eye, Schwann cells, amphicytes, and the adrenal medullae.

Adrenal medulla

Autonomic ganglia

Melanocytes

7 WEEKS
(Distribution of neural crest cells)

Polysynaptic Reflexes

Polysynaptic reflexes can produce far more complicated responses than can monosynaptic reflexes. One reason is that the interneurons involved can control several different muscle groups. Moreover, these interneurons may produce either excitatory or inhibitory postsynaptic potentials (EPSPs or IPSPs) at CNS motor nuclei, so the response can involve the stimulation of some muscles and the inhibition of others.

The Tendon Reflex

The stretch reflex regulates the length of a skeletal muscle. The **tendon reflex** monitors the tension produced during a muscular contraction and prevents tearing or breaking of the tendons. The sensory receptors for this reflex are nerve endings called **Golgi tendon organs.** These receptors are located among the collagen fibers of the tendon. The neuron is stimulated when the collagen fibers are stretched. In other words, these receptors monitor the external tension in the skeletal muscle. ∞ *[p. 293]* Within the spinal cord, these neurons stimulate inhibitory interneurons that innervate the motor neurons controlling the skeletal muscle. The greater the tension in the tendon, the greater the inhibitory effect on the motor neurons. As a result, skeletal muscles are generally prevented from developing enough tension to break their tendons.

Withdrawal Reflexes

Withdrawal reflexes move affected portions of the body away from a source of stimulation. The strongest withdrawal reflexes are triggered by painful stimuli, but these reflexes can sometimes be initiated by the stimulation of touch or pressure receptors.

There are several types of withdrawal reflexes. The **flexor reflex** (Figure 13-17a●) is a withdrawal reflex affecting the muscles of a limb. As you will recall from Chapters 9 and 11, *flexion* is a reduction in the angle between two articulating bones, and the contractions of *flexor muscles* perform this movement. ∞ *[pp. 259, 322]* When you step on a tack, a dramatic flexor reflex is produced in the affected leg. When the pain receptors in your foot are stimulated, the sensory neurons activate interneurons in the spinal cord that stimulate motor neurons in the anterior gray horns. The result is a contraction of flexor muscles that yanks your foot off the ground.

When a specific muscle contracts, opposing muscles must relax to permit the movement.

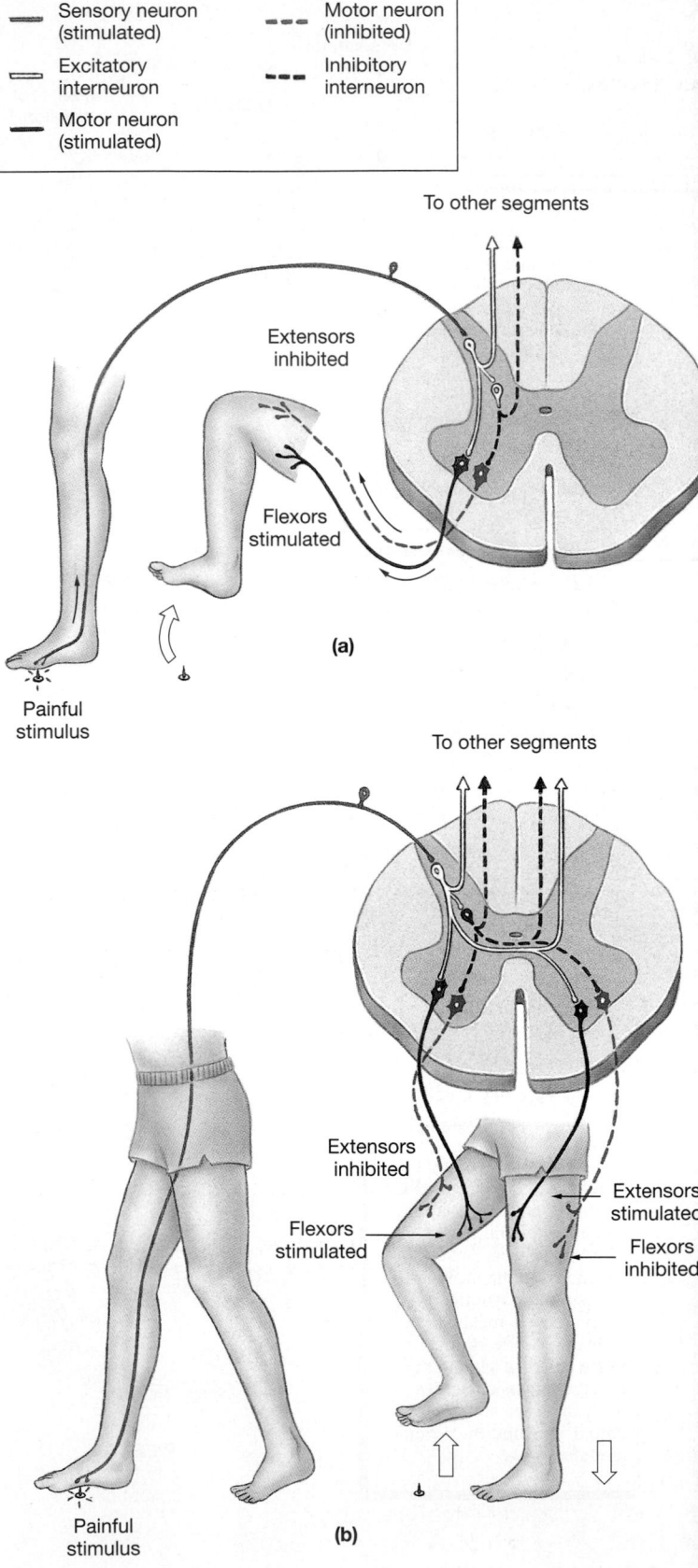

●**FIGURE 13-17** The Flexor and Crossed Extensor Reflexes. **(a)** The flexor reflex, an example of a withdrawal reflex. **(b)** The crossed extensor reflex.

For example, the flexor muscles that bend the leg are opposed by extensor muscles that straighten it out. A potential conflict exists here: Contraction of a flexor muscle should theoretically trigger a stretch reflex in the extensors that would cause them to contract, opposing the movement that is under way. Interneurons in the spinal cord prevent such competition through **reciprocal inhibition.** When one set of motor neurons is stimulated, those neurons that control antagonistic muscles are inhibited. The term *reciprocal* refers to the fact that the system works both ways. When the flexors contract, the extensors relax; when the extensors contract, the flexors relax.

Withdrawal reflexes are much more complex than any monosynaptic reflex. They also show tremendous versatility, because the sensory neurons activate many pools of interneurons. If the stimuli are strong, interneurons will carry excitatory and inhibitory impulses up and down the spinal cord, affecting motor neurons in many different segments. The end result is always the same: a coordinated movement away from the source of stimulation. But the distribution of the effects and the strength and character of the motor responses depend on the intensity and location of the stimulus. When you step on something sharp, mild discomfort might provoke a brief contraction in muscles of your ankle and foot. More powerful stimuli would produce coordinated muscular contractions affecting the positions of your ankle, foot, and leg. Severe pain would also stimulate contractions of your shoulder, trunk, and arm muscles. These contractions could persist for several seconds due to the activation of reverberating circuits. ∞ *[p. 409]* In contrast, monosynaptic reflexes are relatively invariable and brief; the entire knee jerk reflex is completed in roughly 20 msec.

Crossed Extensor Reflexes

The stretch, tendon, and withdrawal reflexes are **ipsilateral reflex arcs** (*ipsi,* same + *lateral,* side), in which the sensory stimulus and the motor response occur on the same side of the body. The **crossed extensor reflex** (Figure 13-17b●) is called a **contralateral reflex arc** (*contra,* opposite) because the motor response occurs on the side opposite the stimulus.

The crossed extensor reflex complements the flexor reflex, and the two occur simultaneously. When you step on a tack, while the flexor reflex pulls that foot away from the ground, the crossed extensor reflex stiffens your other leg to support your body weight. In the crossed extensor reflex, the axons of interneurons responding to the pain cross to the other side of the spinal cord and stimulate motor neurons that control the extensor muscles of the uninjured leg. Your opposite leg straightens to support the shifting weight. Reverberating circuits use positive feedback to ensure that the movement lasts long enough to be effective, despite the absence of motor commands from higher centers of the brain.

The flexor and crossed extensor reflexes are not the most complicated spinal reflexes. Postural and motor reflexes associated with standing, walking, and running are also spinal reflexes. But all polysynaptic reflexes share the same basic characteristics:

1. *They involve pools of interneurons.* Processing occurs in pools of interneurons before motor neurons are activated. The result may be excitation or inhibition; the tendon reflex produces inhibition of motor neurons, whereas the flexor and crossed extensor reflexes direct specific muscle contractions.

2. *They are intersegmental in distribution.* The interneuron pools extend across spinal segments, and they may activate muscle groups in many parts of the body.

3. *They involve reciprocal innervation.* Reciprocal innervation coordinates muscular contractions and reduces resistance to movement. In the flexor and crossed extensor reflexes, the contraction of one muscle group is associated with the inhibition of opposing muscles.

4. *They have reverberating circuits that prolong the reflexive motor response.* Positive feedback between interneurons that innervate motor neurons and the processing pool maintains the stimulation even after the initial stimulus has faded.

5. *Several reflexes may cooperate to produce a coordinated, controlled response.* As a reflex movement is under way, any antagonistic reflexes are inhibited. There may also be interaction between monosynaptic and polysynaptic reflexes. For example, during the stretch reflex, antagonistic muscles are inhibited; during the tendon reflex, antagonistic muscles are stimulated.

INTEGRATION AND CONTROL OF SPINAL REFLEXES

Reflex motor behaviors occur automatically, without instructions from higher centers. However, higher centers can have a profound effect on reflex performance. Processing centers in the brain can facilitate or inhibit reflex motor patterns based in the spinal cord. Descending tracts originating in the brain synapse on interneurons and motor neurons throughout the spinal cord. These synapses are continuously active, producing EPSPs or IPSPs at the postsynaptic membrane.

Reinforcement and Inhibition

Synapses producing EPSPs make postsynaptic neurons more sensitive to other excitatory stimuli. The resulting level of generalized facilitation rises and falls, depending on the activity in higher centers. For example, a voluntary effort to pull apart clasped hands elevates the general state of facilitation along the spinal cord. This elevated facilitation leads to **reinforcement,** or an enhancement of spinal reflexes.

Other descending fibers have an inhibitory effect on spinal reflexes. In adults, stroking the foot on the side of the sole produces a curling of the toes, called a **plantar reflex,** or *negative Babinski reflex,* after about a 1-second delay (Figure 13-18a●). Stroking an infant's

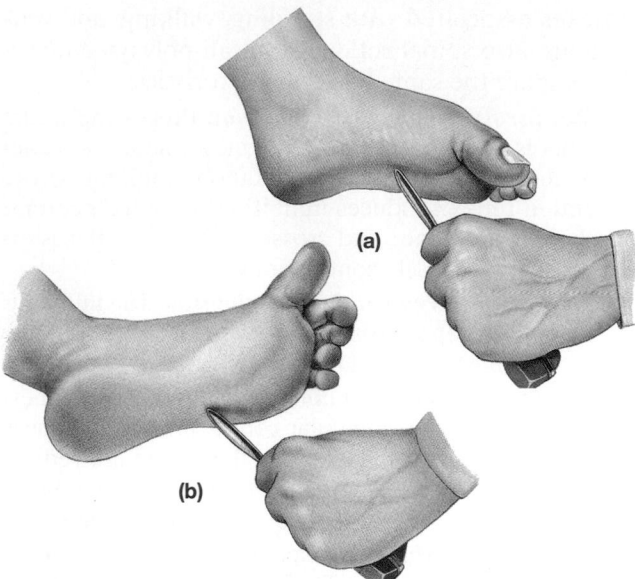

•FIGURE 13-18 The Babinski Reflexes. (a) The plantar reflex (negative Babinski reflex), a curling of the toes, is seen in normal adults. **(b)** The Babinski sign (positive Babinski reflex) occurs in the absence of descending inhibition. It is normal in infants but pathological in adults.

foot on the side of the sole produces a fanning of the toes known as the **Babinski sign,** or *positive Babinski reflex*. This response disappears as descending motor pathways develop. If either the higher centers or the descending tracts are damaged, the Babinski sign will reappear in an adult (Figure 13-18b•). As a result, this reflex is often tested if CNS injury is suspected.

Many reflexes can be assessed through careful observation and the use of simple tools. The procedures are easy to perform, and the results can provide valuable information about damage to the spinal cord or spinal nerves. By testing a series of spinal and cranial reflexes, a physician can assess the function of sensory pathways and motor centers throughout the spinal cord and brain. AM *Reflexes and Diagnostic Testing*

Voluntary Movements and Reflex Motor Patterns

Centers in the brain can interact with the stereotyped motor patterns programmed into the spinal cord. By making use of preexisting patterns, a relatively small number of descending fibers can control complex motor functions. For example, the motor patterns for walking, running, and jumping are primarily directed by neuronal pools in the spinal cord. The descending pathways provide appropriate facilitation, inhibition, or "fine-tuning" of the established patterns.

When complicated voluntary movements are under way, the neurons involved with spinal reflexes assist with muscular coordination and control. For example, the descending tracts that stimulate motor neurons controlling the biceps brachii, brachialis, and other elbow flexors also (1) stimulate the gamma efferents to the intrafusal fibers in these muscles and (2) stimulate inhibitory interneurons to relax antagonistic muscle groups, such as the triceps brachii. The gamma efferents regulate the sensitivity of the stretch reflex, and the interneurons are the same as those activated in the withdrawal and crossed extensor reflexes. As the muscle contraction proceeds, tendon reflexes keep tension production within tolerable limits.

Motor control therefore involves a series of interacting levels. At the lowest level are monosynaptic reflexes that are rapid but stereotyped and relatively inflexible. At the highest level are centers in the brain that can modulate or build on reflexive motor patterns.

☑ A weight lifter is straining to lift a 200-kg barbell. Shortly after he lifts it to chest height, his muscles appear to relax, and he drops the barbell. What reflex does this response demonstrate?

☑ During a withdrawal reflex, what happens to the limb on the side opposite the stimulus? What is this response called?

☑ After suffering an injury to her back, Tina exhibits a positive Babinski reflex. What does this imply about Tina's injury?

SELECTED CLINICAL TERMINOLOGY

Terms Discussed in This Chapter

Babinski sign (*positive Babinski reflex*): A spinal reflex in infants, consisting of a fanning of the toes and produced by stroking the foot on the side of the sole; in adults, a sign of CNS injury. *(p. 440)*

epidural block: Injection of anesthetic into the epidural space to eliminate sensory and motor innervation via spinal nerves in the area of injection. *(p. 420 and AM)*

lumbar puncture: A spinal tap performed between adjacent lumbar vertebrae inferior to the level of the conus medullaris. *(p. 421)*

meningitis: Inflammation of the meninges involving the spinal cord (*spinal meningitis*) and/or brain (*cerebral meningitis*). Bacterial or viral pathogens are generally responsible. *(p. 420 and AM)*

myelography: A diagnostic procedure in which a radiopaque dye is introduced into the cerebrospinal fluid to obtain an X-ray of the spinal cord and cauda equina. *(p. 421)*

nerve growth factor: A peptide factor that promotes the growth and maintenance of neurons. Other factors impor-

tant to neuron growth and repair include BDNF, NT-3, NT-4, GAP-43, and a variety of other compounds. *(p. 427 and AM)*

palsies: Regional losses of sensory and motor function as a result of nerve trauma or compression; also called *peripheral neuropathies*. Common palsies include *radial nerve palsy, ulnar palsy, sciatica,* and *peroneal palsy. (p. 426 and AM)*

paraplegia: Paralysis involving loss of motor control of the lower but not the upper limbs. *(p. 423)*

patellar reflex: The knee jerk reflex resulting from the stimulation of stretch receptors in the quadriceps muscles. *(p. 434)*
plantar reflex *(negative Babinski reflex)*: A spinal reflex in adults, consisting of a curling of the toes and produced by stroking the foot on the side of the sole. *(p. 439)*

quadriplegia: Paralysis involving loss of sensation and motor control of the upper and lower limbs. *(p. 423 and AM)*
shingles: A condition caused by infection of neurons in dorsal root ganglia by the virus *Herpes varicella-zoster*. The primary symptom is a painful rash along the

sensory distribution of the affected spinal nerves. *(p. 425 and AM)*
spinal tap: A procedure in which cerebrospinal fluid is removed from the subarachnoid space through a needle, generally inserted between the lumbar vertebrae. *(p. 420)*

AM Additional Terms Discussed in the *Applications Manual*

---

abdominal reflex: A reflexive twitch in abdominal muscles that moves the navel toward a stimulus; often used to provide information about descending tracts in the spinal cord.
areflexia (ā-re-FLEK-sē-uh): A condition in which normal reflexes fail to appear, even with reinforcement.
caudal anesthesia: The introduction of anesthetics into the epidural space of the sacrum to paralyze lower abdominal and perineal structures.
clonus (KLŌ-nus). Oscillations that develop between opposing stretch reflexes in individuals with hyperreflexia.

Hansen's disease *(leprosy)*: A bacterial infection that begins in sensory nerves of the skin and gradually progresses to a motor paralysis of the same regions.
hyperreflexia: A condition resulting from a high degree of facilitation along the spinal cord; reflexes are easily triggered, and the responses may be grossly exaggerated.
hyporeflexia: A condition in which normal reflexes are weak but demonstrable with reinforcement.
mass reflex: Sudden, abnormal, and prolonged activation of somatic and visceral reflexes throughout the spinal cord,

resulting from an extreme state of motor neuron facilitation.
paresthesia: An abnormal tingling sensation, usually descibed as "pins and needles," that accompanies sensory return after a temporary palsy.
sciatica: (sī-AT-i-kuh): The painful result of compression of the roots of the sciatic nerve.
spinal shock: A period of depressed sensory and motor function following any severe injury to the spinal cord. Spinal shock may accompany *spinal concussion, spinal contusion, spinal laceration, spinal compression,* or *spinal transection.*

CHAPTER REVIEW

 On-line resources for this chapter are on our World Wide Web site at:
http://www.prenhall.com/martini/fap

STUDY OUTLINE

INTRODUCTION, p. 417

1. In addition to relaying information to and from the brain, the spinal cord integrates and processes information on its own.

GROSS ANATOMY OF THE SPINAL CORD, p. 417

1. The adult spinal cord includes localized **enlargements** that provide innervation to the limbs. The spinal cord has 31 segments, each associated with a pair of **dorsal roots** and a pair of **ventral roots.** *(Figure 13-1)*
2. The **filum terminale** (a strand of fibrous tissue) that originates at the **conus medullaris** ultimately becomes part of the coccygeal ligament. *(Figure 13-1)*
3. **Spinal nerves** are **mixed nerves**: They contain both afferent (sensory) and efferent (motor) fibers.

Spinal Meninges, p. 419

4. The **spinal meninges** provide physical stability and shock absorption for neural tissues of the spinal cord; the **cranial meninges** surround the brain. *(Figure 13-2)*
5. The **dura mater** covers the spinal cord; caudally it tapers into the **coccygeal ligament.** The **epidural space** separates the dura mater from the walls of the vertebral canal. *(Figure 13-2)*
6. Interior to the inner surface of the dura mater are the **subdural space,** the **arachnoid** (the second meningeal layer), and the **subarachnoid space.** The subarachnoid space contains **cerebrospinal fluid,** which acts as a shock absorber and a diffusion medium for dissolved gases, nutrients, chemical messengers, and waste products. *(Figure 13-2)*
7. The **pia mater,** a meshwork of elastic and collagen fibers, is the innermost meningeal layer. Unlike more superficial meninges, it is bound to the underlying neural tissue. *(Figure 13-2)*

Sectional Anatomy of the Spinal Cord, p. 420

8. The *white matter* contains myelinated and unmyelinated axons, whereas the *gray matter* contains cell bodies of neurons and glial cells and unmyelinated axons. The projections of gray matter toward the outer surface of the spinal cord are called **horns.** *(Figure 13-4)*
9. The **posterior gray horns** contain somatic and visceral sensory nuclei; nuclei in the **anterior gray horns** deal with somatic motor control. The **lateral gray horns** contain visceral motor neurons. The **gray commissures** contain axons that cross from one side of the cord to the other. *(Figure 13-4)*
10. The white matter can be divided into six **columns** *(funiculi)*, each of which contains **tracts** *(fasciculi)*. **Ascending tracts** relay information from the spinal cord to the brain, and **descending tracts** carry information from the brain to the spinal cord. *(Figure 13-4)*

SPINAL NERVES, p. 423

1. There are 31 pairs of spinal nerves. Each has an **epineurium** (outermost layer), **perineurium,** and **endoneurium** (innermost layer). *(Figure 13-5)*

Peripheral Distribution of Spinal Nerves, p. 424

2. A typical spinal nerve has a **white ramus** (containing myelinated axons), a **gray ramus** (containing unmyelinated fibers that innervate glands and smooth muscles in the body wall or limbs), a **dorsal ramus** (providing sensory and motor innervation to the skin and muscles of the back), and a **ventral ramus** (supplying the ventrolateral body surface, structures in the body wall, and the limbs). Each pair of nerves monitors a region of the body surface called a **dermatome.** *(Figures 13-6, 13-7)*

Nerve Plexuses, p. 425

3. A complex, interwoven network of nerves is a **nerve plexus.** The three large plexuses are the **cervical plexus,** the **brachial plexus,** and the **lumbosacral plexus.** The latter can be divided into the **lumbar plexus** and the **sacral plexus.** *(Figures 13-8–13-11; Tables 13-1–13-3)*

AN INTRODUCTION TO REFLEXES, p. 430

1. **Reflexes** are rapid, automatic responses to stimuli. A *neural reflex* is an automatic, involuntary motor response by the nervous system that helps preserve homeostasis by rapidly adjusting the functions of organs or organ systems.

The Reflex Arc, p. 431

2. A **reflex arc** is the neural "wiring" of a single reflex. *(Figure 13-12)*

3. There are five steps involved in a neural reflex: (1) arrival of a stimulus and activation of a receptor, (2) activation of a sensory neuron, (3) information processing, (4) activation of a motor neuron, and (5) response by an effector. *(Figure 13-12)*

4. A *receptor* is a specialized cell that monitors conditions in the body or external environment. Each receptor has a characteristic range of sensitivity.

Classification of Reflexes, p. 432

5. Reflexes are classified according to (1) their development, (2) the site of information processing, (3) the nature of the resulting motor response, and (4) the complexity of the neural circuit.

6. **Innate reflexes** result from the connections that form between neurons during development. **Acquired reflexes** are learned and typically are more complex. *(Figure 13-13)*

7. Reflexes processed in the brain are **cranial reflexes.** In a **spinal reflex,** the important interconnections and processing events occur inside the spinal cord. *(Figure 13-13)*

8. **Somatic reflexes** control skeletal muscles, and **visceral reflexes** (*autonomic reflexes*) control the activities of other systems.

9. In a **monosynaptic reflex**, the simplest reflex arc, a sensory neuron synapses directly on a motor neuron, which acts as the processing center. **Polysynaptic reflexes**, which have at least one interneuron between the sensory afferent and the motor efferent, have a longer delay between stimulus and response. *(Figure 13-14)*

SPINAL REFLEXES, p. 433

1. Spinal reflexes range from simple monosynaptic reflexes to more complex polysynaptic reflexes, in which many segments interact to produce a coordinated motor response.

Monosynaptic Reflexes, p. 433

2. The **stretch reflex** (such as the **patellar,** or **knee jerk**, **reflex**) is a monosynaptic reflex that automatically regulates skeletal muscle length and muscle tone. The sensory receptors involved are **muscle spindles.** *(Figures 13-15, 13-16)*

3. A **postural reflex** maintains normal upright posture.

Polysynaptic Reflexes, p. 438

4. Polysynaptic reflexes can produce more complicated responses than can monosynaptic reflexes. Examples include the **tendon reflex** (which monitors the tension produced during muscular contractions and prevents damage to tendons) and **withdrawal reflexes** (which move affected portions of the body away from a source of stimulation). The **flexor reflex** is a withdrawal reflex affecting the muscles of a limb; the **crossed extensor reflex** complements the withdrawal reflex. *(Figure 13-17)*

5. In an **ipsilateral reflex arc,** the sensory stimulus and motor response occur on the same side of the body. In a **contralateral reflex arc,** the motor response occurs on the side opposite the stimulus. *(Figure 13–17)*

6. All polysynaptic reflexes (1) involve pools of interneurons, (2) are intersegmental in distribution, (3) involve reciprocal innervation, and (4) have reverberating circuits that prolong the reflexive motor response; (5) several reflexes may cooperate to produce a coordinated response.

INTEGRATION AND CONTROL OF SPINAL REFLEXES, p. 439

1. The brain can facilitate or inhibit reflex motor patterns based in the spinal cord.

Reinforcement and Inhibition, p. 439

2. The enhancement of spinal reflexes is called **reinforcement.** Spinal reflexes may instead be inhibited, as in the **plantar reflex** in adults versus the **Babinski sign** in infants. *(Figure 13-18)*

Voluntary Movements and Reflex Motor Patterns, p. 440

3. Motor control involves a series of interacting levels. Monosynaptic reflexes form the lowest level; at the highest level are the centers in the brain that can modulate or build on reflexive motor patterns.

REVIEW QUESTIONS

Level 1 Reviewing Facts and Terms

1. The expanded area of the spinal cord that supplies nerves to the shoulder girdle and arms is the
   (a) conus medullaris
   (b) filum terminale
   (c) lumbar enlargement
   (d) cervical enlargement

2. The dorsal roots of each spinal segment
   (a) bring sensory information into the spinal cord
   (b) control peripheral effectors
   (c) contain both somatic motor and visceral motor neurons
   (d) a, b, and c are correct

3. Spinal nerves are called mixed nerves because
   (a) they contain sensory and motor fibers
   (b) they exit at intervertebral foramina
   (c) they are associated with a pair of dorsal root ganglia
   (d) they are associated with dorsal and ventral roots

4. The spinal cord continues to enlarge and elongate until an individual is
   (a) born   (b) 1 year old
   (c) 4 years old   (d) 10 years old

5. The tough, fibrous outermost covering of the spinal cord is the
   (a) arachnoid   (b) pia mater
   (c) dura mater   (d) epidural block

6. The gray matter of the spinal cord is dominated by
  (a) myelinated axons only
  (b) unmyelinated axons only
  (c) Schwann cells and satellite cells
  (d) cell bodies of neurons, glial cells, and unmyelinated axons

7. The outermost layer of connective tissue that surrounds each spinal nerve is the
  (a) perineurium   **(b)** epineurium
  (c) endoneurium   **(d)** epimysium

8. A sensory region monitored by the dorsal rami of a single spinal segment is
  (a) a ganglion   **(b)** a fascicle
  (c) a dermatome   **(d)** a ramus

9. The major nerve of the cervical plexus that innervates the diaphragm is the
  (a) median nerve   **(b)** axillary nerve
  (c) phrenic nerve   **(d)** peroneal nerve

10. The genitofemoral, femoral, and lateral femoral cutaneous nerves are major nerves of the
  (a) lumbar plexus   **(b)** sacral plexus
  (c) brachial plexus **(d)** cervical plexus

11. The final step in a neural reflex is
  (a) activation of a receptor
  (b) response by an effector
  (c) activation of a motor neuron
  (d) information processing

12. The reflexes that control the most rapid, stereotyped motor responses of the nervous system to stimuli are
  (a) monosynaptic reflexes
  (b) polysynaptic reflexes
  (c) tendon reflexes

13. An example of a stretch reflex triggered by passive muscle movement is the
  (a) tendon jerk   **(b)** knee jerk
  (c) flexor reflex   **(d)** ipsilateral reflex

14. The contraction of flexor muscles and the relaxation of extensor muscles illustrates the principle of
  (a) reverberating circuitry **(b)** generalized facilitation
  (c) reciprocal inhibition   **(d)** reinforcement

15. Reflex arcs in which the sensory stimulus and the motor response occur on the same side of the body are
  (a) contralateral   **(b)** paraesthetic
  (c) ipsilateral   **(d)** monosynaptic

16. All polysynaptic reflexes
  (a) have reverberating circuits that prolong the reflexive motor response
  (b) involve pools of interneurons
  (c) involve reciprocal innervation
  (d) a, b, and c are correct

17. Which three meningeal layers provide physical stability and shock absorption for the spinal cord?

18. Beginning with the outermost layer, list the connective tissue layers that surround each spinal nerve.

Level 2   Reviewing Concepts

19. Polysynaptic reflexes can produce far more complicated responses than can monosynaptic reflexes because
  (a) the response time is quicker
  (b) the response is initiated by highly sensitive receptors
  (c) motor neurons carry impulses at a faster rate than do sensory neurons
  (d) the interneurons involved can control several different muscle groups

20. Why do cervical nerves outnumber cervical vertebrae?

21. If the anterior gray horns of the spinal cord were damaged, what type of control would be affected?

22. Proceeding inward from the outermost layer, number the following in the correct sequence:
  __ walls of vertebral canal   __ subdural space
  __ pia mater   __ subarachnoid space
  __ dura mater   __ epidural space
  __ arachnoid membrane   __ spinal cord

23. In which of the structures in Question 22 is cerebrospinal fluid (CSF) located? What are the functions of the CSF?

24. What five characteristics are common to all polysynaptic reflexes?

25. Which ramus would be involved if you felt something brushing against your
  (a) abdomen   **(b)** left forearm
  (c) upper back   **(d)** right ankle

26. Which plexus has primary responsibility for innervating the following muscles?
  (a) diaphragm   **(b)** pelvic diaphragm
  (c) deltoid   **(d)** gastrocnemius

27. Why is response time in a monosynaptic reflex much faster than the response time in a polysynaptic reflex?

28. List, in sequence, the five steps involved in a neural reflex.

29. What would happen if the ventral root of a spinal nerve were damaged or transected?

30. How does the stimulation of a sensory neuron that innervates an extrafusal muscle fiber affect muscle tone?

31. Why is it important that a spinal tap be done between the third and fourth lumbar vertebrae?

Level 3   Critical Thinking and Clinical Applications

32. Mary complains that when she wakes up in the morning, her thumb and forefinger are always "asleep." She mentions this condition to her physician, who asks if she sleeps with her arm above her head. She replies that she does. Her physician tells her that sleeping in that position compresses a portion of one of the nerves of the brachial plexus, producing her symptoms. Which nerve is involved?

33. Improper use of crutches can produce a condition known as "crutch paralysis," which is characterized by lack of response by

the extensor muscles of the arm and a condition known as wrist drop. Which nerve is involved?

34. During childbirth, some women are given a local anesthetic that temporarily deadens sensory neurons in the region of the genitals. Which nerve does this anesthetic affect?

35. While playing football, Ramon is tackled hard and suffers an injury to his left leg. As he tries to get up, he cannot flex his left thigh or extend the leg. Which nerve is damaged, and how would this damage affect sensory perception in his left leg?

14

The Brain and Cranial Nerves

*O*ur perceptions of the world around us depend on thousands of interactions
among neurons within the central nervous system. We seldom realize how complex these
processes are unless they go wrong in some way. This child has dyslexia, a condition characterized
by difficulties with the recognition and use of words. Although the cause of dyslexia remains a mystery,
there is general agreement that it results from problems with the integration and processing of visual or
auditory information. There is much that we still do not understand about such activities, which ultimately
create our consciousness and our unique personalities. This chapter introduces the brain regions involved in
our conscious and subconscious thought processes and considers complex neural functions, such as
memory and learning.

CHAPTER OUTLINE AND OBJECTIVES

The brain is probably the most fascinating organ in the body, yet we know relatively little about its structural and functional complexities. What we do know is that all our dreams, passions, plans, and memories result from brain activity. The brain contains tens of billions of neurons organized into hundreds of neuronal pools; it has a complex three-dimensional structure and performs a bewildering array of functions. If an individual's heart, liver, lung, or kidney stops working, heroic measures, including organ transplantation, can be taken to preserve the person's life. But if the brain permanently stops working, the person is classified as "brain dead" and for all intents and purposes ceases to exist.

The brain is far more complex than the spinal cord, and it can respond to stimuli with greater versatility. That versatility results from (1) the tremendous number of neurons and neuronal pools in the brain and (2) the complexity of the interconnections between those neurons and neuronal pools. Because the interconnections are complex and the pools are large, the response can be varied to meet changing circumstances. But this versatility has a price: A response cannot be immediate, precise, and adaptable all at the same time. Adaptability requires multiple processing steps, and every synapse adds to the delay between stimulus and response. One of the major functions of spinal reflexes is to provide an *immediate* response that can be fine-tuned or elaborated on by more versatile—but slower—processing centers in the brain. For example, the flexor reflex will pull your hand from a hot stove before you become consciously aware that something is wrong. In effect, the spinal reflex prevents further injury and gives the brain time to decide what to do next. ∞ *[p. 438]*

In this chapter, we consider the brain and the cranial nerves and their primary functional roles. The chapter organization, which parallels that of Chapter 13, will make it easy for you to compare the functional organization of the brain with that of the spinal cord.

AN INTRODUCTION TO THE ORGANIZATION OF THE BRAIN

The adult human brain contains almost 98 percent of the body's neural tissue. A "typical" brain weighs 1.4 kg (3 lb) and has a volume of 1200 cc (71 in.^{3}). There is considerable individual variation in brain size, and the brains of males average about 10 percent larger than those of females, owing to differences in average body size. No correlation exists between brain size and intelligence. Individuals with the smallest brains (750 cc) and the largest brains (2100 cc) are functionally normal.

Embryology of the Brain

To introduce the organization of the adult brain, we will consider its embryological origins. The development of the brain is detailed in the Embryology Summary on pages 454–455. The CNS begins as a hollow *neural tube* with a fluid-filled internal cavity called the *neurocoel*. In the cephalic portion of the neural tube, three areas enlarge rapidly through expansion of the neurocoel. This enlargement creates three prominent divisions called **primary brain vesicles.** The primary brain vesicles are named for their relative positions: the **prosencephalon** (prō-zen-SEF-a-lon; *proso,* forward + *enkephalos,* brain), or "forebrain"; the **mesencephalon** (*mesos,* middle), or "midbrain"; and the **rhombencephalon** (rom-ben-SEF-a-lon), or "hindbrain."

The fate of the three primary divisions of the brain is summarized in Table 14-1. The prosencephalon and rhombencephalon are subdivided further, forming **secondary brain vesicles.** The prosencephalon forms the **telencephalon** (tel-en-SEF-a-lon; *telos,* end) and the **diencephalon** (dī-en-SEF-a-lon; *dia,* through). The telencephalon will ultimately form the *cerebrum* of the adult brain. The mesencephalon thickens, and the neurocoel becomes a relatively narrow passageway comparable to the central canal of the spinal cord. The portion of the rhombencephalon adjacent to the mesencephalon forms the **metencephalon** (met-en-SEF-a-lon; *meta,* after). The dorsal portion of the metencephalon will become the *cerebellum,* and the ventral portion will develop into the *pons.* The portion of the rhombencephalon closer to the spinal cord forms the **myelencephalon** (mī-el-en-SEF-a-lon; *myelon,* spinal cord), which will become the *medulla oblongata.*

A Preview of Major Regions and Landmarks

The adult brain is dominated in size by the cerebrum:

- *Cerebrum.* Viewed from the anterior and superior surfaces (Figure 14-1a,b●, p. 447), the **cerebrum** (SER-e-brum, or se-RĒ-brum) of the adult brain can be divided into large, paired **cerebral hemispheres.** Conscious thoughts, sensations, intellect, memory, and complex movements originate in the cerebrum. The surfaces of the cerebral hemispheres and cerebellum are highly folded and covered by **neural cortex** (*cortex,* rind or bark), a superficial layer of gray matter. The term *cerebral cortex* refers to the neural cortex of the cerebral hemispheres, as opposed to the *cerebellar cortex* of the cerebellar hemispheres.

Second in size to the cerebrum is the cerebellum:

- *Cerebellum.* The hemispheres of the **cerebellum** (ser-e-BEL-um) are partially hidden by the cerebral hemispheres (Figure 14-1b–d●). The cerebellum adjusts ongoing movements on the basis of comparisons

TABLE 14-1 Development of the Human Brain

Primary Brain Vesicles (3 weeks)	Secondary Brain Vesicles (6 weeks)	Brain Regions at Birth
Prosencephalon	Telencephalon	Cerebrum
	Diencephalon	Diencephalon
Mesencephalon	Mesencephalon	Mesencephalon
Rhombencephalon	Metencephalon	Cerebellum
		Pons
	Myelencephalon	Medulla oblongata

between arriving sensations and sensations experienced previously, allowing you to perform the same movements.

The other major regions of the brain can best be examined after the cerebral hemispheres have been removed (Figure 14-2•, p. 448).

- *Diencephalon.* The walls of the diencephalon are composed of the **left** and **right thalamus** (THAL-a-mus). Each thalamus contains relay and processing centers for sensory information. A narrow stalk, the *infundibulum,* connects the **hypothalamus** (*hypo-,* below), or floor of the diencephalon, to the **pituitary gland,** a component of the endocrine system. The hypothalamus contains centers involved with emotions, autonomic function, and hormone production. As we shall see in Chapter 18, it is the primary link between the nervous and endocrine systems.

The diencephalon is a structural and functional link between the cerebral hemispheres and the components of the *brain stem.* The **brain stem** includes the mesencephalon, pons, and medulla oblongata.[1] It contains a variety of important processing centers and nuclei that relay information headed to or from the cerebrum or cerebellum:

- *Mesencephalon.* Sensory nuclei in the mesencephalon, or midbrain, process visual and auditory information

and control reflexes triggered by these stimuli. For example, your immediate, reflexive responses to a loud, unexpected noise (eye movements and head turning) are directed by nuclei in the midbrain. This region also contains centers involved with the maintenance of consciousness.

- *Pons.* The term *pons* is Latin for "bridge"; the **pons** of the brain connects the cerebellum to the brain stem. In addition to tracts and relay centers, the pons also contains nuclei involved with somatic and visceral motor control.

- *Medulla oblongata.* The spinal cord connects to the brain at the **medulla oblongata**. Near the pons, the roof of the medulla oblongata is thin and membranous. The caudal portion of the medulla oblongata resembles the spinal cord in that it has a narrow central canal. The medulla oblongata relays sensory information to the thalamus and to centers in other portions of the brain stem. The medulla oblongata also contains major centers concerned with the regulation of autonomic function, such as heart rate, blood pressure, and digestion.

The boundaries and general functions of the diencephalon and brain stem are listed in Figure 14-2•.

☑ What are the three primary brain vesicles, and what does each contribute to the structure of the adult brain?

☑ In response to a loud noise, your head automatically turns toward the source of the sound. What part of the brain directs this response?

---

[1]Some sources consider the brain stem to include the diencephalon. We will use the more restrictive definition here.

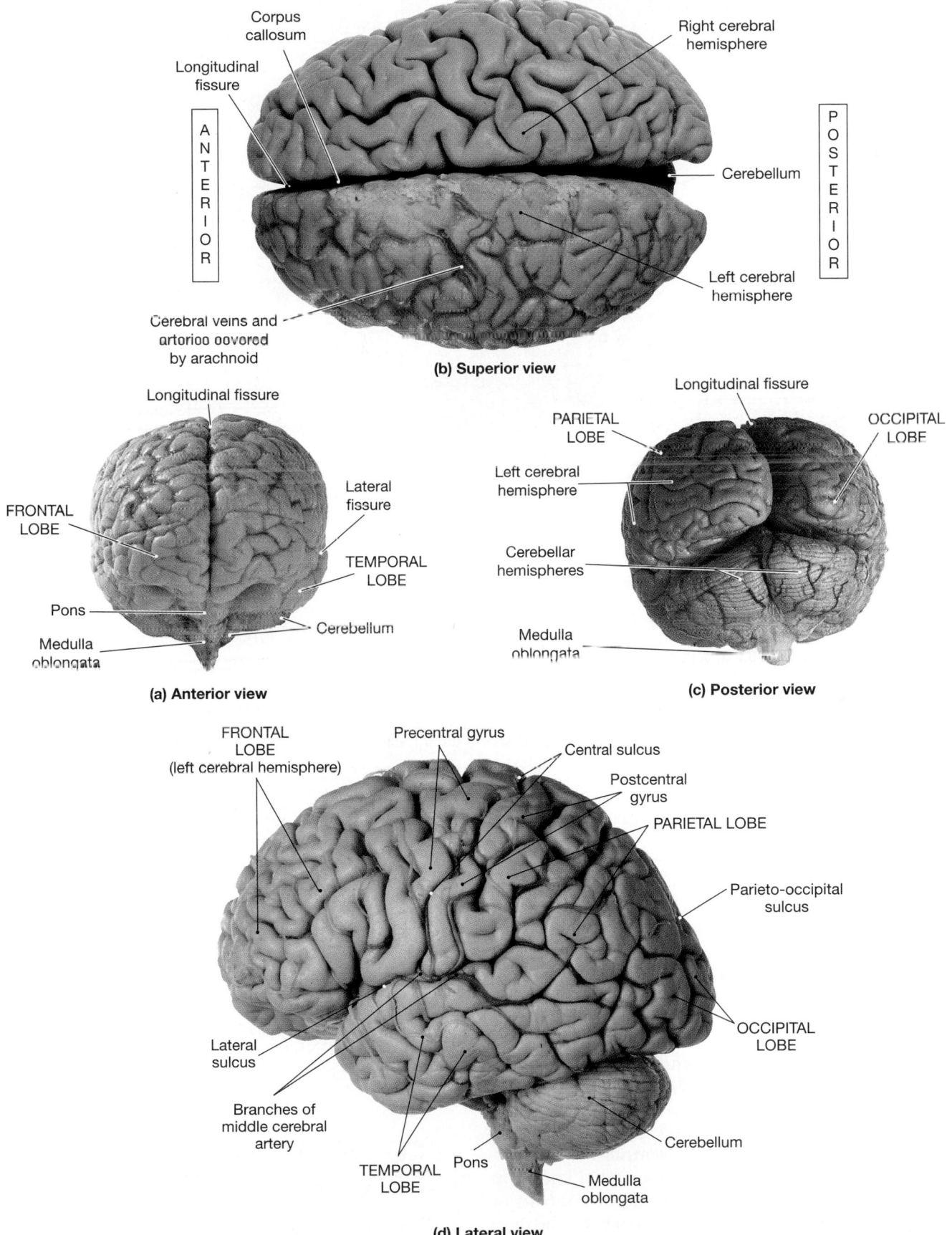

•**FIGURE 14-1** **The Human Brain.** **(a)** Anterior view of the brain. **(b)** Superior view, which is dominated by the paired cerebral hemispheres. **(c)** Posterior view. **(d)** Lateral view.

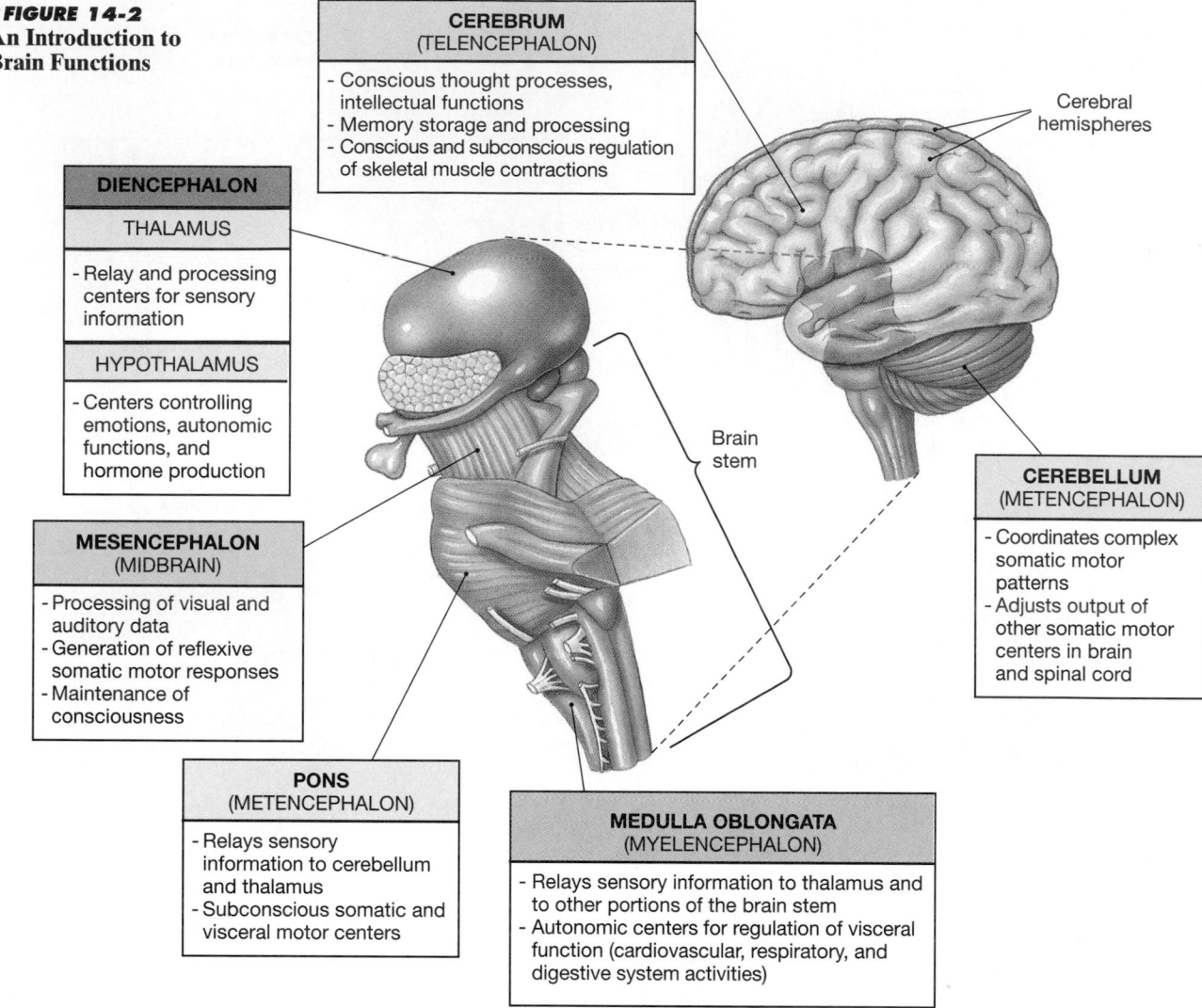

•FIGURE 14-2
An Introduction to Brain Functions

CEREBRUM (TELENCEPHALON)
- Conscious thought processes, intellectual functions
- Memory storage and processing
- Conscious and subconscious regulation of skeletal muscle contractions

DIENCEPHALON

THALAMUS
- Relay and processing centers for sensory information

HYPOTHALAMUS
- Centers controlling emotions, autonomic functions, and hormone production

MESENCEPHALON (MIDBRAIN)
- Processing of visual and auditory data
- Generation of reflexive somatic motor responses
- Maintenance of consciousness

PONS (METENCEPHALON)
- Relays sensory information to cerebellum and thalamus
- Subconscious somatic and visceral motor centers

MEDULLA OBLONGATA (MYELENCEPHALON)
- Relays sensory information to thalamus and to other portions of the brain stem
- Autonomic centers for regulation of visceral function (cardiovascular, respiratory, and digestive system activities)

CEREBELLUM (METENCEPHALON)
- Coordinates complex somatic motor patterns
- Adjusts output of other somatic motor centers in brain and spinal cord

Cerebral hemispheres

Brain stem

Ventricles of the Brain

During development, the neurocoel within the cerebral hemispheres, diencephalon, metencephalon, and medulla oblongata expands to form chambers called **ventricles** (VEN-tri-kls). The ventricles are lined by cells of the *ependyma.* ∞ *[p. 376]*

Each cerebral hemisphere contains an enlarged ventricle. A thin medial partition, the **septum pellucidum,** separates this pair of **lateral ventricles.** There is no direct connection between the two lateral ventricles, but each communicates with the ventricle of the diencephalon through an **interventricular foramen** (*foramen of Monro*) (Figure 14-3•). Because there are two lateral ventricles (first and second), the one in the diencephalon is called the **third ventricle.**

The mesencephalon has a slender canal known as the **mesencephalic aqueduct** (*aqueduct of Sylvius* or *cerebral aqueduct*). This passageway connects the third ventricle with the **fourth ventricle.** The superior portion of the fourth ventricle lies between the posterior surface of the pons and the anterior surface of the

cerebellum. The fourth ventricle extends into the superior portion of the medulla oblongata (Figure 14-3a•). This ventricle then narrows and becomes continuous with the central canal of the spinal cord (Figure 14-3a,c•).

The ventricles are filled with cerebrospinal fluid. There is a continuous circulation of cerebrospinal fluid (CSF) from the ventricles and central canal into the *subarachnoid space* of the meninges that surround the CNS. The CSF passes between the interior and exterior of the CNS through foramina in the roof of the fourth ventricle.

PROTECTION AND SUPPORT OF THE BRAIN

The delicate tissues of the brain are protected from mechanical forces by (1) the bones of the cranium, (2) the *cranial meninges,* and (3) cerebrospinal fluid. In addition, the neural tissue of the brain is biochemi-

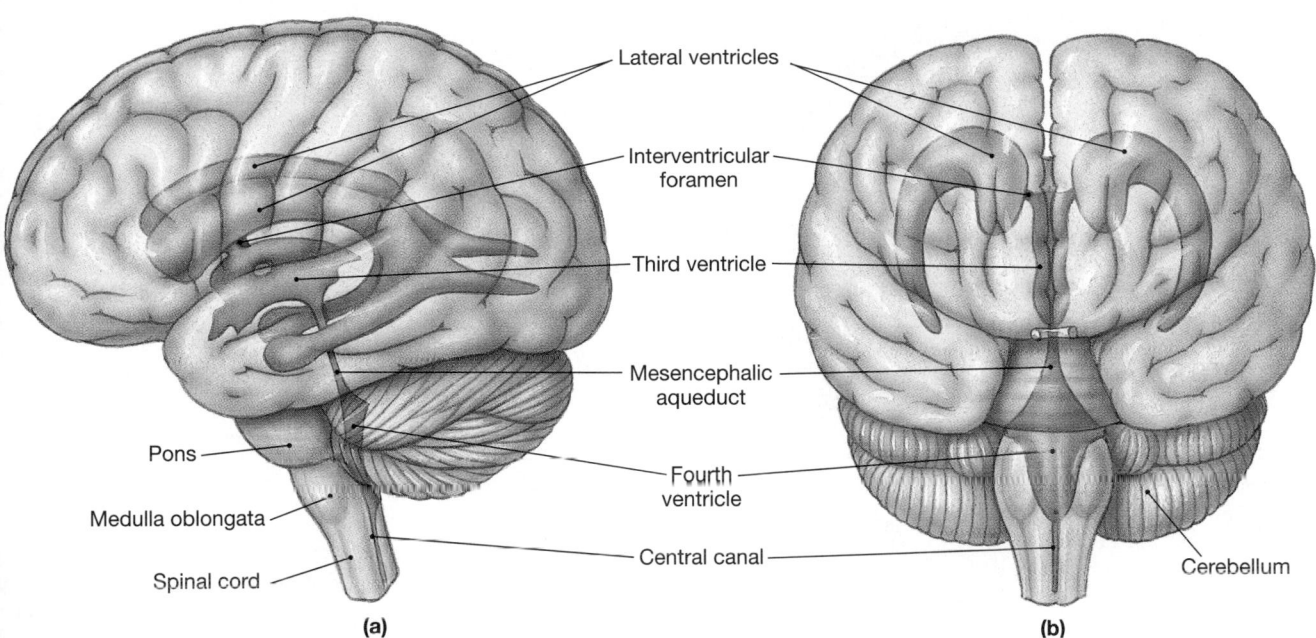

(a)

(b)

•FIGURE 14-3 Ventricles of the Brain. Orientation and extent of the ventricles as they would appear if the brain were transparent. **(a)** Lateral view. **(b)** Anterior view, showing the relationships among the lateral ventricles and the third ventricle. **(c)** A diagrammatic view of the CNS in frontal section, showing the ventricles and central canal. If you compare this diagram with (a), you will find it impossible to make a straight cut that would show all the connections between ventricles. *(See Figure 14-12b, p. 466.)* [AM] *Scans 1, 2*

cally isolated from the general circulation by the *blood–brain barrier.* Refer to Figures 7-3 and 7-4• (pp. 201–203) for a review of the bones of the cranium. We will discuss the other protective factors here.

The Cranial Meninges

The layers that make up the cranial meninges—the *dura mater, arachnoid,* and *pia mater*—are continuous with those of the spinal cord. ∞ *[p. 419]* However, the cranial meninges have distinctive anatomical and functional characteristics.

- The *dura mater* consists of outer and inner fibrous layers. The outer layer is fused to the periosteum of the cranial bones. As a result, there is no *epidural space* comparable to that surrounding the spinal cord. The outer, or *endosteal,* and inner, or *meningeal,* layers are separated by a slender gap that contains tissue fluids and blood vessels, including large venous channels known as **dural sinuses.** The veins of the brain open into these sinuses, which deliver the venous blood to the *internal jugular veins* of the neck.
- The *arachnoid* consists of the *arachnoid membrane* and the cells and fibers of the subarachnoid space. The arachnoid membrane covers the brain, providing a smooth surface that does not follow the brain's underlying folds. The arachnoid membrane, an epithelial

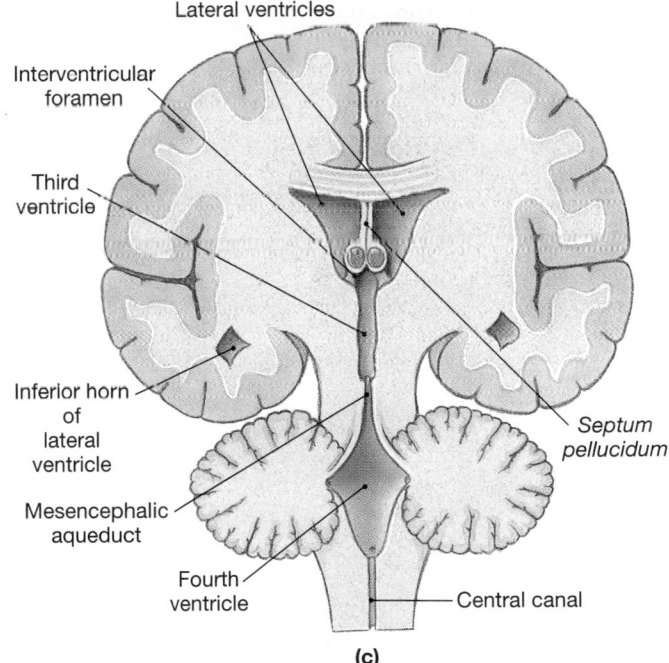

(c)

layer, is in contact with the inner epithelial layer of the dura mater. The subarachnoid space extends between the arachnoid membrane and the pia mater.

- The *pia mater* sticks to the surface of the brain, anchored by the processes of astrocytes. It extends into every fold and curve and accompanies the branches of cerebral blood vessels as they penetrate the surface of the brain to reach internal structures.

Functions of the Cranial Meninges

The brain is cradled within the cranium. There is an obvious correspondence between the shape of the brain and that of the cranial cavity (Figure 14-4a•).

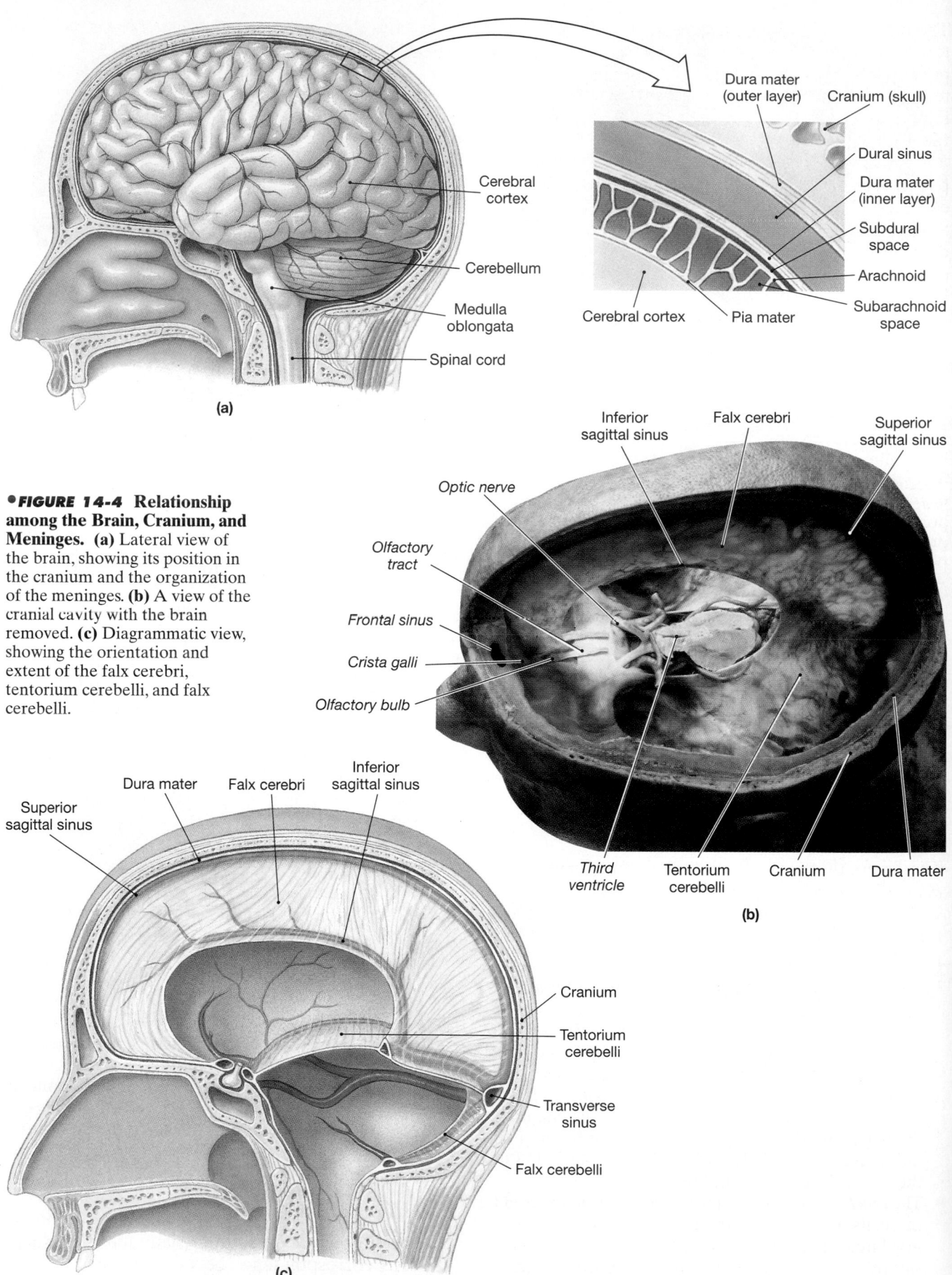

•*FIGURE 14-4* **Relationship among the Brain, Cranium, and Meninges.** **(a)** Lateral view of the brain, showing its position in the cranium and the organization of the meninges. **(b)** A view of the cranial cavity with the brain removed. **(c)** Diagrammatic view, showing the orientation and extent of the falx cerebri, tentorium cerebelli, and falx cerebelli.

(a)

Cerebral cortex
Cerebellum
Medulla oblongata
Spinal cord

Dura mater (outer layer)
Cranium (skull)
Dural sinus
Dura mater (inner layer)
Subdural space
Arachnoid
Subarachnoid space
Cerebral cortex
Pia mater

Inferior sagittal sinus
Falx cerebri
Superior sagittal sinus
Optic nerve
Olfactory tract
Frontal sinus
Crista galli
Olfactory bulb
Third ventricle
Tentorium cerebelli
Cranium
Dura mater

(b)

Superior sagittal sinus
Dura mater
Falx cerebri
Inferior sagittal sinus
Cranium
Tentorium cerebelli
Transverse sinus
Falx cerebelli

(c)

The massive cranial bones provide mechanical protection, but they also pose a threat. The brain is like a person driving a car. If the car hits a tree, the car protects the driver from contact with the tree, but serious injury will occur unless a seat belt or airbag protects the driver from contact with the car.

Cranial trauma is a head injury resulting from impact with another object. There are roughly 8 million cases of cranial trauma each year in the United States, but only 1 case in 8 results in serious brain damage. The percentage is relatively low because the cranial meninges provide effective protection for the brain. Tough, fibrous *dural folds* act like safety belts that hold the brain in position. The cerebrospinal fluid contained in the subarachnoid space acts like an airbag that cushions sudden jolts and shocks. |AM|
Cranial Trauma

DURAL FOLDS In several locations, the inner layer of the dura mater extends into the cranial cavity, forming a sheet that dips inward and then returns. These **dural folds** provide additional stabilization and support to the brain. Dural sinuses may be found between the two layers of a dural fold. The three largest dural folds are called the *falx cerebri,* the *tentorium cerebelli,* and the *falx cerebelli* (Figure 14-4b•):

1. The **falx cerebri** (FALKS ser-Ē brē; *falx,* curving or sickle-shaped) is a fold of dura mater that projects between the cerebral hemispheres in the longitudinal fissure. Its inferior portions attach anteriorly to the crista galli and posteriorly to the *internal occipital crest* (see Figure 7-4b•, p. 203). Two large venous sinuses, the **superior sagittal sinus** and the **inferior sagittal sinus,** travel within this dural fold. The posterior margin of the falx cerebri intersects the *tentorium cerebelli.*

2. The **tentorium cerebelli** (ten-TOR-ē-um ser-e-BEL-ē; *tentorium,* a covering) separates and protects the cerebellar hemispheres from those of the cerebrum. It extends across the cranium at right angles to the falx cerebri. The **transverse sinus** lies within the tentorium cerebelli.

3. The **falx cerebelli** divides the two cerebellar hemispheres along the midsagittal line inferior to the tentorium cerebelli.

Cerebrospinal Fluid

Cerebrospinal fluid (CSF) completely surrounds and bathes the exposed surfaces of the CNS. The CSF has several important functions, including the following:

- *Cushioning delicate neural structures.*
- *Supporting the brain.* In essence, the brain is suspended inside the cranium and floats in the cerebrospinal fluid. A human brain weighs about 1400 g in air but only about 50 g when supported by the cerebrospinal fluid.

- *Transporting nutrients, chemical messengers, and waste products.* Except at the *choroid plexus,* where CSF is produced, the ependymal lining is freely permeable, and the CSF is in constant chemical communication with the interstitial fluid of the CNS.

Because free exchange occurs between the interstitial fluid and CSF, changes in CNS function may produce changes in the composition of the CSF. As we noted in Chapter 13, a *spinal tap* can provide useful clinical information concerning CNS injury, infection, or disease. ∞ *[pp. 420, 421]*

The Formation of CSF

The **choroid plexus** (*choroid,* vascular coat, *plexus,* network) consists of a combination of specialized ependymal cells and permeable capillaries for the production of cerebrospinal fluid. Two extensive folds of the choroid plexus originate in the roof of the third ventricle and extend through the interventricular foramina. These folds cover the floors of the lateral ventricles (Figure 14-5a•). In the lower brain stem, a region of the choroid plexus in the roof of the fourth ventricle projects between the cerebellum and pons.

Specialized ependymal cells, interconnected by tight junctions, surround the capillaries of the choroid plexus. The ependymal cells secrete CSF into the ventricles; they also remove waste products from the CSF and adjust its composition over time. The differences in composition between CSF and blood plasma (blood with the cellular elements removed) are quite pronounced. For example, the blood contains high concentrations of soluble proteins, but the CSF does not. There are also differences in the concentrations of individual ions and in the levels of amino acids, lipids, and waste products.

Circulation of CSF

The choroid plexus produces CSF at a rate of about 500 ml/day. The total volume of CSF at any given moment is approximately 150 ml; thus, the entire volume of CSF is replaced roughly every 8 hours. Despite this rapid turnover, the composition of CSF is closely regulated, and the rate of removal normally keeps pace with the rate of production.

The CSF circulates from the choroid plexus through the ventricles and the central canal of the spinal cord (Figure 14-5a•). As the CSF circulates, there is unrestricted diffusion between it and the interstitial fluid of the CNS between and across the ependymal cells. The CSF reaches the subarachnoid space through two **lateral apertures** and a single **median aperture** in the roof of the fourth ventricle. Cerebrospinal fluid then flows through the subarachnoid space surrounding the brain, spinal cord, and cauda equina.

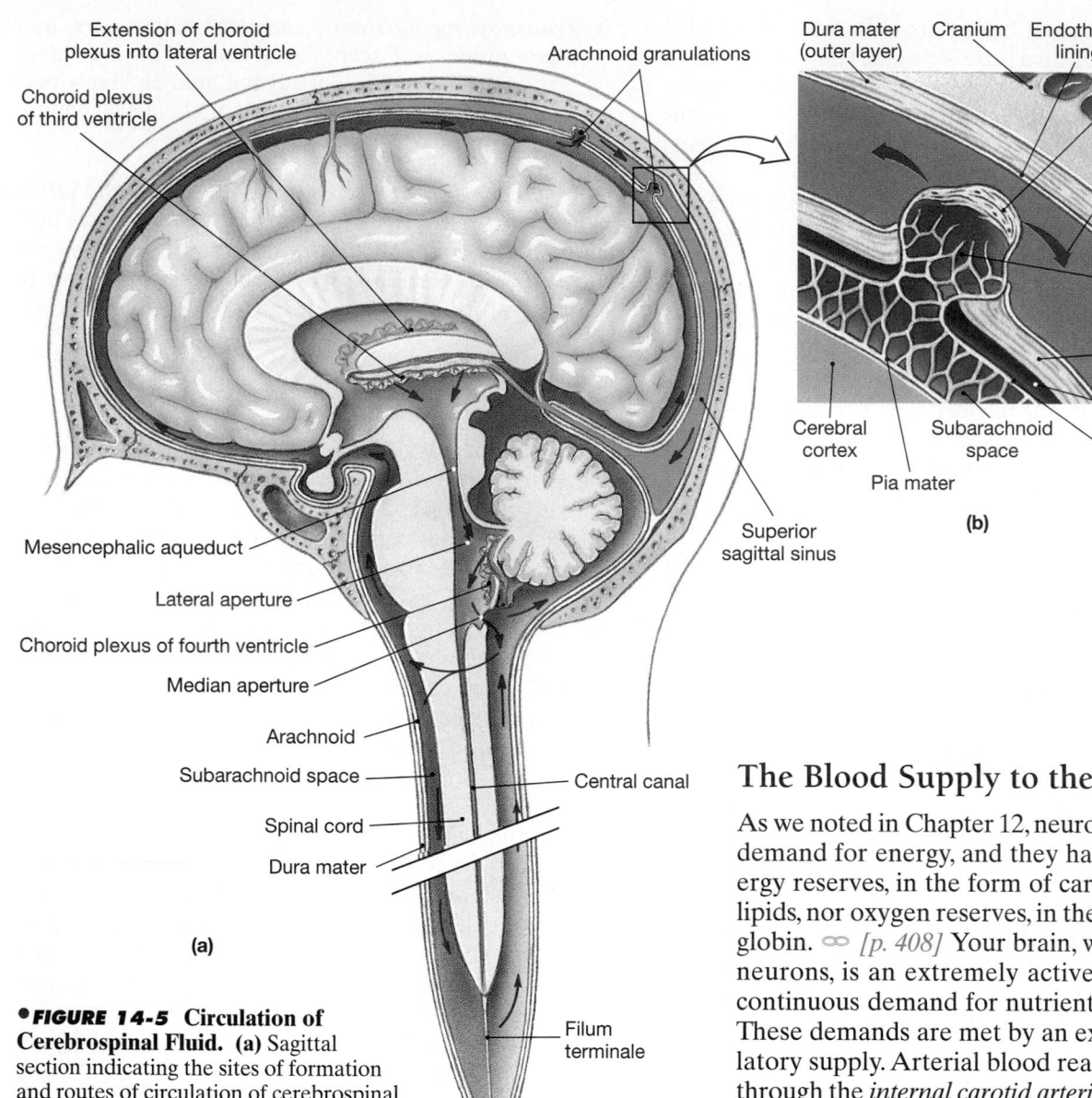

Extension of choroid plexus into lateral ventricle

Choroid plexus of third ventricle

Arachnoid granulations

Mesencephalic aqueduct

Lateral aperture

Choroid plexus of fourth ventricle

Median aperture

Arachnoid

Subarachnoid space

Spinal cord

Dura mater

Central canal

Superior sagittal sinus

Filum terminale

(a)

Dura mater (outer layer)   Cranium   Endothelial lining   Epithelial cap

Fluid movement

Superior sagittal sinus

Arachnoid granulation

Dura mater (inner layer)

Subdural space

Arachnoid

Cerebral cortex   Subarachnoid space

Pia mater

(b)

•FIGURE 14-5 **Circulation of Cerebrospinal Fluid.** **(a)** Sagittal section indicating the sites of formation and routes of circulation of cerebrospinal fluid (red arrows). **(b)** Orientation of the arachnoid granulations. AM *Scan 2*

Along the axis of the superior sagittal sinus, fingerlike extensions of the arachnoid membrane, called the *arachnoid villi*, penetrate the dura mater. In adults, clusters of villi form large **arachnoid granulations** (Figure 14-5b•). Cerebrospinal fluid is absorbed into the venous circulation at the arachnoid granulations. If the normal circulation or reabsorption of CSF is interrupted, a variety of clinical problems may appear. For example, a problem with the reabsorption of CSF in infancy is responsible for symptoms of *hydrocephalus,* or "water on the brain." Infants with this condition have enormously expanded skulls due to the presence of an abnormally large volume of CSF. In an adult, failure of reabsorption or blockage of CSF circulation can cause distortion and damage to the brain.

The Blood Supply to the Brain

As we noted in Chapter 12, neurons have a high demand for energy, and they have neither energy reserves, in the form of carbohydrates or lipids, nor oxygen reserves, in the form of myoglobin. ∞ *[p. 408]* Your brain, with billions of neurons, is an extremely active organ with a continuous demand for nutrients and oxygen. These demands are met by an extensive circulatory supply. Arterial blood reaches the brain through the *internal carotid arteries* and the *vertebral arteries.* Most of the venous blood from the brain leaves the cranium in the *internal jugular veins,* which drain the dural sinuses. A head injury that damages cerebral blood vessels may cause bleeding into dura mater, either near the dural epithelium or between the outer layer of the dura mater and the bones of the skull. These are serious conditions because the blood entering these spaces compresses and distorts the relatively soft tissues of the brain.

Cerebrovascular diseases are circulatory disorders that interfere with the normal circulatory supply to the brain. The particular distribution of the vessel involved will determine the symptoms, and the degree of oxygen or nutrient starvation will determine their severity. A *stroke,* or **cerebrovascular accident (CVA),** occurs when the blood supply to a portion of the brain is shut off. Affected neurons begin to die in a matter of minutes.

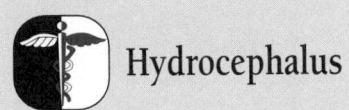 Hydrocephalus

The adult brain is surrounded by the relatively inflexible bones of the cranium. The enclosed cranial cavity contains two fluids—blood and cerebrospinal fluid—and the soft tissues of the brain. Because the total volume cannot change, when the volume of blood or CSF increases, the volume of the brain must decrease. For example, if a dural or subarachnoid blood vessel ruptures, the fluid volume increases as blood collects within the cranial cavity. The rising intracranial pressure compresses the brain, leading to neural dysfunction that commonly ends in unconsciousness and ultimately death.

Any alteration in the rate of CSF production is normally matched by an increase in the rate of removal at the arachnoid granulations. If this equilibrium is disturbed, clinical problems appear as the intracranial pressure changes. The volume of cerebrospinal fluid will increase if the rate of formation accelerates or the rate of removal decreases. In either event, the increased fluid volume leads to compression and distortion of the brain. Increased rates of formation may accompany head injuries, but the most common problems arise from masses, such as tumors or abscesses, or from developmental abnormalities. These conditions have the same effect: They restrict the normal circulation and reabsorption of CSF. Because CSF production continues, the ventricles gradually expand, distorting the surrounding neural tissues and causing brain function to deteriorate.

Infants are especially sensitive to alterations in intracranial pressure, because the arachnoid granulations do not appear until roughly 3 years of age. (Until that time, CSF is reabsorbed into small vessels within the subarachnoid space and beneath the ependyma lining the ventricles.) As in an adult, if intracranial pressure becomes abnormally high, the ventricles will expand. But in an infant, the cranial sutures have yet to fuse, and the skull can enlarge to accommodate the extra fluid volume. This enlargement produces an enormously expanded skull, a condition called **hydrocephalus**, or "water on the brain." Infant hydrocephalus (Figure 14-6●) typically results from some interference with normal CSF circulation, such as blockage of the mesencephalic aqueduct or constriction of the connection between the subarachnoid spaces of the cranial and spinal meninges. Untreated infants commonly suffer some degree of mental retardation. Successful treatment generally involves the installation of a **shunt**, a bypass that either avoids the blockage site or drains the excess cerebrospinal fluid. In either case, the goal is reduction of the intracranial pressure. The shunt may be removed if (1) further growth of the brain eliminates the blockage or (2) the intracranial pressure decreases after the arachnoid granulations develop when the child reaches 3 years of age.

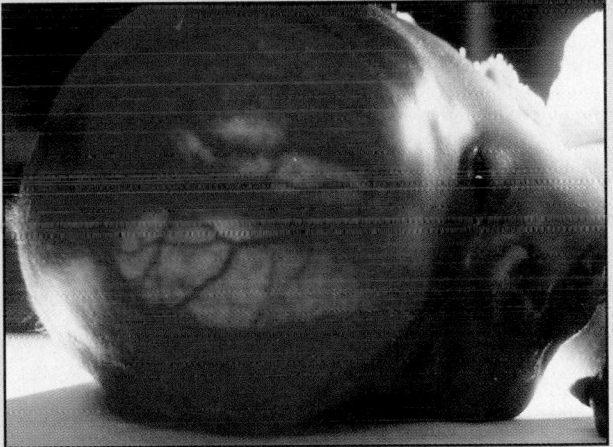

●**FIGURE 14-6** **Hydrocephalus.** This infant has hydrocephalus, a condition generally caused by impaired circulation and impaired removal of cerebrospinal fluid. The buildup of CSF leads to distortion of the brain and enlargement of the cranium.

EPIDURAL AND SUBDURAL HEMORRHAGES
A severe head injury may damage meningeal vessels and cause bleeding into the cranial cavity. The most serious cases involve an arterial break, because arterial blood pressure is relatively high. If blood is forced between the endosteal layer and the cranium, the condition is known as an *epidural hemorrhage*. The elevated fluid pressure then distorts the underlying tissues of the brain. The individual loses consciousness from minutes to hours after the injury, and death follows in untreated cases. An epidural hemorrhage involving a damaged vein does not produce massive symptoms immediately, and the indiviudal may become unconscious from several hours to several days or even weeks after the original injury. As a result, the problem may not be noticed until the nervous tissue has been severely damaged by distortion, compression, and secondary hemorrhaging. Epidural hemorrhages are rare, occurring in fewer than 1 percent of head injuries. However, the mortality rate is 100 percent in untreated cases and over 50 percent even after the blood pool has been removed and the damaged vessels have been closed.

The term *subdural hemorrhage* may be misleading, because in many cases blood enters the meningeal layer of the dura mater, flowing beneath the epithelium that contacts the arachnoid membrane, rather than between the dura mater and the arachnoid. Subdural hemorrhages are roughly twice as common as epidural hemorrhages. The most common source of blood is a small vein or one of the dural sinuses. Because the individual's blood pressure is lower than in epidural arterial hemorrhages, the extent of distortion produced and the effects on brain function can be quite variable.

Development of the Brain and Cranial Nerves

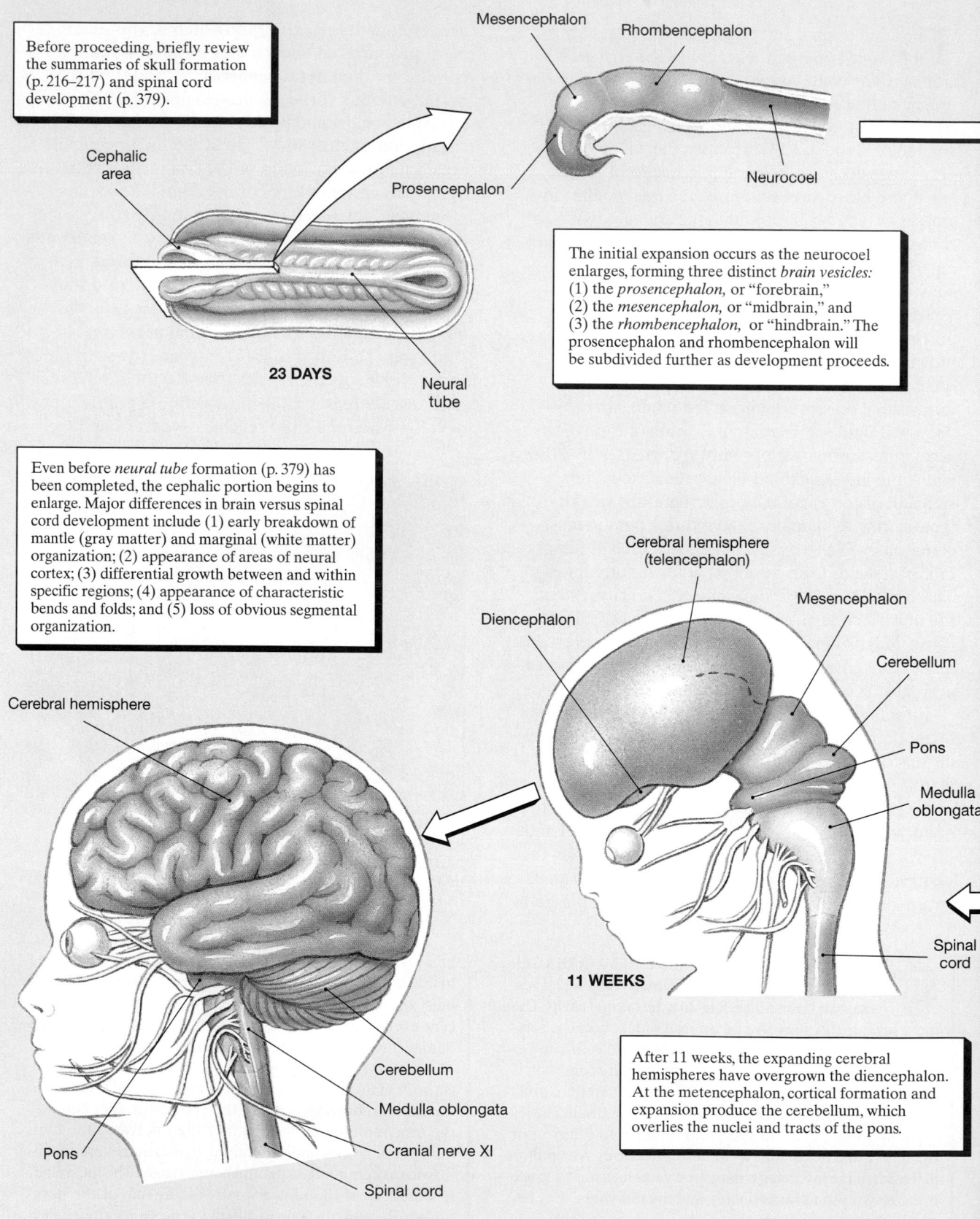

Before proceeding, briefly review the summaries of skull formation (p. 216–217) and spinal cord development (p. 379).

Cephalic area

Prosencephalon

Mesencephalon

Rhombencephalon

Neurocoel

23 DAYS

Neural tube

The initial expansion occurs as the neurocoel enlarges, forming three distinct *brain vesicles:* (1) the *prosencephalon,* or "forebrain," (2) the *mesencephalon,* or "midbrain," and (3) the *rhombencephalon,* or "hindbrain." The prosencephalon and rhombencephalon will be subdivided further as development proceeds.

Even before *neural tube* formation (p. 379) has been completed, the cephalic portion begins to enlarge. Major differences in brain versus spinal cord development include (1) early breakdown of mantle (gray matter) and marginal (white matter) organization; (2) appearance of areas of neural cortex; (3) differential growth between and within specific regions; (4) appearance of characteristic bends and folds; and (5) loss of obvious segmental organization.

Cerebral hemisphere (telencephalon)

Mesencephalon

Diencephalon

Cerebellum

Pons

Medulla oblongata

Cerebral hemisphere

Spinal cord

11 WEEKS

Cerebellum

Medulla oblongata

Cranial nerve XI

Pons

Spinal cord

CHILD

After 11 weeks, the expanding cerebral hemispheres have overgrown the diencephalon. At the metencephalon, cortical formation and expansion produce the cerebellum, which overlies the nuclei and tracts of the pons.

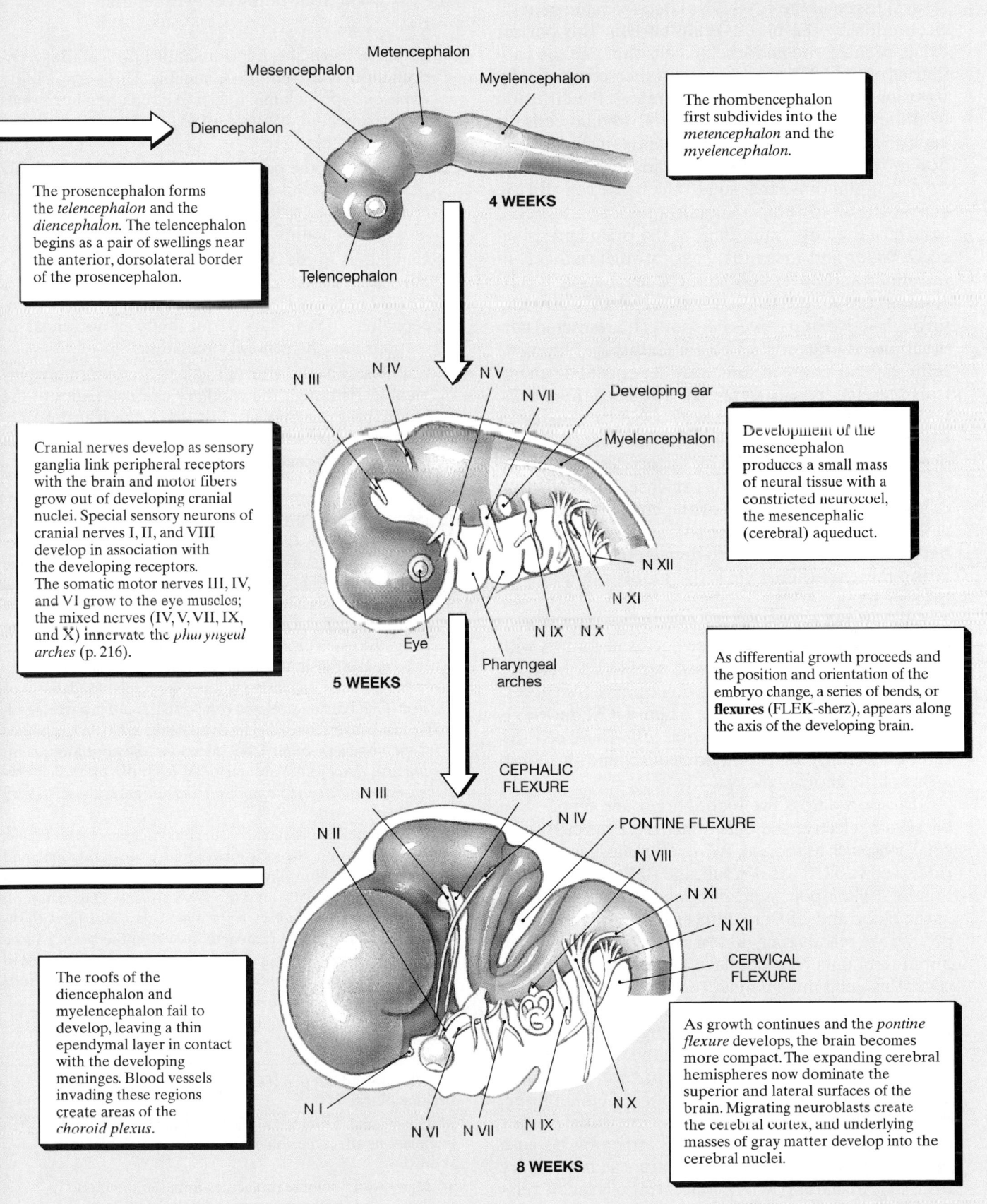

Mesencephalon

Metencephalon

Diencephalon

Myelencephalon

4 WEEKS

Telencephalon

The prosencephalon forms the *telencephalon* and the *diencephalon*. The telencephalon begins as a pair of swellings near the anterior, dorsolateral border of the prosencephalon.

The rhombencephalon first subdivides into the *metencephalon* and the *myelencephalon*.

N III   N IV   N V

N VII

Developing ear

Myelencephalon

Cranial nerves develop as sensory ganglia link peripheral receptors with the brain and motor fibers grow out of developing cranial nuclei. Special sensory neurons of cranial nerves I, II, and VIII develop in association with the developing receptors. The somatic motor nerves III, IV, and VI grow to the eye muscles; the mixed nerves (IV, V, VII, IX, and X) innervate the *pharyngeal arches* (p. 216).

Development of the mesencephalon produces a small mass of neural tissue with a constricted neurocoel, the mesencephalic (cerebral) aqueduct.

N XII

N XI

Eye

N IX   N X

Pharyngeal arches

5 WEEKS

As differential growth proceeds and the position and orientation of the embryo change, a series of bends, or **flexures** (FLEK-sherz), appears along the axis of the developing brain.

CEPHALIC FLEXURE

N III

N II

N IV

PONTINE FLEXURE

N VIII

N XI

N XII

CERVICAL FLEXURE

The roofs of the diencephalon and myelencephalon fail to develop, leaving a thin ependymal layer in contact with the developing meninges. Blood vessels invading these regions create areas of the *choroid plexus.*

N I

N VI   N VII   N IX

N X

8 WEEKS

As growth continues and the *pontine flexure* develops, the brain becomes more compact. The expanding cerebral hemispheres now dominate the superior and lateral surfaces of the brain. Migrating neuroblasts create the cerebral cortex, and underlying masses of gray matter develop into the cerebral nuclei.

The Blood–Brain Barrier

Neural tissue in the CNS is isolated from the general circulation by the **blood–brain barrier.** This barrier exists because the endothelial cells that line the capillaries of the CNS are extensively interconnected by tight junctions. These junctions prevent the diffusion of materials between adjacent endothelial cells. In general, only lipid-soluble compounds (including carbon dioxide, oxygen, ammonia, lipids, such as steroids or prostaglandins, and small alcohols) can diffuse across the lipid bilayer membranes of endothelial cells into the interstitial fluid of the brain and spinal cord. Water and ions must pass through channels in the inner and outer cell membranes. Larger water-soluble compounds can cross the capillary walls only through active or passive transport. The restricted permeability characteristics of the endothelial lining of brain capillaries are in some way dependent on chemicals secreted by astrocytes. We described these cells, which are in close contact with CNS capillaries, in Chapter 12. ∞ *[p. 376]* The outer surfaces of the endothelial cells are covered by the processes of astrocytes. Because the astrocytes release chemicals that control the permeabilities of the endothelium, these cells play a key supporting role in the blood–brain barrier. If the astrocytes are damaged or stop stimulating the endothelial cells, the blood–brain barrier disappears.

The choroid plexus lies outside the neural tissue of the brain, and there are no astrocytes in contact with the endothelial cells. As a result, capillaries there are highly permeable. Substances do not have free access to the CNS, however, because a **blood–CSF barrier** is created by specialized ependymal cells. These cells, interconnected by tight junctions, surround the capillaries of the choroid plexus.

Transport across the blood–brain and blood–CSF barriers is selective and directional. Even the passage of small ions, such as sodium, hydrogen, potassium, or chloride, is controlled. As a result, the pH and concentrations of sodium, potassium, calcium, and magnesium ions in the blood and CSF are different. Some organic compounds are readily transported, and others cross only in minute amounts. Neurons have a constant need for glucose. This need must be met regardless of the relative concentrations in the blood and interstitial fluid. Even when circulating glucose levels are low, endothelial cells continue to transport glucose from the blood to the interstitial fluid of the brain. In contrast, only trace amounts of circulating norepinephrine, epinephrine, dopamine, or serotonin pass into the interstitial fluid or CSF of the brain. This limitation is important because these compounds are neurotransmitters, and their entry from the circulation (where concentrations can be relatively high) could result in the uncontrolled stimulation of neurons throughout the brain.

The blood–brain barrier remains intact throughout the CNS, with four noteworthy exceptions:

1. In portions of the hypothalamus, the capillary endothelium is extremely permeable. This permeability exposes hypothalamic nuclei to circulating hormones and permits the diffusion of hypothalamic hormones into the circulation.

2. Capillaries in the posterior pituitary gland are highly permeable. At this site, the hormones *ADH* and *oxytocin*, produced by hypothalamic neurons, are released into the circulation.

3. Capillaries in the *pineal gland* are also very permeable. The pineal gland, an endocrine structure, is located in the posterior, superior surface of the diencephalon. The capillary permeability allows pineal secretions into the general circulation.

4. Capillaries at the choroid plexus are extremely permeable. Although the capillary characteristics of the blood–brain barrier are lost there, the transport activities of specialized ependymal cells within the choroid plexus maintain the blood–CSF barrier.

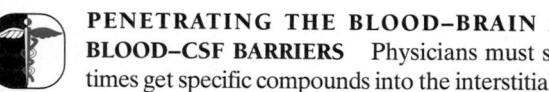 **PENETRATING THE BLOOD–BRAIN AND BLOOD–CSF BARRIERS** Physicians must sometimes get specific compounds into the interstitial fluid of the brain to fight CNS infections or to treat other neural disorders. To do this, they must understand the limitations of the blood–brain and blood–CSF barriers. For example, *Parkinson's disease,* discussed on page 471, results from inadequate release of the neurotransmitter dopamine in a nucleus in the cerebrum. Although dopamine will not cross the blood–brain or blood–CSF barrier, a related compound, *L-dopa,* passes readily and is converted to dopamine inside the brain. In treatments for meningitis or other CNS infections, the antibiotics *penicillin* and *tetracycline* are excluded from the brain, but *erythromycin, sulfisoxazole,* and *sulfadiazine* enter the CNS very rapidly.

When the CNS is damaged, the blood–brain barrier breaks down. As a result, the locations of injury sites, infections, and tumors can be determined by injecting into the bloodstream tracers that cannot enter the CNS tissues. One common method is to use *albumin,* a plasma protein, labeled with radioactive iodine. Any radioactivity within the brain will be easily detected in a brain scan, using procedures described in Section 2 of the *Applications Manual.* [AM] *Medical Importance of Radioisotopes*

☑ What would happen if an interventricular foramen became blocked?

☑ How would decreased diffusion across the arachnoid granulations affect the volume of cerebrospinal fluid in the ventricles?

☑ Many water-soluble molecules found in the blood in relatively large amounts occur in small amounts or not at all in the extracellular fluid of the brain. Why?

THE CEREBRUM

The cerebrum is the largest region of the brain. Conscious thoughts and all intellectual functions originate in the cerebral hemispheres. Much of the cerebrum is involved in the processing of somatic sensory and motor information.

Figure 14-1b,d●, p. 447, provides perspective on the cerebrum and its relationships with other regions of the brain. Gray matter in the cerebrum is found in the *cerebral cortex* and in deeper *cerebral nuclei*. The *central white matter* lies beneath the neural cortex and around the cerebral nuclei (Figure 14-7●).

The Cerebral Cortex

A blanket of neural cortex covers the paired cerebral hemispheres, which dominate the superior and lateral surfaces of the cerebrum (Figure 14-1●, p. 447). The cortical surface forms a series of elevated ridges, or **gyri** (JĪ-rī; singular, *gyrus*). The gyri are separated by shallow depressions called **sulci** (SUL-sī) or by deeper grooves called **fissures.** Gyri increase the surface area of the cerebral hemispheres. The greater the area of neural cortex, the larger the number of neurons. The total surface area of the cerebral hemispheres is roughly equivalent to 2200 cm^2 (2.5 ft^2) of flat surface. An area that large can be packed into the skull only when folded like a crumpled piece of paper. Complex analytical and integrative functions require large numbers of neurons. The entire brain has enlarged in the course of human evolution, but the cerebral hemispheres have enlarged at a much faster rate than has the rest of the brain.

Cerebral Landmarks and Lobes

The two cerebral hemispheres are almost completely separated by a deep **longitudinal fissure** (Figure 14-1a●, p. 447). They remain connected by a thick band of white matter called the *corpus callosum.* Each cerebral hemisphere can be divided into lobes named after the overlying bones of the skull. Your brain has a unique pattern of sulci and gyri, as individual as a fingerprint, but the boundaries between lobes are reliable landmarks. On each hemisphere, the **central sulcus,** a deep groove, divides the anterior **frontal lobe** from the more posterior **parietal lobe.** The roughly horizontal **lateral sulcus** separates the frontal lobe from the **temporal lobe.** Pushing the temporal lobe to the side exposes the **insula** (IN-sū-luh; *insula,* island), an "island" of cortex that is otherwise invisible. The **parieto-occipital sulcus** separates the parietal lobe from the **occipital lobe.**

Each lobe contains functional regions whose boundaries are less clearly defined. Some of these regions deal with sensory information and others with motor commands. Keep in mind three points about the cerebral lobes:

1. *Each cerebral hemisphere receives sensory information from and sends motor commands to the opposite side of the body.* For example, the motor areas of the left cerebral hemisphere control muscles on the right side, and the right cerebral hemisphere controls muscles on the left side. This crossing over has no known functional significance.

2. *The two hemispheres have different functions, even though they look almost identical.* We will discuss these differences in a later section.

3. *The assignment of a specific function to a specific region of the cerebral cortex is imprecise.* Because the boundaries are indistinct and have considerable overlap, one region may have several different functions. Some aspects of cortical function, such as consciousness, cannot easily be assigned to any single region. However, it is clear that normal individuals use all portions of the brain.

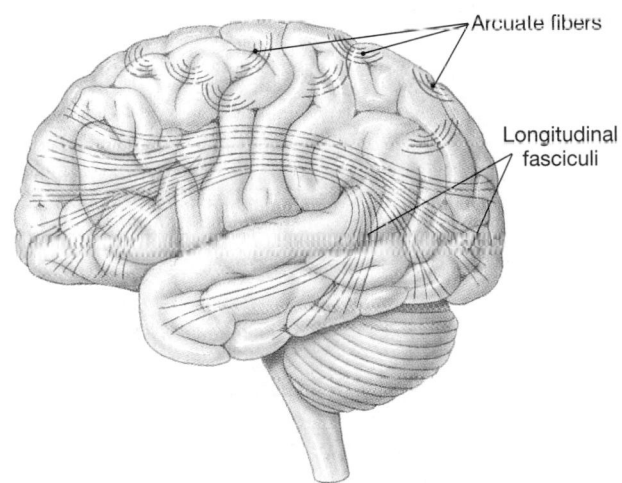

(a) Lateral view

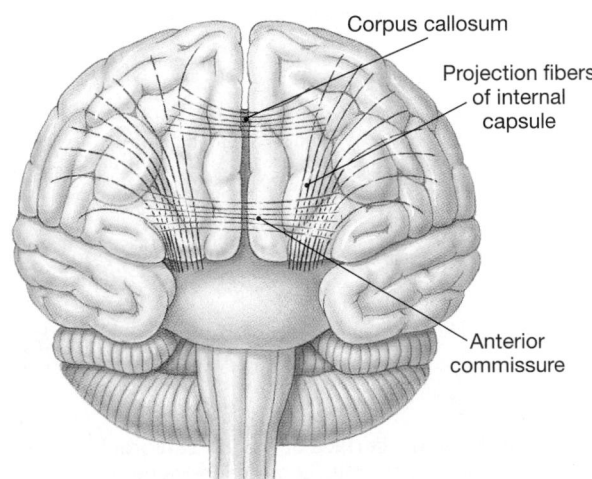

(b) Anterior view

●*FIGURE 14-7* **The Central White Matter**

The Central White Matter

The **central white matter** contains three major groups of axons: *association fibers, commissural fibers,* and *projection fibers* (Figure 14-7●):

1. **Association fibers** interconnect areas of neural cortex within a single cerebral hemisphere. The shorter association fibers are called **arcuate** (AR-kū-āt) **fibers,** because they curve in an arc to pass from one gyrus to another. The longer association fibers are organized into discrete bundles, or *fasciculi.* The **longitudinal fasciculi** connect the frontal lobe to the other lobes of the same hemisphere.

2. **Commissural** (kom-MIS-ū-rul) **fibers** (*commissura,* crossing over) interconnect and permit communication between the two cerebral hemispheres. Prominent bands of commissural fibers linking the cerebral hemispheres include the **corpus callosum** and the **anterior commissure.** The corpus callosum alone contains more than 200 million axons, carrying an estimated 4 billion impulses per second!

3. **Projection fibers** link the cerebral cortex to the diencephalon, brain stem, cerebellum, and spinal cord. All projection fibers must pass through the diencephalon, where axons heading to sensory areas of the cerebral cortex pass among the axons descending from motor areas of the cortex. In gross dissection, the ascending fibers and descending fibers look alike. The entire collection of fibers is known as the **internal capsule.**

Motor and Sensory Areas of the Cortex

The major motor and sensory regions of the cerebral cortex are detailed in Figure 14-8● and Table 14-2. The central sulcus separates the motor and sensory areas of the cortex. The **precentral gyrus** of the frontal lobe forms the anterior border of the central sulcus. The surface of this gyrus is the **primary motor cortex.** Neu-

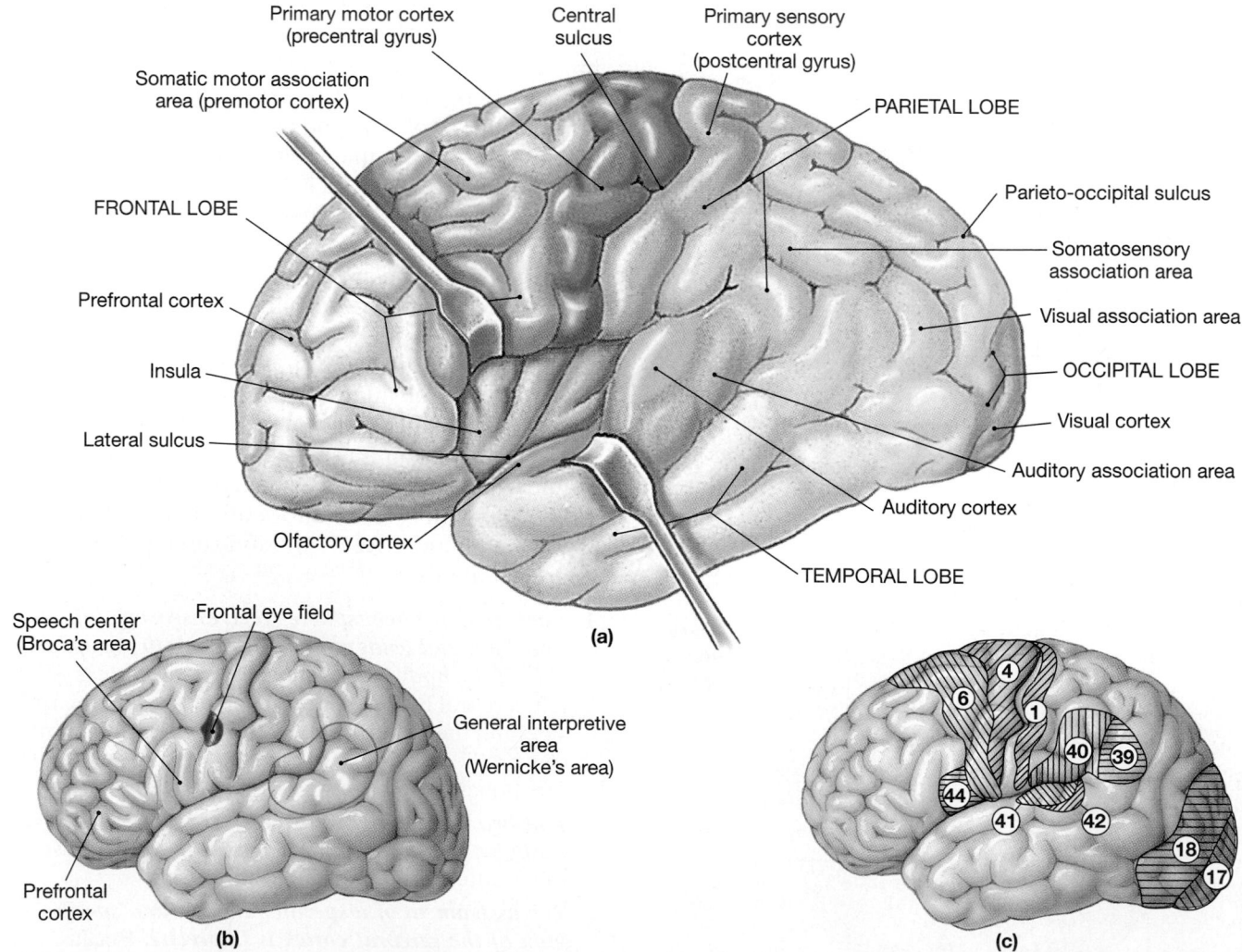

●**FIGURE 14-8 The Cerebral Hemispheres. (a)** Major anatomical landmarks on the surface of the left cerebral hemisphere. The lateral sulcus has been pulled apart to expose the insula. **(b)** The left hemisphere generally contains the general interpretive area and the speech center. The prefrontal cortex of each hemisphere is involved with conscious intellectual functions. **(c)** Regions of the cerebral cortex as determined by histological analysis. Several of the 47 regions described by Brodmann are shown for comparison with the results of functional mapping.

TABLE 14-2 The Cerebral Cortex

Region/Nucleus	Function
Frontal lobe Primary motor cortex	Voluntary control of skeletal muscles
Parietal lobe Primary sensory cortex	Conscious perception of touch, pressure, vibration, pain, temperature, and taste
Occipital lobe Visual cortex	Conscious perception of visual stimuli
Temporal lobe Auditory cortex and olfactory cortex	Conscious perception of auditory and olfactory stimuli
All lobes Association areas	Integration and processing of sensory data; processing and initiation of motor activities

rons of the primary motor cortex direct voluntary movements by controlling somatic motor neurons in the brain stem and spinal cord. These cortical neurons are called **pyramidal cells,** and the pathway that provides voluntary motor control is known as the **pyramidal system.** The primary motor cortex is like the keyboard of a piano. If you strike a specific piano key, you produce a specific sound; if you stimulate a specific motor neuron in the primary motor cortex, you generate a contraction in a specific skeletal muscle.

The sensory areas of the cerebral cortex are like the gauges in the dashboard of a car: They report key information. At each location, sensory information is reported in the pattern of neuron activity in the cortex. The **postcentral gyrus** of the parietal lobe forms the posterior border of the central sulcus, and its surface contains the **primary sensory cortex.** Neurons in this region receive somatic sensory information from touch, pressure, pain, vibration, taste, and temperature receptors. We are consciously aware of these sensations only when nuclei in the thalamus relay the information to the primary sensory cortex.

Sensory information concerning sensations of sight, sound, and smell arrive at other portions of the cerebral cortex (Figure 14-8a●). The **visual cortex** of the occipital lobe receives visual information, and the **auditory cortex** and **olfactory cortex** of the temporal lobe receive information about hearing and smell, respectively.

Association Areas

The sensory and motor regions of the cortex are connected to nearby **association areas,** which interpret in-

coming data or coordinate a motor response (Figure 14-8a●). Like the information provided by the gauges in a car, the arriving information by itself accomplishes nothing; it must be noticed and interpreted before it can be acted on. The **sensory association areas** monitor and interpret the information that arrives at the sensory areas of the cortex. The **somatosensory association area** monitors activity in the primary sensory cortex. It is the somatosensory association area that in summer allows you to recognize a touch as light as the arrival of a mosquito (and gives you a chance to swat the mosquito before it bites).

The special senses of smell, sight, and hearing involve separate areas of the sensory cortex, and each has its own association area. These areas monitor and interpret arriving sensations. For example, the **visual association area** monitors the patterns of activity in the visual cortex and interprets the results. You see the symbols c, a, and r when the stimulation of receptors in your eyes leads to the stimulation of neurons in your visual cortex. Your visual association area recognizes that these are letters and that $c + a + r = car$. An individual with a damaged visual association area could scan the lines of a printed page and see rows of symbols that are clear but would perceive no meaning from the symbols.

The **somatic motor association area,** or **premotor cortex,** is responsible for the coordination of learned movements. The primary motor cortex does nothing on its own, any more than a piano keyboard can play itself. The neurons in the primary motor cortex must be stimulated by neurons in other parts of the cerebrum. When you perform a voluntary movement, the instructions are relayed to the primary motor cortex by the premotor cortex. With repetition, the proper pattern of stimulation becomes stored in your premotor cortex. You can then perform the movement smoothly and easily by triggering the *pattern* rather than by controlling the individual neurons. This principle applies to any learned movement, from something as simple as picking up a glass to something as complex as playing the piano. One area of the premotor cortex, the *frontal eye field,* controls learned eye movements, such as when you scan these lines of type. Someone with damage to the frontal eye field can understand written letters and words but cannot read because his or her eyes cannot follow the lines on a printed page.

Integrative Centers

Major integrative centers receive information from many different association areas and direct extremely complex motor activities. These centers also perform complicated analytical functions. For example, the *prefrontal cortex* of the frontal lobe (Figure 14-8b●) integrates information from sensory association areas and performs abstract intellectual functions, such as predicting the consequences of possible responses.

Integrative centers are located on the lobes and cortical areas of both cerebral hemispheres. Integrative centers concerned with the performance of complex processes, such as speech, writing, mathematical computation, or understanding spatial relationships, are restricted to either the left or the right hemisphere. These centers include the *general interpretive area* and the *speech center.* The corresponding regions on the opposite hemisphere are also active, but their functions are less well-defined.

THE GENERAL INTERPRETIVE AREA   The **general interpretive area,** or *Wernicke's area* (Figure 14-8b●), also called the *gnostic area,* receives information from all the sensory association areas. This analytical center is present in only one hemisphere (typically the left). This region plays an essential role in your personality by integrating sensory information and coordinating access to complex visual and auditory memories. Damage to the general interpretive area affects the ability to interpret what is seen or heard, even though the words are understood as individual entities. For example, if your general intrerpretive area were damaged, you might still understand the meaning of the spoken words *sit* and *here,* because word recognition occurs in the **auditory association areas.** But you would be totally bewildered by the request *sit here.* Damage to another portion of the general interpretive area might leave you able to see a chair clearly and to know that you recognize it, but you would be unable to name it because the connection to your visual association area has been disrupted.

Aphasia (*a-,* without + *phasia,* speech) is a disorder affecting the ability to speak or read. **Global aphasia** results from extensive damage to the general interpretive area or to the associated sensory tracts. Affected individuals are totally unable to speak, to read, or to understand the speech of others. Global aphasia often accompanies a severe stroke or tumor that affects a large area of cortex including the speech and language areas. Recovery is possible when the condition results from edema or hemorrhage, but the process often takes months or even years. [AM] *Aphasia*

Dyslexia (*lexis,* diction) is a disorder affecting the comprehension and use of words. **Developmental dyslexia** affects children; there are estimates that up to 15 percent of children in the United States suffer from some degree of dyslexia. For unknown reasons, the problem is much more common among left-handed children than among right-handed children. Children with dyslexia have difficulty reading and writing, although their other intellectual functions may be normal or above normal. Their writing looks un-

even and disorganized; letters are typically written in the wrong order (*dig* becomes *gid*) or reversed (*E* becomes Ǝ). Recent evidence suggests that at least some forms of dyslexia result from problems in processing, sorting, and integrating visual or auditory information.

THE SPEECH CENTER   Some of the neurons in the general interpretive area innervate the **speech center,** also called the *motor speech area,* or *Broca's area* (Figure 14-8b●). This center lies along the edge of the premotor cortex in the same hemisphere as the general interpretive area (usually the left). The speech center regulates the patterns of breathing and vocalization needed for normal speech. This regulation involves coordinating the activities of the respiratory muscles, the laryngeal and pharyngeal muscles, and the muscles of the tongue, cheeks, lips, and jaws. A person with a damaged motor speech area can make sounds but not words.

The motor commands issued by the motor speech area are adjusted by feedback from the auditory association area, also called the *receptive speech area.* Damage to the related sensory areas can cause a variety of speech-related problems. Some individuals have difficulty speaking, although they know exactly what words to use; others talk constantly but use all the wrong words.

THE PREFRONTAL CORTEX   The **prefrontal cortex** of the frontal lobe coordinates information from the secondary and special association areas of the entire cortex. In doing so, it performs such abstract intellectual functions as predicting the future consequences of events or actions. Damage to the prefrontal cortex leads to difficulties in estimating the temporal relationships between events. Questions such as "How long ago did this happen?" or "What happened first?" become difficult to answer.

The prefrontal cortex has extensive connections with other cortical areas and with other portions of the brain. Feelings of frustration, tension, and anxiety are generated at the prefrontal cortex as it interprets ongoing events and makes predictions about future situations or consequences. If the connections between the prefrontal cortex and other brain regions are severed, the frustrations, tensions, and anxieties are removed. Earlier in this century, this rather drastic procedure, called **prefrontal lobotomy,** was used to "cure" a variety of mental illnesses, especially those associated with violent or antisocial behavior. After a lobotomy, the patient would no longer be concerned about what had previously been a major problem, whether psychological (hallucinations) or physical (severe pain). However, the individual was often equally unconcerned about tact, decorum, and toilet training. Now that drugs have been developed to target specific pathways and regions of the CNS, lobotomies are no longer used to change behavior.

BRODMANN'S AREAS Early in the twentieth century, several attempts were made to describe and classify regional differences in the histological organization of the cerebral cotex. It was hoped that the patterns of cellular organization could be correlated to specific functions. By 1919, at least 200 different patterns had been described, but most of the classification schemes have since been abandoned. However, the cortical map prepared by Korbinian Brodmann in 1909 has proved useful to neuroanatomists. Brodmann, a German neurologist, described 47 patterns of cellular organization in the cerebral cortex. Several of these *Brodmann's areas* are shown in Figure 14-8c●. Some correspond to known functional areas. For example, Brodmann's area 44 corresponds to the speech center, and area 4 follows the contours of the primary motor cortex. In other cases, the correspondence is less precise. For example, Brodmann's areas 1, 2, and 3 all lie within the primary sensory cortex.

Hemispheric Specialization

Figure 14-9● indicates the major functional differences between the hemispheres. In most people, the left hemisphere contains the general interpretive and speech centers, and this is the hemisphere responsible for language-based skills. For example, reading, writing, and speaking are dependent on processing done in the left cerebral hemisphere. In addition, the premotor cortex involved with the control of hand movements is larger on the left side for right-handed than for left-handed individuals. The left hemisphere is also important in performing analytical tasks, such as mathematical calculations and logical decision making. For those reasons, the left hemisphere was formerly called the *dominant hemisphere*. A more appropriate term is the **categorical hemisphere,** because the right hemisphere also has many important functions.

The right cerebral hemisphere analyzes sensory information and relates the body to the sensory environment. Interpretive centers in this hemisphere permit you to identify familiar objects by touch, smell, sight, taste, or feel. For example, the right hemisphere plays a dominant role in recognizing faces and in understanding three-dimensional relationships. It is also important in analyzing the emotional context of a conversation—for example, determining whether the phrase "get lost" was intended as a threat or a question. In-

dividuals with a damaged right hemisphere may be unable to add emotional inflections to their own words. Because the right hemisphere deals with spatial relationships and analyses, the term **representational hemisphere** is used to refer to it.

The left hemisphere is the categorical hemisphere in 96 percent of right-handed individuals; the right hemisphere is categorical in the other 4 percent. Left-handed persons represent 9 percent of the human population. The left hemisphere is the categorical hemisphere in 70 percent of left-handed persons, and the right hemisphere in 15 percent; in the remaining 15 percent, there is no apparent specialization. Interestingly, an unusually high percentage of musicians and artists are left-handed. The complex motor activities performed by these individuals are directed by the primary motor cortex and association areas on the right (representational) hemisphere.

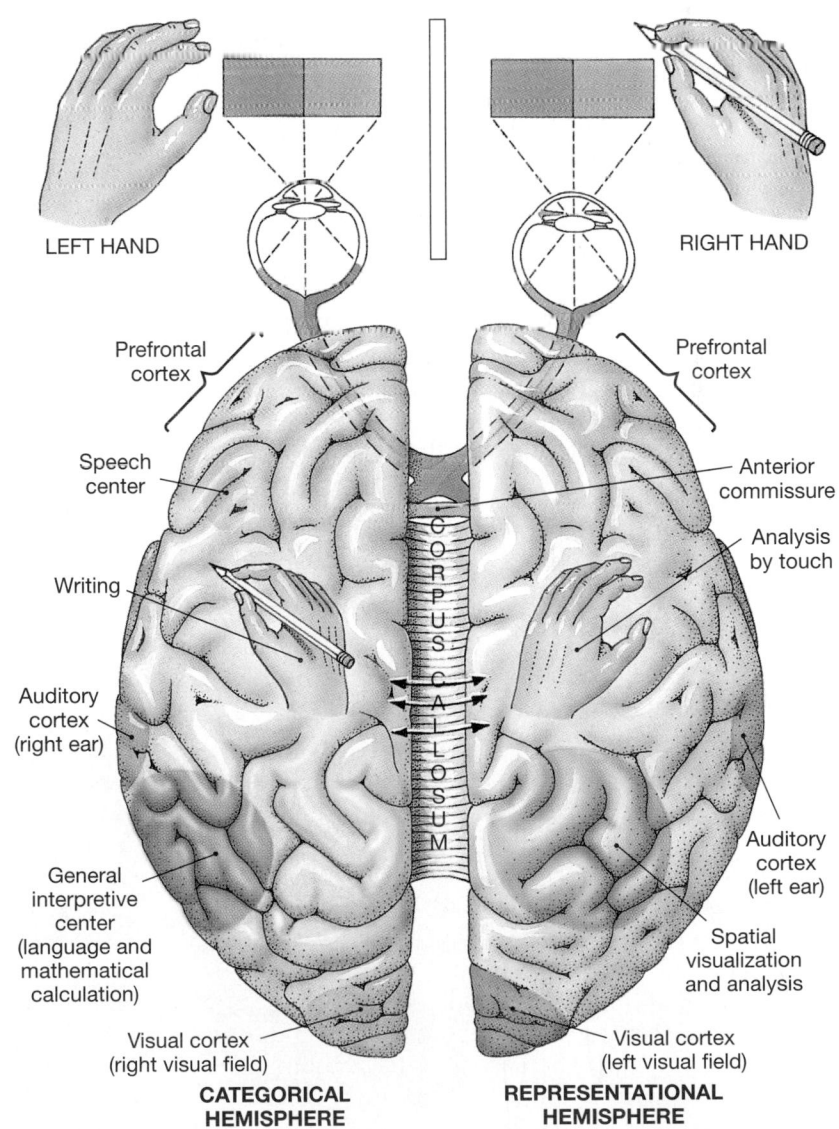

LEFT HAND

RIGHT HAND

Prefrontal cortex

Prefrontal cortex

Speech center

Anterior commissure

Analysis by touch

Writing

CORPUS CALLOSUM

Auditory cortex (right ear)

General interpretive center (language and mathematical calculation)

Auditory cortex (left ear)

Spatial visualization and analysis

Visual cortex (right visual field)

Visual cortex (left visual field)

CATEGORICAL HEMISPHERE

REPRESENTATIONAL HEMISPHERE

●**FIGURE 14-9 Hemispheric Specialization.** Functional differences between the left and right cerebral hemispheres.

DISCONNECTION SYNDROME The functional differences between the hemispheres become apparent if the corpus callosum is cut, a procedure sometimes performed to treat epileptic seizures that cannot be controlled by other methods. This surgery produces symptoms of **disconnection syndrome.** In this condition, the two hemispheres function independently, each "unaware" of stimuli or motor commands that involve its counterpart. The result is some rather interesting changes in the individual's abilities. For example, objects touched by the left hand can be recognized but not verbally identified, because the sensory information arrives at the right hemisphere but the speech center is on the left. The object can be verbally identified if felt with the right hand, but the person will not be able to say whether it is the same object previously touched with the left hand. Sensory information from the left side of the body arrives at the right hemisphere and cannot reach the general interpretive area. Thus, conscious decisions are made without regard for sensations from the left side. Two years after a surgical sectioning of the corpus callosum, the most striking behavioral abnormalities have disappeared, and the person may test normally. In addition, individuals born without a functional corpus callosum do not have sensory, motor, or intellectual problems. In some way, the CNS adapts to the situation, probably by increasing the amount of information transferred across the anterior commissure.

☑ Shelly suffers a head injury that damages her primary motor cortex. Where is this area located?

☑ What senses would be affected by damage to the temporal lobes of the cerebrum?

☑ After suffering a stroke, Jake is unable to speak. He can understand what is said to him, and he can understand written messages, but he cannot express himself verbally. What part of his brain has been affected by the stroke?

The Cerebral Nuclei

While your cerebral cortex is consciously directing a complex movement or solving some intellectual puzzle, other centers of your cerebrum, diencephalon, and brain stem are processing sensory information and issuing motor commands outside your conscious awareness. Many of these activities, which occur at the *subconscious* level, are directed by the cerebral nuclei. The **cerebral nuclei** (Figure 14-10•) are masses of gray matter that lie within each hemisphere be-

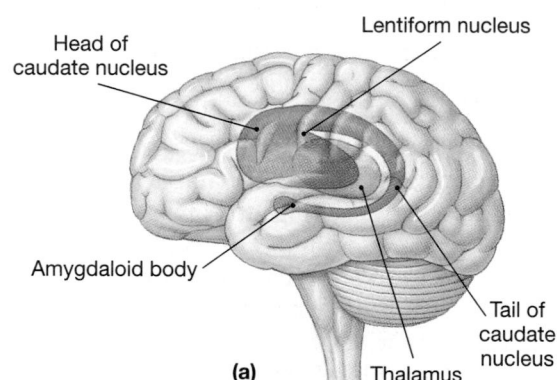

(a)

•*FIGURE 14-10* **The Cerebral Nuclei.** **(a)** The relative positions of the cerebral nuclei in the intact brain. **(b)** The cerebral nuclei seen in frontal section. **(c)** The cerebral nuclei in horizontal section; compare this with **(d)** the view in dissection.

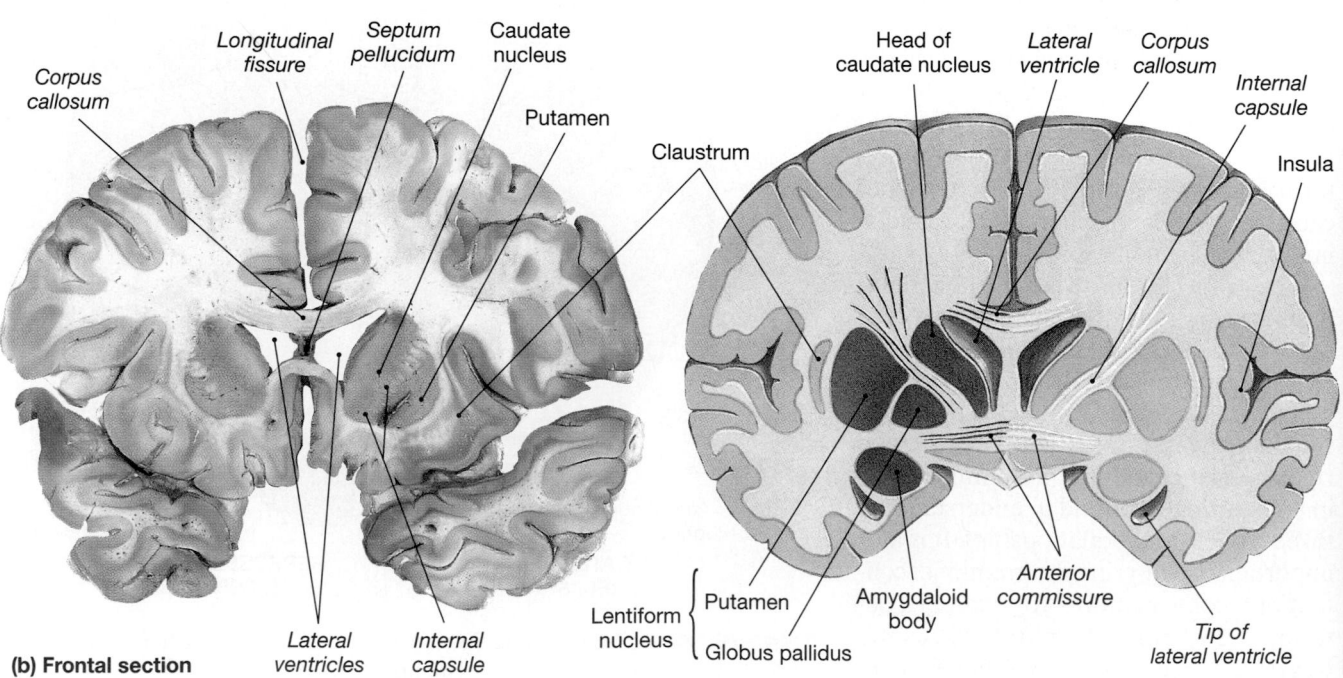

(b) **Frontal section**

neath the floor of the lateral ventricle. They are embedded within the central white matter, and the radiating projection fibers and commissural fibers travel around or between these nuclei. The cerebral nuclei are components of a functional group known as the **basal nuclei,** or *basal ganglia.*[2] The latter term has persisted despite the fact that ganglia are otherwise restricted to the PNS.

The **caudate nucleus** has a massive head and a slender, curving tail that follows the curve of the lateral ventricle. The head of the caudate nucleus lies superior to the **lentiform** (lens-shaped) **nucleus,** which consists of a medial **globus pallidus** (GLŌ-bus PAL-i-dus; pale globe) and a lateral **putamen** (pū-TĀ-men). The term *corpus striatum* (striated body) has been used to refer to the caudate and lentiform nuclei, or to the caudate and putamen. The name refers to the striated (striped) appearance of the internal capsule as it passes among these nuclei. The **claustrum** (KLAWS-trum) is a small nucleus that lies between the lentiform nucleus and the surface of the insula. The **amygdaloid** (ah-MIG-da-loyd) **body** (*amygdale,* almond) is a separate nucleus located anterior to the tail of the caudate and inferior to the lentiform nucleus. The

amygdaloid body is a component of the *limbic system;* we will be consider it further in a later section.

Functions of the Cerebral Nuclei

The cerebral nuclei are components of the **extrapyramidal system,** a motor system that provides subconscious control of skeletal muscle tone and coordinates learned movement patterns and other somatic motor activities. Under normal conditions, these nuclei do not initiate particular movements, but once a movement is under way the cerebral nuclei provide the general pattern and rhythm, especially for movements of the trunk and proximal limb muscles.

Information arrives at the caudate nucleus and putamen from sensory, motor, and association areas of the cerebral cortex. Processing and integration occur in these nuclei and in the adjacent globus pallidus. Most of the output of the cerebral nuclei leaves the globus pallidus and synapses in the thalamus. Nuclei in the thalamus then project the information to appropriate areas of the cerebral cortex. The cerebral nuclei alter the motor commands issued by the cerebral cortex through this feedback loop. For example:

- When you walk, your cerebral nuclei control the cycles of arm and thigh movements that occur between the time you decide to "start" walking and the time you give the "stop" order.
- As you begin a voluntary movement, your cerebral nuclei control and adjust muscle tone, particularly in the

[2] The term *basal ganglia*, like the term *brain stem*, has historically been used in several different ways. In some sources, *basal ganglia* is synonymous with cerebral nuclei, whereas others use the term to refer to a functional group that consists of the cerebral nuclei and nuclei in the diencephalon and mesencephalon.

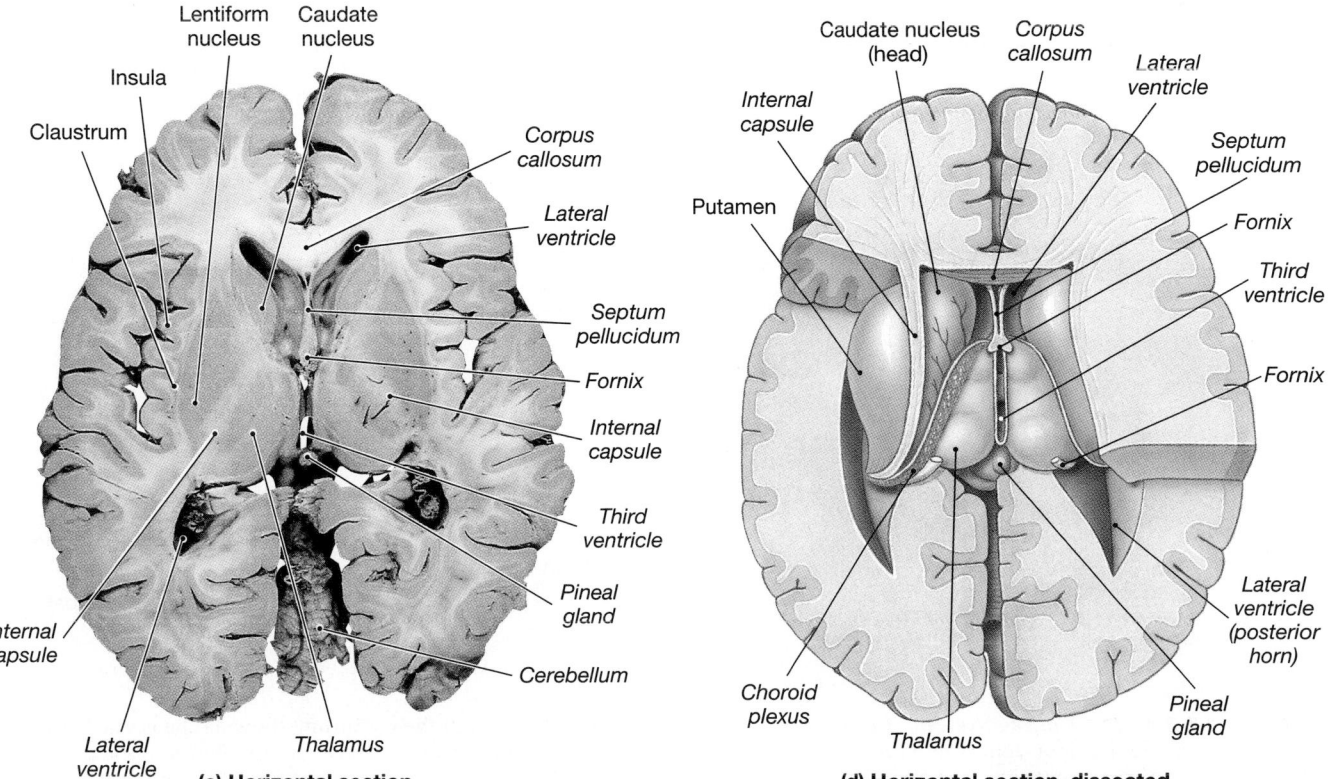

(c) Horizontal section

(d) Horizontal section, dissected

appendicular muscles, to set your body position. When you decide to pick up a pencil, you consciously reach and grasp with your forearm, wrist, and hand, while the cerebral nuclei position your shoulder and stabilize your arm.

THE LIMBIC SYSTEM

The **limbic** (LIM-bik) **system** (*limbus,* border) includes nuclei and tracts along the border between the cerebrum and diencephalon. This system is a functional grouping rather than an anatomical one and consists of components of the cerebrum, diencephalon, and mesencephalon. The functions of the limbic system include (1) establishing emotional states and related behavioral drives; (2) linking the conscious, intellectual functions of the cerebral cortex with the unconscious and autonomic functions of the brain stem; and (3) facilitating memory storage and retrieval.

Figure 14-11● focuses on major components of the limbic system. This system is a functional grouping rather than an anatomical one, and the limbic system includes components of the cerebrum, diencephalon, and mesencephalon. We have already described the amygdaloid body at the tail end of the caudate nucleus as a cerebral contribution to the limbic system (p. 463). The amygdaloid body appears to act as an interface between the limbic system, the cerebrum, and various sensory systems. It plays a role in the regulation of heart rate, in the control of the "flight or fight" response, and in linking emotions with specific memories. The **limbic lobe** of the cerebral hemisphere consists of two gyri that curve along the corpus callosum and onto the medial surface of the temporal lobe (Figure 14-11a●). The **cingulate** (SIN-gū-lāt) **gyrus** (*cingulum,* girdle or belt) sits superior to the corpus callosum. The **dentate gyrus** and the adjacent **parahippocampal** (pa-ra-hip-ō-KAM-pal) **gyrus** conceal the **hippocampus,** a folded area of cortex that lies inferior to the floor of the lateral ventricle. Early anatomists thought this structure resembled a sea horse (*hippocampus*); it appears to be important in learning, especially in the storage and retrieval of new long-term memories.

The **fornix** (FOR-niks) (Figures 14-10c,d and 14-11●) is a tract of white matter that connects the hippocampus with the hypothalamus. From the hippocampus, the fornix curves medially, meeting its counterpart from the opposing hemisphere. The fornix proceeds anteriorly, inferior to the corpus callosum and septum pellucidum, before curving toward the hypothalamus. Many of the fibers end in the **mamillary** (MAM-i-lar-ē) **bodies** (*mamilla,* or *mammilla,* breast or nipple), prominent nuclei in the floor of the hypothalamus (Figures 14-11 and 14-12●). The mamillary bodies process sensory information, including olfactory sensations. They also contain motor nuclei that control reflex movements associated with eating, such as chewing, licking, and swallowing.

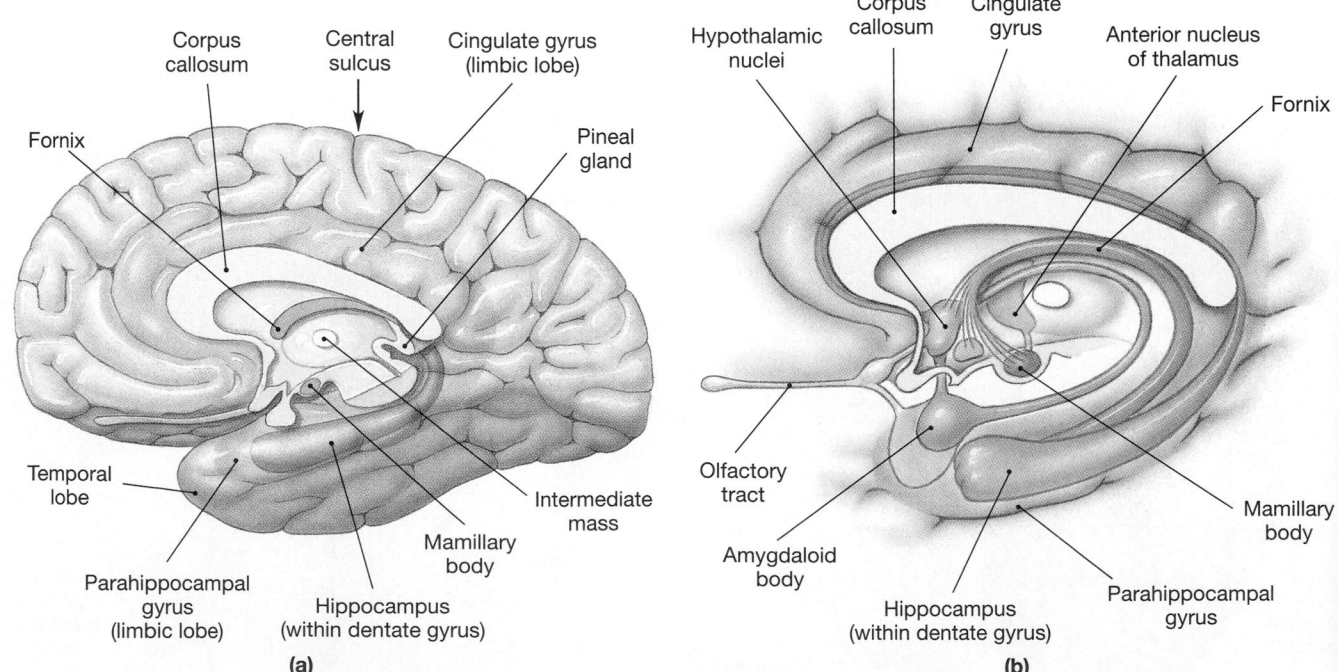

●**FIGURE 14-11** **The Limbic System.** **(a)** A diagrammatic sagittal section through the cerebrum, showing the cortical areas associated with the limbic system. The parahippocampal gyrus is shown as if transparent so that deeper limbic components can be seen. **(b)** Three-dimensional reconstruction of the limbic system, showing the relationships among the major components.

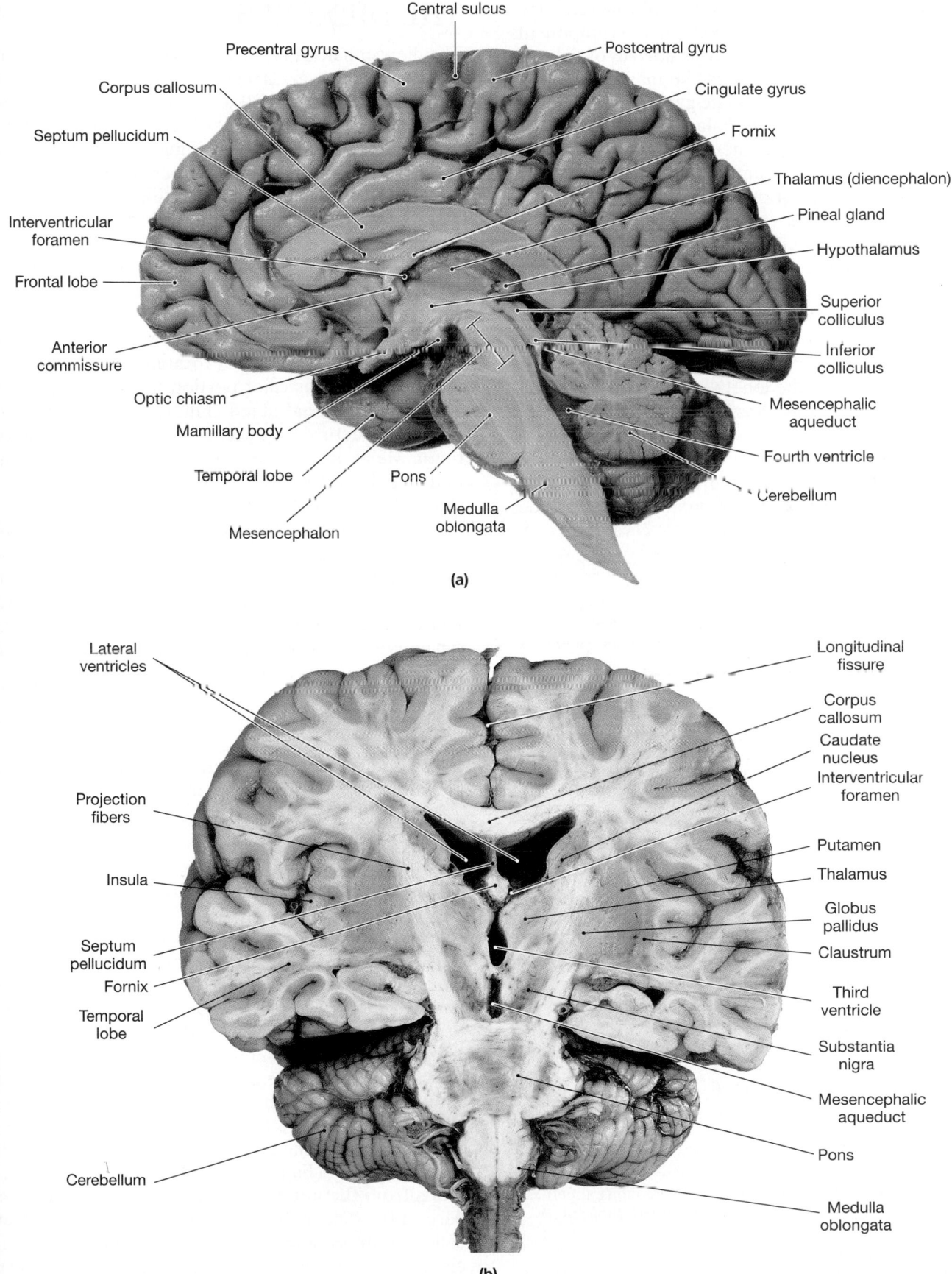

Central sulcus

Precentral gyrus

Postcentral gyrus

Corpus callosum

Cingulate gyrus

Septum pellucidum

Fornix

Thalamus (diencephalon)

Interventricular foramen

Pineal gland

Hypothalamus

Frontal lobe

Superior colliculus

Anterior commissure

Inferior colliculus

Optic chiasm

Mesencephalic aqueduct

Mamillary body

Fourth ventricle

Temporal lobe

Pons

Cerebellum

Mesencephalon

Medulla oblongata

(a)

Lateral ventricles

Longitudinal fissure

Corpus callosum

Caudate nucleus

Interventricular foramen

Projection fibers

Putamen

Insula

Thalamus

Globus pallidus

Septum pellucidum

Claustrum

Fornix

Third ventricle

Temporal lobe

Substantia nigra

Mesencephalic aqueduct

Pons

Cerebellum

Medulla oblongata

(b)

•**FIGURE 14-12** **The Brain in Section. (a)** Midsagittal section. **(b)** Frontal section. AM *Plate 3.1; Scan 1e*

Several other nuclei in the wall (thalamus) and floor (hypothalamus) of the diencephalon are components of the limbic system. The **anterior nucleus** of the thalamus relays information from the mamillary bodies (hypothalamus) to the cingulate gyrus. The boundaries between some of the hypothalamic nuclei tend to be poorly defined, but experimental stimulation has outlined a number of important hypothalamic centers responsible for the emotions of rage, fear, pain, sexual arousal, and pleasure.

Stimulation of specific regions of the hypothalamus can also produce heightened alertness and a generalized excitement or generalized lethargy and sleep. These responses are caused by stimulation or inhibition of the reticular formation. The **reticular formation** is an interconnected network of neurons and nuclei that extends the length of the brain stem. This network has its headquarters in the mesencephalon.

Table 14-3 summarizes the organization and functions of the limbic system. Figure 14-12●, which shows the brain in midsagittal and frontal sections, includes many of the limbic structures of the cerebrum, diencephalon, and brain stem.

☑ What symptoms would you expect to observe in an individual who has damage to the extrapyramidal system?

☑ Paul is having a difficult time remembering facts and recalling long-term memories. What part of his cerebrum is probably involved?

☑ After suffering a head injury in an automobile accident, Terri is having trouble chewing and swallowing. This difficulty suggests that Terri may have sustained damage to what part of her brain?

TABLE 14-3 The Limbic System

FUNCTION
Processing of memories, creation of emotional states, drives, and associated behaviors

CEREBRAL COMPONENTS
Cortical areas: limbic lobe (cingulate gyrus, dentate gyrus, and parahippocampal gyrus)
Nuclei: hippocampus, amygdaloid body
Tracts: fornix

DIENCEPHALIC COMPONENTS
Thalamus: anterior nuclear group
Hypothalamus: centers concerned with emotions, appetites (thirst, hunger), and related behaviors (*see Table 14-5*)

OTHER COMPONENTS
Reticular formation: network of interconnected nuclei throughout brain stem

THE DIENCEPHALON

The diencephalon plays a vital role in integrating conscious and unconscious sensory information and motor commands. Figures 14-10c,d through 14-13● show the position of the diencephalon and its relationship to landmarks on the brain stem. Figure 14-13a● includes the origins of 11 of the 12 pairs of cranial nerves. The individual cranial nerves are identified by Roman numerals. We will introduce the full names of these nerves in a later section (pp. 474–484).

The anterior portion of the diencephalic roof, or *epithalamus,* forms the membranous roof of the third ventricle. The anterior portion of the epithalamus contains an extensive area of choroid plexus that extends through the interventricular foramina into the lateral ventricles. The posterior portion of the epithalamus contains the **pineal gland** (Figure 14-13b●), an endocrine structure that secretes the hormone **melatonin.** Melatonin may be important in the regulation of day/night cycles and of reproductive functions. (We shall describe the role of melatonin in Chapter 18.)

Most of the neural tissue in the diencephalon is concentrated in two structures: (1) the *thalami* (plural of *thalamus;* wall), and (2) the *hypothalamus* (floor). Ascending sensory information from the spinal cord and cranial nerves (other than the olfactory tract) synapses in a nucleus in the left or right thalamus before reaching the cerebral cortex and our conscious awareness. The hypothalamus contains centers involved with emotions and visceral processes that affect the cerebrum as well as other components of the brain stem. It also controls a variety of autonomic functions and forms the link between the nervous and endocrine systems.

The Thalamus

The thalamus is the final relay point for ascending sensory information that will be projected to the primary sensory cortex. It acts as a filter, passing on only a small portion of the arriving sensory information. The thalamus also coordinates the activities of the pyramidal and extrapyramidal systems by relaying information between the cerebral nuclei and the cerebral cortex.

The left thalamus and right thalamus are separated by the third ventricle. Each thalamus consists of a rounded mass of *thalamic nuclei* (Figure 14-14●). The lateral border of the thalamus is established by the fibers of the internal capsule (p. 458). Viewed in midsagittal section (Figure 14-12a●), each thalamus extends from the anterior commissure to the inferior base of the pineal gland. A projection of gray matter called an **intermediate mass** extends into the ventricle from the thalamus on either side (Figure 14-11a●). In roughly 70 percent of the population, the two thalami fuse in the midline.

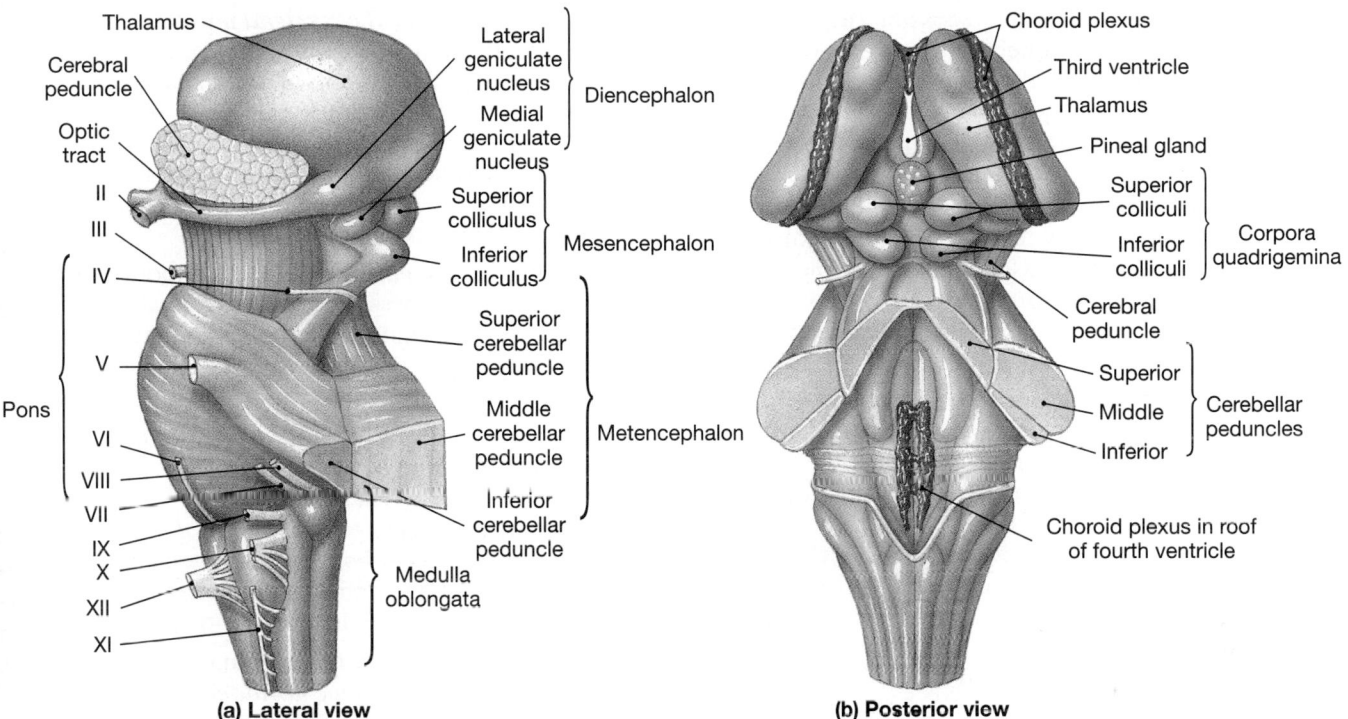

(a) Lateral view

(b) Posterior view

•FIGURE 14-13 The Diencephalon and Brain Stem. **(a)** Lateral view, as seen from the left side. **(b)** Posterior view.

Functions of Thalamic Nuclei

The thalamic nuclei deal primarily with the relay of sensory information to the cerebral nuclei and cerebral cortex. The four major groups of thalamic nuclei, detailed in Table 14-4 and Figure 14-14•, are the *anterior group,* the *medial group,* the *ventral group,* and the *posterior group:*

1. The **anterior nuclei** are part of the limbic system, discussed previously (p. 465).

TABLE 14-4 The Thalamus

Group/Nuclei	Function
Anterior group	Part of the limbic system
Medial group	Integrates sensory information for projection to the frontal lobes
Ventral group	Projects sensory information to the primary sensory cortex; relays information from cerebellum and cerebral nuclei to motor area of cerebral cortex
Posterior group	
Pulvinar	Integrates sensory information for projection to association areas of cerebral cortex
Lateral geniculate nuclei	Project visual information to the visual cortex
Medial geniculate nuclei	Project auditory information to the auditory cortex

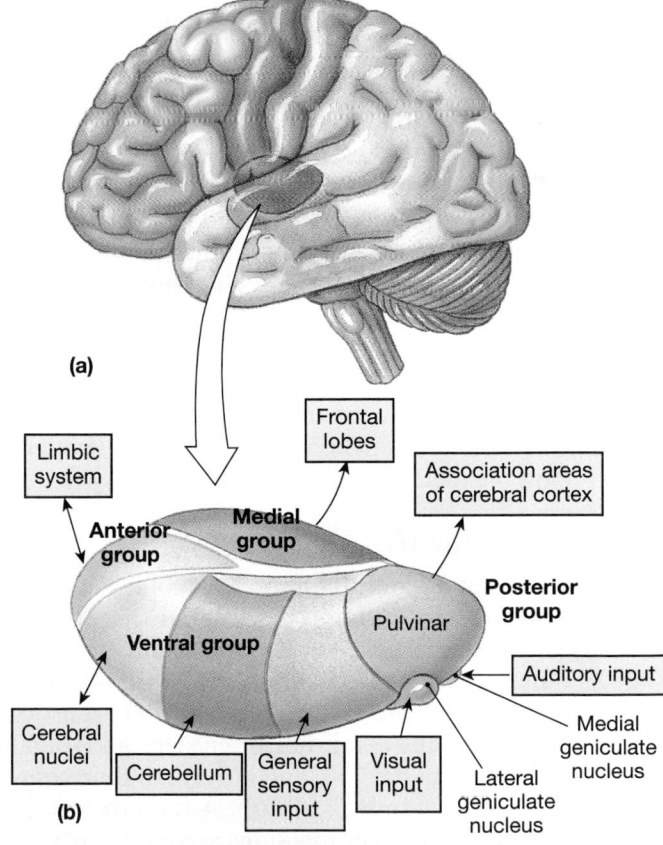

•FIGURE 14-14 The Thalamus. **(a)** Lateral view of the brain, showing the positions of the thalamic nuclei. The colored areas of the cerebral cortex are regions that receive input from the thalamus. **(b)** Enlarged view of the thalamic nuclei of the left side.

2. The **medial nuclei** provide a conscious awareness of emotional states by connecting emotional centers in the hypothalamus with the frontal lobes of the cerebral hemispheres. The medial nuclei also receive and relay sensory information from other portions of the thalamus.

3. The **ventral nuclei** relay information from the cerebral nuclei and cerebellum to the primary motor cortex and related motor association areas. The ventral nuclei also relay sensory information concerning touch, pressure, pain, temperature, proprioception, and taste to the primary sensory cortex.

4. The **posterior nuclei** include the *pulvinar* and the *geniculate nuclei.* The **pulvinar** integrates sensory information for projection to the association areas of the cerebral cortex. The **lateral geniculate** (je-NIK-ū-lāt) **nucleus** (*genicula,* little knee) of each thalamus receives visual information from the eyes over the **optic tract.** The output of the lateral geniculate nucleus goes to the visual cortex and to the mesencephalon. The **medial geniculate nucleus** relays auditory information to the auditory cortex from the specialized receptors of the inner ear.

The Hypothalamus

The hypothalamus extends from the area superior to the *optic chiasm,* where the *optic tracts* from the eyes arrive at the brain, to the posterior margins of the mamillary bodies (Figure 14-12a●). Posterior to the optic chiasm, the **infundibulum** (in-fun-DIB-ū-lum; *infundibulum,* funnel) extends inferiorly, connecting the floor of the hypothalamus to the pituitary gland.

Viewed in sagittal section (Figure 14-15●), the floor of the hypothalamus between the infundibulum and the mamillary bodies is the **tuber cinereum** (sin-Ē-rē-um; *tuber,* swelling + *cinereus,* ashen color). The tuber cinereum is a mass of gray matter whose neurons are involved with the control of pituitary gland function. The anterior portion of the tuber cinereum adjacent to the infundibulum is thickened and slightly elevated; this region is called the **median eminence.**

Functions of the Hypothalamus

The hypothalamus contains important control and integrative centers, in addition to those associated with the limbic system. These centers and their functions are summarized in Figure 14-15a● and Table 14-5. Hypothalamic centers may be stimulated by (1) sensory information from the cerebrum, brain stem, and spinal cord; (2) changes in the CSF and interstitial fluid composition; or (3) chemical stimuli in the circulating blood that enter the hypothalamus across highly permeable capillaries; there is no blood–brain barrier in the hypothalamus.

The following functions are performed by the hypothalamus:

1. *Subconscious control of skeletal muscle contractions.* The hypothalamus directs somatic motor patterns associated with the emotions of rage, pleasure, pain, and sexual arousal by stimulating centers in other portions of the brain. For example, the changes in facial expression that accompany rage and the basic movements associated with sexual activity are controlled by hypothalamic centers.

2. *Control of autonomic function.* The hypothalamus adjusts and coordinates the activities of autonomic centers in the pons and medulla oblongata that regulate heart rate, blood pressure, respiration, and digestive functions.

3. *Coordination of activities of the nervous and endocrine systems.* The hypothalamus coordinates neural and endocrine activities by inhibiting or stimulating endocrine cells in the pituitary gland through the production of *regulatory hormones.* These are produced at the tuber cinereum and released into the local capillaries for transport to the anterior pituitary gland.

4. *Secretion of hormones.* The hypothalamus secretes two hormones: (1) *antidiuretic hormone* (ADH), which is produced by the **supraoptic nucleus** and restricts water loss at the kidneys, and (2) *oxytocin,* which is produced by the **paraventricular nucleus** and stimulates smooth muscle contractions in the uterus and mammary glands of females and the prostate gland of males. These hormones are transported along axons that pass through the infundibulum to the posterior pituitary gland. There the hormones are released into the local circulation for distribution throughout the body.

5. *Production of emotions and behavioral drives.* Specific hypothalamic centers produce sensations that lead to conscious or subconscious changes in behavior. These unfocused "impressions" originating in the hypothalamus are called **drives.** For example, stimulation of the **feeding center** produces the sensation of hunger, and stimulation of the **thirst center** produces the sensation of thirst. The conscious sensations are only part of the hypothalamic response. For example, the thirst center also orders the release of ADH by neurons in the supraoptic nucleus.

6. *Coordination between voluntary and autonomic functions.* When you think about a dangerous or stressful situation, your heart rate and respiratory rate go up and your body prepares for an emergency. These autonomic adjustments are made by the hypothalamus.

7. *Regulation of body temperature.* The **preoptic area** of the hypothalamus coordinates the activities of other CNS centers and regulates other physiological systems to maintain normal body temperature. If body temperature falls, the preoptic area sends instructions to the *vasomotor center* in the medulla, an autonomic center that controls blood flow by regulating the diameter of peripheral blood vessels. In response, the vasomotor center decreases the blood supply to the skin, reducing the rate of heat loss.

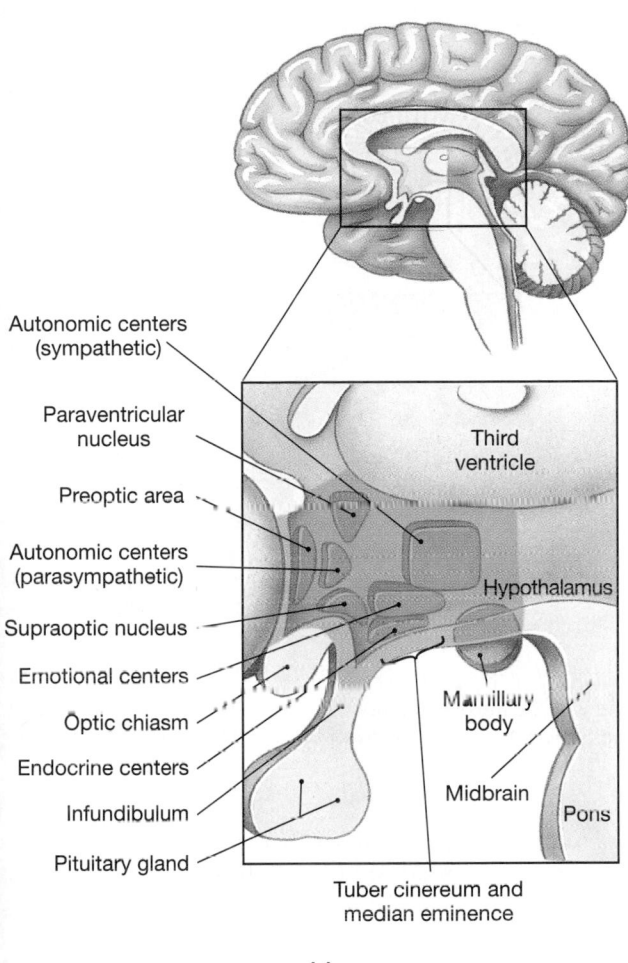

Autonomic centers
(sympathetic)

Paraventricular
nucleus

Preoptic area

Autonomic centers
(parasympathetic)

Supraoptic nucleus

Emotional centers

Optic chiasm

Endocrine centers

Infundibulum

Pituitary gland

Third
ventricle

Hypothalamus

Mamillary
body

Midbrain

Pons

Tuber cinereum and
median eminence

(a)

☑ Damage to the lateral geniculate nuclei of the thalamus would interfere with the functions of which of the senses?

☑ What area of the diencephalon would be stimulated by changes in body temperature?

TABLE 14-5 **Components and Functions of the Hypothalamus**

Region/Nucleus	Function
Supraoptic nucleus	Secretes ADH, restricting water loss at the kidneys
Paraventricular nucleus	Secretes oxytocin
Preoptic areas	Regulate body temperature
Tuber cinereum and median eminence	Releases hormones that control endocrine cells of the anterior pituitary
Autonomic centers	Control medullary nuclei that regulate heart rate and blood pressure
Mamillary bodies	Control feeding reflexes (licking, swallowing, etc.)

•**FIGURE 14-15** **The Hypothalamus in Sagittal Section.** **(a)** Diagrammatic view of the hypothalamus, showing the locations of major nuclei and centers. **(b)** Photograph of the hypothalamus and adjacent portions of the brain.

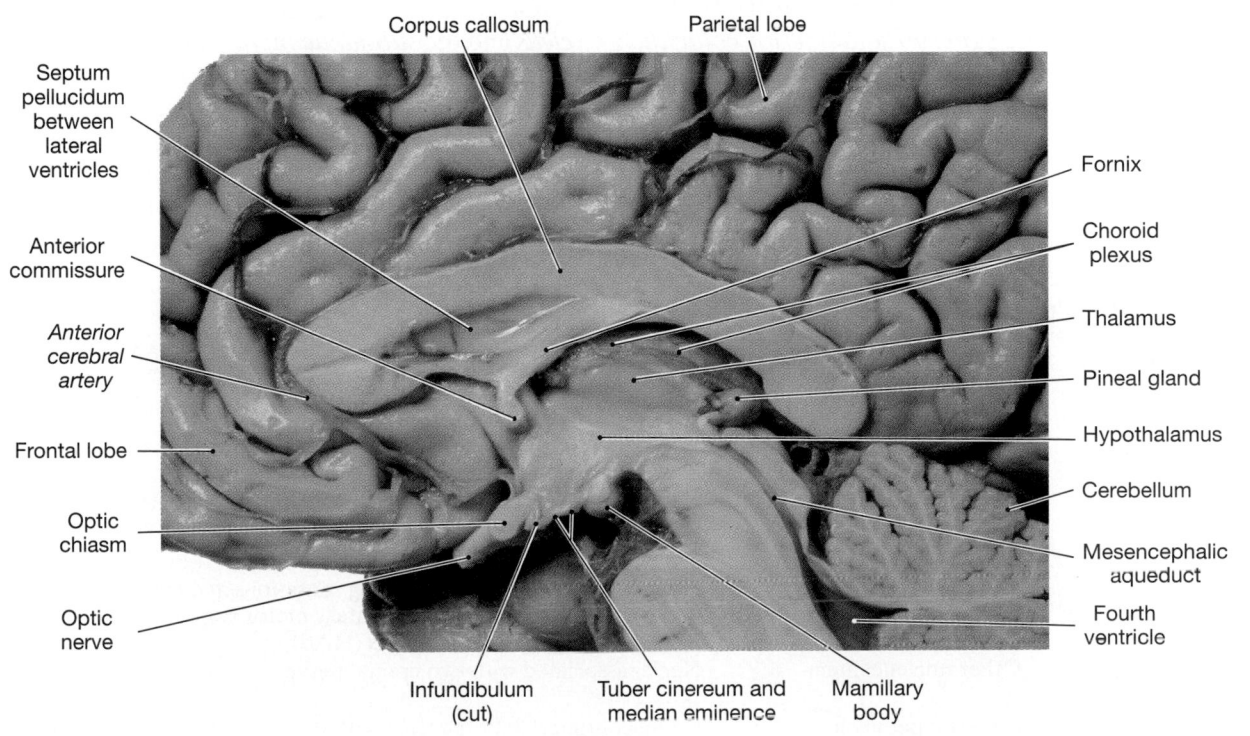

Corpus callosum

Parietal lobe

Septum
pellucidum
between
lateral
ventricles

Anterior
commissure

*Anterior
cerebral
artery*

Frontal lobe

Optic
chiasm

Optic
nerve

Fornix

Choroid
plexus

Thalamus

Pineal gland

Hypothalamus

Cerebellum

Mesencephalic
aqueduct

Fourth
ventricle

Infundibulum
(cut)

Tuber cinereum and
median eminence

Mamillary
body

(b)

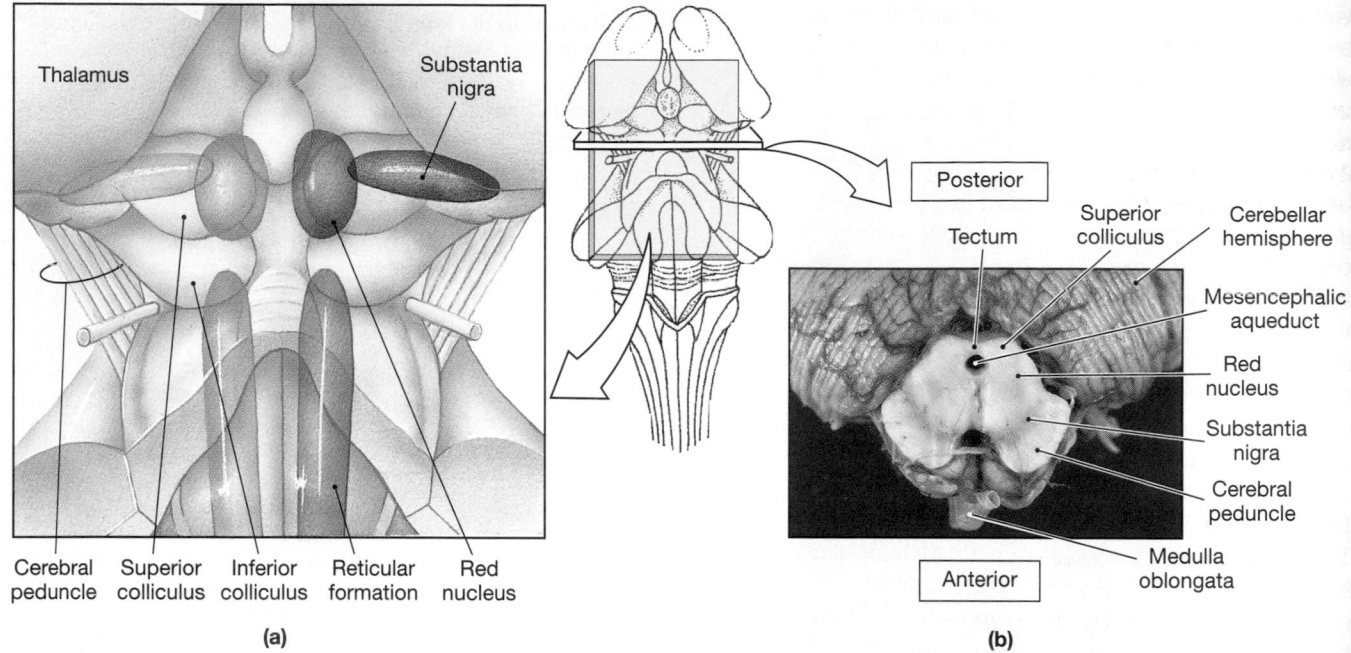

•FIGURE 14-16 **The Mesencephalon.** **(a)** Diagrammatic posterior view of the mesencephalon. Superficial structures are labeled on the left, and underlying nuclei are colored and labeled on the right. **(b)** Superior view of a section through the brain at the level of the mesencephalon.

THE MESENCEPHALON

The external anatomy of the mesencephalon, or midbrain, is shown in Figure 14-13•, and the major nuclei are detailed in Figure 14-16• and Table 14-6. The roof, or **tectum,** of the mesencephalon contains two pairs of sensory nuclei known collectively as the **corpora quadrigemina** (KOR-po-ra quad-ri-JEM-i-nuh). These nuclei, the *superior* and *inferior colliculi,* process visual and auditory sensations. Each **superior colliculus** (kol-IK-ū-lus; *colliculus,* a small hill) receives visual inputs from the lateral geniculate of the thalamus on that side. The **inferior colliculus** receives auditory data from nuclei in the medulla oblongata

and pons. Some of this information may be forwarded to the medial geniculate on the same side. The superior colliculi control the reflex movements of the eyes, head, and neck in response to visual stimuli, such as a bright light. The inferior colliculi control reflex movements of the head, neck, and trunk in response to auditory stimuli, such as a loud noise.

On each side, the mesencephalon contains a *red nucleus* and the *substantia nigra.* The **red nucleus** contains numerous blood vessels that give it a rich red color. This nucleus receives information from the cerebrum and cerebellum and issues subconscious motor commands concerned with the maintenance of muscle tone and posture. The **substantia nigra** (NĪ-gruh; black) lies lateral

TABLE 14-6 **Components and Functions of the Mesencephalon**

Subdivision	Region/Nucleus	Function
Gray matter		
Tectum (roof)	Superior colliculi	Integrate visual information with other sensory inputs; initiate involuntary motor responses
	Inferior colliculi	Relay auditory information to medial geniculate nuclei
Walls and floor	Red nuclei	Involuntary control of muscle tone and posture
	Substantia nigra	Regulates activity in the cerebral nuclei
	Reticular formation (headquarters)	Automatic processing of incoming sensations and outgoing motor commands; can initiate involuntary motor responses to stimuli; maintenance of consciousness (RAS)
	Other nuclei/centers	Nuclei associated with two cranial nerves (N III, N IV)
White matter	Cerebral peduncles	Connect primary motor cortex with motor neurons in brain and spinal cord; carry ascending sensory information to thalamus

to the red nucleus. The gray matter in this region contains darkly pigmented cells, giving it a black color.

Neurons in the substantia nigra inhibit the activity of the cerebral nuclei by releasing the neurotransmitter *dopamine.* ∞ *[p. 399]* If the substantia nigra is damaged or the neurons secrete less dopamine, the cerebral nuclei become more active. The result is a gradual, generalized increase in muscle tone and the appearance of symptoms characteristic of **Parkinson's disease.** Persons with Parkinson's disease have difficulty starting voluntary movements, because opposing muscle groups do not relax—they must be overpowered. Once a movement is under way, every aspect must be voluntarily controlled through intense effort and concentration.

[AM] *The Cerebral Nuclei and Parkinson's Disease*

The nerve fiber bundles on the ventrolateral surfaces of the mesencephalon (Figures 14-13, p. 467, and 14-16●) are the **cerebral peduncles** (*peduncles,* little feet). They contain (1) descending fibers that go to the cerebellum by way of the pons and (2) descending fibers that carry voluntary motor commands from the primary motor cortex of each cerebral hemisphere.

The mesencephalon also contains the headquarters of the **reticular activating system** (RAS), a specialized component of the reticular formation. Stimulation of the mesencephalic portion of the RAS makes you more alert and attentive. We shall consider the role of the RAS in the maintenance of consciousness in Chapter 15.

THE CEREBELLUM

The cerebellum (Table 14-7 and Figure 14-17●) is an automatic processing center. It has two primary functions:

1. *Adjusting the postural muscles of the body.* The cerebellum coordinates rapid, automatic adjustments that maintain balance and equilibrium. These alterations in muscle tone and position are made by modifying the activities of the red nuclei.

2. *Programming and fine-tuning movements controlled at the conscious and subconscious levels.* The cerebellum refines learned movement patterns. This func-

tion is performed indirectly by regulating activity along both pyramidal and extrapyramidal motor pathways at the cerebral cortex, cerebral nuclei, and motor centers in the brain stem. The cerebellum compares the motor commands with proprioceptive information (position sense) and performs any adjustments needed to make the movement smooth.

The cerebellum has a complex, highly convoluted surface composed of neural cortex. The folds, or **folia** (FŌ-lē-uh; leaves), of the surface are less prominent than the gyri of the cerebral hemispheres. The **anterior** and **posterior lobes** are separated by the **primary fissure** (Figure 14-17a●). Along the midline, a narrow band of cortex known as the **vermis** (VER-mis; worm) separates the **cerebellar hemispheres**. The slender **flocculonodular** (flok-ū-lō-NOD-ū-ler) **lobe** lies between the roof of the fourth ventricle and the cerebellar hemispheres and vermis (Figure 14-17b●).

The cerebellar cortex contains huge, highly branched **Purkinje** (pur-KIN-jē) **cells.** The extensive dendrites of each cell receive input from up to 200,000 synapses. Internally, the white matter of the cerebellum forms a branching array that in sectional view resembles a tree. Anatomists call it the **arbor vitae,** or "tree of life." The cerebellum receives proprioceptive information from the spinal cord and monitors all proprioceptive, visual, tactile, balance, and auditory sensations received by the brain. Information concerning the motor commands issued at the conscious and subconscious levels reaches the cerebellum indirectly, relayed by nuclei in the pons. A relatively small portion of the afferent fibers synapse in **cerebellar nuclei** before projecting to the cerebellar cortex. Most axons that carry sensory information do not synapse in the cerebellar nuclei but pass through the deeper layers of the cerebellum on their way to the cerebellar cortex.

Tracts that link the cerebellum with the brain stem, cerebrum, and spinal cord leave the cerebellar hemispheres as the *superior, middle,* and *inferior cerebellar peduncles.* The **superior cerebellar peduncles** link the cerebellum with nuclei in the midbrain, diencephalon, and cerebrum. The **middle cerebellar peduncles** are

TABLE 14-7 Components of the Cerebellum

Subdivision	Region/Nucleus	Function
Gray matter	Cerebellar cortex	Involuntary coordination and control of ongoing movements of body parts
	Cerebellar nuclei	Same as for cerebellar cortex
White matter	Arbor vitae	Connects cerebellar cortex and nuclei with cerebellar peduncles
	Cerebellar peduncles	
	Superior	Link the cerebellum with mesencephalon, diencephalon, and cerebrum
	Middle	Contain transverse fibers and carry communications between the cerebellum and pons
	Inferior	Link the cerebellum with the medulla oblongata and spinal cord
	Transverse fibers	Interconnect pontine nuclei with the cerebellar hemispheres on the opposite side

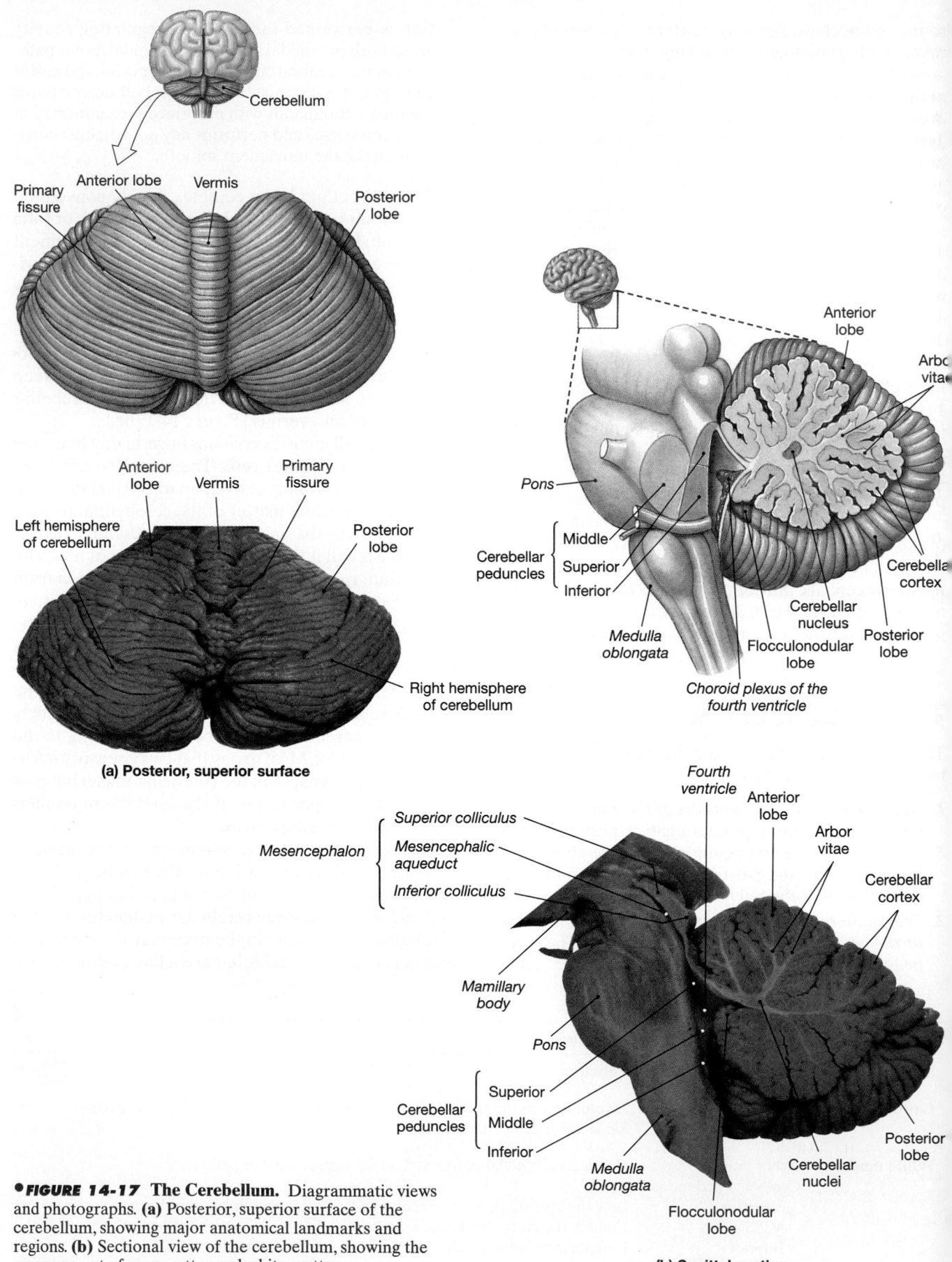

(a) Posterior, superior surface

•FIGURE 14-17 The Cerebellum. Diagrammatic views and photographs. **(a)** Posterior, superior surface of the cerebellum, showing major anatomical landmarks and regions. **(b)** Sectional view of the cerebellum, showing the arrangement of gray matter and white matter.

(b) Sagittal section

connected to a broad band of fibers that cross the ventral surface of the pons at right angles to the axis of the brain stem. The middle cerebellar peduncles also connect the cerebellar hemispheres with sensory and motor nuclei in the pons. The **inferior cerebellar peduncles** permit communication between the cerebellum and nuclei in the medulla oblongata and carry ascending and descending cerebellar tracts from the spinal cord.

The cerebellum may be permanently damaged by trauma or stroke or temporarily affected by drugs such as alcohol. These alterations can produce **ataxia** (a-TAK-sē-uh; *ataxia,* lack of order), a disturbance in balance. In severe ataxia, the individual may be unable to sit or stand without assistance. AM
Cerebellar Dysfunction

THE PONS

The pons links the cerebellar hemispheres with the mesencephalon, diencephalon, cerebrum, and spinal cord. Important features and regions of the pons are indicated in Figures 14-13, p. 467, and 14-18a●. The pons contains the following:

- *Sensory and motor nuclei for four cranial nerves* (N V, N VI, N VII, and N VIII). These cranial nerves innervate the jaw muscles, the anterior surface of the face, one of the extrinsic eye muscles (the lateral rectus), and the sense organs of the inner ear (the *vestibular* and *cochlear nuclei*).

- *Nuclei dealing with the control of respiration.* On each side of the brain, the reticular formation in this region contains two respiratory centers—the *apneustic center* and the *pneumotaxic center*. These centers modify the activity of the *respiratory rhythmicity center* in the medulla oblongata.

- *Nuclei and tracts that process and relay information heading to or from the cerebellum.* The pons links the cerebellum with the brain stem, cerebrum, and spinal cord.

- *Ascending, descending, and transverse tracts.* Longitudinal tracts interconnect other portions of the CNS. The middle cerebellar peduncles are connected to the **transverse fibers** that cross the anterior surface of the pons. These fibers are axons that link nuclei of the pons (*pontine nuclei*) with the cerebellar hemisphere of the opposite side.

THE MEDULLA OBLONGATA

The medulla oblongata is continuous with the spinal cord. Figure 14-13●, p. 467, details the external surface of the medulla oblongata; Figure 14-18b● and Table 14-8 summarize its major components.

The medulla oblongata connects the brain with the spinal cord, and many of its functions are directly re-

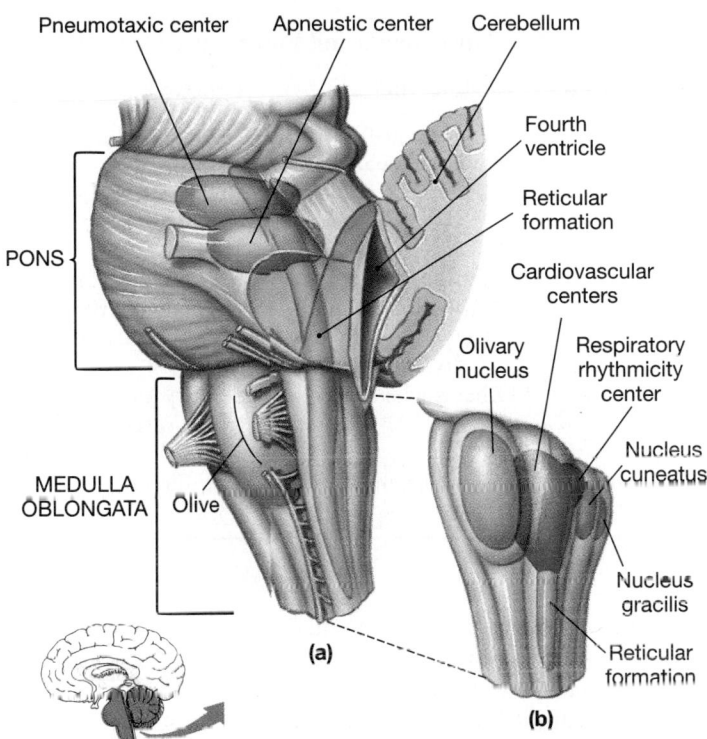

●**FIGURE 14-18** The Pons and Medulla Oblongata.
(a) Nuclei in the pons. **(b)** Nuclei in the medulla oblongata.

lated to that fact. For example, all communication between the brain and spinal cord involves tracts that ascend or descend through the medulla oblongata.

Nuclei in the medulla oblongata include (1) nuclei associated with the autonomic control of visceral activities, (2) sensory or motor nuclei associated with cranial nerves, and (3) relay stations along sensory or motor pathways. We will focus on three types of nuclei that we will encounter in later chapters:

1. *Autonomic nuclei.* The reticular formation in the medulla oblongata contains nuclei and centers responsible for the regulation of vital autonomic functions. These **reflex centers** receive inputs from cranial nerves, the cerebral cortex, and the brain stem. The output of these centers controls or adjusts the activities of one or more peripheral systems. Major centers regulating autonomic functions include the following:

- The **cardiovascular centers,** which adjust heart rate, the strength of cardiac contractions, and the flow of blood through peripheral tissues. In terms of function, the cardiovascular centers are subdivided into **cardiac** (*kardia,* heart) and **vasomotor** (*vas,* canal) **centers,** but their anatomical boundaries are difficult to determine.

- The **respiratory rhythmicity centers,** which set the basic pace for respiratory movements. Their activity is regulated by inputs from the apneustic and pneumotaxic centers of the pons.

TABLE 14-8 Components and Functions of the Medulla Oblongata

Subdivision	Region/Nucleus	Function
Gray matter	Nucleus gracilis Nucleus cuneatus	Relay somatic sensory information to the thalamus
	Olivary nuclei	Relay information from the red nucleus, other midbrain nuclei, and the cerebral cortex to the cerebellum
	Reflex centers	
	Cardiac centers	Regulate heart rate and force of contraction
	Vasomotor centers	Regulate distribution of blood flow
	Respiratory rhythmicity centers	Set the pace of respiratory movements
	Other nuclei/centers	Sensory and motor nuclei of five cranial nerves Nuclei relaying ascending sensory information from the spinal cord to higher centers
White matter	Ascending and descending tracts	Link the brain with the spinal cord

2. *Nuclei of cranial nerves.* The medulla oblongata contains sensory and motor nuclei associated with five of the cranial nerves (N VIII, N IX, N X, N XI, and N XII). These cranial nerves provide motor commands to muscles of the pharynx, neck, and back as well as to the visceral organs of the thoracic and peritoneal cavities. N VIII carries sensory information from receptors in the inner ear to the vestibular and cochlear nuclei, which extend from the pons into the medulla.

3. *Relay stations.* The **nucleus gracilis** and the **nucleus cuneatus** pass somatic sensory information to the thalamus, and the **olivary nuclei** relay information from the spinal cord, the cerebral cortex, and brain stem to the cerebellar cortex. The bulk of these nuclei create the **olives,** prominent bulges along the ventrolateral surface

of the medulla oblongata. (The olives of the medulla oblongata got their name because they are shaped like the olives found in salads.) The tracts leaving these nuclei cross to the opposite side of the brain before reaching their destinations. This crossing over is called *decussation* (dē-kuh-SĀ-shun; *decussatio*, crossing over).

☑ In what part of the brain would you find a worm (vermis) and a tree (arbor vitae)?

☑ The medulla oblongata is one of the smallest sections of the brain, yet damage there can cause death, whereas similar damage in the cerebrum might go unnoticed. Why?

☑ How would decreased release of neurotransmitter from cells of the substantia nigra affect cerebral neurons activity?

FOCUS   Cranial Nerves

Cranial nerves are PNS components that connect to the brain rather than to the spinal cord. Twelve pairs of cranial nerves are located on the ventrolateral surface of the brain (Figure 14-19●), each with a name related to its appearance or function.

The number assigned to a cranial nerve roughly corresponds to its position along the longitudinal axis of the brain, beginning at the cerebrum. Roman numerals are usually used, either alone or with the prefix N or CN. We will use the abbreviation N, which is generally preferred by neuroanatomists and clinical neurologists. CN, an equally valid abbreviation, is preferred by comparative anatomists.

Each cranial nerve attaches to the brain near the associated sensory or motor nuclei. The sensory nuclei act as switching centers, with the postsynaptic neurons relaying the information to other nuclei or

to processing centers within the cerebral or cerebellar cortex. In a similar fashion, the motor nuclei receive convergent inputs from higher centers or from other nuclei along the brain stem.

In this section, we classify cranial nerves as primarily sensory, special sensory, motor, or mixed (sensory and motor). In this classification, sensory nerves carry somatosensory information, including touch, pressure, vibration, temperature, and pain. Special sensory nerves carry the sensations of smell, sight, hearing, and balance. Motor nerves are dominated by the axons of somatic motor neurons; mixed nerves have a mixture of sensory and motor fibers. This is a useful method of classification, but it is based on the primary function, and a cranial nerve can have important secondary functions. Three examples are worth noting:

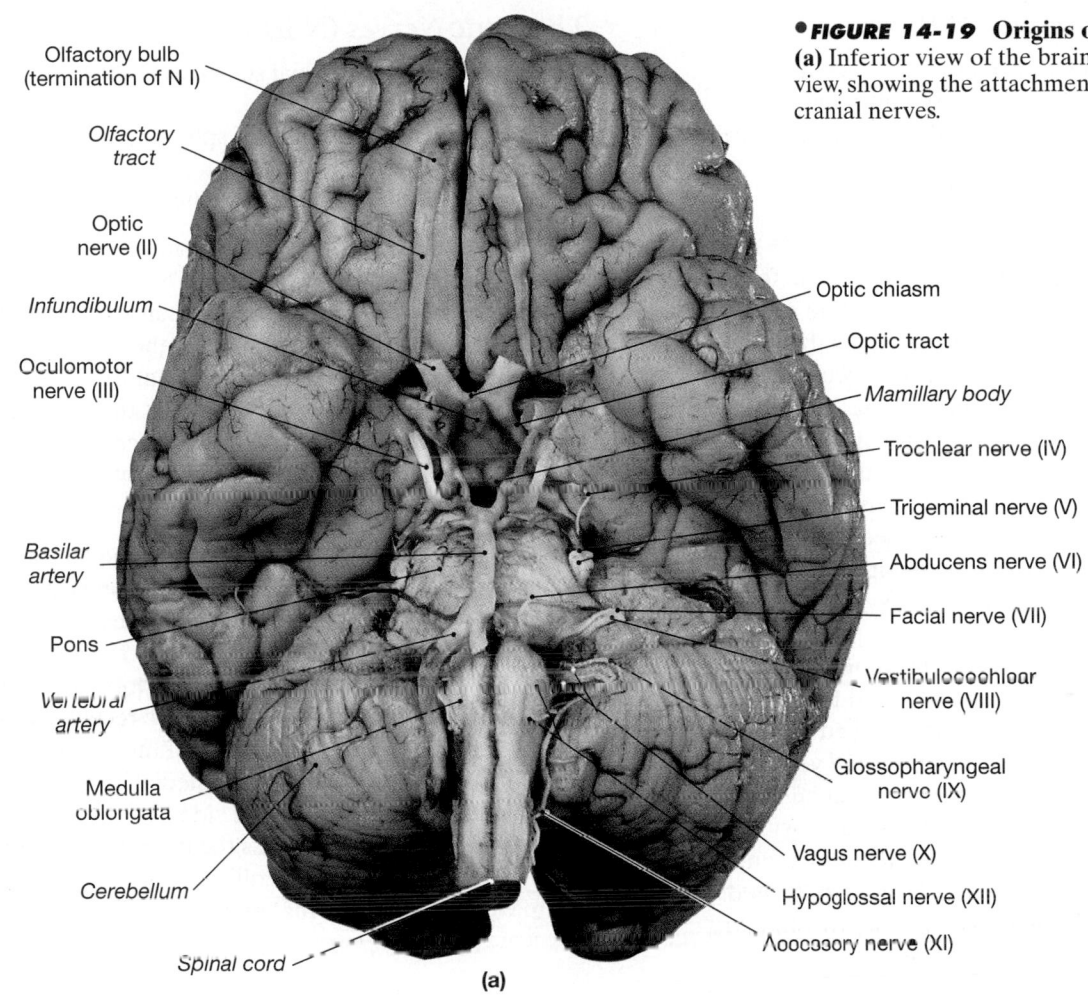

Olfactory bulb
(termination of N I)

*Olfactory
tract*

Optic
nerve (II)

Infundibulum

Oculomotor
nerve (III)

*Basilar
artery*

Pons

*Vertebral
artery*

Medulla
oblongata

Cerebellum

Spinal cord

Optic chiasm

Optic tract

Mamillary body

Trochlear nerve (IV)

Trigeminal nerve (V)

Abducens nerve (VI)

Facial nerve (VII)

Vestibulocochlear
nerve (VIII)

Glossopharyngeal
nerve (IX)

Vagus nerve (X)

Hypoglossal nerve (XII)

Accessory nerve (XI)

(a)

• *FIGURE 14-19* **Origins of the Cranial Nerves.**
(a) Inferior view of the brain. **(b)** Diagrammatic
view, showing the attachment of the 12 pairs of
cranial nerves.

1. The olfactory receptors, the visual receptors, and the receptors of the inner ear are innervated by cranial nerves that are dedicated almost entirely to carrying special sensory information. The sensation of taste, considered to be one of the special senses, is carried by axons that form only a small part of large cranial nerves that have other primary functions.

2. As elsewhere in the PNS, a nerve containing tens of thousands of motor fibers that lead to a skeletal muscle will also contain sensory fibers from muscle spindles and tendon organs in that muscle. We assume that these sensory fibers are present but ignore them in the classification of the nerve.

3. Regardless of their other functions, several cranial nerves (N III, VII, IX, and X) distribute autonomic fibers to peripheral ganglia, just as spinal nerves deliver them to ganglia along the spinal cord. We will note the presence of small numbers of autonomic fibers (and shall discuss them further in Chapter 16) but ignore them in the classification of the nerve.

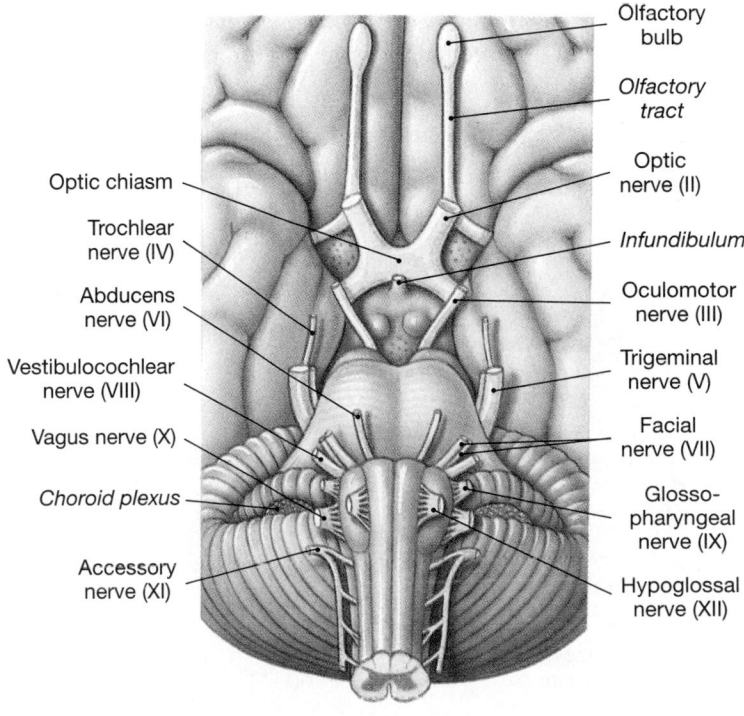

Optic chiasm

Trochlear
nerve (IV)

Abducens
nerve (VI)

Vestibulocochlear
nerve (VIII)

Vagus nerve (X)

Choroid plexus

Accessory
nerve (XI)

Olfactory
bulb

*Olfactory
tract*

Optic
nerve (II)

Infundibulum

Oculomotor
nerve (III)

Trigeminal
nerve (V)

Facial
nerve (VII)

Glosso-
pharyngeal
nerve (IX)

Hypoglossal
nerve (XII)

(b)

The Olfactory Nerves (N I)

Primary function: Special sensory (smell)
Origin: Receptors of olfactory epithelium
Pass through: Cribriform plate of ethmoid bone
∞ *[pp. 203, 208]*
Destination: Olfactory bulbs

The first pair of cranial nerves (Figure 14-19•) carries special sensory information responsible for the sense of smell. The olfactory receptors are specialized neurons in the epithelium covering the roof of the nasal cavity, the superior nasal conchae, and the superior portions of the nasal septum. Axons from these sensory neurons collect to form 20 or more bundles that penetrate the cribriform plate of the ethmoid bone. These bundles are components of the **olfactory nerve** (N I). Almost at once they enter the **olfactory bulbs,** neural masses on either side of the crista galli. The olfactory afferents synapse within the olfactory bulbs, and the axons of the postsynaptic neurons proceed to the cerebrum along the slender **olfactory tracts** (Figures 14-19 and 14-20•).

Because the olfactory tracts looked like typical peripheral nerves, anatomists a hundred years ago misidentified these tracts as the first cranial nerve. Later studies demonstrated that the olfactory tracts and bulbs are part of the cerebrum, but by then the numbering system was already firmly established. Anatomists were left with a forest of tiny olfactory nerve bundles lumped together as N I.

The olfactory nerves are the only cranial nerves attached directly to the cerebrum. The rest originate or terminate within nuclei of the diencephalon or brain stem, and the ascending sensory information synapses in the thalamus before reaching the cerebrum.

The Optic Nerves (N II)

Primary function: Special sensory (vision)
Origin: Retina of eye
Pass through: Optic canals of sphenoid bone
∞ *[pp. 203, 207]*
Destination: Diencephalon via the optic chiasm

The **optic nerves** (N II) carry visual information from special sensory ganglia in the eyes. These nerves (Figure 14-21•) contain about 1 million sensory nerve fibers. The optic nerves pass through the optic canals of the sphenoid before they converge at the ventral, anterior margin of the diencephalon, at the **optic chiasm** (*chiasma,* a crossing). At the optic chiasm, fibers from the nasal half of each retina cross over to the opposite side of the brain.

The reorganized axons continue toward the lateral geniculate nuclei of the thalamus as the **optic tracts** (Figures 14-19 and 14-21•). After synapsing in the lateral geniculates, projection fibers deliver the information to the visual cortex of the occipital lobe. With this arrangement, each cerebral hemisphere receives visual information from the lateral half of the retina of the eye on that side and from the medial half of the retina of the eye of the opposite side. Relatively few axons in the optic tracts bypass the lateral geniculates and synapse in the superior colliculus of the midbrain. We shall consider that pathway in Chapter 17.

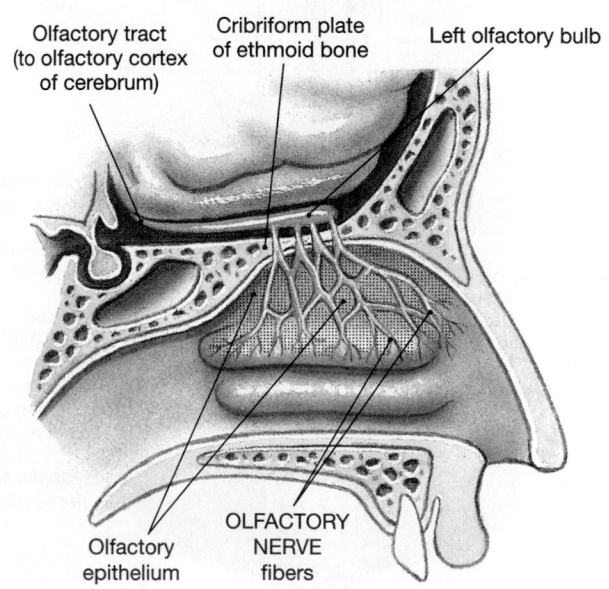

•*FIGURE 14-20* **The Olfactory Nerve**

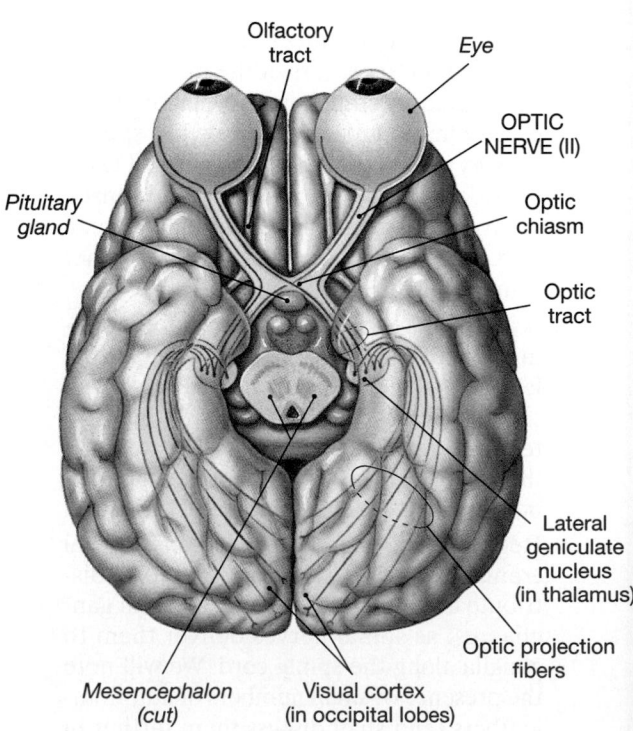

•*FIGURE 14-21* **The Optic Nerve**

The Oculomotor Nerves (N III)

Primary function: Motor (eye movements)
Origin: Mesencephalon
Pass through: Superior orbital fissures of sphenoid bone ∞ *[pp. 202, 207, 210]*
Destination: Somatic motor: superior, inferior, and medial rectus muscles; inferior oblique muscle; true levator palpebrae superioris muscle. ∞ *[p. 327]*
Visceral motor: intrinsic eye muscles

The mesencephalon contains the motor nuclei controlling the third and fourth cranial nerves. Each **oculomotor nerve** (N III) innervates four of the six extrinsic eye muscles and the levator palpebrae superioris, the muscle that raises the upper eyelid (Figure 14-22●). On each side of the brain, N III emerges from the ventral surface of the mesencephalon and penetrates the posterior orbital wall at the superior orbital fissure. Individuals with damage to this nerve often complain of pain over the eye and double vision, because the movements of the left and right eyes cannot be coordinated properly.

The oculomotor nerve also delivers preganglionic autonomic fibers to neurons of the **ciliary ganglion.** The ganglionic neurons control intrinsic eye muscles. These muscles change the diameter of the pupil, adjusting the amount of light entering the eye, and change the shape of the lens to focus images on the retina.

The Trochlear Nerves (N IV)

Primary function: Motor (eye movements)
Origin: Mesencephalon
Pass through: Superior orbital fissures of sphenoid bone ∞ *[pp. 202, 207, 210]*
Destination: Superior oblique muscle ∞ *[p. 327]*

The **trochlear** (TRŌK-lē-ar) **nerve** (*trochlea,* a pulley), smallest of the cranial nerves, innervates the superior oblique muscle of each eye (Figure 14-22●). The name "trochlear nerve" should remind you that the innervated muscle passes through a ligamentous pulley, or sling, on its way to its insertion on the surface of the eye. An individual with damage to this nerve or to its nucleus will have difficulty looking upward.

The Abducens Nerves (N VI)

Primary function: Motor (eye movements)
Origin: Pons
Pass through: Superior orbital fissures of sphenoid bone ∞ *[pp. 202, 207, 210]*
Destination: Lateral rectus muscle ∞ *[p. 327]*

The **abducens** (ab-DŪ senz) **nerve** innervates the lateral rectus, the sixth of the extrinsic eye muscles. Innervation of this muscle makes lateral movements of the eyeball possible. Each abducens nerve emerges from the inferior surface of the brain stem at the border between the pons and the medulla oblongata. Along with the oculomotor and trochlear nerves from that side, it reaches the orbit through the superior orbital fissure (Figure 14-22●).

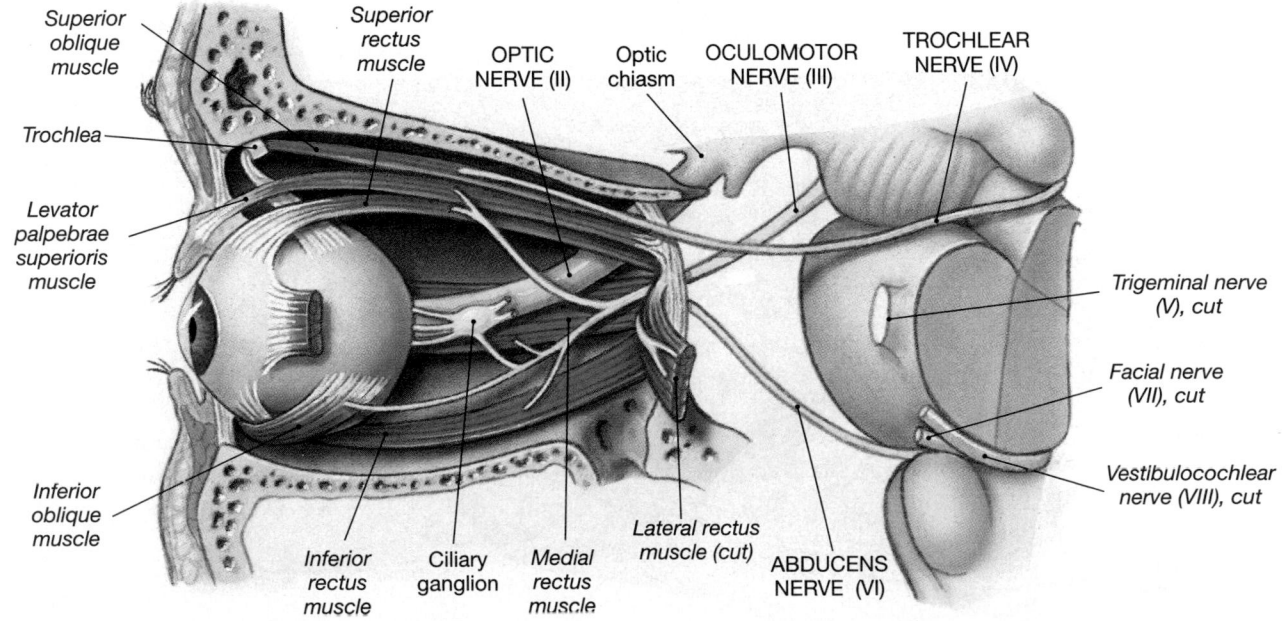

●**FIGURE 14-22** **Cranial Nerves Controlling the Extrinsic Eye Muscles**

The Trigeminal Nerves (N V)

Primary function: Mixed (sensory and motor) to face

Origin: *Ophthalmic branch* (sensory): orbital structures, nasal cavity, skin of forehead, upper eyelid, eyebrow, nose (part). *Maxillary branch* (sensory): lower eyelid, upper lip, gums, and teeth; cheek; nose, palate, and pharynx (part). *Mandibular branch* (mixed): sensory from lower gums, teeth, and lips; palate and tongue (part); motor from motor nuclei of pons

Pass through (on each side): Ophthalmic branch through superior orbital fissure, maxillary branch through foramen rotundum, mandibular branch through foramen ovale ∞ *[pp. 203, 207, 210]*

Destination: Ophthalmic and maxillary branches to sensory nuclei in pons; mandibular branch innervates muscles of mastication ∞ *[p. 328]*

The pons contains the nuclei associated with three cranial nerves (N V, VI, and VII) and contributes to a fourth (N VIII). The **trigeminal** (trī-JEM-i-nal) **nerve** (Figure 14-23●), the largest of the cranial nerves, is a mixed nerve. It provides both somatic sensory information from the head and face and motor control over the muscles of mastication. Sensory (dorsal) and motor (ventral) roots originate on the lateral surface of the pons. The sensory branch is larger, and the enormous **semilunar ganglion** contains the cell bodies of the sensory neurons. As the name implies, the trigeminal has three major branches; the relatively small motor root contributes to only one of the three. **Tic douloureux** (doo-loo-ROO; *douloureux,* painful) is a painful condition affecting the area innervated by the maxillary and mandibular branches of the trigeminal nerve. Sufferers complain of severe, almost totally debilitating pain triggered by contact with the lip, tongue, or gums. The cause of the condition is unknown.   **AM** *Tic Douloureux*

The trigeminal nerve branches are associated with the *ciliary, sphenopalatine, submandibular,* and *otic ganglia.* These are autonomic (parasympathetic) ganglia whose neurons innervate structures of the face. However, although its nerve fibers may pass around or through these ganglia, the trigeminal nerve does not contain visceral motor fibers. We discussed the ciliary ganglion previously (p. 477) and will detail the other ganglia next, with the branches of the *facial nerve* (N VII).

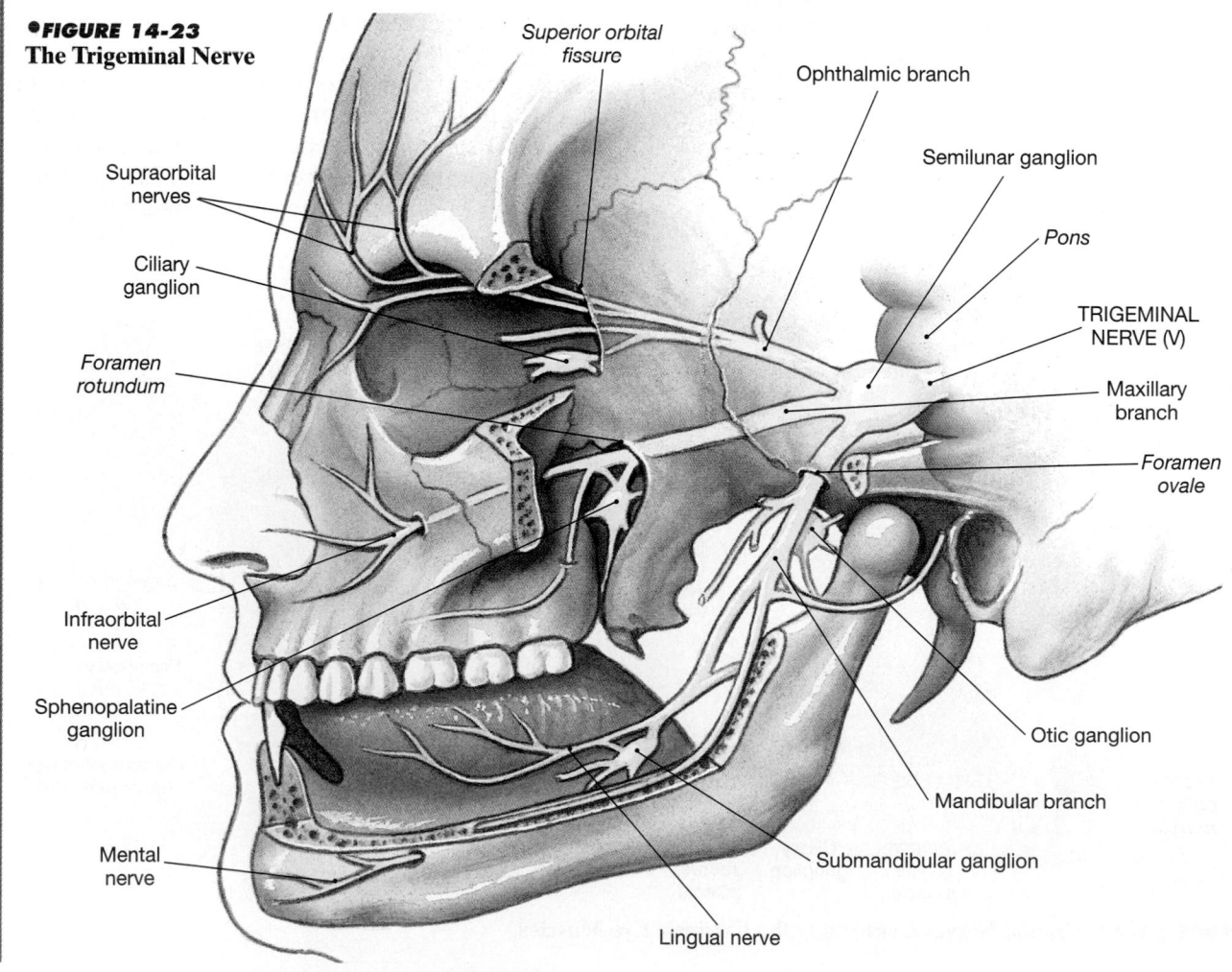

●FIGURE 14-23
The Trigeminal Nerve

Superior orbital fissure

Ophthalmic branch

Supraorbital nerves

Semilunar ganglion

Ciliary ganglion

Pons

Foramen rotundum

TRIGEMINAL NERVE (V)

Maxillary branch

Foramen ovale

Infraorbital nerve

Sphenopalatine ganglion

Otic ganglion

Mental nerve

Mandibular branch

Submandibular ganglion

Lingual nerve

The Facial Nerves (N VII)

Primary function: Mixed (sensory and motor) to face
Origin: Sensory from taste receptors on anterior two thirds of tongue; motor from motor nuclei of pons
Pass through: Internal acoustic canals to the *facial canals*, which end at the stylomastoid foramina ∞ *[pp. 203, 206]*
Destination: *Sensory* to sensory nuclei of pons. *Somatic motor:* muscles of facial expression. ∞ *[p. 325] Visceral motor:* lacrimal (tear) gland and nasal mucous glands via sphenopalatine ganglion; submandibular and sublingual salivary glands via submandibular ganglion

The **facial nerve** is a mixed nerve. The cell bodies of the sensory neurons are located in the **geniculate ganglion,** and the motor nuclei are in the pons. The sensory and motor roots emerge from the side of the pons and enter the internal acoustic canal of the temporal bone. Each facial nerve then passes through the facial canal to reach the face by way of the stylomastoid foramen (Figure 14-24). The sensory neurons monitor proprioceptors in the facial muscles, provide deep pressure sensations over the face, and receive taste information from receptors along the anterior two-thirds of the tongue. Somatic motor fibers control the superficial muscles of the scalp and face and deep muscles near the ear.

The facial nerve carries preganglionic autonomic fibers to the *sphenopalatine* and *submandibular ganglia.* Postganglionic fibers from the **sphenopalatine ganglion** innervate the lacrimal gland and small glands of the nasal cavity and pharynx. The **submandibular ganglion** innervates the *submandibular* and *sublingual* (*sub-*, under, + *lingua*, tongue) salivary glands.

Bell's palsy is a cranial nerve disorder that results from an inflammation of the facial nerve. The condition is probably due to a viral infection. Symptoms include paralysis of facial muscles on the affected side and loss of taste sensations from the anterior two-thirds of the tongue. The condition is usually painless, and in most cases, Bell's palsy goes away after a few weeks or months.

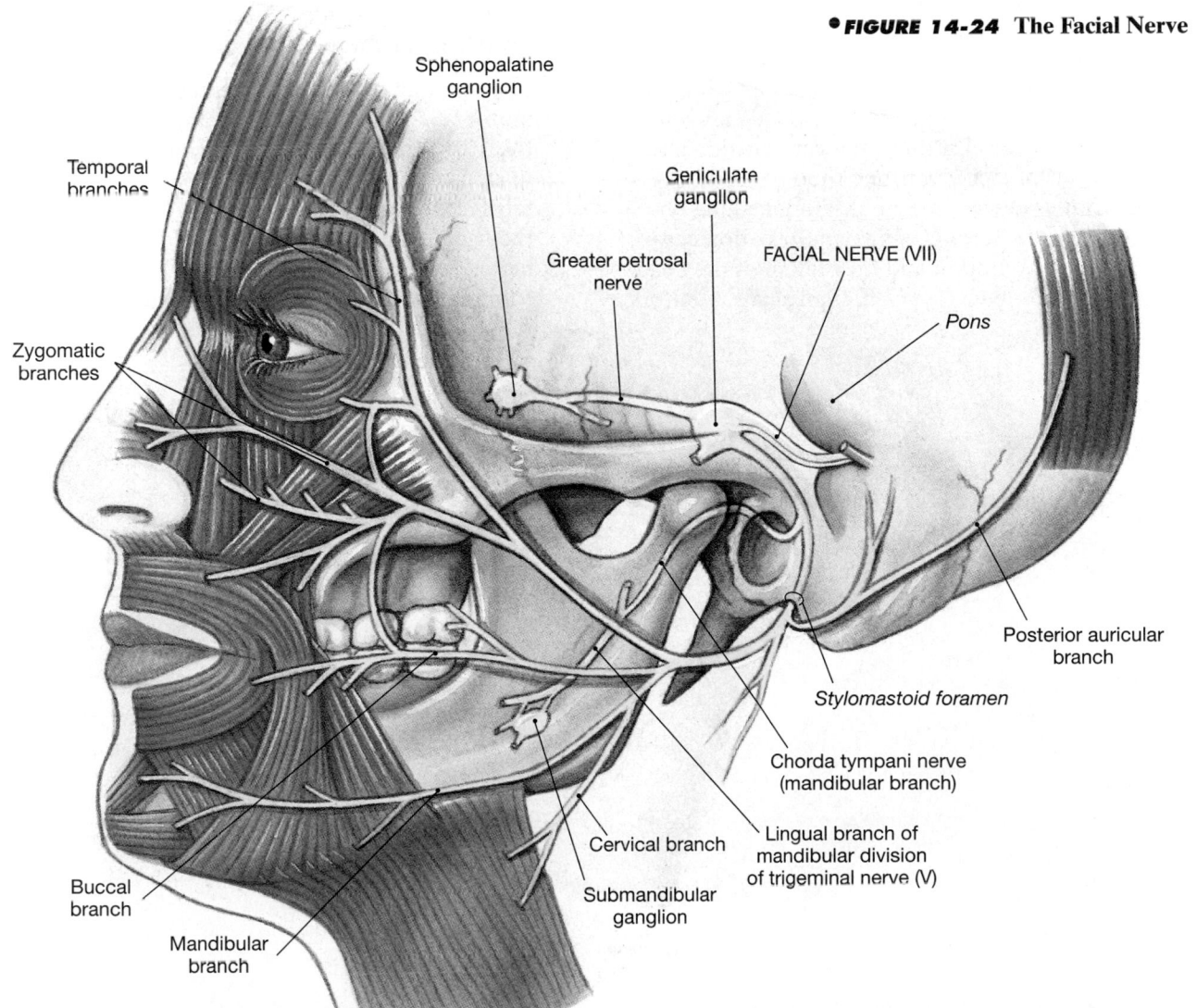

●*FIGURE 14-24* **The Facial Nerve**

Sphenopalatine ganglion

Geniculate ganglion

Greater petrosal nerve

FACIAL NERVE (VII)

Pons

Temporal branches

Zygomatic branches

Posterior auricular branch

Stylomastoid foramen

Chorda tympani nerve (mandibular branch)

Lingual branch of mandibular division of trigeminal nerve (V)

Cervical branch

Submandibular ganglion

Buccal branch

Mandibular branch

The Vestibulocochlear Nerves (N VIII)

Primary function: Special sensory: balance and equilibrium (vestibular branch) and hearing (cochlear branch)
Origin: Monitor receptors of the inner ear (vestibule and cochlea)
Pass through: Internal acoustic canals of temporal bones ∞ *[pp. 203, 206]*
Destination: Vestibular and cochlear nuclei of pons and medulla oblongata

The **vestibulocochlear nerve** is also known as the *acoustic nerve,* the *auditory nerve,* and the *stato-acoustic nerve.* We will use the term *vestibulo-cochlear,* because it indicates the names of the nerve's two major branches: the *vestibular branch* and the *cochlear branch.* Each vestibulocochlear nerve lies posterior to the origin of the facial nerve, straddling the boundary between the pons and the medulla oblongata (Figure 14-25●). This nerve reaches the sensory receptors of the inner ear by entering the internal acoustic canal in company with the facial nerve. There are two distinct bundles of sensory fibers within the vestibulocochlear nerve. The **vestibular branch** (*vestibulum,* cavity) originates at the receptors of the *vestibule,* the portion of the inner ear concerned with balance sensations. The sensory neurons are located within an adjacent sensory ganglion, and their axons target the **vestibular nuclei** of the pons and medulla oblongata. These afferents convey information concerning the orientation and movement of the head. The **cochlear branch** (*cochlea,* snail shell) monitors the receptors in the cochlea that provide the sense of hearing. The neurons are located within a peripheral ganglion (the *spiral ganglion*), and their axons synapse within the **cochlear nuclei** of the pons and medulla oblongata. Axons leaving the vestibular and cochlear nuclei relay the sensory information to other centers or initiate reflexive motor responses. We shall discuss balance and the sense of hearing in Chapter 17.

The Glossopharyngeal Nerves (N IX)

Primary function: Mixed (sensory and motor) to head and neck
Origin: Sensory from posterior one-third of the tongue, part of the pharynx and palate, carotid arteries of the neck; motor from motor nuclei of medulla oblongata
Pass through: Jugular foramina between the occipital bone and the temporal bones ∞ *[pp. 203, 206]*
Destination: *Sensory* to sensory nuclei of medulla oblongata. *Somatic motor:* pharyngeal muscles involved in swallowing. *Visceral motor:* parotid salivary gland by way of the otic ganglion

The medulla oblongata contains the sensory and motor nuclei of the ninth, tenth, eleventh, and twelfth cranial nerves, in addition to the vestibular nucleus of N VIII. The **glossopharyngeal** (glos-ō-fah-RIN-jē-al) **nerve** (*glossum,* tongue) innervates the tongue and pharynx. Each glossopharyngeal nerve penetrates the cranium within the jugular foramen, with N X and N XI (Figure 14-26●).

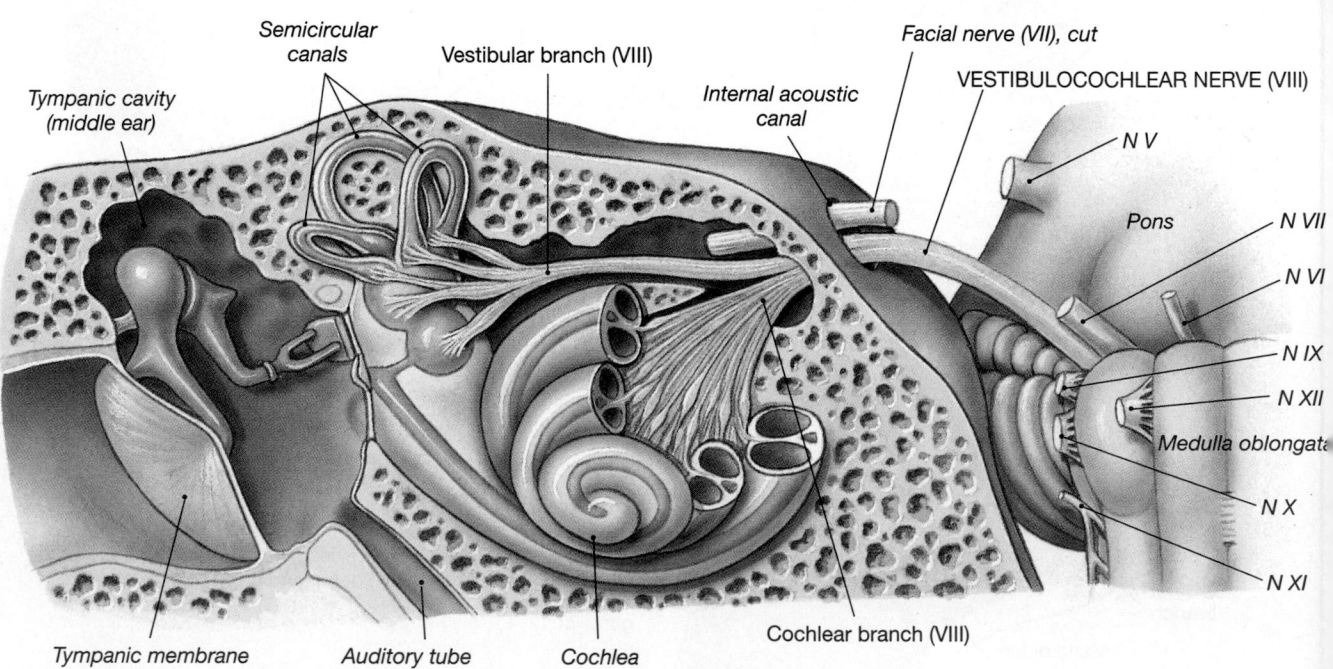

●**FIGURE 14-25** The Vestibulocochlear Nerve

The glossopharyngeal is a mixed nerve, but sensory fibers are most abundant. The sensory neurons are in the **superior** (*jugular*) **ganglion** and **inferior** (*petrosal*) **ganglion.** The sensory fibers carry general sensory information from the lining of the pharynx and the soft palate to a nucleus in the medulla oblongata. This nerve also provides taste sensations from the posterior third of the tongue and has special receptors that monitor the blood pressure and dissolved gas concentrations within major blood vessels.

The somatic motor fibers control the pharyngeal muscles involved in swallowing. Visceral motor fibers synapse in the *otic ganglion*, and postganglionic fibers innervate the parotid salivary gland of the cheek.

The Vagus Nerves (N X)

Primary function: Mixed (sensory and motor), widely distributed in the thorax and abdomen
Origin: Sensory from pharynx (part), pinna and external auditory canal, diaphragm, and visceral organs in thoracic and abdominopelvic cavities; motor from motor nuclei in medulla oblongata
Pass through: Jugular foramina between the occipital bone and the temporal bones ∞ *[pp. 203, 206]*
Destination: Sensory fibers to sensory nuclei and autonomic centers of medulla oblongata; visceral motor fibers to muscles of the palate, pharynx, digestive, respiratory, and cardiovascular systems in the thoracic and abdominal cavities

The **vagus** (VĀ-gus) **nerve** arises immediately posterior to the glossopharyngeal. Many small rootlets contribute to its formation, and developmental studies indicate that this nerve probably represents the fusion of several smaller cranial nerves during our evolutionary history. As its name suggests (*vagus,* wanderer), the vagus nerve branches and radiates extensively. Figure 14-27● shows only the general pattern of distribution.

Sensory neurons are located within the **jugular** and **nodose** (NŌ-dōs) **ganglia** (*node,* knot). The vagus provides somatic sensory information concerning the external auditory canal, a portion of the ear, and the diaphragm and special sensory information from pharyngeal taste receptors. But most of the vagal afferents carry visceral sensory information from receptors along the esophagus, respiratory tract, and abdominal viscera as distant as the last portions of the large intestine. This visceral sensory information is vital to the autonomic control of visceral function.

The motor components of the vagus are equally diverse. The vagus nerve carries preganglionic autonomic (parasympathetic) fibers that affect the heart and control smooth muscles and glands within the areas monitored by its sensory fibers, including the stomach, intestines, and gallbladder. Difficulty in swallowing is one of the most common signs of damage to either nerve IX or nerve X, because damage to either one prevents coordination of the swallowing reflex.

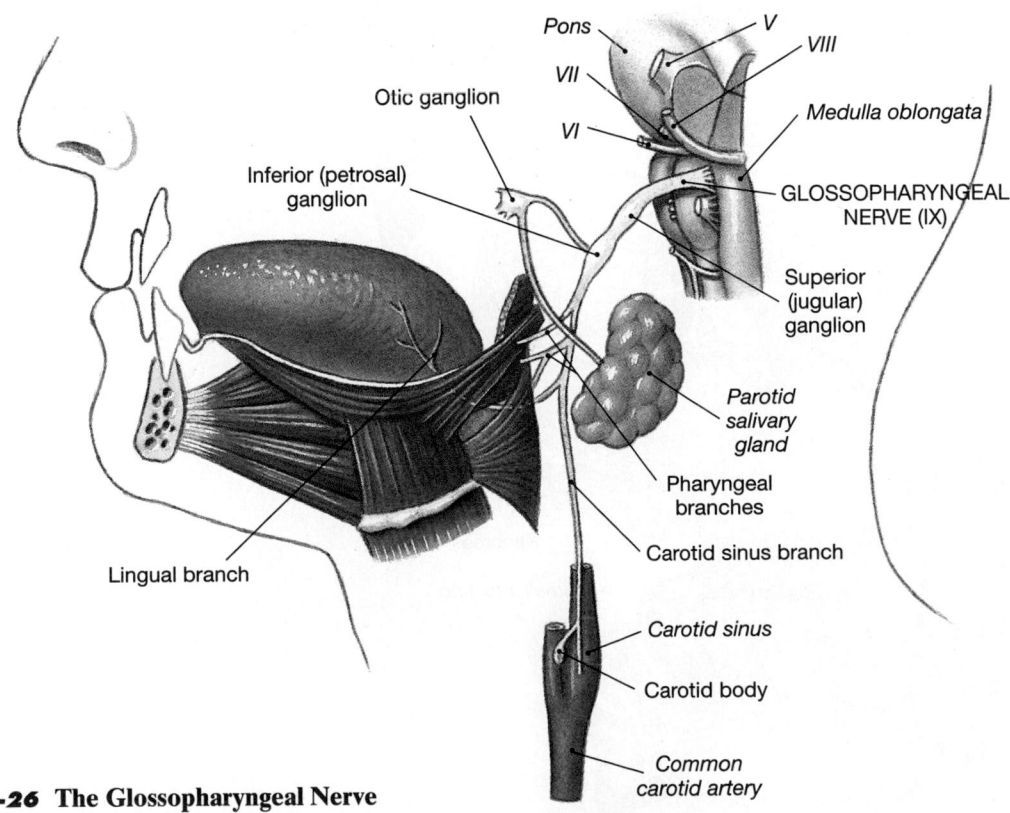

●*FIGURE 14-26* **The Glossopharyngeal Nerve**

•FIGURE 14-27 **The Vagus Nerve**

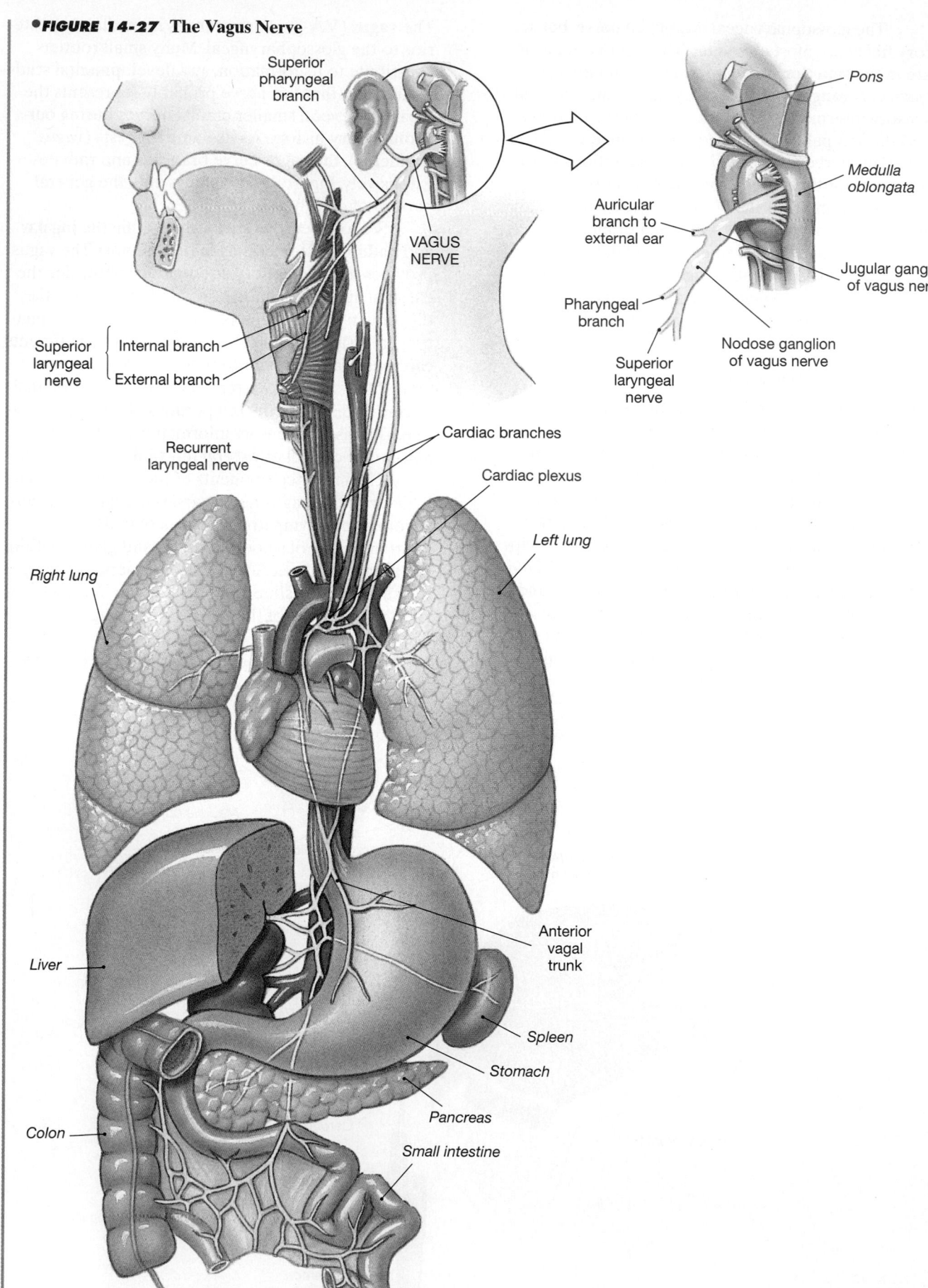

Superior pharyngeal branch

VAGUS NERVE

Pons

Medulla oblongata

Auricular branch to external ear

Pharyngeal branch

Superior laryngeal nerve

Jugular ganglion of vagus nerve

Nodose ganglion of vagus nerve

Superior laryngeal nerve

Internal branch

External branch

Recurrent laryngeal nerve

Cardiac branches

Cardiac plexus

Right lung

Left lung

Liver

Anterior vagal trunk

Spleen

Stomach

Pancreas

Colon

Small intestine

The Accessory Nerves (N XI)

Primary function: Motor to muscles of the neck and upper back

Origin: Motor nuclei of spinal cord and medulla oblongata

Pass through: Jugular foramina between the occipital bone and the temporal bones ∞ *[pp. 202, 203]*

Destination: Medullary branches innervate voluntary muscles of palate, pharynx, and larynx; spinal branches control sternocleidomastoid and trapezius muscles

The **accessory nerve** is also known as the *spinal accessory nerve* or the *spinoaccessory nerve*. Unlike other cranial nerves, each accessory nerve has some motor fibers that originate in the lateral gray horns of the first five cervical segments of the spinal cord (Figure 14-28●). These somatic motor fibers enter the cranium through the foramen magnum, join the motor fibers from a nucleus in the medulla oblongata, and leave the cranium through the jugular foramen. Each accessory nerve consists of two branches:

1. The **medullary branch** joins the vagus nerve and innervates the voluntary swallowing muscles of the soft palate and pharynx and the intrinsic muscles that control the vocal cords.

2. The **spinal branch** controls the sternocleidomastoid and trapezius muscles of the neck and back. ∞ *[pp. 330, 339]* The motor fibers of this branch originate in the lateral gray horns of C_1 to C_5.

The Hypoglossal Nerves (N XII)

Primary function: Motor (tongue movements)

Origin: Motor nuclei of medulla oblongata

Pass through: Hypoglossal canals of occipital bone ∞ *[pp. 202, 204]*

Destination: Muscles of the tongue ∞ *[p. 329]*

Each **hypoglossal** (hī-pō-GLOS-al) **nerve** leaves the cranium through the hypoglossal canal of the occipital bone (Figure 14-28●). It then follows a curving path to reach the skeletal muscles of the tongue. This nerve provides voluntary motor control over movements of the tongue. Its condition is checked by sticking out the tongue; damage to the nerve or its nuclei will cause the tongue to veer toward the affected side.

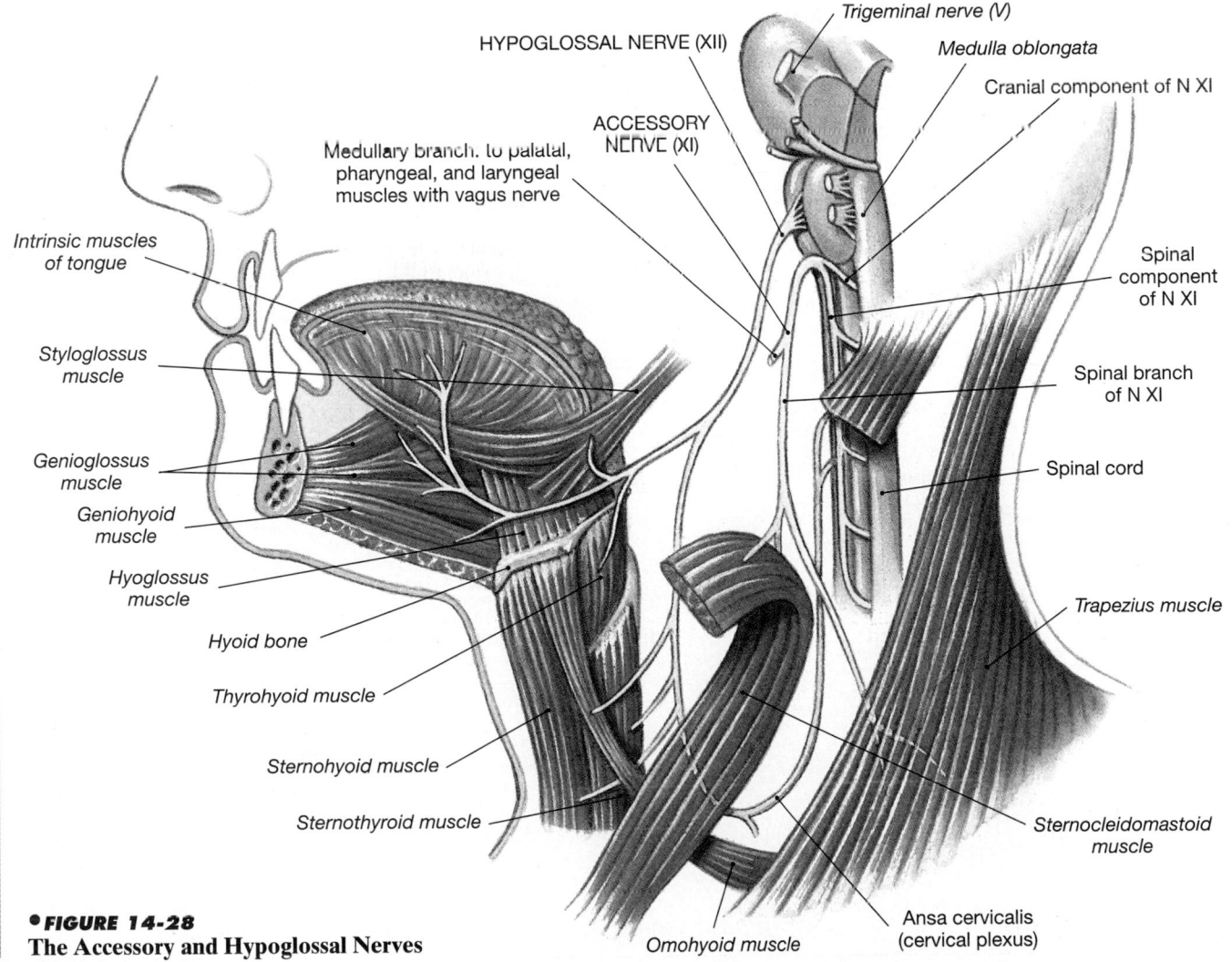

HYPOGLOSSAL NERVE (XII)

Trigeminal nerve (V)

Medulla oblongata

Cranial component of N XI

ACCESSORY NERVE (XI)

Medullary branch to palatal, pharyngeal, and laryngeal muscles with vagus nerve

Intrinsic muscles of tongue

Styloglossus muscle

Genioglossus muscle

Geniohyoid muscle

Hyoglossus muscle

Hyoid bone

Thyrohyoid muscle

Sternohyoid muscle

Sternothyroid muscle

Spinal component of N XI

Spinal branch of N XI

Spinal cord

Trapezius muscle

Sternocleidomastoid muscle

Ansa cervicalis (cervical plexus)

Omohyoid muscle

●**FIGURE 14-28**
The Accessory and Hypoglossal Nerves

A Summary of Cranial Nerve Branches and Functions

Few people are able to remember the names, numbers, and functions of the cranial nerves without a struggle. Mnemonic devices may prove useful. The best known is *On Old Olympus's Towering Top, A Finn And German Viewed Some Hops. (And* refers to the acoustic nerve, an alternative name for N VIII, and *Some* is for spinal accessory [N XI].) A more modern mnemonic device, *Oh, Once One Takes The Anatomy Final, Very Good Vacations Are Heavenly*, may be a bit easier to remember. Table 14-9 provides a detailed summary of the basic distribution and function of each cranial nerve.

TABLE 14-9 The Cranial Nerves

Cranial Nerve (number)	Sensory Ganglion	Branch	Primary Function	Foramen	Innervation
Olfactory (I)			Special sensory	Cribriform plate of ethmoid	Olfactory epithelium
Optic (II)			Special sensory	Optic canal	Retina of eye
Oculomotor (III)			Motor	Superior orbital fissure	Inferior, medial, superior rectus, inferior oblique and levator palpebrae muscles; intrinsic eye muscles
Trochlear (IV)			Motor	Superior orbital fissure	Superior oblique muscle
Trigeminal (V)	Semilunar		Mixed		Areas associated with the jaws
		Ophthalmic	Sensory	Superior orbital fissure	Orbital structures, nasal cavity, skin of forehead, upper eyelid, eyebrows, nose (part)
		Maxillary	Sensory	Foramen rotundum	Lower eyelid; upper lip, gums, and teeth; cheek, nose (part), palate and pharynx (part)
		Mandibular	Mixed	Foramen ovale	*Sensory* to lower gums, teeth, lips; palate (part) and tongue (part) *Motor* to muscles of mastication
Abducens (VI)			Motor	Superior orbital fissure	Lateral rectus muscle
Facial (VII)	Geniculate		Mixed	Internal acoustic canal to facial canal; exits at stylomastoid foramen	*Sensory* to taste receptors on anterior 2/3 of tongue *Motor* to muscles of facial expression, lacrimal gland, submandibular gland, sublingual salivary glands
Vestibulocochlear (Acoustic) (VIII)			Special sensory	Internal acoustic canal	
		Cochlear Vestibular			Cochlea (receptors for hearing) Vestibule (receptors for motion and balance)
Glossopharyngeal (IX)	Superior (jugular) and inferior (petrosal)		Mixed	Jugular foramen	*Sensory* from posterior 1/3 of tongue; pharynx and palate (part); receptors for blood pressure, pH, oxygen, and carbon dioxide concentrations *Motor* to pharyngeal muscles and parotid salivary gland
Vagus (X)	Jugular and nodose		Mixed	Jugular foramen	*Sensory* from pharynx; pinna and external auditory canal; diaphragm; visceral organs in thoracic and abdominopelvic cavities *Motor* to palatal and pharyngeal muscles, and to visceral organs in thoracic and abdominopelvic cavities
Accessory (XI)		Medullary	Motor	Jugular foramen	Skeletal muscles of palate, pharynx, and larynx (with vagus nerve)
		Spinal	Motor	Jugular foramen	Sternocleidomastoid and trapezius muscles
Hypoglossal (XII)			Motor	Hypoglossal canal	Tongue musculature

CRANIAL REFLEXES

Cranial reflexes are reflex arcs that involve the sensory and motor fibers of cranial nerves. We shall discuss examples of cranial reflexes in later chapters; in this section, we will simply provide an overview and general introduction.

Table 14-10 lists representative examples of cranial reflexes and their functions. These reflexes are clinically important because they provide a quick and easy method for observing the condition of cranial nerves and specific nuclei and tracts in the brain. Thus, they help localize the site of damage or disease.

Cranial somatic reflexes are seldom more complex than the somatic reflexes of the spinal cord. Table 14-10 includes four somatic reflexes: the *corneal reflex,* the *tympanic reflex,* the *auditory reflexes,* and the *vestibuloocular reflexes.* The normal functions of these reflexes are indicated in the table. In many instances, these reflexes are used to check for damage to the cranial nerves or processing centers involved. The brain stem contains many reflex centers that control visceral motor activity. Many of these reflex centers are located in the medulla oblongata, and they can direct very complex visceral motor responses to stimuli. These reflexes are essential to the control of respiratory, digestive, and cardiovascular functions. In Chapter 16, we shall consider them in greater detail. [AM] *Cranial Nerve Tests*

TABLE 14-10   Cranial Reflexes

Reflex	Stimulus	Afferents	Central Synapse	Efferents	Response
Somatic reflexes					
Corneal reflex	Contact with corneal surface	N V (trigeminal)	Motor nucleus for N VII (facial)	N VII	Blinking of eyelids
Tympanic reflex	Loud noise	N VIII (vestibulocochlear)	Inferior colliculus (midbrain)	N VII	Reduced movement of auditory ossicles
Auditory reflexes	Loud noise	N VIII	Motor nuclei of brain stem and spinal cord	N III, IV, VI, VII, X, cervical nerves	Eye and/or head movements triggered by sudden sounds
Vestibulo-ocular reflexes	Rotation of head	N VIII	Motor nuclei controlling eye muscles	N III, IV, VI	Opposite movement of eyes to stabilize field of vision
Visceral reflexes					
Direct light reflex	Light striking photoreceptors	N II (optic)	Superior colliculus (midbrain)	N III (oculomotor)	Constriction of ipsilateral pupil
Consensual light reflex	Light striking photoreceptors	N II	Superior colliculus	N III	Constriction of contralateral pupil

---

SELECTED CLINICAL TERMINOLOGY

Terms Discussed in This Chapter

ataxia: A disturbance of balance that in severe cases leaves the individual unable to stand without assistance. It is caused by problems affecting the cerebellum. *(p. 473)*

Bell's palsy: A condition resulting from an inflammation of the facial nerve; symptoms include paralysis of facial muscles on the affected side and loss of taste sensations from the anterior two-thirds of the tongue. *(p. 479)*

cerebrovascular accident (CVA): Also known as stroke; occurs when the blood supply to a portion of the brain is blocked off. *(p. 452)*

cranial trauma: A head injury resulting from violent contact with another object.

Cranial trauma may cause a **concussion,** a condition characterized by a temporary loss of consciousness and a variable period of amnesia. *(p. 451 and AM)*

dyslexia: A disorder affecting the comprehension and use of words. *(p. 460)*

epidural hemorrhage: A condition involving bleeding into the epidural space, generally resulting from cranial trauma. *(p. 453)*

hydrocephalus: A condition resulting from interference with the normal circulation and/or reabsorption of cerebrospinal fluid. Also known as "water on the brain." *(p. 453 and AM)*

Parkinson's disease *(paralysis agitans):* A condition characterized by a pronounced increase in muscle tone, resulting from excitation of neurons in the cerebral nuclei. *(p. 471 and AM)*

subdural hemorrhage: A condition in which blood accumulates under the dural epithelium in contact with the arachnoid membrane. *(p. 453)*

tic douloureux (doo-loo-ROO), or **trigeminal neuralgia:** A disorder of the maxillary and mandibular branches of N V characterized by severe, almost totally debilitating pain triggered by contact with the lip, tongue, or gums. *(p. 478 and AM)*

Additional Terms Discussed in the *Applications Manual*

decerebrate rigidity: A generalized state of muscular contraction resulting from loss of CNS inhibitory control.

dysmetria (dis-MET-rē-a): An inability to stop a movement at a precise, predetermined position; it commonly leads to an **intention tremor** in the affected individual. Dysmetria generally reflects cerebellar dysfunction.

spasticity: A condition characterized by hesitant, jerky voluntary movements, increased muscle tone, and hyperactive stretch reflexes.

tremor: Cyclic oscillations in limb position resulting from a "tug of war" between antagonistic muscle groups.

CHAPTER REVIEW

On-line resources for this chapter are on our World Wide Web site at:
http://www.prenhall.com/martini/fap

STUDY OUTLINE

INTRODUCTION, p. 445

1. The brain is far more complex and adaptable than the spinal cord.

AN INTRODUCTION TO THE ORGANIZATION OF THE BRAIN, p. 445

Embryology of the Brain, p. 445

1. The brain forms from three swellings at the superior tip of the developing neural tube: the **prosencephalon,** the **mesencephalon,** and the **rhombencephalon.** The prosencephalon forms the **telencephalon** (which becomes the *cerebrum*) and diencephalon; the rhombencephalon forms the **metencephalon** (*cerebellum* and *pons*) and **myelencephalon** (*medulla oblongata*). (*Table 14-1*)

A Preview of Major Regions and Landmarks, p. 445

2. There are six regions in the adult brain: cerebrum, diencephalon, mesencephalon (midbrain), pons, cerebellum, and medulla oblongata. (*Figures 14-1, 14-2; Table 14-1*)

3. Conscious thought, intellectual functions, memory, and complex motor patterns originate in the **cerebrum.** The **cerebellum** adjusts ongoing motor activities on the basis of sensory data and stored memories. (*Figure 14-2*)

4. Each wall of the diencephalon forms a **thalamus** that contains relay and processing centers for sensory data. The **hypothalamus** contains centers involved with emotions, autonomic function, and hormone production. (*Figure 14-2*)

5. The **mesencephalon** processes visual and auditory information and generates reflexive somatic motor responses. (*Figure 14-2*)

6. The **pons** (which, along with the cerebellum, forms from the metencephalon) connects the cerebellum to the **brain stem** and is involved with somatic and visceral motor control. The spinal cord connects to the brain at the **medulla oblongata,** which relays sensory information and regulates autonomic functions. (*Figure 14-2*)

7. The brain contains extensive areas of **neural cortex,** a layer of gray matter on the surfaces of the cerebrum and cerebellum.

Ventricles of the Brain, p. 448

8. The central passageway of the brain expands to form chambers called **ventricles.** Cerebrospinal fluid continuously circulates from the ventricles and central canal of the spinal cord into the subarachnoid space of the meninges that surround the CNS. (*Figure 14-3*)

PROTECTION AND SUPPORT OF THE BRAIN, p. 448

The Cranial Meninges, p. 449

1. The cranial meninges (the **dura mater, arachnoid,** and **pia mater**) are continuous with those of the spinal cord but have anatomical and functional differences.

2. Folds of dura mater, including the **falx cerebri, tentorium cerebelli,** and **falx cerebelli,** stabilize the position of the brain. (*Figure 14-4*)

Cerebrospinal Fluid, p. 451

3. Cerebrospinal fluid (CSF) is produced at the **choroid plexus.**

4. Cerebrospinal fluid (1) cushions delicate neural structures, (2) supports the brain, and (3) transports nutrients, chemical messengers, and waste products. Cerebrospinal fluid reaches the subarachnoid space through the **lateral apertures** and a **median aperture.** Diffusion across the **arachnoid granulations** into the **superior sagittal sinus** returns CSF to the venous circulation. (*Figure 14-5*)

The Blood Supply to the Brain, p. 452

5. The **blood–brain barrier** isolates neural tissue from the general circulation.

6. The blood–brain barrier is intact throughout the CNS except in parts of the hypothalamus, the pineal gland, and at the choroid plexus in the diencephalon and medulla oblongata.

THE CEREBRUM, p. 457

The Cerebral Cortex, p. 457

1. The cortical surface contains **gyri** (elevated ridges) separated by **sulci** (shallow depressions) or **fissures** (deeper grooves). The **longitudinal fissure** separates the two **cerebral hemispheres.** The **central sulcus** separates the **frontal** and **parietal lobes.** Other sulci form the boundaries of the **temporal** and **occipital lobes.** (*Figures 14-1, 14-8*)

2. Each cerebral hemisphere receives sensory information and generates motor commands that concern the opposite side of the body. There are significant functional differences between the two hemispheres.

3. The **central white matter** contains **association fibers, commissural fibers,** and **projection fibers.** (*Figure 14-7*)

4. The **primary motor cortex** of the **precentral gyrus** directs voluntary movements by means of the **pyramidal system.** The **primary sensory cortex** of the **postcentral gyrus** receives somatic sensory information from touch, pressure, pain, taste, and temperature receptors. (*Figure 14-8; Table 14-2*)

5. **Association areas,** such as the **visual association area** and **somatic motor association area (premotor cortex),** control

our ability to understand sensory information and coordinate a motor response. *(Figure 14-8; Table 14-2)*

6. The **general interpretive area** receives information from all the sensory association areas. It is present in only one hemisphere—generally the left. *(Figure 14-8)*

7. The **speech center** regulates the patterns of breathing and vocalization needed for normal speech. *(Figure 14-8)*

8. The **prefrontal cortex** coordinates information from the secondary and special association areas of the entire cortex and performs abstract intellectual functions. *(Figure 14-8)*

9. The left hemisphere is ordinarily the **categorical hemisphere;** it contains the general interpretive and speech centers and is responsible for language-based skills. The right hemisphere, or **representational hemisphere,** is responsible for spatial relationships and analyses. *(Figure 14-9)*

The Cerebral Nuclei, p. 462

10. The **cerebral nuclei** in the central white matter include the **caudate nucleus, amygdaloid body, claustrum, globus pallidus,** and **putamen**—parts of the **extrapyramidal system,** which controls muscle tone and coordinates learned movement patterns and other somatic motor activities. *(Figure 14-10)*

THE LIMBIC SYSTEM, p. 464

1. The **limbic system** includes the amygdaloid body, **cingulate gyrus, parahippocampal gyrus, hippocampus, mamillary bodies,** and the **fornix.** The functions of the limbic system involve emotional states and related behavioral drives. *(Figures 14-10–14-12; Table 14-3)*

THE DIENCEPHALON, p. 465

1. The diencephalon provides the switching and relay centers necessary to integrate the conscious and subconscious sensory and motor pathways. *(Figures 14-11–14-13)*

The Thalamus, p. 465

2. The thalamus is the final relay point for ascending sensory information and coordinates the pyramidal and extrapyramidal systems. *(Figures 14-12–14-14; Table 14-4)*

The Hypothalamus, p. 468

3. The hypothalamus contains important control and integrative centers. It can (1) control somatic motor activities at the subconscious level, (2) control autonomic function, (3) coordinate activities of the nervous and endocrine systems, (4) secrete hormones, (5) produce emotions and behavioral drives, (6) coordinate voluntary and autonomic functions, and (7) regulate body temperature. *(Figure 14-15; Table 14-5)*

THE MESENCEPHALON, p. 470

1. The **tectum** (roof) of the mesencephalon contains two pairs of sensory nuclei, the **corpora quadrigemina.** On each side, the **superior colliculus** receives visual inputs from the thalamus, and the **inferior colliculus** receives auditory data from the medulla oblongata. The **red nucleus** integrates information from the cerebrum and issues motor commands related to muscle tone and posture. The **substantia nigra** regulates the motor output of the cerebral nuclei. The **cerebral peduncles** contain ascending fibers headed for thalamic nuclei and descending pyramidal fibers that carry voluntary motor commands from the primary motor cortex. *(Figures 14-13, 14-16; Table 14-6)*

THE CEREBELLUM, p. 471

1. The cerebellum adjusts postural muscles and programs and tunes ongoing movements. The **cerebellar hemispheres** consist of neural cortex formed into folds, or **folia.** The surface consists of the **anterior** and **posterior lobes,** the **vermis,** and the **flocculonodular lobe.** *(Figure 14-17; Table 14-7)*

THE PONS, p. 473

1. The pons contains (1) sensory and motor nuclei for four cranial nerves; (2) nuclei concerned with the control of respiration; (3) nuclei and tracts linking the cerebellum with the brain stem, cerebrum, and spinal cord; and (4) ascending and descending tracts. *(Figures 14-13, 14-18)*

THE MEDULLA OBLONGATA, p. 473

1. The medulla oblongata connects the brain and spinal cord. It contains **olivary nuclei,** which relay information from the spinal cord, cerebral cortex, and brain stem to the cerebellar cortex. Its **reflex centers,** including the **cardiovascular and respiratory rhythmicity centers,** control or adjust activities of peripheral systems. *(Figures 14-13, 14-18; Table 14-8)*

FOCUS: Cranial Nerves, p. 474

1. There are 12 pairs of cranial nerves. Each nerve attaches to the brain's ventrolateral surface near the associated sensory or motor nuclei. *(Figure 14-19)*

2. The **olfactory nerves** (N I) carry sensory information responsible for the sense of smell. The olfactory afferents synapse within the **olfactory bulbs.** *(Figures 14-19, 14-20)*

3. The **optic nerves** (N II) carry visual information from special sensory receptors in the eyes. *(Figures 14-19, 14-21)*

4. The **oculomotor nerves** (N III) are the primary source of innervation for four of the extrinsic oculomotor muscles. *(Figure 14-22)*

5. The **trochlear nerves** (N IV), the smallest cranial nerves, innervate the eyes' superior oblique muscles. *(Figure 14-22)*

6. The **trigeminal nerves** (N V), the largest cranial nerves, are mixed nerves with **ophthalmic, maxillary,** and **mandibular branches.** *(Figure 14-23)*

7. The **abducens nerves** (N VI) innervate the lateral rectus muscles. *(Figure 14-22)*

8. The **facial nerves** (N VII) are mixed nerves that control muscles of the scalp and face. They provide pressure sensations over the face and receive taste information from the tongue. *(Figure 14-24)*

9. The **vestibulocochlear nerves** (N VIII) contain the **vestibular nerves,** which monitor sensations of balance, position, and movement; and **cochlear nerves,** which monitor hearing receptors. *(Figure 14-25)*

10. The **glossopharyngeal nerves** (N IX) are mixed nerves that innervate the tongue and pharynx and control the action of swallowing. *(Figure 14-26)*

11. The **vagus nerves** (N X) are mixed nerves that are vital to the autonomic control of visceral function. *(Figure 14-27)*

12. The **accessory nerves** (N XI) have **medullary branches,** which innervate voluntary swallowing muscles of the soft palate and pharynx, and **spinal branches,** which control muscles associated with the pectoral girdle. *(Figure 14-28)*

13. The **hypoglossal nerves** (N XII) provide voluntary motor control over tongue movements. *(Figure 14-28)*

A Summary of Cranial Nerve Branches and Functions, p. 484

14. The branches and functions of the cranial nerves are summarized in Table 14-9.

CRANIAL REFLEXES, p. 485

1. **Cranial reflexes** are reflex arcs that involve the sensory and motor fibers of cranial nerves. *(Table 14-10)*

REVIEW QUESTIONS

Level 1 — Reviewing Facts and Terms

Match each numbered item with the most closely related lettered item. Use letters for answers in the spaces provided.

Set 1: The Brain

___ 1. prosencephalon
___ 2. rhombencephalon
___ 3. brain stem
___ 4. ventricle in diencephalon
___ 5. dural fold
___ 6. choroid plexus
___ 7. Broca's area
___ 8. Wernicke's area
___ 9. limbic system
___ 10. pineal gland
___ 11. supraoptic nucleus
___ 12. paraventricular nucleus
___ 13. substantia nigra
___ 14. red nucleus
___ 15. Purkinje cells
___ 16. arbor vitae
___ 17. ataxia
___ 18. reticular formation
___ 19. nucleus gracilis
___ 20. hypothalamus

a. third ventricle
b. oxytocin
c. auditory association area
d. maintains muscle tone and posture
e. melatonin
f. white matter of cerebellum
g. mesencephalon, pons, medulla oblongata
h. link between nervous and endocrine systems
i. releases the neurotransmitter dopamine
j. forebrain
k. cerebellar cortex
l. falx cerebri
m. a disturbance in balance
n. apneustic, pneumotaxic centers
o. hindbrain
p. amygdaloid body
q. medulla oblongata
r. production of CSF
s. antidiuretic hormone
t. speech center

Set 2: The Cranial Nerves

___ 21. facial nerve
___ 22. vagus nerve
___ 23. olfactory nerve
___ 24. trochlear nerve
___ 25. hypoglossal nerve
___ 26. accessory nerve
___ 27. trigeminal nerve
___ 28. optic nerve
___ 29. glossopharyngeal nerve
___ 30. oculomotor nerve
___ 31. vestibulocochlear nerve
___ 32. abducens nerve

a. sensory, smell
b. N V
c. motor, tongue movements
d. sensory, vision
e. N X
f. N III
g. equilibrium, hearing
h. N VII
i. N VI
j. N IX
k. motor, eye movements
l. N XI

33. The structural and functional link between the cerebral hemispheres and the components of the brain stem is the
  (a) neural cortex
  (b) medulla oblongata
  (c) mesencephalon
  (d) diencephalon

34. The primary link between the nervous and endocrine systems is the
  (a) hypothalamus
  (b) mesencephalon
  (c) pons
  (d) medulla oblongata

35. Immediate reflexive responses to a loud, unexpected noise are directed by nuclei in the
  (a) pons
  (b) diencephalon
  (c) mesencephalon
  (d) medulla oblongata

36. Regulation of autonomic function, such as heart rate and blood pressure, originates in the
  (a) cerebrum
  (b) cerebellum
  (c) diencephalon
  (d) medulla oblongata

37. The ventricles in the brain are filled with
  (a) blood
  (b) cerebrospinal fluid
  (c) air
  (d) neural tissue

38. The smooth surface that covers the brain but does not follow the underlying neural convolutions or sulci is the
  (a) neural cortex
  (b) dura mater
  (c) pia mater
  (d) arachnoid membrane

39. The meningeal layer that adheres to the surface contour of the brain, extending into every fold and curve, is the
  (a) pia mater
  (b) dura mater
  (c) arachnoid membrane
  (d) neural cortex

40. The dural fold that divides the two cerebellar hemispheres is the
  (a) transverse sinus
  (b) falx cerebri
  (c) tentorium cerebelli
  (d) falx cerebelli

41. Production and secretion of cerebrospinal fluid occur in the
  (a) hypothalamus
  (b) choroid plexus
  (c) medulla oblongata
  (d) crista galli

42. The primary purpose of the blood–brain barrier is to
  (a) provide the brain with oxygenated blood
  (b) drain venous blood via the internal jugular veins
  (c) isolate neural tissue in the CNS from the general circulation
  (d) a, b, and c are correct

43. Conscious thought processes and all intellectual functions originate in the
  (a) cerebellum
  (b) corpus callosum
  (c) cerebral hemispheres
  (d) medulla oblongata

44. The two cerebral hemispheres are functionally different even though anatomically they appear the same.
  (a) true
  (b) false

45. The inability to speak, to read, or to understand the speech of others is
  (a) dyslexia
  (b) global aphasia
  (c) tic douloureux
  (d) ataxia

46. The area of the brain that generates feelings of frustration, tension, and anxiety as it interprets ongoing events and makes predictions about future situations is the
  (a) Broca's area
  (b) postcentral gyrus
  (c) prefrontal cortex
  (d) temporal lobes

47. Reading, writing, and speaking are dependent on processing in the
  (a) right cerebral hemisphere
  (b) left cerebral hemisphere
  (c) prefrontal cortex
  (d) postcentral gyrus

48. Tracts in the central white matter that connect the two cerebral hemispheres are
  (a) integrative fibers
  (b) projection fibers
  (c) association fibers
  (d) commissural fibers

49. The motor system that controls skeletal muscle tone and coordinates learned movement patterns and other somatic activities is the
- (a) limbic system
- (b) extrapyramidal system
- (c) reticular system
- (d) choroid plexus

50. Establishment of emotional states and related behavioral drives are functions of the
- (a) limbic system
- (b) tectum
- (c) mamillary bodies
- (d) thalamus

51. The final relay point for ascending sensory information that will be projected to the primary sensory cortex is the
- (a) hypothalamus
- (b) thalamus
- (c) spinal cord
- (d) medulla oblongata

52. The two hormones secreted by the hypothalamus are
- (a) epinephrine and norepinephrine
- (b) antidiuretic hormone and oxytocin
- (c) melatonin and serotonin
- (d) FSH and ATP

53. Symptoms characteristic of Parkinson's disease appear if there is damage to the
- (a) red nucleus
- (b) cerebral peduncles
- (c) corpora quadrigemina
- (d) substantia nigra

54. The part of the brain that coordinates rapid, automatic adjustments that maintain balance and equilibrium is the
- (a) cerebrum
- (b) cerebellum
- (c) pons
- (d) medulla oblongata

55. The centers in the pons that modify the activity of the respiratory rhythmicity centers in the medulla oblongata are the
- (a) apneustic and pneumotaxic centers
- (b) inferior and superior peduncles
- (c) cardiac and vasomotor centers
- (d) nucleus gracilis and nucleus cuneatus

56. What are the primary functions of the cerebrum?

57. Briefly summarize the overall function of the cerebellum.

58. What role does the hypothalamus play in the body?

59. What three layers make up the cranial meninges?

60. What are the three important functions of the CSF?

61. Which three areas in the brain are not isolated from the general circulation by the blood–brain barrier?

62. What are pyramidal cells, and what is their function?

63. What functional properties are associated with the neurons in the primary sensory cortex?

64. What is the extrapyramidal system, and what is its function?

65. Where is the limbic system located, and what are its three functions?

66. What three kinds of nuclei occur in the medulla oblongata?

67. Using the mnemonic device "Oh, Once One Takes The Anatomy Final, Very Good Vacations Are Heavenly," list the 12 pairs of cranial nerves and their functions.

Level 2 Reviewing Concepts

68. A person with a damaged visual association area may be
- (a) unable to scan the lines of a page or see rows of clear symbols
- (b) declared legally blind
- (c) unable to see letters but able to identify words and their meanings
- (d) able to see letters quite clearly but unable to recognize or interpret them

69. If the corpus callosum is cut,
- (a) cross-referencing of sensory information is inhibited
- (b) the individual talks constantly but uses the wrong words
- (c) objects touched with the left hand cannot be recognized but can be verbally identified
- (d) objects touched with the right hand cannot be verbally identified

70. What structure in the brain would your A & P instructor be referring to when talking about a nucleus that resembles a sea horse and that appears to be important in the storage and retrieval of long-term memories? In what functional system of the brain would it be found?

71. Stimulation of what part of the brain would produce sensations of hunger and thirst?

72. Damage to the corpora quadrigemina would interfere with
- (a) control of autonomic function
- (b) regulation of body temperature
- (c) processing of visual and auditory sensations
- (d) conscious control of skeletal muscles

73. If symptoms characteristic of Parkinson's disease appear, what part of the mesencephalon is inhibited from secreting a neurotransmitter? Which neurotransmitter?

74. What are the principal functional differences between the categorical hemisphere and the representational hemisphere?

75. What major part of the brain is associated with identifying the clinical signs of life?

76. Why can the brain respond to stimuli with greater versatility than the spinal cord can?

77. What kinds of problems are associated with the presence of lesions in Wernicke's area and Broca's area?

Level 3 Critical Thinking and Clinical Applications

78. Smelling salts can sometimes help restore consciousness after a person has fainted. The active ingredient of smelling salts is ammonia, and it acts by irritating the lining of the nasal cavity. Propose a mechanism by which smelling salts would raise a person from the unconscious state to the conscious state.

79. A police officer has just stopped Bill on suspicion of driving while intoxicated. The officer asks Bill to walk the yellow line on the road and then to place the tip of his index finger on the tip of his nose. How would these activities indicate Bill's level of sobriety? What part of the brain is being tested by these activities?

80. While having some dental work performed, Tyler is given an injection of local anesthetic in his lower jaw. His dentist tells him not to eat until the anesthetic wears off, not because of his teeth but because of his tongue. Why is the dentist giving him this advice?

CHAPTER 15

Integrative Functions

*T*here is a saying that "big cities never sleep." In New York City or Los Angeles at 3:00 A.M., shops are open, deliveries are made, people are on the street, and traffic moves briskly. The central nervous system (CNS) is much more complex than any city and far busier. There is a continuous flow of information between the brain, spinal cord, and peripheral nerves. At any moment, millions of sensory neurons are delivering information to processing centers in the CNS, and millions of motor neurons are controlling or adjusting the activities of peripheral effectors. This process continues 24 hours a day, whether we are awake or asleep. We may sleep soundly, but our nervous system does not; many brain stem centers are active throughout our lives, performing vital autonomic functions. This chapter examines the flow of sensory information and motor commands between the PNS and the CNS. It also examines complex aspects of human consciousness, such as memory and learning, that involve multiple regions of the brain.

CHAPTER OUTLINE AND OBJECTIVES

Chapters 13 and 14 introduced the major components and functions of the CNS and PNS. Those chapters, however, could not tell the entire story, for the CNS is more than a collection of independent parts. Think of the nervous system as a symphony orchestra. Thus far, you have studied the appearance and sound of the individual cellos, clarinets, and other instruments. Although essential to an understanding of an orchestra, this information does not tell you how the orchestra performs or what a symphony sounds like. While you are reading this text and trying to memorize important information, you are also breathing, balancing upright, controlling your body temperature, and possibly listening to the radio or TV. To accomplish these tasks, many centers in your CNS must work together in a coordinated and integrated fashion. In this chapter, we consider the neural "orchestra" as it performs complex intellectual functions. In doing so, we examine the integration of processing centers in the brain and spinal cord as well as functional relationships between the CNS and PNS.

The essential communication between the CNS and the PNS occurs over tracts and nuclei that relay sensory information and motor commands. **Pathways** consist of the nuclei and tracts that link the brain with the rest of the body. Sensory and motor pathways involve a series of synapses, one after the other. For example, an ascending axon of a sensory neuron may synapse in the *nucleus gracilis* in the medulla oblongata. There the axon activates a neuron that synapses in the thalamus. Thalamic neurons relay the sensory information to the cerebral cortex. At each step, there are opportunities for facilitation or inhibition that will magnify or suppress the sensations involved. There are also opportunities for divergence and the activation of additional neuronal pools that operate outside our conscious awareness. For example, while your cerebral cortex struggles to decide on a voluntary response to a stumble, collateral processing within your brain stem and cerebellum may already be issuing the commands that prevent a fall.

Many subtle forms of interaction, feedback, and regulation link higher centers with the various components of the brain stem. Only a few are understood in any detail. The following sections focus on basic patterns and principles in the sensory, motor, and higher-order functions of the brain.

SENSORY AND MOTOR PATHWAYS

We will begin by considering the pathways that utilize the major ascending (sensory) and descending (motor) tracts of the spinal cord. It may help you to remember that the tract names indicate the origins and destinations of the axons:

- If the name of a tract begins with *spino-*, the tract must *start* in the spinal cord and *end* in the brain. The rest of the name indicates the tract's destination. For example, a *spinothalamic tract* begins in the spinal cord and ends in the thalamus.
- If the name of a tract ends in *-spinal*, the tract's axons must *start* in a higher center and *end* in the spinal cord. The rest of the name indicates the specific nucleus or cortical area of the brain where the tract originates. For example, a *corticospinal tract* carries motor commands from the cerebral cortex to the spinal cord.

Sensory Pathways

Sensory receptors represent the interface between the nervous system and the internal and external environments. A **sensory receptor** is a specialized cell or dendritic process that monitors conditions in the body or the external environment. When stimulated, a receptor passes information to the CNS in the form of action potentials along the axon of a sensory neuron. The arriving information is called a **sensation.** We introduced three general categories of sensory receptors in Chapter 12—*exteroceptors, proprioceptors*, and *interoceptors.* ∞ *[p. 375]* These categories reflect the source of the stimulation: Exteroceptors provide information about the external environment; proprioceptors track the positions of bones and joints; and interoceptors monitor internal tissues and organs.

The term **general senses** refers to sensations of temperature, pain, touch, pressure, vibration, and proprioception. General sensory receptors are distributed throughout the body, and they are relatively simple in structure. Some of the information they send to the CNS reaches the primary sensory cortex and our conscious awareness. As we noted in Chapter 12, sensory information is interpreted on the basis of the frequency of arriving action potentials. ∞ *[p. 405]* For example, when pressure sensations are arriving, the harder the pressure, the higher the frequency of action potentials. The conscious awareness of a sensation is called a **perception.**

In Chapter 17, we will consider the origins of sensations and the pathways involved in relaying the information to conscious and subconscious processing centers in the CNS. Most of the processing occurs in centers along the sensory pathways in the spinal cord or brain stem. Only about 1 percent of the information provided by afferent fibers reaches the cerebral cortex and our conscious awareness. For example, we usually do not feel the clothes we wear or hear the hum of the engine in our car.

Receptors for the General Senses

Receptors for the general senses are classified according to the nature of the stimulus that excites them:

1. **Nociceptors** (nō-sē-SEP-torz; *noceo,* hurt) respond to a variety of stimuli generally associated with tissue damage. Receptor activation causes the sensation of pain.

2. **Thermoreceptors** respond to temperature changes.

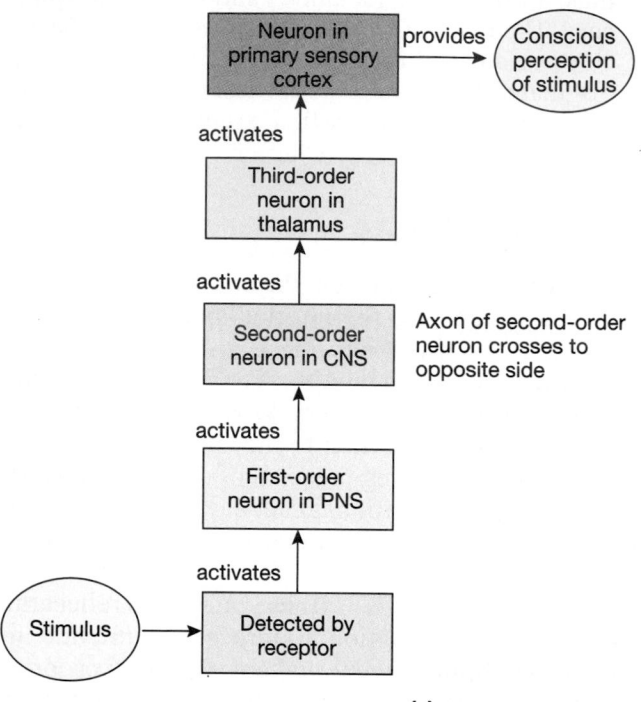

(a)

3. **Mechanoreceptors** are stimulated or inhibited by physical distortion, contact, or pressure on their cell membranes. There are many different types of mechanoreceptors. Examples introduced in earlier chapters include *proprioceptors,* such as muscle spindles and Golgi tendon organs, and *tactile receptors,* such as touch receptors in the skin.

4. **Chemoreceptors** monitor the chemical composition of body fluids and respond to the presence of specific molecules. Chemoreceptors are important components of several visceral reflexes.

In the next section, we discuss ascending pathways that carry somatic sensory information to the primary sensory cortex of the cerebral hemispheres. Sensations arriving from visceral organs are generally routed to processing centers in the brain stem and diencephalon rather than to the cerebrum. As a result, those sensations seldom reach our conscious awareness.

The Organization of Sensory Pathways

The sensory neuron that delivers the sensations to the CNS is often called a **first-order neuron** (Figure 15-1a●). The soma of a first-order sensory neuron is located in a dorsal root ganglion. Inside the CNS, the axon of that sensory neuron synapses on an interneuron known as a **second-order neuron,** which may be located in the spinal cord or brain stem. If the sensation is to reach our conscious awareness, the second-order neuron will synapse on a **third-**

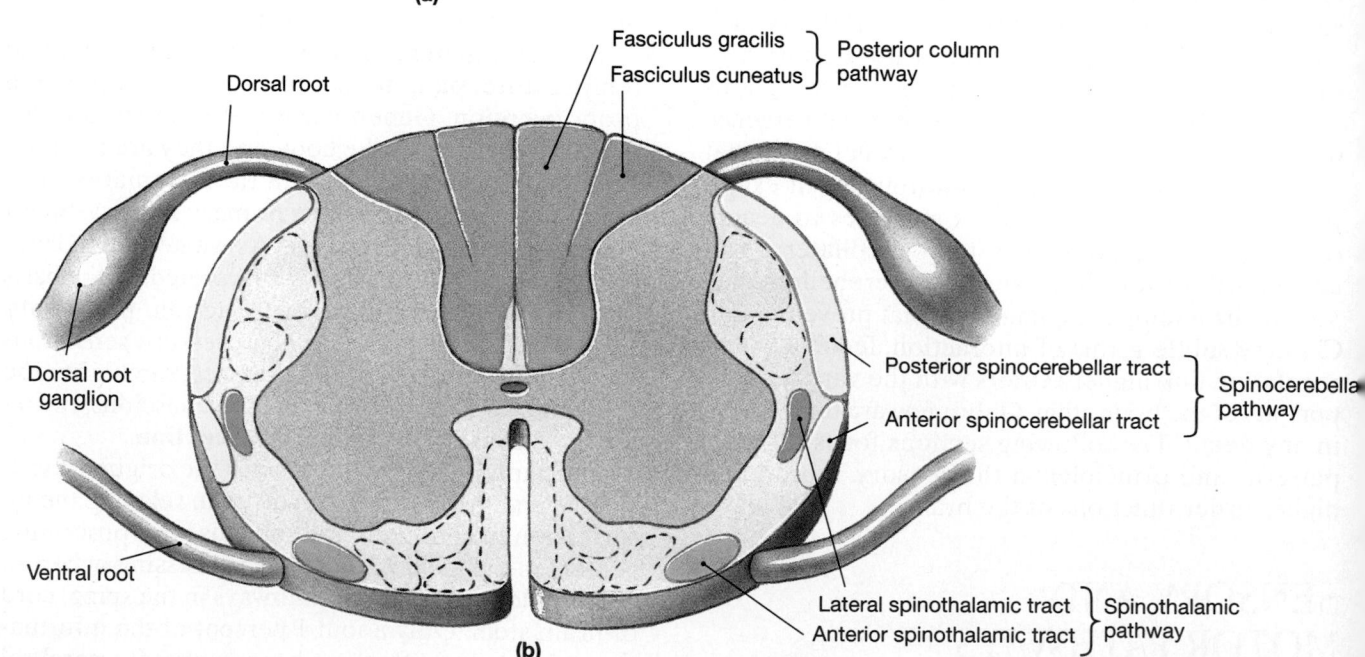

(b)

●**FIGURE 15-1** **Ascending Sensory Pathways and Tracts in the Spinal Cord.** **(a)** Steps in a sensory pathway. **(b)** A cross-sectional view indicating the locations of the major ascending sensory tracts in the spinal cord. For information about these tracts, see Table 15-1. Descending motor tracts (identified in Figure 15-5) are shown in dashed outline.

order neuron in the thalamus. Somewhere along its length, the axon of the second-order neuron crosses over to the opposite side of the body. As a result, the right side of the thalamus receives sensory information from the left side of the body, and vice versa.

The axons of the third-order neurons synapse on neurons of the primary sensory cortex of the cerebral hemisphere. Because the axons leaving the thalamus ascend on the same side of the brain, the right cerebral hemisphere receives sensory information from the left side of the body, and the left cerebral hemisphere receives sensations from the right side.

There are three major somatic sensory pathways, or **somatosensory pathways:** (1) the *posterior column pathway*, (2) the *spinothalamic pathway*, and (3) the *spinocerebellar pathway*. Two pairs of spinal tracts are associated with each pathway, one pair on each side of the spinal cord. All the axons within a tract share a common origin and destination. Figure 15-1b● indicates the relative positions of the spinal tracts involved.

The Posterior Column Pathway

The **posterior column pathway** carries sensations of highly localized ("fine") touch, pressure, vibration, and proprioception (position sense). This pathway begins at a peripheral receptor and ends at the primary sensory cortex of the cerebral hemispheres. Sensations that originate on the right side of the body will ultimately arrive at the primary sensory cortex of the left cerebral hemisphere. Sensations entering the posterior column pathway from the left side of the body will ultimately arrive at the right cerebral hemisphere. The spinal tracts involved are the left and right **fasciculus gracilis** (*gracilis*, delicate) and the left and right **fasciculus cuneatus** (*cuneus*, wedge-shaped). On each side of the posterior median sulcus, the fasciculus gracilis is medial to the fasciculus cuneatus.

The axons of the first-order neurons reach the CNS within the dorsal roots of spinal nerves and the sensory roots of cranial nerves. The axons ascending within the posterior column are organized according to the region innervated. Axons carrying sensations from the lower half of the body ascend within the fasciculus gracilis and synapse in the *nucleus gracilis* of the medulla oblongata. Axons carrying sensations from the upper half of the trunk, upper limbs, and neck ascend in the fasciculus cuneatus and synapse in the *nucleus cuneatus*. We introduced those nuclei in Chapter 14. ∞ *[p. 474]* Axons of the second-order neurons then ascend to the thalamus.

As the axons of the second-order neurons leave the nucleus, they cross over to the opposite side of the brain stem. Movement of an axon from the left side to the right side, or vice versa, is called **decussation**. Once on the opposite side of the brain, the axons enter a tract called the **medial lemniscus** (*lemniskos*, ribbon). Before reaching the thalamus, the medial lemniscus receives sensory information collected by cranial nerves V, VII, IX, and X.

The axons in the medial lemniscus synapse on third-order neurons in the thalamus. Axons that carry sensory information from the entire body, excluding the face, synapse in the **ventral posterolateral nucleus** (VPL). Axons carrying sensory information from the face synapse in the **ventral posteromedial nucleus** (VPM). These nuclei sort the arriving information according to (1) the nature of the stimulus and (2) the region of the body involved. Processing in the thalamus determines whether we perceive a given sensation as fine touch rather than as pressure or vibration.

Our ability to localize the sensation—to determine precisely where on the body a specific stimulus originated—depends on the projection of information from the ventral posterolateral nucleus to the primary sensory cortex. Sensory information from the toes arrives at one end of the primary sensory cortex, and information from the head arrives at the other. When neurons in one portion of your primary sensory cortex are stimulated, you become consciously aware of sensations originating at a specific location. If your primary sensory cortex were damaged or the projection fibers were cut, you could detect a light touch but would be unable to determine its source.

The same sensations are reported whether the cortical neurons are activated by axons ascending from the thalamus or by direct electrical stimulation. Researchers have electrically stimulated the primary sensory cortex in awake individuals during brain surgery and asked the subjects where they thought the stimulus originated. The results were used to create a functional map of the primary sensory cortex. This map, seen in Figure 15-2a●, is called a **sensory homunculus** ("little man").

The proportions of the sensory homunculus are very different from those of any individual. For example, the face is huge and distorted, with enormous lips and tongue, whereas the back is relatively tiny. These distortions occur because the area of sensory cortex devoted to a particular region is proportional not to its absolute size but rather to the *number of sensory receptors* it contains. In other words, many more cortical neurons are required to process sensory information arriving from the tongue, which has tens of thousands of taste and touch receptors, than to analyze sensations originating on the back, where touch receptors are few and far between.

The Spinothalamic Pathway

The **spinothalamic pathway** provides conscious sensations of poorly localized ("crude") touch, pressure, pain, and temperature. This pathway begins at peripheral receptors and ends at the primary sensory cortex of the cerebral hemispheres.

The axons of first-order sensory neurons enter the spinal cord and synapse on second-order neurons within the posterior gray horns. The axons of these interneurons cross to the opposite side of the spinal cord and ascend within the anterior or lateral spinothalamic tracts. The **anterior spinothalamic tract** carries crude touch and pressure sensations (Figure 15-2b●), whereas the **lateral spinothalamic tract** carries pain and temperature sensations (Figure 15-2c●). On each side of the brain, the anterior and lateral spinothalamic tracts end at third-order neurons in

●**FIGURE 15-2** **The Posterior Column and Spinothalamic Pathways.** For clarity, this figure shows only the pathway for sensations originating on the right side of the body. **(a)** The posterior column pathway delivers fine touch, vibration, and proprioception information to the primary sensory cortex of the cerebral hemisphere on the opposite side of the body. **(b)** The anterior spinothalamic tracts carry sensations of crude touch and pressure to the primary sensory cortex of the opposite cerebral hemisphere. **(c)** The lateral spinothalamic tracts carry sensations of pain and temperature to the primary sensory cortex on the opposite side of the body.

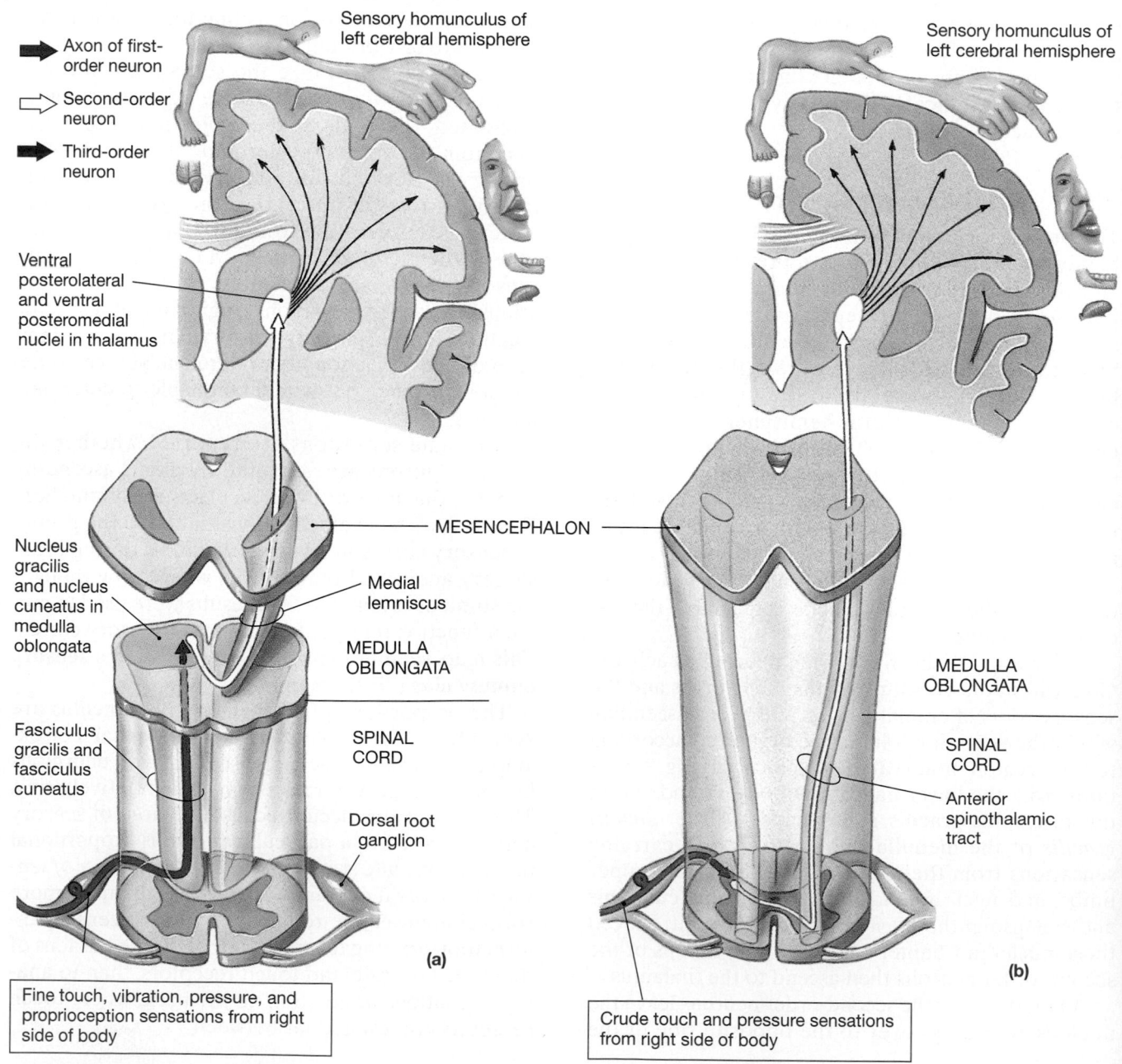

Axon of first-order neuron

Second-order neuron

Third-order neuron

Sensory homunculus of left cerebral hemisphere

Ventral posterolateral and ventral posteromedial nuclei in thalamus

Nucleus gracilis and nucleus cuneatus in medulla oblongata

Medial lemniscus

MESENCEPHALON

MEDULLA OBLONGATA

Fasciculus gracilis and fasciculus cuneatus

SPINAL CORD

Dorsal root ganglion

(a)

Fine touch, vibration, pressure, and proprioception sensations from right side of body

Sensory homunculus of left cerebral hemisphere

MEDULLA OBLONGATA

SPINAL CORD

Anterior spinothalamic tract

(b)

Crude touch and pressure sensations from right side of body

the ventral posterolateral nucleus of the thalamus. After they have been sorted and processed, the sensations are relayed to the primary sensory cortex.

As in the posterior column pathway, the determination that an arriving stimulus is painful rather than cold, hot, or vibrating is determined by the projection of the sensation to different populations of second-order and third-order neurons. The ability to localize that stimulus depends on stimulation of an appropriate area of the primary sensory cortex. Any abnormality along the pathway can result in inappropriate sensations. Consider these examples:

- An individual can experience painful sensations that arc not real. For example, a person may continue to experience pain in an amputated limb. This condition, called *phantom limb pain,* is caused by activity in the sensory neurons or interneurons along the spinothalamic pathway. The neurons involved once monitored conditions in the intact limb.

- An individual can feel pain in an uninjured part of the body when the pain actually originates at another location. For example, strong visceral pain sensations arriving at a segment of the spinal cord can stimulate interneurons that are normally part of the spinothalamic pathway. Activity in these interneurons leads to stimulation of the primary sensory cortex, so the individual feels pain in a specific part of the body surface. This phenomenon is called **referred pain.** Two familiar examples are (1) the pain of a heart attack, which is typically felt in the left arm, and (2) the pain of appendicitis, which is generally felt in the right lower quadrant. Additional examples are illustrated in Figure 15-3•. We will consider the origins and distribution of pain sensations in greater detail in Chapter 17.

Sensory homunculus of left cerebral hemisphere

MESENCEPHALON

Lateral spinothalamic tract

MEDULLA OBLONGATA

SPINAL CORD

(c)

Pain and temperature sensations from right side of body

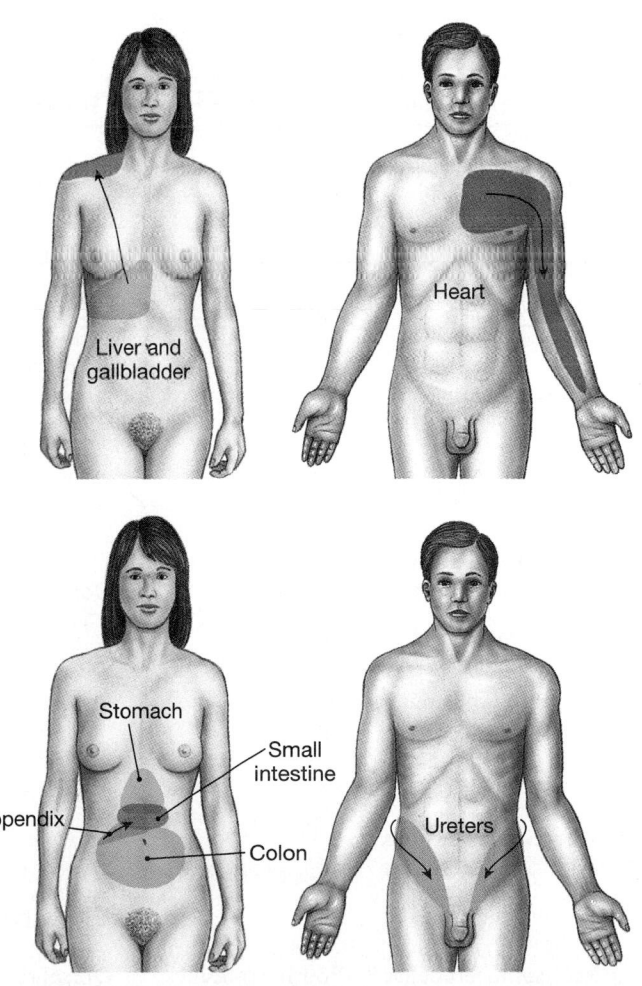

•*FIGURE 15-3* **Referred Pain.** Pain sensations originating in visceral organs are often perceived as involving specific regions of the body surface innervated by the same spinal segments. Each surface region of perceived pain is labeled according to the organ at which the pain originates.

Liver and gallbladder

Heart

Stomach

Small intestine

Appendix

Colon

Ureters

The Spinocerebellar Pathway

The cerebellum receives proprioceptive information concerning the position of skeletal muscles, tendons, and joints via the **spinocerebellar pathway** (Figure 15-4●). This information does not reach our conscious awareness. The axons of first-order sensory neurons synapse on interneurons in the dorsal gray horns of the spinal cord. The axons of these second-order neurons ascend in one of the *spinocerebellar tracts*. Axons that enter the left or right **posterior spinocerebellar tract** do not cross over to the opposite side of the spinal cord. These axons reach the cerebellar cortex via the inferior cerebellar peduncle of that side. Axons that do cross over to the opposite side of the spinal cord enter the left or right **anterior spinocerebellar tract**.[1] These axons reach the cerebellar cortex via the superior cerebellar peduncle.

[1]Each anterior spinocerebellar tract also contains uncrossed axons from interneurons on the same side of the spinal cord as the stimulus.

Information carried by the axons of the spinocerebellar tracts reaches the *Purkinje cells* of the cerebellar cortex. These cells are complex and highly branched. Proprioceptive information from each part of the body is relayed to a specific portion of the cerebellar cortex. We will consider the integration of proprioceptive information and the role of the cerebellum in somatic motor control in a later section.

Table 15-1 reviews the three major sensory pathways we discussed in this section.

☑ As a result of pressure on her spinal cord, Jill cannot feel touch or pressure on her lower limbs. Which spinal tract is being compressed?

☑ Which spinal tract carries action potentials generated by nociceptors?

☑ Which cerebral hemisphere receives impulses conducted by the right fasciculus gracilis?

TABLE 15-1 Principal Ascending (Sensory) Tracts in the Spinal Cord and the Sensory Information They Provide

Pathway/Tract	Sensations	Location of Neurons		
		First-Order	Second-Order	Third-Order
Posterior Column Pathway				
Fasciculus gracilis	Proprioception and fine touch, pressure, and vibration	Dorsal root ganglia of lower body; axons enter CNS in dorsal roots and join fasciculus gracilis	Nucleus gracilis of medulla oblongata; axons cross over before entering medial lemniscus	Ventral posterolateral nucleus of thalamus
Fasciculus cuneatus	Proprioception and fine touch, pressure, and vibration	Dorsal root ganglia of upper body; axons enter CNS in dorsal roots and join fasciculus cuneatus	Nucleus cuneatus of medulla oblongata; axons cross over before entering medial lemniscus	Ventral posterolateral and ventral posteromedial nuclei of thalamus
Spinothalamic Pathway				
Lateral spinothalamic tracts	Pain and temperature	Dorsal root ganglia; axons enter CNS in dorsal roots	Interneurons in posterior gray horn; axons enter lateral spinothalamic tract on opposite side	Ventral posterolateral nucleus of thalamus
Anterior spinothalamic tracts	Crude touch and pressure	Dorsal root ganglia; axons enter CNS in dorsal roots	Interneurons in posterior gray horn; axons enter anterior spinothalamic tract on opposite side	Ventral posterolateral nucleus of thalamus
Spinocerebellar Pathway				
Posterior spinocerebellar tracts	Proprioception	Dorsal root ganglia; axons enter CNS in dorsal roots	Interneurons in posterior gray horn; axons enter posterior spinocerebellar tract on same side	Not present
Anterior spinocerebellar tracts	Proprioception	Dorsal root ganglia; axons enter CNS in dorsal roots	Interneurons in same spinal segment; axons enter anterior spinocerebellar tract on same or opposite side	Not present

Motor Pathways

Motor commands issued by the CNS are distributed by the somatic nervous system and the autonomic nervous system (ANS). ∞ *[p. 369]* The somatic nervous system, also called the *somatic motor system*, controls the contractions of skeletal muscles. The output of the somatic nervous system is under voluntary control. The autonomic nervous system, or *visceral motor system*, controls visceral effectors, such as smooth muscle, cardiac muscle, and glands. In Chapter 16, we will examine the organization of the ANS; our interest here is the structure of the somatic motor system. Throughout this discussion we will use the terms *motor neuron* and *motor control* to refer specifically to somatic motor neurons and pathways that control skeletal muscles.

Somatic motor pathways always involve at least two motor neurons: an **upper motor neuron,** whose soma lies in a CNS processing center, and a **lower motor neuron** whose soma lies in a nucleus of the brain stem or spinal cord. The upper motor neuron synapses on the lower motor neuron, which in turn innervates a single motor unit in a skeletal muscle. Activity in the upper motor neuron may facilitate or inhibit the lower motor neuron. Activation of the lower motor neuron triggers a contraction in the innervated muscle. Only the axon of the lower motor neuron extends outside the CNS. Destruction or damage to a lower motor neuron invariably produces a flaccid paralysis in the innervated motor unit.

Conscious and subconscious motor commands control skeletal muscles by traveling over two integrated motor pathways, the *pyramidal system* and the *extrapyramidal system.* These systems were thought

Final Destination	Site of Cross-Over
Primary sensory cortex on side opposite stimulus	Axons of second-order neurons before joining medial lemniscus
Primary sensory cortex on side opposite stimulus	Axons of second-order neurons before joining medial lemniscus
Primary sensory cortex on side opposite stimulus	Axons of second-order neurons at level of entry
Primary sensory cortex on side opposite stimulus	Axons of second-order neurons at level of entry
Cerebellar cortex on side of stimulus	None
Cerebellar cortex on side opposite stimulus (and to side of stimulus)	Axons of most second-order neurons cross before entering tract

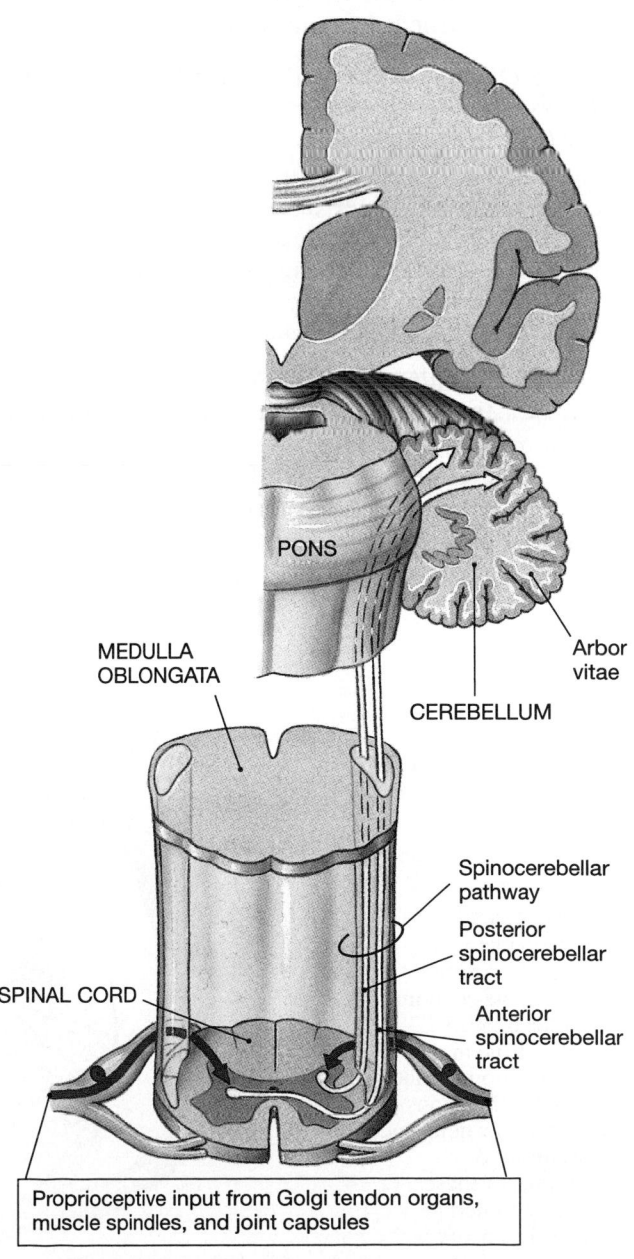

Proprioceptive input from Golgi tendon organs, muscle spindles, and joint capsules

•FIGURE 15-4 The Spinocerebellar Pathway

to be completely separate, with the pyramidal system providing conscious control and the extrapyramidal system working at the subconscious level. In fact, these systems are extensively integrated. Figure 15-5● indicates the positions of the pyramidal and extrapyramidal tracts descending in the spinal cord.

The Pyramidal System

The **pyramidal system** (Figure 15-6●) provides voluntary control over skeletal muscles. This system begins at the **pyramidal cells** of the primary motor cortex. These cells are upper motor neurons whose cell bodies are shaped like pyramids. The axons of pyramidal cells descend into the brain stem and spinal cord to synapse on lower motor neurons. The pyramidal system is direct—there are no synapses between the pyramidal cells and the lower motor neurons.

TRACTS OF THE PYRAMIDAL SYSTEM The pyramidal system contains three pairs of descending tracts: (1) the *corticobulbar tracts,* (2) the *lateral corticospinal tracts,* and (3) the *anterior corticospinal tracts.* These tracts enter the white matter of the internal capsule, descend into the brain stem, and emerge on either side of the mesencephalon as the *cerebral peduncles* (Figures 15-6, 14-13, and 14-16●, pp. 467, 470).

The Corticobulbar Tracts. Axons in the **corticobulbar** (kor-ti-kō-BUL-bar) **tracts** (*bulbar,* brain stem) synapse on lower motor neurons in the motor nuclei of eight cranial nerves (N III, IV, V, VI, VII, IX, XI, and XII). The corticobulbar tracts provide conscious control over skeletal muscles that move the eye, jaw, and face and control some muscles of the neck and pharynx as well.

The Corticospinal Tracts. Axons in the **corticospinal tracts** synapse on lower motor neurons in the anterior gray horns of the spinal cord. As they descend, the corticospinal tracts are visible along the ventral surface of the medulla oblongata as a pair of thick bands, the **pyramids.** Along the length of the pyramids roughly 85 percent of the axons cross the midline *(decussate)* to enter the descending **lateral corticospinal tracts** on the opposite side of the spinal cord. The other 15 percent continue uncrossed along the spinal cord as the **anterior corticospinal tracts.** At the spinal segment it targets, an axon in the anterior corticospinal tract crosses over to the opposite side of the spinal cord in the anterior white commissure before synapsing on lower motor neurons in the anterior gray horns.

THE MOTOR HOMUNCULUS Activity of pyramidal cells in a specific portion of the primary motor cortex will result in the contraction of peripheral muscles. The identities of the stimulated muscles depend on the region of motor cortex that is active. As in the case of the primary sensory cortex, there is a fine point-to-point correspondence between the primary motor cortex and specific regions of the body. The cortical areas have been mapped out in diagrammatic form, creating a **motor homunculus.** Figure 15-6● shows the motor homunculus of the left cerebral hemisphere and identifies the pyramidal system components involved in controlling the right side of the body.

The proportions of the motor homunculus are quite different from those of the actual body (Figure 15-6●), because the motor area devoted to a specific region of the cortex is proportional to the number of

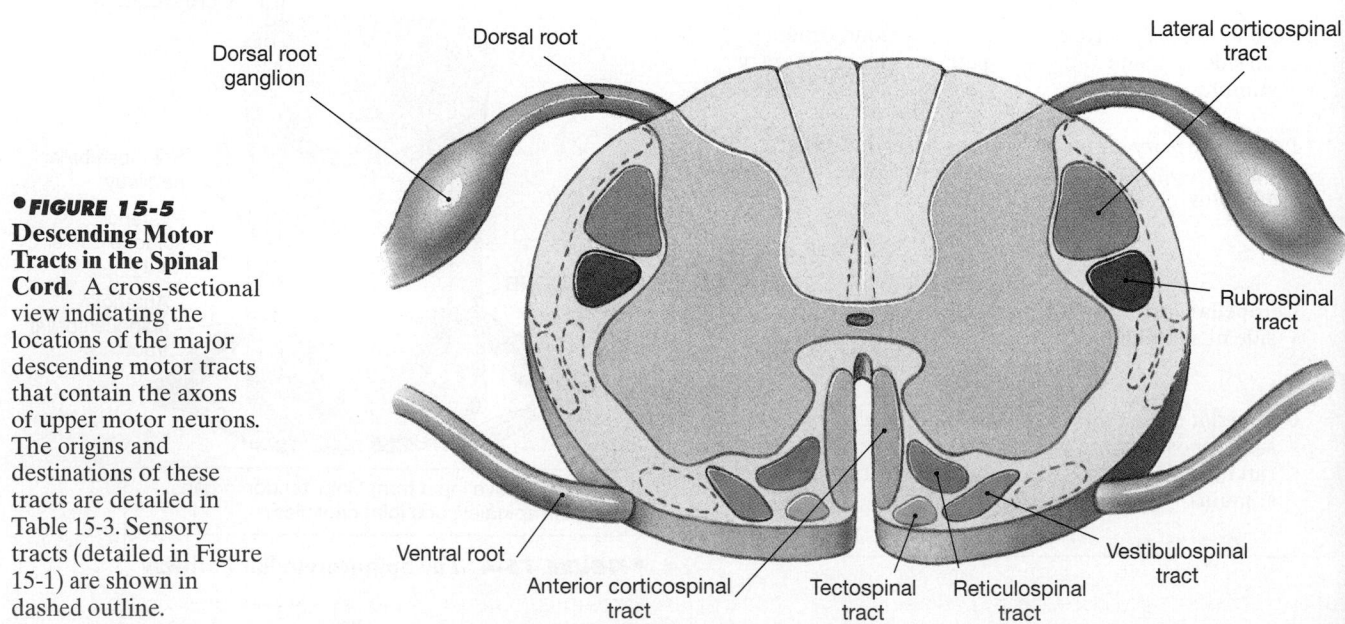

●**FIGURE 15-5**
Descending Motor Tracts in the Spinal Cord. A cross-sectional view indicating the locations of the major descending motor tracts that contain the axons of upper motor neurons. The origins and destinations of these tracts are detailed in Table 15-3. Sensory tracts (detailed in Figure 15-1) are shown in dashed outline.

Dorsal root ganglion

Dorsal root

Lateral corticospinal tract

Rubrospinal tract

Ventral root

Anterior corticospinal tract

Tectospinal tract

Reticulospinal tract

Vestibulospinal tract

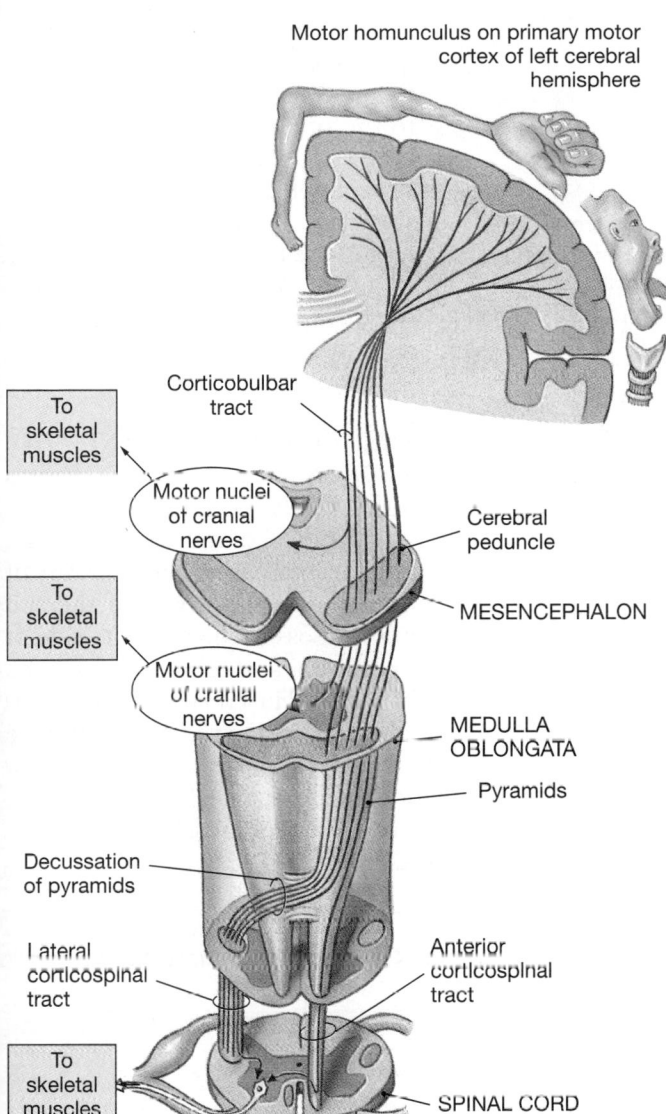

Motor homunculus on primary motor cortex of left cerebral hemisphere

Corticobulbar tract

To skeletal muscles

Motor nuclei of cranial nerves

Cerebral peduncle

MESENCEPHALON

To skeletal muscles

Motor nuclei of cranial nerves

MEDULLA OBLONGATA

Pyramids

Decussation of pyramids

Lateral corticospinal tract

Anterior corticospinal tract

To skeletal muscles

SPINAL CORD

•**FIGURE 15-6** **The Pyramidal System.** The pyramidal system originates at the primary motor cortex. The cortico-bulbar tracts end at the motor nuclei of cranial nerves on the opposite side of the brain. Most of the pyramidal fibers cross over in the medulla and enter the lateral corticospinal tracts; the rest descend in the anterior corticospinal tracts and cross over on reaching target segments in the spinal cord.

motor units involved in its control rather than to its actual size. As a result, the homunculus provides an indication of the degree of fine motor control available. For example, the hands, face, and tongue, all of which are capable of varied and complex movements, appear very large, whereas the trunk is relatively small. These proportions are similar to those of the sensory homunculus (Figure 15-2•, pp. 494–495). The sensory and motor homunculi differ in other respects because some highly sensitive regions, such as the sole of the foot, contain few motor units, and some areas with an abundance of motor units, such as the eye muscles, are not particularly sensitive.

CEREBRAL PALSY The term **cerebral palsy** refers to a number of disorders that affect voluntary motor performance; they appear during infancy or childhood and persist throughout the life of the affected individual. The cause may be trauma associated with premature or unusually stressful birth, maternal exposure to drugs, including alcohol, or a genetic defect that causes improper development of the motor pathways. Problems with labor and delivery result from compression or interruption of placental circulation or oxygen supplies. If the oxygen concentration in the fetal blood declines significantly for as little as 5–10 minutes, CNS function may be permanently impaired. The cerebral cortex, cerebellum, cerebral nuclei, hippocampus, and thalamus are likely targets, producing abnormalities in motor skills, posture and balance, memory, speech, and learning abilities.

The Extrapyramidal System

Several centers in the cerebrum, diencephalon, and brain stem may issue somatic motor commands as a result of processing performed at a subconscious level. These centers and their associated tracts form the **extrapyramidal system (EPS).** The extrapyramidal system can modify or direct skeletal muscle contractions by stimulating, facilitating, or inhibiting lower motor neurons. In addition, feedback from the cerebral nuclei and cerebellum to the primary motor cortex can modify the commands issued by the pyramidal system.

The components of the extrapyramidal system, introduced in Chapter 14, are spread throughout the brain (Table 15-2). The upper motor neurons of the extrapyramidal system are located in the *vestibular nuclei,* the *superior* and *inferior colliculi,* the *red nucleus,* and the *reticular formation.* The *cerebral nuclei* and *cerebellum* form a higher level of control. Their output either (1) stimulates or inhibits the activity of other extrapyramidal nuclei or (2) stimulates or inhibits the activity of pyramidal cells in the primary motor cortex.

The axons of upper motor neurons in the extrapyramidal system synapse on the same lower motor neurons innervated by the pyramidal system. In the brain stem, they synapse in the motor nuclei of cranial nerves. Axons controlling lower motor neurons of the spinal cord descend within the *vestibulospinal, tectospinal, rubrospinal,* and *reticulospinal tracts* (Table 15-3). Figure 15-5• indicates the location of these tracts in the spinal cord.

THE VESTIBULAR NUCLEI AND VESTIBULOSPINAL TRACTS \The vestibular nuclei receive information, via the eighth cranial nerve, from receptors in the inner ear that monitor the position and movement of the head. These nuclei respond to changes in the orientation of the head, sending motor commands that alter the muscle tone, extension, and position of the neck, eyes, head, and limbs. The primary goal is to maintain posture and balance. The descending fibers within the spinal cord constitute the **vestibulospinal tracts.**

TABLE 15-2 Components of the Extrapyramidal System

Center	Location	Primary Function
Cerebellar nuclei and cortex ∞ *[p. 471]*	Cerebellum	Coordination of movements, integration of sensory feedback
Cerebral nuclei ∞ *[p. 462]*	Cerebrum	Preparation and coordination of limb and trunk movements
Vestibular nuclei ∞ *[pp. 473, 474]*	Pons and medulla oblongata	Processing of equilibrium sensations and control of associated reflexes
Superior colliculi ∞ *[p. 470]*	Mesencephalon	Processing of visual information and control of associated reflexes
Inferior colliculi ∞ *[p. 470]*	Mesencephalon	Processing of auditory information and control of associated reflexes
Red nucleus ∞ *[p. 470]*	Mesencephalon	Control of skeletal muscle tone
Reticular formation ∞ *[p. 465]*	Mesencephalon and other brain stem areas	Interconnected network of neurons and centers with varied functions
Thalamus ∞ *[p. 465]*	Diencephalon	Relay station for somatic sensory information and for feedback from cerebral nuclei and cerebellum

THE COLLICULI AND THE TECTOSPINAL TRACTS
The superior and inferior colliculi are located in the roof, or *tectum*, of the mesencephalon (Figure 14-16•, p. 470). The colliculi receive visual (superior) and auditory (inferior) sensations. Axons of upper motor neurons in the superior and inferior colliculi descend in the **tectospinal tracts**. These axons cross to the opposite side immediately, before descending to synapse on lower motor neurons in the brain stem or spinal cord. The axons within the tectospinal tracts direct reflexive changes in the position of the head, neck, and upper limbs in response to bright lights, sudden movements, or loud noises.

THE RED NUCLEI AND THE RUBROSPINAL TRACTS The red nuclei of the mesencephalon control the level of muscle tone in the body's skeletal muscles. Axons of upper motor neurons in the red nuclei cross to the opposite side of the brain and descend into the spinal cord in the **rubrospinal tracts** (*ruber*, red). Motor neurons in the red nuclei may be stimulated or inhibited by input from the primary motor cortex, cerebral nuclei, cerebellum, or reticular formation. Stimulation increases muscle tone, and inhibition decreases it.

TABLE 15-3 Principal Descending (Motor) Tracts in the Spinal Cord and the General Functions of the Associated Nuclei in the Brain

Tract	Location of Upper Motor Neurons	Destination	Site of Cross-Over	Action
Pyramidal Tracts				
Corticobulbar tracts	Primary motor cortex (cerebral hemisphere)	Lower motor neurons of cranial nerve nuclei in brain stem	Brain stem	Conscious motor control of skeletal muscles
Lateral corticospinal tracts	As above	Lower motor neurons of anterior gray horns of spinal cord	Pyramids of medulla oblongata	As above
Anterior corticospinal tracts	As above	As above	Level of lower motor neuron	As above
Extrapyramidal Tracts				
Rubrospinal tracts	Red nuclei of mesencephalon	Lower motor neurons of anterior gray horns	Brain stem (mesencephalon)	Subconscious regulation of posture and muscle tone
Reticulospinal tracts	Reticular formation (network of nuclei in brain stem)	As above	Uncrossed	Subconscious regulation of reflex activity
Vestibulospinal tracts	Vestibular nuclei (at border of pons and medulla oblongata)	As above	Uncrossed	Subconscious regulation of balance and muscle tone
Tectospinal tracts	Tectum (mesencephalon: superior and inferior colliculi)	Lower motor neurons of anterior gray horns of cervical spinal cord	Brain stem (mesencephalon)	Subconscious regulation of eye, head, neck, and upper limb position in response to visual and auditory stimuli

THE RETICULAR FORMATION AND THE RETICULOSPINAL TRACTS The reticular formation is a loosely organized network of neurons that extends throughout the brain stem. The reticular formation receives collateral input from almost every ascending and descending pathway. It also has extensive interconnections with the cerebrum, the cerebellum, and brain stem nuclei. Axons of upper motor neurons in the reticular formation descend into the **reticulospinal tracts** without crossing to the opposite side. The effects of reticular formation stimulation are determined by the region stimulated. For example, stimulation of upper motor neurons in one portion of the reticular formation produces eye movements, whereas stimulation of another region of the reticular formation activates respiratory muscles.

THE CEREBRAL NUCLEI The cerebral nuclei blur the distinctions between conscious and subconscious motor control. In effect, they provide the background patterns of movement involved in the performance of voluntary motor activities. These nuclei do not exert direct control over lower motor neurons; instead, they adjust the activities of upper motor neurons in both the extrapyramidal and pyramidal systems. The cerebral nuclei receive input from all portions of the cerebral cortex as well as from the substantia nigra.

The cerebral nuclei use three major pathways to adjust or establish patterns of movement:

1. One group of axons synapses on thalamic neurons, which then send their axons to the premotor cortex, the motor association area that directs activities of the primary motor cortex. This arrangement creates a feedback loop that changes the sensitivity of the pyramidal cells and alters the pattern of instructions carried by the corticospinal tracts.

2. A second group of axons innervates the red nucleus and alters the activity in the rubrospinal tracts, thereby changing muscle tone and the degree of resistance to specific movements.

3. A third group of axons synapses in the reticular formation, altering the output of the reticulospinal tracts.

Two distinct populations of neurons exist: one that stimulates neurons by releasing acetylcholine (ACh) and another that inhibits neurons by the release of gamma-aminobutyric acid (GABA). Under normal conditions, the excitatory interneurons are kept inactive, and under ordinary circumstances the tracts leaving the cerebral nuclei have an inhibitory effect on upper motor neurons. In *Parkinson's disease* (Chapter 14), the excitatory neurons become more active, leading to problems with the voluntary control of movement. ∞ *[p. 471]*

If the primary motor cortex is damaged, the individual loses the ability to exert fine control over skeletal muscles. However, some voluntary movements can still be controlled by the cerebral nuclei. In effect, the extrapyramidal system functions as it usually does, but the pyramidal system cannot fine-tune the movements. For example, after damage to the primary motor cortex, the cerebral nuclei could still receive information about planned movements from the prefrontal cortex and perform preparatory movements of the trunk and limbs. But, because the pyramidal system is inoperative, precise movements of the forearms, wrists, and hands cannot occur. An individual in this condition could stand, maintain balance, and even walk, but all movements would be hesitant, awkward, and poorly controlled.

THE CEREBELLUM The cerebellum monitors proprioceptive (position) sensations, visual information from the eyes, and vestibular (balance) sensations from the inner ear as movements are under way. Axons relaying proprioceptive information reach the cerebellar cortex in the spinocerebellar tracts. Visual information is relayed from the superior colliculi, and balance information is relayed from the vestibular nuclei. The output of the cerebellum affects upper motor neuron activity in both the extrapyramidal and pyramidal systems.

Both pyramidal and extrapyramidal centers send information to the cerebellum when motor commands are issued. As the movement proceeds, the cerebellum monitors proprioceptive and vestibular information and compares the arriving sensations with those experienced during previous movements. It then adjusts the activities of the upper motor neurons involved. In general, any voluntary movement begins with the activation of far more motor units than are actually required—or even desirable. The cerebellum provides the necessary inhibition, reducing the number of motor commands to an efficient minimum. As the movement proceeds, the pattern and degree of inhibition change, producing the desired result.

The patterns of cerebellar activity are learned by trial and error, over many repetitions. Many of the basic patterns are established early in life; examples include the fine balancing adjustments you make while standing and walking. The ability to fine-tune a complex pattern of movement improves with practice, until the movements become fluid and automatic. For example, consider the relaxed, smooth movements of acrobats, dancers, golfers, tennis players, and sushi chefs. These people move without consciously thinking about the details of their movements. This is important, because when you concentrate on voluntary control, the rhythm and pattern of the movement usually fall apart as your pyramidal system starts overriding the commands of the cerebral nuclei the and cerebellum.

AMYOTROPHIC LATERAL SCLEROSIS Amyotrophic lateral sclerosis (ALS), formerly known as *Lou Gehrig's disease*, is a progressive, degenerative disorder that affects motor neurons in the spinal cord, brain stem, and cerebral hemispheres. Most people with ALS do not develop symptoms until after age 50, although there are inherited forms that affect individuals in their early teens. Affected individuals typically die within 5 years after symptoms appear.

The degeneration affects both upper and lower motor neurons of the pyramidal and extrapyramidal systems. Because a motor neuron and its dependent muscle fibers are so intimately related, the destruction of CNS neurons causes atrophy of the associated skeletal muscles. The specificity of this disease is remarkable: Motor neurons are destroyed, but sensory neurons, sensory pathways, and intellectual functions remain unaffected. Roughly 90 percent of ALS cases have no known cause. About 10 percent of cases can be linked to genetic factors, but the nature of the genetic abnormality is uncertain. There is no effective treatment. [AM] *Amyotrophic Lateral Sclerosis*

☑ For what anatomical reason does the left side of the brain control motor function on the right side of the body?

☑ An injury involving the superior portion of the motor cortex would affect what region of the body?

☑ What effect would increased stimulation of the motor neurons of the red nucleus have on muscle tone?

Levels of Processing and Motor Control

When you touch a hot stove, you withdraw your hand before you are consciously aware that you have burned yourself. Voluntary motor responses, such as shaking your hand, stepping back, and crying out, occur somewhat later. This is a general pattern: Spinal and cranial reflexes provide rapid, automatic, preprogrammed responses that preserve homeostasis over the short term. Voluntary responses are more complex, but they require more time to prepare and execute.

Figure 15-7● reviews the primary sites of somatic motor control. Cranial and spinal reflexes control the most basic motor activities. Integrative centers in the brain perform more elaborate processing, and as we move from the medulla oblongata to the cerebral cortex, the motor patterns become increasingly complex and variable. The most complex and variable motor activities are directed by the primary motor cortex of the cerebral hemispheres.

During development, the spinal and cranial reflexes are the first to appear. More complex reflexes and motor patterns develop as CNS neurons multiply, enlarge, and interconnect. The process proceeds relatively slowly, as billions of neurons estab-

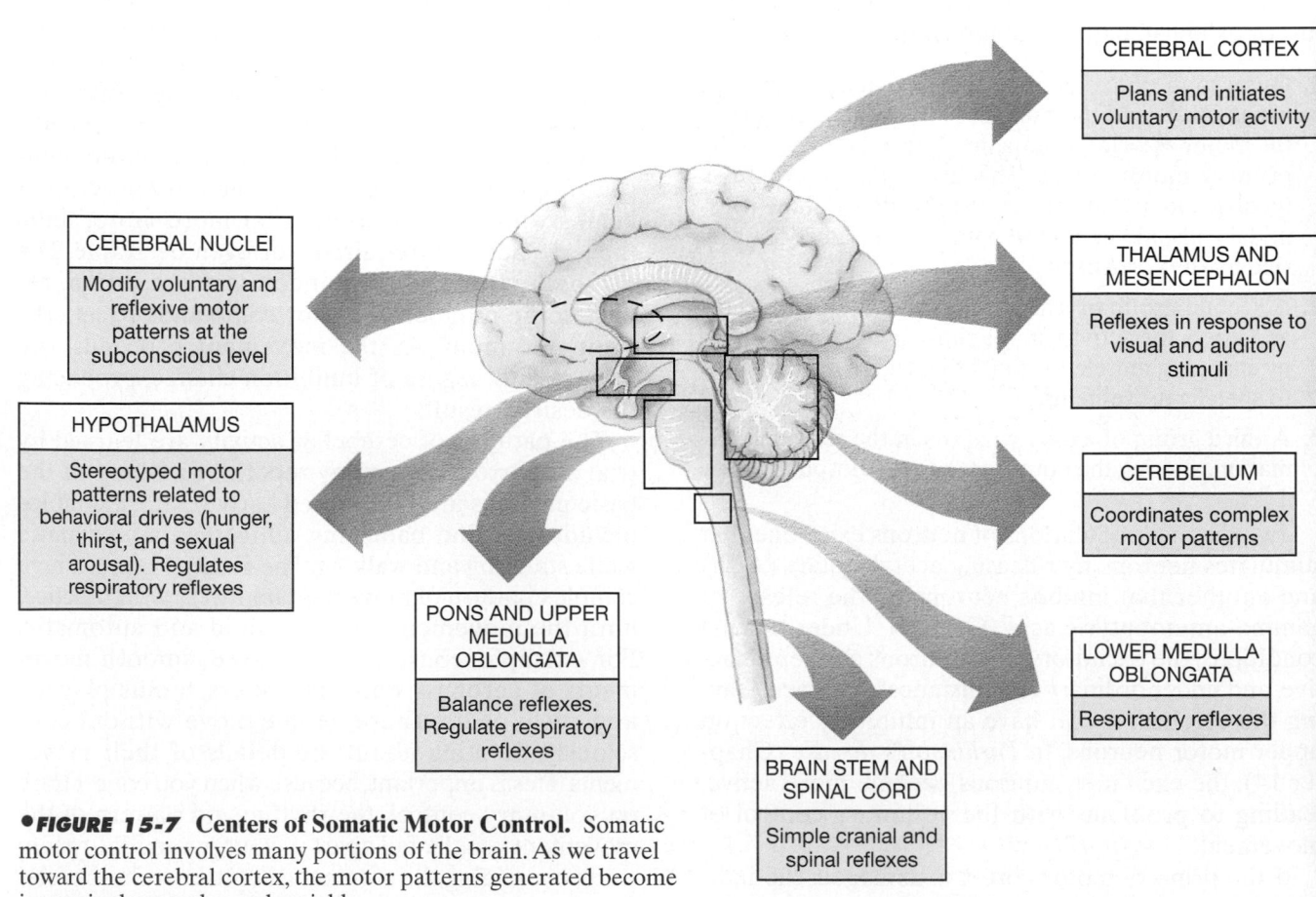

●**FIGURE 15-7** **Centers of Somatic Motor Control.** Somatic motor control involves many portions of the brain. As we travel toward the cerebral cortex, the motor patterns generated become increasingly complex and variable.

lish trillions of synaptic connections. At birth, neither the cerebral nor the cerebellar cortex is fully functional, and the behavior of newborn infants is directed primarily by centers in the diencephalon and brain stem.

Among the anatomical factors that contribute to the maturation of the CNS and the refinement of motor skills are the following:

- The number of neurons in the cerebral cortex continues to increase until at least age 1.
- The brain as a whole grows in size and complexity until at least age 4.
- Myelination of CNS axons continues at least until age 1–2, and peripheral myelination may continue through puberty. Myelination reduces the delay between stimulus and response and improves motor control.

As these events occur, cortical neurons continue to establish new synaptic interconnections that will have a long-term effect on mental capabilities.

ANENCEPHALY Although it may sound strange, physicians generally take a newborn infant into a dark room and shine a light against the skull. They are checking for **anencephaly** (an-en-SEF-uh-lē), a rare condition in which the brain fails to develop at levels above the mesencephalon or lower diencephalon. In most such cases, the cranium also fails to develop, and diagnosis is easy, but in some cases, a normal skull forms. In such instances, the cranium is empty and translucent enough to transmit light. Unless the condition is discovered right away, the parents may take the infant home, totally unaware of the problem. All the normal behavior patterns expected of a newborn are present, including suckling, stretching, yawning, crying, kicking, sticking fingers in the mouth, and tracking movements with the eyes. However, death will occur naturally over a period of days to months. This tragic condition provides a striking demonstration of the role of the brain stem in controlling complex motor patterns. During normal development, these patterns become incorporated into variable and versatile behaviors as control and analytical centers appear in the cerebral cortex.

MONITORING BRAIN ACTIVITY: THE ELECTROENCEPHALOGRAM

The primary sensory cortex and the primary motor cortex have been mapped by direct stimulation in patients undergoing brain surgery. The functions of other regions of the cerebrum can be revealed by the behavioral changes that follow localized injuries or strokes, and the activities of specific regions can be examined by a PET scan or sequential MRI scans.

The electrical activity of the brain is commonly monitored to assess brain activity. Neural function depends on electrical events within the cell membrane of neurons. The brain contains billions of neurons, and their activity generates an electrical field that can be measured by placing electrodes on the brain or the outer surface of the skull. The electrical activity changes constantly, as nuclei and cortical areas are stimulated or quiet down. A printed report on the electrical activity of the brain is called an **electroencephalogram (EEG)**. The electrical patterns observed are called *brain waves*.

Typical brain waves are shown in Figure 15-8a–d•. **Alpha waves** occur in the brains of healthy, awake adults who are resting with their eyes closed (Figure 15-8e•). Alpha waves disappear during sleep, but they also vanish when the individual begins to concentrate on some specific task. During attention to stimuli or tasks, alpha waves are replaced by higher-frequency **beta waves.** Beta waves are typical of individuals who are either concentrating on a task, under stress, or in a state of psychological tension. **Theta waves** may appear transiently during sleep in normal adults but are most often observed in children and in intensely frustrated adults. The presence of theta waves under other circumstances may indicate the presence of a brain disorder, such as a tumor. **Delta waves** are very large-amplitude, low-frequency waves. They are normally seen during deep sleep in individuals of all ages. Delta waves are also seen in the brains of infants (in whom cortical development is still incomplete) and in the awake adult when a tumor, vascular blockage, or inflammation has damaged portions of the brain.

Electrical activity in the two hemispheres is generally synchronized by a "pacemaker" mechanism that appears to involve the thalamus. Asynchrony between the hemispheres can therefore be used to detect localized damage or other cerebral abnormalities. For example, a tumor or injury affecting one hemisphere typically changes the pattern in that hemisphere, and the patterns of the two hemispheres are no longer aligned. A **seizure** is a temporary cerebral disorder accompanied by abnormal movements, unusual sensations, inappropriate behavior, or some combination of these symptoms. Clinical conditions characterized by seizures are known as seizure disorders, or *epilepsies*. Seizures of all kinds are accompanied by a marked change in the pattern of electrical activity monitored in an electroencephalogram. The alteration begins in one portion of the cerebral cortex but may subsequently spread across the entire cortical surface, like a wave on the surface of a pond.

The nature of the symptoms produced depends on the region of the cortex involved. If a seizure affects the primary motor cortex, movements will occur; if it affects the auditory cortex, the individual will hear strange sounds. [AM] *Seizures and Epilepsies*

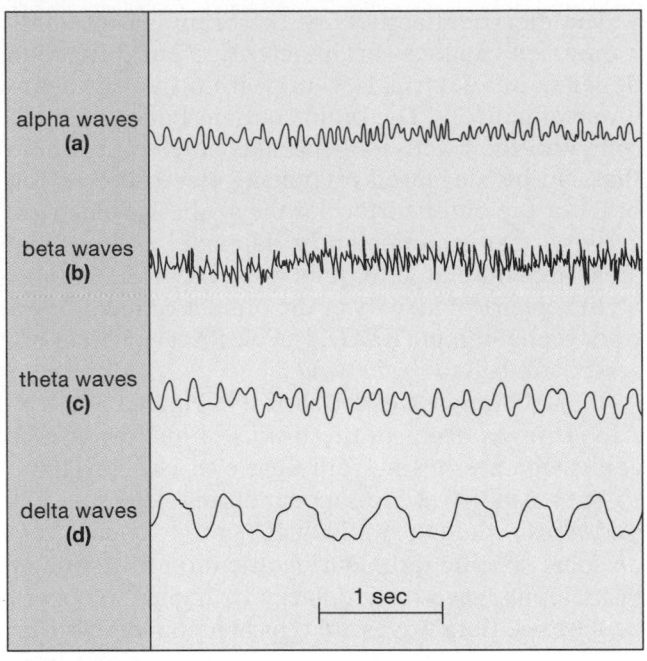

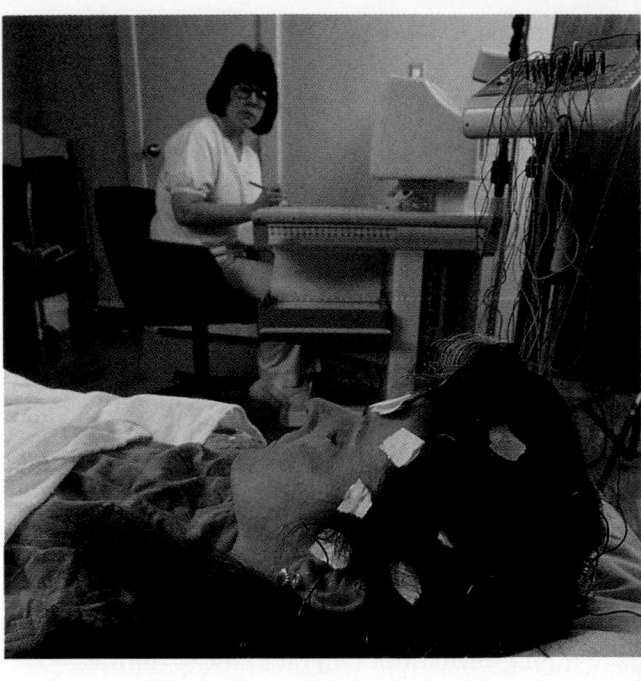

(e)

●*FIGURE 15-8* **Brain Waves.** **(a)** Alpha waves are characteristic of healthy resting adults. **(b)** Beta waves typically accompany intense concentration. **(c)** Theta waves are seen in children and in frustrated adults. **(d)** Delta waves occur during deep sleep and in certain pathological states. (Note: These wave patterns are not drawn to the same scale.) **(e)** A patient wired for EEG monitoring.

HIGHER-ORDER FUNCTIONS

Higher-order functions share the following characteristics:

- The cerebral cortex is required for their performance, and they involve complex interactions among areas of the cortex and between the cerebral cortex and other areas of the brain.
- They involve both conscious and unconscious information processing.
- They are not part of the programmed "wiring" of the brain; therefore, the functions are subject to modification and adjustment over time.

In Chapter 14, we considered functional areas of the cerebral cortex and the regional specializations of the left and right cerebral hemispheres. ∞ *[pp. 461–462]* In this section, we consider the mechanisms of memory and learning and detail the neural interactions responsible for consciousness, sleep, and arousal. In the next section, we will provide an overview of brain chemistry and its effects on behavior and personality.

Memory

What was the topic of the last sentence you read? What do your parents look like? What is your Social Security number? When did Columbus reach the New World?

What does a red traffic light mean? What does a hot dog taste like? Answering these questions involves accessing *memories,* stored bits of information gathered through prior experience. **Fact memories** are specific bits of information, such as the color of a stop sign or the smell of a perfume. **Skill memories** are learned motor behaviors. For example, you can probably remember how to light a match or open a screw-top jar. With repetition, skill memories become incorporated at the unconscious level. Examples include the complex motor patterns involved in skiing, playing the violin, and similar activities. Skill memories related to programmed behaviors, such as eating, are stored in appropriate portions of the brain stem. Complex skill memories involve integration of motor patterns in the cerebral nuclei, cerebral cortex, and cerebellum.

There appear to be two classes of memories. **Short-term memories,** or *primary memories,* do not last long, but while they persist, the information can be recalled immediately. Primary memories contain small bits of information, such as a person's name or a telephone number. Repeating a phone number or other bit of information reinforces the original short-term memory and helps ensure its conversion to a long-term memory. That conversion is called **memory consolidation.** **Long-term memories** last much longer, in some cases for an entire lifetime. There are two types of long-term memory: (1) *Secondary memories* are long-term memories that fade with time and may require considerable effort to recall. (2) *Tertiary memories* are long-term memories that seem to be part of consciousness, such as your name or the contours of your own body. Proposed relationships among these memory classes are diagrammed in Figure 15-9●.

Brain Regions Involved in Memory Consolidation and Access

Two components of the limbic system, the amygdaloid body and the hippocampus (Figure 14-11●, p. 464), are essential to memory consolidation. Damage to the hippocampus leads to an inability to convert short-term memories to new long-term memories, although existing long-term memories remain intact and accessible. Tracts leading from the amygdaloid body to the hypothalamus may link memories to specific emotions.

The **nucleus basalis,** a cerebral nucleus near the diencephalon, plays an uncertain role in memory storage and retrieval. Tracts connect this nucleus with the hippocampus, amygdaloid body, and all areas of the cerebral cortex. Damage to this nucleus is associated with changes in emotional states, memory, and intellectual function (as we will see in the discussion of Alzheimer's disease later in this chapter).

Most long-term memories are stored in the cerebral cortex. Conscious motor and sensory memories are referred to the appropriate association areas. For example, visual memories are stored in the visual association area, and memories of voluntary motor activity in the premotor cortex. Special portions of the occipital and temporal lobes are crucial to the memories of faces, voices, and words. In at least some cases, a specific memory probably depends on the activity of a single neuron. For example, in one portion of the temporal lobe an individual neuron responds to the sound of one word and ignores others. A specific neuron may also be activated by the proper combination of sensory stimuli associated with a particular individual, such as your grandmother. As a result, these neurons are called "grandmother cells."

Information on one subject is parceled out to many different regions of the brain. Your memories of cows are stored in the visual association area (what a cow looks like, how the letters *c-o-w* mean "cow"), the auditory association area (the "moo" sound and how the word *cow* sounds), the speech center (how to say the word *cow*), and the frontal lobes (how big cows are, what they eat). Related information, such as how you feel about cows and what milk tastes like, is stored in other locations. If one of those storage areas is damaged, your memory will be incomplete in some way. How these memories are accessed and assembled on demand remains a mystery.

Cellular Mechanisms of Memory Formation and Storage

Memory consolidation at the cellular level involves anatomical and physiological changes in neurons and synapses. For legal, ethical, and logistical reasons, it is not possible to perform much research on these mechanisms with human subjects. Research on other animals, commonly those with relatively simple nervous systems, has indicated that the following mechanisms may be involved:

- **Increased neurotransmitter release.** A synapse that is frequently active increases the amount of neurotransmitter that it stores, and it releases more on each stimulation. The greater the amount of neurotransmitter released, the greater the effect on the postsynaptic neuron.

- **Facilitation at synapses.** When a neural circuit is repeatedly activated, the synaptic terminals begin releasing neurotransmitter in small quantities continuously. The neurotransmitter binds to receptors on the postsynaptic membrane, producing a graded depolarization

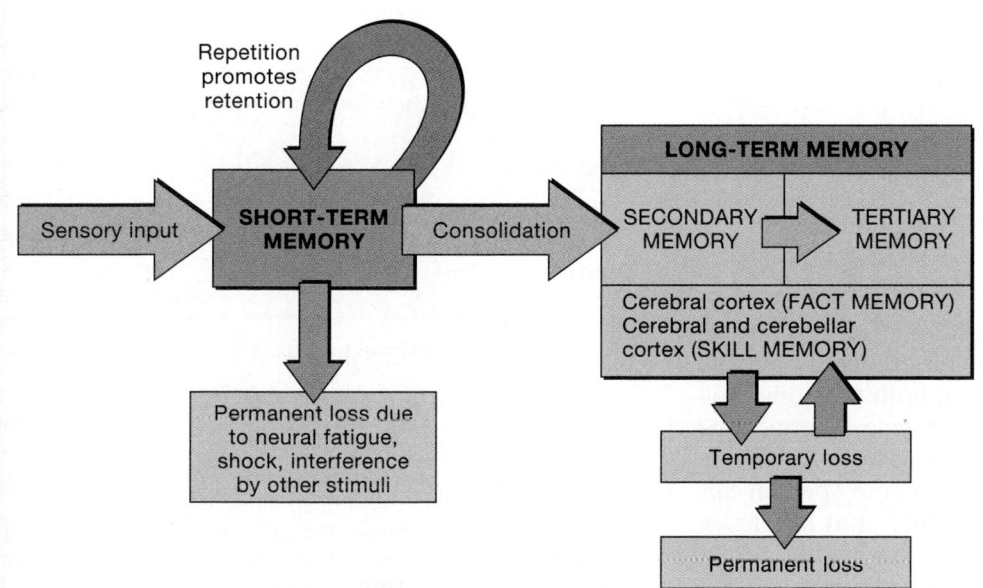

●**FIGURE 15-9** Memory Storage. Steps in the storage of memories and the conversion from short-term to long-term memories.

that brings the membrane closer to threshold. The facilitation that results affects all the neurons in the circuit.

- ***Formation of additional synaptic connections.*** There is evidence that when one neuron repeatedly communicates with another, the axon tip branches and forms additional synapses on the postsynaptic neuron. As a result, the presynaptic neuron will have a greater effect on the transmembrane potential of the postsynaptic neuron.

These processes create anatomical changes that facilitate communication along a specific neural circuit. This facilitated communication is thought to be the basic method of memory storage. A single circuit that corresponds to a single memory has been called a **memory engram.** This definition is based on function rather than structure; we know too little about the organization and storage of memories to be able to describe the neural circuits involved.

Efficient conversion of a short-term memory into a memory engram takes time, usually at least an hour and often longer. Whether that conversion will occur depends on several factors, including the nature, intensity, and frequency of the original stimulus. Very strong, repeated, or exceedingly pleasant or unpleasant events are most likely to be converted to long-term memories. Drugs that stimulate the CNS, such as caffeine and nicotine, may enhance memory consolidation through facilitation; we discussed the membrane effects of those drugs in Chapter 12. ∞ *[p. 404]*

Amnesia refers to the loss of memory from disease or trauma. The type of memory loss depends on the specific regions of the brain affected. For example, damage to the auditory association areas may make it difficult to remember sounds. Damage to thalamic and limbic structures, especially the hippocampus, will affect new memory storage and consolidation. AM *Amnesia*

☑ Toward the end of her A & P test, Tina begins to feel a great deal of stress. What type of brain wave pattern would you expect her to exhibit?

☑ After suffering a head injury in an automobile accident, David has difficulty comprehending what he hears or reads. This symptom might indicate damage to which portion of his brain?

☑ As you recall facts while you take your A & P test, which type of memory are you using?

Consciousness

A conscious individual is alert and attentive; an unconscious individual is not. The difference is obvious, but there are many gradations of both the conscious and unconscious states. For example, a healthy conscious person can be almost asleep, wide awake, or high-strung and jumpy; a healthy sleeping person can be dozing lightly or so deeply asleep that he or she cannot easily be awakened. Healthy individuals cycle between the alert, conscious state and the asleep state

each day. The degree of wakefulness at any given moment is an indication of the level of ongoing CNS activity. When CNS function becomes abnormal or depressed, the state of wakefulness can be affected. As a result, clinicians are quick to note changes in the responsiveness of their patients. AM *Altered States*

Sleep

Conscious implies a state of awareness of and attention to external events and stimuli. *Unconscious* can imply a number of different conditions, ranging from the deep, unresponsive state induced by anesthesia before major surgery to the light, drifting "nod" that occasionally plagues students who are reading anatomy and physiology textbooks. You are considered to be asleep when you are unconscious but can still be awakened by normal sensory stimuli.

There are two levels of sleep, each typified by characteristic patterns of brain wave activity (Figure 15-10●):

1. In **deep sleep,** also called *slow wave* or *non-REM (NREM) sleep,* your entire body relaxes, and activity at the cerebral cortex is at a minimum. Heart rate, blood pressure, respiratory rate, and energy utilization decline by up to 30 percent.

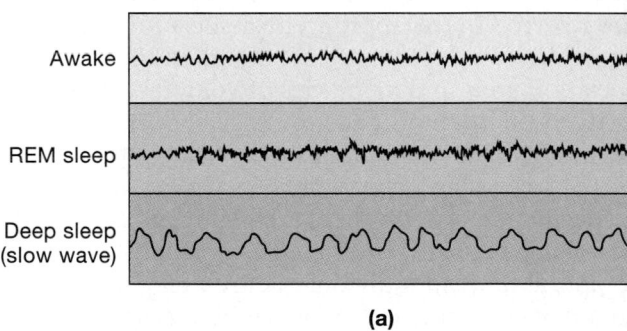

(a)

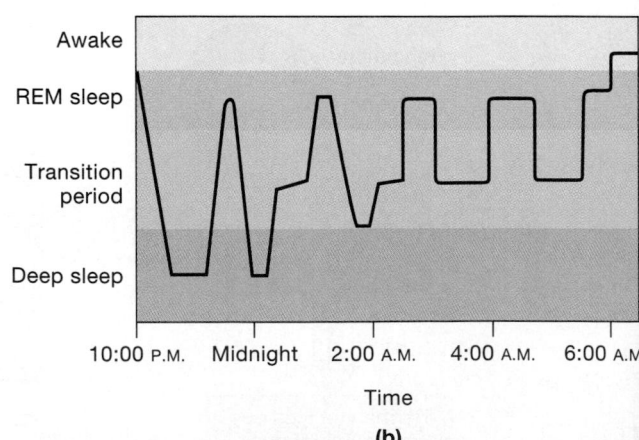

(b)

●FIGURE 15-10 Levels of Sleep. (a) EEG from the awake, REM, and deep (slow wave) sleep states. The EEG pattern during REM sleep resembles the alpha waves typical of awake adults. **(b)** Typical pattern of oscillation between sleep stages of a healthy young adult during a single night's sleep.

2. During **rapid eye movement (REM) sleep,** active dreaming occurs, accompanied by alterations in your blood pressure and respiratory rates. Although the EEG shows traces resembling those of the awake state, you become even less receptive to outside stimuli than in deep sleep, and muscle tone decreases markedly. Intense inhibition of somatic motor neurons probably prevents you from physically producing the responses you envision during the dream sequence. The oculomotor neurons escape this inhibitory influence, and your eyes move rapidly as the imaginary events unfold.

Periods of REM and deep sleep alternate throughout the night, beginning with a period of deep sleep that lasts about an hour and a half. REM periods initially average about 5 minutes in length, but they gradually increase to about 20 minutes over an 8-hour night. Although each night we probably spend less than 2 hours dreaming, there are many sources of variation. For example, children devote more time to REM sleep than do adults, and extremely tired individuals have very short and infrequent REM periods.

Sleep produces only minor alterations in the physiological activities of other organs and systems, and none of these alterations appear to be essential to normal function. The significance of sleep must lie in its impact on the CNS, but the physiological or biochemical basis remains to be determined. We do know that protein synthesis in neurons increases during sleep. Extended periods without sleep will lead to a variety of disturbances in mental function. Roughly 25 percent of the U.S. population experiences some form of a *sleep disorder.* Examples of such disorders include abnormal patterns or duration of REM sleep, or unusual behaviors performed while sleeping, such as sleepwalking. In some cases, these problems begin to affect the individual's conscious activities. Slowed reaction times, irritability, and behavioral changes may result. AM
Sleep Disorders

Arousal

Arousal appears to be one of the functions of the reticular formation. The reticular formation is especially well-suited for providing "watchdog" services, for it has extensive interconnections with the sensory, motor, and integrative nuclei and pathways all along the brain stem.

THE RETICULAR ACTIVATING SYSTEM Your state of consciousness is determined by complex interactions between your brain stem and cerebral cortex. One of the most important brain stem components is a diffuse network in the reticular formation known as the **reticular activating system (RAS).** ∞ *[p. 471]*

This network, diagrammed in Figure 15-11•, extends from the midbrain to the medulla oblongata. The output of the RAS projects to thalamic nuclei that influence large areas of the cerebral cortex. When the RAS is inactive, so is the cerebral cortex; stimulation of the RAS produces a widespread activation of the cerebral cortex.

The mesencephalic portion of the RAS appears to be the center of the system. Stimulation of this area produces the most pronounced and long-lasting effects on the cerebral cortex. Stimulating other portions of the RAS seems to have an effect only to the degree that it changes the activity of the mesencephalic region. The greater the stimulation to the mesencephalic region of the RAS, the more alert and attentive the individual will be to incoming sensory information. The thalamic nuclei associated with the RAS may also play an important role in focusing attention on specific mental processes.

Sleep may be ended by any stimulus sufficient to activate the reticular formation and RAS. Arousal occurs rapidly, but the effects of a single stimulation of the RAS last less than a minute. Thereafter, consciousness can be maintained by positive feedback, because activity in the cerebral cortex, cerebral nu-

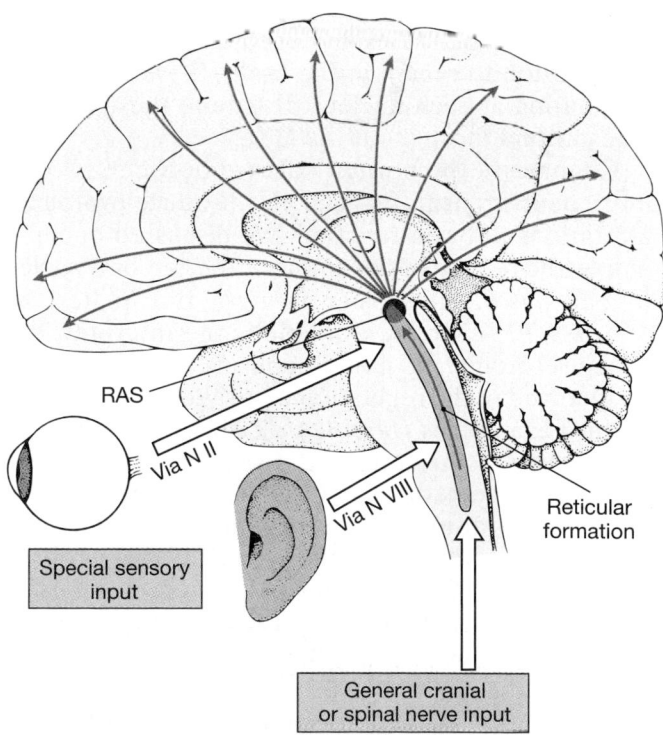

•**FIGURE 15-11** **The Reticular Activating System.** The mesencephalic headquarters of the reticular formation receives collateral inputs from a variety of sensory pathways. Stimulation of this region produces arousal and heightened states of attentiveness.

clei, and sensory and motor pathways will continue to stimulate the RAS.

After many hours of activity, the reticular formation becomes less responsive to stimulation. The individual becomes less alert and more lethargic. The precise mechanism remains unknown, but neural fatigue probably plays a relatively minor role in the reduction of RAS activity. Evidence suggests that the regulation of awake/asleep cycles involves an interplay between brain stem nuclei that use different neurotransmitters. One group of nuclei stimulates the RAS with norepinephrine and maintains the awake, alert state. The other group, which depresses RAS activity with serotonin, promotes deep sleep. These "dueling" nuclei are located in the brain stem.

BRAIN CHEMISTRY AND BEHAVIOR

We discussed the general distribution of neurotransmitters in the CNS in Chapter 12. ∞ *[p. 401]* Of the roughly 50 known neurotransmitters, the most important and widespread are acetylcholine (ACh), norepinephrine (NE), glutamate, aspartate, glycine, gamma aminobutyric acid (GABA), dopamine, serotonin, histamine, Substance P, and the enkephalins and endorphins (see Table 12-5, p. 399). The distribution of neurons releasing each compound varies from region to region in the brain. However, tracts originating at each nucleus distribute these neurotransmitters throughout the CNS.

Changes in the normal balance between two or more neurotransmitters can also produce profound alterations in brain function. We discussed one example, the release of dopamine at the cerebral nuclei by neurons of the substantia nigra, in Chapter 14. ∞ *[p. 470]* A second example is the interplay between serotonin and norepinephrine that appears to be involved in the regulation of awake/asleep cycles. A third example is *Huntington's disease*. The primary problem in this inherited disease is the destruction of ACh-secreting and GABA-secreting neurons in the cerebral nuclei. The reason for this destruction is unknown. Symptoms appear as the cerebral nuclei and frontal lobes slowly degenerate. An individual with Huntington's disease has difficulty controlling movements, and there is a gradual decline in intellectual abilities. AM *Huntington's Disease*

In many cases, the importance of a specific neurotransmitter has been revealed in seeking a mechanism for the effects of administered drugs. Three examples demonstrate patterns that are emerging:

1. An extensive network of tracts delivers serotonin to nuclei and higher centers throughout the brain.

Compounds that enhance the effects of serotonin produce hallucinations; for example, *LSD (lysergic acid diethylamide)* is a powerful hallucinogenic drug that activates serotonin receptors in the brain stem, hypothalamus, limbic system, and spinal cord. Compounds that inhibit serotonin production or block its action cause severe depression and anxiety. The most effective antidepressive drug now in use, *fluoxetine (Prozac®),* slows the removal of serotonin at synapses, causing an increase in serotonin concentrations at the postsynaptic membrane.

2. Norepinephrine is another important neurotransmitter with pathways throughout the brain. Drugs that stimulate NE release cause exhilaration, and those that depress NE release cause depression. One inherited form of depression has been linked to a defective enzyme involved in NE synthesis.

3. Disturbances in dopamine transmission have been linked to several neurological disorders. Inadequate dopamine production causes the motor problems of Parkinson's disease. Excessive production of dopamine may be associated with **schizophrenia,** a psychological disorder marked by pronounced disturbances of mood, thought patterns, and behavior. Amphetamine, or "speed," stimulates dopamine secretion and in large doses can produce symptoms resembling those of schizophrenia. (It also affects other neurotransmitter systems, producing a variety of changes in CNS function.) The *Applications Manual* contains additional information concerning the effects and classification of CNS-active drugs. AM *Pharmacology and Drug Abuse*

Personality and Other Mysteries

The basic pathways and components of the CNS have been traced, and memory acquisition, learning, and associated mental processes can now be manipulated and explored in animal experiments. But the truly remarkable human characteristics of self-awareness and personality remain phenomena without discrete anatomical foundations. That is not to say that an anatomical basis does not exist, just that self-awareness and personality are more characteristics of the brain as an integrated system than functions of any specific nucleus or region. As a result, the origins of human consciousness and personality may remain a mystery for the immediate future. But our knowledge of the anatomy and biochemistry of specific tracts, nuclei, and regions will continue to provide useful clinical information.

In recent years, significant advances have been made by correlating anatomical or physiological deficits with observed behavioral disorders. In some cases, such as the use of L-dopa to treat Parkinson's disease, this procedure has led to new and effective forms of treatment.

AGING AND THE NERVOUS SYSTEM

The aging process affects all bodily systems, and the nervous system is no exception. Anatomical and physiological changes begin shortly after maturity (probably by age 30) and accumulate over time. Although an estimated 85 percent of people above age 65 lead relatively normal lives, they exhibit noticeable changes in mental performance and CNS functioning.

Age-Related Anatomical and Functional Changes

Common age-related anatomical changes in the nervous system include the following:

- *A reduction in brain size and weight.* This reduction results primarily from a decrease in the volume of the cerebral cortex. The brains of elderly individuals have narrower gyri and wider sulci than those of young persons, and the subarachnoid space is enlarged.

- *A reduction in the number of neurons.* Brain shrinkage has been linked to a loss of cortical neurons, although evidence exists that neuronal loss does not occur (at least to the same degree) in brain stem nuclei.

- *A decrease in blood flow to the brain.* With age, fatty deposits gradually accumulate in the walls of blood vessels. Like a kink in a garden hose or a clog in a drain, these deposits reduce the rate of blood flow through arteries. (This process, called *arteriosclerosis*, affects arteries throughout the body; we shall discuss it further in Chapter 21.) The reduction in blood flow may not cause an acute cerebral crisis, but it does increase the chances that the individual will suffer a stroke.

- *Changes in synaptic organization of the brain.* In many areas, the number of dendritic branchings and interconnections appears to decrease. Synaptic connections are lost, and the rate of neurotransmitter production declines.

- *Intracellular and extracellular changes in CNS neurons.* Many neurons in the brain accumulate abnormal intracellular deposits, including *lipofuscin* and *neurofibrillary tangles.* **Lipofuscin** is a granular pigment that has no known function. **Neurofibrillary tangles** are masses of neurofibrils that form dense mats inside the soma and axon. **Plaques** are extracellular accumulations of **amyloid,** an unusual fibrillar protein, surrounded by abnormal dendrites and axons. The significance of these cellular and extracellular abnormalities is unknown. There is evidence that they appear in all aging brains, but when present in excess, they seem to be associated with clinical abnormalities.

These anatomical changes are linked to a series of functional alterations. In general, neural processing becomes less efficient with age. For example, memory consolidation typically becomes more difficult, and secondary memories, especially those of the recent past, become harder to access. The sensory systems of the elderly—notably, hearing, balance, vision, smell, and taste—become less acute. Lights must be brighter, sounds louder, and smells stronger before they are perceived. Reaction times are slowed, and reflexes—even some withdrawal reflexes—become weaker or disappear. There is a decrease in the precision of motor control, and it takes longer to perform a given motor pattern than it did 20 years earlier.

For roughly 85 percent of the elderly population, these changes do not interfere with their abilities to function in society. But for as yet unknown reasons, some elderly individuals become incapacitated by progressive CNS changes. By far the most common such incapacitating condition is *Alzheimer's disease.*

Alzheimer's Disease

Alzheimer's disease is a progressive disorder characterized by the loss of higher cerebral functions. It is the most common cause of **senile dementia,** or senility. The first symptoms generally appear at 50–60 years of age, although the disease occasionally affects younger individuals. Alzheimer's disease has widespread impact. An estimated 2 million people in the United States, including roughly 15 percent of those over age 65, suffer from some form of the condition, and it causes approximately 100,000 deaths each year.

The link remains uncertain, but the areas containing concentrations of plaques and neurofibrillary tangles are the same regions involved with memory, emotions, and intellectual function. It remains to be determined whether these deposits cause Alzheimer's or are secondary signs of ongoing metabolic alterations with an environmental, hereditary, or infectious basis.

Genetic factors certainly play a major role. The late-onset form of Alzheimer's disease has been traced to a gene on chromosome 19 that codes for proteins involved in cholesterol transport. Less than 5 percent of people with Alzheimer's disease have the early-onset form; these individuals develop the condition before age 50. The early-onset form of Alzheimer's disease has been linked to genes on chromosomes 21 and 14. Interestingly, the majority of individuals with *Down syndrome* develop Alzheimer's disease relatively early in life. (Down syndrome results from an extra copy of chromosome 21; we shall discuss this condition further in Chapter 29.) AM *Alzheimer's Disease*

☑ You are asleep. What would happen to you if your reticular activating system (RAS) were suddenly stimulated?

☑ What would you expect to be the effect of a drug that substantially increases the amount of serotonin released in the brain?

☑ One of the problems associated with aging is difficulty in recalling things or even a total loss of memory. What are some possible reasons for these changes?

SELECTED CLINICAL TERMINOLOGY

Terms Discussed in This Chapter

Alzheimer's disease: A progressive disorder marked by the loss of higher cerebral functions. *(p. 509 and AM)*

amnesia: A temporary or permanent loss of memory. *(p. 506 and AM)*

amyotrophic lateral sclerosis (ALS): A progressive, degenerative disorder affecting motor neurons of the spinal cord, brain stem, and cerebral hemispheres. *(p. 502 and AM)*

anencephaly (an-en-SEF-a-lē): A rare condition in which the brain fails to develop at levels above the mesencephalon or lower diencephalon. *(p. 503)*

cerebral palsy: A disorder that affects voluntary motor performance and arises in infancy or childhood as a result of pre-

natal trauma, drug exposure, or a congenital defect. *(p. 499)*

electroencephalogram (EEG): A printed record of the brain's electrical activity over time. *(p. 503)*

epilepsy: One of more than 40 different conditions characterized by a recurring pattern of seizures over extended periods. *(p. 503 and AM)*

Huntington's disease: An inherited disease marked by a progressive deterioration of mental abilities and by motor disturbances. *(p. 508 and AM)*

schizophrenia: A psychological disorder marked by pronounced disturbances of mood, thought patterns, and behavior. *(p. 508)*

seizure: A temporary disorder of cerebral function, accompanied by abnormal movements, unusual sensations, and/or inappropriate behavior; includes *focal seizures, temporal lobe seizures, convulsive seizures, generalized seizures, grand mal seizures,* and *petit mal seizures.* *(p. 503 and AM)*

senile dementia: A progressive loss of memory, spatial orientation, language, and personality as a consequence of aging. *(p. 509)*

AM Additional Terms Discussed in the *Applications Manual*

anterograde amnesia: Loss of the ability to store new memories after an injury, although earlier memories remain intact and accessible.

aphasia: A disorder affecting the ability to speak or read.

delirium: A state characterized by wild oscillations in the level of wakefulness, little or no grasp of reality, and possible hallucinations. Also called an *acute confusional state* (ACS).

dementia: A stable, chronic state of consciousness characterized by deficits in memory, spatial orientation, language, or personality.

hypersomnia: Unusually long periods of sleep that is otherwise normal.

insomnia: Shortened sleeping periods and difficulty in getting to sleep.

parasomnias: Abnormal behaviors performed during sleep.

post-traumatic amnesia (PTA): Amnesia caused by head injury, in which the individual can neither remember the past nor consolidate memories of the present.

retrograde amnesia (*retro-,* behind): The loss of memories of events that occurred prior to an injury.

sleep apnea: A condition in which a sleeping individual stops breathing for short intervals throughout the night.

CHAPTER REVIEW

 On-line resources for this chapter are on our World Wide Web site at:
http://www.prenhall.com/martini/fap

STUDY OUTLINE

INTRODUCTION, p. 491

1. There is continuous communication among the brain, spinal cord, and peripheral nerves.

SENSORY AND MOTOR PATHWAYS, p. 491

Sensory Pathways, p. 491

1. A **sensation** arrives in the form of action potentials in an afferent fiber. The **posterior column pathway** carries fine touch, pressure, and proprioceptive sensations. The axons ascend within the **fasciculus gracilis** and **fasciculus cuneatus** and relay information to the thalamus via the **medial lemniscus.** Before the axons enter the medial lemniscus, they *decussate* (cross over) to the opposite side of the brain stem. *(Figures 15-1, 15-2; Table 15-1)*

2. The **spinothalamic pathway** carries poorly localized sensations of touch, pressure, pain, and temperature. The axons involved decussate in the spinal cord and ascend within the **anterior** and **lateral spinothalamic tracts** to the **ventral posterolateral** and **ventral posteromedial nuclei** of the thalamus. *(Figures 15-2, 15-3; Table 15-1)*

3. The **spinocerebellar pathway,** including the **posterior** and **anterior spinocerebellar tracts,** carries sensations to the cerebellum concerning the position of muscles, tendons, and joints. *(Figure 15-4; Table 15-1)*

Motor Pathways, p. 497

4. Somatic motor pathways always involve an **upper motor neuron** (whose soma lies in a CNS processing center) and a **lower motor neuron** (whose soma is located in a motor nucleus of the brain stem or spinal cord). *(Figure 15-5)*

5. The neurons of the primary motor cortex are **pyramidal cells.** The **pyramidal system** provides voluntary skeletal muscle control. The **corticobulbar tracts** terminate at the cranial nerve nuclei; the **corticospinal tracts** synapse on motor neurons in the anterior gray horns of the spinal cord. The corticospinal tracts are visible along the medulla as a pair of thick bands, the **pyramids,** where most of the axons decussate to enter the descending **lateral corticospinal tracts.** Those that do not cross over enter the **anterior corticospinal tracts.** The pyramidal system provides a rapid, direct mechanism for controlling skeletal muscles. *(Figures 15-5, 15-6)*

6. The **extrapyramidal system (EPS)** consists of several other centers that may issue motor commands as a result of processing performed at a subconscious level. Its outputs descend via the *vestibulospinal, tectospinal, rubrospinal,* and *reticulospinal tracts. (Figure 15-5; Tables 15-2, 15-3)*

7. The **vestibulospinal tracts** carry information related to maintaining balance and posture. Commands carried by the **tectospinal tracts** change the position of the eyes, head, neck, and arms in response to bright lights, sudden movements, or loud noises. The **rubrospinal tracts** carry motor responses to spinal motor neurons. Motor commands carried by the **reticulospinal tracts** vary according to the region stimulated.

8. The cerebral nuclei (the most important and complex components of the extrapyramidal system) adjust the motor commands issued in other processing centers through synapses in the thalamus.

9. The cerebellum regulates the activity along both the pyramidal and the extrapyramidal motor pathways. The integrative activities performed by neurons in the cerebellar cortex and cerebellar nuclei are essential for precise control of movements. *(Figure 15-7)*

MONITORING BRAIN ACTIVITY: THE ELECTROENCEPHALOGRAM, p. 503

1. An **electroencephalogram (EEG)** is a printed report of the electrical activity of the brain over time. An EEG can be used to check for a variety of clinical conditions. *(Figure 15-8)*

HIGHER-ORDER FUNCTIONS, p. 504

1. Higher-order functions (1) are performed by the cerebral cortex and involve complex interactions among areas of the cerebral cortex and between the cortex and other areas of the brain, (2) involve conscious and unconscious information processing, and (3) are subject to modification and adjustment over time.

Memory, p. 504

2. Memories can be classified into short-term or long-term. **Short-term,** or *primary*, **memories** are short-lived, but while they persist, the information can be recalled immediately. **Long-term memories** last much longer.

3. Long-term memories can in turn be subdivided into *secondary memories* (which fade with time and may be difficult to recall) and *tertiary memories* (which seem to be part of consciousness). *(Figure 15-9)*

4. Conversion from short-term to long-term memory is **memory consolidation.** The amygdaloid body and the hippocampus are essential to memory consolidation.

5. Cellular mechanisms that seem to be involved in memory formation and storage include increased neurotransmitter release, facilitation at synapses, and formation of additional synaptic connections.

Consciousness, p. 506

6. In **deep sleep** (*slow wave* or *non-REM sleep*), the body relaxes and cerebral cortex activity is low. During **rapid eye movement (REM) sleep**, active dreaming occurs. Periods of REM and deep sleep alternate throughout the night. *(Figure 15-10)*

7. Consciousness is determined by interactions between the brain stem and cerebral cortex. One of the most important brain stem components is a network in the reticular formation called the **reticular activating system (RAS).** *(Figure 15-11)*

BRAIN CHEMISTRY AND BEHAVIOR, p. 508

1. Changes in the normal balance between two or more neurotransmitters can profoundly alter brain function. The most important and widespread neurotransmitters are acetylcholine (ACh), norepinephrine (NE), gamma aminobutyric acid (GABA), dopamine, serotonin, histamine, Substance P, and the enkephalins and endorphins.

Personality and Other Mysteries, p. 508

2. The human characteristics of self-awareness and personality remain a mystery. They appear to be characteristic of the brain as an integrated system rather than a function of any specific component.

AGING AND THE NERVOUS SYSTEM, p. 509

Age-Related Anatomical and Functional Changes, p. 509

1. Age-related changes in the nervous system include reduction in brain size and weight, reduction in the number of neurons, decrease in blood flow to the brain, changes in synaptic organization of the brain, and intracellular and extracellular changes in CNS neurons.

Alzheimer's Disease, p. 509

2. **Alzheimer's disease** is a progressive disorder characterized by the loss of higher cerebral functions. It is the most common cause of **senile dementia.**

REVIEW QUESTIONS

Level 1   Reviewing Facts and Terms

1. The corticospinal tract
   (a) carries motor commands from the cerebral cortex to the spinal cord
   (b) carries sensory information from the spinal cord to the brain
   (c) starts in the spinal cord and ends in the brain
   (d) a, b, and c are correct

2. The ability to localize a specific stimulus depends on the organized distribution of sensory information to the
   (a) spinothalamic pathway
   (b) posterior column tract
   (c) spinocerebellar pathway
   (d) primary sensory cortex

3. Destruction or damage to the lower motor neuron in the somatic nervous system produces
  (a) the inability to localize a stimulus
  (b) subconscious response to stimulation
  (c) a flaccid paralysis in the innervated motor unit
  (d) stimulation of the innervated muscle

4. A progressive disorder characterized by the loss of higher cerebral functions is
  (a) Parkinson's disease
  (b) parasomnia
  (c) Huntington's disease
  (d) Alzheimer's disease

5. What four kinds of sensory receptors for the general senses are found in the body? What is the nature of the stimuli that excite each one?

6. What are the three major somatic sensory pathways in the body, and what is the function of each pathway?

7. What three pairs of descending tracts make up the pyramidal system?

8. What four motor pathways make up the extrapyramidal system?

9. What are the two primary functional roles of the cerebellum?

10. What three anatomical factors contribute to the maturation of the CNS and the refinement of motor skills?

11. What three characteristics are shared by higher-order functions?

12. As a result of animal studies, what cellular mechanisms are thought to be involved in memory formation and storage?

13. What physiological activities distinguish non-REM sleep from REM sleep?

14. What anatomical and functional changes in the brain are linked to alterations that occur with aging?

Level 2   Reviewing Concepts

15. Damage to the hippocampus, a component of the limbic system, leads to
  (a) a loss of emotion due to forgetfulness
  (b) a loss of consciousness
  (c) a loss of long-term memory
  (d) an immediate loss of short-term memory

16. Describe the relationship among first-, second-, and third-order neurons.

17. How is the autonomic nervous system (ANS) *functionally* different from the somatic nervous system (SNS)?

18. What is a motor homunculus? How does it differ from a sensory homunculus?

19. What effect does injury to the primary motor cortex have on peripheral muscles?

20. By which structures and in which part of the brain is the level of muscle tone in the body's skeletal muscles controlled? How is this control exerted?

21. One patient suffers a cerebrovascular accident (CVA) in the left hemisphere; another suffers a CVA in the right hemisphere. What functions could be affected in each case?

Level 3   Critical Thinking and Clinical Applications

22. Kelly is having difficulty controlling movements of her eyes, and she notices that she has lost some control of her facial muscles. After an examination and testing, her physician tells Kelly that her cranial nerves are perfectly normal but that a small tumor is putting pressure on certain fiber tracts in her brain. This pressure is the cause of her symptoms. Where is the tumor most likely located?

23. Doris develops a clot that blocks the right branch of the middle cerebral artery, a blood vessel that serves the anterior portion of the right cerebral hemisphere. What symptoms would you expect to observe as a result of this blockage?

24. Researchers studying an illicit drug find that it causes individuals to see strange shapes and hear sounds that aren't actually present. Perceptions of color are exaggerated, and individuals taking the drug report increased sexual appetites. Which neurotransmitter is this drug mimicking, and which part or parts of the brain are involved?

16

The Autonomic Nervous System

*T*his is the control center of a nuclear power station. The generation and harnessing of nuclear energy are complex, potentially dangerous processes. This power station operates reliably because multiple systems—reactor, cooling, and electrical—are working smoothly and in balance. The systems involved are so complex, and the responses must be so swift and precise, that no one individual can manage a single system unaided, let alone the entire facility. Instead, computers are programmed to make adjustments in system operation, and the people look at the "big picture." The human body is an extremely complex biological mechanism. As in the power station, the individual systems are regulated automatically, and our conscious minds are aware of only the general patterns and results of those regulatory activities. In this chapter, we will examine the regulatory mechanism and explore the structure and function of the autonomic nervous system.

CHAPTER OUTLINE AND OBJECTIVES

Your conscious thoughts, plans, and actions represent a tiny fraction of the activities of the nervous system. In practical terms, your conscious thoughts and the somatic motor system, which operates under conscious control, seldom have a direct effect on your long-term survival. Of course, the somatic motor system can be important in moving you out of the way of a speeding bus or pulling your hand from a hot stove—but it was your conscious movements that put you in jeopardy in the first place. If all consciousness were eliminated, vital physiological processes would continue virtually unchanged; a night's sleep is not a life-threatening event. Longer, deeper states of unconsciousness are not necessarily more dangerous, as long as nourishment is provided. People who have suffered severe brain injuries can survive in a coma for decades.

Survival under these conditions is possible because routine homeostatic adjustments in physiological systems are made by the *autonomic nervous system (ANS)*. It is the ANS that coordinates cardiovascular, respiratory, digestive, excretory, and reproductive functions. In doing so, the ANS adjusts internal water, electrolyte, nutrient, and dissolved gas concentrations in body fluids—and it does so without instructions or interference from the conscious mind.

In this chapter, we examine the anatomical and physiological divisions of the ANS. Our understanding of this system has had a profound effect on the practice of medicine. For example, in 1960, the 5-year survival rate for patients surviving their first heart attack was very low, primarily because it was difficult and sometimes impossible to control high blood pressure. Nearly forty years later, many survivors of heart attacks lead normal lives. This dramatic change occurred as we learned to manipulate the ANS with specific drugs and procedures.

AN OVERVIEW OF THE ANS

It will be useful to compare the organization of the ANS with that of the somatic nervous system, which controls our skeletal muscles. We will focus on (1) the neural interactions that direct motor output and (2) the subdivisions of the ANS, based on structural and functional patterns of peripheral innervation.

Figure 16-1● compares the organization of the somatic and autonomic nervous systems. In the somatic system, lower motor neurons exert direct control over skeletal muscles (Figure 16-1a●). In the ANS (Figure 16-1b●), *there is always a second visceral motor neuron interposed between the CNS and the peripheral effector.* The visceral motor neurons in the CNS are known as **preganglionic neurons.** The axons of these neurons are called *preganglionic fibers.* The

preganglionic fibers leave the CNS and synapse on neurons in **autonomic ganglia**. These are called **ganglionic neurons.** The ganglionic neurons control peripheral effectors such as cardiac muscle, smooth muscles, glands, and adipose tissues. The axons of ganglionic neurons are called **postganglionic fibers,** because they begin at the ganglia and extend to the peripheral target organs.[1]

Subdivisions of the ANS

The ANS contains two subdivisions, the *sympathetic division* and the *parasympathetic division*. Most often, the two divisions have opposing effects; if the sympathetic division causes excitation, the parasympathetic causes inhibition. However, this is not always the case, because (1) the two divisions may work independently, with some structures innervated by only one division or the other, and (2) the two divisions may work together, each controlling one stage of a complex process.

One way to recall the general functions of the sympathetic and parasympathetic divisions is to imagine a situation dominated by one system or the other. Although we will consider in detail each of the responses described in the following scenarios later in the chapter, this exercise will give you a sense of the overall *pattern* of response.

The **sympathetic division** prepares the body for heightened levels of somatic activity. When fully activated, this division produces what is known as the "fight or flight" response, which prepares the body for a crisis that may require sudden, intense physical activity. To understand the nature of this response, imagine walking down a long, dark alley and hearing strange noises in the darkness ahead. Your body responds immediately, and you become more alert and aware of your surroundings. Your metabolic rate rises quickly, to as much as twice the resting level. Your digestive and urinary activities are suspended temporarily, and blood flow to your skeletal muscles increases. You begin breathing more quickly and more deeply. Both your heart rate and blood pressure increase, circulating your blood more rapidly. You feel warm and begin to perspire. The general pattern can be summarized as follows: (1) heightened mental alertness, (2) increased metabolic rate, (3) reduced digestive and urinary function, (4) activation of energy reserves, (5) increased respiratory rate and dilation of respiratory passageways, (6) increased heart rate and blood pressure, and (7) activation of sweat glands.

---

[1]Ganglionic neurons are sometimes called *postganglionic neurons*. This term is potentially confusing because (1) only the axons extend beyond the ganglia, and (2) the adrenal medulla contains ganglionic neurons that lack postganglionic fibers.

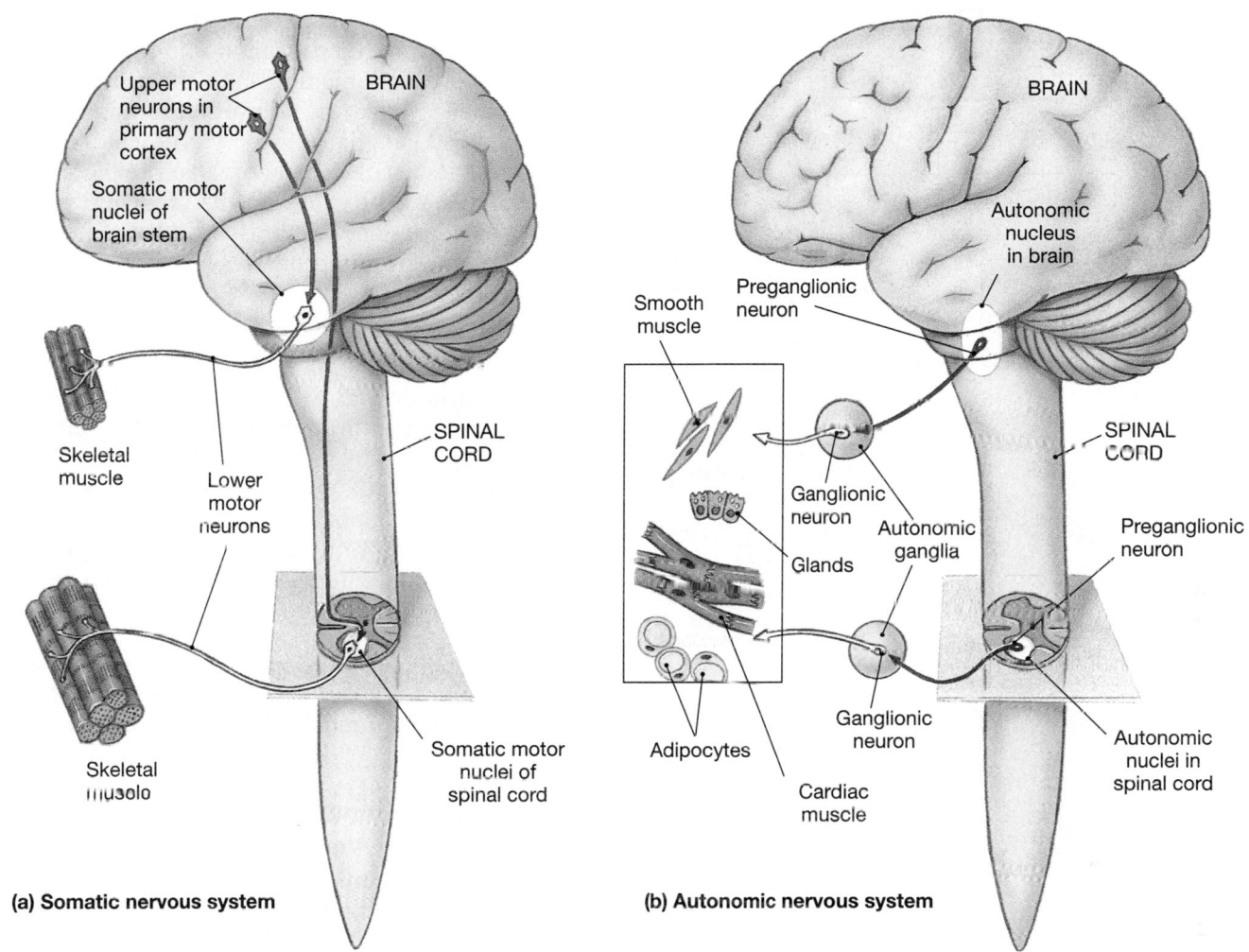

•FIGURE 16-1 Organization of the Somatic and Autonomic Nervous Systems. (a) In the SNS, an upper motor neuron in the CNS controls a lower motor neuron in the brain stem or spinal cord. The axon of the lower motor neuron has direct control over skeletal muscle fibers. Stimulation of the lower motor neuron always has an excitatory effect on the skeletal muscle fibers. **(b)** In the ANS, the axon of a preganglionic neuron in the CNS controls ganglionic neurons in the periphery. Stimulation of the ganglionic neurons may lead to excitation or inhibition of the visceral effector innervated.

The **parasympathetic division** stimulates visceral activity. For example, it is responsible for the state of "rest and repose" that follows a big dinner. Your body relaxes, energy demands are minimal, and both your heart rate and blood pressure are relatively low. Meanwhile, your digestive organs are highly stimulated. Your salivary glands and other secretory glands are active, your stomach is contracting, and smooth muscle contractions move materials along your digestive tract. This movement promotes defecation; at the same time, smooth muscle contractions along your urinary tract promote urination. The overall pattern is as follows: (1) decreased metabolic rate, (2) decreased heart rate and blood pressure, (3) increased secretion by salivary and digestive glands, (4) increased motility and blood flow in the digestive tract, and (5) stimulation of urination and defecation.

The two systems are anatomically as well as functionally distinct. Figure 16-2• provides an introduction to the major differences. We will now take a closer look at the structural and functional organization of each system.

Sympathetic (Thoracolumbar) Division

Preganglionic fibers from the thoracic and upper lumbar spinal cord segments synapse in ganglia near the spinal cord. These axons and ganglia are part of the sympathetic division, or *thoracolumbar* (tho-ra-kō-LUM-bar) *division*, of the ANS (Figure 16-2•). In the sympathetic division, the preganglionic fibers are short and the postganglionic fibers are long. An increase in sympathetic activity generally stimulates tissue metabolism, increases alertness, and prepares the body to deal with emergencies.

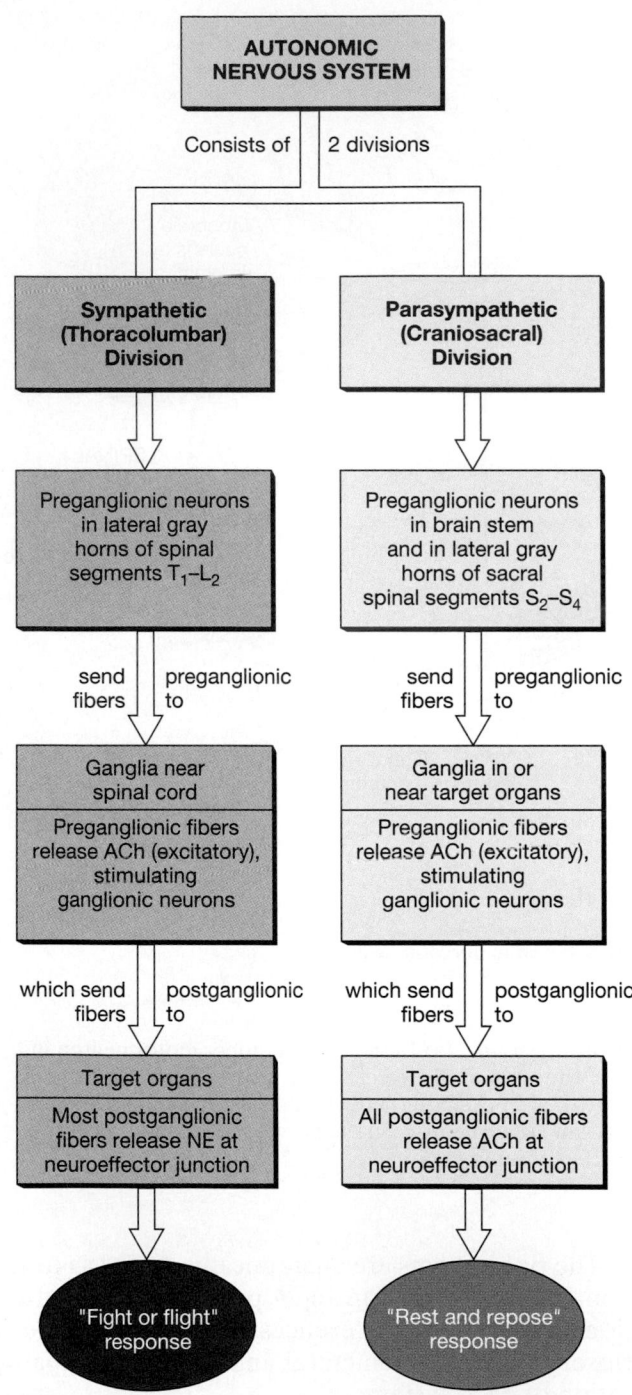

•FIGURE 16-2 **Divisions of the Autonomic Nervous System (ANS).** At the thoracic and lumbar levels, the visceral efferent fibers that emerge form the sympathetic, or thoracolumbar, division. At the cranial and sacral levels, the visceral efferent fibers from the CNS make up the parasympathetic, or craniosacral, division.

Parasympathetic (Craniosacral) Division

Preganglionic fibers originating in the brain and the sacral spinal cord segments are part of the parasympathetic division, or *craniosacral* (krā-nē-ō-SĀ-krul)

division, of the ANS (Figure 16-2•). In the parasympathetic division, the preganglionic fibers are long and the postganglionic fibers are short. The preganglionic fibers synapse on neurons of **terminal ganglia** or **intramural ganglia** (*murus,* wall) located near or within the tissues of visceral organs. The parasympathetic division is often called the "rest and repose" system because general parasympathetic activation conserves energy and promotes sedentary activities, such as digestion.

Innervation Patterns

The sympathetic and parasympathetic divisions affect target organs through the controlled release of neurotransmitters by postganglionic fibers. Whether the result is a stimulation or an inhibition of activity depends on the response of the membrane receptor to the presence of the neurotransmitter. We can make three general statements about the neurotransmitters and their effects:

1. All preganglionic autonomic fibers (sympathetic and parasympathetic) release *acetylcholine (ACh)* at their synaptic terminals. The effects are always excitatory.
2. Postganglionic parasympathetic fibers also release ACh, but the effects may be excitatory or inhibitory, depending on the nature of the receptor.
3. Most postganglionic sympathetic terminals release the neurotransmitter *norepinephrine (NE)*. The effects are generally excitatory, but the response depends on the nature of the postsynaptic receptor.

☑ How many motor neurons are required to conduct an action potential from the spinal cord to smooth muscles in the wall of the intestine?

☑ While out for a brisk walk, Julie is suddenly confronted by an angry dog. Which division of the autonomic nervous system is responsible for the physiological changes that occur in Julie as she turns and runs from the animal?

☑ How could you distinguish the sympathetic division from the parasympathetic division of the autonomic nervous system on the basis of anatomy?

THE SYMPATHETIC DIVISION

The sympathetic division (Figure 16-3•) consists of *preganglionic neurons located between segments T_1 and L_2 of the spinal cord* and *ganglionic neurons located in ganglia near the vertebral column.* The preganglionic neurons are situated in the lateral gray horns, and their axons enter the ventral roots of these segments. The ganglionic neurons occur in three different locations:

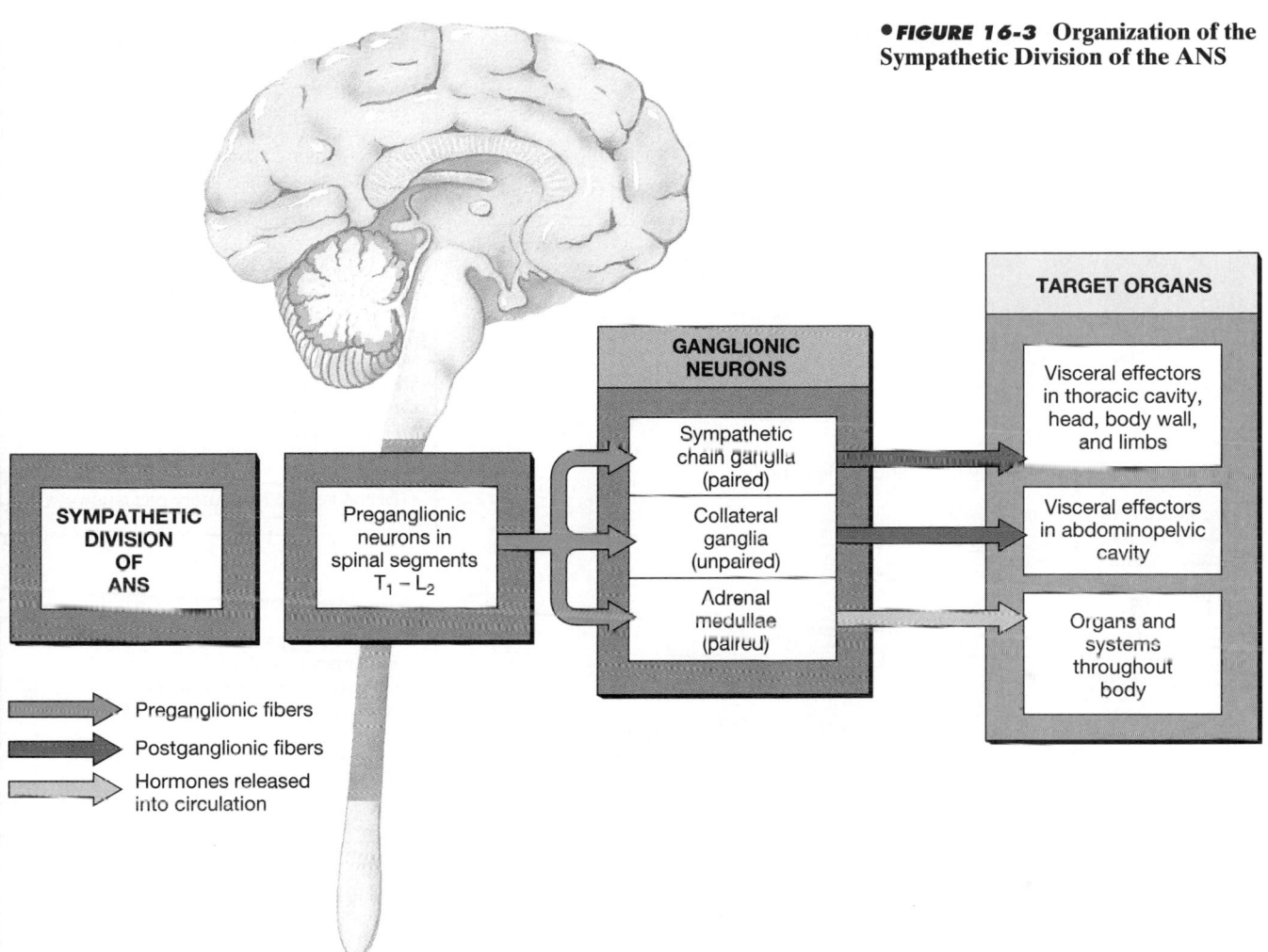

•*FIGURE 16-3* Organization of the Sympathetic Division of the ANS

1. *Sympathetic chain ganglia.* **Sympathetic chain ganglia,** also called *paravertebral ganglia* or *lateral ganglia,* lie on either side of the vertebral column. Neurons in these ganglia control effectors in the body wall and inside the thoracic cavity (Figure 16-4a•).

2. *Collateral ganglia.* **Collateral ganglia,** also known as *prevertebral ganglia,* are anterior to the vertebral bodies (Figure 16-4b•). Collateral ganglia contain ganglionic neurons that innervate tissues and organs in the abdominopelvic cavity.

3. *The adrenal medullae.* The center of each adrenal gland, an area known as the *adrenal medulla,* is a modified sympathetic ganglion. The ganglionic neurons of the adrenal medullae have very short axons; when stimulated, they release their neurotransmitters into the bloodstream (Figure 16-4c•). This change in the release site—from a synapse to a capillary—allows the neurotransmitters to function as hormones that affect target cells throughout the body.

The preganglionic fibers are relatively short, because the ganglia are located relatively near the spinal cord. In contrast, the postganglionic fibers are relatively long, except at the adrenal medullae.

The Sympathetic Chain

The ventral roots of spinal segments T_1 to L_2 contain sympathetic preganglionic fibers. The basic pattern of sympathetic innervation in these regions was described in Figure 13-6a•, p. 425. After passing through the intervertebral foramen, each ventral root gives rise to a myelinated *white ramus,* or *white ramus communicans,* that carries preganglionic fibers into a nearby sympathetic chain ganglion. These fibers may synapse within the sympathetic chain ganglia (Figure 16-4a•), at one of the collateral ganglia (Figure 16-4b•), or in the adrenal medullae (Figure 16-4c•). Extensive divergence occurs, with one preganglionic fiber synapsing on two dozen or more ganglionic neurons. Preganglionic fibers running between the sympathetic chain ganglia interconnect them, making the chain resemble a string of beads. Each ganglion in the sympathetic chain innervates a particular body segment or group of segments.

If a preganglionic fiber carries motor commands that target structures in the body wall or the thoracic cavity, it will synapse in one or more of the sympathetic chain ganglia (Figure 16-4a•). The paths of the unmyelinated postganglionic fibers differ depending on whether their

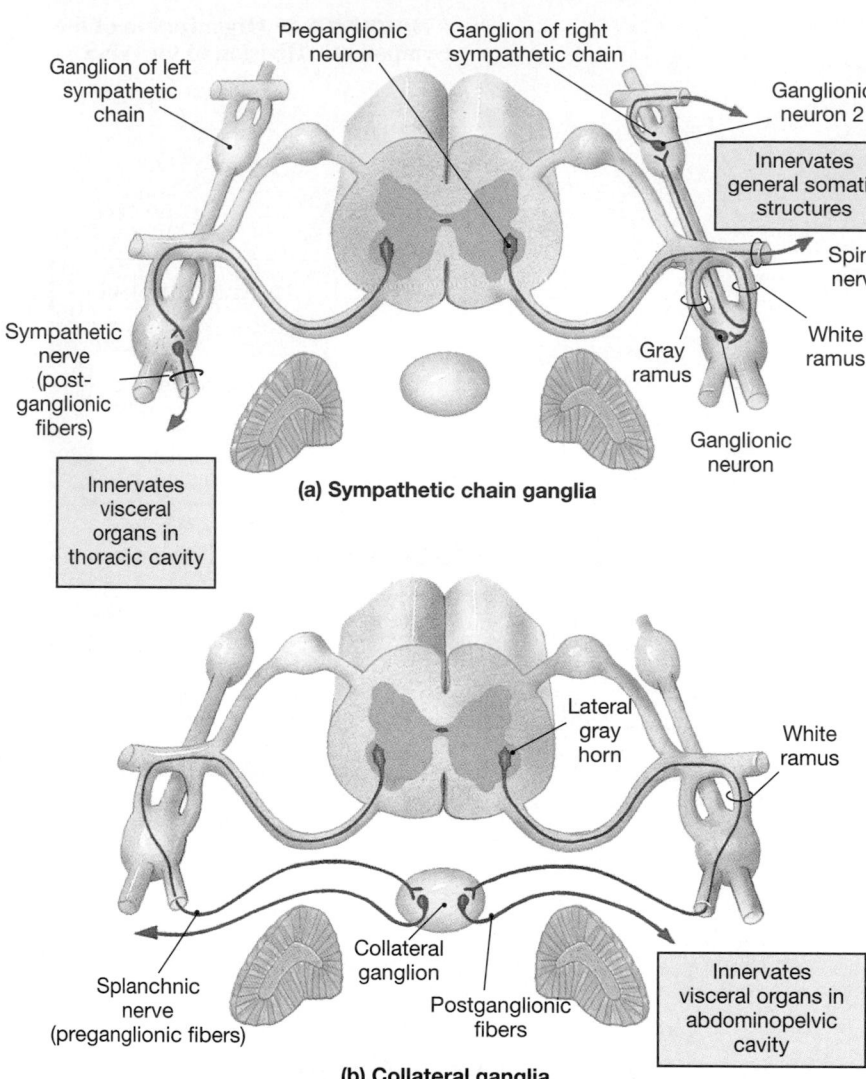

Ganglion of left
sympathetic
chain

Preganglionic
neuron

Ganglion of right
sympathetic chain

Ganglionic
neuron 2

Innervates
general somatic
structures

Spinal
nerve

White
ramus

Gray
ramus

Sympathetic
nerve (post-
ganglionic
fibers)

Ganglionic
neuron

Innervates
visceral
organs in
thoracic
cavity

(a) Sympathetic chain ganglia

Major effects produced by sympathetic
postganglionic fibers in spinal nerves:
• Constriction of cutaneous blood vessels,
  reduction in circulation to the skin and to
  most other organs in the body wall
• Acceleration of blood flow to skeletal
  muscles and brain
• Stimulation of energy production and use
  by skeletal muscle tissue
• Release of stored lipids from subcutaneous
  adipose tissue
• Stimulation of secretion by sweat glands
• Dilation of the pupils and focusing for
  distant objects
Major effects produced by postganglionic
fibers entering the thoracic cavity in
sympathetic nerves:
• Acceleration of the heart rate and
  increasing the strength of cardiac
  contractions
• Dilation of the respiratory passageways

Lateral
gray
horn

White
ramus

Splanchnic
nerve
(preganglionic fibers)

Collateral
ganglion

Postganglionic
fibers

Innervates
visceral organs in
abdominopelvic
cavity

(b) Collateral ganglia

Major effects produced by preganglionic
fibers innervating the collateral ganglia:
• Constriction of small arteries and
  reduction in the flow of blood to visceral
  organs
• Decrease in the activity of digestive
  glands and organs
• Stimulation of the release of glucose from
  glycogen reserves in the liver
• Stimulation of the release of lipids from
  adipose tissue
• Relaxation of the smooth muscle in the
  wall of the urinary bladder
• Reduction of the rate of urine formation at
  the kidneys
• Control of some aspects of sexual
  function, such as ejaculation in males

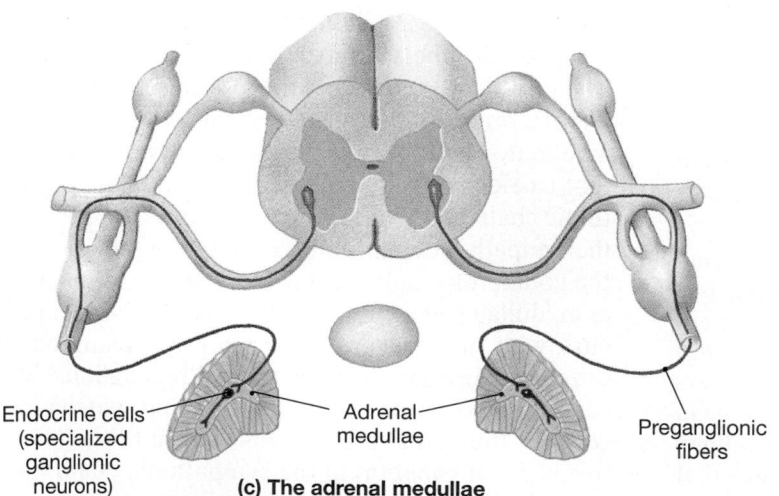

Endocrine cells
(specialized
ganglionic
neurons)

Adrenal
medullae

Preganglionic
fibers

(c) The adrenal medullae

Major effect produced by preganglionic
fibers innervating the adrenal medullae:
• Release of epinephrine and norepinephrine
  into the general circulation

•**FIGURE 16-4** **Sympathetic Pathways**

targets lie in the body wall or within the thoracic cavity. Postganglionic fibers that control visceral effectors in the body wall, such as the sweat glands of the skin or the smooth muscles in superficial blood vessels, enter the *gray ramus (gray ramus communicans)* and return to the spinal nerve for subsequent distribution. However, spinal nerves do not innervate structures in the ventral body cavities. Postganglionic fibers targeting structures in the thoracic cavity, such as the heart and lungs, form **sympathetic nerves** that proceed directly to their peripheral targets. Although Figure 16-4a• shows sympathetic nerves on the left side and spinal nerve distribution on the right, in reality *both* innervation patterns are found on *each* side of the body.

Postganglionic fibers leaving the sympathetic chain reach their peripheral targets by way of spinal nerves and sympathetic nerves. Figure 16-4a• summarizes the primary results of increased activity in the postganglionic fibers leaving the sympathetic

chain ganglia in spinal nerves and sympathetic nerves. How these effects are brought about will be the focus of a later section.

Anatomy of the Sympathetic Chain

Figure 16-5• provides a more detailed diagram of the structure of the sympathetic division. The left side represents the distribution to the skin and to skeletal muscles and other tissues of the body wall; the right side depicts the innervation of visceral structures.

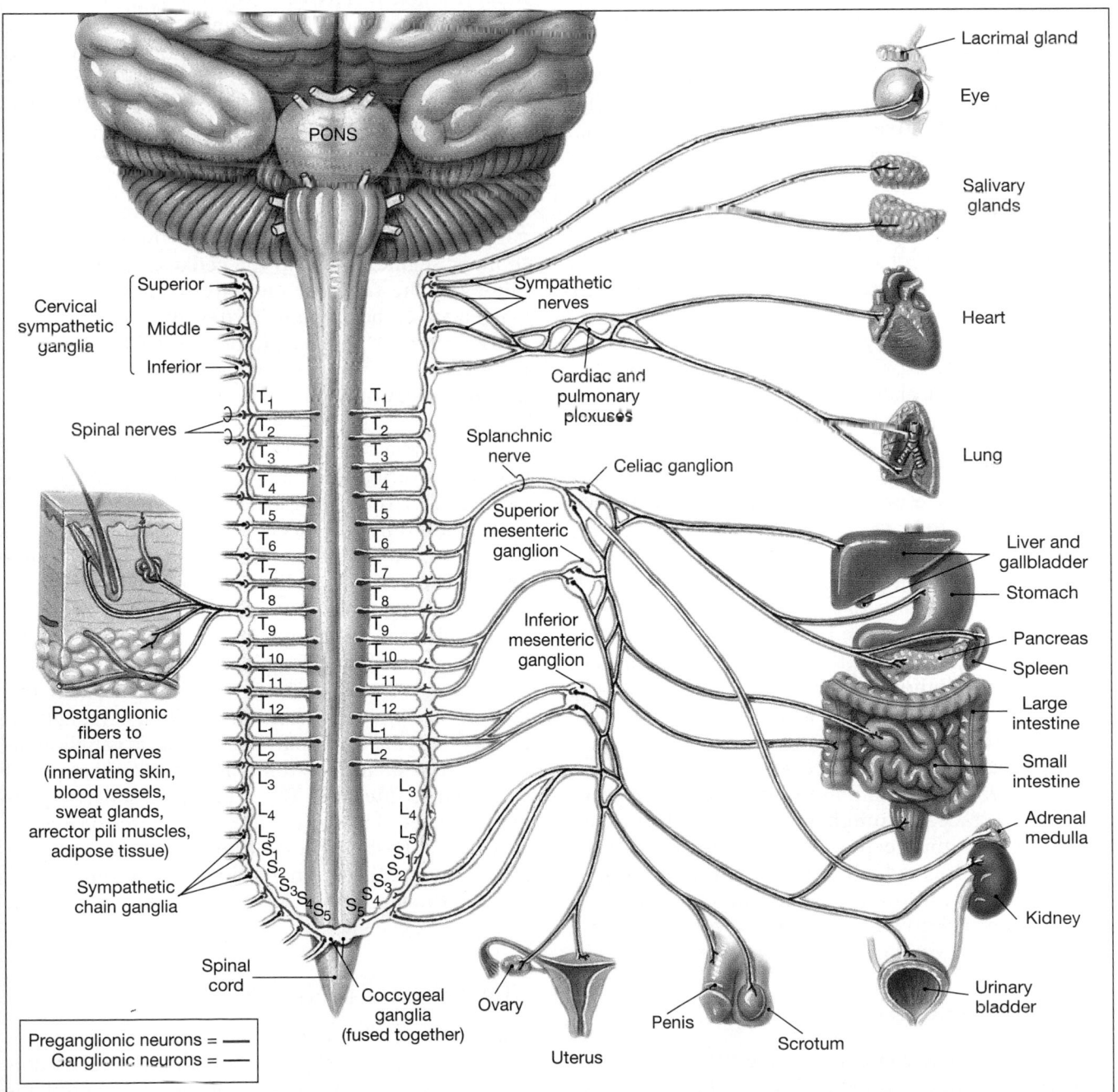

• **FIGURE 16-5** **The Sympathetic Division.** The distribution of sympathetic fibers is the same on both sides of the body. For clarity, the innervation of somatic structures is shown to the left, and the innervation of visceral structures to the right.

In each chain, there are 3 cervical, 10–12 thoracic, 4–5 lumbar, and 4–5 sacral sympathetic ganglia and 1 coccygeal sympathetic ganglion. Preganglionic neurons are limited to spinal cord segments T_1–L_2, and these spinal nerves have both white rami (myelinated preganglionic fibers) and gray rami (unmyelinated postganglionic fibers). The neurons in the cervical, lower lumbar, and sacral sympathetic chain ganglia are innervated by preganglionic fibers that run along the axis of the chain. In turn, these chain ganglia provide postganglionic fibers, via gray rami, to the cervical, lumbar, and sacral spinal nerves. As a result, although only spinal nerves T_1–L_2 have white rami, *every spinal nerve has a gray ramus* that carries sympathetic postganglionic fibers for distribution in the body wall.

About 8 percent of the axons in each spinal nerve are sympathetic postganglionic fibers. As a result, the dorsal and ventral rami of the spinal nerves, which provide somatic motor innervation to skeletal muscles of the body wall and limbs, also distribute sympathetic postganglionic fibers. In the head, postganglionic sympathetic fibers leaving the cervical chain ganglia supply the regions and structures innervated by cranial nerves III, VII, IX, and X (Figure 16-5●). ∞ *[pp. 477, 479–481]*

In summary: (1) Only the thoracic and upper lumbar ganglia receive preganglionic fibers from white rami; (2) the cervical, lower lumbar, and sacral chain ganglia receive preganglionic innervation through collateral fibers; and (3) every spinal nerve receives a gray ramus from a ganglion of the sympathetic chain.

This anatomical arrangement means that if the ventral roots of thoracic spinal nerves are damaged, there will be no sympathetic motor function on the affected side of the head, neck, and trunk. Yet damage to the ventral roots of cervical spinal nerves will produce voluntary muscle paralysis on the affected side but will leave sympathetic function intact, because the preganglionic fibers innervating the cervical ganglia originate in the white rami of thoracic segments, which are undamaged.

In contrast, damage to the cervical ganglia or thoracic segments can eliminate sympathetic innervation to the face, although sensation and muscle control remain unaffected. The affected side of the face becomes flushed, although sweating does not occur, and the pupil constricts. This combination of symptoms is known as *Horner's syndrome.* AM
Hypersensitivity and Sympathetic Function

Collateral Ganglia

The abdominopelvic viscera receive sympathetic innervation by way of preganglionic fibers that pass through the sympathetic chain without synapsing. These fibers originate at preganglionic neurons in the lower thoracic and upper lumbar segments. They synapse within separate *collateral ganglia* (Figure 16-4b●). Preganglionic fibers that innervate the collateral ganglia form the **splanchnic** (SPLANK-nik) **nerves,** which lie in the dorsal wall of the abdominal cavity. Splanchnic nerves from both sides of the body converge on these ganglia. Although there are two sympathetic chains, one on each side of the vertebral column, most collateral ganglia are single rather than paired.

Postganglionic fibers leaving the collateral ganglia extend throughout the abdominopelvic cavity, innervating a variety of visceral tissues and organs. A summary of the effects of increased sympathetic activity along these postganglionic fibers is included in Figure 16-4b●. The general pattern is (1) a reduction of blood flow and energy use by visceral organs that are not important to short-term survival, such as the digestive tract, and (2) the release of stored energy reserves.

The splanchnic nerves innervate three collateral ganglia. Preganglionic fibers from the seven lower thoracic segments end at the **celiac** (SĒ-lē-ak) **ganglion** and the **superior mesenteric ganglion.** These ganglia are embedded in an extensive network of autonomic nerves. Preganglionic fibers from the lumbar segments form splanchnic nerves that end at the **inferior mesenteric ganglion.** These ganglia are diagrammed in Figure 16-5●.

The ganglia are named by their association with adjacent arteries. For example, the celiac ganglion is named after the *celiac artery.* The celiac ganglion most commonly consists of a pair of interconnected masses of gray matter situated at the base of that artery. The celiac ganglion may also form a single mass or many small, interwoven masses. Postganglionic fibers from this ganglion innervate the stomach, liver, pancreas, and spleen.

The superior mesenteric ganglion sits near the base of the *superior mesenteric artery.* Postganglionic fibers leaving the superior mesenteric ganglion innervate the small intestine and the initial segments of the large intestine. The inferior mesenteric ganglion is located near the base of the *inferior mesenteric artery.* Postganglionic fibers from this ganglion provide sympathetic innervation to the terminal portions of the large intestine, the kidney and bladder, and the sex organs.

The Adrenal Medullae

Preganglionic fibers entering an adrenal gland proceed to its center, a region called the **adrenal medulla** (Figures 16-4c and 16-5●). The adrenal medulla is a modified sympathetic ganglion. Within the medulla, preganglionic fibers synapse on *neuroendocrine* *cells,* specialized neurons that release the neurotransmitters *epinephrine (E)* and *norepinephrine (NE)* into the general circulation. Epinephrine, also called *adrenaline,* accounts for 75–80 percent of the secretory output; the rest is NE.

The bloodstream then carries the neurotransmitters throughout the body, causing changes in the metabolic activities of many different cells. In general, these effects resemble those produced by the stimulation of sympathetic postganglionic fibers. They differ, however, in two respects: (1) Cells not innervated by sympathetic postganglionic fibers are affected as well, and (2) the effects last much longer than those produced by direct sympathetic innervation.

Sympathetic Activation

The sympathetic division can change tissue and organ activities by releasing NE at peripheral synapses and by distributing E and NE throughout the body in the bloodstream. The visceral motor fibers that target specific effectors, such as smooth muscle fibers in blood vessels of the skin, can be activated in reflexes that do not involve other peripheral effectors. In a crisis, however, the entire division responds. This event is called **sympathetic activation.** Sympathetic activation is controlled by sympathetic centers in the hypothalamus. The effects are not limited to peripheral tissues; sympathetic activation also alters CNS activity. When sympathetic activation occurs, an individual experiences the following:

- Increased alertness, via stimulation of the reticular activating system, causing the individual to feel "on edge."
- A feeling of energy and euphoria, often associated with a disregard for danger and a temporary insensitivity to painful stimuli.
- Increased activity in the cardiovascular and respiratory centers of the pons and medulla oblongata, leading to elevations in blood pressure, heart rate, breathing rate, and depth of respiration.
- A general elevation in muscle tone through stimulation of the extrapyramidal system, so the person *looks* tense and may begin to shiver.

These changes, plus the peripheral changes already noted, complete the preparations necessary for the individual to cope with stressful situations.

Neurotransmitters and Sympathetic Function

We have examined the distribution of sympathetic impulses and the general effects of sympathetic activation. We will now consider the cellular basis of these effects on peripheral organs. On stimulation, sympathetic preganglionic fibers release ACh at synapses that innervate ganglionic neurons. Synapses that use ACh as a transmitter are called *cholinergic.* The effect on the ganglionic neurons is always excitatory.

Stimulation of the ganglionic neurons in the sympathetic division leads to the release of neurotransmitter at postganglionic neuroeffector junctions. (Recall from Chapter 12 that neuroeffector junctions are synapses between a neuron and another cell type.) ∞ *[p. 373]* The synaptic terminals are typically different from the neuroeffector junctions of the somatic nervous system. Instead of forming individual synaptic knobs, the telodendria form a network or chain of **varicosities,** swollen segments that either contact the target cells or end in the adjacent connective tissues (Figure 16-6●).

Most sympathetic ganglionic neurons release NE at their varicosities. As we learned in Chapter 12, neurons that use NE as a neurotransmitter are called *adrenergic.* ∞ *[p 399]* The sympathetic division also contains a small but significant number of ganglionic neurons that release ACh rather than

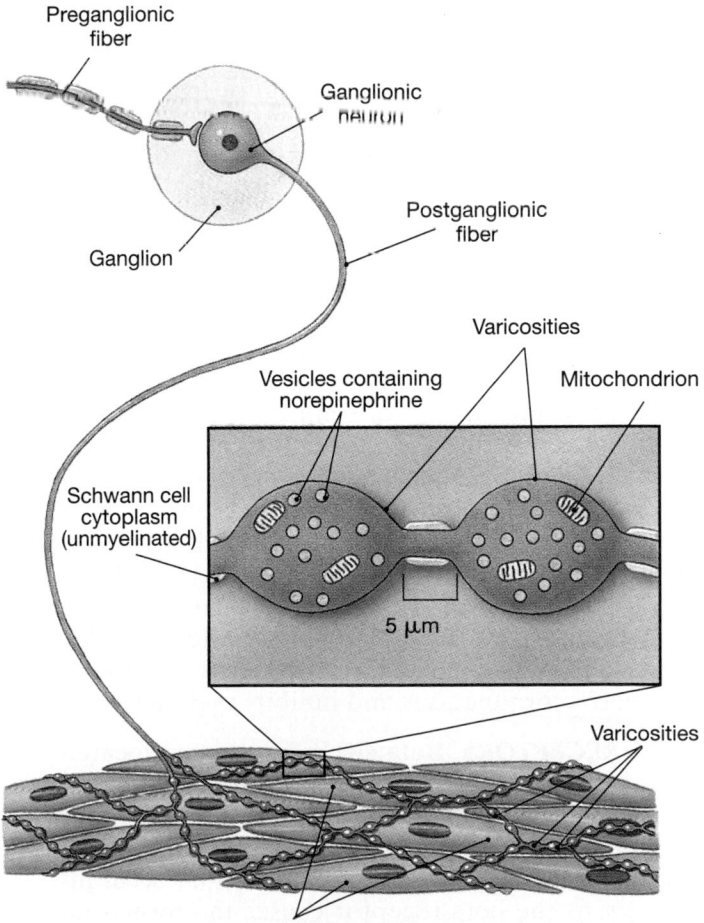

Preganglionic fiber

Ganglionic neuron

Ganglion

Postganglionic fiber

Varicosities

Mitochondrion

Vesicles containing norepinephrine

Schwann cell cytoplasm (unmyelinated)

5 μm

Varicosities

Smooth muscle cells

●**FIGURE 16-6** **Sympathetic Postganglionic Nerve Endings**

NE. The varicosities involved with ACh release are located in the body wall, the skin, and in skeletal muscles.

The NE released by varicosities affects its targets for just a few seconds before it is inactivated by enzymes. (As usual, the specific effects on the target cells vary with the nature of the receptor on the postsynaptic membrane.) Because your bloodstream does not contain the enzymes that break down NE or E, and because most tissues contain relatively low concentrations of those enzymes, the effects of E or NE released by the adrenal medullae last much longer. Tissue concentrations of epinephrine throughout the body may remain elevated for as long as 30 seconds, and the effects may persist for several minutes.

Membrane Receptors

The effects of sympathetic stimulation result primarily from interactions with membrane receptors sensitive to NE and E. There are two classes of sympathetic receptors: *alpha receptors* and *beta receptors*. In general, norepinephrine stimulates alpha receptors more than it does beta receptors, whereas epinephrine stimulates both classes of receptors.

ALPHA RECEPTORS Stimulation of **alpha (α) receptors** activates enzymes on the inside of the cell membrane. There are two types of alpha receptors: alpha-1 (α_1) and alpha-2 (α_2) (Figure 16-7a●). The result for the most common type of alpha receptor, α_1, is the release of intracellular calcium ions from reserves in the endoplasmic reticulum. This response follows the release of *second messengers* inside the target cell. ∞ *[p. 400]* The release of calcium ions generally has an excitatory effect on the target cell. For example, the stimulation of α_1 receptors on the surfaces of smooth muscle cells is responsible for the constriction of peripheral blood vessels and the closure of sphincters along the digestive tract. Alpha-2 receptors are less common; their stimulation results in a lowering of cyclic-AMP (cAMP) levels in the cytoplasm. This reduction generally has an inhibitory effect on the cell. The presence of α_2 receptors within the parasympathetic division helps coordinate sympathetic and parasympathetic activities. When the sympathetic division is active, the NE released binds to parasympathetic neuroeffector junctions and inhibits their activity.

BETA RECEPTORS Beta (β) **receptors** are located in many organs, including skeletal muscles, the lungs, the heart, and the liver. Stimulation of beta receptors at these sites triggers changes in the metabolic activity of the target cell. These alterations occur indirectly, as the beta receptor causes the formation of a second messenger, cAMP, which activates or inactivates key enzymes (Figure 16-7b●).

There are two major types of beta receptors: beta-1 (β_1) and beta-2 (β_2). Stimulation of β_1 receptors leads to an increase in metabolic activity. For example, stimulation of β_1 receptors on skeletal muscles accelerates the metabolic activities of the muscles. Stimulation of β_1 receptors in the heart causes an in-

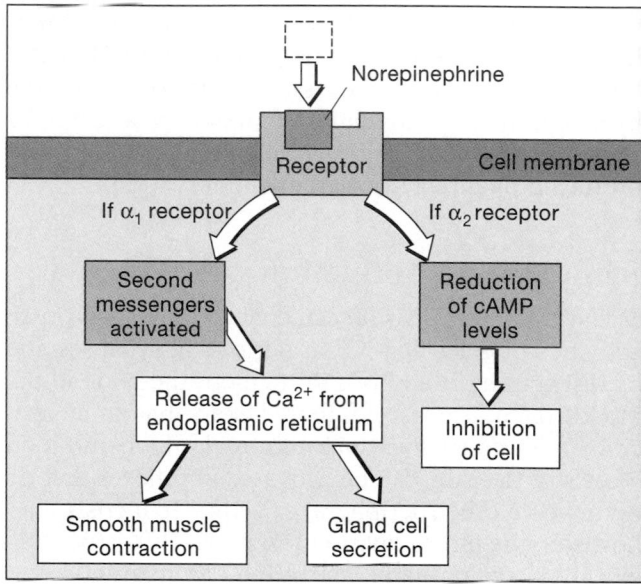

(a) Alpha-receptor stimulation

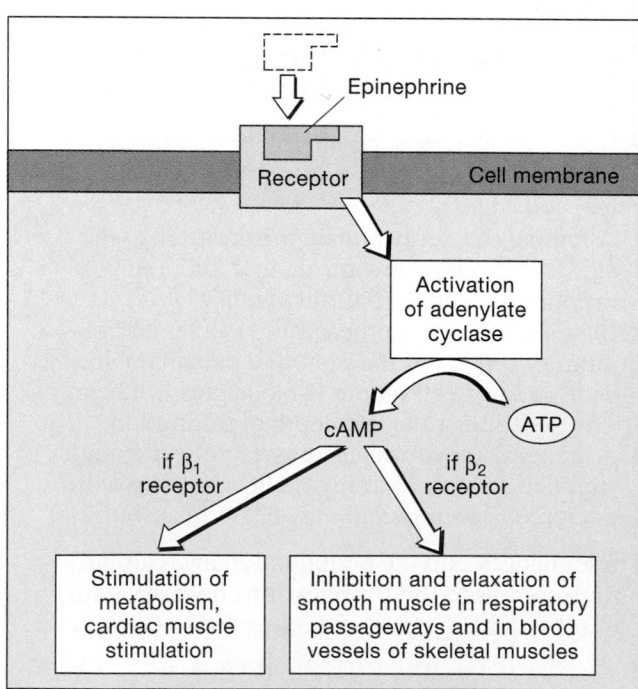

(b) Beta-receptor stimulation

●**FIGURE 16-7** **Sympathetic Receptor Classification.**
(a) Stimulation of an alpha-1 receptor causes release of calcium ions into the cytoplasm. Stimulation of an alpha-2 receptor causes a reduction in the concentration of cyclic-AMP in the cytoplasm. **(b)** Beta receptor stimulation may lead to excitation or inhibition of the target cell.

crease in heart rate and in the force of contraction. Stimulation of β_2 receptors causes inhibition. When stimulated, these receptors trigger a relaxation of smooth muscles along the respiratory tract, increasing the diameter of the respiratory passageways and making breathing easier. This response accounts for the effectiveness of the inhalers used to treat asthma.

The effects of NE on the postsynaptic membrane last longer than those of ACh, because the NE is removed relatively slowly. From 50 to 80 percent of the NE is re-absorbed by the varicosities and either re-used or broken down by the enzyme *monoamine oxidase (MAO)*. The rest of the NE diffuses out of the area or is broken down by the enzyme *catechol-O-methyltransferase (COMT)* in surrounding tissues.

Sympathetic Stimulation, ACh, and NO

Although the vast majority of sympathetic postganglionic fibers are adrenergic, releasing NE, a few postganglionic fibers are cholinergic. These postganglionic fibers innervate sweat glands of the skin and the blood vessels to skeletal muscles and the brain. Activation of these sympathetic fibers stimulates sweat gland secretion and dilates the blood vessels.

It may seem strange that sympathetic terminals release ACh, which is the neurotransmitter used by the parasympathetic nervous system. However, (1) ACh stimulates sweat gland secretion much more than does NE; (2) NE release causes constriction of most peripheral arteries; and (3) neither the body wall nor skeletal muscles are innervated by the parasympathetic division. The distribution of cholinergic fibers via the sympathetic division provides a method of stimulating sweat gland secretion and selectively enhancing blood flow to skeletal muscles while the adrenergic terminals reduce the blood flow to other tissues in the body wall.

The sympathetic division also includes *nitroxidergic* synapses, which release *nitric oxide (NO)* as a neurotransmitter. As we mentioned in Chapter 12, such synapses occur where neurons innervate smooth muscles in the walls of blood vessels in many regions, notably in skeletal muscles and the brain. ∞ *[p. 400]* Activity of these synapses promotes immediate vasodilation and increased blood flow through the region.

A Summary of the Sympathetic Division

To summarize our discussion of the sympathetic division:

1. The sympathetic division of the ANS includes two segmentally arranged sympathetic chains, one on each side of the spinal column; three collateral ganglia anterior to the spinal column; and two adrenal medullae.

2. The preganglionic fibers are short, because the ganglia are close to the spinal cord. The postganglionic fibers are relatively long and extend a considerable distance before reaching their target organs. (In the case of the adrenal medullae, very short axons end at capillaries that carry their secretions to the bloodstream.)

3. The sympathetic division shows extensive divergence, and a single preganglionic fiber may innervate two dozen or more ganglionic neurons in different ganglia. As a result, a single sympathetic motor neuron inside the CNS can control a variety of peripheral effectors and produce a complex and coordinated response.

4. All preganglionic neurons release ACh at their synapses with ganglionic neurons. Most postganglionic fibers release NE, but a few release ACh or NO.

5. The effector response depends on the nature of the channels or enzymes activated when NE or E binds to alpha or beta receptors.

☑ Where do the nerves that synapse in the collateral ganglia originate?

☑ How would a drug that stimulates acetylcholine receptors affect the sympathetic nervous system?

☑ An individual with high blood pressure may be given a medication that blocks beta receptors. How would this medication help that person's condition?

THE PARASYMPATHETIC DIVISION

The parasympathetic division of the ANS (Figure 16-8●) consists of the following:

1. ***Preganglionic neurons in the brain stem and in sacral segments of the spinal cord.*** In the brain, the mesencephalon, pons, and medulla oblongata contain autonomic nuclei associated with cranial nerves III, VII, IX, and X. In the sacral segments of the spinal cord, the autonomic nuclei lie in the lateral gray horns of spinal segments S_2–S_4.

2. ***Ganglionic neurons in peripheral ganglia located within or adjacent to the target organs.*** The preganglionic fibers of the parasympathetic division do not diverge as extensively as do those of the sympathetic division. A typical preganglionic fiber synapses on six to eight ganglionic neurons. In contrast to the pattern in the sympathetic division, all these ganglionic neurons are located in the same ganglion and their postganglionic fibers influence the same target organ. As a result, the effects of parasympathetic stimulation are more specific and localized than those of the sympathetic division.

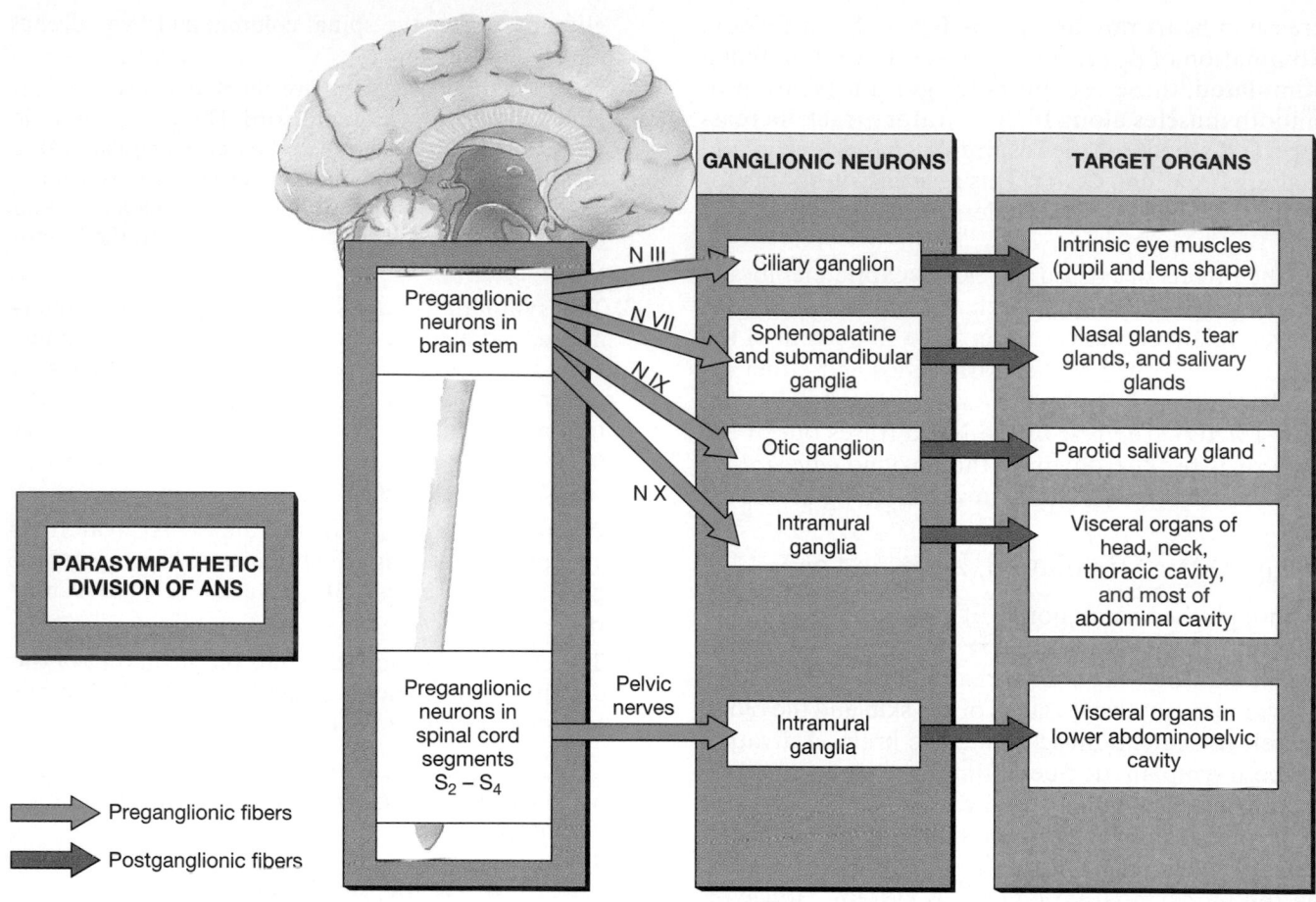

●**FIGURE 16-8** Organization of the Parasympathetic Division of the ANS

Organization and Anatomy of the Parasympathetic Division

Parasympathetic preganglionic fibers leave the brain as components of cranial nerves III (oculomotor), VII (facial), IX (glossopharyngeal), and X (vagus) (Figure 16-9●). These fibers carry the cranial parasympathetic output. Parasympathetic fibers in the oculomotor, facial, and glossopharyngeal nerves control visceral structures in the head. These fibers synapse in the *ciliary, sphenopalatine, submandibular*, and *otic ganglia.* ∞ *[pp. 478–479]* Short postganglionic fibers then continue to their peripheral targets. The vagus nerve provides preganglionic parasympathetic innervation to structures in the thoracic and abdominopelvic cavity as distant as the last segments of the large intestine. The vagus nerve alone provides roughly 75 percent of all parasympathetic outflow.

The preganglionic fibers in the sacral segments of the spinal cord carry the sacral parasympathetic output. These fibers do not join the ventral roots of the spinal nerves. Instead, the preganglionic fibers form distinct **pelvic nerves,** which innervate intramural ganglia in the kidney and urinary bladder, the terminal portions of the large intestine, and the sex organs.

General Functions of the Parasympathetic Division

The following is a partial listing of the major effects produced by the parasympathetic division:

- Constriction of the pupils, to restrict the amount of light that enters the eyes, and focusing the eyes on nearby objects.
- Secretion by digestive glands, including salivary glands, gastric glands, duodenal glands, intestinal glands, pancreas, and liver.
- Secretion of hormones that promote the absorption and utilization of nutrients by peripheral cells.
- Increased smooth muscle activity along the digestive tract.
- Stimulation and coordination of defecation.
- Contraction of the urinary bladder during urination.
- Constriction of the respiratory passageways.
- Reduction in heart rate and in the force of contraction.
- Sexual arousal and stimulation of sexual glands in both genders.

These functions center on relaxation, food processing, and energy absorption. The parasympathetic division has been called the *anabolic system* because its

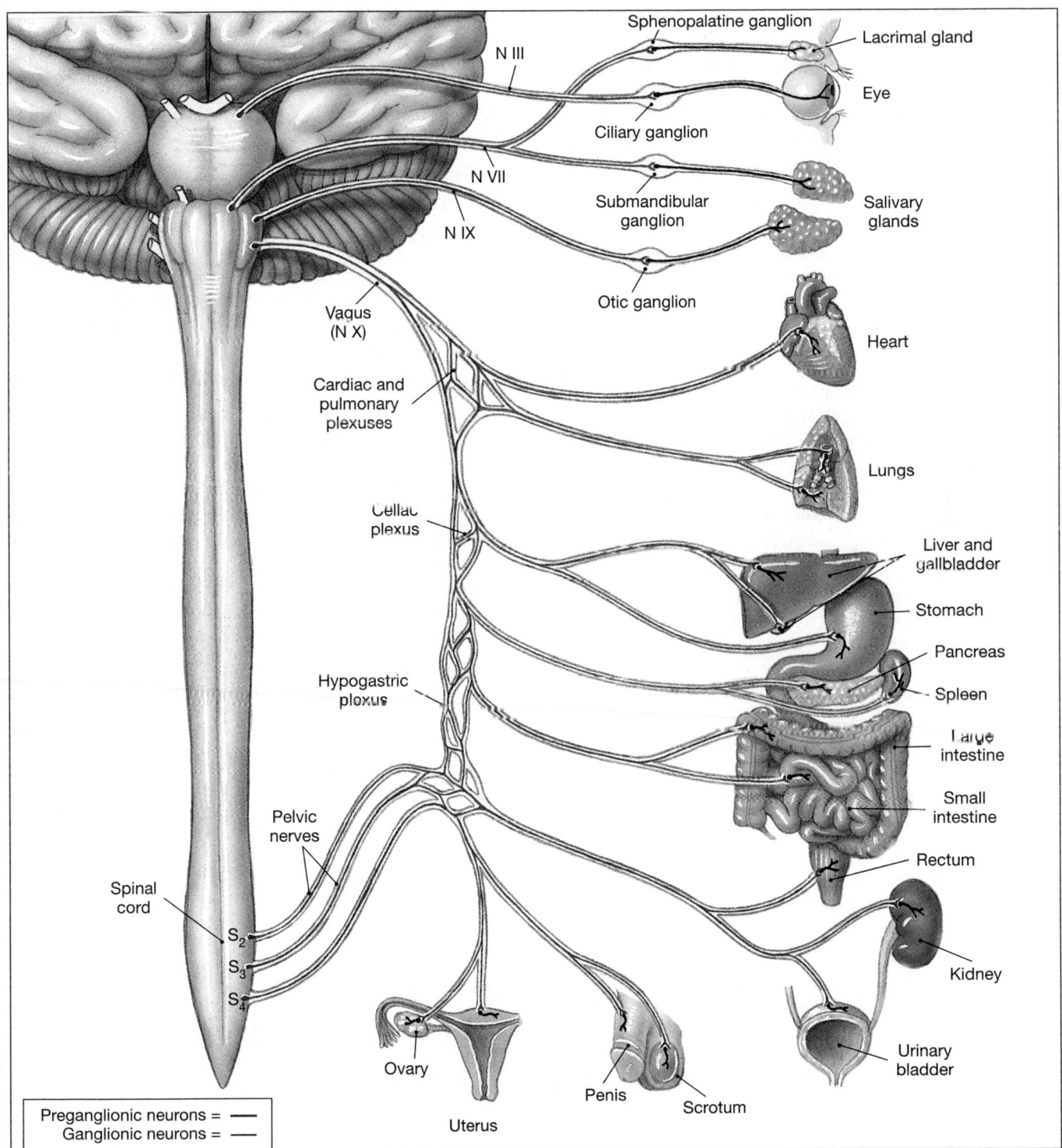

•FIGURE 16-9 Distribution of Parasympathetic Innervation

stimulation leads to a general increase in the nutrient content of the blood. (*Anabolic* comes from the Greek word *anabole*, which means "a rising up." ∞ *[p. 39]*) Cells throughout the body respond to this increase by absorbing nutrients and using them to support growth, cell division, and the creation of energy reserves in the form of lipids or glycogen.

Parasympathetic Activation and Neurotransmitter Release

All the preganglionic and postganglionic fibers in the parasympathetic division release ACh at synapses and neuroeffector junctions. The neuroeffector junctions are small and have narrow synaptic clefts. The effects of stimulation are short-lived, because most of the ACh released is inactivated by acetylcholinesterase (AChE) within the synapse. Any ACh diffusing into the surrounding tissues will be inactivated by the enzyme *tissue cholinesterase.* As a result, the effects of parasympathetic stimulation are quite localized, and they last a few seconds at most.

Membrane Receptors and Responses

Although all the synapses (neuron to neuron) and neuroeffector junctions (neuron to effector) of the parasympathetic division use the same transmitter, ACh, two different types of ACh receptors occur on the postsynaptic membranes:

1. **Nicotinic** (nik-ō-TIN-ik) **receptors** are located on the surfaces of ganglion cells of both the parasympathetic and sympathetic divisions as well as at neuromuscular junctions of the somatic motor system. Exposure to ACh always causes excitation of the ganglionic neuron or muscle fiber via the opening of membrane ion channels.

2. **Muscarinic** (mus-kar-IN-ik) **receptors** are located at cholinergic neuroeffector junctions in the parasympathetic division as well as at the few cholinergic neuroeffector junctions in the sympathetic division. Stimulation of muscarinic receptors produces longer-lasting effects than does stimulation of nicotinic receptors. The response, which reflects the activation or inactivation of specific enzymes, may be excitatory or inhibitory.

The names *nicotinic* and *muscarinic* indicate the chemical compounds that stimulate these receptor sites. Nicotinic receptors bind *nicotine,* a powerful toxin that can be obtained from a variety of sources, including tobacco leaves. Muscarinic receptors are stimulated by *muscarine,* a toxin produced by some poisonous mushrooms.

These compounds have discrete actions, targeting either the autonomic ganglia and skeletal neuromuscular junctions (nicotine) or the parasympathetic neuroeffector junctions (muscarine). They produce dangerously exaggerated, uncontrolled responses that parallel those produced by normal receptor stimulation. For example, nicotine poisoning occurs if as little as 50 mg of the compound is ingested or absorbed through the skin. The symptoms reflect widespread autonomic activation—vomiting, diarrhea, high blood pressure, rapid heart rate, sweating, and profuse salivation. Because the neuromuscular junctions of the somatic motor system are stimulated, convulsions occur. In severe cases, stimulation of nicotinic receptors inside the CNS may lead to coma and death within minutes. The symptoms of muscarine poisoning are almost entirely restricted to the parasympathetic division: salivation, nausea, vomiting, diarrhea, constric-

tion of respiratory passages, low blood pressure, and an abnormally slow heart rate.

Table 16-1 summarizes details about the adrenergic and cholinergic receptors of the ANS.

A Summary of the Parasympathetic Division

To summarize our discussion of the parasympathetic division:

1. The parasympathetic division includes visceral motor nuclei associated with four cranial nerves (III, VII, IX, and X) and with sacral segments S_2–S_4.

2. The ganglionic neurons are located within or next to their target organs.

3. The parasympathetic division innervates areas serviced by the cranial nerves and organs in the thoracic and abdominopelvic cavities.

4. All parasympathetic neurons are cholinergic. Ganglionic neurons have nicotinic receptors, which are excited by ACh. Muscarinic receptors present at neuroeffector junctions may produce either excitation or inhibition, depending on the nature of the enzymes activated when ACh binds to the receptor.

5. The effects of parasympathetic stimulation are generally brief and restricted to specific organs and sites.

☑ Which nerve is responsible for parasympathetic innervation of the lungs, heart, stomach, liver, pancreas, and parts of the small and large intestines?

☑ What effect would stimulation of muscarinic receptors in cardiac muscle have on the heart?

☑ Why is the parasympathetic division of the ANS sometimes referred to as the anabolic system?

INTERACTIONS OF THE SYMPATHETIC AND PARASYMPATHETIC DIVISIONS

The sympathetic division has widespread impact, reaching visceral organs and tissues throughout the body. The parasympathetic division innervates only visceral structures that are serviced by the cranial nerves or that lie within the abdominopelvic cavity. Although some organs are innervated by one division or the other, most vital organs receive **dual innervation**—that is, they receive instructions from both the sympathetic and the parasympathetic divisions. Where dual innervation exists, the two divisions commonly have opposing effects. Dual innervation with op-

TABLE 16-1 Adrenergic and Cholinergic Receptors of the ANS

Receptor	Location	Response	Mechanism
Adrenergic Receptors			
α_1	Widespread, found in most tissues	Excitation, stimulation of metabolism	Activation of enzymes, release of intracellular Ca^{2+}
α_2	Sympathetic neuroeffector junctions	Inhibition of effector cell	Reduction of cAMP concentrations
	Parasympathetic neuroeffector junctions	Inhibition of neurotransmitter release	Reduction of cAMP concentrations
β_1	Heart, kidneys, liver, adipose tissue*	Stimulation, increased energy consumption	Enzyme activation
β_2	Smooth muscle in vessels of heart and skeletal muscle; smooth muscle layers in the intestines, lungs, and bronchi	Inhibition, relaxation	Enzyme activation
Cholinergic Receptors			
Nicotinic	All autonomic synapses between preganglionic and ganglionic neurons; neuromuscular junctions of the SNS	Stimulation, excitation; muscular contraction	Opening of chemically regulated Na^+ channels
Muscarinic	All parasympathetic neuroeffector junctions; cholinergic sympathetic neuroeffector junctions	Variable	Enzyme activation causing changes in membrane permeability to K^+

*Adipocytes also contain an additional receptor type, β_3, not found in other tissues.

posing effects is most evident in the digestive tract, the heart, and the lungs. At other sites, the responses may be divergent or complementary. Secretory control of the salivary glands and the sexual functions of the male reproductive tract are other examples.

Anatomy of Dual Innervation

In the head, parasympathetic postganglionic fibers from the ciliary, sphenopalatine, submandibular, and otic ganglia accompany the cranial nerves to their peripheral destinations. The sympathetic innervation reaches the same structures by traveling directly from the superior cervical ganglia of the sympathetic chain.

In the thoracic and abdominopelvic cavities, the sympathetic postganglionic fibers mingle with parasympathetic preganglionic fibers at a series of *autonomic plexuses*: the *cardiac plexus,* the *pulmonary plexus,* the *esophageal plexus,* the *celiac plexus,* the *inferior mesenteric plexus,* and the *hypogastric plexus* (Figure 16-10•). Nerves leaving these networks travel with the blood vessels and lymphatics that supply visceral organs.

Autonomic fibers entering the thoracic cavity intersect at the **cardiac plexus** and the **pulmonary plexus.** These plexuses contain sympathetic and parasympathetic fibers bound for the heart and lungs, respectively, as well as the parasympathetic ganglia whose output

affects those organs. The **esophageal plexus** contains descending branches of the vagus nerve and splanchnic nerves leaving the sympathetic chain on either side.

Parasympathetic preganglionic fibers of the vagus nerve enter the abdominopelvic cavity with the esophagus. There the fibers join the network of the **celiac plexus,** also known as the *solar plexus.* The celiac plexus and associated smaller plexuses, such as the **inferior mesenteric plexus,** innervate viscera down to the initial segments of the large intestine. The **hypogastric plexus** contains the parasympathetic outflow of the pelvic nerves, sympathetic postganglionic fibers from the inferior mesenteric ganglion, and splanchnic nerves from the sacral sympathetic chain. This plexus innervates the digestive, urinary, and reproductive organs of the pelvic cavity.

A Comparison of the Sympathetic and Parasympathetic Divisions

Figure 16-11• and Table 16-2 (p. 529) compare key features of the sympathetic and parasympathetic divisions of the ANS. The distinctions have physiological and functional correlates. Table 16-3 (p. 530) provides a more detailed comparison, taking into account the effect of sympathetic or parasympathetic activity on specific organs and systems.

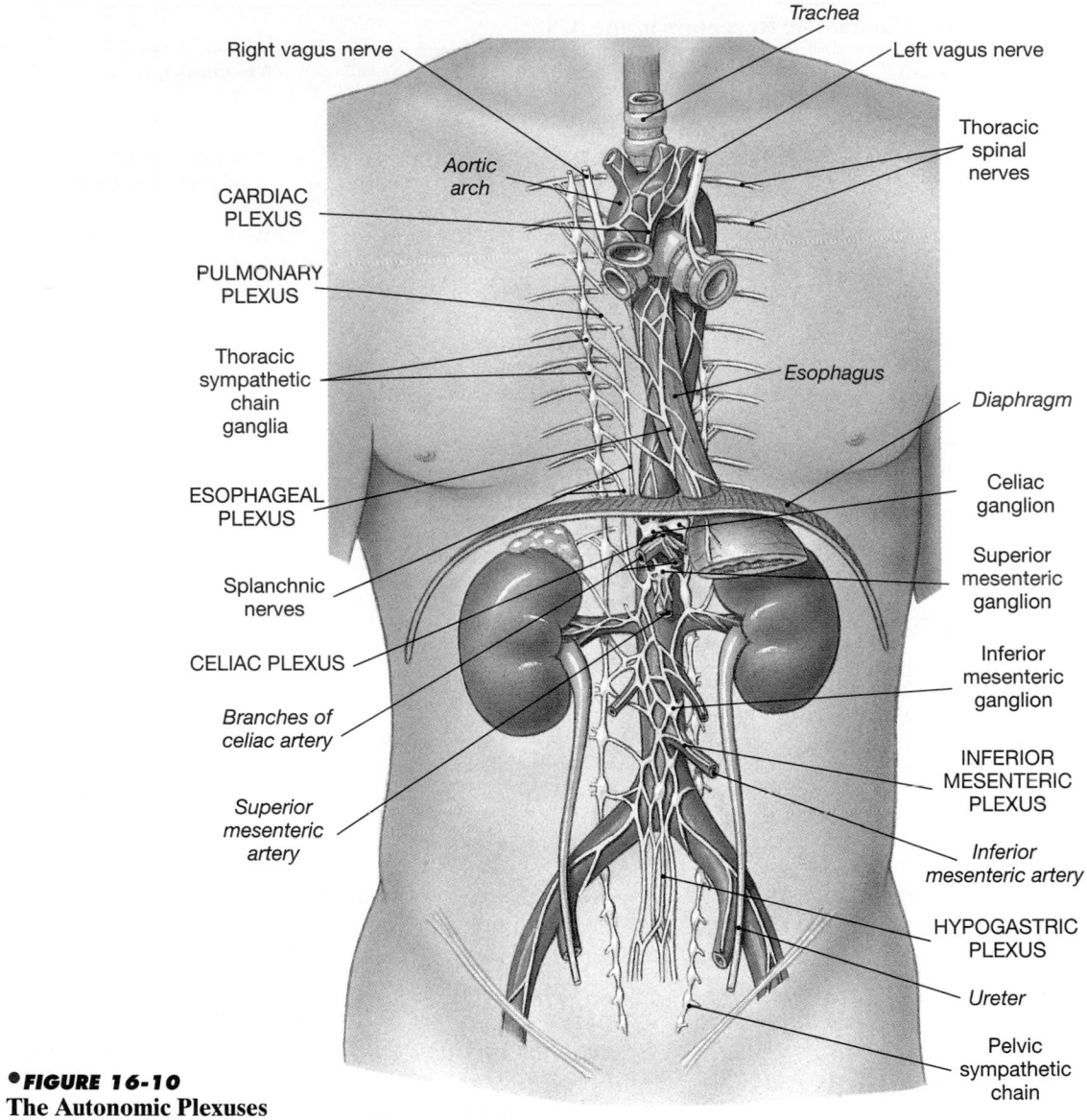

●**FIGURE 16-10**
The Autonomic Plexuses

Autonomic Tone

Even in the absence of stimuli, autonomic motor neurons show a resting level of spontaneous activity. The level of activation determines an individual's **autonomic tone**. Autonomic tone is an important aspect of ANS function, just as muscle tone is a key aspect of muscular function. If a nerve is absolutely inactive under normal conditions, then all it can do is increase its activity on demand. But if the nerve maintains a background level of activity, it may either increase or decrease its activity, providing a range of control options.

Autonomic tone is significant where dual innervation occurs and the ANS divisions have opposing effects. It is even more important in situations in which dual innervation does not occur. To demonstrate how autonomic tone affects ANS function, we will consider one example of each arrangement.

Autonomic Tone in the Presence of Dual Innervation

The heart is an organ that receives dual innervation. The two autonomic divisions have opposing effects on heart function. Acetylcholine released by postganglionic fibers of the parasympathetic division causes a reduction in heart rate, whereas NE released by varicosities of the sympathetic division accelerates the heart rate. Because autonomic tone exists, small amounts of both of these neurotransmitters are released continuously. By means of small adjustments in parasympathetic stimulation versus sympathetic stimulation, the heart rate can be controlled very precisely. In a crisis, the stimulation of the sympathetic innervation and the inhibition of the parasympathetic innervation accelerate the heart rate to the maximum extent possible.

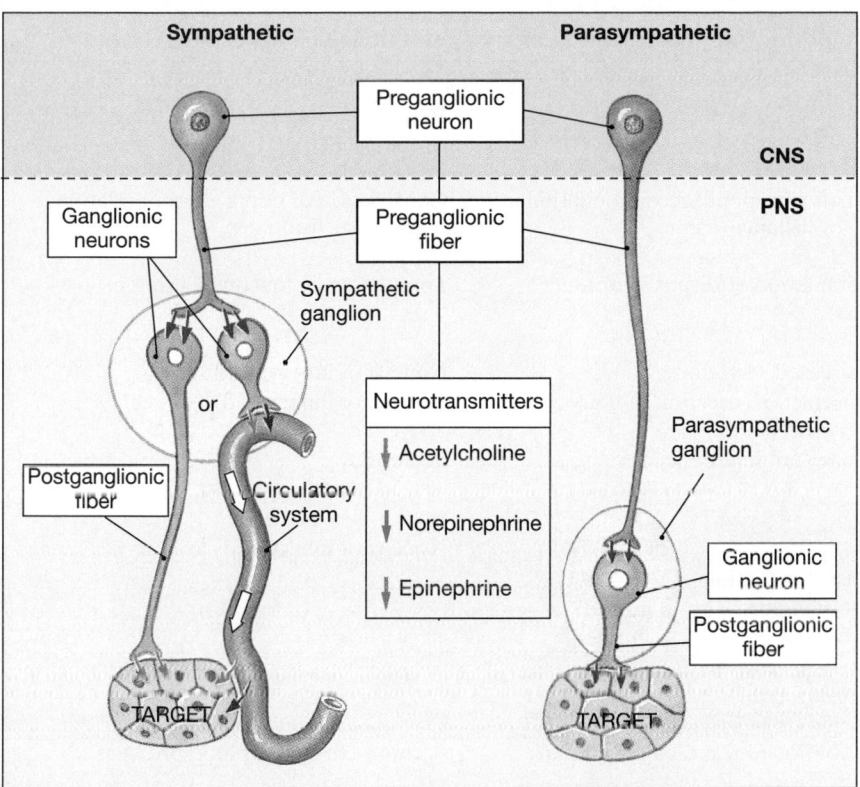

FIGURE 16-11 A Diagrammatic Summary of the Anatomical Differences between the Sympathetic and Parasympathetic Divisions

TABLE 16-2 A Structural Comparison of the Sympathetic and Parasympathetic Divisions of the ANS

Characteristic	Sympathetic Division	Parasympathetic Division
Location of CNS visceral motor neuron	Lateral gray horns and spinal segments T_1–L_2	Brain stem and spinal segments S_2–S_4
Location of PNS ganglia	Near spinal column	Typically intramural
Preganglionic fibers		
Length	Relatively short	Relatively long
Neurotransmitter released	Acetylcholine	Acetylcholine
Postganglionic fibers		
Length	Relatively long	Relatively short
Neurotransmitter released	Generally norepinephrine	Always acetylcholine
Neuroeffector junction	Varicosities and enlarged terminal knobs that release transmitter near target cells	Neuroeffector junctions that release transmitter to special receptor surface
Degree of divergence from CNS to ganglion cells	Approximately 1:32	Approximately 1:6
General function(s)	Stimulate metabolism; increase alertness; prepare for emergency ("fight or flight")	Promote relaxation, nutrient uptake, energy storage ("rest and repose")

Autonomic Tone in the Absence of Dual Innervation

Several organs are innervated by one division only. For example, most structures in the body wall, such as blood vessels, glands, adipose tissue, and skeletal muscles, receive only sympathetic innervation. Conversely, the constrictor muscles of the pupil, the tear glands, and nasal glands are innervated only by the parasympathetic division.

The sympathetic control of blood vessel diameter demonstrates how autonomic tone allows fine adjustment of peripheral activities when the target organ is not innervated by both ANS divisions. Sympathetic postganglionic fibers innervate the smooth muscle cells in the walls of peripheral vessels. The background sympathetic tone keeps these muscles partially contracted, so the vessels are ordinarily at roughly half their maximum diameter. Because the normal diameter is maintained by sympathetic tone, increasing or decreasing sympathetic stimulation provides precise control of vessel diameter over its entire range. When the vessel dilates, blood flow to the region increases; when the diameter decreases, blood flow is reduced.

TABLE 16-3 A Functional Comparison of the Sympathetic and Parasympathetic Divisions of the ANS

Structure	Sympathetic Receptor Type	Sympathetic Innervation Effect	Parasympathetic Innervation Effect (all muscarinic receptors)
Eye	α_1	Dilation of pupil; accommodation for distance vision	Constriction of pupil; accommodation for close vision
Salivary Glands	α_1, β_1	Stimulation of serous secretion	Stimulation of mucous secretion
Skin			
Sweat glands	α_1	Increased secretion	None (not innervated)
Arrector pili	α_1	Contraction, erection of hairs	None (not innervated)
Tear Glands		None (not innervated)	Secretion
Cardiovascular System			
Blood vessels			None (not innervated)
To integument	α_1	Vasoconstriction	
To skeletal muscles	β_2	Vasodilation	
To heart	β_2	Vasodilation	
To digestive viscera	α_1	Vasoconstriction	
Veins	α_1, β_1	Constriction	
Heart	α_1, β_1	Increased heart rate, force of contraction, and blood pressure	Decreased heart rate, force of contraction, and blood pressure
Adrenal Gland		Secretion of epinephrine, norepinephrine by medulla	None (not innervated)
Posterior Pituitary	β_1	Secretion of ADH	None (not innervated)
Respiratory System			
Airways	β_2	Increased airway diameter	Decreased airway diameter
Digestive System			
Sphincters	α_1	Constriction	Dilation
General level of activity	α_2, β_2	Decreased	Increased
Secretory glands	α_2	Inhibition	Stimulation
Liver	α_1, β_2	Glycogen breakdown, glucose synthesis and release	Glycogen synthesis
Pancreas	α_1	Decreased exocrine secretion	Increased exocrine secretion
	α_2	Decreased hormone (insulin) secretion	Increased hormone (insulin) secretion
Skeletal Muscles	β_2	Increased force of contraction, glycogen breakdown	None (not innervated)
	α_2	Facilitation of ACh release at neuromuscular junction	None (not innervated)
Adipose Tissue	β_1, β_3	Lipolysis, fatty acid release	
Urinary System			
Kidneys	β_2	Decreased urine production	Increased urine production
Urinary bladder	α_1, β_2	Constriction of internal sphincter, relaxation of urinary bladder	Tensing of urinary bladder, relaxation of internal sphincter to eliminate urine
Male Reproductive System	α_1	Increased glandular secretion and ejaculation	Erection
Female Reproductive System	α_1	Increased glandular secretion; contraction of pregnant uterus	Variable (depending on hormones present)
	β_2	Relaxation of nonpregnant uterus	Variable (depending on hormones present)

INTEGRATION AND CONTROL OF AUTONOMIC FUNCTIONS

Figure 15-7•, p. 502, diagrammed the relationships among centers involved in somatic motor control. The lowest level of regulatory control consists of the lower motor neurons involved in cranial and spinal reflex arcs. The highest level consists of the pyramidal motor neurons of the primary motor cortex, operating with the assistance of extrapyramidal and cerebellar nuclei.

The ANS is also organized into a series of interacting levels. At the bottom are visceral motor neurons in the lower brain stem and spinal cord that are involved in cranial and spinal visceral reflexes. **Visceral reflexes** provide automatic motor responses that can be modified, facilitated, or inhibited by higher centers, especially those of the hypothalamus.

For example, when a light is shined in one of your eyes, a visceral reflex constricts the pupils of *both* eyes (the *consensual light reflex,* described in Chapter 14). ∞ *[p. 485]* The visceral motor commands are distributed by parasympathetic fibers. In darkness, your pupils dilate; this *pupillary reflex* is directed by sympathetic postganglionic fibers. However, the motor nuclei directing pupillary constriction or dilation are also controlled by hypothalamic centers concerned with emotional states. When you are queasy or nauseated, your pupils constrict; when you are sexually aroused, your pupils dilate.

Visceral Reflexes

Each **visceral reflex arc** consists of a receptor, a sensory neuron, a processing center (interneuron or visceral motor neuron), and two visceral motor neurons (Figure 16-12•). All visceral reflexes are polysynaptic. **Long reflexes** are the autonomic equivalents of the spinal reflexes introduced in Chapter 13. ∞ *[p. 433]* In a long reflex, the processing steps occur within the CNS. Short reflexes involve interneurons within autonomic ganglia. Short reflexes control very simple motor responses with localized effects. For example, short reflexes control patterns of smooth muscle contraction that affect small segments of the digestive tract, whereas long reflexes coordinate the activities of entire organs. We will consider long and short reflexes in more detail when we examine the control of digestive and urinary functions in Chapters 24 and 26.

Visceral sensory neurons deliver information to the CNS along the dorsal roots of spinal nerves, within the sensory branches of cranial nerves, and within the autonomic nerves that innervate peripheral effectors. Figure 13-6b•, p. 425, diagrammed the distribution of visceral sensory fibers in a representative spinal nerve.

As we examine other body systems in later chapters, we will encounter many examples of autonomic reflexes involved in respiration, cardiovascular function, and other visceral activities. Some of the most important are previewed in Table 16-4. Note that the parasympathetic division participates in reflexes that affect individual or-

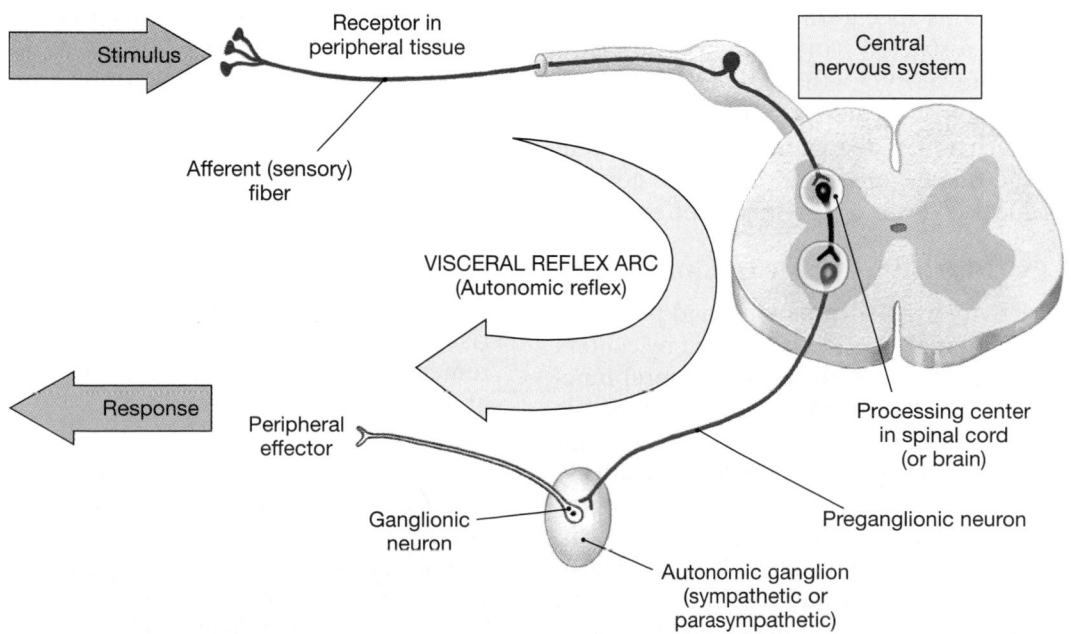

•**FIGURE 16-12 Visceral Reflexes.** Visceral reflexes have the same basic components as somatic reflexes, but all visceral reflexes are polysynaptic.

TABLE 16-4 **Representative Visceral Reflexes**

Reflex	Stimulus	Response	Comments
Parasympathetic Reflexes			
Gastric and intestinal reflexes (*Chapter 24*)	Pressure and physical contact with materials	Smooth muscle contractions that propel materials and mix with secretions	Via vagus nerve
Defecation (*Chapter 24*)	Distention of rectum	Relaxation of internal anal sphincter	Requires voluntary relaxation of external sphincter
Urination (*Chapter 26*)	Distention of urinary bladder	Contraction of the walls of the urinary bladder, relaxation of internal urethral sphincter	Urination requires voluntary relaxation of external urethral sphincter
Direct light and consensual light reflexes (*Chapter 14*)	Bright light shining in eye(s)	Constriction of pupils of both eyes	
Swallowing reflex (*Chapter 24*)	Movement of material into upper pharynx	Smooth muscle and skeletal muscle contractions	Coordinated by medullary swallowing center
Coughing reflex (*Chapter 23*)	Irritation of respiratory tract	Sudden explosive ejection of air	Coordinated by medullary coughing center
Baroreceptor reflex (*Chapters 17, 20, 21*)	Sudden rise in carotid blood pressure	Reduction in heart rate and force of contraction	Coordinated in cardiac center in medulla oblongata
Sexual arousal (*Chapter 28*)	Erotic stimuli (visual or tactile)	Increased glandular secretions, sensitivity	
Sympathetic Reflexes			
Cardioacceleratory reflex	Sudden decline in blood pressure in carotid artery	Increase in heart rate and force of contraction	Coordinated in cardiac center in medulla oblongata
Vasomotor reflexes (*Chapter 21*)	Changes in blood pressure in major arteries	Changes in diameter of peripheral vessels	Coordinated in vasomotor center in medulla oblongata
Pupillary reflex (*Chapter 17*)	Low light level reaching visual receptors	Dilation of pupil	
Ejaculation (in males) (*Chapter 28*)	Erotic stimuli (tactile)	Skeletal muscle contractions ejecting semen	

gans and systems. This specialization reflects the relatively specific and restricted pattern of innervation. In contrast, there are fewer sympathetic reflexes. The sympathetic division is typically activated as a whole, in part because it has such a high degree of divergence and in part because the release of hormones by the adrenal medullae produces widespread peripheral effects.

Higher Levels of Autonomic Control

The levels of activity in the sympathetic and parasympathetic divisions of the ANS are controlled by centers in the brain stem that deal with specific visceral functions. Figure 16-13● diagrams the levels of autonomic control. As in the somatic motor system, in the ANS simple reflexes based in the spinal cord provide relatively rapid and automatic responses to stimuli. More complex sympathetic and parasympathetic reflexes are coordinated by processing centers in the medulla oblongata. In addition to the cardiovascular and respiratory centers, the medulla contains centers and nuclei involved with salivation, digestive secretions, peristalsis, and urinary function. These medullary centers are in turn subject to regulation by the hypothalamus. ∞ *[pp. 468, 473]*

The term *autonomic* was originally applied to the visceral motor system because the regulatory centers were thought to function without regard for other CNS activities. This view has been drastically revised in light of subsequent research. Because the hypothalamus interacts with all other portions of the brain, activity in the limbic system, thalamus, or cerebral cortex can have dramatic effects on autonomic function. For example, when you become angry, your heart rate accelerates, your blood pressure rises, and your respiratory rate increases; when you remember your last big dinner, your stomach "growls" and your mouth waters.

☑ What effect would loss of sympathetic tone have on blood flow to a tissue?

☑ What physiological changes would you expect to observe in a patient who is about to have a root canal performed and who is quite anxious about the procedure?

☑ Harry has a brain tumor that is interfering with the function of his hypothalamus. Would you expect this tumor to interfere with autonomic function? Why or why not?

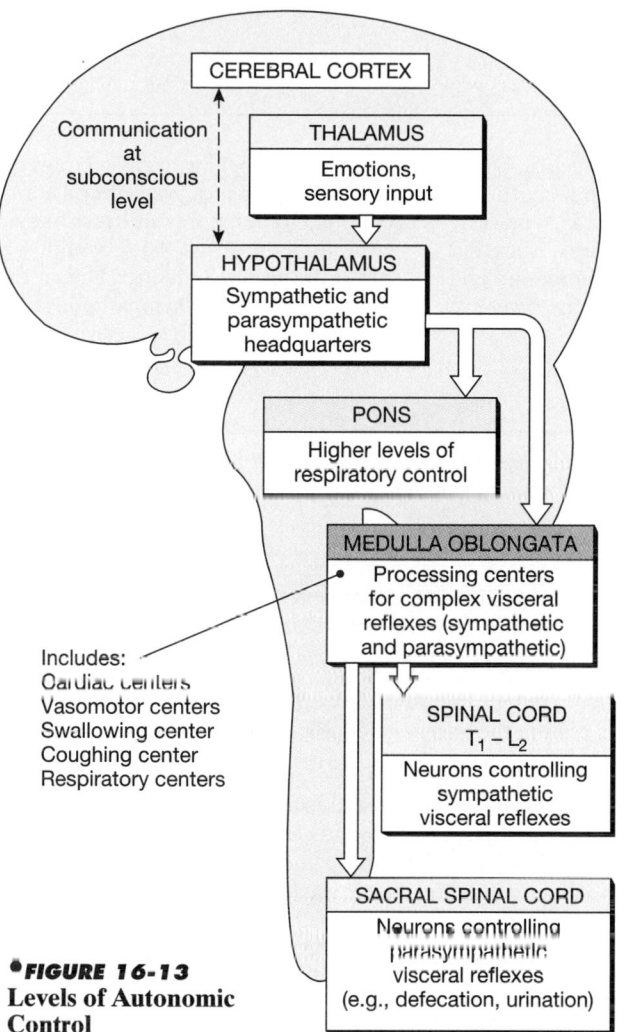

•FIGURE 16-13
Levels of Autonomic Control

 **MODIFYING AUTONOMIC NERVOUS SYSTEM FUNCTION** Drugs may be administered to counteract or reduce symptoms caused or aggravated by autonomic activities. These drugs are called **mimetic** if they mimic the activity of one of the normal autonomic transmitters. Drugs that reduce the effects of autonomic stimulation by keeping the neurotransmitter from affecting the postsynaptic membranes are known as **blocking agents**. Mimetic drugs commonly have advantages over natural neurotransmitters. For example, norepinephrine and epinephrine must be administered into the bloodstream by injection or infusion, and the side effects are short-lived. In contrast, **sympathomimetic drugs** may survive oral administration, produce longer-lasting effects, or have more-specific actions than does E or NE. Examples include drugs applied topically to reduce hemorrhaging, by spray to reduce nasal congestion, by inhalation to dilate the respiratory passageways, or in drops to dilate the pupils.

Sympathetic blocking agents bind to the receptor sites and prevent a normal response to the presence of neurotransmitters or sympathomimetic drugs. **Alpha-blockers** eliminate the peripheral vasoconstriction that accompanies sympathetic stimulation, and **beta-blockers** are effective and clinically useful for treating chronic high blood pressure and other forms of cardiovascular disease.

Parasympathomimetic drugs may also be used to increase the activity along the digestive tract and to encourage defecation and urination. **Parasympathetic blocking agents** target the muscarinic receptors at the neuroeffector junctions. These drugs have diverse effects, but they are often used to control the diarrhea and cramps associated with various forms of food poisoning. For information on specific mimetic drugs and blocking agents and details on their modes of action, see the *Applications Manual*. **AM** *Pharmacology and the Autonomic Nervous System*

Biofeedback

Although conscious thought processes affect the ANS, you normally do not perceive the effect because visceral sensory information does not reach your cerebral cortex. Even when a conscious mental process triggers a physiological shift, such as a change in blood pressure, sweat gland activity, skin temperature, or muscle tone, the information never arrives at your sensory cortex. **Biofeedback** is an attempt to bridge this gap. In this technique, a person's physiological processes are monitored, and a visual or auditory signal is used to alert the subject when a particular change takes place. These signals let the subject know when ongoing conscious thought processes have triggered a desirable change in autonomic function. For example, when biofeedback is used to regulate blood pressure, a light or tone informs the subject

when blood pressure drops. With practice, some people can learn to re-create the mood or thought pattern that will lower their blood pressure.

Biofeedback techniques have been used to promote conscious control of blood pressure, heart rate, circulatory pattern, skin temperature, brain waves, and so forth. By reducing stress, lowering blood pressure, and improving circulation, these techniques can reduce the severity of clinical symptoms. In the process, they also lower the risks for serious complications, such as heart attacks or strokes, in patients with high blood pressure. Unfortunately, not everyone can learn to influence their autonomic functions, and the combination of variable results and expensive equipment makes biofeedback unsuitable for widespread application.

SELECTED CLINICAL TERMINOLOGY

Terms Discussed in This Chapter

alpha-blockers: Drugs that eliminate the peripheral vasoconstriction that accompanies sympathetic stimulation. *(p. 533 and AM)*

beta-blockers: Drugs that decrease heart rate and force of contraction, lowering peripheral blood pressure. *(p. 533 and AM)*

parasympathetic blocking agents: Drugs that target the muscarinic receptors at neuroeffector junctions. *(p. 533 and AM)*

parasympathomimetic drugs: Drugs that mimic parasympathetic stimulation and increase the activity along the digestive tract. *(p. 533 and AM)*

sympathetic blocking agents: Drugs that bind to receptor sites, preventing a normal response to neurotransmitters or sympathomimetic drugs. *(p. 533 and AM)*

sympathomimetic drugs: Drugs that mimic the effects of sympathetic stimulation. *(p. 533 and AM)*

CHAPTER REVIEW

 On-line resources for this chapter are on our World Wide Web site at: http://www.prenhall.com/martini/fap

STUDY OUTLINE

INTRODUCTION, p. 514

1. The *autonomic nervous system (ANS)* coordinates cardiovascular, respiratory, digestive, excretory, and reproductive functions.

AN OVERVIEW OF THE ANS, p. 514

1. **Preganglionic neurons** in the CNS send axons to synapse on **ganglionic neurons** in **autonomic ganglia** outside the CNS. *(Figure 16-1)*

Subdivisions of the ANS, p. 514

2. Visceral efferents from the thoracic and lumbar segments form the *thoracolumbar division,* or **sympathetic division** ("fight or flight" system) of the ANS. Visceral efferents leaving the brain and sacral segments form the *craniosacral division,* or **parasympathetic division** ("rest and repose" system). *(Figure 16-2)*

THE SYMPATHETIC DIVISION, p. 516

1. The sympathetic division consists of preganglionic neurons between segments T_1 and L_2, ganglionic neurons in ganglia near the vertebral column, and specialized neurons inside the adrenal glands. *(Figures 16-3–16-5)*
2. There are two types of sympathetic ganglia: **sympathetic chain ganglia** *(paravertebral ganglia)* and **collateral ganglia** *(prevertebral ganglia). (Figures 16-3, 16-4)*

The Sympathetic Chain, p. 517

3. **Postganglionic fibers** targeting structures in the body wall and limbs rejoin the spinal nerves and reach their destinations by way of the dorsal and ventral rami. *(Figure 16-4)*
4. Postganglionic fibers targeting structures in the thoracic cavity form **sympathetic nerves** that go directly to their visceral destination. Preganglionic fibers run between the sympathetic chain ganglia and interconnect them. *(Figure 16-4)*

Collateral Ganglia, p. 520

5. The abdominopelvic viscera receive sympathetic innervation via preganglionic fibers that synapse within collateral ganglia. The preganglionic fibers that innervate the collateral ganglia form the **splanchnic nerves.** *(Figure 16-4)*
6. The **celiac ganglion** innervates the stomach, liver, pancreas, and spleen; the **superior mesenteric ganglion** innervates the small intestine and initial segments of the large intestine; and the **inferior mesenteric ganglion** innervates the kidney,

urinary bladder, sex organs, and terminal portions of the large intestine. *(Figures 16-5, 16-10)*

The Adrenal Medullae, p. 520

7. Preganglionic fibers entering an adrenal gland synapse within the **adrenal medulla.** *(Figures 16-4, 16-5)*

Sympathetic Activation, p. 521

8. In a crisis, the entire division responds, an event called **sympathetic activation.** Its effects include increased alertness, a feeling of energy and euphoria, increased cardiovascular and respiratory activity, and general elevation in muscle tone.

Neurotransmitters and Sympathetic Function, p. 521

9. Stimulation of the sympathetic division has two distinctive results: the release of *norepinephrine (NE)* at specific locations and secretion of *epinephrine (E)* and norepinephrine into the general circulation. *(Figure 16-6)*
10. There are two types of sympathetic receptors: **alpha receptors** (which respond to NE or E by depolarizing the membrane) and **beta receptors** (which are particularly sensitive to E). *(Figure 16-7)*
11. Most postganglionic fibers are *adrenergic;* a few are *cholinergic* or *nitroxidergic.* Postganglionic fibers innervating sweat glands and blood vessels to skeletal muscles release ACh. Postganglionic fibers innervating smooth muscle in the walls of small arteries in skeletal muscles and the brain release *nitric oxide (NO),* promoting vasodilation.

A Summary of the Sympathetic Division, p. 523

12. Preganglionic sympathetic fibers are relatively short. Except for those of the adrenal medullae, postganglionic fibers are quite long. Each fiber typically synapses with many ganglionic neurons in different ganglia. A single effector cell may have more than one type of receptor, so that the result of sympathetic stimulation depends on interaction among the receptors.

THE PARASYMPATHETIC DIVISION, p. 523

Organization and Anatomy of the Parasympathetic Division, p. 524

1. The parasympathetic division includes preganglionic neurons in the brain stem and sacral segments of the spinal cord and ganglionic neurons in peripheral ganglia located within or next to target organs. *(Figure 16-8)*

2. Preganglionic fibers leaving the sacral segments form **pelvic nerves.** *(Figure 16-9)*

General Functions of the Parasympathetic Division, p. 524

3. The effects produced by the parasympathetic division center on relaxation, food processing, and energy absorption.

Parasympathetic Activation and Neurotransmitter Release, p. 525

4. All parasympathetic preganglionic and postganglionic fibers release ACh at synapses and neuroeffector junctions. The effects are short-lived because of the actions of cholinesterase.

5. Two different ACh receptors are found in postsynaptic membranes. Stimulation of **muscarinic receptors** produces a longer-lasting effect than does stimulation of **nicotinic receptors.** *(Table 16-1)*

A Summary of the Parasympathetic Division, p. 526

6. The parasympathetic division innervates areas serviced by cranial nerves and organs in the thoracic and abdominopelvic cavities. All preganglionic and postganglionic parasympathetic neurons are cholinergic, and the effects of stimulation are usually brief and restricted to specific sites.

INTERACTIONS OF THE SYMPATHETIC AND PARASYMPATHETIC DIVISIONS, p. 526

1. The sympathetic division has widespread influence on visceral and somatic structures.

2. The parasympathetic division innervates only visceral structures serviced by cranial nerves or lying within the abdominopelvic cavity. Organs with **dual innervation** receive input from both divisions.

Anatomy of Dual Innervation, p. 527

3. In body cavities, the parasympathetic and sympathetic nerves intermingle to form a series of characteristic nerve plexuses (nerve networks), which include the **cardiac, pulmonary, esophageal, celiac, inferior mesenteric,** and **hypogastric plexuses.** *(Figure 16-10)*

A Comparison of the Sympathetic and Parasympathetic Divisions, p. 527

4. There are important physiological and functional differences between the sympathetic and parasympathetic divisions. *(Figure 16-11; Tables 16-2, 16-3)*

5. Parasympathetic effects include decreased metabolic rate, heart rate, and blood pressure; salivary and digestive gland secretion; increased digestive tract activity; and urination and defecation.

6. Sympathetic effects include mental alertness; increased metabolic rate; suspended digestive and urinary tract function; activation of energy reserves; increased respiratory rate and efficiency; increased heart rate and blood pressure; and activation of sweat glands.

Autonomic Tone, p. 528

7. Even when stimuli are absent, autonomic motor neurons show a resting level of activation, the **autonomic tone.**

INTEGRATION AND CONTROL OF AUTONOMIC FUNCTIONS, p. 531

Visceral Reflexes, p. 531

1. **Visceral reflexes** are the simplest function of the ANS. *(Figure 16-12)*

2. Parasympathetic reflexes include (1) gastric and intestinal reflexes, (2) defecation, (3) urination, (4) direct light and consensual light reflexes, (5) swallowing reflex, (6) coughing reflex, (7) baroreceptor reflex, and (8) sexual arousal. Sympathetic reflexes include (1) cardioacceleratory reflex, (2) vasomotor reflexes, (3) pupillary reflex, and (4) ejaculation (in males). *(Table 16-4)*

Higher Levels of Autonomic Control, p. 532

3. The hypothalamus receives input from many other portions of the brain, including the cerebral cortex. Centers in the hypothalamus are concerned with the coordination of sympathetic and parasympathetic function. The regulatory activities occur outside conscious awareness and control. *(Figure 16-13)*

REVIEW QUESTIONS

Level 1   Reviewing Facts and Terms

1. There is always a synapse between the CNS and the peripheral effector in the
  (a) autonomic nervous system
  (b) somatic nervous system
  (c) reflex arc
  (d) a, b, and c are correct

2. The preganglionic fibers of the sympathetic division of the ANS originate in the
  (a) cerebral cortex of the brain
  (b) medulla oblongata
  (c) craniosacral region of the brain stem and spinal cord
  (d) thoracolumbar region of the spinal cord

3. All preganglionic autonomic fibers release _____ at their synaptic terminals, and the effects are always _____:
  (a) norepinephrine; inhibitory
  (b) norepinephrine; excitatory
  (c) acetylcholine; excitatory
  (d) acetylcholine; inhibitory

4. Ganglionic neurons that innervate tissues and organs in the abdominopelvic cavity are located in
  (a) sympathetic chain ganglia
  (b) collateral ganglia
  (c) paravertebral ganglia
  (d) lateral ganglia

5. In the dorsal wall of the abdominal cavity, preganglionic fibers that innervate the collateral ganglia form the
  (a) splanchnic nerve
  (b) phrenic nerve
  (c) sciatic nerve
  (d) adrenal medulla

6. Postganglionic fibers from the celiac ganglion innervate the
  (a) heart, lungs, thymus, and diaphragm
  (b) cerebrum, pons, medulla, and cerebellum
  (c) stomach, liver, pancreas, and spleen
  (d) large intestine, kidney, urinary bladder, and sex organs

7. The effect of the neurotransmitter on the target cell depends on the nature of the
  (a) neurotransmitter
  (b) receptor on the presynaptic membrane
  (c) receptor on the postsynaptic membrane
  (d) a, b, and c are correct

8. Approximately 75 percent of parasympathetic outflow is provided by the
  (a) vagus nerve
  (b) sciatic nerve
  (c) glossopharyngeal nerves
  (d) pelvic nerves

9. The neurotransmitter at all synapses and neuroeffector junctions in the parasympathetic division of the ANS is
  (a) epinephrine   (b) norepinephrine
  (c) cyclic-AMP   (d) acetylcholine

10. How does the emergence of sympathetic fibers from the spinal cord differ from the emergence of parasympathetic fibers?

11. Starting in the spinal cord, trace an impulse through the sympathetic division of the ANS until it reaches a target organ in the abdominopelvic region.

12. Which three collateral ganglia serve as origins for ganglionic neurons that innervate organs or tissues within the abdominopelvic region?

13. What two distinctive results are produced by the stimulation of sympathetic ganglionic neurons?

14. What is the difference between a cholinergic synapse and an adrenergic synapse?

15. Which four pairs of cranial nerves are associated with the cranial segment of the parasympathetic division of the ANS?

16. Which four ganglia serve as origins for postganglionic fibers that deal with the control of visceral structures in the head?

17. Which six plexuses in the thorax and abdominopelvic cavities innervate visceral organs, and what are the effects of sympathetic versus parasympathetic stimulation?

18. What are the components of a visceral reflex arc?

19. Which control centers in the brain stem influence the levels of activity in the sympathetic and parasympathetic divisions of the ANS?

Level 2   Reviewing Concepts

20. Autonomic tone in autonomic motor neurons exists because
  (a) ANS neurons from both divisions commonly innervate the same organ
  (b) ANS neurons rarely innervate the same organs as somatic motor neurons
  (c) ANS neurons are inactive unless stimulated by higher centers
  (d) ANS neurons are always active to some degree

21. Dual innervation refers to situations in which
  (a) vital organs receive instructions from sympathetic and parasympathetic fibers
  (b) the atria and ventricles of the heart receive autonomic stimulation from the same nerves
  (c) sympathetic and parasympathetic fibers have similar effects
  (d) a, b, and c are correct

22. Compare the general effects of the sympathetic and parasympathetic divisions of the ANS.

23. Why does sympathetic function remain intact even when there is damage to the ventral roots of the cervical spinal nerves?

24. Why is the adrenal medulla considered to be a modified sympathetic ganglion?

25. How does the result of stimulating alpha receptors differ from that of stimulating beta receptors?

26. Why are the effects of parasympathetic stimulation quite localized and of short duration?

27. Why is autonomic tone a significant part of ANS function?

28. You are alone in your home late at night when you hear what sounds like breaking glass. What physiological effects would this experience probably produce, and what would be their cause?

Level 3   Critical Thinking and Clinical Applications

29. In some very severe cases of stomach ulcers, the branches of the vagus nerve (N X) that lead to the stomach are surgically severed. How would this procedure help control the ulcers?

30. Mr. Martin is suffering from a condition known as ventricular tachycardia, in which his heart beats too quickly. Would you suggest an alpha-blocker or a beta-blocker to help alleviate his problem? Why?

31. Little Billy is stung on his cheek by a wasp. Because he is allergic to wasp venom, his throat begins to swell and his

respiratory passages constrict. Would acetylcholine or epinephrine be more helpful in relieving his symptoms? Why?

32. While studying the activity of smooth muscle in blood vessels, Shelly discovers that, when applied to muscle membrane, a molecule chemically similar to a neurotransmitter triggers an increase in intracellular calcium ions. Which neurotransmitter is the molecule mimicking, and to which receptors is it binding?

17

Sensory Function

*O*ur comprehension of the world around us is based on information provided by
our senses. For example, if we can see something, it must be there, because seeing is believing . . .
or is it? The more you study this image, the more confusing it becomes. We are so used to trusting what
we see that when we see something that is physically impossible, it shocks us; some people become dizzy
at the sight of an Escher drawing. This chapter examines the way receptors provide sensory information
as well as the sensory pathways that distribute this information to provide us with our senses of smell, taste,
vision, equilibrium, and hearing.

CHAPTER OUTLINE AND OBJECTIVES

Our knowledge of the world around us is limited to those characteristics that stimulate our sensory receptors. Although we may not realize it, our picture of the environment is incomplete. Colors we cannot distinguish guide insects to flowers; sounds we cannot hear and smells we cannot detect provide dolphins, dogs, and cats with important information about their surroundings.

What we *do* perceive varies considerably with the state of our nervous systems. For example, when sympathetic activation occurs, we experience heightened awareness of sensory information and hear sounds that would normally escape our notice. Yet when concentrating on a difficult problem, we may remain unaware of relatively loud noises. Finally, our perception of any stimulus reflects activity in the cerebral cortex, and that activity can be inappropriate. In cases of phantom limb pain, a person feels pain in a missing limb; during an epileptic seizure, an individual may experience sights, sounds, or smells that have no physical basis. ∞ *[pp. 495, 503]*

We begin this chapter by examining receptor function and basic concepts in sensory processing. We will then apply that information to each of the general and special senses.

RECEPTORS

Sensory receptors are specialized cells or cell processes that provide your central nervous system with information about conditions inside or outside your body. A sensory receptor detects an arriving stimulus and translates it into an action potential that can be conducted to the CNS. This translation process is called **transduction.**

Sensory receptors are the interface between your nervous system and your internal and external environments. The **general senses** of pain, temperature, touch, pressure, vibration, and proprioception (position sense) are distributed throughout the body. We introduced the ascending pathways that provide conscious perception of those sensations in Chapter 15. ∞ *[pp. 492–495]* The **special senses** are **olfaction** (smell), **vision** (sight), **gustation** (taste), **equilibrium** (balance), and **hearing.** These sensations are provided by receptors that are structurally more complex than those of the general senses. Special sensory receptors are located in **sense organs** such as the eye or ear, where they are protected by surrounding tissues. The information provided by these receptors is distributed to specific areas of the cerebral cortex (the auditory cortex, the visual cortex, and so forth) and to centers throughout the brain stem. AM *Analyzing Sensory Disorders*

Specificity

Each receptor has a characteristic sensitivity. For example, a touch receptor is very sensitive to pressure but relatively insensitive to chemical stimuli, whereas a taste receptor is sensitive to dissolved chemicals but insensitive to pressure. This concept is called **receptor specificity.**

Specificity may result from the structure of the receptor cell or from the presence of accessory cells or structures that shield the receptor cell from other stimuli. The simplest receptors are the dendrites of sensory neurons. The dendritic processes, called **free nerve endings,** are not protected by accessory structures. They can be stimulated by many different stimuli (Figure 17-1a●). For example, free nerve endings that respond to tissue damage by providing pain sensations may be stimulated by chemical stimulation, pressure, temperature changes, or trauma. Complex receptors, such as the eye's visual receptors, are protected by accessory cells and connective tissue layers. These cells are seldom exposed to any stimulus other than light and so provide very specific information.

The area monitored by a single receptor cell is its **receptive field** (Figure 17-1b●). Whenever a sufficiently strong stimulus arrives in the receptive field, the CNS receives the information "stimulus arriving at receptor X." The larger the receptive field, the poorer your ability to localize a stimulus. For example, a touch receptor on the general body surface may have a receptive field 7 cm (2.5 in.) in diameter. As a result, you can describe a light touch there as affecting only a general area, not an exact spot. On the tongue or fingertips, where the receptive fields are less than a millimeter in diameter, you can be very precise about the location of a stimulus.

An arriving stimulus can take many different forms. It may be a physical force (such as pressure), a dissolved chemical, a sound, or a light. However, regardless of the nature of the stimulus, sensory information must be sent to the CNS in the form of action potentials, which are electrical events.

Transduction

Transduction can be divided into three steps:

STEP 1: *An arriving stimulus alters the transmembrane potential of the receptor membrane.*

The nature of the interaction between stimulus and receptor determines the receptor's specificity. Regardless of the nature of the interaction, however, the result is always the same: The transmembrane potential of the receptor cell changes. The change in the transmembrane potential that accompanies receptor stimulation is called a **receptor potential.**

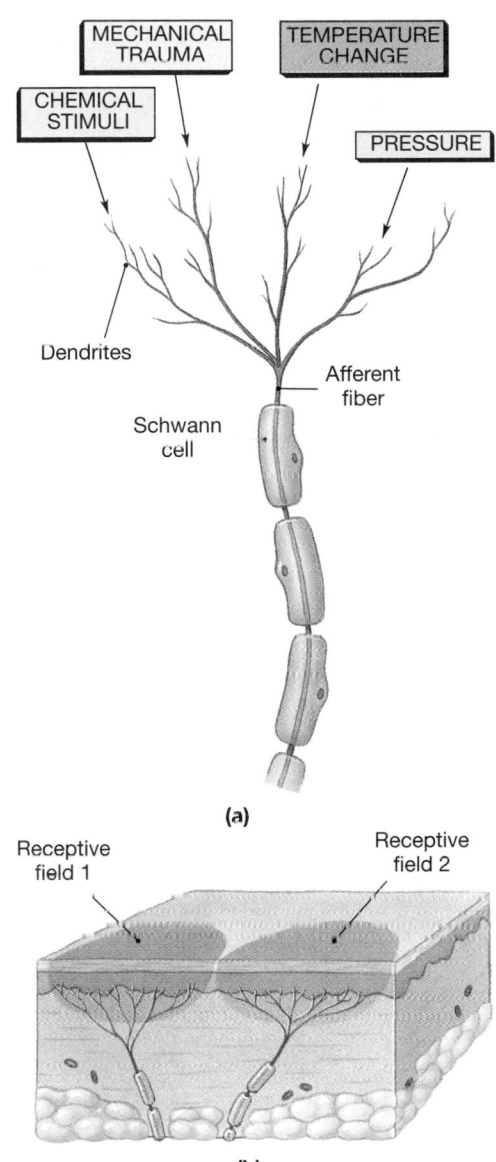

(a)

(b)

●**FIGURE 17-1** Receptors and Receptive Fields. (a) A free nerve ending consists of sensory dendrites that may be stimulated by a variety of stimuli. (b) Each receptor monitors a specific area known as the receptive field.

The receptor potential may be a depolarization or a hyperpolarization. ∞ *[p. 386]* It is a graded potential change, and the stronger the stimulus, the larger the receptor potential.

STEP 2: *The receptor potential directly or indirectly affects a sensory neuron.*

A membrane depolarization that leads to an action potential in a sensory neuron is called a **generator potential.** The typical receptors for the general senses are the dendrites of sensory neurons (Figure 17-2a●). Receptor potentials in the dendrites spread to and summate at the initial segment, producing a generator

potential that can be large enough to trigger an action potential in the axon. When a sensory neuron acts as a receptor, the terms *receptor potential* and *generator potential* can be used interchangeably.

Sensations of taste, hearing, equilibrium, and vision are provided by specialized receptor cells that communicate with sensory neurons across chemical synapses. The receptor cells develop graded receptor potentials in response to stimulation, and the change in membrane potential alters the rate of neurotransmitter release at the synapse. In the case of taste receptors (Figure 17-2b●), stimulation triggers the release of neurotransmitters that depolarize the membrane of the sensory neuron. The larger the receptor potential, the more neurotransmitter released, and the greater the depolarization. When specialized receptor cells are involved, the receptor potential and the generator potential are distinct in both location and timing. The receptor potential appears in the receptor cell when the stimulus arrives. The generator potential develops later, in the sensory neuron, following the arrival of neurotransmitter at the postsynaptic membrane.

STEP 3: *Action potentials travel to the CNS along an afferent fiber.*

When a generator potential appears, action potentials develop in the afferent fiber. For reasons that we discussed in Chapter 12, the greater the depolarization produced by the generator potential, the higher the frequency of action potentials in the afferent fiber. ∞ *[p. 405]* The arriving information is then processed and interpreted by the CNS at the conscious and subconscious levels.

Interpretation of Sensory Information

Sensory information that arrives at the CNS is routed according to the location and nature of the stimulus. In previous chapters, we emphasized the fact that axons in the CNS are organized in bundles with specific origins and destinations. Along sensory pathways, axons relay information from one point (the receptor) to another point (a neuron at a specific site in the cerebral cortex). For example, touch, pressure, pain, temperature, and taste sensations arrive at the primary sensory cortex; visual, auditory, and olfactory sensations reach the visual, auditory, and olfactory regions of the cortex.

The neural link between receptor and cortical neuron is called a **labeled line.** Each labeled line consists of axons carrying information about one **modality**, or type of stimulus (touch, pressure, light, sound, and so forth). A labeled line begins at receptors in a specific part of the body. The CNS interprets the modality entirely on the basis of the labeled line over which it ar-

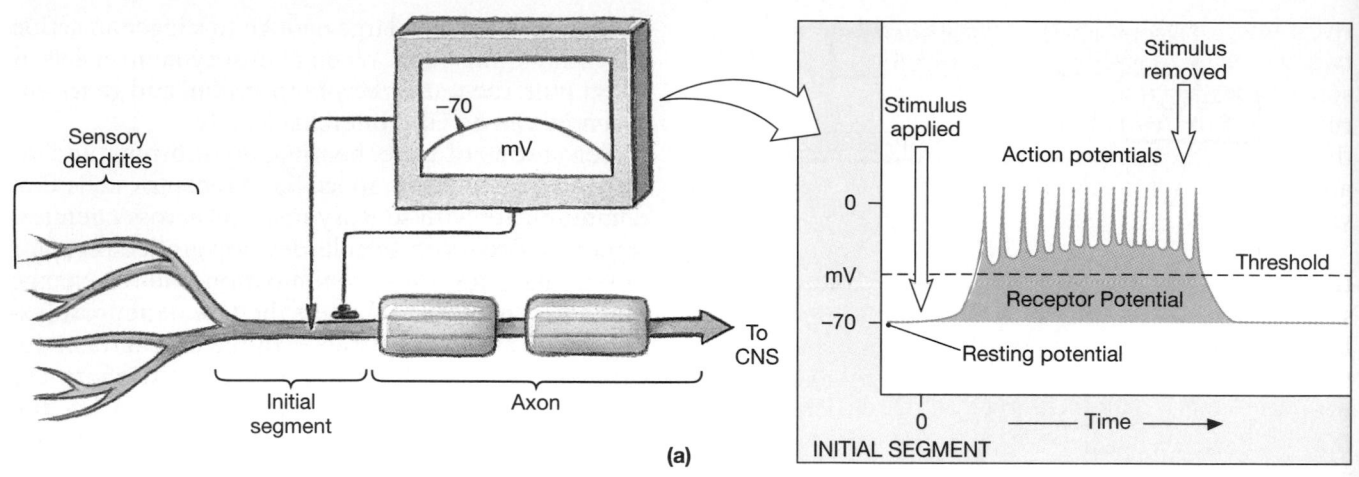

(a)

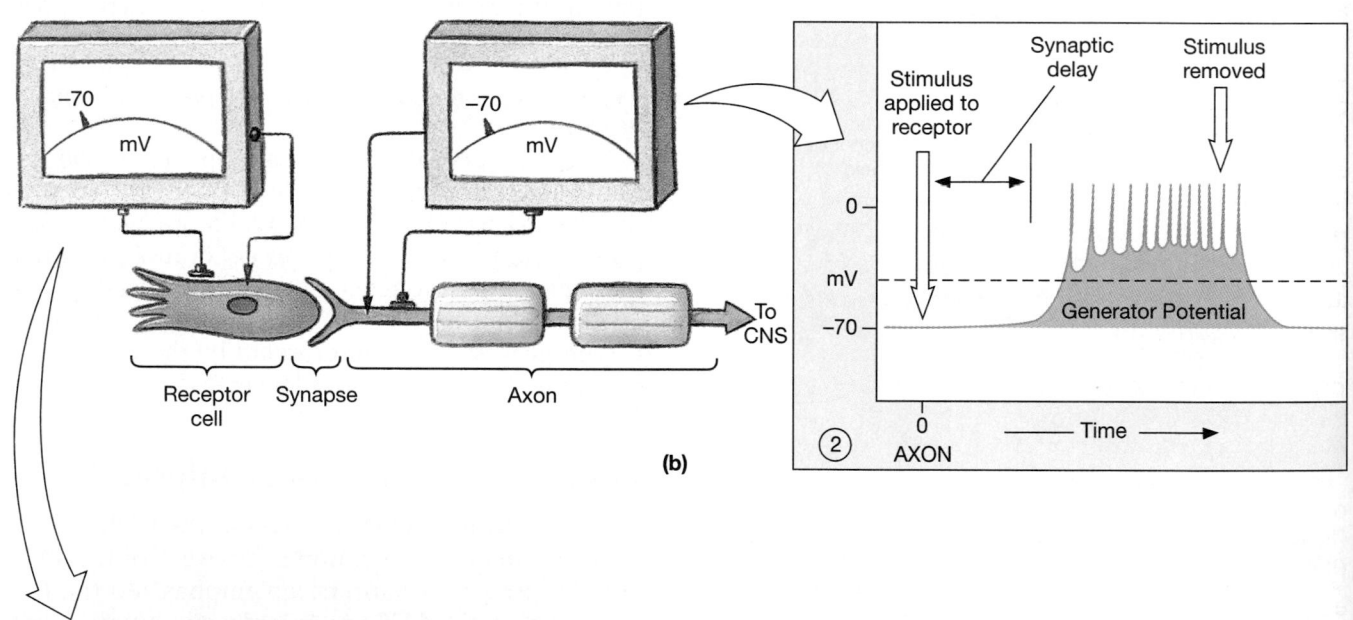

(b)

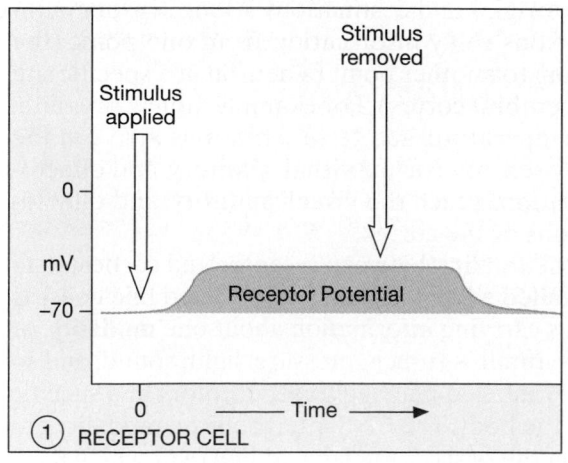

●*FIGURE 17-2* **Receptor and Generator Potentials.** **(a)** When the sensory neuron acts as the receptor, a stimulus that depolarizes the dendrites may bring the initial segment of the axon to threshold. The receptor and the neuron are the same cell, so the receptor potential is a generator potential. **(b)** In the special senses of taste, equilibrium, hearing, and vision, the receptor cells are specialized cells that communicate with neurons across chemical synapses. The receptor cell shows a receptor potential in response to stimulation (1). In this example, the receptor potential is a depolarization that accelerates neurotransmitter release, and the neurotransmitter produces a generator potential in the postsynaptic membrane (2).

rives. As a result, you cannot tell the difference between a true sensation and a false one generated somewhere along the line. For example, when you rub your eyes, you commonly see flashes of light. Although the stimulus is mechanical rather than visual, any activity along the optic nerve is projected to the visual cortex and experienced as a visual perception.

The identity of the active labeled line indicates the type of the stimulus. Where it arrives within the sensory cortex determines its perceived location. For example, if activity in a labeled line that carries touch sensations stimulates the facial region of your primary sensory cortex, you perceive a touch on the face. *All other characteristics* of the stimulus—its strength, duration, and variation—are conveyed by the frequency and pattern of action potentials. The translation of complex sensory information into meaningful patterns of action potentials is called *sensory coding.*

Some sensory neurons, called **tonic receptors,** are always active. The frequency with which they generate action potentials indicates the background level of stimulation. When the stimulus increases or decreases, their rate of action potential generation changes accordingly. Other receptors are normally inactive but become active for a short time whenever there is a change in the conditions they are monitoring. These receptors, called **phasic receptors,** provide information on the intensity and rate of change of a stimulus. Receptors that combine phasic and tonic coding can convey extremely complicated sensory information.

Central Processing and Adaptation

Adaptation is a reduction in sensitivity in the presence of a constant stimulus. **Peripheral adaptation** occurs when the receptors or sensory neurons alter their levels of activity. The receptor responds strongly at first, but thereafter the activity along the afferent fiber gradually declines, in part because the size of the generator potential gradually decreases. This response is characteristic of phasic receptors, which are also called **fast-adapting receptors.** Temperature receptors (*thermoreceptors*) are phasic receptors; you seldom notice room temperature unless it changes suddenly. Tonic receptors show little peripheral adaptation and so are called **slow-adapting receptors.** Pain receptors (*nociceptors*) are slow-adapting receptors, and that's one reason why pain sensations remind you of an injury long after the initial damage has occurred.

Adaptation also occurs inside the CNS along the sensory pathways. For example, a few seconds after you are exposed to a new smell, conscious awareness of the stimulus virtually disappears, although the sensory neurons are still quite active. This process is known as **central adaptation.** Central adaptation generally involves the inhibition of nuclei along a sensory pathway.

Peripheral adaptation reduces the amount of information that reaches the CNS. Central adaptation at the subconscious level further restricts the amount of detail that arrives at the cerebral cortex. Most of the incoming sensory information is processed in centers along the spinal cord or brain stem, potentially triggering reflex responses, such as the withdrawal reflex. Only about 1 percent of the information provided by afferent fibers reaches the cerebral cortex and our conscious awareness.

The output from higher centers can increase receptor sensitivity or facilitate transmission along a sensory pathway. The reticular activating system ∞ *[pp. 471, 505]* in the mesencephalon helps focus our attention and thus heightens or reduces our awareness of arriving sensations. This adjustment of sensitivity can occur under conscious or subconscious direction. When you "listen carefully," your sensitivity and awareness of auditory stimuli increase. Output from higher centers can also inhibit transmission along a sensory pathway. This inhibition occurs when you enter a noisy factory or walk along a crowded city street, as you automatically tune out the high level of background noise.

In this discussion, we have introduced basic concepts of receptor function and sensory processing. We will next consider how those concepts apply to the general senses.

THE GENERAL SENSES

Receptors for the general senses are scattered throughout the body and are relatively simple in structure. The simple classification scheme introduced in Chapter 12 divides them into *exteroceptors, proprioceptors,* and *interoceptors.* ∞ *[p. 375]* **Exteroceptors** provide information about the external environment, **proprioceptors** report the positions of skeletal muscles and joints, and **interoceptors** monitor visceral organs and functions.

A more detailed classification system divides the general sensory receptors into four types according to the nature of the stimulus that excites them: *nociceptors* (pain), *thermoreceptors* (temperature), *mechanore-ceptors* (touch, pressure, proprioception), and *chemoreceptors* (chemical sense). Each class of receptors has distinct structural and functional characteristics.

Nociceptors

Pain receptors, or **nociceptors,** are especially common in the superficial portions of the skin, in joint capsules, within the periostea of bones, and around the walls of blood vessels. There are few nociceptors in other deep tissues or in most visceral organs. Pain receptors are free nerve endings with large receptive fields (Figure 17-3a●). As a consequence, it is often difficult to determine the exact source of a painful sensation.

There are three different populations of nociceptors: (1) those sensitive to extremes of temperature, (2) those sensitive to mechanical damage, and (3) those sensitive to dissolved chemicals, such as chemicals released by injured cells. Very strong stimuli, however, will excite all three receptor types. For that reason, people describing very painful sensations—whether caused by acids, heat, or a deep cut—use similar descriptive terms, such as "burning."

Stimulation of the dendrites of a nociceptor causes depolarization. When the initial segment of the axon reaches threshold, an action potential heads toward the CNS. The sensory neurons use the amino acid *glutamate* and the neuropeptide *Substance P* as neurotransmitters. We detailed the distribution of painful sensations in Chapter 15: The axon synapses on an interneuron, whose axon crosses the spinal cord to ascend within the *lateral spinothalamic tract.* After a synapse in the thalamus, the information is relayed to the primary sensory cortex. ∞ *[p. 494]*

Two types of axons—type A and type C fibers—carry painful sensations. Myelinated Type A fibers (∞ *[p. 394]*) carry sensations of **fast pain,** or *prickling pain.* An injection or a deep cut produces this type of pain. These sensations very quickly reach the CNS, where they often trigger somatic reflexes. They are also relayed to the primary sensory cortex and so receive conscious attention. In most cases, the arriving information permits localization of the stimulus to an area several inches in diameter.

Slower Type C fibers (∞ *[p. 394]*) carry sensations of **slow pain,** or *burning and aching pain.* These sensations cause a generalized activation of the reticular formation and thalamus. The individual becomes aware of the pain but has only a general idea of the area affected.

Pain receptors exhibit a tonic response to stimulation. Significant peripheral adaptation does not occur, and the receptors continue to respond as long as the painful stimulus remains. Painful sensations cease only after tissue damage has ended. However, central adaptation may reduce *perception* of the pain while the pain receptors are still stimulated. This effect involves the inhibition of centers in the thalamus, reticular formation, lower brain stem, and spinal cord.

An understanding of the origins of pain sensations and an ability to control or reduce pain levels have always been among the most important aspects of medical treatment. After all, it is usually pain that induces someone to seek treatment; conditions that are not painful are typically ignored or tolerated. Although we often use the term *pain pathways*, it is becoming clear that pain distribution and perception are extremely complex—more so than had previously been imagined. The neurotransmitters and mechanisms involved are detailed in the *Applications Manual.* **AM** *A Closer Look: Pain Mechanisms, Pathways, and Control*

Thermoreceptors

Temperature receptors, or **thermoreceptors,** are free nerve endings located in the dermis of the skin, in skeletal muscles, in the liver, and in the hypothalamus. Cold receptors are three or four times more numerous than warm receptors. There are no known structural differences between warm and cold thermoreceptors.

Temperature sensations are conducted along the same pathways that carry pain sensations. They are sent to the reticular formation, the thalamus, and (to a lesser extent) the primary sensory cortex. Thermoreceptors are phasic receptors: They are very active when the temperature is changing, but they quickly adapt to a stable temperature. When you enter an air-conditioned classroom on a hot summer day or a warm lecture hall on a brisk fall evening, the temperature seems unpleasant at first, but your discomfort fades as adaptation occurs.

Mechanoreceptors

Mechanoreceptors are sensitive to stimuli that distort their cell membranes. These membranes contain *mechanically regulated ion channels* whose gates open or close in response to stretching, compression, twisting, or other distortions of the membrane. There are three classes of mechanoreceptors:

1. **Tactile receptors,** which provide sensations of touch, pressure, and vibration. These sensations are closely related. Touch sensations provide information about shape or texture, whereas pressure sensations indicate the degree of mechanical distortion. Vibration sensations indicate a pulsing or oscillating pressure. The receptors involved may be specialized in some way. For example, rapidly adapting tactile receptors are best suited for detecting vibration. But your interpretation of a sensation as touch rather than pressure is typically a matter of the degree of stimulation and not of differences in the type of receptor stimulated.

2. **Baroreceptors** (bar-ō-rē-SEP-torz; *baro-,* pressure), which detect pressure changes in the walls of blood vessels and in portions of the digestive, reproductive, and urinary tracts.

3. *Proprioceptors,* which monitor the positions of joints and muscles. These are the most structurally and functionally complex of the general sensory receptors.

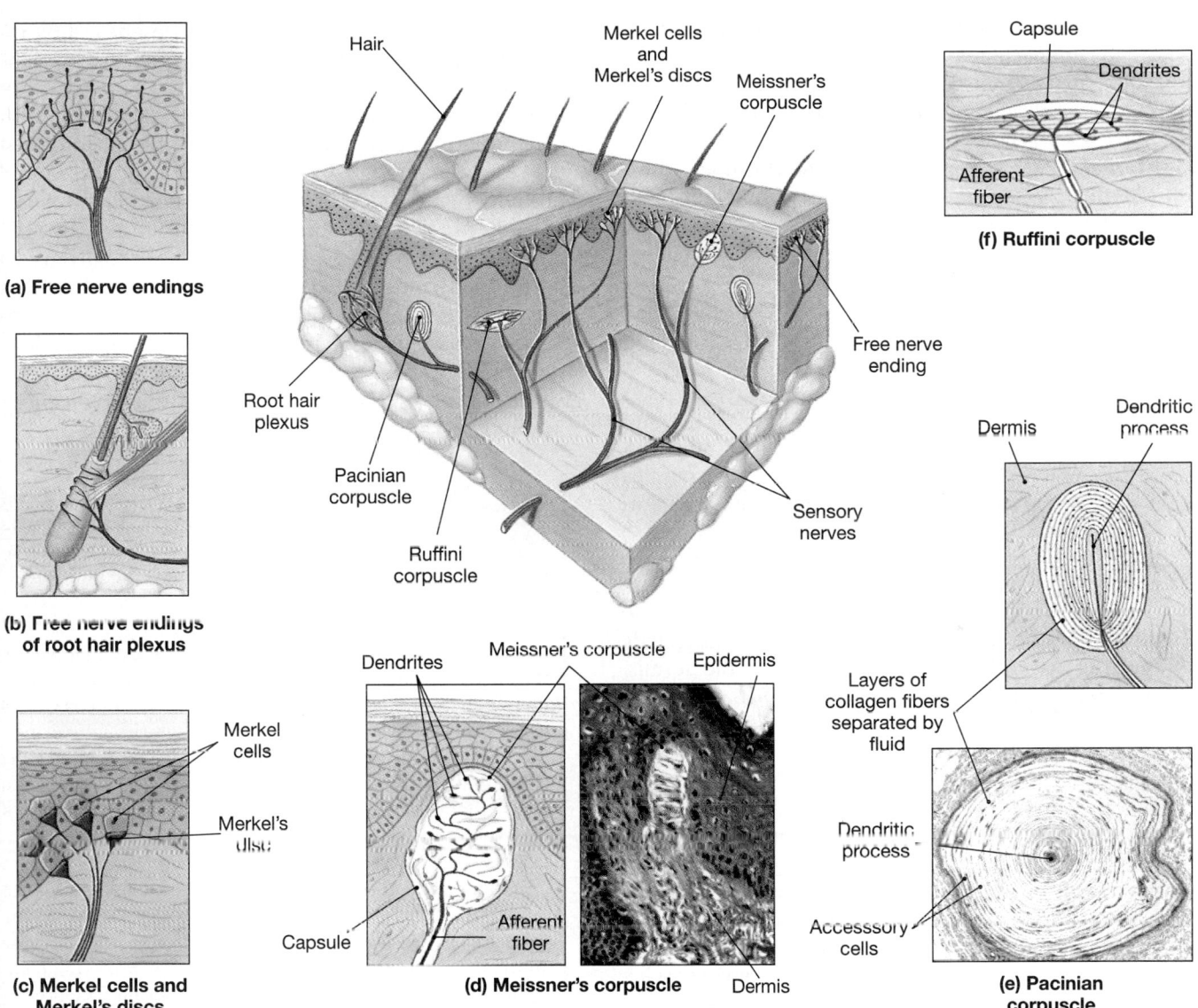

(a) Free nerve endings

(b) Free nerve endings of root hair plexus

Root hair plexus

Pacinian corpuscle

Ruffini corpuscle

Hair

Merkel cells and Merkel's discs

Meissner's corpuscle

Free nerve ending

Sensory nerves

(f) Ruffini corpuscle

Capsule

Dendrites

Afferent fiber

(c) Merkel cells and Merkel's discs

Merkel cells

Merkel's disc

Dendrites

Meissner's corpuscle

Epidermis

Capsule

Afferent fiber

(d) Meissner's corpuscle

Dermis

Dermis

Dendritic process

Layers of collagen fibers separated by fluid

Dendritic process

Accesssory cells

(e) Pacinian corpuscle

•**FIGURE 17-3** **Tactile Receptors in the Skin.** (a) Free nerve endings. (b) Root hair plexus. (c) Merkel cells and Merkel's discs. (d) Meissner's corpuscle. (e) Pacinian corpuscle. (f) Ruffini corpuscle.

Tactile Receptors

Fine touch and pressure receptors provide detailed information about a source of stimulation, including its exact location, shape, size, texture, and movement. These receptors are extremely sensitive and have relatively narrow receptive fields. The sensory information reaches our conscious awareness through the *posterior column pathway.* ∞ *[p. 493]* **Crude touch and pressure receptors** provide poor localization and, because of relatively large receptive fields, give little additional information about the stimulus. These sensations ascend within the *anterior spinothalamic tract,* and the thalamus relays the information to appropriate areas of the primary sensory cortex. ∞ *[p. 494]*

Tactile receptors range in complexity from free nerve endings to specialized sensory complexes with accessory cells and supporting structures. Figure 17-3• shows six types of tactile receptors in the skin:

1. *Free nerve endings* sensitive to touch and pressure are situated between epidermal cells (Figure 17-3a•). There are no apparent structural differences between these receptors and the free nerve endings that provide temperature or pain sensations. These are the only sensory receptors on the corneal surface of the eye, but in other portions of the body surface, more-specialized tactile receptors are probably more important. Free nerve endings that provide touch sensations are tonic receptors with small receptive fields.

2. Wherever hairs are located, the nerve endings of the **root hair plexus** (Figure 17-3b•) monitor distortions and movements across the body surface. When a hair is displaced, the movement of the follicle distorts the sensory dendrites and produces action potentials.

These receptors adapt rapidly, so they are best at detecting initial contact and subsequent movements. For example, you generally feel your clothing only when you move or when you consciously focus on tactile sensations from the skin.

3. **Merkel's** (MER-kelz) **discs** are fine touch and pressure receptors (Figure 17-3c●). They are tonically active, extremely sensitive, and have very small receptive fields. The dendritic processes of a single myelinated afferent fiber make close contact with unusually large epithelial cells in the stratum germinativum of the skin; we described these *Merkel cells* in Chapter 5. ∞ [p. 149]

4. **Meissner's** (MĪS-nerz) **corpuscles** (Figure 17-3d●) perceive sensations of fine touch and pressure and low-frequency vibration. They adapt to stimulation within a second after contact. Meissner's corpuscles are fairly large structures, measuring roughly 100 μm in length and 50 μm in width. These receptors are most abundant in the eyelids, lips, fingertips, nipples, and external genitalia. The dendrites are highly coiled and interwoven, and they are surrounded by modified Schwann cells. A fibrous capsule surrounds the entire complex and anchors it within the dermis.

5. **Pacinian** (pa-SIN-ē-an) **corpuscles,** or *lamellated corpuscles,* are sensitive to deep pressure. Because they are fast-adapting receptors, they are most sensitive to pulsing or high-frequency vibrating stimuli. A single dendritic process lies within a series of concentric layers of collagen fibers and supporting cells (specialized fibroblasts). The entire corpuscle may reach 4 mm in length and 1 mm in diameter (Figure 17-3e●). The concentric layers, separated by interstitial fluid, shield the dendrite from virtually every source of stimulation other than direct pressure. Pacinian corpuscles adapt quickly because distortion of the capsule soon relieves pressure on the sensory process. These receptors are scattered throughout the dermis, notably in the fingers, mammary glands, and external genitalia. They occur also in the superficial and deep fasciae, in joint capsules, in mesenteries, in the pancreas, and in the walls of the urethra and urinary bladder.

6. **Ruffini** (roo-FĒ-nē) **corpuscles** are also sensitive to pressure and distortion of the skin, but they are located in the reticular (deep) dermis. These receptors are tonically active and show little if any adaptation. The capsule surrounds a core of collagen fibers that are continuous with those of the surrounding dermis (Figure 17-3f●). Inside the capsule, a network of dendrites is intertwined with the collagen fibers. Any tension or distortion of the dermis tugs or twists the capsular fibers, stretching or compressing the attached dendrites and altering the activity in the myelinated afferent fiber.

In Chapter 15, we considered the central distribution of tactile sensations through the posterior column and spinothalamic pathways. ∞ [pp. 493–495] Our sensitivity to tactile sensations may be altered by infection, disease, or damage to sensory neurons or pathways. As a result, mapping tactile responses can sometimes aid clinical assessment. Sensory losses with clear regional boundaries indicate trauma to spinal nerves. For example, sensory loss along a dermatomal boundary (described in Chapter 13) can permit a reasonably precise determination of the affected spinal nerve or nerves. ∞ [p. 425] Regional sensitivity to light touch can be checked by gentle contact with a fingertip or a slender wisp of cotton. Vibration receptors are tested by applying the base of a tuning fork to the skin. We discuss more-detailed procedures, such as the *two-point discrimination test,* in the *Applications Manual.* AM *Assessment of Tactile Sensitivities*

TICKLING AND ITCHING Tickle and itch sensations are closely related to the sensations of touch and pain. The receptors involved are free nerve endings, and the information is carried by unmyelinated Type C fibers. Tickle sensations, which are usually (but not always) described as pleasurable, are produced by a light touch that moves across the skin. Psychological factors are involved in the interpretation of tickle sensations, and there are tremendous individual differences in tickle sensitivity. Little is known about the labeled line involved, but the sensations travel over the spinothalamic pathway. Itching is probably produced by stimulation of the same receptors. There are specific "itch spots" that can be mapped in the skin, the inner surfaces of the eyelids, and the mucous membrane of the nose. Itch sensations are absent from other mucous membranes and from deep tissues and viscera. Itching is extremely unpleasant, even more unpleasant than pain. Individuals with extreme itching will scratch even when pain is the result. Itch receptors can be stimulated by the injection of histamine or proteolytic enzymes into the epidermis and superficial dermis. The precise receptor mechanism is unknown.

Baroreceptors

Baroreceptors monitor changes in pressure. A baroreceptor consists of free nerve endings that branch within the elastic tissues in the wall of a distensible organ, such as a blood vessel or a portion of the respiratory, digestive, or urinary tract. When the pressure changes, the elastic walls of the tract recoil or expand. This movement distorts the dendritic branches and alters the rate of action potential generation. Baroreceptors respond immediately to a change in pressure, but they adapt rapidly, and the output along the afferent fibers gradually returns to normal. Figure 17-4● provides an overview of baroreceptor distribution and function.

Baroreceptors monitor blood pressure in the walls of major vessels, including the carotid artery (at the *carotid sinus*) and the aorta (at the *aortic sinus*). The information plays a major role in regulating cardiac function and adjusting blood flow to vital tissues. Baroreceptors in the lungs monitor the degree of lung

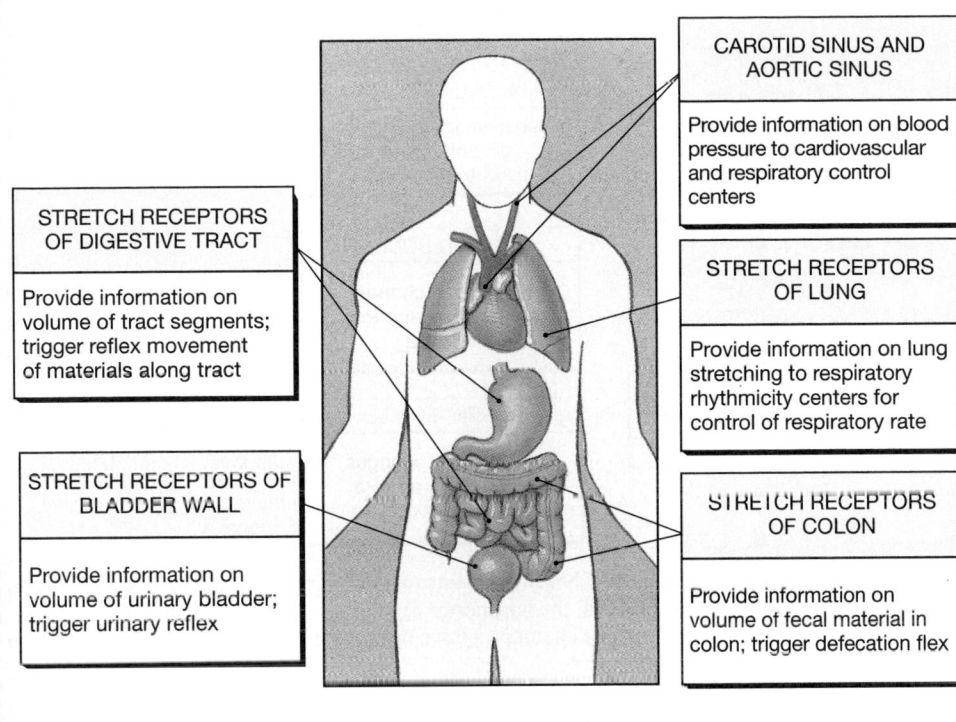

CAROTID SINUS AND AORTIC SINUS

Provide information on blood pressure to cardiovascular and respiratory control centers

STRETCH RECEPTORS OF DIGESTIVE TRACT

Provide information on volume of tract segments; trigger reflex movement of materials along tract

STRETCH RECEPTORS OF LUNG

Provide information on lung stretching to respiratory rhythmicity centers for control of respiratory rate

STRETCH RECEPTORS OF BLADDER WALL

Provide information on volume of urinary bladder; trigger urinary reflex

STRETCH RECEPTORS OF COLON

Provide information on volume of fecal material in colon; trigger defecation flex

•*FIGURE 17-4* **Baroreceptors and the Regulation of Autonomic Functions.** Baroreceptors provide information essential to the regulation of autonomic activities, including cardiovascular function, urination, defecation, and respiration.

expansion. This information is relayed to the respiratory rhythmicity centers, which set the pace of respiration. Comparable stretch receptors in the digestive and urinary tracts trigger a variety of visceral reflexes, including those of urination and defecation. We shall detail those baroreceptor reflexes in chapters that deal with specific physiological systems.

Proprioceptors

Proprioceptors monitor the position of joints, the tension in tendons and ligaments, and the state of muscular contraction. We described two representative examples in earlier chapters: *Golgi tendon organs,* which monitor the strain on a tendon, and *muscle spindles,* which monitor the length of a skeletal muscle.

In a Golgi tendon organ, dendrites branch repeatedly and wind around the densely packed collagen fibers in a tendon. Tension or distortion of the tendon stimulates the receptor, and excessive stimulation triggers a reduction in the strength of skeletal muscle contractions. We discussed this *tendon reflex* in Chapter 13. ∞ *[p. 438]* In Chapters 10 and 13, we described muscle spindles and considered their role in regulating skeletal muscle tone and maintaining normal posture and balance. ∞ *[pp. 295, 434–435]*

Proprioceptors do not adapt to constant stimulation, and each receptor continuously sends information to the CNS. A relatively small proportion of the arriving proprioceptive information reaches your conscious awareness over the *posterior column pathway.* Most of the proprioceptive information is processed at subconscious levels. The *spinocerebellar pathway* delivers proprioceptive information to the cerebellum outside your conscious awareness. ∞ *[p. 496]*

Chemoreceptors

Specialized chemoreceptive neurons can detect small changes in the concentration of specific chemicals or compounds. In general, **chemoreceptors** respond only to water-soluble and lipid-soluble substances that are dissolved in the surrounding fluid. The receptors show a pronounced peripheral adaptation over a period of seconds, and central adaptation may also occur. Figure 17-5• provides an overview of the locations and functions of chemoreceptors.

The chemoreceptors included in the general senses do not send information to the primary sensory cortex, and we are not consciously aware of the sensations they provide. The arriving sensory information is routed to brain stem centers that deal with the autonomic control of respiratory and cardiovascular function. Neurons within the respiratory centers of the brain respond to the concentration of hydrogen ions and carbon dioxide molecules in the cerebrospinal fluid. Chemoreceptive neurons are also located within the **carotid bodies,** near the origin of the internal carotid arteries on each side of the neck, and in the **aortic bodies** between the major branches of the aortic arch. These receptors monitor the carbon dioxide and oxygen concentrations of arterial blood. The afferent fibers leaving the carotid and aortic bodies reach the respiratory centers by traveling within cranial nerves IX (glossopharyngeal) and X (vagus).

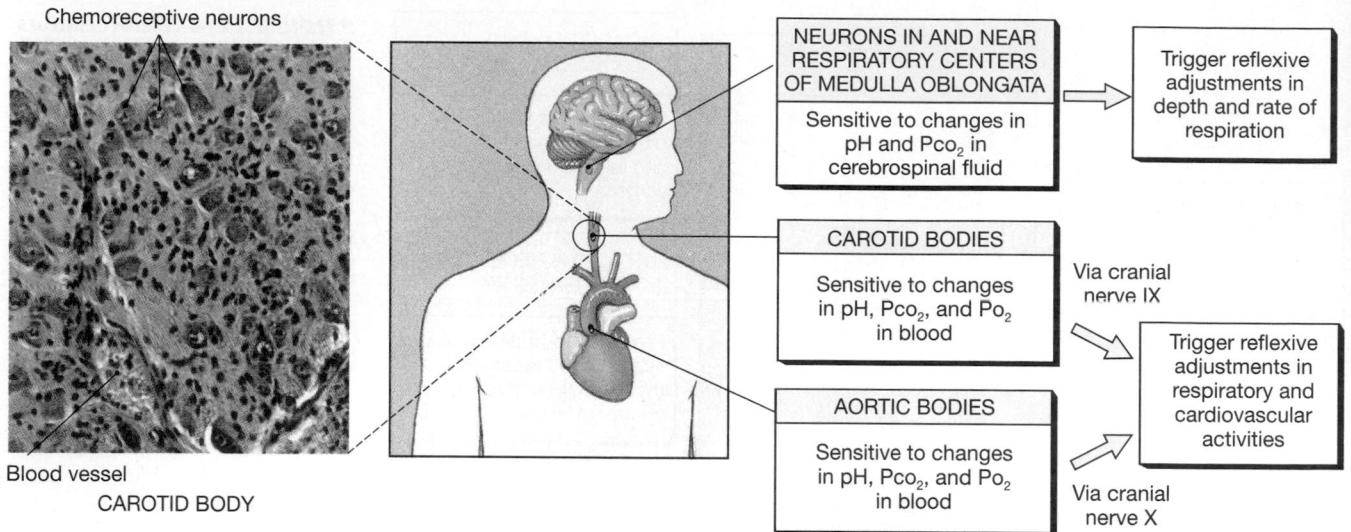

Chemoreceptive neurons

Blood vessel
CAROTID BODY

NEURONS IN AND NEAR RESPIRATORY CENTERS OF MEDULLA OBLONGATA

Sensitive to changes in pH and P_{CO_2} in cerebrospinal fluid

Trigger reflexive adjustments in depth and rate of respiration

CAROTID BODIES

Sensitive to changes in pH, P_{CO_2}, and P_{O_2} in blood

Via cranial nerve IX

AORTIC BODIES

Sensitive to changes in pH, P_{CO_2}, and P_{O_2} in blood

Via cranial nerve X

Trigger reflexive adjustments in respiratory and cardiovascular activities

•**FIGURE 17-5** **Chemoreceptors.** Chemoreceptors are located inside the CNS, on the ventrolateral surfaces of the medulla oblongata, and in the aortic and carotid bodies. These receptors are involved in the autonomic regulation of cardiovascular and respiratory function. The micrograph shows the histological appearance of the chemoreceptive neurons in the carotid body. (LM × 1150)

☑ Receptor A has a circular receptive field with a diameter of 2.5 cm. Receptor B has a circular receptive field 7.0 cm in diameter. Which receptor will provide more-precise sensory information?

☑ When the nociceptors in your hand are stimulated, what sensation do you perceive?

☑ What would happen to you if the information from proprioceptors in your legs were blocked from reaching the CNS?

We will next turn our attention to the five *special senses:* olfaction, gustation, vision, equilibrium, and hearing. Although the sense organs involved are structurally more complex than those of the general senses, the same basic principles of receptor function apply.

OLFACTION

The sense of smell, more precisely called *olfaction,* is provided by paired **olfactory organs.** These organs are located in the nasal cavity on either side of the nasal septum (Figure 17-6a•). The olfactory organs contain the following two layers (Figure 17-6b•):

1. An **olfactory epithelium,** which contains the **olfactory receptors,** supporting cells, and **basal cells** (*stem cells*). The olfactory epithelium covers the inferior surface of the cribriform plate, the superior portion of the nasal septum, and the superior nasal conchae.

2. The *lamina propria,* an underlying layer of loose connective tissue. This layer contains **olfactory glands,** also called *Bowman's glands,* whose secretions absorb water and form a thick, pigmented mucus. It also contains numerous blood vessels and nerves.

When you draw air in through your nose, the air swirls and eddies within the nasal cavity. This turbulence brings airborne compounds to your olfactory organs. A normal, relaxed inspiration carries a small sample (about 2 percent) of the inspired air to the olfactory organs. Sniffing repeatedly increases the flow of air across the olfactory epithelium, intensifying the stimulation of the receptors. Once compounds have reached the olfactory organs, water-soluble and lipid-soluble materials must diffuse into the mucus before they can stimulate the olfactory receptors.

Olfactory Receptors

The olfactory receptors are highly modified neurons. The exposed tip of each receptor cell forms a prominent knob that projects beyond the epithelial surface (Figure 17-6b•). That projection provides a base for up to 20 cilia that extend into the surrounding mucus. These cilia lie parallel to the epithelial surface, exposing their considerable surface area to dissolved chemical compounds.

Olfactory reception occurs on the surfaces of the olfactory cilia as dissolved chemicals interact with receptors, called *odorant binding proteins*, on the membrane surface. *Odorants* are chemicals that stimulate olfactory receptors. In general, odorants are small organic molecules; the strongest smells are associated with molecules of high water and lipid solubilities. The binding of an odorant to its binding protein leads to the activation of adenylate cyclase, the enzyme that converts ATP to cyclic-AMP (cAMP). The cAMP then opens sodium channels in the membrane, which, as a result, begins to depolar-

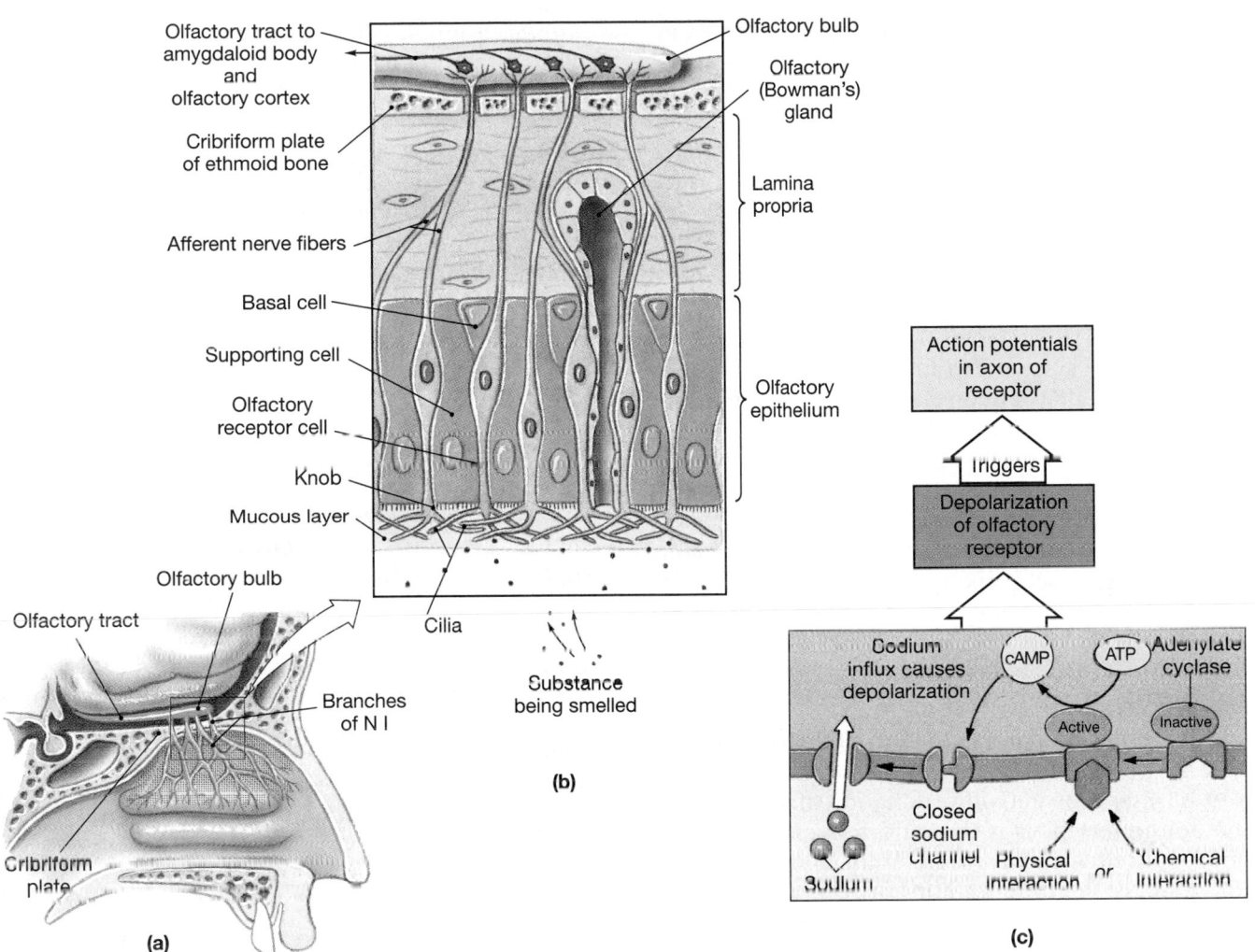

●**FIGURE 17-6** **The Olfactory Organs.** **(a)** The structure of the olfactory organ on the left side of the nasal septum. **(b)** An olfactory receptor is a modified neuron with multiple cilia extending from its free surface. **(c)** Steps in the transduction process.

ize. If sufficient depolarization occurs, an action potential is triggered in the axon, and the information is relayed to the CNS (Figure 17-6c●).

Somewhere between 10 and 20 million olfactory receptors are packed into an area of roughly 5 cm² (0.8 in.²). If we take into account the exposed ciliary surfaces, the actual sensory area probably approaches that of the entire body surface. Nevertheless, our olfactory sensitivities cannot compare with those of other vertebrates such as dogs, cats, or fishes. A German shepherd sniffing for smuggled drugs or explosives has an olfactory receptor surface 72 times greater than that of the nearby customs inspector.

Olfactory Pathways

The olfactory system is very sensitive. As few as four molecules of an odorous substance can activate an olfactory receptor. However, the activation of an afferent fiber does not guarantee a conscious awareness of the stimulus. Considerable convergence ∞ *[p. 408]* occurs along the olfactory pathway, and inhibition at the intervening synapses can prevent the sensations from reaching the *olfactory cortex* of the cerebral hemispheres. (The olfactory cortex is located on the inferior surface of each temporal lobe.) The olfactory receptors adapt very little to a persistent stimulus. It is central adaptation that ensures that you quickly lose awareness of a new smell but retain sensitivity to others.

Axons leaving the olfactory epithelium collect into 20 or more bundles that penetrate the cribriform plate of the ethmoid bone to reach the *olfactory bulbs* of the cerebrum (Figure 17-6a●), where the first synapse occurs. Efferent fibers from nuclei elsewhere in the brain also innervate neurons of the olfactory bulbs. This arrangement provides a mechanism for central adaptation or facilitation of olfactory sensitivity. Axons leaving the olfactory bulb travel along the olfactory tract to reach the olfactory cortex, the hypothalamus, and portions of the limbic system.

Olfactory stimulation is the only type of sensory information that reaches the cerebral cortex without first synapsing in the thalamus. The extensive limbic and hypothalamic connections help explain the profound emotional and behavioral responses that can be produced by certain smells. The perfume industry, which understands the practical implications of these connections, expends considerable effort to develop odors that trigger sexual responses.

Olfactory Discrimination

The olfactory system can make subtle distinctions among 2000–4000 chemical stimuli. No apparent structural differences exist among the olfactory cells, but the epithelium as a whole contains receptor populations with distinct sensitivities. There are at least 50 different "primary smells," and our language does not allow us to describe these sensory impressions effectively. It appears likely that the CNS interprets each smell on the basis of the overall pattern of receptor activity.

▧ Aging and Olfactory Sensitivity

The olfactory receptor population shows considerable turnover, with new receptor cells being produced by division and differentiation of basal cells in the epithelium. This is the only proven example of neuronal replacement in the adult human. Despite this process, the total number of receptors declines with age, and the remaining receptors become less sensitive. As a result, elderly individuals have difficulty detecting odors in low concentrations. This decline in the number of receptors accounts for Grandmother's tendency to apply perfume in excessive quantities and explains why Grandfather's aftershave seems so overdone: They must apply more to be able to smell it.

GUSTATION

Gustation, or taste, provides information about the foods and liquids we consume. The **gustatory** (GUS-ta-tor-ē), or **taste, receptors** are distributed over the superior surface of the tongue and adjacent portions of the pharynx and larynx. By the time we reach adulthood, the taste receptors on the pharynx and larynx have decreased in importance and abundance. The most important taste receptors are on the tongue, although a few remain on the epiglottis and adjacent areas of the pharynx. Taste receptors and specialized epithelial cells form sensory structures called **taste buds**. An adult has about 3000 taste buds.

The superior surface of the tongue bears epithelial projections called **lingual papillae** (pa-PIL-lē;

papilla, a nipple-shaped mound). There are three types of lingual papillae on the human tongue: (1) **filiform** (*filum,* thread) **papillae,** (2) **fungiform** (*fungus,* mushroom) **papillae,** and (3) **circumvallate** (sir-kum-VAL-āt) **papillae** (*circum-,* around + *vallum,* wall). There are regional differences in the distribution of the lingual papillae (Figure 17-7a●). Filiform papillae provide friction against objects in the mouth. They do not contain taste buds. Each small fungiform papilla contains about five taste buds; each large circumvallate papilla contains as many as 100 taste buds. The circumvallate papillae form a V near the posterior margin of the tongue.

Gustatory Receptors

Each taste bud (Figure 17-7b,c●) contains about 40 slender receptors called **gustatory cells** and many supporting cells. Taste buds are recessed into the surrounding epithelium and isolated from the relatively unprocessed oral contents. Each gustatory cell extends slender microvilli, sometimes called *taste hairs,* into the surrounding fluids through a narrow opening, the **taste pore.** Despite this relatively protected position, it's still a hard life: A typical gustatory cell survives for only about 10 days before it is replaced by the division of nearby epithelial cells.

Gustatory Pathways

Taste buds are monitored by cranial nerves VII (facial), IX (glossopharyngeal), and X (vagus). The facial nerve monitors all the taste buds located on the anterior two-thirds of the tongue, from the tip to the line of circumvallate papillae. The circumvallate papillae and the posterior one-third of the tongue are innervated by the glossopharyngeal nerve. The vagus nerve innervates taste buds scattered on the surface of the epiglottis. The sensory afferents carried by these cranial nerves synapse within the **nucleus solitarius,** or *solitary nucleus*, of the medulla oblongata, and the axons of the postsynaptic neurons enter the medial lemniscus. There the neurons join axons that carry somatic sensory information on touch, pressure, and proprioception. After another synapse in the thalamus, the information is projected to the appropriate portions of the primary sensory cortex.

A conscious perception of taste is produced as the information received from the taste buds is correlated with other sensory data. Information concerning the general texture of the food, together with taste-related sensations such as "peppery" or "burning hot," is provided by sensory afferents in the trigeminal nerve (N V). In addition, the level of stimulation from the olfactory receptors plays an

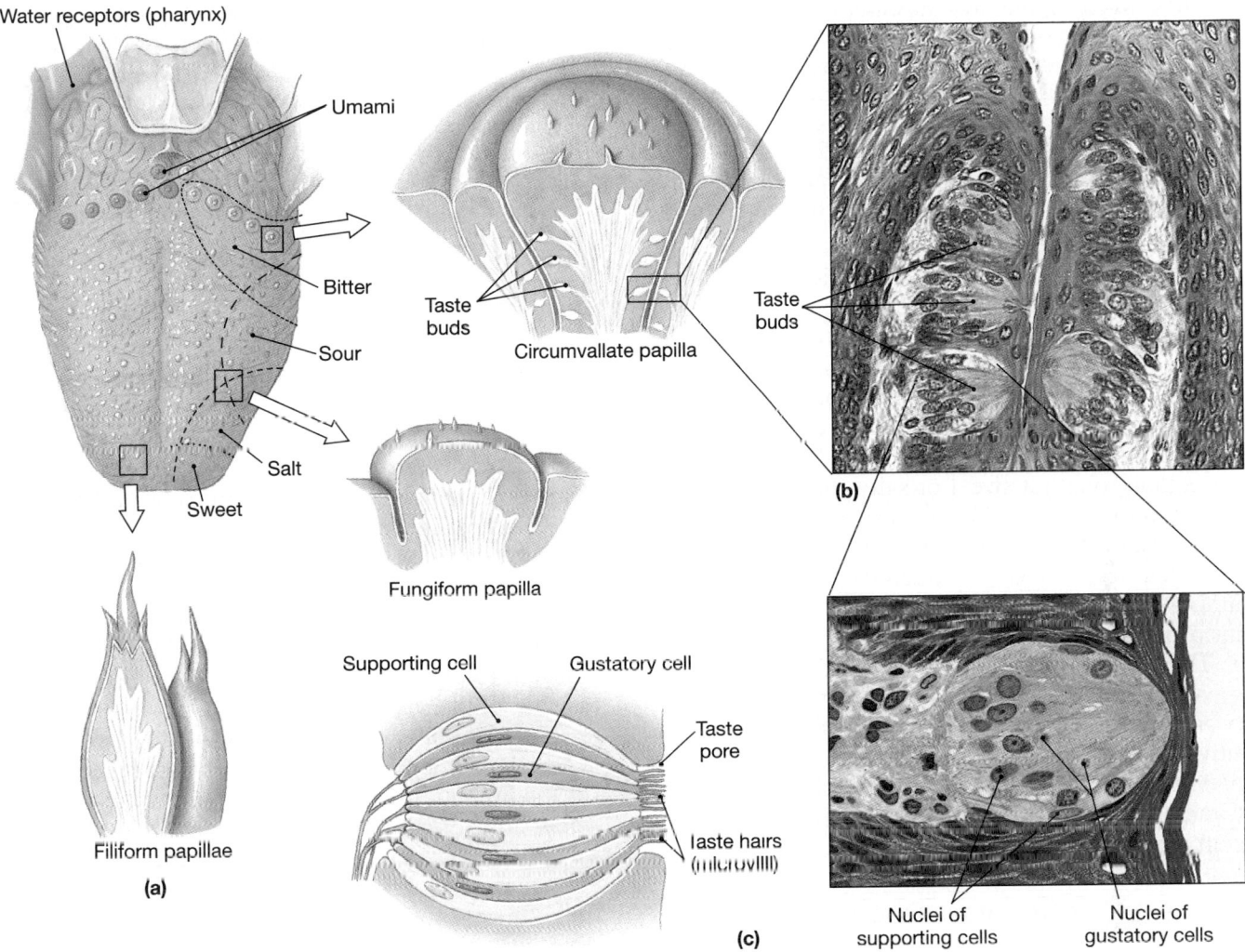

●*FIGURE 17-7* **Gustatory Reception.** **(a)** Gustatory receptors are located in taste buds that form pockets in the epithelium of the fungiform and circumvallate papillae. **(b)** Photomicrograph of taste buds in a circumvallate papilla. (LM × 280) **(c)** A taste bud, showing receptor (gustatory) cells and supporting cells. The diagrammatic view shows details of the taste pore that are not visible in the micrograph. (LM × 650)

overwhelming role in taste perception. You are several thousand times more sensitive to "tastes" when your olfactory organs are fully functional. When you have a cold, airborne molecules cannot reach your olfactory receptors, and meals taste dull and unappealing. This reduction in taste perception will occur even though the taste buds may be responding normally.

Gustatory Discrimination

The mechanism behind gustatory reception resembles that of olfaction. Dissolved chemicals contacting the taste hairs bind to receptor proteins of the gustatory cell. The different tastes involve different receptor mechanisms. For example, salt receptors depolarize after Na^+ channels are opened, whereas sweet receptors depolarize after K^+ channels are closed. The end result of gustatory receptor stimu-

lation is the release of neurotransmitter by the receptor cell. The dendrites of the sensory afferents are tightly wrapped by folds of the receptor cell membrane, and neurotransmitter release leads to the generation of action potentials in the afferent fiber. Gustatory receptors adapt slowly, but central adaptation quickly reduces your sensitivity to a new taste.

You are probably already familiar with the four **primary taste sensations:** sweet, salt, sour, and bitter. At least two additional tastes have been discovered in humans:

- *Umami.* Umami (oo-MAH-mē) is a pleasant taste that is characteristic of beef broth and chicken broth. This taste is produced by receptors sensitive to the presence of amino acids, especially glutamate, small peptides, and nucleotides. The distribution of these receptors is not known in detail, but they are present in taste buds of the circumvallate papillae.

- *Water.* Most people say that water has no taste. However, research on humans and other vertebrates has demonstrated the presence of **water receptors**, especially in the pharynx. Their sensory output is processed in the hypothalamus and affects several systems that deal with water balance and regulation of blood volume. For example, a mouthful of distilled water held for 20 minutes will inhibit ADH secretion and promote water loss at the kidneys.

A sensory map of the tongue indicates that there are regional differences in primary sensitivity (Figure 17-7a●). The threshold for receptor stimulation varies for each of the primary taste sensations, and the taste receptors respond more readily to unpleasant than to attractive stimuli. For example, we are almost a thousand times more sensitive to acids, which give a sour taste, than to either sweet or salty chemicals, and we are a hundred times more sensitive to bitter compounds than to acids. This sensitivity has survival value, for acids can damage the mucous membranes of the mouth and pharynx, and many potent biological toxins produce an extremely bitter taste.

There are significant individual differences in taste sensitivity. Many conditions related to taste sensitivity are inherited. The best-known example involves sensitivity to the compound *phenylthiourea*, also known as *phenylthiocarbamide*, or **PTC.** Roughly 70 percent of Caucasians can taste this substance; the other 30 percent are unable to detect it.

⧖ Aging and Gustatory Discrimination

Our tasting abilities change with age. We begin life with more than 10,000 taste buds, but the number begins declining dramatically by age 50. The sensory loss becomes especially significant because, as we noted, aging individuals also experience a decline in the number of olfactory receptors. As a result, many elderly people find that their food tastes bland and unappetizing, whereas children tend to find the same foods too spicy.

☑ When you first enter the A & P lab for dissection, you are very aware of the odor of preservatives. By the end of the lab period, the smell doesn't seem to be nearly as strong. Why?

☑ If you completely dry the surface of your tongue and then place salt or sugar crystals on it, you can't taste them. Why not?

☑ Your grandfather tells you that he can't understand why foods he used to enjoy just don't taste the same anymore. What will you tell him?

VISION

We rely more on vision than on any other special sense. Our visual receptors are contained in the eyes, elaborate structures that enable us not only to detect light but also to create detailed visual images. We will begin our discussion of these fascinating organs by considering the *accessory structures* of the eye that provide protection, lubrication, and support.

Accessory Structures of the Eye

The **accessory structures** of the eye include the eyelids, the superficial epithelium of the eye, and the structures associated with the production, secretion, and removal of tears. Figure 17-8● details the superficial anatomy of the eye and its accessory structures.

Eyelids

The **eyelids**, or **palpebrae** (pal-PĒ-brē), are a continuation of the skin. The eyelids act like windshield wipers: Their continual blinking movements keep the surface lubricated and free from dust and debris. They can also close firmly to protect the delicate surface of the eye. The free margins of the upper and lower eyelids are separated by the **palpebral fissure,** but the two are connected at the **medial canthus** (KAN-thus) and the **lateral canthus** (Figure 17-8●). The **eyelashes** along the palpebral margins are very robust hairs. These help prevent foreign matter and insects from reaching the surface of the eye.

The eyelashes are associated with the *glands of Zeis* (ZĪS), large sebaceous glands. Along the inner margin of the lid, modified sebaceous glands called **Meibomian** (mī-BŌ-mē-an) **glands** secrete a lipid-rich product that helps keep the eyelids from sticking together. At the medial canthus, the **lacrimal caruncle** (KAR-ung-kul), a soft tissue mass, contains glands producing the thick secretions that contribute to the gritty deposits sometimes found after a good night's sleep. These various glands are subject to occasional invasion and infection by bacteria. A cyst, or **chalazion** (kah-LĀ-zē-on; small lump) generally results from the infection of a Meibomian gland. An infection in a sebaceous gland of one of the eyelashes, a Meibomian gland, or one of the many sweat glands that open to the surface between the follicles produces a painful localized swelling known as a **sty.**

The skin covering the visible surface of the eyelid is very thin. Beneath the skin lie the muscle fibers of the *orbicularis oculi* and the *levator palpebrae superioris.* ∞ *[p. 325, 477]* These skeletal muscles are responsible for closing the eyelids and raising the upper eyelid, respectively.

The epithelium covering the inner surfaces of the eyelids and the outer surface of the eye is called the **conjunctiva** (kon-junk-TĪ-vuh) (Figure 17-9a●, p. 552). It is a mucous membrane covered by a specialized stratified squamous epithelium. The **palpebral conjunctiva** covers the inner surface of the eyelids, and the **ocular conjunctiva,** or *bulbar conjunctiva,* covers the anterior surface of the eye, extending to the edges of the **cornea** (KOR-nē-uh), a transparent, fibrous layer. The **cornea** is covered

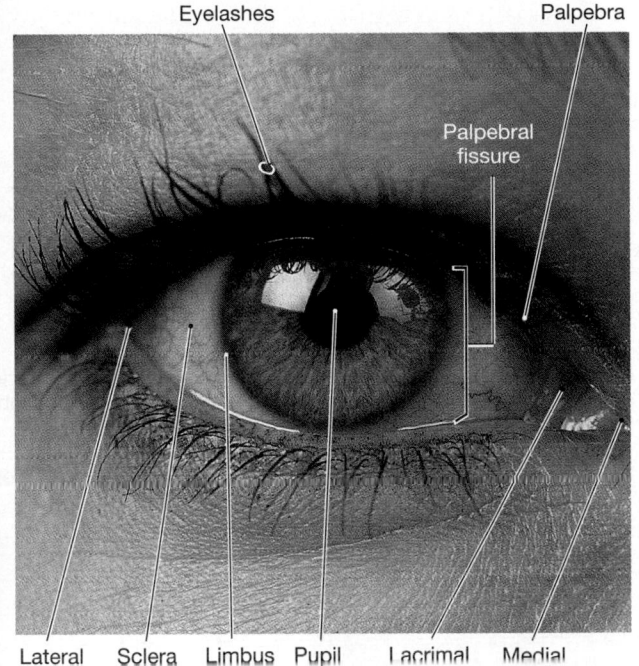

Eyelashes

Palpebra

Palpebral fissure

Lateral canthus   Sclera   Limbus   Pupil   Lacrimal caruncle   Medial canthus

(a)

by a very delicate squamous *corneal epithelium*, 5–7 cells thick, that is continuous with the ocular conjunctiva. A constant supply of fluid washes over the surface of the eyeball, keeping the ocular conjunctiva and cornea moist and clean. Goblet cells within the epithelium assist the various accessory glands in providing a superficial lubricant that prevents friction and drying of the opposing conjunctival surfaces.

Conjunctivitis, or pinkeye, results from damage to and irritation of the conjunctival surface. The most obvious symptom, reddening, is due to dilation of the blood vessels beneath the conjunctival epithelium. This condition may be caused by pathogenic infection or by physical or chemical irritation of the conjunctival surface.

The Lacrimal Apparatus

A constant flow of tears keeps conjunctival surfaces moist and clean. Tears reduce friction, remove debris, prevent bacterial infection, and provide nutrients and oxygen to portions of the conjunctival epithelium. The **lacrimal apparatus** produces, distributes, and removes tears. The lacrimal apparatus of each eye consists of (1) a *lacrimal gland* with associated ducts, (2) *lacrimal canals*, (3) a *lacrimal sac*, and (4) a *nasolacrimal duct* (Figure 17-8b●).

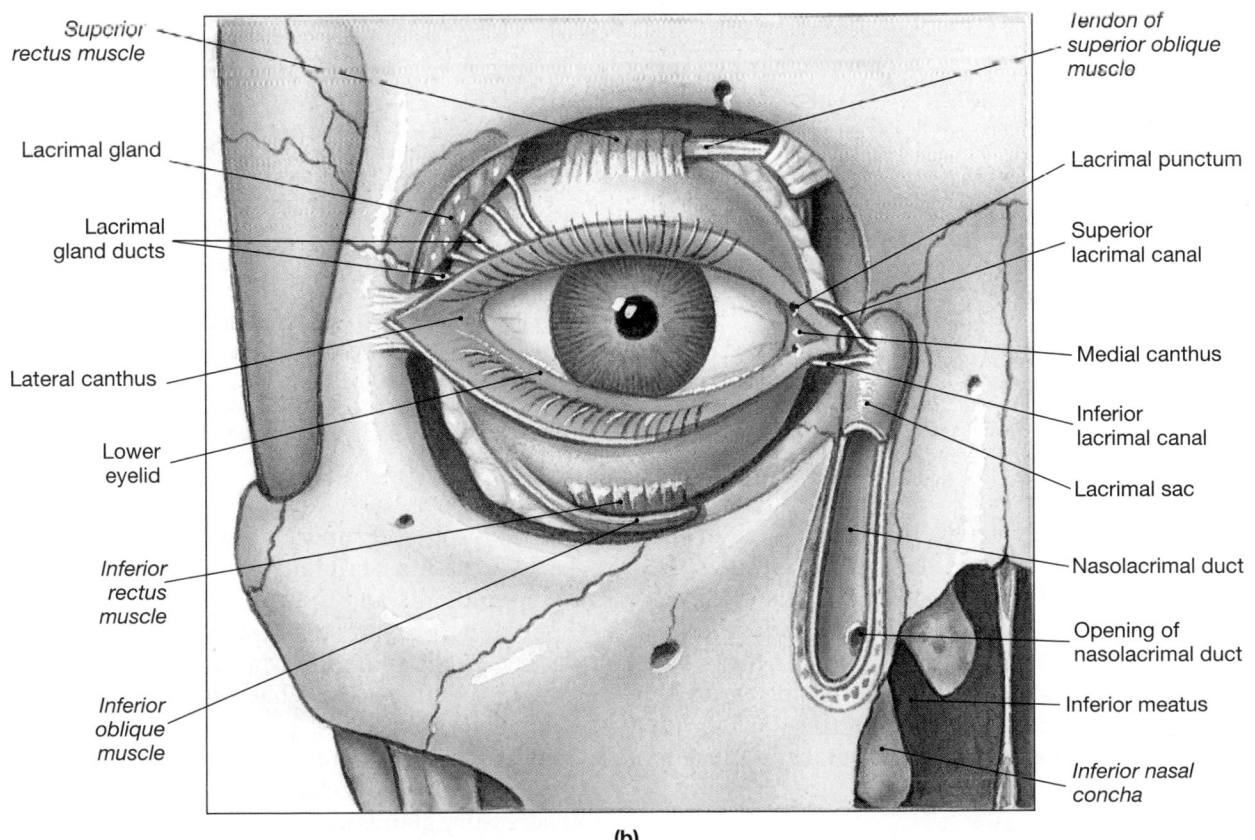

Superior rectus muscle

Lacrimal gland

Lacrimal gland ducts

Lateral canthus

Lower eyelid

Inferior rectus muscle

Inferior oblique muscle

Tendon of superior oblique muscle

Lacrimal punctum

Superior lacrimal canal

Medial canthus

Inferior lacrimal canal

Lacrimal sac

Nasolacrimal duct

Opening of nasolacrimal duct

Inferior meatus

Inferior nasal concha

(b)

●**FIGURE 17-8 External Features and Accessory Structures of the Eye. (a)** Gross and superficial anatomy of the accessory structures. **(b)** Details of the organization of the lacrimal apparatus.

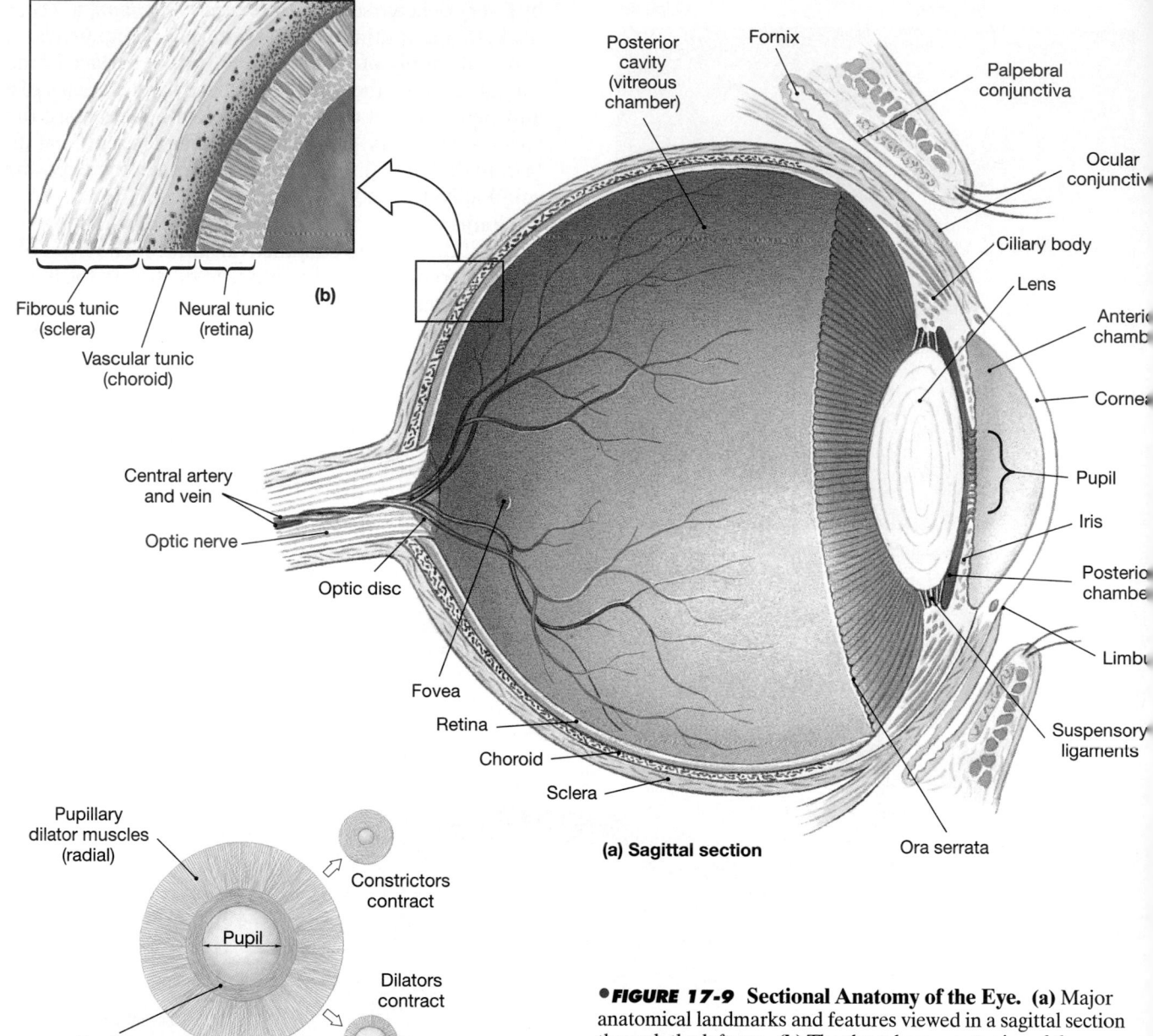

(b)

Fibrous tunic (sclera)
Vascular tunic (choroid)
Neural tunic (retina)

Posterior cavity (vitreous chamber)
Fornix
Palpebral conjunctiva
Ocular conjunctiva
Ciliary body
Lens
Anterior chamber
Cornea
Pupil
Iris
Posterior chamber
Limbus
Suspensory ligaments
Ora serrata

Central artery and vein
Optic nerve
Optic disc
Fovea
Retina
Choroid
Sclera

(a) Sagittal section

Pupillary dilator muscles (radial)
Constrictors contract
Dilators contract
Pupil
Pupillary constrictor muscles (sphincter)
(c)

•FIGURE 17-9 **Sectional Anatomy of the Eye. (a)** Major anatomical landmarks and features viewed in a sagittal section through the left eye. **(b)** The three layers, or tunics, of the eye. **(c)** The pupillary muscles. **(d)** Landmarks and features viewed in a horizontal section through the right eye.

The pocket created where the conjunctiva of the eyelid connects with that of the eye is known as the **fornix** (FOR-niks) (Figure 17-9a●).[1] The lateral portion of the superior fornix receives 10–12 ducts from the **lacrimal gland,** or tear gland (Figure 17-8b●). The lacrimal gland is about the size and shape of an almond, measuring roughly 12–20 mm (0.5–0.75 in.). It nestles within a depression in the frontal bone, just in-

side the orbit and superior and lateral to the eyeball. ∞ *[p. 205]* The lacrimal gland normally provides the key ingredients and most of the volume of the tears that bathe the conjunctival surfaces. Its secretions are watery, slightly alkaline, and contain the enzyme **lysozyme,** which attacks bacteria.

The lacrimal gland produces tears at a rate of around 1 ml/day. Once the lacrimal secretions have reached the ocular surface, they mix with the products of accessory glands and the oily secretions of the Meibomian glands. The result is a superficial "oil slick" that assists in lubrication and slows evaporation.

Blinking sweeps the tears across the ocular surface, and they accumulate at the medial canthus in

---

[1]*Fornix*, like *diaphragm* and *membrane*, is a term that refers to structures in several systems. Regardless of which system is involved, each fornix has an arched or curved shape. We have already discussed the fornix of the brain, a curving component of the limbic system.

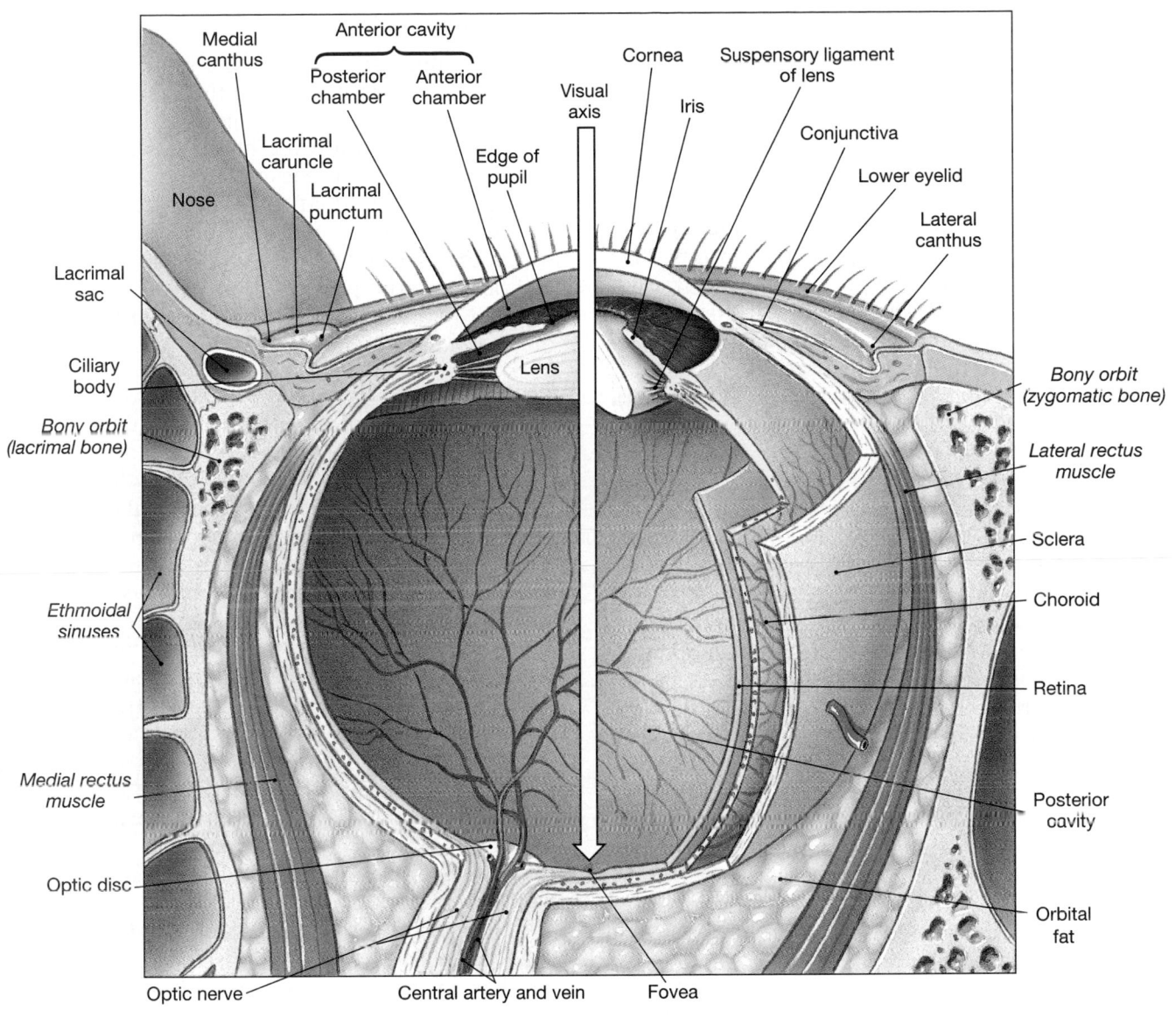

(d) Horizontal section

an area known as the **lacrimal lake** *(lacus lacrimalis),* or "lake of tears." The lacrimal lake covers the lacrimal caruncle, which bulges anteriorly. Two small pores, the **lacrimal puncta** (singular, *punctum*), drain the lacrimal lake. They empty into the **lacrimal canals,** which in turn lead to the **lacrimal sac,** which nestles within the lacrimal groove of the orbit. ∞ *[p. 210]* From the inferior portion of the lacrimal sac, the **nasolacrimal duct** passes through the *nasolacrimal canal,* formed by the lacrimal bone and the maxillary bone. This duct delivers the tears to the *inferior meatus* of the nasal cavity.

The Eye

The eyes are extremely sophisticated visual instruments—more versatile and adaptable than the most expensive cameras, yet light, compact, and durable. Each eye is a slightly irregular spheroid with an average diameter of 24 mm (almost 1 in.), a little smaller than a Ping-Pong ball, and a weight of about 8 g (0.28 oz). Within the orbit, the eyeball shares space with the extrinsic eye muscles, the lacrimal gland, and the cranial nerves and blood vessels that supply the eye and adjacent portions of the orbit and face. A mass of **orbital fat** provides padding and insulation.

The wall of the eye contains three distinct layers, or *tunics* (Figure 17-9b●): an outer *fibrous tunic,* an intermediate *vascular tunic,* and an inner *neural tunic.* The visual receptors, or *photoreceptors*, are located in the neural tunic. The eyeball is hollow; its interior can be divided into two *cavities* (Figure 17-9a,d●). The large **posterior cavity** is also called the *vitreous cham-*

ber, because it contains the gelatinous *vitreous body.* The smaller **anterior cavity** is subdivided into two *chambers,* anterior and posterior. The shape of the eye is stabilized in part by the vitreous body and the clear *aqueous humor* that fills the anterior cavity.

The Fibrous Tunic

The **fibrous tunic,** the outermost layer of the eye, consists of the *sclera* and the *cornea.* The fibrous tunic (1) provides mechanical support and some degree of physical protection, (2) serves as an attachment site for the extrinsic eye muscles, and (3) contains structures that assist in the focusing process.

Most of the ocular surface is covered by the **sclera** (SKLER-uh) (Figure 17-9a,d●). The sclera, or "white of the eye," consists of a dense fibrous connective tissue containing both collagen and elastic fibers. This layer is thickest over the posterior surface of the eye, near the exit of the optic nerve, and thinnest over the anterior surface. The six extrinsic eye muscles insert on the sclera, blending their collagen fibers with those of the outer tunic. ∞ *[p. 327]*

The surface of the sclera contains small blood vessels and nerves that penetrate the sclera to reach internal structures. The network of small vessels interior to the ocular conjunctiva generally does not carry enough blood to lend an obvious color to the sclera, but on close inspection the vessels are visible as red lines against the white background of collagen fibers.

The transparent cornea is structurally continuous with the sclera; the **limbus** (Figures 17-8a, 17-9a●) is the border between the two. Beneath the delicate corneal epithelium, the cornea consists primarily of a dense matrix containing multiple layers of collagen fibers. These fibers are organized into a series of layers that do not interfere with the passage of light. There are no blood vessels in the cornea. The superficial epithelial cells must obtain oxygen and nutrients from the tears that flow across their free surfaces. There are also numerous free nerve endings in the cornea, and it is the most sensitive portion of the eye.

CORNEAL TRANSPLANTS Corneal damage will cause blindness even though the rest of the eye—including the photoreceptors—may be perfectly normal. The cornea has a very restricted ability to repair itself, so corneal injuries must be treated immediately to prevent serious vision losses. Restoring vision after corneal scarring generally requires the replacement of the cornea through a corneal transplant. Corneal replacement is probably the most common form of transplant surgery. Such transplants can be performed between unrelated individuals, because there are no blood vessels to carry white blood cells, which attack foreign tissues, into the area. Corneal grafts are obtained from the eyes of donors who have died from illness or accident. For best results, the tissues must be removed within 5 hours after the donor's death.

The Vascular Tunic (Uvea)

The **vascular tunic,** or **uvea,** contains numerous blood vessels, lymphatics, and the intrinsic eye muscles. The functions of this middle layer include (1) providing a route for blood vessels and lymphatics that supply tissues of the eye; (2) regulating the amount of light that enters the eye; (3) secreting and reabsorbing the *aqueous humor* that circulates within the eye; and (4) controlling the shape of the lens, an essential part of the focusing process. The vascular tunic includes the *iris,* the *ciliary body,* and the *choroid* (Figure 17-9a,d●).

THE IRIS The **iris,** which we can see through the transparent corneal surface, contains blood vessels, pigment cells, and two layers of smooth muscle fibers (Figure 17-9c●). When these muscles contract, they change the diameter of the central opening, or **pupil,** of the iris. One group of smooth muscle fibers, the **pupillary constrictor muscles,** forms a series of concentric circles around the pupil. When these sphincter muscles contract, the diameter of the pupil decreases. A second group of smooth muscles, the **pupillary dilator muscles,** extends radially away from the edge of the pupil. Contraction of these muscles enlarges the pupil. Both muscle groups are controlled by the autonomic nervous system. For example, parasympathetic activation in response to bright light causes the pupils to constrict (the *consensual light reflex*), and sympathetic activation in response to dim light causes the pupils to dilate (the *pupillary reflex*).

The body of the iris consists of a highly vascular, pigmented, loose connective tissue. The anterior surface has no epithelial covering, although there is an incomplete layer of fibroblasts and melanocytes. Melanocytes are also scattered within the body of the iris. The posterior surface is covered by a pigmented epithelium that contains melanin granules. This pigmented epithelium is part of the neural tunic. Your eye color is determined by the density and distribution of melanocytes and by the density of the pigmented epithelium. When there are few melanocytes in the connective tissue of the iris, light passes through it and bounces off the pigmented epithelium. The eye then appears blue. Individuals with gray, brown, or black eyes have increasing numbers of melanocytes in the body and surface of the iris. The eyes of human albinos appear a very pale gray or blue-gray.

THE CILIARY BODY At its periphery, the iris attaches to the anterior portion of the ciliary body. The **ciliary body** begins at the junction between the cornea and the sclera (Figure 17-9a,d●). It extends posteriorly to the **ora serrata** (Ō-ra ser-RA-tuh; serrated mouth), the serrated anterior edge of the *neural retina,* which contains the visual receptors. The bulk of the ciliary body consists of the **ciliary muscle**, a smooth muscular ring that projects into the interior of the eye. Its epithelium is thrown into numerous folds called **ciliary processes**. The **suspensory ligaments** of the lens attach to these processes. These connective tissue fibers hold the lens posterior to the iris and centered on the pupil. As a result, any light passing through the pupil and headed for the photoreceptors will pass through the lens.

THE CHOROID The **choroid** is a vascular layer that separates the fibrous and neural tunics posterior to the ora serrata (Figure 17-9a,b,d●). It is covered by the sclera and attached to the outermost layer of the retina. The choroid contains an extensive capillary network that delivers oxygen and nutrients to the retina. It also contains melanocytes, which are especially numerous near the sclera.

The Neural Tunic

The **neural tunic,** or **retina,** is the innermost layer of the eye. It consists of a thin, outer **pigmented layer** and a thick inner layer, the **neural retina,** which contains the visual receptors and associated neurons (Figure 17-10a●). The pigmented layer absorbs light that passes through the neural retina, and the pigment cells have important biochemical interactions with the retinal photoreceptors. The neural retina contains (1) the *photoreceptors* that respond to light, (2) supporting cells and neurons that perform preliminary processing and integration of visual information, and (3) blood vessels supplying tissues that line the posterior cavity.

The two retinal layers are normally very close together but not tightly interconnected. The pigmented layer of the retina continues over the ciliary body and iris, although the neural retina extends anteriorly only as far as the ora serrata. The neural retina thus forms a cup that establishes the posterior and lateral boundaries of the posterior cavity (Figures 17-9a,d and 17-10b●).

RETINAL ORGANIZATION In sectional view, the retina contains several layers of cells (Figure 17-10a●). The outermost layer, closest to the pigmented layer, contains the **photoreceptors**, or cells that detect light.

The photoreceptors are entirely dependent on the diffusion of oxygen and nutrients from blood vessels in the choroid. In a **detached retina,** the neural retina becomes separated from the pigmented layer. This condition can result from a sudden impact to the eye or from a variety of other factors. Unless the two layers of the neural tunic are reattached, the photoreceptors will degenerate and vision will be lost. The reattachment is generally performed by "welding" the

two layers together by means of laser beams focused through the cornea.

There are two types of photoreceptors: rods and cones. **Rods** do not discriminate among different colors of light. They are very light-sensitive and enable us to see in dimly lit rooms, at twilight, and in pale moonlight. **Cones** provide us with color vision. There are three types of cones, and their stimulation in various combinations provides the perception of different colors. Cones give us sharper, clearer images than rods do, but cones require more-intense light. If you sit outside at sunset, you will probably be able to tell when your visual system shifts from cone-based vision (clear images in full color) to rod-based vision (relatively grainy images in black and white).

Rods and cones are not evenly distributed across the outer surface of the retina. Approximately 125 million rods form a broad band around the periphery of the retina. Roughly 6 million cones span the posterior retinal surface. Most of these are concentrated in the area where a visual image arrives after it passes through the cornea and lens. There are no rods in this region, which is known as the **macula lutea** (LOO-tē-uh; yellow spot). The highest concentration of cones occurs in the central portion of the macula lutea, an area called the **fovea** (FŌ-vē-uh; shallow depression), or *fovea centralis* (Figure 17-10c●). The fovea is the site of sharpest vision: When you look directly at an object, its image falls on this portion of the retina (Figure 17-9d●). A line drawn from the center of that object through the center of the lens to the fovea establishes the **visual axis** of the eye.

You are probably already aware of the visual consequences of this distribution. When you look directly at an object, you are focusing its image on your fovea, the center of color vision. You see a very good image as long as there is enough light to stimulate the cones. In very dim light, cones cannot function. For example, when you try to stare at a dim star, you are unable to see it. But if you look a little to one side rather than directly at the star, you will see it quite clearly. Shifting your gaze moves the image of the star from the fovea, where it does not provide enough light to stimulate the cones, to the periphery, where it can affect the more sensitive rods.

The rods and cones synapse with roughly 6 million **bipolar cells** (Figure 17-10a●). Bipolar cells in turn synapse within the layer of **ganglion cells** adjacent to the posterior cavity. A network of **horizontal cells** extends across the outer portion of the retina at the level of the synapses between photoreceptors and bipolar cells. A comparable layer of **amacrine** (AM-a-krin) **cells** occurs where bipolar cells synapse with ganglion cells. Horizontal and amacrine cells can facilitate or inhibit communication between photoreceptors and ganglion cells, adjusting the sensitivity of the retina.

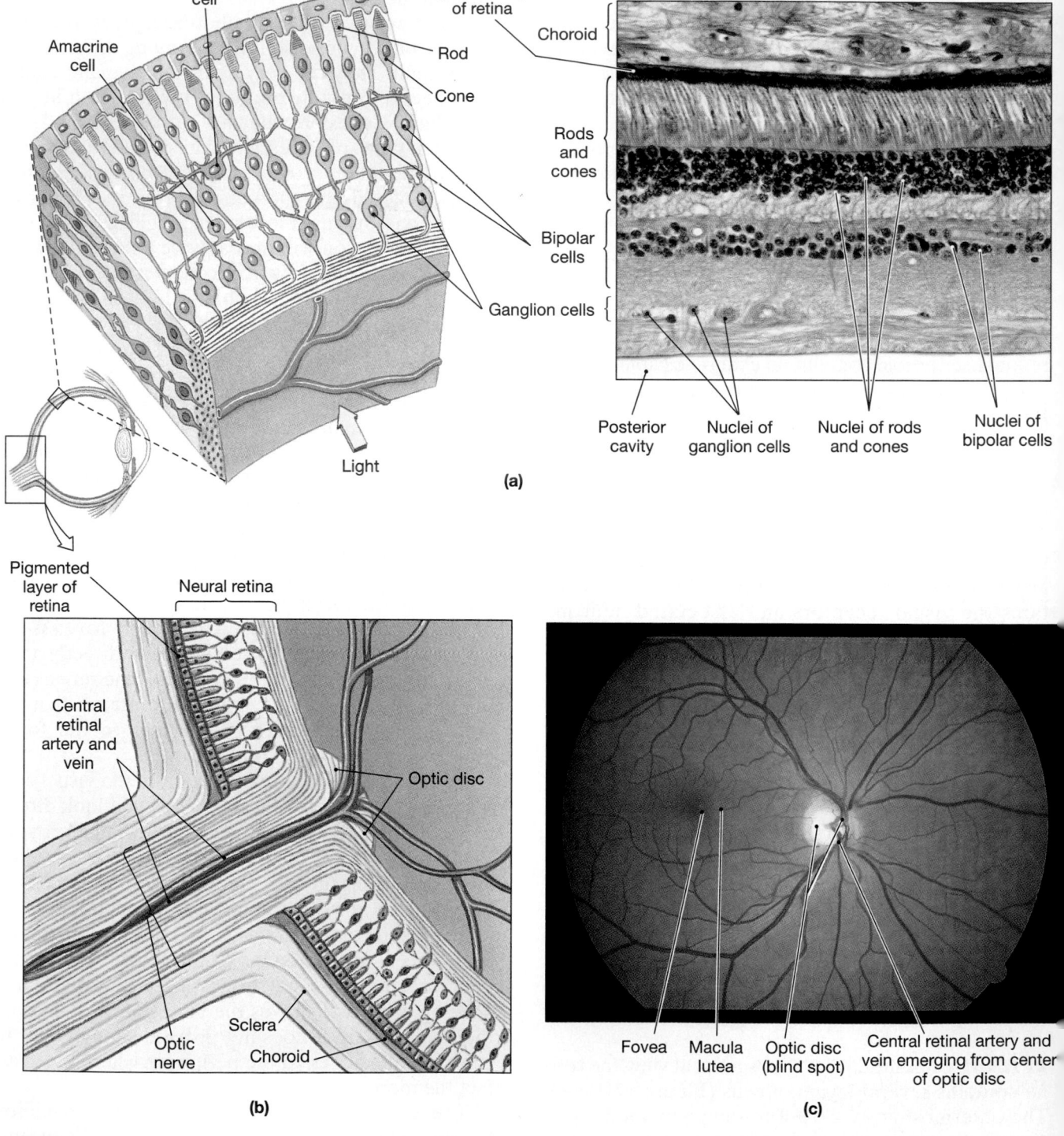

●**FIGURE 17-10 Retinal Organization.** **(a)** Cellular organization of the retina. Note that the photoreceptors are located closest to the choroid rather than near the posterior cavity (vitreous chamber). (LM × 290) **(b)** The optic disc in diagrammatic horizontal section. **(c)** A photograph of the retina as seen through the pupil of the eye.

The effect could be compared to adjusting the "contrast" setting on a television. The activities of these cells play an important role in the eye's adjustment to dim or brightly lit environments.

THE OPTIC DISC Axons from an estimated 1 million ganglion cells converge on the **optic disc**, a circular region just medial to the fovea. The optic disc is the origin of the optic nerve (N II). From this point, the axons turn, penetrate the wall of the eye, and proceed toward the diencephalon (Figure 17-10b•). The *central retinal artery* and *central retinal vein,* which supply the retina, pass through the center of the optic nerve and emerge on the surface of the optic disc (Figure 17-10b,c•). There are no photoreceptors or other retinal structures at the optic disc. Because light striking this area goes unnoticed, the optic disc is commonly called the **blind spot.** You do not notice a blank spot in your visual field primarily because involuntary eye movements keep the visual image moving and allow your brain to fill in the missing information. A simple experiment, shown in Figure 17-11•, will demonstrate the presence and location of the blind spot within your visual field.

> **DIABETIC RETINOPATHY** A *retinopathy* is a disease of the retina. *Diabetic retinopathy* develops in most individuals with *diabetes mellitus,* an endocrine disorder that affects primarily glucose metabolism. Many different systems are affected, but serious circulatory problems are very common. Diabetic retinopathy develops over a period of years due to degeneration and rupture of retinal blood vessels. Visual acuity is lost, and, over time, the photoreceptors are destroyed because of oxygen starvation.

The Chambers of the Eye

The ciliary body and lens divide the interior of the eye into a large *posterior cavity,* also called the *vitreous*

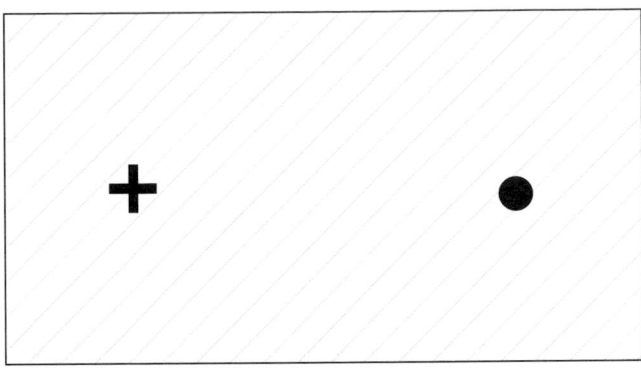

•*FIGURE 17-11* **The Optic Disc.** Close your left eye and stare at the cross with your right eye, keeping the cross in the center of your field of vision. Begin with the page a few inches away from your eye, and gradually increase the distance. The dot will disappear when its image falls on the blind spot, at your optic disc. To check the blind spot in your left eye, close your right eye and repeat this sequence while you stare at the dot.

chamber, and a smaller *anterior cavity.* The anterior cavity is subdivided into the **anterior chamber,** which extends from the cornea to the iris, and a **posterior chamber** between the iris and the ciliary body and lens. The anterior and posterior chambers are filled with the fluid *aqueous humor.* Aqueous humor circulates within the anterior cavity, passing from the posterior to the anterior chamber through the pupil (Figure 17-12•). The posterior cavity is filled with a gelatinous substance known as the *vitreous body,* or *vitreous humor.*

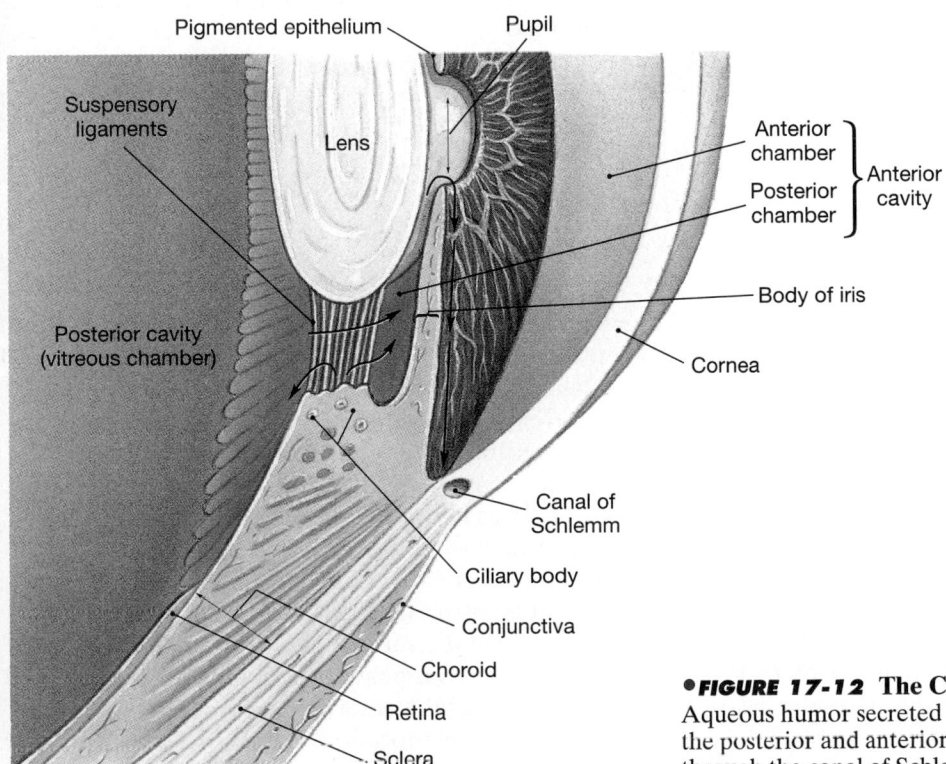

•*FIGURE 17-12* **The Circulation of Aqueous Humor.** Aqueous humor secreted at the ciliary body circulates through the posterior and anterior chambers before it is reabsorbed through the canal of Schlemm.

AQUEOUS HUMOR **Aqueous humor** forms through active secretion by epithelial cells of the ciliary body's ciliary processes (Figure 17-12●). It is secreted to the posterior chamber at a rate of 1–2 μl per minute. The epithelial cells regulate its composition, which resembles that of cerebrospinal fluid. The aqueous humor circulates, so in addition to forming a fluid cushion, it provides an important route for nutrient and waste transport.

The eye is filled with fluid, and fluid pressure in the aqueous humor helps retain the eye's shape. Fluid pressure also stabilizes the position of the retina, pressing the photoreceptor layer against the pigmented layer. In effect, the aqueous humor acts like the air inside a balloon. The **intraocular pressure** can be measured in the anterior chamber, where the fluid pushes against the inner surface of the cornea. Intraocular pressure is most often checked by bouncing a tiny blast of air off the surface of the eye and then measuring the deflection produced. Normal intraocular pressures range from 12 to 21 mm Hg.

Aqueous humor returns to the circulation from the anterior chamber. After filtering through a network of connective tissues located near the base of the iris, aqueous humor enters the **canal of Schlemm,** a passageway that extends completely around the eye at the level of the limbus. Collecting channels deliver the aqueous humor from the canal to veins in the sclera. The rate of removal normally keeps pace with the rate of generation at the ciliary processes, and aqueous humor is removed and recycled within a few hours of its formation.

THE VITREOUS BODY The posterior cavity of the eye contains the **vitreous body** (or *vitreous humor*), a gelatinous mass. The vitreous body helps stabilize the shape of the eye and gives additional physical support to the retina. Specialized cells embedded in the vitreous body produce the collagen fibers and proteoglycans that account for the consistency of this mass. Unlike the aqueous humor, the vitreous body is not replaced. Aqueous humor produced in the posterior chamber freely diffuses through the vitreous body and across the surface of the retina (Figure 17-12●).

The Lens

The **lens** lies posterior to the cornea, held in place by the suspensory ligaments that originate on the ciliary body of the choroid (Figures 17-9, pp. 552–553, and 17-12●). The primary function of the lens is to focus the visual image on the retinal photoreceptors. It does so by changing its shape.

The lens consists of concentric layers of cells that are precisely organized. A dense fibrous capsule covers the entire lens. Many of the capsular fibers are elastic. Unless an outside force is applied, they will contract and make the lens spherical. Around the edges of the lens, the capsular fibers intermingle with those of the suspensory ligaments. The cells in the interior of the lens are called **lens fibers**. These highly specialized cells have lost their nuclei and other organelles. They are slender and elongate and are filled with transparent proteins called **crystallins**.

 Glaucoma

If aqueous humor cannot enter the canal of Schlemm, the condition of **glaucoma** results. Although drainage is impaired, the production of aqueous humor continues, and the intraocular pressure begins to rise. The fibrous scleral coat cannot expand significantly, so the increasing pressure begins to distort soft tissues within the eye. The tough sclera cannot enlarge like an inflating balloon, but it does have one weak point, where the optic nerve penetrates the wall of the eye. Because the optic nerve must penetrate all three tunics, it is not wrapped in connective tissue. When intraocular pressures have risen to roughly twice normal levels, the distortion of the nerve fibers begins to block action potential propagation, and vision begins to deteriorate. If this condition is not corrected, blindness eventually results.

Glaucoma affects roughly 2 percent of the population over age 35. In most cases, the primary factors responsible cannot be determined. Because glaucoma is a relatively common condition—over 2 million cases in the United States alone—most eye exams include a test of intraocular pressure. Glaucoma may be treated by the application of drugs that constrict the pupil and tense the edge of the iris, making the surface more permeable to aqueous humor. Surgical correction involves perforating the wall of the anterior chamber to encourage drainage. This procedure is now performed by laser surgery on an outpatient basis.

CATARACTS The transparency of the lens depends on a precise combination of structural and biochemical characteristics. When that balance becomes disturbed, the lens loses its transparency, and the abnormal lens is known as a **cataract**. Cataracts may result from drug reactions, injuries, or radiation, but **senile cataracts**, a natural consequence of aging, are the most common form.

Over time, the lens takes on a yellowish hue, and eventually it begins to lose its transparency. As the lens becomes "cloudy," the individual needs brighter and brighter light for reading, and visual clarity begins to fade. If the lens becomes completely opaque, the person will be functionally blind, even though the retinal receptors are normal. Modern surgical procedures involve removing the lens, either intact or in pieces, after it has been shattered with high-frequency sound. The missing lens can be replaced by an artificial substitute, and vision can then be fine-tuned with glasses or contact lenses.

REFRACTION There are approximately 130 million photoreceptors in the retina, each monitoring a specific location. A visual image results from the processing of information provided by the entire receptor population. The eye has often been compared to a camera. Like a camera lens, the lens of the eye must focus the arriving image if it is to provide any useful information. "In focus" means that the rays of light arriving from an object strike the sensitive surface of the film or retina in precise order so as to form a miniature image of the original. If the rays are not perfectly focused, the image will be blurry. Focusing in the eye normally occurs in two steps, as light passes through (1) the cornea and (2) the lens.

Light is bent, or **refracted,** when it passes from a medium with one density to a medium with a different density. You can see refraction clearly by sticking a pencil into a glass of water. Because refraction occurs as the light passes into the air from the much denser water, the pencil shaft appears to bend sharply at the air–water interface.

In the human eye, the greatest amount of refraction occurs when light passes from the air into the corneal tissues, which have a density close to that of water. When you open your eyes underwater, you cannot see clearly because the air–water interface has been eliminated; light is passing from one watery medium to another.

Additional refraction takes place when the light passes from the aqueous humor into the relatively dense lens. The lens provides the extra refraction you need to focus the light rays from an object toward a specific point of intersection, called a **focal point,** on the retina. The distance between the center of the lens and its focal point is the **focal distance.** The focal distance of a lens, whether the lens of your eye or the lens in your camera, is determined by the following two factors:

1. *The distance of the object from the lens.* The focal distance increases as an object moves closer to the lens (Figure 17-13a●).
2. *The shape of the lens.* The rounder the lens, the more refraction occurs, so a very round lens has a shorter focal distance than a flatter one (Figure 17-13b●).

ACCOMMODATION Using a camera, we can focus an image by moving the lens toward or away from the film. This method of focusing cannot work in our eyes, because the distance from the lens to the macula lutea cannot change. We focus images on the retina by changing the shape of the lens so as to keep the focal length constant. This process is called **accommodation.** During accommodation, the lens becomes rounder to focus the image of a nearby object on the retina, and it flattens when we focus on a distant object, as we noted in number 2 above.

The lens is held in place by the suspensory ligaments that originate at the ciliary body. Smooth muscle fibers in the ciliary body act like sphincter muscles. When the ciliary muscle contracts, the ciliary body moves toward the lens. This movement reduces the tension in the suspensory ligaments, and the elastic

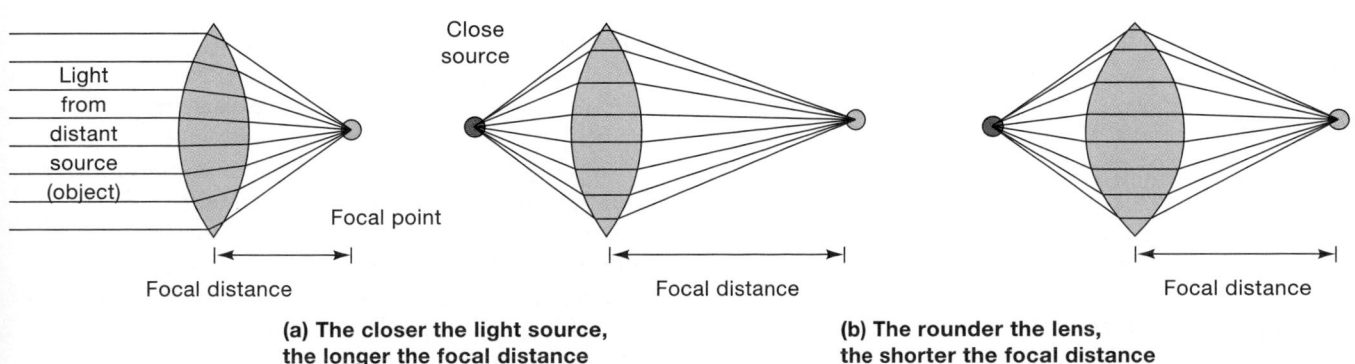

(a) **The closer the light source, the longer the focal distance**

(b) **The rounder the lens, the shorter the focal distance**

●*FIGURE 17-13* **Principles of Image Formation.** Light rays from a given source are refracted when they reach the lens of the eye. From the lens, the rays are focused onto a single focal point. **(a)** The focal distance increases as the object nears the lens. **(b)** A rounder lens has a shorter focal distance than a flatter lens does.

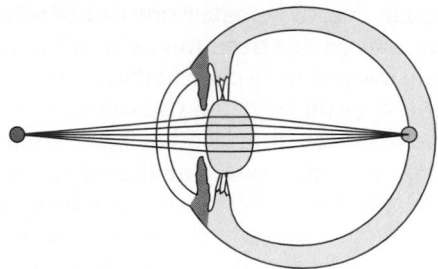

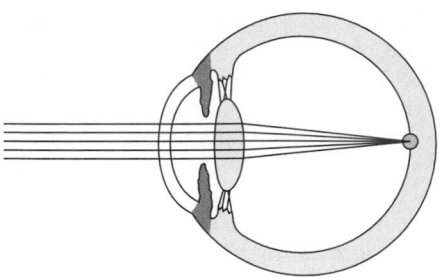

(a) Ciliary muscle contracted, lens rounded for close vision

(b) Ciliary muscle relaxed, lens flattened for distant vision

●*FIGURE 17-14* **Accommodation.** For the eye to form a sharp image, the focal distance must equal the distance between the center of the lens and the retina. **(a)** When the ciliary muscle contracts, the ligaments allow the lens to round up. **(b)** When the ciliary muscle relaxes, the suspensory ligaments pull against the margins of the lens and flatten it.

capsule pulls the lens into a more spherical shape. The rounder shape increases the refractive power of the lens, enabling it to bring light from nearby objects into focus on the retina (Figure 17-14a●). When the ciliary muscle relaxes, the suspensory ligaments pull at the circumference of the lens, making the lens relatively flat (Figure 17-14b●).

The greatest amount of refraction is required to view objects that are very close to the lens. The inner limit of clear vision, known as the *near point of vision,* is determined by the degree of elasticity in the lens. Children can usually focus on something 7–9 cm from the eye, but over time the lens tends to become stiffer and less responsive. A young adult can usually focus on objects 15–20 cm

 ## Accommodation Problems

In the healthy eye, when the ciliary muscles are relaxed and the lens is flattened, a distant image will be focused on the retinal surface (Figure 17-15a●). This condition is called **emmetropia** (*emmetro-,* proper), or normal vision.

Figure 17-15b,c● diagrams two common problems with the accommodation mechanism. If the eyeball is too deep or the resting curvature of the lens too great, the image of a distant object will form in front of the retina (Figure 17-15b●). The individual will see distant objects as blurry and out of focus. Vision at close range will be normal, because the lens will be able to round up as needed to focus the image on the retina. As a result, such individuals are said to be nearsighted. Their condition is more formally termed **myopia** (*myein,* to shut + *ops,* eye). Myopia can be treated by placing a *diverging* lens in front of the eye (Figure 17-15d●). This lens shape, typical of the lenses used in prescription corrective glasses, spreads the light rays apart as if the object were closer to the viewer.

If the eyeball is too shallow or the lens too flat, **hyperopia** results (Figure 17-15c●). The ciliary muscles must contract to focus even a distant object on the retina, and at close range the lens cannot provide enough refraction to focus an image on the retina. Individuals with this problem are said to be farsighted,

because they can see distant objects most clearly. Older individuals become farsighted as their lenses lose elasticity; this form of hyperopia is called **presbyopia** (*presbys,* old man). Hyperopia can be corrected by placing a *converging* lens in front of the eye. This lens provides the additional refraction needed to bring nearby objects into focus (Figure 17-15e●).

Variable success at correcting myopia and hyperopia has been achieved by surgically reshaping the cornea to alter its refractive powers. This procedure, called **radial keratotomy,** remains controversial. Although roughly two-thirds of patients are satisfied with the results, corneal healing takes several years. Many ophthalmological surgeons have expressed concerns that the scarring that develops after radial keratotomy creates weak points in the cornea that may increase the chances for a dangerous corneal rupture.

Another controversial procedure is **photorefractive keratectomy (PRK),** in which a computer-guided laser shapes the cornea to exact specifications. Tissue is removed only to a depth of 10–20 μm—no more than about 10 percent of the cornea's thickness. The entire procedure can be done in less than a minute. Each year, an estimated 100,000 people undergo PRK therapy in the United States. Advances in laser design and improvements in accuracy may make the procedure more popular during the next decade.

away. As aging proceeds, this distance gradually increases; the near point at age 60 is typically about 83 cm.

If light passing through the cornea and lens fails to refract properly, the visual image will be distorted. In the condition called **astigmatism,** the degree of curvature in the cornea or lens varies from one axis to another. For example, the cornea may be more strongly curved in the vertical axis than it is in the horizontal axis. Astigmatism can be corrected by glasses or special contact lenses. Minor astigmatism is very common; the image distortion may be so minimal that people are unaware of the condition.

Image Reversal. Thus far, we have considered light that originates at a single source, either near or away from the viewer. An object in view is a complex light source that can be treated as a large number of individual points. Light from each point is focused on the retina as indicated in Figure 17-16a,b•. The result is the creation of a miniature image of the original, but the image arrives upside down and backward.

To understand why, consider Figure 17-16c•, a sagittal section through an eye that is looking at a telephone pole. The image of the top of the pole lands at the bottom of the retina, and the image of the bottom hits the top of the retina. Now consider Figure 17-16d•, a horizontal section through an eye that is looking at a picket fence. The image of the left edge of the fence falls on the right side of the retina, and the image of the right edge falls on the left side of the retina. The brain compensates for this image reversal, and we are not consciously aware of any difference between the orientation of the image on the retina and that of the object.

VISUAL ACUITY The clarity of vision, or **visual acuity,** is rated against a "normal" standard. A person whose vision is rated 20/20 can see details at a distance of 20 feet as clearly as would an individual with normal vision. Vision noted as 20/15 is better than average, for at 20 feet the person is able to see details that would be clear to a normal eye only at a distance of 15 feet. Conversely, a person with 20/30 vision must be 20 feet from an object to discern details that a person with normal vision could make out at a distance of 30 feet.

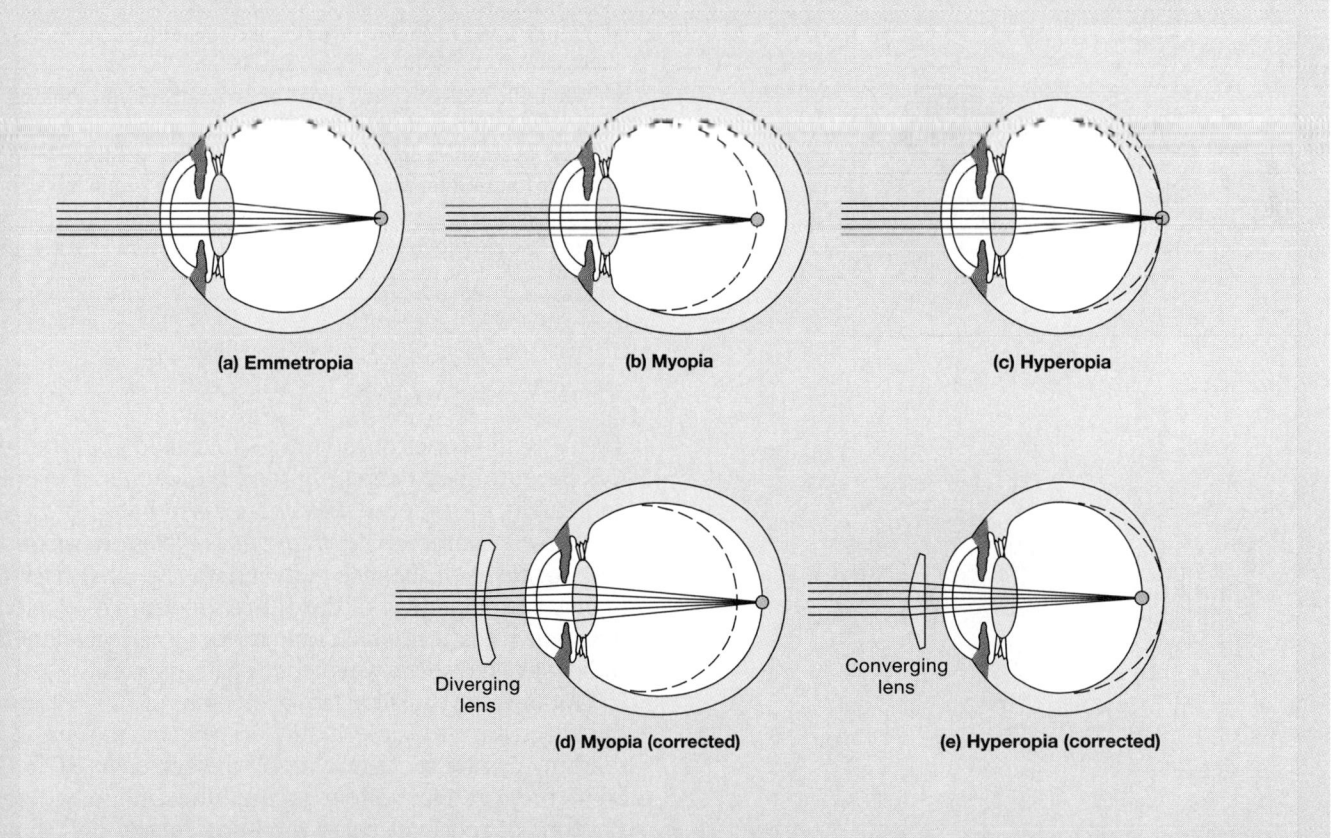

•*FIGURE 17-15* **Visual Abnormalities.** **(a)** In normal vision, the lens focuses the visual image on the retina. Common problems with the accommodation mechanism involve **(b)** myopia, an inability to lengthen the focal distance enough to focus the image of a distant object on the retina, and **(c)** hyperopia, an inability to shorten the focal distance adequately for nearby objects. These conditions can be corrected by placing appropriately shaped lenses in front of the eyes. **(d)** A diverging lens is used to correct myopia, and **(e)** a converging lens is used to correct hyperopia.

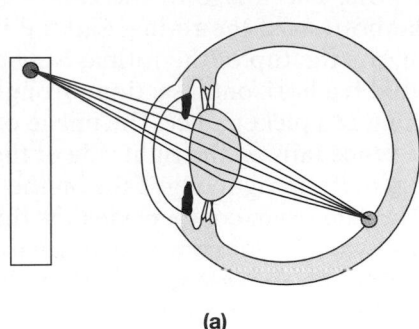

(a)

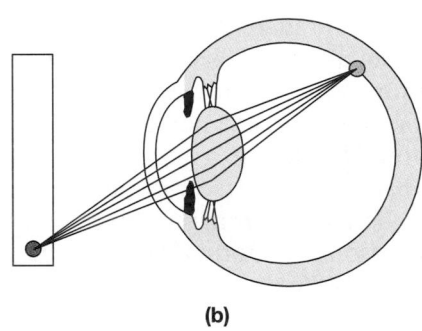

(b)

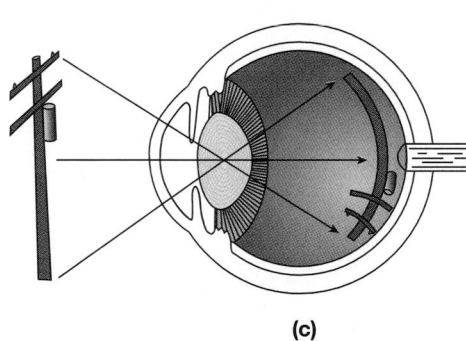

(c)

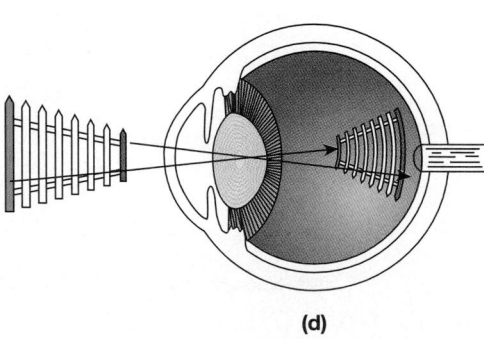

(d)

•*FIGURE 17-16* **Image Formation. (a,b)** Light from each portion of an object is focused on a different part of the retina. The resulting image arrives **(c, d)** upside down and backward.

When visual acuity falls below 20/200, even with the help of glasses or contact lenses, the individual is considered to be legally blind. There are probably fewer than 400,000 legally blind people in the United States; more than half are over 65 years of age. The term *blindness* implies a total absence of vision due to damage to the eyes or to the optic pathways. Common causes of blindness include diabetes mellitus, cataracts, glaucoma, corneal scarring, retinal detachment, accidental injuries, and hereditary factors that are as yet poorly understood.

SCOTOMAS Abnormal blind spots, or **scotomas** (skō-TŌ-muhz), may appear in the visual field at positions other than at the optic disc. Scotomas are permanent abnormalities that are fixed in position. They may result from compression of the optic nerve, damage to photoreceptors, or central damage along the visual pathway. You are probably familiar with *floaters,* small spots that drift across the field of vision. Floaters are generally temporary phenomena that result from blood cells or cellular debris within the vitreous body. They can be detected by staring at a blank wall or a white sheet of paper.

✔ Which layer of the eye would be affected first by inadequate tear production?

✔ When the lens of your eye is very round, are you looking at an object that is close to you or far from you?

✔ As Renee enters a dark room, most of the available light becomes focused on the fovea of her eye. Will she be able to see very clearly?

✔ How would blockage of the canal of Schlemm affect your vision?

Visual Physiology

The rods and cones of the retina are called *photoreceptors* because they detect *photons,* basic units of visible light. Light, a form of *radiant energy,* is radiated in waves that have a characteristic *frequency* (cycles per second) and *wavelength* (distance between wave peaks). Visible light is one small part of the entire spectrum of electromagnetic radiation, which ranges from long-wavelength radio waves to short-wavelength gamma rays.

Our eyes are sensitive to wavelengths of 700–400 nm, the spectrum of visible light. This spectrum, seen in a rainbow, can be remembered by the acronym ROY G. BIV (red, orange, yellow, green, blue, indigo, violet). Photons of red light carry the least energy, and those from the violet portion of the spectrum contain the most. Rods provide the CNS with information on the presence or absence of photons, without regard to wavelength. Cones provide information on the wavelength of arriving photons, giving us a perception of color.

Anatomy of Rods and Cones

Figure 17-17a● compares the structures of rods and cones. The elongate **outer segment** of a photoreceptor contains hundreds to thousands of flattened membranous plates, or **discs.** The names *rod* and *cone* refer to the outer segment's shape. In a rod, each disc is an independent entity and the outer segment forms an elongate cylinder. In cones, the discs are infoldings of the cell membrane and the outer segment tapers to a blunt tip.

A narrow connecting stalk attaches the outer segment to the **inner segment,** a region that contains all the usual cellular organelles. The inner segment makes synaptic contact with other cells, and it is here that neurotransmitters are released.

VISUAL PIGMENTS The discs of the outer segment in both rods and cones contain special organic compounds called **visual pigments.** The absorption of photons by visual pigments is the first key step in the process of photoreception. Visual pigments are derivatives of the compound **rhodopsin** (rō-DOP-sin), or *visual purple,* the visual pigment found in rods. Rhodopsin consists of a protein, **opsin,** bound to the pigment **retinal** (RET-i-nal). Retinal, also called *retinene,* is synthesized from **vitamin A**. One form of opsin is characteristic of all rods. Figure 17-17b● diagrams the structure of a rhodopsin molecule.

Cones contain the same retinal pigment that rods do, but in cones the retinal is attached to other forms of opsin. The type of opsin present determines the

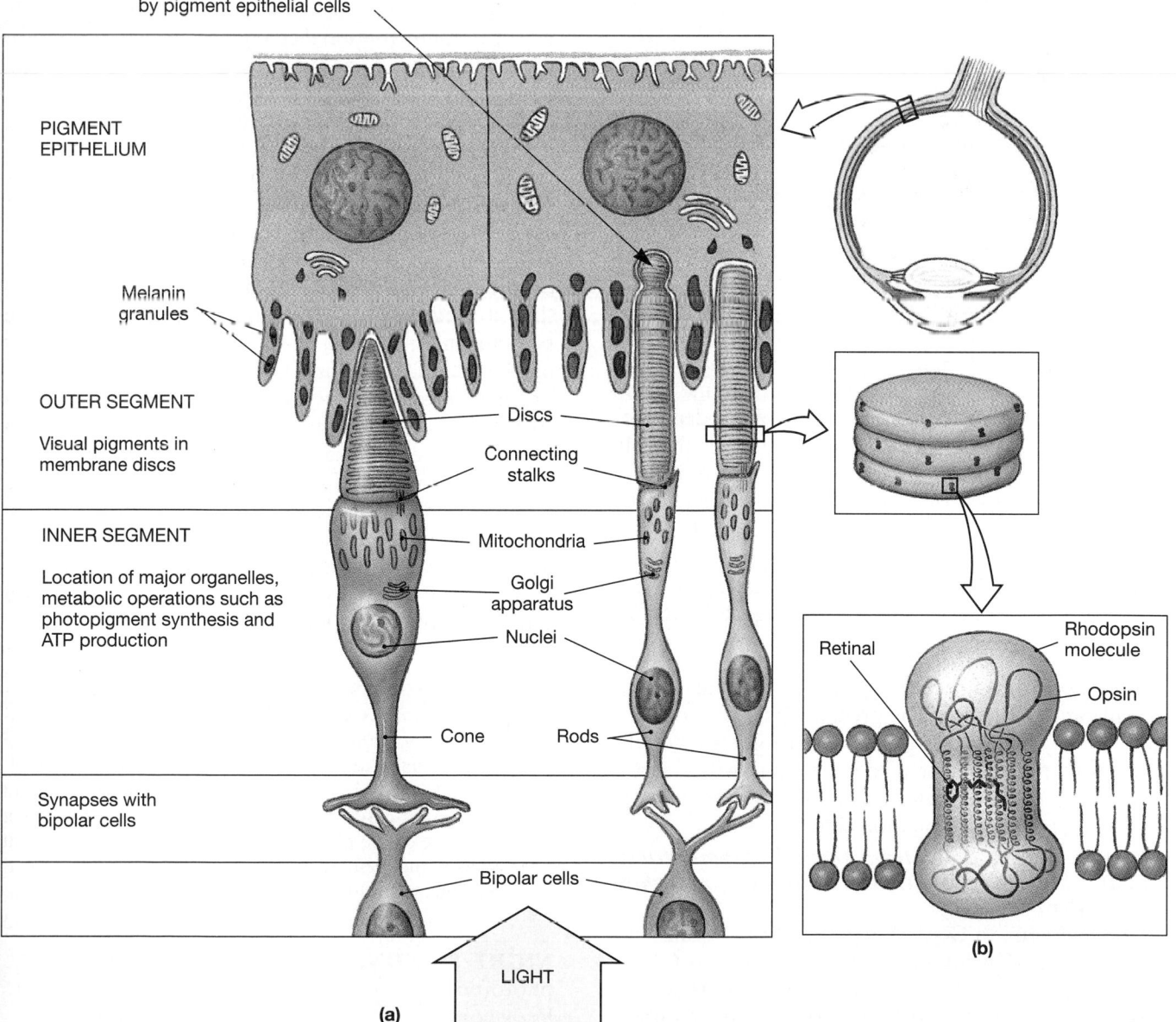

●FIGURE 17-17 Rods and Cones. (a) The structure of rods and cones. **(b)** The structure of a rhodopsin molecule.

wavelength of light that can be absorbed by the retinal pigment. Differential stimulation of these cone populations is the basis of color vision.

New discs containing visual pigment are continuously assembled at the base of the outer segment. A completed disc then moves toward the tip of the outer segment. After about 10 days, it will be shed in a small droplet of cytoplasm. These droplets are absorbed by the pigment cells, which break down the membrane components and reconvert the retinal to vitamin A. The vitamin A is then stored within the pigment cells for subsequent transfer to the photoreceptors.

Photoreception

We will now follow the steps in rhodopsin-based photoreception. In darkness, the cell membrane in the outer segment of the photoreceptor contains open, chemically regulated sodium ion channels. These gated channels are kept open in the presence of *cyclic-GMP* (*cyclic guanosine monophosphate*, or *cGMP*), a derivative of the high-energy compound *guanosine triphosphate* (GTP). Because these channels are open, the transmembrane potential is approximately –40 mV rather than the –70 mV typical of resting neurons. ∞ *[p. 382]* The inner segment continuously pumps sodium ions out of the cytoplasm. The movement of Na^+ into the outer segment, to the inner segment, and out of the cell is known as the *dark current* (Figure 17-18●).

At this transmembrane potential, the photoreceptor is continuously releasing neurotransmitters (glutamate) across synapses at the inner segment. The arrival of a photon reduces the dark current. As the membrane hyperpolarizes, the rate of neurotransmitter release decreases.

The process begins when a photon strikes the retinal portion of a rhodopsin molecule. There are two possible configurations for the bound retinal molecule: the **11-*cis*** form and the **11-*trans*** form. It is normally in the 11-*cis* form; on absorbing light, it adopts the more linear 11-*trans* form. This change in shape triggers a chain of enzymatic steps (Figure 17-18●):

STEP 1: *Opsin activation occurs.*

Rhodopsin is a receptor protein embedded in the membrane of the disc. The protein opsin is an enzyme that is inactive in the dark. When light strikes the rhodopsin molecule, the pigment retinal switches from the 11-*cis* to the 11-*trans* form. This switch activates the opsin portion of the molecule.

STEP 2: *Opsin activates a second enzyme, transducin, which in turn activates a third enzyme, phosphodiesterase (PDE).*

Transducin is a **G-protein,** a membrane-bound enzyme activated by interaction with receptor proteins in the cell membranes. In this case, transducin is activated by opsin, and transducin in turn activates **phosphodiesterase (PDE)**. Phosphodiesterase is an enzyme that breaks down cGMP.

STEP 3: *Cyclic-GMP levels decline, and gated sodium ion channels close.*

The removal of cGMP from the gated sodium channels results in their inactivation. The rate of Na^+ entry into the cytoplasm then decreases.

STEP 4: *The rate of neurotransmitter release declines.*

Because active transport continues to remove Na^+ from the cytoplasm, when the sodium channels close, the transmembrane potential drops toward –70 mV. As the membrane hyperpolarizes, the rate of neurotransmitter release decreases. This decrease indicates to the adjacent bipolar cell that the photoreceptor has absorbed a photon.

RECOVERY AFTER STIMULATION After absorbing a photon, retinal does not spontaneously convert back to the 11-*cis* form. Instead, the entire rhodopsin molecule must be broken down and reassembled. Shortly after the conformational change occurs, the rhodopsin molecule begins to break down into retinal and opsin, a process known as *bleaching* (Figure 17-19●, p. 566). Before it can recombine with opsin, the retinal must be enzymatically converted to the 11-*cis* form. This conversion requires energy in the form of ATP (adenosine triphosphate), and it takes time. Bleaching contributes to the lingering visual impression you have after you see a flashbulb go off. After an intense exposure to light, a photoreceptor cannot respond to further stimulation until its rhodopsin molecules have been regenerated. As a result, a "ghost" image remains on the retina. Bleaching is seldom noticeable under ordinary circumstances, because the eyes are constantly making small, involuntary changes in position that move the image across the retinal surface.

While the rhodopsin molecule is being reassembled, membrane permeability is returning to normal. Opsin is inactivated when bleaching occurs, and the breakdown of cGMP halts as a result. As other enzymes generate cGMP in the cytoplasm, the chemically gated sodium channels reopen.

RETINITIS PIGMENTOSA The term *retinitis pigmentosa* (RP) refers to a collection of inherited retinopathies. Together they are the most common inherited visual pathology, affecting approximately 1 individual in 3000. The visual receptors gradually deteriorate, and blindness eventually results. The mutations responsible affect the structure of the photoreceptors—specifically, the visual pigments of the membrane discs. It is not known how the altered pigments lead to photoreceptor destruction.

NIGHT BLINDNESS The visual pigments of the photoreceptors are synthesized from vitamin A. The body contains vitamin A reserves sufficient for several months, and a significant amount is stored in the cells of the pigmented layer of the neural tunic. If

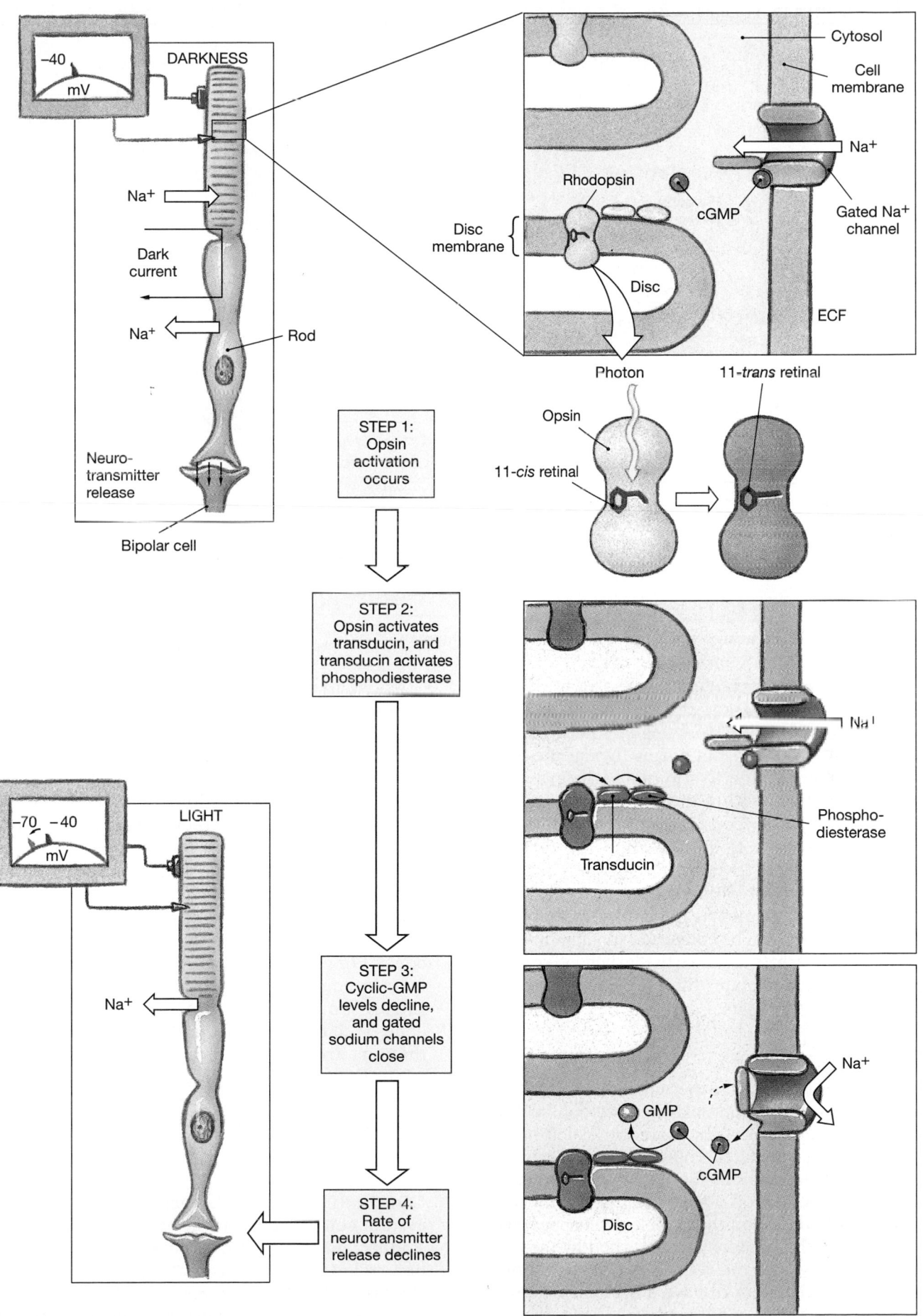

●**FIGURE 17-18** Photoreception

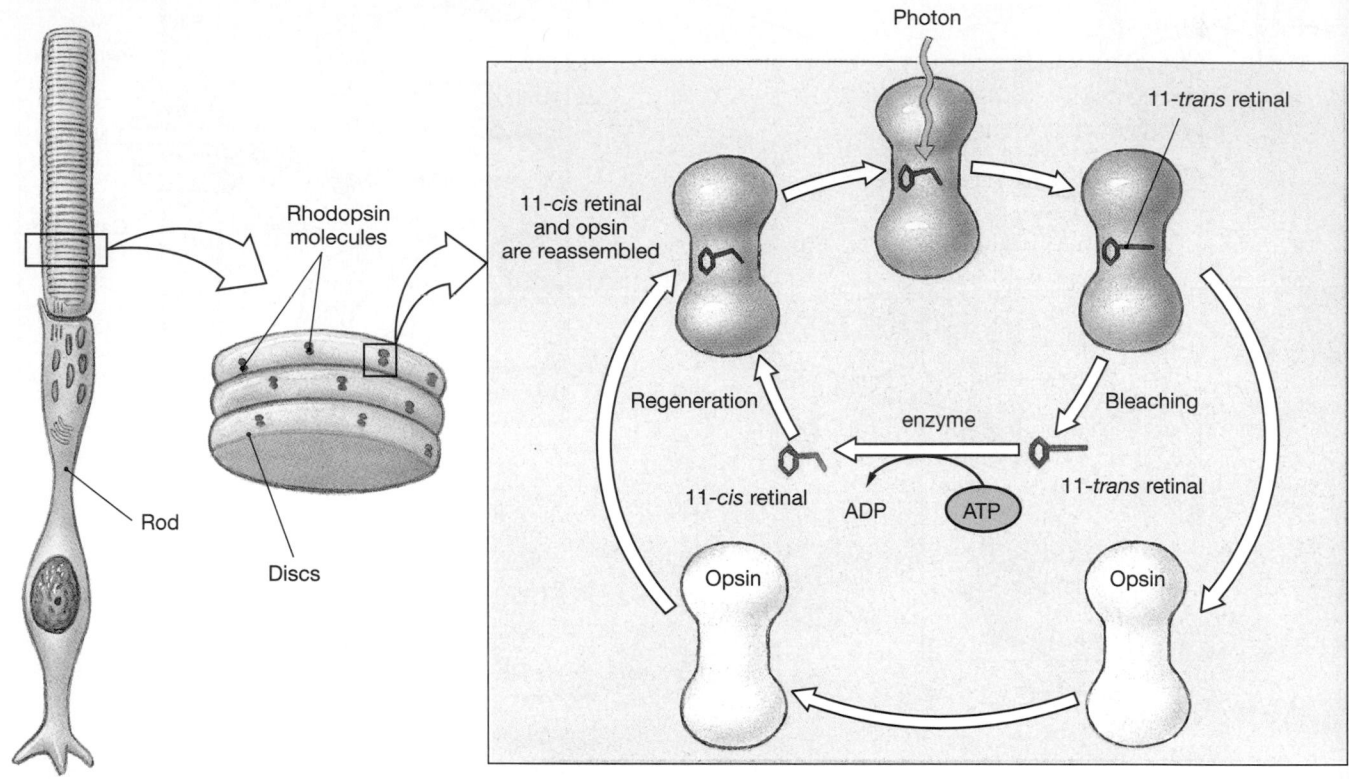

•*FIGURE 17-19* **Bleaching and Regeneration of Visual Pigments**

dietary sources are inadequate, these reserves are gradually exhausted, and the amount of visual pigment in the photoreceptors begins to decline. Daylight vision is affected, but during the day the light is usually bright enough to stimulate whatever visual pigments remain within the densely packed cone population of the fovea. As a result, the problem first becomes apparent at night, when the dim light proves insufficient to activate the rods. This condition, known as **night blindness**, can be treated by administration of vitamin A. The carotene pigments in many vegetables can be converted to vitamin A within the body. Carrots are a particularly good source of carotene, which explains the old adage that carrots are good for your eyes.

Color Vision

An ordinary lightbulb or the sun emits photons of all wavelengths. These photons stimulate both rods and cones. When all types of cones are stimulated, or when rods alone are stimulated, you see a "white" light. Your eyes also detect photons that reach your retina after the photons bounce off objects around you. If photons of all colors bounce off the object, it will appear white to you; if all the photons are absorbed by the object (and so none reach the retina), it will appear black. An object will appear to have a particular color if it reflects (or transmits) photons from one portion of the visible spectrum and absorbs the rest.

There are three types of cones: **blue cones, green cones,** and **red cones.** Each type has a different form of opsin and a sensitivity to a different range of wavelengths. Their stimulation in various combinations is the basis for color vision. In a normal individual, the cone population consists of 16 percent blue cones, 10 percent green cones, and 74 percent red cones. Although their sensitivities overlap, each type is most sensitive to a specific portion of the visual spectrum (Figure 17-20a•).

Color discrimination occurs through the integration of information arriving from all three types of cones. For example, the perception of yellow results from a combination of inputs from green cones (highly stimulated), red cones (stimulated), and blue cones (relatively unaffected). If all three cone populations are stimulated, we perceive the color as white. Because we also perceive white if rods, rather than cones, are stimulated, everything appears to be black and white when we enter dimly lit surroundings or walk by starlight.

COLOR BLINDNESS Persons unable to distinguish certain colors have a form of **color blindness**. The standard tests for color vision involve picking numbers or letters out of a complex and colorful picture (Figure 17-20b•). Color blindness occurs when one or more classes of cones are nonfunctional. The cones may be absent entirely, or they may be present

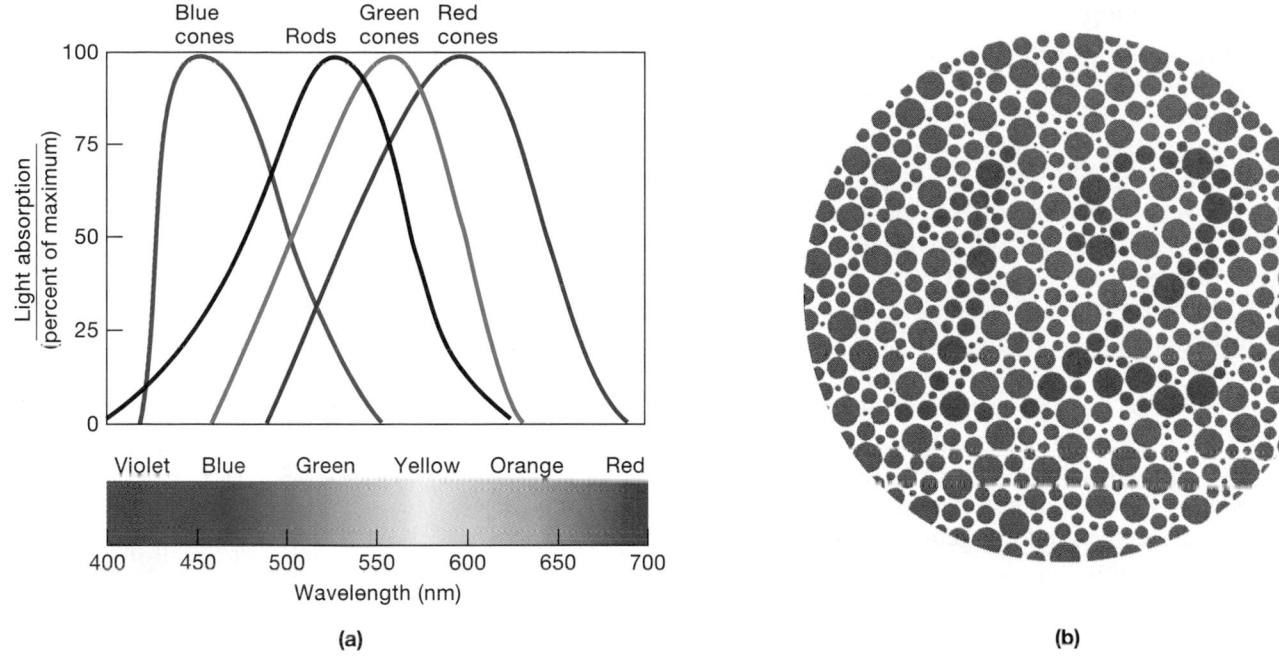

(a)

(b)

•**FIGURE 17-20 Cones and Color Vision.** (a) A graph comparing the absorptive characteristics of blue, green, and red cones with those of typical rods. Notice that the rod sensitivities overlap those of the cones and that the various cone types have overlapping sensitivity curves. **(b)** Part of a standard test for color vision. If you lack one or more populations of cones, you will be unable to distinguish the patterned image (the number 12).

but unable to manufacture the necessary visual pigments. In the most common type of color blindness (red–green color blindness), the red cones are missing, and the individual cannot distinguish red light from green light.

Inherited color blindness involving one or two cone pigments is not unusual. The genes for the red and green cone pigments are located on the X chromosome. As we shall see in Chapter 29, women have two copies of this chromosome (XX), whereas men have one X chromosome paired with a Y chromosome (XY). As a result, men are much more likely to be color blind than women, who must have the same abnormal cone pigment genes on both X chromosomes to exhibit color blindness. Ten percent of all men show some color blindness, whereas the incidence among women is only about 0.67 percent. Total color blindness is extremely rare; only 1 person in 300,000 fails to manufacture any cone pigments.

Light and Dark Adaptation

The sensitivity of your visual system varies with the intensity of illumination. After 30 minutes or more in the dark, almost all visual pigments will be fully receptive to stimulation. This is the **dark-adapted state.** When dark-adapted, the visual system is extremely sensitive. For example, a single rod will hyperpolarize in response to a single photon of light. Even more remarkable, if as few as seven rods absorb photons at one time, you will see a flash of light.

When the lights come on, at first they seem almost unbearably bright, but over the next few minutes your sensitivity decreases as bleaching occurs. Eventually, the rate of visual pigment breakdown is balanced by the rate of reformation. This condition is the **light-adapted state.** If you moved from the depths of a cave to the full light of midday, your receptor sensitivity would decrease by about 25,000 times.

A variety of central responses further adjust light sensitivity. Constriction of the pupil, via the pupillary constrictor reflex, reduces the amount of light entering your eye to 1/30 the maximum dark-adapted levels. Dilating the pupil fully can produce a 30-fold increase in the amount of light entering the eye, and facilitating some of the synapses along the visual pathway can perhaps triple its sensitivity. Hence, the entire system may increase its efficiency by a factor of more than 1 million.

The Visual Pathway

The visual pathway begins at the photoreceptors and ends at the visual cortex of the cerebral hemispheres. In other sensory pathways we have examined, there is at most one synapse between a receptor and a sensory neuron that delivers information to the CNS. In the visual pathway, the message must cross two synapses (photoreceptor to bipolar cell and bipolar cell to ganglion cell) before it heads toward the brain. The extra synapse adds to the synaptic delay, but it provides an opportunity for the processing and integration of visual information before it leaves the retina.

Retinal Processing

Each photoreceptor in the retina monitors a specific receptive field. The retina contains about 130 million photoreceptors, 6 million bipolar cells, and 1 million ganglion cells. There is thus a considerable amount of convergence at the start of the visual pathway. The degree of convergence differs between rods and cones. Regardless of the amount of convergence, each ganglion cell monitors a specific portion of the visual field.

As many as a thousand rods may pass information via their bipolar cells to a single ganglion cell. The ganglion cells that monitor rods are relatively large, and they are called **M cells** (*magnocells*). M cells provide information on general form, motion, and shadows under dim light conditions. Because there is so much convergence, when an M cell becomes active, it indicates that light has arrived in a general area rather than at a specific location.

The loss of specificity due to convergence is partially overcome by the fact that ganglion cells vary their activity according to the pattern of activity in their sensory field. The sensory field is generally circular (Figure 17-21●), and a ganglion cell typically responds differently to stimuli that arrive in the center than at the edges of its receptive field. Some ganglion cells **(on-center neurons)** are excited by light arriving in the center of their sensory field and are inhibited when light strikes the edges of their receptive field. Others **(off-center neurons)** are inhibited by light in the central zone but stimulated by illumination at the edges. On-center and off-center neurons provide information about which portion of their sensory field is illuminated.

Cones typically show very little convergence; in the fovea, the ratio of cones to ganglion cells is 1:1. The ganglion cells that monitor cones, called **P cells** (*parvo cells*), are smaller and more numerous than M cells. P cells are active in bright light, and they provide information on edges, fine detail, and color. Because little convergence occurs, the activation of a P cell means that light has arrived at one specific location. As a result, cones provide more-precise information about a visual image than do rods. In photographic terms, pictures formed by rods have a coarse, grainy appearance that blurs the details, whereas those produced by cones are fine-grained, sharp, and clear.

Central Processing of Visual Information

Axons from the entire population of ganglion cells converge on the optic disc, penetrate the wall of the eye, and proceed toward the diencephalon as the optic nerve (N II). The two optic nerves, one from each eye, reach the diencephalon at the optic chiasm (Figure 17-22●). From this point, approximately half the fibers proceed toward the lateral geniculate nu-

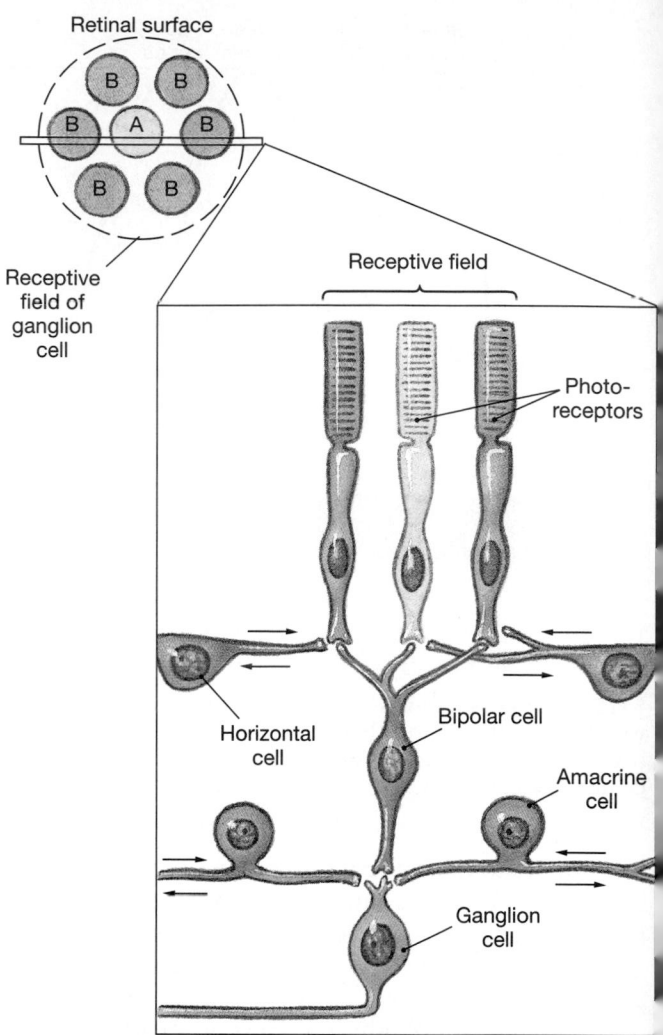

●**FIGURE 17-21** **Convergence and Ganglion Cell Function.** Photoreceptors are organized in groups within the visual field. Each ganglion cell monitors a well-defined portion of the visual field. Some ganglion cells (on-center neurons) respond strongly to light arriving at the center of their receptive field (Receptor A). Others (off-center neurons) respond most strongly to illumination of the edges of their receptive field (Receptors B).

cleus of the same side of the brain, whereas the other half cross over to reach the lateral geniculate nucleus of the opposite side. ∞ [p. 468] The lateral geniculate nuclei act as switching and processing centers that relay visual information to reflex centers in the brain stem as well as to the cerebral cortex. For example, the pupillary reflexes and reflexes that control eye movement are triggered by information relayed to the superior colliculi by the lateral geniculate nuclei.

From each lateral geniculate, visual information travels to the occipital cortex of the cerebral hemisphere on that side. The bundle of projection fibers linking the lateral geniculate with the visual cortex is known as the **optic radiation.**

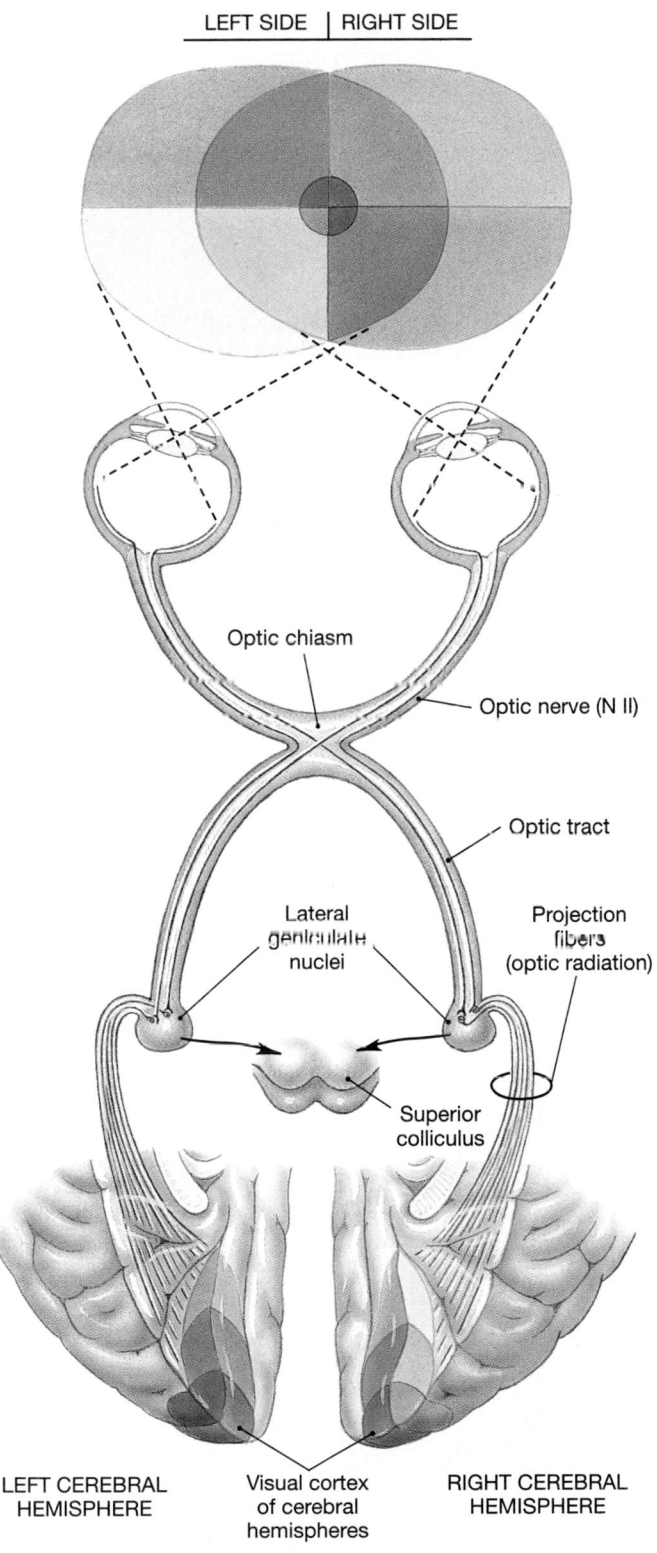

LEFT SIDE | RIGHT SIDE

Optic chiasm

Optic nerve (N II)

Optic tract

Lateral geniculate nuclei

Projection fibers (optic radiation)

Superior colliculus

LEFT CEREBRAL HEMISPHERE

Visual cortex of cerebral hemispheres

RIGHT CEREBRAL HEMISPHERE

●*FIGURE 17-22* **The Visual Pathways.** At the optic chiasm, a partial crossover of nerve fibers occurs. As a result, each hemisphere receives visual information from the medial half of the visual field of the eye on that side and from the lateral half of the visual field of the eye on the opposite side. Visual association areas integrate this information to develop a composite picture of the entire visual field.

THE VISUAL FIELD The perception of a visual image reflects the integration of information that arrives at the visual cortex of the occipital lobes. Each eye receives a slightly different visual image, because (1) the foveas are 5–7.5 cm apart, and (2) the nose and eye socket block the view of the opposite side. Depth perception, an interpretation of the three-dimensional relationships among objects in view, is obtained by comparing the relative positions of objects within the images received by the left and right eyes.

When you look straight ahead, the visual images from your left and right eyes overlap (Figure 17-22●). The image received by the fovea of each eye lies in the center of the region of overlap. A vertical line drawn through the center of the region of overlap marks the division of visual information at the optic chiasm. Visual information from the left half of the combined visual field will reach the visual cortex of your right occipital lobe; visual information from the right half of the combined visual field will arrive at your left visual cortex.

The cerebral hemispheres thus contain a map of the entire field of vision. As in the case of the primary sensory cortex, the map does not faithfully duplicate the relative areas within the sensory field. For example, the area assigned to the macula lutea and fovea covers about 35 times the surface it would cover if the map were proportionally accurate. The map is also upside down and backward, duplicating the orientation of the visual image at the retina.

THE BRAIN STEM AND VISUAL PROCESSING Many centers in the brain stem receive visual information, either from the lateral geniculates or via collaterals from the optic tracts. Collaterals that bypass the lateral geniculates synapse in the superior colliculi or hypothalamus. The superior colliculus of the mesencephalon issues motor commands that control unconscious eye, head, or neck movements in response to visual stimuli. Visual inputs to the **suprachiasmatic** (soo-pra-kī-az-MA-tik) **nucleus** of the hypothalamus affect the function of other brain stem nuclei. This nucleus and the *pineal gland* of the epithalamus establish a daily pattern of visceral activity that is tied to the day/night cycle. This **circadian rhythm** (*circa,* about + *dies,* day) affects your metabolic rate, endocrine function, blood pressure, digestive activities, awake/asleep cycle, and other physiological and behavioral processes that we discussed in Chapters 14–16.

☑ If you had been born without cones in your eyes, would you still be able to see? Explain.

☑ How could a diet deficient in vitamin A affect vision?

☑ What effect would a decrease in phosphodiesterase activity in photoreceptor cells have on vision?

EQUILIBRIUM AND HEARING

The *inner ear,* a receptor complex located in the petrous portion of the temporal bone of the skull, provides two senses: equilibrium and hearing. ∞ *[p. 206]* *Equilibrium* sensations inform us of the position of the head in space by monitoring gravity, linear acceleration, and rotation. *Hearing* enables us to detect and interpret sound waves. The basic receptor mechanism for both senses is the same. The receptors, or *hair cells,* are simple mechanoreceptors. The complex structure of the inner ear and the different arrangement of accessory structures account for the abilities of the hair cells to respond to different stimuli and thus to provide the input for the two different senses.

Anatomy of the Ear

The ear is divided into three anatomical regions: the external ear, the middle ear, and the inner ear (Figure 17-23●). The *external ear,* the visible portion of the ear,

collects and directs sound waves to the *eardrum.* The *middle ear* is a chamber located within the petrous portion of the temporal bone. Structures of the middle ear collect sound waves and transmit them to an appropriate portion of the inner ear. The *inner ear* contains the sensory organs for hearing and equilibrium.

The External Ear

The **external ear** includes the fleshy flap and cartilaginous **pinna,** or *auricle,* that surrounds the **external auditory canal,** or **ear canal.** The pinna protects the opening of the canal and provides directional sensitivity to the ear. Sounds coming from behind the head are blocked by the pinna; sounds coming from the side or front are collected and channeled into the external auditory canal of the temporal bone. (When you "cup" your ear with your hand to hear a faint sound more clearly, you are exaggerating this effect.) The external auditory canal is a passageway that ends at the **tympanic membrane,** also called the *tympanum,* or

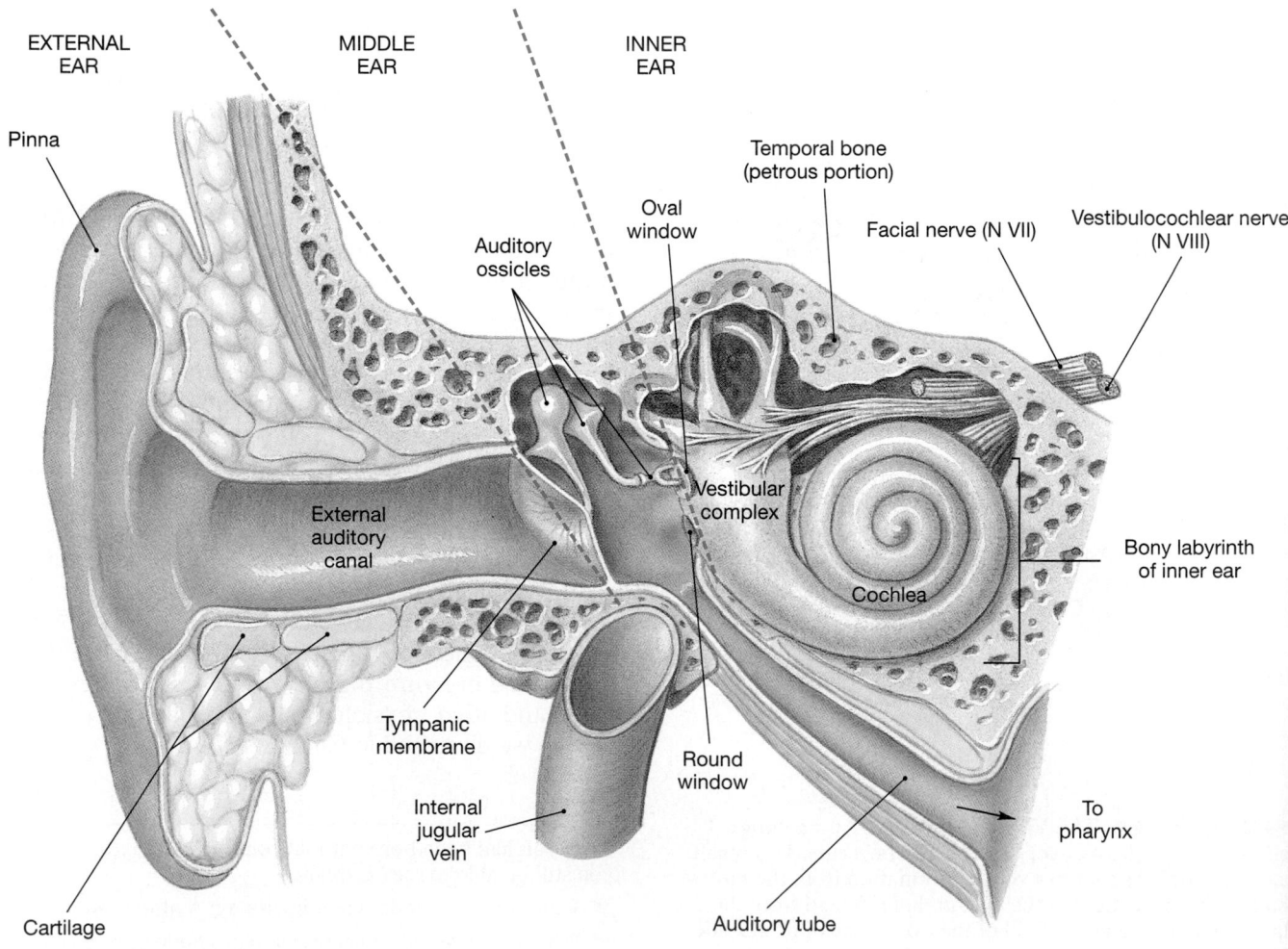

●**FIGURE 17-23 Anatomy of the Ear.** The orientation of the external, middle, and inner ears.

eardrum. The tympanic membrane is a thin, semi-transparent sheet that separates the external ear from the middle ear (Figure 17-23●).

The tympanic membrane is very delicate. The pinna and the narrow external auditory canal provide some protection from accidental injury. In addition, **ceruminous glands,** integumentary glands along the external auditory canal, secrete a waxy material that helps deny access to foreign objects or insects, as do many small, outwardly projecting hairs. The waxy secretion of the ceruminous glands, called **cerumen,** also slows the growth of microorganisms in the external auditory canal and reduces the chances of infection.

The Middle Ear

The **middle ear,** or **tympanic cavity,** is filled with air. It is separated from the external auditory canal by the tympanic membrane, but it communicates with the *nasopharynx* (the superior portion of the pharynx) through the **auditory tube** and with the mastoid air cells through a number of small and variable connections (Figure 17-23●). ∞ *[p. 206]* The auditory tube is also called the *pharyngotympanic tube* or the *Eustachian tube*. This tube, about 4 cm long, consists of two portions. The portion near the connection to the middle ear is relatively narrow and is supported by cartilage (Figure 17-24●). The portion near the opening into the nasopharynx is relatively broad and funnel-shaped. The auditory tube serves to equalize the pressure inside the eardrum with that outside the eardrum. Unfortunately, it can also allow microorganisms to travel from the nasopharynx into the tympanic cavity. Invasion by microorganisms can lead to an unpleasant middle ear infection known as *otitis media*. AM *Otitis Media and Mastoiditis*

THE AUDITORY OSSICLES The middle ear contains three tiny ear bones, collectively called **auditory ossicles**. The ear bones connect the tympanic membrane with the receptor complex of the inner ear (Figures 17-23 and 17-24●). The three auditory ossicles are the *malleus*, the *incus*, and the *stapes*. The **malleus** (*malleus*, hammer) attaches at three points to the interior surface of the tympanic membrane. The middle bone, the **incus** (*incus*, anvil), attaches the malleus to the inner **stapes** (*stapes*, stirrup). The base of the stapes is bound to a ligamentous sheet that spans the *oval window*, an opening in the bone that surrounds the inner ear.

Vibration of the tympanic membrane converts arriving sound waves into mechanical movements. The auditory ossicles act as levers that conduct those vibrations to the fluid-filled chamber of the inner ear. The ossicles are connected in such a way that an in–out movement of the tympanic membrane produces a rocking motion of the stapes. The tympanic membrane is 22 times larger and heavier than the oval window, so a 1-μm movement of the tympanic membrane produces a 22-μm deflection of the base of the stapes. Thus, the amount of movement increases markedly from tympanic membrane to oval window.

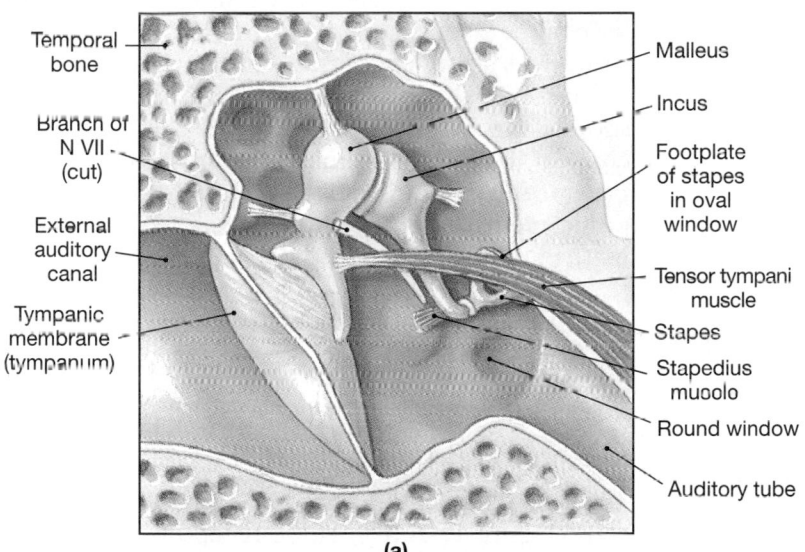

(a)

Temporal bone
Branch of N VII (cut)
External auditory canal
Tympanic membrane (tympanum)
Malleus
Incus
Footplate of stapes in oval window
Tensor tympani muscle
Stapes
Stapedius muscle
Round window
Auditory tube

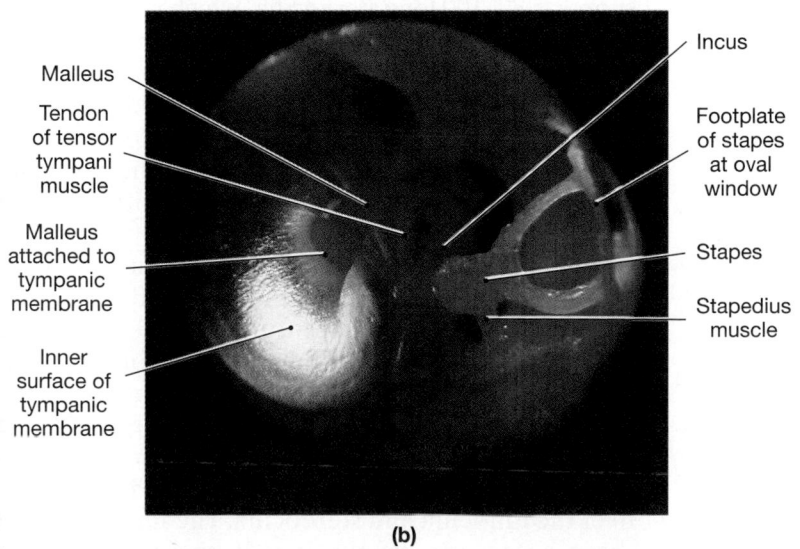

(b)

Malleus
Tendon of tensor tympani muscle
Malleus attached to tympanic membrane
Inner surface of tympanic membrane
Incus
Footplate of stapes at oval window
Stapes
Stapedius muscle

●*FIGURE 17-24* **The Middle Ear. (a)** Detail of the structures of the middle ear. **(b)** Tympanic membrane and auditory ossicles.

Because this amplification occurs, we can hear very faint sounds. But that degree of amplification can be a problem when we are exposed to very loud noises. Within the tympanic cavity, two small muscles serve to protect the eardrum and ossicles from violent movements under very noisy conditions:

1. The **tensor tympani** (TEN-sor tim-PAN-ē) **muscle** is a short ribbon of muscle whose origin is the petrous portion of the temporal bone and the auditory tube and whose insertion is on the "handle" of the malleus. When the tensor tympani contracts, the malleus is pulled medially, stiffening the tympanic membrane. This increased stiffness reduces the amount of possible movement. The tympani muscle is innervated by motor fibers of the mandibular branch of the trigeminal nerve (N V).

2. The **stapedius** (sta-PĒ-dē-us) **muscle,** innervated by the facial nerve (N VII), originates from the posterior wall of the tympanic cavity and inserts on the stapes. Contraction of the stapedius pulls the stapes, reducing movement of the stapes at the oval window.

The Inner Ear

The senses of equilibrium and hearing are provided by the receptors of the **inner ear** (Figures 17-23 and 17-25a●). The receptors lie within a collection of fluid-filled tubes and chambers known as the **membranous labyrinth** (*labyrinthos,* network of canals). The membranous labyrinth contains **endolymph** (EN-dō-limf), a fluid with electrolyte concentrations different from those of typical body fluids. (See Appendix VI for a chemical analysis of perilymph, endolymph, and other body fluids.)

The **bony labyrinth** is a shell of dense bone that surrounds and protects the membranous labyrinth. Its inner contours closely follow the contours of the membranous labyrinth, and its outer walls are fused with the surrounding temporal bone. Between the bony and membranous labyrinths flows the **perilymph** (PER-i-limf), a liquid whose properties closely resemble those of cerebrospinal fluid.

The bony labyrinth can be subdivided into the *vestibule,* the *semicircular canals,* and the *cochlea* (Figure 17-25a●).

VESTIBULE The **vestibule** (VES-ti-būl) includes a pair of membranous sacs: the **saccule** (SAK-ūl) and the **utricle** (Ū-tri-kul), or the *sacculus* and *utriculus.* Receptors in the saccule and utricle provide sensations of gravity and linear acceleration.

SEMICIRCULAR CANALS The **semicircular canals** enclose slender **semicircular ducts.** Receptors in the semicircular ducts are stimulated by rotation of the head. The combination of vestibule and semicircular canals is called the *vestibular complex;* the fluid-filled chambers within the vestibule are broadly continuous with those of the semicircular canals.

COCHLEA The **cochlea** (KOK-lē-uh; *cochlea,* a snail shell) is a spiral-shaped, bony chamber that contains the **cochlear duct** of the membranous labyrinth. Receptors within the cochlear duct provide the sense of hearing. The cochlear duct is sandwiched between a pair of perilymph-filled chambers. The entire complex makes $2^1/_2$ turns around a central bony hub. In sectional view, the spiral arrangement resembles that of a snail shell.

The walls of the bony labyrinth consist of dense bone everywhere except at two small areas near the base of the cochlear spiral (Figure 17-23●) The **round window** is a thin, membranous partition that separates the perilymph of the cochlear chambers from the air spaces of the middle ear. Collagen fibers connect the bony margins of the **oval window** to the base of the stapes. When a sound vibrates the tympanic membrane, the movements are conducted by the auditory ossicles to the stapes. Movement of the stapes ultimately leads to the stimulation of receptors within the cochlear duct, and we hear the sound.

Receptor Function in the Inner Ear

The sensory receptors of the inner ear are called **hair cells** (Figure 17-25b●). These receptor cells are surrounded by **supporting cells** and are monitored by sensory afferent fibers. The free surface of each hair cell supports 80–100 long stereocilia, which resemble very long microvilli. ∞ *[p. 110]* Each hair cell in the vestibule also contains a **kinocilium,** a single large cilium. Hair cells do not actively move their kinocilia and stereocilia. However, when an external force pushes against these processes, the distortion of the cell membrane alters the rate of chemical transmitter release by the hair cell.

Hair cells provide information about the direction and strength of mechanical stimuli. The stimuli involved, however, are quite varied: gravity or acceleration in the vestibule, rotation in the semicircular canals, and sound in the cochlea. The sensitivities of the hair cells differ, because each of these regions has different accessory structures that determine which stimulus will provide the force to deflect the kinocilia and stereocilia. The importance of these accessory structures will become apparent when we consider the way sensations of equilibrium and hearing are produced.

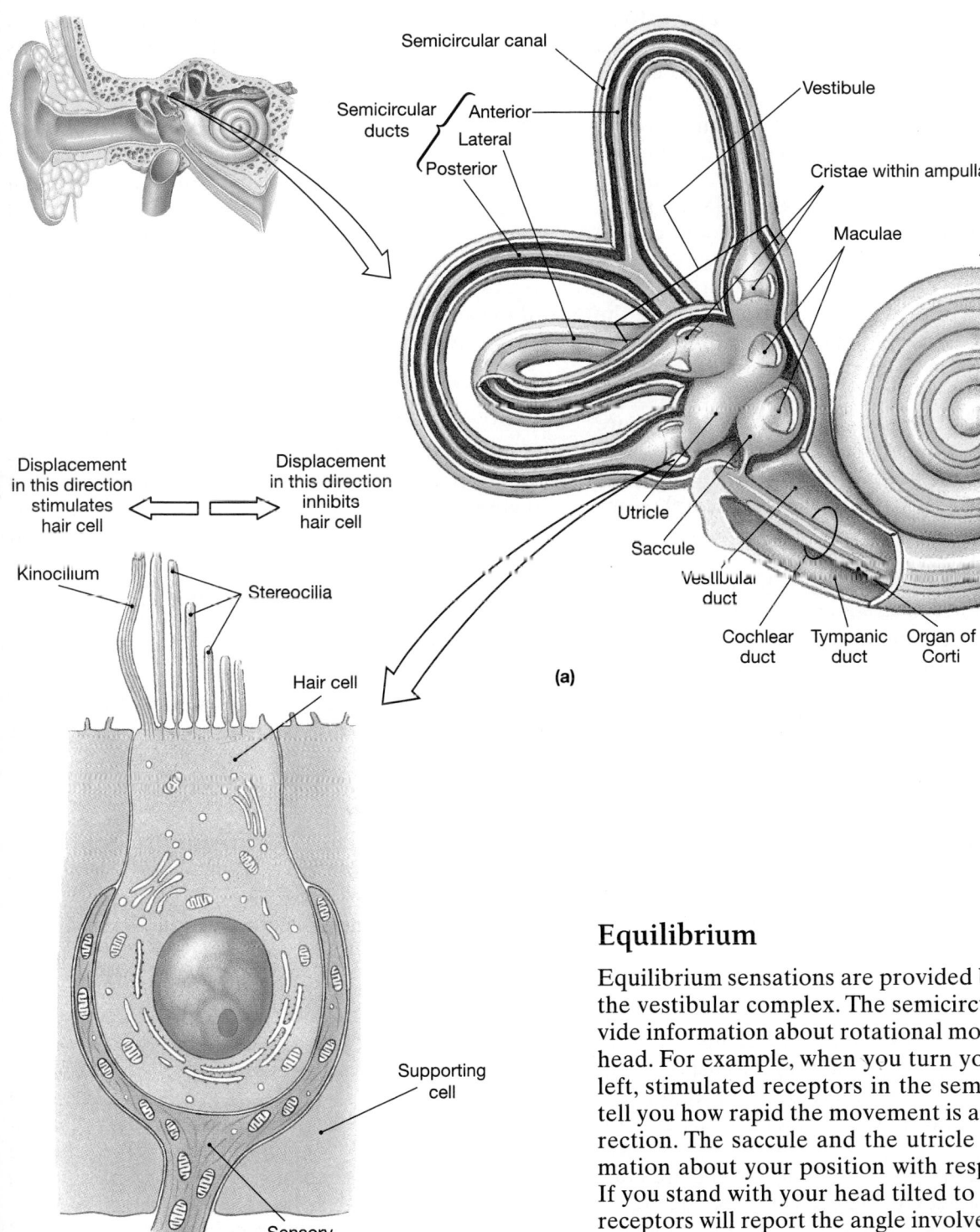

Semicircular canal

Semicircular ducts
Anterior
Lateral
Posterior

Vestibule

Cristae within ampullae

Maculae

Cochlea

Utricle

Saccule

Vestibular duct

Cochlear duct **Tympanic duct** **Organ of Corti**

(a)

Displacement in this direction stimulates hair cell

Displacement in this direction inhibits hair cell

Kinocilium

Stereocilia

Hair cell

Supporting cell

Sensory nerve ending

(b) Hair cell

•*FIGURE 17-25* **The Inner Ear. (a)** The bony and membranous labyrinths. Areas of the membranous labyrinth containing sensory receptors (cristae, maculae, and the organ of Corti) are highlighted. **(b)** A representative hair cell (receptor) from the vestibular complex. Bending the stereocilia toward the kinocilium depolarizes the cell and stimulates the sensory neuron. Displacement in the opposite direction inhibits the sensory neuron.

Equilibrium

Equilibrium sensations are provided by receptors of the vestibular complex. The semicircular ducts provide information about rotational movements of the head. For example, when you turn your head to the left, stimulated receptors in the semicircular ducts tell you how rapid the movement is and in which direction. The saccule and the utricle provide information about your position with respect to gravity. If you stand with your head tilted to one side, these receptors will report the angle involved and whether your head tilts forward or backward. These receptors are also stimulated by sudden acceleration. When your car accelerates from a stop, the saccular and utricular receptors give you the impression of increasing speed.

The Semicircular Ducts

Receptors in the semicircular ducts respond to rotational movements of the head. These hair cells are active during a movement but quiet when the body is motionless. The **anterior, posterior,** and **lateral semicircular ducts** are continuous with the utricle (Figures 17-25a and 17-26a●). Each semicircular duct contains an **ampulla,** an expanded region that contains the sensory receptors. Hair cells attached to the wall of the ampulla form a raised structure known as a **crista** (Figure 17-26b●). The kinocilia and stereocilia of the hair cells are embedded in the **cupula** (KŪ-pū-luh), a gelatinous structure. Because the cupula has a density very close to that of the surrounding endolymph, it essentially floats above the receptor surface, nearly filling the ampulla. When your head rotates in the plane of the duct, movement of the endolymph along the canal axis pushes the cupula and distorts the receptor processes (Figure 17-26c●). Fluid movement in one direction stimulates the hair cells, and movement in the opposite direction inhibits them. When the endolymph stops moving, the elastic nature of the cupula makes it bounce back to its normal position.

Even the most complex movement can be analyzed in terms of motion in three rotational planes. Each semicircular duct responds to one of these rotational movements. A horizontal rotation, as in shaking your head "no," stimulates the hair cells of the lateral semicircular duct. Nodding "yes" excites the anterior duct, and tilting your head from side to side activates the receptors in the posterior duct.

The Utricle and Saccule

The utricle and saccule provide equilibrium sensations whether or not the body is moving. These chambers are connected by a slender passageway continuous with the narrow **endolymphatic duct** (Figure 17-26a●). The endolymphatic duct ends in a blind pouch, the **endolymphatic sac,** that projects through the dura mater lining the temporal bone into the subdural space. Portions of the cochlear duct continuously secrete endolymph, and at the endolymphatic sac excess fluids return to the general circulation.

The hair cells of the utricle and saccule are clustered in oval **maculae** (MAK-ū-lē; *macula,* spot) (Figure 17-26a●). As they are in the ampullae, the hair cell processes are embedded in a gelatinous mass, but the macular receptors lie under a thin layer containing densely packed calcium carbonate crystals, called **otoliths** ("ear stones"), or *otoconia* (*oto,* ear + *conia,* dust).

The macula of the saccule is diagrammed in Figure 17-26d●. When your head is in the normal, upright position the otoliths sit atop the macula. Their weight presses down on the macular surfaces, pushing the hair cell processes down rather than to one

side or another. When your head is tilted, the pull of gravity on the otoliths shifts the mass to the side. This shift distorts the hair cell processes, and the change in receptor activity tells the CNS that your head is no longer level (Figure 17-26e●).

A similar mechanism accounts for your perception of linear acceleration when you are in a car that speeds up suddenly. The otoliths lag behind, and the effect on the hair cells is comparable to tilting your head back. Under normal circumstances, your nervous system distinguishes between the sensations of tilting and linear acceleration by integrating vestibular sensations with visual information.

Pathways for Equilibrium Sensations

Hair cells of the vestibule and semicircular ducts are monitored by sensory neurons located in adjacent **vestibular ganglia.** Sensory fibers from each ganglion form the **vestibular branch** of the vestibulocochlear nerve (N VIII). These fibers innervate neurons within the **vestibular nuclei** at the boundary between the pons and medulla oblongata. The two vestibular nuclei have the following four functions:

1. Integrating the sensory information, regarding balance and equilibrium, that arrives from both sides of the head.
2. Relaying information from the vestibular apparatus to the cerebellum.
3. Relaying information from the vestibular apparatus to the cerebral cortex, providing a conscious sense of head position and movement.
4. Sending commands to motor nuclei in the brain stem and spinal cord.

The reflexive motor commands issued by the vestibular nuclei are distributed to the motor nuclei for cranial nerves involved with eye, head, and neck movements (N III, IV, VI, and XI). Instructions descending in the *vestibulospinal tracts* of the spinal cord adjust peripheral muscle tone and complement the reflexive movements of the head or neck. These pathways are indicated in Figure 17-27●(p. 576); we considered the role of the vestibulospinal tracts in the extrapyramidal system in Chapter 15. ∞ *[p. 499]*

The automatic eye movements that occur in response to sensations of motion are directed by the *superior colliculus* of the mesencephalon. ∞ *[pp. 470, 500]* These movements attempt to keep your gaze focused on a specific point in space despite changes in body position and orientation. If your body is turning or spinning rapidly, your eyes will fix on one point for a moment, then jump ahead to another in a series of short, jerky movements. This type

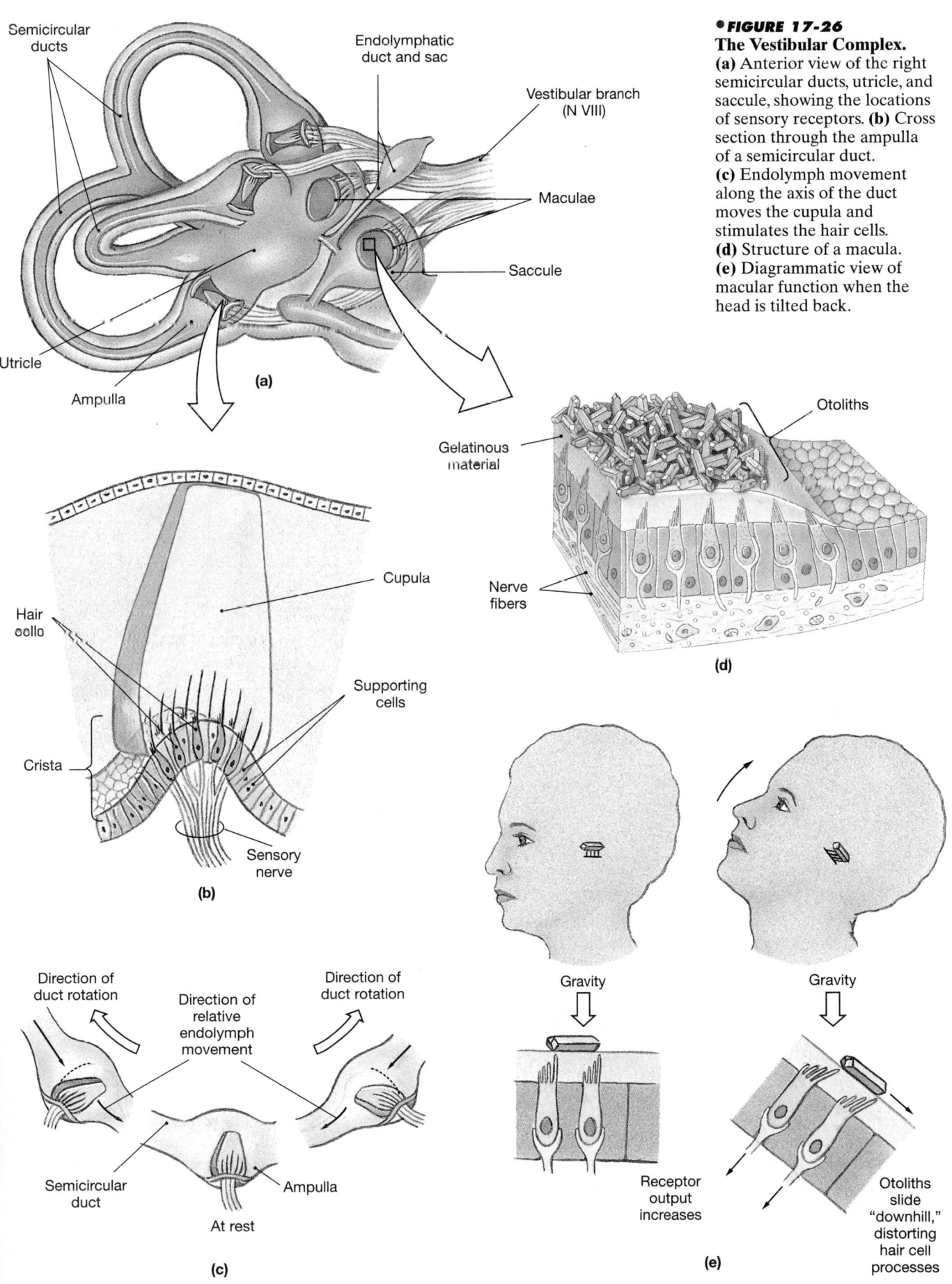

The Vestibular Complex.
(a) Anterior view of the right semicircular ducts, utricle, and saccule, showing the locations of sensory receptors. **(b)** Cross section through the ampulla of a semicircular duct. **(c)** Endolymph movement along the axis of the duct moves the cupula and stimulates the hair cells. **(d)** Structure of a macula. **(e)** Diagrammatic view of macular function when the head is tilted back.

Semicircular ducts

Endolymphatic duct and sac

Vestibular branch (N VIII)

Maculae

Saccule

Utricle

Ampulla

(a)

Gelatinous material

Otoliths

Nerve fibers

(d)

Cupula

Hair cells

Supporting cells

Crista

Sensory nerve

(b)

Direction of duct rotation

Direction of relative endolymph movement

Direction of duct rotation

Semicircular duct

Ampulla

At rest

(c)

Gravity

Gravity

Receptor output increases

Otoliths slide "downhill," distorting hair cell processes

(e)

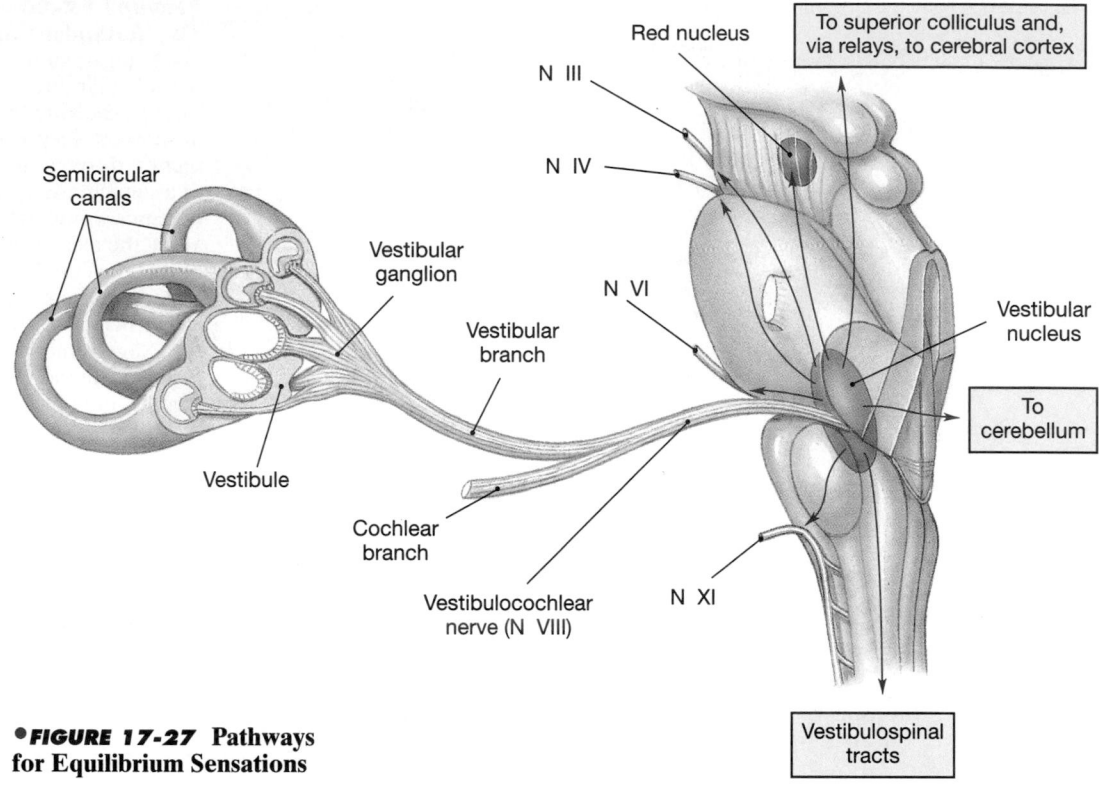

●*FIGURE 17-27* Pathways for Equilibrium Sensations

of eye movement can occur even when the body is stationary if either the brain stem or the inner ear is damaged. An individual with this condition, which is called **nystagmus,** has trouble controlling eye movements. Physicians commonly check for nystagmus by asking the patient to watch a small penlight as it is moved across the field of vision. AM *Vertigo and Ménière's Disease*

 ## Motion Sickness

The exceedingly unpleasant symptoms of **motion sickness** include headache, sweating, flushing of the face, nausea, vomiting, and various changes in mental perspective. (Sufferers may go from a state of giddy excitement to almost suicidal despair in a matter of moments.) It has been suggested that the condition results when central processing stations, such as the mesencephalic tectum, receive conflicting sensory information. Why and how these conflicting reports result in nausea, vomiting, and other symptoms are not known. Sitting below the deck on a moving boat or reading in a car or airplane tends to provide the necessary conditions. Your eyes (which are tracking lines on a page) report that your position in space is not changing, but your labyrinthine receptors report that your body is lurching and turning. As a result, seasick sailors watch the horizon rather than their immediate surroundings so that their eyes will provide visual confirmation of the movements detected by their inner ears. It is not known why some individuals are almost immune to motion sickness, whereas others find travel by boat or plane almost impossible.

Drugs commonly administered to prevent motion sickness include dimenhydrinate *(Dramamine)*, scopolamine, and promethazine. These compounds appear to depress activity at the vestibular nuclei. Sedatives, such as prochlorperazine *(Compazine)*, may also be effective. Scopolamine can be administered across the skin surface by using an adhesive patch *(Transderm-Scōp™)*. (However, this product is currently unavailable in the United States.)

Hearing

The receptors of the cochlear duct provide us with a sense of hearing that enables us to detect the quietest whisper yet remain functional in a crowded, noisy room. The receptors responsible for auditory sensations are hair cells similar to those of the vestibular complex. However, their placement within the cochlear duct and the organization of the surrounding accessory structures shield them from stimuli other than sound. In conveying vibrations from the tympanic membrane to the oval window, the auditory ossicles convert pressure waves in air to pressure pulses in the perilymph of the cochlea. These pressure pulses stimulate hair cells along the cochlear spiral. The *frequency* of the perceived sound is determined by *which part* of the cochlear duct is stimulated. The *intensity* (volume) of the perceived sound is determined by *how many* of the hair cells at that location are stimulated. We will now consider the mechanics of this remarkably elegant process.

The Cochlear Duct

In sectional view (Figure 17-28a,b,c●), the cochlear duct, or *scala media,* lies between a pair of perilym-

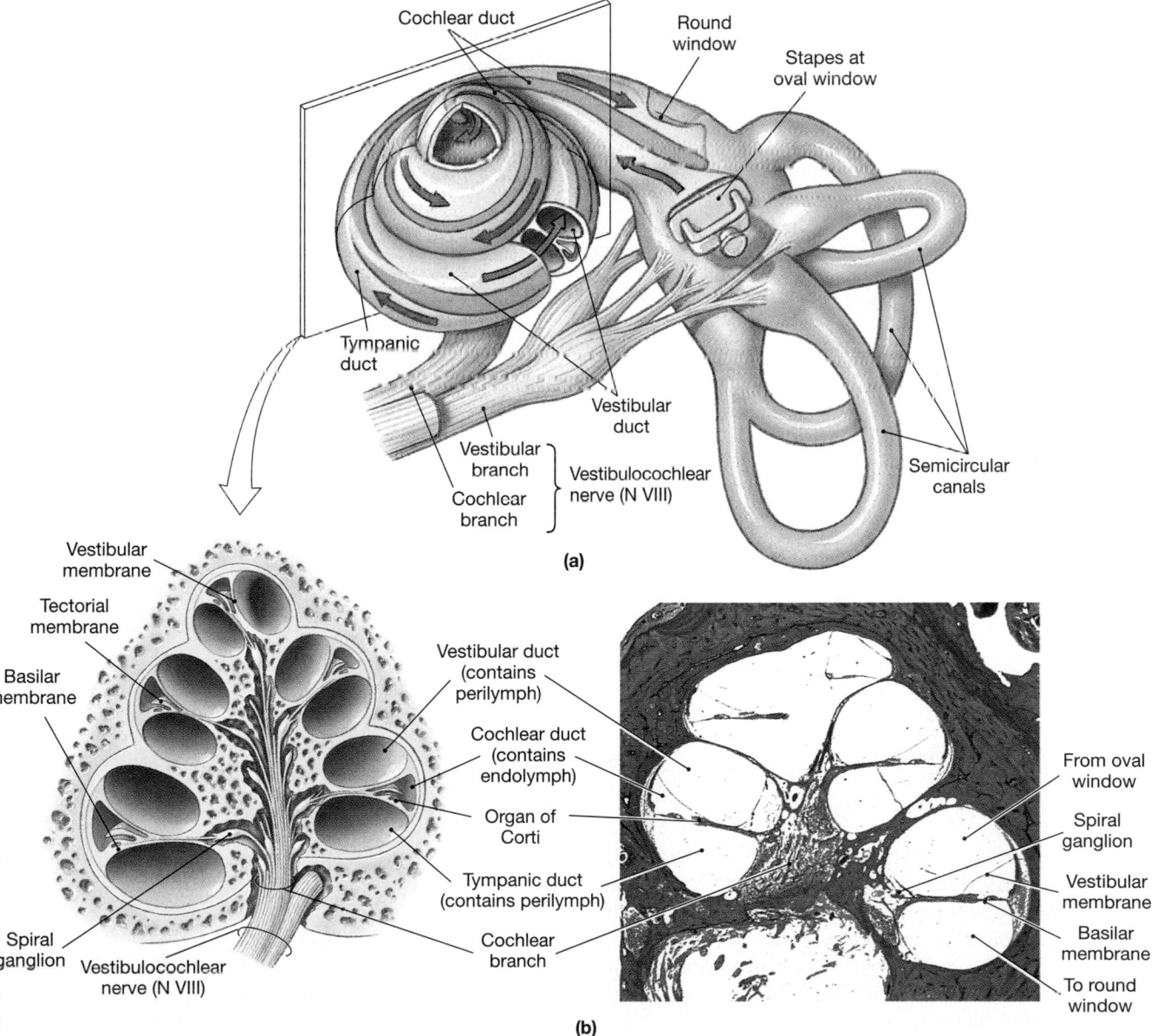

(a)

(b)

●*FIGURE 17-28* **The Cochlea.** **(a)** Structure of the cochlea. **(b)** Diagrammatic and sectional views of the cochlear spiral. **(c)** Three-dimensional section showing details of the compartments, tectorial membrane, and organ of Corti. **(d)** Diagrammatic and sectional views of the receptor hair cell complex of the organ of Corti. (LM × 1233)

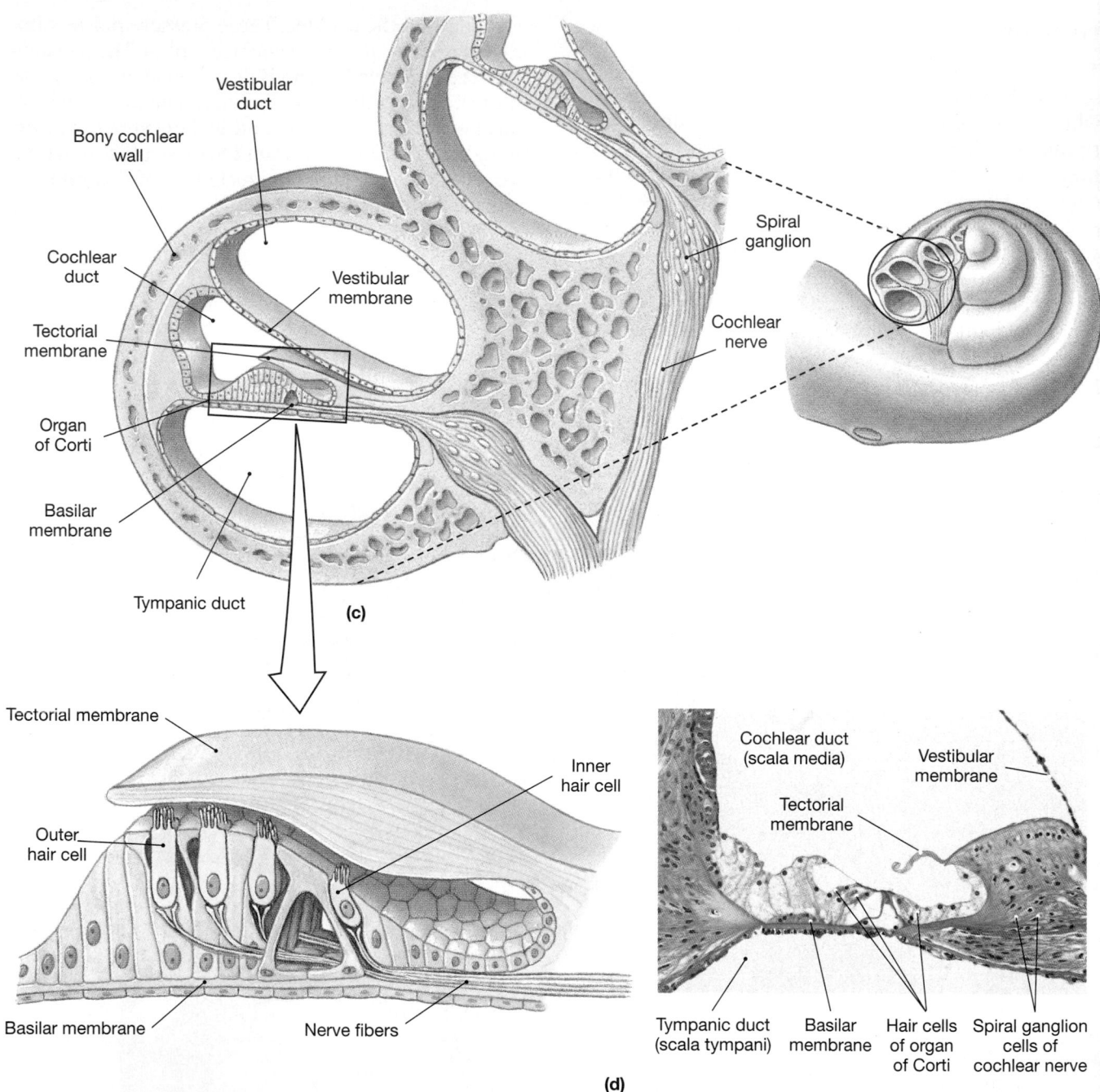

(c)

Vestibular duct

Bony cochlear wall

Cochlear duct

Tectorial membrane

Organ of Corti

Basilar membrane

Tympanic duct

Vestibular membrane

Spiral ganglion

Cochlear nerve

Tectorial membrane

Outer hair cell

Inner hair cell

Basilar membrane

Nerve fibers

Cochlear duct (scala media)

Tectorial membrane

Vestibular membrane

Tympanic duct (scala tympani)

Basilar membrane

Hair cells of organ of Corti

Spiral ganglion cells of cochlear nerve

(d)

•*FIGURE 17-28 (continued)*

phatic chambers: the **vestibular duct** (*scala vestibuli*) and the **tympanic duct** (*scala tympani*). The vestibular and tympanic ducts are interconnected at the tip of the cochlear spiral. The outer surfaces of these ducts are encased by the bony labyrinth everywhere except at the oval window (base of the vestibular duct) and the round window (base of the tympanic duct).

The hair cells of the cochlear duct are located in a structure called the **organ of Corti** (Figure 17-28b,c,d•). This sensory structure sits above the **basilar membrane**, a membrane that separates the

cochlear duct from the tympanic duct. The hair cells are arranged in a series of longitudinal rows. They lack kinocilia, and their stereocilia are in contact with the overlying **tectorial** (tek-TOR-ē-al) **membrane** (*tectum,* roof). This membrane is firmly attached to the inner wall of the cochlear duct. When a portion of the basilar membrane bounces up and down, the stereocilia of the hair cells are distorted. The basilar membrane moves in response to pressure waves within the perilymph. These waves are produced when sounds arrive at the tympanic membrane. To understand how these waves develop, we must consider the basic properties of sound.

An Introduction to Sound

Hearing is the detection of sound, which consists of pressure waves conducted through a medium such as air or water. In air, each pressure wave consists of a region where the air molecules are crowded together and an adjacent zone where they are relatively far apart (Figure 17-29a•). These waves travel through the air at approximately 1235 km/h (768 mph). Physicists use the term **cycles** rather than *waves;* the number of cycles per second (cps), or **hertz (Hz),** represents the frequency of the sound. What we perceive as the **pitch** of a sound is our sensory response to its frequency. A *high-frequency* sound (high pitch) might have a frequency of 15,000 Hz or more; a very *low-frequency* sound (low pitch) could have a frequency of 100 Hz or less.

It takes energy to produce sound waves. When you strike a tuning fork, it vibrates and pushes against the surrounding air, producing sound waves whose frequency depends on the frequency of vibration. The harder you strike the tuning fork, the more energy you provide and the louder the sound. The loudness increases because the sound waves carry more energy away with them. The energy content, or *power*, of a sound determines its **intensity** (volume), which is reported in **decibels** (DES-i-

belz). Table 17-1 indicates the decibel levels of familiar sounds.

When sound waves strike an object, their energy is a physical pressure. You may have seen windows move in a room in which a stereo is blasting. The more flexible the object, the easier it will respond to sound pressure. Even soft stereo music will vibrate a sheet of paper held in front of the speaker. Given the right combination of frequencies and intensities, an object will begin to vibrate at the same frequency as the sound, a phenomenon called *resonance*. The greater the sound intensity, the greater the amount of vibration produced. For you to be able to hear, your thin, flexible tympanic membrane must vibrate in resonance with sound waves.

The Hearing Process

The process of hearing can be divided into six basic steps, diagrammed in Figure 17-29b• and summarized in Table 17-2 (p. 581):

STEP 1: *Sound waves arrive at the tympanic membrane.* Sound waves enter the external auditory canal and travel toward the tympanic membrane. The orientation of the canal provides some directional sensitivity. Sound waves approaching the side of the head have direct access to the tympanic membrane on that side, whereas sounds arriving from another direction must bend around corners or pass through the pinna or other body tissues.

TABLE 17-1 Sound Intensities

Typical Decibel Level	Example	Dangerous Time Exposure
0	Lowest audible sound	
30	Quiet library; soft whisper	
40	Quiet office; living room; bedroom away from traffic	
50	Light traffic at a distance; refrigerator; gentle breeze	
60	Air conditioner at 20 feet; conversation; sewing machine in operation	
70	Busy traffic; noisy restaurant	Some damage if continuous
80	Subway; heavy city traffic; alarm clock at 2 feet; factory noise	More than 8 hours
90	Truck traffic; noisy home appliances; shop tools; gas lawnmower	Less than 8 hours
100	Chain saw; boiler shop; pneumatic drill	2 hours
120	"Heavy metal" rock concert; sandblasting; thunderclap nearby	Immediate danger
140	Gunshot; jet plane	Immediate danger
160	Rocket launching pad	Hearing loss inevitable

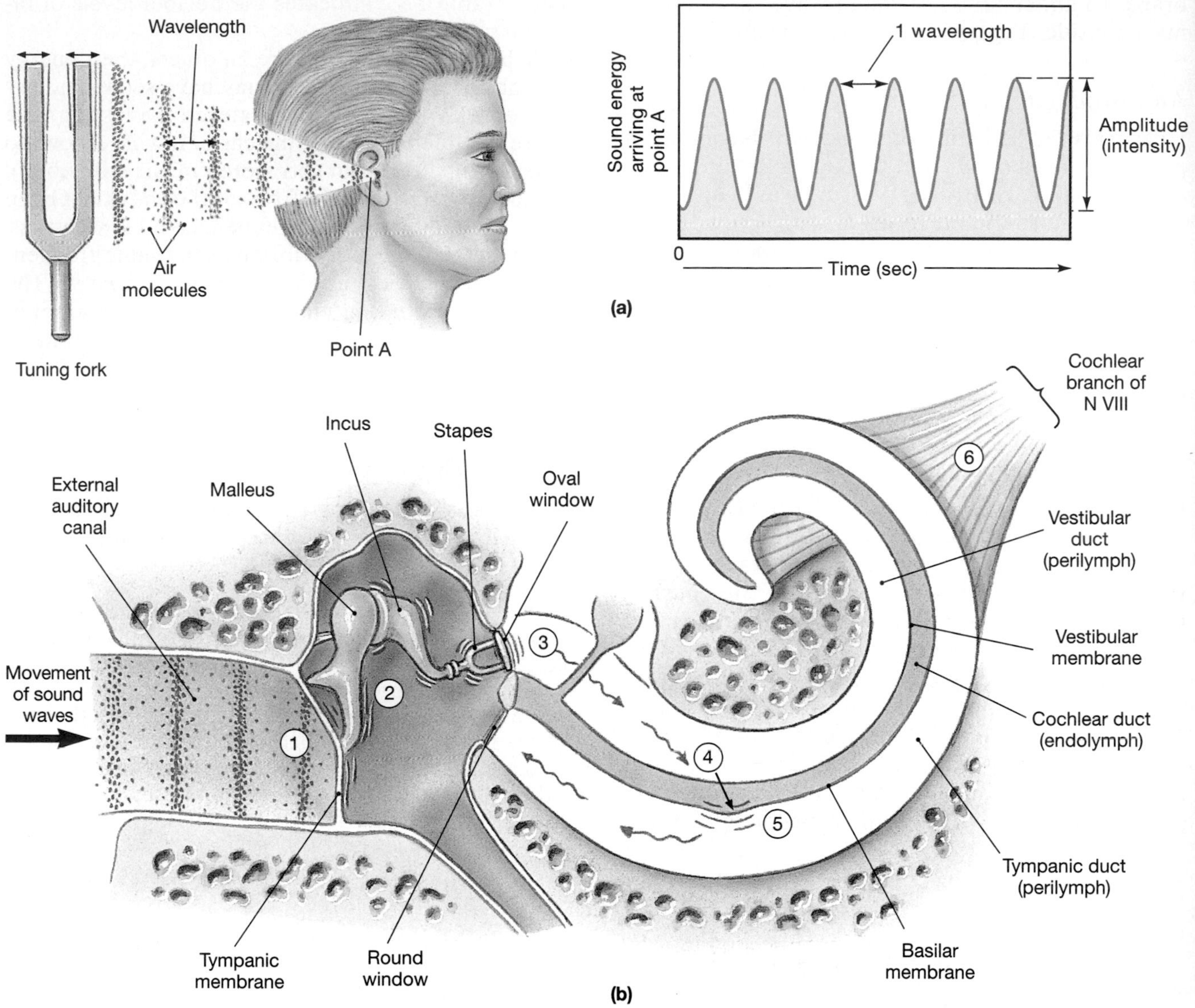

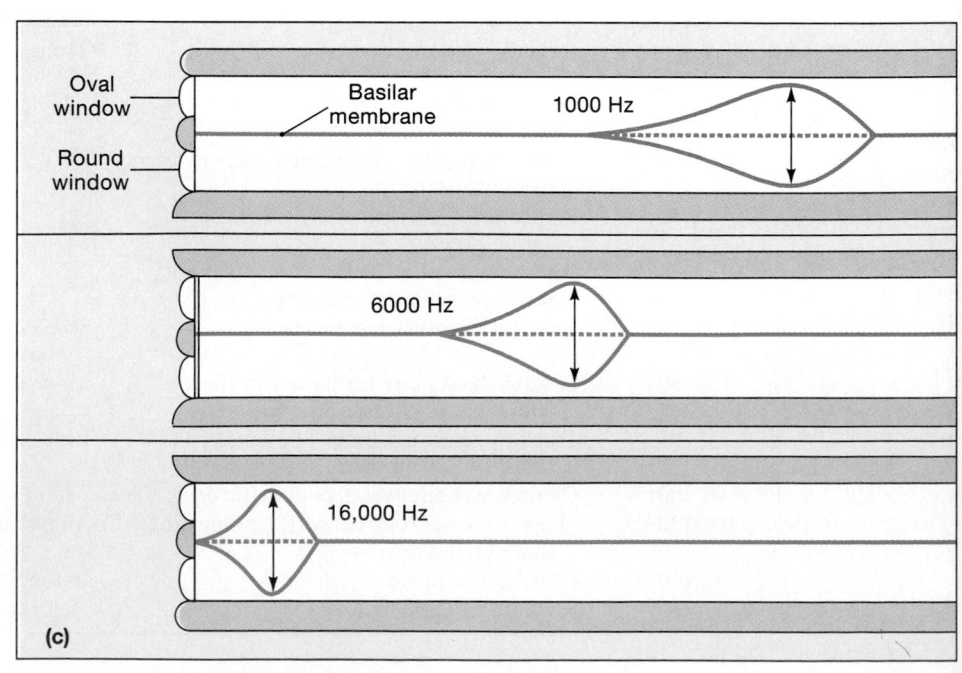

Wavelength

Tuning fork

Air molecules

Point A

(a)

Sound energy arriving at point A

1 wavelength

Amplitude (intensity)

0

Time (sec)

Incus

Stapes

Malleus

Oval window

External auditory canal

Cochlear branch of N VIII

Movement of sound waves

Vestibular duct (perilymph)

Vestibular membrane

Cochlear duct (endolymph)

Tympanic duct (perilymph)

Basilar membrane

Tympanic membrane

Round window

(b)

Oval window

Round window

Basilar membrane

1000 Hz

6000 Hz

16,000 Hz

(c)

•*FIGURE 17-29* **Sound and Hearing.** **(a)** Sound waves generated by a tuning fork travel through the air as pressure waves. The frequency of the sound wave is the number of wavelengths that pass a fixed reference point each second. Frequencies are reported in terms of cycles per second (cps), or hertz (Hz). **(b)** Steps in the reception and transduction of sound energy. (See text and Table 17-2 for steps 1–6.) **(c)** The location of distortion in the basilar membrane shifts toward the oval window as the frequency of the sound increases.

TABLE 17-2 Steps in the Production of an Auditory Sensation

1. Sound waves arrive at the tympanic membrane.
2. Movement of the tympanic membrane causes displacement of the auditory ossicles.
3. Movement of the stapes at the oval window establishes pressure waves in the perilymph of the vestibular duct.
4. The pressure waves distort the basilar membrane on their way to the round window of the tympanic duct.
5. Vibration of the basilar membrane causes vibration of hair cells against the tectorial membrane.
6. Information about the region and intensity of stimulation is relayed to the CNS over the cochlear branch of N VIII.

STEP 2: *Movement of the tympanic membrane causes displacement of the auditory ossicles.* The tympanic membrane provides the surface for sound collection, and it vibrates in resonance to sound waves with frequencies between approximately 20 and 20,000 Hz (in a young child). When the tympanic membrane vibrates, so do the malleus and, through their articulations, the incus and stapes, and the sound is amplified.

STEP 3: *Movement of the stapes at the oval window establishes pressure waves in the perilymph of the vestibular duct.* Liquids are incompressible; if you push down on *one part* of a water bed, the water bed bulges *somewhere else.* Because the rest of the cochlea is sheathed in bone, pressure applied at the oval window can be relieved only at the round window. When the stapes moves inward, the round window bulges outward. As the stapes moves in and out, vibrating at the frequency of the sound arriving at the tympanic membrane, it creates pressure pulses, or *pressure waves,* within the perilymph.

STEP 4: *The pressure waves distort the basilar membrane on their way to the round window of the tympanic duct.* The pressure waves established by movement of the stapes travel through the perilymph of the vestibular and tympanic ducts to reach the round window. In doing so, these pressure waves distort the basilar membrane. The location of maximum distortion varies with the frequency of the sound. High-frequency sounds, which have a very short wavelength, vibrate the basilar membrane near the oval window. The lower the frequency of the sound, the longer the wavelength, and the farther from the oval window the area of maximum distortion will be. Thus, *frequency* information is translated into *position* information.

The *amount* of movement at a given location will depend on the amount of force applied by the stapes. The amount of force is a function of the intensity of the sound. The louder the sound, the greater the movement of the basilar membrane.

STEP 5: *Vibration of the basilar membrane causes vibration of hair cells against the tectorial membrane.* Vibration of the affected region of the basilar membrane moves hair cells against the tectorial membrane. This movement leads to the displacement of the stereocilia, which in turn opens ion channels in their surfaces. A subsequent inrush of ions depolarizes the hair cells. The depolarization leads to neurotransmitter release and thus to the stimulation of sensory neurons.

The hair cells of the organ of Corti are arranged in several rows. A very soft sound may stimulate only a few hair cells in a portion of one row. As the intensity of a sound increases, not only do these hair cells become more active, but additional hair cells—at first in the same row, and then in adjacent rows—are stimulated as well. The number of hair cells responding in a given region of the organ of Corti thus provides information on the intensity of the sound.

STEP 6: *Information about the region and intensity of stimulation is relayed to the CNS over the cochlear branch of the vestibulocochlear nerve (N VIII).* The cell bodies of the bipolar sensory neurons that monitor the cochlear hair cells are located at the center of the bony cochlea (Figure 17-28c,d●) in the **spiral ganglion.** From there, the information is carried to the cochlear nuclei of the medulla oblongata for subsequent distribution to other centers in the brain.

Auditory Pathways

Hair cell stimulation activates sensory neurons whose cell bodies are in the adjacent spiral ganglion (Figure 17-28●). Their afferent fibers form the **cochlear branch** of the vestibulocochlear nerve (N VIII) (Figure 17-30●). These axons enter the medulla oblongata and synapse at the **cochlear nucleus.** From there, the information crosses to the opposite side of the brain and ascends to the inferior colliculus of the mesencephalon. This processing center coordinates a number of responses to acoustic stimuli, including auditory reflexes that involve skeletal muscles of the head, face, and trunk. These reflexes automatically change the position of your head in response to a sudden loud noise; you usually turn your head and your eyes toward the source of the sound.

Before reaching the cerebral cortex and your conscious awareness, ascending auditory sensations synapse in the medial geniculate nucleus of the thalamus. Projection fibers then deliver the information to the auditory cortex of the temporal lobe. Information travels to the cortex over labeled lines: High-frequency sounds activate one portion of the cortex, and low-frequency sounds activate another. In ef-

Development of Special Sense Organs

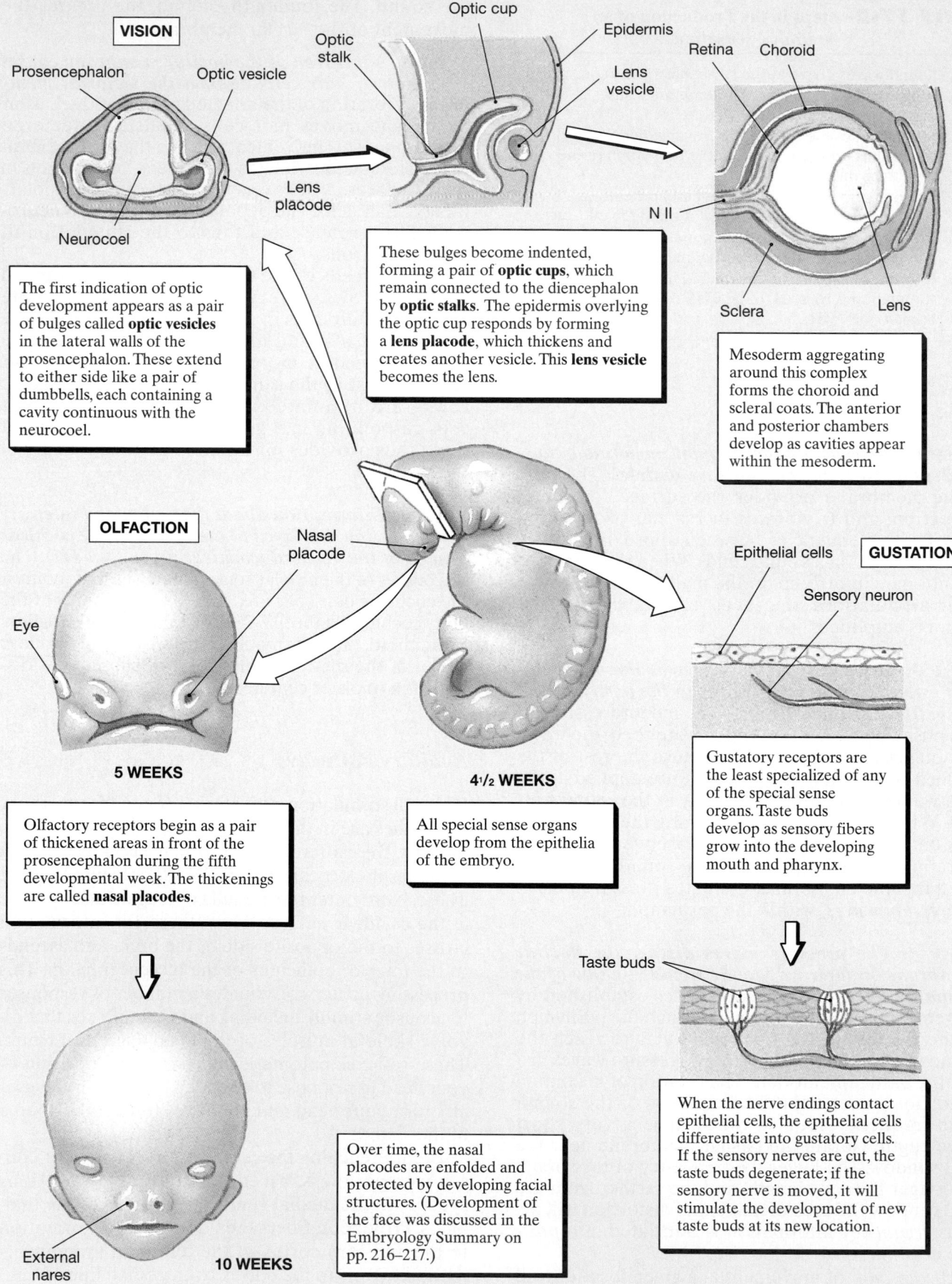

VISION

Prosencephalon
Optic vesicle
Lens placode
Neurocoel

Optic stalk
Optic cup
Epidermis
Lens vesicle

Retina
Choroid
Sclera
Lens
N II

The first indication of optic development appears as a pair of bulges called **optic vesicles** in the lateral walls of the prosencephalon. These extend to either side like a pair of dumbbells, each containing a cavity continuous with the neurocoel.

These bulges become indented, forming a pair of **optic cups**, which remain connected to the diencephalon by **optic stalks**. The epidermis overlying the optic cup responds by forming a **lens placode**, which thickens and creates another vesicle. This **lens vesicle** becomes the lens.

Mesoderm aggregating around this complex forms the choroid and scleral coats. The anterior and posterior chambers develop as cavities appear within the mesoderm.

OLFACTION

Nasal placode

Eye

5 WEEKS

4 1/2 WEEKS

Epithelial cells
GUSTATION
Sensory neuron

All special sense organs develop from the epithelia of the embryo.

Gustatory receptors are the least specialized of any of the special sense organs. Taste buds develop as sensory fibers grow into the developing mouth and pharynx.

Olfactory receptors begin as a pair of thickened areas in front of the prosencephalon during the fifth developmental week. The thickenings are called **nasal placodes**.

Taste buds

External nares
10 WEEKS

Over time, the nasal placodes are enfolded and protected by developing facial structures. (Development of the face was discussed in the Embryology Summary on pp. 216–217.)

When the nerve endings contact epithelial cells, the epithelial cells differentiate into gustatory cells. If the sensory nerves are cut, the taste buds degenerate; if the sensory nerve is moved, it will stimulate the development of new taste buds at its new location.

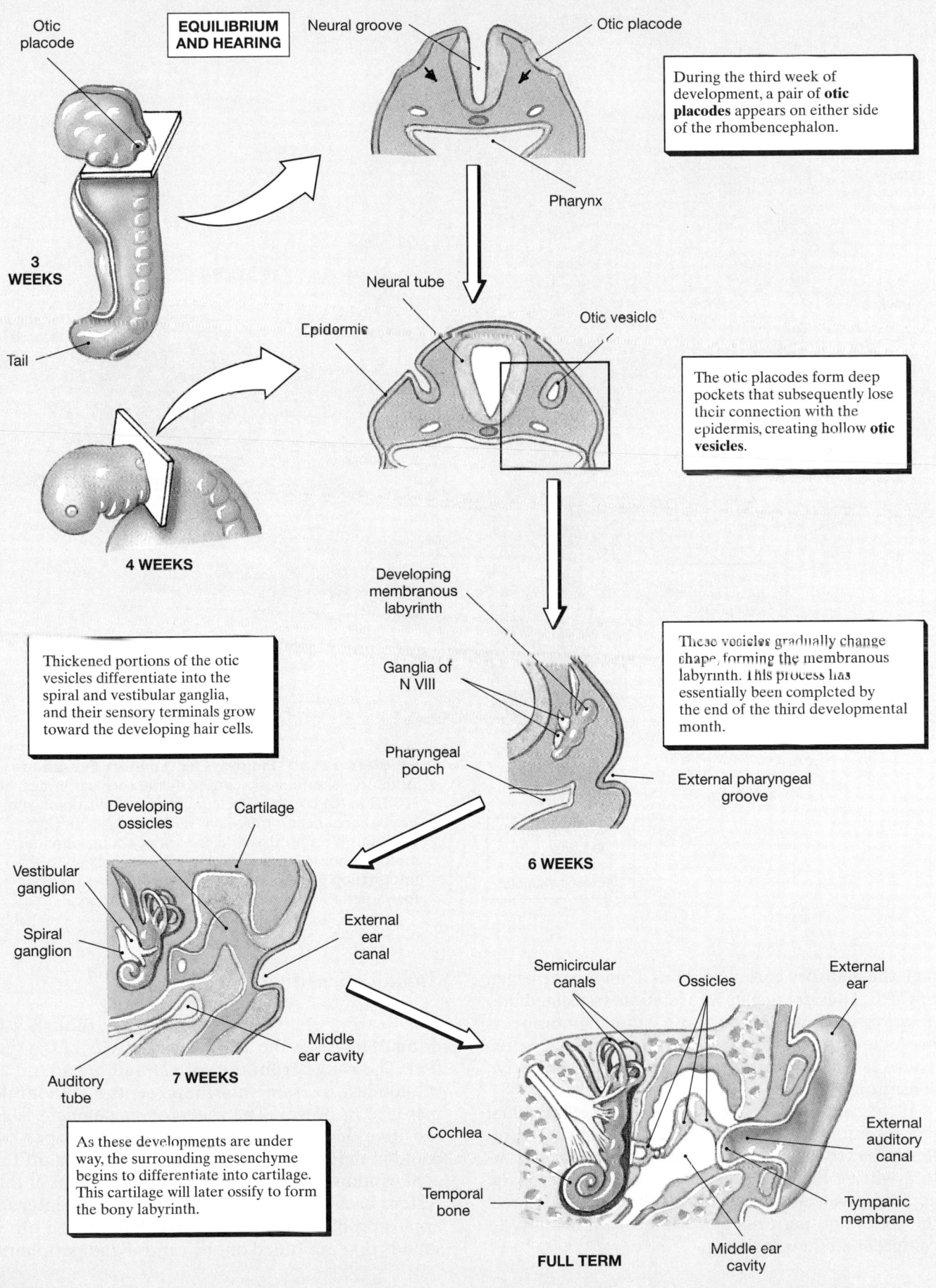

Otic
placode

**EQUILIBRIUM
AND HEARING**

Neural groove

Otic placode

During the third week of
development, a pair of **otic
placodes** appears on either side
of the rhombencephalon.

Pharynx

**3
WEEKS**

Tail

Neural tube

Epidermis

Otic vesicle

The otic placodes form deep
pockets that subsequently lose
their connection with the
epidermis, creating hollow **otic
vesicles**.

4 WEEKS

Developing
membranous
labyrinth

Ganglia of
N VIII

These vesicles gradually change
shape, forming the membranous
labyrinth. This process has
essentially been completed by
the end of the third developmental
month.

Pharyngeal
pouch

External pharyngeal
groove

Thickened portions of the otic
vesicles differentiate into the
spiral and vestibular ganglia,
and their sensory terminals grow
toward the developing hair cells.

6 WEEKS

Developing
ossicles

Cartilage

Vestibular
ganglion

Spiral
ganglion

External
ear
canal

Semicircular
canals

Ossicles

External
ear

Middle
ear cavity

7 WEEKS

Auditory
tube

Cochlea

External
auditory
canal

As these developments are under
way, the surrounding mesenchyme
begins to differentiate into cartilage.
This cartilage will later ossify to form
the bony labyrinth.

Temporal
bone

Tympanic
membrane

Middle ear
cavity

FULL TERM

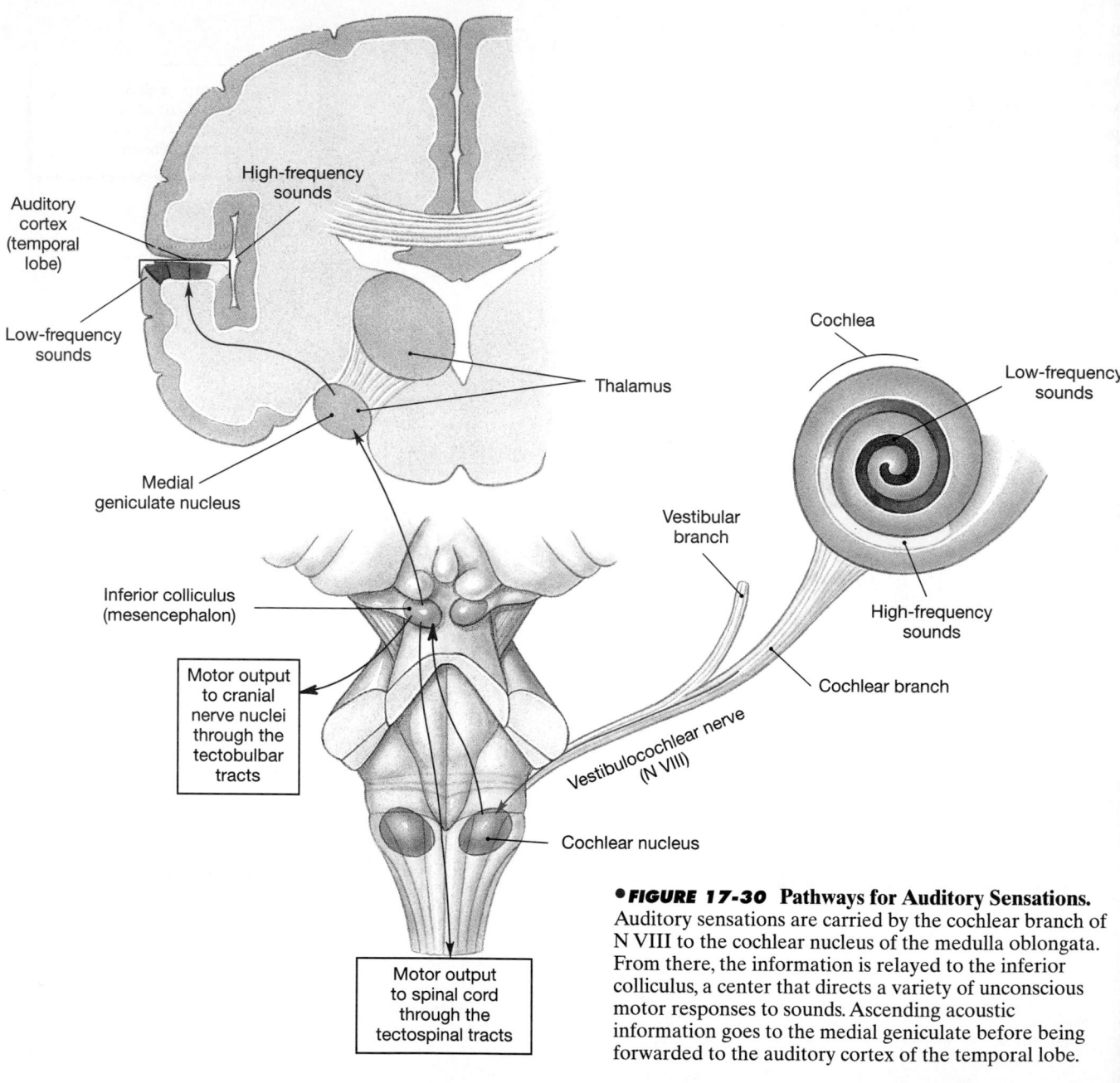

•FIGURE 17-30 Pathways for Auditory Sensations.
Auditory sensations are carried by the cochlear branch of
N VIII to the cochlear nucleus of the medulla oblongata.
From there, the information is relayed to the inferior
colliculus, a center that directs a variety of unconscious
motor responses to sounds. Ascending acoustic
information goes to the medial geniculate before being
forwarded to the auditory cortex of the temporal lobe.

fect, the auditory cortex contains a map of the organ
of Corti. Thus, *frequency* information, translated into
position information on the basilar membrane, is
projected in this form onto the auditory cortex.
There it is interpreted to produce your subjective
sensation of pitch.

If the auditory cortex is damaged, the individual
will respond to sounds and have normal acoustic reflexes, but sound interpretation and pattern recognition will be difficult or impossible. Damage to the
adjacent association area leaves the ability to detect
the tones and patterns but produces an inability to
comprehend their meaning.

Auditory Sensitivity

Our hearing abilities are remarkable, though it is
difficult to assess the absolute sensitivity of the system. The range from the softest audible sound to
the loudest tolerable blast represents a trillionfold
increase in power. The receptor mechanism is so
sensitive that, if we were to remove the stapes, we
could in theory hear air molecules bouncing off the
oval window. We never use the full potential of this
system because body movements and our internal
organs produce squeaks, groans, thumps, and other
sounds that are tuned out by central and peripheral

adaptation. When other environmental noises fade away, the level of adaptation drops, and the system becomes increasingly sensitive. For example, when you relax in a quiet room, your heartbeat seems to get louder and louder as the auditory system adjusts to the level of background noise.

Young children have the greatest hearing range: They can detect sounds ranging from a 20-Hz buzz to a 20,000-Hz whine. With age, damage due to loud noises or other injuries accumulates. The eardrum gets less flexible, the articulations between the ossicles stiffen, and the round window may begin to ossify. As a result, older individuals show some degree of hearing loss.

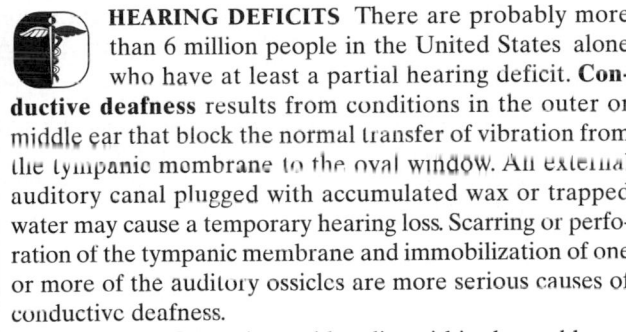 **HEARING DEFICITS** There are probably more than 6 million people in the United States alone who have at least a partial hearing deficit. **Conductive deafness** results from conditions in the outer or middle ear that block the normal transfer of vibration from the tympanic membrane to the oval window. An external auditory canal plugged with accumulated wax or trapped water may cause a temporary hearing loss. Scarring or perforation of the tympanic membrane and immobilization of one or more of the auditory ossicles are more serious causes of conductive deafness.

In **nerve deafness,** the problem lies within the cochlea or somewhere along the auditory pathway. The vibrations reach the oval window and enter the perilymph, but the receptors either cannot respond or their response cannot reach its central destinations. For example:

- Very loud (high-intensity) sounds can produce nerve deafness by breaking stereocilia off the surfaces of the hair cells. (The reflex contraction of the tensor tympani and stapedius muscles in response to a dangerously loud noise occurs in less than 0.1 second, but this may not be fast enough.)

- Drugs such as the aminoglycoside antibiotics (*neomycin* or *gentamicin*) may diffuse into the endolymph and kill the hair cells. Because hair cells and sensory nerves can also be damaged by bacterial infection, the potential side effects must be balanced against the severity of infection.

There are many treatment options for conductive deafness; treatment options for nerve deafness are relatively limited. Because many of these problems become progressively worse, early diagnosis improves the chances of successful treatment. AM *Testing and Treating Hearing Deficits*

☑ If the round window were not able to bulge out with increased pressure in the perilymph, how would sound perception be affected?

☑ How would loss of stereocilia from the hair cells of the organ of Corti affect hearing?

☑ Why does blockage of the auditory tube produce an earache?

SELECTED CLINICAL TERMINOLOGY

Terms Discussed in This Chapter

cataract: An abnormal lens that has lost its transparency. *(p. 559)*
color blindness: A condition in which a person is unable to distinguish certain colors. *(p. 566)*
conductive deafness: Deafness resulting from conditions in the outer or middle ear that block the transfer of vibrations from the tympanic membrane to the oval window. *(p. 585)*
glaucoma: A condition characterized by increased intraocular pressure due to impaired reabsorption of aqueous humor; may result in blindness. *(p. 558)*
hyperopia: Farsightedness—a condition in which nearby objects are blurry but distant objects are clear. *(p. 560)*
myopia: Nearsightedness—a condition in which vision at close range is normal but distant objects appear blurry. *(p. 560)*
nerve deafness: Deafness resulting from problems within the cochlea or along the auditory pathway. *(p. 585)*
nystagmus: Abnormal eye movements that may appear after the brain stem or inner ear is damaged. *(p. 576)*
presbyopia: A type of hyperopia that develops with age as lenses become less elastic. *(p. 560)*
retinitis pigmentosa: A group of inherited retinopathies characterized by the progressive deterioration of photoreceptors, eventually resulting in blindness. *(p. 564)*

AM **Additional Terms Discussed in the** *Applications Manual*

anesthesia: A total loss of sensation.
hypesthesia: A reduction in sensitivity.
Ménière's disease: A condition in which high fluid pressures rupture the walls of the membranous labyrinth, resulting in acute vertigo and inappropriate auditory sensations.
paresthesia: Abnormal sensations, such as a "pins and needles" sensation.

CHAPTER REVIEW

On-line resources for this chapter are on our World Wide Web site at:
http://www.prenhall.com/martini/fap

STUDY OUTLINE

INTRODUCTION, p. 538

RECEPTORS, p. 538

1. The **general senses** are pain, temperature, touch, pressure, vibration, and proprioception. Receptors for those sensations are distributed throughout the body. The **special senses** are **olfaction** (smell), **gustation** (taste), **vision** (sight), **equilibrium** (balance), and **hearing**. Receptors for these senses are located in specialized areas called **sense organs.**
2. A *sensory receptor* is a specialized cell that, when stimulated, sends a *sensation* to the CNS.

Specificity, p. 538

3. **Receptor specificity** allows each receptor to respond to particular stimuli. The simplest receptors are **free nerve endings.** The area monitored by a single receptor cell is the **receptive field.** *(Figure 17-1)*

Transduction, p. 538

4. **Transduction** is the translation of a stimulus into an action potential. This process involves the development of **receptor potentials** that can summate to produce a **generator potential** in an afferent fiber. The resulting action potential travels to the CNS. *(Figure 17-2)*

Interpretation of Sensory Information, p. 539

5. The identity of a sensation is indicated by the **labeled line** that carries the action potentials into the CNS. The intensity of the stimulus is indicated by the frequency and pattern of action potentials. This phenomenon is called *sensory coding*.
6. **Tonic receptors** are always sending signals to the CNS; **phasic receptors** become active only when the conditions they monitor change.

Central Processing and Adaptation, p. 541

7. **Adaptation** (a reduction in sensitivity in the presence of a constant stimulus) may involve changes in receptor sensitivity **(peripheral adaptation)** or inhibition along the sensory pathways **(central adaptation). Fast-adapting receptors** are phasic; **slow-adapting receptors** are tonic.

THE GENERAL SENSES, p. 541

Nociceptors, p. 541

1. **Nociceptors** respond to a variety of stimuli, generally ones associated with tissue damage. There are two types of these painful sensations: **fast** *(prickling)* **pain** and **slow** *(burning and aching)* **pain.** *(Figure 17-3)*

Thermoreceptors, p. 542

2. **Thermoreceptors** respond to changes in temperature.

Mechanoreceptors, p. 542

3. **Mechanoreceptors** respond to physical distortion, contact, or pressure on their cell membranes. The three types of mechanoreceptors are **tactile receptors, baroreceptors,** and **proprioceptors**.
4. Tactile receptors respond to touch, pressure, and vibration and can be classified by the fineness of touch they detect. **Fine touch and pressure receptors** provide detailed infor-

mation about a source of stimulation; **crude touch and pressure receptors** are poorly localized. The six types of tactile receptors are free nerve endings, the **root hair plexus, Merkel's discs, Meissner's corpuscles, Pacinian corpuscles,** and **Ruffini corpuscles.** *(Figure 17-3)*
5. **Baroreceptors** monitor changes in blood pressure in the walls of major arteries and veins. Baroreceptors along the digestive tract help coordinate reflex activities of digestion, and others monitor pressure changes in the reproductive and urinary tracts. These mechanoreceptors respond immediately but adapt rapidly. *(Figure 17-4)*
6. **Proprioceptors** monitor the position of joints, tension in tendons and ligaments, and the state of muscular contraction. Proprioceptors include tendon organs and muscle spindles.

Chemoreceptors, p. 545

7. In general, **chemoreceptors** respond to water-soluble and lipid-soluble substances dissolved in the surrounding fluid. These receptors monitor the chemical composition of body fluids. *(Figure 17-5)*

OLFACTION, p. 546

Olfactory Receptors, p. 546

1. The **olfactory organs** contain the **olfactory epithelium** with **olfactory receptors** (neurons sensitive to chemicals dissolved in the overlying mucus), *supporting cells,* and **basal** *(stem)* **cells.** Their surfaces are coated with the secretions of the **olfactory glands.** *(Figure 17-6)*
2. The olfactory receptors are modified neurons.

Olfactory Pathways, p. 547

3. The highly sensitive olfactory system has extensive limbic and hypothalamic connections.

Olfactory Discrimination, p. 548

4. The olfactory system can distinguish thousands of chemical stimuli. The CNS interprets smells by the pattern of receptor activity.

Aging and Olfactory Sensitivity, p. 548

5. The olfactory receptor population shows considerable turnover. The total number of olfactory receptors declines with age.

GUSTATION, p. 548

1. **Gustatory (taste) receptors** are clustered in **taste buds**.
2. Taste buds are associated with epithelial projections on the dorsal surface of the tongue **(lingual papillae).** *(Figure 17-7)*

Gustatory Receptors, p. 548

3. Each taste bud contains **gustatory cells,** which extend *taste hairs* through a narrow **taste pore.** *(Figure 17-7)*

Gustatory Pathways, p. 548

4. The taste buds are monitored by cranial nerves that synapse within the **nucleus solitarius.**

Gustatory Discrimination, p. 549

5. The **primary taste sensations** are sweet, salt, sour, and bitter. *(Figure 17-7)*

6. There are significant individual differences in taste sensitivity, some of which are inherited.
7. The number of taste buds declines with age.

VISION, p. 550

Accessory Structures of the Eye, p. 550

1. The **accessory structures** of the eye include the **palpebrae (eyelids),** which are separated by the **palpebral fissure.** The **eyelashes** line the palpebral margins. Along the inner margin of the lid are **Meibomian glands,** which secrete a lipid-rich product. Glands at the **lacrimal caruncle** produce other secretions. *(Figure 17-8)*
2. An epithelium called the **conjunctiva** covers most of the exposed surface of the eye. The **ocular** *(bulbar)* **conjunctiva** covers the anterior surface of the eye, and the **palpebral conjunctiva** lines the inner surface of the eyelids. The **cornea** is transparent. *(Figure 17-9)*
3. The secretions of the **lacrimal gland** bathe the conjunctiva; these secretions are slightly alkaline and contain **lysozyme** (an enzyme that attacks bacteria). Tears collect in the **lacrimal lake.** The tears reach the inferior meatus of the nose after they pass through the **lacrimal puncta,** the **lacrimal canals,** the **lacrimal sac,** and the **nasolacrimal duct.** *(Figure 17-8)*

The Eye, p. 553

4. The eye has three layers: an outer **fibrous tunic,** a **vascular tunic,** and an inner **neural tunic.**
5. The fibrous tunic consists of the **sclera** (a dense fibrous connective tissue that covers most of the ocular surface), the **cornea,** and the **limbus** (the border between the sclera and the cornea). *(Figure 17-9)*
6. The vascular tunic, or **uvea,** includes the **iris,** the **ciliary body,** and the **choroid.** The iris contains blood vessels, pigment cells, and muscle fibers that change the diameter of the **pupil.** The ciliary body contains the **ciliary muscle** and the **ciliary processes,** which attach to the **suspensory ligaments** of the lens. *(Figure 17-9)*
7. The neural tunic, or **retina,** consists of an outer **pigmented layer** and an inner **neural retina;** the latter contains visual receptors and associated neurons. *(Figures 17-9, 17-10)*
8. The direct line to the CNS proceeds from the **photoreceptors** to **bipolar cells,** then to **ganglion cells,** and to the brain via the optic nerve. Axons of ganglion cells converge at the **optic disc,** or **blind spot. Horizontal cells** and **amacrine cells** modify the signals passed among other retinal components. *(Figures 17-10, 17-11)*
9. The ciliary body and lens divide the interior of the eye into a large **posterior cavity,** or *vitreous chamber,* and a smaller **anterior cavity.** The anterior cavity is subdivided into the **anterior chamber,** which extends from the cornea to the iris, and a **posterior chamber** between the iris and the ciliary body and lens. *(Figure 17-12)*
10. The fluid **aqueous humor** circulates within the eye and reenters the circulation after diffusing through the walls of the anterior chamber and into the **canal of Schlemm.** *(Figure 17-12)*
11. The **lens,** held in place by the suspensory ligaments, lies posterior to the cornea and forms the anterior boundary of the posterior cavity. This cavity contains the **vitreous body,** a gelatinous mass that helps stabilize the shape of the eye and support the retina. *(Figures 17-9, 17-12)*
12. The lens focuses a visual image on the retinal receptors. A lens that has lost its transparency is a **cataract.**

13. Light is **refracted** (bent) when it passes through the cornea and lens. During **accommodation,** the shape of the lens changes to focus an image on the retina. "Normal" **visual acuity** is rated 20/20. *(Figures 17-13–17-16)*

Visual Physiology, p. 562

14. There are two types of photoreceptors (visual receptors of the retina): **rods,** which respond to almost any photon, regardless of its energy content; and **cones,** which have characteristic ranges of sensitivity. Many cones are densely packed within the **fovea** (the central portion of the **macula lutea**), the site of sharpest vision.
15. Each photoreceptor contains an **outer segment** with membranous **discs.** A narrow stalk connects the outer to the **inner segment.** Light absorption occurs in the **visual pigments,** which are derivatives of **rhodopsin** (**opsin** plus the pigment **retinal,** which is synthesized from **vitamin A**). The retinal molecule can be in either the normal **11-*cis*** or (after absorbing light) the **11-*trans*** form. *(Figures 17-17–17-19)*
16. **Color blindness** is the inability to detect certain colors. *(Figure 17-20)*
17. In the **dark-adapted state,** almost all visual pigments will be fully receptive to stimulation. The **light-adapted state** is characterized by constriction of the pupil and bleaching of the visual pigments.

The Visual Pathway, p. 567

18. Each photoreceptor monitors a specific receptive field. The ganglion cells that monitor rods, called **M cells** *(magnocells),* are relatively large. The ganglion cells that monitor cones, called **P cells** *(parvo cells),* are smaller and more numerous. *(Figure 17-21)*
19. Visual information from the left half of the combined visual field arrives at the visual cortex of the right occipital lobe; information from the right half of the combined visual field arrives at the visual cortex of the left occipital lobe. *(Figure 17-22)*
20. Visual inputs to the **suprachiasmatic nucleus** of the hypothalamus affect the function of other brain stem nuclei. This nucleus establishes a visceral **circadian rhythm** that is tied to the day/night cycle and affects other metabolic processes.

EQUILIBRIUM AND HEARING, p. 570

1. The senses of equilibrium and hearing are provided by the receptors of the inner ear.

Anatomy of the Ear, p. 570

2. The ear is divided into the **external ear, middle ear,** and **inner ear.** *(Figure 17-23)*
3. The external ear includes the **pinna,** which surrounds the entrance to the **external auditory canal.** That canal ends at the **tympanic membrane** *(eardrum). (Figure 17-23)*
4. The middle ear communicates with the nasopharynx via the **auditory** *(pharyngotympanic)* **tube.** The middle ear encloses and protects the **auditory ossicles,** which connect the tympanic membrane with the receptors of the inner ear. *(Figures 17-23, 17-24)*
5. The **membranous labyrinth** (the chambers and tubes) of the inner ear contain the fluid **endolymph.** The **bony labyrinth** surrounds and protects the membranous labyrinth. The bony labyrinth can be subdivided into the **vestibule,** the **semicircular canals,** and the **cochlea.** The structures and air spaces of the external and middle ears help capture and transmit sound to the cochlea. *(Figure 17-25)*

6. The vestibule of the inner ear encloses a pair of membranous sacs, the **saccule** and **utricle,** whose receptors provide sensations of gravity and linear acceleration. The semicircular canals contain the **semicircular ducts,** whose receptors provide sensations of rotation. The cochlea contains the **cochlear duct,** an elongated portion of the membranous labyrinth. *(Figure 17-25)*

7. The basic receptors of the inner ear are **hair cells.** The surfaces of each hair cell support **stereocilia** and (except in the cochlear duct) a single **kinocilium.** Hair cells provide information about the direction and strength of varied mechanical stimuli. *(Figure 17-25)*

Equilibrium, p. 573

8. The **anterior, posterior,** and **lateral semicircular ducts** are continuous with the utricle. Each duct contains an **ampulla** with sensory receptors. Here the cilia contact a gelatinous **cupula.** *(Figures 17-25, 17-26)*

9. The saccule and utricle are connected by a passageway continuous with the **endolymphatic duct,** which terminates in the **endolymphatic sac.** In the saccule and utricle, hair cells cluster within **maculae,** where their cilia contact **otoliths** (densely packed mineral crystals). When the head tilts, the pile of otoliths shifts, and the resulting distortion in the hair cell processes signals the CNS. *(Figure 17-26)*

10. The vestibular receptors activate sensory neurons of the **vestibular ganglia.** The axons form the **vestibular branch** of the vestibulocochlear nerve (N VIII), synapsing within the **vestibular nuclei.** *(Figure 17-27)*

Hearing, p. 577

11. The energy content of a sound determines its **intensity,** measured in **decibels.** Sound waves travel toward the tympanic membrane, which vibrates; the auditory ossicles conduct these vibrations to the inner ear. Movement at the **oval window** applies pressure to the **perilymph** of the **vestibular duct.** *(Tables 17-1, 17-2; Figures 17-28, 17-29)*

12. Pressure waves distort the **basilar membrane** and push the hair cells of the **organ of Corti** against the **tectorial membrane.** The **tensor tympani** and **stapedius muscles** contract to reduce the amount of motion when very loud sounds arrive. *(Figures 17-28, 17-29)*

13. The sensory neurons are located in the **spiral ganglion** of the cochlea. Their afferent fibers form the **cochlear branch** of the vestibulocochlear nerve (N VIII), synapsing at the **cochlear nucleus.** *(Figure 17-30)*

REVIEW QUESTIONS

Level 1   Reviewing Facts and Terms

1. Regardless of the nature of a stimulus, sensory information must be sent to the CNS in the form of
   (a) dendritic processes
   (b) action potentials
   (c) neurotransmitter molecules
   (d) generator potentials

2. A membrane depolarization that leads to an action potential in a sensory neuron is
   (a) a labeled line   (b) a sensation
   (c) a neurotransmitter   (d) a generator potential

3. A reduction in sensitivity in the presence of a constant stimulus is
   (a) transduction   (b) sensory coding
   (c) line labeling   (d) adaptation

4. Sensations produced by an injection or a deep cut would reach the CNS very quickly via
   (a) Type A fibers   (b) Type C fibers
   (c) tactile receptors   (d) proprioceptors

5. Mechanoreceptors that detect pressure changes in the walls of blood vessels and in portions of the digestive, reproductive, and urinary tracts are
   (a) tactile receptors   (b) baroreceptors
   (c) proprioceptors   (d) free nerve endings

6. Pacinian corpuscles in the skin detect
   (a) deep pressure   (b) fine touch
   (c) muscle stretch   (d) temperature changes

7. Examples of proprioceptors that monitor the position of joints and the state of muscular contraction are
   (a) Pacinian and Meissner's corpuscles
   (b) carotid and aortic sinuses
   (c) Merkel's discs and Ruffini corpuscles
   (d) Golgi tendon organs and muscle spindles

8. When chemicals dissolve in the nasal cavity, they stimulate
   (a) gustatory cells   (b) olfactory hairs
   (c) rod cells   (d) tactile receptors

9. The taste sensation of sweetness is experienced on the
   (a) posterior part of the tongue
   (b) anterior part of the tongue
   (c) right and left lateral sides of the tongue
   (d) the middle part of the tongue

10. The anterior, transparent part of the fibrous tunic is known as the
   (a) cornea   (b) iris
   (c) sclera   (d) retina

11. Meibomian glands secrete a lipid-rich product that helps
   (a) produce and secrete tears in the lacrimal apparatus
   (b) keep the cornea moist and clean
   (c) dilate the blood vessels beneath the conjunctival epithelium
   (d) keep the eyelids from sticking together

12. Tears produced by the lacrimal apparatus
   (a) keep conjunctival surfaces moist and clean
   (b) reduce friction and remove debris from the eye
   (c) provide nutrients and oxygen to the conjunctional epithelium
   (d) a, b, and c are correct

13. The thick, gel-like fluid that helps support the structure of the eyeball is the
   (a) vitreous humor   (b) aqueous humor
   (c) ora serrata   (d) perilymph

14. Pupillary muscle groups are controlled by the ANS. Parasympathetic activation causes pupillary _____, and sympathetic activation causes _____:
   (a) dilation, constriction
   (b) dilation, dilation
   (c) constriction, dilation
   (d) constriction, constriction

15. The retina is the
   (a) vascular tunic   (b) fibrous tunic
   (c) neural tunic   (d) a, b, and c are correct

16. At sunset or sunrise your visual system adapts to
   (a) fovea vision   (b) rod-based vision
   (c) macular vision   (d) cone-based vision
17. The primary function of the lens of the eye is to
   (a) stabilize the shape of the eye
   (b) facilitate communication between photoreceptors and ganglion cells
   (c) focus the visual image on the retinal photoreceptors
   (d) a, b, and c are correct
18. A better-than-average visual acuity rating is
   (a) 20/20   (b) 20/30
   (c) 15/20   (d) 20/15
19. Ceruminous glands along the external auditory canal are responsible for
   (a) slowing the growth of microorganisms in the exterior auditory canal
   (b) reducing chances of infection in the ear
   (c) helping deny access to foreign objects
   (d) a, b, and c are correct
20. The malleus, incus, and stapes are the tiny ear bones located in the
   (a) outer ear   (b) middle ear
   (c) inner ear   (d) membranous labyrinth
21. Receptors in the saccule and utricle provide sensations of
   (a) balance and equilibrium
   (b) hearing
   (c) vibration
   (d) gravity and linear acceleration

22. The organ of Corti is located within the _____ of the inner ear:
   (a) utricle   (b) bony labyrinth
   (c) vestibule   (d) cochlea
23. Auditory information about the region and intensity of stimulation is relayed to the CNS over the cochlear branch of cranial nerve
   (a) N IV   (b) N VI
   (c) N VIII   (d) N X
24. What three steps are necessary for the process of transduction to occur?
25. What three types of mechanoreceptors respond to stretching, compression, twisting, or other distortions of the cell membrane?
26. Identify six types of tactile receptors located in the skin, and describe their sensitivities.
27. Trace the olfactory pathway from the time an odor reaches the olfactory epithelium until it reaches its final destination in the brain.
28. What are the three types of papillae on the human tongue?
29. (a) What structures make up the fibrous tunic of the eye?
   (b) What are the functions of the fibrous tunic?
30. What structures make up the vascular tunic of the eye?
31. Trace the pathway of a nerve impulse from the optic nerve to the visual cortex.
32. What are the three auditory ossicles in the middle ear, and what are their functions?
33. What six basic steps are involved in the process of hearing?

Level 2   Reviewing Concepts

34. The CNS interprets sensory information entirely on the basis of the
   (a) strength of the action potential
   (b) number of generator potentials
   (c) line over which it arrives
   (d) a, b, and c are correct
35. If the auditory cortex is damaged, the individual will respond to sounds and have normal acoustic reflexes, but
   (a) the sounds may produce nerve deafness
   (b) the auditory ossicle may be immobilized
   (c) sound interpretation and pattern recognition may be impossible
   (d) normal transfer of vibration to the oval window is inhibited

36. Distinguish between the general senses and the special senses in the human body.
37. In what form does the CNS receive a stimulus detected by a sensory receptor?
38. What type of information about a stimulus does sensory coding provide?
39. Why are olfactory sensations long-lasting and an important part of our memories and emotions?
40. What is the usual result if a sebaceous gland of an eyelash or a Meibomian gland becomes infected?
41. Jane makes an appointment with the optometrist for a vision test. Her test results are reported as 20/15. What does this test result mean? Is a rating of 20/20 better or worse?

Level 3   Critical Thinking and Clinical Applications

42. You are at a park watching some deer 35 feet away from you. A friend taps you on the shoulder to ask a question. As you turn to look at your friend, who is standing just 2 feet away, what changes would your eyes undergo?
43. Sally's driver's license indicates that she must wear glasses when she drives. She does not need glasses to read or to see objects that are close. What type of lenses are in Sally's glasses?
44. After attending a Fourth of July fireworks extravaganza, Millie finds it difficult to hear normal conversation and her ears keep "ringing." What is causing her hearing problems?

45. For a few seconds after you ride the express elevator from the twentieth floor to the ground floor, you still feel as if you are descending, even though you have come to a stop. Why?
46. Juan has a disorder involving the saccule and the utricle. He is asked to stand with his feet together and arms extended forward. As long as he keeps his eyes open, he exhibits very little movement. But when he closes his eyes, his body begins to sway a great deal, and his arms tend to drift in the direction of the impaired vestibular receptors. Why does this occur?

The Endocrine System

Gheorghe Muresan (left) and Tony Massenburg differ in height by about 0.3 m—about 1 foot. Genetically programmed levels of hormones, especially growth hormone, were probably responsible for this difference. Hormone levels during childhood and adolescence provided Muresan with a height advantage, but both players have the right blend of coordination, balance, strength, speed, and intelligence. This is a common pattern: Hormones direct long-term changes (such as those that determine body form); moment-to-moment adjustments (which make basketball games exciting) are controlled by the nervous system. This chapter introduces the major hormones and the endocrine organs that produce them.

CHAPTER OUTLINE AND OBJECTIVES

To function effectively, every cell in the body must communicate with its neighbors and with cells and tissues in distant portions of the body. Table 18-1 provides an overview of the ways our cells and tissues communicate with one another. In a few specialized tissues, cellular activities are coordinated by the exchange of ions and molecules from one cell to the next across gap junctions. This **direct communication** occurs between cells of the same type, and the two cells must be in extensive physical contact. The two cells are communicating so closely that they function as a single entity. For example, gap junctions (1) coordinate ciliary movement among epithelial cells, (2) coordinate the contractions of cardiac muscle cells, and (3) facilitate the propagation of action potentials from one neuron to the next at an electrical synapse. ∞ *[pp. 112, 306, 397]*

Direct communication is highly specialized and relatively rare. Most of the communication between living cells involves the release and receipt of chemical messages. Each living cell is continuously "talking" to its neighbors by releasing chemicals, often called *cytokines*, into the extracellular fluid. These chemicals tell cells what their neighbors are doing at any given moment. The result is the coordination of tissue function at the local level. Examples of these *local hormones* include the prostaglandins, introduced in Chapter 2, and the various *growth factors* we discussed in Chapter 3. ∞ *[pp. 49, 102]* The use of chemical messengers to transfer information from cell to cell within a single tissue is called **paracrine communication**, and the chemicals involved are called **paracrine factors**.

Paracrine factors enter the circulation, but the concentrations are usually so low that distant cells and tissues are not affected. However, some paracrine factors, including several of the prostaglandins and related chemicals, have primary effects in their tissues of origin and secondary effects in other tissues and organs. When secondary effects occur, the paracrine factors are also acting as hormones. **Hormones** are chemical messengers that are released in one tissue and transported within the circulation to reach certain cells in other tissues. Whereas all cells release paracrine factors, typical hormones are produced only by specialized cells. In intercellular communication, hormones are letters, and the circulatory system is the postal service. A hormone released into the circulation will be distributed throughout the body. Each hormone has *target cells*, specific cells that will respond to its presence. These cells possess the receptors needed to bind and "read" the hormonal message. The general pattern is indicated in Table 18-1. Although every cell in the body is exposed to the mixture of hormones in circulation at any given moment, each individual cell will respond to only a few of the hormones present. The other hormones are treated like junk mail and ignored, because the cell lacks the receptors to read the messages they contain. The use of hormones to coordinate cellular activities in tissues in distant portions of the body is called **endocrine communication**.

TABLE 18-1 Mechanisms of Intercellular Communication

Mechanism	Transmission	Chemical Mediators	Distribution of Effects
Direct communication	Through gap junctions from cytoplasm to cytoplasm	Ions, small solutes, lipid-soluble materials	Limited to adjacent cells that are directly interconnected by connexons
Paracrine communication	Through extracellular fluid	Paracrine factors	Primarily limited to local area, where concentrations are relatively high; target cells must have appropriate receptors
Endocrine communication	Through the circulatory system	Hormones	Target cells are primarily in other tissues and organs and must have appropriate receptors
Synaptic communication	Across synaptic clefts	Neurotransmitters	Limited to very specific area; target cells must have appropriate receptors

Because the target cells can be anywhere in the body, a single hormone can alter the metabolic activities of multiple tissues and organs simultaneously. These effects may be slow to appear, but they typically persist for days. Consequently, hormones are effective in coordinating cell, tissue, and organ activities on a sustained, long-term basis. For example, circulating hormones keep body water content and levels of electrolytes and organic nutrients within normal limits 24 hours a day throughout our entire lives.

While the effects of a single hormone persist, a cell may receive additional instructions from other hormones. The result will be a further modification of cellular operations. Gradual changes in the quantities and identities of circulating hormones can produce complex changes in physical structure and physiological capabilities. Examples include the processes of embryological and fetal development, growth, and puberty.

The nervous system also relies primarily on chemical communication, but it does not use the circulation for message delivery. Instead, as we saw in Chapter 12, neurons release a neurotransmitter very close to target cells that bear the appropriate receptors. ∞ *[p. 397]* The command to release the neurotransmitter rapidly travels from one location to another in the form of action potentials propagated along axons. The nervous system thus acts like a telephone company, carrying high-speed "messages" from one location in the body to another and delivering them to a specific destination. The effects of neural stimulation are generally short-lived, and they tend to be restricted to specific target cells—primarily because the neurotransmitter is rapidly broken down or recycled. This form of **synaptic communication** is ideal for crisis management: If you are in danger of being hit by a speeding bus, the nervous system can coordinate and direct your leap to safety. Once the crisis is over and the neural circuit quiets down, things soon return to normal.

When viewed from a general perspective, the differences between the nervous and endocrine systems seem relatively clear. In fact, these broad organizational and functional distinctions are the basis for treating them as two separate systems. Yet, when we consider them in detail, the two systems are organized along parallel lines. For example:

- Both systems rely on the release of chemicals that bind to specific receptors on their target cells.
- The two systems share many chemical messengers; for example, norepinephrine and epinephrine are called *hormones* when released into the circulation but *neurotransmitters* when released across synapses.
- Both systems are regulated primarily by negative feedback control mechanisms.
- The two systems share a common goal: to preserve homeostasis by coordinating and regulating the activities of other cells, tissues, organs, and systems.

In this chapter, we introduce the components and functions of the endocrine system and explore the interactions between the nervous and endocrine systems. In later chapters, we shall consider specific endocrine organs, hormones, and functions in detail.

AN OVERVIEW OF THE ENDOCRINE SYSTEM

The endocrine system includes all the endocrine cells and tissues of the body. As we noted in Chapter 4, *endocrine cells* are glandular secretory cells that release their secretions into the extracellular fluid. This characteristic distinguishes them from *exocrine cells,* which secrete their products onto epithelial surfaces, generally by way of ducts. ∞ *[p. 117]* The chemicals released by endocrine cells may affect only adjacent cells, as in the case of most paracrine factors, or they may affect cells throughout the body.

The components of the endocrine system are introduced in Figure 18-1•. This figure also lists the major hormones produced in each endocrine tissue and organ. Some of these organs, such as the pituitary gland, have endocrine secretion as a primary function; others, such as the pancreas, have many other functions in addition to endocrine secretion. We shall consider the structure and functions of these endocrine organs in detail in chapters dealing with other systems. Examples include the hypothalamus (Chapter 14), the adrenal medullae (Chapter 16), the heart (Chapter 20), the thymus (Chapter 22), the pancreas and digestive tract (Chapter 24), the kidneys (Chapter 26), the reproductive organs (Chapter 28), and the placenta (Chapter 29).

Hormone Structure

Hormones can be divided into three groups on the basis of chemical structure: (1) *amino acid derivatives,* (2) *peptide hormones,* and (3) *lipid derivatives.*

Amino Acid Derivatives

Several hormones are relatively small molecules that are structurally similar to amino acids, the building blocks of proteins. ∞ *[p. 52]* This group of hormones, sometimes known as the *biogenic amines,* includes *epinephrine (E), norepinephrine (NE), dopamine,* the *thyroid hormones*, and the pineal hormone *melatonin.* Epinephrine, norepinephrine, and dopamine are structurally similar; these compounds are sometimes called **catecholamines** (kat-e-KŌ-la-mēnz). As we saw in Chapter 16, E and NE are secreted by the adrenal medullae during sympathetic activation. ∞ *[p. 521]* **Thyroid hormones** are released by the thyroid gland, and **melatonin** is produced by the pineal gland.

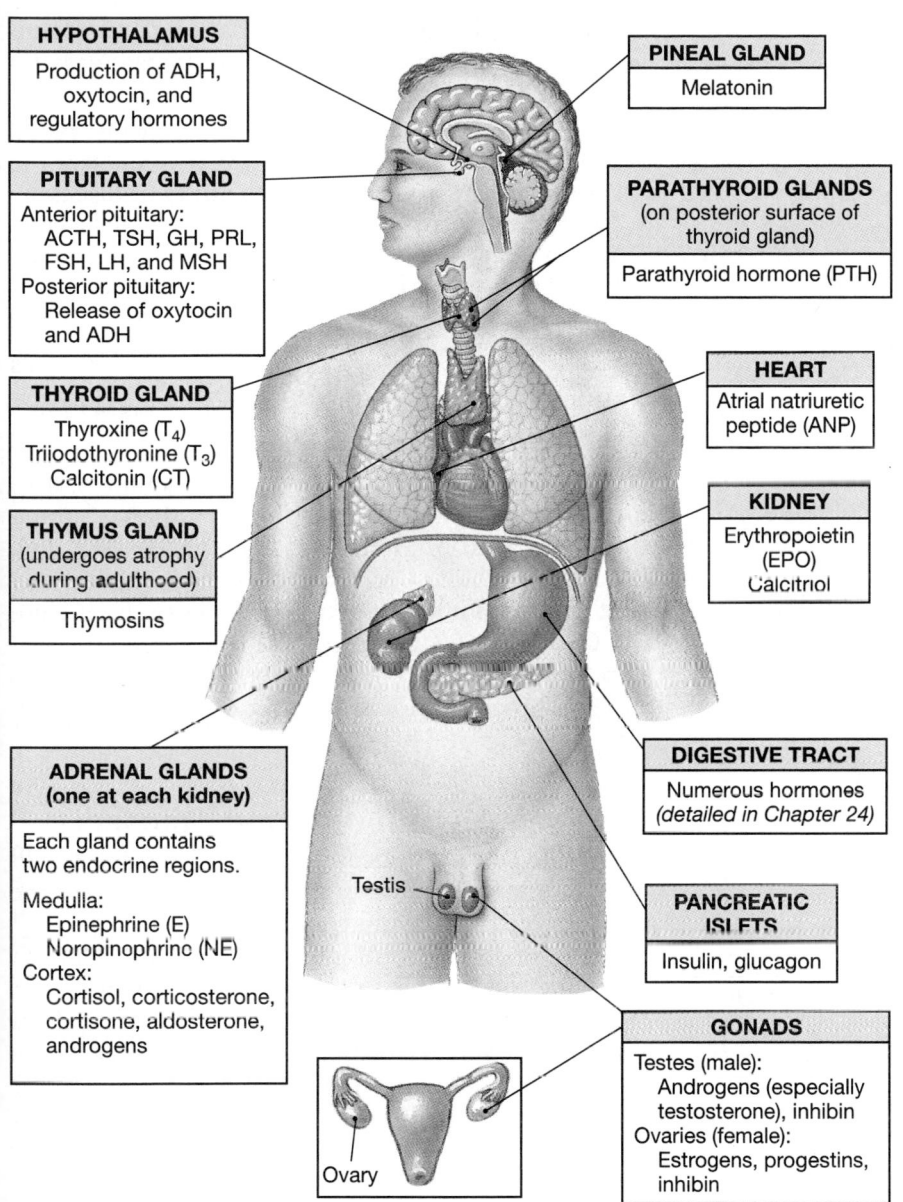

HYPOTHALAMUS
Production of ADH, oxytocin, and regulatory hormones

PITUITARY GLAND
Anterior pituitary: ACTH, TSH, GH, PRL, FSH, LH, and MSH
Posterior pituitary: Release of oxytocin and ADH

THYROID GLAND
Thyroxine (T_4)
Triiodothyronine (T_3)
Calcitonin (CT)

THYMUS GLAND
(undergoes atrophy during adulthood)
Thymosins

ADRENAL GLANDS
(one at each kidney)
Each gland contains two endocrine regions.
Medulla:
    Epinephrine (E)
    Norepinephrine (NE)
Cortex:
    Cortisol, corticosterone, cortisone, aldosterone, androgens

PINEAL GLAND
Melatonin

PARATHYROID GLANDS
(on posterior surface of thyroid gland)
Parathyroid hormone (PTH)

HEART
Atrial natriuretic peptide (ANP)

KIDNEY
Erythropoietin (EPO)
Calcitriol

DIGESTIVE TRACT
Numerous hormones
(detailed in Chapter 24)

PANCREATIC ISLETS
Insulin, glucagon

GONADS
Testes (male):
    Androgens (especially testosterone), inhibin
Ovaries (female):
    Estrogens, progestins, inhibin

Testis

Ovary

•*FIGURE 18-1* **The Endocrine System**

Catecholamines and thyroid hormones are synthesized from molecules of the amino acid *tyrosine* (TĪ-rō-sēn). Melatonin is manufactured from molecules of the amino acid *tryptophan.*

Peptide Hormones

Peptide hormones are chains of amino acids. They can be divided into two groups. One large and diverse group includes hormones that range from short peptide chains, such as antidiuretic hormone (*ADH*) and *oxytocin* (9 amino acids apiece), to small proteins, such as growth hormone (*GH*; 191 amino acids) and *prolactin* (*PRL*; 198 amino acids). This group includes all the hormones secreted by the hypothalamus, posterior pituitary gland, heart, thymus, digestive tract, and pan-

creas and most of the hormones of the anterior pituitary gland.

The second group of peptide hormones consists of glycoproteins. ∞ *[p. 57]* These proteins are more than 200 amino acids long and have carbohydrate side chains. The glycoproteins include thyroid-stimulating hormone (*TSH*), luteinizing hormone (*LH*), and follicle-stimulating hormone (*FSH*) from the anterior pituitary gland as well as several hormones produced in other organs.

Figure 18-2• illustrates the stages in the synthesis of peptide hormones. In general, the process follows the pattern of protein synthesis and secretion that we discussed in Chapter 3. ∞ *[p. 89]* After mRNA transcription has occurred in the nucleus, translation and peptide synthesis begin at ribosomes of the rough endoplasmic reticulum. The peptide chains produced at the RER are called **prehormones,** because they are structurally distinct from the final hormone product. For example, the prehormone may contain extra amino acids, or it may be missing important carbohydrate components.

Modification in the endoplasmic reticulum converts the prehormone to a **prohormone,** an inactive form of the hormone. Final conversion to an active hormone may occur prior to release, as the prohormone passes through the Golgi apparatus and secretory vesicles, or after release, within the bloodstream.

Lipid Derivatives

There are two classes of lipid-based hormones: (1) *steroid hormones,* derived from cholesterol, and (2) *eicosanoids* (ī-KŌ-sa-noydz), derived from *arachidonic acid,* a 20-carbon fatty acid.

STEROID HORMONES **Steroid hormones** are lipids structurally similar to cholesterol (Figure 2-15•, p. 51). Steroid hormones are released by male and female reproductive organs (*androgens* by the testes, *estrogens* and *progestins* by the ovaries), the adrenal glands (*corticosteroids*), and the kidneys (*calcitriol*). The individ-

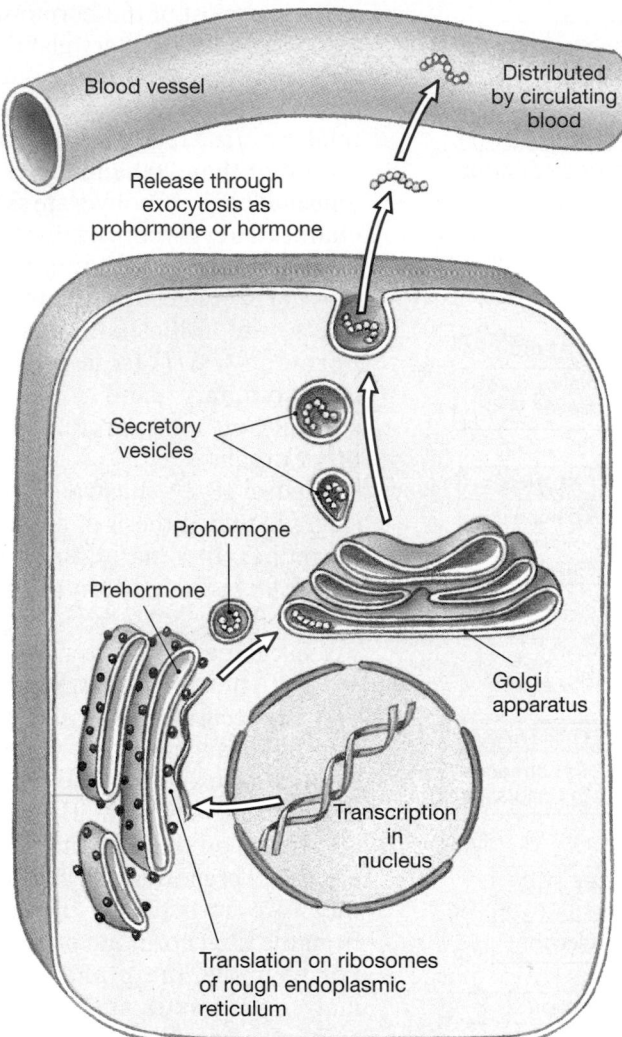

Blood vessel

Distributed by circulating blood

Release through exocytosis as prohormone or hormone

Secretory vesicles

Prohormone

Prehormone

Golgi apparatus

Transcription in nucleus

Translation on ribosomes of rough endoplasmic reticulum

•**FIGURE 18-2** **Peptide Hormone Production.**
Activated genes in the nucleus are transcribed, producing mRNA that travels to the rough endoplasmic reticulum (RER). Translation on ribosomes of the RER then produces molecules of a prehormone. These molecules undergo modification in the RER, forming a prohormone. Conversion of the prohormone to an active hormone may occur in the Golgi apparatus, in the secretory vesicles, or after release through exocytosis.

ual hormones differ in the side chains attached to the central ring structure.

In the blood, steroid hormones are bound to specific transport proteins in the plasma. For this reason, they remain in circulation longer than do secreted peptide hormones. The liver gradually absorbs these steroids and converts them to a soluble form that can be excreted in the bile or urine.

EICOSANOIDS **Eicosanoids** are small molecules with a five-carbon ring at one end. These compounds are important paracrine factors that coordinate cellular activities and affect enzymatic processes (such as blood clotting) that occur in extracellular fluids.

Some of the eicosanoids also have secondary roles as hormones. Examples of important eicosanoids include the following:

- **Leukotrienes** are released by activated white blood cells, or *leukocytes.* Leukotrienes are important in coordinating tissue responses to injury or disease.
- **Prostaglandins** are produced in most tissues of the body. Within each tissue, the prostaglandins released are involved primarily in coordinating local cellular activities. Circulating platelets convert prostaglandins to **thromboxanes** (throm-BOX-ānz) and **prostacyclins** (pros-ta-SĪ-klinz) that are released when blood clotting occurs. We introduced platelets, normal components of the circulating blood, in Chapter 4. ∞ *[p. 126]*

Our focus in this chapter is on circulating hormones whose primary functions are the coordination of activities in many different tissues and organs. We shall consider eicosanoids in chapters that deal with individual tissues and organs, including Chapters 19 (the blood), 22 (the lymphatic system), and 28 (the reproductive system). Examples of each group of hormones discussed here are included in Figure 18-3•.

Hormone Distribution and Transport

Hormone release typically occurs where capillaries are abundant, and the hormones quickly enter the circulation for distribution throughout the body. Within the blood, hormones may circulate freely, or they may be bound to special carrier proteins. A freely circulating hormone remains functional for less than 1 hour, and sometimes for as little as 2 minutes. It is inactivated when (1) it diffuses out of the bloodstream and binds to receptors in target tissues, (2) it is absorbed and broken down by cells of the liver or kidneys, or (3) it is broken down by enzymes in the plasma or interstitial fluids.

Thyroid hormones and steroid hormones remain in the circulation much longer, because when these hormones enter the bloodstream, almost all of them become attached to special transport proteins. Less than 1 percent of these hormones circulate freely. An equilibrium state exists between the free hormones and the bound hormones; as the free hormones are removed and inactivated, they are replaced by the release of bound hormones. Thus the circulation contains a substantial reserve (several weeks' supply) of these hormones at any time.

Hormone Function and Mechanisms of Action

All cellular structures and functions depend on proteins. Structural proteins determine the general shape and internal structure of a cell, and enzymes affect its metabolic activities. ∞ *[p. 51]* Hormones alter cellular operations by changing the *types, ac-*

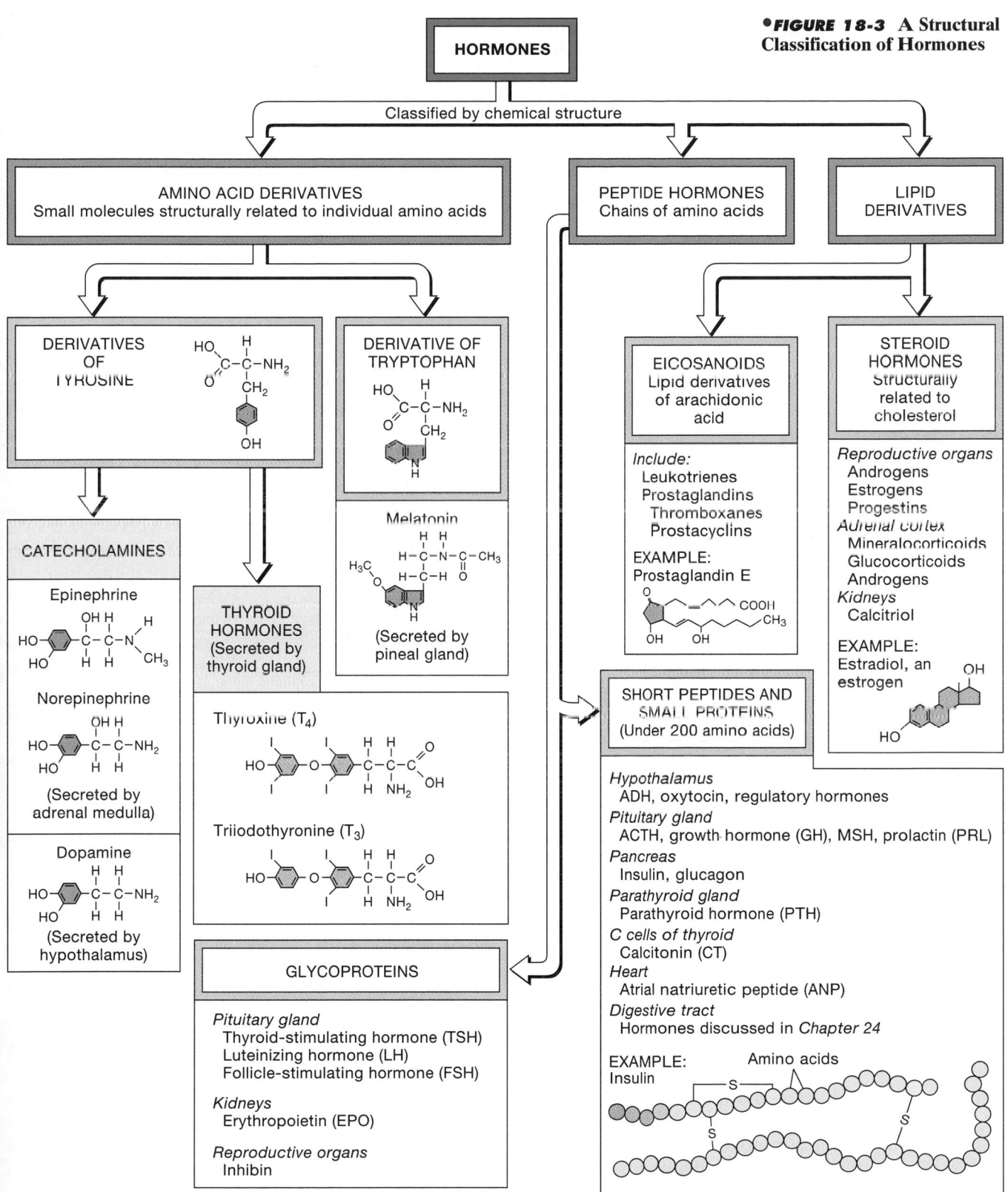

•*FIGURE 18-3* A Structural Classification of Hormones

tivities, or *quantities* of important enzymes and structural proteins. In other words, a hormone may:

- Stimulate the synthesis of an enzyme or a structural protein not already present in the cytoplasm by activating appropriate genes in the nucleus.

- Turn an existing enzyme "on" or "off" by changing its shape or structure.

- Increase or decrease the rate of synthesis of a particular enzyme or other protein by changing the rate of transcription or translation.

Through one or more of these mechanisms, a hormone can modify the physical structure or biochemical properties of its target cells.

To affect a target cell, a hormone must first interact with an appropriate receptor. Each cell has the receptors needed to respond to several different hormones, but cells in different tissues have different combinations of receptors. This arrangement accounts for the differential effects of hormones on specific tissues. For every cell, the presence or absence of a specific receptor determines the cell's hormonal sensitivities. If the cell has a receptor that will bind a particular hormone, that cell will respond to the hormone's presence. If the cell lacks the proper receptor for that hormone, the hormone will have no effect.

Hormone receptors are located either (1) on the cell membrane or (2) inside the cell. Using a few specific examples, we will now introduce the basic mechanisms involved.

Hormones and the Cell Membrane

The receptors for catecholamines (E, NE, and dopamine), peptide hormones, and eicosanoids are in the cell membranes of their respective target cells. Because catecholamines and peptide hormones are not lipid-soluble, they are unable to penetrate a cell membrane. Instead, these hormones bind to receptor proteins at the outer surface of the cell membrane. Conversely, eicosanoids, which are lipid-soluble, diffuse across the membrane to reach receptor proteins on the inner membrane surface.

Any hormone that binds to receptors in the cell membrane does not have direct effects on the target cell. For example, such a hormone does not begin building a protein or catalyzing a specific reaction. The hormone acts as a **first messenger** that causes the appearance of a **second messenger** in the cytoplasm. The second messenger may function as an enzyme activator, inhibitor, or cofactor, but the net result is a change in the rates of various metabolic reactions. The most important second messengers are (1) *cyclic-AMP (cAMP),* a derivative of ATP; (2) *cyclic-GMP (cGMP),* a derivative of GTP, another high-energy compound; and (3) calcium ions.

The binding of a small number of hormone molecules to membrane receptors may lead to the release of thousands of second messengers within a cell. This process, which magnifies the effect of a hormone on the target cell, is called **amplification.** In addition, the arrival of a single hormone may promote the release of more than one type of second messenger. Thus, the hormone can alter many aspects of cell function at the same time.

G-PROTEINS The link between the first messenger and the second messenger generally involves a **G-protein**, an integral membrane protein that interacts with the membrane receptor. There are several different types of G-proteins, but in each case the G-protein is activated when a hormone binds to its receptor at the membrane surface. Figure 18-4● diagrams important mechanisms that may be set in motion by the activation of a G-protein. Each sequence affects levels of a second messenger in the cytoplasm:

1. Activation of adenylate cyclase. **Adenylate cyclase**, also called *adenyl cyclase* or *adenylyl cyclase*, converts ATP to a ring-shaped molecule of **cyclic-AMP (cAMP)**. Cyclic-AMP then functions as the second messenger. The first step typically involves the activation of a kinase. A *kinase* is an enzyme that performs *phosphorylation,* the attachment of a phosphate group (PO_4^{3-}) to another molecule. Cyclic-AMP activates kinases that phosphorylate proteins. The effect on the target cell depends on the nature of the proteins affected. The phosphorylation of membrane proteins can open ion channels, and in the cytoplasm many important enzymes can be activated only by phosphorylation. The hormones calcitonin, parathyroid hormone, ADH, ACTH, epinephrine, FSH, LH, TSH, and glucagon produce their effects by increasing intracellular concentrations of cyclic-AMP. The increase is usually short-lived, because the cytoplasm contains another enzyme, **phosphodiesterase (PDE),** that inactivates cyclic-AMP by converting it to AMP (adenosine monophosphate).

2. Entry and/or release of calcium ions. An activated G-protein can trigger the opening of calcium ion channels in the membrane or the release of calcium ions from intracellular stores. The G-protein first activates the enzyme **phospholipase C** (PLC). This enzyme generates the second messengers **diacylglycerol** (DAG) and **inositol triphosphate** (IP_3) from membrane phospholipids known as *phosphatidylinositols* (fos-fa-tī-dil-i-NOS-i-tols). The mechanism proceeds as follows:

STEP 1. Inositol triphosphate diffuses into the cytoplasm and triggers the release of Ca^{2+} from intracellular reserves, such as those in the smooth endoplasmic reticulum of many cells.

STEP 2. The combination of diacylglycerol and intracellular calcium ions activates another membrane protein, **protein kinase C** (PKC). Activation of PKC leads to the phosphorylation of calcium channel proteins. This opens the channels and permits the entry of extracellular Ca^{2+}.

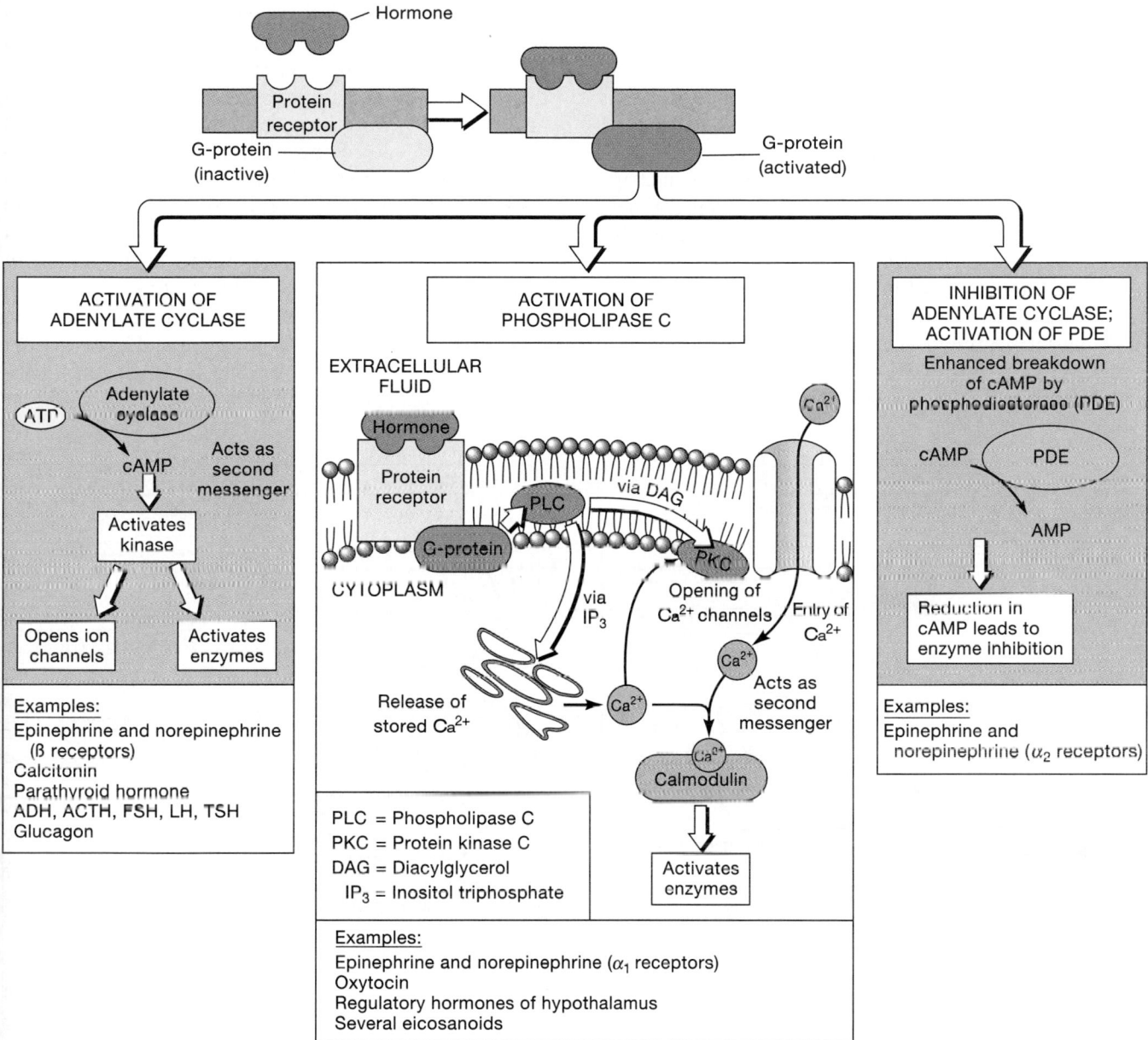

•FIGURE 18-4 **Mechanisms of Hormone Activity.** Peptide hormones, catecholamines, and eicosanoids act by binding to membrane receptors and activating G-proteins. Three possible results are shown, each affecting the concentration of second messengers in the cytoplasm. G-protein activation may lead to (1) activation of adenylate cyclase and subsequent formation of cyclic-AMP, (2) opening of calcium ion channels and calcium release into the cytoplasm, or (3) inhibition of cellular activities by reduction of second messenger concentrations.

STEP 3. The calcium ions then serve as second messengers, generally in combination with an intracellular protein called **calmodulin.** Once it has bound calcium ions, calmodulin can activate specific cytoplasmic enzymes. This chain of events is responsible for the stimulatory effects that follow the activation of α_1 receptors by epinephrine or norepinephrine. ∞ *[p. 522]* Calmodulin activation is also involved in the responses to oxytocin and to several regulatory hormones secreted by the hypothalamus.

3. Inhibition of cellular activities. There are also inhibitory G-proteins whose activation leads to a reduction in cAMP levels inside the cell. In the example shown in Figure 18-4•, the activated G-protein inhibits adenylate cyclase activity and stimulates PDE activity. Levels of cAMP then decline. This decline has an inhibitory effect on the cell because without phosphorylation, key enzymes remain inactive. This is the mechanism responsible for the inhibitory effects that follow the stimulation of α_2 receptors, as we discussed in Chapter 16. ∞ *[p. 522]*

Hormones and Intracellular Receptors

Steroid hormones diffuse across the lipid part of the cell membrane and bind to receptors in the cytoplasm or nucleus (Figure 18-5a●). The hormone-receptor complexes then bind to DNA segments called **hormone-responsive elements** (HREs). This event triggers the activation or inactivation of specific genes. By this mechanism, steroid hormones can alter the rate of DNA transcription in the nucleus, thereby changing the pattern of protein synthesis. In turn, the cell structure or, if enzymes are involved, the metabolic activity of the target cell may be altered. For example, the sex hormone *testosterone* stimulates the production of enzymes and structural proteins in skeletal muscle fibers, causing an increase in muscle size and strength.

Thyroid hormones cross the cell membrane by diffusion or by an as-yet unidentified carrier mechanism. Once within the cytosol, these hormones bind to receptors within the nucleus and on mitochondria (Figure 18-5b●). The hormone-receptor complexes in the nucleus bind to HREs, activating specific genes or increasing or decreasing transcription rates. This alteration affects the metabolic activities of the cell by ordering the synthesis of specific enzymes or by increasing or decreasing the concentration of specific enzymes. Thyroid hormones bound to mitochondria increase the mitochondria's rates of ATP production.

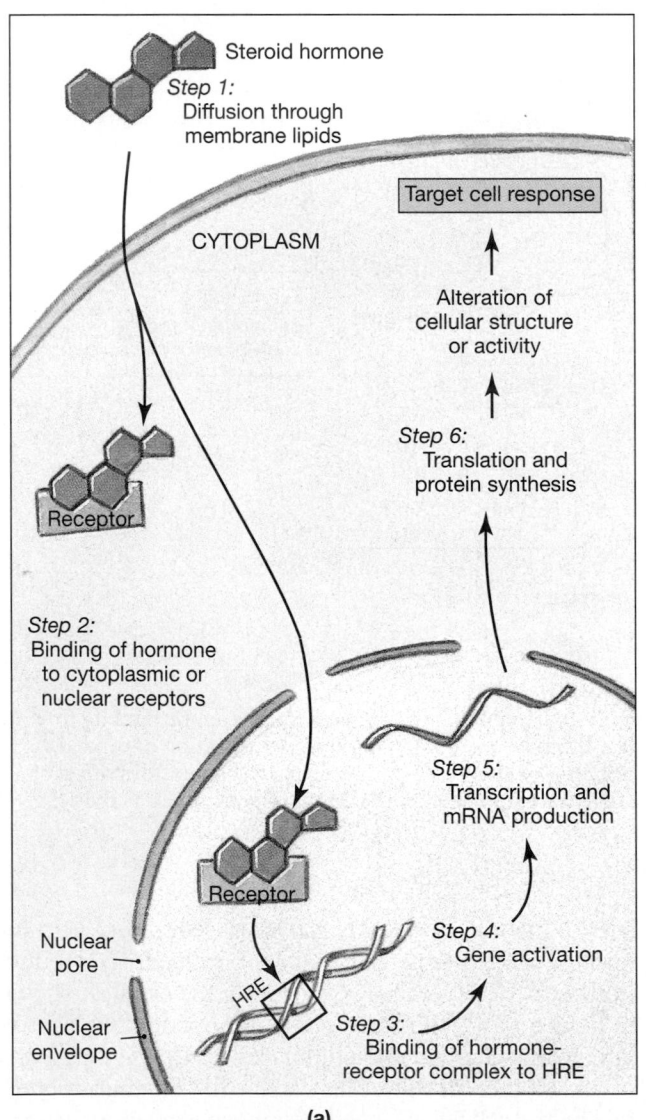

(a)

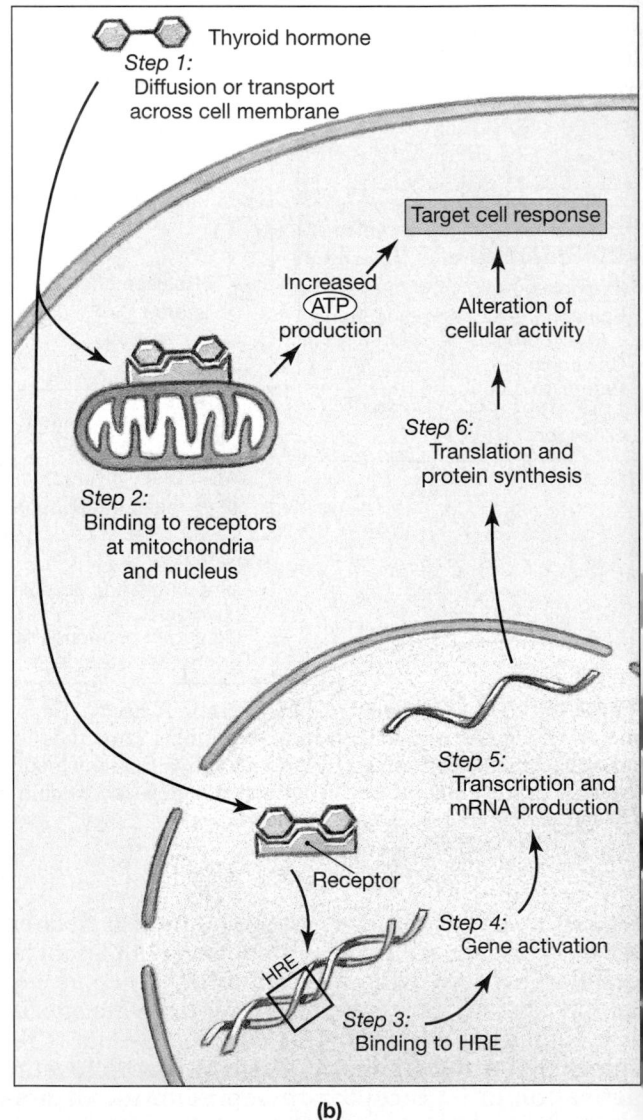

(b)

●**FIGURE 18-5** **Hormone Effects on Gene Activity.** **(a)** Steroid hormones diffuse through the membrane lipids and bind to receptors in the cytoplasm or nucleus. The complex then binds to HREs in the nucleus and activates specific genes. **(b)** Thyroid hormones enter the cytoplasm and bind to receptors in the nucleus to activate specific genes. They also bind to receptors on mitochondria and accelerate ATP production.

Control of Endocrine Activity

As we noted earlier, there are many functional parallels between the organization of the nervous and endocrine systems. In Chapter 13, we considered the basic operation of neural reflex arcs, the simplest organizational units in the nervous system. ∞ *[p. 431]* The most direct arrangement was a monosynaptic reflex, such as the stretch reflex diagrammed in Figure 18-6a•. Polysynaptic reflexes provide more complex and variable responses to stimuli, and higher centers, which integrate multiple inputs, can facilitate or inhibit these reflexes as needed.

Endocrine Reflexes

Endocrine reflexes are the functional counterparts of neural reflexes. They can be triggered by (1) *humoral stimuli* (changes in the composition of the extracellular fluid), (2) *hormonal stimuli* (the arrival or removal of a specific hormone), or (3) *neural stimuli* (the arrival of neurotransmitter at neuroglandular junctions). In most cases, endocrine reflexes are controlled by negative feedback mechanisms: A stimulus triggers the production of a hormone whose direct or indirect effects reduce the intensity of the stimulus.

Endocrine cells in a simple endocrine reflex (Figure 18-6b•) involve only one hormone. The endocrine cells involved respond directly to changes in the composition of the extracellular fluid. The secreted hormone adjusts the activities of target cells and restores homeostasis. Simple endocrine reflexes control hormone secretion by the heart, pancreas, parathyroid gland, and digestive tract.

More-complex endocrine reflexes involve one or more intermediary steps and two or more hormones. The hypothalamus, the highest level of endocrine control, integrates the activities of the nervous and endocrine systems. This integration involves three mechanisms, summarized in Figure 18-7• (p. 601):

1. The hypothalamus secretes **regulatory hormones,** or *regulatory factors,* special hormones that control endocrine cells in the pituitary gland. By convention, the use of *hormone* in the name indicates that the substance's identity is known; if the identity is not known, the term *factor* is used instead. The basic pattern of regulation is detailed in Figure 18-6c•. The hypothalamic regulatory factors control the secretory activities of endocrine cells in the anterior portion of the pituitary gland. The hormones released by the anterior pituitary, in turn, control the activities of endocrine cells in the thyroid, adrenal cortex, and reproductive organs. This arrangement closely parallels that of a complex neural reflex: Cells in the hypothalamus act as receptors, the endocrine organs are processing centers, and the target cells are effectors.

The primary difference is that endocrine reflexes use hormones rather than axons to carry information and instructions from place to place.

2. The hypothalamus contains autonomic centers that exert direct neural control over the endocrine cells of the adrenal medullae. When the sympathetic division is activated, the adrenal medullae release hormones into the bloodstream (Figure 18-6d•).

3. The hypothalamus acts as an endocrine organ by releasing hormones into the circulation at the posterior pituitary. We introduced two of these hormones, ADH and oxytocin, in Chapter 14. ∞ *[p. 468]*

The hypothalamus secretes ADH and regulatory hormones in response to changes in the composition of the circulating blood. The pathways leading to secretion of E, NE, and oxytocin involve both neural and hormonal components. For example, the commands issued to the adrenal medullae that trigger secretion of E and NE arrive in the form of action potentials along efferent fibers rather than as circulating hormones. Such pathways are called **neuroendocrine reflexes** (Figure 18-6d,e•), because they include both neural and endocrine components. We will consider these reflex patterns in more detail as we deal with specific endocrine tissues and organs.

In our discussion of sensory coding in Chapter 17, we noted that receptors provide complex information by varying the frequency and pattern of action potentials in a sensory neuron. In the endocrine system, complex commands are issued by changing the amount of hormone secreted and the pattern of hormone release. In a simple endocrine reflex, hormones are released continuously, but the rate of secretion rises and falls in response to humoral stimuli. For example, when blood glucose levels climb, the pancreas increases its secretion of *insulin*, a hormone that stimulates glucose absorption and utilization. As insulin levels rise, glucose levels decline; in turn, the stimulation of the insulin-secreting cells is reduced. As glucose levels return to normal, the rate of insulin secretion reaches resting levels. (We discussed the same pattern in Chapter 1 when we considered the negative feedback control of body temperature.) ∞ *[p. 13]*

In this example, the responses of the target cells change over time, because the impact of insulin is proportional to its concentration. However, the relationship between hormone concentration and target cell response is not always so predictable. For instance, a hormone can have one effect at low concentrations and additional effects—or even different effects—at high concentrations. (We will consider specific examples later in the chapter.)

Several hormones of the hypothalamus and pituitary gland are released in sudden bursts called *pulses* rather

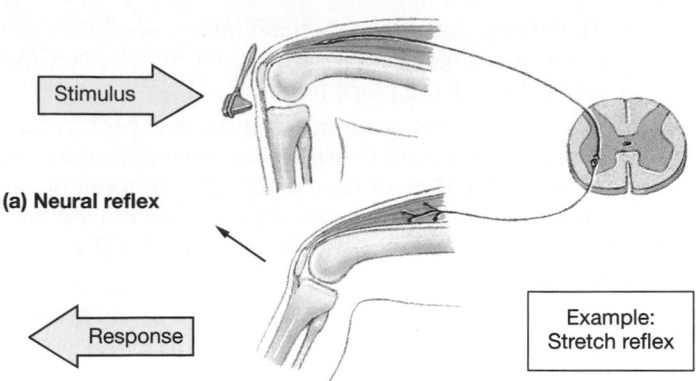

(a) Neural reflex

Stimulus

Response

Example: Stretch reflex

● **FIGURE 18-6 Patterns of Endocrine Activity and Control.** (a) A neural reflex, for comparison with endocrine control mechanisms. (b) A simple endocrine reflex. The endocrine cell responds to a change in the composition of the extracellular fluid, and the hormone released stimulates a target cell to respond in a way that restores homeostasis. (c) A more complex endocrine reflex, involving the hypothalamus. In this case, the stimulus may be a chemical change in the blood or some change in CNS activity. In response, the hypothalamus produces a regulatory hormone that controls the hormonal output of the pituitary gland. In a neuroendocrine reflex, a sensory or motor axon is part of the reflex arc, as in (d) the control of secretion by the adrenal medullae and in (e) milk ejection at the mammary glands.

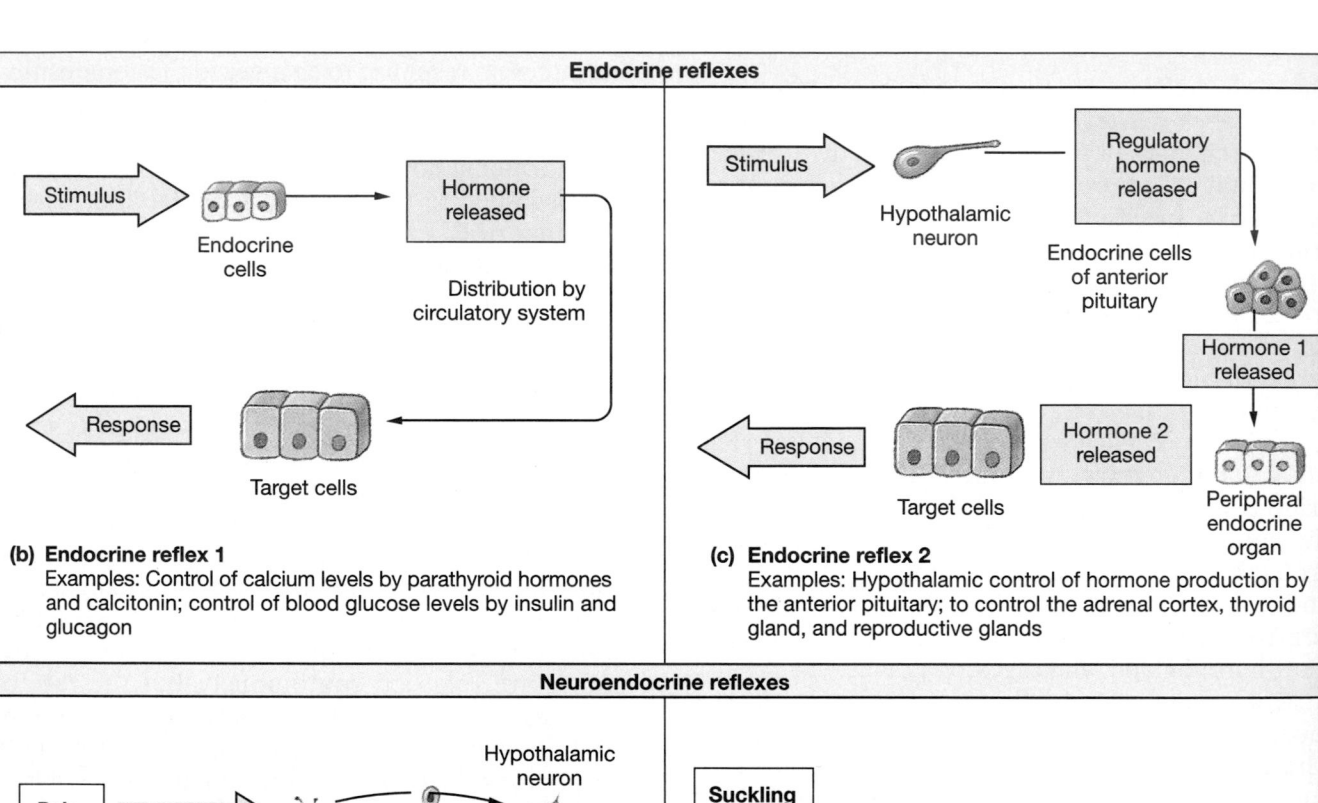

Endocrine reflexes

(b) Endocrine reflex 1
Examples: Control of calcium levels by parathyroid hormones and calcitonin; control of blood glucose levels by insulin and glucagon

(c) Endocrine reflex 2
Examples: Hypothalamic control of hormone production by the anterior pituitary; to control the adrenal cortex, thyroid gland, and reproductive glands

Neuroendocrine reflexes

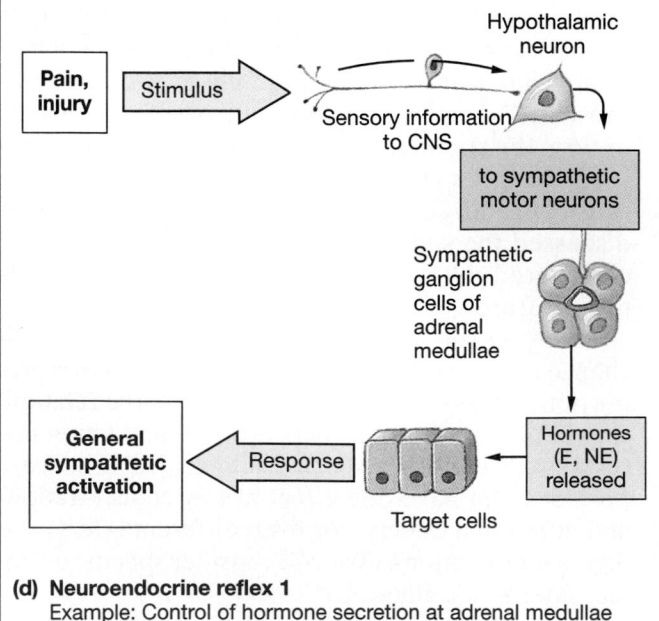

(d) Neuroendocrine reflex 1
Example: Control of hormone secretion at adrenal medullae during sympathetic activation

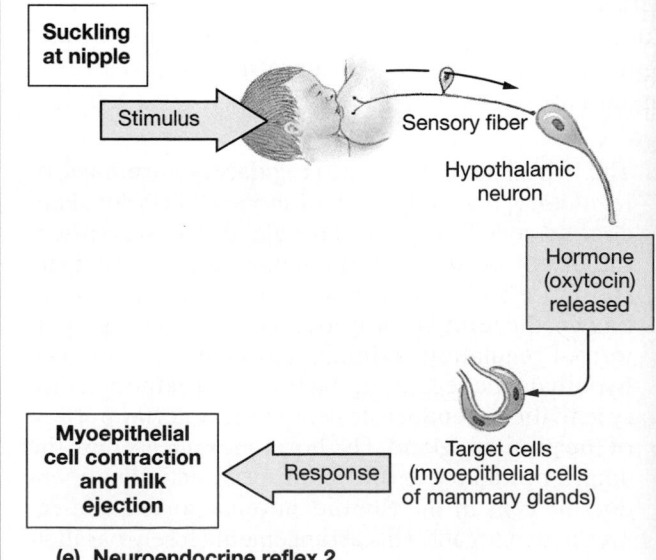

(e) Neuroendocrine reflex 2
Example: Control of milk ejection by mammary glands

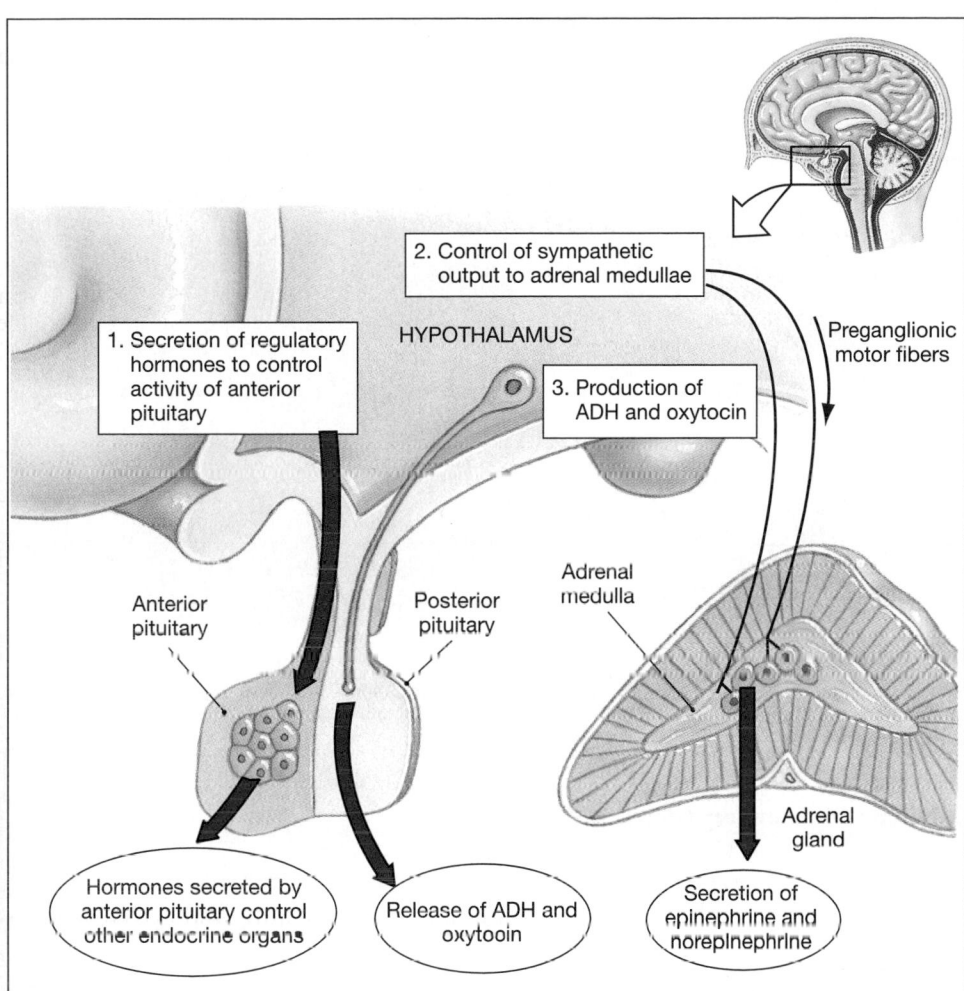

2. Control of sympathetic output to adrenal medullae

HYPOTHALAMUS

1. Secretion of regulatory hormones to control activity of anterior pituitary

3. Production of ADH and oxytocin

Preganglionic motor fibers

Anterior pituitary

Posterior pituitary

Adrenal medulla

Adrenal gland

Hormones secreted by anterior pituitary control other endocrine organs

Release of ADH and oxytocin

Secretion of epinephrine and norepinephrine

than continuously. When hormones arrive in pulses, target cells may vary their response with the frequency of pulses. For example, the target cell response to one pulse every 3 hours can differ from the response when pulses arrive every 30 minutes. The most complicated hormonal instructions issued by the hypothalamus involve changes in the frequency of pulses *and* in the amount secreted in each pulse.

☑ How could you distinguish between a neural response and an endocrine response on the basis of response time and response duration?

☑ How would the presence of a molecule that blocks adenylate cyclase affect the activity of a hormone that produces its cellular effects by way of the second messenger cAMP?

☑ What primary factor determines each cell's hormonal sensitivities?

THE PITUITARY GLAND

Figure 18-8● details the anatomical organization of the **pituitary gland**, or **hypophysis** (hī-POF-i-sis).

This small, oval gland lies nestled within the *sella turcica,* a depression in the sphenoid bone (Figure 7-8●, p. 207). The pituitary gland hangs inferior to the hypothalamus, connected by the slender **infundibulum** (in-fun-DIB-ū-lum; funnel). The base of the infundibulum lies between the optic chiasm and the mamillary bodies. The pituitary gland is cradled by the sella turcica and held in position by the *diaphragma sellae,* a dural sheet that encircles the infundibulum. The diaphragma sellae locks the pituitary in position and isolates it from the cranial cavity.

The pituitary gland can be divided into posterior and anterior divisions on the basis of function and developmental anatomy (see the Embryology Summary on pp. 626–627). Nine important peptide hormones are released by the pituitary gland—seven by the anterior pituitary and two by the posterior pituitary. All nine hormones bind to membrane receptors, and all nine use cAMP as a second messenger. Table 18-2 summarizes key information about these hormones. Refer to this table as our discussion proceeds.

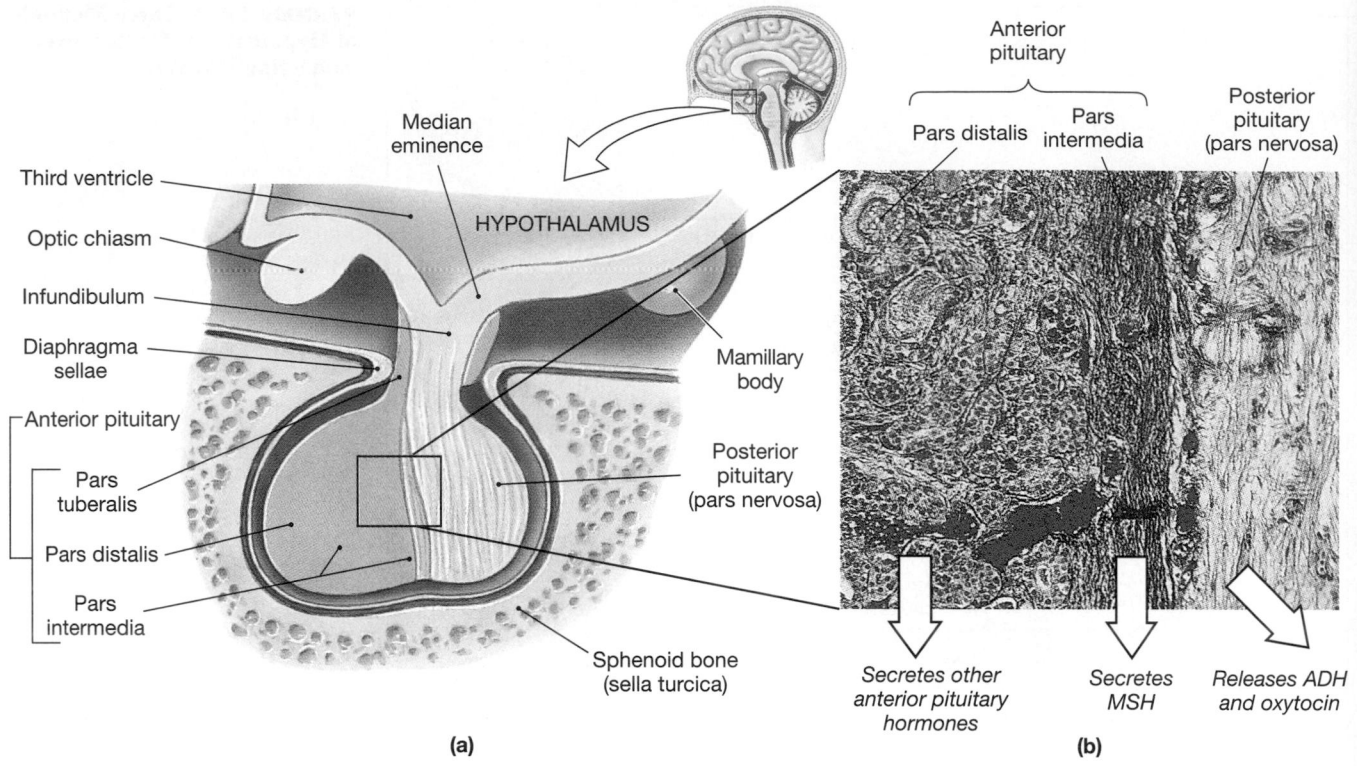

°FIGURE 18-8 Anatomy and Orientation of the Pituitary Gland. (LM × 62)

The Anterior Pituitary

The **anterior pituitary,** or **adenohypophysis** (ad-ē-nō-hī-POF-i-sis), contains a variety of endocrine cell types. The anterior pituitary can be subdivided into three regions: (1) a **pars distalis** (dis-TAL-is; distal part), which is the largest portion of the entire pituitary gland; (2) an extension called the **pars tuberalis**, which wraps around the adjacent portion of the infundibulum; and (3) a slender **pars intermedia** (in-ter-MĒ-dē-uh; intermediate part), which forms a narrow band bordering the posterior pituitary (Figure 18-8●). An extensive capillary network radiates through these regions, so every endocrine cell has immediate access to the circulatory system.

The Hypophyseal Portal System

The production of hormones in the anterior pituitary is controlled by the hypothalamus through the secretion of specific regulatory factors. At the median eminence, a swelling near the attachment of the infundibulum, hypothalamic neurons release regulatory factors into the surrounding interstitial fluids. Their secretions enter the circulation quite easily, because the endothelial cells lining the capillaries in this region are unusually permeable. These **fenestrated** (FEN-es-trā-ted) **capillaries** (*fenestra,* window) allow relatively large molecules to enter or leave the circulatory system. The capillary networks

in the median eminence are supplied by the *superior hypophyseal artery.*

Before leaving the hypothalamus, the capillary networks unite to form a series of larger vessels that spiral around the infundibulum to reach the anterior pituitary gland. Once within the anterior pituitary, these vessels form a second capillary network that branches among the endocrine cells (Figure 18-9●).

This is an unusual vascular arrangement. A typical artery conducts blood from the heart to a capillary network, and a typical vein carries blood from a capillary network back to the heart. The vessels between the median eminence and the anterior pituitary, however, carry blood from one capillary network to another. Blood vessels that link two capillary networks are called **portal vessels,** and the entire complex is termed a **portal system.**

Portal systems provide an efficient means of chemical communication by ensuring that all the blood entering the portal vessels will reach the intended target cells before it returns to the general circulation. The communication is strictly one-way, however, because any chemicals released by the cells "downstream" must do a complete tour of the circulatory system before they reach the capillaries at the start of the portal system. Portal vessels are named after their destinations, so this particular network of vessels is known as the **hypophyseal portal system.**

TABLE 18-2 The Pituitary Hormones

Region/Area	Hormones	Targets	Hormonal Effects	Hypothalamic Regulatory Hormones
Anterior pituitary (Adenohypophysis)				
Pars distalis	Thyroid-stimulating hormone (TSH)	Thyroid gland	Secretion of thyroid hormones	Thyrotropin-releasing hormone (TRH)
	Adrenocorticotropic hormone (ACTH)	Adrenal cortex (zona fasciculata)	Glucocorticoid secretion	Corticotropin-releasing hormone (CRH)
	Gonadotropic hormones:			
	Follicle-stimulating hormone (FSH)	Follicle cells of ovaries	Estrogen secretion, follicle development	Gonadotropin-releasing hormone (GnRH)
		Sustentacular cells of testes	Stimulation of sperm maturation	As above
	Luteinizing hormone (LH)	Follicle cells of ovaries	Ovulation, formation of corpus luteum, progesterone secretion	As above
		Interstitial cells of testes	Testosterone secretion	As above
	Prolactin (PRL)	Mammary glands	Production of milk	Prolactin-inhibiting hormone (PIH) Prolactin-releasing factor (PRF)
	Growth hormone (GH)	All cells	Growth, protein synthesis, lipid mobilization and catabolism	Growth hormone–releasing hormone (GH–RH) Growth hormone–inhibiting hormone (GH–IH)
Pars intermedia (not active in normal adults)	Melanocyte-stimulating hormone (MSH)	Melanocytes	Increased melanin synthesis in epidermis	Melanocyte-stimulating hormone–inhibiting hormone (MSH–IH)
Posterior pituitary (Neurohypophysis or Pars Nervosa)	Antidiuretic hormone (ADH)	Kidneys	Reabsorption of water, elevation of blood volume and pressure	Transported along axons from supraoptic nucleus to posterior pituitary gland
	Oxytocin (OT)	Uterus, mammary glands (females)	Labor contractions, milk ejection	Transported along axons from paraventricular nucleus to posterior pituitary gland
		Ductus deferens and prostate gland (males)	Contractions of ductus deferens and prostate	

Hypothalamic Control of the Anterior Pituitary

There are two classes of regulatory hormones: (1) *releasing hormones* and (2) *inhibiting hormones.* A **releasing hormone (RH)** stimulates synthesis and secretion of one or more hormones at the anterior pituitary, whereas an **inhibiting hormone (IH)** prevents the synthesis and secretion of hormones from the anterior pituitary. An endocrine cell in the anterior pituitary may be controlled by releasing hormones, inhibiting hormones, or some combination of the two. The regulatory hormones released at the hypothalamus are transported directly to the anterior pituitary by the hypophyseal portal system.

The rate of regulatory hormone secretion by the hypothalamus is controlled by negative feedback. The primary regulatory patterns are diagrammed in Figure 18-10● (p. 605); we will reference them as we examine specific pituitary hormones.

Hormones of the Anterior Pituitary

We will discuss seven hormones whose functions and control mechanisms are reasonably well understood: *thyroid-stimulating hormone, adrenocorticotropic hormone, follicle-stimulating hormone, luteinizing hormone, prolactin, growth hormone*, and *melanocyte-stimulating hormone.* Of the six hor-

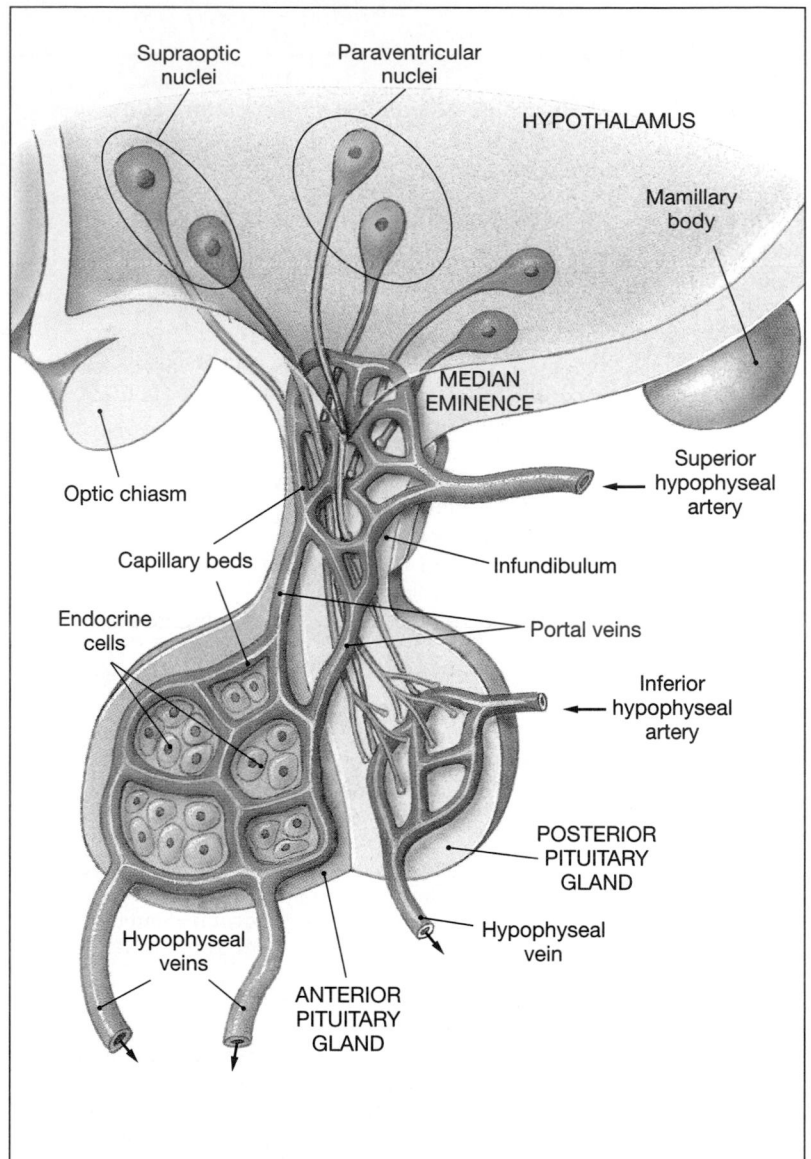

•FIGURE 18-9 The Hypophyseal Portal System

Labels in figure:

Supraoptic nuclei

Paraventricular nuclei

HYPOTHALAMUS

Mamillary body

MEDIAN EMINENCE

Optic chiasm

Superior hypophyseal artery

Capillary beds

Infundibulum

Endocrine cells

Portal veins

Inferior hypophyseal artery

POSTERIOR PITUITARY GLAND

Hypophyseal veins

Hypophyseal vein

ANTERIOR PITUITARY GLAND

mones produced by the pars distalis, four regulate the production of hormones by other endocrine glands. The names of these hormones indicate their activities, but many of the phrases are so long that abbreviations are often used instead.

The hormones of the anterior pituitary are also called *tropic hormones* (*tropé,* a turning), because they turn on endocrine glands or support the functions of other organs. (Some references use *trophic hormones* [*trophé,* nourishment] to refer to these hormones.)

THYROID-STIMULATING HORMONE (TSH) **Thyroid-stimulating hormone (TSH)**, or *thyrotropin,* targets the thyroid gland and triggers the release of thyroid hormones. TSH is released in response to *thy-*

rotropin-releasing hormone (TRH) from the hypothalamus. As circulating concentrations of thyroid hormones rise, the rates of TRH and TSH production decline (Figure 18-10a•).

ADRENOCORTICOTROPIC HORMONE (ACTH) **Adrenocorticotropic hormone (ACTH)**, also known as *corticotropin,* stimulates the release of steroid hormones by the *adrenal cortex,* the outer portion of the adrenal gland. ACTH specifically targets cells that produce hormones called *glucocorticoids* (gloo-kō-KOR-ti-koyds), which affect glucose metabolism. ACTH release occurs under the stimulation of **corticotropin-releasing hormone (CRH)** from the hypothalamus. As glucocorticoid levels increase, the rates of CRH release and ACTH release decline (Figure 18-10a•).

THE GONADOTROPINS Follicle-stimulating hormone and luteinizing hormone are called **gonadotropins** (gō-nad-ō-TRŌ-pinz), because they regulate the activities of the male and female gonads (testes and ovaries, respectively). The production of gonadotropins occurs under stimulation by **gonadotropin-releasing hormone (GnRH)** from the hypothalamus.

Follicle-Stimulating Hormone (FSH). **Follicle-stimulating hormone (FSH)**, or *follitropin,* promotes follicle development in women and, in combination with luteinizing hormone, stimulates the secretion of *estrogens* (ES-trō-jenz) by ovarian cells. *Estradiol* is the most important estrogen. In men, FSH stimulates *sustentacular cells,* specialized cells in the tubules where sperm differentiate. In response, the sustentacular cells promote the physical maturation of developing sperm. FSH production is inhibited by *inhibin,* a peptide hormone released by cells in the testes and ovaries (Figure 18-10a•). (Disagreement exists as to whether inhibin suppresses the release of GnRH as well as FSH.)

Luteinizing Hormone (LH). **Luteinizing** (LOO-tē-in-ī-zing) **hormone (LH),** or *lutropin,* induces ovulation in women and promotes the ovarian secretion of estrogens and the *progestins* (such as *progesterone*), which prepare the body for possible pregnancy. In men, LH is sometimes called *interstitial cell–stimulating hormone* (ICSH), because it stimulates the pro-

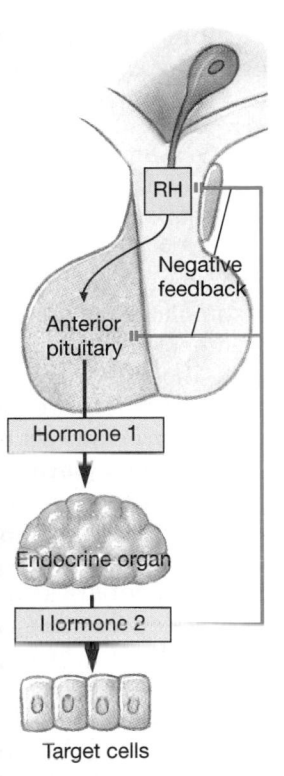

Regulatory Hormone	Pituitary Hormone 1	Endocrine Target	Hormone 2
TRH	TSH	Thyroid gland	Thyroid hormones
CRH	ACTH	Adrenal cortex	Gluco-corticoids
GnRH	FSH	Testes	Inhibin
		Ovaries	{ Inhibin Estrogens
	LH	Ovaries	{ Progestins Estrogens
		Testes	Androgens

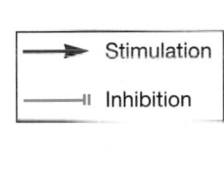

→ Stimulation

–‖ Inhibition

(a)

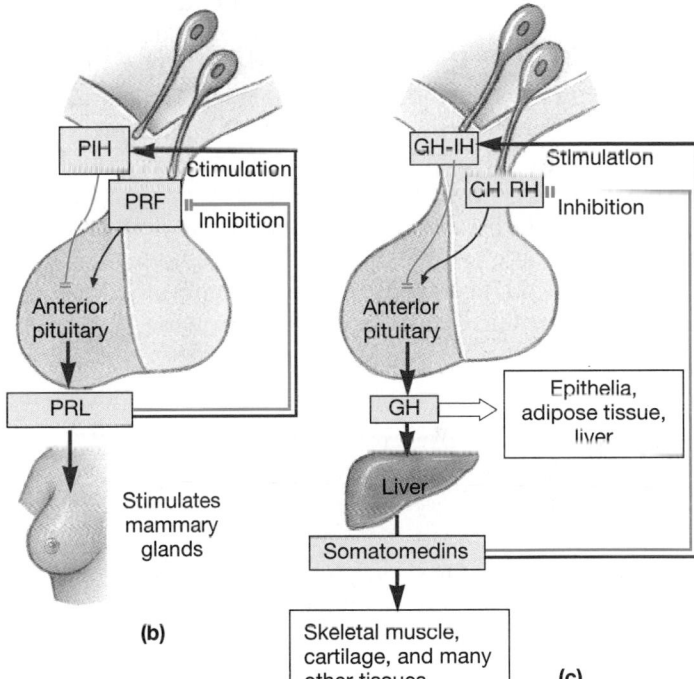

(b)

(c)

•FIGURE 18-10 **Feedback Control of Endocrine Secretion. (a)** Typical pattern of regulation when multiple endocrine organs are involved. The hypothalamus produces a releasing hormone (RH) to stimulate hormone production by other glands, and control is via negative feedback. **(b)** Regulation of prolactin (PRL) production by the anterior pituitary. In this case, the hypothalamus produces both a releasing factor (PRF) and an inhibiting hormone (PIH); when one is stimulated, the other is inhibited. **(c)** Regulation of growth hormone (GH) production by the anterior pituitary; when GH-RH release is inhibited, GH-IH release is stimulated.

duction of sex hormones by the *interstitial cells* of the testes. These sex hormones are called **androgens** (AN-drō-jenz; *andros,* man), the most important of which is *testosterone.* LH production, like FSH production, is stimulated by GnRH from the hypothalamus. GnRH production is inhibited by estrogens, progestins, and androgens (Figure 18-10a•).

PROLACTIN (PRL) Prolactin (prō-LAK-tin; *pro-,* before + *lac,* milk) **(PRL),** or *mammotropin,* works with other hormones to stimulate mammary gland development. In pregnancy and during the nursing period that follows delivery, PRL also stimulates milk production by the mammary glands. The functions of PRL in males are poorly understood, but there is evidence that PRL has an indirect role in the regulation of androgen production. (PRL appears to make interstitial cells more sensitive to LH.)

Prolactin production is inhibited by **prolactin-inhibiting hormone (PIH).** This hormone is identical to the neurotransmitter *dopamine.* The hypothalamus also secretes a prolactin-releasing hormone, but the identity of this *prolactin-releasing factor (PRF)* is a mystery. Circulating PRL stimulates PIH release and inhibits the secretion of prolactin-releasing factor (Figure 18-10b•).

Although PRL exerts the dominant effect on the glandular cells, normal development of the mammary glands is regulated by the interaction of several hormones. Prolactin, estrogens, progesterone, glucocorticoids, pancreatic hormones, and hormones produced by the placenta cooperate in preparing the mammary glands for secretion, and milk ejection occurs only in response to oxytocin release at the posterior pituitary. We shall detail the functional development of the mammary glands.

GROWTH HORMONE **Growth hormone (GH),** or **somatotropin** (*soma,* body), stimulates cell growth and replication by accelerating the rate of protein synthesis. Although virtually every tissue responds to some degree, skeletal muscle cells and chondrocytes (cartilage cells) are particularly sensitive to GH levels.

The stimulation of growth by GH involves two different mechanisms. The primary mechanism, which is indirect, is best understood. Liver cells respond to the presence of GH by synthesizing and releasing **insulin-like growth factors (IGFs),** or **somatomedins,** which are peptide hormones that bind to receptor sites on a variety of cell membranes (Figure 18-10c●). In skeletal muscle fibers, cartilage cells, and other target cells, somatomedins increase the rate of uptake of amino acids and their incorporation into new proteins. These effects develop almost immediately after GH release occurs; they are particularly important after a meal, when the blood contains high concentrations of glucose and amino acids. In functional terms, cells can now obtain ATP easily through the aerobic metabolism of glucose, and amino acids are readily available for protein synthesis. Under these conditions, GH, acting through the somatomedins, stimulates protein synthesis and cell growth.

The direct actions of GH are more selective and tend not to appear until after blood glucose and amino acid concentrations have returned to normal levels:

- In epithelia and connective tissues, GH stimulates stem cell divisions and the differentiation of daughter cells. The subsequent growth of these daughter cells will be stimulated by somatomedins.

- GH also has metabolic effects in adipose tissue and in the liver. In adipose tissue, GH stimulates the breakdown of stored triglycerides by adipocytes (fat cells), which then release fatty acids into the blood. As circulating fatty acid levels rise, many tissues stop breaking down glucose and start breaking down fatty acids to generate ATP. This process is termed a **glucose-sparing effect.** In the liver, GH stimulates the breakdown of glycogen reserves by liver cells. These cells then release glucose into the circulation. Because most other tissues are now metabolizing fatty acids rather than glucose, blood glucose concentrations begin to climb, perhaps to levels significantly higher than normal. The elevation of blood glucose levels by GH has been called a **diabetogenic effect,** because *diabetes mellitus,* an endocrine disorder we will consider later in the chapter, is characterized by abnormally high blood glucose concentrations.

 ## Growth Hormone Abnormalities

Growth hormone stimulates muscular and skeletal development. If it is administered before the epiphyseal plates have closed, it will cause an increase in height, weight, and muscle mass; in extreme cases, *gigantism* can result. ∞ *[p. 186]* In **acromegaly** (*akron,* extremity + *megale,* great), an excessive amount of GH is released after puberty, when most of the epiphyseal plates have already closed. Cartilages and small bones respond to the hormone, however, resulting in abnormal growth at the hands, feet, lower jaw, skull, and clavicle. Figure 18-23a● (p. 634) shows a typical acromegalic individual.

Children unable to produce adequate concentrations of GH suffer from *pituitary growth failure,* a condition we introduced in Chapter 6. ∞ *[p. 186]* The steady growth and maturation that typically precede and accompany puberty do not occur in these individuals, who have short stature, slow epiphyseal growth, and larger-than-normal adipose tissue reserves.

Normal growth patterns can be restored by the administration of GH. Before the advent of gene splicing and recombinant DNA techniques, GH had to be carefully extracted and purified from the pituitaries of cadavers at considerable expense. It is

now possible to use genetically manipulated bacteria to produce human GH in commercial quantities.

The current availability of purified human growth hormone has led to its use under medically questionable circumstances. For example, it is now being praised as an "anti-aging" miracle cure. Although GH supplements do slow or even reverse the losses in bone and muscle mass that accompany aging, little is known about adverse side effects that may accompany long-term use in mature adults. GH is also being sought by some parents of short but otherwise healthy children. These parents view short stature as a handicap that merits treatment by a physician. Whether we are considering GH treatment of adults or children, it is important to remember that GH and the somatomedins affect many different tissues and have widespread metabolic impacts. For example, children exposed to GH may grow faster, but their body fat content declines drastically and sexual maturation is delayed. This decline is associated with metabolic changes in many organs. The range and significance of these metabolic side effects are now the subject of long-term studies.

Control of Growth Hormone Production. The production of GH is regulated by **growth hormone–releasing hormone (GH-RH,** or *somatocrinin*) and **growth hormone–inhibiting hormone (GH-IH,** or *somatostatin*) from the hypothalamus. Somatomedins stimulate GH-IH and inhibit GH-RH (Figure 18-10c●).

MELANOCYTE-STIMULATING HORMONE The pars intermedia may secrete two forms of **melanocyte-stimulating hormone (MSH),** or *melanotropin.* As the name indicates, MSH stimulates the melanocytes of the skin, increasing their production of melanin, a brown, black, or yellow-brown pigment. MSH release is inhibited by an inhibiting hormone now known to be dopamine. MSH is important in the control of skin pigmentation in fishes, amphibians, reptiles, and many mammals other than primates. The pars intermedia in adult humans is virtually nonfunctional, and the circulating blood usually does not contain MSH. However, MSH is secreted by the human pars intermedia (1) during fetal development, (2) in very young children, (3) in pregnant women, and (4) in some disease states. The functional significance of MSH secretion under these circumstances is not known. Administration of a synthetic form of MSH causes darkening of the skin, and it has been suggested as a means of obtaining a "sunless tan."

The Posterior Pituitary

The **posterior pituitary** is also called the **neurohypophysis** (noo-rō-hī-POF-i-sis), or *pars nervosa* (nervous part), because it contains the axons of hypothalamic neurons. Neurons of the **supraoptic** and **paraventricular nuclei** manufacture antidiuretic hormone (ADH) and oxytocin, respectively. These products move by axoplasmic transport ∞ *[p. 373]* along axons in the infundibulum to the basement membranes of capillaries in the posterior pituitary gland.

Antidiuretic Hormone

Antidiuretic hormone (ADH), also known as *vasopressin* or *arginine vasopressin (AVP),* is released in response to a variety of stimuli, most notably a rise in the electrolyte concentration in the blood or a fall in blood volume or blood pressure. A rise in the electrolyte concentration stimulates the secretory neurons directly. Because they respond to a change in the osmotic concentration of body fluids, these neurons are called *osmoreceptors.* ADH secretion after a fall in blood volume or pressure occurs under the stimulation of another hormone, *angiotensin II* (*angeion,* vessel + *teinein,* to stretch). (We will detail that pathway on pp. 619–620.)

The primary function of ADH is to decrease the amount of water lost at the kidneys. With losses minimized, any water absorbed from the digestive tract will be retained, reducing the concentration of electrolytes in the extracellular fluid. In high concentrations, ADH also causes *vasoconstriction,* a constriction of peripheral blood vessels that helps elevate blood pressure. ADH release is inhibited by alcohol, which explains the increased fluid excretion that follows the consumption of alcoholic beverages.

DISORDERS OF ADH PRODUCTION There are several forms of **diabetes,** all characterized by excessive urine production (**polyuria**). Although diabetes can be caused by physical damage to the kidneys, most forms are the result of endocrine abnormalities. The two most prevalent forms are *diabetes insipidus* and *diabetes mellitus.* Diabetes mellitus is described on page 623. **Diabetes insipidus** generally develops because the posterior pituitary no longer releases adequate amounts of ADH. Water conservation at the kidneys is impaired, and excessive amounts of water are lost in the urine. As a result, the individual is constantly thirsty, but the fluids consumed are not retained by the body.

Mild cases of diabetes insipidus may not require treatment as long as fluid and electrolyte intake keep pace with urinary losses. In severe cases, the fluid losses can reach 10 liters per day, and fatal dehydration and electrolyte imbalances will occur unless treatment is provided. In one innovative treatment, *desmopressin acetate* (DDAVP), a synthetic form of ADH, is administered in a nasal spray. The drug enters the bloodstream after diffusing through the nasal epithelium. This method is not effective in treating *nephrogenic diabetes insipidus,* which results from an inability to concentrate urine in the kidneys. In this condition, the circulating levels of ADH are already elevated, but the kidney cells are unable to respond appropriately to the hormonal stimulation.

Oxytocin

In women, **oxytocin** (*oxy-,* quick + *tokos,* childbirth), or **OT,** stimulates smooth muscle tissue in the wall of the uterus, promoting labor and delivery. After delivery, oxytocin stimulates the contraction of myoepithelial cells around the secretory alveoli and the ducts of the mammary glands, promoting the ejection of milk.

Until the last stages of pregnancy, the uterine smooth muscles are relatively insensitive to oxytocin, but sensitivity becomes more pronounced as the time of delivery approaches. The trigger for normal labor and delivery is probably a sudden rise in oxytocin levels at the uterus. There is good evidence, however, that the oxytocin released by the posterior pituitary plays only a supporting role and that most of the oxytocin involved is secreted by the uterus and fetus.

Oxytocin secretion and milk ejection are part of the neuroendocrine reflex diagrammed in Figure 18-6e●, p. 600. The stimulus is an infant suckling at the breast, and sensory nerves innervating the nipples relay the information to the hypothalamus. Oxytocin is then released into the circulation at the posterior pituitary, and the myoepithelial cells respond by squeezing milk from the secretory alveoli into large collecting ducts. This "milk let-down" reflex can be modified by any factor that affects the hypothalamus. For example, anxiety, stress, and other factors can prevent the flow of milk, even when the mammary glands are fully functional. In contrast, nursing mothers can become conditioned to associate a baby's crying with suckling. These women may begin milk let-down as soon as they hear a baby cry.

Although the functions of oxytocin in male and female sexual activity remain uncertain, it is known that circulating concentrations of oxytocin in both genders rise during sexual arousal and peak at orgasm. There is evidence that in men oxytocin stimulates smooth muscle contractions in the walls of the sperm duct (*ductus deferens*) and prostate gland. These actions may be important in *emission*, the ejection of prostatic secretions, sperm, and the secretions of other glands into the male reproductive tract before ejaculation. There are indications that the oxytocin released during intercourse in females may stimulate smooth muscle contractions in the uterus and vagina that promote sperm transport toward the uterine tubes.

Figure 18-11● and Table 18-2 summarize important information concerning the hormonal products of the pituitary gland. Review these carefully before considering the structure and function of other endocrine organs.

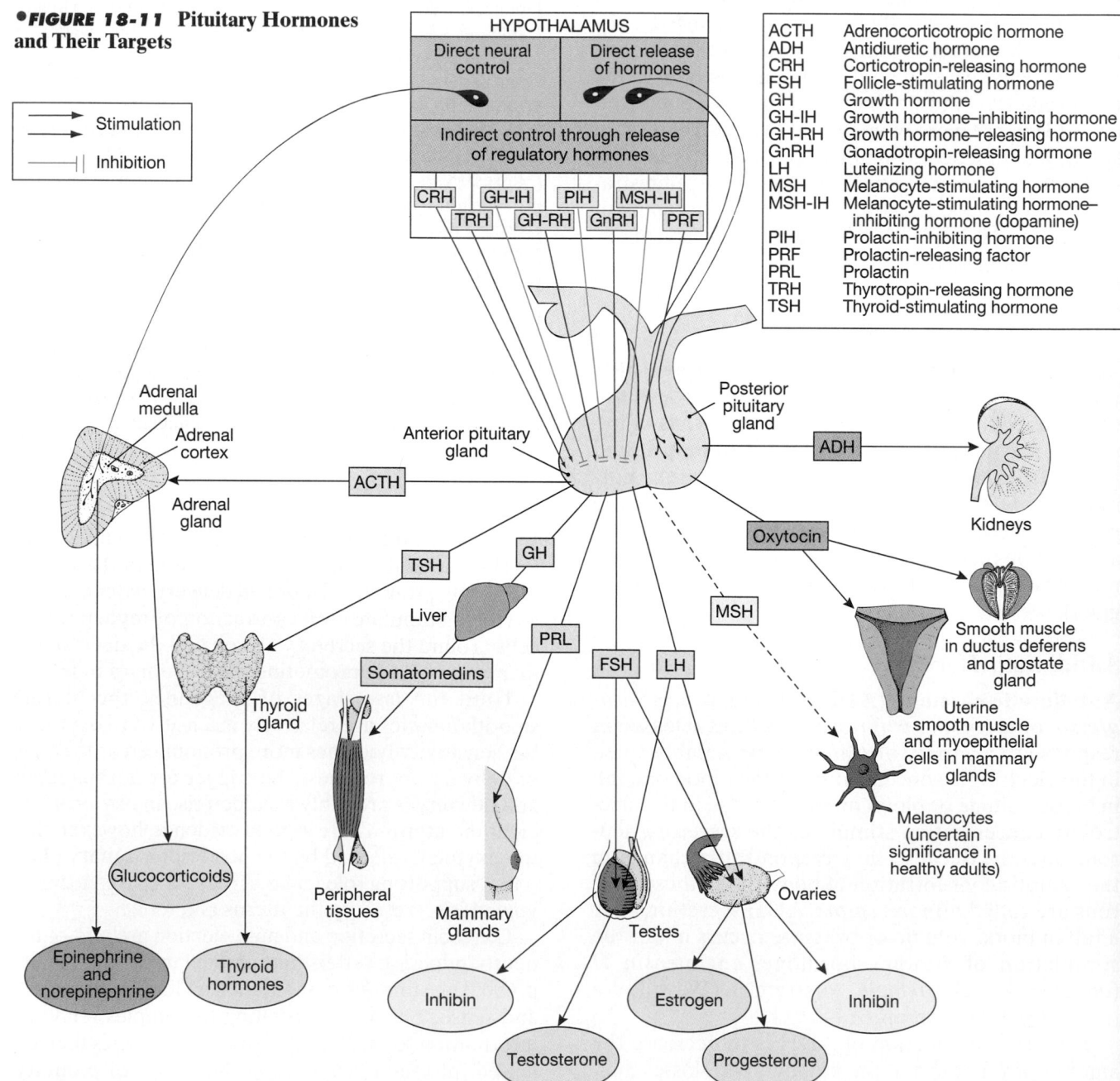

•*FIGURE 18-11* **Pituitary Hormones and Their Targets**

ACTH	Adrenocorticotropic hormone
ADH	Antidiuretic hormone
CRH	Corticotropin-releasing hormone
FSH	Follicle-stimulating hormone
GH	Growth hormone
GH-IH	Growth hormone–inhibiting hormone
GH-RH	Growth hormone–releasing hormone
GnRH	Gonadotropin-releasing hormone
LH	Luteinizing hormone
MSH	Melanocyte-stimulating hormone
MSH-IH	Melanocyte-stimulating hormone–inhibiting hormone (dopamine)
PIH	Prolactin-inhibiting hormone
PRF	Prolactin-releasing factor
PRL	Prolactin
TRH	Thyrotropin-releasing hormone
TSH	Thyroid-stimulating hormone

☑ If a person were suffering from dehydration, how would this condition affect the level of ADH released by the posterior pituitary gland?

☑ A blood sample shows elevated levels of somatomedins. What pituitary hormone would you expect to be elevated as well?

☑ What effect would elevated circulating levels of cortisol, a steroid hormone from the adrenal cortex, have on the pituitary secretion of ACTH?

THE THYROID GLAND

The **thyroid gland** curves across the anterior surface of the trachea just inferior to the *thyroid* ("shield-shaped") *cartilage,* which forms most of the anterior surface of the larynx (Figure 18-12a•). The two **lobes** of the thyroid gland are united by a slender connection, the **isthmus** (IS-mus). You can easily feel the gland with your fingers. When something goes

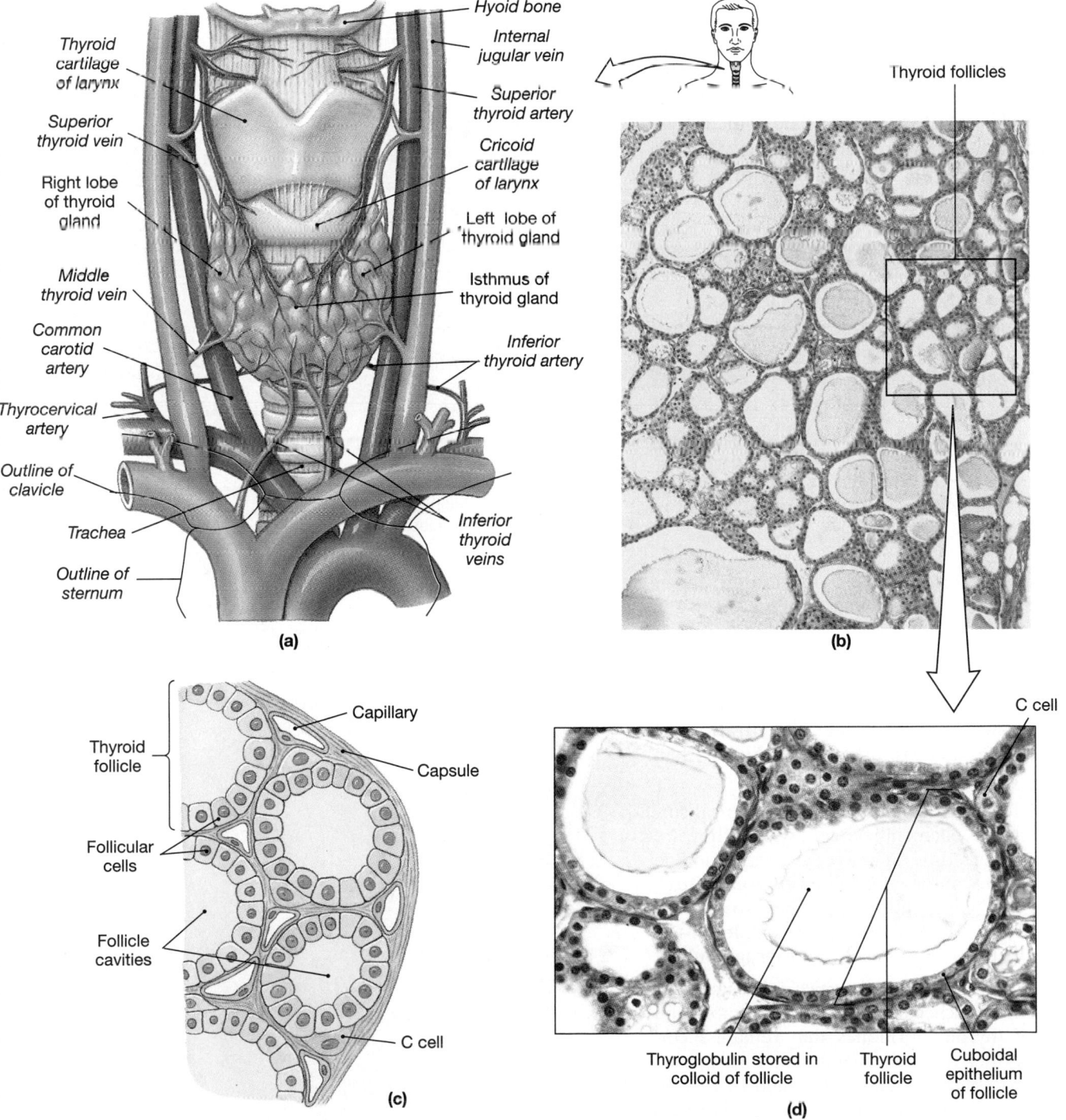

•*FIGURE 18-12* **The Thyroid Gland. (a)** Location and anatomy of the thyroid. **(b)** Histological organization. (LM × 99) **(c)** Diagrammatic view of a section through the wall of the thyroid gland. **(d)** Histological details, showing thyroid follicles. (LM × 211)

wrong with it, the thyroid gland typically becomes visible as it swells and distorts the surface of the neck. The size of the gland is quite variable, depending on heredity and environmental and nutritional factors, but its average weight is about 34 g (1.2 oz). An extensive blood supply gives the thyroid gland a deep red color.

Thyroid Follicles and Thyroid Hormones

The thyroid gland contains large numbers of **thyroid follicles.** Individual follicles are spheres lined by a simple cuboidal epithelium (Figure 18-12b,c,d●). The follicle cells surround a **follicle cavity.** This cavity holds a viscous *colloid,* a fluid containing large quantities of suspended proteins. A network of capillaries surrounds each follicle, delivering nutrients and regulatory hormones to the glandular cells and accepting their secretory products and metabolic wastes.

Follicular cells synthesize a globular protein called **thyroglobulin** (thī-rō-GLOB-ū-lin) and secrete it into the colloid of the thyroid follicles (Figures 18-12d and 18-13a●). Each thyroglobulin molecule contains the amino acid *tyrosine,* the building block of thyroid hormones. The formation of thyroid hormones involves three basic steps:

1. Iodide ions are absorbed from the diet at the digestive tract and delivered to the thyroid gland by the circulation. Carrier proteins in the basal membrane of the follicle cells transport iodide ions (I^-) into the cytoplasm. The follicle cells normally maintain intracellular concentrations of iodide that are many times higher than those in the extracellular fluid.

2. The iodide ions diffuse to the apical surface of each follicle cell, where they are converted to an activated form of iodide (I^+) by an enzyme called *thyroid peroxidase.* This reaction sequence also attaches either one or two of these iodide ions to the tyrosine molecules of thyroglobulin.

3. Tyrosine molecules to which iodide ions have been attached are paired, forming molecules of thyroid hormones that remain incorporated into thyroglobulin. The pairing process is probably performed by thyroid peroxidase. The hormone **thyroxine** (thī-ROKS-ēn), also known as *tetraiodothyro-*

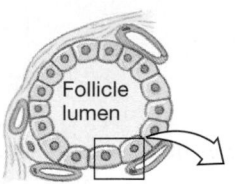

●*FIGURE 18-13* **The Functions of the Thyroid Follicles. (a)** The synthesis, storage, and secretion of thyroid hormones. For a detailed explanation of the numbered events, see the text. **(b)** The regulation of thyroid secretion.

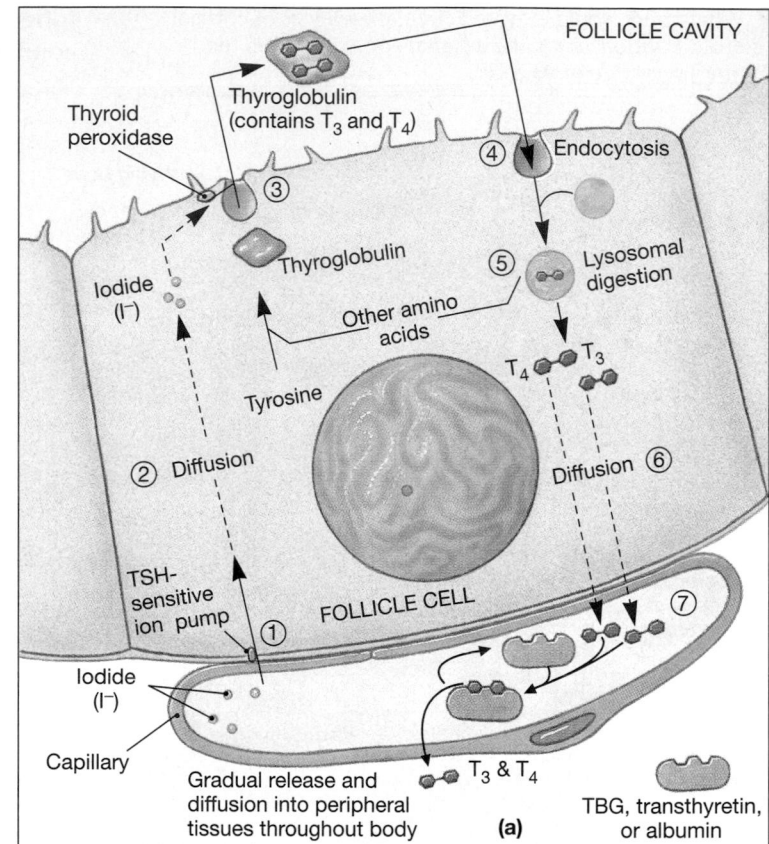

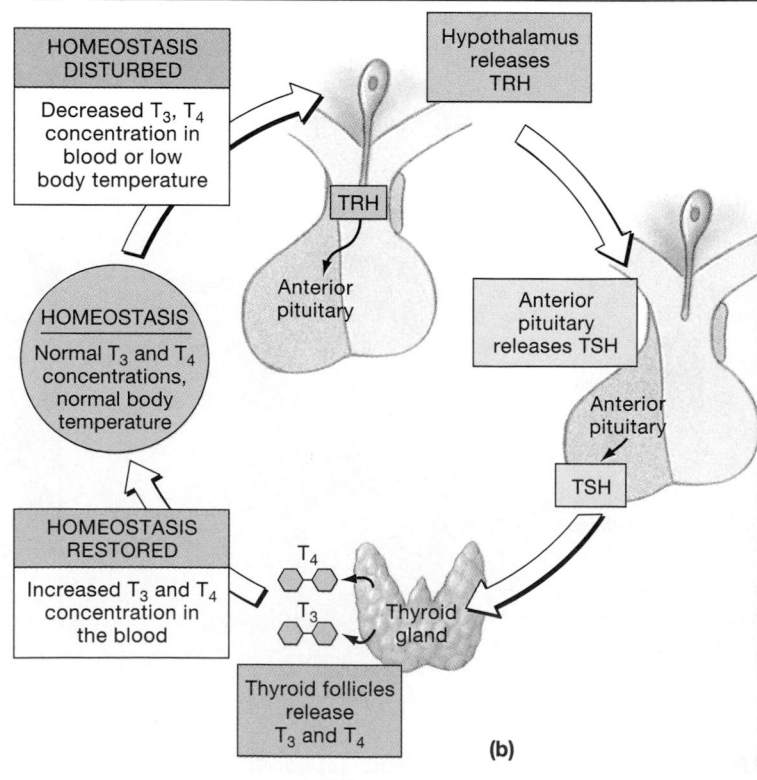

nine, or simply **T₄**, contains four iodide ions. **Triiodothyronine,** or **T₃,** is a related molecule containing three iodide ions. Eventually, each molecule of thyroglobulin contains 4–8 molecules of T_3 or T_4 hormones or both.

The major factor controlling the rate of thyroid hormone release is the concentration of TSH in the circulating blood (Figure 18-13b•). TSH stimulates iodide transport into the follicle cells and stimulates the production of thyroglobulin and thyroid peroxidase. TSH also stimulates the release of thyroid hormones. Under the influence of TSH, the following steps occur:

4. Follicle cells remove thyroglobulin from the follicles through endocytosis.

5. Lysosomal enzymes then break the protein down, and the amino acids and thyroid hormones enter the cytoplasm. The amino acids are recycled and used to synthesize thyroglobulin.

6. The released molecules of T_3 and T_4 diffuse across the basement membrane and enter the circulation. Thyroxine (T_4) accounts for roughly 90 percent of all thyroid secretions, and triiodothyronine (T_3) is secreted in comparatively small amounts.

7. Roughly 75 percent of the T_4 and 70 percent of the T_3 molecules entering the circulation become attached to transport proteins called **thyroid-binding globulins**

(TBGs). Most of the rest of the T_4 and T_3 in the circulation is attached to **transthyretin,** also known as *thyroid-binding prealbumin (TBPA),* or to *albumin,* one of the plasma proteins. Only the relatively small quantities of thyroid hormones that remain unbound—roughly 0.3 percent of the circulating T_3 and 0.03 percent of the circulating T_4—are free to diffuse into peripheral tissues.

An equilibrium exists between the bound and unbound thyroid hormones. At any moment, free thyroid hormones are being bound to carriers at the same rate at which bound hormones are being released. When unbound thyroid hormones diffuse out of the circulation and into other tissues, the equilibrium is disturbed. The carrier proteins then release additional thyroid hormones until a new equilibrium is reached. The bound thyroid hormones thus represent a substantial reserve, and the bloodstream normally contains more than a week's supply of thyroid hormones.

TSH plays a key role in both the synthesis and the release of thyroid hormones. In the absence of TSH, the thyroid follicles become inactive, and neither synthesis nor secretion occurs. TSH binds to membrane receptors and, by stimulating adenylate cyclase, activates key enzymes (Figure 18-4•, p. 597).

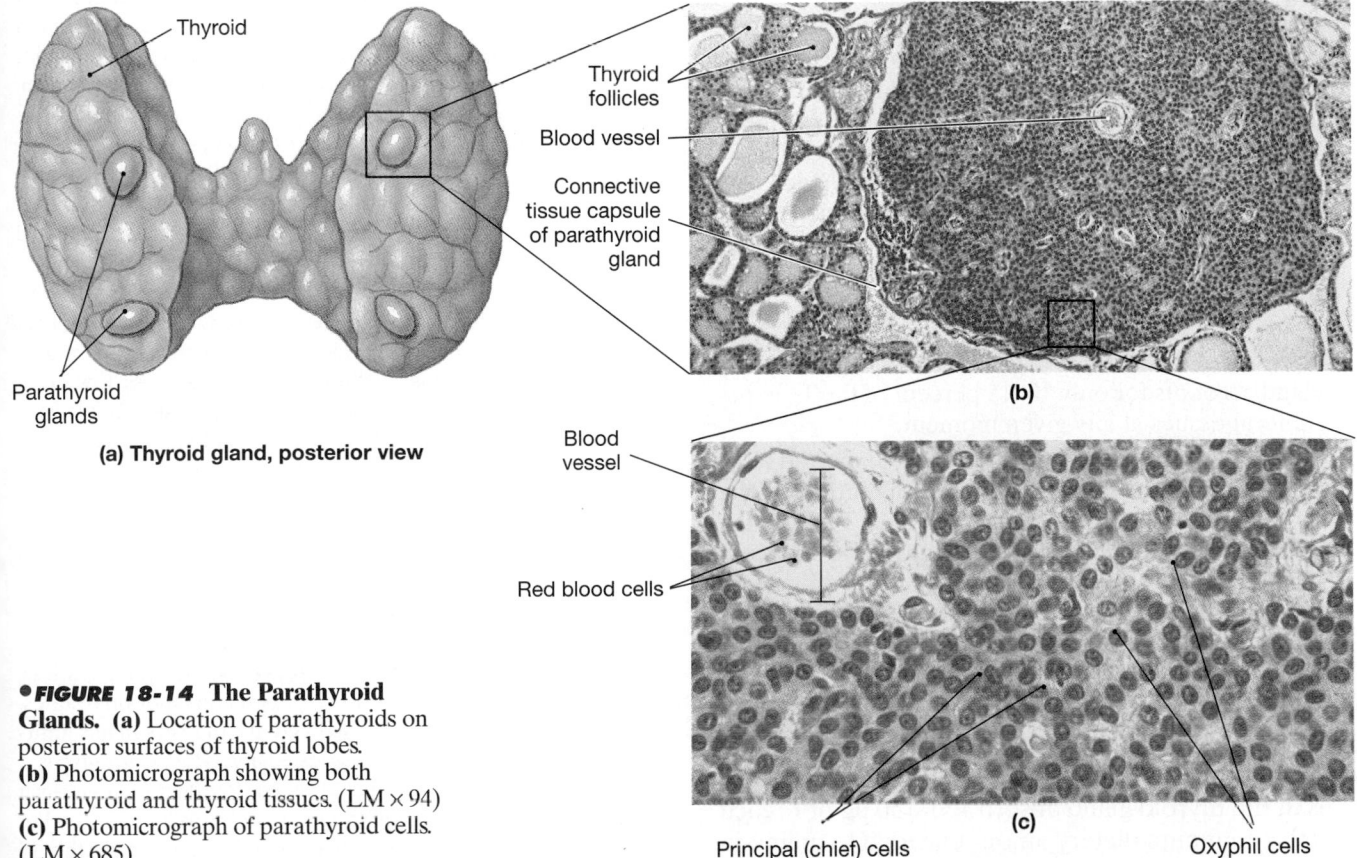

•*FIGURE 18-14* The Parathyroid Glands. (a) Location of parathyroids on posterior surfaces of thyroid lobes. **(b)** Photomicrograph showing both parathyroid and thyroid tissues. (LM × 94) **(c)** Photomicrograph of parathyroid cells. (LM × 685)

Thyroid

Parathyroid glands

(a) Thyroid gland, posterior view

Thyroid follicles

Blood vessel

Connective tissue capsule of parathyroid gland

(b)

Blood vessel

Red blood cells

Principal (chief) cells

Oxyphil cells

(c)

Functions of Thyroid Hormones

Thyroid hormones readily cross cell membranes, and they affect almost every cell in the body. Inside a cell, they bind to (1) receptors in the nucleus, (2) receptors on the surfaces of mitochondria, and (3) receptors in the cytoplasm. Thyroid hormones bound to cytoplasmic receptors are essentially held in storage. If intracellular levels of thyroid hormones decline, the bound thyroid hormones are released into the cytoplasm.

The thyroid hormones binding to mitochondria increase the rates of mitochondrial ATP production. The binding to receptors in the nucleus activates genes that control the synthesis of enzymes involved with energy transformation and utilization. One specific effect is accelerated production of sodium–potassium ATPase, the membrane protein responsible for the ejection of intracellular sodium and the recovery of extracellular potassium. As we noted in Chapter 3, this exchange pump consumes large amounts of ATP. ∞ *[p. 78]*

Thyroid hormones also activate genes that code for the synthesis of enzymes involved in glycolysis and ATP production. This effect, coupled with the direct effect of thyroid hormones on mitochondria, increases the metabolic rate of the cell. Because the cell consumes more energy and because energy use is measured in *calories,* the effect is called the **calorigenic effect** of thyroid hormones. When the metabolic rate increases, more heat is generated. In young children, TSH production increases in cold weather; the calorigenic effect may help them adapt to cold climates. (This response does not occur in adults.) In growing children, thyroid hormones are also essential to normal development of the skeletal, muscular, and nervous systems.

T_3 VERSUS T_4 The thyroid gland produces large amounts of T_4, but T_3 is primarily responsible for the observed effects of thyroid hormones: a strong, immediate, short-lived increase in the rate of cellular metabolism. There are two sources of T_3 in peripheral tissues:

1. **T_3 released by the thyroid gland.** T_3 from the thyroid gland accounts for only 10–15 percent of the T_3 in peripheral tissues at any given moment.

2. **The conversion of T_4 to T_3.** T_4 can be converted to T_3 by enzymes in the liver, kidneys, and other tissues. The thyroid gland releases much more T_4 than T_3, although T_3 is the primary active form. Roughly 85–90 percent of the T_3 that reaches the target cells is produced by the conversion of T_4 within peripheral tissues.

Table 18-3 summarizes the effects of thyroid hormones on major organs and systems.

IODINE AND THYROID HORMONES Iodine in the diet is absorbed at the digestive tract as I^-. The follicle cells in the thyroid gland absorb 120–150 μg of I^- each day, the minimum dietary amount needed to maintain normal thyroid function. The iodide ions are actively transported into the thyroid follicle cells, so the concentration of I^- inside thyroid follicle cells is generally about 30 times higher than that in the plasma. If plasma I^- levels rise, so do levels inside the follicle cells.

The thyroid follicles contain most of the iodide reserves in the body. The active transport mechanism for iodide is stimulated by TSH, and the increased movement of iodide into the cytoplasm accelerates the formation of thyroid hormones.

The typical diet in the United States provides approximately 500 μg of iodide per day, roughly three times the minimum daily requirement. Much of the excess is due to the addition of I^- to the table salt sold in grocery stores as "iodized salt." Thus, iodide deficiency is seldom responsible for limiting the rate of thyroid hormone production. (This is not necessarily the case in other countries.) Excess I^- is filtered out of the blood at the kidneys, and each day a small amount of I^- (about 20 μg) is excreted by the liver into the bile. The losses in the bile, which continue even if the diet contains less than the minimum iodide requirement, can gradually deplete the iodide reserves in the thyroid. Thyroid hormone production then declines, regardless of the circulating levels of TSH.

THYROID GLAND DISORDERS Normal production of thyroid hormones establishes the background rates of cellular metabolism. These hormones exert their primary effects on metabolically active tissues and organs, including skeletal muscles, the liver, and the kidneys. Inadequate production of thyroid hormones is called **hypothyroidism**. Hypothyroidism in an infant produces a condition known as *cretinism,* marked by inadequate skeletal and nervous development and a metabolic rate as much as 40 percent below normal levels. This condition affects approximately 1 birth out of every 5000. Cretinism developing later in childhood will retard growth and mental development and delay puberty.

Adults with hypothyroidism are lethargic and unable to tolerate cold temperatures. The symptoms of adult hypothyroidism, collectively known as *myxedema* (miks-e-DĒ-muh), include subcutaneous swelling, dry skin, hair loss, low body temperature, muscular weakness, and slowed reflexes. Hypothyroidism and myxedema due to inadequate dietary iodide are very rare in the United States but may be relatively common in many poorer countries.

Hyperthyroidism, or *thyrotoxicosis,* occurs when thyroid hormones are produced in excessive quantities. The metabolic rate climbs, and the skin becomes flushed and moist with perspiration. Blood pressure and heart rate increase, and the heartbeat may become irregular as circulatory demands escalate. The effects on the CNS make the individual restless, excitable, and subject to shifts in mood and emotional states. Despite the drive for increased activity, the person has limited energy reserves and fatigues easily. *Graves' disease* is a form of hyperthyroidism that afflicted President George Bush and Barbara Bush during their stay in the White House. We consider both hypothyroidism and hyperthyroidism further in the *Applications Manual.* **AM** *Thyroid Gland Disorders*

TABLE 18-3 Effects of Thyroid Hormones on Peripheral Tissues

1. Elevate oxygen consumption and energy consumption rate; in children, may cause a rise in body temperature
2. Increase heart rate and force of contraction; generally cause a rise in blood pressure
3. Increase sensitivity to sympathetic stimulation
4. Maintain normal sensitivity of respiratory centers to changes in oxygen and carbon dioxide concentrations
5. Stimulate formation of red blood cells and thereby enhance oxygen delivery
6. Stimulate activity of other endocrine tissues
7. Accelerate turnover of minerals in bone

The C Cells of the Thyroid Gland: Calcitonin

A second population of endocrine cells lies sandwiched between the cuboidal follicle cells and their basement membrane. These cells are larger than those of the follicular epithelium, and they do not stain as clearly. They are the **C (clear) cells,** or *parafollicular cells,* which produce the hormone **calcitonin (CT).** Calcitonin aids in the regulation of Ca^{2+} concentrations in body fluids. We introduced the functions of this hormone in Chapter 6. ∞ *[p. 187]* The net effect of calcitonin release is a drop in the Ca^{2+} concentration in body fluids. This reduction is accomplished by (1) the inhibition of osteoclasts, which slows the rate of Ca^{2+} release from bone, and (2) the stimulation of Ca^{2+} excretion at the kidneys.

The control of calcitonin secretion is an example of direct endocrine regulation (Figure 18-6b•, p. 600), because the C cells respond directly to elevations in the Ca^{2+} concentration of the blood. When those concentrations rise, calcitonin secretion increases. This increase lowers the Ca^{2+} concentrations, eliminating the stimulus and "turning off" the C cells.

Calcitonin is probably most important during childhood, when it stimulates bone growth and mineral deposition in the skeleton. It also appears to be important in reducing the loss of bone mass (1) during prolonged starvation and (2) in the late stages of pregnancy, when the maternal skeleton competes with the developing fetus for calcium ions absorbed by the digestive tract. The role of calcitonin in the healthy nonpregnant adult is uncertain.

In several chapters, we have considered the importance of Ca^{2+} in controlling muscle cell and neuron activities. Calcium ion concentrations also affect the sodium permeabilities of excitable membranes. At high Ca^{2+} concentrations, sodium permeability decreases and membranes become less responsive. Such problems are relatively rare. Problems caused by lower-than-normal Ca^{2+} concentrations are equally dangerous and much more common. When calcium ion concentrations decline, sodium permeabilities increase, and cells become

extremely excitable. If calcium levels fall too far, convulsions or muscular spasms may result. Maintenance of adequate calcium levels involves the *parathyroid glands* and *parathyroid hormone.*

THE PARATHYROID GLANDS

Two pairs of **parathyroid glands** are embedded in the posterior surfaces of the thyroid gland (Figure 18-14a•). The gland cells are separated by the dense capsular fibers of the thyroid. All together the four parathyroid glands weigh a mere 1.6 g (0.06 oz). The histological appearance of a parathyroid gland is shown in Figure 18-14b,c•. There are at least two different cell populations in the parathyroid. The **chief cells** produce parathyroid hormone; the functions of the other cells, called *oxyphils*, are unknown.

Parathyroid Hormone Secretion

Like the C cells of the thyroid, the chief cells monitor the circulating concentration of calcium ions. When the Ca^{2+} concentration of the blood falls below normal, the chief cells secrete **parathyroid hormone (PTH),** or *parathormone.* The net result of PTH secretion is an increase in Ca^{2+} concentration in body fluids. Parathyroid hormone has four major effects:

1. It stimulates osteoclasts, accelerating mineral turnover and the release of Ca^{2+} from bone.
2. It inhibits osteoblasts, reducing the rate of calcium deposition in bone.
3. It reduces urinary excretion of Ca^{2+}.
4. It stimulates the formation and secretion of *calcitriol* at the kidneys. In general, the effects of calcitriol complement or enhance those of PTH, but one major effect of calcitriol is the enhancement of Ca^{2+} and PO_4^{3-} absorption by the digestive tract. We will consider calcitriol in a later section (p. 618).

Figure 18-15• compares the functions of calcitonin and PTH. It is likely that PTH, aided by calcitriol, is the primary regulator of circulating calcium ion concentrations in healthy adults.

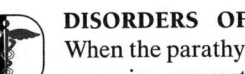 **DISORDERS OF PARATHYROID FUNCTION** When the parathyroid gland secretes inadequate or excessive amounts of parathyroid hormone, Ca^{2+} concentrations move outside normal homeostatic limits. Inadequate parathyroid hormone production, a condition called **hypoparathyroidism,** leads to low Ca^{2+} concentrations in body fluids. The most obvious symptoms involve neural and muscle tissues: The nervous system becomes more excitable, and the affected individual may experience *hypocalcemic tetany,* prolonged muscle spasms affecting the limbs and face.

In **hyperparathyroidism,** Ca^{2+} concentrations become abnormally high. Bones grow thin and brittle, skeletal muscles become weak, CNS function is depressed, nausea and vomiting occur. In severe cases, the patient may become comatose.

AM *Disorders of Parathyroid Function*

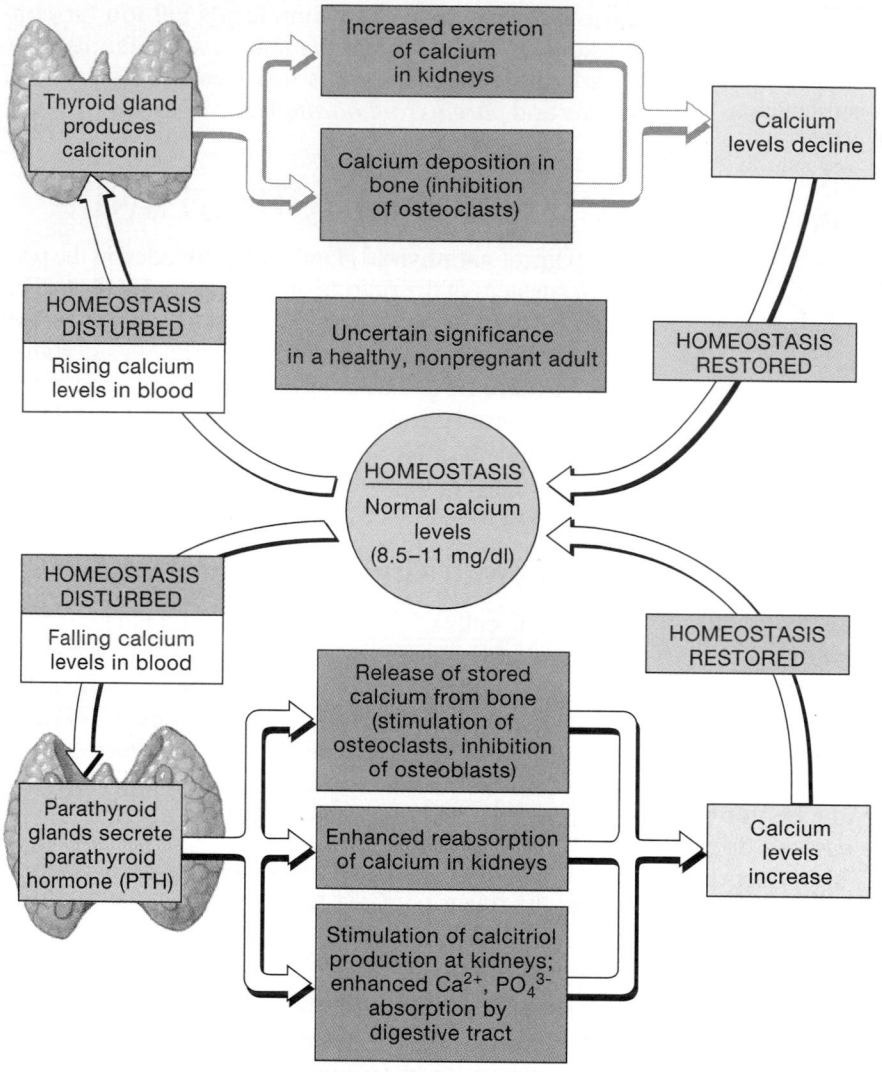

●**FIGURE 18-15** The Homeostatic Regulation of Calcium Ion Concentrations

The thymus produces several hormones that are important to the development and maintenance of normal immunological defenses (Table 18-4). **Thymosin** (THĪ-mō-sin) is the name originally given to a thymic extract that promotes the development and maturation of *lymphocytes,* the white blood cells responsible for immunity. The thymic extract actually contains a blend of several different, complementary hormones—*thymosin-α, thymosin-β, thymosin V, thymopoietin, thymulin,* and several others. The term *thymosins* is now used to refer to all thymic hormones.

Information concerning the hormones of the thyroid, parathyroid, and thymus glands is summarized in Table 18-4. We will further consider the histological organization of the thymus and the functions of the thymosins in Chapter 22.

THE ADRENAL GLANDS

A yellow, pyramid-shaped **adrenal gland,** or *suprarenal* (soo-pra-RĒ-nal) *gland* (supra-, above + renes, kidneys), sits on the superior border of each kidney (Figure 18-16a●). Each adrenal gland lies at roughly the level of the twelfth rib and is firmly attached to the superior portion of each kidney by a dense fibrous capsule. The adrenal gland on each side nestles between the kidney, the diaphragm, and the major arteries and veins that run along the dorsal wall of the abdominopelvic cavity. The adrenal glands project into the peritoneal cavity, and their anterior surfaces are covered by a layer of parietal peritoneum. Like other endocrine glands, the adrenal glands are highly vascularized.

A typical adrenal gland weighs about 7.5 g (0.19 oz), but adrenal size can vary greatly as secretory demands change. The adrenal gland is divided into two parts: a superficial **adrenal cortex** and an inner **adrenal medulla** (Figure 18-16b●).

The Adrenal Cortex

The yellowish color of the adrenal cortex is due to the presence of stored lipids, especially cholesterol and various fatty acids. The adrenal cortex produces more than two dozen different steroid hormones, collec-

THE THYMUS

The **thymus** is located in the mediastinum, generally just posterior to the sternum. ∞ *[p. 24]* In newborn infants and young children, the thymus is relatively large, commonly extending from the base of the neck to the superior border of the heart. Although the thymus continues to increase in size throughout childhood, the body as a whole grows even faster, so the size of the thymus relative to that of the other organs in the mediastinum gradually decreases. The thymus reaches its maximum absolute size, at a weight of about 40 g (1.4 oz), just before puberty. After puberty, it gradually diminishes in size; by the time an individual reaches age 50, the thymus may weigh less than 12 g (0.3 oz). It has been suggested that the gradual decrease in the size and secretory abilities of the thymus may make the elderly more susceptible to disease.

TABLE 18-4 Hormones of the Thyroid Gland, Parathyroid Glands, and Thymus

Gland/Cells	Hormone(s)	Targets	Hormonal Effects	Regulatory Control
Thyroid				
Follicular epithelium	Thyroxine (T_4), triiodothyronine (T_3)	Most cells	Increase energy utilization, oxygen consumption, growth, and development (*see Table 18-3*)	Stimulated by TSH from anterior pituitary (*see Figure 18-10a*)
C cells	Calcitonin (CT)	Bone, kidneys	Decreases Ca^{2+} concentrations in body fluids (*see Figure 18-15*)	Stimulated by elevated blood Ca^{2+} levels; actions opposed by PTH
Parathyroids				
Chief cells	Parathyroid hormone (PTH)	Bone, kidneys	Increases Ca^{2+} concentrations in body fluids (*see Figure 18-15*)	Stimulated by low blood Ca^{2+} levels; PTH effects enhanced by calcitriol and opposed by calcitonin
Thymus	Thymosins (thymosin-α, -β, thymosin V, thymopoietin, thymulin, thymolympho-tropin, thymic-factor X)	Lymphocytes	Stimulate development and maturation of immune response	Unknown

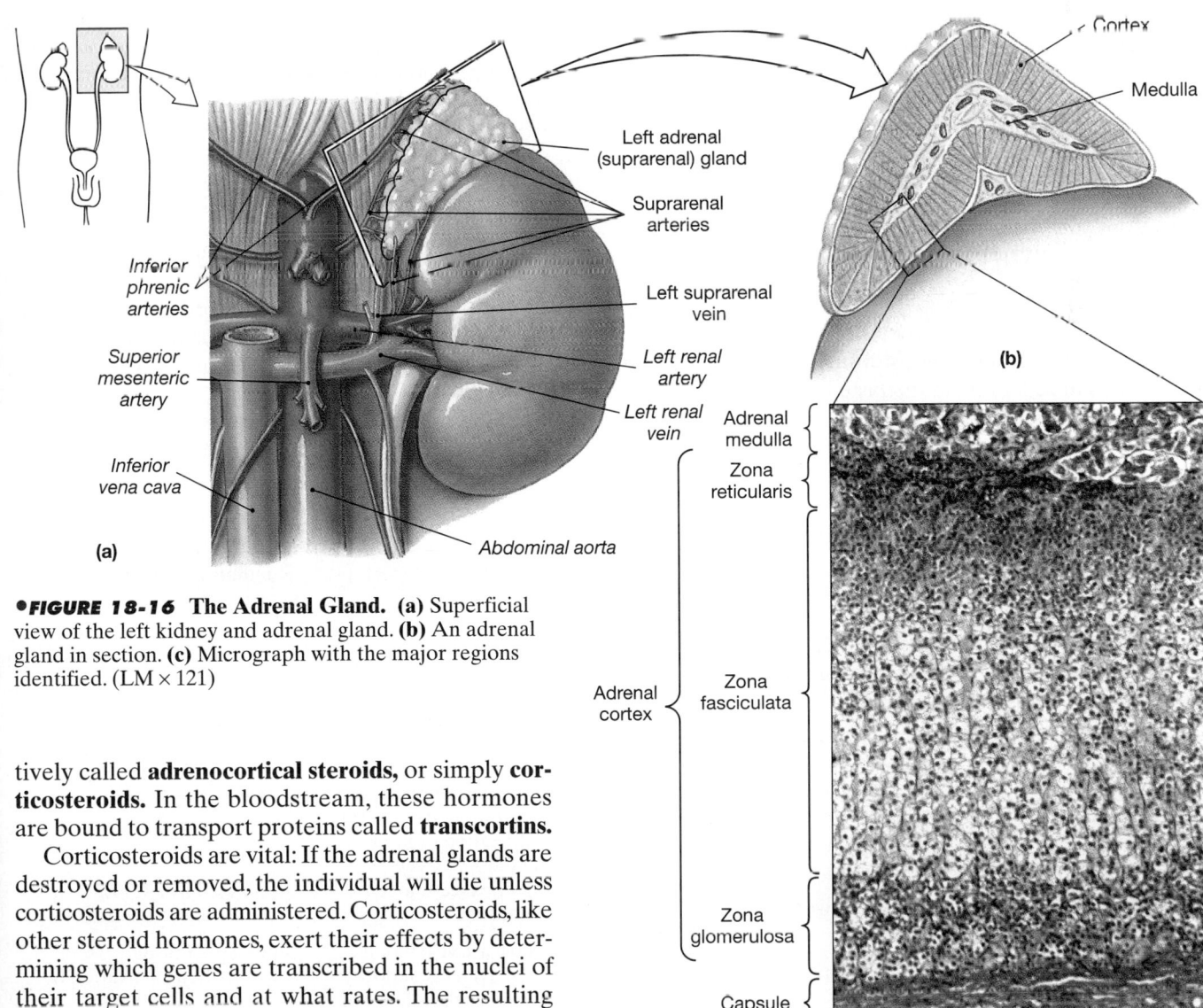

•**FIGURE 18-16** The Adrenal Gland. **(a)** Superficial view of the left kidney and adrenal gland. **(b)** An adrenal gland in section. **(c)** Micrograph with the major regions identified. (LM × 121)

tively called **adrenocortical steroids,** or simply **corticosteroids.** In the bloodstream, these hormones are bound to transport proteins called **transcortins.**

Corticosteroids are vital: If the adrenal glands are destroyed or removed, the individual will die unless corticosteroids are administered. Corticosteroids, like other steroid hormones, exert their effects by determining which genes are transcribed in the nuclei of their target cells and at what rates. The resulting changes in the nature and concentration of enzymes in the cytoplasm affect cellular metabolism.

Beneath the adrenal capsule are three distinct regions, or *zones,* in the adrenal cortex: (1) an outer, *zona glomerulosa;* (2) a middle, *zona fasciculata;* and (3) an inner, *zona reticularis* (Figure 18-16c●). Each zone synthesizes specific steroid hormones, as detailed in Table 18-5.

The Zona Glomerulosa

The **zona glomerulosa** (glō-mer-ū-LŌ-suh) produces **mineralocorticoids (MCs),** steroid hormones that affect the electrolyte composition of body fluids. **Aldosterone** is the principal mineralocorticoid produced by the human adrenal cortex.

The zona glomerulosa is the outer region of the adrenal cortex, accounting for about 15 percent of the cortical volume (Figure 18-16c●). A *glomerulus* is a little ball; as the term *zona glomerulosa* implies, the endocrine cells in this region form small, dense knots or clusters. This zone extends from the capsule to the radiating cords of the underlying zona fasciculata.

ALDOSTERONE Aldosterone secretion stimulates the conservation of sodium ions and the elimination of potassium ions. This hormone targets cells that regulate the ionic composition of excreted fluids. It causes the retention of sodium ions at the kidneys, sweat glands, salivary glands, and pancreas, preventing Na^+ loss in urine, sweat, saliva, and digestive secretions. The retention of Na^+ is accompanied by a loss of K^+. As a secondary effect, the reabsorption of Na^+ enhances the osmotic reabsorption of water at the kidneys, sweat glands, salivary glands, and pancreas. The effect at the kidneys is most dramatic when normal levels of ADH are present. In addition, aldosterone increases the sensitivity of salt receptors in the taste buds of the tongue. As a result, there is an increased interest in (and consumption of) salty food.

Aldosterone secretion occurs in response to a drop in blood Na^+ content, blood volume, or blood pressure or to a rise in the blood K^+ concentration. Changes in Na^+ or K^+ concentration have a direct effect on the zona glomerulosa, stimulating aldosterone release. Aldosterone release also occurs in response to *angiotensin II.* We will discuss this hormone, part of the *renin–angiotensin system,* in a later section.

DISORDERS OF ALDOSTERONE SECRETION In **hypoaldosteronism**, the zona glomerulosa fails to produce enough aldosterone, generally because the kidneys are not releasing adequate amounts of renin. Affected individuals lose excessive amounts of water and Na^+ at the kidneys; the water loss leads to low blood volume and low blood pressure. Changes in electrolyte concentrations affect transmembrane potentials, eventually disrupting neural and muscular tissues throughout the body.

Hypersecretion of aldosterone results in the condition of **aldosteronism.** Under continued aldosterone stimulation, the kidneys retain sodium ions very effectively, but potassium ions are lost in large quantities. A crisis eventually develops when low extracellular K^+ concentrations disrupt normal cardiac, neural, and kidney cell functions. AM *Disorders of the Adrenal Cortex*

The Zona Fasciculata

The **zona fasciculata** (fa-sik-ū-LA-tuh; *fasciculus,* little bundle) produces steroid hormones collectively known as **glucocorticoids (GCs)** because of their effects on glucose metabolism. This zone begins at the inner border of the zona glomerulosa and extends toward the adrenal medulla. It contributes about 78 percent of the cortical volume. The endocrine cells are larger and contain more lipid than do those of the zona

TABLE 18-5 The Adrenal Hormones

Region/Zone	Hormone(s)	Targets	Hormonal Effects	Regulatory Control
Cortex				
Zona glomerulosa	Mineralocorticoids (MC), primarily aldosterone	Kidneys	Increase renal reabsorption of Na^+ and water (especially in the presence of ADH) and accelerate urinary loss of K^+	Stimulated by angiotensin II; inhibited by ANP *(see Figure 18-18)*
Zona fasciculata	Glucocorticoids (GC): cortisol (hydrocortisone), corticosterone, cortisone	Most cells	Release of amino acids from skeletal muscles, and lipids from adipose tissues; promote liver formation; promote peripheral glycogen utilization of lipids; anti-inflammatory effects	Stimulated by ACTH from anterior pituitary gland *(see Figure 18-10a)*
Zona reticularis	Androgens		Uncertain significance under normal conditions	Stimulated by ACTH; significance uncertain
Medulla	Epinephrine, norepinephrine	Most cells	Increase cardiac activity, blood pressure, glycogen breakdown, blood glucose levels; release of lipids by adipose tissue	Stimulated during sympathetic activation by sympathetic preganglionic fibers *(see Chapter 16)*

glomerulosa, and the lipid droplets give the cytoplasm a pale, foamy appearance. The cells of the zona fasciculata form individual cords composed of stacks of cells. Adjacent cords are separated by flattened blood vessels with fenestrated walls.

THE GLUCOCORTICOIDS When stimulated by ACTH from the anterior pituitary, the zona fasciculata secretes primarily **cortisol** (KOR-ti-sol), also called *hydrocortisone*, along with smaller amounts of the related steroids **corticosterone** (kor-ti-KOS-te-rōn) and **cortisone**. Glucocorticoid secretion is regulated by negative feedback: The glucocorticoids released have an inhibitory effect on the production of corticotropin-releasing hormone (CRH) in the hypothalamus and of ACTH in the anterior pituitary. This relationship was diagrammed in Figure 18-10a•, p. 605.

EFFECTS OF GLUCOCORTICOIDS Glucocorticoids accelerate the rates of glucose synthesis and glycogen formation, especially within the liver. Adipose tissue responds by releasing fatty acids into the blood, and other tissues begin to break down fatty acids and proteins instead of glucose. This process is another example of a *glucose-sparing effect.*

Glucocorticoids also show **anti-inflammatory** effects by inhibiting the activities of white blood cells and other components of the immune system. "Steroid creams" are commonly used to control irritating allergic rashes, such as those produced by poison ivy, and injections of glucocorticoids may be used to control more severe allergic reactions. Glucocorticoids slow the migration of phagocytic cells into an injury site and cause phagocytic cells already in the area to become less active. In addition, mast cells exposed to these steroids are less likely to release histamine and other chemicals that promote inflammation. ∞ *[p. 140]* As a result, swelling and further irritation are dramatically reduced. On the negative side, the rate of wound healing decreases, and the weakening of the region's defenses makes it an easy target for infectious organisms. For that reason, topical steroids are used to treat superficial rashes but are never applied to open wounds.

ABNORMAL GLUCOCORTICOID PRODUCTION
Addison's disease is caused primarily by inadequate glucocorticoid production (see Figure 18-23d•, p. 634). The usual cause is destruction of the zona fasciculata, either as the result of an *autoimmune response,* in which the body's defenses mistakenly attack healthy tissues, or after infection by the bacteria responsible for *tuberculosis.* Affected individuals become weak and lose weight; they cannot effectively use their lipid reserves to generate ATP, and blood glucose concentrations fall sharply within hours after a meal. Because in most cases the zona glomerulosa is affected as well, symptoms similar to those of hypoaldosteronism may also appear.

Cushing's disease results from overproduction of glucocorticoids, generally because of hypersecretion of ACTH. The symptoms resemble those of a protracted and exaggerated response to stress (see Figure 18-23e•, p. 634). Glucose metabolism is suppressed, lipid reserves are mobilized, and peripheral proteins are broken down. Lipid reserves are redistributed, and the distribution of body fat changes. Individuals develop a "moon-faced" appearance due to the deposition of lipids in the subcutaneous tissues of the face. Skeletal muscles become weak as contractile proteins break down. **AM** *Disorders of the Adrenal Cortex*

The Zona Reticularis

The **zona reticularis** (re-tik-ū-LAR-is; *reticulum,* network) forms a narrow band bordering each adrenal medulla (Figure 18-16c•). In total, the zona reticularis accounts for only about 7 percent of the total volume of the adrenal cortex. The endocrine cells of the zona reticularis form a folded, branching network, and fenestrated blood vessels wind between the cells.

The zona reticularis normally produces small quantities of androgens, the sex hormones produced in large quantities by the male testes. Once in the circulation, some of the androgens released by the zona reticularis are converted to estrogens, the dominant sex hormone in females. When secreted in normal amounts, neither the androgens nor the estrogens have an effect on sexual characteristics, and the significance of adrenal sex hormone production remains uncertain. ACTH stimulates the zona reticularis to a slight degree, but the effects are generally insignificant.

ABNORMAL ACTIVITY IN THE ZONA RETICULARIS The zona reticularis ordinarily produces a negligible amount of androgens. If a tumor develops in this portion of the adrenal cortex, androgen secretion may increase dramatically, producing symptoms of the **adrenogenital syndrome**. In women, this condition leads to the gradual development of male secondary sexual characteristics, including body and facial hair patterns, adipose tissue distribution, and muscular development. Tumors affecting the zona reticularis of males may sometimes result in the production of large quantities of estrogens. In this condition, the affected male develops enlarged breasts and, in some cases, other female secondary sexual characteristics. This array of symptoms is called **gynecomastia** (*gynaikos,* woman + *mastos,* breast).

The Adrenal Medullae

The boundary between the adrenal cortex and the adrenal medulla does not form a straight line, and the supporting connective tissues and blood vessels are extensively interconnected. The adrenal medulla has a reddish brown coloration due in part to the many blood vessels in this area. It contains large, rounded cells— similar to those in sympathetic ganglia that are innervated by preganglionic sympathetic fibers. The secretory activities of the adrenal medullae are controlled by the sympathetic division of the ANS, as we detailed in Chapter 16. ∞ *[p. 520]*

The adrenal medullae contain two populations of secretory cells: one producing epinephrine (adrenaline) and the other norepinephrine (noradrenaline). There is evidence that the two cell types are distributed in different areas of the medulla and that their secretory activities can be independently controlled. The secretions are packaged in vesicles that form dense clusters just inside the cell membranes. The hormones within these vesicles are continuously being released at low levels, through exocytosis. Sympathetic stimulation dramatically accelerates the rate of exocytosis and hormone release.

Epinephrine and Norepinephrine

Epinephrine makes up 75–80 percent of the secretions from the medullae, the rest being norepinephrine. We detailed the peripheral effects of these hormones, which result from interaction with alpha and beta receptors on cell membranes, in Chapter 16. ∞ *[p. 522]* Stimulation of α_1 and β_1 receptors, the most common types, accelerates cellular energy utilization and the mobilization of energy reserves.

When the adrenal medulla is activated:

- Epinephrine and norepinephrine trigger a mobilization of glycogen reserves in skeletal muscles and accelerate the breakdown of glucose to provide ATP. This combination increases muscular strength and endurance.
- In adipose tissue, stored fats are broken down to fatty acids, which are released into the circulation for use by other tissues.
- In the liver, glycogen molecules are broken down. The resulting glucose molecules are released into the circulation, primarily for use by neural tissues, which cannot shift to fatty acid metabolism.
- In the heart, stimulation of β_1 receptors triggers an increase in the rate and force of cardiac muscle contraction.

The metabolic changes that follow catecholamine release are at their peak 30 seconds after adrenal stimulation, and they linger for several minutes thereafter. As a result, the effects produced by stimulation of the adrenal medullae outlast the other signs of sympathetic activation. [AM] *Disorders of the Adrenal Medullae*

☑ What symptoms would you expect to see in an individual whose diet lacks iodine?

☑ When a person's thyroid gland is removed, signs of decreased thyroid hormone concentration do not appear until about one week later. Why?

☑ Removal of the parathyroid glands would result in a decrease in the blood concentration of which important mineral?

☑ What effect would elevated cortisol levels have on the level of glucose in the blood?

THE KIDNEYS

The kidneys release the steroid hormone *calcitriol*, the peptide hormone *erythropoietin*, and an enzyme, *renin*. Calcitriol is important for calcium ion homeostasis. Erythropoietin and renin are involved in the regulation of blood volume and blood pressure.

Calcitriol

Calcitriol is a steroid hormone secreted by the kidney in response to the presence of parathyroid hormone (PTH) (Figure 18-17a●). Calcitriol synthesis is dependent on the availability of *cholecalciferol* (vitamin D_3), a related steroid that may be synthesized in the skin or absorbed from the diet. During transport in the bloodstream, cholecalciferol from either source is bound to *transcalciferin*, a transport protein. Cholecalciferol is absorbed from the bloodstream by the liver and converted to an intermediary product (*25-hydroxy-D_3*). This intermediary is released into the circulation and absorbed by the kidneys for conversion to calcitriol. The term *vitamin D* applies to the entire group of related steroids, including calcitriol, cholecalciferol, and various intermediaries.

The best-known function of calcitriol is the stimulation of calcium and phosphate ion absorption along the digestive tract. The effects of PTH on Ca^{2+} absorption result primarily through stimulation of calcitriol release. Calcitriol's other effects on calcium metabolism include (1) stimulating formation and differentiation of osteoprogenitor cells and osteoclasts, (2) stimulating bone resorption by osteoclasts, (3) stimulating Ca^{2+} reabsorption at the kidneys, and (4) suppressing PTH production. There is now good evidence that calcitriol has many other effects that have little to do with regulating calcium levels in body fluids. For example, calcitriol appears to stimulate muscle contractility, insulin secretion, and phagocytosis, and it may depress sperm production and the secretion of testosterone. The importance of these effects in a healthy individual remains to be determined.

Erythropoietin

Erythropoietin (e-rith-rō-poy-Ē-tin; *erythros*, red + *poiesis*, making), or **EPO,** is a peptide hormone released by the kidney in response to low oxygen levels in kidney tissues (Figure 18-17b●). The reduced oxygen concentrations may result from (1) reduced renal blood flow caused by blood loss, low blood pressure, or blockage of renal vessels; (2) a reduction in the number or the oxygen-carrying capacity of circulating red blood cells; (3) a reduction in the oxygen content of the air in the lungs, due to respiratory problems or increased altitude; or (4) problems at the respiratory surfaces of the lungs.

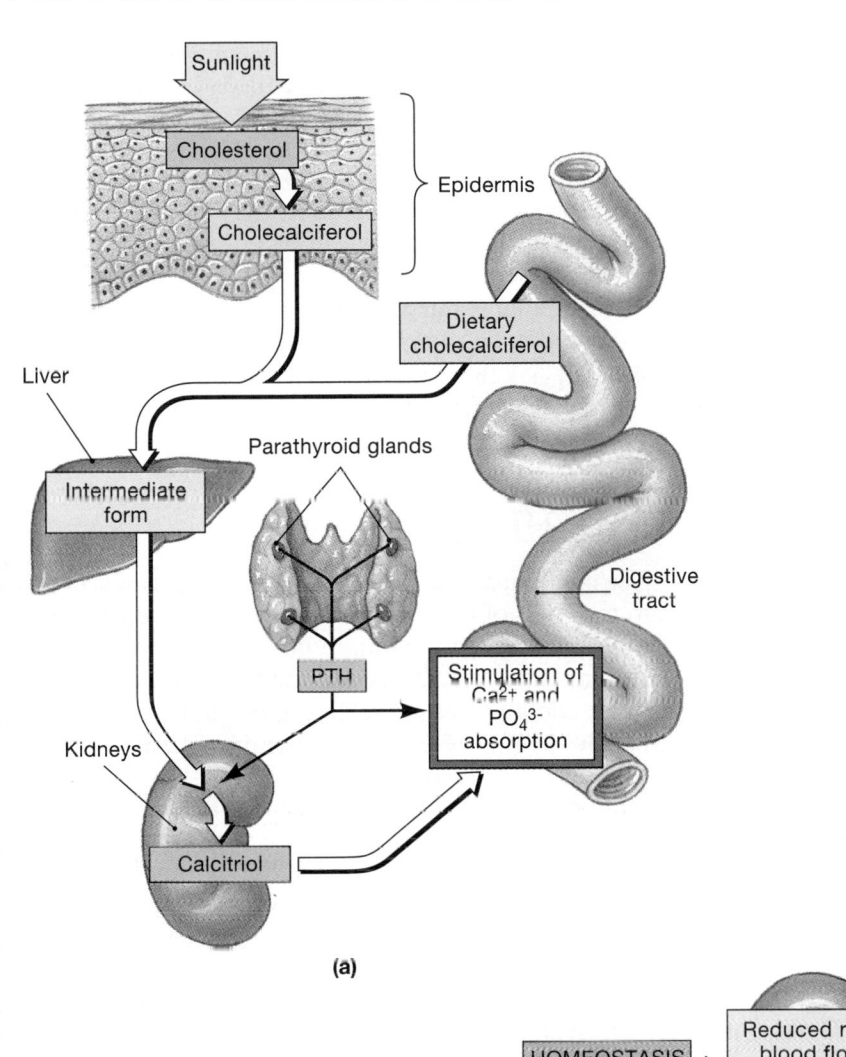

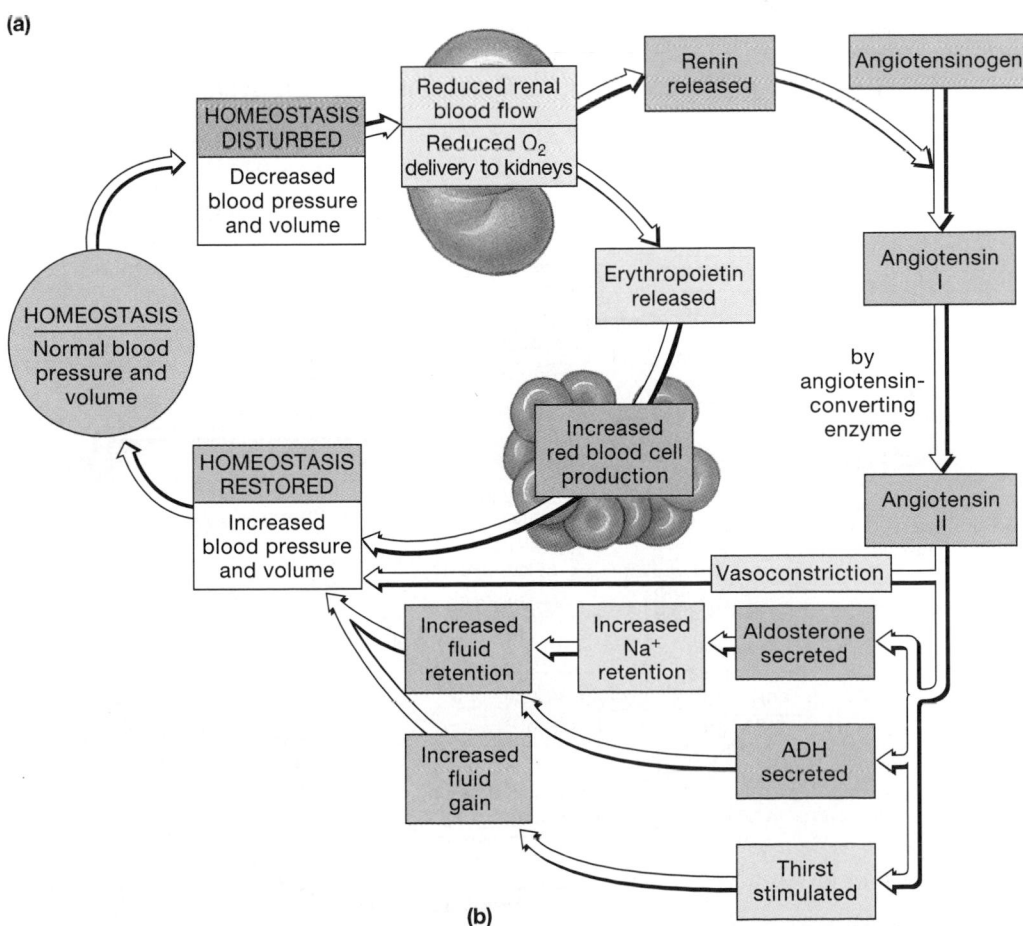

•*FIGURE 18-17* **The Endocrine Functions of the Kidneys.** **(a)** The production of calcitriol. **(b)** The renin–angiotensin system assists in the elevation of blood volume and blood pressure; erythropoietin enhances red blood cell production.

EPO stimulates the production of red blood cells by the bone marrow. The increase in the number of erythrocytes elevates blood volume to some degree. Because these cells transport oxygen, the increase in their number improves oxygen delivery to peripheral tissues. We shall consider EPO in greater detail when we discuss the formation of blood cells in Chapter 19.

Renin

Renin plays a key role in the renin–angiotensin system. As a result, many physiological and endocrinological references consider renin among the hormones. Renin levels rise in response to (1) sympathetic stimulation or (2) a decline in renal blood flow. Reduced renal blood flow may be caused by a fall in blood volume, blood pressure, or both as a result of hemorrhage, dehydration, or other factors.

Renin is released by specialized cells of the *juxtaglomerular apparatus* of the kidney; those cells also release EPO. Renin is secreted as *prorenin*, an inactive form. Before entering the circulation and leaving the kidneys, most of the prorenin is converted to renin. Once in the bloodstream, renin functions as an enzyme that starts an enzymatic chain reaction (Figure 18-17b•). The initial step occurs when renin converts **angiotensinogen,** a plasma protein produced by the liver, to **angiotensin I.** In the capillaries of the lungs, *angiotensin-converting enzyme* (ACE) then modifies angiotensin I to **angiotensin II,** an active hormone with diverse effects. Angiotensin II has four major effects:

1. It stimulates the adrenal production of aldosterone, causing Na^+ retention and K^+ loss at the kidneys.
2. It stimulates the secretion of ADH, in turn stimulating water reabsorption at the kidneys and complementing the effects of aldosterone. The combination of increased Na^+ reabsorption and water retention prevents further reductions in blood volume due to urinary water losses.
3. It stimulates thirst, resulting in increased fluid consumption and an elevation of blood volume.
4. It triggers the constriction of arterioles; this vasoconstriction elevates systemic blood pressure. The effect is 4–8 times greater than that produced by norepinephrine.

THE HEART

The endocrine cells in the heart are cardiac muscle cells in the walls of the *atria,* chambers that receive venous blood. If the blood volume becomes too great, these cardiac muscle cells are excessively stretched. Under these conditions, they release a hormone called **atrial natriuretic** (nā-trē-ū-RET-ik) **peptide (ANP)** (*natrium,* sodium + *ouresis,* making water). ANP has four effects that oppose those of angiotensin II (Figure 18-18•):

1. It promotes the loss of Na^+ and water at the kidneys.
2. It inhibits renin release and the secretion of water-conserving hormones, such as ADH and aldosterone.
3. It suppresses thirst.
4. It blocks the action of hormones such as angiotensin II or norepinephrine on arterioles. The resulting relaxation of the blood vessels helps reduce blood pressure.

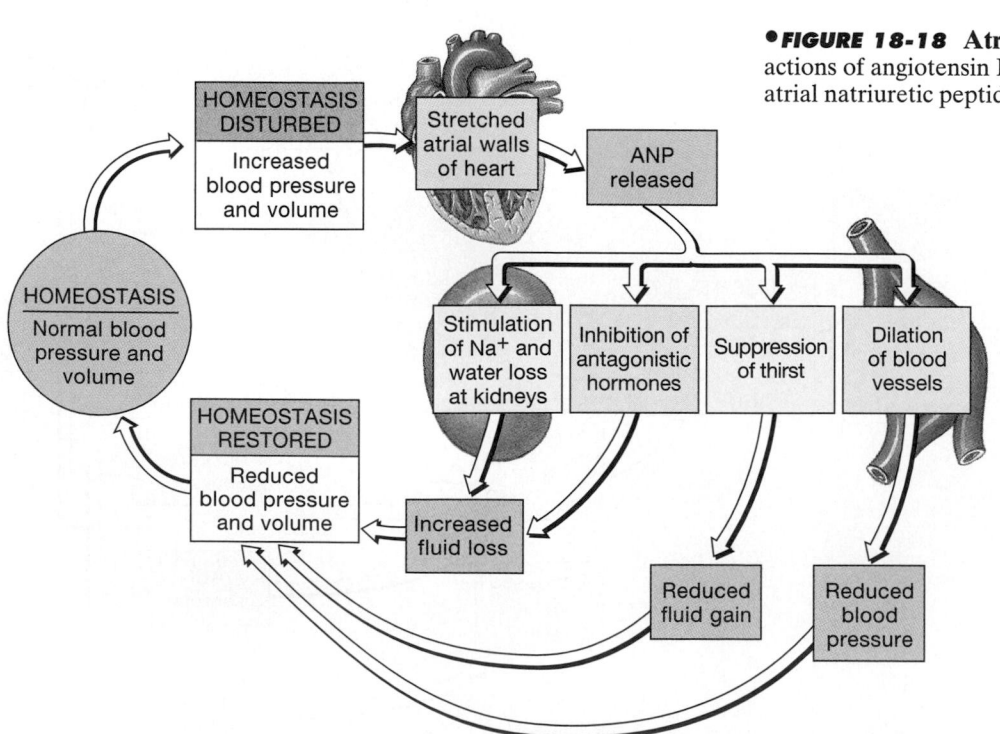

•FIGURE 18-18 Atrial Natriuretic Peptide. The actions of angiotensin II are opposed by those of atrial natriuretic peptide.

This combination lowers blood volume and reduces the stretching of the cardiac muscle cells in the atrial walls. We will consider the actions of this hormone in greater detail when we discuss the control of blood pressure and volume in Chapter 21.

☑ How would an increase in the amount of atrial natriuretic peptide affect the Na^+ concentration in the urine?

ENDOCRINE TISSUES OF THE DIGESTIVE SYSTEM

The linings of the digestive tract, the liver, and the pancreas produce a variety of exocrine secretions that are essential to the normal breakdown and absorption of food. Although the pace of digestive activities can be affected by the ANS, most digestive processes are controlled locally. The various components of the digestive tract communicate with one another by means of hormones that we will consider in Chapter 24. One digestive organ, the pancreas, produces two hormones with widespread effects.

The Pancreas

The **pancreas** lies within the abdominopelvic cavity in the J-shaped loop between the stomach and the small intestine (Figure 18-19a●). It is a slender, pale organ with a nodular (lumpy) consistency. The adult pancreas is 20–25 cm (8–10 in.) long and weighs about 80 g (2.8 oz). We shall consider the detailed anatomy of the pancreas in Chapter 24, because it is primarily a digestive organ that manufactures digestive enzymes. The **exocrine pancreas,** roughly 99 percent of the pancreatic volume, consists of clusters of gland cells, called *pancreatic acini*, and their attached ducts. Together the gland and duct cells secrete large quantities of an alkaline, enzyme-rich fluid. This secretion reaches the lumen of the digestive tract by traveling along a network of secretory ducts.

The **endocrine pancreas** consists of small groups of cells scattered among the exocrine cells. The endocrine clusters are known as **pancreatic islets,** or the *islets of Langerhans* (LAN-ger-hanz) (Figure 18-19b●). Pancreatic islets account for only about 1 percent of the pancreatic cell population. Nevertheless, a typical pancreas contains roughly 2 million pancreatic islets.

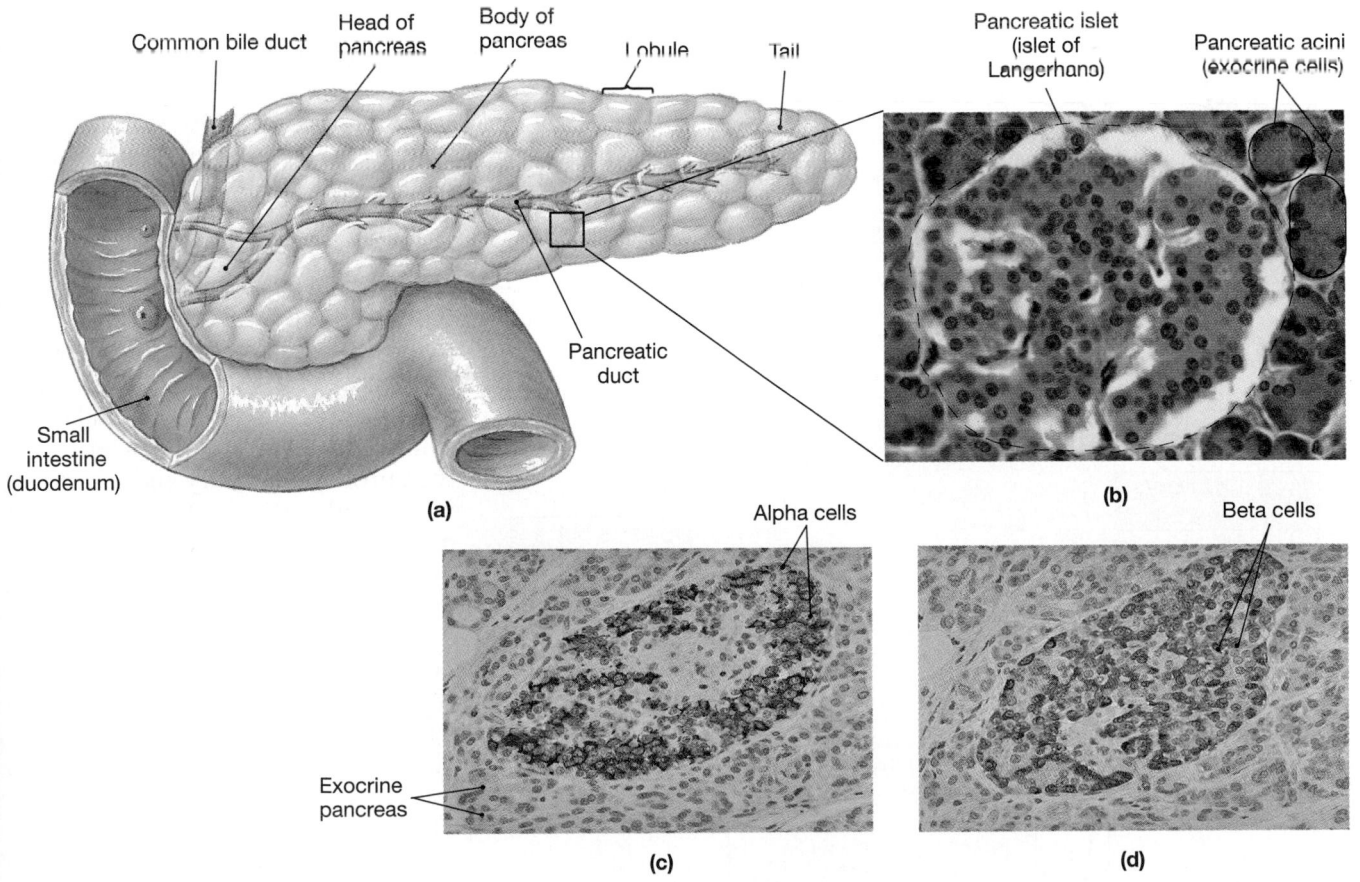

●**FIGURE 18-19** **The Endocrine Pancreas.** **(a)** Gross anatomy of the pancreas. **(b)** A pancreatic islet surrounded by exocrine cells. (LM × 276) Special staining techniques can be used to differentiate between **(c)** alpha cells and **(d)** beta cells in pancreatic islets.

Like other endocrine tissue, the pancreatic islets are surrounded by an extensive, fenestrated capillary network that carries its hormones into the circulation. Each islet contains four different cell types:

1. **Alpha cells** (Figure 18-19c●) produce the hormone **glucagon** (GLOO-ka-gon). Glucagon raises blood glucose levels by increasing the rates of glycogen breakdown and glucose release by the liver.

2. **Beta cells** (Figure 18-19d●) produce the hormone **insulin** (IN-suh-lin). Insulin lowers blood glucose by increasing the rate of glucose uptake and utilization by most body cells and increasing glycogen synthesis in skeletal muscles and the liver. Beta cells also secrete *amylin,* a recently discovered peptide hormone whose role is uncertain.

3. **Delta cells** produce a peptide hormone identical to **somatostatin** (GH-IH), a hypothalamic regulatory hormone. Somatostatin produced in the pancreas suppresses glucagon and insulin release by other islet cells and slows the rates of food absorption and enzyme secretion along the digestive tract.

4. **F cells** produce the hormone **pancreatic polypeptide (PP).** Little is known about the specific functions of pancreatic polypeptide. It inhibits gallbladder contractions and regulates the production of some pancreatic enzymes; it may help control the rate of nutrient absorption by the digestive tract.

We will focus on insulin and glucagon, the hormones responsible for the regulation of blood glucose concentrations. These hormones interact to control blood glucose levels. When blood glucose levels rise, beta cells secrete insulin, which then stimulates the transport of glucose across cell membranes. When blood glucose levels decline, alpha cells secrete glucagon, which stimulates glucose release by the liver. This relationship is diagrammed in Figure 18-20●.

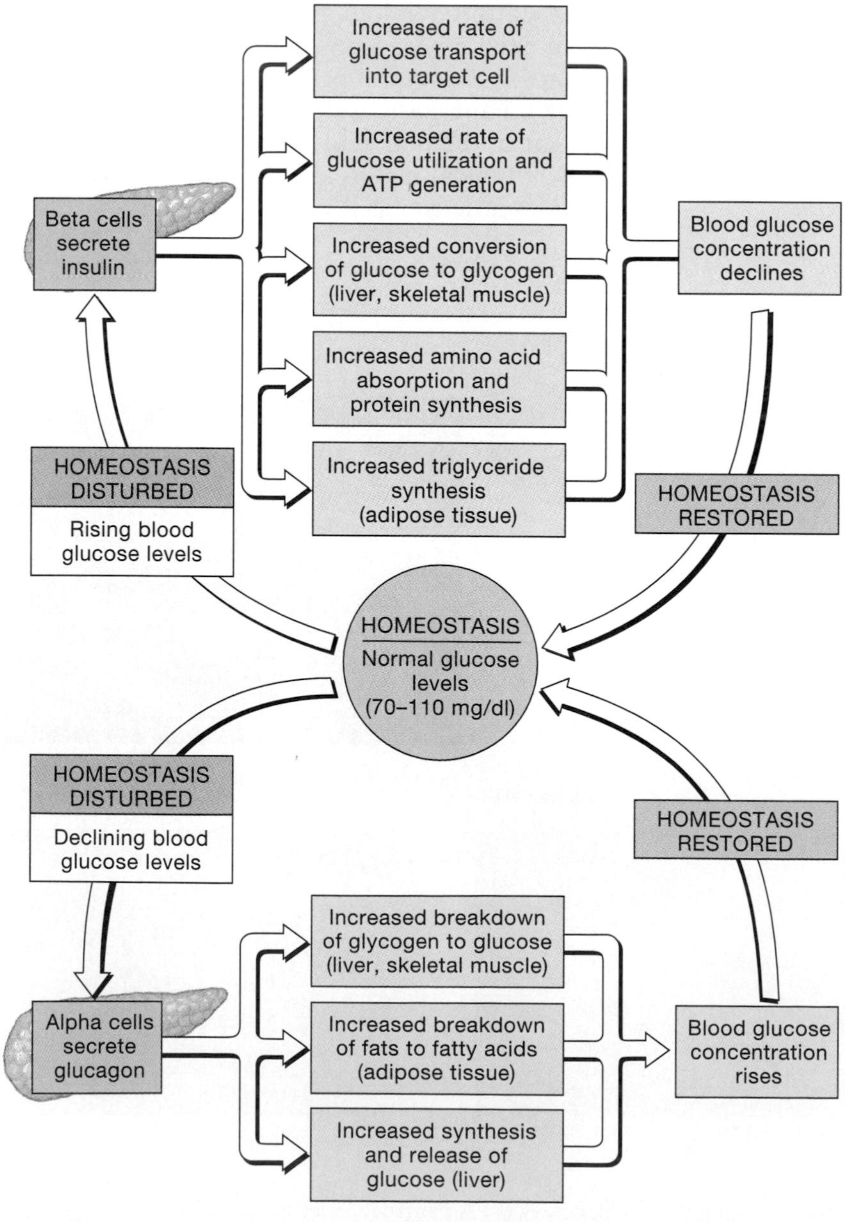

Insulin

Insulin is a peptide hormone released by beta cells when glucose levels rise above normal levels (70–110 mg/dl). Insulin secretion is also stimulated by elevated levels of some amino acids, including arginine and leucine. Insulin exerts its effects on cellular metabolism in a series of steps that begins when insulin binds to receptor proteins on the cell membrane. Binding leads to the activation of the receptor, which functions as a kinase and attaches phosphate groups to intracellular enzymes. Phosphorylation of enzymes then produces primary and secondary effects within the cell; the biochemical details remain unresolved.

One of the most important effects is the enhancement of glucose absorption and utilization. Insulin receptors are present in most cell membranes; such cells are called *insulin-dependent.* However, cells in the brain and kidneys, cells in the lining of the digestive tract, and red blood cells lack insulin receptors. These cells are called *insulin-independent,* because they can absorb and utilize glucose without insulin stimulation.

The effects of insulin on its target cells include:

●*FIGURE 18-20* **The Regulation of Blood Glucose Concentrations**

- *Acceleration of glucose uptake (all target cells).* This effect results from an increase in the number of glucose transport proteins in the cell membrane. These proteins transport glucose into the cell by facilitated diffusion. This movement follows the concentration gradient for glucose, and ATP is not required.
- *Acceleration of glucose utilization (all target cells) and enhanced ATP production.* This effect occurs for two reasons: (1) The rate of glucose use is proportional to its availability; when more glucose enters the cell, more is used. (2) Second messengers activate a key enzyme involved in the initial steps of glycolysis.
- *Stimulation of glycogen formation (skeletal muscles and liver cells).* When excess glucose enters these cells, it is stored in the form of glycogen.
- *Stimulation of amino acid absorption and protein synthesis.*
- *Stimulation of triglyceride formation in adipose tissues.* Insulin stimulates the absorption of fatty acids and glyc-

erol by adipocytes. The adipose cells then store these components as triglycerides. Adipocytes also increase their absorption of glucose; excess glucose is used in the synthesis of additional triglycerides.

In summary, insulin is secreted when glucose is abundant, and this hormone stimulates glucose utilization to support growth and the establishment of carbohydrate (glycogen) and lipid (triglyceride) reserves. The accelerated use of glucose soon brings circulating glucose levels within normal limits.

Glucagon

When glucose concentrations fall below normal, alpha cells release glucagon, and energy reserves are mobilized. When glucagon binds to a receptor in the cell membrane, it activates adenylate cyclase, and cAMP acts as a second messenger that activates cy-

---

 ## Diabetes Mellitus

Blood glucose levels are usually very closely regulated by insulin and glucagon. Whether glucose is absorbed across the digestive tract or manufactured and released by the liver, very little glucose leaves the body intact once it has entered the circulation. Glucose does get filtered out of the blood at the kidneys, but the cells lining the kidney tubules usually reabsorb virtually all of it, and urinary glucose losses are negligible.

Diabetes mellitus (mel-Ī-tus; *mellitum,* honey) is characterized by glucose concentrations that are high enough to overwhelm the reabsorption capabilities of the kidneys. Glucose appears in the urine **(glycosuria)** (glī-kō-SOO-rē-a), and urine production generally becomes excessive (*polyuria*). Other metabolic products, such as fatty acids and other lipids, are also present in abnormal concentrations.

Diabetes mellitus may be caused by genetic abnormalities, and some of the genes responsible have been identified. For example, genes that result in inadequate insulin production, the synthesis of abnormal insulin molecules, or the production of defective receptor proteins will produce the same basic symptoms. Diabetes mellitus may also appear as the result of other pathological conditions, injuries, immune disorders, or hormonal imbalances. There are two major types of diabetes mellitus: *insulin-dependent (Type I) diabetes* and *non-insulin-dependent (Type II) diabetes.* We discuss both in greater detail in the *Applications Manual.* **AM** *Diabetes Mellitus*

Probably because glucose levels cannot be stabilized adequately, even with treatment, persons with diabetes mellitus commonly develop chronic medical problems. In general, these problems are related to circulatory system abnormalities. The most common examples include the following:

- Vascular changes at the retina, including proliferation of capillaries and hemorrhaging, typically cause partial or complete blindness. This condition is called *diabetic retinopathy.*
- Changes occur in the clarity of the lens of the eye, producing cataracts.
- Small hemorrhages and inflammation at the kidneys cause degenerative changes that can lead to kidney failure. This condition is called **diabetic nephropathy.** Treatment with drugs that improve blood flow to the kidneys can slow the progression to kidney failure.
- A variety of neural problems appear, including *peripheral neuropathies* and abnormal autonomic function. ∞ *[p. 426]* These disorders, collectively termed **diabetic neuropathy,** are probably related to disturbances in the blood supply to neural tissues.
- Degenerative changes in cardiac circulation can lead to early heart attacks. For a given age group, heart attacks are 3–5 times more likely in diabetic individuals than in nondiabetic persons.
- Other peripheral changes in the vascular system can disrupt normal circulation to the extremities. For example, a reduction in blood flow to the feet can lead to tissue death, ulceration, infection, and loss of toes or a major portion of one or both feet.

toplasmic enzymes. The primary effects of glucagon are as follows:

- *Stimulation of glycogen breakdown in skeletal muscle and liver cells.* The glucose molecules released will either be metabolized for energy (skeletal muscle fibers) or released into the bloodstream (liver cells).
- *Stimulation of triglyceride breakdown in adipose tissues.* The adipocytes then release the fatty acids into the circulation for use by other tissues.
- *Stimulation of glucose production at the liver.* Liver cells absorb amino acids from the bloodstream, convert them to glucose, and release the glucose into the circulation. This process of glucose synthesis in the liver is called *gluconeogenesis* (gloo-kō-nē-ō-JEN-e-sis).

The results are a reduction in glucose use and the release of more glucose into the bloodstream. Consequently, blood glucose concentrations soon rise toward normal levels.

Pancreatic alpha cells and beta cells monitor blood glucose concentrations, and the secretion of glucagon and insulin occur without endocrine or nervous instructions. Yet, because the alpha cells and beta cells are very sensitive to changes in blood glucose levels, any hormone that affects blood glucose concentrations will indirectly affect the production of both insulin and glucagon. Insulin production is also influenced by autonomic activity. Parasympathetic stimulation enhances insulin release, and sympathetic stimulation inhibits it.

Information concerning insulin, glucagon, and other pancreatic hormones is summarized in Table 18-6.

ENDOCRINE TISSUES OF THE REPRODUCTIVE SYSTEM

The endocrine tissues of the reproductive system are restricted primarily to the male and female reproductive organs, the testes and ovaries. We shall consider details of the anatomy of the reproductive organs and the endocrinological control of reproductive function in Chapter 28. The primary regulatory hormones involved are FSH and LH from the anterior pituitary. Abnormally low production of either of these hormones will produce **hypogonadism.** Children with this condition will not undergo sexual maturation, and adults with hypogonadism will not produce and support functional sperm or oocytes.

The Testes

In the male, the **interstitial cells** (*Leydig cells*) of the testes produce the male hormones known as androgens. **Testosterone** (tes-TOS-ter-ōn) is the most important androgen. This hormone promotes the production of functional sperm, maintains the secretory glands of the male reproductive tract, stimulates growth, and determines male secondary sexual characteristics, such as the distribution of facial hair and body fat. Testosterone also affects metabolic operations throughout the body; it stimulates protein synthesis and muscle growth, and it produces aggressive behavioral responses. During embryonic development, the production of testosterone affects the de-

TABLE 18-6 Hormones of the Pancreas

Structure/ Cells	Hormone	Primary Targets	Hormonal Effects	Regulatory Control
Pancreatic islets				
Alpha cells	Glucagon	Liver, adipose tissues	Mobilizes lipid reserves; promotes glucose synthesis and glycogen breakdown in liver; elevates blood glucose concentrations	Stimulated by low blood glucose concentrations; inhibited by somatostatin from delta cells
Beta cells	Insulin	Most cells	Facilitates uptake of glucose by target cells; stimulates lipid and glycogen formation and storage	Stimulated by high blood glucose concentrations, parasympathetic stimulation, and high levels of some amino acids; inhibited by somatostatin from delta cells and by sympathetic activation
Delta cells	Somatostatin (GH-IH)	Other islet cells, digestive epithelium	Inhibits insulin and glucagon secretion; slows rates of nutrient absorption and enzyme secretion along digestive tract	Stimulated by protein-rich meal; mechanism uncertain
F cells	Pancreatic polypeptide	Digestive organs	Inhibits gallbladder contraction; regulates production of pancreatic enzymes; influences rate of nutrient absorption by digestive tract	Stimulated by protein-rich meal and by parasympathetic stimulation

velopment of CNS structures, including hypothalamic nuclei that will later influence sexual behaviors.

Sustentacular cells (*Sertoli cells*) in the testes support the differentiation and physical maturation of spermatozoa. Under FSH stimulation, these cells secrete the hormone **inhibin,** which inhibits the secretion of FSH at the anterior pituitary and perhaps suppresses GnRH release at the hypothalamus.

The Ovaries

Immature female gametes, called *oocytes*, are produced in specialized structures called **follicles.** Follicles develop under FSH stimulation. The follicle cells surrounding each developing oocyte produce **estrogens** under FSH and LH stimulation. Estrogens are steroid hormones that support the maturation of the oocytes and stimulate the growth of the uterine lining. **Estradiol** is the principal estrogen. Under FSH stimulation, follicle cells secrete inhibin, which suppresses FSH release through a feedback mechanism comparable to that in males.

A surge in LH secretion during the ovarian cycle triggers ovulation, and LH then causes the remaining follicular cells to reorganize into a **corpus luteum** (LOO-tē-um; "yellow body"). These follicular cells release a mixture of estrogens and **progestins. Progesterone** (prō-JES-ter-ōn), the principal progestin, has several important functions, including the following:

- It prepares the uterus for the arrival of a developing embryo.

- It accelerates the movement of the oocyte or, if fertilization occurs, the developing embryo along the uterine tube.

- It causes an enlargement of the mammary glands, working in combination with other hormones, such as estradiol, growth hormone, and prolactin.

Relaxin

During pregnancy, the corpus luteum, placenta, and uterus produce an additional hormone, **relaxin.** Relaxin levels are elevated throughout pregnancy. This hormone loosens the articulation at the pubic symphysis, permitting expansion of the lower portion of the uterus and the vagina during delivery. In combination with many other hormones, it also stimulates functional development of the mammary glands. Early in a pregnancy, relaxin production occurs under stimulation by LH from the anterior pituitary. Later in pregnancy, additional stimulation is provided by a placental hormone, *human chorionic gonadotropin (hCG).* Relaxin secretion stops after delivery, because the placenta is expelled, and in the absence of placental hormones, the corpus luteum degenerates.

During pregnancy, additional hormones produced by the placenta interact with those produced by the ovaries and the pituitary gland to promote normal fetal development and delivery. We shall consider the endocrinological aspects of pregnancy in Chapter 29.

Information concerning the reproductive hormones of the testes and ovaries is summarized in Table 18-7.

TABLE 18-7 Hormones of the Reproductive System

Structure/Cells	Hormone(s)	Primary Targets	Hormonal Effects	Regulatory Control
Testes				
Interstitial cells	Androgens	Most cells	Support functional maturation of sperm, protein synthesis in skeletal muscles, male secondary sexual characteristics, and associated behaviors	Stimulated by LH from anterior pituitary gland (*see Figure 18-10a*)
Sustentacular cells	Inhibin	Anterior pituitary gland	Inhibits secretion of FSH	Stimulated by FSH from anterior pituitary gland (*see Figure 18-10a*)
Ovaries				
Follicular cells	Estrogens	Most cells	Support follicle maturation, female secondary sexual characteristics, and associated behaviors	Stimulated by FSH and LH from anterior pituitary gland (*see Figure 18-10a*)
	Inhibin	Anterior pituitary gland	Inhibits secretion of FSH	Stimulated by FSH from anterior pituitary gland (*see Figure 18-10a*)
Corpus luteum	Progestins	Uterus, mammary glands	Prepare uterus for implantation; prepare mammary glands for secretory activity	Stimulated by LH from anterior pituitary gland (*see Figure 18-10a*)
	Relaxin	Pubic symphysis, uterus, mammary glands	Loosens pubic symphysis, relaxes uterine (cervical) muscles, stimulates mammary gland development	Stimulated by LH from the anterior pituitary gland and by placental hCG (human chorionic gonadotropin)

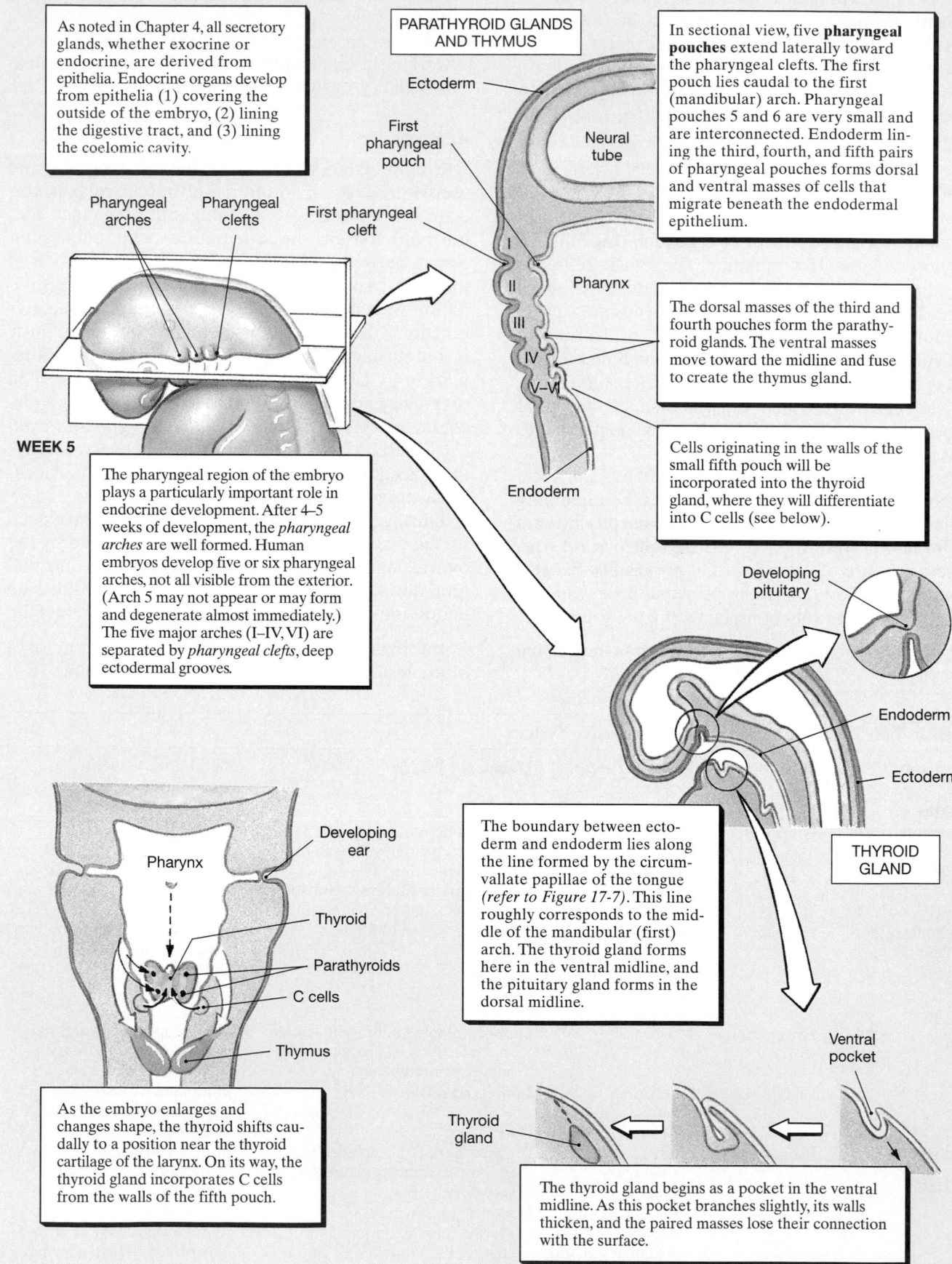

As noted in Chapter 4, all secretory glands, whether exocrine or endocrine, are derived from epithelia. Endocrine organs develop from epithelia (1) covering the outside of the embryo, (2) lining the digestive tract, and (3) lining the coelomic cavity.

PARATHYROID GLANDS AND THYMUS

In sectional view, five **pharyngeal pouches** extend laterally toward the pharyngeal clefts. The first pouch lies caudal to the first (mandibular) arch. Pharyngeal pouches 5 and 6 are very small and are interconnected. Endoderm lining the third, fourth, and fifth pairs of pharyngeal pouches forms dorsal and ventral masses of cells that migrate beneath the endodermal epithelium.

Ectoderm

First pharyngeal pouch

First pharyngeal cleft

Pharyngeal arches

Pharyngeal clefts

Neural tube

Pharynx

I

II

III

IV

V–VI

Endoderm

The dorsal masses of the third and fourth pouches form the parathyroid glands. The ventral masses move toward the midline and fuse to create the thymus gland.

Cells originating in the walls of the small fifth pouch will be incorporated into the thyroid gland, where they will differentiate into C cells (see below).

WEEK 5

The pharyngeal region of the embryo plays a particularly important role in endocrine development. After 4–5 weeks of development, the *pharyngeal arches* are well formed. Human embryos develop five or six pharyngeal arches, not all visible from the exterior. (Arch 5 may not appear or may form and degenerate almost immediately.) The five major arches (I–IV, VI) are separated by *pharyngeal clefts*, deep ectodermal grooves.

Developing pituitary

Endoderm

Ectoderm

THYROID GLAND

Developing ear

Pharynx

Thyroid

Parathyroids

C cells

Thymus

The boundary between ectoderm and endoderm lies along the line formed by the circumvallate papillae of the tongue *(refer to Figure 17-7)*. This line roughly corresponds to the middle of the mandibular (first) arch. The thyroid gland forms here in the ventral midline, and the pituitary gland forms in the dorsal midline.

As the embryo enlarges and changes shape, the thyroid shifts caudally to a position near the thyroid cartilage of the larynx. On its way, the thyroid gland incorporates C cells from the walls of the fifth pouch.

Ventral pocket

Thyroid gland

The thyroid gland begins as a pocket in the ventral midline. As this pocket branches slightly, its walls thicken, and the paired masses lose their connection with the surface.

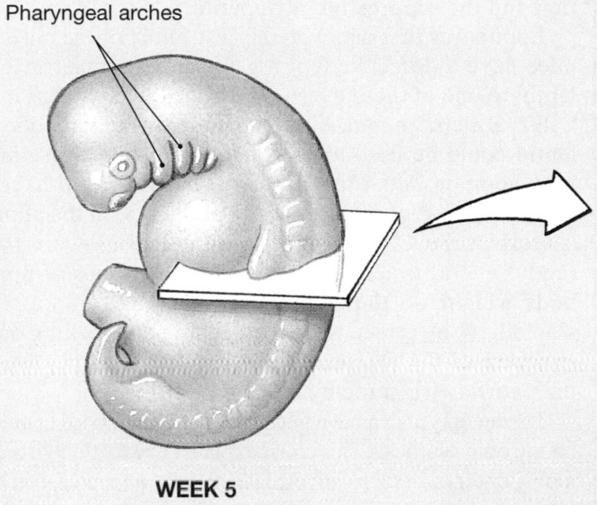

Pharyngeal arches

WEEK 5

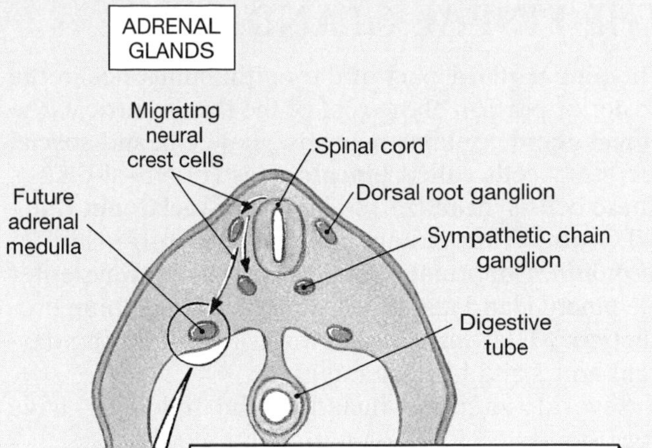

ADRENAL
GLANDS

Migrating
neural
crest cells

Spinal cord

Dorsal root ganglion

Future
adrenal
medulla

Sympathetic chain
ganglion

Digestive
tube

Each adrenal gland also has a
compound origin. Shortly after the
formation of the *neural tube*, neural
crest cells migrate away from the CNS.
This migration leads to the formation
of the dorsal root ganglia and
autonomic ganglia. On each side of the
coelomic cavity, neural crest cells
aggregate in a mass that will become
an adrenal medulla.

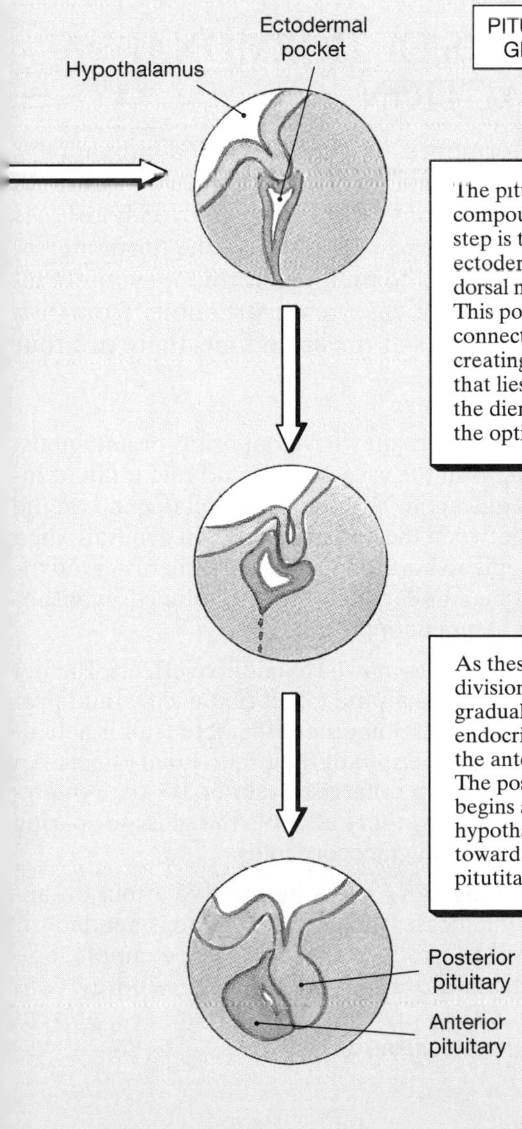

Hypothalamus

Ectodermal
pocket

PITUITARY
GLAND

The pituitary gland has a
compound origin. The first
step is the formation of an
ectodermal pocket in the
dorsal midline of the pharynx.
This pocket loses its
connection to the pharynx,
creating a hollow ball of cells
that lies inferior to the floor of
the diencephalon posterior to
the optic chiasm.

Neural crest cell mass

Lining of
coelomic cavity

Sympathetic
preganglionic fibers

Mesothelium

Adrenal medulla

Adrenal cortex

As these cells undergo
division, the central chamber
gradually disappears. This
endocrine mass will become
the anterior pituitary gland.
The posterior pituitary gland
begins as a depression in the
hypothalamic floor and grows
toward the developing anterior
pitutitary.

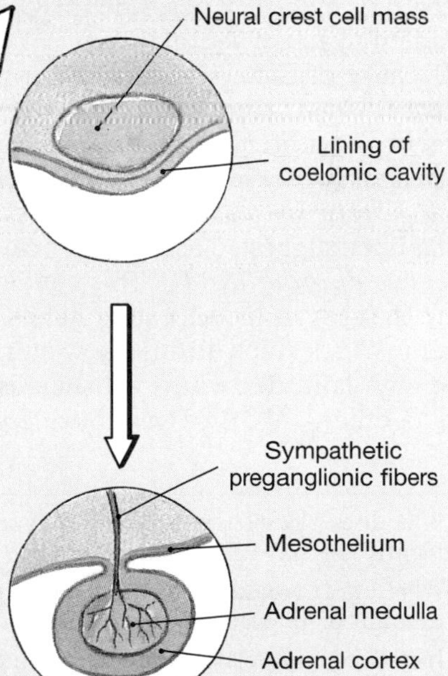

Overlying epithelial cells respond by
undergoing division, and the daughter cells
surround the neural crest cells to form a
thick adrenal cortex.

Posterior
pituitary

Anterior
pituitary

For additional details concerning the development of other
endocrine organs, refer to the Embryology Summaries in
Chapters 22, 24, 25, 26, and 27.

THE PINEAL GLAND

The **pineal gland**, part of the epithalamus, lies in the posterior portion of the roof of the third ventricle. The pineal gland contains neurons, glial cells, and special secretory cells called **pinealocytes** (pi-NĒ-al-ō-sīts). These cells synthesize the hormone **melatonin** (mel-a-TŌ-nin) from molecules of the neurotransmitter *serotonin.* Collaterals from the visual pathways enter the pineal gland and affect the rate of melatonin production. Melatonin production is lowest during daylight hours and highest at night.

Several functions, including the following, have been suggested for melatonin in humans:

- In a variety of other mammals, melatonin slows the maturation of sperm, eggs, and reproductive organs by reducing the rate of GnRH secretion. The significance of this effect remains uncertain, but there is circumstantial evidence that melatonin may play a role in the timing of human sexual maturation. For example, melatonin levels in the blood decline at puberty, and pineal tumors that eliminate melatonin production will cause premature puberty in young children.

- Melatonin is a very effective *antioxidant;* it may protect CNS neurons from free radicals, such as nitric oxide (NO) or hydrogen peroxide (H_2O_2), that may be generated in active neural tissue.

- Because of the cyclical nature of pineal activity, the pineal gland may also be involved with the establishment or maintenance of basic *circadian rhythms,* daily changes in physiological processes that follow a regular pattern. Increased melatonin secretion in darkness has been suggested as a primary cause for *seasonal affective disorder (SAD).* This condition, characterized by changes in mood, eating habits, and sleeping patterns, can develop during the winter in people who live at high latitudes, where sunshine is scarce or lacking altogether. ☐AM ☐ *Light and Behavior*

✔ Why does a person with Type I or Type II diabetes urinate frequently and have a pronounced thirst?

✔ What effect would increased levels of glucagon have on the amount of glycogen stored in the liver?

✔ Increased amounts of light would inhibit the production of which hormone?

LEPTIN: A NEW HORMONE Since 1990, several new hormones have been identified. In most cases, their structures and modes of action remain to be determined. One of the most interesting of these hormones is **leptin**. Leptin is a peptide hormone secreted by adipose tissues throughout the body. It has several functions, the best known being the feedback control of appetite. When you eat, adipose tissues absorb glucose and lipids and synthesize triglycerides for storage. At the same time, they release leptin into the circulation. Leptin binds to neurons in the CNS that are concerned with emotion and the control of appetite. The result is a sense of satiation and the suppression of appetite.

Leptin was first discovered in a strain of obese mice. The mice have a defective leptin gene, and administration of leptin to one of these overweight mice quickly turns it into a slim, athletic animal. Although there was initial hope that leptin could be used to treat human obesity, those hopes were soon dashed. Most obese people appear to have defective leptin receptors (or leptin pathways) in the appetite centers of the CNS. Their circulating leptin levels are already several times higher than in individuals of normal body weight, so the administration of additional leptin would have no effect. Researchers are now working on the structure of the receptor protein and the biochemistry of the pathway triggered by leptin binding.

Leptin may also have a permissive effect on GnRH and gonadotropin synthesis. This could explain (1) why thin girls commonly enter puberty relatively late, (2) why an increase in body fat content can improve fertility, and (3) why women stop menstruating when their body fat content becomes very low.

PATTERNS OF HORMONAL INTERACTION

Although hormones are usually studied individually, the extracellular fluids contain a mixture of hormones whose concentrations change daily or even hourly. As a result, cells never respond to only one hormone; instead, they respond to multiple hormones simultaneously. When a cell receives instructions from two different hormones at the same time, there are four possible results:

1. The two hormones may have opposing, or **antagonistic,** effects, as in the case of PTH and calcitonin or insulin and glucagon. The net result will depend on the balance between the two hormones. In general, when an antagonistic hormone is present, the observed effects will be smaller than those produced by either hormone acting unopposed.

2. The two hormones may have additive effects. The net result is greater than the effect that each would produce acting alone. Sometimes the net result is actually greater than the *sum* of their individual effects. An example of such a **synergistic** (sin-er-JIS-tik; *synairesis,* a drawing together) effect is the glucose-sparing action of GH and glucocorticoids.

3. One hormone can have a **permissive** effect on another. In such cases, the first hormone is needed for the second to produce its effect. For example, epinephrine has no apparent effect on energy consumption unless thyroid hormones are also present in normal concentrations.

4. Finally, hormones may produce different but complementary results in specific tissues and organs. These **integrative** effects are important in coordinating the activities of diverse physiological systems. The differing effects of calcitriol and parathyroid hormone on tissues involved in calcium metabolism are an example.

In this section, we shall present three examples of the ways hormones interact to produce complex, well-coordinated results. The patterns introduced will provide the background needed to understand the more detailed discussions found in chapters that deal with cardiovascular function, metabolism, excretion, and reproduction.

Hormones and Growth

Normal growth requires the cooperation of several endocrine organs. Six different hormones are especially important, although many others have secondary effects on growth. The circulating concentrations of these hormones are regulated independently. Every time the hormonal mixture changes, metabolic operations are modified to some degree. The alterations vary in duration and intensity, producing unique individual growth patterns. We will consider GH, thyroid hormones, insulin, PTH, calcitriol, and gonadal hormones:

1. *Growth hormone.* Growth hormone assists in the maintenance of normal blood glucose concentrations and the mobilization of lipid reserves stored in adipose tissues. It is not the primary hormone involved, however, and an adult with a GH deficiency but normal levels of thyroxine (T_4), insulin, and glucocorticoids will have no physiological problems.

   The effects of GH on protein synthesis and cellular growth are most apparent in children, in whom GH supports muscular and skeletal development. Undersecretion or oversecretion of GH can lead to pituitary growth failure or gigantism, disorders considered in Chapter 6. ∞ *[p. 186]*
2. *Thyroid hormones.* Normal growth also requires appropriate levels of thyroid hormones. If these hormones are absent for the first year after birth, the nervous system will fail to develop normally, and mental retardation and other signs of cretinism will result. If T_4 concentrations decline later in life but before puberty, normal skeletal development will not continue.
3. *Insulin.* Growing cells need adequate supplies of energy and nutrients. Without insulin, produced by the pancreatic islets, the passage of glucose and amino acids across cell membranes will be drastically reduced or eliminated.
4, 5. *Parathyroid hormone and calcitriol.* Parathyroid hormone and calcitriol promote the absorption of calcium salts for subsequent deposition in bone. Without adequate levels of both hormones, bones can still enlarge, but they will be poorly mineralized, weak, and flexible. For example, in *rickets,* a condition caused by inadequate calcitriol production in growing children, the limb bones are so weak that they bend under the weight of the body. ∞ *[p. 188]*
6. *Gonadal hormones.* The activity of osteoblasts in key locations and the growth of specific cell populations are affected by the presence or absence of sexual hormones (androgens in males, estrogens in females). Sex hormones stimulate cell growth and differentiation in their target tissues. The targets differ for androgens and estrogens, and the differential growth induced by these hormones accounts for gender-related differences in skeletal proportions and secondary sexual characteristics.

Hormones and Stress

Any condition within the body that threatens homeostasis is a form of **stress.** Stresses are produced by the action of *stressors,* factors that may be (1) physical, such as illness or injury; (2) emotional, such as depression or anxiety; (3) environmental, such as extreme heat or cold; or (4) metabolic, such as acute starvation. The stresses produced may be opposed by specific homeostatic adjustments. For example, a decline in body temperature will result in responses such as shivering or changes in the pattern of circulation in an attempt to restore normal body temperature.

In addition, the body has a *general* response to stress that can occur while other, more specific, responses are under way. Exposure to a wide variety of stressors will produce the same general pattern of hormonal and physiological adjustments. These responses are part of the **stress response,** also known as the **general adaptation syndrome (GAS).** The GAS, first described by Hans Selye in 1936, can be divided into three phases: (1) the *alarm phase,* (2) the *resistance phase,* and (3) the *exhaustion phase.*

The Alarm Phase

During the **alarm phase,** there is an immediate response to the stress. This response is directed by the sympathetic division of the ANS. In the alarm phase, (1) energy reserves are mobilized, mainly in the form of glucose, and (2) the body prepares to deal with the stressor by "fight or flight" responses, detailed in Chapter 16. ∞ *[p. 514]*

Epinephrine is the dominant hormone of the alarm phase, and its secretion accompanies a generalized sympathetic activation. The characteristics of the alarm phase include:

- Increased mental alertness.
- Increased energy consumption by skeletal muscles and other peripheral tissues.

- The mobilization of energy reserves (glycogen and lipids).
- Increased blood flow to skeletal muscles.
- Decreased blood flow to the skin, kidneys, and digestive organs.
- A drastic reduction in digestion and urine production.
- Increased sweat gland secretion.
- Increased blood pressure, heart rate, and respiratory rate.

Although the effects of epinephrine are most apparent during the alarm phase, other hormones play supporting roles. Sympathetic activation also triggers the release of renin, leading to the activation of angiotensin II. The reduction of water losses by circulatory changes, ADH production, and aldosterone secretion may be very important if the stress(es) involve a loss of blood.

The Resistance Phase

The temporary adjustments of the alarm phase are often sufficient to remove or overcome the stress. But some stresses, such as starvation, acute illness, or severe anxiety, can persist for hours, days, or weeks. If a stress lasts longer than a few hours, the individual will enter the **resistance phase** of the GAS.

Glucocorticoids are the dominant hormones of the resistance phase. Epinephrine, GH, and thyroid hormones play supporting roles. Energy demands in the resistance phase remain higher than normal due to the combined effects of these hormones.

Neural tissue has a high demand for energy, and neurons must have a reliable supply of glucose. If blood glucose concentrations fall too far, neural function will deteriorate. Glycogen reserves are adequate to maintain normal glucose concentrations during the alarm phase but are nearly exhausted after several hours. The endocrine secretions of the resistance phase are coordinated to achieve four integrated results:

1. *The mobilization of lipid and protein reserves.* The hypothalamus produces GH-RH and CRH, stimulating the release of GH and, by means of ACTH, the secretion of glucocorticoids. Adipose tissues respond to GH and glucocorticoids by releasing stored fatty acids. Skeletal muscles respond to glucocorticoids by breaking down proteins and releasing amino acids into the circulation.
2. *The conservation of glucose for neural tissues.* Glucocorticoids and GH from the anterior pituitary stimulate lipid metabolism in peripheral tissues. These glucose-sparing effects maintain normal blood glucose levels even after long periods of starvation. Neural tissues do not alter their metabolic activities, however, and they continue to use glucose as an energy source.

3. *The elevation and stabilization of blood glucose concentrations.* As blood glucose levels decline, glucagon and glucocorticoids stimulate the liver to manufacture glucose from other carbohydrates, glycerol from triglycerides, and amino acids provided by skeletal muscles. The glucose molecules are then released into the circulation, and blood glucose concentrations return to normal levels.
4. *The conservation of salts and water and the loss of K^+ and H^+.* Blood volume is conserved through the actions of ADH and aldosterone. As Na^+ is conserved, K^+ and H^+ are lost.

The lipid reserves of the body are sufficient to maintain the resistance phase for a period of weeks or even months. (These reserves account for the ability to endure lengthy periods of starvation.) But the resistance phase cannot be sustained indefinitely. If starvation is the primary stress, the resistance phase ends as lipid reserves are exhausted and structural proteins become the primary energy source. If starvation is not the problem, the resistance phase will end due to complications brought about by hormonal side effects. For example:

- Although the metabolic effects of glucocorticoids are essential to normal resistance, their anti-inflammatory action slows wound healing and increases susceptibility to infection.
- The continued conservation of fluids under the influence of ADH and aldosterone stresses the cardiovascular system by producing elevated blood volumes and higher-than-normal blood pressures.
- The adrenal cortex may become unable to continue producing glucocorticoids, quickly eliminating the ability to maintain acceptable blood glucose concentrations.

Poor nutrition, emotional or physical trauma, chronic illness, or damage to key organs such as the heart, liver, or kidneys will hasten the end of the resistance phase.

The Exhaustion Phase

When the resistance phase ends, homeostatic regulation breaks down and the **exhaustion phase** begins. Unless corrective actions are taken almost immediately, the failure of one or more organ systems will prove fatal.

Problems with mineral balance contribute to the existing problems with major systems. The production of aldosterone throughout the resistance phase results in a conservation of Na^+ at the expense of K^+. As the body's K^+ content declines, a variety of cells—notably neurons and muscle fibers—begin to malfunction. Although a single cause, such as heart failure, may be listed as the cause of death, the underlying problem is the inability to sustain the endocrine and metabolic adjustments of the resistance phase. Figure 18-21• summarizes the three phases of GAS.

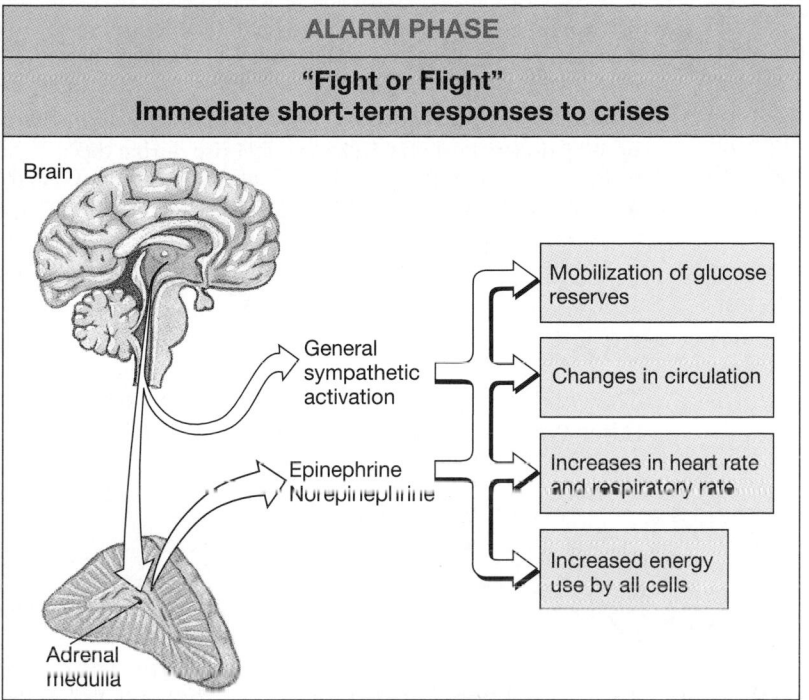

ALARM PHASE

"Fight or Flight"
Immediate short-term responses to crises

Brain

General sympathetic activation

Epinephrine
Norepinephrine

Adrenal medulla

- Mobilization of glucose reserves
- Changes in circulation
- Increases in heart rate and respiratory rate
- Increased energy use by all cells

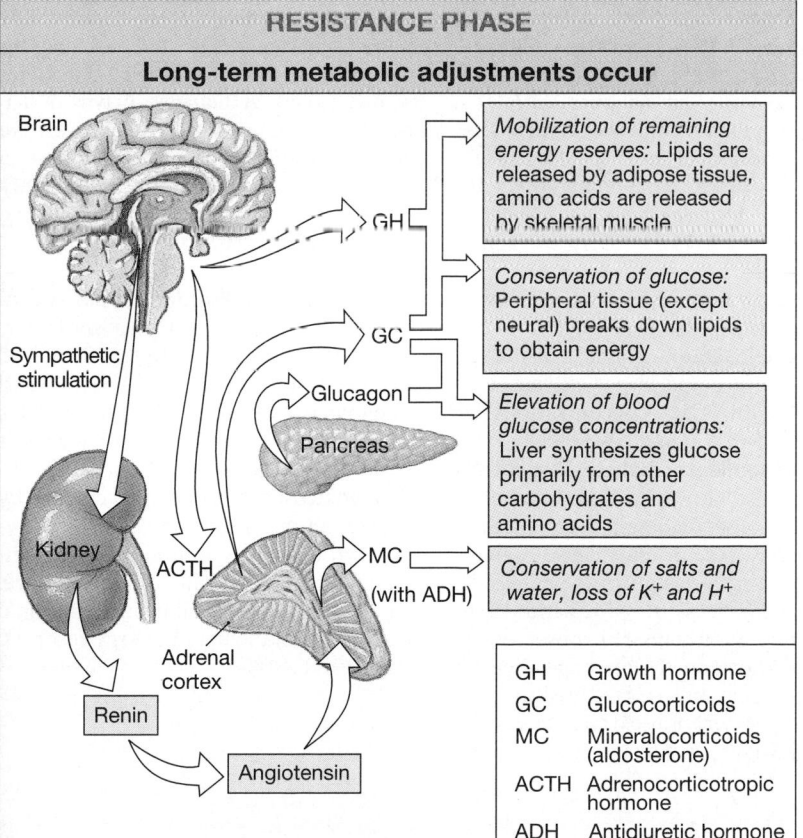

RESISTANCE PHASE

Long-term metabolic adjustments occur

Brain

Sympathetic stimulation

Kidney

ACTH

Renin

Angiotensin

Adrenal cortex

GH

GC

Glucagon

Pancreas

MC (with ADH)

- Mobilization of remaining energy reserves: Lipids are released by adipose tissue, amino acids are released by skeletal muscle
- Conservation of glucose: Peripheral tissue (except neural) breaks down lipids to obtain energy
- Elevation of blood glucose concentrations: Liver synthesizes glucose primarily from other carbohydrates and amino acids
- Conservation of salts and water, loss of K⁺ and H⁺

GH	Growth hormone
GC	Glucocorticoids
MC	Mineralocorticoids (aldosterone)
ACTH	Adrenocorticotropic hormone
ADH	Antidiuretic hormone

EXHAUSTION PHASE

Collapse of vital systems

Causes may include:
— Exhaustion of lipid reserves
— Inability to produce glucocorticoids
— Failure of electrolyte balance
— Cumulative structural or functional damage to vital organs

•*FIGURE 18-21* The General Adaptation Syndrome

Hormones and Behavior

As we have seen, many endocrine functions are regulated by the hypothalamus, and hypothalamic neurons monitor the levels of many circulating hormones. Other portions of the CNS are also quite sensitive to hormonal stimulation.

The clearest demonstrations of the effects of specific hormones involve individuals whose endocrine glands are oversecreting or undersecreting. But even normal changes in circulating hormone levels can cause behavioral changes. In *precocious* (premature) *puberty,* sex hormones are produced at an inappropriate time, perhaps as early as 5 or 6 years of age. The affected children not only begin to develop adult secondary sexual characteristics but also undergo significant behavioral changes. The "nice little kid" disappears, and the child becomes aggressive and assertive. These behavioral alterations represent the effects of sex hormones on CNS function. Thus behaviors that in normal teenagers are usually attributed to environmental stimuli, such as peer pressure, actually have some physiological basis as well. In the adult, changes in the mixture of hormones reaching the CNS can have significant effects on intellectual capabilities, memory, learning, and emotional states.

⧗ Aging and Hormone Production

The endocrine system shows relatively few functional changes with age. The most dramatic exception is the decline in the concentration of reproductive hormones. We noted the effects of these hormonal changes on the skeletal system in Chapter 6; we shall continue the discussion in Chapter 29.

Blood and tissue concentrations of many other hormones, including TSH, thyroid hormones, ADH, PTH, prolactin, and glucocorticoids, remain unchanged with advancing age. Although circulating hormone levels may remain within normal limits, however, some endocrine tissues become less responsive to stimulation. For example, in elderly individuals smaller amounts of GH and insulin are secreted after a carbohydrate-rich meal or in a glucose tolerance test. The reduction in levels of GH and other tropic hormones affects tissues throughout the body; these hormonal effects are associated with the reductions in bone density and muscle mass noted in earlier chapters.

Finally, it should be noted that age-related changes in other tissues affect their abilities to respond to hormonal stimulation. As a result, peripheral tissues may become less responsive to some hormones. This loss of sensitivity has been documented for glucocorticoids and ADH.

☑ Insulin lowers the level of glucose in the blood, and glucagon causes glucose levels to rise. What is this type of hormonal interaction called?

☑ The lack of which hormones would inhibit skeletal formation?

☑ Why do levels of GH-RH and CRH rise during the resistance phase of the general adaptation syndrome (GAS)?

INTEGRATION WITH OTHER SYSTEMS

The relationships between the endocrine system and other systems are summarized in Figure 18-22●.

SELECTED CLINICAL TERMINOLOGY

Terms Discussed in This Chapter

Addison's disease: A condition caused by hyposecretion of glucocorticoids and mineralocorticoids; characterized by an inability to mobilize energy reserves and maintain normal blood glucose levels. *(p. 617* and *AM)*

cretinism: A condition caused by hypothyroidism in infancy; marked by inadequate skeletal and nervous development, and a metabolic rate as much as 40 percent below normal levels. *(p. 612)*

Cushing's disease: A condition caused by hypersecretion of glucocorticoids; characterized by excessive breakdown and relocation of lipid reserves and proteins. *(p. 617* and *AM)*

diabetes insipidus: A disorder that develops when the posterior pituitary no longer releases adequate amounts of ADH or when the kidneys cannot respond to ADH. *(p. 607)*

diabetes mellitus (mel-Ī-tus)**:** A disorder characterized by glucose concentrations high enough to overwhelm the kidneys' reabsorption capabilities. *(p. 623* and *AM)*

diabetic retinopathy, nephropathy, neuropathy: Disorders of the retina, kidneys,

and peripheral nerves related to diabetes mellitus; the conditions most often afflict middle-aged or older diabetics. *(p. 623)*

general adaptation syndrome (GAS): The pattern of hormonal and physiological adjustments with which the human body responds to all forms of stress. *(p. 629)*

glycosuria: The presence of glucose in the urine. *(p. 623)*

goiter: An abnormal enlargement of the thyroid gland. *(p. 634* and *AM)*

hypocalcemic tetany: Muscle spasms affecting the face and upper extremities; caused by low Ca^{2+} concentrations in body fluids. *(p. 613* and *AM)*

insulin-dependent diabetes mellitus (IDDM), also known as **Type I diabetes** or **juvenile-onset diabetes:** A type of diabetes mellitus; the primary cause is inadequate insulin production by the beta cells of the pancreatic islets. *(p. 623* and *AM)*

myxedema: Symptoms of hyposecretion of thyroid hormones, including subcutaneous swelling, hair loss, dry skin, low body temperature, muscle weakness, and slowed reflexes. *(p. 612* and *AM)*

non-insulin-dependent diabetes mellitus (NIDDM), also known as **Type II diabetes** or **maturity-onset diabetes:** A type of diabetes mellitus in which insulin levels are normal or elevated but peripheral tissues no longer respond normally. *(p. 623* and *AM)*

polyuria: The production of excessive amounts of urine; a symptom of diabetes. *(p. 607)*

seasonal affective disorder (SAD): A condition characterized by depression, lethargy, an inability to concentrate, and altered sleep and eating habits. It has been linked to enhanced melatonin levels in subjects living in areas or seasons that have short periods of daylight. *(p. 628* and *AM)*

thyrotoxicosis: A condition caused by oversecretion of thyroid hormones (hyperthyroidism). Symptoms include increased metabolic rate, blood pressure, and heart rate; excitability and emotional instability; and lowered energy reserves. *(p. 612* and *AM)*

AM Additional Terms Discussed in the *Applications Manual*

exophthalmos: Protrusion of the eyes from their sockets; commonly a symptom of hypersecretion of thyroid hormones.

ketoacidosis: A condition in which large numbers of ketone bodies in the blood lead to a dangerously low blood pH.

ketone bodies: Organic acids produced during the breakdown of fatty acids.

thyrotoxic crisis: An acute state of hyperthyroidism, characterized by high

fever, rapid heart rate, and the malfunctioning of multiple organs.

INTEGUMENTARY SYSTEM

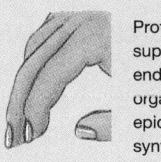

Protects superficial endocrine organs; epidermis synthesizes cholecalciferol

Sex hormones stimulate sebaceous gland activity, influence hair growth, fat distribution, and apocrine sweat gland activity; PRL stimulates development of mammary glands; adrenal hormones alter dermal blood flow, stimulate release of lipids from adipocytes; MSH stimulates melanocyte activity

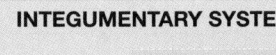

Male

Female

SKELETAL SYSTEM

Protects endocrine organs, especially in brain, chest and pelvic cavity

Skeletal growth regulated by several hormones; calcium mobilization regulated by parathyroid hormone and calcitonin; sex hormones speed growth and closure of epiphyseal plates at puberty and help maintain bone mass in adults

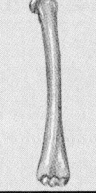

MUSCULAR SYSTEM

Skeletal muscles provide protection for some endocrine organs

Hormones adjust muscle metabolism, energy production, and growth; regulate Ca^{2+} and PO_4^{3-} levels in body fluids; speed skeletal muscle growth

NERVOUS SYSTEM

Hypothalamic hormones directly control pituitary and indirectly control secretions of other endocrine organs; controls adrenal medulla; secretes ADH and oxytocin

Several hormones affect neural metabolism; hormones help regulate fluid and electrolyte balance; reproductive hormones influence CNS development and behaviors

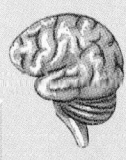

CARDIOVASCULAR SYSTEM

Circulatory system distributes hormones throughout the body; heart secretes ANP

Erythropoietin regulates production of RBCs; several hormones elevate blood pressure; epinephrine elevates heart rate and contractile force

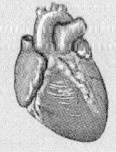

LYMPHATIC SYSTEM

Lymphocytes provide defense against infection and, with other WBCs, assist in repair after injury

Glucocorticoids have anti-inflammatory effects; thymosins stimulate development of lymphocytes; many hormones affect immune function

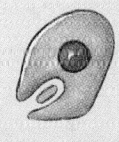

RESPIRATORY SYSTEM

Provides O_2 and eliminates CO_2 generated by endocrine cells; converting enzyme in lung capillaries converts angiotensin I to angiotensin II

Epinephrine and norepinephrine stimulate respiratory activity and dilate respiratory passageways

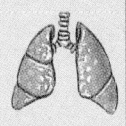

DIGESTIVE SYSTEM

Provides nutrients and substrates to endocrine cells; endocrine cells of pancreas secrete insulin and glucagon; liver produces and releases angiotensinogen

E and NE stimulate constriction of sphincters and depress activity along digestive tract; digestive tract hormones coordinate secretory activities along tract

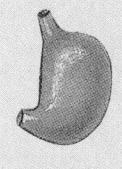

URINARY SYSTEM

Kidney cells (1) release renin and erythropoietin when local blood pressure declines and (2) produce calcitriol

Aldosterone, ADH, and ANP adjust rates of fluid and electrolyte reabsorption in kidneys

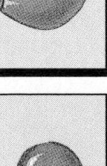

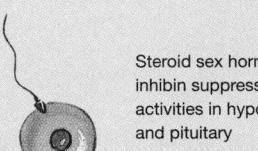

Steroid sex hormones and inhibin suppress secretory activities in hypothalamus and pituitary

Hypothalamic factors and pituitary hormones regulate sexual development and function; oxytocin stimulates uterine and mammary gland smooth muscle contractions

REPRODUCTIVE SYSTEM

● **FIGURE 18-22** **Functional Relationships between the Endocrine System and Other Systems**

Endocrine Disorders

Endocrine disorders fall into two basic categories: They reflect either abnormal hormone production or abnormal cellular sensitivity. The symptoms are interesting because they highlight the significance of normally "silent" hormonal contributions. The characteristic features of many of these conditions are illustrated in Figure 18-23●, and an abbreviated summary is presented in Table 18-8.

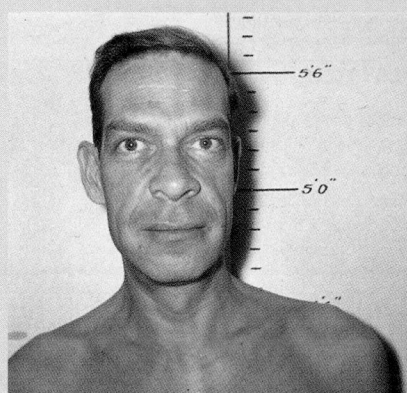

(a) *Acromegaly* results from the over-production of growth hormone after the epiphyseal plates have fused. Bone shapes change, and cartilaginous areas of the skeleton enlarge. Note the broad facial features and the enlarged lower jaw.

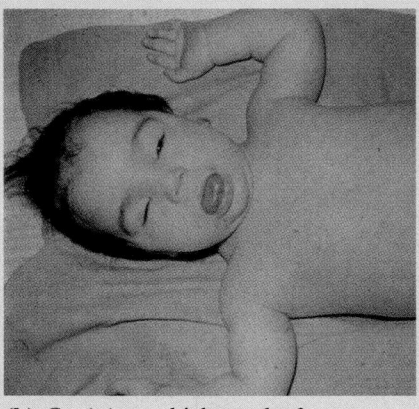

(b) *Cretinism*, which results from thyroid hormone insufficiency in infancy.

(c) An enlarged thyroid gland, or *goiter*, is usually associated with thyroid hyposecretion due to iodine insufficiency.

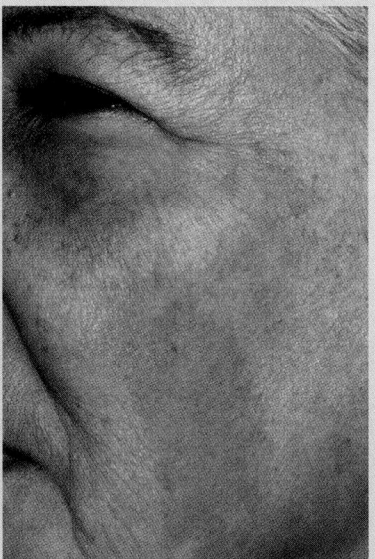

(d) *Addison's disease* is caused by hyposecretion of corticosteroids, especially glucocorticoids. Pigment changes result from stimulation of melanocytes by ACTH, which is structurally similar to MSH.

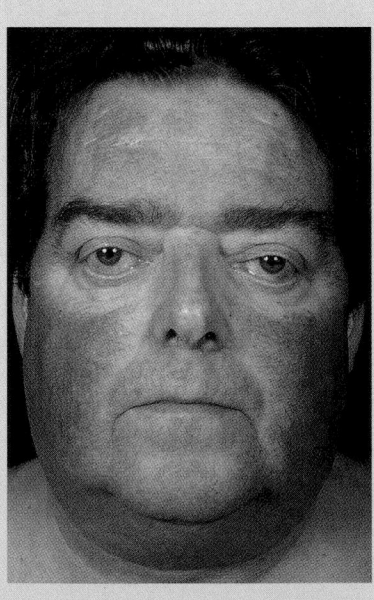

(e) *Cushing's disease* is caused by hypersecretion of glucocorticoids. Lipid reserves are mobilized, and adipose tissue accumulates in the cheeks and at the base of the neck.

●**FIGURE 18-23** **Endocrine Abnormalities**

TABLE 18-8 **Clinical Implications of Endocrine Malfunctions**

Hormone	Underproduction Syndrome	Principal Symptoms	Overproduction Syndrome	Principal Symptoms
Growth hormone (GH)	Pituitary growth failure *(pp. 606, 629)*	Retarded growth, abnormal fat distribution, low blood glucose hours after a meal	Gigantism, acromegaly *(pp. 629, 606 and AM)*	Excessive growth
Antidiuretic hormone (ADH) or vasopressin (AVP)	Diabetes insipidus *(p. 607)*	Polyuria, dehydration, thirst	SIADH (syndrome of inappropriate ADH secretion) *(AM)*	Increased body weight and water content
Thyroxine (T_4), triiodothyronine (T_3)	Myxedema, cretinism *(p. 612 and AM)*	Low metabolic rate, body temperature; impaired physical and mental development	Hyperthyroidism, Graves' disease *(p. 612 and AM)*	High metabolic rate and body temperature
Parathyroid hormone (PTH)	Hypopara-thyroidism *(p. 613 and AM)*	Muscular weakness, neurological problems, formation of dense bones, tetany due to low blood Ca^{2+} concentrations	Hyperparathyroidism *(p. 613 and AM)*	Neurological, mental, muscular problems due to high blood Ca^{2+} concentrations; weak and brittle bones
Insulin	Diabetes mellitus (Type I) *(p. 623)*	High blood glucose, impaired glucose utilization, dependence on lipids for energy; glycosuria	Excess insulin production or administration	Low blood glucose levels, possibly causing coma
Mineralocorticoids (MCs)	Hypoaldosteronism *(p. 616 and AM)*	Polyuria, low blood volume, high blood K^+, low blood Na^+ concentrations	Aldosteronism *(p. 616 and AM)*	Increased body weight due to Na^+ and water retention; low blood K^+ concentration
Glucocorticoids (GCs)	Addison's disease *(p. 617 and AM)*	Inability to tolerate stress, mobilize energy reserves, or maintain normal blood glucose concentrations	Cushing's disease *(p. 617 and AM)*	Excessive breakdown of tissue proteins and lipid reserves; impaired glucose metabolism
Epinephrine (E), norepinephrine (NE)	None identified		Pheochromocytoma *(AM)*	High metabolic rate, body temperature, and heart rate; elevated blood glucose levels
Estrogens (females)	Hypogonadism *(p. 624)*	Sterility; lack of secondary sexual characteristics	Adrenogenital syndrome *(p. 617)*	Overproduction of androgens by zona reticularis of adrenal cortex leads to masculinization
			Precocious puberty *(p. 631)*	Premature sexual maturation and related behavioral changes
Androgens (males)	Hypogonadism *(p. 624)*	Sterility, lack of secondary sexual characteristics	Adrenogenital syndrome (gynecomastia) *(p. 617)*	Abnormal production of estrogen, sometimes due to adrenal or interstitial cell tumors; leads to breast enlargement
			Precocious puberty *(p. 631)*	Premature sexual maturation and related behavioral changes

Endocrinology and Athletic Performance

Medical research involving humans is generally subject to tight ethical constraints and meticulous scientific scrutiny. Yet a clandestine, unscientific, and potentially quite dangerous program of "experimentation" with hormones is being pursued today by athletes in many countries. The use of hormones to improve performance is banned by the International Olympic Committee, the U.S. Olympic Committee, the National Collegiate Athletic Association, and the National Football League, and it is condemned by the American Medical Association and the American College of Sports Medicine. A significant number of amateur and professional athletes, however, persist in this dangerous practice. Synthetic forms of testosterone are used most often, but athletes may use any combination of testosterone, GH, EPO, and a variety of synthetic hormones.

Androgen Abuse. The use of androgens, or *anabolic steroids*, has become popular with many athletes, both amateur and professional. The goal of steroid use is to increase muscle mass, endurance, and "competitive spirit." It has been suggested that as many as 30 percent of college and professional athletes and 10–20 percent of male high school athletes may be using anabolic steroids (with or without GH) to improve their performance. Among bodybuilders, the proportion who use steroids in the United States may be as high as 80 percent.

One supposed justification for this practice has been the unfounded opinion that compounds manufactured in the body are not only safe but are good for you. In reality, the administration of natural or synthetic androgens in abnormal amounts carries unacceptable health risks. Androgens affect many tissues in a variety of ways. Known complications include (1) premature epiphyseal closure, (2) various liver dysfunctions (including jaundice and hepatic tumors), (3) prostate enlargement and urinary tract obstruction, and (4) testicular atrophy and infertility. A link to heart attacks, impaired cardiac function, and strokes has also been suggested.

Moreover, the normal regulation of androgen production involves a feedback mechanism comparable to that described for adrenal steroids earlier in this chapter. GnRH stimulates the production of LH, and LH stimulates the secretion of testosterone and other androgens by the interstitial cells of the testes. Circulating androgens, in turn, inhibit the production of both GnRH and LH. Thus, when synthetic androgens are administered in high doses, they can (1) suppress the normal production of testosterone and (2) depress the manufacture of GnRH by the hypothalamus. *This suppression of GnRH release may be permanent.*

The use of androgenic "bulking agents" by female bodybuilders may not only add muscle mass but may alter muscular proportions and secondary sexual characteristics as well. For example, women taking steroids can develop irregular menstrual periods and changes in body hair distribution (including baldness). Finally, androgen abuse may cause a generalized depression of the immune system.

EPO Abuse. Because it is now being synthesized by recombinant DNA techniques, EPO is readily available. Athletes engaged in endurance sports, such as cycling or marathon running, may use it to boost the number of oxygen-carrying red blood cells in circulation. Although this effect improves the oxygen content of the blood, it also makes the blood denser, and the heart must work harder to push it around the circulatory system. Between 1991 and 1994 the deaths of 18 young, otherwise healthy European cyclists were attributed to heart failure related to EPO abuse.

GHB and Clenbuterol. Androgens and EPO are known hormones with reasonably well understood effects. Because drug testing is now widespread in amateur and professional sports, people interested in "getting an edge" are experimenting with drugs not easily detected by standard tests. The long-term and short-term effects of these drugs are difficult to predict. Two examples are the recent use of *GHB* and *clenbuterol* by amateur athletes.

Gamma-hydroxybutyrate (GHB) was tested for use as an anesthetic 30 years ago. It was rejected, in part because it was linked to petit mal and grand mal seizures. In 1990, this drug appeared in health-food stores, where it was sold as an anabolic agent and diet aid. During a 5-month period in 1990, 16 cases of severe reaction to GHB were treated in San Francisco alone. Symptoms included confusion, hallucinations, seizures, and coma at doses from 0.25 teaspoons to 4 tablespoons.

Clenbuterol, sometimes used to treat asthma, mimics epinephrine and stimulates β_2 receptors. This effect increases the diameter of the respiratory passageways and accelerates blood flow through active skeletal muscles. Although it is also rumored to have anabolic properties comparable to those of androgens, there is no evidence to support that notion. Clenbuterol abuse is reportedly widespread, although exact numbers are difficult to obtain. Heavy usage can cause severe headaches, tremors, insomnia, and potentially dangerous abnormal heartbeats. During the 1992 Olympics in Barcelona, Spain, two American athletes were disqualified because they tested positive for this drug.

CHAPTER REVIEW

On-line resources for this chapter are on our World Wide Web site at:
http://www.prenhall.com/martini/fap

STUDY OUTLINE

INTRODUCTION, p. 591

1. In general, the nervous system performs short-term "crisis management," whereas the endocrine system regulates longer-term, ongoing metabolic processes. Endocrine cells release chemicals called **hormones,** which alter the metabolic activities of many different tissues and organs simultaneously. *(Figure 18-1; Table 18-1)*

AN OVERVIEW OF THE ENDOCRINE SYSTEM, p. 592

Hormone Structure, p. 592

1. Hormones can be divided into three groups on the basis of chemical structure: *amino acid derivatives* are structurally similar to amino acids; **peptide hormones** are chains of amino acids; and *lipid derivatives* include **steroid hormones** and **eicosanoids.** *(Figures 18-2, 18-3)*

Hormone Distribution and Transport, p. 594

2. Hormones may circulate freely or bound to transport proteins. Free hormones are rapidly removed from the circulation.

Hormone Function and Mechanisms of Action, p. 594

3. Hormones exert their effects by modifying the activities of *target cells*. Receptors for **catecholamines** and peptide hormones are located on the cell membranes of target cells; the hormone acts as a **first messenger,** which causes a **second messenger** to appear in the cytoplasm. Thyroid and steroid hormones cross the cell membrane and bind to receptors in the cytoplasm or nucleus. The hormone–receptor complex activates or inactivates specific genes. *(Figures 18-4, 18-5)*

Control of Endocrine Activity, p. 599

4. **Endocrine reflexes** are the functional counterparts of neural reflexes. *(Figures 18-6, 18-7)*
5. The hypothalamus regulates the activities of the nervous and endocrine systems by three mechanisms: (1) It secretes **regulatory hormones,** which control the activities of endocrine cells in the pituitary gland; (2) its autonomic centers exert direct neural control over the endocrine cells of the adrenal medullae; and (3) it acts as an endocrine organ by releasing hormones into the circulation at the posterior pituitary. *(Figure 18-7)*

THE PITUITARY GLAND, p. 601

1. The **pituitary gland** releases nine important peptide hormones; all bind to membrane receptors and use **cyclic-AMP** as a second messenger. *(Figure 18-8; Table 18-2)*

The Anterior Pituitary, p. 602

2. The **anterior pituitary gland (adenohypophysis)** can be subdivided into the **pars distalis,** the **pars intermedia,** and the **pars tuberalis.** *(Figure 18-8)*
3. At the *median eminence*, neurons release regulatory factors into the surrounding interstitial fluids. Their secretions enter the circulation easily, because the capillaries in this area are **fenestrated.** *(Figure 18-9)*
4. The **hypophyseal portal system** ensures that all the blood entering the portal vessels will reach the intended target cells before it returns to the general circulation. *(Figure 18-9)*

5. **Thyroid-stimulating hormone (TSH)** triggers the release of thyroid hormones. *Thyrotropin-releasing hormone (TRH)* promotes the secretion of TSH.
6. **Adrenocorticotropic hormone (ACTH)** stimulates the release of *glucocorticoids* by the adrenal gland. **Corticotropin-releasing hormone (CRH)** causes the secretion of ACTH.
7. **Follicle-stimulating hormone (FSH)** stimulates follicle development and estrogen secretion in women and sperm production in men. **Luteinizing hormone (LH)** causes ovulation and **progestin** production in women and **androgen** production in men. **Gonadotropin-releasing hormone (GnRH)** promotes the secretion of both FSH and LH.
8. **Prolactin (PRL),** with other hormones, stimulates both the development of the mammary glands and milk production. Prolactin secretion is inhibited by **prolactin-inhibiting hormone (PIH)** and stimulated by a *prolactin-releasing factor.* *(Figure 18-10)*
9. **Growth hormone (GH, or somatotropin)** stimulates cell growth and replication through the release of **somatomedins** from liver cells and has direct metabolic effects. The production of GH is regulated by **growth hormone–releasing hormone (GH-RH)** and **growth hormone–inhibiting hormone (GH-IH).** *(Figure 18-10)*
10. In many nonhuman vertebrates, **melanocyte-stimulating hormone (MSH),** released by the pars intermedia, stimulates melanocytes to produce melanin, but MSH is not normally secreted by the nonpregnant adult human.

The Posterior Pituitary, p. 607

11. The **posterior pituitary (neurohypophysis)** contains the axons of hypothalamic neurons. Neurons of the **supraoptic** and **paraventricular nuclei** manufacture **antidiuretic hormone (ADH)** and **oxytocin**, respectively. ADH decreases the amount of water lost at the kidneys and in higher concentrations elevates blood pressure through vasoconstriction. In women, **oxytocin** stimulates contractile cells in the mammary glands and has a stimulatory effect on uterine and prostatic smooth muscles. *(Figure 18-11)*

THE THYROID GLAND, p. 609

1. The **thyroid gland** lies near the *thyroid cartilage* of the larynx and consists of two **lobes** connected by a narrow **isthmus.** *(Figure 18-12)*

Thyroid Follicles and Thyroid Hormones, p. 610

2. The thyroid gland contains numerous **thyroid follicles.** Thyroid follicles release several hormones, including **thyroxine (T_4)** and **triiodothyronine (T_3).** *(Figures 18-12, 18-13; Table 18-4)*
3. The follicle cells manufacture **thyroglobulin** and store it as a colloid. *(Figure 18-13)*
4. Most of the thyroid hormones entering the bloodstream are attached to special **thyroid-binding globulins (TBGs),** and the rest are attached to **transthyretin** or to albumin. The unbound hormones affect peripheral tissues immediately; the binding proteins gradually release their hormones over a week or more. *(Figure 18-13; Table 18-3)*
5. Thyroid hormones exert a **calorigenic effect,** which enables us to adapt to cold temperatures.

The C Cells of the Thyroid Gland: Calcitonin, p. 613

6. The **C cells** of the follicles produce **calcitonin (CT),** which helps regulate Ca^{2+} concentrations in body fluids, especially during childhood and pregnancy. *(Figures 18-6, 18-12, 18-15; Table 18-4)*

THE PARATHYROID GLANDS, p. 613

Parathyroid Hormone Secretion, p. 613

1. Four **parathyroid glands** are embedded in the posterior surface of the thyroid gland. The **chief cells** of the parathyroid produce **parathyroid hormone (PTH)** in response to lower-than-normal concentrations of Ca^{2+}. The parathyroid glands, aided by calcitriol from the kidneys, are the primary regulators of Ca^{2+} levels in healthy adults. *(Figures 18-14, 18-15; Table 18-4)*

THE THYMUS, p. 614

1. The **thymus** produces several hormones, collectively known as **thymosins.** The thymosins play a role in developing and maintaining normal immunological defenses. *(Table 18-4)*

THE ADRENAL GLANDS, p. 614

1. A single **adrenal** *(suprarenal)* **gland** lies along the superior border of each kidney. Each gland is surrounded by a fibrous capsule. The gland can be subdivided into the superficial **adrenal cortex** and the inner **adrenal medulla.** *(Figure 18-16)*

The Adrenal Cortex, p. 614

2. The adrenal cortex manufactures steroid hormones called **adrenocortical steroids (corticosteroids).** The cortex can be subdivided into three areas: (1) the **zona glomerulosa,** which releases **mineralocorticoids (MCs),** principally **aldosterone,** which restrict Na^+ losses at the kidneys, sweat glands, digestive tract, and salivary glands; (2) the **zona fasciculata,** which produces **glucocorticoids (GCs),** notably **cortisol, corticosterone,** and **cortisone** (all of which exert a *glucose-sparing effect* on peripheral tissues); and (3) the **zona reticularis,** which produces androgens of uncertain significance. *(Figure 18-16; Table 18-5)*

The Adrenal Medullae, p. 617

3. The adrenal medullae produce epinephrine (75–80 percent of medullary secretions) and norepinephrine (20–25 percent.) *(Figure 18-16; Table 18-5)*

THE KIDNEYS, p. 618

1. Endocrine cells in the kidneys produce the hormones calcitriol and erythropoietin and the enzyme renin. These secretions are important for the regulation of calcium metabolism, blood volume, and blood pressure.

Calcitriol, p. 618

2. **Calcitriol** stimulates calcium (Ca^{2+}) and phosphate ($PO_4{}^{3-}$) ion absorption along the digestive tract. *(Figure 18-17)*

Erythropoietin, p. 618

3. **Erythropoietin (EPO)** stimulates red blood cell production by the bone marrow. *(Figure 18-17)*

Renin, p. 620

4. **Renin** converts **angiotensinogen** to **angiotensin I.** In the capillaries of the lungs, *angiotensin-converting enzyme* (ACE) converts this compound to **angiotensin II,** the hormone that (1) stimulates the adrenal production of aldos-terone and the pituitary release of ADH, (2) promotes thirst, and (3) produces vasoconstriction. *(Figure 18-17)*

THE HEART, p. 620

1. Specialized muscle cells in the heart produce **atrial natriuretic peptide (ANP)** when blood volume becomes excessive. *(Figure 18-18)*

ENDOCRINE TISSUES OF THE DIGESTIVE SYSTEM, p. 621

1. The linings of the digestive tract, the liver, and the pancreas produce exocrine secretions that are essential to the normal breakdown and absorption of food. The digestive tract also produces numerous hormones important to the coordination of digestive activities.

The Pancreas, p. 621

2. The **pancreas** contains both exocrine and endocrine cells. The **exocrine pancreas** secretes an enzyme-rich fluid that travels to the lumen of the digestive tract. Cells of the **endocrine pancreas** form clusters called **pancreatic islets** *(islets of Langerhans).* These islets contain **alpha cells** (which secrete the hormone **glucagon), beta cells** (which secrete **insulin), delta cells** (which secrete **somatostatin),** and **F cells** (which secrete **pancreatic polypeptide).** *(Figure 18-19; Table 18-6)*

3. Insulin lowers blood glucose by increasing the rate of glucose uptake and utilization; glucagon raises blood glucose by increasing the rates of glycogen breakdown and glucose manufacture in the liver. Somatostatin suppresses glucagon and insulin release and slows the rates of food absorption and enzyme secretion in the digestive tract. Pancreatic polypeptide inhibits gallbladder contractions and regulates the production of some pancreatic enzymes. *(Figure 18-20)*

ENDOCRINE TISSUES OF THE REPRODUCTIVE SYSTEM, p. 624

The Testes, p. 624

1. The **interstitial cells** of the testes produce androgens. **Testosterone** is the most important sex hormone in males.

The Ovaries, p. 625

2. In women, *oocytes* develop in **follicles;** follicle cells produce **estrogens,** especially **estradiol.** After ovulation, the remaining follicle cells reorganize into a **corpus luteum.** Those follicle cells release a mixture of estrogens and **progestins,** especially **progesterone.** *(Table 18-7)*

THE PINEAL GLAND, p. 628

1. The **pineal gland** contains **pinealocytes,** which synthesize **melatonin.** Although research and debate continue, melatonin appears to (1) slow the maturation of sperm, oocytes, and reproductive organs by reducing the rate of GnRH secretion and (2) establish daily *circadian rhythms.*

PATTERNS OF HORMONAL INTERACTION, p. 628

1. The endocrine system functions as an integrated unit, and hormones often interact. These interactions may exert several effects: (1) **antagonistic** (opposing) effects; (2) **synergistic** (additive) effects; (3) **permissive** effects, in which one hormone is necessary for another to produce its effect; or (4) **integrative** effects, in which hormones produce different but complementary results.

Hormones and Growth, p. 629

2. Normal growth requires the cooperation of several endocrine organs. Six hormones are especially important: GH, thyroid hormones, insulin, PTH, calcitriol, and gonadal hormones.

Hormones and Stress, p. 629

3. Any condition that threatens homeostasis represents a **stress.** Our bodies respond to a variety of *stressors* by the **stress response,** or **general adaptation syndrome (GAS).** The GAS can be divided into three phases: (1) the **alarm phase** (an immediate, "fight or flight" response, under the direction of the sympathetic division of the ANS); (2) the **resistance phase,** dominated by glucocorticoids (mobilization of lipid and protein reserves, elevation and stabilization of blood glucose concentrations, and conservation of glu-

cose for neural tissues); and (3) the **exhaustion phase,** the eventual breakdown of homeostatic regulation and failure of one or more organ systems. *(Figure 18-21)*

Hormones and Behavior, p. 631

4. Many hormones affect the nervous system, producing changes in mood, emotional states, and behavior.

Aging and Hormone Production, p. 631

5. The endocrine system shows few functional changes with advanced age. The chief change is a decline in the concentration of reproductive hormones.

INTEGRATION WITH OTHER SYSTEMS, p. 632

1. The endocrine system interacts with each of the other body systems. *(Figure 18-22)*

REVIEW QUESTIONS

Level 1   Reviewing Facts and Terms

1. Peptide chains synthesized at ribosomes of the RER are
   - (a) prehormones
   - (b) prohormones
   - (c) hormones
   - (d) liposomes

2. Cyclic-AMP functions as a second messenger to
   - (a) build proteins and catalyze specific reactions
   - (b) activate adenylate cyclase
   - (c) open ion channels and activate key enzymes in the cytoplasm
   - (d) bind the hormone-receptor complex to DNA segments

3. Adrenocorticotropic hormone (ACTH) stimulates the release of
   - (a) thyroid hormones by the hypothalamus
   - (b) gonadotropins by the adrenal glands
   - (c) somatotropins by the hypothalamus
   - (d) steroid hormones by the adrenal glands

4. FSH production in males supports
   - (a) maturation of sperm by stimulating sustentacular cells
   - (b) development of muscles and strength
   - (c) production of male sex hormones
   - (d) increased desire for sexual activity

5. The hormone that induces ovulation in women and promotes the ovarian secretion of progesterone is
   - (a) interstitial cell–stimulating hormone
   - (b) estradiol
   - (c) luteinizing hormone
   - (d) prolactin

6. The two hormones released by the posterior pituitary are
   - (a) somatotropin and gonadotropin
   - (b) estrogen and progesterone
   - (c) GH and prolactin
   - (d) ADH and oxytocin

7. The primary function of ADH is to
   - (a) increase the amount of water lost at the kidneys
   - (b) decrease the amount of water lost at the kidneys
   - (c) dilate peripheral blood vessels to decrease blood pressure
   - (d) increase absorption along the digestive tract

8. The element required for normal thyroid function is
   - (a) magnesium
   - (b) calcium
   - (c) potassium
   - (d) iodine

9. Reduced fluid losses in the urine due to retention of Na^+ and water is a result of the action of
   - (a) antidiuretic hormone
   - (b) calcitonin
   - (c) aldosterone
   - (d) cortisone

10. The adrenal medulla produces the hormones
    - (a) cortisol and cortisone
    - (b) epinephrine and norepinephrine
    - (c) corticosterone and testosterone
    - (d) androgens and progesterone

11. Hormones released by the kidneys include
    - (a) calcitriol and erythropoietin
    - (b) ADH and aldosterone
    - (c) epinephrine and norepinephrine
    - (d) cortisol and cortisone

12. What four events are set in motion by activation of a G-protein due to the binding of a hormone to its receptor site?

13. What three higher-level mechanisms are involved in integrating activities of the nervous and endocrine systems?

14. Which seven hormones are released by the anterior pituitary gland?

15. Which two hormones are released by the posterior pituitary gland?

16. What five primary effects result from the action of thyroid hormones?

17. What effects do calcitonin and parathyroid hormone have on blood calcium levels?

18. What three zones make up the adrenal cortex, and what kinds of hormones are produced by each zone?

19. What peripheral effects does activation of the adrenal medulla have on alpha and beta receptors?

20. Which two hormones are released by the kidneys, and what is the importance of each hormone?

21. What are the three opposing effects of atrial natriuretic peptide and angiotensin II?

22. What four cell populations make up the endocrine pancreas? What hormone does each cell type produce?

23. (a) What three phases of the general adapation syndrome (GAS) constitute the body's response to stress?
    (b) Which endocrine secretions play dominant roles in the alarm and resistance phases?

Level 2 | Reviewing Concepts

24. What is the primary difference in the way the nervous and endocrine systems communicate with their target cells?

25. How can a hormone modify the activities of its target cells?

26. What possible results occur when a cell receives instructions from two different hormones at the same time?

27. What is an endocrine reflex? Compare endocrine and neural reflexes.

28. How would blocking the activity of phosphodiesterase affect a cell that responds to hormonal stimulation by the cAMP second messenger system?

Level 3 | Critical Thinking and Clinical Applications

29. Roger has been suffering from extreme thirst; he drinks numerous glasses of water every day and urinates a great deal. Name two disorders that could produce these symptoms. What test could a clinician perform to determine which disorder is present?

30. Julie is pregnant and is not receiving prenatal care. She has a poor diet consisting mostly of fast food. She drinks no milk, preferring colas instead. How would this situation affect Julie's level of parathyroid hormone?

31. Sherry tells her physician that she has been restless and irritable lately. Sherry has a hard time sleeping and complains of diarrhea and shortness of breath. During the examination, her physician notices a higher-than-normal heart rate and decreased circulating levels of cholesterol. What tests would you suggest to help the physician make a positive diagnosis of Sherry's condition?

32. Patients receiving steroid hormone frequently retain large amounts of water. Why?

19

Blood

O *ut of sight, out of mind. That old saying certainly applies to blood. As long as we don't see*
it, we don't think or worry much about it. But looking at vials or bags of it in a hospital—or, worse,
watching even a drop of our own oozing from a cut or scrape—makes most of us feel distinctly uneasy, some
people faint at the sight. This anxiety probably comes from being reminded of how dependent we are on our
relatively small supply of the precious fluid. If you have three half-gallon milk containers in your refrigerator
at home, you probably have more milk than you do blood. You can easily spare a drop of blood or even
donate a pint, but if you suddenly lose more than about 15 to 20 percent of your blood supply, you will need
to get more in a hurry. Fortunately, blood transfusions are now fairly routine matters—but, as we will see in
this chapter, not just anyone's blood will do. We'll also explore the many functions of this vital substance.

CHAPTER OUTLINE AND OBJECTIVES

Small embryos don't need cardiovascular systems, because diffusion across their exposed surfaces provides adequate oxygen and removes waste products as rapidly as they are generated. By the time a human embryo reaches a few millimeters in length, however, developing tissues are consuming oxygen and nutrients faster than they can be provided by simple diffusion. At that stage, the cardiovascular system must begin functioning to provide a rapid-transport system for oxygen, nutrients, and waste products. It is the first system to become fully operational: The heart begins beating at the end of the third week of embryonic life, when many other systems have barely begun their development.

When the heart starts beating, blood begins circulating. The embryo can now make more efficient use of the nutrients obtained from the maternal bloodstream, and its size doubles in the next week. In the adult, circulating blood provides nutrients, oxygen, chemical instructions, and a mechanism for waste removal to each of the roughly 75 trillion individual cells in the human body. The blood also transports specialized cells that defend peripheral tissues from infection and disease. These services are absolutely essential, and a region completely deprived of circulation may die in a matter of minutes.

We can compare the cardiovascular system to the cooling system of a car. The basic components include a circulating fluid (blood), a pump (the heart), and an assortment of conducting pipes (the blood vessels). Although the cardiovascular system is far more complicated and versatile, both mechanical and biological systems can suffer from fluid losses, pump failures, or damaged pipes.

In the next three chapters, we shall examine the individual components of the cardiovascular system. This chapter deals with the nature of the circulating blood, Chapter 20 covers the structure and function of the heart, and Chapter 21 examines the organization of blood vessels and the integrated functioning of the cardiovascular system.

The cardiovascular system is intimately bound up with the *lymphatic system*. The two systems are interconnected and interdependent. Fluid leaves the bloodstream, enters the tissues, and returns to the blood within the vessels of the lymphatic system. For this reason, the cardiovascular and lymphatic systems are often said to be components of a single *circulatory system*. The bloodstream carries cells, antibodies, and cytokines of the lymphatic system throughout the body, and the cardiovascular system assists the lymphatic system in defending the body against invasion by pathogens. In Chapter 22, we detail the components of the lymphatic system and consider the array of defenses that protect the body from internal and external hazards.

FUNCTIONS OF BLOOD

In this chapter, we examine the structure and functions of blood, a specialized connective tissue that contains cells suspended in a fluid matrix. ∞ *[p. 126]* The functions of blood include the following:

- *The transportation of dissolved gases, nutrients, hormones, and metabolic wastes.* Blood carries oxygen from the lungs to the tissues and carbon dioxide from the tissues to the lungs. Blood distributes the nutrients absorbed at the digestive tract or released from storage in adipose tissue or the liver. It carries hormones from endocrine glands toward their target tissues, and the wastes produced by tissue cells are absorbed by the blood and carried to the kidneys for excretion.
- *The regulation of the pH and electrolyte composition of interstitial fluids throughout the body.* The blood absorbs and neutralizes the acids generated by active tissues, such as the lactic acid produced by skeletal muscles.
- *The restriction of fluid losses through damaged vessels or at other injury sites.* Blood contains enzymes and other substances that respond to breaks in the vessel walls by initiating the process of *blood clotting*. A blood clot acts as a temporary patch and prevents further blood loss.
- *Defense against toxins and pathogens.* Blood transports *white blood cells,* specialized cells that migrate into peripheral tissues to fight infections or remove debris. It also delivers *antibodies,* special proteins that attack invading organisms or foreign compounds.
- *Stabilization of body temperature.* Blood absorbs the heat generated by active skeletal muscles and redistributes it to other tissues. When body temperature is already high, that heat will be lost across the surface of the skin. If body temperature is too low, the warm blood is directed to the brain and to other temperature-sensitive organs.

COMPOSITION OF BLOOD

Blood has a characteristic and unique composition (Figure 19-1•). It is a fluid connective tissue with an extracellular matrix called **plasma** (PLAZ-muh). Plasma consists of *plasma proteins* and a ground substance called *serum*. The plasma proteins are in solution rather than forming the insoluble fibers found in other connective tissues, such as loose connective tissue or cartilage. Because the plasma proteins are in solution, plasma is slightly denser than water.

Formed elements are blood cells and cell fragments that are suspended in the plasma. **Red blood cells (RBCs),** or **erythrocytes** (e-RITH-rō-sīts; *erythros,* red + *cyte,* cell) are the most abundant blood cells. These specialized cells are essential for the transport of oxygen in the blood. The less numerous **white blood cells**

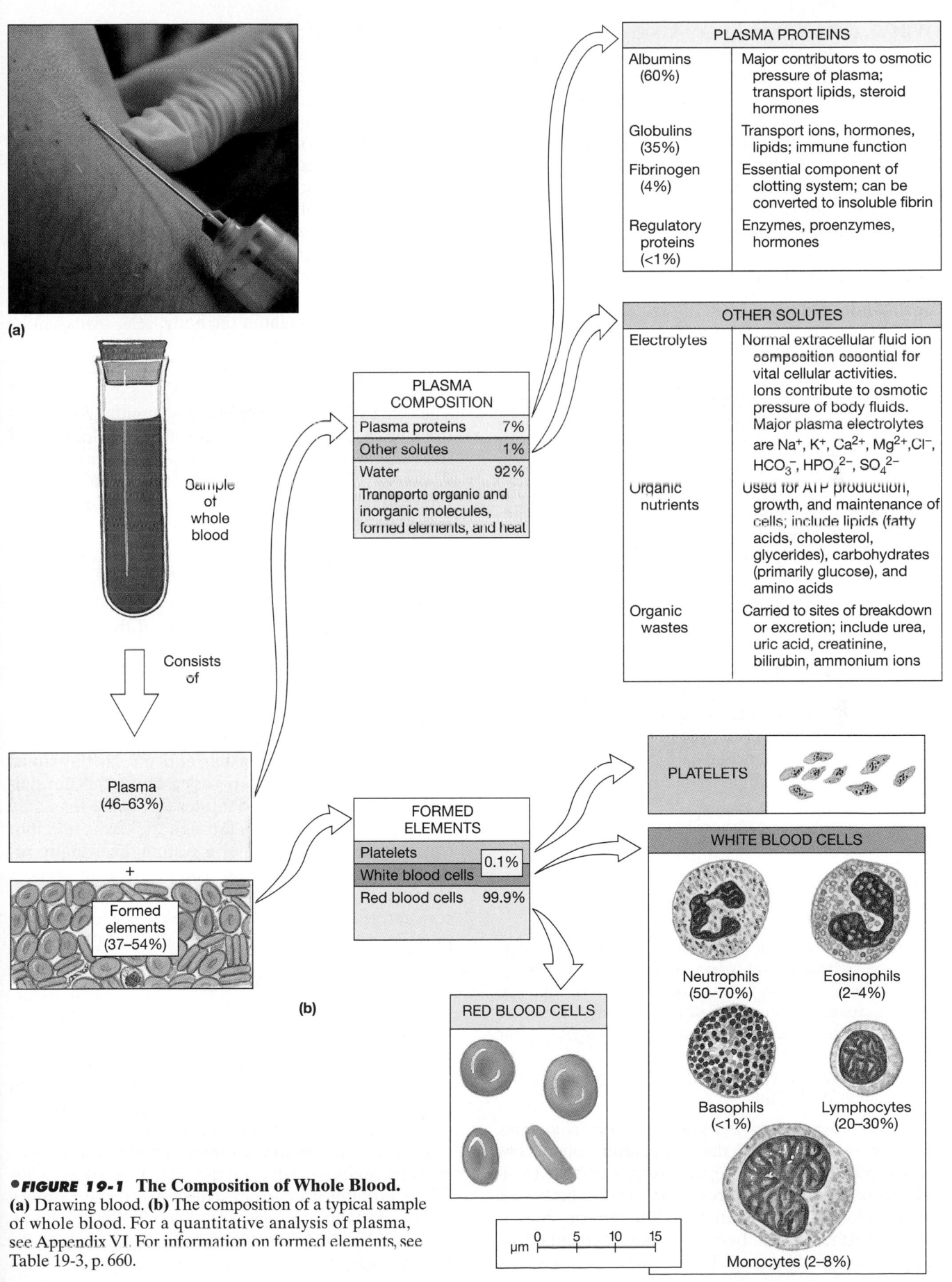

•FIGURE 19-1 **The Composition of Whole Blood.**
(a) Drawing blood. **(b)** The composition of a typical sample
of whole blood. For a quantitative analysis of plasma,
see Appendix VI. For information on formed elements, see
Table 19-3, p. 660.

(WBCs), or **leukocytes** (LOO-kō-sīts; *leukos,* white) are cells involved with the body's defense mechanisms. **Platelets** are small, membrane-bound packets of cytoplasm that contain enzymes and other substances important to the process of blood clotting.

Together, the plasma and formed elements constitute **whole blood.** These components may be **fractionated,** or separated, for analytical or clinical purposes; we shall encounter examples of uses for fractionated blood later in the chapter.

Blood Collection and Analysis

Fresh whole blood is generally collected from a superficial vein, such as the *median cubital vein* on the anterior surface of the elbow (Figure 19-1a●). This procedure is called **venipuncture** (VĒN-i-punk-chur; *vena,* vein + *punctura,* a piercing). Venipuncture is a common sampling technique because (1) superficial veins are easy to locate; (2) the walls of veins are thinner than those of comparably sized arteries; and (3) blood pressure in the venous system is relatively low, so the puncture wound seals quickly. The most common clinical procedures examine venous blood.

Blood from peripheral capillaries can be obtained by puncturing the tip of a finger, the lobe of an ear, or (in infants) the great toe or heel. A small drop of capillary blood can be used to prepare a *blood smear,* a thin film of blood on the surface of a microscope slide. The blood smear can then be stained with special dyes to show the different types of formed elements.

An **arterial puncture,** or "arterial stick," may be used for checking the efficiency of gas exchange at the lungs. Samples are generally drawn from the *radial artery* at the wrist or the *brachial artery* at the elbow.

Whole blood from any of these sources—venous blood, blood from peripheral capillaries, or arterial blood—has the same basic physical characteristics:

- *Temperature.* Blood temperature is roughly 38°C (100.4°F), slightly higher than normal body temperature.
- *Viscosity.* Blood is five times as viscous as water; in other words, blood is five times stickier, more cohesive, and resistant to flow than water. The high viscosity results from interactions among dissolved proteins, formed elements, and the surrounding water molecules in plasma.
- *pH.* Blood pH ranges from 7.35 to 7.45, averaging 7.4—slightly alkaline.

The cardiovascular system of an adult man contains 5–6 liters of whole blood; that of an adult woman contains 4–5 liters. The gender differences in blood volume primarily reflect differences in average body size. Blood volume in liters can be estimated for an individual of either gender by calculating 7 percent of the body weight in kilograms. For example, a 75-kg individual would have a blood volume of approximately 5.25 liters, or about 1.4 gallons. More precise determinations of blood volume can be obtained by using special dyes or radioisotopes, such as phosphorus-32. Clinicians use the terms **hypovolemic** (hī-pō-vō-LĒ-mik), **normovolemic** (nor-mō-vō-LĒ-mik), and **hypervolemic** (hī-per-vō-LĒ-mik) to refer to low, normal, and excessive blood volumes, respectively. These conditions are potentially dangerous because variations in blood volume affect other components of the cardiovascular system. For example, an abnormally large blood volume can place a severe stress on the heart, which must keep the extra fluid circulating through the lungs and throughout the body. [AM] *Transfusions*

PLASMA

The composition of whole blood is summarized in Figure 19-1b●. Plasma contributes some 46–63 percent of the volume of whole blood, with water accounting for 92 percent of the plasma volume. Together, plasma and interstitial fluid account for most of the volume of extracellular fluid (ECF) in the body.

Differences between Plasma and Interstitial Fluid

In many respects, the composition of the plasma resembles that of interstitial fluid. The concentrations of the major plasma ions, for example, are similar to those of interstitial fluid and differ markedly from those inside living cells. This similarity is understandable, as there is a continuous exchange of water, ions, and small solutes between the plasma and interstitial fluids across the walls of capillaries. The capillaries normally deliver more liquid and solutes to a tissue than they remove. The excess flows through the tissue and into the vessels of the lymphatic system, eventually returning to the bloodstream. The primary differences between plasma and interstitial fluid involve (1) the concentrations of dissolved proteins, because plasma proteins cannot cross capillary walls, and (2) the levels of respiratory gases (oxygen and carbon dioxide), due to the respiratory activities of tissue cells.

Plasma Proteins

Plasma contains significant quantities of dissolved proteins. On average, there are 7.6 g (0.3 oz) of protein in each 100 ml of plasma, almost five times the concentration in interstitial fluid. The large size and globular shapes of most blood proteins prevent them from crossing capillary walls, and they remain trapped within the circulatory system. There are three primary classes of plasma proteins: (1) *albumins* (al-BŪ-minz), (2) *globulins* (GLOB-ū-linz), and (3) *fibrinogen* (fī-BRIN-ō-jen).

Albumins, Globulins, and Fibrinogen

Albumins constitute roughly 60 percent of the plasma proteins. As the most abundant plasma proteins, they are major contributors to the osmotic pressure of the plasma. Albumins are also important in the transport of fatty acids, thryoid hormones, some steroid hormones, and other substances.

Globulins account for approximately 35 percent of the protein population. Examples of important plasma globulins include *immunoglobulins* and *transport globulins*. **Immunoglobulins** (i-mū-nō-GLOB-ū-linz), also called **antibodies,** attack foreign proteins and pathogens. We shall describe several different classes of immunoglobulins in Chapter 22.

Transport globulins bind small ions, hormones, or compounds that might otherwise be lost at the kidneys or have very low solubility in water. Important examples of transport globulins include the following:

- *Thyroid-binding globulin* and *transthyretin,* which transport thyroid hormones; *transcortin,* which transports ACTH; and *transcalciferin,* which transports calcitriol.
- *Metalloproteins* (me-tal-ō-PRŌ-tēnz), which transport metal ions. *Transferrin,* for example, is a metalloprotein that transports iron (Fe^{2+}).
- *Apolipoproteins,* which carry triglycerides and other lipids in the blood. When bound to lipids, an apolipoprotein becomes a **lipoprotein** (lī-pō-PRŌ-tēn).
- *Steroid binding proteins,* which transport steroid hormones in the blood. For example, *testosterone-binding globulin* (TeBG) binds and transports testosterone.

The third type of plasma protein, **fibrinogen,** functions in blood clotting. Fibrinogen normally accounts for roughly 4 percent of plasma proteins. Under certain conditions, fibrinogen molecules interact, forming large, insoluble strands of **fibrin** (FĪ-brin). These fibers provide the basic framework for a blood clot. If steps are not taken to prevent clotting, the conversion of fibrinogen to fibrin will occur in a sample of plasma. This conversion removes the clotting proteins, leaving a fluid known as **serum.** The clotting process also removes calcium ions and other materials from solution. As a result, there are several significant differences between plasma and serum. Because the normal range of values may differ, the results of a blood test generally indicate whether the sample source is plasma (P) or serum (S).

Other Plasma Proteins

Peptide hormones, including insulin, prolactin (PRL), and the glycoproteins thyroid-stimulating hormone (TSH), follicle-stimulating hormone (FSH), and luteinizing hormone (LH), are normally present in the circulating blood. ∞ *[p. 593]* Their plasma concentrations rise and fall from day to day or even hour to hour.

Origins of the Plasma Proteins

The liver synthesizes and releases more than 90 percent of the plasma proteins, including all of the albumins and fibrinogen, most of the globulins, and the prohormone *angiotensinogen.* Immunoglobulins (antibodies) are produced by *plasma cells.* Plasma cells are derived from lymphocytes, the primary cells of the lymphatic system. Peptide hormones are produced in a variety of endocrine organs, as we detailed in Chapter 18.

Because the liver is the primary source of plasma proteins, liver disorders can alter the composition and functional properties of the blood. For example, some forms of liver disease can lead to uncontrolled bleeding due to inadequate synthesis of fibrinogen and other proteins involved in blood clotting.

PLASMA EXPANDERS *Plasma expanders* can be used to increase blood volume temporarily, over a period of hours, while preliminary lab work is under way to determine blood type. These solutions contain large carbohydrate molecules, rather than proteins, to maintain proper osmolarity. (We noted the emergency use of the carbohydrate dextran in sodium chloride solutions in Chapter 3. ∞ *[p. 76]*) Although these carbohydrates are not metabolized, they are gradually removed from the circulation by phagocytes, and the blood volume steadily declines. Plasma expanders are easily stored, and their sterile preparation ensures that there are no problems with viral or bacterial contamination. Isotonic electrolyte solutions such as normal saline may also be used, alone or in combination with concentrated purified albumin. Although they provide a temporary solution to hypovolemia, plasma expanders do not increase the amount of oxygen carried by the blood, as that function is performed by red blood cells.

☑ Why is venipuncture a common technique for obtaining a blood sample?

☑ What would be the effects of a decrease in the amount of plasma proteins?

☑ Which plasma protein would you expect to be elevated during a viral infection?

FORMED ELEMENTS

As we noted earlier in this chapter, the major cellular components of blood are red blood cells and white blood cells. Blood also contains noncellular formed elements called *platelets,* small, membrane-enclosed packets of cytoplasm that function in blood clotting.

Hemopoiesis

Formed elements are produced through the process of **hemopoiesis** (hēm-ō-poy-Ē-sis), or *hematopoiesis.* Blood cells appear in the circulation during the third week of embryonic development. These cells divide repeatedly, increasing in number. The vessels of the embryonic yolk sac are the primary site of blood formation for the first 8 weeks of development. As other organ systems appear, some of the embryonic blood cells move out of the circulation and into the liver, spleen, thymus, and bone marrow. These embryonic cells differentiate into **stem cells,** which produce blood cells by their divisions. The liver and spleen are the primary sites of hemopoiesis from the second to fifth months of development. As the skeleton enlarges, the bone marrow becomes increasingly important, and it predominates after the fifth developmental month. In the adult, it is the only site of red blood cell production as well as the primary site of white blood cell formation.

Hemocytoblasts, or *pleuripotent stem cells*, divide to produce *myeloid stem cells* and *lymphoid stem cells*. These stem cells remain capable of division, but the daughter cells they generate are involved with the production of specific types of blood cells. Lymphoid stem cells are responsible for lymphocyte production, whereas myeloid stem cells are responsible for the production of all other kinds of formed elements. We will consider the fates of the myeloid and lymphoid stem cells as we discuss the formation of each type of formed element.

Red Blood Cells

Red blood cells (RBCs) contain the red pigment *hemoglobin*, which binds and transports oxygen and carbon dioxide. RBCs are the most abundant blood cells, accounting for 99.9 percent of the formed elements. They give whole blood its deep red color.

Abundance of RBCs

A standard blood test reports the number of RBCs per **microliter** (μl) of whole blood. One microliter, or *cubic millimeter* (mm^3), of whole blood from a man contains 4.5–6.3 million erythrocytes; a microliter from a woman contains 4.2–5.5 million. There are approximately 260 million RBCs in a single drop of whole blood, and 25 trillion in the blood of an average adult. The number of erythrocytes thus accounts for roughly one-third of the cells in the human body.

The **hematocrit** (he-MA-tō-krit) is the percentage of whole blood occupied by cellular elements. The normal hematocrit in an adult man averages 46 (range: 40–54); the average for an adult woman is 42 (range: 37–47). The difference in hematocrit between males and females reflects the fact that androgens stimulate red blood cell production, whereas estrogens do not.

The hematocrit is determined by centrifuging a blood sample so that all the formed elements will come out of suspension. Whole blood contains roughly 1000 red blood cells for each white blood cell. After centrifugation, the white blood cells and platelets form a very thin *buffy coat* above a thick layer of RBCs. Because the hematocrit value is due almost entirely to the volume of erythrocytes, hematocrits are commonly reported as the **volume of packed red cells (VPRC)** or simply the **packed cell volume (PCV).**

Many different factors can alter the hematocrit. For example, the hematocrit increases in cases of dehydration, owing to a reduction in plasma volume, or after *EPO (erythropoietin)* stimulation. (We introduced EPO in Chapter 18.) ∞ *[p. 618]* The hematocrit may decrease due to internal bleeding or to problems with RBC formation. As a result, the hematocrit alone does not provide specific diagnostic information. Yet a change in hematocrit is an indication that other, more specific tests are needed. (We will consider some of those tests later in the chapter.) AM *Polycythemia*

Structure of RBCs

Red blood cells (Figure 19-2•) are among the most specialized cells of the body. A red blood cell is very different from the "typical cell" we discussed in Chapter 3. Each red blood cell is a biconcave disc with a thin central region and a thicker outer margin. An average erythrocyte has a diameter of 7.8 μm and a maximum thickness of 2.6 μm, although the center narrows to about 0.8 μm. (Normal ranges of measurement are given in Figure 19-2d•.)

This unusual shape has three important effects on RBC function:

1. It gives each RBC a relatively large surface area. A large surface area is important because the RBC carries oxygen bound to intracellular proteins. That oxygen must be absorbed or released quickly as the RBC passes through the capillaries of the lungs or peripheral tissues. The larger the surface area, the faster the exchange between the interior of the cell and the surrounding plasma. The total surface area of the red blood cells in the blood of a typical adult is approximately 3800 square meters, roughly 2000 times the total surface area of the body.

2. It enables them to form stacks, like dinner plates, that smooth the flow through narrow blood vessels. These stacks, called **rouleaux** (roo-LŌ; little rolls; singular, *rouleau*), form and dissociate repeatedly without affecting the cells involved. An entire rouleau can pass along a blood vessel only slightly larger than the diameter of a single erythrocyte, whereas individual cells would bump the walls, bang together, and form logjams that could restrict or prevent blood flow.

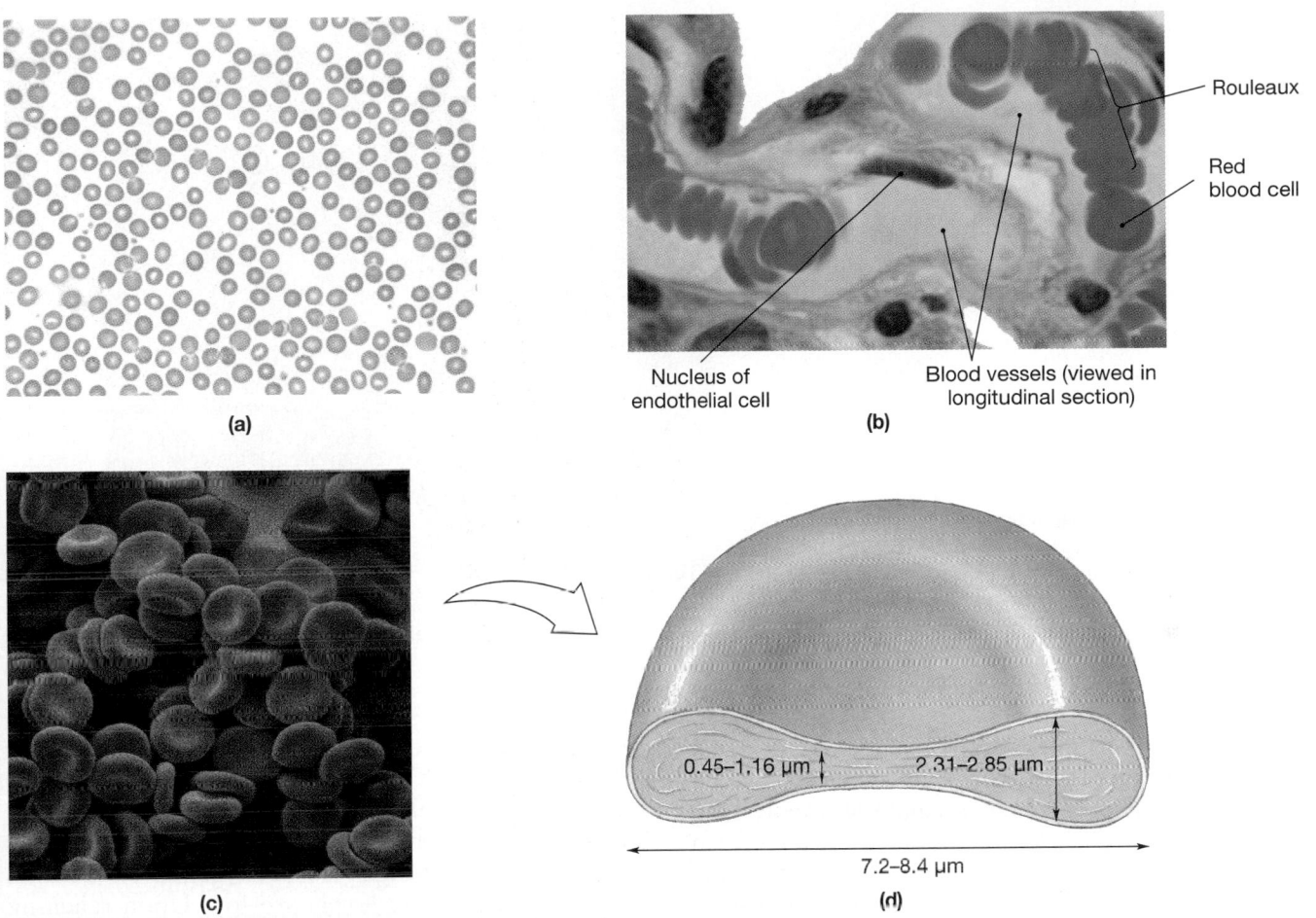

Rouleaux

Red
blood cell

Nucleus of
endothelial cell

Blood vessels (viewed in
longitudinal section)

(a)

(b)

(c)

0.45–1.16 µm ↕   2.31–2.85 µm

7.2–8.4 µm

(d)

● *FIGURE 19-2* **Anatomy of Red Blood Cells.** **(a)** When viewed in a standard blood smear, red blood cells appear as two-dimensional objects, because they are flattened against the surface of the slide. (LM × 320) **(b)** When traveling through relatively narrow capillaries, erythrocytes may stack like dinner plates, forming rouleaux. (LM × 1072) **(c)** A scanning electron micrograph of red blood cells reveals their three-dimensional structure quite clearly. (× 1195) **(d)** A sectional view of a mature red blood cell, showing average dimensions.

3. It enables them to bend and flex when entering small capillaries and branches. RBCs are very flexible, and, by changing shape, individual red blood cells can squeeze through capillaries as narrow as 4 µm.

During their differentiation, the red blood cells of humans and other mammals lose most of their organelles, including nuclei; these cells retain only the cytoskeletal elements. (The RBCs of nonmammalian vertebrates are nucleated cells.) Because they lack nuclei and ribosomes, your circulating RBCs are incapable of dividing and of synthesizing structural proteins or enzymes. As a result, RBCs cannot perform repairs, and their life span is relatively short—normally less than 120 days. Their energy demands are low. In the absence of mitochondria, they obtain the energy they need through the anaerobic metabolism of glucose absorbed from the surrounding plasma. The lack of mitochondria ensures that absorbed oxygen will be carried to peripheral tissues, not "stolen" by mitochondria in the cell.

Hemoglobin

In effect, a developing erythrocyte loses any intracellular component not directly associated with its primary function—the transport of respiratory gases. Molecules of **hemoglobin** (HĒ-mō-glō-bin) **(Hb)** account for over 95 percent of the intracellular proteins. The hemoglobin content of whole blood is reported in terms of grams of Hb per 100 ml of whole blood (g/dl). Normal ranges are 14–18 g/dl in males and 12–16 g/dl in females. Hemoglobin is responsible for the cell's ability to transport oxygen and carbon dioxide.

HEMOGLOBIN STRUCTURE Each Hb molecule has a complex quaternary shape (Figure 19-3●). There are two *alpha (α) chains* and two *beta (β) chains* of polypeptides in each Hb molecule. Each individual chain is a globular protein subunit that resembles the myoglobin in skeletal and cardiac muscle cells. Like myoglobin, each of the Hb chains contains a single molecule of

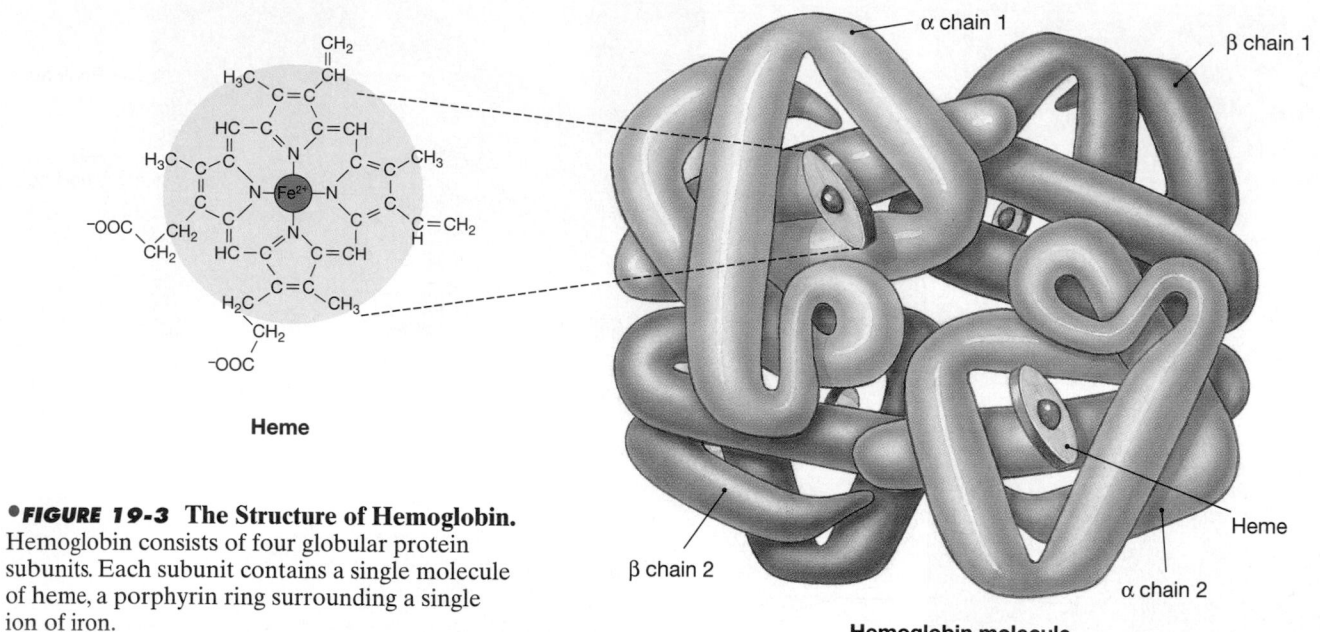

Heme

•FIGURE 19-3 **The Structure of Hemoglobin.**
Hemoglobin consists of four globular protein
subunits. Each subunit contains a single molecule
of heme, a porphyrin ring surrounding a single
ion of iron.

Hemoglobin molecule

heme, a pigment complex shown in Figure 19-3•. Heme
is a **porphyrin**, an organic compound generally associated with metal ions. Each heme unit holds an iron ion
in such a way that the iron can interact with an oxygen
molecule, forming *oxyhemoglobin*, HbO_2. The iron–
oxygen interaction is very weak; the two can easily be
separated without damaging the heme unit or the oxygen molecule. The binding of an oxygen molecule to the
iron in a heme unit is therefore completely reversible.

The red blood cells of an embryo or fetus contain a
different form of hemoglogin, known as *fetal hemoglobin,* which has a higher affinity for oxygen than does the
hemoglobin of adults. Because fetal hemoglobin binds
oxygen more readily than does adult hemoglobin, the
developing fetus can "steal" oxygen from the maternal
bloodstream at the placenta. The conversion from fetal
hemoglobin to the adult form begins shortly before birth
and continues over the next year. The production of fetal
hemoglobin can be stimulated in adults by the administration of drugs such as *hydroxyurea* or *butyrate*. This is
one method of treatment for conditions, such as *sickle
cell anemia* or *thalassemia*, that result from the production of abnormal forms of adult hemoglobin.

HEMOGLOBIN FUNCTION There are approximately 280 million Hb molecules in each red blood cell.
Because a Hb molecule contains four heme units, each
erythrocyte can potentially carry more than a billion
molecules of oxygen. Roughly 98.5 percent of the oxygen carried by the blood travels through the circulation bound to Hb molecules inside red blood cells.

The amount of oxygen bound to hemoglobin depends
primarily on the oxygen content of the plasma. When
plasma oxygen levels are low, hemoglobin releases

oxygen. Under these conditions, typical of peripheral
capillaries, plasma carbon dioxide levels are elevated.
The alpha and beta chains of hemoglobin then bind
carbon dioxide, forming **carbaminohemoglobin.** In the
capillaries of the lungs, plasma oxygen levels are high
and carbon dioxide levels are low. Upon reaching
these capillaries, RBCs absorb oxygen, which is then
bound to hemoglobin, and release carbon dioxide.
(These processes are detailed in Chapter 23.)

Normal activity levels can be sustained only when
tissue oxygen levels are kept within normal limits. If
the hematocrit is low or the hemoglobin content of
the RBCs is reduced, the condition of **anemia** exists.
Anemia produces clinical symptoms because it interferes with oxygen delivery to peripheral tissues.
Every system is affected as organ function deteriorates due to oxygen starvation. Anemic individuals
become weak, lethargic, and often confused as the
brain is affected. There are many different forms of
anemia; we will consider specific examples both in
this chapter and in the *Applications Manual.*

RBC Life Span and Circulation

An erythrocyte is exposed to severe mechanical
stresses. A single round-trip from the heart, through
the peripheral tissues, and back to the heart usually
takes less than a minute. In that time, a red blood
cell gets pumped out of the heart and forced along
vessels, where it bounces off the walls and collides
with other RBCs. It stacks in rouleaux, contorts and
squeezes through tiny capillaries, and then joins its
comrades in a headlong rush back to the heart for
another round.

Abnormal Hemoglobin

There are several inherited disorders character-ized by the production of abnormal hemoglo-bin. Two of the best known are *thalassemia* and *sickle cell anemia (SCA)*.

The various forms of **thalassemia** (thal-ah-SĒ-mē-uh) result from an inability to produce adequate amounts of alpha or beta chains. As a result, the rate of RBC production is slowed, and the mature RBCs are fragile and short-lived. The scarcity of healthy RBCs reduces the oxygen-carrying capacity of the blood and leads to problems with the development and growth of systems throughout the body. Individuals with severe thalassemia must periodically undergo *transfusions* —the administration of blood components—to keep adequate numbers of RBCs in the circulation. AM *Thalassemia; Transfusions*

Sickle cell anemia results from a mutation affecting the amino acid sequence of the beta chains of the Hb molecule. When the blood contains abundant oxygen, the Hb molecules and the erythrocytes that carry them appear normal (Figure 19-4a●). But when the defective hemoglobin gives up enough of its bound oxygen, the adjacent Hb molecules interact, and the cells change shape, becoming stiff and curved (Figure 19-4b●). This "sickling" makes the erythrocytes fragile and easily damaged. Moreover, when an RBC that has folded to squeeze into a narrow capillary delivers its oxygen to the surrounding tissue, the cell can become stuck as sickling occurs. A circulatory blockage results, and nearby tissues become oxygen-starved. (More information on this condition and new treatment options are included in the *Applications Manual*.) AM *Sickle Cell Anemia*

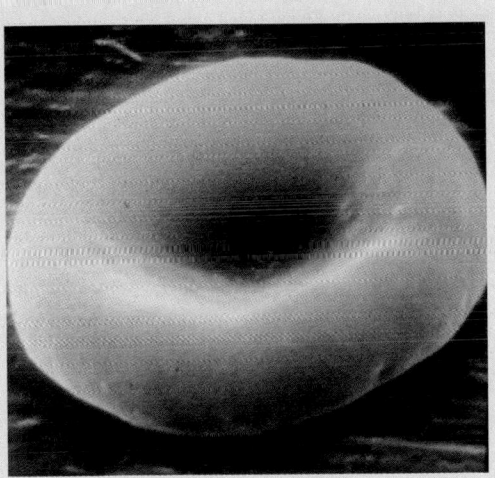
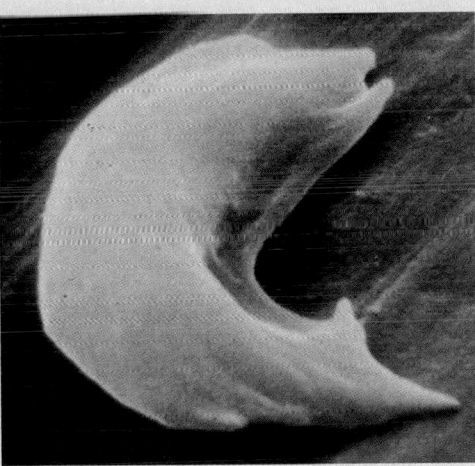

(a) Normal RBC　　**(b) Sickled RBC**

●**FIGURE 19-4** Sickling in Red Blood Cells. **(a)** When fully oxygenated, the cells of an individual with the sickling trait appear relatively normal. **(b)** At lower oxygen concentrations, the RBCs change shape, becoming relatively rigid and sharply curved. (SEM × 6750)

With all this wear and tear and no repair mecha-nisms, a typical RBC has a relatively short life span. After traveling about 700 miles in 120 days, either the cell membrane ruptures or the damage is detected by phagocytic cells, and the cell is engulfed. The continu-ous elimination of RBCs is usually unnoticed because new erythrocytes are entering the circulation at a com-parable rate. About 1 percent of the circulating ery-throcytes are replaced each day, and in the process approximately 3 million new erythrocytes enter the circulation *each second!*

HEMOGLOBIN CONSERVATION AND RECYCLING
Phagocytic cells of the liver, spleen, and bone marrow monitor the condition of circulating erythrocytes; the cells generally recognize and engulf erythrocytes be-fore they **hemolyze**, or rupture. These phagocytes also detect and remove Hb molecules and cell fragments from the relatively small number of RBCs that he-molyze in the bloodstream (about 10 percent of the total recycled each day).

If the hemoglobin released by hemolysis is not phagocytized, its components will not be recycled. Hemoglobin remains intact only under the condi-tions found inside RBCs. When hemolysis occurs, the Hb molecule breaks down, and the alpha and beta chains are filtered by the kidneys and excreted in the urine. When abnormally large numbers of eryth-rocytes break down in the circulation, the urine may develop a reddish or brown coloration. This condition is called **hemoglobinuria. Hematuria** (hē-ma-TOO-rē-uh), the presence of intact RBCs in the urine, occurs only after kidney damage or damage to vessels along the urinary tract.

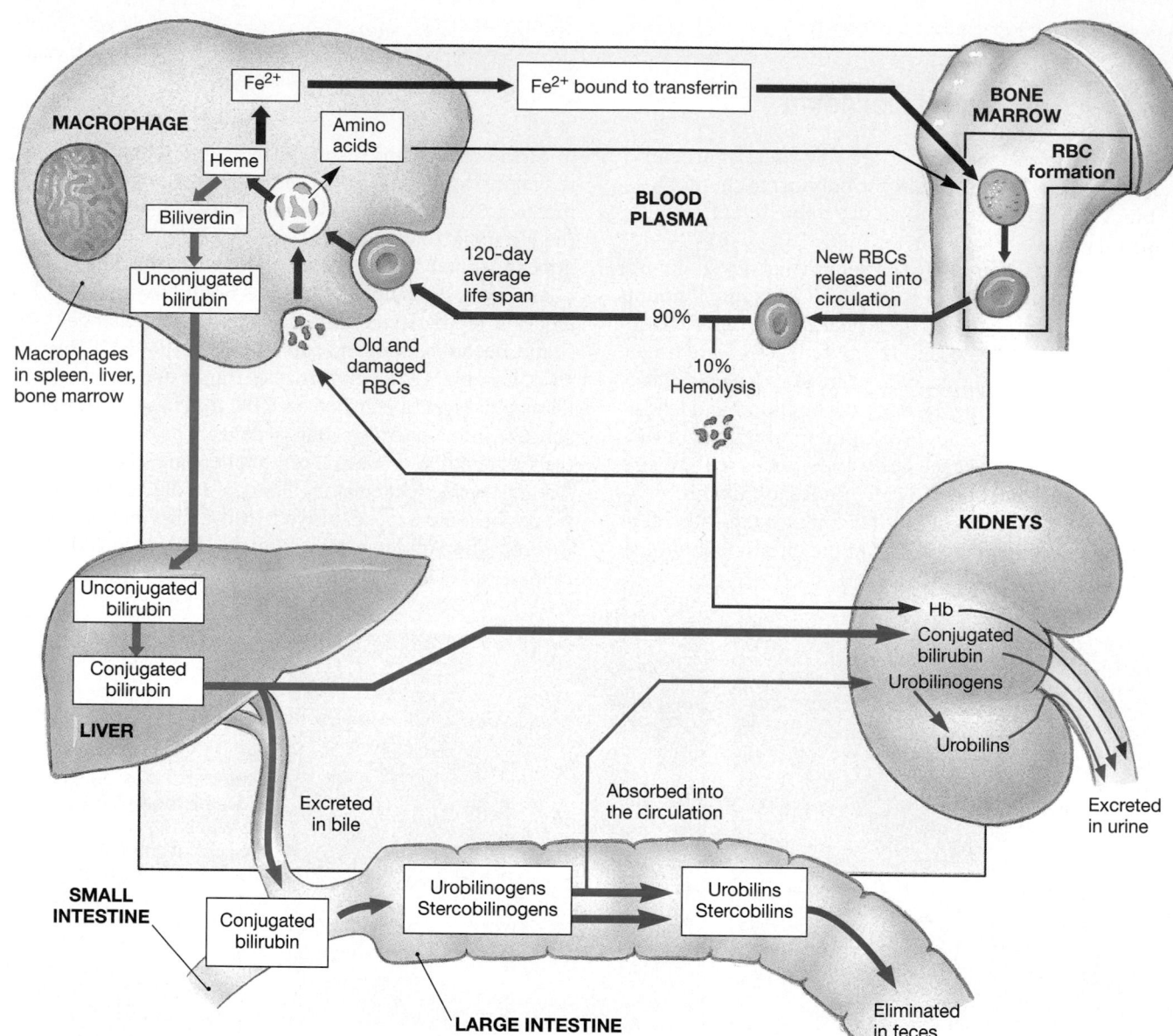

● *FIGURE 19-5* **Red Blood Cell Turnover.** This diagram shows the normal pathways for recycling amino acids and iron from aging or damaged red blood cells. The amino acids are absorbed, especially by developing cells in the bone marrow. The iron is stored in many different sites. The porphyrins of the heme units are converted to bilirubin, absorbed by the liver, and subsequently excreted in the bile or urine; there is some recirculation of the breakdown products produced in the large intestine.

Once a red blood cell has been engulfed and broken down by a phagocytic cell, each component of the Hb molecule has a different fate (Figure 19-5●). The globular proteins are disassembled into their component amino acids. These amino acids are either metabolized by the cell or released into the circulation for use by other cells. Each heme unit is stripped of its iron and converted to **biliverdin** (bil-ē-VER-din), a porphyrin derivative with a green color. (Bad bruises commonly develop a green tint due to biliverdin formation in the blood-filled tissues.) Biliverdin is then converted to **bilirubin** (bil-ē-ROO-bin), which has an orange-yellow color, and released into the circulation. There, the bilirubin binds to albumin and is transported to the liver.

Liver cells absorb the bilirubin and convert it to **conjugated bilirubin,** or *bilirubin diglucuronide.* Although similar in color and structure, conjugated bilirubin is much more soluble than the (unconjugated) bilirubin absorbed by the liver. This distinction is important because most of the conjugated bilirubin will be excreted in the bile, at concentrations at which unconjugated bilirubin would form insoluble crystals. A small amount of conjugated bilirubin reenters the circulation; under normal circumstances, it will be removed at the kidneys and eliminated in the urine. Urinary concentrations of conjugated bilirubin are very low, and unconjugated bilirubin does not normally enter the urine at all.

If the bile ducts are blocked or the liver cannot absorb or excrete bilirubin, circulating levels of conjugated and unconjugated bilirubin climb rapidly. Bilirubin then diffuses into peripheral tissues, giving them a yellow color that is most apparent in the skin and over the sclera of the eyes. This combination of signs (yellow skin and eyes) is called **jaundice** (JAWN-dis). AM *Bilirubin Tests and Jaundice*

Inside the large intestine, bacteria convert the conjugated bilirubin to *urobilinogens* and *stercobilinogens*. Some of the urobilinogens are absorbed into the circulation. These will subsequently be excreted in the urine. On exposure to oxygen, some of the urobilinogens and stercobilinogens convert to **urobilins** and **stercobilins.** Urobilins in the urine give it a yellow color; the brownish color of feces is due to urobilins and stercobilins.

Iron. Large quantities of free iron are toxic to cells, and iron within the body is generally bound to transport or storage proteins. Iron extracted from the heme molecules may be bound and stored within the phagocytic cell or released into the bloodstream, where it binds to **transferrin** (tranz-FER-in), a plasma protein. Red blood cells developing in the bone marrow absorb the amino acids and transferrins from the circulation and use them to synthesize new Hb molecules. Excess transferrins are removed in the liver and spleen, and the iron is stored in two special protein–iron complexes, **ferritin** (FER-I-tin) and **hemosiderin** (hē-mō-SID-e-rin).

This recycling system is remarkably efficient; although roughly 26 mg of iron are incorporated into Hb molecules each day, a dietary supply of 1–2 mg will keep pace with the incidental losses that occur at the kidney and the digestive tract.

Any impairment in iron uptake or metabolism can cause serious clinical problems due to impairment of RBC formation. *Iron-deficiency anemia*, which results from a lack of iron in the diet or from problems with iron absorption, is one example. Too much iron can also cause problems due to excessive buildup in secondary storage sites, such as the liver and cardiac muscle tissue. Excessive iron deposition in cardiac muscle cells has also been linked to heart disease. AM *Iron Deficiencies and Excesses*

Red Blood Cell Formation

Red blood cell formation, or **erythropoiesis** (e-rith-rō-poy-Ē-sis), in the adult occurs in the red bone marrow, or **myeloid** (MĪ-e-loyd) **tissue** (*myelos,* marrow). **Red marrow,** where active blood cell production occurs, is located in portions of the vertebrae, sternum, ribs, skull, scapulae, pelvis, and proximal limb bones. Other marrow areas contain a fatty tissue known as **yellow marrow.** ∞ *[p. 174]* Under extreme stimulation, such

as severe and sustained blood loss, areas of yellow marrow can convert to red marrow, increasing the rate of RBC formation.

STAGES IN RBC MATURATION During its maturation, a red blood cell passes through a series of stages. **Hematologists** (hē-ma-TOL-o-jists), specialists in blood formation and function, have given specific names to key stages. Divisions of hemocytoblasts in bone marrow produce (1) **myeloid stem cells**, which in turn divide to produce red blood cells and several classes of white blood cells, and (2) **lymphoid stem cells**, which divide to produce the various classes of lymphocytes. Cells destined to become RBCs first differentiate into **proerythroblasts** and then proceed through the various stages of **erythroblasts** (Figure 19-6•). Erythroblasts actively synthesize hemoglobin. There are several stages of erythroblasts, categorized on the basis of total size, the amount of hemoglobin present, and the size and appearance of the nucleus.

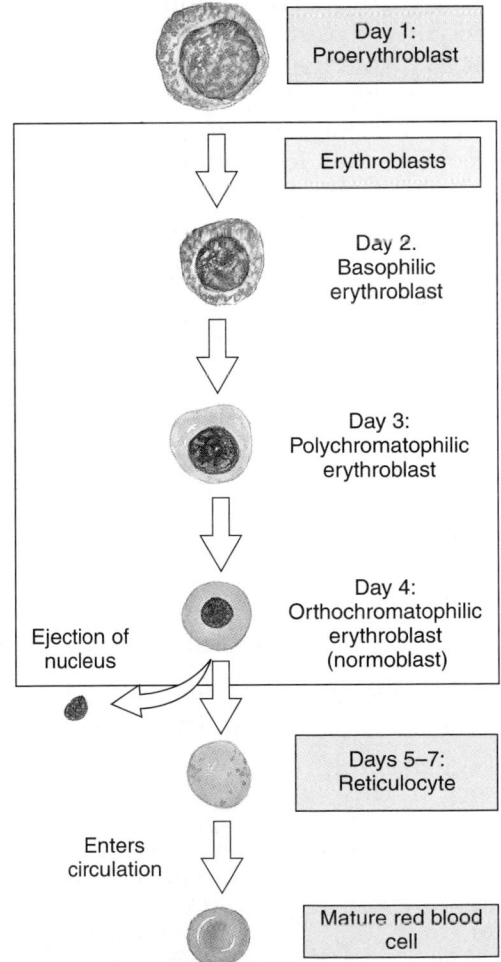

Day 1: Proerythroblast

Erythroblasts

Day 2. Basophilic erythroblast

Day 3: Polychromatophilic erythroblast

Day 4: Orthochromatophilic erythroblast (normoblast)

Ejection of nucleus

Days 5–7: Reticulocyte

Enters circulation

Mature red blood cell

•***FIGURE 19-6*** **Stages in RBC Maturation.** RBC production occurs in the red bone marrow. The color density indicates the abundance of hemoglobin in the cytoplasm. Note the reduction in the size of the cell and the size of the nucleus before a reticulocyte is formed.

After roughly 4 days of differentiation, the erythroblast, now called a *normoblast*, sheds its nucleus and becomes a **reticulocyte** (re-TIK-ū-lō-sīt). A reticulocyte contains 80 percent of the hemoglobin of a mature RBC, and hemoglobin synthesis continues for 2–3 more days. Over this period, while the cells are synthesizing proteins, their cytoplasm still contains RNA, which can be visualized with appropriate stains. After 2 days in the bone marrow, reticulocytes enter the circulation. At this time, they can still be detected by staining, and reticulocytes normally account for about 0.8 percent of the erythrocyte population in the blood. After 24 hours in circulation, the reticulocytes complete their maturation and become indistinguishable from other mature RBCs.

REGULATION OF ERYTHROPOIESIS For erythropoiesis to proceed normally, the myeloid tissues must receive adequate supplies of amino acids, iron, and vitamins (including B_{12}, B_6, and folic acid) required for protein synthesis. For example, if vitamin B_{12} is not obtained from the diet, normal stem cell divisions cannot occur, and *pernicious anemia* results. We obtain **vitamin B_{12}** from dairy products and meat, and its absorption requires the presence of *intrinsic factor* produced in the stomach. Thus, pernicious anemia may be caused by a diet deficient in vitamin B_{12}, a problem with gastric production of intrinsic factor, or a problem with B_{12} complex absorption.

Erythropoiesis is stimulated directly by erythropoietin and indirectly by several hormones, including thyroxine, androgens, and growth hormone. As we noted above, estrogens do not have a stimulatory effect on erythropoiesis.

Erythropoietin, also called **EPO** or **erythropoiesis-stimulating hormone,** is a glycoprotein that appears in the plasma when peripheral tissues, especially the kidneys, are exposed to low oxygen concentrations (Figure 18-17•, p. 619). The state of low tissue oxygen levels is called **hypoxia** (hī-POKS-ē-uh; *hypo-,* below + *ox-,* presence of oxygen). EPO is released (1) during anemia, (2) when blood flow to the kidneys declines, (3) during adaptation to high altitudes, and (4) when the respiratory surfaces of the lungs are damaged. ∞ *[p. 618]* Once in the circulation, EPO travels to areas of red bone marrow, where it stimulates stem cells and developing erythrocytes.

Erythropoietin has two major effects: (1) It stimulates increased rates of cell divisions in erythroblasts and in the stem cells that produce erythroblasts; and (2) it speeds up the maturation of red blood cells, primarily by accelerating the rate of Hb synthesis. Under maximum EPO stimulation, the bone marrow can increase the rate of RBC formation tenfold, to about 30 million per second.

This reserve is important to a person recovering from a severe blood loss. However, if EPO is administered to a healthy individual, as in the case of the cyclists mentioned in Chapter 18, the hematocrit may rise to 65 or more. ∞ *[p. 636]* This increase can place an intolerable strain on the heart. Comparable problems can occur after **blood doping.** In this practice, athletes reinfuse packed RBCs removed at an earlier date and stored, with the goal of increasing hematocrit to improve performance. ▲M *Erythrocytosis and Blood Doping*

BLOOD TESTS AND RBCS Blood tests provide information about the general health of an individual, usually with a minimum of trouble and expense. Several common blood tests focus on red blood cells, the most abundant formed elements. These tests assess the number, size, shape, and maturity of circulating RBCs. Such assessment provides an indication of the erythropoietic activities under way, and it can also be useful in detecting problems, such as internal bleeding, that may not produce other obvious symptoms. Table 19-1 lists examples of important blood tests and related terms. (A more detailed discussion, including sample calculations, can be found in the *Applications Manual.*) ▲M *Blood Tests and RBCs*

TABLE 19-1 RBC Tests and Related Terminology

Test	Determines	Terms Associated with Abnormal Values	
		Elevated	**Depressed**
Hematocrit (Hct)	Percentage of formed elements in whole blood Normal = 37–54	Polycythemia (may reflect erythrocytosis or leukocytosis)	Anemia
Reticulocyte count (Retic.)	Circulating percentage of reticulocytes Normal = 0.8%	Reticulocytosis	
Hemoglobin concentration (Hb)	Concentration of hemoglobin in blood Normal = 12–18 g/dl		Anemia
RBC count	Number of RBCs per μl of whole blood Normal = 4.2–6.3 million/μl	Erythrocytosis/ polycythemia	Anemia
Mean corpuscular volume (MCV)	Average volume of single RBC Normal = 82–101 μm^3 (normocytic)	Macrocytic	Microcytic
Mean corpuscular hemoglobin concentration (MCHC)	Average amount of Hb in one RBC Normal = 27–34 pg/μl (normochromic)	Hyperchromic	Hypochromic

☑ How would the hematocrit change after an individual suffered a hemorrhage?

☑ Dave develops a blockage in his renal arteries that restricts blood flow to the kidneys. Will his hematocrit change?

☑ How would the level of bilirubin in the blood be affected by a disease that causes damage to the liver?

Blood Types

Antigens are materials that can trigger an *immune response*, a defense mechanism that protects you from infection. Most antigens are proteins, although some other types of organic molecules are antigens as well. Your cell membranes contain **surface antigens** that your immune defenses recognize as "normal." In other words, these substances are ignored rather than attacked by your own immune system as "foreign." Your **blood type** is determined by the presence or absence of specific surface antigens in the RBC cell membrane. The surface antigens involved are integral glycoproteins or glycolipids whose characteristics are genetically determined. The surface antigens of RBCs are often called **agglutinogens** (a-gloo-TIN-ō-jenz). Red blood cells have at least 50 different kinds of surface antigens. Three of particular importance have been designated as surface antigens **A, B**, and **Rh (D).**

The RBCs of an individual may have (1) either A or B surface antigens, (2) both A and B surface antigens, or (3) neither A nor B surface antigens (Figure 19-7•). **Type A** blood has antigen A only, **Type B** has antigen B only, **Type AB** has both A and B, and **Type O** has neither A nor B. These blood types are not evenly distributed throughout the population. The average values for the U.S. population are: Type O, 46 percent; Type A, 40 percent; Type B, 10 percent; and Type AB, 4 percent.

The term **Rh-positive** (Rh⁺) indicates the presence of the Rh surface antigen, sometimes called the *Rh factor*. The absence of this antigen is indicated as **Rh-negative** (Rh⁻). When the complete blood type is recorded, the term *Rh* is usually omitted, and the data are reported as O-negative (O⁻), A-positive (A⁺), and so forth. As in the distribution of A and B surface antigens, there are racial and regional differences in Rh type (Table 19-2).

You are probably aware that your blood type must be checked before you can give or receive blood. Your immune system ignores the surface antigens on your own RBCs. However, your plasma contains antibodies, sometimes called **agglutinins** (a-GLOO-ti-ninz), that will attack the antigens on "foreign" RBCs. When the agglutinins attack, the foreign cells clump together, or **agglutinate** (a-GLOO-ti-nāt); the process is called **agglutination** (a-gloo-ti-NĀ-shun). If the difference between agglutinins and agglutinogens confuses you, remember that *agglutinins* cause *agglutination* but that *agglutinogens* are antigens.

The plasma of a Type A, Type B, or Type O individual always contains either anti-A or anti-B antibodies, or both, even if the person has never been exposed to red blood cells that carry foreign surface antigens. If you have Type A blood, your plasma contains anti-B antibodies, which will attack Type B surface antigens. If you have Type B blood, your plasma contains anti-A antibodies. The RBCs of an individual with Type O blood have neither A nor B surface antigens, and that person's plasma contains both anti-A and anti-B antibodies. A Type AB individual has RBCs with both A and B surface antigens, and the plasma does not contain anti-A or anti-B antibodies.

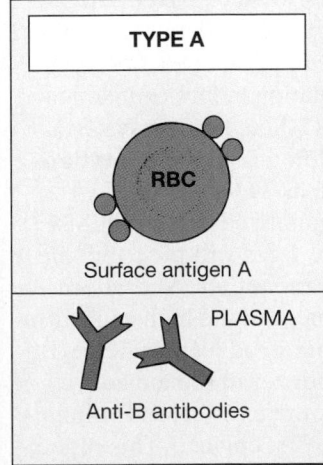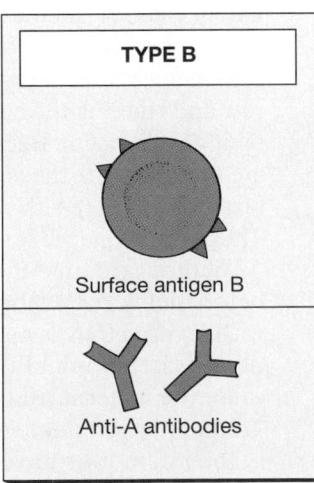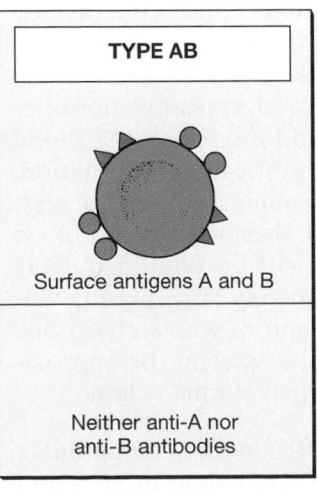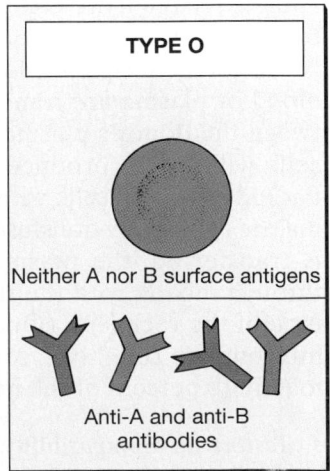

•FIGURE 19-7 **Blood-Typing.** The blood type depends on the presence of surface antigens (agglutinogens) on RBC surfaces. The plasma contains antibodies (agglutinins) that will react with foreign surface antigens. The relative frequencies of each blood type in the U.S. population are indicated in Table 19-2.

TABLE 19-2 Differences in Blood Group Distribution

Population	Percentage with Each Blood Type				
	O	A	B	AB	Rh⁺
U.S. (average)	46	40	10	4	85
Caucasian	45	40	11	4	85
African American	49	27	20	4	95
Chinese American	42	27	25	6	100
Japanese American	31	39	21	10	100
Korean American	32	28	30	10	100
Filipino American	44	22	29	6	100
Hawaiian	46	46	5	3	100
Native North American	79	16	4	<1	100
Native South American	100	0	0	0	100
Australian Aborigines	44	56	0	0	100

In contrast to the situation with surface antigens A and B, the plasma of an Rh-negative individual does not necessarily contain anti-Rh antibodies. These antibodies are present only if the individual has been **sensitized** by previous exposure to Rh-positive erythrocytes. Such exposure may occur accidentally during a transfusion, but it may also accompany a seemingly normal pregnancy involving an Rh-negative mother and an Rh-positive fetus. (See the box "Hemolytic Disease of the Newborn.")

CROSS-REACTIONS When an antibody meets its specific antigen, a **cross-reaction** occurs. Initially, the RBCs agglutinate, and they may also hemolyze. Clumps and fragments of RBCs under attack form drifting masses that can plug small vessels in the kidneys, lungs, heart, or brain, damaging or destroying dependent tissues. Such reactions can be prevented by taking care to ensure that the blood types of the donor and the recipient are **compatible**—that is, the donor's blood cells will not undergo cross-reaction with the plasma of the recipient. Unless large volumes of whole blood or plasma are transferred, cross-reactions between the donor's plasma and the recipient's blood cells will fail to produce significant agglutination. Packed red blood cells, with a minimal amount of plasma, are commonly transfused. Even when whole blood is transferred, the plasma will be diluted quickly through mixing with the relatively large plasma volume of the recipient. (One unit of whole blood, 500 ml, contains roughly 275 ml of plasma, or approximately 10 percent of the normal plasma volume.)

Testing for Compatibility. Testing for compatibility normally involves two steps: (1) a determination of blood type and (2) a *cross-match test*. At least 50 surface antigens have been identified on RBCs, but the standard test for blood type considers only the three

Hemolytic Disease of the Newborn

Genes controlling the presence or absence of any surface antigen in the erythrocyte membrane are provided by both parents, so a child may have a blood type different from that of either parent. During pregnancy, when fetal and maternal circulatory systems are closely intertwined, the mother's antibodies may cross the placental barrier, attacking and destroying fetal red blood cells. The resulting condition is called **hemolytic disease of the newborn (HDN)**.

There are many forms of HDN, some so mild as to remain undetected, but those involving the Rh surface antigen are potentially quite dangerous. Because HDN results from the passage of maternal antibodies across the placental barrier, an Rh-positive mother (who lacks anti-Rh antibodies) can carry an Rh-negative fetus without difficulty. Potential problems appear when an Rh-negative woman carries an Rh-positive fetus. Sensitization generally occurs at delivery, when bleeding takes place at the placenta and uterus. This event can trigger the maternal production of anti-Rh antibodies. Within 6 months of delivery, roughly 20 percent of Rh-negative mothers who carried Rh-positive children have become sensitized.

Because the anti-Rh antibodies are not produced in significant amounts until after delivery, the first infant will not be affected. (Some fetal red blood cells cross into the maternal circulation during pregnancy but generally not in numbers sufficient to stimulate antibody production.) But if a second pregnancy occurs involving an Rh-positive fetus, maternal anti-Rh antibodies produced after the first delivery cross the placenta and enter the fetal circulation. These antibodies destroy fetal RBCs and produce a dangerous anemia. The fetal demand for blood cells increases, and they begin leaving the bone marrow and entering the circulation before completing their development. Because these immature RBCs are erythroblasts, the condition is known as **erythroblastosis fetalis** (e-rith-rō-blas-TŌ-sis fē-TAL-is). The entire sequence is summarized in Figure 19-8●.

Without treatment, the fetus will probably die before delivery or shortly thereafter. A newborn with severe HDN is anemic, and the high concentration of circulating bilirubin produces jaundice. Because the maternal antibodies will remain active for 1–2 months after delivery, the entire blood volume of the infant may have to be replaced. This blood replacement removes most of the maternal anti-Rh antibodies as well as the affected erythrocytes, reducing the complications and the chance of death.

When there is a danger that the fetus may not survive to full term, premature delivery may be induced after 7–8 months of development. In a severe case affecting a fetus at an earlier stage, one or more transfusions can be given while the fetus continues to develop within the uterus.

The maternal production of anti-Rh antibodies can be prevented by administering anti-Rh antibodies (available under the name *RhoGam*) to the mother in the last 3 months of pregnancy and during and after delivery. (It is also given after a miscarriage or abortion.) These antibodies will destroy any fetal RBCs that cross the placental barrier before they can stimulate an immune response in the mother. Because sensitization does not occur, anti-Rh antibodies are not produced. This relatively simple procedure could almost entirely prevent HDN mortality caused by Rh incompatibilities.

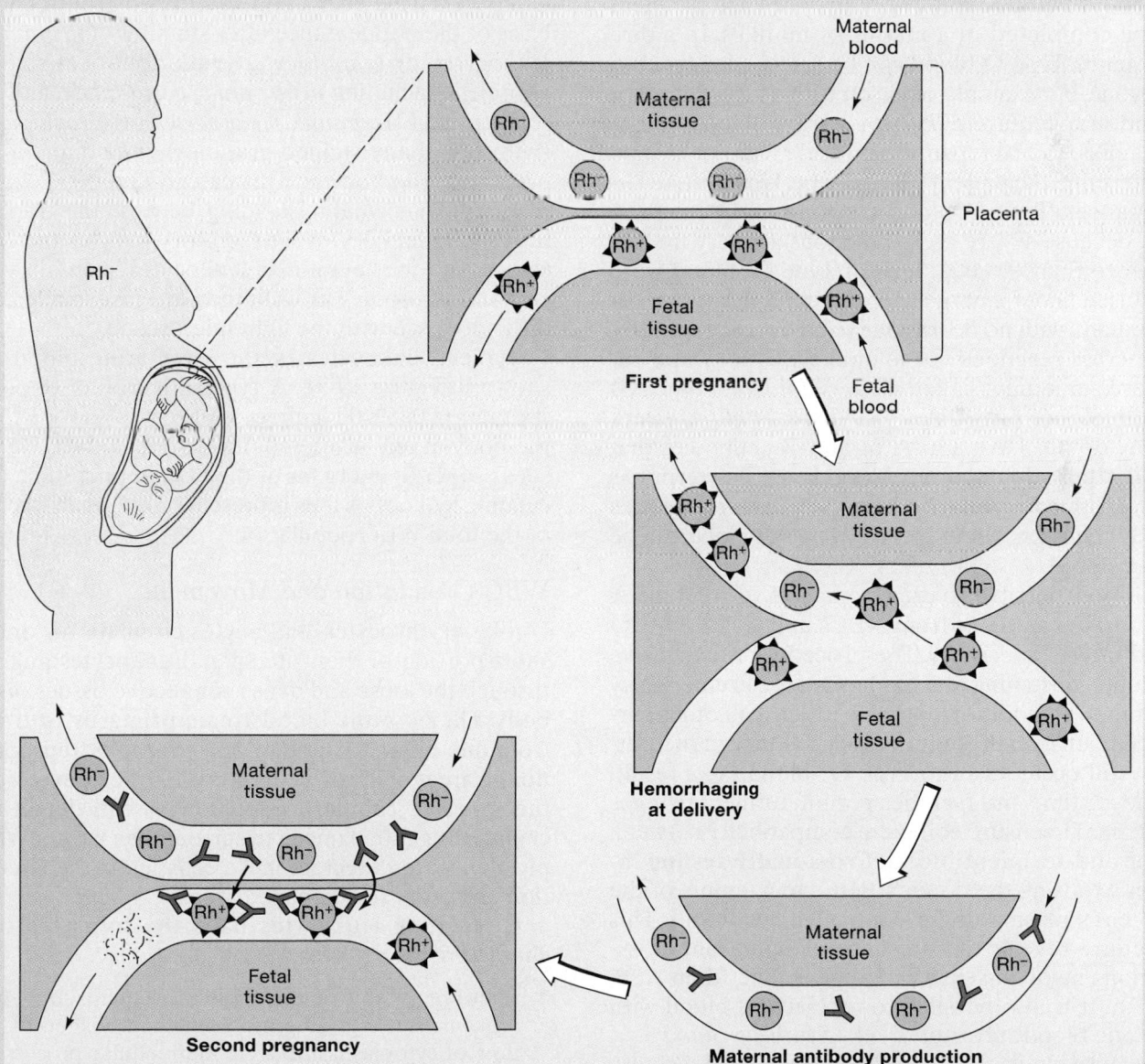

•**FIGURE 19-8 Rh Factors and Pregnancy.** When an Rh-negative woman has her first Rh-positive child, mixing of fetal and maternal blood occurs at delivery when the placental connection breaks down. The appearance of Rh-positive blood cells in the maternal circulation sensitizes the mother, stimulating the production of anti-Rh antibodies. If another pregnancy occurs with an Rh-positive fetus, maternal antibodies can cross the placental barrier and attack fetal blood cells, producing symptoms of hemolytic disease of the newborn (HDN).

most likely to produce dangerous cross-reactions. The test involves taking drops of blood and mixing them separately with solutions containing anti-A, anti-B, and anti-Rh (anti-D) antibodies. Any cross-reactions are then recorded. For example, if an individual's RBCs clump together when exposed to anti-A and to anti-B, the individual has Type AB blood. If no reactions occur after exposure, that person must have Type O blood. The presence or absence of the Rh agglutinogen is also noted, and the individual is classified as Rh-positive or Rh-negative on that basis. Type O$^+$ is the most common blood type. Although the RBCs of Type O$^+$ individuals do not have surface antigens A or B, these cells do have the Rh antigen.

Standard blood-typing of both donor and recipient can be completed in a matter of minutes. In a dire emergency, Type O blood can be safely administered to anyone. For example, a person with a severe gunshot wound may require 5 *liters* or more of blood before the damage can be repaired, and there may not be time to determine the patient's blood type. Under these circumstances, Type O blood (preferably O$^-$) will be used, because severe cross-reactions are very unlikely. (More than 100,000 units of Type O$^-$ blood were transfused under emergency conditions during the war in Vietnam, with no deaths due to cross-reactions.) Because cross-reactions are so unlikely, Type O individuals are sometimes called *universal donors*. Type AB individuals are sometimes called *universal recipients*, as they do not have anti-A or anti-B antibodies that would attack donated red blood cells. This term has been dropped, primarily because if the recipient's blood type is known to be AB, Type AB blood can be administered.

It is now possible to use enzymes to strip off the A or B surface antigens from RBCs and create Type O blood in the laboratory. The procedure is expensive and time-consuming and has limited use in emergency treatment. In addition, because at least 48 other surface antigens are present, cross-reactions can theoretically still occur, even to Type O$^-$ blood. As a result, whenever time and facilities permit, further testing is performed to ensure complete compatibility between donor and recipient blood. **Cross-match testing** involves exposing the donor's RBCs to a sample of the recipient's plasma under controlled conditions. This procedure reveals the presence of significant cross-reactions involving surface antigens other than A, B, and Rh. It is also possible to replace lost blood with synthetic blood substitutes. **AM** *Synthetic Blood*

Because blood groups are inherited, blood tests are also used as paternity tests and in crime detection. The blood collected cannot prove that a particular individual *is* the father or *is* guilty of a specific crime, but it can prove that the individual is *not* involved. For example, it is impossible for an adult with Type AB blood to be the parent of an infant with Type O blood.

☑ Which blood type(s) can be transfused into a person with Type O blood?

☑ Why can't a person with Type A blood safely receive blood from a person with Type B blood?

☑ What are surface antigens on RBCs?

White Blood Cells

Unlike red blood cells, white blood cells, or *leukocytes,* have nuclei and other organelles, and these cells lack hemoglobin. White blood cells help defend the body against invasion by pathogens, and they remove toxins, wastes, and abnormal or damaged cells. Traditionally, leukocytes have been divided into two groups on the basis of their appearance after staining:[1] (1) *granular leukocytes,* or *granulocytes* (with abundant stained granules), including *neutrophils*, *eosinophils*, and *basophils*; and (2) *agranular leukocytes,* or *agranulocytes* (with few if any stained granules), including *monocytes* and *lymphocytes*. This categorization is convenient but somewhat misleading, because the granules in granular leukocytes are secretory vesicles and lysosomes, and the "agranular leukocytes" contain vesicles and lysosomes as well; they are just smaller and difficult to see with the light microscope.

Typical leukocytes in the circulating blood are shown in Figure 19-9●. A typical microliter of blood contains 6000–9000 leukocytes. Most of the WBCs in the body at any given moment are in connective tissues proper or in organs of the lymphatic system. Circulating leukocytes thus represent only a small fraction of the total WBC population.

WBC Circulation and Movement

Unlike erythrocytes, leukocytes circulate for only a short portion of their life span. Leukocytes migrate through the loose and dense connective tissues of the body. They use the bloodstream primarily to travel from one organ to another and for rapid transportation to areas of invasion or injury. As they travel along the miles of capillaries, leukocytes can detect the chemical signs of damage to surrounding tissues. When problems are detected, these cells leave the circulation and enter the damaged area.

Circulating leukocytes have the following four characteristics:

1. They are capable of amoeboid movement. *Amoeboid movement* is a gliding motion accomplished by the flow of cytoplasm into a slender cellular process extended in front of the cell. The mechanism is not fully understood, but (1) it involves the continuous rearrangement of bonds between actin filaments in the

---

[1]The stains used in standard blood work are *Wright's stain* and the *Giemsa stain*.

cytoskeleton, and (2) it requires calcium ions and ATP. This mobility allows WBCs to move along the walls of blood vessels and, outside the bloodstream, through surrounding tissues.

2. They can migrate out of the bloodstream by squeezing between adjacent endothelial cells in the capillary wall, a process known as **diapedesis** (dī-a-pe-DĒ-sis).

3. They are attracted to specific chemical stimuli. This characteristic, called **positive chemotaxis** (kē-mō-TAK-sis), guides them to invading pathogens, damaged tissues, and active WBCs.

4. Neutrophils, monocytes, and eosinophils are capable of phagocytosis. These cells may engulf pathogens, cell debris, or other materials. Before becoming phagocytic, monocytes first leave the bloodstream and then differentiate into the connective tissue cells known as *macrophages*. ∞ *[p. 120]* Neutrophils and eosinophils are sometimes called *microphages,* to distinguish them from the larger macrophages.

General Functions

Neutrophils, eosinophils, basophils, and monocytes contribute to the body's *nonspecific defenses.* These defenses are activated by a variety of stimuli, but they do not discriminate between one type of threat and another. Lymphocytes, in contrast, are responsible for *specific immunity:* the body's ability to mount a counterattack against particular invading pathogens or foreign proteins on an individual basis. We will discuss the interactions among WBCs and the relationships between specific and nonspecific defenses in Chapter 22.

Neutrophils

Fifty to seventy percent of the circulating WBCs are **neutrophils** (NOO-trō-filz). This name reflects the fact that their granules are chemically neutral and thus are difficult to stain with either acidic or basic dyes. A mature neutrophil (Figure 19-9a●) has a very dense, segmented nucleus that forms two to five lobes resembling beads on a string. This characteristic has given neutrophils another name, **polymorphonuclear** (pol-ē-mōr-fō-NOO-klē-ar) **leukocytes** (*poly,* many + *morphe,* form), or **PMNs.** "Polymorphs," or "polys," as they are often called, are roughly 12 μm in diameter. Their cytoplasm is packed with pale granules containing lysosomal enzymes and bactericidal (bacteria-killing) compounds.

Neutrophils are highly mobile, and they are generally the first of the WBCs to arrive at an injury site. They are very active cells that specialize in attacking and digesting bacteria that have been "marked" with antibodies or with *complement proteins,* plasma proteins involved in tissue defenses. (We shall detail the complement system in Chapter 22.)

Upon encountering a bacterium, the neutrophil quickly engulfs it, and the metabolic rate of the neutrophil increases dramatically. This *respiratory burst*

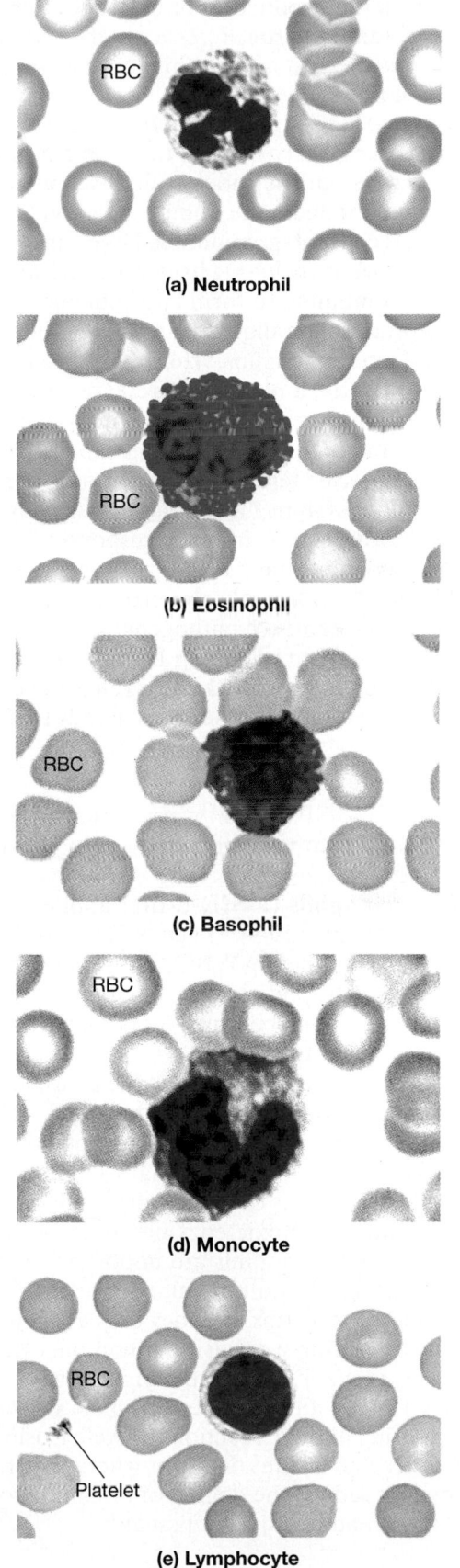

●*FIGURE 19-9* **White Blood Cells. (a)** Neutrophil. **(b)** Eosinophil. **(c)** Basophil. **(d)** Monocyte. **(e)** Lymphocyte. (LMs × 1500)

accompanies the production of destructive chemical agents, including *hydrogen peroxide* (H_2O_2) and *superoxide anions* (O_2^-). These highly reactive chemicals can kill bacteria.

Meanwhile, the vesicle containing the engulfed pathogen fuses with lysosomes that contain digestive enzymes and small peptides called **defensins.** This process, which reduces the number of granules in the cytoplasm, is called **degranulation.** Defensins kill a variety of pathogens, including bacteria, fungi, and some viruses, by combining to form large channels in their cell membranes. The digestive enzymes then break down the bacterial remains. While actively engaged in attacking bacteria, a neutrophil releases prostaglandins and leukotrienes. The prostaglandins contribute to local inflammation by increasing capillary permeability in the affected region. Leukotrienes are hormones of the immune system that attract other phagocytes and help coordinate the immune response.

Most neutrophils have a short life span, surviving in the bloodstream for only about 10 hours. When actively engulfing debris or pathogens, they may last 30 minutes or less. After engulfing 1–2 dozen bacteria, a neutrophil dies, but its breakdown releases additional chemicals that attract other neutrophils to the site.

Eosinophils

Eosinophils (ē-ō-SIN-ō-filz) were so named because their granules stain darkly with *eosin,* a red dye. The granules will also stain with other acid dyes, so the name **acidophils** (a-SID-ō-filz) applies as well. Eosinophils (Figure 19-9b●) generally represent 2–4 percent of the circulating WBCs. These cells are similar in size to neutrophils, but the combination of deep red granules and a bilobed (two-lobed) nucleus makes an eosinophil easy to identify.

Eosinophils attack objects that have already been coated with antibodies. They are phagocytic cells and will engulf antibody-marked bacteria, protozoa, or cellular debris. However, their primary mode of attack involves the exocytosis of toxic compounds, including nitric oxide and cytotoxic enzymes, onto the surface of their targets. Eosinophils are important in the defense against large multicellular parasites, such as flukes or parasitic worms, and they increase in number dramatically during a parasitic infection.[2] Because they are also sensitive to circulating *allergens* (materials that trigger allergies), eosinophils increase in number during allergic reactions as well. Eosinophils are also attracted to sites of injury, where they release enzymes that reduce the degree of inflammation and control its spread to adjacent tissues.

[2]For a review of major classes of pathogens, see "The Nature of Pathogens" in Section 2 of the *Applications Manual.*

Basophils

Basophils (BĀ-sō-filz) have numerous granules that stain darkly with basic dyes. In a standard blood smear, the inclusions are deep purple or blue (Figure 19-9c●). Basophils are smaller than neutrophils or eosinophils, measuring 8–10 μm in diameter. They are also relatively rare, accounting for less than 1 percent of the circulating leukocyte population.

Basophils migrate to sites of injury and cross the capillary endothelium to accumulate within the damaged tissues. There they discharge their granules into the interstitial fluids. The granules contain histamine and *heparin,* a chemical that prevents blood clotting. Stimulated basophils release these chemicals into the interstitial fluids, and their arrival enhances the local inflammation initiated by mast cells. ∞ *[p. 140]* Although the same compounds are released by mast cells in damaged connective tissues, mast cells and basophils are distinct populations with separate origins. Other chemicals released by stimulated basophils attract eosinophils and other basophils to the area.

Monocytes

Monocytes (MON-ō-sīts) in the blood are spherical cells that may exceed 15 μm in diameter, nearly twice the diameter of a typical erythrocyte (Figure 19-9d●). When flattened in a blood smear, they look even larger, so monocytes are relatively easy to identify. The nucleus is large and tends to be oval or kidney bean–shaped. Monocytes normally account for 2–8 percent of the circulating leukocytes.

An individual monocyte uses the bloodstream as a highway, remaining in circulation for only about 24 hours before entering peripheral tissues to become a tissue macrophage. Macrophages are aggressive phagocytes, often attempting to engulf items as large as or larger than themselves. Upon encountering a foreign object too large for a single cell to engulf, several macrophages may fuse together to create a single **phagocytic giant cell** big enough to do the job. While fusing, they release chemicals that attract and stimulate neutrophils, monocytes, and other phagocytic cells. Active macrophages also secrete substances that lure fibroblasts into the region. The fibroblasts then begin producing the scar tissue that will wall off the injured area.

Lymphocytes

Typical **lymphocytes** (LIM-fō-sīts) are slightly larger than RBCs and lack abundant, deeply stained granules. In fact, when you see a lymphocyte in a blood smear, you generally see just a thin halo of cytoplasm around a relatively large nucleus (Figure 19-9e●).

Lymphocytes account for 20–30 percent of the leukocyte population of the blood. Lymphocytes continuously migrate from the bloodstream, through peripheral tissues, and back to the bloodstream. Circulating lymphocytes represent only a minute fraction of the entire lymphocyte population, for at any moment most of your body's lymphocytes are in other connective tissues and in organs of the lymphatic system.

The circulating blood contains three classes of lymphocytes, although they cannot be distinguished with a light microscope:

1. *T cells.* **T cells** are responsible for *cellular immunity,* a defense mechanism against invading foreign cells and tissues, and for the coordination of the immune response. Activated *cytotoxic T cells* enter peripheral tissues and destroy foreign cells directly by physical and chemical attack. *Regulatory T cells,* including *helper T cells* and *suppressor T cells,* stimulate or inhibit the activities of other lymphocytes.

2. *B cells.* **B cells** are responsible for *humoral immunity,* a defense mechanism that involves the production and distribution of antibodies that can attack foreign antigens throughout the body. Activated B cells differentiate into **plasma cells,** which are specialized for the synthesis and secretion of antibodies. Whereas the T cells responsible for cellular immunity must migrate to their targets, the antibodies produced by plasma cells in one location can destroy antigens almost anywhere in the body.

3. *NK cells.* **NK cells,** sometimes known as *large granular lymphocytes,* are responsible for *immune surveillance,* the detection and subsequent destruction of abnormal tissue cells. These cells are important in preventing cancer.

The Differential Count and Changes in WBC Profiles

A variety of disorders, including pathogenic infection, inflammation, and allergic reactions, cause characteristic changes in circulating populations of WBCs. By examining a stained blood smear, we can obtain a **differential count** of the WBC population. The values reported indicate the number of each type of cell in a sample of 100 WBCs.

The normal range for each cell type is indicated in Table 19-3. The term **leukopenia** (loo-kō-PĒ-nē-uh; *penia,* poverty) indicates inadequate numbers of WBCs. **Leukocytosis** (loo-kō-sī-TŌ-sis) refers to excessive numbers of WBCs. A modest leukocytosis is normal during an infection. Extreme leukocytosis (>100,000/μl or more) generally indicates the presence of some form of **leukemia** (loo-KĒ-mē-uh). There are many different types of leukemias. Treatment helps in some cases; unless treated, all are fatal. The endings *-penia* and *-osis* can also indicate low or high numbers

of specific types of WBCs. For example, *lymphopenia* means too few lymphocytes, and *lymphocytosis* means too many. AM *The Leukemias*

White Blood Cell Production

Stem cells responsible for the production of white blood cells originate in the bone marrow, with the divisions of hemocytoblasts. As we noted earlier, hemocytoblast divisions produce myeloid stem cells and lymphoid stem cells. Myeloid stem cell division creates **progenitor cells,** which give rise to all the formed elements except lymphocytes. The divisions of one type of progenitor cell produce RBCs, as we detailed on pages 651–652, and the divisions of a second type is responsible for the production of platelets.

Neutrophils, eosinophils, basophils, and monocytes are produced by the divisions of a third type of progenitor cell. All but the monocytes complete their development in the bone marrow. Monocytes begin their differentiation in the bone marrow, enter the circulation, and complete their development when they become free macrophages in peripheral tissues. Each of the other cell types goes through a characteristic series of maturational stages, proceeding from *blast cells* to *myelocytes* to *band cells* before becoming mature WBCs. For example, a cell differentiating into a neutrophil goes from a myeloblast to a *neutrophilic myelocyte* and then becomes a *neutrophilic band cell.* Some band cells enter the circulation before completing their maturation; normally 3–5 percent of all circulating WBCs are band cells. Figure 19-10● (p. 661) summarizes the relationships among the various WBC populations and compares the formation of WBCs and RBCs.

Many of the lymphoid stem cells responsible for the production of lymphocytes, a process called **lymphopoiesis,** migrate from the bone marrow to peripheral **lymphoid tissues,** including the thymus, spleen, and lymph nodes. As a result, lymphocytes are produced in these organs as well as in the bone marrow.

REGULATION OF WBC PRODUCTION Factors that regulate lymphocyte maturation are as yet incompletely understood. Prior to maturity, the "thymosins" produced by the thymus promote the differentiation and maintenance of T cell populations. ∞ *[p. 614]* The importance of the thymus in adulthood, especially in aging, remains controversial. In the adult, production of B and T lymphocytes is regulated primarily by exposure to antigens (foreign proteins, cells, or toxins). When antigens appear, lymphocyte production escalates. We shall describe the control mechanisms in Chapter 22.

Several hormones are involved in the regulation of other white blood cell populations. Figure 19-10● diagrams the targets of these hormones, called **colony-stimulating factors (CSFs),** and the target of eryth-

TABLE 19-3 A Review of the Formed Elements of the Blood

Cell	Abundance (average number per μl)	Appearance in a Stained Blood Smear	Functions	Remarks
Red Blood Cells	5.2 million (range: 4.4–6.0 million)	Flattened, circular cell; no nucleus, mitochondria, or ribosomes; red	Transport oxygen from lungs to tissues and carbon dioxide from tissues to lungs	Remain in circulation; 120-day life expectancy; amino acids and iron recycled; produced in bone marrow
White Blood Cells	7000 (range: 6000–9000)			
Neutrophils	4150 (range: 1800–7300) Differential count: 50–70%	Round cell; nucleus lobed and may resemble a string of beads; cytoplasm contains large, pale inclusions	Phagocytic: Engulf pathogens or debris in tissues, release cytotoxic enzymes and chemicals	Move into tissues after several hours; may survive minutes to days, depending on tissue activity; produced in bone marrow
Eosinophils	165 (range: 0–700) Differential count: 2–4%	Round cell; nucleus generally in two lobes; cytoplasm contains large granules that generally stain bright red	Phagocytic: Engulf antibody-labeled materials, release cytotoxic enzymes, reduce inflammation	Move into tissues after several hours; survive minutes to days, depending on tissue activity; produced in bone marrow
Basophils	44 (range: 0–150) Differential count: <1%	Round cell; nucleus generally cannot be seen through dense, blue-stained granules in cytoplasm	Enter damaged tissues and release histamine and other chemicals that promote inflammation	Survival time unknown; assist mast cells of tissues in producing inflammation; produced in bone marrow
Monocytes	456 (range: 200–950) Differential count: 2–8%	Very large cell; kidney bean–shaped nucleus; abundant pale cytoplasm	Enter tissues to become macrophages; engulf pathogens or debris	Move into tissues after 1–2 days; survive for months or longer; primarily produced in bone marrow
Lymphocytes	2185 (range: 1500–4000) Differential count: 20–30%	Generally round cell, slightly larger than RBC; round nucleus; very little cytoplasm	Cells of lymphatic system, providing defense against specific pathogens or toxins	Survive for months to decades; circulate from blood to tissues and back; produced in bone marrow and lymphoid tissues
Platelets	350,000 (range: 150,000–500,000)	Round to spindle-shaped cytoplasmic fragment; contains enzymes and proenzymes; no nucleus	Hemostasis: Clump together and stick to vessel wall (platelet phase); activate intrinsic pathway of coagulation phase	Remain in circulation or in vascular organs; remain intact for 7–12 days; produced by megakaryocytes in bone marrow

ropoietin (EPO), discussed on page 652. Four CSFs have been identified, each stimulating WBC or WBC and RBC formation. The abbreviation for each factor indicates its target:

1. **M-CSF** stimulates activity along the monocyte/macrophage line.
2. **G-CSF** stimulates production of granulocytes (neutrophils, eosinophils, and basophils). A genetically engineered form of G-CSF, sold under the name of *filgrastim (Neupogen),* is now used to stimulate pro-

duction of neutrophils in patients undergoing cancer chemotherapy.
3. **GM-CSF** stimulates production of both granulocytes and monocytes.
4. **Multi-CSF** accelerates production of granulocytes, monocytes, platelets, and erythrocytes.

Chemical communication between lymphocytes and other leukocytes assists in the coordination of the immune response. For example, active macrophages release chemicals that make lymphocytes more sensitive to antigens and accelerate the development of specific immunity. In turn, active lymphocytes release multi-CSF and GM-CSF, reinforcing nonspecific defenses. The molecular structure of many of the stimulating factors has been determined, and several can be produced by genetic engineering. The U.S. Food and Drug Administration approved the administration of synthesized forms of erythropoietin, G-CSF, and GM-CSF to stimulate the production of specific blood cell lines. As of early 1997, clinical trials were under way for M-CSF and six other stimulatory chemicals, including some novel compounds not found in the body.

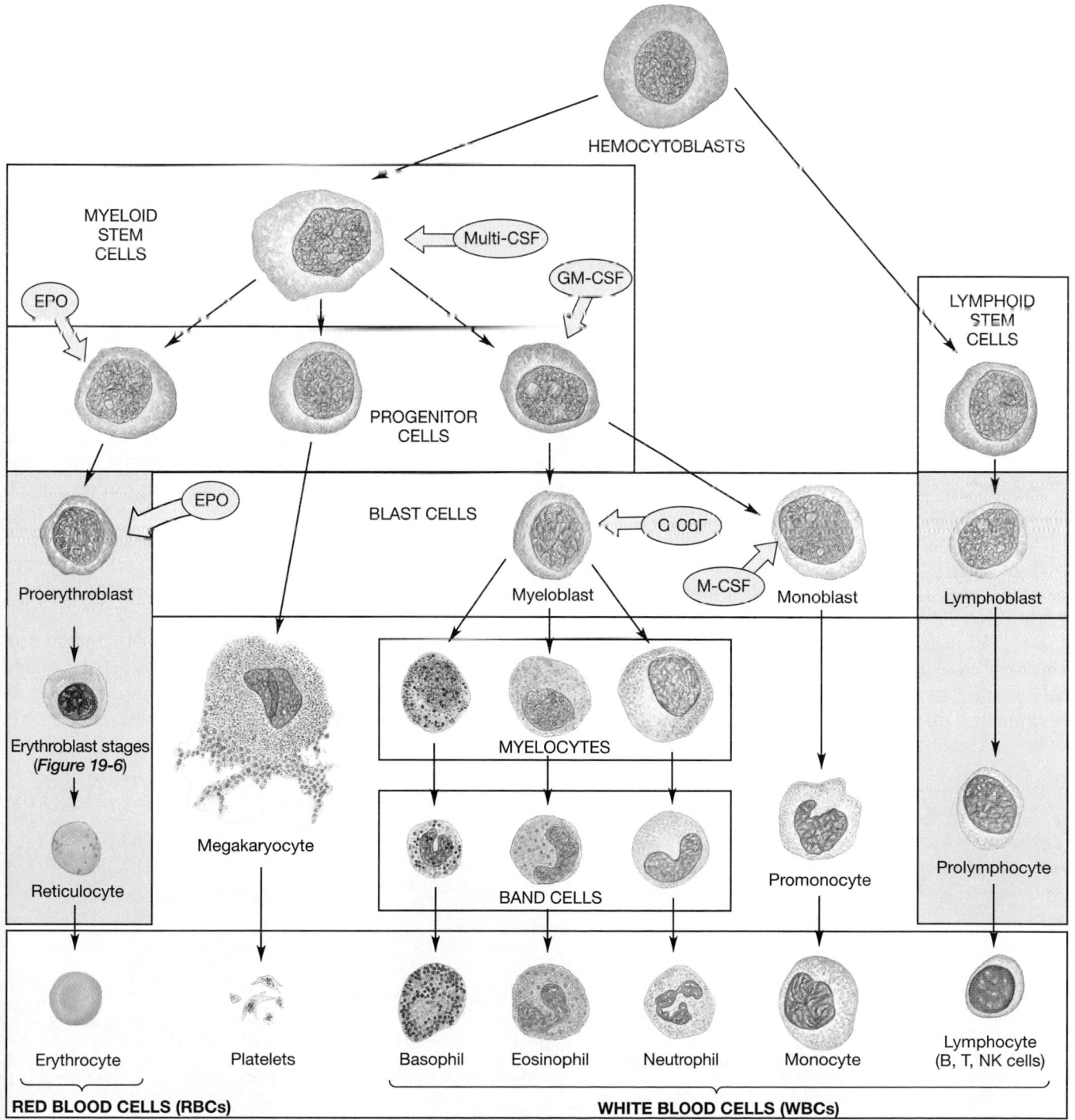

•*FIGURE 19-10* **The Origins and Differentiation of Blood Cells.** Hemocytoblast divisions give rise to myeloid and lymphoid stem cells. Lymphoid stem cells produce the various classes of lymphocytes. Myeloid stem cells produce progenitor cells with more restricted fates; these divide to produce the other classes of blood cells. The targets of EPO and the four colony-stimulating factors are indicated.

☑ Which type of white blood cell would you expect to find in the greatest numbers in an infected cut?

☑ Which cell type would you expect to find in elevated numbers in a person who is producing large amounts of circulating antibodies to combat a virus?

☑ How do basophils respond during inflammation?

Platelets

Platelets (PLĀT-lets) are flattened discs, round when viewed from above and spindle-shaped when seen in section or in a blood smear (Figure 19-9e●). They average about 4 μm in diameter and are roughly 1 μm thick. Platelets in nonmammalian vertebrates are nucleated cells called **thrombocytes** (THROM-bō-sīts; *thrombos,* clot). Because in humans they are cell fragments rather than individual cells, the term *platelet* is preferred when referring to our blood.

Platelets are continuously replaced. Each platelet circulates for 9–12 days before being removed by phagocytes, mainly in the spleen. There are 150,000–500,000 platelets in each microliter of circulating blood; 350,000/μl is the average concentration. Roughly one-third of the platelets in the body at any moment are held in the spleen and other vascular organs rather than in the circulation. These reserves are mobilized during a circulatory crisis, such as severe bleeding.

An abnormally low platelet count (80,000/μl or less) is known as **thrombocytopenia** (throm-bō-sī-tō-PĒ-nē-uh). Thrombocytopenia generally indicates excessive platelet destruction or inadequate platelet production. Symptoms include bleeding along the digestive tract, within the skin, and occasionally inside the CNS. Platelet counts in **thrombocytosis** (throm-bō-sī-TŌ-sis) may exceed 1,000,000/μl. Thrombocytosis generally results from accelerated platelet formation in response to infection, inflammation, or cancer.

Platelet Functions

Platelets are one participant in a vascular *clotting system* that also includes plasma proteins and the cells and tissues of the circulatory network. The functions of platelets include:

- **Transport of chemicals important to the clotting process.** By releasing enzymes and other factors at the appropriate times, platelets help initiate and control the clotting process.
- **Formation of a temporary patch in the walls of damaged blood vessels.** Platelets clump together at an injury site, forming a *platelet plug,* which can slow the rate of blood loss while clotting occurs.
- **Active contraction after clot formation has occurred.** Platelets contain filaments of actin and myosin. After a blood clot has formed, the contraction of platelet filaments shrinks the clot and reduces the size of the break in the vessel wall.

Platelet Production

Platelet production, or **thrombocytopoiesis,** occurs in the bone marrow. Normal bone marrow contains a number of **megakaryocytes** (meg-a-KAR-ē-ō-sīts; *mega-,* big + *karyon,* nucleus + -*cyte,* cell). As the name suggests, these very unusual cells are enormous (up to 160 μm in diameter) and have large nuclei. The dense nucleus of a megakaryocyte may be lobed or ring-shaped, and the surrounding cytoplasm contains Golgi apparatus, ribosomes, and mitochondria in abundance. The cell membrane communicates with an extensive membrane network that radiates throughout the peripheral cytoplasm (Figure 19-11●).

During their development and growth, megakaryocytes manufacture structural proteins, enzymes, and membranes. They then begin shedding cytoplasm in small membrane-enclosed packets. These packets are

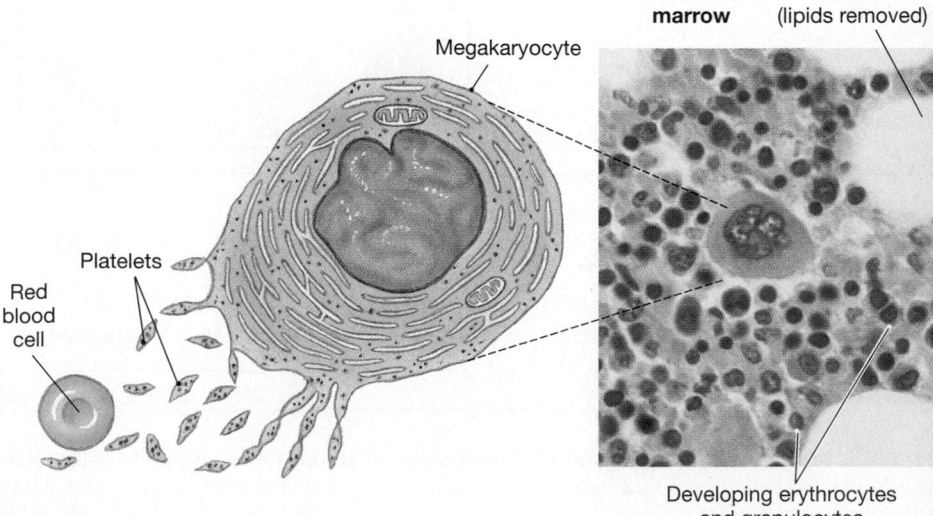

●**FIGURE 19-11**
Megakaryocytes and Platelet Formation. Megakaryocytes stand out in bone marrow sections because they are enormous and have a nucleus with an unusual shape. These cells continuously shed chunks of cytoplasm that enter the circulation as platelets. (LM × 673)

Red bone marrow

Fat cell (lipids removed)

Megakaryocyte

Platelets

Red blood cell

Developing erythrocytes and granulocytes

the platelets that enter the circulation. A mature megakaryocyte gradually loses all of its cytoplasm, producing about 4000 platelets before the nucleus is engulfed by phagocytes and broken down for recycling.

The rate of megakaryocyte activity and platelet formation is regulated by three factors:

1. **Thrombopoietin** (TPO), or *thromobocyte-stimulating factor,* is a peptide hormone that accelerates platelet formation and stimulates the production of megakaryocytes. The result is a rapid rise in the platelet count. TPO is produced in the kidneys and perhaps other sites as well. Very little is known about this hormone, other than its existence. For example, we do not know its amino acid structure, the feedback control mechanism involved, or the kidney cells responsible for its production.

2. **Interleukin-6** (Il-6), a hormone of the immune system, has a stimulatory effect on platelet formation. Current evidence indicates that the effects of Il-6 are distinct from, and less pronounced than, the effects of TPO.

3. Multi-CSF stimulates platelet production by promoting the formation and growth of megakaryocytes.

HEMOSTASIS

The process of **hemostasis** (*haima,* blood + *stasis,* halt) prevents the loss of blood through the walls of damaged vessels. At the same time, it establishes a framework for tissue repairs. Although hemostasis can be analyzed as a series of steps, it is more like a chain reaction. In essence, each step modifies the events already under way. As a result, it is easier to say when a particular step begins than when it ends.

The Vascular Phase

Cutting the wall of a blood vessel triggers a contraction in the smooth muscle fibers of the vessel wall (Figure 19-12●). This contraction produces a local **vascular spasm** that decreases the diameter of the vessel at the site of injury. Such a constriction can slow or even stop the loss of blood through the wall of a small vessel. The period of local vasoconstriction, called the **vascular phase** of hemostasis, lasts about 30 minutes.

During the vascular phase, there are changes in the endothelium at the injury site:

- The endothelial cells contract and expose the underlying basement membrane to the bloodstream.
- The endothelial cells begin releasing chemical factors and local hormones. We shall discuss several of these factors, including *ADP, tissue factor,* and *prostacyclin,* in later sections. Endothelial cells also release **endothelins**, peptide hormones that (1) stimulate

smooth muscle contraction and promote vascular spasms and (2) stimulate the division of endothelial cells, smooth muscle fibers, and fibroblasts to accelerate the repair process.

- The endothelial cell membranes become "sticky," and in small capillaries endothelial cells on opposite sides of the vessel may stick together and close off the passageway.

The Platelet Phase

Platelets now begin to attach to sticky endothelial surfaces, to the basement membrane, and to exposed collagen fibers. This attachment marks the start of the **platelet phase** of hemostasis (Figure 19-12●). The attachment of platelets to exposed surfaces is called **platelet adhesion.** As more and more platelets arrive, they begin sticking to one another as well. This process, called **platelet aggregation,** forms a **platelet plug** that may close the break in the vascular lining, if the damage is not severe or the vessel is relatively small. Platelet aggregation begins within 15 seconds after an injury occurs.

As they arrive, platelets become activated. The first sign of activation is that they change shape, becoming more spherical and developing cytoplasmic processes that extend toward adjacent platelets. At this time, the platelets begin synthesizing and releasing a wide variety of compounds, including the following:

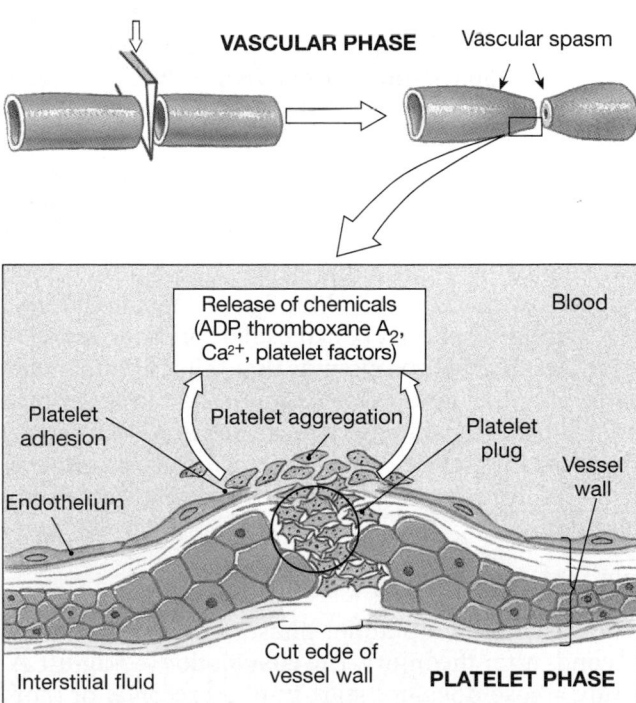

● **FIGURE 19-12** **The Vascular and Platelet Phases of Hemostasis**

- *ADP.* ADP, the primary stimulus for platelet aggregation, shape changes, and platelet secretion, is released by endothelial cells at the injury site during the vascular phase. This release in turn leads to ADP release by activated platelets. ADP causes platelet aggregation by binding to **aggregin,** a receptor protein on the platelet membrane. This binding activates adenylate cyclase, and cAMP then activates a variety of cytoplasmic enzymes that lead to changes in cell shape and to the synthesis and secretion of ADP and other compounds.
- *Thromboxane A_2.* **Thromboxane A_2,** a paracrine factor related to prostaglandins, is released by activated platelets. It causes (1) platelet aggregation and the stimulation of secretory activities by individual platelets and (2) smooth muscle contractions in the vessel walls, enhancing the vascular spasms.
- *Serotonin.* Serotonin released by activated platelets and by mast cells in the surrounding connective tissue assists thromboxane A_2 in stimulating local vasoconstriction.
- *Platelet factors.* Important platelet factors include proteins called *procoagulants,* which play a role in blood coagulation, and *platelet-derived growth factor* (PDGF), a peptide that promotes vessel repair by stimulating the division of endothelial cells, smooth muscle cells, and fibroblasts.
- In addition, calcium ions released by activated platelets supplement the local plasma supply of Ca^{2+}. Calcium ions are required for platelet aggregation and by several steps in the clotting process.

Regulation of the Platelet Phase

The platelet phase proceeds rapidly because each arriving platelet releases ADP, thromboxane, and calcium ions that stimulate further aggregation. This positive feedback loop ultimately produces a platelet plug that will be reinforced as clotting occurs. However, platelet aggregation must be controlled and restricted to the area of the injury. Several key factors limit the growth of the platelet plug: (1) **prostacyclin,** a prostaglandin that inhibits platelet aggregation and is released by endothelial cells; (2) inhibitory compounds released by white blood cells entering the area; (3) circulating plasma enzymes that break down ADP near the plug; (4) compounds that, when abundant, inhibit plug formation (for example, serotonin at high concentrations will block the action of ADP); and (5) the development of a blood clot, which reinforces the platelet plug but separates it from the general circulation.

The Coagulation Phase

The vascular and platelet phases begin within a few seconds after the injury. The **coagulation** (cō-ag-ū-LĀ-shun) **phase** does not start until 30 seconds or more after the vessel has been damaged. Coagulation, or *blood clotting,* involves a complex sequence of steps leading to the conversion of circulating fibrinogen into the insoluble protein fibrin. As the fibrin network grows, it covers the surface of the platelet plug. Passing blood cells and additional platelets are trapped within the fibrous tangle, forming a **blood clot** that effectively seals off the damaged portion of the vessel. Figure 19-13● shows the structure of a blood clot as viewed with a scanning electron microscope. Key stages of the coagulation phase are summarized in Figure 19-14●.

Clotting Factors

Normal coagulation cannot occur unless the plasma contains the necessary **clotting factors.** Important clotting factors, called **procoagulants,** include Ca^{2+} and 11 different proteins. Many of the proteins are **proenzymes** that, when converted to active enzymes, direct essential reactions in the clotting response.

For reference, specific clotting factors are identified in Table 19-4. Many are identified by Roman numerals; Ca^{2+}, for example, is also known as clotting Factor IV. All but three of the procoagulants (Factors III, IV, and VIII) are synthesized and released by the liver and are always present in the circulation. Activated platelets release five procoagulants (Factors III, IV, V, VIII, and XIII) during the platelet phase. During the coagulation phase, enzymes and proenzymes interact. The activation of one proenzyme commonly creates an enzyme that activates a second proenzyme, and so on in a chain reaction, or *cascade.*

Figure 19-14● provides an overview of the cascades involved in the *extrinsic, intrinsic,* and *common pathways.* The extrinsic pathway begins outside the blood-

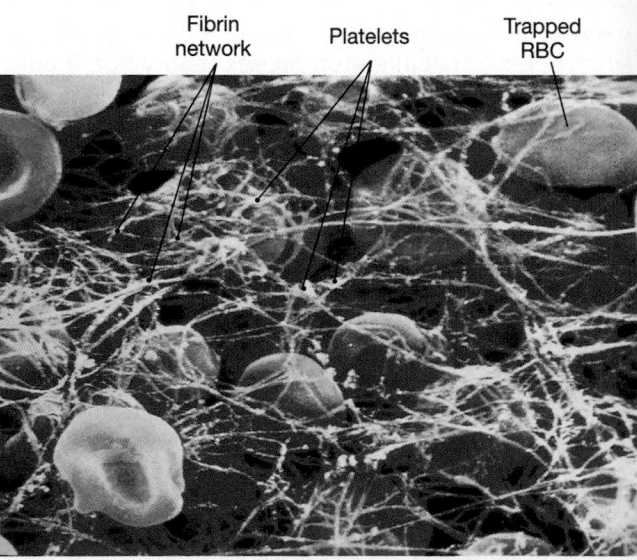

●*FIGURE 19-13* **Structure of a Blood Clot.** A color-enhanced micrograph showing the network of fibrin that forms the framework of a clot. (SEM × 3561) Red blood cells trapped in those fibers add to the mass of the blood clot and give it a red color. Platelets that stick to the fibrin strands gradually contract, shrinking the clot and tightly packing the RBCs.

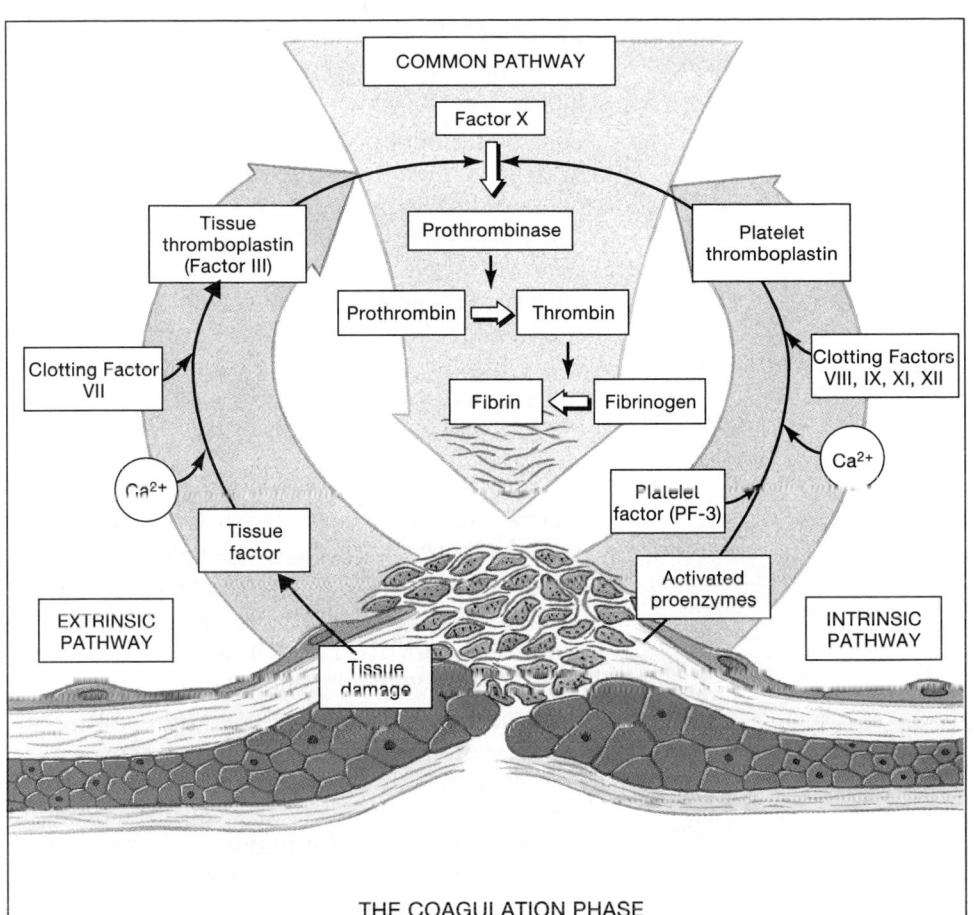

THE COAGULATION PHASE

TABLE 19-4 Procoagulants

Factor	Structure	Name	Source	Concentration in Plasma (μg/ml)	Pathway
I	Protein	Fibrinogen	Liver	2500–3500	Common
II	Protein	Prothrombin	Liver, requires vitamin K	100	Common
III	Lipoprotein	Thromboplastin (*Tissue factor*)	Damaged tissue, activated platelets	0	Extrinsic and intrinsic
IV	Ion	Calcium ions	Bone, intestines, platelets	100	Entire process
V	Protein	Proaccelerin	Liver, platelets	10	Extrinsic and intrinsic
VI	(No longer used)				
VII	Protein	Proconvertin	Liver, requires vitamin K	0.5	Extrinsic
VIII	Protein	Antihemophilic factor (AHF)	Platelets, endothelial cells	15	Intrinsic
IX	Protein	Christmas factor	Liver, requires vitamin K	3	Intrinsic
X	Protein	Stuart–Prower factor	Liver, requires vitamin K	10	Extrinsic and intrinsic
XI	Protein	Plasma thromboplastin antecedent (PTA)	Liver	<5	Intrinsic
XII	Protein	Hageman factor	Liver	<5	Intrinsic; also activates plasmin
XIII	Protein	Fibrin-stabilizing factor (FSF)	Liver, platelets	20	Stabilizes fibrin, slows fibrinolysis

stream; the intrinsic pathway begins inside the bloodstream. The extrinsic and intrinsic pathways converge at the common pathway. If you are interested in specific details concerning the steps in these pathways, consult the *Applications Manual*. AM *A Closer Look: The Clotting System*

The Extrinsic Pathway

The **extrinsic pathway** begins with the release of **tissue factor** (TF) by damaged endothelial cells or peripheral tissues. The greater the damage, the more tissue factor is released and the faster clotting occurs. Tissue factor then combines with Ca^{2+} and another procoagulant (Factor VII) to form an enzyme called **tissue thromboplastin** (Factor III).

The Intrinsic Pathway

The **intrinsic pathway** begins with the activation of proenzymes exposed to collagen fibers at the injury site. This pathway proceeds with the assistance of a platelet factor **(PF-3)** released by aggregating platelets. Platelets also release a variety of other factors that accelerate the reactions of the intrinsic pathway. After a series of linked reactions, several activated proenzymes, Ca^{2+}, and PF-3 combine to form a complex called **platelet thromboplastin.**

The Common Pathway

The **common pathway** begins when thromboplastin from either the extrinsic or the intrinsic pathway appears in the plasma. The first step involves the activation of a clotting factor (Factor X) and the formation of the enzyme **prothrombinase.** Prothrombinase converts the proenzyme prothrombin into the enzyme **thrombin** (THROM-bin). Thrombin then completes the coagulation process by converting fibrinogen, a plasma protein, to insoluble strands of fibrin.

Interactions among the Pathways

When a blood vessel is damaged, both the extrinsic and the intrinsic pathways respond, and clotting usually begins in about 15 seconds. The extrinsic pathway is shorter and faster than the intrinsic pathway, and it initiates the clotting process. In essence, the extrinsic pathway produces a small amount of thrombin very quickly. This quick patch is reinforced by the intrinsic pathway, which produces greater amounts of thrombin, but somewhat later. Once the intrinsic pathway begins generating platelet thromboplastin, it produces a larger, more effective clot than can the extrinsic pathway alone.

The time required to complete clot formation varies with the site and the nature of the injury. In tests of the clotting system, blood held in fine glass tubes normally clots in 8–18 minutes (the *coagulation time*), and a small puncture wound typically stops bleeding in 1–4 minutes (the *bleeding time*).

Feedback Control of Coagulation

Thrombin generated in the common pathway stimulates the process of coagulation by (1) stimulating formation of tissue thromboplastin and (2) stimulating the release of PF-3 by platelets. Thus, the activity of the common pathway has a stimulatory effect on both the intrinsic and extrinsic pathways. This positive feedback loop accelerates the clotting process, and speed can be very important in reducing blood loss after a severe injury.

The coagulation process is restricted by factors that either inactivate or remove procoagulants and other stimulatory agents from the blood. Examples include the following:

- Normal plasma contains several **anticoagulants,** enzymes that inhibit coagulation. One, **antithrombin-III,** inhibits several procoagulants, including thrombin.

- **Heparin,** a compound released by basophils and mast cells, is a cofactor that accelerates the activation of antithrombin-III. Heparin is used clinically to impede or prevent coagulation.

- **Thrombomodulin** is released by endothelial cells. This protein binds to thrombin and converts it to an enzyme that activates protein C. **Protein C** is a plasma protein that inactivates several clotting factors and stimulates the formation of *plasmin,* an enzyme that breaks down fibrin strands.

- *Prostacyclin* released during the platelet phase inhibits platelet aggregation and opposes the stimulatory action of thrombin, ADP, and other factors.

- Other plasma proteins with anticoagulant properties include *alpha-2-macroglobulin,* which inhibits thrombin, and *C_1 inactivator,* which inhibits several procoagulants involved in the intrinsic pathway.

Calcium Ions, Vitamin K, and Blood Coagulation

Calcium ions and **vitamin K** affect almost every aspect of the clotting process. All three pathways (intrinsic, extrinsic, and common) require the presence of Ca^{2+}, and a disorder that lowers plasma Ca^{2+} concentrations will impair blood coagulation.

Adequate amounts of vitamin K must be present for the liver to be able to synthesize four of the clotting factors, including prothrombin. Vitamin K is a fat-soluble vitamin, present in green vegetables, grain, and organ meats, that is absorbed with dietary lipids. It is obtained from the diet, and a significant amount (roughly half the daily requirement) is manufactured by bacteria within the large intestine. A diet inadequate in fats or vitamin K, or a disorder that affects fat digestion and absorption (such as problems with bile production) will lead to a vitamin K deficiency. This condition will cause the eventual breakdown of the common pathway due to a lack of procoagulants, and the entire clotting system will be inactivated.

 Abnormal Hemostasis

The coagulation process involves a complex chain of events, and a disorder that affects any individual clotting factor may disrupt the entire process. As a result, there are many different clinical conditions involving the clotting system. AM *Testing the Clotting System*

Excessive or Abnormal Coagulation

If the clotting response is inadequately controlled, clot formation will begin in the circulation rather than at an injury site. These blood clots do not stick to the wall of the vessel but continue to drift around until either plasmin digests them or they become stuck in a small blood vessel. A drifting blood clot is a type of **embolus** (EM-bō-lus; *embolos,* plug), an abnormal mass within the bloodstream. An embolus that becomes stuck in a blood vessel blocks circulation to the area downstream, killing the affected tissues. The blockage is called an **embolism**, and the tissue damage caused by the circulatory interruption is an **infarct**.

An embolus in the arterial system may get stuck in capillaries in the brain, causing a *stroke.* An embolus that forms in the venous system will probably become lodged in one of the capillaries of the lungs, causing a condition known as a *pulmonary embolism.*

A **thrombus** (*thrombos,* clot) begins to form when platelets stick to the wall of an intact blood vessel. Often, the platelets are attracted to areas called **plaques,** where endothelial and smooth muscle cells contain large quantities of lipids. The blood clot gradually enlarges, projecting into the lumen of the vessel and reducing its diameter. Eventually, the vessel may be completely blocked, or a large chunk of the clot may break off, creating an equally dangerous embolus.

Treatment of these conditions must be prompt to prevent irreparable damage to the tissues whose vessels have been restricted or blocked by emboli or thrombi. The clots may be surgically removed, or they may be attacked by enzymes such as *streptokinase* or by plasmin stimulated by administered *t-PA (tissue plasminogen activator).* The benefits of t-PA outweigh those of streptokinase or urokinase, despite the relatively high cost of t-PA.

Inadequate Coagulation

Hemophilia (hē-mō-FĒL-ē-uh) is one of many inherited disorders characterized by inadequate production of clotting factors. This condition affects about 1 in 10,000 people, with males accounting for 80–90 percent of those affected. In hemophilia, production of a single clotting factor (most commonly Factor VIII) is inadequate; the severity of the condition depends on the degree of underproduciton. In severe cases, extensive bleeding accompanies the slightest mechanical stress, and hemorrhages occur spontaneously at joints and around muscles.

In many cases, transfusions of clotting factors can reduce or control the symptoms of hemophilia, but plasma samples from many individuals must be pooled (combined) to obtain adequate amounts of clotting factors. This procedure makes the treatment very expensive and increases the risk of blood-borne infections such as hepatitis or AIDS. Gene-splicing techniques have been used to manufacture clotting Factor VIII, an essential component of the intrinsic clotting pathway. Although supplies are now limited, this procedure should eventually provide a safer and cheaper method of treatment.

DISSEMINATED INTRAVASCULAR COAGULATION The clotting process is complex and normally is precisely regulated. In *disseminated intravascular coagulation (DIC),* bacterial toxins activate thrombin, which then converts fibrinogen to fibrin within the circulating blood. Much of the fibrin is removed by phagocytes or dissolved by plasmin, but small clots may block small vessels and damage the dependent tissues. If the liver cannot keep pace with the demand for fibrinogen, clotting abilities gradually decline and uncontrolled bleeding may occur. DIC is one of the complicating factors of *septicemia,* a dangerous infection of the bloodstream that spreads bacteria and bacterial toxins throughout the body.

Clot Retraction

Once the fibrin meshwork has appeared, platelets and RBCs stick to the fibrin strands. The platelets then contract, and the entire clot begins to undergo **clot retraction,** or **syneresis** (sin-ER-ē-sis; "a drawing together"). Clot retraction occurs over a period of 30–60 minutes. It is important because (1) it pulls the torn edges of the vessel closer together, reducing residual bleeding and stabilizing the injury site; and (2) it reduces the size of the damaged area, making it easier for fibroblasts, smooth muscle cells, and endothelial cells to complete repairs.

Fibrinolysis

As the repairs proceed, the clot gradually dissolves. This process, called **fibrinolysis** (fī-bri-NOL-i-sis), begins with the activation of the proenzyme **plasminogen** (plaz-MIN-ō-jen) by two enzymes: thrombin, produced by the common pathway, and **tissue plasminogen activator (t-PA),** released by damaged tissues at the injury site. The activation of plasminogen by thrombin or t-PA produces the enzyme **plasmin** (PLAZ-min), which begins digesting the fibrin strands and eroding the foundation of the clot.

Manipulation of Hemostasis

Clinicians may attempt to prevent unwanted clotting by administering drugs that either depress the clotting response or dissolve clots already present. Important anticoagulant drugs include:

- Heparin, which activates antithrombin-III.
- **Coumadin** (COO-ma-din), or *warfarin* (WAR-fa-rin), and **dicumarol** (dī-KŪ-ma-rol), which depress the synthesis of several clotting factors by blocking the action of vitamin K.
- t-PA synthesized by recombinant DNA techniques, which stimulates plasmin formation.
- **Streptokinase** (strep-tō-KĪ-nās) or **urokinase** (ū-rō-KĪ-nās), enzymes that convert plasminogen to plasmin.
- Aspirin, which inactivates platelet enzymes involved with the production of thromboxanes and prostaglandins and inhibits the endothelial cell production of prostacyclin. Daily ingestion of small quantities of aspirin reduces the sensitivity of the clotting process. This method has been proven to be effective in preventing heart attacks in people with significant heart disease.

These compounds have widespread use in the treatment of chronic or acute circulatory blockages. We shall consider the use of the "clot-busting" agents t-PA, streptokinase, and urokinase in the treatment of heart attacks in Chapter 20.

Blood samples may be stabilized temporarily by adding heparin or *EDTA* (*e*thylene*d*iamine*t*etroacetic *a*cid) to the sample. EDTA removes Ca^{2+} from plasma, effectively preventing coagulation. In units of whole blood held for extended periods in a blood bank, *citratephosphate dextrose* (CPD) is typically added. CPD, like EDTA, ties up plasma Ca^{2+}.

☑ A sample of bone marrow has unusually few megakaryocytes. What body process would you expect to be impaired as a result?

☑ About half of our vitamin K is produced by bacteria in the large intestine. How would you expect extended use of antibiotics to affect blood clotting?

☑ Unless chemically treated, blood will coagulate in a test tube. The process begins when Factor XII becomes activated. Which clotting pathway is involved in this process?

To perform its vital functions, blood must be kept in motion. If the circulatory supply is cut off, dependent tissues may be destroyed in a matter of minutes. Individual RBCs complete two trips around the circulatory system each minute. The circulation of blood begins in the third week of embryonic development and continues throughout life. In Chapter 20, we shall examine the structure and function of the heart, the pump that maintains this vital blood flow.

SELECTED CLINICAL TERMINOLOGY

Terms Discussed in This Chapter

anemia (a-NĒ-mē-uh): A condition in which the oxygen-carrying capacity of the blood is reduced owing to low hematocrit or low blood hemoglobin concentrations. (*p. 648* and *AM*)

embolism: A condition in which a drifting blood clot becomes stuck in a blood vessel, blocking circulation to the area downstream. (*p. 667*)

hematocrit (hē-MA-tō-krit): The value that indicates the percentage of whole blood occupied by cellular elements. (*p. 646*)

hematuria (hē-ma-TOO-rē-uh): The presence of blood cells in the urine. (*p. 649*)

hemoglobinuria: The presence of hemoglobin in the urine. (*p. 649*)

hemolytic disease of the newborn (HDN): A condition in which fetal red blood cells have been destroyed by maternal antibodies. (*p. 654*)

hemophilia (hē-mō-FĒL-ē-uh): Inherited disorders characterized by inadequate production of clotting factors. (*p. 667*)

hypervolemic (hī-per-vō-LĒ-mik): Having an excessive blood volume. (*p. 644*)

hypovolemic (hī-pō-vō-LĒ-mik): Having a low blood volume. (*p. 644*)

hypoxia (hī-POKS-ē-uh): Low tissue oxygen levels. (*p. 652* and *AM*)

jaundice: A condition characterized by yellow skin and eyes, caused by abnormally high levels of plasma bilirubin; examples include *hemolytic jaundice* and *obstructive jaundice*. (*p. 651* and *AM*)

leukemia (loo-KĒ-mē-uh): A condition characterized by extremely elevated levels of circulating white blood cells; includes both *myeloid* and *lymphoid* forms. (*p. 659* and *AM*)

leukocytosis (loo-kō-sī-TŌ-sis): Excessive numbers of white blood cells in the circulation. (*p. 659*)

leukopenia (loo-kō-PĒ-nē-uh): Inadequate numbers of white blood cells in the circulation. (*p. 659*)

normochromic: The condition in which red blood cells contain normal amounts of hemoglobin. *(p. 652 and AM)*
normocytic: A term referring to cells of normal size. *(p. 652 and AM)*
normovolemic (nōr-mō-vō-LĒ-mik): Having a normal blood volume. *(p. 644)*
plaque: An abnormal area within a blood vessel where large quantities of lipids accumulate. *(p. 667)*
septicemia: A dangerous infection of the bloodstream that distributes bacteria and bacterial toxins throughout the body; may cause *disseminated intravascular coagulation* (DIC). *(p. 667)*

sickle cell anemia: An anemia resulting from the production of hemoglobin S rather than normal hemoglobin; this hemo-globin form causes red blood cell sickling at low oxygen levels. *(p. 649 and AM)*
tests of RBC status: These tests include a *reticulocyte count, hematocrit, hemoglobin concentration, RBC count, mean corpuscular volume,* and *mean corpuscular hemoglobin concentration. (p. 652 and AM)*
thalassemia: A disorder resulting from production of an abnormal form of hemoglobin. *(p. 649 and AM)*

thrombus: A blood clot that forms at the lumenal surface of a blood vessel. *(p. 667)*
transfusion: A procedure in which blood components are given to someone whose blood volume has been reduced or whose blood is defective. *(p. 649 and AM)*
venipuncture (VEN-i-punk-chur): The puncturing of a vein for any purpose, including the withdrawal of blood or the administration of medication. *(p. 644)*

AM Additional Terms Discussed in the *Applications Manual*

---

autologous marrow transplant: Reinfusion of an individual's own bone marrow collected before chemotherapy or radiation treatment.
erythrocytosis (e-rith-rō-sī-TŌ-sis)· A polycythemia affecting only red blood cells.
heterologous marrow transplant: Transplantation of bone marrow from one individual to another to replace bone

marrow destroyed during cancer therapy.
packed red cells: Red blood cells from which most of the plasma has been removed.
polycythemia (po-lē-sī-THĒ-mē-uh): A descriptive term indicating an elevated hematocrit with a normal blood volume.
tests of the clotting system: These tests include *bleeding time, coagulation time,*

partial thromboplastin time, and *plasma prothrombin time.*
thrombocytopenic purpura: An immune system disorder resulting in the production of antibodies that attack platelets. The platelet count is low, and the individual bruises easily because the remaining platelets are very fragile.

CHAPTER REVIEW

On-line resources for this chapter are on our World Wide Web site at:
http://www.prenhall.com/martini/fap

STUDY OUTLINE

INTRODUCTION, p. 642

1. The cardiovascular system provides a mechanism for the rapid transport of nutrients, waste products, respiratory gases, and cells within the body.

FUNCTIONS OF BLOOD, p. 642

1. Blood is a specialized connective tissue. Its functions include: (1) transporting dissolved gases, nutrients, hormones, and metabolic wastes; (2) regulating the pH and electrolyte composition of interstitial fluids; (3) restricting fluid losses through damaged vessels or at other injury sites; (4) defending the body against toxins and pathogens; and (5) regulating body temperature by absorbing and redistributing heat.

COMPOSITION OF BLOOD, p. 642

1. Blood contains **plasma, red blood cells (RBCs), white blood cells (WBCs),** and **platelets.** The plasma and **formed elements** constitute **whole blood,** which can be **fractionated** for analytical or clinical purposes. *(Figure 19-1; Table 19-3)*

Blood Collection and Analysis, p. 644

2. Whole blood from any region of the body has roughly the same temperature, viscosity, and pH.

PLASMA, p. 644

1. Plasma accounts for about 55 percent of the volume of blood; roughly 92 percent of plasma is water. *(Figure 19-1)*

Differences between Plasma and Interstitial Fluid, p. 644

2. Compared with interstitial fluid, plasma has a higher dissolved oxygen concentration and more dissolved proteins.

Plasma Proteins, p. 644

3. There are three primary classes of plasma proteins: *albumins, globulins,* and *fibrinogen.*
4. **Albumins** constitute about 60 percent of plasma proteins. **Globulins** constitute roughly 35 percent of plasma proteins: They include **immunoglobulins (antibodies),** which attack foreign proteins and pathogens, and **transport globulins,** which bind ions, hormones, and other compounds. **Fibrinogen** molecules are converted to **fibrin** in the clotting reaction. The removal of fibrinogen from plasma leaves a fluid called **serum.**

FORMED ELEMENTS, p. 645

Hemopoiesis, p. 645

1. **Hemopoiesis** is the process of blood cell formation. Circulating **stem cells** called **hemocytoblasts** divide to form all types of blood cells.

Red Blood Cells, p. 646

2. Red blood cells **(erythrocytes)** account for slightly less than half the blood volume and 99.9 percent of the formed elements. The **hematocrit** value indicates the percentage of whole blood occupied by formed elements.

3. RBCs transport oxygen and carbon dioxide within the bloodstream. They are highly specialized cells with a relatively large surface area. RBCs typically degenerate after about 120 days in the circulation. *(Figure 19-2)*

4. Molecules of **hemoglobin (Hb)** account for over 95 percent of the RBCs proteins. Hemoglobin is a globular protein formed from two pairs of polypeptide subunits. Each subunit contains a single molecule of **heme** (a **porphyrin**) and can reversibly bind an oxygen molecule. Damaged or dead RBCs are recycled by phagocytes. *(Figures 19-3, 19-5)*

5. **Erythropoiesis,** the formation of erythrocytes, occurs mainly within the **myeloid tissue** (red bone marrow) in adults. RBC formation increases under stimulation by **erythropoiesis-stimulating hormone (erythropoietin, EPO)**. Stages in RBC development include **erythroblasts** and **reticulocytes.** *(Figures 19-6, 19-10)*

6. **Blood type** is determined by the presence or absence of specific **surface antigens (agglutinogens)** in the RBC cell membranes: antigens **A, B,** and **Rh (D)**. Antibodies **(agglutinins)** within the plasma will react with RBCs bearing different surface antigens. *(Figure 19-7; Table 19-2)*

White Blood Cells, p. 656

7. White blood cells **(leukocytes)** defend the body against pathogens and remove toxins, wastes, and abnormal or damaged cells. *(Figure 19-9)*

8. Leukocytes exhibit **diapedesis** (the ability to move through vessel walls) and **chemotaxis** (attraction to specific chemicals).

9. *Granular leukocytes* (*granulocytes*) are subdivided into **neutrophils, eosinophils,** and **basophils.** Fifty to seventy percent of circulating WBCs are neutrophils, which are highly mobile phagocytes. The much less common eosinophils are phagocytes attracted to foreign compounds that have reacted with circulating antibodies. The relatively rare basophils migrate to damaged tissues and release histamine and **heparin**, aiding the inflammation response. *(Figure 19-9)*

10. *Agranular leukocytes* (*agranulocytes*) are subdivided into **monocytes** and **lymphocytes.** Monocytes migrating into peripheral tissues become tissue macrophages. Lymphocytes, the primary cells of the lymphatic system, include **T cells** (which enter peripheral tissues and attack foreign cells directly), **B cells** (which produce antibodies), and **NK cells** (which destroy abnormal tissue cells). *(Figure 19-9; Table 19-3)*

11. Granulocytes and monocytes are produced by stem cells in the bone marrow. Stem cells responsible for **lymphopoiesis** (production of lymphocytes) also originate in the bone marrow, but many migrate to peripheral **lymphoid tissues.** *(Figure 19-10)*

12. Factors that regulate lymphocyte maturation are not completely understood. Several **colony-stimulating factors (CSFs)** are involved in regulating other WBC populations and coordinating RBC and WBC production. *(Figure 19-10)*

Platelets, p. 662

13. **Megakaryocytes** in the bone marrow release packets of cytoplasm (platelets) into the circulating blood. The functions of platelets include (1) transporting chemicals important to the clotting process; (2) forming a temporary patch in the walls of damaged blood vessels; and (3) contracting after a clot has formed in order to reduce the size of the break in the vessel wall. *(Figure 19-11)*

HEMOSTASIS, p. 663

1. The process of **hemostasis** prevents the loss of blood through the walls of damaged vessels.

The Vascular Phase, p. 663

2. The **vascular phase** is a period of local vasoconstriction resulting from **vascular spasm** at the injury site. *(Figure 19-12)*

The Platelet Phase, p. 663

3. The **platelet phase** follows as platelets are activated, aggregate at the site, and adhere to the damaged surfaces. *(Figure 19-12)*

The Coagulation Phase, p. 664

4. The **coagulation phase** occurs as factors released by platelets and endothelial cells interact with **clotting factors** to form a **blood clot**. In this reaction sequence, suspended fibrinogen is converted to large, insoluble fibers of fibrin. *(Figures 19-13, 19-14; Table 19-4)*

Clot Retraction, p. 667

5. During **clot retraction,** platelets contract and pull the torn edges of the damaged vessel closer together.

Fibrinolysis, p. 668

6. During **fibrinolysis,** the clot gradually dissolves through the action of **plasmin**, the activated form of circulating **plasminogen**.

Manipulation of Hemostasis, p. 668

7. Clotting may be prevented by administering drugs that depress the clotting response or dissolve existing clots. Important anticoagulant drugs include heparin, **coumadin, dicumarol, t-PA, streptokinase, urokinase,** and aspirin.

REVIEW QUESTIONS

Level 1 Reviewing Facts and Terms

1. The formed elements of the blood include
   (a) plasma, fibrin, serum
   (b) albumins, globulins, fibrinogen
   (c) WBCs, RBCs, platelets
   (d) a, b, and c are correct

2. Blood temperature is approximately _____, and the blood pH averages _____:
   (a) 98.6°F, 7.0   (b) 104°F, 7.8
   (c) 100.4°F, 7.4   (d) 96.8°F, 7.0

3. Plasma contributes approximately _____ percent of the volume of whole blood, and water accounts for _____ percent of the plasma volume:

(a) 55, 92
(b) 25, 55
(c) 92, 55
(d) 35, 72

4. When the clotting proteins are removed from plasma, _____ remains:

(a) fibrinogen
(b) fibrin
(c) serum
(d) heme

5. In an adult, the only site of red blood cell production, and the primary site of white blood cell formation, is the

(a) liver
(b) spleen
(c) thymus
(d) red bone marrow

6. The most numerous WBCs found in a differential count of a healthy individual are

(a) neutrophils
(b) basophils
(c) lymphocytes
(d) monocytes

7. The differential count of a person who has an allergy would show a high number of

(a) neutrophils
(b) eosinophils
(c) basophils
(d) monocytes

8. Stem cells responsible for the process of lymphopoiesis are located in the

(a) thymus and spleen
(b) lymph nodes
(c) red bone marrow
(d) a, b, and c are correct

9. The first step in the process of hemostasis is

(a) coagulation
(b) the platelet phase
(c) fibrinolysis
(d) vascular spasm

10. The complex sequence of steps leading to the conversion of fibrinogen to fibrin is

(a) fibrinolysis
(b) coagulation
(c) retraction
(d) the platelet phase

11. What five major functions are performed by the blood?

12. What three primary classes of plasma proteins are found in the blood? What is the major function of each?

13. Which type of antibodies does the plasma contain for each of the following blood types?

(a) Type A
(b) Type B
(c) Type AB
(d) Type O

14. What four processes facilitate the movement of WBCs to areas of invasion or injury?

15. Which kinds of WBCs contribute to the body's nonspecific defenses?

16. Which three classes of lymphocytes are the primary cells of the lymphatic system? What are the functions of each class?

17. Which kinds of WBCs are produced by each of the four colony-stimulating factors (CSFs)?

18. What are the three functions of platelets during the clotting process?

19. What three factors regulate the rate of megakaryocyte activity and platelet formation?

20. What five steps are necessary for hemostasis to be completed after damage to the wall of a blood vessel?

21. What contribution from the intrinsic and the extrinsic pathways is necessary for the common pathway to begin?

22. Distinguish between an embolus and a thrombus.

Level 2   Reviewing Concepts

23. Dehydration would cause

(a) an increase in the hematocrit
(b) a decrease in the hematocrit
(c) no effect on the hematocrit
(d) an increase in plasma volume

24. Erythropoietin directly stimulates RBC formation by

(a) increasing rates of mitotic divisions in erythroblasts
(b) speeding up the maturation of red blood cells
(c) accelerating the rate of hemoglobin synthesis
(d) a, b, and c are correct

25. A person with Type A blood has

(a) antigen A in the plasma
(b) anti-B antibodies in the plasma
(c) anti-A antibodies on the red blood cells
(d) antigen B on the red blood cell membranes

26. Hemolytic disease of the newborn may result if

(a) the woman is Rh-positive and the man is Rh-negative

(b) both the man and the woman are Rh-negative
(c) both the man and the woman are Rh-positive
(d) an Rh-negative woman carries an Rh-positive fetus

27. How do red blood cells differ from typical cells in the body?

28. How does the blood defend against toxins and pathogens in the body?

29. What is the role of blood in the stabilization and maintenance of body temperature?

30. You are a respiratory therapist, and you need blood to check the efficiency of gas exchange at a patient's lungs. Which type of "stick" will you use, and from which blood vessels might you draw the sample?

31. Linda was given RhoGam during and after the delivery of her baby. What does the administration of RhoGam imply, and what effect does it have on the mother?

32. Why is aspirin sometimes prescribed for the prevention of vascular problems?

Level 3   Critical Thinking and Clinical Applications

33. A test for prothrombin time is used to determine deficiencies in the extrinsic clotting system and is prolonged if any of the factors are deficient. A test for activated partial thromboplastin time is used in a similar fashion to detect deficiencies in the intrinsic clotting system. Which factor would be deficient if a person had a prolonged prothrombin time but a normal partial thromboplastin time?

34. Which of the formed elements would you expect to see increase after the donation of a pint of blood?

35. Mary took the antibiotic cephalosporin, and her platelet

count dropped to 50,000/μl. What signs and symptoms would you expect her to exhibit?

36. Why do people suffering from advanced kidney disease commonly become anemic?

37. After Randy was diagnosed with stomach cancer, nearly all of his stomach had to be removed. Postoperative treatment included weekly injections of vitamin B_{12}. Why was this vitamin prescribed, and why was this mode of administration specified?

38. How would you expect extended use of antibiotics to affect blood clotting?

20

The Heart

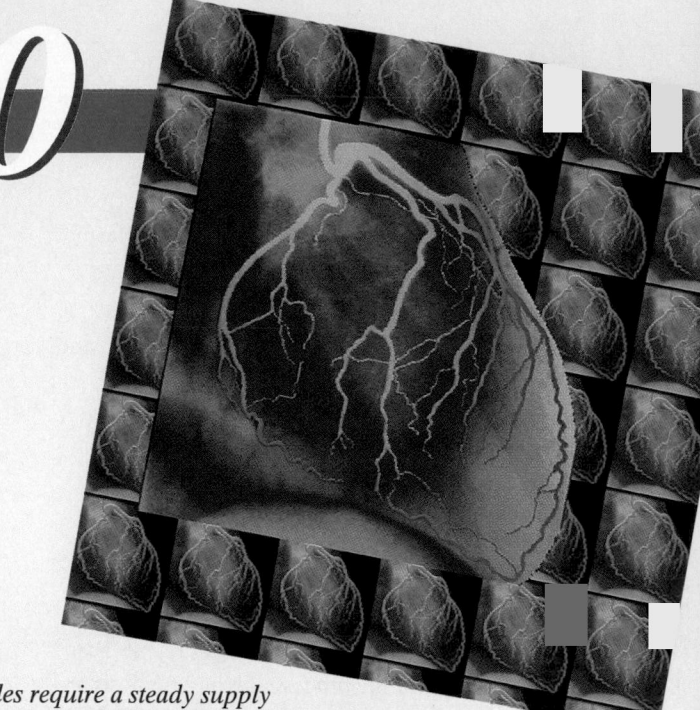

*Y*ou know from previous chapters that hard-working muscles require a steady supply
of blood to provide them with nutrients and oxygen. This is especially true of the hardest-working
muscle of all—your heart. Unlike most other muscles, the heart never rests. Not surprisingly, then, any
substantial interruption or reduction in the flow of blood to this organ has grave consequences: what we
commonly call a heart attack. Such attacks typically occur when there is an obstruction in one of the arteries
that supply the heart muscle. Physicians can check the status of these vessels using scanning techniques that
produce computer-enhanced images like these. In this chapter, we shall see how the heart supplies blood, not
only to itself but to all other tissues and organs as well. We shall also see how its activities are adjusted so that
the blood supply to active tissues—including the heart—can meet the body's constantly changing demands.

C H A P T E R   O U T L I N E   A N D   O B J E C T I V E S

Every living cell relies on the surrounding interstitial fluid for oxygen, nutrients, and waste disposal. The composition of the interstitial fluid in tissues throughout the body is kept stable through continuous exchange between the peripheral tissues and the bloodstream. Yet the blood can help maintain homeostasis only as long as it stays in motion. If blood remains stationary, its oxygen and nutrient supplies are quickly exhausted, its capacity to absorb wastes is soon saturated, and neither hormones nor white blood cells can reach their intended targets. Thus all the functions of the cardiovascular system ultimately depend on the heart. This muscular organ beats approximately 100,000 times *each day,* pumping roughly 8000 liters of blood—enough to fill 40 55-gallon drums, or 8800 quart-sized milk cartons.

We begin this chapter by examining the structural features that enable the heart to perform so reliably.

We will then consider the physiological mechanisms that regulate cardiac activity to meet changing circumstances.

AN OVERVIEW OF THE CARDIOVASCULAR SYSTEM

Blood flows through a network of blood vessels that extend between the heart and peripheral tissues. Those blood vessels can be subdivided into a **pulmonary circuit,** which carries blood to and from the gas exchange surfaces of the lungs, and a **systemic circuit,** which transports blood to and from the rest of the body. Each circuit begins and ends at the heart (Figure 20-1●), and blood travels through these circuits in sequence. For example, blood returning to the heart from the systemic circuit must complete the pulmonary circuit before reentering the systemic circuit.

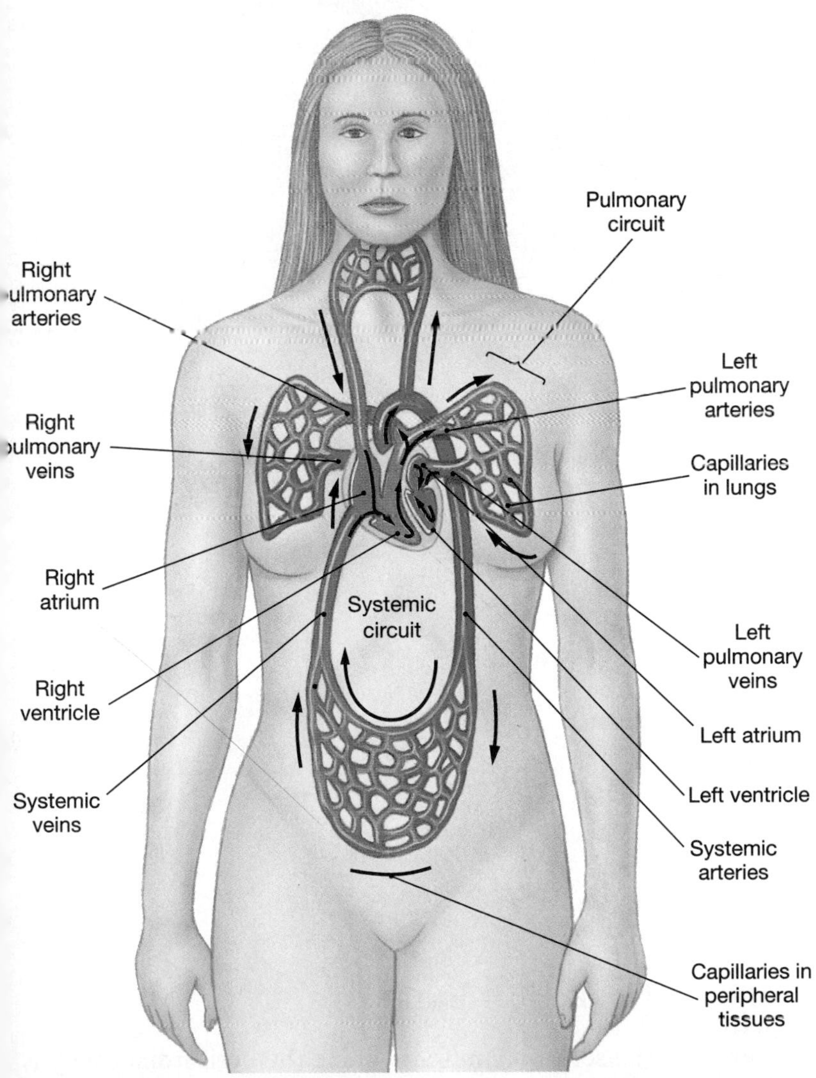

Arteries, or *efferent vessels,* carry blood away from the heart; **veins,** or *afferent vessels,* return blood to the heart. **Capillaries** are small, thin-walled vessels between the smallest arteries and veins. Capillaries are called **exchange vessels,** because their thin walls permit the exchange of nutrients, dissolved gases, and waste products between the blood and surrounding tissues.

Despite its impressive workload, the heart is a small organ, roughly the size of a clenched fist. The heart contains four muscular chambers, two associated with each circuit. The **right atrium** (Ā-trē-um; chamber; plural, *atria*) receives blood from the systemic circuit and passes it to the **right ventricle** (VEN-tri-kul; little belly). The right ventricle discharges blood into the pulmonary circuit. The **left atrium** collects blood from the pulmonary circuit and empties it into the **left ventricle.** Contraction of the left ventricle ejects blood into the systemic circuit. When the heart beats, the atria contract first, followed by the ventricles. The two ventricles contract at the same time and eject equal volumes of blood into the pulmonary and systemic circuits.

ANATOMY OF THE HEART

The heart is located near the anterior chest wall, directly posterior to the sternum (Figure 20-2a●). A midsagittal section through the trunk would not divide the heart into two equal halves because the heart (1) lies slightly to the left of the midline, (2) sits at an angle

●**FIGURE 20-1 An Overview of the Cardiovascular System.** Blood flows through separate pulmonary and systemic circuits, driven by the pumping of the heart. Each circuit begins and ends at the heart and contains arteries, capillaries, and veins.

Labels on figure:
- Right pulmonary arteries
- Right pulmonary veins
- Right atrium
- Right ventricle
- Systemic veins
- Pulmonary circuit
- Left pulmonary arteries
- Capillaries in lungs
- Left pulmonary veins
- Left atrium
- Left ventricle
- Systemic arteries
- Capillaries in peripheral tissues
- Systemic circuit

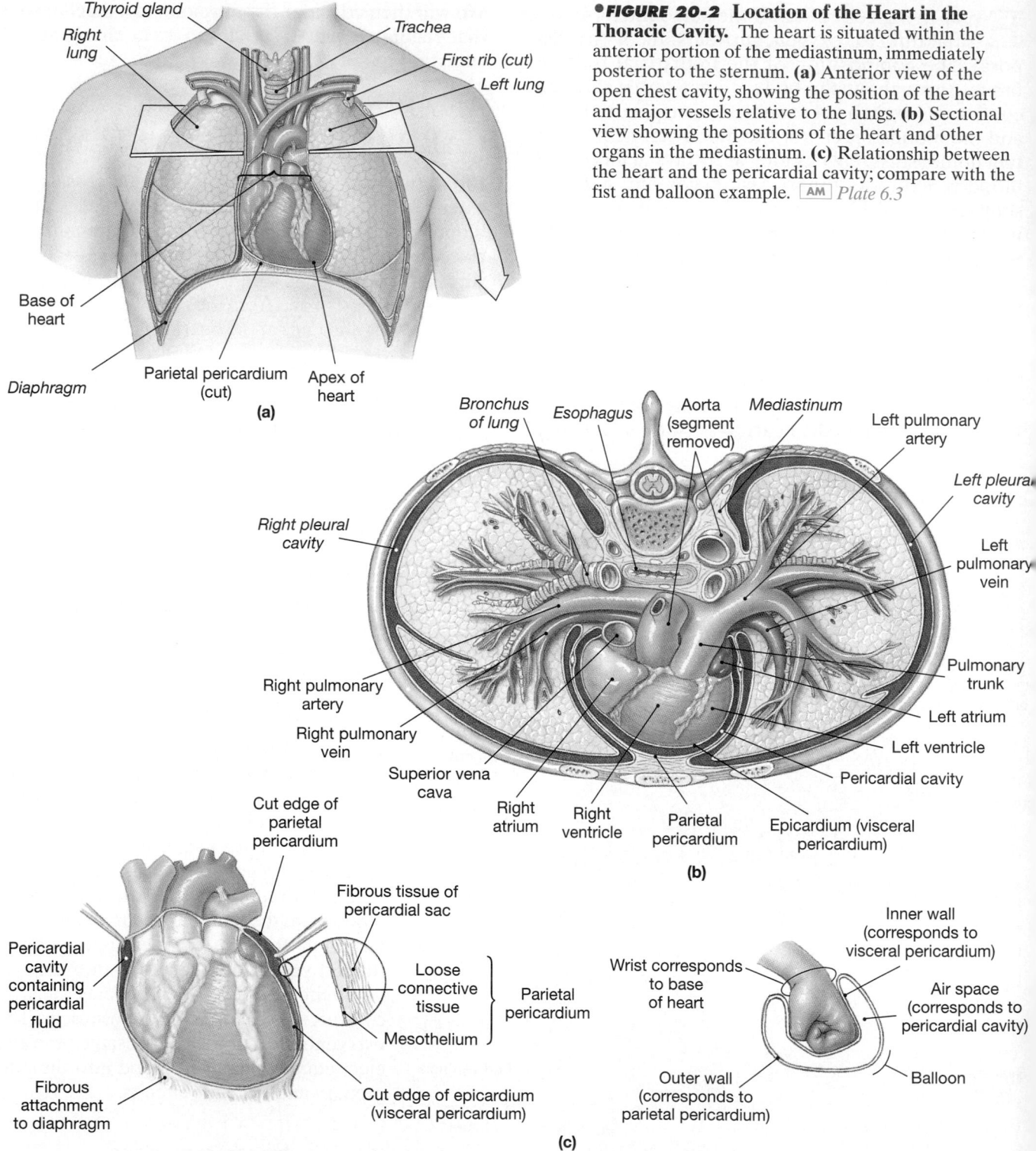

Thyroid gland
Right lung
Trachea
First rib (cut)
Left lung
Base of heart
Diaphragm
Parietal pericardium (cut)
Apex of heart

(a)

●**FIGURE 20-2** **Location of the Heart in the Thoracic Cavity.** The heart is situated within the anterior portion of the mediastinum, immediately posterior to the sternum. **(a)** Anterior view of the open chest cavity, showing the position of the heart and major vessels relative to the lungs. **(b)** Sectional view showing the positions of the heart and other organs in the mediastinum. **(c)** Relationship between the heart and the pericardial cavity; compare with the fist and balloon example. AM *Plate 6.3*

Bronchus of lung
Esophagus
Aorta (segment removed)
Mediastinum
Left pulmonary artery
Left pleural cavity
Right pleural cavity
Left pulmonary vein
Pulmonary trunk
Right pulmonary artery
Left atrium
Right pulmonary vein
Left ventricle
Superior vena cava
Pericardial cavity
Right atrium
Right ventricle
Parietal pericardium
Epicardium (visceral pericardium)

(b)

Cut edge of parietal pericardium
Fibrous tissue of pericardial sac
Inner wall (corresponds to visceral pericardium)
Pericardial cavity containing pericardial fluid
Wrist corresponds to base of heart
Air space (corresponds to pericardial cavity)
Loose connective tissue
Parietal pericardium
Mesothelium
Fibrous attachment to diaphragm
Cut edge of epicardium (visceral pericardium)
Outer wall (corresponds to parietal pericardium)
Balloon

(c)

to the longitudinal axis of the body, and (3) is rotated toward the left side. The heart is surrounded by the **pericardial** (per-i-KAR-dē-al) **cavity,** located in the anterior portion of the mediastinum. The mediastinum, which separates the two pleural cavities, also contains the thymus, esophagus, and trachea. ∞ *[p. 24]* Figure 20-2b● is a sectional view that illustrates the position of the heart relative to other structures in the mediastinum.

The Pericardium

The serous membrane lining the pericardial cavity is called the **pericardium** ∞ *[pp. 24, 132]* To visualize the relationship between the heart and the pericardial cavity, imagine pushing your fist toward the center of a large balloon (Figure 20-2c●). The balloon represents the pericardium, and your fist is the heart. Your wrist, where the balloon folds back on itself, corresponds to the base of the heart, where the *great vessels*, the largest veins and arteries in the body, are attached to the heart. The space inside the balloon is the pericardial cavity.

The pericardium can be subdivided into the *visceral pericardium* and the *parietal pericardium*. The **visceral pericardium**, or *epicardium*, covers the outer surface of the heart; the **parietal pericardium** lines the inner surface of the **pericardial sac,** which surrounds the heart (Figure 20-2c●). The pericardial sac, which is reinforced by a dense network of collagen fibers, stabilizes the position of the heart and associated vessels within the mediastinum.

The space between the opposing parietal and visceral surfaces is the pericardial cavity. This cavity normally contains 10–20 ml of **pericardial fluid** secreted by the pericardial membranes. Pericardial fluid acts as a lubricant, reducing friction between the opposing surfaces as the heart beats.

PERICARDITIS A variety of pathogens may infect the pericardium, producing a condition known as **pericarditis.** The inflamed pericardial surfaces rub against one another, producing a distinctive scratching sound. In addition, the pericardial irritation and inflammation commonly result in an increased production of pericardial fluid. Fluid then collects in the pericardial cavity, restricting the movement of the heart. This condition is called **cardiac tamponade** (tam-pō-NĀD; *tampon*, plug). [AM] *Infection and Inflammation of the Heart*

Superficial Anatomy of the Heart

The four cardiac chambers can easily be identified in a superficial view of the heart (Figure 20-3●). The two atria have relatively thin muscular walls, and they are highly expandable. When not filled with blood, the outer portion of each atrium deflates and becomes a lumpy, wrinkled flap. This expandable extension of an atrium is called an **auricle** (AW-ri-kul; *auris,* ear), because it reminded early anatomists of the external ear, or an *atrial appendage* (Figure 20-3a●). The **coronary sulcus,** a deep groove, marks the border between the atria and the ventricles. The **anterior interventricular sulcus** and the **posterior interventricular sulcus,** shallower depressions, mark the boundary line between the left and right ventricles (Figure 20-3a,b●).

The connective tissue of the epicardium at the coronary and interventricular sulci generally contains substantial amounts of fat. In fresh or preserved hearts, this fat must be stripped away to expose the underlying grooves. These sulci also contain the arteries and veins that supply blood to the cardiac muscle of the heart.

The heart has an attached *base* and a free *apex*. The great veins and arteries of the circulatory system are connected to the superior end of the heart at the **base.** The base sits posterior to the sternum at the level of the third costal cartilage, centered about 1.2 cm (0.5 in.) to the left side (Figure 20-3c●). The inferior, pointed tip of the heart is the **apex** (Ā-peks). A typical adult heart measures approximately 12.5 cm (5 in.) from the attached base to the apex. The apex reaches the fifth intercostal space approximately 7.5 cm (3 in.) to the left of the midline.

Internal Anatomy and Organization

The right atrium communicates with the right ventricle, and the left atrium with the left ventricle. The two atria are separated by the **interatrial septum** (*septum,* wall), and the two ventricles are separated by the much thicker **interventricular septum** (Figure 20-4a,c●, p. 677). Each septum is a muscular partition. **Atrioventricular (AV) valves,** folds of fibrous tissue, extend into the openings between the atria and ventricles. These valves permit blood flow in one direction only: from the atria into the ventricles.

The Right Atrium

The right atrium receives blood from the systemic circuit through the two great veins, the **superior vena cava** (VĒ-na KĀ-vuh; plural, *venae cavae*) and the **inferior vena cava.** The superior vena cava delivers blood to the right atrium from the head, neck, upper limbs, and chest. The superior vena cava opens into the posterior and superior portion of the right atrium. The inferior vena cava carries blood to the right atrium from the rest of the trunk, the viscera, and the lower limbs. The inferior vena cava opens into the posterior and inferior portion of the right atrium. The *coronary veins* of the heart return blood to the **coronary sinus,** which opens into the right atrium inferior to the connection with the inferior vena cava.

Prominent muscular ridges, the **pectinate muscles** (*pectin,* comb), or *musculi pectinati,* run along the inner surface of the auricle and across the adjacent anterior atrial wall (Figure 20-4a,c●). The interatrial septum separates the right atrium from the left atrium. From the fifth week of embryonic development until birth, the **foramen ovale,** an oval opening, penetrates the septum and connects the two atria. The foramen ovale permits blood flow from the right atrium to the left atrium while the lungs are developing. At birth, the foramen ovale closes; after 48 hours, the opening is permanently sealed. A small depression, the **fossa ovalis,** persists at this site in the adult heart (Figure 20-4a,c●). If the foramen ovale does not close, blood will flow from the left atrium into the right atrium rather than the opposite way, because after birth, blood pressure in the pulmonary circuit is lower than that in the systemic circuit. We will consider the physiological effects of this condition in Chapter 21.

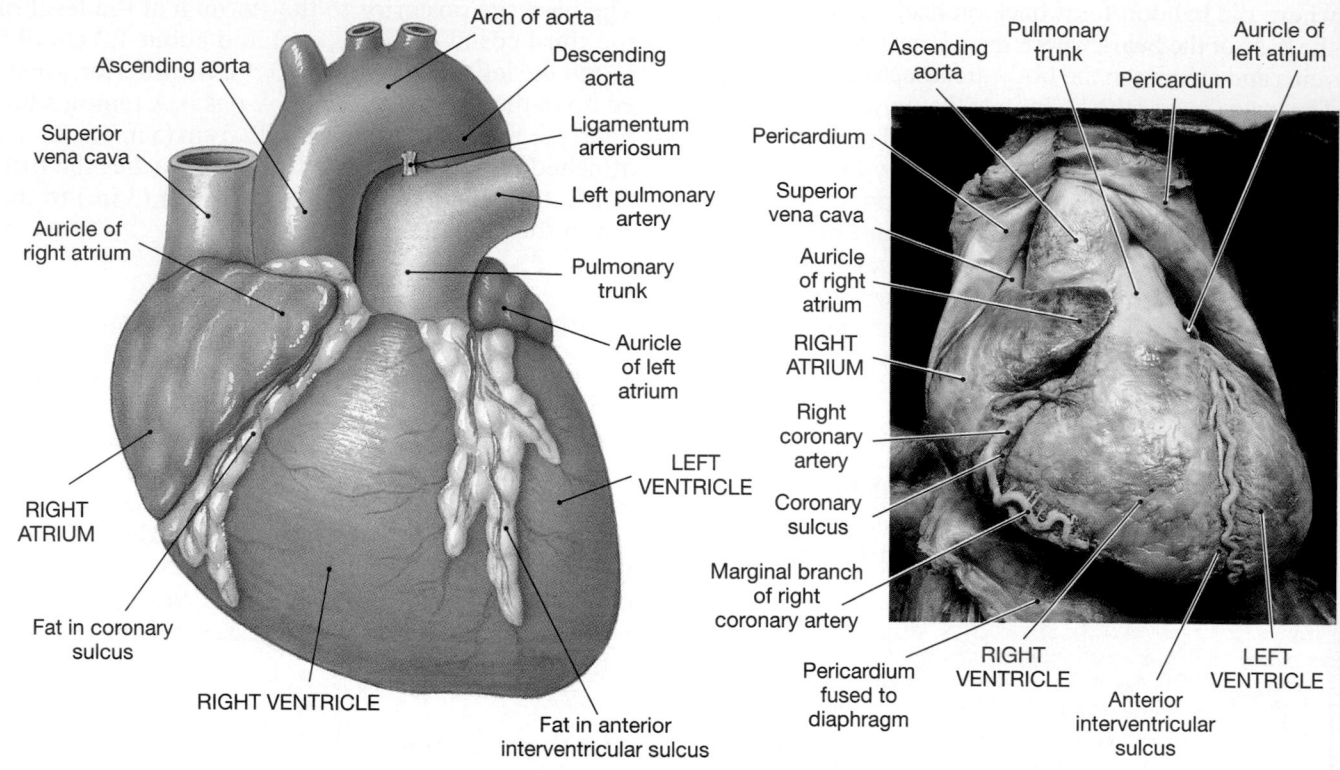

(a) Anterior (sternocostal) surface

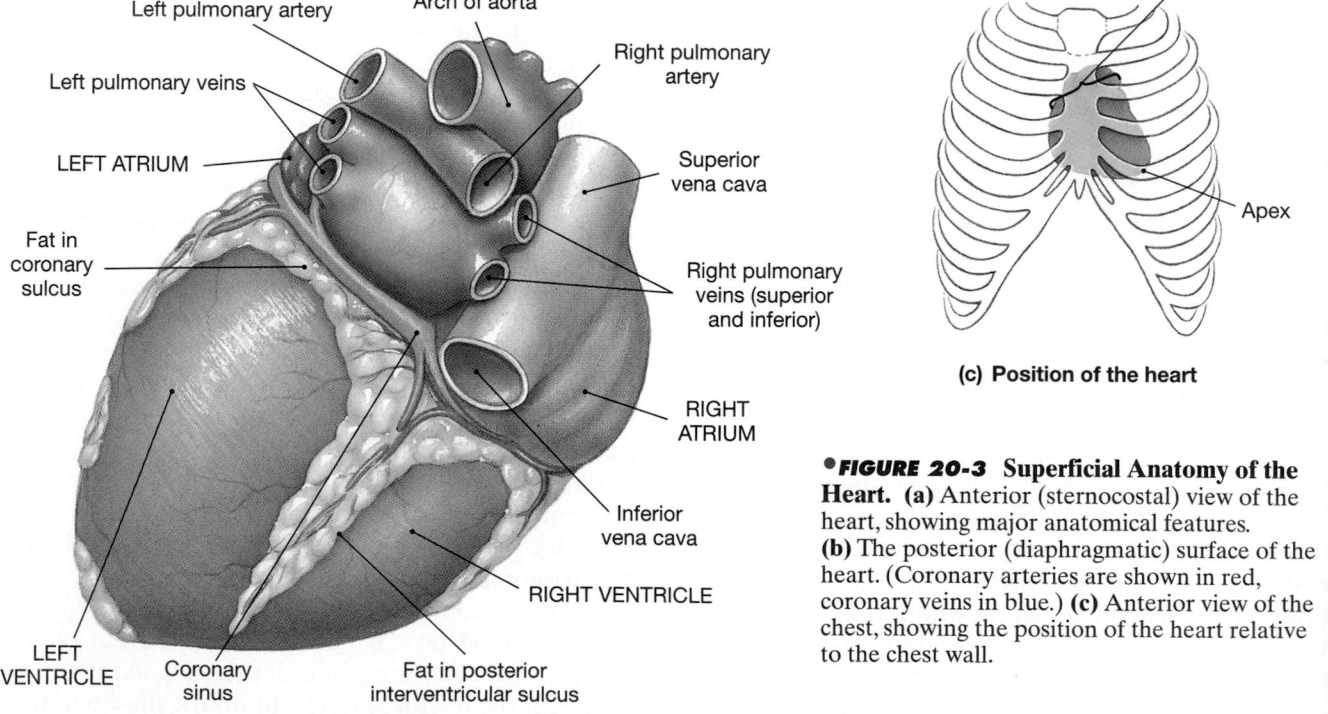

(b) Posterior (diaphragmatic) surface

(c) Position of the heart

•FIGURE 20-3 **Superficial Anatomy of the Heart.** **(a)** Anterior (sternocostal) view of the heart, showing major anatomical features. **(b)** The posterior (diaphragmatic) surface of the heart. (Coronary arteries are shown in red, coronary veins in blue.) **(c)** Anterior view of the chest, showing the position of the heart relative to the chest wall.

The Right Ventricle

Blood travels from the right atrium into the right ventricle through a broad opening bounded by three fibrous flaps. These flaps, or **cusps,** are part of the **right atrioventricular (AV) valve,** also known as the **tricuspid** (trī-KUS-pid; *tri,* three) **valve**. The free edge of each cusp is attached to tendinous connective tissue fibers called the **chordae tendineae** (KOR-dē TEN-di-nē-ē; tendinous cords). These

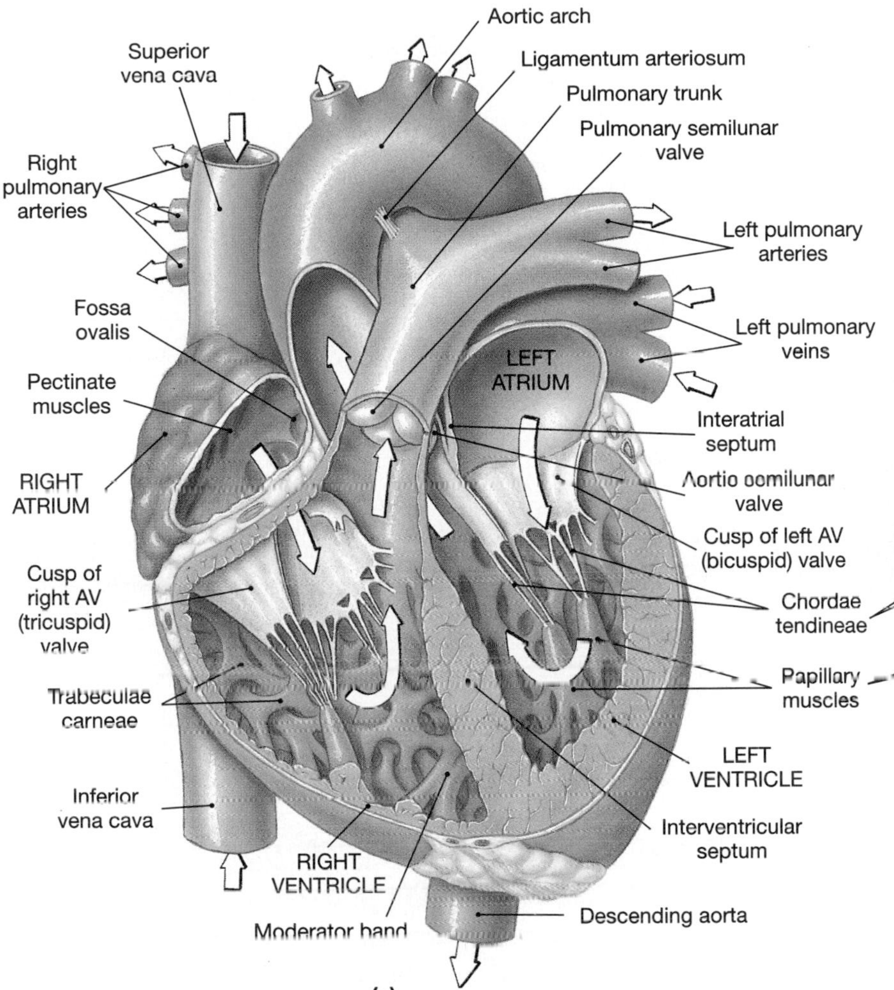

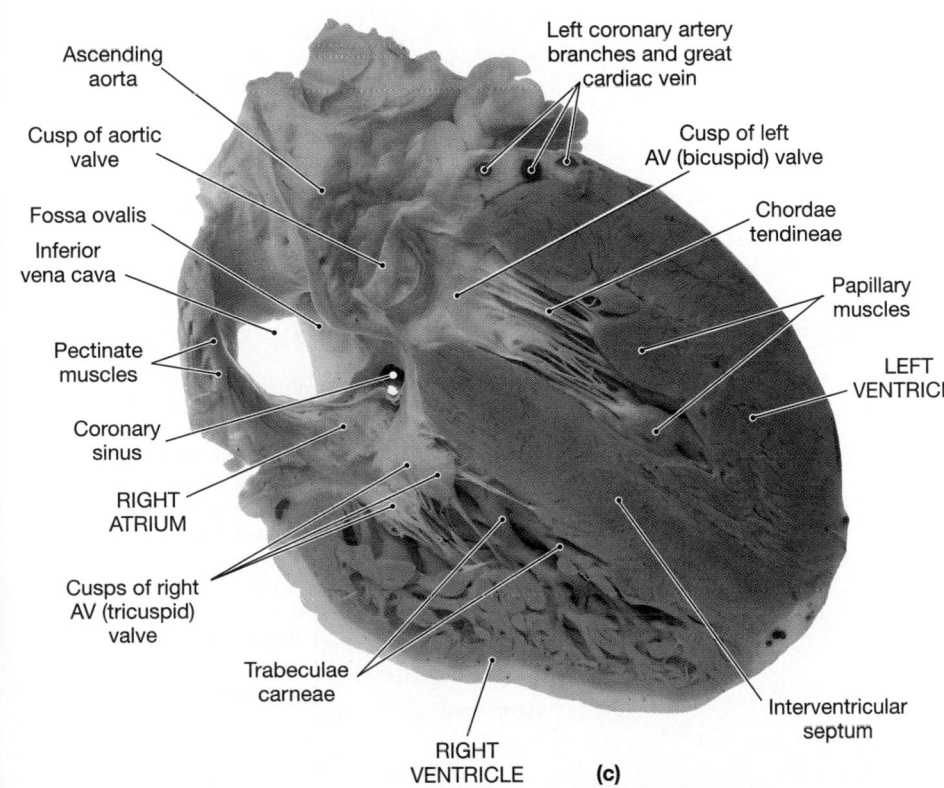

Superior vena cava

Aortic arch

Ligamentum arteriosum

Pulmonary trunk

Pulmonary semilunar valve

Right pulmonary arteries

Left pulmonary arteries

Fossa ovalis

LEFT ATRIUM

Left pulmonary veins

Pectinate muscles

Interatrial septum

Aortic semilunar valve

RIGHT ATRIUM

Cusp of left AV (bicuspid) valve

Cusp of right AV (tricuspid) valve

Chordae tendineae

Trabeculae carneae

Papillary muscles

Inferior vena cava

LEFT VENTRICLE

Interventricular septum

RIGHT VENTRICLE

Moderator band

Descending aorta

(a)

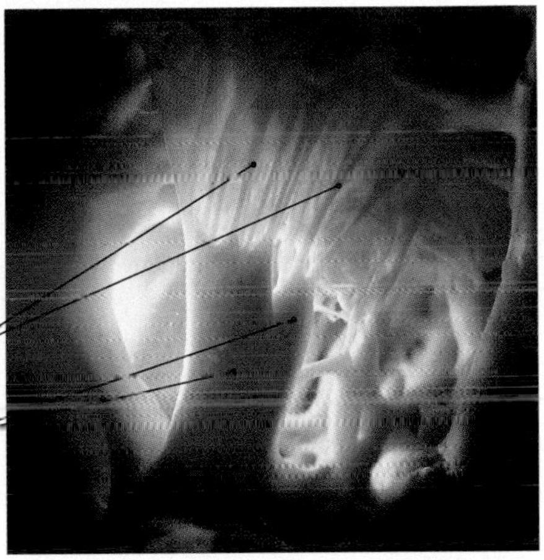

(b)

Ascending aorta

Left coronary artery branches and great cardiac vein

Cusp of aortic valve

Cusp of left AV (bicuspid) valve

Fossa ovalis

Chordae tendineae

Inferior vena cava

Papillary muscles

Pectinate muscles

LEFT VENTRICLE

Coronary sinus

RIGHT ATRIUM

Cusps of right AV (tricuspid) valve

Trabeculae carneae

Interventricular septum

RIGHT VENTRICLE **(c)**

● **FIGURE 20-4** Sectional **Anatomy of the Heart. (a)** A diagrammatic frontal section through the heart, showing major landmarks and the path of blood flow through the atria, ventricles, and associated vessels. **(b)** Photograph of papillary muscles and chordae tendineae supporting the right AV (tricuspid) valve. The picture was taken inside the right ventricle, looking toward a light shining from the right atrium. **(c)** Sectional view of the heart.

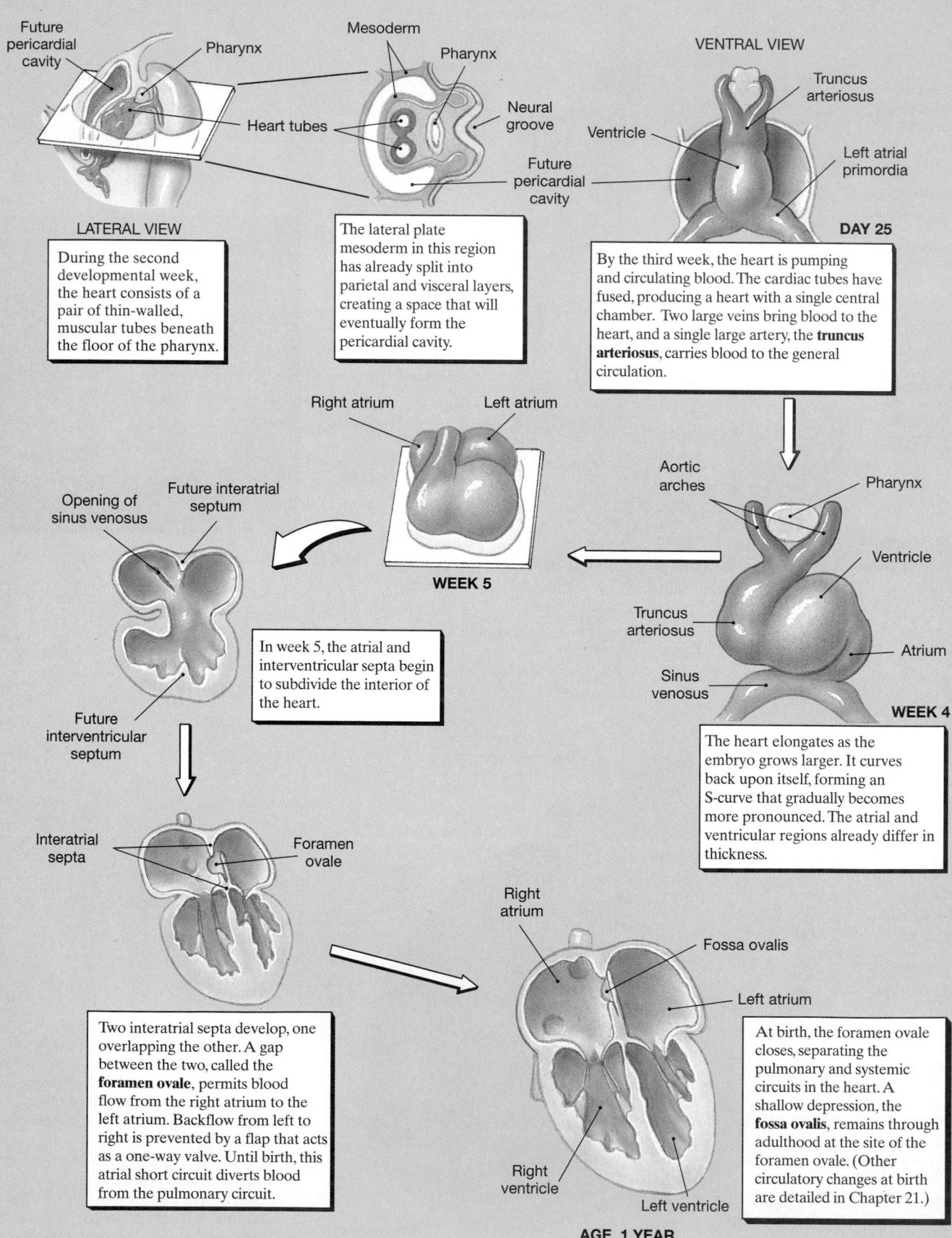

LATERAL VIEW

Future pericardial cavity

Pharynx

Heart tubes

During the second developmental week, the heart consists of a pair of thin-walled, muscular tubes beneath the floor of the pharynx.

Mesoderm

Pharynx

Neural groove

Future pericardial cavity

The lateral plate mesoderm in this region has already split into parietal and visceral layers, creating a space that will eventually form the pericardial cavity.

VENTRAL VIEW

Truncus arteriosus

Ventricle

Left atrial primordia

DAY 25

By the third week, the heart is pumping and circulating blood. The cardiac tubes have fused, producing a heart with a single central chamber. Two large veins bring blood to the heart, and a single large artery, the **truncus arteriosus**, carries blood to the general circulation.

Right atrium

Left atrium

WEEK 5

Opening of sinus venosus

Future interatrial septum

Future interventricular septum

In week 5, the atrial and interventricular septa begin to subdivide the interior of the heart.

Aortic arches

Pharynx

Ventricle

Truncus arteriosus

Atrium

Sinus venosus

WEEK 4

The heart elongates as the embryo grows larger. It curves back upon itself, forming an S-curve that gradually becomes more pronounced. The atrial and ventricular regions already differ in thickness.

Interatrial septa

Foramen ovale

Two interatrial septa develop, one overlapping the other. A gap between the two, called the **foramen ovale**, permits blood flow from the right atrium to the left atrium. Backflow from left to right is prevented by a flap that acts as a one-way valve. Until birth, this atrial short circuit diverts blood from the pulmonary circuit.

Right atrium

Fossa ovalis

Left atrium

Right ventricle

Left ventricle

AGE 1 YEAR

At birth, the foramen ovale closes, separating the pulmonary and systemic circuits in the heart. A shallow depression, the **fossa ovalis**, remains through adulthood at the site of the foramen ovale. (Other circulatory changes at birth are detailed in Chapter 21.)

fibers originate at the **papillary** (PAP-i-ler-ē) **muscles,** conical muscular projections that arise from the inner surface of the right ventricle (Figure 20-4b●). The valve closes when the right ventricle contracts, preventing the backflow of blood into the right atrium. Without the chordae tendineae, the cusps would be like swinging doors that permitted blood flow in both directions.

The internal surface of the ventricle also contains a series of muscular ridges, the **trabeculae carneae** (tra-BEK-ū-lē CAR-nē-ē; *carneus,* fleshy). The **moderator band** is a muscular ridge that extends horizontally from the inferior portion of the interventricular septum and connects to the anterior papillary muscle. The moderator band is variable in size in humans. It is noteworthy because it contains a portion of the *conducting system,* an internal network that coordinates the contractions of cardiac muscle cells. The moderator band delivers the contraction stimulus to the papillary muscles so that they begin tensing the chordae tendineae before the rest of the ventricle contracts.

The superior end of the right ventricle tapers to a conical pouch, the **conus arteriosus,** which ends at the **pulmonary semilunar valve.** The pulmonary semilunar valve consists of three semilunar (half-moon–shaped) cusps of thick connective tissue. Blood flowing from the right ventricle passes through this valve to enter the **pulmonary trunk,** the start of the pulmonary circuit. The arrangement of cusps prevents backflow as the right ventricle relaxes. Once within the pulmonary trunk, blood flows into the **left pulmonary arteries** and the **right pulmonary arteries.** These vessels branch repeatedly within the lungs before supplying the capillaries where gas exchange occurs.

The Left Atrium

From the respiratory capillaries, blood collects into small veins that ultimately unite to form the four *pulmonary veins.* The posterior wall of the left atrium receives blood from two **left** and two **right pulmonary veins.** Like the right atrium, the left atrium has an auricle and a valve, the **left atrioventricular (AV) valve,** or **bicuspid** (bī-KUS-pid) **valve** (Figure 20-4a,c●). As the name *bicuspid* implies, the left AV valve contains a pair, not a trio, of cusps. Clinicians often use the term **mitral** (MĪ-tral; *mitre,* a bishop's hat) when referring to this valve. The left AV valve permits the flow of blood from the left atrium into the left ventricle.

The Left Ventricle

The right and left ventricles contain equal amounts of blood, but the left ventricle is much larger than the right because it has thicker walls. The thick, muscular wall enables the left ventricle to develop pressure

sufficient to push blood through the large systemic circuit; the right ventricle needs to pump blood, at lower pressure, only about 15 cm (6 in.) to and from the lungs. The internal organization of the left ventricle resembles that of the right ventricle (Figure 20-4a,c●). The trabeculae carneae are prominent, and a pair of large papillary muscles tense the chordae tendineae that brace the cusps of the AV valve and prevent backflow of blood into the left atrium.

Blood leaves the left ventricle by passing through the **aortic semilunar valve** into the **ascending aorta.** The arrangement of cusps in the aortic semilunar valve is the same as that in the pulmonary semilunar valve. Saclike dilations of the base of the ascending aorta occur adjacent to each cusp. These sacs, called **aortic sinuses,** prevent the individual cusps from sticking to the wall of the aorta when the valve opens. Once the blood has been pumped out of the heart and into the systemic circuit, the aortic semilunar valve prevents backflow into the left ventricle. From the ascending aorta, blood flows on through the **aortic arch** and into the **descending aorta** (Figure 20-4a●). The pulmonary trunk is attached to the aortic arch by the *ligamentum arteriosum,* which marks the path of an important fetal blood vessel that linked the pulmonary and systemic circuits.

Structural Differences between the Left and Right Ventricles

The function of an atrium is to collect blood that is returning to the heart and deliver it to the attached ventricle. The functional demands on the right and left atria are very similar, and the two chambers look almost identical. The demands on the right and left ventricles, however, are very different, and there are significant structural differences between the two.

Anatomical differences between the left and right ventricles are best seen in a three-dimensional view (Figure 20-5a●). The lungs are close to the heart, and the pulmonary blood vessels are relatively short and wide. Thus the right ventricle normally does not need to push very hard to propel blood through the pulmonary circuit. The wall of the right ventricle is relatively thin, and in sectional view it resembles a pouch attached to the massive wall of the left ventricle. When it contracts, the right ventricle acts like a bellows pump, squeezing the blood against the mass of the left ventricle. This mechanism moves blood very efficiently with minimal effort, but it develops relatively low pressures.

A comparable pumping arrangement would not be suitable for the left ventricle, because six to seven times as much force must be exerted to push blood around the systemic circuit. The left ventricle has an extremely thick muscular wall, and it is round in cross

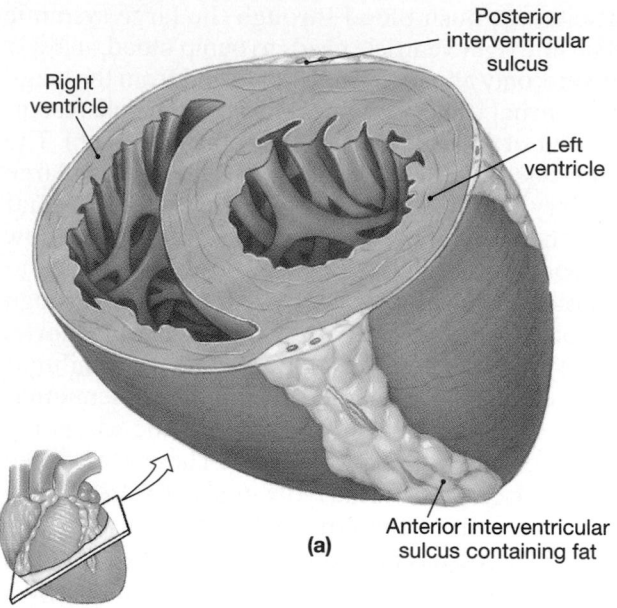

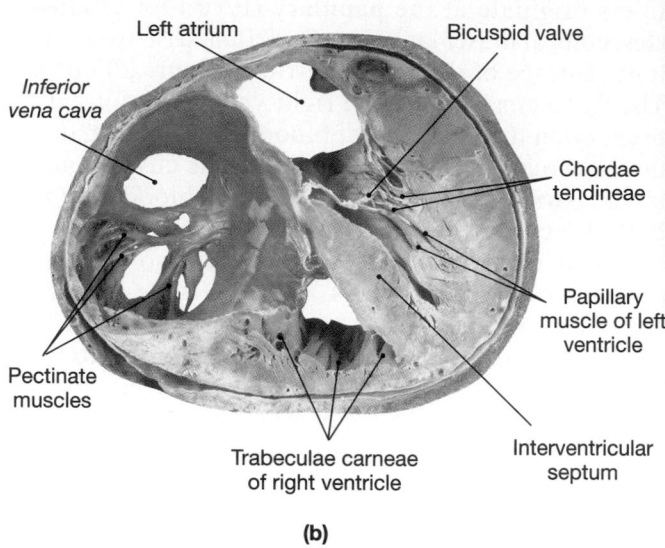

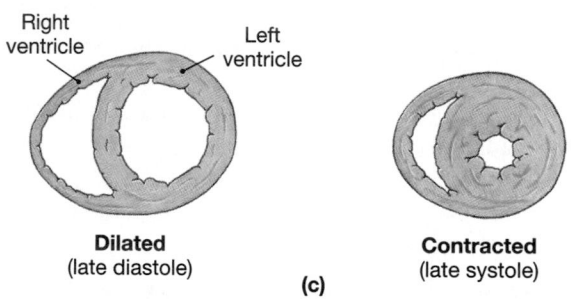

●FIGURE 20-5 **Structural Differences between the Left and Right Ventricles.** **(a)** Diagrammatic sectional view through the heart, showing the relative thicknesses of the two ventricles. Note the pouchlike shape of the right ventricle and the mass of the left ventricle. **(b)** Horizontal section through the heart at the level of vertebra T$_8$. **(c)** Diagrammatic views of the ventricles just before (left) and just after a contraction.

section (Figure 20-5●). When this ventricle contracts, two things happen: (1) The distance between the base and apex decreases, and (2) the diameter of the ventricular chamber decreases. If you imagine the effects of simultaneously squeezing and rolling up the end of a toothpaste tube, you will get the idea. The forces generated are quite powerful, more than enough to open the semilunar valve and eject blood into the ascending aorta. As the powerful left ventricle contracts, it also bulges into the right ventricular cavity (Figure 20-5c●). This dual action improves the efficiency of the right ventricle's efforts. Individuals whose right ventricular musculature has been severely damaged may survive because the contraction of the left ventricle helps push blood into the pulmonary circuit.

AM *The Cardiomyopathies*

The Heart Valves

We will now detail the structure and function of the various heart valves.

THE ATRIOVENTRICULAR VALVES The atrioventricular valves prevent the backflow of blood from the ventricles to the atria when the ventricles are contracting. The chordae tendineae and papillary muscles play an important role in the normal function of the AV valves. During the period known as *ventricular diastole* (dī-AS-tō-lē), the ventricles are relaxed. As each relaxed ventricle fills with blood, the chordae tendineae are loose and the AV valves offer no resistance to the flow of blood from the atria to the ventricles (Figure 20-6a●). The ventricles contract during the period of *ventricular systole* (SIS-tō-lē). As the ventricles begin to contract, blood moving back toward the atria swings the cusps together, closing the valves (Figure 20-6b●). At the same time, the contraction of the papillary muscles tenses the chordae tendineae and stops the cusps before they swing into the atria. If the chordae tendineae are cut or the papillary muscles damaged, the valves act like swinging doors, and there is backflow, or **regurgitation,** of blood into the atria each time the ventricles contract.

THE SEMILUNAR VALVES The pulmonary and aortic semilunar valves prevent the backflow of blood from the pulmonary trunk and aorta into the right and left ventricles. The semilunar valves do not require muscular braces because the arterial walls do not contract, and the relative positions of the cusps are stable. When these valves close, the three symmetrical cusps support one another like the legs of a tripod (Figure 20-6a,c●).

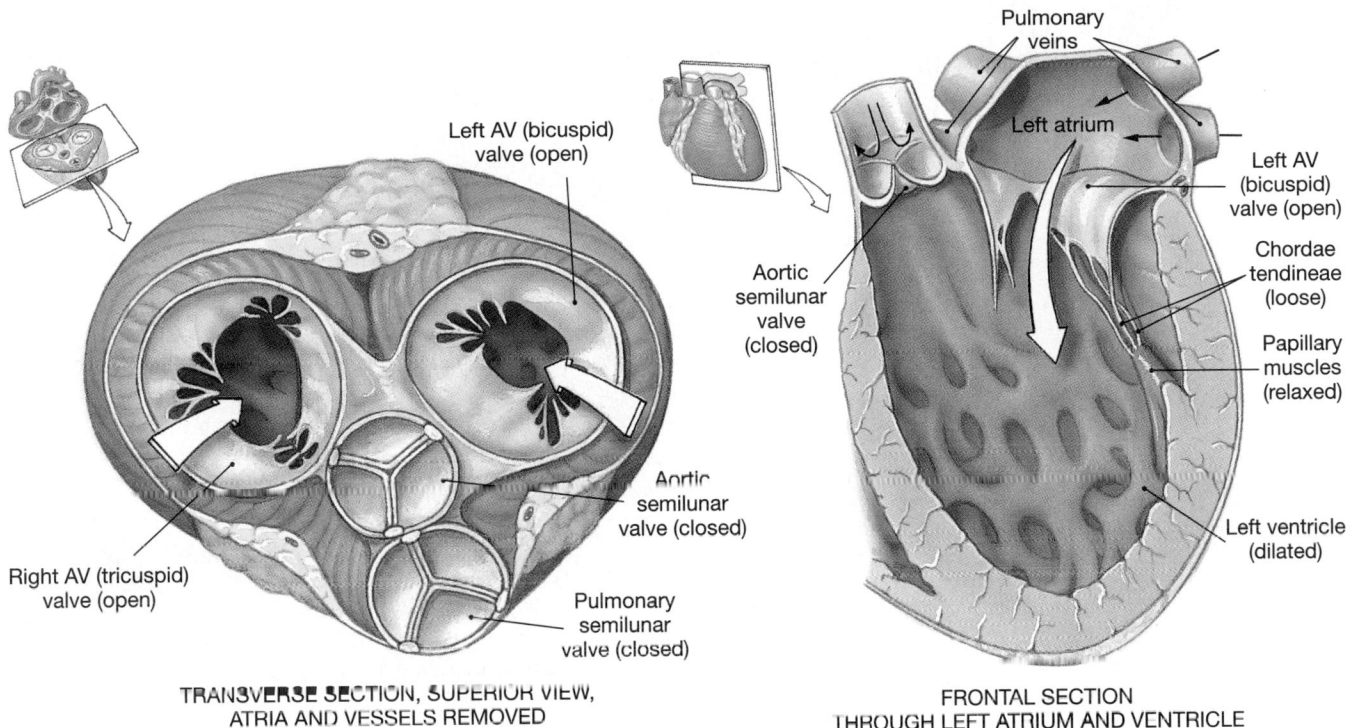

(a) Ventricular diastole (relaxation)

Left AV (bicuspid) valve (open)

Aortic semilunar valve (closed)

Right AV (tricuspid) valve (open)

Aortic semilunar valve (closed)

Pulmonary semilunar valve (closed)

TRANSVERSE SECTION, SUPERIOR VIEW, ATRIA AND VESSELS REMOVED

Pulmonary veins

Left atrium

Left AV (bicuspid) valve (open)

Chordae tendineae (loose)

Papillary muscles (relaxed)

Left ventricle (dilated)

FRONTAL SECTION THROUGH LEFT ATRIUM AND VENTRICLE

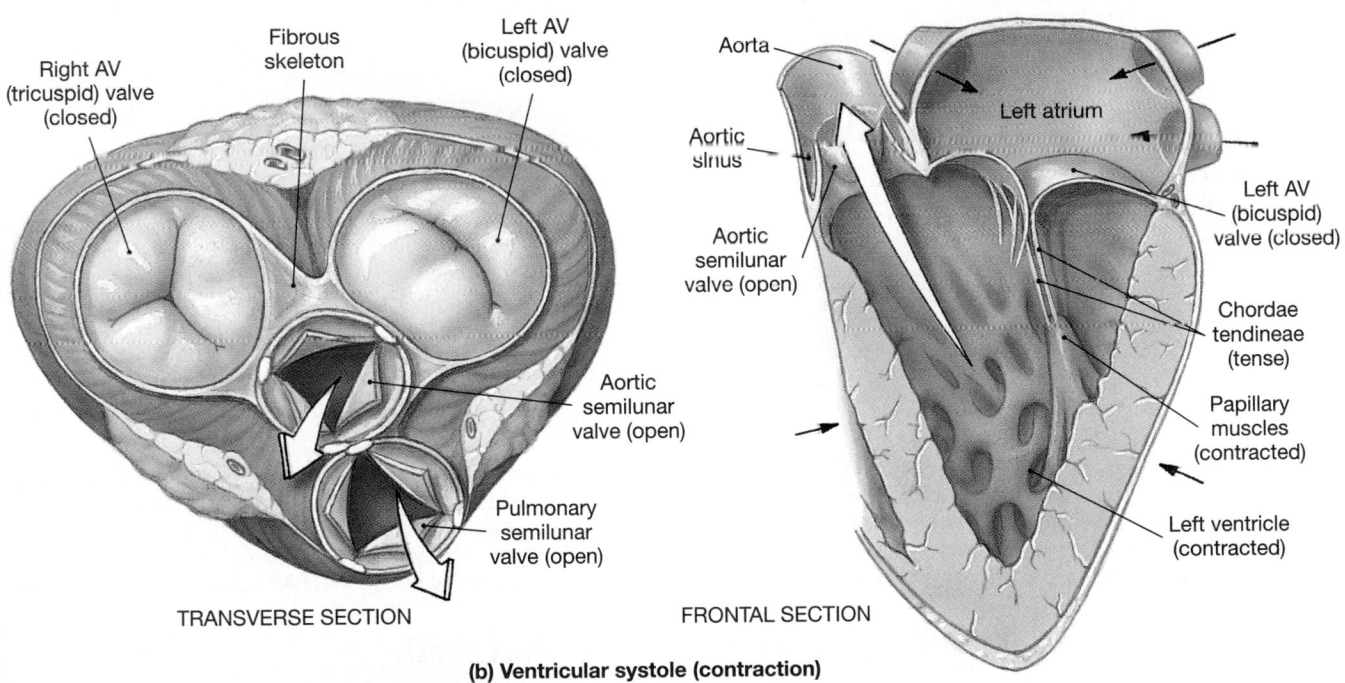

(b) Ventricular systole (contraction)

Right AV (tricuspid) valve (closed)

Fibrous skeleton

Left AV (bicuspid) valve (closed)

Aortic semilunar valve (open)

Pulmonary semilunar valve (open)

TRANSVERSE SECTION

Aorta

Aortic sinus

Aortic semilunar valve (open)

Left atrium

Left AV (bicuspid) valve (closed)

Chordae tendineae (tense)

Papillary muscles (contracted)

Left ventricle (contracted)

FRONTAL SECTION

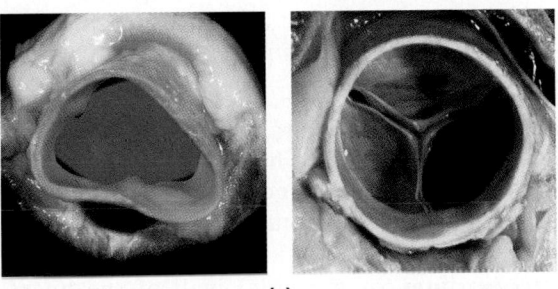

(c)

•*FIGURE 20-6* **Valves of the Heart.** **(a)** Valve position during ventricular diastole (relaxation), when the AV valves are open and the semilunar valves are closed. Note that the chordae tendineae are loose and the papillary muscles are relaxed. **(b)** The appearance of the cardiac valves during ventricular systole (contraction), when the AV valves are closed and the semilunar valves are open. In the frontal section, note the attachment of the left AV valve to the chordae tendineae and papillary muscles. **(c)** The aortic semilunar valve in the open (left) and closed (right) positions. Note how the individual cusps brace one another in the closed position.

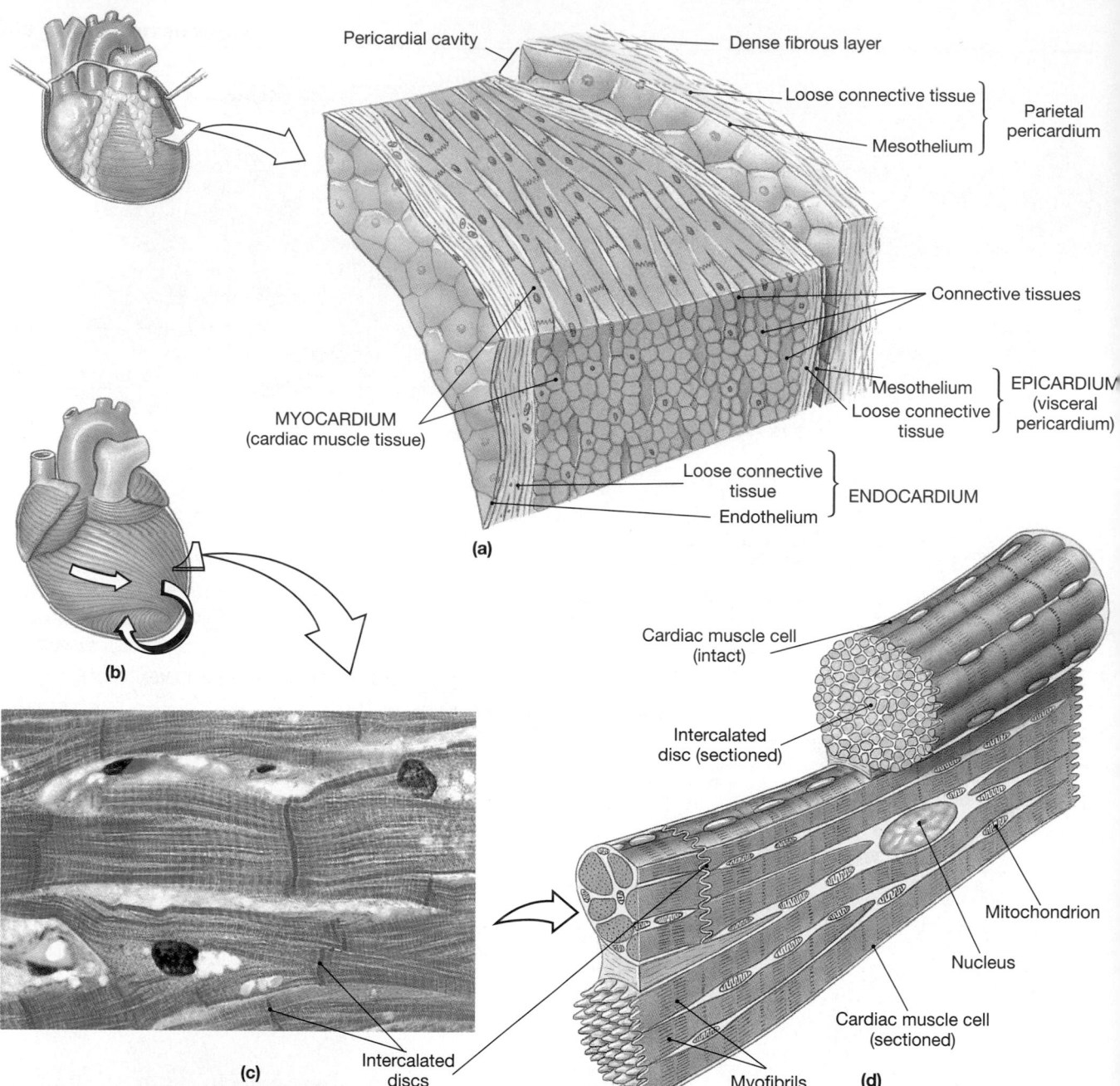

•FIGURE 20-7 The Heart Wall. **(a)** A diagrammatic section through the heart wall, showing the relative positions of the epicardium, myocardium, and endocardium. **(b)** Cardiac muscle tissue in the heart forms concentric layers that wrap around the atria and spiral within the walls of the ventricles. **(c)** Sectional and **(d)** diagrammatic views of cardiac muscle tissue. Histological characteristics of cardiac muscle cells compared with those of skeletal muscle fibers include (1) small size, (2) single, centrally placed nucleus, (3) branching interconnections between cells, and (4) presence of intercalated discs. (LM × 575)

VALVULAR HEART DISEASE Serious valve problems can interfere with cardiac function. If valve function deteriorates to the point at which the heart cannot maintain adequate circulatory flow, symptoms of **valvular heart disease (VHD)** appear. Congenital malformations may be responsible, but in many cases the condition develops after *carditis,* an inflammation of the heart, occurs.

One relatively common cause of carditis is **rheumatic** (roo-MA-tik) **fever,** an inflammatory condition that may develop after infection by streptococcal bacteria. Valve problems serious enough to reduce cardiac function may not appear until 10–20 years after the initial infection. The resulting clinical disorder is known as *rheumatic heart disease (RHD).* AM
RHD and Valvular Stenosis

The Heart Wall

A section through the wall of the heart (Figure 20-7a•) reveals three distinct layers: (1) an outer *epicardium,* (2) a middle *myocardium,* and (3) an inner *endocardium.*

- The **epicardium** is the visceral pericardium that covers the outer surface of the heart. This serous membrane consists of an exposed mesothelium and an underlying layer of loose connective tissue that is attached to the myocardium.

- The **myocardium,** or muscular wall of the heart, forms both atria and ventricles. The myocardium contains cardiac muscle tissue, blood vessels, and nerves. The myocardium consists of concentric layers of cardiac muscle tissue. The atrial myocardium contains muscle bundles that wrap around the atria and form figure-eights that pass through the interatrial septum. Superficial ventricular muscles wrap around both ventricles; deeper muscle layers spiral around and between the ventricles toward the apex (Figure 20-7b●).

- The inner surfaces of the heart, including those of the heart valves, are covered by the **endocardium** (en-dō-KAR-dē-um; *endo-,* inside). The endocardium is simple squamous epithelium that is continuous with the endothelium of the attached blood vessels.

Cardiac Muscle Tissue

Recall from Chapter 4 that **cardiac muscle cells** are interconnected by **intercalated discs.** ∞ *[p. 135]* These discs convey the force of contraction from cell to cell and propagate action potentials. In Chapter 10, we briefly compared the properties of cardiac muscle tissue with the properties of other muscle types. ∞ *[pp. 306, 308]* Figure 20-7c,d● and Table 20-1 provide a quick review of the structural and functional differences between cardiac muscle cells and skeletal muscle fibers that we noted in those earlier discussions.

Connective Tissues and the Fibrous Skeleton

The connective tissues of the heart include large numbers of collagen and elastic fibers. Each cardiac muscle cell is wrapped in a strong but elastic sheath, and adjacent cells are tied together by fibrous crosslinks, or "struts." These fibers are in turn interwoven into sheets that separate the superficial and deep muscle layers. These connective tissue fibers (1) provide physical support for the cardiac muscle fibers, blood vessels, and nerves of the myocardium; (2) help distribute the forces of contraction; (3) add strength and prevent overexpansion of the heart; and (4) provide elasticity that helps return the heart to its original size and shape after a contraction.

The **fibrous skeleton** of the heart consists of four dense bands of fibroelastic tissue that encircle the bases of the pulmonary trunk and aorta and the heart valves (Figure 20-6●). These bands stabilize the positions of the heart valves and ventricular muscle cells and physically isolate the ventricular cells from the atrial cells.

The Blood Supply to the Heart

The heart works continuously, and cardiac muscle cells require reliable supplies of oxygen and nutrients. The

---

TABLE 20-1 Structural and Functional Differences between Cardiac Muscle Cells and Skeletal Muscle Fibers

Feature	Cardiac Muscle Cells	Skeletal Muscle Fibers
Size	10–20 μm × 50–100 μm	100 μm × up to 40 cm
Nuclei	Typically 1 (rarely 2–5)	Multiple (hundreds)
Contractile proteins	Sarcomeres along myofibrils	Sarcomeres along myofibrils
Internal membranes	Short T tubules; no triads formed with sarcoplasmic reticulum	Long T tubules form triads with cisternae of the sarcoplasmic reticulum
Mitochondria	Abundant (25% of cell volume)	Relatively scarce
Inclusions	Myoglobin, lipids, glycogen	Extensive glycogen reserves
Circulatory supply	Very extensive	More extensive than in most connective tissues but sparse compared with supply to cardiac muscle cells
Metabolism (resting)	Aerobic	Aerobic
Metabolism (active)	Aerobic, using any available substrate	Anaerobic, breakdown of glycogen reserves
Contractions	Twitches with brief relaxation periods; long refractory period prevents tetanic contractions	Usually sustained tetanic contractions
Stimulus for contraction	Autorhythmicity of pacemaker cells generates action potentials	Activity of somatic motor neuron generates action potentials in sarcolemma
Trigger for contraction	Calcium entry from the ECF and calcium release from the sarcoplasmic reticulum	Calcium release from the sarcoplasmic reticulum
Intercellular connections	Branching network with cell membranes locked together at intercalated discs; connective tissue fibers tie adjacent layers together	Adjacent fibers tied together by connective tissue fibers

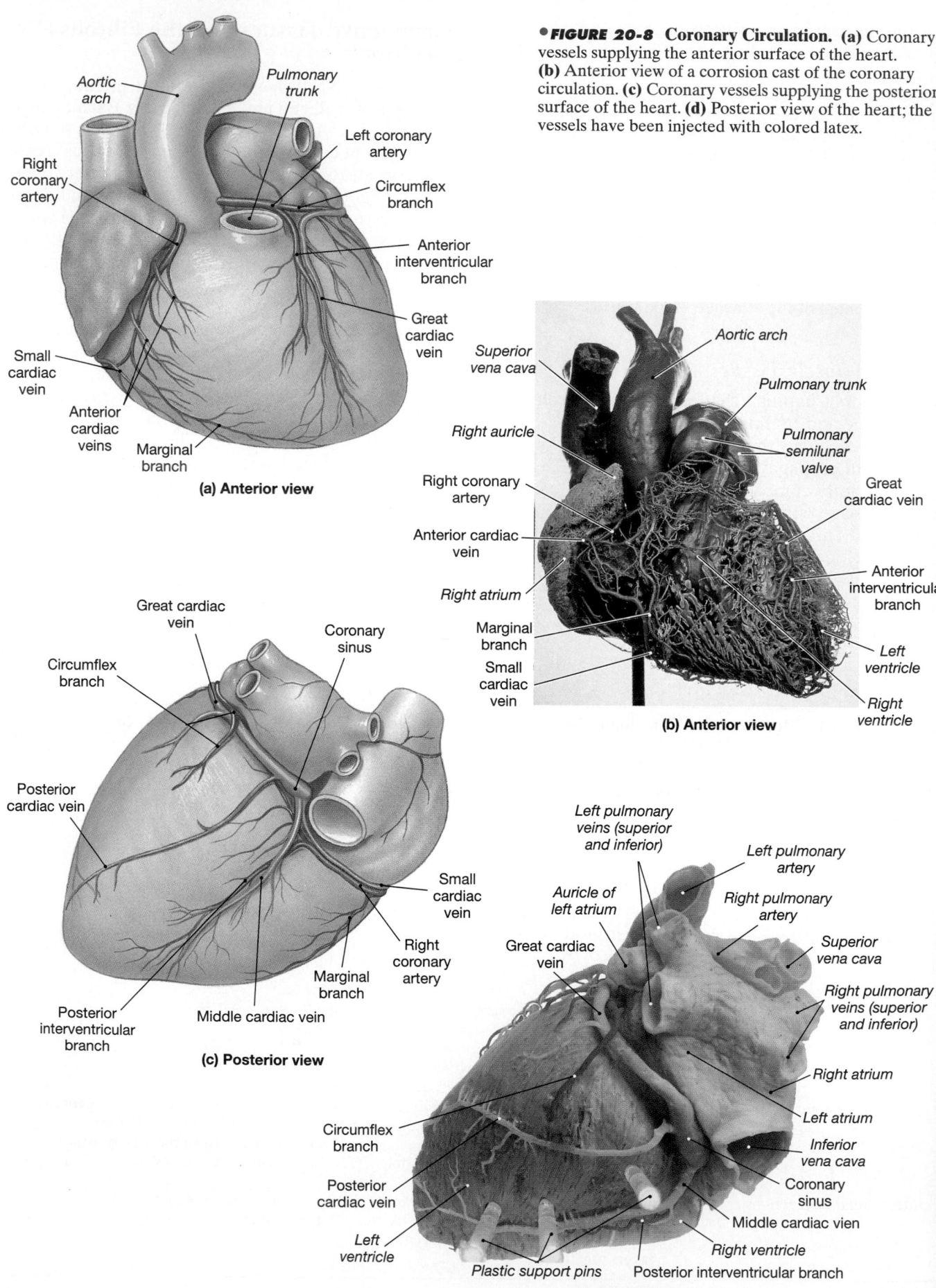

FIGURE 20-8 Coronary Circulation. (a) Coronary vessels supplying the anterior surface of the heart. **(b)** Anterior view of a corrosion cast of the coronary circulation. **(c)** Coronary vessels supplying the posterior surface of the heart. **(d)** Posterior view of the heart; the vessels have been injected with colored latex.

Aortic arch

Pulmonary trunk

Left coronary artery

Right coronary artery

Circumflex branch

Anterior interventricular branch

Great cardiac vein

Small cardiac vein

Anterior cardiac veins

Marginal branch

(a) Anterior view

Superior vena cava

Aortic arch

Pulmonary trunk

Right auricle

Pulmonary semilunar valve

Right coronary artery

Great cardiac vein

Anterior cardiac vein

Right atrium

Anterior interventricular branch

Marginal branch

Small cardiac vein

Left ventricle

Right ventricle

(b) Anterior view

Great cardiac vein

Coronary sinus

Circumflex branch

Posterior cardiac vein

Small cardiac vein

Right coronary artery

Posterior interventricular branch

Middle cardiac vein

Marginal branch

(c) Posterior view

Left pulmonary veins (superior and inferior)

Left pulmonary artery

Auricle of left atrium

Right pulmonary artery

Great cardiac vein

Superior vena cava

Right pulmonary veins (superior and inferior)

Right atrium

Left atrium

Inferior vena cava

Coronary sinus

Middle cardiac vien

Right ventricle

Posterior interventricular branch

Circumflex branch

Posterior cardiac vein

Left ventricle

Plastic support pins

(d) Posterior view

coronary circulation supplies blood to the muscles of the heart. During maximum exertion, the oxygen demand rises considerably, and the blood flow to the heart may increase to nine times that of resting levels. The coronary circulation (Figure 20-8●) includes an extensive network of coronary blood vessels.

The Coronary Arteries

The left and right **coronary arteries** originate at the base of the ascending aorta (Figure 20-8a,b●). Blood pressure here is the highest in the systemic circuit, and this pressure ensures a continuous flow of blood to meet the demands of active cardiac muscle tissue.

THE RIGHT CORONARY ARTERY The **right coronary artery,** which follows the coronary sulcus around the heart, supplies blood to (1) the right atrium, (2) portions of both ventricles, and (3) portions of the conducting system of the heart, including the *SA* (sinoatrial) and *AV nodes*. The cells of the SA node and AV node are essential to establishing the normal heart rate. We will focus on their functions and their part in regulation of the heart rate in a later section. Inferior to the right atrium, the right coronary artery generally gives rise to one or more **marginal branches,** which extend across the ventricular surface (Figure 20-8a,b,c●). It then continues across the posterior

surface of the heart, supplying the **posterior interventricular branch,** or *posterior descending artery,* which runs toward the apex within the posterior interventricular sulcus. The posterior interventricular branch supplies blood to the interventricular septum and adjacent portions of the ventricles.

THE LEFT CORONARY ARTERY The **left coronary artery** supplies blood to the left ventricle, left atrium, and the interventricular septum. As it reaches the anterior surface of the heart, it gives rise to a *circumflex branch* and an *anterior interventricular branch* (Figure 20-8a,b,c●). The **circumflex branch** curves to the left around the coronary sulcus, eventually meeting and fusing with small branches of the right coronary artery. The much larger **anterior interventricular branch,** or *left anterior descending artery,* swings around the pulmonary trunk and runs along the anterior surface within the anterior interventricular sulcus. This branch supplies small tributaries continuous with those of the posterior interventricular branch of the right coronary artery. Such interconnections between arteries are called **anastomoses** (a-nas-tō-MŌ-sēz; *anastomosis,* outlet). Because the arteries are interconnected in this way, the blood supply to the cardiac muscle remains relatively constant despite pressure fluctuations in the left and right coronary arteries as the heart beats.

Coronary Artery Disease

The term **coronary artery disease (CAD)** refers to degenerative changes in the coronary circulation. Cardiac muscle cells need a constant supply of oxygen and nutrients, and any reduction in coronary circulation produces a corresponding reduction in cardiac performance. Such reduced circulatory supply, known as **coronary ischemia** (is-KĒ-mē-uh), generally results from partial or complete blockage of the coronary arteries. The usual cause is the formation of a fatty deposit, or *plaque,* in the wall of a coronary vessel. The plaque or an associated thrombus then narrows the passageway and reduces blood flow; spasms in the smooth muscles of the vessel wall can further decrease or even stop blood flow. The plaques may be visible through *angiography* or high-resolution ultrasound, and the metabolic effects can be detected in PET scans of the heart (Figure 20-9a,b●). We shall consider plaque development and growth in Chapter 21.

One of the first symptoms of CAD is commonly **angina pectoris** (an-JĪ-nuh PEK-tor-is; *angina,* pain spasm + *pectoris,* of the chest). In the most common form of angina, a temporary ischemia develops

when the workload of the heart increases. Although the individual may feel comfortable at rest, any unusual exertion or emotional stress can produce a sensation of pressure, chest constriction, and pain that may radiate from the sternal area to the arms, back, and neck.

Angina can typically be controlled by a combination of drug treatment and changes in lifestyle. Lifestyle changes to combat angina include (1) limiting activities known to trigger angina attacks, such as strenuous exercise, and avoiding stressful situations; (2) stopping smoking; and (3) lowering fat consumption. Medications useful for controlling angina include drugs that block sympathetic stimulation (*propranolol* or *metoprolol*); vasodilators such as *nitroglycerin* (nī-trō-GLIS-er-in) and *atrial natriuretic peptide (ANP)*; and drugs that block calcium movement into the cardiac muscle cells *(calcium channel blockers)*.

Angina can also be treated surgically. A single, soft plaque may be reduced with the aid of a long, slender **catheter** (KATH-e-ter). The catheter, a

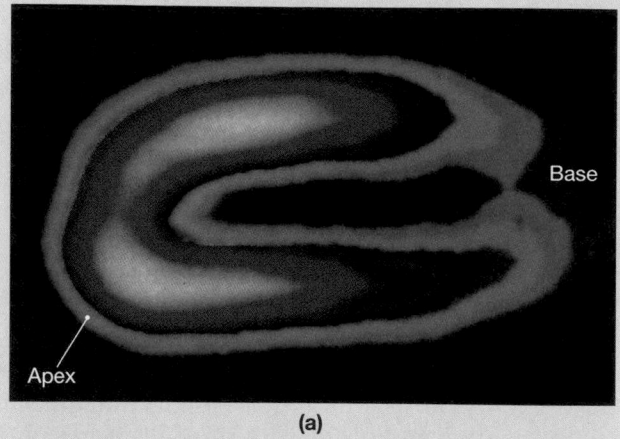

(a)

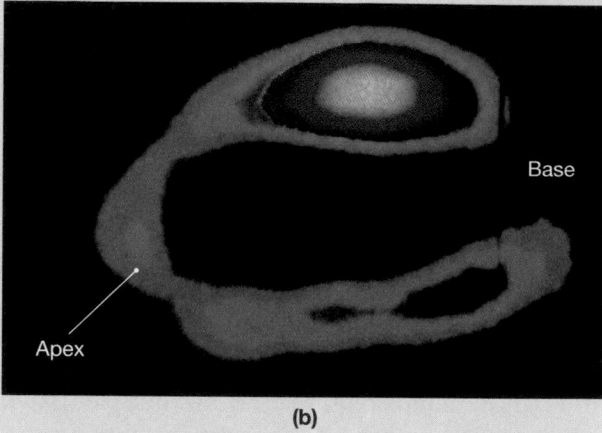

(b)

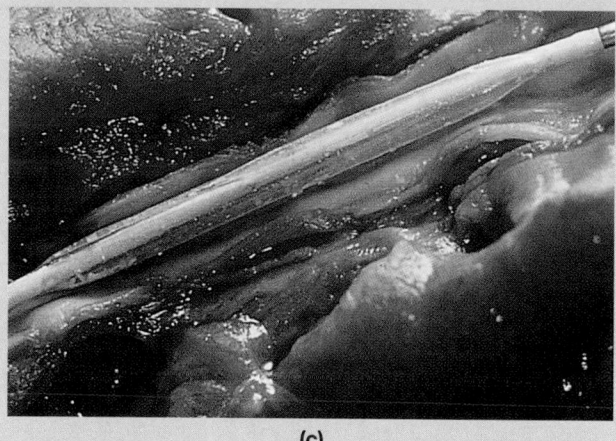

(c)

•FIGURE 20-9 **Coronary Circulation and Clinical Testing.** **(a)** PET scan of a healthy heart after the introduction of an imaging dye that contains the radioisotope technetium-99. The brighter the color, the greater the blood flow through the tissue. (The atria are not shown.) The ventricular walls have an extensive circulatory supply. **(b)** PET scan of a damaged heart. Most of the ventricular myocardium is deprived of circulation. **(c)** Balloon angioplasty can in some cases be used to remove a circulatory blockage. The catheter is guided through the coronary arteries to the site of blockage and inflated to press the soft plaque against the vessel wall.

small-diameter tube, is inserted into a large artery and guided into a coronary artery to the plaque. A variety of surgical tools can be slid into the catheter, and the plaque can then be removed with laser beams or chewed to pieces by a miniature version of the Roto-Rooter® drain-cleaning machine. Debris created during plaque destruction is sucked up by the catheter, preventing blockage of smaller vessels.

In **balloon angioplasty** (AN-jē-ō-plas-tē; *angeion,* vessel), the catheter tip contains an inflatable balloon (Figure 20-9c•). Once in position, the balloon is inflated, pressing the plaque against the vessel walls. This procedure works best in small (under 10 mm), soft plaques. Several factors make this a highly attractive treatment: (1) The mortality rate during surgery is only about 1 percent; (2) the success rate is over 90 percent; and (3) it can be performed on an outpatient basis. Although in about 20 percent of patients the plaque deposit returns to its original size within 6 months, the process can be repeated as needed. Unfortunately, only about 10 percent of severe angina patients have isolated problems suitable for balloon angioplasty.

In a **coronary artery bypass graft (CABG),** a small section is removed from either a small artery (commonly the *internal thoracic artery*) or a peripheral vein (such as the *great saphenous vein* of the leg) and used to create a detour around the obstructed portion of a coronary artery. As many as four coronary arteries can be rerouted this way during a single operation. The procedures are named according to the number of vessels repaired, so we speak of single, double, triple, or quadruple coronary bypass operations. The mortality rate during surgery for operations performed before significant heart damage has occurred is relatively low (1–2 percent). Under these conditions, the procedure completely eliminates the angina symptoms in 70 percent of the cases and provides partial relief in another 20 percent.

Although coronary bypass surgery does offer certain advantages, recent studies have shown that for mild angina, this surgery does not yield significantly better results than drug therapy. Current recommendations are that coronary bypass surgery be reserved for cases of severe angina that do not respond to other treatment.

THE CARDIAC VEINS The **great cardiac vein** begins on the anterior surface of the ventricles, along the interventricular sulcus. This vein drains blood from the region supplied by the anterior interventricular branch of the left coronary artery. The great cardiac vein reaches the level of the atria and then curves around the left side of the heart within the coronary sulcus. The vein empties into the **coronary sinus,** a large, thin-walled vein that lies in the posterior portion of the coronary sulcus. The coronary sinus communicates with the right atrium near the base of the inferior vena cava. The other cardiac veins, which empty into the great cardiac vein or the coronary sinus, include (1) the **posterior cardiac vein**, draining the area served by the circumflex branch of the left coronary artery; (2) the **middle cardiac vein,** draining the area supplied by the posterior interventricular branch of the right coronary artery; and (3) the **small cardiac vein** and **anterior cardiac veins,** draining the other regions supplied by the right coronary artery and its tributaries (Figure 20-8a,c,d●).

☑ Damage to the semilunar valves on the right side of the heart would interfere with blood flow to which vessel?

☑ What prevents the AV valves from opening back into the atria?

☑ Why is the left ventricle more muscular than the right ventricle?

Innervation of the Heart

The sympathetic and parasympathetic divisions of the ANS provide innervation to the heart through the *cardiac plexus* (Figure 20-10●). Postganglionic sympathetic neurons are located in the cervical and upper thoracic ganglia. The vagus nerve (N X) carries parasympathetic preganglionic fibers to small ganglia in the cardiac plexus. Both ANS divisions innervate the SA and AV nodes and the atrial muscle cells. Although the ventricular muscle cells are innervated by both divisions, sympathetic fibers far outnumber parasympathetic fibers.

The *cardiac centers* of the medulla oblongata contain the autonomic

headquarters for cardiac control. (We introduced these centers in Chapter 14.) ∞ *[p. 473]* Stimulation of the **cardioacceleratory center** activates the necessary sympathetic neurons; the nearby **cardioinhibitory center** governs the activities of the parasympathetic neurons. The cardiac centers receive input from higher centers, especially from the parasympathetic and sympathetic headquarters in the hypothalamus. Information about the status of the cardiovascular system arrives over visceral sensory fibers accompanying the vagus nerve and the sympathetic nerves of the cardiac plexus.

The cardiac centers monitor baroreceptors and chemoreceptors innervated by the glossopharyngeal

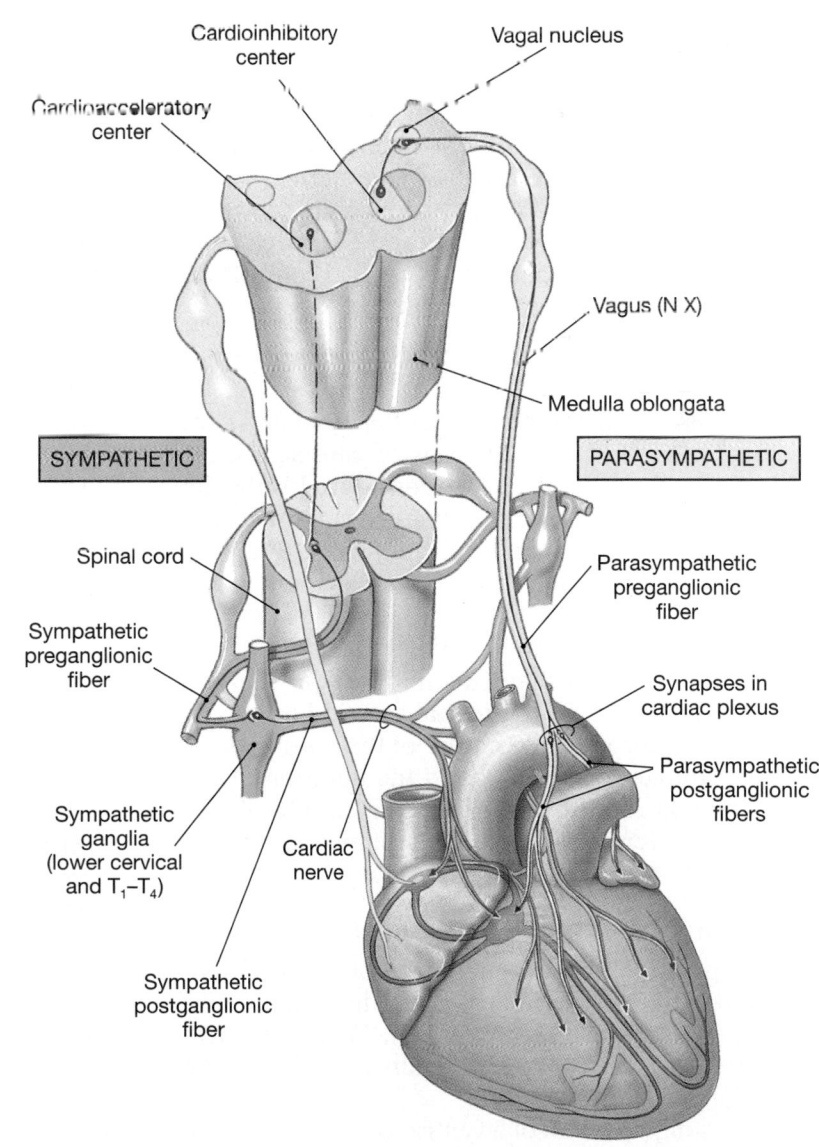

●***FIGURE 20-10*** **Autonomic Innervation of the Heart**

(N IX) and vagus nerves (N X). ∞ *[pp 480–482]* On the basis of this information, they adjust cardiac performance to maintain adequate circulation to vital organs, such as the brain. These centers respond to changes in blood pressure and in the arterial concentrations of dissolved oxygen and carbon dioxide. For example, a decline in blood pressure or oxygen concentrations or an increase in carbon dioxide levels generally indicates that the heart must work harder to meet the demands of peripheral tissues. The cardiac centers then call for an increase in cardiac activity. We shall detail these reflexes and their effects on the heart and peripheral vessels in Chapter 21.

THE HEARTBEAT

In a single **heartbeat**, the entire heart—atria and ventricles—contracts in a coordinated manner so that blood flows in the right direction at the proper time. Each time your heart beats, the contractions of individual cardiac muscle cells within the atria and ventricles must occur in a specific sequence. Two types of cardiac muscle cells are involved in a normal heartbeat: (1) *contractile cells* that produce the powerful contractions that propel blood, and (2) specialized muscle cells of the *conducting system* that control and coordinate the activities of the contractile cells.

Contractile Cells

Contractile cells form the bulk of the atrial and ventricular walls. In the discussions of cardiac muscle tissue in earlier chapters, we considered only the structure of contractile cells, which account for roughly 99 percent of the muscle cells in the heart. In both cardiac muscle cells and skeletal muscle fibers, (1) an action potential leads to the appearance of Ca^{2+} among the myofibrils, and (2) the binding of Ca^{2+} to troponin on the thin filaments initiates the contraction (see Table 20-1). But skeletal and cardiac muscle cells differ in terms of the nature of the action potential, the source of the Ca^{2+}, and the duration of the resulting contraction. We noted the structural and functional differences between these muscle types in Chapter 10. ∞ *[p. 309]*

The Action Potential in Cardiac Muscle Cells

We compared the basic appearance of action potentials in skeletal muscle fibers and cardiac muscle cells in Chapter 12 (Figure 12-19●, p. 396). We will now take a closer look at the origin and conduction of an action potential in a contractile cell.

The resting potential of a ventricular contractile cell is approximately –90 mV, comparable to that of a resting skeletal muscle fiber (–85 mV). (The resting potential of atrial contractile cells is about –80 mV, but the basic principles described here apply to atrial cells as well.) Although their resting potentials are similar, an action potential in a ventricular contractile cell lasts 250–300 msec, 30 times as long as a typical action potential in a skeletal muscle fiber (Figure 20-11a,b●).

An action potential begins when the membrane of the ventricular muscle cell is brought to threshold, usually about –75 mV. This normally occurs in a portion of the membrane next to an intercalated disc. The typical stimulus is the excitation of an adjacent muscle cell. Once threshold has been reached, the action potential proceeds in three basic steps:

STEP 1: *Rapid depolarization.* The stage of rapid depolarization in a cardiac muscle cell resembles that of a skeletal muscle fiber. At threshold, voltage-regulated sodium channels open, and the membrane suddenly becomes permeable to Na^+. The result is rapid depolarization of the sarcolemma. The channels involved are called **fast channels,** because they open quickly and remain open for only a few milliseconds.

STEP 2: *The plateau.* As the transmembrane potential approaches +30 mV, the voltage-regulated sodium channels close. They will remain closed and inactivated until the membrane potential reaches –60 mV. As the sodium channels are closing, voltage-regulated calcium channels are opening. The calcium channels are called **slow channels,** because they open slowly and remain open for a relatively long period—roughly 175 msec. As a result, the transmembrane potential remains near 0 mV for an extended period. This portion of the action potential curve is called the *plateau. The presence of a plateau is the major difference between action potentials in cardiac muscle cells and skeletal muscle fibers.* In a skeletal muscle fiber, rapid depolarization is immediately followed by a period of rapid repolarization.

STEP 3: *Repolarization.* As the plateau continues, slow calcium channels begin closing and **slow potassium channels** begin opening. The result is a period of rapid repolarization that restores the resting potential.

THE REFRACTORY PERIOD For some time after an action potential begins, the membrane will not respond normally to a second stimulus. This time is called the *refractory period.* In the *absolute refractory period,* the membrane cannot respond at all, because the sodium channels are either already open or closed and inactivated. In a ventricular muscle cell, the absolute refractory period lasts approximately 200 msec, spanning the duration of the plateau and the initial period of rapid repolarization.

The absolute refractory period is followed by a shorter (50 msec) *relative refractory period.* During this period, the voltage-regulated sodium channels are closed but capable of opening. The membrane will respond to a stronger-than-normal stimulus by initiating another action potential.

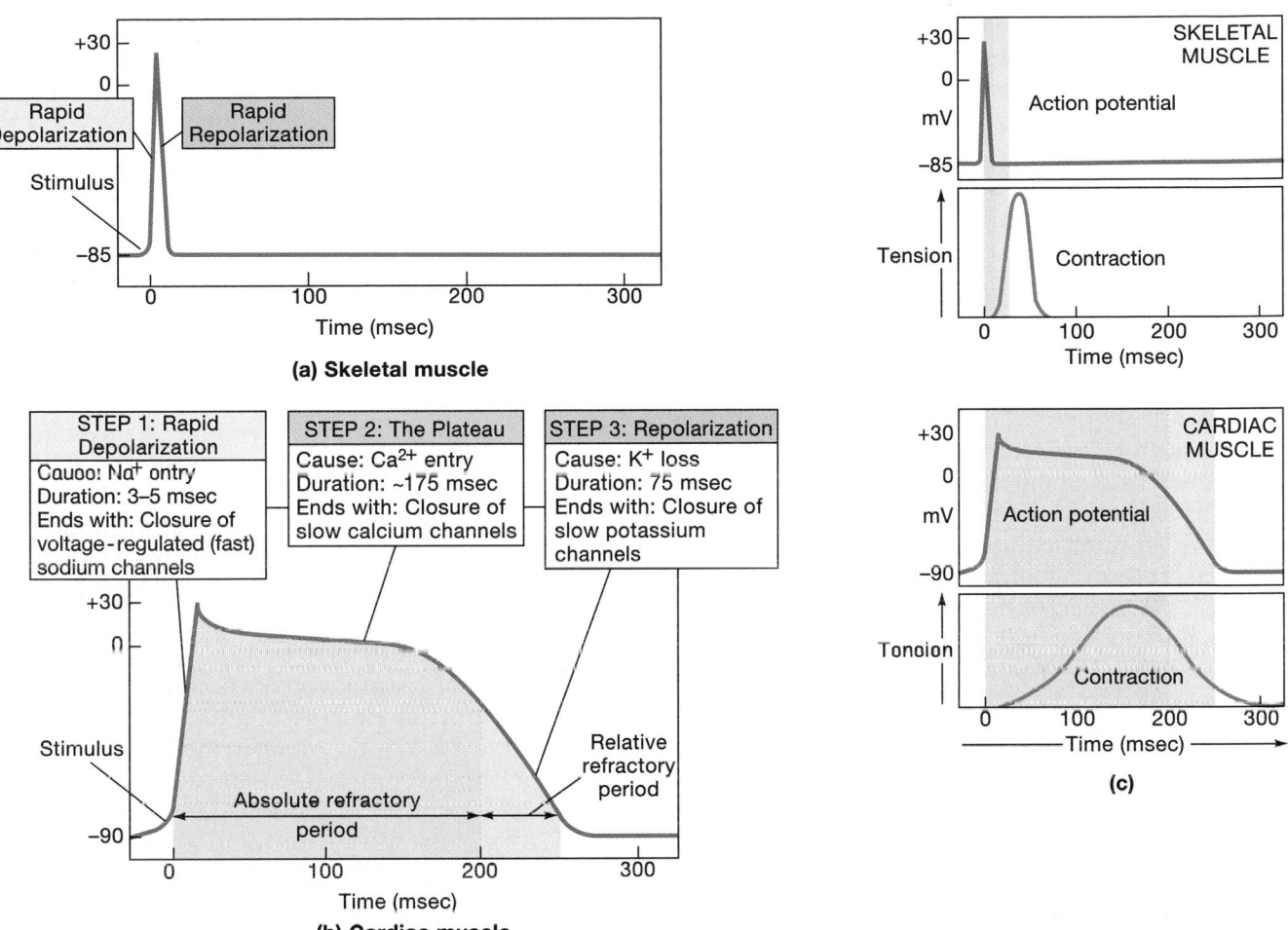

FIGURE 20-11 **The Action Potential in Skeletal and Cardiac Muscle.** **(a)** Action potential in a representative skeletal muscle fiber; for more details, see Figure 10-10, p. 292. **(b)** Action potential in a ventricular muscle cell. **(c)** The relationship between an action potential and a twitch contraction in skeletal and cardiac muscle.

Calcium Ions and Cardiac Contractions

The appearance of an action potential in the cardiac muscle cell membrane produces a contraction by causing an increase in the concentration of Ca^{2+} around the myofibrils. This process occurs in two steps:

1. Calcium ions entering the cell membrane during the plateau phase of the action potential provide roughly 20 percent of the Ca^{2+} required for a contraction.

2. The arrival of extracellular Ca^{2+} is the trigger for the release of additional Ca^{2+} from reserves in the sarcoplasmic reticulum (SR).

Extracellular calcium ions thus have direct and indirect effects on cardiac muscle cell contraction. For this reason, cardiac muscle tissue is very sensitive to changes in the Ca^{2+} concentration of the extracellular fluid.

In a skeletal muscle fiber, the action potential is relatively brief and ends as the related twitch contraction begins (Figure 20-11c●). The twitch con-

traction is short and ends as the SR reclaims the Ca^{2+} it released. In a cardiac muscle cell, as we have seen, the action potential is prolonged and calcium ions continue to enter the cell throughout the plateau. As a result, the period of active muscle cell contraction continues until the plateau ends. As the slow calcium channels close, the intracellular calcium ions are absorbed by the SR or pumped out of the cell and the muscle cell relaxes.

In skeletal muscle fibers, the refractory period ends before the muscle fiber develops peak tension. As a result, twitches can summate and tetanus can occur. In cardiac muscle cells, the absolute refractory period continues until relaxation is under way. Because summation is not possible, tetanic contractions cannot occur in a normal cardiac muscle cell, regardless of the frequency or intensity of stimulation. This feature is absolutely vital: A heart in tetany could not pump blood. With a single twitch lasting 250 msec or longer, a normal cardiac muscle cell can increase its rate of contraction to 300–400 per minute under maximum stimulation. This rate is not reached in a normal heart due to limitations imposed by the conducting system, which we consider next.

The Conducting System

In contrast to skeletal muscle, cardiac muscle tissue contracts on its own, in the absence of neural or hormonal stimulation. This property is called *automaticity,* or *autorhythmicity.* The cells responsible for initiating and distributing the stimulus to contract are part of the *conducting system* of the heart. The **conducting system** is a network of specialized cardiac muscle cells that initiates and distributes electrical impulses. The actual contraction lags behind the passage of an electrical impulse, as excitation–contraction coupling occurs and cross-bridge interactions take place. The basic mechanical process of contraction in a cardiac muscle cell resembles that which we already described for skeletal muscle fibers in Chapter 10. ∞ *[pp. 288–289]*

The conducting system (Figure 20-12a●) includes:

- The *sinoatrial (SA) node,* located in the wall of the right atrium.
- The *atrioventricular (AV) node,* located at the junction between the atria and ventricles.
- *Conducting cells*, which interconnect the two nodes and distribute the contractile stimulus throughout the myocardium. Conducting cells in the atria are found in the *internodal pathways,* which distribute the contractile stimulus to atrial muscle cells as the impulse travels from the SA node to the AV node. (The importance of these pathways in relaying the signal to the AV node remains in dispute, because an impulse spread from contractile cell to contractile cell reaches the AV node at about the same time as an impulse that traverses an internodal pathway.) The ventricular conducting cells include those in the *AV bundle* and the *bundle branches* as well as the *Purkinje* (pur-KIN-jē) *fibers,* which distribute the stimulus to the ventricular myocardium.

Most of the cells of the conducting system are smaller than the contractile cells of the myocardium and contain very few myofibrils. These cells share another important characteristic: They cannot maintain a stable resting potential. Each time repolarization occurs, the membrane again gradually drifts toward threshold. This gradual depolarization is called a **prepotential.**

The rate of spontaneous depolarization varies in different portions of the conducting system. It is fastest at the SA node, which in the absence of neural or hormonal stimulation will generate action potentials at a rate of 80–100 per minute (Figure 20-12b●). Isolated cells of the AV node depolarize more slowly, generating 40–60 action potentials per

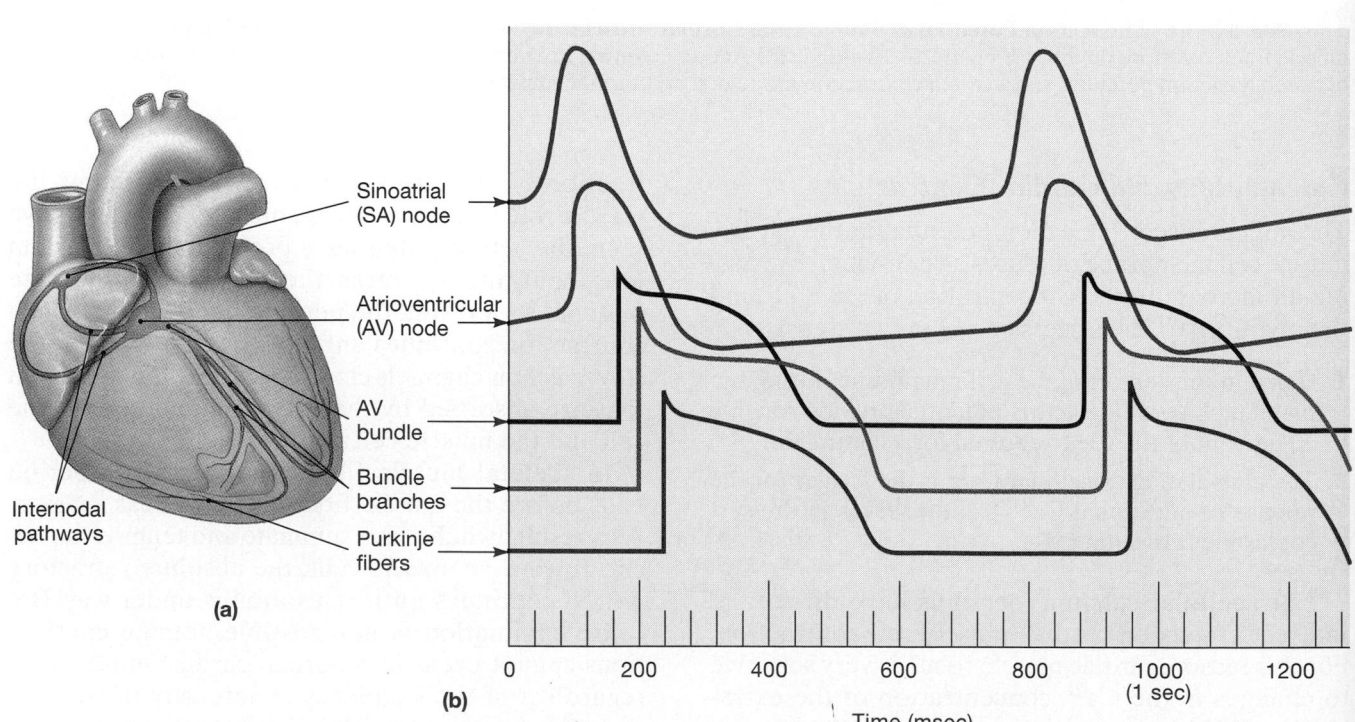

(a)

(b)

Sinoatrial (SA) node

Atrioventricular (AV) node

AV bundle

Bundle branches

Purkinje fibers

Internodal pathways

Time (msec)

●***FIGURE 20-12*** **The Conducting System of the Heart.** **(a)** Components of the conducting system. **(b)** This graph follows the spontaneous changes in the membrane potentials of nodal cells and the resulting action potentials in conducting cells. (The vertical axis is not at the same scale for each curve; the graphs are spread apart for clarity.) The rate of spontaneous depolarization is highest at the SA node. Because these components are interconnected in the intact heart, the SA node is the cardiac pacemaker.

minute. Under normal conditions, most cells of the AV bundle, the bundle branches, and Purkinje fibers do not depolarize spontaneously.

In the normal heart, the cells of the conducting system are interconnected. An action potential appearing at any one location will be conducted across the entire network. Because the SA node reaches threshold first, it establishes the heart rate.

We shall now trace the path of an impulse from its initiation at the SA node and consider its effects on the surrounding myocardium.

The Sinoatrial (SA) Node

The **sinoatrial** (sī-nō-Ā-trē-al) **node (SA node)** is embedded in the posterior wall of the right atrium, near the entrance of the superior vena cava (Figures 20-12a and 20-13 [Step 1]●). The SA node contains **pacemaker cells**, which establish the heart rate. As a result, the SA node is also known as the *cardiac pacemaker* or the *natural pacemaker*. It is connected to the larger AV node by the internodal pathways in the atrial walls. It takes roughly 50 msec for an action potential to travel from the SA node to the AV node along these pathways. Along the way, the conducting cells pass the contractile stimulus to cardiac muscle cells of the right and left atria. The action potential then spreads across the atrial surfaces through cell-to-cell contact (Figure 20-13 [Step 2]●). The stimulus affects only the atria, because the fibrous skeleton separates the atrial myocardium from the ventricular myocardium.

The Atrioventricular (AV) Node

The relatively large **atrioventricular (AV) node** sits within the floor of the right atrium near the opening of the coronary sinus. The rate of propagation slows as the impulse leaves the internodal pathways and enters the AV node, because the nodal cells are smaller in diameter than the conducting cells. (Chapter 12 included a discussion of the relationship between diameter and propagation speed. ∞ *[p. 394]*) In addition, the connections between nodal cells are less efficient than those between conducting cells at relaying the impulse from one cell to another. As a result, it takes about 100 msec for the impulse to pass through the AV node and enter the AV bundle (Figure 20-13 [Step 3]●).

The delay at the AV node is important because the atria must contract before the ventricles; once the ventricles start contracting, the AV valves close. If all the chambers contracted at the same time, the AV valves would prevent blood flow from the atria and into the ventricles. Because there is a delay at the AV node, the atrial myocardium completes its contraction before ventricular contraction begins.

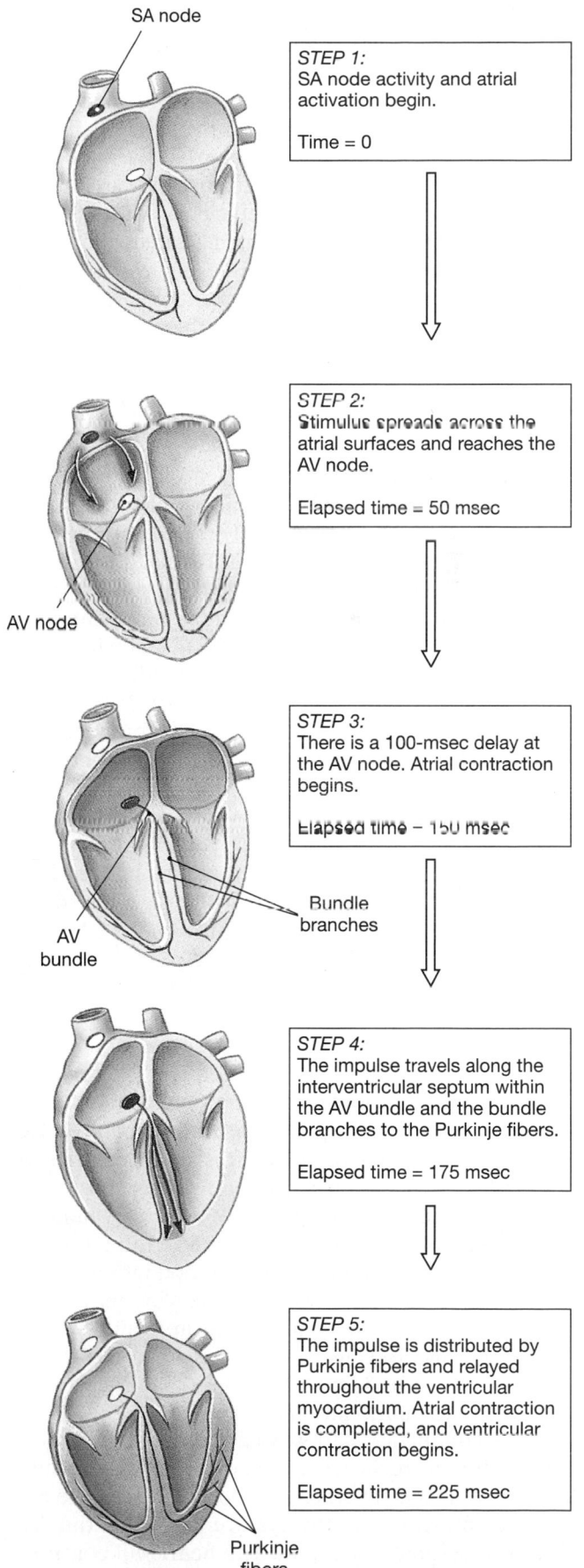

SA node

STEP 1:
SA node activity and atrial activation begin.

Time = 0

STEP 2:
Stimulus spreads across the atrial surfaces and reaches the AV node.

Elapsed time = 50 msec

AV node

STEP 3:
There is a 100-msec delay at the AV node. Atrial contraction begins.

Elapsed time = 150 msec

AV bundle

Bundle branches

STEP 4:
The impulse travels along the interventricular septum within the AV bundle and the bundle branches to the Purkinje fibers.

Elapsed time = 175 msec

STEP 5:
The impulse is distributed by Purkinje fibers and relayed throughout the ventricular myocardium. Atrial contraction is completed, and ventricular contraction begins.

Elapsed time = 225 msec

Purkinje fibers

●*FIGURE 20-13* **Impulse Conduction through the Heart**

The cells of the AV node can conduct impulses at a maximum rate of 230 per minute. Because each impulse will result in a ventricular contraction, this value is the maximum normal heart rate. Even if the SA node generates impulses at a faster rate, the ventricles will still contract at 230 beats per minute (bpm). This limitation is important because mechanical factors, discussed later, begin to decrease the pumping efficiency of the heart at rates above approximately 180 bpm. At rates above 230 bpm, pumping efficiency becomes dangerously low. Ventricular rates increase toward their theoretical limit of 300–400 bpm only when the heart or the conducting system has been damaged or affected by stimulatory drugs.

The AV Bundle, Bundle Branches, and Purkinje Fibers

The connection between the AV node and the **AV bundle,** or *bundle of His,* is the only electrical connection between the atria and the ventricles. Once an impulse enters the AV bundle, it travels to the interventricular septum and enters the **right** and **left bundle branches.** The left bundle branch, which supplies the relatively massive left ventricle, is much larger than the right bundle branch. Both bundle branches extend toward the apex of the heart, turn, and fan out beneath the endocardial surface. As the branches diverge, they conduct the impulse to **Purkinje fibers** and to papillary muscles.

The Purkinje fibers conduct action potentials very rapidly, as fast as small myelinated axons. Within about 75 msec, the signal to begin a contraction has reached all the ventricular cardiac muscle cells. The entire process, from generation of an impulse at the SA node to the complete depolarization of the ventricular myocardium, normally takes around 225 msec. The atria have now completed their contractions, and ventricular contraction can safely occur.

Because the bundle branches deliver the impulse across the moderator band to the papillary muscles directly rather than via Purkinje fibers, the papillary muscles begin contracting before the rest of the ventricular musculature. This contraction applies tension to the chordae tendineae, bracing the AV valves. By limiting the movement of the cusps, tension in the papillary muscles prevents the backflow of blood into the atria when the ventricles contract. Ventricular contraction proceeds in a wave that begins at the apex and spreads toward the base. Because the ventricles contract in this way, blood is pushed toward the base of the heart, into the aortic and pulmonary trunks.

If the conducting pathways are damaged, the normal rhythm of the heart will be disturbed. The resulting problems are called *conduction deficits.* If the SA node or internodal pathways are damaged, the AV node will assume command. The heart will continue beating normally, although at a slower rate. If an ab-

normal conducting cell or ventricular muscle cell begins generating action potentials at a more rapid rate, the impulses can override those of the SA or AV node. The origin of these abnormal signals is called an **ectopic** (ek-TOP-ik; out of place) **pacemaker.** The activity of an ectopic pacemaker partially or completely bypasses the conducting system, disrupting the timing of ventricular contraction. The result is a dangerous reduction in the efficiency of the heart. Such conditions are commonly diagnosed with the aid of an *electrocardiogram.* (For a more detailed discussion of conduction deficits and other problems, consult the *Applications Manual.*)

AM *Diagnosing Abnormal Heartbeats*

☑ If the cells of the SA node failed to function, what effect would this failure have on heart rate?

☑ Why is it important for impulses from the atria to be delayed at the AV node before they pass into the ventricles?

☑ If the cardioinhibitory center of the medulla oblongata were damaged, which part of the autonomic nervous system would be affected, and what effect would this damage have on the heart?

☑ How would an increase in the extracellular concentration of calcium ions affect the strength of a cardiac contraction?

The Electrocardiogram

The electrical events occurring in the heart are powerful enough to be detected by electrodes on the body surface. A recording of these electrical activities constitutes an **electrocardiogram** (ē-lek-trō-KAR-dē-ō-gram), also called an **ECG** or **EKG.** Each time the heart beats, a wave of depolarization radiates through the atria, reaches the AV node, travels down the interventricular septum to the apex, turns, and spreads through the ventricular myocardium toward the base (Figure 20-13●).

By comparing the information obtained from electrodes placed at different locations, a clinician can monitor the electrical activity of the heart, which is directly related to the performance of specific nodal, conducting, and contractile components. For example, when a portion of the heart has been damaged, the affected muscle cells will no longer conduct action potentials. An ECG will reveal an abnormal pattern of impulse conduction.

The appearance of the ECG recording varies with the placement of the monitoring electrodes, or *leads.* Figure 20-14a● shows the leads in one of the standard configurations. Figure 20-14b● depicts the important features of an ECG obtained with that configuration. Note the following ECG features:

1. The small **P wave** accompanies the depolarization of the atria. The atria begin contracting about 100 msec after the start of the P wave.

2. The **QRS complex** appears as the ventricles depolarize. This is a relatively strong electrical signal, because

the mass of the ventricular muscle is much larger than that of the atria. The ventricles begin contracting shortly after the peak of the **R wave**.

3. The smaller **T wave** indicates ventricular repolarization. You do not see a deflection corresponding to atrial repolarization, because it occurs while the ventricles are depolarizing and the electrical events are masked by the QRS complex.

ECG ANALYSIS To analyze an ECG, you must measure the size of the voltage changes and determine the durations and temporal relationships of the various components. Attention usually focuses on the amount of depolarization occurring during the P wave and the QRS complex. For example, a smaller-than-normal electrical signal may mean that the mass of the heart muscle has decreased; an excessively large QRS complex may indicate that the heart has become enlarged.

The size and shape of the T wave may also be affected by any condition that slows ventricular repolarization. For example, starvation and low cardiac energy reserves, coronary ischemia, or abnormal ion concentrations will reduce the size of the T wave.

You must also take measurements of the time between waves. The values are reported as *segments* or *intervals*. The terms used are indicated in Figure 20-14b●, and you will find that the names do not always seem to fit. For example:

■ The **P–R interval** extends from the start of atrial depolarization to the start of the QRS complex (ventricular depolarization) rather than to R, because in abnormal ECGs, the peak can be difficult to determine. Extension of the P–R interval to more than 0.2 seconds can indicate damage to the conducting pathways or AV node.

■ The **Q–T interval** indicates the time required for the ventricles to undergo a single cycle of depolarization and repolarization. It is usually measured from the end of the P–R interval rather than from the bottom of the Q wave. The Q–T interval can be lengthened by conduction problems, coronary ischemia, or myocardial damage. A congenital heart defect that can cause sudden death without warning may be detected by a prolonged Q–T interval.

Despite the variety of sophisticated equipment available to assess or visualize cardiac function, in the majority of cases the ECG provides the most important diagnostic information. ECG analysis is especially useful in detecting and diagnosing **cardiac arrhythmias** (ā-RITH-mē-az), abnormal patterns of cardiac electrical activity. Momentary arrhythmias are not inherently dangerous; about 5 percent of the healthy population experiences a few abnormal heartbeats each day. Clinical problems appear when the arrhythmias reduce the pumping efficiency of the heart. Serious arrhythmias may indicate damage to the myocardial musculature, injuries to the pacemakers or conduction pathways, exposure to drugs, or variations in the electrolyte composition of extracellular fluids. (For a discussion of the most common types of arrhythmias detected with the ECG, see the *Applications Manual*.) AM *Diagnosing Abnormal Heartbeats; Monitoring the Heart*

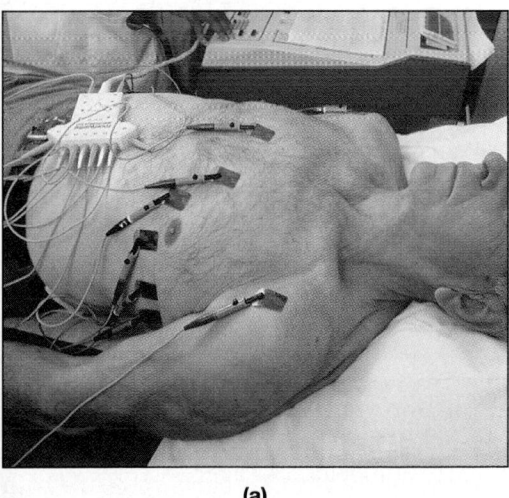

(a)

●**FIGURE 20-14** An Electrocardiogram.
(a) Electrode placement for recording a standard ECG. **(b)** An ECG printout is a strip of graph paper containing a record of the electrical events monitored by electrodes attached to the body surface. The placement of electrodes affects the size and shape of the waves recorded. This is an example of a normal ECG; the enlarged section indicates the major components of the ECG and the measurements most often taken during clinical analysis.

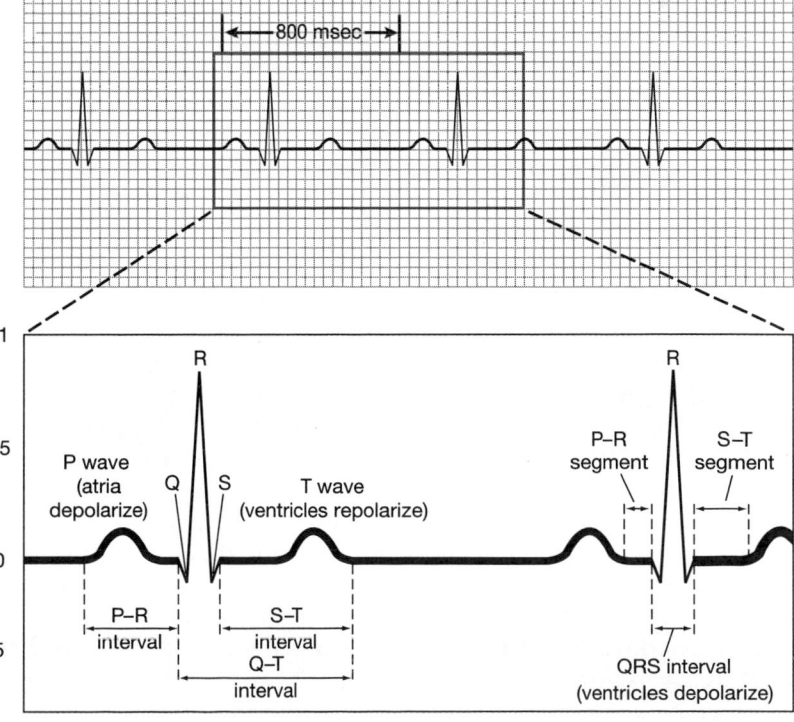

(b)

Heart Attacks

In a **myocardial** (mī-ō-KAR-dē-al) **infarction (MI),** or *heart attack,* the coronary circulation becomes blocked and the cardiac muscle cells die from lack of oxygen. The affected tissue then degenerates, creating a nonfunctional area known as an *infarct.* Heart attacks most commonly result from severe coronary artery disease (CAD). The consequences depend on the site and nature of the circulatory blockage. If it occurs near the base of one of the coronary arteries, the damage will be widespread and the heart will probably stop beating. If the blockage involves one of the smaller arterial branches, the individual may survive the immediate crisis, but there are many potential complications, all unpleasant. As scar tissue forms in the damaged area, the heartbeat may become irregular, and other vessels can become constricted, creating additional circulatory problems.

Myocardial infarctions are generally associated with fixed blockages, such as those seen in CAD. When the crisis develops because of thrombus (clot) formation at a plaque, the condition is called **coronary thrombosis.** A vessel already narrowed by plaque formation may also become blocked by a sudden spasm in the smooth muscles of the vascular wall. The individual then may experience intense pain, similar to that of an angina attack but persisting even at rest. However, pain does not always accompany a heart attack. These *silent heart attacks* may be even more dangerous, because the condition may not be diagnosed and treated before a fatal MI occurs. Roughly 25 percent of heart attacks are not recognized when they occur.

The cytoplasm of a damaged cardiac muscle cell differs from that of a normal muscle cell. As the supply of oxygen decreases, the cells become more dependent on anaerobic metabolism to meet their energy needs. ⊂⊃ *[p. 299]* Over time, the cytoplasm accumulates large numbers of enzymes involved with anaerobic energy production.

As the cardiac muscle cell membranes deteriorate, these enzymes enter the surrounding intercellular fluids. The appearance of such enzymes in the circulation thus indicates that an infarct has occurred. The enzymes tested for in a diagnostic blood test include **lactate dehydrogenase (LDH), serum glutamic oxaloacetic transaminase (SGOT,** also called *aspartate aminotransferase*), **creatine phosphokinase (CPK,** or **CK),** and a special form of creatine phosphokinase that occurs only in cardiac muscle **(CK-MB).**

Roughly 25 percent of MI patients die before obtaining medical assistance, and 65 percent of MI deaths among those under age 50 occur within an hour after the initial infarct. The goals of treatment are to limit the size of the infarct and prevent additional complications by preventing irregular contractions, improving circulation with vasodilators, providing additional oxygen, reducing the cardiac workload, and, if possible, eliminating the cause of the circulatory blockage. Anticoagulants may help prevent the formation of additional thrombi, and clot-dissolving enzymes may reduce the extent of the damage if they are administered within 6 hours after the MI has occurred. Current evidence suggests that t-PA, which is relatively expensive, is more beneficial than other fibrinolytic agents, such as urokinase or streptokinase. A procedure that combines small amounts of t-PA with streptokinase has been proposed as a workable compromise that produces effects equivalent to high doses of t-PA alone. Follow-up treatment with heparin, aspirin, or both is recommended; without further treatment, the circulatory blockages will reappear in roughly 20 percent of patients.

Roughly 1.3 million MIs occur in the United States each year, and half the victims die within a year of the incident. The following 10 factors appear to increase the risk of a heart attack: (1) smoking, (2) high blood pressure, (3) high blood cholesterol levels, (4) high circulating levels of low-density lipoproteins (LDL), (5) diabetes, (6) male gender (below age 70), (7) severe emotional stress, (8) obesity, (9) genetic predisposition, and (10) a sedentary lifestyle. We shall consider the role of lipoproteins and cholesterol in plaque formation and heart disease in Chapter 21. Although the heart attack rate of women under age 70 is lower than that of men, the mortality rate for women is actually higher—perhaps because heart disease in women is neither diagnosed as early nor treated as aggressively as is heart disease in men.

The presence of two risk factors more than doubles the risk of heart attack, so eliminating as many risk factors as possible will improve one's chances of preventing or surviving a heart attack. Changing the diet to limit cholesterol, exercising to lower weight, and seeking treatment for high blood pressure are steps in the right direction. It has been estimated that reduction of coronary risk factors could prevent 150,000 deaths each year in the United States alone.

The Cardiac Cycle

The period between the start of one heartbeat and the beginning of the next is a single **cardiac cycle.** The cardiac cycle therefore includes alternating periods of contraction and relaxation. For any one chamber in the heart, the cardiac cycle can be divided into two phases: (1) *systole* and (2) *diastole*. During **systole** (SIS-tō-lē), or contraction, the chamber contracts and pushes blood into an adjacent chamber or into an arterial trunk. Systole is followed by the second phase, **diastole** (dī-AS-tō-lē), or relaxation. During diastole, the chamber fills with blood and prepares for the next cardiac cycle.

Fluids tend to move from an area of high pressure to one of lower pressure. In the course of the cardiac cycle, the pressure within each chamber rises in systole and falls in diastole. Valves between adjacent chambers help ensure that blood flows in the desired direction, but the mere presence of valves is not enough. Blood will flow from one chamber to another only if the pressure in the first chamber exceeds that in the second. This basic principle governs the movement of blood between atria and ventricles, between ventricles and arterial trunks, and between the major veins and the atria.

The correct pressure relationships are dependent on the careful timing of contractions. For example, blood movement could not occur in the desired direction if an atrium and its attached ventricle contracted at precisely the same moment. The elaborate pacemaking and conducting systems normally provide the required spacing between atrial and ventricular systoles. As a result, *atrial systole* and *ventricular systole* do not occur at the same time, and *atrial diastole* and *ventricular diastole* differ in duration. Figure 20-15● shows the duration and timing of systole and diastole for a representative heart rate of 75 bpm. At this heart rate, a sequence of systole and diastole in either the atria or the ventricles lasts 800 msec. For convenience, we will consider that the cardiac cycle is determined by the atria and that it includes one cycle of atrial systole and atrial diastole. This convention follows our description of the conducting system and the propagation of the stimulus for contraction

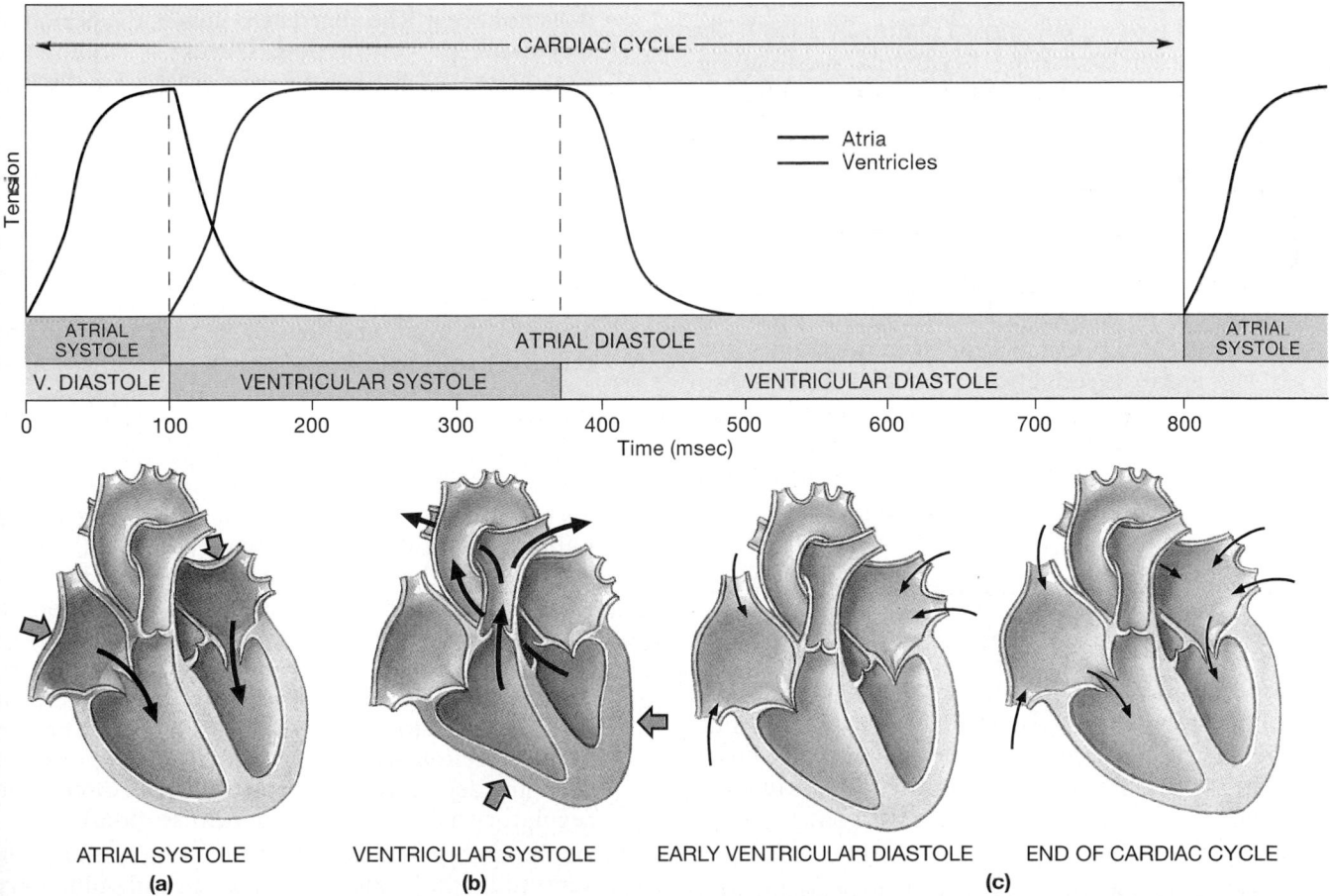

●FIGURE 20-15 **Phases of the Cardiac Cycle.** The timing of the repeated cycles of systole and diastole differs between the atria and ventricles. A cardiac cycle consists of one cycle of systole and diastole in the atria or the ventricles; we will consider a cardiac cycle as determined by the state of the atria for a heart rate of 75 bpm. **(a)** Atrial systole. The atria contract and complete ventricular filling. **(b)** Ventricular systole. Blood is ejected into the pulmonary and aortic trunks. **(c)** Ventricular diastole. Once the AV valves open, passive filling of the ventricles occurs through the end of this cardiac cycle, when atrial systole begins the next cycle.

Phases of the Cardiac Cycle

The phases of atrial systole, ventricular systole, and ventricular diastole are diagrammed in Figure 20-15a,b,c● for a heart rate of 75 bpm. As this cardiac cycle begins, the chambers are relaxed and the ventricles are partially filled with blood. During atrial systole, the atria contract and the ventricles become completely filled with blood (Figure 20-15a●). Atrial systole lasts 100 msec, and over this period, blood cannot flow into the atria because atrial pressure exceeds venous pressure. Although there are no valves at the connections with the venous system, very little backflow occurs, because blood takes the path of least resistance. There is less resistance to blood flow through the broad AV connections and into the ventricles than through the smaller, angled openings of the large veins, which are connected to miles of smaller vessels.

The atria next enter a period of atrial diastole, which continues until the start of the next cardiac cycle. As atrial diastole begins, ventricular systole occurs. Ventricular systole lasts 270 msec. During this period, blood is pushed through the systemic and pulmonary circuits and toward the atria (Figure 20-15b●). The heart then enters a period of ventricular diastole (Figure 20-15c●), which lasts 530 msec (the 430 msec remaining in this cardiac cycle, plus the first 100 msec of the next). For the rest of this cardiac cycle, filling occurs passively and both the atria and the ventricles are relaxed. We have now arrived at the start of the next cardiac cycle, which begins with atrial systole and the completion of ventricular filling.

When the heart rate increases, all the phases of the cardiac cycle become shorter in duration. However, the greatest reduction occurs in the length of time spent in diastole. When the heart rate climbs from 75 bpm to 200 bpm, the time spent in systole drops by less than 40 percent but the duration of diastole is reduced by almost 75 percent.

Pressure and Volume Changes in the Cardiac Cycle

We will consider the pressure and volume changes that occur during the cardiac cycle, using reference numbers that match those in Figure 20-16●. The narrative description applies to both sides of the heart. Although pressures are lower in the right atrium and right ventricle, both sides of the heart contract at the same time, and they eject equal volumes of blood.

ATRIAL SYSTOLE The cardiac cycle begins with atrial systole, which lasts about 100 msec in a resting adult:

1. As the atria contract, rising atrial pressures push blood into the ventricles through the open right and left AV valves.

2. At the start of atrial systole, the ventricles are already filled to about 70 percent of capacity, and atrial systole essentially "tops them off" by providing the additional 30 percent. (The ventricles filled passively during the previous cardiac cycle.)

3. At the end of atrial systole, each ventricle contains the maximum amount of blood that it will hold in this cardiac cycle. That quantity is called the **end-diastolic volume (EDV).** The end-diastolic volume in a person standing at rest is typically about 130 ml.

VENTRICULAR SYSTOLE As atrial systole ends, ventricular systole begins. This period lasts for approximately 270 msec in a resting adult. As the pressures inside the ventricles rise above those in the atria, the AV valves swing shut.

4. During this stage of ventricular systole, the ventricles are contracting. Blood flow has yet to occur, because ventricular pressures are not high enough to force open the semilunar valves and push blood into the pulmonary or aortic trunks. Over this period, the ventricles contract *isometrically;* they generate tension, and ventricular pressures rise, but blood flow does not occur. The ventricle is now in the period of **isovolumetric contraction.** During isovolumetric contraction, all the heart valves are closed, the volume of the ventricle remains constant, and ventricular pressure rises.

5. Once pressure in the ventricle exceeds that in the arterial trunks, the semilunar valves open and blood flows into the pulmonary and aortic trunks. This point marks the beginning of the period of **ventricular ejection.** The ventricles now contract *isotonically;* the muscle cells shorten, and tension production remains relatively constant. (For a review of isotonic versus isometric contractions, see Figure 10-14●, p. 296.)

   After reaching a peak, ventricular pressures gradually decline near the end of ventricular systole. Figure 20-16● shows values for the left ventricle and aorta. Although pressures in the right ventricle and pulmonary trunk are much lower, the right ventricle also goes through periods of isovolumetric contraction and ventricular ejection. During this period, each ventricle will eject approximately 80 ml of blood. This is the **stroke volume (SV)** of the heart. The stroke volume in this case is roughly 60 percent of the end-diastolic volume. This percentage, known as the *ejection fraction,* can vary in response to changing demands on the heart. (We will discuss the regulatory mechanisms in the next section.)

6. At the end of ventricular systole, ventricular pressures fall rapidly. Blood in the aorta and pulmonary trunks now starts to flow back toward the ventricles, and this movement closes the semilunar valves. As the backflow begins, pressure decreases in the aorta. When the semilunar valves close, pressure rises again as the elastic arterial walls recoil. This

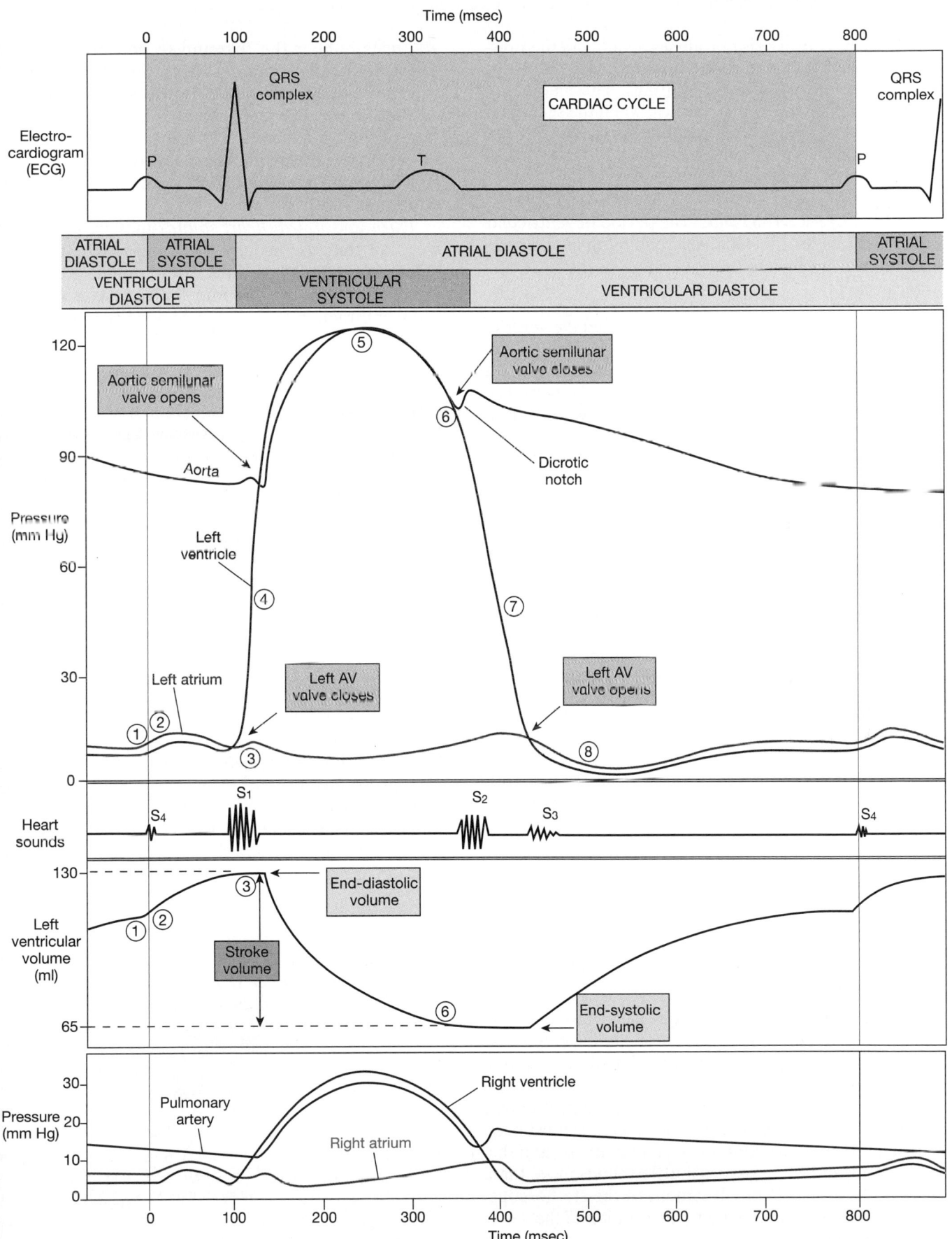

● **FIGURE 20-16** **Pressure and Volume Relationships in the Cardiac Cycle.** Major features of the cardiac cycle are shown for a heart rate of 75 bpm. The circled numbers correspond to the numbered list on pages 696 and 698.

small, temporary rise produces a valley in the pressure tracing that is called a *dicrotic* (dī-KRO-tik) *notch* (*dikrotos,* double-beating). The amount of blood remaining in the ventricle when the semilunar valve closes is the **end-systolic volume (ESV).** The end-systolic volume in this case is 50 ml, about 40 percent of the end-diastolic volume.

VENTRICULAR DIASTOLE The period of ventricular diastole lasts for the 430 msec remaining in this cycle and continues through atrial systole in the next cardiac cycle.

7. All the heart valves are now closed, and the ventricular myocardium is relaxing. Because ventricular pressure is still higher than atrial pressure, blood cannot flow into the ventricles. This is the period of **isovolumetric relaxation.** Ventricular pressures drop rapidly over this period, because the elasticity of the connective tissues of the heart and fibrous skeleton helps reexpand the ventricles toward their resting dimensions.

8. When ventricular pressures fall below those of the atria, the atrial pressure forces the AV valves open. Blood now flows from the atria into the ventricles. Both the atria and the ventricles are now in diastole, but the ventricular pressures continue to fall as the ventricular chambers expand. Throughout this period, pressures in the ventricles are so far below those in the major veins that blood pours through the relaxed atria and into the ventricles. This passive mechanism is the primary method of ventricular filling. The ventricles will be nearly three-quarters full before the cardiac cycle ends.

The relatively minor contribution that atrial systole makes to ventricular volume explains why individuals can survive quite normally when their atria have been so severely damaged that they can no longer function. By contrast, damage to one or both ventricles can leave the heart unable to maintain adequate blood flow through peripheral tissues and organs. A condition of **heart failure** then exists. AM *Heart Failure*

Heart Sounds

Listening to the heart, a technique called *auscultation,* is a simple and effective method of cardiac diagnosis. Physicians use an instrument called a **stethoscope** to listen to normal and abnormal heart sounds. Where to place the stethoscope depends on which valve is under examination (Figure 20-17●). Valve sounds must pass through the pericardium, surrounding tissues, and the chest wall, and some tissues muffle sounds more than others. As a result, the placement of the stethoscope does not always correspond to the position of the valve under review.

There are four **heart sounds**, designated as S_1–S_4 (Figure 20-16●). When you listen to your own heart,

you usually hear the *first* and *second heart sounds.* These sounds accompany the action of your heart valves. The first heart sound, known as "lubb" (S_1) lasts a little longer than the second. S_1, which marks the start of ventricular contraction, is produced as the AV valves close. The second heart sound, "dupp" (S_2) occurs at the beginning of ventricular filling, when the semilunar valves close.

Third and *fourth heart sounds* may be audible as well, but they are usually very faint and seldom are detectable in healthy adults. These sounds are associated with blood flowing into the ventricles (S_3) and atrial contraction (S_4) rather than valve action.

 MITRAL VALVE PROLAPSE Minor abnormalities in valve shape are relatively common. For example, an estimated 10 percent of healthy individuals between 14 and 30 years of age have some degree of **mitral valve prolapse.** In this condition, the mitral valve cusps do not close properly. The problem may involve abnormally long (or short) chordae tendineae or malfunctioning papillary muscles. Because the valve does not work perfectly, some regurgitation occurs during left ventricular systole. The surges, swirls, and eddies that occur during regurgitation create a rushing, gurgling sound known as a **heart murmur.** Most individuals with this condition are completely asymptomatic and live a normal life, unaware of any circulatory malfunction.

The Energy for Cardiac Contractions

When a normal heart is beating, the energy required is obtained by the mitochondrial breakdown of fatty acids (stored as lipid droplets) and glucose (stored as glycogen). These aerobic reactions can occur only when oxygen is readily available. In addition to obtaining oxygen from the coronary circulation, cardiac muscle cells maintain sizable reserves of oxygen bound

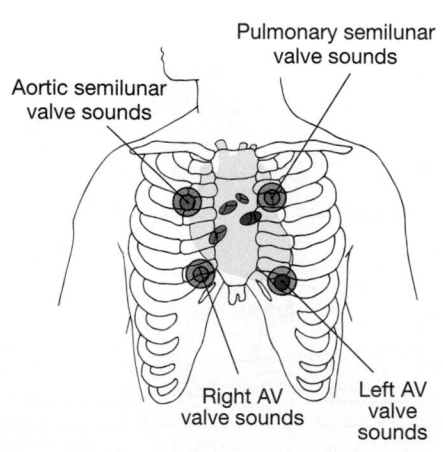

●**FIGURE 20-17 Detecting Heart Sounds.** Placements of a stethoscope for listening to the different sounds produced by individual valves.

to the heme units of myoglobin molecules. (We detailed the structure of this globular protein in Chapter 2 and discussed its function in slow skeletal muscle fibers in Chapter 10.) ∞ *[pp. 55, 302]* Under normal circumstances, the combination of circulatory supplies plus myoglobin reserves is enough to meet the oxygen demands of your heart, even when it is working at maximum capacity.

☑ Is the heart always pumping blood when pressure in the left ventricle is rising? Explain.

☑ What could cause an increase in the size of the QRS complex of an ECG recording?

CARDIODYNAMICS

The term **cardiodynamics** refers to the movements and forces generated during cardiac contractions. Each time the heart beats, the two ventricles eject equal amounts of blood. Earlier we introduced the following important terms:

- *End-diastolic volume (EDV).* The amount of blood in each ventricle at the end of ventricular diastole (the start of ventricular systole).
- *End-systolic volume (ESV).* The amount of blood remaining in each ventricle at the end of ventricular systole (the start of ventricular diastole).
- *Stroke volume (SV).* The amount of blood pumped out of each ventricle during a single beat, which can be expressed as EDV – ESV = SV.
- *Ejection fraction.* The percentage of the EDV represented by the SV.

Stroke volume is the most important factor in an examination of a single cardiac cycle. When considering cardiac function over time, physicians generally are most interested in the **cardiac output (CO),** the amount of blood pumped by each ventricle in 1 minute.

The cardiac output provides a useful indication of ventricular efficiency over time. We can calculate it by multiplying the average stroke volume (SV) by the heart rate (HR):

CO	=	SV	×	HR
cardiac		stroke		heart
output		volume		rate
(ml/min)		(ml/beat)		(beats/min)

For example, if the stroke volume is 80 ml and the heart rate is 75 bpm, the cardiac output will be

$$CO = 80 \text{ ml} \times 75 \text{ bpm}$$
$$= 6000 \text{ ml/min (6 l/min)}$$

Cardiac output is precisely adjusted so that peripheral tissues receive an adequate circulatory supply under a variety of conditions. When necessary, stroke volume in a normal heart can almost double, and the heart rate can increase by 250 percent. In most healthy people, increasing both the stroke volume and the heart rate, as during heavy exercise, can raise the cardiac output by 300–500 percent, to 18–30 l/min. Trained athletes exercising at maximal levels may increase cardiac output by nearly 700 percent, to 40 l/min. The difference between resting and maximal cardiac output is the **cardiac reserve.**

For convenience, we will consider the regulation of stroke volume and heart rate separately, although alterations in cardiac output generally reflect changes in both aspects of cardiac function.

Factors Controlling Stroke Volume

The stroke volume is the difference between the end-diastolic volume and the end-systolic volume. Thus, changes in either the EDV or the ESV can alter the stroke volume and change cardiac output.

The EDV

The EDV indicates the amount of blood a ventricle contains at the end of diastole, just before a contraction begins. This volume is affected by two factors: the **filling time** (the duration of ventricular diastole) and the **venous return** (the rate of blood flow over this period):

1. Filling time depends entirely on the heart rate. The faster the heart rate, the shorter the available filling time.
2. Venous return changes in response to alterations in cardiac output, blood volume, patterns of peripheral circulation, skeletal muscle activity, and other factors that affect the rate of blood flow through the venae cavae. (We shall explore these factors in Chapter 21.)

The ESV

After the ventricle contracts and the stroke volume has been ejected, the ESV is the amount of blood that remains in the ventricle at the end of ventricular systole. Three factors that influence the ESV are (1) the *preload*, (2) the *contractility* of the ventricle, and (3) the *afterload*.

PRELOAD The degree of stretching experienced during ventricular diastole is called the **preload.** The preload is directly proportional to the EDV: The greater the EDV, the larger the preload. Preload is significant because it affects the ability of muscle cells to produce tension. In Chapter 10, we considered the length–tension relationship for skeletal muscle fibers (Figure 10-8●, pp. 288–289); the same principles apply to cardiac muscle cells. If a muscle cell is stimulated when the resting length is very short, a powerful contraction cannot occur, because the sarcomeres are already at minimum size. The degree of shortening and the amount of tension produced during a contraction increase as the resting length increases. There is a relatively narrow range of optimal lengths. As a muscle cell stretches *beyond* optimal length, its abilities to shorten and produce tension decline again. If stretched to the point at which thick and thin filaments no longer overlap, a muscle cell cannot contract.

The amount of preload varies with the demands placed on the heart. When you are standing at rest, your EDV is low; the ventricular muscle is stretched very little, and the sarcomeres are relatively short. During ventricular systole, the cardiac muscle cells develop little power, and the ESV is relatively high because the muscle cells contract only a short distance. If you begin exercising, venous return increases, and more blood flows into your heart. Your EDV increases, and the myocardium stretches further. As the sarcomeres approach optimal lengths, the ventricular muscle cells can contract more efficiently and produce more forceful contractions. They also shorten more, and more blood is pumped out of your heart.

In general, the greater the EDV, the larger the stroke volume. Stretching *past* optimal length, which would reduce the force of contraction, does not normally occur because ventricular expansion is limited by myocardial connective tissues, the fibrous skeleton, and the pericardial sac.

The Frank–Starling Principle. The relationship between the amount of ventricular stretching and contractile force means that within normal physiological limits, increasing the EDV results in a corresponding increase in the stroke volume. This general rule of "more in = more out" was first proposed by Ernest H. Starling on the basis of an analysis of research performed by Otto Frank. The relationship is therefore known as the **Frank–Starling principle,** or *Starling's law of the heart.*

Autonomic adjustments to cardiac output make it difficult to see the effects of the Frank–Starling principle. However, it can be demonstrated effectively in individuals who have received heart transplants, because the implanted hearts are not innervated by the ANS. The most obvious effect of the Frank–Starling principle in these hearts is that the outputs of the left and right ventricles remain balanced under a variety of conditions. For example, consider a situation in which an individual is at rest and the two ventricles are ejecting equal volumes of blood. The ventricles operate in series: When the heart contracts, blood leaves the right ventricle; during the next period of ventricular diastole, a comparable volume of blood will pass through the left atrium, to be ejected by the left ventricle at the next contraction. If the venous return decreases, the EDV of the right ventricle will decline. During ventricular systole it will then pump less blood through the pulmonary circuit to the left atrium. In the next

cardiac cycle, the EDV of the left ventricle will be reduced, and it will eject a smaller volume of blood. The output of the two ventricles will again be in balance, but both will have smaller stroke volumes.

CONTRACTILITY **Contractility** is the amount of force produced during a contraction, at a given preload. Under normal circumstances, contractility may be altered by autonomic innervation or circulating hormones. Under special circumstances, contractility can be altered by administered drugs or as the result of abnormal changes in ion concentrations in the extracellular fluid.

Factors that increase contractility are said to have a *positive inotropic action;* those that decrease contractility have a *negative inotropic action.* Positive inotropic agents typically stimulate Ca^{2+} entry into cardiac muscle cells, thus increasing the force and duration of ventricular contractions. Negative inotropic agents may block Ca^{2+} movement or depress cardiac muscle metabolism in some way. Positive and negative inotropic factors include ANS activity, hormones, and changes in extracelluar ion concentrations.

Autonomic Activity. Autonomic activity alters the degree of contraction and changes the ESV in the following ways:

- Sympathetic stimulation has a positive inotropic effect. It causes the release of norepinephrine (NE) by postganglionic fibers and the secretion of epinephrine (E) and NE by the adrenal medullae. In addition to their effects on heart rate, discussed below, these compounds stimulate cardiac muscle cell metabolism and increase the force and degree of contraction. These effects are produced by stimulating alpha and beta receptors in cardiac muscle cell membranes. ∞ *[p. 597]* The net effect is that the ventricles contract more forcefully, increasing the ejection fraction and decreasing the ESV.

- Parasympathetic stimulation from the vagus nerve has a negative inotropic effect. The primary effect of acetylcholine (ACh) is at the membrane surface, where it produces hyperpolarization and inhibition. The primary result is a decrease in heart rate through effects on the SA and AV nodes. There is also a reduction in the force of cardiac contractions; because the ventricles are not extensively innervated by the parasympathetic division, the atria show the greatest changes in contractile force. However, under strong parasympathetic stimulation or after the administration of drugs that mimic the actions of ACh, the ventricles contract less forcefully, the ejection fraction decreases, and the ESV enlarges.

Hormones. Several hormones, including the following, affect the contractility of the heart:

- Epinephrine and norepinephrine have positive inotropic effects.

- Glucagon has a positive inotropic effect. Before synthetic inotropic agents were available, glucagon was widely used to stimulate cardiac function. It is still used in cardiac emergencies and to treat some forms of heart disease.

- Thyroid hormones have positive inotropic effects. Thyroid gland disorders that result in hypersecretion of thyroid hormones commonly produce excessive stimulation of cardiac muscle tissue, leading to cardiac arrhythmias. Former President George Bush experienced such problems during his term in the White House.

Changes in Ion Concentration. Changes in extracellular Ca^{2+} concentrations affect primarily the strength and duration of cardiac contractions. If the extracellular concentration of Ca^{2+} is elevated, a condition termed **hypercalcemia** (hī-per-kal-SĒ-mē-uh), cardiac muscle cells become extremely excitable, and their contractions become powerful and prolonged. In extreme cases, the heart goes into an extended state of contraction that is generally fatal. When the Ca^{2+} concentration is abnormally low, a condition termed **hypocalcemia** (hī-pō-kal-SĒ-mē-uh), the contractions become very weak and may cease altogether.

Changes in the extracellular K^+ concentration affect primarily the transmembrane potential and the rates of depolarization and repolarization. When K^+ concentrations are high, a condition termed **hyperkalemia** (hī-per-ka-LĒ-mē-uh), the muscle cells depolarize, and repolarzation is inhibited. The cardiac contractions then become weak and irregular, and in severe cases the heart eventually stops in diastole. When K^+ concentrations are abnormally low, a condition termed **hypokalemia** (hī-pō-ka-LĒ-mē-uh), the membranes of cardiac muscle cells hyperpolarize, and their rate of contraction declines. The heart rate decreases, and blood pressure falls. In severe cases, the heart eventually stops in systole. Severe hyperkalemia and hypokalemia are life-threatening conditions that require immediate corrective action.

EFFECTS OF DRUGS ON CONTRACTILITY The drugs *isoproterenol, dopamine,* and *dobutamine* mimic the action of E and NE by stimulating beta-1 receptors on cardiac muscle cells. Dopamine (at high doses) and dobutamine also stimulate Ca^{2+} entry through alpha-1 receptor stimulation. *Digitalis* and related drugs also elevate intracellular Ca^{2+} concentrations, but by a different mechanism. These drugs interfere with the removal of Ca^{2+} from the sarcoplasm of cardiac muscle cells.

Many of the drugs used to treat hypertension (high blood pressure) have a negative inotropic action. Drugs such as *propranolol, timolol, metoprolol, atenolol,* several *barbiturates,* and *labetalol* block beta receptors, alpha receptors, or both and prevent sympathetic stimulation of the heart.

AFTERLOAD The **afterload** is the amount of tension the contracting ventricle must produce to force open the semilunar valve and eject blood. The greater the afterload, the longer the period of isovolumetric contraction, the shorter the duration of ventricular ejection, and the larger the ESV. In other words, as the afterload increases, the stroke volume decreases.

Afterload is increased by any factor that restricts blood flow through the arterial system. For example, constriction of peripheral blood vessels or a circulatory blockage will elevate arterial blood pressure and increase afterload. If the afterload is too great, the ventricle will be unable to eject blood. Afterload that high is rare in the normal heart, but damage to the heart muscle can weaken the myocardium enough that even a modest rise in arterial blood pressure can reduce stroke volume to dangerously low levels, producing symptoms of *heart failure.*

☑ Why is it a potential problem if the heart beats too rapidly?

☑ What effect would stimulating the acetylcholine receptors of the heart have on cardiac output?

☑ What effect would an increased venous return have on the stroke volume?

☑ How would increased sympathetic stimulation of the heart affect the end-systolic volume?

☑ Joe's end-systolic volume is 40 ml, and his end-diastolic volume is 125 ml. What is Joe's stroke volume?

Factors Affecting the Heart Rate

Autonomic activity, hormones and various drugs, changes in ion concentrations, and alterations in body temperature may alter the basic rhythm of contraction established by the SA node. Under normal circumstances, autonomic activity and circulating hormones are responsible for making delicate adjustments to the heart rate as circulatory demands change. Significant changes in ion concentration or body temperature are unusual and may indicate problems with homeostatic control mechanisms. All these factors act by modifying the natural rhythm of the heart. Even a heart removed for a heart transplant will continue to beat unless steps are taken to prevent it.

A number of clinical problems are the result of abnormal pacemaker function. **Bradycardia** (brā-dē-KAR-dē-uh; *bradys,* slow) is the term used to indicate a heart rate that is slower than normal, whereas **tachycardia** (tak-ē-KAR-dē-uh; *tachys,* swift) indicates a faster-than-normal heart rate. These are relative terms, and in clinical practice the definitions vary with the normal resting heart rate of the individual. **AM** *Problems with Pacemaker Function*

Autonomic Innervation

We presented the anatomical details concerning cardiac innervation in Chapter 16 and diagrammed them in Figure 20-10•, p. 687. ∞ *[p. 527]* As do other organs with dual innervation, the heart has a resting autonomic tone. Both autonomic divisions are normally active at a steady background level, releasing ACh and NE both at the nodes and into the myocardium. Thus, cutting the vagus nerves increases the heart rate, and sympathetic blocking agents slow the heart rate.

In a healthy, resting individual, parasympathetic effects dominate. The heart rate in the absence of autonomic innervation is established by the pacemaker cells of the SA node. Such a heart beats at a rate of 80–100 bpm. At rest, a typical adult heart with intact innervation beats at 70–80 bpm due to activity in the parasympathetic nerves innervating the SA node. If parasympathetic activity increases, the heart rate declines further; the heart rate will increase if either parasympathetic activity decreases or sympathetic activation occurs. Through dual innervation and adjustments in autonomic tone, the ANS can make very delicate adjustments in cardiovascular function to meet the demands of other systems.

EFFECTS ON THE SA NODE The sympathetic and parasympathetic divisions alter the heart rate by changing the permeabilities of cells in the conducting system. The most dramatic effects are seen at the SA node, where there are changes in the rate of impulse generation that affect the heart rate. As a secondary effect, autonomic activity also increases or decreases the rate of impulse propagation along the conducting system. We will ignore the changes in propagation rate and focus on the SA node and the modification of the heart rate.

Consider the SA node of a resting individual whose heart is beating at 75 bpm (Figure 20-18a•). Any factor that changes the rate of spontaneous depolarization or the duration of repolarization will alter the heart rate by changing the time required to reach threshold. Acetylcholine released by parasympathetic neurons opens chemically regulated K^+ channels in the cell membrane. This opening dramatically slows the rate of spontaneous depolarization and also slightly extends the duration of repolarization (Figure 20-18b•). The result is a decline in heart rate.

The NE released by sympathetic neurons binds to beta-1 receptors; this binding leads to the opening of calcium ion channels. The influx of Ca^{2+} increases the rate of depolarization and shortens the period of repolarization. The nodal cells reach threshold more quickly, and the heart rate increases (Figure 20-18c•).

THE ATRIAL REFLEX The **atrial reflex,** or *Bainbridge reflex,* involves adjustments in heart rate in response to an increase in the venous return. When the walls of the right atrium are stretched, the stimulation of stretch receptors in the atrial walls triggers a reflexive increase in heart rate caused by increased sympathetic activity (Figure 20-18c•).

Hormones

Just as epinephrine, norepinephrine, and thyroid hormones increase contractility, these hormones also cause an increase in the heart rate. The effects of epinephrine on the SA node are similar to those of norepinephrine. Epinephrine also affects the contractile cells; after massive sympathetic stimulation of the adrenal medullae, the myocardium may become so excitable that abnormal contractions occur.

EFFECTS OF DRUGS ON HEART RATE In Chapter 12, we noted that several drugs, including caffeine and nicotine, have a stimulatory effect on excitable membranes in the nervous system. ∞ *[p. 404]* These drugs also cause an increase in heart rate. Caffeine acts directly on the conducting system and increases the rate of depolarization at the SA node. Nicotine acts indirectly by stimulating the activity of sympathetic neurons that innervate the heart.

Changes in Ion Concentration

The pace the SA node establishes depends on (1) the resting membrane potential of the nodal cells and (2) the rate of spontaneous depolarization. Changes in the ionic composition of the extracellular fluid that affect either the resting potential or the rate of spontaneous depolarization will therefore alter the heart rate. The most obvious and clinically important examples involve K^+ and Ca^{2+}. Changes in extracellular K^+ concentrations will alter the resting potential at the SA node and thereby change the heart rate. When the extracellular concentration of potassium ions declines, the rate of K^+ diffusion out of the nodal cell increases. As a result, a decrease in extracellular potassium ion concentration leads to membrane hyperpolarization and a reduction in heart rate. Changes in the concentration of Ca^{2+} affect the SA node and conducting system, but these effects are overshadowed by the effects on the contractility of the myocardium (as we described in an earlier section).

Changes in Body Temperature

Temperature changes affect metabolic operations throughout the body. For example, a reduction in temperature slows the rate of depolarization at the SA node, lowers the heart rate, and reduces the

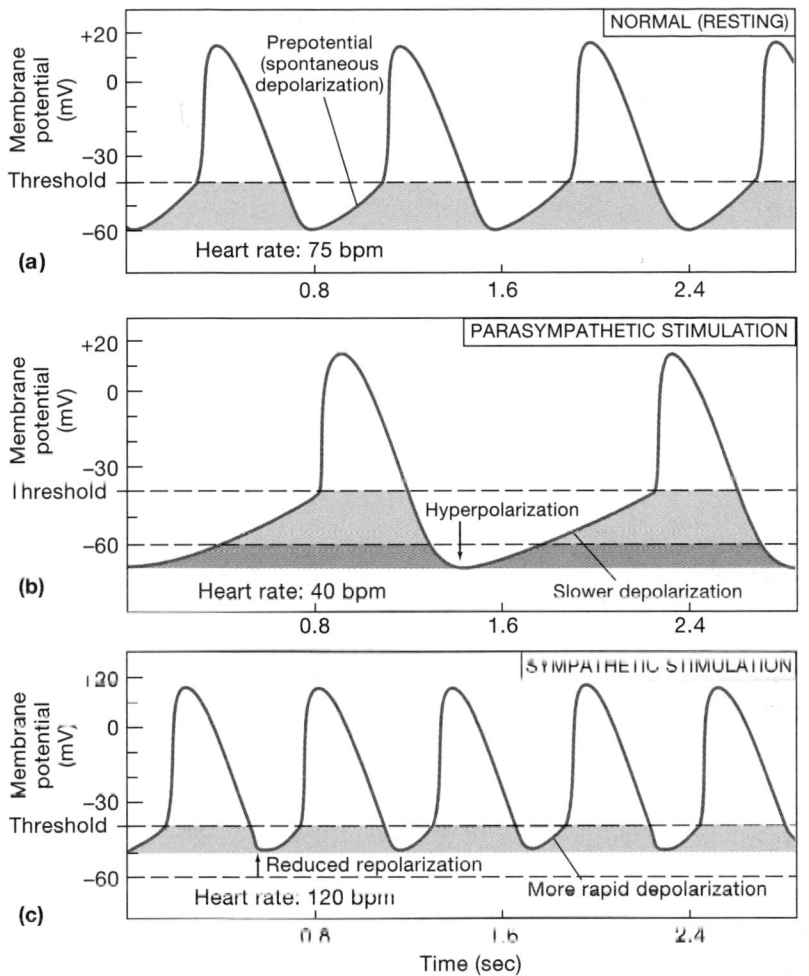

(a)

(b)

(c)

●FIGURE 20-18 Pacemaker Function.
(a) Pacemaker cells have membrane potentials slightly lower than those of other cardiac muscle cells (–60 mV versus –80 to –90 mV). Their cell membranes undergo spontaneous depolarization to threshold, producing action potentials at a frequency determined by (1) the resting membrane potential and (2) the slope of the prepotential (the gradual depolarization). **(b)** Parasympathetic stimulation releases ACh, which extends repolarization and decreases the rate of spontaneous depolarization. The heart rate slows. **(c)** Sympathetic stimulation releases NE, which shortens repolarization and accelerates the rate of spontaneous depolarization. The heart rate increases as a result.

strength of cardiac contractions. (In open-heart surgery, the exposed heart may be chilled until it stops beating.) An elevated body temperature accelerates the heart rate and the contractile force. That is one reason why your heart seems to race and pound whenever you have a fever.

Cardiac Output and Heart Rate

Figure 20-19● summarizes factors involved in the regulation of heart rate and stroke volume—the factors that interact to determine cardiac output. Cardiac output cannot increase indefinitely, primarily

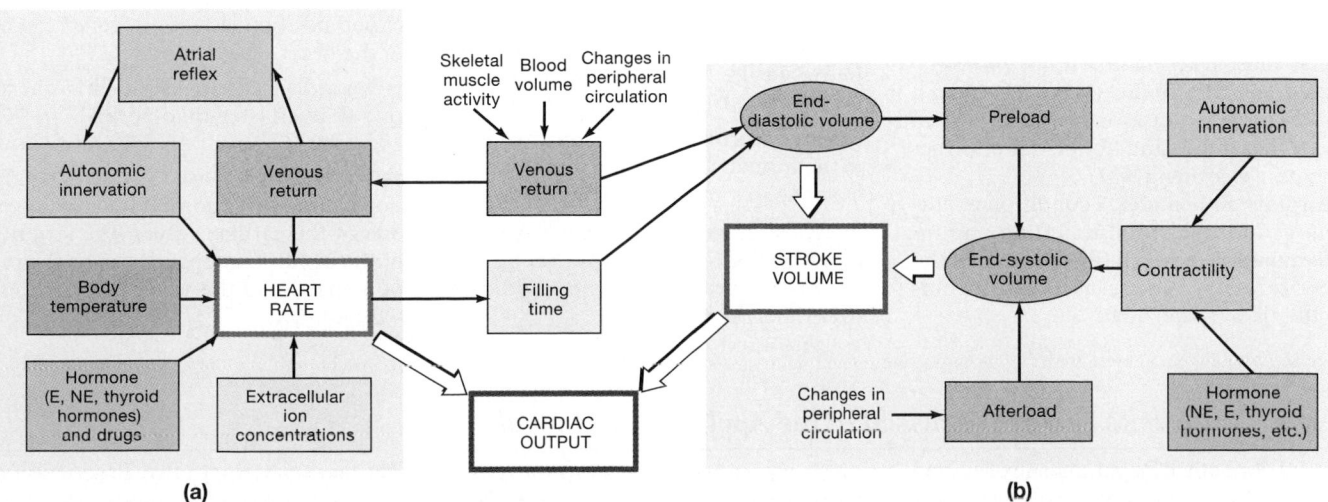

(a)

(b)

●FIGURE 20-19 Factors Affecting Cardiac Output. **(a)** Factors affecting heart rate. **(b)** Factors affecting stroke volume.

because the available filling time becomes shorter and shorter as the heart rate increases. At heart rates up to 160–180 bpm, the combination of an increased rate of venous return and increased contractility compensates for the reduction in filling time. Over this range, cardiac output and heart rate increase together. But if the heart rate continues to climb, the stroke volume begins to drop, and cardiac output first plateaus and then declines.

THE HEART AND THE CARDIOVASCULAR SYSTEM

The goal of cardiovascular regulation is to maintain adequate blood flow to all body tissues. The heart cannot accomplish this goal by itself, and it does not work in isolation. For example, when blood pressure changes, car-

diovascular centers not only adjust the heart rate but also alter the diameters of peripheral blood vessels. These adjustments work together to keep blood pressure within normal limits and to maintain circulation to vital tissues and organs. Chapter 21 begins with the functional anatomy of blood vessels and then details the cardiovascular responses to changing activity patterns and circulatory emergencies. It then completes our discussion of the cardiovascular system by examining the anatomy of the pulmonary and systemic circuits.

☑ What effect would drinking large amounts of caffeinated coffee have on the heart?

☑ If a person suffers from bradycardia, is cardiac output likely to be increased or decreased? Explain.

☑ How would a drug that increases the length of time required for repolarization of pacemaker cells affect a person's heart rate?

SELECTED CLINICAL TERMINOLOGY

Terms Discussed in This Chapter

angina pectoris (an-JĪ-nuh PEK-tor-is): A condition in which exertion or stress can produce severe chest pain, resulting from temporary insufficiency and ischemia when the heart's workload increases. *(p. 685)*

balloon angioplasty: A technique for reducing the size of a coronary plaque by compressing it against the arterial walls, using a catheter with an inflatable collar. *(p. 686)*

bradycardia (brā-dē-KAR-dē-uh): A heart rate that is slower than normal. *(p. 701 and AM)*

calcium channel blockers: Drugs administered to reduce contractility of the heart by slowing the influx of calcium ions during the plateau phase of the cardiac muscle action potential. *(p. 685)*

cardiac arrhythmias (ā-RITH-mē-az): Abnormal patterns of cardiac electrical activity, indicating abnormal contractions. *(p. 693 and AM)*

cardiac tamponade: A condition resulting from pericardial irritation and inflammation in which fluid collects in the pericardial sac and restricts cardiac output. *(p. 675 and AM)*

carditis (kar-DĪ-tis): A general term indicating inflammation of the heart *(p. 682 and AM)*

conduction deficit: An abnormality in the conducting system of the heart that affects the timing and pacing of cardiac contractions. *(p. 692 and AM)*

coronary artery bypass graft (CABG): Replacement of a coronary artery or one of its branches by a vessel transplanted from another part of the body. *(p. 686)*

coronary artery disease (CAD): Degenerative changes in the coronary circulation. *(p. 685)*

coronary ischemia: Restriction of the circulatory supply to the heart, potentially causing tissue damage and a reduction in cardiac efficiency. *(p. 685)*

coronary thrombosis: A blockage due to the formation of a clot (thrombus) at a plaque in a coronary artery. *(p. 694)*

electrocardiogram (ē-lek-trō-KAR-dē-ō-gram) **(ECG or EKG):** A recording of the electrical activities of the heart over time. *(p. 692)*

heart failure: A condition in which the heart weakens and peripheral tissues suffer from oxygen and nutrient deprivation. *(p. 698 and AM)*

myocardial (mī-ō-KAR-dē-al) **infarction (MI):** A condition in which the coronary circulation becomes blocked and the cardiac muscle cells die from oxygen starvation; also called a *heart attack*. *(p. 694)*

pericarditis: Inflammation of the pericardium. *(p. 675)*

rheumatic heart disease (RHD): A disorder in which the heart valves become thickened and stiffen into a partially closed position, affecting the efficiency of the heart. *(p. 682 and AM)*

tachycardia (tak-ē-KAR-dē-uh): A heart rate that is faster than normal. *(p. 701 and AM)*

valvular heart disease (VHD): A condition caused by abnormal functioning of one of the cardiac valves. The severity of the condition depends on the degree of damage and the valve involved. *(p. 682 and AM)*

AM Additional Terms Discussed in the *Applications Manual*

atrial fibrillation: Rapid, uncontrolled, and uncoordinated spasms in the atrial walls.

atrial flutter: Rapid contractions of the atria that are ineffective in propelling blood.

cardiomyopathies (kar-dē-ō-mī-OP-a-thēz): A group of diseases characterized by the progressive, irreversible degeneration of the myocardium.

coronary arteriography: Production of an X-ray image of coronary circulation after the introduction of a radiopaque dye into one of the coronary arteries via a catheter; the resulting image is a *coronary angiogram.*

defibrillator: A device used to eliminate ventricular fibrillation and restore normal cardiac rhythm.

echocardiography: Ultrasound analysis of the heart and the blood flow through major vessels.

endocarditis: Inflammation of the endocardial lining.

heart block: A condition affecting the normal rhythm of the heart; characterized by impaired communication between the SA node, the AV node, and the ventricular myocardium resulting from damage to conduction pathways caused by mechanical distortion, ischemia, infection, or inflammation.

myocarditis: Inflammation of the myocardium.

paroxysmal atrial tachycardia (PAT): Bursts of rapid atrial depolarization, leading to multiple atrial contractions with little pause between them.

premature atrial contractions (PACs): Atrial contractions that occur in advance of the rhythm established by the SA node.

premature ventricular contractions (PVCs): Premature ventricular beats that are directed by an ectopic pacemaker.

valvular stenosis (ste-NŌ-sis)**:** A condition in which the opening between thickened heart valves becomes narrower.

ventricular escape: The initiation of ven-

tricular contractions that are controlled by the AV node, Purkinje fibers, or other parts of the ventricular conducting system rather than by the SA node.

ventricular fibrillation (VF): The most serious type of cardiac arrhythmia, in which the normal rhythm becomes disrupted because the cardiac muscle fibers are stimulating one another at such a rapid rate. Ventricular contractions are rapid, uncontrolled, and uncoordinated; the ventricles are incapable of pumping blood. Also known as *cardiac arrest.*

ventricular tachycardia (VT): Rapid ventricular beats that are not coordinated with atrial activity.

CHAPTER REVIEW

On-line resources for this chapter are on our World Wide Web site at:
http://www.prenhall.com/martini/fap

STUDY OUTLINE

INTRODUCTION, p. 673

AN OVERVIEW OF THE CARDIOVASCULAR SYSTEM, p. 673

1. The circulatory pathways can be subdivided into the **pulmonary circuit** (which carries blood to and from the lungs) and the **systemic circuit** (which transports blood to and from the rest of the body). **Arteries** carry blood away from the heart; **veins** return blood to the heart. **Capillaries** are narrow-diameter vessels that connect the smallest arteries and veins. *(Figure 20-1)*

2. The heart has four chambers: the **right atrium** and **right ventricle** and the **left atrium** and **left ventricle.**

ANATOMY OF THE HEART, p. 673

1. The heart is surrounded by the **pericardial cavity** and lies within the anterior portion of the mediastinum, which separates the two pleural cavities.

The Pericardium, p. 674

2. The pericardial cavity is lined by the **pericardium**. The **visceral pericardium (epicardium)** covers the heart's outer surface, and the **parietal pericardium** lines the inner surface of the **pericardial sac,** which surrounds the heart. *(Figure 20-2)*

Superficial Anatomy of the Heart, p. 675

3. A deep groove, the **coronary sulcus,** marks the boundary between the atria and the ventricles. Other surface markings also provide useful reference points in describing the heart and associated structures. *(Figure 20-3)*

Internal Anatomy and Organization, p. 675

4. The atria are separated by the **interatrial septum,** and the ventricles are divided by the **interventricular septum.** The right atrium receives blood from the systemic circuit via two large veins, the **superior vena cava** and the **inferior vena cava.** The atrial walls contain prominent muscular ridges, the **pectinate muscles.** *(Figure 20-4)*

5. Blood flows from the right atrium into the right ventricle via the **right atrioventricular (AV) valve (tricuspid valve).** This opening is bounded by three **cusps** of fibrous tissue braced by the tendinous **chordae tendineae,** which are connected to **papillary muscles.**

6. Blood leaving the right ventricle enters the **pulmonary trunk** after passing through the **pulmonary semilunar valve.** The pulmonary trunk divides to form the **left** and **right pulmonary arteries.** The **left** and **right pulmonary veins** return blood to the left atrium. Blood leaving the left atrium flows into the left ventricle via the **left atrioventricular (AV) valve (bicuspid,** or **mitral, valve).** Blood leaving the left ventricle passes through the **aortic semilunar valve** and into the systemic circuit via the **ascending aorta.**

7. Anatomical differences between the ventricles reflect the functional demands placed on them. The wall of the right ventricle is relatively thin, whereas the left ventricle has a massive muscular wall. *(Figure 20-5)*

8. Valves normally permit blood flow in only one direction, preventing the **regurgitation** (backflow) of blood. *(Figure 20-6)*

The Heart Wall, p. 682

9. The bulk of the heart consists of the muscular **myocardium.** The **endocardium** lines the inner surfaces of the heart. *(Figure 20-7)*

10. **Cardiac muscle cells** are interconnected by **intercalated discs,** which convey the force of contraction from cell to cell and conduct action potentials. *(Table 20-1)*

Connective Tissues and the Fibrous Skeleton, p. 683

11. The connective tissues of the heart (mainly collagen and elastic fibers) and the **fibrous skeleton** support the heart's contractile cells and valves. *(Figure 20-6).*

The Blood Supply to the Heart, p. 683

12. The **coronary circulation** meets the high oxygen and nutrient demands of cardiac muscle cells. The **coronary arteries** originate at the base of the ascending aorta. Interconnec-

tions between arteries called **anastomoses** ensure a constant blood supply. The **great, posterior, small, anterior,** and **middle cardiac veins** carry blood from the coronary capillaries to the **coronary sinus.** *(Figure 20-8)*

13. In **coronary artery disease (CAD)**, the coronary circulation undergoes degenerative changes. *(Figure 20-9)*

Innervation of the Heart, p. 687

14. The **cardioacceleratory center** in the medulla oblongata activates sympathetic neurons; the **cardioinhibitory center** governs the activities of the parasympathetic neurons. The cardiac centers receive inputs from higher centers and from receptors monitoring blood pressure and the concentrations of dissolved gases. *(Figure 20-10)*

THE HEARTBEAT, p. 688

1. Two general classes of cardiac muscle cells are involved in the normal **heartbeat: contractile cells** and cells of the **conducting system**.

Contractile Cells, p. 688

2. Cardiac muscle cells have a long refractory period, so rapid stimulation produces isolated rather than tetanic contractions. *(Figure 20-11)*

The Conducting System, p. 690

3. The **conducting system** is composed of the *sinoatrial node*, the *atrioventricular node*, and *conducting cells*. This system initiates and distributes electrical impulses within the heart. Nodal cells establish the rate of cardiac contraction, and conducting cells distribute the contractile stimulus from node to node (along the *internodal pathways*) and from the nodes to the general myocardium. *(Figure 20-12)*

4. Unlike skeletal muscle, cardiac muscle contracts without neural or hormonal stimulation. **Pacemaker cells** in the **sinoatrial (SA) node (cardiac pacemaker)** normally establish the rate of contraction. From the SA node, the stimulus travels to the **atrioventricular (AV) node,** then to the **AV bundle,** which divides into **bundle branches.** From there, **Purkinje fibers** convey the impulses to the ventricular myocardium. *(Figure 20-13)*

The Electrocardiogram, p. 692

5. A recording of electrical activities in the heart is an **electrocardiogram (ECG or EKG).** Important landmarks of an ECG include the **P wave** (atrial depolarization), the **QRS complex** (ventricular depolarization), and the **T wave** (ventricular repolarization). *(Figure 20-14)*

The Cardiac Cycle, p. 695

6. The **cardiac cycle** contains periods of **atrial** and **ventricular systole** (contraction) and **diastole** (relaxation). *(Figure 20-15)*

7. When the heart beats, the two ventricles eject equal volumes of blood. *(Figure 20-16)*

8. The closing of valves and rushing of blood through the heart cause characteristic **heart sounds** that can be heard during *auscultation*. *(Figures 20-16, 20-17)*

CARDIODYNAMICS, p. 699

1. The amount of blood ejected by a ventricle during a single beat is the **stroke volume (SV);** the amount of blood pumped by a ventricle each minute is the **cardiac output (CO).** The difference between resting and maximal cardiac output is the **cardiac reserve.**

Factors Controlling Stroke Volume, p. 699

2. The stroke volume is the difference between the **end-diastolic volume (EDV)** and the **end-systolic volume (ESV).** The **filling time** and **venous return** interact to determine the end-diastolic volume. Normally, the greater the EDV, the more powerful the succeeding contraction (the **Frank–Starling principle**).

Factors Affecting the Heart Rate, p. 701

3. The basic heart rate is established by the pacemaker cells of the SA node, but it can be modified by the ANS. The **atrial reflex** accelerates the heart rate when the walls of the right atrium are stretched. *(Figure 20-18)*

4. Sympathetic activity produces more powerful contractions that reduce the ESV, and parasympathetic stimulation slows the heart rate, reduces the contractile strength, and raises the ESV.

5. Cardiac output is affected by various factors, including autonomic activity, hormones, drugs, ion concentrations, and temperature changes. *(Figure 20-19)*

THE HEART AND THE CARDIOVASCULAR SYSTEM, p. 704

1. The heart does not work in isolation in maintaining adequate blood flow to all tissues.

REVIEW QUESTIONS

Level 1   Reviewing Facts and Terms

1. The blood supply to the muscles of the heart is provided by the
- **(a)** systemic circulation
- **(b)** pulmonary circulation
- **(c)** coronary circulation
- **(d)** coronary portal system

2. The autonomic centers for cardiac function are located in the
- **(a)** myocardial tissue of the heart
- **(b)** cardiac centers of the medulla oblongata
- **(c)** cerebral cortex
- **(d)** a, b, and c are correct

3. The serous membrane covering the outer surface of the heart is the
- **(a)** parietal pericardium   **(b)** visceral pericardium
- **(c)** myocardium   **(d)** endocardium

4. The simple squamous epithelium covering the valves of the heart constitutes the
- **(a)** epicardium   **(b)** endocardium
- **(c)** myocardium   **(d)** fibrous skeleton

5. The structure that permits blood flow from the right atrium to the left atrium while the lungs are developing is the
- **(a)** foramen ovale   **(b)** interatrial septum
- **(c)** coronary sinus   **(d)** fossa ovalis

6. Blood leaves the left ventricle by passing through the
  (a) aortic semilunar valve
  (b) pulmonary semilunar valve
  (c) mitral valve
  (d) tricuspid valve

7. The QRS complex of the ECG appears as the
  (a) atria depolarize   **(b)** ventricles depolarize
  (c) ventricles repolarize **(d)** atria repolarize

8. In the cardiac cycle, during diastole the chambers of the heart
  (a) relax and fill with blood
  (b) contract and push blood into an adjacent chamber
  (c) experience a sharp increase in pressure
  (d) reach a pressure of approximately 120 mm Hg

9. During the cardiac cycle, the amount of blood remaining in the ventricle when the semilunar valve closes is the
  (a) stroke volume (SV)
  (b) end-diastolic volume (EDV)
  (c) end-systolic volume (ESV)
  (d) cardiac output (CO)

10. What role do the chordae tendineae and papillary muscles have in the normal function of the AV valves?

11. What is the purpose of each of the three distinct layers that make up the wall of the heart?

12. What are the principal valves found in the heart, and what is the function of each?

13. Trace the normal pathway of an electrical impulse through the conducting system of the heart.

14. (a) What is the cardiac cycle?
  (b) What phases and events are necessary to complete the cardiac cycle?

15. What three important factors regulate stroke volume to ensure that the left and right ventricles pump equal volumes of blood?

Level 2   Reviewing Concepts

16. During the absolute refractory period, a cell membrane cannot respond at all because
  (a) sodium channels are either open or closed and inactivated
  (b) potassium channels are temporarily closed
  (c) the membrane is hyperpolarized
  (d) a, b, and c are correct

17. Tetanic muscle contractions cannot occur in a normal cardiac muscle cell because
  (a) cardiac muscle tissue contracts on its own
  (b) there is no neural or hormonal stimulation
  (c) the refractory period lasts until the muscle cell relaxes
  (d) the refractory period ends before the muscle cell reaches peak tension

18. The amount of blood that is forced out of the heart depends on
  (a) the degree of stretching at the end of ventricular diastole
  (b) the contractility of the ventricle
  (c) the amount of pressure required to eject blood
  (d) a, b, and c are correct

19. The cardiac output cannot increase indefinitely because
  (a) available filling time becomes shorter as the heart rate increases
  (b) cardiovascular centers adjust the heart rate
  (c) the rate of spontaneous depolarization decreases
  (d) the ion concentrations of pacemaker cell membranes decrease

20. Describe the association of the four muscular chambers of the heart with the pulmonary and systemic circuits.

21. What are the source and significance of the heart sounds?

22. (a) Differentiate between stroke volume and cardiac output.
  (b) How is cardiac output calculated?

23. What factors influence cardiac output?

24. What is the significance of preload to cardiac muscle cells?

25. (a) What effect does sympathetic stimulation have on the heart?
  (b) What effect does parasympathetic stimulation have on the heart?

26. Describe the effects that epinephrine, norepinephrine, glucagon, and thyroid hormones have on the contractility of the heart.

Level 3   Critical Thinking and Clinical Applications

27. Most of the ATP produced in cardiac muscle is derived from the metabolism of fatty acids. During times of exertion, cardiac muscle cells can use lactic acid as an energy source. Why would this adaptation be advantageous to cardiac function?

28. A patient's ECG recording shows a consistent pattern of two P waves followed by a normal QRS complex and T wave. What is the cause of this abnormal wave pattern?

29. The following measurements were made on two individuals. (The values recorded remained stable for 1 hour.)
  Person 1: heart rate, 75 bpm; stroke volume, 60 ml
  Person 2: heart rate, 90 bpm; stroke volume, 95 ml
Which person has the greater venous return? Which person has the longer ventricular filling time?

30. Karen is taking the medication *verapamil*, a drug that blocks the calcium channels in cardiac muscle cells. What effect would you expect this medication to have on Karen's stroke volume?

31. After a myocardial infarction, the cells surrounding the damaged tissue frequently become hyperexcitable and act as ectopic pacemakers. This condition can lead to abnormal ventricular rhythms, with fatal consequences. What would cause the excitability of the uninjured cells?

32. Preventricular contractions (PVCs) occur when a Purkinje cell or contractile cell in the ventricle depolarizes to threshold, triggering a premature contraction. A person experiencing a PVC may feel that her heart has "skipped a beat." If the PVC causes an extra contraction, why does the person feel that a beat has been skipped?

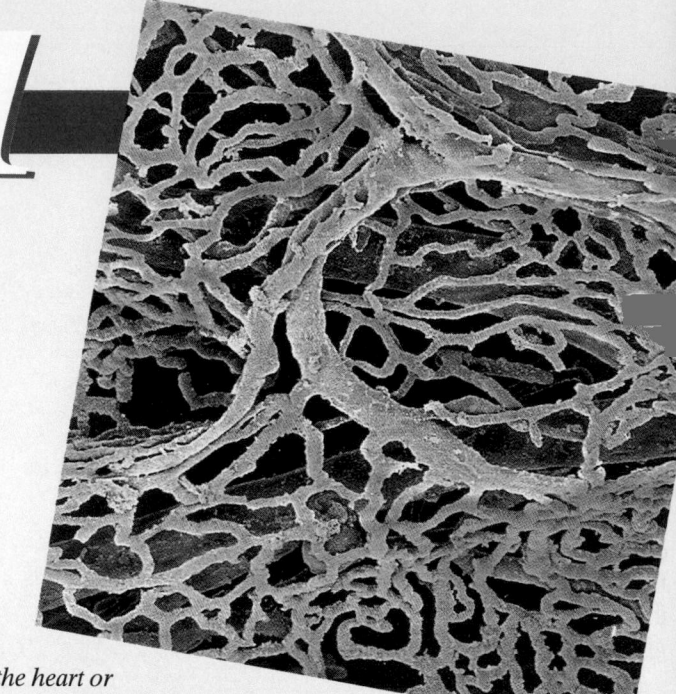

CHAPTER 21

Blood Vessels and Circulation

When we think of the cardiovascular system, we think first of the heart or of the great blood vessels that leave it and return to it. But the real work of the cardiovascular system is done in microscopic vessels that permeate most tissues. This is a network of capillaries, delicate vessels that permit diffusion between the blood and the interstitial fluid. The cardiovascular system performs all its major functions within such networks. Capillaries weave throughout active tissues, forming intricate networks that surround muscle fibers, radiate through connective tissues, and branch beneath the basement membranes of epithelia. Capillaries provide oxygen and nutrients and remove carbon dioxide and wastes. Homeostatic mechanisms operating at the local, regional, and systemic levels adjust blood flow through the capillaries to meet the demands of peripheral tissues. In this chapter, we shall consider these mechanisms and the structure and distribution of capillaries and other blood vessels.

CHAPTER OUTLINE AND OBJECTIVES

In the last two chapters, we examined the composition of blood and the structure and function of the heart, whose pumping action keeps the blood in motion. We shall now consider the vessels that carry blood to peripheral tissues and the nature of the exchange that occurs between the blood and interstitial fluids of the body.

Blood leaves the heart in the pulmonary and aortic trunks, each with an internal diameter of about 2.5 cm (1 in.). The pulmonary arteries that branch from the pulmonary trunk carry blood to the lungs. The systemic arteries that branch from the aorta distribute blood to all other organs. Within these organs, further branching occurs, creating several hundred million tiny arteries that provide blood to more than 10 billion capillaries. These capillaries, barely the diameter of a single red blood cell, form extensive, branching networks. If all the capillaries in the body were placed end to end, they would circle the globe with a combined length of over 25,000 miles.

The vital functions of the cardiovascular system depend entirely on events at the capillary level: *All chemical and gaseous exchange between the blood and interstitial fluid takes place across capillary walls.* Tissue cells rely on capillary diffusion to obtain nutrients and oxygen and to remove metabolic wastes, such as carbon dioxide and urea. Diffusion occurs very rapidly, because the distances involved are very small; few living cells lie farther than 125 μm (0.005 in.) from a capillary.

As blood flows through peripheral tissues, blood pressure forces water and solutes out of the plasma, across capillary walls. About 3.6 liters of water and solutes flows through peripheral tissues each day and enters the *lymphatic system,* which empties into the bloodstream (Figure 21-1•). The continuous movement of water out of the capillaries, through peripheral tissues, then back to the bloodstream via the lymphatic system has four important functions:

1. It ensures that the plasma and the interstitial fluid, two major components of the extracellular fluid, are in constant communication.

2. It accelerates the distribution of nutrients, hormones, and dissolved gases throughout the tissue.

3. It assists in the transport of insoluble lipids and tissue proteins that cannot enter the circulation by crossing the capillary walls.

4. It has a flushing action that carries bacterial toxins and other chemical stimuli to lymphoid tissues and organs responsible for providing immunity from disease.

This chapter begins with a description of the histological organization of arteries, capillaries, and veins. We then consider the functions of these vessels and basic principles of cardiovascular regulation. In the final section, we examine the distribution of major blood vessels in the body. We will then be ready to consider the organization and function of the lymphatic system, the focus of Chapter 22.

THE ANATOMY OF BLOOD VESSELS

Arteries carry blood away from the heart toward **capillaries**, and **veins** return blood from capillaries to the heart. As they approach the capillaries, arteries branch repeatedly, and the branches decrease in diameter. The smallest arterial branches are called **arterioles** (ar-TE-rē-ōlz). From the arterioles, blood moves into the capillaries. By regulating smooth muscle tone in the walls of the arterioles, the *cardiovascular centers* in the brain ensure that the volume, pressure, and speed of blood movement through peripheral capillaries remain within acceptable limits. ∞ *[p. 473]*

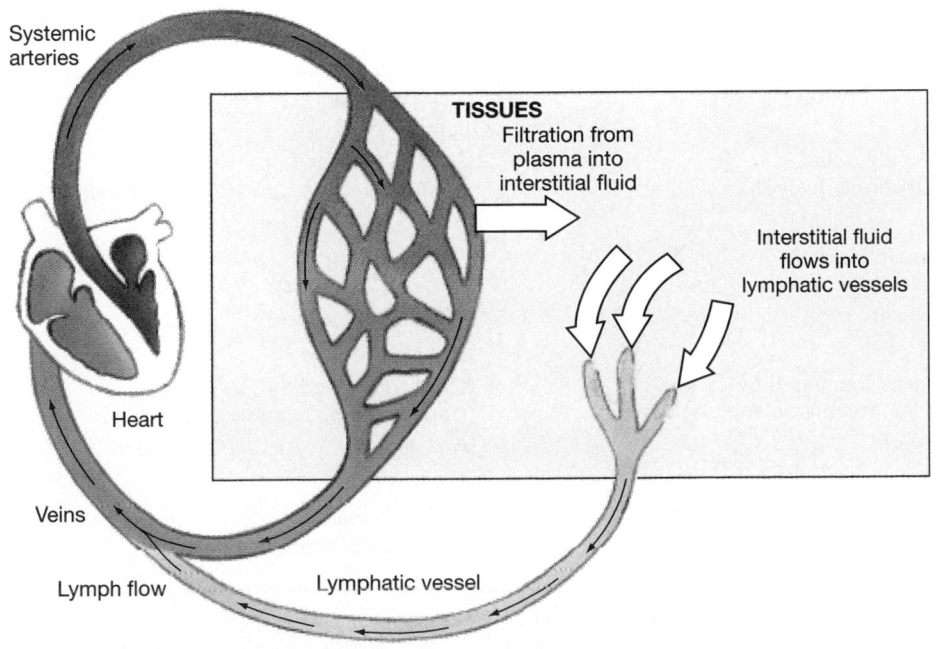

Systemic arteries

TISSUES
Filtration from plasma into interstitial fluid

Interstitial fluid flows into lymphatic vessels

Heart

Veins

Lymph flow

Lymphatic vessel

•*FIGURE 21-1* The **Circulation of Extracellular Fluid.** There is continuous movement of fluid from the plasma into the interstitial fluid at capillaries and back to the plasma via the lymphatic system.

The Structure of Vessel Walls

The walls of arteries and veins contain three distinct layers: a *tunica interna* (in-TER-nuh) on the inside, a *tunica media* in the middle, and a *tunica externa* (eks-TER-nuh) on the outside (Figure 21-2●):

1. The **tunica interna,** or *tunica intima,* is the innermost layer of a blood vessel. This layer includes the endothelial lining and an underlying layer of connective tissue with a variable number of elastic fibers. In arteries, the outer margin of the tunica interna contains a thick layer of elastic fibers called the **internal elastic membrane.**

2. The **tunica media,** the middle layer, contains concentric sheets of smooth muscle tissue in a framework of loose connective tissue. The collagen fibers bind the tunica media to the tunica interna and tunica externa. The tunica media is commonly the thickest layer in the wall of a small artery. It is separated from the surrounding tunica externa by the **external elastic membrane,** a thin band of elastic fibers. The smooth muscle cells of the tunica media encircle the endothelium lining the lumen of the blood vessel. When these smooth muscles contract, the vessel decreases in diameter; when they relax, the diameter increases. Large arteries also contain layers of longitudinally arranged smooth muscle cells.

3. The **tunica externa,** or *tunica adventitia* (ad-ven-TISH-ē-uh), is the outermost layer and forms a connective tissue sheath around the vessel. In arteries, this layer contains collagen fibers with scattered bands of elastic fibers. In veins, this layer, which is

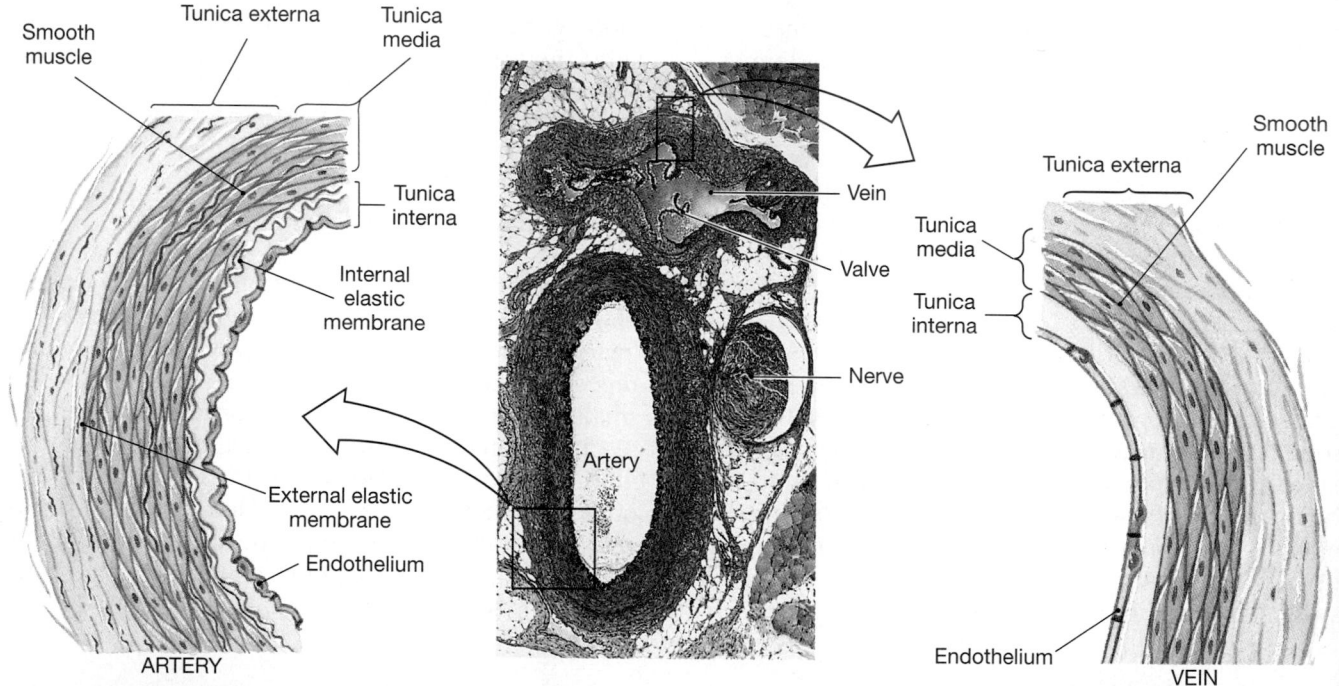

Feature	Typical Artery	Typical Vein
General appearance in sectional view	Usually round, with relatively thick wall	Usually flattened or collapased, with relatively thin wall
Tunica interna Endothelium	Usually rippled due to vessel constriction	Often smooth
Internal elastic membrane	Present	Absent
Tunica media	Thick, dominated by smooth muscle and elastic fibers	Thin, dominated by smooth muscle and collagen fibers
External elastic membrane	Present	Absent
Tunica externa	Collagen and elastic fibers	Collagen, elastic, and smooth muscle fibers

●**FIGURE 21-2 A Comparison of a Typical Artery and a Typical Vein.** (LM × 74)

generally thicker than the tunica media, contains networks of elastic fibers and bundles of smooth muscle cells. The connective tissue fibers of the tunica externa typically blend into those of adjacent tissues, stabilizing and anchoring the blood vessel.

Their layered walls give arteries and veins considerable strength. The muscular and elastic components also permit controlled alterations in diameter as blood pressure or blood volume changes. However, the walls of arteries and veins are too thick to allow diffusion between the bloodstream and surrounding tissues, or even between the blood and the tissues of the vessel itself. For this reason, the walls of large vessels contain small arteries and veins that supply the smooth muscle cells and fibroblasts of the tunica media and tunica externa. These blood vessels are called the **vasa vasorum** ("vessels of vessels").

Differences between Arteries and Veins

Arteries and veins supplying the same region typically lie side by side (Figure 21-2●). In sectional view, arteries and veins may be distinguished by the following characteristics:

- In general, the walls of arteries are thicker than those of veins. The tunica media of an artery contains more smooth muscle and elastic fibers than does that of a vein. These contractile and elastic components resist the pressure generated by the heart as it forces blood into the circuit.
- When not opposed by blood pressure, the elastic fibers in the arterial walls recoil, constricting the lumen. Thus, when seen on dissection or in sectional view, the lumen of an artery looks smaller than that of the corresponding vein. Because the walls of arteries are relatively thick and strong, they retain their circular shape in section (Figure 21-2●). Cut veins tend to collapse, and in section they often look flattened or grossly distorted.
- The endothelial lining of an artery cannot contract, so when an artery constricts, its endothelium is thrown into folds that give arterial sections a pleated appearance. The lining of a vein lacks these folds.

In gross dissection, arteries and veins can generally be distinguished because:

- The thicker walls of the arteries can be felt if the vessels are compressed.
- Arteries usually retain their cylindrical shape, whereas veins often collapse.
- Arteries are more resilient; when stretched, they keep their shape and elongate, and when released they snap back. A small vein cannot tolerate as much distortion without collapsing or tearing.
- Veins typically contain *valves,* internal structures that prevent the backflow of blood toward the capillaries.

In an intact vein, the location of each valve is marked by a slight distension of the vessel wall. (We will consider valve structure in a later section.)

Arteries

Their relatively thick, muscular walls make arteries *elastic* and *contractile.* Elasticity permits passive changes in vessel diameter in response to alterations in blood pressure. It allows arteries to absorb the pressure pulses that accompany the contractions of the ventricles.

The contractility of the arterial walls gives them the ability to change in diameter actively, primarily under the control of the sympathetic division of the ANS. When stimulated, arterial smooth muscles contract and thereby constrict the artery, a process called **vasoconstriction.** Relaxation of the smooth muscles causes an increase in the diameter of the lumen, a process called **vasodilation.** Vasoconstriction and vasodilation affect (1) the afterload on the heart, (2) peripheral blood pressure, and (3) capillary blood flow. We will explore these effects in a later section. Contractility is also important during the *vascular phase* of hemostasis, when contraction of a damaged vessel wall helps reduce bleeding. ∞ *[p. 663]*

In traveling from the heart to peripheral capillaries, blood passes through *elastic arteries, muscular arteries,* and *arterioles* (Figure 21-3●).

Elastic Arteries

Elastic arteries, or *conducting arteries,* are large vessels with a diameter up to 2.5 cm (1 in.). These vessels transport large volumes of blood away from the heart. The pulmonary and aortic trunks and their major arterial branches (the *pulmonary, common carotid, subclavian,* and *common iliac arteries*) are elastic arteries.

The walls of elastic arteries (Figure 21-3●) are extremely resilient. The tunica media of these vessels contains a high density of elastic fibers and relatively few smooth muscle cells. As a result, elastic arteries are able to tolerate the pressure changes that occur during the cardiac cycle. When ventricular systole occurs, pressures rise rapidly and the elastic arteries expand. During ventricular diastole, blood pressure within the arterial system falls, and the elastic fibers recoil to their original dimensions. Their expansion cushions the sudden rise in pressure during ventricular systole, and their recoil slows the drop in pressure during ventricular diastole. This feature is important because blood pressure is the driving force behind blood flow; the greater the pressure oscillations, the greater the changes in blood flow. The elasticity of the arterial system dampens the pressure peaks and valleys that accompany the heartbeat. By the time blood reaches the arterioles, the pressure

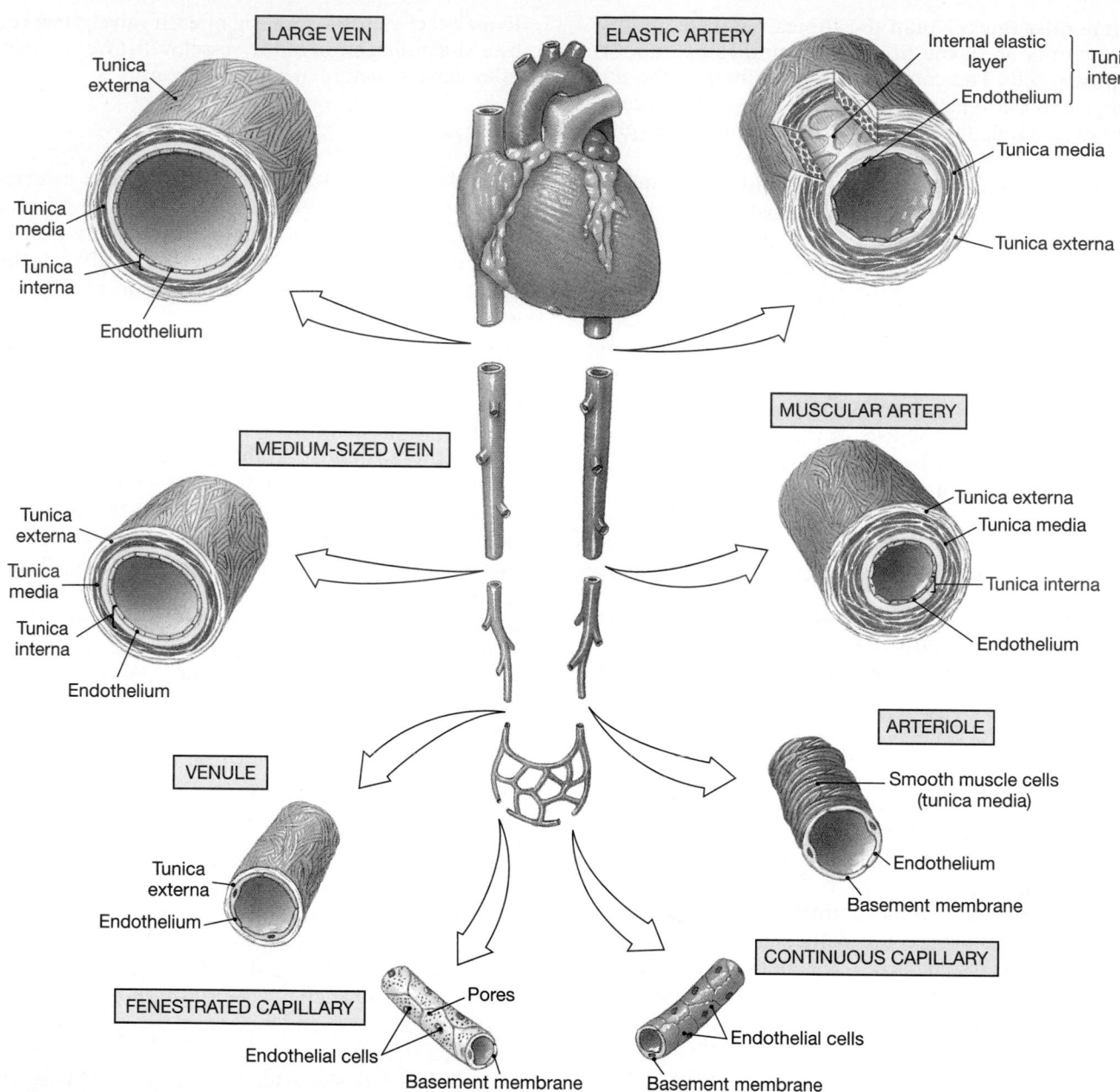

● **FIGURE 21-3** **Histological Structure of Blood Vessels.** Representative diagrammatic cross-sectional views of the walls of arteries, veins, and capillaries. Note the differences in relative size of the layers in these vessels.

oscillations have disappeared and blood flow is continuous.

Muscular Arteries

Muscular arteries, also known as *medium-sized arteries* or *distribution arteries,* distribute blood to the body's skeletal muscle and internal organs. A typical muscular artery has a lumen diameter of approximately 0.4 cm (0.15 in.). Muscular arteries are characterized by a thick tunica media that contains more smooth muscle cells than does the tunica media of elastic arteries (Figures 21-2 and 21-3●). The *external carotid arteries* of

the neck, the *brachial arteries* of the arms, and the *femoral arteries* of the thighs are muscular arteries.

Arterioles

Arterioles, with an internal diameter of 30 μm or less, are considerably smaller than muscular arteries. Arterioles have a poorly defined tunica externa, and the tunica media in the larger arterioles consists of one or two layers of smooth muscle cells (Figure 21-3●). The tunica media of the smallest arterioles contains scattered smooth muscle cells that do not form a complete layer.

The diameters of smaller muscular arteries and arterioles change in response to local conditions or to sympathetic or endocrine stimulation. For example, arterioles in most tissues vasodilate when oxygen levels are low and, as we saw in Chapter 16, vasoconstrict under sympathetic stimulation. ∞ *[p. 522]* Changes in their diameter affect the amount of force required to push blood around the cardiovascular system; more pressure is required to push blood through a constricted vessel than through a dilated one. The force opposing blood flow is called **resistance (R),** and arterioles are therefore called **resistance vessels.**

Elastic and muscular arteries are interconnected, and vessel characteristics change gradually as we travel away from the heart. For example, the largest muscular arteries contain a considerable amount of elastic tissue, whereas the smallest resemble heavily muscled arterioles.

Arteriosclerosis

Arteriosclerosis (ar-tē-rē-ō-skle-RŌ-sis) is a thickening and toughening of arterial walls. Although this condition may not sound life-threatening, complications related to arteriosclerosis account for roughly half of all deaths in the United States. The effects of arteriosclerosis are varied; for example, arteriosclerosis of coronary vessels is responsible for *coronary artery disease (CAD),* and arteriosclerosis of arteries supplying the brain can lead to strokes. ∞ *[p. 452]*

There are two major forms of arteriosclerosis:

1. **Focal calcification** is the gradual degeneration of smooth muscle in the tunica media and the subsequent deposition of calcium salts. This process typically involves arteries of the limbs and genital organs. Some focal calcification occurs as part of the aging process, and it may develop in association with atherosclerosis. Rapid and severe calcification may occur as a complication of diabetes mellitus, an endocrine disorder we considered in Chapter 18. ∞ *[p. 607]*

2. **Atherosclerosis** (ath-er-ō-skle-RŌ-sis) is associated with damage to the endothelial lining and the formation of lipid deposits in the tunica media. This is the most common form of arteriosclerosis.

Many factors may be involved in the development of atherosclerosis. One major factor is lipid levels in the blood. Atherosclerosis tends to develop in persons whose blood contains elevated levels of plasma lipids, specifically cholesterol. Circulating cholesterol is transported to peripheral tissues in lipoproteins, protein-lipid complexes. (We shall discuss the various types of lipoproteins and their interrelationships in Chapter 25.) Recent evidence indicates that many forms of atherosclerosis are associated with either (1) low levels of *apolipoprotein-E (ApoE),* a transport protein whose lipids are quickly removed by peripheral tissues, or (2) high levels of *lipoprotein(a),* a *low-density lipoprotein (LDL)* that is removed at a much slower rate.

When ApoE levels are low or lipoprotein(a) levels are high, cholesterol-rich lipoproteins remain in circulation for an extended period. Circulating monocytes then begin removing them from the bloodstream. Eventually, the monocytes become filled with lipid droplets. Now called *foam cells,* they attach themselves to the endothelial walls of blood vessels, where they release growth factors. These cytokines stimulate the divisions of smooth muscle cells near the tunica interna, thickening the vessel wall.

Other monocytes then invade the area, migrating between the endothelial cells. As these changes occur, the monocytes, smooth muscle cells, and endothelial cells begin phagocytizing lipids as well. The result is a **plaque,** a fatty mass of tissue that projects into the lumen of the vessel. At this point, the plaque has a relatively simple structure, and there is evidence that the process can be reversed if appropriate dietary adjustments are made.

If the conditions persist, the endothelial cells become swollen with lipids, and gaps appear in the endothelial lining. Platelets now begin sticking to the exposed collagen fibers, and the combination of platelet adhesion and aggregation leads to the formation of a localized blood clot that will further restrict blood flow through the artery. The structure of the plaque is now relatively complex. Plaque growth may be halted, but the structural changes are generally permanent.

Typical plaques are shown in Figure 21-4•. Elderly individuals, especially elderly men, are most likely to develop atherosclerotic plaques. There is evidence that estrogens may slow plaque formation; this may account for the lower incidence of CAD, myocardial infarctions (MIs), and strokes in women. After menopause, when estrogen production declines, the risk of CAD, MIs, and strokes in women increases markedly.

In addition to advanced age and male gender, other important risk factors include high blood cholesterol levels, high blood pressure, and cigarette smoking. Roughly 20 percent of middle-aged men have all three of these risk factors; these individuals are four times more likely to experience an MI or cardiac arrest than are other men in their age group. Although fewer women develop this condition, elderly women smokers with high blood cholesterol and high blood pressure are at much greater risk than other women. Other factors that may promote development of atherosclerosis in both men and women include diabetes mellitus, obesity, and stress. There is also evidence that at least some forms of atherosclerosis may be linked to chronic infection with *Chlamydia pneumoniae,* a bacterium responsible for several types of respiratory infections, including some forms of pneumonia.

We discussed potential treatments for atherosclerotic plaques, such as catheterization, balloon angioplasty, and bypass surgery, in Chapter 20. ∞ *[p. 667]* In cases in which dietary modifications do not lower circulating LDL levels sufficiently, there are drug therapies that can bring them under control. Genetic engineering techniques have recently been used to treat an inherited form of *hypercholesterolemia* (high blood cholesterol) linked to extensive plaque formation. (Individuals with this condition are unable to absorb and recycle cholesterol in the liver.) In this experimental procedure, circulating cholesterol levels declined after copies of appropriate genes were inserted into some of the individual's liver cells.

Without question, the best approach to atherosclerosis is to try to avoid it by eliminating or reducing associated risk factors. Suggestions include (1) reducing the amount of dietary cholesterol and saturated fats by restricting consumption of fatty meats (such as beef, lamb, and pork), egg yolks, and cream; (2) not smoking; (3) checking your blood pressure and taking steps to lower it if necessary; (4) having your blood cholesterol levels checked at annual physical examinations; (5) controlling your weight; and (6) exercising regularly.

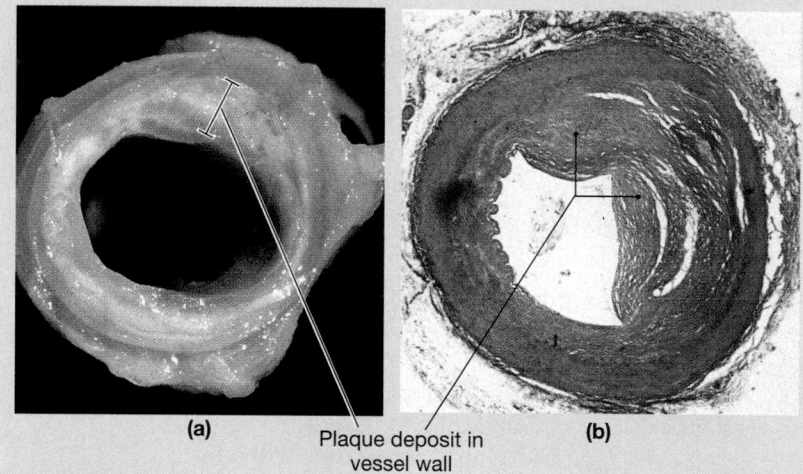

(a)   Plaque deposit in vessel wall   **(b)**

•*FIGURE 21-4* **A Plaque Blocking a Peripheral Artery.** **(a)** A section of a coronary artery narrowed by plaque formation. **(b)** Sectional view of a large plaque. (LM × 18)

ANEURYSMS An **aneurysm** (AN-ū-rizm) is a bulge in the weakened wall of an artery. This bulge resembles a bubble in the wall of a tire; like a bad tire, the affected artery may suffer a catastrophic blowout. The most dangerous aneurysms are those involving arteries of the brain, where they cause strokes, and of the aorta, where a ruptured aneurysm will cause fatal bleeding in a matter of minutes.

Aneurysms most commonly occur in individuals with *arteriosclerosis.* Over time, arteriosclerosis causes vessel walls to become less elastic, and a weak spot may develop. Aneurysms may also be associated with other conditions that weaken arterial walls, such as arterial inflammation or infection, and with *Marfan's syndrome,* a connective tissue disorder we introduced in Chapter 4. ∞ *[p. 123]* As you might expect, individuals with high blood pressure are most likely to develop dangerous aneurysms, because the elevated arterial pressures place great stresses on the vessel walls. Unfortunately, because they are often painless, aneurysms are likely to go undetected. Treatment options for aneurysms are discussed in the *Applications Manual.* [AM] *Aneurysms*

Capillaries

Capillaries are the only blood vessels whose walls permit exchange between the blood and the surrounding interstitial fluids. Because capillary walls are relatively thin, the diffusion distances are small, and exchange can occur quickly. In addition, blood flows through capillaries relatively slowly, allowing sufficient time for diffusion or active transport of materials across the capillary walls. Thus, the histological structure of capillaries permits a two-way exchange of substances between blood and interstitial fluid.

A typical capillary consists of an endothelial tube inside a delicate basement membrane. There is neither a tunica media nor a tunica externa. The average diameter of a capillary is a mere 8 μm, very close to that of a single red blood cell. There are two major types of capillaries: (1) *continuous capillaries* and (2) *fenestrated capillaries* (Figure 21-5•)

Continuous Capillaries

Most regions of the body are supplied by **continuous capillaries**. In a continuous capillary, the capillary endothelium is a complete lining. A cross section through a large continuous capillary will cut across several endothelial cells (Figure 21-5a●). In a small continuous capillary, a single endothelial cell may wrap all the way around the lumen, as your hand wraps around a small glass.

Continuous capillaries are located in all tissues except epithelia and cartilage. Continuous capillaries permit the diffusion of water, small solutes, and lipid-soluble materials into the surrounding interstitial fluid but prevent the loss of blood cells and plasma proteins. In addition, some transport may occur between the blood and the interstitial fluid through the movement of vesicles that form at the inner endothelial surface. We introduced this form of vesicular transport in Chapter 3. ∞ *[p. 80]*

The endothelial cells in specialized continuous capillaries throughout most of the CNS and in the thymus gland are bound together by tight junctions. These capillaries have very restricted permeability characteristics. We discussed one example, the capillaries responsible for the *blood–brain barrier,* in Chapter 14. ∞ *[p. 456]*

Fenestrated Capillaries

Fenestrated (FEN-es-trā-ted) **capillaries** (*fenestra,* window) are capillaries that contain "windows," or pores, that span the endothelial lining (Figure 21-5b●). The pores permit the rapid exchange of water and solutes as large as small peptides between the plasma and interstitial fluid. Examples of fenestrated capillaries noted in earlier chapters include the *choroid plexus* of the brain and the blood vessels in a variety of endocrine organs, including the hypothalamus, the pituitary, the pineal gland, and the thyroid gland. Fenestrated capillaries are also located along absorptive areas of the intestinal tract and at filtration sites in the kidneys. Both the number of pores and their permeability characteristics may vary from one region of the capillary to another.

SINUSOIDS Sinusoids (SĪ-nus-oydz) are specialized fenestrated capillaries that are flattened and irregular. In contrast to other fenestrated capillaries, sinusoids commonly have gaps between adjacent endothelial cells, and the basement membrane may be incomplete or absent. As a result, sinusoids permit the free exchange of water and solutes as large as plasma proteins.

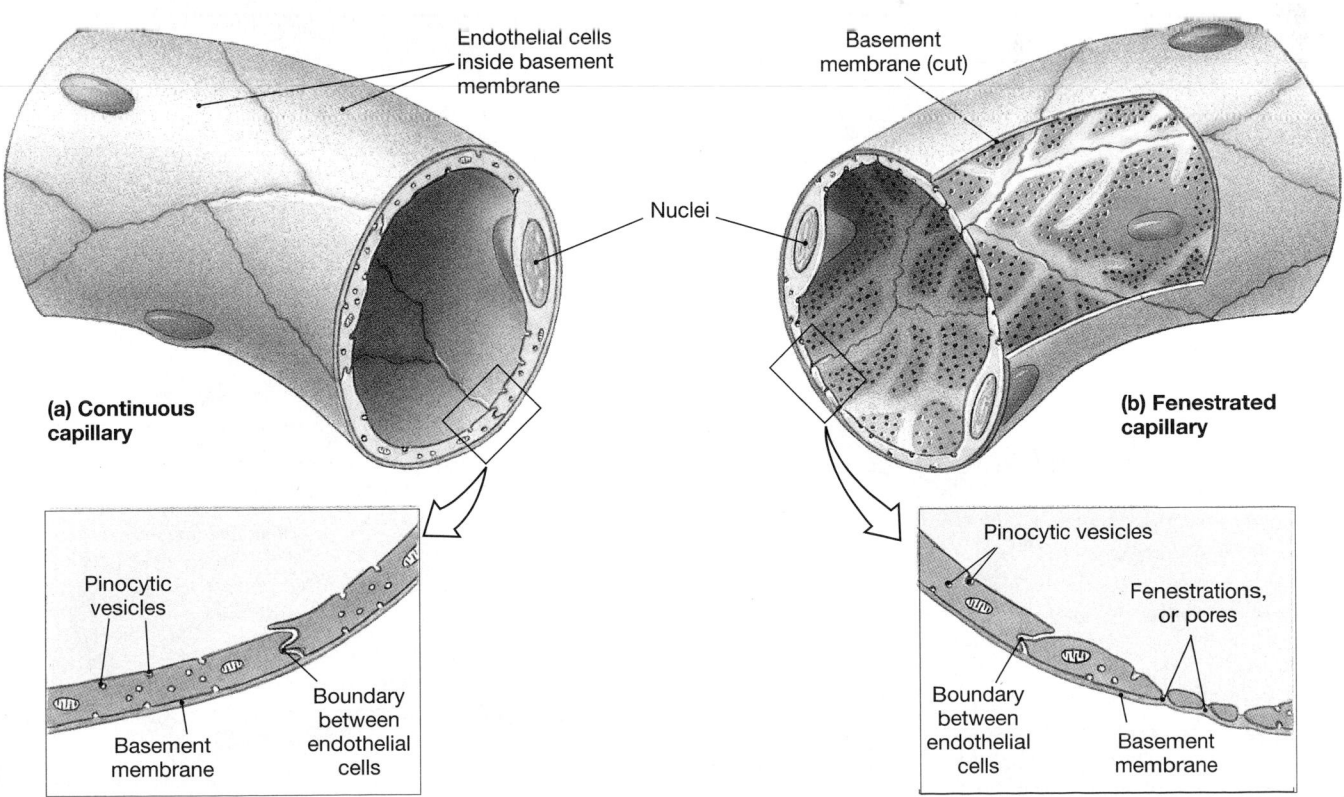

●**FIGURE 21-5** Capillary Structure. **(a)** A continuous capillary, showing routes for the diffusion of water and solutes. **(b)** A fenestrated capillary, showing the pores that facilitate diffusion across the endothelial lining.

Blood moves through sinusoids relatively slowly, maximizing the time available for absorption and secretion across the sinusoidal walls. Sinusoids occur in the liver, bone marrow, the anterior pituitary gland, and adrenal glands. At the liver sinusoids, plasma proteins secreted by the liver cells enter the circulation. Along sinusoids of the liver and bone marrow, phagocytic cells monitor the passing blood, engulfing damaged red blood cells, pathogens, and cellular debris.

Capillary Beds

Capillaries do not function as individual units but as part of an interconnected network called a **capillary plexus,** or **capillary bed** (Figure 21-6•). A single arteriole generally gives rise to dozens of capillaries that empty into several *venules,* the smallest vessels of the venous system. The entrance to each capillary is guarded by a band of smooth muscle called a **precapillary sphincter.** Contraction of the smooth muscle cells constricts and narrows the diameter of the capillary entrance, thereby reducing the flow of blood. Relaxation of the sphincter dilates the opening, allowing blood to enter the capillary at a faster rate.

Within the capillary bed, **preferred channels** provide a relatively direct means of communication between arterioles and venules. The arteriolar segment of the channel contains smooth muscles capable of altering its diameter. This segment is often called a **metarteriole** (met-ar-TĒ-rē-ōl) (Figure 21-6a,b•). The rest of the preferred channel, which resembles a typical capillary in structure, is called a **thoroughfare channel.**

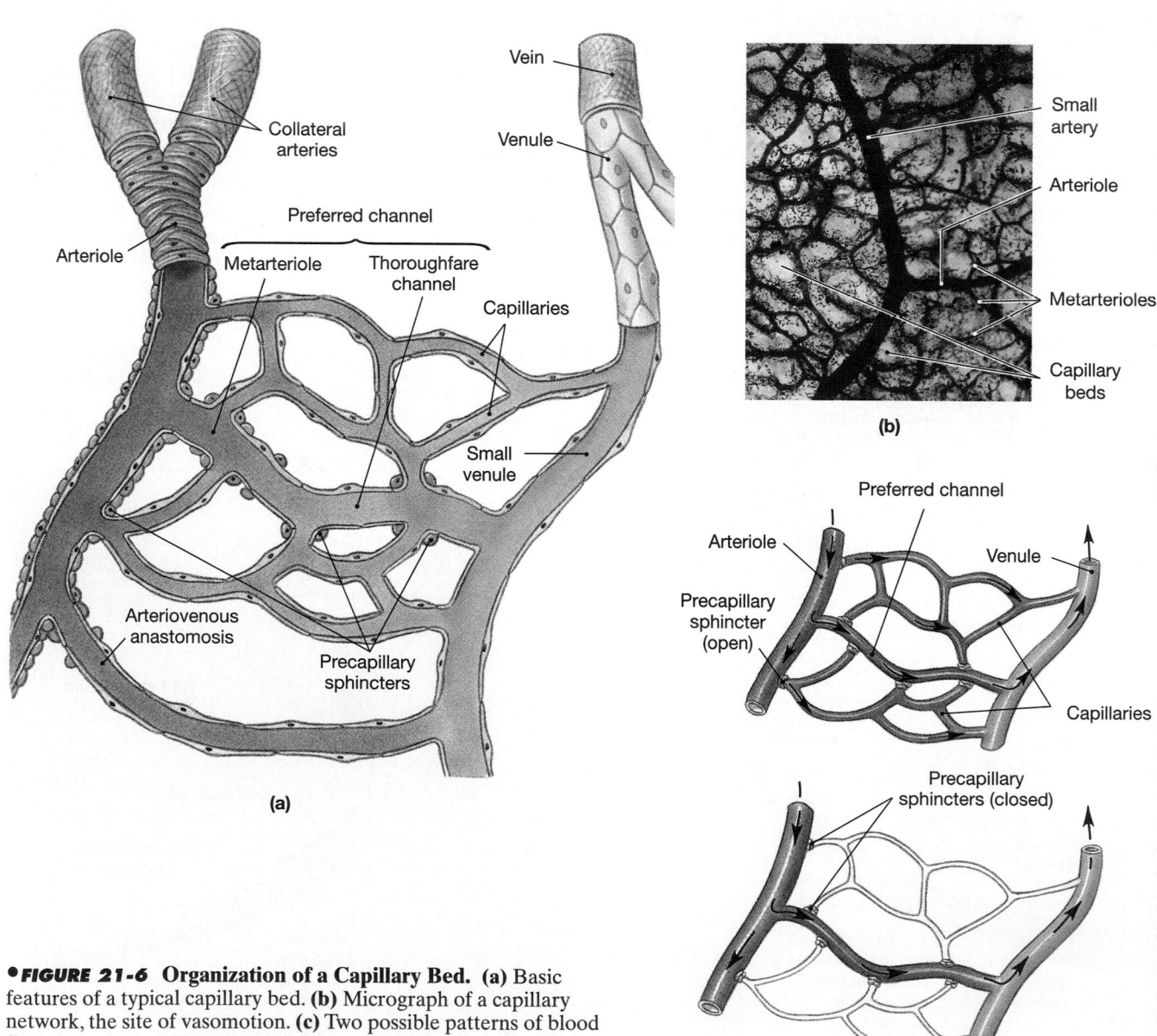

•**FIGURE 21-6** **Organization of a Capillary Bed.** (a) Basic features of a typical capillary bed. (b) Micrograph of a capillary network, the site of vasomotion. (c) Two possible patterns of blood flow through a capillary network. The actual pattern varies between these extremes, changing continuously in response to local alterations in tissue oxygen demand.

A single capillary bed may receive blood from more than one artery. The multiple arteries, called **collaterals,** enter the region and fuse before giving rise to arterioles. The fusion of two collateral arteries that supply a capillary bed is an example of an **arterial anastomosis.** The interconnections between the *anterior* and *posterior interventricular branch* arteries of the heart are arterial anastomoses. ∞ *[p. 685]* An arterial anastomosis acts like an insurance policy: If one artery is compressed or blocked, capillary circulation will continue.

Arteriovenous (ar-tē-rē-ō-VĒ-nus) **anastomoses** are direct connections between arterioles and venules. When an arteriovenous anastomosis is dilated, blood will bypass the capillary bed and flow directly into the venous circulation. The pattern of blood flow through an arteriovenous anastomosis is regulated primarily by sympathetic innervation under the control of the cardiovascular centers of the medulla oblongata.

Vasomotion

Although blood normally flows from the arterioles to the venules at a constant rate, the flow within each capillary is quite variable. Each precapillary sphincter goes through cycles of alternately contracting and relaxing, perhaps a dozen times per minute. As a result, the blood flow within any one capillary occurs in a series of pulses rather than as a steady and constant stream. The net effect is that blood may reach the venules by one route now and by a quite different route later (Figure 21-6c●). The cycling of contraction and relaxation of smooth muscles that causes alteration of blood flow through capillary beds is called **vasomotion.**

Vasomotion is controlled at the local level by changes in the concentrations of chemicals and dissolved gases within the interstitial fluids. For example, when dissolved oxygen concentrations decline within a tissue, the capillary sphincters relax, and blood flow to the area increases (Figure 21-6c●). This process, an example of capillary *autoregulation,* will be the focus of a later section.

When you are at rest, blood flows through roughly 25 percent of the vessels within a capillary bed. Your cardiovascular system does not contain enough blood to maintain adequate blood flow to all the capillaries in all the capillary beds in your body at the same time. As a result, when many tissues become active, the blood flow through capillary beds must be coordinated. We shall detail mechanisms by which the cardiovascular centers perform this coordination later in the chapter.

Veins

Veins collect blood from all tissues and organs and return it to the heart. The walls of veins are thinner than those of corresponding arteries. Venous walls need not be as thick as arterial walls because the blood pressure in veins is lower than that in the arteries. Veins are classified on the basis of their size. Even though their walls are thinner, in general, veins are larger in diameter than their corresponding arteries. Review Figures 21-2, p. 710, and 21-3●, p.712, to compare typical arteries and veins.

Venules

Venules, which collect blood from capillary beds, vary widely in size and character. An average venule has an internal diameter of roughly 20 μm. Venules smaller than 50 μm lack a tunica media altogether, and the smallest venules resemble expanded capillaries.

Medium-Sized Veins

Medium-sized veins range from 2 to 9 mm in internal diameter and are thus comparable in size to muscular arteries. In these veins, the tunica media is thin and contains relatively few smooth muscle cells. The thickest layer of a medium-sized vein is the tunica externa, which contains longitudinal bundles of elastic and collagen fibers.

Large Veins

Large veins include the superior and inferior venae cavae and their tributaries within the abdominopelvic and thoracic cavities. All the tunica layers are present in large veins. The slender tunica media is surrounded by a thick tunica externa composed of a mixture of elastic and collagen fibers.

Venous Valves

The arterial system is a high-pressure system: It takes almost all the force developed by the heart to push blood through the network of arteries and across miles of capillaries. Blood pressure within a peripheral venule is only about 10 percent of that in the ascending aorta, and pressures continue to fall along the venous system.

The blood pressure in venules and medium-sized veins is so low that it cannot oppose the force of gravity. In the limbs, veins of this size contain valves (Figure 21-7●). **Valves** are folds of the tunica interna that project from the vessel wall and point in the direction of blood flow. These valves act like the valves in the heart. They permit blood flow in one direction only and prevent the backflow of blood toward the capillaries.

As long as the valves function normally, any movement that distorts or compresses a vein will push blood toward the heart. This mechanism is particularly important when you are standing, because

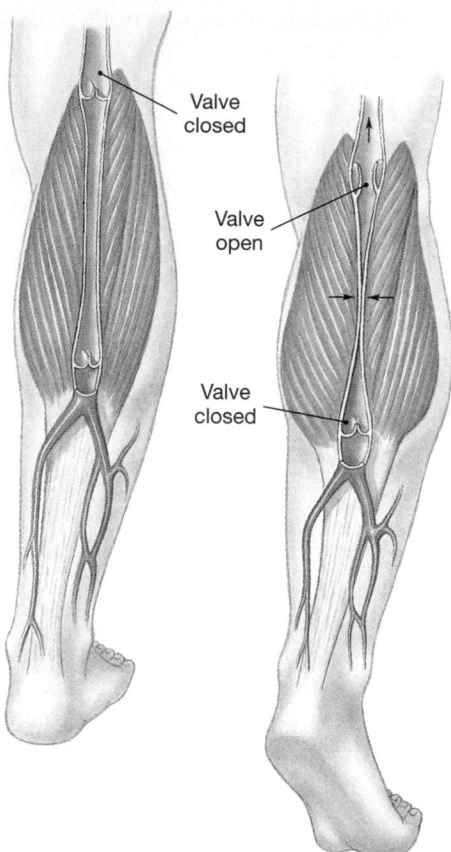

• **FIGURE 21-7** **Function of Valves in the Venous System.**
Valves in the walls of medium-sized veins prevent the
backflow of blood. Venous compression caused by the
contraction of adjacent skeletal muscles assists in maintaining
venous blood flow.

blood returning from your feet must overcome the
pull of gravity to ascend to the heart. Valves com-
partmentalize the blood within the veins, thereby
dividing the weight of the blood between the compartments. Any movement in the surrounding skeletal muscles squeezes the blood toward the heart.
Although you are probably not aware of it, when
you are standing, rapid cycles of contraction and relaxation are occurring within your leg muscles. These
contractions help push blood toward the trunk.
When you are lying down, venous valves have much
less impact on venous return, because your heart and
major vessels are at the same level.

If the walls of the veins near the valves weaken or
become stretched and distorted, the valves may not
work properly. Blood then pools in the veins, and the
vessels become grossly distended. The effects range
from mild discomfort and a cosmetic problem, as in
superficial *varicose veins* in the thighs and legs, to
painful distortion of adjacent tissues, as in *hemorrhoids*. We discuss these conditions and potential treatments in the *Applications Manual.* AM *Problems with
Venous Valve Function*

The Distribution of Blood

The total blood volume is unevenly distributed
among arteries, veins, and capillaries (Figure 21-8•).
The heart, arteries, and capillaries normally contain
30–35 percent of the blood volume (roughly 1.5 liters
of whole blood), and the venous system contains the
rest (65–70 percent, or about 3.5 liters). Of the blood
in the venous system, roughly one-third (about a
liter) is circulating within the liver, bone marrow, and
skin. These organs contain extensive venous networks that at any given moment contain large volumes of blood.

Because their walls are thinner and contain a
lower proportion of smooth muscle, veins are much
more distensible than arteries. For a given rise in
blood pressure, a typical vein will stretch about eight
times as much as a corresponding artery. The *capacitance* of a blood vessel is the relationship between
the volume of blood it contains and the blood pressure. The more expandable the vessel, the larger the
capacitance. Veins are called **capacitance vessels,** because they are easily distensible, and large changes
in blood volume have little effect on blood pressure.
If the blood volume rises or falls, the elastic walls
stretch or recoil, changing the volume of blood in
the venous system.

If serious hemorrhaging occurs, the *vasomotor
centers* of the medulla oblongata stimulate sympathetic nerves that innervate smooth muscle cells in
the walls of medium-sized veins. This activity has two
major effects:

1. Systemic veins constrict, and this **venoconstriction**
(vē-nō-kon-STRIK-shun) reduces the volume of the
venous system. Reducing the amount of blood in the
venous system maintains the volume within the arterial system and capillaries at near-normal levels
despite a significant blood loss.

2. The constriction of veins in the liver, skin, and lungs
redistributes a significant proportion of the total
blood volume. As a result, blood flow to delicate organs, such as the brain, and to active skeletal muscles
can be increased or maintained after a blood loss.
The amount of blood shifted from these organs to
the general circulation is called the **venous reserve.**
The venous reserve normally amounts to about 20
percent of the total blood volume.

☑ Examination of a cross section of tissue shows several
small, thin-walled vessels with very little smooth muscle tissue
in the tunica media. Which type of vessels are these?

☑ Why are valves located in veins but not in arteries?

☑ Where in the body would you expect to find fenestrated
capillaries?

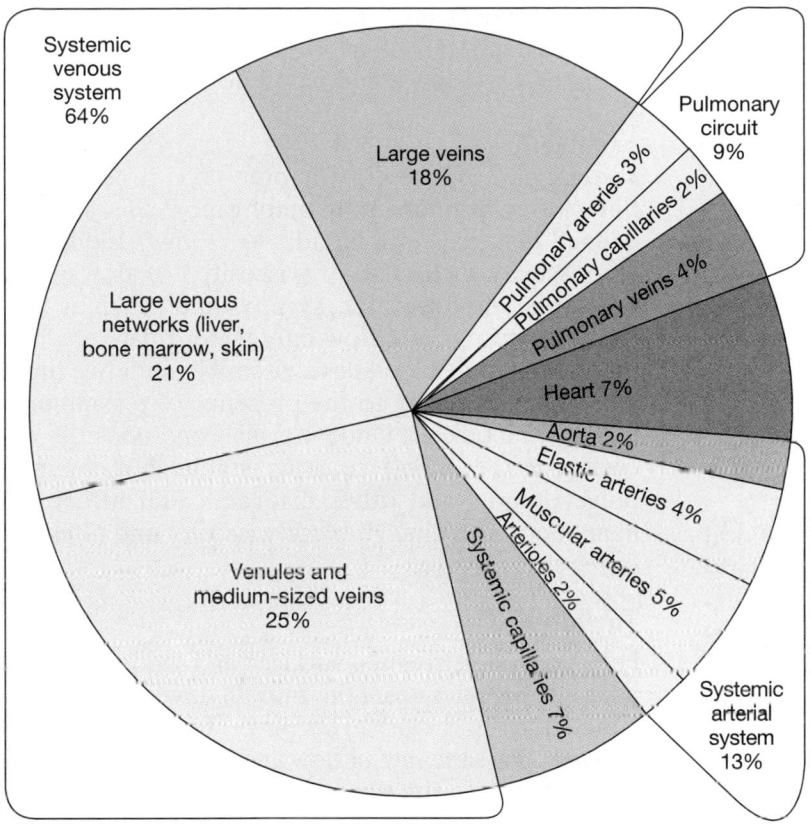

Systemic
venous
system
64%

Large veins
18%

Pulmonary
circuit
9%

Pulmonary arteries 3%

Pulmonary capillaries 2%

Pulmonary veins 4%

Large venous
networks (liver,
bone marrow, skin)
21%

Heart 7%

Aorta 2%

Elastic arteries 4%

Muscular arteries 5%

Arterioles 2%

Venules and
medium-sized veins
25%

Systemic capillaries 7%

Systemic
arterial
system
13%

•**FIGURE 21-8** The Distribution of Blood in the Cardiovascular System

CARDIOVASCULAR PHYSIOLOGY

The goal of cardiovascular regulation is the maintenance of adequate **blood flow** through peripheral tissues and organs. Under normal circumstances, blood flow is equal to cardiac output. When cardiac output goes up, so does the blood flow through capillary beds; when cardiac output declines, capillary blood flow is reduced. We considered the primary factors involved in the regulation of cardiac output in Chapter 20. ∞ *[p. 703]* The afterload of the heart is determined by the interplay between *pressure* and *resistance* in the circulatory network. If there were no resistance to blood flow in the cardiovascular system, the heart would not have to generate pressure to force blood around the pulmonary and systemic circuits.

Pressure

Liquids, including blood, are incompressible. A force exerted against a liquid generates **hydrostatic pressure (HP)** that is conducted in all directions. If a pressure gradient exists, hydrostatic pressure will push a liquid from an area of higher pressure toward an area of lower pressure. The hydrostatic pressure in the water pipes of your home is higher than atmospheric pressure; when you open

a faucet, water flows out. The greater the water pressure, the faster the water flow. In other words, the flow rate is directly proportional to the pressure gradient.

In the systemic circuit of the cardiovascular system, the pressure gradient is the **circulatory pressure,** the pressure difference between the base of the ascending aorta and the entrance to the right atrium. Circulatory pressures average about 100 mm Hg. This relatively high pressure is needed primarily to force blood through the arterioles—*resistance vessels*—and into peripheral capillaries. For convenience, the circulatory pressure is often divided into three components:

1. *Blood pressure.* When referring to arterial pressure, we will use the term **blood pressure (BP)** to distinguish it from the total circulatory pressure. Capillary blood flow is directly proportional to blood pressure, which is closely regulated by a combination of neural and hormonal mechanisms. Blood pressure in the systemic arterial system ranges from an average of 100 mm Hg to roughly 35 mm Hg at the start of a capillary network

2. *Capillary pressure.* **Capillary pressure** is the pressure within capillary beds. Along the length of a typical capillary, pressures decline from roughly 35 mm Hg to about 18 mm Hg.

3. *Venous pressure.* **Venous pressure** is the pressure within the venous system. Venous pressure is quite low; the pressure gradient from the venules to the right atrium is only about 18 mm Hg.

Resistance

A *resistance* is any force that opposes movement. If you put a kink in a garden hose or cover the mouth of the hose with your finger, you will increase the resistance and decrease the water flow. With the tap closed at the kitchen sink, there is sufficent resistance to stop the water flow. The more open the tap, the lower the resistance and the faster the water flow.

The resistance of the cardiovascular system opposes the movement of blood. The greater the resistance, the slower the blood flow. For circulation to occur, the pressure gradient must be great enough to overcome the **total peripheral resistance,** the resistance of the entire cardiovascular system. Because the resistance of the venous system is very low, for reasons we will detail below, attention focuses on the **peripheral resistance (PR),** the resistance of the arterial system. For blood to flow into peripheral capillaries, the pressure gradient must be great enough to overcome the peripheral resistance: The higher the peripheral resistance, the lower the rate of blood flow: $F \propto \Delta P/R$. In words, the flow, F, is directly proportional to the pressure gradient, ΔP, and inversely proportional to resistance, R. Sources of peripheral resistance include *vascular resistance, viscosity,* and *turbulence.*

Vascular Resistance

Vascular resistance is the resistance of the blood vessels; it is the largest component of peripheral resistance. *The most important factor in vascular resistance is friction between the blood and the vessel walls.* The amount of friction depends on the length and diameter of the vessel.

VESSEL LENGTH Increasing the length of a blood vessel increases friction: The longer the vessel, the larger the surface area in contact with the blood. For example, you can easily blow the water out of a snorkel that is 2.5 cm (1 in.) in diameter and 25 cm (10 in.) long, but you cannot blow the water out of a 15-m-long garden hose, because the total friction is too great. The most dramatic changes in blood vessel length occur between birth and maturity, as growth occurs. In an adult, vessel length can increase or decrease gradually when the individual gains or loses weight, but on a day-to-day basis this component of vascular resistance can be considered constant.

VESSEL DIAMETER Friction affects the blood primarily in a narrow zone closest to the vessel wall. In a small-diameter vessel, nearly all the blood will be slowed down by friction with the walls. Resistance will therefore be relatively high. Blood near the center of a large-diameter vessel will not encounter resistance from friction with the walls, and the resistance is therefore relatively low.

Differences in diameter have much more significant effects on resistance than do differences in length. If there are two vessels of equal diameter, one twice as long as the other, the longer vessel will offer twice as much resistance to blood flow. But with two vessels of equal length, one twice the diameter of the other, the narrower one will offer 16 times as much resistance to blood flow. This relationship, expressed in terms of the vessel radius, r, and resistance, R, can be summarized as $R \propto 1/r^4$.

More significantly, vessel length is constant, but vessel diameter can change. Most of the peripheral resistance occurs in the arterioles, the smallest vessels of the arterial system. As we noted earlier in the chapter, arterioles are extremely muscular; the wall of an arteriole 30 μm in diameter may contain a 20-μm-thick layer of smooth muscle. When these smooth muscles contract or relax, peripheral resistance increases or decreases. Because a small change in diameter produces a large change in resistance, mechanisms that alter the diameters of arterioles provide control over peripheral resistance and blood flow.

Viscosity

Viscosity, introduced in Chapter 19, is resistance to flow caused by interactions among molecules and suspended materials in a liquid. ∞ *[p. 644]* Liquids of low viscosity, such as water (viscosity 1.0), flow at low pressures, whereas thick, syrupy fluids, such as molasses (viscosity 300), flow only under relatively high pressures. Whole blood has a viscosity about five times that of water, owing to the presence of plasma proteins and blood cells. Under normal conditions, the viscosity of the blood remains stable, but anemia, polycythemia, and other disorders that affect the hematocrit also change blood viscosity and alter peripheral resistance.

Turbulence

High flow rates, irregular surfaces, and sudden changes in vessel diameter upset the smooth flow of blood, creating eddies and swirls. This phenomenon, called **turbulence,** slows the rate of flow and increases resistance.

Turbulence normally occurs when blood flows between the atria and ventricles and between the ventricles and the aortic and pulmonary trunks. In addition to increasing resistance, this turbulence generates the *third* and *fourth heart sounds* that can sometimes be heard through a stethoscope. Turbulent blood flow across damaged or malformed heart valves is responsible for the sound of *heart murmurs.* ∞ *[p. 698]* Turbulence also develops in large arteries, such as the aorta, when cardiac output and arterial flow rates are very high.

Turbulence seldom occurs in smaller vessels unless their walls are damaged. For example, scar tissue formation at an injury site or the development of an atherosclerotic plaque will create abnormal turbulence and restrict blood flow. Because of the distinctive sound, or *bruit* (broo-Ē), produced by this turbulence, the presence of plaques in large blood vessels can often be detected with a stethoscope.

Table 21-1 provides a quick review of the terms and relationships we discussed in this section.

An Overview of Circulatory Pressures

The graphs in Figure 21-9● provide an overview of the vessel diameters, areas, pressures, and velocity of blood flow in the systemic circuit. Notice the following:

- As you proceed from the aorta toward the capillaries, the arteries branch repeatedly; each branch is smaller in diameter than the preceding one (Curve 1). As you proceed from the capillaries toward the venae cavae,

TABLE 21-1 Some of the Key Terms and Relationships Pertaining to Blood Circulation

Blood flow (F): The volume of blood flowing per unit of time through a vessel or group of vessels; may refer to circulation through a capillary, a tissue, an organ, or the entire vascular network. Total blood flow is equal to cardiac output.

Blood pressure (BP): The hydrostatic pressure in the arterial system that pushes blood through capillary beds.

Circulatory pressure: The pressure difference between the base of the ascending aorta and the entrance to the right atrium.

Hydrostatic pressure: A pressure exerted by a liquid in response to an applied force.

Peripheral resistance (PR): The resistance of the arterial system. Factors that affect peripheral resistance include vascular resistance, viscosity, and turbulence.

Resistance (R): A force that opposes movement (in this case, blood flow).

Total peripheral resistance: The resistance of the entire cardiovascular system.

Turbulence: Resistance due to irregular, swirling movement of blood at high flow rates or exposure to irregular surfaces.

Vascular resistance: Resistance due to friction within the vessel, primarily between the blood and the walls of the blood vessels. Increases with increasing length or decreasing diameter; vessel length is constant, but vessel diameter may change.

Venous pressure: The hydrostatic pressure in the venous system.

Viscosity: Resistance to flow due to interactions among molecules within a liquid.

Relationships among these terms:

F ∝ ΔP (flow is proportional to the pressure gradient)

F ∝ 1/R (flow is inversely proportional to resistance)

F ∝ ΔP/R (flow is directly proportional to the pressure gradient and inversely proportional to resistance)

F ∝ BP/PR (flow is directly proportional to blood pressure and inversely proportional to peripheral resistance)

$R \propto 1/r^4$ (resistance is inversely proportional to the fourth power of the vessel radius)

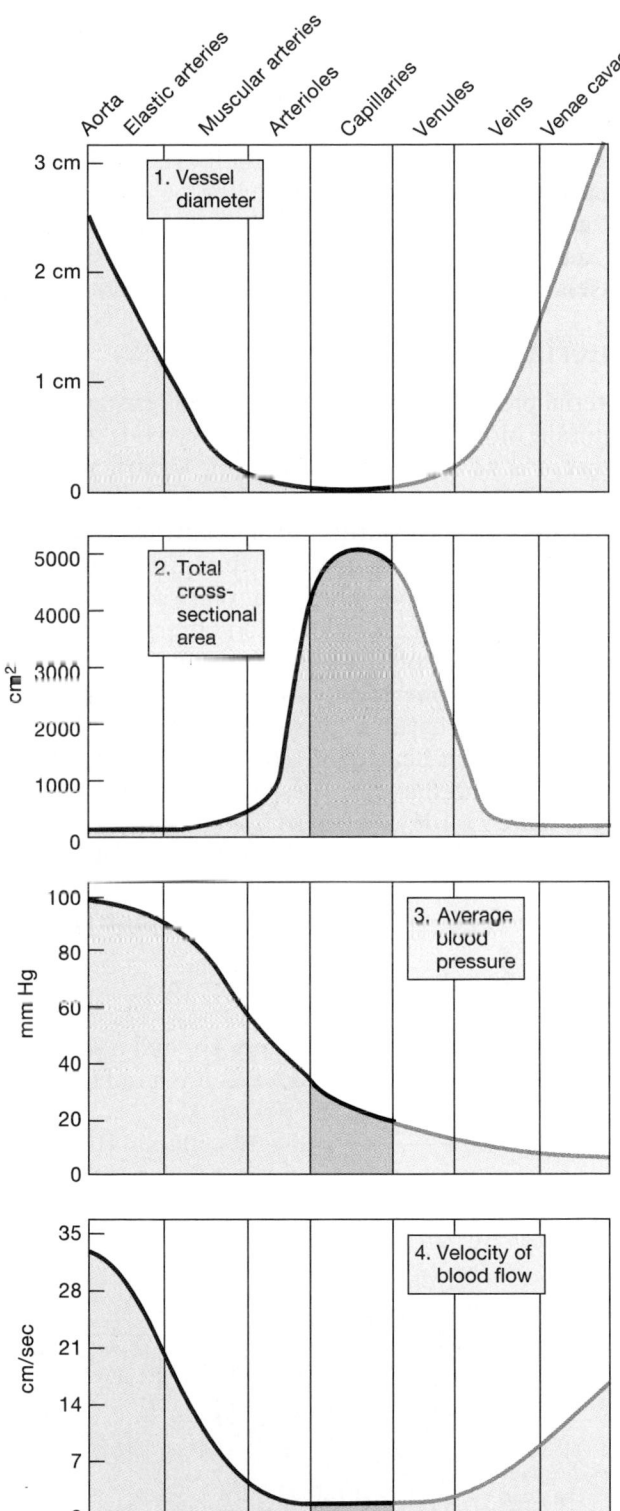

●**FIGURE 21-9** Relationships among Vessel Diameter, Cross-Sectional Area, Blood Pressure, and Blood Velocity

convergence occurs and the diameters increase as venules combine to form small and medium-sized veins.

- Although the arterioles, capillaries, and venules are small in diameter, there are a great many of them in the body (Curve 2). What is important is not the cross-sectional area of each individual vessel but their *combined* cross-sectional area. The aorta has a cross-sectional area of 4.5 cm², and all the blood entering the systemic circuit flows through it. That systemic blood also flows through the peripheral capillaries, which have a total cross-sectional area of 5000 cm².

- As branching proceeds, blood pressure falls rapidly. Most of the decline occurs in the small arteries and arterioles of the arterial system; venous pressures are relatively low (Curve 3).

- Like a fast-flowing river delivering water to a floodplain, the blood flow decreases in velocity as the total cross-sectional area of the vessels increases from the aorta toward the capillaries; it then rises in velocity as the cross-sectional area drops from the capillaries toward the venae cavae (Curve 4).

Figure 21-10● details the blood pressure throughout the cardiovascular system. Systemic pressures are highest in the aorta, peaking at about 120 mm Hg, and reach a minimum at the entrance to the right atrium. Pressures in the pulmonary circuit are much lower than those in the systemic circuit; the right ventricle is not a high-pressure pump, and the pulmonary vessels are much shorter and more distensible than systemic vessels and thus provide less resistance to blood flow.

Arterial Blood Pressure

Arterial pressures overcome peripheral resistance and maintain blood flow through capillary beds. Arterial pressure is not stable; it rises during ventricular systole and falls during ventricular diastole. The peak blood pressure measured during ventricular systole is called **systolic pressure,** and the minimum blood pressure at the end of ventricular diastole is called **diastolic pressure.** In recording blood pressure, we separate systolic and diastolic pressures by a slash mark, as in "120/80" ("one-twenty over eighty") or "110/75."

A *pulse* is a rhythmic pressure oscillation that accompanies each heartbeat. The difference between the systolic and diastolic pressures is the **pulse pressure** (Figure 21-10●). To report a single blood pressure value, we use the **mean arterial pressure (MAP)**. The mean arterial pressure is calculated by adding one third of the pulse pressure to the diastolic pressure (P_{dia}):

$$MAP = P_{dia} + \frac{pulse\ pressure}{3}$$

For a systolic pressure of 120 mm Hg and a diastolic pressure of 90 mm Hg, the MAP is 100 mm Hg:

$$MAP = 90 + \frac{120 - 90}{3} = 90 + 10 = 100\ mm\ Hg$$

Elastic Rebound

As systolic pressure climbs, the arterial walls stretch, just as an extra puff of air expands a partially inflated balloon. This expansion allows the arterial system to accommodate some of the blood provided by ventricular systole. When diastole begins and blood pressures fall, the arteries recoil to their original dimensions. Because the aortic semilunar valve prevents the return of blood to the heart, the arterial recoil pushes blood toward the capillaries. This phenomenon is called **elastic rebound.**

Pressures in Small Arteries and Arterioles

The mean arterial pressure and the pulse pressure become smaller as the distance from the heart increases (Figure 21-10●):

- The mean arterial pressure declines as the arterial branches become smaller and more numerous. In essence, the blood pressure decreases as it produces blood flow and overcomes friction.

- The pulse pressure fades as a result of the cumulative effects of elastic rebound. Each arterial segment reduces the magnitude of the pressure change that is experienced by its downstream neighbors. The effect is like that of a loud shout creating a series of ever-softer echoes. Each time an echo is produced, the reflecting surface absorbs some of the sound energy. Eventually, the echo disappears. The pressure surge accompanying ventricular ejection is the shout, and it is reflected by the wall of the aorta, echoing down the arterial system until it finally disappears at the level of the small arterioles. By the time blood reaches a precapillary sphincter, there are no pressure oscillations and the blood pressure remains steady at approximately 35 mm Hg.

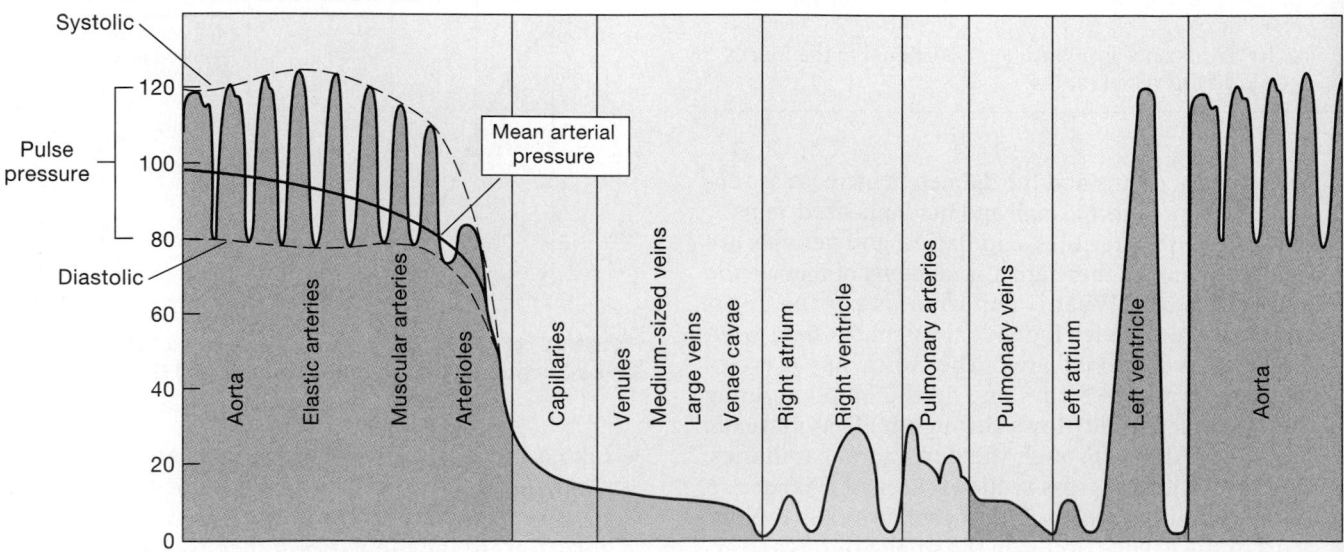

●**FIGURE 21-10** **Pressures within the Cardiovascular System.** Notice the general reduction in circulatory pressure within the systemic circuit and the elimination of the pulse pressure within the arterioles.

HYPERTENSION AND HYPOTENSION Hypertension is the presence of abnormally high blood pressure. The usual criterion for hypertension in an adult is a blood pressure greater than 150/90. One study estimated that 20 percent of the U.S. Caucasian population has blood pressures greater than 160/95 and that another 25 percent is on the borderline, with pressures above 140/90. The figures for other racial groups vary; the incidence of hypertension among African Americans is roughly twice that of Caucasian Americans. The extent to which these data reflect genetic rather than environmental factors is not known.

Hypertension significantly increases the workload on the heart, and the left ventricle gradually enlarges. More muscle mass means a greater oxygen demand. When the coronary circulation cannot keep pace, symptoms of coronary ischemia ap-

pear. ∞ *[p. 685]* Increased arterial pressures also place a physical stress on the walls of blood vessels throughout the body. This stress promotes or accelerates the development of arteriosclerosis and increases the risk of aneurysms, heart attacks, and strokes. **Hypotension,** or low blood pressure, is most often seen in patients who have received overly aggressive treatment for hypertension. AM *Hypertension and Hypotension*

Capillary Exchange

Capillary walls are very thin and delicate (Figure 21-3•, p. 712); they consist of a single squamous endothelial cell, generally supported by a basement membrane. This arrangement minimizes the distance

---

Checking the Pulse and Blood Pressure

You can feel your pulse within any of the large or medium-sized arteries. The usual procedure involves using your fingertips to squeeze an artery against a relatively solid mass, preferably a bone. When the vessel is compressed, you feel your pulse as a pressure against your fingertips. The inside of the wrist is commonly used, because the *radial artery* can easily be pressed against the distal portion of the radius (Figure 21-23•, p. 742). Other accessible arteries include the *external carotid, brachial, temporal, facial, femoral,* and *popliteal arteries* (Figures 21-23, 21-24, and 21-28•, pp. 742, 744, and 749). Firm pressure exerted at these locations, called **pressure points,** can reduce or eliminate arterial bleeding distal to the site.

Blood pressure not only forces blood through the circulatory system, but it also pushes outward against the walls of the containing vessels, just as air pushes against the walls of an inflated balloon. As a result, we can measure blood pressure indirectly by determining how forcefully the blood presses against the vascular walls.

The instrument used to measure blood pressure is called a **sphygmomanometer** (sfig-mō-ma-NOM-e-ter; *sphygmos,* pulse + *manometer,* device for measuring pressure). An inflatable cuff is placed around the arm in such a position that its inflation compresses the brachial artery (Figure 21-11•). A stethoscope is placed over the artery distal to the cuff, and the cuff is then inflated. A tube connects the cuff to a pressure gauge that reports the cuff pressure in millimeters of mercury (mm Hg). Inflation continues until cuff pressure is roughly 30 mm Hg above the pressure sufficient to collapse the brachial artery completely, stop the flow of blood, and eliminate the sound of the pulse.

The investigator then slowly lets the air out of the cuff. When the pressure in the cuff falls below systolic pressure, blood can again enter the artery. At first, blood enters only at peak systolic pressures, and the stethoscope picks up the sound of blood pulsing through the artery. As the pressure falls further, the sound changes, because the artery is remaining open for longer and longer periods. When the cuff pressure falls below diastolic pressure, blood flow becomes continuous and the sound of the pulse becomes muffled or disappears. Thus the pressure at which the pulse appears corresponds to the peak systolic pressure; when the pulse fades, the pressure has reached diastolic levels. The distinctive sounds heard during this test are called *sounds of Korotkoff* (sometimes spelled *Korotkov* or *Korotkow*). These sounds are produced by turbulence as blood flows past the constricted portion of the artery.

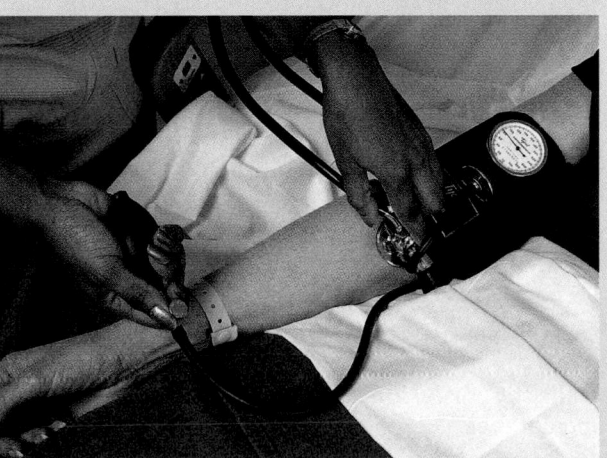

•**FIGURE 21-11** Using a Sphygmomanometer to Measure Blood Pressure

between the blood and the interstitial fluid, facilitating rapid diffusion or transport of materials into or out of the circulation. In Chapter 3, we introduced the forces responsible for the movement of water and solutes across membranes, and in this section we shall assume some familiarity with basic concepts. If necessary, return to Chapter 3 for a more detailed treatment of each mechanism. ∞ *[pp. 72–80]*

The most important processes involved in the movement of materials across typical capillary walls are *diffusion, filtration,* and *reabsorption.* We shall discuss these mechanisms in detail next. In specialized capillaries, such as the continuous capillaries of the CNS, carrier-mediated transport and vesicular transport are also important.

Diffusion

As we saw in Chapter 3, *diffusion* is the net movement of ions or molecules from an area where their concentration is relatively high to an area where their concentration is relatively low. ∞ *[p. 72]* The difference between the high and low concentrations represents a *concentration gradient,* and diffusion tends to eliminate that gradient. Diffusion occurs most rapidly when (1) the distances involved are small, (2) the concentration gradient is large, and (3) the ions or molecules involved are small.

Diffusion across capillary walls can occur by several different routes:

1. Water, ions, and small organic molecules, such as glucose, amino acids, and urea, can usually enter or leave the circulation by diffusion between adjacent endothelial cells or through the pores of fenestrated capillaries. In the CNS and thymus, the endothelial cells of continuous capillaries are locked together, and there are no intervening gaps that permit passive diffusion of water, ions, and solutes. This arrangement occurs at the blood–brain barrier ∞ *[p. 456]*, and at the blood–thymus barrier (Chapter 22).

2. Many ions, including sodium, potassium, calcium, and chloride, can diffuse across the endothelial cells by passing through channels in the cell membranes.

3. Large water-soluble compounds are unable to enter or leave the circulation except at fenestrated capillaries, such as those of the hypothalamus, the kidneys, many endocrine organs, and the intestinal tract.

4. Lipids, such as fatty acids and steroids, and lipid-soluble materials, including soluble gases such as oxygen and carbon dioxide, can cross the capillary walls by diffusion through the endothelial cell membranes.

5. Plasma proteins are normally unable to cross the endothelial lining anywhere except in sinusoids, such as those of the liver, where plasma proteins enter the circulation.

Filtration

The driving force for filtration is hydrostatic pressure. As we saw earlier, hydrostatic pressure pushes water from an area of high pressure to an area of lower pressure. Blood pressure is really the hydrostatic pressure of the blood *(BHP)*.

In *capillary filtration,* water is forced across a capillary wall and small solute molecules travel with the water (Figure 21-12●). The solute molecules must be small enough to pass between adjacent endothelial cells or through the pores in a fenestrated capillary; larger solutes and suspended proteins are filtered out and remain in the circulation.

Along the length of a typical capillary, blood pressure gradually falls from about 35 mm Hg to roughly 18 mm Hg, the pressure at the start of the venous system. Filtration occurs primarily at the arterial end of a capillary, where BHP is highest.

Reabsorption

Reabsorption occurs as the result of osmosis. *Osmosis* is a special term used to refer to the diffusion of water across a selectively permeable membrane separating two solutions of differing solute concentrations. Water molecules tend to diffuse across a membrane *toward*

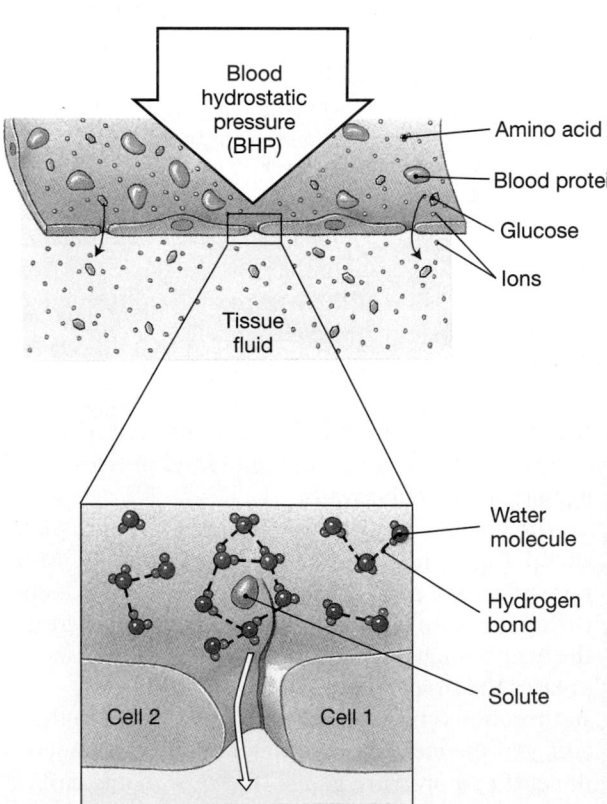

●FIGURE 21-12 Capillary Filtration. Blood hydrostatic pressure forces water and solutes across capillary walls through the gaps between adjacent endothelial cells in continuous capillaries. The size of the solutes that move across the vessel wall is determined primarily by the dimensions of the gaps.

the solution containing a higher solute concentration, because in so doing they are moving down the concentration gradient for water (Figure 3-7●, p. 75).

The **osmotic pressure (OP)** of a solution is an indication of the force of water movement resulting from its solute concentration. The higher the solute concentration of a solution, the greater the solution's osmotic pressure. The osmotic pressure of the blood is also called **blood colloid osmotic pressure (BCOP),** because only the suspended proteins are unable to cross the capillary walls.[1] Osmotic water movement will continue until either the solute concentrations are equalized or the movement is prevented by an opposing hydrostatic pressure.

We will now consider the interplay between filtration and reabsorption along the length of a typical capillary. As the discussion proceeds, remember that hydrostatic pressure forces water *out of* a solution, whereas osmotic pressure draws water *into* a solution.

The Interplay between Filtration and Reabsorption

The rates of filtration and reabsorption gradually change as blood passes along the length of a capillary. The factors involved are diagrammed in Figure 21-13●.

The *net hydrostatic pressure* tends to push water and solutes into the interstitial fluid. The net hydrostatic pressure is the difference between

1. the **blood hydrostatic pressure (BHP),** which ranges from 35 mm Hg at the arterial end of a capillary to 18 mm Hg at the venous end, and
2. the **hydrostatic pressure of the interstitial fluid (IHP).** Measurements of IHP have yielded very small values that differ from tissue to tissue—from +6 mm Hg in the brain to –6 mm Hg in subcutaneous tissues. A positive IHP opposes BHP, and the tissue hydrostatic pressure must be overcome before fluid can move out of the capillary. A negative IHP assists BHP, and additional fluid will be pulled out of the capillary.

The *net colloid osmotic pressure* tends to pull water and solutes into the capillary. The net colloid osmotic pressure is the difference between

1. the **blood colloid osmotic pressure (BCOP),** which is roughly 25 mm Hg, and
2. the **interstitial fluid colloid osmotic pressure (ICOP).** The ICOP is as variable and low as the IHP, because the interstitial fluid in most tissues contains negligible quantities of suspended proteins. Reported values of ICOP are from 0 to 5 mm Hg, within the range of pressures recorded for the IHP.

[1]Clinicians often use the term *oncotic pressure* (*onkos,* a swelling) when referring to the colloid osmotic pressure of body fluids. The two terms are equivalent.

Because (1) the ICOP and the IHP oppose one another, (2) these pressures are of comparable size, and (3) they are difficult to measure accurately, both values have been set at 0 mm Hg in our model. This value is acceptable because BHP and BOP are the primary forces acting across a capillary wall under normal circumstances. However, the method of calculation described here would still apply, regardless of the values selected for IHP and ICOP.

The **net filtration pressure (NFP)** is the difference between the net hydrostatic pressure and the net osmotic pressure. In terms of the factors listed above, this means that

$$NFP = (BHP - IHP) - (BCOP - ICOP)$$

At the arterial end of the capillary, the net filtration pressure is +10 mm Hg:

$$NFP = (35 - 0) - (25 - 0) = 35 - 25 = 10 \text{ mm Hg}$$

Because this value is positive, it indicates that fluid will tend to move *out of* the capillary and into the interstitial fluid. At the venous end of the capillary, the net filtration pressure is –7 mm Hg:

$$NFP = (18 - 0) - (25 - 0) = 18 - 25 = -7 \text{ mm Hg}$$

The minus sign indicates that there is a net movement of fluid *into* the capillary; that is, reabsorption is occurring.

The transition between filtration and reabsorption occurs where the BHP is 25 mm Hg, because at that point the hydrostatic and osmotic forces are equal, and the NFP is 0 mm Hg. If the maximum filtration pressure at the arterial end of the capillary were equal to the maximum reabsorption pressure at the venous end, this transition point would lie midway along the length of the capillary. Under these circumstances, filtration would occur along the first half of the capillary, and an identical amount of reabsorption would occur along the second half. However, the maximum filtration pressure is higher than the maximum reabsorption pressure, so the transition point between filtration and reabsorption normally lies closer to the venous end of the capillary than to the arterial end. As a result, more filtration than reabsorption occurs along the length of the capillary. Of the roughly 24 liters of fluid that moves out of the plasma and into the interstitial fluid each day, 85 percent is reabsorbed. The remainder (3.6 liters) flows through the tissues and into lymphatic vessels, for eventual return to the venous system.

Any condition that affects hydrostatic or osmotic pressures in the blood or tissues will shift the balance between hydrostatic and osmotic forces. We can then predict the effects on the basis of an understanding of capillary dynamics. For example:

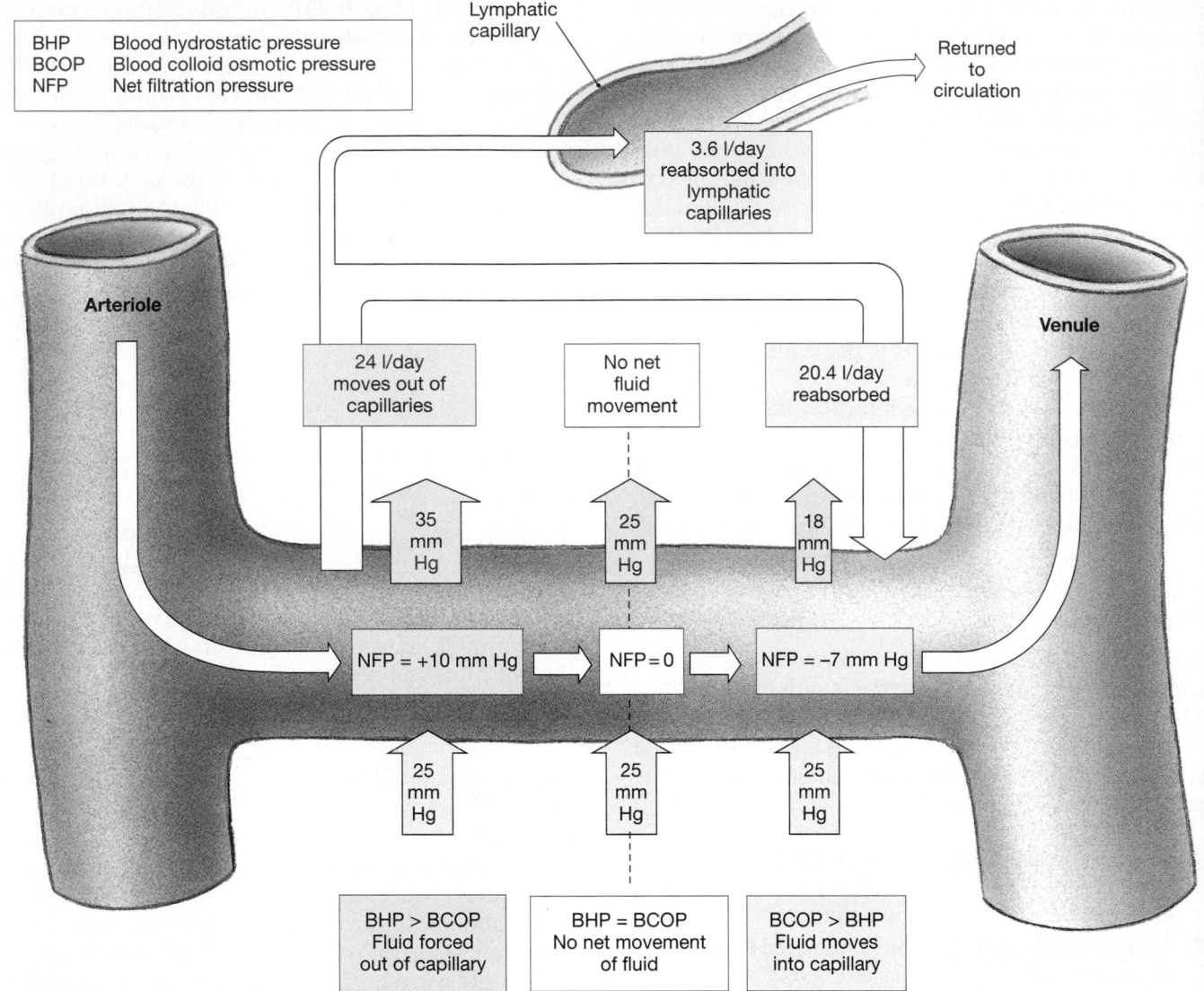

BHP	Blood hydrostatic pressure
BCOP	Blood colloid osmotic pressure
NFP	Net filtration pressure

Lymphatic capillary

Returned to circulation

3.6 l/day reabsorbed into lymphatic capillaries

Arteriole

Venule

24 l/day moves out of capillaries

No net fluid movement

20.4 l/day reabsorbed

35 mm Hg

25 mm Hg

18 mm Hg

NFP = +10 mm Hg → NFP = 0 → NFP = −7 mm Hg

25 mm Hg

25 mm Hg

25 mm Hg

| BHP > BCOP Fluid forced out of capillary | BHP = BCOP No net movement of fluid | BCOP > BHP Fluid moves into capillary |

•FIGURE 21-13 **Forces Acting across Capillary Walls.** At the arterial end of the capillary, blood hydrostatic pressure (BHP) is stronger than blood colloid osmotic pressure (BCOP), and fluid moves out of the capillary. Near the venule, BHP is lower than BCOP, and fluid moves into the capillary. In this model, interstitial fluid osmotic pressure (ICOP) and interstitial fluid hydrostatic pressure (IHP) are assumed to be 0 mm Hg.

- If hemorrhaging occurs, both blood volume and blood pressure decline. This reduction in BHP lowers the NFP and increases the amount of reabsorption. The result is a reduction in the volume of interstitial fluid and an increase in the circulating plasma volume. This process is known as a *recall of fluids.*
- If dehydration occurs, the plasma volume decreases owing to water loss, and the concentration of plasma proteins increases. The increase in BCOP accelerates reabsorption and a recall of fluids that delays the onset and severity of clinical symptoms.
- If BHP rises or BCOP declines, fluid moves out of the blood and builds up within peripheral tissues, a condition called *edema.*

EDEMA Edema (e-DĒ-muh) is an abnormal accumulation of interstitial fluid. There are many different causes of edema, and we will encounter specific examples in later chapters. The underlying problem in all types of edema is a disturbance in the normal balance between hydrostatic and osmotic forces at the capillary level. For example:

- When a capillary is damaged, plasma proteins can cross the capillary wall and enter the interstitial fluid. The resulting elevation of the ICOP will reduce the rate of capillary reabsorption and produce a localized edema. This is why you usually have swelling at a bruise.
- In acute starvation, the liver cannot synthesize enough plasma proteins to maintain normal concentrations in the blood. BCOP declines, and fluids begin moving from the blood

into peripheral tissues. In children, fluid accumulates in the abdominopelvic cavity, producing the swollen bellies typical of starvation victims. A reduction in BCOP is also seen after severe burns and in several types of liver or kidney diseases.

- In the U.S. population, serious cases of edema most commonly result from an increase in the arterial blood pressure, the venous pressure, or the total circulatory pressure. The increase may result from heart problems, such as heart failure, venous blood clots that elevate venous pressures, or other circulatory abnormalities. The net result is an increase in BHP that accelerates the movement of fluid into the tissues.

Edema can also result from problems with other systems, such as the blockage of lymphatic vessels or impaired urine formation·

- If the lymphatic vessels in a region become blocked, the volume of interstitial fluid will continue to rise, and the IHP will gradually increase until capillary filtration ceases. In *filariasis,* a condition we shall consider in Chapter 22, parasites can block lymphatic vessels and cause a massive regional edema known as *elephantiasis.*
- If the kidneys are unable to produce urine but the individual continues to drink liquids, the blood volume will rise. This situation ultimately leads to elevated BHP and enhances fluid movement into the peripheral tissues.

Venous Pressure and Venous Return

Venous pressure, although low, determines the *venous return,* which has a direct impact on cardiac output and peripheral blood flow. Although pressure at the start of the venous system is only about one-tenth that at the start of the arterial system, before returning to the heart the blood must still travel through a vascular network as complex as the arterial system.

Pressures at the entrance to the right atrium fluctuate, but they average about 2 mm Hg. Thus the effective pressure in the venous system is roughly 16 mm Hg (from 18 mm Hg in the venules to 2 mm Hg in the venae cavae), as compared with 65 mm Hg in the arterial system (from 100 mm Hg at the aorta to 35 mm Hg at the capillaries). Yet, although venous pressures are low, the veins offer comparatively little resistance, and pressure declines very slowly as blood moves through the venous system. As blood continues toward the heart, the veins become larger, resistance drops, and the velocity of blood flow increases (Figure 21-9•, p. 721).

When you are standing, the venous blood returning from your body inferior to the heart must overcome gravity as it ascends within the inferior vena cava. Two factors cooperate to assist the relatively low venous pressures in propelling blood toward your heart: (1) *muscular compression* of peripheral veins and (2) the *respiratory pump.*

Muscular Compression

The contractions of skeletal muscles near a vein compress it, helping push blood toward the heart. The valves in small and medium-sized veins ensure that blood flow occurs in one direction only (Figure 21-7•, p. 718). During normal standing and walking, the cycles of contraction and relaxation that accompany normal movements assist venous return. If you stand at attention, with knees locked and leg muscles immobilized, that assistance is lost. The reduction in venous return then leads to a fall in cardiac output, which reduces the blood supply to the brain. This decline is sometimes enough to cause **fainting,** a temporary loss of consciousness. You would then collapse, and in the horizontal position both venous return and cardiac output return to normal.

The Respiratory Pump

As you inhale, your thoracic cavity expands, and pressures within the pleural cavities decline. This drop in pressure pulls air into your lungs. At the same time, blood is pulled into the inferior vena cava and right atrium from the smaller veins of your abdominal cavity and lower body. The effect on venous return from the superior vena cava is less pronounced, as blood in that vessel normally flows "downhill" and so has the assistance of gravity.

As you exhale, your thoracic cavity decreases in size. Internal pressures then rise, forcing air out of your lungs and pushing venous blood into the right atrium. This mechanism is called the **respiratory pump,** or *thoracoabdominal pump.* The importance of this pumping action increases during heavy exercise, when respirations are deep and frequent.

☑ In a healthy individual, where would you expect the blood pressure to be greater—at the aorta or at the inferior vena cava? Explain.

☑ While standing in the hot sun, Sally begins to feel lightheaded and faints. Explain.

☑ Terry's blood pressure is 125/70. At what pressure did the nurse taking his blood pressure first hear the sounds of Korotkoff?

CARDIOVASCULAR REGULATION

Homeostatic mechanisms regulate cardiovascular activity to ensure that tissue blood flow, also called **tissue perfusion,** meets the demand for oxygen and nutrients. The three variable factors are (1) cardiac output, (2) peripheral resistance, and (3) blood pressure. We discussed cardiac output in Chapter 20 and considered peripheral resistance and blood pressure earlier in this chapter. ∞ *[p. 699]*

Most cells are relatively close to capillaries. When a group of cells becomes active, the circulation to that region must increase to deliver the necessary oxygen and nutrients and to carry away the waste products and carbon dioxide they generate. The goal of cardiovascular regulation is to ensure that these blood flow changes occur (1) at an appropriate time, (2) in the right area, and (3) without drastically altering blood pressure and blood flow to any vital organs.

Factors involved in the regulation of cardiovascular function include:

- *Local factors.* Local factors change the pattern of blood flow within capillary beds in response to chemical changes in the interstitial fluids. This is an example of *autoregulation* at the tissue level. Autoregulation causes immediate, localized homeostatic adjustments. If autoregulation fails to normalize conditions at the tissue level, central and endocrine mechanisms are activated (Figure 21-14●).
- *Central mechanisms.* Central mechanisms respond to changes in arterial pressure or blood gas levels at specific sites. When those changes occur, the cardiovascular centers of the ANS adjust cardiac output and peripheral resistance to maintain blood pressure and ensure adequate blood flow.

- *Endocrine factors.* The endocrine system releases hormones that enhance short-term adjustments and direct long-term changes in cardiovascular performance.

We shall next consider each of these regulatory mechanisms in greater detail.

Autoregulation of Blood Flow within Tissues

Under normal resting conditions, cardiac output remains stable, and peripheral resistance within individual tissues is adjusted to control local blood flow.

Local Vasodilators

Factors that promote dilation of precapillary sphincters are called **vasodilators. Local vasodilators** are produced at the tissue level and accelerate blood flow through the tissue of origin. Examples include:

- Decreased tissue oxygen levels or increased CO_2 levels.
- The generation of lactic acid or other acids by tissue cells.
- The release of nitric oxide (NO), formerly known as *endothelium-derived relaxation factor (EDRF),* from endothelial cells.
- Rising concentrations of potassium ions or hydrogen ions in the interstitial fluid.
- Chemicals released during local inflammation, including histamine and nitric oxide.
- Elevated local temperatures.

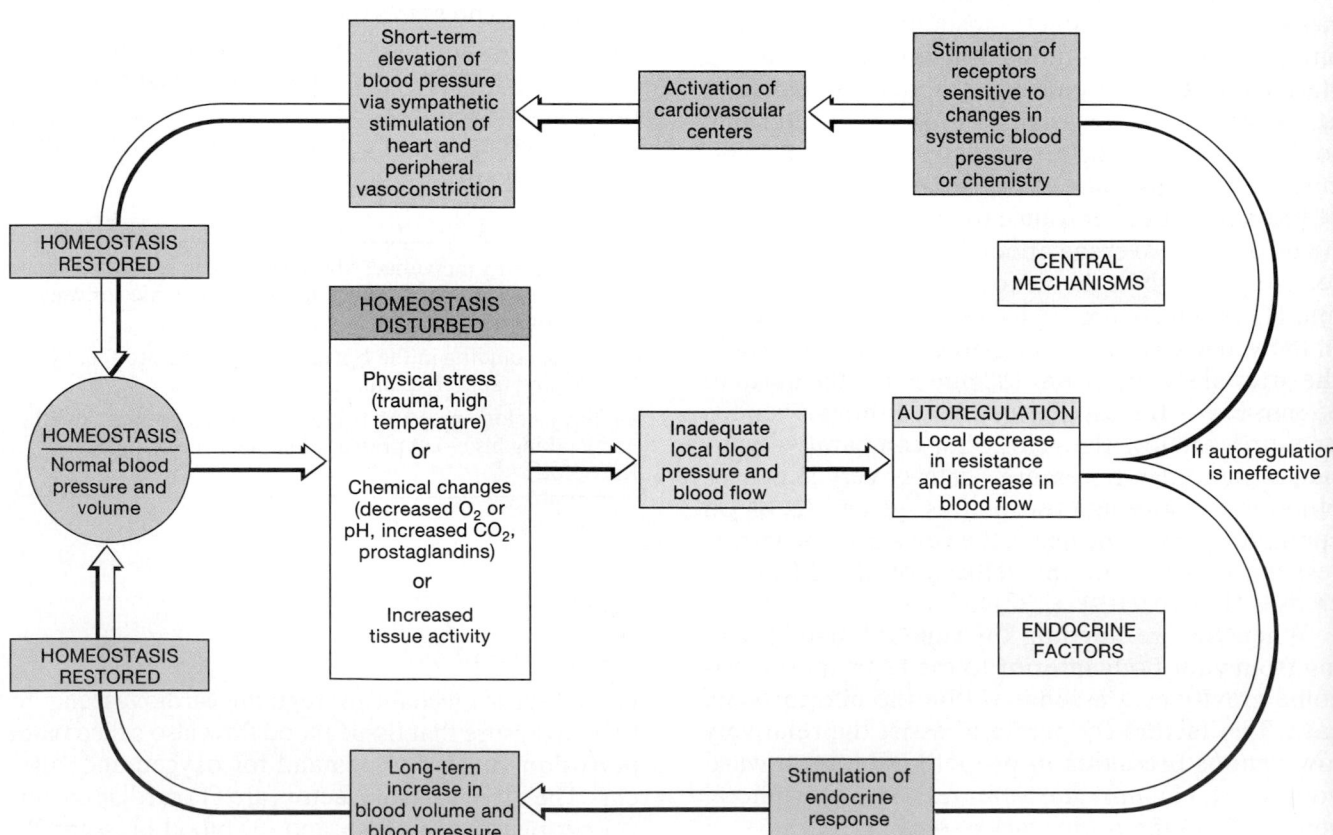

●FIGURE 21-14 Homeostatic Adjustments that Maintain Blood Pressure and Blood Flow

These factors work by stimulating the relaxation of the smooth muscle cells of the precapillary sphincters. All these factors indicate that conditions in the tissue are abnormal in one way or another. An improvement in blood flow, which will bring oxygen, nutrients, and buffers, may be sufficient to restore homeostasis.

Local Vasoconstrictors

As we noted in Chapter 19, aggregating platelets and damaged tissues produce compounds that stimulate the constriction of precapillary sphincters. These compounds are **local vasoconstrictors.** Examples include prostaglandins and thromboxanes released by activated platelets and white blood cells and the endothelins released by damaged endothelial cells. ∞ *[p. 664]*

Local vasodilators and vasoconstrictors control blood flow within a single capillary bed (Figure 21-6•, p. 716). When present in high concentrations, these factors also affect arterioles, increasing or decreasing blood flow to all the capillary beds in a given area.

The Neural Control of Blood Pressure and Blood Flow

The nervous system is responsible for adjusting cardiac output and peripheral resistance to maintain adequate blood flow to vital tissues and organs. Centers responsible for these regulatory activities include the *cardiac centers* and the *vasomotor centers* of the medulla oblongata. ∞ *[p. 473]* It is difficult to distinguish the cardiac and vasomotor centers anatomically, and they are often considered to form complex **cardiovascular (CV) centers.** In functional terms, however, the cardiac and vasomotor centers often act independently.

As we noted in Chapter 20, each cardiac center consists of a *cardioacceleratory center,* which increases cardiac output through sympathetic innervation, and a *cardioinhibitory center,* which reduces cardiac output through parasympathetic innervation. ∞ *[p. 687]*

The vasomotor centers contain two populations of neurons: (1) a very large group responsible for widespread vasoconstriction and (2) a relatively small group responsible for the vasodilation of arterioles in skeletal muscles and the brain. The vasomotor centers exert their effects by controlling the activity of sympathetic motor neurons:

1. *Control of vasoconstriction.* The neurons innervating peripheral blood vessels in most tissues are *adrenergic,* releasing norepinephrine. The response to NE release is the stimulation of smooth muscle in the walls of arterioles, producing vasoconstriction.
2. *Control of vasodilation.* Vasodilator neurons innervate blood vessels in skeletal muscles and in the brain. Stimulation of these neurons will relax smooth muscle cells in the walls of arterioles, producing vasodilation. Relaxation of smooth muscle cells is triggered by the

appearance of NO in their surroundings. The vasomotor centers may control NO release indirectly or directly. The most common vasodilator synapses are *cholinergic,* and their synaptic knobs release ACh. ACh stimulates the release of NO by endothelial cells in the area; the NO then causes local vasodilation. Another population of vasodilator synapses is *nitroxidergic*; the synaptic knobs release NO as a neurotransmitter. Nitric oxide has an immediate and direct relaxing effect on the vascular smooth muscle cells in the area.

Vasomotor Tone

In Chapter 16, we discussed the significance of autonomic tone in setting a background level of neural activity that can increase or decrease on demand. ∞ *[p. 528]* The sympathetic vasoconstrictor nerves are chronically active, producing a significant **vasomotor tone.** Vasoconstrictor activity is normally sufficient to keep the arterioles partially constricted. Under maximal stimulation, arterioles constrict to about half their resting diameter, whereas a fully dilated arteriole increases its resting diameter by roughly 1.5 times. Constriction has a significant effect on resistance, because, as we saw earlier, the resistance increases sharply as the diameter decreases. The resistance of a maximally constricted arteriole is roughly *80 times* that of a fully dilated arteriole. Because blood pressure varies directly with peripheral resistance, the vasomotor centers can provide very effective control over arterial blood pressure by making modest adjustments in vessel diameter. Extreme stimulation of the vasomotor centers will also produce venoconstriction and a mobilization of the venous reserve.

Reflex Control of Cardiovascular Function

The cardiovascular centers detect changes in tissue demand by monitoring arterial blood, with particular attention to blood pressure and to the pH and dissolved gas concentrations. The *baroreceptor reflexes* respond to changes in blood pressure, and the *chemoreceptor reflexes* monitor changes in the chemical composition of arterial blood. These reflexes are regulated through a negative feedback loop. Stimulation of a receptor by an abnormal condition leads to a response that counteracts the stimulus and restores normal conditions.

BARORECEPTOR REFLEXES *Baroreceptors* are specialized receptors that monitor the degree of stretch in the walls of expandable organs. ∞ *[p. 544]* The baroreceptors involved with cardiovascular regulation are located in the walls of (1) the **carotid sinuses,** expanded chambers near the bases of the *internal carotid arteries* of the neck (Figure 21-24•, p. 744), (2) the **aortic sinuses,** pockets in the walls of the ascending aorta adjacent to the heart (Figure 20-6•, p. 681), and (3) the wall of the right atrium. These receptors are components of the **baroreceptor reflexes,** which adjust cardiac output and peripheral resistance to maintain normal arterial pressures.

Baroreceptor Reflexes of the Carotid and Aortic Sinuses. Aortic baroreceptors monitor blood pressure within the ascending aorta. The **aortic reflex** adjusts blood pressure in response to changes in pressure at this location. The goal is to maintain adequate blood pressure and blood flow through the systemic circuit. Carotid sinus baroreceptors respond to changes in blood pressure at the carotid sinus. These receptors trigger reflexes that maintain adequate blood flow to the brain. The blood flow to the brain must remain constant, and the carotid sinus receptors are extremely sensitive.

Figure 21-15● presents the basic organization of the baroreceptor reflexes triggered by changes in blood pressure at these locations.

When blood pressure climbs, the increased output from the baroreceptors alters activity in the CV centers and produces two major effects (Figure 21-15a●):

1. *A decrease in the cardiac output,* due to parasympathetic stimulation and inhibition of sympathetic activity.
2. *Widespread peripheral vasodilation,* due to the inhibition of excitatory neurons in the vasomotor centers.

The decrease in cardiac output reflects primarily a reduction in heart rate due to ACh release at the SA node. ∞ *[p. 691]* The widespread vasodilation lowers peripheral resistance, and this effect, combined with a reduction in cardiac output, leads to a decline in blood pressure to normal levels.

When blood pressure falls below normal, there is a corresponding reduction in baroreceptor output (Figure 21-15b●). This change has two major effects:

1. *An increase in cardiac output* through stimulation of the sympathetic innervation to the heart. This results from stimulation of the cardioacceleratory centers and is accompanied by inhibition of the cardioinhibitory centers.
2. *Widespread peripheral vasoconstriction,* caused by stimulation of sympathetic vasoconstrictor neurons by the vasomotor centers.

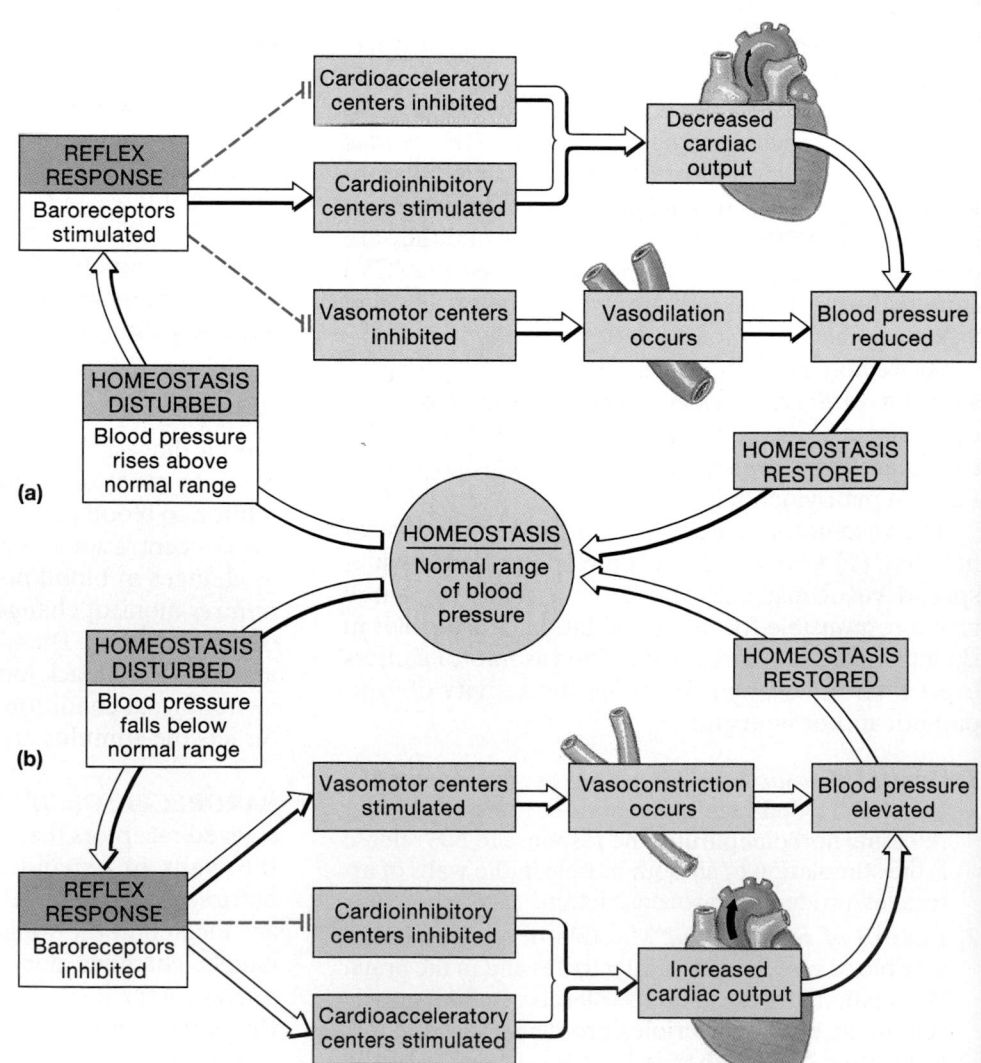

●**FIGURE 21-15** The Carotid and Aortic Sinus Baroreceptor Reflexes

The effects on the heart result from the release of NE by sympathetic neurons innervating the SA node, the AV node, and the general myocardium. In a crisis situation, sympathetic activation occurs, and its effects will be enhanced by the release of both E and NE from the adrenal medullae. The net effect is an immediate increase in heart rate and stroke volume and a corresponding rise in cardiac output. The vasoconstriction, which also results from NE release by sympathetic neurons, increases peripheral resistance. These adjustments, increased cardiac output and increased peripheral resistance, work together to elevate blood pressure.

Atrial Baroreceptors. **Atrial baroreceptors** monitor blood pressure at the end of the systemic circuit—at the venae cavae and the right atrium. The **atrial reflex,** introduced in Chapter 20, responds to stretching of the wall of the right atrium. ∞ *[p. 702]*

Under normal circumstances the heart pumps blood into the aorta at the same rate that it is arriving at the right atrium. When blood pressure rises at the right atrium, blood is arriving at the heart faster than it is being pumped out. The atrial baroreceptors solve the problem by stimulating the CV centers and increasing cardiac output until the backlog of venous blood is removed. Atrial pressure then returns to normal.

CHEMORECEPTOR REFLEXES The **chemoreceptor reflexes** (Figure 21-16●) respond to changes in the carbon dioxide, oxygen, or pH levels in the blood and cerebrospinal fluid. The chemoreceptors involved are sensory neurons located in the **carotid bodies,** situated in the neck near the carotid sinus, and the **aortic bodies,** near the arch of the aorta. These receptors monitor the composition of the arterial blood. Additional chemoreceptors are located on the ventrolateral surfaces of the medulla oblongata. These receptors monitor the composition of the cerebrospinal fluid.

When chemoreceptors in the carotid bodies or the aortic bodies detect a rise in the carbon dioxide content or a fall in the pH of the arterial blood, the cardioacceleratory centers are stimulated and the cardioinhibitory centers are inhibited (Figure 21-16●). The primary result is an elevation in arterial pressure via stimulation of the vasomotor center. A drop in the oxygen level at the aortic bodies will have the same effects. Strong stimulation of the carotid or aortic chemoreceptors causes a more widespread sympathetic activation, increasing heart rate and cardiac output.

The medullary chemoreceptors are primarily involved with the control of respiratory function and secondarily with regulating blood flow to the brain. For example, a steep rise in CSF carbon dioxide levels will trigger the vasodilation of cerebral vessels but will produce vasoconstriction in most other organs. The result is increased blood flow—and hence increased oxygen delivery—to the brain.

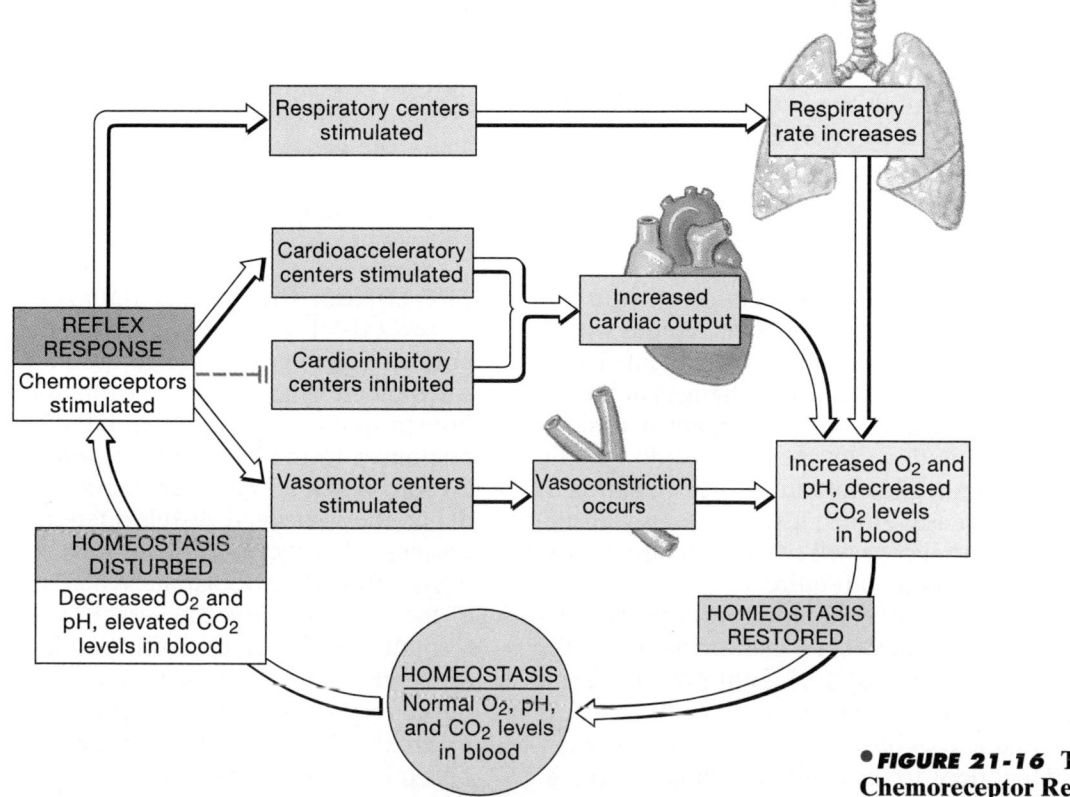

●*FIGURE 21-16* The Chemoreceptor Reflexes

Arterial CO_2 levels can be reduced and O_2 levels increased most effectively by coordinating cardiovascular and respiratory activity. Chemoreceptor stimulation also stimulates the respiratory centers, and the rise in cardiac output and blood pressure is associated with an increased respiratory rate. Coordination of cardiovascular and respiratory activity is vital, because accelerating tissue blood flow is useful only if the circulating blood contains adequate oxygen. In addition, a rise in the respiratory rate accelerates venous return through the action of the respiratory pump. (We shall consider other aspects of chemoreceptor activity and respiratory control in Chapter 23.)

Special Circulation

The vasoconstriction that occurs in response to a fall in blood pressure or a rise in CO_2 levels affects multiple tissues and organs simultaneously. The term *special circulation* refers to the circulation through organs where blood flow is controlled by separate mechanisms. We will note three important examples: the circulation to the brain, the heart, and the lungs.

THE BRAIN In Chapter 14, we noted the existence of the blood–brain barrier, which isolates most CNS tissue from the general circulation. ∞ *[p. 456]* The brain has a very high demand for oxygen and receives a substantial supply of blood. Under a variety of conditions, the blood flow to the brain remains steady at about 750 ml/min. This supply is roughly 12 percent of the cardiac output, delivered to an organ that represents less than 2 percent of the body weight. Neurons do not maintain significant energy reserves, and in functional terms most of the adjustments made by the cardiovascular system treat circulation to the brain as the number one priority. Even when a circulatory crisis is under way, blood flow through the brain remains as near normal as possible. While the cardiovascular centers are calling for widespread peripheral vasoconstriction, the cerebral vessels are told to dilate.

Although total blood flow to the brain remains relatively constant, circulation to specific regions within the brain changes from moment to moment. These changes occur in response to local changes in the interstitial fluid composition that accompany neural activity. When you read, write, speak, or walk, specific regions of your brain become active. The circulation to those regions increases almost instantaneously, ensuring that the active neurons will continue to receive the oxygen and nutrients they require.

The brain receives arterial blood from four different arteries. Because these arteries form anastomoses inside the cranium, interruption of any one vessel will not compromise the circulatory supply to the brain. If a plaque or blood clot blocks an artery or if an artery ruptures, dependent tissues will be injured or killed.

Symptoms of a *stroke,* or *cerebrovascular accident (CVA),* then appear. The *Applications Manual* contains a discussion of current views on the causes and treatment of strokes. [AM] *The Causes and Treatment of Cerebrovascular Disease*

THE HEART We described the anatomy of the coronary circulation in Chapter 20. ∞ *[p. 683]* The coronary arteries arise at the base of the ascending aorta, where systemic pressures are highest. Each time the heart contracts, it squeezes the coronary vessels, and blood flow is reduced. In the left ventricle, systolic pressures are high enough that blood can flow into the myocardium only during diastole. Normal cardiac muscle cells can tolerate these brief circulatory interruptions because they have substantial oxygen reserves.

When you are at rest, your coronary blood flow is about 250 ml/min. When the workload on your heart increases, local factors, such as reduced O_2 levels and lactic acid production, dilate the coronary vessels and increase blood flow. Epinephrine released during sympathetic stimulation promotes vasodilation of coronary vessels while increasing your heart rate and the strength of cardiac contractions. As a result, coronary blood flow increases at the same time that vasoconstriction occurs in other tissues.

For uncertain reasons, some individuals experience *coronary spasms* that can temporarily restrict coronary circulation and produce symptoms of angina. Permanent restriction or blockage of coronary vessels, as in coronary artery disease (CAD), and tissue damage, as caused by a myocardial infarction (MI), can limit the heart's ability to increase cardiac output, even under maximal stimulation. These individuals will experience symptoms of heart failure when the cardiac workload increases much above resting levels. The causes and treatments of heart failure are detailed in the *Applications Manual.* [AM] *Heart Failure*

THE LUNGS The lungs contain roughly 300 million *alveoli* (al-VĒ-ō-lī; *alveolus,* sac), delicate epithelial pockets where gas exchange occurs. Each alveolus is surrounded by an extensive capillary network. Blood flow through the lungs is regulated primarily by local responses to the levels of oxygen within individual alveoli. When an alveolus contains oxygen in abundance, the associated vessels dilate, and blood flow increases. This increase promotes the absorption of oxygen from the alveolar air. When the oxygen content of the air is very low, the vessels constrict, and blood is shunted to alveoli that still contain significant levels of oxygen. This mechanism maximizes the efficiency of the respiratory system, because there is no benefit to circulating blood through the capillaries of an alveolus unless it contains oxygen.

This mechanism is precisely the opposite of the situation in other tissues, where a decline in oxygen levels causes local vasodilation rather than vasoconstriction. The difference makes functional sense, but its physiological basis remains a mystery.

Blood pressure in the pulmonary capillaries, averaging 10 mm Hg, is lower than the pressure in systemic capillaries. The BCOP (25 mm Hg) is the same as elsewhere in the circulation. As a result, reabsorption exceeds filtration in pulmonary capillaries. There is a continuous movement of fluid into the pulmonary capillaries across the alveolar surfaces. This movement prevents fluid buildup in the alveoli that could interfere with the diffusion of respiratory gases. If the blood pressure in pulmonary capillaries rises above 25 mm Hg, fluid will enter the alveoli, causing *pulmonary edema,* a dangerous condition that is commonly seen in *congestive heart failure* (CHF). (CHF is discussed in the *Applications Manual.)*

CNS Activities and the Cardiovascular Centers

The output of the cardiovascular centers can also be influenced by activities in other areas of the brain. For example, activation of either division of the ANS will affect output from the cardiovascular centers:

- The cardioacceleratory and vasomotor centers are stimulated when a general sympathetic activation occurs. The result is an increase in cardiac output and blood pressure.
- When the parasympathetic division is activated, the cardioinhibitory centers are stimulated, producing a reduction in cardiac output. Parasympathetic activity does not directly affect the vasomotor centers, but vasodilation occurs as sympathetic activity declines.

The activities of higher brain centers can also affect blood pressure. Our thought processes or emotional states can produce significant changes in blood pressure by influencing cardiac output and vasomotor tone. For example, strong emotions of anxiety, fear, or rage are accompanied by an elevation in blood pressure, caused by cardiac stimulation and vasoconstriction.

Hormones and Cardiovascular Regulation

The endocrine system provides both short-term and long-term regulation of cardiovascular performance. As we have seen, E and NE from the adrenal medullae stimulate cardiac output and peripheral vasoconstriction. Other hormones important in regulating cardiovascular function include (1) antidiuretic hormone (ADH), (2) angiotensin II, (3) erythropoietin (EPO), and (4) atrial natriuretic peptide (ANP). We described those hormones and their functions in Chapter 18 and will provide only an overview here. ∞ *[pp. 607, 618, 620]*

Although ADH and angiotensin II affect blood pressure, all four hormones are concerned primarily with the long-term regulation of blood volume (Figure 21-17●).

Antidiuretic Hormone

ADH is released at the posterior pituitary in response to a decrease in blood volume or an increase in the osmotic concentration of the plasma or, secondarily, in response to circulating angiotensin II. The immediate result is a peripheral vasoconstriction that elevates blood pressure. This hormone also stimulates the conservation of water at the kidneys, thus preventing a reduction in blood volume that would further reduce blood pressure (Figure 21-17a●).

Angiotensin II

Angiotensin II appears in the blood following the release of renin by specialized kidney cells in response to a fall in renal blood pressure. Renin starts a chain reaction that ultimately converts angiotensinogen, an inactive protein, to the hormone angiotensin II (Figure 18-17b●, p. 619).

Angiotensin II has two short-term effects: (1) an extremely powerful vasoconstriction that elevates blood pressure and (2) a positive inotropic action on the heart, which elevates cardiac output. Two other effects of angiotension II have long-term significance: (1) *It stimulates the secretion of ADH by the pituitary and of aldosterone by the adrenal cortex* (Figure 21-17a●). Aldosterone stimulates the reabsorption of sodium ions at the kidneys, reducing sodium ion and water losses in the urine. (2) *It stimulates thirst.* The presence of ADH and aldosterone ensures that the additional water consumed will be retained, elevating the plasma volume.

Erythropoietin

Erythropoietin (EPO) is released at the kidneys if the blood pressure declines or if the oxygen content of the blood becomes abnormally low (Figure 21-17a●). This hormone stimulates red blood cell production and maturation, elevating the blood volume and improving the oxygen-carrying capacity of the blood (Figure 18-17b●, p. 619).

Atrial Natriuretic Peptide

Atrial natriuretic peptide (ANP) is produced by cardiac muscle cells in the wall of the right atrium in response to excessive stretching during diastole. This peptide hormone reduces blood volume and blood pressure by (1) increasing sodium ion excretion at the kidneys, (2) promoting water losses by increasing the volume of urine produced; (3) reducing thirst; (4) blocking the release of ADH, aldosterone, epinephrine, and norepinephrine; and (5) stimulating peripheral vasodilation (Figure 21-17b●). As blood volume and blood pressure decline, the stress on the atrial walls is removed, and ANP production ceases (Figure 18-18●, p. 620).

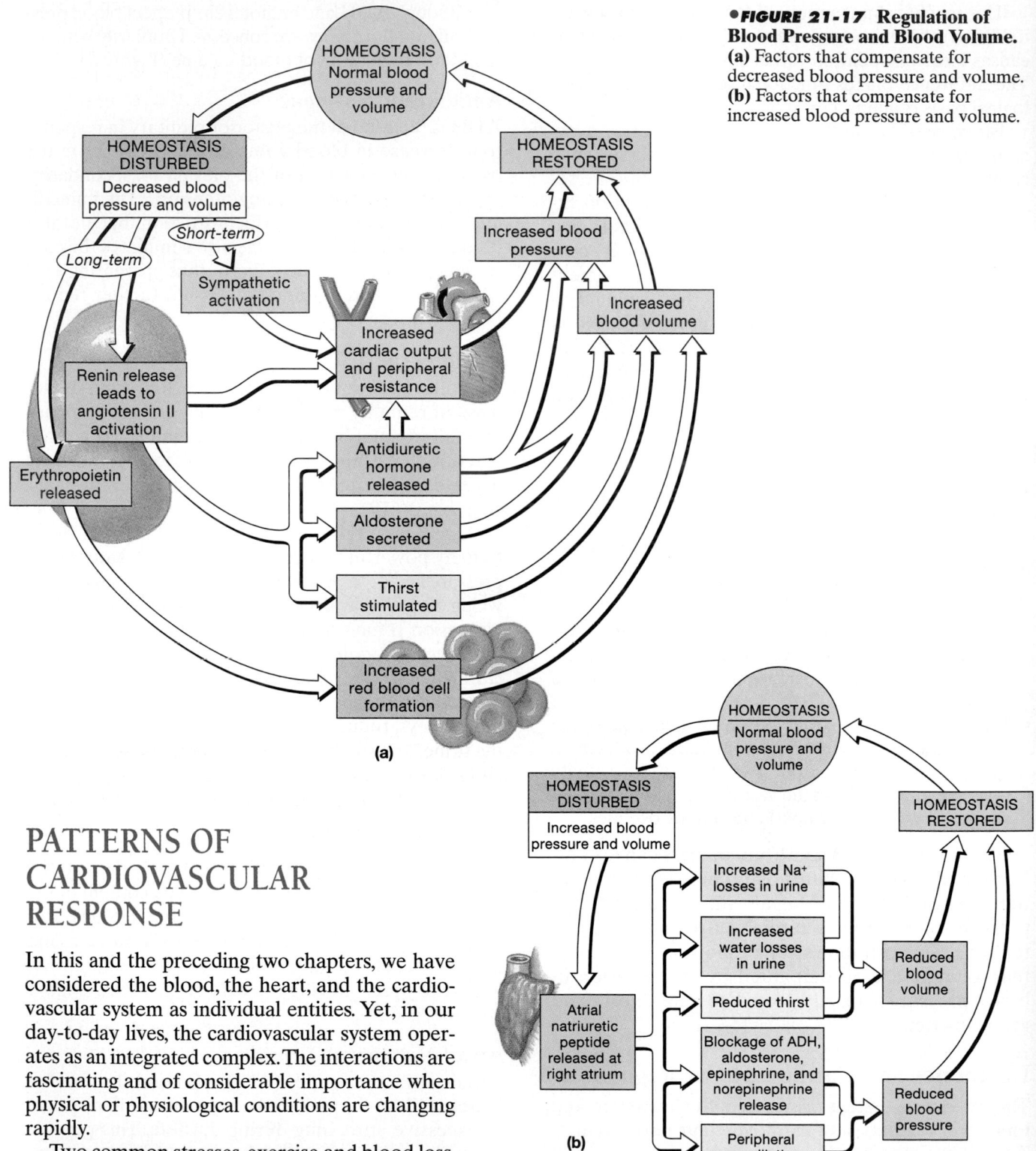

•**FIGURE 21-17** Regulation of **Blood Pressure and Blood Volume.** (a) Factors that compensate for decreased blood pressure and volume. (b) Factors that compensate for increased blood pressure and volume.

PATTERNS OF CARDIOVASCULAR RESPONSE

In this and the preceding two chapters, we have considered the blood, the heart, and the cardiovascular system as individual entities. Yet, in our day-to-day lives, the cardiovascular system operates as an integrated complex. The interactions are fascinating and of considerable importance when physical or physiological conditions are changing rapidly.

Two common stresses, exercise and blood loss, provide examples of the adaptability of this system and its ability to maintain homeostasis. The homeostatic responses involve an interplay among the cardiovascular system, the endocrine system, and other systems, and the central mechanisms are aided by autoregulation at the tissue level. We shall also consider the physiological mechanisms involved in shock, an important homeostatic disorder.

Exercise and the Cardiovascular System

At rest, cardiac output averages about 5.6 liters per minute. That value changes dramatically during exercise. In addition, the pattern of blood distribution changes markedly, as detailed in Table 21-2.

TABLE 21-2 Distribution of Blood During Exercise

	Tissue Blood Flow (ml/min)		
	Rest	Light Exercise	Strenuous Exercise
Muscle	1200	4500	12,500
Heart	250	350	750
Brain	750	750	750
Skin	500	1500	1900
Kidney	1100	900	600
Abdominal viscera	1400	1100	600
Miscellaneous	600	400	400
Total cardiac output	5800	9500	17,500

Light Exercise

Before you begin to exercise, your heart rate increases slightly because there is a general rise in sympathetic activity as you think about the workout ahead. As you begin light exercise, three interrelated changes occur:

1. **Extensive vasodilation occurs** as the rate of skeletal muscle oxygen consumption increases. Peripheral resistance drops, blood flow through the capillaries increases, and blood enters the venous system at an accelerated rate.

2. **The venous return increases,** as skeletal muscle contractions squeeze blood along the peripheral veins and an increased breathing rate pulls blood into the venae cavae via the respiratory pump.

3. **Cardiac output rises,** primarily in response to (1) the rise in venous return (the Frank–Starling principle ∞ [p. 700]) and (2) atrial stretching (the atrial reflex). There is some sympathetic stimulation, leading to increases in heart rate and contractility, but massive sympathetic activation does not occur. The increased cardiac output keeps pace with the elevated demand, and arterial pressures are maintained despite the drop in peripheral resistance.

This regulation by venous feedback produces a gradual increase in cardiac output to about double resting levels. Over this range, there are increases in the blood flow to skeletal muscle, cardiac muscle, and the skin. The increased flow to the muscles reflects the dilation of arterioles and precapillary sphincters in response to local factors; the increased blood flow to the skin occurs in response to the rise in body temperature.

Heavy Exercise

At higher levels of exertion, other physiological adjustments occur as the cardiac and vasomotor centers call for the general activation of the sympathetic nervous system. Cardiac output increases toward maximal levels, and there are major changes in the peripheral distribution of blood, facilitating the blood flow to active skeletal muscles.

Under massive sympathetic stimulation, your cardioacceleratory centers can increase cardiac output to levels as high as 20–25 liters per minute. Even at these rates, the increased circulatory demands of the skeletal muscles can be met only if the vasomotor centers severely restrict the blood flow to "nonessential" organs, such as those of your digestive system. During exercise at maximal levels, your blood essentially races between the skeletal muscles and the lungs and heart. Although most tissues experience reductions in blood flow, skin perfusion increases further, because body temperature continues to climb. Only the blood supply to the brain remains unaffected.

Exercise, Cardiovascular Fitness, and Health

Cardiovascular performance improves significantly with training. Table 21-3 compares the cardiac performance of athletes with that of nonathletes. Trained athletes have bigger hearts and larger stroke volumes than do nonathletes, and these are important functional differences.

Recall that cardiac output is equal to the stroke volume times the heart rate; for the same cardiac output, an individual with a larger stroke volume will have a slower heart rate. A professional athlete at rest can maintain normal blood flow to peripheral tissues at a heart rate as low as 50 bpm (beats per

TABLE 21-3 Effects of Training on Cardiovascular Performance

Subject	Heart Weight (g)	Stroke Volume (ml)	Heart Rate (bpm)	Cardiac Output (l/min)	Blood Pressure (systolic/diastolic)
Nonathlete (rest)	300	60	83	5.0	120/80
Nonathlete (maximum)		104	192	19.9	187/75
Trained athlete (rest)	500	100	53	5.3	120/80
Trained athlete (maximum)		167	182	30.4	200/90*

*Diastolic pressures in athletes during maximal activity have not been accurately measured.

minute), and when necessary the cardiac output can increase to levels 50 percent higher than those of nonathletes. Thus, a trained athlete can tolerate sustained levels of activity that are well outside the capabilities of nonathletes.

Exercise and Cardiovascular Disease

Regular exercise has several beneficial effects. Even a moderate exercise routine (jogging 5 miles a week, for example) can lower total blood cholesterol levels. A high cholesterol level is one of the major risk factors for atherosclerosis, leading to cardiovascular disease and strokes. In addition, a healthy lifestyle—regular exercise, a balanced diet, weight control, and not smoking—reduces stress, lowers blood pressure, and slows plaque formation.

Regular moderate exercise may cut the incidence of heart attacks almost in half. However, at present only an estimated 8 percent of adults in the United States exercise at recommended levels. Exercise is also beneficial in accelerating recovery after a heart attack. Regular light-to-moderate exercise, such as walking, jogging, or bicycling, coupled with a low-fat diet and a low-stress lifestyle, not only reduces symptoms of CAD, such as angina, but also improves both mood and the overall quality of life. However, exercise does not remove the underlying medical problem, and atherosclerotic plaques do not disappear and seldom grow smaller with exercise.

There is no evidence that *intense* athletic training lowers the incidence of cardiovascular disease. On the contrary, the strains placed on all physiological systems, including the cardiovascular system, during an ultramarathon, iron-man triathlon, or other athletic extreme can be severe. Individuals with congenital aneurysms, cardiomyopathy, or cardiovascular disease risk fatal circulatory problems, such as an arrhythmia or heart attack, during severe exercise. Even healthy individuals can develop acute physiological disorders, such as kidney failure, after extreme exercise. We will detail the effects of exercise on other systems in later chapters.

Cardiovascular Response to Hemorrhaging

In Chapter 19, we considered the local circulatory reaction to a break in the wall of a blood vessel. ∞ *[p. 663]* When hemostasis fails to prevent a significant blood loss, the entire cardiovascular system begins making adjustments to maintain blood pressure and restore blood volume (Figure 21-18●). The immediate problem is the maintenance of adequate blood pressure and peripheral blood flow. The long-term problem is the restoration of normal blood volume.

Short-Term Elevation of Blood Pressure

Almost as soon as the pressures start to decline, short-term adjustments begin. The steps include the following:

- The carotid and aortic reflexes increase cardiac output and cause peripheral vasoconstriction (pp. 730–731). With the blood volume reduced, cardiac output is maintained by increasing the heart rate, typically to rates of 180–200 bpm.
- Sympathetic activation provides assistance by increasing vasomotor tone, constricting the arterioles, and elevating blood pressure (p. 729). At the same time, the small muscular arteries, arterioles, and veins decrease in diameter.
- The venoconstriction demanded by the vasomotor centers mobilizes the venous reserve and quickly improves venous return (p. 718).
- Hormonal adjustments occur. (1) Sympathetic activation causes the secretion of E and NE by the adrenal medulla, increasing cardiac output and extending peripheral vasoconstriction. (2) The release of ADH by the posterior pituitary and the production of angiotensin II enhance vasoconstriction while participating in the long-term response.

This combination of adjustments elevates blood pressure and improves peripheral blood flow. After blood losses of up to 20 percent of the total blood volume, these short-term adjustments can restore normal arterial pressures and peripheral blood flow. For example, these adjustments are more than sufficient to compensate for the blood loss experienced when you donate blood. (Most blood banks collect 500 ml of whole blood, roughly 10 percent of the total blood volume.)

Long-Term Restoration of Blood Volume

Short-term compensation involves adjusting to the loss in blood volume. Long-term compensation involves restoring normal blood volume. After a serious hemorrhage, this process may take several days. The steps include the following:

- The decline in capillary blood pressure triggers the recall of fluids from the interstitial spaces (p. 726).
- ADH and aldosterone promote fluid retention and reabsorption at the kidneys, preventing further reductions in blood volume.
- Thirst increases, and additional water is obtained by absorption across the digestive tract. This intake of fluid elevates plasma volume and ultimately replaces the interstitial fluids "borrowed" at the capillaries.
- Erythropoietin targets the bone marrow, stimulating the maturation of red blood cells, which increase the blood volume and improve oxygen delivery to peripheral tissues.

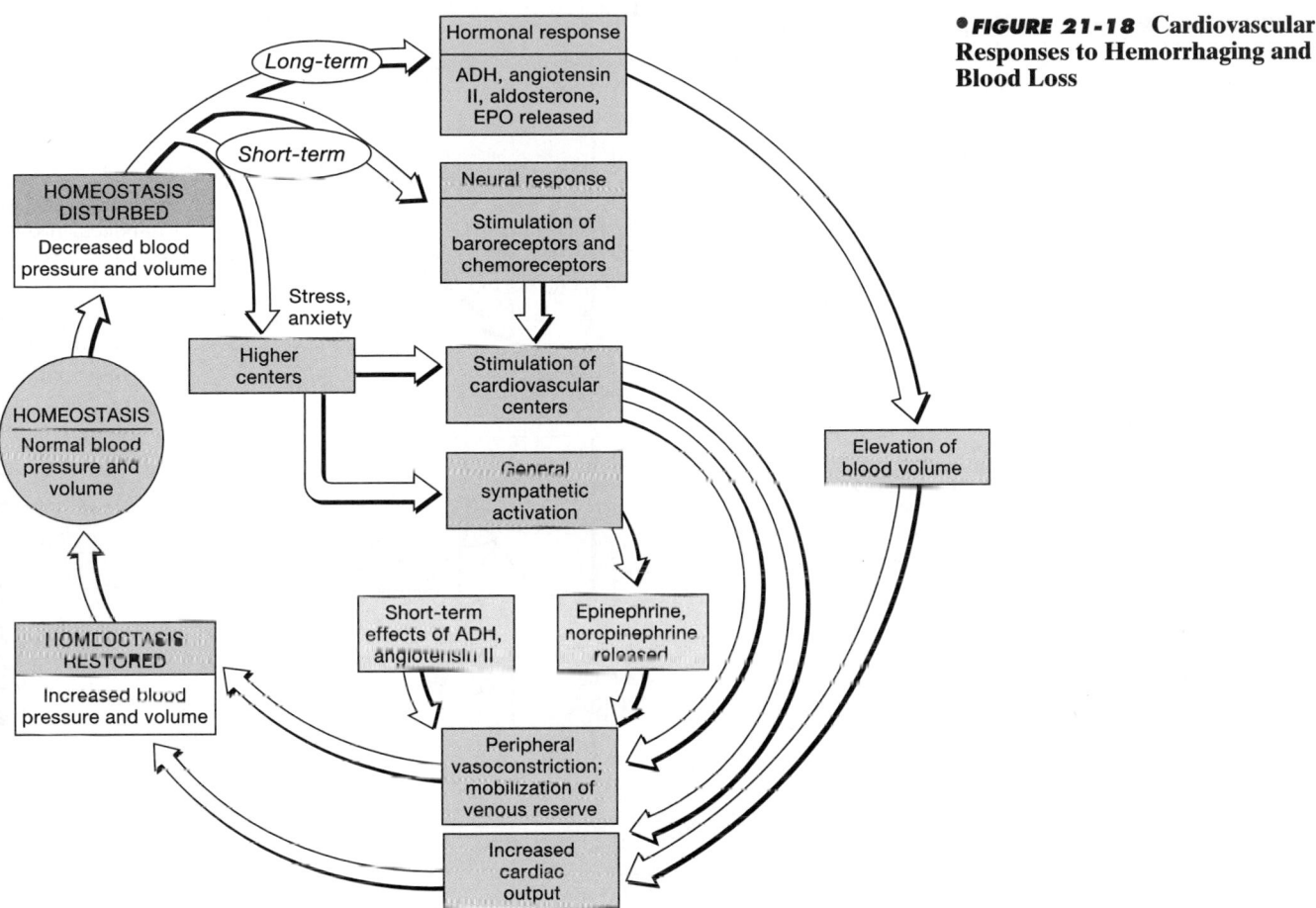

•FIGURE 21-18 Cardiovascular Responses to Hemorrhaging and Blood Loss

Shock

Shock is an acute circulatory crisis marked by low blood pressure (hypotension) and inadequate peripheral blood flow. Severe and potentially fatal symptoms develop as vital tissues become starved for oxygen and nutrients. Common causes of shock are (1) a drop in cardiac output after hemorrhaging or other fluid losses, (2) damage to the heart, (3) external pressure on the heart, and (4) extensive peripheral vasodilation. We shall focus on the cause, symptoms, and treatment of circulatory shock. To read about other types of shock, consult the *Applications Manual.* AM *Other Types of Shock*

Circulatory Shock

A severe reduction in blood volume produces symptoms of **circulatory shock.** Symptoms of circulatory shock appear after fluid losses of about 30 percent of the total blood volume. The cause can be hemorrhaging or fluid losses to the environment, as in dehydration or after severe burns. All cases of circulatory shock share the same basic symptoms:

1. Hypotension, with systolic pressures below 90 mm Hg.
2. Pale, cool, and moist ("clammy") skin. The skin is pale and cool due to peripheral vasoconstriction; the moisture reflects sympathetic activation of the sweat glands.

3. Frequent confusion and disorientation, due to a drop in blood pressure at the brain.
4. A rise in heart rate and a rapid, weak pulse.
5. Cessation of urination, because the reduced blood flow to the kidneys slows or stops urine production.
6. A drop in blood pH (*acidosis*), due to lactic acid generation in oxygen-deprived tissues.

Circulatory shock is often divided into *compensated, progressive,* and *irreversible* stages.

THE COMPENSATED STAGE (STAGE I) During the **compensated stage,** homeostatic adjustments can cope with the situation. The short-term and long-term responses detailed in Figure 21-18• are part of the compensation process. During the period of compensation, peripheral blood flow is reduced but remains within tolerable limits.

THE PROGRESSIVE STAGE (STAGE II) When blood volume declines by more than 35 percent, the individual enters the **progressive stage** of circulatory shock. Homeostatic mechanisms are now unable to cope with the situation. Despite sustained vasoconstriction and the mobilization of the venous reserve, blood pressure remains abnormally low, venous return is reduced, and cardiac output is inadequate (Figure 21-19a•). A vicious

●**FIGURE 21-19** Shock. (a) The progressive stage is characterized by a gradual decline in systemic blood pressure, tissue blood flow, and cardiac output. (b) The irreversible stage involves a series of integrated chain reactions leading to a rapid decline in cardiac output, a dramatic and irreversible fall in blood pressure, circulatory collapse, and eventual death.

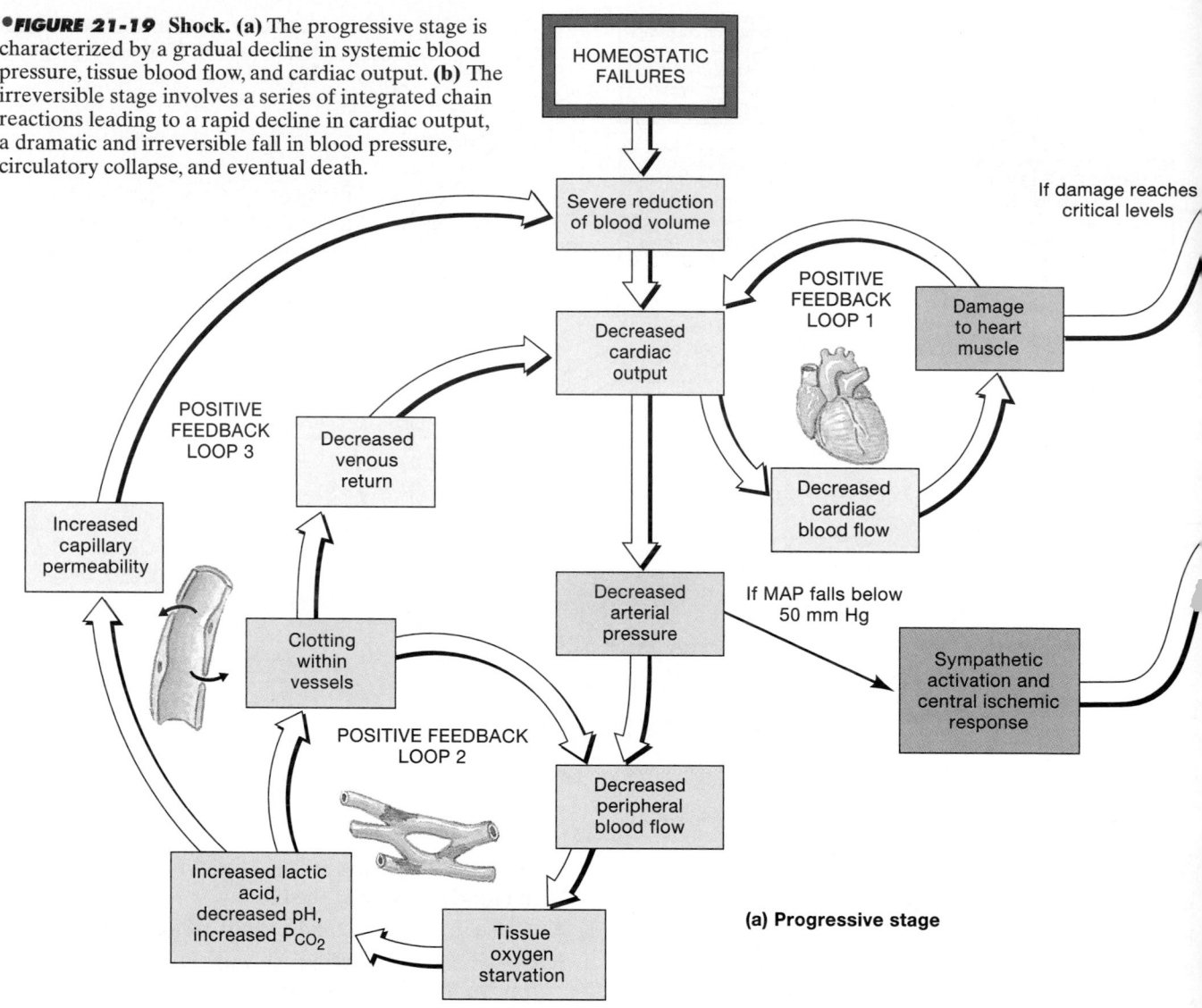

(a) Progressive stage

cycle begins when the low cardiac output causes myocardial damage. This damage leads to a further reduction in cardiac output and subsequent reductions in blood pressure and venous return. As blood flow to the brain decreases, the individual becomes disoriented and confused.

When the mean arterial blood pressure falls to about 50 mm Hg, carotid sinus baroreceptors trigger a massive activation of the vasomotor centers. In essence, the goal now is to preserve the circulation to the brain at any cost. The sympathetic output now causes a sustained and maximal vasoconstriction. This reflex, called the **central ischemic response,** reduces peripheral circulation to an absolute minimum, but it elevates blood pressure to about 70 mm Hg and improves blood flow to cerebral vessels.

The central ischemic response is a last-ditch effort that maintains adequate blood flow to the brain at the expense of other tissues. Unless prompt treatment is provided, the condition will soon prove fatal. Treatment

must concentrate on (1) preventing further fluid losses and (2) giving a transfusion of whole blood, plasma expanders, or blood substitutes.

THE IRREVERSIBLE STAGE (STAGE III) In the absence of treatment, progressive shock will soon turn into **irreversible shock** (Figure 21-19b●). At this point, conditions in the heart, liver, kidneys, and CNS are rapidly deteriorating to the point at which death will occur, even *with* medical treatment.

Irreversible shock begins when conditions in the tissues become so abnormal that the arteriolar smooth muscles and precapillary sphincters become unable to contract, despite the commands of the vasomotor centers. The result is a widespread peripheral vasodilation and an immediate and fatal decline in blood pressure. This event is called **circulatory collapse.** The blood pressure in many tissues then falls so low that the capillaries collapse like deflating balloons. Blood flow through these capillary beds then stops completely, and the cells

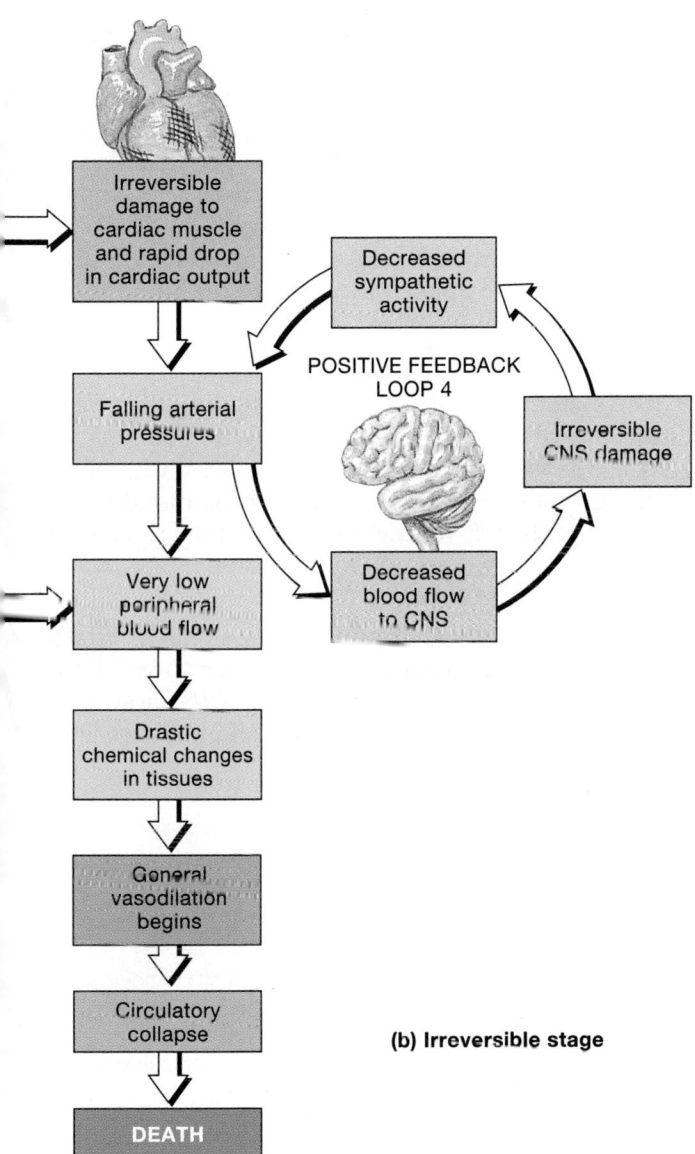

(b) Irreversible stage

the lungs. This circuit begins at the right ventricle and ends at the left atrium. From the left ventricle, the arteries of the systemic circuit transport oxygenated blood and nutrients to all organs and tissues, ultimately returning deoxygenated blood to the right atrium. Figure 21-20• summarizes the primary circulatory routes within the pulmonary and systemic circuits.

In the pages that follow, we shall examine the vessels of the pulmonary and systemic circuits in detail. Two general functional patterns are worth noting at the outset:

1. The peripheral distributions of arteries and veins on the left and right sides are generally identical except near the heart, where the largest vessels connect to the atria or ventricles. For example, there are *left* and *right subclavian, axillary, brachial,* and *radial arteries* whose distribution parallels that of the *left* and *right subclavian, axillary, brachial,* and *radial veins,* respectively.

2. A single vessel may have several different names as it crosses specific anatomical boundaries, making accurate anatomical descriptions possible when the vessel extends far into the periphery. For example, the *external iliac artery* becomes the *femoral artery* as it leaves the trunk and enters the lower limb.

in the affected tissues die. The dying cells release more abnormal chemicals, and the effect quickly spreads throughout the body.

☑ Why does blood pressure increase during exercise?

☑ How would applying a small pressure to the common carotid artery affect your heart rate?

☑ What effect would vasoconstriction of the renal artery have on blood pressure and blood volume?

☑ Why does a person suffering from circulatory shock have a rapid and weak pulse?

THE BLOOD VESSELS

You already know that the cardiovascular system is divided into the *pulmonary circuit* and the *systemic circuit.* The pulmonary circuit is composed of arteries and veins that transport blood between the heart and

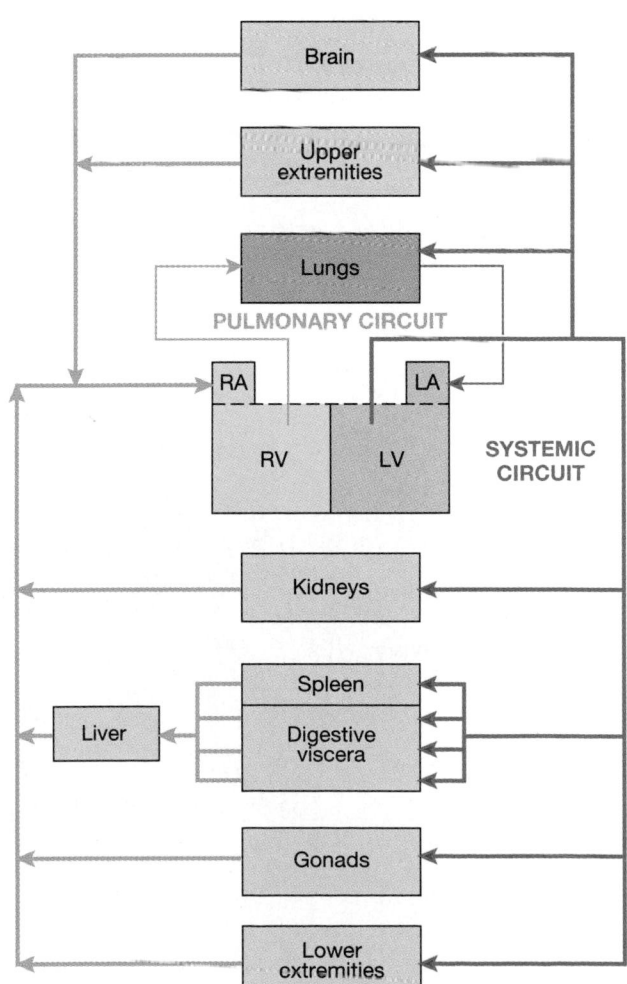

•*FIGURE 21-20* **An Overview of the Pattern of Circulation**

The Pulmonary Circulation

Blood entering the right atrium has just returned from the peripheral capillary beds, where oxygen was released and carbon dioxide was absorbed. After traveling through the right atrium and ventricle, blood enters the pulmonary trunk, the start of the pulmonary circuit (Figure 21-21 ●). At the lungs, oxygen will be replenished, carbon dioxide will be released, and the oxygenated blood will be returned to the heart for distribution via the systemic circuit. Compared with the systemic circuit, the pulmonary circuit is relatively short: The base of the pulmonary trunk and the lungs are only about 15 cm (6 in.) apart.

The arteries of the pulmonary circuit differ from those of the systemic circuit in that they carry deoxygenated blood. (For this reason, most color-coded diagrams show the pulmonary arteries in blue, the same color as systemic veins.) As it curves over the superior border of the heart, the pulmonary trunk gives rise to the **left** and **right pulmonary arteries.** These large arteries enter the lungs before branching repeatedly, giving rise to smaller and smaller arteries. The smallest

branches, the *pulmonary arterioles,* provide blood to capillary networks that surround *alveoli.* The walls of these small air pockets are thin enough for gas exchange to occur between the capillary blood and inspired air. As it leaves the alveolar capillaries, oxygenated blood enters venules that in turn unite to form larger vessels carrying blood toward the **pulmonary veins.** These four veins, two from each lung, empty into the left atrium, completing the pulmonary circuit.

The Systemic Circulation

The systemic circulation supplies the capillary beds in all parts of the body not serviced by the pulmonary circuit. The systemic circuit, which at any given moment contains about 84 percent of the total blood volume, begins at the left ventricle and ends at the right atrium.

Systemic Arteries

Figure 21-22 ● is an overview of the systemic arterial system. This figure indicates the relative locations of major systemic arteries. Figures 21-23 to 21-29 ● show the detailed distribution of these vessels and their branches.

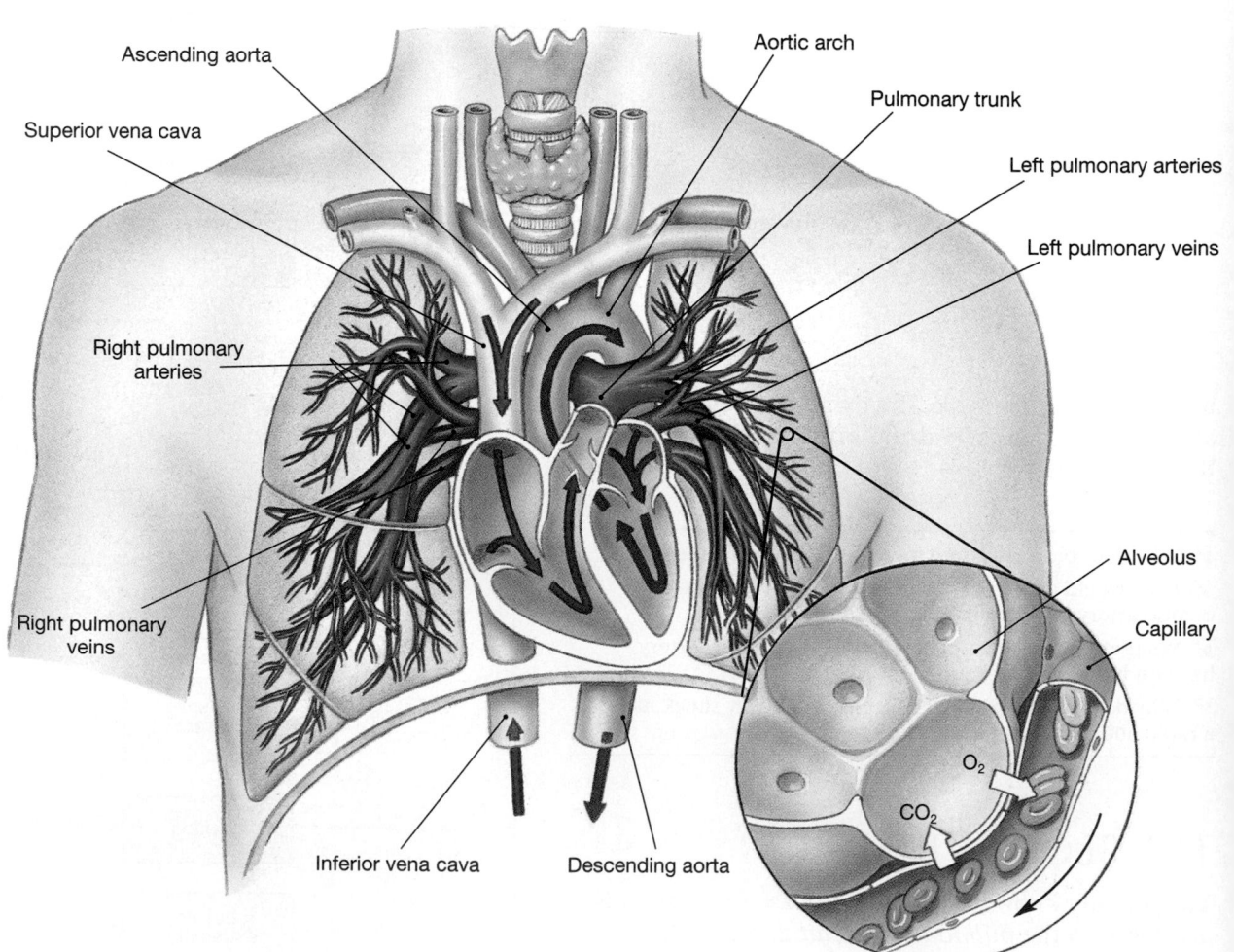

●**FIGURE 21-21** **The Pulmonary Circuit.** [AM] *Plate 6.3b; Scan 10*

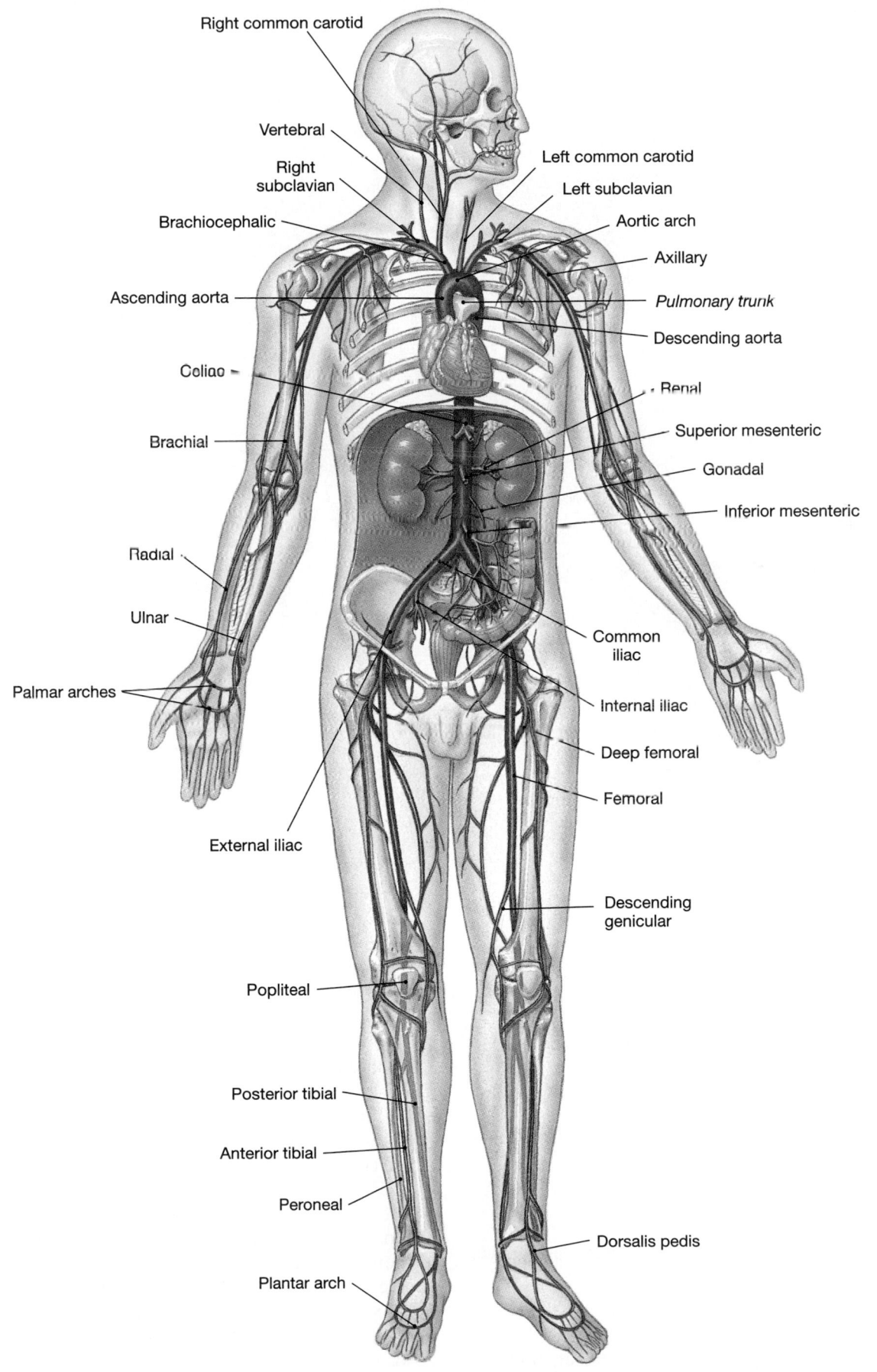

Right common carotid

Vertebral

Right subclavian

Brachiocephalic

Ascending aorta

Celiac

Brachial

Radial

Ulnar

Palmar arches

External iliac

Popliteal

Posterior tibial

Anterior tibial

Peroneal

Plantar arch

Left common carotid

Left subclavian

Aortic arch

Axillary

Pulmonary trunk

Descending aorta

Renal

Superior mesenteric

Gonadal

Inferior mesenteric

Common iliac

Internal iliac

Deep femoral

Femoral

Descending genicular

Dorsalis pedis

●*FIGURE 21-22* **An Overview of the Systemic Arterial System**

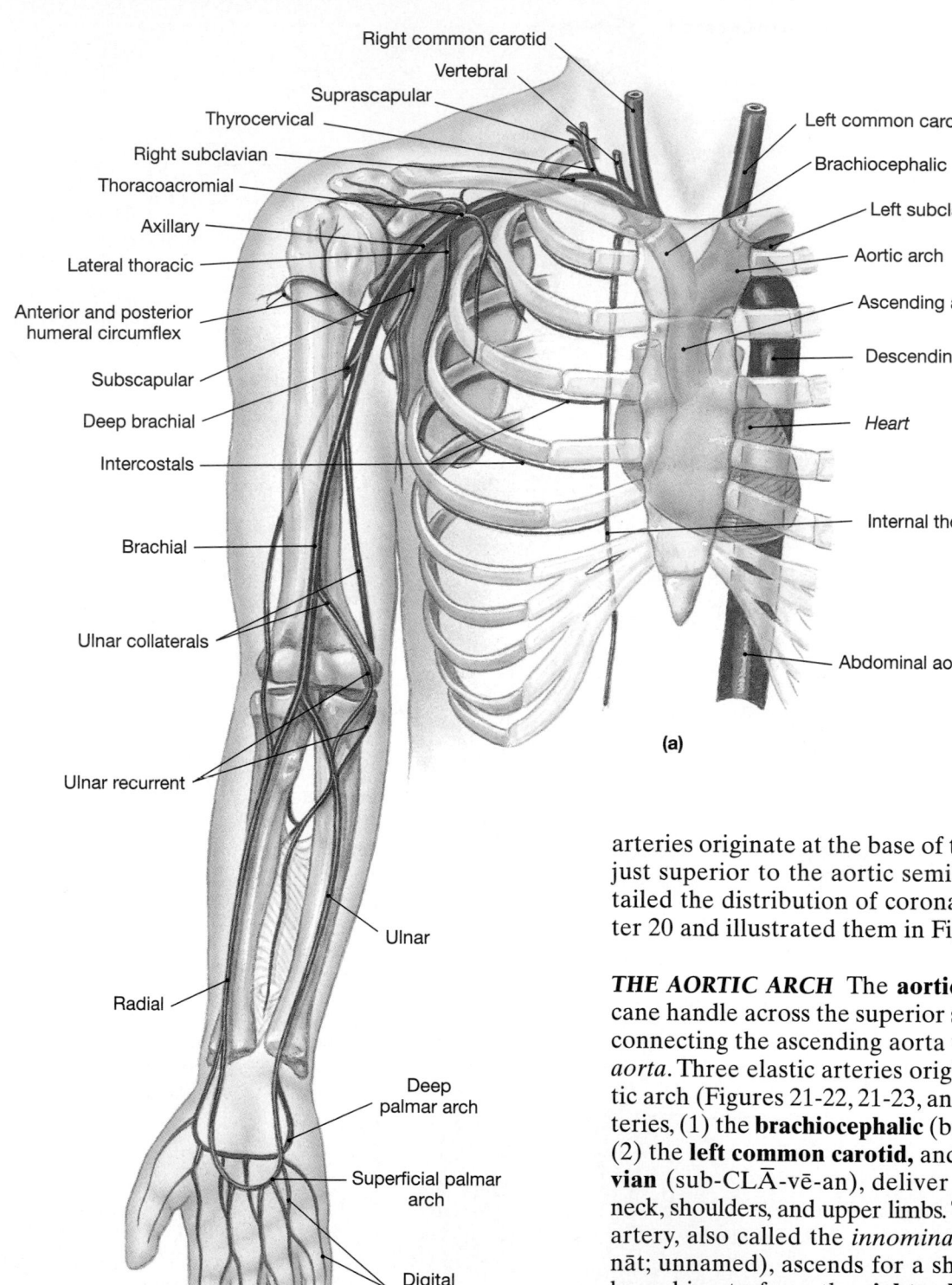

Right common carotid
Vertebral
Suprascapular
Thyrocervical
Right subclavian
Thoracoacromial
Axillary
Lateral thoracic
Anterior and posterior humeral circumflex
Subscapular
Deep brachial
Intercostals
Brachial
Ulnar collaterals
Ulnar recurrent

Left common carotid
Brachiocephalic
Left subclavian
Aortic arch
Ascending aorta
Descending (thoracic) aorta
Heart
Internal thoracic
Abdominal aorta

Ulnar
Radial
Deep palmar arch
Superficial palmar arch
Digital arteries

(a)

•FIGURE 21-23 **Arteries of the Chest and Upper Limb.** **(a)** Diagrammatic view. **(b)** Flowchart. AM *Plates 5.1, 5.2; Scan 11*

THE ASCENDING AORTA The **ascending aorta** begins at the aortic semilunar valve of the left ventricle (Figure 21-23•). The left and right coronary

arteries originate at the base of the ascending aorta, just superior to the aortic semilunar valve. We detailed the distribution of coronary vessels in Chapter 20 and illustrated them in Figure 20-8•, p. 684.

THE AORTIC ARCH The **aortic arch** curves like a cane handle across the superior surface of the heart, connecting the ascending aorta with the *descending aorta.* Three elastic arteries originate along the aortic arch (Figures 21-22, 21-23, and 21-24•). These arteries, (1) the **brachiocephalic** (brā-kē-ō-se-FAL-ik), (2) the **left common carotid,** and (3) the **left subclavian** (sub-CLĀ-vē-an), deliver blood to the head, neck, shoulders, and upper limbs. The brachiocephalic artery, also called the *innominate artery* (i-NOM-i-nāt; unnamed), ascends for a short distance before branching to form the **right subclavian artery** and the **right common carotid artery.**

There is only one brachiocephalic artery, and the left common carotid and left subclavian arteries arise separately from the aortic arch. However, in terms of their peripheral distribution, the vessels on the left side are mirror images of those on the right side. Because the descriptions that follow focus on major branches found on both sides of the body, for clarity we will not use the terms *right* and *left* in the following discussion. Figures 21-23 and 21-24• illustrate the major branches of these arteries.

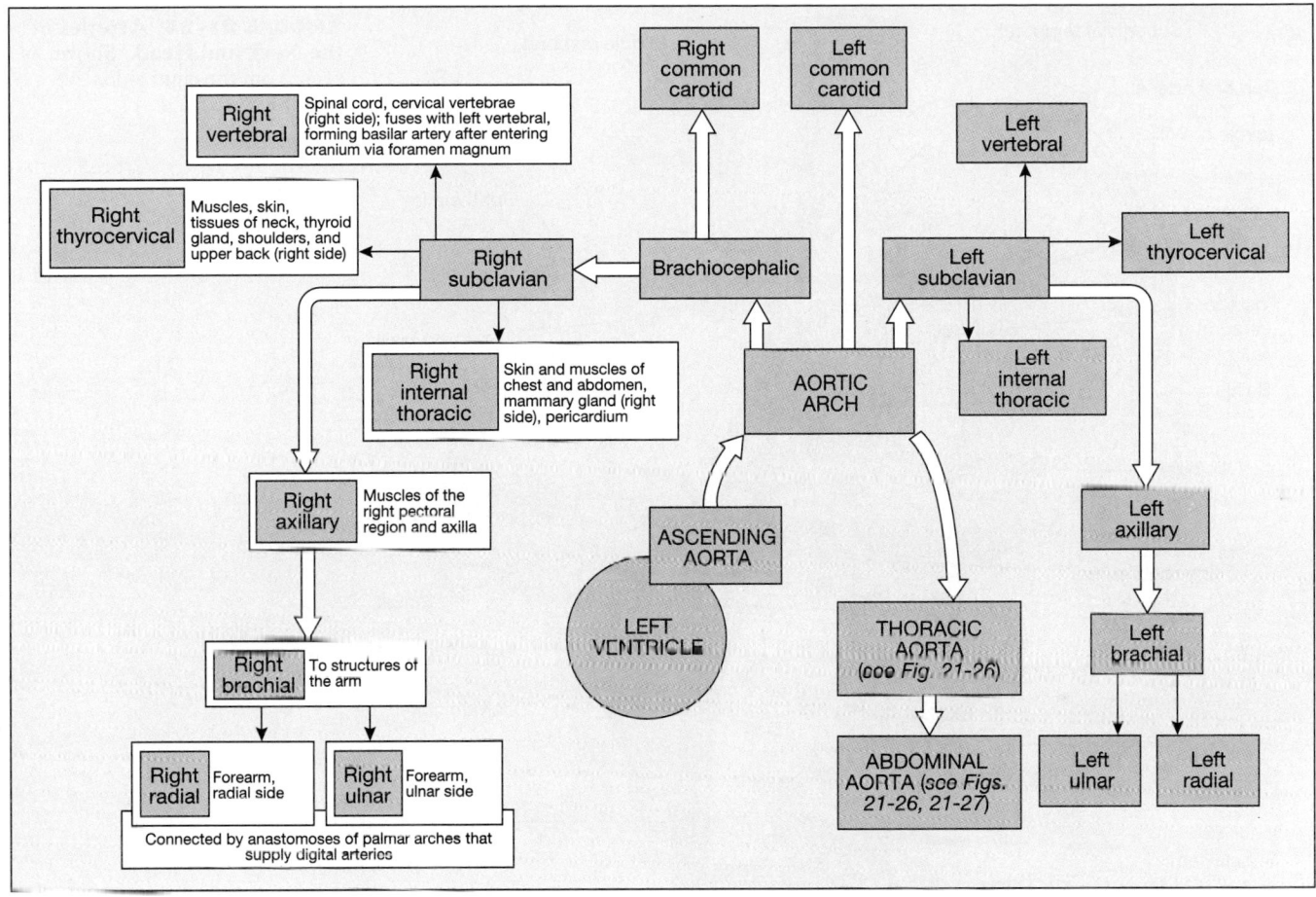

(b)

The Subclavian Arteries. The subclavian arteries supply blood to the arms, chest wall, shoulders, back, and CNS (Figures 21-22 and 21-23●). Three major branches arise before a subclavian artery leaves the thoracic cavity: (1) the **thyrocervical artery,** which provides blood to muscles and other tissues of the neck, shoulder, and upper back; (2) the **internal thoracic artery,** supplying the pericardium and anterior wall of the chest; and (3) the **vertebral artery,** which provides blood to the brain and spinal cord.

After leaving the thoracic cavity and passing across the superior border of the first rib, the subclavian is called the **axillary artery.** The axillary artery crosses the axilla to enter the arm, where it becomes the **brachial artery.** The brachial artery supplies blood to the upper extremity. At the antecubital fossa, the brachial artery divides into the **radial artery,** which follows the radius, and the **ulnar artery,** which follows the ulna to the wrist. These arteries supply blood to the forearm. At the wrist, they anastomose to form the **superficial palmar arch** and the **deep palmar arch,** which supply blood to the hand and to the **digital arteries** of the thumb and fingers.

The Carotid Artery and the Blood Supply to the Brain. The common carotid arteries ascend deep in the tissues of the neck. You can usually locate the

carotid artery by pressing gently along either side of the windpipe (trachea) until you feel a strong pulse.

Each common carotid artery divides into an **external carotid** and an **internal carotid artery** (Figure 21-24●). The carotid sinus, located at the base of the internal carotid, may extend along a portion of the common carotid. The external carotids supply blood to the structures of the neck, esophagus, pharynx, larynx, lower jaw, and face. The internal carotids enter the skull through the *carotid canals* of the temporal bones, delivering blood to the brain (see Figures 7-3 and 7-4●, pp. 201–203).

The internal carotids ascend to the level of the optic nerves, where each divides into three branches: (1) an **ophthalmic artery,** which supplies the eyes; (2) an **anterior cerebral artery,** which supplies the frontal and parietal lobes of the brain; and (3) a **middle cerebral artery,** which supplies the mesencephalon and lateral surfaces of the cerebral hemispheres (Figures 21-24 and 21-25●, p. 745).

The brain is extremely sensitive to changes in its circulatory supply. An interruption of circulation for several seconds will produce unconsciousness, and after 4 minutes there may be some permanent neural damage. Such circulatory crises are rare, because blood reaches the brain through the vertebral arteries as well as by way of the internal carotids. The left

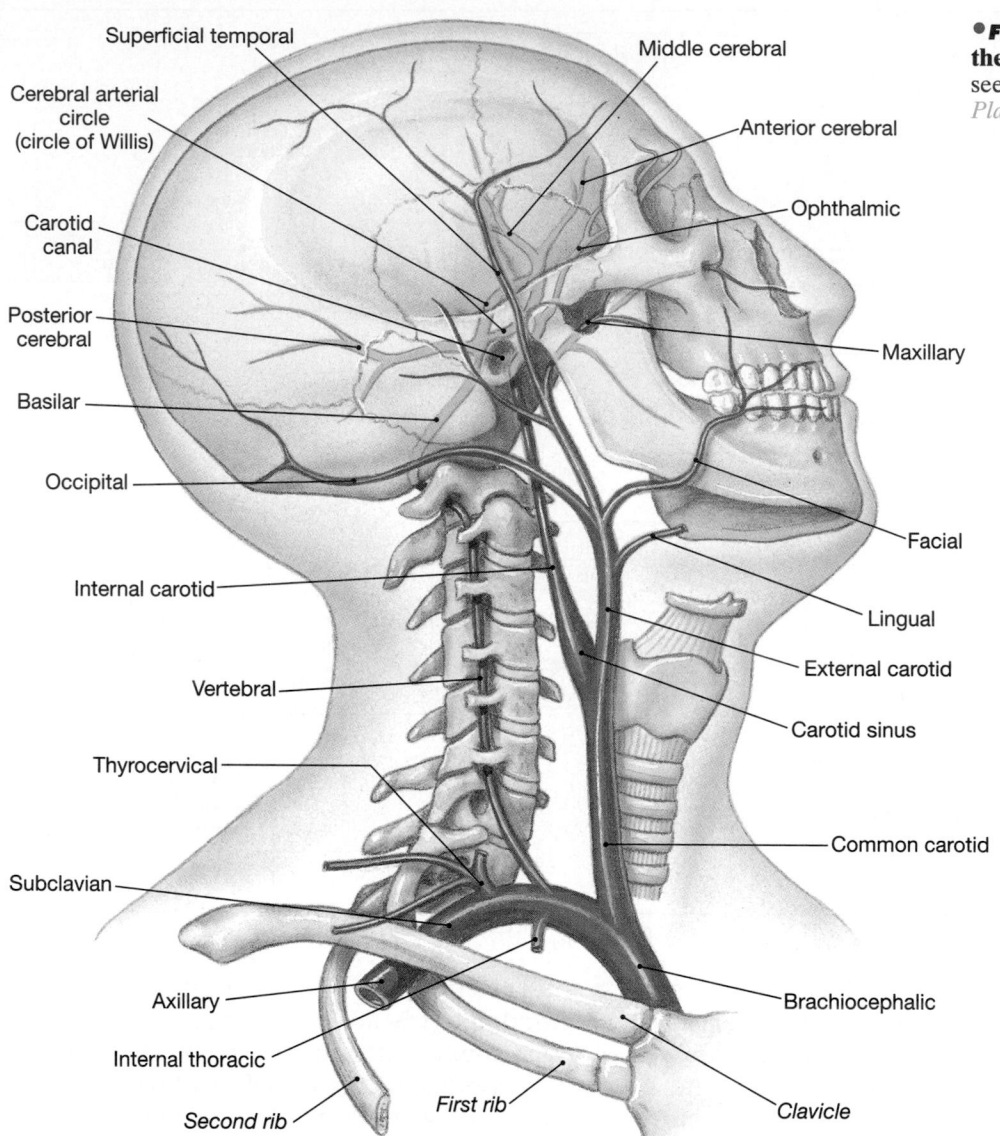

Superficial temporal

Cerebral arterial circle (circle of Willis)

Carotid canal

Posterior cerebral

Basilar

Occipital

Internal carotid

Vertebral

Thyrocervical

Subclavian

Axillary

Internal thoracic

Second rib

First rib

Middle cerebral

Anterior cerebral

Ophthalmic

Maxillary

Facial

Lingual

External carotid

Carotid sinus

Common carotid

Brachiocephalic

Clavicle

and right vertebral arteries arise from the subclavian arteries and ascend within the *transverse foramina* of the cervical vertebrae (see Figure 7-19•, p. 221). The vertebral arteries enter the cranium at the foramen magnum, where they fuse along the ventral surface of the medulla oblongata to form the **basilar artery.** The basilar artery continues on the ventral surface along the pons, branching many times before dividing into the **posterior cerebral arteries.** The **posterior communicating arteries** branch off the posterior cerebral arteries (Figure 21-25•).

The internal carotids normally supply the arteries of the anterior half of the cerebrum, and the rest of the brain receives blood from the vertebral arteries. But this circulatory pattern can easily change, because the internal carotids and the basilar artery are interconnected in a ring-shaped anastomosis called the **cerebral arterial circle,** or *circle of Willis,* which encircles the infundibulum of the pituitary gland (Figure 21-25•). With

this arrangement, the brain can receive blood from either the carotids or the vertebrals, and the chances for a serious interruption of circulation are reduced.

STROKES *Strokes,* or *cerebrovascular accidents (CVAs),* are interruptions of the vascular supply to a portion of the brain. The *middle cerebral artery,* a major branch of the cerebral arterial circle, is the most common site of a stroke. Superficial branches deliver blood to the temporal lobe and large portions of the frontal and parietal lobes; deep branches supply the cerebral nuclei and portions of the thalamus. If a stroke blocks the middle cerebral artery on the left side of the brain, aphasia and a sensory and motor paralysis of the right side result. In a stroke affecting the middle cerebral on the right side, the individual experiences a loss of sensation and motor control over the left side and has difficulty drawing or interpreting spatial relationships. Strokes affecting vessels supplying the brain stem also produce distinctive symptoms; those affecting the lower brain stem are commonly fatal.

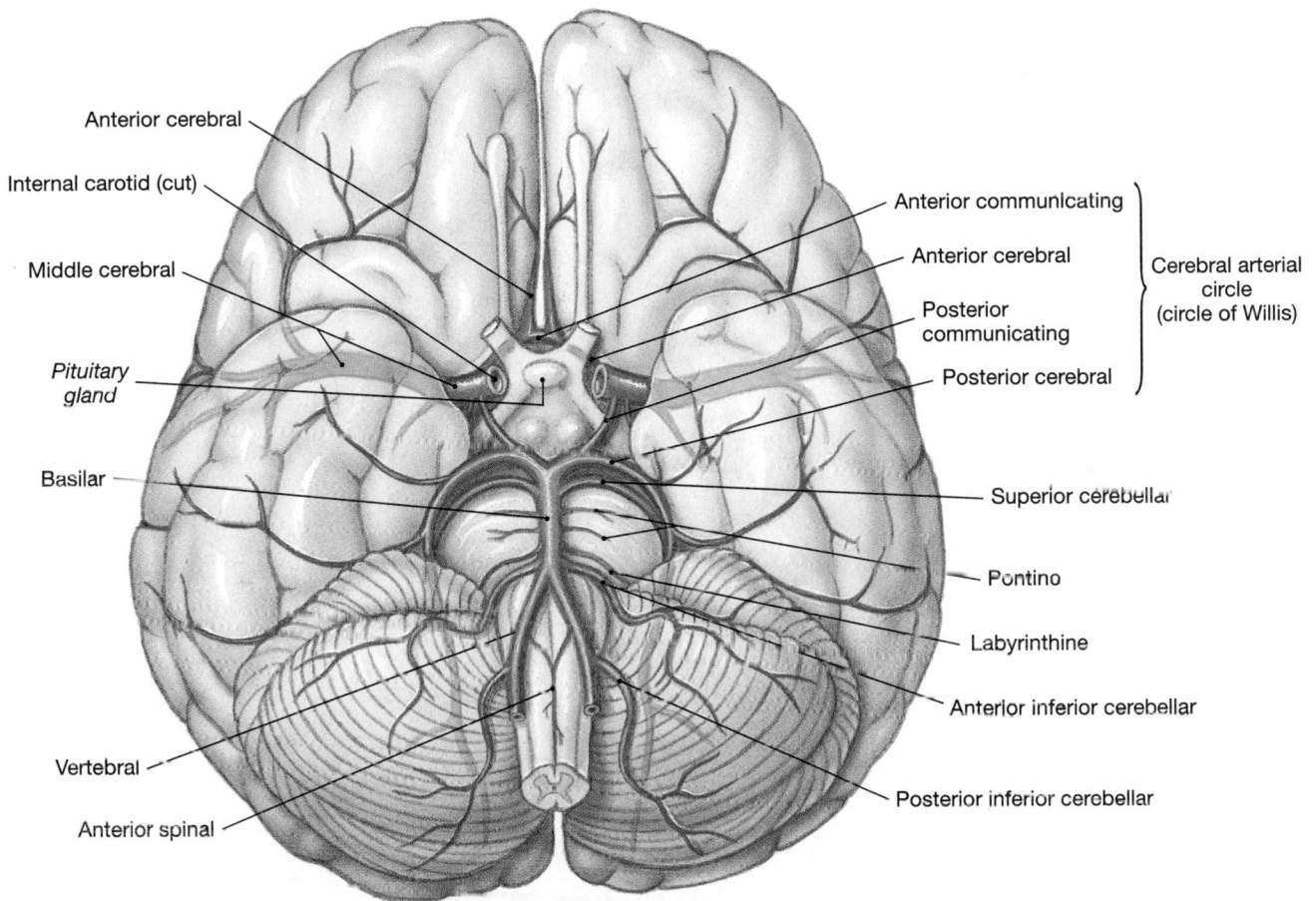

Anterior cerebral

Internal carotid (cut)

Middle cerebral

Pituitary gland

Basilar

Vertebral

Anterior spinal

Anterior communicating

Anterior cerebral

Posterior communicating

Posterior cerebral

Cerebral arterial circle (circle of Willis)

Superior cerebellar

Pontine

Labyrinthine

Anterior inferior cerebellar

Posterior inferior cerebellar

•**FIGURE 21-25** Arteries of the Brain. *Inferior surface of the brain, showing the arterial distribution.* [AM] *Plate 3.5a; Scan 12*

THE DESCENDING AORTA The **descending aorta** is continuous with the aortic arch. The diaphragm divides the descending aorta into a superior **thoracic aorta** and an inferior **abdominal aorta** (Figures 21-26, 21-27, p. 748, and 21-28•, p. 749.

The Thoracic Aorta. The thoracic aorta begins at the level of vertebra T_5 and penetrates the diaphragm at the level of vertebra T_{12}. The thoracic aorta travels within the mediastinum, on the dorsal thoracic wall, slightly to the left of the vertebral column. It supplies blood to branches servicing the tissues and organs of the mediastinum, the muscles of the chest and the diaphragm, and the thoracic spinal cord.

The branches of the thoracic aorta are anatomically grouped as either *visceral branches* or *parietal branches.* Visceral branches supply the organs of the chest: The **bronchial arteries** supply the nonrespiratory tissues of the lungs, the **pericardial arteries** supply the pericardium, the **esophageal arteries** supply the esophagus, and the **mediastinal arteries** supply the tissues of the mediastinum. The parietal branches supply the chest wall: The **intercostal arteries** supply the chest muscles and the vertebral column area, and the **superior phrenic** (FREN-ik) **arteries** deliver

blood to the superior surface of the diaphragm, which separates the thoracic and abdominopelvic cavities. The branches of the thoracic aorta are detailed in Figure 21-26•.

The Abdominal Aorta. The abdominal aorta, which begins immediately inferior to the diaphragm, is a continuation of the thoracic aorta (Figure 21-26a•). The abdominal aorta descends slightly to the left of the vertebral column but posterior to the peritoneal cavity. It is commonly surrounded by a cushion of adipose tissue. At the level of vertebra L_4, the abdominal aorta splits into two major arteries—the *left* and *right common iliac arteries*—that supply deep pelvic structures and the lower limbs. The region where the aorta splits is called the *terminal segment of the aorta.*

The abdominal aorta delivers blood to all the abdominopelvic organs and structures. The major branches to visceral organs are unpaired, and they arise on the anterior surface of the abdominal aorta and extend into the mesenteries. Branches to the body wall, the kidneys, the urinary bladder, and other structures outside the abdominopelvic cavity are paired, and they originate along the lateral surfaces of the abdominal

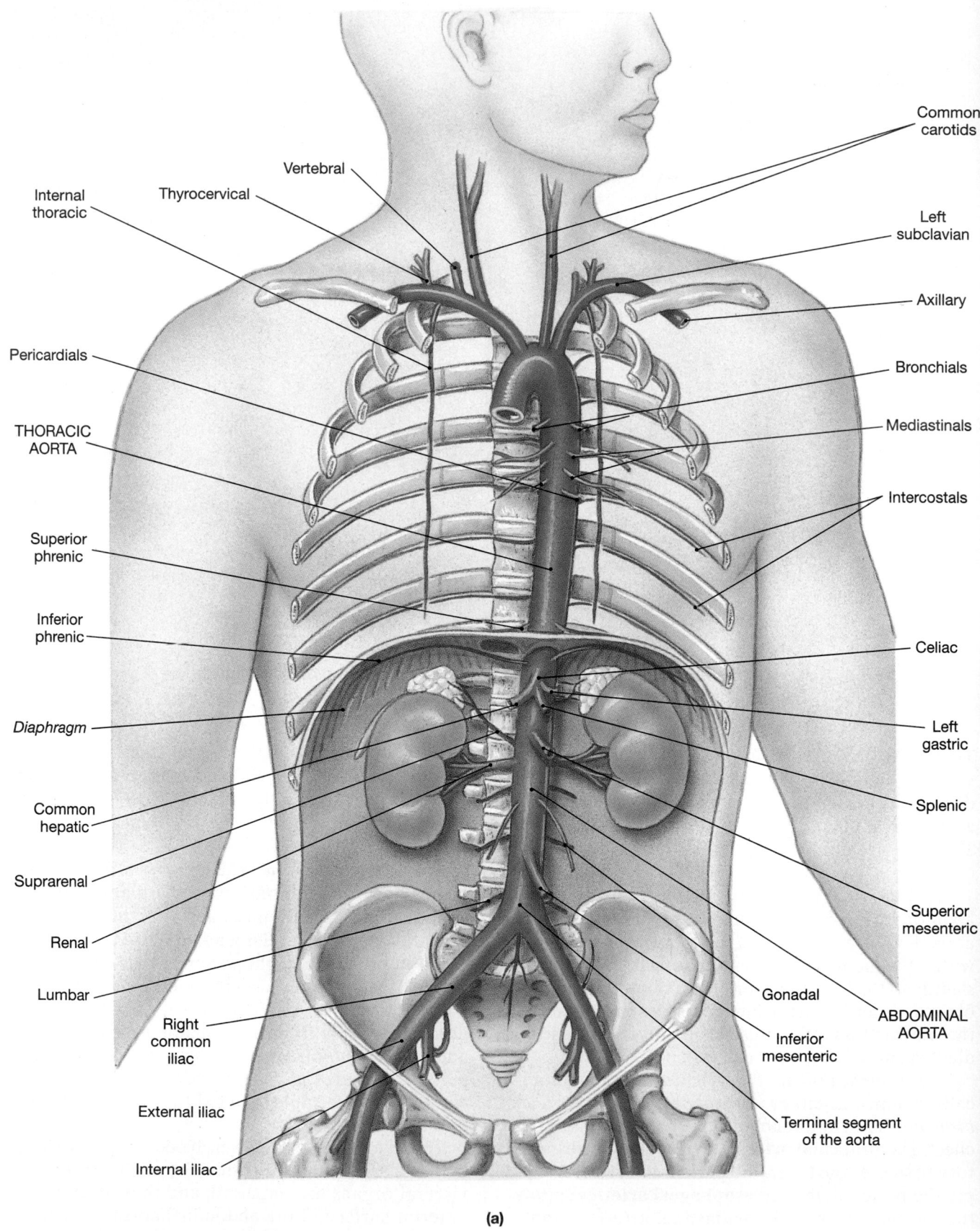

(a)

•FIGURE 21-26 Major Arteries of the Trunk. (a) Diagrammatic view, with most of the thoracic and abdominal organs removed. **(b)** Flowchart. AM *Plate 6.5k, 6.8, 7.1b; Scan 13*

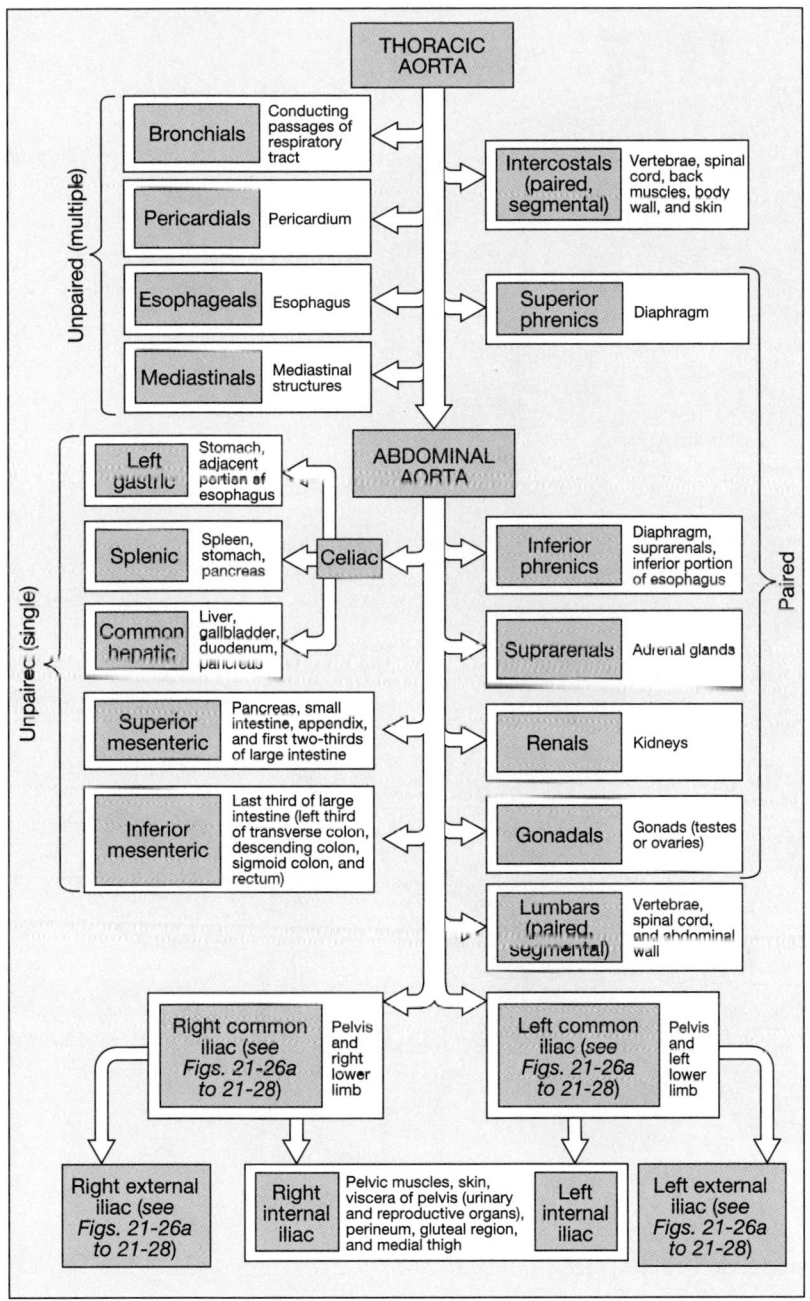

(b)

b. The **splenic artery,** which supplies the spleen and arteries to the stomach *(left gastroepiploic)* and pancreas *(pancreatic).*

c. The **common hepatic artery,** which supplies arteries to the liver *(hepatic),* stomach *(right gastric),* gallbladder *(cystic),* and duodenal area *(gastroduodenal, right gastroepiploic,* and *superior pancreaticoduodenal artery).*

2. The **superior mesenteric** (mez-en-TER-ik) **artery** arises about 2.5 cm inferior to the celiac artery to supply arteries to the pancreas and duodenum *(pancreaticoduodenal),* small intestine *(intestinal),* and most of the large intestine *(right* and *middle colic* and the *ileocolic).*

3. The **inferior mesenteric artery** arises about 5 cm superior to the terminal aorta and delivers blood to the terminal portions of the colon *(left colic* and *sigmoid)* and the rectum *(rectal).*

The abdominal aorta also gives rise to five paired arteries:

1. The **inferior phrenics,** which supply the inferior surface of the diaphragm.

2. The **suprarenal arteries,** which originate on either side of the aorta near the base of the superior mesenteric artery. Each suprarenal artery supplies one of the adrenal glands that cap the superior portion of the kidneys.

3. The short (about 7.5 cm) **renal arteries,** which arise along the posterolateral surface of the abdominal aorta, about 2.5 cm (1 in.) inferior to the superior mesenteric artery, and travel posterior to the peritoneal lining to reach the adrenal glands and kidneys. We shall consider the branches of the renal arteries in Chapter 26.

aorta. Figure 21-26● shows the major arteries of the trunk after removal of most of the thoracic and abdominal organs. Figure 21-27● gives the distribution of those arteries to abdominopelvic organs.

The abdominal aorta gives rise to three unpaired arteries (Figures 21-26 and 21-27●):

1. The **celiac** (SĒ-lē-ak) **artery** delivers blood to the liver, stomach, and spleen. The celiac divides into three branches:

a. The **left gastric artery,** which supplies the stomach and inferior portion of the esophagus.

4. The **gonadal** (gō-NAD-al) **arteries,** which originate between the superior and inferior mesenteric arteries. In males, they are called *testicular arteries* and are long, thin arteries that supply blood to the testes and scrotum. In females, they are termed *ovarian arteries* and supply blood to the ovaries, uterine tubes, and uterus. The distribution of gonadal vessels (both arteries and veins) differs in males and females; we shall describe the differences in Chapter 28.

5. Small **lumbar arteries,** which arise on the posterior surface of the aorta and supply the vertebrae, spinal cord, and abdominal wall.

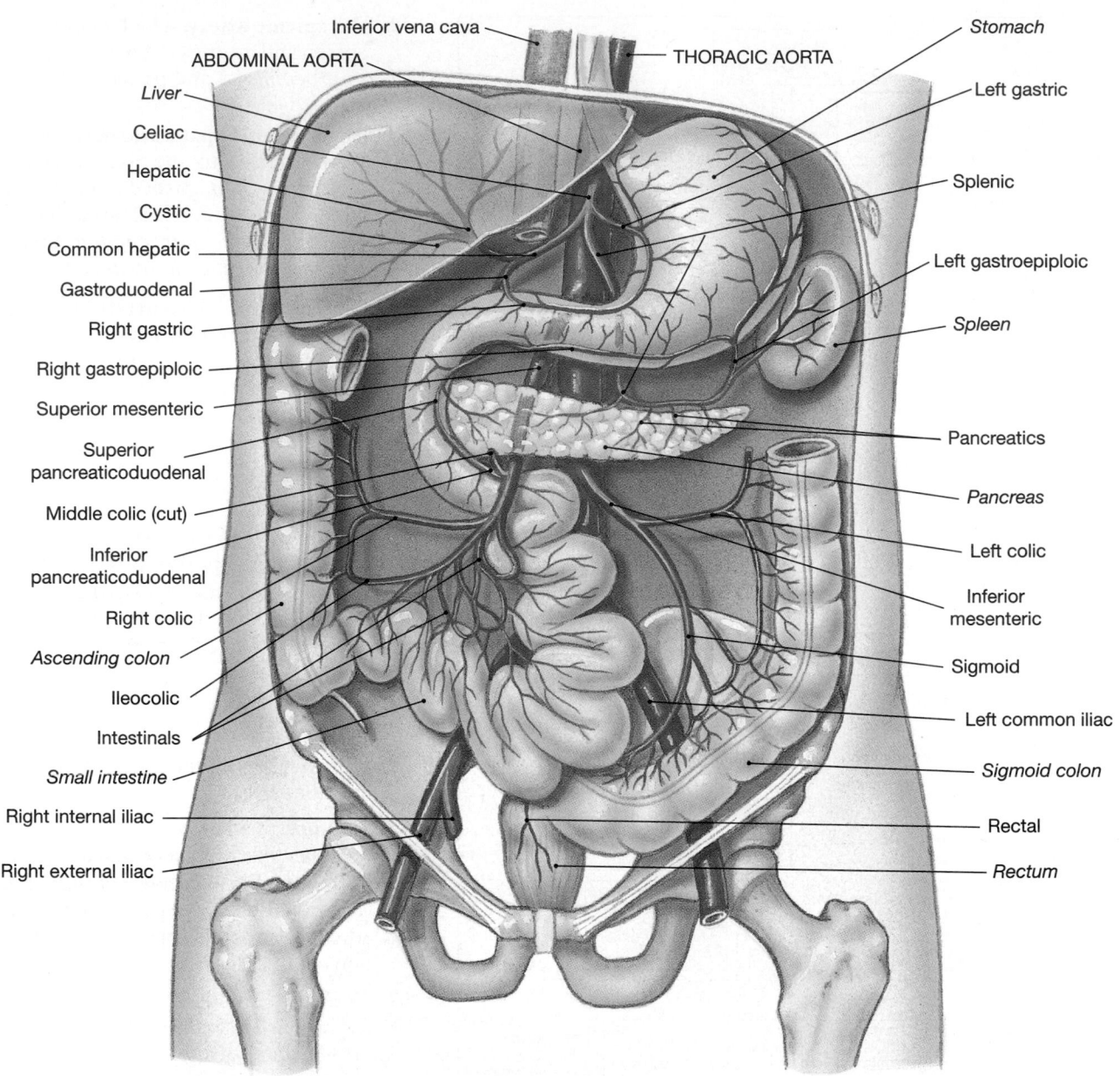

●**FIGURE 21-27** **Arteries Supplying the Abdominopelvic Organs.** *(See also Figure 24-18, p. 737.)* [AM] *Plates 6.5i–6.5k*

ARTERIES OF THE PELVIS AND LOWER LIMBS

Near the level of vertebra L_4, the terminal segment of the abdominal aorta divides to form a pair of elastic arteries, the **right** and **left common iliac** (IL-ē-ak) **arteries** (Figure 21-26b●). These arteries carry blood to the pelvis and lower limbs (Figures 21-28 and 21-29●). As these arteries travel along the inner surface of the ilium, they descend posterior to the cecum and sigmoid colon. At the level of the lumbosacral joint, each common iliac divides to form an **internal iliac artery** and an **external iliac artery** (Figure 21-27●). The internal iliac arteries enter the pelvic cavity to supply the urinary bladder, the internal and external walls of the pelvis, the external genitalia, the medial side of the thigh, and, in females, the uterus and

vagina. The external iliac arteries supply blood to the lower limbs, and they are much larger in diameter than the internal iliac arteries.

Arteries of the Thigh and Leg. The external iliac artery crosses the surface of the iliopsoas muscle and penetrates the abdominal wall midway between the anterior superior iliac spine and the pubic symphysis. It emerges on the anteromedial surface of the thigh as the **femoral artery.** Roughly 5 cm distal to the emergence of the femoral artery, the **deep femoral artery** branches off its lateral surface (Figure 21-28a,b●). The deep femoral artery, which gives rise to the *medial* and *lateral circumflex arteries*, supplies blood to the ventral and lateral regions of the skin and deep muscles of the thigh.

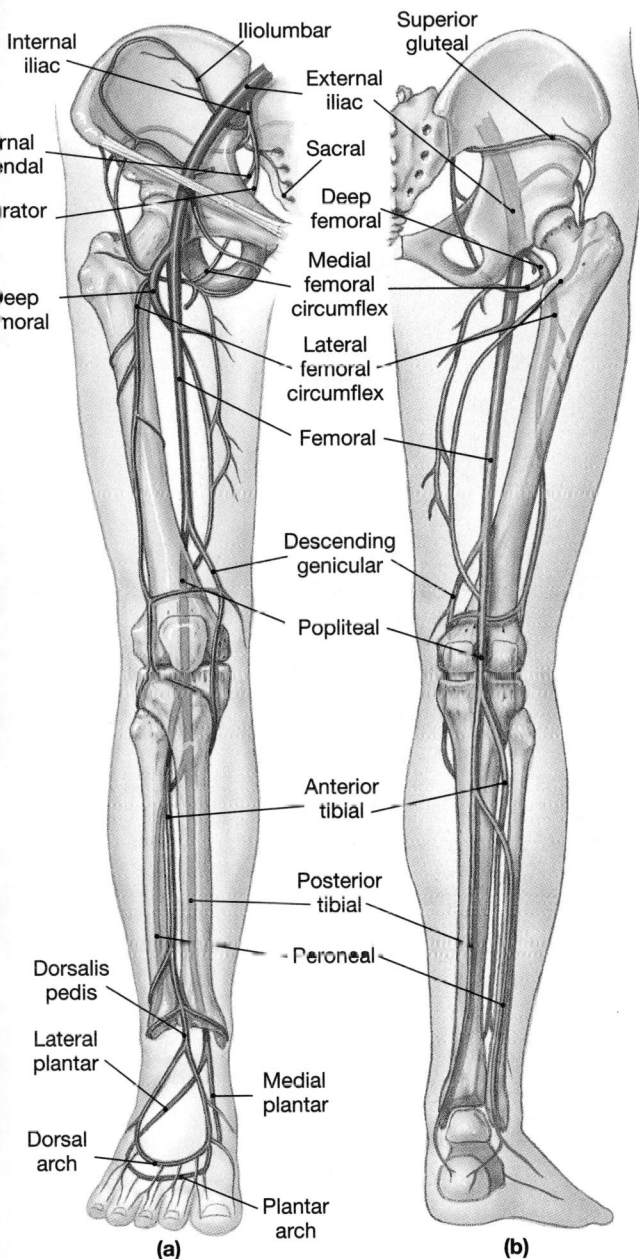

Internal iliac — Iliolumbar — Superior gluteal

External iliac

Sacral

Deep femoral

Medial femoral circumflex

Lateral femoral circumflex

Femoral

Descending genicular

Popliteal

Anterior tibial

Posterior tibial

Peroneal

Dorsalis pedis

Lateral plantar

Dorsal arch

Medial plantar

Plantar arch

(a) **(b)**

●**FIGURE 21-28** **Arteries of the Lower Limb.**
(a) Anterior view. **(b)** Posterior view. AM *Plate 7.2; Scans 5, 6*

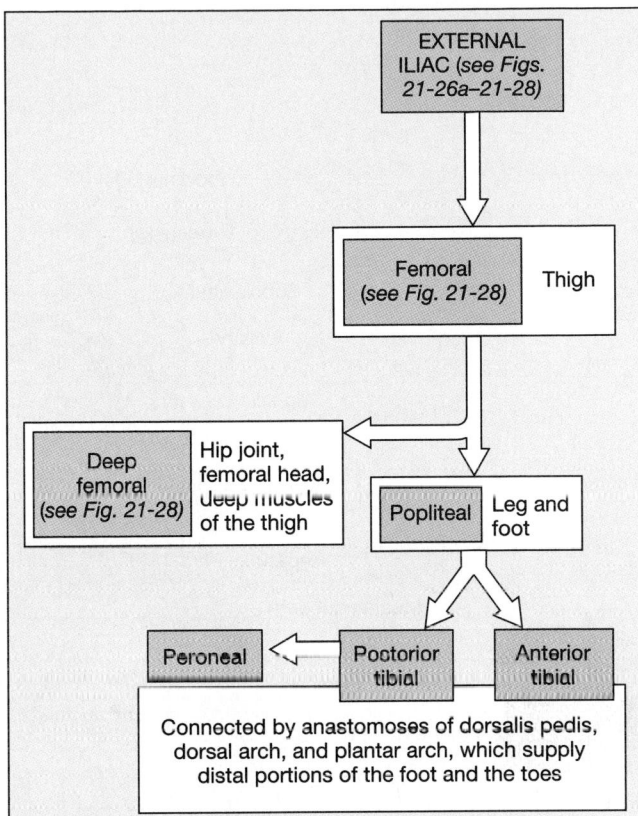

EXTERNAL ILIAC (*see Figs. 21-26a–21-28*)

Femoral (*see Fig. 21-28*) Thigh

Deep femoral (*see Fig. 21-28*) Hip joint, femoral head, deep muscles of the thigh

Popliteal Leg and foot

Peroneal ← Posterior tibial Anterior tibial

Connected by anastomoses of dorsalis pedis, dorsal arch, and plantar arch, which supply distal portions of the foot and the toes

●**FIGURE 21-29** **Flowchart for Blood Flow to a Lower Limb**

Arteries of the Foot. When it reaches the ankle, the anterior tibial artery becomes the **dorsalis pedis artery.** The dorsalis pedis branches repeatedly, supplying the ankle and dorsal portion of the foot (Figure 21-28●).

As it reaches the ankle, the posterior tibial artery divides to form the **medial** and **lateral plantar arteries,** which supply blood to the plantar surface of the foot. The medial and lateral plantar arteries are connected to the dorsalis pedis artery through a pair of anastomoses. This arrangement produces a **dorsal arch** *(arcuate arch)* and a **plantar arch.** Small arteries branching off these arches supply the distal portions of the foot and the toes.

☑ Blockage of which branch from the aortic arch would interfere with the blood flow to the left arm?

☑ Why would compression of the common carotid arteries cause a person to lose consciousness?

☑ Grace is in an automobile accident and ruptures her celiac trunk. Which organs will be affected most directly by this injury?

Systemic Veins

Veins collect blood from the body's tissues and organs by means of an elaborate venous network that drains into the right atrium of the heart via the superior and inferior venae cavae (Figure 21-30●). The branching pattern of peripheral veins is much more variable than that of arteries. The discussion that follows is based on

The femoral artery continues inferiorly and posterior to the femur. At the *popliteal fossa,* the femoral artery becomes the **popliteal** (pop-LIT-ē-al) **artery.** The popliteal artery crosses the popliteal fossa before branching to form the **posterior** and **anterior tibial arteries.** The posterior tibial artery gives rise to the **peroneal artery** and continues inferiorly along the posterior surface of the tibia. The anterior tibial artery passes between the tibia and fibula, emerging on the anterior surface of the tibia. As it descends toward the foot, the anterior tibial provides blood to the skin and muscles of the anterior portion of the leg.

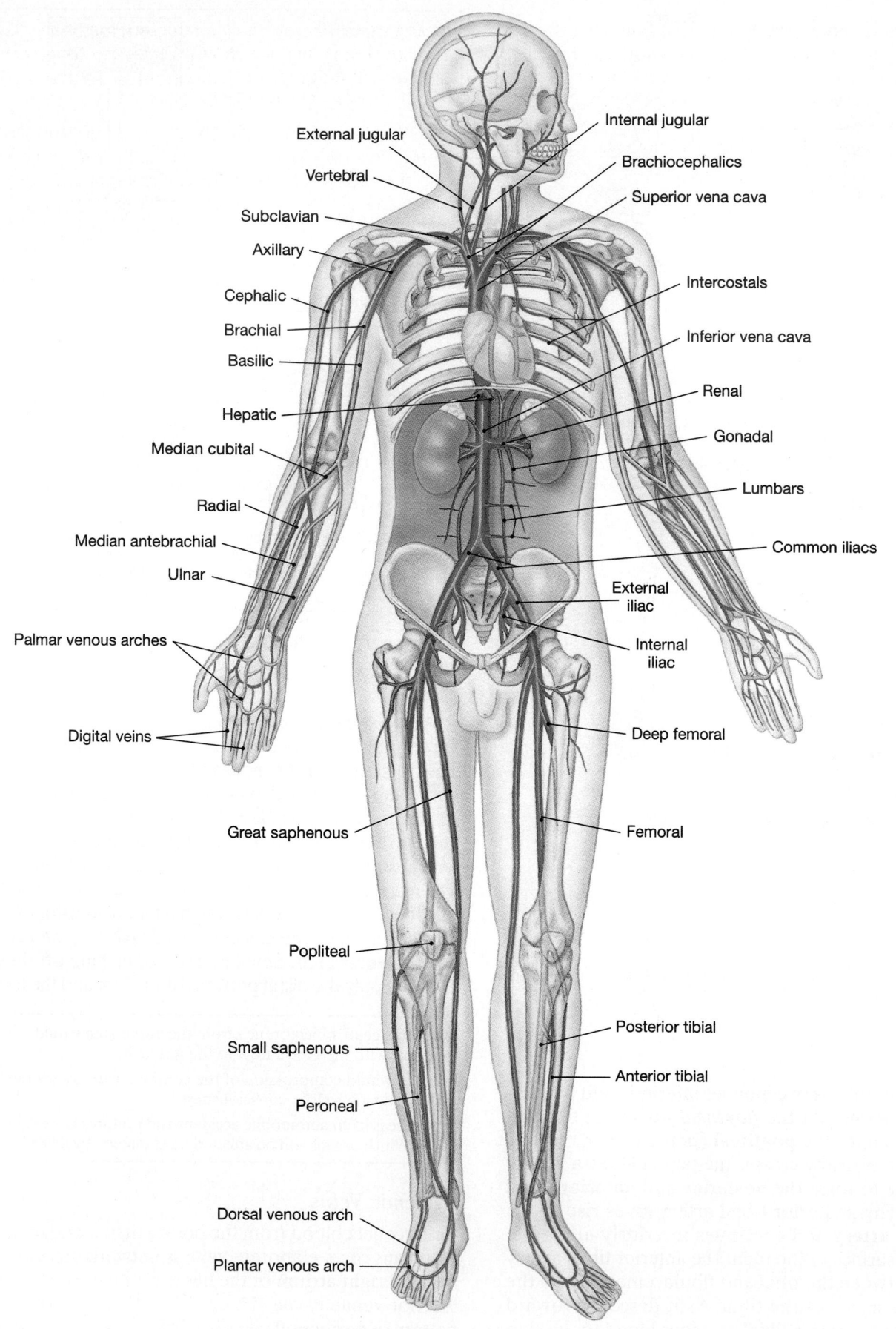

• **FIGURE 21-30** An Overview of the Systemic Venous System

the most common arrangement of veins. Complementary arteries and veins commonly run side by side, and in many cases they have comparable names.

One significant difference between the arterial and venous systems concerns the distribution of major veins in the neck and limbs. Arteries in these areas are located deep beneath the skin, protected by bones and surrounding soft tissues. In contrast, the neck and limbs generally have two sets of peripheral veins, one superficial and the other deep. This dual venous drainage is important for controlling body temperature. In hot weather, venous blood flows in superficial veins, where heat loss can occur; in cold weather, blood is routed to the deep veins to minimize heat loss.

THE SUPERIOR VENA CAVA All the body's systemic veins (except the cardiac veins) drain into ei-

ther the *superior vena cava* or the *inferior vena cava.* The **superior vena cava (SVC)** receives blood from the tissues and organs of the head, neck, chest, shoulders, and upper limbs (Figure 21-30●).

Venous Return from the Cranium. Numerous *superficial cerebral veins* and *internal cerebral veins* drain the cerebral hemispheres. The **superficial cerebral veins** and small veins of the brain stem, such as the *pontal* and *petrosal veins,* empty into a network of dural sinuses. These sinuses include the *superior* and *inferior sagittal sinuses,* the *petrosal sinuses,* the *occipital sinus,* the *left* and *right transverse sinuses,* and the *straight sinus* (Figure 21-31●). The largest sinus, the **superior sagittal sinus,** is in the falx cerebri (see Figure 14-4●, p. 450). The majority of the **internal cerebral veins** collect inside the brain to form

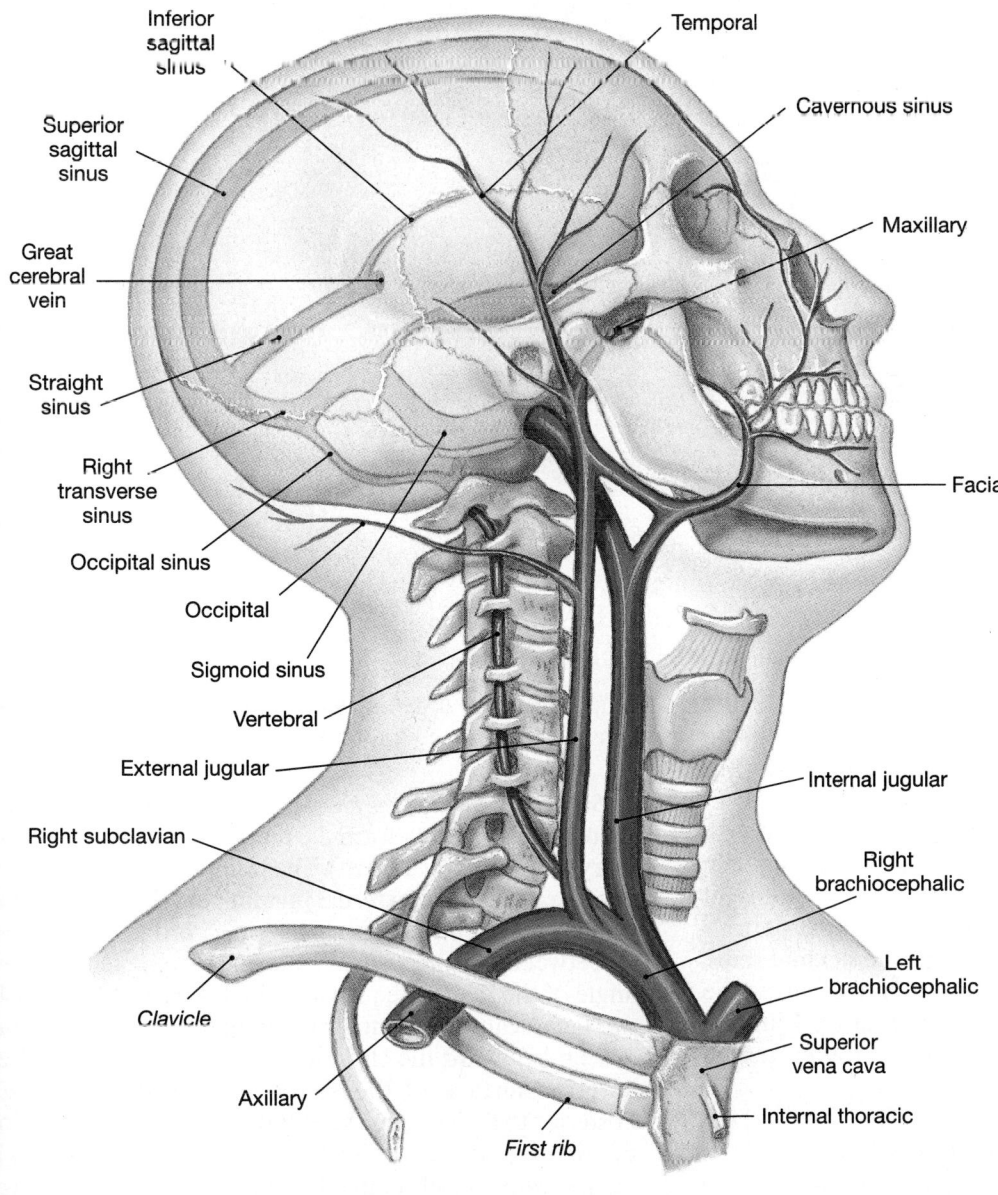

Inferior sagittal sinus

Superior sagittal sinus

Great cerebral vein

Straight sinus

Right transverse sinus

Occipital sinus

Occipital

Sigmoid sinus

Vertebral

External jugular

Right subclavian

Clavicle

Axillary

First rib

Temporal

Cavernous sinus

Maxillary

Facial

Internal jugular

Right brachiocephalic

Left brachiocephalic

Superior vena cava

Internal thoracic

(a)

●FIGURE 21-31 Major Veins of the Head, Neck, and Brain. (a) Veins draining superficial and deep portions of the head and neck. **(b)** Inferior surface of the brain, showing the venous distribution. For the relationship of these veins to meningeal layers, see Figure 14-4, p. 450.
AM *Plates 4.2b, 4.3b*

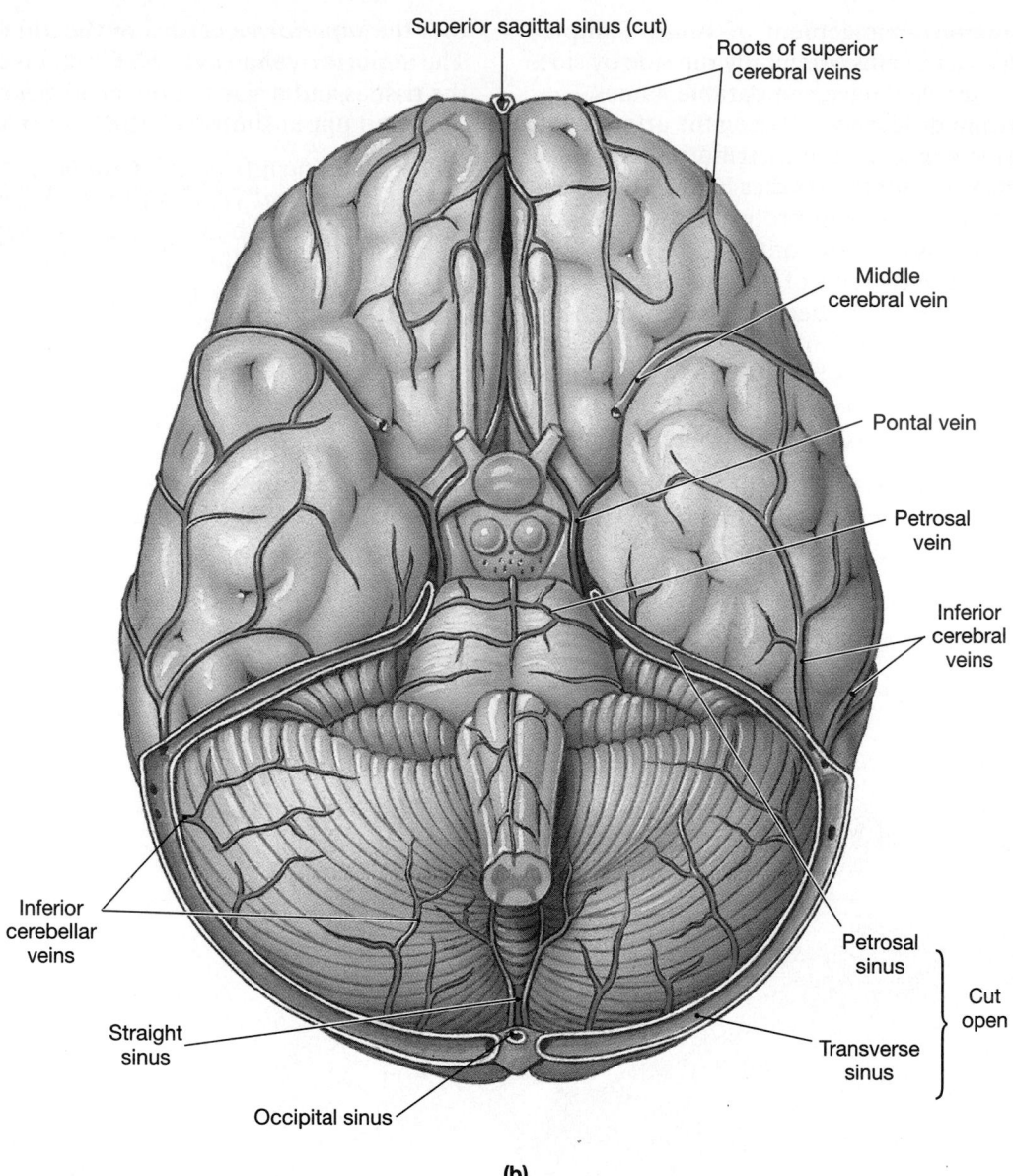

Superior sagittal sinus (cut)

Roots of superior cerebral veins

Middle cerebral vein

Pontal vein

Petrosal vein

Inferior cerebral veins

Petrosal sinus

Cut open

Transverse sinus

Inferior cerebellar veins

Straight sinus

Occipital sinus

(b)

● *FIGURE 21-31 (continued)*

the **great cerebral vein,** which collects blood from the interior of the cerebral hemispheres and the choroid plexus and delivers it to the **straight sinus.** Other cerebral veins drain into the **cavernous sinus** in company with numerous small veins from the orbit.

The venous sinuses converge within the dura mater in the region of the lambdoidal suture. The left and right transverse sinuses converge at the base of the petrous portion of the temporal bone, forming the **sigmoid sinus,** which penetrates the jugular foramen and leaves the skull as the **internal jugular vein.** The internal jugular vein descends parallel to the common carotid artery in the neck (pp. 762–763).

Vertebral veins drain the cervical spinal cord and the posterior surface of the skull. These vessels descend within the transverse foramina of the cervical vertebrae, in company with the vertebral arteries. The vertebral veins

empty into the *brachiocephalic veins* of the chest (discussed later in the chapter).

Superficial Veins of the Head and Neck. Superficial veins of the head collect to form the **temporal, facial,** and **maxillary veins** (Figure 21-31●). The temporal and maxillary veins drain into the **external jugular vein.** The facial vein drains into the internal jugular vein. A broad anastomosis between the external and internal jugular veins at the angle of the mandible provides dual venous drainage of the face, scalp, and cranium. The external jugular vein descends toward the chest just beneath the skin on the anterior surface of the sternocleidomastoid muscle. Posterior to the clavicle, the external jugular empties into the *subclavian vein.* In healthy individuals, the external jugular vein is easily palpable, and a *jugular venous pulse (JVP)* can sometimes be seen at the base of the neck.

Venous Return from the Upper Limbs. The **digital veins** empty into the **superficial** and **deep palmar veins** of the hand, which are interconnected to form the **palmar venous arches.** The superficial arch empties into the **cephalic vein,** which ascends along the radial side of the forearm, the **median antebrachial vein,** and the **basilic vein,** which ascends on the ulnar side (Figure 21-32•). Anterior to the elbow is the superficial **median cubital vein.** This vein passes from the cephalic vein, medially and at an oblique angle, to connect to the basilic vein. (The median cubital is the vein from which venous blood samples are typically collected.) From the elbow, the basilic vein passes superiorly along the median surface of the biceps brachii.

The deep palmar veins drain into the **radial vein** and the **ulnar vein.** After crossing the elbow, these veins fuse to form the **brachial vein.** The brachial vein lies parallel to the brachial artery. As the brachial vein continues toward the trunk, it receives blood from the basilic vein before entering the axilla as the **axillary vein.**

Formation of the Superior Vena Cava. The cephalic vein joins the axillary vein on the lateral surface of the first rib, forming the **subclavian vein,** which continues into the chest. The subclavian vein passes superior to the first rib and along the superior margin of the clavicle. The vein then meets and merges with the external and internal jugular veins of that side. This fusion creates the

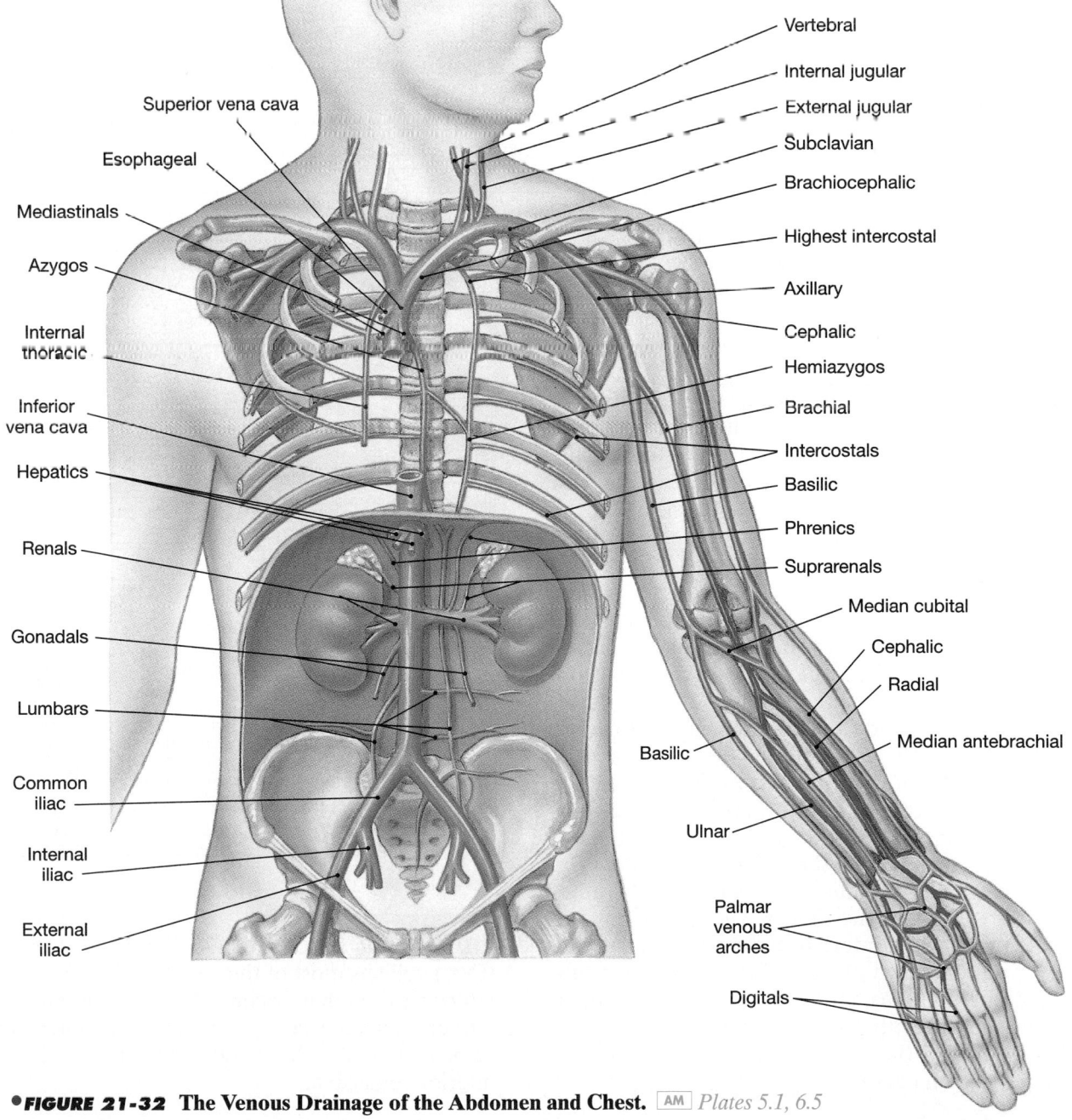

Vertebral
Internal jugular
External jugular
Subclavian
Brachiocephalic
Highest intercostal
Axillary
Cephalic
Hemiazygos
Brachial
Intercostals
Basilic
Phrenics
Suprarenals
Median cubital
Cephalic
Radial
Median antebrachial

Superior vena cava
Esophageal
Mediastinals
Azygos
Internal thoracic
Inferior vena cava
Hepatics
Renals
Gonadals
Lumbars
Common iliac
Internal iliac
External iliac

Basilic
Ulnar
Palmar venous arches
Digitals

•**FIGURE 21-32** **The Venous Drainage of the Abdomen and Chest.** [AM] *Plates 5.1, 6.5*

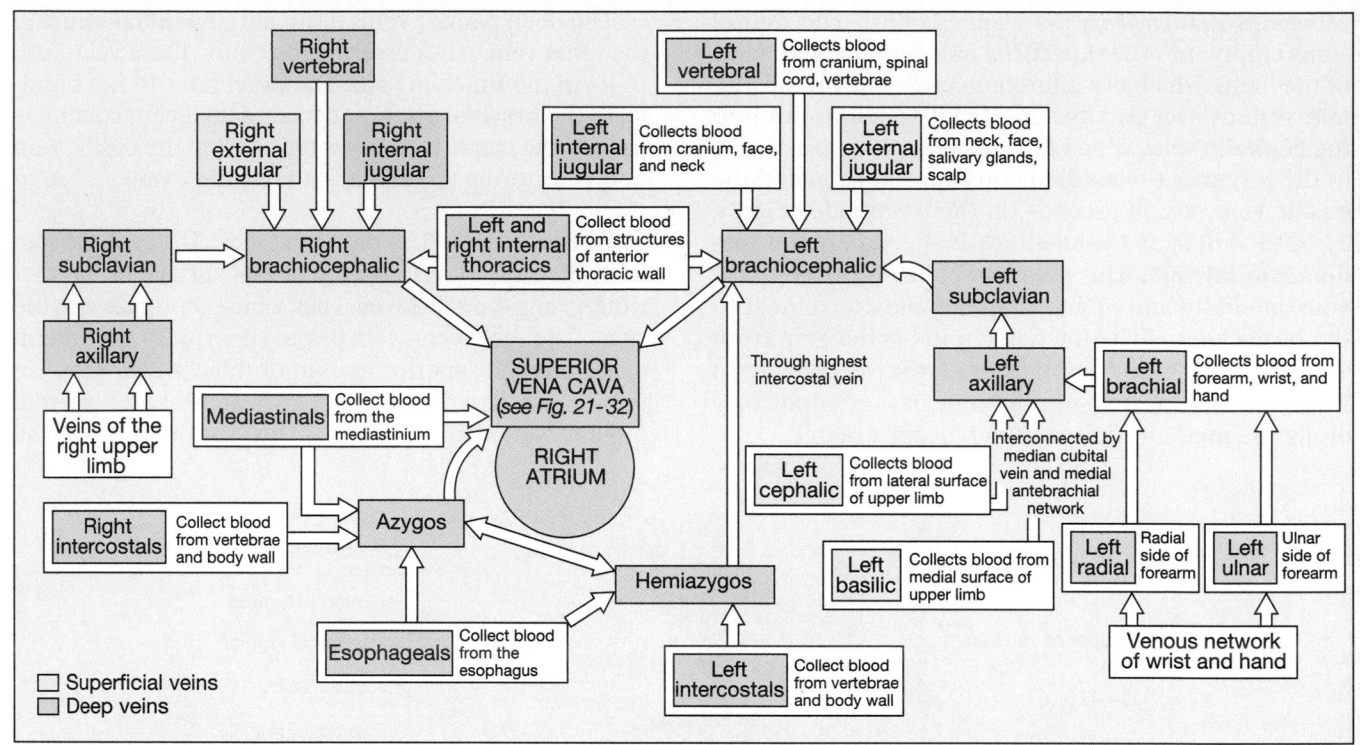

(a)

•*FIGURE 21-33* **Flowchart of Circulation to the Superior and Inferior Venae Cavae.** **(a)** Tributaries of the SVC. **(b)** Tributaries of the IVC.

brachiocephalic vein, or *innominate vein*, which penetrates the body wall and enters the thoracic cavity.

Each brachiocephalic vein receives blood from the **vertebral vein** of the same side, which drains the back of the skull and spinal cord. Near the heart, at the level of the first and second ribs, the left and right brachiocephalic veins combine, creating the superior vena cava. Close to the point of fusion, the **internal thoracic vein** empties into the brachiocephalic vein.

The **azygos** (AZ-i-gos) **vein** is the major tributary of the superior vena cava. This vessel ascends from the lumbar region over the right side of the vertebral column to invade the thoracic cavity through the diaphragm. The azygos joins the superior vena cava at the level of vertebra T_2. On the left side, the azygos receives blood from the smaller **hemiazygos vein,** which in many people also drains into the left brachiocephalic through the *highest intercostal vein.*

The azygos and hemiazygos veins are the chief collecting vessels of the thorax. They receive blood from (1) numerous **intercostal veins,** which receive blood from the chest muscles, (2) **esophageal veins,** which drain blood from the esophagus, and (3) smaller veins draining other mediastinal structures.

Figure 21-33• diagrams the venous tributaries of the superior vena cava.

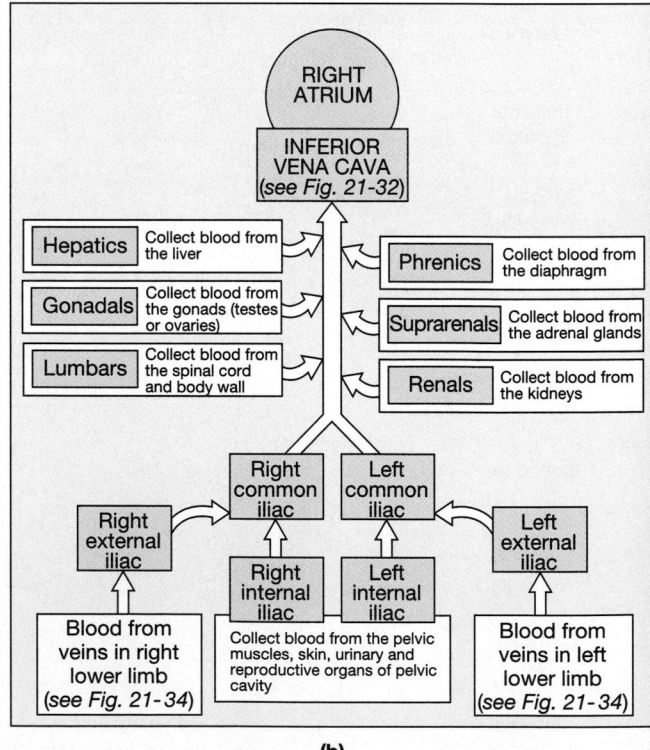

(b)

THE INFERIOR VENA CAVA The **inferior vena cava (IVC)** collects most of the venous blood from organs inferior to the diaphragm. (A small amount reaches the superior vena cava via the azygos and hemiazygos veins.) Figure 21-33b• diagrams the tributaries of the inferior vena cava.

Veins Draining the Lower Limb. Blood leaving capillaries in the sole of each foot collects into a network of **plantar veins,** which supply the **plantar venous arch.** The plantar network provides blood to the deep veins of the leg: the **anterior tibial vein,** the **posterior tibial vein,** and the **peroneal vein** (Figure 21-34a●). The **dorsal venous arch** collects blood from capillaries on the superior surface of the foot and the **digital veins** of the toes. There are extensive interconnections between the plantar arch and the dorsal arch, and the path of blood flow can easily shift from superficial to deep veins.

The dorsal venous arch is drained by two superficial veins, the **great saphenous** (sa-FĒ-nus) **vein** (*saphenes,* prominent) and the **small saphenous vein.** The great saphenous vein ascends along the medial aspect of the leg and thigh, draining into the *femoral vein* near the hip joint. The small saphenous vein arises from the dorsal venous arch and ascends along the posterior and lateral aspect of the calf. This vein then enters the popliteal fossa, where it meets the **popliteal vein,** formed by the union of the tibial and peroneal veins. The popliteal vein is easily palpated in the popliteal fossa adjacent to the adductor magnus muscle (Figure 21-34b●). Once it reaches the femur, the popliteal vein becomes the **femoral vein,** which ascends along the thigh, next to the femoral artery. Immediately before penetrating the abdominal wall, the femoral vein receives blood from the great saphenous vein and the **deep femoral vein,** which collects blood from the thigh. The femoral vein then penetrates the body wall and emerges in the pelvic cavity as the **external iliac vein.**

Veins Draining the Pelvis. The external iliac veins receive blood from the lower limbs, pelvis, and lower abdomen. As the left and right external iliacs travel across the inner surface of the ilium, they are joined by the **internal iliac veins,** which drain the pelvic organs. The union

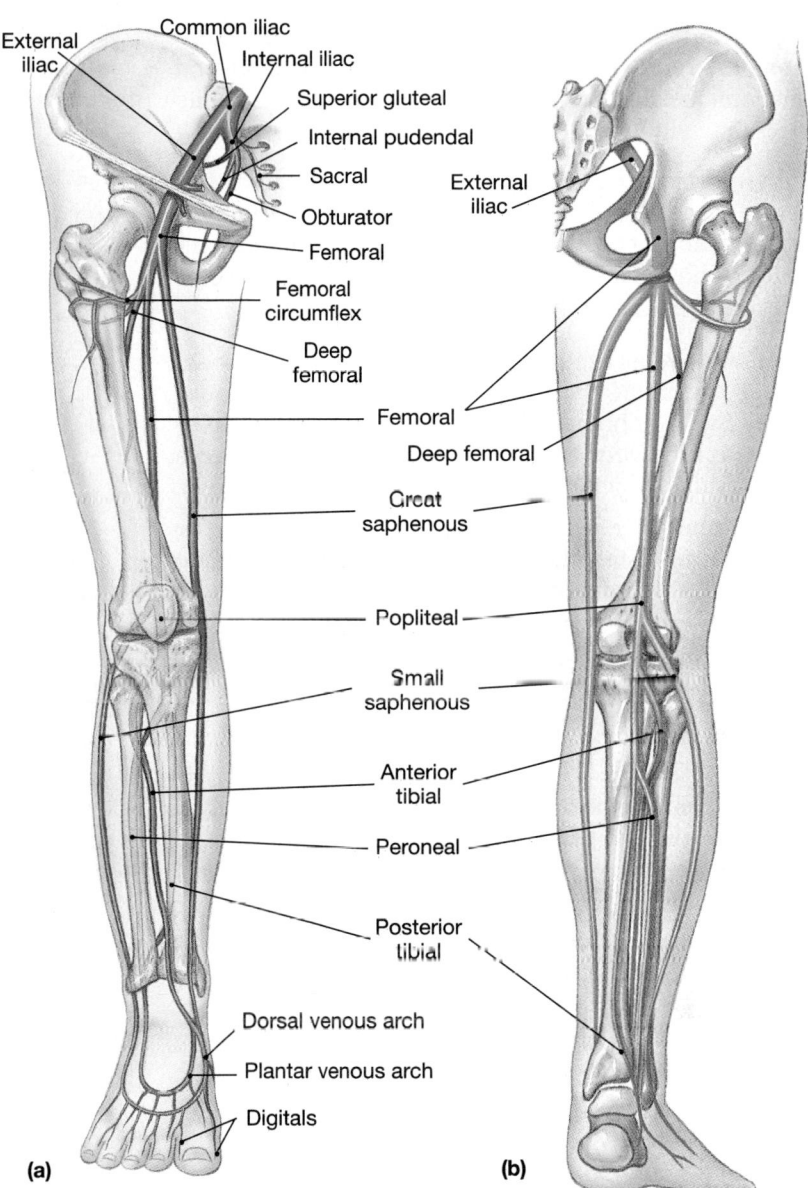

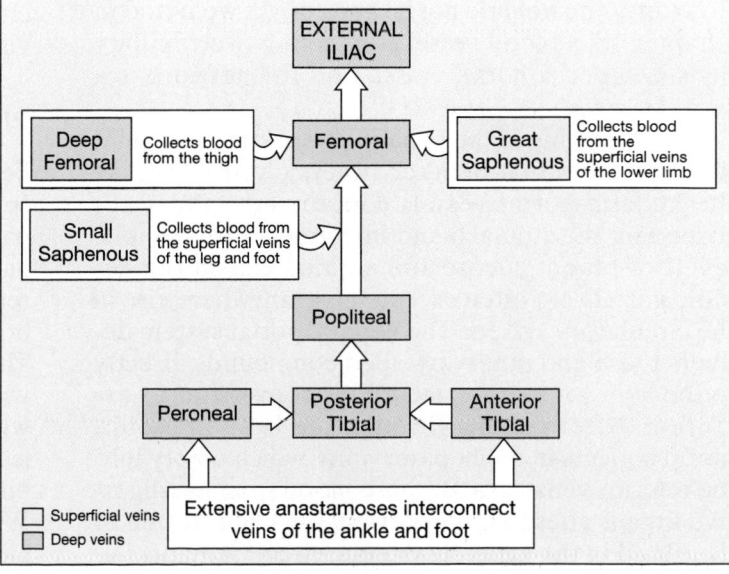

●**FIGURE 21-34** **Venous Drainage from the Lower Limb.** **(a)** Anterior view. **(b)** Posterior view. **(c)** Flowchart of venous circulation to a lower limb. AM *Plate 7.2; Scans 5, 6*

of external and internal iliac veins results in the **common iliac vein.** The right and left common iliacs ascend at an oblique angle, and anterior to vertebra L_5 they unite to form the inferior vena cava (Figure 21-32●).

Veins Draining the Abdomen. The inferior vena cava ascends posterior to the peritoneal cavity, parallel to the aorta. The abdominal portion of the inferior vena cava collects blood from six major veins (Figures 21-32 and 21-33●):

1. **Lumbar veins** drain the lumbar portion of the abdomen, including the spinal cord and body wall muscles. Superior branches of these veins are connected to the azygos vein (right side) and hemiazygos vein (left side), which empty into the superior vena cava.

2. **Gonadal** *(ovarian* or *testicular)* **veins** drain the ovaries or testes. The right gonadal vein empties into the inferior vena cava; the left gonadal generally drains into the left renal vein.

3. **Hepatic veins** leave the liver and empty into the inferior vena cava at the level of vertebra T_{10}.

4. **Renal veins** collect blood from the kidneys. These are the largest tributaries of the inferior vena cava.

5. **Suprarenal veins** drain the adrenal glands. Generally, only the right suprarenal vein drains into the inferior vena cava, and the left drains into the left renal vein.

6. **Phrenic veins** drain the diaphragm. Only the right phrenic vein drains into the inferior vena cava; the left drains into the left renal vein.

THE HEPATIC PORTAL SYSTEM The liver is the only digestive organ drained by the inferior vena cava. Instead of traveling directly to the inferior vena cava, blood leaving the capillaries supplied by the celiac, superior, and inferior mesenteric arteries flows into the **hepatic portal system.** As we noted in Chapter 18, a blood vessel connecting two capillary beds is called a *portal vessel,* and the network is a *portal system.* ∞ *[p. 602]*

Blood flowing in the hepatic portal system is quite different from that in other systemic veins, because the hepatic portal vessels contain substances absorbed by the stomach and intestines. For example, levels of blood glucose and amino acids in the hepatic portal vein often exceed those anywhere else in the circulatory system. The hepatic portal system delivers these and other absorbed compounds directly to the liver for storage, metabolic conversion, or excretion. After passing through the liver sinusoids, blood collects in the hepatic veins, which empty into the inferior vena cava. Because blood from the digestive organs goes to the liver first, the composition of the blood in the general systemic circuit remains relatively stable, regardless of any digestive activities.

The hepatic portal system begins in the capillaries of the digestive organs and ends as the hepatic portal vein discharges blood into sinusoids in the liver. The tributaries of the hepatic portal vein (Figure 21-35●) include the following:

- The **inferior mesenteric vein,** which collects blood from capillaries along the lower portion of the large intestine. Its tributaries include the *left colic vein* and the *superior rectal veins,* which drain the descending colon, sigmoid colon, and rectum.
- The **splenic vein,** formed by the union of the inferior mesenteric vein and veins from the spleen, the lateral border of the stomach *(left gastroepiploic),* and the pancreas *(pancreatic).*
- The **superior mesenteric vein,** which collects blood from veins draining the stomach *(right gastroepiploic),* the small intestine *(intestinal),* and two-thirds of the large intestine *(ileocolic, right colic,* and *middle colic).*

The hepatic portal vein forms through the fusion of the superior mesenteric and splenic veins. Of the two, the superior mesenteric normally contributes the greater volume of blood and most of the nutrients. As it proceeds toward the liver, the hepatic portal receives blood from the **gastric veins,** which drain the medial border of the stomach, and the **cystic vein** from the gallbladder.

Fetal Circulation

There are significant differences between the fetal and adult circulatory systems that reflect differing sources of respiratory and nutritional support. The embryonic lungs are collapsed and nonfunctional, and the digestive tract has nothing to digest. All the embryonic nutritional and respiratory needs are provided by diffusion across the placenta.

Placental Blood Supply

Fetal circulation is diagrammed in Figure 21-36● (p. 758). Blood flow to the placenta is provided by a pair of **umbilical arteries,** which arise from the internal iliac arteries and enter the umbilical cord. Blood returns from the placenta in the single **umbilical vein,** bringing oxygen and nutrients to the developing fetus. The umbilical vein drains into the **ductus venosus,** which is connected to an intricate network of veins within the developing liver. The ductus venosus collects blood from the veins of the liver and from the umbilical vein and empties into the inferior vena cava. When the placental connection is broken at birth, blood flow ceases along the umbilical vessels, and they soon degenerate.

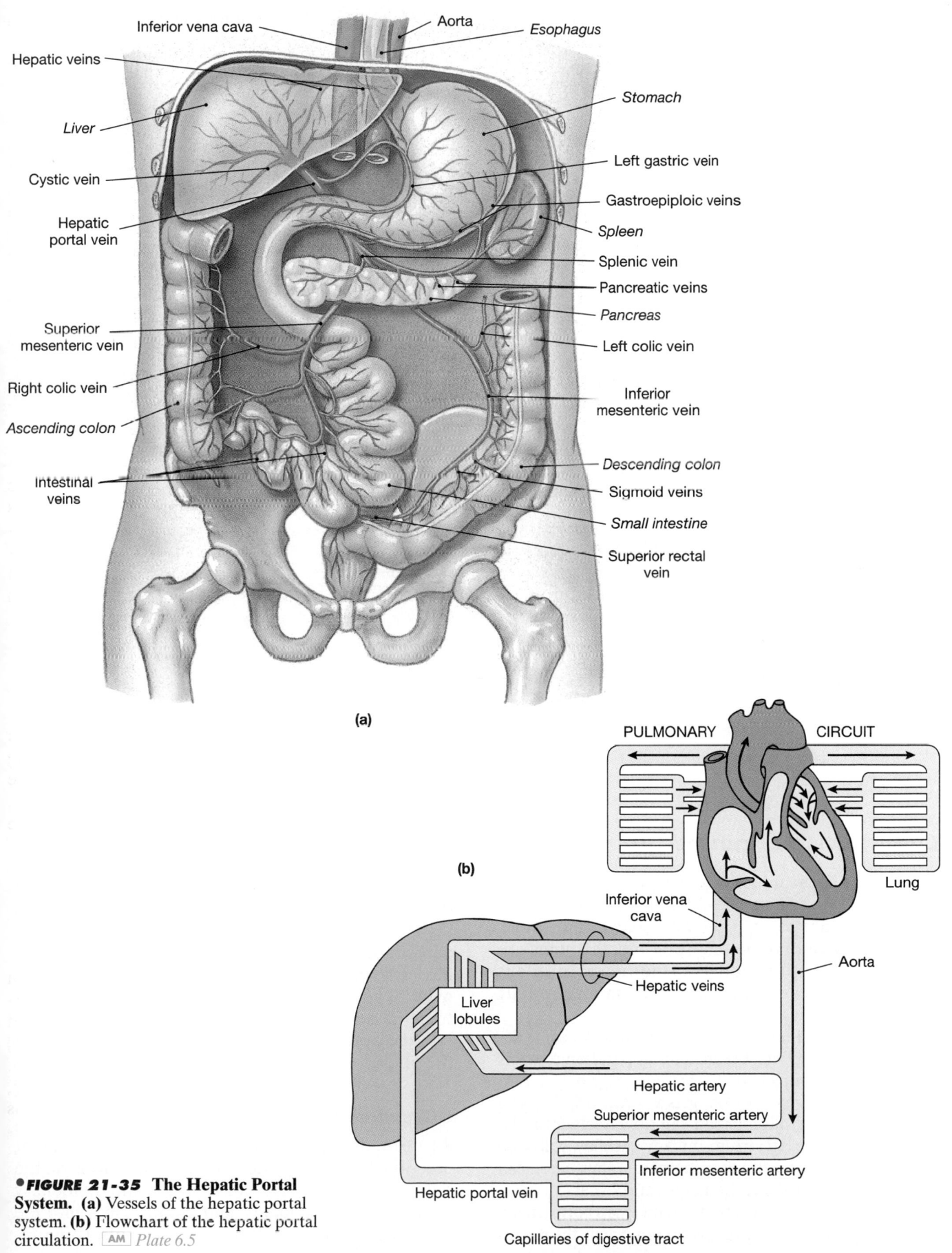

•FIGURE 21-35 **The Hepatic Portal System.** (a) Vessels of the hepatic portal system. (b) Flowchart of the hepatic portal circulation. AM *Plate 6.5*

•**FIGURE 21-36** Fetal Circulation. (a) Blood flow to and from the placenta. (b) Blood flow through the neonatal (newborn) heart.

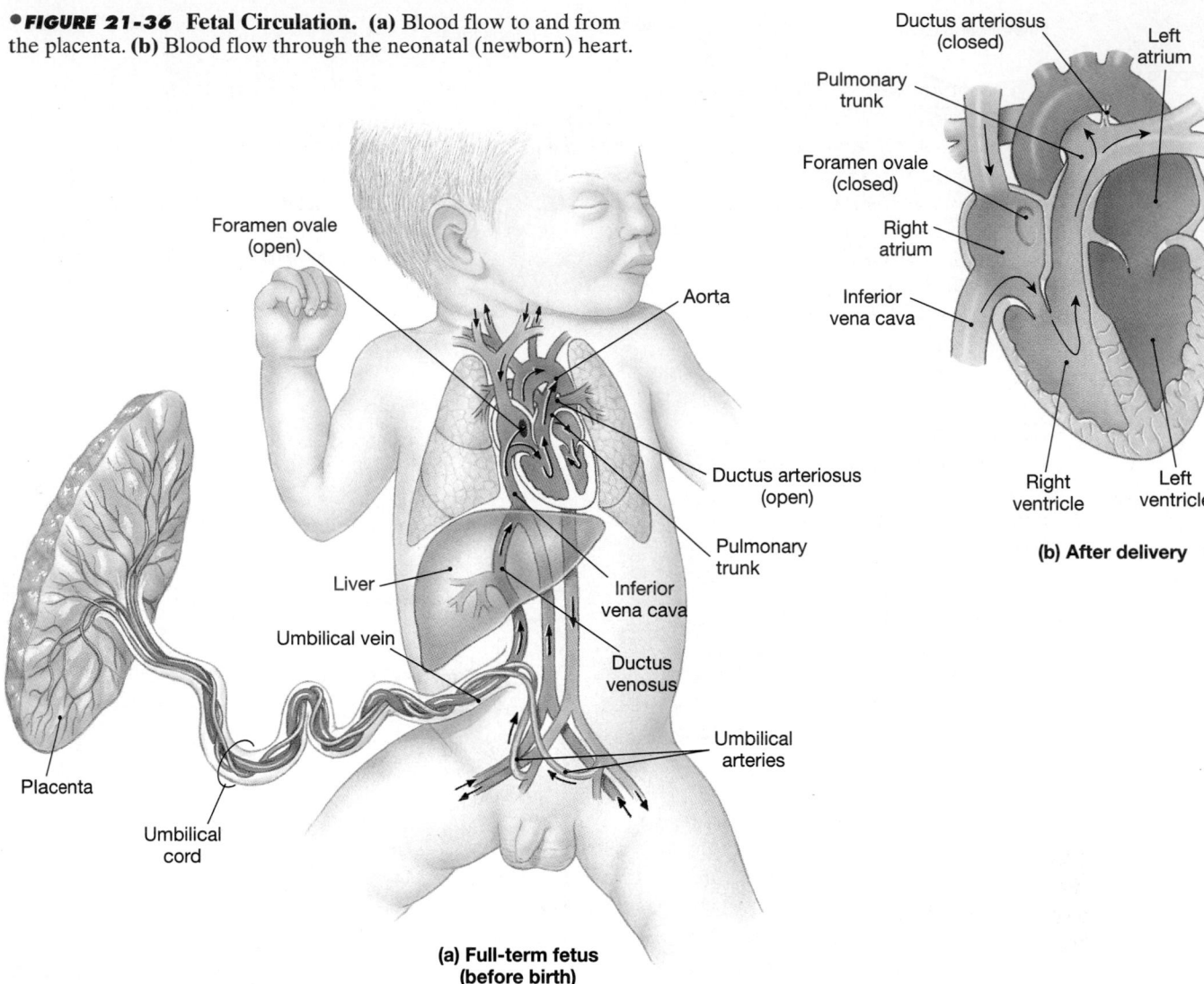

(a) Full-term fetus (before birth)

(b) After delivery

Circulation in the Heart and Great Vessels

One of the most interesting aspects of circulatory development reflects the differences between the life of an embryo or fetus and that of an infant. Throughout embryonic and fetal life, the lungs are collapsed; yet, after delivery, the newborn infant must be able to extract oxygen from inspired air rather than across the placenta.

Although the interatrial and interventricular septa develop early in fetal life, the interatrial partition remains functionally incomplete up to the time of birth. The *foramen ovale,* or *interatrial opening,* is associated with a long flap that acts as a valve. Blood can flow freely from the right atrium to the left atrium, but any backflow will close the valve and isolate the two chambers. Thus blood can enter the heart at the right atrium and bypass the pulmonary circuit. A second short-circuit exists between the pulmonary and aortic trunks. This connection, the **ductus arteriosus,** consists of a short, muscular vessel.

With the lungs collapsed, the capillaries are compressed, and little blood flows through the lungs. During diastole, blood enters the right atrium and flows into the right ventricle, but it also passes into the left atrium through the foramen ovale. About 25 percent of the blood arriving at the right atrium bypasses the pulmonary circuit in this way. In addition, over 90 percent of the blood leaving the right ventricle passes through the ductus arteriosus and enters the systemic circuit rather than continuing to the lungs.

Circulatory Changes at Birth

At birth, dramatic changes occur. When the infant takes the first breath, the lungs expand, and so do the pulmonary vessels. The resistance in the pulmonary circuit declines suddenly, and blood rushes into the pulmonary vessels. Within a few seconds, rising O_2 levels stimulate constriction of the ductus arteriosus,

 Congential Circulatory Problems

Minor individual variations in the circulatory network are quite common. For example, very few individuals have identical patterns of venous distribution. Congenital circulatory problems serious enough to represent a threat to homeostasis are relatively rare. They generally reflect abnormal formation of the heart or problems with the interconnections between the heart and the great vessels. Several examples of congenital circulatory defects are illustrated in Figure 21-37●. All these conditions can be surgically corrected, although multiple surgeries may be required.

The incomplete closure of the foramen ovale or ductus arteriosus (Figure 21-37a●) results in similar types of problems. If the foramen ovale remains open, or *patent*, blood recirculates into the pulmonary circuit without entering the left ventricle. The movement, driven by the relatively high systemic pressure, is called a "left-to-right shunt." Arterial oxyen content is normal, but the left ventricle must work much harder than usual to provide adequate blood flow through the systemic circuit, and the pressures rise in the pulmonary circuit. Although the abnormality may not be immediately apparent, pulmonary hypertension,

pulmonary edema, and cardiac enlargement are the eventual results. If the ductus arteriosus remains open, the same basic problems develop as blood ejected by the left ventricle reenters the pulmonary circuit. If, as happens in some cases, valve defects, constricted pulmonary vessels, or other abnormalities occur as well, pulmonary pressures can rise enough to force blood into the systemic circuit through the ductus arteriosus. This movement is called a "right-to-left shunt." Because normal blood oxygenation does not occur, the circulating blood develops a deep red color. The skin then develops the blue tones typical of *cyanosis,* a condition we noted in Chapter 5, and the infant is known as a "blue baby."

Ventricular septal defects (Figure 21-37b●) are the most common congenital heart problems, affecting 0.12 percent of newborn infants. The opening between the left and right ventricles has the reverse effect of a connection between the atria: When the more powerful left ventricle beats, it ejects blood into the right ventricle and pulmonary circuit. The end results are the same as for a patent foramen ovale—a left-to-right shunt, with eventual pulmonary hypertension, pulmonary edema, and cardiac enlargement.

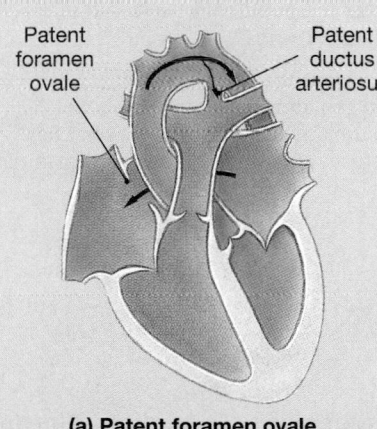

(a) Patent foramen ovale and ductus arteriosus

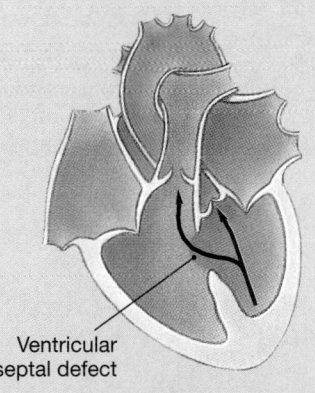

(b) Ventricular septal defect

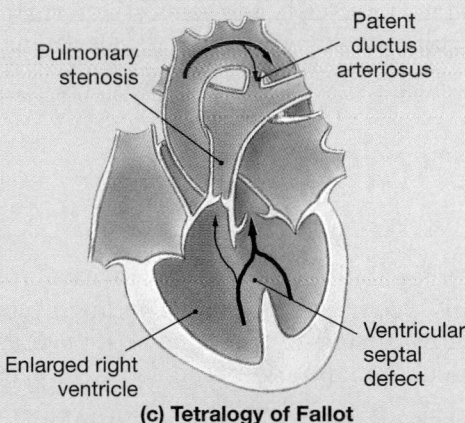

(c) Tetralogy of Fallot

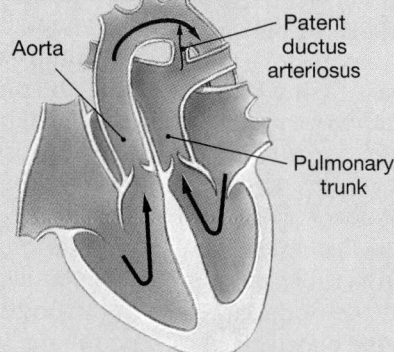

(d) Transposition of great vessels

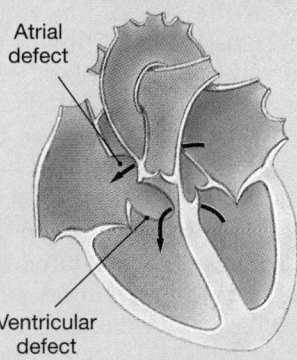

(e) Atrioventricular septal defect

●*FIGURE 21-37* Congenital Circulatory Problems

The *tetralogy of Fallot* (fa-LŌ) (Figure 21-37c●) is a complex group of heart and circulatory defects that affect 0.10 percent of newborn infants. In this condition, (1) the pulmonary trunk is abnormally narrow, (2) the interventricular septum is incomplete, (3) the aorta originates where the interventricular septum normally ends, and (4) the right ventricle is enlarged and both ventricles are thickened due to increased workloads.

In the *transposition of great vessels,* the aorta is connected to the right ventricle and the pulmonary artery is connected to the left ventricle. (Figure 21-37d●), This malformation affects 0.05 percent of newborn infants.

In an *atrioventricular septal defect*, the atria and ventricles are incompletely separated (Figure 21-37e●). The results are quite variable, depending on the extent of the defect and the effects on the atrioventricular valves. This type of defect most commonly affects infants with *Down syndrome*, a disorder caused by the presence of an extra copy of chromosome 21.

isolating the pulmonary and aortic trunks. As pressures rise in the left atrium, the valvular flap closes the foramen ovale and completes the circulatory remodeling. In the adult, the interatrial septum bears a shallow depression, the *fossa ovalis,* that marks the site of the foramen ovale (see Figure 20-4●, p. 677). The remnants of the ductus arteriosus persist as a fibrous cord, the *ligamentum arteriosum.*

If the proper circulatory changes do not occur at birth or shortly thereafter, problems will eventually develop. The severity of the problem depends on which connection remains open and on the size of the opening. Treatment may involve surgical closure of the foramen ovale, the ductus arteriosus, or both. Other forms of congenital heart defects result from abnormal cardiac development or inappropriate connections between the heart and major arteries and veins. (For further discussion of circulatory changes during development, see the Embryology Summary on pp. 762–763.)

⊠ AGING AND THE CARDIOVASCULAR SYSTEM

The capabilities of your cardiovascular system gradually decline as you age. The major changes are as follows, listed in the same sequence as the cardiovascular chapters—blood, heart, and vessels.

- *Age-related changes in the blood.* Age-related changes in the blood may include (1) decreased hematocrit; (2) constriction or blockage of peripheral veins by a *thrombus* (stationary blood clot); the thrombus can become detached, pass through the heart, and become wedged in a small artery, commonly in the lungs, causing a *pulmonary embolism;* and (3) pooling of blood in the veins of the legs because valves are not working effectively.

- *The aging heart.* Age-related changes in the heart include (1) a reduction in the maximum cardiac output; (2) changes in the activities of the nodal and conducting cells; (3) a reduction in the elasticity of the fibrous skeleton; (4) a progressive atherosclerosis

that can restrict coronary circulation; and (5) replacement of damaged cardiac muscle cells by scar tissue.

- *Aging and blood vessels.* Age-related changes in blood vessels are typically related to arteriosclerosis. For example, (1) the inelastic walls of arteries become less tolerant of sudden pressure increases, which may lead to an *aneurysm,* whose subsequent rupture may cause a stroke, myocardial infarction, or massive blood loss (depending on the vessel involved); (2) calcium salts can be deposited on weakened vascular walls, increasing the risk of a stroke or myocardial infarction; and (3) thrombi can form at atherosclerotic plaques.

✔ Whenever Tim gets angry, a large vein bulges in the lateral region of his neck. Which vein is this?

✔ A thrombus that blocks the popliteal vein would interfere with blood flow in which other veins?

✔ A blood sample taken from the umbilical cord shows a high concentration of oxygen and nutrients and a low concentration of carbon dioxide and waste products. Did this blood sample come from an umbilical artery or from the umbilical vein? Explain.

INTEGRATION WITH OTHER SYSTEMS

The cardiovascular system is both anatomically and functionally linked to all other systems. In the section on vessel distribution, we demonstrated the extent of the anatomical connections. Figure 21-38● summarizes the physiological relationships between the cardiovascular system and other organ systems.

The most extensive communication occurs between the cardiovascular and lymphatic systems. Not only are the two systems physically interconnected, but cell populations of the lymphatic system use the cardiovascular system as a highway to move from one part of the body to another. In Chapter 22, we examine the lymphatic system in detail and consider its role in the immune response.

INTEGUMENTARY SYSTEM

Mast cell stimulation produces localized changes in blood flow and capillary permeability

Delivers immune system cells to injury sites; clotting response seals breaks in skin surface; carries away toxins from sites of infection; provides heat

SKELETAL SYSTEM

Provides calcium needed for normal cardiac muscle contraction; protects blood cells developing in bone marrow

Provides Ca^{2+} and PO_4^{3-} for bone deposition; delivers EPO to bone marrow, parathyroid hormone and calcitonin to osteoblasts and osteoclasts

MUSCULAR SYSTEM

Skeletal muscle contractions assist in moving blood through veins; protects superficial blood vessels, especially in neck and limbs

Delivers oxygen and nutrients, removes carbon dioxide, lactic acid, and heat during skeletal muscle activity

THE CARDIOVASCULAR SYSTEM

NERVOUS SYSTEM

Controls patterns of circulation in peripheral tissues; modifies heart rate and regulates blood pressure; releases ADH

Endothelial cells maintain blood–brain barrier, help generate CSF

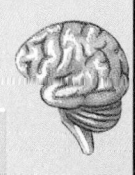

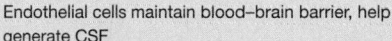

ENDOCRINE SYSTEM

Erythropoietin regulates production of RBCs; several hormones elevate blood pressure; epinephrine stimulates cardiac muscle, elevating heart rate and contractile force

Distributes hormones throughout the body; heart secretes ANP

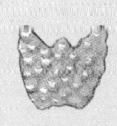

LYMPHATIC SYSTEM

Defends against pathogens or toxins in blood; fights infections of cardiovascular organs; returns tissue fluid to circulation

Distributes WBCs; carries antibodies that attack pathogens; clotting response assists in restricting spread of pathogens; granulocytes and lymphocytes produced in bone marrow

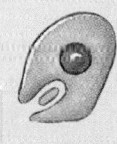

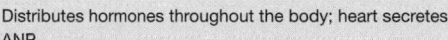

FOR ALL SYSTEMS
Delivers oxygen, hormones, nutrients, and WBCs; removes carbon dioxide and metabolic wastes; transfers heat

RESPIRATORY SYSTEM

Provides oxygen to cardiovascular organs and removes carbon dioxide; enzyme in lung capillaries converts angiotensin I to angiotensin II

RBCs transport oxygen and carbon dioxide between lungs and peripheral tissues

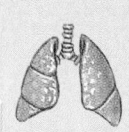

DIGESTIVE SYSTEM

Provides nutrients to cardiovascular organs; absorbs water and ions essential to maintenance of normal blood volume

Distributes digestive tract hormones; carries nutrients, water, and ions away from sites of absorption; delivers nutrients and toxins to liver

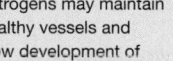

Estrogens may maintain healthy vessels and slow development of atherosclerosis with age

Distributes reproductive hormones; provides nutrients, oxygen, and waste removal for developing fetus; local blood pressure changes responsible for physical changes during sexual arousal

REPRODUCTIVE SYSTEM

URINARY SYSTEM

Releases renin to elevate blood pressure and erythropoietin to accelerate red blood cell production

Delivers blood to capillaries, where filtration occurs; accepts fluids and solutes reabsorbed during urine production

•**FIGURE 21-38** Functional Relationships between the Cardiovascular System and Other Systems

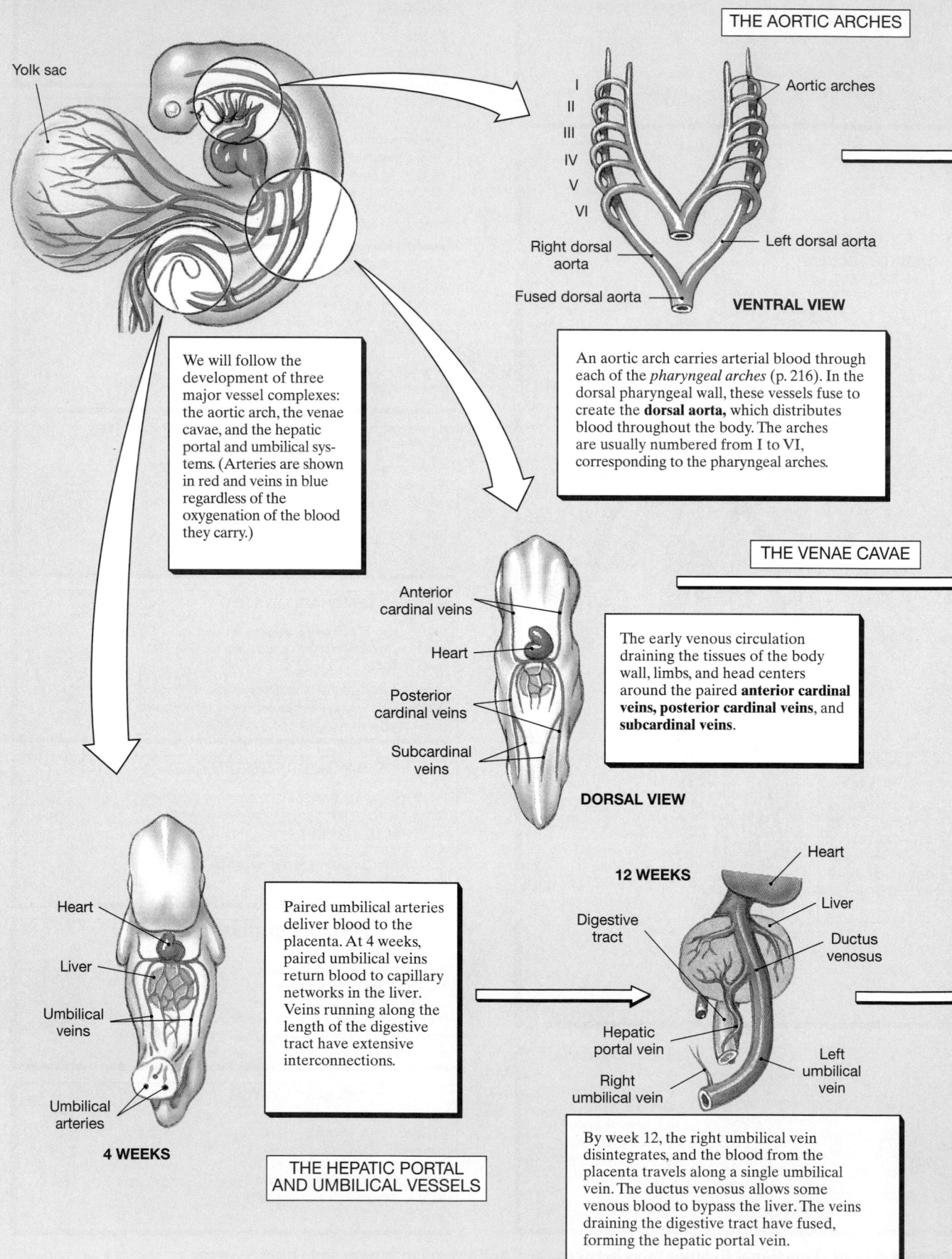

Yolk sac

THE AORTIC ARCHES

I
II
III
IV
V
VI

Aortic arches

Right dorsal aorta

Left dorsal aorta

Fused dorsal aorta

VENTRAL VIEW

We will follow the development of three major vessel complexes: the aortic arch, the venae cavae, and the hepatic portal and umbilical systems. (Arteries are shown in red and veins in blue regardless of the oxygenation of the blood they carry.)

An aortic arch carries arterial blood through each of the *pharyngeal arches* (p. 216). In the dorsal pharyngeal wall, these vessels fuse to create the **dorsal aorta,** which distributes blood throughout the body. The arches are usually numbered from I to VI, corresponding to the pharyngeal arches.

THE VENAE CAVAE

Anterior cardinal veins

Heart

Posterior cardinal veins

Subcardinal veins

The early venous circulation draining the tissues of the body wall, limbs, and head centers around the paired **anterior cardinal veins, posterior cardinal veins**, and **subcardinal veins**.

DORSAL VIEW

Heart

Liver

Umbilical veins

Umbilical arteries

4 WEEKS

Paired umbilical arteries deliver blood to the placenta. At 4 weeks, paired umbilical veins return blood to capillary networks in the liver. Veins running along the length of the digestive tract have extensive interconnections.

12 WEEKS

Heart

Liver

Ductus venosus

Digestive tract

Hepatic portal vein

Right umbilical vein

Left umbilical vein

By week 12, the right umbilical vein disintegrates, and the blood from the placenta travels along a single umbilical vein. The ductus venosus allows some venous blood to bypass the liver. The veins draining the digestive tract have fused, forming the hepatic portal vein.

THE HEPATIC PORTAL AND UMBILICAL VESSELS

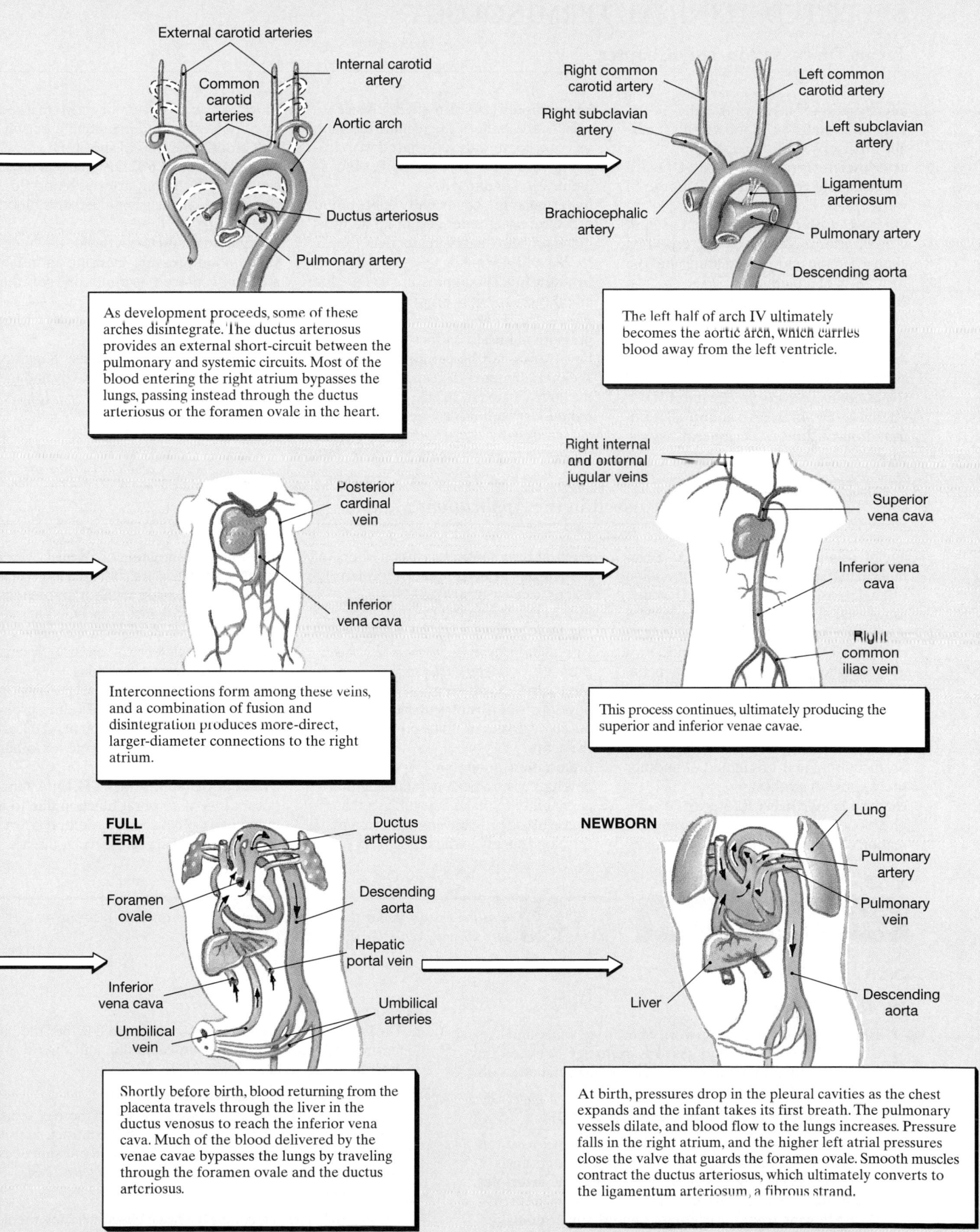

External carotid arteries

Common carotid arteries

Internal carotid artery

Aortic arch

Ductus arteriosus

Pulmonary artery

As development proceeds, some of these arches disintegrate. The ductus arteriosus provides an external short-circuit between the pulmonary and systemic circuits. Most of the blood entering the right atrium bypasses the lungs, passing instead through the ductus arteriosus or the foramen ovale in the heart.

Right common carotid artery

Left common carotid artery

Right subclavian artery

Left subclavian artery

Ligamentum arteriosum

Brachiocephalic artery

Pulmonary artery

Descending aorta

The left half of arch IV ultimately becomes the aortic arch, which carries blood away from the left ventricle.

Posterior cardinal vein

Inferior vena cava

Interconnections form among these veins, and a combination of fusion and disintegration produces more-direct, larger-diameter connections to the right atrium.

Right internal and external jugular veins

Superior vena cava

Inferior vena cava

Right common iliac vein

This process continues, ultimately producing the superior and inferior venae cavae.

FULL TERM

Ductus arteriosus

Foramen ovale

Descending aorta

Hepatic portal vein

Inferior vena cava

Umbilical vein

Umbilical arteries

Shortly before birth, blood returning from the placenta travels through the liver in the ductus venosus to reach the inferior vena cava. Much of the blood delivered by the venae cavae bypasses the lungs by traveling through the foramen ovale and the ductus arteriosus.

NEWBORN

Lung

Pulmonary artery

Pulmonary vein

Liver

Descending aorta

At birth, pressures drop in the pleural cavities as the chest expands and the infant takes its first breath. The pulmonary vessels dilate, and blood flow to the lungs increases. Pressure falls in the right atrium, and the higher left atrial pressures close the valve that guards the foramen ovale. Smooth muscles contract the ductus arteriosus, which ultimately converts to the ligamentum arteriosum, a fibrous strand.

SELECTED CLINICAL TERMINOLOGY

Terms Discussed in This Chapter

aneurysm (AN-ū-rizm): A bulge in the weakened wall of a blood vessel, generally an artery. *(p. 714 and AM)*

arteriosclerosis (ar-tē-rē-ō-skle-RŌ-sis): A thickening and toughening of arterial walls. *(p. 713)*

atherosclerosis (ath-er-ō-skle-RŌ-sis): A type of arteriosclerosis characterized by changes in the endothelial lining and the formation of a plaque. *(p. 713)*

congestive heart failure: A condition that develops when the left ventricle can no longer keep pace with the right ventricle and blood backs up into the pulmonary circuit. Elevated pulmonary pressures lead to *pulmonary edema. (p. 733 and AM)*

edema (e-DĒ-muh): An abnormal accumulation of fluid in peripheral tissues. *(p. 726)*

hemorrhoids (HEM-o-roydz): Varicose veins in the walls of the rectum, the anus, or both; commonly associated with frequent straining to force bowel movements. *(p. 718 and AM)*

hypertension: Abnormally high blood pressure; usually defined in an adult as blood pressure that is greater than 150/90. *(p. 723 and AM)*

hypotension: Blood pressure so low that circulation to vital organs may be impaired. *(p. 723 and AM)*

pressure points: Locations where muscular arteries can be compressed against skeletal elements to restrict or prevent the flow of blood in an emergency. *(p. 723)*

pulmonary embolism: Circulatory blockage caused by the trapping of a freed thrombus in a pulmonary artery. *(p. 760)*

shock: An acute circulatory crisis marked by hypotension and inadequate peripheral blood flow. *(p. 737 and AM)*

sounds of Korotkoff: Distinctive sounds caused by turbulent arterial blood flow heard while measuring a person's blood pressure. *(p. 723)*

sphygmomanometer: A device that measures blood pressure by using an inflatable cuff placed around one of the extremities. *(p. 723)*

thrombus: A stationary blood clot within a blood vessel. *(p. 760)*

varicose (VAR-i-kōs) **veins:** Sagging, swollen veins distorted by gravity and the failure of the venous valves. *(p. 718)*

AM Additional Terms Discussed in the *Applications Manual*

angiotensin-converting enzyme (ACE) inhibitors: Drugs that block the conversion of angiotensin I to angiotensin II; sometimes used in the treatment of chronic hypertension and congestive heart failure.

cardiogenic (kar-dē-ō-JEN-ik) **shock:** Progressive shock that develops when the heart becomes unable to maintain normal cardiac output.

cerebral embolism: Occlusion of one of the cerebral arteries by a blood clot, fatty mass, or air bubble that originated at another site; a cause of strokes.

cerebral hemorrhage: Rupture of a cerebral blood vessel, commonly following formation of an aneurysm; a cause of strokes.

cerebral thrombosis: Occlusion of one of the cerebral arteries by a clot formed at a plaque; a cause of strokes.

distributive shock: Progressive shock that results from a widespread, uncontrolled vasodilation, as in *neurogenic shock, septic shock,* or *anaphylactic shock.*

obstructive shock: Progressive shock that develops when ventricular output is reduced by tissues or fluids pressing against the heart.

orthostatic hypotension: Hypotension that develops when a person is standing, due to an inability to make adjustments that increase blood pressure enough to maintain the blood supply to the brain.

phlebitis: Inflammation of a vein.

primary hypertension (essential hypertension): Hypertension without an obvious cause. Known risk factors include a hereditary history of hypertension, male gender, high plasma cholesterol, obesity, chronic stresses, and cigarette smoking.

secondary hypertension: Hypertension that develops as the result of kidney problems or abnormal production of ADH, aldosterone, renin, epinephrine, or other hormones.

transient ischemic attack (TIA): A temporary loss of cerebral function due to a momentary blockage in a cerebral artery; an indication of cerebrovascular disease.

CHAPTER REVIEW

On-line resources for this chapter are on our World Wide Web site at:
http://www.prenhall.com/martini/fap

STUDY OUTLINE

INTRODUCTION, p. 709

1. Blood flows through a network of arteries, veins, and capillaries. All chemical and gaseous exchange between the blood and interstitial fluid takes place across capillary walls. *(Figure 21-1)*

THE ANATOMY OF BLOOD VESSELS, p. 709

1. **Arteries** and **veins** form an internal distribution system through which the heart propels blood. Arteries branch repeatedly, decreasing in size until they become **arterioles.** From the arterioles, blood enters the **capillary** networks. Blood flowing from the capillaries enters small **venules** before entering larger veins.

The Structure of Vessel Walls, p. 710

2. The walls of arteries and veins contain three layers: the innermost **tunica interna,** the **tunica media,** and the outermost **tunica externa.** *(Figure 21-2)*

Differences between Arteries and Veins, p. 711

3. In general, the walls of arteries are thicker than those of veins. The endothelial lining of an artery cannot contract, so it is thrown into folds. Arteries constrict when blood pressure does not distend them; veins constrict very little. *(Figure 21-2)*

Arteries, p. 711

4. The arterial system includes the large **elastic arteries,** medium-sized **muscular arteries,** and smaller arterioles. As we

proceed toward the capillaries, the number of vessels increases, but the diameter of the individual vessels decreases and the walls become thinner. *(Figure 21-3)*

5. **Atherosclerosis,** a type of **arteriosclerosis,** is associated with changes in the endothelial lining of arteries. Fatty masses of tissue called **plaques** typically develop during atherosclerosis. *(Figure 21-4)*

Capillaries, p. 714

6. Capillaries are the only blood vessels whose walls permit exchange between blood and interstitial fluid. Capillaries may be **continuous** or **fenestrated. Sinusoids** have fenestrated walls and form elaborate networks that allow very slow blood flow. Sinusoids are located in the liver and in various endocrine organs. *(Figure 21-5)*

7. Capillaries form interconnected networks called **capillary plexuses (capillary beds).** A **precapillary sphincter** (a band of smooth muscle) adjusts the blood flow into each capillary. Blood flow within a capillary changes as **vasomotion** occurs. The entire capillary plexus may be bypassed by blood flow through **arteriovenous anastomoses** or through **preferred channels** within the capillary plexus. *(Figure 21-6)*

Veins, p. 717

8. Venules collect blood from the capillaries and merge into **medium-sized veins** and then **large veins.** The arterial system is a high-pressure system; blood pressure in veins is much lower. **Valves** in these vessels prevent the backflow of blood. *(Figures 21-2, 21-3, 21-7)*

The Distribution of Blood, p. 718

9. Peripheral **venoconstriction** helps maintain adequate blood volume in the arterial system after a hemorrhage. The **venous reserve** normally accounts for about 20 percent of the total blood volume. *(Figure 21-8)*

CARDIOVASCULAR PHYSIOLOGY, p. 719

Pressure, p. 719

1. Flow is proportional to pressure difference; blood will flow from an area of higher pressure to one of lower pressure. The **circulatory pressure** is the pressure gradient across the systemic circuit.

Resistance, p. 719

2. The resistance (R) determines the rate of blood flow through the systemic circuit. The major determinant of blood flow rate is the **peripheral resistance (PR),** the resistance of the arterial system. Neural and hormonal control mechanisms regulate blood pressure and peripheral resistance.

3. The most important determinant of peripheral resistance is the diameter of arterioles.

An Overview of Circulatory Pressures, p. 720

4. The high arterial pressures overcome peripheral resistance and maintain blood flow through peripheral tissues. Capillary pressures are normally low; small changes in capillary pressure determine the rate of fluid movement into or out of the bloodstream. Venous pressure, normally low, determines venous return and affects cardiac output and peripheral blood flow. *(Figure 21-9; Table 21-1)*

Arterial Blood Pressure, p. 722

5. Arterial blood pressure rises during ventricular systole and falls during ventricular diastole. The difference between

these two blood pressures is the **pulse pressure.** *(Figures 21-10, 21-11)*

Capillary Exchange, p. 723

6. At the capillaries, solute molecules diffuse across the capillary lining, and water-soluble materials diffuse through small spaces between endothelial cells. Water will move when driven by either **hydrostatic** or **osmotic pressure.** The direction of water movement is determined by the balance between these two opposing pressures. Any change in hydrostatic or osmotic pressure in the blood or tissues will shift the point where the two forces are equal and determine whether there is a net loss or a net gain of fluid along the length of the capillary. *(Figures 21-12, 21-13)*

Venous Pressure and Venous Return, p. 727

7. Valves, muscular compression, and the **respiratory pump** *(thoracoabdominal pump)* help the relatively low venous pressures propel blood toward the heart. *(Figure 21-7)*

CARDIOVASCULAR REGULATION, p. 727

1. Homeostatic mechanisms ensure that **tissue perfusion** (blood flow) delivers adequate oxygen and nutrients.

2. Local, autonomic, and endocrine factors influence the coordinated regulation of cardiovascular function. Local factors change the pattern of blood flow within capillary beds in response to chemical changes in interstitial fluids. Central mechanisms respond to changes in arterial pressure or blood gas levels. Hormones can assist in short-term adjustments (changes in cardiac output and peripheral resistance) and long-term adjustments (changes in blood volume that affect cardiac output and gas transport). *(Figure 21-14)*

Autoregulation of Blood Flow within Tissues, p. 728

3. Peripheral resistance is adjusted at the tissues by local factors that result in the dilation or constriction of precapillary sphincters.

The Neural Control of Blood Pressure and Blood Flow, p. 729

4. **Baroreceptor reflexes** are autonomic reflexes that adjust cardiac output and peripheral resistance to maintain normal arterial pressures. *Baroreceptors* are located in the **aortic sinuses,** the **carotid sinuses,** and the right atrium. *(Figure 21-15)*

5. **Chemoreceptor reflexes** respond to changes in the oxygen or carbon dioxide levels in the blood and cerebrospinal fluid. Stimulation of the *cardioacceleratory* and *vasomotor centers* leads to the stimulation of the heart rate and blood pressure through sympathetic activation; the *cardioinhibitory centers* exert their effects over the parasympathetic division. Epinephrine and norepinephrine stimulate cardiac output and peripheral vasoconstriction. *(Figure 21-16)*

Hormones and Cardiovascular Regulation, p. 733

6. The endocrine system provides short-term regulation of cardiac output and peripheral resistance with epinephrine and norepinephrine from the adrenal medullae. Hormones involved in the long-term regulation of blood pressure and volume are antidiuretic hormone (ADH), angiotensin II, aldosterone, erythropoietin (EPO), and atrial natriuretic peptide (ANP). *(Figure 21-17)*

7. ADH and angiotensin II also promote peripheral vasoconstriction in addition to their other functions. ADH and aldosterone promote water and electrolyte retention and stimulate thirst. EPO stimulates red blood cell production.

ANP encourages fluid loss, reduces blood pressure, inhibits thirst, and lowers peripheral resistance.

PATTERNS OF CARDIOVASCULAR RESPONSE, p. 734

Exercise and the Cardiovascular System, p. 734

1. During exercise, blood flow to skeletal muscles increases at the expense of circulation to nonessential organs, and cardiac output rises. Cardiovascular performance improves with training. Athletes have larger stroke volumes, slower resting heart rates, and larger cardiac reserves than do nonathletes. *(Tables 21-2, 21-3)*

Cardiovascular Response to Hemorrhaging, p. 736

2. Blood loss lowers blood volume and venous return and decreases cardiac output. Compensatory mechanisms include an increase in cardiac output, mobilization of venous reserves, peripheral vasoconstriction, and the liberation of hormones that promote fluid retention and the manufacture of erythrocytes. *(Figure 21-18)*

Shock, p. 737

3. **Shock** is an acute circulatory crisis marked by hypotension and inadequate peripheral blood flow. A severe drop in blood volume produces symptoms of **circulatory shock.** Causes of fluid loss may include hemorrhaging, dehydration, and severe burns. *(Figure 21-19)*

THE BLOOD VESSELS, p. 739

1. The peripheral distributions of arteries and veins are generally identical on both sides of the body, except near the heart. *(Figure 21-20)*

The Pulmonary Circulation, p. 740

2. The pulmonary circuit includes the pulmonary trunk, the **left** and **right pulmonary arteries,** and the **pulmonary veins,** which empty into the left atrium. *(Figure 21-21)*

The Systemic Circulation, p. 740

3. The **ascending aorta** gives rise to the coronary circulation. The **aortic arch** communicates with the **descending aorta.** *(Figures 21-22–21-29)*

4. Arteries in the neck and limbs are deep beneath the skin; in contrast, there are generally two sets of peripheral veins, one superficial and one deep. This dual venous drainage is important for controlling body temperature.

5. The **superior vena cava** receives blood from the head, neck, chest, shoulders, and arms. *(Figures 21-30–21-33)* The **inferior vena cava** collects most of the venous blood from organs inferior to the diaphragm. *(Figures 21-33, 21-34)*

6. The **hepatic portal system** directs blood from the other digestive organs to the liver before the blood returns to the heart. *(Figures 21-35)*

Fetal Circulation, p. 756

7. Major circulatory changes occur at birth as the lungs expand and the pulmonary circuit is activated. *(Figure 21-36)*

8. Congenital circulatory problems generally reflect abnormalities of the heart or of interconnections between the heart and great vessels. *(Figure 21-37)*

AGING AND THE CARDIOVASCULAR SYSTEM, p. 760

1. Age-related changes in the blood can include (1) decreased hematocrit; (2) constriction or blockage of peripheral veins by a *thrombus* (stationary blood clot); and (3) pooling of blood in the veins of the legs because valves are not working effectively.

2. Age-related changes in the heart include (1) a reduction in the maximum cardiac output; (2) changes in the activities of the nodal and conducting cells; (3) a reduction in the elasticity of the fibrous skeleton; (4) a progressive atherosclerosis that can restrict coronary circulation; (5) replacement of damaged cardiac muscle cells by scar tissue.

3. Age-related changes in blood vessels, commonly related to arteriosclerosis, include (1) a weakening in the walls of arteries, potentially leading to *aneurysm* formation; (2) calcium salt deposition on weakened vascular walls, increasing the risk of a stroke or myocardial infarction; and (3) thrombus formation at atherosclerotic *plaques.*

INTEGRATION WITH OTHER SYSTEMS, p. 760

1. The most extensive communication occurs between the cardiovascular and lymphatic systems. *(Figure 21-38)*

REVIEW QUESTIONS

Level 1   Reviewing Facts and Terms

Match each numbered item with the most closely related lettered item. Use letters for answers in the spaces provided:

__ **1.** veins	**a.** specialized fenestrated capillaries	__ **11.** blood pressure	**k.** capacitance vessels
__ **2.** vasa vasorum	**b.** minimum blood pressure	__ **12.** arterioles	**l.** drains the liver
__ **3.** conducting arteries	**c.** vasomotor centers	__ **13.** cholinergic	**m.** largest artery in body
__ **4.** muscular arteries	**d.** drains the kidney	__ **14.** adrenergic	**n.** stationary blood clot
__ **5.** sinusoids	**e.** vasoconstrictor fibers	__ **15.** baroreceptors	**o.** distribution arteries
__ **6.** precapillary sphincter	**f.** ventricular stretching	__ **16.** Frank–Starling principle	**p.** foramen ovale
__ **7.** medulla oblongata	**g.** elastic arteries	__ **17.** aorta	**q.** vasomotion
__ **8.** vascular resistance	**h.** blood supply to pelvis	__ **18.** internal iliac artery	**r.** resistance vessels
__ **9.** systolic pressure	**i.** vasodilator fibers	__ **19.** renal vein	**s.** migrating blood clot
__ **10.** diastolic pressure	**j.** largest superficial vein in body	__ **20.** hepatic vein	**t.** carotid sinus
		__ **21.** great saphenous	**u.** friction
		__ **22.** interatrial opening	**v.** sounds of Korotkoff
		__ **23.** thrombus	**w.** "vessels of vessels"
		__ **24.** embolus	**x.** peak blood pressure

25. Blood vessels that carry blood away from the heart are
    (a) veins   (b) arterioles
    (c) venules   (d) arteries
26. The layer of the arteriole wall that provides the properties of contractility and elasticity is the
    (a) tunica adventitia   (b) tunica media
    (c) tunica intima   (d) tunica externa
27. Blood vessels that supply the walls of arteries and veins with blood are
    (a) coronary vessels   (b) capillaries
    (c) vasa vasorum   (d) metarterioles
28. Of the following arteries, the one that is an elastic artery is the
    (a) subclavian artery
    (b) external carotid artery
    (c) brachial artery
    (d) femoral artery
29. The two-way exchange of substances between blood and body cells occurs only through
    (a) arterioles
    (b) capillaries
    (c) venules
    (d) a, b, and c are correct
30. Large molecules such as peptides and proteins move into and out of the circulation by way of
    (a) continuous capillaries
    (b) fenestrated capillaries
    (c) preferred channels
    (d) metarterioles
31. The alteration of blood flow due to the action of precapillary sphincters is
    (a) vasomotion   (b) autoregulation
    (c) selective resistance   (d) turbulence
32. The blood vessels that collect blood from all tissues and organs and return it to the heart are the
    (a) veins   (b) arteries
    (c) capillaries   (d) arterioles
33. Blood is compartmentalized within the veins because of the presence of
    (a) venous reservoirs   (b) muscular walls
    (c) clots   (d) valves
34. The most important factor in vascular resistance is
    (a) the viscosity of the blood
    (b) friction between the blood and the vessel walls
    (c) turbulence due to irregular surfaces of blood vessels
    (d) the length of the blood vessels
35. Plasma proteins are normally unable to cross the endothelial linings anywhere except in the
    (a) continuous capillaries
    (b) fenestrated capillaries
    (c) sinusoids
    (d) arteriovenous anastomoses
36. Hydrostatic pressure forces water _____ a solution; osmotic pressure forces water _____ a solution:
    (a) into, out of
    (b) out of, into
    (c) out of, out of
    (d) a, b, and c are incorrect
37. The two arteries formed by the division of the brachiocephalic artery are the
    (a) aorta and internal carotid
    (b) axillary and brachial
    (c) external and internal carotid
    (d) common carotid and subclavian

38. The unpaired arteries supplying blood to the visceral organs include the
    (a) suprarenal, renal, lumbar
    (b) iliac, gonadal, femoral
    (c) celiac, superior and inferior mesenteric
    (d) a, b, and c are correct
39. The paired arteries supplying blood to the body wall and other structures outside the abdominopelvic cavity include the
    (a) left gastric, hepatic, splenic, phrenic
    (b) suprarenals, renals, lumbars, gonadals
    (c) iliacs, femorals, gonadals, ileocecals
    (d) celiac, left gastric, superior and inferior mesenteric
40. The vein that drains the dural sinuses of the brain is the
    (a) cephalic
    (b) great saphenous
    (c) internal jugular
    (d) superior vena cava
41. The vein that drains the thorax and empties into the superior vena cava is the
    (a) azygos   (b) basilic
    (c) cardiac   (d) cephalic
42. The vein that collects most of the venous blood from below the diaphragm is the
    (a) superior vena cava
    (b) great saphenous
    (c) inferior vena cava
    (d) azygos
43. The tributaries of the hepatic portal vein include the
    (a) lumbar, gonadal, renal, and suprarenal veins
    (b) left colic, splenic, inferior and superior mesenteric veins
    (c) phrenic, hepatic, renal, and suprarenal veins
    (d) peroneal, iliac, saphenous, and femoral veins
44. Differentiate among the following types of pressure:
    (a) systolic
    (b) diastolic
    (c) pulse
    (d) mean arterial
45. (a) What are the primary forces that cause fluid to move out of a capillary and into the interstitial fluid at its arterial end?
    (b) What are the primary forces that cause fluid to move into a capillary from the interstitial fluid at its venous end?
46. What two factors assist relatively low venous pressures in propelling blood toward the heart?
47. Give at least five examples of local vasodilators produced at the tissue level that accelerate blood flow through the tissue.
48. What two effects occur when the baroreceptor response to elevated blood pressure is triggered?
49. What factors affect the activity of chemoreceptors in the carotid and aortic bodies?
50. List six hormones important in the regulation of cardiovascular function.
51. What interrelated cardiovascular changes occur before and during light exercise?
52. What criteria are used to identify the basic symptoms of all forms of shock?
53. What circulatory changes occur at birth?
54. What age-related changes take place in the blood, heart, and blood vessels?

Level 2   Reviewing Concepts

55. When dehydration occurs, there is
    (a) accelerated reabsorption of water at the kidneys
    (b) a recall of fluids
    (c) an increase in the blood colloidal osmotic pressure
    (d) a, b, and c are correct

56. Increased CO_2 levels in tissues would promote
    (a) constriction of precapillary sphincters
    (b) an increase in the pH of the blood
    (c) dilation of precapillary sphincters
    (d) decreased blood flow to tissues

57. Elevated levels of the hormones ADH and angiotensin II will produce
    (a) increased peripheral vasodilation
    (b) increased peripheral vasoconstriction
    (c) increased peripheral blood flow
    (d) increased venous return

58. Secretion of ADH and aldosterone are typical of the body's long-term compensation following
    (a) a heart attack
    (b) hypertension
    (c) a serious hemorrhage
    (d) heavy exercise

59. Relate the anatomical differences between arteries and veins to their functions.

60. What causes variations in the blood flow in individual capillaries?

61. Why do capillaries permit the diffusion of materials whereas arteries and veins do not?

62. How is blood pressure maintained in veins to cope with the force of gravity?

63. How do pressure and resistance affect cardiac output and peripheral blood flow?

64. Why is blood flow to the brain relatively continuous and constant?

65. Compare the effects of the cardioacceleratory and cardioinhibitory centers on cardiac output and blood pressure.

66. A nurse practitioner tells Mrs. B. that her blood pressure is 150/90. Explain what these numbers represent, and calculate Mrs. B.'s mean arterial pressure (MAP).

67. An accident victim displays the following symptoms: hypotension; pale, cool, moist skin; confusion and disorientation. Identify her condition, and explain why these symptoms occur. If you took her pulse, what would you probably find?

68. Emilio has been diagnosed with congestive heart failure. Because of this condition, his ankles and feet appear to be swollen. What is the relationship between congestive heart failure and accumulation of fluid in the feet and ankles?

Level 3   Critical Thinking and Clinical Applications

69. Bob is sitting outside on a warm day and is sweating profusely. Mary wants to practice taking blood pressures, and he agrees to play patient. Mary finds that Bob's blood pressure is elevated, even though he is resting and has lost fluid from sweating. (She reasons that fluid loss should lower blood volume and thus blood pressure.) Mary asks you why Bob's blood pressure is high instead of low. What will you tell her?

70. The most common site of varicose veins is the greater saphenous vein of the leg. Why do you think this is the case?

71. Would you expect a resting athlete or a resting person who never exercises and has a sedentary job to have a higher pulse pressure? Why?

72. People who suffer from allergies frequently take antihistamines with decongestants to relieve their symptoms. The container warns that the medication should not be taken by individuals who are being treated for high blood pressure. Why not?

73. Jolene awakens suddenly to the sound of her alarm clock. Realizing she is late for class, she jumps to her feet, feels lightheaded, and falls back on her bed. What probably caused this to happen? Why doesn't this happen all the time?

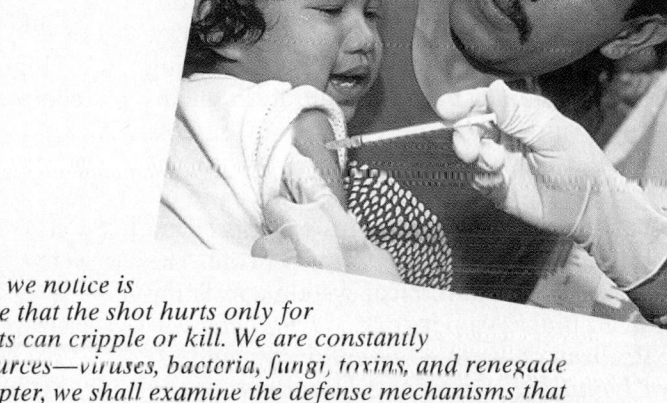

CHAPTER 22

The Lymphatic System and Immunity

*T*his is every child's nightmare! As children, all we notice is
the sting of a vaccination. As adults, we realize that the shot hurts only for
an instant (if at all), whereas the disease it prevents can cripple or kill. We are constantly
besieged by threats to our health from various sources—viruses, bacteria, fungi, toxins, and renegade
body cells that have turned malignant. In this chapter, we shall examine the defense mechanisms that
the body deploys to fight off these threats. We shall also see how modern medicine can manipulate the
immune system to enhance the body's defenses.

CHAPTER OUTLINE AND OBJECTIVES

The world is not always kind to the human body. Accidental bumps, cuts, and scrapes, chemical and thermal burns, extreme cold, and ultraviolet radiation are just a few of the hazards in our physical environment. Making matters worse, the world around us contains an assortment of *viruses, bacteria, fungi,* and *parasites* capable of not only surviving but thriving inside our bodies—and potentially causing us great harm in the process. These organisms, called **pathogens,** are responsible for many human diseases. Each has a different lifestyle and attacks the body in a specific way. For example, viruses spend most of their time hiding within cells, whereas many bacteria multiply in the interstitial fluids, and the largest parasites burrow through internal organs. [AM] *The Nature of Pathogens*

Many organs and systems work together in an effort to keep us alive and healthy. In this ongoing struggle, the **lymphatic system** plays a central role. The lymphatic system is an anatomical system consisting of (1) *lymph,* a fluid; (2) a network of *lymphatic vessels;* (3) specialized cells called *lymphocytes;* and (4) an array of *lymphoid tissues* and *lymphoid organs* scattered throughout the body.

AN OVERVIEW OF THE LYMPHATIC SYSTEM

We introduced lymphocytes, the primary cells of the lymphatic system, in Chapters 4 and 19. ∞ *[pp. 126, 658]* These cells are vital to our ability to resist or overcome infection and disease. Lymphocytes respond to the presence of (1) invading pathogens, such as bacteria or viruses, (2) abnormal body cells, such as virus-infected cells or cancer cells, and (3) foreign proteins, such as the toxins released by some bacteria. They attempt to eliminate these threats or render them harmless by a combination of physical and chemical attacks.

Lymphocytes respond to specific threats. If bacteria invade peripheral tissues, lymphocytes organize a defense against that particular type of bacterium. For this reason, lymphocytes are said to provide a *specific defense,* known as the **immune response. Immunity** is the ability to resist infection and disease through the activation of specific defenses. All the cells and tissues involved with the production of immunity are sometimes considered to be part of an *immune system,* a physiological system that includes not only the lymphatic system but also components of the integumentary, cardiovascular, respiratory, digestive, and other systems. For example, interactions between lymphocytes and Langerhans cells of the skin are important in mobilizing specific defenses against skin infections. We begin this chapter by examining the organization of the lymphatic system. We shall then consider how the lymphatic system interacts with cells and tissues of other systems to defend the body against infection and disease.

ORGANIZATION OF THE LYMPHATIC SYSTEM

The lymphatic system consists of the following:

1. A network of **lymphatic vessels** that begin in peripheral tissues and end at connections to the venous system.
2. **Lymph,** a fluid that resembles plasma but contains a much lower concentration of suspended proteins.
3. **Lymphoid organs** connected to the lymphatic vessels and containing large numbers of lymphocytes.

Figure 22-1● provides a general overview of the components of this system.

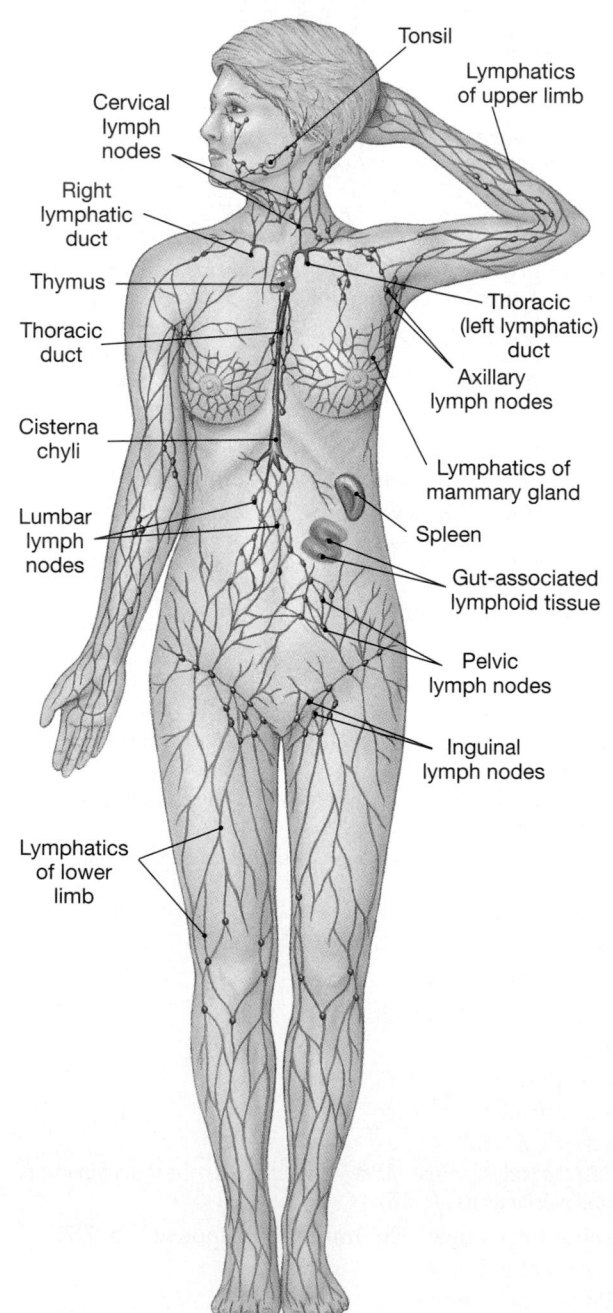

●*FIGURE 22-1* **Components of the Lymphatic System**

Functions of the Lymphatic System

The three primary functions of the lymphatic system are:

1. *The production, maintenance, and distribution of lymphocytes.* Lymphocytes are produced and stored within (1) lymphoid tissues and organs, such as the spleen and thymus, and (2) areas of red bone marrow.

2. *The return of fluid and solutes from peripheral tissues to the blood.* Capillaries normally deliver more fluid to the tissues than they carry away. ∞ *[p. 725]* The return of tissue fluids through lymphatic vessels maintains normal blood volume and eliminates local variations in the composition of the interstitial fluid.

3. *The distribution of hormones, nutrients, and waste products from their tissues of origin to the general circulation.* Substances that originate in the tissues but are for some reason unable to enter the bloodstream directly may do so by way of the lymphatic vessels. For example, lipids absorbed by the digestive tract commonly fail to enter the circulation at the capillary level. They reach the bloodstream only after they have traveled along lymphatic vessels (a process we shall explore further in Chapter 24).

Lymphatic Vessels

Lymphatic vessels, often called **lymphatics,** carry lymph from peripheral tissues to the venous system. The smallest lymphatic vessels are called *lymphatic capillaries.*

Lymphatic Capillaries

The lymphatic network begins with the **lymphatic capillaries,** or *terminal lymphatics,* that branch through peripheral tissues. They differ from blood capillaries in that lymphatic capillaries (1) originate as blind pockets, (2) are larger in diameter, (3) have thinner walls, and (4) in sectional view typically have a flattened or irregular outline (Figure 22-2●). Although lined by endothelial cells, they have no underlying basement membrane. The endothelial cells of a lymphatic capillary are not tightly bound together, but they do overlap. The region of overlap acts as a one-way valve. It permits the entry of fluids and solutes, even those as large as proteins, but it prevents their return to the intercellular spaces (Figure 22-2●).

Lymphatic capillaries are present in almost every tissue and organ in the body. Prominent lymphatic capillaries in the small intestine are called *lacteals;* these are important in the transport of lipids absorbed by the digestive tract. Lymphatic capillaries are absent in areas that lack a blood supply, such as the cornea of the eye. There are also no lymphatics in the CNS.

Valves of Lymphatic Vessels

From the lymphatic capillaries, lymph flows into larger lymphatic vessels that lead toward the trunk. The walls of these lymphatic vessels contain layers comparable to

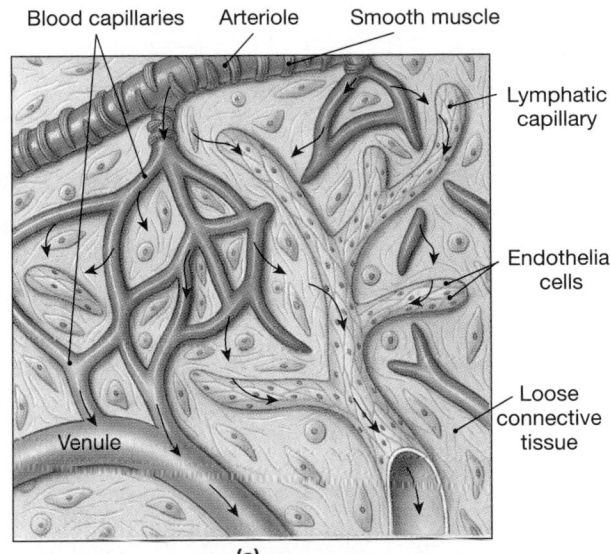

(a)

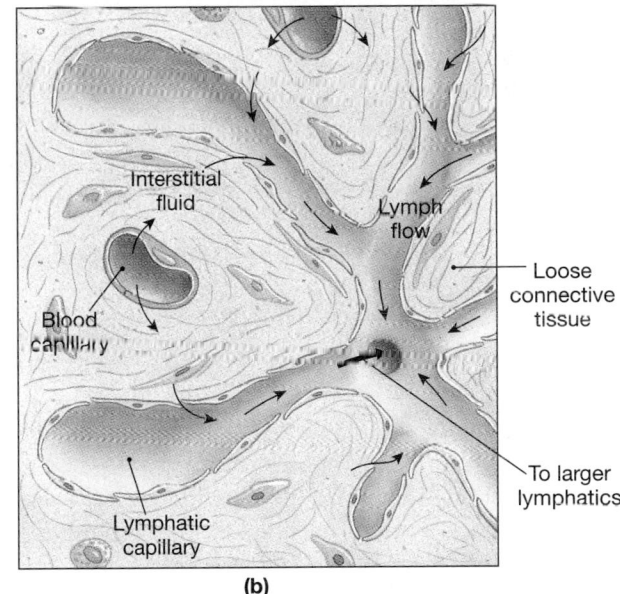

(b)

●**FIGURE 22-2 Lymphatic Capillaries. (a)** Diagrammatic view of the interwoven network formed by blood capillaries and lymphatic capillaries. Arrows show the net direction of interstitial fluid and lymph movement. **(b)** Sectional view showing the path of fluid movement from the plasma through the interstitial fluid and into the lymphatic system.

those of veins, and, like veins, the larger lymphatic vessels contain valves (Figure 22-3a,b●). The valves are quite close together, and at each valve the lymphatic vessel bulges noticeably. As a result, large lymphatics have a beaded appearance (Figure 22-3a●). The valves prevent the backflow of lymph within lymph vessels, especially those of the limbs. Pressures within the lymphatic system are minimal, and the valves are essential to maintaining normal lymph flow toward the thoracic cavity.

Lymphatic vessels commonly occur in association with blood vessels (Figure 22-3a●). Note the differences in relative size, general appearance, and branching pattern that distinguish the lymphatic vessels from

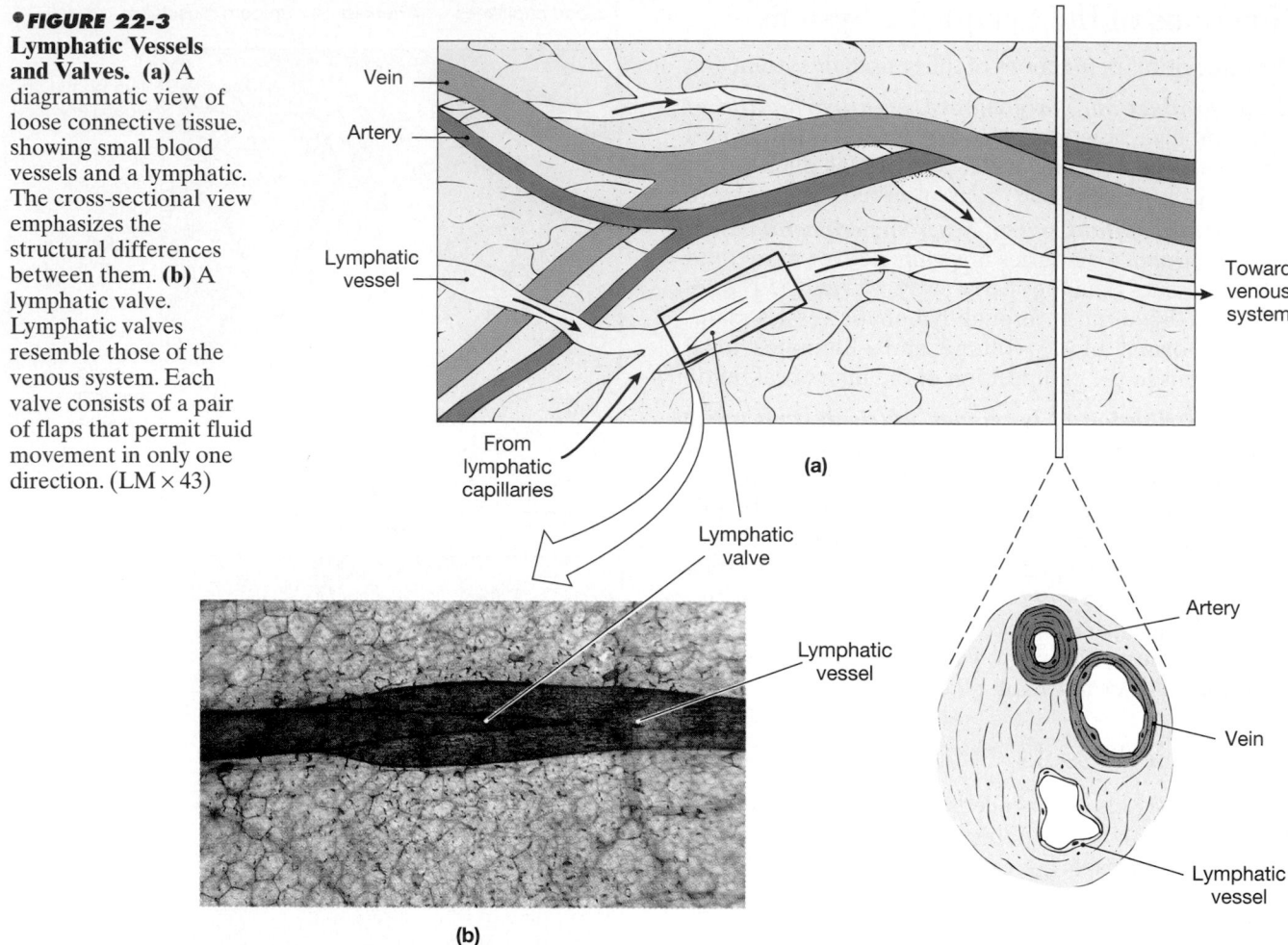

•*FIGURE 22-3*
Lymphatic Vessels and Valves. **(a)** A diagrammatic view of loose connective tissue, showing small blood vessels and a lymphatic. The cross-sectional view emphasizes the structural differences between them. **(b)** A lymphatic valve. Lymphatic valves resemble those of the venous system. Each valve consists of a pair of flaps that permit fluid movement in only one direction. (LM × 43)

Vein

Artery

Lymphatic vessel

From lymphatic capillaries

(a)

Toward venous system

Lymphatic valve

Lymphatic vessel

Artery

Vein

Lymphatic vessel

(b)

arteries and veins. There are also characteristic color differences that are apparent on examining living tissues. Most arteries are bright red, veins dark red, and lymphatic vessels a pale golden color. In general, a tissue will contain many more lymphatics than veins, but the lymphatics are much smaller.

Major Lymph-Collecting Vessels

Two sets of lymphatic vessels collect blood from the lymphatic capillaries: *superficial lymphatics* and *deep lymphatics*. **Superficial lymphatics** are located in the subcutaneous layer beneath the skin; in the loose connective tissues of the mucous membranes lining the digestive, respiratory, urinary, and reproductive tracts; and in the loose connective tissues of the serous membranes lining the pleural, pericardial, and peritoneal cavities. **Deep lymphatics** are larger lymph vessels. They accompany deep arteries and veins supplying skeletal muscles and other organs of the neck, limbs, and trunk and the walls of visceral organs.

Superficial and deep lymphatics converge to form larger vessels called *lymphatic trunks* that in turn empty into two large collecting vessels: the *thoracic duct* and the *right lymphatic duct*. The **thoracic duct** collects lymph from the body inferior to the diaphragm and from the left side of the body superior to the diaphragm. The smaller **right lymphatic duct** collects lymph from the right side of the body superior to the diaphragm.

THE THORACIC DUCT The thoracic duct is formed inferior to the diaphragm at the level of vertebra L_2. The base of the thoracic duct is an expanded, saclike chamber called the **cisterna chyli** (KĪ-lē) (Figure 22-4•). The cisterna chyli receives lymph from the lower abdomen, pelvis, and lower limbs via the *right* and *left lumbar trunks* and the *intestinal trunk*.

The inferior segment of the thoracic duct lies anterior to the vertebral column. From the second lumbar vertebra, it penetrates the diaphragm at the *aortic hiatus* and ascends along the left side of the vertebral column to the level of the left clavicle. After collecting lymph from the *left bronchomediastinal trunk,* the *left subclavian trunk,* and the *left jugular trunk,* it empties into the left subclavian vein near the left internal jugular vein (Figure 22-4•). Lymph collected from the left side of the head, neck, and thorax, as well as lymph from the entire body inferior to the diaphragm, reenters the venous system in this way.

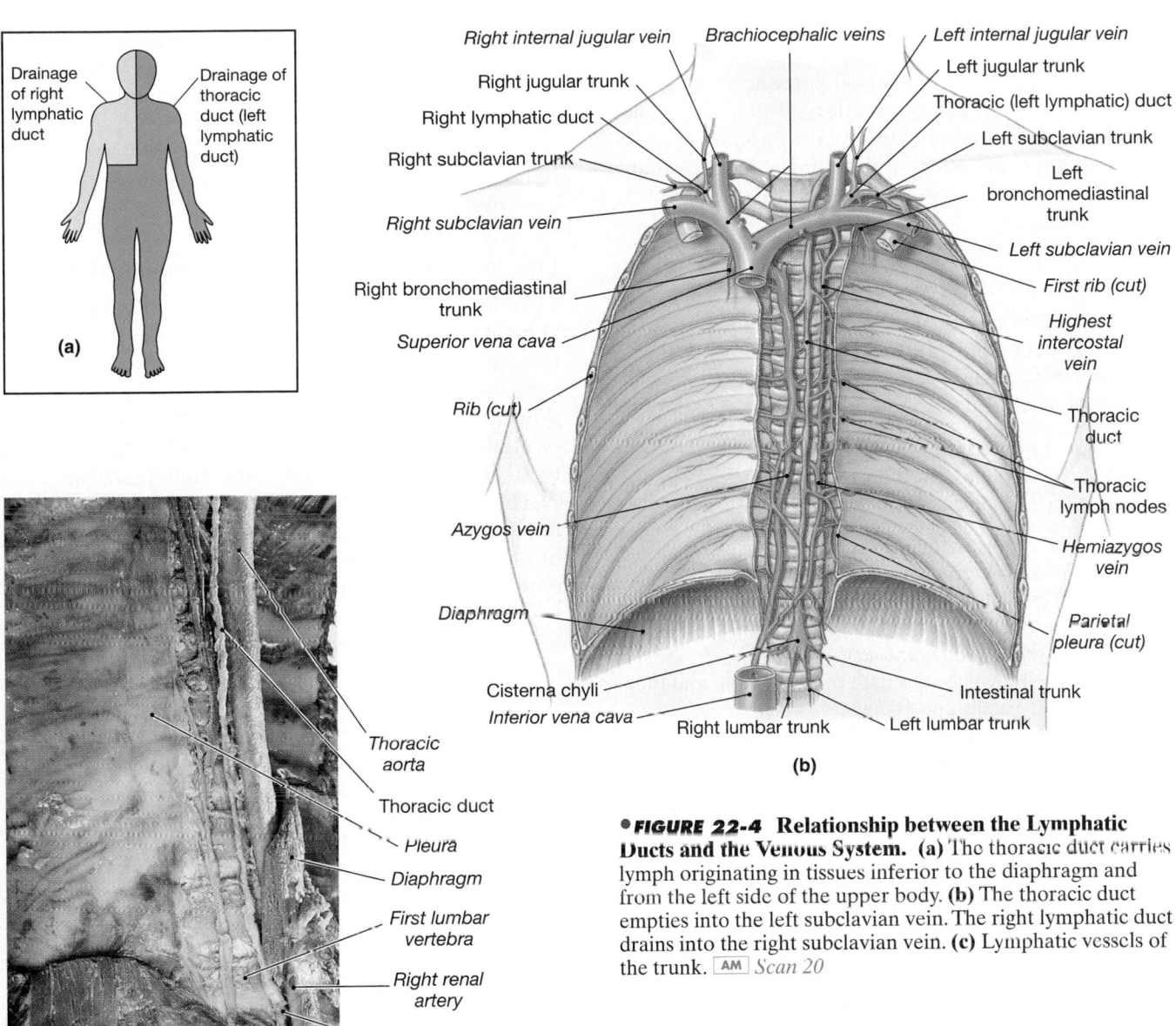

Figure labels: Right internal jugular vein, Brachiocephalic veins, Left internal jugular vein, Left jugular trunk, Thoracic (left lymphatic) duct, Left subclavian trunk, Left bronchomediastinal trunk, Left subclavian vein, First rib (cut), Highest intercostal vein, Thoracic duct, Thoracic lymph nodes, Hemiazygos vein, Parietal pleura (cut), Intestinal trunk, Left lumbar trunk, Right lumbar trunk, Inferior vena cava, Cisterna chyli, Diaphragm, Azygos vein, Rib (cut), Superior vena cava, Right bronchomediastinal trunk, Right subclavian vein, Right subclavian trunk, Right lymphatic duct, Right jugular trunk

(a) Drainage of right lymphatic duct. Drainage of thoracic duct (left lymphatic duct).

(c) Thoracic aorta, Thoracic duct, Pleura, Diaphragm, First lumbar vertebra, Right renal artery, Cisterna chyli, Abdominal aorta

•FIGURE 22-4 **Relationship between the Lymphatic Ducts and the Venous System. (a)** The thoracic duct carries lymph originating in tissues inferior to the diaphragm and from the left side of the upper body. **(b)** The thoracic duct empties into the left subclavian vein. The right lymphatic duct drains into the right subclavian vein. **(c)** Lymphatic vessels of the trunk. [AM] *Scan 20*

THE RIGHT LYMPHATIC DUCT The right lymphatic duct is formed by the merging of the *right jugular, right subclavian,* and *right bronchomediastinal trunks* in the area near the right clavicle. This duct empties into the right subclavian vein, delivering lymph from the right side of the body superior to the diaphragm (Figure 22-4•).

LYMPHEDEMA Blockage of the lymphatic drainage from a limb produces **lymphedema** (lim-fe-DĒ-muh). In this painless condition, interstitial fluids accumulate, and the limb gradually becomes swollen and grossly distended. If the condition persists, the connective tissues lose their elasticity, and the swelling becomes permanent. Lymphedema by itself does not pose a major threat to life. The danger comes from the constant risk that an uncon-

trolled infection will develop in the affected area. Because the interstitial fluids are essentially stagnant, toxins and pathogens can accumulate and overwhelm the local defenses without fully activating the immune system.

Temporary lymphedema may result from tight clothing. Chronic lymphedema can result from scar tissue formation or from parasitic infections. In **filariasis** (fil-a-RĪ-a-sis), larvae of a parasitic nematode (roundworm), generally *Wuchereria bancrofti,* is transmitted by mosquitoes or blackflies. The adult worms form massive colonies within lymphatic vessels and lymph nodes. Repeated scarring of the passageways eventually blocks lymphatic drainage and produces extreme lymphedema with permanent distension of tissues. The limbs or external genitalia typically become grossly distended, a condition known as **elephantiasis** (el-e-fan-TĪ-a-sis).

Therapy for chronic lymphedema consists of treating infections by the administration of antibiotics and (when possible) reducing the swelling. One possible treatment involves the application of elastic wrappings that squeeze the tissue. This external compression elevates the hydrostatic pressure of the interstitial fluids and opposes the entry of additional fluid from the capillaries.

Lymphocytes

Lymphocytes account for 20–30 percent of the circulating white blood cell population. ∞ *[p. 659]* However, circulating lymphocytes are only a small fraction of the total lymphocyte population. The body contains some 10^{12} lymphocytes, with a combined weight of over a kilogram.

Types of Lymphocytes

There are three classes of lymphocytes in the blood: (1) **T** (**t**hymus-dependent) **cells**, **B** (**b**one marrow–derived) **cells**, and **NK** (**n**atural **k**iller) **cells**. Each type has distinctive biochemical and functional characteristics.

T CELLS Approximately 80 percent of circulating lymphocytes are classified as T cells. There are many different types of T cells, including the following:

- **Cytotoxic T cells,** which attack foreign cells or body cells infected by viruses. Their attack commonly involves direct contact. These lymphocytes are the primary cells involved in the production of *cell-mediated immunity,* or *cellular immunity.*
- **Helper T cells,** which stimulate the activation and function of both T cells and B cells.
- **Suppressor T cells,** which inhibit the activation and function of both T cells and B cells.

The interplay between suppressor and helper T cells helps establish and control the sensitivity of the immune response. For this reason, these cells are also known as *regulatory T cells.*

These are the T cells that we will examine in the course of this chapter. It is not a complete list, however; there are other types of T cells that participate in the immune response. For example, *inflammatory T cells* stimulate regional inflammation and local defenses in an injured tissue, and *suppressor/inducer T cells* suppress B cell activity but stimulate other T cells.

B CELLS B cells account for 10–15 percent of circulating lymphocytes. When stimulated, B cells can differentiate into **plasma cells.** Plasma cells, introduced in Chapter 4, are responsible for the production and secretion of *antibodies,* soluble proteins that are also known as *immunoglobulins.* ∞ *[p. 121]* These proteins react with specific chemical targets called **antigens.** Most antigens are pathogens, parts or products of pathogens, or other foreign compounds. Most antigens are proteins, but some lipids, polysaccharides, and nucleic acids can also stimulate antibody production. When an antibody binds to its target antigen, a chain of events begins that leads to the destruction of the target compound or organism. B cells are responsible for *antibody-mediated immunity,* which is also known as *humoral* ("liquid") *immunity* because antibodies occur in body fluids.

NK CELLS The remaining 5–10 percent of circulating lymphocytes are NK cells, also known as **large granular lymphocytes.** These lymphocytes will attack foreign cells, normal cells infected with viruses, and cancer cells that appear in normal tissues. Their continuous policing of peripheral tissues has been called *immunological surveillance.*

Life Span and Circulation of Lymphocytes

Lymphocytes are not evenly distributed in the blood, bone marrow, spleen, thymus, and peripheral lymphoid tissues. The ratio of B cells to T cells varies with the tissue or organ considered. For example, B cells are seldom found in the thymus, and in the blood T cells outnumber B cells by a ratio of 8:1. This ratio changes to 1:1 in the spleen and 1:3 in the bone marrow.

The lymphocytes within these organs are visitors, not residents. All types of lymphocytes move throughout the body. They wander through a tissue and then enter a blood vessel or lymphatic vessel for transport to another site.

T cells move relatively quickly. For example, a wandering T cell may spend about 30 minutes in the blood, 5–6 hours in the spleen, and 15–20 hours in a lymph node. B cells, which are responsible for antibody production, move more slowly. A typical B cell spends about 30 hours in a lymph node before moving on.

Lymphocytes have relatively long life spans. Roughly 80 percent survive for 4 years, and some last 20 years or more. Throughout your life, you maintain normal lymphocyte populations by producing new lymphocytes in your bone marrow and lymphatic tissues.

Lymphocyte Production

In Chapter 19, we discussed *hemopoiesis,* the formation of the cellular elements of blood. ∞ *[p. 645]* Erythropoiesis (red blood cell formation) in adults is normally confined to the bone marrow, but lymphocyte production, or **lymphopoiesis** (lim-fō-poy-Ē-sis), involves the bone marrow, thymus, and peripheral lymphoid tissues. The relationships are diagrammed in Figure 22-5●.

The bone marrow plays the primary role in the maintenance of normal lymphocyte populations. Hemocytoblast divisions in the bone marrow of an adult generate the lymphoid stem cells responsible for the production of all types of lymphocytes. Two distinct populations of lymphoid stem cells are produced in the bone marrow.

One group of lymphoid stem cells remains in the bone marrow (Figure 22-5a●). Divisions of these cells produce immature B cells and NK cells. B cell development involves intimate contact with large **stromal cells** (*stroma,* a bed) in the bone marrow. The cytoplasmic extensions of stromal cells contact or even wrap around the developing B cells. The stromal cells produce an immune sys-

●FIGURE 22-5 **Derivation and Distribution of Lymphocytes.** Hemocytoblast divisions produce lymphocytic stem cells with two different fates. **(a)** One group remains in the bone marrow, producing daughter cells that mature into B cells and NK cells. **(b)** The other group migrates to the thymus, where subsequent divisions produce daughter cells that mature into T cells. The mature B cells, NK cells, and T cells circulate throughout the body in the bloodstream **(c)** and then leave the circulation to take temporary residence in peripheral tissues.

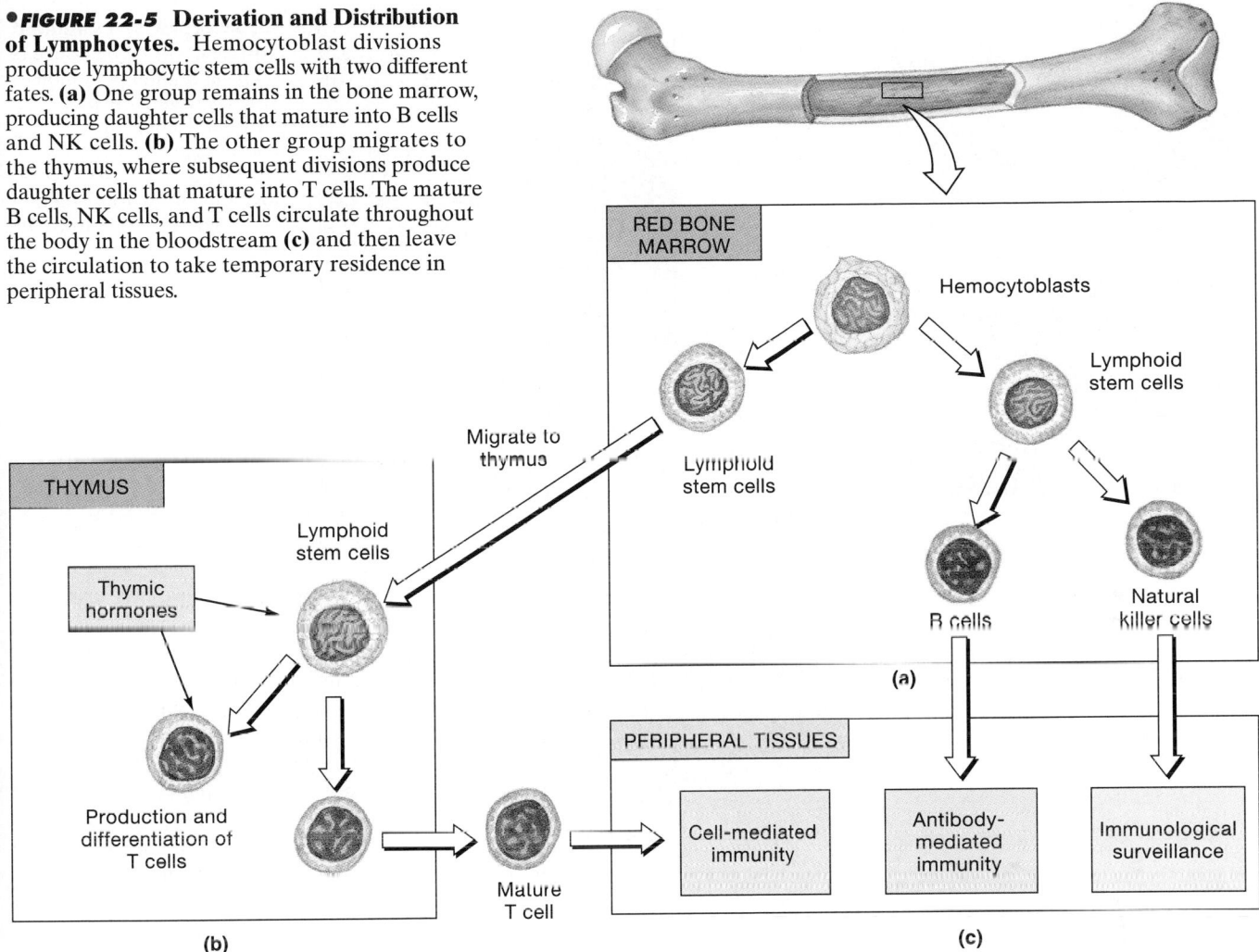

tem hormone, or *cytokine*, called *interleukin-7*, which promotes the differentiation of B cells. (We will consider cytokines and their varied effects in a later section.)

As they mature, B cells and NK cells enter the circulation and migrate to peripheral tissues (Figure 22-5c●). Most of the B cells move into lymph nodes, the spleen, or other lymphoid tissues. The NK cells migrate throughout the body, moving through peripheral tissues in their search for abnormal cells.

The second group of lymphoid stem cells migrates to the thymus (Figure 22-5b●). While in the thymus, these cells and their descendants develop further in an environment that is isolated from the general circulation by the **blood–thymus barrier.** Under the influence of thymic hormones collectively known as **thymosins,** the lymphoid stem cells divide repeatedly, producing the various kinds of T cells. At least seven thymosins have been identified, including *thymosin-α, thymosin-β, thymosin V, thymopoietin, thymulin, thymolymphotropin,* and *thymic-factor X.* Their precise functions and interactions have yet to be determined.

When their development is nearing completion, T cells reenter the circulation and return to the bone

marrow. They also travel to lymphoid tissues and organs, such as the spleen (Figure 22-5c●).

The T cells and B cells that migrate from their sites of origin retain the ability to divide. Their divisions produce daughter cells of the same type; for example, a dividing B cell produces other B cells, not T cells or NK cells. As we shall see, the ability to increase the number of lymphocytes of a specific type is important to the success of the immune response.

Lymphoid Tissues

Lymphoid tissues are connective tissues dominated by lymphocytes. In a **lymphoid nodule,** or *lymphatic nodule,* the lymphocytes are densely packed in an area of loose connective tissue (Figure 22-6●). Lymphoid nodules are found in the connective tissue beneath the epithelia lining the respiratory, digestive, and urinary tracts. Typical nodules average about a millimeter in diameter, but the boundaries are not distinct, because no fibrous capsule surrounds them. They commonly have a central zone called a **germinal center,** which contains dividing lymphocytes (Figure 22-6b●).

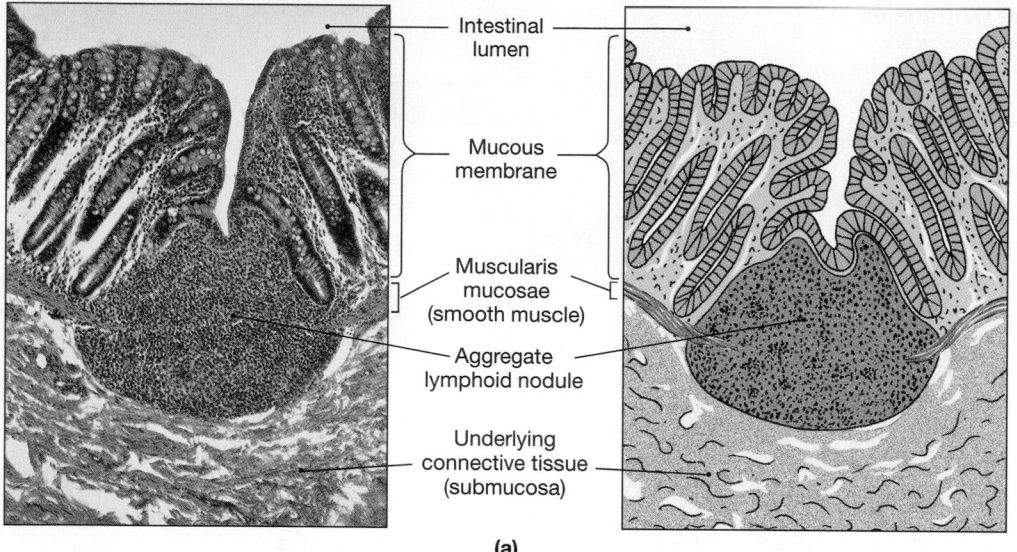

Intestinal lumen

Mucous membrane

Muscularis mucosae (smooth muscle)

Aggregate lymphoid nodule

Underlying connective tissue (submucosa)

(a)

●*FIGURE 22-6* **Lymphoid Nodules. (a)** Appearance of a typical nodule in section. Note the relatively pale germinal center, where lymphocyte cell divisions occur. (LM × 17) **(b)** The positions of the tonsils and the appearance of a tonsil in section.

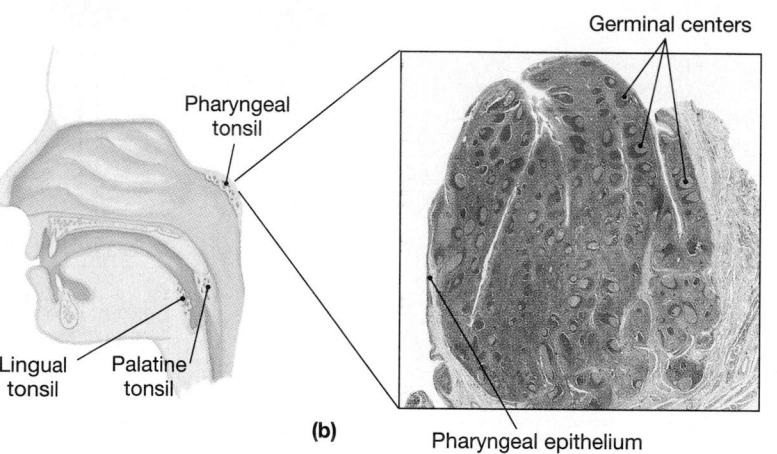

Germinal centers

Pharyngeal tonsil

Lingual tonsil

Palatine tonsil

(b)

Pharyngeal epithelium

INFECTED LYMPHOID NODULES The lymphocytes in a lymphoid nodule are not always able to destroy bacterial or viral invaders that have crossed the adjacent epithelium. If pathogens become established in a lymphoid nodule, an infection develops. Two examples are probably familiar to you: *tonsillitis,* an infection of one of the tonsils (generally the pharyngeal or palatine), and *appendicitis,* an infection of the appendix that begins in the lymphoid nodules. Treatment commonly consists of antibiotic therapy, sometimes combined with surgical removal of the infected tissues. AM

Infected Lymphoid Nodules

GALT

The extensive collection of lymphoid tissues linked with the digestive system is called the **gut-associated lymphoid tissue (GALT).** Clusters of lymphoid nodules beneath the epithelial lining of the intestine are known as **aggregate lymphoid nodules,** or *Peyer's patches* (Figure 22-6a●). In addition, the walls of the appendix, a blind pouch that originates near the junction between the small and large intestines, contain a mass of fused lymphoid nodules.

Tonsils

Large nodules in the walls of the pharynx are called **tonsils** (Figure 22-6b●). Most people have five tonsils:

1. A single **pharyngeal tonsil,** often called the *adenoids,* located in the posterior superior wall of the nasopharynx.
2,3. A pair of **palatine tonsils,** found at the posterior margin of the oral cavity, along the boundary with the pharynx.
4,5. A pair of **lingual tonsils,** which are usually not visible because they are located under the attached base of the tongue.

Lymphoid Organs

Lymphoid organs are separated from surrounding tissues by a fibrous connective tissue capsule. Important lymphoid organs include the *lymph nodes,* the *thymus,* and the *spleen.*

Lymph Nodes

Lymph nodes are small, oval lymphoid organs ranging in diameter from 1 to 25 mm (to about 1 in.). Figure 22-1●, p. 770, shows the general pattern of lymph node distribution in the body. Each lymph node is covered by a capsule of dense connective tissue. Bundles of collagen fibers extend from the capsule into the interior of the node. These fibrous partitions are called **trabeculae** (*trabecula,* a wall).

The shape of a typical lymph node resembles that of a kidney bean. Blood vessels and nerves attach to the lymph node at the indentation, or **hilus** (Figure 22-7●). Two sets of lymphatic vessels are connected to each lymph node: *efferent lymphatics* and *afferent lymphatics.*

1. **Efferent lymphatics** are attached to the lymph node at the hilus. These lymphatic vessels carry lymph away from the lymph node and toward the venous system.
2. **Afferent lymphatics** bring lymph to the lymph node from peripheral tissues. The afferent lymphatics penetrate the capsule of the lymph node on the side opposite the hilus.

LYMPH FLOW Lymph delivered by the afferent lymphatics flows through the lymph node within a network of sinuses, open passageways with incomplete walls. Lymph first enters a *subcapsular sinus,* which contains a meshwork of branching reticular fibers, macrophages, and **dendritic cells.** Dendritic cells are involved in the initiation of the immune response, and we shall consider their role in a later section. After passing through the subcapsular sinus, lymph flows through the **outer cortex** of the node. The outer cortex contains B cells within germinal centers that resemble those of lymphoid nodules.

Lymph then continues through lymph sinuses in the **deep cortex** (*paracortical area*). Lymphocytes leave the circulation and enter the lymph node by crossing the walls of blood vessels within the deep cortex. The deep cortical area is dominated by T cells.

After flowing through the sinuses of the deep cortex, lymph continues into the core, or **medulla,** of the lymph node. The medulla contains B cells and plasma cells organized into elongate masses known as **medullary cords.** Lymph enters the efferent lymphatics at the hilus after passing through a network of sinuses in the medulla.

LYMPH NODE FUNCTION A lymph node functions like a kitchen water filter: It filters and purifies lymph before that fluid reaches the venous system. As lymph flows through a lymph node, at least 99 percent of the antigens in the arriving lymph will be removed. Fixed macrophages in the walls of the lymphatic sinuses engulf debris or pathogens in the lymph as it flows past. Antigens removed in this way are then processed by the macrophages and "presented" to nearby lymphocytes. Other antigens bind to receptors on the surfaces of dendritic cells, where they can stimulate lymphocyte activity. This process, called *antigen presentation,* is generally the first step in the activation of the immune response.

In addition to filtering, lymph nodes provide an early-warning system. Any infection or other abnormality in a peripheral tissue will introduce abnormal antigens into the interstitial fluid and thus into the lymph leaving the area. These antigens will then stimulate macrophages and lymphocytes in nearby lymph nodes.

If you wanted to protect a house against intrusion, you might guard all entrances and exits or place traps by the windows and doors. The distribution of lymphatic tissues and lymph nodes follows such a pattern. The largest lymph nodes are found where peripheral lymphatics con-

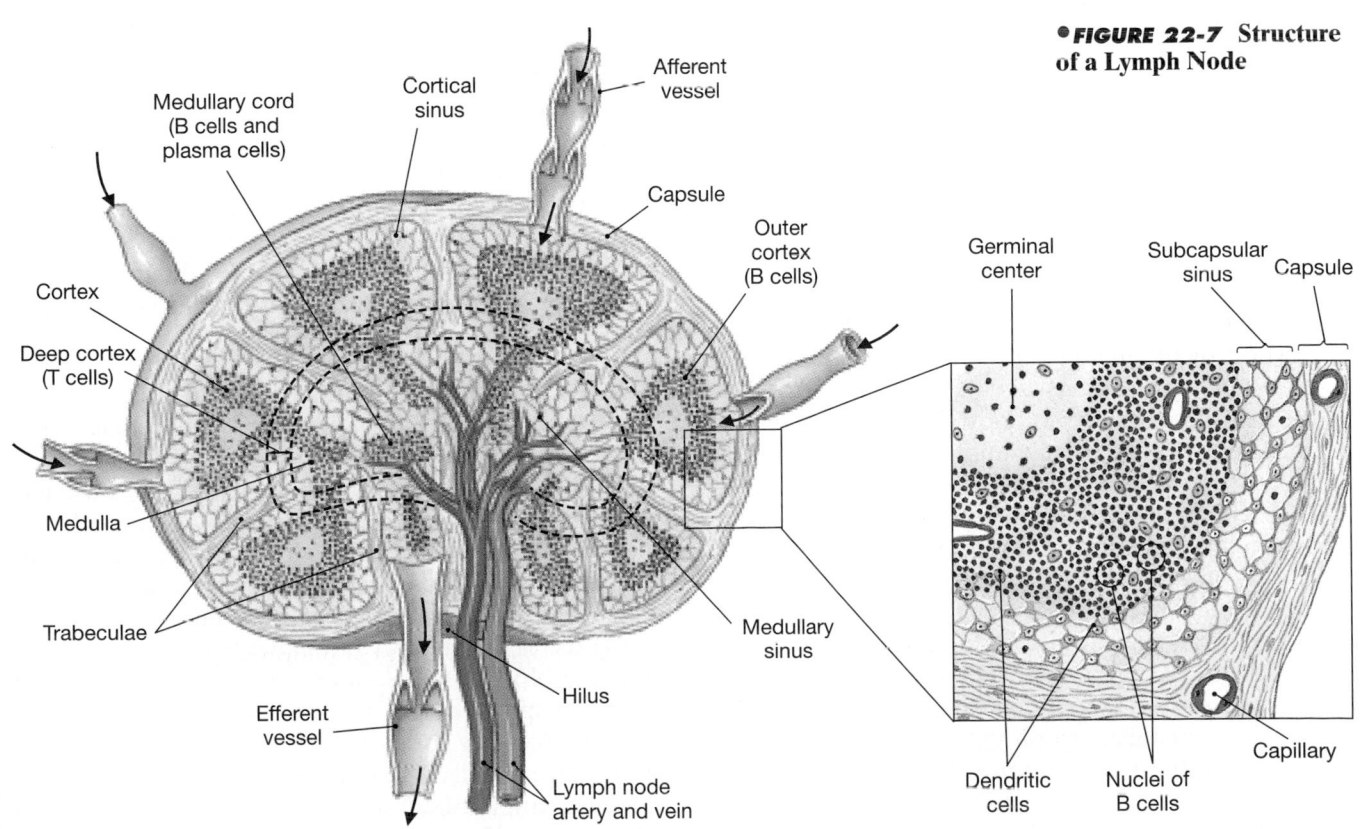

•FIGURE 22-7 Structure of a Lymph Node

nect with the trunk, in regions such as the groin, the axillae, and the base of the neck. These nodes are often called *lymph glands*. Because lymph is monitored in the cervical, inguinal, or axillary lymph nodes, potential problems can be detected and dealt with before they affect the vital organs of the trunk. Aggregations of lymph nodes also exist within the mesenteries of the gut, near the trachea and passageways leading to the lungs, and in association with the thoracic duct (Figure 22-4●, p. 773). These lymph nodes protect against pathogens and other antigens within the digestive and respiratory systems.

A minor injury commonly produces a slight enlargement of the nodes along the lymphatics draining the region. This symptom, often called "swollen glands," typically indicates inflammation or infection of peripheral structures. The enlargement generally results from an increase in the number of lymphocytes and phagocytes in the node in response to a minor, localized infection. Chronic or excessive enlargement of lymph nodes constitutes **lymphadenopathy** (lim-fad-e-NOP-a-thē). This condition may occur in response to bacterial or viral infections, endocrine disorders, or cancer.

CANCER AND THE LYMPHATIC SYSTEM Lymphatics are located in almost all portions of the body except the CNS, and the lymphatic capillaries offer little resistance to the passage of cancer cells. As a result, metastasizing cancer cells commonly spread along the lymphatics. Under these circumstances, the lymph nodes serve as way stations for migrating cancer cells. Thus, an analysis of lymph nodes can provide information on the spread of the cancer cells, and such information has a direct influence on the selection of appropriate therapies. In Chapter 29, we shall detail one example: identifying the stages of breast cancer by the degree of nodal involvement. *Lymphomas*, one group of cancers originating within the lymphatic system, are discussed in the *Applications Manual*. [AM] *Lymphomas*

The Thymus

The **thymus** lies posterior to the sternum, in the anterior portion of the mediastinum. It has a pinkish coloration and a grainy consistency. The thymus reaches its greatest size (relative to body size) in the first year or two after birth and its maximum absolute size during puberty, when it weighs between 30 and 40 g. Thereafter the thymus gradually decreases in size and becomes increasingly fibrous, a process called *involution*.

The capsule that covers the thymus divides it into two **thymic lobes** (Figure 22-8a,b●). Fibrous partitions, or **septae,** from the capsule divide the lobes into **lobules** averaging 2 mm in width (Figure 22-8b,c●). Each lobule consists of a densely packed outer **cortex** and a paler, central **medulla.** Lymphocytes in the cortex are dividing, and as the T cells mature, they migrate into the medulla. After roughly 3 weeks, these T cells leave the thymus by entering one of the medullary blood vessels.

THE CORTEX Lymphocytes in the cortex are arranged in clusters that are completely surrounded by **reticular epithelial cells.** These cells, which developed from epithelial cells of the embryo, also encircle the blood vessels of the cortex. The reticular epithelial cells (1) maintain the blood–thymus barrier and (2) secrete the thymic hormones (thymosins) that stimulate stem cell divisions and T cell differentiation.

THE MEDULLA As they mature, T cells leave the cortex and enter the medulla of the thymus. There is no blood–thymus barrier in the medulla. The reticular epithelial cells in the medulla cluster together in concentric layers, forming distinctive structures known as **Hassall's corpuscles** (Figure 22-8d●). Despite their imposing appearance, the function of Hassall's corpuscles remains unknown. T cells within the medulla can enter or leave the circulation via the blood vessels in this region or within one of the efferent lymphatics that collect lymph from the thymus.

The Spleen

The adult **spleen** contains the largest collection of lymphoid tissue in the body. It is about 12 cm (5 in.) long and weighs on average nearly 160 g. The spleen lies along the curving lateral border of the stomach, extending between the ninth and eleventh ribs on the left side. It is attached to the lateral border of the stomach by a broad mesenterial band, the **gastrosplenic ligament** (Figure 22-9a●, p. 780).

FUNCTIONS OF THE SPLEEN On gross dissection, the spleen has a deep red color due to the blood it contains. In essence, the spleen performs the same functions for the blood that lymph nodes perform for lymph. Splenic functions can be summarized as (1) the removal of abnormal blood cells and other blood components through phagocytosis, (2) the storage of iron from recycled red blood cells, and (3) the initiation of immune responses by B cells and T cells in response to antigens in the circulating blood.

SURFACES OF THE SPLEEN The spleen has a soft consistency, and its shape primarily reflects its association with the structures around it. It is in contact with the stomach, the left kidney, and the muscular diaphragm. The *diaphragmatic surface* is smooth and convex, conforming to the shape of the diaphragm and body wall. The *visceral surface* contains indentations that record the shape of the stomach (the *gastric area*) and of the kidney (the *renal area*) (Figure 22-9b●). Splenic blood vessels and lymphatics communicate with the spleen on the visceral surface at the **hilus,** a groove marking the border between the gastric and renal depressions. The **splenic artery,** the **splenic vein,** and the lymphatics draining the spleen are attached at the hilus.

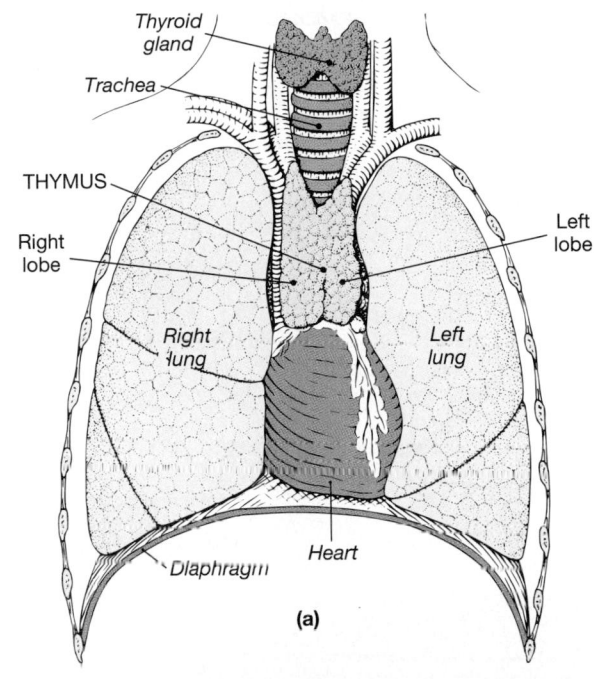

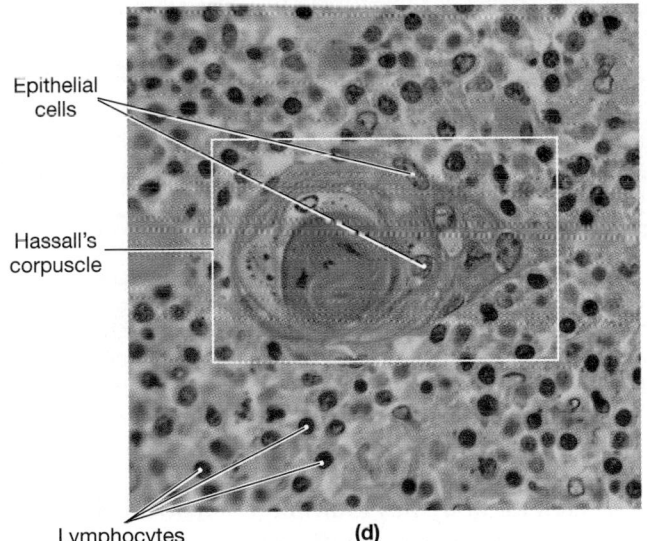

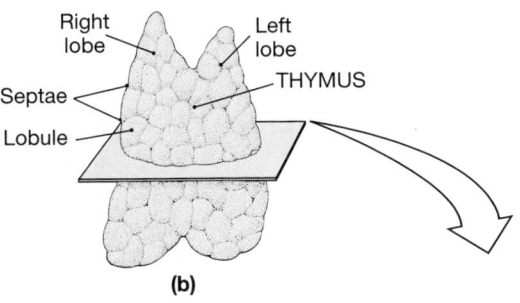

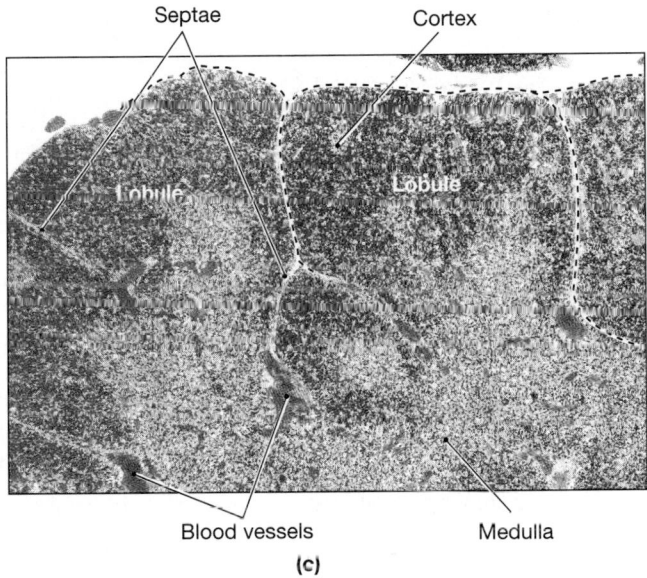

•FIGURE 22-8 The Thymus. (a) Appearance and position of the thymus in relation to other organs in the chest. **(b)** Anatomical landmarks on the thymus. **(c)** A low-power micrograph of the thymus. Note the fibrous septae that divide the thymic tissue into lobules resembling interconnected lymphatic nodules. (LM × 40) **(d)** At higher magnification, we can see the unusual structure of Hassall's corpuscles. The small cells in view are lymphocytes in various stages of development. (LM × 532)

HISTOLOGY OF THE SPLEEN The spleen is surrounded by a capsule containing collagen and elastic fibers.[1] The cellular components within constitute the **pulp** of the spleen (Figure 22-9c•). Areas of **red pulp** contain large quantities of red blood cells, whereas areas of **white pulp** resemble lymphoid nodules.

The splenic artery enters at the hilus and branches to produce a number of arteries that radiate outward toward the capsule. These **trabecular arteries** branch extensively, and their finer branches are surrounded by areas of white pulp. Capillaries then discharge the blood into the red pulp.

The cell population of the red pulp includes all the normal components of the circulating blood, plus fixed and free macrophages. The structural framework of the red pulp consists of a network of reticular fibers. The blood passes through this meshwork and enters large sinusoids, also lined by fixed macrophages. The sinusoids empty into small veins, and these ultimately collect into **trabecular veins** that continue toward the hilus.

This circulatory arrangement gives the phagocytes of the spleen an opportunity to identify and engulf any damaged or infected cells in the circulating blood. Lymphocytes are scattered throughout the red pulp, and the *marginal zone* surrounding each area of white pulp has a high concentration of macrophages and dendritic cells. Thus any microorganism or other antigen in the blood will quickly come to the attention of the splenic lymphocytes.

[1]The spleens of dogs and cats have extensive layers of smooth muscle that can contract to eject blood into the circulation; the human spleen lacks those muscle layers and cannot contract.

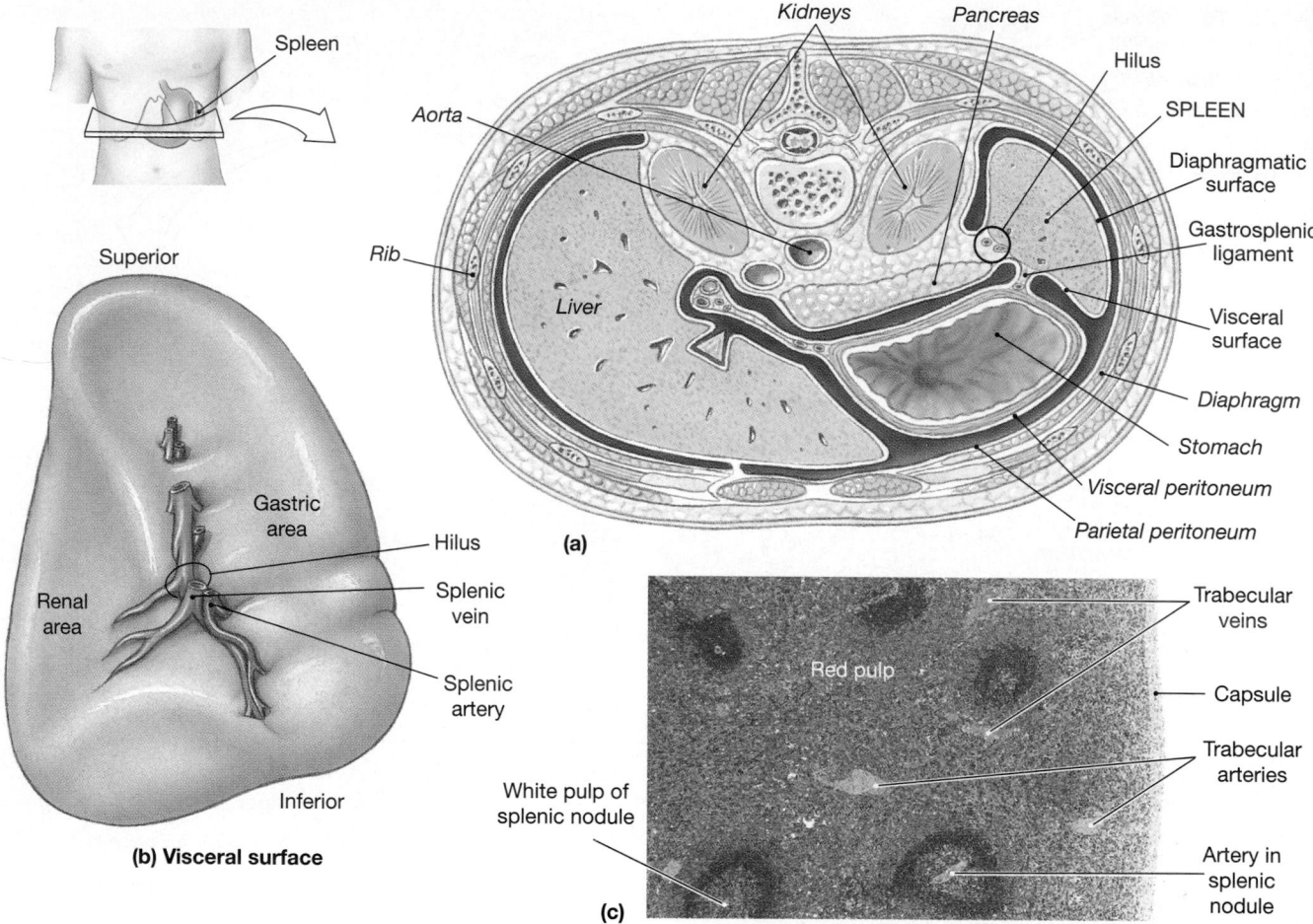

FIGURE 22-9 The Spleen. **(a)** The shape of the spleen roughly conforms to the shapes of adjacent organs. This transverse section through the trunk shows the typical position of the spleen within the abdominopelvic cavity. **(b)** External appearance of the intact spleen, showing major anatomical landmarks. Compare this view with that of part **(a)**. **(c)** Histological appearance of the spleen. Areas of white pulp are dominated by lymphocytes; they appear blue because the nuclei of lymphocytes stain very darkly. Areas of red pulp contain a preponderance of red blood cells. (LM × 38). AM *Plate 6.5*

INJURY TO THE SPLEEN An impact to the left side of the abdomen can distort or damage the spleen. Such injuries are known risks of contact sports, such as football and hockey, and more solitary athletic activities, such as skiing and sledding. However, the spleen tears so easily that a seemingly minor blow to the side may rupture the capsule. The result is serious internal bleeding and eventual circulatory shock.

Because the spleen is relatively fragile, it is very difficult to repair surgically. (Sutures typically tear out before they have been tensed enough to stop the bleeding.) Treatment for a severely ruptured spleen involves its complete removal, a process called a **splenectomy** (sple-NEK-to-mē).

The spleen may also be damaged through infection, inflammation, or invasion by cancer cells. These conditions and related symptoms are considered in the *Applications Manual*. AM *Disorders of the Spleen*

☑ How would blockage of the thoracic duct affect the circulation of lymph?

☑ If the thymus failed to produce thymosins, what particular population of lymphocytes would be affected?

☑ Why do lymph nodes enlarge during some infections?

The Lymphatic System and Body Defenses

The human body has multiple defense mechanisms, but these can be sorted into two general categories:

1. **Nonspecific defenses** do not discriminate between one threat and another. These defenses, which are present at birth, include *physical barriers, phagocytic cells, immunological surveillance, interferons, complement, inflammation,* and *fever.* They provide the body with a defensive capability known as *nonspecific resistance.*

2. **Specific defenses** protect against particular threats. For example, a specific defense may protect against infection by one type of bacteria but ignore other bacteria and viruses. Many specific defenses develop after birth, as a result of accidental or deliberate exposure to environmental hazards. *Specific defenses are dependent on the activities of lymphocytes.* The body's specific defenses produce a state of protection known as immunity, or **specific resistance.**

Nonspecific and specific resistances are complementary, and both must function normally to provide adequate resistance to infection and disease.

NONSPECIFIC DEFENSES

Nonspecific defenses prevent the approach, deny the entrance, or limit the spread of microorganisms or other environmental hazards. The seven major categories of nonspecific defenses are summarized in Figure 22-10●:

1. *Physical barriers* keep hazardous organisms and materials outside the body. For example, a mosquito that lands on your head may be unable to reach the surface of the scalp if you have a full head of hair.
2. *Phagocytes* are cells that engulf pathogens and cell debris. Examples of phagocytes are the macrophages of peripheral tissues and the microphages of the blood.
3. *Immunological surveillance* is the destruction of abnormal cells by NK cells in peripheral tissues.
4. *Interferons* are chemical messengers that coordinate the defenses against viral infection.
5. *Complement* is a system of circulating proteins that assist antibodies in the destruction of pathogens.
6. The *inflammatory response* is a local response to injury or infection that is directed at the tissue level. Inflammation tends to restrict the spread of an injury as well as combat an infection.
7. *Fever* is an elevation in body temperature that accelerates tissue metabolism and defenses.

Physical Barriers

To cause trouble, an antigenic compound or pathogen must enter the body tissues, and that means crossing an epithelium. The epithelial covering of the skin, described in Chapter 5, has multiple layers, a keratin coating, and a network of desmosomes that lock adjacent cells together. ∞ *[pp. 149–152]* These barriers provide very effective protection for underlying tissues. Even along the more delicate internal passageways of the respiratory, digestive, and urinary tracts, the epithelial cells are tied together by tight junctions and generally are supported by a dense and fibrous basement membrane.

In addition to the barriers posed by the epithelial cells, most epithelia are protected by specialized accessory structures and secretions. The hairs found in most areas of your body surface provide some protection against mechanical abrasion (especially on the scalp), and they often prevent hazardous materials or insects from contacting your skin surface. The epidermal surface also receives the secretions of sebaceous and sweat glands. These secretions flush the surface, washing away microorganisms and chemical agents. The secretions may also contain bactericidal chemicals, destructive enzymes (*lysozymes*), and antibodies.

The epithelia lining the digestive, respiratory, urinary, and reproductive tracts are more delicate, but they are equally well defended. Mucus bathes most surfaces of your digestive tract, and your stomach contains a powerful acid that can destroy many potential pathogens. Mucus moves across the lining of the respiratory tract, urine flushes the urinary passageways, and glandular secretions do the same for the reproductive tract. Special enzymes, antibodies, and an acidic pH may add to the effectiveness of these secretions.

Phagocytes

Phagocytes perform janitorial and police services in peripheral tissues, removing cellular debris and responding to invasion by foreign compounds or pathogenic organisms. These cells represent the "first line" of cellular defense against pathogenic invasion. Many phagocytes attack and remove the microorganisms before lymphocytes become aware of the incident. There are two general classes of phagocytic cells in the human body: *microphages* and *macrophages*.

Microphages

Microphages are the neutrophils and eosinophils that normally circulate in the blood. These phagocytic cells leave the bloodstream and enter peripheral tissues subjected to injury or infection. As we noted in Chapter 19, neutrophils are abundant, mobile, and quick to phagocytize cellular debris or invading bacteria. ∞ *[p. 657]* Eosinophils, which are less abundant, target foreign compounds or pathogens that have been coated with antibodies.

Macrophages

Macrophages are large, actively phagocytic cells. Your body contains several different types of macrophages, and most are derived from the monocytes of the circulating blood. Although there are no purely phagocytic organs or tissues, almost every tissue in the body shelters resident or visiting macrophages. This relatively diffuse collection of phagocytic cells has been called the **monocyte–macrophage system,** or the *reticuloendothelial system.*

An activated macrophage may respond to a pathogen in several different ways. For example:

- It may engulf a pathogen or other foreign object and destroy it with lysosomal enzymes.
- It may bind or remove a pathogen from the interstitial fluid but be unable to destroy it until assisted by other cells.
- It may destroy its target by releasing toxic chemicals, such as *tumor necrosis factor,* nitric oxide, or hydrogen peroxide, into the interstitial fluid.

We shall consider those responses in more detail in a later section.

FIXED MACROPHAGES **Fixed macrophages** are permanent residents of specific tissues and organs. These cells are normally incapable of movement, so the ob-

•FIGURE 22-10 Nonspecific Defenses

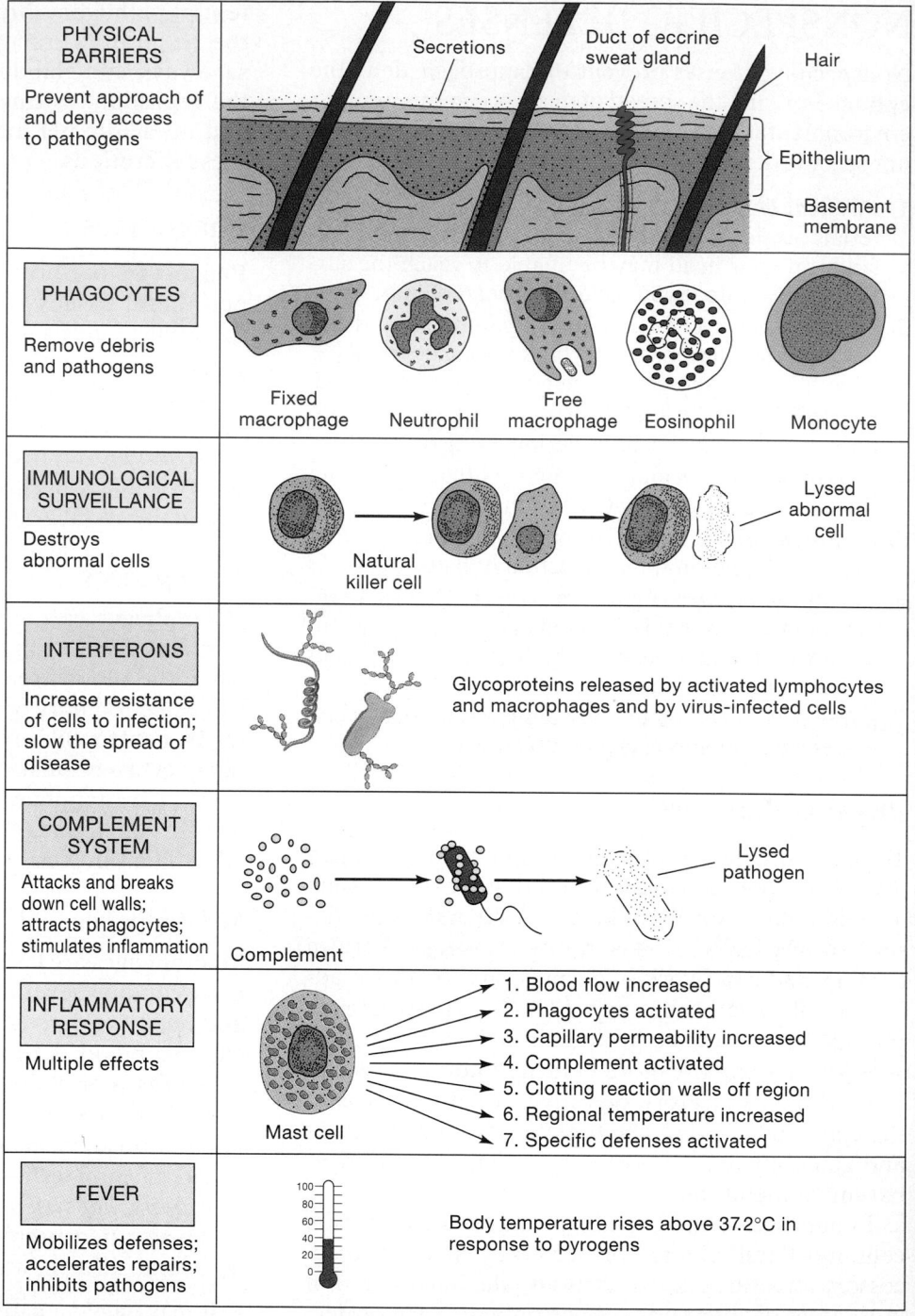

PHYSICAL BARRIERS	
Prevent approach of and deny access to pathogens	Secretions · Duct of eccrine sweat gland · Hair · Epithelium · Basement membrane
PHAGOCYTES	
Remove debris and pathogens	Fixed macrophage · Neutrophil · Free macrophage · Eosinophil · Monocyte
IMMUNOLOGICAL SURVEILLANCE	
Destroys abnormal cells	Natural killer cell · Lysed abnormal cell
INTERFERONS	
Increase resistance of cells to infection; slow the spread of disease	Glycoproteins released by activated lymphocytes and macrophages and by virus-infected cells
COMPLEMENT SYSTEM	
Attacks and breaks down cell walls; attracts phagocytes; stimulates inflammation	Complement · Lysed pathogen
INFLAMMATORY RESPONSE	
Multiple effects	Mast cell · 1. Blood flow increased 2. Phagocytes activated 3. Capillary permeability increased 4. Complement activated 5. Clotting reaction walls off region 6. Regional temperature increased 7. Specific defenses activated
FEVER	
Mobilizes defenses; accelerates repairs; inhibits pathogens	Body temperature rises above 37.2°C in response to pyrogens

jects of their phagocytic attention must diffuse or otherwise move through the surrounding tissue until they are within range. Fixed macrophages are scattered among connective tissues, usually in close association with collagen or reticular fibers. Their presence has been noted in the papillary and reticular layers of the dermis, in the subarachnoid space of the meninges, and in bone marrow. In some organs, the fixed macrophages have special names: **Microglia** are macrophages inside the CNS, and **Kupffer cells** are macrophages located in and around the liver sinusoids.

FREE MACROPHAGES Free macrophages, or *mobile macrophages,* travel throughout the body, arriving at an injury site by migration through adjacent tissues or by recruitment from the circulating blood. In the skeletal system, the fusion of macrophages produces the osteoclasts that dissolve and recycle the mineral content of bone. Some tissues contain free macrophages with distinctive characteristics. For example, the exchange surfaces of the lungs are patrolled by **alveolar macrophages,** also known as *phagocytic dust cells.*

Both classes of macrophages are derived from the monocytes of the blood, and the primary difference is that fixed macrophages are always located within a given tissue, but the free macrophages come and go. During an infection, this distinction commonly breaks down, for the fixed macrophages may lose their attachments and begin roaming around the damaged tissue.

Orientation and Phagocytosis

Free macrophages and microphages share a number of functional characteristics:

- Both can move through capillary walls by squeezing between adjacent endothelial cells, a process known as **diapedesis.** The endothelial cells in an injured area develop membrane "markers" that let passing blood cells know that something is wrong. The cells then attach to the endothelial lining and migrate into the surrounding tissues.
- Both may be attracted or repelled by chemicals in the surrounding fluids, a phenomenon called **chemotaxis.** They are particularly sensitive to cytokines released by other body cells and to the chemicals released by pathogens.
- Phagocytosis begins with **adhesion** of the phagocyte to its target. Adhesion depends on binding of the surface of the target to receptors on the cell membrane of the phagocyte. Adhesion is followed by the formation of a vesicle containing the bound target, as detailed in Figure 3-11a•, p. 80. The contents of the vesicle are then digested when the vesicle fuses with lysosomes or peroxisomes.

All phagocytic cells function in much the same way, although the items selected for phagocytosis may differ from one cell type to another. The life span of an actively phagocytic cell can be rather brief. For example, most neutrophils expire before they have engulfed more than 25 bacteria, and in an infection a neutrophil may attack that many in an hour.

Immunological Surveillance

The immune system generally ignores the body's own cells unless they become abnormal in some way. NK (natural killer) cells are responsible for recognizing and destroying abnormal cells when they appear in peripheral tissues. The constant monitoring of normal tissues by NK cells is called **immunological surveillance.**

The cell membrane of an abnormal cell generally contains antigens not found on the membranes of normal cells. NK cells recognize an abnormal cell by detecting the presence of those antigens. The recognition mechanism differs from that used by T cells or B cells. A T cell or B cell can be activated only by exposure to a *specific* antigen at a *specific* site on a cell membrane. An NK cell responds to a variety of abnormal antigens that may appear anywhere on a cell membrane. NK cells are therefore much less selective about their targets than are other lymphocytes; if a membrane contains abnormal antigens, it will be attacked. As a result, NK cells are very versatile:

A single NK cell can attack bacteria in the interstitial fluid, body cells infected with virus, or cancer cells.

NK cells also respond much more rapidly than T cells or B cells. The activation of T cells and B cells involves a relatively complex and time-consuming sequence of events. NK cells respond immediately on contact with an abnormal cell.

NK Cell Activation

An NK cell must first be activated, and then it reacts in a predictable way:

1. If a cell has unusual surface proteins or other components in its cell membrane, an NK cell recognizes that other cell as abnormal. Activation then occurs, and the NK cell adheres to its target cell.

2. The Golgi apparatus moves around the nucleus until the maturing face points directly toward the abnormal cell. This process might be compared to the rotation of a tank turret to point the cannon toward the enemy. A flood of secretory vesicles is then produced at the Golgi apparatus. These vesicles travel through the cytoplasm, to be released at the cell surface through exocytosis. The secretory vesicles contain proteins called **perforins.**

3. The vesicles are then released through exocytosis; the perforins diffuse across the gap between the NK cell and its target.

4. On reaching the opposing cell membrane, perforin molecules interact with one another and with the membrane. The result is the creation of a network of pores in the cell membrane (Figure 22-11•). The pores created by perforin are large enough to permit the free passage of ions, proteins, and other intracellular materials. The target cell can no longer maintain its internal environment, and it quickly disintegrates.

It is not clear why perforin does not affect the membrane of the NK cell itself. It has been proposed that the NK cell membranes contain a second protein, called *protectin,* that binds and inactivates perforin.

NK CELLS AND CANCER NK cells attack cancer cells and cells infected with viruses. Cancer cells probably appear throughout life, but their cell membranes generally contain unusual proteins called **tumor-specific antigens,** which NK cells recognize as abnormal. The affected cells are then destroyed, preserving tissue integrity. Unfortunately, some cancer cells avoid detection, perhaps because they lack tumor-specific antigens or because these antigens are covered in some way. Other cancer cells are able to destroy the NK cells that detect them. This process of avoiding detection or neutralizing body defenses is called **immunological escape.** Once immunological escape has occurred, cancer cells can multiply and spread without interference by NK cells.

NK CELLS AND VIRAL INFECTIONS NK cells are also important in fighting viral infections. Viruses reproduce

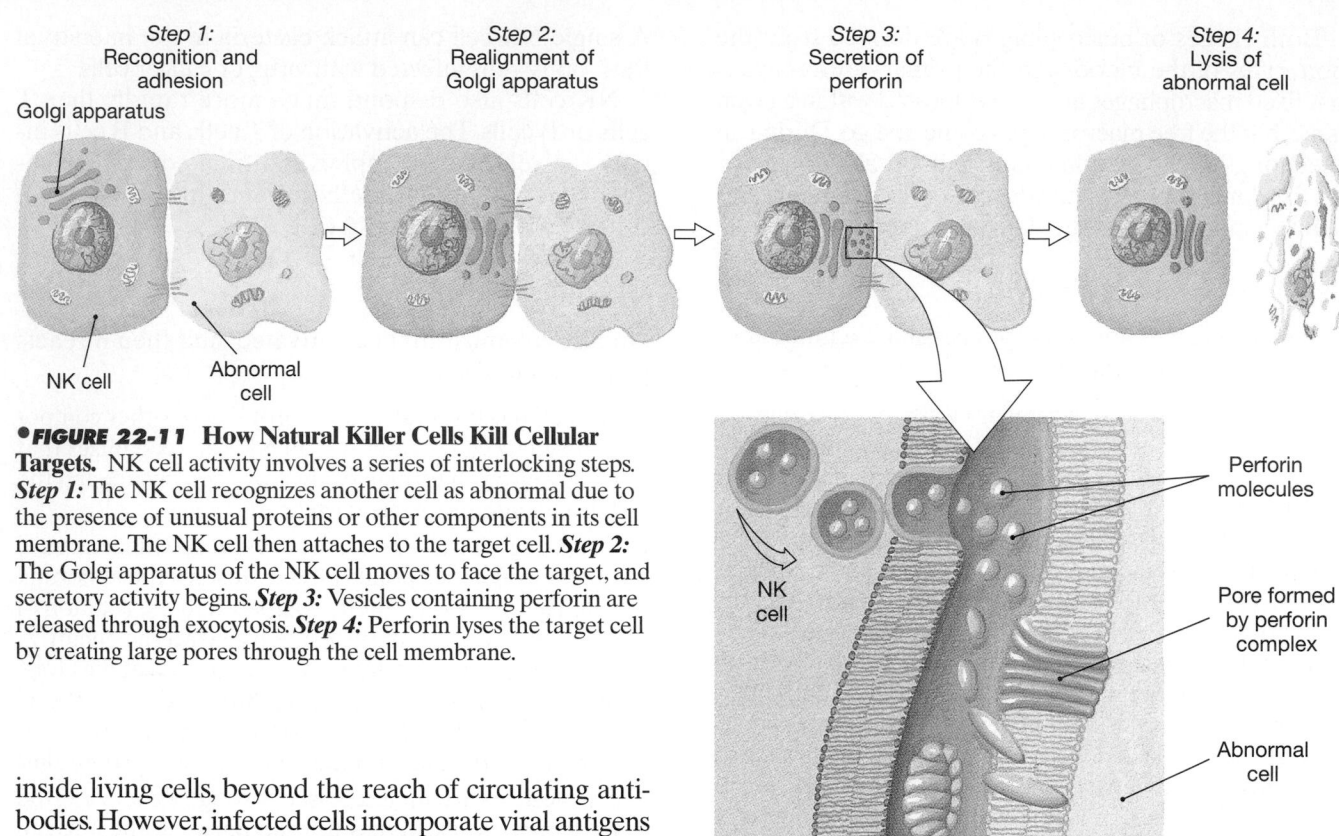

Step 1:
Recognition and
adhesion

Step 2:
Realignment of
Golgi apparatus

Step 3:
Secretion of
perforin

Step 4:
Lysis of
abnormal cell

Golgi apparatus

NK cell

Abnormal
cell

NK
cell

Perforin
molecules

Pore formed
by perforin
complex

Abnormal
cell

•*FIGURE 22-11* **How Natural Killer Cells Kill Cellular Targets.** NK cell activity involves a series of interlocking steps. *Step 1:* The NK cell recognizes another cell as abnormal due to the presence of unusual proteins or other components in its cell membrane. The NK cell then attaches to the target cell. *Step 2:* The Golgi apparatus of the NK cell moves to face the target, and secretory activity begins. *Step 3:* Vesicles containing perforin are released through exocytosis. *Step 4:* Perforin lyses the target cell by creating large pores through the cell membrane.

inside living cells, beyond the reach of circulating antibodies. However, infected cells incorporate viral antigens into their cell membranes, and NK cells recognize these infected cells as abnormal. By destroying them, NK cells can slow or prevent the spread of a viral infection.

Interferons

Interferons (in-ter-FĒR-onz) are small proteins released by activated lymphocytes and macrophages and by tissue cells infected with viruses. When an interferon reaches the membrane of a normal cell, it binds to surface receptors on the cell and, via second messengers, triggers the production of **antiviral proteins** in the cytoplasm. Antiviral proteins do not interfere with the entry of viruses, but they interfere with viral replication inside the cell. In addition to their role in slowing the spread of viral infections, interferons stimulate the activities of macrophages and NK cells.

There are at least three different interferons, each of which has additional specialized functions:

- *Alpha- (α) interferon*, produced by several different leukocytes, attracts and stimulates NK cells.
- *Beta- (β) interferon*, secreted by fibroblasts, slows inflammation in a damaged area.
- *Gamma- (γ) interferon*, secreted by T cells and NK cells, stimulates macrophage activity.

Most cell types other than lymphocytes and macrophages will respond to viral infection by secreting beta-interferon.

Interferons are examples of **cytokines** (SĪ-tō-kīnz), chemical messengers released by tissue cells to coordinate local activities. ∞ *[p. 591]* Cytokines produced by most cells are used only for paracrine communi-

cation. But the cytokines released by cellular defenders also act as hormones; they alter the activities of cells and tissues throughout the body. We shall discuss their role in the regulation of specific defenses in a later section.

Complement

Your plasma contains 11 special **complement proteins (C),** which form the **complement system.** The term *complement* refers to the fact that this system complements, or supplements, the action of antibodies.

The complement proteins interact with one another in chain reactions, or *cascades*, comparable to those of the clotting system. ∞ *[p. 664]* Figure 22-12• provides an overview of the complement system. If you are interested in further details, refer to the *Applications Manual*, which contains additional information about the reactions involved in the complement system. AM *A Closer Look: The Complement System*

Known effects of complement activation include:

- *Destruction of target cell membranes.* Five of the interacting complement proteins bind to the cell membrane. They form a group known as the **membrane attack complex (MAC),** which creates a pore in the membrane. The pores formed in this way are comparable to those produced by perforin and have the same effect. The target cell is soon destroyed.

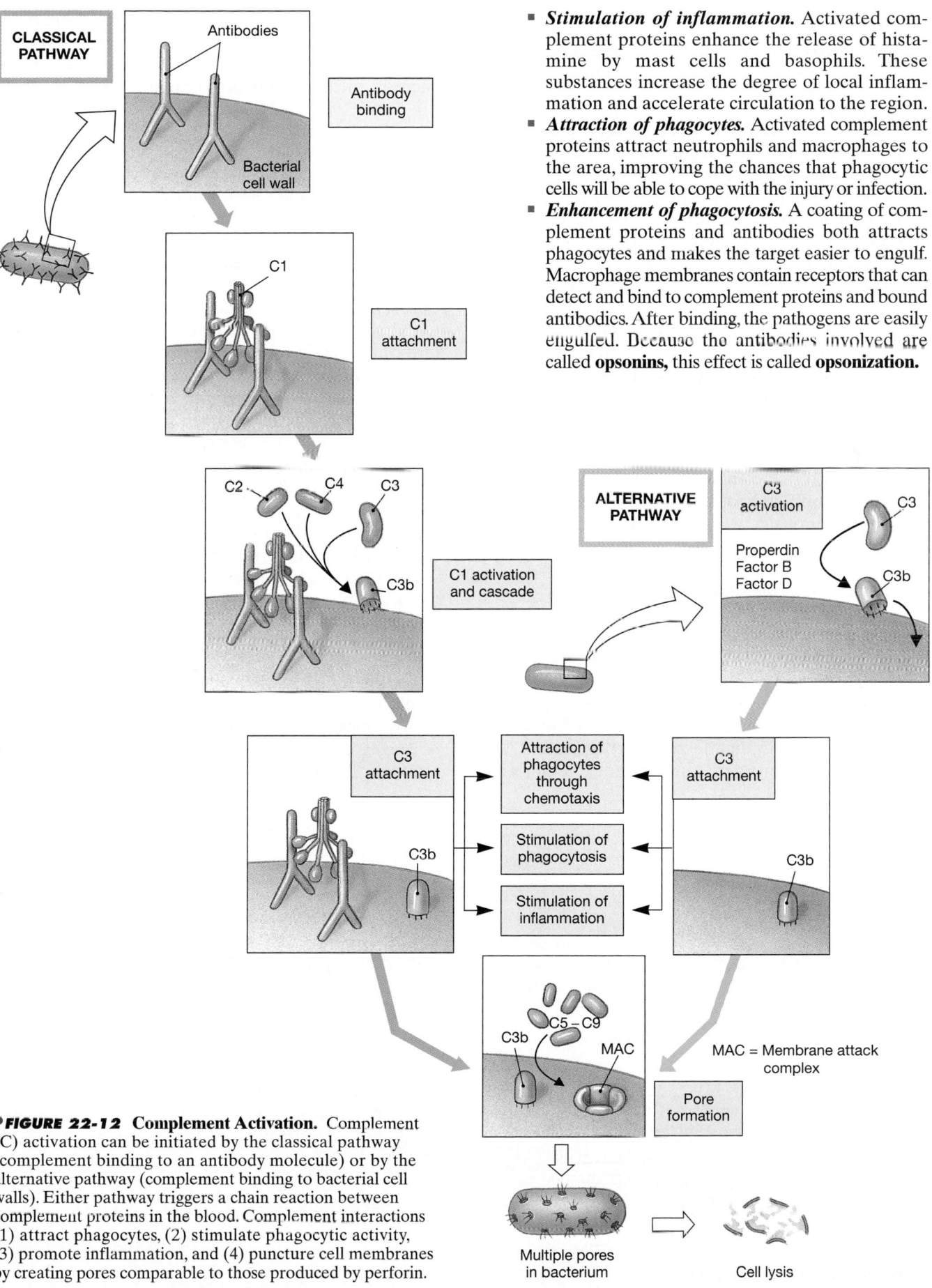

- ■ *Stimulation of inflammation.* Activated complement proteins enhance the release of histamine by mast cells and basophils. These substances increase the degree of local inflammation and accelerate circulation to the region.
- ■ *Attraction of phagocytes.* Activated complement proteins attract neutrophils and macrophages to the area, improving the chances that phagocytic cells will be able to cope with the injury or infection.
- ■ *Enhancement of phagocytosis.* A coating of complement proteins and antibodies both attracts phagocytes and makes the target easier to engulf. Macrophage membranes contain receptors that can detect and bind to complement proteins and bound antibodies. After binding, the pathogens are easily engulfed. Because the antibodies involved are called **opsonins,** this effect is called **opsonization.**

●FIGURE 22-12 Complement Activation. Complement (C) activation can be initiated by the classical pathway (complement binding to an antibody molecule) or by the alternative pathway (complement binding to bacterial cell walls). Either pathway triggers a chain reaction between complement proteins in the blood. Complement interactions (1) attract phagocytes, (2) stimulate phagocytic activity, (3) promote inflammation, and (4) puncture cell membranes by creating pores comparable to those produced by perforin.

Complement Activation

There are two routes of complement activation: (1) the *classical pathway* and (2) the *alternative pathway* (Figure 22-12●).

THE CLASSICAL PATHWAY The most rapid and effective activation of the complement system occurs via the **classical pathway** (Figure 22-12●). The process begins when one of the complement proteins (C1) binds to an antibody molecule already attached to its specific antigen. The bound complement protein then acts as an enzyme, catalyzing a series of reactions involving other complement proteins. The classical pathway ends with the conversion of an inactive complement protein, **C3,** to an active form, **C3b.** C3b binds the surface of the antigen and subsequently stimulates phagocytosis and promotes inflammation. It also triggers additional reactions that end with the creation of the membrane attack complex.

THE ALTERNATIVE PATHWAY A less effective, slower activation of the complement system occurs in the absence of antibody molecules. This **alternative pathway,** or *properdin pathway,* is important in the defense against bacteria, some parasites, and virus-infected cells. The pathway begins when several complement proteins, including **properdin,** or *factor P,* interact in the plasma (Figure 22-12●). This interaction can be triggered by exposure to foreign materials, such as the capsule of a bacterial cell. Like the classical pathway, the alternative pathway ends with the conversion of C3 to C3b, the stimulation of phagocytosis, and the subsequent formation of the membrane attack complex.

Some bacteria are unaffected by complement, because the complement proteins cannot interact with their cell membranes. These bacteria coat themselves in a protective capsule of carbohydrates. However, the classical pathway still provides a defense against these bacteria, once antibodies have bound to the capsule.

Inflammation

Inflammation, a localized tissue response to injury (introduced in Chapter 4), produces local sensations of swelling, redness, heat, and pain. ∞ *[p. 140]* Many stimuli can produce inflammation, including impact, abrasion, distortion, chemical irritation, infection by pathogenic organisms, or extreme temperatures (hot or cold). The common factor is that each of these stimuli kills cells, damages connective tissue fibers, or injures the tissue in some other way. These changes alter the chemical composition of the interstitial fluid. Damaged cells release prostaglandins, proteins, and potassium ions, and the injury itself may have introduced foreign proteins or pathogens. The changes in the interstitial environment trigger a complex process called *inflammation,* or the *inflammatory response.*

Figure 22-13● contains a summary of the events that occur during inflammation of the skin. Comparable events will occur in almost any tissue subjected to physical damage or infection. The three goals of inflammation are:

1. To perform a temporary repair at the injury site and prevent the access of additional pathogens.
2. To slow the spread of pathogens away from the injury site.
3. To mobilize local, regional, and systemic defenses that can overcome the pathogens and facilitate permanent repairs. The repair process is called *regeneration.*

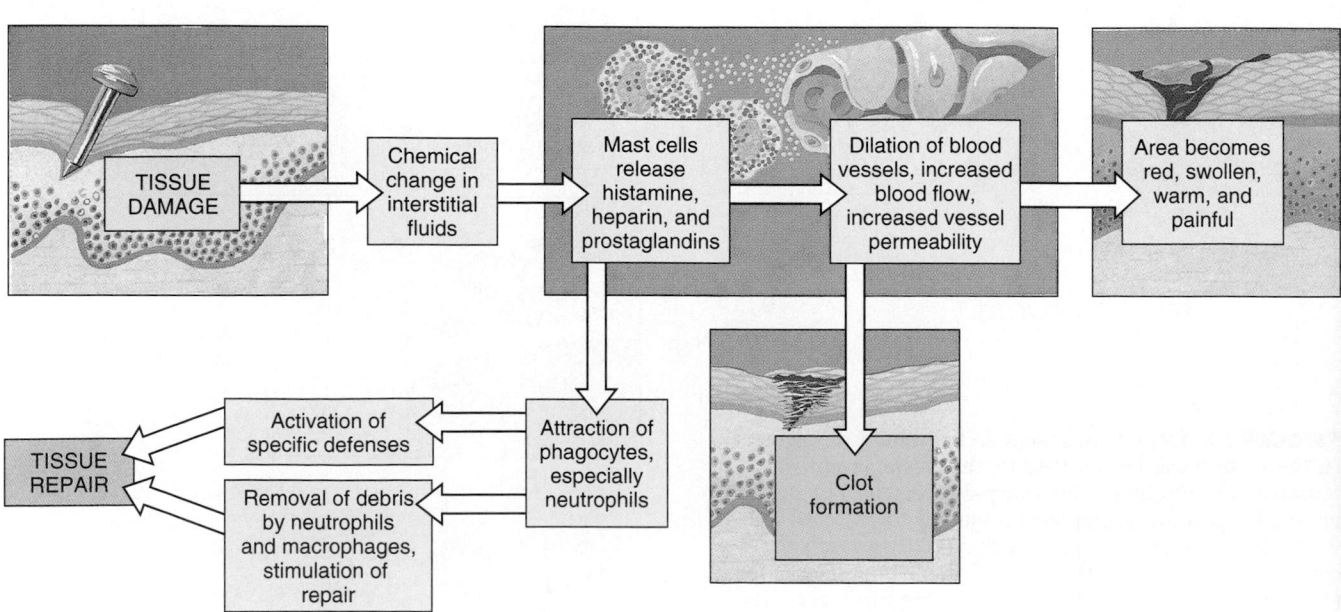

●*FIGURE 22-13* Inflammation

The Response to Injury

Mast cells play a pivotal role in the inflammatory process. When stimulated by mechanical stress or chemical changes in the local environment, these cells release histamine, heparin, prostaglandins, and other chemicals into the interstitial fluid. Events then proceed in a series of integrated steps:

1. The released histamine increases capillary permeability and accelerates blood flow through the area. The increased blow flow brings more cellular defenders to the site and carries away toxins and debris, diluting them and reducing their local impact.

2. Clotting factors and complement proteins leave the bloodstream and enter the injured or infected area. Clotting does not occur at the actual site of injury, due to the presence of heparin. However, a clot soon forms around the damaged area, and this clot both isolates the region and slows the spread of the chemical or pathogen into healthy tissues. Meanwhile, complement activation through the alternative pathway breaks down bacterial cell walls and attracts phagocytes.

3. The increased blood flow elevates the local temperature, increasing the rate of enzymatic reactions and accelerating the activity of phagocytes. The rise in temperatures may also denature foreign proteins or vital enzymes of invading microorganisms.

4. Debris and bacteria are attacked by neutrophils drawn to the area by chemotaxis. As they circulate through a blood vessel in an injured area, neutrophils undergo *activation*. In this process, (1) they stick to the side of the vessel and move into the tissue by diapedesis; (2) their metabolic rate goes up dramatically, and, while this *respiratory burst* continues, they generate reactive compounds, such as nitric oxide and hydrogen peroxide, that can destroy engulfed pathogens; and (3) they secrete cytokines that attract other neutrophils and macrophages to the area.

5. Fixed and free macrophages are also actively engulfing pathogens and cell debris. At first, these cells are outnumbered by neutrophils, but as the macrophages and neutrophils continue to secrete cytokines, the number of macrophages increases rapidly. Eosinophils may become involved if the foreign materials become coated with antibodies.

6. Other cytokines released by active phagocytes stimulate fibroblasts to begin barricading the area with scar tissue, reinforcing the clot and further slowing the invasion of adjacent tissues.

7. The combination of abnormal tissue conditions and chemicals released by mast cells stimulates local sensory neurons, producing sensations of pain. The individual is consciously aware of these sensations and may take steps to limit the damage, such as removing a splinter or cleaning a wound.

8. As inflammation is under way, the foreign proteins, toxins, microorganisms, and active phagocytes in the area activate the body's specific defenses.

 NECROSIS In the period following an injury, tissue conditions generally become even more abnormal before they begin to improve. **Necrosis** (ne-KRŌ-sis) refers to the tissue degeneration that occurs after cells have been injured or destroyed. The process begins several hours after the initial event, and the damage is caused by lysosomal enzymes. Lysosomes, described in Chapter 3, break down through autolysis, releasing digestive enzymes that first destroy the injured cells and then attack surrounding tissues. ∞ *[p. 90]*

As local inflammation continues, debris, fluid, dead and dying cells, and necrotic tissue components accumulate at the injury site. This viscous fluid mixture is known as **pus**. An accumulation of pus in an enclosed tissue space is called an **abscess**. [AM] *Complications of Inflammation*

Fever

Fever is the maintenance of a body temperature greater than 37.2°C (99°F). In Chapter 14, we noted the presence of a temperature-regulating center in the preoptic area of the hypothalamus. ∞ *[p. 468]* Circulating proteins called **pyrogens** (PĪ-rō-jenz) can reset the thermostat and cause a rise in body temperature. A variety of stimuli, including pathogens, bacterial toxins, and antigen–antibody complexes, either act as pyrogens themselves or stimulate the release of pyrogens by macrophages. The pyrogen released by active macrophages is a cytokine called **endogenous pyrogen**, or **interleukin-1** (in-ter-LOO-kin), abbreviated **Il-1.**

Within limits, a fever may be beneficial. High body temperatures may inhibit some viruses and bacteria, but the most likely beneficial effect is on body metabolism. For each 1°C rise in temperature, your metabolic rate jumps by 10 percent. Your cells can move more quickly, and enzymatic reactions occur at a faster rate. The net results may be faster mobilization of tissue defenses and an accelerated repair process.

☑ What types of cells would be affected by a decrease in the monocyte-forming cells in the bone marrow?

☑ A rise in the level of interferon in the body would suggest what kind of infection?

☑ What effects do pyrogens have in the body?

SPECIFIC RESISTANCE: THE IMMUNE RESPONSE

Specific resistance, or immunity, is provided by the coordinated activities of T cells and B cells, which respond to the presence of specific antigens. The basic functional relationship can be summarized as follows:

1. *T cells* are responsible for **cell-mediated immunity** (or **cellular immunity**), our defense against abnormal cells and pathogens inside living cells.

2. *B cells* provide **antibody-mediated immunity** (also called **humoral immunity**), our defense against antigens and pathogenic organisms in body fluids.

Both mechanisms are important, because they come into play under different circumstances. Activated T cells do not respond to antigenic materials in solution, and the antibodies produced by activated B cells cannot cross cell membranes. Moreover, helper T cells play a crucial role in antibody-mediated immunity by stimulating the activity of B cells.

Our understanding of immunity has greatly improved in the past decade or so, and a comprehensive discussion would involve hundreds of pages and thousands of details. The discussion that follows emphasizes important patterns and introduces general principles that will provide a foundation for future courses in microbiology and immunology.

Forms of Immunity

Immunity may be either *innate* or *acquired* (Figure 22-14●).

Innate Immunity

Innate immunity is genetically determined; it is present at birth and has no relationship to previous exposure to the antigen involved. For example, people are not subject to the same diseases as goldfish. Innate immunity breaks down only in the case of *AIDS* or other conditions that depress all aspects of specific resistance.

Acquired Immunity

Acquired immunity is not present at birth; you acquire immunity to a specific antigen only when you have been exposed to that antigen. Acquired immunity can be active or passive. **Active immunity** appears after exposure to an antigen, as a consequence of the immune response. The immune system is *capable* of defending against a large number of antigens. However, the appropriate defenses are mobilized only after you encounter a particular antigen. Active acquired immunity may develop as a result of (1) natural exposure to an antigen in the environment (naturally acquired immunity) or (2) from deliberate exposure to an antigen (induced active immunity). **Naturally acquired immunity** normally begins to develop after birth, and it is continually enhanced as you encounter "new" pathogens or other antigens. You might compare this process to vocabulary development; a child begins with a few basic common words and learns new ones on an as-needed basis. The purpose of **induced active immunity** is to stimulate antibody production under controlled conditions so that you will be able to overcome natural exposure to the pathogen at some time in the future. This is the basic principle behind immunization to prevent disease.

Passive immunity is produced by the transfer of antibodies from another individual. **Natural passive immunity** results when antibodies produced by the mother provide protection against infections during development (across the placenta) or in early infancy (through breast milk). In **induced passive immunity,** antibodies are administered to fight infection or prevent disease.

●**FIGURE 22-14** Types of Immunity

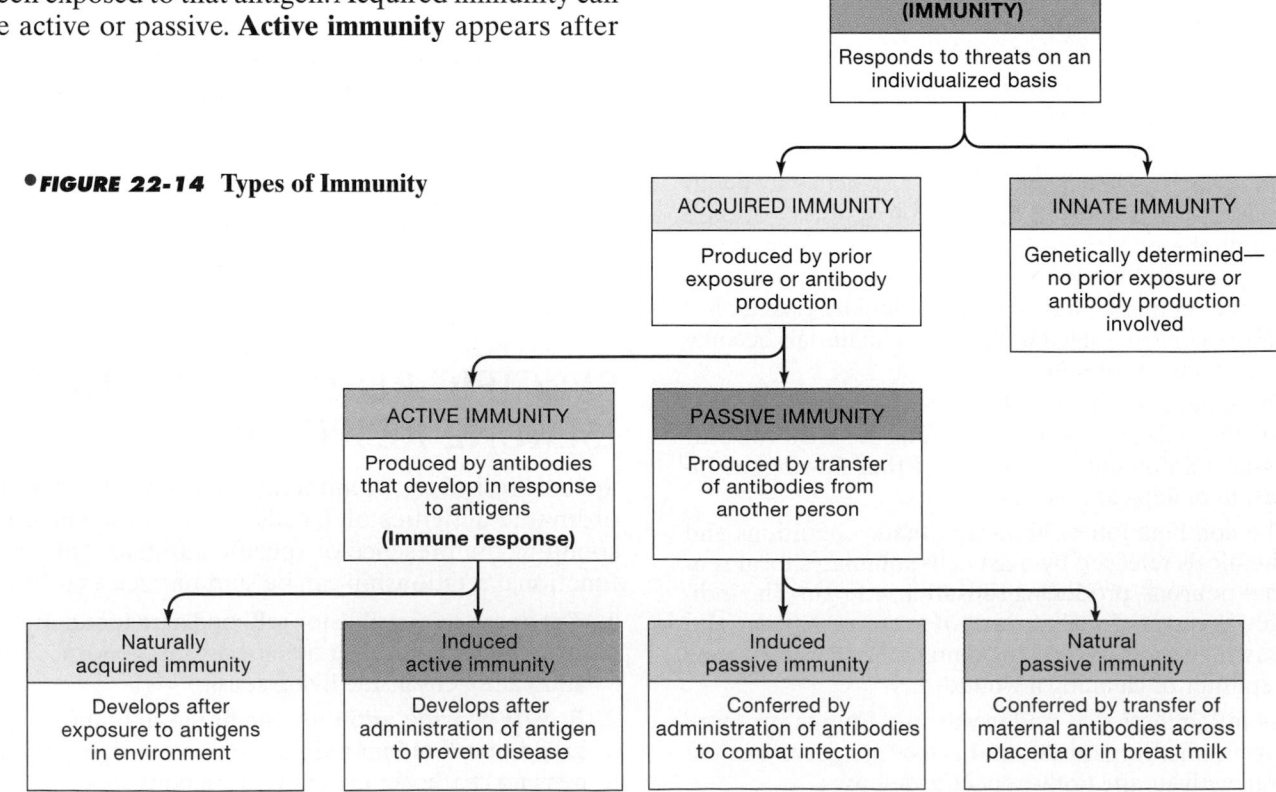

Properties of Immunity

We shall begin our discussion of immunity by noting four general properties:

1. *Specificity.* A specific defense is activated by an antigen, and the response targets that particular antigen and no others. **Specificity** results from the activation of appropriate lymphocytes and the production of antibodies with targeted effects.

2. *Versatility.* In the course of a normal lifetime, an individual encounters tens of thousands of antigens. The immune system can differentiate among them, producing appropriate and specific responses to each. **Versatility** results in part from the large diversity of lymphocytes present in the body and in part from variability in the structure of synthesized antibodies.

3. *Memory.* The immune system "remembers" antigens that it encounters. As a result of this **memory**, the immune response that occurs after a second exposure to an antigen is stronger and lasts longer than the response to the first exposure.

4. *Tolerance.* **Tolerance** is said to exist when the immune system does not respond to a particular antigen. For example, all cells and tissues in the body contain antigens that normally fail to stimulate an immune response.

Specificity

Specificity occurs because T cells and B cells respond to the molecular structure of an antigen. The antigen shape and size determine which lymphocytes will respond to its presence. Each T cell or B cell has receptors that will bind to one specific antigen, ignoring all others. The response of an activated T cell or B cell is equally specific. Either lymphocyte will destroy or inactivate that antigen without affecting other antigens or normal tissues.

Versatility

Millions of antigens in the environment may pose a threat to health. In your lifetime, only a relatively small fraction of those antigens will enter your body tissues. Your immune system, however, has no way of anticipating which antigens it will encounter. It must be ready to confront *any* antigen at *any* time.

During development, differentiation of cells in the lymphatic system produces an enormous number of lymphocytes with varied antigen sensitivities. The trillion or more T cells and B cells in the human body include millions of different lymphocyte populations. Each population consists of several thousand cells with receptors in their membranes that differ from those of other lymphocyte populations. As a result, each group of lymphocytes will respond to the presence of a different antigen. Representatives of each population are distributed throughout the body.

A few thousand lymphocytes are not enough to overcome a pathogenic invasion. When activated in the presence of an appropriate antigen, however, a lymphocyte begins to divide, producing more lymphocytes with the same specificity. Thus, whenever an antigen is encountered, more lymphocytes are produced. The term **clone** refers to all the cells produced by the division of an activated lymphocyte. *All the members of a clone are sensitive to the same specific antigen.*

To understand the basic principle, you might think about running a snack shop with only samples on display. When customers come in and make a selection, you prepare the food on the spot. This is an efficient way to do business. You can offer a wide selection, because the samples don't take up much space, and you do not have to expend energy preparing food that will not be eaten.

Memory

During the initial response to an antigen, lymphocytes sensitive to its presence undergo repeated cycles of cell division. Two kinds of cells are produced: some that attack the invader and others that remain inactive unless they are exposed to the same antigen at a later date. These inactive *memory cells* enable your immune system to "remember" antigens it has previously encountered and to launch a faster, stronger counterattack if one of them ever appears again.

Tolerance

The immune response targets foreign cells and compounds, but it generally ignores normal tissues. During their differentiation in the bone marrow (B cells) and thymus (T cells), cells that react to antigens normally present in the body are destroyed. As a result, mature B and T cells will ignore normal, or *self,* antigens and attack foreign, or *nonself,* antigens. Tolerance can also develop over time in response to chronic exposure to an antigen in the environment. Such tolerance lasts only as long as the exposure continues.

We shall now consider the origins and functions of T cells and B cells in greater detail.

T Cells and Cell-Mediated Immunity

T cells play a key role in the initiation, maintenance, and control of the immune response. We have already introduced three major types of T cells:

1. *Cytotoxic T cells* (T_C) are responsible for cell-mediated immunity. These cells enter peripheral tissues and subject antigens to direct physical and chemical attack.

2. *Helper T cells* (T_H) stimulate the responses of both T cells and B cells. They are absolutely vital to the immune response, because B cells must be activated by helper T cells before they will begin producing antibodies. The reduction in the helper T cell population that occurs in AIDS is largely responsible for the loss of immunity. (We shall discuss AIDS on p. 806.)

3. *Suppressor T cells* (T_S) inhibit T cell and B cell activities and moderate the immune response.

T Cell Activation

Before an immune response can begin, T cells must be activated by exposure to an antigen. This activation seldom happens by direct lymphocyte–antigen interaction, and foreign compounds or pathogens entering a tissue commonly fail to stimulate an immediate immune response.

ANTIGEN PRESENTATION T cells recognize antigens when they are bound to glycoproteins in cell membranes. **Antigen presentation** occurs when an antigen–glycoprotein combination capable of activating T cells appears in a cell membrane. Glycoproteins are integral membrane components introduced in Chapters 2, 3, and 19. ∞ *[pp. 50, 70, 653]* The structure of these glycoproteins is genetically determined. The genes controlling their synthesis are located along one portion of chromosome 6, in a region called the **major histocompatibility complex (MHC).** These membrane glycoproteins are called **MHC proteins,** or *human leukocyte antigens (HLAs).*

The amino acid sequences and shapes of MHC proteins differ from individual to individual. Each MHC molecule has a distinct three-dimensional shape with a relatively narrow central groove. An antigen that fits into this groove can be held in position by hydrogen bonding.

Two major classes of MHC proteins are known: *Class I* and *Class II.*

MHC Class I. **Class I** MHC proteins are in the membranes of all nucleated cells. These proteins are continuously being synthesized and exported to the cell membrane in vesicles created at the Golgi apparatus. As they form, Class I proteins pick up small peptides from the surrounding cytoplasm and carry them to the cell surface. If the cell is healthy and the peptides are normal, T

cells will ignore them. If the cytoplasm contains abnormal peptides, they will soon appear in the cell membrane, and T cells will be activated. Ultimately, their activation will lead to the destruction of the abnormal cell.

Viruses entering body tissues enter their target cells and begin replicating. In this process, viral proteins appear in the cell membranes of infected cells (Figure 22-15a●), bound to Class I proteins. T cells activated by contact with these antigens play an important role in the defense against viral infection.

MHC Class II. **Class II** MHC proteins are present only in the membranes of antigen-presenting cells and lymphocytes. **Antigen-presenting cells (APCs)** are spe-

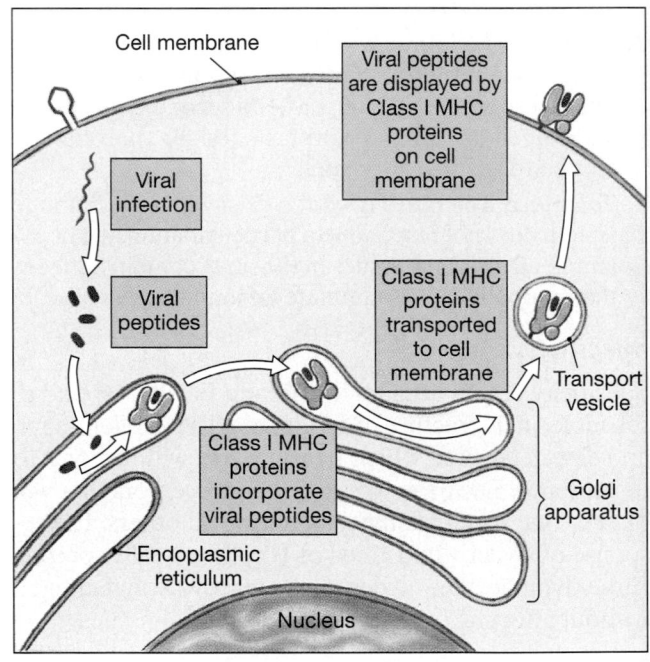

(a) Virus-infected cell

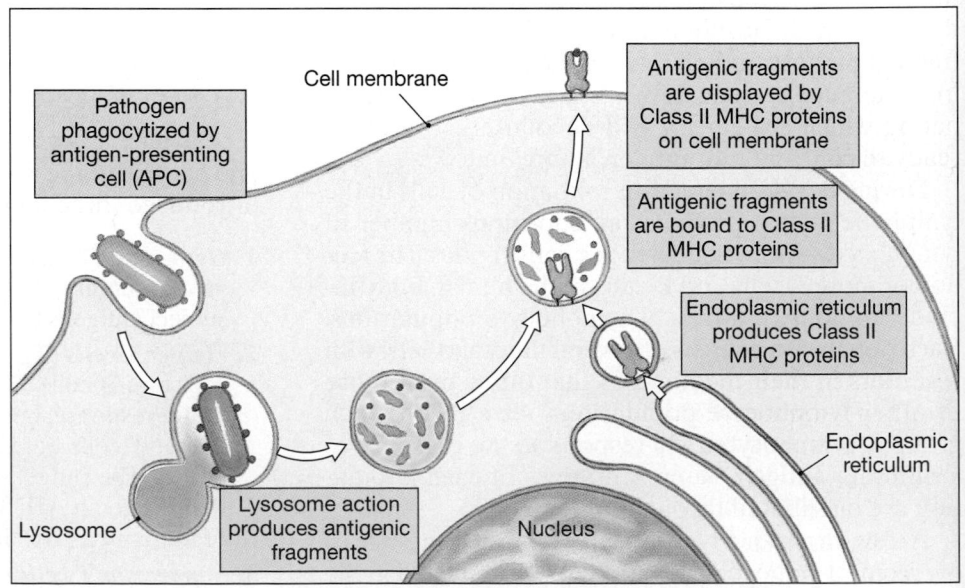

(b) Phagocytic antigen-presenting cell

●**FIGURE 22-15** **Antigens and MHC Proteins. (a)** Viral or other foreign antigens appear in cell membranes bound to Class I MHC proteins. **(b)** Processed antigens appear on the surfaces of antigen-presenting cells bound to Class II MHC proteins.

cialized cells responsible for activating T cell defenses against foreign cells (including bacteria) and foreign proteins. Antigen-presenting cells include all the phagocytic cells of the monocyte–macrophage group discussed in other chapters, including (1) the free and fixed macrophages in connective tissues (Chapter 4); (2) the Kupffer cells of the liver, and (3) the microglia in the CNS (Chapter 12). ∞ *[pp. 120, 378]* The Langerhans cells of the skin and the dendritic cells of the lymph nodes and spleen are APCs that are not phagocytic. ∞ *[p. 150]*

Phagocytic APCs engulf and break down pathogens or foreign antigens. This **antigen processing** creates antigenic fragments, which are then bound to Class II MHC proteins and inserted into the cell membrane (Figure 22-15b•). *Class II MHC proteins appear in the cell membrane only when the cell is processing antigens.* Exposure to an APC membrane containing processed antigen can stimulate appropriate T cells.

The Langerhans cells and dendritic cells remove antigenic materials from their surroundings by pinocytosis rather than by phagocytosis. The cell membrane of these cells present antigens that are bound to Class II MHC proteins.

ANTIGEN RECOGNITION Inactive T cells have receptors that recognize Class I or Class II MHC proteins. The receptors also have binding sites that will detect the presence of specific bound antigens. If an MHC protein contains any antigen other than its specific target, the T cell will remain inactive. If it contains the antigen that the T cell is programmed to detect, binding will occur. This process is called **antigen recognition**, because the T cell recognizes that it has found an appropriate target.

Some T cells can recognize antigens bound to Class I MHC proteins, whereas others can recognize antigens bound to Class II MHC proteins. Whether a T cell responds to antigens held by Class I or Class II proteins depends on the structure of the T cell membrane. The membrane proteins involved are members of a larger class of proteins called **CD** (*cluster designation*) **markers.**

CD Markers. Lymphocytes, macrophages, and other related cells have CD markers. There are over 50 different types of CD markers, each designated by an identifying number. We shall consider only the two CD markers associated with the MHC receptor complex:

1. **CD8** markers are found on cytotoxic T cells and suppressor T cells. These cells are often called *CD8 T cells,* or *CD8+ T cells.* CD8 T cells can be activated by exposure to antigens presented by Class I MHC proteins (Figure 22-16•).

2. **CD4** markers are found on helper T cells. These cells are often called *CD4 T cells,* or *CD4+ T cells.* Helper T cells respond to antigens presented by Class II MHC proteins (Figure 22-17•).

COSTIMULATION A T cell does not become active immediately after recognizing an antigen. Antigen recognition simply prepares the cell for activation. Before activation can occur, a T cell must bind to the stimulating cell at a second site. This vital secondary binding process is called *costimulation.* Costimulation essentially confirms the "OK to activate" signal. Appropriate costimulation proteins appear in the presenting cell only if that cell has engulfed antigens or is

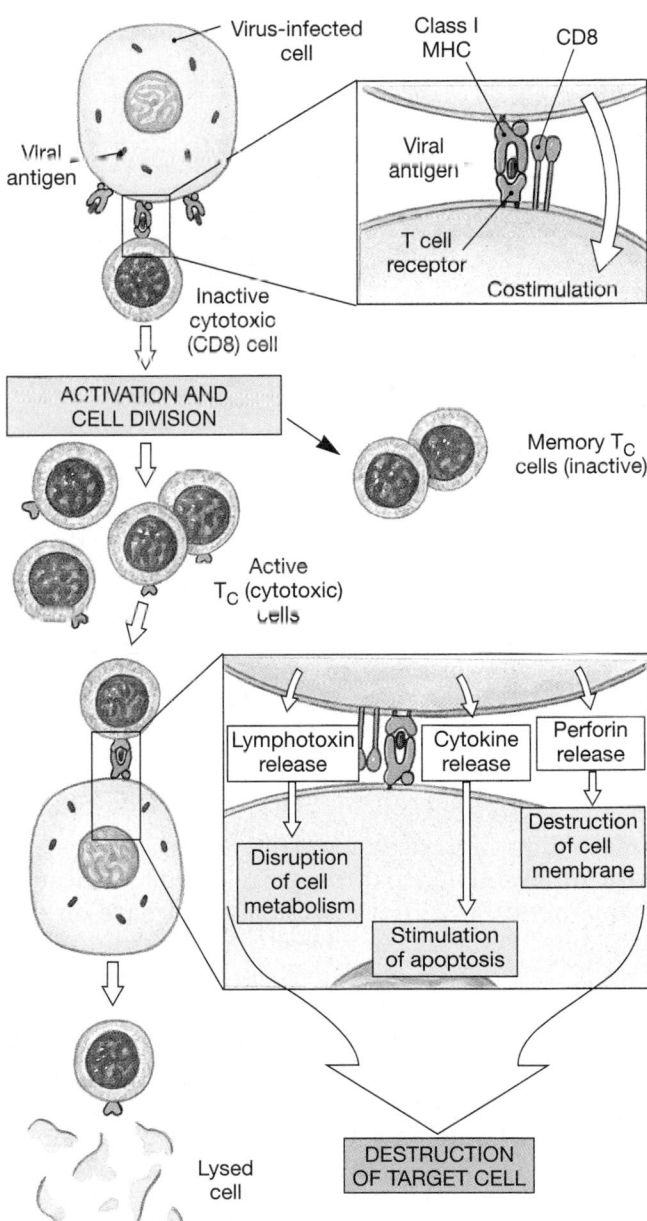

•FIGURE 22-16 **Antigen Recognition and the Activation of Cytotoxic T Cells.** An inactive cytotoxic T cell must encounter an appropriate antigen bound to Class I MHC proteins. It must also receive costimulation from the membrane it contacts. It is then activated and undergoes divisions that produce memory T_C cells and active T_C cells. When one of these active cells encounters a membrane displaying the target antigen, the T_C cell will use one of several methods to destroy the cell.

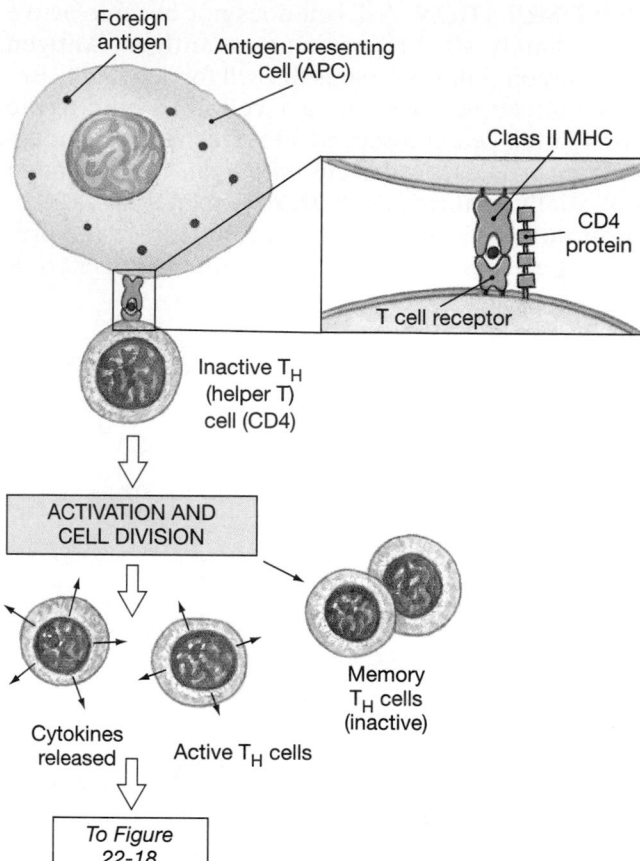

•FIGURE 22-17 **Antigen Recognition and Activation of Helper T Cells.** Inactive helper T cells must be exposed to appropriate antigens bound to Class II MHC proteins. The T_H cells then undergo activation, dividing to produce memory T_H cells and active T_H cells. Active T_H cells secrete cytokines that stimulate cell-mediated and antibody-mediated immunities. They also interact with sensitized B cells, as Figure 22-18 shows.

infected by viruses; many costimulation proteins are structurally related to cytokines released by activated lymphocytes. The effects these proteins have on the exposed T cell vary but typically include the stimulation of transcription at the nucleus, thereby promoting cell division and differentiation.

Costimulation is like the safety on a gun: It helps prevent T cells from mistakenly attacking normal tissues. If a cell displays an unusual antigen but does not display the "I am an active phagocyte" or "I am infected" signal, T cell activation will not occur. Costimulation is important only in determining whether a T cell will become activated. Once activation has occurred, the safety is off and the T cell will attack any cells that carry the target antigens.

We shall now consider the functions of CD8 and CD4 T cells in the immune response.

Cytotoxic T Cells

When activated, CD8 T cells undergo a series of divisions that generate cytotoxic T cells, or T_C cells, and

memory T_C cells. The cytotoxic T cells, also called **killer T cells,** roam through the injured tissue. When a cytotoxic T cell encounters its target antigens bound to Class I MHC proteins of another cell, it immediately destroys that cell. A cytotoxic T cell may accomplish this destruction in three ways (Figure 22-16•):

1. By rupturing the antigenic cell membrane through the release of perforin.
2. By killing the target cell by secreting a poisonous **lymphotoxin** (lim-fō-TOK-sin).
3. By activating genes within the cell's nucleus that tell the cell to die. We introduced the process of genetically programmed cell death, called *apoptosis* (ap-op-TŌ-sis; *apo,* away + *ptosis,* a falling), in Chapter 3. ∞ *[p. 98]*

The entire sequence of events, from appearance of the antigen in a tissue to cell destruction by cytotoxic T cells, takes a significant amount of time. After first exposure to an antigen, 2 days or more may pass before the concentration of cytotoxic T cells reaches effective levels at the site of an injury or infection. Over this period, the damage or infection may spread, making it more difficult to control.

MEMORY T_C CELLS Memory T_C cells ensure that there will not be a delay in the response if the antigen reappears. These cells do not differentiate further the first time the antigen triggers an immune response, although thousands of them are produced. Instead, they remain in reserve. If the same antigen appears a second time, these cells will *immediately* differentiate into cytotoxic T cells, producing a swift, effective cellular response that can overwhelm an invading organism before it becomes well established in the tissues.

Suppressor T Cells

Suppressor T cells, or T_S cells, depress the responses of other T cells and B cells by secreting *suppression factors,* inhibitory cytokines of unknown structure. This suppression does not occur immediately: Suppressor T cell activation takes much longer than the activation of other types of T cells. In addition, most of the CD8 T cells in circulation will on activation produce cytotoxic T cells rather than suppressor T cells. As a result, suppressor T cells act *after* the initial immune response. In effect, these cells put on the brakes and limit the degree of immune system activation from a single stimulus.

Helper T Cells

Upon activation, CD4 T cells undergo a series of cell divisions that produce helper T cells, or T_H cells, and **memory T_H cells** (Figure 22-17•). The memory T_H cells remain in reserve, whereas the helper T cells secrete a variety of cytokines that (1) coordinate specific and nonspecific defenses and (2) stimulate cell-mediated and antibody-mediated immunities. For example, activated helper T cells secrete cytokines that:

1. Stimulate the T cell divisions that produce memory T cells and accelerate maturation of cytotoxic T cells.
2. Enhance nonspecific defenses by attracting macrophages to the area, preventing their departure, and stimulating their phagocytic activity and effectiveness.
3. Attract and stimulate the activity of NK cells, providing another mechanism for the destruction of abnormal cells and pathogens.
4. Promote B cell division, plasma cell maturation, and antibody production.

In addition, activated helper T cells participate in the direct activation of B cells sensitive to the antigen involved. We shall consider this aspect of helper T cell function in the next section.

GRAFT REJECTION AND IMMUNOSUPPRESSION Organ transplantation is a treatment option for patients with severe disorders of the kidneys, liver, heart, lungs, or pancreas. Finding a suitable donor is the first major problem. In the United States, each day 6 people die while awaiting an organ transplant, and another 48 people are added to the transplant waiting list. After surgery has been performed, the major problem is **graft rejection.** In graft rejection, T cells are activated by contact with MHC proteins on cell membranes in the donated tissues. The cytotoxic T cells that develop then attack and destroy the foreign cells.

Significant improvements in transplant success can be made by reducing the sensitivity of the immune system. Until recently, the drugs used to produce this **immunosuppression** did not selectively target the immune response. For example, **prednisone** (PRED-ni-sōn), a corticosteroid, was used because it has anti-inflammatory effects that reduce the number of circulating white blood cells and depress the immune response. However, corticosteroid use also caused undesirable changes in glucose metabolism and other side effects. ∞ *[p. 617]*

An understanding of the communication among T cells, macrophages, and B cells has now led to the development of drugs with more selective effects. **Cyclosporin A (CsA),** a compound derived from a fungus, was the most important *immunosuppressive drug* developed in the 1980s. This compound depresses all aspects of the immune response, primarily by suppressing helper T cell activity while leaving suppressor T cells relatively unaffected. [AM] *Transplants and Immunosuppressive Drugs*

B Cells and Antibody-Mediated Immunity

B cells are responsible for launching a chemical attack on antigens through the production of appropriate *antibodies*. B cell activation proceeds in a series of steps diagrammed in Figure 22-18•.

B Cell Sensitization

As we noted earlier, there are millions of different populations of B cells. Each B cell carries its antibody molecules in its cell membrane. If antigens that will bind to those antibodies appear in the interstitial fluid,

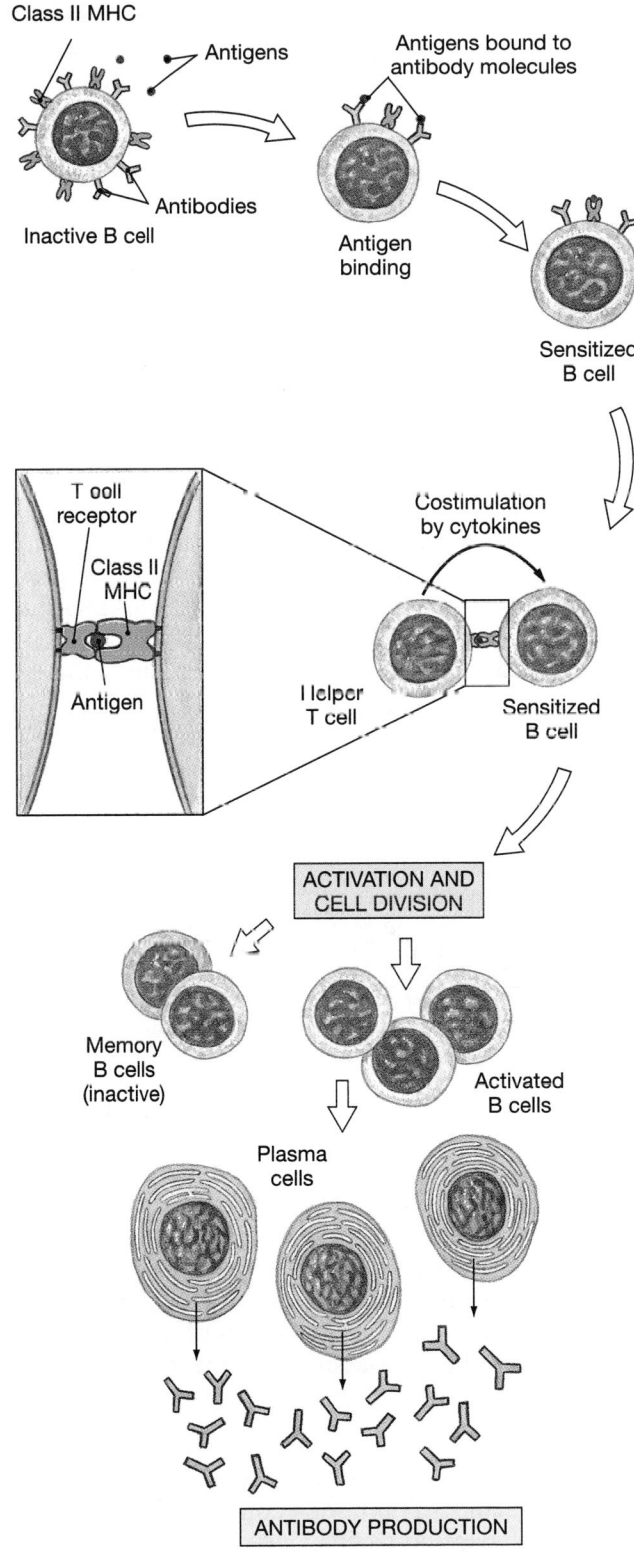

•*FIGURE 22-18* **Sensitization and Activation of B Cells.** A B cell is sensitized through exposure to antigens that bind to antibodies contained in the B cell membrane. The B cell then displays those antigens in its cell membrane. Activated helper T cells encountering those antigens release cytokines that trigger the activation of the B cell. The activated B cell then goes through cycles of division producing memory B cells and B cells that differentiate into plasma cells. The plasma cells secrete antibodies.

the B cell becomes **sensitized.** Because B cells migrate throughout the body, pausing briefly in one lymphoid tissue or another, sensitization typically occurs at the lymph node closest to the site of infection or injury.

During sensitization, antigenic materials enter the cell through endocytosis, and they then appear on the surface of the B cell, bound to Class II MHC proteins. A sensitized B cell is now ready to go, but it generally will not undergo activation unless it receives the OK from a helper T cell.[2] The need for activation by a helper T cell helps prevent inappropriate activation, the same way that costimulation acts as a "safety" for cell-mediated immunity.

B Cell Activation

A sensitized B cell next encounters a helper T cell already activated by antigen presentation. The helper T cell binds to the MHC complex, recognizes the presence of an antigen, and begins secreting cytokines that (1) promote B cell activation, (2) stimulate B cell division, (3) accelerate plasma cell formation, and (4) enhance antibody production.

Figure 22-18 • diagrams the events that occur when a B cell is activated by a helper T cell. The activated B cell typically divides several times, producing daughter cells that differentiate into plasma cells and **memory B cells.** The plasma cells begin synthesizing and secreting large numbers of antibodies into the interstitial fluid. These antibodies have the same target as the antibodies on the surface of the sensitized B cell. When stimulated by cytokines from helper T cells, a plasma cell can secrete up to 100 million antibody molecules each hour.

MEMORY B CELLS Memory B cells perform the same role for antibody-mediated immunity that memory T cells perform for cell-mediated immunity. Memory B cells do not respond to a threat on first exposure. Instead, they remain in reserve to deal with subsequent injuries or infections that involve the same antigens. On subsequent exposure, the memory B cells respond by dividing and differentiating into plasma cells that secrete antibodies in massive quantities.

Antibody Structure

Figure 22-19 • presents different views of a single antibody molecule. An antibody consists of two parallel pairs of polypeptide chains: one pair of **heavy chains** and one pair of **light chains**. Each chain contains *constant* and *variable segments.*

---

[2]Under certain circumstances, complex antigens can directly stimulate B cells. The B cell must bind multiple antigenic molecules simultaneously in a single, localized portion of its cell membrane. This process, called *capping*, is relatively rare.

THE CONSTANT SEGMENTS The constant segments of the heavy chains form the base of the antibody molecule (Figure 22-19a,b•). B cells produce only five different types of constant segments. These form the basis of a classification scheme that identifies antibodies as *IgG, IgE, IgD, IgM,* or *IgA,* as we shall discuss in the next section. The structure of the constant segments of the heavy chains determines the way the antibody is secreted and how it is distributed within the body. For example, antibodies in one class circulate in body fluids, whereas another type binds to the membranes of basophils and mast cells.

The heavy chain constant segments, which are bound to constant segments of the light chains, also contain binding sites that can activate the complement system. These binding sites are covered when the antibody is secreted, but they become exposed when the antibody binds to an antigen.

THE VARIABLE SEGMENTS The specificity of the antibody molecule depends on the structure of the variable segments of the light and heavy chains. The free tips of the two variable segments contain the **antigen binding sites** of the antibody molecule (Figure 22-19a•). These sites can interact with an antigen in the same way that the active site of an enzyme interacts with a substrate molecule. ∞ *[pp. 54–56]*

Small differences in the amino acid sequence of the variable segments affect the precise shape of the antigen binding site. These differences account for differences in specificity between the antibodies produced by different B cells. The distinctions are the result of minor genetic variations that occur during the production, division, and differentiation of B cells. The normal adult body contains roughly 10 trillion B cells. It has been estimated that these cells can produce 100 million different types of antibodies, each with a different specificity.

The Antigen–Antibody Complex. When an antibody molecule binds to its corresponding antigen, an **antigen–antibody complex** is formed. Once the two molecules are in position, hydrogen bonding and other weak chemical forces lock them together.

Antibodies do not bind to the entire antigen. They bind to certain portions of its exposed surface, regions called **antigenic determinant sites** (Figure 22-19c•). The specificity of that binding depends initially on the three-dimensional "fit" between the variable segments of the antibody molecule and the corresponding antigenic determinant sites. A **complete antigen** is an antigen with at least two antigenic determinant sites, one for each of the antigen binding sites on an antibody molecule. Exposure to a complete antigen can lead to B cell sensitization and a subsequent immune response. Most environmental antigens have multiple antigenic determinant sites; entire microorganisms may have thousands.

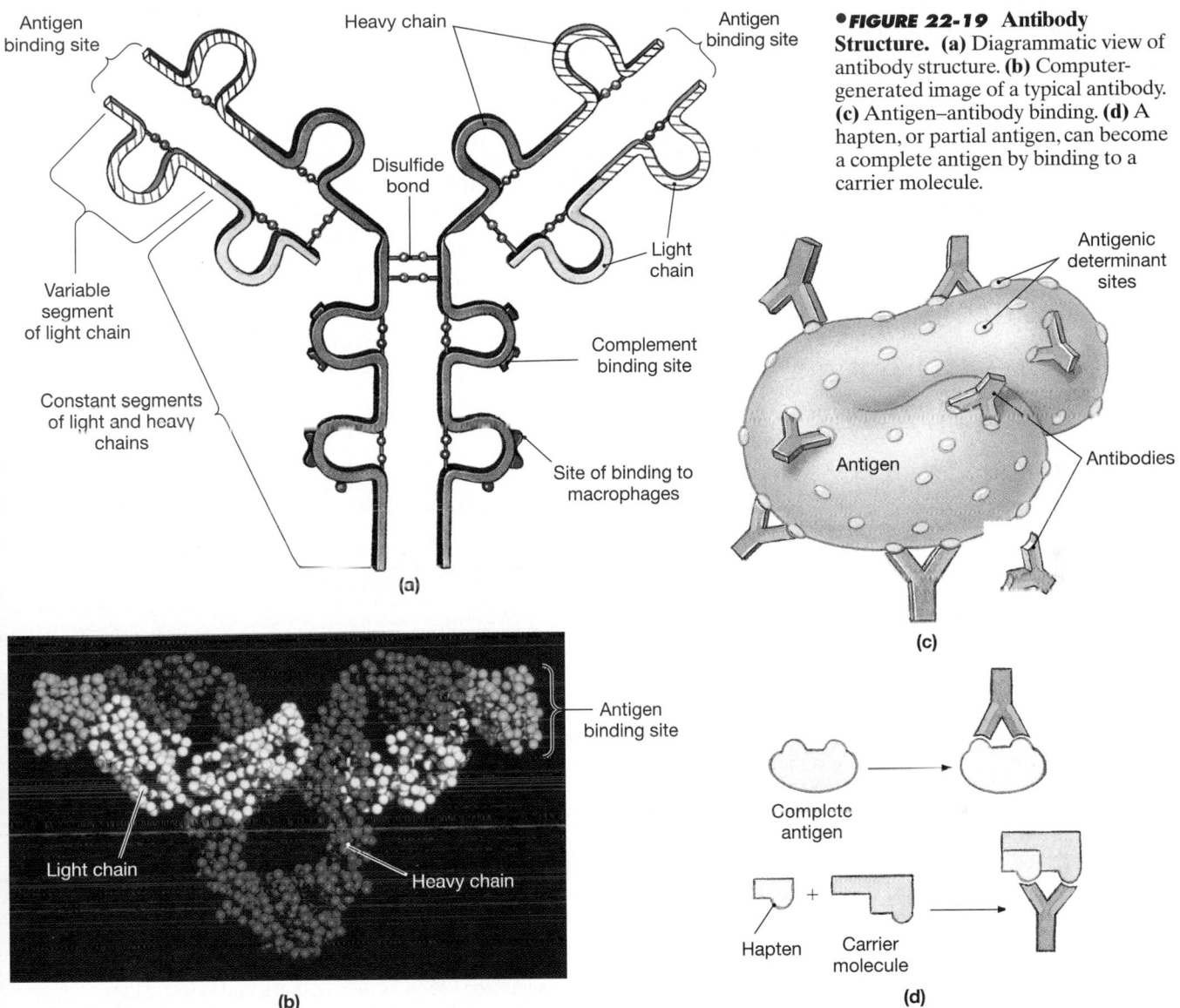

• *FIGURE 22-19* **Antibody Structure.** **(a)** Diagrammatic view of antibody structure. **(b)** Computer-generated image of a typical antibody. **(c)** Antigen–antibody binding. **(d)** A hapten, or partial antigen, can become a complete antigen by binding to a carrier molecule.

DRUG REACTIONS **Haptens,** or **partial antigens,** do not ordinarily cause B cell activation and antibody production. Haptens include short peptide chains, steroids and other lipids, and several drugs, including antibiotics such as *penicillin*. Haptens may, however, become attached to carrier molecules, forming combinations that can function as complete antigens (Figure 22-19d •). In some cases, the carrier contributes an antigenic determinant site. The antibodies produced will attack both the hapten and the carrier molecule. If the carrier molecule is normally present in the tissues, the antibodies may begin attacking and destroying normal cells. This process is the basis for several drug reactions, including penicillin allergies.

Classes and Actions of Antibodies

There are five classes of antibodies, or **immunoglobulins (Ig),** in body fluids: *IgG, IgE, IgD, IgM,* and *IgA* (Table 22-1). The classes are determined by variations in the structure of the heavy chain constant segments and so

have no effect on the specificity of the antibody. The formation of an antigen–antibody complex may cause the elimination of the antigen in seven different ways:

1. *Neutralization.* Both viruses and bacterial toxins have specific sites that must bind to target regions on body cells in order to enter or injure them. Antibodies may bind to those sites, making the virus or toxin incapable of attaching itself to a cell. This mechanism is known as **neutralization.**

2. *Agglutination and precipitation.* Each antibody molecule has two antigen binding sites, and most antigens have many antigenic determinant sites. If separate antigens (such as macromolecules or bacterial cells) are far apart, an antibody molecule will necessarily bind to two antigenic sites on the same antigen. If antigens are close together, however, an antibody can bind to antigenic determinant sites on two different antigens. In this way, antibodies can form "bridges" that tie

TABLE 22-1 Classes of Immunoglobulins

Structure	Description
IgG	**IgG** is the largest and most diverse class of antibodies. There are several types of IgG, but each occur as an individual molecule. Together they account for 80 percent of all immunoglobulins. The IgG antibodies are responsible for resistance against many viruses, bacteria, and bacterial toxins. IgG antibodies can cross the placenta, and during gestational development maternal IgG provides passive immunity to the developing fetus. However, the anti-Rh (anti-D) antibodies produced by Rh-negative mothers sensitized to Rh surface antigens are also IgG antibodies. These antibodies can cross the placenta and attack fetal Rh-positive red blood cells, producing the *hemolytic disease of the newborn* discussed in Chapter 19. ∞ *[p. 654]*
IgE	**IgE** attaches as an individual molecule to the exposed surfaces of basophils and mast cells. When a suitable antigen appears and binds to the IgE molecules, the cell is stimulated to release histamine and other chemicals that accelerate inflammation in the immediate area. IgE is also important in the allergic response.
IgD	**IgD** is an individual molecule on the surfaces of B cells, where it can bind antigens in the extracellular fluid. This binding can play a role in the activation of the B cell involved.
IgM	**IgM** is the first antibody type secreted after the arrival of an antigen. The concentration of IgM then declines as IgG production accelerates. Although plasma cells secrete individual IgM molecules, IgM circulates as a five-antibody starburst. This configuration makes these antibodies particularly effective in forming immune complexes. The anti-A and anti-B antibodies responsible for the agglutination of cross-matched blood are IgM antibodies. ∞ *[p. 654]* IgM antibodies may also attack bacteria that are insensitive to IgG.
IgA (Secretory piece)	**IgA** is found primarily in glandular secretions such as mucus, tears, and saliva. These antibodies attack pathogens before they gain access to internal tissues. IgA antibodies circulate in the blood as individual molecules or in pairs. Epithelial cells absorb them from the blood and attach a *secretory piece*, which confers solubility, before secreting the IgA molecules onto the epithelial surface.

large numbers of antigens together (Figure 22-20•). The three-dimensional structure created in this way is known as an **immune complex.** When the antigen is a soluble molecule, such as a toxin, this process may create complexes too large to remain in solution. The formation of insoluble immune complexes is called **precipitation.** When the antigenic target is on the surface of a cell or virus, the formation of large complexes is called **agglutination.** The clumping of red blood cells that occurs when incompatible blood types are mixed (Chapter 19) is an agglutination reaction. ∞ *[p. 653]*

3. *Activation of complement.* Upon binding to an antigen, portions of the antibody molecule change shape, exposing areas that bind complement proteins. The bound complement molecules then activate the complement system, destroying the antigen as we detailed previously.

4. *Attraction of phagocytes.* Antigens covered with antibodies attract eosinophils, neutrophils, and macrophages—cells that phagocytize pathogens and destroy foreign or abnormal cell membranes.

5. *Opsonization.* A coating of antibodies and complement proteins increases the effectiveness of phagocy-

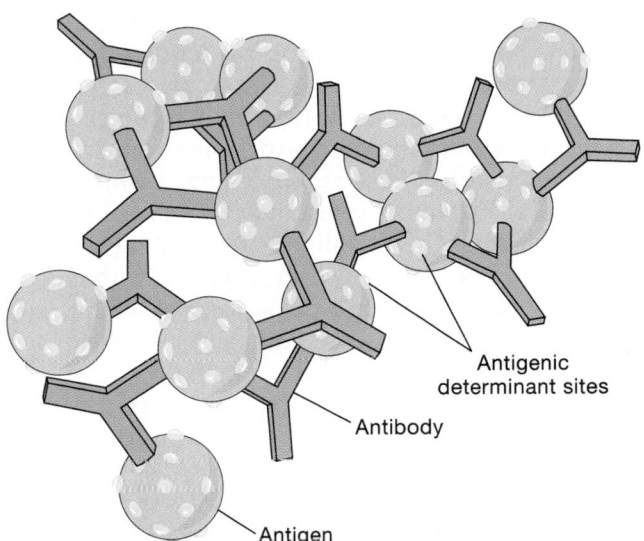

• FIGURE 22-20 An Immune Complex. An immune complex forms when large numbers of antibodies and antigens interact. The antibodies bind to antigenic determinant sites on two different antigens, forming a complex structure.

tosis. Microorganisms such as bacteria have slick cell membranes or capsules, and a phagocyte must be able to hang onto its prey before it can engulf the prey. Phagocytes can bind more easily to antibodies and complement proteins on the surface of a pathogen than they can to the bare surface.

6. ***Stimulation of inflammation.*** Antibodies may promote inflammation through the stimulation of basophils and mast cells.

7. ***Prevention of bacterial and viral adhesion.*** Antibodies dissolved in saliva, mucus, and perspiration coat epithelia and provide an additional layer of defense. A covering of antibodies makes it difficult for pathogens to attach to body surfaces and penetrate their defenses.

☑ How can the presence of an abnormal peptide in the cytoplasm of a cell initiate an immune response?

☑ A decrease in the number of cytotoxic T cells would affect which type of immunity?

☑ How would a lack of helper T cells affect the antibody-mediated immune response?

☑ A sample of lymph contains an elevated number of plasma cells. Would you expect the amount of antibodies in the blood to be increasing or decreasing? Why?

Primary and Secondary Responses to Antigen Exposure

The initial response to antigen exposure is called the **primary response.** When an antigen appears a second time, it triggers a more extensive and prolonged **secondary response.** The secondary response reflects the presence of large numbers of memory cells that are already primed for the arrival of the antigen. Primary and secondary responses are characteristic of both cell-mediated and antibody-mediated immunities. The differences between these responses are most easily demonstrated by following the production of antibodies over time.

The Primary Response

Because the antigen must activate the appropriate B cells, which must then respond by differentiating into plasma cells, the primary response does not appear immediately (Figure 22-21a•). As plasma cells differentiate and begin secreting, there is a gradual, sustained rise in the concentration of circulating antibodies.

During the primary response, the antibody activity, or **antibody titer** ("standard"), in the plasma does not peak until 1–2 weeks after the initial exposure. If the individual is no longer exposed to the antigen, the

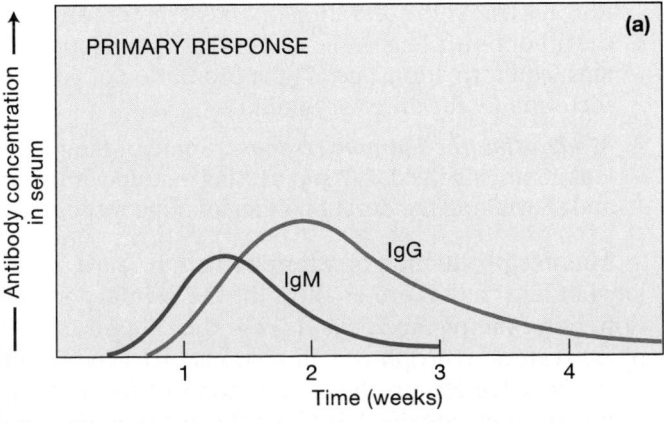

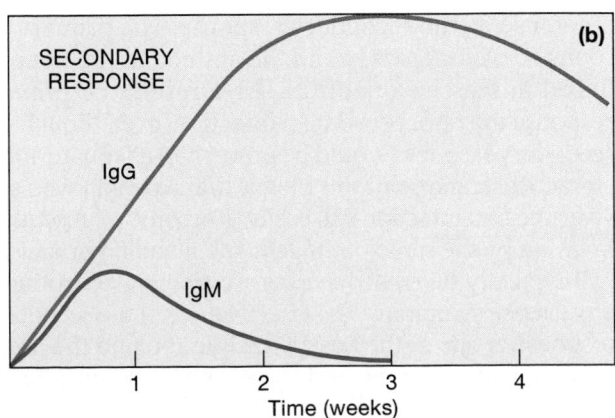

• FIGURE 22-21 The Primary and Secondary Immune Responses. (a) The primary response takes about 2 weeks to develop peak antibody titers, and IgM and IgG antibody concentrations do not remain elevated. **(b)** The secondary response is characterized by a very rapid increase in IgG antibody titer, to levels much higher than those of the primary response. Antibody activity remains elevated for an extended period after the second exposure to the antigen.

antibody concentration thereafter declines. This reduction in antibody production occurs because (1) plasma cells have very high metabolic rates and survive for only a short time, and (2) further production of plasma cells is inhibited by suppression factors released by suppressor T cells. However, suppressor T cell activity does not begin immediately after antigen exposure, and under normal conditions helper cells outnumber suppressors by more than 3 to 1. As a result, many B cells are activated before suppressor T cell activity has a noticeable effect.

Two types of antibodies are involved in the primary response. Molecules of *immunoglobulin M, or IgM,* are the first to appear in the circulation. IgM is secreted by the plasma cells that form immediately after B cell activation. These lymphocytes do not pause to produce memory cells. Levels of *immunoglobulin G, or IgG,* rise more slowly, because the stimulated lymphocytes undergo repeated cell divisions and generate large numbers of memory cells as well as plasma cells. In effect, IgM provides an immediate but limited defense that fights the infection until massive quantities of IgG can be produced.

The Secondary Response

Memory B cells do not differentiate into plasma cells unless they are exposed to the same antigen a second time. If and when that exposure occurs, these cells respond immediately, dividing and differentiating into plasma cells that secrete antibodies in massive quantities. This is the secondary response, or **anamnestic response** (an-am-NES-tik; *anamnesis,* a memory), to antigen exposure.

During the secondary response, antibody titers increase more rapidly and reach levels many times higher than they did in the primary response (Figure 22-21b•). The secondary response appears even if the second exposure occurs years after the first, for memory cells are long-lived: They may survive for 20 years or more.

The primary response to an antigen is not as effective a defense as the secondary response. The primary response develops slowly, and the antibodies are not produced in massive quantities. As a result, the primary response may not prevent an infection, even though the secondary response would be more than adequate. From a practical standpoint this means that a person who survives the first infection will easily overcome a subsequent invasion by the same pathogen. The invading organisms will typically be destroyed before they have produced any disease symptoms. The effectiveness of the secondary response is one of the basic principles behind the use of immunization to prevent disease. [AM] *Immunization*

Hormones of the Immune System

The specific and nonspecific defenses of the body are coordinated by physical interaction and by the release of chemical messengers. One example of physical interaction is antigen presentation by activated macrophages and helper T cells. An example of chemical messenger release is the release of cytokines by many of the cells involved in the immune response. Cytokines can be classified according to their origins: *Lymphokines* are secreted by lymphocytes, and *monokines* by active macrophages and other antigen-presenting cells. These terms are misleading, however, because lymphocytes and macrophages may secrete the same cytokines, and cytokines can also be secreted by cells involved with nonspecific defenses and tissue repair.

Table 22-2 contains a brief summary of the cytokines that have been identified to date. Five subgroups merit special attention: (1) *interleukins,* (2) *interferons,* (3) *tumor necrosis factors,* (4) chemicals that regulate phagocytic activities, and (5) *colony-stimulating factors.*

Interleukins

Interleukins are probably the most diverse and important chemical messengers in the immune system. More than a dozen different types of interleukins have now been identified (Table 22-2). Lymphocytes and macrophages are the primary sources of interleukins, but specific interleukins, such as interleukin-1 (Il-1), are also produced by endothelial cells, fibroblasts, and astrocytes.

Interleukins have the following general functions:

1. *Increasing T cell sensitivity to antigens exposed on macrophage membranes.* This heightened sensitivity accelerates the production of cytotoxic and regulatory T cells.

2. *Stimulating B cell activity, plasma cell formation, and antibody production.* These events promote the production of antibodies and the development of antibody-mediated immunity.

3. *Enhancing nonspecific defenses.* Known effects of interleukin production include (1) the stimulation of inflammation, (2) the formation of scar tissue by fibroblasts, (3) the elevation of body temperature via the preoptic nucleus of the CNS, (4) the stimulation of mast cell formation, and (5) the promotion of ACTH secretion by the anterior pituitary.

4. *Moderating the immune response.* Some of the interleukins are involved with suppressing immune function and shortening the duration of an immune response.

Massive production of interleukins can cause problems at least as severe as those of the primary infection. For example, in *Lyme disease* the release of Il-1 by activated macrophages in response to a localized bacterial infection produces symptoms of fever, pain, skin rash, and arthritis that affect the entire body. [AM] *Lyme Disease*

Some interleukins enhance the immune response, whereas others suppress it. The relative quantities se-

TABLE 22-2   Chemical Mediators (Cytokines) of the Immune Response

Compound	Functions
Interleukins	
Interleukin-1 (Il-1)	Stimulates T cells to produce Il-2, promotes inflammation; causes fever
Interleukin-2 (Il-2)	Stimulates growth and activation of other T cells and NK cells
Interleukin-3 (Il-3)	Stimulates production of mast cells and other blood cells
Il-4 (B cell differentiating factor); Il-5 (B cell growth factor); Il-6, Il-7 (B cell stimulating factors); Il-10; Il-11	Promote differentiation and growth of B cells and stimulate plasma cell formation and antibody production; each has somewhat different effects on macrophage and microphage activities
Il-8	Stimulates blood vessel formation (angiogenesis)
Il-9	Stimulates myeloid cell production (RBCs, platelets, granulocytes, monocytes)
Il-12	Stimulates T cell activity and cell-mediated immunity
Il-13	Suppresses production of several other cytokines (Il-1, Il-8, TNF); stimulates Class II MHC antigen presentation
Interferons (alpha, beta, gamma)	Activate other cells to prevent viral entry; inhibit viral replication; stimulate NK cells and macrophages
Tumor Necrosis Factors (TNFs)	Kill tumor cells; slow tumor growth; stimulate activities of T cells and eosinophils; inhibit parasites and viruses
Monocyte-Chemotactic Factor (MCF)	Attracts monocytes; activates them to macrophages
Migration-Inhibitory Factor (MIF)	Prevents macrophage migration from the area
Macrophage-Activating Factor (MAF)	Makes macrophages more active and aggressive
Microphage-Chemotactic Factor	Attracts microphages from the blood
Colony-Stimulating Factors (CSFs)	Stimulate RBC and WBC production
Defensins	Kill cells by piercing cell membrane
Growth-Inhibitory Factor (GIF)	Reduces/inhibits replication of target cells
Hemopoiesis-Stimulating Factor	Promotes blood cell production in bone marrow
Leukotrienes	Stimulate regional inflammation
Lymphotoxins	Kill cells; damage tissue; promote inflammation
Perforin	Destroys cell membranes by creating large pores
Transforming Growth Factor-β (TGF-β)	Stimulates production of IgA and of matrix proteins in connective tissues; inhibits macrophage activation and T_C maturation
Suppression Factors	Inhibit T_C cell and B cell activity; depress immune response
Transfer Factor	Sensitizes other T cells to same antigen

creted at any given moment therefore have a significant effect on the nature and intensity of the response to an antigen. For example, in the course of a typical infection, the pattern of interleukin secretion is constantly changing. Whether the individual succeeds in overcoming the infection is determined in part by whether stimulatory or suppressive interleukins predominate. As a result, interleukins and their interactions are now the focus of an intensive research effort.

Interferons

Interferons make the cell that synthesizes them, and that cell's neighbors, resistant to viral infection, thereby slowing the spread of the virus. These compounds may have other beneficial effects in addition to their antiviral activity. For example, **alpha-interferons** and **gamma-interferons** attract and stimulate NK cells, and **beta-interferons** slow the progress of inflammation associated with viral infection. Gamma-interferons also stimulate macrophages, making them more effective at killing bacterial or fungal pathogens.

Because they stimulate NK cell activity, interferons can be used to fight some cancers. For example, alpha-interferons have been used in the treatment of malignant melanoma, bladder cancer, ovarian cancer, and two forms of leukemia. Alpha- or gamma-interferons may be used to treat Kaposi's sarcoma, a cancer that typically develops in AIDS patients. [AM] *AIDS*

Tumor Necrosis Factors

Tumor necrosis factors (TNFs) slow tumor growth and kill sensitive tumor cells. Activated macrophages secrete one type of TNF and carry the molecules in their cell membranes. Cytotoxic T cells produce a different type of TNF. In addition to their effects on tumor cells, tumor necrosis factors stimulate granular leukocyte production, promote eosinophil activity, cause fever, and increase T cell sensitivity to interleukins.

MANIPULATING THE IMMUNE RESPONSE As our understanding of the immune system grows, complex therapies involving combinations of cytokines, including interleukins, and *monoclonal antibodies*[3] are appearing. For example, in one procedure, cytotoxic T cells were removed from patients with *malignant melanoma,* a particularly dangerous type of skin cancer. These lymphocytes were able to recognize tumor cells and had migrated to the tumor, but for some reason they appeared to be unable to kill the tumor cells.

The extracted lymphocytes were cultured, and viruses were used to insert multiple copies of the genes responsible for the production of tumor necrosis factor. The patients were then given periodic infusions of these "supercharged" T cells. To enhance T cell activity further, the researchers also administered doses of interleukin-2. Initial results are promising. It is clear that the ability to manipulate the immune response will revolutionize the treatment of many serious diseases.

[3]Monoclonal antibodies are produced by a single clone of B cells under laboratory conditions. For a detailed discussion of the process, see the *Applications Manual.* AM *Technology, Immunity, and Disease*

Chemicals Regulating Phagocytic Activities

Several cytokines coordinate the specific and nonspecific defenses by adjusting the activities of phagocytic cells. These cytokines include factors that attract free macrophages and microphages to the area and prevent their premature departure.

Colony-Stimulating Factors

We introduced **colony-stimulating factors (CSFs)** and their functions in Chapter 19. ∞ *[p. 660]* These factors are produced by active T cells, cells of the monocyte–macrophage group, endothelial cells, and fibroblasts. CSFs stimulate production of blood cells in bone marrow and lymphocytes in lymphoid tissues and organs.

PATTERNS OF IMMUNE RESPONSE

We have now detailed the basic chemical and cellular interactions that follow the appearance of a foreign antigen. Figure 22-22● presents a broader, integrated view of the immune response and its relationship to nonspecific defenses. Figure 22-23● provides an overview of the time course of the events responsible for overcoming a bacterial infection.

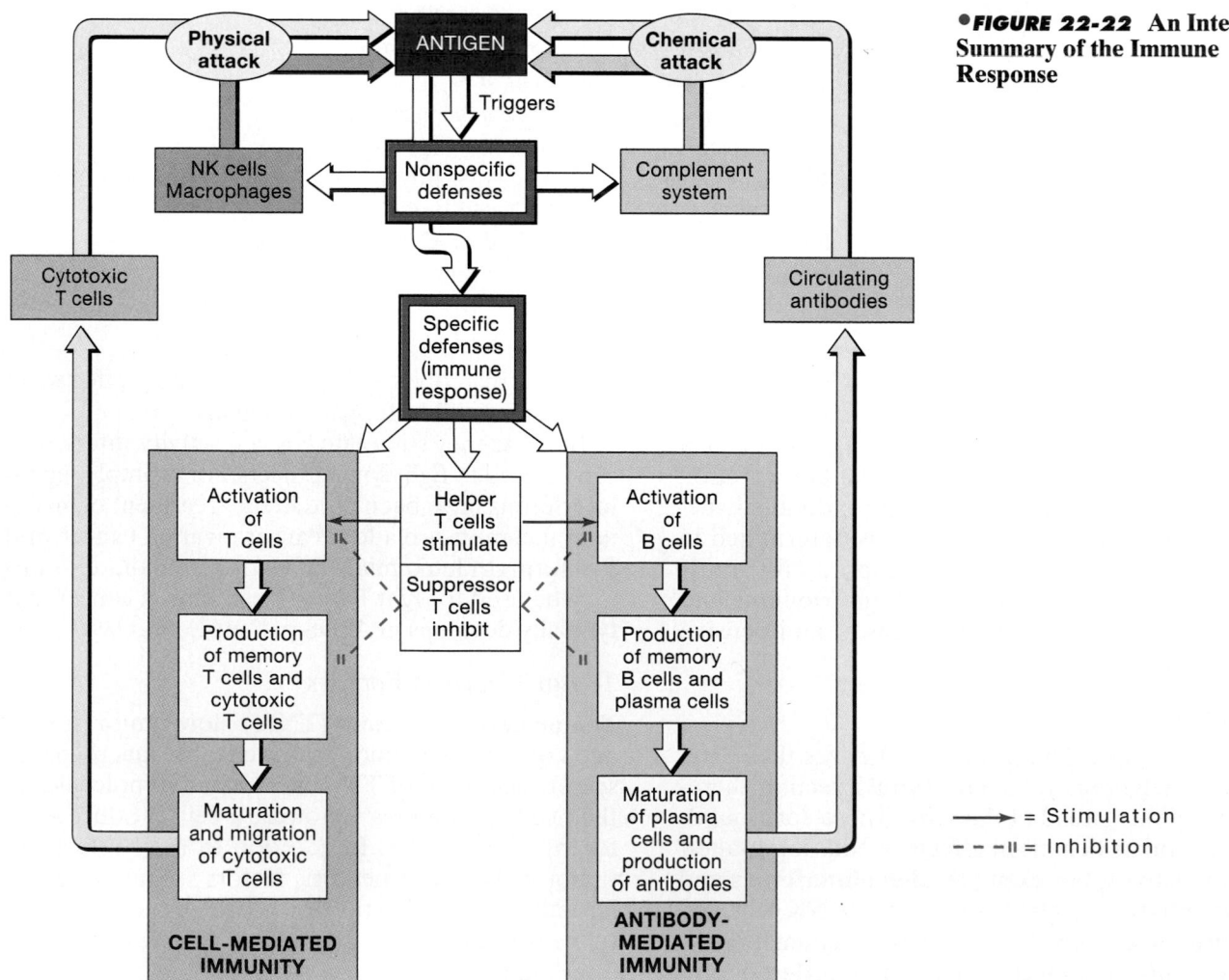

●*FIGURE 22-22* An Integrated Summary of the Immune Response

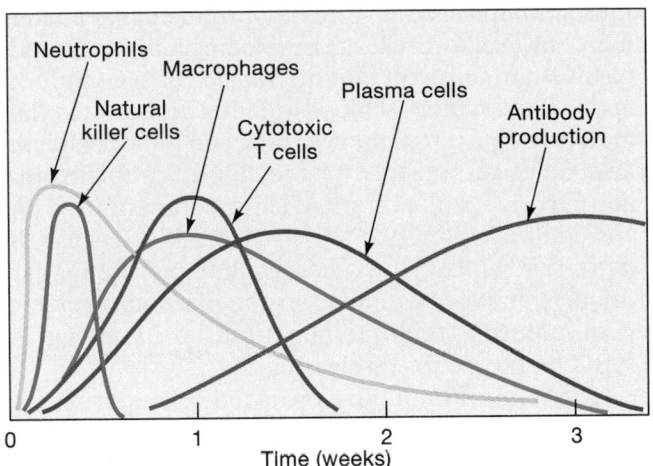

• **FIGURE 22-23** The Time Course of the Body's Response to Bacterial Infection. An outline of the basic sequence of events that begins with the appearance of pathogens in peripheral tissues.

In the early stages of infection, before antigen processing has occurred, neutrophils and NK cells migrate into the threatened area and destroy bacterial cells. Over time, cytokines draw increasing numbers of phago-

cytes to the region (Figure 22-23•). Cytotoxic T cells appear as arriving T cells are activated by antigen presentation. Last of all, the population of plasma cells rises as activated B cells differentiate. This rise is followed by a gradual, sustained increase in the level of circulating antibodies.

The basic sequence of events is very similar when a viral infection occurs. The initial steps are different, however, because cytotoxic T cells and NK cells can be activated by contact with infected cells. Figure 22-24• contrasts the steps involved in defense against bacteria with the defense against infection by viruses. Table 22-3 completes our summary by reviewing the cells involved.

THE DEVELOPMENT OF RESISTANCE

The ability to demonstrate an immune response upon exposure to an antigen is **immunological competence.** Cell-mediated immunity can be demonstrated as early as the third month of fetal development, but active antibody-mediated immunity does not appear until later.

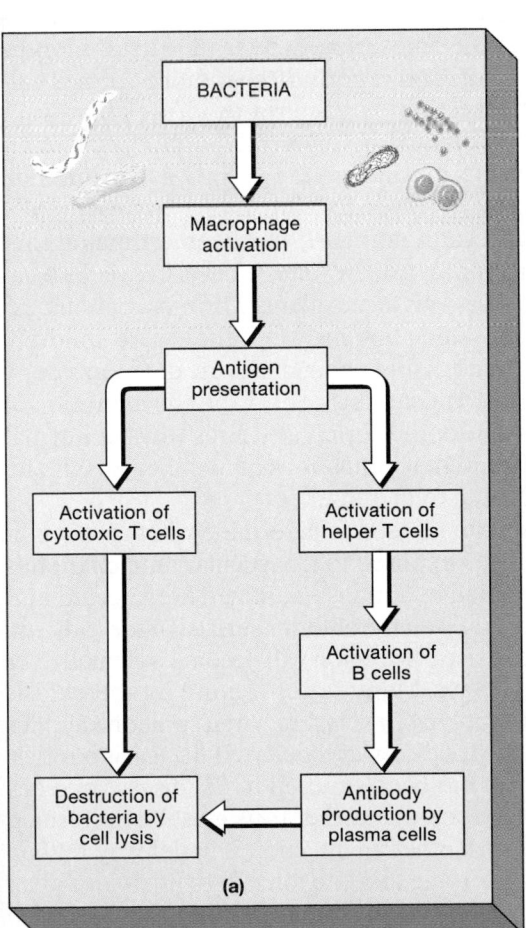

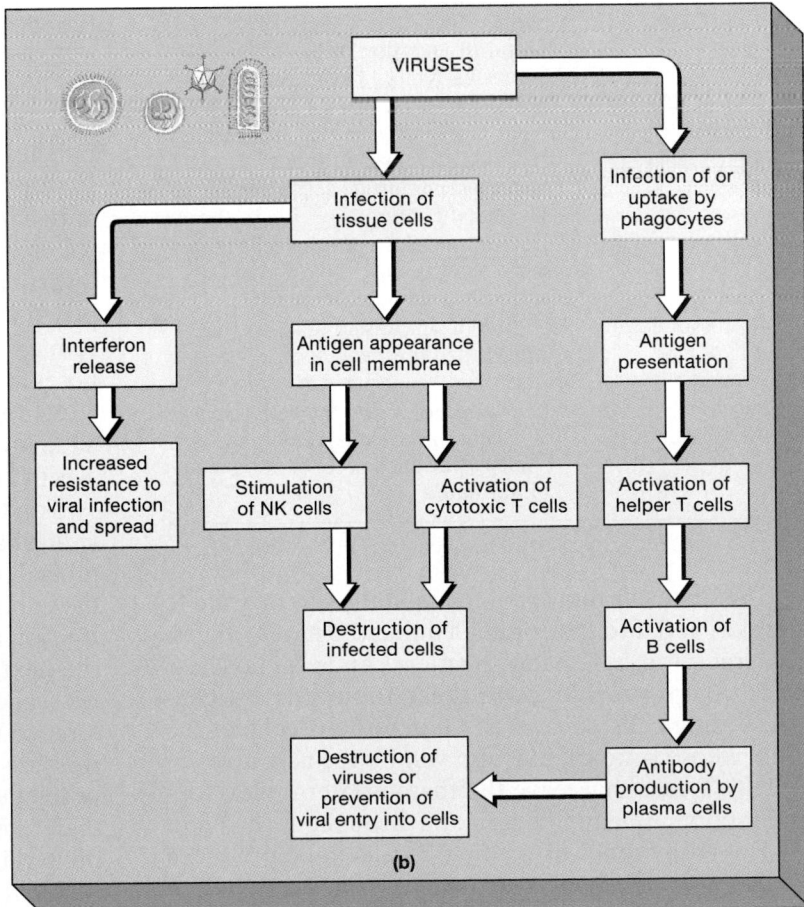

• **FIGURE 22-24** Defenses against Bacterial and Viral Pathogens. (a) Defenses against bacteria are usually initiated by active macrophages. (b) Defenses against viruses are usually activated after infection of normal cells.

TABLE 22-3 Cells That Participate in Tissue Defenses

Cell	Functions
Neutrophils	Phagocytosis; stimulation of inflammation
Eosinophils	Phagocytosis of antigen–antibody complexes; suppression of inflammation; involved in allergic response
Mast cells and basophils	Stimulation and coordination of inflammation by release of histamine, heparin, leukotrienes, prostaglandins
Antigen-Presenting Cells	
Macrophages (free and fixed macrophages, Kupffer cells, microglia, etc.)	Phagocytosis; antigen processing; antigen presentation with Class II MHC proteins; secretion of cytokines, especially interleukins and interferons
Dendritic cells, Langerhans cells	Antigen presentation bound to Class II MHC proteins
Lymphocytes	
NK cells	Destruction of cell membranes containing abnormal antigens
Cytotoxic T cells (T_C, CD8 marker)	Lysis of cell membranes containing antigens bound to Class I MHC proteins; secretion of perforins, defensins, lymphotoxins, and other cytokines
Helper T cells (T_H, CD4 marker)	Secretion of cytokines that stimulate cell-mediated and antibody-mediated immunity; activation of sensitized B cells
B cells	Differentiation into plasma cells that secrete antibodies and provide antibody-mediated immunity
Suppressor T cells (T_S, CD8 marker)	Secretion of suppression factors that inhibit the immune response
Memory cells (T_C, T_H, B)	Cells produced during the activation of T cells and B cells; remain in tissues awaiting arrival of antigen at a later date

The first cells that leave the fetal thymus migrate to the skin and into the epithelia lining the mouth, digestive tract, and, in females, the uterus and vagina. These cells take up residence in these tissues as antigen-presenting cells, such as the Langerhans cells of the skin, whose primary function will be the activation of T cells. T cells that leave the thymus later in development populate lymphoid organs throughout the body.

The membranes of the first B cells to be produced in the liver and bone marrow carry IgM antibodies. Sometime after the fourth developmental month, the fetus may, if exposed to specific pathogens, produce IgM antibodies. Fetal antibody production is uncommon, however, because the developing fetus has natural passive immunity due to the transfer of IgG antibodies from the maternal bloodstream. These are the only antibodies that cross the placenta, and they include the antibodies responsible for the clinical problems that accompany fetal–maternal Rh incompatibility, discussed in Chapter 19. ∞ *[pp. 654–655]* Because the anti-A and anti-B antibodies (agglutinins) are IgM antibodies, and IgM cannot cross the placenta, problems with maternal–fetal incompatibilities involving the ABO blood groups rarely occur.

The natural immunity provided by maternal IgG may not be enough to protect the fetus if the maternal defenses are overrun by a bacterial or viral infection. For example, the microorganisms responsible for syphilis and rubella ("German measles") can cross from the maternal to the fetal bloodstream, producing a congenital infection that leads to the production of fetal antibodies. IgM provides only a partial defense, and these infections can result in severe developmental problems for the fetus. AM *Fetal Infections*

Delivery eliminates the maternal supply of IgG. Although the mother provides IgA antibodies in the breast milk, the infant gradually loses its passive immunity. The amount of maternal IgG in the infant's bloodstream declines rapidly over the first 2 months after birth. During this period, the infant becomes vulnerable to infection by bacteria or viruses that were previously overcome by maternal antibodies. The infant also begins producing its own IgG, as the immune system begins to respond to infections, environmental antigens, and vaccinations. It has been estimated that children encounter a "new" antigen every 6 weeks from birth to age 12. (This fact explains why most parents, exposed to the same antigens when they were children, remain healthy while their children develop runny noses and colds.) Over this period, the concentration of circulating antibodies gradually rises toward normal adult levels, and the populations of memory B cells and T cells continue to increase.

Skin tests can be used to determine whether an individual has been exposed to a particular antigen. In this procedure, small quantities of antigen are injected into the skin, generally on the anterior surface of the forearm. If resistance exists, the region will become inflamed over the next 2–4 days. Many states require a tuberculosis test, called a *tuberculin skin test,* when someone applies for a job in the food-service industry. If the test is positive, the individual has been exposed to the disease and has developed antibodies. Further tests must be performed to determine whether an infection is under way. (When preparing or serving food, anyone with tuberculosis can accidentally transmit the bacteria.) Skin tests are also used to check for allergies to environmental antigens. AM *Delayed Hypersensitivity and Skin Tests*

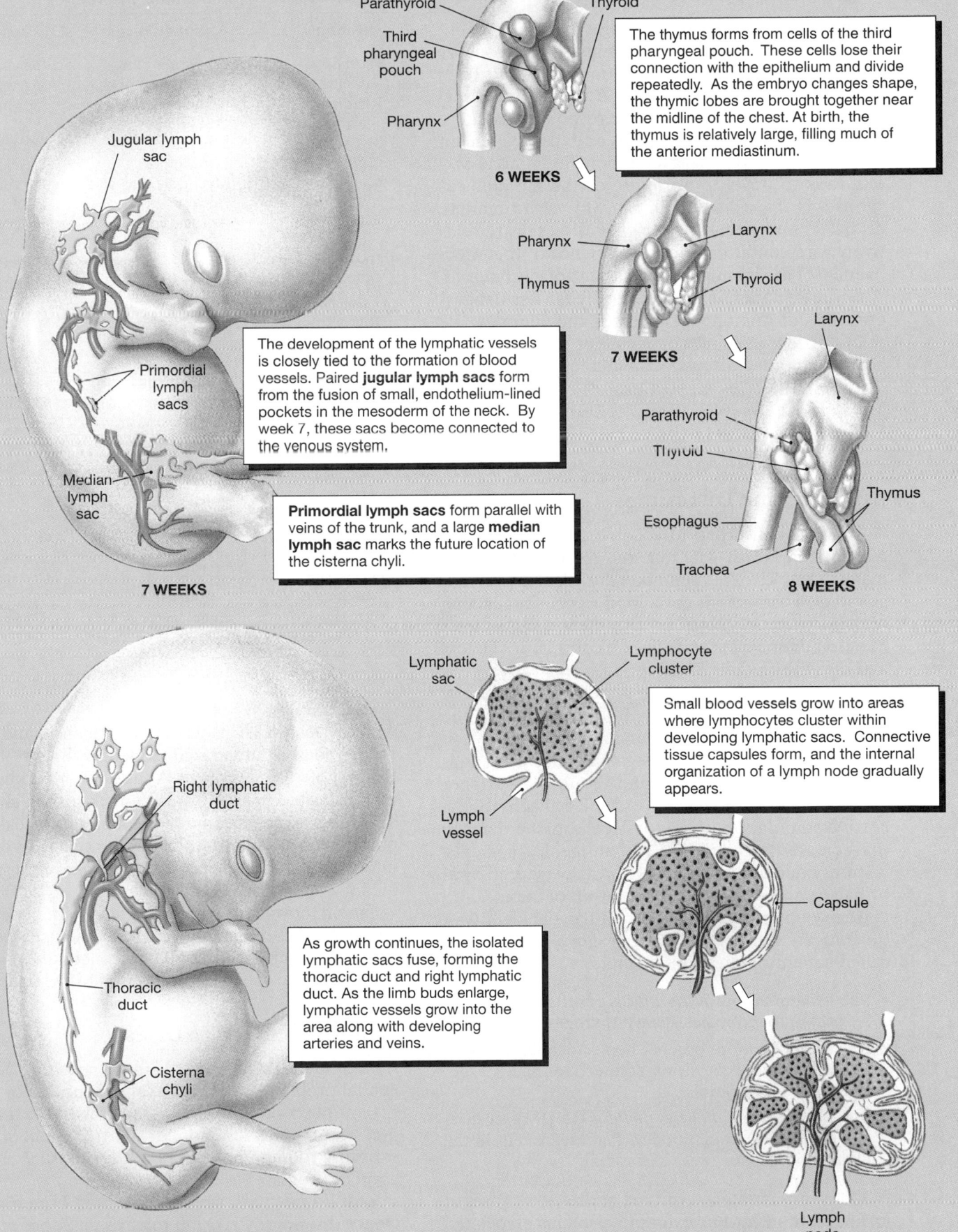

EMBRYOLOGY SUMMARY

Development of the Lymphatic System

Parathyroid — **Thyroid**
Third pharyngeal pouch
Pharynx

The thymus forms from cells of the third pharyngeal pouch. These cells lose their connection with the epithelium and divide repeatedly. As the embryo changes shape, the thymic lobes are brought together near the midline of the chest. At birth, the thymus is relatively large, filling much of the anterior mediastinum.

6 WEEKS

Pharynx — **Larynx**
Thymus — **Thyroid**

7 WEEKS

Larynx
Parathyroid
Thyroid
Esophagus
Trachea
Thymus

8 WEEKS

Jugular lymph sac

Primordial lymph sacs

Median lymph sac

The development of the lymphatic vessels is closely tied to the formation of blood vessels. Paired **jugular lymph sacs** form from the fusion of small, endothelium-lined pockets in the mesoderm of the neck. By week 7, these sacs become connected to the venous system.

Primordial lymph sacs form parallel with veins of the trunk, and a large **median lymph sac** marks the future location of the cisterna chyli.

7 WEEKS

Right lymphatic duct

Thoracic duct

Cisterna chyli

As growth continues, the isolated lymphatic sacs fuse, forming the thoracic duct and right lymphatic duct. As the limb buds enlarge, lymphatic vessels grow into the area along with developing arteries and veins.

8 WEEKS

Lymphatic sac — **Lymphocyte cluster**
Lymph vessel

Small blood vessels grow into areas where lymphocytes cluster within developing lymphatic sacs. Connective tissue capsules form, and the internal organization of a lymph node gradually appears.

Capsule

Lymph node

IMMUNE DISORDERS

Because the immune response is so complex, there are many opportunities for things to go wrong. A great variety of clinical conditions may result from disorders of immune function. **Autoimmune disorders** develop when the immune response mistakenly targets normal body cells and tissues. In an **immunodeficiency disease,** either the immune system fails to develop normally or the immune response is blocked in some way. Autoimmune disorders and immunodeficiency diseases are relatively rare conditions—clear evidence of the effectiveness of the immune system's control mechanisms. A far more common, and generally far less dangerous, class of immune disorders is the **allergies.** We shall consider examples of each type of disorder. For a more extended discussion, including possible therapies, see the relevant sections of the *Applications Manual.* [AM] *Immune Complex Disorders; Systemic Lupus Erythematosus*

Autoimmune Disorders

Autoimmune disorders affect an estimated 5 percent of adults in North America and Europe. We noted many examples of autoimmune disorders in previous chapters in discussions of their effects on the function of major systems. Table 22-4 lists the major examples of autoimmune disorders that we consider in the text and the *Applications Manual.*

In most cases, the immune system recognizes and ignores the antigens normally found in the body. The recognition system can malfunction, however. When it does, the activated B cells begin manufacturing antibodies against other cells and tissues. The trigger may be a reduction in suppressor T cell activity, excessive stimulation of helper T cells, tissue damage that releases large quantities of antigenic fragments, haptens bound to compounds normally ignored, viral or bacterial toxins, or a combination of factors.

The symptoms produced depend on the identity of the antigen attacked by these misguided antibodies, called **autoantibodies.** For example:

- The inflammation of *thyroiditis* results from the release of autoantibodies against thyroglobulin.

- *Rheumatoid arthritis* occurs when autoantibodies form immune complexes within connective tissues, especially around the joints.

- *Insulin-dependent diabetes mellitus* (IDDM) is generally caused by autoantibodies that attack cells in the pancreatic islets.

Many autoimmune disorders appear to be cases of mistaken identity. For example, proteins associated

TABLE 22-4 Autoimmune Disorders

Disorder	Antibody Target	Discussion
Psoriasis	Epidermis of skin	*p. 162* and *AM*
Vitiligo	Melanocytes of skin	*p. 153* and *AM*
Rheumatoid arthritis	Connective tissues at joints	*p. 256* and *AM*
Myasthenia gravis	Synaptic ACh receptors	*p. 287* and *AM*
Multiple sclerosis	Myelin sheaths of axons	*p. 380* and *AM*
Addison's disease	Adrenal cortex	*p. 617* and *AM*
Graves' disease	Thyroid follicles	*p. 612* and *AM*
Hypoparathyroidism	Chief cells of parathyroid	*p. 613* and *AM*
Thyroiditis	Thyroid-binding globulin	*AM*
Type I diabetes	Beta cells of pancreatic islets	*p. 623* and *AM*
Rheumatic fever	Myocardium and heart valves	*p. 682* and *AM*
Systemic lupus erythematosus	DNA, cytoskeletal proteins, other tissue components	*AM*
Ménière's disease	Collagen	*AM*
Thrombocytopenic purpura	Platelets	*AM*
Pernicious anemia	Parietal cells of stomach	*AM*
Chronic hepatitis	Hepatocytes of liver	*p. 894* and *AM*

with the measles, Epstein–Barr, influenza, and other viruses contain amino acid sequences that are similar to those of myelin proteins. As a result, antibodies that target these viruses may also attack myelin sheaths. This mechanism accounts for the neurological complications that sometimes follow a vaccination or viral infection. It is also the probable mechanism responsible for *multiple sclerosis.*

For unknown reasons, the risk of autoimmune problems increases if an individual has an unusual type of MHC proteins. At least 50 clinical conditions have been linked to specific variations in MHC structure.

Immunodeficiency Diseases

Immunodeficiency diseases are the result of (1) congenital problems with the development of immunity; (2) an infection with a virus, such as HIV, that depresses immune function; or (3) treatment with or exposure to immunosuppressive agents, such as radiation or drugs.

Individuals with **severe combined immunodeficiency disease (SCID)** fail to develop either cellular or antibody-mediated immunity. Lymphocyte populations are low, and normal B and T cells are not present. Such individuals are unable to provide an immune defense, and even a mild infection can prove fatal. Total isolation offers protection at great cost and with severe restrictions on lifestyle. Bone marrow transplants from compatible donors, normally a close relative, have been used to colonize lymphoid tissues with functional lymphocytes. Gene-splicing techniques have led to therapies that can treat at least one form of SCID. AM *Genetic Engineering and Gene Therapy*

AIDS, an immunodeficiency disease that we consider on page 806, is the result of a viral infection that targets primarily helper T cells. As the number of T cells declines, the normal immune control mechanism breaks down. When an infection occurs, suppressor factors released by suppressor T cells inhibit an immune response before the few surviving helper T cells can stimulate the formation of cytotoxic T cells or plasma cells in adequate numbers.

Immunosuppressive agents may destroy stem cells and lymphocytes, leading to a complete immunological failure. This outcome is one of the potentially fatal consequences of radiation exposure. Immunosuppressive drugs have been used for many years to prevent graft rejection after transplant surgery.

Allergies

Allergies are inappropriate or excessive immune responses to antigens. The sudden increase in cellular activity or antibody titers can have a number of unpleasant side effects. For example, neutrophils or cytotoxic T cells may destroy normal cells while attacking the antigen, or the antigen–antibody complex may trigger a massive inflammatory response. Antigens that trigger allergic reactions are often called **allergens**.

There are several types of allergies. A complete classification recognizes four categories: *immediate hypersensitivity (Type I), cytotoxic reactions (Type II), immune complex disorders (Type III),* and *delayed hypersensitivity (Type IV).* We shall consider immediate hypersensitivity at this time, as it is probably the most common form of allergy. One form, *allergic rhinitis,* includes "hay fever" and environmental allergies that may affect 15 percent of the U.S. population. We discussed one example of a cytotoxic (Type II) reaction, the cross-reactions that occur following the transfusion of an incompatible blood type, in Chapter 19. ∞ *[p. 654]* Other types of allergies are detailed in the *Applications Manual.* AM *Immune Complex Disorders; Delayed Hypersensitivity and Skin Tests*

Immediate Hypersensitivity

Immediate hypersensitivity begins with the process of **sensitization.** Sensitization is the initial exposure to an allergen that leads to the production of antibodies, specifically large quantities of IgE. The tendency to produce IgE antibodies in response to specific allergens may be genetically determined. In drug reactions, such as penicillin allergies, IgE antibodies are produced in response to a hapten (partial antigen) bound to a larger molecule inside the body.

Due to the lag time needed to activate B cells, produce plasma cells, and synthesize antibodies, the first exposure to an allergen does not produce symptoms. It merely sets the stage for the next encounter. After sensitization, the IgE antibodies become attached to the cell membranes of basophils and mast cells throughout the body. When exposed to the same allergen at a later date, the bound antibodies stimulate these cells to release histamine, heparin, several cytokines, prostaglandins, and other chemicals into the surrounding tissues. The result is a sudden, massive inflammation of the affected tissues.

Basophils, eosinophils, T cells, and macrophages are drawn to the area by the cytokines and other secretions of the mast cells. These cells release chemicals of their own, extending and exaggerating the responses initiated by mast cells.

The severity of the allergic reaction depends on the individual's sensitivity and on the location involved. If allergen exposure occurs at the body surface, the response may be restricted to that area. If the allergen enters the systemic circulation, the response may be more dramatic and perhaps lethal.

ANAPHYLAXIS In **anaphylaxis** (a-na-fi-LAK-sis; *ana-,* again + *phylaxis,* protection), a circulating allergen affects mast cells throughout the body (Figure 22-25•, p. 807). The entire range of symptoms can develop within minutes. Changes in capillary permeabilities produce swelling and edema in the dermis, and raised welts, or *hives,* appear on the skin surface. Smooth muscles along the respiratory passageways contract, and the narrowed passages make breathing extremely difficult. In severe cases of anaphylaxis, an extensive peripheral vasodilation occurs, producing a fall in blood pressure that may lead to a circulatory collapse. This response has been called **anaphylactic shock.**

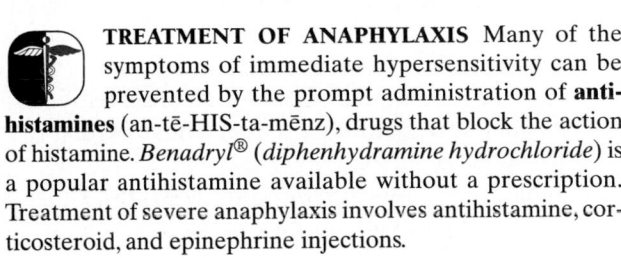

TREATMENT OF ANAPHYLAXIS Many of the symptoms of immediate hypersensitivity can be prevented by the prompt administration of **antihistamines** (an-tē-HIS-ta-mēnz), drugs that block the action of histamine. *Benadryl®* (*diphenhydramine hydrochloride*) is a popular antihistamine available without a prescription. Treatment of severe anaphylaxis involves antihistamine, corticosteroid, and epinephrine injections.

 AIDS

Acquired immune deficiency syndrome (AIDS), or *late-stage HIV disease,* is caused by a virus known as the **human immunodeficiency virus (HIV).** HIV is a *retrovirus,* a virus that carries its genetic information in RNA rather than in DNA. The virus enters the cell through receptor-mediated endocytosis. In the human body, the virus binds to CD4, the membrane protein characteristic of helper T cells. Several types of antigen-presenting cells, including those of the monocyte–macrophage line, are also infected by HIV, but it is the infection of helper T cells that leads to clinical problems.

Once the virus is inside a cell, *reverse transcriptase,* a viral enzyme, synthesizes a complementary strand of DNA, which is then incorporated into the genome of the cell. When these inserted viral genes are activated, the infected cell begins synthesizing viral proteins. In effect, the introduced viral genes take over the synthetic machinery of the cell and force it to produce additional viruses. These new viruses are then shed at the cell surface. (For a more complete discussion, refer to Section 2 of the *Applications Manual.*) AM *The Nature of Pathogens*

Cells infected with HIV are ultimately killed. Several mechanisms may be responsible for cell destruction, including (1) the formation of pores in the cell membrane as the viruses are shed, (2) cessation of cell maintenance due to the continuing synthesis of viral components, (3) autolysis, and (4) stimulation of apoptosis.

The gradual destruction of helper T cells by itself impairs the immune response, because these cells play a central role in coordinating cell-mediated and antibody-mediated responses to antigens. To make matters worse, suppressor T cells are relatively unaffected by the virus, and over time the excess of suppressing factors "turns off" the normal immune response. Circulating antibody levels decline, cell-mediated immunity is reduced, and the body is left without defenses against a wide variety of microbial invaders. With immune function so reduced, ordinarily harmless pathogens can initiate lethal infections known as *opportunistic infections.* Because immune surveillance is also depressed, the risk of cancer increases.

Infection with HIV occurs through intimate contact with the body fluids of infected individuals. Although all body fluids carry the virus, the major routes of transmission involve contact with blood, semen, or vaginal secretions. Most AIDS patients become infected through sexual contact with an HIV-infected person (who may *not* necessarily be exhibiting the clinical symptoms of AIDS). The next largest group of patients consists of intravenous drug users who shared contaminated needles. A relatively small number of individuals have become infected with the virus after receiving a transfusion of contaminated blood or blood products. Finally, an increasing number of infants are born with the disease, having acquired it from infected mothers.

AIDS is a public health problem of massive proportions. By the end of 1996, an estimated 350,000 people had died from AIDS in the United States alone; over 100,000 are living with AIDS. Estimates of the number of individuals infected with HIV in the United States range from 1 to 2 million. The virus has spread throughout the population. For example, a study performed in 1990 indicated that the incidence of infection at U.S. colleges and universities was 1 student in 500. The numbers worldwide are even more frightening: The World Health Organization estimates that 20–30 million people are already infected, and the number is increasing rapidly in Africa, Asia, and South America. At the current rate of increase, there may be 40 to 110 million infected individuals by the year 2000. *Every 15 seconds, another person becomes infected with the HIV virus.*

The best defense against AIDS consists of avoiding sexual contact with infected individuals. *All forms of sexual intercourse carry the risk of viral transmission.* The use of synthetic condoms greatly reduces the chance of infection (although it does not completely eliminate it). Condoms that are not made of synthetic materials are effective in preventing pregnancy but do not block the passage of viruses.

Clinical symptoms of AIDS may not appear for 5–10 years or more following infection. When they do appear, they are commonly mild, consisting of lymphadenopathy and chronic but nonfatal infections. So far as is known, however, AIDS is almost always fatal, and most people who carry the virus will eventually die of the disease. (A handful of infected individuals have been able to tolerate the virus without apparent illness. For details, see the *Applications Manual.*)

Despite intensive efforts, a vaccine has yet to be developed that will provide immunity from HIV infection. While efforts to prevent the spread of HIV continue, the survival rate for AIDS patients has been steadily increasing, because new drugs are available that slow the progression of the disease and improved antibiotic therapies help combat secondary infections. This combination is extending the life span of patients while the search for more effective treatment continues. For more information on the distribution of HIV infection, current and future drug therapies, and additional details on HIV disease, consult the *Applications Manual.* AM *AIDS*

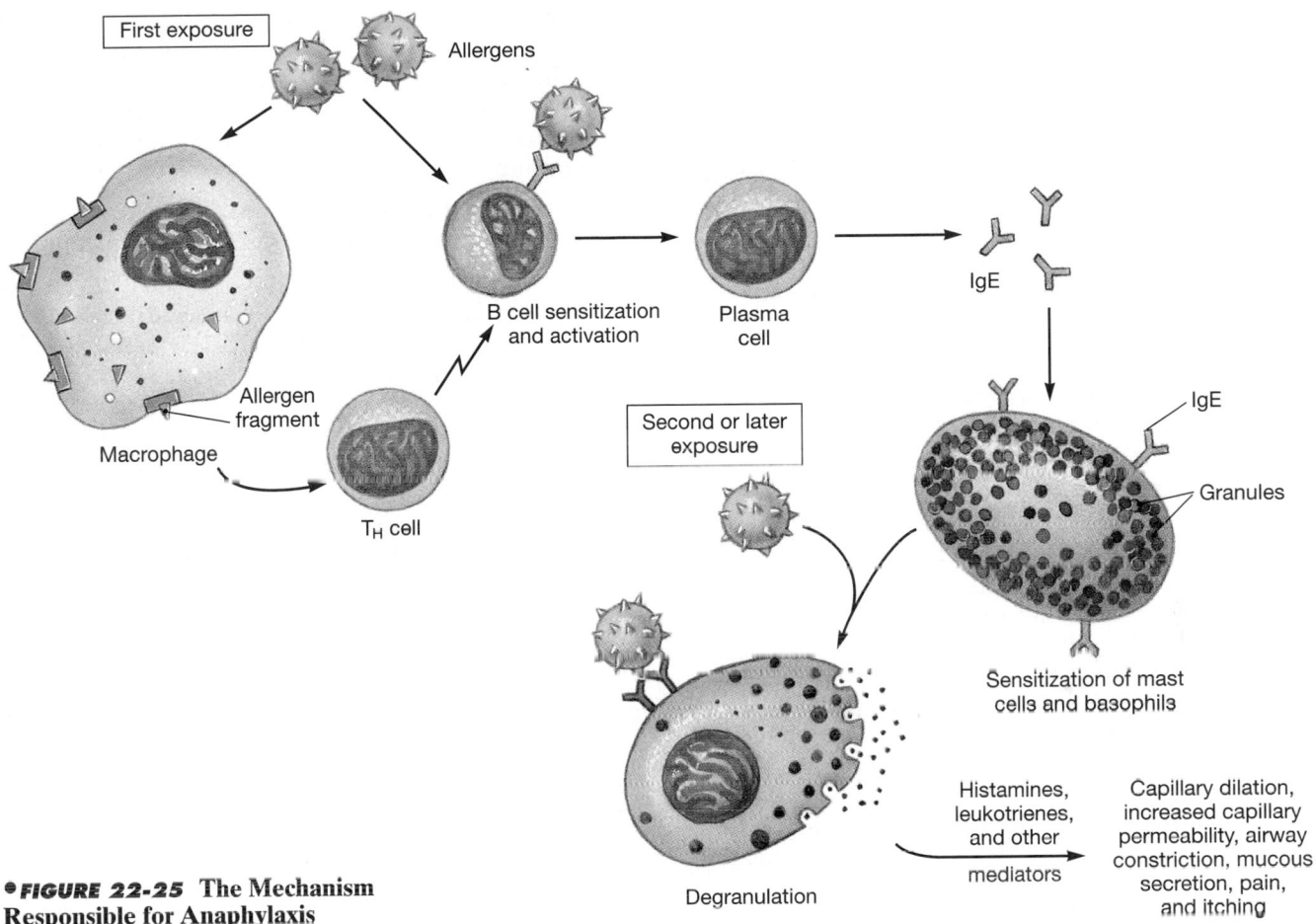

● **FIGURE 22-25** The Mechanism Responsible for Anaphylaxis

STRESS AND THE IMMUNE RESPONSE

One of the normal effects of interleukin-1 secretion is the stimulation of ACTH production by the anterior pituitary. ACTH production in turn leads to the secretion of glucocorticoids by the adrenal cortex. (We described the functions of ACTH and glucocorticoids in Chapter 18. ∞ *[p. 617]*) The anti-inflammatory effects of the glucocorticoids may help control the size of the immune response. However, the long-term secretion of glucocorticoids, as in the resistance phase of the *general adaptation syndrome,* can inhibit the immune response and lower resistance to disease. ∞ *[p. 629]* Several of the tissue responses to glucocorticoids alter the effectiveness of specific and nonspecific defenses. Examples include the following:

■ *Depression of the inflammatory response.* Glucocorticoids inhibit mast cells and make capillaries less permeable. Inflammation is therefore less likely. When inflammation does occur, the reduced permeability of the capillaries slows the entry of fibrinogen, complement, and cellular defenders.

■ *Reduction in the activities and numbers of phagocytes in peripheral tissues.* This reduction further impairs nonspecific defense mechanisms and interferes with the processing and presentation of antigens to lymphocytes.

■ *Inhibition of interleukin secretion.* A reduction in interleukin production depresses the lymphocytic response, even to antigens bound to MHC proteins.

The mechanisms responsible for these changes are still under investigation, but it should be noted that immune system depression due to chronic stress can represent a serious threat to health.

⊠ AGING AND THE IMMUNE RESPONSE

With advancing age, your immune system becomes less effective at combating disease. T cells become less responsive to antigens; as a result, fewer cytotoxic T cells respond to an infection. This effect may, at least in part, be associated with the gradual involution of the thymus and a reduction in circulating levels of thymosins. Because the number of helper T cells is also reduced, B cells are less responsive, and antibody levels do not rise as quickly after antigen exposure. The net result is an increased susceptibility to viral and bacterial infection. For this reason, vaccinations for acute viral diseases, such as the flu (influenza), are strongly recommended for elderly individuals. The increased incidence of cancer in the elderly reflects the fact that immune surveillance declines, so tumor cells are not eliminated as effectively.

INTEGRATION WITH OTHER SYSTEMS

Figure 22-26● summarizes the interactions between the lymphatic system and other physiological systems. The relationships among the cells of the immune response and the nervous and endocrine systems are now the focus of intense research. It is apparent that there is substantial interaction between these systems.

For example, the cells of the immune system can influence CNS and endocrine activity in the following ways:

- The thymus secretes oxytocin, ADH, and endorphins as well as the thymic hormones. The effects of these hormones on the CNS are not known, but removal or destruction of the thymus leads to a decrease in brain endorphin levels.
- Thymic hormones as well as cytokines are involved in establishing the normal level of corticotropin-releasing hormone (CRH) produced by the hypothalamus. Removal or destruction of the thymus leads to a decrease in ACTH production and a decline in circulating adrenocortical hormones. Thymic hormones also stimulate the production of thyroid hormone–releasing hormone (TRH), leading to thyroid-stimulating hormone (TSH) secretion at the anterior pituitary. As a result,

circulating thyroid hormone levels increase when an immune response is under way. The thyroid hormones further stimulate cell and tissue metabolism.

- Other thymic hormones affect the anterior pituitary directly, stimulating secretion of prolactin and growth hormone.

Conversely, the nervous system can apparently adjust the sensitivity of the immune response. For example:

- The CNS innervates dendritic cells in the lymph nodes and spleen, Langerhans cells in the skin, and other antigen-presenting cells. The nerve endings release neurotransmitters that exaggerate and enhance local immune responses. For this reason, some skin conditions, such as *psoriasis,* become worse when a person is under stress.
- Glial cells in the CNS produce cytokines that promote an immune response.
- A sudden decline in immune function can occur after even a brief period of emotional distress. Recent studies indicate that a heated argument can produce a steep reduction in circulating numbers of white blood cells over the following 24 hours.

The mechanisms involved and the functional significance of these interactions have yet to be determined. An explosive growth in our understanding of the immune response occurred during the last decade; that growth will continue in the decades to come.

☑ Would the primary response or the secondary response be more affected by a lack of memory B cells for a particular antigen?

☑ Which kind of immunity protects the developing fetus, and how is that immunity produced?

☑ How does increased stress decrease the effectiveness of the immune response?

SELECTED CLINICAL TERMINOLOGY

Terms Discussed in This Chapter

acquired immune deficiency syndrome (AIDS): A disorder that develops following HIV infection, characterized by reduced circulating antibody levels and depressed cell-mediated immunity. *(p. 806 and AM)*

allergen: An antigen capable of triggering an allergic reaction. *(p. 805)*

allergy: An inappropriate or excessive immune response to antigens, triggered by the stimulation of mast cells bound to IgE. *(p. 804 and AM)*

anaphylactic shock: A drop in blood pressure that may lead to circulatory collapse, resulting from a severe case of anaphylaxis. *(p. 805)*

anaphylaxis (a-na-fi-LAK-sis): A type of allergy in which a circulating allergen affects mast cells throughout the body,

producing numerous symptoms very quickly. *(p. 805)*

appendicitis: Infection and inflammation of the aggregate lymphoid nodules in the appendix. *(p. 776 and AM)*

autoimmune disorder: A disorder that develops when the immune response mistakenly targets normal body cells and tissues. *(p. 804)*

bacteria: Prokaryotic cells (cells lacking nuclei and other membranous organelles) that may be extracellular or intracellular pathogens. *(p. 770 and AM)*

filariasis: An infection by larval stages of a parasitic nematode; the adult worms may scar and block lymphatic vessels, causing acute lymphedema, commonly affecting the external genitalia and lower limbs **(elephantiasis)**. *(p. 773)*

fungi (singular, *fungus*): Eukaryotic organisms that absorb organic materials from the remains of dead cells; some fungi are pathogenic. *(p. 770 and AM)*

human immunodeficiency virus (HIV): The virus responsible for AIDS and related immunodeficiency disorders. *(p. 806 and AM)*

immune complex disorder: A disorder caused by the precipitation of immune complexes at sites such as the kidneys, where their presence disrupts normal tissue function. *(p. 805 and AM)*

immunodeficiency disease: A disease in which either the immune system fails to develop normally or the immune response is somehow blocked. *(p. 804)*

immunosuppression: A reduction in the sensitivity of the immune system. *(p. 793)*

INTEGUMENTARY SYSTEM

Provides physical barriers to pathogen entry; Langerhans cells in epidermis and macrophages in dermis resist infection and present antigens to trigger immune response; mast cells trigger inflammation, mobilize cells of lymphatic system

Provides IgA for secretion onto integumentary surfaces

THE LYMPHATIC SYSTEM

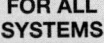

FOR ALL SYSTEMS
Provides specific defenses against infection; immune surveillance eliminates cancer cells; returns tissue fluid to circulation

---

SKELETAL SYSTEM

Lymphocytes and other cells involved in the immune response are produced and stored in bone marrow

Assists in repair of bone after injuries; macrophages fuse to become osteoclasts

---

MUSCULAR SYSTEM

Protects superficial lymph nodes and the lymphatic vessels in the abdominopelvic cavity; muscle contractions help propel lymph along lymphatic vessels

Assists in repair after injuries

NERVOUS SYSTEM

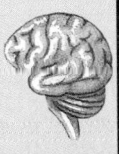

Microglia present antigens that stimulate specific defenses; glial cells secrete cytokines; innervation stimulates APCs

Cytokines affect hypothalamic production of CRH and TRH

ENDOCRINE SYSTEM

Glucocorticoids have anti-inflammatory effects; thymosins stimulate development and maturation of lymphocytes; many hormones affect immune function

Thymus secretes thymosins; cytokines affect cells throughout the body

---

CARDIOVASCULAR SYSTEM

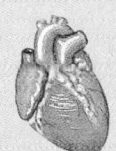

Distributes WBCs; carries antibodies that attack pathogens; clotting response helps restrict spread of pathogens; granulocytes and lymphocytes produced in bone marrow

Fights infections of cardiovascular organs; returns tissue fluid to circulation

---

RESPIRATORY SYSTEM

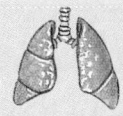

Alveolar phagocytes present antigens and trigger specific defenses; provides O_2 required by lymphocytes and eliminates CO_2 generated during their metabolic activities

Tonsils protect against infection at entrance to respiratory tract

---

DIGESTIVE SYSTEM

Provides nutrients required by lymphatic tissues, digestive acids, and enzymes; provides nonspecific defense against pathogens

Tonsils and GALT defend against infection and toxins absorbed from tract; lymphatics carry absorbed lipids to venous system

---

URINARY SYSTEM

Eliminates metabolic wastes generated by cellular activity; acid pH of urine provides nonspecific defense against urinary tract infection

---

Lysozymes and bactericidal chemicals in secretions provide nonspecific defense against reproductive tract infections

Provides IgA for secretion by epithelial glands

REPRODUCTIVE SYSTEM

•*FIGURE 22-26* **Functional Relationships between the Lymphatic System and Other Systems**

immunosuppressive drugs: Drugs administered to inhibit the immune response; examples include prednisone, cyclophosphamide, azathioprine, cyclosporin, and FK506. *(p. 793 and AM)*

lymphadenopathy (lim-fad-e-NOP-a-thē): Chronic or excessive enlargement of lymph nodes. *(p. 778)*

lymphedema: A painless accumulation of lymph in a region whose lymphatic drainage has been blocked. *(p. 773)*

lymphomas: Malignant cancers consisting of abnormal lymphocytes or lymphoid stem cells; examples include *Hodgkin's disease* and *non-Hodgkin's lymphoma. (p. 778 and AM)*

severe combined immunodeficiency disease (SCID): A congenital disorder resulting from the failure to develop cell-mediated or antibody-mediated immunity. *(p. 804 and AM)*

tonsillitis: Infection of one or more tonsils; symptoms include a sore throat, high fever, and leukocytosis (an abnormally high white blood cell count). *(p. 776 and AM)*

viruses: Noncellular pathogens that replicate by directing the synthesis of virus-specific proteins and nucleic acids inside tissue cells. *(p. 770 and AM)*

AM ## Additional Terms Discussed in the *Applications Manual*

active immunization: An immunization that intentionally stimulates a primary response to a pathogen before it is encountered in the environment.

bone marrow transplantation: Infusion of bone marrow from a compatible donor after the destruction of host marrow through radiation or chemotherapy; a treatment option for acute, late-stage lymphoma.

graft-versus-host disease (GVH): A condition that results when T cells in donor tissues, such as bone marrow, attack the tissues of the recipient.

hybridoma (hī-bri-DŌ-muh): A cell formed by the fusion of a cancer cell with a plasma cell; used to produce large amounts of a single antibody.

hypersplenism (hī-per-SPLĒN-ism): A condition caused by an overactive spleen; symptoms include anemia, leukopenia, and thrombocytopenia.

hyposplenism (hī-pō-SPLĒN-ism): A condition resulting from the absence or nonfunctional state of the spleen; there are few symptoms.

monoclonal (mo-nō-KLŌ-nal) **antibody:** An antibody produced in large quantities by a population of genetically identical cells.

mononucleosis: A condition resulting from chronic infection by the *Epstein–Barr virus (EBV);* symptoms include splenic enlargement, fever, sore throat, widespread swelling of lymph nodes, increased numbers of lymphocytes in the blood, and the presence of circulating antibodies to the virus.

passive immunization: An immunization in which the patient receives a dose of antibodies that will attack a pathogen immediately without involving the host's immune system.

protozoa: Unicellular eukaryotic organisms that are abundant in soil and water; some protozoa are pathogenic.

splenomegaly (splen-ō-MEG-a-lē): Enlargement of the spleen.

systemic lupus erythematosus (LOO-pus e-rith-ē-ma-TŌ-sis) **(SLE):** An autoimmune disorder resulting from a breakdown in the antigen recognition mechanism, leading to the production of antibodies that destroy healthy cells and tissues.

tonsillectomy: The removal of an inflamed tonsil.

vaccine (vak-SĒN): A preparation of antigens derived from a specific pathogen; administered during immunization.

CHAPTER REVIEW

On-line resources for this chapter are on our World Wide Web site at:
http://www.prenhall.com/martini/fap

STUDY OUTLINE

INTRODUCTION, p. 770

1. The cells, tissues, and organs of the **lymphatic system** play a central role in the body's defenses against a variety of **pathogens,** or disease-causing organisms.

AN OVERVIEW OF THE LYMPHATIC SYSTEM, p. 770

1. Lymphocytes, the primary cells of the lymphatic system, provide an **immune response** to specific threats to the body. **Immunity** is the ability to resist infection and disease through the activation of specific defenses.

ORGANIZATION OF THE LYMPHATIC SYSTEM, p. 770

1. The lymphatic system includes a network of **lymphatic vessels** that carry **lymph** (a fluid similar to plasma but with a lower concentration of proteins). A series of **lymphoid organs** is connected to the lymphatic vessels. *(Figure 22-1)*

Functions of the Lymphatic System, p. 771

2. The lymphatic system produces, maintains, and distributes lymphocytes (cells that attack invading organisms, abnor-

mal cells, and foreign proteins). The system also helps maintain blood volume and eliminate local variations in the composition of the interstitial fluid.

Lymphatic Vessels, p. 771

3. Lymph flows along a network of **lymphatics,** the smallest of which are the **lymphatic capillaries** *(terminal lymphatics).* The lymphatic vessels empty into the **thoracic duct** and the **right lymphatic duct.** *(Figures 22-1–22-4)*

Lymphocytes, p. 774

4. There are three classes of lymphocytes: **T cells** (thymus-dependent), **B cells** (bone marrow–derived), and **NK cells** (natural killer).

5. **Cytotoxic T cells** attack foreign cells or body cells infected by viruses and provide **cell-mediated (cellular) immunity. Regulatory T cells** regulate and coordinate the immune response.

6. B cells can differentiate into **plasma cells,** which produce and secrete *antibodies* that react with specific chemical tar-

gets called **antigens.** Antibodies in body fluids are called *immunoglobulins.* B cells are responsible for **antibody-mediated (humoral) immunity.**

7. NK cells (also called **large granular lymphocytes**) attack foreign cells, normal cells infected with viruses, and cancer cells. NK cells provide **immunological surveillance.**

8. Lymphocytes continuously migrate into and out of the blood through the lymphatic tissues and organs. **Lymphopoiesis** (lymphocyte production) involves the bone marrow, thymus, and peripheral lymphoid tissues. *(Figure 22-5)*

Lymphoid Tissues, p. 775

9. **Lymphoid tissues** are connective tissues dominated by lymphocytes. In a **lymphoid nodule,** the lymphocytes are densely packed in an area of loose connective tissue. *(Figure 22-6)*

Lymphoid Organs, p. 776

10. Important lymphoid organs include the **lymph nodes,** the **thymus,** and the **spleen.** Lymphoid tissues and organs are distributed in areas especially vulnerable to injury or invasion.

11. Lymph nodes are encapsulated masses of lymphatic tissue. The **deep cortex** is dominated by T cells; the **outer cortex** and **medulla** contain B cells. *(Figure 22-7)*

12. The thymus lies behind the sternum, in the anterior mediastinum. **Reticular epithelial cells** scattered among the lymphocytes produce thymic hormones. *(Figure 22-8)*

13. The adult spleen contains the largest mass of lymphoid tissue in the body. The cellular components form the **pulp** of the spleen. **Red pulp** contains large numbers of red blood cells, and areas of **white pulp** resemble lymphoid nodules. *(Figure 22-9)*

The Lymphatic System and Body Defenses, p. 780

14. The lymphatic system is a major component of the body's defenses. These defenses are either (1) **nonspecific defenses,** which do not discriminate between one threat and another, or (2) **specific defenses,** which protect against threats on an individual basis.

NONSPECIFIC DEFENSES, p. 781

1. Nonspecific defenses prevent the approach, deny the entrance, or limit the spread of living or nonliving hazards. *(Figure 22-10)*

Physical Barriers, p. 781

2. Physical barriers include hair, epithelia, and various secretions of the integumentary and digestive systems.

Phagocytes, p. 781

3. There are two types of phagocytic cells: **microphages** and **macrophages** (cells of the **monocyte–macrophage system**). Microphages are the neutrophils and eosinophils in circulating blood.

4. **Phagocytes** move among cells by **diapedesis** and exhibit **chemotaxis** (sensitivity and orientation to chemical stimuli).

Immunological Surveillance, p. 783

5. Immunological surveillance involves constant monitoring of normal tissues by NK cells sensitive to abnormal antigens on the surfaces of otherwise normal cells. Cancer cells with **tumor-specific antigens** on their surfaces are killed. *(Figure 22-11)*

Interferons, p. 784

6. **Interferons,** small proteins released by cells infected with viruses, trigger the production of **antiviral proteins** that interfere with viral replication inside the cell. Interferons are **cytokines,** chemical messengers released by tissue cells to coordinate local activities.

Complement, p. 784

7. At least 11 **complement proteins** make up the **complement system.** They interact with each other in chain reactions to destroy target cell membranes, stimulate inflammation, attract phagocytes, and/or enhance phagocytosis. *(Figure 22-12)*

Inflammation, p. 786

8. Inflammation represents a coordinated nonspecific response to tissue injury. *(Figure 22-13)*

Fever, p. 787

9. A **fever** (body temperature greater than 37.2°C, or 99°F) can inhibit pathogens and accelerate metabolic processes.

SPECIFIC RESISTANCE: THE IMMUNE RESPONSE, p. 787

Forms of Immunity, p. 788

1. Specific immunity may involve **innate immunity** (genetically determined and present at birth) or **acquired immunity.** The two types of acquired immunity are **active immunity** (which appears after exposure to an antigen) and **passive immunity** (produced by the transfer of antibodies from another source). *(Figure 22-14)*

Properties of Immunity, p. 789

2. Lymphocytes provide specific immunity, which has four general characteristics: **specificity, versatility, memory,** and **tolerance.** *Memory cells* enable the immune system to "remember" previous target antigens. Tolerance refers to the ability of the immune system to ignore some antigens, such as those of body cells.

T Cells and Cell-Mediated Immunity, p. 789

3. Foreign antigens must generally be processed by other cells (typically macrophages) and incorporated into cell membranes **(antigen presentation)** before the antigens can activate lymphocytes.

4. All body cells have membrane glycoproteins. The genes controlling their synthesis make up a chromosomal region called the **major histocompatibility complex (MHC).** The membrane glycoproteins are called **MHC proteins.** Lymphocytes are not activated by lone antigens but will respond to an antigen bound to either a **Class I** or a **Class II** MHC protein.

5. Class I MHC proteins are in all nucleated body cells. Class II MHC proteins are only in **antigen-presenting cells (APCs)** and lymphocytes. *(Figure 22-15)*

6. Whether a T cell responds to antigens held in Class I or II MHC proteins depends on the structure of the T *cell membrane.* T cell membranes contain proteins called **CD** (*cluster designation*) **markers. CD8 markers** are found on cytotoxic and suppressor T cells. **CD4 markers** are found on all helper T cells.

7. Cell-mediated immunity (cellular immunity) results from the activation of CD8 T cells by antigens bound to Class I MHCs. When activated, most of these T cells divide to generate **cytotoxic T cells** and **memory T cells,** which remain on reserve to guard against future such attacks. **Suppressor T cells** depress the responses of other T and B cells. *(Figure 22-16)*

8. **Helper,** or **CD4, T cells** respond to antigens presented by Class II MHC proteins. When activated, they secrete lymphokines that help coordinate specific and nonspecific defenses and regulate cell-mediated and antibody-mediated immunity. *(Figure 22-17)*

B Cells and Antibody-Mediated Immunity, p. 793

9. B cells become **sensitized** when antigens bind to their membrane antibody molecules. The antigens are then displayed

on the Class II MHC proteins of the B cells, which become activated by helper T cells activated by the same antigen.

10. An active B cell may differentiate into a plasma cell or produce daughter cells that differentiate into plasma cells and **memory B cells.** Antibodies are produced by the plasma cells. *(Figure 22-18)*

11. An antibody molecule consists of two parallel pairs of polypeptide chains containing *constant* and *variable segments. (Figure 22-19)*

12. When antibody molecules bind to an antigen, they form an **antigen–antibody complex.** Effects that appear after binding include **neutralization; precipitation** (formation of an insoluble **immune complex**) and **agglutination** (formation of large complexes); **opsonization;** stimulation of **inflammation;** and prevention of bacterial or viral adhesion. *(Figure 22-20)*

13. There are five classes of antibodies in body fluids: **immunoglobulin** (1) **G (IgG),** responsible for resistance against many viruses, bacteria, and bacterial toxins; (2) **E (IgE),** which releases chemicals that accelerate local inflammation; (3) **D (IgD),** located on the surfaces of B cells; (4) **M (IgM),** the first antibody type secreted after an antigen arrives; and (5) **A (IgA),** found in glandular secretions. *(Table 22-1)*

Primary and Secondary Responses to Antigen Exposure, p. 797

14. The antibodies first produced by plasma cells are the agents of the **primary response.** The maximum **antibody titer** appears during the **secondary (anamnestic) response** to antigen exposure. *(Figure 22-21)*

Hormones of the Immune System, p. 798

15. **Interleukins** increase T cell sensitivity to antigens exposed on macrophage membranes; stimulate B cell activity, plasma cell formation, and antibody production; and enhance nonspecific defenses.

16. Interferons slow the spread of a virus by making the synthesizing cell and its neighbors resistant to viral infections.

17. **Tumor necrosis factors** slow tumor growth and kill tumor cells.

18. Several lymphokines adjust the activities of phagocytic cells in order to coordinate specific and nonspecific defenses. *(Table 22-2)*

PATTERNS OF IMMUNE RESPONSE, p. 800

1. The initial steps in the immune response to viral and bacterial infections differ. *(Figures 22-22– 22-24; Table 22-3)*

THE DEVELOPMENT OF RESISTANCE, p. 801

1. **Immunological competence** is the ability to demonstrate an immune response upon exposure to an antigen. The developing fetus receives passive immunity from the maternal bloodstream. After delivery, the infant begins developing acquired immunity following exposure to environmental antigens.

IMMUNE DISORDERS, p. 804

Autoimmune Disorders, p. 804

1. **Autoimmune disorders** develop when the immune response mistakenly targets normal body cells and tissues. *(Table 22-4)*

Immunodeficiency Diseases, p. 804

2. In an **immunodeficiency disease,** either the immune system does not develop normally or the immune response is somehow blocked.

Allergies, p. 805

3. **Allergies** are inappropriate or excessive immune responses to **allergens** (antigens that trigger allergic reactions). There are four types of allergies: **immediate hypersensitivity** *(Type I), cytotoxic reactions (Type II), immune complex disorders (Type III),* and *delayed hypersensitivity (Type IV).*

4. In **anaphylaxis,** a circulating allergen affects mast cells throughout the body. *(Figure 22-25)*

STRESS AND THE IMMUNE RESPONSE, p. 807

1. **Interleukin-1** released by active macrophages triggers the release of ACTH by the anterior pituitary. Glucocorticoids produced by the adrenal cortex moderate the immune response, but their long-term secretion can lower resistance to disease.

AGING AND THE IMMUNE RESPONSE, p. 807

1. With aging, the immune system becomes less effective at combating disease.

INTEGRATION WITH OTHER SYSTEMS, p. 808

1. There is substantial interaction between the cells and tissues involved with the immune response and the nervous and endocrine systems. *(Figure 22-26)*

REVIEW QUESTIONS

Level 1   Reviewing Facts and Terms

1. Lymph from the lower abdomen, pelvis, and lower limbs is received by the
- (a) right lymphatic duct
- (b) cisterna chyli
- (c) right thoracic duct
- (d) aorta

2. Lymphoid stem cells that can form all types of lymphocytes occur in the
- (a) bloodstream
- (b) thymus
- (c) bone marrow
- (d) spleen

3. Lymphatics are located in all portions of the body except the
- (a) lower limbs
- (b) central nervous system
- (c) integument
- (d) digestive tract

4. The body's largest collection of lymphoid tissue is in the
- (a) adult spleen
- (b) adult thymus
- (c) bone marrow
- (d) tonsils

5. Red blood cells that are damaged or defective are removed from the circulation by the
- (a) thymus
- (b) lymph nodes
- (c) spleen
- (d) tonsils

6. Phagocytes move through capillary walls by squeezing between adjacent endothelial cells, a process known as
- (a) diapedesis
- (b) chemotaxis
- (c) adhesion
- (d) perforation

7. Perforins and protectin are proteins associated with the activity of
- (a) T cells
- (b) B cells
- (c) NK cells
- (d) plasma cells

8. Complement activation
- (a) stimulates inflammation
- (b) attracts phagocytes
- (c) enhances phagocytosis
- (d) a, b, and c are correct

9. Inflammation
   (a) aids in temporary repair at an injury site
   (b) slows the spread of pathogens
   (c) facilitates permanent repair
   (d) a, b, and c are correct
10. CD4 markers are associated with
   (a) cytotoxic T cells   **(b)** suppressor T cells
   (c) helper T cells   **(d)** a, b, and c are correct
11. Which two large collecting vessels are responsible for returning lymph to the veins of the circulatory system? What areas of the body does each serve?
12. Give a function for each of the following:
   (a) cytotoxic T cells
   (b) helper T cells
   (c) suppressor (regulatory) T cells

   (d) plasma cells
   (e) NK cells
   (f) stromal cells
   (g) reticular epithelial cells
   (h) interferons
   (i) pyrogens
   (j) B cells
   (k) plasma cells
   (l) interleukins
   (m) tumor necrosis factor
   (n) colony-stimulating factors
13. What are the three classes of lymphocytes, and where does each class originate?
14. What seven defenses, present at birth, provide the body with the defensive capability known as nonspecific resistance?

Level 2   Reviewing Concepts

15. Compared with nonspecific defenses, *specific defenses*
   (a) do not discriminate between one threat and another
   (b) are always present at birth
   (c) provide protection against threats on an individual basis
   (d) deny entrance of pathogens to the body
16. T cells and B cells can be activated only by
   (a) pathogenic organisms
   (b) interleukins, interferons, and colony-stimulating factors
   (c) cells infected with viruses, bacterial cells, or cancer cells
   (d) exposure to a specific antigen at a specific site on a cell membrane
17. Class II MHC proteins appear in the cell membrane only when
   (a) the plasma cells are releasing antibodies
   (b) the cell is processing antigens
   (c) cytotoxic T cells are inhibited
   (d) NK cells are activated

18. List the four general properties of immunity; give an explanation of each.
19. How does a cytotoxic T cell destroy another cell displaying antigens bound to Class I MHC proteins?
20. How does the formation of an antigen–antibody complex cause elimination of an antigen?
21. What effects follow the activation of the complement system?
22. An anesthesia technician is advised that she should be vaccinated against hepatitis B, which is caused by a virus. She is given one injection and is told to come back for a second injection in a month and a third injection after 6 months. Why is this series of injections necessary?

Level 3   Critical Thinking and Clinical Applications

23. An investigator at a crime scene discovers some body fluid on the victim's clothing. The investigator carefully takes a sample and sends it to the crime lab for analysis. On the basis of analysis of immunoglobulins, could the crime lab determine whether the sample is blood plasma or semen? Explain.
24. Ted finds out that he has been exposed to the measles. He is concerned that he might have contracted the disease, so he goes to see his physician. The physician takes a blood sample and sends it to a lab for antibody titers. The results show an elevated level of IgM antibodies to rubella (measles) virus but very few IgG antibodies to the virus. Has Ted contracted the disease?
25. While walking along the street, you and your friend see an elderly lady whose left arm appears to be swollen to several

times its normal size. Your friend remarks that she must have been in the tropics and contracted a filarial disease that produces elephantiasis. You disagree, saying that it is more likely that the woman had a radical mastectomy (removal of a breast because of cancer). Explain the rationale behind your answer.
26. Tilly has T cells that are capable of responding to antigen A. Tilly's friend Harry has been exposed to antigen A, and Tilly offers to have some of her T cells transfused so that Harry will not come down with an infection. You overhear their conversation and observe that such a transfusion would not work. Why not?
27. You are a researcher interested in studying cells that can respond to the hormone FSH. How could you use antibodies to help you locate FSH-responsive cells?

23

The Respiratory System

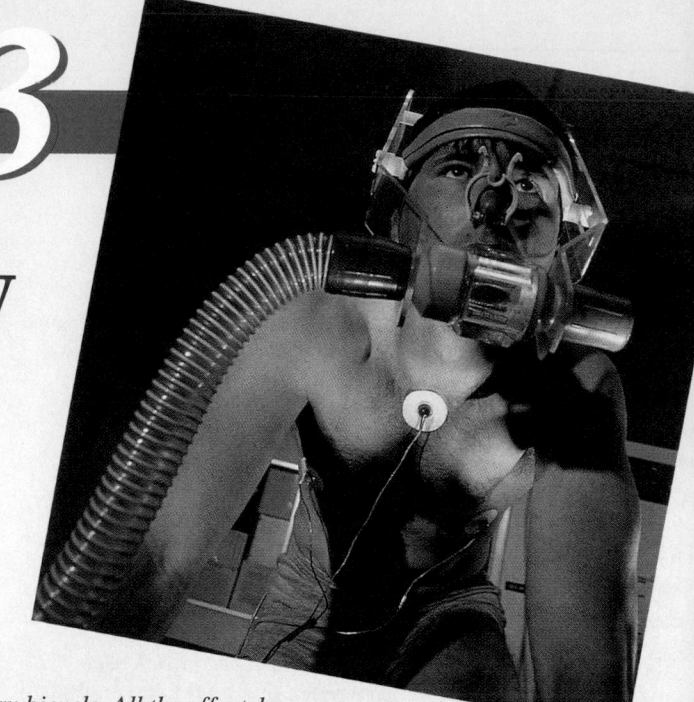

*T*his is one way to go nowhere fast—pedaling on a stationary bicycle. All the effort, however, is not wasted. The awkward-looking array of hoses and wires attached to this rider is designed to monitor respiratory and cardiovascular performance during exercise. Similar equipment can be used in a clinical setting to assess respiratory function in resting or active individuals. The respiratory and cardiovascular systems are closely linked, both anatomically and functionally. In this chapter, we shall see how they cooperate to supply our tissues with vital oxygen and to protect them from the carbon dioxide they generate. We shall also examine the homeostatic mechanisms responsible for regulating respiratory activities as levels of physical activity change.

CHAPTER OUTLINE AND OBJECTIVES

Living cells need energy for maintenance, growth, defense, and replication. Our cells obtain that energy through aerobic mechanisms that require oxygen and produce carbon dioxide. ∞ *[p. 87]* Many aquatic organisms can obtain oxygen and excrete carbon dioxide by diffusion across the surface of the skin or in specialized structures, such as the gills of a fish. Such arrangements are poorly suited for life on land, because the exchange surfaces must be very thin and relatively delicate to permit rapid diffusion. In air, the exposed membranes collapse, evaporation and dehydration reduce blood volume, and the delicate surfaces become vulnerable to attack by pathogenic organisms. Our respiratory exchange surfaces are just as delicate as those of an aquatic organism, but they are confined to the inside of the lungs—in a warm, moist, protected environment. Under these conditions, diffusion can occur between the air and the blood.

The cardiovascular system provides the link between your interstitial fluids and the exchange surfaces of your lungs. Your circulating blood carries oxygen from the lungs to peripheral tissues; it also accepts and transports the carbon dioxide generated by those tissues, delivering it to the lungs.

Diffusion between the blood and air occurs at the **alveoli** (al-VĒ-ō-lī) of the lungs. The distance between the blood in an alveolar capillary and the air inside an alveolus is generally less than 1 µm and in some cases as small as 0.1 µm. To meet the metabolic requirements of peripheral tissues, the exchange surfaces of the lungs must be very large. In fact, the total surface area for gas exchange in the adult lungs is at least 35 times the surface area of the body. It is difficult to measure the exchange surfaces with precision; estimates of the surface area involved range from 70 m^2 to 140 m^2.

Our discussion of the respiratory system begins by following the air as it travels toward the alveoli. We shall then consider the mechanics of breathing and the physiology of respiration.

FUNCTIONS OF THE RESPIRATORY SYSTEM

The **respiratory system** has five basic functions:

1. Providing an extensive area for gas exchange between the air and the circulating blood.
2. Moving air to and from the exchange surfaces of the lungs.
3. Protecting respiratory surfaces from dehydration, temperature changes, or other environmental variations and defending the respiratory system and other tissues from invasion by pathogens.
4. Producing sounds involved in speaking, singing, and nonverbal communication.
5. Providing olfactory sensations to the CNS from the olfactory epithelium in the superior portions of the nasal cavity.

In addition, the capillaries of the lungs indirectly assist in the regulation of blood volume and blood pressure, through the conversion of angiotensin I to angiotensin II. ∞ *[pp. 620, 733]*

ORGANIZATION OF THE RESPIRATORY SYSTEM

We can divide the respiratory system (Figure 23-1●) into an *upper respiratory system* and a *lower respiratory system*. The **upper respiratory system** consists of the nose, nasal cavity, paranasal sinuses, and pharynx. These passageways filter, warm, and humidify the incoming air—protecting the more delicate surfaces of the lower respiratory system—and cool and dehumidify outgoing air. The **lower respiratory system** includes the larynx (voice box), trachea (windpipe), bronchi, bronchioles, and alveoli of the lungs.

Your **respiratory tract** consists of the airways that carry air to and from the exchange surfaces of your lungs. The respiratory tract can be divided into a *conducting portion* and a *respiratory portion*. The conducting portion begins at the entrance to the nasal cavity and extends through the pharynx and larynx and along the trachea, bronchi, and bronchioles to the *terminal bronchioles*. The respiratory portion of the tract includes the delicate *respiratory bronchioles* and the sites of gas exchange, the alveoli.

Filtering, warming, and humidification of the inspired air begin at the entrance to the upper respiratory system and continue throughout the rest of the conducting system. By the time air reaches the alveoli, most foreign particles and pathogens have been removed, and the humidity and temperature are within acceptable limits. The success of this "conditioning process" is due primarily to the properties of the *respiratory mucosa*.

The Respiratory Mucosa

The **respiratory mucosa** (mū-KŌ-suh) lines the conducting portion of the respiratory system. A *mucosa* is a *mucous membrane*, one of the four types of membranes introduced in Chapter 4. It consists of an epithelium and an underlying layer of loose connective tissue. ∞ *[p. 132]*

The Respiratory Epithelium

A pseudostratified, ciliated, columnar epithelium with numerous goblet cells (Figure 23-2●, p. 817) lines the nasal cavity and the superior portion of the pharynx. ∞ *[p. 117]* The structure of the respiratory epithelium changes as you proceed along the respiratory tract. The epithelium lining inferior portions of the pharynx is a stratified squamous epithelium similar to that of the oral cavity. These portions of the pharynx, which conduct air to the lower respiratory tract, also convey food to the

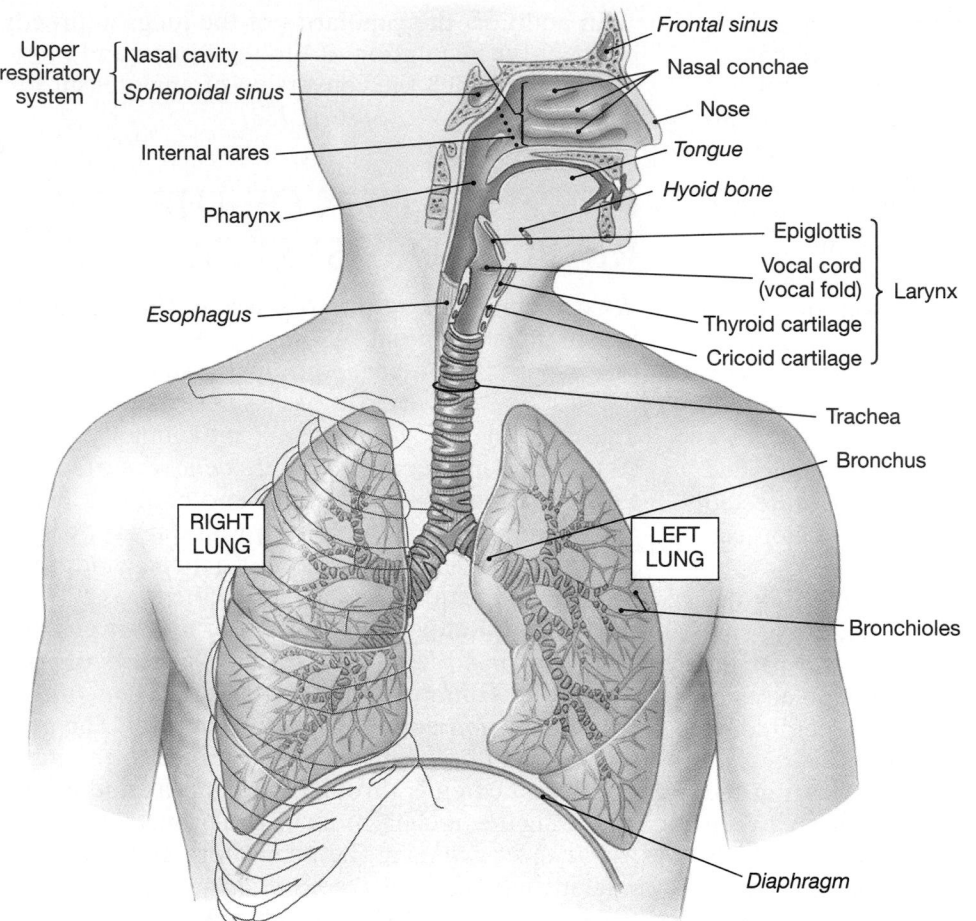

Upper respiratory system {
Nasal cavity
Sphenoidal sinus

Internal nares

Pharynx

Esophagus

Frontal sinus

Nasal conchae

Nose

Tongue

Hyoid bone

Epiglottis
Vocal cord (vocal fold) } Larynx
Thyroid cartilage
Cricoid cartilage

Trachea

Bronchus

RIGHT LUNG

LEFT LUNG

Bronchioles

Diaphragm

•*FIGURE 23-1* **Components of the Respiratory System.** The conducting portion of the respiratory system is shown here. At this scale, the alveoli are not visible. [AM] *Plates 3.1, 4.4, 6.3*

esophagus. The pharyngeal epithelium must therefore provide protection from abrasion and chemical attack.

At the beginning of the lower respiratory tract is a pseudostratified ciliated columnar epithelium comparable to that of the nasal cavity. In the smaller bronchioles, this pseudostratified epithelium is replaced by a cuboidal epithelium with scattered cilia. The exchange surfaces of the alveoli are lined by a very delicate simple squamous epithelium. Other, more specialized cells are scattered within the alveolar epithelium.

The Lamina Propria

The **lamina propria** (LA-mi-nuh PRŌ-prē-uh) is the underlying layer of loose connective tissue that supports the respiratory epithelium. In the upper respiratory system and in the trachea and bronchi, the lamina propria contains mucous glands that discharge their secretions onto the epithelial surface. The lamina propria in the conducting portions of the lower respiratory system contains bundles of smooth muscle cells. At the level of the bronchioles, the smooth muscles form relatively thick bands that encircle or spiral around the lumen.

The Respiratory Defense System

The delicate exchange surfaces of the respiratory system can be severely damaged if the inspired air becomes contaminated with debris or pathogens. Such contami-

nation is prevented by a series of filtration mechanisms that together make up the **respiratory defense system.**

Along much of the length of the respiratory tract, goblet cells in the epithelium and mucous glands in the lamina propria produce a sticky mucus that bathes exposed surfaces. In the nasal cavity, cilia sweep that mucus and any trapped debris or microorganisms toward the pharynx, where it will be swallowed and exposed to the acids and enzymes of the stomach. In the lower respiratory system, the cilia also beat toward the pharynx, moving a carpet of mucus toward the pharynx and cleaning the respiratory surfaces. This process is often described as a *mucus escalator* (Figure 23-2c•).

Filtration in the nasal cavity removes virtually all particles larger than about 10 μm from the inspired air. Smaller particles may be trapped by the mucus of the nasopharynx or secretions of the pharynx before proceeding farther along the conducting system. Exposure to unpleasant stimuli, such as noxious vapors, large quantities of dust and debris, allergens, or pathogens, generally causes a rapid increase in the rate of mucus production in the nasal cavity and paranasal sinuses. (The familiar symptoms of the "common cold" result from the invasion of this respiratory epithelium by any of more than 200 viruses.)

Most particles 1–5 μm in diameter are trapped in the mucus coating the respiratory bronchioles or in

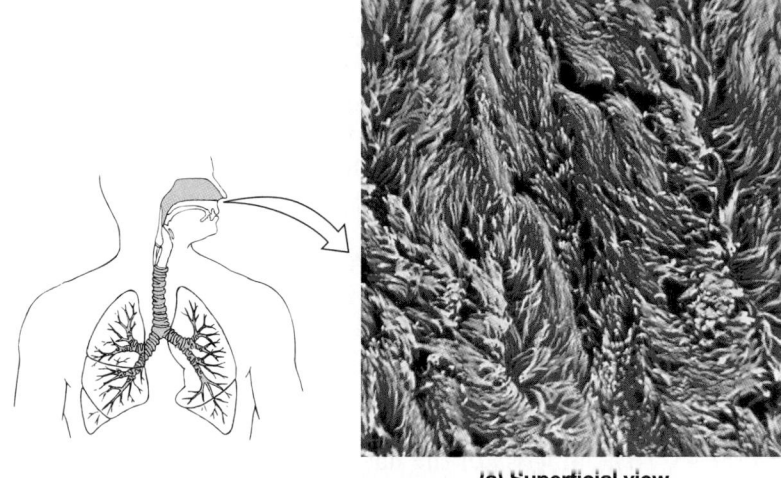

(a) Superficial view

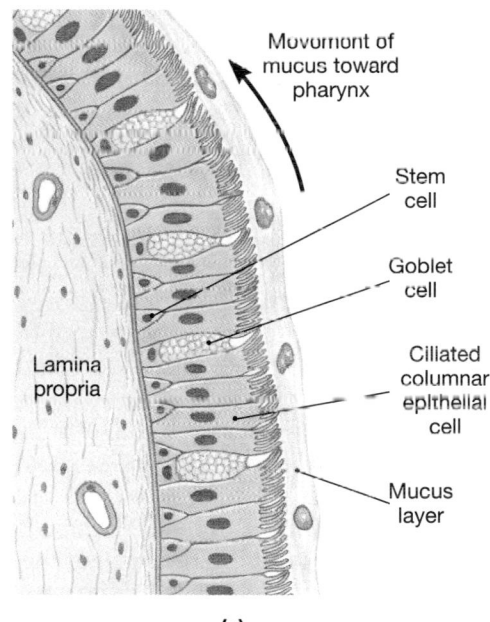

●FIGURE 23-2 The Respiratory Epithelium of the Nasal Cavity and Conducting System. **(a)** A surface view of the epithelium, as seen with the scanning electron microscope. The cilia of the epithelial cells form a dense layer that resembles a shag carpet. The movement of these cilia propels mucus across the epithelial surface. (× 1614) **(b)** Light micrograph showing the sectional appearance of the respiratory epithelium. (× 1062) **(c)** Diagrammatic view of the respiratory epithelium of the trachea, showing the mechanism of mucus transport.

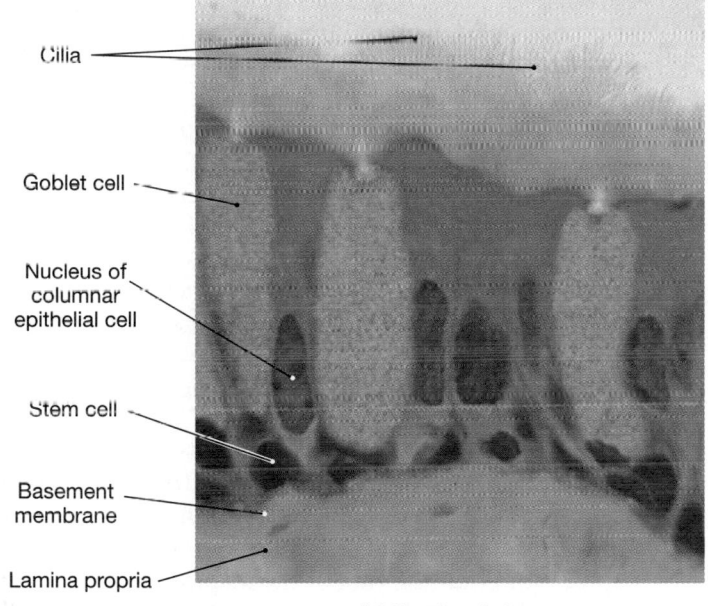

Cilia

Goblet cell

Nucleus of columnar epithelial cell

Stem cell

Basement membrane

Lamina propria

(b) Sectional view

Movement of mucus toward pharynx

Stem cell

Goblet cell

Ciliated columnar epithelial cell

Mucus layer

Lamina propria

(c)

the liquid covering the alveolar surfaces. These areas are outside the boundaries of the mucus escalator, but the foreign particles can be engulfed by alveolar macrophages. Most particles smaller than about 0.5 μm remain suspended in the air.

Large quantities of airborne particles may overload the respiratory defenses and produce a variety of illnesses. For example, the presence of irritants in the lining of the conducting passageways can provoke the formation of abscesses that block airflow and reduce pulmonary function, and damage to the epithelium in the affected area may allow irritants to enter the surrounding tissues of the lung. The irritants then produce local inflammation, and there is a strong link between airborne irritants and the development of lung cancer (p. 857). (We noted smoking-induced changes induced in the respiratory epithelium in Chapter 4, Figure 4-20●, p. 142.) [AM] *Overloading the Respiratory Defenses*

CYSTIC FIBROSIS Cystic fibrosis (CF) is the most common lethal inherited disease affecting Caucasians of Northern European descent, occurring at a frequency of 1 birth in 2500. It occurs with less frequency in people of Southern European ancestry, in the Ashkenazi Jewish population, and in African Americans. The condition results from a defective gene located on chromosome 7. Individuals with CF seldom survive past age 30; death is generally the result of a massive bacterial infection of the lungs and associated heart failure.

The most serious symptoms appear because the respiratory mucosa in these individuals produces a dense, viscous mucus that cannot be transported by the respiratory defense system. The mucus escalator stops working, and mucus blocks the smaller respiratory passageways. This blockage reduces the diameter of the airways, making breathing difficult, and the inactivation of the normal respiratory defenses leads to frequent bacterial infections. For information on the genetic basis of this disorder and current strategies for treatment, refer to the *Applications Manual.* [AM] *Cystic Fibrosis*

THE UPPER RESPIRATORY SYSTEM

The upper respiratory system consists of the nose, nasal cavity, paranasal sinuses, and pharynx (Figures 23-1 and 23-3●).

The Nose and Nasal Cavity

The nose is the primary passageway for air entering the respiratory system. Air normally enters the respiratory system through the paired **external nares** (NĀR-ēz), or *nostrils* (Figure 23-3a●), which open into the **nasal cavity.** The **vestibule** (VES-ti-būl) is the space contained within the flexible tissues of the nose (Figure 23-3c●). The epithelium of the vestibule contains coarse hairs that extend across the external nares. Large airborne particles, such as sand, sawdust, or even insects, are trapped in these hairs and are thereby prevented from entering the nasal cavity.

The *nasal septum* divides the nasal cavity into left and right portions (Figure 23-3b●). The bony portion of the nasal septum is formed by the fusion of the perpendicular plate of the ethmoid bone and the plate of the vomer (Figure 7-3d●, p. 202). The anterior portion of the nasal septum is formed of hyaline cartilage. This cartilaginous plate supports the bridge, or *dorsum nasi* (DOR-sum NĀ-zī), and *apex* (tip) of the nose.

The maxillary, nasal, frontal, ethmoid, and sphenoid bones form the lateral and superior walls of the nasal cavity. The mucous secretions produced in the associated *paranasal sinuses* (Figure 7-11●, p. 210), aided by the tears draining through the nasolacrimal ducts, help keep the surfaces of the nasal cavity moist and clean. The *olfactory region,* or superior portion of the nasal cavity, includes the areas lined by olfactory epithelium: (1) the inferior surface of the cribriform plate, (2) the superior portion of the nasal septum, and (3) the superior nasal conchae. Receptors in the olfactory epithelium provide your sense of smell. ∞ *[p. 546]*

The *superior, middle,* and *inferior nasal conchae* project toward the nasal septum from the lateral walls of the nasal cavity. ∞ *[pp. 208, 210]* To pass from the vestibule to the internal nares, air tends to flow between adjacent conchae, through the **superior, middle,** and **inferior meatuses** (mē-Ā-tus; *meatus,* a passage) (Figure 23-3b●). These are narrow grooves rather than open passageways, and the incoming air bounces off the conchal surfaces and churns around like a stream flowing over rapids. This turbulence serves a purpose: As the air eddies and swirls, small airborne particles are likely to come into contact with the mucus that coats the lining of the nasal cavity. In addition to promoting filtration, the turbulence allows extra time for warming and humidifying the incoming air. It also creates eddy currents that bring olfactory stimuli to the olfactory receptors.

A bony **hard palate,** formed by portions of the maxillary and palatine bones, forms the floor of the nasal cavity and separates the oral and nasal cavities. A fleshy **soft palate** extends posterior to the hard palate, marking the boundary between the superior **nasopharynx** (nā-zō-FAR-inks) and the rest of the pharynx. The nasal cavity opens into the nasopharynx at the **internal nares.**

The Nasal Mucosa

The mucosa of the nasal cavity prepares the air you breathe for arrival at your lower respiratory system. Throughout much of the nasal cavity, the lamina propria contains an abundance of arteries, veins, and capillaries that bring nutrients and water to the secretory cells. The lamina propria of the nasal conchae also contains an extensive network of large and highly expandable veins. This extensive vascularization provides a mechanism for warming and humidifying the incoming air (as well as for cooling and dehumidifying the outgoing air). As cool, dry air passes inward over the exposed surfaces of the nasal cavity, the warm epithelium radiates heat and the water in the mucus evaporates. Air moving from your nasal cavity to your lungs has been heated almost to body temperature, and it is nearly saturated with water vapor. This mechanism protects more delicate respiratory surfaces from chilling or drying out—two potentially disastrous events. Breathing through your mouth eliminates much of the preliminary filtration, heating, and humidifying of the inspired air. To avoid alveolar damage, patients breathing on a respirator, which utilizes a tube to provide air directly into the trachea, must receive air that has been externally filtered and humidified.

As air moves out of the respiratory tract, it again passes across the epithelium of the nasal cavity. This air is warmer and more humid than the air that enters; it warms the nasal mucosa, and moisture condenses on the epithelial surfaces. Thus breathing through your nose also helps prevent heat loss and water loss to your environment.

NOSEBLEEDS The extensive vascularization of the nasal cavity and the relatively vulnerable position of the nose make a nosebleed, or **epistaxis** (ep-i-STAK-sis), a fairly common event. Bleeding generally involves vessels of the mucosa covering the cartilaginous portion of the septum. A variety of factors may be responsible, including trauma, such as a punch in the nose, drying, infections, allergies, and clotting disorders. Hypertension may also provoke a nosebleed by rupturing small vessels of the lamina propria.

•**FIGURE 23-3** **The Nose, Nasal Cavity, and Pharynx.** **(a)** The nasal cartilages and external landmarks on the nose. **(b)** The meatuses and the maxillary and ethmoidal sinuses. **(c)** The nasal cavity and pharynx, as seen in sagittal section with the nasal septum removed. AM *Plates 3.1, 4.4; Scan 2*

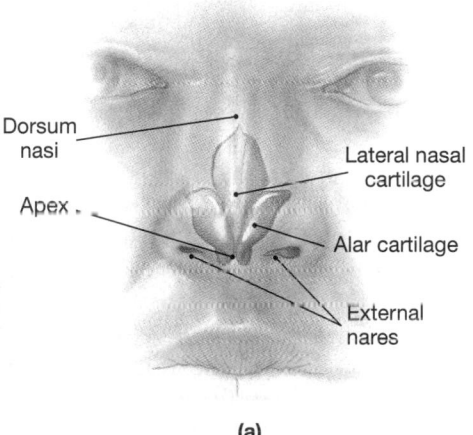

Dorsum nasi
Apex
Lateral nasal cartilage
Alar cartilage
External nares

(a)

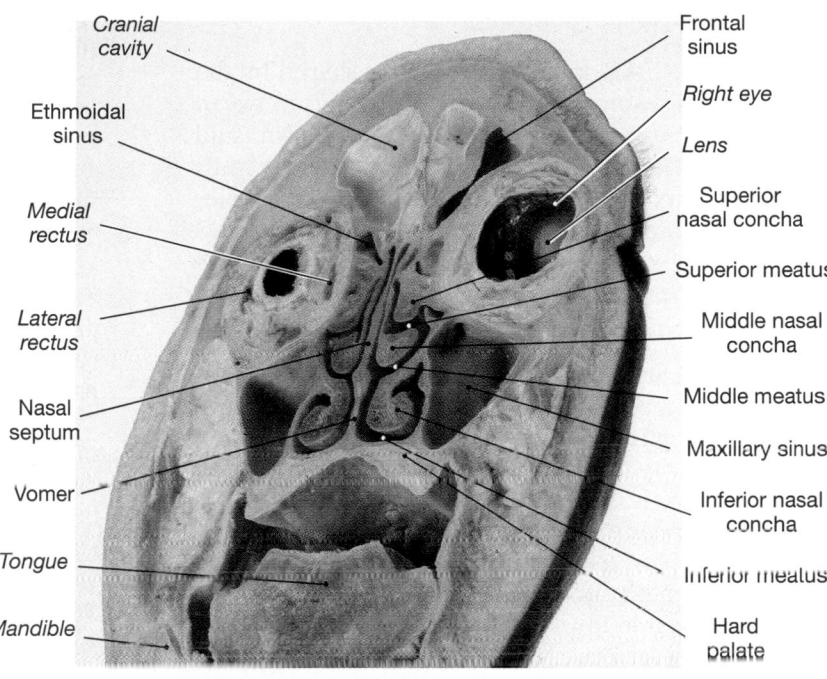

Cranial cavity
Ethmoidal sinus
Medial rectus
Lateral rectus
Nasal septum
Vomer
Tongue
Mandible

Frontal sinus
Right eye
Lens
Superior nasal concha
Superior meatus
Middle nasal concha
Middle meatus
Maxillary sinus
Inferior nasal concha
Inferior meatus
Hard palate

(b)

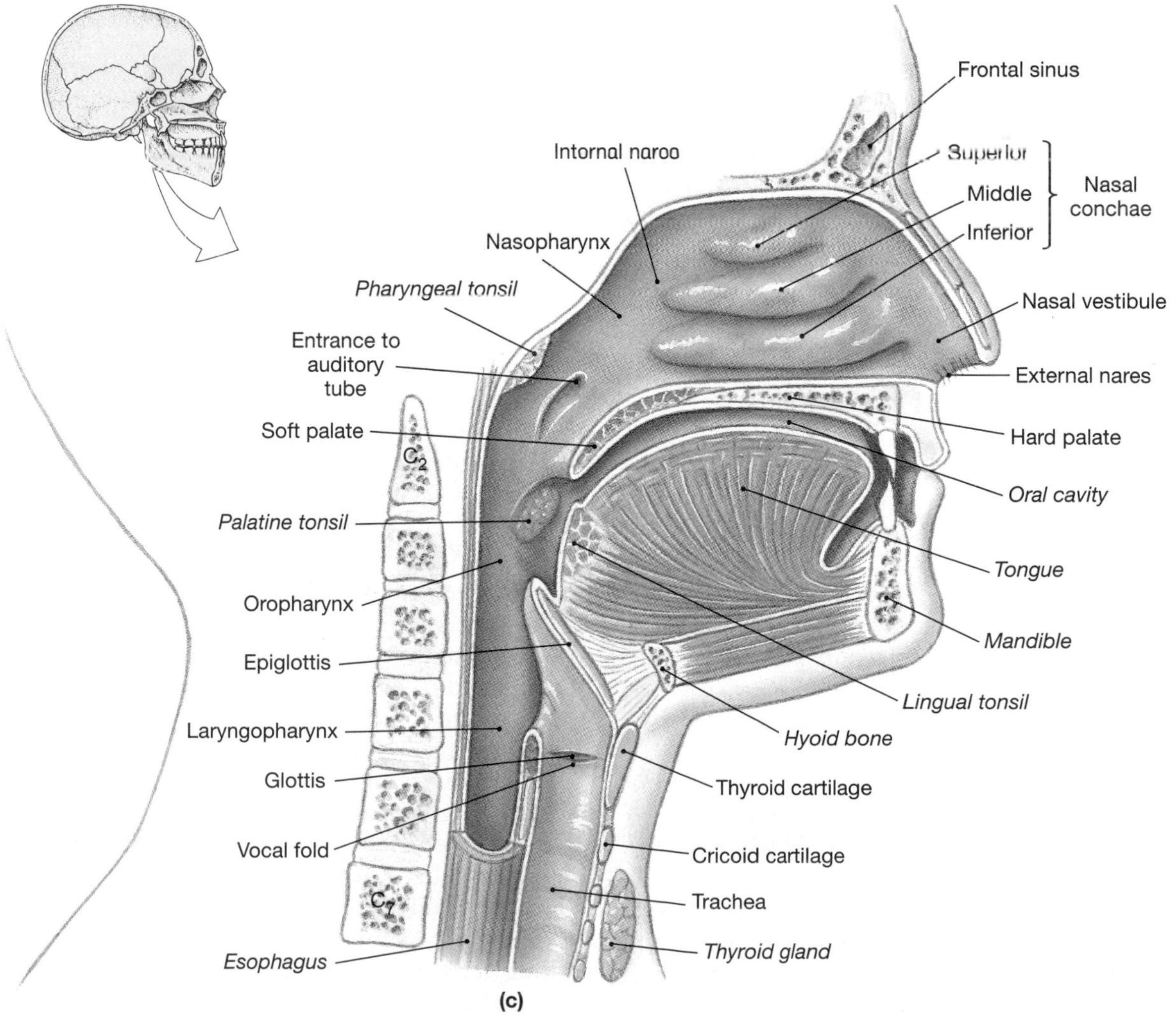

Frontal sinus
Internal nares
Nasopharynx
Pharyngeal tonsil
Entrance to auditory tube
Soft palate
Palatine tonsil
Oropharynx
Epiglottis
Laryngopharynx
Glottis
Vocal fold
Esophagus

Superior
Middle
Inferior
Nasal conchae
Nasal vestibule
External nares
Hard palate
Oral cavity
Tongue
Mandible
Lingual tonsil
Hyoid bone
Thyroid cartilage
Cricoid cartilage
Trachea
Thyroid gland

C₂
C₇

(c)

The Pharynx

The **pharynx** (FAR-inks) is a chamber shared by the digestive and respiratory systems. It extends between the internal nares and the entrances to the larynx and esophagus. The curving superior and posterior walls of the pharynx are closely bound to the axial skeleton, but the lateral walls are flexible and muscular.

The pharynx is divided into three regions (Figure 23-3c●): the *nasopharynx,* the *oropharynx,* and the *laryngopharynx:*

1. The **nasopharynx** is the superior portion of the pharynx. It is connected to the posterior portion of the nasal cavity through the internal nares and is separated from the oral cavity by the soft palate (Figure 23-3c●). The nasopharynx is lined by the same pseudostratified ciliated columnar epithelium as that in the nasal cavity. The *pharyngeal tonsil* is located on the posterior wall of the nasopharynx; on each side, one of the *auditory tubes* opens into the nasopharynx. ∞ *[p. 776, 571]*

2. The **oropharynx** (*oris,* mouth) extends between the soft palate and the base of the tongue at the level of the hyoid bone. The posterior portion of the oral cavity communicates directly with the oropharynx, as does the posterior inferior portion of the nasopharynx. At the boundary between the nasopharynx and the oropharynx, the epithelium changes from a pseudostratified columnar epithelium to a stratified squamous epithelium.

3. The narrow **laryngopharynx** (la-rin-gō-FAR-inks), the inferior portion of the pharynx, includes that portion of the pharynx that lies between the hyoid bone and the entrance to the larynx and esophagus (Figure 23-3c●). Like the oropharynx, it is lined by a stratified squamous epithelium that can resist mechanical abrasion, chemical attack, and pathogenic invasion.

THE LARYNX

Inspired (inhaled) air leaves the pharynx by passing through the **glottis** (GLOT-is), a narrow opening. The **larynx** (LAR-inks) surrounds and protects the glottis. The larynx begins at the level of vertebra C_4 or C_5 and ends at the level of vertebra C_6. The larynx is essentially a cylinder whose incomplete cartilaginous walls are stabilized by ligaments and skeletal muscles (Figure 23-4●).

Cartilages of the Larynx

Three large, unpaired cartilages form the body of the larynx: the *thyroid cartilage,* the *cricoid cartilage,* and the *epiglottis* (Figure 23-4●):

1. The **thyroid cartilage** (*thyroid,* shield-shaped) is the largest laryngeal cartilage. Consisting of hyaline cartilage, it forms most of the anterior and lateral walls of the larynx. The thyroid cartilage in section is U-shaped;

it is incomplete posteriorly. The prominent anterior surface of this cartilage, which you can easily see and feel, is commonly called the *Adam's apple.* The inferior surface of the thyroid cartilage articulates with the cricoid cartilage. The superior surface has ligamentous attachments to the hyoid bone and to the epiglottis and smaller laryngeal cartilages.

2. The thyroid cartilage sits superior to the **cricoid** (KRĪ-koyd; ring-shaped) **cartilage,** another hyaline cartilage. The posterior portion of the cricoid is greatly expanded, providing support in the absence of the thyroid cartilage. The cricoid and thyroid cartilages protect the glottis and the entrance to the trachea, and their broad surfaces provide sites for the attachment of important laryngeal muscles and ligaments. Ligaments attach the inferior surface of the cricoid cartilage to the first cartilage of the trachea. The superior surface of the cricoid cartilage articulates with the small, paired *arytenoid cartilages.*

3. The shoehorn-shaped **epiglottis** (ep-i-GLOT-is) projects superior to the glottis. Composed of elastic cartilage, it has ligamentous attachments to the anterior and superior borders of the thyroid cartilage and the hyoid bone. During swallowing, the larynx is elevated and the epiglottis folds back over the glottis, preventing the entry of liquids or solid food into the respiratory passageways.

The larynx also contains three pairs of smaller hyaline cartilages: the *arytenoid, corniculate,* and *cuneiform cartilages:*

■ The **arytenoid** (ar-i-TĒ-noyd; ladle-shaped) **cartilages** articulate with the superior border of the enlarged portion of the cricoid cartilage.

■ The **corniculate** (kor-NIK-ū-lāt; horn-shaped) **cartilages** articulate with the arytenoid cartilages. The corniculate and arytenoid cartilages are involved with the opening and closing of the glottis and the production of sound.

■ Elongate, curving **cuneiform** (kū-NĒ-i-form; wedge-shaped) **cartilages** lie within folds of tissue (the *aryepiglottic folds*) that extend between the lateral aspect of each arytenoid cartilage and the epiglottis.

Laryngeal Ligaments

Intrinsic ligaments bind all nine cartilages together to form the larynx (Figure 23-4a,b●). **Extrinsic ligaments** attach the thyroid cartilage to the hyoid bone and the cricoid cartilage to the trachea. The **ventricular ligaments** and the **vocal ligaments** extend between the thyroid cartilage and the arytenoids.

The ventricular and vocal ligaments are covered by folds of laryngeal epithelium that project into the glottis. The ventricular ligaments lie within the superior pair of folds, known as the **ventricular folds** (Figure 23-4b–e●). The ventricular folds, which are relatively inelastic, help prevent foreign objects from entering the glottis and provide protection for the more delicate **vocal folds** (Figure 23-4b–e●).

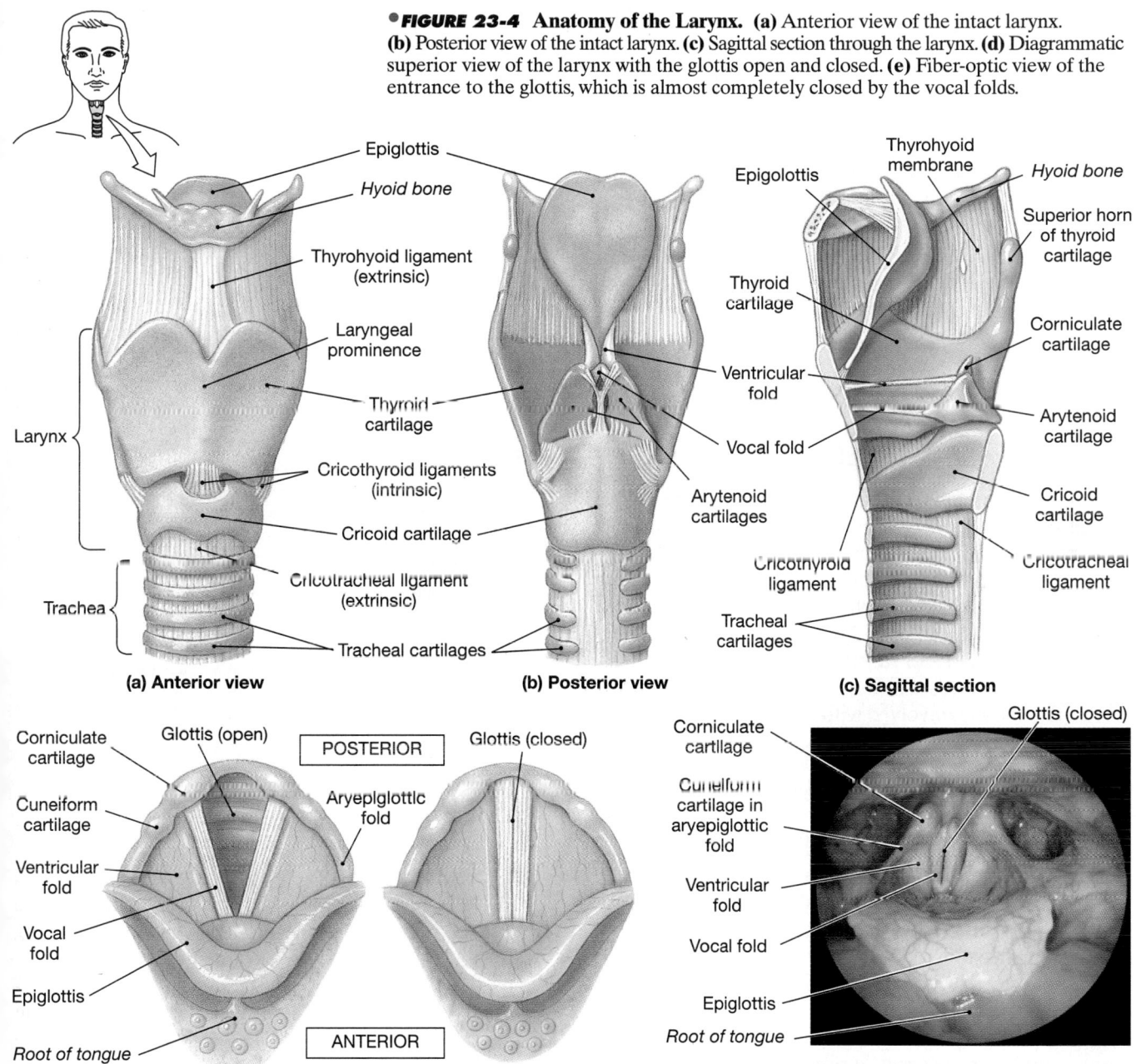

•FIGURE 23-4 Anatomy of the Larynx. (a) Anterior view of the intact larynx. **(b)** Posterior view of the intact larynx. **(c)** Sagittal section through the larynx. **(d)** Diagrammatic superior view of the larynx with the glottis open and closed. **(e)** Fiber-optic view of the entrance to the glottis, which is almost completely closed by the vocal folds.

The vocal folds guard the entrance to the glottis. They are located inferior to the ventricular folds. The vocal folds are highly elastic, because they contain bands of elastic tissue called the *vocal ligaments.* The vocal folds are involved with the production of sounds, and for this reason they are known as the **true vocal cords.** Because the ventricular folds play no part in sound production, they are often called the **false vocal cords.**

Sound Production

Air passing through the glottis vibrates the vocal folds and produces sound waves. The pitch of the sound produced depends on the diameter, length, and tension in the vocal folds. The diameter and length are directly related to the size of the larynx. The tension is controlled by the contraction of voluntary muscles that change the position of the arytenoid cartilages relative to that of the thyroid cartilage. When the distance increases, the vocal folds tense and the pitch rises; when the distance decreases, the vocal folds relax and the pitch falls.

Anatomically, children of both genders have slender, short vocal folds, and their voices tend to be high-pitched. At puberty, the larynx of a male enlarges considerably more than that of a female. The true vocal cords of an adult male are thicker and longer, and they produce lower tones, than those of an adult female.

Sound production at the larynx is called **phonation** (fō-NĀ-shun; *phone,* voice). Phonation is one component of speech production, but clear speech also requires **articulation** (ar-tik-ū-LĀ-shun), the modification of those sounds by other structures. In a stringed instrument, such as a guitar, the quality of the sound produced does not depend solely on the nature of the vibrating string. The entire instrument becomes involved as the walls vibrate and the composite sound echoes within the hollow body. Similar amplification and resonance occur within your pharynx, oral cavity, nasal cavity, and paranasal sinuses. The combination determines the particular and distinctive sound of your voice. The final production of distinct words further depends on voluntary movements of the tongue, lips, and cheeks.

The Laryngeal Musculature

The larynx is associated with two groups of muscles: (1) the *extrinsic laryngeal muscles* and (2) the *intrinsic laryngeal muscles.* The extrinsic laryngeal musculature, which includes muscles of the neck and pharynx, positions and stabilizes the larynx. We considered these muscles in Chapter 11. ∞ *[p. 330]* The **intrinsic laryngeal muscles** have two major functions. One set regulates tension in the vocal folds; a second set opens and closes the glottis. The muscles involved with the vocal folds insert on the thyroid, arytenoid, and corniculate cartilages. The opening or closing of the glottis involves rotational movements of the arytenoids that move the vocal folds apart or together.

When you swallow, both extrinsic and intrinsic muscles cooperate to prevent food or drink from entering the glottis. Before the material is swallowed, it is crushed and chewed into a pasty mass known as a *bolus.* Extrinsic muscles then elevate the larynx, bending the epiglottis over the entrance to the glottis, so that the bolus can glide across the epiglottis rather than falling into the larynx (Figure 23-5 ●). While this movement is under way, intrinsic muscles close the glottis. Food particles or liquids that touch the surfaces of the ventricular or vocal folds will trigger the *coughing reflex.* In a cough, the glottis is kept closed while the expiratory muscles contract, elevating intrapulmonary pressure. When the glottis is opened suddenly, the resulting blast of air from the trachea generally ejects any material that blocks the entrance to the glottis.

LARYNGEAL PROBLEMS Infection or inflammation of the larynx is known as **laryngitis** (lar-in-JĪ-tis). This condition commonly affects the vibrational qualities of the vocal cords; hoarseness is the most familiar symptom. Mild cases are temporary and seldom serious. However, bacterial or viral infection of the epiglottis can be very dangerous, because the resulting swelling may close the glottis and cause suffocation. This condition, acute **epiglottitis** (ep-i-glot-TĪ-tis), can develop relatively rapidly after a bacterial infection of the throat. Young children are most likely to be affected.

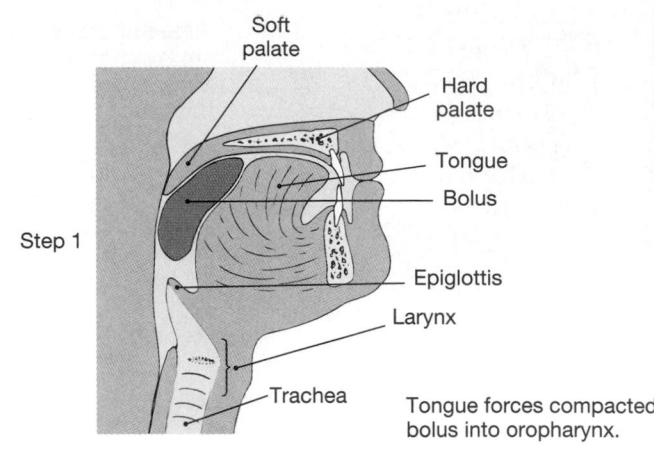

Step 1

Soft palate
Hard palate
Tongue
Bolus
Epiglottis
Larynx
Trachea

Tongue forces compacted bolus into oropharynx.

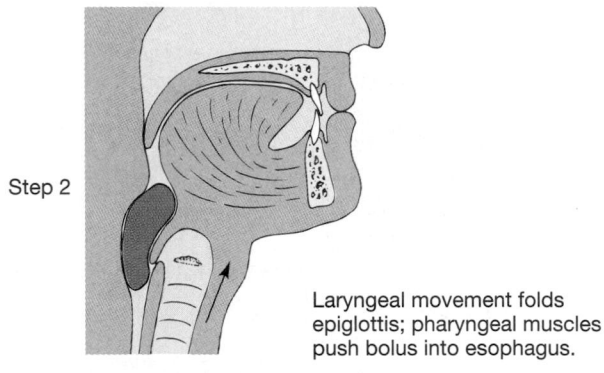

Step 2

Laryngeal movement folds epiglottis; pharyngeal muscles push bolus into esophagus.

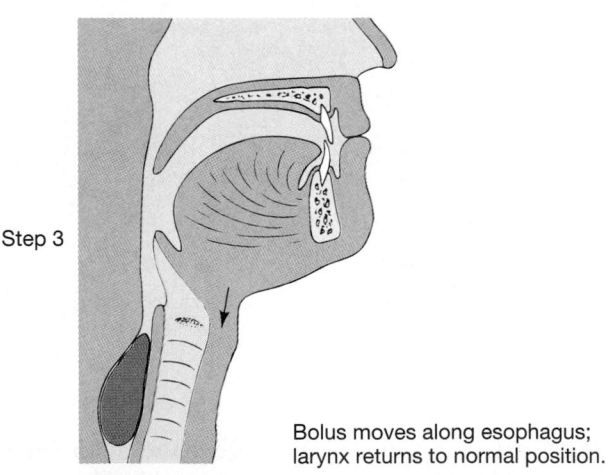

Step 3

Bolus moves along esophagus; larynx returns to normal position.

●**FIGURE 23-5** **Movements of the Larynx During Swallowing.** During swallowing, the elevation of the larynx folds the epiglottis over the glottis so that materials slide into the esophagus rather than into the trachea.

☑ Why is the vascularization of the nasal cavity important?

☑ Why is the lining of the nasopharynx different from that of the oropharynx and laryngopharynx?

☑ When the tension in your vocal cords increases, what happens to the pitch of your voice?

THE TRACHEA

The epithelium of the larynx is continuous with that of the **trachea** (TRĀ-kē-uh), or windpipe. The trachea is a tough, flexible tube with a diameter of about 2.5 cm (1 in.) and a length of approximately 11 cm (4.25 in.) (Figure 23-6●). The trachea begins anterior to vertebra C_6 in a ligamentous attachment to the cricoid cartilage. It ends in the mediastinum, at the level of vertebra T_5, where it branches to form the *right* and *left primary bronchi*.

The mucosa of the trachea resembles that of the nasal cavity and nasopharynx (Figure 23-2a●, p. 817). The **submucosa** (sub-mū-KŌ-suh), a thick layer of connective tissue, surrounds the mucosa. The submucosa contains mucous glands that communicate with the epithelial surface through a number of secretory ducts. The trachea contains 15–20 **tracheal cartilages** (Figure 23-6a●). Each tracheal cartilage is bound to neighboring cartilages by elastic **annular ligaments.** The tracheal cartilages stiffen the tracheal walls and protect the air-way. They also prevent its collapse or overexpansion as pressures change in the respiratory system.

Each tracheal cartilage is C-shaped. The closed portion of the C protects the anterior and lateral surfaces of the trachea. The open portion of the C faces posteriorly, toward the esophagus. Because the tracheal cartilages do not continue around the trachea, the posterior tracheal wall can easily distort when you swallow, permitting the passage of large masses of food through the esophagus.

An elastic ligament and the **trachealis,** a band of smooth muscle, connect the ends of each tracheal cartilage (Figure 23-6b,c●). Contraction of the trachealis muscle alters the diameter of the trachea, changing the trachea's resistance to airflow. The normal diameter of the trachea changes from moment to moment, primarily under the control of the sympathetic division of the ANS. Sympathetic stimulation increases the diameter of the trachea and makes it easier to move large volumes of air along the respiratory passageways.

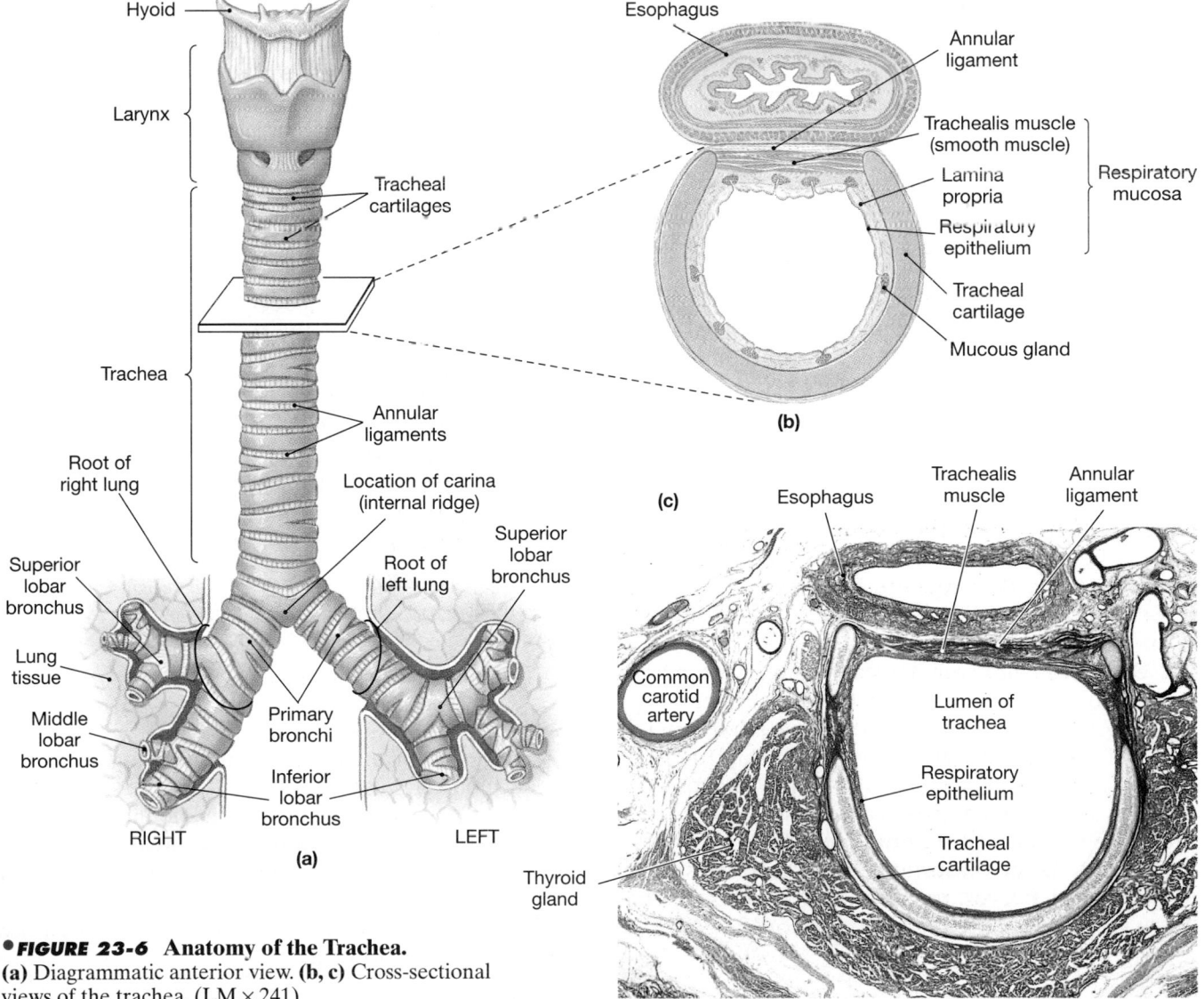

●**FIGURE 23-6   Anatomy of the Trachea.**
(a) Diagrammatic anterior view. (b, c) Cross-sectional views of the trachea. (LM × 241)

TRACHEAL BLOCKAGE We sometimes inadvertently breathe in foreign objects; this process is called *aspiration*. Foreign objects that become lodged in the larynx or trachea are generally expelled by coughing. If the individual can speak or make a sound, the airway is still open, and no emergency measures should be taken. If the person can neither breathe nor speak, an immediate threat to life exists. Unfortunately, many victims become acutely embarrassed by this situation; instead of seeking assistance, they run to the nearest rest room and die there.

In the **Heimlich** (HĪM-lik) **maneuver,** or *abdominal thrust,* a rescuer applies compression to the victim's abdomen just inferior to the diaphragm. This action elevates the diaphragm forcefully and may generate enough pressure to remove the blockage. The maneuver must be performed properly to avoid damage to internal organs. Many organizations such as the American Red Cross, local fire departments, and other charitable groups periodically hold brief training sessions in the proper performance of the Heimlich maneuver.

If blockage results from a swelling of the epiglottis or tissues surrounding the glottis, a professionally qualified rescuer may insert a curved tube through the pharynx and glottis to permit airflow. This procedure is called an *intubation*. If the blockage is immovable or the larynx has been crushed, a **tracheostomy** (trā-kē-OS-to-mē) may be performed. In this procedure, an incision is made through the anterior tracheal wall, and a tube is inserted. The tube bypasses the larynx and permits air to flow directly into the trachea.

THE PRIMARY BRONCHI

The trachea branches within the mediastinum, giving rise to the **right** and **left primary bronchi** (BRONG-kī). A ridge called the **carina** (ka-RĪ-nuh) marks the line of separation between the two bronchi (Figure 23-6a●). The histological organization of the primary bronchi is the same as that of the trachea, with cartilaginous C-shaped supporting rings. The right primary bronchus supplies the right lung, and the left supplies the left lung. The right primary bronchus is larger in diameter, and descends toward the lung at a steeper angle, than the left. Thus most foreign objects that enter the trachea find their way into the right bronchus rather than the left.

Before branching further, each primary bronchus travels to a groove along the medial surface of its lung. This groove, the **hilus** of the lung, also provides access for entry to pulmonary vessels and nerves (Figure 23-7●). The entire array is firmly anchored in a meshwork of dense connective tissue. This complex, known as the **root** of the lung (Figure 23-6a●), attaches to the mediastinum and fixes the positions of the major nerves, vessels, and lymphatics. The roots of the lungs are located anterior to vertebrae T_5 (right) and T_6 (left).

THE LUNGS

The left and right lungs (Figure 23-7●) are situated in the left and right pleural cavities. Each lung is a blunt cone, with the tip, or **apex,** pointing superiorly. The apex on each side extends into the base of the neck superior to the first rib. The broad concave inferior portion, or **base,** of each lung rests on the superior surface of the diaphragm.

Lobes and Surfaces of the Lungs

The lungs have distinct **lobes** separated by deep fissures (Figure 23-7●). The right lung has three lobes: *superior, middle,* and *inferior*, separated by the *horizontal* and *oblique fissures*. The left lung has only two lobes: *superior* and *inferior*, separated by the *oblique fissure*. The right lung is broader than the left, because most of the heart and great vessels project into the left thoracic cavity. However, the left lung is longer than the right lung, because the diaphragm rises on the right side to accommodate the mass of the liver.

The curving anterior portion of the lung that follows the inner contours of the rib cage is the **costal surface.** The **mediastinal surface,** containing the hilus, has a more irregular shape. The mediastinal surfaces of both lungs bear grooves that mark the passage of the great vessels and of the **cardiac impressions,** concavities that conform to the shape of the pericardium (Figures 23-7 and 23-8●, p. 826). The cardiac impression of the left lung is deeper than that of the right lung. In anterior view, the medial edge of the right lung forms a vertical line, whereas the margin of the left lung is indented at the **cardiac notch.**

TUBERCULOSIS Tuberculosis (too-ber-kū-LŌ-sis), or TB, results from a bacterial infection of the lungs; other organs may be invaded as well. The bacterium *Mycobacterium tuberculosis* may colonize the respiratory passageways, the interstitial spaces, the alveoli, or a combination of the three. Symptoms are variable but generally include coughing and chest pain, with fever, night sweats, fatigue, and weight loss. To learn about the incidence and treatment of TB, see the *Applications Manual.* [AM] *Tuberculosis*

The Bronchi

The primary bronchi and their branches form the **bronchial tree** (Figure 23-9●, p. 827). Because the left and right primary bronchi are outside the lungs, they are also called **extrapulmonary bronchi.** As the primary bronchi enter the lungs, they divide to form smaller passageways (Figures 23-6, 23-9, and 23-10a●, p. 828). Those branches are collectively called the **intrapulmonary bronchi.**

●**FIGURE 23-7** Gross Anatomy of the Lungs. [AM] *Plate 6.3*

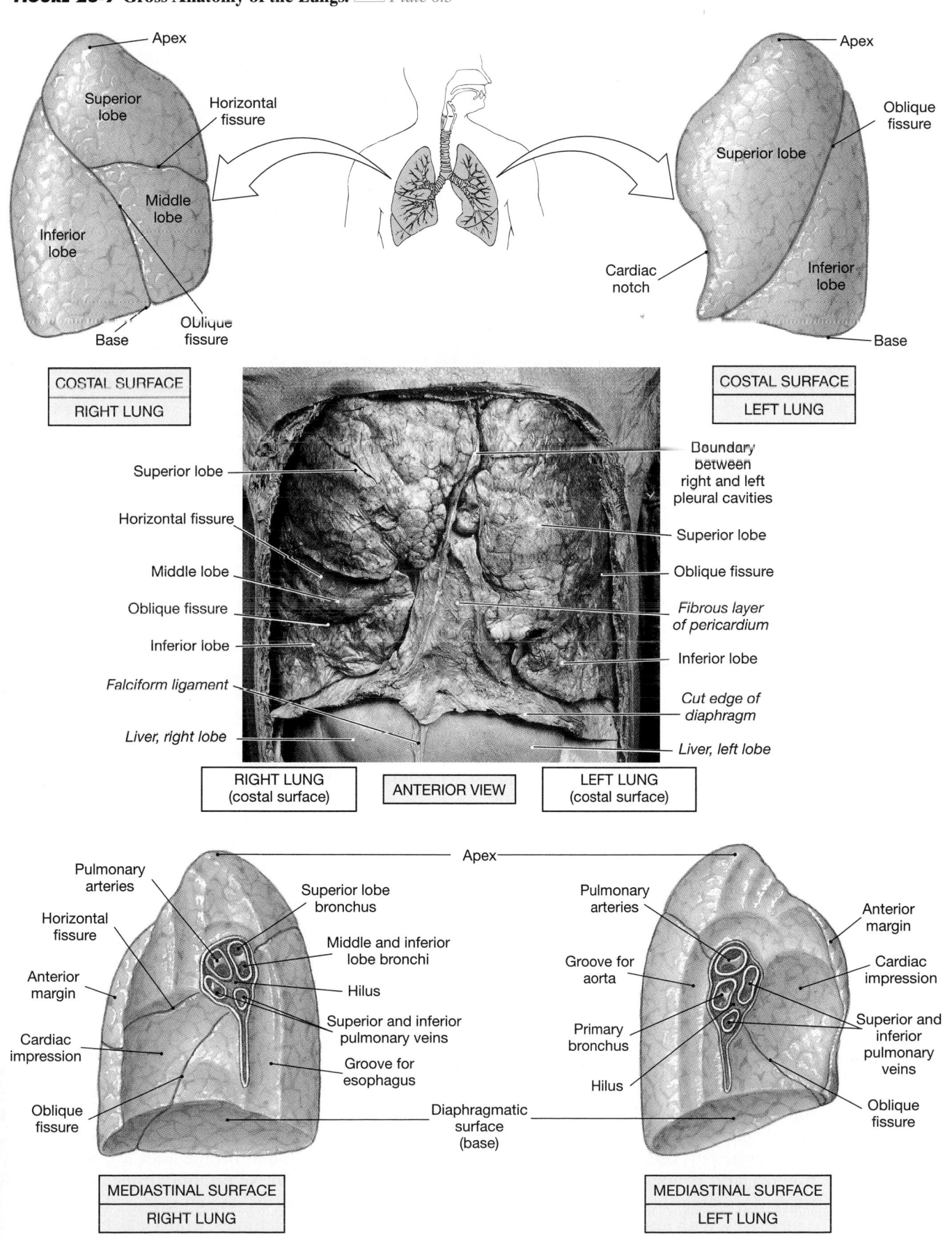

Apex

Superior
lobe

Horizontal
fissure

Middle
lobe

Inferior
lobe

Base

Oblique
fissure

Apex

Oblique
fissure

Superior lobe

Cardiac
notch

Inferior
lobe

Base

COSTAL SURFACE
RIGHT LUNG

COSTAL SURFACE
LEFT LUNG

Superior lobe

Horizontal fissure

Middle lobe

Oblique fissure

Inferior lobe

Falciform ligament

Liver, right lobe

Boundary
between
right and left
pleural cavities

Superior lobe

Oblique fissure

*Fibrous layer
of pericardium*

Inferior lobe

*Cut edge of
diaphragm*

Liver, left lobe

RIGHT LUNG
(costal surface)

ANTERIOR VIEW

LEFT LUNG
(costal surface)

Apex

Pulmonary
arteries

Horizontal
fissure

Anterior
margin

Cardiac
impression

Oblique
fissure

Superior lobe
bronchus

Middle and inferior
lobe bronchi

Hilus

Superior and inferior
pulmonary veins

Groove for
esophagus

Pulmonary
arteries

Groove for
aorta

Primary
bronchus

Hilus

Diaphragmatic
surface
(base)

Anterior
margin

Cardiac
impression

Superior and
inferior
pulmonary
veins

Oblique
fissure

MEDIASTINAL SURFACE
RIGHT LUNG

MEDIASTINAL SURFACE
LEFT LUNG

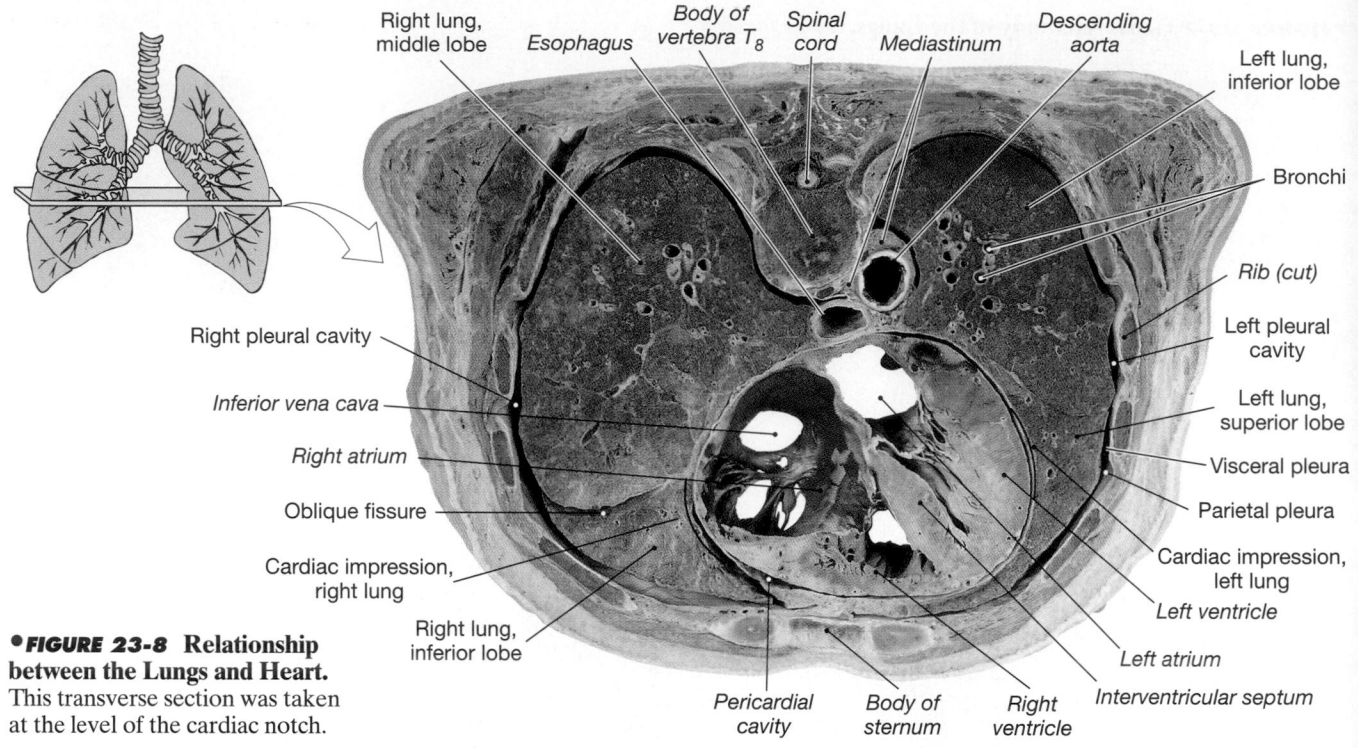

Right lung, middle lobe
Esophagus
Body of vertebra T$_8$
Spinal cord
Mediastinum
Descending aorta
Left lung, inferior lobe
Bronchi
Rib (cut)
Left pleural cavity
Left lung, superior lobe
Visceral pleura
Parietal pleura
Cardiac impression, left lung
Left ventricle
Left atrium
Interventricular septum

Right pleural cavity
Inferior vena cava
Right atrium
Oblique fissure
Cardiac impression, right lung
Right lung, inferior lobe
Pericardial cavity
Body of sternum
Right ventricle

•**FIGURE 23-8** Relationship between the Lungs and Heart. This transverse section was taken at the level of the cardiac notch.

Each primary bronchus divides to form **secondary bronchi,** also known as *lobar bronchi.* The right lung has three lobes, and the right primary bronchus divides into three secondary bronchi: (1) a *superior lobar bronchus,* (2) a *middle lobar bronchus,* and (3) an *inferior lobar bronchus.* The left lung has two lobes, and the left primary bronchus divides into two secondary bronchi: (1) a *superior lobar bronchus* and (2) an *inferior lobar bronchus.*

Figure 23-10• (p. 828) follows the branching pattern of the left primary bronchus as it enters the lung. (The number of branches have been reduced for clarity.) Within each lung, the secondary bronchi branch to form **tertiary bronchi,** or *segmental bronchi.* The branching pattern differs between the two lungs, but each tertiary bronchus ultimately supplies air to a single **bronchopulmonary segment,** a specific region of one lung. There are 10 bronchopulmonary segments in the right lung (Figure 23-9b•). During development, the left lung also has 10 segments, but subsequent fusion of adjacent tertiary bronchi generally reduces that number to eight or nine.

The walls of the primary, secondary, and tertiary bronchi contain progressively lesser amounts of cartilage. In the secondary and tertiary bronchi, the cartilages form plates arranged around the lumen. These cartilages serve the same purpose as the rings of cartilage in the trachea and primary bronchi. As the amount of cartilage decreases, the relative amount of smooth muscle increases. With less cartilaginous support, the amount of tension in those smooth muscles

has a greater effect on bronchial diameter and the resistance to airflow.

BRONCHOSCOPY During a respiratory infection, the bronchi and bronchioles may become inflamed. In this condition, called **bronchitis,** the smaller passageways may become greatly constricted, leading to difficulties in breathing. One method of investigating the status of the respiratory passageways is the use of a bronchoscope. A **bronchoscope** is a fiber-optic bundle small enough to be inserted into the trachea and steered along the conducting passageways to the level of the smaller bronchioles. This procedure is called **bronchoscopy** (brong-KOS-ko-pē). In addition to permitting direct visualization of bronchial structures, the bronchoscope can collect tissue or mucus samples from the respiratory tract. In **bronchography** (brong-KOG-ra-fē), a bronchoscope or catheter introduces a radiopaque material into the bronchi. This technique can permit detailed X-ray analysis of bronchial masses, such as tumors, or other obstructions along the bronchial tree (Figure 23-9a•).

The Bronchioles

Each tertiary bronchus branches several times within the bronchopulmonary segment, giving rise to multiple **bronchioles**. These branch further into the finest conducting branches, called **terminal bronchioles.** Roughly 6500 terminal bronchioles are supplied by each tertiary bronchus. Terminal bronchioles have a lumenal diameter of 0.3–0.5 mm.

The walls of bronchioles, which lack cartilaginous supports, are dominated by smooth muscle tissue

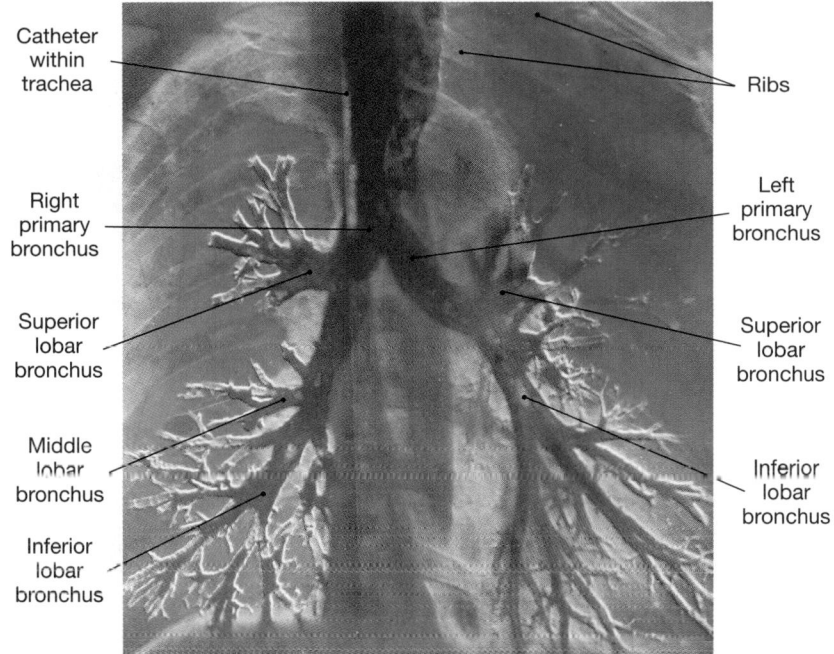

Catheter
within
trachea

Right
primary
bronchus

Superior
lobar
bronchus

Middle
lobar
bronchus

Inferior
lobar
bronchus

Ribs

Left
primary
bronchus

Superior
lobar
bronchus

Inferior
lobar
bronchus

(a)

•FIGURE 23-9 **The Bronchial Tree. (a)** An X-ray showing the major branches of the bronchial tree. **(b)** A cast of the bronchial tree. The finer branches have been painted to indicate the regions supplied by different bronchopulmonary segments. AM *Plate 6.5*

(Figure 23-10●). In functional terms, the bronchioles are to the respiratory system what the arterioles are to the cardio-vascular system. Varying the diameter of the bronchioles provides control over the amount of resistance to airflow and the distribution of air within the lungs.

The ANS regulates the activity in this smooth muscle layer and thereby controls the diameter of the bronchioles. Sympathetic activation leads to enlargement of the airway diameter, or **bronchodilation.** Parasympathetic stimulation leads to **bronchoconstriction,** a reduction in the diameter of the airways. Bronchoconstriction also occurs during allergic reactions such as anaphylaxis (Chapter 22), in response to histamine released by activated mast cells and basophils. ∞ *[p. 805]*

Bronchodilation and bronchoconstriction alter the resistance to airflow toward or away from the respiratory exchange surfaces. Tension in the smooth muscles commonly throws the bronchiolar mucosa into a series of folds, limiting airflow; excessive stimulation, as in *asthma* (AZ-muh), can almost completely prevent airflow along the terminal bronchioles. AM *Asthma*

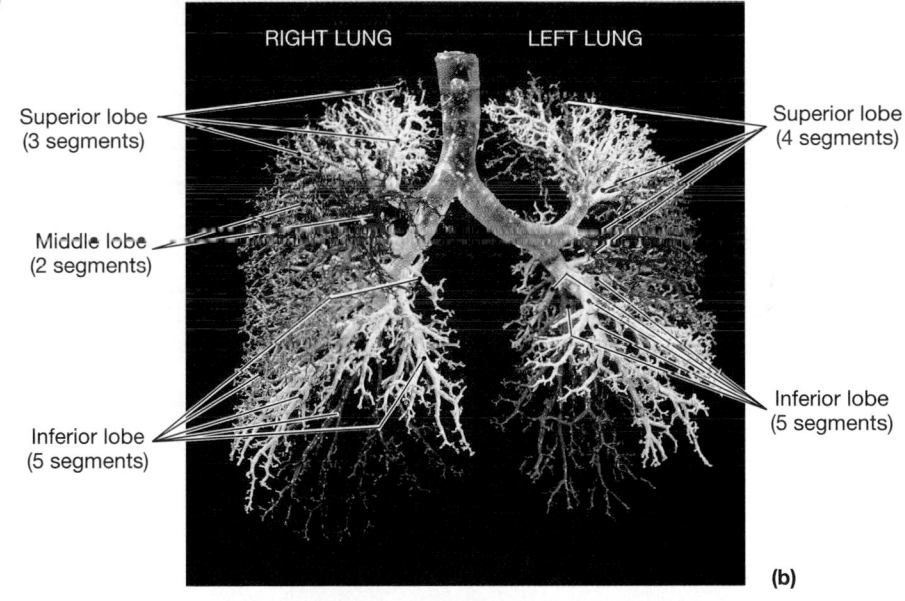

RIGHT LUNG   LEFT LUNG

Superior lobe
(3 segments)

Middle lobe
(2 segments)

Inferior lobe
(5 segments)

Superior lobe
(4 segments)

Inferior lobe
(5 segments)

(b)

Pulmonary Lobules

The connective tissues of the root of each lung extend into the lung's parenchyma. The fibrous partitions, or trabeculae, contain elastic fibers, smooth muscles, and lymphatic vessels. The trabeculae branch repeatedly, dividing the lobes into ever smaller compartments. The branches of the conducting passageways, pulmonary vessels, and nerves of the lungs follow these trabeculae. The finest partitions, or **interlobular septa** (*septum,* a wall) divide the lung into **pulmonary lobules** (LOB-ūlz), each supplied by branches of the pulmonary arteries, pulmonary veins, and respiratory passageways (Figure 23-10●). The connective tissues of the septa are in turn continuous with those of the *visceral pleura*, the serous membrane covering the lungs.

●*FIGURE 23-10* The Bronchi and Lobules of the Lung.
(a) Branching pattern of bronchi in the left lung, simplified.
(b) The structure of a single lobule, part of a
bronchopulmonary segment.

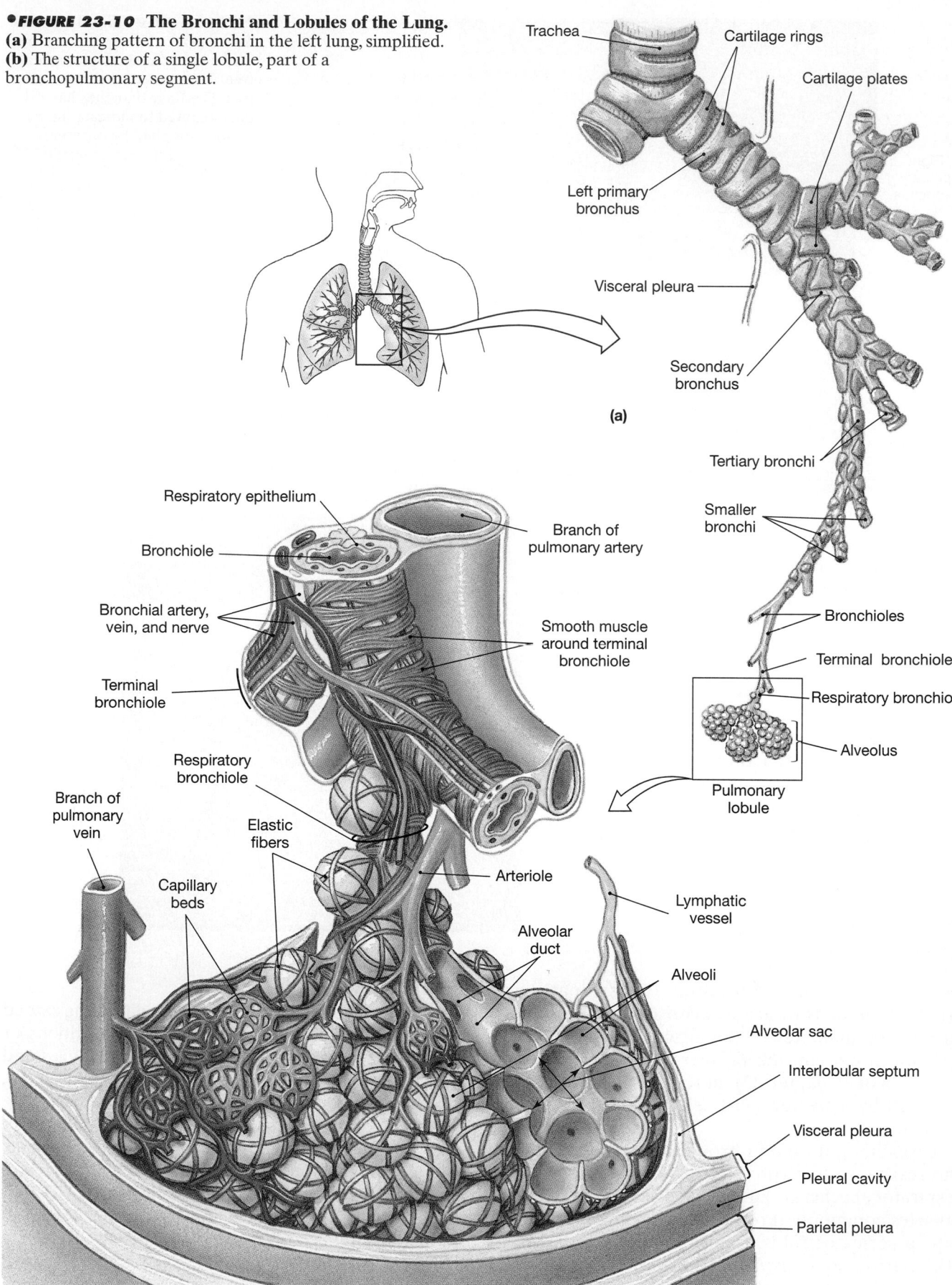

Trachea

Cartilage rings

Cartilage plates

Left primary bronchus

Visceral pleura

Secondary bronchus

(a)

Tertiary bronchi

Smaller bronchi

Bronchioles

Terminal bronchiole

Respiratory bronchiole

Alveolus

Pulmonary lobule

Respiratory epithelium

Bronchiole

Branch of pulmonary artery

Bronchial artery, vein, and nerve

Smooth muscle around terminal bronchiole

Terminal bronchiole

Respiratory bronchiole

Branch of pulmonary vein

Elastic fibers

Capillary beds

Arteriole

Lymphatic vessel

Alveolar duct

Alveoli

Alveolar sac

Interlobular septum

Visceral pleura

Pleural cavity

Parietal pleura

(b)

Each terminal bronchiole delivers air to a single pulmonary lobule. Within the lobule, the terminal bronchiole branches to form several **respiratory bronchioles.** These are the thinnest and most delicate branches of the bronchial tree. They deliver air to the exchange surfaces of the lungs.

The preliminary filtration and humidification of the incoming air are completed before air leaves the terminal bronchioles. The epithelial cells of the terminal bronchioles and respiratory bronchioles are cuboidal, with only scattered cilia, and there are no goblet cells or underlying mucous glands.

Alveolar Ducts and Alveoli

Respiratory bronchioles are connected to individual alveoli and to multiple alveoli along regions called **alveolar ducts** (Figures 23-10 and 23-11●). These passageways end at **alveolar sacs,** common chambers connected to multiple individual alveoli. Each lung contains about 150 million alveoli, and their abundance gives the lung an open, spongy appearance. An extensive network of capillaries is associated with each alveolus (Figure 23-12a●); the capillaries are surrounded by a network of elastic fibers. This elastic tissue helps maintain the relative positions of the alveoli and respiratory bronchioles. Recoil of these fibers during exhalation reduces the size of the alveoli and helps push air out of the lungs.

The Alveolus and the Respiratory Membrane

The alveolar epithelium consists primarily of simple squamous epithelium (Figure 23-12b●). The squamous epithelial cells, called *Type I cells,* are unusually thin and delicate. Roaming **alveolar macrophages** (*dust cells*) patrol the epithelium, phagocytizing any particulate matter that has eluded the respiratory defenses and reached the alveolar surfaces. **Septal cells,** also called **surfactant** (sur-FAK-tant) **cells** or *Type II cells,* are scattered among the squamous cells. These large cells produce an oily secretion, or **surfactant,** containing a mixture of phospholipids and proteins. Surfactant is secreted onto the alveolar surfaces, where it forms a superficial coating over a thin layer of water.

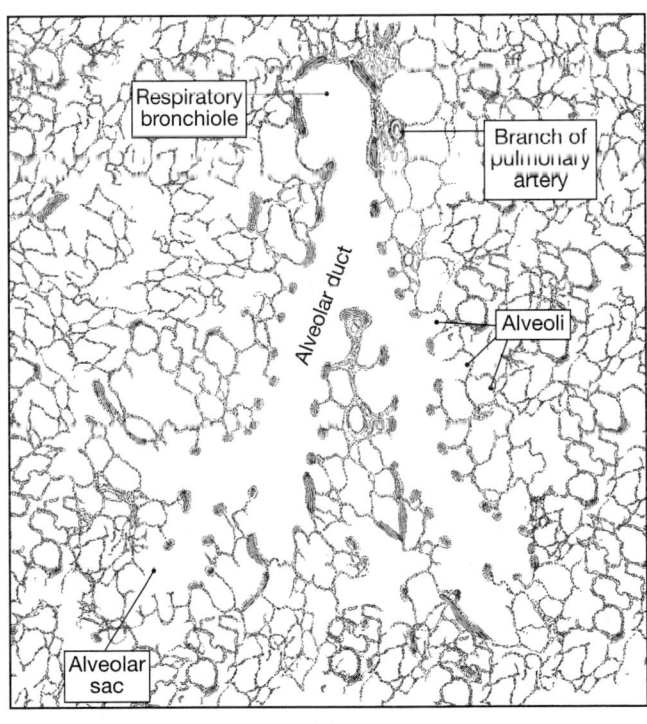

(a)

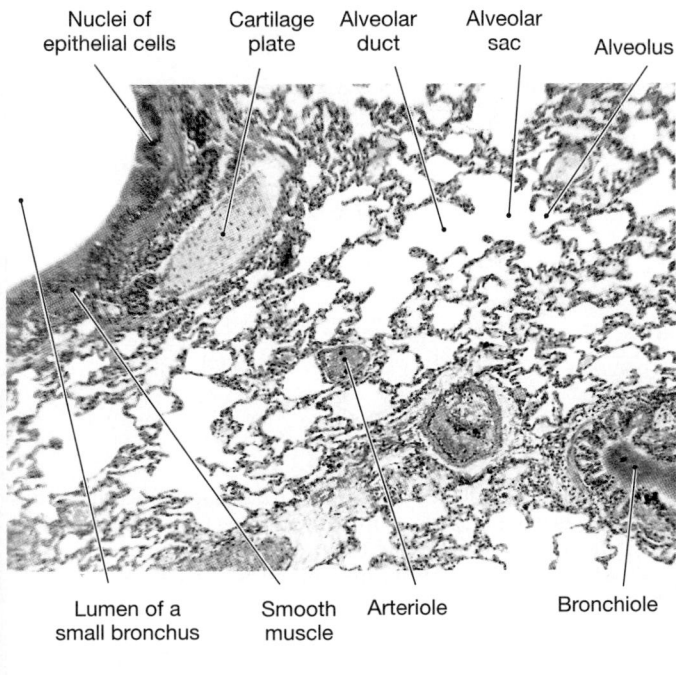

(b)

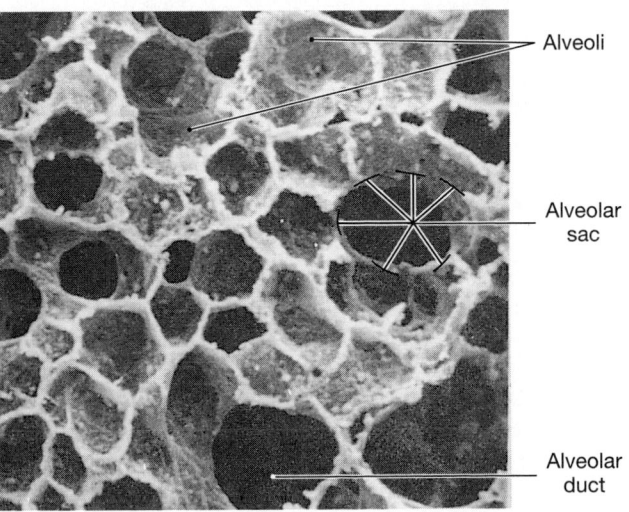

(c)

●*FIGURE 23-11* **The Bronchioles. (a)** Distribution of a respiratory bronchiole supplying a portion of a lobule. **(b)** The alveolar sacs and alveoli. (LM × 42) **(c)** An SEM of the lung. Note the open, spongy appearance of the lung tissue. Compare with Figure 23-10b.

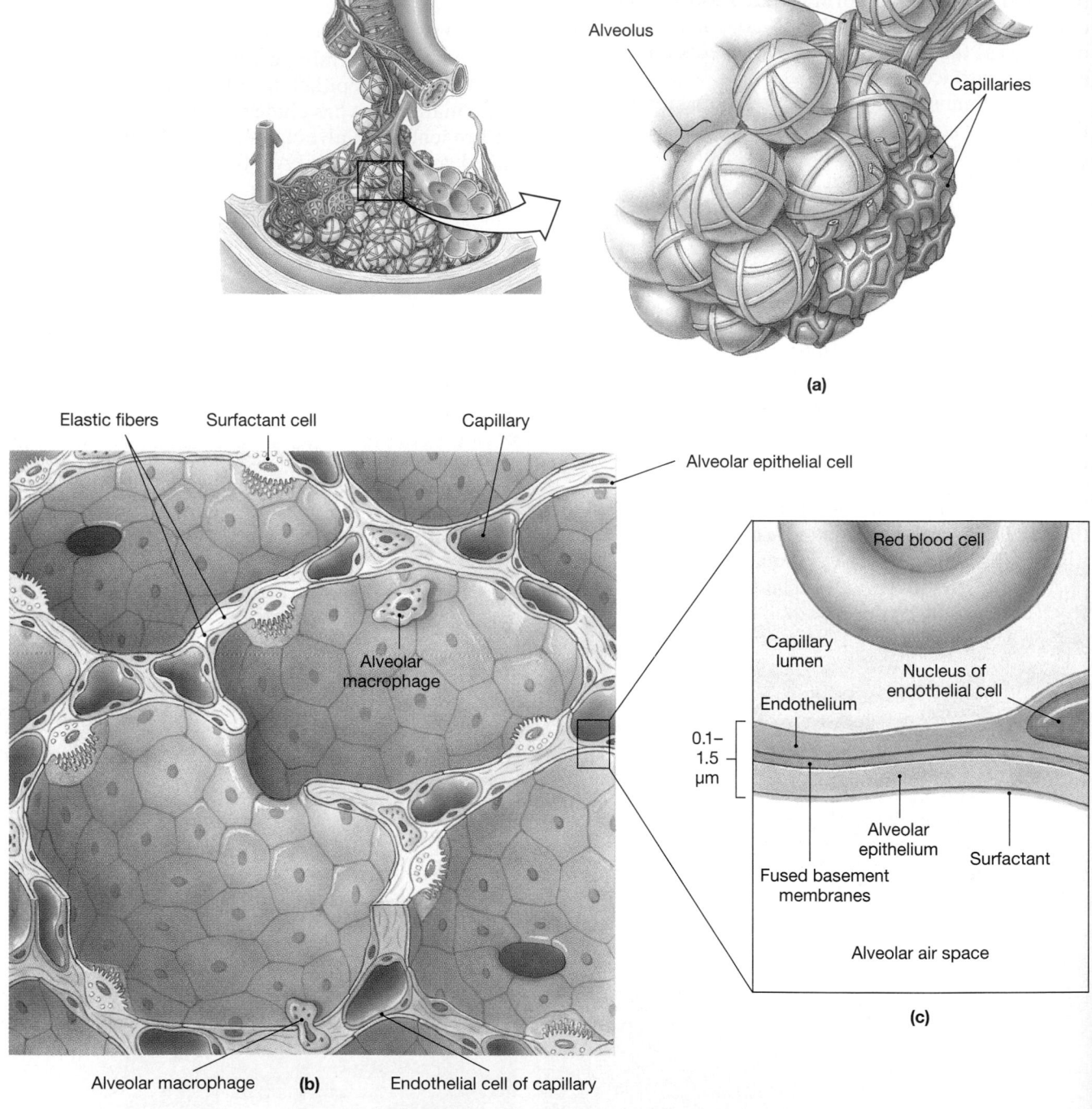

•FIGURE 23-12 Alveolar Organization. **(a)** Basic structure of a portion of a single lobule. A network of capillaries surrounds each alveolus. These capillaries are supported by elastic fibers. Respiratory bronchioles also contain wrappings of smooth muscle that can vary the diameter of these airways. **(b)** Diagrammatic view of alveolar structure. Note that a single capillary may be involved with gas exchange across several different alveoli simultaneously. **(c)** The respiratory membrane, which consists of an alveolar epithelial cell, a capillary endothelial cell, and their fused basement membranes.

Surfactant is important because it reduces surface tension in the liquid coating the alveolar surface. As we saw in Chapter 2, surface tension results from the attraction between water molecules at an air–water boundary. ∞ *[p. 37]* The alveolar walls are very delicate, and without surfactant the surface tension would be so high that the alveoli would collapse. The surfactant forms a thin surface layer that interacts with the water molecules, reducing the surface tension and keeping the alveoli open.

If surfactant cells produce inadequate amounts of surfactant due to injury or genetic abnormalities, the alveoli will collapse, and respiration will become difficult. On each breath, the inhalation must be forceful enough to pop open the alveoli. A person with this problem, *respiratory distress syndrome*, is soon exhausted by the effort required to keep inflating and deflating the lungs. AM *Respiratory Distress Syndrome (RDS)*

THE RESPIRATORY MEMBRANE Gas exchange occurs across the **respiratory membrane** of the alveoli. The respiratory membrane (Figure 23-12c•) is a composite structure consisting of three parts:

1. The squamous epithelial cell lining the alveolus.
2. The endothelial cell lining an adjacent capillary.
3. The fused basement membranes that lie between the alvcolar and cndothelial cells.

At the respiratory membrane, the total distance separating the alveolar air and the blood can be as little as 0.1 μm. Diffusion across the respiratory membrane proceeds very rapidly, because (1) the distance is small and (2) both oxygen and carbon dioxide are lipid-soluble. The membranes of the epithelial and endothelial cells thus do not pose a barrier to the movement of oxygen and carbon dioxide between the blood and alveolar air spaces.

PNEUMONIA Pneumonia (noo-MŌ-nē-uh) develops from a pathogenic infection or any other stimulus that causes inflammation of the lobules of the lung. As inflammation occurs, fluids leak into the alveoli and there is swelling and constriction of the respiratory bronchioles. Respiratory function deteriorates as a result. When bacteria are involved, they are generally species that normally inhabit the mouth and pharynx but have somehow managed to evade the respiratory defenses. Pneumonia becomes more likely when the respiratory defenses have already been compromised by other factors, such as epithelial damage from smoking or the breakdown of the immune system in AIDS. The most common pneumonia that develops in AIDS patients results from infection by the fungus *Pneumocystis carinii*. These organisms are normally found in the alveoli, but in healthy individuals the respiratory defenses are able to prevent infection and tissue damage.

☑ Why are the cartilages that reinforce the trachea C-shaped?

☑ What would happen to the alveoli if surfactant were not produced?

☑ What path does air take in flowing from the glottis to the respiratory membrane?

The Blood Supply to the Lungs

Your respiratory exchange surfaces receive blood from arteries of your pulmonary circuit. The pulmonary arteries enter the lungs at the hilus and branch with the bronchi as they approach the lobules. Each lobule re-

ceives an arteriole and a venule, and a network of capillaries surrounds each alveolus directly beneath the respiratory membrane. In addition to providing a mechanism for gas exchange, the endothelial cells of the alveolar capillaries are the primary source of *angiotensin-converting enzyme (ACE)*. This enzyme, which converts circulating angiotensin I to angiotensin II, plays an important role in the regulation of blood volume and blood pressure (as discussed in Chapters 18 and 21). ∞ *[pp. 620, 733]*

Blood from the alveolar capillaries passes through the pulmonary venules and then enters the pulmonary veins, which deliver it to the left atrium. The conducting portions of your respiratory tract receive blood from the *external carotid arteries* (nasal passages and larynx), the *thyrocervical arteries* (the inferior larynx and trachea), and the *bronchial arteries* (the bronchi and bronchioles). (See Figures 21-23, 21-24, and 21-26, pp. 742, 744, 746.) The capillaries supplied by the bronchial arteries provide oxygen and nutrients to the conducting passageways of your lungs. The venous blood flows into the pulmonary veins, bypassing the rest of the systemic circuit and diluting the oxygenated blood leaving the alveoli.

PULMONARY EMBOLISM Blood pressure in the pulmonary circuit is usually relatively low, with systemic pressures of 30 mm Hg or less. With pressures that low, pulmonary vessels can easily become blocked by small blood clots, fat masses, or air bubbles in the pulmonary arteries. Because the lungs receive the entire cardiac output, any drifting masses in the blood are likely to cause problems almost at once. The blockage of a branch of a pulmonary artery will stop blood flow to a group of lobules or alveoli. This condition is called a **pulmonary embolism.** *Atherosclerosis,* described in Chapter 21 (∞ *[p. 713]*), and *venous thrombosis* (the blockage of a vein by a blood clot) can promote the development of a pulmonary embolism, because in both cases there is a tendency for small blood clots to form, break loose, and drift in the circulation. If a pulmonary embolism remains in place for several hours, the alveoli will permanently collapse. If the blockage occurs in a major pulmonary vessel, rather than a minor tributary, pulmonary resistance increases. This resistance places extra strain on the right ventricle, which may be unable to maintain cardiac output. Congestive heart failure may be the result.

THE PLEURAL CAVITIES AND PLEURAL MEMBRANES

The thoracic cavity has the shape of a broad cone. Its walls are the rib cage, and the muscular diaphragm forms the floor. The two **pleural cavities** are separated by the mediastinum (Figure 23-8•, p. 826). Each lung occupies a single pleural cavity, which is lined by a serous membrane called the **pleura** (PLOO-ra; plur-

Development of the Respiratory System

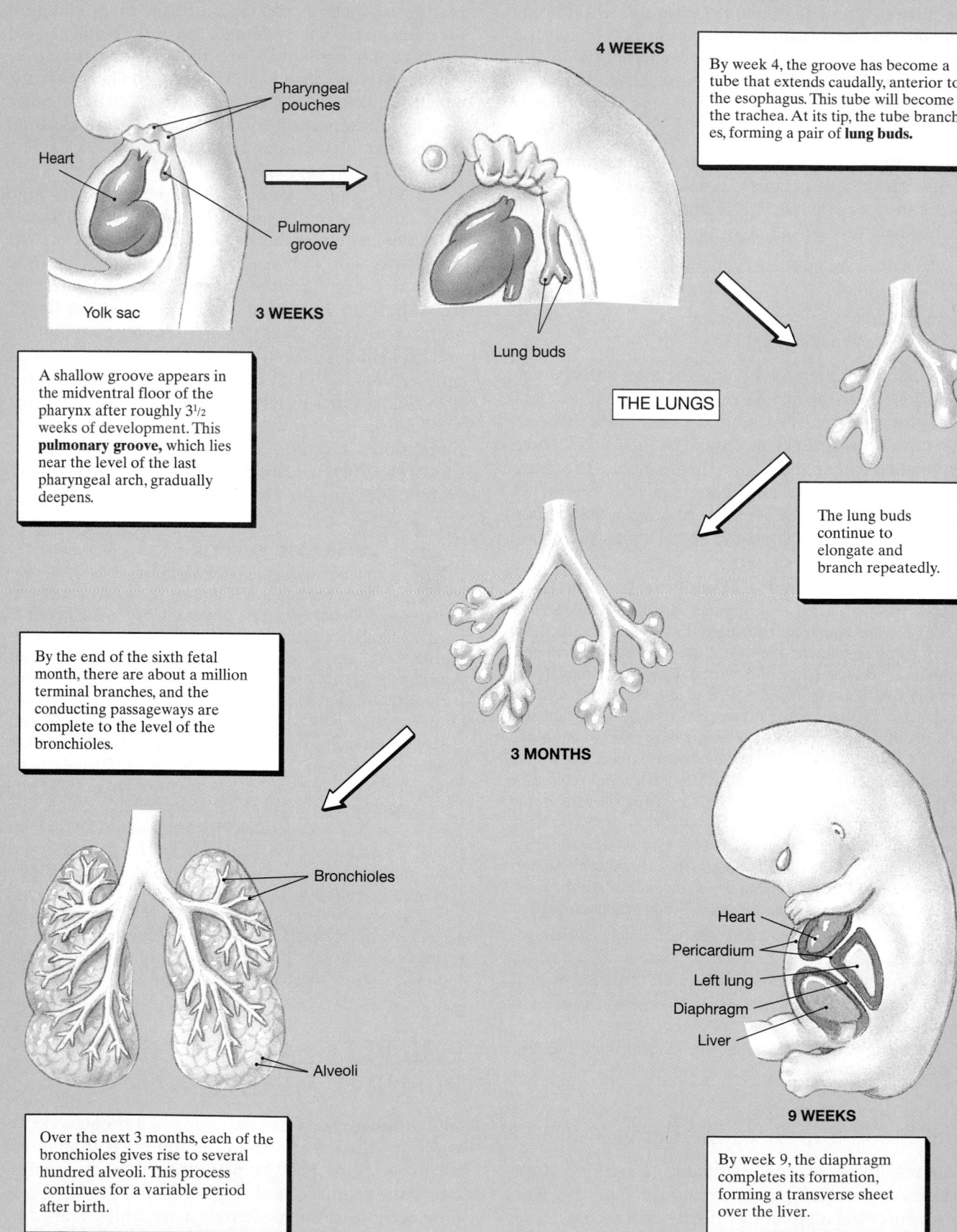

4 WEEKS

By week 4, the groove has become a tube that extends caudally, anterior to the esophagus. This tube will become the trachea. At its tip, the tube branches, forming a pair of **lung buds.**

Pharyngeal pouches

Heart

Pulmonary groove

Yolk sac

3 WEEKS

Lung buds

THE LUNGS

A shallow groove appears in the midventral floor of the pharynx after roughly 3$^1/_2$ weeks of development. This **pulmonary groove,** which lies near the level of the last pharyngeal arch, gradually deepens.

The lung buds continue to elongate and branch repeatedly.

By the end of the sixth fetal month, there are about a million terminal branches, and the conducting passageways are complete to the level of the bronchioles.

3 MONTHS

Bronchioles

Heart

Pericardium

Left lung

Diaphragm

Liver

Alveoli

9 WEEKS

Over the next 3 months, each of the bronchioles gives rise to several hundred alveoli. This process continues for a variable period after birth.

By week 9, the diaphragm completes its formation, forming a transverse sheet over the liver.

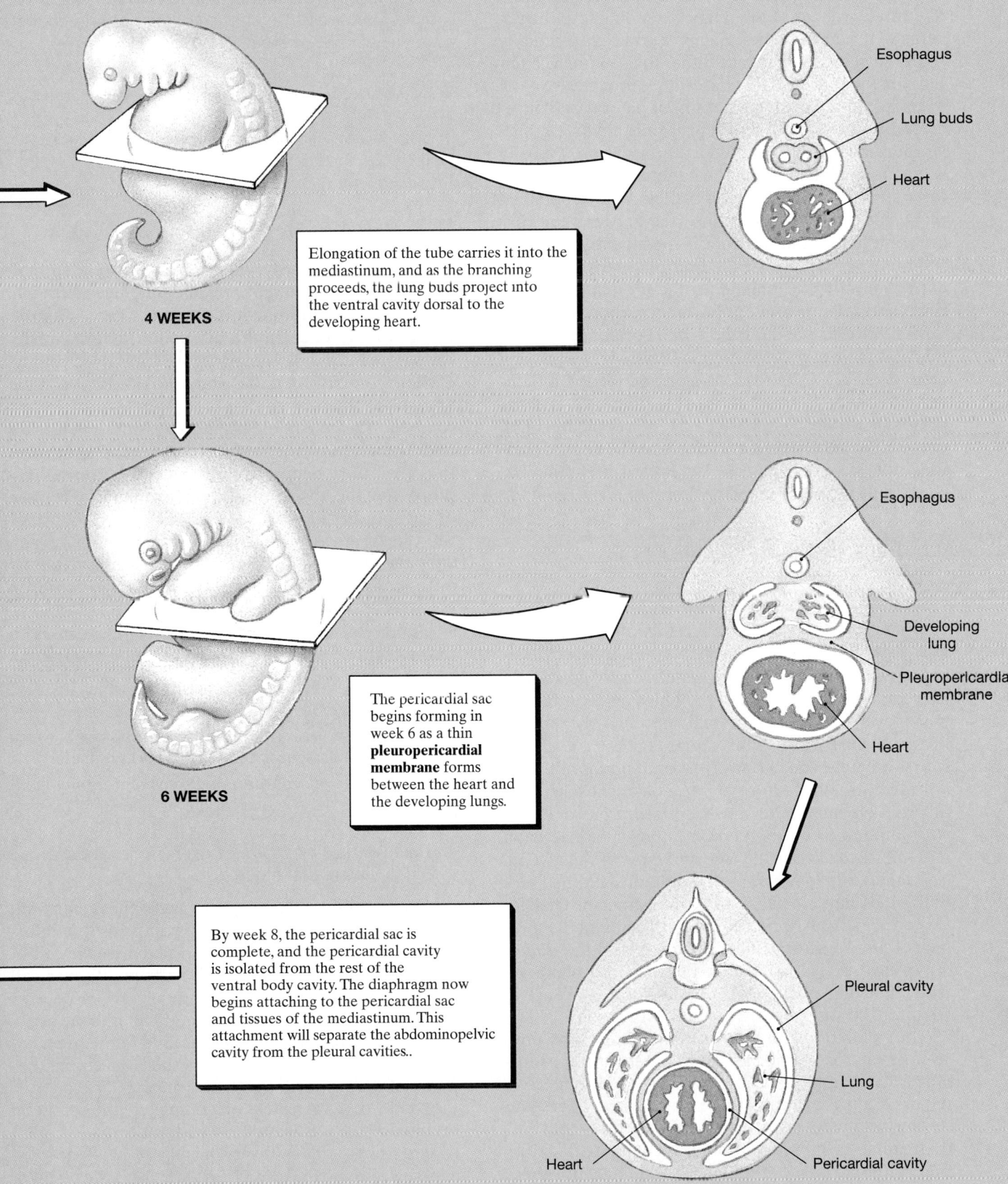

THE PLEURAL CAVITIES

4 WEEKS

Elongation of the tube carries it into the mediastinum, and as the branching proceeds, the lung buds project into the ventral cavity dorsal to the developing heart.

Esophagus

Lung buds

Heart

6 WEEKS

The pericardial sac begins forming in week 6 as a thin **pleuropericardial membrane** forms between the heart and the developing lungs.

Esophagus

Developing lung

Pleuropericardial membrane

Heart

By week 8, the pericardial sac is complete, and the pericardial cavity is isolated from the rest of the ventral body cavity. The diaphragm now begins attaching to the pericardial sac and tissues of the mediastinum. This attachment will separate the abdominopelvic cavity from the pleural cavities..

Pleural cavity

Lung

Heart

Pericardial cavity

8 WEEKS

al, *pleurae*). The **parietal pleura** covers the inner surface of the thoracic wall and extends over the diaphragm and mediastinum. The **visceral pleura** covers the outer surfaces of the lungs, extending into the fissures between the lobes. Each pleural cavity actually represents a potential space rather than an open chamber, for the parietal and visceral pleurae are usually in close contact. A small amount of **pleural fluid** is secreted by both pleurae. Pleural fluid gives a moist, slippery coating that provides lubrication, thereby reducing friction between the parietal and visceral surfaces as you breathe. Samples of pleural fluid, obtained by means of a long needle inserted between the ribs, are sometimes obtained for diagnostic purposes. This sampling procedure is called *thoracentesis*. The fluid extracted is then examined for the presence of bacteria, blood cells, or other abnormal components.

In some disease states, the normal coating of pleural fluid is unable to prevent friction between the opposing pleural surfaces. The result is pain and pleural inflammation, a condition called **pleurisy.** When pleurisy develops, there may be excessive secretion of pleural fluid, or the inflamed pleurae may adhere to one another, limiting relative movement. In either case, breathing becomes difficult, and prompt medical attention is required.

Changes in the Respiratory System at Birth

There are several important differences between the respiratory system of a fetus and that of a newborn infant. Prior to delivery, pulmonary arterial resistance is high, because the pulmonary vessels are collapsed. The rib cage is compressed, and the lungs and conducting passageways contain only small amounts of fluid and no air. At birth, the newborn infant takes a truly heroic first breath through powerful contractions of the diaphragmatic and external intercostal muscles. The inspired air must enter the respiratory passageways with enough force to overcome the surface tension and inflate the bronchial tree and most of the alveoli. The same drop in pressure that pulls air into the lungs pulls blood into the pulmonary circulation; the changes in blood flow that occur lead to the closure of the *foramen ovale,* an interatrial connection, and the *ductus arteriosus,* the fetal connection between the pulmonary trunk and the aorta. (We detailed these events in Chapters 20 and 21.) ∞ *[pp. 675, 758)*

The exhalation that follows fails to empty the lungs completely, for the rib cage does not return to its former, fully compressed state. Cartilages and connective tissues keep the conducting passageways open, and the surfactant covering the alveolar surfaces prevents their collapse. Subsequent breaths complete the inflation of the alveoli. Pathologists sometimes use these physical changes to determine whether a newborn infant died before delivery or shortly thereafter. Prior to the first breath, the lungs are completely filled with fluid, and they will sink if placed in water. After the infant's first breath, even the collapsed lungs contain enough air to keep them afloat.

☑ Which arteries supply blood to the conducting portions and respiratory exchange surfaces of the lungs?

☑ What are the functions of the pleura, and what does it secrete?

RESPIRATORY PHYSIOLOGY

The general term *respiration* refers to two integrated processes: *external respiration* and *internal respiration.* The precise definitions of these terms vary from reference to reference. In this discussion, **external respiration** includes all the processes involved in the exchange of oxygen and carbon dioxide between the interstitial fluids of the body and the external environment. The goal of external respiration, and the primary function of the respiratory system, is to meet the respiratory demands of living cells. **Internal respiration** is the absorption of oxygen and the release of carbon dioxide by those cells. We shall consider the biochemical pathways responsible for oxygen consumption and carbon dioxide generation by mitochondria, often called *cellular respiration,* in Chapter 25.

Our discussion of respiratory physiology focuses on four integrated steps involved in external respiration:

1. *Pulmonary ventilation,* or breathing, which involves the physical movement of air into and out of the lungs.
2. *Gas diffusion across the respiratory membrane* between the alveolar air spaces and the alveolar capillaries.
3. *The storage and transport of oxygen and carbon dioxide* between the alveolar capillaries and capillary beds in other tissues.
4. *The exchange of dissolved gases* between the blood and the interstitial fluids.

Abnormalities affecting any one of these steps will ultimately affect the gas concentrations of the interstitial fluids and thereby cellular activities as well. If the oxygen content declines, the affected tissues will become oxygen-starved. **Hypoxia**, or low tissue oxygen levels, places severe limits on the metabolic activities of the affected area. For example, the effects of *coronary ischemia* (Chapter 20) result from chronic hypoxia affecting cardiac muscle cells. ∞ *[p. 686]* If the supply of oxygen is cut off completely, the condition of **anoxia** (a-NOKS-ē-uh; *a-*, without + *ox-*, oxygen) results. Anoxia kills cells very quickly. Much of the damage caused by strokes and heart attacks is the result of localized anoxia.

Pulmonary Ventilation

Pulmonary ventilation is the physical movement of air into and out of the respiratory tract. The primary function of pulmonary ventilation is to maintain adequate *alveolar ventilation,* air movement into and out of the alveoli. Alveolar ventilation prevents the buildup of carbon dioxide in the alveoli and ensures a continuous supply of oxygen that keeps pace with absorption by the bloodstream. To understand this mechanical process, we need to take a look at basic physical principles governing the movement of air.

Boyle's Law: Gas Pressure and Volume

The primary differences between liquids and gases such as air reflect the interactions between individual molecules. Although the molecules in a liquid are in constant motion, they are held closely together by weak interactions. In a polar liquid such as water, these interactions include hydrogen bonding between adjacent atoms. But, because the electrons of adjacent atoms tend to repel one another, liquids tend to resist compression. If you squeeze a balloon filled with water, it will distort into a different shape, but the volume of the new shape will be the same as that of the original.

In a gas, the molecules bounce around as independent entities. At normal atmospheric pressures, gas molecules are much farther apart than the molecules in a liquid, so the density of air is relatively low. The forces acting between gas molecules are minimal—the molecules are too far apart for weak interactions to occur—so an applied pressure can push them closer together. Consider a sealed container of air at atmospheric pressure (Figure 23-13●). The pressure exerted by the enclosed gas results from the collision of gas molecules with the walls of the container. The greater the number of collisions, the higher the pressure.

You can change the gas pressure within a container by altering the volume of the container, thereby giving the gas molecules more or less room in which to bounce around. If you decrease the volume of the container, the collisions will occur more frequently over a given time period, elevating the pressure of the gas (Figure 23-13a●). If you increase the volume, there will be fewer collisions per unit time, because it will take longer for a gas molecule to travel from one wall to another. As a result, the gas pressure inside the container will decline (Figure 23-13b●).

An inverse relationship thus exists between the pressure and volume of a gas in a closed container: Gas pressure is inversely proportional to volume. *If you decrease the volume of a gas, its pressure rises; if you increase the volume of a gas, its pressure falls.* In particular, the relationship between pressure and volume is reciprocal: If you double the external pressure on a flexible container, its volume will drop by half. If you reduce the

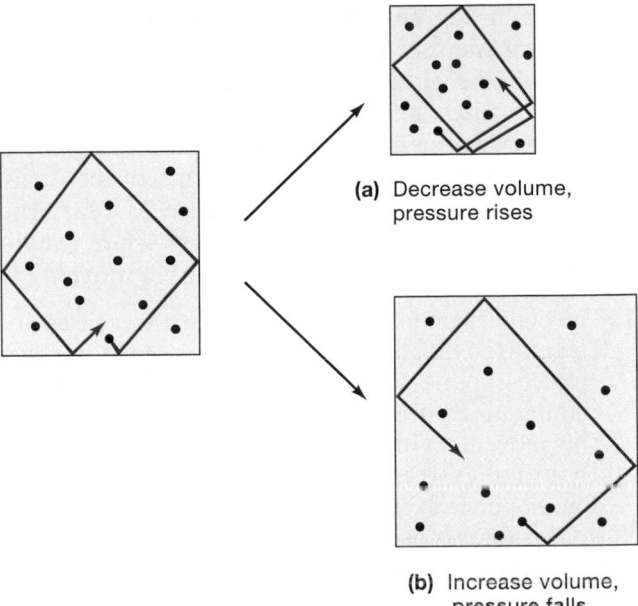

(a) Decrease volume, pressure rises

(b) Increase volume, pressure falls

●**FIGURE 23-13 Pressure and Volume Relationships.** Gas molecules within a container bounce off the walls and off one another, traveling the distance indicated in a given period of time. **(a)** If the volume decreases, each molecule will travel the same distance in that same time period but will strike the walls more frequently. The pressure exerted by the gas increases. **(b)** Alternatively, if the volume of the container increases, the gas molecules will strike the walls less often, lowering the pressure in the container.

external pressure by half, the volume of the container will double. This relationship, P = 1/V, first recognized by Robert Boyle in the 1600s, is called **Boyle's law.** AM
Boyle's Law and Air Overexpansion Syndrome

Pressure and Airflow to the Lungs

Air will flow from an area of higher pressure to an area of lower pressure. This tendency for directed airflow plus the pressure/volume relationships of Boyle's law provide the basis for pulmonary ventilation. A single respiratory cycle consists of an *inspiration,* or inhalation, and an *expiration,* or exhalation. Inhalation and exhalation involve changes in the volume of the lungs. These changes create pressure gradients that move air into or out of the respiratory tract.

Each lung lies within a pleural cavity. The parietal and visceral pleurae are separated by only a thin film of pleural fluid. Although the two membranes can slide across one another, they are held together by that fluid film. You encounter the same principle whenever you set a wet glass on a smooth tabletop. You can slide the glass quite easily, but when you try to lift it, you encounter considerable resistance. As you pull the glass away from the tabletop, you create a powerful suction. The only way to defeat it is to tilt the glass so that air is pulled between the glass and the table, breaking the fluid bond.

A comparable fluid bond exists between the parietal pleura and the visceral pleura covering the lungs. As a result, the surface of each lung sticks to the inner wall of the chest and the superior surface of the diaphragm. Movements of the chest wall or diaphragm thus have a direct effect on the volume of the lungs. The basic principle is shown in Figure 23-14a●. The volume of the thoracic cavity changes when the diaphragm changes position or the rib cage moves:

1. The diaphragm forms the floor of the thoracic cavity. The relaxed diaphragm has the shape of a dome that projects superiorly into the thoracic cavity. When the diaphragm contracts, it tenses and moves inferiorly. This movement increases the volume of the thoracic cavity and exerts pressure on the contents of the abdominopelvic cavity. When the diaphragm relaxes, it returns to its original position, and the volume of the thoracic cavity decreases.

2. Due to the nature of the articulations between the ribs and the vertebrae, superior movement of the rib cage increases the depth and width of the thoracic cavity. Inferior movement of the rib cage reverses the process and reduces the volume of the thoracic cavity.

At the start of a breath, pressures inside and outside the thoracic cavity are identical, and there is no movement of air into or out of the lungs (Figure 23-14b●). When the thoracic cavity enlarges, the pleural cavities and lungs expand to fill the additional space (Figure 23-14c●). This expansion lowers the pressure inside the lungs. Air now enters the respiratory passageways, because the pressure inside the lungs (P_{inside}) is lower than atmospheric pressure (pressure outside, or $P_{outside}$). Air continues to enter until the volume stops increasing and the internal pressure is the same as that outside. When the thoracic cavity decreases in volume, pressures rise inside the lungs, forcing air out of the respiratory tract (Figure 23-14d●).

COMPLIANCE The **compliance** of the lungs is an indication of their expandability. The lower the compliance, the greater the force required to fill and empty the lungs. Factors affecting compliance include the following:

- **The connective tissue structure of the lungs.** The loss of supporting tissues resulting from alveolar damage, as in *emphysema,* increases compliance (p. 851).
- **The level of surfactant production.** The collapse of alveoli on expiration, due to inadequate surfactant, as in respiratory distress syndrome, reduces compliance (p. 831).
- **The mobility of the thoracic cage.** Arthritis or other skeletal disorders that affect the articulations of the ribs or spinal column will also reduce compliance.

At rest, the muscular activity involved in pulmonary ventilation accounts for 3–5 percent of the resting energy demand. If compliance is reduced, that figure climbs dramatically, and an individual may become exhausted simply trying to continue breathing.

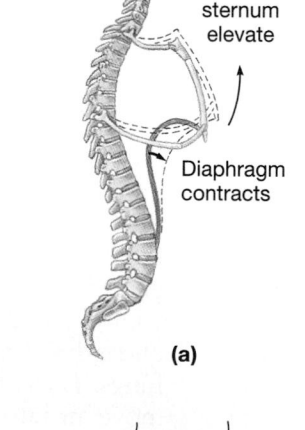

Ribs and sternum elevate

Diaphragm contracts

(a)

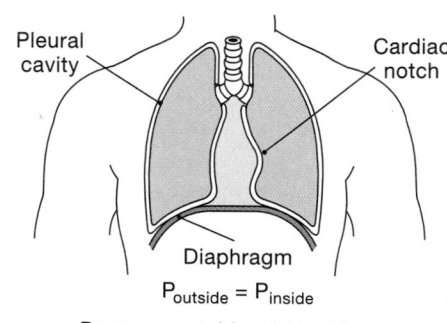

Pleural cavity

Cardiac notch

Diaphragm

$P_{outside} = P_{inside}$

Pressure outside and inside are equal, so no air movement occurs

(b)

●**FIGURE 23-14** **Mechanisms of Pulmonary Ventilation.** **(a)** As the ribs are elevated or the diaphragm is depressed, the volume of the thoracic cavity increases. **(b)** Anterior view at rest, with no air movement. **(c)** *Inhalation:* Elevation of the rib cage and contraction of the diaphragm increase the size of the thoracic cavity. Pressure decreases, and air flows into the lungs. **(d)** *Exhalation:* When the rib cage returns to its original position, the volume of the thoracic cavity decreases. Pressure rises, and air moves out of the lungs.

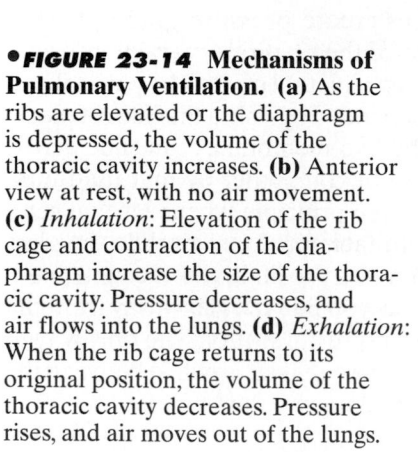

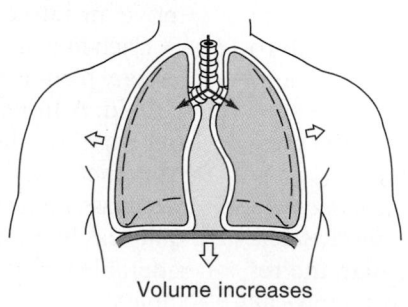

Volume increases

$P_{outside} > P_{inside}$

Pressure inside falls, so air flows in

(c)

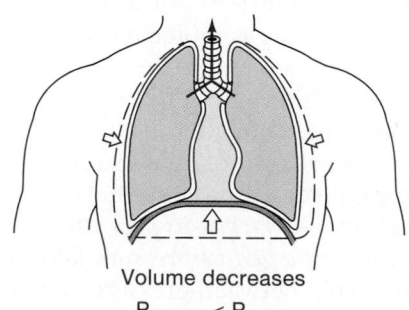

Volume decreases

$P_{outside} < P_{inside}$

Pressure inside rises, so air flows out

(d)

Pressure Changes during Inhalation and Exhalation

To understand the mechanics of respiration and the principles of gas exchange, we need to be familiar with the actual pressures recorded inside and outside the respiratory tract. We can report the pressure readings in several different ways (Table 23-1), but we will use millimeters of mercury (mm Hg), as we did when considering blood pressure.

Although we are seldom reminded of the fact, the body is under considerable pressure due to the weight of Earth's atmosphere. This **atmospheric pressure** pushes against everything around us. One atmosphere of pressure **(1 atm)** is equivalent to 760 mm Hg.

THE INTRAPULMONARY PRESSURE   The direction of airflow is determined by the relationship between atmospheric pressure and *intrapulmonary pressure*. The **intrapulmonary** (in-tra-PUL-mo-ner-ē) **pressure,** or **intra-alveolar** (in-tra-al-VĒ-ō-lar) **pressure,** is the pressure measured inside the respiratory tract, at the alveoli.

When you are relaxed and breathing quietly, the pressure difference between atmospheric and intrapulmonary pressures is relatively small. On inhalation, your lungs expand, and the intrapulmonary pressure drops to about 759 mm Hg. Because the intrapulmonary pressure is 1 mm Hg below atmospheric, it is generally reported as –1 mm Hg. On exhalation, your lungs recoil, and the intrapulmonary pressure rises to 761 mm Hg, or +1 mm Hg (Figure 23-15a•).

The size of the pressure gradient increases when you breathe heavily. When a trained athlete breathes at maximum capacity, the pressure differentials can reach –30 mm Hg during inspiration and +100 mm Hg if the individual is straining with the glottis kept closed. This is one reason you are told to exhale while lifting weights; exhaling keeps your intrapulmonary pressures and peritoneal pressure from climbing so high that an alveolar rupture or hernia could occur. Exhaling less forcefully against a closed glottis, a procedure known as the *Valsalva maneuver,* causes reflexive changes in blood pressure and cardiac output due to compression of the aorta and venae cavae. When abdominal and intrapleural pressures rise, the venae cavae collapse and the venous return decreases. The resulting fall in cardiac output and blood pressure stimulates the aortic and carotid baroreceptors. Their stimulation causes a reflexive increase in heart rate and peripheral vasoconstriction. When the glottis opens and pressures return to normal, venous return rises, and so does cardiac output. Because vasoconstriction has occurred, blood pressure then rises sharply. The rise in blood pressure inhibits the baroreceptors, and cardiac output, heart rate, and blood pressure quickly return to normal levels. The Valsalva maneuver is thus a very simple way to check the cardiovascular responses to changes in arterial pressure and venous return.

THE INTRAPLEURAL PRESSURE   The **intrapleural pressure** is the pressure measured in the space between the parietal and visceral pleurae. The intrapleural pressure averages about –4 mm Hg and reaches –18 mm Hg during a powerful inspiration (Figure 23-15b•). The pressure is below atmospheric pressure due to the relationship between the lungs and the body wall. We noted earlier that the lungs are highly elastic. In fact, they would collapse to about 5 percent of their normal resting volume if the elastic fibers could recoil completely. The elastic fibers cannot recoil significantly, however, because they are not strong enough to overcome the fluid bond between the parietal and visceral pleurae.

The elastic fibers are continuously fighting that bond and pulling the lungs away from the chest wall and diaphragm. This pull lowers the intrapleural pressure slightly. Because the elastic fibers remain stretched even after a full exhalation, the intrapleural pressure remains below atmospheric pressure throughout normal cycles of inhalation and exhalation. The cyclical changes in the intrapleural pressure are responsible for the *respiratory pump,* discussed in Chapter 21, that assists the venous return to the heart. ∞ *[p. 727]*

The Respiratory Cycle

A **respiratory cycle** is a single cycle of inhalation and exhalation. The **tidal volume** is the amount of air you move into or out of your lungs during a single respiratory cycle. The graphs in Figure 23-15a,b• follow the intrapleural and intrapulmonary pressures during a single respiratory cycle of an individual at rest, and they relate those changes to the tidal volume (Figure 23-15c•).

TABLE 23-1   **The Four Most Common Methods of Reporting Gas Pressures**

- *millimeters of mercury* (mm Hg): This is the most common method of reporting blood pressure and gas pressures. Normal atmospheric pressure is approximately 760 mm Hg.
- *torr:* This unit of measurement is preferred by many respiratory therapists; it is also commonly used in Europe and in some technical journals. One torr is equivalent to 1 mm Hg; in other words, normal atmospheric pressure is equal to 760 torr.
- *centimeters of water* (cm H_2O): In a hospital setting, anesthetic gas pressures and oxygen pressures are commonly reported in terms of centimeters of water. One cm H_2O is equivalent to 0.735 mm Hg; normal atmospheric pressure is 1033.6 cm H_2O.
- *pounds per square inch* (psi): Pressures in compressed gas cylinders and other industrial applications are generally reported in terms of psi. Normal atmospheric pressure at sea level is approximately 15 psi.

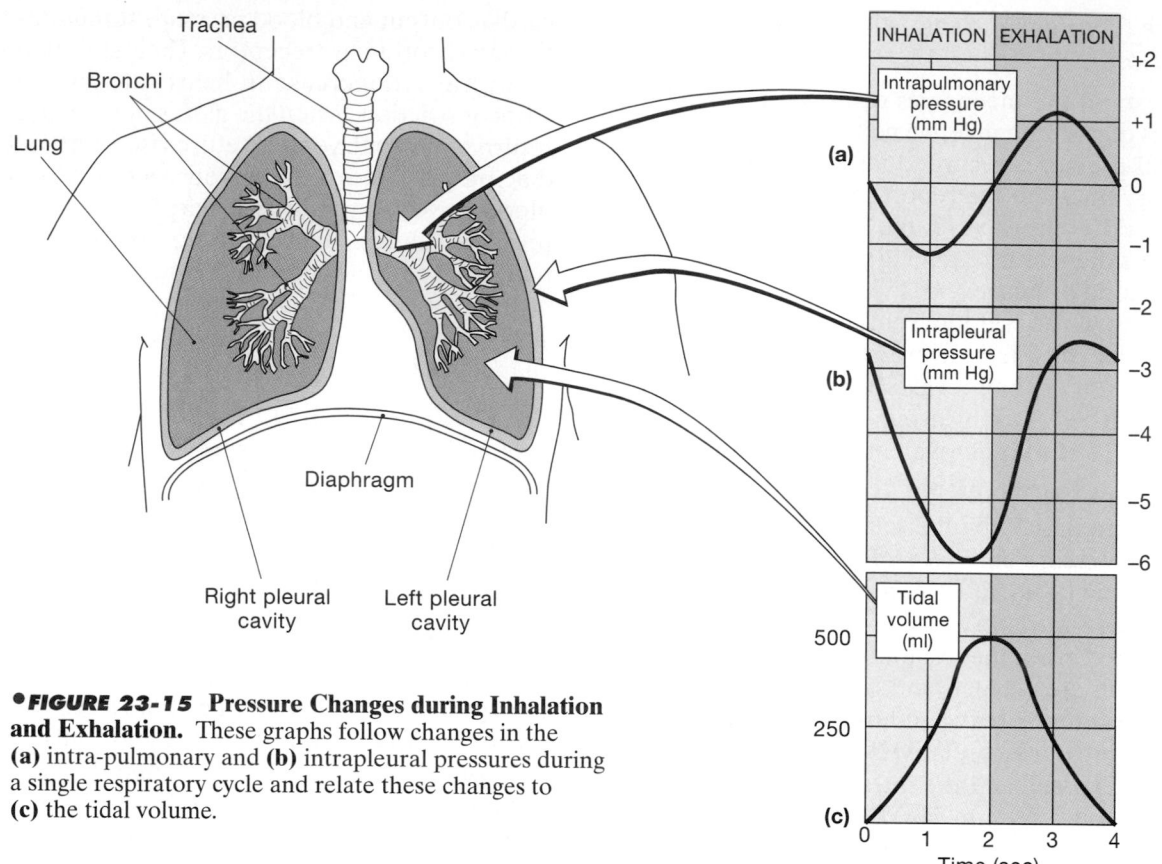

•**FIGURE 23-15** **Pressure Changes during Inhalation and Exhalation.** These graphs follow changes in the **(a)** intra-pulmonary and **(b)** intrapleural pressures during a single respiratory cycle and relate these changes to **(c)** the tidal volume.

At the start of the cycle, the intrapulmonary and atmospheric pressures are equal, and no air movement is occurring. Inhalation begins with the fall of intrapleural pressure that accompanies the expansion of the thoracic cavity. The intrapleural pressure gradually falls to approximately –6 mm Hg. Over this period, the intrapulmonary pressure drops to just under –1 mm Hg; it then begins to rise as air flows into the lungs. When exhalation begins, intrapleural and intrapulmonary pressures rise rapidly, forcing air out of the lungs. At the end of expiration, air movement again ceases when the pressure difference between intrapulmonary and atmospheric pressure has been eliminated. The amount of air moved into the lungs during inhalation is equal to the amount moved out of the lungs during exhalation. That amount is the tidal volume.

PNEUMOTHORAX An injury to the chest wall that penetrates the parietal pleura or damages the alveoli and the visceral pleura can allow air into the pleural cavity. This **pneumothorax** (noo-mō-THO-raks) breaks the fluid bond between the pleurae and allows the elastic fibers to recoil. The result is called a collapsed lung, or **atelectasis** (at-e-LEK-ta-sis; *ateles*, imperfect + *ektasis*, expansion). Treatment for a collapsed lung involves the removal of as much of the air as possible from the affected pleural cavity before the opening is sealed. This treatment lowers the intrapleural pressure and reinflates the lung.

Respiratory Muscles

In Chapter 11, we introduced the skeletal muscles involved in respiratory movements. Of those, the most important respiratory muscles are the *diaphragm* (Figure 11-12•, p. 337) and the *external intercostals* (Figures 11-11, 11-14, and 11-15•, pp. 336, 340–341, and 343). These muscles are involved in normal breathing at rest. The **accessory respiratory muscles** become active when the depth and frequency of respiration must be increased markedly. These muscles include the *internal intercostals, sternocleidomastoid,* the *serratus anterior,* the *pectoralis minor,* the *scalenes,* the *transversus thoracis,* the *transversus abdominis,* the *external* and *internal obliques,* and the *rectus abdominis.* These muscles are diagrammed in Figures 11-11, 11-12, and 23-16•.

MUSCLES USED IN INHALATION Inhalation is an active process involving the contraction of one or more of these muscles:

- Contraction of the diaphragm increases the volume of the thoracic cavity by tensing and flattening its floor, and this increase draws air into the lungs. Diaphragmatic contraction is responsible for roughly 75 percent of the air movement in normal *quiet breathing* (breathing at rest).
- The external intercostals assist in inhalation by elevating the ribs. This action contributes roughly 25 percent to the volume of air in the lungs.

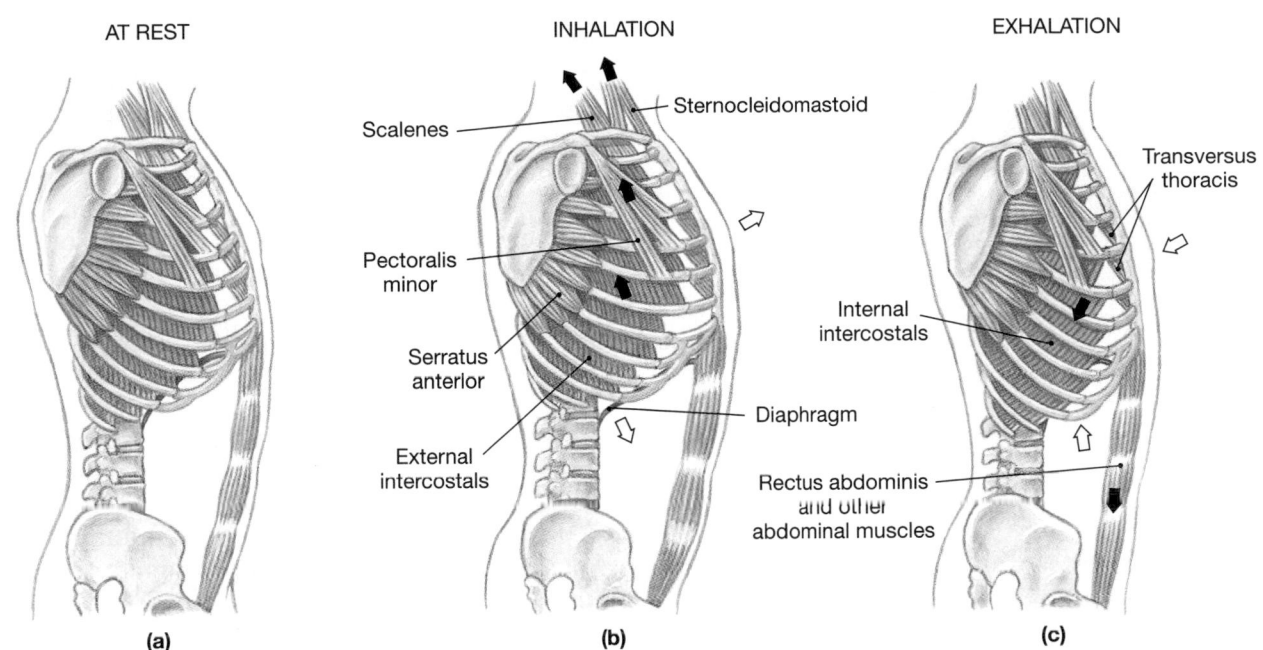

AT REST   INHALATION   EXHALATION

Scalenes — Sternocleidomastoid

Transversus thoracis

Pectoralis minor

Internal intercostals

Serratus anterior

Diaphragm

External intercostals

Rectus abdominis and other abdominal muscles

(a)   (b)   (c)

•FIGURE 20-16 The Respiratory Muscles. The abdominal muscles that act as accessory expiratory muscles are represented by a single muscle (the rectus abdominis). Diaphragmatic movements were also illustrated in Figure 23-14. **(a)** Lateral view at rest, with no air movement. **(b)** Inhalation, showing the muscles that elevate the ribs. **(c)** Exhalation, showing the muscles that depress the ribs.

- Accessory muscles, including the sternocleidomastoid, the serratus anterior, the pectoralis minor, and the scalenes, can assist the external intercostals in elevating the ribs. These muscles increase the speed and amount of rib movement.

MUSCLES USED IN EXHALATION Exhalation may be passive or active, depending on the level of respiratory activity. When exhalation is active, it may involve one or more of the following muscles:

- The internal intercostals and transversus thoracis depress the ribs and reduce the width and depth of the thoracic cavity.
- The abdominal muscles, including the external and internal obliques, the transversus abdominis, and the rectus abdominis, can assist the internal intercostals in expiration by compressing the abdomen and forcing the diaphragm upward.

ARTIFICIAL RESPIRATION Artificial respiration is a technique to provide air to an individual whose respiratory muscles are no longer functioning. In **mouth-to-mouth resuscitation,** a rescuer provides ventilation by exhaling into the mouth or mouth and nose of the victim. After each breath, contact is broken to permit passive exhalation by the victim. Air provided in this way contains adequate oxygen to meet the needs of the victim. If the victim's cardiovascular system is nonfunctional as well, a technique called *cardiopulmonary resuscitation (CPR)* is required to maintain adequate blood flow and tissue oxygenation. [AM] *CPR*

Modes of Breathing

The respiratory muscles may be used in various combinations depending on the volume of air that must be moved into or out of the system. Respiratory movements are usually classified as *quiet breathing* or *forced breathing,* depending on the pattern of muscle activity in the course of a single respiratory cycle. Relationships between these categories are indicated in Figure 23-17•.

QUIET BREATHING In **quiet breathing,** or **eupnea** (ŪP-nē-uh), inhalation involves muscular contractions, but exhalation is a passive process. Inhalation usually involves the contraction of both the diaphragm and the external intercostals. The relative contributions of these muscles can vary, depending on the circumstances:

- During **diaphragmatic breathing,** or **deep breathing,** contraction of the diaphragm provides the necessary change in thoracic volume. Air is drawn into the lungs as the diaphragm contracts, and exhalation occurs passively when the diaphragm relaxes.
- In **costal breathing,** or **shallow breathing,** the thoracic volume changes because the rib cage changes shape. Inhalation occurs when contractions of the external intercostals elevate the ribs and enlarge the thoracic cavity. Exhalation occurs passively when these muscles relax.

During quiet breathing, expansion of the lungs stretches their elastic fibers. In addition, elevation of the rib cage stretches opposing skeletal muscles and

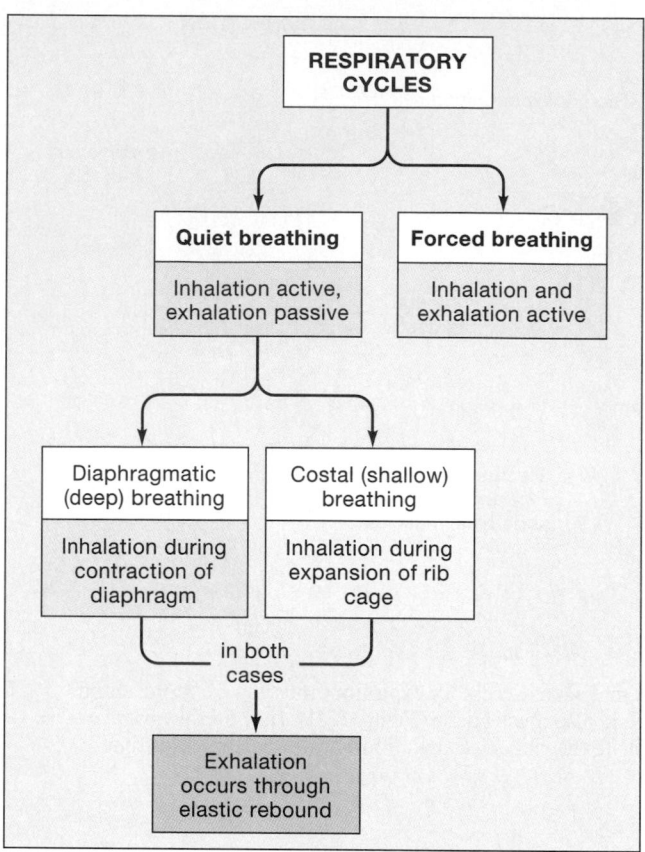

RESPIRATORY CYCLES

Quiet breathing
Inhalation active, exhalation passive

Forced breathing
Inhalation and exhalation active

Diaphragmatic (deep) breathing
Inhalation during contraction of diaphragm

Costal (shallow) breathing
Inhalation during expansion of rib cage

in both cases

Exhalation occurs through elastic rebound

•*FIGURE 23-17* **A Classification of Respiratory Activity**

elastic fibers in the connective tissues of the body wall. When the inspiratory muscles relax, these elastic components recoil, returning the diaphragm, the rib cage, or both to their original positions. This phenomenon is called **elastic rebound.**

At minimal levels of activity, eupnea involves primarily diaphragmatic movement with little costal motion. As increased volumes of air are required, inspiratory movements become larger and the contribution of costal movement increases. Even when you are at rest, costal breathing can predominate when abdominal pressures, fluids, or masses restrict diaphragmatic movements. For example, pregnant women increasingly rely on costal breathing as the uterus enlarges and pushes the abdominal viscera against the diaphragm.

FORCED BREATHING **Forced breathing**, or **hyperpnea** (hī-PERP-nē-uh), involves active inspiratory and expiratory movements. Forced breathing calls on the accessory muscles to assist with inspiration, and expiration involves contraction of the internal intercostals. At absolute maximum levels of forced breathing, the abdominal muscles are used in exhalation. Their contraction compresses the abdominal contents, pushing them up against the diaphragm and further reducing the volume of the thoracic cavity.

Respiratory Rates

As you read this, you are probably breathing quietly, but when necessary your rate of respiration can change dramatically. Your **respiratory rate** is the number of breaths you take per minute. The normal respiratory rate of a resting adult ranges from 12 to 18 breaths per minute. Children breathe more rapidly, at rates of about 18–20 breaths per minute.

The Respiratory Minute Volume

We can calculate the amount of air moved each minute by multiplying the respiratory rate (f) by the tidal volume (V_T). This value is called the **respiratory minute volume,** $\dot{V}_E$. The tidal volume at rest varies from individual to individual, but it averages around 500 ml per breath. Therefore the respiratory minute volume at rest, at 12 breaths per minute, will be approximately 6 liters per minute:

$$\dot{V}_E \text{ (volume of air moved each minute)} =$$

$$f \text{ (breaths per minute)} \times V_T \text{ (tidal volume)} =$$

$$12 \times 500 \text{ ml per minute} = 6000 \text{ ml per minute}$$

$$= 6.0 \text{ liters per minute}$$

Alveolar Ventilation

The respiratory minute volume measures pulmonary ventilation and provides an indication of how much air is moving into and out of the respiratory tract. Only a portion of the inspired air reaches the alveolar exchange surfaces. A typical tidal inspiration pulls about 500 ml of air into your respiratory system. The first 350 ml inspired travels along the conducting passageways and enters the alveolar spaces. The last 150 ml of inspired air never gets farther than the conducting passageways and does not participate in gas exchange with the blood. The volume of air in the conducting passages is known as the **anatomic dead space** (V_D). **Alveolar ventilation** ($\dot{V}_A$), is the amount of air reaching the alveoli each minute. Alveolar ventilation is less than the respiratory minute volume, because some of the air never reaches the alveoli but remains in the dead space of the lungs. We can calculate alveolar ventilation by subtracting the dead space from the tidal volume, using the following formula:

$$\dot{V}_A = f \times (V_T - V_D)$$

At rest, alveolar ventilation rates are approximately 4.2 liters per minute (12×350 ml). However, the gas arriving in the alveoli is significantly different from that of the surrounding atmosphere, because the air entering the respiratory tract on inhalation mixes with air in the anatomic dead space before reaching the exchange surfaces. The air in the alveoli thus contains less oxygen and more carbon dioxide than atmospheric air.

Relationships among V_T, $\dot{V}_E$, and $\dot{V}_A$

The respiratory minute volume can be increased by (1) increasing the tidal volume or (2) increasing the respiratory rate. Under maximum stimulation, the tidal volume can increase to roughly 4.8 liters. At peak respiratory rates of 40–50 breaths per minute and maximum cycles of inspiration and expiration, the respiratory minute volume can approach 200 liters (about 55 gal) per minute.

In functional terms, the alveolar ventilation rate is more important than the respiratory minute volume, because it determines the rate of oxygen delivery to the alveoli:

- For a given respiratory rate, increasing the tidal volume will increase the alveolar ventilation rate.
- For a given tidal volume, increasing the respiratory rate will increase the alveolar ventilation rate.

It is important to note that *the alveolar ventilation rate can change independently of the respiratory minute volume.* In our previous example, the respiratory minute volume at rest was 6 liters, and the alveolar ventilation rate was 4.2 l/min. If the respiratory rate increases to 20 breaths per minute but the tidal volume drops to 300 ml, the respiratory minute volume remains the same (20 × 300 = 6000). However, the alveolar ventilation rate drops to only 3 l/min (20 × [300 − 150] = 3000). Thus, whenever the demand for oxygen increases, both tidal volume *and* respiratory rate must be regulated closely. The regulatory mechanisms involved will be the focus of a later section.

Respiratory Performance and Volume Relationships

Only a small proportion of the air in the lungs is exchanged during a single quiet respiratory cycle; the tidal volume can be increased by inhaling more vigorously and exhaling more completely. We can divide the total volume of the lungs (Figure 23-18●) into *volumes* and *capacities*. These values can be experimentally determined, and the values obtained are useful in diagnosing problems with pulmonary ventilation.

Pulmonary volumes include the following:

- The **resting tidal volume** is the amount of air you move into or out of your lungs during a single quiet respiratory cycle. The resting tidal volume averages about 500 ml in both males and females. Many other respiratory volumes and capacities differ between the genders, because adult females on average have smaller bodies, and thus smaller lung volumes, than males. Representative values for both genders are indicated in Figure 23-18●.

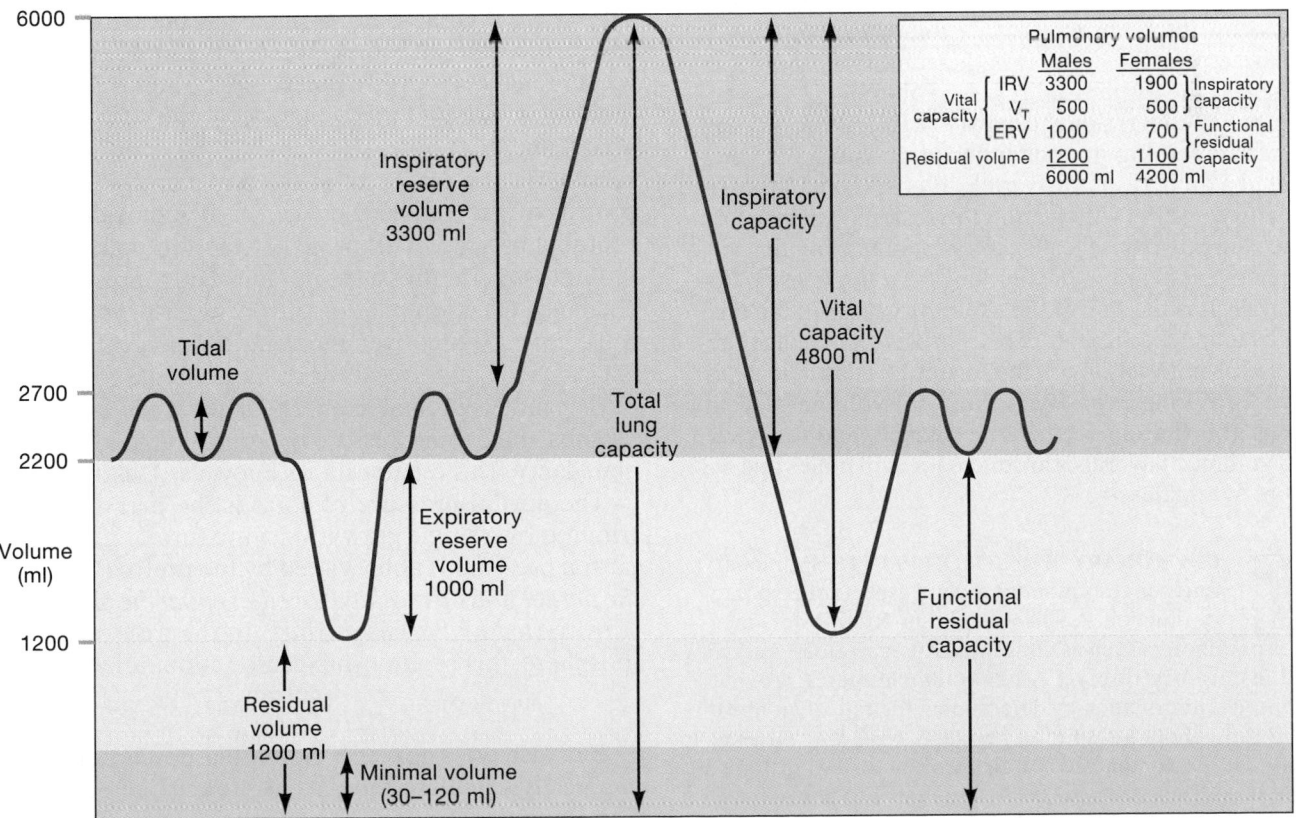

●**FIGURE 23-18** **Respiratory Volumes and Capacities.** The graph diagrams the relationships between the respiratory volumes and capacities of an average male.

- The **expiratory reserve volume** (ERV) is the amount of air that you can voluntarily expel after you have completed a normal quiet respiratory cycle. If, with maximum use of the accessory muscles, you can expel an additional 1000 ml of air, your expiratory reserve volume is 1000 ml.

- The **residual volume** is the amount of air that remains in your lungs even after a maximal exhalation—typically about 1200 ml in males and 1100 ml in females. The **minimal volume,** a component of the residual volume, is the amount of air that would remain in your lungs if they were allowed to collapse. The minimal volume ranges from 30 to 120 ml, but unlike other volumes it cannot be measured in a healthy person. The minimal volume and the residual volume are very different, because the fluid bond between the lungs and the chest wall prevents the recoil of the elastic fibers. Some air remains in the lungs, even at minimal volume, because the surfactant coating the alveolar surfaces prevents their collapse.

- The **inspiratory reserve volume** (IRV) is the amount of air that you can take in over and above the tidal volume. There are significant differences between the inspiratory reserve volumes of males and females, because on average the lungs of males are larger. The inspiratory reserve volume of males averages 3300 ml versus 1900 ml in females.

We can determine respiratory capacities by adding the values of various volumes. Examples include the following:

- The **inspiratory capacity** is the amount of air that you can draw into your lungs after you have completed a quiet respiratory cycle. It is the sum of the tidal volume and the inspiratory reserve volume.

- The **functional residual capacity** (FRC) is the amount of air remaining in your lungs after you have completed a quiet respiratory cycle. It is the sum of the expiratory reserve volume and the residual volume.

- The **vital capacity** is the maximum amount of air that you can move into or out of your lungs in a single respiratory cycle. It is the sum of the expiratory reserve, the tidal volume, and the inspiratory reserve. The vital capacity averages around 4.8 liters in males and 3.1 liters in females.

- The **total lung capacity** is the total volume of your lungs. It is the sum of the vital capacity and the residual volume; it averages around 6 liters in males and 4.2 liters in females.

PULMONARY FUNCTION TESTS Pulmonary function tests monitor several aspects of respiratory function. A **spirometer** (spī-ROM-e-ter) measures parameters such as vital capacity, expiratory reserve, and inspiratory reserve. A **pneumotachometer** provides additional information by determining the rate of air movement throughout a respiratory cycle. A *peak flow meter* records the maximum rate of air movement during forced expiration. These tests are relatively simple to perform, and they have considerable diagnostic significance. For example, in people with asthma, the constricted airways tend to close before an exhalation is completed. As a result, pulmonary function tests show a reduction in vital capacity, expiratory

reserve volume, and peak flow rate. Air whistling through the constricted airways produces the characteristic "wheezing" that accompanies an asthmatic attack. [AM] *Asthma*

Conditions that restrict the maximum distensibility of the lungs have the same effect on vital capacity, because they lower the inspiratory reserve. However, because the airways are not affected, the expiratory reserve and expiratory flow rates are relatively normal.

☑ Mark breaks a rib that punctures the chest wall on his left side. What do you expect will happen to his left lung as a result?

☑ In pneumonia, fluid accumulates in the alveoli of the lungs. How would this accumulation affect vital capacity?

Gas Exchange at the Respiratory Membrane

Pulmonary ventilation ensures that your alveoli are supplied with oxygen, and it removes the carbon dioxide arriving from your bloodstream. The actual process of gas exchange occurs between the blood and alveolar air across the respiratory membrane. To understand these events, we consider (1) the *partial pressures* of the gases involved and (2) the diffusion of molecules between a gas and a liquid.

Dalton's Law and Partial Pressures

The air we breathe is not a single gas but a mixture of gases. Nitrogen molecules (N_2) are the most abundant, accounting for about 78.6 percent of the atmospheric gas molecules. Oxygen molecules (O_2), the second most abundant, account for roughly 20.9 percent of the atmospheric content. Most of the remaining 0.5 percent consists of water molecules, with carbon dioxide (CO_2) contributing a mere 0.04 percent to the atmospheric total.

Atmospheric pressure, 760 mm Hg, represents the combined effects of collisions involving each type of molecule in air. At any given moment, 78.6 percent of those collisions will involve nitrogen molecules, 20.9 percent oxygen molecules, and so on. Thus each of the gases contributes to the total pressure in proportion to its relative abundance. This relationship is known as **Dalton's law.**

The **partial pressure** of a gas is the pressure contributed by a single gas within a mixture of gases. The partial pressure is abbreviated by the prefix *P* or *p*. *All the partial pressures added together equal the total pressure exerted by the gas mixture.* In the case of the atmosphere, this relationship can be summarized as

$$P_{N_2} + P_{O_2} + P_{H_2O} + P_{CO_2} = 760 \text{ mm Hg}$$

Because we know the individual percentages, we can easily calculate the partial pressure of each gas. For example, the partial pressure of oxygen, P_{O_2}, is 20.9 percent of 760 mm Hg, or roughly 159 mm Hg. The partial pressures for other atmospheric gases are included in Table 23-2.

TABLE 23-2 Partial Pressures (mm Hg) and Normal Gas Concentrations (%) in Air

Source of Sample	Nitrogen (N_2)	Oxygen (O_2)	Carbon Dioxide (CO_2)	Water Vapor (H_2O)
Inspired air (dry)	597 (78.6%)	159 (20.9%)	0.3 (0.04%)	3.7 (0.5%)
Alveolar air (saturated)	573 (75.4%)	100 (13.2%)	40 (5.2%)	47 (6.2%)
Expired air (saturated)	569 (74.8%)	116 (15.3%)	28 (3.7%)	47 (6.2%)

Henry's Law: Diffusion between Liquids and Gases

Differences in pressure move gas molecules from one place to another. Pressure differences also affect the movement of gas molecules into and out of solution. At a given temperature, _the amount of a particular gas in solution is directly proportional to the partial pressure of that gas._ This principle is known as **Henry's law**.

When a gas under pressure contacts a liquid, the pressure tends to force gas molecules into solution. At a given pressure, the number of dissolved gas molecules will rise until an equilibrium becomes established. At equilibrium, gas molecules are diffusing out of the liquid as quickly as they are entering, and the total number of gas molecules in solution remains constant. If the partial pressure goes up, more gas molecules will go into solution; if the partial pressure goes down, gas molecules will come out of solution. These relationships are diagrammed in Figure 23-19●.

You see Henry's law in action whenever you open a can of soda. The soda was put into the can under pressure, and the gas (carbon dioxide) is in solution.

When you open the can, the pressure falls, and the gas molecules begin coming out of solution. The process will theoretically continue until an equilibrium develops between the air and the gas in solution. In fact, the volume of the can is so small, and the volume of the atmosphere so great, that over the course of a half-hour or so, virtually all of the carbon dioxide comes out of solution. You are then left with a "flat" soda.

The actual _amount_ of a gas in solution at a given partial pressure and temperature depends on the solubility of the gas in that particular liquid. Carbon dioxide is very soluble, oxygen is somewhat less soluble, and nitrogen has very limited solubility in body fluids. The dissolved gas content is usually reported in terms of milliliters of gas per 100 ml (1 dl) of solution. To see the differences in relative solubility, we can compare the amount in solution with the partial pressure of each gas. For example, the plasma in a pulmonary vein generally contains 2.62 ml/dl of dissolved CO_2 (P_{CO_2} = 40 mm Hg), 0.29 ml/dl of dissolved O_2 (P_{O_2} = 100 mm Hg), and 1.25 ml/dl of dissolved N_2 (P_{N_2} = 573 mm Hg).

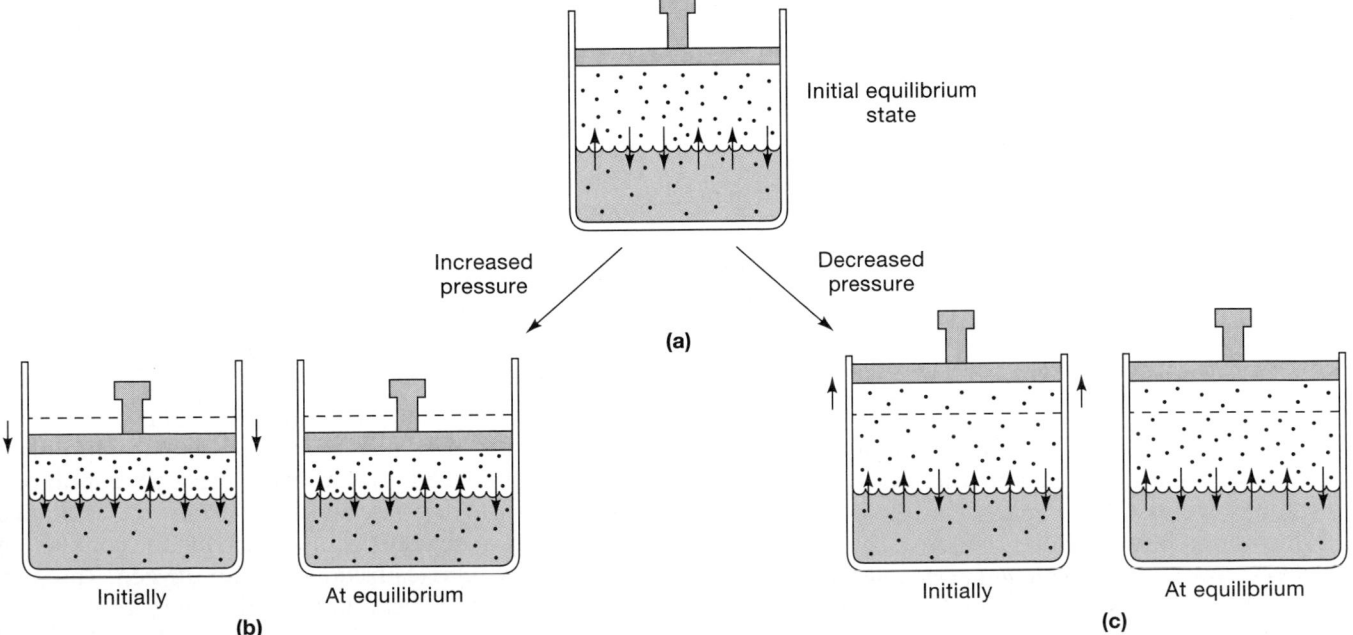

Initial equilibrium state

Increased pressure

Decreased pressure

(a)

Initially

At equilibrium

(b)

Initially

At equilibrium

(c)

●**_FIGURE 23-19_** **Henry's Law and the Relationship between Solubility and Pressure.** **(a)** A solution containing dissolved gas molecules at equilibrium with air under a given pressure. **(b)** Increasing the pressure drives additional gas molecules into solution until a new equilibrium becomes established. **(c)** When the pressure decreases, some of the dissolved gas molecules leave the solution until equilibrium is restored.

DECOMPRESSION SICKNESS Decompression **sickness** is a painful condition that develops when a person is suddenly exposed to a drop in atmospheric pressure. Nitrogen is the gas responsible for the problems experienced, due to its high partial pressure. When the pressure drops, nitrogen comes out of solution, forming bubbles like those in a shaken can of soda. The bubbles may form in joint cavities, in the bloodstream, and in the cerebrospinal fluid (CSF). The victims typically curl up from the pain in affected joints; this reaction accounts for the common name for the condition, *the bends.* Decompression sickness most commonly affects scuba divers, who while submerged are breathing air under greater-than-normal pressures. It also may affect people in airplanes subject to sudden losses of cabin pressure. [AM] *Decompression Sickness*

The Composition of Alveolar Air

As soon as air enters the respiratory tract, its characteristics begin to change. In passing through the nasal cavity, the air becomes warmer, and the amount of water vapor increases. Humidification and filtration continue as the air travels through the pharynx, trachea, and bronchial passageways. On reaching the alveoli, the incoming air mixes with air that had remained in the alveoli from the previous respiratory cycle. The alveolar air thus contains more carbon dioxide and less oxygen than does atmospheric air.

The last 150 ml of inspired air never gets farther than the conducting passageways and remains in the anatomic dead space of the lungs. During the subsequent expiration, the departing alveolar air mixes with air in the dead space to produce yet another mixture that differs from both atmospheric and alveolar samples. The differences in composition between atmospheric (inspired) and alveolar air are given in Table 23-2.

Diffusion at the Respiratory Membrane

Gas exchange at the respiratory membrane is efficient for the following five reasons:

1. **The differences in partial pressure across the respiratory membrane are substantial.** This fact is important because the greater the difference in partial pressure, the faster the rate of gas diffusion. Conversely, if the P_{O_2} in the alveoli decreases, the rate of oxygen diffusion into the blood will drop. This is why many people feel light-headed at altitudes of 3000 m or more—the partial pressure of oxygen in their alveoli has dropped low enough that the rate of oxygen absorption is significantly reduced.

2. **The distances involved in gas exchange are small.** The fusion of capillary and alveolar basement membranes reduces the distance to an average of 0.5 μm. Inflammation of the lung tissue or fluid buildup inside the alveoli will increase the diffusion distance and impair alveolar gas exchange.

3. **The gases are lipid-soluble.** Both oxygen and carbon dioxide diffuse readily through the surfactant layer and the alveolar and endothelial cell membranes.

4. **The total surface area is large.** The combined alveolar surface area at peak inspiration may approach 140 m². Damage to alveolar surfaces, as occurs in emphysema, reduces the available surface area and the efficiency of gas transfer.

5. **Blood flow and airflow are coordinated.** This arrangement improves the efficiency of both pulmonary ventilation and pulmonary circulation. For example, blood flow is greatest around alveoli with the highest P_{O_2} values, where oxygen uptake can proceed with maximum efficiency. If the normal circulation is impaired, as occurs in a *pulmonary embolism,* or the normal airflow is interrupted, as in various forms of *pulmonary obstruction,* this coordination is lost, and respiratory efficiency decreases. [AM] *Bronchitis, Emphysema, and COPD*

Partial Pressures in the Alveolar Air and Alveolar Capillaries

Figure 23-20● details the partial pressures of oxygen and carbon dioxide in the pulmonary and systemic circuits. Blood arriving in the pulmonary arteries has a lower P_{O_2} and a higher P_{CO_2} than does alveolar air. Diffusion between the alveolar mixture and the pulmonary capillaries thus elevates the P_{O_2} of the blood while lowering its P_{CO_2}. By the time the blood enters the pulmonary venules, it has reached equilibrium with the alveolar air, so it departs the alveoli with a P_{O_2} of about 100 mm Hg and a P_{CO_2} of roughly 40 mm Hg (Figure 23-20a●).

Diffusion between the blood in the pulmonary capillaries and the alveolar air occurs very rapidly. When you are at rest, a red blood cell moves through one of your pulmonary capillaries in about 0.75 second; when you exercise, that passage takes less than 0.3 second. This amount of time is usually sufficient to reach an equilibrium between the alveolar air and the blood.

Partial Pressures in the Systemic Circuit

The oxygenated blood now leaves the alveolar capillaries and returns to the heart, to be discharged into the systemic circuit. As it enters the pulmonary veins, the oxygenated blood from the alveolar capillaries mixes with blood that flowed through capillaries around conducting passageways. Because gas exchange occurs only at alveoli, the blood leaving the conducting passageways carries relatively little oxygen. The partial pressure of oxygen in the pulmonary veins therefore drops to about 95 mm Hg. This is the P_{O_2} in the blood that enters the systemic circuit, and no further changes in partial pressure occur until the blood reaches the peripheral capillaries (Figure 23-20b●).

Normal interstitial fluid has a P_{O_2} of 40 mm Hg. As a result, oxygen diffuses out of the capillaries and carbon dioxide diffuses in, until the capillary partial pressures are the same as those in the adjacent tissues. At a normal tissue P_{O_2}, the blood entering the venous system still contains about 75 percent of its total oxygen content.

Blood entering the systemic circuit normally has a P_{CO_2} of 40 mm Hg. Inactive peripheral tissues normally have a P_{CO_2} of about 45 mm Hg. As a result, carbon dioxide diffuses into the blood as oxygen diffuses out (Figure 23-20b●).

BLOOD GAS ANALYSIS Blood samples may be analyzed to determine the concentration of dissolved gases. The usual tests include determination of pH, P_{CO_2}, and P_{O_2} in an arterial sample. This analysis can be very helpful in monitoring patients after a heart attack or in chronic respiratory conditions such as obstructive disorders or asthma. A blood sample provides information about the degree of oxygenation in peripheral tissues. For example, if the P_{CO_2} is very high and the P_{O_2} very low, tissues are not receiving adequate oxygen. This condition may be corrected by providing a gas mixture that has a high P_{O_2} (or even pure oxygen, with a P_{O_2} of 760 mm Hg). In addition, blood gas measurements determine the efficiency of gas exchange at the lungs. If the arterial P_{O_2} remains low despite oxygen administration or the P_{CO_2} is very high, pulmonary exchange problems, such as pulmonary edema, emphysema, or pneumonia, must exist.

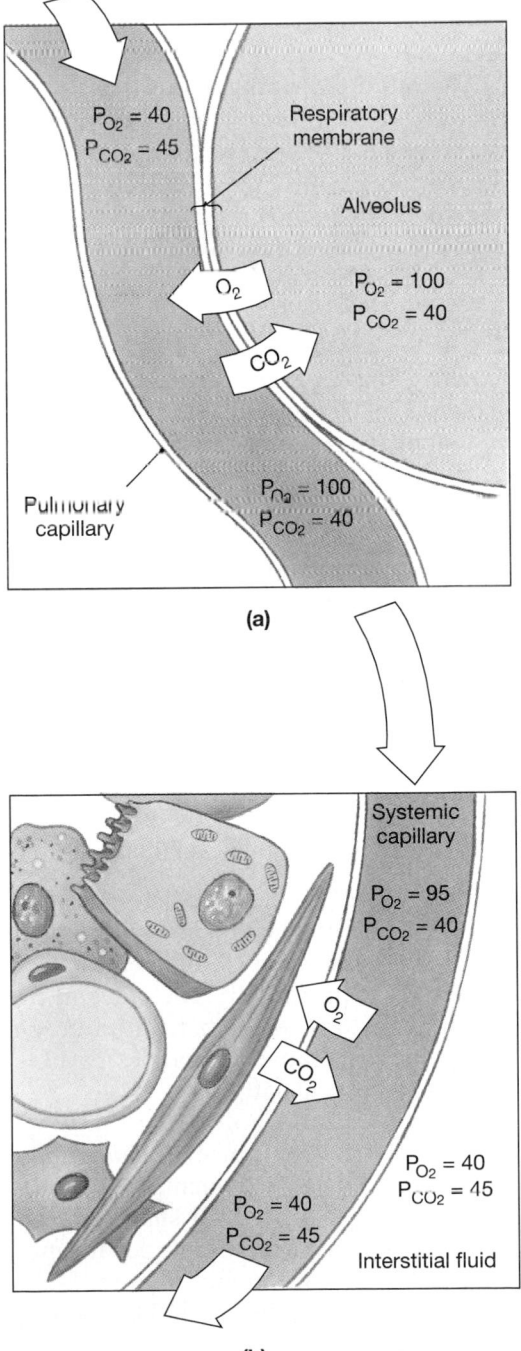

(a)

(b)

●*FIGURE 23-20* **An Overview of Respiratory Processes and Partial Pressures in Respiration.** **(a)** Partial pressures and diffusion at the respiratory membrane. **(b)** Partial pressures and diffusion in other tissues.

Gas Pickup and Delivery

Oxygen and carbon dioxide have limited solubilities in blood plasma. For example, at the normal alveolar P_{O_2}, 100 ml of plasma will absorb only about 0.3 ml of oxygen. The limited solubilities of these gases are a problem because peripheral tissues need more oxygen and generate more carbon dioxide than the plasma can absorb and transport.

The problem is solved by the red blood cells, which remove dissolved oxygen and carbon dioxide molecules and either bind them (in the case of oxygen) or use them to manufacture soluble compounds (in the case of carbon dioxide). Because these reactions remove dissolved gas from the plasma, gas will continue to diffuse into the blood but will never reach equilibrium.

The important thing about these reactions is that they are (1) *temporary* and (2) *completely reversible*. When plasma oxygen or carbon dioxide concentrations are high, the excess molecules are removed by the RBCs. When the plasma concentrations are falling, the RBCs release their stored reserves.

Oxygen Transport

The blood leaving the alveolar capillaries carries away roughly 20 ml of oxygen per 100 ml. Of this amount, only about 0.3 ml (1.5 percent) consists of oxygen molecules in solution. The rest are bound to *hemoglobin (Hb) molecules,* specifically to the iron ions in the center of heme units. We detailed the structure of hemoglobin in Chapter 19. ∞ *[p. 648]* A hemoglobin molecule consists of four globular protein subunits, each containing a heme unit. Thus each hemoglobin molecule can bind four molecules of oxygen, forming oxyhemoglobin. This is a reversible reaction that can be summarized as

$$Hb + O_2 \rightleftharpoons HbO_2$$

There are approximately 280 million molecules of hemoglobin in each red blood cell. Because a hemoglobin molecule contains four heme units, each RBC potentially can carry more than a billion molecules of oxygen.

HEMOGLOBIN SATURATION **Hemoglobin saturation** is the percentage of heme units containing bound oxygen. At most, each Hb molecule can carry four O_2 molecules. If all the Hb molecules in the blood are fully loaded, there is 100 percent saturation. If, on average, each Hb molecule carries two O_2 molecules, there is 50 percent saturation.

In Chapter 2, we noted that the shape and functional properties of a protein change in response to alterations in its environment. ∞ *[p. 54]* Hemoglobin is no exception, and any shape changes that occur can affect oxygen binding. Under normal conditions, the most important environmental factors affecting hemoglobin are (1) the P_{O_2} of the blood, (2) blood pH, (3) temperature, and (4) ongoing metabolic activity within RBCs.

Hemoglobin and P_{O_2}. An **oxygen–hemoglobin saturation curve** (Figure 23-21●), or *oxygen–hemoglobin dissociation curve,* is a graph that relates the saturation of hemoglobin to the partial pressure of oxygen. The binding and dissociation of oxygen to hemoglobin is a typical reversible reaction. At equilibrium, oxygen molecules are binding to heme at the same rate as other oxygen molecules are being released. If you increase the P_{O_2}, you shift the reaction to the right and more oxygen gets bound to hemoglobin. If you decrease the P_{O_2}, the reaction shifts to the left and more oxygen is released by hemoglobin. The graph of this relationship is a curve rather than a straight line, because the shape of a hemoglobin molecule changes slightly each time it binds an oxygen molecule, and the change affects its ability to bind *another* oxygen molecule. In other words, the attachment of the first oxygen molecule makes it easier to bind the second, binding second promotes binding of the third, and binding of the third enhances binding of the fourth.

Because each arriving oxygen molecule increases the affinity of hemoglobin for the *next* oxygen molecule, the saturation curve takes the form shown in Figure 23-21●. The slope is gradual until the first oxygen molecule binds to the hemoglobin; then the slope rises rapidly, with a prolonged plateau near 100 percent saturation. Over the steep initial slope, a very small decrease in the plasma P_{O_2} will result in a large change in the amount of oxygen bound to or released from HbO_2. Because the curve rises quickly, hemoglobin will be more than 90 percent saturated if exposed to an alveolar P_{O_2} above 60 mm Hg. This means that near-normal oxygen transport can continue despite a de-

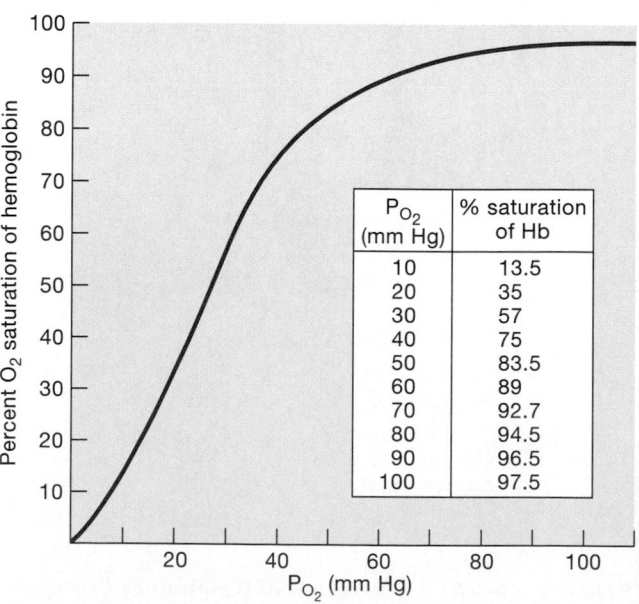

P_{O_2} (mm Hg)	% saturation of Hb
10	13.5
20	35
30	57
40	75
50	83.5
60	89
70	92.7
80	94.5
90	96.5
100	97.5

●**FIGURE 23-21** **The Oxygen–Hemoglobin Saturation Curve.** This curve indicates the normal saturation characteristics of hemoglobin at various partial pressures of oxygen.

crease in the oxygen content of alveolar air. Without this ability, you would be unable to survive at high altitudes, and conditions significantly reducing pulmonary ventilation would be immediately fatal.

At normal alveolar pressures (P_{O_2} = 100 mm Hg), the percent saturation is very high (97.5 percent), although complete saturation does not occur until the P_{O_2} reaches excessively high levels (about 250 mm Hg). In functional terms, the maximum percent saturation is not as important as the ability of hemoglobin to provide oxygen over the normal P_{O_2} range in body tissues. Over that range, from 100 mm Hg at the alveoli to perhaps 15 mm Hg in active tissues, the percent saturation drops from 97.5 percent to less than 20 percent, and a small change in P_{O_2} makes a big difference in terms of the amount of oxygen bound by hemoglobin.

Note that the relationship between P_{O_2} and percent saturation remains valid whether the P_{O_2} is rising or falling. *If the P_{O_2} increases, the percent saturation goes up and hemoglobin stores oxygen. If the P_{O_2} decreases, hemoglobin releases oxygen into its surroundings.* When oxygenated blood arrives in the peripheral capillaries, the blood P_{O_2} declines rapidly as a result of gas exchange with the interstitial fluid. As the P_{O_2} falls, hemoglobin gives up its oxygen.

The relationship between P_{O_2} and hemoglobin saturation provides a mechanism for automatic regulation of oxygen delivery. Inactive tissues have little demand for oxygen, and the local P_{O_2} is usually about 40 mm Hg. Under these conditions, hemoglobin will not release much oxygen. As it passes through the capillaries, it will go from 97 percent saturation (P_{O_2} = 95 mm Hg) to 75 percent saturation (P_{O_2} = 40 mm Hg). Because it still retains three-quarters of its oxygen, the venous blood has a relatively large oxygen reserve. This reserve is important, because it can be mobilized if tissue oxygen demands increase.

Active tissues consume oxygen at an accelerated rate, and the P_{O_2} may drop to 15–20 mm Hg. Hemoglobin passing through these capillaries will then go from 97 percent saturation to about 20 percent saturation. In practical terms, this means that as blood circulates through peripheral capillaries, active tissues will receive 3.5 times as much oxygen as will inactive tissues.

CARBON MONOXIDE POISONING Murder or suicide victims who died in their cars inside a locked garage are popular characters for mystery writers. In real life, entire families are killed each winter by leaky furnaces or space heaters. The cause of death is carbon monoxide poisoning. The exhaust of automobiles and other petroleum-burning engines, of oil lamps, and of fuel-fired space heaters contains *carbon monoxide* (CO). Carbon monoxide competes with oxygen molecules for the binding sites on heme units. Unfortunately, the carbon monoxide usually wins, for it has a much stronger affinity for hemoglobin at very low partial pressures. The bond formed is extremely durable, and the attachment of a carbon monoxide molecule

essentially inactivates that heme unit for respiratory purposes. Carbon monoxide will bind to hemoglobin at very low partial pressures. If carbon monoxide molecules make up just 0.1 percent of the components of inspired air, enough hemoglobin will be affected that survival will become impossible without medical assistance. Treatment may include (1) administration of pure oxygen, for at sufficiently high partial pressures the oxygen molecules will gradually replace carbon monoxide at the hemoglobin molecules, and, if necessary, (2) the transfusion of compatible red blood cells.

Hemoglobin and pH. The oxygen–hemoglobin saturation curve in Figure 23-21● was determined in normal blood, with a pH of 7.4 and a temperature of 37°C. In addition to consuming oxygen, active tissues generate acids that lower the pH of the interstitial fluid. When the pH drops, there is a change in the shape of hemoglobin molecules. As a result of this change, the slope of the hemoglobin saturation curve changes (Figure 23-22a●), and the hemoglobin molecules release their oxygen reserves more readily. In other words, the saturation percentage declines. Thus, at a tissue P_{O_2} of 40 mm Hg, hemoglobin molecules will release 15 percent more oxygen at a pH of 7.2 than they would at a pH of 7.4. This effect of pH on the hemoglobin saturation curve is called the **Bohr effect.**

Carbon dioxide is the primary compound responsible for the Bohr effect. When CO_2 diffuses into the blood, it rapidly diffuses into red blood cells. There, an enzyme called **carbonic anhydrase** (an-HĪ-dras) catalyzes its reaction with water molecules:

$$CO_2 + H_2O \overset{\text{carbonic}}{\underset{\text{anhydrase}}{\rightleftharpoons}} H_2CO_3 \rightleftharpoons H^+ + HCO_3^-$$

The product of this enzymatic reaction, H_2CO_3, is called *carbonic acid,* because it dissociates into a hydrogen ion (H^+) and a bicarbonate ion (HCO_3^-). The rate of carbonic acid formation depends on the amount of carbon dioxide in solution, which as we noted earlier depends on the P_{CO_2}. When the P_{CO_2} rises, the reaction proceeds from left to right, and the rate of carbonic acid formation accelerates. The hydrogen ions generated diffuse out of the red blood cells, and the pH of the plasma drops. When the P_{CO_2} declines, the reaction proceeds from right to left; hydrogen ions then diffuse into the RBCs, and the pH of the plasma rises.

Hemoglobin and Temperature. Temperature changes also affect the slope of the hemoglobin saturation curve, as indicated in Figure 23-22b●. As the temperature rises, hemoglobin releases more oxygen; as the temperature declines, hemoglobin holds oxygen more tightly. Temperature effects are significant only in active tissues where large amounts of heat are being generated. For example, active skeletal muscles generate heat, and the heat warms blood that flows through these organs. As the blood warms, the hemoglobin

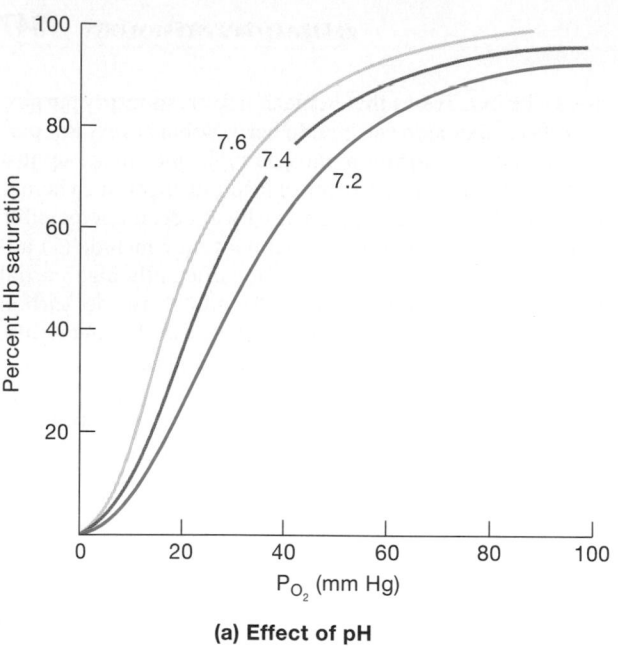

(a) Effect of pH

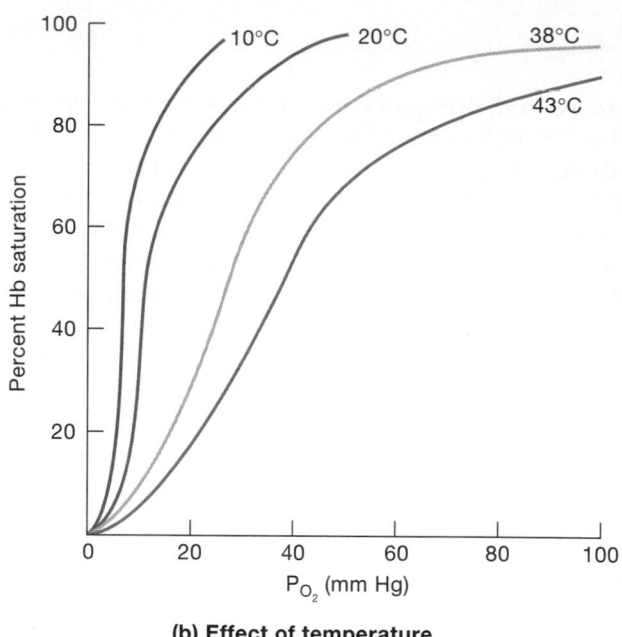

(b) Effect of temperature

•FIGURE 23-22 **Effects of pH and Temperature on Hemoglobin Saturation.** **(a)** When the pH drops below normal levels, more oxygen is released and the hemoglobin saturation curve shifts to the right. If the pH increases, the curve shifts to the left as less oxygen is released. **(b)** When the temperature rises, the saturation curve shifts to the right.

molecules release more oxygen than can be used by the active muscle fibers.

Hemoglobin and BPG. Red blood cells can produce ATP only through glycolysis, which, as we saw in Chapter 10, involves the formation of lactic acid. ∞ *[p. 299]* The metabolic pathways involved in glycolysis in an RBC also generate a compound called **2,3-bisphosphoglycerate** (bis-fos-fō-GLIS-e-rāt), or **BPG,** also known as *2,3-diphosphoglycerate,* or *DPG.* BPG is always present in the normal red blood cell and has a direct effect on oxygen binding and release. For any partial pressure of oxygen, the higher the concentration of BPG, the more oxygen will be released by the hemoglobin molecules.

The concentration of BPG can be increased by thyroid hormones, growth hormone, epinephrine, androgens, and a high blood pH. These factors improve oxygen delivery to the tissues, because when BPG levels are elevated, hemoglobin will release about 10 percent more oxygen at a given P_{O_2} than it would otherwise. BPG synthesis and the Bohr effect improve oxygen delivery when the pH changes: BPG levels rise when the pH increases, and the Bohr effect appears when the pH decreases.

BPG production decreases as RBCs age. Thus, the level of BPG can determine how long a blood bank can hold fresh whole blood. When BPG levels get too low, hemoglobin becomes firmly bound to the available oxygen. The blood is then useless for transfusions, because the RBCs will no longer release oxygen to peripheral tissues, even at a disastrously low P_{O_2}.

FETAL HEMOGLOBIN The RBCs of a developing fetus contain **fetal hemoglobin.** The structure of fetal

hemoglobin, which differs from that of adult hemoglobin, gives it a much higher affinity for oxygen. Therefore, at the same P_{O_2}, fetal hemoglobin will bind more oxygen than will adult hemoglobin (Figure 23-23•). This characteristic is important in transferring oxygen across the placenta.

A developing fetus obtains oxygen from the maternal bloodstream. The maternal blood at the placenta has a relatively low P_{O_2}, ranging from 35 to 50 mm Hg. If maternal blood arrives at the placenta with a P_{O_2} of 40 mm Hg, hemoglobin saturation will be roughly 75 percent. The fetal blood arriving at the placenta has a P_{O_2} close to 20 mm Hg. However, because fetal hemoglobin has a higher affinity for oxygen, it is still 58 percent saturated.

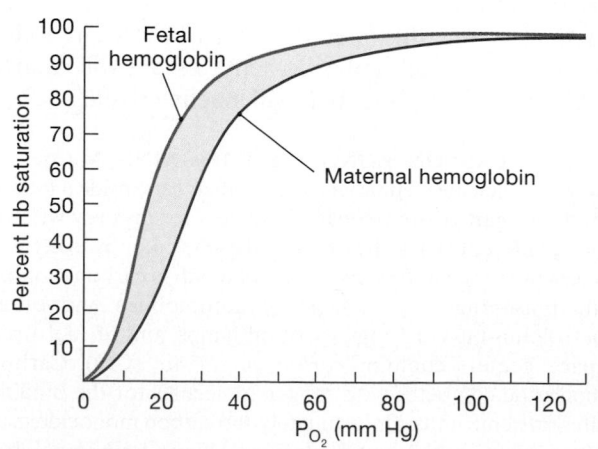

•FIGURE 23-23 **A Functional Comparison of Fetal and Maternal Hemoglobins**

As diffusion occurs between fetal and maternal blood, oxygen enters the fetal circulation until the P_{O_2} equilibrates at 30 mm Hg. At this P_{O_2} the maternal hemoglobin will be less than 60 percent saturated, but the fetal hemoglobin will be over 80 percent saturated.

When the fetal RBCs arrive in the peripheral tissues, the steep slope of the saturation curve for fetal hemoglobin means that it will release a large amount of oxygen in response to a very small change in P_{O_2}.

The functional difference between adult and fetal hemoglobin becomes apparent when you compare the uptake and delivery of oxygen in the adult and fetal circulations. Adult hemoglobin leaves the alveoli at a P_{O_2} of 100 mm Hg with 97.5 percent saturation, and it returns to the lungs with a P_{O_2} of 40 mm Hg and a 75 percent saturation. Thus it releases about 22 percent of its oxygen reserves over a 60 mm Hg drop in P_{O_2}. Fetal hemoglobin leaves the placenta at a P_{O_2} of 30 mm Hg with an 80 percent saturation and returns to the placenta with a P_{O_2} of 20 mm Hg and 58 percent saturation. It has also released 22 percent of its oxygen reserves but over a drop of just 10 mm Hg.

ADAPTATIONS TO HIGH ALTITUDE Atmospheric pressure decreases with increasing altitude, and so do the partial pressures of the component gases, including oxygen. People living in Denver or Mexico City function normally with alveolar oxygen pressures in the 80–90 mm Hg range. At higher elevations, the alveolar P_{O_2} continues to decline. At 3300 meters, an altitude familiar to many hikers and skiers, the alveolar P_{O_2} falls to around 60 mm Hg. Despite the low alveolar P_{O_2}, millions of people live and work at altitudes this high or higher. Important physiological adjustments include an increased respiratory rate, an increased heart rate, and an elevated hematocrit. Thus, even though the hemoglobin is not fully saturated, there is more of it in circulation, and the round-trip between the lungs and the peripheral tissues takes less time. These responses represent an excellent example of the functional interplay between the respiratory and cardiovascular systems. However, most of these adaptations take days to weeks to appear. As a result, athletes planning to compete in events held at high altitude must begin training well in advance.

Not everyone can tolerate high-altitude conditions. Roughly 20 percent of people who ascend to 2600 meters or higher experience symptoms of *altitude sickness,* or *mountain sickness.* Symptoms may include headache, disorientation, and fatal pulmonary or cerebral edema. AM *Mountain Sickness*

Carbon Dioxide Transport

Carbon dioxide is generated by aerobic metabolism in peripheral tissues. After entering the bloodstream, a CO_2 molecule may be (1) converted to a molecule of carbonic acid, (2) bound to the protein portion of hemoglobin molecules within RBCs, or (3) dissolved in the plasma (Figure 23-24•). All three are completely reversible reactions. We shall consider the events under way as blood enters peripheral tissues in which the P_{CO_2} is 45 mm Hg.

CARBONIC ACID FORMATION Most of the carbon dioxide absorbed by the blood (roughly 70 percent of the total) will be transported as molecules of carbonic acid. Carbon dioxide is converted to carbonic acid through the activity of the enzyme carbonic anhydrase within RBCs. The carbonic acid molecules immediately dissociate into a hydrogen ion and a bicarbonate ion. The entire reaction sequence can be summarized as

$$CO_2 + H_2O \overset{\text{carbonic}}{\underset{\text{anhydrase}}{\rightleftharpoons}} H_2CO_3 \rightleftharpoons H^+ + HCO_3^-$$

The conversion of CO_2 to H^+ and HCO_3^- occurs very rapidly, and it is completely reversible. Because the carbonic acid dissociates into bicarbonate and hydrogen ions, we can ignore the intermediate step and summarize the reaction as

$$CO_2 + H_2O \overset{\text{carbonic}}{\underset{\text{anhydrase}}{\rightleftharpoons}} H^+ + HCO_3^-$$

In peripheral capillaries this reaction proceeds vigorously, tying up large numbers of carbon dioxide molecules. The reaction continues as carbon dioxide diffuses out of the interstitial fluids.

The hydrogen ions and bicarbonate ions have different fates. Most of the hydrogen ions bind to hemoglobin molecules, forming Hb • H^+. The hemoglobin molecules thus function as buffers, tying up the released hydrogen ions before the ions leave the RBCs and affect the plasma pH. The bicarbonate ions move into the surrounding plasma with the aid of a countertransport

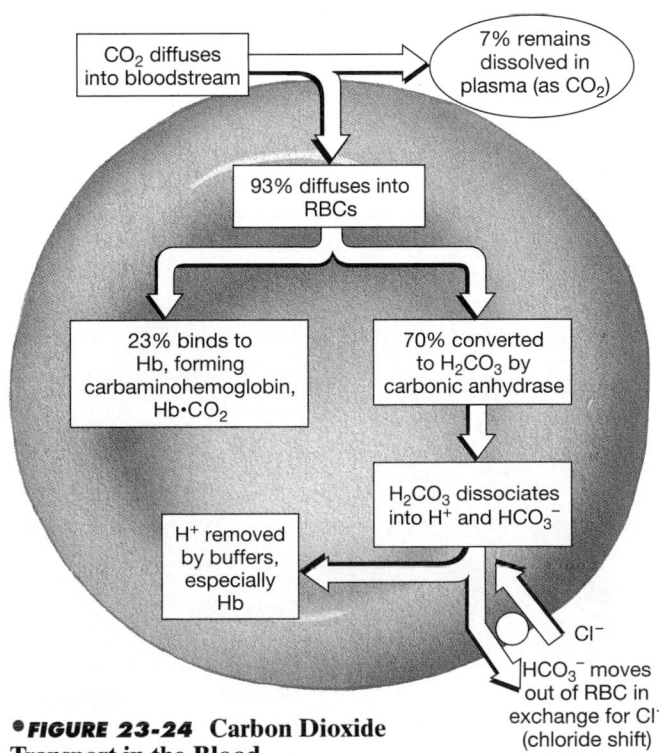

•**FIGURE 23-24** Carbon Dioxide Transport in the Blood

mechanism that exchanges intracellular bicarbonate ions for extracellular chloride ions. This exchange results in a mass movement of chloride ions into the RBCs, an event known as the **chloride shift.**

HEMOGLOBIN BINDING Roughly 23 percent of the carbon dioxide carried by your blood will be bound to the globular protein portions of the Hb molecules inside RBCs. These carbon dioxide molecules are attached to exposed amino groups ($-NH_2$) of the Hb molecules. The result is called **carbaminohemoglobin** (kar-ba-mē-nō-hē-mō-GLŌ-bin), $Hb \bullet CO_2$. This is a reversible reaction that is summarized as

$$CO_2 + HbNH_2 \rightleftharpoons HbNHCOOH$$

This reaction is abbreviated without the amino groups as

$$CO_2 + Hb \rightleftharpoons Hb \bullet CO_2$$

PLASMA TRANSPORT Plasma becomes saturated with carbon dioxide quite rapidly, and only about 7 percent of the carbon dioxide absorbed by peripheral capillaries is transported in the form of dissolved gas molecules. The rest is absorbed by the RBCs for conversion by carbonic anhydrase or storage as carbaminohemoglobin.

Gas Transport: A Summary

Figure 23-25● summarizes the transportation of oxygen and carbon dioxide in the respiratory and cardiovascular systems. This transport is dynamic, capable of varying its responses to meet changing circumstances. Some of the responses are automatic and result from the basic chemistry of the transport mechanisms. Other responses require coordinated adjustments in the activities of the cardiovascular and respiratory systems. We shall now consider these levels of control and regulation.

☑ As you exercise, hemoglobin releases more oxygen to your active skeletal muscles than it does when the muscles are at rest. Why?

☑ How would blockage of the trachea affect the blood pH?

CONTROL OF RESPIRATION

Peripheral cells are continuously absorbing oxygen from the interstitial fluids and generating carbon dioxide. Under normal conditions, the cellular rates of absorption and generation are matched by the capillary

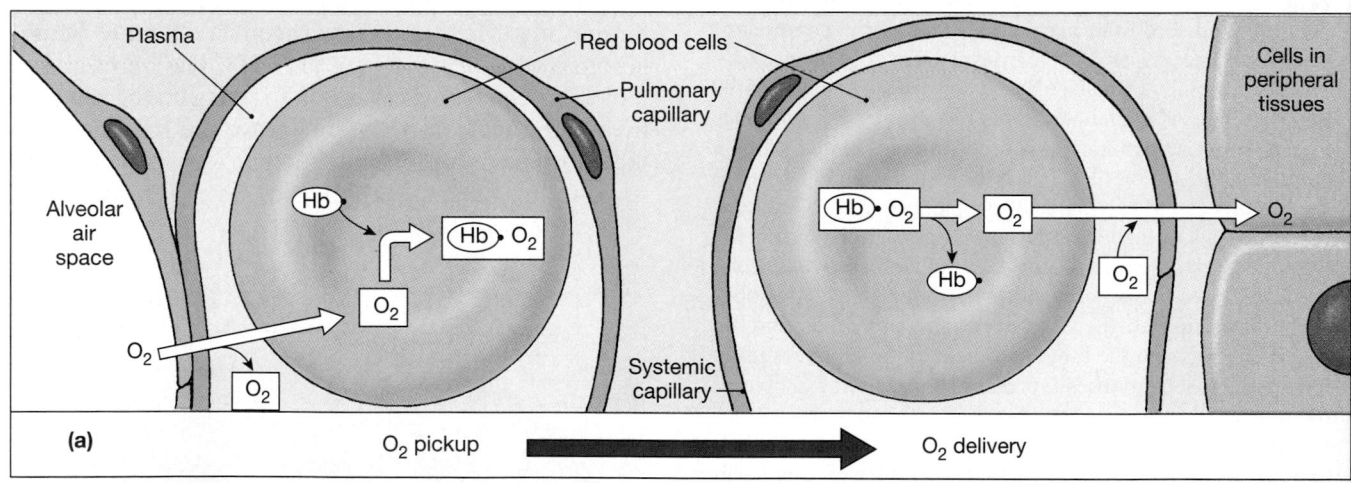

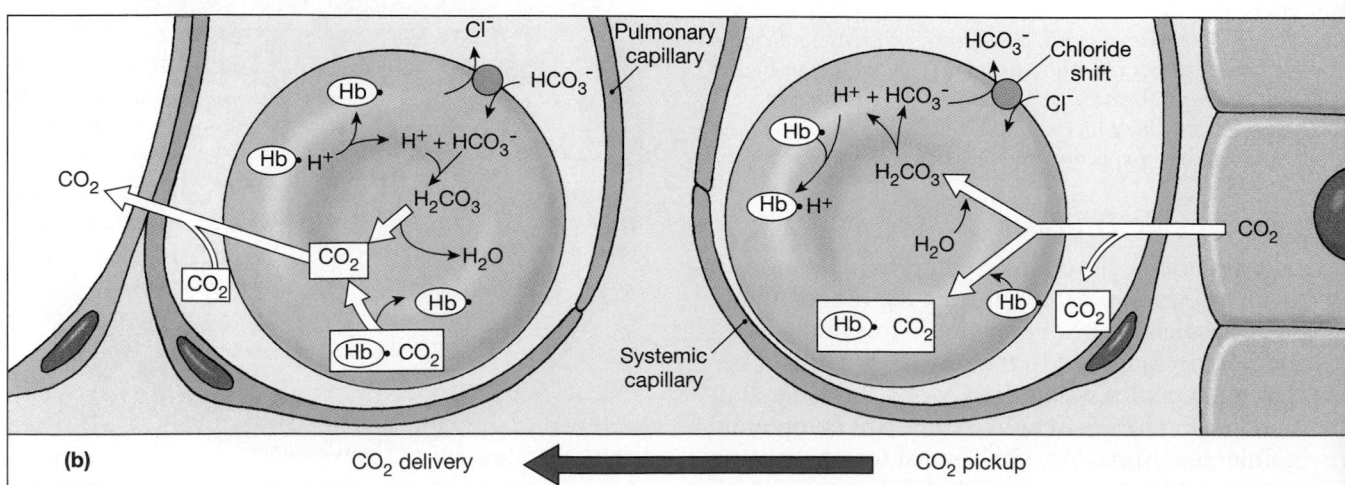

●FIGURE 23-25 **A Summary of the Primary Gas Transport Mechanisms.** **(a)** Oxygen transport. **(b)** Carbon dioxide transport.

rates of delivery and removal. These rates are identical to those of oxygen absorption and carbon dioxide excretion at the lungs. If these rates become unbalanced, homeostatic mechanisms intervene to restore equilibrium. Those mechanisms involve (1) changes in blood flow and oxygen delivery that are regulated at the local level and (2) variations in the depth and rate of respiration under the control of the brain's respiratory centers. The activities of the respiratory centers are coordinated with changes in cardiovascular function, such as variations in blood pressure and cardiac output.

Local Regulation of Gas Transport and Alveolar Function

The rate of oxygen delivery in each tissue and the efficiency of oxygen pickup at the lungs are largely regulated at the local level. For example, if a peripheral tissue becomes more active, the interstitial P_{O_2} falls and the P_{CO_2} rises. This change increases the difference between the partial pressures in the tissues and in the arriving blood, so more oxygen is delivered and more carbon dioxide is carried away. In addition, the rising P_{CO_2} levels cause the relaxation of smooth muscles in the walls of arterioles and capillaries in the area, increasing local blood flow. ∞ *[p. 728]*

Local factors regulate blood flow, or *perfusion*, and airflow, or *ventilation*, over a wide range of conditions and activity levels.

Changes in Lung Perfusion

As blood flows toward the alveolar capillaries, it is directed toward lobules in which the P_{O_2} is relatively high. This movement occurs because alveolar capillaries constrict when the local P_{O_2} is low. (We noted this response, the opposite of that seen in peripheral tissues, in Chapter 21.) ∞ *[p. 732]* Such a shift in circulation tends to eliminate temporary differences in the oxygen and carbon dioxide content of alveoli, lobules, or groups of lobules that could otherwise result from minor variations in local blood flow.

Changes in Alveolar Ventilation

Smooth muscles in the walls of bronchioles are sensitive to the P_{CO_2} of the air they contain. When the P_{CO_2} goes up, the bronchioles increase in diameter (bronchodilation). When the P_{CO_2} declines, the bronchioles constrict (bronchoconstriction). Airflow is therefore directed to lobules in which the P_{CO_2} is high. Because the carbon dioxide is obtained from the blood, these lobules are actively engaged in gas exchange. This response is especially important because the improvement of airflow to functional alveoli can at least partially compensate for damage to pulmonary lobules.

By directing blood flow to alveoli with low levels of carbon dioxide and improving airflow to alveoli with high levels of carbon dioxide, local adjustments improve the efficiency of gas transport. When activity levels increase and the demand for oxygen rises, the cardiac output and respiratory rates increase under neural control, but the adjustments in alveolar blood flow and bronchiolar diameter occur automatically.

EMPHYSEMA Emphysema (em-fi-SĒ-muh) is a chronic, progressive condition characterized by shortness of breath and an inability to tolerate physical exertion. The underlying problem is the destruction of alveolar surfaces and inadequate surface area for oxygen and carbon dioxide exchange. In essence, respiratory bronchioles and alveoli are functionally eliminated. The alveoli gradually expand, their walls become incomplete, and adjacent alveoli merge to form larger air spaces supported by fibrous tissue without alveolar capillary networks. As connective tissues are eliminated, compliance increases; air moves into and out of the lungs more easily than before. However, the loss of respiratory surface area restricts oxygen absorption, so the individual becomes short of breath.

Emphysema has been linked to the inhalation of air that contains fine particulate matter or toxic vapors, such as those found in cigarette smoke. There are also genetic factors that predispose individuals to this condition. Some degree of emphysema is a normal consequence of aging, and an estimated 66 percent of adult males and 25 percent of adult females have detectable areas of emphysema in their lungs. [AM] *Bronchitis, Emphysema, and COPD*

The Respiratory Centers of the Brain

Respiratory control has both involuntary and voluntary components. Your brain's involuntary centers regulate the activities of the respiratory muscles and control the respiratory minute volume by adjusting the frequency and depth of pulmonary ventilation. They do so in response to sensory information arriving from your lungs and other portions of the respiratory tract as well as from a variety of other sites.

The voluntary control of respiration reflects activity in the cerebral cortex that affects either the output of the respiratory centers in the medulla oblongata and pons or motor neurons in the spinal cord that control respiratory muscles. The **respiratory centers** are three pairs of nuclei in the reticular formation of the medulla (the *respiratory rhythmicity centers*) and pons (the *apneustic centers* and *pneumotaxic centers*). The motor neurons in the spinal cord are generally controlled by *respiratory reflexes*, but they can also be controlled voluntarily through commands delivered by the pyramidal system.

Respiratory Centers in the Medulla Oblongata

We introduced the *respiratory rhythmicity centers* of the medulla oblongata in Chapter 14. ∞ *[p. 473]* These paired centers set the pace for respiration. Each center can be subdivided into a **dorsal respiratory group (DRG)** and a **ventral respiratory group (VRG).** The DRG's *in-*

spiratory center contains neurons that control lower motor neurons innervating the external intercostal muscles and the diaphragm. This group functions in every respiratory cycle, whether quiet or forced. The VRG functions only during forced respiration. It includes neurons that innervate lower motor neurons controlling accessory respiratory muscles involved in active exhalation (an *expiratory center*) and maximal inhalation (an *inspiratory center*).

There is reciprocal inhibition between the neurons involved with inhalation and exhalation. When the inspiratory neurons are active, the expiratory neurons are inhibited, and vice versa. The pattern of interaction between these groups differs for quiet and forced respirations. During quiet respiration (Figure 23-26a●):

- Activity in the DRG increases over a period of about 2 seconds, providing stimulation to the inspiratory muscles. Over this period, inspiration occurs.
- After 2 seconds, the DRG neurons become inactive. They remain quiet for the next 3 seconds and allow the inspiratory muscles to relax. Over this period, passive expiration occurs.

During forced respiration (Figure 23-26b●):

- As the level of activity in the DRG increases, it in some way stimulates neurons of the VRG that activate the accessory muscles involved in inspiration.
- At the end of each inspiration, active expiration occurs as the neurons of the expiratory center stimulate the appropriate accessory muscles.

The basic pattern of respiration thus reflects a cyclic interaction between the DRG and VRG. The pace of this interaction is thought to be established by pacemaker cells that spontaneously undergo rhythmic patterns of activity. Attempts to locate the pacemaker, however, have been unsuccessful.

Central nervous system stimulants, such as amphetamines or even caffeine, will increase your respiratory rate by facilitating the pacemaker neurons and respiratory centers. These actions can be opposed by CNS depressants, such as barbiturates or opiates.

The Apneustic and Pneumotaxic Centers

The **apneustic** (ap-NOO-stik) **centers** and the **pneumotaxic** (noo-mō-TAKS-ik) **centers** of the pons are paired nucei that adjust the output of the respiratory rhythmicity centers. Their activities adjust the respiratory rate and the depth of respiration in response to sensory stimuli or input from other centers in the brain. Each apneustic center provides continuous stimulation to the DRG on that side of the brain stem. During quiet respiration, stimulation from the apneustic center helps increase the intensity of inspiration over the following 2 seconds. Under normal conditions, after 2 seconds the apneustic center is inhibited by signals from the pneumotaxic center on that side. During forced respirations, the apneustic centers also respond to sensory input from

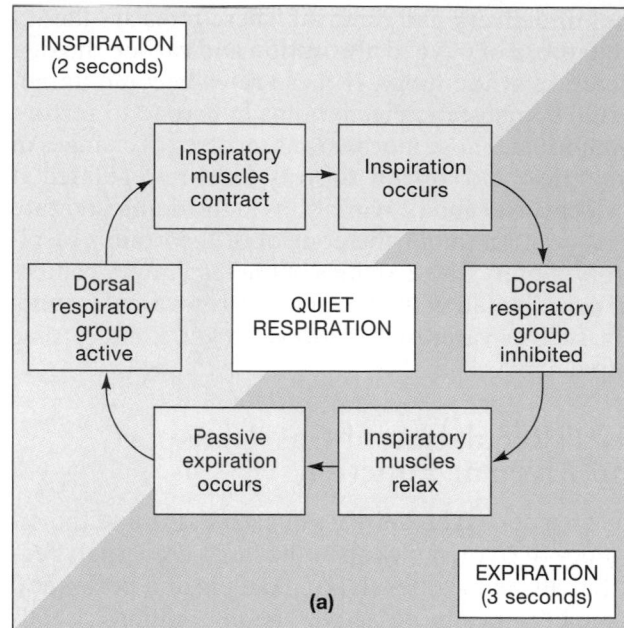

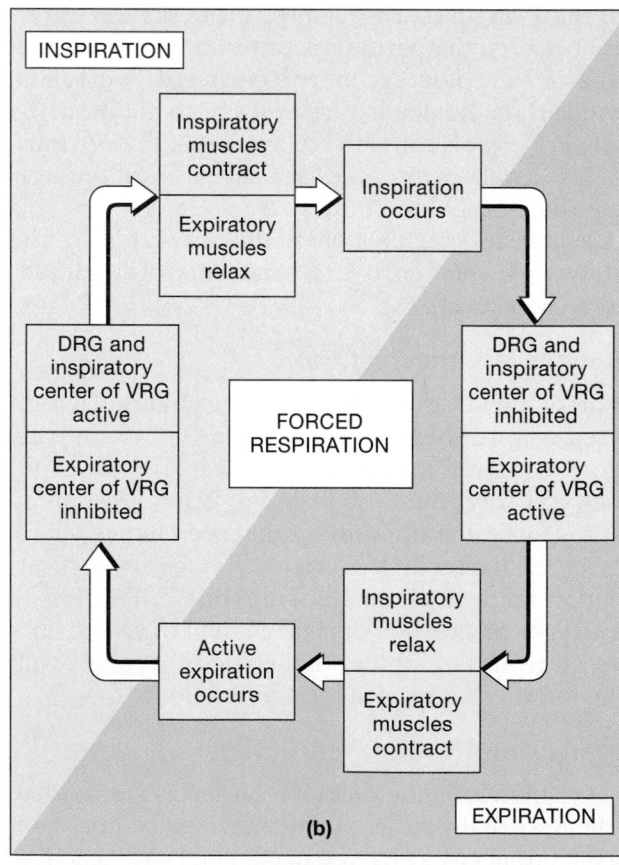

●*FIGURE 23-26* **Basic Regulatory Patterns.** **(a)** Quiet respiration. **(b)** Forced respiration.

the vagus nerves regarding the amount of lung inflation.

The pneumotaxic centers inhibit the apneustic centers and promote passive or active exhalation. Centers in the hypothalamus and cerebrum can alter the activity of the pneumotaxic centers and change the respiratory rate and depth. However, essentially normal respiratory cycles continue even if the brain stem above the pons has been severely damaged. If the inhibitory output of the pneumotaxic centers is cut off by a stroke or other damage to the brain stem and if sensory innervation from the lungs is eliminated by cutting the vagus nerves, the person inhales to maximum capacity and maintains that state for 10–20 seconds at a time. Intervening exhalations are brief, and little pulmonary ventilation occurs.

The CNS regions involved with respiratory control are diagrammed in Figure 23-27•. Interactions between the DRG and VRG establish the basic pace and depth of respiration. The pneumotaxic centers modify that pace; an increase in pneumotaxic output quickens the pace of respiration by shortening the duration of each inhalation. A decrease in pneumotaxic output slows the respiratory pace but increases the depth of respiration, because the apneustic centers are more active.

Respiratory Reflexes

The activities of the respiratory centers are modified by sensory information from:

1. *Chemoreceptors sensitive to the P_{CO_2}, pH, and/or P_{O_2} of the blood or CSF.*

2. *Changes in blood pressure in the aorta or carotid sinuses.*

3. *Stretch receptors that respond to changes in the volume of the lungs.*

4. *Irritating physical or chemical stimuli in the nasal cavity, larynx, or bronchial tree.*

5. *Other sensations,* including pain, changes in body temperature, and abnormal visceral sensations.

Information from these receptors alters the pattern of respiration. The induced changes have been called *respiratory reflexes.*

 SIDS Sudden infant death syndrome (SIDS), also known as *crib death*, kills an estimated 10,000 infants each year in the United States alone. Most crib deaths occur between midnight and 9:00 A.M., in the late fall or winter, and involve infants 2–4 months old. Eyewitness accounts indicate that the sleeping infant suddenly stops breathing, turns blue, and relaxes. Genetic factors appear to be involved, but controversy remains as to the relative importance of other factors, such as laryngeal spasms, cardiac arrhythmias, upper respiratory tract infections, viral infections, and CNS malfunctions. The age at the time of death corresponds with a period when the pacemaker complex and respiratory centers are establishing connections with other portions of the brain. It has recently been proposed that SIDS results from a problem in the interconnection process that disrupts the reflexive respiratory pattern.

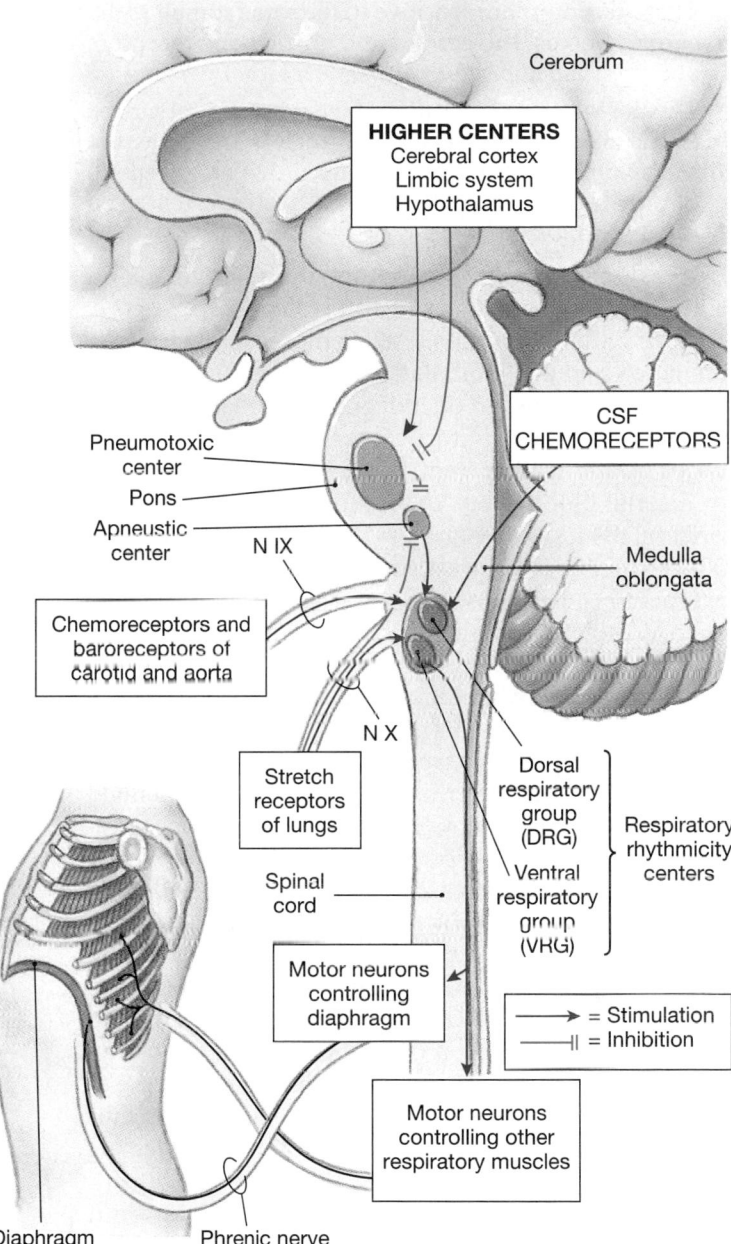

•**FIGURE 23-27** **Respiratory Centers and Reflex Controls.** The positions of the major respiratory centers and other factors important to the reflex control of respiration. Pathways for conscious control over respiratory muscles are not shown.

The Chemoreceptor Reflexes

The respiratory centers are strongly influenced by chemoreceptor inputs from the ninth and tenth cranial nerves and from receptors that monitor CSF composition:

- The glossopharyngeal nerve (N IX) carries chemoceptive information from the carotid bodies, adjacent to the carotid sinus. ∞ *[p. 480]* The carotid bodies are stimulated by a decrease in the pH or P_{O_2} of the blood. Because changes in P_{CO_2} affect pH, these receptors are indirectly stimulated by a rise in the P_{CO_2}.

- The vagus nerve (N X) monitors chemoreceptors in the aortic bodies, near the aortic arch. ∞ *[p. 481]*

These receptors are sensitive to the same stimuli as the carotid bodies. The carotid and aortic body receptors are often called *peripheral chemoreceptors.*

■ Chemoreceptors are located on the ventrolateral surface of the medulla oblongata in a region known as the *chemosensitive area.* The neurons in that area respond only to the P_{CO_2} and pH of the CSF and are often called *central chemoreceptors.*

We discussed chemoreceptors and their effects on cardiovascular function in Chapters 17 and 21. ∞ *[pp. 545, 731]* Stimulation of these chemoreceptors leads to an increase in the depth and rate of respiration. Under normal conditions, a reduction in the arterial P_{O_2} has little effect on the respiratory centers until the arterial P_{O_2} drops by about 40 percent, below 60 mm Hg. If the P_{O_2} of arterial blood drops to 40 mm Hg, the level in peripheral tissues, the respiratory rate will increase by only 50–70 percent. In contrast, a rise of just 10 percent in the arterial P_{CO_2} will cause the respiratory rate to double, even if the P_{O_2} remains completely normal. Carbon dioxide levels are therefore responsible for regulating respiratory activity under normal conditions.

Although the receptors monitoring carbon dioxide levels are more sensitive, oxygen and carbon dioxide receptors work together in a crisis. Carbon dioxide is generated during oxygen consumption, so when oxygen concentrations are falling rapidly, carbon dioxide levels are usually increasing. This cooperation breaks down only under unusual circumstances. For example, you can hold your breath longer if you prepare by taking deep, full breaths, but the practice is very dangerous. The danger lies in the fact that your increased ability is due not to extra oxygen but to the loss of carbon dioxide. If the P_{CO_2} is driven down far enough, breath-holding ability may increase to the point at which the individual becomes unconscious from oxygen starvation in the brain without ever feeling the urge to breathe. AM *Shallow Water Blackout*

The chemoreceptors are subject to adaptation if the P_{O_2} or P_{CO_2} remains abnormal for an extended period; this adaption can complicate the treatment of chronic respiratory disorders. For example, if the P_{O_2} remains low for an extended period and the P_{CO_2} remains chronically elevated, the chemoreceptors will reset to those values. They will accept those values as normal and oppose any attempts to return the partial pressures to the proper range. AM *Chemoreceptor Accommodation and Opposition*

Because the chemoreceptors monitoring carbon dioxide levels are also sensitive to pH, any condition altering the pH of the blood or CSF will affect respiratory performance. For example, the rise in lactic acid levels after exercise causes a drop in pH that helps stimulate respiratory activity.

HYPERCAPNIA AND HYPOCAPNIA   **Hypercapnia** is an increase in the P_{CO_2} of arterial blood. Figure 23-28● diagrams the central response to hypercapnia, which

is triggered by the stimulation of chemoreceptors in the carotid and aortic bodies and reinforced by stimulation of CNS chemoreceptors. Carbon dioxide crosses the blood–brain barrier quite rapidly, so a rise in arterial P_{CO_2} almost immediately elevates CSF CO_2 levels, lowering the pH of the CSF and stimulating the chemoreceptive neurons of the medulla oblongata.

These receptors stimulate your respiratory centers to increase the rate and depth of respiration. Your breathing becomes more rapid, and more air moves into and out of your lungs with each breath. Because more air moves into and out of the alveoli each minute, alveolar concentrations of carbon dioxide decline, accelerating the diffusion of carbon dioxide from the alveolar capillaries. Thus homeostasis is restored.

If the rate and depth of respiration exceed the demands for oxygen delivery and carbon dioxide removal, the condition of **hyperventilation** exists. Hyperventilation will gradually lead to **hypocapnia,** an abnormally low P_{CO_2}. If the arterial P_{CO_2} drops below normal levels (Figure 23-28●), chemoreceptor activity decreases and the respiratory rate falls. This situation continues until the P_{CO_2} returns to normal and homeostasis is restored.

The most common cause of hypercapnia is hypoventilation. In **hypoventilation,** the respiratory rate remains abnormally low and is insufficient to meet the demands for normal oxygen delivery and carbon dioxide removal. Carbon dioxide then accumulates in the blood.

The Baroreceptor Reflexes

We described the effects of carotid and aortic baroreceptor stimulation on systemic blood pressure in Chapter 21. ∞ *[pp. 729–731]* The output from these baroreceptors also affects the respiratory centers. When blood pressure falls, the respiratory rate increases; when blood pressure rises, the respiratory rate declines. This adjustment results from stimulation or inhibition of the respiratory centers by sensory fibers in the glossopharyngeal (IX) and vagus (X) nerves.

The Hering–Breuer Reflexes

The **Hering–Breuer reflexes** are named after the physiologists who described them in 1865. The sensory information from these reflexes is distributed to the apneustic centers and the VRG. The Hering–Breuer reflexes are not involved in normal quiet breathing or in tidal volumes under 1000 ml. There are two such reflexes—the *inflation reflex* and the *deflation reflex:*

1. The **inflation reflex** prevents overexpansion of the lungs during forced breathing. The stretch receptors are located in the smooth muscle tissue around bronchioles and are stimulated by lung expansion. Sensory fibers leaving the stretch receptors of each lung reach the respiratory rhythmicity center on that side over the vagus nerve. As

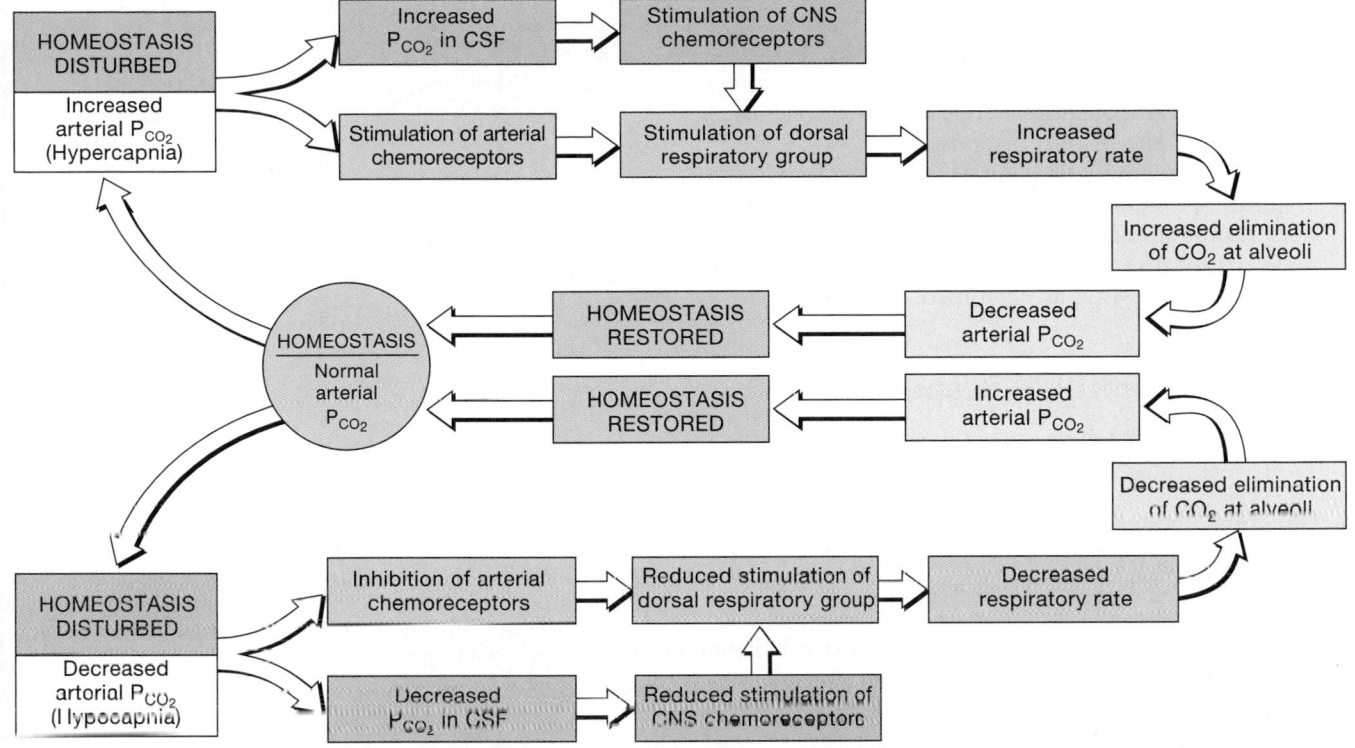

•FIGURE 23-28 Chemoreceptor Response to Changes in P$_{CO_2}$. A rise in arterial P$_{CO_2}$ (top) stimulates chemoreceptors that accelerate breathing cycles at the inspiratory center. This change increases the respiratory rate, encourages CO_2 loss at the lungs, and lowers arterial P$_{CO_2}$. A drop in arterial P$_{CO_2}$ (bottom) inhibits these chemoreceptors. In the absence of stimulation, the rate of respiration decreases, slowing the rate of CO_2 loss and elevating arterial P$_{CO_2}$.

lung volume increases, the dorsal respiratory group is gradually inhibited, and the expiratory center of the VRG is stimulated. Inspiration then stops as the lungs near maximum volume, and active expiration then begins.

2. The **deflation reflex** inhibits the expiratory centers and stimulates the inspiratory centers when the lungs are deflating. The receptors, which are distinct from those of the inflation reflex, are located in the alveolar wall near the alveolar capillary network. The smaller the volume of the lungs, the greater the degree of inhibition, until expiration stops and inspiration begins. This reflex normally functions only during forced expiration, when both the inspiratory and expiratory centers are active.

Protective Reflexes

Protective reflexes operate when you are exposed to toxic vapors, chemical irritants, or mechanical stimulation of the respiratory tract. The receptors involved are located within the epithelium of the respiratory tract. Examples of protective reflexes include sneezing, coughing, and *laryngeal spasms*.

Sneezing is triggered by irritation of the wall of your nasal cavity. Coughing is triggered by irritation of your larynx, trachea, or bronchi. Both reflexes involve a temporary period of **apnea** (AP-nē-uh), in which respiration is suspended. They are usually followed by a forceful expulsion of air intended to remove the offending stimulus. The glottis is forcibly closed while the lungs are still relatively full. The abdominal and internal intercostal muscles then contract suddenly, creating pressures that will blast air out of your respiratory passageways when the glottis reopens. Air leaving the larynx may travel at 160 kph, carrying mucus, foreign particles, and irritating gases out of the respiratory tract via the nose or mouth.

Laryngeal spasms result from the entry of chemical irritants, foreign objects, or fluids into the area around the glottis. This reflex generally closes your airway temporarily. A very strong stimulus, such as a toxic gas, could close the glottis so powerfully that you could lose consciousness and die without taking another breath. Fine chicken or fish bones that pierce the laryngeal walls may also stimulate laryngeal spasms, swelling, or both, restricting the airway.

Other Sensations That Affect Respiratory Function

Several other sensory stimuli can affect the activities of the respiratory centers. In some cases, the mechanism involved is not known. Examples include the following:

- Sudden pain or immersion in cold water can produce a temporary apnea. Chronic pain stimulates the sympathetic division of the ANS, leading to an increase in the respiratory rate.
- Vomiting and swallowing involve automatic adjustments in respiratory activity to prevent entrance of foreign objects into the trachea.

- Fever or an increase in body temperature due to exertion or overheating will cause an increase in the respiratory rate. A reduction in body temperature leads to a decrease in the respiratory rate.
- Curiously, stretching the anal sphincter stimulates the respiratory centers and increases the rate of respiration. Although this reflex is occasionally used to stimulate respiration in an emergency situation, it is not clear which pathways are involved.

Voluntary Control of Respiration

Activity of your cerebral cortex has an indirect effect on your respiratory centers. For example:

- Conscious thought processes tied to strong emotions, such as rage or fear, affect the respiratory rate by stimulating centers in the hypothalamus.
- Emotional states can affect respiration through activation of the sympathetic or parasympathetic division of the ANS. Sympathetic activation causes bronchodilation and increases the respiratory rate, and parasympathetic stimulation has the opposite effects.
- Anticipation of strenuous exercise can trigger an automatic increase in the respiratory rate, along with increased cardiac output, by sympathetic stimulation.

We also have conscious control over our respiratory activities. This control may bypass the respiratory centers completely, using pyramidal fibers that innervate the same lower motor neurons controlled by the DRG and VRG. This control mechanism is an essential part of speaking, singing, and swimming, when respiratory activities must be precisely timed. Higher centers may also have an inhibitory effect on the apneustic centers and on the DRG and VRG, and this effect is important when you hold your breath.

There are limits to our abilities to override the respiratory centers, however. The chemoreceptor reflexes are extremely powerful respiratory stimulators, and they cannot be consciously suppressed. For example, you can hold your breath before you dive into a swimming pool and thereby keep yourself from inhaling water. But you cannot kill yourself by holding your breath "till you turn blue." Once the P_{CO_2} rises to critical levels, you will be forced to take a breath.

⊠AGING AND THE RESPIRATORY SYSTEM

Many factors interact to reduce the efficiency of the respiratory system in elderly individuals. Three examples are particularly noteworthy:

1. As age increases, elastic tissue deteriorates throughout the body. This deterioration reduces the compliance of the lungs, lowering the vital capacity.

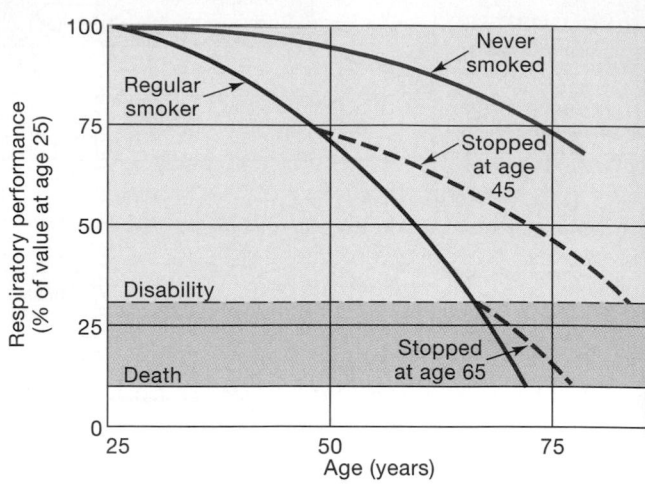

•FIGURE 23-29 Aging and the Decline in Respiratory Performance. Graph comparing the relative respiratory performance of individuals who have never smoked, individuals who quit smoking at age 45, individuals who quit smoking at age 65, and lifelong smokers.

2. Chest movements are restricted by arthritic changes in the rib articulations and by decreased flexibility at the costal cartilages. Along with the changes noted in item 1, the stiffening and reduction in chest movement effectively limit the respiratory minute volume. This restriction contributes to the reduction in exercise performance and capabilities with increasing age.

3. Some degree of emphysema is normally found in individuals over age 50. However, the extent varies widely with the lifetime exposure to cigarette smoke and other respiratory irritants. Figure 23-29• compares the respiratory performance of individuals who have never smoked with individuals who have smoked for varying periods of time. The message is quite clear: Although some decrease in respiratory performance is inevitable, you can prevent serious respiratory deterioration by stopping smoking or never starting.

☑ What effect would exciting the pneumotaxic centers have on respiration?

☑ Are peripheral chemoreceptors as sensitive to the levels of carbon dioxide as they are to the levels of oxygen?

☑ Johnny is angry with his mother, so he tells her that he will hold his breath until he turns blue and dies. Should Johnny's mother worry?

INTEGRATION WITH OTHER SYSTEMS

The respiratory system has extensive anatomical connections to the cardiovascular system, and we have noted examples throughout this chapter. The functional ties are just as extensive.

The goal of respiratory activity is to maintain homeostatic oxygen and carbon dioxide levels in peripheral tissues. Changes in respiratory activity alone are seldom sufficient to accomplish this. For example, during exercise, when tissue oxygen demands are high, alveolar ventilation increases through stimulation of the respiratory muscles. This response, which actually increases the demand for oxygen, serves no purpose unless cardiac output accelerates simultaneously, delivering more oxygen to active tissues.

We shall consider four examples of the integration between the respiratory and cardiovascular systems:

1. At the local level, changes in lung perfusion in response to variations in alveolar P_{O_2} improve the efficiency of gas exchange within or among lobules.

2. Chemoreceptor stimulation not only increases respiratory drive, it also causes an elevation in blood pressure, through stimulation of the vasomotor centers, and increased cardiac output. We noted these responses in Chapter 21. *[pp. 731–732]*

3. The stimulation of baroreceptors in the lungs has secondary effects on cardiovascular function. For example, the stimulation of airway stretch receptors not only triggers the inflation reflex but also increases the heart rate. Thus, as the lungs fill, cardiac output rises, and more blood flows through the alveolar capillaries.

4. Stimulation of baroreceptors in systemic blood vessels has secondary effects on respiratory function. When systemic blood pressure falls, stimulating carotid and aortic baroreceptors, the respiratory rate increases; when systemic blood pressure rises, the respiratory rate declines.

The respiratory system is functionally linked to all other systems as well. Figure 23-30● details these interrelationships.

LUNG CANCER   Lung cancer, or *pleuropulmonary neoplasm,* is an aggressive class of malignancies originating in the bronchial passageways or alveoli. These cancers affect the epithelial cells that line conducting passageways, mucous glands, or alveoli. Symptoms generally do not appear until the condition has progressed to the point at which the tumor masses are restricting airflow or compressing adjacent mediastinal structures. Chest pain, shortness of breath, a cough or wheeze, and weight loss are common symptoms. Treatment programs vary depending on the cellular organization of the tumor and whether metastasis (cancer cell migration) has occurred, but surgery, radiation exposure, or chemotherapy may be involved.

Deaths from lung cancer were rare at the turn of the twentieth century, but there were 29,000 in 1956, 105,000 in 1978, and 160,400 in 1997 in the United States. These figures continue to rise, with the number of diagnosed cases doubling every 15 years. Each year, 22 percent of new cancers detected are lung cancers, and in 1997 98,300 men and 79,800 women were diagnosed with this condition. The rate is increasing markedly among women; the lung cancer rates in 1989 were 101,000 men and 54,000 women. For additional information about the diagnosis and treatment of lung cancer, see the *Applications Manual.* [AM] *Lung Cancer*

SELECTED CLINICAL TERMINOLOGY

Terms Discussed in This Chapter

anoxia (a-NOKS-ē-uh): A condition of tissue oxygen starvation caused by (1) circulatory blockage, (2) respiratory problems, or (3) cardiovascular problems. *(p. 834)*

asthma (AZ-muh): An acute respiratory disorder characterized by unusually sensitive, irritable conducting passageways. *(p. 827 and AM)*

atelectasis (at-e-LEK-ta-sis): A collapsed lung. *(p. 838)*

bronchitis (brong-KĪ-tis): An inflammation of the bronchial lining. *(p. 826 and AM)*

bronchodilation (brong-kō-dī-LĀ-shun): Enlargement of the respiratory passageways. *(p. 827)*

bronchography (brong-KOG-ra-fē): A procedure in which radiopaque materials are introduced into the airways to improve X-ray imaging of the bronchial tree. *(p. 826)*

bronchoscope: A fiber-optic bundle small enough to be inserted into the trachea and finer airways; the procedure is called *bronchoscopy. (p. 826)*

cardiopulmonary resuscitation (CPR): The application of cycles of compression to the rib cage and mouth-to-mouth breathing to maintain cardiovascular and respiratory function. *(p. 839 and AM)*

cystic fibrosis (CF): A relatively common lethal inherited disease caused by an abnormal membrane channel protein; a major symptom is that mucous secretions become too thick to be transported easily, leading to respiratory problems. *(p. 817 and AM)*

decompression sickness: The bends, a condition caused by a drop in atmospheric pressure and the resulting formation of nitrogen gas bubbles in body fluids, tissues, and organs. *(p. 844 and AM)*

emphysema (em-fi-SĒ-muh): A chronic, progressive condition characterized by shortness of breath and an inability to tolerate physical exertion. *(p. 851 and AM)*

epistaxis (ep-i-STAK-sis): A nosebleed. *(p. 818)*

Heimlich (HĪM-lik) **maneuver,** or *abdominal thrust*: Compression applied to the abdomen just inferior to the diaphragm, to force air out of the lungs and clear a blocked trachea or larynx. *(p. 824)*

hypercapnia (hī-per-KAP-nē-uh): An increase in the P_{CO_2} of arterial blood. *(p. 854)*

hypocapnia: An abnormally low arterial P_{CO_2}. *(p. 854)*

hypoxia (hī-POKS-ē-uh): A condition of reduced tissue P_{O_2}. *(p. 834)*

lung cancer (*pleuropulmonary neoplasm*): A class of aggressive malignancies originating in the bronchial passageways or alveoli. *(p. 857 and AM)*

mountain sickness: An acute disorder resulting from CNS effects of the low gas partial pressures that occur at high altitudes. *(p. 849 and AM)*

pleurisy: Inflammation of the pleurae and secretion of excess amounts of pleural fluid. *(p. 834)*

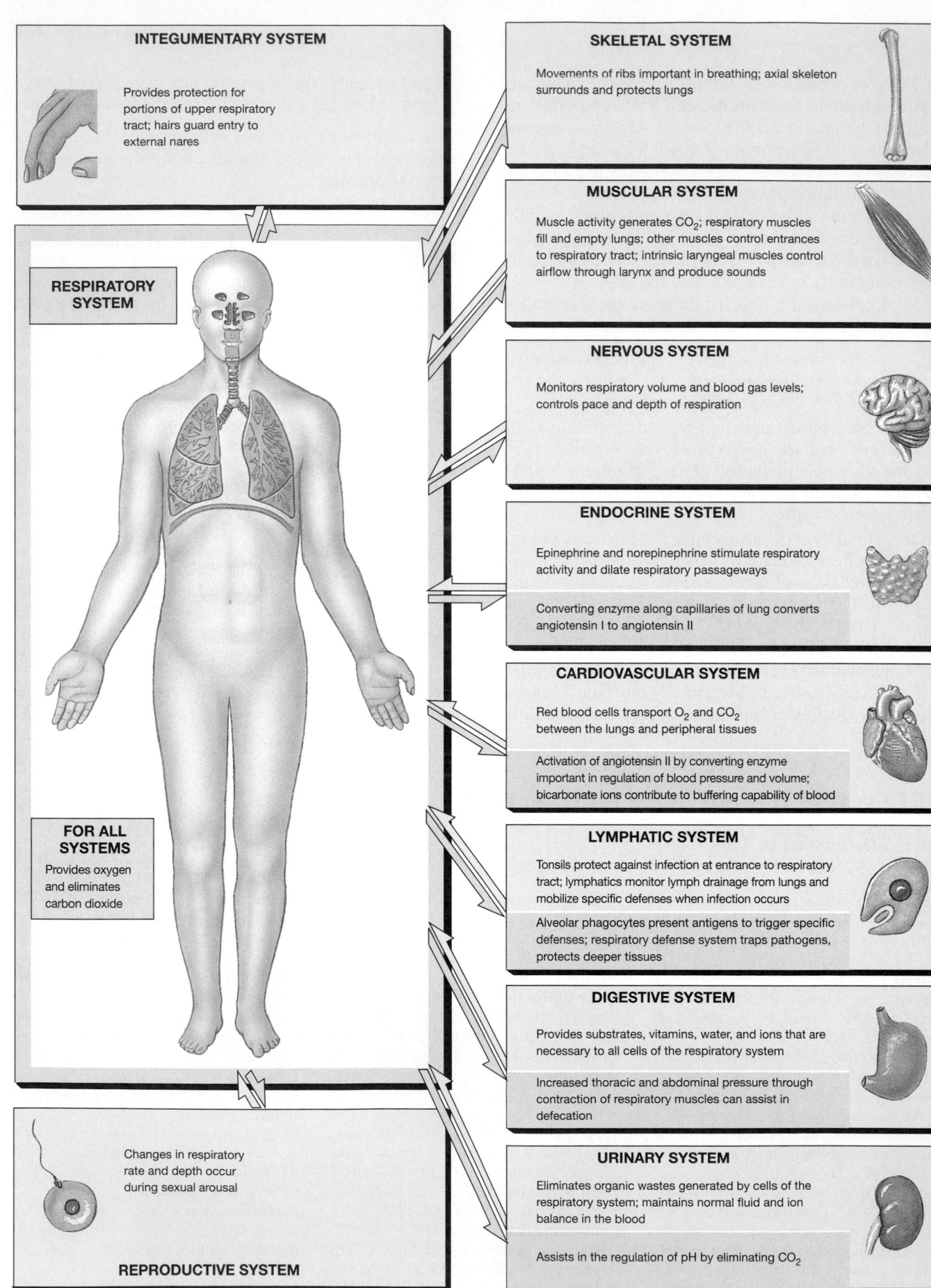

INTEGUMENTARY SYSTEM

Provides protection for portions of upper respiratory tract; hairs guard entry to external nares

RESPIRATORY SYSTEM

FOR ALL SYSTEMS

Provides oxygen and eliminates carbon dioxide

SKELETAL SYSTEM

Movements of ribs important in breathing; axial skeleton surrounds and protects lungs

MUSCULAR SYSTEM

Muscle activity generates CO_2; respiratory muscles fill and empty lungs; other muscles control entrances to respiratory tract; intrinsic laryngeal muscles control airflow through larynx and produce sounds

NERVOUS SYSTEM

Monitors respiratory volume and blood gas levels; controls pace and depth of respiration

ENDOCRINE SYSTEM

Epinephrine and norepinephrine stimulate respiratory activity and dilate respiratory passageways

Converting enzyme along capillaries of lung converts angiotensin I to angiotensin II

CARDIOVASCULAR SYSTEM

Red blood cells transport O_2 and CO_2 between the lungs and peripheral tissues

Activation of angiotensin II by converting enzyme important in regulation of blood pressure and volume; bicarbonate ions contribute to buffering capability of blood

LYMPHATIC SYSTEM

Tonsils protect against infection at entrance to respiratory tract; lymphatics monitor lymph drainage from lungs and mobilize specific defenses when infection occurs

Alveolar phagocytes present antigens to trigger specific defenses; respiratory defense system traps pathogens, protects deeper tissues

DIGESTIVE SYSTEM

Provides substrates, vitamins, water, and ions that are necessary to all cells of the respiratory system

Increased thoracic and abdominal pressure through contraction of respiratory muscles can assist in defecation

URINARY SYSTEM

Eliminates organic wastes generated by cells of the respiratory system; maintains normal fluid and ion balance in the blood

Assists in the regulation of pH by eliminating CO_2

Changes in respiratory rate and depth occur during sexual arousal

REPRODUCTIVE SYSTEM

● **FIGURE 23-30** **Functional Relationships between the Respiratory System and Other Systems**

pneumonia (noo-MŌ-nē-uh): A respiratory disorder characterized by fluid leakage into the alveoli and/or swelling and constriction of the respiratory bronchioles. *(p. 831)*
pneumothorax (noo-mō-THO-raks): The entry of air into the pleural cavity. *(p. 838)*
pulmonary embolism: A blockage of a branch of a pulmonary artery, with interruption of blood flow to a group of lobules and/or alveoli. *(p. 831)*

respiratory distress syndrome: A condition resulting from inadequate surfactant production and associated alveolar collapse. *(p. 831 and AM)*
sudden infant death syndrome (SIDS): Crib death, the death of an infant due to respiratory arrest; the cause remains uncertain. *(p. 853)*

tracheostomy (trā-kē-OS-to-mē): Insertion of a tube directly into the trachea to bypass a blocked or damaged larynx. *(p. 824)*
tuberculosis: A respiratory disorder caused by infection of the lungs by the bacterium *Mycobacterium tuberculosis.* *(p. 824 and AM)*

AM Additional Terms Discussed in the *Applications Manual*

---

chronic obstructive pulmonary disease (COPD): A condition characterized by *chronic bronchitis* and *chronic airways obstruction.*

shallow water blackout: A potential risk for individuals who hyperventilate before swimming underwater; the CNS shuts down suddenly due to hypoxia.

CHAPTER REVIEW

 On-line resources for this chapter are on our World Wide Web site at: http://www.prenhall.com/martini/fap

STUDY OUTLINE

INTRODUCTION, p. 815

1. Body cells must obtain oxygen and eliminate carbon dioxide. Gas exchange occurs at the **alveoli** of the lungs.

FUNCTIONS OF THE RESPIRATORY SYSTEM, p. 815

1. The functions of the **respiratory system** include (1) providing an area for gas exchange between air and circulating blood; (2) moving air to and from exchange surfaces; (3) protecting respiratory surfaces from environmental variations and defending the respiratory system and other tissues from invasion by pathogens; (4) permitting vocal communication; and (5) providing olfactory sensations to the CNS.

ORGANIZATION OF THE RESPIRATORY SYSTEM, p. 815

1. The respiratory system includes the nose, nasal cavity, paranasal sinuses, pharynx, larynx, trachea, bronchi, bronchioles, and alveoli of the lungs. *(Figure 23-1)*
2. The **respiratory tract** consists of the conducting passageways that carry air to and from the alveoli. The passageways of the **upper respiratory system** filter and humidify the incoming air. The **lower respiratory system** includes delicate conduction passages and the alveolar exchange surfaces.

The Respiratory Mucosa, p. 815

3. The **respiratory mucosa** (respiratory epithelium and underlying connective tissue) lines the upper and lower respiratory tracts.
4. The respiratory epithelium changes in structure along the respiratory tract. It is supported by a layer of loose connective tissue, the **lamina propria.** *(Figure 23-2)*

THE UPPER RESPIRATORY SYSTEM, p. 818

1. The components of the upper respiratory system consists of the nose, nasal cavity, paranasal sinuses, and pharynx. *(Figures 23-1, 23-3)*

The Nose and Nasal Cavity, p. 818

2. Air normally enters the respiratory system through the **external nares,** which open into the **nasal cavity.** The **vestibule** (entryway) is guarded by hairs that screen out large particles.

3. Incoming air flows through the **superior, middle,** and **inferior meatuses** (narrow grooves) and bounces off the conchal surfaces.
4. The **hard palate** separates the oral and nasal cavities. The **soft palate** separates the superior nasopharynx from the rest of the pharynx. The connections between the nasal cavity and nasopharynx are the **internal nares.**
5. The nasal mucosa traps particles, warms and humidifies incoming air, and cools and dehumidifies outgoing air.

The Pharynx, p. 820

6. The **pharynx** is a chamber shared by the digestive and respiratory systems. The **nasopharynx** is the superior part of the pharynx. The **oropharynx** is continuous with the oral cavity. The **laryngopharynx** includes the narrow zone between the hyoid and the entrance to the esophagus. *(Figure 23-3)*

THE LARYNX, p. 820

1. Inspired air passes through the **glottis** en route to the lungs; the **larynx** surrounds and protects the glottis. *(Figure 23-4)*

Cartilages of the Larynx, p. 820

2. The cylindrical larynx is composed of three large cartilages (the **thyroid, cricoid,** and **epiglottis**) and three smaller pairs of cartilages (the **arytenoid, corniculate,** and **cuneiform**). The epiglottis projects into the pharynx.
3. Two pairs of folds span the glottis: the inelastic **ventricular folds** and the more delicate **vocal folds.** Air passing through the glottis vibrates the vocal folds, producing sound.

The Laryngeal Musculature, p. 822

4. The **extrinsic laryngeal muscles** position and stabilize the larynx. The **intrinsic laryngeal muscles** regulate tension in the vocal folds and open and close the glottis. During swallowing, both sets of muscles help prevent particles from entering the glottis. *(Figure 23-5)*

THE TRACHEA, p. 823

1. The **trachea** extends from the sixth cervical vertebra to the fifth thoracic vertebra. The **submucosa** contains C-shaped **tracheal**

cartilages, which stiffen the tracheal walls and protect the airway. The posterior tracheal wall can distort to permit large masses of food to pass through the esophagus. *(Figure 23-6)*

THE PRIMARY BRONCHI, p. 824

1. The trachea branches within the mediastinum to form the **right** and **left primary bronchi.** Each bronchus enters a lung at the **hilus** (a groove). The **root** of the lung is a connective tissue mass including the bronchus, pulmonary vessels, and nerves.

THE LUNGS, p. 824

Lobes and Surfaces of the Lungs, p. 824

1. The **lobes** of the lungs are separated by fissures; the right lung has three lobes, and the left lung has two. *(Figure 23-7)*
2. The anterior **costal surfaces** follow the inner contours of the rib cage. The concavity of the **mediastinal surface** of the left lung is the **cardiac notch,** which conforms to the shape of the pericardium. *(Figure 23-8)*

The Bronchi, p. 824

3. The primary bronchi and their branches form the **bronchial tree.** The **secondary** and **tertiary bronchi** are branches within the lungs. As they branch, the amount of cartilage in their walls decreases and the amount of smooth muscle increases. *(Figure 23-9)*
4. Each tertiary bronchus supplies air to a single **bronchopulmonary segment.** *(Figure 23-10)*

The Bronchioles, p. 826

5. **Bronchioles** within the bronchopulmonary segments ultimately branch into **terminal bronchioles.** Each terminal bronchiole delivers air to a single **pulmonary lobule.** Within the lobule, the terminal bronchiole branches into **respiratory bronchioles.** The connective tissues of the root extend into the parenchyma of the lung as a series of trabeculae (partitions). These branch to form **septa** that divide the lung into lobules. *(Figures 23-10, 23-11)*

Alveolar Ducts and Alveoli, p. 829

6. The respiratory bronchioles open into **alveolar ducts;** many alveoli are interconnected at each duct. The respiratory exchange surfaces are extensively connected to the circulatory system via the vessels of the pulmonary circuit.
7. The **respiratory membrane** consists of a simple squamous epithelium; **surfactant cells** scattered in it produce an oily secretion that keeps the alveoli from collapsing. **Alveolar macrophages** patrol the epithelium and engulf foreign particles. *(Figure 23-12)*

The Blood Supply to the Lungs, p. 831

8. The conducting portions of the respiratory tract receive blood from the external carotid arteries, the thyrocervical trunks, and the bronchial arteries. The venous blood flows into the pulmonary veins, bypassing the rest of the systemic circuit and diluting the oxygenated blood leaving the alveoli.

THE PLEURAL CAVITIES AND PLEURAL MEMBRANES, p. 831

1. Each lung occupies a single pleural cavity lined by a **pleura** (serous membrane). There are two pleurae: a **parietal pleura**, covering the inner surface of the thoracic wall, and a **visceral pleura**, covering the lungs.

Changes in the Respiratory System at Birth, p. 834

2. Before delivery, the fetal lungs are fluid-filled and collapsed.

At the first breath, the lungs inflate and never collapse completely thereafter.

RESPIRATORY PHYSIOLOGY, p. 834

1. Respiratory physiology focuses on a series of integrated processes: **external respiration** (the exchange of oxygen and carbon dioxide between the interstitial fluids of the body and the external environment), which includes **pulmonary ventilation** (breathing), and **internal respiration** (the exchange of oxygen and carbon dioxide between the interstitial fluid and living cells). If oxygen content declines, the affected tissues will suffer from **hypoxia;** if the oxygen supply is completely shut off, **anoxia** and tissue death result.

Pulmonary Ventilation, p. 835

2. As pressure on a gas decreases, its volume expands; as pressure increases, gas volume contracts; this inverse relationship is known as **Boyle's law.** *(Figure 23-13; Table 23-1)*
3. The relationship between **intrapulmonary pressure** (the pressure inside the respiratory tract) and **atmospheric pressure (atm)** determines the direction of airflow. The **intrapleural pressure** is the pressure in the space between the parietal and visceral pleurae. *(Figures 23-14, 23-15)*
4. The diaphragm and the external and internal intercostals are involved in normal **quiet breathing,** or eupnea. Accessory muscles become active during the active inspiratory and expiratory movements of **forced breathing,** or hyperpnea. *(Figures 23-16, 23-17)*
5. **Alveolar ventilation** is the amount of air reaching the alveoli each minute. The **vital capacity** includes the **tidal volume** plus the **expiratory** and **inspiratory reserve volumes.** The air left in the lungs at the end of maximum expiration is the **residual volume.** *(Figure 23-18)*

Gas Exchange at the Respiratory Membrane, p. 842

6. In a mixed gas, the individual gases exert a pressure proportional to their abundance in the mixture **(Dalton's law).** The pressure contributed by a single gas is its **partial pressure.**
7. The amount of a gas in solution is directly proportional to the partial pressure of that gas **(Henry's law).** *(Figure 23-19)*
8. Alveolar and atmospheric air differ in composition. *(Figure 23-20; Table 23-2)*

Gas Pickup and Delivery, p. 846

9. Blood entering peripheral capillaries delivers oxygen and absorbs carbon dioxide. The transport of oxygen and carbon dioxide in the blood involves reactions that are completely reversible.
10. Over the range of oxygen pressures normally present in the body, a small change in plasma P_{O_2} will lead to a large change in the amount of oxygen bound or released. At alveolar P_{O_2}, hemoglobin is almost fully saturated; at the P_{O_2} of peripheral tissues, it retains a substantial oxygen reserve. When low plasma P_{O_2} continues for extended periods, red blood cells generate more **2,3-bisphosphoglycerate (BPG),** which reduces hemoglobin's affinity for oxygen. *(Figures 23-21, 23-22)*
11. **Fetal hemoglobin** has a stronger affinity for oxygen than does adult hemoglobin, aiding the removal of oxygen from the maternal blood. *(Figure 23-23)*
12. Aerobic metabolism in peripheral tissues generates CO_2. About 7 percent of the CO_2 transported in the blood is dissolved in the plasma; 23 percent is bound as **carbaminohemoglobin;** the rest is converted to carbonic acid, which dissociates into H^+ and HCO_3^-. *(Figures 23-24, 23-25)*

CONTROL OF RESPIRATION, p. 850

1. Local regulation compensates for small oscillations affecting individual tissues and organs; large-scale or extended changes require the integration of cardiovascular and respiratory responses.

Local Regulation of Gas Transport and Alveolar Function, p. 851

2. Local factors regulate blood flow (*perfusion*) and airflow (*ventilation*). Alveolar capillaries constrict under conditions of low oxygen, and bronchioles dilate under conditions of high carbon dioxide.

The Respiratory Centers of the Brain, p. 851

3. The **respiratory centers** include three pairs of nuclei in the reticular formation of the pons and medulla oblongata. The *respiratory rhythmicity centers* set the pace for respiration. The **apneustic centers** cause strong, sustained inspiratory movements, and the **pneumotaxic centers** inhibit the apneustic centers and promote exhalation. (*Figure 23-26*)

4. The **inflation reflex** prevents overexpansion of the lungs during forced breathing. The **deflation reflex** stimulates inspiration when the lungs are collapsing. Chemoreceptor reflexes respond to changes in the P_{O_2} and P_{CO_2} of the blood and cerebrospinal fluid. (*Figures 23-27, 23-28*)

Voluntary Control of Respiration, p. 856

5. Conscious and unconscious thought processes can affect respiration by affecting the respiratory centers.

AGING AND THE RESPIRATORY SYSTEM, p. 856

1. The respiratory system is generally less efficient in the elderly because (1) elastic tissue deteriorates, lowering the vital capacity of the lungs, (2) movements of the chest are restricted by arthritic changes and decreased flexibility of costal cartilages, and (3) some degree of emphysema is generally present. (*Figure 23-29*)

INTEGRATION WITH OTHER SYSTEMS, p. 856

1. The respiratory system has extensive anatomical and functional connections to the cardiovascular system. (*Figure 23-30*)

REVIEW QUESTIONS

Level 1   Reviewing Facts and Terms

1. The C shape of the tracheal cartilages is important because
   - (a) large masses of food can pass through the esophagus during swallowing
   - (b) large masses of air can pass through the trachea
   - (c) it allows greater tracheal elasticity and flexibility
   - (d) a, b, and c are correct

2. Control over the amount of resistance to airflow and the distribution of air within the lungs is provided by the
   - (a) diaphragm   (b) trachea
   - (c) bronchioles   (d) alveoli

3. Pulmonary ventilation involves
   - (a) internal respiration   (b) external respiration
   - (c) cellular respiration   (d) air movement

4. The presence of an abnormally low oxygen content in tissue fluids is
   - (a) anoxia   (b) edema
   - (c) hypoxia   (d) emphysema

5. During inhalation, the lungs expand and the intrapulmonary pressure
   - (a) rises to about 761 mm Hg
   - (b) remains at 760 mm Hg
   - (c) drops to about 759 mm Hg
   - (d) does not change

6. During a normal inhalation, the intrapleural pressure is approximately
   - (a) −1 mm Hg   (b) +6 mm Hg
   - (c) +1 mm Hg   (d) −6 mm Hg

7. According to Henry's law, if the partial pressure of a gas increases,
   - (a) gas molecules will come out of solution
   - (b) more gas molecules will enter solution
   - (c) the solubility of the gas will decrease
   - (d) the volume of the gas will decrease

8. At a normal tissue partial pressure of oxygen, the blood entering the venous system contains about _____ of its total oxygen content:
   - (a) 25 percent   (b) 50 percent
   - (c) 75 percent   (d) 90 percent

9. Approximately 70 percent of the carbon dioxide absorbed by the blood will be transported
   - (a) as bicarbonate ions
   - (b) bound to hemoglobin
   - (c) in the form of dissolved gas molecules
   - (d) bound to oxygen molecules

10. The apneustic centers of the pons provide
   - (a) inhibition of the pneumotaxic and inspiratory centers
   - (b) continuous stimulation to the inspiratory center
   - (c) monitoring of blood gas levels
   - (d) alterations in chemoreceptor sensitivity

11. All the following provide chemoreceptor input to the medullary respiratory centers except the
   - (a) olfactory epithelium
   - (b) medullary chemoreceptors
   - (c) aortic body
   - (d) carotid body

12. Sneezing and coughing are classic examples of
   - (a) inflation reflexes   (b) deflation reflexes
   - (c) protective reflexes   (d) Hering–Breuer reflexes

13. What are the five primary functions of the respiratory system?

14. Distinguish between the structures of the upper respiratory system and the lower respiratory system.

15. What defense mechanisms protect the respiratory system from becoming contaminated with debris or pathogens?

16. What are the three regions of the pharynx called, and where is each located?

17. What prevents you from swallowing food or liquids while you are breathing?

18. How does the parietal pleura differ from the visceral pleura?

19. What four integrated steps are involved in external respiration?

20. What important physiological adjustments are necessary to adapt to and tolerate high-altitude conditions?

21. By what three mechanisms is carbon dioxide transported in the bloodstream?

22. What effect does P_{CO_2} have on smooth muscles in the walls of bronchioles?

Level 2   Reviewing Concepts

23. Parasympathetic stimulation to the smooth muscle tissue layer in the bronchioles causes
- **(a)** bronchoconstriction
- **(b)** bronchodilation
- **(c)** a relaxation of muscle tone
- **(d)** an increase in tidal volume

24. If you have a respiration rate of 15 breaths per minute and a tidal volume of 500 ml of air, your respiratory minute volume is
- **(a)** 7.5 l/min
- **(b)** 75 l/min
- **(c)** 750 l/min
- **(d)** 7500 l/min

25. If a tidal inspiration pulls in 1000 ml of air, the amount of air reaching the alveolar spaces is about
- **(a)** 300 ml
- **(b)** 850 ml
- **(c)** 150 ml
- **(d)** 700 ml

26. Gas exchange at the respiratory membrane is efficient because
- **(a)** the differences in partial pressure are substantial
- **(b)** the gases are lipid-soluble
- **(c)** the total surface area is large
- **(d)** a, b, and c are correct

27. For any partial pressure of oxygen, if the concentration of 2,3-bisphosphoglycerate (BPG) increases,
- **(a)** the amount of oxygen released by hemoglobin will decrease
- **(b)** the oxygen levels in hemoglobin will be unaffected
- **(c)** the amount of oxygen released by hemoglobin will increase
- **(d)** the amount of carbon dioxide carried by hemoglobin will increase

28. The primary physiological adjustment(s) necessary for an athlete to compete at high altitudes is (are)
- **(a)** increased respiratory rate
- **(b)** increased heart rate
- **(c)** elevated hematocrit
- **(d)** a, b, and c are correct

29. An increase in the partial pressure of carbon dioxide in arterial blood causes the chemoreceptors to stimulate the respiratory centers, causing
- **(a)** decreased respiratory rate
- **(b)** increased respiratory rate
- **(c)** hypocapnia
- **(d)** hypercapnia

30. What is the functional significance of the decrease in the amount of cartilage and increase in the amount of smooth muscle in the lower respiratory passageways?

31. Why can't you swallow solid food or liquid properly while you are talking?

32. Why is breathing through the nasal cavity more desirable than breathing through the mouth?

33. How would you justify the statement, "the bronchioles are to the respiratory system what the arterioles are to the cardiovascular system"?

34. How are septal cells involved with keeping the alveoli from collapsing?

35. Why is diffusion across the respiratory membrane rapid and efficient?

36. How does pulmonary ventilation differ from alveolar ventilation, and what is the function of each type of ventilation?

37. What is the significance of **(a)** Boyle's law, **(b)** Dalton's law, and **(c)** Henry's law to the process of respiration?

38. If the pleural cavity is penetrated due to an injury, why does the lung collapse?

39. What happens to the process of respiration when a person is sneezing or coughing?

40. What are the differences between pulmonary volumes and respiratory capacities? How are pulmonary volumes and respiratory capacities determined?

41. What are the primary effects of pH, temperature, and 2,3-bisphosphoglycerate (BPG) on hemoglobin saturation?

42. What is the functional difference between the dorsal respiratory group (DRG) and the ventral respiratory group (VRG) of the medulla oblongata?

43. What types of sensory information modify the activities of the respiratory centers?

Level 3   Critical Thinking and Clinical Applications

44. Billy's normal alveolar ventilation rate (AVR) during mild exercise is 6.0 l/min. While at the beach on a warm summer day, he decides to go snorkeling. The snorkel has a volume of 50 ml. Assuming that the water is not too cold and that snorkeling for Billy is mild exercise, what would his respiratory rate have to be in order for him to maintain an AVR of 6.0 l/min while snorkeling? (Assume a constant tidal volume of 500 ml and an anatomic dead space of 150 ml.)

45. A decrease in blood pressure will trigger a baroreceptor reflex that leads to increased ventilation. What is the possible advantage of this reflex?

46. Mr. B. has suffered from chronic advanced emphysema for 15 years. While hospitalized with a respiratory infection, he goes into respiratory distress. Without thinking, his nurse immediately administers pure oxygen, which causes Mr. B. to stop breathing. Why did this occur?

47. You spend the night at a friend's house during the winter. Your friend's home is quite old, and the hot-air furnace lacks a humidifier. When you wake up in the morning, you have a fair amount of nasal congestion and decide you might be coming down with a cold. After a steamy shower and some juice for breakfast, the nasal congestion disappears. Explain.

48. Why would a person with kyphosis (see Chapter 7, p. 218) exhibit a lower-than-normal vital capacity?

49. Why do patients suffering from anemia generally not exhibit an increase in respiratory rate or tidal volume even though their blood is not carrying enough oxygen?

50. Doris has an obstruction of her right primary bronchus. As a result, how would you expect the oxygen–hemoglobin saturation curve for her right lung to compare with that for her left?

The Digestive System

A *lthough Giuseppe Arcimboldo may have taken the notion a bit too literally,*
modern physiology has confirmed that, in a sense, "you are what you eat." We all need a
regular supply of nutrients in our diet. Unfortunately, the nutrients in food are not ready for use by our
cells. Without processing by the digestive system, food would be of no more use to us than a lump of coal in the
gas tank of a car. This chapter introduces the organs of the digestive system and their varied functions in digestion.

CHAPTER OUTLINE AND OBJECTIVES

Few people give any serious thought to the digestive system unless it malfunctions. Yet we spend hours of conscious effort filling and emptying it. References to this system are part of our everyday language. Think of how often we use expressions relating to the digestive system: We may "have a gut feeling," "want to chew on" something, or find someone's opinions "hard to swallow." (Other, less polite remarks may be heard almost as often.) When something does go wrong with the digestive system, even something minor, most of us seek treatment immediately. For this reason, television advertisements promote toothpaste and mouthwash, diet supplements, antacids, and laxatives on an hourly basis.

Cells perform metabolic reactions that provide energy for the synthesis of ATP. These reactions require two essential ingredients: (1) oxygen and (2) organic molecules that can be broken down by intracellular enzymes. The respiratory system, working with the cardiovascular system, provides the necessary oxygen. The digestive system, working with the cardiovascular and lymphatic systems, provides the organic molecules. In effect, the digestive system provides both the fuel that keeps all the body's cells running and the building blocks needed for cell growth and repair.

The digestive system consists of a muscular tube, the **digestive tract,** and various **accessory organs.** The *oral cavity* (mouth), *pharynx, esophagus, stomach, small intestine,* and *large intestine* make up the digestive tract. Accessory digestive organs include the teeth, tongue, and various *glandular organs,* such as the salivary glands, liver, and pancreas, that secrete into ducts emptying into the digestive tract. Food enters the digestive tract and passes along its length. On the way, the secretions of the glandular organs, which contain water, enzymes, buffers, and other components, assist in preparing organic and inorganic nutrients for absorption across the epithelium of the digestive tract.

We can consider digestive functions to be a series of integrated steps:

1. *Ingestion* occurs when materials enter the digestive tract via the mouth. Ingestion is an active process involving conscious choice and decision making.

2. *Mechanical processing* is physical manipulation and distortion that makes materials easier to propel along the digestive tract and increases the surface area for enzymatic attack. Mechanical processing may or may not be required before ingestion; you can swallow liquids immediately but must process most solids first. Tearing and crushing with the teeth, followed by squashing and compaction by the tongue, are examples of preliminary mechanical processing. Swirling, mixing, and churning motions of the stomach and intestines provide mechanical processing after ingestion.

3. *Digestion* refers to the chemical breakdown of food into small organic fragments suitable for absorption by the digestive epithelium. Simple molecules in food, such as glucose, can be absorbed intact, but epithelial cells have no way to deal with molecules the size and complexity of proteins, polysaccharides, or triglycerides. These molecules must be disassembled by digestive enzymes prior to absorption. For example, the starches in a potato are of no value until enzymes have broken them down to simple sugars that you can absorb and distribute to your cells.

4. *Secretion* is the release of water, acids, enzymes, buffers, and salts by the epithelium of the digestive tract and by glandular organs.

5. *Absorption* is the movement of organic substrates, electrolytes (inorganic ions), vitamins, and water across the digestive epithelium and into the interstitial fluid of the digestive tract.

6. *Excretion* is the elimination of waste products from the body. The digestive tract and glandular organs secrete waste products in secretions discharged into the lumen of the tract. Most of these waste products, after mixing with the indigestible residue of the digestive process, will leave the body. The ejection of materials from the digestive tract, a process called **defecation** (def-e-KĀ-shun), or *egestion,* eliminates materials as **feces**.

The lining of the digestive tract also plays a protective role by safeguarding surrounding tissues against (1) the corrosive effects of digestive acids and enzymes, (2) mechanical stresses, such as abrasion, and (3) bacteria that either are swallowed with food or reside inside the digestive tract. The digestive epithelium and its secretions provide a nonspecific defense against these bacteria. Bacteria reaching the underlying tissues are attacked by macrophages and other cells of the immune system.

AN OVERVIEW OF THE STRUCTURE AND FUNCTION OF THE DIGESTIVE TRACT

Figure 24-1● shows the major components of the digestive system. The digestive tract begins at the oral cavity, continues through the pharynx, esophagus, stomach, small intestine, and large intestine, and ends at the anus. These structures have overlapping functions, but each has certain areas of specialization and shows distinctive histological characteristics.

Histological Organization of the Digestive Tract

The major layers of the digestive tract include (1) the *mucosa,* (2) the *submucosa,* (3) the *muscularis externa,* and (4) the *serosa.* Sectional, diagrammatic views

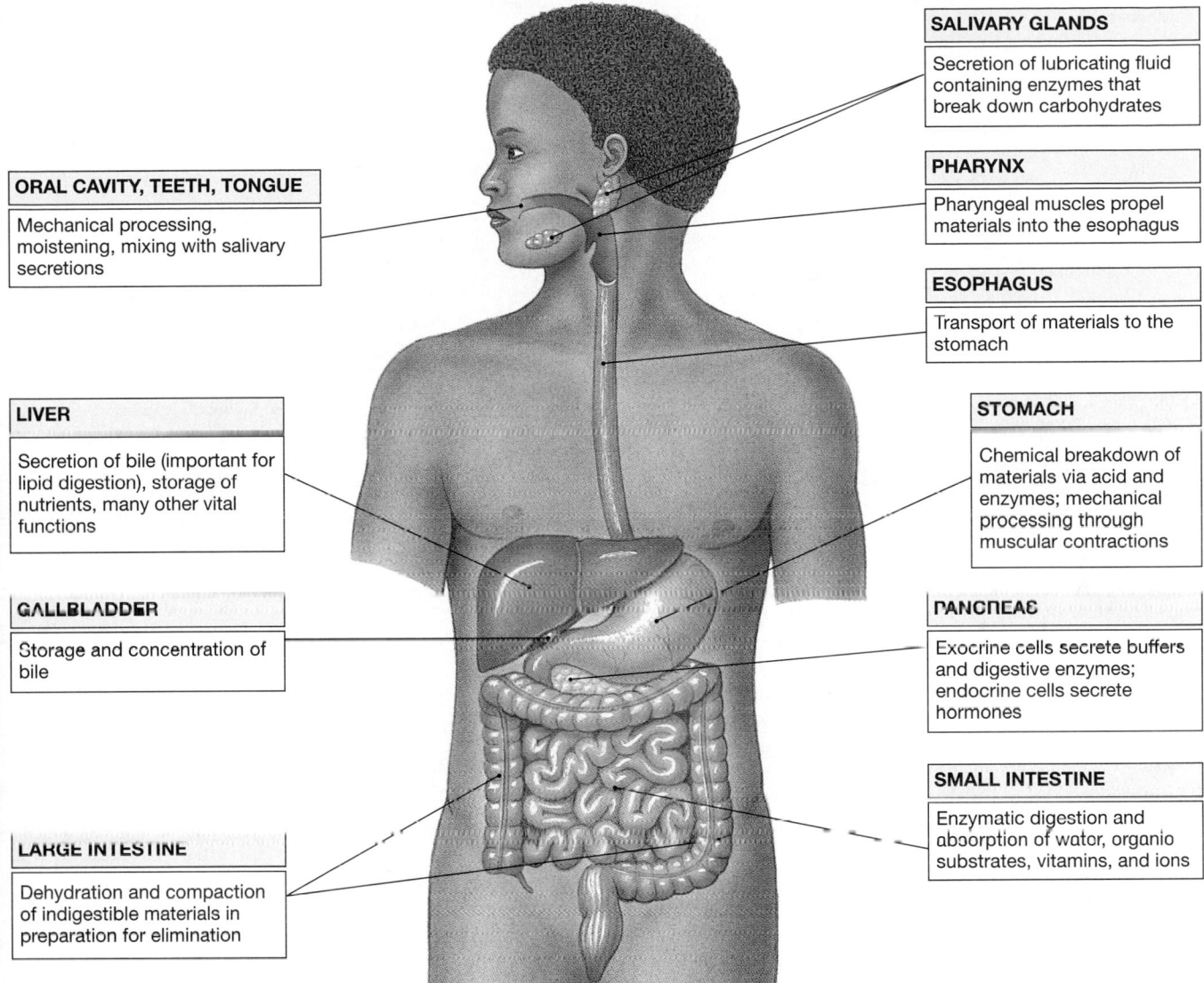

SALIVARY GLANDS

Secretion of lubricating fluid containing enzymes that break down carbohydrates

PHARYNX

Pharyngeal muscles propel materials into the esophagus

ESOPHAGUS

Transport of materials to the stomach

STOMACH

Chemical breakdown of materials via acid and enzymes; mechanical processing through muscular contractions

PANCREAS

Exocrine cells secrete buffers and digestive enzymes; endocrine cells secrete hormones

SMALL INTESTINE

Enzymatic digestion and absorption of water, organic substrates, vitamins, and ions

ORAL CAVITY, TEETH, TONGUE

Mechanical processing, moistening, mixing with salivary secretions

LIVER

Secretion of bile (important for lipid digestion), storage of nutrients, many other vital functions

GALLBLADDER

Storage and concentration of bile

LARGE INTESTINE

Dehydration and compaction of indigestible materials in preparation for elimination

•*FIGURE 24-1* **Components of the Digestive System.** The major regions and accessory organs of the digestive tract, together with their primary functions.

of these layers are presented in Figure 24-2•. There are regional variations in the structure of the layers; Figure 24-2 is a composite view that most closely resembles the appearance of the small intestine, the longest segment of the digestive tract.

The Mucosa

The inner lining, or **mucosa,** of the digestive tract is a *mucous membrane.* Mucous membranes, introduced in Chapter 4, consist of a layer of an epithelium moistened by glandular secretions and an underlying layer of loose connective tissue, the *lamina propria.* ∞ *[p. 132]*

THE DIGESTIVE EPITHELIUM The mucosal epithelium may be either simple or stratified, depending on the location and the stresses to which it is most often subjected. For example, the oral cavity, pharynx, and esophagus (where mechanical stresses are most severe) are lined by a stratified squamous epithelium, whereas the stomach, the small intestine, and almost the entire length of the large intestine (where absorption occurs) have a simple columnar epithelium that contains goblet cells. Scattered among the columnar cells are **enteroendocrine cells.** These cells secrete hormones that coordinate the activities of the digestive tract and the accessory glands.

The lining of the digestive tract is often thrown into longitudinal folds, which disappear as the tract fills, and permanent transverse folds, or *plicae* (PLĪ - sē; singular, *plica* [PLĪ - ka]) (Figure 24-2•). The folding dramatically increases the surface area available for absorption. The secretions of gland cells located in the mucosa and submucosa—or within accessory glandular organs—are carried to the epithelial surfaces by ducts.

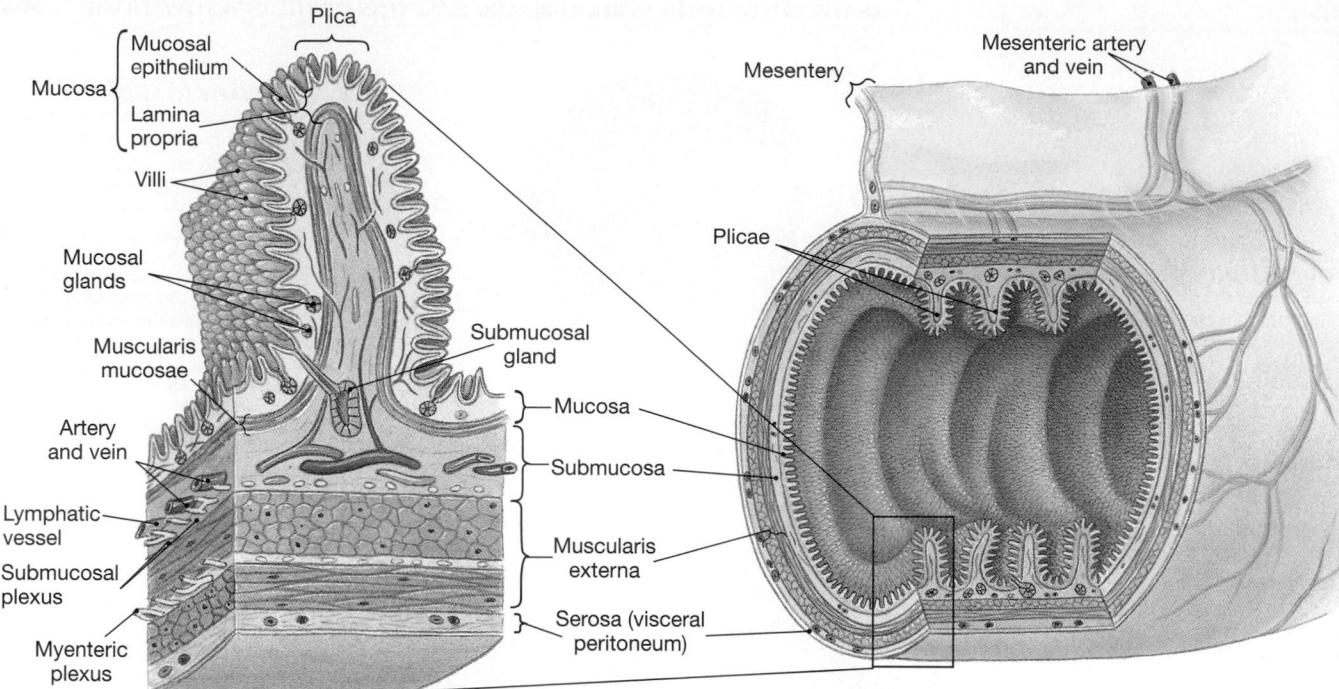

•FIGURE 24-2 Structure of the Digestive Tract. Diagrammatic view of a representative portion of the digestive tract. The features illustrated are those of the small intestine.

EPITHELIAL RENEWAL AND REPAIR The life span of a typical epithelial cell varies from 2 to 3 days in the esophagus to 6 days in the large intestine. The lining of the entire digestive tract is therefore continuously being renewed through the divisions of epithelial stem cells, keeping pace with the rate of cell destruction and loss at epithelial surfaces. This high rate of cell division explains why radiation and anticancer drugs that inhibit mitosis have drastic effects on the digestive tract. Lost epithelial cells are no longer replaced, and the cumulative damage to the epithelial lining quickly leads to problems in nutrient absorption. In addition, the exposure of the lamina propria to digestive enzymes can cause internal bleeding and other serious problems.

THE LAMINA PROPRIA The lamina propria consists of a layer of loose connective tissue that also contains blood vessels, sensory nerve endings, lymphatic vessels, smooth muscle cells, and scattered areas of lymphoid tissue. In the oral cavity, pharynx, esophagus, stomach, and *duodenum* (the proximal portion of the small intestine), the lamina propria also contains the secretory cells of mucous glands.

In most areas of the digestive tract, the outer portion of the lamina propria contains a narrow band of smooth muscle and elastic fibers. This band is called the **muscularis** (mus-kū-LAR-is) **mucosae.** The smooth muscle cells in the muscularis mucosae are arranged in two concentric layers (Figure 24-2•). The inner layer encircles the lumen (the *circular muscle*), and the outer layer contains muscle cells oriented parallel to the long axis of the tract (the *longitudinal layer*). Contractions in these layers alter the shape of the lumen and move the epithelial pleats and folds.

The Submucosa

The **submucosa** is a layer of dense connective tissue that surrounds the muscularis mucosae. This layer has large blood vessels and lymphatics, and in some regions the submucosa also contains exocrine glands that secrete buffers and enzymes into the lumen of the digestive tract. Along its outer margin, the submucosa contains a network of nerve fibers and scattered neurons. This **submucosal plexus,** or *plexus of Meissner,* contains sensory neurons, parasympathetic ganglionic neurons, and sympathetic postganglionic fibers that innervate the mucosa and submucosa (Figure 24-2•).

The Muscularis Externa

The submucosal plexus lies along the inner border of the **muscularis externa,** a region dominated by smooth muscle cells. As in the muscularis mucosae, the smooth muscle cells are arranged in an inner, circular layer and an outer, longitudinal layer. These layers play an essential role in mechanical processing and in the movement of materials along the digestive tract. These movements are coordinated primarily by neurons of the **myenteric** (mī-en-TER-ik) **plexus** (*mys,* muscle + *enteron,* intestine), or *plexus of Auerbach.* This network of parasympathetic ganglia and sympathetic postganglionic fibers lies sandwiched between the circular and longitudinal muscle layers. Parasympathetic stimulation increases muscle tone and activity, and sympathetic stimulation promotes muscular inhibition and relaxation.

The Serosa

Along most portions of the digestive tract inside the peritoneal cavity, the muscularis externa is covered by a *serous membrane* known as the **serosa** (Figure 24-2•). There is no serosa covering the muscularis externa of the oral cavity, pharynx, esophagus, and rectum, where a dense network of collagen fibers firmly attaches the digestive tract to adjacent structures. This fibrous sheath is called an *adventitia*.

The Movement of Digestive Materials

Your digestive tract contains layers of *visceral smooth muscle tissue,* a type of smooth muscle introduced in Chapter 10. ∾ *[p. 311]* The smooth muscle along the digestive tract shows rhythmic cycles of activity due to the presence of *pacesetter cells.* These smooth muscle cells undergo spontaneous depolarization, and their contraction triggers a wave of contraction that spreads through the entire muscular sheet. Pacesetter cells are located in the muscularis mucosae and muscularis externa, whose layers surround the lumen of the digestive tract. The coordinated contractions of the muscularis externa play a vital role in the movement of materials along the tract, through *peristalsis,* and in mechanical processing, through *segmentation*.

Peristalsis

The muscularis externa propels materials from one portion of the digestive tract to another through the contractions known as **peristalsis** (per-i-STAL-sis). Peristalsis consists of waves of muscular contractions that move along the length of the digestive tract. During a peristaltic movement, the circular muscles contract behind the digestive contents. Longitudinal muscles contract next, shortening adjacent segments. A wave of contraction in the circular muscles then forces the materials in the desired direction (Figure 24-3•).

Segmentation

Most areas of the small intestine and some portions of the large intestine undergo **segmentation.** These movements churn and fragment the digestive materials, mixing the contents with intestinal secretions. Because they do not follow a set pattern, segmentation movements do not produce directional movement of materials along the tract.

The Control of Digestive Function

The activities of the digestive system are regulated by neural, hormonal, and local mechanisms (Figure 24-4•).

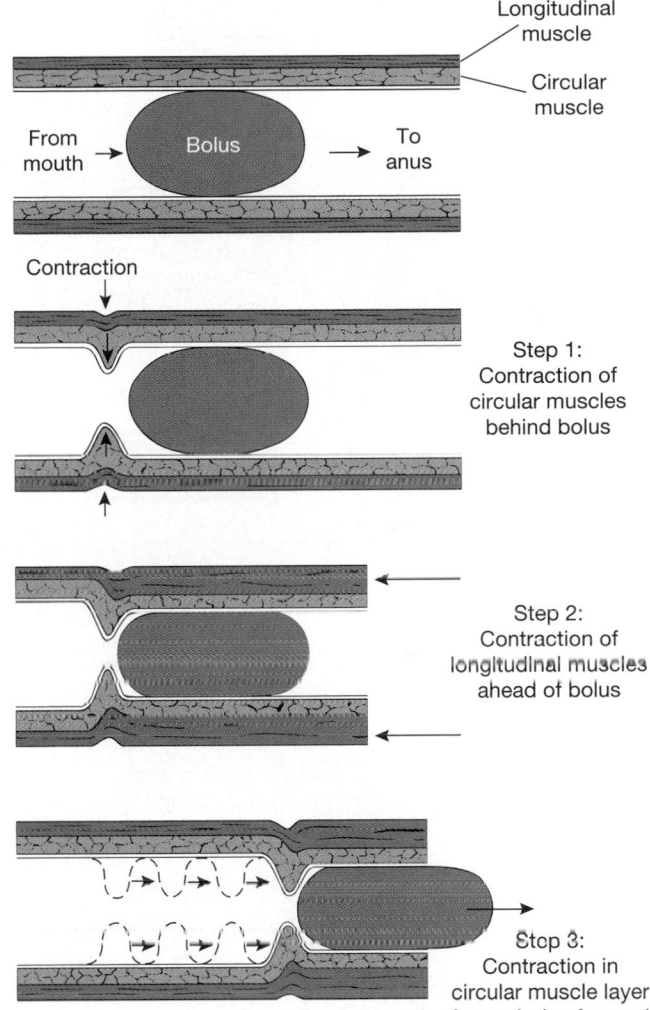

•**FIGURE 24-3** **Peristalsis.** Peristalsis propels materials along the length of the digestive tract.

Neural Mechanisms

The movement of materials along your digestive tract and many secretory functions are controlled primarily by neural mechanisms. For example, peristaltic movements limited to a few centimeters of the digestive tract are triggered by sensory receptors in the walls of the digestive tract. The motor neurons that control smooth muscle contraction and glandular secretion are located in the myenteric plexus. These neurons are usually considered to be parasympathetic, because they are innervated by parasympathetic preganglionic fibers. However, the sensory neurons and interneurons that are responsible for local reflexes are also found in the myenteric plexus rather than in the CNS. The reflexes controlled by these myenteric neurons are called **short reflexes,** or *myenteric reflexes.* The term *enteric nervous system* is often used to refer to the neural network within the digestive tract.

Sensory information from receptors in the digestive tract is also distributed to the CNS, where it can

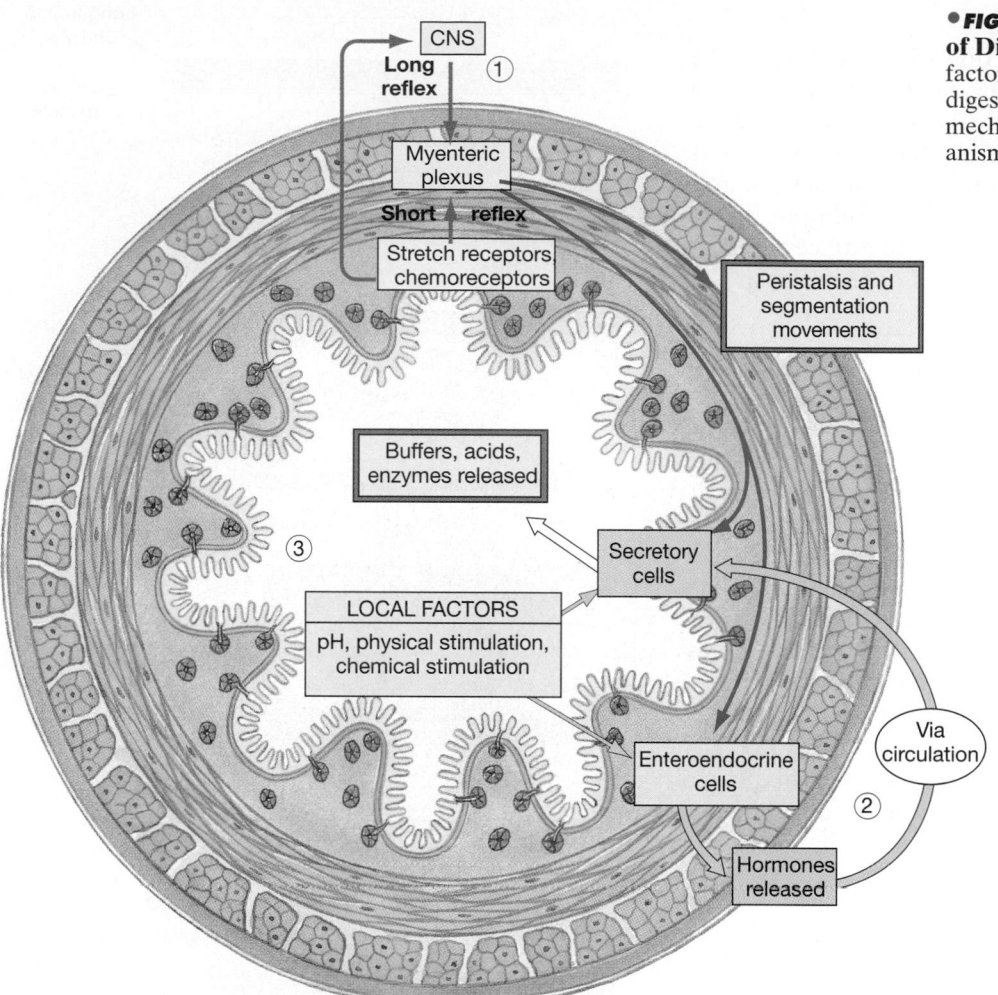

•*FIGURE 24-4* **The Regulation of Digestive Activities.** The major factors responsible for regulating digestive activities: ① neural mechanisms, ② hormonal mechanisms, and ③ local mechanisms.

trigger *long reflexes.* Long reflexes provide a higher level of control that coordinates activities along the length of the digestive tract. These reflexes generally control large-scale peristaltic waves that move materials from one region of the digestive tract to another. Long reflexes may involve motor fibers in the glossopharyngeal, vagus, or pelvic nerves that synapse in the myenteric plexus. The sensitivity of the smooth muscle cells to neural commands can be enhanced or inhibited by digestive hormones.

Hormonal Mechanisms

Your digestive tract is a hormone factory that produces at least 18 different hormones. These hormones affect almost every aspect of digestive function, and some of them also affect the activities of other systems. The hormones (*gastrin, secretin,* and others) are peptides produced by enteroendocrine cells in the digestive tract. They reach their target organs by distribution in the circulatory system. We shall consider each of these hormones in more detail as we proceed down the length of the digestive tract.

Local Mechanisms

Prostaglandins, histamine, and other chemicals released into the interstitial fluid may affect cells within a small segment of the digestive tract. These local messengers are important in coordinating a response to conditions (such as a change in the local pH or other chemical or physical stimulation) that affect only a portion of the digestive tract. For example, histamine release in the lamina propria of the stomach stimulates the secretion of acid by cells in the adjacent epithelium.

The Peritoneum

The abdominopelvic cavity contains the *peritoneal cavity.* The peritoneal cavity is lined by a serous membrane that consists of a superficial mesothelium covering a layer of loose connective tissue. We can divide the serous membrane into the serosa, or *visceral peritoneum,* which covers organs within the peritoneal cavity, and the *parietal peritoneum,* which lines the inner surfaces of the body wall. ∞ *[p. 24]*

The serous membrane lining the peritoneal cavity continuously produces peritoneal fluid that lubricates the peritoneal surfaces. About 7 liters of fluid is secreted and reabsorbed each day, although the volume within the peritoneal cavity at any one time is very small. Liver disease, kidney disease, and heart failure can cause an increase in the rate of fluid movement into the peritoneal cavity. The accumulation of fluid creates a characteristic abdominal swelling that is called **ascites** (a-SĪ-tēz). Distortion of internal organs by the contained fluid can result in a variety of symptoms; heartburn, indigestion, and low back pain are common complaints.

Mesenteries

Portions of your digestive tract are suspended within the peritoneal cavity by sheets of serous membrane that connect the parietal peritoneum with the visceral peritoneum. These **mesenteries** (MEZ-en-ter-ēz) are double sheets of peritoneal membrane (Figure 24-2•). The loose connective tissue between the mesothelial surfaces provides an access route for the passage of the blood vessels, nerves, and lymphatics to and from the digestive tract. Mesenteries also stabilize the positions of the attached organs and prevent your intestines from becoming entangled during digestive movements or sudden changes in body position.

During development, the digestive tract and accessory organs are suspended within the peritoneal cavity by *dorsal* and *ventral mesenteries* (Figure 24-5a•). The ventral mesentery later disappears along most of the digestive tract, persisting in the adult only (1) on the ventral surface of the stomach, between the stomach and liver (the *lesser omentum*), and (2) between the liver and the anterior abdominal wall (the *falciform* [FAL-si-form] *ligament*) (Figure 24-5b,c,d•). The **lesser omentum** (ō-MEN-tum; *omentum,* fat skin) stabilizes the position of the stomach and provides an access route for blood vessels and other structures entering or leaving the liver. The **falciform ligament** helps stabilize the position of the liver relative to the diaphragm and abdominal wall. (For more about the development of the digestive tract, accessory organs, and associated mesenteries, see the Embryology Summary on pp. 910–911.)

As the digestive tract elongates, it twists and turns within the relatively crowded peritoneal cavity. The dorsal mesentery of the stomach becomes greatly enlarged and forms an enormous pouch that extends inferiorly between the body wall and the anterior surface of the small intestine. This pouch is the **greater omentum** (Figure 24-5b,d•), which hangs like an apron from the lateral and inferior borders of the stomach. Adipose tissue in the greater omentum conforms to the shapes of the surrounding organs, providing padding and protection across the anterior and lateral surfaces of the abdomen.

The lipids in the adipose tissue are an important energy reserve, and the greater omentum provides insulation that reduces heat loss across the anterior abdominal wall.

All but the first 25 cm of the small intestine is suspended by the **mesentery proper**, a thick mesenterial sheet that provides stability but permits a degree of independent movement. The mesentery associated with the initial portion of the small intestine (the *duodenum*) and the pancreas fuses with the posterior abdominal wall, locking those structures in position. Their anterior surfaces are covered by peritoneum, but the rest of the organs lie outside the peritoneal cavity. Organs that lie posterior to, rather than within, the peritoneal cavity are called **retroperitoneal** (*retro,* behind).

A **mesocolon** is a mesentery associated with a portion of the large intestine. During normal development, the mesocolon of the *ascending colon,* the *descending colon,* and the *rectum* of the large intestine fuse to the dorsal body wall and are thus locked in place. Thereafter these organs are retroperitoneal, with the visceral peritoneum covering only their anterior surfaces and portions of their lateral surfaces (Figure 24-5b,c•). The **transverse mesentery,** which supports the transverse colon, and the **sigmoid mesocolon,** which supports the sigmoid colon, are all that remains of the original mesocolon.

PERITONITIS Inflammation of the peritoneal membrane produces symptoms of **peritonitis** (per i tō NĪ tis), a painful condition that interferes with the normal functioning of the affected organs. Physical damage, chemical irritation, and bacterial invasion of the peritoneum can lead to severe and even fatal cases of peritonitis. Peritonitis may be caused in untreated appendicitis by the rupturing of the appendix and the subsequent release of bacteria into the peritoneal cavity. Peritonitis is always a potential complication of any surgery in which the peritoneal cavity is opened.

☑ What is the importance of the mesenteries?

☑ Which would be more efficient in propelling intestinal contents from one place to another—peristalsis or segmentation?

☑ What effect would a drug that blocks parasympathetic stimulation of the digestive tract have on peristalsis?

THE ORAL CAVITY

Our exploration of the digestive tract will follow the path of food from the mouth to the anus. The mouth opens into the *oral cavity*. We can summarize the functions of the oral cavity as follows: (1) *analysis* of material before swallowing; (2) *mechanical processing* through the actions of the teeth, tongue, and palatal surfaces; (3) *lubrication* by mixing with mucus and salivary secretions; and (4) limited *digestion* of carbohydrates and lipids.

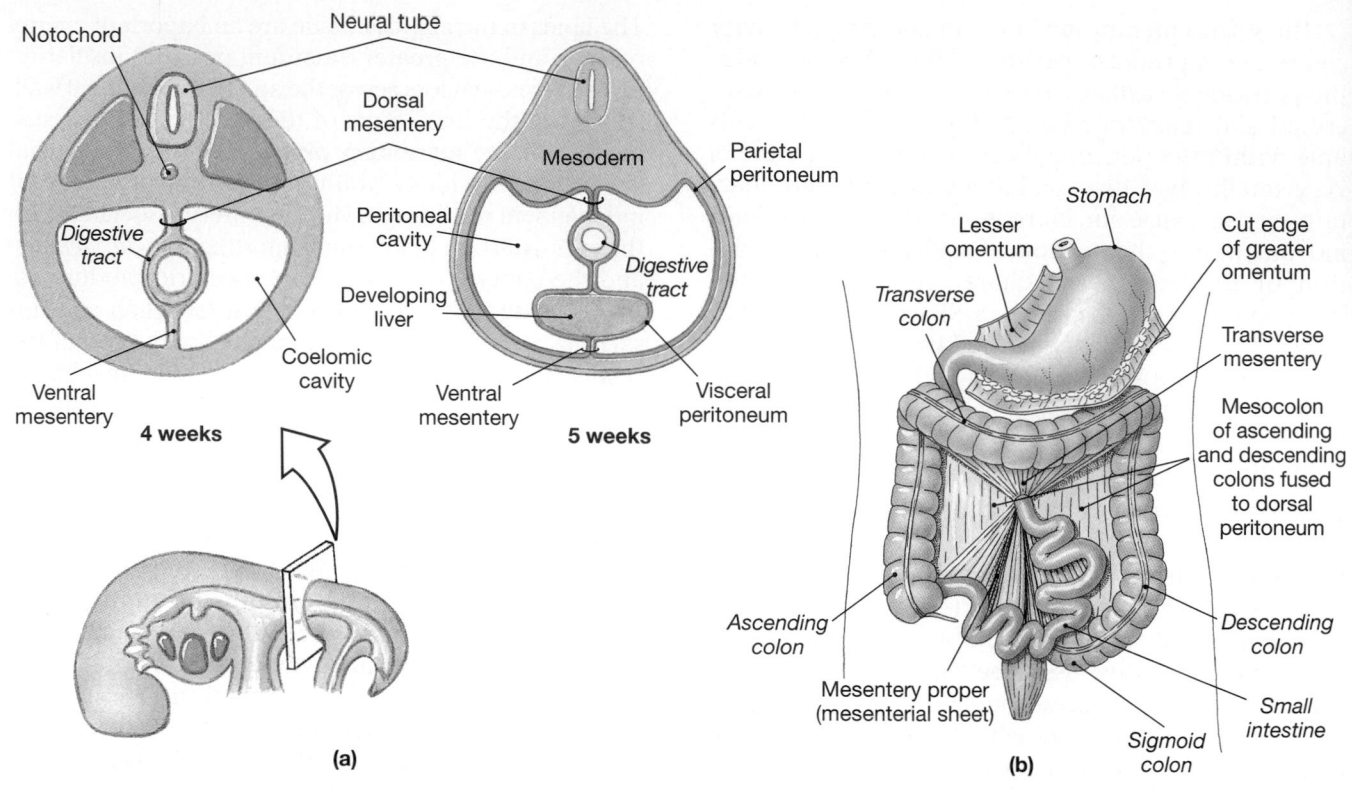

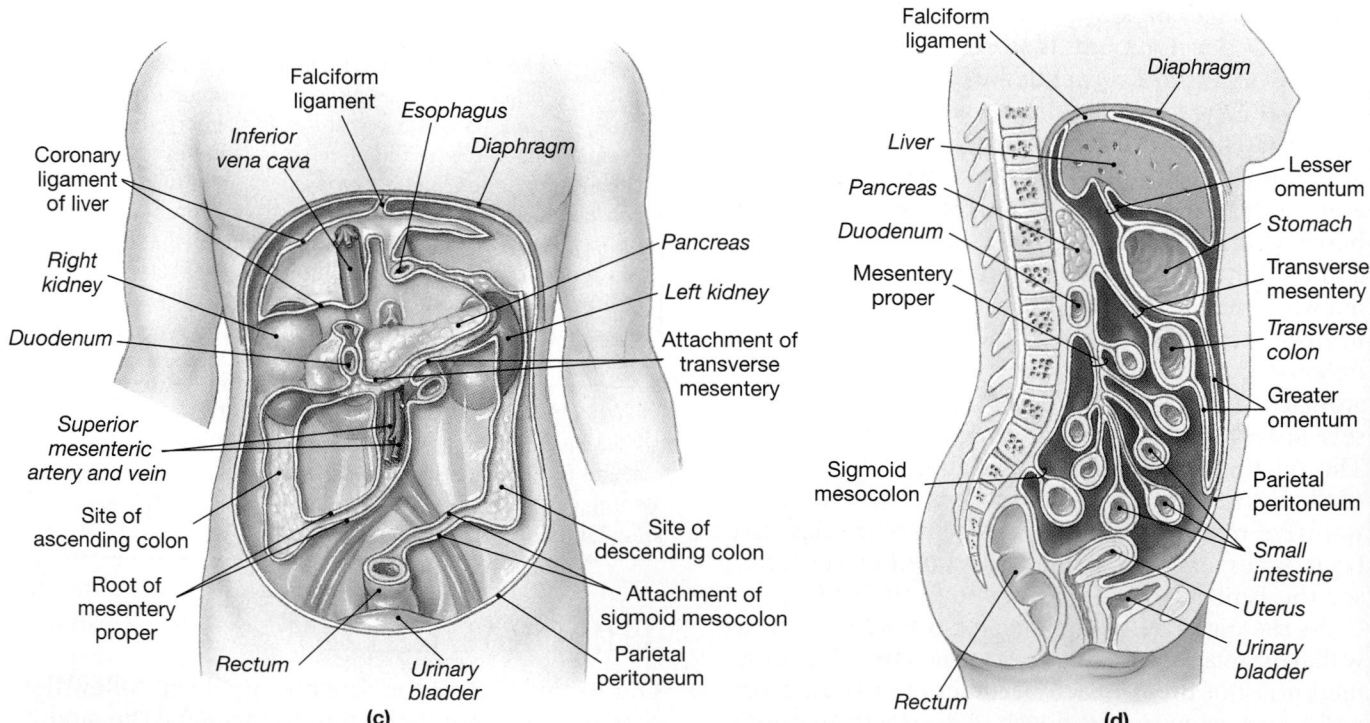

● *FIGURE 24-5* **Mesenteries.** **(a)** During development, the digestive tube is initially suspended by dorsal and ventral mesenteries. The ventral mesentery in the adult is lost except where it connects the stomach to the liver *(lesser omentum)* and the liver to the anterior body wall and diaphragm *(falciform ligament).* **(b)** A diagrammatic view of the organization of mesenteries in the adult. As the digestive tract enlarges, mesenteries associated with the proximal portion of the small intestine, the pancreas, and the ascending and descending portions of the colon fuse to the body wall. **(c)** Anterior view of the empty peritoneal cavity, showing attachment sites where fusion occurs. **(d)** Mesenteries of the adult as seen in sagittal section. Note that the pancreas, duodenum, and rectum are retroperitoneal.

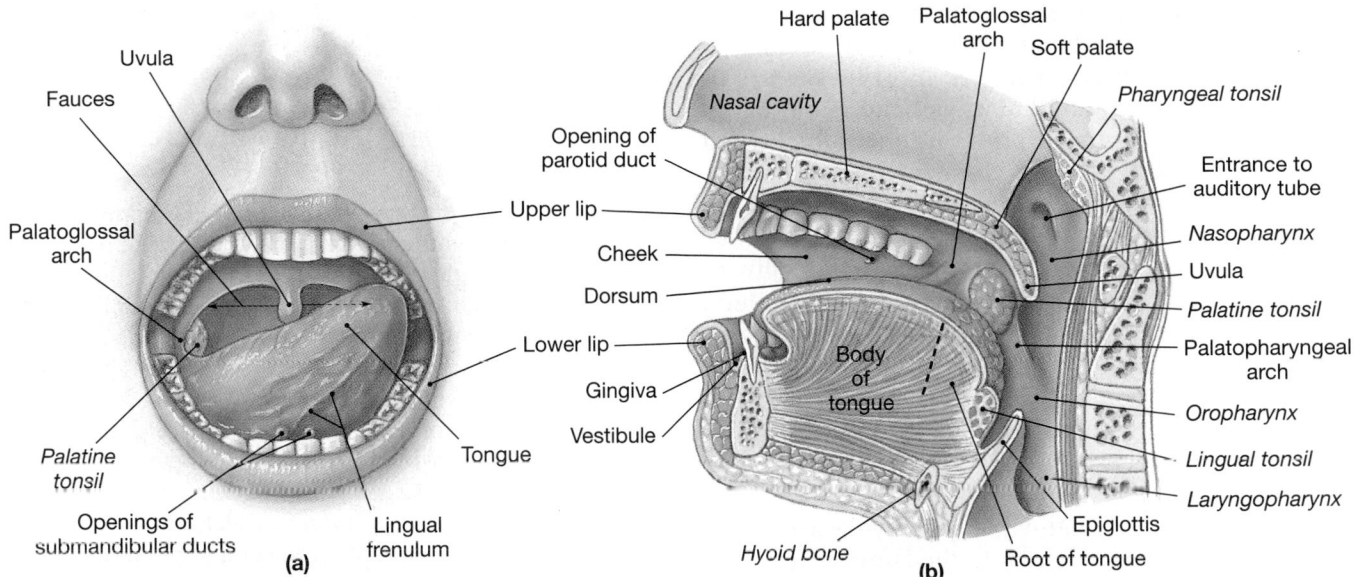

●*FIGURE 24-6* **The Oral Cavity.** (a) An anterior view of the oral cavity, as seen through the open mouth. (b) The oral cavity as seen in sagittal section. AM *Plate 4.4*

The oral cavity, or **buccal** (BUK-al) **cavity** (Figure 24-6●), is lined by the **oral mucosa,** which has a stratified squamous epithelium. Only the regions exposed to severe abrasion, such as the superior surface of the tongue and the opposing surfaces of the hard palate, are covered by a layer of keratinized cells. The epithelial lining of the cheeks, lips, and undersurface of the tongue is relatively thin, nonkeratinized, and delicate. Although nutrient absorption does not occur in the oral cavity, the mucosa inferior to the tongue is thin enough and vascular enough to permit the rapid absorption of lipid-soluble drugs. *Nitroglycerin* is sometimes administered via this route to treat acute angina attacks.

The mucosae of the **cheeks,** or lateral walls of the oral cavity, are supported by pads of fat and the *buccinator muscles.* ∞ *[p. 324]* Anteriorly, the mucosa of each cheek is continuous with that of the lips, or **labia** (LĀ-bē-uh; singular, *labium*). The **vestibule** is the space between the cheeks (or lips) and the teeth. Ridges of oral mucosa, the *gums,* or **gingivae** (JIN-ji-vē), surround the base of each tooth on the alveolar processes of the maxillary bones and mandible. ∞ *[pp. 209, 211]* The gingivae in most regions are firmly bound to the peri-ostea of the underlying bones.

The roof of the oral cavity is formed by the hard and soft palates; the tongue dominates its floor (Figure 24-6b●). ∞ *[p. 818]* The floor of the mouth inferior to the tongue receives extra support from the *mylohyoid muscle.* ∞ *[p. 330]* The hard palate is formed by the palatine processes of the maxillary bones and the horizontal plates of the palatine bones. A prominent central ridge, or *raphe,* extends along the midline of the hard palate. The mucosa lateral and anterior to the raphe is thick, with complex ridges. When your tongue compresses food against your hard palate, these ridges provide traction. The soft palate lies posterior to the hard palate. A thinner and more-delicate mucosa covers the posterior margin of the hard palate and extends onto the soft palate.

The posterior margin of the soft palate supports the dangling **uvula** (Ū-vū-luh) and two pairs of muscular **pharyngeal arches** (Figure 24-6●). On either side, a palatine tonsil lies between an anterior **palatoglossal** (pal-a-tō-GLOS-al) **arch** and a posterior **palatopharyngeal** (pal-a-tō-fa-RIN-jē-al) **arch.** A curving line that connects the palatoglossal arches and uvula forms the boundaries of the **fauces** (FAW-sēz), the passageway between the oral cavity and the oropharynx.

The Tongue

The tongue (Figure 24-6●) manipulates materials inside the mouth and is occasionally used to bring foods (such as ice cream) into the oral cavity. The primary functions of the tongue are (1) mechanical processing by compression, abrasion, and distortion; (2) manipulation to assist in chewing and to prepare the material for swallowing; (3) sensory analysis by touch, temperature, and taste receptors, and (4) secretion of mucus and an enzyme, *lingual lipase.*

We can divide the tongue into an anterior **body,** or *oral portion,* and a posterior **root,** or *pharyngeal portion.* The superior surface, or *dorsum,* of the body contains a forest of fine lingual papillae. We described the structure of the *lingual papillae* in Chapter 17, in our discussion of taste buds and taste sensations. ∞ *[p. 548]* The thickened epithelium covering each papilla assists in the movement of materials by the tongue. A V-shaped line of circumvallate papillae roughly indicates the boundary between the body and the root of the tongue, which is situated in the pharynx (Figure 24-6b●).

The epithelium covering the inferior surface of the tongue is thinner and more delicate than that of the dorsum. Along the inferior midline is the **lingual frenulum** (FREN-ū-lum; *frenulum,* a small bridle), a thin fold of mucous membrane that connects the body of the tongue to the mucosa covering the floor of the oral cavity. Ducts from two pairs of salivary glands open on either side of the lingual frenulum. The lingual frenulum prevents extreme movements of the tongue. However, if your lingual frenulum is *too* restrictive, you cannot eat or speak normally. When properly diagnosed, this condition, called **ankyloglossia** (ang-ki-lō-GLOS-ē-uh), can be corrected surgically.

The tongue's epithelium is flushed by the secretions of small glands that extend into the lamina propria of the tongue. These secretions contain water, mucus, and the enzyme **lingual lipase.** Lingual lipase begins the enzymatic breakdown of lipids, specifically triglycerides, before you have swallowed the food.

Your tongue contains two groups of skeletal muscles: (1) **intrinsic tongue muscles** and (2) **extrinsic tongue muscles.** All gross movements of the tongue are performed by the relatively large extrinsic muscles, which we detailed in Chapter 11. ∞ *[p. 329]* The smaller intrinsic muscles alter the shape of your tongue and assist the extrinsic muscles during precise movements, as in speech. Both intrinsic and extrinsic tongue muscles are under the control of the hypoglossal nerve (N XII).

Salivary Glands

Three pairs of salivary glands (Figure 24-7a●) secrete into the oral cavity. Each pair of salivary glands has a distinctive cellular organization and produces saliva with slightly different properties:

1. The large **parotid** (pa-ROT-id) **salivary glands** lie inferior to the zygomatic arch beneath the skin that covers the lateral and posterior surface of the mandible. Each gland has an irregular shape, extending from the mastoid process of the temporal bone across the outer surface of the masseter muscle. ∞ *[p. 328]* The parotid salivary glands produce a thick, serous secretion containing large amounts of *salivary amylase,* an enzyme that breaks down starches (complex carbohydrates). The secretions of each parotid gland are drained by a **parotid duct** (*Stensen's duct*), which empties into the vestibule at the level of the second upper molar.

2. The **sublingual** (sub-LING-gwal) **salivary glands** are covered by the mucous membrane of the floor of the mouth. These glands produce a watery, mucous secretion that acts as a buffer and lubricant. Numerous **sublingual ducts** (*Rivinus' ducts*) open along either side of the lingual frenulum.

3. The **submandibular salivary glands** are situated in the floor of the mouth along the inner surfaces of the mandible within a depression (∞ *[p. 211]*) called the *mandibular groove.* The submandibular glands (Figure 24-7b●) secrete a mixture of buffers, glycoproteins called *mucins,* and salivary amylase. The **submandibular ducts** (*Wharton's ducts*) open into the mouth on either side of the lingual frenulum immediately posterior to the teeth (Figure 24-6a●).

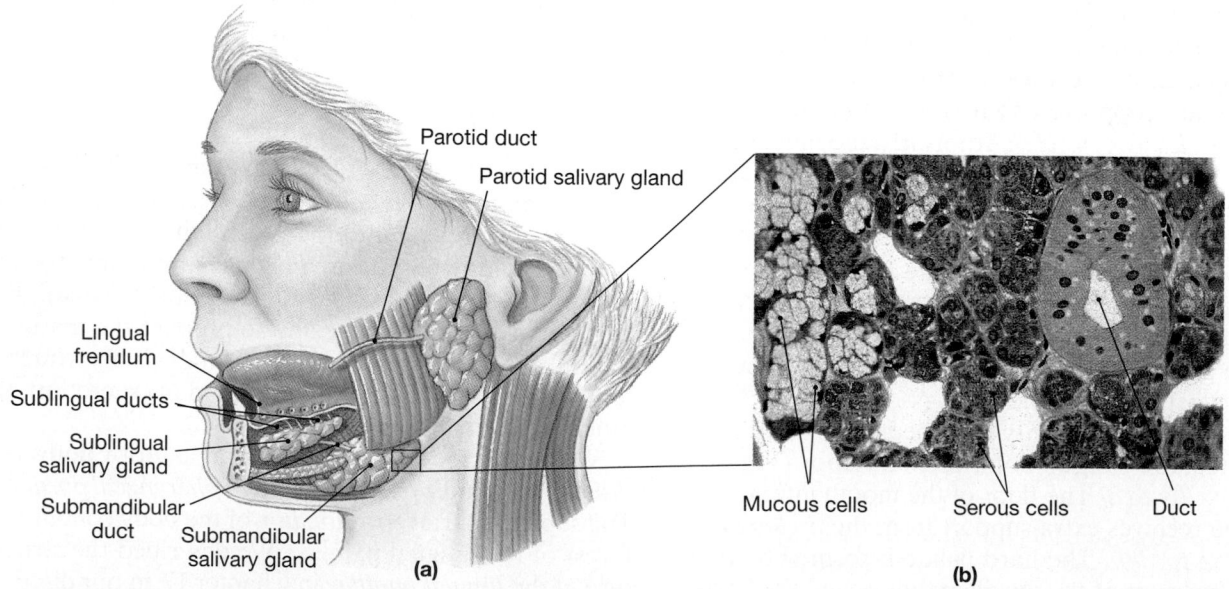

●FIGURE 24-7 The Salivary Glands. (a) Lateral view, showing the relative positions of the salivary glands and ducts on the left side of the head. For clarity, the left ramus and a portion of the right ramus of the mandible have been removed. For the positions of the ducts inside the oral cavity, see Figure 24-6. **(b)** The submandibular gland secretes a mixture of mucins, produced by mucous cells, and enzymes, produced by serous cells. (LM × 303) AM *Plate 4.3*

Saliva

Your salivary glands produce 1.0–1.5 liters of saliva each day. Saliva is 99.4 percent water, and the remaining 0.6 percent includes an assortment of electrolytes (principally Na^+, Cl^-, and HCO_3^-), buffers, glycoproteins, antibodies, enzymes, and waste products. The glycoproteins, called **mucins,** are primarily responsible for the lubricating action of saliva. ∞ *[p. 118]* Saliva is a mixture of glandular secretions; about 70 percent of the saliva originates in the submandibular salivary glands, 25 percent in the parotids, and the remaining 5 percent in the sublingual salivary glands.

A continuous background level of secretion flushes the oral surfaces, helping keep them clean. Buffers within the saliva keep the pH of your mouth near 7.0 and prevent the buildup of acids produced through bacterial action. In addition, saliva contains immunoglobulins (IgA) and lysozymes that help control populations of oral bacteria. A reduction or elimination of salivary secretions, caused by radiation exposure, emotional distress, or other factors, triggers a bacterial population explosion in the oral cavity. This proliferation rapidly leads to recurring infections and the progressive erosion of the teeth and gums.

The saliva produced when you eat has a variety of functions, including:

- Lubricating the mouth.
- Moistening and lubricating materials in the mouth.
- Dissolving chemicals that can stimulate the taste buds and provide sensory information about the material.
- Initiating the digestion of complex carbohydrates before the material is swallowed. The enzyme involved is **salivary amylase,** which is also known as *ptyalin* or *alpha-amylase.* Although the digestive process begins in the oral cavity, it is not completed there, and no absorption of nutrients occurs across the lining of the oral cavity.

MUMPS The **mumps virus** most often targets the salivary glands, especially the parotid salivary glands, although other organs may also become infected. Infection typically occurs at 5–9 years of age. The first exposure stimulates antibody production and in most cases confers permanent immunity; active immunity can be conferred by immunization. In postadolescent males, the mumps virus may also infect the testes and cause sterility. Infection of the pancreas by the mumps virus may produce temporary or permanent diabetes; other organ systems, including the CNS, may be affected in severe cases. An effective mumps vaccine is available. Widespread distribution of that vaccine has reduced the incidence of the disease in the United States.

Control of Salivary Secretion

Salivary secretions are normally controlled by the ANS. Each gland receives parasympathetic and sympathetic innervation. The parasympathetic outflow originates in the **salivatory nuclei** of the medulla oblongata and synapses within the submandibular and otic ganglia. ∞ *[p. 478]* Any object placed in your mouth can trigger a salivary reflex by stimulating receptors monitored by the trigeminal nerve or by stimulating taste buds innervated by N VII, IX, or X. Parasympathetic stimulation accelerates secretion by all the salivary glands, resulting in the production of large amounts of saliva. The role of the sympathetic innervation remains uncertain; evidence suggests that it provokes the secretion of small amounts of very thick saliva.

The salivatory nuclei are also influenced by other brain stem nuclei as well as by the activities of higher centers. For example, chewing with an empty mouth, the smell of food, or even thinking about food will initiate an increase in salivary secretion rates. The presence of irritating stimuli in the esophagus, stomach, or intestines will also accelerate saliva production, as will the sensation of nausea. In functional terms, increased saliva production in response to unpleasant stimuli helps reduce the magnitude of the stimulus by dilution, a rinsing action, or by buffering strong acids or bases.

The Teeth

Movements of the tongue are important in passing food across the opposing surfaces of the teeth. The **occlusal,** or opposing, **surfaces** of your teeth perform chewing, or **mastication** (mas-ti-KĀ-shun), of food. Mastication breaks down tough connective tissues in meat and the plant fibers in vegetable matter, and it helps saturate the materials with salivary secretions and enzymes.

Figure 24-8a• is a sectional view through an adult tooth. The bulk of each tooth consists of a mineralized matrix similar to that of bone. This material, called **dentin** (DEN-tin), differs from bone in that it does not contain living cells. Instead, cytoplasmic processes extend into the dentin from cells in the central **pulp cavity.** The pulp cavity receives blood vessels and nerves from the **root canal,** a narrow tunnel located at the base, or **root,** of the tooth. Blood vessels and nerves enter the root canal through the **apical foramen** to supply the pulp cavity.

The root of each tooth sits within a bony socket called an *alveolus.* Collagen fibers of the **periodontal ligament** extend from the dentin of the root to the alveolar bone, creating a strong articulation known as a *gomphosis.* ∞ *[p. 255]* A layer of **cementum** (se-MEN-tum) covers the dentin of the root, providing protection and firmly anchoring the periodontal ligament. Cementum is very similar in histological structure to bone, and it is less resistant to erosion than is dentin.

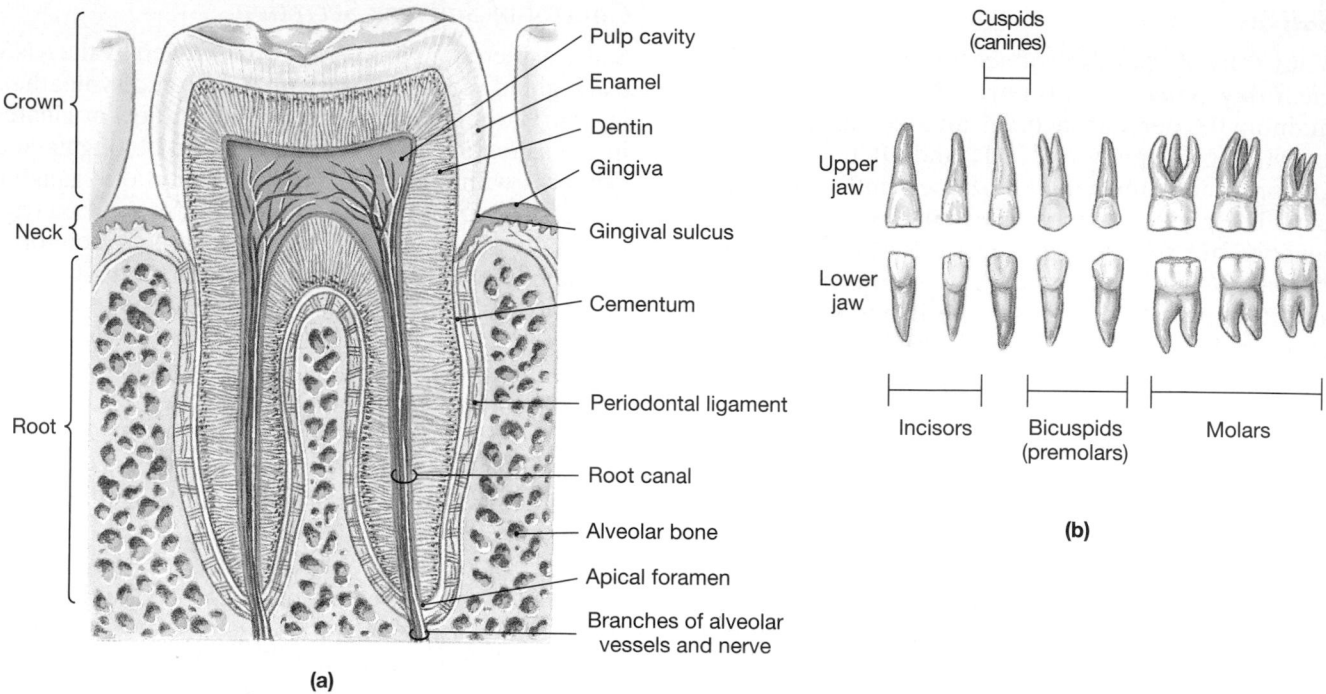

●*FIGURE 24-8* **Teeth.** (a) Diagrammatic section through a typical adult tooth. (b) The adult teeth.

The **neck** of the tooth marks the boundary between the root and the **crown.** The crown is the exposed portion of the tooth. A shallow **gingival sulcus** surrounds the neck of each tooth. The mucosa of the gingival sulcus is very thin, and it is not tightly bound to the periosteum. The epithelium is bound to the tooth over an extensive area. This epithelial attachment prevents bacterial access to the lamina propria of the gingiva and the relatively soft cementum of the root. When you brush and massage your gums, you stimulate the epithelial cells and strengthen the attachment. If the epithelial attachment breaks down, bacterial infection of the gingivae can occur, a condition called **gingivitis.**

The dentin of the crown is covered by a layer of **enamel.** Enamel contains calcium phosphate in a crystalline form; it is the hardest biologically manufactured substance. Adequate amounts of calcium, phosphates, and vitamin D during childhood are essential if the enamel coating is to be complete and resistant to decay.

Tooth decay generally results from the action of bacteria that inhabit your mouth. Bacteria adhering to the surfaces of the teeth produce a sticky matrix that traps food particles and creates deposits of **plaque.** Over time, this organic material can become calcified, forming a hard layer of *tartar,* or *dental calculus,* which can be difficult to remove. Tartar deposits most commonly develop at or near the gingival sulcus, where brushing cannot remove the relatively soft plaque deposits.

DENTAL PROBLEMS The mass of a plaque deposit protects the oral bacteria from salivary secretions. As the bacteria digest nutrients, these pathogens generate acids that erode the structure of the teeth. The result is *dental caries,* otherwise known as cavities. *Streptococcus mutans* is the most abundant bacterium at these sites, and vaccines are now being developed to promote resistance to it and thereby prevent dental caries. If *S. mutans* (or another bacterium) reaches the pulp and infects it, *pulpitis* (pul-PĪ-tis) results. Treatment generally involves the complete removal of the pulp tissue, especially the sensory innervation and all areas of decay; the pulp cavity is then packed with appropriate materials. This procedure is called a *root canal.*

Brushing the exposed surfaces of your teeth after you eat helps prevent the settling of bacteria and the entrapment of food particles, but bacteria between your teeth, in the region known as the *interproximal space,* and within the gingival sulcus may elude the brush. Dentists therefore recommend the daily use of dental floss to clean these spaces and stimulate the gingival epithelium. If bacteria and plaque remain within the gingival sulcus for extended periods, the acids generated will begin eroding the connections between the neck of the tooth and the gingiva. The gums appear to recede from the teeth, and *periodontal disease* develops. As the disease progresses, the bacteria attack the cementum, eventually destroying the periodontal ligament and eroding the alveolar bone. This deterioration loosens the tooth, and it falls out or must be pulled. Periodontal disease is the most common cause for the loss of teeth.

Types of Teeth

The alveolar processes of the maxillary bones and the mandible form the *upper* and *lower dental arches,* respectively. There are four types of teeth within these arches, each with specific functions (Figure 24-8b●):

1. **Incisors** (in-SĪ-zerz) are blade-shaped teeth located at the front of the mouth. Incisors are useful for clipping or cutting, as when you nip off the tip of a carrot stick. These teeth have a single root.

2. The **cuspids** (KUS-pidz), or *canines,* are conical, with a sharp ridgeline and a pointed tip. They are used for tearing or slashing. You might weaken a tough piece of celery by the clipping action of the incisors but then take advantage of the shearing action provided by the cuspids. Cuspids have a single root.

3. **Bicuspids** (bī-KUS-pidz), or *premolars,* have flattened crowns with prominent ridges. They are used for crushing, mashing, and grinding. Bicuspids have one or two roots.

4. **Molars** have very large, flattened crowns with prominent ridges. They excel at crushing and grinding. You can usually shift a tough nut or sparerib to your bicuspids and molars for successful crunching. Molars typically have three or more roots.

Dental Succession

During development, two sets of teeth begin to form. The first to appear are the **primary dentition: deciduous teeth** (de-SID-ū-us; *deciduus,* falling off), also known as *primary teeth, milk teeth,* or *baby teeth.* Most children have 20 deciduous teeth—five on each side of the upper and lower jaws (Figure 24-9a●). These teeth will later be replaced by the adult **secondary dentition,** or *permanent dentition* (Figure 24-9b●). Adult jaws are larger and can accommodate more than 20 permanent teeth. Three additional molars appear on each side of the upper and lower jaws as the individual ages. These teeth extend the length of the tooth rows posteriorly and bring the permanent tooth count to 32.

On each side of the upper or lower jaw, the primary dentition consists of two incisors, one cuspid, and a pair of deciduous molars. These deciduous teeth are gradually replaced by the permanent dentition.

As replacement proceeds, the periodontal ligaments and roots of the primary teeth are eroded until eventually the deciduous teeth either fall out or are pushed aside by the **eruption,** or emergence, of the secondary teeth. The adult premolars take the place of the deciduous molars, and the definitive adult molars extend the tooth row as the jaw enlarges. The third molars, or *wisdom teeth,* may not erupt before age 21. When wisdom teeth fail to erupt, it is because they either develop in

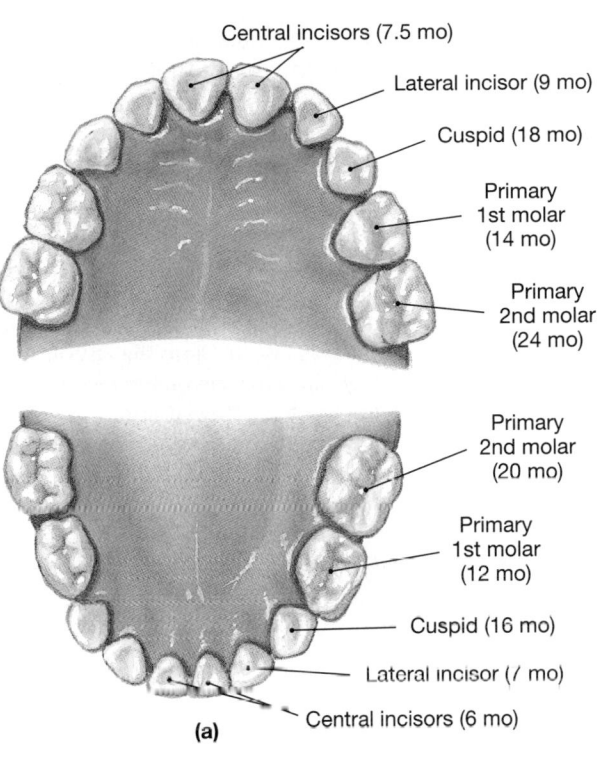

Central incisors (7.5 mo)
Lateral incisor (9 mo)
Cuspid (18 mo)
Primary 1st molar (14 mo)
Primary 2nd molar (24 mo)
Primary 2nd molar (20 mo)
Primary 1st molar (12 mo)
Cuspid (16 mo)
Lateral incisor (7 mo)
Central incisors (6 mo)

(a)

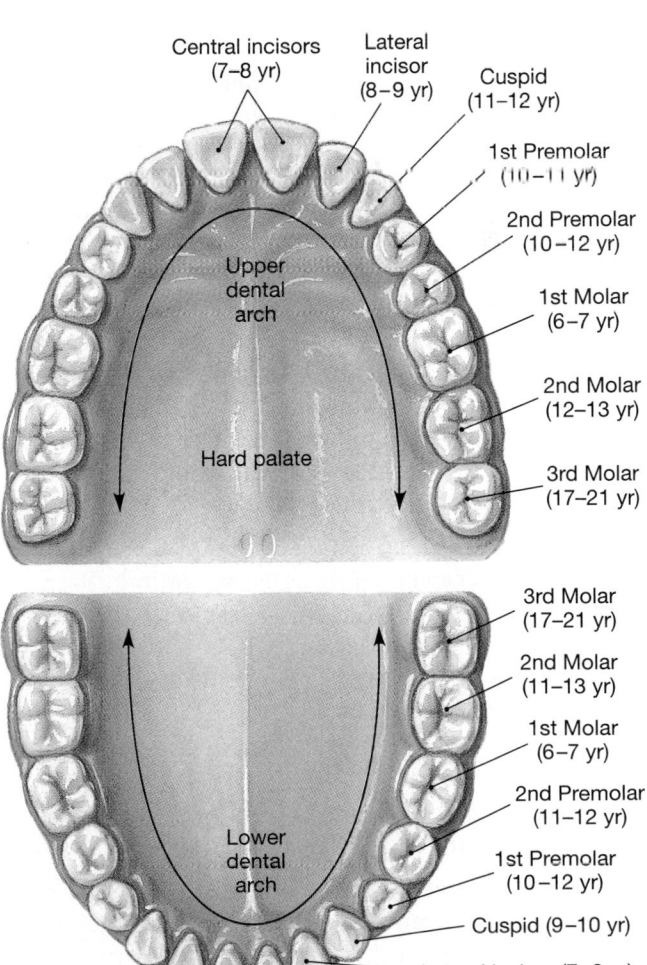

Central incisors (7–8 yr)
Lateral incisor (8–9 yr)
Cuspid (11–12 yr)
1st Premolar (10–11 yr)
2nd Premolar (10–12 yr)
1st Molar (6–7 yr)
2nd Molar (12–13 yr)
3rd Molar (17–21 yr)
Upper dental arch
Hard palate

3rd Molar (17–21 yr)
2nd Molar (11–13 yr)
1st Molar (6–7 yr)
2nd Premolar (11–12 yr)
1st Premolar (10–12 yr)
Cuspid (9–10 yr)
Lateral incisor (7–8 yr)
Central incisors (6–7 yr)
Lower dental arch

(b)

●FIGURE 24-9 Primary and Secondary Teeth. **(a)** The primary teeth, with the age at eruption given in months. **(b)** The adult teeth, with the age at eruption given in years.

inappropriate positions or because there is inadequate space on the dental arch. Any teeth that develop in locations that do not permit their eruption are called *impacted teeth.* Impacted wisdom teeth can be surgically removed to prevent formation of abscesses in later life.

DENTAL IMPLANTS Lost or broken teeth have commonly been replaced by "false teeth" attached to a plate or frame inserted into the mouth. In the 1980s, an alternative was developed that uses *dental implants.* A ridged titanium cylinder is inserted into the alveolus, and osteoblasts lock the ridges into the surrounding bone. After 4–6 months, an artificial tooth is screwed into the cylinder. Dental implants are not suitable for everyone. For example, there must be enough alveolar bone present to provide a firm attachment, and there may be complications during or after surgery. Nevertheless, as the technique evolves, dental implants will become increasingly important. Roughly 42 percent of individuals over age 65 have lost all their teeth; the rest have lost an average of 10 teeth.

Mastication

The *muscles of mastication* close your jaws and slide or rock your lower jaw from side to side. ∞ *[p. 328]* Chewing is not a simple process, as it can involve any combination of mandibular elevation, depression, protraction, retraction, and medial–lateral movement. (Try classifying the movements involved the next time you eat.)

During mastication, you force food from the oral cavity to the vestibule and back, crossing and recrossing the occlusal surfaces. This movement results in part from the action of the masticatory muscles, but control would be impossible without the aid of the buccal, labial, and lingual muscles. ∞ *[pp. 324–326, 329]* Once you have shredded or torn the material to a satisfactory consistency and moistened it with salivary secretions, your tongue begins compacting the debris into a **bolus,** or small oval mass. You can swallow a compact, moist, cohesive bolus relatively easily.

☑ Which type of epithelium lines the oral cavity?

☑ The digestion of which nutrient would be affected by damage to the parotid glands?

☑ Which type of tooth is most useful for chopping off bits of relatively rigid foods?

THE PHARYNX

The pharynx serves as a common passageway for solid food, liquids, and air. We described the epithelial lining and divisions of the pharynx—the nasopharynx, the oropharynx, and the laryngopharynx—in Chapter 23. ∞ *[p. 820]* Food normally passes through the oropharynx and laryngopharynx on its way to the esophagus. Both of these divisions have a stratified squamous epithelium similar to that of the oral cavity, and the lamina propria contains scattered mucous glands and the lymphoid tissue of the pharyngeal, palatal, and lingual tonsils. Beneath the lamina propria lies a dense layer of elastic fibers, bound to the underlying skeletal muscles.

We detailed the specific pharyngeal muscles involved in swallowing in Chapter 11. ∞ *[pp. 328–331]* In summary:

- The *pharyngeal constrictors* push the bolus toward the esophagus.
- The *palatopharyngeus* and *stylopharyngeus* elevate the larynx.
- The *palatal muscles* elevate the soft palate and adjacent portions of the pharyngeal wall.

These muscles cooperate with muscles of the oral cavity and esophagus to initiate the process of swallowing, which pushes the bolus along the esophagus and into the stomach.

THE ESOPHAGUS

The esophagus (Figure 24-10●) is a hollow muscular tube with a length of approximately 25 cm (1 ft) and a diameter of about 2 cm (0.75 in.) at its widest point. The primary function of the esophagus is to carry solid food and liquids to the stomach.

The esophagus begins posterior to the cricoid cartilage, at the level of vertebra C_6. From this point, the narrowest portion of the esophagus, it descends toward the thoracic cavity posterior to the trachea, passes inferiorly along the dorsal wall of the mediastinum, and enters the abdominopelvic cavity through the **esophageal hiatus** (hī-Ā-tus), an opening in the diaphragm. The esophagus then empties into the stomach anterior to vertebra T_7.

Within the neck, the esophagus receives blood from the *external carotid* and *thyrocervical arteries*. In the mediastinum, it is supplied by the *esophageal arteries* and branches of the *bronchial arteries*. As it passes through the esophageal hiatus, the esophagus receives blood from the *inferior phrenic arteries*, and the portion adjacent to the stomach is supplied by the *left gastric artery*. Blood from esophageal capillaries collects into the *esophageal, inferior thyroid, azygos,* and *gastric veins.* ∞ *[pp. 747, 754, 756]*

The esophagus is innervated by parasympathetic and sympathetic fibers from the *esophageal plexus.* ∞ *[p. 527]* Resting muscle tone in the circular muscle layer in the upper 3 cm (1 in.) of the esophagus normally prevents air from entering your esophagus.

•FIGURE 24-10 The Esophagus.
(a) Low-power view of a section through the esophagus. (b) The esophageal mucosa. (LM × 77) (c) The transition between the esophageal and gastric mucosae at the lower esophageal sphincter. (LM × 94)

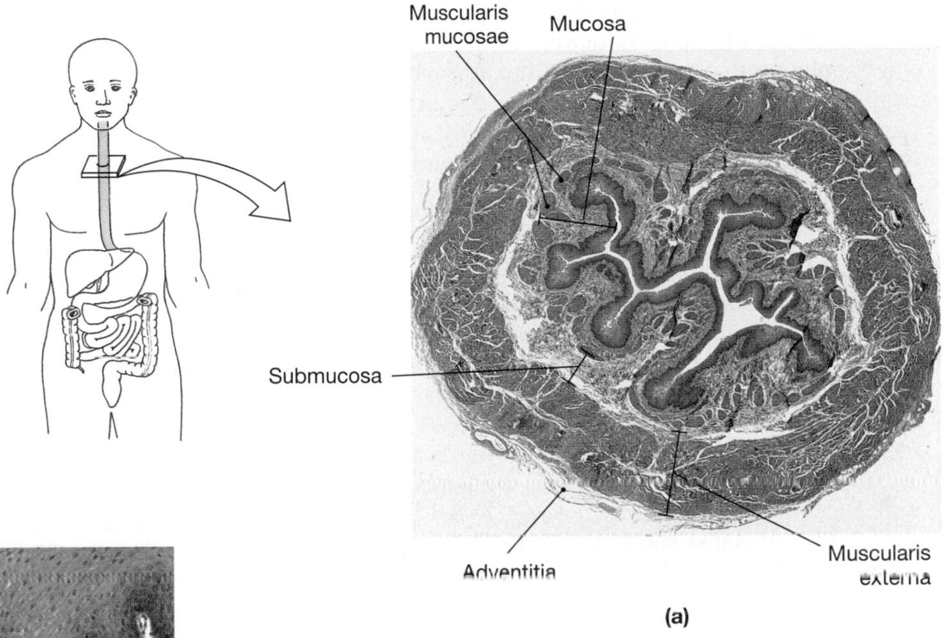

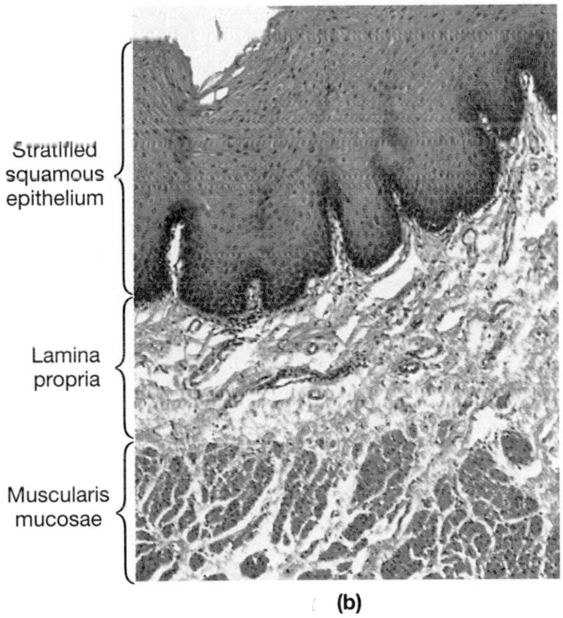

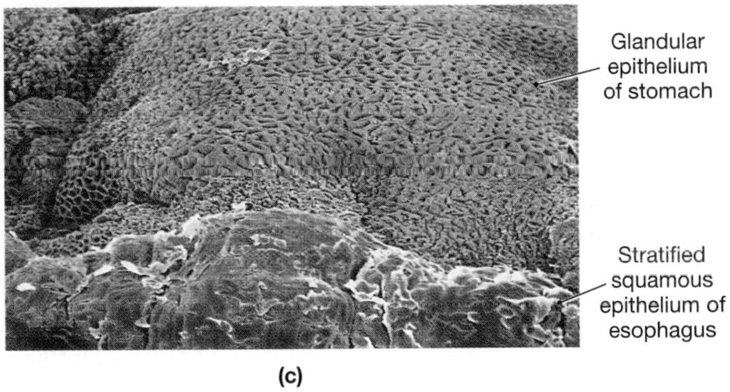

(a)

(b)

(c)

A comparable zone at the inferior end of the esophagus normally remains in a state of active contraction. This condition prevents the backflow of materials from the stomach into the esophagus. Neither region contains a well-defined sphincter muscle comparable to those located elsewhere along the digestive tract. Nevertheless, the terms *upper esophageal sphincter* and *lower esophageal sphincter (cardiac sphincter)* are often used in recognition of the similarity in function.

Histology of the Esophagus

The wall of the esophagus contains mucosal, submucosal, and muscularis layers comparable to those described in Figure 24-2•, p. 864. Distinctive features of the esophageal wall (Figure 24-10•) include the following:

- The mucosa of the esophagus contains a nonkeratinized stratified squamous epithelium similar to that of the pharynx and oral cavity.
- The mucosa and submucosa are thrown into large folds that extend the length of the esophagus. These folds allow for expansion during the passage of a large bolus; except when you swallow, muscle tone in the walls keeps the lumen closed.
- The muscularis mucosae consists of an irregular layer of smooth muscle.
- The submucosa contains scattered *esophageal glands* that produce a mucous secretion that reduces friction between the bolus and the esophageal lining.
- The muscularis externa has inner circular and outer longitudinal layers. In the upper third of the esophagus, these layers contain skeletal muscle fibers; in the middle third, there is a mixture of skeletal and smooth muscle tissue; along the lower third, there is only smooth muscle.

- There is no serosa, but an adventitia of connective tissue outside the muscularis externa anchors the esophagus in position against the dorsal body wall. Over the 1–2 cm between the diaphragm and stomach, the esophagus is retroperitoneal, with peritoneum covering the anterior and left lateral surfaces.

ESOPHAGEAL VARICES The veins draining the inferior portion of the esophagus empty into tributaries of the hepatic portal vein. If the venous pressure in the hepatic portal vein becomes abnormally high, due to liver damage or constriction of the vessels, blood will pool in the submucosal veins of the esophagus. The veins become grossly distended and create bulges in the esophageal wall. These distorted *esophageal varices* (VAR-i-sēz; *varices,* dilated veins) may constrict the esophageal passageway. They may also rupture, bleeding into the submucosal tissues or onto the epithelial surface. Esophageal varices commonly develop in individuals with advanced *cirrhosis,* a chronic liver disorder that restricts hepatic blood flow.

Swallowing

Swallowing, or **deglutition**, is a complex process whose initiation can be voluntarily controlled, but it proceeds automatically once it begins. Although you are consciously aware of, and voluntarily control, swallowing when you eat or drink, swallowing can also occur unconsciously, as saliva collects at the back of the mouth. Each day you swallow approximately 2400 times. We can divide swallowing into *buccal, pharyngeal,* and *esophageal phases:*

1. The **buccal phase** begins with the compression of the bolus against the hard palate. Subsequent retraction of the tongue then forces the bolus into the pharynx and assists in the elevation of the soft palate, thereby isolating the nasopharynx (Figure 24-11a,b•). The buccal phase is strictly voluntary. Once the bolus enters the oropharynx, reflex responses are initiated, and the bolus is moved toward the stomach.

2. The **pharyngeal phase** begins as the bolus comes into contact with the palatoglossal and palatopharyngeal arches and the posterior pharyngeal wall (Figure 24-11c,d•). The **swallowing reflex** begins as tactile receptors on the palatal arches and uvula are stimulated by the passage of the bolus. The information is relayed to the **swallowing center** of the medulla oblongata over the trigeminal and glossopharyngeal nerves. Motor commands originating at this center then target the pharyngeal musculature, producing a

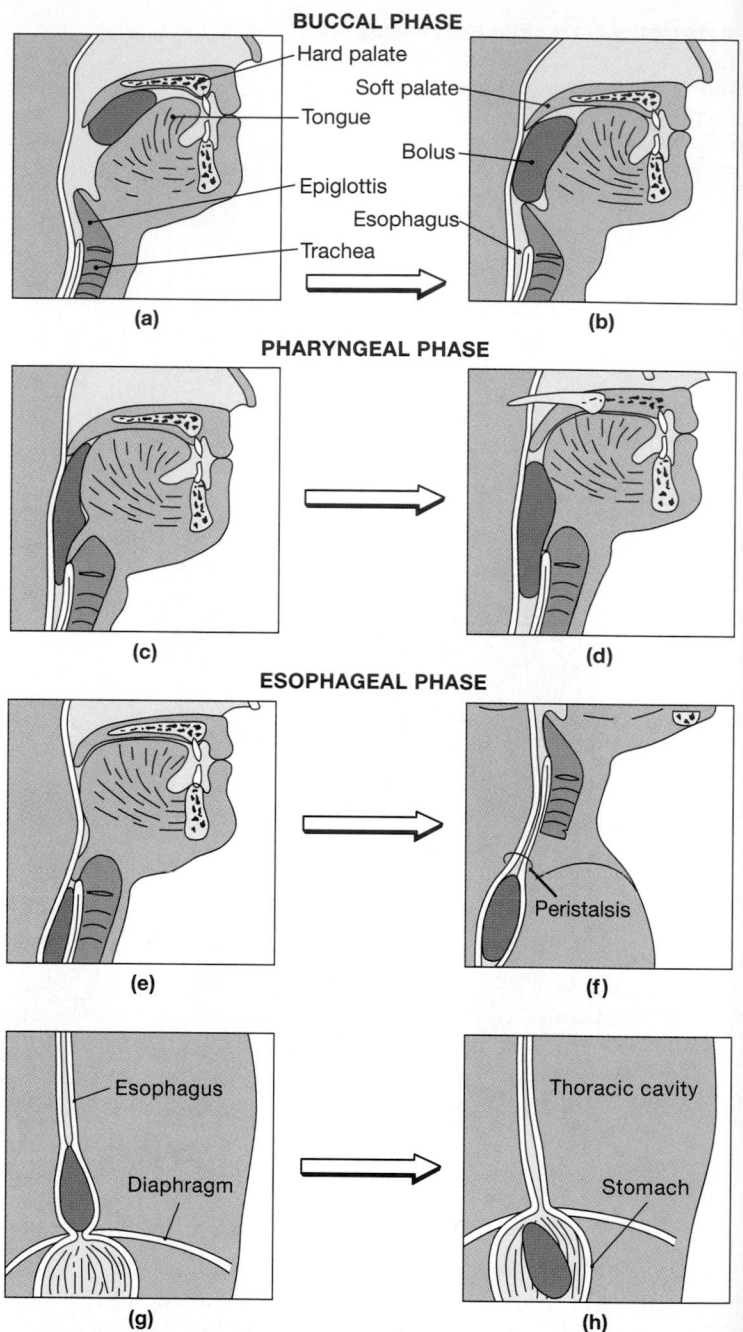

BUCCAL PHASE

Hard palate
Soft palate
Tongue
Bolus
Epiglottis
Esophagus
Trachea

(a)   (b)

PHARYNGEAL PHASE

(c)   (d)

ESOPHAGEAL PHASE

Peristalsis

(e)   (f)

Esophagus   Thoracic cavity

Diaphragm   Stomach

(g)   (h)

•**FIGURE 24-11** **The Swallowing Process.** This sequence, based on a series of X-rays, shows the stages of swallowing and the movement of materials from the mouth to the stomach.

coordinated and stereotyped pattern of muscle contraction. Elevation of the larynx and folding of the epiglottis direct the bolus past the closed glottis, while the uvula and soft palate block passage back to the nasopharynx. It takes less than a second for the pharyngeal muscles to propel the bolus into the esophagus. During this period the respiratory centers are inhibited, and breathing stops.

3. The **esophageal phase** of swallowing begins as the contraction of pharyngeal muscles forces the bolus

through the entrance to the esophagus (Figure 24-11e–h●). Once within the esophagus, the bolus is pushed toward the stomach by a peristaltic wave. The approach of the bolus triggers the opening of the lower esophageal sphincter, and the bolus then continues into the stomach.

Primary peristaltic contractions are coordinated by afferent and efferent fibers within the glossopharyngeal and vagus nerves. For a typical bolus, the entire trip takes about 9 seconds to complete. Liquids may make the journey in a few seconds, arriving ahead of the peristaltic contractions with the assistance of gravity.

A relatively dry or poorly lubricated bolus travels much more slowly, and a series of **secondary peristaltic waves** may be required to push it all the way to the stomach. Secondary peristaltic waves are local reflexes triggered by the stimulation of sensory receptors in the esophageal walls. These receptors relay information by way of the submucosal and myenteric plexuses, producing peristaltic contractions in the absence of CNS instructions. You cannot swallow a completely dry bolus, because friction with the walls of the esophagus will make peristalsis ineffective. (For this reason, you cannot swallow an entire slice of processed white bread without taking a drink.)

ACHALASIA AND ESOPHAGITIS In the condition known as *achalasia* (ak-a-LĀ-zē-uh), a bolus descends relatively slowly, due to abnormally weak peristaltic waves, and its arrival does not trigger the opening of the lower esophageal sphincter. Materials then accumulate at the base of the esophagus like cars at a stop light. Secondary peristaltic waves may occur repeatedly, adding to the individual's discomfort. The most successful treatment involves cutting the circular muscle layer at the base of the esophagus or expanding a balloon in the lower esophagus until the muscle layer tears.

A weakened or permanently relaxed sphincter can cause inflammation of the esophagus, or *esophagitis* (ē-sof-a-JĪ-tis), as powerful stomach acids enter the lower esophagus. The esophageal epithelium has few defenses against acid and enzyme attack, and inflammation, epithelial erosion, and intense discomfort are the result. Occasional incidents of reflux, or backflow, from the stomach are responsible for the symptoms of *heartburn*. This relatively common problem supports a multimillion dollar industry devoted to producing and promoting antacids.

✔ What is unusual about the muscularis externa of the esophagus?

✔ Where in the human body would you find the fauces?

✔ What is occurring when the soft palate and larynx elevate and the glottis closes?

THE STOMACH

The stomach performs four major functions: (1) the bulk storage of ingested food, (2) the mechanical breakdown of ingested food, (3) the disruption of chemical bonds in food material through the action of acids and enzymes, and (4) the production of *intrinsic factor,* a glycoprotein whose presence in the digestive tract is required for the absorption of vitamin B_{12}. The mixing of ingested substances with the secretions of the glands of the stomach produces a viscous, highly acidic, soupy mixture of partially digested food. This material is called **chyme** (kīm).

Anatomy of the Stomach

The stomach has the shape of an expanded J (Figure 24-12●). A short **lesser curvature** forms the **medial surface** of the organ, and a long **greater curvature** forms the **lateral surface.** The **anterior** and **posterior surfaces** are smoothly rounded. The shape and size of the stomach are extremely variable from individual to individual and even from one meal to the next. In an "average" stomach, the lesser curvature has a length of approximately 10 cm (4 in.), and the greater curvature measures about 40 cm (16 in.). The stomach typically extends between the levels of vertebrae T_7 and L_3.

We can divide the stomach into four regions (Figure 24-12●):

1. *The cardia.* The **cardia** (KAR-dē-uh) is the smallest part of the stomach. It consists of the superior, medial portion of the stomach within 3 cm (1.2 in.) of the junction with the esophagus. The cardia contains abundant mucous glands whose secretions coat the connection with the esophagus and help protect that tube from the acids and enzymes of the stomach.

2. *The fundus.* The portion of the stomach superior to the junction between the stomach and esophagus is the **fundus** (FUN-dus). The fundus contacts the inferior and posterior surface of the diaphragm (Figure 24-12a●).

3. *The body.* The area between the fundus and the curve of the J is the **body** of the stomach. The body is the largest region of the stomach, and it functions as a mixing tank for ingested food and secretions produced within the stomach. *Gastric glands* (*gaster,* stomach) in the fundus and body secrete most of the acids and enzymes involved in gastric digestion.

4. *The pylorus.* The **pylorus** (pī-LOR-us) is the curve of the J. The pylorus is divided into a **pyloric antrum** (*antron,* cavity), which is connected to the body, and a **pyloric canal** that empties into the *duodenum,* the proximal segment of the small intestine. As mixing movements occur during digestion, the pylorus frequently changes shape. A muscular **pyloric sphincter** regulates the release of chyme into the duodenum. Glands in the pylorus secrete mucus and important digestive hormones, including *gastrin,* a hormone that stimulates the activity of gastric glands.

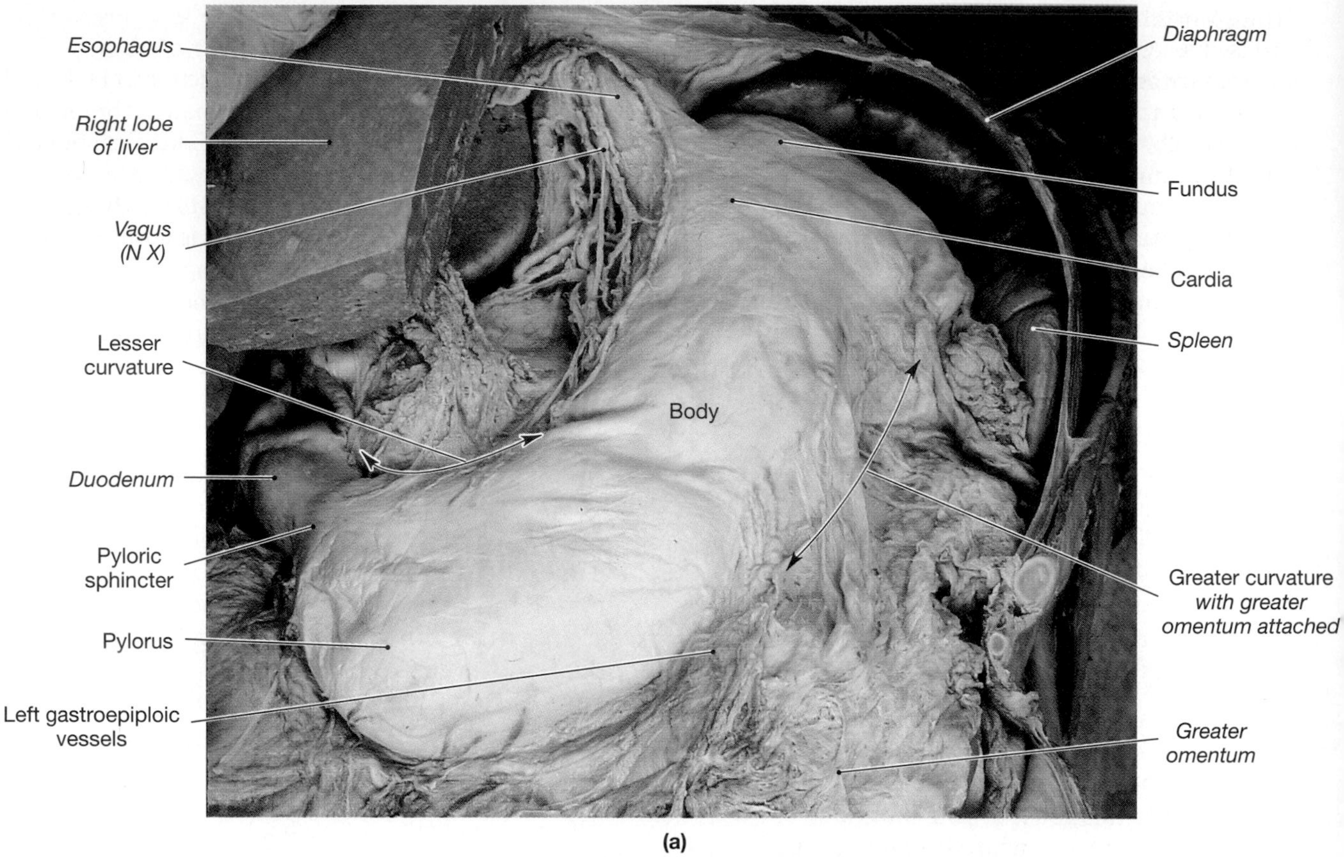

Esophagus

Right lobe of liver

Vagus (N X)

Lesser curvature

Duodenum

Pyloric sphincter

Pylorus

Left gastroepiploic vessels

Diaphragm

Fundus

Cardia

Spleen

Body

Greater curvature with greater omentum attached

Greater omentum

(a)

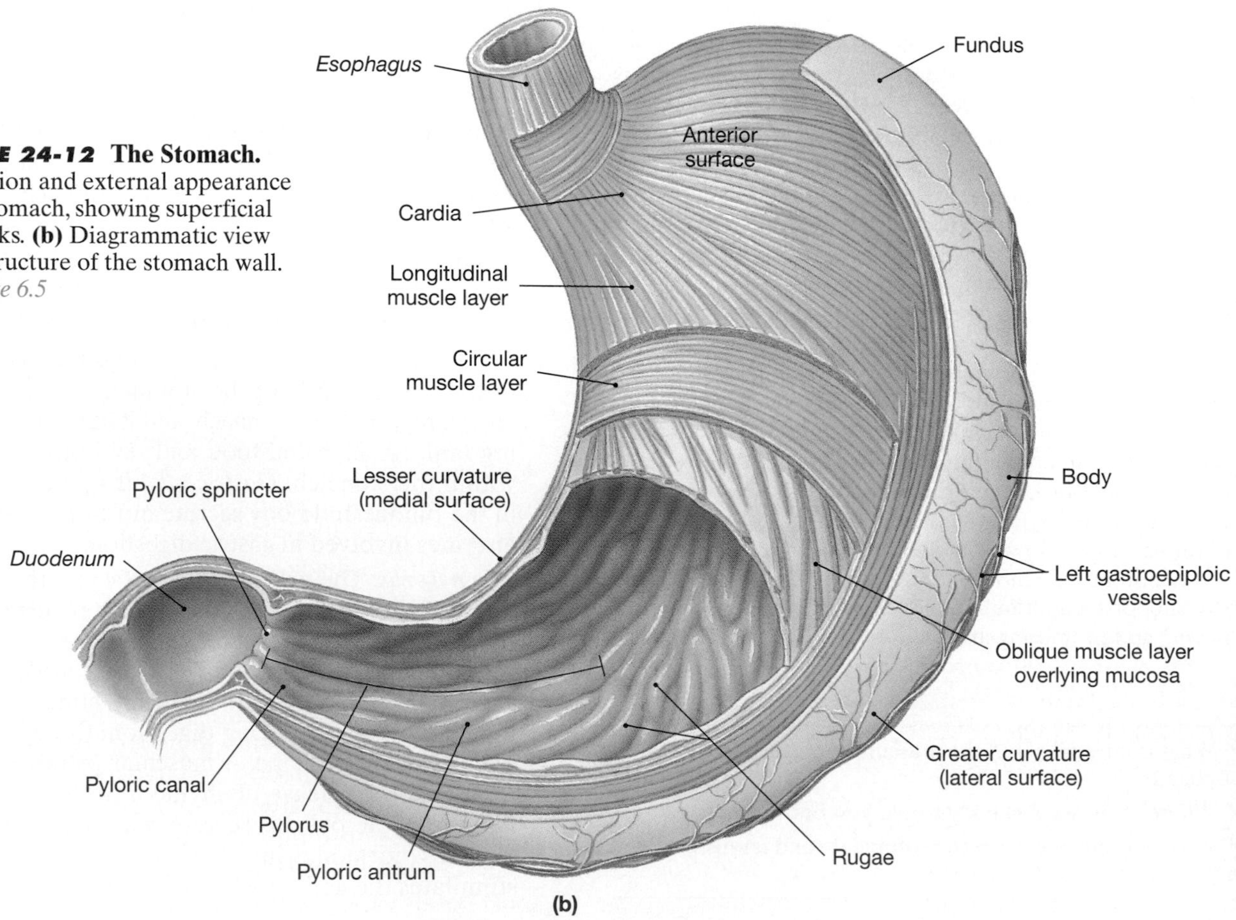

•FIGURE 24-12 **The Stomach.**
(a) Position and external appearance of the stomach, showing superficial landmarks. **(b)** Diagrammatic view of the structure of the stomach wall.
AM Plate 6.5

Esophagus

Fundus

Anterior surface

Cardia

Longitudinal muscle layer

Circular muscle layer

Pyloric sphincter

Duodenum

Lesser curvature (medial surface)

Body

Left gastroepiploic vessels

Oblique muscle layer overlying mucosa

Greater curvature (lateral surface)

Pyloric canal

Pylorus

Pyloric antrum

Rugae

(b)

The stomach's volume increases at mealtime, then decreases as chyme enters the small intestine. When your stomach is relaxed (empty), the mucosa is thrown into prominent folds called **rugae** (ROO-gē; wrinkles) (Figure 24-12b•). Rugae are temporary features that let the gastric lumen expand. As your stomach fills, the rugae flatten out. When your stomach is full, the rugae almost disappear. When empty, your stomach resembles a muscular tube with a narrow, constricted lumen. When full, it can expand to contain 1–1.5 liters of material.

Musculature of the Stomach

The muscularis mucosae and muscularis externa of the stomach contain extra layers of smooth muscle cells in addition to the usual circular and longitudinal layers. The muscularis mucosae generally contains an outer, circular layer of muscle cells. The muscularis externa has an inner, **oblique layer** of smooth muscle (Figure 24-12b•). The extra layer of smooth muscle strengthens the stomach wall and assists in the mixing and churning activities essential to the formation of chyme.

Histology of the Stomach

A simple columnar epithelium lines all portions of the stomach (Figure 24-13a•). The epithelium is a *secretory sheet* that produces a carpet of mucus that covers the interior surfaces of the stomach. The alkaline mucous layer protects epithelial cells against the acids and enzymes in the gastric lumen.

The stomach receives blood from (1) the *left gastric artery;* (2) the *splenic artery,* which supplies the *left gastroepiploic artery;* and (3) the *common hepatic artery,* which supplies the *right gastric, gastroduodenal,* and *right gastroepiploic arteries.* We detailed those vessels in Chapter 21. ∞ *[pp. 745–747]*

Shallow depressions called **gastric pits** open onto the gastric surface (Figure 24-13b,c,d•). The mucous cells at the base, or *neck,* of each gastric pit are actively dividing, replacing superficial cells that are shed into the chyme. The continuous replacement of epithelial cells provides an additional defense against the acidic gastric contents. A typical epithelial cell has a life span of 3–7 days, but exposure to strong alcohol or other chemicals can increase the rate of cell turnover.

GASTRIC GLANDS In the fundus and body of the stomach, each gastric pit communicates with several **gastric glands** that extend deep into the underlying lamina propria. Gastric glands (Figure 24-13b,c,d•) are dominated by two types of secretory cells: (1) *parietal cells* and (2) *chief cells.* Together they secrete about 1500 ml of **gastric juice** each day.

Parietal Cells. **Parietal cells** are especially common along the proximal portions of each gastric gland. These cells secrete *intrinsic factor* and *hydrochloric acid* (HCl). **Intrinsic factor** is a glycoprotein that fa-

cilitates the absorption of **vitamin B$_{12}$** across the intestinal lining. We introduced this vitamin, essential for normal erythropoiesis, in Chapter 19. ∞ *[p. 652]*

The parietal cells do not produce HCl in the cytoplasm, because it is such a strong acid that it would erode a secretory vesicle and destroy the cell. Instead, H$^+$ and Cl$^-$, the two ions that form HCl, are transported independently by different mechanisms (Figure 24-14•, p. 881). Hydrogen ions are generated inside the cell as the enzyme carbonic anhydrase converts carbon dioxide and water to carbonic acid. The carbonic acid promptly dissociates into hydrogen ions and bicarbonate ions. The hydrogen ions are actively transported into the lumen of the gastric gland. The bicarbonate ions are ejected into the interstitial fluid by a countertransport mechanism that exchanges intracellular bicarbonate ions for extracellular chloride ions. The chloride ions then diffuse across the cell and through open chloride channels in the cell membrane to the lumen of the gastric gland.

The bicarbonate ions released by the parietal cell diffuse through the interstitial fluid into the bloodstream. When gastric glands are actively secreting, enough bicarbonate ions enter the circulation to increase the pH of the blood significantly. This sudden influx of bicarbonate ions has been called the *alkaline tide.*

The secretory activities of the parietal cells can keep the stomach contents at a pH of 1.5–2.0. This highly acidic environment does not by itself digest the chyme but has four important functions:

1. The low pH of gastric juice kills most of the microorganisms ingested with food.

2. The low pH denatures proteins and inactivates most of the enzymes in food.

3. The acid helps break down plant cell walls and the connective tissues in meat.

4. An acidic environment is essential for the activation and function of *pepsin,* a protease (protein-digesting enzyme) secreted by the chief cells.

Chief Cells. **Chief cells** are most abundant near the base of a gastric gland. These cells secrete **pepsinogen** (pep-SIN-ō-jen), an inactive proenzyme. Pepsinogen is converted by the acid in the gastric lumen to **pepsin**, an active *proteolytic* (protein-digesting) enzyme. Pepsin functions most effectively at a strongly acidic pH of 1.5–2.0. The stomachs of newborn infants (but not of adults) also produce **rennin**, also known as *chymosin,* and **gastric lipase**, enzymes important for the digestion of milk. Rennin coagulates milk proteins; gastric lipase initiates the digestion of milk fats.

PYLORIC GLANDS Glands in the pylorus produce primarily a mucous secretion rather than enzymes or acid. In addition, several different types of enteroendocrine cells are scattered among the mucus-secreting cells. These cells produce at least seven different hormones, most notably the hormone **gastrin** (GAS-trin). Gastrin

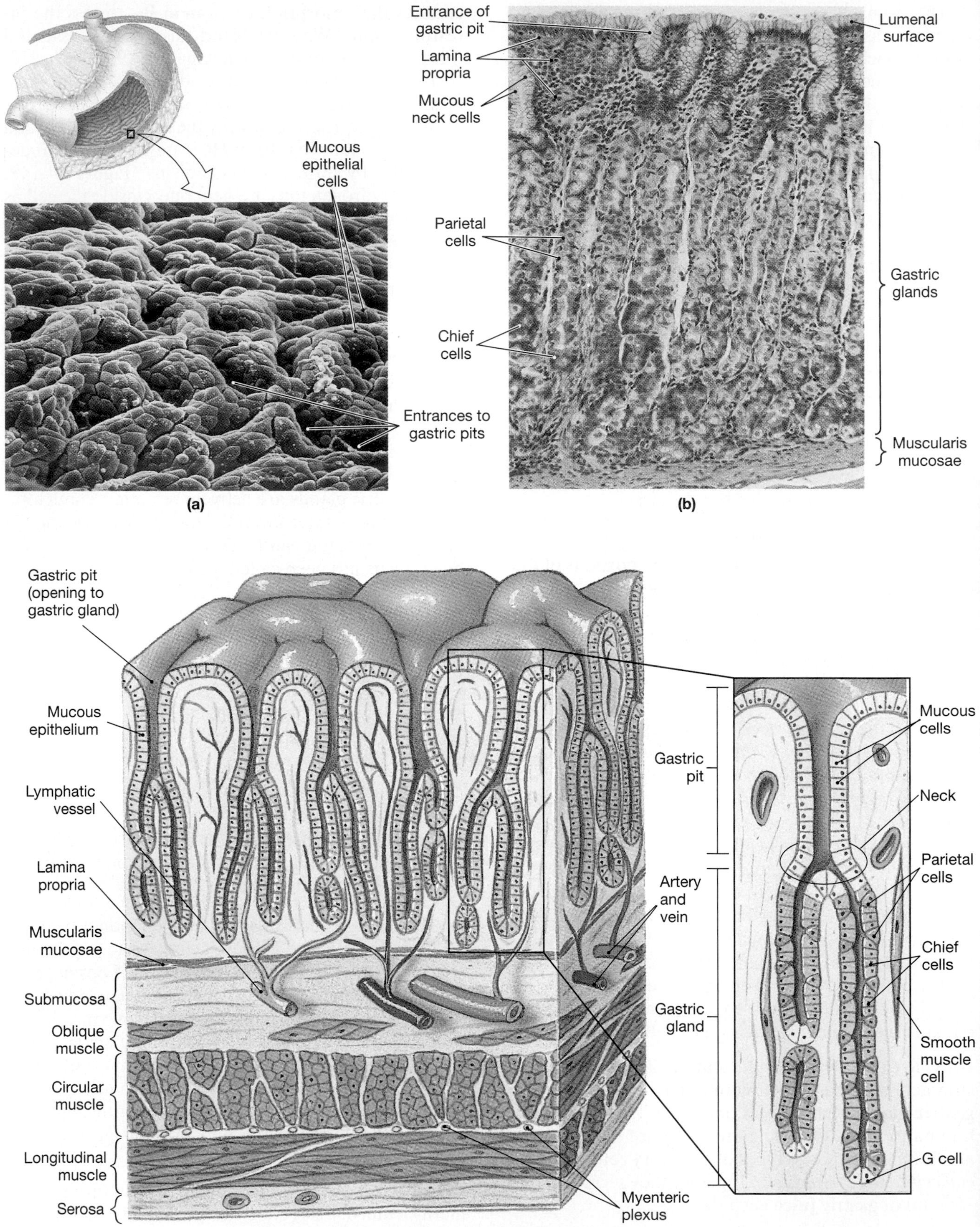

(a)

Mucous epithelial cells

Entrances to gastric pits

Entrance of gastric pit

Lamina propria

Mucous neck cells

Parietal cells

Chief cells

Lumenal surface

Gastric glands

Muscularis mucosae

(b)

Gastric pit (opening to gastric gland)

Mucous epithelium

Lymphatic vessel

Lamina propria

Muscularis mucosae

Submucosa

Oblique muscle

Circular muscle

Longitudinal muscle

Serosa

Artery and vein

Gastric pit

Gastric gland

Myenteric plexus

(c)

Mucous cells

Neck

Parietal cells

Chief cells

Smooth muscle cell

G cell

(d)

•*FIGURE 24-13* **The Stomach Lining. (a)** Surface view of the gastric mucosa of the full stomach, showing the entrances to the gastric pits. (SEM × 35) **(b)** A portion of a gastric pit and gastric gland. (LM × 300) **(c)** Diagrammatic view of the organization of the stomach wall. **(d)** A gastric gland.

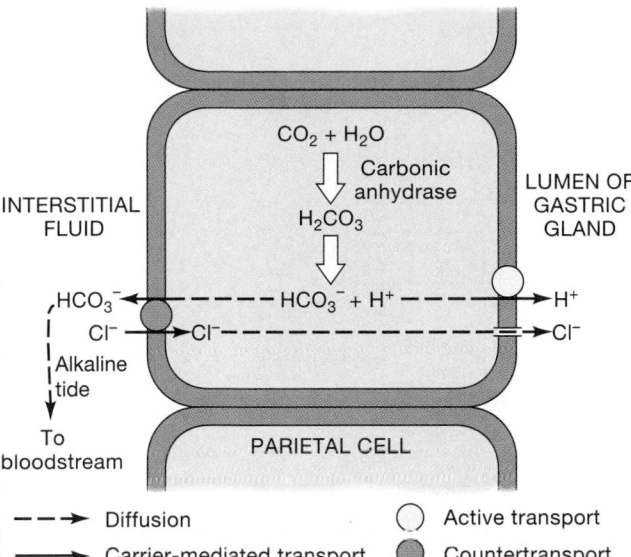

•FIGURE 24-14 Gastric Acid Secretion. An active parietal cell generates H^+ by the dissociation of carbonic acid within the cell. The bicarbonate is exchanged for Cl^- in the interstitial fluid; the chloride ions diffuse into the lumen of the gastric gland as the hydrogen ions are transported out of the cell.

is produced by *G cells.* G cells are most abundant in the gastric pits of the pyloric antrum. Gastrin stimulates (1) the secretion of both parietal and chief cells and (2) contractions of the gastric wall that mix and stir the gastric contents. The pyloric glands also contain *D cells,* which release **somatostatin,** a hormone that inhibits gastrin release. D cells are continuously releasing their secretions into the interstitial fluid adjacent to the G cells. This inhibition of gastrin production can be overpowered by neural and hormonal stimuli when the stomach is preparing for digestion or is already engaged in the digestion of food.

GASTRITIS AND PEPTIC ULCERS Inflammation of the gastric mucosa is called **gastritis** (gas-TRI-tis). This condition may develop after a person has swallowed drugs, including alcohol and aspirin. Gastritis may also appear after severe emotional or physical stress, bacterial infection of the gastric wall, or the ingestion of strongly acidic or alkaline chemicals.

A **peptic ulcer** develops when the digestive acids and enzymes manage to erode their way through the defenses of the stomach lining or proximal portions of the small intestine. The specific locations are indicated by the terms **gastric ulcer** (stomach) or **duodenal ulcer** (duodenum). Peptic ulcers result from the excessive production of acid or the inadequate production of the alkaline mucus that provides an epithelial defense. Since the late 1970s, drugs such as *cimetidine (Tagamet®)* have been used to inhibit acid production by parietal cells. Infections involving the bacterium *Helicobacter pylori* are now thought to be responsible for at least 80 percent of peptic ulcers, and treatment for gastric ulcers today commonly involves the administration of antibiotic drugs. **AM** *Peptic Ulcers*

Regulation of Gastric Activity

The production of acid and enzymes by the gastric mucosa can be (1) controlled by the CNS, (2) regulated by short reflexes coordinated in the wall of the stomach, and (3) regulated by digestive tract hormones. Several phases of gastric control can be identified, although considerable overlap exists among them. These phases are summarized in Figure 24-15•.

The Cephalic Phase

The **cephalic phase** of gastric secretion (Figure 24-15a•) begins when you see, smell, taste, or think of food. This stage, which is directed by the CNS, prepares your stomach to receive food. The neural output proceeds via the parasympathetic division of the ANS and reaches the stomach by means of the vagus nerves. The output then synapses in the submucosal plexus. Next, postganglionic parasympathetic fibers innervate mucous cells, chief cells, parietal cells, and G cells of the stomach. In response to stimulation, the production of gastric juice accelerates, reaching rates of about 500 ml/h. This phase generally lasts for a relatively brief period before the *gastric phase* commences. Emotional states can exaggerate or inhibit the cephalic phase. For example, anger or hostility leads to excessive gastric secretion, whereas anxiety, stress, or fear decreases gastric secretion and motility.

The Gastric Phase

The **gastric phase** (Figure 24-15b•) begins with the arrival of food in the stomach and builds on the stimulation provided during the cephalic phase. The stimuli that initiate the gastric phase are (1) distension of the stomach, (2) an increase in the pH of the gastric contents, and (3) the presence of undigested materials in the stomach, especially proteins and peptides. The gastric phase consists of the following responses:

1. *Neural response.* The stimulation of stretch receptors in the stomach wall and chemoreceptors in the mucosa triggers short reflexes coordinated in the submucosal and myenteric plexuses. The postganglionic fibers leaving the submucosal plexus innervate parietal cells and chief cells, and the release of ACh stimulates their secretion. Proteins, alcohol in small doses, and caffeine enhance gastric secretion markedly by stimulating chemoreceptors in the gastric lining. Stimulation of the myenteric plexus produces mixing waves in the muscularis externa.

2. *Hormonal response.* Neural stimulation and the presence of peptides and amino acids in the chyme stimulate the secretion of gastrin, primarily by G cells of the pyloric antrum. Gastrin entering the interstitial fluid of the stomach must enter capillaries and complete a round-trip of the circulation before

(a) The Cephalic Phase   Sight, smell, taste, or thoughts of food

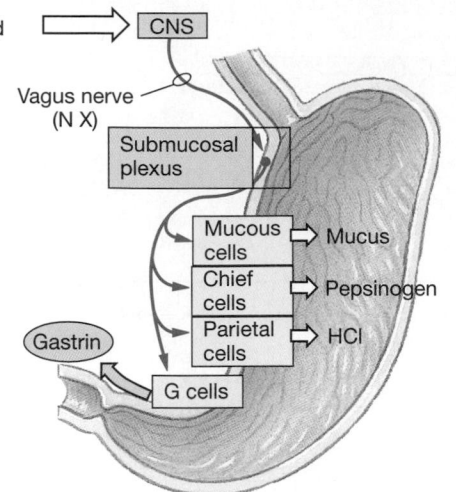

Function:
  Prepare stomach for arrival of food
Duration:
  Short (minutes)
Mechanism:
  Neural, via preganglionic fibers in vagus nerve and
    synapses in submucosal plexus
Actions:
  Primary: increased volume of gastric juice by
    stimulating mucus, enzyme, and acid production
  Secondary: stimulation of gastrin release by G cells

(b) The Gastric Phase

Functions:
  Enhance secretion started in
    cephalic stage; homogenize
    and acidify chyme;
    initiate digestion of proteins by pepsin
Duration:
  Long (3–4 hours)
Mechanisms:
  Neural: short reflexes triggered by
    (1) stimulation of stretch receptors as stomach fills
    (2) stimulation of chemoreceptors as pH increases
  Hormonal: stimulation of gastrin
    release by G cells by parasympathetic
    activity and presence of peptides
    and amino acids in chyme
  Local: release of histamine by mast cells
    as stomach fills (not shown)
Actions:
  Increased acid and pepsinogen production;
    increased motility and initiation of mixing waves

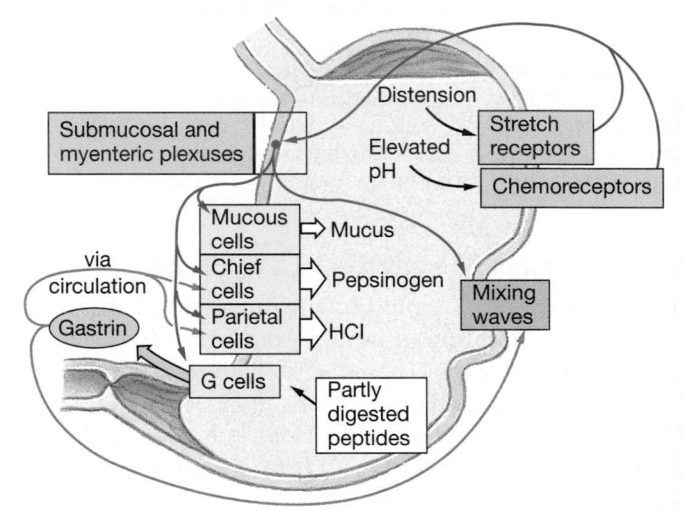

(c) The Intestinal Phase

Function:
  Control rate of chyme entry
    into duodenum
Duration:
  Long (hours)
Mechanisms:
  Neural: short reflexes (enterogastric
    reflex) triggered by extension of
    duodenum
  Hormonal:
  Primary: stimulation of CCK, GIP, and
    secretin release by presence of acid,
    carbohydrates, and lipids
  Secondary: release of gastrin stimulated
    by presence of undigested proteins
    and peptides (not shown)
Actions:
  Feedback inhibition of gastric acid
    and pepsinogen production; reduction
    of gastric motility

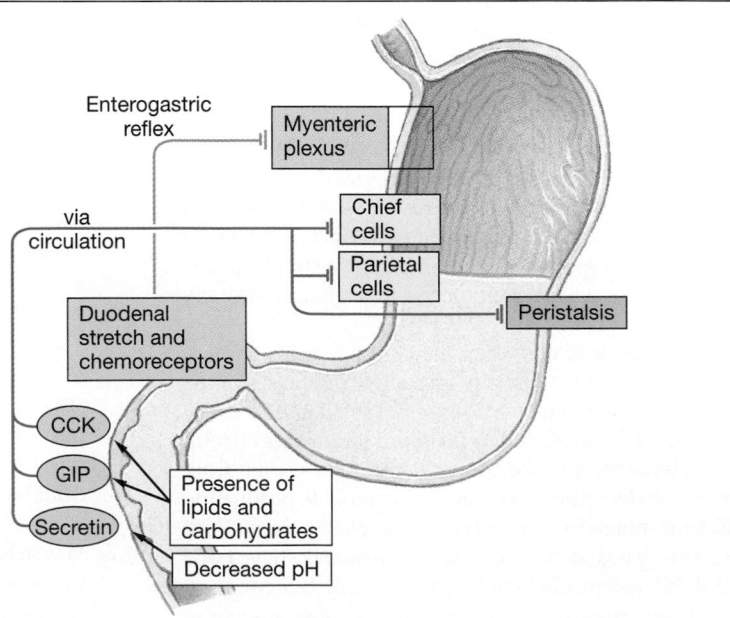

•*FIGURE 24-15* **The Phases of Gastric Secretion**

the hormone stimulates parietal and chief cells of the fundus and body. Both parietal and chief cells respond to the presence of gastrin by accelerating their secretion rates. The effect on the parietal cells is the most pronounced, and the pH of the gastric juice declines as a result. Gastrin also stimulates gastric motility.

3. *Local response.* Distortion of the gastric wall also stimulates the release of histamine in the lamina propria. The source of the histamine is thought to be mast cells within the connective tissue of this layer. Histamine binds to receptors on the parietal cells and stimulates acid secretion.

DURATION OF THE GASTRIC PHASE The gastric phase may continue for several hours while the ingested materials are processed by the acids and enzymes. During this period, gastrin stimulates contractions in the muscularis externa of the stomach and intestinal tract. The effects are strongest in the stomach, where stretch receptors are stimulated as well. The initial contractions are weak pulsations in the gastric walls. These *mixing waves* occur several times per minute, and they gradually increase in intensity. After an hour, the material within the stomach is churning like the clothing in a washing machine.

At the time the contractions begin, the pH of the gastric contents is high, and only the material in contact with the gastric epithelium is exposed to undiluted digestive acids and enzymes. As mixing occurs, the acid is diluted, and the pH remains elevated until a large volume of gastric juice has been secreted and the contents thoroughly mixed. This process generally takes several hours. As the pH throughout the chyme reaches 1.5–2.0, and the amount of undigested protein decreases, gastrin production declines, and so do the rates of acid and enzyme secretion by parietal and chief cells.

The Intestinal Phase

The **intestinal phase** of gastric secretion (Figure 24-15c•) begins when chyme starts to enter the small intestine. The intestinal phase generally starts after several hours of mixing contractions, when waves of contraction begin sweeping down the length of the stomach. Each time the pylorus contracts, a small quantity of chyme squirts through the pyloric sphincter. The purpose of the intestinal phase is to control the rate of gastric emptying and ensure that the secretory, digestive, and absorptive functions of the small intestine can proceed with reasonable efficiency. Although we shall consider the intestinal phase now as it affects stomach activity, the arrival of chyme in the small intestine also triggers other neural and hormonal events that coordinate the activities of the intestinal tract and the pancreas, liver, and gallbladder.

The intestinal phase involves a combination of neural and hormonal responses.

1. *Neural response.* Chyme leaving the stomach relieves some of the distension in the stomach wall, thus reducing the stimulation of stretch receptors. At the same time, distension of the duodenum by acidic chyme stimulates stretch receptors and chemoreceptors that trigger the **enterogastric reflex.** The enterogastric reflex temporarily inhibits both central and local stimulation of gastrin production and gastric contractions. It also stimulates contraction of the pyloric sphincter. The net result is that immediately after chyme enters the small intestine, gastric contractions decrease in strength and frequency, and further discharge of chyme is prevented. This result gives the duodenum time to deal with the arriving acids before the next wave of gastric contraction occurs. At the same time, local reflexes at the duodenum stimulate mucus production that helps protect the intestinal lining from the arriving acids and enzymes.

2. *Hormonal response.* Several hormonal responses are triggered by the arrival of chyme in the duodenum.

   a. The arrival of lipids (especially triglycerides and fatty acids) and carbohydrates in the duodenum stimulates the secretion of the hormones *cholecystokinin* (kō-lē-sis-tō-KĪ-nin), or CCK, and *gastric inhibitory peptide* (GIP). CCK has multiple effects on the digestive system, including inhibition of gastric secretion of acids and enzymes. GIP, which also targets the pancreas, inhibits gastric secretion and reduces the rate and force of gastric contractions. As a result, a meal high in fats stays in your stomach longer, and enters the duodenum at a more leisurely pace, than a low-fat meal. This delay allows more time for lipids to be digested and absorbed in the small intestine.

   b. A drop in pH below 4.5 stimulates secretion of *secretin* (se-KRĒ-tin) by enteroendocrine cells of the duodenum. Secretin inhibits parietal cell and chief cell activity in the stomach. It also targets (1) the pancreas, where it stimulates the production of buffers that will protect the duodenum by neutralizing the acid in chyme, and (2) the liver, where it stimulates bile secretion.

   c. The arrival of partially digested proteins in the duodenum stimulates G cells in the duodenal wall. These cells secrete gastrin, which circulates to the stomach and accelerates acid and enzyme production. In effect, this is a feedback mechanism that regulates the amount of gastric processing to meet the requirements of a specific meal.

In general, the rate of chyme movement into the small intestine is highest when your stomach is greatly distended and the meal contains relatively little protein. A large meal containing small amounts of protein, large amounts of carbohydrates (such as rice or pasta), wine (alcohol), and after-dinner coffee (caffeine) will leave your stomach extremely quickly because both alcohol and caffeine stimulate gastric secretion and motility.

Digestion and Absorption in the Stomach

The stomach performs preliminary digestion of proteins by pepsin and, for a variable period, permits the digestion of carbohydrates and lipids by salivary amylase and lingual lipase. Until the pH throughout the material in the stomach falls below 4.5, the salivary amylase and lingual lipase continue to digest carbohydrates and lipids in the meal. These enzymes generally remain active 1–2 hours after a meal.

As the stomach contents become more fluid and the pH approaches 2.0, pepsin activity increases and protein disassembly begins. Protein digestion is not completed in the stomach, because time is limited and pepsin attacks only specific types of peptide bonds, not all peptide bonds. However, there is generally enough time for pepsin to break down complex proteins into smaller peptide and polypeptide chains before the chyme enters the duodenum.

Although digestion does occur in the stomach, there is no nutrient absorption there because (1) the epithelial cells are covered by a blanket of alkaline mucus and are not directly exposed to the chyme, (2) the epithelial cells lack the specialized transport mechanisms of cells that line the small intestine, (3) the gastric lining is relatively impermeable to water, and (4) digestion has not proceeded to completion by the time chyme leaves the stomach. At this stage, most carbohydrates, lipids, and proteins are only partially broken down.

Some drugs can be absorbed in the stomach. For example, ethyl alcohol can diffuse through the mucous barrier and penetrate the lipid membranes of the epithelial cells. As a result, you absorb alcohol in your stomach before any nutrients in a meal reach the circulation. Meals containing large amounts of fat will slow the rate of alcohol absorption, because alcohol is lipid-soluble, and some of it will be dissolved in fat droplets within the chyme. Aspirin is another lipid-soluble drug that can enter the circulation across the gastric mucosa. Aspirin and related drugs alter the properties of the mucous layer and can promote epithelial damage by stomach acids and enzymes. Prolonged aspirin use can cause gastric bleeding, and individuals with stomach ulcers usually avoid aspirin.

STOMACH CANCER *Stomach,* or *gastric, cancer* is one of the most common lethal cancers, responsible for roughly 14,000 deaths in the United States each year. Because the symptoms may resemble those of gastric ulcers, the condition may not be reported in its early stages. Diagnosis generally involves X-rays of the stomach at various degrees of distension. The mucosa can also be visually inspected by using a flexible instrument called a *gastroscope.* Attachments permit the collection of tissue samples for histological analysis. Treatment of gastric cancer involves the surgical removal of part or all of the stomach. People can survive even a total *gastrectomy* (gas-TREK-to-mē), because the

only absolutely vital function of the stomach is the secretion of intrinsic factor. Protein breakdown can still be performed by the small intestine, although at reduced efficiency, and the loss of gastric functions such as food storage and acid production is not life-threatening.

☑ How would a large meal affect the pH of the blood that leaves the stomach?

☑ When a person suffers from chronic ulcers in the stomach, the branches of the vagus nerve that serve the stomach are sometimes severed. Why?

THE SMALL INTESTINE AND ASSOCIATED GLANDULAR ORGANS

Your stomach is a holding tank where food is saturated with gastric juices and exposed to stomach acids and the digestive effects of pepsin. These are preliminary steps, for most of the important digestive and absorptive functions occur in your small intestine, where the products of digestion are absorbed. The mucosa of the small intestine produces only a few of the enzymes involved. The pancreas provides digestive enzymes as well as buffers that assist in the neutralization of acidic chyme. The liver and gallbladder provide *bile,* a solution that contains additional buffers and *bile salts,* compounds that facilitate the digestion and absorption of lipids.

The Small Intestine

The small intestine plays the primary role in the digestion and absorption of nutrients. The small intestine averages 6 m (20 ft) in length (range: 4.5–7.5 m; 14.8–24.6 ft) and has a diameter ranging from 4 cm (1.6 in.) at the stomach to about 2.5 cm (1 in.) at the junction with the large intestine. It occupies all abdominal regions except the right and left hypochondriac and epigastric regions (Figure 1-8●, p. 18). Ninety percent of nutrient absorption occurs in the small intestine, and most of the rest occurs in the large intestine. The small intestine has three subdivisions: (1) the *duodenum,* (2) the *jejunum,* and (3) the *ileum.*

The **duodenum** (doo-AH-de-num), 25 cm (10 in.) in length, is the section closest to the stomach. This portion of the small intestine is a "mixing bowl" that receives chyme from the stomach and digestive secretions from the pancreas and liver. From its connection with the stomach, the duodenum curves in a C that encloses the pancreas. Except for the proximal 2.5 cm (1 in.), the duodenum is in a retroperitoneal position between vertebrae L_1 and L_4 (Figure 24-5●, p. 868).

A rather abrupt bend marks the boundary between the duodenum and the **jejunum** (je-JOO-num). At this junction, the small intestine reenters the peritoneal cavity, supported by a sheet of mesentery. The jejunum is about 2.5 meters (8 ft) long. The bulk of chemical digestion and nutrient absorption occurs in the jejunum.

The **ileum** (IL-ē-um) is the third and last segment of the small intestine. It is also the longest, averaging 3.5 meters (12 ft) in length. The ileum ends at a sphincter, the **ileocecal valve**, which controls the flow of materials from the ileum into the *cecum* of the large intestine.

The small intestine fills much of the peritoneal cavity, and its position is stabilized by a broad mesentery attached to the dorsal body wall (Figure 24-5•, p. 868). Movement of the small intestine during digestion is restricted by the stomach, the large intestine, the abdominal wall, and the pelvic girdle. Blood vessels, lymphatics, and nerves reach these segments of the small intestine within the connective tissue of the mesentery. The primary blood vessels involved are branches of the *superior mesenteric artery* and the *superior mesenteric vein.* ∞ *[pp. 747, 756]*

Histology of the Small Intestine

The intestinal lining bears a series of transverse folds called **plicae**, or *plicae circulares* (Figure 24-16a,b•). Unlike the rugae in the stomach, each plica is a permanent feature that does not disappear when the small intestine fills with chyme. There are roughly 800 plicae along the length of the small intestine, and their presence greatly increases the surface area available for absorption.

Figure 24-16c,d,e• presents a more detailed view of the intestinal wall.

INTESTINAL VILLI The mucosa of the small intestine is thrown into a series of fingerlike projections, the **intestinal villi.** The intestinal villi are covered by a simple columnar epithelium that is carpeted with microvilli. Because the microvilli project from the epithelium like the bristles on a brush, these cells are said to have a *brush border* (Figure 24-16c,d,e•).

If the small intestine were a simple tube with smooth walls, it would have a total absorptive area of roughly 3300 cm^2 (3.6 ft^2). Instead, the mucosa contains plicae circulares, each plica supports a forest of villi, and each villus is covered by epithelial cells whose exposed surfaces contain microvilli. This arrangement increases the total area for absorption by a factor of more than 600, to approximately 2 million square centimeters, or more than 2200 square feet.

The lamina propria of each villus contains an extensive network of capillaries. These capillaries originate in a vascular network within the submucosa. They transport respiratory gases and carry absorbed nutrients to the hepatic portal circulation for delivery to the liver. The liver adjusts the nutrient concentrations of the blood before it reaches the general systemic circulation.

In addition to capillaries and nerve endings, each villus contains a lymphatic capillary called a **lacteal** (LAK-tē-al; *lacteus,* milky) (Figure 24-16d,e•). Lacteals transport materials that are unable to enter blood capillaries. For example, absorbed fatty acids are assembled into protein-lipid packages too large to diffuse into the bloodstream. These packets, called *chylomicrons,* reach the venous circulation by way of the thoracic duct, which delivers lymph into the left subclavian vein. ∞ *[p. 772]* The name *lacteal* refers to the pale, cloudy appearance of lymph that contains large quantities of lipids.

Contractions of the muscularis mucosae and smooth muscle cells within the villi move the villi back and forth, exposing the epithelial surfaces to the liquefied intestinal contents. This movement improves the efficiency of absorption, because local differences in the nutrient concentration of the chyme will be quickly eliminated. Movements of the villi also squeeze the lacteals, thus assisting in the movement of lymph out of the villi.

INTESTINAL CRYPTS Between the columnar epithelial cells, goblet cells eject mucins onto the intestinal surfaces. ∞ *[p. 118]* At the bases of the villi are the entrances to the **intestinal crypts** (also known as **intestinal glands** or *crypts of Lieberkuhn*). These glandular pockets extend deep into the underlying lamina propria (Figure 24-16c•). Near the base of each intestinal gland, stem cell divisions produce new generations of epithelial cells. These new cells are continuously displaced toward the intestinal surface. Within a few days, they will have reached the tip of a villus, where they are shed into the intestinal lumen. This ongoing process renews the epithelial surface and the subsequent disintegration of the shed cells adds enzymes to the chyme.

Several important *brush border enzymes* enter the intestinal lumen in this way. Brush border enzymes are integral membrane proteins located on the surfaces of intestinal microvilli. The enzymes have important digestive functions: Materials in contact with the brush border are attacked by these enzymes, and the breakdown products are absorbed by the epithelial cells. Once the epithelial cells are shed, they disintegrate within the lumen, and the intracellular and brush border enzymes enter the chyme. There they continue to function until proteolytic enzymes break them apart. *Enterokinase*, also called *enteropeptidase*, is a brush border enzyme that enters the lumen in this way. Enterokinase does not directly participate in digestion, but it activates proenzymes secreted by the pancreas. (We shall consider the functions of enterokinase and other brush border enzymes in a later section.) Intestinal crypts also contain enteroendocrine cells responsible for the production of several intestinal hormones, including gastrin, cholecystokinin, and secretin.

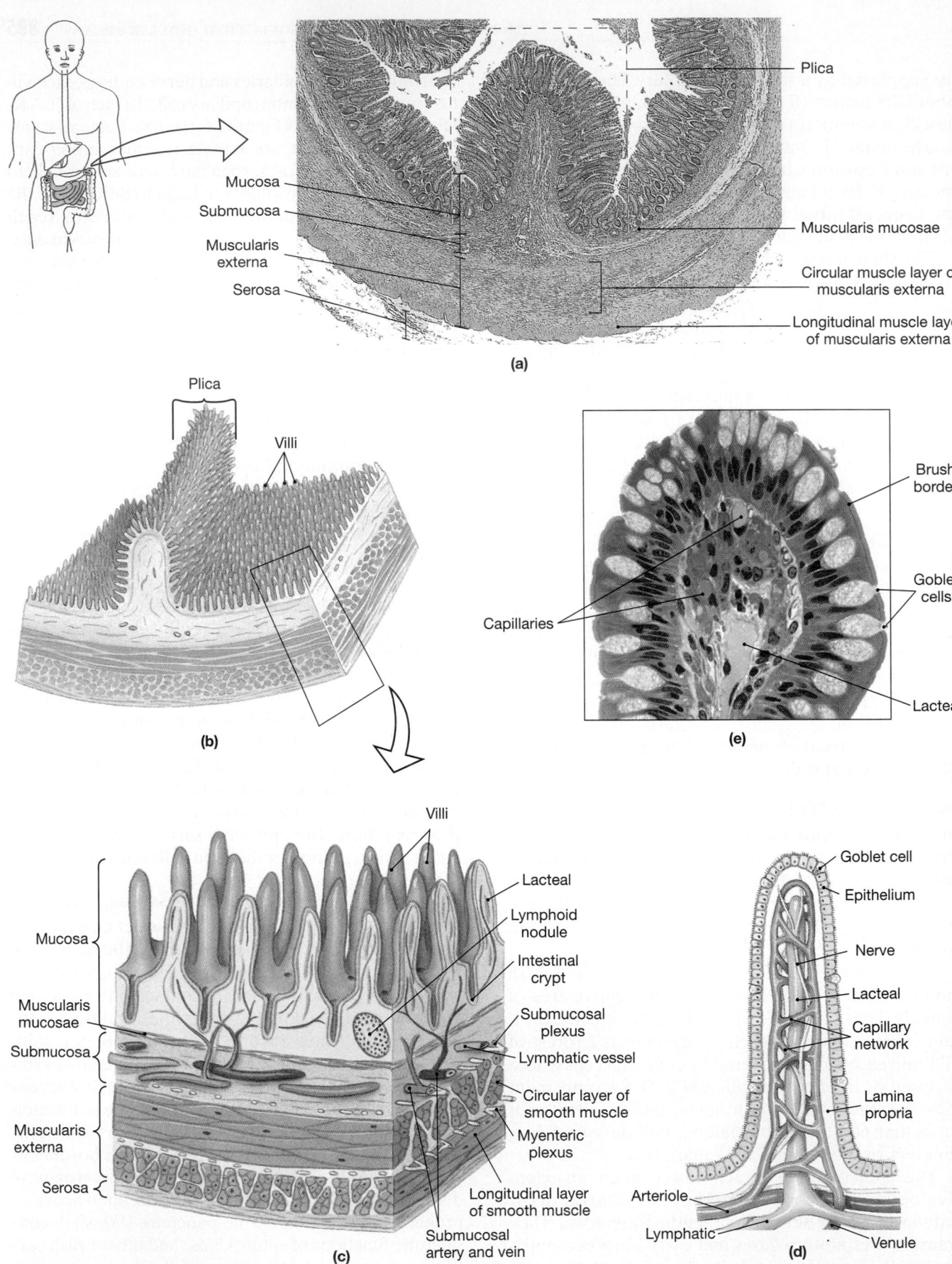

Mucosa
Submucosa
Muscularis externa
Serosa

Plica

Muscularis mucosae

Circular muscle layer of muscularis externa

Longitudinal muscle layer of muscularis externa

(a)

Plica

Villi

(b)

Brush border

Capillaries

Goblet cells

Lacteal

(e)

Villi

Mucosa

Muscularis mucosae

Submucosa

Muscularis externa

Serosa

Lacteal

Lymphoid nodule

Intestinal crypt

Submucosal plexus

Lymphatic vessel

Circular layer of smooth muscle

Myenteric plexus

Longitudinal layer of smooth muscle

Submucosal artery and vein

(c)

Goblet cell

Epithelium

Nerve

Lacteal

Capillary network

Lamina propria

Arteriole

Lymphatic

Venule

(d)

•**FIGURE 24-16** **The Intestinal Wall.** **(a)** A view of the intestine in section. (LM × 2.5) **(b)** A single plica and multiple villi. **(c)** Diagrammatic view of the organization of the intestinal wall. **(d)** Internal structures in a single villus, showing the capillary and lymphatic supply. **(e)** A villus in sectional view. (LM × 252)

REGIONAL SPECIALIZATIONS The regions of the small intestine have histological specializations related to their primary functions. The locations of these segments in the peritoneal cavity are indicated in Figure 24-17a●. Representative sections and photographs of each region of the small intestine are presented in Figure 24-17b,c●.

Duodenum. The duodenum contains few plicae; the villi are numerous but shorter and stumpier than those of the jejunum. There are numerous mucous glands in the duodenum, both within the epithelium and beneath it. In addition to the intestinal crypts, the submucosa contains **submucosal glands,** or *Brunner's glands,* which produce copious quantities of mucus when chyme arrives from the stomach. Mucus produced by these glands protects the epithelium from the acidic chyme. It also contains buffers that help elevate the pH of the chyme. Along the length of the duodenum, the pH of the chyme goes from 1–2 to 7–8. The submucosal glands also secrete the hormone **urogastrone,** which inhibits gastric acid production. Urogastrone, or *epidermal growth factor* (EGF), stimulates the division of epithelial cells along the digestive tract as well as stem cell activity in other areas.

Jejunum. Plicae and villi are prominent over the proximal half of the jejunum. As materials approach the ileum, the plicae and villi become smaller and continue to diminish in size to the end of the ileum. This reduction parallels the reduction in absorptive activity; most nutrient absorption has occurred before ingested materials reach the ileum. One rather drastic surgical method of promoting weight loss is the removal of a significant portion of the jejunum. The reduction in absorptive area causes a marked weight loss, but the side effects can be very troublesome. AM *Drastic Weight-Loss Techniques*

Ileum. The ileum adjacent to the large intestine lacks plicae altogether, and the scattered villi are stumpy and conical. The ileum also contains 20–30 masses of lymphoid tissue called *aggregate lymphoid nodules,* or *Peyer's patches,* which we described in Chapter 22. ∞ *[p. 776]* The lymphocytes in these nodules protect the small intestine from bacteria that are normal inhabitants of the large intestine. Lymphoid nodules are most abundant in the terminal portion of the ileum, near the entrance to the large intestine.

Intestinal Secretions

Roughly 1.8 liters of watery **intestinal juice** enter your intestinal lumen each day. Intestinal juice moistens the chyme, assists in buffering acids, and liquefies both the digestive enzymes provided by the pancreas and the products of digestion. Much of this fluid volume arrives through osmosis, as water flows out of the mucosa and into the relatively concentrated chyme. The rest is provided by intestinal glands, stimulated by the activation of touch and stretch receptors in the intestinal walls.

The submucosal glands help protect the duodenal epithelium from gastric acids and enzymes. These glands increase their secretory activities in response to (1) local reflexes, (2) the release of the hormone *enterocrinin* by duodenal enteroendocrine cells, and (3) parasympathetic stimulation carried by the vagus nerves. Mechanisms 1 and 2 operate only after chyme arrives in the duodenum. However, because vagal activity triggers their secretion, the submucosal glands begin secreting during the cephalic phase of gastric secretion, long before the chyme reaches the pyloric sphincter. Thus the duodenal lining has protection in advance.

Sympathetic stimulation will inhibit the activation of the submucosal glands, leaving the duodenal lining relatively unprepared for the arrival of acidic chyme. This fact probably accounts for the common observation that duodenal ulcers can be caused by chronic stress or other factors that promote sympathetic activation.

Intestinal Movements

After chyme has arrived in the duodenum, weak peristaltic contractions move it slowly toward the jejunum. These contractions are myenteric reflexes not under CNS control. Their effects are limited to within a few centimeters of the site of the original stimulus. These short reflexes are controlled by motor neurons in the submucosal and myenteric plexuses. In addition, some of the smooth muscle cells contract periodically, even without stimulation, establishing a basic contractile rhythm that then spreads from cell to cell.

The stimulation of the parasympathetic system increases the sensitivity of these myenteric reflexes and accelerates both local peristalsis and segmentation. More-elaborate reflexes coordinate activities along the entire length of the small intestine. Two examples are triggered by the stimulation of stretch receptors in the stomach as it fills. The **gastroenteric reflex** stimulates motility and secretion along the entire length of the small intestine; the **gastroileal** (gas-trō-IL-ē-al) **reflex** triggers the relaxation of the ileocecal valve. The net result is that materials pass from the small intestine into the large intestine. Thus the gastroenteric and gastroileal reflexes accelerate movement along the small intestine—the opposite effect of the enterogastric reflex.

Hormones released by the digestive tract can enhance or suppress reflex responses. For example, the gastroileal reflex is triggered by stretch receptor stimulation, but the degree of ileocecal valve relaxation is enhanced by gastrin, which is secreted in large quantities when food enters the stomach.

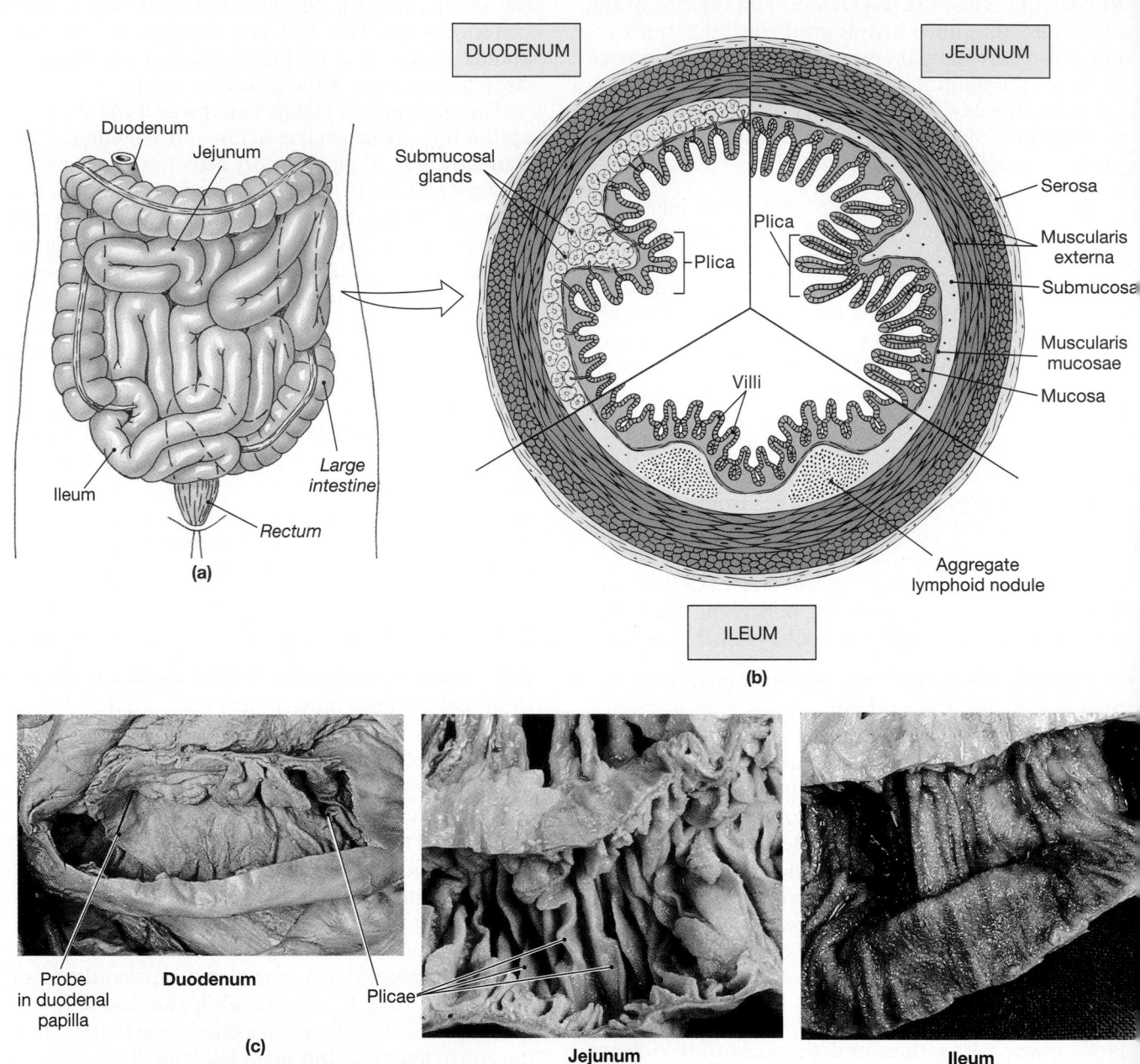

DUODENUM

JEJUNUM

Duodenum

Jejunum

Submucosal
glands

Serosa

Plica

Muscularis
externa

Plica

Submucosa

Plica

Muscularis
mucosae

Villi

Mucosa

Ileum

Large
intestine

Rectum

(a)

Aggregate
lymphoid nodule

ILEUM

(b)

Probe
in duodenal
papilla

Duodenum

Plicae

(c)

Jejunum

Ileum

•*FIGURE 24-17* **Regions of the Small Intestine. (a)** Positions of the duodenum, jejunum, and ileum within the abdominopelvic cavity. **(b)** Diagrammatic comparison of the organization of the intestinal wall in each of these regions. **(c)** Representative views of the three regions of the small intestine.

VOMITING The responses of the digestive tract to chemical or mechanical irritation are rather predictable. Gastric secretion accelerates all along the digestive tract, and the intestinal contents are eliminated as quickly as possible. The *vomiting reflex* occurs in response to irritation of the fauces, pharynx, esophagus, stomach, or proximal portions of the small intestine. These sensations are relayed to the vomiting center of the medulla oblongata, which coordinates the motor responses. During the preparatory phase, the pylorus relaxes, and the contents of the duodenum and proximal jejunum are discharged into the stomach by strong peristaltic waves that travel toward the stomach rather than toward the ileum. Vomiting, or *emesis* (EM-e-sis), then occurs as the stomach regurgitates its contents through the esophagus and pharynx. As regurgitation occurs, the uvula and soft palate block the entrance to the nasopharynx. Increased salivary secretion assists in buffering the stomach acids, thereby preventing erosion of the teeth. In conditions marked by repeated vomiting, severe tooth damage can occur; we shall discuss one example, the eating disorder *bulimia,* in Chapter 25. Most of the force of vomiting comes from expiratory movements that elevate intra-abdominal pressures and force the stomach against the tensed diaphragm.

The Pancreas

Your pancreas lies posterior to your stomach, extending laterally from the duodenum toward the spleen (Figure 24-18a●). The pancreas is an elongate, pinkish gray organ with a length of approximately 15 cm (6 in.) and a weight of about 80 g (3 oz). The broad **head** of the pancreas lies within the loop formed by the duodenum as it leaves the pylorus. The slender **body** extends transversely toward the spleen, and the **tail** is short and bluntly rounded. The pancreas is retroperitoneal and is firmly bound to the posterior wall of the abdominal cavity.

The surface of the pancreas has a lumpy, lobular texture. A thin, transparent connective tissue capsule wraps the entire organ. You can see the pancreatic lobules, associated blood vessels, and excretory ducts through the anterior capsule and the overlying layer of peritoneum. Arterial blood reaches the pancreas by way of branches of the *splenic, superior mesenteric,* and *common hepatic arteries.* The *pancreatic arteries* and *pancreaticoduodenal arteries* are the major branches from these vessels. The *splenic vein* and its branches drain the pancreas.

The pancreas is primarily an exocrine organ, producing digestive enzymes and buffers. The large **pancreatic duct** *(duct of Wirsung)* delivers these secretions to the duodenum. A small **accessory duct,** or *duct of Santorini,* may branch from the pancreatic duct. The pancreatic duct extends within the attached mesentery to reach the duodenum, where it meets the *common bile duct* from the liver and gallbladder. The two ducts then empty into the *duodenal ampulla,* a chamber located roughly halfway along the length of the

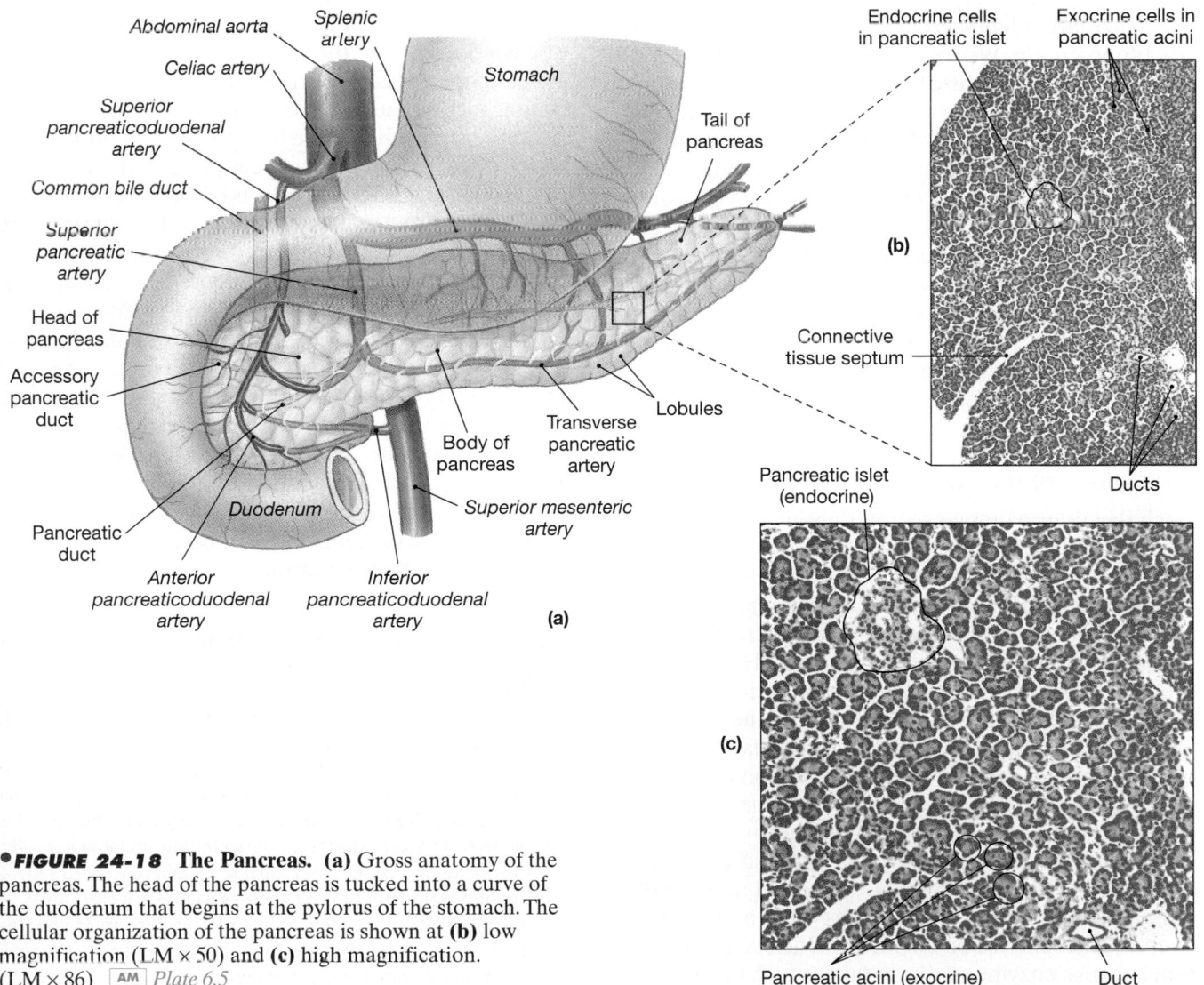

●**FIGURE 24-18** **The Pancreas.** **(a)** Gross anatomy of the pancreas. The head of the pancreas is tucked into a curve of the duodenum that begins at the pylorus of the stomach. The cellular organization of the pancreas is shown at **(b)** low magnification (LM × 50) and **(c)** high magnification. (LM × 86) **AM** *Plate 6.5*

duodenum (Figure 24-21b, ● p. 895). When present, the accessory duct generally empties into the duodenum independently, outside the duodenal ampulla.

Histological Organization

Partitions of connective tissue divide the pancreatic tissue into distinct lobules. The blood vessels and tributaries of the pancreatic ducts are situated within these connective tissue septa (Figure 24-18b●). The pancreas is an example of a *compound tubuloacinar gland,* a gland structure that we described in Chapter 4. ∞ *[p. 119]* Within each lobule, the ducts branch repeatedly before ending in blind pockets called the **pancreatic acini** (AS-i-nī). Each pancreatic acinus is lined by a simple cuboidal epithelium. *Pancreatic islets,* the endocrine tissues of the pancreas, are scattered among the acini (Figure 24-18b,c●). The islets account for only about 1 percent of the cellular population of the pancreas.

The pancreas has two distinct functions, one endocrine and the other exocrine. The endocrine cells of the pancreatic islets secrete insulin and glucagon into the bloodstream. We described those hormones and their actions in Chapter 18. ∞ *[p. 622]* The exocrine cells include the acinar cells and the epithelial cells that line the duct system. Together they secrete an alkaline **pancreatic juice** into the small intestine. Pancreatic juice is a mixture of digestive enzymes, water, and ions. Pancreatic enzymes are secreted by the acinar cells. These enzymes do most of the digestive work in the small intestine, breaking down ingested materials into small molecules suitable for absorption. The water and ions, secreted primarily by the cells lining the pancreatic ducts, assist in diluting and buffering the acids in the chyme.

Physiology of the Pancreas

Each day your pancreas secretes about 1000 ml (1 qt) of pancreatic juice. The secretory activities are controlled primarily by hormones from the duodenum. When acid chyme arrives in the duodenum, *secretin* is released. This hormone triggers the pancreatic secretion of a watery buffer solution with a pH of 7.5–8.8. Among its other components, this secretion contains bicarbonate and phosphate buffers that help elevate the pH of the chyme. A different duodenal hormone, *cholecystokinin,* stimulates the production and secretion of pancreatic enzymes. Pancreatic enzyme secretion also increases under stimulation by the vagus nerves. As we noted earlier, this stimulation occurs during the cephalic phase of gastric regulation, so the pancreas starts to synthesize enzymes before food even reaches the stomach. Such a head start is important because enzyme synthesis takes much longer than

buffer production. By starting early, the pancreatic cells will be ready to meet the demand when chyme arrives in the duodenum.

The specific pancreatic enzymes involved include the following:

- **Pancreatic alpha-amylase** is a **carbohydrase** (kar-bō-HĪ-drās), an enzyme that breaks down certain starches. Pancreatic alpha-amylase is almost identical to salivary amylase.
- **Pancreatic lipase** breaks down certain complex lipids, releasing fatty acids and other products that can be easily absorbed.
- **Nucleases** break down nucleic acids.
- **Proteolytic enzymes** break certain proteins apart. The proteolytic enzymes of the pancreas include **proteases** and **peptidases;** proteases break apart large protein complexes, whereas peptidases break small peptide chains into individual amino acids.

Proteolytic enzymes account for about 70 percent of the total pancreatic enzyme production. The enzymes are secreted as inactive *proenzymes* that are activated only after they reach the small intestine. Proenzymes discussed earlier in the text include pepsinogen, angiotensinogen, plasminogen, fibrinogen, and many of the clotting factors and enzymes of the complement system. As in the stomach, release of a proenzyme rather than an active enzyme in the pancreas protects the secretory cells from the destructive effects of their own products. Among the proenzymes secreted by the pancreas are **trypsinogen** (trip-SIN-ō-jen), **chymotrypsinogen** (kī-mō-trip-SIN-ō-jen), **procarboxypeptidase** (prō-kar-bok-sē-PEP-ti-dās), and **proelastase** (pro-ē-LAS-tās).

Once inside the duodenum, enterokinase located on the brush border and in the lumen triggers the conversion of trypsinogen to **trypsin,** an active protease. Trypsin then activates the other proenzymes, producing **chymotrypsin, carboxypeptidase,** and **elastase.** Each enzyme attacks peptide bonds linking specific amino acids and ignores others. Together, they break down complex proteins into a mixture of dipeptides, tripeptides, and amino acids.

PANCREATITIS Pancreatitis (pan-krē-a-TĪ-tis) is an inflammation of the pancreas. Blockage of the excretory ducts, bacterial or viral infections, ischemia, and drug reactions, especially those involving alcohol, are among the factors that may produce this condition. These stimuli provoke a crisis by injuring exocrine cells in at least a portion of the organ. Lysosomes within the damaged cells then activate the proenzymes, and autodigestion begins. The proteolytic enzymes digest the surrounding, undamaged cells, activating their enzymes and starting a chain reaction. In most cases, only a portion of the pancreas will be affected, and the condition subsides in a few days. In 10–15 percent of pancreatitis cases, the process does not subside, and the enzymes may ultimately destroy the pancreas.

The Liver

The liver, the largest visceral organ, is one of the most versatile organs in the body. Most of its mass lies within the right hypochondriac and epigastric regions, but it may extend into the left hypochondriac and umbilical regions as well. The liver weighs about 1.5 kg (3.3 lb). This large, firm, reddish brown organ provides essential metabolic and synthetic services that fall into three general categories: (1) *metabolic regulation,* (2) *hematological regulation,* and (3) *bile production.* We shall detail those general functions after we examine the anatomy and organization of the liver.

Anatomy of the Liver

The liver is wrapped in a tough fibrous capsule and covered by a layer of visceral peritoneum. On the anterior surface, the **falciform ligament** marks the division between the **left lobe** and the **right lobe** of the liver (Figure 24-19a,b●). A thickening in the posterior margin of the falciform ligament is the **round ligament,** or *ligamentum teres,* a fibrous band that marks the path of the fetal umbilical vein.

On the posterior surface of the liver, the impression left by the inferior vena cava marks the division between the right lobe and the small **caudate** (KAW-dāt) **lobe** (Figure 24-19c●). Inferior to the caudate lobe lies the **quadrate lobe,** sandwiched between the left lobe and the gallbladder. Afferent blood vessels and other structures reach the liver by traveling within the connective tissue of the lesser omentum. They converge at the **hilus** of the liver, a region known as the *porta hepatis* ("doorway to the liver").

The *gallbladder* is a muscular sac that stores and concentrates bile prior to its excretion into the small intestine. The gallbladder is located in a recess, or fossa, in the posterior surface of the liver's right lobe. We shall describe the gallbladder and associated structures in a later section.

THE BLOOD SUPPLY TO THE LIVER
We detailed the circulation to the liver in Chapter 21 and summarized that circulation pattern in Figures 21-27 and 21-35●, pp. 748, 757. Roughly one-third of the blood supply to the liver is arterial blood from the *hepatic artery.* The remainder consists of venous blood from the *hepatic portal vein,* which begins in the capillaries of the esophagus, stomach, small intestine, and most of the large intestine. We described the distribution and major tributaries of the hepatic portal vein in Chapter 21. ∞ *[p. 756]* Liver cells, called **hepatocytes** (he-PAT-ō-sīts), adjust circulating levels of nutrients by selective absorption and secretion. Blood leaving the liver returns to the systemic circuit through the *hepatic veins,* which open into the inferior vena cava.

PORTAL HYPERTENSION Pressures in the hepatic portal system are usually low, averaging 10 mm Hg or less. This pressure can increase markedly if blood flow through the liver becomes restricted as a result of liver damage or a blood clot. A rise in portal pressure is called **portal hypertension.** As pressures rise, small peripheral veins and capillaries in the portal system become distended and are likely to rupture, and intestinal bleeding can occur. Under these conditions, esophageal varices (p. 876) may develop, and there may also be leakage of fluid into the peritoneal cavity across the serosal surfaces of the liver and viscera, producing ascites (p. 867).

Histological Organization of the Liver

Each lobe of the liver is divided by connective tissue into approximately 100,000 **liver lobules,** the basic functional units of the liver. The histological organization and structure of a typical liver lobule are shown in Figure 24-20● (p. 893).

THE LIVER LOBULE
Adjacent lobules are separated from each other by an *interlobular septum.* The hepatocytes in a liver lobule form a series of irregular plates arranged like the spokes of a wheel (Figure 24-20a,b●). The plates are only one cell thick, and exposed hepatocyte surfaces are covered with short microvilli. Within a lobule, sinusoids between adjacent plates empty into the **central vein.** (We introduced sinusoids in Chapter 21. ∞ *[p. 715]*) The liver sinusoids lack a basement membrane, so large openings between the endothelial cells allow solutes—even those as large as plasma proteins—to pass out of the circulation and into the spaces surrounding the hepatocytes.

In addition to typical endothelial cells, the sinusoidal lining includes a large number of **Kupffer** (KOOP-fer) **cells,** also known as *stellate reticuloendothelial cells.* ∞ *[p. 782]* These phagocytic cells, part of the monocyte–macrophage system, engulf pathogens, cell debris, and damaged blood cells. Kupffer cells are also responsible for storing (1) iron, (2) some lipids, and (3) heavy metals, such as tin or mercury, that are absorbed by the digestive tract.

Blood enters the liver sinusoids from small branches of the portal vein and hepatic artery. A typical lobule has a hexagonal shape in cross section (Figure 24-20a,b●). There are six **portal areas,** or *hepatic triads,* one at each corner of the lobule. A portal area contains three structures: (1) a branch of the hepatic portal vein, (2) a branch of the hepatic artery, and (3) a small branch of the bile duct (Figure 24-20c,d●).

Branches from the arteries and veins deliver blood to the sinusoids of adjacent liver lobules (Figure 24-20a,b●). As blood flows through the sinusoids, hepatocytes absorb solutes from the plasma and secrete materials such as plasma proteins. Blood then leaves the sinusoids and enters the central vein of the lobule.

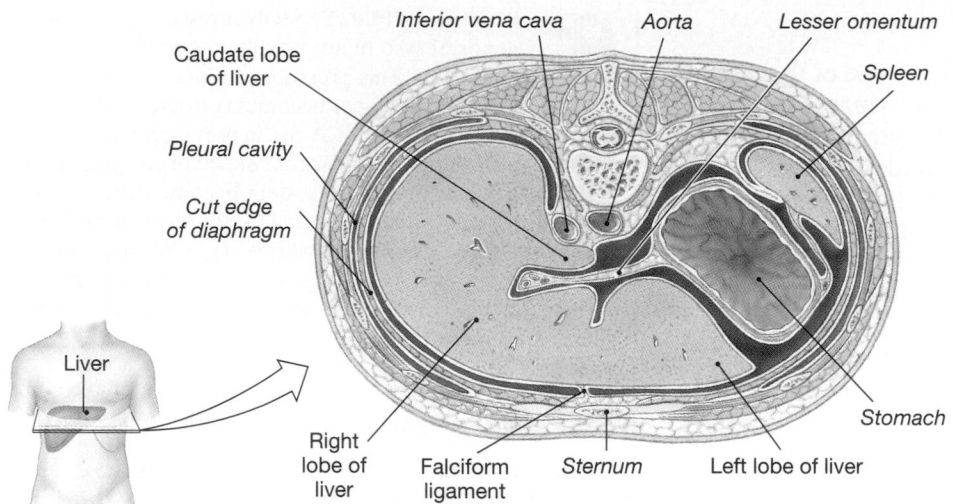

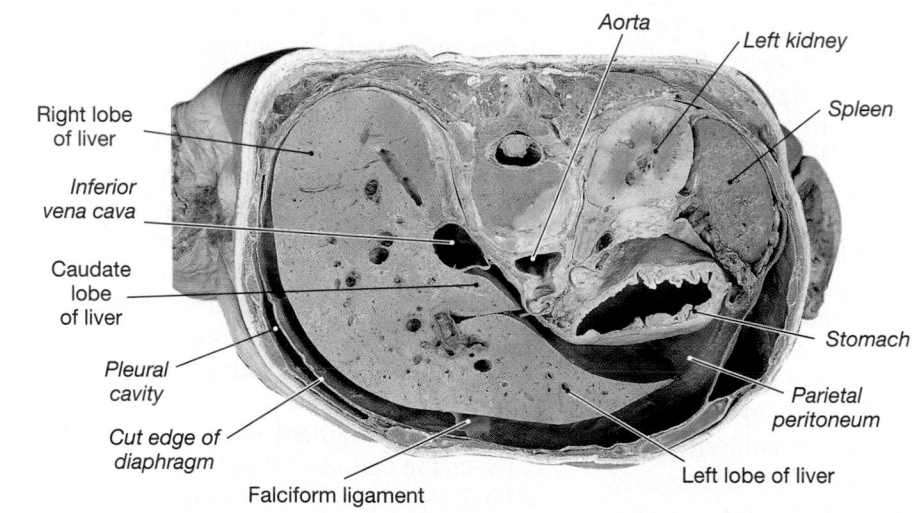

(a) Horizontal section

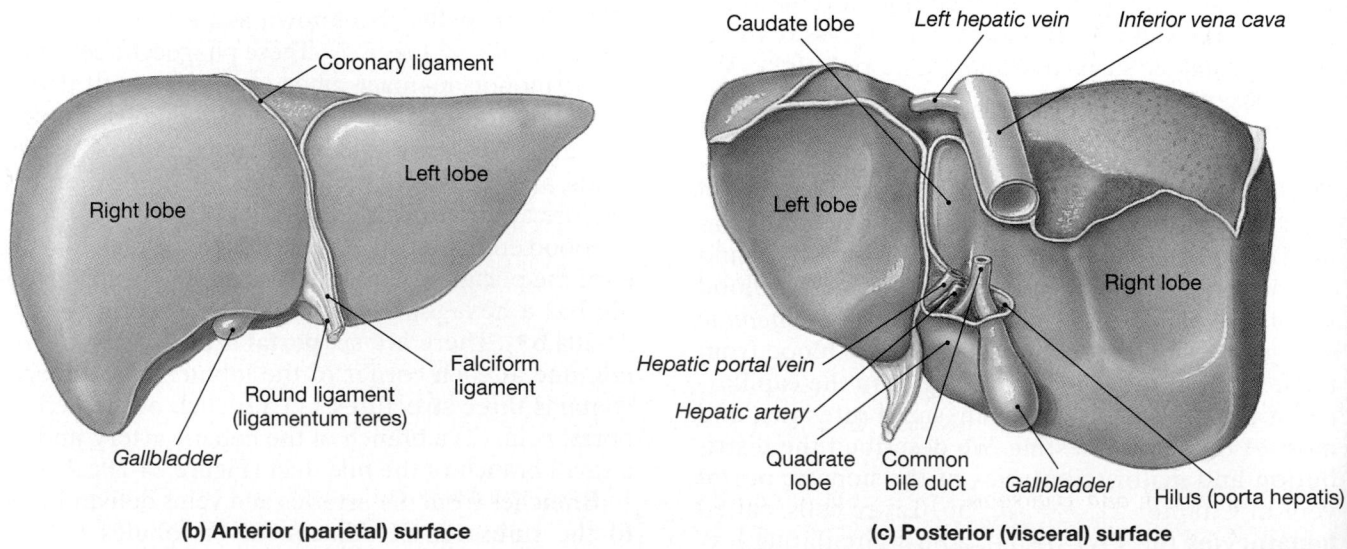

(b) Anterior (parietal) surface

(c) Posterior (visceral) surface

●*FIGURE 24-19*
Anatomy of the Liver.
(a) Horizontal sectional views (diagrammatic and actual) through the upper abdomen. **(b)** The anterior surface of the liver. **(c)** The posterior surface of the liver.

The central veins ultimately merge to form the hepatic veins, which then empty into the inferior vena cava. Liver diseases, such as the various forms of *hepatitis,* and conditions such as alcoholism can lead to degenerative changes in the liver tissue and constriction of the circulatory supply.   AM *Liver Disease*

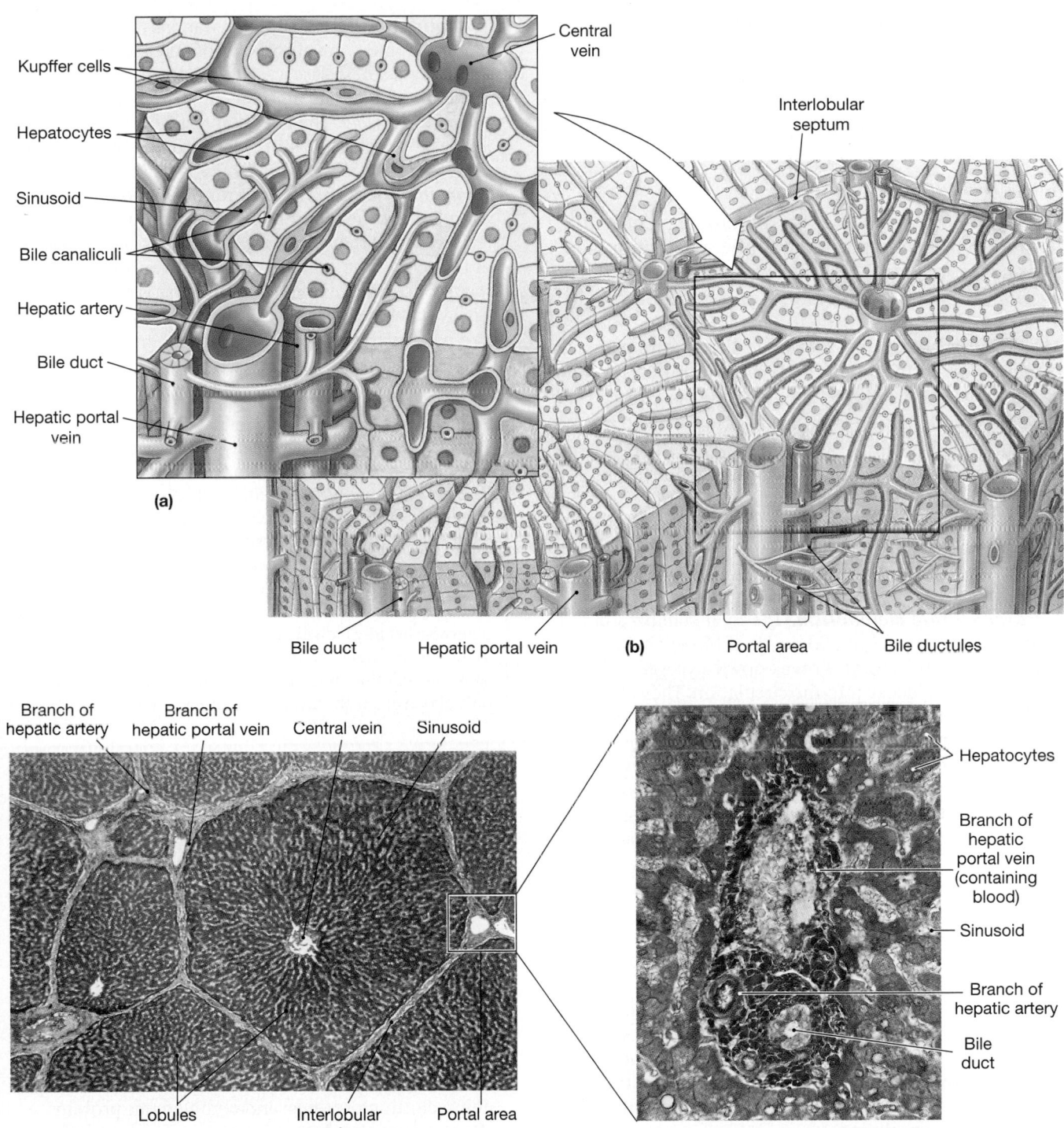

•FIGURE 24-20 Liver Histology. (a) Diagrammatic view of a single liver lobule and its cellular components. **(b)** Diagrammatic view of liver structure, showing relationships among lobules. **(c)** Micrograph showing a section through liver lobules from a pig liver. (LM × 38) (The interlobular septa in a human liver are very difficult to see at comparable magnification.) **(d)** A portal area. (LM × 31)

Bile Secretion and Transport. Bile is secreted into a network of narrow channels between the opposing membranes of adjacent liver cells. These passageways, called **bile canaliculi,** extend outward, away from the central vein. Eventually they connect with fine **bile ductules** (DUK-tūlz), which carry bile to bile ducts in the nearest portal area. The **right** and **left hepatic ducts** collect bile from all the bile ducts of the liver lobes. These ducts unite to form the **common hepatic duct,** which leaves the liver (Figure 24-21•). The bile within the common hepatic duct may either (1) flow into the *common bile duct,* which empties into the duodenal ampulla, or (2) enter the *cystic duct,* which leads to the gallbladder.

The **common bile duct** is formed by the union of the **cystic duct** and the common hepatic duct. The common bile duct passes within the lesser omentum toward the stomach, turns, and penetrates the wall of the duodenum to meet the pancreatic duct at the duodenal ampulla.

The Physiology of the Liver

The liver is responsible for metabolic regulation, hematological regulation, and bile production. The liver has more than 200 different functions; in this discussion, we shall provide only a general overview.

METABOLIC REGULATION The liver is the primary organ involved in regulating the composition of your circulating blood. All blood leaving the absorptive surfaces of the digestive tract enters the hepatic portal system and flows into the liver. Liver cells can thus extract absorbed nutrients or toxins from the blood before it reaches the systemic circulation through the hepatic veins. Excess nutrients are removed and stored, and deficiencies are corrected by mobilizing stored reserves or performing synthetic activities. For example:

- **Carbohydrate metabolism.** The liver stabilizes blood glucose levels at about 90 mg/dl. If blood glucose levels drop, the hepatocytes break down glycogen reserves and release glucose into the circulation. They also synthesize glucose from other carbohydrates or from available amino acids. The synthesis of glucose from other compounds is a process called *gluconeogenesis*. If blood glucose levels climb, liver cells remove glucose from the circulation and either store it as glycogen or use it to synthesize lipids that can be stored in the liver or other tissues. These metabolic activities are regulated by circulating hormones, such as insulin and glucagon, as we noted in Chapter 18. ∞ *[pp. 622–624]*
- **Lipid metabolism.** The liver regulates circulating levels of triglycerides, fatty acids, and cholesterol. When those levels decline, the liver breaks down its lipid reserves and releases them into the circulation. When the levels are high, the lipids are removed for storage. However, because most lipids absorbed by the digestive tract bypass the hepatic portal circulation, this regulation occurs only after lipid levels have risen within the general circulation.
- **Amino acid metabolism.** The liver removes excess amino acids from the circulation. These amino acids may be used to synthesize proteins, or they may be converted to lipids or glucose for storage.
- **Removal of waste products.** When converting amino acids to lipids or carbohydrates, or when breaking down amino acids to get energy, the liver strips off the amino groups, a process called *deamination*. This process produces ammonia, a toxic waste product the liver neutralizes by conversion to *urea,* a relatively harmless compound excreted at the kidneys. Other waste products, circulating toxins, and drugs are also removed from the blood for inactivation, storage, or excretion.

- **Vitamin storage.** Fat-soluble vitamins (A, D, E, and K) and vitamin B_{12} are absorbed from the blood and stored in the liver. These reserves are called on when your diet contains inadequate amounts of those vitamins.
- **Mineral storage.** The liver converts the body's iron reserves to ferritin and stores this protein–iron complex, as we learned in Chapter 19. ∞ *[p. 651]*
- **Drug inactivation.** The liver removes and breaks down circulating drugs, thereby limiting the duration of their effects. When they prescribe drugs, physicians must take into account the rate at which the liver removes a particular drug. For example, a drug that is absorbed relatively quickly must be administered every few hours to keep the plasma concentrations at therapeutic levels.

LIVER DISEASE Any condition that severely damages the liver represents a serious threat to life. The liver has a limited ability to regenerate itself after injury, but liver function will not fully recover unless the normal vascular pattern returns. Examples of important types of liver disease include cirrhosis, which is characterized by the replacement of lobules by fibrous tissue, and various forms of hepatitis as a result of viral infections. Liver transplants are in some cases used to combat liver disease, but the supply of suitable donor tissue is limited, and the success rate is highest in young, otherwise healthy individuals. Clinical trials are now under way to test an artificial liver known as *ELAD* (extracorporeal *l*iver *a*ssist *d*evice) that may prove suitable for the long-term support of persons with chronic liver disease. **AM** *Liver Disease*

HEMATOLOGICAL REGULATION The liver, the largest blood reservoir in your body, receives about 25 percent of the cardiac output. As blood passes by, the liver performs the following functions:

- **Phagocytosis and antigen presentation.** Kupffer cells in the liver sinusoids engulf old or damaged RBCs, cellular debris, and pathogens from the circulation. Kupffer cells are antigen-presenting cells that can stimulate an immune response. ∞ *[p. 791]*
- **Plasma protein synthesis.** The hepatocytes synthesize and release most of the plasma proteins. These include the albumins, which contribute to the osmotic concentration of the blood; the various types of transport proteins; clotting proteins; and complement proteins.
- **Removal of circulating hormones.** The liver is the primary site for the absorption and recycling of epinephrine, norepinephrine, insulin, thyroid hormones, and steroid hormones such as the sex hormones (estrogens and androgens) and corticosteroids. The liver also absorbs cholecalciferol (vitamin D_3) from the blood. Liver cells then convert cholecalciferol, which may be synthesized in the skin or absorbed in the diet, into an intermediary product, 25-hydroxy-D_3, that is released back into the circulation. The intermediary is absorbed by the kidneys and used to generate calcitriol, a hormone important to Ca^{2+} metabolism. ∞ *[p. 618]*
- **Removal of antibodies.** The liver absorbs and breaks down antibodies, releasing amino acids to be recycled.

■ *Removal or storage of toxins.* Lipid-soluble toxins in the diet, such as DDT, are absorbed by the liver and stored in lipid deposits, where they do not disrupt cellular functions. Other toxins are removed from the circulation and are either broken down or excreted in the bile.

■ *Synthesis and secretion of bile.* **Bile** is synthesized in the liver and excreted into the lumen of the duodenum. Bile consists mostly of water, with minor amounts of ions, *bilirubin* (a pigment derived from hemoglobin), cholesterol, and an assortment of lipids collectively known as the **bile salts.** The water and ions assist in the dilution and buffering of acids in chyme as it enters the small intestine.

Bile salts are synthesized from cholesterol in the liver. Several related compounds are involved; the most abundant are derivatives of the steroids *cholate* and *chenodeoxycholate.*

THE FUNCTIONS OF BILE Most dietary lipids are not water-soluble. Mechanical processing in the stomach creates large drops containing a variety of lipids. Pancreatic lipase is not lipid-soluble, so the enzymes can interact with lipids only at the surface of a lipid drop. The larger the droplet, the more lipids are inside, isolated and protected from these enzymes. Bile salts break the droplets apart, a process called **emulsification** (ē-mul-si-fi-KĀ-shun).

Emulsification creates tiny *emulsion droplets* with a superficial coating of bile salts. The formation of tiny droplets increases the surface area available for enzymatic attack. In addition, the layer of bile salts facilitates interaction between the lipids and lipid-digesting enzymes supplied by the pancreas. After lipid digestion has been completed, bile salts promote absorption of lipids by the intestinal epithelium. More than 90 percent of the bile salts are themselves reabsorbed, primarily in the ileum, as lipid digestion is completed. The reabsorbed bile salts enter the hepatic portal circulation and are collected and recycled by the liver. The cycling of bile salts from the liver to the small intestine and back is called the **enterohepatic circulation of bile.**

The Gallbladder

The **gallbladder** is a hollow, pear-shaped, muscular organ. It is divided into three regions: (1) the **fundus,** (2) the **body,** and (3) the **neck** (Figure 24-21a●). The cystic duct leads from the gallbladder toward its union with the common hepatic duct to form the common bile duct. At the duodenum, the common bile duct meets the pancreatic duct before emptying into the **duodenal ampulla** (am-PUL-a) (Figure 24-21b●). The

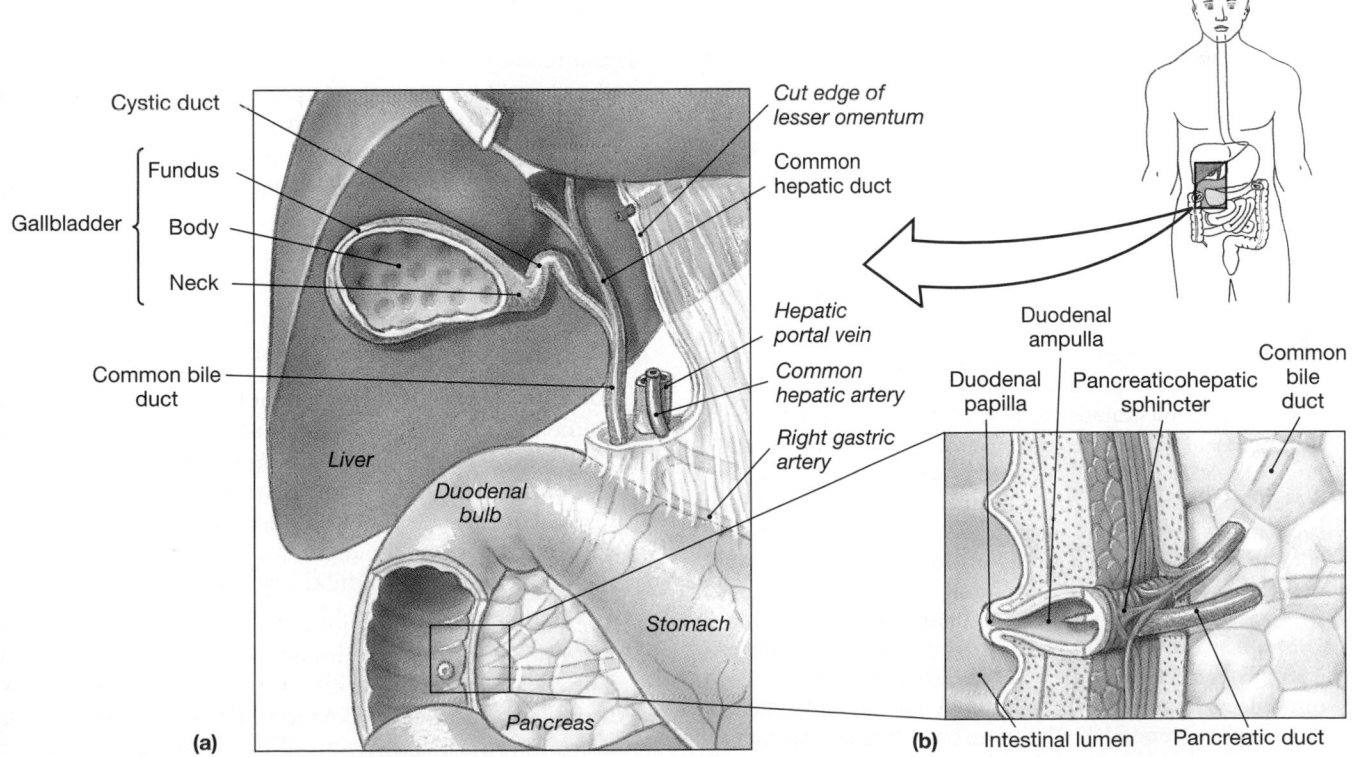

●FIGURE 24-21 The Gallbladder. (a) A view of the inferior surface of the liver, showing the position of the gallbladder and ducts that transport bile from the liver to the gallbladder and duodenum. A portion of the lesser omentum has been cut away to make it easier to see the relationship of the common bile duct to the common hepatic duct and cystic duct. **(b)** Interior view of the duodenum, showing the duodenal ampulla and related structures. AM *Plate 6.5; Scan 16*

duodenal ampulla receives buffers and enzymes from the pancreas and bile from the liver and gallbladder. It opens into the duodenum at a small mound, the **duodenal papilla.**

The **pancreaticohepatic sphincter** (*sphincter of Oddi*), a muscular sphincter, encircles the lumen of the common bile duct and generally the pancreatic duct and ampulla as well. This sphincter remains contracted unless stimulated by the intestinal hormone *cholecystokinin.*

Physiology of the Gallbladder

The gallbladder has two major functions (1) *bile storage* and (2) *bile modification.* Liver cells produce roughly 1 liter of bile each day, but the pancreaticohepatic sphincter remains closed until chyme enters the duodenum. In the meantime, when bile cannot flow along the common bile duct, it enters the cystic duct for storage within the expandable gallbladder. When full, the gallbladder contains 40–70 ml of bile. The composition of bile gradually changes as it remains in the gallbladder. Much of the water is absorbed, and the bile salts and other components of bile become increasingly concentrated.

Bile secretion occurs continuously, but bile release into the duodenum occurs only under stimulation of cholecystokinin. In the absence of CCK, the pancreaticohepatic sphincter remains closed and bile exiting the liver in the common hepatic duct reaches the gallbladder by way of the cystic duct. Cholecystokinin (1) relaxes the pancreaticohepatic sphincter and (2) stimulates contractions in the walls of the gallbladder. These contractions push bile into the small intestine. Cholecystokinin release occurs whenever acid chyme enters the duodenum, but the amount secreted increases markedly when the chyme contains large amounts of lipids.

PROBLEMS WITH BILE STORAGE AND EXCRETION If bile becomes too concentrated, crystals of insoluble minerals and salts begin to appear. These deposits are called **gallstones.** Merely having them, a condition termed **cholelithiasis** (kō-lē-li-THĪ-a-sis; *chole,* bile), does not represent a problem as long as the stones remain small. Small stones are normally flushed down the bile duct and excreted. In *cholecystitis,* the gallstones are so large that they can damage the wall of the gallbladder or block the cystic or common bile duct. A recent therapy for cholecystitis involves immersing the individual in water and shattering the stones with focused sound waves. The apparatus used is called a *lithotripter.* The particles produced are then small enough to pass through the bile duct without difficulty. In severe cases of cholecystitis, the gallbladder may become infected, inflamed, or perforated. Under these conditions, it may be surgically removed, in a procedure known as a *cholecystectomy.* This loss does not seriously impair digestion, for bile production continues at normal levels. However, the bile is more dilute, and its entry into the small intestine is not as closely tied to the arrival of food in the duodenum. [AM] *Cholecystitis*

The Coordination of Secretion and Absorption

A combination of neural and hormonal mechanisms coordinates the activities of the digestive glands. These regulatory mechanisms are centered around the duodenum, for it is there that the acids must be neutralized and the appropriate enzymes added.

Neural mechanisms involving the CNS are concerned with (1) preparing the digestive tract for activity (parasympathetic innervation) or inhibiting gastrointestinal activity (sympathetic innervation) and (2) coordinating the movement of materials along the length of the digestive tract (the enterogastric, gastroenteric, and gastroileal reflexes).

In addition, motor neurons synapsing within the digestive tract release a variety of neurotransmitters. Many of these chemicals are also released inside the CNS, but in general their functions are poorly understood. Examples of potentially important neurotransmitters include Substance P, enkephalins, and endorphins.

We introduced hormones important to the regulation of intestinal and glandular function in the course of our discussion. We will now summarize the introductory information and consider some additional details about the regulatory mechanisms involved. Information on these hormones is summarized in Table 24-1.

Intestinal Hormones

The intestinal tract secretes a variety of hormones, but its hormonal secretions are poorly understood. It has proved very difficult to determine the primary effects of these hormones, largely because all are peptide hormones with similar chemical structures. Careful analyses have led to a marked increase in the number of identified intestinal hormones, but their specific functions have yet to be sorted out to everyone's satisfaction.

Duodenal enteroendocrine cells produce the following hormones known to coordinate digestive functions:

- *Enterocrinin.* **Enterocrinin,** a hormone released when acidic chyme enters the small intestine, stimulates the submucosal glands of the duodenum. The effects of enterocrinin are relatively straightforward, but other intestinal hormones have multiple effects that may target several regions of the digestive tract as well as affecting the accessory glandular organs.

- *Secretin.* **Secretin** is released when chyme arrives in the duodenum. The primary effect of secretin is to cause an increase in the secretion of bile and buffers by the liver and pancreas. Among its secondary effects, secretin reduces gastric motility and secretory rates.

- *Cholecystokinin.* **Cholecystokinin (CCK)** is secreted when chyme arrives in the duodenum, especially when the chyme contains lipids and partially digested proteins. In the pancreas, CCK accelerates the produc-

TABLE 24-1 **Important Gastrointestinal Hormones and Their Primary Effects**

Hormone	Stimulus	Origin	Target	Effects
Cholecystokinin (CCK)	Arrival of chyme containing lipids and partially digested proteins	Duodenum	Pancreas	Stimulates production of pancreatic enzymes
			Gallbladder	Stimulates contraction of gallbladder
			Duodenum	Causes relaxation of pancreaticohepatic sphincter
			Stomach	Inhibits gastric secretion and motion
			CNS	May reduce hunger
Enterocrinin	Arrival of chyme in the duodenum	Duodenum	Submucosal (Brunner's) glands	Stimulates production of alkaline mucus
Gastric inhibitory peptide (GIP)	Arrival of chyme containing large quantities of fats and glucose	Duodenum	Pancreas	Stimulates release of insulin by pancreatic islets
			Stomach	Inhibits gastric secretion and motility
			Adipose tissue	Stimulates lipid synthesis
			Skeletal muscle	Stimulates glucose use
Gastrin	Vagus nerve stimulation or arrival of food in the stomach	Stomach	Stomach	Stimulates production of acids and enzymes; increases motility
	Arrival of chyme containing large quantities of undigested proteins	Duodenum	Stomach	As above
Secretin	Arrival of chyme in the duodenum	Duodenum	Pancreas	Stimulates production of alkaline buffers
			Stomach	Inhibits gastric secretion and motility
			Liver	Increases rate of bile secretion
Vasoactive intestinal peptide (VIP)	Arrival of chyme in the duodenum	Duodenum	Duodenal glands, stomach	Stimulates buffer secretion; inhibits acid production; dilates intestinal capillaries

tion and secretion of all types of digestive enzymes. It also causes relaxation of the pancreaticohepatic sphincter and contraction of the gallbladder, resulting in the ejection of bile and pancreatic juice into the duodenum. The net effects of CCK are thus (1) to increase the secretion of pancreatic enzymes and (2) to push pancreatic secretions and bile into the duodenum. The presence of CCK in high concentrations has two additional effects: (1) It inhibits gastric activity, and (2) it appears to have CNS effects that reduce the sensation of hunger.

- ***Gastric inhibitory peptide.*** **Gastric inhibitory peptide (GIP)** is secreted when fats and carbohydrates, especially glucose, enter the small intestine. The inhibition of gastric activity is accompanied by the stimulation of insulin release at the pancreatic islets, so GIP is also known as *glucose-dependent insulinotropic peptide.* Like secretin, GIP stimulates the activity of the duodenal submucosal glands. It also affects a variety of other tissues. For example, GIP works with insulin to

stimulate lipid synthesis in adipose tissue, and this peptide increases glucose use in skeletal muscle.

- ***Vasoactive intestinal peptide.*** **Vasoactive intestinal peptide (VIP)** stimulates the secretion of intestinal glands, dilates regional capillaries, and inhibits acid production in the stomach. By dilating capillaries in active areas of the intestinal tract, VIP provides an efficient mechanism for removing absorbed nutrients.

- ***Gastrin.*** Gastrin is secreted by G cells in the duodenum when they are exposed to large quantities of incompletely digested proteins.

- ***Other intestinal hormones.*** Several other hormones are produced in relatively small quantities. Examples include *motilin,* which stimulates intestinal contractions, *villikinin,* which promotes movement of villi and associated lymph flow, and *somatostatin,* which inhibits gastric secretion.

Functional interactions among gastrin, secretin, CCK, GIP, and VIP are diagrammed in Figure 24-22●.

•**FIGURE 24-22** The Activities of Major
Digestive Tract Hormones. The primary
actions of gastrin, GIP, secretin, CCK, and
VIP.

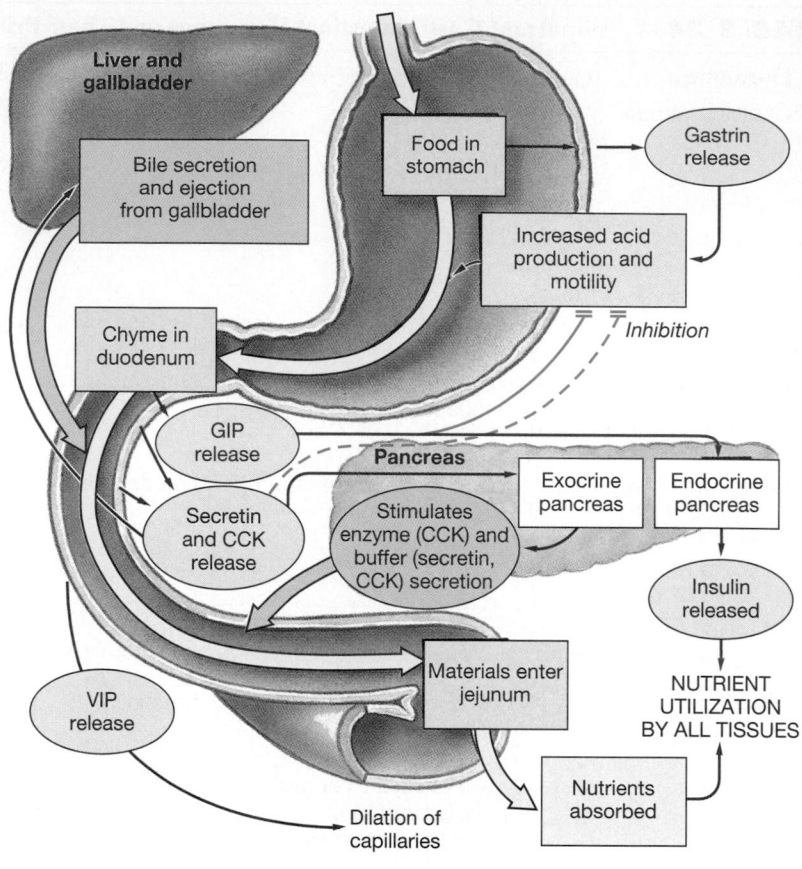

•**FIGURE 24-22** The Activities of Major
Digestive Tract Hormones. The primary
actions of gastrin, GIP, secretin, CCK, and
VIP.

Intestinal Absorption

On average it takes about 5 hours for materials to pass from your duodenum to the end of your ileum, so the first of the materials to enter the duodenum after you eat breakfast may leave the small intestine at lunchtime. Along the way, absorptive effectiveness is enhanced by the fact that so much of the mucosa is movable. The microvilli can be moved by their supporting microfilaments, the individual villi by smooth muscle cells, groups of villi by the muscularis mucosae, and the plicae by the muscularis mucosae and the muscularis externa. All these movements tend to stir and mix the intestinal contents. As a result, the environment around each epithelial cell changes from moment to moment.

☑ How is the small intestine adapted for the absorption of nutrients?

☑ How would a meal that is high in fat affect the level of cholecystokinin (CCK) in the blood?

☑ How would the pH of intestinal contents be affected if the small intestine did not secrete the hormone secretin?

☑ The digestion of which nutrient would be most impaired by damage to the exocrine pancreas?

THE LARGE INTESTINE

The horseshoe-shaped large intestine begins at the end of the ileum and ends at the anus. The large intestine lies inferior to the stomach and liver and almost completely frames the small intestine (Figure 24-17a•, p. 888). The major functions of the large intestine include (1) the reabsorption of water and compaction of intestinal contents into feces, (2) the absorption of important vitamins liberated by bacterial action, and (3) the storing of fecal material prior to defecation.

The large intestine, or the **large bowel,** has an average length of about 1.5 meters (5 ft) and a width of 7.5 cm (3 in.). We can divide it into three parts: (1) the pouchlike *cecum,* the first portion of the large intestine; (2) the *colon,* the largest portion; and (3) the *rectum,* the last 15 cm (6 in.) of the large intestine and the end of the digestive tract (Figure 24-23•, pp. 900–901).

The Cecum

Material arriving from the ileum first enters an expanded pouch called the **cecum** (SĒ-kum). The ileum attaches to the medial surface of the cecum and opens into the cecum at the **ileocecal** (il-ē-ō-SĒ-kal) **valve** (Figure 24-23a•). The cecum collects and stores chyme and begins the process of compaction. The slender, hollow **vermiform appendix** (*vermis,* worm), or simply *appendix*, is attached to the posteromedial surface of the cecum. The appendix is generally about 9 cm (3.5 in.) long, but its size and shape are quite variable. A small mesentery called the **mesoappendix** connects the appendix to the ileum and cecum. The mucosa and submucosa of the appendix are dominated by lymphoid nodules, and the appendix's primary function is as an organ of the lymphatic system. Inflammation of the appendix is known as *appendicitis*. ∞ *[p. 776]*

The Colon

The **colon** has a larger diameter and a thinner wall than the small intestine. Distinctive features of the colon (Figure 24-23●) include the following:

- The wall of the colon forms a series of pouches, or **haustra** (HAWS-truh; singular, *haustrum*). Cutting into the intestinal lumen reveals that the creases between the haustra affect the mucosal lining as well, producing a series of internal folds. Haustra permit the expansion and elongation of the colon rather like the bellows that allow an accordion to lengthen.

- Three separate longitudinal ribbons of smooth muscle—the **taenia coli** (TĒ-nē-a KŌ-lī), are visible on the outer surfaces of the colon just beneath the serosa. These bands correspond to the outer layer of the muscularis externa in other portions of the digestive tract. Muscle tone within these bands creates the haustra.

- The serosa of the colon contains numerous teardrop-shaped sacs of fat called **epiploic** (e-pip-LŌ-ik) **appendages.**

Regions of the Colon

We can subdivide the colon into four regions: the *ascending colon,* the *transverse colon,* the *descending colon,* and the *sigmoid colon* (Figure 24-23a●):

1. The **ascending colon** begins at the superior border of the cecum and ascends along the right lateral and posterior wall of the peritoneal cavity to the inferior surface of the liver. At this point, the colon makes a sharp bend to the left at the **right colic flexure,** or *hepatic flexure.* This flexure marks the end of the ascending colon and the beginning of the *transverse colon.* The ascending colon is retroperitoneal, and only its lateral and anterior surfaces are covered by the peritoneum (Figure 24-5●, p. 868).

2. The **transverse colon** curves anteriorly from the right colic flexure and crosses the abdomen from right to left. It is supported by the transverse mesentery and is separated from the anterior abdominal wall by the layers of the greater omentum. As the transverse colon reaches the left side of the body, it passes inferior to the greater curvature of the stomach. Near the spleen, the colon makes a 90° turn at the **left colic flexure,** or *splenic flexure,* and becomes the *descending colon.*

3. The **descending colon** proceeds inferiorly along the left side until reaching the iliac fossa. The descending colon is retroperitoneal and firmly attached to the abdominal wall. At the iliac fossa, the descending colon curves at the **sigmoid flexure** and becomes the *sigmoid colon.*

4. The sigmoid flexure is the start of the **sigmoid** (SIG-moyd) **colon** (*sigmeidos,* the Greek letter *S*), an S-shaped segment that is only about 15 cm (6 in.) long. It lies posterior to the urinary bladder, suspended from the sigmoid mesocolon (Figure 24-5●, p. 868). The sigmoid colon empties into the *rectum.*

The large intestine receives blood from tributaries of the *superior mesenteric* and *inferior mesenteric arteries,* and venous blood is collected by the *superior mesenteric* and *inferior mesenteric veins.* We detailed those vessels in Chapter 21. ∞ *[p. 756]*

The Rectum

The **rectum** (REK-tum) forms the last 15 cm (6 in.) of the digestive tract (Figure 24-23a,b●). The rectum is an expandable organ for the temporary storage of fecal material. Movement of fecal materials into the rectum triggers the urge to defecate.

The last portion of the rectum, the **anorectal** (ā-nō-REK-tal) **canal,** contains small longitudinal folds, the **rectal columns.** The distal margins of the rectal columns are joined by transverse folds that mark the boundary between the columnar epithelium of the proximal rectum and a stratified squamous epithelium like that in the oral cavity. Very close to the **anus,** or **anal orifice** (the exit of the anorectal canal), the epidermis becomes keratinized and identical to the surface of the skin.

There is a network of veins in the lamina propria and submucosa of the anorectal canal. If venous pressures there rise too high due to straining during defecation, the veins may become distended, producing *hemorrhoids.* The circular muscle layer of the muscularis externa in this region forms the **internal anal sphincter.** The smooth muscle cells of the internal anal sphincter are not under voluntary control. The **external anal sphincter** guards the anus. This sphincter, which consists of a ring of skeletal muscle fibers, is under voluntary control.

COLON CANCER Colon cancers are relatively common. Approximately 94,100 cases are diagnosed in the United States each year (in addition to 37,100 cases of rectal cancers). In 1997 there were an estimated 46,600 deaths from colon cancers (and another 8300 deaths from rectal cancers). The mortality rate for these cancers remains high, and the best defense appears to be early detection and prompt treatment. The standard screening test involves checking the feces for blood. This is a simple procedure that can easily be performed on a stool (fecal) sample in the course of a routine physical. AM
Colon Inspection and Cancer

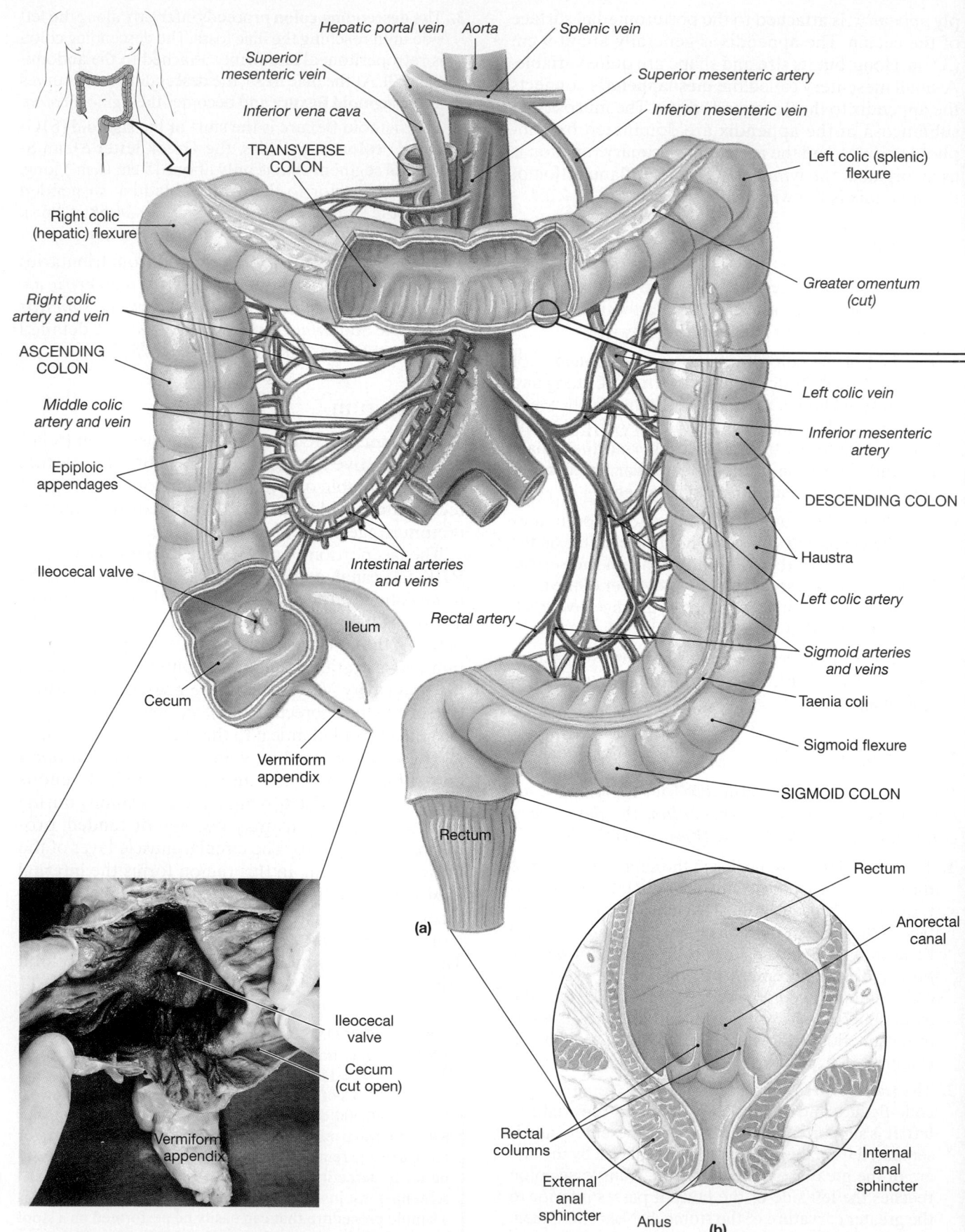

•*FIGURE 24-23* **The Large Intestine.** (a) Gross anatomy and regions of the large intestine. (b) Detailed anatomy of the rectum and anus. (c) The mucosa and glands of the colon. (LM × 114) [AM] *Scan 19*

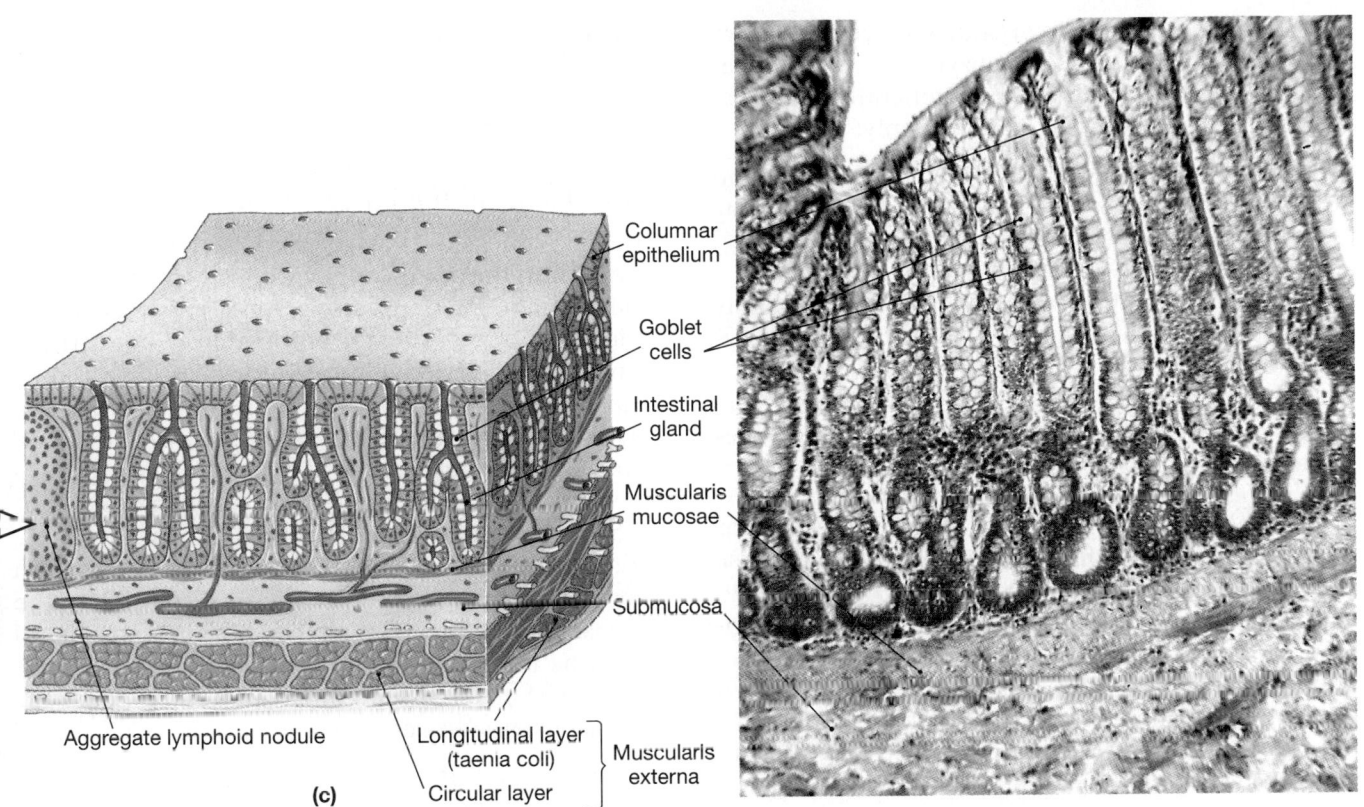

Columnar epithelium

Goblet cells

Intestinal gland

Muscularis mucosae

Submucosa

Aggregate lymphoid nodule

Longitudinal layer (taenia coli)

Circular layer

Muscularis externa

(c)

Histology of the Large Intestine

Although the diameter of the colon is roughly three times that of the small intestine, its wall is much thinner. The major characteristics of the colon are the lack of villi, the abundance of goblet cells, and the presence of distinctive intestinal glands (Figure 24-23c●). The glands in the large intestine are deeper than those of the small intestine, and they are dominated by goblet cells. The mucosa of the large intestine does not produce enzymes; any digestion that occurs results from enzymes introduced in the small intestine or from bacterial action. The mucus is important in providing lubrication as the fecal material becomes less moist and more compact. Mucous secretion occurs as local stimuli, such as friction or exposure to harsh chemicals, trigger short reflexes involving local nerve plexuses. Large lymphoid nodules are scattered throughout the lamina propria and submucosa.

The muscularis externa of the large intestine is unusual because the longitudinal layer has been reduced to the muscular bands of the taenia coli. However, the mixing and propulsive contractions of the colon resemble those of the small intestine.

DIVERTICULOSIS In *diverticulosis* (dī-ver-tik-ū-LŌ-sis), pockets (*diverticula*) form in the mucosa, generally in the sigmoid colon. These pockets get forced outward, probably by the pressures generated during defecation. If the pockets push through weak points in the muscularis externa, they form semi-isolated chambers that are subject to recurrent infection and inflammation. The infections cause pain and occasional bleeding, a condition known as *diverticulitis* (dī-ver-tik-ū-LĪ-tis). Inflammation of other portions of the colon is called *colitis* (kō-LĪ-tis). AM *Inflammatory Bowel Disease*

Physiology of the Large Intestine

Less than 10 percent of the absorption under way in the digestive tract occurs in the large intestine. Nevertheless, the absorptive operations in this segment of the digestive tract are important. The large intestine also prepares the fecal material for ejection from the body.

Absorption in the Large Intestine

The reabsorption of water is an important function of the large intestine. Although roughly 1500 ml of material enters your colon each day, only about 200 ml of feces is ejected. The remarkable efficiency of digestion can best be appreciated by considering the average composition of fecal wastes: 75 percent water, 5 percent bacteria, and the rest a mixture of indigestible materials, small quantities of inorganic matter, and the remains of epithelial cells.

In addition to reabsorbing water, the large intestine absorbs a number of other substances that remain in the fecal material or that were secreted into the digestive tract along its length:

VITAMINS Vitamins are organic molecules that are important as cofactors or coenzymes in many metabolic pathways. Colonic bacteria generate three vitamins that supplement the dietary supply:

1. *Vitamin K,* a fat-soluble vitamin that the liver needs to enable it to synthesize four clotting factors, including prothrombin. Intestinal bacteria produce roughly half of our daily vitamin K requirements.

2. *Biotin,* a water-soluble vitamin important to a variety of reactions, notably those involved with glucose metabolism.

3. *Vitamin B_5* (pantothenic acid), a water-soluble vitamin required in the manufacture of steroid hormones and some neurotransmitters.

Disorders resulting from deficiencies of biotin or vitamin B_5 are extremely rare after infancy because the intestinal bacteria produce sufficient amounts to supplement any shortage in the diet. Vitamin K deficiencies, which lead to impaired blood clotting, can result from (1) a deficiency of lipids in the diet, which impairs the absorption of all fat-soluble vitamins, or (2) problems affecting lipid processing and absorption, such as inadequate bile production or chronic diarrhea.

UROBILINOGEN In Chapter 19, we discussed the fate of conjugated bilirubin, a breakdown product of heme. ∞ *[p. 650]* Inside the large intestine, bacteria convert the conjugated bilirubin to *urobilinogens* and *stercobilinogens.* Some of the urobilinogens are absorbed into the circulation and excreted in the urine. On exposure to oxygen, some of the urobilinogens and stercobilinogens are converted to **urobilins** and **stercobilins.** These pigments give feces a brown coloration.

BILE SALTS Some of the bile salts remaining in the feces will be reabsorbed in the cecum.

TOXINS Bacterial action breaks down peptides that remain in the feces and generates (1) ammonia, in the form of soluble *ammonium ions* (NH_4^+), (2) *indole* and *skatole,* two nitrogen-containing compounds that are primarily responsible for the odor of feces, and (3) hydrogen sulfide (H_2S), a gas that produces a "rotten egg" odor. Significant amounts of ammonia and smaller amounts of other toxins cross the colonic epithelium and enter the hepatic portal circulation. These are removed by the liver and converted to relatively nontoxic compounds that can be released into the blood and excreted at the kidneys.

Indigestible carbohydrates are not altered by intestinal enzymes, and they arrive in the colon virtually intact. These complex polysaccharides provide a reliable nutrient source for colonic bacteria, whose metabolic activities are responsible for the small quantities of intestinal gas, or **flatus,** in the large intestine.

Meals containing large numbers of indigestible carbohydrates (such as franks and beans) stimulate bacterial gas production, leading to colonic distension, cramps, and the frequent discharge of intestinal gases.

Movements of the Large Intestine

The gastroileal and gastroenteric reflexes move materials into the cecum at mealtimes. Movement from the cecum to the transverse colon occurs very slowly, allowing hours for water absorption to convert the already thick material into a sludgy paste. Peristaltic waves move material along the length of the colon, and segmentation movements, sometimes called *haustral churning,* mix the contents of adjacent haustra. Movement from the transverse colon through the rest of the large intestine results from powerful peristaltic contractions called **mass movements,** which occur a few times each day. The stimulus is distension of the stomach and duodenum, and the commands are relayed over the intestinal nerve plexuses. The contractions force fecal materials into the rectum and produce the conscious urge to defecate.

DEFECATION The rectal chamber is usually empty except when one of those powerful peristaltic contractions forces fecal materials out of the sigmoid colon. Distension of the rectal wall then triggers the defecation reflex. The **defecation reflex** involves two positive feedback loops (Figure 24-24●); both are triggered by stretch receptor stimulation in the walls of the rectum. The first loop is a short reflex that triggers a series of peristaltic contractions in the rectum that move feces toward the anus (Steps 1, 2, 3a, and 4a). The second loop is a long reflex coordinated by the sacral parasympathetic system. This reflex stimulates mass movements that push fecal materials toward the rectum from the descending colon and sigmoid colon (Steps 3b and 4b).

Rectal stretch receptors also trigger two reflexes important to the *voluntary* control of defecation. One is a visceral reflex mediated by parasympathetic innervation within the pelvic nerves. This reflex causes the relaxation of the internal anal sphincter, a smooth muscle sphincter that controls the movement of feces into the anorectal canal. The second (Step 3c in Figure 24-24●) is a somatic reflex that stimulates the immediate contraction of the external anal sphincter, a skeletal muscle introduced in Chapter 11. ∞ *[p. 338]* The motor commands are carried by the pudendal nerves.

The elimination of feces requires that both the internal and external anal sphincters be relaxed, but these reflexes open the internal sphincter and close the external sphincter. The actual release of feces requires conscious effort to open the external sphincter. In addition to opening the external sphincter, consciously

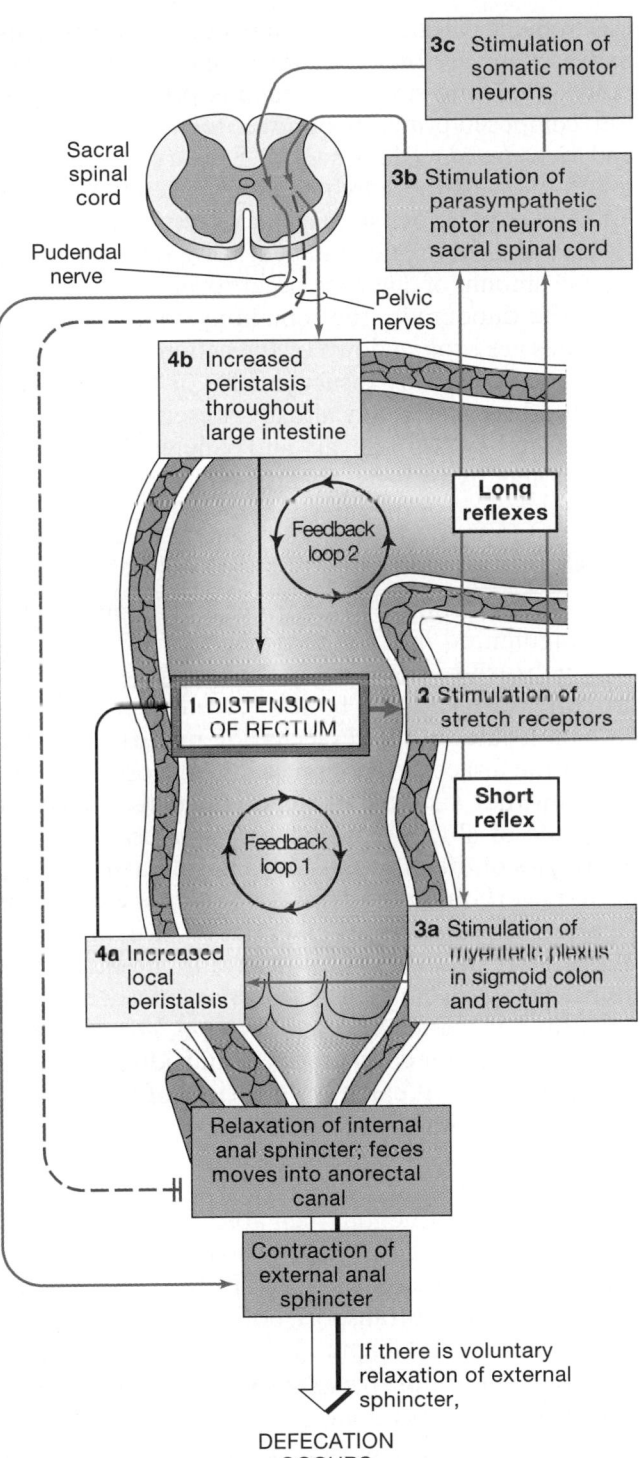

Sacral spinal cord

Pudendal nerve

3c Stimulation of somatic motor neurons

3b Stimulation of parasympathetic motor neurons in sacral spinal cord

Pelvic nerves

4b Increased peristalsis throughout large intestine

Long reflexes

Feedback loop 2

1 DISTENSION OF RECTUM

2 Stimulation of stretch receptors

Short reflex

Feedback loop 1

4a Increased local peristalsis

3a Stimulation of myenteric plexus in sigmoid colon and rectum

Relaxation of internal anal sphincter; feces moves into anorectal canal

Contraction of external anal sphincter

If there is voluntary relaxation of external sphincter,

DEFECATION OCCURS

● **FIGURE 24-24** The Defecation Reflex. Short and long reflexes (Steps 1, 2, 3a, and 4a and Steps 3b and 4b, respectively) promote movement of fecal material toward the anus. Another long reflex triggered by rectal stretch receptor stimulation (Step 3c) prevents involuntary defecation.

directed activities such as tensing the abdominal muscles or making expiratory movements while closing the glottis elevate intra-abdominal pressures and help force fecal materials out of the rectum.

If the external sphincter remains constricted, the peristaltic contractions cease until additional rectal expansion triggers the defecation reflex a second time. The urge to defecate usually develops when rectal pressure reaches about 15 mm Hg. If pressure inside the rectum exceeds 55 mm Hg, the external sphincter will relax, and defecation will occur. This mechanism regulates defecation in infants and in adults with severe spinal cord injuries.

DIARRHEA AND CONSTIPATION Diarrhea (dī-a-RĒ-uh) exists when an individual has frequent, watery bowel movements. This condition results when the colonic mucosa becomes unable to maintain normal levels of absorption or the rate of fluid entry into the colon exceeds its maximum reabsorptive capacity. Bacterial, viral, or protozoan infection of the colon or small intestine may cause acute bouts of diarrhea lasting several days. Severe diarrhea can be life-threatening due to cumulative losses of fluids and ions. In *cholera* (KOL-e-ruh), bacteria bound to the intestinal lining release toxins that stimulate a massive fluid secretion across the intestinal epithelium. Without treatment, a person with cholera may die of acute dehydration in a matter of hours. |AM| *Diarrhea*

Constipation is infrequent defecation, generally involving dry, hard feces. Constipation occurs when fecal materials move through the colon so slowly that excessive water reabsorption occurs. The feces then become extremely compact, difficult to move, and highly abrasive. Inadequate dietary fiber and fluids, coupled with a lack of exercise, are the common causes. Constipation can generally be treated by oral administration of stool softeners, such as *Colace®*, laxatives, or **cathartics** (ka-THAR-tiks), which promote defecation. These compounds promote water movement into the feces, increase fecal mass, or irritate the lining of the colon to stimulate peristalsis. For example, indigestible fiber adds bulk to the feces, retaining moisture and stimulating stretch receptors that promote peristalsis. The promotion of peristalsis is one of the benefits of high-fiber cereals. Active movement during exercise also assists in the movement of fecal materials through the colon.

DIGESTION AND ABSORPTION

A typical meal contains a mixture of carbohydrates, proteins, lipids, water, electrolytes, and vitamins. Your digestive system handles each of those components differently. Large organic molecules must be broken down through digestion before absorption can occur. Water, electrolytes, and vitamins can be absorbed without preliminary processing, but special transport mechanisms are commonly involved.

The Processing and Absorption of Nutrients

Food contains large organic molecules, many of them insoluble. The digestive system first breaks down the physical structure of the ingested material and then proceeds to disassemble the component molecules into smaller fragments. This disassembly eliminates any antigenic properties, making the fragments suitable for absorption. The molecules released into the bloodstream will be absorbed by cells and either (1) broken down to provide energy for the synthesis of ATP or (2) used to synthesize carbohydrates, proteins, and lipids. This section will focus on the mechanics of digestion and absorption; the fate of the compounds inside cells will be the focus of Chapter 25.

Most ingested organic materials are complex chains of simpler molecules. In a typical dietary carbohydrate, the basic molecules are simple sugars. In a protein, the building blocks are amino acids, and in lipids, they are generally fatty acids. Digestive enzymes break the bonds between the component molecules in a process called *hydrolysis.* (We detailed the hydrolysis of carbohydrates, lipids, and proteins in Chapter 2.) ∞ *[p. 40]*

The classes of digestive enzymes differ with respect to their specific targets. *Carbohydrases* break the bonds between sugars, *proteases* split the linkages between amino acids, and *lipases* separate the fatty acids from glycerides. Specific enzymes in each class may be even more selective, breaking bonds between specific molecules. For example, a particular carbohydrase might break bonds between glucose molecules but not those between glucose and another simple sugar.

The enzymes involved in digestion are localized in two different sites. The enzymes secreted by the salivary glands, tongue, stomach, and pancreas are mixed into the ingested material as it passes along the digestive tract. These enzymes break down large proteins, lipids, and carbohydrates into smaller fragments, which in turn must typically be broken down further before absorption can occur. The final enzymatic steps involve brush border enzymes, which are attached to the exposed surfaces of microvilli.

Figure 24-25● summarizes information concerning the digestive fates of carbohydrates, lipids, and proteins, and Table 24-2 (p. 906) reviews the major digestive enzymes and their functions. We shall now take a closer look at the digestion and absorption of carbohydrates, lipids, and proteins.

Carbohydrate Digestion and Absorption

The digestion of complex carbohydrates (simple polysaccharides and starches) proceeds in two steps. One step involves carbohydrases produced by the salivary glands and pancreas; the other involves brush border enzymes.

Salivary and Pancreatic Enzymes

The digestion of complex carbohydrates involves two enzymes—salivary amylase and pancreatic alpha-amylase (Figure 24-25a●), that function effectively at a pH of 6.7–7.5. Carbohydrate digestion begins in the mouth during mastication, by the action of salivary amy-

lase from the parotid and submandibular salivary glands. Salivary amylase breaks down starches (complex carbohydrates) into smaller fragments, producing a mixture composed primarily of *disaccharides* (two sugars) and *trisaccharides* (three sugars). Salivary amylase continues to digest the starches and glycogen in the meal for 1–2 hours before stomach acids render it inactive. Because the enzymatic content of saliva is not high, only a small amount of digestion occurs over this period.

In the duodenum, the remaining complex carbohydrates are broken down by the action of pancreatic alpha-amylase. Any disaccharides or trisaccharides produced, as well as any already present in the food, are ignored by both salivary and pancreatic amylases. Additional hydrolysis does not occur until these molecules contact the intestinal mucosa.

Brush Border Enzymes

Prior to absorption, disaccharides and trisaccharides are fragmented into *monosaccharides* (simple sugars) by brush border enzymes of the intestinal microvilli. **Maltase** splits bonds between the two glucose molecules of the disaccharide **maltose**. **Sucrase** breaks the disaccharide **sucrose** into glucose and *fructose,* another six-carbon sugar. **Lactase** releases a molecule of glucose and one of *galactose* from the hydrolysis of the disaccharide **lactose.** Lactose is the primary carbohydrate in milk, so by breaking down lactose, lactase provides essential services throughout infancy and early childhood. If the intestinal mucosa stops producing lactase by adolescence, the individual becomes *lactose-intolerant.* After ingesting milk and other dairy products, individuals who are lactose-intolerant can have a variety of unpleasant digestive problems. AM *Lactose Intolerance*

Absorption of Monosaccharides

The intestinal epithelium then absorbs the monosaccharides by *facilitated diffusion* and *cotransport mechanisms.* We detailed facilitated diffusion (see Figure 3-9a●, p. 77) and cotransport, both of which involve a carrier protein, in Chapter 3. ∞ *[pp. 77–78]* There are three major differences between facilitated diffusion and cotransport:

1. *Facilitated diffusion moves only one molecule or ion through the cell membrane; cotransport moves more than one molecule or ion through the membrane at the same time.* In cotransport, the transported materials move in the same direction: down the concentration gradient for at least one of the transported substances.

2. *Facilitated diffusion does not require ATP.* Although cotransport by itself does not consume ATP, the cell must often expend ATP to preserve homeostasis. For example, the process may introduce sodium ions that must later be pumped out of the cell.

3. *Facilitated diffusion will not occur if there is an opposing concentration gradient for the particular mole-*

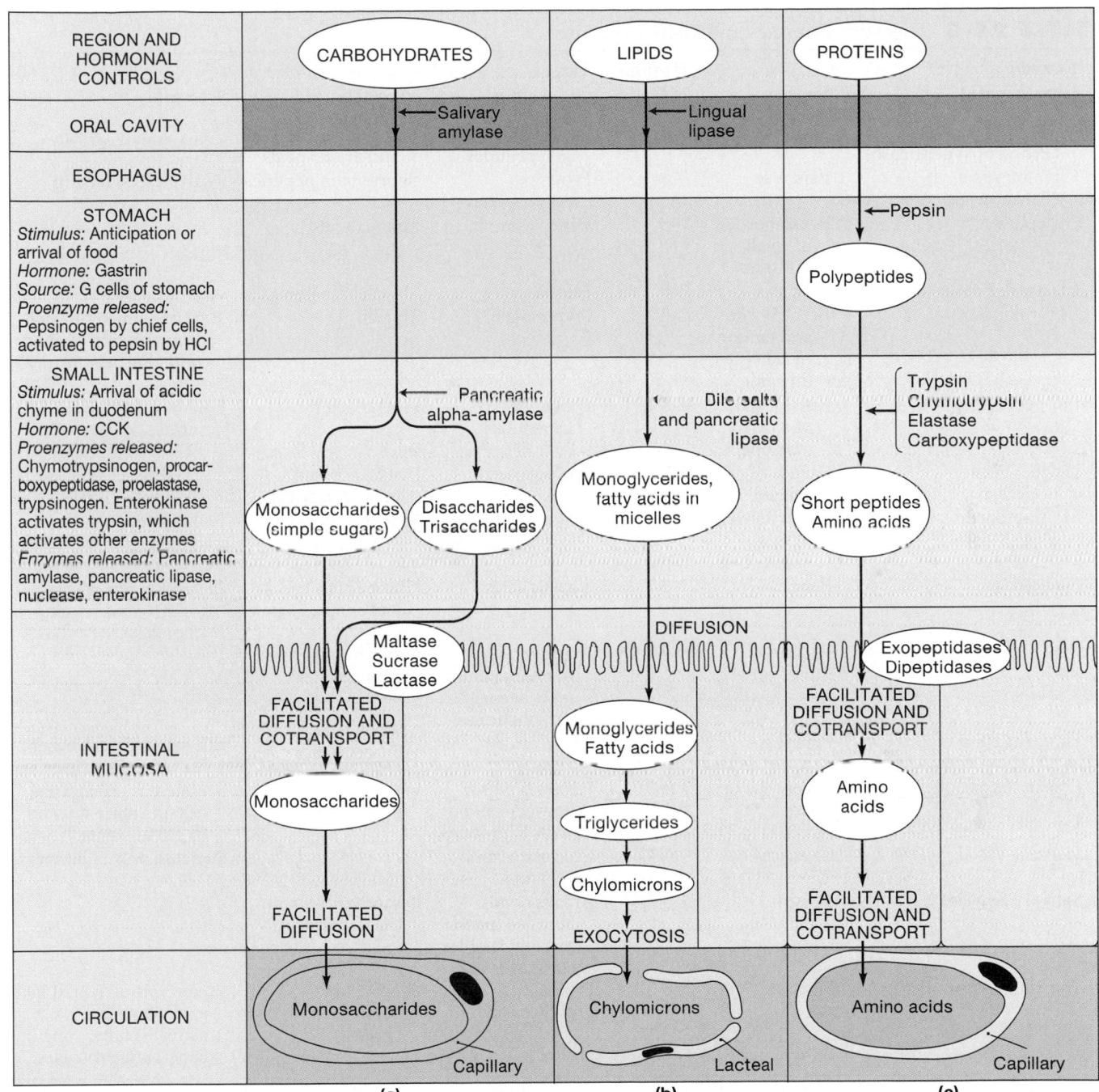

• **FIGURE 24-25** **A Summary of the Chemical Events in Digestion.** For further details on the enzymes involved, see Table 24-2.

cule or ion. Cotransport can occur despite an opposing concentration gradient for one of the transported substances. For example, cells lining the small intestine will continue to absorb glucose when glucose concentrations inside the cells are much higher than they are in the intestinal contents.

The cotransport system responsible for the uptake of glucose also brings sodium ions into the cell. This passive process resembles facilitated diffusion except that both a sodium ion and a glucose molecule must bind to the carrier protein before they can move into the cell. Glu-

cose cotransport is an example of *sodium-linked cotransport.* Comparable cotransport mechanisms exist for other simple sugars and for some amino acids. Although these mechanisms deliver valuable nutrients to the cytoplasm, they also bring in sodium ions that must be ejected by the sodium–potassium exchange pump.

The simple sugars entering the cell diffuse by the cytoplasm and reach the interstitial fluid by facilitated diffusion at the base of the cell. Once in the interstitial fluid, these monosaccharides diffuse into the capillaries of the villus for eventual transport to the liver in the hepatic portal vein.

TABLE 24-2 Digestive Enzymes and Their Functions

Enzyme (Proenzyme)	Source	Optimal pH	Target and Action	Products	Remarks
Carboxypeptidase (Procarboxypeptidase)	Pancreas	7–8	Proteins, polypeptides	Short-chain peptides and amino acids	Activated by trypsin
Chymotrypsin (Chymotrypsinogen)	Pancreas	7–8	Proteins, polypeptides	Short-chain peptides	Activated by trypsin
Dipeptidases	Brush border of small intestine	7–8	Dipeptides	Amino acids	
Elastase (Proelastase)	Pancreas	7–8	Elastin	Short-chain peptides	Activated by trypsin
Enterokinase	Brush border and lumen of small intestine	7–8	Trypsinogen	Trypsin	Reaches lumen through disintegration of shed epithelial cells
Exopeptidases	Brush border of small intestine	7–8	Dipeptides, tripeptides	Amino acids	Found in membrane surface of microvilli
Lingual lipase	Glands of tongue	6.7–7.5	Triglycerides	Fatty acids and monoglycerides	
Maltase, sucrase, lactase	Brush border of small intestine	7–8	Maltose, sucrose, lactose	Monosaccharides	Found in membrane surface of microvilli
Nuclease	Pancreas	7–8	Nucleic acids	Nitrogenous bases and simple sugars	Includes ribonuclease for RNA and deoxyribonuclease for DNA
Pancreatic alpha-amylase	Pancreas	6.7–7.5	Breaks bonds between monomers in complex carbohydrates	Disaccharides and trisaccharides	
Pancreatic lipase	Pancreas	7–8	Triglycerides	Fatty acids and monoglycerides	Bile salts must be present for efficient action
Pepsin (Pepsinogen)	Chief cells of stomach	1.5–2.0	Breaks bonds between amino acids in proteins	Short-chain polypeptides	Secreted as proenzyme, pepsinogen; activated by H^+ in stomach acid
Rennin	Stomach	3.5–4.0	Coagulates milk proteins		Secreted only in infants
Salivary amylase	Salivary glands	6.7–7.5	Breaks bonds between monomers in complex carbohydrates	Disaccharides and trisaccharides	
Trypsin (Trypsinogen)	Pancreas	7–8	Proteins, polypeptides	Short-chain peptides	Proenzyme activated by enterokinase; activates other pancreatic proteases

Lipid Digestion and Absorption

Lipid digestion involves lingual lipase from glands of the tongue and pancreatic lipase from the pancreas (Figure 24-25b●). The most important and abundant dietary lipids are *triglycerides,* which consist of three fatty acids attached to a single molecule of glycerol (see Figure 2-14●, p. 50). The lingual and pancreatic lipases break off two of the fatty acids, leaving *monoglycerides.*

Lipases are water-soluble enzymes, and lipids tend to form large drops that exclude water molecules. As a result, lipases can attack only the exposed surfaces of the lipid drops. Lingual lipase begins breaking down triglycerides in the mouth and stomach (until inacti-

vated), but the lipid drops are so large, and the available time so short, that very little lipid digestion has occurred by the time the chyme enters the duodenum.

Bile salts improve chemical digestion by emulsifying the lipid drops into tiny emulsion droplets, thus providing better access for pancreatic lipase. The emulsification occurs only after the chyme has been mixed with bile in the duodenum. Pancreatic lipase then breaks apart the triglycerides to form a mixture of fatty acids and monoglycerides. As these molecules are released, they interact with bile salts in the surrounding chyme to form small lipid-bile salt complexes called **micelles** (mī-SELZ). A micelle is only about 2.5 nm (0.0025 μm) in diameter.

When a micelle contacts the intestinal epithelium, the lipids diffuse across the cell membrane and enter the cytoplasm. The intestinal cells synthesize new triglycerides from the monoglycerides and fatty acids. These triglycerides, in company with absorbed steroids and phospholipids, are then coated with proteins, creating complexes known as **chylomicrons** (kī-lō-MĪ-kronz; *chylos,* milky lymph + *mikros,* small).

The intestinal cells then secrete the chylomicrons into the interstitial fluids by exocytosis. Their superficial coating of proteins keeps the chylomicrons suspended in the interstitial fluids but generally prevents them from diffusing into capillaries. Most of the chylomicrons released diffuse into the intestinal lacteals, which lack basement membranes and have large gaps between adjacent endothelial cells. From the lacteals they proceed along the lymphatics and through the thoracic duct, finally entering the circulation at the left subclavian vein.

Protein Digestion and Absorption

Proteins have very complex structures, so protein digestion is both complex and time-consuming. The first problem to overcome is the disruption of the three-dimensional organization of the food so that proteolytic enzymes can attack individual proteins. This step involves mechanical processing in the oral cavity, through mastication, and chemical processing in the stomach, through the action of hydrochloric acid. Exposure of the bolus to a strongly acidic environment breaks down plant cell walls and the connective tissues in animal products and has the extra benefit of killing most pathogenic microorganisms.

The acidic contents of the stomach also provide the proper environment for the activity of pepsin, the proteolytic enzyme secreted by chief cells of the stomach (Figure 24-25c•). Pepsin, which works effectively at a pH of 1.5–2.0, breaks peptide bonds within a polypeptide chain. A protease with this kind of activity is called an **endopeptidase.** For instance, pepsin might take a polypeptide 500 amino acids long and break it into two smaller polypeptides, one containing 200 amino acids and the other 300. In the few hours chyme spends in the stomach, pepsin has time to reduce the relatively huge proteins (10,000–100,000 amino acids in length) of the chyme into smaller polypeptide fragments.

When the chyme enters the duodenum, enterokinase produced in the small intestine triggers the conversion of trypsinogen to **trypsin,** and the pH is adjusted to 7–8. Pancreatic proteases can now begin working. Trypsin, chymotrypsin, and elastase are endopeptidases that, like pepsin, break peptide bonds within a polypeptide. Each enzyme has a different specialty. For example, trypsin breaks peptide bonds involving the amino acids *arginine* or *lysine,* whereas chymotrypsin targets peptide bonds involving *tyrosine* or *phenylalanine.* Carboxypeptidase is called an **exopeptidase,** because it simply chops off the last amino acid of a polypeptide chain, ignoring the identities of the amino acids involved. Thus, while the endopeptidases are generating peptide fragments of varying length, carboxypeptidase is producing free amino acids.

Absorption of Amino Acids

The epithelial surfaces of the small intestine contain exopeptidases and **dipeptidases** that break short peptide chains into individual amino acids. (Dipeptidases break apart *dipeptides.*) These amino acids, as well as those produced by the pancreatic enzymes, are absorbed through both facilitated diffusion and cotransport mechanisms.

After diffusing to the opposite end of the cell, the amino acids are released into the interstitial fluids by facilitated diffusion and cotransport. Once within the interstitial fluids, the amino acids diffuse into intestinal capillaries for transport to the liver within the hepatic portal vein.

Water Absorption

Our cells are unable to absorb or secrete water actively. Thus all water movements across the lining of the digestive tract and the production of glandular secretions, such as saliva, involve passive water flow along osmotic gradients. When two fluids are separated by a selectively permeable membrane, water will tend to flow into the solution that contains the higher concentration of solutes. Osmotic movements are relatively rapid, so the interstitial fluids and the fluids in the intestinal lumen always have the same osmolarity (concentration of solutes).

Our epithelial cells are continuously absorbing nutrients and ions from the intestinal contents; these activities gradually lower the solute concentration in the chyme. As the solute concentration lowers, water moves into the surrounding tissues, maintaining osmotic equilibrium.

Each day 2–2.5 liters of water enters your digestive tract in the form of food or drink. The salivary, gastric, intestinal, pancreatic, and bile secretions provide another 6–7 liters. Of that total, only about 150 ml is lost in the fecal wastes. The sites of secretion and absorption of water are indicated in Figure 24-26• (p. 910).

Ion Absorption

For osmotic purposes, it really does not matter which ions are being transported. But for metabolic purposes, the particular ions matter a great deal, and each ion is handled individually (Figure 24-27•). Many of the

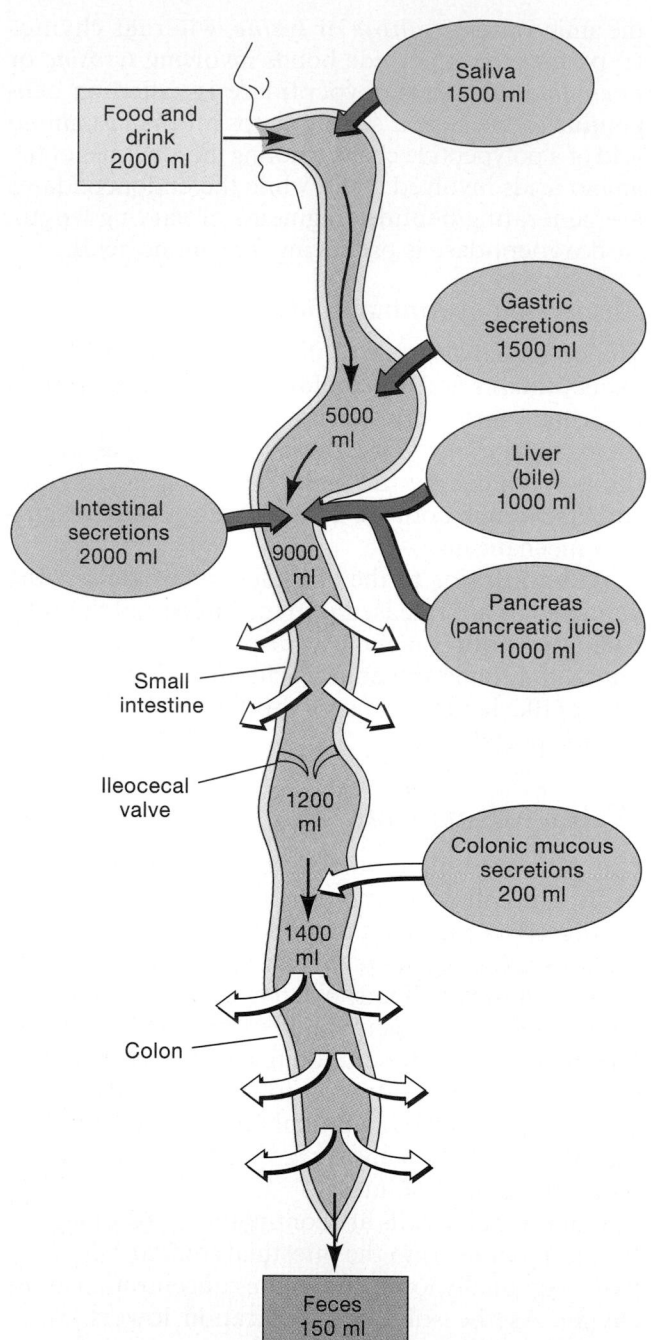

•FIGURE 24-26 Secretion and Water Absorption along the Digestive Tract

regulatory mechanisms controlling the rates of absorption are poorly understood.

Sodium ions, usually the most abundant cations in the chyme, may enter intestinal cells through diffusion or cotransport with another nutrient. These ions are then pumped into the interstitial fluid across the base of the cell. The rate of Na$^+$ uptake from the lumen is generally proportional to the concentration of Na$^+$ in the intestinal contents. As a result, eating heavily salted meals leads to increased sodium ion absorption and an associated gain of water through osmosis. The rate of sodium ion absorption by the digestive tract is increased by aldosterone, a steroid hormone from the adrenal cortex that was discussed in Chapter 18. ∞ *[p. 616]*

Calcium ion absorption involves active transport at the epithelial surface. The rate of transport is accelerated by parathyroid hormone and calcitriol. We considered the roles of those hormones in calcium ion homeostasis in Chapters 6 and 18. ∞ *[pp. 187, 612–613, 618]*

As other solutes move out of the chyme, the concentration of potassium ions increases. These ions can diffuse into the epithelial cells following the concentration gradient. The absorption of magnesium, iron, and other cations involves specific carrier proteins, and the cell must use ATP to obtain and transport these ions to the interstitial fluid. Regulatory factors controlling the absorption of these cations are poorly understood.

The anions chloride, iodide, bicarbonate, and nitrate may be absorbed through diffusion or carrier-mediated transport. Phosphate and sulfate ions enter epithelial cells only through active transport.

Vitamin Absorption

Vitamins are organic compounds required in very small quantities. There are two major groups of vitamins. Vitamins A, D, E, and K are **fat-soluble vitamins;** their structure allows them to dissolve in lipids. The nine **water-soluble vitamins** include the B vitamins, common in milk and meats, and vitamin C, found in citrus fruits. In Chapter 25, we shall consider the functions of vitamins and associated nutritional problems.

All but one of the water-soluble vitamins are easily absorbed by diffusion across the digestive epithelium. Vitamin B$_{12}$ cannot be absorbed by the intestinal mucosa in normal amounts unless it has been bound to *intrinsic factor,* a glycoprotein secreted by the parietal cells of the stomach. The combination is then absorbed via active transport.

Fat-soluble vitamins in the diet enter the duodenum in fat droplets, mixed with triglycerides. The vitamins remain in association with these lipids as they form emulsion droplets and, after further digestion, micelles. The fat-soluble vitamins are then absorbed from the micelles along with the fatty acids and monoglycerides. Vitamin K produced in the colon is absorbed with other lipids released through bacterial action. Taking supplements of fat-soluble vitamins while you have an empty stomach, are fasting, or are on a low-fat diet will be relatively ineffective, as proper absorption requires the presence of other lipids.

Figure 24-27• summarizes the digestive fates of the major electrolytes and vitamins. Review this figure carefully to become familiar with the outlined mechanisms and events. In Chapter 25, we shall examine the interplay between the respiratory and digestive systems.

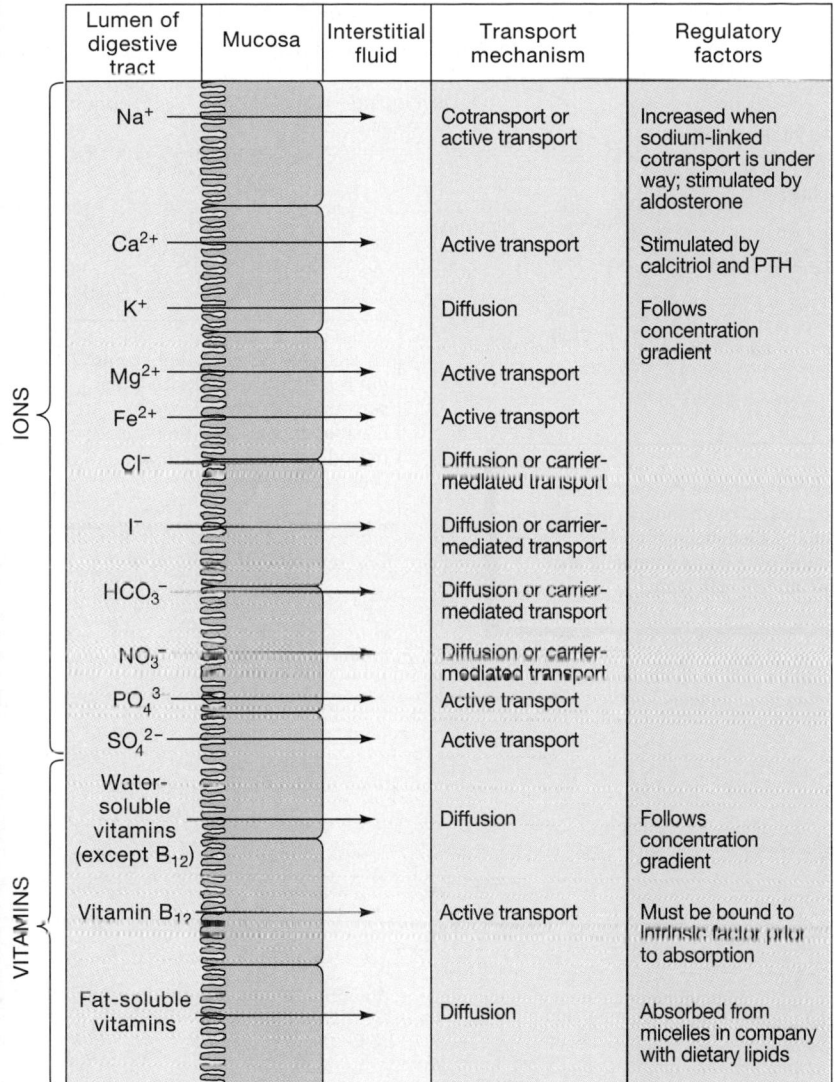

Lumen of digestive tract	Mucosa	Interstitial fluid	Transport mechanism	Regulatory factors
Na⁺			Cotransport or active transport	Increased when sodium-linked cotransport is under way; stimulated by aldosterone
Ca²⁺			Active transport	Stimulated by calcitriol and PTH
K⁺			Diffusion	Follows concentration gradient
Mg²⁺			Active transport	
Fe²⁺			Active transport	
Cl⁻			Diffusion or carrier-mediated transport	
I⁻			Diffusion or carrier-mediated transport	
HCO₃⁻			Diffusion or carrier-mediated transport	
NO₃⁻			Diffusion or carrier-mediated transport	
PO₄³⁻			Active transport	
SO₄²⁻			Active transport	
Water-soluble vitamins (except B₁₂)			Diffusion	Follows concentration gradient
Vitamin B₁₂			Active transport	Must be bound to intrinsic factor prior to absorption
Fat-soluble vitamins			Diffusion	Absorbed from micelles in company with dietary lipids

IONS — Na⁺, Ca²⁺, K⁺, Mg²⁺, Fe²⁺, Cl⁻, I⁻, HCO₃⁻, NO₃⁻, PO₄³⁻, SO₄²⁻
VITAMINS — Water-soluble vitamins, Vitamin B₁₂, Fat-soluble vitamins

•FIGURE 24-27 Ion and Vitamin Absorption by the Digestive Tract

MALABSORPTION SYNDROMES Difficulties in the absorption of all classes of compounds will result from damage to the accessory glands or the intestinal mucosa. If the accessory organs are functioning normally but their secretions cannot reach the duodenum, the condition is called *biliary obstruction* (bile duct blockage) or *pancreatic obstruction* (pancreatic duct blockage). Alternatively, the ducts may remain open but the glandular cells may be damaged and unable to continue normal secretory activities. We noted two examples, pancreatitis and cirrhosis, earlier in this chapter.

Even with the normal enzymes present in the lumen, absorption will not occur if the mucosa cannot function properly. A genetic inability to manufacture specific enzymes will result in discrete patterns of malabsorption; *lactose intolerance* is a good example. Mucosal damage due to ischemia, radiation exposure, toxic compounds, or infection will adversely affect absorption and will deplete nutrient and fluid reserves as a result.

⧖AGING AND THE DIGESTIVE SYSTEM

Essentially normal digestion and absorption occur in elderly individuals. However, there are many changes in the digestive system that parallel age-related changes we have already described for other systems:

- **The rate of epithelial stem cell division declines.** The digestive epithelium becomes more susceptible to damage by abrasion, acids, or enzymes. Peptic ulcers therefore become more likely. In the mouth, esophagus, and anus, the stratified epithelium becomes thinner and more fragile.

- **Smooth muscle tone decreases.** General motility decreases, and peristaltic contractions are weaker. These changes slow the rate of fecal movement and promote constipation. Sagging and inflammation of the haustra in the colon can produce symptoms of *diverticulosis*. Straining to eliminate compacted fecal materials can stress the less-resilient walls of blood vessels, producing hemorrhoids. Problems are not restricted to the lower digestive tract. For example, weakening of muscular sphincters can lead to esophageal reflux and frequent bouts of "heartburn."

- **The effects of cumulative damage become apparent.** A familiar example is the gradual loss of teeth due to dental caries or gingivitis. Cumulative damage can involve internal organs as well. Toxins such as alcohol and other injurious chemicals that are absorbed by the digestive tract are transported to the liver for processing. The cells of the liver are not immune to these toxic compounds, and chronic exposure can lead to cirrhosis or other types of liver disease.

- **Cancer rates increase.** Not surprisingly, cancers are most common in organs in which stem cells divide to maintain epithelial cell populations. Rates of colon cancer and stomach cancer rise in the elderly; oral and pharyngeal cancers are particularly common among elderly individuals who smoke.

- **Changes in other systems have direct or indirect effects on the digestive system.** For example, the reduction in bone mass and calcium content in the skeleton is associated with erosion of the tooth sockets and eventual tooth loss. The decline in olfactory and gustatory sensitivity with age can lead to dietary changes that affect the entire body.

EMBRYOLOGY SUMMARY

Development of the Digestive System

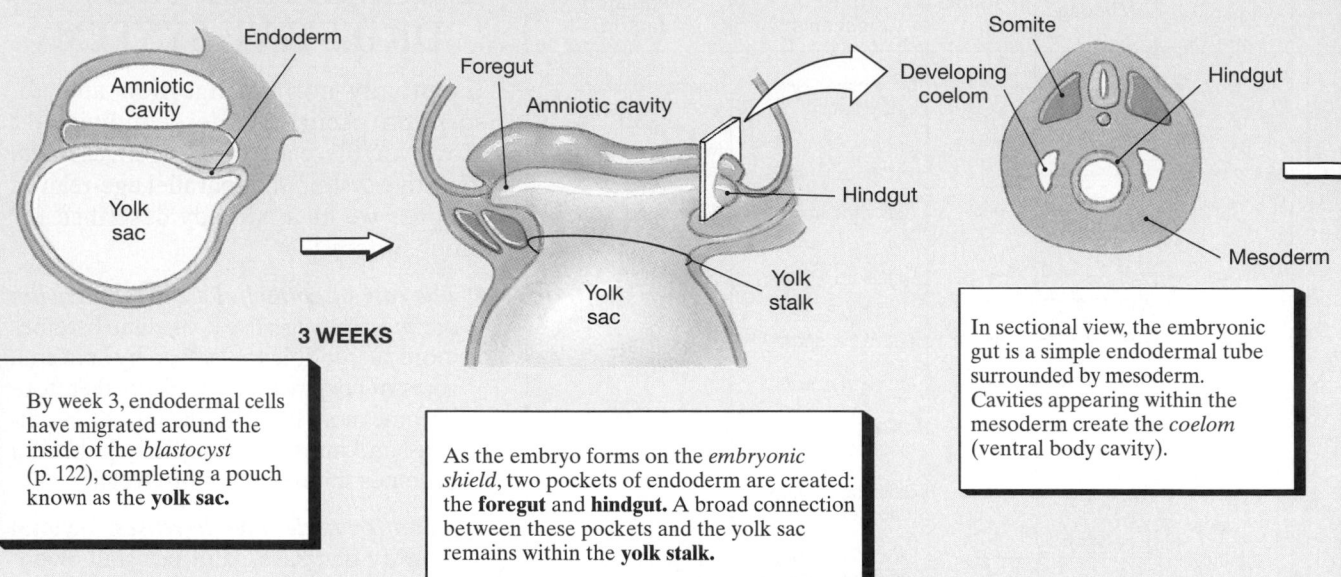

3 WEEKS

By week 3, endodermal cells have migrated around the inside of the *blastocyst* (p. 122), completing a pouch known as the **yolk sac.**

As the embryo forms on the *embryonic shield*, two pockets of endoderm are created: the **foregut** and **hindgut.** A broad connection between these pockets and the yolk sac remains within the **yolk stalk.**

In sectional view, the embryonic gut is a simple endodermal tube surrounded by mesoderm. Cavities appearing within the mesoderm create the *coelom* (ventral body cavity).

8 WEEKS

A partition grows across the cloaca, dividing it into a posterior rectum and an anterior **urogenital sinus** that retains a connection to the allantois.

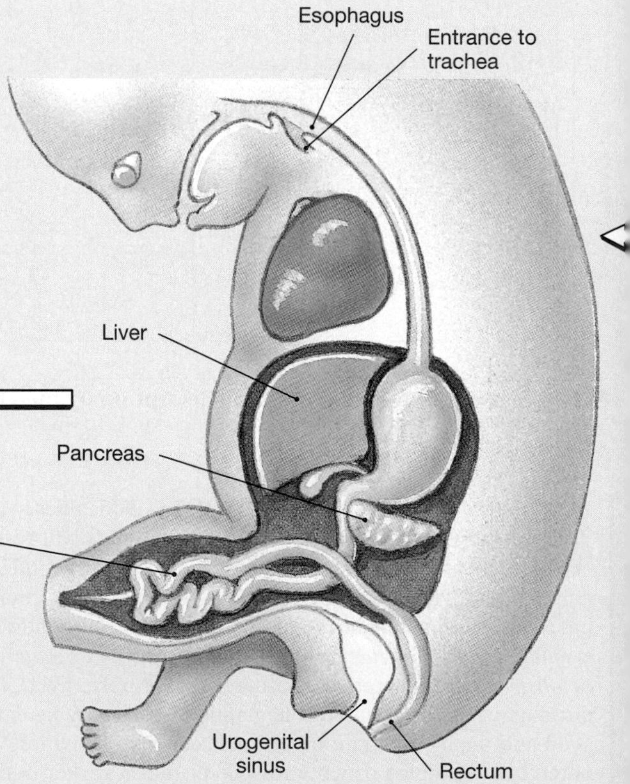

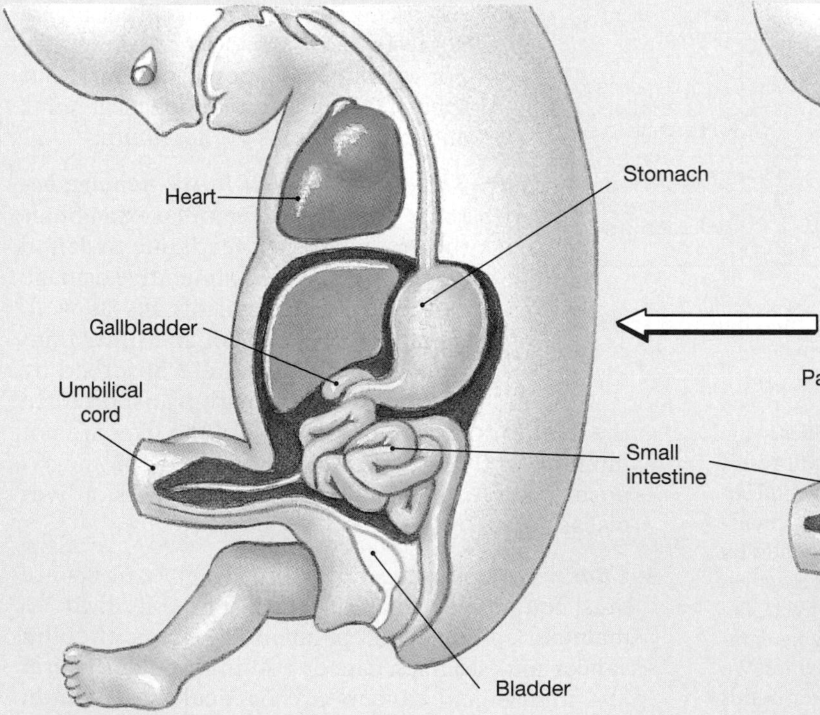

10 WEEKS

By week 10, the intestines have begun moving back into the coelomic cavity, although they continue to grow longer.

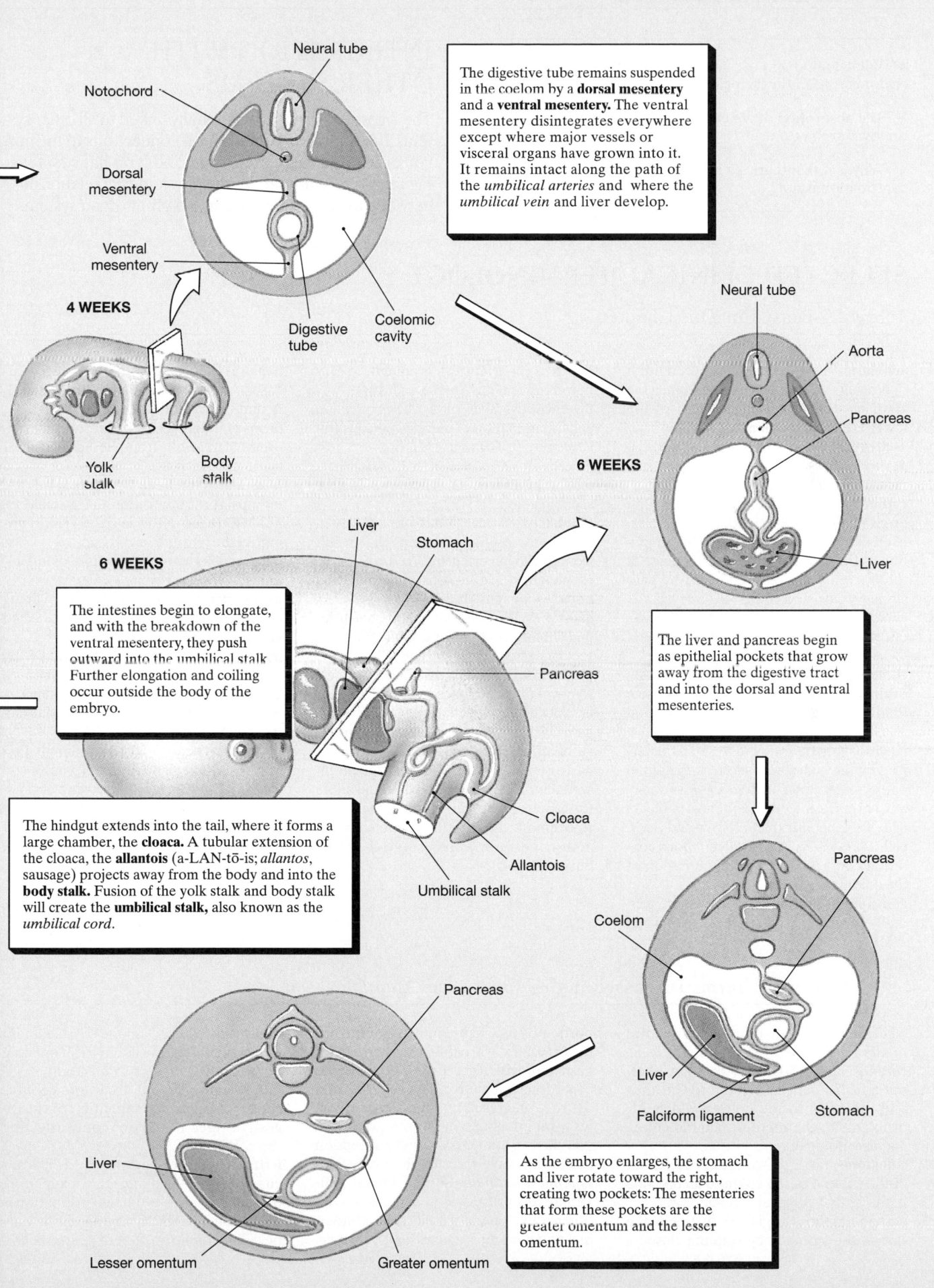

Neural tube

Notochord

Dorsal mesentery

Ventral mesentery

4 WEEKS

Digestive tube

Coelomic cavity

The digestive tube remains suspended in the coelom by a **dorsal mesentery** and a **ventral mesentery.** The ventral mesentery disintegrates everywhere except where major vessels or visceral organs have grown into it. It remains intact along the path of the *umbilical arteries* and where the *umbilical vein* and liver develop.

Yolk stalk

Body stalk

Neural tube

Aorta

Pancreas

6 WEEKS

Liver

The liver and pancreas begin as epithelial pockets that grow away from the digestive tract and into the dorsal and ventral mesenteries.

Liver

Stomach

6 WEEKS

The intestines begin to elongate, and with the breakdown of the ventral mesentery, they push outward into the umbilical stalk. Further elongation and coiling occur outside the body of the embryo.

Pancreas

Cloaca

Allantois

Umbilical stalk

The hindgut extends into the tail, where it forms a large chamber, the **cloaca.** A tubular extension of the cloaca, the **allantois** (a-LAN-tō-is; *allantos,* sausage) projects away from the body and into the **body stalk.** Fusion of the yolk stalk and body stalk will create the **umbilical stalk,** also known as the *umbilical cord.*

Pancreas

Coelom

Pancreas

Liver

Falciform ligament

Stomach

As the embryo enlarges, the stomach and liver rotate toward the right, creating two pockets: The mesenteries that form these pockets are the greater omentum and the lesser omentum.

Liver

Lesser omentum

Greater omentum

☑ What component of a meal would increase the number of chylomicrons in the lacteals?

☑ The absorption of which vitamin would be impaired by removal of the stomach?

☑ Why is it that diarrhea is potentially life-threatening but constipation is not?

INTEGRATION WITH OTHER SYSTEMS

The digestive system is functionally linked to all other systems. It has extensive anatomical connections to the nervous, cardiovascular, endocrine, and lymphatic systems. Figure 24-28● summarizes the physiological relationships between the digestive system and other organ systems.

SELECTED CLINICAL TERMINOLOGY

Terms Discussed in This Chapter

achalasia (ak-a-LĀ-zē-uh): A condition that results when a bolus cannot reach the stomach due to the constriction of the lower esophageal sphincter. *(p. 877)*

ascites (a-SĪ-tēz): Fluid leakage into the peritoneal cavity across the serosal surfaces of the liver and viscera. *(p. 867)*

cathartics (ka-THAR-tiks): Drugs that promote defecation. *(p. 903)*

cholecystitis (kō-lē-sis-TĪ-tis): Inflammation of the gallbladder due to blockage of the cystic or common bile duct by gallstones. *(p. 896 and AM)*

cholelithiasis (kō-lē-li-THĪ-a-sis): The presence of gallstones in the gallbladder. *(p. 896 and AM)*

cholera (KOL-e-ruh): A bacterial infection of the digestive tract that causes massive fluid losses through diarrhea. *(p. 903 and AM)*

cirrhosis (sir-Ō-sis): A disease characterized by the widespread destruction of hepatocytes by exposure to drugs (especially alcohol), viral infection, ischemia, or blockage of the hepatic ducts. *(p. 876 and AM)*

colitis (kō-LĪ-tis): A general term for a condition characterized by inflammation of the colon. *(p. 901 and AM)*

constipation: Infrequent defecation, generally involving dry, hard feces. *(p. 903 and AM)*

diarrhea (dī-a-RĒ-uh): Frequent, watery bowel movements. *(p. 903 and AM)*

diverticulitis (dī-ver-tik-ū-LĪ-tis): Infection and inflammation of mucosal pockets (diverticula). *(p. 901 and AM)*

diverticulosis (dī-ver-tik-ū-LŌ-sis): Formation of diverticula, generally along the sigmoid colon. *(p. 901)*

esophageal varices (VAR-i-sēz): Swollen and fragile esophageal veins that result from portal hypertension. *(p. 876)*

esophagitis (ē-sof-a-JĪ-tis): Inflammation of the esophagus. *(p. 877)*

gallstones: Deposits of minerals, bile salts, and cholesterol that form if bile becomes too concentrated. *(p. 896 and AM)*

gastrectomy (gas-TREK-to-mē): Surgical removal of the stomach, generally to treat advanced stomach cancer. *(p. 884)*

gastritis (gas-TRĪ-tis): Inflammation of the gastric mucosa. *(p. 881)*

gastroscope: A fiber-optic instrument inserted into the mouth and directed along the esophagus and into the stomach; used to examine the interior of the stomach and to perform minor surgical procedures. *(p. 884)*

hepatitis (hep-a-TĪ-tis): Virus-induced disease of the liver; forms include *hepatitis A, B,* and *C. (p. 892 and AM)*

lactose intolerance: A malabsorption syndrome that results from the lack of the enzyme *lactase* at the brush border of the intestinal epithelium. *(p. 909 and AM)*

mumps: A viral infection that in children tends to focus in the salivary glands, primarily the parotid. *(p. 871)*

pancreatitis (pan-krē-a-TĪ-tis): An inflammation of the pancreas. *(p. 890)*

peptic ulcer: Erosion of the gastric lining or duodenal lining by stomach acids and enzymes. *(p. 881 and AM)*

periodontal disease: A loosening of the teeth within the alveolar sockets caused by erosion of the periodontal ligaments by acids produced through bacterial action. *(p. 872)*

peritonitis (per-i-tō-NĪ-tis): Inflammation of the peritoneal membrane. *(p. 867)*

portal hypertension: High venous pressures in the hepatic portal system. *(p. 891)*

pulpitis (pul-PĪ-tis): Infection of the pulp of a tooth; may be treated by a *root canal* procedure. *(p. 872)*

AM Additional Terms Discussed in the *Applications Manual*

colectomy (kō-LEK-to-mē): The removal of all or a portion of the colon.

colonoscope (kō-LON-ō-skōp): A fiber-optic instrument inserted into the anus and moved into the colon to look for problems in the rectum, sigmoid colon, and ascending colon.

colostomy (kō-LOS-to-mē): Attachment of the cut end of the colon to an opening in the body wall after a colectomy.

gastric stapling: Surgical reduction in the size of the stomach by stapling closed a portion of the gastric lumen; an attempt to correct an overeating problem.

gastroenteritis (gas-trō-en-ter-Ī-tis): A condition characterized by vomiting and diarrhea and resulting from bacterial toxins, viral infections, or various poisons.

giardiasis (jē-ar-DĪ-a-sis): An infection of the digestive tract by the protozoan *Giardia intestinalis.* Some infected individuals do not develop symptoms, but others develop violent diarrhea, abdominal pains, cramps, nausea, and vomiting.

ileostomy (il-ē-OS-to-mē): Attachment of the cut end of the ileum to an opening in the body wall after a colectomy.

inflammatory bowel disease (*ulcerative colitis):* Chronic inflammation of the digestive tract, most commonly affecting the colon.

irritable bowel syndrome: A disorder characterized by diarrhea, constipation, or both alternately. When constipation is the primary problem, this condition may be called a *spastic colon* or *spastic colitis.*

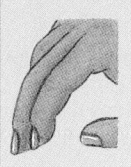

INTEGUMENTARY SYSTEM

Provides vitamin D₃ needed for the absorption of calcium and phosphorus

Provides lipids for storage by adipocytes in subcutaneous layer

SKELETAL SYSTEM

Skull, ribs, vertebrae, and pelvic girdle support and protect parts of digestive tract; teeth important in mechanical processing of food

Absorbs calcium and phosphate ions for incorporation into bone matrix; provides lipids for storage in yellow marrow

MUSCULAR SYSTEM

Protects and supports digestive organs in abdominal cavity; controls entrances and exits to digestive tract; liver metabolizes lactic acid from active muscles

Liver regulates blood glucose and fatty acid levels, removes lactic acid from circulation

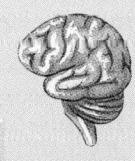

NERVOUS SYSTEM

ANS regulates movement and secretion; reflexes coordinate passage of materials along tract; control over skeletal muscles regulates ingestion and defecation; hypothalamic centers control hunger, satiation, and feeding behaviors

Provides substrates essential for neurotransmitter synthesis

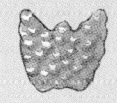

ENDOCRINE SYSTEM

Epinephrine and norepinephrine stimulate constriction of sphincters and depress digestive activity; hormones coordinate activity along tract

Provides nutrients and substrates to endocrine cells; endocrine cells of pancreas secrete insulin and glucagon; liver produces angiotensinogen

DIGESTIVE SYSTEM

CARDIOVASCULAR SYSTEM

Distributes hormones of the digestive tract; carries nutrients, water, and ions from sites of absorption; delivers nutrients and toxins to liver

Absorbs fluid to maintain normal blood volume; absorbs vitamin K; liver excretes heme (as conjugated bilirubin), synthesizes coagulation proteins

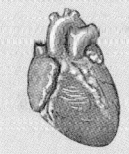

LYMPHATIC SYSTEM

Tonsils and GALT defend against infection; GALT provides defense against toxins absorbed from the tract; lymphatics carry absorbed lipids to venous system

Secretions of digestive tract (acids and enzymes) provide nonspecific defense against pathogens

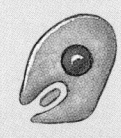

FOR ALL SYSTEMS

Absorbs organic substrates, vitamins, ions, and water required by all living cells

RESPIRATORY SYSTEM

Increased thoracic and abdominal pressure through contraction of respiratory muscles can assist in defecation

Pressure of digestive organs against the diaphragm can assist exhalation and limit inhalation

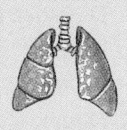

URINARY SYSTEM

Excretes toxins absorbed by the digestive epithelium; excretes some conjugated bilirubin produced by liver

Absorbs water needed to excrete waste products at the kidneys; absorbs ions needed to maintain normal body fluid concentrations

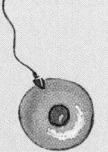

Provides additional nutrients required to support gamete production and (in pregnant women) embryonic and fetal development

REPRODUCTIVE SYSTEM

•FIGURE 24-28 Functional Relationships between the Digestive System and Other Systems

laparoscopy (lap-a-ROS-ko-pē): Use of a flexible fiber-optic instrument introduced through the abdominal wall to permit direct visualization of the viscera, tissue sampling, and limited surgical procedures.
liver biopsy: A sample of liver tissue, generally taken through the anterior abdominal wall by means of a long needle.

liver scans: Radiological procedures that involve the injection of radioisotope-labeled compounds that will be selectively absorbed by Kupffer cells, liver cells, or abnormal liver tissues.
perforated ulcer: A particularly dangerous ulcer in which the gastric acids erode

through the wall of the digestive tract and enter the peritoneal cavity.
polyps (PO-lips): Small mucosal tumors that grow from the intestinal wall.

CHAPTER REVIEW

 On-line resources for this chapter are on our World Wide Web site at:
http://www.prenhall.com/martini/fap

STUDY OUTLINE

INTRODUCTION, p. 862

1. The digestive system consists of the muscular **digestive tract** and various **accessory organs.**
2. Digestive functions are *ingestion, mechanical processing, digestion, secretion, absorption,* and *excretion.*

AN OVERVIEW OF THE STRUCTURE AND FUNCTION OF THE DIGESTIVE TRACT, p. 862

1. The digestive tract consists of the oral cavity, pharynx, esophagus, stomach, small intestine, large intestine, rectum, and anus. *(Figure 24-1)*

Histological Organization of the Digestive Tract, p. 862

2. The digestive tract is lined by a mucous epithelium moistened by glandular secretions of the epithelial and accessory organs.
3. The *lamina propria* and epithelium form the **mucosa** (mucous membrane) of the digestive tract. Proceeding outward, we encounter the **submucosa,** the **muscularis externa,** and a layer of loose connective tissue called the *adventitia.* Within the peritoneal cavity, the muscularis externa is covered by a serous membrane called the **serosa.** *(Figure 24-2)*

The Movement of Digestive Materials, p. 865

4. Many smooth muscle cells are not innervated by motor neurons, and the neurons that do innervate smooth muscles are not under voluntary control.
5. The muscularis externa propels materials through the digestive tract by the contractions of **peristalsis. Segmentation** movements in areas of the small intestine churn digestive materials. *(Figure 24-3)*

The Control of Digestive Function, p. 865

6. Digestive tract activities are controlled by neural, hormonal, and local mechanisms. *(Figure 24-4)*

The Peritoneum, p. 866

7. Double sheets of peritoneal membrane called **mesenteries** suspend the digestive tract. The **greater omentum** lies in front of the abdominal viscera. Its adipose tissue provides padding, protection, insulation, and an energy reserve. *(Figure 24-5)*

THE ORAL CAVITY, p. 867

1. The functions of the oral cavity are (1) *analysis* of foods; (2) *mechanical processing* by the teeth, tongue, and palatal surfaces; (3) *lubrication* by mixing with mucus and salivary secretions; and (4) initiating the *digestion* of carbohydrates and lipids.

2. The oral cavity, or **buccal cavity,** is lined by **oral mucosa.** The **hard** and **soft palates** form its roof, and the tongue forms its floor. *(Figure 24-6)*

The Tongue, p. 869

3. **Intrinsic** and **extrinsic tongue muscles** are controlled by the hypoglossal nerve. *(Figure 24-6)*

Salivary Glands, p. 870

4. The **parotid, sublingual,** and **submandibular salivary glands** discharge their secretions into the oral cavity. *(Figure 24-7)*

The Teeth, p. 871

5. **Mastication** (chewing) of the **bolus** occurs through the contact of the **occlusal** (opposing) **surfaces** of the teeth. The **periodontal ligament** anchors each tooth in an *alveolus,* or bony socket. **Dentin** forms the basic structure of a tooth. The **crown** is coated with **enamel,** and the **root** with **cementum.** *(Figure 24-8)*
6. The 20 primary teeth, or **deciduous teeth,** are replaced by the 32 teeth of the **secondary dentition** during development. *(Figure 24-9)*

THE PHARYNX, p. 874

1. Propulsion of the bolus results from the contractions of the *pharyngeal constrictors* and the *palatal muscles* and from elevation of the larynx.

THE ESOPHAGUS, p. 874

1. The esophagus carries solids and liquids from the pharynx to the stomach through the **esophageal hiatus,** an opening in the diaphragm.

Histology of the Esophagus, p. 875

2. The esophageal mucosa consists of a stratified epithelium. Mucous secretion by esophageal glands of the submucosa reduces friction during the passage of foods. The proportions of skeletal and smooth muscle of the muscularis externa change from the pharynx to the stomach. *(Figure 24-10)*

Swallowing, p. 876

3. Swallowing (**deglutition**) can be divided into **buccal, pharyngeal,** and **esophageal phases.** Swallowing begins with the compaction of a bolus and its movement into the pharynx, followed by the elevation of the larynx, reflection of the epiglottis, and closure of the glottis. After the *upper esophageal sphincter* is opened, peristalsis moves the bolus down the esophagus to the *lower esophageal sphincter.* *(Figure 24-11)*

THE STOMACH, p. 877

1. The stomach has four major functions: (1) bulk storage of ingested food, (2) mechanical breakdown of food, (3) disruption of chemical bonds via acids and enzymes, and (4) production of *intrinsic factor*.

Anatomy of the Stomach, p. 877

2. The four regions of the stomach are the **cardia, fundus, body,** and **pylorus.** The **pyloric sphincter** guards the exit from the stomach. In a relaxed state, the stomach lining contains numerous **rugae** (ridges and folds). *(Figure 24-12)*

3. Within the **gastric glands, parietal cells** secrete **intrinsic factor** and *hydrochloric acid.* **Chief cells** secrete **pepsinogen,** which is converted by acids in the gastric lumen to the enzyme **pepsin. Enteroendocrine** cells of the stomach secrete several compounds, notably the hormone **gastrin.** *(Figures 24-13, 24-14)*

Regulation of Gastric Activity, p. 881

4. Gastric secretion consists of (1) the **cephalic phase,** which prepares the stomach to receive ingested materials; (2) the **gastric phase,** which begins with the arrival of food in the stomach; and (3) the **intestinal phase,** which controls the rate of gastric emptying. *(Figure 24-15)*

Digestion and Absorption in the Stomach, p. 884

5. The preliminary digestion of proteins by pepsin begins in the stomach. Very little nutrient absorption occurs in the stomach.

THE SMALL INTESTINE AND ASSOCIATED GLANDULAR ORGANS, p. 884

1. Most of the important digestive and absorptive functions occur in the small intestine. Digestive secretions and buffers are provided by the pancreas, liver, and gallbladder.

The Small Intestine, p. 884

2. The small intestine consists of the **duodenum,** the **jejunum,** and the **ileum.** A sphincter, the **ileocecal valve,** marks the transition between the small and large intestines.

3. The intestinal mucosa bears transverse folds called **plicae circulares** and small projections called **intestinal villi.** These folds and projections increase the surface area for absorption. Each villus contains a terminal lymphatic called a **lacteal.** Pockets called **intestinal crypts** are lined by enteroendocrine, goblet, and stem cells. *(Figure 24-16)*

4. The **submucosal** *(Brunner's)* **glands** of the duodenum produce mucus, buffers, and the hormone **urogastrone.** The ileum contains masses of lymphoid tissue called *Peyer's patches* near the entrance to the large intestine. *(Figure 24-17)*

5. **Intestinal juice** moistens the chyme, helps buffer acids, and holds digestive enzymes and the products of digestion in solution.

Intestinal Movements, p. 887

6. Some of the smooth muscle cells contract periodically, without stimulation, to produce a pattern of peristaltic contractions that move the intestinal contents.

The Pancreas, p. 889

7. The **pancreatic duct** penetrates the wall of the duodenum. Within each pancreatic lobule, ducts branch repeatedly before ending in the **pancreatic acini** (blind pockets). *(Figure 24-18)*

8. The pancreas has two functions: endocrine (secreting insulin and glucagon into the blood) and exocrine (secreting water, ions, and digestive enzymes into the small intestine). Pancreatic enzymes include **carbohydrases, lipases, nucleases,** and **proteolytic enzymes.**

The Liver, p. 891

9. The liver performs metabolic and hematological regulation and produces **bile.** The bile ducts from all the **liver lobules** unite to form the **common hepatic duct,** which meets the **cystic duct** to form the **common bile duct,** which empties into the duodenum. *(Figure 24-19)*

10. The liver lobule is the organ's basic functional unit. **Hepatocytes** form irregular plates arranged as wheel spokes. **Bile canaliculi** carry bile to the **bile ductules,** which lead to **portal areas.** *(Figure 24-20)*

11. In **emulsification, bile salts** break apart large drops of lipids, making them accessible to lipases secreted by the pancreas.

The Gallbladder, p. 895

12. The **gallbladder** stores and concentrates bile. *(Figure 24-21)*

The Coordination of Secretion and Absorption, p. 896

13. Neural and hormonal mechanisms coordinate the activities of the digestive glands. Gastrointestinal activity is stimulated by parasympathetic innervation and inhibited by sympathetic innervation. The **enterogastric, gastroenteric,** and **gastroileal reflexes** coordinate movement from the stomach to the large intestine.

14. Intestinal hormones include **enterocrinin, secretin, cholecystokinin (CCK), gastric inhibitory peptide (GIP), vasoactive intestinal peptide (VIP),** and gastrin. *(Figure 24-22; Table 24-1)*

THE LARGE INTESTINE, p. 898

1. The main functions of the large intestine are to (1) reabsorb water and compact materials into feces, (2) absorb vitamins produced by bacteria, and (3) store fecal material prior to defecation. *(Figure 24-23)*

The Cecum, p. 898

2. The **cecum** collects and stores material from the ileum and begins the process of compaction. The **vermiform appendix** is attached to the cecum.

The Colon, p. 899

3. The **colon** has a larger diameter and a thinner wall than the small intestine. It bears **haustra** (pouches) and the **taenia coli** (longitudinal bands of muscle).

The Rectum, p. 899

4. The **rectum** terminates in the **anorectal canal** leading to the **anus.** *(Figure 24-23)*

Histology of the Large Intestine, p. 901

5. Characteristics of the colon include an absence of villi and the presence of goblet cells and deep intestinal glands. *(Figure 24-23)*

Physiology of the Large Intestine, p. 901

6. The large intestine reabsorbs water and other substances such as vitamins, urobilinogen, bile salts, and toxins. Bacteria are responsible for intestinal gas, or **flatus.**

7. The gastroileal reflex moves chyme into the cecum at mealtimes. Distension of the stomach and duodenum stimulates **mass movements** of materials from the transverse colon through the rest of the large intestine and into the rectum.

Muscular sphincters control the passage of fecal materal to the anus. Distension of the rectal wall triggers the **defecation reflex**. *(Figure 24-24)*

DIGESTION AND ABSORPTION, p. 903

The Processing and Absorption of Nutrients, p. 903

1. The digestive system breaks down the physical structure of the ingested material and then disassembles the component molecules into smaller fragments through *hydrolysis*. *(Figure 24-25; Table 24-2)*

Carbohydrate Digestion and Absorption, p. 904

2. Salivary and pancreatic amylases break down complex carbohydrates into disaccharides and trisaccharides. These are broken down into monosaccharides by enzymes at the epithelial surface. The monosaccharides are then absorbed by the intestinal epithelium through facilitated diffusion and cotransport.

Lipid Digestion and Absorption, p. 906

3. Triglycerides are emulsified into lipid droplets. The resulting fatty acids and monoglycerides interact with bile salts to form **micelles,** from which they diffuse across the intestinal epithelium.

Protein Digestion and Absorption, p. 907

4. Protein digestion involves the gastric enzyme pepsin and the various pancreatic proteases. Peptidases liberate amino acids that are absorbed and exported to the interstitial fluids.

Water Absorption, p. 907

5. About 2–2.5 liters of water is ingested each day, and digestive secretions provide 6–7 liters. Nearly all is reabsorbed by osmosis. *(Figure 24-26)*

Ion Absorption, p. 907

6. Various processes such as diffusion, cotransport, and carrier-mediated and active transport are responsible for the movements of cations (sodium, calcium, potassium, and so on) and anions (chloride, iodide, bicarbonate, and so on) into epithelial cells. *(Figure 24-27)*

Vitamin Absorption, p. 908

7. The **water-soluble vitamins** (except for B_{12}) easily diffuse across the digestive epithelium. **Fat-soluble vitamins** are enclosed within fat droplets and are absorbed with the products of lipid digestion. *(Figure 24-27)*

AGING AND THE DIGESTIVE SYSTEM, p. 909

1. Age-related changes include a thinner and more fragile epithelium due to a reduction in epithelial stem cell division, weaker peristaltic contractions as smooth muscle tone decreases, effects of cumulative damage, increased cancer rates, and related changes in other systems.

INTEGRATION WITH OTHER SYSTEMS, p. 912

1. The digestive system has extensive anatomical connections to the nervous, cardiovascular, endocrine, and lymphatic systems. *(Figure 24-28)*

REVIEW QUESTIONS

Level 1	Reviewing Facts and Terms

1. The enzymatic breakdown of large molecules into their basic building blocks is
    (a) absorption
    (b) secretion
    (c) mechanical digestion
    (d) chemical digestion

2. The inner layer of the gastrointestinal tract is the
    (a) mucosa   (b) submucosa
    (c) serosa   (d) muscularis

3. The muscularis externa propels materials from one portion of the digestive tract to another through the contractions of
    (a) segmentation   (b) propulsion
    (c) mass movements   (d) peristalsis

4. The activities of the digestive system are regulated by
    (a) hormonal mechanisms
    (b) local mechanisms
    (c) neural mechanisms
    (d) a, b, and c are correct

5. Double sheets of peritoneum that provide support and stability for the organs of the peritoneal cavity are
    (a) mediastinums   (b) mucous membranes
    (c) omenta   (d) mesenteries

6. The layer of the peritoneum that lines the inner surfaces of the body wall is the
    (a) visceral peritoneum   (b) parietal peritoneum
    (c) greater omentum   (d) lesser omentum

7. The peritoneal fold that stabilizes and supports the small intestine is the
    (a) serosa   (b) lesser omentum
    (c) mesentery   (d) parietal peritoneum

8. Intrinsic factor and hydrochloric acid are secreted by cells in the stomach wall called
    (a) parietal cells   (b) chief cells
    (c) acinar cells   (d) G cells

9. Protein digestion in the stomach results primarily from secretions released by
    (a) G cells   (b) parietal cells
    (c) chief cells   (d) D cells

10. The part of the gastrointestinal tract that plays the primary role in the digestion and absorption of nutrients is the
    (a) large intestine   (b) small intestine
    (c) stomach   (d) cecum and colon

11. The duodenal hormone that stimulates the production and secretion of pancreatic enzymes is
    (a) enterokinase   (b) gastrin
    (c) secretin   (d) cholecystokinin

12. The essential metabolic or synthetic service provided by the liver is
    (a) metabolic regulation
    (b) hematological regulation
    (c) bile production
    (d) a, b, and c are correct

13. Bile release from the gallbladder into the duodenum occurs only under the stimulation of
- **(a)** cholecystokinin
- **(b)** secretin
- **(c)** gastrin
- **(d)** enterokinase

14. The major function(s) of the large intestine is (are)
- **(a)** reabsorption of water and compaction of chyme into feces
- **(b)** absorption of vitamins liberated by bacterial action
- **(c)** storage of fecal material prior to defecation
- **(d)** a, b, and c are correct

15. The part of the colon that empties into the rectum is the
- **(a)** ascending colon
- **(b)** descending colon
- **(c)** transverse colon
- **(d)** sigmoid colon

16. Three vitamins generated by colonic bacteria are
- **(a)** vitamins A, D, and E
- **(b)** B complex vitamins and vitamin C
- **(c)** vitamin K, biotin, and pantothenic acid
- **(d)** niacin, thiamine, and riboflavin

17. The final enzymatic steps in the digestive process are accomplished by
- **(a)** brush border enzymes of the microvilli
- **(b)** enzymes secreted by the stomach
- **(c)** enzymes secreted by the pancreas
- **(d)** the action of bile from the gallbladder

18. What are the six steps involved in digestion?

19. What is the purpose of the transverse or longitudinal folds of the epithelium in the digestive tract?

20. Name and describe the layers of the digestive tract, proceeding from the innermost to the outermost layer.

21. What three basic mechanisms regulate the activities of the digestive tract?

22. What is the relationship among the parietal peritoneum, visceral peritoneum, and mesenteries?

23. What are the four primary functions of the oral (buccal) cavity?

24. List the three pairs of salivary glands, and identify the ducts used by their glandular secretions.

25. What are the functions of saliva, and how are salivary secretions controlled?

26. What specific function does each of the four types of teeth perform in the oral cavity?

27. What role do the pharyngeal muscles play in swallowing?

28. What are the three phases of swallowing, and how are they controlled?

29. What contributions do the gastric and pyloric glands of the stomach make to the digestive process?

30. What three subdivisions of the small intestine are involved in the digestion and absorption of food?

31. What are the primary functions of the pancreas, liver, and gallbladder in the digestive process?

32. Which hormones produced by duodenal enteroendocrine cells effectively coordinate digestive functions?

33. What are the three primary functions of the large intestine?

34. What two positive feedback loops are involved in the defecation reflex?

35. What are the five age-related changes that occur in the digestive system?

Level 2   Reviewing Concepts

36. If the lingual frenulum is too restrictive, an individual
- **(a)** has difficulty tasting food
- **(b)** cannot swallow properly
- **(c)** cannot control movements of the tongue
- **(d)** cannot eat or speak normally

37. Increased secretion by all the salivary glands results from
- **(a)** sympathetic stimulation
- **(b)** hormonal stimulation
- **(c)** parasympathetic stimulation
- **(d)** myenteric reflexes

38. The production of acid and enzymes by the gastric mucosa is controlled and regulated by
- **(a)** the central nervous system
- **(b)** short reflexes coordinated in the stomach wall
- **(c)** digestive tract hormones
- **(d)** a, b, and c are correct

39. The gastric phase of secretion is initiated by
- **(a)** distension of the stomach
- **(b)** an increase in the pH of the gastric contents
- **(c)** the presence of undigested materials in the stomach
- **(d)** a, b, and c are correct

40. Chyme reaches the small intestine most quickly when
- **(a)** the stomach is greatly distended
- **(b)** a large amount of gastrin is released
- **(c)** the meal contains a relatively small quantity of proteins
- **(d)** a and c are correct

41. A drop in pH below 4.5 in the duodenum stimulates the secretion of
- **(a)** secretin
- **(b)** cholecystokinin
- **(c)** gastrin
- **(d)** a, b, and c are correct

42. If you wanted to nip the tip off a carrot stick, which teeth would you most likely use?

43. Differentiate between the action and outcome of peristalsis and segmentation.

44. How does the stomach facilitate and assist in digestion?

45. How do the three phases of gastric secretion promote and facilitate gastric control?

46. Why should a person who has stomach ulcers avoid aspirin?

47. Nutritionists have found that after a heavy meal, there is a slight increase in the pH of the blood, especially in the veins that carry blood away from the stomach. What causes this "postenteric alkaline tide" to occur?

Level 3   Critical Thinking and Clinical Applications

48. Some patients who have gallstones develop pancreatitis. How could this occur?

49. Barb suffers from Crohn's disease, a regional inflammation of the intestine. This disease is thought to have some genetic basis, but the actual cause is as yet unknown. When the disease flares up, she experiences abdominal pain, weight loss, and anemia. Which part(s) of the intestine is (are) probably involved, and what is the cause of her symptoms?

50. What symptoms would you expect to observe in a patient who suffers from a blockage of the small intestine at the level of the jejunum?

CHAPTER 25

Metabolism and Energetics

*H*uman beings are remarkably adaptable animals; we can maintain homeostasis near
the equator, where the year-round temperature averages 27°C (80.6°F), and near the poles, where
the temperature remains below 0°C (32°F) most of the time. The key to survival under such varied
environmental conditions is the preservation of normal body temperature. In part, we can accomplish this
by shedding or adding layers of clothing, but hormonal and metabolic adjustments are also crucial. Thus,
our metabolic rate rises in cold climates, generating additional heat that warms the body. This chapter will
examine several aspects of metabolism, including nutrition, energy use, heat production, and the regulation
of body temperature under normal and abnormal conditions.

CHAPTER OUTLINE AND OBJECTIVES

Living cells are chemical factories that break down organic molecules to obtain energy; this energy can be used to generate ATP. As we noted in Chapter 3, reactions within mitochondria provide most of the energy needed by a typical cell. ∞ *[p. 87]* To carry out these metabolic reactions, our cells must have a reliable supply of oxygen and nutrients, including water, vitamins, mineral ions, and organic substrates. (Recall from Chapter 2 that *substrates* are the reactants in enzymatic reactions.) Oxygen is absorbed at the lungs; the other substances are obtained by absorption at the digestive tract. The cardiovascular system ensures prompt distribution of all those substances to interstitial fluids throughout the body, where they become accessible to our cells.

The energy released inside the cell supports growth, cell division, contraction, secretion, and a variety of other special functions that vary from cell to cell and tissue to tissue. Because different tissue types contain different populations of cells, the energy and nutrient requirements of any two tissues, such as loose connective tissue and cardiac muscle, are quite different. When cells, tissues, and organs modify their patterns of activity, the metabolic needs of the body change. There are short-term and long-term changes. For example, nutrient requirements vary from moment to moment (resting versus active), hour to hour (asleep versus awake), and year to year (growing child versus adult).

Differentiation creates tissues and organs that are specialized to store excess nutrients; the storage of lipids in adipose tissue is one familiar example. When nutrients are abundant, energy reserves are built up. These reserves, generally glycogen deposits or lipid inclusions, can then be called on whenever the digestive tract cannot provide the right quantity or quality of nutrients. The endocrine system, with the assistance of the nervous system, adjusts and coordinates the metabolic activities of the body's tissues and controls the storage and mobilization of nutrient reserves.

AN OVERVIEW OF AEROBIC METABOLISM

Figure 25-1● provides an overview of the ways cells use organic nutrients absorbed from the interstitial fluid. Amino acids, lipids, and simple sugars cross the cell membrane to join nutrients already in the cytoplasm. All the cell's organic building blocks collectively form a *nutrient pool*.

Catabolism breaks down organic substrates, releasing energy that can be used to synthesize ATP or other high-energy compounds. Catabolism proceeds in a series of steps. In general, preliminary processing occurs in the cytosol, where enzymes break down large or-

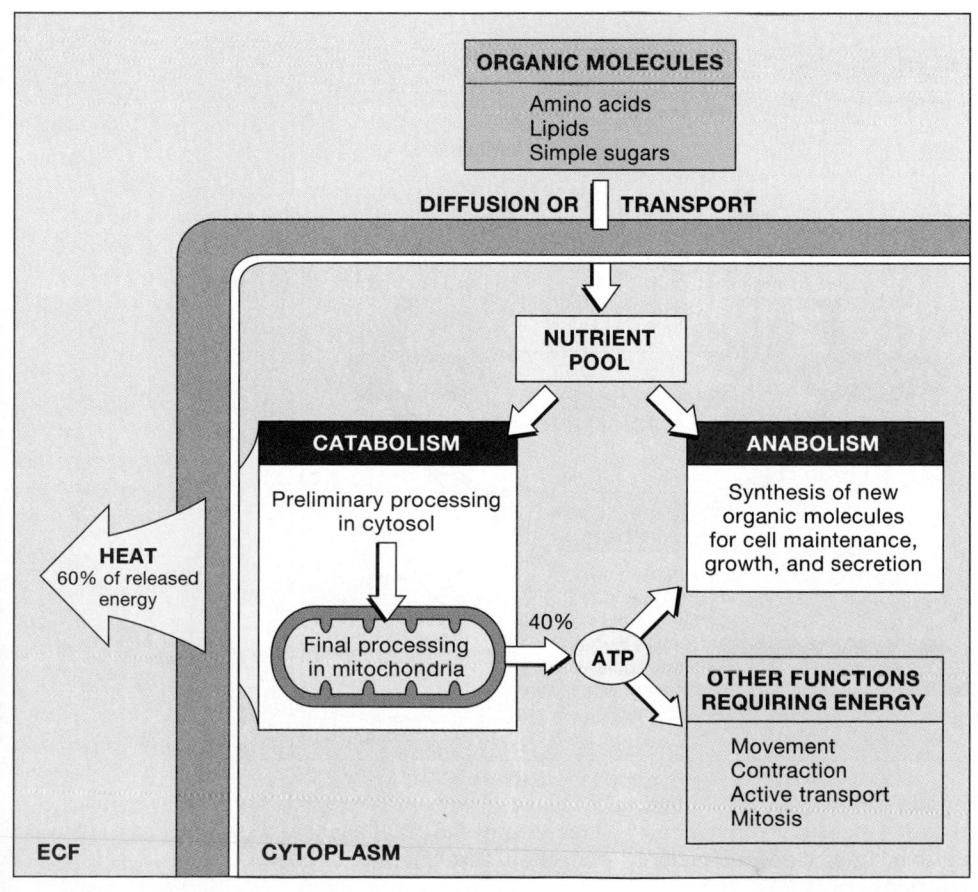

●**FIGURE 25-1** **Aerobic Metabolism.** The cell obtains organic molecules from the extracellular fluid (ECF) and breaks them down to obtain ATP. Only about 40 percent of the energy released through catabolism is captured in ATP; the rest is radiated as heat. The ATP generated through catabolism provides energy for all vital cellular activities, including anabolism.

ganic molecules into smaller fragments. For example, carbohydrates are broken down to short carbon chains, triglycerides are split into fatty acids and glycerol, and proteins are broken down to individual amino acids.

Relatively little ATP is produced during these preparatory steps. However, the simple molecules formed can be absorbed and processed by mitochondria, and the mitochondrial steps release significant amounts of energy. As mitochondrial enzymes break the covalent bonds that hold these molecules together, they capture roughly 40 percent of the energy released. The captured energy is used to convert ADP to ATP. The rest escapes as heat that warms the interior of the cell and the surrounding tissues.

The ATP produced by mitochondria provides energy to support *anabolism,* the synthesis of new organic molecules, as well as other cell functions. Those additional functions, such as ciliary or cell movement, contraction, active transport, and cell division, vary from one cell to another. For example, muscle fibers need ATP to provide energy for contraction, and gland cells need ATP to synthesize and transport their secretions. We have considered such specialized functions in other chapters, so we shall restrict our focus here to anabolic processes.

Anabolism is an "uphill" process that involves the formation of new chemical bonds. Cells synthesize new organic components for three basic reasons (Figure 25-2•):

1. ***To perform structural maintenance or repairs.*** All cells must expend energy to perform ongoing maintenance and repairs, because most structures in the cell are temporary rather than permanent. Their removal and replacement are part of the process of *metabolic turnover,* described in Chapter 2. ∞ *[p. 61]*

2. ***To support growth.*** Cells preparing for division increase in size and synthesize additional proteins and organelles.

3. ***To produce secretions.*** Secretory cells must synthesize their products and deliver them to the interstitial fluid.

The nutrient pool is the source of the substrates for both catabolism and anabolism. As you might expect, the cell tends to conserve the materials needed to build new compounds and breaks down the rest. The cell is continuously replacing membranes, organelles, enzymes, and structural proteins. These anabolic activities require more amino acids than lipids and few carbohydrates. *In general, when a cell with excess carbohydrates, lipids, and amino acids needs energy, it will break down carbohydrates first.* Lipids are a second choice, and amino acids are seldom broken down if other energy sources are available.

In the next section, we shall detail the most important catabolic and anabolic reactions that occur within our cells.

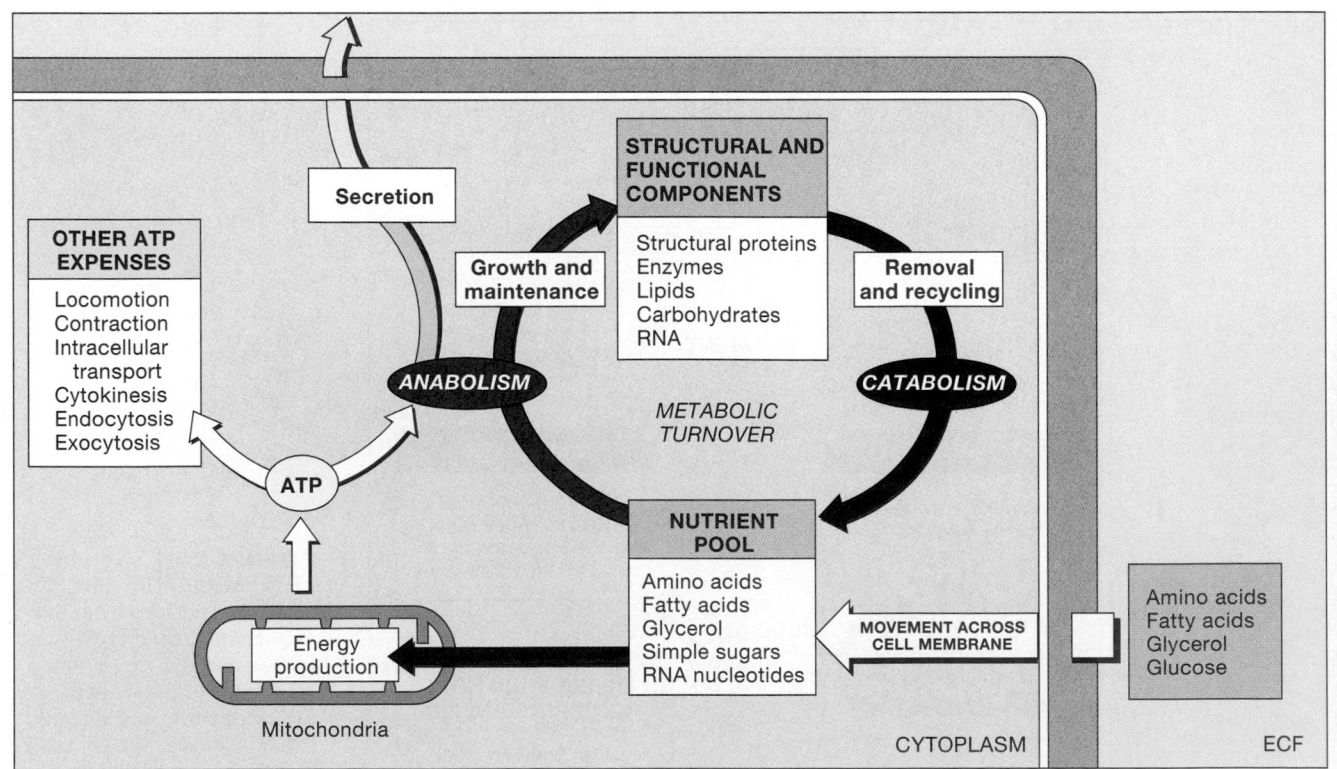

•FIGURE 25-2 Metabolic Turnover and Cellular ATP Production. Metabolic turnover is integrated into other cellular functions. Organic molecules released during metabolic turnover may be recycled by means of anabolic pathways and used to build other large molecules. They can also be broken down for energy production. Through catabolism, the cell must provide ATP for anabolism and for other cellular functions.

CARBOHYDRATE METABOLISM

Most cells generate ATP and other high-energy compounds through the breakdown of carbohydrates, especially glucose. The complete reaction sequence can be summarized as follows:

$$C_6H_{12}O_6 + 6\,O_2 \rightarrow 6\,CO_2 + 6\,H_2O$$

glucose   oxygen   carbon   water
                              dioxide

The breakdown occurs in a series of small steps, and several of the steps release sufficient energy to support the conversion of ADP to ATP. The complete catabolism of one molecule of glucose will provide a typical cell in the body with a net gain of 36 molecules of ATP.

Although most of the actual ATP production occurs inside mitochondria, the first steps take place in the cytosol. We outlined the steps in Chapter 10; because those steps do not require oxygen, they are said to be *anaerobic*. The subsequent reactions, which occur in mitochondria, consume oxygen and are called *aerobic*. ∞ *[pp. 298–299]* The mitochondrial activity responsible for ATP production is called *cellular respiration*, or *aerobic metabolism*.

Glycolysis

Glycolysis (glī-KOL-i-sis; *glykus,* sweet + *lysis,* dissolution) is the breakdown of glucose to **pyruvic acid**. In this process, a series of enzymatic steps breaks the six-carbon glucose molecule ($C_6H_{12}O_6$) into two three-carbon molecules of pyruvic acid (CH_3—CO—COOH). At the normal pH inside the cell, each pyruvic acid molecule loses a hydrogen ion and exists as the negatively charged ion CH_3—CO—COO$^-$. This ionized form is usually called *pyruvate* rather than pyruvic acid.

Glycolysis requires (1) glucose molecules, (2) appropriate cytoplasmic enzymes, (3) ATP and ADP, (4) inorganic phosphates, and (5) **NAD** (*nicotinamide adenine dinucleotide*), a coenzyme that removes hydrogen atoms during one of the enzymatic reactions. If any of these participants is missing, glycolysis cannot take place.

The steps in glycolysis are diagrammed in Figure 25-3•. This is an overview rather than a detailed description of the pathway involved. Glycolysis begins when an enzyme *phosphorylates*—that is, attaches a phosphate group—to the last (sixth) carbon atom on a glucose molecule, creating **glucose-6-phosphate.** Although this step "costs" the cell one ATP molecule, it has two important results: (1) It traps the glucose molecule within the cell, because phosphorylated glucose cannot cross the cell membrane; and (2) it prepares the glucose molecule for further biochemical reactions.

A second phosphorylation occurs in the cytosol before the six-carbon chain is broken into two three-carbon fragments. Energy benefits begin to appear as these fragments are converted to pyruvic acid. Two of the steps release enough energy to generate ATP from ADP and inorganic phosphate (PO_4^{3-}, or P_i). In addition, two molecules of NAD are converted to NADH. The net reaction looks like this:

$$\text{Glucose} + 2\text{ NAD} + 2\text{ ADP} + 2\text{ P}_i \rightarrow$$
(6-carbon)

$$\text{2 Pyruvic acid} + 2\text{ NADH} + 2\text{ ATP}$$
(3-carbon)

This anaerobic reaction sequence provides for the cell a net gain of two molecules of ATP for each glucose molecule converted to two pyruvic acid molecules. A few highly specialized cells, such as red blood cells, lack mitochondria and derive all their ATP through glycolysis. Skeletal muscle fibers rely on glycolysis for ATP production during periods of active contraction; using the ATP provided by glycolysis alone, most cells can survive for brief periods. However, when oxygen is readily available, mitochondrial activity provides most of the ATP required by our cells.

Mitochondrial ATP Production

Glycolysis yields for the cell an immediate net gain of two ATP molecules for each glucose molecule. However, a great deal of additional energy is still locked in the chemical bonds of pyruvic acid. The ability to capture that energy depends on the availability of oxygen. If oxygen supplies are adequate, mitochondria absorb the pyruvic acid molecules and break them down. The hydrogen atoms of each pyruvic acid molecule (CH_3—CO—COOH) are removed by coenzymes and will ultimately be the source of most of the energy gain for the cell. The carbon and oxygen atoms are removed and released as carbon dioxide, a process called **decarboxylation** (dē-kar-boks-i-LĀ-shun).

Each mitochondrion has a double layer of membrane around it. The *outer mitochondrial membrane* contains large-diameter pores that are permeable to ions and small organic molecules such as pyruvic acid. The ions and molecules easily enter the *intermembrane space* that separates the outer mitochondrial membrane from the *inner mitochondrial membrane*. The inner mitochondrial membrane contains a carrier protein that moves pyruvic acid into the mitochondrial matrix.

Once inside the mitochondrion, a pyruvic acid molecule participates in a complex reaction involving NAD and an additional coenzyme called **coenzyme A (CoA).** This reaction yields one molecule of carbon dioxide, one molecule of NADH, and one molecule of **acetyl-CoA** (as-Ē-til-KŌ-ā). Acetyl-CoA consists of a two-carbon **acetyl group** (CH_3CO) bound to coenzyme A. Next, the acetyl group is transferred from CoA to a four-carbon molecule of *oxaloacetic acid*, producing *citric acid*.

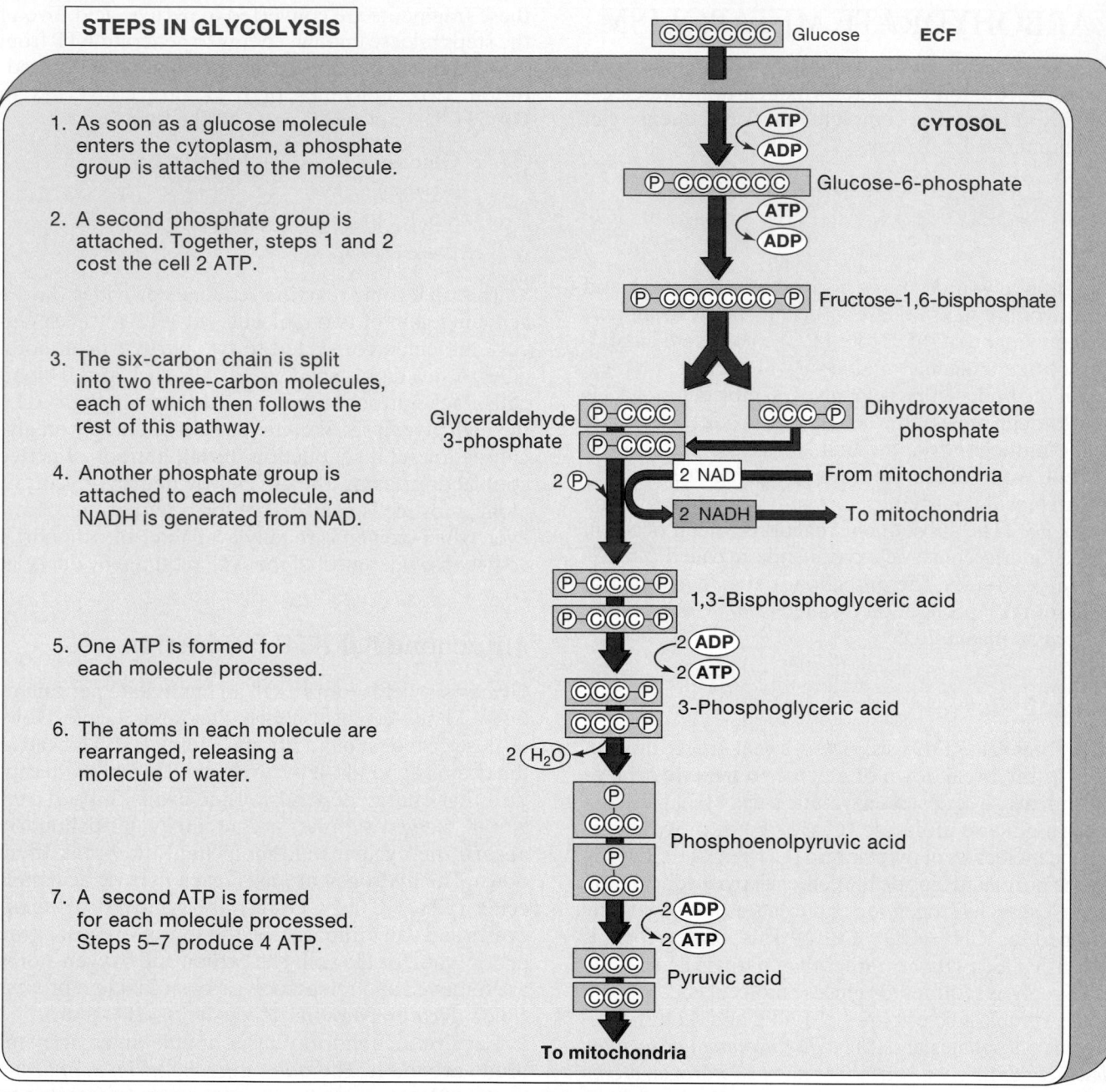

STEPS IN GLYCOLYSIS

1. As soon as a glucose molecule enters the cytoplasm, a phosphate group is attached to the molecule.

2. A second phosphate group is attached. Together, steps 1 and 2 cost the cell 2 ATP.

3. The six-carbon chain is split into two three-carbon molecules, each of which then follows the rest of this pathway.

4. Another phosphate group is attached to each molecule, and NADH is generated from NAD.

5. One ATP is formed for each molecule processed.

6. The atoms in each molecule are rearranged, releasing a molecule of water.

7. A second ATP is formed for each molecule processed. Steps 5–7 produce 4 ATP.

•**FIGURE 25-3** **Glycolysis.** Glycolysis breaks down a six-carbon glucose molecule into two three-carbon molecules of pyruvic acid through a series of enzymatic steps. This diagram follows the fate of the carbon chain. There is a net gain of two ATP molecules for each glucose molecule converted to pyruvic acid. In addition, two molecules of the coenzyme NAD are converted to NADH. Once transferred to mitochondria, both the pyruvic acid and the NADH can still yield a great deal more energy. The further catabolism of pyruvic acid begins with its entry into a mitochondrion (see Figure 25-4).

The TCA Cycle

The formation of citric acid is the first step in a sequence of enzymatic reactions called the **tricarboxylic** (trī-kar-bok-SIL-ik) **acid (TCA) cycle,** or **citric acid cycle.** This reaction sequence is sometimes called the *Krebs cycle* in honor of Hans Krebs, the biochemist who described these reactions in 1937. The function of the cycle is to remove hydrogen atoms from organic molecules and transfer them to coenzymes. The overall pattern of the cycle is shown in Figure 25-4•;

if you are interested in the biochemical details of glycolysis and the TCA cycle, consult the *Applications Manual.* AM *A Closer Look: Aerobic Metabolism*

At the start of the TCA cycle, the two-carbon acetyl group carried by CoA is attached to a four-carbon oxaloacetic acid molecule to make the six-carbon compound citric acid. Coenzyme A is released intact and can thus bind another acetyl group. A complete revolution of the TCA cycle removes two carbon atoms, regenerating the four-carbon chain. (This is why the reaction sequence is called a *cycle.*)

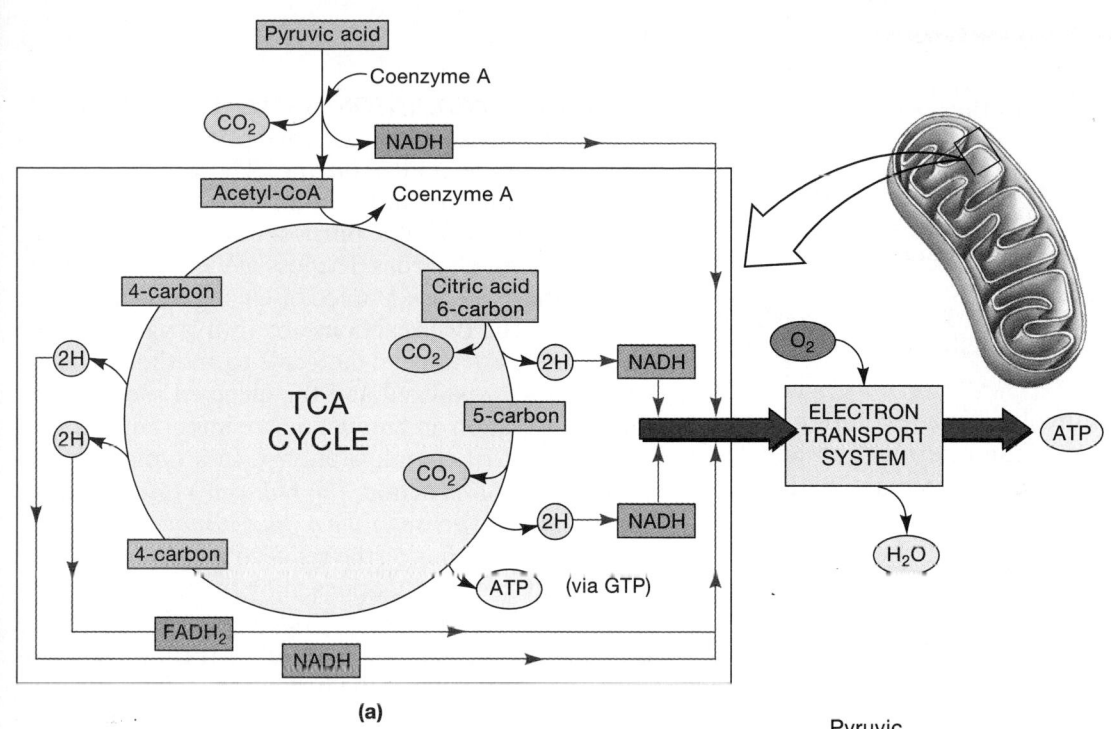

(a)

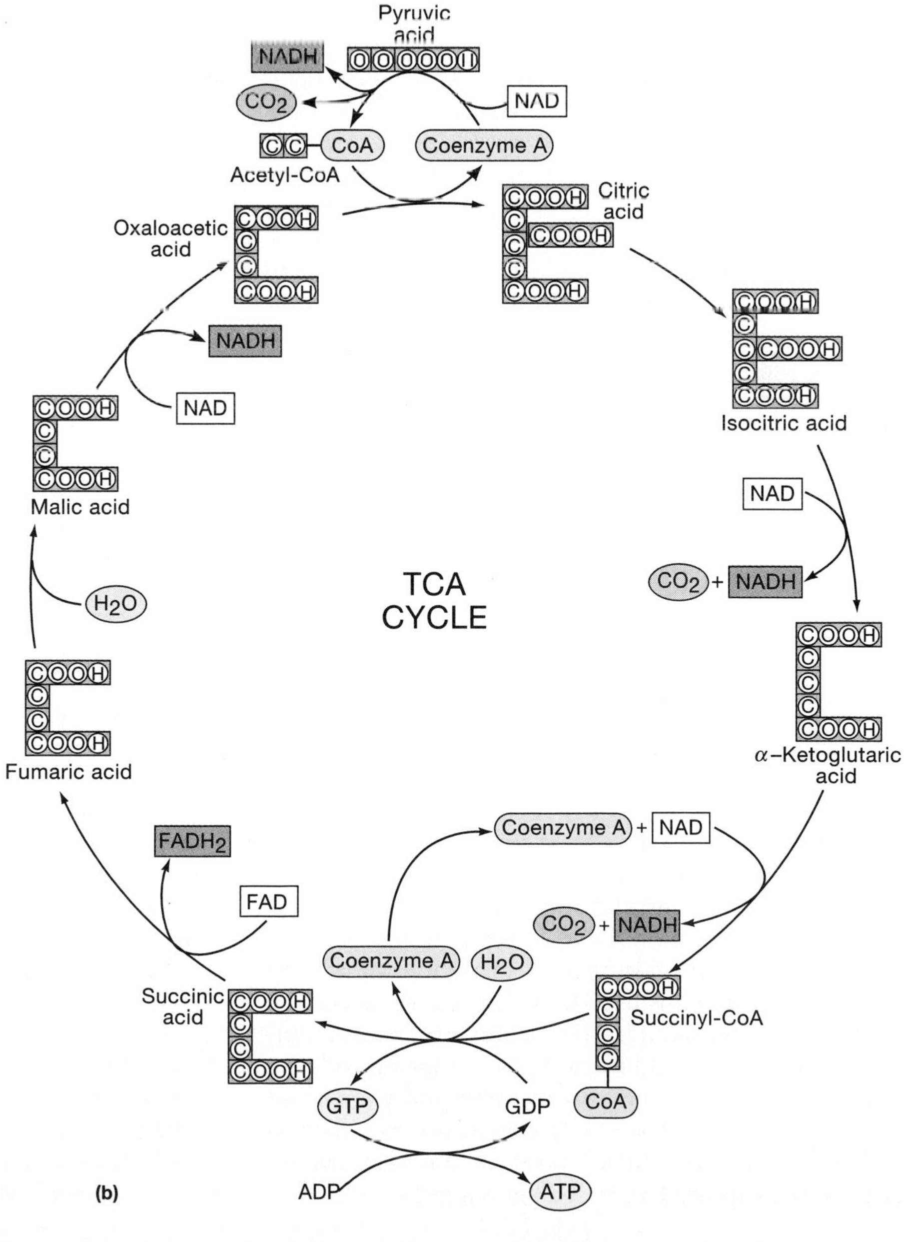

TCA CYCLE

Pyruvic acid — Coenzyme A — CO_2 — NADH — Acetyl-CoA — Coenzyme A — 4-carbon — Citric acid 6-carbon — CO_2 — 2H — NADH — 5-carbon — CO_2 — 2H — NADH — 4-carbon — 2H — ATP (via GTP) — $FADH_2$ — NADH

O_2 — ELECTRON TRANSPORT SYSTEM — ATP — H_2O

•**FIGURE 25-4** **The TCA Cycle.**
The TCA cycle completes the breakdown of organic molecules begun by glycolysis and other catabolic pathways. **(a)** An overview of the TCA cycle and the distribution of carbon, hydrogen, and oxygen atoms. **(b)** A more complete diagrammatic view of the TCA cycle, showing the fate of the carbon chains. The TCA cycle begins with the transfer of an acetyl group from coenzyme A to oxaloacetic acid, a four-carbon molecule in the mitochondrial matrix. In a series of enzymatic reactions, the two added carbon atoms, together with oxygen atoms, are eliminated as carbon dioxide. One molecule of ATP is produced indirectly (via GTP) for each turn of the cycle, but much more ATP will ultimately be obtained from the hydrogen atoms that are removed by the coenzymes NAD and FAD.

Pyruvic acid — NADH — CO_2 — NAD — CoA — Coenzyme A — Acetyl-CoA — COOH Citric acid — Oxaloacetic acid COOH — NADH — NAD — Malic acid COOH — Isocitric acid COOH — NAD — CO_2 + NADH — H_2O — Fumaric acid COOH — α–Ketoglutaric acid COOH — $FADH_2$ — FAD — Coenzyme A + NAD — CO_2 + NADH — Coenzyme A H_2O — Succinyl-CoA COOH — Succinic acid COOH — CoA — GTP — GDP — ATP — ADP

TCA CYCLE

(b)

We can summarize the fate of the atoms in the acetyl group as follows:

- The two carbon atoms are removed in enzymatic reactions that incorporate four oxygen atoms and generate two molecules of carbon dioxide, a metabolic waste product.
- The hydrogen atoms are removed by the coenzyme NAD or a related coenzyme called **FAD** (*flavin adenine dinucleotide*).

Several of the steps involved in a revolution of the TCA cycle involve more than one reaction and require more than one enzyme. Water molecules are tied up in two of those steps. The entire sequence can be summarized as

$$CH_3CO—CoA + 3\,NAD + FAD + GDP + P_i + 2\,H_2O \rightarrow$$
$$CoA + 2\,CO_2 + 3\,NADH + FADH_2 + 2\,H^+ + GTP$$

The only immediate energy benefit of one revolution of the TCA cycle is the formation of a single molecule of GTP (*guanosine triphosphate*). In practical terms, GTP is the equivalent of ATP, because GTP readily transfers a phosphate group to ADP, producing ATP: $GTP + ADP \rightarrow GDP + ATP$.

The formation of GTP from GDP (*guanosine diphosphate*) in the TCA cycle is an example of **substrate-level phosphorylation.** In this process, an enzyme uses the energy released by a chemical reaction to transfer a phosphate group to a suitable acceptor molecule. Although GTP is formed in the TCA cycle, many reaction pathways in the cytosol phosphorylate ADP and form ATP directly. For example, the ATP produced during glycolysis is generated through substrate-level phosphorylation. Normally, however, substrate-level phosphorylation provides a relatively small amount of energy compared with *oxidative phosphorylation*.

Oxidative Phosphorylation

Oxidative phosphorylation is the generation of ATP within mitochondria in a reaction sequence that requires coenzymes and consumes oxygen. This process produces over 90 percent of the ATP used by our cells. The foundation of this reaction sequence is very simple:

$$2\,H_2 + O_2 \rightarrow 2\,H_2O$$

Living cells can easily obtain the ingredients for this reaction. Hydrogen is a component of all organic molecules, and oxygen is an atmospheric gas. The only problem is that the reaction releases a tremendous amount of energy all at once. In fact, this reaction releases so much energy that it is used to launch the space shuttle into orbit. Cells cannot handle energy explosions; energy release must be gradual, and it is gradual in oxidative phosphorylation. During oxidative phosphorylation, this powerful reaction proceeds in a series of small, enzymatically controlled steps. Under these conditions, energy can be captured and ATP generated.

OXIDATION, REDUCTION, AND ENERGY TRANSFER

The enzymatic steps of oxidative phosphorylation involve *oxidation* and *reduction*. There are different types of oxidation and reduction reactions, but the most important for our purposes are those involving the transfer of electrons. The loss of electrons is a form of **oxidation**; the acceptance of electrons is a form of **reduction**. The two reactions are always paired. When electrons pass from one molecule to another, the electron donor is oxidized and the electron recipient reduced. Oxidation and reduction are important because electrons carry chemical energy. In a typical oxidation–reduction reaction, *the reduced molecule gains energy at the expense of the oxidized molecule.*

In such an exchange, the reduced molecule does not acquire all the energy released by the oxidized molecule. Some energy is always released as heat, and additional energy may be used to perform physical or chemical work, such as the formation of ATP. By leading the electrons through a series of oxidation–reduction reactions before they ultimately combine with oxygen atoms, cells can capture and use much of the energy released in the formation of water.

Coenzymes play a key role in this process. A coenzyme acts as an intermediary that accepts electrons from one molecule and transfers them to another molecule. In the TCA cycle, NAD and FAD remove hydrogen atoms from organic substrates. Each hydrogen atom consists of an electron (e^-) and a proton (a hydrogen ion, H^+), so when a coenzyme accepts hydrogen atoms, the coenzyme is reduced and gains energy. The donor molecule loses electrons and energy as it gives up its hydrogen atoms.

NADH and $FADH_2$, the reduced forms of NAD and FAD, then transfer their hydrogen atoms to other coenzymes. The protons are subsequently released, and the electrons, which carry the chemical energy, enter a sequence of oxidation–reduction reactions. This sequence ends with the electrons' transfer to oxygen and the formation of a water molecule. At several steps along the oxidation–reduction sequence, enough energy is released to support the synthesis of ATP from ADP. We shall now consider that reaction sequence in greater detail.

FAD accepts two hydrogen atoms from the TCA cycle and in doing so gains two electrons. NAD also gains two electrons as two hydrogen atoms are removed from the donor molecule. However, NAD binds one electron as part of a hydrogen atom, forming NADH; it removes the electron from the other hydrogen atom and releases the proton (H^+). For this reason, the reduced form of NAD is often described as "NADH + H^+."

Other coenzymes involved in the initial steps of oxidative phosphorylation are **FMN** (*flavin mononucleotide*) and **coenzyme Q** (*ubiquinone*). These coenzymes are either free in the mitochondrial matrix (NAD) or attached to the inner mitochondrial membrane (FAD, FMN, coenzyme Q).

THE ELECTRON TRANSPORT SYSTEM The **electron transport system (ETS)**, or *respiratory chain,* is a sequence of metalloproteins called **cytochromes** (SĪ-tō-krōmz; *cyto-,* cell + *chroma,* color). Each cytochrome has two components: a protein and a pigment. The protein, embedded in the inner mitochondrial membrane, surrounds the pigment complex, which contains a metal ion (either iron, Fe^{3+}, or copper, Cu^{2+}). We shall be concerned with four cytochromes: *b, c, a,* and *a₃*.

Figure 25-5● summarizes the major steps in oxidative phosphorylation. We shall first consider the path taken by the electrons that are captured and delivered by coenzymes (Figure 25-5a●):

STEP 1: *A coenzyme strips a pair of hydrogen atoms from a substrate molecule.* As we have seen, different coenzymes are used for different substrate molecules. During glycolysis, which occurs in the cytoplasm, NAD is reduced to NADH. Within mitochondria, both NAD and FAD are reduced through reactions of the TCA cycle, producing NADH and $FADH_2$, respectively.

STEP 2: *NADH and $FADH_2$ deliver hydrogen atoms to coenzymes embedded in the inner mitochondrial membrane.* It is the electrons that carry the energy, and the protons that accompany them will be released before the electrons are transferred to the electron transport system. As indicated in Figure 25-5a●, the path taken to the ETS depends on whether the donor is NADH or $FADH_2$.

STEP 3: *Coenzyme Q accepts hydrogen atoms from $FMNH_2$ and $FADH_2$ and passes electrons to cytochrome* b.

STEP 4: *Electrons are passed along the electron transport system, losing energy in a series of small steps.* The sequence is cytochrome *b* to *c* to *a* to *a₃*.

STEP 5: *At the end of the electron transport system, an oxygen atom accepts the electrons, creating an*

oxygen ion (O⁻). This ion has a very strong affinity for hydrogen ions (H^+); it quickly combines with two hydrogen ions in the mitochondrial matrix, forming a molecule of water.

Notice that this reaction sequence started with the removal of two hydrogen atoms from a substrate molecule, and it ended with the formation of water from two hydrogen ions and one oxygen ion. This is the reaction that we described initially as releasing too much energy if performed in a single step. Because the reaction has occurred in a series of small steps, however, the combination of hydrogen and oxygen can take place quietly rather than explosively.

The electron transport system does not produce ATP directly. Instead, it creates the conditions necessary for ATP production. The electrons release energy as they pass from coenzyme to cytochrome and from cytochrome to cytochrome. The energy released at several steps drives hydrogen ion pumps that move hydrogen ions from the mitochondrial matrix into the space between the inner and outer mitochondrial membranes. These hydrogen ion pumps create a large concentration gradient for hydrogen ions across the inner mitochondrial membrane. This concentration gradient is a form of potential energy, and it is used to provide the energy to convert ADP to ATP (Figure 25-5b●). Despite the concentration gradient, the hydrogen ions cannot diffuse into the matrix because they are not lipid-soluble. However, there are hydrogen ion channels in the inner mitochondrial membrane that will permit the diffusion of H^+ into the matrix. The entire complex of an ion channel and associated proteins looks like a miniature Tootsie Roll® Pop. These ion channels and their attached *coupling factors* in some way use the kinetic energy of passing hydrogen ions to generate ATP. This process is called **chemiosmosis** (kem-ē-oz-MŌ-sis), or *chemiosmotic phosphorylation.*

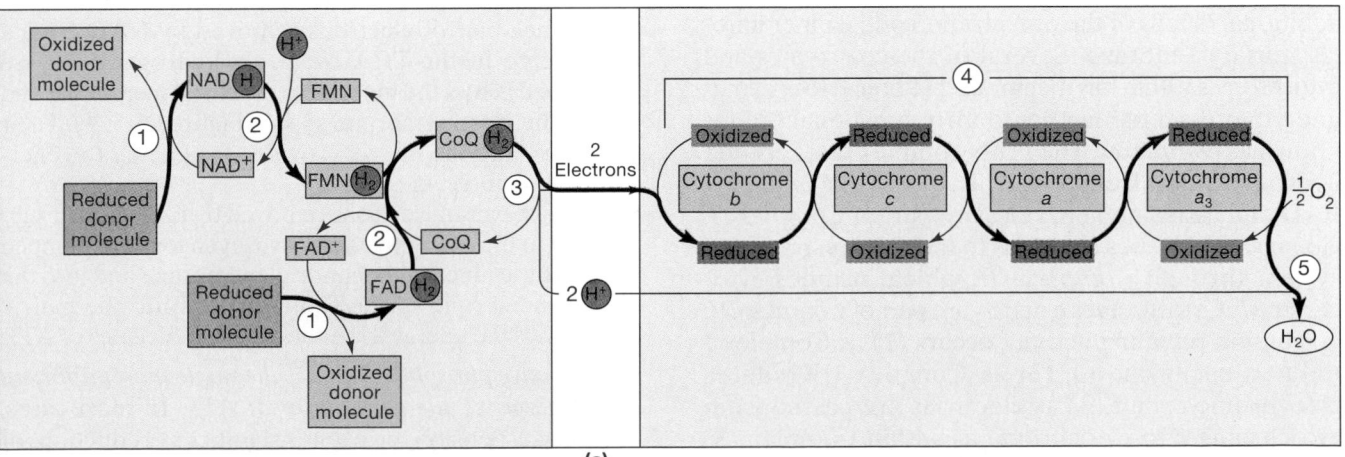

●**FIGURE 25-5** Oxidative Phosphorylation. **(a)** The sequence of oxidation–reduction reactions involved.

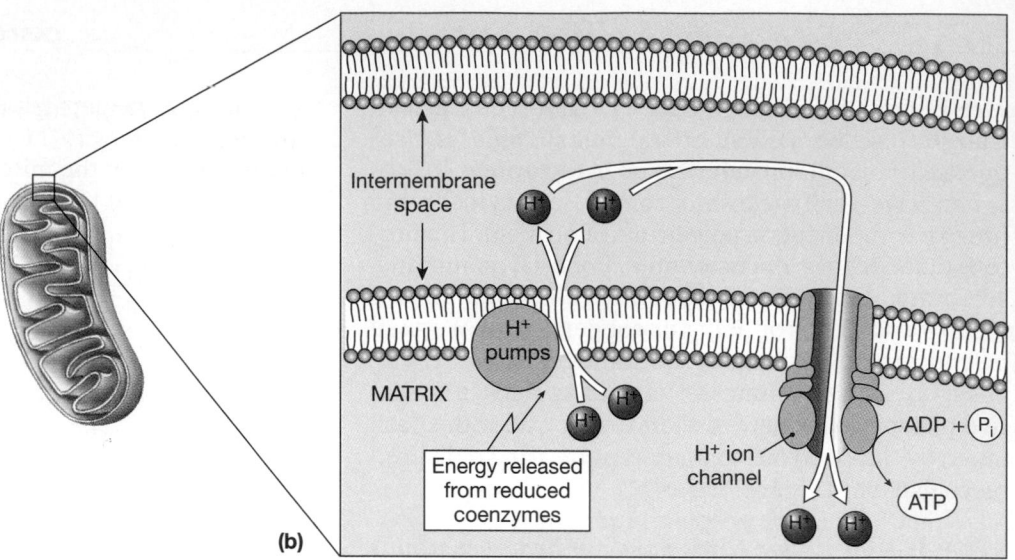

●**FIGURE 25-5 (continued).**
(b) An overview of the basic principle of chemiosmosis.
(c) Location of the coenzymes and the electron transport system. Note the sites where hydrogen ions are pumped into the intermembrane space, providing the concentration gradient essential to the generation of ATP.

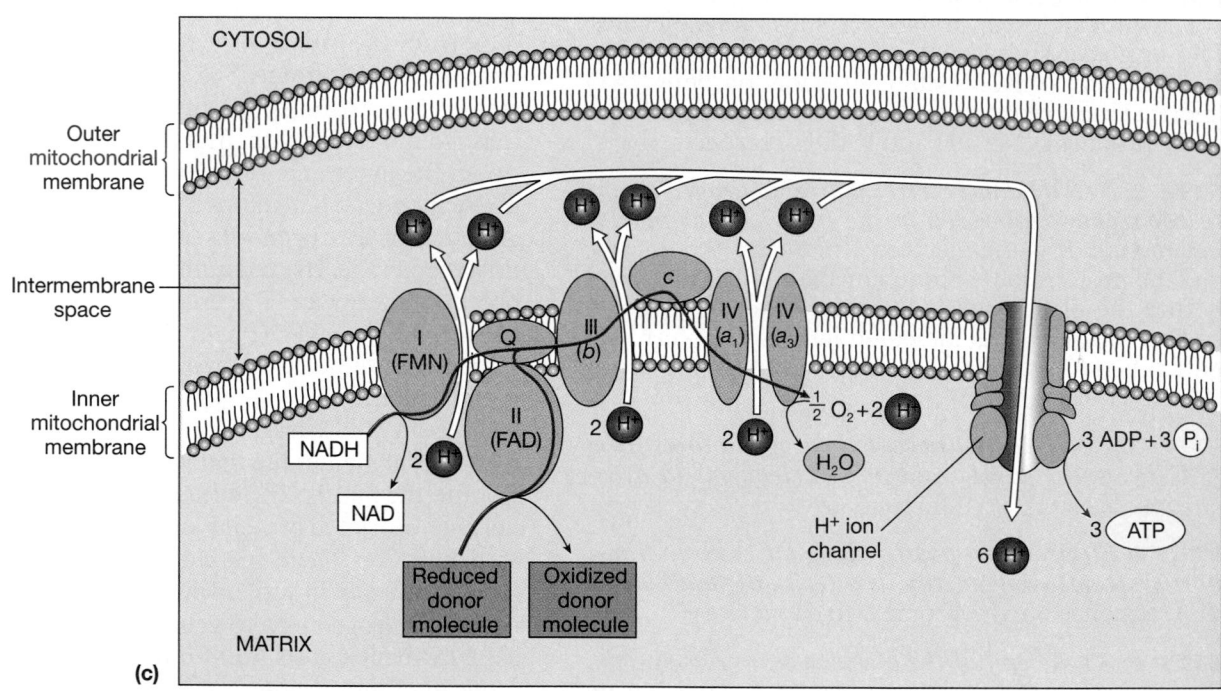

ATP GENERATION AND THE ETS Figure 25-5c● diagrams the mechanism of ATP generation and shows additional details of the organization of the inner mitochondrial membrane. Several of the coenzymes and cytochromes within this membrane are bound to enzymes and other essential factors to form functional groups known as *complexes*. The FMN complex that receives hydrogen atoms from NADH is known as *Complex I*; FAD and associated enzymes are part of *Complex II*. Coenzyme Q passes electrons to the electron transport system through *Complex III*, which includes cytochrome *b*. Cytochromes *a* and *a₃* are part of *Complex IV*.

The ion pumping activity occurs (1) as Complex I reduces coenzyme Q, (2) as Complex III reduces cytochrome *c*, and (3) as electrons are passed from cytochrome *a* to cytochrome *a₃* within Complex IV. The amount of ATP produced depends on how many hydrogen ions move across the membrane, which in turn depends on how many are pumped into the intermembrane space by the electron transport chain.

- For each pair of electrons removed by NAD from a substrate in the TCA cycle, six hydrogen ions are pumped across the inner mitochondrial membrane and into the intermembrane space (Figure 25-5c●). Their reentry into the matrix provides the energy to generate three molecules of ATP.

- For each pair of electrons removed by FAD from a substrate in the TCA cycle, four hydrogen ions are pumped across the inner mitochondrial membrane and into the intermembrane space. Their reentry into the matrix provides the energy to generate two molecules of ATP.

Oxidative phosphorylation is the single most important mechanism for the generation of ATP. In most cases, chronic suspension, or even a significant reduction, of the rate of oxidative phosphorylation will kill the cell. If many cells are affected, the individual may die. Oxidative phosphorylation requires both oxygen and electrons; the rate of ATP generation can thus be limited by the availability of either oxygen or electrons.

Cells obtain oxygen by diffusion from the extracellular fluid. If the supply of oxygen is cut off, mitochondrial ATP production will cease, because reduced cytochrome a_3 has no acceptor for its electrons. With the last reaction stopped, the entire ETS comes to a halt, like cars at a washed-out bridge. Oxidative phosphorylation can no longer take place, and cells quickly succumb to energy starvation. Because neurons have a high demand for energy, the brain is one of the first organs to be affected. Hydrogen cyanide gas is sometimes used as a pesticide to kill rats or mice; in some states where capital punishment is legal, the gas is used to execute criminals. The cyanide ion (CN^-)

binds to cytochrome a_3 and prevents the transfer of electrons to oxygen. As a result, cells die from energy starvation.

Energy Yield of Glycolysis and Cellular Respiration

For most cells, the complete reaction pathway that begins with glucose and ends with carbon dioxide and water is the primary method of generating ATP. Figure 25-6• reviews the entire process in terms of energy production:

- **Glycolysis.** During glycolysis, the cell gains 2 molecules of ATP directly for each glucose molecule broken down anaerobically to pyruvic acid. Two molecules of

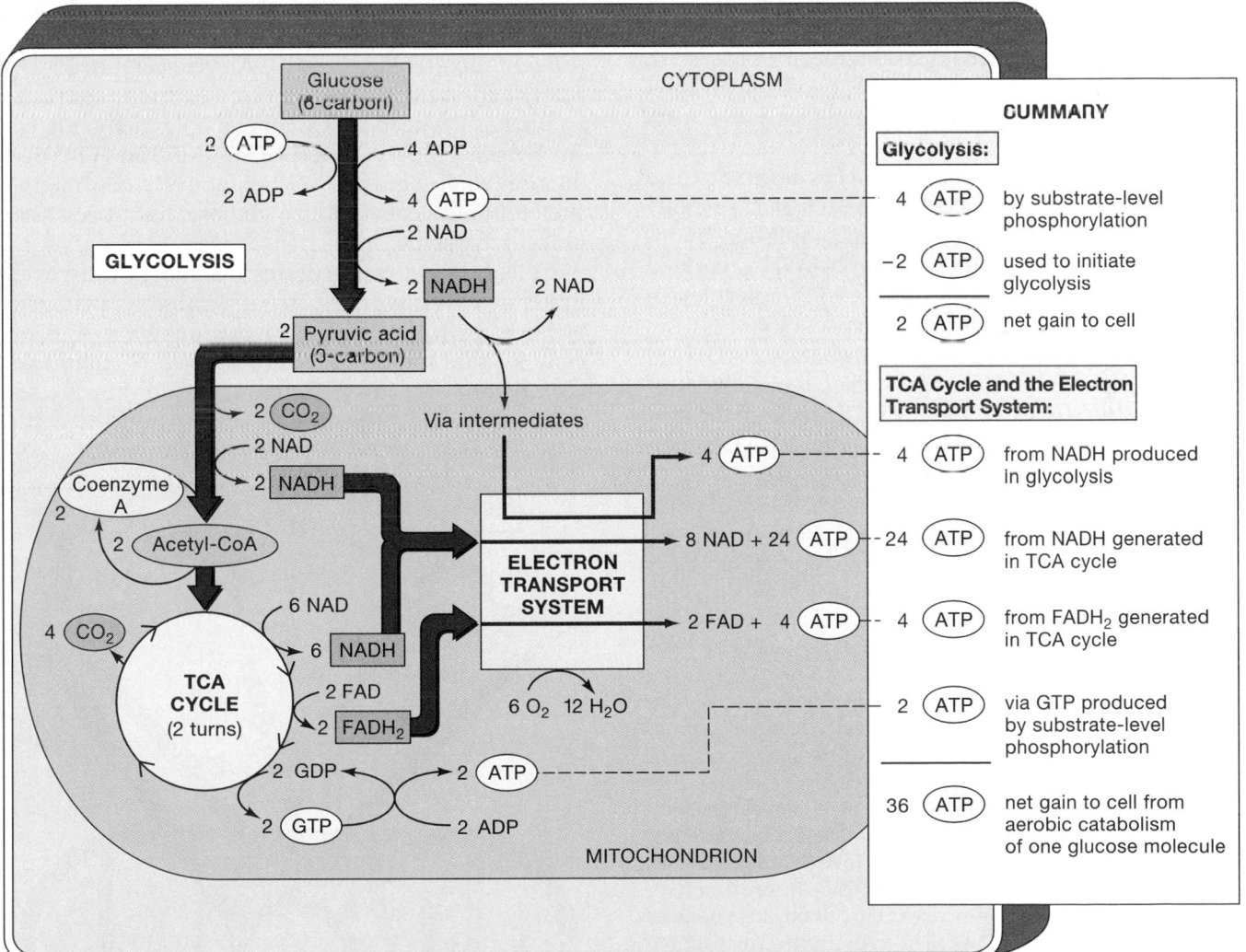

•**FIGURE 25-6** **A Summary of the Energy Yield of Aerobic Metabolism.** For each glucose molecule broken down via glycolysis, only two molecules of ATP (net) are produced. The TCA cycle generates an additional two ATP molecules in the form of GTP. However, glycolysis, the formation of acetyl-CoA, and the TCA cycle all yield molecules of reduced coenzymes (NADH and/or $FADH_2$). Many additional ATP molecules are produced when electrons from these coenzymes pass through the electron transport system. Each of the eight NADH molecules produced inside the mitochondrion yields three ATP molecules. In most cells, each of the two NADH molecules produced in glycolysis provides only two ATP molecules, the same amount gained from each of the two $FADH_2$ molecules generated within the mitochondrion.

NADH are also produced. In most cells, electrons are passed from NADH to FAD via an intermediate in the intermembrane space and thence to CoQ and the electron transport system. This sequence of events will ultimately provide an additional 4 ATP molecules.

- **The TCA cycle.** The 2 pyruvic acid molecules derived from each glucose molecule are fully broken down aerobically in mitochondria. Two revolutions of the TCA cycle, each yielding 1 molecule of ATP by way of GTP, provide a total gain of 2 more molecules of ATP. These revolutions also generate molecules of reduced coenzymes (8 NADH and 2 FADH$_2$).

- **The electron transport system.** The NADH and FADH$_2$ feed electrons into the electron transport chain. Through the ETS, each of the 8 molecules of NADH yields 3 ATP and 1 water molecule; each of the 2 FADH$_2$ molecules yields 2 ATP and 1 water molecule. The total yield of electron transport for each glucose molecule is an additional 28 molecules of ATP (Figure 25-6●).

Summing up, for each glucose molecule processed, the cell gains 36 molecules of ATP: two from glycolysis, 4 from the NADH generated in glycolysis, 2 from the TCA cycle, and 28 from the ETS.

Your heart muscle cells and liver cells are able to gain an additional 2 ATP molecules for each glucose molecule broken down. This gain is accomplished by increasing the energy yield from the NADH generated during glycolysis from 4 ATP to 6 ATP molecules. In these cells, each NADH molecule passes electrons to an intermediate that generates NADH, not than FADH$_2$, in the mitochondrial matrix. The subsequent transfer of electrons to FMN, CoQ, and the ETS results in the production of 6 ATP molecules, just as if the 2 NADH molecules had been generated in the TCA cycle.

CARBOHYDRATE LOADING Although other nutrients can be broken down to provide substrates for the TCA cycle, carbohydrates require the least processing and preparation. It is not surprising, therefore, that athletes have tried to devise ways of exploiting these compounds as ready sources of energy.

Eating carbohydrates *just before* you exercise does not improve your performance and may actually decrease your endurance by slowing the mobilization of existing energy reserves. Runners or swimmers preparing for lengthy endurance contests, such as a marathon or 5-km swim, do not eat immediately before competing, and for 2 hours before the race they limit their intake to drinking water. However, these athletes often eat carbohydrate-rich meals for 3 days before the event. This **carbohydrate loading** increases the carbohydrate reserves of muscle tissue that will be called on during the competition.

You can obtain maximum effects by exercising to exhaustion for 3 days before you start a high-carbohydrate diet; this practice is called *carbohydrate depletion/loading*. There are a number of potentially unpleasant side effects to carbohydrate depletion/loading, including muscle and kidney damage. Sports physiologists recommend that athletes use this routine fewer than three times a year.

☑ What is the primary role of the TCA cycle in the production of ATP?

☑ The NADH produced by glycolysis in skeletal muscle fibers leads to the production of 2 ATP molecules in the mitochondria, but the NADH produced by glycolysis in cardiac muscle cells leads to the production of 3 ATP molecules. Why?

☑ How would a decrease in the level of cytoplasmic NAD affect ATP production in mitochondria?

Other Catabolic Pathways

Aerobic metabolism is relatively efficient and capable of generating large amounts of ATP. It is the cornerstone of normal metabolism, but it has one obvious limitation: The cell must have adequate supplies of both oxygen and glucose. Cells can survive only for brief periods without adequate oxygen. Low glucose concentrations have a much lesser effect on most cells, because they can break down other nutrients to provide substrates for the TCA cycle (Figure 25-7●). Many cells can switch from one nutrient source to another as the need arises. For example, many cells can shift from glucose-based to lipid-based ATP production when necessary. When actively contracting, skeletal muscles catabolize glucose, but at rest they utilize fatty acids.

Cells break down proteins for energy only when lipids or carbohydrates are unavailable, primarily because proteins make up the enzymes and organelles that the cell needs to survive. Nucleic acids are present only in small amounts, and they are seldom catabolized for energy, even when the cell is

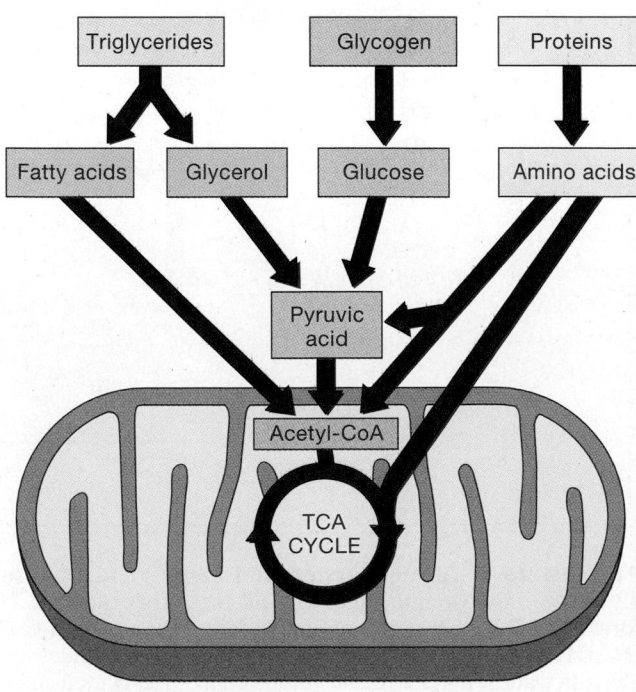

●FIGURE 25-7 Alternative Catabolic Pathways

dying of acute starvation. This restraint makes sense, because it is the DNA in the nucleus that determines all the structural and functional characteristics of the cell. We shall consider the catabolism of other compounds in later sections as we discuss lipid, protein, and nucleic acid metabolism.

Gluconeogenesis

Because some of the steps in glycolysis— glucose breakdown—are essentially irreversible, cells cannot generate glucose by performing glycolysis in reverse, using the same enzymes. Therefore glucose production involves a different set of regulatory enzymes, and glucose breakdown and synthesis are independently regulated. Key regulatory enzymes are indicated in Figure 25-8•. Pyruvic acid or other three-carbon molecules can be used to synthesize glucose; as a result, a cell can create glucose molecules from other carbohydrates, lactic acid, glycerol, or some amino acids. However, acetyl-CoA cannot be used to make glucose, because the decarboxylation step between pyruvic acid and acetyl-CoA cannot be reversed. Fatty acids and many amino acids cannot be used for *gluconeogenesis*, because their catabolic pathways produce acetyl-CoA. **Gluconeogenesis** (gloo-kō-nē-ō-JEN-e-sis) is the synthesis of glucose from noncarbohydrate precursors, such as lactic acid, glycerol, or amino acids.

Glucose molecules created by gluconeogenesis can be used to manufacture other simple sugars, complex carbohydrates, proteoglycans, or nucleic acids. In the liver and in skeletal muscle, glucose molecules are stored as glycogen. The process of glycogen formation from glucose is known as **glycogenesis.** It is a complex process that involves several steps and requires the high-energy compound UTP (*uridine triphosphate*). Glycogen is an important energy reserve that can be broken down when the cell cannot obtain enough glucose from the interstitial fluid. The process of glycogen breakdown is called **glycogenolysis.** Glycogenolysis occurs quickly and involves a single enzymatic step. Although glycogen molecules are large, glycogen reserves take up very little space because they form compact, insoluble granules.

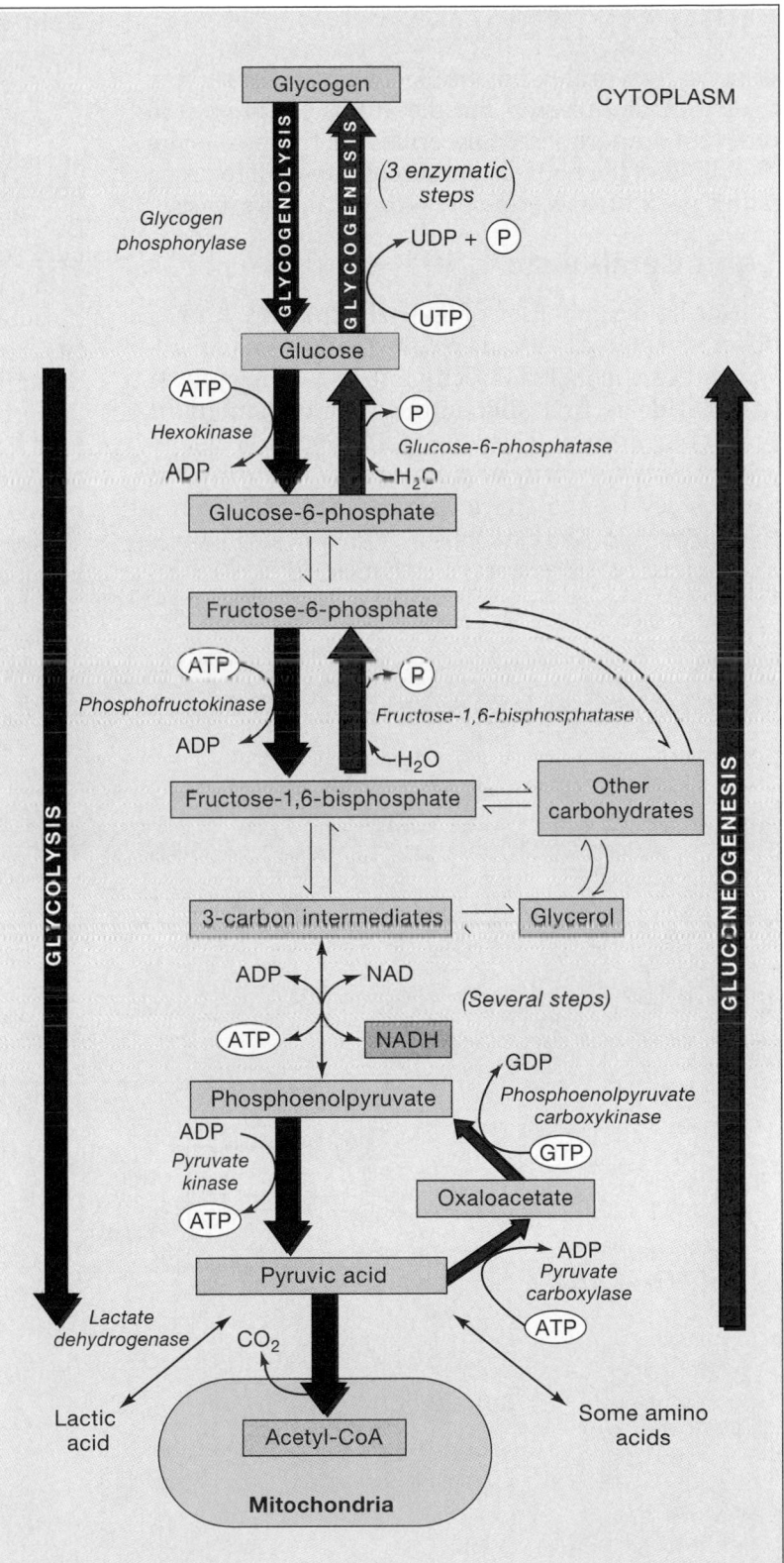

•**FIGURE 25-8 Carbohydrate Synthesis.** A flowchart of the pathways for glycolysis and gluconeogenesis. Many of the reactions are freely reversible, but separate regulatory enzymes control the key steps, which are indicated by colored arrows. Some amino acids, carbohydrates other than glucose, and glycerol can be converted to glucose. Notice that the enzymatic reaction that converts pyruvic acid to acetyl-CoA cannot be reversed.

LIPID METABOLISM

Like carbohydrates, lipid molecules contain carbon, hydrogen, and oxygen, but the atoms are present in different proportions. Triglycerides are the most abundant lipid in the body, so our discussion shall focus on pathways for triglyceride breakdown and synthesis.

Lipid Catabolism

During lipid catabolism, or **lipolysis,** lipids are broken down into pieces that can be either converted to pyruvic acid or channeled directly into the TCA cycle. A triglyceride is first split into its component parts through hydrolysis. This step yields one molecule of glycerol and three fatty acid molecules. Glycerol enters the TCA cycle after enzymes in the cytosol convert it to pyruvic acid. The catabolism of fatty acids involves a completely different set of enzymes.

Beta-Oxidation

Fatty acid molecules are broken down into two-carbon fragments, in a sequence of reactions known as **beta-oxidation.** This process occurs inside mitochondria, so the carbon chains can enter the TCA cycle immediately. Figure 25-9● diagrams one step in the process of beta-oxidation. Each step generates molecules of acetyl-CoA, NADH, and $FADH_2$, leaving a shorter carbon chain bound to coenzyme A.

Beta-oxidation provides substantial energy benefits. For each two-carbon fragment removed from the fatty acid, the cell gains 12 ATP molecules from the processing of acetyl-CoA in the TCA cycle, plus 5 ATP molecules from the NADH and $FADH_2$. The cell can therefore gain 144 ATP molecules from the breakdown of one 18-carbon fatty acid molecule. This result is almost 1.5 times the energy obtained by the complete breakdown of three six-carbon glucose molecules. The catabolism of other lipids follows similar patterns, generally ending with the formation of acetyl-CoA.

Lipids and Energy Production

Lipids are important as an energy reserve because they can provide large amounts of ATP. Because they are insoluble, lipids can be stored in compact droplets in the cytosol. This storage method saves space, but when the lipid droplets are large, it is difficult for water-soluble enzymes to get at them. Lipid reserves are therefore more difficult to access than carbohydrate reserves. Also, most lipids are processed inside mitochondria, and mitochondrial activity is limited by the availability of oxygen. The net result is that lipids cannot provide large amounts of ATP in a short time. However, cells with modest energy demands can shift to lipid-based energy production when glucose supplies are limited. Skeletal muscle fibers normally cycle between lipid and carbohydrate metabolism. When at rest (when energy demands are low), these cells break down fatty acids. When active (when energy demands are large and immediate), skeletal muscle fibers shift to glucose metabolism.

Lipid Synthesis

The synthesis of lipids is known as **lipogenesis** (li-pō-JEN-e-sis). Figure 25-10● (p. 932) shows the major pathways of lipogenesis. Glycerol is synthesized from *dihydroxyacetone phosphate,* one of the intermediate products of glycolysis. The synthesis of most other types of lipids, in-

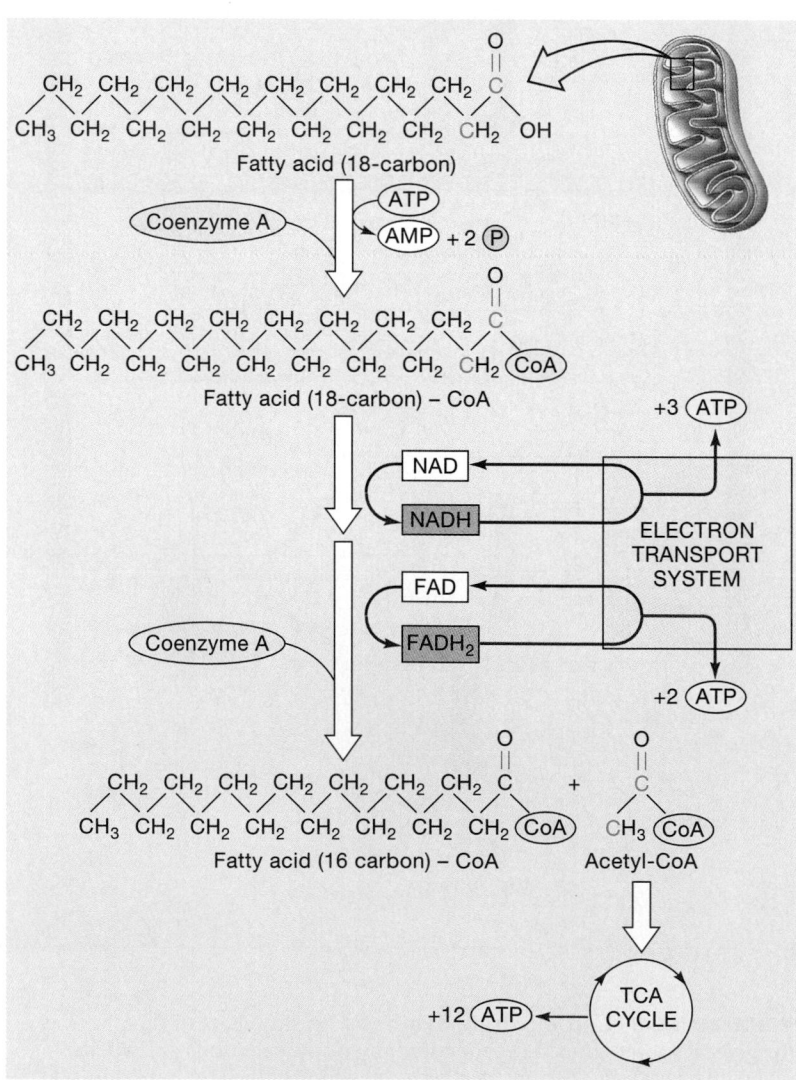

●**FIGURE 25-9** **Beta-Oxidation.** During beta-oxidation, the carbon chains of fatty acids are broken down to yield molecules of acetyl-CoA, which can be used in the TCA cycle. The reaction also donates hydrogen atoms to coenzymes that deliver them to the electron transport system.

Dietary Fats and Cholesterol

Elevated cholesterol levels are associated with the development of atherosclerosis (Chapter 21) and coronary artery disease (CAD; Chapter 20). ∞ *[pp. 713, 685]* Current nutritional advice suggests that you reduce cholesterol intake to under 300 mg per day. This amount represents a 40 percent reduction for the average American adult. As a result of rising concerns about cholesterol, such phrases as "low-cholesterol," "contains no cholesterol," and "cholesterol-free" are now widely used in the advertising and packaging of foods. Cholesterol content alone does not tell the entire story; we must consider some basic information about cholesterol and about lipid metabolism in general:

- **Cholesterol has many vital functions in the human body.** It serves as a waterproofing for the epidermis, a lipid component of all cell membranes, a key constituent of bile, and the precursor of several steroid hormones and one vitamin (vitamin D_3). Because cholesterol is so important, dietary restrictions should have the goal of keeping cholesterol levels within acceptable limits. The goal is not to eliminate cholesterol from the diet or from the circulating blood.

- **The cholesterol content of the diet is not the only source of circulating cholesterol.** The human body can manufacture cholesterol from the acetyl-CoA obtained through glycolysis or the beta-oxidation of other lipids. Dietary cholesterol probably accounts for only about 20 percent of the cholesterol in circulation; the rest is the result of metabolism of saturated fats in the diet. If the diet contains an abundance of saturated fats, serum cholesterol levels will rise, because excess lipids are broken down to acetyl-CoA and used to synthesize cholesterol. Consequently, a person trying to lower serum cholesterol by dietary control must restrict other lipids—especially saturated fats—as well.

- **Genetic factors affect each individual's cholesterol level.** If you reduce your dietary supply of cholesterol, your body will synthesize more to maintain "acceptable" concentrations in the blood. The acceptable level depends on your genetic programming. Because individuals differ in genetic makeup, their cholesterol levels can vary even on similar diets. In virtually all instances, however, dietary restrictions can lower blood cholesterol significantly.

- **Cholesterol levels vary with age and physical condition.** At age 19, three out of four males have fasting cholesterol levels below 170 mg/dl. Cholesterol levels in females of this age are slightly higher, typically at or below 175 mg/dl. As age increases, the cholesterol values gradually climb; over age 70 the values are 230 mg/dl (males) and 250 mg/dl (females). Cholesterol levels are considered unhealthy if they are higher than those of 90 percent of the

population in that age group. For males, this value ranges from 185 mg/dl at age 19 to 250 mg/dl at age 70. For females, the comparable values are 190 mg/dl and 275 mg/dl, respectively.

To determine whether you may need to do anything about your cholesterol level, just remember three simple rules:

1. Individuals of any age with total cholesterol values below 200 mg/dl probably do not need to change their lifestyle unless they have a family history of CAD and atherosclerosis.

2. Those with cholesterol levels between 200 and 239 mg/dl should modify their diet, lose weight (if overweight), and have annual checkups.

3. Cholesterol levels over 240 mg/dl warrant drastic changes in dietary lipid consumption, perhaps coupled with drug treatment. Drug therapies are always recommended in cases in which the serum cholesterol level exceeds 350 mg/dl. Examples of drugs used to lower cholesterol levels include *cholestyramine, colestipol,* and *lovastatin.*

Most physicians, when ordering a blood test for cholesterol, also request information on circulating triglycerides. In fasting individuals, triglycerides are usually present at levels of 40–150 mg/dl. (After a person has consumed a fatty meal, triglyceride levels may be temporarily elevated.)

When cholesterol levels are high, or when an individual has a family history of atherosclerosis or CAD, further tests and calculations may be performed. The HDL level is measured, and the LDL level is calculated as

$$LDL = \frac{Cholesterol - HDL - Triglycerides}{5}$$

A high total cholesterol value linked to a high LDL level spells trouble. In effect, an unusually large amount of cholesterol is being exported to peripheral tissues. Problems can also exist if the individual has high total cholesterol—or even normal total cholesterol—but low HDL levels (below 35 mg/dl). In this case, excess cholesterol delivered to the tissues cannot easily be returned to the liver for excretion. In either event, the amount of cholesterol in peripheral tissues, and especially in arterial walls, is likely to increase. For years, LDL:HDL ratios were taken to be valid predictors of the risk of developing atherosclerosis. Risk-factor analysis and LDL levels are now thought to be more-accurate indicators. For a male with more than one risk factor, many clinicians recommend dietary and drug therapy if LDL levels exceed 130 mg/dl, regardless of the total cholesterol or HDL levels.

●**FIGURE 25-10** Lipid
Synthesis. Pathways of lipid
synthesis begin with acetyl-CoA.
Molecules of acetyl-CoA can be
strung together in the cytosol,
yielding fatty acids. Those fatty
acids can be used to synthesize
glycerides or other lipid
molecules. Lipids can be
synthesized from amino acids or
carbohydrates by using acetyl-
CoA as an intermediate.

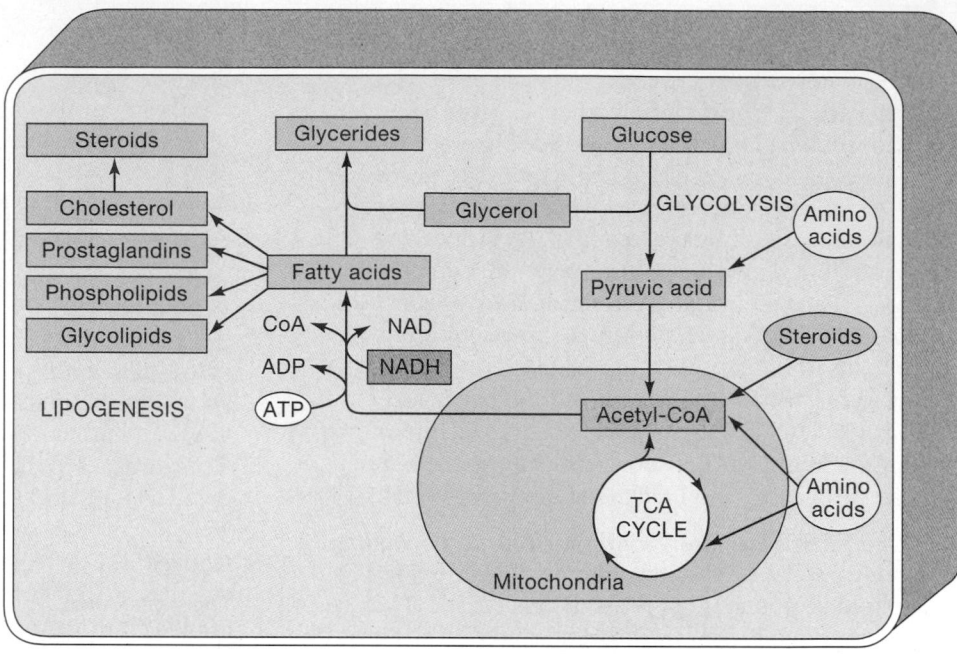

cluding nonessential fatty acids and steroids, begins
with acetyl-CoA. Lipogenesis can use almost any
organic substrate, because lipids, amino acids, and
carbohydrates can be converted to acetyl-CoA.

Fatty acid synthesis involves a reaction sequence
quite distinct from that of beta-oxidation. Your cells
cannot *build* every fatty acid they can break down.
Linoleic acid and **linolenic acid,** both 18-carbon unsaturated fatty acids, cannot be synthesized. These are
called **essential fatty acids,** because they must be
included in your diet. They are synthesized by plants.
Essential fatty acid deficiency generally occurs only
among hospitalized individuals receiving nutrients in
an intravenous solution. A diet poor in linoleic acid
slows growth and alters the appearance of the skin.
These fatty acids are also needed to synthesize
prostaglandins and some of the phospholipids incorporated into cell membranes throughout the body.

Lipid Transport and Distribution

Like glucose, lipids are needed throughout the body. All
cells need lipids to maintain their cell membranes, and
there are important steroid hormones that must reach
target cells in many different tissues. Because most lipids
are not soluble in water, special transport mechanisms
are required to carry them from one region of the body
to another. Most lipids circulate through the bloodstream as *free fatty acids* or *lipoproteins*.

Free Fatty Acids

Free fatty acids (FFAs) are lipids that can easily diffuse
across cell membranes. Most free fatty acids in the

blood are bound to albumin, the most abundant plasma protein.

Sources of free fatty acids in the blood include the
following:

■ Fatty acids that are not used in the synthesis of triglycerides diffuse out of the intestinal epithelium and into
the blood.

■ Fatty acids diffuse out of lipid stores, such as the liver
and adipose tissue, when triglycerides are broken down.

Liver cells, cardiac muscle cells, skeletal muscle
fibers, and many other body cells can metabolize free
fatty acids. These fatty acids are an important energy
source during periods of starvation, when glucose supplies are limited.

Lipoproteins

Lipoproteins are lipid–protein complexes that contain large insoluble glycerides and cholesterol with
a superficial coating of phospholipids and proteins.
The proteins and phospholipids make the entire
complex soluble, and exposed proteins, which can
bind to specific membrane receptors, determine
which cells absorb the associated lipids.

Lipoproteins are usually classified according to
size and the relative proportions of lipid versus protein. The following five major groups are recognized:

1. **Chylomicrons.** Roughly 95 percent of the weight of
a chylomicron consists of triglycerides. Chylomicrons
are the largest lipoproteins, ranging in diameter from
0.03 to 0.5 μm. They are produced by intestinal epithelial cells, as we described in Chapter 24. ∞ *[p.
907]* Chylomicrons carry absorbed lipids from the

intestinal tract to the circulation. The other lipoproteins shuttle lipids among various tissues in the body, such as between the liver and adipose tissue. The liver is the primary source of all other types of lipoproteins.

2. **Very low-density lipoproteins (VLDLs).** Very low-density lipoproteins contain triglycerides manufactured by the liver plus small amounts of phospholipids and cholesterol. The primary function of VLDLs is the transport of these triglycerides to peripheral tissues. VLDLs range in diameter from 25 to 75 nm (0.025–0.075 μm).

3. **Intermediate-density lipoproteins (IDLs).** Intermediate-density lipoproteins are intermediate in size and lipid composition between VLDLs and low-density lipoproteins. They contain smaller amounts of triglycerides than do VLDLs and relatively more phospholipids and cholesterol than do LDLs.

4. **Low-density lipoproteins (LDLs).** Low-density lipoproteins contain cholesterol, lesser amounts of phospholipids, and very few triglycerides. These lipoproteins, which are about 25 nm in diameter, deliver cholesterol to peripheral tissues. Because this cholesterol may wind up in arterial plaques, LDL cholesterol is often called bad cholesterol.

5. **High-density lipoproteins (HDLs).** High-density lipoproteins, about 10 nm in diameter, have roughly equal amounts of lipid and protein. The lipids are largely cholesterol and phospholipids. The primary function of HDLs is transporting excess cholesterol from peripheral tissues back to the liver for storage or excretion in the bile. Because HDL cholesterol is returning from peripheral tissues and will not cause circulatory problems, it is sometimes called good cholesterol. Actually, applying the terms *good* and *bad* to cholesterol can be misleading, for cholesterol metabolism is complex and variable. (For more details, see the discussion "Dietary Fats and Cholesterol" on p. 931.)

Figure 25-11● diagrams the probable relationships among these lipoproteins. Chylomicrons produced in the intestinal tract reach the venous circulation by entering lymphatic capillaries and traveling through the thoracic duct (Figure 25-11a●). Although chylomicrons are too large to diffuse across capillary walls, the endothelial lining of capillaries in skeletal muscle, cardiac muscle, adipose tissue, and the liver contains the enzyme **lipoprotein lipase,** which breaks down complex lipids. When chylomicrons contact these endothelial walls, enzymatic activity releases fatty acids and monoglycerides that can diffuse across the endothelium and into the interstitial fluid.

The liver controls the distribution of other lipoproteins (Figure 25-11b●):

STEP 1: Liver cells synthesize VLDL for discharge into the circulation.

STEP 2: On arrival in peripheral capillaries, lipoprotein lipase removes many of the triglycerides from VLDL, leaving an IDL; the triglycerides are broken down into fatty acids and monoglycerides.

STEP 3: When the IDL reaches the liver, additional triglycerides are removed, and the protein content is altered. This process creates an LDL that then returns to peripheral tissues to deliver cholesterol.

STEP 4: The LDLs leave the circulation through capillary pores or cross the endothelium by means of vesicular transport.

STEP 5: Once in peripheral tissues, the LDLs are absorbed by means of *receptor-mediated endocytosis*. The amino acids and cholesterol then enter the cytoplasm.

STEP 6: The cholesterol not used by the cell in the synthesis of lipid membranes or other products diffuses out of the cell.

STEP 7: It then reenters the circulation, where it is absorbed by HDLs and returned to the liver.

STEP 8: On arrival in the liver, the HDLs are absorbed and their cholesterol is extracted. Some of the cholesterol recovered will be used in the synthesis of LDL, and the rest will be excreted in bile salts.

STEP 9: The HDLs stripped of their cholesterol are released into the circulation to travel into peripheral tissues and absorb additional cholesterol.

PROTEIN METABOLISM

There are roughly 100,000 different proteins in the human body, with various forms, functions, and structures. Each protein contains some combination of the same 20 amino acids. (Appendix IV details the structures of these amino acids.) Under normal conditions, there is a continuous turnover of cellular proteins. Peptide bonds are broken, and the free amino acids are used to manufacture new proteins. This recycling occurs in the cytosol.

If other energy sources are inadequate, mitochondria can break down amino acids in the TCA cycle to generate ATP. Not all amino acids enter the TCA cycle at the same point, so the ATP benefits vary. However, the average ATP yield per gram is comparable to that of carbohydrate catabolism.

Amino Acid Catabolism

The first step in amino acid catabolism is the removal of the amino group. This process requires a coenzyme derivative of **vitamin B$_6$** (*pyridoxine*). The amino group may be removed by *transamination* (trans-am-i-NĀ-shun) or *deamination* (dē-am-i-NĀ-

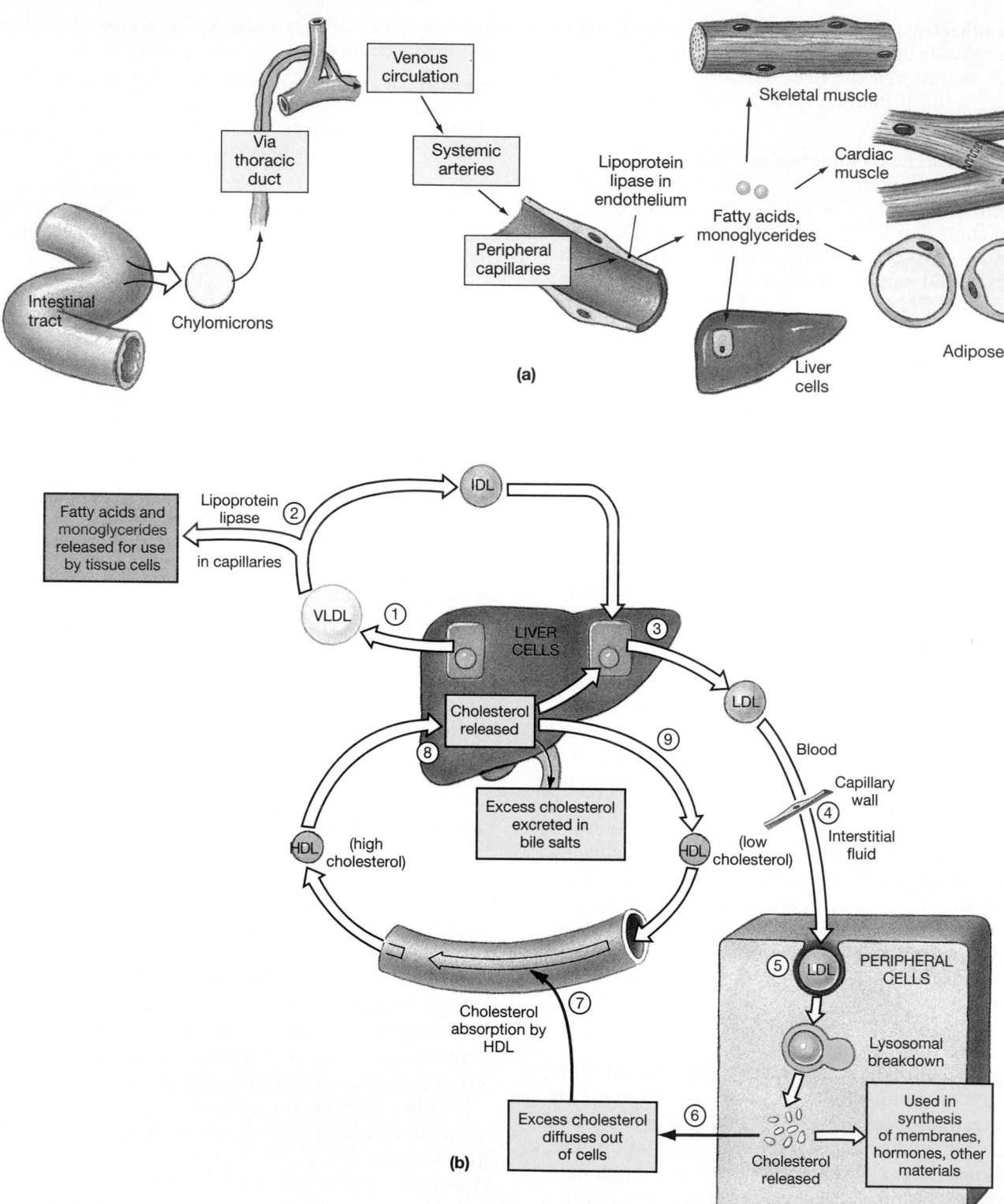

•*FIGURE 25-11* **Lipid Transport and Utilization.** **(a)** Chylomicrons synthesized at the intestinal epithelium reach the circulation via the thoracic duct. They are broken down by lipoprotein lipase in capillaries that supply blood to skeletal muscle, cardiac muscle, adipose tissue, and the liver. **(b)** ① Liver cells synthesize a VLDL that delivers triglycerides to peripheral tissues. ② Lipoprotein lipase in endothelial cells breaks down these triglycerides and releases fatty acids and monoglycerides that diffuse into the surrounding tissues. ③ The IDL that remains returns to the liver, where it is absorbed and converted to an LDL that contains cholesterol. ④ The LDL circulates to peripheral tissues, crosses the endothelium, and is absorbed by cells through endocytosis. ⑤ Some of the cholesterol is used in cellular processes. ⑥ The excess diffuses back into the circulation. ⑦ In the plasma, the cholesterol is absorbed by an HDL produced by the liver. ⑧ On returning to the liver, the HDL is absorbed and the cholesterol is extracted. Some of the cholesterol will be exported once again, in an LDL; excess cholesterol is excreted in bile salts. ⑨ The HDLs stripped of their cholesterol are released into the circulation.

shun). We shall consider other details of amino acid catabolism in a later section.

Transamination

Transamination (Figure 25-12a●) attaches the amino group of an amino acid to a **keto acid.** A keto acid resembles an amino acid except that the second carbon binds an oxygen atom rather than an amino group. Transamination produces a "new" amino acid that can enter the cytosol and be used for protein synthesis. It also converts the original amino acid to a keto acid that can enter the TCA cycle. Many different tissues perform transaminations. These reactions enable a cell to synthesize many of the amino acids needed for protein synthesis. Cells of the liver,

skeletal muscles, heart, lung, kidney, and brain, which are particularly active in protein synthesis, perform large numbers of transaminations.

Deamination

Deamination (Figure 25-12b●) is performed in preparing an amino acid for breakdown in the TCA cycle. Deamination is the removal of an amino group and a hydrogen atom in a reaction that generates an ammonia (NH_3) molecule or an ammonium ion (NH_4^+). Ammonia molecules are highly toxic, even in low concentrations. Your liver, the primary site of deamination, has the enzymes needed to deal with the problem of ammonia generation. Liver cells convert the ammonia to **urea,** a relatively harmless, water-soluble compound excreted in the urine. The **urea cycle** is the reaction sequence involved in urea production (Figure 25-12c●).

When glucose supplies are low and lipid reserves are inadequate, liver cells break down internal proteins and absorb additional amino acids from the blood. The amino acids are deaminated, and the carbon chains are broken down to provide ATP.

Several inherited metabolic disorders result from an inability to produce specific enzymes involved with amino acid metabolism. *Phenylketonuria* (fen-il-kē-tō-NOO-rē-uh), or PKU, is an example. Individuals with PKU cannot convert phenylalanine to tyrosine, due to a defect in the enzyme *phenylalanine hydroxylase*. This reaction is an essential step in the synthesis of norepinephrine, epinephrine, dopamine, and melanin. If PKU is not detected in infancy, CNS development is inhibited, and severe brain damage results. [AM] *Phenylketonuria*

Proteins and ATP Production

Three factors make protein catabolism an impractical source of quick energy:

1. Proteins are more difficult to break apart than are complex carbohydrates or lipids.

2. One of the byproducts, ammonia, is a toxin that can damage cells.

3. Proteins form the most important structural and functional components of any cell. Extensive protein catabolism therefore threatens homeostasis at the cellular and systems levels.

Protein Synthesis

We detailed the basic mechanism for protein synthesis in Chapter 3 (Figures 3-22

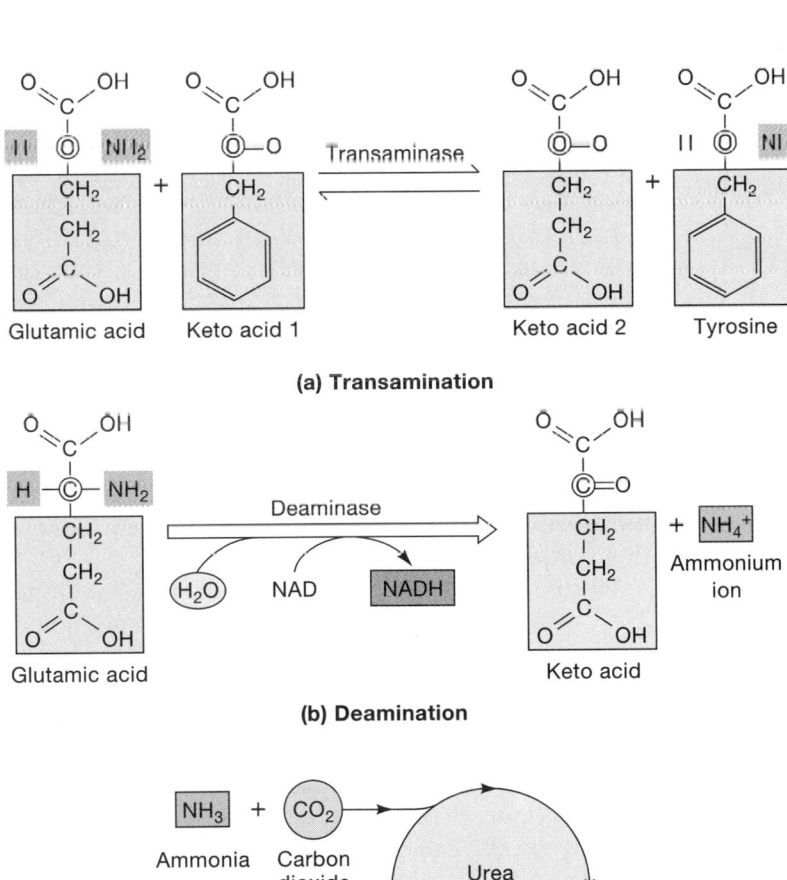

(a) Transamination

(b) Deamination

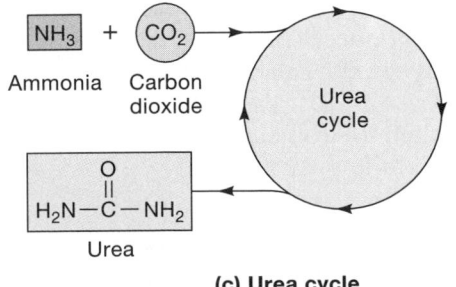

(c) Urea cycle

●**FIGURE 25-12 Amino Acid Catabolism. (a)** During transamination, an enzyme removes the amino group (—NH_2) from one molecule and attaches it to a keto acid. **(b)** During deamination, an enzyme strips the amino group and a hydrogen atom from an amino acid and produces a keto acid and ammonia. **(c)** The urea cycle takes two metabolic waste products, carbon dioxide and ammonia, and produces a molecule of urea. Urea is a relatively harmless, soluble compound that is excreted in the urine.

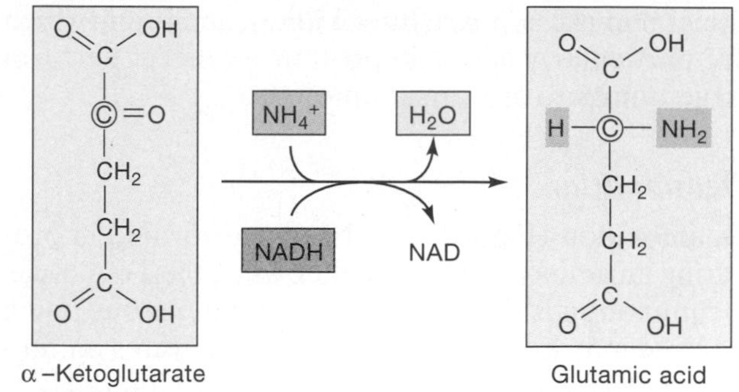

•**FIGURE 25-13** **Amination.** Amination attaches an amino group to a keto acid. This is an important step in the synthesis of nonessential amino acids. Amino groups can also be attached through transamination (see Figure 25-12).

and 3-23•, pp. 95, 97). Your body can synthesize roughly half of the various amino acids needed to build proteins. There are 10 **essential amino acids.** You cannot synthesize eight of them (*isoleucine, leucine, lysine, threonine, tryptophan, phenylalanine, valine,* and *methionine*); the other two (*arginine* and *histidine*) can be synthesized but in amounts that are insufficient for growing children. Because the body can make other amino acids on demand, they are called the **nonessential amino acids.** Your body can readily synthesize the carbon frameworks of the nonessential amino acids, and a nitrogen group can be attached through transamination or through amination. **Amination** is the attachment of an amino group (Figure 25-13•).

Protein deficiency diseases develop when an individual does not consume adequate amounts of all essential amino acids. All amino acids must be available if protein synthesis is to occur. Every transfer RNA molecule must appear at the active ribosome at the proper time bearing its individual amino acid. As soon as the amino acid called for by a particular codon is missing, the entire process comes to a halt. Regardless of its energy content, if the diet is deficient in essential amino acids, the individual will be malnourished to some degree. Examples of protein deficiency diseases include *marasmus* and *kwashiorkor.* More than 100 million children worldwide have symptoms of these disorders, although neither condition is common in the United States today. [AM]
Protein Deficiency Diseases

NUCLEIC ACID METABOLISM

Living cells contain both DNA and RNA. In Chapter 3, we considered the replication of DNA, the mechanics of cell division, and the importance of DNA in regulating the structural and functional characteristics of the cell. ∞ *[pp. 91–101]* The genetic infor-

mation contained in the DNA of the nucleus is essential to the cell's long-term survival. As a result, the DNA in the nucleus is never catabolized for energy, even if the cell is dying of starvation. But the RNA in the cell is involved in protein synthesis, and RNA molecules are broken down and replaced regularly.

RNA Catabolism

In the breakdown of a strand of RNA, the bonds between nucleotides are broken, and the molecule is disassembled into individual nucleotides. These nucleotides are usually recycled into new nucleic acids. However, they can be catabolized to simple sugars and nitrogenous bases.

RNA and Energy Production

RNA catabolism makes a relatively insignificant contribution to the total energy budget of the cell. Proteins account for 30 percent of the weight of the cell, and much more energy can be provided through the catabolism of nonessential proteins. Even when RNA is broken down, only the sugars and pyrimidines provide energy. The sugars can enter the glycolytic pathways. The pyrimidines (cytosine and uracil, in RNA) are converted to acetyl-CoA and metabolized through the TCA cycle. The purines (adenine and guanine) cannot be catabolized. Instead, they are deaminated and excreted as **uric acid.** Like urea, uric acid is a relatively nontoxic waste product, but it is far less soluble than urea. Urea and uric acid are called **nitrogenous wastes,** because they are waste products that contain nitrogen atoms.

Normal plasma uric acid concentrations average 2.7–7.4 mg/dl, depending on gender and age. When plasma concentrations exceed 7.4 mg/dl, *hyperuricemia* (hī-per-ū-ri-SĒ-mē-uh) exists. This condition may affect 18 percent of the U.S. population. At concentrations over 7.4 mg/dl, body fluids are saturated with uric acid. Although symptoms may not appear at once, uric acid crystals may begin to form in body fluids. The condition that then develops is called *gout.* Most cases of hyperuricemia and gout are linked to problems with the excretion of uric acid by the kidneys. [AM] *Gout*

Nucleic Acid Synthesis

Most cells synthesize RNA, but DNA synthesis occurs only in cells that are preparing for mitosis and cell division or meiosis (nuclear events involved in gamete production). We described the process of DNA replication in Chapter 3. ∞ *[p. 98]* Messenger RNA (mRNA), transfer RNA (tRNA), and ribosomal RNA (rRNA) are transcribed by different forms of RNA polymerase:

- Messenger RNA is manufactured only when specific genes are activated; a strand of mRNA has a life span measured in minutes or hours.
- Ribosomal RNA and tRNA are more durable than mRNA strands. For example, the average life span of a strand of rRNA is just over 5 days. However, because a typical cell contains roughly 100,000 ribosomes and many times that number of tRNA molecules, their replacement involves a considerable amount of synthetic activity.

☑ How would a diet that is deficient in pyridoxine (vitamin B₆) affect protein metabolism?

☑ Elevated levels of uric acid in the blood could be an indicator of increased metabolism of which organic compound?

☑ Why are high-density lipoproteins (HDLs) considered to be beneficial?

METABOLIC INTERACTIONS

Figure 25-14• summarizes the metabolic pathways for lipids, carbohydrates, and proteins. The diagram follows the reactions in a "typical" cell. Yet no one cell can perform all the anabolic and catabolic operations and interconversions required by the body as a whole. As differentiation proceeds, each cell type develops its own complement of enzymes, and this enzyme complement determines the cell's metabolic capabilities. In the presence of such cellular diversity, homeostasis can be preserved only when the metabolic activities of tissues, organs, and organ systems are coordinated.

The nutrient requirements of each tissue vary with the types and quantities of enzymes present in the cytoplasm. From a metabolic standpoint, we can consider the body in terms of five distinctive components: the *liver, adipose tissue, skeletal muscle, neural tissue,* and *other peripheral tissues:*

1. **The liver.** The liver is the focal point for metabolic regulation and control. Liver cells contain a great diversity of enzymes, and they can break down or synthesize most of the carbohydrates, lipids, and amino acids needed by other cells in the body. Because they have an extensive circulatory supply, liver cells are in an excellent position to monitor and adjust the nutrient composition of the circulating blood. The liver also contains significant energy reserves in the form of glycogen deposits.

2. **Adipose tissue.** Adipose tissue stores lipids, primarily as triglycerides. Adipocytes are located in many areas

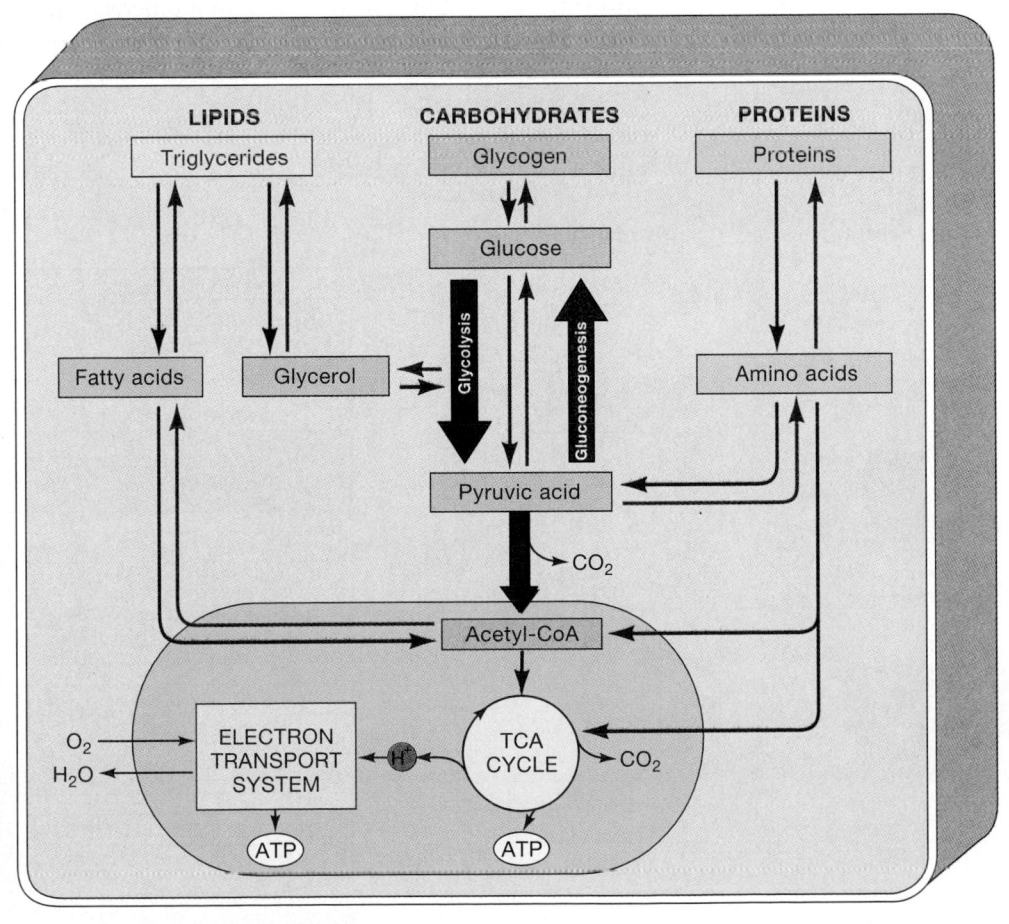

•FIGURE 25-14 A **Summary of the Pathways of Catabolism and Anabolism.** An overview of major catabolic and anabolic pathways for lipids, carbohydrates, and proteins.

of the body; in previous chapters, we noted the presence of fat cells in loose connective tissue, in mesenteries, within red and yellow marrows, in the epicardium, and around the eyes.

3. *Skeletal muscle.* Skeletal muscle accounts for almost half of an individual's body weight, and these cells maintain substantial glycogen reserves. In addition, their contractile proteins can be broken down and the amino acids used as an energy source if other nutrients are unavailable.

4. *Neural tissue.* Neural tissue has a high demand for energy, but the cells do not maintain reserves of carbohydrates, lipids, or proteins. Neurons must be provided with a reliable supply of glucose, because they are generally unable to metabolize other molecules. If blood glucose becomes too low, neural tissue in the CNS cannot continue to function, and the individual becomes unconscious.

5. *Other peripheral tissues.* Other peripheral tissues do not maintain large metabolic reserves, but they are able to metabolize glucose, fatty acids, or other substrates. Their preferred source of energy varies with the instructions provided by the endocrine system.

The interrelationships among these five components can best be understood by considering the events that occur over a 24-hour period. During this time, the body experiences two broad patterns of metabolic activity: (1) the *absorptive state* and (2) the *postabsorptive state*.

The Absorptive State

The **absorptive state** is the period following a meal, when nutrient absorption is under way. After a typical meal, the absorptive state continues for about 4 hours. If you are fortunate enough to eat three meals a day, you spend 12 out of every 24 hours in the absorptive state.

A typical meal contains proteins, lipids, and carbohydrates in varying proportions; while you are in the absorptive state, your intestinal mucosa busily absorbs these nutrients. Glucose and amino acids enter the circulation, and the hepatic portal vein carries them to the liver. Most of the absorbed fatty acids are packaged in chylomicrons, which enter the lacteals.

Some of the carbohydrates, lipids, and amino acids will be broken down at once to provide the energy needed to support cellular operations. The remainder will be stored, lessening the impact of future shortages. Insulin is the primary hormone of the absorptive state, although various other hormones stimulate amino acid uptake (growth hormone, or GH) and protein synthesis (GH, androgens, and estrogens). We shall now consider the activities under way in specific sites, with particular reference to Figure 25-15● and Table 25-1.

The Liver

The liver regulates the levels of glucose and amino acids in the blood of the hepatic portal vein before that blood reaches the inferior vena cava. Despite the continuous absorption of glucose at the intestinal mucosa, blood glucose levels do not skyrocket, because the liver cells, under insulin stimulation, remove glucose from the hepatic portal circulation. Blood glucose levels do rise, but only from about 90 mg/dl to perhaps 150 mg/dl, even after a meal rich in carbohydrates. The liver uses some

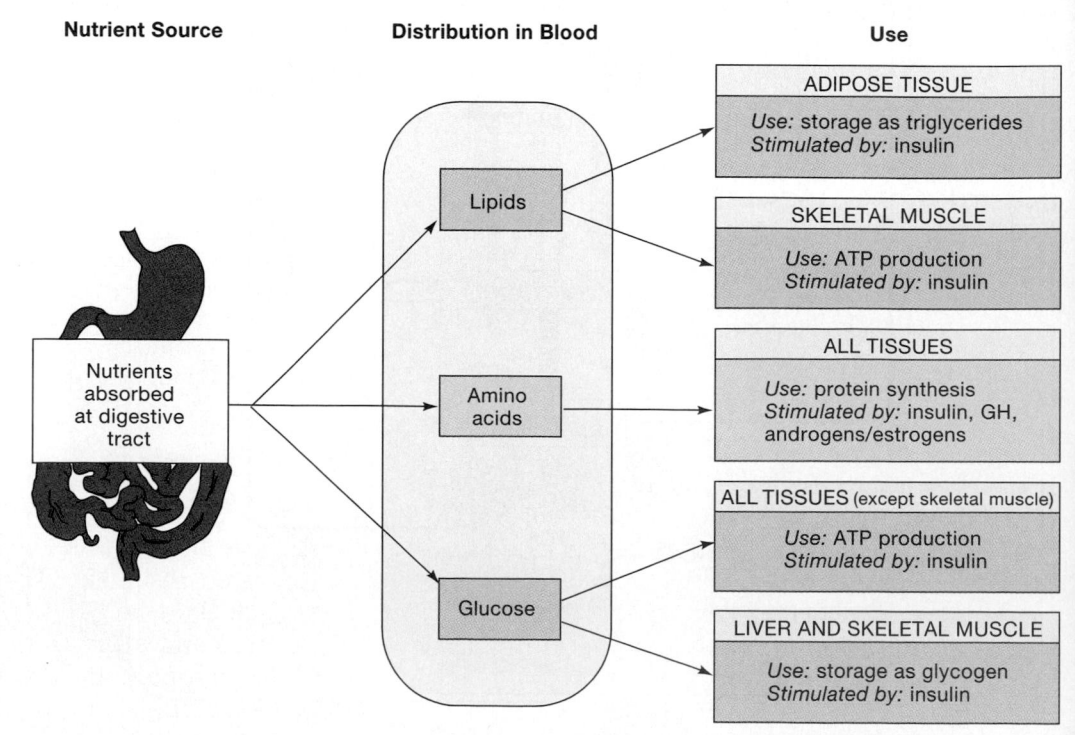

●**FIGURE 25-15**
The Absorptive State. During the absorptive state, the primary metabolic goal is anabolic activity, especially growth and the storage of energy reserves.

TABLE 25-1 Regulatory Hormones and Their Effects on Peripheral Metabolism

State/Hormone	Effect on General Peripheral Tissues	Selective Effects on Target Tissues
Absorptive		
Insulin	Increased glucose uptake and utilization	*Liver:* Glycogenesis *Adipose tissue:* Lipogenesis *Skeletal muscle:* Glycogenesis
Insulin and growth hormone	Increased amino acid uptake and protein synthesis	*Skeletal muscle:* Fatty acid catabolism
Androgens, estrogens	Increased amino acid use in protein synthesis	
Postabsorptive		
Glucagon		*Liver:* Glycogenolysis
Epinephrine		*Liver:* Glycogenolysis *Adipose tissue:* Lipolysis
Glucocorticoids	Decreased use of glucose; increased reliance on ketone bodies and fatty acids	*Liver:* Glycogenolysis *Adipose tissue:* Lipolysis, gluconeogenesis *Skeletal muscle:* Glycogenolysis, protein breakdown, amino acid release
Growth hormone	Complements effects of glucocorticoids	

of the absorbed glucose to generate the ATP required to perform synthetic operations, such as glycogen formation, plasma protein synthesis, or the manufacture of proenzymes. Glycogenesis (glycogen formation) continues until glycogen accounts for about 5 percent of the total liver weight. If excess glucose still remains in the circulation, the hepatocytes use glucose to synthesize triglycerides. Although small quantities of lipids are normally stored in the liver, most of the synthesized triglycerides are bound to transport proteins and released into the bloodstream as VLDLs. Peripheral tissues, primarily adipose tissues, then absorb these lipids for storage.

The liver does not control circulating levels of amino acids as precisely as it does glucose concentrations. Plasma amino acid levels normally range between 35 and 65 mg/dl, but they may become elevated after a protein-rich meal. The absorbed amino acids are used to support the synthesis of proteins, including plasma proteins and the proenzymes of the clotting system. Liver cells can also synthesize many amino acids, and an amino acid present in abundance may be converted to another, less common type and released into the circulation.

Most of the lipids absorbed by the digestive tract do not reach the liver. Triglycerides, cholesterol, and large fatty acids reach the general venous circulation in chylomicrons that are transported in the thoracic duct. Most of these lipids will be absorbed by other tissues, and relatively few will reach the liver.

Adipose Tissue

During the absorptive state, adipocytes remove fatty acids and glycerol from the circulation. Removal of lipids from the blood continues for 4–6 hours after you have eaten a fatty meal. Over this period, the presence of chylomicrons may give the plasma a milky appearance, a characteristic called **lipemia** (lip-Ē-mē-uh).

Adipocytes are particularly active in absorbing these lipids and in synthesizing new triglycerides for storage. At normal blood glucose concentrations, any glucose entering these cells will be catabolized to provide the energy needed for lipogenesis (lipid synthesis). Adipocytes also absorb amino acids as required for protein synthesis. Although these cells can use glucose or amino acids to manufacture triglycerides, they do so only if circulating concentrations are unusually high.

If on a daily basis you take in more nutrients during the absorptive state than you catabolize during the postabsorptive state, the fat deposits in your adipose tissue will enlarge. Most of the increase represents an increase in the size of individual adipocytes. An increase in the total number of adipocytes does not ordinarily occur, except in children before puberty and in extremely obese adults.

OBESITY A useful definition of *obesity* is 20 percent over ideal weight, because this is the point at which serious health risks appear. On the basis of that criterion, 20–30 percent of men and 30–40 percent of women in the United States can be considered obese. Simply stated, obese individuals are taking in more food energy than they are using. Unfortunately, there is very little agreement as to the underlying cause of this situation. The two major categories of obesity, *regulatory obesity* and *metabolic obesity*, are detailed in the *Applications Manual*. **AM** *Obesity*

Skeletal Muscle, Neural Tissue, and Other Peripheral Tissues

When blood glucose and amino acid concentrations are elevated, insulin is released from the pancreatic islets, and all the body's tissues increase their rates of absorption and utilization. Glucose molecules are catabolized for energy, and the amino acids are used to build proteins.

Glucose is normally retained in the body, because the kidneys prevent the loss of glucose molecules in the urine. The kidneys' ability to conserve glucose breaks down only when blood glucose concentrations are extraordinarily high, somewhere in excess of 180 mg/dl. The amino acid content of the urine is not as carefully regulated; amino acids commonly appear in the urine after a protein-rich meal.

When blood glucose levels are elevated, most cells ignore the circulating lipids, and the adipocytes have little competition. In resting skeletal muscles, a significant portion of the metabolic demand is met through the catabolism of fatty acids. Glucose molecules are used to build glycogen reserves, which may account for 0.5–1 percent of the weight of each muscle fiber.

The Postabsorptive State

The **postabsorptive state** is the period when nutrient absorption is not under way and your body must rely on internal energy reserves to continue meeting its energy demands. You spend roughly 12 hours each day in the postabsorptive state, although a person who is skipping meals can extend it considerably. Metabolic activity in the postabsorptive state is focused on the mobilization of energy reserves and the maintenance of normal blood glucose levels. These activities are coordinated by several hormones, including glucagon, epinephrine, the glucocorticoids, and growth hormone (Table 25-1).

The metabolic reserves of a typical 70-kg (154-lb) individual include carbohydrates, lipids, and proteins (Figure 25-16a●). Due to its high energy content, the adipose tissue represents a disproportionate percentage of the total reserve in the form of triglycerides. Most of the available protein reserve is located in the contractile proteins of skeletal muscle. Carbohydrate reserves are relatively small and sufficient for only a few hours or, at most, overnight.

We shall now examine the events under way in specific tissues during this period, as diagrammed in Figure 25-16b●.

The Liver

As the absorptive state ends, your intestinal cells stop providing glucose to the portal circulation. At first, the peripheral tissues continue to remove glucose from the blood, and blood glucose levels begin to decline. The liver responds by reducing its synthetic activities. When plasma concentrations fall below 80 mg/dl, liver cells begin breaking down glycogen reserves and releasing glucose into the circulation. This glycogenolysis occurs in response to a rise in circulating levels of glucagon and epinephrine. The liver contains 75–100 g of glycogen that is readily available, and this reserve is adequate to maintain blood glucose levels for about 4 hours.

As glycogen reserves decline and plasma glucose levels fall to about 70 mg/dl, liver cells begin to manufacture glucose in an attempt to stabilize blood glucose concentrations. The shift from glycogenolysis to gluconeogenesis occurs under stimulation by *glucocorticoids,* steroid hormones from the adrenal cortex.

GLUCONEOGENESIS Through gluconeogenesis, liver cells synthesize glucose molecules from smaller carbon fragments. In effect, any carbon fragment that can be converted to pyruvic acid or one of the three-carbon compounds involved in the cytoplasmic reactions of glycolysis can be used to synthesize glucose. (We discussed the conversion of lactic acid to glucose in the liver in Chapter 10. (∞ *[p. 301]*) With glucose already in short supply, lipids and amino acids must be catabolized to provide the ATP molecules needed for these synthetic operations.

UTILIZATION OF LIPIDS During the postabsorptive state, your liver absorbs fatty acids and glycerol from the blood. The three-carbon glycerol molecules are converted to glucose. Fatty acids are broken down through beta-oxidation to produce large quantities of acetyl-CoA. However, because the enzymatic reaction that converts pyruvic acid to acetyl-CoA cannot be reversed, acetyl-CoA cannot be used to synthesize glucose. Instead,

- Some of the acetyl-CoA molecules deliver their two-carbon acetyl fragments to the TCA cycle, where they are broken down as we described previously. The ATP generated can then be used to support gluconeogenesis.

- In addition, some of the molecules of acetyl-CoA are converted to special compounds known as **ketone bodies,** which can be utilized by peripheral tissues. Ketone bodies are organic acids that are also produced during the catabolism of amino acids.

UTILIZATION OF AMINO ACIDS Before an amino acid can be used for either gluconeogenesis or energy production via the TCA cycle, the amino group ($-NH_2$) must be removed. The structure of the remaining carbon chain determines its subsequent fate. After deamination, some amino acids can be converted to molecules of pyruvic acid or to one of the intermediary molecules of the TCA cycle. These amino acids, which can be used for gluconeogenesis, are known as the **glucogenic amino acids.** Other amino acids can be converted only to acetyl-CoA and must be either broken down further or converted to ketone bodies. These are called the **ketogenic amino acids.**

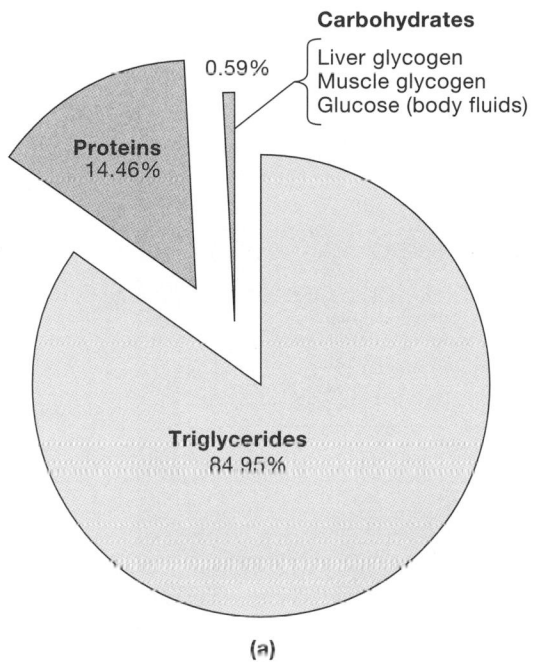

Carbohydrates
Liver glycogen
Muscle glycogen
Glucose (body fluids)

0.59%

Proteins
14.46%

Triglycerides
84.95%

(a)

Most of the essential amino acids are ketogenic. The metabolic fates of glucogenic and ketogenic amino acids are diagrammed in Figure 25-17•.

In the liver, glucogenic and ketogenic amino acids are broken down, and the ammonia generated by the deamination reactions is converted to urea. This relatively harmless, water-soluble compound is later excreted in the urine. The urea concentration in the blood rises during the postabsorptive period, because the rate of amino acid catabolism increases.

KETONE BODIES During the postabsorptive state, liver cells conserve glucose and break down lipids and amino acids. Both lipid catabolism and amino acid catabolism generate acetyl-CoA. As the concentration of acetyl-CoA rises, ketone bodies begin to form. There are three ketone bodies: (1) **aceto-acetate** (as-ē-tō-AS-e-tāt), (2) **acetone** (AS-e-tōn), and (3) **betahydroxybutyrate** (bā-ta-hī-droks-ē-BŪ-te-rāt). Liver cells do not metabolize any of the ketone bodies, and these compounds diffuse through

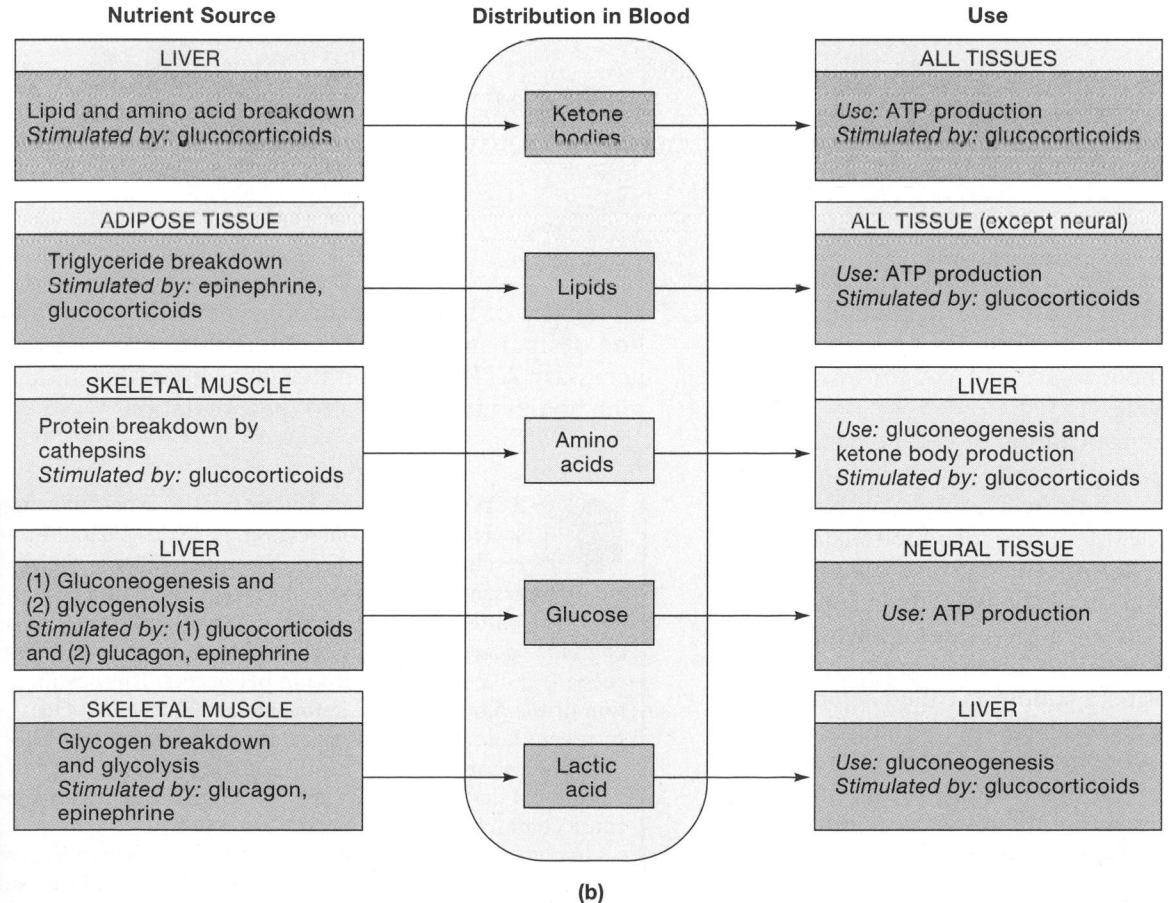

(b)

•***FIGURE 25-16*** **The Postabsorptive State.** **(a)** The distribution of the estimated metabolic reserves of a 70-kg individual. **(b)** In the postabsorptive state, energy reserves are mobilized, and peripheral tissues (except neural tissues) shift from glucose catabolism to fatty acid or ketone body catabolism to obtain energy.

•FIGURE 25-17 Routes of Amino
Acid Catabolism

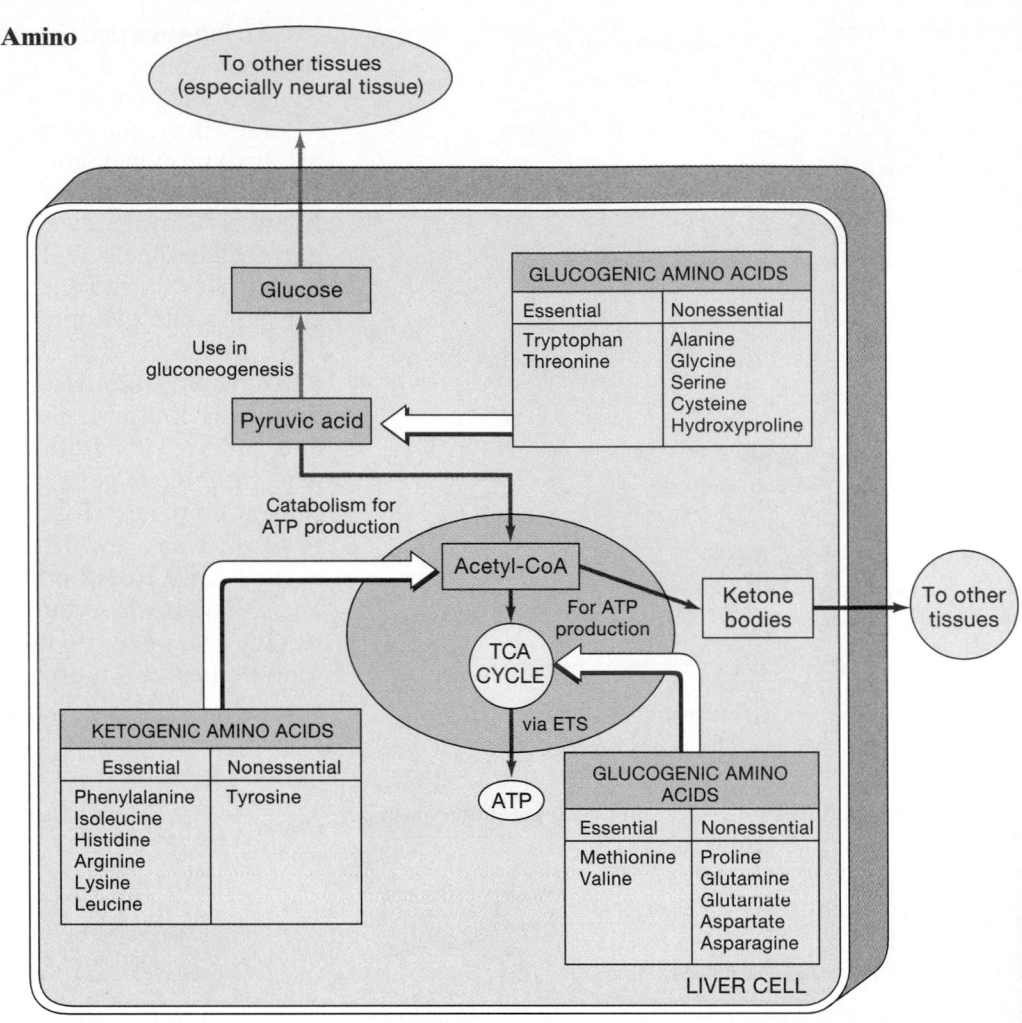

the cytoplasm and into the general circulation. Cells
in peripheral tissues then absorb the ketone bodies
and reconvert them to acetyl-CoA for introduction
into the TCA cycle.

The normal concentration of ketone bodies in the
blood is about 30 mg/dl, and very few of these compounds appear in the urine. During even a brief period of fasting, the increased production of ketone
bodies results in **ketosis** (kē-TŌ-sis), a high concentration of ketone bodies in body fluids. In ketosis, there
is an elevated concentration of ketone bodies in the
blood, a condition called **ketonemia** (kē-tō-NĒ-mē-
uh), and in the urine, a condition called **ketonuria** (kē-
tō-NOO-rē-uh). Ketonemia and ketonuria are clear
indications that protein and lipid catabolism are under
way. Acetone, which diffuses out of the pulmonary
capillaries and into the alveoli very readily, may be
smelled on the individual's breath.

SUMMARY In summary, during the postabsorptive
state your liver attempts to stabilize blood glucose concentrations, first by the breakdown of glycogen reserves and later by gluconeogenesis. Over the

remainder of the postabsorptive state, the combination of lipid and amino acid catabolism provides the
necessary ATP and generates large quantities of ketone bodies that diffuse into the circulation.

KETOACIDOSIS A ketone body is an acid that dissociates in solution, releasing a hydrogen ion. As a
result, the appearance of ketone bodies in the circulation presents a threat to the plasma pH that must be controlled by buffers. During prolonged starvation, ketone levels
continue to rise. Eventually, the buffering capacities are exceeded, and a dangerous drop in pH occurs. This acidification of the blood is called **ketoacidosis** (kē-tō-as-i-DŌ-sis).
In severe ketoacidosis, the circulating concentration of ketone bodies can reach 200 mg/dl, and the pH may fall below
7.05. A pH that low can disrupt normal tissue activities and
cause coma, cardiac arrhythmias, and death.

In *diabetes mellitus,* most peripheral tissues cannot utilize
glucose, because they lack insulin. ∞ *[p. 623]* Under these
circumstances, cells survive by catabolizing lipids and proteins. The result is the production of large numbers of ketone
bodies. This condition leads to *diabetic ketoacidosis,* the most
common and life-threatening form of ketoacidosis.

Adipose Tissue

Adipose tissue contains a tremendous storehouse of energy in the form of triglycerides. Fat accounts for approximately 15 percent of the body weight of the average individual—enough to provide a 1–2 month reserve of ATP. Although some areas, including the eyelids, nose, and the backs of the hands and feet, rarely contain adipose tissue, other regions are preferential sites of deposition. Typically, an individual's adipose tissue is distributed among the hypodermis (50 percent), the greater omentum (10–15 percent), between muscles (5–8 percent), and packed around the kidneys (12 percent) and reproductive organs (15–20 percent).

As blood glucose levels decline, the rate of triglyceride synthesis falls. Under the stimulation of epinephrine, glucocorticoids, and growth hormone, the adipocytes soon begin breaking down their lipid reserves, releasing fatty acids and glycerol into the bloodstream. This process, called *fat mobilization,* continues for the duration of the postabsorptive state. A normal individual retains about 2 months' supply of energy in the triglycerides of adipose tissue. The evolutionary advantages are obvious: The retention of an energy reserve provides a buffer against daily, monthly, and even seasonal changes in the available food supply.

Skeletal Muscle

At the start of the postabsorptive state, skeletal muscles obtain energy by breaking down their glycogen reserves and catabolizing the glucose released. As the concentrations of fatty acids and ketone bodies in the circulation increase, these substrates become increasingly important as an energy source.

Skeletal muscle as a whole contains twice as much glycogen as the liver, but it is distributed throughout your muscular system. These glucose reserves are not directly available to other tissues, because the lack of a key enzyme makes skeletal muscle cells unable to release glucose into the circulation. Once skeletal muscle fibers are metabolizing fatty acids as an energy source, they continue to break down their glycogen reserves and convert the pyruvic acid molecules to lactic acid. Lactic acid then diffuses out of the muscle fibers and into the circulation. However, even if all of the available glycogen reserves in your body were mobilized as glucose or as lactic acid, the energy provided would be only enough to get you through a good night's sleep. If the postabsorptive state continues for an unusually long period, long enough that lipid reserves are being depleted, muscle proteins will be broken down by special enzymes called **cathepsins** (ka-THEP-sinz). The amino acids released by cathepsins diffuse into the blood for use by the liver in gluconeogenesis.

Other Peripheral Tissues

With rising plasma concentrations of lipids and ketone bodies and falling blood glucose levels, peripheral tissues gradually decrease their reliance on glucose. Circulating ketone bodies and fatty acids are absorbed and converted to acetyl-CoA for entry into the TCA cycle.

Neural Tissue

Neurons are unusual in that they continue "business as usual" during the postabsorptive state. Neurons are dependent on a reliable supply of glucose, and the alterations in the activity of liver, adipose tissue, skeletal muscle, and other peripheral tissues are intended to ensure that the supply of glucose to the nervous system continues unaffected, despite daily or even weekly changes in the availability of nutrients. Only after a prolonged period of starvation will neural tissue begin to metabolize ketone bodies and lactic acid molecules as well as glucose.

You should now take the time to compare Figures 25-15, p. 938, and 25-16•, p. 941, until you clearly understand the functional relationships in each state. The adaptations to starvation are exaggerations of the basic themes seen each day in the postabsorptive state. For a discussion of the physiological adaptations to prolonged starvation, consult the *Applications Manual.* **AM** *Adaptations to Starvation*

☑ What process in the liver would you expect to increase after you have eaten a meal that is high in carbohydrates?

☑ Why is there an increase in the amount of urea in the blood during the postabsorptive state?

☑ If a cell accumulates more acetyl-CoA than it can metabolize by way of the TCA cycle, what products are likely to be formed?

DIET AND NUTRITION

The postabsorptive state can be maintained for a considerable period, but homeostasis can be maintained indefinitely only if your digestive tract absorbs fluids, organic substrates, minerals, and vitamins on a regular basis, keeping pace with cellular demands. The absorption of nutrients from food is called **nutrition.**

The individual requirement for each nutrient varies from day to day and from person to person. *Nutritionists* attempt to analyze a diet in terms of its ability to meet the needs of a specific individual. A **balanced diet** contains all the ingredients necessary to maintain homeostasis, including adequate substrates for energy generation, essential amino acids and fatty acids, minerals, and vitamins. In addition, your diet must include enough water to replace losses in urine, feces, and evaporation. A balanced diet prevents **malnutrition,** an unhealthy state resulting from inadequate or excessive intake of one or more nutrients. **AM** *Nutrition and Nutritionists*

Food Groups and Food Pyramids

For several decades, the traditional American method of avoiding malnutrition was to include members of each of the four **basic food groups** in the diet. These were (1) the *dairy group,* (2) the *meat group,* (3) the *vegetable and fruit group,* and (4) the *bread and cereal group.* Each group differs from the others in the typical balance of proteins, carbohydrates, and lipids contained as well as in the amount and identity of vitamins and minerals.

In an attempt to provide more guidance as to how much you should rely on each food source, nutritionists have recently increased the four groups to six by separating out the fruit group and establishing a *fats, oils, and sweets group.* The six groups are now arranged in a *food pyramid* with the bread and cereal group at the bottom (Figure 25-18●). The aim of this display is to emphasize the need to restrict dietary fats, oils, and sugar and to increase your consumption of breads and cereals, which are rich in complex carbohydrates (polysaccharides such as starch).

These are rather artificial groupings at best, and they are downright misleading at worst. For instance, many processed foods belong in a combination of these groups. What is important is that you obtain nutrients in sufficient *quantity* (adequate to meet your energy needs) and *quality* (including essential amino acids, fatty acids, vitamins, and minerals). How these nutrients are packaged is a secondary concern. There is nothing magical about the number six; since 1940, the U.S. government has advocated 11, 7, 4, or 6 food groups at various times. The key is making intelligent choices about what you eat. The wrong choices can lead to malnutrition even if all six groups are represented.

For example, consider the case of the essential amino acids. Your liver cannot synthesize any of these ketogenic amino acids, and you must obtain them from your diet. Some members of the meat and milk groups, such as beef, fish, poultry, eggs, and milk, contain all the essential amino acids in sufficient quantities. They are said to contain **complete proteins.** Many plants also contain adequate *amounts* of protein, but these are **incomplete**

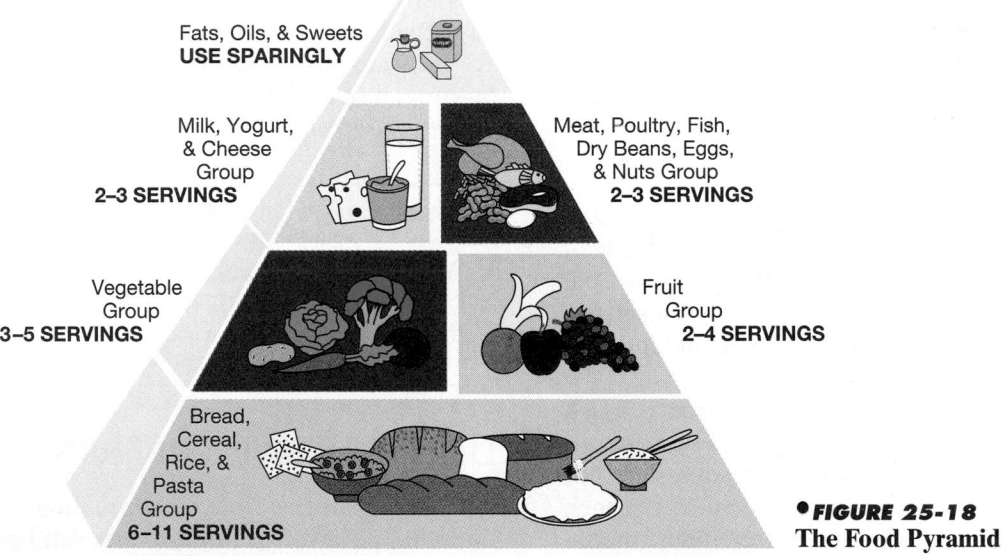

●**FIGURE 25-18**
The Food Pyramid

A Guide to Daily Food Choices

Nutrient Group	Provides	Deficiencies
Fats, oils, sweets	Calories	The majority are deficient in most minerals and vitamins
Milk, yogurt, cheese	Complete proteins; fats; carbohydrates; calcium; potassium; magnesium; sodium; phosphorus; vitamins A, B$_{12}$, pantothenic acid, thiamine, riboflavin	Dietary fiber, vitamin C
Meat, poultry, fish, dry beans, eggs, nuts	Complete proteins; fats; potassium; phosphorus; iron; zinc; vitamins E, thiamine, B$_6$	Carbohydrates, dietary fiber, several vitamins
Fruits	Carbohydrates; vitamins A, C, E, folacin; dietary fiber; potassium	Many are low in fats, calories, and protein
Vegetables	Carbohydrates; vitamins A, C, E, folacin; dietary fiber; potassium	Many are low in fats, calories, and protein
Bread, cereal, rice, pasta	Carbohydrates; vitamins E, thiamine, niacin, folacin; calcium; phosphorus; iron; sodium; dietary fiber	Fats

proteins, which are deficient in one or more of the essential amino acids. Vegetarians, who restrict themselves to the fruit and vegetable groups (with or without the bread and cereal group), must become adept at juggling the constituents of their meals to include a combination of ingredients that will meet all their amino acid requirements. Even with a proper balance of amino acids, vegetarians face a significant problem, because vitamin B_{12} is obtained only from animal products or from fortified cereals or tofu. (Although some health-food products, such as *Spirulina,* are marketed as sources of this vitamin, the B_{12} present is in a form that humans cannot utilize.)

Nitrogen Balance

A variety of important compounds in the body contain nitrogen atoms. These **N compounds** include:

- Amino acids, which are part of the framework of all proteins and protein derivatives, such as glycoproteins and lipoproteins.
- Purines and pyrimidines, the nitrogenous bases of RNA and DNA.
- *Creatine*, important in energy storage in muscle tissue (as creatine phosphate).
- *Porphyrins*, complex ring-shaped molecules that bind metal ions and are essential to the function of hemoglobin, myoglobin, and the cytochromes.

Nitrogenous compounds are essential components of living systems, for they play key roles in determining the direction and rates of intracellular processes, and they form structural proteins. When you consume an adequate diet of fats, carbohydrates, and proteins, you catabolize the fats and carbohydrates and incorporate the amino acids into proteins or convert them to other N compounds. As a result, dietary fats and carbohydrates are often called **protein sparers.**

Despite the importance of nitrogen to these compounds, your body neither stores nitrogen nor maintains reserves of N compounds as it does carbohydrates (glycogen) and lipids (triglycerides). Your body can synthesize the carbon chains of the N compounds, but you must obtain nitrogen atoms either by recycling N compounds already in the body or by absorbing nitrogen from your diet. You are in **nitrogen balance** if the amount of nitrogen you absorb from the diet balances the amount you lose in the urine and feces. This is the normal condition, and it means that the rates of N compound synthesis and breakdown are equivalent.

Growing children, athletes, persons recovering from an illness or injury, and pregnant or lactating women are actively synthesizing N compounds, so they must absorb more nitrogen than they excrete. Such individuals are in a state of **positive nitrogen balance.** When excretion exceeds ingestion, a **negative nitrogen balance** exists.

This is an extremely unsatisfactory situation: The body contains only about a kilogram of nitrogen tied up in N compounds, and a decrease of one-third can be fatal. Even when energy reserves are mobilized, as during starvation, carbohydrates and lipid reserves are broken down first and N compounds are conserved.

Such conservation is relative, not absolute. Proteins are the most abundant organic constituents of living cells, accounting for roughly 20 percent of total body weight. If periods of energy shortage are prolonged, protein catabolism becomes increasingly important as other reserves are exhausted. For example, during the first week of starvation, a person may catabolize 1–1.5 kg of proteins, but in the eighth week the same person may be catabolizing that amount each *day*.

Minerals and Vitamins

Minerals and vitamins are essential components of the diet. Your body cannot synthesize minerals, and your cells can generate only a small quantity of very few vitamins.

Minerals

Minerals are inorganic ions released through the dissociation of electrolytes. Minerals are important for three reasons:

1. *Ions such as sodium and chloride determine the osmolarities of body fluids.* Potassium is important in maintaining the osmolarity of the cytoplasm inside body cells.

2. *Ions in various combinations play major roles in important physiological processes,* as we have discussed. These processes include

   - The maintenance of transmembrane potentials (Chapters 3, 10, and 12).
   - Action potential generation (Chapter 12).
   - Neurotransmitter release (Chapters 10 and 12).
   - Muscle contraction (Chapters 10 and 20).
   - The construction and maintenance of the skeleton (Chapter 6).
   - The transport of respiratory gases (Chapter 23).
   - Buffer systems (Chapters 2 and 27).
   - Fluid absorption (Chapter 24).
   - Waste removal (Chapters 26 and 27).

3. *Ions are essential cofactors in a variety of enzymatic reactions.* For example, the calcium-dependent ATPase in skeletal muscle also requires the presence of magnesium ions, and another ATPase required for the conversion of glucose to pyruvic acid needs both potassium and magnesium ions. Carbonic anhydrase, important in CO_2 transport, buffering systems, and gastric acid secretion, requires the presence of zinc ions. Finally, each component of the electron transport system requires an iron atom, and the terminal cytochrome (a_3) must bind a copper ion as well.

TABLE 25-2 Minerals and Mineral Reserves*

Mineral	Significance	Total Body Content	Primary Route of Excretion	Recommended Daily Intake
Bulk Minerals				
Sodium	Major cation in body fluids; essential for normal membrane function	110 g, primarily in body fluids	Urine, sweat, feces	0.5–1.0 g
Potassium	Major cation in cytoplasm; essential for normal membrane function	140 g, primarily in cytoplasm	Urine	1.9–5.6 g
Chloride	Major anion in body fluids	89 g, primarily in body fluids	Urine, sweat	0.7–1.4 g
Calcium	Essential for normal muscle and neuron function, and normal bone structure	1.36 kg, primarily in skeleton	Urine, feces	0.8–1.2 g
Phosphorus	As phosphate in high-energy compounds, nucleic acids, and bone matrix	744 g, primarily in skeleton	Urine, feces	0.8–1.2 g
Magnesium	Cofactor of enzymes, required for normal membrane functions	29 g (skeleton, 17 g; cytoplasm and body fluids, 12 g)	Urine	0.3–0.4 g
Trace Minerals				
Iron	Component of hemoglobin, myoglobin, cytochromes	3.9 g (1.6 g stored as ferritin or hemosiderin)	Urine (traces)	10–18 mg
Zinc	Cofactor of enzyme systems, notably carbonic anhydrase	2 g	Urine, hair (traces)	15 mg
Copper	Required as cofactor for hemoglobin synthesis	127 mg	Urine, feces (traces)	2–3 mg
Manganese	Cofactor for some enzymes	11 mg	Feces, urine (traces)	2.5–5 mg

*For information on the effects of deficiencies and excesses, see Table 27-2, p. 1017.

The major minerals and a summary of their functional roles are given in Table 25-2. Your body contains significant reserves of several important minerals; these reserves help reduce the effects of variations in dietary supply. The reserves are often small, however, and chronic dietary reductions can lead to a variety of clinical problems. Alternatively, because storage capabilities are limited, a dietary excess of mineral ions can be equally dangerous.

Problems involving iron are particularly common. The body of a normal man contains about 3.5 g of iron in the ionic form Fe^{2+}. Of that amount, 2.5 g is bound to the hemoglobin of circulating red blood cells, and the rest is stored in the liver and bone marrow. In women, the total body iron content averages 2.4 g, with roughly 1.9 g incorporated into red blood cells. Thus a woman's iron reserves consist of only 0.5 g, half that of a typical man. If the diet contains inadequate amounts of iron, women are therefore more likely to develop signs of iron deficiency than are men.

Vitamins

Vitamins can be assigned to either of two groups, depending on their chemical structure and characteristics: (1) *fat-soluble vitamins* or (2) *water-soluble vitamins.*

FAT-SOLUBLE VITAMINS Vitamins A, D, E, and K are the **fat-soluble vitamins.** These vitamins are absorbed primarily from your digestive tract along with the lipid contents of micelles. However, when exposed to sunlight, your skin can synthesize small amounts of vitamin D, and intestinal bacteria produce some vitamin K. There is considerable uncertainty over the mode of action of these vitamins. Vitamin A has long been recognized as a structural component of the visual pigment retinal, but its more general metabolic effects are not well understood. Vitamin D is ultimately converted to calcitriol, which binds to cytoplasmic receptors within the intestinal epithelium and promotes an increase in the rate of intestinal calcium and phosphorus absorption. Vitamin E is thought to stabilize intracellular membranes. Vitamin K is a necessary participant in a reaction essential to the synthesis of several proteins, including at least three of the clotting factors. Current information concerning the fat-soluble vitamins is summarized in Table 25-3.

Because they dissolve in lipids, fat-soluble vitamins normally diffuse into cell membranes and other lipids in the body, including the lipid inclusions in the liver and adipose tissue. Your body therefore contains a significant reserve of these vitamins, and normal metabolic operations can continue for several months after

TABLE 25-3 The Fat-Soluble Vitamins

Vitamin	Significance	Sources	Daily Requirement	Effects of Deficiency	Effects of Excess
A	Maintains epithelia; required for synthesis of visual pigments	Leafy green and yellow vegetables	1 mg	Retarded growth, night blindness, deterioration of epithelial membranes	Liver damage, skin peeling, CNS effects (nausea, anorexia)
D (steroids including cholecalciferol, or D_3)	Required for normal bone growth, calcium and phosphorus absorption at gut and retention at kidneys	Synthesized in skin exposed to sunlight	None*	Rickets, skeletal deterioration	Calcium deposits in many tissues, disrupting functions
E (tocopherols)	Prevents breakdown of vitamin A and fatty acids	Meat, milk, vegetables	12 mg	Anemia, other problems suspected	None reported
K	Essential for liver synthesis of prothrombin and other clotting factors	Vegetables; production by intestinal bacteria	0.7–0.14 mg	Bleeding disorders	Liver dysfunction, jaundice

*Unless sunlight exposure is inadequate for extended periods and alternative sources (fortified milk products) are unavailable.

dietary sources have been cut off. As Table 25-3 points out, *too much* of a vitamin may produce effects just as unpleasant as *too little*. **Hypervitaminosis** (hī-per-vī-ta-min-Ō-sis) occurs when the dietary intake exceeds the abilities to store, utilize, or excrete a particular vitamin. This condition most commonly involves one of the fat-soluble vitamins, because the excess is retained and stored in body lipids.

HYPERVITAMINOSIS "If a little is good, a lot must be better" is a common but dangerously incorrect attitude about vitamins. When the dietary supply of fat-soluble vitamins is excessive, the tissue lipids absorb the additional vitamins. Because these vitamins will later diffuse back into circulation, once the symptoms of hypervitaminosis appear, they are likely to persist.

When absorbed in massive amounts (from ten to thousands of times the recommended daily allowance), fat-soluble vitamins can produce acute symptoms of *vitamin toxicity*. Vitamin A toxicity is the most common condition; it afflicts some children whose parents are overanxious about proper nutrition and vitamins. A single enormous overdose can produce nausea, vomiting, headache, dizziness, lethargy, and even death. Chronic overdose can lead to hair loss, joint pain, hypertension, weight loss, and liver enlargement.

At least 19 cases of vitamin D toxicity were reported in the Boston area during 1992. Symptoms included fatigue, weight loss, and potentially severe damage to the kidneys and cardiovascular system. The problems resulted from drinking milk fortified with vitamin D. Due to problems at one dairy, some of the milk sold had over 230,000 units of vitamin D per quart instead of the usual 400 units per quart. The incident highlighted the need for quality control in production, and care in the consumption, of vitamin supplements.

WATER-SOLUBLE VITAMINS Most of the **water-soluble vitamins** (Table 25-4) are components of coenzymes. For example, NAD is derived from niacin, FAD from vitamin B_2 (riboflavin), and coenzyme A from vitamin B_5 (pantothenic acid).

Water-soluble vitamins are rapidly exchanged between the fluid compartments of the digestive tract and the circulating blood, and excessive amounts are readily excreted in the urine. For this reason, hypervitaminosis involving water-soluble vitamins is relatively uncommon except among individuals taking large doses of vitamin supplements. Only vitamins B_{12} and C are stored in significant quantities, and insufficient intake of other water-soluble vitamins may lead to initial symptoms of vitamin deficiency within a period of days to weeks. The condition that results is termed a **deficiency disease,** or **avitaminosis** (ā-vī-ta-min-Ō-sis). Avitaminosis involving either fat-soluble or water-soluble vitamins can be caused by a variety of factors other than dietary deficiencies. An inability to absorb a vitamin from the digestive tract, inadequate storage, or excessive demand may each produce the same result.

The bacterial inhabitants of our intestines help prevent deficiency diseases by producing small amounts of five of the nine water-soluble vitamins, in addition to fat-soluble vitamin K. Your intestinal epithelium can easily absorb all the water-soluble vitamins except B_{12}. The B_{12} molecule is large, and, as you will recall from Chapter 24, it must be bound to the *intrinsic factor* from the gastric mucosa before absorption can occur. ∞ *[p. 908]*

TABLE 25-4 The Water-Soluble Vitamins

Vitamin	Significance	Sources	Daily Require-ment	Effects of Deficiency	Effects of Excess
B₁ (thiamine)	Coenzyme in decarboxylation reactions	Milk, meat, bread	1.9 mg	Muscle weakness, CNS and cardiovascular problems, including heart disease; called *beriberi*	Hypotension
B₂ (riboflavin)	Part of FMN and FAD	Milk, meat	1.5 mg	Epithelial and mucosal deterioration	Itching, tingling sensations
Niacin (nicotinic acid)	Part of NAD	Meat, bread, potatoes	14.6 mg	CNS, GI, epithelial, and mucosal deterioration; called *pellagra*	Itching, burning sensations; vaso-dilation; death after large dose
B₅ (pantothenic acid)	Part of acetyl-CoA	Milk, meat	4.7 mg	Retarded growth, CNS disturbances	None reported
B₆ (pyridoxine)	Coenzyme in amino acid and lipid metabolism	Meat	1.42 mg	Retarded growth, anemia, convulsions, epithelial changes	CNS alterations, perhaps fatal
Folacin (folic acid)	Coenzyme in amino acid and nucleic acid metabolism	Vegetables, cereal, bread	0.1 mg	Retarded growth, anemia, gastrointestinal disorders, developmental abormalities*	Few noted except at massive doses
B₁₂ (cobalamin)	Coenzyme in nucleic acid metabolism	Milk, meat	4.5 μg	Impaired RBC production, causing *pernicious anemia*	Polycythemia
Biotin	Coenzyme in decarboxylation reactions	Eggs, meat, vegetables	0.1–0.2 mg	Fatigue, muscular pain, nausea, dermatitis	None reported
C (ascorbic acid)	Coenzyme; delivers hydrogen ions, antioxidant	Citrus fruits	60 mg	Epithelial and mucosal deterioration; called *scurvy*	Kidney stones

*Folic acid deficiency during pregnancy can cause neural tube defects (NTDs). ∞ *[p. 436]*

Diet and Disease

Diet has a profound influence on general health. We have already considered the effects of too many or too few nutrients, above- or below-normal concentrations of minerals, and hypervitaminosis or avitaminosis. More subtle, long-term problems may occur when the diet includes the wrong proportions or combinations of nutrients. The average American diet contains too many calories, and too great a proportion of those calories is provided by lipids. This diet increases the incidence of obesity, heart disease, atherosclerosis, hypertension, and diabetes in the U.S. population. **AM** *Perspectives on Dieting*

☑ Would an athlete in intensive training try to maintain a positive or a negative nitrogen balance?

☑ How would a decrease in the amount of bile salts in the bile affect the amount of vitamin A in the body?

BIOENERGETICS

The study of *bioenergetics* examines the acquisition and use of energy by organisms. When chemical bonds are broken, energy is released. Inside cells, a significant amount of energy may be used to synthesize ATP, but much of it is lost to the environment as heat. The process of **calorimetry** (kal-o-RIM-e-trē) determines the total amount of energy released when the bonds of organic molecules are broken. The unit of measurement is the **calorie** (KAL-o-rē), defined as the amount of energy required to raise the temperature of 1 g of water one degree centigrade. One gram of water is not a very practical measure when you are interested in the metabolic operations that keep a 70-kg human alive, so the **kilocalorie** (KIL-ō-kal-o-rē) (kc), or **Calorie** (with a capital *C*), also known as "large calorie," is used instead. Each Calorie represents the amount of energy needed to raise the temperature of 1 *kilo*gram of water one degree centigrade. The numbers in a dieting guide that give the caloric value of various foods indicate Calories, not calories.

Food and Energy

In living cells, organic molecules are oxidized to carbon dioxide and water. Oxidation also occurs when something burns, and this process can be experimentally controlled. A known amount of material is placed in a

 ## Alcohol: A Risky Diversion

Alcohol production and sales are big business throughout the Western world. Beer commercials on television, billboards advertising various brands of liquor, TV or movie characters enjoying a drink—all demonstrate the significance of alcohol in our society. Most people are unaware of the medical consequences of this cultural fondness for alcohol. Problems with alcohol are usually divided into those stemming from alcohol abuse and those involving alcoholism. The boundary between these conditions is rather hazy. *Alcohol abuse* is the general term for overuse and the resulting behavioral and physical effects of overindulgence. *Alcoholism* is chronic alcohol abuse with the physiological changes associated with addiction to other CNS-active drugs. Alcoholism has received the most attention in recent years, although alcohol abuse—especially when combined with driving an automobile—is also in the limelight.

Consider the following statistics:

- Alcoholism affects more than 10 million people in the United States alone. The lifetime risk of developing alcoholism is estimated at 10 percent.
- Alcoholism is probably the most expensive health problem today, with an annual estimated direct cost of more than $136 billion. Indirect costs, in terms of damage to automobiles, property, and innocent accident victims, are unknown.
- An estimated 25–40 percent of U.S. hospital patients are undergoing treatment related to alcohol consumption. There are approximately 200,000 deaths annually due to alcohol-related medical conditions. Some major clinical conditions are caused almost entirely by alcohol consumption. For example, alcohol is responsible for 60–90 percent of all liver disease in the United States.
- Alcohol affects all physiological systems. Major clinical symptoms of alcoholism include (1) disorientation and confusion (nervous system); (2) ulcers, diarrhea, and cirrhosis (digestive system); (3) cardiac arrhythmias, cardiomyopathy, and anemia (cardiovascular system); (4) depressed sexual drive and testosterone levels (reproductive system); and (5) itching and angiomas (integumentary system).
- The toll on newborn infants has risen steadily since the 1960s as the number of women drinkers has increased. Women consuming 1 ounce of alcohol per day during pregnancy have a higher rate of spontaneous abortion and produce children with lower birthweights than do women who consume no alcohol. Women who drink heavily may bear children with *fetal alcohol syndrome (FAS)*. This condition is marked by characteristic facial abnormalities, a small head, slow growth, and mental retardation.
- Perhaps most disturbing of all, the problem of alcohol abuse is considerably more widespread than alcoholism. Although the medical effects are less well-documented, they are certainly significant.

Several factors interact to produce alcoholism. The primary risk factors are gender (males are more likely to become alcoholics than are females) and a family history of alcoholism. There does appear to be a genetic component; a gene on chromosome 11 has been implicated in some inherited forms. The relative importance of genes versus social environment has been difficult to assess. It is likely that alcohol abuse and alcoholism can result from a variety of factors. Treatment may consist of counseling and behavior modification. To be successful, treatment must include the total avoidance of alcohol. Supporting groups, such as Alcoholics Anonymous, can be very helpful in providing a social framework for abstinence. Use of the drug *disulfiram (Antabuse)* has not proved to be as successful as originally anticipated. Antabuse sensitizes the individual to alcohol so that a drink produces intense nausea; it was anticipated that this would be an effective deterrent. Clinical tests indicated that it could increase the time between drinks but could not prevent drinking altogether.

chamber called a **calorimeter** (kal-o-RIM-e-ter) that is filled with oxygen and surrounded by a known volume of water. Once the material is inside, the chamber is sealed, and the substance is electrically ignited. When it has completely oxidized and only ash remains in the chamber, the number of Calories released can be determined by comparing the water temperatures before and after the test. The energy potential of food is usually expressed in Calories per gram (C/g). The catabolism of lipids entails the release of a considerable amount of energy, roughly 9.46 C/g. The catabolism of carbohydrates or proteins is not as rewarding, because many of the carbon and hydrogen atoms are already bound to oxygen. Their average yields are comparable: 4.18 C/g for carbohydrate and 4.32 C/g for protein. Most foods are mixtures of fats, proteins, and carbohydrates, and the values in a "Calorie counter" vary as a result.

 Eating Disorders

Eating disorders are psychological problems that result in either inadequate or excessive food consumption. The most common conditions are *anorexia nervosa,* characterized by self-induced starvation, and *bulimia,* characterized by feeding binges followed by vomiting, laxative use, or both. Adolescent females account for most cases. These conditions are less common in males, who account for only 5–10 percent of anorexia or bulimia cases. A common thread in the two conditions is an obsessive concern about food and body weight.

According to current estimates, the incidence of **anorexia nervosa** in the United States ranges from 0.4 to 1.5 per 100,000 population. The incidence among Caucasian women ages 12–18 is estimated to be 1 percent. A typical person with this condition is an adolescent Caucasian woman whose weight is roughly 30 percent below normal levels. Although underweight, she is convinced that she is still too fat and refuses to eat normal amounts of food.

The psychological factors responsible for anorexia are complex. Young women with this condition tend to be high achievers who are attempting to reach an "ideal" weight that will be envied and admired and thereby achieve a sense of security and accomplishment. The factors thus tend to be a combination of their view of society ("thin is desirable or demanded"), their view of themselves ("I am not yet thin enough"), and a desire to be able to control their fate ("I can decide when to eat"). Female models, figure skaters, gymnasts, and theater and arts majors of any age may feel forced to drop weight to remain competitive. The few male anorexics diagnosed typically face comparable stresses. They tend to be athletes, such as jockeys or wrestlers, who need to maintain a minimal weight to succeed in their careers.

Young anorexic women will often continue to starve themselves down to a weight of 30–35 kg (66–77 lb). Dry skin, peripheral edema, an abnormally low heart rate and blood pressure, a reduction in bone and muscle mass, and a cessation of menstrual cycles are relatively common symptoms. Some of the changes, especially in bone mass, may be permanent. Treatment is difficult, and only 50–60 percent of anorexics who regain normal weight stay there for 5 years or more. Death rates from severe anorexia nervosa range from 10 to 15 percent.

Bulimia is more common than anorexia. In this condition, the individual goes on an "eating binge" that may involve a meal that lasts 1–2 hours and may include 20,000 or more calories. The meal is followed by induced vomiting, commonly accompanied by the use of laxatives (to promote movement of the material through the digestive tract) and diuretics (drugs that promote fluid loss in the urine). These often expensive binges may occur several times each week, separated by periods of either normal eating or fasting.

Bulimia generally involves women of the same age group as anorexia nervosa. The actual incidence is difficult to determine; published estimates for young college-age women range from 5 to 18 percent. However, many bulimics are not diagnosed until they are age 30–40. Because they ingest plenty of food, bulimics may have normal body weight, and therefore the condition is harder to diagnose than anorexia nervosa. The health risks of bulimia result from (1) cumulative damage to the stomach, esophagus, oral cavity, and teeth by repeated exposure to stomach acids; (2) electrolyte imbalances resulting from the loss of sodium and potassium ions in the gastric juices, diarrhea, and urine; (3) edema; and (4) cardiac arrhythmias.

The underlying cause of bulimia remains uncertain. Societal factors are certainly involved, but bulimia has also been strongly correlated with depression and with elevated levels of ADH in the cerebrospinal fluid.

Metabolic Rate

Clinicians can examine your metabolic state and determine how many Calories you are utilizing. The result can be expressed as Calories per hour, Calories per day, or Calories per unit of body weight per day; what is actually measured is the sum of all the varied anabolic and catabolic processes occurring in your body. This value represents your **metabolic rate** at that time. The metabolic rate will change accord-ing to the activity under way—sprinting and sleeping measurements are quite different. In an attempt to reduce the variations, clinicians standardize the testing conditions so as to determine the **basal metabolic rate (BMR).** Ideally, the BMR would represent the minimum, resting energy expenditures of an awake, alert person.

A direct method of determining the BMR simply monitors respiratory activity, for in resting subjects energy utilization is proportional to oxygen consumption. If we assume that average amounts of carbohydrates, lipids, and proteins are being catabolized, the ratio gives 4.825 Calories per liter of oxygen consumed.

An average individual has a BMR of 70 C per hour or about 1680 C per day. Although the test conditions are standardized, there are many uncontrollable factors that can influence the BMR. These include age, gender, physical condition, body weight, and genetic differences such as variations among ethnic groups.

Because the BMR is technically difficult to measure, and because circulating thyroid hormone levels have a profound effect on the BMR, clinicians usually monitor the concentration of thyroid hormones rather than the actual metabolic rate. The results are then compared with normal values, to obtain an index of metabolic activity. One such test, the **T₄ assay,** measures the amount of thyroxine in the blood.

Daily energy expenditures for a given individual vary widely depending on the activities undertaken. For example, a person leading a sedentary life may have near-basal energy demands, but a single hour of swimming can increase the daily caloric requirements by 500 C or more. If your daily energy intake exceeds your total energy demands, you will store the excess, primarily as triglycerides in adipose tissue. If your daily caloric expenditures exceed your dietary supply, there will be a net reduction in your body's energy reserves and a corresponding loss in weight. This relationship accounts for the significance of calorie counting and exercise in a weight-control program.

The control of appetite is poorly understood. Stretch receptors along the digestive tract, especially in the stomach, do play a role, but other factors are probably more important. Social factors, psychological pressures, and dietary habits are all important. There is also evidence that complex hormonal stimuli interact to affect appetite. For example, the hormones *cholecystokinin* and *ACTH* will suppress appetite. The hormone *leptin,* released by adipose tissues, also plays a role. During the absorptive state, adipose tissues release leptin into the circulation as they synthesize triglycerides. Leptin binds to neurons in the CNS that deal with emotion and the control of appetite. The result is a sense of satiation and the suppression of appetite.

Thermoregulation

The BMR estimates the rate of energy use by the body. The energy not captured and harnessed by living cells is released as heat. This heat loss serves an important homeostatic purpose. Humans are subject to vast changes in environmental temperatures, but our complex biochemical systems have a major limitation: The enzyme systems will operate over only a relatively narrow range of temperatures. Our bodies have anatomical and physiological mechanisms that keep body temperatures within acceptable limits, regardless of the environmental conditions. This homeostatic process is called **thermoregulation.** Failure to control body temperature can result in a series of physiological changes. For example, a body temperature below 36°C (97°F) or above 40°C (104°F) can cause disorientation, and a temperature above 42°C (108°F) can cause convulsions and permanent cell damage.

We are continuously producing heat as a byproduct of metabolism; that heat must be lost to the environment at the same rate if body temperature is to remain constant. When the environmental conditions vary from "ideal," becoming too warm or too cold, the body must control the gains or losses to maintain homeostasis.

Mechanisms of Heat Transfer

Heat exchange with the environment involves four basic processes: *radiation, conduction, convection,* and *evaporation.* These mechanisms are illustrated in Figure 25-19●.

RADIATION Warm objects lose heat energy as infrared **radiation.** When you feel the heat from the sun, you are experiencing radiant heat. Your body loses heat the same way but in proportionately smaller amounts. Over half the heat you lose is attributable to radiation. The exact amount varies with both body temperature and skin temperature.

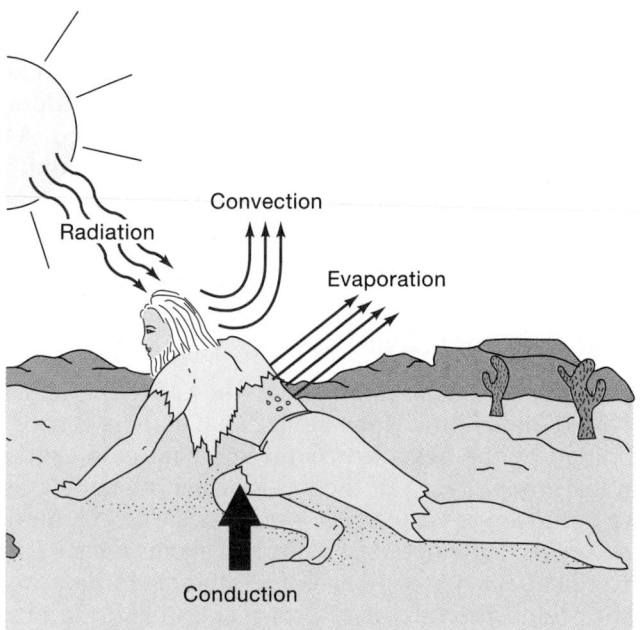

●**FIGURE 25-19** **Routes of Heat Gain and Loss**

CONDUCTION **Conduction** refers to the direct transfer of energy through physical contact. When you arrive in an air-conditioned classroom and sit down on a cold plastic chair, you are immediately aware of this process. Conduction is generally not an effective mechanism for you to gain or lose heat.

CONVECTION **Convection** is the result of conductive heat loss to the air that overlies the surface of the body. Warm air rises, because it is lighter than cool air. As your body conducts heat to the air next to your skin, that air warms and rises, moving away from the skin surface. Cooler air replaces it; as it in turn becomes warmed, the pattern repeats. Convection accounts for roughly 15 percent of your heat loss; convection is insignificant as a mechanism for heat gain.

EVAPORATION When water evaporates, it changes from a liquid to a vapor. **Evaporation** absorbs energy, roughly 0.58 C per gram of water evaporated. The rate of evaporation occurring at your skin is highly variable. Each hour, 20–25 ml of water crosses epithelia and evaporates from the alveolar surfaces and the surface of the skin. This insensible perspiration remains relatively constant; at rest it accounts for roughly one-fifth of your average heat loss. The sweat glands responsible for sensible perspiration have a tremendous scope of activity, ranging from virtual inactivity to secretory rates of 2–4 liters per hour.

Mechanisms of Heat Distribution

Biochemical reactions produce heat, and that heat is retained in the water that accounts for nearly 66 percent of your body weight. Water is an excellent conductor of heat, so the heat produced in one region of the body is rapidly distributed by diffusion as well as via the circulation. The more biochemical reactions are under way, the more heat will be produced. As a result, heat production in the body can be quite variable. The maintenance of a constant body temperature requires a finely tuned and adaptable homeostatic mechanism.

Heat Gain and Heat Loss

Heat loss and heat gain involve the coordinated activity of many different systems. That activity is coordinated by the **heat-loss center** and **heat-gain center** in the *preoptic area* of the anterior hypothalamus, as we discussed in Chapter 14. ∞ *[p. 468]* These centers modify the activities of other hypothalamic nuclei. The overall effect is to control temperature by influencing two events: the rate of heat production and the rate of heat loss to the environment. These may be further supported by behavioral modifications.

MECHANISMS FOR INCREASING HEAT LOSS When the temperature at the preoptic nucleus exceeds its thermostat setting, the heat-loss center is stimulated. Stimulation of this center has three major effects:

1. Inhibition of the vasomotor center causes peripheral vasodilation, and warm blood flows to the surface of the body. The skin takes on a reddish color, skin temperatures rise, and radiational and convective losses increase.

2. As integumentary blood flow increases, sweat glands are stimulated to increase their secretory output. The perspiration flows across the body surface, and evaporative losses accelerate. A maximal secretion rate would, if it were completely evaporated, remove 2320 C per hour.

3. The respiratory centers are stimulated, and the depth of respiration increases. Often the individual begins respiring through an open mouth rather than through the nasal passageways, increasing evaporative losses through the lungs.

HEAT EXHAUSTION AND HEAT STROKE *Heat exhaustion* and *heat stroke* represent malfunctions of the thermoregulatory system. In **heat exhaustion**, also known as *heat prostration*, the individual experiences difficulties with the maintenance of blood volume. The heat-loss center is stimulating sweat glands, whose secretions moisten the surface of the skin to provide evaporative cooling. As fluid losses mount, blood volume decreases. The resultant decline in blood pressure is not countered by peripheral vasoconstriction, because the heat-loss center is actively stimulating peripheral vasodilation. As blood flow to the brain declines, headache, nausea, and eventual collapse follow. Treatment is straightforward: Provide fluids, salts, and a cooler environment.

Heat stroke is more serious and may follow an untreated case of heat exhaustion. Predisposing factors include any pre-existing condition, such as heart disease or diabetes, that affects peripheral circulation. The thermoregulatory center ceases to function, the sweat glands are inactive, and the skin becomes hot and dry. Unless the situation is recognized in time, body temperature may climb to 41°–45°C (106°–113°F). Temperatures in this range will quickly disrupt a variety of vital physiological systems and destroy brain, liver, skeletal muscle, and kidney cells. Proper treatment involves lowering body temperature as rapidly as possible.

MECHANISMS FOR PROMOTING HEAT GAIN The function of the heat-gain center of the brain is to prevent **hypothermia** (hī-pō-THER-mē-uh), or below-normal body temperature. When the temperature at the preoptic nucleus drops below acceptable levels, the heat-loss center is inhibited, and the heat-gain center is activated.

Heat Conservation. The sympathetic vasomotor center decreases blood flow to the dermis of the skin, thus reducing radiational, convective, and conductive losses. The skin cools, and with the circulation

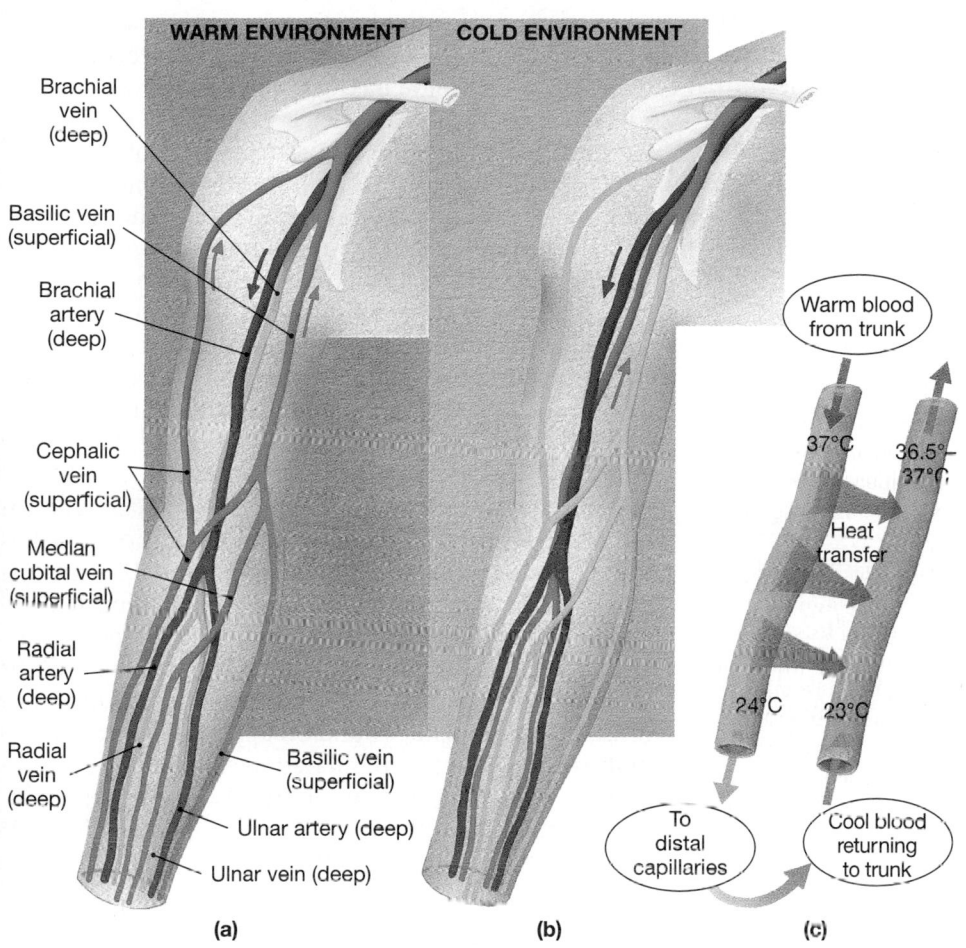

WARM ENVIRONMENT

COLD ENVIRONMENT

Brachial
vein
(deep)

Basilic vein
(superficial)

Brachial
artery
(deep)

Cephalic
vein
(superficial)

Median
cubital vein
(superficial)

Radial
artery
(deep)

Radial
vein
(deep)

Basilic vein
(superficial)

Ulnar artery (deep)

Ulnar vein (deep)

(a)

(b)

Warm blood
from trunk

37°C

36.5°-
37°C

Heat
transfer

24°C

23°C

To
distal
capillaries

Cool blood
returning
to trunk

(c)

•**FIGURE 25-20**
Countercurrent Heat Exchange. (a) Circulation through the blood vessels of the forearm in a warm environment. Blood enters the limb in a deep artery and returns to the trunk in a network of superficial veins. These veins radiate heat into the environment through the overlying skin. **(b)** Circulation through the blood vessels of the forearm in a cold environment. Blood now returns to the trunk via a network of deep veins that flow around the artery. The amount of heat loss is reduced, as indicated in (c). **(c)** Countercurrent heat exchange occurs as heat radiates from the warm arterial blood into the cooler venous blood flowing in the opposite direction. By the time the arterial blood reaches distal capillaries, where most of the heat loss to the environment occurs, it is already 13°C cooler than it was when it left the trunk. This mechanism reduces the rate of heat loss while conserving body heat. In effect, the countercurrent exchange traps heat near the trunk.

restricted, it may take on a bluish or pale coloration. The epithelial cells are not damaged, because they can tolerate extended periods at temperatures as low as 25°C (77°F) or as high as 49°C (120°F).

In addition, blood returning from the extremities is shunted into a network of deep veins that we introduced in Chapter 21. ∞ *[p. 751]* Figure 25-20• shows the changes in circulation that occur. In warm weather, blood flows in a superficial venous network (Figure 25-20a•). In cold weather, blood is diverted to a network of deep veins that lie beneath an insulating layer of subcutaneous fat. This venous network wraps around the deep arteries (Figure 25-20b•). Heat diffuses from the warm blood flowing outward to the limbs into the cooler blood returning from the periphery. This arrangement traps the heat close to the body core and restricts heat loss by reducing the temperature gradient between the arterial blood and the outside world. Diffusion between fluids that are moving in opposite directions is called *countercurrent exchange* (Figure 25-20c•). (We shall return to this topic in Chapter 26.)

Heat Generation. The mechanisms available to generate heat can be divided into two broad categories: (1) *shivering thermogenesis* and (2) *nonshiv-*

ering thermogenesis. In **shivering thermogenesis** (ther-mō-JEN-e-sis), a gradual increase in muscle tone increases the energy consumption of skeletal muscle tissue throughout your body. Both agonists and antagonists are involved, and the degree of stimulation varies with the demand.

If the heat-gain center is extremely active, muscle tone increases to the point at which stretch receptor stimulation will produce brief, oscillatory contractions of antagonistic muscles. In other words, you begin to **shiver.** Shivering increases the workload of the muscles and further elevates oxygen and energy consumption. The heat that is produced warms the deep vessels, to which blood has been shunted by the sympathetic vasomotor center. Shivering can elevate body temperature quite effectively because it increases the rate of heat generation by as much as 400 percent.

In **nonshivering thermogenesis,** hormones are released that increase the metabolic activity of all tissues:

- The heat-gain center stimulates the adrenal medullae, via the sympathetic division of the ANS, and epinephrine is released. Epinephrine increases the rates of glycogenolysis in liver and skeletal muscle and the metabolic rate of most tissues. These effects are immediate.

- The preoptic nucleus controls the production of TRH by the hypothalamus. In children, when temperatures are below normal, additional TRH is released, stimulating the release of TSH by the anterior pituitary gland. The thyroid gland responds by increasing the rate of thyroxine release into the blood. Thyroxine increases not only the rate of carbohydrate catabolism but also the rate of catabolism of all other nutrients. These effects develop gradually, over a period of days to weeks.

INDUCED HYPOTHERMIA   Hypothermia may be intentionally produced during surgery to reduce the metabolic rate of a particular organ or of the entire body. In controlled hypothermia, the individual is first anesthetized to prevent the shivering that would otherwise fight the process.

During open-heart surgery the body is typically cooled to 25°–32°C (79°–89°F). This cooling reduces the metabolic demands of the body, which will be receiving blood from an external pump or oxygenator. The heart must be stopped completely during the operation, and it cannot be well supplied with blood over this period. So the heart is exposed to an *arresting solution* at 0°–4°C (32°–39°F) and maintained at a temperature below 15°C (60°F) for the duration of the operation. At these temperatures, the cardiac muscle can tolerate several hours of ischemia without damage.

When cardiac surgery is performed on infants, a deep hypothermia may be produced by cooling the entire body to temperatures as low as 11°C (52°F) for an hour or more. In effect, this procedure duplicates the conditions experienced by the accidental drowning victims discussed in the *Applications Manual*.   [AM] *Accidental Hypothermia*

Sources of Individual Variation

The timing of thermoregulatory responses may differ from individual to individual. A person may undergo **acclimatization** (a-klī-ma-ti-ZĀ-shun), making physiological adjustment to a particular environment over time. For example, natives of Tierra del Fuego once lived naked in the snow, but Hawaii residents unpack their sweaters when the temperature drops below 22°C (72°F).

Another interesting source of variation is body size. Although heat *production* occurs within the mass of the body, heat *loss* must occur across a body surface. The relationship between heat production and heat loss is thus linked to what is called the *surface-to-volume ratio*. As an object (or person) gets larger, its surface area increases at a much slower rate than does its total volume.

For example, consider a cube that measures 1 meter in width, depth, and height. Each of its six sides has a surface area of 1 m^2 (1×1), so the cube has a total surface area of 6 m^2. Its volume is 1 m^3 ($1 \times 1 \times 1$), and it has a surface-to-volume ratio of 6 to 1. Now consider a cube that is twice as big, measuring 2 meters in each dimension. Each side has 4 m^2 of surface area (2×2), so the cube has a total surface area of 24 m^2. Its volume is 8 m^3 ($2 \times 2 \times 2$), and it has a surface-to-volume ratio of 3 to 1. In doubling the volume, the surface-to-volume ratio was halved. In general, the smaller the object, the larger the surface-to-volume ratio.

This ratio affects thermoregulation, because heat is generated by the "volume" (that is, by internal tissues) and lost at the body surface. Because they have larger surface-to-volume ratios, small individuals lose heat more readily than do large individuals.

THERMOREGULATORY PROBLEMS OF INFANTS
Infants have problems with thermoregulation due to their relatively high surface-to-volume ratios. During embryonic development, temperature regulation is no concern of theirs, as the maternal surroundings are at normal body temperature. At birth, the temperature-regulating mechanisms are not fully functional. With such high surface-to-volume ratios, newborns must be dried quickly and kept bundled up; for those born prematurely, a thermally regulated incubator is required. Infants' body temperatures are also more unstable than those of adults. Their metabolic rates decline when they are sleeping, then rise after arousal.

Infants cannot shiver, but they have a different mechanism for raising body temperature rapidly. The adipose tissue between the shoulder blades, around the neck, and possibly elsewhere in the upper body is histologically and functionally different from most of the adipose tissue in the adult. The tissue is highly vascularized, and the individual adipocytes contain numerous mitochondria. Together these characteristics give the tissue a deep, rich color responsible for the name **brown fat**. ∞ *[p. 124]* The individual adipocytes are innervated by sympathetic autonomic fibers. When these nerves are stimulated, lipolysis accelerates in the adipocytes. The cells do not capture the energy released through fatty acid catabolism, and it radiates into the surrounding tissues as heat. This heat quickly warms the blood that passes through the surrounding network of vessels, and it is then distributed throughout the body. In this way, an infant can accelerate metabolic heat generation by 100 percent very quickly, whereas nonshivering thermogenesis in the adult will raise heat production by only 10–15 percent after a period of weeks.

With increasing age and size, body temperature becomes more stable, and the importance of this thermoregulatory mechanism declines. There is little if any brown fat in the adult; with increased body size, skeletal muscle mass, and insulation, shivering thermogenesis is significantly more effective in elevating body temperature.

THERMOREGULATORY VARIATIONS AMONG ADULTS

Adults of the same body weight may differ in their thermal responses if their weight is distributed differently. For instance, they may have different surface-to-volume ratios; consider two 70-kg individuals, one 2 meters tall, and the other 1.5 meters tall. Which tissues account for their weight is also a factor. Adipose tissue is an excellent insulator, conducting heat at only about one-third the rate of other tissues. As a result, individuals with a more substantial layer of subcutaneous fat may not begin to shiver until long after their more slender companions.

In addition to hormone levels, environmental acclimatization, body weight, age, tissue distribution, and surface-to-volume ratios, our hypothalamic thermostats also affect our thermoregulatory responses. Two otherwise similar individuals may differ in their response to temperature changes because their hypothalamic thermostats are at different settings. There are daily oscillations in body temperature, with temperatures falling 1°–2°C (equivalent to 1.8°–3.6°F) at night and peaking sometime during the day or early evening. Individuals vary in terms of their time of maximum temperature setting, and some have a series of peaks, with an afternoon low. The origin of these patterns is uncertain. It is not the result of daily activity regimens—people who work at night still show their temperature peaks over the same range of times as the rest of the population.

Fevers

An elevated body temperature, or **pyrexia** (pī-REK-sē-uh), may occur for a variety of reasons, not all of them pathological. In young children, transient fevers with no ill effects may result from exercise in warm weather. Similar exercise-related elevations were rarely encountered in adults until running marathons became a national pastime. Temperatures ranging from 39° to 41°C (103° to 106°F) may result, and it is for this reason that competitions are usually held when the air temperature is below 28°C (82°F).

We discussed fevers when we examined nonspecific defenses in Chapter 22. ∞ *[p. 787]* A fever is the maintenance of a body temperature greater than 37.2°C (99°F). Fevers may result from such factors as the following:

- Abnormalities affecting the entire thermoregulatory mechanism, such as heat exhaustion or heat stroke.
- Clinical problems that restrict circulation, such as congestive heart failure.
- Conditions that impair sweat gland activity, such as drug reactions and some skin conditions.
- The resetting of the hypothalamic thermostat by circulating *pyrogens,* most notably interleukin-1.

We can classify fevers as *chronic* or *acute.* The classification and treatment of fevers are discussed in the *Applications Manual.* **AM** *Fevers*

☑ How would the BMR (basal metabolic rate) of a pregnant woman compare with her BMR in the nonpregnant state?

☑ What effect would vasoconstriction of peripheral blood vessels have on body temperature on a hot day?

☑ Why do infants have greater problems with thermoregulation than do adults?

SELECTED CLINICAL TERMINOLOGY

Terms Discussed in This Chapter

avitaminosis (ā-vī-ta-min-Ō-sis): A vitamin deficiency disease. *(p. 947)*

carbohydrate loading: Eating large quantities of carbohydrates in the days preceding an athletic competition to increase endurance. *(p. 928)*

eating disorders: Psychological problems that result in inadequate or excessive food consumption. Examples include anorexia nervosa and bulimia. *(p. 950)*

heat exhaustion: A malfunction of the thermoregulatory system caused by excessive fluid loss in perspiration. *(p. 952)*

heat stroke: A condition in which the thermoregulatory center stops functioning and body temperature rises uncontrollably. *(p. 952)*

hyperuricemia (hī-per-ū-ri-SĒ-mē-uh): Levels of plasma uric acid above 7.4 mg/dl; may result in the condition called *gout. (p. 936)*

hypervitaminosis (hī-per-vī-ta-min-Ō-sis): A disorder caused by ingestion of excessive quantities of one or more vitamins. *(p. 947)*

hypothermia (hī-pō-THER-mē-uh): Below-normal body temperature. *(p. 952 and AM)*

ketoacidosis (kē-to-ās-i-DŌ-sis): Acidification of the blood due to the presence of ketone bodies. *(p. 942 and AM)*

ketonemia (kē-tō-NE-mē-uh): Elevated levels of ketone bodies in the blood. *(p. 942)*

ketonuria (kē-tō-NOO-rē-uh): The presence of ketone bodies in the urine. *(p. 942)*

ketosis (kē-TŌ-sis): Abnormally high concentration of ketone bodies in body fluids. *(p. 942)*

obesity: A body weight more than 20 percent above the ideal weight for a given individual. *(p. 939 and AM)*

phenylketonuria (fen-il-kē-tō-NOO-rē-uh): An inherited metabolic disorder resulting from an inability to convert phenylalanine to tyrosine. *(p. 935 and AM)*

protein deficiency diseases: Nutritional disorders resulting from a lack of essential amino acids. *(p. 936 and AM)*

CHAPTER REVIEW

 On-line resources for this chapter are on our World Wide Web site at: http://www.prenhall.com/martini/fap

STUDY OUTLINE

INTRODUCTION, p. 919

1. Cells in the human body are chemical factories that break down organic substrates to obtain energy.

AN OVERVIEW OF AEROBIC METABOLISM, p. 919

1. In general, cells will break down excess carbohydrates first, then lipids, while conserving amino acids. Only about 40 percent of the energy released through catabolism is captured in ATP; the rest is released as heat. *(Figure 25-1)*
2. Cells synthesize new compounds: (1) to perform structural maintenance or repairs, (2) to support growth, and (3) to produce secretions. *(Figure 25-2)*

CARBOHYDRATE METABOLISM, p. 921

1. Most cells generate ATP and other high-energy compounds through the breakdown of carbohydrates.

Glycolysis, p. 921

2. **Glycolysis** and *aerobic metabolism* provide most of the ATP used by typical cells. Glycogen can be broken down to glucose molecules. In glycolysis, each molecule of glucose yields 2 molecules of pyruvic acid (as pyruvate ions), a net of 2 molecules of ATP, and 2 NADH. *(Figure 25-3)*

Mitochondrial ATP Production, p. 921

3. In the presence of oxygen, the pyruvic acid molecules enter mitochondria, where they are broken down completely in the **tricarboxylic acid (TCA) cycle**. Carbon and oxygen atoms are lost as carbon dioxide; hydrogen atoms are passed to coenzymes, which initiate the oxygen-consuming and ATP-generating reaction **oxidative phosphorylation.** *(Figure 25-4)*
4. **Cytochromes** pass electrons along the respiratory chain of the **electron transport system,** which eventually generates ATP and water as the electrons and hydrogen ions combine with oxygen to form water. *(Figure 25-5)*

Energy Yield of Glycolysis and Cellular Respiration, p. 927

5. For each glucose molecule processed through glycolysis and aerobic metabolism, most cells gain 36 molecules of ATP. *(Figure 25-6)*

Other Catabolic Pathways, p. 928

6. Cells can break down other nutrients to provide substrates for the TCA cycle if supplies of glucose are limited. *(Figure 25-7)*

Gluconeogenesis, p. 929

7. **Gluconeogenesis,** the synthesis of glucose from noncarbohydrate precursors, such as lactic acid, glycerol, or amino acids, enables a liver cell to synthesize glucose molecules when carbohydrate reserves are depleted. **Glycogenesis** is the process of glycogen formation. Glycogen is an important energy reserve when the cell cannot obtain enough glucose from the extracellular fluid. *(Figure 25-8)*

LIPID METABOLISM, p. 930

Lipid Catabolism, p. 930

1. During **lipolysis** (lipid catabolism), lipids are broken down into pieces that can be converted into pyruvic acid or channeled into the TCA cycle.

2. Triglycerides, the most abundant lipids in the body, are split into glycerol and fatty acids. The glycerol enters the glycolytic pathways, and the fatty acids enter the mitochondria.
3. **Beta-oxidation** is the breakdown of a fatty acid molecule into two-carbon fragments that can be used in the TCA cycle. The steps of beta-oxidation cannot be reversed, and the body cannot manufacture all the fatty acids needed for normal metabolic operations. *(Figure 25-9)*
4. Lipids cannot provide large amounts of ATP in a short amount of time. However, cells can shift to lipid-based energy production when glucose reserves are limited.

Lipid Synthesis, p. 930

5. In **lipogenesis,** the synthesis of lipids, almost any organic substrate can be used to form glycerol. **Essential fatty acids** cannot be synthesized and must be included in the diet. *(Figure 25-10)*

Lipid Transport and Distribution, p. 932

6. Lipids circulate as **free fatty acids (FFA)** (water-soluble lipids that can easily diffuse across cell membranes) and as **lipoproteins** (lipid–protein complexes that contain large glycerides and cholesterol). The largest lipoproteins, chylomicrons, carry absorbed lipids from the intestinal tract to the circulation. All other lipoproteins are derived from the liver and carry lipids to and from various tissues of the body. *(Figure 25-11)*
7. Capillary walls of adipose tissue, skeletal muscle, cardiac muscle, and the liver contain **lipoprotein lipase,** an enzyme that breaks down complex lipids, releasing a mixture of fatty acids and monoglycerides. *(Figure 25-11)*

PROTEIN METABOLISM, p. 933

Amino Acid Catabolism, p. 933

1. If other energy sources are inadequate, mitochondria can break down amino acids in the TCA cycle to generate ATP. In the mitochondria, the amino group may be removed by **transamination** or **deamination,** and the carbon skeleton is converted to one of the compounds involved in glycolysis or oxidative respiration. *(Figure 25-12)*
2. Protein catabolism is an impractical source for quick energy.

Protein Synthesis, p. 935

3. Roughly half the amino acids needed to build proteins can be synthesized. There are 10 **essential amino acids** that must be acquired through the diet. **Amination,** the attachment of an amino acid group to a carbon framework, is an important step in the synthesis of **nonessential amino acids.** *(Figure 25-13)*

NUCLEIC ACID METABOLISM, p. 936

1. DNA in the nucleus is never catabolized for energy.

RNA Catabolism, p. 936

2. RNA molecules are broken down and replaced regularly. They are generally recycled as new nucleic acids, but the nucleotides can be catabolized to simple sugars and nitrogenous bases. In general, nucleic acids do not contribute significantly to the cell's energy reserves.

Nucleic Acid Synthesis, p. 936

3. Most cells synthesize RNA, but DNA synthesis occurs only in cells preparing for mitosis or meiosis.

METABOLIC INTERACTIONS, p. 937

1. No one cell of a human can perform all the anabolic and catabolic operations necessary to support life. Homeostasis can be preserved only when metabolic activities of different tissues are coordinated. *(Figure 25-14)*
2. The body has five metabolic components: the liver, adipose tissue, skeletal muscle, neural tissue, and other peripheral tissues. The liver is the focal point for metabolic regulation and control. Adipose tissue stores lipids, primarily in the form of triglycerides. Skeletal muscle contains substantial glycogen reserves, and the contractile proteins can be mobilized and the amino acids used as an energy source. Neural tissue, which does not contain energy reserves, depends on aerobic metabolism for energy production. Other peripheral tissues are able to metabolize glucose, fatty acids, or other substrates under the direction of the endocrine system.

The Absorptive State, p. 938

3. For about 4 hours after a meal, nutrients enter the blood as intestinal absorption proceeds. *(Figure 25-15; Table 25-1)*
4. The liver closely regulates the glucose content of blood and the circulating levels of amino acids.
5. **Lipemia** (milky appearance of the plasma due to the presence of lipids) commonly marks the **absorptive state.** Adipocytes remove fatty acids and glycerol from the circulation and synthesize new triglycerides to be stored for later use.
6. During the absorptive state, glucose molecules are catabolized and amino acids are used to build proteins. Skeletal muscles may also catabolize circulating fatty acids, and the energy is used to increase glycogen reserves.

The Postabsorptive State, p. 940

7. The **postabsorptive state** extends from the end of the absorptive state to the next meal. *(Figure 25-16)*
8. When blood glucose falls, the liver begins breaking down glycogen reserves and releasing the glucose into the circulation. As the duration of the fast increases, liver cells synthesize glucose molecules from smaller carbon fragments and from glycerol molecules. Fatty acids undergo beta-oxidation; the fragments enter the TCA cycle or combine to form **ketone bodies.** *(Table 25-1)*
9. **Glucogenic amino acids** can be converted to pyruvic acid and used for gluconeogenesis. **Ketogenic amino acids** can be converted to acetyl-CoA and catabolized or converted to ketone bodies. *(Figure 25-17)*
10. The average individual carries a 1–2 month energy reserve in adipose tissue. During the postabsorptive state, lipolysis increases and the fatty acids are released into the circulation for catabolism.
11. Skeletal muscles metabolize ketone bodies and fatty acids. Their glycogen reserves are broken down to yield lactic acid, which diffuses into the bloodstream. After a prolonged fast, **cathepsins** (proteolytic enzymes) begin breaking down contractile proteins.
12. Neural tissue continues to be supplied with glucose as an energy source until blood glucose levels become extremely low.

DIET AND NUTRITION, p. 943

1. **Nutrition** is the absorption of nutrients from food. A **balanced diet** contains all the ingredients necessary to maintain homeostasis; it prevents **malnutrition.**

Food Groups and Food Pyramids, p. 944

2. The six **basic food groups** are milk; meat; vegetable; fruit; fats, oils, and sweets; and bread and cereal. These are arranged in a food pyramid with the bread and cereal group forming the base. *(Figure 25-18)*

Nitrogen Balance, p. 945

3. Amino acids, purines, pyrimidines, creatine, and porphyrins are **N compounds,** which contain nitrogen atoms. An adequate dietary supply of nitrogen is essential, because the body does not maintain large nitrogen reserves.

Minerals and Vitamins, p. 945

4. **Minerals** act as cofactors in a variety of enzymatic reactions. They also contribute to the osmolarity of body fluids and play a role in transmembrane potentials, action potentials, neurotransmitter release, muscle contraction, construction and maintenance of the skeleton, transport of gases, buffer systems, fluid absorption, and waste removal. *(Table 25-2)*
5. Vitamins are needed in very small amounts. Vitamins A, D, E, and K are the **fat-soluble vitamins;** taken in excess, they can lead to **hypervitaminosis. Water-soluble vitamins** are not stored in the body; lack of adequate dietary supplies may lead to **deficiency disease (avitaminosis).** *(Tables 25-3, 25-4)*

Diet and Disease, p. 948

6. A balanced diet can improve general health.

BIOENERGETICS, p. 948

1. The energy content of food is usually expressed as **Calories** per gram (C/g). Our cells can capture less than half the energy content of glucose or any other nutrient.

Food and Energy, p. 948

2. The catabolism of lipids releases 9.46 C/g, about twice the amount released by equivalent weights of carbohydrates and proteins.

Metabolic Rate, p. 950

3. The total of all the anabolic and catabolic processes under way is an individual's **metabolic rate.** The **basal metabolic rate (BMR)** is the rate of energy utilization at rest.

Thermoregulation, p. 951

4. The homeostatic regulation of body temperature is **thermoregulation.** Heat exchange with the environment involves four processes: **radiation, conduction, convection, and evaporation.** *(Figure 25-19)*
5. The *preoptic area* of the hypothalamus acts as the body's thermostat, affecting the **heat-loss center** and the **heat-gain center.**
6. Mechanisms for increasing heat loss include both physiological mechanisms (peripheral vasodilation, increased perspiration, and increased respiration) and behavioral modifications.
7. Responses that conserve heat include decreased blood flow to the dermis and *countercurrent exchange.* *(Figure 25-20)*
8. Heat may be generated by **shivering thermogenesis** and **nonshivering thermogenesis.**
9. Thermoregulatory responses differ between individuals. One important source of variation is **acclimatization** (adjusting physiologically to an environment over time).
10. **Pyrexia** (fever), a body temperature above 37.2°C (99°F), can result from problems with the thermoregulatory mechanism, circulation, or sweat gland activity or from the resetting of the hypothalamic thermostat by circulating pyrogens.

REVIEW QUESTIONS

Level 1   Reviewing Facts and Terms

1. Cells synthesize new organic components to
   (a) perform structural maintenance and repairs
   (b) support growth
   (c) produce secretions
   (d) a, b, and c are correct

2. During the complete catabolism of one molecule of glucose, a typical cell gains
   (a) 4 ATP   **(b)** 18 ATP
   (c) 36 ATP   **(d)** 144 ATP

3. The breakdown of glucose to pyruvic acid is
   (a) glycolysis
   (b) gluconeogenesis
   (c) cellular respiration
   (d) oxidative phosphorylation

4. Glycolysis yields an *immediate* net gain of _____ for the cell:
   (a) 1 ATP   **(b)** 2 ATP
   (c) 4 ATP   **(d)** 36 ATP

5. The only *immediate* energy benefit of one turn of the TCA cycle is the formation of a single molecule of
   (a) GTP   **(b)** NAD
   (c) FAD   **(d)** CoA

6. The process that produces over 90 percent of the ATP used by our cells is
   (a) glycolysis
   (b) the TCA cycle
   (c) substrate-level phosphorylation
   (d) oxidative phosphorylation

7. The electron transport system and phosphorylation associated with it generally yield a total of _____ molecules of ATP in the complete catabolism of one glucose molecule:
   (a) 2   **(b)** 4
   (c) 32   **(d)** 36

8. The synthesis of glucose from nonglucose precursors is
   (a) glycolysis
   (b) glycogenesis
   (c) gluconeogenesis
   (d) beta-oxidation

9. The sequence of reactions responsible for the breakdown of fatty acid molecules is
   (a) beta-oxidation   **(b)** the TCA cycle
   (c) lipogenesis   **(d)** a, b, and c are correct

10. The essential fatty acids that cannot be synthesized by the body but must be included in the diet are
   (a) linoleic and linolenic
   (b) leucine and lysine
   (c) cholesterol and glycerol
   (d) HDLs and LDLs

11. The lipoproteins that transport excess cholesterol from peripheral tissues back to the liver for storage or excretion in the bile are the
   (a) chylomicrons   **(b)** VLDLs
   (c) LDLs   **(d)** HDLs

12. The removal of an amino group in a reaction that generates an ammonia molecule is
   (a) ketoacidosis
   (b) transamination
   (c) deamination
   (d) denaturation

13. The part of the RNA molecule that *cannot* be catabolized to provide energy is the
   (a) phosphate   **(b)** sugar
   (c) pyrimidine   **(d)** purine

14. The focal point of metabolic regulation and control is the
   (a) brain   **(b)** liver
   (c) heart   **(d)** kidneys

15. When the body is relying on internal energy reserves to continue meeting its energy demands, it is in the
   (a) postabsorptive state
   (b) absorptive state
   (c) starvation state
   (d) deprivation state

16. A complete protein contains
   (a) the proper balance of amino acids
   (b) all the essential amino acids in sufficient quantities
   (c) a combination of nutrients selected from the food pyramid
   (d) N compounds produced by the body

17. All minerals and most vitamins
   (a) are fat-soluble
   (b) cannot be stored by the body
   (c) cannot be synthesized by the body
   (d) must be synthesized by the body because they are not present in adequate amounts in the diet

18. The vitamins generally associated with vitamin toxicity are
   (a) fat-soluble vitamins
   (b) water-soluble vitamins
   (c) the B complex vitamins
   (d) vitamins C and B_{12}

19. The basal metabolic rate (BMR) represents the
   (a) maximum energy expenditure when exercising
   (b) minimum, resting energy expenditure of an awake, alert person
   (c) minimum amount of energy expenditure during light exercise
   (d) muscular energy expenditure added to the resting energy expenditure

20. The abuse of laxatives and diuretics is commonly associated with
   (a) avitaminosis
   (b) kwashiorkor
   (c) bulimia
   (d) marasmus

21. Over half the heat loss from our bodies is attributable to
   (a) radiation   **(b)** conduction
   (c) convection   **(d)** evaporation

22. The most effective mechanism for elevating body temperature in an adult is
   (a) lipolysis in brown fat
   (b) carbohydrate loading
   (c) fatty acid catabolism
   (d) shivering thermogenesis

23. The resetting of the hypothalamic thermostat by circulating pyrogens such as interleukin-1 produces
   (a) hypothermia
   (b) fever
   (c) maintenance of normal body temperature
   (d) shivering

24. Define the terms *metabolism, anabolism,* and *catabolism.*

25. Write the complete reaction sequence for carbohydrate metabolism.

26. What is a lipoprotein? What are the major groups of lipoproteins, and how do they differ?

27. Why are vitamins and minerals essential components of the diet?

28. What energy yields, in Calories per gram, are associated with the catabolism of carbohydrates, lipids, and proteins?

29. What is the basal metabolic rate (BMR)?

30. What four basic mechanisms are involved in the body's thermoregulation and the environment?

31. What causative factors or conditions are associated with the production of fever?

Level 2   Reviewing Concepts

32. The function of the TCA cycle is to
  (a) produce energy during periods of active muscle contraction
  (b) break six carbon chains into three carbon fragments
  (c) prepare the glucose molecule for further reactions
  (d) remove hydrogen atoms from organic molecules and transfer them to coenzymes

33. During periods of fasting or starvation, the presence of ketone bodies in the circulation causes
  (a) an increase in the pH
  (b) a decrease in the pH
  (c) lipemia
  (d) diabetes insipidus

34. What happens during the process of glycolysis? What conditions are necessary for this process to take place?

35. Why is the TCA cycle called a cycle? What substance(s) enter(s) the cycle, and what substances leave it?

36. What is oxidative phosphorylation? Explain how the electron transport system is involved in this process.

37. How are lipids catabolized in the body? How is beta-oxidation involved with lipid catabolism?

38. How is RNA catabolized to produce energy?

39. How do the absorptive and postabsorptive states maintain normal blood glucose levels?

40. Why is the liver the focal point for metabolic regulation and control?

41. How can the food pyramid be used as a tool to obtain nutrients in sufficient quantity and quality? Why are the dietary fats, oils, and sugars at the top of the pyramid and the breads and cereals at the bottom?

42. How is the brain involved in the regulation of body temperature?

43. Some articles in popular magazines refer to "good cholesterol" and "bad cholesterol." To what types and functions of cholesterol might these terms refer? Explain your answer.

Level 3   Critical Thinking and Clinical Applications

44. While resting and alert, Mary has a respiratory rate of 10 breaths per minute and a tidal volume of 300 ml. The air she is breathing contains 18 percent oxygen. Estimate Mary's basal metabolic rate for a 24-hour period.

45. Why is a starving person more susceptible to infectious disease than one who is well nourished?

46. Individuals suffering from anorexia nervosa typically exhibit bradycardia, hypotension, and decreased heart size. These

problems can eventually lead to death from heart failure. How does anorexia cause these symptoms?

47. The drug *colestipol* binds bile salts in the intestine, forming complexes that cannot be absorbed. How would this drug affect cholesterol levels in the blood?

CHAPTER

26

The Urinary System

*U*rinating in public is considered bad form. But this whimsical fountain in Brussels, Belgium, depicts a very natural and necessary process. After our digestive system assimilates a meal and our body harvests and utilizes the nutrients it needs, we must somehow discharge the resulting waste products. The urinary system performs this service while efficiently conserving water and other valuable substances. It also, as we shall see, carries out a variety of other vital but less obvious functions (none of which has been immortalized in sculpture, so far as we know).

CHAPTER OUTLINE AND OBJECTIVES

The human body contains trillions of cells bathed in extracellular fluid. In previous chapters, we compared these cells to factories that burn nutrients to obtain energy. Imagine what would happen if *real* factories were built as close together as cells in the body. What a mess they would make! Each would generate piles of garbage, and the smoke they produced (while depleting the oxygen supply) would drastically reduce air quality. In short, there would be a serious pollution problem.

The coordinated activities of the digestive, cardiovascular, respiratory, and urinary systems prevent the development of similar pollution problems inside the body. The digestive tract absorbs nutrients from food, and the liver adjusts the nutrient concentration of the circulating blood. The cardiovascular system delivers these nutrients and oxygen from the respiratory system to peripheral tissues. As blood leaves these tissues, it carries the carbon dioxide and waste products to sites of excretion. The carbon dioxide is eliminated at the lungs. Most of the organic waste products are removed and excreted by the urinary system, the focus of this chapter.

The urinary system performs vital excretory functions and eliminates the organic waste products generated by cells throughout the body. It also has other essential functions that are often overlooked. A more complete list of urinary system functions includes the following:

- Regulating blood volume and blood pressure by (1) adjusting the volume of water lost in the urine, (2) releasing erythropoietin, and (3) releasing renin. ∞ *[p. 733]*
- Regulating plasma concentrations of sodium, potassium, chloride, and other ions by controlling the quantities lost in the urine and controlling calcium ion levels by the synthesis of calcitriol. ∞ *[p. 618]*
- Contributing to the stabilization of blood pH by controlling the loss of hydrogen ions and bicarbonate ions in the urine.
- Conserving valuable nutrients by preventing their excretion in the urine while eliminating organic waste products, especially nitrogenous wastes, such as *urea* and *uric acid.*
- Assisting the liver in detoxifying poisons and, during starvation, deaminating amino acids so that they can be broken down by other tissues.

These activities are carefully regulated to keep the composition of the blood within acceptable limits. A disruption of any one of these functions will have immediate and potentially fatal consequences.

In this chapter, we shall consider the functional organization of the urinary system and describe the major regulatory mechanisms that control urine production and concentration.

ORGANIZATION OF THE URINARY SYSTEM

The urinary system (Figure 26-1a●) includes the *kidneys, ureters, urinary bladder,* and *urethra.* The excretory functions of the urinary system are performed by the two **kidneys.** These organs produce **urine,** a fluid containing water, ions, and small soluble compounds. Urine leaving the kidneys travels along the paired **ureters** (ū-RĒ-terz) to the **urinary bladder** for temporary storage. Urine excretion, a process called **urination,** or **micturition** (mik-tū-RI-shun), occurs when the contraction of the muscular urinary bladder forces urine through the **urethra** and out of the body.

THE KIDNEYS

The kidneys are located on either side of the vertebral column between vertebrae T_{12} and L_3. The left kidney lies slightly superior to the right kidney (Figure 26-1a,b●).

On gross dissection, the anterior surface of the right kidney is covered by the liver, the right colic (hepatic) flexure of the colon, and the duodenum. The anterior surface of the left kidney is covered by the stomach, pancreas, jejunum, and left colic (splenic) flexure of the colon. The superior surface of each kidney is capped by an adrenal gland (Figures 26-1a,b and 26-2●). The kidneys and adrenal glands lie between the muscles of the dorsal body wall and the parietal peritoneum in a retroperitoneal position (Figure 26-1c●).

The position of the kidneys in the abdominal cavity is maintained by (1) the overlying peritoneum, (2) contact with adjacent visceral organs, and (3) supporting connective tissues. Each kidney is protected and stabilized by three concentric layers of connective tissue (Figure 26-1c●):

1. The **renal capsule** is a layer of collagen fibers that covers the outer surface of the entire organ. This layer is also known as the *fibrous tunic* of the kidney.
2. The **adipose capsule,** a layer of adipose tissue, surrounds the renal capsule. This layer can be quite thick, and on dissection it generally obscures the outline of the kidney.
3. The **renal fascia** is a dense outer layer. Collagen fibers extend outward from the renal capsule through the adipose capsule to this layer. The renal fascia anchors the kidney to surrounding structures. Posteriorly, the renal fascia fuses with the deep fascia surrounding the muscles of the body wall. Anteriorly, the renal fascia forms a thick fibrous layer that fuses with the peritoneum.

In effect, each kidney hangs suspended by collagen fibers from the renal fascia and packed in a soft cushion of adipose tissue. This arrangement prevents the jolts and

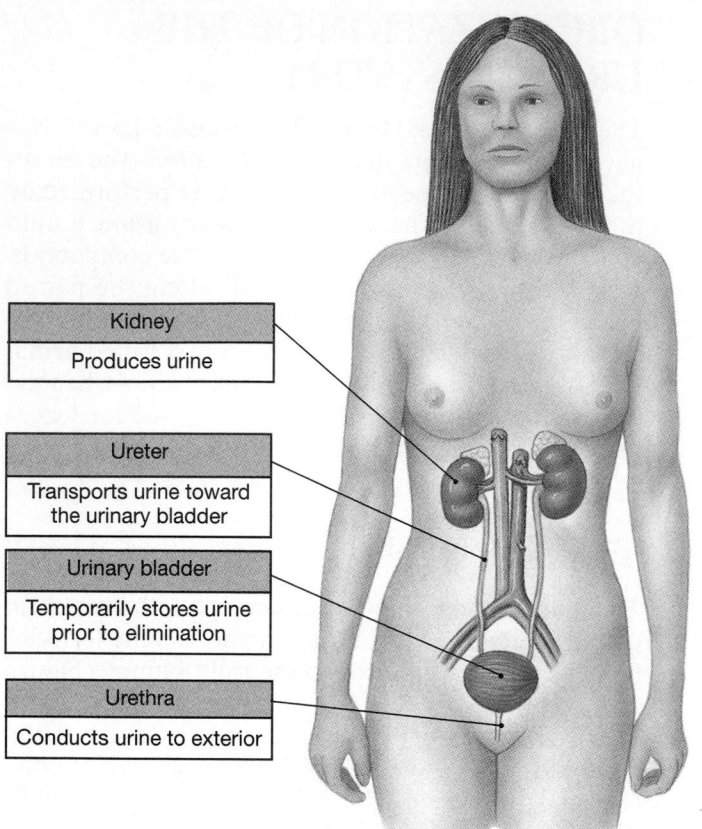

Kidney

Produces urine

Ureter

Transports urine toward the urinary bladder

Urinary bladder

Temporarily stores urine prior to elimination

Urethra

Conducts urine to exterior

(a) Anterior view

●**FIGURE 26-1** **An Introduction to the Urinary System.** (a) Anterior view of the urinary system, showing the positions of the kidneys and other components. (b) Posterior view of the trunk. (c) Sectional view at the level indicated in (b).

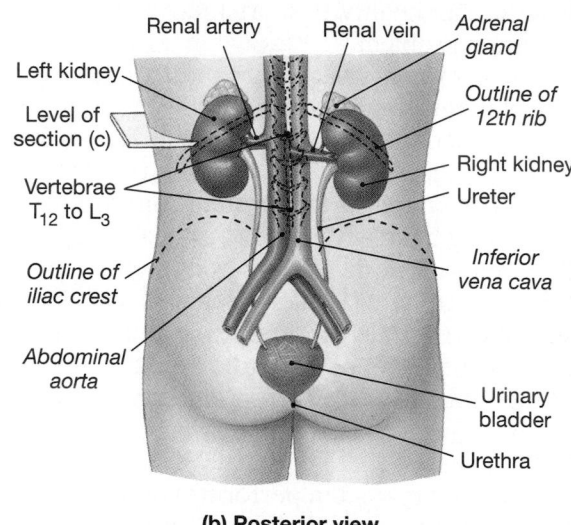

(b) Posterior view

shocks of day-to-day existence from disturbing normal kidney function. If the suspensory fibers break or become detached, a slight bump or jar may displace the kidney and stress the attached vessels and ureter. This condition, called a *floating kidney*, can be especially dangerous, because the ureters or renal blood vessels may become twisted or kinked during movement.

Superficial Anatomy of the Kidneys

Each reddish brown kidney has the shape of a kidney bean. A typical adult kidney (Figures 26-2 and 26-3●, p. 964) is about 10 cm (4 in.) in length, 5.5 cm (2.2 in.) in width, and 3 cm (1.2 in.) in thickness. Each kidney weighs about 150 g (5.25 oz). The **hilus,** a prominent medial indentation, is the point of entry for the *renal artery* and *renal nerves* and the point of exit for the *renal vein* and *ureter*.

Sectional Anatomy of the Kidneys

The fibrous renal capsule has inner and outer layers. In sectional view (Figure 26-3a●), the inner layer folds inward at the hilus and lines an internal cavity, the

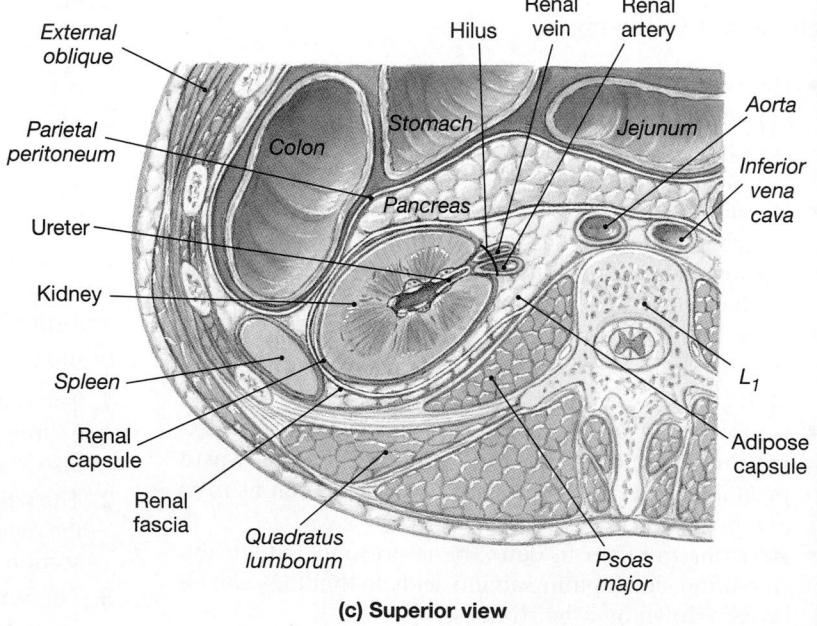

(c) Superior view

renal sinus. Renal blood vessels and the ureter draining the kidney pass through the hilus and branch within the renal sinus. A thickened, outer layer of the capsule extends across the hilus and stabilizes the position of these structures.

The renal **cortex** is the outer layer of the kidney in contact with the capsule. The cortex is reddish

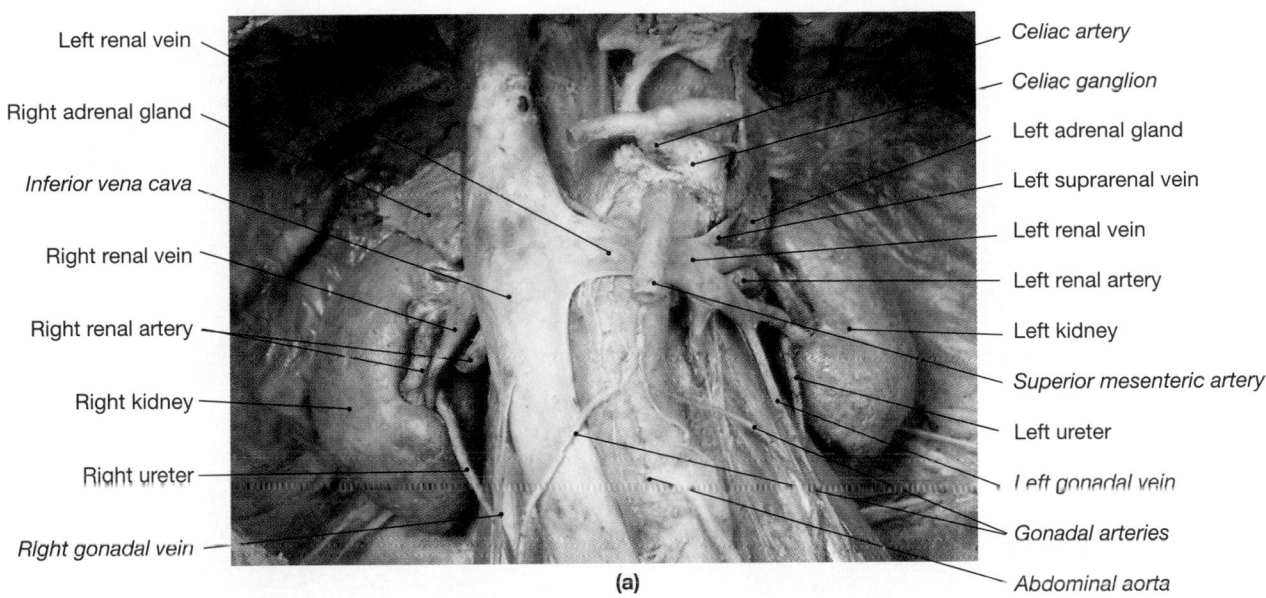

Left renal vein
Right adrenal gland
Inferior vena cava
Right renal vein
Right renal artery
Right kidney
Right ureter
Right gonadal vein

Celiac artery
Celiac ganglion
Left adrenal gland
Left suprarenal vein
Left renal vein
Left renal artery
Left kidney
Superior mesenteric artery
Left ureter
Left gonadal vein
Gonadal arteries
Abdominal aorta

(a)

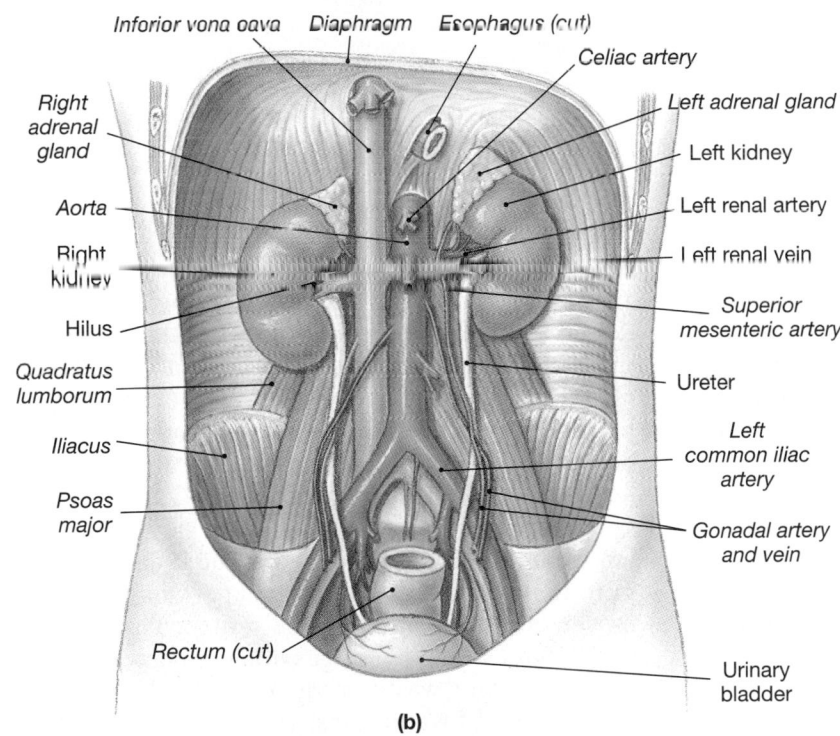

Inferior vena cava   Diaphragm   Esophagus (cut)
Celiac artery
Right adrenal gland
Left adrenal gland
Left kidney
Aorta
Left renal artery
Right kidney
Left renal vein
Hilus
Superior mesenteric artery
Quadratus lumborum
Ureter
Iliacus
Left common iliac artery
Psoas major
Gonadal artery and vein
Rectum (cut)
Urinary bladder

(b)

● *FIGURE 26-2* **The Urinary System in Gross Dissection. (a)** Anterior view of the kidneys after removal of the abdominal organs. **(b)** Diagrammatic view of the abdominopelvic cavity, showing the kidneys, ureters, urinary bladder, and the blood supply to the urinary structures.

brown and granular in texture. The renal **medulla** consists of 6 to 18 distinct conical or triangular structures called **renal pyramids.** The base of each pyramid faces the cortex, and the tip of each pyramid, a region known as the **renal papilla,** projects into the renal sinus. Each pyramid has a series of fine grooves that converge at the papilla. Adjacent renal pyramids

are separated by bands of cortical tissue called **renal columns**, which extend into the medulla. The columns have a distinctly granular texture, similar to that of the cortex. A **renal lobe** consists of a renal pyramid, the overlying area of renal cortex, and adjacent tissues of the renal columns.

Urine production occurs in the renal lobes. Ducts within each renal papilla discharge urine into a cup-shaped drain called a **minor calyx** (KĀ-liks). Four or five minor calyces (KĀL-i-sēz) merge to form a **major calyx,** and two or three major calyces combine to form the **renal pelvis,** a large, funnel-shaped chamber. The renal pelvis, which fills most of the renal sinus, is connected to the ureter at the hilus of the kidney.

Urine production begins in microscopic structures called **nephrons** (NEF-ronz) in the cortex of each renal lobe. There are roughly 1.25 million nephrons in each kidney, with a combined length of about 145 km (85 miles).

The Nephron

Each nephron consists of a *renal corpuscle* and a *renal tubule* roughly 50 mm long. The **renal tubule** begins at the **renal corpuscle** (KOR-pus-ul), a cup-shaped chamber (Figure 26-4●). The renal corpuscle is approximately 200 μm (0.2 mm) in diameter. It contains a capillary

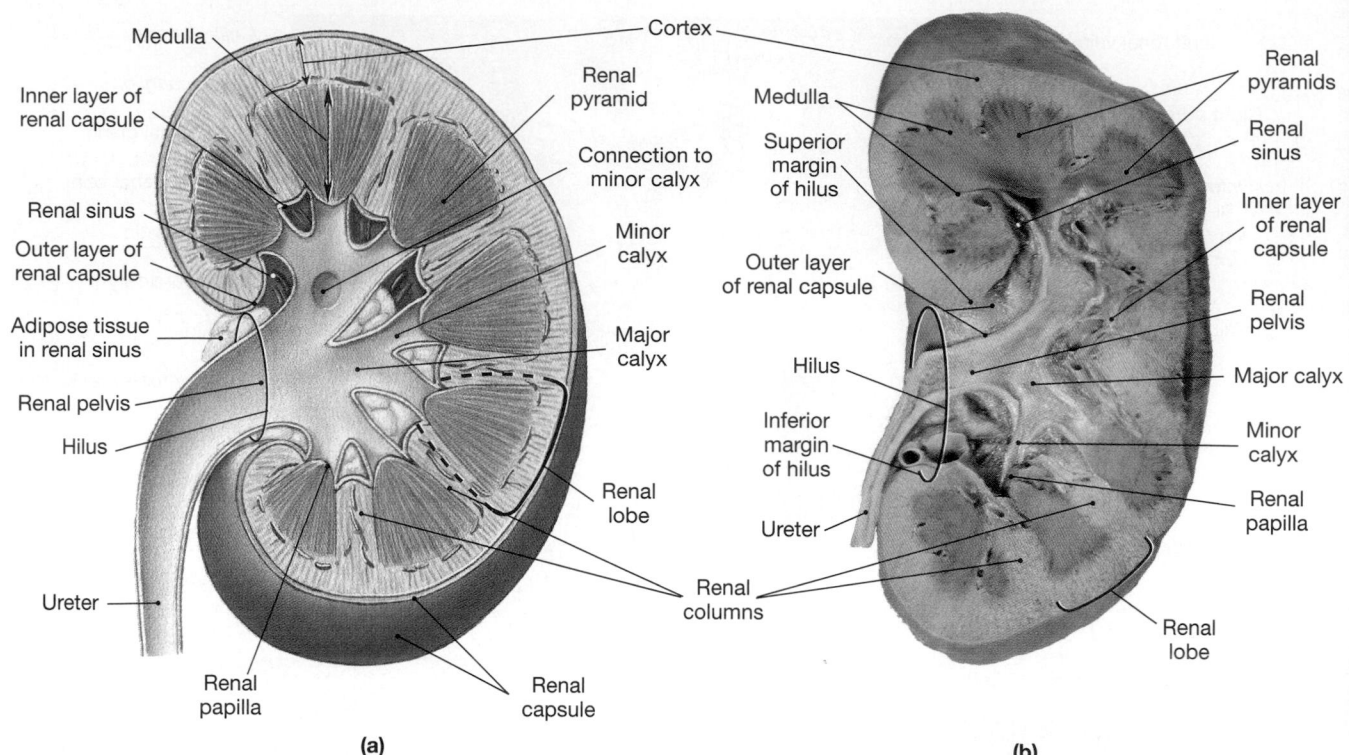

•FIGURE 26-3 Structure of the Kidney. (a) Diagrammatic view of a frontal section through the left kidney, showing major structures. **(b)** Frontal section of the left kidney.

network called the **glomerulus** (glo-MER-ū-lus; plural, *glomeruli*), which consists of about 50 intertwining capillaries. Blood arrives at the glomerulus via the *afferent arteriole* and departs in the *efferent arteriole*. Filtration occurs in the renal corpuscle as blood pressure forces fluid and dissolved solutes out of the glomerular capillaries and into the *capsular space*. Filtration produces an essentially protein-free solution, known as a **filtrate,** that is otherwise very similar to blood plasma.

From the renal corpuscle, the filtrate enters a long tubular passageway. The renal tubule has two convoluted (coiled or twisted) segments—the *proximal convoluted tubule* (PCT) and the *distal convoluted tubule* (DCT)—separated by a simple U-shaped tube, the *loop of Henle* (HEN-lē). The convoluted segments are in the cortex, and the loop extends partially or completely into the medulla. For clarity, the nephron diagrammed in Figure 26-4• has been shortened and straightened. The regions of the nephron vary in their structural and functional characteristics. As it travels along the tubule, the filtrate, now called **tubular fluid**, gradually changes in composition. The changes that occur and the characteristics of the urine that results vary with the activities under way in each segment of the nephron; Figure 26-4• provides an overview of the regional specializations.

Each nephron empties into the **collecting system.** A *connecting tubule* carries the tubular fluid from the distal convoluted tubule to a nearby *collecting duct*. The collecting duct, which receives tubular fluid from many different nephrons, leaves the cortex and descends into the medulla, carrying fluid to a *papillary duct* that drains into a minor calyx.

The urine arriving at the renal pelvis is very different from the filtrate produced at the renal corpuscle. Filtration is a passive process that permits or prevents movement across a barrier solely on the basis of solute size. A filter with pores large enough to permit the passage of organic waste products is unable to prevent the passage of water, ions, and other organic molecules, such as glucose, fatty acids, and amino acids. These useful substances must be reclaimed and the waste products excreted. The segments of the nephron distal to the renal corpuscle are responsible for:

- Reabsorbing all the useful organic substrates that enter the renal tubule.
- Reabsorbing over 90 percent of the water present in the filtrate.
- Secreting into the tubular fluid any waste products that were missed by the filtration process.

Additional water and salts will be removed from the tubular fluid in the collecting system before the

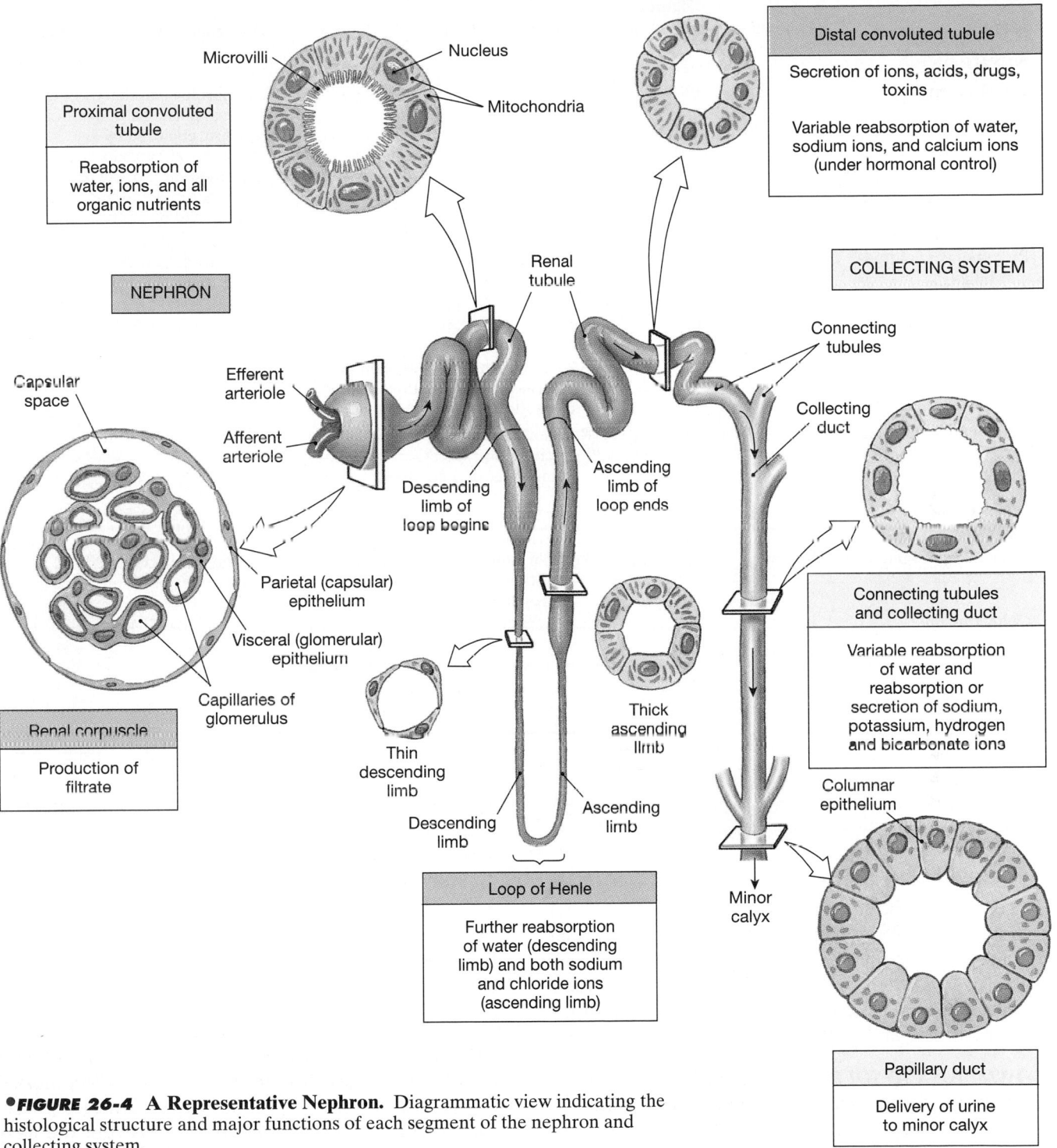

•FIGURE 26-4 A Representative Nephron. Diagrammatic view indicating the histological structure and major functions of each segment of the nephron and collecting system.

fluid is released into the renal sinus as urine. Table 26-1 gives an overview of important information concerning the regions of the nephron and collecting system.

Nephrons differ slightly in structure, depending on their location. Roughly 85 percent of all nephrons are **cortical nephrons**; they are located in the superficial cortex of the kidney. The remaining 15 percent of nephrons, termed **juxtamedullary** (juks-ta-MED-ū-lar-ē) **nephrons** (*juxta*, near), are located closer to the medulla. Because they are more numerous than juxtamedullary nephrons, cortical nephrons perform most of the reabsorptive and secretory functions of the kidneys. However, the juxtamedullary nephrons are responsible for the ability to produce a concentrated urine.

TABLE 26-1 The Organization of the Nephron and Collecting System in the Kidney

Region	Length	Diameter (μm)	Primary Function	Histological Characteristics
NEPHRON				
Renal corpuscle	Spherical	150–250	Filtration of plasma	Glomerulus (capillary knot), mesangial cells, and lamina densa, enclosed by Bowman's capsule; visceral epithelium (podocytes) and parietal epithelium separated by capsular space
Renal tubule				
Proximal convoluted tubule (PCT)	14	60	Reabsorption of ions, organic molecules, vitamins, water; secretion of drugs, toxins, acids	Cuboidal cells with microvilli
Loop of Henle	30	15	Descending limb: reabsorption of water from tubular fluid	Low cuboidal or squamous cells
		30	Ascending limb: reabsorption of ions; assists in creation of a concentration gradient in the medulla	
Distal convoluted tubule (DCT)	5	30–50	Reabsorption of sodium ions and calcium ions; secretion of acids, ammonia, drugs, toxins	Cuboidal cells with few if any microvilli
COLLECTING SYSTEM				
Connecting tubule	Variable	50	Reabsorption of water, sodium ions; secretion or reabsorption of hydrogen ions or bicarbonate ions	Cuboidal cells without microvilli; pale compared with DCT
Collecting duct	15	50–100	Reabsorption of water, sodium ions; secretion or reabsorption of bicarbonate ions or hydrogen ions	Cuboidal to columnar cells
Papillary duct	5	100–200	Conduction of tubular fluid to minor calyx; contributes to concentration gradient of the medulla	Columnar cells

We shall now examine the structure of each segment of a representative nephron.

The Renal Corpuscle

The renal corpuscle (Figure 26-5a,b,c●) has a diameter averaging 150–250 μm. It includes (1) the glomerular capillary network and (2) a region known as **Bowman's capsule.** Connected to the initial segment of the renal tubule, Bowman's capsule forms the outer wall of the renal corpuscle and covers the glomerular capillaries.

BOWMAN'S CAPSULE The glomerulus projects into Bowman's capsule much as the heart projects into the pericardial cavity (Figure 26-5c●). The outer wall of the capsule is lined by a simple squamous **parietal epithelium** (capsular epithelium). This layer is continuous with the **visceral epithelium** (glomerular epithelium) that covers the glomerular capillaries. The visceral epithelium consists of large cells with complex processes, or "feet," that wrap around the **lamina densa,** the specialized basement membrane of the glomerular capillaries (Figure 26-5c,d,e●). These unusual cells are called **podocytes** (PŌ-do-sīts; podos, foot + -cyte, cell). The podocyte feet are known as **pedicels.** Materials passing out of the blood at the glomerulus must be small enough to pass between the narrow gaps, or **filtration slits,** between adjacent pedicels. These slits are small enough to prevent the loss of all but the smallest plasma proteins.

The **capsular space** separates the visceral and parietal epithelia (Figures 26-4, 26-5b,c, and 26-6a●). The connection between the two epithelial layers lies at the **vascular pole** of the renal corpuscle. At the vascular pole, blood flows into and out of the glomerular capillaries. Blood arrives in an **afferent arteriole** and departs in an **efferent arteriole.**

THE GLOMERULAR CAPILLARIES The glomerular capillaries (Figure 26-5d●) are *fenestrated capillaries* whose endothelium contains large-diameter pores. The openings are small enough to prevent the passage of blood cells but too large to restrict the diffusion of dissolved or suspended compounds, even those the size of plasma proteins.

The endothelial cells lining the capillaries are surrounded by the lamina densa (Figure 26-5e●). During filtration, the lamina densa restricts the passage of large plasma proteins but permits the movement of smaller molecules, including albumin, many organic nutrients, and ions. Unlike basement membranes elsewhere, the lamina densa may encircle two or more capillaries. When it does, **mesangial cells** are situated between the capillaries. Mesangial cells have several important functions:

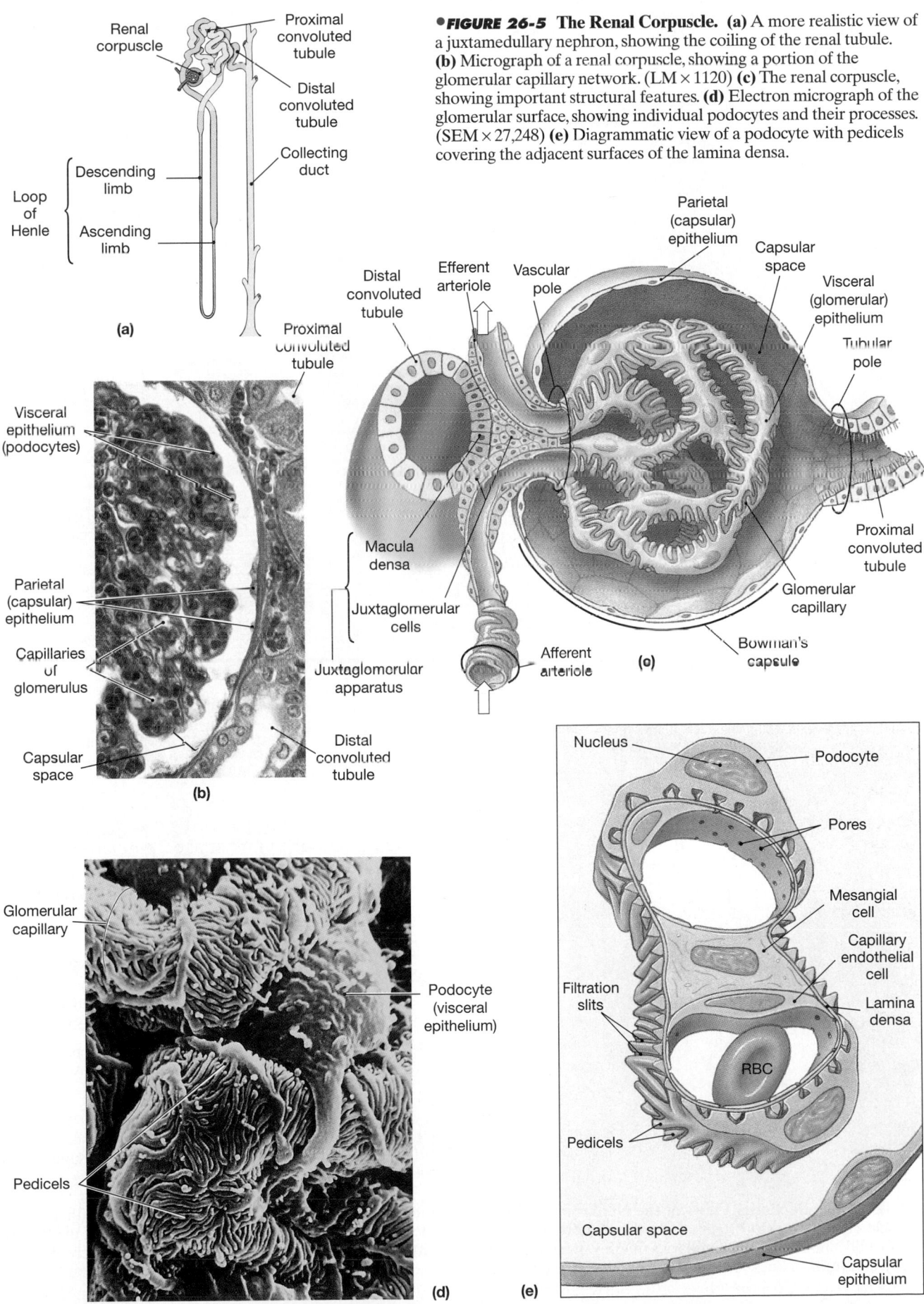

•**FIGURE 26-5** **The Renal Corpuscle.** **(a)** A more realistic view of a juxtamedullary nephron, showing the coiling of the renal tubule. **(b)** Micrograph of a renal corpuscle, showing a portion of the glomerular capillary network. (LM × 1120) **(c)** The renal corpuscle, showing important structural features. **(d)** Electron micrograph of the glomerular surface, showing individual podocytes and their processes. (SEM × 27,248) **(e)** Diagrammatic view of a podocyte with pedicels covering the adjacent surfaces of the lamina densa.

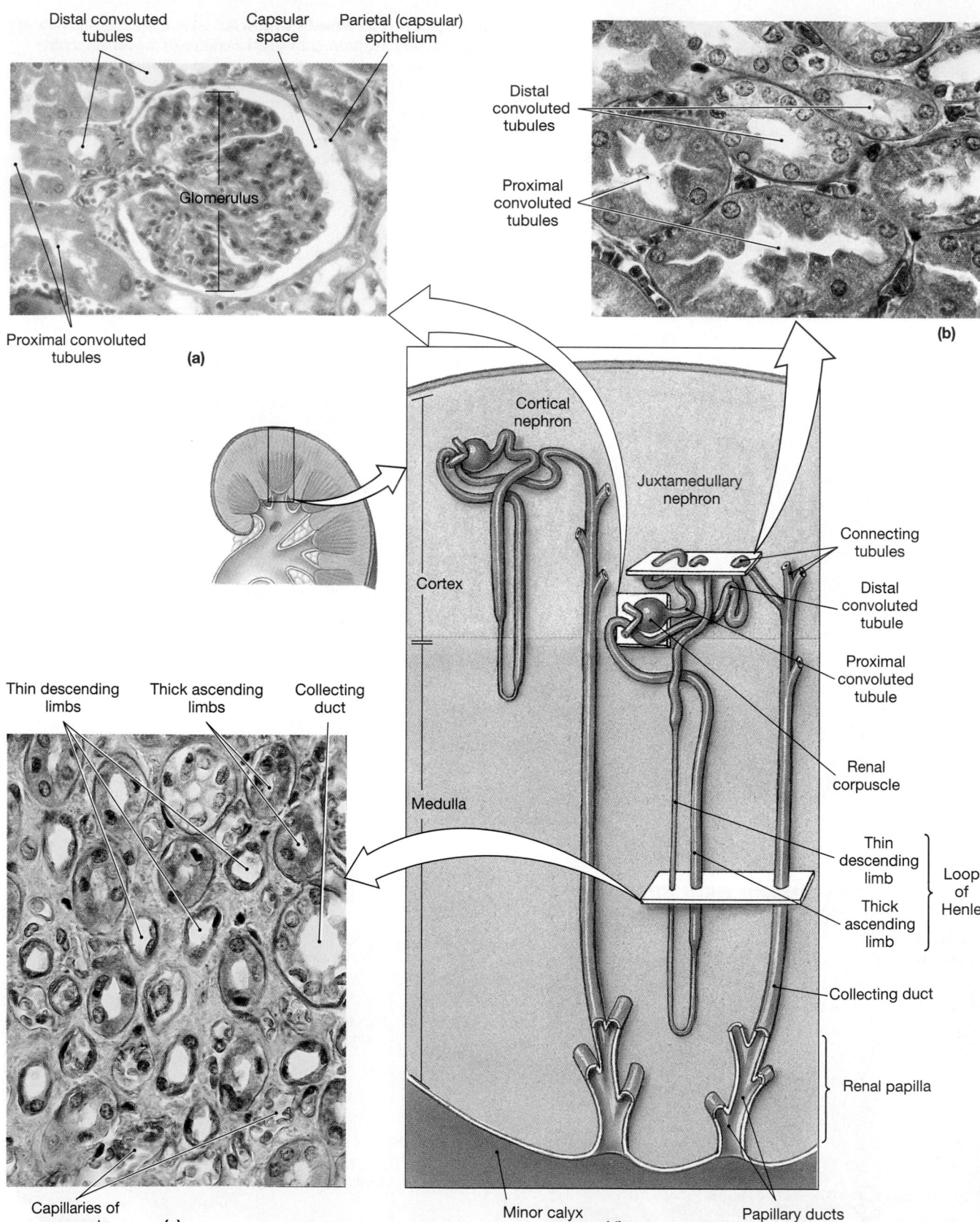

●**FIGURE 26-6 Sectional Views of the Nephron. (a)** The association of proximal and distal convoluted tubules with a renal corpuscle. (LM × 370) **(b)** Proximal and distal convoluted tubules. (LM × 560) **(c)** Descending and ascending limbs of the loop of Henle, a collecting duct, and capillaries of the vasa recta. (LM × 500) **(d)** Approximate locations of the sections in micrographs (a), (b), and (c).

- Providing physical support to the capillaries.
- Engulfing organic materials that might otherwise clog the filter at the lamina densa.
- Contracting or relaxing to change the diameter of the glomerular capillaries and to change the filtration rate.

Together, the fenestrated endothelium, the lamina densa, and the filtration slits form the *filtration membrane.* During filtration, blood pressure forces water and small solutes across this membrane and into the capsular space. The larger solutes, especially plasma proteins, are excluded. Filtration at the renal corpuscle is both effective and passive, but it has one major limitation: In addition to metabolic wastes and excess ions, compounds such as glucose, free fatty acids, amino acids, vitamins, and other solutes enter the capsular space. These potentially useful materials are recaptured before the filtrate leaves the kidneys, with much of the reabsorption occurring in the proximal convoluted tubule.

GLOMERULONEPHRITIS *Glomerulonephritis* (glo-mer-ū-lō-nef-RĪ-tis) is an inflammation of the renal cortex that affects the filtration mechanism of the kidneys. This condition, which may develop after an infection involving *Streptococcus* bacteria, is an *immune complex disorder,* a class of diseases we introduced in Chapter 22. ∞ *[p. 805]* The primary infection may not occur in or near the kidneys. However, as the immune system responds to the infection, the number of circulating antigen–antibody complexes skyrockets. These complexes are small enough to pass through the lamina densa but too large to fit between the slit pores of the filtration membrane. As a result, the filtration mechanism clogs up, and filtrate production drops. Any condition that leads to a massive immune response can cause glomerulonephritis, including viral infections and autoimmune disorders.

The Proximal Convoluted Tubule

The entrance to the **proximal convoluted tubule (PCT)** lies almost directly opposite the vascular pole, at the **tubular pole** of the renal corpuscle (Figure 26-5c●). The lining of the PCT consists of a simple cuboidal epithelium whose exposed surfaces are blanketed with microvilli (Figures 26-4 and 26-6b●). The cuboidal tubular cells actively absorb organic nutrients, ions, and plasma proteins (if any) from the tubular fluid and release them into the **peritubular fluid,** the interstitial fluid surrounding the renal tubule. As these solutes are absorbed and transported, osmotic forces pull water across the wall of the PCT and into the peritubular fluid. Although reabsorption is the primary function of the PCT, the epithelial cells can also secrete substances into the lumen.

The Loop of Henle

The PCT makes an acute bend that turns the renal tubule toward the renal medulla. This turn marks the start of the **loop of Henle** (Figures 26-4b and 26-5a●).

The loop of Henle can be divided into a **descending limb** and an **ascending limb.** Fluid in the descending limb travels toward the renal pelvis, and that in the ascending limb travels toward the renal cortex. Each limb contains a **thick segment** and a **thin segment.** (The terms *thick* and *thin* refer to the height of the epithelium, not to the diameter of the lumen.)

The thick segments have a cuboidal epithelium, whereas a thin squamous epithelium lines the thin segments (Figures 26-4 and 26-6c●). The thick descending limb has functions similar to those of the PCT. The thick ascending limb pumps sodium and chloride ions out of the tubular fluid. The effect of this pumping is most noticeable in the medulla, where the long ascending limbs of juxtamedullary nephrons create unusually high solute concentrations in the peritubular fluid.

The Distal Convoluted Tubule

The thick ascending limb of the loop of Henle ends where it forms a sharp angle near the vascular pole of the renal corpuscle. The **distal convoluted tubule (DCT)** begins there. The initial portion of the DCT passes between the afferent and efferent arterioles (Figure 26-5c●).

In sectional view (Figures 26-4 and 26-6b●), the DCT differs from the PCT in that (1) the DCT has a smaller diameter, (2) the epithelial cells of the DCT lack microvilli, and (3) the boundaries between the epithelial cells in the DCT are distinct. The DCT is an important site for:

- The active secretion of ions, acids, and other materials.
- The selective reabsorption of sodium ions and calcium ions from the tubular fluid.
- The selective reabsorption of water, which assists in concentrating the tubular fluid.

THE JUXTAGLOMERULAR APPARATUS The epithelial cells of the DCT near the vascular pole of the renal corpuscle are taller than those elsewhere along the DCT, and their nuclei are clustered together. This region, detailed in Figure 26-5c●, is called the **macula densa** (MAK-ū-la DEN-sa). The cells of the macula densa are closely associated with unusual smooth muscle fibers in the wall of the afferent arteriole. These fibers are known as **juxtaglomerular cells.** Together, the macula densa and juxtaglomerular cells form the **juxtaglomerular apparatus (JGA).** The juxtaglomerular apparatus is an endocrine structure that secretes the *erythropoietin* and *renin,* as we described in Chapter 18. ∞ *[p. 733]*

The Collecting System

The DCT, the last segment of the nephron, opens into the collecting system. The collecting system consists of *connecting tubules, collecting ducts,* and *papillary ducts*

(Figure 26-4•, p. 965). Individual **connecting tubules** connect each nephron to a nearby **collecting duct** (Figure 26-6c,d•). Each collecting duct receives tubular fluid from many connecting tubules. Several collecting ducts converge to empty into a larger **papillary duct,** which in turn empties into a minor calyx. The epithelium lining the collecting system begins with simple cuboidal cells in the connecting tubules and changes to a columnar epithelium in the collecting and papillary ducts.

In addition to transporting tubular fluid from the nephron to the renal pelvis, the collecting system adjusts its composition and determines the final osmotic concentration and volume of the urine.

POLYCYSTIC KIDNEY DISEASE *Polycystic* (po-lē-SIS-tik) *kidney disease* is an inherited condition affecting the structure of kidney tubules. Swellings (cysts) develop along the length of the tubules, some growing large enough to compress adjacent nephrons and vessels. Kidney function deteriorates, and the nephrons may eventually become nonfunctional. The process is so gradual that serious problems seldom appear before the individual is 30–40 years of age. Common symptoms include sharp pain in the sides, recurrent urinary infections, and the presence of blood in the urine. Treatment is symptomatic, focusing on prevention of infection and reduction of pain with analgesics. In severe cases, hemodialysis or kidney transplantation may be required.

The Blood Supply to the Kidneys

Your kidneys receive 20–25 percent of your total cardiac output. In normal individuals, about 1200 ml of blood flows through the kidneys each minute. That is a phenomenal amount of blood for organs with a combined weight of less than 300 g (10.5 oz)!

Each kidney receives blood from a *renal artery* that originates along the lateral surface of the abdominal aorta near the level of the superior mesenteric artery (Figure 21-26a•, p. 746). As it enters the renal sinus, the renal artery provides blood to the **segmental arteries** (Figure 26-7a•). Segmental arteries further divide into a series of **interlobar arteries** that radiate outward through the renal columns between the renal pyramids. The interlobar arteries supply blood to the **arcuate** (AR-kū-āt) **arteries,** which arch along the boundary between the cortex and medulla of the kidney. Each arcuate artery gives rise to a number of **interlobular arteries,** which supply parts of the adjacent renal lobe. Branching from each interlobular artery are numerous afferent arterioles (Figure 26-7b•).

Blood reaches the vascular pole of each glomerulus through an afferent arteriole and leaves in an efferent arteriole (Figures 26-5c and 26-7c•). Blood travels from the efferent arteriole to form a capillary plexus, a network of **peritubular capillaries** that supplies the PCT and DCT. The peritubular capillaries provide a route for the pickup or delivery of substances that are reabsorbed or secreted by these portions of the nephron.

In juxtamedullary nephrons, the efferent arterioles and peritubular capillaries are connected to a series of long, slender capillaries that accompany the loops of Henle into the medulla (Figure 26-7d•). These capillaries, known as the **vasa recta** (*rectus,* straight), absorb and transport solutes and water reabsorbed into the medulla from tubular fluid in the loops of Henle and collecting ducts. Under normal conditions, the removal of solutes and water by the vasa recta balances the rates of solute and water reabsorption in the medulla.

From the peritubular capillaries and vasa recta, blood enters a network of venules and small veins that converge on the **interlobular veins.** In a mirror image of the arterial distribution, the interlobular veins deliver blood to **arcuate veins,** which empty into **interlobar veins.** The interlobar veins drain into the **segmental veins,** which merge to form a *renal vein*. Many of the blood vessels just described are visible in the corrosion cast of the kidneys shown in Figure 26-8• (p. 972).

ANALYSIS OF RENAL BLOOD FLOW The rate of blood flow through the kidneys can be determined by administering the compound *para-aminohippuric acid (PAH)*, which is removed by filtration at the glomeruli and by active secretion along the proximal and distal convoluted tubules. Virtually all the PAH contained in the blood that arrives at the kidneys will be removed before the blood departs in the renal veins. Renal blood flow can be calculated by comparing plasma concentrations of PAH with the amount secreted in the urine. The *Applications Manual* describes the calculations involved. [AM] *PAH and the Calculation of Renal Blood Flow*

Innervation of the Kidneys

The kidneys and ureters are innervated by **renal nerves.** Most of the nerve fibers involved are sympathetic postganglionic fibers from the superior mesenteric ganglion. A renal nerve enters each kidney at the hilus and follows the tributaries of the renal arteries to reach individual nephrons. The sympathetic innervation targets (1) the juxtaglomerular apparatus, (2) the smooth muscles in the walls of the afferent and efferent arterioles, and (3) mesangial cells. Known functions of sympathetic innervation include the following:

- Regulation of glomerular blood flow and pressure, through control of the diameters of the afferent and efferent arterioles and the glomerular capillaries.
- Stimulation of renin release from the juxtaglomerular apparatus.
- Direct stimulation of water and sodium ion reabsorption.

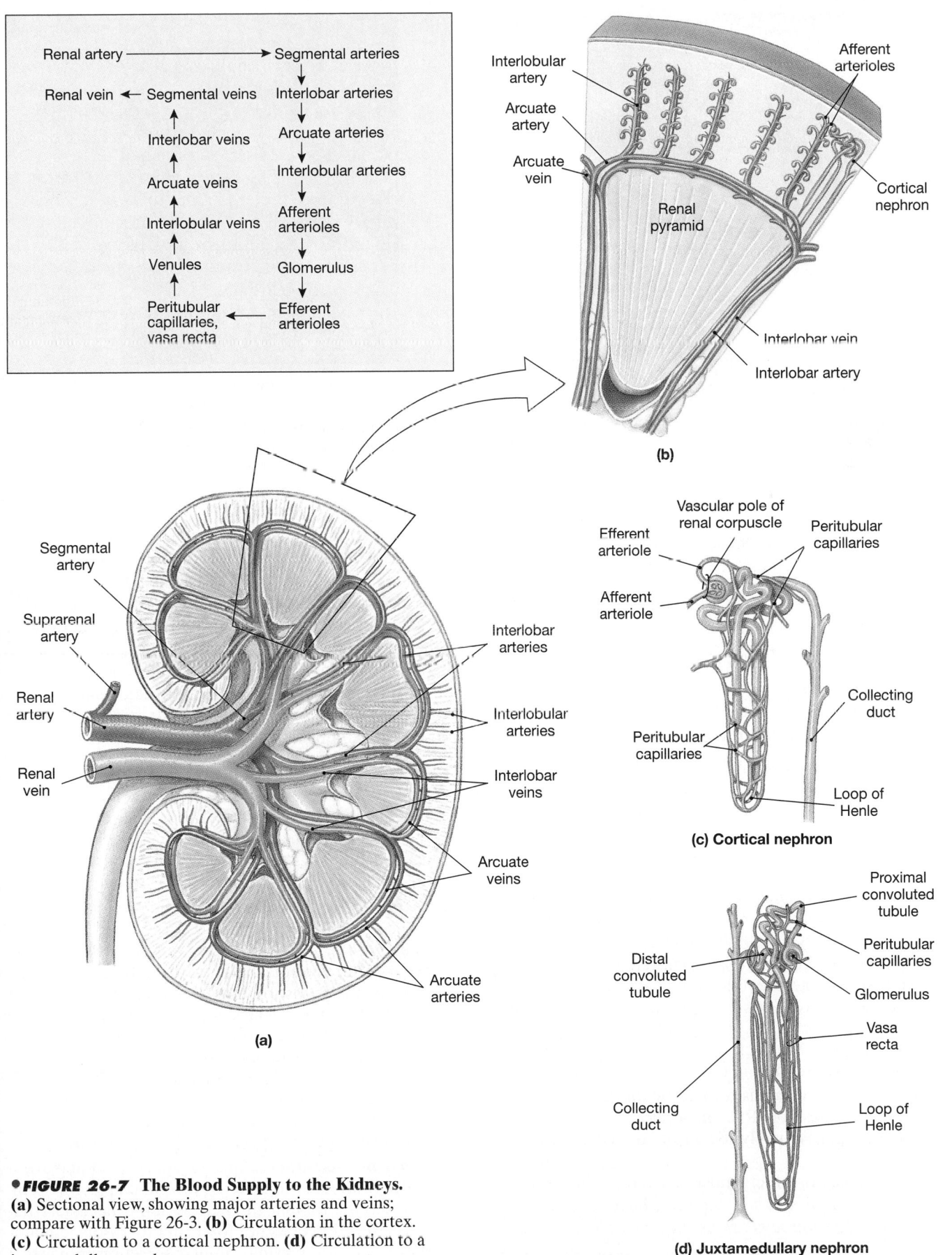

•FIGURE 26-7 The Blood Supply to the Kidneys.
(a) Sectional view, showing major arteries and veins; compare with Figure 26-3. **(b)** Circulation in the cortex. **(c)** Circulation to a cortical nephron. **(d)** Circulation to a juxtamedullary nephron.

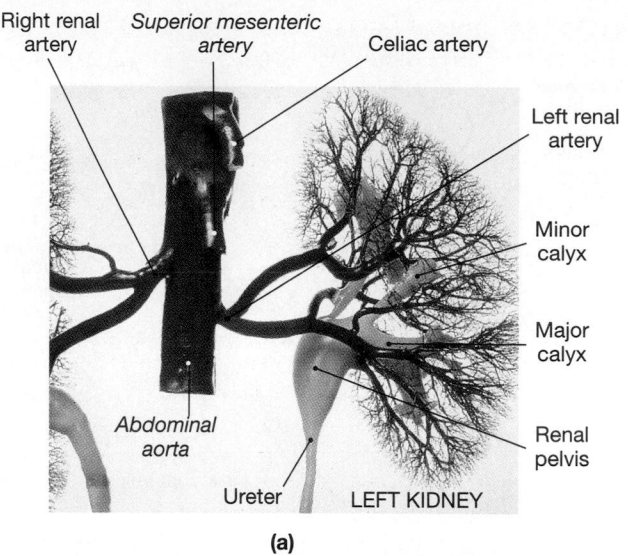

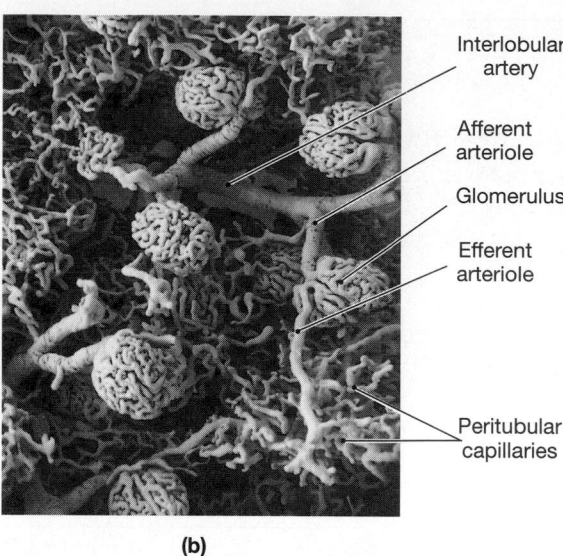

•FIGURE 26-8 The Blood Supply to the Kidneys. (a) A corrosion cast of the circulation and conducting passageways of the kidneys. **(b)** The arteriolar and glomerular vessels. (SEM × 75) Reproduced from R. G. Kessel and R. H. Kardon, "Tissues and Organs: A Text-Atlas of Scanning Electron Microscopy," W. H. Freeman & Co., 1979. ☐AM☐ *Scans 13–15*

☑ What portions of the nephron are located in the renal cortex?

☑ Why don't plasma proteins pass into the capsular space under normal circumstances?

☑ Damage to what part of the nephron would interfere with the control of blood pressure?

RENAL PHYSIOLOGY

Basic Principles of Urine Formation

The goal in urine production is to maintain homeostasis by regulating the volume and composition of the blood. This process involves the excretion and elimination of solutes, specifically metabolic waste products. There are three noteworthy organic waste products:

1. *Urea.* Urea is the most abundant organic waste. ∞ *[p. 935]* You generate roughly 21 g of urea each day. Most of it is produced during the breakdown of amino acids.

2. *Creatinine.* Creatinine is generated in skeletal muscle tissue by the breakdown of *creatine phosphate,* a high-energy compound that plays an important role in muscle contraction. ∞ *[p. 298]* Your body generates roughly 1.8 g of creatinine each day, and virtually all of it is excreted in the urine.

3. *Uric acid.* Uric acid is formed by the recycling of nitrogenous bases from RNA molecules. ∞ *[p. 934]* You produce approximately 480 mg of uric acid each day.

These waste products must be excreted in solution, so their elimination is accompanied by an unavoidable water loss. The kidneys are usually capable of producing a concentrated urine with an osmotic concentration of 1200–1400 mOsm/l, more than four times that of plasma. If the kidneys were not able to concentrate the filtrate produced by glomerular filtration, fluid losses would lead to fatal dehydration in a matter of hours. At the same time, your kidneys ensure that the fluid that *is* lost does not contain potentially useful organic substrates, such as sugars or amino acids, that are found in blood plasma. These valuable materials must be reabsorbed and retained for use by other tissues.

To accomplish these goals, your kidneys rely on three distinct processes:

1. *Filtration.* In **filtration,** blood pressure forces water across a filtration membrane. At the kidneys, the filtration membrane includes the glomerular endothelium, the lamina densa, and the filtration slits. Solute molecules small enough to pass through this filtration complex are carried along by the surrounding water molecules.

2. *Reabsorption.* **Reabsorption** is the removal of water and solutes from the filtrate, across the tubular epithelium, and into the peritubular fluid. Reabsorption occurs after the filtrate has left the renal corpuscle. Most of the reabsorbed materials are nutrients that your body can use. Whereas filtration occurs solely on the basis of size, reabsorption is a selective process. Solute reabsorption may involve simple diffusion or the activity of carrier proteins in the tubular epithelium. The reabsorbed substances pass into the peritubular fluid and eventually reenter the blood. Water reabsorption occurs passively, through osmosis.

3. *Secretion.* **Secretion** is the transport of solutes from the peritubular fluid, across the tubular epithelium, and into the tubular fluid. Secretion is necessary because filtration does not force all the dissolved materials out of the plasma. Tubular secretion provides a backup for the filtration process that can further lower

the plasma concentration of undesirable materials. Secretion can be the primary method of excretion for some compounds, including many drugs.

Together, these processes create a fluid that is very different from other body fluids. Table 26-2 provides an indication of the efficiency of the renal system by comparing the concentrations of some of the substances present in urine and plasma.

All segments of the nephron and collecting system participate in the process of urine formation. Most regions of the nephron perform a combination of reabsorption and secretion, but the balance between the two shifts from one region to another:

- Filtration occurs exclusively in the renal corpuscle, across a filtration membrane consisting of the glomerular endothelium, the lamina densa, and the filtration slits.
- Nutrient reabsorption occurs primarily along the proximal convoluted tubules, but it also occurs elsewhere along the renal tubule and within the collecting system.
- Active secretion occurs primarily at the proximal and distal convoluted tubules.
- The loops of Henle, especially the long loops of the juxtamedullary nephrons, and the collecting system interact to regulate the final volume and solute concentration of the urine.

Normal kidney function can continue only as long as the processes of filtration, reabsorption, and secretion function within relatively narrow limits. A disruption in kidney function has immediate effects on the composition of the circulating blood. If both kidneys are affected, death will occur within a few days unless medical assistance is provided.

 EXPRESSING SOLUTE CONCENTRATIONS The osmotic concentration, or *osmolarity,* of a solution is the total number of solute particles in each liter. Osmolarity is usually expressed in terms of **osmoles** per liter (Osm/l) or **milliosmoles** per liter (mOsm/l). If each liter of a fluid contains one mole of dissolved particles, the solute concentration is 1 Osm/l, or 1000 mOsm/l. Body fluids have an osmotic concentration of about 300 mOsm/l. In comparison, seawater has an osmolarity of about 1000 mOsm/l, and fresh water about 5 mOsm/l. Ion concentrations are often reported in terms of *millequivalents* per liter (mEq/l), whereas the concentrations of large organic molecules are usually reported in terms of grams or milligrams per unit volume of solution (typically mg or g per dl). For a more detailed explanation of the various methods of reporting solute concentrations, see the *Applications Manual.* ▧ *Solutions and Concentrations*

Filtration

Filtration occurs within the renal corpuscle, as fluids move across the wall of the glomerulus and into the capsular space. The process of **glomerular filtration** involves passage across three physical barriers (Figures 26-5e, p. 967, and 26-9a●):

1. *The capillary endothelium.* The glomerular capillaries are fenestrated capillaries with pores ranging from 60 to 100 nm (0.06 to 0.1 μm) in diameter. These openings are small enough to prevent the passage of blood cells, but they are too large to restrict the diffusion of solutes, even those the size of plasma proteins.
2. *The basement membrane.* The lamina densa that surrounds the capillary endothelium has several times the density and thickness of a typical basement membrane. This layer restricts the passage of the larger plasma proteins but permits the movement of smaller plasma proteins, nutrients, and ions.
3. *The filtration slits.* The filtration slits between the pedicels of podocytes are only 6–9 nm wide, small enough to block the passage of most of the smaller protein molecules. As a result, under normal circumstances none of the larger plasma proteins and very few albumin molecules (average diameter 7 nm) enter the capsular space. The filtrate contains dissolved ions and small organic molecules in roughly the same concentrations as in plasma.

Filtration Pressures

We discussed the major forces that act across capillary walls in Chapters 21 and 22. (You may find it helpful to review Figures 21-12 and 21-13● before you proceed.) ∞ *[pp. 724, 726]* The primary factor involved in glomerular filtration is basically the same as that governing fluid and solute movement across capillaries throughout the body: the balance between hydrostatic and colloid os-

TABLE 26-2 Significant Differences between Urinary and Plasma Solute Concentrations

Component	Urine	Plasma
Ions (mEq/l)		
Sodium (Na$^+$)	147.5	138.4
Potassium (K$^+$)	47.5	4.4
Chloride (Cl$^-$)	153.3	106
Bicarbonate (HCO$_3^-$)	1.9	27
Metabolites and nutrients (mg/dl)		
Glucose	0.009	90
Lipids	0.002	600
Amino acids	0.188	4.2
Proteins	0.000	7.5 g/dl
Nitrogenous wastes (mg/dl)		
Urea	1800	10–20
Creatinine	150	1–1.5
Ammonia	60	<0.1
Uric acid	40	3

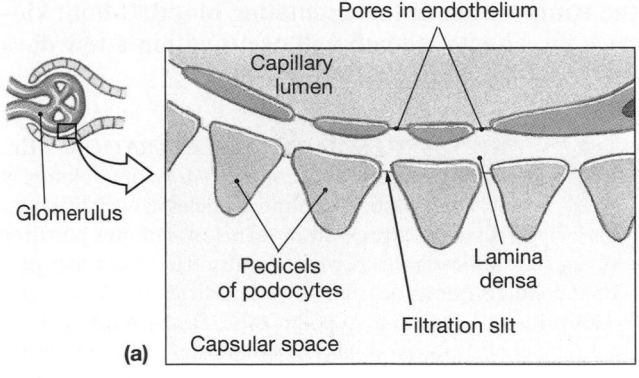

(a)

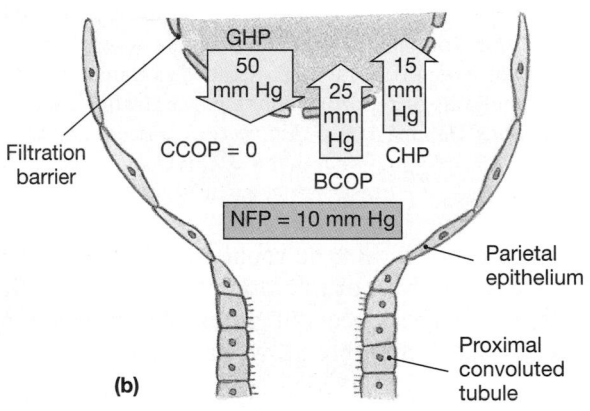

(b)

NFP = (GHP – CHP) – (BCOP – CCOP)
   = (50 – 15) – (25 – 0)
   = 10 mm Hg

GHP = Glomerular (blood) hydrostatic pressure
BCOP = Blood colloid osmotic pressure
CHP = Capsular hydrostatic pressure
NFP = Net filtration pressure
CCOP = Capsular colloid osmotic pressure

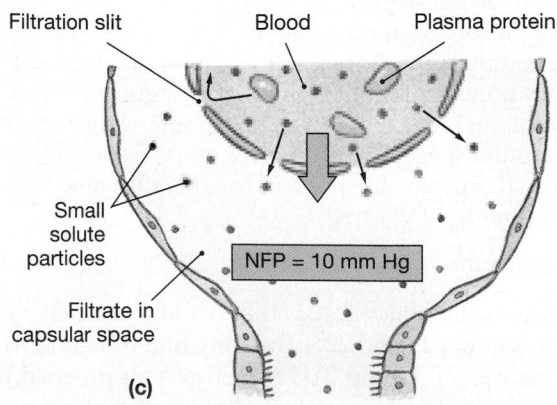

(c)

●**FIGURE 26-9 Glomerular Filtration. (a)** The filtration membrane. **(b)** Forces acting across the glomerular wall. **(c)** The net filtration pressure (NFP).

motic pressures on either side of the capillary walls. These forces are diagrammed in Figure 26-9b,c●.

HYDROSTATIC PRESSURE The *net hydrostatic pressure* tends to drive water and solutes out of the plasma and into the intercapsular space. The net hydrostatic pressure is made up of two opposing components: the *glomerular hydrostatic pressure* and the *capsular hydrostatic pressure.*

Glomerular Hydrostatic Pressure. The **glomerular hydrostatic pressure (GHP)** is the blood pressure in the glomerular capillaries. It tends to push water and solute molecules out of the plasma and into the filtrate. The GHP is significantly higher than capillary pressures elsewhere in the systemic circuit due to the arrangement of vessels at the glomerulus.

Blood pressure is low in typical systemic capillaries, because the capillary blood flows into the venous system, where resistance is relatively low. However, at the glomerulus, blood leaving the glomerular capillaries flows into an efferent arteriole, whose diameter is *smaller* than that of the afferent arteriole. The efferent arteriole offers considerable resistance, and relatively high pressures are needed to force blood into it. As a result, glomerular pressures are similar to those of small arteries, averaging 45–55 mm Hg instead of the 35 mm Hg typical of peripheral capillaries.

Capsular Hydrostatic Pressure. Glomerular hydrostatic pressure is opposed by the **capsular hydrostatic pressure (CHP).** The capsular hydrostatic pressure tends to push water and solutes out of the filtrate and into the plasma. The CHP results from the resistance to flow along the nephron and conducting system. (Before additional filtrate can enter the capsule, some of the filtrate already present must be forced into the proximal convoluted tubule.) The capsular hydrostatic pressure averages about 15 mm Hg.

The *net hydrostatic pressure (NHP)* is the difference between the glomerular hydrostatic pressure and the capsular hydrostatic pressure:

NHP = GHP – CHP = 50 mm Hg –15 mm Hg = 35 mm Hg

COLLOID OSMOTIC PRESSURE As you may recall from Chapter 21, the colloid osmotic pressure of a solution is the osmotic pressure resulting from the presence of suspended proteins. ∞ *[p. 725]* The *net colloid osmotic pressure* has two components: the *blood colloid osmotic pressure* and the *capsular colloid osmotic pressure.*

Blood Colloid Osmotic Pressure. The **blood colloid osmotic pressure (BCOP)** tends to draw water out of the filtrate and into the plasma. It thus opposes filtration. Over the entire length of the glomerular capillary bed, the BCOP averages about 25 mm Hg.

Capsular Colloid Osmotic Pressure. Under normal conditions, very few plasma proteins enter the capsular space, and the **capsular colloid osmotic pressure (CCOP)** is negligible. However, if the glomeruli are damaged by disease or injury and blood proteins begin passing into the capsular space, the CCOP increases. When present, the CCOP tends to draw water out of the plasma and into the filtrate. It thus promotes filtration.

The *net colloid osmotic pressure (NCOP)* is the difference between the blood colloid osmotic pressure and the capsular colloid osmotic pressure:

$$NCOP = BCOP - CCOP = 25 \text{ mm Hg} - 0 \text{ mm Hg}$$
$$= 25 \text{ mm Hg}$$

NET FILTRATION PRESSURE The **net filtration pressure (NFP)** at the glomerulus is the difference between the net hydrostatic pressure and the net colloid osmotic pressure. Under normal circumstances, this relationship can be summarized as

$$NFP = NHP - NCOP = 35 \text{ mm Hg} - 25 \text{ mm Hg}$$
$$= 10 \text{ mm Hg}$$

This value represents the average pressure forcing water and dissolved materials out of the glomerular capillaries and into the capsular spaces.

Abnormal changes in the net filtration pressure can result in significant alterations in kidney function. For example:

- Damage to the glomerulus that increases the CCOP will enhance the movement of water and solutes into the capsular space.
- Interference with normal fluid movement along the tubules or urine flow along the ureters will elevate CHP and reduce or eliminate the net filtration pressure. **AM** *Conditions Affecting Filtration*

Glomerular Filtration Rate

The **glomerular filtration rate (GFR)** is the amount of filtrate your kidneys produce each minute. Each kidney contains about 6 m^2 of filtration surface, and the GFR averages an astounding *125 ml per minute.* This means that roughly 10 percent of the fluid delivered to your kidneys by the renal arteries leaves the bloodstream and enters the capsular spaces.

A *creatinine clearance test* is often used to estimate the GFR. Creatinine, which results from the breakdown of creatine phosphate in muscle tissue, is normally excreted in the urine. Creatinine enters the filtrate at the glomerulus, and it is not reabsorbed in significant amounts. By monitoring the concentrations in the blood and the amount excreted in the urine, a clinician can easily estimate the GFR. For example, consider a person who excretes 84 mg of creatinine each hour and has a plasma creatinine concentration of 1.4 mg/dl. The filtration rate is equal to the amount secreted, divided by the plasma concentration: (84 mg/h)/(1.4 mg/dl) = 60 dl/h, or 100 ml/min. The GFR is usually reported in terms of milliliters per minute.

This value is only an approximation of the GFR, because up to 15 percent of creatinine enters the urine through active tubular secretion. When necessary, a more accurate GFR determination can be performed by using the complex carbohydrate *inulin*, which is not metabolized in the body and is neither reabsorbed nor secreted by the kidney tubules.

In the course of a single day, your glomeruli generate about 180 liters (50 gal) of filtrate, roughly 70 times the total plasma volume. But as the filtrate passes through the renal tubules, about 99 percent of it is reabsorbed. You should now appreciate the significance of tubular reabsorption!

BLOOD PRESSURE AND GLOMERULAR FILTRATION The glomerular filtration rate depends on the filtration pressure acting across the glomerular capillaries. Any factor that alters the filtration pressure will therefore alter the GFR and affect kidney function. One of the most significant factors is a drop in renal blood pressure.

Despite the large volume of filtrate generated, the filtration pressure is relatively low. Whenever the mean arterial pressure falls by 10 percent (from a normal value of about 100 mm Hg to 90 mm Hg), the GFR is severely restricted. If blood pressure at the glomeruli drops by 20 percent, from 50 mm Hg to 40 mm Hg, kidney filtration will cease altogether, because the net filtration pressure will be 0 mm Hg. The kidneys are therefore sensitive to changes in blood pressure that have little or no effect on other organs. Hemorrhaging, shock, and dehydration are relatively common clinical conditions that can cause a dangerous decline in the GFR.

Controlling the GFR

Glomerular filtration is the vital first step essential to all other kidney functions. If filtration does not occur, waste products are not excreted, pH control is jeopardized, and an important mechanism of blood volume regulation is eliminated. It should be no surprise to find that a variety of regulatory mechanisms exist to ensure that your GFR remains within normal limits.

Filtration depends on adequate circulation to the glomerulus and the maintenance of normal filtration pressures. Three interacting levels of control stabilize your GFR: (1) *autoregulation* at the local level, (2) *hormonal regulation* initiated by the kidneys, and (3) *autonomic regulation*, primarily by the sympathetic division of the ANS.

AUTOREGULATION OF THE GFR The goal of autoregulation is to maintain an adequate GFR despite changes in local blood pressure and blood flow. This goal is accomplished by changing the diameters of the afferent arterioles, the efferent arterioles, and the glomerular capillaries. The most important regulatory mechanisms stabilize the GFR when systemic blood pressure declines. A reduction in blood flow and a decline in glomerular blood pressure trigger (1) the dilation of the afferent arteriole, (2) the relaxation of mesangial cells and dilation of the glomerular capillaries, and (3) the constriction of the efferent arteriole. This combination increases blood flow and elevates glomerular blood pressure to normal levels. As a result, filtration rates remain relatively constant. The GFR also remains relatively constant when systemic blood pressure rises. A rise in renal blood pressure stretches the walls of the afferent arterioles, and the smooth muscle cells respond by contracting. The reduction in the diameter of the afferent arterioles decreases glomerular blood flow and keeps the GFR within normal limits.

HORMONAL REGULATION OF THE GFR The GFR is regulated by the hormones *renin* and *atrial natriuretic peptide (ANP)*. We introduced those hormones and their actions in Chapters 18 and 21. ∞ *[pp. 618–620, 733]*

The JGA and Renin Secretion. Renin is an enzyme released when (1) glomerular blood pressure drops or (2) the osmolarity of the tubular fluid that reaches the DCT decreases.

Renin is released by the juxtaglomerular apparatus (JGA) when renal blood flow declines as a result of a decrease in blood volume, a fall in systemic pressures, or a blockage in the renal artery or its tributaries. Renin is also released in response to a reduction in the osmolarity of the tubular fluid reaching the JGA. When the rate of glomerular filtration decreases, so does the rate of fluid movement along the nephron. Because the tubular fluid then spends more time in the ascending limb of the loop of Henle, the concentration of sodium and chloride ions becomes abnormally low.

We introduced the functions of the renin–angiotensin system in Chapter 18. ∞ *[p. 620]* Renin converts the inactive protein angiotensinogen to angiotensin I, and the angiotensin-converting enzyme in the capillaries of the lungs then converts angiotensin I to the active form, angiotensin II. Figure 26-10● diagrams the primary effects of angiotensin II. They include the following:

- *In peripheral capillary beds,* angiotensin II causes a brief but powerful vasoconstriction of arterioles and precapillary sphincters, elevating pressures in the renal arteries and their tributaries.
- *At the nephron,* angiotensin II causes the constriction of the efferent arteriole, further elevating glomerular pressures and filtration rates. It also (1) directly stimulates the reabsorption of sodium ions and water at the proximal convoluted tubule and (2) indirectly stimulates the reabsorption of sodium by the DCT and collecting system by triggering aldosterone secretion.
- *In the CNS,* angiotensin II (1) triggers the release of ADH, stimulating the reabsorption of water in the distal portion of the DCT and the collecting system, (2) causes

the sensation of thirst, and (3) increases sympathetic motor tone, further stimulating peripheral vasoconstriction.
- *At the adrenal glands,* angiotensin II stimulates the secretion of aldosterone by the adrenal cortex. The aldosterone accelerates sodium reabsorption in the DCT and cortical portion of the collecting system.

The final results are an increase in systemic blood volume and blood pressure and the restoration of normal GFR.

ANP and the GFR. Atrial natriuretic peptide is released in response to the stretching of the atrial walls of the heart by increased blood volume or blood pressure. Among its other effects, ANP triggers the dilation of the afferent arteriole and constriction of the efferent arteriole. This effect elevates glomerular pressures and increases the GFR, leading to increased urine production and decreased blood volume and pressure.

AUTONOMIC REGULATION OF THE GFR Most of the autonomic innervation of the kidneys consists of sympathetic postganglionic fibers. (The role of the few parasympathetic fibers in regulating kidney function is not known.) Sympathetic activation has one direct effect on the GFR: It produces a powerful vasoconstriction of the afferent arterioles, decreasing the GFR and slowing the production of filtrate. The sympathetic activation triggered by an acute fall in blood pressure or a heart attack will override the local regulatory mechanisms that act to stabilize the GFR. As the crisis passes and sympathetic tone decreases, the filtration rate gradually returns to normal levels.

When the sympathetic division alters regional patterns of blood circulation, blood flow to the kidneys is often affected. For example, the dilation of superficial vessels in warm weather shunts blood away from your kidneys, and glomerular filtration declines temporarily. The effect becomes especially pronounced during periods of strenuous exercise. As the blood flow to your skin and skeletal muscles increases, kidney perfusion gradually decreases. These changes may be opposed, with variable success, by autoregulation at the local level.

At maximal levels of exertion, renal blood flow may be less than 25 percent of your normal resting levels. This reduction can create problems for distance swimmers and marathon runners, because metabolic wastes build up over the course of a long competition. *Proteinuria* (protein loss in the urine) commonly occurs after such events, because the glomerular cells have been injured by prolonged hypoxia (low oxygen levels). If the damage is substantial, *hematuria* (blood loss in the urine) will occur. Hematuria develops in roughly 18 percent of marathon runners. Proteinuria and hematuria generally disappear within 48 hours, as the glomerular tissues are repaired. However, a small number of marathon and ultramarathon runners experience *renal failure,* with permanent impairment of kidney function.

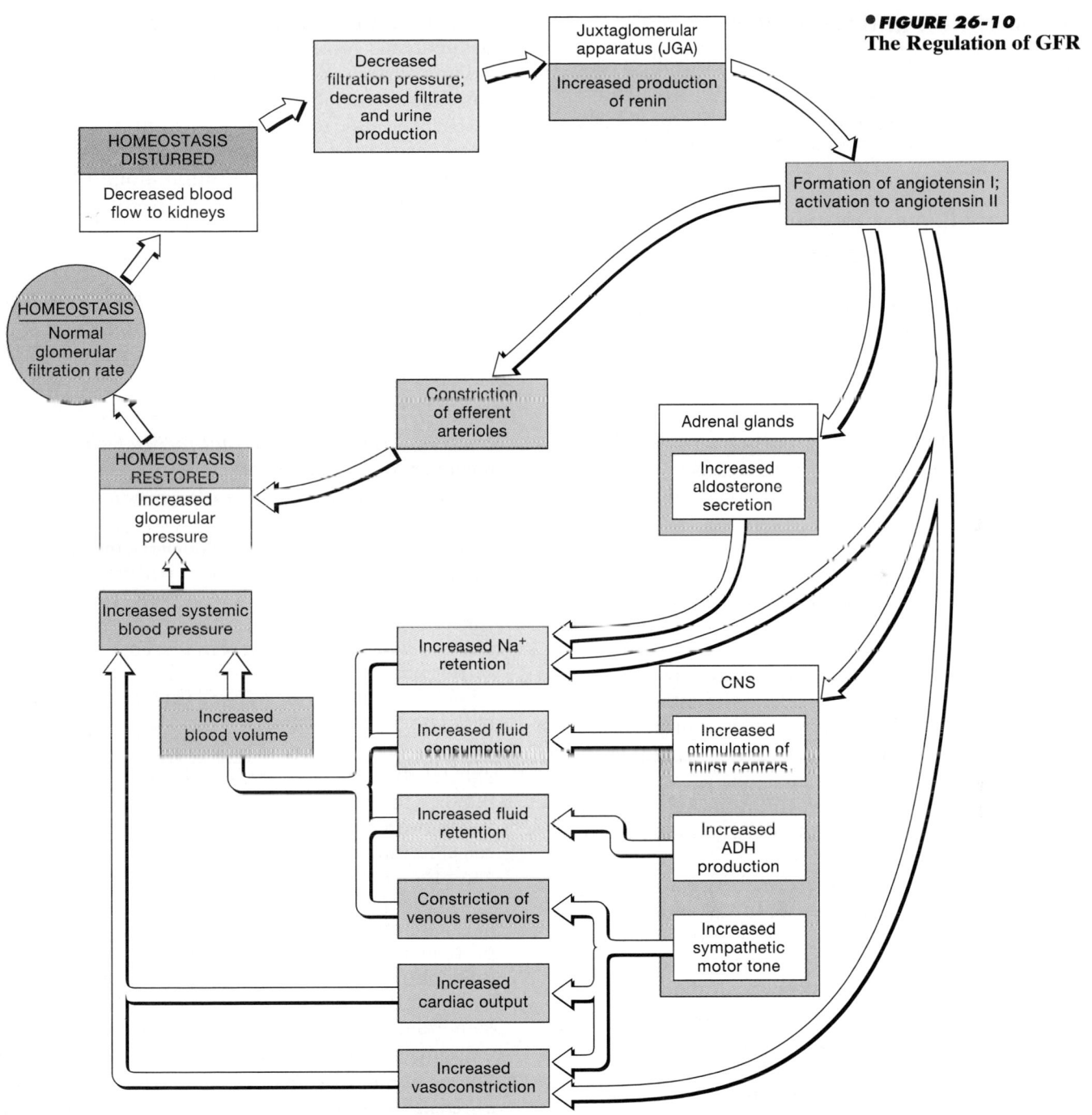

●FIGURE 26-10
The Regulation of GFR

RENAL FAILURE Renal failure occurs when the kidneys become unable to perform the excretory functions needed to maintain homeostasis. When kidney filtration slows for any reason, urine production declines. As the decline continues, symptoms of renal failure appear because water, ions, and metabolic wastes are retained. Virtually all systems in the body are affected. For example, fluid balance, pH, muscular contraction, metabolism, and digestive function are disturbed. The individual generally becomes hypertensive, anemia develops due to a decline in erythropoietin production, and CNS problems may lead to sleeplessness, seizures, delirium, and even coma.

Acute renal failure occurs when exposure to toxic drugs, renal ischemia, urinary obstruction, or trauma causes filtration to slow suddenly or to stop altogether. The reduction in kidney function occurs over a period of a few days and persists for weeks. Sensitized individuals may also develop acute renal failure after exposure to antibiotics or anesthetics. Individuals in acute renal failure may recover if they survive the incident. (With supportive treatment, the mortality rate is approximately 50 percent.) In *chronic renal failure*, kidney function deteriorates gradually, and the associated problems accumulate over time. The condition generally cannot be reversed, only prolonged, and symptoms of acute renal failure eventually develop. The symptoms of chronic and acute renal failure can be relieved by the use of dialysis equipment, but this treatment is not a cure. In some cases, kidney transplantation, an option discussed later in the chapter, is the best long-term solution.

Reabsorption and Secretion

Reabsorption and secretion at the kidneys involve a combination of diffusion, osmosis, and carrier-mediated transport. We have considered diffusion and osmosis in many chapters, so we shall pause at this time only for a brief review of carrier-mediated transport mechanisms.

Carrier-Mediated Transport

In previous chapters, we considered four major types of *carrier-mediated transport:* (1) *facilitated diffusion,* (2) *active transport,* (3) *cotransport,* and (4) *countertransport.*

In facilitated diffusion (Figure 3-9a●, p. 77) a carrier protein transports a molecule across the cell membrane without an expenditure of energy. The transport always follows the concentration gradient for the ion or molecule transported. Facilitated diffusion is important in the reabsorption of glucose and amino acids when their concentrations in the tubular fluid are relatively high.

Active transport is driven by the hydrolysis of ATP to ADP on the inner membrane surface. The sodium–potassium exchange pump (Figure 3-9b●, p. 77), which ejects three intracellular sodium ions in exchange for two extracellular potassium ions, is a familiar example. Exchange pumps are just one of several types of ion pumps active along the kidney tubules. Many other carrier proteins transport only one type of ion across the membrane. Examples include the active transport mechanisms for calcium, magnesium, chloride, iodide, and iron. Active transport mechanisms can operate despite an opposing concentration gradient.

In cotransport, carrier protein activity is not directly linked to the hydrolysis of ATP. In this process, two substrates (ions, molecules, or both) cross the membrane while they are bound to the carrier protein. The movement always follows the concentration gradient of *at least one of the transported substances.* In most instances, a sodium ion is one of the substrates transported. This fact is significant, because there is always a concentration gradient for sodium ions across the cell membrane. This concentration gradient then "drives" the transport of some other ion or molecule. Cotransport mechanisms are responsible for the reabsorption of glucose and other simple sugars, amino acids, lactic acid, phosphate, chloride ions, and hydrogen ions from the tubular fluid.

Countertransport resembles cotransport in all respects except that the two transported ions move in opposite directions. Most cells in the body use sodium–calcium countertransport to keep intracellular calcium concentrations low. Along the kidney tubules, carriers on the lumenal surface bring calcium ions *into the cell,* whereas those on the basal surface (attached to the basement membrane) pump them *into the peritubular fluid.* Parathyroid hormone and calcitriol stimulate these carrier proteins and reduce the urinary loss of calcium ions. The DCT and collecting system contain countertransport mechanisms that exchange chloride ions for bicarbonate ions. Bicarbonate ions are important buffers, and their reabsorption can help prevent a decline in blood pH. Hydrogen ion secretion along the DCT and collecting system occurs by means of a countertransport mechanism that reabsorbs sodium ions from the tubular fluid.

CHARACTERISTICS OF CARRIER-MEDIATED TRANSPORT All four carrier-mediated processes share features that are important for an understanding of kidney function:

1. *In carrier-mediated transport, a specific substrate binds to a carrier protein that facilitates its movement across the membrane.*

2. *A given carrier protein normally works in one direction only.* In facilitated diffusion, that direction is determined by the concentration gradient of the substance being transported. In active transport, cotransport, and countertransport, the location and orientation of the carrier proteins determine whether a particular substance will be reabsorbed or secreted. For example, the carrier protein that transports amino acids from the tubular fluid to the cytoplasm will not carry amino acids back into the tubular fluid.

3. *The distribution of carrier proteins can vary from one portion of the cell surface to another.* Transport between the tubular fluid and the interstitial fluid involves two steps, because the material must enter the cell at its lumenal surface and then leave the cell and enter the peritubular fluid. Each step involves a different set of carrier proteins. For example, the lumenal membranes of cells along the PCT contain carrier proteins that bring amino acids, glucose, and many other nutrients into these cells by sodium-linked cotransport, whereas the surfaces near the basement membrane contain carrier proteins that move those nutrients out of the cell by facilitated diffusion.

4. *The membrane of a single tubular cell contains many different types of carrier protein.* Each cell can have multiple functions, and the same cell that reabsorbs one compound can secrete another.

5. *Carrier proteins, like enzymes, can be saturated.* As you may recall from Chapter 2, when an enzyme is *saturated,* further increases in substrate concentration have no effect on the rate of reaction. ∞ *[p. 56]* When a carrier protein is saturated, further increases in substrate concentration have no effect on the rate of transport across the cell membrane. For any substance, the concentration at saturation is called the **transport maximum (T_m),** or *tubular maximum.* The saturation of carrier proteins involved with tubular secretion seldom occurs in healthy individuals, but carriers involved with tubular reabsorption are often at risk of saturation, especially during the absorptive period.

THE T_M ***AND RENAL THRESHOLD*** Normally, any plasma proteins and nutrients, such as amino acids and glucose, are removed from the tubular fluid through cotransport or facilitated diffusion. If the concentrations of these nutrients rise in the tubular fluid, the rates of reabsorption increase until the carrier proteins are saturated. *A concentration higher than the tubular maximum will exceed the reabsorptive abilities of the nephron, and some of the material will remain in the tubular fluid and appear in the urine.* The transport maximum thus determines the **renal threshold,** the plasma concentration at which a specific compound or ion will begin appearing in the urine.

The renal threshold varies with the substance involved. The renal threshold for glucose is approximately 180 mg/dl. When plasma glucose concentrations remain higher than 180 mg/dl, glucose concentrations in the tubular fluid will exceed the T_m of the tubular cells, and glucose will appear in the urine. The presence of glucose in the urine is a condition called *glycosuria.* After you have eaten a meal rich in carbohydrates, your plasma glucose levels may exceed the T_m for a brief period. Your liver will quickly lower circulating glucose levels, and very little glucose will be lost in the urine. Chronically elevated plasma and urinary glucose concentrations are very unusual. (Glycosuria is one of the key symptoms of diabetes mellitus, which we described in Chapter 18. ∞ *[p. 623]*)

The renal threshold for amino acids is lower than that for glucose; amino acids appear in the urine when plasma concentrations exceed 65 mg/dl. Your plasma amino acid levels commonly exceed the renal threshold after you have eaten a protein-rich meal, causing some amino acids to appear in the urine. This condition is termed *aminoaciduria.*

The T_m values for water-soluble vitamins are relatively low, and as a result, excess quantities of these vitamins are excreted in the urine. (This is typically the fate of daily water-soluble vitamin supplements.) The renal tubular cells ignore a number of other compounds in the tubular fluid. As water and other compounds are removed, the concentration of those materials gradually rises. Table 26-3 contains a partial listing of substances actively reabsorbed, secreted, or ignored by the renal tubules.

We shall now proceed along the nephron and consider the changes in the composition and concentration of the filtrate that is produced at the renal corpuscle. Most of what follows applies equally to cortical and juxtamedullary nephrons. The major differences between the two types of nephrons are that the loop of Henle of a cortical nephron is shorter and does not extend as far into the medulla as the loop of Henle of a juxtamedullary nephron (Figure 26-6d•, p. 968). The long loop of Henle in a juxtamedullary nephron extends deep into the renal pyramids, where it plays a vital role in water conservation and the formation of concentrated urine. Because this process is so important, affecting the tubular fluid pro-

TABLE 26-3 Tubular Reabsorption and Secretion

Reabsorbed	Secreted	No Transport Mechanism
Ions	**Ions**	Urea
Na^+, Cl^-, K^+,	K^+, H^+, Ca^{2+},	Water
Ca^{2+}, Mg^{2+},	PO_4^{3-}	Urobilinogen
SO_4^{2-},		Bilirubin
HCO_3^-	**Wastes**	
	Creatinine	
Metabolites	Ammonia	
Glucose	Organic acids	
Amino acids	and bases	
Proteins		
Vitamins	**Miscellaneous**	
	Neurotransmitters (ACh,	
	NE, E, dopamine)	
	Histamine	
	Drugs (penicillin, atropine,	
	morphine, many others)	

duced by every nephron in the kidney, and because the functions of the renal corpuscle, PCT, and DCT are the same in all nephrons, we shall use a juxtamedullary nephron as our example.

The Proximal Convoluted Tubule

The cells of the PCT normally reabsorb 60–70 percent of the volume of the filtrate produced in the renal corpuscle. The reabsorbed materials enter the peritubular fluid and diffuse into the peritubular capillaries.

The PCT has five major functions:

1. ***Reabsorption of organic nutrients.*** Under normal circumstances, before the tubular fluid enters the loop of Henle, the PCT reabsorbs more than 99 percent of the glucose, amino acids, and other organic nutrients. This reabsorption involves a combination of facilitated transport and cotransport.

2. ***Active reabsorption of ions.*** The PCT actively transports ions, including sodium, potassium, magnesium, bicarbonate, phosphate, and sulfate ions. The ion pumps involved are individually regulated and may be influenced by circulating ion or hormone levels. For example, angiotensin II stimulates Na^+ reabsorption along the PCT. By absorbing carbon dioxide, the PCT indirectly recaptures roughly 90 percent of the bicarbonate ions from the tubular fluid (Figure 26-11•). Bicarbonate is important in stabilizing blood pH, a process we shall examine further in Chapter 27.

3. ***Reabsorption of water.*** The reabsorptive processes have a direct effect on the solute concentrations inside and outside the tubules. The filtrate entering the PCT has the same osmotic concentration as the peritubular fluid. As transport activities proceed, the solute concentration of the tubular fluid decreases and that of the peritubular fluid and adjacent capillaries increases. Osmosis then pulls water out of the tubular fluid and

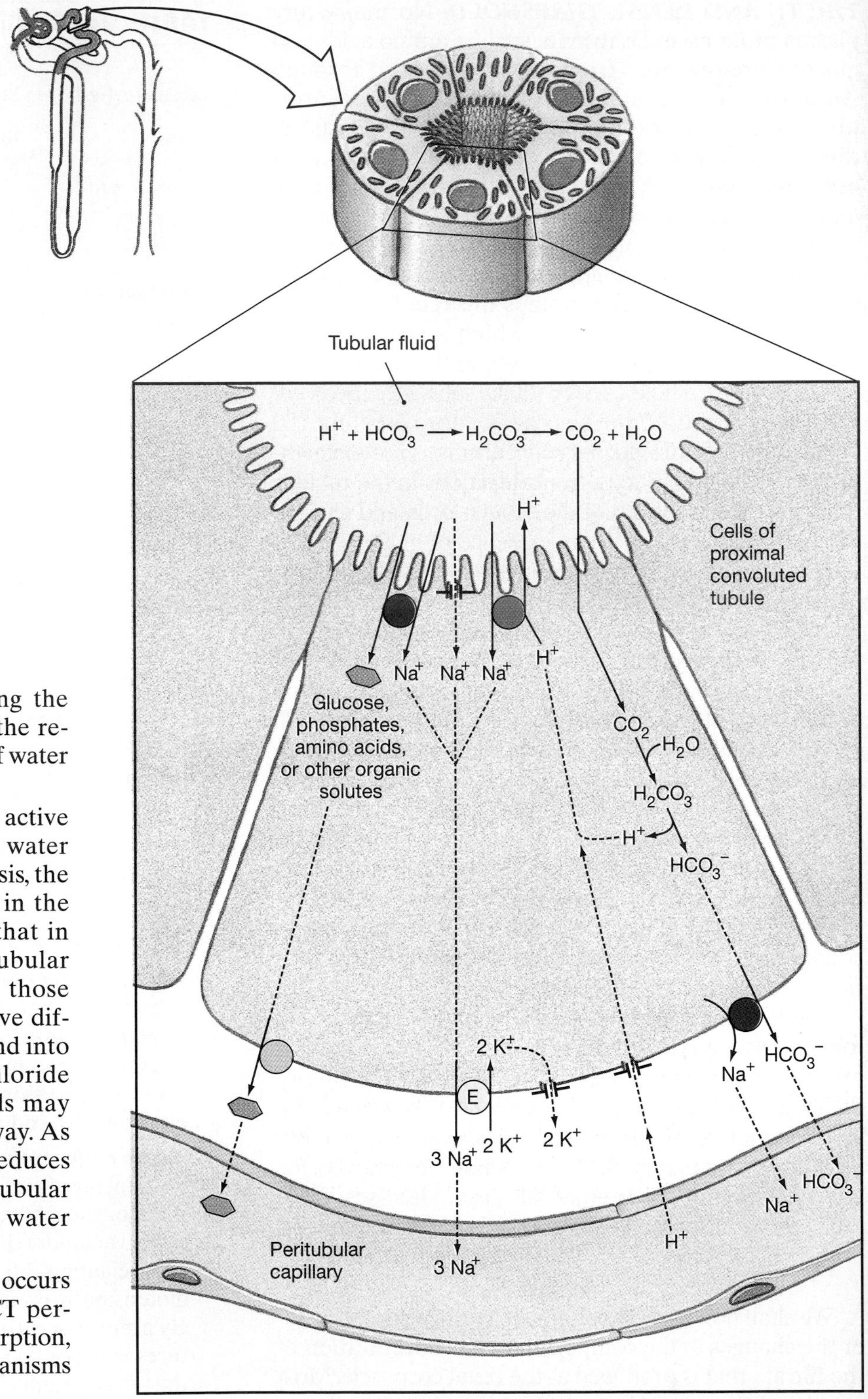

•*FIGURE 26-11* **Transport Activities at the PCT.** Sodium ions may enter the tubular cell from the filtrate by diffusion, cotransport, or countertransport mechanisms. The sodium ions are then pumped into the peritubular fluid via the sodium–potassium exchange pump. Other reabsorbed solutes may be ejected into the peritubular fluid by separate active transport mechanisms.

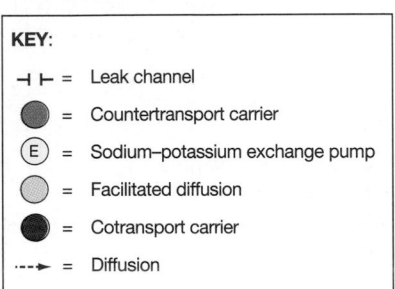

KEY:

⊣ ⊢ = Leak channel

⬤ = Countertransport carrier

Ⓔ = Sodium–potassium exchange pump

◯ = Facilitated diffusion

● = Cotransport carrier

---▶ = Diffusion

into the peritubular fluid. Along the PCT, this mechanism results in the reabsorption of roughly 108 liters of water each day.

4. *Passive reabsorption of ions.* As active reabsorption of ions occurs and water leaves the tubular fluid by osmosis, the concentration of other solutes in the tubular fluid increases above that in the peritubular fluid. If the tubular cells are permeable to them, those solutes will move through passive diffusion across the tubular cells and into the peritubular fluid. Urea, chloride ions, and lipid-soluble materials may diffuse out of the PCT in this way. As this diffusion occurs, it further reduces the solute concentration of the tubular fluid and promotes additional water reabsorption through osmosis.

5. *Secretion.* Active secretion also occurs along the PCT. Because the DCT performs comparatively little reabsorption, we will consider secretory mechanisms when we discuss the DCT.

Sodium ion reabsorption plays an important role in these processes (Figure 26-11•). Sodium ions may enter the tubular cells (1) by diffusion through sodium leak channels, (2) in the sodium-linked cotransport of glucose, amino acids, or other organic solutes, and (3) through countertransport for hydrogen ions. Once inside the tubular cell, the sodium ions diffuse toward the basement membrane. The cell membrane in this region

contains sodium–potassium exchange pumps that eject sodium ions in exchange for extracellular potassium ions. The ejected sodium ions then diffuse into the adjacent peritubular capillaries. Most of the potassium ions reclaimed by the cell will diffuse back out of the cell and into the peritubular fluid through K⁺ leak channels.

The reabsorption of ions and compounds along the PCT involves many different carrier proteins. Some

---

Figure labels (within diagram):

Tubular fluid

$H^+ + HCO_3^- \longrightarrow H_2CO_3 \longrightarrow CO_2 + H_2O$

H^+

Cells of proximal convoluted tubule

Na^+ Na^+ Na^+ H^+

Glucose, phosphates, amino acids, or other organic solutes

CO_2 H_2O

H_2CO_3

H^+

HCO_3^-

$2 K^+$

Ⓔ

$3 Na^+$ $2 K^+$ $2 K^+$

Na^+ HCO_3^-

HCO_3^-

Na^+

Peritubular capillary

$3 Na^+$

H^+

people have an inherited inability to manufacture one or more of these carrier proteins and are therefore unable to recover specific solutes from the tubular fluid. For example, in *renal glycosuria* (glī-cō-SOO-rē-uh), a defective carrier protein makes it impossible for the PCT to reabsorb glucose from the tubular fluid. `AM` *Inherited Problems with Tubular Function*

☑ How does the reabsorption of calcium ions by the kidney affect the sodium ion concentration of the filtrate?

☑ What occurs when the plasma concentration of a substance exceeds its tubular maximum?

☑ How would a decrease in blood pressure affect the glomerular filtration rate (GFR)?

The Loop of Henle and Countercurrent Exchange

Roughly 60–70 percent of the volume of filtrate produced at the glomerulus has been reabsorbed before the tubular fluid reaches the loop of Henle. In the process, the useful organic substrates have been reclaimed, along with many mineral ions. The loop of Henle will reabsorb roughly half of the water as well as two-thirds of the sodium and chloride ions remaining in the tubular fluid.

COUNTERCURRENT MULTIPLICATION The loop of Henle accomplishes this reabsorption in a very efficient way. It uses the principle of countercurrent exchange, which we introduced in Chapter 25 in our discussion of heat conservation techniques (Figure 25-20c•, p. 953).

The thin descending limb and the thick ascending limb of the loop of Henle are very close together, separated by peritubular fluid. The exchange that occurs between these segments is called **countercurrent multiplication.** *Countercurrent* refers to the fact that the exchange occurs between fluids moving in opposite directions; the tubular fluid in the descending limb is moving toward the renal pelvis, whereas the tubular fluid in the ascending limb is moving toward the cortex. *Multiplication* refers to the fact that the effect of the exchange increases as fluid movement continues.

The two parallel segments of the loop of Henle have very different permeability characteristics. The thin descending limb is permeable to water but relatively impermeable to solutes. The thick ascending limb, which is relatively impermeable to both water and solutes, contains active transport mechanisms that pump sodium and chloride ions from the tubular fluid into the peritubular fluid of the medulla. The basic concept involved in countercurrent multiplication is as follows:

■ *Sodium and chloride are pumped out of the thick ascending limb and into the peritubular fluid.*

■ *This pumping elevates the osmotic concentration in the peritubular fluid around the thin descending limb.*

■ *The result is an osmotic flow of water out of the thin descending limb and into the peritubular fluid, increasing the solute concentration inside the thin descending limb.*

■ *The arrival of the highly concentrated solution in the thick ascending limb accelerates the transport of sodium and chloride ions into the peritubular fluid of the medulla.*

Note that this arrangement is a simple positive feedback loop: Solute pumping at the ascending limb leads to higher solute concentrations in the descending limb, which then result in accelerated solute pumping in the ascending limb.

Figure 26-12a• diagrams ion transport across the epithelium of the thick ascending limb. Active transport at the lumenal surface moves sodium, potassium, and chloride ions out of the tubular fluid. The carrier is called a **Na^+–K^+/2 Cl^- transporter,** because each cycle of the pump carries a sodium ion, a potassium ion, and two chloride ions into the tubular cell. Potassium and chloride ions are pumped across the basal membrane of the cell, but the potassium ions are recovered as the sodium–potassium exchange pump ejects sodium ions from the tubular cell. The reclaimed potassium ions then diffuse back into the lumen of the tubule through potassium leak channels. The net result is that Na^+ and Cl^- enter the peritubular fluid of the renal medulla.

The removal of sodium and chloride ions from the tubular fluid in the ascending limb elevates the osmotic concentration of the peritubular fluid around the thin descending limb (Figure 26-12b•). The thin descending limb is permeable to water but impermeable to solutes. Thus, as the tubular fluid travels into the medulla along the thin descending limb, water moves out of the tubule by osmosis. Solutes remain behind, so the tubular fluid reaching the turn of the loop has a higher osmolarity than it did when it entered the medulla.

The pumping mechanism of the thick ascending limb is very effective: Almost two-thirds of the sodium and chloride ions that enter the thick ascending limb are pumped out of the tubular fluid before it reaches the DCT. In other tissues, differences in solute concentration are quickly eliminated through osmosis. Osmosis cannot occur across the wall of the thick ascending limb, however, because the tubular epithelium is impermeable to water. Thus, as Na^+ and Cl^- are removed, the solute concentration in the tubular fluid declines.

The rate of ion transport by the thick ascending limb is proportional to the concentration in the tubular fluid. As a result, more sodium and chloride ions are pumped into the medulla at the start of the thick ascending limb, where NaCl concentrations are highest, than near the cortex, where the concentrations are

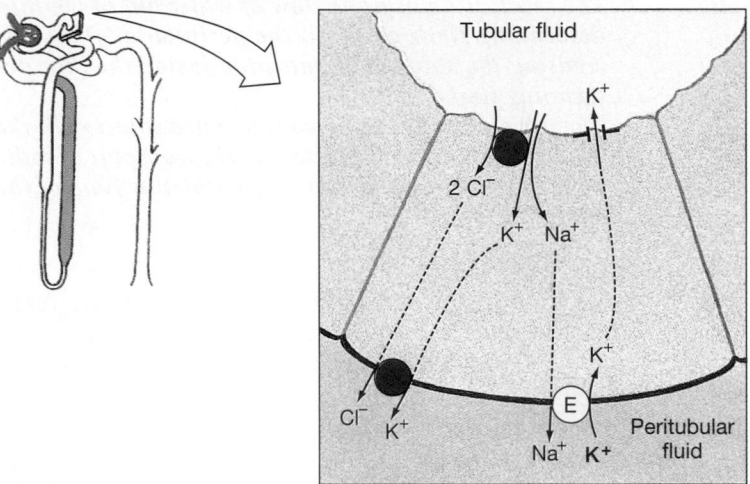

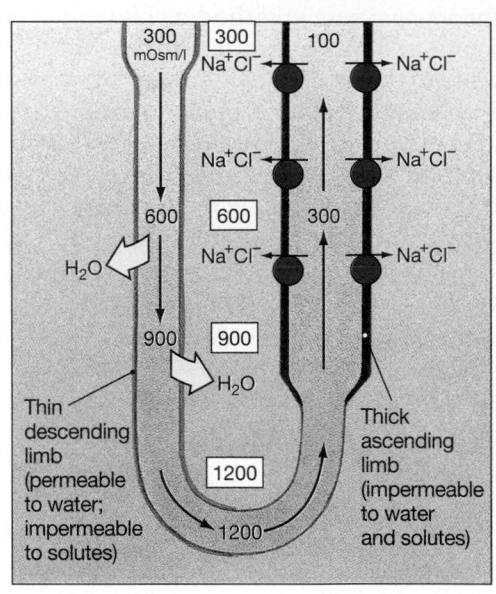

(a) The mechanism of sodium and chloride ion transport involves (1) a Na^+-K^+/ 2 Cl^- carrier at the apical surface and (2) two carriers, a potassium–chloride cotransport pump and a sodium–potassium exchange pump, at the basal surface of the tubular cell. Because the potassium ions removed from the lumen of the tubule ultimately diffuse back in through leak channels, the net result is the transport of sodium and chloride ions into the peritubular fluid.

(b) Active transport of NaCl along the ascending thick limb results in the movement of water from the descending limb.

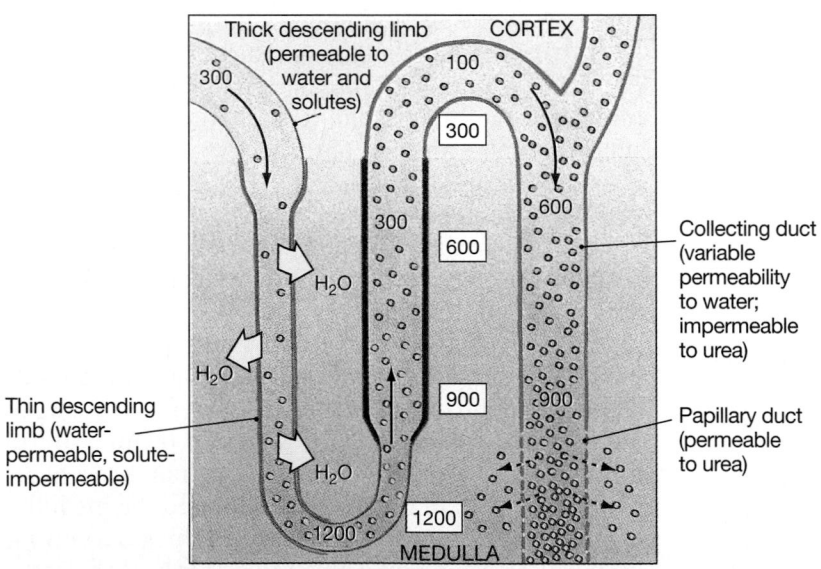

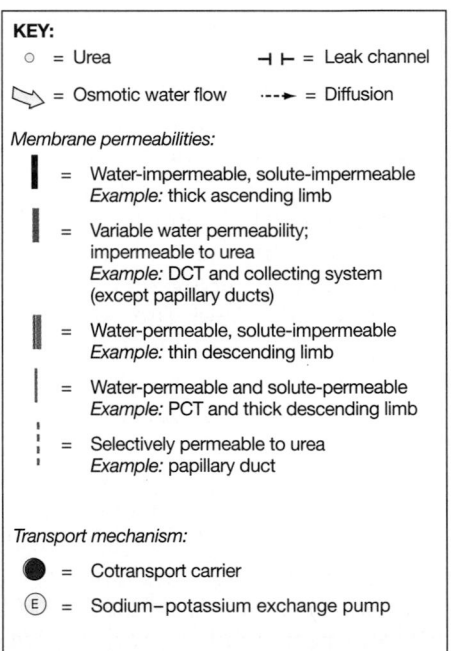

(c) The permeability characteristics of the loop and the collecting duct tend to concentrate urea in the tubular fluid and in the medulla. The ascending thick limb, DCT, and collecting duct are impermeable to urea. As water reabsorption occurs, the urea concentration rises. The papillary ducts' permeability to urea accounts for roughly one-third of the solutes in the deepest portions of the medulla.

•**FIGURE 26-12** **Countercurrent Multiplication and Concentration of Urine**

relatively low. This regional difference in the rate of ion transport is the basis of the concentration gradient within the medulla. The tubular fluid arrives at the DCT with an osmotic concentration of only about 100 mOsm/l, one-third of the concentration of the peritubular fluid of the renal cortex.

THE CONCENTRATION GRADIENT OF THE MEDULLA In a normal kidney, the maximum solute concentration of the peritubular fluid near the turn of the loop is about 1200 mOsm/l. Sodium and chloride ions pumped out of the ascending limb of the loop of Henle account for roughly two-thirds of that gradient (750 mOsm/l). The rest of the concentration gradient results from the presence of urea. To understand how the urea arrived in the medulla, we must look ahead to events in the last segments of the collecting system (Figure 26-12c●). The thick ascending limb of the loop of Henle and the DCT, connecting tubules, and the collecting ducts are impermeable to urea. As water reabsorption occurs, the concentration of urea gradually rises, and the tubular fluid reaching the papillary duct typically contains urea at a concentration of about 450 mOsm/l. Because the papillary ducts are permeable to urea, the urea concentration in the deepest portions of the medulla also averages 450 mOsm/l.

BENEFITS OF COUNTERCURRENT MULTIPLICA-TION The countercurrent mechanism performs two services for the kidneys:

1. It is an efficient way to reabsorb solutes and water before the tubular fluid reaches the DCT and collecting system.
2. It establishes a concentration gradient that will permit the passive reabsorption of water from the tubular fluid in the collecting system. This reabsorption is regulated by circulating levels of antidiuretic hormone (ADH).

The tubular fluid arriving at the descending limb of the loop of Henle has an osmolarity of roughly 300 mOsm/l, primarily due to the presence of ions such as Na^+ and Cl^-. The concentration of organic wastes, such as urea, is relatively low. Roughly half of the tubular fluid entering the loop of Henle is reabsorbed along the thin descending limb, and two-thirds of the Na^+ and Cl^- is reabsorbed along the thick ascending limb. As a result, the DCT receives a reduced volume of tubular fluid with an osmolarity of about 100 mOsm/l. Urea and other organic wastes, which were not pumped out of the thick ascending limb, now represent a significant percentage of the dissolved solutes.

The Distal Convoluted Tubule

The composition and volume of the tubular fluid change dramatically as it travels from the capsular space to the DCT. Only 15–20 percent of the initial filtrate volume reaches the DCT, and the concentrations of electrolytes and organic wastes in the arriving tubular fluid no longer resemble the concentrations in blood plasma. Selective reabsorption or secretion, primarily along the DCT, makes the final adjustments in the solute composition and volume of the tubular fluid.

REABSORPTION Throughout most of the DCT, the tubular cells actively transport Na^+ and Cl^- out of the tubular fluid (Figure 26-13a●). Tubular cells along the distal portions of the DCT also contain ion pumps that reabsorb tubular Na^+ in exchange for another cation (usually K^+) (Figure 26-13b●). The ion pump and the Na^+ channels involved are controlled by the hormone

aldosterone, produced by the adrenal cortex. Aldosterone stimulates the synthesis and incorporation of sodium ion pumps and sodium channels in cell membranes along the DCT, the connecting tubule, and the collecting duct. The net result is a reduction in the number of sodium ions lost in the urine. However, sodium ion conservation is associated with potassium ion loss (Figure 26-13b●). Prolonged aldosterone stimulation can therefore produce *hypokalemia,* a dangerous reduction in the plasma K^+ concentration.

As we noted in Chapter 3, sodium–potassium exchange pumps eject three sodium ions from the cell in exchange for two potassium ions. ∞ *[p. 77]* Thus the reabsorption of sodium along the DCT and collecting system cause a net reduction in the osmolarity of the tubular fluid. This reduction results in a small osmotic flow of water out of the tubular fluid. (We shall discuss other mechanisms involved in the reabsorption of water in a later section.)

The DCT is also the primary site of Ca^{2+} reabsorption. This process is regulated by circulating levels of parathyroid hormone and calcitriol. ∞ *[p. 613, 619]*

SECRETION Filtration does not force all the dissolved materials out of the plasma, and blood entering the peritubular capillaries still contains a number of potentially undesirable substances. In most cases, the concentrations of these materials are too low to cause physiological problems. However, any ions or compounds present in the peritubular capillaries will diffuse into the peritubular fluid. If those concentrations become too high, the tubular cells may absorb these materials from the peritubular fluid and secrete them into the tubular fluid. Table 26-3, p. 979, gives a partial listing of substances secreted into the tubular fluid by the DCT and PCT.

The rate of K^+ and H^+ secretion rises or falls in response to changes in their concentrations in the peritubular capillaries. The higher the concentration in the peritubular fluid, the higher the rate of secretion. Potassium and hydrogen ions merit special attention because their concentrations in body fluids must be maintained within relatively narrow limits.

Potassium Ion Secretion. Figure 26-13a,b● diagrams the mechanism of K^+ secretion. Potassium ions are removed from the peritubular fluid in exchange for sodium ions obtained from the tubular fluid. At the lumenal surface of the tubule, these potassium ions diffuse into the lumen through potassium channels. In effect, the tubular cells are trading sodium ions in the tubular fluid for excess potassium ions in the body fluids.

Hydrogen Ion Secretion. Hydrogen ion secretion is also associated with the reabsorption of sodium. Figure 26-13c● shows two secretory routes. Both involve a familiar reaction sequence—the generation of carbonic

•FIGURE 26-13 Tubular Secretion and Solute Reabsorption at the DCT. **(a)** Basic pattern of the absorption of sodium and chloride ions and the secretion of potassium ions. **(b)** Aldosterone-regulated absorption of sodium ions, linked to the passive loss of potassium ions. **(c)** Hydrogen ion secretion and the acidification of the urine occur by two different routes. The central theme is the exchange of hydrogen ions in the cytoplasm for sodium ions in the tubular fluid and the reabsorption of the bicarbonate ions generated in the process.

KEY:

▬▬▬	=	ADH-regulated water permeability
⊣ ⊢	=	Leak channel
⊣ ʌ	=	Aldosterone-sensitive leak channel
Ⓔ	=	Sodium–potassium exchange pump
Ⓐ	=	Aldosterone-sensitive exchange pump
⬤	=	Cotransport carrier
⬤	=	Countertransport carrier
---▶	=	Diffusion

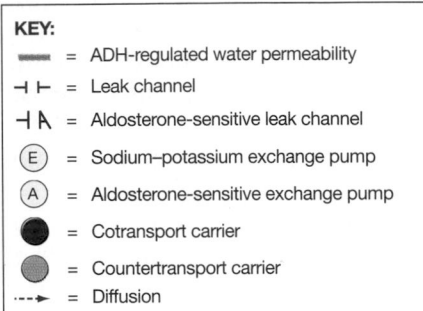

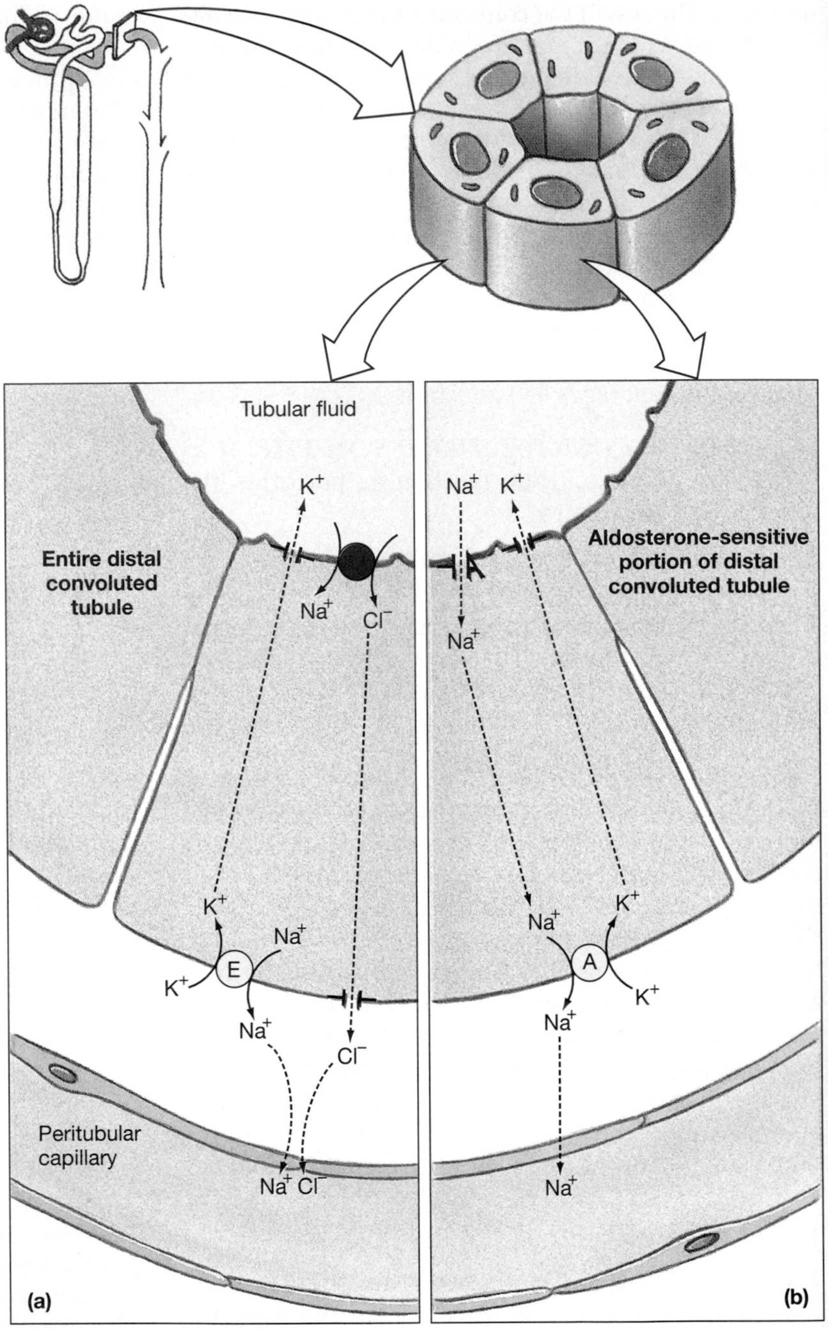

acid by the enzyme *carbonic anhydrase.* ∞ *[p. 847, 849]* Hydrogen ions generated by the dissociation of the carbonic acid are secreted by means of sodium-linked countertransport in exchange for Na$^+$ in the tubular fluid. The bicarbonate ions diffuse into the peritubular fluids and then into the bloodstream, where they help prevent changes in plasma pH.

Hydrogen ion secretion acidifies the tubular fluid while elevating the pH of the blood. Hydrogen ion secretion accelerates when the pH of the blood falls, as in *lactic acidosis,* which can develop after exhaustive muscle activity (∞ *[p. 300]*) or *ketoacidosis.* ∞ *[p. 942]* The combination of H$^+$ removal and HCO$_3^-$ production at the kidneys plays an important role in the control of blood pH. Because one of the secretory pathways is aldosterone-sensitive, aldosterone stimulates H$^+$ secretion. Prolonged aldosterone stimulation can cause *alkalosis.*

In Chapter 25, we noted that the production of lactic acid and ketone bodies during the postabsorptive state can cause acidosis. Under these conditions, the PCT and DCT will deaminate amino acids in reactions that strip off the amino groups (—NH$_2$). The reaction sequence ties up H$^+$ and yields **ammonium ions,** NH$_4^+$, and HCO$_3^-$. As indicated in Figure 26-13c•, the ammonium ions are then pumped into the tubular fluid by sodium-linked countertransport, and the bicarbonate ions enter the bloodstream by way of the peritubular fluid.

Tubular deamination thus has two major benefits: (1) It provides carbon chains suitable for catabolism, and (2) it generates bicarbonate ions that add to the buffering capabilities of the plasma.

The Collecting System

The collecting ducts receive tubular fluid from many nephrons and carry it toward the renal sinus through the concentration gradient in the medulla. The normal amount of water and solute loss in the collecting system is regulated in two ways: (1) by aldosterone, which controls sodium ion pumps in the distal DCT and the cortical portion of the collecting system, and (2) by ADH, which controls the permeability of the DCT and

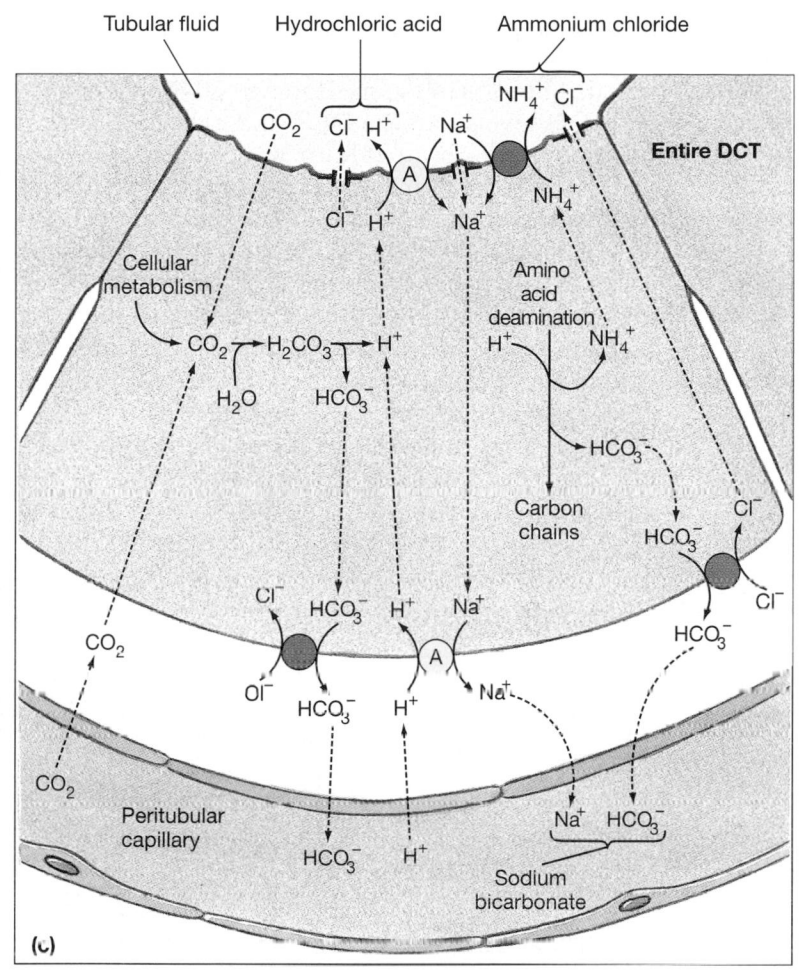

Tubular fluid   Hydrochloric acid   Ammonium chloride

Entire DCT

Cellular metabolism

Amino acid deamination

Carbon chains

Cellular metabolism

Peritubular capillary

Sodium bicarbonate

(c)

concentration in this region usually averages 450 mOsm/l, roughly one-third of the total osmolarity.

SECRETION The collecting system is an important site for the control of body fluid pH through the secretion of hydrogen or bicarbonate ions. If the pH of the peritubular fluid declines, carrier proteins pump hydrogen ions into the tubular fluid and reabsorb bicarbonate ions that help restore normal pH. If the pH of the peritubular fluid rises (a much less common event), the collecting system secretes bicarbonate ions and pumps hydrogen ions into the peritubular fluid. The net result is that the body eliminates a buffer and gains hydrogen ions that will lower the pH. We shall examine those responses in more detail in Chapter 27 when we consider acid–base balance.

The Control of Urine Volume and Osmolarity

Urine volume and osmolarity are regulated by controlling water reabsorption. Water reabsorption by osmosis occurs in the PCT and the descending limb of the loop of Henle. The water permeabilities of these regions cannot be adjusted, and water reabsorption occurs whenever the osmotic concentration of the peritubular fluid exceeds that of the tubular fluid. The ascending limb of the loop of Henle is impermeable to water, but 1–2 percent of the volume of water in the original filtrate is recovered during sodium ion reabsorption in the DCT and collecting system, as we noted on page 983. Because these water movements cannot be prevented, they represent the **obligatory water reabsorption.** Obligatory water reabsorption usually recovers 85 percent of the volume of filtrate produced.

The volume of water lost in the urine depends on how much of the water in the remaining tubular fluid (15 percent of the filtrate volume, or roughly 27 liters per day) is reabsorbed along the DCT and the collecting system. The amount can be precisely controlled through a process called **facultative water reabsorption.** Precise control is possible because these segments are relatively impermeable to water *except in the presence of ADH.* ADH causes the appearance of special *water channels* in the apical membranes. Water channels dramatically enhance the rate of osmotic water movement. The higher the circulating levels of ADH, the greater the number of water channels and the greater the water permeability of these segments.

As we noted earlier in this chapter, the tubular fluid arriving at the DCT has an osmolarity of only about

collecting system to water. The collecting system also has other reabsorptive and secretory functions, many of which are important to the control of body fluid pH.

REABSORPTION Important examples of solute reabsorption in the collecting system include the following:

- **Sodium ion reabsorption.** The cortical region of the collecting system contains aldosterone-sensitive ion pumps that exchange Na$^+$ in the tubular fluid for K$^+$ in the peritubular fluid (Figure 26-13a,b●).
- **Bicarbonate reabsorption.** Bicarbonate ions are reabsorbed in exchange for chloride ions in the peritubular fluid, as we detailed in Figure 26-13c●.
- **Urea reabsorption.** The concentration of urea in the tubular fluid entering the collecting duct is relatively high. The epithelia lining the thick ascending limb of the loop of Henle, the DCT, the connecting tubules, and the collecting ducts are relatively impermeable to urea. When water reabsorption occurs along these segments, the urea concentration rises. The fluid entering the papillary duct generally has the same osmotic concentration as the interstitial fluid of the medulla—about 1200 mOsm/l—but contains a much higher concentration of urea. As a result, urea tends to diffuse out of the tubular fluid and into the peritubular fluid of the deepest portion of the medulla. The urea

100 mOsm/l. In the presence of ADH, osmosis occurs and water moves out of the DCT until the osmolarity of the tubular fluid is the same as that of the surrounding cortex (roughly 300 mOsm/l). The tubular fluid then travels along the collecting duct, which passes through the concentration gradient of the medulla. Additional water is then reabsorbed, and the urine reaching the minor calyx will have an osmolarity closer to 1200 mOsm/l. Just how closely the osmolarity approaches 1200 mOsml/l depends on how much ADH is present.

Figure 26-14• diagrams the effects of ADH on the DCT and collecting system. In the absence of ADH, water reabsorption does not occur in these segments, and all the fluid reaching the DCT will be eliminated in the urine. The individual then produces large amounts of very dilute urine (Figure 26-14a•). That

is what happens in cases of *diabetes insipidus* (∞ *[p. 607]*), when urinary water losses may reach 24 liters per day, and urine osmolarity is 30–400 mOsm/l. As ADH levels rise, the DCT and collecting system become more permeable to water, the amount of water reabsorbed increases, and urine osmolarity climbs. Under maximum ADH stimulation (Figure 26-14b•), the DCT and collecting system become so permeable to water that the osmolarity of the urine is equal to that of the deepest portion of the medulla. Note that the concentration of the urine can never *exceed* that of the medulla, because the concentrating mechanism relies on osmosis.

You are continuously secreting ADH at low levels, and your DCT and collecting system always have a significant degree of water permeability. The DCT normally reabsorbs roughly 9 liters per day, or

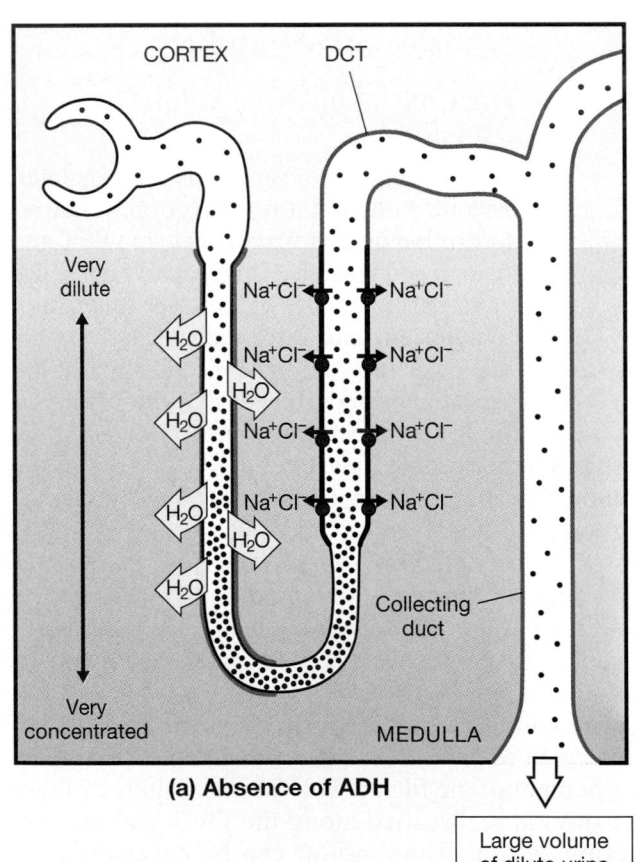

(a) Absence of ADH

Large volume of dilute urine

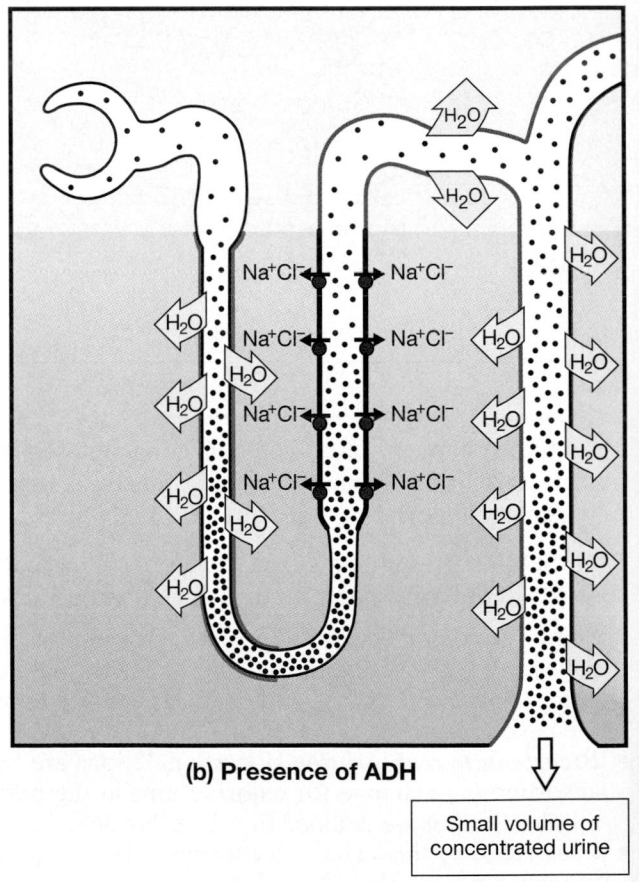

(b) Presence of ADH

Small volume of concentrated urine

•*FIGURE 26-14* **The Effects of ADH on the DCT and Collecting Duct. (a)** Tubule permeabilities and urine osmolarity without ADH. **(b)** Tubule permeabilities and urine osmolarity at maximal ADH levels.

KEY:

•.• = Solutes H_2O = Osmotic water flow   ● = Cotransport carrier

Transport mechanism:

Membrane permeabilities:

| = Water-impermeable and solute-impermeable
Example: thick ascending limb

| = Variable water permeability and Na⁺ transport; impermeable to urea
Example: DCT and collecting system (except papillary ducts)

| = Water-permeable and solute-impermeable
Example: thin descending limb

| = Water-permeable and solute-permeable
Example: PCT and thick descending limb

about 5 percent of the original volume of filtrate produced by your glomeruli. At normal levels of ADH, the collecting system will reabsorb roughly 16.8 liters per day, or about 9.3 percent of the original volume of filtrate. A healthy adult typically produces 1200 ml of urine per day (about 0.6 percent of the filtrate volume), with an osmolarity of 800–1000 mOsm/l.

The effects of ADH are opposed by those of atrial natriuretic peptide (ANP). ANP is released by cardiac muscle cells in the atrial walls when they are excessively stretched. This hormone (1) elevates glomerular pressures and increases the GFR, as we noted on page 976; (2) suppresses Na^+ reabsorption along the DCT and collecting system; (3) blocks the release of aldosterone and ADH; and (4) blocks the responses of the DCT and collecting system to aldosterone and ADH already in the circulation. The net result is that ANP stimulates the production of a large volume of relatively dilute urine. This fluid loss soon restores normal plasma volume.

The Function of the Vasa Recta

The solutes and water reabsorbed in the medulla must be returned to the general circulation without disrupting the concentration gradient. This return is

the function of the vasa recta. As illustrated in Figure 26-15•, for an individual nephron, blood flow through the vasa recta and tubular fluid movement through the loop of Henle may occur in opposite directions. This arrangement has no special significance, however. In the functioning kidney, there is no exclusive 1:1 relationship (1 vasa recta:1 loop of Henle); the medulla is literally packed with loops of Henle, vasa recta, and collecting ducts, without obvious organization (see Figure 26-6c•, p. 968). Thus the mechanism of vasa recta function is independent of the direction of flow within the renal tubule or collecting system.

Blood enters the vasa recta with an osmolarity of approximately 300 mOsm/l (Figure 26-15•). The blood descending into the medulla gradually increases in osmolarity as the solute concentration in the peritubular fluid rises. This increase in blood osmolarity involves both solute absorption and water loss, but solute absorption predominates, because the plasma proteins limit osmotic flow out of the blood. ∞ *[p. 725]* Blood flowing toward the cortex gradually decreases in osmolarity as the solute concentration of the peritubular fluid declines. Again, this decrease involves a combination of solute diffusion and osmosis, but in this case osmosis predominates, because the presence of plasma proteins does not

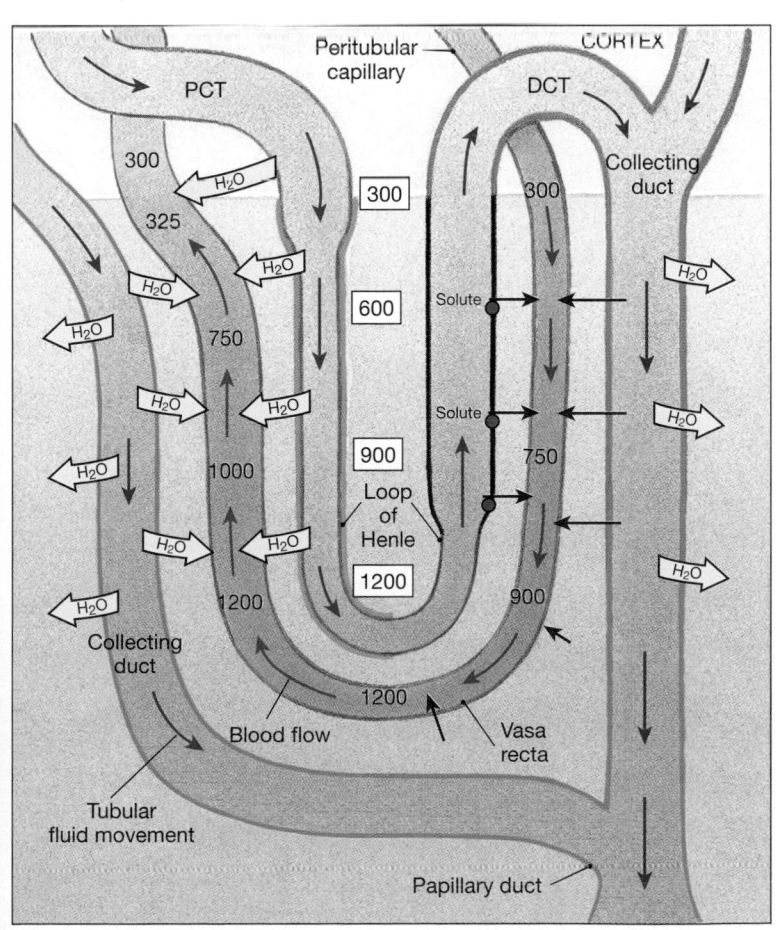
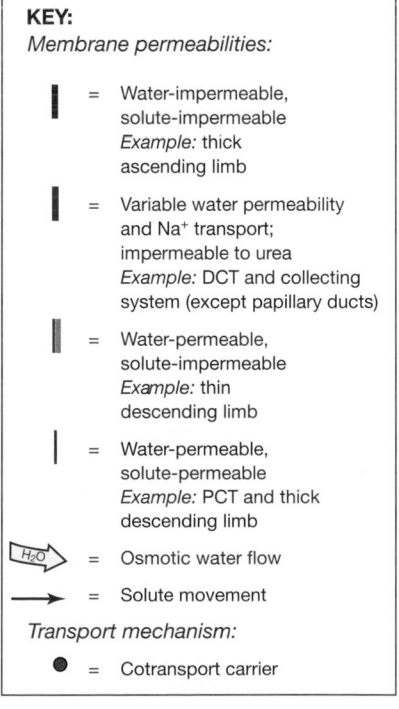

•FIGURE 26-15 **The Vasa Recta.** Capillaries of the vasa recta maintain the concentration gradient in the medulla by removing the water and solutes reclaimed from the tubular fluid in the nephron and collecting duct.

oppose the osmotic flow of water into the blood. The net results are that (1) some of the solutes absorbed in the descending portion of the vasa recta do not diffuse out in the ascending portion, and (2) more water moves into the ascending portion of the vasa recta than is moved out in the descending portion. Thus the vasa recta carries both water and solutes out of the medulla. Under normal conditions, the re-

moval of solutes and water by the vasa recta precisely balances the rates of solute reabsorption and osmosis in the medulla.

A Summary of Renal Function

Table 26-4 details the functions of each segment of the nephron and collecting system. Figure 26-16● provides a functional overview that summarizes the

TABLE 26-4 Renal Segments and Their Functions

Segment	General Functions	Specific Functions	Mechanisms
Renal corpuscle	*Filtration of plasma;* generates approximately 180 liters/day of filtrate similar in composition to blood plasma without plasma proteins	*Filtration* of water, inorganic and organic solutes from plasma; retention of plasma proteins and blood cells	Glomerular hydrostatic (blood) pressure working across capillary endothelium, lamina densa, and filtration slits
Proximal convoluted tubule	*Reabsorption* of 60–70% of the water (approximately 180 liters/day), 99–100% of the organic substrates, and 60–70% of the sodium and chloride ions in the original filtrate	*Reabsorption:* Active: glucose, other simple sugars, amino acids, vitamins, ions (including sodium, potassium, magnesium, phosphate, and bicarbonate) Passive: urea, chloride ions, lipid-soluble materials, water *Secretion:* Hydrogen ions, ammonium ions, creatinine, drugs, and toxins (as at DCT)	Carrier-mediated transort, including facilitated transport (glucose, amino acids), cotransport (glucose, ions), or countertransport (with secretion of H^+) Diffusion (solutes) or osmosis (water) Countertransport with sodium ions
Loop of Henle	*Reabsorption* of 25% of the water (45 liters/day) and 20–25% of the sodium and chloride ions present in the original filtrate; creation of the concentration gradient in the medulla	*Reabsorption:* Sodium and chloride ions Water	Active transport via Na^+–K^+/2 Cl^- transporter Osmosis
Distal convoluted tubule	*Reabsorption* of a variable amount of water (usually 5%, or 9 liters/day), under ADH stimulation, and a variable amount of sodium ions, under aldosterone stimulation	*Reabsorption:* Sodium and chloride ions Sodium ions (variable) Calcium ions (variable) Water (variable) *Secretion:* Hydrogen ions, ammonium ions Creatinine, drugs, toxins	Cotransport Countertransport with potassium ions; aldosterone-regulated Carrier-mediated transport stimulated by parathyroid hormone and calcitriol Osmosis; ADH-regulated Countertransport with sodium ions Carrier-mediated transport
Collecting system	*Reabsorption* of a variable amount of water (usually 9.3%, or 16.8 liters/day), under ADH stimulation, and a variable amount of sodium ions, under aldosterone stimulation	*Reabsorption:* Sodium ions (variable) Bicarbonate ions (variable) Water (variable) Urea (distal portions only) *Secretion:* Potassium and hydrogen ions (variable)	Countertransport with potassium or hydrogen ions; aldosterone-regulated Diffusion, generated within tubular cells Osmosis; ADH-regulated Diffusion Carrier-mediated transport
Peritubular capillaries	*Redistribution* of water and solutes reabsorbed in the cortex	Return of water and solutes to the general circulation	Osmosis and diffusion
Vasa recta	*Redistribution* of water and solutes reabsorbed in the medulla and stabilization of the concentration gradient of the medulla	Return of water and solutes to the general circulation	Osmosis and diffusion

major steps involved in the reabsorption of water and the production of a concentrated urine:

STEP 1. The filtrate produced at the renal corpuscle has the same osmolarity as the plasma, about 300 mOsm/l.

STEP 2. In the PCT, the active removal of ions and organic substrates produces a continuous osmotic flow of water out of the tubular fluid. This process reduces the volume of filtrate but keeps the solutions inside and outside the tubule isotonic. Between 60 and 70 percent of the filtrate volume has been reabsorbed before the tubular fluid reaches the descending limb of the loop of Henle.

STEP 3. In the PCT and descending limb of the loop of Henle, water moves into the surrounding peri-tubular fluids, leaving a small volume (15–20 percent of the original filtrate) of highly concentrated tubular fluid. The volume reduction has occurred through obligatory water reabsorption.

STEP 4. The thick ascending limb is impermeable to water and solutes. The tubular cells actively transport Na^+ and Cl^- out of the tubular fluid. This transport lowers the osmotic concentration of the tubular fluid *without affecting the volume of the tubular fluid.* The tubular fluid reaching the DCT is hypotonic relative to the peritubular fluid, with an osmolarity of only about 100 mOsm/l. Because only Na^+ and Cl^- are removed, urea now accounts for a significantly higher proportion of the total osmotic concentration at the end of the loop than it did at the start.

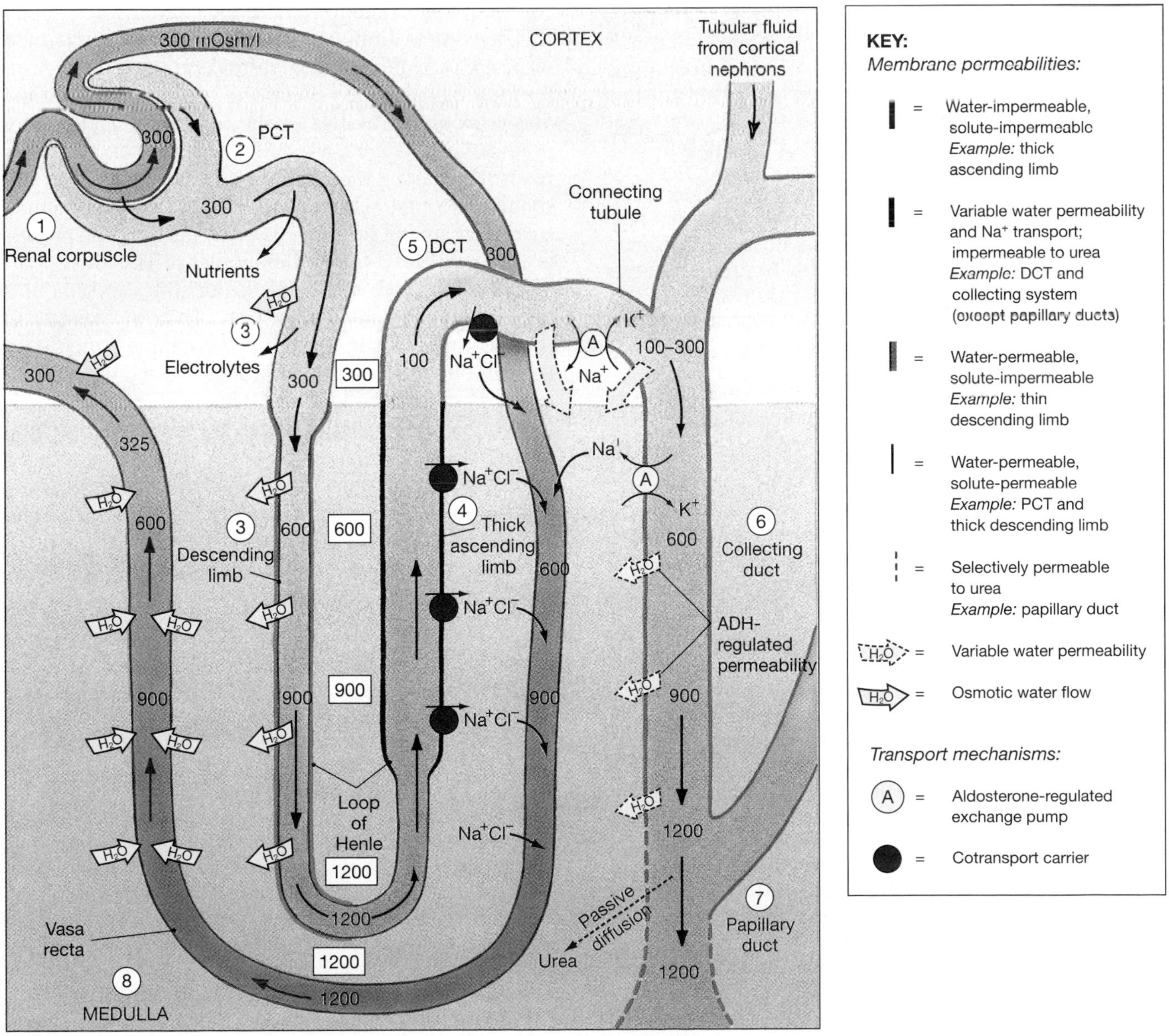

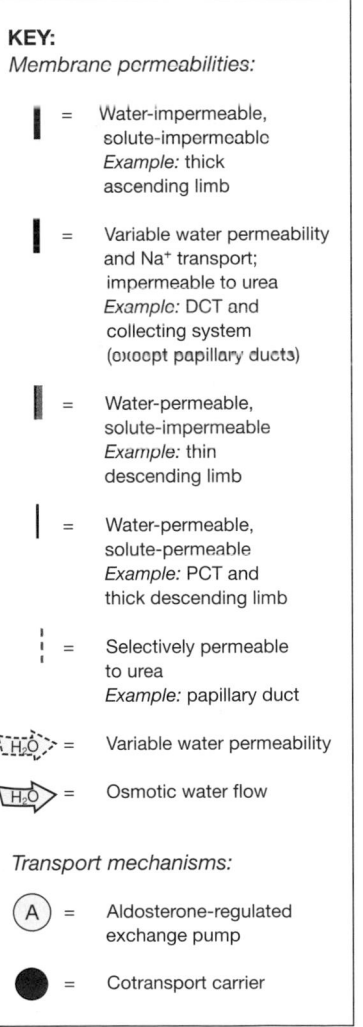

• **FIGURE 26-16** **Major Steps in the Production of a Concentrated Urine**

STEP 5. The final adjustments in the composition of the tubular fluid are made in the DCT and the collecting system. Although the DCT, connecting tubule, and collecting duct are generally impermeable to solutes, the osmolarity of the tubular fluid can be adjusted through active transport (reabsorption or secretion). Sodium–chloride cotransport and the aldosterone-stimulated reabsorption of Na^+ and secretion of K^+ are shown in Figure 26-13● (pp. 984–985).

STEP 6. The final adjustments in the volume and osmotic concentration of the tubular fluid are made by controlling the water permeabilities of the distal portions of the DCT and the collecting system. These segments are relatively impermeable to water unless exposed to ADH. At normal concentrations of ADH, the distal portions of the DCT, the connecting tubules, and the collecting ducts are somewhat permeable to water, and there is an osmotic flow of water out of the tubular fluid. Under these conditions, the urine entering the minor calyx has an osmolarity below 1200 mOsm/l. Under maximum ADH stimulation, urine volume is at a minimum and urine osmolarity is equal to that of the peritubular fluid in the deepest portion of the medulla (roughly 1200 mOsm/l).

STEP 7. As the tubular fluid becomes increasingly concentrated, the urea concentration rises accordingly. The papillary ducts are permeable to urea, and urea molecules diffuse out of the tubular fluid and into the peritubular fluid. The urea entering in this way contributes to the concentration gradient of the medulla.

STEP 8. The vasa recta absorbs solutes and water reabsorbed by the loop of Henle and the collecting ducts and thereby maintains the concentration gradient of the medulla.

DIURETICS The term **diuresis** (dī-ū-RĒ-sis; *dia*, through + *ouresis*, urination) refers to urine excretion. It typically indicates the production of a large volume of urine. **Diuretics** (dī-ū-RET-iks) are drugs that promote the loss of water in the urine. The usual goal in diuretic administration is a reduction of blood volume, blood pressure, or both. The ability to control renal water losses with relatively safe and effective diuretics has saved the lives of many patients, especially those with high blood pressure or congestive heart failure.

Diuretics have many different mechanisms of action, but all affect transport activities or water reabsorption along the nephron and collecting system. For example, a class of diuretics called *thiazides* (THĪ-a-zīdz) promotes water loss by reducing sodium and chloride ion transport in the proximal and distal convoluted tubules. For a detailed discussion of the major classes of diuretics and their modes of action, refer to the *Applications Manual*. [AM] *Diuretics*

Diuretic use for nonclinical reasons is currently on the rise. For example, large doses of diuretics may be taken by bodybuilders to improve muscle definition and by fashion models or jockeys to reduce body weight. This is an extremely dangerous practice that has caused several deaths due to severe dehydration and cardiac arrest.

☑ What effect would increased amounts of aldosterone have on the K^+ concentration of the urine?

☑ What effect would a decrease in the Na^+ concentration of the filtrate have on the pH of the tubular fluid?

☑ How would the lack of juxtamedullary nephrons affect the volume and osmolarity of the urine produced by the kidneys?

☑ Why does a decrease in the amount of Na^+ in the distal convoluted tubule lead to an increase in blood pressure?

The Composition of Normal Urine

More than 99 percent of the 180 liters or more of filtrate produced each day by the glomeruli is reabsorbed and never reaches the renal pelvis. General characteristics of normal urine are listed in Table 26-5. However, the composition of the excreted 1.2 liters of urine varies with the metabolic and hormonal events under way.

The composition and concentration of the excreted urine are two integrated but distinct properties. The *composition* of the urine reflects the filtration, absorption, and secretion activities of the nephrons. Some compounds, such as urea, are neither actively excreted nor reabsorbed along the nephron. In contrast, organic nutrients are completely reabsorbed, and other compounds, such as creatinine, that are missed by the filtration process are actively secreted into the tubular fluid.

These processes determine the identities and amounts of materials eliminated in the urine. The *concentration* of these components in a given urine sample depends on the osmotic movement of water across the walls of the tubules and collecting ducts. Because the composition and concentration of the urine vary independently, you can produce a small quantity of concentrated urine or a large quantity of dilute urine and still excrete the same amount of dissolved materials. As a result, physicians interested in a detailed assessment of renal function often rely on a 24-hour urine collection rather than the analysis of a single urine sample.

URINALYSIS Normal urine is a clear, sterile solution with a yellow color. The color results from the presence of urobilin, generated in the kidneys from the urobilinogens produced by intestinal bacteria and absorbed in the colon (see Figure 19-5●, p. 650). The evaporation of small molecules, such as ammonia, accounts for the characteristic odor of urine. Other substances not normally present, such as acetone or other ketone bodies, may also impart a distinctive smell. The analysis of a urine sample is a diagnostic tool of considerable importance, even in modern high-technology medicine. A standard **urinalysis** involves an assessment of urine color and appearance, two characteristics that can be determined without specialized equipment. In the seventeenth century, physicians classified the taste of the urine as sweet, salty, and so on, but quantitative chemical tests have long since replaced the taste bud assay. Average values for urinalysis are presented in Table 26-6. [AM] *Urinalysis*

TABLE 26-5 General Characteristics of Normal Urine

Characteristic	Normal Range
pH	6.0 (range: 4.5–8)
Specific gravity	1.003–1.030
Osmolarity	855–1335 mOsm/l
Water content	93–97 percent
Volume	1200 ml/day
Color	Clear yellow
Odor	Varies with composition
Bacterial content	Sterile

URINE TRANSPORT, STORAGE, AND ELIMINATION

Filtrate modification and urine production end when the fluid enters the renal pelvis. The remaining parts of the urinary system (the *ureters, urinary bladder,* and *urethra*) are responsible for the transport, storage, and elimination of urine. A *pyelogram* (PĪ-el-ō-gram) is an image of the urinary system, obtained by taking an X-ray of the kidneys after a radiopaque compound has been administered. Such an image provides an orientation to the relative sizes and positions of these organs (Figure 26-18●). The sizes of the minor and

TABLE 26-6 Typical Values from Standard Urine Testing

Compound	Primary Source	Daily Excretion*	Concentration	Remarks
Nitrogenous wastes				
Urea	Deamination of amino acids at liver and kidneys	21 g	1.8 g/dl	Rises if negative nitrogen balance exists
Creatinine	Breakdown of creatinine phosphate in skeletal muscle	1.8 g	150 mg/dl	Proportional to muscle mass; decreases during atrophy or muscle disease
Ammonia	Deamination by liver and kidney, absorption from intestinal tract	0.68 g	60 mg/dl	
Uric acid	Breakdown of purines	0.53 g	40 mg/dl	Increases in gout, liver diseases
Hippuric acid	Breakdown of dietary toxins	4.2 mg	350 µg/dl	
Urobilin	Urobilinogens absorbed at colon	1.5 mg	125 µg/dl	Gives urine its yellow color
Bilirubin	Hemoglobin breakdown product	0.3 mg	20 µg/dl	Increase may indicate problem with liver excretion or excess production; causes yellow skin color in jaundice
Nutrients and metabolites				
Carbohydrates		0.11 g	9 µg/dl	Primarily glucose; *glycosuria* develops if T_m is exceeded
Ketone bodies		0.21 g	17 µg/dl	Ketonuria may occur during postabsorptive state
Lipids		0.02 g	1.6 µg/dl	May increase in some kidney diseases
Amino acids		2.25 g	287.5 µg/dl	Note relatively high loss compared with other metabolites due to low T_m; excess (*aminoaciduria*) indicates T_m problem
Ions				
Sodium		4.0 g	333 mg/dl	Varies with diet, urine pH, hormones, etc.
Chloride		6.4 g	533 mg/dl	
Potassium		2.0 g	166 mg/dl	Varies with diet, urine pH, hormones, etc.
Calcium		0.2 g	17 mg/dl	Hormonally regulated (PTH/CT)
Magnesium		0.15 g	13 mg/dl	
Blood cells[†]				
RBCs		130,000/day	100/ml	Excess (*hematuria*) indicates vascular damage
WBCs		650,000/day	500/ml	Excess (*pyuria*) indicates renal infection or inflammation

*Representative values for a 70-kg male. †Usually estimated by counting the cells in a sample of sediment after urine centrifugation.

Advances in the Treatment of Renal Failure

Many conditions can result in renal failure. Management of chronic renal failure typically involves restricting water and salt intake and reducing protein intake to a minimum, with few dietary proteins. This combination reduces strain on the urinary system by (1) minimizing the volume of urine produced and (2) preventing the generation of large quantities of nitrogenous wastes. Acidosis, a common problem in patients with renal failure, can be countered by the ingestion of bicarbonate ions.

If drugs and dietary controls cannot stabilize the composition of the blood, more drastic measures are taken. In **hemodialysis** (hē-mō-dī-AL-i-sis), an artificial membrane is used to regulate the composition of the blood by means of a *dialysis machine* (Figure 26-17a●). The basic principle involved in this process, called **dialysis,** involves passive diffusion across a selectively permeable membrane. The patient's blood flows past an artificial *dialysis membrane* that contains pores large enough to permit the diffusion of small ions but small enough to prevent the loss of plasma proteins. On the other side of the membrane flows a special **dialysis fluid.**

The composition of typical dialysis fluid is indicated in Table 26-7. As diffusion takes place across the membrane, the composition of the blood changes. Potassium ions, phosphate ions, sulfate ions, urea,

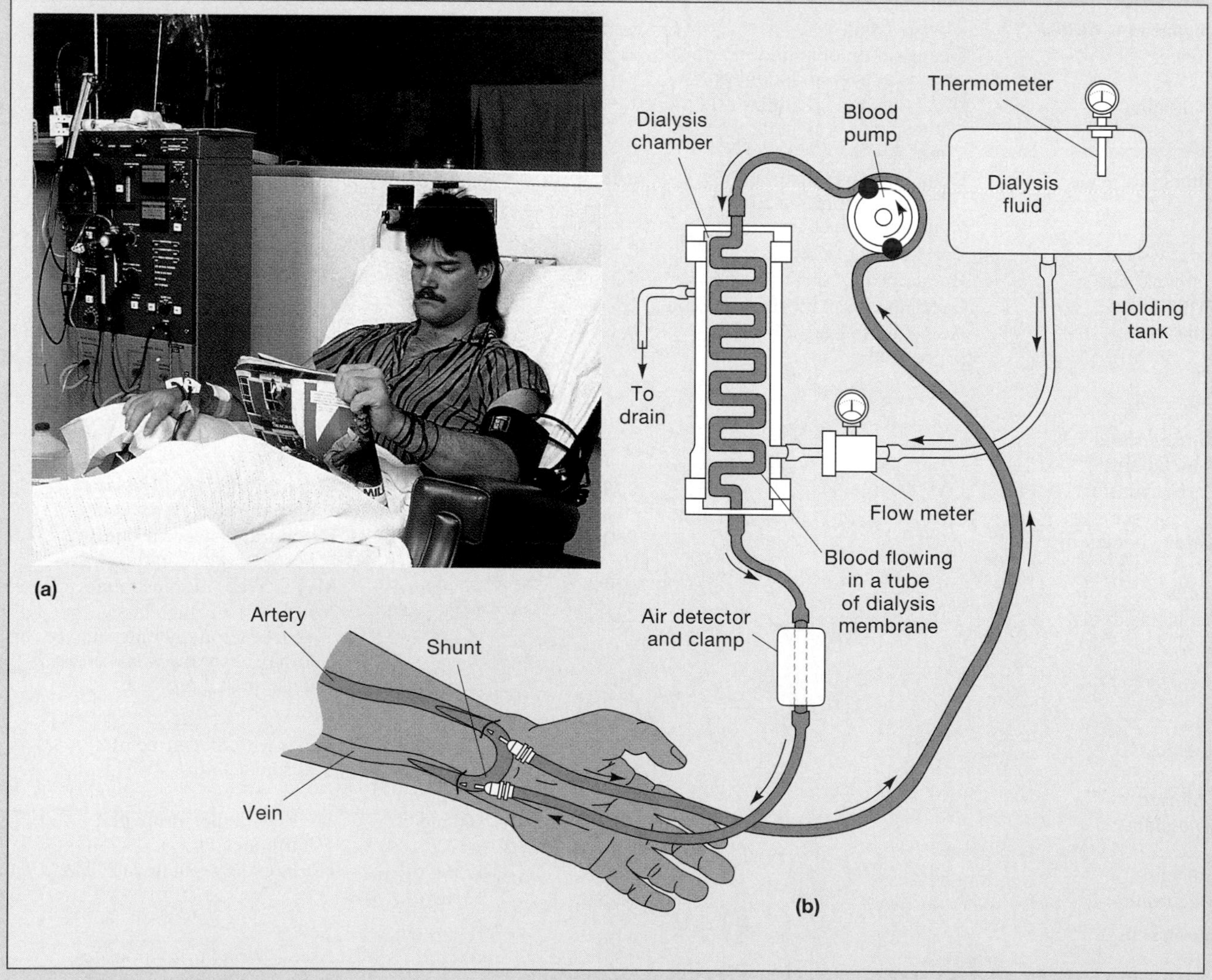

(a)

Dialysis chamber

Blood pump

Thermometer

Dialysis fluid

Holding tank

To drain

Flow meter

Blood flowing in a tube of dialysis membrane

Air detector and clamp

Artery

Shunt

Vein

(b)

●*FIGURE 26-17* Hemodialysis. (a) A patient is hooked up to a dialysis machine. **(b)** A diagrammatic view of the dialysis procedure. Preparation for hemodialysis typically involves the implantation of a pair of shunts connected by a loop that permits normal blood flow when the patient is not hooked up to the machine.

TABLE 26-7 The Composition of Dialysis Fluid

Component	Plasma	Dialysis Fluid
Electrolytes (mEq/l)		
Potassium	4	3
Bicarbonate	27	36
Phosphate	3	0
Sulfate	0.5	0
Nutrients (mg/dl)		
Glucose	80–100	125
Nitrogenous wastes (mg/dl)		
Urea	20	0
Creatinine	1	0
Uric acid	3	0

Note: Only the significant variations are given; values for other electrolytes are usually similar. Although these values are representative, the precise composition can be tailored to meet the specific clinical needs. For example, if plasma potassium levels are too low, the dialysis fluid concentration can be elevated to remedy the situation. Changes in the osmolarity of the dialysis fluid can also be used to adjust an individual's blood volume, generally by adjusting the glucose content of the dialysis fluid.

creatinine, and uric acid diffuse across the membrane into the dialysis fluid. Bicarbonate ions and glucose diffuse into the bloodstream. In effect, diffusion across the dialysis membrane takes the place of normal glomerular filtration, and the characteristics of the dialysis fluid ensure that important metabolites remain in the circulation rather than diffusing across the membrane.

In practice, silicone rubber tubes called *shunts* are inserted into a medium-sized artery and vein. (The typical location is in the forearm, although the lower leg is sometimes used.) The two shunts are then connected as shown in Figure 26-17b●. The connection acts like a short-circuit that does not impede blood flow, and the shunts can be used like taps in a wine barrel, to draw a blood sample or to connect the individual to a dialysis machine. For long-term dialysis, a surgically created arteriovenous anastomosis provides access.

When connected to the dialysis machine, the individual sits quietly while blood circulates from the arterial shunt, through the machine, and back through the venous shunt. Inside the machine, the blood flows within a tube composed of dialysis membrane, and diffusion occurs between the blood and the surrounding dialysis fluid.

Use of a dialysis machine is suggested when a patient's *BUN (blood urea nitrogen)* exceeds 100 mg/dl (the normal value is up to 30 mg/dl). Dialysis techniques can maintain patients who are awaiting a transplant or those whose kidney function has been temporarily disrupted. Hemodialysis does have drawbacks, however: (1) The patient must sit by the machine about 15 hours a week; (2) between treatments, the symptoms of uremia will gradually develop; (3) hypotension can develop as a result of fluid loss during dialysis; (4) air bubbles in the tubing can cause an embolism to form in the bloodstream; (5) anemia commonly develops; and (6) the shunts can serve as sites of recurring infections.

One alternative to the use of a dialysis machine is peritoneal dialysis. In **peritoneal dialysis,** the peritoneal lining is used as a dialysis membrane. Dialysis fluid is introduced into the peritoneum through a catheter in the abdominal wall, and at intervals the fluid is removed and replaced. For example, one procedure involves cycling 2 liters of fluid in an hour—15 minutes for infusion, 30 minutes for exchange, and 15 minutes for fluid reclamation. This process is usually performed in a hospital. An interesting variation on this procedure is called **continuous ambulatory peritoneal dialysis (CAPD).** In this procedure, the patient administers 2 liters of dialysis fluid through the catheter and then continues with life as usual until 4–6 hours later, when the fluid is removed and replaced.

Probably the most satisfactory solution, in terms of overall quality of life, is *kidney transplantation.* This procedure involves the implantation of a new kidney obtained from a living donor or from a cadaver. One-third of the estimated 11,000 kidneys transplanted in 1997 were obtained from living, related donors. In most cases, the damaged kidney is removed, and its blood supply is connected to the transplant. When the original kidney is left in place, an arterial graft is inserted to carry blood from the iliac artery or the aorta to the transplant, which is placed in the pelvis or lower abdomen.

The success rate for kidney transplantation varies, depending on how aggressively the recipient's T cells attack the donated organ and whether an infection develops. The 1-year success rate for implantation is now 85–95 percent. The use of kidneys taken from close relatives significantly improves the chances that the transplant will succeed. Immunosuppressive drugs are administered to reduce tissue rejection, but unfortunately this treatment also lowers the individual's resistance to infection.

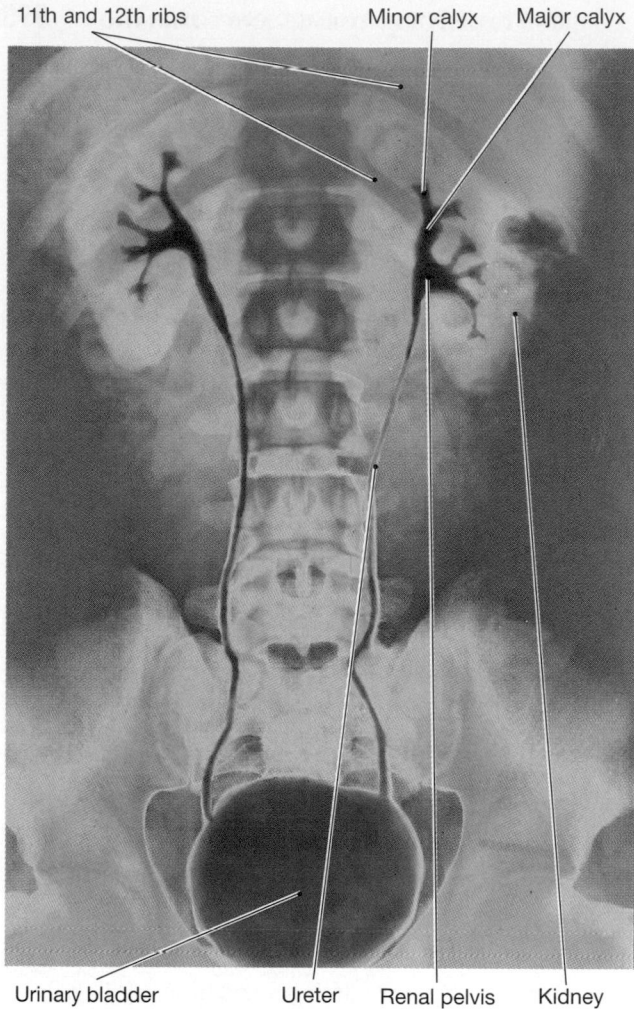

11th and 12th ribs | Minor calyx | Major calyx

Urinary bladder | Ureter | Renal pelvis | Kidney

• **FIGURE 26-18** A Radiographic View of the Urinary System. This pyelogram (a posterior view) has been color-enhanced.

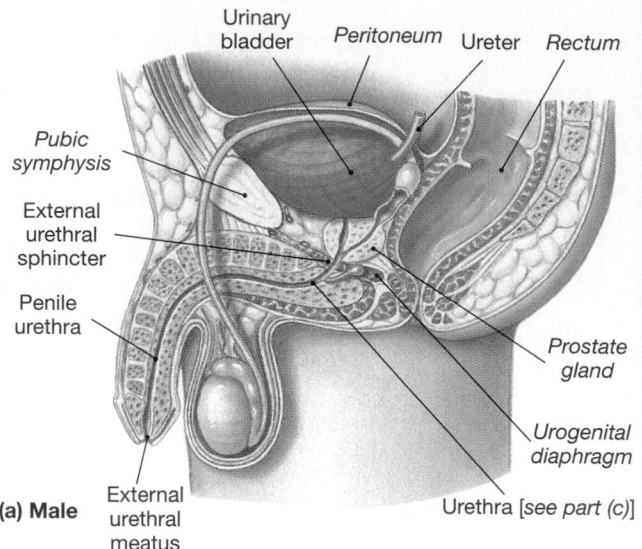

Urinary bladder | Peritoneum | Ureter | Rectum

Pubic symphysis

External urethral sphincter

Penile urethra

Prostate gland

Urogenital diaphragm

Urethra [see part (c)]

(a) Male | External urethral meatus

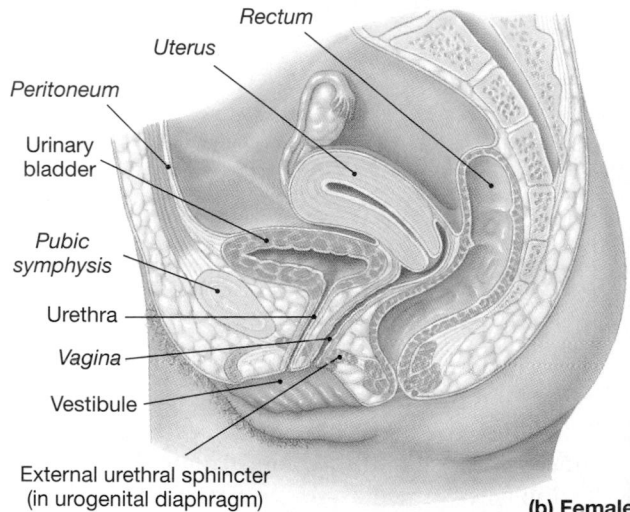

Rectum

Uterus

Peritoneum

Urinary bladder

Pubic symphysis

Urethra

Vagina

Vestibule

External urethral sphincter (in urogenital diaphragm)

(b) Female

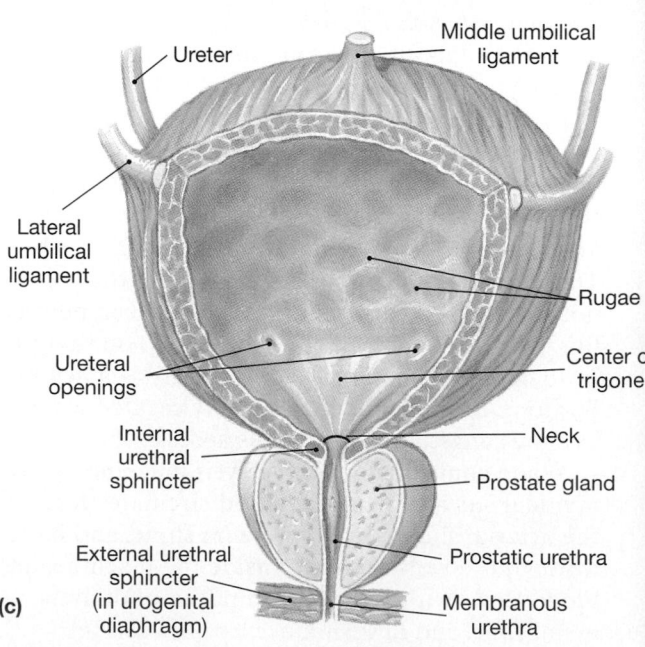

Ureter | Middle umbilical ligament

Lateral umbilical ligament

Ureteral openings

Internal urethral sphincter

External urethral sphincter (in urogenital diaphragm)

(c)

Rugae

Center of trigone

Neck

Prostate gland

Prostatic urethra

Membranous urethra

• **FIGURE 26-19** Organs for the Conduction and Storage of Urine. **(a)** The ureter, urinary bladder, and urethra in the male and **(b)** in the female. **(c)** The urinary bladder of a male.
AM *Plates 6.6, 6.7*

major calyces, the renal pelvis, the ureters, the urinary bladder, and the proximal portion of the urethra are somewhat variable, because these regions are lined by a *transitional epithelium* that can tolerate cycles of distension and contraction without damage. ∞ *[p. 114]*

We shall now examine these components of the urinary system.

The Ureters

The ureters are a pair of muscular tubes that extend inferiorly from the kidneys for about 30 cm (12 in.) before reaching the urinary bladder. Each ureter begins at the funnel-shaped renal pelvis (Figure 26-3•, p. 964). The ureters extend inferiorly and medially, passing over the anterior surfaces of the *psoas major* muscles (Figures 26-1c, p. 962, and 26-2b•, p. 963). The ureters are retroperitoneal and are firmly attached to the posterior abdominal wall. The paths taken by the ureters in men and women are different due to variations in the nature, size, and position of the reproductive organs. As Figure 26-19a• shows, in males, the base of the urinary bladder lies between the rectum and the pubic symphysis; in females, the base of the urinary bladder sits inferior to the uterus and anterior to the vagina (Figure 26-19b●).

The ureters penetrate the posterior wall of the urinary bladder without entering the peritoneal cavity. They pass through the bladder wall at an oblique angle, and the **ureteral openings** are slit-like rather than rounded (Figure 26-19c●). This shape helps prevent the backflow of urine toward the ureter and kidneys when the urinary bladder contracts.

Histology of the Ureters

The wall of each ureter consists of three layers: (1) an inner mucosa covered by a transitional epithelium, (2) a middle muscular layer made up of longitudinal and circular bands of smooth muscle, and (3) an outer connective tissue layer that is continuous with the fibrous renal capsule and peritoneum (Figure 26-20a●). About every 30 seconds, a peristaltic contraction begins at the renal pelvis and sweeps along the ureter, forcing urine toward the urinary bladder.

 PROBLEMS WITH THE CONDUCTING SYSTEM
Local blockages of the connecting tubules, collecting ducts, or ureter may result from the formation of *casts*—small blood clots, epithelial cells, lipids, or other materials. Casts are commonly excreted in the urine and are visible in microscopic analysis of urine samples. **Calculi** (KAL-kū-lī), or kidney stones, form within the urinary tract from calcium deposits, magnesium salts, or crystals of uric acid. This condition is called *nephrolithiasis* (nef-rō-li-THĪ-a-sis). The blockage of the urinary passage by a stone or by other factors, such as external compression, results in **urinary obstruction.** Urinary obstruction is a serious problem because, in addition to causing pain, it will reduce or eliminate filtration in the affected kidney by elevating the capsular hydrostatic pressure.

Kidney stones are generally visible on an X-ray. If peristalsis and fluid pressures are insufficient to dislodge them, they must be surgically removed or destroyed. One interesting nonsurgical procedure involves breaking kidney stones apart with a *lithotripter*, the same apparatus used to destroy gallstones. ∞ *[p. 896]* Another nonsurgical approach entails the insertion of a catheter that is armed with a laser that can shatter kidney stones with intense light beams.

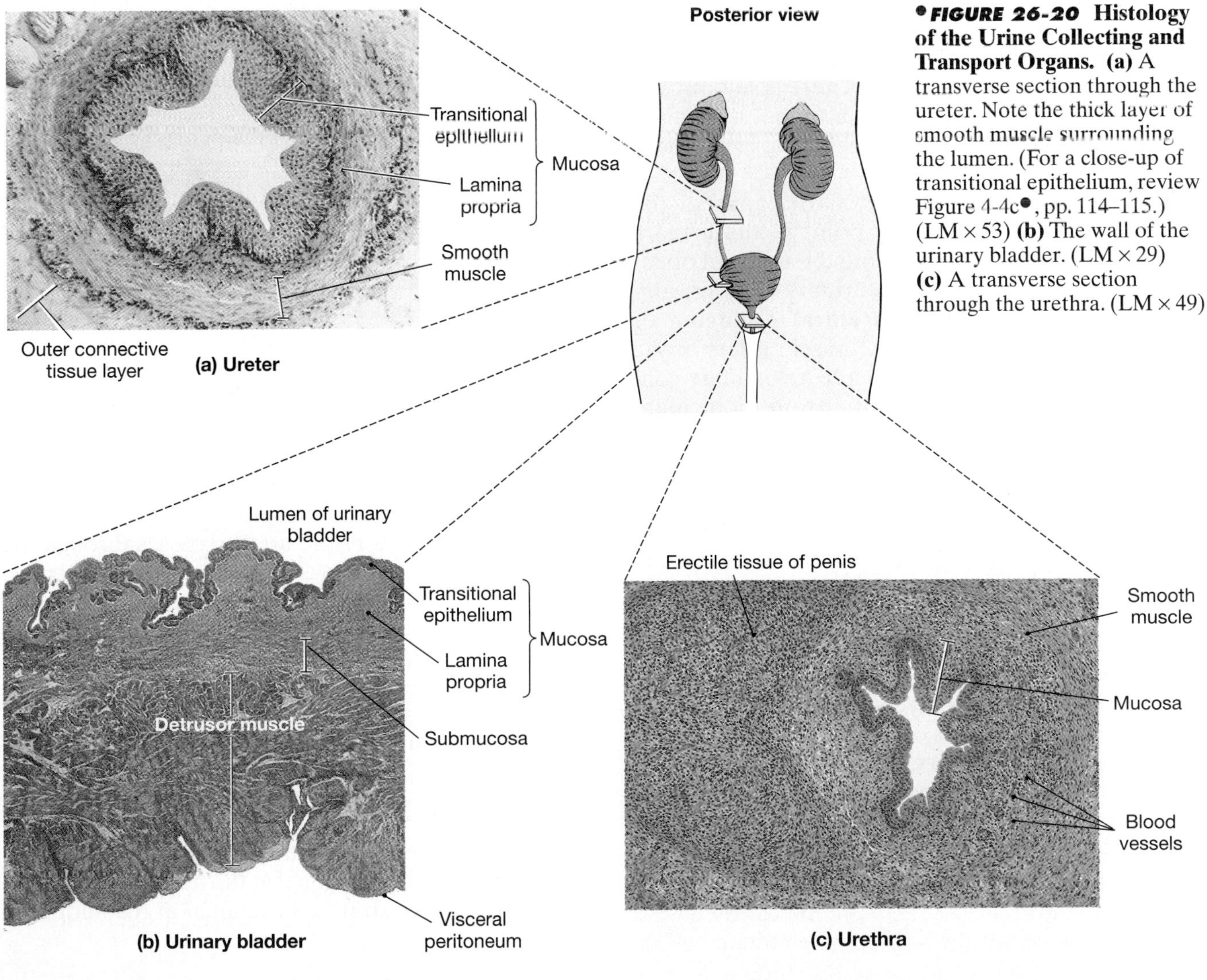

Posterior view

●*FIGURE 26-20* **Histology of the Urine Collecting and Transport Organs. (a)** A transverse section through the ureter. Note the thick layer of smooth muscle surrounding the lumen. (For a close-up of transitional epithelium, review Figure 4-4c●, pp. 114–115.) (LM × 53) **(b)** The wall of the urinary bladder. (LM × 29) **(c)** A transverse section through the urethra. (LM × 49)

Transitional epithelium
Lamina propria
} Mucosa
Smooth muscle

Outer connective tissue layer **(a) Ureter**

Lumen of urinary bladder
Transitional epithelium
Lamina propria
} Mucosa
Submucosa
Detrusor muscle
Visceral peritoneum
(b) Urinary bladder

Erectile tissue of penis
Smooth muscle
Mucosa
Blood vessels
(c) Urethra

The Urinary Bladder

The urinary bladder is a hollow, muscular organ that functions as a temporary storage reservoir for urine. The dimensions of the urinary bladder vary with the state of distension, but the full urinary bladder can contain about a liter of urine.

The superior surfaces of the urinary bladder are covered by a layer of peritoneum, and several peritoneal folds assist in stabilizing its position (Figure 26-19c•). The **middle umbilical ligament** extends from the anterior and superior border toward the umbilicus (navel). The **lateral umbilical ligaments** pass along the sides of the bladder and also reach the umbilicus. These fibrous cords contain the vestiges of the two *umbilical arteries*, which supplied blood to the placenta during embryonic and fetal development. ∞ *[p. 758]* The urinary bladder's posterior, inferior, and anterior surfaces lie outside the peritoneal cavity. In these areas, tough ligamentous bands anchor the urinary bladder to the pelvic and pubic bones.

In sectional view, the mucosa lining the urinary bladder is usually thrown into folds, or **rugae,** that disappear as the bladder fills. The triangular area bounded by the ureteral openings and the entrance to the urethra constitutes the **trigone** (TRĪ-gōn) of the urinary bladder. The mucosa here is smooth and very thick. The trigone acts as a funnel that channels urine into the urethra when the urinary bladder contracts.

The urethral entrance lies at the apex of the trigone, at the most inferior point in the urinary bladder. The region surrounding the urethral opening, known as the **neck** of the urinary bladder, contains a muscular **internal urethral sphincter,** or *sphincter vesicae.* The smooth muscle fibers of the internal urethral sphincter provide involuntary control over the discharge of urine from the urinary bladder. The urinary bladder is innervated by postganglionic fibers from ganglia in the hypogastric plexus and by parasympathetic fibers from intramural ganglia that are controlled by branches of the pelvic nerves.

Histology of the Urinary Bladder

The wall of the urinary bladder contains mucosa, submucosa, and muscularis layers (Figure 26-20b•). The muscularis layer consists of inner and outer longitudinal smooth muscle layers, with a circular layer sandwiched between. Collectively, these layers form the powerful **detrusor** (de-TROO-sor) **muscle** of the urinary bladder. Contraction of this muscle compresses the urinary bladder and expels its contents into the urethra.

 BLADDER CANCER Each year in the United States, approximately 52,000 new cases of *bladder cancer* are diagnosed, and there are 9500 deaths from this condition. The incidence among males is three times that among females, and most patients are age 60–70. Environmental factors, especially exposure to *2-naphthylamine* or related compounds, are responsible for most bladder cancers. For this reason, the bladder cancer rate is highest among cigarette smokers and employees of chemical and rubber companies. The mechanism appears to involve damage to tumor suppressor genes, such as *p53*, that regulate cell division. The prognosis is reasonably good for localized superficial cancers (88 percent 5-year survival), but it is poor for persons with severe metastatic bladder cancer (9 percent 5-year survival). Treatment of metastatic bladder cancer is very difficult, because the cancer spreads rapidly through adjacent lymphatics and through the bone marrow of the pelvis.

The Urethra

The urethra extends from the neck of the urinary bladder (Figure 26-19c•) to the exterior. The female and male urethrae differ in length and in function. In females, the urethra is very short, extending 3–5 cm (1–2 in.) from the bladder to the vestibule (Figure 26-19b•). The external urethral opening, or **external urethral meatus,** is situated near the anterior wall of the vagina.

In males, the urethra extends from the neck of the urinary bladder to the tip of the penis, a distance that may be 18–20 cm (7–8 in.). The male urethra can be subdivided into three portions (Figure 26-19a,c•): (1) the *prostatic urethra,* (2) the *membranous urethra,* and (3) the *penile urethra.*

The **prostatic urethra** passes through the center of the prostate gland (Figure 26-19c•). The **membranous urethra** includes the short segment that penetrates the *urogenital diaphragm,* the muscular floor of the pelvic cavity. The **penile** (PĒ-nīl) **urethra** extends from the distal border of the urogenital diaphragm to the external urethral meatus at the tip of the penis (Figure 26-19a•). We shall consider the functional differences among these regions in Chapter 28.

In both genders, as the urethra passes through the urogenital diaphragm, a circular band of skeletal muscle forms the **external urethral sphincter.** This muscular band acts as a valve. The external urethral sphincter is under voluntary control, via the perineal branch of the pudendal nerve. ∞ *[p. 430]* This sphincter has a resting muscle tone and must be voluntarily relaxed to permit micturition.

Histology of the Urethra

The urethral lining consists of a stratified epithelium that varies from transitional at the neck of the urinary bladder, through stratified columnar at the midpoint, to stratified squamous near the external urethral meatus. The lamina propria is thick and elastic, and the mucous membrane is thrown into longitudinal folds (Figure 26-20c●). Mucin-secreting cells are located in the epithelial pockets. In males, the epithelial mucous glands may form tubules that extend into the lamina propria. Connective tissues of the lamina propria anchor the urethra to surrounding structures. In females, the lamina propria contains an extensive network of veins, and the entire complex is surrounded by concentric layers of smooth muscle.

URINARY TRACT INFECTIONS Urinary tract infections (UTIs) result from the colonization of the urinary tract by bacterial or fungal invaders. The intestinal bacterium *Escherichia coli* is most commonly involved. Women are particularly susceptible to urinary tract infections, because the external urethral orifice is so close to the anus. Sexual intercourse may push bacteria into the urethra, and, because the female urethra is relatively short, the urinary bladder may become infected.

The condition may be asymptomatic (without symptoms), but it can be detected by the presence of bacteria and blood cells in the urine. If inflammation of the urethral wall occurs, the condition may be termed *urethritis;* inflammation of the lining of the bladder is called *cystitis.* Many infections, including sexually transmitted diseases (STDs) such as gonorrhea, cause a combination of urethritis and cystitis. These conditions cause painful urination, a symptom known as *dysuria* (dis-ū-rē-uh), and the bladder becomes tender and sensitive to pressure. Despite the discomfort produced, the individual feels the urge to urinate frequently. Urinary tract infections generally respond to antibiotic therapies, although reinfections may occur.

In untreated cases, the bacteria may proceed along the ureters to the renal pelvis. The resulting inflammation of the walls of the renal pelvis produces *pyelitis* (pī-e-LĪ-tis). If the bacteria invade the renal cortex and medulla as well, *pyelonephritis* (pī-e-lō-nef-RĪ-tis) results. Signs and symptoms of pyelonephritis include a high fever, intense pain on the affected side, vomiting, diarrhea, and the presence of blood cells and pus in the urine.

The Micturition Reflex and Urination

Urine reaches the urinary bladder by the peristaltic contractions of the ureters. The process of urination is coordinated by the **micturition reflex.** Components of this reflex are diagrammed in Figure 26-21●.

Stretch receptors in the wall of the urinary bladder are stimulated as the bladder fills with urine. Afferent fibers in the pelvic nerves carry the impulses generated to the sacral spinal cord. Their increased level of activity (1) facilitates parasympathetic motor neurons in the sacral spinal cord and (2) stimulates interneurons that relay sensations to the thalamus and on to the cerebral cortex. As a result, you become consciously aware of the fluid pressure in your urinary bladder.

The urge to urinate generally appears when the bladder contains about 200 ml of urine. The micturition reflex begins to function when the stretch receptors have provided adequate stimulation to parasympathetic preganglionic motor neurons. Action potentials carried by efferent fibers within the pelvic nerves then stimulate ganglionic neurons in the wall of the urinary bladder. These neurons in turn stimulate sustained contraction of the detrusor muscle.

This contraction elevates fluid pressures in the urinary bladder, but urine ejection does not occur unless both the internal and external urethral sphincters are relaxed. Relaxation of the external urethral sphincter occurs under voluntary control. When the external urethral sphincter relaxes, so does the internal sphincter. If the external urethral sphincter does not relax, the internal sphincter remains closed and the urinary bladder gradually relaxes.

A further increase in bladder volume begins the cycle again, usually within an hour. Each increase in urinary volume leads to an increase in stretch receptor stimulation that makes the sensation more acute. Once the volume of the urinary bladder exceeds 500 ml, the micturition reflex may generate enough pressure to force open the internal urethral sphincter. This opening leads to a reflexive relaxation of the external sphincter, and urination occurs despite voluntary opposition or potential inconvenience. At the end of a normal micturition, less than 10 ml of urine remains in the bladder.

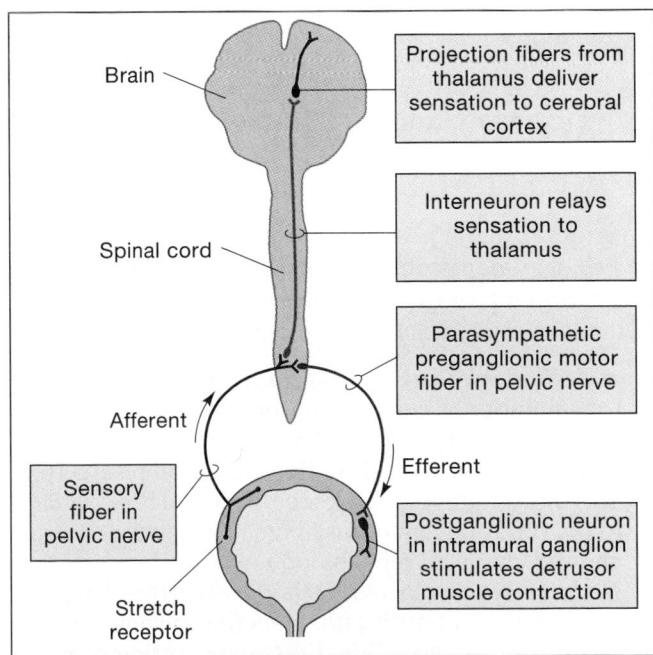

●*FIGURE 26-21* **The Micturition Reflex.** The components of the reflex arc that stimulates smooth muscle contractions in the urinary bladder. Micturition occurs after voluntary relaxation of the external urethral sphincter.

Brain

Spinal cord

Afferent

Sensory fiber in pelvic nerve

Stretch receptor

Projection fibers from thalamus deliver sensation to cerebral cortex

Interneuron relays sensation to thalamus

Parasympathetic preganglionic motor fiber in pelvic nerve

Efferent

Postganglionic neuron in intramural ganglion stimulates detrusor muscle contraction

Infants lack voluntary control over urination, because the necessary corticospinal connections have yet to be established. Toilet training before age 2 typically involves training the parent to anticipate the timing of the reflex rather than training the child to exert conscious control.

INCONTINENCE Incontinence (in-KON-ti-nens) is the inability to control urination voluntarily. Trauma to the internal or external sphincter may contribute to incontinence in otherwise healthy adults. For example, childbirth can stretch and damage the sphincter muscles, and some women then develop *stress incontinence.* In this condition, elevated intra-abdominal pressures, caused for example by a cough or sneeze, can overwhelm the sphincter muscles, causing urine to leak out. Incontinence may also develop in older individuals due to a general loss of muscle tone.

Damage to the CNS, the spinal cord, or the nerve supply to the urinary bladder or external urethral sphincter may also produce incontinence. For example, incontinence commonly accompanies Alzheimer's disease, and it may also result from a stroke or spinal cord injury. In most cases, the affected individual develops an *automatic bladder.* The micturition reflex remains intact, but voluntary control of the external sphincter is lost, and the individual cannot prevent the reflexive emptying of the urinary bladder. Damage to the pelvic nerves can eliminate the micturition reflex entirely, because those nerves carry both afferent and efferent fibers of this reflex arc. The urinary bladder becomes greatly distended with urine. It remains filled to capacity, and the excess trickles into the urethra in an uncontrolled stream. Insertion of a catheter is commonly required to facilitate the discharge of urine.

⧗ AGING AND THE URINARY SYSTEM

In general, aging is associated with an increased incidence of kidney problems. We described examples, such as *nephrolithiasis* (kidney stones), earlier in the chapter. Other age-related changes in the urinary system include:

- *A decline in the number of functional nephrons.* The total number of kidney nephrons drops by 30–40 percent between ages 25 and 85.
- *A reduction in the GFR.* This reduction results from decreased numbers of glomeruli, cumulative damage to the filtration apparatus in the remaining glomeruli, and reductions in renal blood flow.
- *Reduced sensitivity to ADH.* With age, the distal portions of the nephron and collecting system become less responsive to ADH. Less reabsorption of water and sodium ions occurs, and more sodium ions are lost in the urine.
- *Problems with the micturition reflex.* The following three factors are involved in such problems:

1. The sphincter muscles lose muscle tone and become less effective at voluntarily retaining urine. *Incontinence* results, commonly involving a slow leakage of urine.
2. The ability to control micturition may be lost after a stroke, Alzheimer's disease, or other CNS problems affecting the cerebral cortex or hypothalamus.
3. In males, **urinary retention** may develop secondary to enlargement of the prostate gland (*prostatic hypertrophy*). In this condition, swelling and distortion of surrounding prostatic tissues compress the urethra, restricting or preventing the flow of urine.

INTEGRATION WITH OTHER SYSTEMS

The urinary system is not the only organ system concerned with excretion. The urinary, integumentary, respiratory, and digestive systems are considered to form an anatomically diverse **excretory system.** Its components perform all the excretory functions of the body that affect the composition of body fluids. Examples of excretory activities discussed in earlier chapters include these:

- *Integumentary system.* Water and electrolyte losses in sensible perspiration can affect plasma volume and composition. The effects are most apparent when losses are extreme, as when you are at peak sweat production. You also excrete small amounts of metabolic wastes, including urea, in perspiration.
- *Respiratory system.* The lungs excrete the carbon dioxide generated by living cells. Small amounts of other compounds, such as acetone and water, evaporate into the alveoli and are eliminated when you exhale.
- *Digestive system.* Your liver excretes small amounts of metabolic waste products in bile, and you lose a variable amount of water in feces.

These excretory activities have an impact on the composition of body fluids. The respiratory system, for example, is the primary site of carbon dioxide excretion. But the excretory functions of these systems are not regulated as closely as are those of the kidneys. Under normal circumstances, the effects of integumentary and digestive excretory activities are minor compared with those of the urinary system.

Figure 26-22● summarizes the functional relationships between the urinary system and other systems. We shall explore many of these relationships further in Chapter 27 when we consider major aspects of fluid, pH, and electrolyte balance.

☑ What effect would a high-protein diet have on the composition of urine?

☑ An obstruction of a ureter by a kidney stone would interfere with the flow of urine between which two points?

☑ The ability to control the micturition reflex depends on your ability to control which muscle?

INTEGUMENTARY SYSTEM

Sweat glands assist in elimination of water and solutes, especially sodium and chloride ions; keratinized epidermis prevents excessive fluid loss through skin surface; epidermis produces vitamin D₃, important for the renal production of calcitriol

SKELETAL SYSTEM

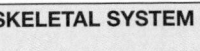

Axial skeleton provides some protection for kidneys and ureters; pelvis protects urinary bladder and proximal urethra

Conserves calcium and phosphate needed for bone growth

MUSCULAR SYSTEM

Sphincter controls urination by closing urethral opening; muscle layers of trunk provide some protection for urinary organs

Removes waste products of protein metabolism; assists in regulation of calcium and phosphate concentrations

NERVOUS SYSTEM

Adjusts renal blood pressure; monitors distension of bladder and controls urination

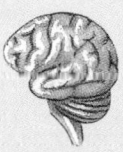

ENDOCRINE SYSTEM

Aldosterone and ADH adjust rates of fluid and electrolyte reabsorption in kidneys

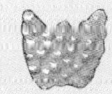

Kidney cells release renin when local blood pressure declines and erythropoietin (EPO) when renal O₂ levels decline

URINARY SYSTEM

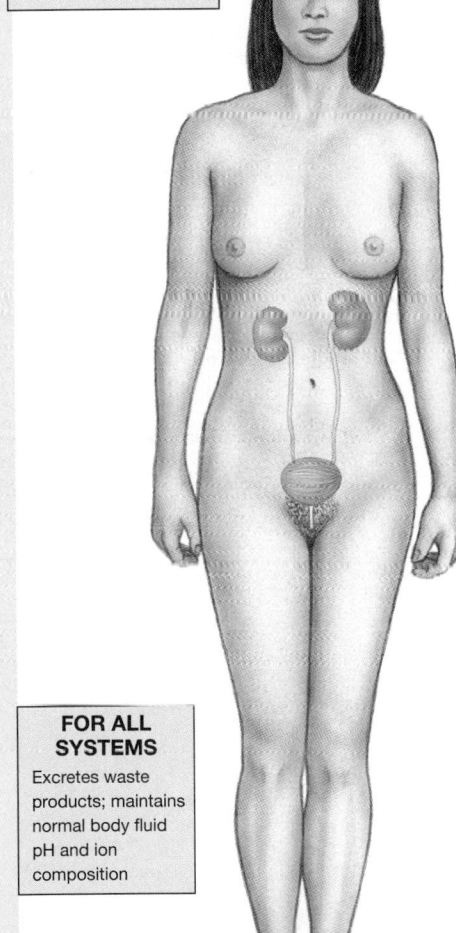

FOR ALL SYSTEMS

Excretes waste products; maintains normal body fluid pH and ion composition

CARDIOVASCULAR SYSTEM

Delivers blood to capillaries, where filtration occurs; accepts fluids and solutes reabsorbed during urine production

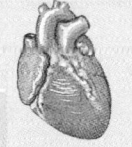

Releases renin to elevate blood pressure and erythropoietin to accelerate red blood cell production

LYMPHATIC SYSTEM

Provides specific defenses against urinary tract infections

Eliminates toxins and wastes generated by cellular activities; acid pH of urine provides nonspecific defense against urinary tract infections

RESPIRATORY SYSTEM

Assists in the regulation of pH by eliminating CO₂

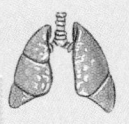

Assists in elimination of CO₂; provides bicarbonate buffers that assist in pH regulation

DIGESTIVE SYSTEM

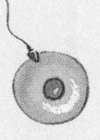

Accessory organ secretions may have antibacterial action that helps prevent urethral infections in males

Urethra in males carries semen to the exterior

Absorbs water needed to excrete wastes at kidneys; absorbs ions needed to maintain normal body fluid concentrations; liver removes conjugated bilirubin

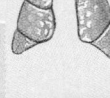

Excretes toxins, such as urobilinogen, absorbed by the digestive epithelium; excretes some conjugated bilirubin and nitrogenous wastes produced by the liver

REPRODUCTIVE SYSTEM

• FIGURE 26-22 Functional Relationships between the Urinary System and Other Systems

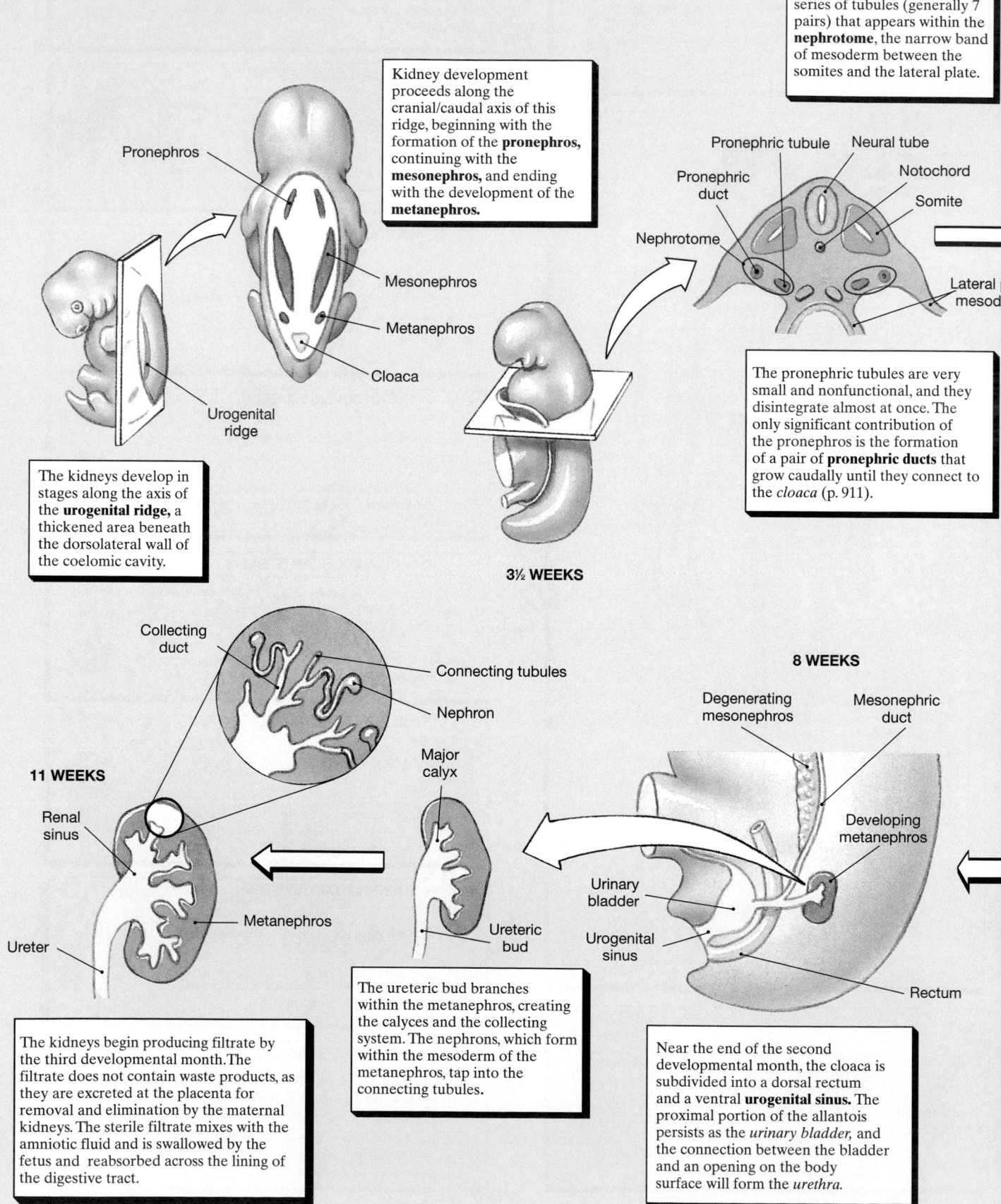

The pronephros consists of a series of tubules (generally 7 pairs) that appears within the **nephrotome**, the narrow band of mesoderm between the somites and the lateral plate.

Kidney development proceeds along the cranial/caudal axis of this ridge, beginning with the formation of the **pronephros,** continuing with the **mesonephros,** and ending with the development of the **metanephros.**

Pronephros

Mesonephros

Metanephros

Cloaca

Urogenital ridge

The kidneys develop in stages along the axis of the **urogenital ridge,** a thickened area beneath the dorsolateral wall of the coelomic cavity.

Pronephric tubule   Neural tube

Pronephric duct   Notochord

Somite

Nephrotome

Lateral p' mesode

The pronephric tubules are very small and nonfunctional, and they disintegrate almost at once. The only significant contribution of the pronephros is the formation of a pair of **pronephric ducts** that grow caudally until they connect to the *cloaca* (p. 911).

3½ WEEKS

Collecting duct

Connecting tubules

Nephron

11 WEEKS

Renal sinus

Ureter

Metanephros

Major calyx

Ureteric bud

The ureteric bud branches within the metanephros, creating the calyces and the collecting system. The nephrons, which form within the mesoderm of the metanephros, tap into the connecting tubules.

The kidneys begin producing filtrate by the third developmental month. The filtrate does not contain waste products, as they are excreted at the placenta for removal and elimination by the maternal kidneys. The sterile filtrate mixes with the amniotic fluid and is swallowed by the fetus and reabsorbed across the lining of the digestive tract.

8 WEEKS

Degenerating mesonephros   Mesonephric duct

Developing metanephros

Urinary bladder

Urogenital sinus

Rectum

Near the end of the second developmental month, the cloaca is subdivided into a dorsal rectum and a ventral **urogenital sinus.** The proximal portion of the allantois persists as the *urinary bladder,* and the connection between the bladder and an opening on the body surface will form the *urethra.*

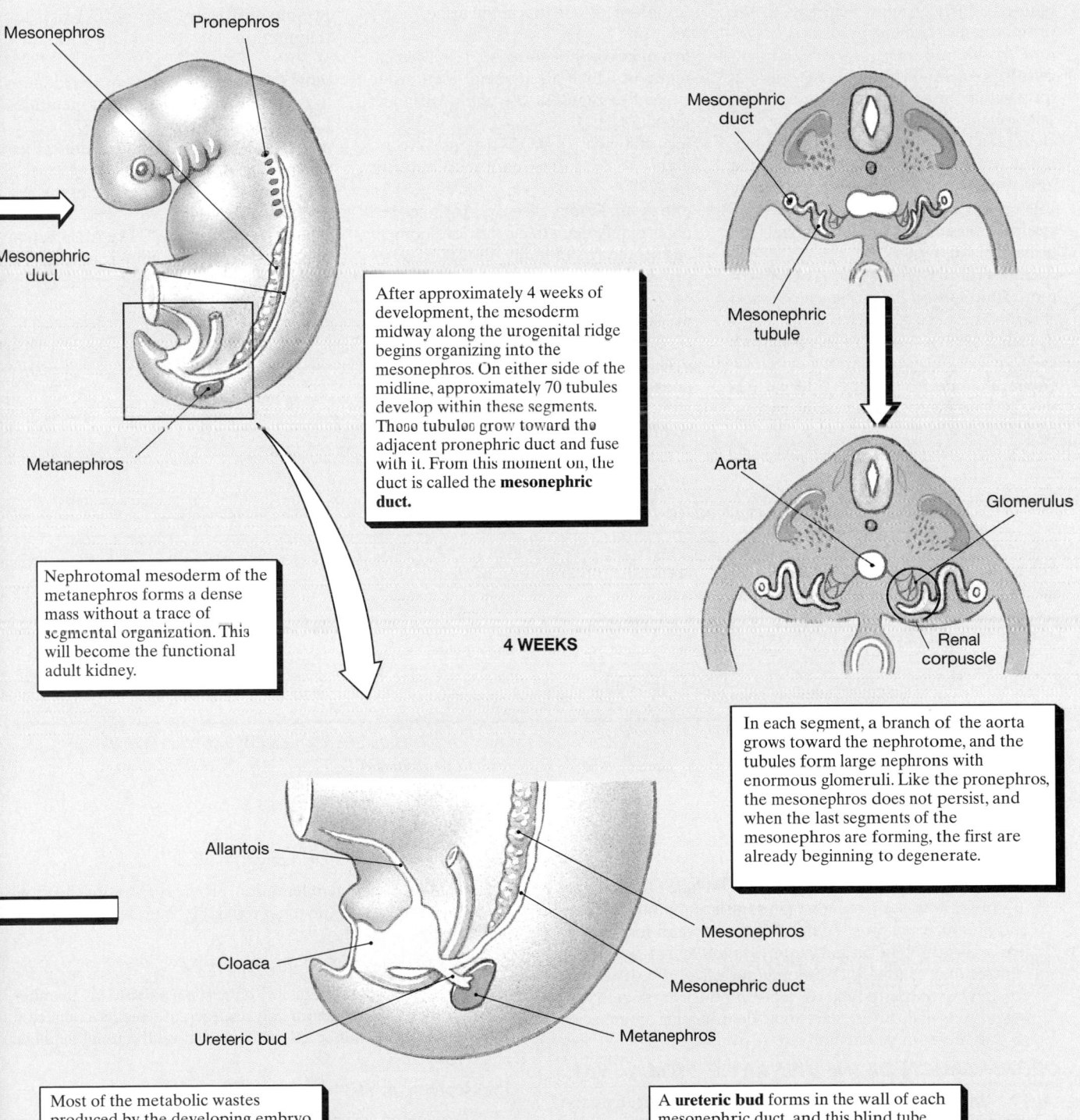

Mesonephros

Pronephros

Mesonephric duct

Mesonephric duct

Metanephros

After approximately 4 weeks of development, the mesoderm midway along the urogenital ridge begins organizing into the mesonephros. On either side of the midline, approximately 70 tubules develop within these segments. These tubules grow toward the adjacent pronephric duct and fuse with it. From this moment on, the duct is called the **mesonephric duct.**

Nephrotomal mesoderm of the metanephros forms a dense mass without a trace of segmental organization. This will become the functional adult kidney.

4 WEEKS

Mesonephric duct

Mesonephric tubule

Aorta

Glomerulus

Renal corpuscle

In each segment, a branch of the aorta grows toward the nephrotome, and the tubules form large nephrons with enormous glomeruli. Like the pronephros, the mesonephros does not persist, and when the last segments of the mesonephros are forming, the first are already beginning to degenerate.

Allantois

Cloaca

Ureteric bud

Mesonephros

Mesonephric duct

Metanephros

Most of the metabolic wastes produced by the developing embryo are passed across the placenta to enter the maternal circulation. The small amount of urine produced by the kidneys accumulates within the cloaca and the *allantois* (p. 911), an endoderm-lined sac that extends into the umbilical stalk.

A **ureteric bud** forms in the wall of each mesonephric duct, and this blind tube elongates and branches within the adjacent metanephros. Tubules developing within the metanephros then connect to the terminal branches.

SELECTED CLINICAL TERMINOLOGY

Terms Discussed in This Chapter

aminoaciduria: Amino acid loss in the urine; the most common form is *cystinuria*. *(p. 979 and AM)*

calculi (KAL-kū-lī): Insoluble deposits that form within the urinary tract from calcium salts, magnesium salts, or uric acid. *(p. 995)*

clearance test: A test that permits calculation of the GFR by monitoring plasma and renal concentrations of a specific solute, such as creatinine. *(p. 975 and AM)*

cystitis: Inflammation of the lining of the urinary bladder. *(p. 997)*

diuretics (dī-ū-RET-iks): Drugs that promote fluid loss in the urine. *(p. 990 and AM)*

dysuria (dis-ū-rē-uh): Painful urination. *(p. 997)*

glomerulonephritis (glo-mer-ū-lō-nef-RĪ-tis): Inflammation of the renal cortex. *(p. 969)*

hematuria: Blood loss in the urine. *(p. 991 and AM)*

hemodialysis (hē-mō-dī-AL-i-sis): A technique in which an artificial membrane is used to regulate the composition of blood. *(p. 992)*

incontinence (in-KON-ti-nens): An inability to control urination voluntarily. *(p. 998)*

polycystic kidney disease: An inherited abnormality that affects the development and structure of kidney tubules. *(p. 970)*

proteinuria: Protein loss in the urine. *(p. 970 and AM)*

pyelitis (pī-e-LĪ-tis): Inflammation of the walls of the renal pelvis. *(p. 997)*

pyelogram (PĪ-el-ō-gram): An image obtained by taking an X-ray of the kidneys after a radiopaque compound has been administered. *(p. 991 and AM)*

pyelonephritis (pī-e-lō-nef-RĪ-tis): Inflammation of the kidney tissues. *(p. 997)*

renal failure: An inability of the kidneys to excrete wastes in sufficient quantities to maintain homeostasis. *(p. 977)*

urinalysis: A physical and chemical assessment of urine. *(p. 990)*

urinary obstruction: Blockage of the urinary tract. *(p. 995)*

urinary tract infection (UTI): An infection that results from the colonization of the urinary tract by bacterial or fungal invaders. *(p. 997)*

AM ## Additional Terms Discussed in the *Applications Manual*

nephritis: Inflammation of the kidneys.
uremia (ū-RĒ-mē-uh): A change in the composition of the blood, indicating that all renal excretory functions are abnormal.

urethritis: Inflammation of the urethral wall.

CHAPTER REVIEW

 On-line resources for this chapter are on our World Wide Web site at:
http://www.prenhall.com/martini/fap

STUDY OUTLINE

INTRODUCTION, p. 961

1. The functions of the urinary system include (1) eliminating organic waste products; (2) regulating blood volume and pressure by adjusting the volume of water lost and releasing erythropoietin and renin; (3) regulating plasma concentrations of ions; (4) helping stabilize blood pH; (5) conserving nutrients; and (6) assisting the liver in detoxifying poisons and, during starvation, deaminating amino acids so that they can be catabolized by other tissues.

ORGANIZATION OF THE URINARY SYSTEM, p. 961

1. The urinary system includes the **kidneys,** the **ureters,** the **urinary bladder,** and the **urethra.** The kidneys produce **urine** (a fluid containing water, ions, and soluble compounds). During **urination (micturition)** urine is forced out of the body. *(Figure 26-1)*

THE KIDNEYS, p. 961

1. The left kidney extends superiorly slightly more than the right kidney. Both kidneys and their adrenal gland caps lie in a retroperitoneal position. *(Figure 26-2)*

Superficial Anatomy of the Kidneys, p. 962

2. The **hilus,** a medial indentation, provides entry for the renal artery and exit for the renal vein and the ureter. *(Figures 26-2, 26-3)*

Sectional Anatomy of the Kidneys, p. 962

3. The ureter communicates with the **renal pelvis.** This chamber branches into two **major calyces,** each of which is connected to four or five **minor calyces** that enclose the **renal papillae.** *(Figure 26-3)*

The Nephron, p. 963

4. The **nephron** (the basic functional unit in the kidney) consists of the **renal corpuscle** and **renal tubule.** The renal tubule is long and narrow and divided into the *proximal convoluted tubule,* the *loop of Henle,* and the *distal convoluted tubule.* **Filtrate** is produced at the renal corpuscle. The nephron empties **tubular fluid** into the **collecting system** through a **connecting tubule,** a tributary of a **collecting duct.** *(Figure 26-4)*

5. Nephrons are responsible for the production of filtrate, the reabsorption of organic nutrients, the reabsorption of water

and ions, and the secretion into the tubular fluid of waste products missed by filtration. (*Table 26-1*)

6. Roughly 85 percent of the nephrons are **cortical nephrons** found in the cortex. The **juxtamedullary nephrons** are closer to the **medulla**, with their loops of Henle extending deep into the **renal pyramids**.

7. The renal tubule begins at the renal corpuscle. It includes a knot of intertwined capillaries called the **glomerulus** surrounded by **Bowman's capsule**. Blood arrives at the glomerulus via the **afferent arteriole** and departs in the **efferent arteriole**. (*Figure 26-5*)

8. At the glomerulus, **podocytes** cover the **lamina densa** of the capillaries that project into the **capsular space**. The **pedicels** of the podocytes are separated by narrow **filtration slits**. (*Figure 26-5*)

9. The **proximal convoluted tubule (PCT)** actively reabsorbs nutrients, plasma proteins, and ions from the filtrate. The **loop of Henle** includes a **descending limb** and an **ascending limb;** each limb contains a **thick segment** and a **thin segment**. The ascending limb delivers fluid to the **distal convoluted tubule (DCT),** which actively secretes ions, toxins, and drugs and reabsorbs sodium ions from the tubular fluid. (*Figures 26-4–26-6*)

The Blood Supply to the Kidneys, p. 970

10. The vasculature of the kidneys includes the **segmental, interlobar, arcuate,** and **interlobular arteries** and the **interlobular, arcuate, interlobar,** and **segmental veins**. Blood travels from the efferent arteriole to the **peritubular capillaries** and the **vasa recta**. Diffusion occurs between the capillaries of the vasa recta and the tubular cells through the **peritubular fluid** that surrounds the nephron. (*Figures 26-7, 26-8*)

Innervation of the Kidneys, p. 970

11. The **renal nerves** that innervate the kidneys and ureters are dominated by sympathetic postganglionic fibers.

RENAL PHYSIOLOGY, p. 972

Basic Principles of Urine Formation, p. 972

1. Urine formation involves **filtration, reabsorption,** and **secretion**. (*Table 26-2*)

Filtration, p. 973

2. **Glomerular filtration** occurs as fluids move across the wall of the glomerulus into the capsular space, in response to the hydrostatic (blood) pressure in the glomerular capillaries—the **glomerular hydrostatic pressure (GHP)**. This movement is opposed by the **capsular hydrostatic pressure (CHP)** and by the **blood colloid osmotic pressure (BCOP)**. The **net filtration pressure (NFP)** at the glomerulus is the difference between the blood pressure and the opposing capsular and osmotic pressures. (*Figure 26-9*)

3. The **glomerular filtration rate (GFR)** is the amount of filtrate produced in the kidneys each minute. Any factor that alters the filtration pressure acting across the glomerular capillaries will change the GFR and affect kidney function. (*Figure 26-10*)

4. A drop in filtration pressures stimulates the **juxtaglomerular apparatus (JGA)** to release renin and erythropoietin.

5. Sympathetic activation (1) produces a powerful vasoconstriction of the afferent arterioles, decreasing the GFR and slowing the production of filtrate, (2) alters the GFR by changing the regional pattern of blood circulation, and (3) stimulates the release of renin by the juxtaglomerular apparatus.

Reabsorption and Secretion, p. 978

6. Four types of *carrier-mediated transport (facilitated diffusion, active transport, cotransport,* and *countertransport)* are involved in modifying the filtrate. The saturation limit of a carrier protein is its **transport maximum**. The transport maximum determines the **renal threshold**, the plasma concentration at which various compounds will appear in the urine. (*Table 26-3*)

7. Glomerular filtration produces a filtrate with a composition similar to blood plasma but with few, if any, plasma proteins.

8. The cells of the PCT normally reabsorb sodium and other ions, water, and almost all the organic nutrients that enter the filtrate. It also secretes various substances into the tubular fluid. (*Figure 26-11*)

9. Water and ions are reclaimed from the tubular fluid by the loop of Henle. A concentration gradient in the medulla encourages the osmotic flow of water out of the tubular fluid. The **countercurrent multiplication** between the ascending and descending limbs of the loop of Henle helps create the osmotic gradient in the medulla. As water is lost by osmosis and the volume of tubular fluid decreases, the urea concentration rises. (*Figure 26-12*)

10. The DCT performs final adjustments by actively secreting or absorbing materials. Sodium ions are actively absorbed, in exchange for potassium or hydrogen ions discharged into the tubular fluid. Aldosterone secretion increases the rate of sodium reabsorption and potassium loss. (*Figure 26-13*)

11. The amount of water and solutes in the tubular fluid of the collecting ducts is further regulated by aldosterone and ADH secretions. (*Figure 26-14*)

The Function of the Vasa Recta, p. 987

12. Normally, the removal of solutes and water by the vasa recta precisely balances the rates of reabsorption and osmosis in the medulla. (*Figure 26-15*)

A Summary of Renal Function, p. 988

13. Each segment of the nephron and collecting system contributes to the production of a concentrated urine. (*Figure 26-16; Table 26-4*)

The Composition of Normal Urine, p. 990

14. More than 99 percent of the filtrate produced each day is reabsorbed before reaching the renal pelvis. (*Tables 26-5, 26-6*)

15. **Hemodialysis** is a technique used to regulate the composition of blood artificially. (*Figure 26-17; Table 26-7*)

URINE TRANSPORT, STORAGE, AND ELIMINATION, p. 991

1. Urine production ends when the tubular fluid enters the renal pelvis. The rest of the urinary system is responsible for transporting, storing, and eliminating the urine. (*Figures 26-18, 26-19*)

The Ureters, p. 994

2. The ureters extend from the renal pelvis to the urinary bladder. Peristaltic contractions by smooth muscles move the urine. (*Figure 26-20*)

The Urinary Bladder, p. 996

3. The urinary bladder is stabilized by the **middle umbilical ligament** and the **lateral umbilical ligaments.** Internal features include the **trigone**, the **neck,** and the **internal urethral sphincter.** The mucosal lining contains prominent **rugae** (folds). Contraction of the **detrusor muscle** compresses the bladder and expels the urine into the urethra. *(Figures 26-19, 26-20)*

The Urethra, p. 996

4. In both genders, as the urethra passes through the *urogenital diaphragm,* a circular band of skeletal muscles forms the **external urethral sphincter,** which is under voluntary control. *(Figures 26-19, 26-20)*

The Micturition Reflex and Urination, p. 997

5. The process of urination is coordinated by the **micturition reflex,** which is initiated by stretch receptors in the bladder wall. Voluntary urination involves coupling this reflex with the voluntary relaxation of the external urethral sphincter, which allows the opening of the **internal urethral sphincter.** *(Figure 26-21)*

AGING AND THE URINARY SYSTEM, p. 998

1. Aging is generally associated with increased kidney problems. Age-related changes in the urinary system include (1) declining numbers of functional nephrons, (2) reduced GFR, (3) reduced sensitivity to ADH, (4) problems with the micturition reflex (**urinary retention** may develop in men whose prostate gland is enlarged).

INTEGRATION WITH OTHER SYSTEMS, p. 998

1. The urinary, integumentary, respiratory, and digestive systems are sometimes considered an anatomically diverse **excretory system,** whose components work together to perform all the excretory functions that affect the composition of body fluids. *(Figure 26–22)*

REVIEW QUESTIONS

Level 1   Reviewing Facts and Terms

1. The point of entry for the renal artery and exit for the renal vein and ureter is a medial indentation called the
   (a) renal column   (b) medulla
   (c) hilus   (d) renal cortex
2. The glomerulus is located within the
   (a) renal corpuscle   (b) renal tubule
   (c) renal pelvis   (d) renal column
3. Specialized cells that engulf organic materials that might otherwise clog the filter at the lamina densa are
   (a) mesangial cells   (b) juxtamedullary cells
   (c) podocytes   (d) macula densa
4. Large cells with complex processes, or "feet," that wrap around the glomerular capillaries are
   (a) vasa recta   (b) podocytes
   (c) astrocytes   (d) mesangial cells
5. After the filtrate leaves the glomerulus, it empties into the
   (a) distal convoluted tubule
   (b) loop of Henle
   (c) proximal convoluted tubule
   (d) collecting duct
6. The distal convoluted tubule is an important site for
   (a) active secretion of ions
   (b) active secretion of acids and other materials
   (c) selective reabsorption of sodium ions from the tubular fluid
   (d) a, b, and c are correct
7. The endocrine structure that secretes renin and erythropoietin is the
   (a) juxtaglomerular apparatus
   (b) vasa recta
   (c) Bowman's capsule
   (d) adrenal gland
8. The primary purpose of the collecting system is to
   (a) transport urine from the bladder to the urethra
   (b) selectively reabsorb sodium ions from tubular fluid
   (c) transport urine from the renal pelvis to the ureters
   (d) make final adjustments to the osmotic concentration and volume of urine

9. Pickup or delivery of substances that are reabsorbed or secreted by the PCT and the DCT is provided by the
   (a) afferent arteriole   (b) peritubular capillaries
   (c) renal artery   (d) efferent arteriole
10. The most abundant organic waste in urine is
    (a) uric acid   (b) creatinine
    (c) urea   (d) creatine phosphate
11. The removal of water and solute molecules from the filtrate after it enters the renal tubules is
    (a) filtration   (b) secretion
    (c) reabsorption   (d) excretion
12. The transport of solutes from the peritubular fluid into the tubular fluid is
    (a) reabsorption   (b) filtration
    (c) excretion   (d) secretion
13. The force that tends to drive water and solutes into the interstitial fluid is the
    (a) glomerular hydrostatic pressure
    (b) net hydrostatic pressure
    (c) capsular hydrostatic pressure
    (d) net colloid osmotic pressure
14. Inflammation of the lining of the urinary bladder is
    (a) urethritis   (b) cystitis
    (c) dysuria   (d) pyelitis
15. What is the primary function of the urinary system?
16. What structures are included as component parts of the urinary system?
17. Trace the pathway of the protein-free filtrate from the time it is produced in the renal corpuscle until it drains into the renal pelvis in the form of urine. (Use arrows to indicate the direction of flow.)
18. Name the segments of the nephron distal to the renal corpuscle, and state the function(s) of each.
19. What role does the lamina densa play in the renal corpuscle?
20. What is the function of the juxtaglomerular apparatus?
21. Using arrows, trace a drop of blood from its entry into the renal artery until its exit via a renal vein.

22. Name and define the three distinct processes involved in the production of urine.

23. What are the primary effects of angiotensin II on kidney function and regulation?

24. Which parts of the urinary system are responsible for the transport, storage, and elimination of urine?

Level 2   Reviewing Concepts

25. The urinary system regulates blood volume and pressure by
  (a) adjusting the volume of water lost in the urine
  (b) releasing erythropoietin
  (c) releasing renin
  (d) a, b, and c are correct

26. The balance of solute and water reabsorption in the medulla is maintained by the
  (a) segmental arterioles and veins
  (b) interlobar arteries and veins
  (c) vasa recta
  (d) arcuate arteries

27. Sympathetic activation of the nerve fibers in the nephron causes
  (a) regulation of glomerular blood flow and pressure
  (b) stimulation of renin release from the juxtaglomerular apparatus
  (c) direct stimulation of water and Na⁺ reabsorption
  (d) a, b, and c are correct

28. An increase in the capsular colloid osmotic pressure caused by damage to the glomerulus would
  (a) decrease the amount of plasma that is delivered to the kidney
  (b) enhance the movement of water and solutes into the capsular space
  (c) cause a decrease in the renal blood pressure
  (d) decrease movement of water and solutes into the capsular space

29. Sodium reabsorption in the DCT and cortical portion of the collecting system is accelerated by secretion of
  (a) ADH   (b) renin
  (c) aldosterone (d) erythropoietin

30. When ADH levels rise,
  (a) the amount of water reabsorbed increases
  (b) the DCT becomes impermeable to water
  (c) the amount of water reabsorbed decreases
  (d) sodium ions are exchanged for potassium ions

31. Control of blood pH by the kidneys involves
  (a) addition of hydrogen ions and removal of bicarbonate ions from the filtrate
  (b) a decrease in the amount of water reabsorbed
  (c) hydrogen ion removal and bicarbonate ion production
  (d) potassium ion secretion

32. How does a cortical nephron differ from a juxtamedullary nephron?

33. How can you determine the net filtration pressure (NFP) at the glomerulus?

34. What interacting controls operate to stabilize the glomerular filtration rate (GFR)?

35. What is the renal threshold?

36. What primary changes occur in the composition and concentration of the filtrate as a result of activity in the proximal convoluted tubule?

37. What two functions does the countercurrent mechanism perform for the kidney?

38. What events in the distal convoluted tubule and collecting system determine the final composition and concentration of the filtrate?

39. Describe the micturition reflex.

Level 3   Critical Thinking and Clinical Applications

40. Why do long-haul trailer truck drivers commonly experience kidney problems?

41. Doctors often ask for urine samples collected over a 24-hour period rather than a single sample. Why?

42. Randy enjoys his four or five cups of coffee a day and his six or seven beers a few nights a week. What are the effects of his caffeine and alcohol consumption on his urinary system?

43. For the past week, Susan has felt a burning sensation in the urethral area when she urinates. She checks her temperature and finds that she has a low-grade fever. What unusual substances are likely to be present in her urine?

44. *Lasix* is a diuretic that acts by decreasing the amounts of sodium and chloride ions actively transported by the ascending limb of the loop of Henle. Why would this medication be given to someone suffering from high blood pressure (hypertension)?

45. Carlos suffers from advanced arteriosclerosis. An analysis of his blood indicates elevated levels of aldosterone and decreased levels of ADH. Explain.

46. *Mannitol* is a sugar that is filtered but not reabsorbed by the kidneys. What effect would drinking a solution of mannitol have on the volume of urine produced?

47. The drug *Diamox* is sometimes used in the treatment of epilepsy. It functions by inhibiting the action of carbonic anhydrase in the proximal convoluted tubule. Polyuria (the elimination of an unusually large volume of urine) is a side effect associated with this medication. Why does this symptom occur?

27

Fluid, Electrolyte, and Acid–Base Balance

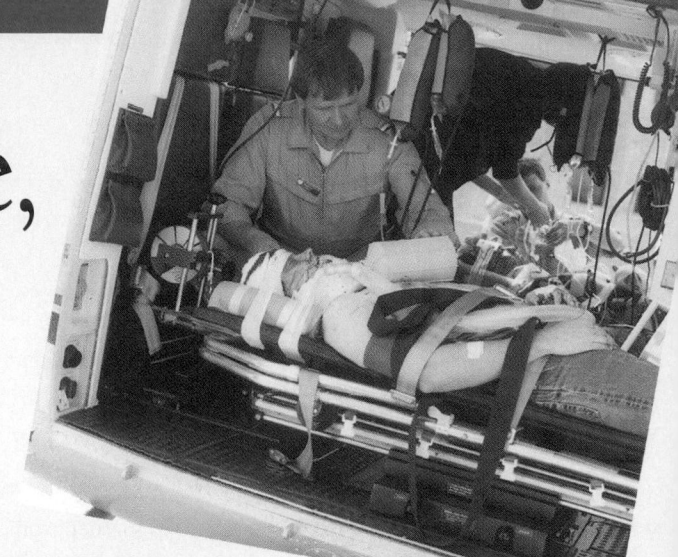

*A*fter a severe accident, getting to a medical facility as quickly as possible can mean the difference between life and death. Emergency measures generally must start before the patient arrives at a hospital even when the trip is made by Med-Evac helicopter. After serious trauma, the body's regulatory mechanisms may be unable to maintain homeostasis without assistance. The patient's internal conditions must therefore be monitored and stabilized while in transit. In virtually all medical emergencies, fluid and electrolyte levels must be adjusted, generally by the administration of intravenous solutions containing water, electrolytes, and buffers. This relatively simple therapy can have multiple benefits: replacing lost blood, providing nutrients, and controlling the pH of body fluids. In this chapter, we shall examine the mechanisms responsible for the maintenance of water, electrolyte, and acid–base balance under normal and abnormal conditions.

CHAPTER OUTLINE AND OBJECTIVES

The next time you see a small pond, take a moment to think about the fish it contains. They live out their lives totally dependent on the quality of that isolated environment. Polluting the pond with toxic substances will, of course, kill the fish, but more subtle changes can have equally grave effects. Changes in the volume of the pond, for example, can be quite important. If evaporation removes too much of the pond water, the fish become overcrowded; oxygen and food supplies run out, and the fish suffocate or starve. The ionic concentration of the pond water is also crucial. Most of the fish in a freshwater pond will die if the water becomes too salty; those in a saltwater pond will die if their environment becomes too dilute. The pH of the pond water, too, is a vital factor; that is one reason acid rain is such a problem.

The cells of your body live in a pond whose shores are the exposed surfaces of your skin. Most of the weight of the human body is water. Water accounts for up to 99 percent of the volume of the fluid outside cells, and it is an essential ingredient of cytoplasm. All of a cell's operations rely on water as a diffusion medium for the distribution of gases, nutrients, and waste products. If the water content of the body changes, cellular activities are jeopardized. For example, when the water content of the body reaches very low levels, proteins denature, enzymes cease functioning, and cells die. Maintenance of normal volume and composition in the **extracellular fluid (ECF)** and **intracellular fluid (ICF)** is essential to our survival.

The ionic concentrations and pH of the body water are as important as its absolute quantity. If concentrations of calcium or potassium ions in the ECF become too high, cardiac arrhythmias develop, and death can result. A pH outside the normal range can lead to a variety of dangerous problems. Low pH is especially dangerous because hydrogen ions break chemical bonds, change the shapes of complex molecules, disrupt cell membranes, and impair tissue functions.

In this chapter, we shall consider the dynamics of exchange among the various body fluids and between the body and the external environment. Stabilizing the volumes, solute concentrations, and pH of the ECF and ICF involves three interrelated processes:

1. *Fluid balance.* You are in **fluid balance** when the amount of water you gain each day is equal to the amount you lose to the environment. The maintenance of normal fluid balance involves regulating body water *content* and *distribution* in the ECF and ICF. The digestive system is the primary source for water gains. A small amount of water is generated by metabolic activity. The urinary system is the primary route for water loss under normal conditions, but as we noted in Chapter 25, sweat gland activity can become important when body temperature is elevated. [p. 952] Because your cells and tissues cannot transport water, fluid balance reflects primarily the control of *electrolyte balance.*

2. *Electrolyte balance.* **Electrolytes** are ions released through the dissociation of inorganic compounds; they are called electrolytes because they will conduct an electrical current in a solution. [p. 42] Each day, your body fluids gain electrolytes from the food and drink you consume and lose electrolytes in urine, sweat, and feces. For each individual ion, daily gains must balance daily losses. For example, if you lose 500 mg of Na^+ in urine and insensible perspiration, you will need to gain 500 mg of Na^+ from food and drink to remain in sodium balance. If the gains and losses for every electrolyte are in balance, you are said to be in **electrolyte balance.** Electrolyte balance primarily involves balancing the rates of absorption across the digestive tract with rates of loss at the kidneys, although excretion at sweat glands and other sites can play a secondary role.

3. *Acid–base balance.* You are in **acid–base balance** when the production of hydrogen ions is precisely offset by their loss. When acid–base balance exists, the pH of body fluids remains within normal limits. Preventing a reduction in pH is the primary problem, because your body generates a variety of acids during normal metabolic operations. Once again, the kidneys play a major role by secreting hydrogen ions into the urine and generating buffers that enter the circulation. As we learned in Chapter 26, this secretion occurs primarily in the distal segments of the distal convoluted tubule (DCT) and along the collecting system. [p. 983] As we shall see, the lungs also play a key role through the elimination of carbon dioxide.

Much of the material in this chapter was introduced in earlier chapters that focused on aspects of fluid, electrolyte, or acid–base balance that affected specific systems. This chapter provides an overview that integrates those discussions to highlight important functional patterns. Few other chapters have such wide-ranging clinical importance: *Treatment of any serious illness affecting the nervous, cardiovascular, respiratory, urinary, or digestive system must include steps to restore normal fluid, electrolyte, and acid–base balance.* Because this chapter builds on information provided in earlier chapters, you will encounter many references to relevant discussions and figures that you should use when you need a quick review.

FLUID AND ELECTROLYTE BALANCE

Figure 27-1a• provides an overview of the composition of the body of a 70-kg individual with a minimum of body fat. This distribution was obtained by averaging values for adult males and adult females. Water accounts for roughly 60 percent of the total body weight of a male and 50 percent of the total body weight of a female. The gender difference primarily reflects the proportionately larger mass of adipose tissue in adult females and the greater average muscle mass in adult males. (Adipose tissue is 10 percent water, whereas skeletal muscle is 75

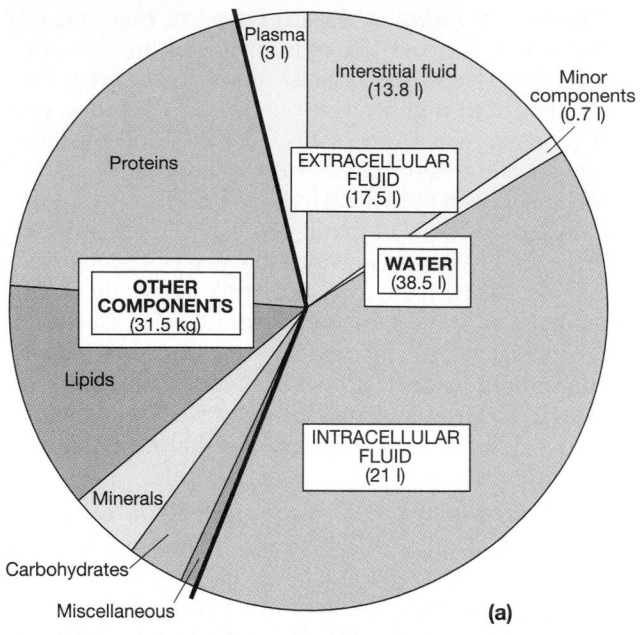

(a)

(b) Body Composition of Adult Males and Females

Material/Compartment	Percentage of Total Body Weight	
	Adult Males	**Adult Females**
Solids (proteins, lipids, minerals, carbohydrates, various organic and inorganic materials)	40	50
Water	60	50
ICF	33	27
ECF	27	23
Interstitial fluid	21.5	18
Plasma	4.5	4.0
Minor components	1.0	1.0

•*FIGURE 27-1* **Composition of the Human Body.**
(a) The body composition and major body fluid compartments of a 70-kg individual. For technical reasons, it is extremely difficult to determine the precise volume of any of these fluid compartments; estimates of their relative sizes vary widely. **(b)** Comparison of the body compositions of adult males and females.

percent water.) In both genders, the intracellular fluid contains more of the total body water than does the extracellular fluid. Exchange between the ICF and the ECF occurs across cell membranes through mechanisms detailed in Chapter 3. (To review the mechanisms involved, see Table 3-2, p. 81.)

The largest subdivisions of the ECF are the *interstitial fluid* of peripheral tissues and the *plasma* of the circulating blood. Minor components of the ECF include lymph, cerebrospinal fluid (CSF), synovial fluid, serous fluids (pleural, pericardial, and peritoneal fluids), aqueous humor, perilymph, and endolymph. Precise measurements of total body water provide additional information on gender differences in the distribution of body water (Figure 27-1b•). The great-

est variation is in the ICF due to differences in the intracellular water content of fat versus muscle. There are less-striking differences in the ECF values due to variations in the interstitial fluid volume of various tissues and the larger blood volume in males.

In clinical situations, it is customary to estimate that two-thirds of the total body water is in the ICF and one-third in the ECF. This ratio underestimates the real volume of the ECF, but that underestimation is appropriate when we consider rapid fluid and solute movements. Portions of the ECF, including the water in bone, in many dense connective tissues, and in many of the minor ECF components, are relatively isolated. Exchange between these fluid volumes and the rest of the ECF occurs more slowly than does exchange between the plasma and other interstitial fluids.

Exchange among the subdivisions of the ECF occurs primarily across the endothelial lining of capillaries. Fluid may also travel from the interstitial spaces to the plasma through lymphatic vessels that drain into the venous system. ∞ *[pp. 725–726]* There are regional variations in the identity and quantity of dissolved electrolytes, proteins, nutrients, and waste products within the ECF. (For a chemical analysis of the composition of ECF compartments, see Appendix VI.) Yet the variations among the segments of the ECF seem minor when compared with the major differences between the ECF and the ICF.

The ECF and ICF are called **fluid compartments,** because they commonly behave as distinct entities. The presence of a cell membrane and active transport at the membrane surface enable cells to maintain internal environments with a composition different from that of their surroundings. The principal ions in the ECF are sodium, chloride, and bicarbonate. The ICF contains an abundance of potassium, magnesium, and phosphate ions, plus large numbers of negatively charged proteins. Figure 27-2• compares the ICF with the two major subdivisions of the ECF.

If the cell membrane were freely permeable, diffusion would continue until these ions were evenly distributed across the membrane. But it does not, because cell membranes are selectively permeable; ions can enter or leave the cell only via specific membrane channels. In addition, there are carrier mechanisms that move specific ions into or out of the cell.

Despite the differences in the concentration of specific substances, the ICF and ECF osmolarities are identical. Osmosis eliminates any minor concentration differences almost at once, because most cell membranes are freely permeable to water. (The only noteworthy exceptions are the lumenal surfaces of the ascending limb of the loop of Henle, the DCT, and the collecting system, as we detailed in Chapter 26.) ∞ *[pp. 988–990]* Because changes in solute concentrations will lead to immediate changes in water distrib-

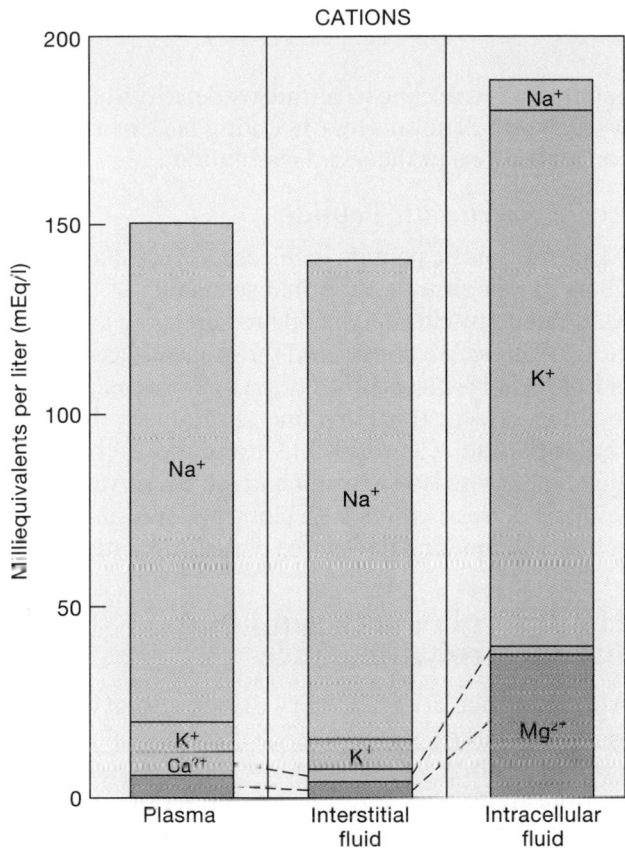

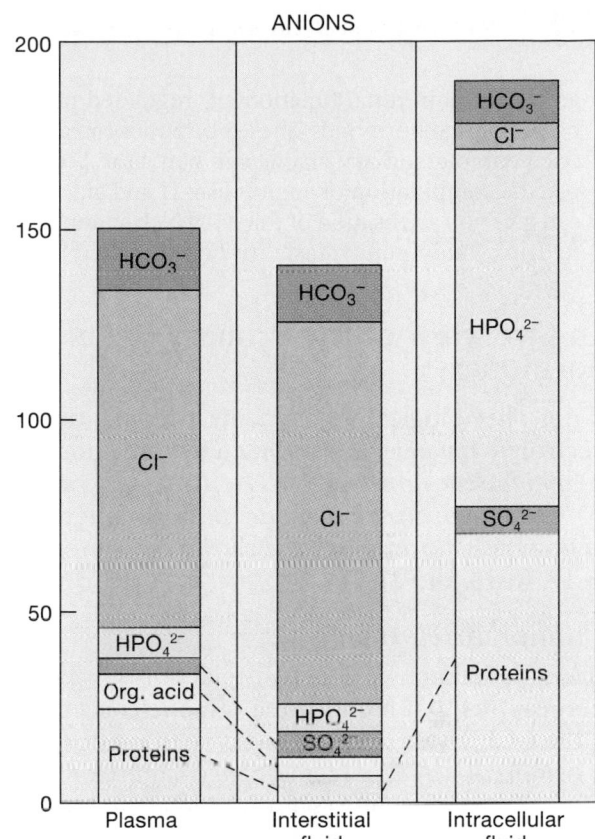

•FIGURE 27-2 Cations and Anions in Body Fluids. Note the differences in cation and anion concentrations in the various body fluid compartments. For information about the composition of other body fluids, see Appendix VI.

ution, the regulation of fluid balance and that of electrolyte balance are tightly intertwined.

Physiologists and clinicians pay particular attention to ionic distributions across membranes and to the electrolyte composition of body fluids. Appendix VI, which contains normal values for the analysis of body fluids, reports values in the units most often encountered in clinical reports.

Basic Concepts Pertaining to Fluid and Electrolyte Regulation

You must understand four basic principles before we can proceed to a discussion of fluid and electrolyte balance:

1. *All the homeostatic mechanisms that monitor and adjust the composition of body fluids respond to changes in the ECF, not in the ICF.* Receptors monitoring the composition of two key components of the ECF, the plasma and the cerebrospinal fluid, detect significant changes in composition or volume and trigger appropriate neural and endocrine responses.

   This arrangement makes functional sense because a change in one ECF component will spread rapidly throughout the extracellular compartment and will affect all the body's cells. In contrast, the ICF is contained within trillions of individual cells physically and chemically isolated from one another by their cell membranes. Thus changes in the ICF in one cell will have no direct effect on the composition of the ICF in distant cells and tissues unless those changes affect the ECF.

2. *There are no receptors that can directly monitor fluid or electrolyte balance.* In other words, our receptors cannot detect how many liters of water or grams of sodium, chloride, or potassium the body contains or count how many liters or grams we gain or lose in the course of a day. But our receptors *can* monitor *plasma volume* and *osmolarity*. Because fluid continuously circulates between the interstitial fluid and plasma and there is exchange between the ECF and ICF, plasma volume and osmolarity are good indicators of the state of fluid and electrolyte balance for the body as a whole.

3. *Our cells cannot move water molecules by active transport.* All water movement across cell membranes and epithelia occurs passively, in response to osmotic gradients. These gradients can be established by the active transport of specific ions, such as sodium and chloride. You may find it useful to remember the simple phrase *water follows salt*. As we saw in earlier chapters, when sodium and chloride ions (or other solutes) are actively transported across a membrane or epithelium, water follows by osmosis. This basic principle accounts for water absorption across the digestive epithelium and water conservation in the kidneys.

4. *The body content of water or electrolytes will rise if dietary gains exceed losses to the environment and will fall if losses exceed gains.* This basic rule is important when you consider the mechanics of fluid and electrolyte balance. Homeostatic adjustments generally affect the balance between urinary excretion and dietary absorption. As we saw in Chapter 26, the physiological

adjustments in renal function are regulated primarily by circulating hormones. These hormones can also produce complementary changes in behavior. For example, the combination of angiotensin II and aldosterone can give you a sensation of thirst—which stimulates you to drink fluids—and a taste for heavily salted foods.

An Overview of the Primary Regulatory Hormones

Major physiological adjustments affecting fluid and electrolyte balance are mediated by three hormones: (1) *antidiuretic hormone (ADH)*, (2) *aldosterone*, and (3) *atrial natriuretic peptide (ANP)*. Interactions among these hormones were shown in Figures 18-16, 18-17, 18-18, and 21-17●. ∞ *[pp. 618, 619, 734]*

Antidiuretic Hormone

The hypothalamus contains special cells known as **osmoreceptors** that monitor the osmotic concentration of the ECF. These cells are sensitive to subtle changes in osmolarity: A 2 percent change (approximately 6 mOsm/l) is sufficient to alter osmoreceptor activity.

The population of osmoreceptors includes neurons that secrete ADH. These neurons are located in the anterior hypothalamus, and their axons release ADH near fenestrated capillaries in the posterior pituitary gland. The rate of ADH release varies directly with the osmolarity: The higher the osmolarity, the greater the amount of ADH released.

Increased release of ADH has two important effects: (1) *It stimulates water conservation at the kidneys, reducing urinary water losses and concentrating the urine;* and (2) *it stimulates the thirst center to promote the drinking of fluids.* As we saw in Chapter 18, the combination of decreased water loss and increased water gain gradually restores normal plasma osmolarity. ∞ *[p. 607]*

Aldosterone

The secretion of aldosterone by the adrenal cortex plays a major role in determining the rate of Na^+ absorption and K^+ loss along the DCT and collecting system of the kidneys. We detailed the mechanism in Chapters 18, 21, and 26. ∞ *[pp. 616, 620, 733–734, 983]* The higher the plasma concentration of aldosterone, the more efficiently the kidneys conserve Na^+. Because "water follows salt," conservation of Na^+ also stimulates water retention. Aldosterone also increases the sensitivity of salt receptors on the tongue. This effect may increase your interest in and consumption of salty foods.

Secretion of aldosterone occurs in response to rising K^+ or falling Na^+ levels in the blood that reaches the adrenal cortex or to activation of the renin–angiotensin system. As we noted in earlier chapters, renin release occurs in response to (1) a drop in plasma volume or blood pressure at the juxtaglomerular apparatus of the

nephron, (2) a decline in filtrate osmolarity at the DCT, or, as we shall soon see, (3) falling Na^+ or rising K^+ concentrations in the renal circulation.

Atrial Natriuretic Peptide

Atrial natriuretic peptide is released by cardiac muscle fibers in response to abnormal stretching of the atrial walls caused by elevated blood pressure or an increase in blood volume. We considered the many effects of ANP in Chapters 18, 21, and 26. ∞ *[pp. 619, 734, 987]* Among its other effects, this hormone (1) reduces thirst and (2) blocks the release of ADH and aldosterone that might otherwise lead to water and salt conservation. The resulting diuresis lowers both blood pressure and plasma volume, eliminating the source of the stimulation.

The Interplay between Fluid and Electrolyte Balance

At first glance, it can be very difficult to distinguish between water balance and electrolyte balance. For example, when you lose body water, your plasma volume decreases and electrolyte concentrations rise. Conversely, when you gain or lose excess electrolytes, there is an associated water gain or loss due to osmosis. However, because the regulatory mechanisms involved are quite different, it is often useful to consider fluid balance and electrolyte balance as distinct entities. This distinction is absolutely vital in a clinical setting, where problems with fluid balance and electrolyte balance must be identified and corrected promptly.

Fluid Balance

Water circulates freely within the ECF compartment. At capillary beds throughout the body, hydrostatic pressure forces water out of the plasma and into interstitial spaces. Some of that water is reabsorbed along the distal portion of the capillary bed, and the rest circulates into lymphatic vessels for transport to the venous circulation. This circulation is diagrammed in Figure 27-3●, which indicates two additional important relationships between components of the ECF:

1. Water moves back and forth across the mesothelial surfaces that line the peritoneal, pleural, and pericardial cavities and through the synovial membranes that line joint capsules. The flow rate is significant; for example, roughly 7 liters of peritoneal fluid is produced and reabsorbed each day.

2. Water also moves between the blood and the cerebrospinal fluid (CSF), the aqueous and vitreous humors of the eye, and the perilymph and endolymph of the inner ear. The volumes involved are very small, and the volume and composition of these fluids are closely regulated. For those reasons, we shall largely ignore them in the discussion that follows.

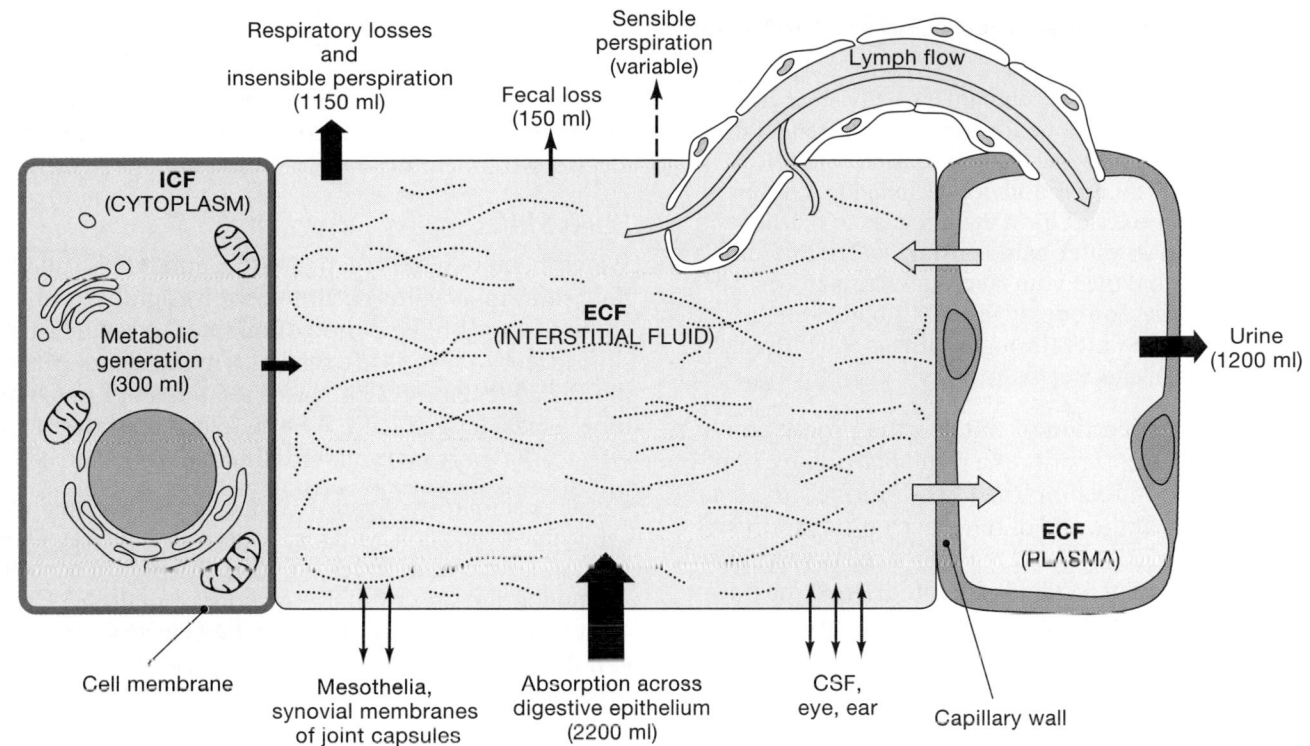

Respiratory losses
and
insensible perspiration
(1150 ml)

Sensible
perspiration
(variable)

Lymph flow

Fecal loss
(150 ml)

ICF
(CYTOPLASM)

Metabolic
generation
(300 ml)

ECF
(INTERSTITIAL FLUID)

Urine
(1200 ml)

ECF
(PLASMA)

Cell membrane

Mesothelia,
synovial membranes
of joint capsules

Absorption across
digestive epithelium
(2200 ml)

CSF,
eye, ear

Capillary wall

•FIGURE 27-3 Fluid Exchanges. Diagrammatic representation of fluid movement between the ICF and ECF and between the ECF and the environment. The volumes are not drawn to scale; functionally, the ICF is twice as large as the ECF.

Fluid Movement within the ECF

In Chapter 21, in the discussion of capillary dynamics, we considered the basic principles that determine fluid movement between the divisions of the ECF. ∞ *[pp. 723–727]* The exchange between the plasma and the interstitial fluid, by far the largest components of the ECF, is determined by the relationship between the net hydrostatic pressure, which tends to push water out of the plasma and into the interstitial fluid, and the net colloid osmotic pressure, which tends to draw water out of the interstitial fluid and into the plasma. The interaction between these opposing forces, diagrammed in Figure 21-13• (p. 726), results in the continuous filtration of fluid from the capillaries into the interstitial fluid. This volume of fluid is then redistributed, returning to the venous system after passing through the channels of the lymphatic system. At any moment, the interstitial fluid and minor fluid compartments contain roughly 80 percent of the ECF volume, and the plasma contains the other 20 percent.

Any factor that affects the net hydrostatic pressure or the net colloid osmotic pressure will alter the distribution of fluid within the ECF. The movement of abnormal amounts of water from the plasma into the interstitial fluids is called *edema*. We detailed the general causes of edema in Chapter 21. ∞ *[pp. 726–727]* For example, pulmonary edema can result from an increase in the blood pressure in pulmonary capillaries, and a generalized edema can result from a decrease in blood colloid osmotic pressure, as in advanced starvation, when plasma protein concentrations decline. Localized edema can result from damage to capillary walls, as in bruising, constriction of the regional venous circu-

lation, or blockage of the lymphatic drainage, as in *lymphedema*, discussed in Chapter 22. ∞ *[p. 773]*

Fluid Exchange with the Environment

Figure 27-3• and Table 27-1 indicate the major routes of exchange with the environment:

- **Water losses.** You lose roughly 2500 ml of water each day through urine, feces, and *insensible perspiration*, the gradual movement of water across the epithelia of the skin and respiratory tract. The losses due to *sensible perspiration*, the secretory activities of the sweat glands, vary with the activities you undertake. Sensible perspiration can cause significant water deficits, with maximum perspiration rates reaching well over 4 liters per hour. We detailed the relationships between sensible and insensible perspiration in Chapter 25. ∞ *[p. 952]*

TABLE 27-1 Water Balance

Daily Input (ml)	
Water content of food	1000
Water consumed as liquid	1200
Metabolic water during catabolism	300
Total	2500

Daily Output (ml)	
Urine	1200
Evaporation at skin	750
Evaporation at lungs	400
Lost in feces	150
Total	2500

- The temperature rise accompanying a fever can also increase water losses. For each degree your temperature rises above normal, your daily insensible water loss increases by 200 ml. So the advice "drink plenty of fluids" for anyone who is sick has a definite physiological basis.
- *Water gains.* A water gain of roughly 2500 ml/day is required to balance your average water losses. This value amounts to roughly 40 ml/kg body weight per day. You obtain water through eating (1000 ml), drinking (1200 ml), and *metabolic generation* (300 ml).

Metabolic generation of water is the production of water within cells, primarily as a result of oxidative phosphorylation in mitochondria. (We described the synthesis of water at the end of the electron transport system in Chapter 25.) ∞ *[p. 925]* When a cell breaks down 1 g of lipid, 1.7 ml of water is generated. Breaking down proteins or carbohydrates results in much lower values (0.41 ml/g protein; 0.55 ml/g carbohydrate). A typical mixed diet in the United States contains 46 percent carbohydrates, 40 percent lipids, and 14 percent protein. This diet would produce roughly 300 ml of water per day, about 12 percent of your average daily requirement.

WATER EXCESS AND WATER DEPLETION The body's water content cannot easily be determined. However, the concentration of Na^+, the most abundant ion in the ECF, provides useful clues to the state of water balance. When the body water content rises enough to reduce the Na^+ concentration of the ECF below 130 mEq/l, the state of *hyponatremia* (*natrium*, sodium) exists. When the body water content declines, the Na^+ concentration rises; when it exceeds 150 mEq/l, *hypernatremia* exists.

Hyponatremia is a sign of **overhydration,** or *water excess.* Hyponatremia may result from (1) ingestion of a large volume of fresh water or infusion of a hypotonic solution; (2) an inability to eliminate excess water in the urine caused by *chronic renal failure* (p. 977), *congestive heart failure* (p. 698), *cirrhosis* (p. 894), or other disorders; or (3) endocrine disorders, such as excessive ADH production. The reduction in Na^+ concentrations leads to a shift of fluid into the ICF. The first signs are the effects on CNS function. In the early stages of hyponatremia, the individual behaves as if drunk on alcohol. This condition, called *water intoxication,* sounds humorous, but is extremely dangerous. Untreated cases can rapidly progress from confusion to hallucinations, convulsions, coma, and death. Treatment of severe water intoxication generally involves diuretics and infusing a concentrated salt solution that elevates Na^+ levels to near-normal levels.

Dehydration, also known as *water depletion* or *volume depletion,* develops when water losses outpace water gains. Plasma osmolarity gradually increases, and hypernatremia results. The loss of body water is associated with severe thirst, dryness and wrinkling of the skin, and a fall in plasma volume and blood pressure. Eventually, circulatory shock develops, generally with fatal consequences. Treatment for dehydration entails administering hy-

potonic fluids by mouth or intravenous infusion. This treatment increases ECF volume and restores normal electrolyte concentrations. [AM] *Water and Weight Loss*

Fluid Shifts

Water movement between the ECF and ICF is called a **fluid shift.** Fluid shifts occur relatively rapidly, reaching equilibrium within a period of minutes to hours. These shifts occur in response to changes in the osmolarity of the ECF. We explored the basic relationships in Chapter 3, especially in Figure 3-8•, p. 75, which showed the effects of hypertonic and hypotonic solutions on cells:

- If the osmolarity of your ECF increases, that fluid will become hypertonic with respect to your ICF. Water will then move from the cells into the ECF until osmotic equilibrium is restored. The osmolarity of the ECF will increase if you lose water but retain electrolytes.
- If the osmolarity of your ECF decreases, that fluid will become hypotonic with respect to your ICF. Water will then move from the ECF into the cells, and the ICF volume will increase. The osmolarity of the ECF will decrease if you gain water but do not gain electrolytes.

In summary, if the osmolarity of the ECF changes, a fluid shift between the ICF and ECF will tend to oppose the change. Because the volume of the ICF is much greater than that of the ECF, the ICF acts as a water reserve. In effect, instead of a large change in the osmolarity of the ECF, there are smaller changes in both the ECF and ICF. Two examples will demonstrate the dynamic exchange of water between the ECF and ICF.

ALLOCATION OF WATER LOSSES When you lose water but retain electrolytes, the osmolarity of the ECF rises. Osmosis then moves water out of the ICF and into the ECF until the two solutions are again isotonic. At this point, both the ECF and ICF will be somewhat more concentrated than normal, and both volumes will be lower than they were before the fluid loss. Because the ICF has roughly twice the functional volume of the ECF, the net change in the ECF is kept relatively small.

Conditions that cause severe water losses include excessive perspiration (exercising in hot weather), inadequate water consumption (being lost in a desert), repeated vomiting, and diarrhea. These disorders promote water losses far in excess of electrolyte losses, so body fluids become increasingly concentrated. Responses that attempt to restore homeostasis include ADH and renin secretion and (as soon as possible) an increase in fluid intake.

DISTRIBUTION OF WATER GAINS When you drink a glass of pure water, or when you are given hypotonic solutions through intravenous infusion, the water content of your body increases without a corresponding increase in the concentration of electrolytes. As a result, the ECF increases in volume but becomes hypotonic with respect to the ICF. A fluid shift then occurs, and the volume of the ICF increases at the expense of the

ECF. Once again, the larger volume of the ICF enables it to limit the amount of osmotic change. After the fluid shift, the ECF and ICF have slightly larger volumes and slightly lower osmolarities than they did originally.

Normally, this situation will be promptly corrected. The reduced plasma osmolarity depresses the secretion of ADH, discouraging fluid intake and increasing water losses in the urine. If the situation is *not* corrected, a variety of clinical problems will develop as intracellular fluid losses disrupt normal cell functions.

Electrolyte Balance

You are in electrolyte balance when the rates of gain and loss are equal for each of the electrolytes in your body. Electrolyte balance is important because:

- *Total electrolyte concentrations have a direct effect on water balance,* as we detailed above.

- *The concentrations of individual electrolytes can affect a cell functions.* We saw many examples in earlier chapters, such as the effect of abnormal Na^+ concentrations on neuron activity and the effects of high or low Ca^{2+} and K^+ concentrations on cardiac muscle tissue.

Two cations, Na^+ and K^+, merit attention, because (1) they are major contributors to the osmolarities of the ECF and ICF, respectively, and (2) they directly affect the normal functioning of all cells. Sodium is the dominant cation in the ECF. More than 90 percent of the osmolarity of the ECF results from the presence of sodium

salts, principally sodium chloride (NaCl) and sodium bicarbonate ($NaHCO_3$), so alterations in the osmolarity of body fluids generally reflect changes in Na^+ concentration. Normal Na^+ concentrations in the ECF average 136–142 mEq/l, versus 10 mEq/l or less in the ICF. Potassium is the dominant cation in the ICF, where concentrations reach 160 mEq/l. Extracellular K^+ concentrations are generally very low, from 3.8 to 5.0 mEq/l.

Two general rules about sodium balance and potassium balance are worth noting:

1. *The most common problems with electrolyte balance are caused by an imbalance between gains and losses of sodium ions.*
2. *Problems with potassium balance are less common but significantly more dangerous than are those related to sodium balance.*

Sodium Balance

The total amount of sodium in the ECF represents a balance between two factors:

1. *Sodium ion uptake across the digestive epithelium.* Sodium ions enter the ECF by crossing the digestive epithelium through diffusion and carrier-mediated transport. The rate of absorption varies directly with the amount of sodium in the diet.
2. *Sodium ion excretion at the kidneys and other sites.* Sodium losses occur primarily by excretion in the urine and through perspiration. The kidneys are the most important sites of Na^+ regulation. We detailed

 ## Athletes and Salt Loss

There are many unfounded notions and rumors about water and salt requirements during exercise. Sweat is a hypotonic solution that contains Na^+ in lower concentration than the ECF. As a result, a person who is sweating profusely loses more water than salt, and this loss leads to a rise in the Na^+ concentration of the ECF. The water content of the ECF decreases as the water loss occurs, and the blood volume drops. As we noted on page 1012, this condition is called *volume depletion*. Because this volume depletion occurs at the same time that blood is being shunted away from the kidneys, waste products accumulate in the blood and kidney function is impaired. We considered this effect in Chapter 26. ∞ *[pp. 976–977]*

To prevent volume depletion, athletes should pause at brief intervals to drink liquids. The primary problem in volume depletion is water loss, and research has shown no basis for the rumor that cramps will result if you drink while exercising. Salt pills and the various sports beverages promoted for "faster

absorption" and "better electrolyte balance" have no apparent benefits despite their relatively high cost. Body reserves of electrolytes are sufficient to tolerate extended periods of strenuous activity, and problems with Na^+ balance are extremely unlikely except during ultramarathons or other activities that involve maximal exertion for more than 6 hours.

Some sports beverages contain sugars and vitamins as well as electrolytes. During endurance events (marathons, ultramarathons, distance cycling), solutions containing less than 10 g/dl of glucose may improve performance when consumed late in the competition, as metabolic reserves are exhausted. However, high sugar concentrations (above 10 g/dl) may cause cramps, diarrhea, and other problems. The benefit of "glucose polymers" in sports drinks has yet to be proved. Drinking beverages "fortified" with vitamins is actively discouraged. Vitamins are not lost during exercise, and the consumption of these beverages in large volumes could, over time, cause hypervitaminosis.

the mechanisms for sodium reabsorption at the kidneys in Chapter 26. ∞ *[p. 981, 985]*

A person in sodium balance typically gains and loses 48–144 mEq (1.1–3.3 g) of Na$^+$ each day. When sodium gains exceed sodium losses, the total Na$^+$ content of the ECF goes up; when losses exceed gains, the Na$^+$ content declines. However, a change in the Na$^+$ *content* of the ECF does not produce a change in Na$^+$ *concentration*. Whenever the rate of sodium intake or output changes, a corresponding gain or loss of water tends to keep the Na$^+$ concentration constant. For example, if you eat a heavily salted meal, the osmolarity of your ECF will not increase. When sodium ions are pumped across your digestive epithelium, the solute concentration in that portion of the ECF increases, whereas that of the intestinal contents decreases. Osmosis then occurs, and additional water enters the ECF from the digestive tract.

When sodium gains exceed sodium losses, the volume of the ECF increases, but its osmolarity remains the same. Thus, persons with high blood pressure are advised to restrict the amount of salt in their diets. If consumed *with* a normal fluid volume, excess salt will result in an elevated blood volume and blood pressure, as water follows salt from the digestive tract into the plasma. If large amounts of salt are consumed *without* adequate fluid, as when you eat heavily salted potato chips without taking a drink, the plasma Na$^+$ concentration will rise temporarily. A change in ECF volume will soon follow, however. Fluid will exit the ICF, increasing ECF volume and lowering Na$^+$ concentrations somewhat. The secretion of ADH restricts water loss and stimulates thirst, promoting additional water consumption. Due to the inhibition of water receptors in the pharynx, ADH secretion begins even before the Na$^+$ absorption occurs; the secretion rate rises further after Na$^+$ absorption due to osmoreceptor stimulation. ∞ *[p. 550]*

When sodium losses exceed gains, the volume of the ECF decreases. This reduction occurs without a significant change in the osmolarity of the ECF. Thus, if you perspire heavily but consume only pure water, you will lose sodium, and the osmolarity of your ECF will drop briefly. However, as soon as the osmolarity drops by 2 percent or more, ADH secretion will decrease, and water losses at your kidneys will increase. As water leaves the ECF, the osmolarity will return to normal.

THE REGULATION OF SODIUM BALANCE As we have seen, under normal circumstances alterations in sodium balance result in the expansion or contraction of the ECF compartment. The regulatory mechanism, diagrammed in Figure 27-4●, changes the ECF volume but keeps the Na$^+$ concentration relatively stable.

Minor changes in the ECF volume do not matter, because they do not cause adverse physiological effects. If the regulation of Na$^+$ concentrations results in a large change in the ECF volume, the situation will be corrected by the same homeostatic mechanisms responsible for regulating blood pressure and blood volume that we detailed in Chapters 18 and 21. ∞ *[pp. 618–620, 733–734]*

Whenever the ECF volume changes, so does plasma volume. When plasma volume changes, it affects blood volume. If the ECF volume rises, blood volume goes up; if the ECF volume drops, blood volume goes down. As we saw in Chapter 21, blood volume has a direct effect on blood pressure. A rise in blood volume elevates blood pressure; a drop in blood volume leads to a fall in blood pressure. The net result is that homeostatic mechanisms can monitor ECF volume indirectly by monitoring blood pressure.

The receptors involved are *baroreceptors* at the carotid sinus, the aortic sinus, and the right atrium. ∞ *[pp. 545, 729–731]* The regulatory steps are detailed

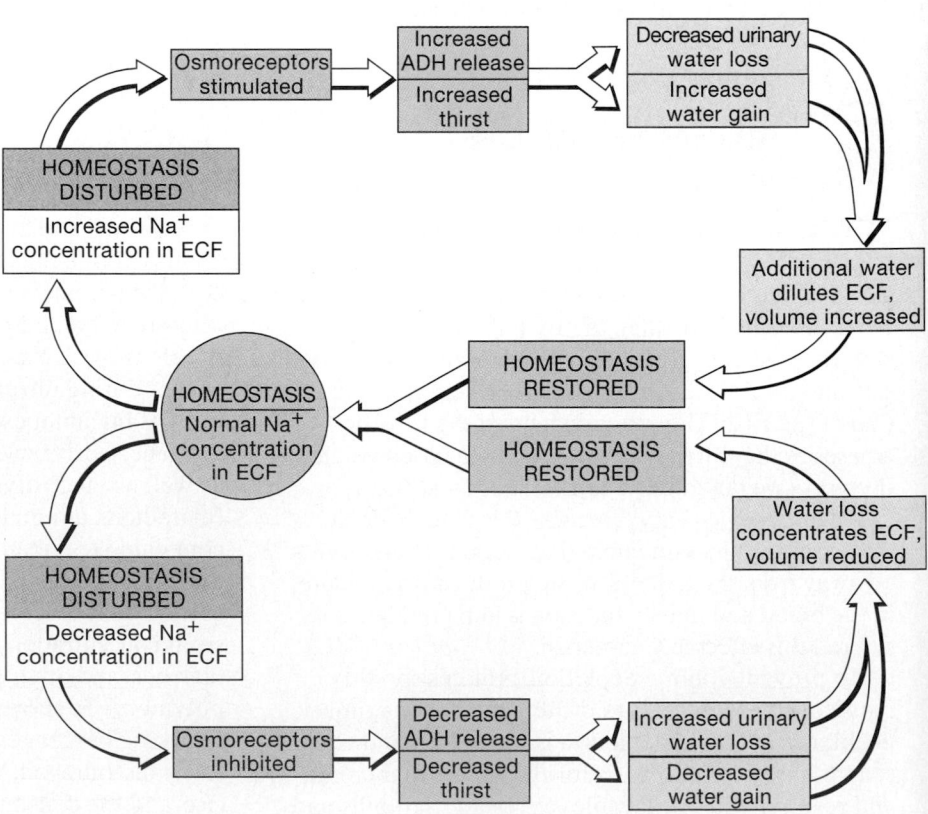

●**FIGURE 27-4** Homeostatic Regulation of Normal Sodium Ion Concentrations in Body Fluids

in Figure 27-5•. If the ECF volume is inadequate, blood volume and blood pressure decline:

- The fall in blood pressure along the afferent arterioles of the kidneys stimulates the release of renin by the juxtaglomerular apparatus of each nephron.

- Renin initiates a chain of events leading to the activation of angiotensin II.

- Angiotensin II produces a coordinated elevation in the ECF volume by stimulating thirst, causing the release of ADH, and triggering the adrenal production and secretion of aldosterone.

- Thirst and ADH secretion lead to increased ingestion and retention of water, whereas aldosterone promotes the retention of Na+ along the distal portions of the kidney nephrons.

Thus, losses of water and Na+ are reduced, and gains of water and Na+ are increased. The net result is that the ECF volume increases. Note that although the *total amount* of Na+ in the ECF is increasing (gains exceed losses), the Na+ *concentration* in the ECF remains unchanged, because Na+ absorption is accompanied by osmotic water movement.

If the plasma volume becomes abnormally large:

- Venous return increases, stretching the atrial wall and stimulating the release of ANP.

- Atrial natriuretic peptide reduces thirst and blocks the secretion of ADH and aldosterone, which together would promote water or salt conservation. This mechanism causes increased salt and water loss at the kidneys, and the volume of the ECF declines.

Potassium Balance

Roughly 98 percent of the potassium content of the human body is in the ICF. Living cells expend energy to recover potassium ions as they diffuse out of the cytoplasm and into the ECF. The K+ concentration outside the cell is relatively low, and the concentration in the ECF at any given moment represents a balance between (1) the rate of entry across the digestive epithelium and (2) the rate of loss into the urine. As we noted in Chapter 26, potassium loss in the urine is regulated by controlling the activities of ion pumps along the distal portions of the nephron and collecting system. ∞ *[pp. 984–985]*

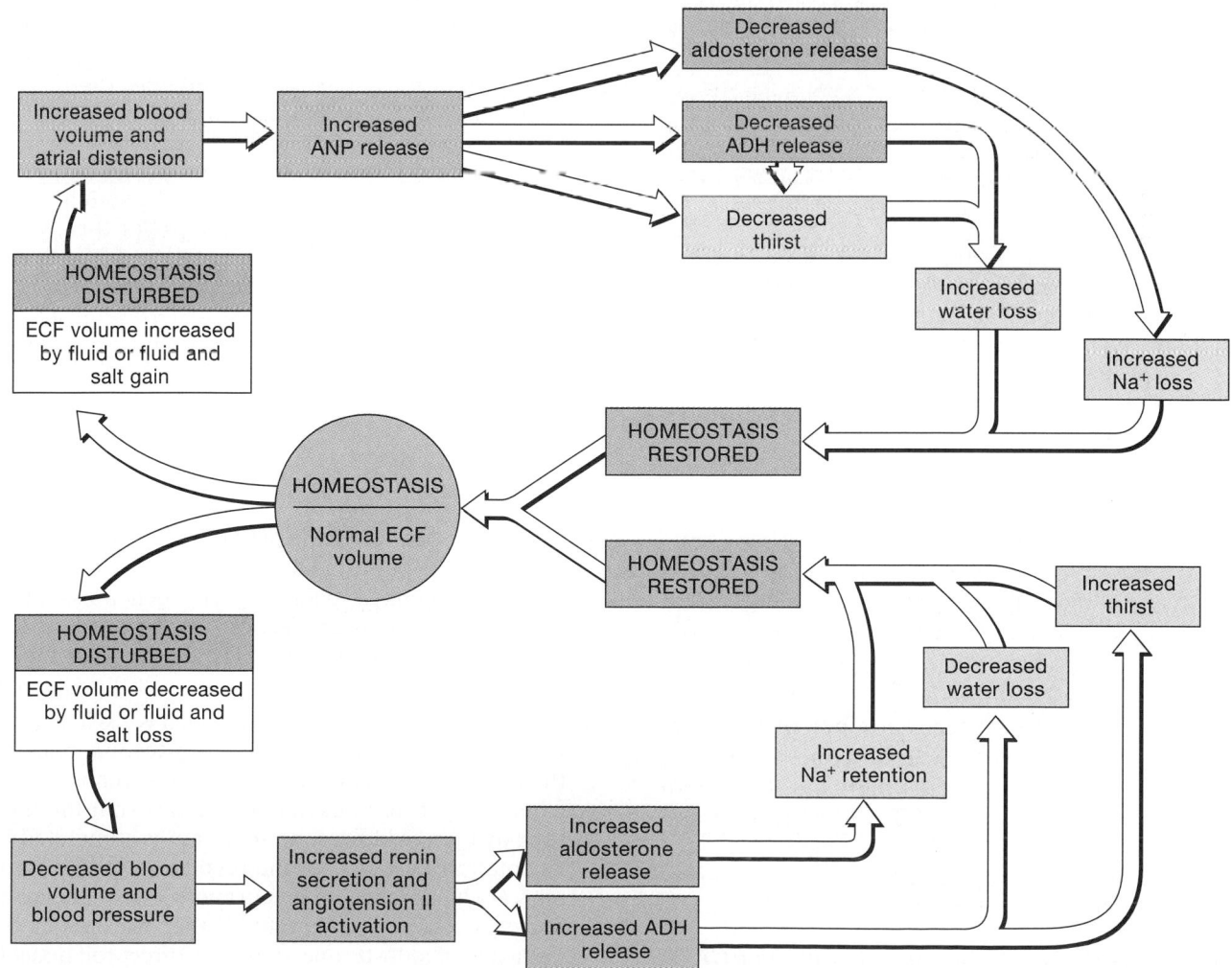

•**FIGURE 27-5** The Integration of Fluid Volume Regulation and Sodium Ion Concentrations in Body Fluids

Hypokalemia and Hyperkalemia

When the plasma concentration of potassium falls below 2 mEq/l, extensive muscular weakness develops, followed by eventual paralysis. We discussed this condition, called *hypokalemia (kalium,* potassium), in Chapter 20 in connection with ion effects on cardiac function. ∞ *[p. 701]* Causes of hypokalemia include:

- *Inadequate dietary K$^+$ intake.* If potassium gains from the diet do not keep pace with the rate of potassium loss in the urine, K$^+$ concentrations in the ECF will drop.

- *Administration of diuretic drugs.* Several diuretics, including *Lasix,* can produce hypokalemia by increasing the volume of urine produced. Although the concentration of K$^+$ in the urine is low, the greater the total volume, the larger the amount of potassium lost.

- *Excessive aldosterone secretion.* The condition of aldosteronism, characterized by excessive aldosterone secretion, results in hypokalemia, because the reabsorption of Na$^+$ is tied to the secretion of K$^+$. We discussed aldosteronism in Chapter 18. ∞ *[p. 616]*

- *An increase in the pH of the ECF.* When the H$^+$ concentration in the ECF declines, cells exchange intracellular H$^+$ for extracellular K$^+$. This ion swap helps stabilize the extracellular pH but gradually lowers the K$^+$ concentration of the ECF.

Treatment for hypokalemia generally includes increasing dietary intake by salting food with potassium salts (KCl) or by taking potassium tablets, such as *Slow-K.* Severe cases are treated by the infusion of a solution containing K$^+$ at a concentration of 40–60 mEq/l.

High K$^+$ concentrations in the ECF produce an equally dangerous condition known as *hyperkalemia.* Severe cardiac arrhythmias appear when the K$^+$ concentration exceeds 8 mEq/l; we discussed the mechanism in Chapter 20. ∞ *[p. 701]* Hyperkalemia may result from factors such as:

- *Renal failure.* Kidney failure due to damage or disease will prevent normal K$^+$ secretion and thereby produce hyperkalemia.

- *The administration of diuretic drugs that block Na$^+$ reabsorption.* Potassium secretion is linked to sodium reabsorption. When sodium reabsorption slows down, so does potassium secretion. Hyperkalemia can result.

- *A decline in the pH of the ECF.* When the pH in the ECF declines, hydrogen ions move into the ICF in exchange for intracellular potassium ions. In addition, potassium secretion at the kidney tubules slows down, because hydrogen ions are secreted instead of potassium ions. The combination of increased K$^+$ entry into the ECF and decreased K$^+$ secretion can produce a dangerous hyperkalemia very rapidly.

Treatment for hyperkalemia includes the (1) elevation of ECF volume with a solution low in potassium; (2) stimulation of urinary potassium loss by using appropriate diuretics, such as *Lasix;* (3) administration of buffers (generally sodium bicarbonate) that can control the pH of the ECF; (4) restriction of dietary potassium intake; and (5) administration of enemas or laxatives that contain compounds, such as *kayexolate,* that promote K$^+$ loss across the digestive lining. In cases resulting from renal failure, kidney dialysis may also be required.

Whenever a sodium ion is reabsorbed from the tubular fluid, it generally is exchanged for a cation, typically a potassium ion, in the peritubular fluid.

Urinary K$^+$ losses are usually limited to the amount gained by absorption across the digestive epithelium, typically 50–150 mEq (1.9–5.8 g) per day. (The potassium losses in feces and perspiration are negligible by comparison.) The K$^+$ concentration in the ECF is controlled by adjustments in the rate of active secretion along the DCT and collecting system of the nephron.

The rate of tubular secretion of K$^+$ changes in response to the following three factors:

1. *Alterations in the K$^+$ concentration of the ECF.* In general, the higher the extracellular concentration of potassium, the higher the rate of secretion.

2. *Changes in pH.* When the pH of the ECF falls, so does the pH of the peritubular fluid. The rate of K$^+$ secretion then declines, because hydrogen ions, rather than potassium ions, are secreted in exchange for sodium ions in the tubular fluid. The mechanisms for H$^+$ secretion were summarized in Figure 26-13c●, p. 985.

3. *Aldosterone levels.* The rate at which K$^+$ is lost in the urine is strongly affected by aldosterone, because the ion pumps sensitive to this hormone reabsorb Na$^+$ from the filtrate in exchange for K$^+$ from the peritubular fluid. Aldosterone secretion is stimulated by angiotensin II as part of the regulation of blood volume. High plasma K$^+$ concentrations also stimulate aldosterone secretion directly. Either way, under the influence of aldosterone, there is a direct relationship between the amount of sodium conserved and the amount of potassium excreted in the urine.

Other Electrolytes

The ECF concentrations of other electrolytes are regulated as well. We shall briefly note the most important ions involved. Additional information concerning these ions is in Table 27-2.

CALCIUM BALANCE Calcium is the most abundant mineral in the human body. A typical body contains 1–2 kg (2.2–4.4 lb) of calcium, with 99 percent of it deposited in the skeleton. In addition to forming the crystalline component of bone, calcium ions play key roles in the control of muscular and neural activity, in blood clotting, as cofactors for enzymatic reactions, and as second messengers in a variety of cell types.

As we noted in Chapters 6 and 18, calcium homeostasis reflects primarily an interplay between the re-

TABLE 27-2 Electrolyte Balance

Ion and Normal ECF Range (mEq/l)	Disorder	Symptoms	Causes	Treatment
Sodium (136–142)	Hypernatremia (>147)	Thirst, dryness and wrinkling of skin, reduced blood volume and pressure, eventual circulatory collapse	Dehydration; loss of hypotonic fluid	Ingestion of water or intravenous infusion of hypotonic solution
	Hyponatremia (<130)	Disturbed CNS function (water intoxication): confusion, hallucinations, convulsions, coma; death in severe cases	Infusion or ingestion of large volumes of hypotonic solution	Diuretic use and infusion of hypertonic salt solution
Potassium (3.8–5.0)	Hyperkalemia (>8))	Severe cardiac arrhythmias	Renal failure; use of diuretics; chronic acidosis	Infusion of hypotonic solution; selection of different diuretics; infusion of buffers; dietary restrictions
	Hypokalemia (<2)	Muscular weakness and paralysis	Low-potassium diet; diuretics; hypersecretion of aldosterone; chronic alkalosis	Increase in dietary K^+ content; ingestion of K^+ tablets or solutions; infusion of potassium solution
Calcium (4.5–5.3)	Hypercalcemia (>11)	Confusion, muscle pain, cardiac arrhythmias, kidney stones, calcification of soft tissues	Hyperparathyroidism; cancer; vitamin D toxicity; calcium supplement overdose	Infusion of hypotonic fluid to lower Ca^{2+} levels; surgery to remove parathyroid gland; administration of calcitonin
	Hypocalcemia (<4)	Muscle spasms, convulsions, intestinal cramps, weak heartbeats, cardiac arrhythmias, osteoporosis	Poor diet; lack of vitamin D; renal failure; hypoparathyroidism; hypomagnesemia	Calcium supplements; administration of vitamin D
Magnesium (1.5–2.5)	Hypermagnesemia (>4)	Confusion, lethargy, respiratory depression, hypotension	Overdose of magnesium supplements or antacids (rare)	Infusion of hypotonic solution to lower plasma concentration
	Hypomagnesemia (<0.8)	Hypocalcemia, muscle weakness, cramps, cardiac arrhythmias, hyptertension	Poor diet; alcoholism; severe diarrhea; kidney disease; malabsorption syndrome; ketoacidosis	Intravenous infusion of solution high in Mg^{2+}
Phosphate (1.8–2.6)	Hyperphosphatemia (>6)	No immediate symptoms; chronic elevation leads to calcification of soft tissues	High dietary phosphate intake; hypoparathyroidism	Dietary reduction
	Hypophosphatemia (<1)	Anorexia, dizziness, muscle weakness, cardiomyopathy, osteoporosis	Poor diet; kidney disease; malabsorption syndrome; hyperparathyroidism; vitamin D deficiency	Dietary improvement; hormone or vitamin supplements
Chloride (100–108)	Hyperchloremia (>112)	Acidosis, hyperkalemia	Dietary excess; increased chloride retention	Infusion of hypotonic solution to lower plasma concentration
	Hypochloremia (<95)	Alkalosis, anorexia, muscle cramps, apathy	Vomiting; hypokalemia	Diuretic use and infusion of hypertonic salt solution

serves in bone, the rate of absorption across the digestive tract, and the rate of loss at the kidneys. The hormones parathyroid hormone (PTH), calcitriol, and, to a lesser degree, calcitonin are responsible for maintaining calcium homeostasis in the ECF. Parathyroid hormone and calcitriol raise Ca^{2+} concentrations; their actions are opposed by calcitonin. ∞ *[pp. 187, 613, 619, 983]*

A relatively small amount of calcium is lost in the bile, and under normal circumstances, very little calcium escapes in the urine or feces. To keep pace with biliary, urinary, and fecal calcium losses, an adult must absorb only 0.8–1.2 g/day of calcium. That represents only about 0.03 percent of the calcium reserve in the skeleton. Calcium absorption at the digestive tract and reabsorption along the distal convoluted tubule are stimulated by PTH from the parathyroid glands and calcitriol from the kidneys.

HYPERCALCEMIA AND HYPOCALCEMIA *Hypercalcemia* exists when the Ca^{2+} concentration of the ECF is greater than 11 mEq/l. The primary cause of hypercalcemia in adults is *hyperparathyroidism*. ∞ *[p. 613]* Less-common causes include malignant cancers of the breast, lung, kidney, or bone marrow. Severe hypercalcemia (12–13 mEq/l) causes a variety of symptoms, including fatigue, confusion, cardiac arrhythmias, and calcification of the kidneys and soft tissues throughout the body.

The condition *hypocalcemia* (with a Ca^{2+} concentration under 4 mEq/l) is much less common than hypercalcemia. Hypoparathyroidism, vitamin D deficiency, or chronic renal failure is typically responsible for hypocalcemia. Symptoms include muscle spasms, sometimes including generalized convulsions, weak heartbeats, cardiac arrhythmias, and osteoporosis.

MAGNESIUM BALANCE The adult body contains about 29 g of magnesium, with almost 60 percent of it deposited in the skeleton. The magnesium in body fluids is contained primarily in the ICF, where the concentration averages about 26 mEq/l. Magnesium is required as a cofactor for several important enzymatic reactions, including the phosphorylation of glucose within cells and the use of ATP by contracting muscle fibers. Magnesium is also important as a structural component of bone.

The Mg^{2+} concentration of the ECF averages 1.5–2.5 mEq/l, considerably lower than levels in the ICF. The PCT reabsorbs magnesium very effectively. Keeping pace with the daily urinary loss requires a minimum dietary intake of only 24–32 mEq (0.3–0.4 g) per day.

PHOSPHATE BALANCE Phosphate ions are required for bone mineralization, and roughly 744 g of PO_4^{3-} is bound up in the mineral salts of the skeleton. In body fluids, the most important functions of PO_4^{3-} involve the ICF, where the ions are required for the formation of high-energy compounds, the activation of enzymes, and the synthesis of nucleic acids.

The PO_4^{3-} concentration of the plasma is usually 1.8–2.6 mEq/l. Phosphate ions are reabsorbed from the tubular fluid along the PCT; urinary and fecal losses of PO_4^{3-} amount to 30–45 mEq (0.8–1.2 g) per day. Phosphate ion reabsorption along the PCT is stimulated by calcitriol.

CHLORIDE BALANCE Chloride ions are the most abundant anions in the ECF. The plasma concentration ranges from 100 to 108 mEq/l. In the ICF, Cl^- concentrations are usually low (3 mEq/l). Chloride ions are absorbed across the digestive tract in company with sodium ions; several carrier proteins along the renal tubules reabsorb Cl^- with Na^+. ∞ *[pp. 982–984]* The rate of Cl^- loss is small; a gain of 48–146 mEq (1.7–5.1 g) per day will keep pace with losses in urine and perspiration.

☑ What effect would drinking a pitcher of distilled water have on your level of ADH?

☑ How would eating a meal high in salt affect the amount of fluid in the ICF?

☑ What effect would being in the desert for a day without water have on your plasma osmolarity?

ACID–BASE BALANCE

In Chapter 2, we introduced the topic of pH and the chemical nature of acids, bases, and buffers. Table 27-3 reviews key terms important to the discussion that follows. If you need a more detailed review, refer to the appropriate sections of Chapter 2 before you proceed. ∞ *[pp. 43–45]*

TABLE 27-3 A Review of Important Terms Relating to Acid–Base Balance

pH	The negative exponent (negative log) of the hydrogen ion concentration
Neutral	A solution with a pH of 7, which contains equal numbers of hydrogen and hydroxide ions
Acidic	A solution with a pH below 7, in which hydrogen ions predominate
Basic, or alkaline	A solution with a pH above 7, in which hydroxide ions predominate
Acid	A substance that dissociates to release hydrogen ions, shifting the pH toward acidity
Base	A substance that dissociates to release hydroxide ions or to tie up hydrogen ions and increase pH
Salt	An ionic compound consisting of a cation other than hydrogen and an anion other than a hydroxide ion
Buffer	A substance that tends to oppose changes in the pH of a solution by removing or replacing hydrogen ions; in body fluids, buffers maintain pH within normal limits (7.35–7.45)

The pH of body fluids may be altered by the introduction of either acids or bases. A general classification of acids and bases sorts them into *strong* versus *weak:*

- **Strong acids and bases.** *Strong acids* and *strong bases* dissociate completely in solution. For example, hydrochloric acid (HCl), a strong acid introduced in Chapter 2, dissociates in solution by the reaction

$$HCl \longrightarrow H^+ + Cl^-$$

- **Weak acids and bases.** When *weak acids* or *weak bases* enter a solution, a significant number of molecules remain intact. This means that if you place molecules of a weak acid in one solution and the same number of molecules of a strong acid in another solution, the weak acid will liberate fewer hydrogen ions and have less of an impact on the pH of the solution than the strong acid will. Carbonic acid is a weak acid we encountered in previous chapters. At the normal pH of the ECF, an equilibrium state exists, and the reaction can be diagrammed as follows:

$$H_2CO_3 \rightleftharpoons H^+ + HCO_3^-$$

The Importance of pH Control

The pH of your body fluids reflects interactions among all the acids, bases, and salts in solution. The pH of the ECF normally remains within relatively narrow limits, usually from 7.35 to 7.45. Any deviation outside the normal range is extremely dangerous, because changes in H$^+$ concentrations disrupt the stability of cell membranes, alter protein structure, and change the activities of important enzymes. You could not survive for long with an ECF pH below 6.8 or above 7.7.

When the pH of the plasma falls below 7.35, **acidemia** exists. The physiological state that results is called **acidosis.** When the pH of the blood rises above 7.45, **alkalemia** exists. The physiological state that results is called **alkalosis.** Acidosis and alkalosis affect virtually all body systems, but the nervous system and cardiovascular system are particularly sensitive to pH fluctuations. For example, severe acidosis (pH below 7.0) can be deadly, because (1) CNS function deteriorates, and the individual becomes comatose; (2) cardiac contractions grow weak and irregular, and symptoms of heart failure develop; and (3) peripheral vasodilation produces a dramatic drop in blood pressure, and circulatory collapse may occur.

The control of pH is therefore an extremely important homeostatic process of great physiological and clinical significance. Although both acidosis and alkalosis are dangerous, in practice, *problems with acidosis are much more common than are problems with alkalosis.* This is the case because several different types of acids, including carbonic acid, are generated by normal cellular activities.

Types of Acids in the Body

There are three general categories of acids in the body: (1) *volatile acids,* (2) *fixed acids,* and (3) *organic acids.*

Volatile Acids

A **volatile acid** is an acid that can leave solution and enter the atmosphere. Carbonic acid (H_2CO_3) is an important volatile acid in body fluids. At the lungs, carbonic acid breaks down into carbon dioxide and water; the carbon dioxide diffuses into the alveoli, a process we described in Chapter 23. ∞ *[pp. 849–850]* In peripheral tissues, carbon dioxide in solution interacts with water to form molecules of carbonic acid, which dissociates to release hydrogen ions and bicarbonate ions. The complete reaction sequence is

$$\underset{\substack{\text{carbon} \\ \text{dioxide}}}{CO_2} + \underset{\text{water}}{H_2O} \rightleftharpoons \underset{\substack{\text{carbonic} \\ \text{acid}}}{H_2CO_3} \rightleftharpoons \underset{}{H^+} + \underset{\substack{\text{bicarbonate} \\ \text{ion}}}{HCO_3^-}$$

This reaction occurs spontaneously in body fluids, but it occurs most rapidly in the presence of *carbonic anhydrase* (CA), an enzyme found in the cytoplasm of red blood cells, liver and kidney cells, parietal cells of the stomach, and in many other cell types.

Because most of the carbon dioxide in solution is converted to carbonic acid, and most of the carbonic acid dissociates, there is an inverse relationship between the partial pressure of carbon dioxide (P_{CO_2}) and the pH (Figure 27-6$\bullet$). When carbon dioxide levels rise, additional hydrogen ions and bicarbonate ions are released, and the pH goes down. (Recall that the pH is a *negative exponent,* so when the concentration of hydrogen ions goes up, the pH value goes down.) The P_{CO_2} is the most important factor affecting the pH in body tissues.

At the alveoli, carbon dioxide diffuses into the atmosphere, the number of hydrogen ions and bicarbonate ions drops, and the pH rises. We shall consider this process, which effectively removes hydrogen ions from solution, in more detail later in the chapter.

Fixed Acids

Fixed acids are acids that do not leave solution; once produced, they remain in body fluids until they are excreted at the kidneys. *Sulfuric acid* and *phosphoric acid* are the most important fixed acids. They are generated in small amounts during the catabolism of amino acids and compounds that contain phosphate groups, including phospholipids and nucleic acids.

Organic Acids

Organic acids are acid participants in or byproducts of aerobic metabolism. Lactic acid, produced by the anaerobic metabolism of pyruvate, and ketone bodies, synthesized from acetyl-CoA, are important organic acids that we considered in Chapters 10 and 25. ∞ *[pp. 299, 941–942]* Under normal conditions, most

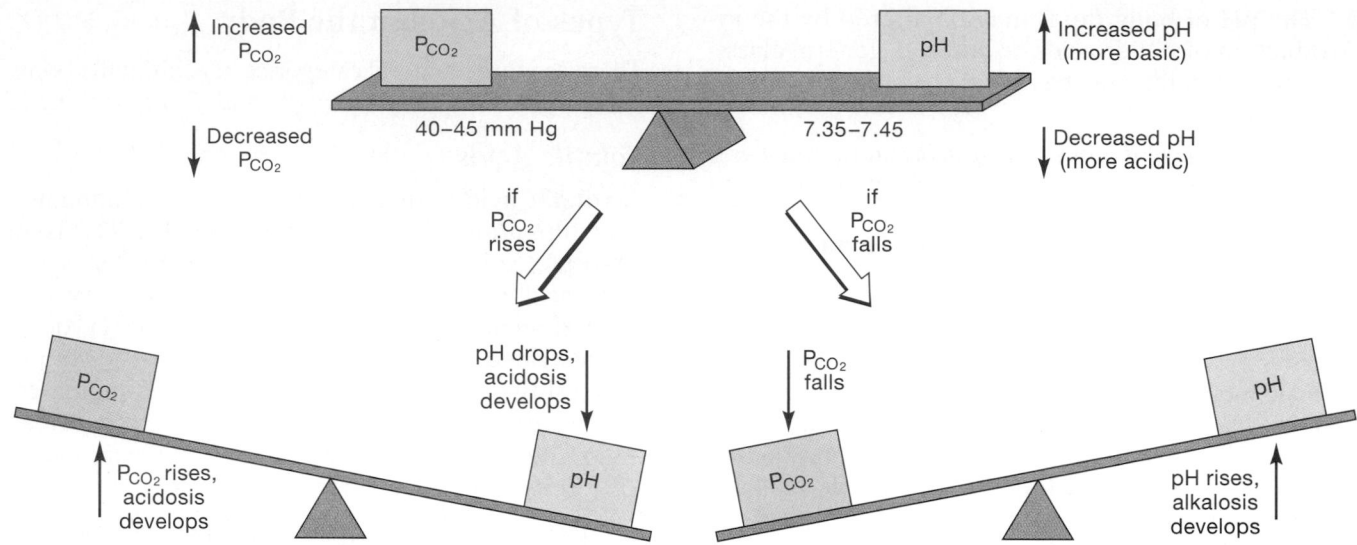

•FIGURE 27-6 **Basic Relationship between P$_{CO_2}$ and Plasma pH.** The P$_{CO_2}$ is inversely related to the pH; when the P$_{CO_2}$ increases, the pH decreases.

organic acids are metabolized rapidly, and significant accumulations do not occur. But relatively large amounts of organic acids are produced (1) during periods of anaerobic metabolism, because lactic acid builds up rapidly; and (2) during starvation or excessive lipid catabolism, because ketone bodies accumulate.

Mechanisms of pH Control

To maintain acid–base balance over long periods of time, your body must balance gains and losses of hydrogen ions. Hydrogen ions are gained at the digestive tract and through metabolic activities within your cells. You must excrete these hydrogen ions at the kidneys, through the secretion of H$^+$ into the urine, and at the lungs, through the formation of water and carbon dioxide from H$^+$ and HCO$_3^-$. The sites of excretion are far removed from the sites of acid production. *Buffer systems* within body fluids neutralize the H$^+$ as these ions travel from their points of origin to the lungs or kidneys, where they will be excreted.

Buffers and Buffer Systems

The acids produced in the course of normal metabolic operations are temporarily neutralized by buffers and buffer systems in body fluids. *Buffers,* introduced in Chapter 2, are dissolved compounds that can provide or remove H$^+$ and thereby stabilize the pH of a solution. ∞ *[p. 45]* Buffers include weak acids that can donate H$^+$ and weak bases that can absorb H$^+$. A **buffer system** in body fluids generally consists of a combination of a weak acid and its dissociation products. The anion released functions as a weak base. Adding H$^+$ to the solution will upset the equilibrium, and the resulting formation of additional molecules of the weak acid will remove some of the H$^+$ from solution.

There are three major buffer systems, each with slightly different characteristics and distribution:

1. ***Protein buffer systems.*** Protein buffer systems contribute to the regulation of pH in the ECF and ICF. These buffer systems interact extensively with the other buffer systems.
2. ***The carbonic acid–bicarbonate buffer system.*** The carbonic acid–bicarbonate buffer system is most important in the ECF.
3. ***The phosphate buffer system.*** The phosphate buffer system has an important role in buffering the pH of the ICF and of the urine.

Figure 27-7• shows the locations of these buffer systems.

PROTEIN BUFFER SYSTEMS **Protein buffer systems** depend on the ability of amino acids to respond to alterations in pH by accepting or releasing H$^+$. The underlying mechanism is shown in Figure 27-8•.

- If the pH climbs, the carboxyl group (—COOH) of the amino acid can dissociate, acting as a weak acid and releasing a hydrogen ion. The carboxyl group then becomes a carboxylate ion (—COO$^-$). At the normal pH of body fluids (7.35–7.45), the carboxyl groups of most amino acids have already given up their hydrogen ions. (Proteins carry negative charges primarily for that reason, as we noted in Chapters 3 and 12. ∞ *[pp. 73, 380]*) However, some amino acids, notably *histidine* and *cysteine*, have side chains (R groups) that will donate hydrogen ions if the pH climbs outside the normal range. Their buffering effects are very important in both the ECF and ICF.
- If the pH drops, the amino group (—NH$_2$) can act as a weak base and accept an additional hydrogen ion, forming an amino ion (—NH$_3^+$). This effect is primarily limited to free amino acids and the last amino acid in a polypeptide chain, because the amino groups in peptide bonds cannot function as buffers.

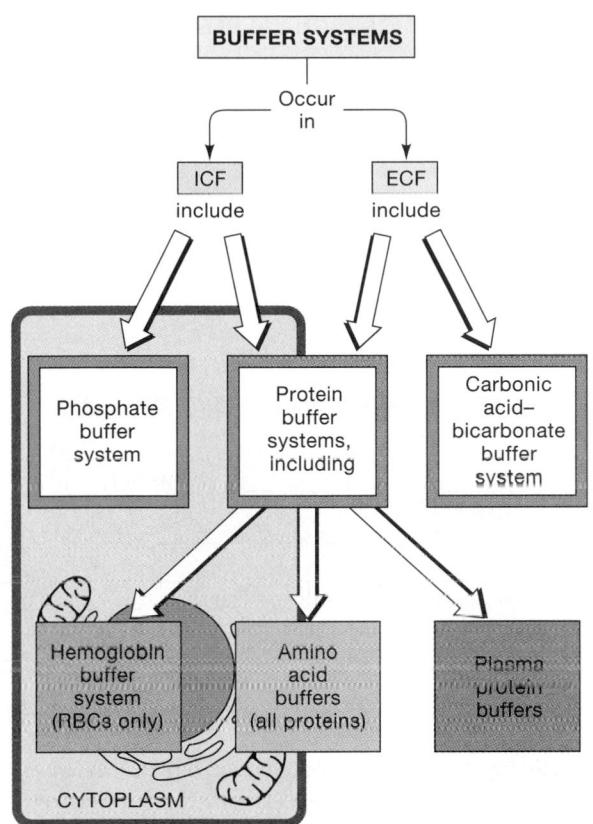

BUFFER SYSTEMS

Occur in

ICF — include

ECF — include

Phosphate buffer system

Protein buffer systems, including

Carbonic acid–bicarbonate buffer system

Hemoglobin buffer system (RBCs only)

Amino acid buffers (all proteins)

Plasma protein buffers

CYTOPLASM

•**FIGURE 27-7** **Buffer Systems in Body Fluids.** Phosphate buffers occur primarily in the ICF, whereas the carbonic acid–bicarbonate buffer system occurs primarily in the ECF. Protein buffer systems are in both the ICF and the ECF. Extensive interactions take place among these buffer systems.

Plasma proteins contribute to the buffering capabilities of the blood. The interstitial fluids contain extracellular protein fibers and dissolved amino acids that also assist in the regulation of pH. In the ICF of active cells, structural and other proteins provide an extensive buffering capability that prevents destructive pH changes when organic acids, such as lactic acid or pyruvic acid, are produced by cellular metabolism.

Because there is an exchange between the ECF and ICF, the protein buffer system can help stabilize the pH of the ECF. For example:

- When the pH of the ECF decreases, cells pump H^+ out of the ECF and into the ICF, where they can be buffered by intracellular proteins.
- When the pH of the ECF rises, exchange pumps in cell membranes exchange H^+ in the ICF for K^+ in the ECF.

These mechanisms can assist in buffering the pH of the ECF, but the process is slow, because hydrogen ions must be individually transported across the cell membrane. As a result, the protein buffer system in most cells cannot make rapid, large-scale adjustments in the pH of the ECF.

The Hemoglobin Buffer System. The situation is somewhat different for red blood cells. Red blood cells, which contain roughly 5.5 percent of the ICF, are normally suspended in the plasma. These cells are densely packed with hemoglobin, and their cytoplasm contains large amounts of carbonic anhydrase. These cells have a significant effect on the pH of the ECF, because they absorb carbon dioxide from the plasma and convert it to carbonic acid. Carbon dioxide can diffuse across the RBC membrane very quickly, and no transport mechanism is needed. As the carbonic acid dissociates, the bicarbonate ions move into the plasma via the *chloride shift*. The hydrogen ions are buffered by hemoglobin molecules. At the lungs, the entire reaction sequence, diagrammed in Figure 23-25•, p. 850, proceeds in reverse. This is known as the **hemoglobin buffer system.**

This is the only intracellular buffer system that can have an immediate effect on the pH of the ECF. *The hemoglobin buffer system helps prevent drastic alterations in pH when the plasma P_{CO_2} is rising or falling.*

THE CARBONIC ACID–BICARBONATE BUFFER SYSTEM With the exception of RBCs, some cancer cells, and tissues temporarily deprived of oxygen, your cells generate carbon dioxide virtually 24 hours a day. As we detailed above, most of the carbon dioxide is converted to carbonic acid, which then dissociates into a hydrogen ion and a bicarbonate ion. The carbonic acid and its dissociation products form the **carbonic acid–bicarbonate buffer system.**

This buffer system (Figure 27-9a•) consists of the following reaction:

$$CO_2 + H_2O \rightleftharpoons H_2CO_3 \rightleftharpoons H^+ + HCO_3^-$$

Because the reaction is freely reversible, changing the concentrations of any participant will affect the concentrations of all other participants. For example, if hydrogen ions are added, most of them will be removed by interactions with HCO_3^-, forming H_2CO_3. In the process, the HCO_3^- acts as a weak base that buffers the excess H^+.

Neutral pH

pH rises

pH falls

In alkaline medium, amino acid acts as an acid

In acidic medium, amino acid acts as a base

•**FIGURE 27-8** **Amino Acid Buffers.** Amino acids can either accept a hydrogen ion or donate one, depending on the pH of their surroundings.

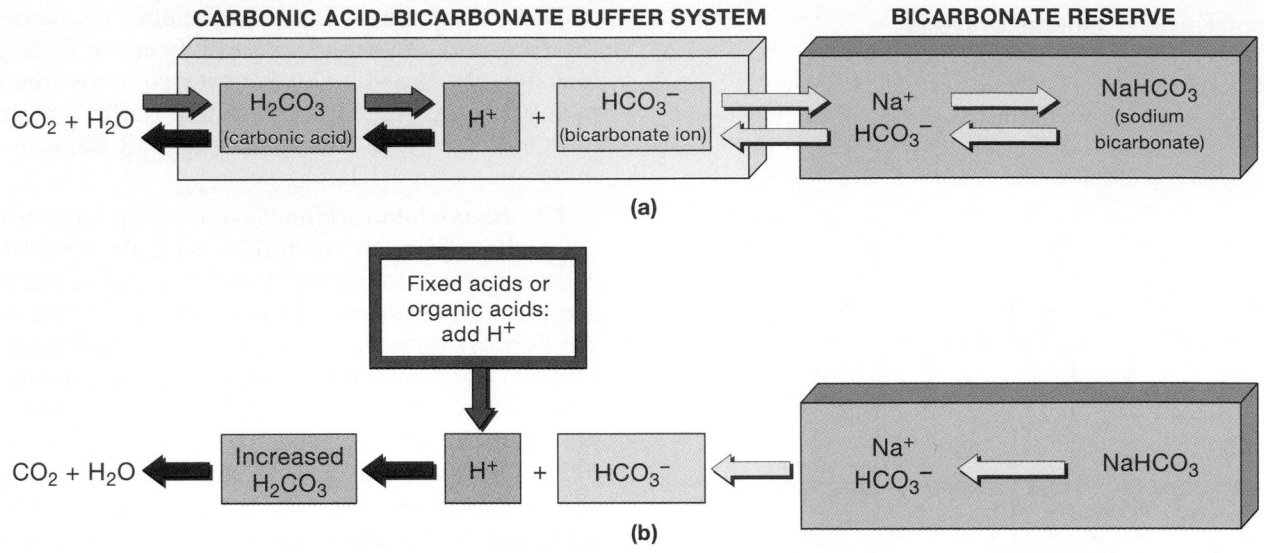

•FIGURE 27-9 **The Carbonic Acid–Bicarbonate Buffer System.** **(a)** Basic components of the carbonic acid–bicarbonate buffer system, showing their relationships to carbon dioxide and the bicarbonate reserve. **(b)** Response of the carbonic acid–bicarbonate buffer system to hydrogen ions generated by fixed or organic acids in body fluids.

The H_2CO_3 formed in this way will in turn dissociate into CO_2 and water (Figure 27-9b•), and the CO_2 can then be excreted at the lungs. In effect, this reaction takes the H^+ released by a strong organic or fixed acid and generates a volatile acid that can easily be eliminated.

This buffer system can also protect against increases in pH, although such changes are relatively rare. If hydrogen ions are removed from the plasma, the reaction is driven to the right: The P_{CO_2} declines, and the dissociation of H_2CO_3 replaces the missing H^+.

The primary role of the carbonic acid–bicarbonate buffer system is to prevent pH changes caused by organic acids and fixed acids in the ECF. The three most important limitations of this buffer system are as follows:

1. *It cannot protect the ECF from pH changes that result from elevated or depressed levels of CO_2.* A buffer system cannot protect against changes in the concentration of its own weak acid. In the case above, the addition of excess H^+ drove the reaction to the left. But consider what would have happened if we had added excess CO_2 instead of excess H^+. The elevated CO_2 would have driven the reaction to the right, forming additional H_2CO_3 that would have dissociated into H^+ and HCO_3^-. This reaction would have reduced the pH of the plasma.

2. *It can function only when the respiratory system and the respiratory control centers are working normally.* Normally, the elevation in P_{CO_2} that occurs when fixed or organic acids are buffered will stimulate an increase in the respiratory rate. This increase accelerates the removal of CO_2 at the lungs. If the respiratory passageways are blocked, circulation to the lungs will be impaired, or if the respiratory centers do not respond normally, the efficiency of the buffer system will be reduced.

3. *The ability to buffer acids is limited by the availability of bicarbonate ions.* Every time a hydrogen ion is removed from the plasma, a bicarbonate ion goes with it. When all the bicarbonate ions have been tied up, buffering capabilities are lost.

Problems due to a lack of bicarbonate ions are rare for several reasons. Body fluids contain a large reserve of HCO_3^-, primarily in the form of dissolved molecules of the weak base *sodium bicarbonate* ($NaHCO_3$). This readily available supply of HCO_3^- is known as the **bicarbonate reserve.** The reaction sequence involved is

$$Na^+ + HCO_3^- \rightleftharpoons NaHCO_3$$

When hydrogen ions enter the ECF, the bicarbonate ions tied up in H_2CO_3 molecules are replaced by HCO_3^- from the bicarbonate reserve. This relationship is diagrammed in Figure 27-9b•.

Additional HCO_3^- can be generated at the kidneys, through mechanisms we introduced in Chapter 26 (Figure 26-13c•, p. 985). In the DCT and collecting system, carbonic anhydrase converts CO_2 within tubular cells into H_2CO_3, which then dissociates. The hydrogen ion is pumped into the tubular fluid in exchange for a sodium ion, and the bicarbonate ion is transported into the peritubular fluid in exchange for a chloride ion. In effect, the tubular cell removes HCl from the peritubular fluid in exchange for $NaHCO_3$.

THE PHOSPHATE BUFFER SYSTEM The **phosphate buffer system** consists of an anion, $H_2PO_4^-$, that is a weak acid. The operation of the phosphate buffer system resembles that of the carbonic acid–bicarbonate buffer system. The reversible reaction involved is

$$H_2PO_4^- \rightleftharpoons H^+ + HPO_4^{2-}$$

The weak acid is *dihydrogen phosphate* ($H_2PO_4^-$), and the anion released is *monohydrogen phosphate* (HPO_4^{2-}). In the ECF, the phosphate buffer system plays only a supporting role in the regulation of pH, primarily because the concentration of HCO_3^- far exceeds that of HPO_4^{2-}. However, the phosphate buffer system is quite important in buffering the pH of the ICF. In addition, cells contain a *phosphate reserve* in the form of the weak base *sodium monohydrogen phosphate* (Na_2HPO_4). The phosphate buffer system is also important in stabilizing the pH of the urine. The dissociation of Na_2HPO_4 provides additional HPO_4^{2-} for use by the buffer system:

$$2\,Na^+ + HPO_4^{2-} \rightleftharpoons Na_2HPO_4$$

Maintenance of Acid–Base Balance

Although buffer systems can tie up excess H^+, they provide only a temporary solution. The hydrogen ions have not been eliminated but merely rendered harmless. For homeostasis to be preserved, the captured H^+ must ultimately be excreted. The problem is that the supply of buffer molecules is limited. Suppose that a buffer molecule prevents a pH change by binding a hydrogen ion that enters the ECF. The buffer will then be tied up, reducing the capacity of the ECF to cope with more H^+. Eventually, all the buffer molecules will be bound to H^+, and pH control will be impossible.

The situation can be resolved only by removing the H^+ from the ECF, thereby freeing the buffer molecules, or by replacing the buffer molecules. Similarly, if a buffer provides a hydrogen ion to maintain normal pH, either another hydrogen ion must be obtained or the buffer must be replaced.

The maintenance of acid–base balance thus includes balancing H^+ gains and losses. This balancing act involves coordinating the actions of buffer systems with pulmonary mechanisms and renal mechanisms. The pulmonary and renal mechanisms support the buffer systems by (1) secreting or absorbing H^+, (2) controlling the excretion of acids and bases, and, when necessary, (3) generating additional buffers. It is the *combination* of buffer systems and these pulmonary and renal mechanisms that maintains your pH within narrow limits.

Respiratory Compensation

Respiratory compensation is a change in the respiratory rate that helps stabilize the pH of the ECF. Respiratory compensation occurs whenever your pH strays outside normal limits. This compensation is effective because respiratory activity has a direct effect on the carbonic acid–bicarbonate buffer system. Increasing or decreasing the rate of respiration alters pH by lowering or raising the P_{CO_2}. When the P_{CO_2} rises, the pH falls; when the P_{CO_2} falls, the pH rises, because the removal of CO_2 drives the carbonic acid–bicarbonate system to the left:

$$CO_2 + H_2O \rightleftharpoons H_2CO_3 \rightleftharpoons H^+ + HCO_3^-$$

We detailed mechanisms responsible for the control of respiratory rate in Chapter 23 and shall present only a brief summary here. (For a more detailed review, refer to Figures 23-27, p. 853, and 23-28●, p. 854.)

1. Chemoreceptors of the carotid and aortic bodies are sensitive to the P_{CO_2} of the circulating blood. Other receptors located on the ventrolateral surfaces of the medulla oblongata monitor the P_{CO_2} of the CSF.

2. A rise in P_{CO_2} stimulates the receptors; a fall in the P_{CO_2} inhibits them.

3. Stimulation of the chemoreceptors leads to an increase in the respiratory rate. As the rate of respiration increases, more CO_2 is lost at the lungs, and the P_{CO_2} returns to normal levels.

4. When the P_{CO_2} of the blood or CSF declines, respiratory activity becomes depressed and the breathing rate decreases. This decrease causes an elevation of the P_{CO_2} in the ECF.

Renal Compensation

Renal compensation is a change in the rates of H^+ and HCO_3^- secretion or reabsorption in response to changes in plasma pH. Under normal conditions, the body generates enough organic and fixed acids to add about 100 mEq of H^+ to the ECF each day. An equivalent number of hydrogen ions must therefore be excreted in the urine to maintain acid–base balance. In addition, the kidneys assist the lungs by eliminating CO_2 that enters the renal tubules during filtration or that diffuses into the tubular fluid as it travels toward the renal pelvis.

Hydrogen ions are secreted into the tubular fluid along the PCT, the DCT, and the collecting system. The basic mechanisms involved were detailed in Figures 26-11 and 26-13●, pp. 980, 985. The ability to eliminate a large number of hydrogen ions in a normal volume of urine depends on the presence of buffers in the urine. Secretion of H^+ can continue only until the pH of the tubular fluid reaches 4.0–4.5. (At that point, the H^+ concentration gradient is so great that hydrogen ions leak out of the tubule as fast as they are pumped in.) If there were no buffers in the tubular fluid, the kidneys could secrete less than 1 percent of the acid produced each day before the pH reached this limit. To maintain acid balance under these conditions, the kidneys would have to produce about 1000 liters of urine each day just to keep pace with the generation of H^+ in the body. Buffers in the tubular fluid are therefore extremely important, because they keep the pH high enough for H^+ secretion to continue. Metabolic acids are continuously being generated; without these buffering mechanisms, the kidneys would be unable to maintain homeostasis.

Figure 27-10● diagrams the primary routes of H^+ secretion and shows the buffering mechanisms that stabilize the pH of the tubular fluid. The three major buffers involved (Figure 27-10a●) are ① the carbonic acid–bicarbonate buffer system, ② the phosphate buffer

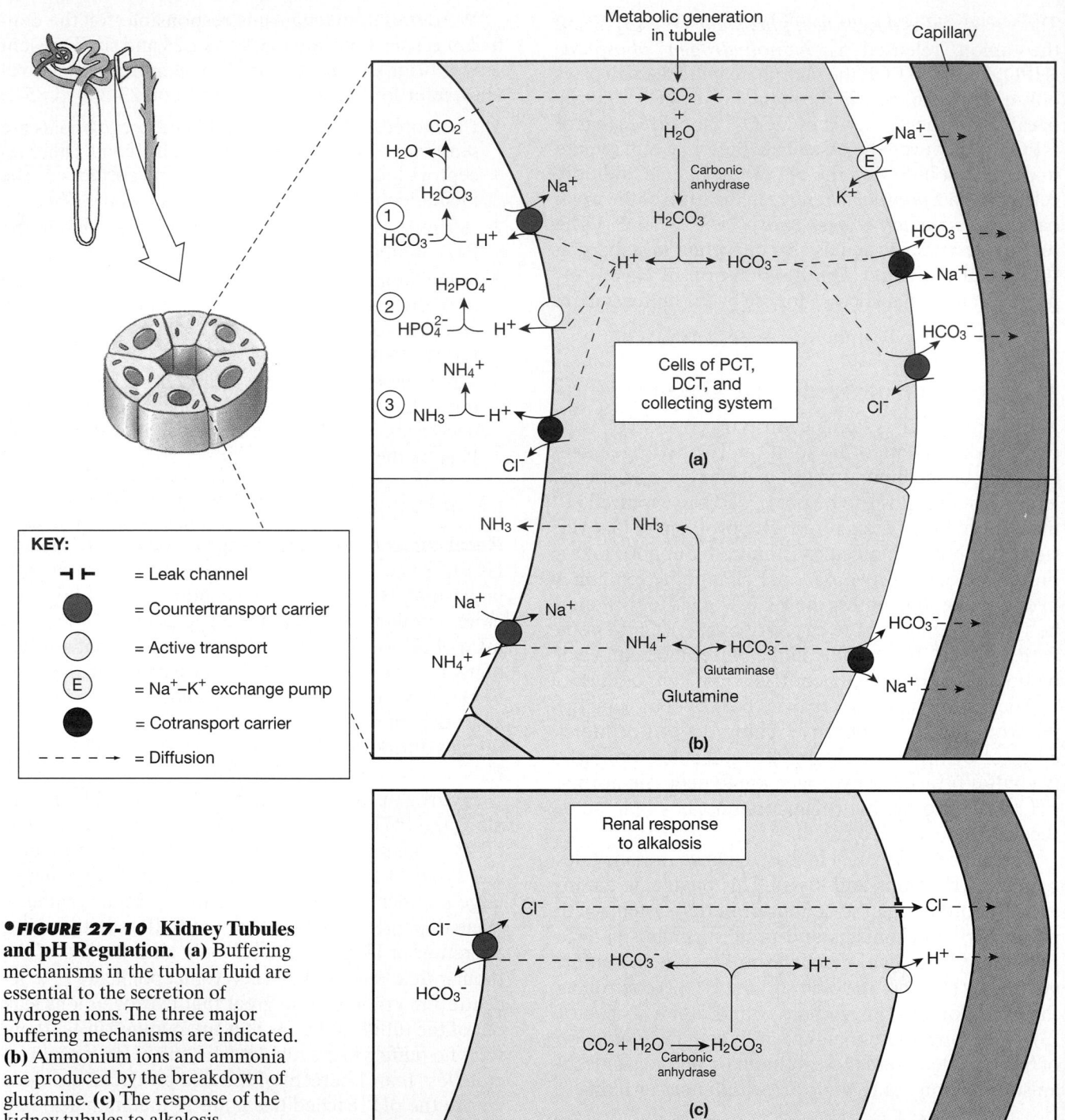

•FIGURE 27-10 Kidney Tubules and pH Regulation. (a) Buffering mechanisms in the tubular fluid are essential to the secretion of hydrogen ions. The three major buffering mechanisms are indicated. (b) Ammonium ions and ammonia are produced by the breakdown of glutamine. (c) The response of the kidney tubules to alkalosis.

system, and ③ ammonia. Glomerular filtration puts components of the carbonic acid–bicarbonate and phosphate buffer systems into the filtrate. The ammonia is generated by tubule cells (primarily those of the PCT).

Figure 27-10a• shows the secretion of H⁺, which relies on carbonic anhydrase (CA) activity within the tubular cells. The hydrogen ions generated may be pumped into the lumen in exchange for sodium ions, individually, or with chloride ions. The net result is the secretion of H⁺, accompanied by the removal of CO_2 (from the tubular fluid, the tubule cells, and the ECF),

and the release of sodium bicarbonate into the ECF. Figure 27-10b• shows the generation of ammonia within the tubules. As the tubule cells use the enzyme *glutaminase* to break down the amino acid *glutamine*, amino groups are released as ammonium ions (NH_4^+) or as ammonia (NH_3). The ammonium ions are transported into the lumen in exchange for Na^+ in the tubular fluid. The NH_3, which is highly volatile and also toxic to cells, rapidly diffuses into the tubular fluid. There it reacts with a hydrogen ion, forming NH_4^+. This reaction buffers the tubular fluid and removes a potentially dangerous compound from body fluids. The carbon chains of the glutamine molecules are ultimately converted to HCO_3^-, which is cotransported with Na^+ into the ECF. The generation of ammonia by tubule cells thus ties up H^+ in the tubular fluid and releases sodium bicarbonate into the ECF, where it contributes to the bicarbonate reserve. These mechanisms of H^+ secretion and buffering are always functioning, but their levels of activity vary widely with the pH of the ECF.

THE RENAL RESPONSES TO ACIDOSIS AND ALKALOSIS Acidosis develops when the normal plasma buffer mechanisms are stressed by excessive numbers of hydrogen ions. The kidney tubules do not distinguish among the various acids that may cause acidosis. Whether the fall in pH results from the production of volatile, fixed, or organic acids, the renal contribution remains limited to (1) secretion of H^+, (2) activity of buffers in the tubular fluid, (3) removal of CO_2, and (4) reabsorption of $NaHCO_3$.

The tubular cells thus bolster the capabilities of the carbonic acid–bicarbonate buffer system by increasing the concentration of bicarbonate ions in the ECF, replacing those already used to remove hydrogen ions from solution. In a starving individual, the tubular cells break down amino acids, yielding ammonium ions that are pumped into the tubular fluid, carbon chains for catabolism, and bicarbonates to help buffer ketone bodies in the blood (see Figure 26-13c•, p. 985).

When alkalosis develops, (1) the rate of H^+ secretion at the kidneys declines, (2) the tubule cells do not reclaim the bicarbonates in the tubular fluid, and (3) the collecting system transports HCO_3^- into the tubular fluid while releasing a stong acid (HCl) into the peritubular fluid. The concentration of HCO_3^- in the plasma decreases, promoting the dissociation of H_2CO_3 and the release of hydrogen ions. The additional H^+ generated at the kidneys helps return the pH to normal levels. The renal response to alkalosis is diagrammed in Figure 27-10c•.

DISTURBANCES OF ACID–BALANCE BALANCE

Figure 27-11• (p. 1027) summarizes the interactions among buffer systems, respiration, and renal function in normal acid–base balance. In combination, these mechanisms are generally able to control pH very precisely, so the pH of the ECF seldom varies more than 0.1 pH units, from 7.35 to 7.45. When buffering mechanisms are severely stressed, the pH wanders outside these limits, producing symptoms of alkalosis or acidosis.

If you are considering a career in any health-related field, an understanding of acid–base dynamics will be essential to you for clinical diagnosis and patient management under a variety of conditions. Temporary alterations in the pH of body fluids occur frequently. Rapid and complete recovery occurs through a combination of buffer system activity and the respiratory and renal responses. More serious and prolonged disturbances of acid–base balance can result from any factor that affects one of the principal regulatory mechanisms:

- Any disorder affecting circulating buffers, respiratory performance, or renal function can disrupt acid–base balance. Several conditions described in earlier chapters, including *emphysema* (Chapter 23) and *renal failure* (Chapter 26), are associated with dangerous changes in pH. ∞ *[pp. 851, 977]*
- Cardiovascular conditions, such as *heart failure* or *hypotension*, can affect the pH of internal fluids by causing fluid shifts, changing glomerular filtration rates, and altering respiratory efficiency.
- Conditions affecting the CNS can disrupt normal acid–base balance. For example, neural damage or CNS disease will affect the respiratory and cardiovascular reflexes that are essential to normal pH regulation.

Serious abnormalities in acid–base balance generally show an initial *acute phase*, in which the pH moves rapidly away from the normal range. If the condition persists, physiological adjustments occur, and the individual enters the *compensated phase*. Unless the underlying problem is corrected, compensation cannot be completed, and blood chemistry will remain abnormal. The pH typically remains outside normal limits, even after compensation has occurred. Even if the pH is within the normal range, the P_{CO_2} or HCO_3^- concentrations may be abnormal.

The primary source of the problem is usually indicated by the name given to the resulting condition. For example:

- ***Respiratory acid–base disorders*** result from a mismatch between carbon dioxide generation in peripheral tissues and carbon dioxide excretion at the lungs. When a respiratory acid–base disorder is present, the carbon dioxide levels in the ECF are abnormal.
- ***Metabolic acid–base disorders*** are caused by the generation of organic or fixed acids or by conditions affecting the concentration of HCO_3^- in the ECF.

Pulmonary compensation alone may restore normal acid–base balance in individuals suffering from respiratory disorders. In contrast, compensation mechanisms for metabolic disorders may be able to stabilize pH, but other aspects of acid–base balance (buffer system function, bicarbonate levels, and P_{CO_2}) remain abnormal until the underlying metabolic cause is corrected.

We can subdivide the respiratory and metabolic categories to create four major classes of acid–base disturbances: (1) *respiratory acidosis,* (2) *respiratory alkalosis,* (3) *metabolic acidosis,* and (4) *metabolic alkalosis.* We shall discuss each class separately.

Fluid, Electrolyte, and Acid–Base Balance in Infants

Fetuses and infants have very different requirements for the maintenance of fluid, electrolyte, and acid–base balance than do adults. A fetus obtains the water, organic nutrients, and electrolytes it needs from the maternal circulation. Buffers in the fetal bloodstream provide short-term pH control, and the maternal kidneys eliminate the generated H^+. The body water content is high; at birth, water accounts for roughly 75 percent of the newborn infant's body weight, as compared with 50–60 percent in adults. Several factors contribute to this difference, including (1) the larger infant blood volume (10–12.5 percent of body weight versus 6–7 percent in adults), (2) a proportionately larger CSF volume, (3) a relatively larger interstitial fluid volume (in part due to existing for 9 months under weightless conditions; body water content also changes in orbiting astronauts), and (4) differences in the water content and in the proportions, in terms of body mass, of organs and tissues; for example, a newborn infant has a proportionately larger heart and brain, which are 75–79 percent water, but less than half as much adipose tissue, which is 10 percent water, than an adult has.

During the first 2–3 days after delivery, roughly 6 percent of the infant's excess water is lost. The loss thereafter is more gradual, with the typical adult body water content appearing after age 2.

At birth, the distribution of body water is also different from that of an adult. In a newborn, the ECF accounts for roughly 35 percent of total body weight, versus 40 percent for the ICF. Because the ECF and ICF in an infant are similar in volume, the ICF is less effective than in an adult at buffering changes in the ECF volume. As a result, less water must be lost from the ECF before the ICF volume is reduced enough to damage cells. As growth occurs, the ICF volume (as a percentage of total body weight) remains relatively stable, whereas the ECF volume gradually decreases. The relative decline in the ECF volume becomes evident after a few months; the adult relationship between the ECF and ICF volumes is reached after approximately 2 years.

Basic aspects of electrolyte balance are the same in a newborn infant as in an adult, but the effects of fluctuations in the diet are much more immediate in a newborn because reserves of minerals and energy sources are much smaller. (At both ages, the bone mass and adipose tissue mass are relatively small.) The problem is compounded by the fact that the metabolic rate (per unit body mass) is twice that of an adult. Thus, infants have an elevated demand for nutrients, and that demand must be met promptly. This is one reason that infants require frequent feedings; another is that they have a much higher demand for water than do adults.

The elevated metabolic rate also means an accelerated production of waste products that must be eliminated at the kidneys. But the kidneys of a newborn are unable to produce urine with an osmolarity above 450 mOsm/l. So a newborn must produce a greater volume of urine to eliminate the metabolic waste products. Water losses at other sites are higher as well. Because the surface-to-volume ratio and the respiratory rate are relatively high, the rate of insensible perspiration is much higher in an infant than in an adult. To keep pace with rates of water loss in urine and insensible perspiration, a newborn must consume, on a proportional basis, seven times as much water as an adult does. If water intake is inadequate, waste products accumulate, and metabolic acidosis may develop. Although the kidneys become effective at concentrating urine after about 1 month, the elevated metabolic rate and water loss remain. As a result, infants continue to consume (and lose) roughly twice as much water as do adults by body weight. Infants are therefore at greater risk of dehydration; they can survive for only 1–3 days without water, whereas adults can survive for a week or more.

The increased metabolic rate of infants also means an accelerated demand for oxygen and more rapid generation of CO_2. The respiratory rate in a newborn is relatively high—roughly 40 breaths per minute. And the breaths are relatively deep; in proportion to body weight, a newborn must move twice as much air as an adult. As a result, the functional residual capacity of the lungs is about half that of adults. ∞ [p. 842] In adults, the functional residual capacity is large relative to the tidal volume, and this proportion helps prevent rapid changes in the P_{O_2} and P_{CO_2} of alveolar air. In infants, as soon as the respiratory rate changes, the alveolar air composition changes. Because changes in the respiratory rate can cause sudden changes in the P_{O_2} and P_{CO_2} of arterial blood, newborns are at greater risk of developing respiratory acidosis or respiratory alkalosis. We will return to acid–base imbalances in newborn infants later in this chapter.

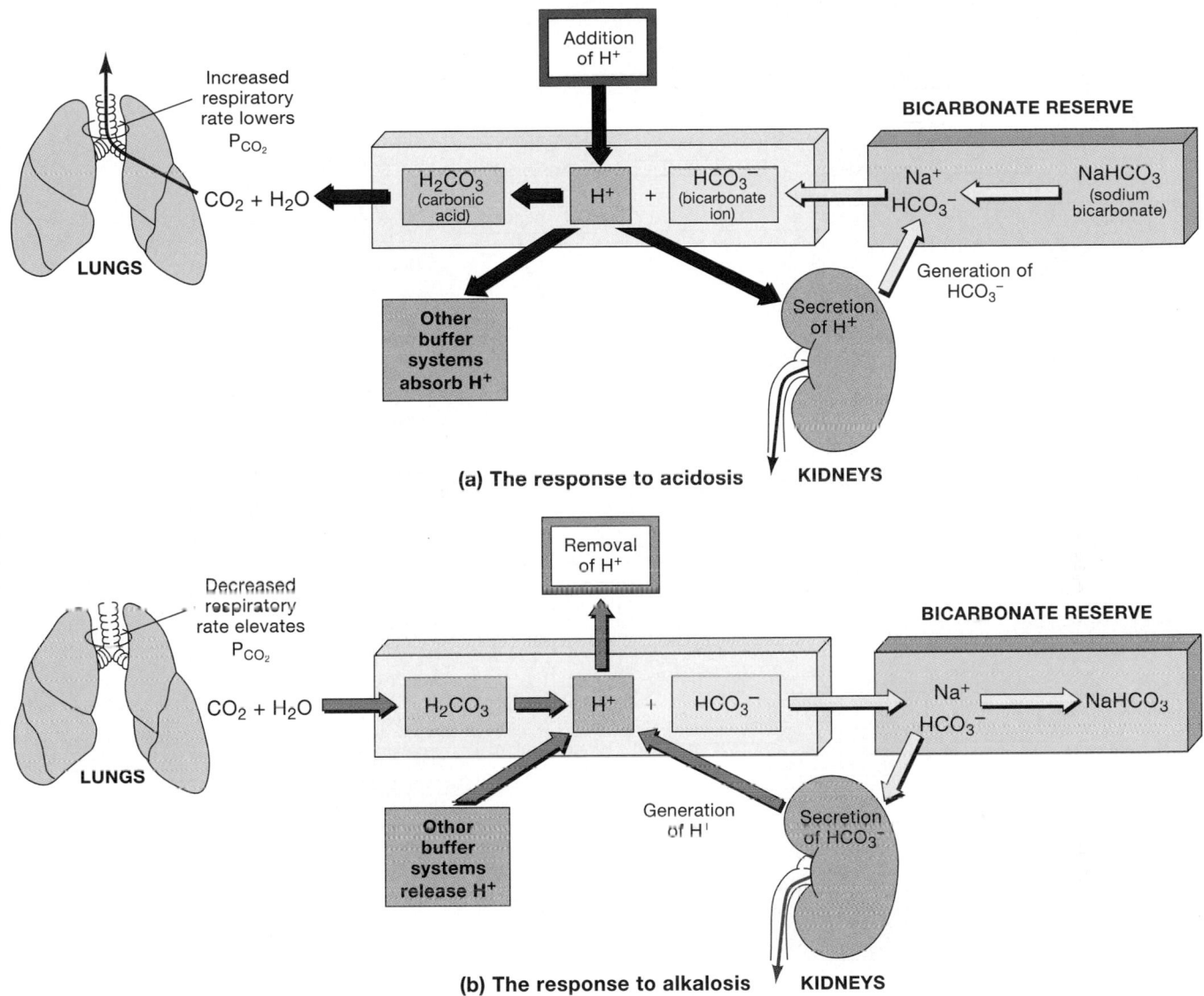

•FIGURE 27-11 The Central Role of the Carbonic Acid–Bicarbonate Buffer System in the Regulation of Plasma pH. Interactions between the carbonic acid–bicarbonate buffer system and other buffer systems and compensation mechanisms. **(a)** The response to a fall in pH caused by the addition of H^+. **(b)** The response to a rise in pH caused by the removal of H^+.

Respiratory Acidosis

Respiratory acidosis develops when the respiratory system cannot eliminate all the carbon dioxide generated by peripheral tissues. The primary symptom is low plasma pH due to **hypercapnia,** an elevated plasma P_{CO_2}. The usual cause is *hypoventilation,* an abnormally low respiratory rate. (We discussed hypoventilation and its effects on P_{CO_2} in Chapter 23.) ∞ *[p. 855]* When the P_{CO_2} in the ECF rises, H^+ and HCO_3^- concentrations also begin rising as H_2CO_3 forms and dissociates. Other buffer systems can tie up some of the H^+, but once the combined buffering capacity has been exceeded, the pH begins to fall rapidly. The effects are diagrammed in Figure 27-12a•.

Respiratory acidosis is the most common challenge to acid–base equilibrium. Your tissues generate carbon dioxide at a rapid rate, and even a few minutes of hypoventilation can cause acidosis, reducing the pH of the ECF to as low as 7.0. Under normal circumstances, the chemoreceptors that monitor the P_{CO_2} of the plasma and the CSF will eliminate the problem by calling for an increase in pulmonary ventilation rates.

Acute Respiratory Acidosis

If the chemoreceptors fail to respond, if pulmonary ventilation cannot be increased, or if the circulatory supply to the lungs is inadequate, the pH will continue to decline. If the decline is severe, **acute respiratory acidosis** will develop. Acute respiratory acidosis is an immediate, life-threatening condition. It is especially dangerous in persons whose tissues are generating large amounts of carbon dioxide or who are incapable of normal respiratory activity. For this reason, the reversal of acute respiratory acidosis is probably the major goal in the resuscitation of

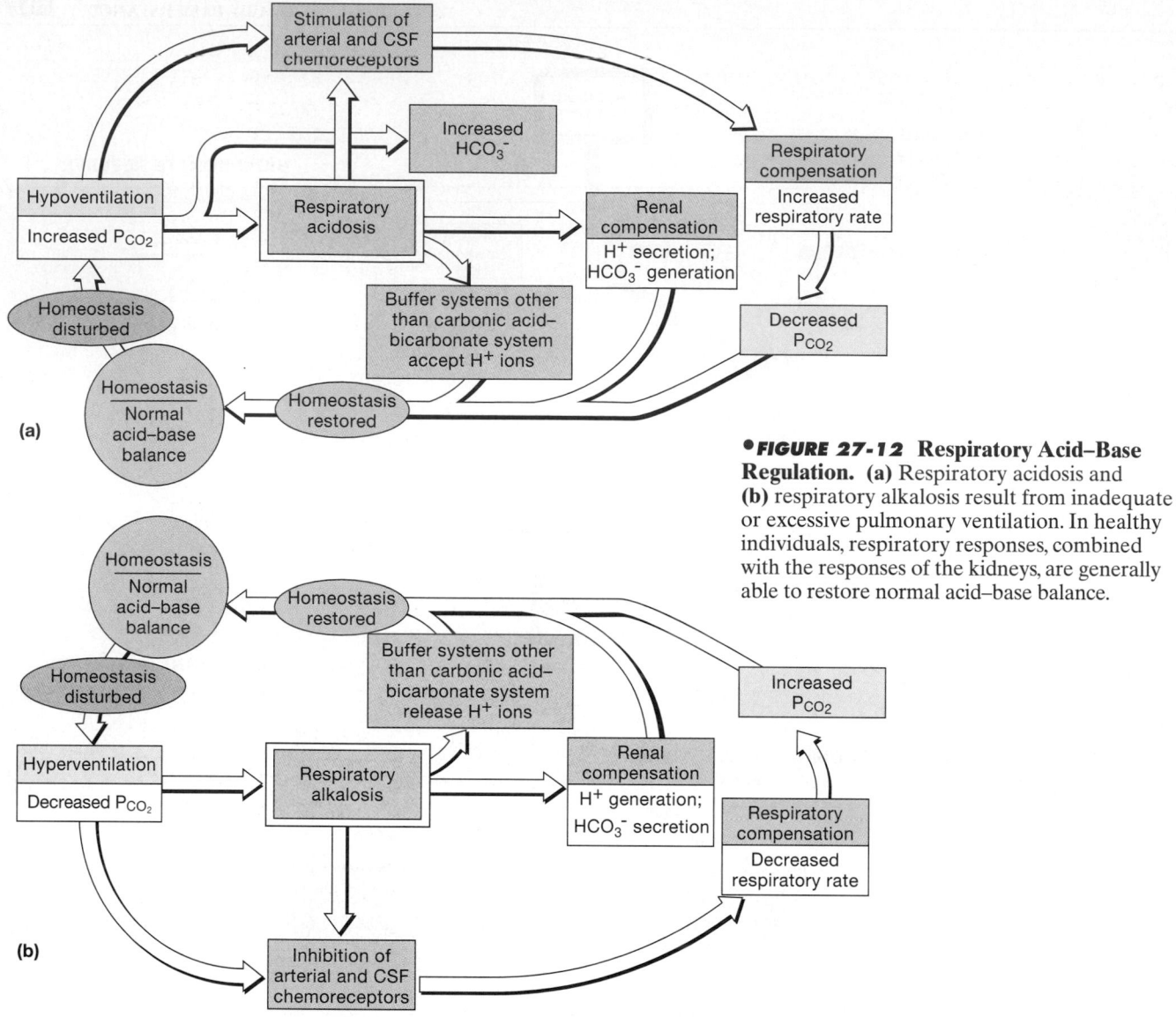

●**FIGURE 27-12 Respiratory Acid–Base Regulation.** **(a)** Respiratory acidosis and **(b)** respiratory alkalosis result from inadequate or excessive pulmonary ventilation. In healthy individuals, respiratory responses, combined with the responses of the kidneys, are generally able to restore normal acid–base balance.

cardiac arrest or drowning victims. Thus first-aid, CPR, and life-saving courses always stress the "ABCs" of emergency care: Airway, Breathing, and Circulation.

Chronic Respiratory Acidosis

Chronic respiratory acidosis develops when normal respiratory function has been compromised but the compensatory mechanisms have not failed completely. Individuals suffering from CNS injuries, or those whose respiratory centers have been desensitized by drugs such as alcohol or barbiturates, do not respond to the warning signals from the chemoreceptors. As a result, these people are prone to developing acidosis due to chronic hypoventilation.

Even when respiratory centers are intact and functional, increased pulmonary exchange may be prevented by damage to some components of the respiratory system. Examples of conditions fostering chronic respiratory acidosis include emphysema, congestive heart failure, and pneumonia (in which alveolar damage or blockage

typically occurs). Pneumothorax and respiratory muscle paralysis have a similar effect, for they too limit the ability to maintain adequate pulmonary ventilation rates.

When a normal pulmonary response does not occur, the kidneys respond by increasing the rate of H⁺ secretion into the tubular fluid. This response slows the rate of pH change. However, renal mechanisms alone cannot return the pH to normal until the underlying respiratory or circulatory problems are corrected.

Treatment of Respiratory Acidosis

The primary problem in respiratory acidosis is that the rate of pulmonary exchange is inadequate to keep the arterial P_{CO_2} within normal limits. The efficiency of pulmonary ventilation can typically be improved temporarily by inducing bronchodilation or using mechanical aids that provide air under positive pressure. If breathing has ceased, artificial respiration or a mechanical ventilator will be required. These measures may be sufficient to restore normal pH if the respiratory acidosis was neither severe nor prolonged. Treatment of acute respiratory acidosis is complicated by the fact that it causes a complementary *metabolic acidosis*, as we shall soon see, due to the generation of lactic acid in oxygen-starved tissues.

Respiratory Alkalosis

Problems with **respiratory alkalosis** (Figure 27-12b●) are relatively uncommon. Respiratory alkalosis develops when respiratory activity lowers plasma P_{CO_2} to below-normal levels, a condition called **hypocapnia.** A temporary hypocapnia can be produced by *hyperventilation*, when increased respiratory activity leads to a reduction in the arterial P_{CO_2}. (We discussed hyperventilation and its effects on P_{CO_2} in Chapter 23.) ∞ *[pp. 855]* Continued hyperventilation can elevate the pH to levels as high as 7.8–8. This condition generally corrects itself, for the reduction in P_{CO_2} removes the stimulation for the chemoreceptors, and the urge to breathe fades until carbon dioxide levels have returned to normal. *Respiratory alkalosis caused by hyperventilation seldom persists long enough to cause a clinical emergency.*

Common causes of hyperventilation include physical stresses, such as pain, or psychological stresses, such as extreme anxiety. Sometimes people will hyperventilate deliberately to prepare for holding their breath; doing this can be dangerous for reasons we noted in Chapter 23. ∞ *[p. 854]* Hyperventilation gradually elevates the pH of the CSF, and CNS function is affect-

ed. The initial symptoms involve tingling sensations in the hands, feet, and lips. A light-headed feeling may also be noted. If the hyperventilation continues, the individual may lose consciousness. When unconsciousness occurs, any contributing psychological stimuli are removed, and the breathing rate declines. The P_{CO_2} then rises until the pH returns to normal.

Treatment of Respiratory Alkalosis

A simple treatment for respiratory alkalosis caused by hyperventilation consists of having the individual breathe and rebreathe the air contained in a small paper bag. As the P_{CO_2} in the bag rises, so do the person's alveolar and arterial CO_2 concentrations. This change eliminates the problem and restores the pH to normal levels. Other problems with respiratory alkalosis are rare and involve primarily (1) individuals adapting to high altitudes, where the low P_{O_2} promotes hyperventilation, (2) patients on mechanical respirators, or (3) individuals with brain stem injuries who are incapable of responding to alterations in plasma CO_2 concentrations.

Metabolic Acidosis

Metabolic acidosis is the second most common type of acid–base imbalance. It has three major causes:

1. The most widespread cause of metabolic acidosis is the production of a large number of fixed or organic acids. As diagrammed in Figure 27-13a●, the hydro-

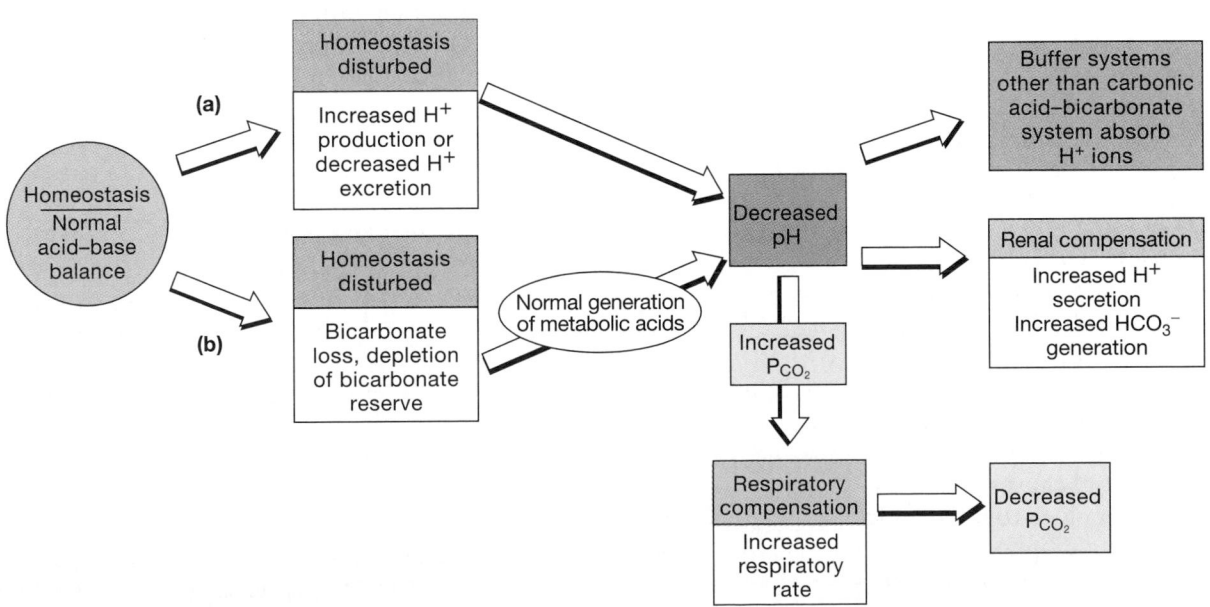

●**FIGURE 27-13** **The Response to Metabolic Acidosis.** Metabolic acidosis can result from either **(a)** increased acid production or decreased acid secretion, leading to a buildup of H^+ in body fluids, or **(b)** a loss of bicarbonate ions that makes the carbonic acid–bicarbonate buffer system incapable of preventing a fall in pH. Pulmonary and renal compensation mechanisms can stabilize pH, but the blood chemistry remains abnormal until the levels of acid production, acid secretion, and bicarbonate ion concentration return to normal.

gen ions liberated by these acids overload the carbonic acid–bicarbonate buffer system, and the pH begins to decline. We introduced two important examples of metabolic acidosis in earlier chapters:

a. **Lactic acidosis** may develop after strenuous exercise or prolonged tissue hypoxia (oxygen starvation) as active cells rely on anaerobic respiration. (For a more detailed discussion, see Chapter 10, especially Figure 10-16●, p. 300.)

b. **Ketoacidosis** results from the generation of large quantities of ketone bodies during the postabsorptive state of metabolism. Ketoacidosis is a problem in starvation and a potential complication of poorly controlled diabetes mellitus. In either case, peripheral tissues are unable to obtain adequate glucose from the circulation and begin metabolizing lipids and ketone bodies. The ketone bodies are generated at the liver during the catabolism of lipids and ketogenic amino acids. (For additional details, see the related discussion in Chapter 25.) ∞ [pp. 940–943]

2. A less common cause of metabolic acidosis is an impaired ability to excrete H^+ at the kidneys. For example, conditions marked by severe kidney damage, such as *glomerulonephritis* (Chapter 26), typically result in a severe metabolic acidosis. ∞ [p. 969] Metabolic acidosis may also be caused by diuretics that turn off the sodium–hydrogen transport system in the kidney tubules. The secretion of H^+ is directly or indirectly linked to the reabsorption of Na^+. When Na^+ reabsorption stops, so does H^+ secretion.

3. Metabolic acidosis can occur after a severe bicarbonate loss (Figure 27-13b●). The carbonic acid–bicarbonate buffer system relies on bicarbonate ions to balance hydrogen ions that threaten pH balance. A drop in the HCO_3^- concentration in the ECF thus reduces the effectiveness of this buffer system, and acidosis soon develops. The most common cause of HCO_3^- depletion is chronic diarrhea. Under normal conditions, most of the bicarbonate ions secreted into the digestive tract in pancreatic, hepatic, and mucous secretions are reabsorbed before the feces are eliminated. In diarrhea, these bicarbonates are lost, and the HCO_3^- concentration of the ECF drops accordingly.

The nature of the problem must be understood before treatment can begin. Because the potential causes are so varied, a clinician must piece together relevant clues to make a diagnosis. In some cases, the diagnosis is very straightforward; for example, a patient in metabolic acidosis after a bicycle race probably has lactic acidosis. In other cases, the clinician must do some detective work. When questions exist, calculation of an *anion gap* can be very helpful. This quan-

tity, the difference between the concentration of Na^+ and the sum of the concentrations of HCO_3^- and Cl^- is discussed in the *Applications Manual*. AM *Diagnostic Classification of Acid–Base Disorders*

Compensation for Metabolic Acidosis

Compensation for metabolic acidosis generally involves a combination of respiratory and renal mechanisms. Hydrogen ions interacting with bicarbonate ions form carbon dioxide molecules that are eliminated at the lungs, whereas the kidneys excrete additional hydrogen ions into the urine and generate bicarbonate ions that are released into the ECF.

Combined Respiratory and Metabolic Acidosis

Metabolic and respiratory acidosis are typically associated, because sustained hypoventilation leads to decreased arterial P_{O_2}, and oxygen-starved tissues generate large quantities of lactic acid. The problem can be particularly serious in cases of near-drowning, in which the body fluids have high P_{CO_2}, low P_{O_2}, and large amounts of lactic acid generated by the muscles of the struggling victim. Prompt emergency treatment is essential. The usual procedure involves some form of artificial or mechanical respiratory assistance coupled with the intravenous infusion of an isotonic solution that contains sodium lactate, sodium gluconate, or sodium bicarbonate.

Metabolic Alkalosis

Metabolic alkalosis occurs when HCO_3^- concentrations become elevated (Figure 27-14●). The bicarbonate ions then interact with hydrogen ions in solution, forming H_2CO_3. The reduction in H^+ concentrations then causes symptoms of alkalosis.

Cases of metabolic alkalosis are relatively rare, but we noted one interesting cause in Chapter 24. ∞ [p. 879]

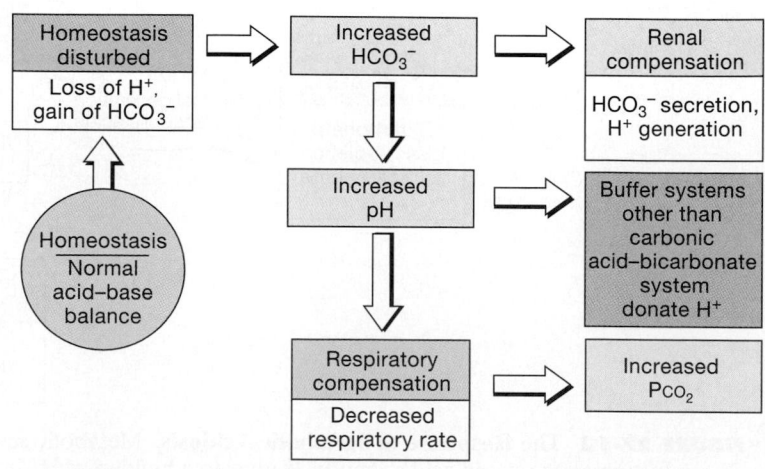

●**FIGURE 27-14** Metabolic Alkalosis. Metabolic alkalosis most commonly results from the loss of acids, especially stomach acid lost through vomiting. As replacement gastric acids are produced, the alkaline tide introduces a great many bicarbonate ions into the bloodstream, and pH increases.

The secretion of hydrochloric acid (HCl) by the gastric mucosa is associated with the influx of large numbers of bicarbonate ions into the ECF. This phenomenon, known as the *alkaline tide,* temporarily elevates the HCO_3^- concentration in the ECF at mealtimes. A person who begins vomiting repeatedly will continue to generate stomach acids to replace those that are lost, so the ECF HCO_3^- concentration will continue to rise.

Compensation for metabolic alkalosis involves a reduction in pulmonary ventilation, coupled with the increased loss of HCO_3^- in the urine. Treatment of mild cases addresses the primary cause, generally by the control of vomiting, and may involve the administration of solutions that contain NaCl or KCl.

Treatment of acute cases of metabolic alkalosis may involve the administration of ammonium chloride (NH_4Cl). Metabolism of the ammonium ion in the liver results in the liberation of a hydrogen ion, so in effect the introduction of NH_4Cl leads to the internal generation of HCl, a strong acid. As the HCl diffuses into the bloodstream, the pH falls toward normal levels.

PROBLEMS WITH ACID–BASE BALANCE IN NEWBORN INFANTS Many newborn infants suffer from some degree of respiratory acidosis caused by the interruption of placental circulation during labor. This condition is known as *fetal stress.* It is therefore important that the newborn begin breathing as soon as possible after delivery so that the excess CO_2 can be eliminated. Most full-term newborns are given oxygen briefly to improve blood oxygenation and to help them eliminate excess CO_2. (Oxygen administration to premature infants may have unwanted side effects.) The pH of infants in fetal stress typically does not fall below 7.25; in most cases, no treatment other than oxygen administration is required.

If the fetus struggles during labor or the delivery is prolonged, the condition of *fetal distress* may develop. Fetal distress results from the combination of respiratory acidosis, due to decreased placental circulation, and metabolic acidosis, due to lactic acid production. The pH can fall as low as 7.0, with potentially fatal effects on the heart and the CNS. Because fetal distress is one of the risks of vaginal delivery, the fetus is commonly monitored during labor and delivery. Early in labor, the heart rate is checked by ultrasound; once delivery is under way, blood samples can be taken from the scalp. If the fetal heart rate becomes too rapid and irregular early in labor, or if the arterial pH drops below 7.2, prompt surgical delivery (by cesarean section) may be recommended. Treatment of a newborn who is suffering from fetal distress typically includes the administration of oxygen and the intravenous infusion of a sodium bicarbonate solution.

The Detection of Acidosis and Alkalosis

Virtually anyone who has a problem that affects the cardiovascular, respiratory, urinary, digestive, or nervous system may develop potentially dangerous problems with acid–base balance. For this reason, most diagnostic blood tests include tests designed to provide information on pH and buffer function. Standard tests include blood pH, P_{CO_2}, and HCO_3^- levels. These data make the recognition of acidosis or alkalosis and the classification of a particular condition as respiratory or metabolic in nature relatively straightforward. Table 27-4 indicates the patterns that characterize the four major categories of acid–base disorders. Additional steps, such as determining the anion gap, plotting blood test results on a graph, or *nomogram,* or using a diagnostic chart may help in identifying possible causes of the problem and in

TABLE 27-4 Changes in Blood Chemistry Associated with the Major Classes of Acid–Base Disorders

Disorder	pH (normal = 7.35–7.45)	HCO_3^- (normal = 24–28 mEq/l)	P_{CO_2} (mm Hg) (normal = 35–45)	Remarks	Treatment
Respiratory acidosis	Decreased (below 7.35)	*Acute:* normal *Compensated:* increased (above 28)	Increased (above 50)	Generally caused by hypoventilation and CO_2 buildup in tissues and blood	Improve ventilation; in some cases, with bronchodilation and mechanical assistance
Metabolic acidosis	Decreased (below 7.35)	Decreased (below 24)	*Acute:* normal *Compensated:* decreased (below 35)	Caused by buildup of organic or fixed acid, impaired H^+ excretion at kidneys, or HCO_3^- loss in urine or feces	Administration of bicarbonate (gradual) with other steps as needed to correct primary cause
Respiratory alkalosis	Increased (above 7.45)	*Acute:* normal *Compensated:* decreased (below 24)	Decreased (below 35)	Generally caused by hyperventilation and reduction in plasma CO_2 levels	Reduce respiratory rate, allow rise in P_{CO_2}
Metabolic alkalosis	Increased (above 7.45)	Increased (above 28)	Increased (above 45)	Generally caused by prolonged vomiting and associated acid loss	pH below 7.55: no treatment; pH above 7.55: may require administration of ammonium chloride

distinguishing compensated from uncompensated conditions. Details and diagnostic exercises are included in the *Applications Manual*.

✔ What effect would a decrease in the pH of the body fluids have on the respiratory rate?

✔ Why does the tubular fluid in nephrons need to be buffered?

✔ How would a prolonged fast affect the body's pH?

✔ Why can prolonged vomiting produce alkalosis?

⧗ AGING AND FLUID, ELECTROLYTE, AND ACID–BASE BALANCE

Aging affects many aspects of fluid, electrolyte, and acid–base balance, including the following:

- Total body water content gradually decreases with age. Between ages 40 and 60, total body water content averages 55 percent for males and 47 percent for females. After age 60, the values decline to roughly 50 percent for males and 45 percent for females. Among other effects, this decrease reduces the dilution of waste products, toxins, and administered drugs.

- A reduction in the GFR and in the number of functional neurons reduces the ability to regulate pH through renal compensation.

- The ability to concentrate urine declines, and more water is lost in the urine. In addition, the rate of insensible perspiration increases as the skin becomes thinner and more delicate. Maintaining fluid balance therefore requires a higher daily water intake. A reduction in ADH and aldosterone sensitivity makes older people less able than younger people to conserve body water when losses exceed gains.

- Many people over age 60 experience a net loss in body mineral content as muscle mass and skeletal mass decrease. This loss can, at least in part, be prevented by a combination of exercise and increased dietary mineral supply.

- The reduction in vital capacity that accompanies aging reduces the ability to perform respiratory compensation, increasing the risk of respiratory acidosis. This problem can be compounded by arthritis, which can reduce vital capacity; and by emphysema, another condition that, to some degree, develops with aging.

- With increasing age, people are likely to have developed conditions that affect fluid, electrolyte, or acid–base balance. Examples mentioned in earlier chapters include myocardial infarction, congestive heart failure, malnutrition, cirrhosis, nephrolithiasis, and urinary obstruction.

SELECTED CLINICAL TERMINOLOGY

Terms Discussed in This Chapter

---

hyperkalemia: Plasma potassium levels above 8 mEq/l. *(p. 1016)*

hypernatremia: Plasma sodium levels above 150 mEq/l. *(p. 1012)*

hypokalemia: Plasma potassium levels below 2 mEq/l. *(p. 1016)*

hyponatremia: Plasma sodium levels below 130 mEq/l. *(p. 1012)*

metabolic acidosis: A type of acidosis caused by the inability to excrete hydro-

gen ions, the production of numerous fixed and/or organic acids, or a severe bicarbonate loss. *(p. 1029 and AM)*

metabolic alkalosis: A rare form of alkalosis resulting from high concentrations of bicarbonate ions in body fluids. *(p. 1030 and AM)*

nomogram: A graph that can be used to assess acid–base balance in a clinical setting. *(p. 1031 and AM)*

respiratory acidosis: Acidosis resulting from inadequate respiratory activity, characterized by elevated levels of carbon dioxide (hypercapnia) in body fluids. *(p. 1027 and AM)*

respiratory alkalosis: Alkalosis due to excessive respiratory activity, which depresses carbon dioxide levels and elevates the pH of body fluids. *(p. 1029 and AM)*

CHAPTER REVIEW

 On-line resources for this chapter are on our World Wide Web site at: http://www.prenhall.com/martini/fap

STUDY OUTLINE

INTRODUCTION, p. 1007

1. The maintenance of normal volume and normal composition of the extracellular and intracellular fluids is vital to life. Three types of homeostasis are involved in this maintenance: **fluid balance, electrolyte balance,** and **acid–base balance.**

FLUID AND ELECTROLYTE BALANCE, p. 1007

1. The **intracellular fluid (ICF)** contains nearly two-thirds of the total body water; the **extracellular fluid (ECF)** contains the rest. Exchange occurs between the ICF and ECF, but the two **fluid compartments** retain their distinctive characteristics. *(Figures 27-1, 27-2)*

Basic Concepts Pertaining to Fluid and Electrolyte Regulation, p. 1009

2. Homeostatic mechanisms that monitor and adjust the composition of body fluids respond to changes in the ECF.
3. No receptors directly monitor fluid or electrolyte balance; receptors involved in fluid and electrolyte balance respond to changes in plasma volume or osmolarity.
4. Our cells cannot move water molecules by active transport; all water movements across cell membranes and epithelia occur passively in response to osmotic gradients.
5. The body content of water or electrolytes will rise if intake exceeds outflow and will fall if losses exceed gains.

An Overview of the Primary Regulatory Hormones, p. 1010

6. ADH encourages water reabsorption at the kidneys and stimulates thirst. Aldosterone increases the rates of sodium reabsorption at the kidneys. ANP opposes those actions and promotes fluid and electrolyte losses in the urine.

The Interplay between Fluid and Electrolyte Balance, p. 1010

7. The regulatory mechanisms are quite different between fluid balance and electrolyte balance, and the distinction is important clinically.

Fluid Balance, p. 1010

8. Water circulates freely within the ECF compartment.
9. Water losses are normally balanced by gains through eating, drinking, and **metabolic generation.** *(Figure 27-3; Table 27-1)*
10. Water movement between the ECF and ICF is called a **fluid shift**. If the ECF becomes hypertonic relative to the ICF, water will move from the ICF into the ECF until osmotic equilibrium has been restored. If the ECF becomes hypotonic relative to the ICF, water will move from the ECF into the cells, and the volume of the ICF will increase.

Electrolyte Balance, p. 1013

11. Problems with electrolyte balance generally result from an imbalance between sodium gains and losses. Problems with potassium balance are less common but more dangerous.
12. The rate of sodium uptake across the digestive epithelium is directly proportional to the amount of sodium in the diet. Sodium losses occur mainly in the urine and through perspiration. *(Figure 27-4)*
13. Alterations in sodium balance result in the expansion or contraction of the ECF. Large variations in the ECF volume are corrected by the homeostatic mechanisms triggered by changes in blood volume. If the volume becomes too low, ADH and aldosterone are secreted; if the volume becomes too high, ANP is secreted. *(Figure 27-5)*
14. Potassium ion concentrations in the ECF are very low and not as closely regulated as are sodium ion concentrations. Potassium excretion increases as ECF concentrations rise, under aldosterone stimulation, and when the pH rises. Potassium retention occurs when the pH falls.
15. The ECF concentrations of other electrolytes such as calcium, magnesium, phosphate, and chloride are also regulated. *(Table 27-2)*

ACID–BASE BALANCE, p. 1018

1. Acids and bases are either *strong* or *weak*. *(Table 27-3)*

The Importance of pH Control, p. 1019

2. The pH of normal body fluids ranges from 7.35 to 7.45; variations outside this relatively narrow range produce symptoms of **acidosis** or **alkalosis.**

Types of Acids in the Body, p. 1019

3. **Volatile acids** can leave solution and enter the atmosphere; **fixed acids** remain in body fluids until excreted at the kidneys; **organic acids** are acid participants in or byproducts of aerobic metabolism.
4. Carbonic acid is the most important factor affecting the pH of the ECF. In solution, CO_2 reacts with water to form carbonic acid. An inverse relationship exists between pH and the concentration of CO_2. *(Figure 27-6)*
5. *Sulfuric acid* and *phosphoric acid*, the most important fixed acids, are generated during the catabolism of amino acids and compounds containing phosphate groups.
6. Organic acids include metabolic products such as lactic acid and ketone bodies.

Mechanisms of pH Control, p. 1020

7. A **buffer system** typically consists of a weak acid and the anion released by its dissociation, which functions as a weak base. There are three major buffer systems: (1) **protein buffer systems** in the ECF and ICF; (2) the **carbonic acid–bicarbonate buffer system,** most important in the ECF; and (3) the **phosphate buffer system** in the ICF and urine. *(Figure 27-7)*
8. In protein buffer systems, the component amino acids respond to changes in pH by accepting or releasing hydrogen ions. The **hemoglobin buffer system** is a protein buffer system that helps prevent drastic changes in pH when the P_{CO_2} is rising or falling. *(Figure 27-8)*
9. The carbonic acid–bicarbonate buffer system prevents pH changes caused by organic acids and fixed acids in the ECF. The readily available supply of bicarbonate ions is the **bicarbonate reserve.** *(Figure 27-9)*
10. The phosphate buffer system plays a supporting role in regulating the pH of the ECF but is important in buffering the pH of the ICF and of the urine.

Maintenance of Acid–Base Balance, p. 1023

11. The lungs help regulate pH by affecting the carbonic acid–bicarbonate buffer system. Changing the respiratory rate can raise or lower the P_{CO_2} of body fluids, affecting the buffering capacity. This process is **respiratory compensation.**
12. In the process of **renal compensation,** the kidneys vary their rates of hydrogen ion secretion and bicarbonate ion reabsorption, depending on the pH of the ECF. *(Figure 27-10)*

DISTURBANCES OF ACID–BASE BALANCE, p. 1025

1. Interactions among buffer systems, respiratory, and renal functions normally maintain the pH of the ECF within a pH range of 7.35–7.45. *(Figure 27-11)*
2. **Respiratory acid–base disorders** result when abnormal respiratory function causes an extreme rise or fall in CO_2 levels in the ECF. **Metabolic acid–base disorders** are caused by the generation of organic or fixed acids or by conditions affecting the concentration of bicarbonate ions in the ECF.

Respiratory Acidosis, p. 1027

3. **Respiratory acidosis** results from excessive levels of CO_2 in body fluids. **Chronic respiratory acidosis** develops when compensatory mechanisms have not completely failed. If normal homeostatic adjustments do not occur, **acute respiratory acidosis** develops. *(Figure 27-12)*

Respiratory Alkalosis, p. 1029

4. **Respiratory alkalosis** is a relatively rare condition associated with hyperventilation. *(Figure 27-12)*

Metabolic Acidosis, p. 1029

5. **Metabolic acidosis** results from the depletion of the bicarbonate reserve. It is caused by an inability to excrete hydrogen ions at the kidneys, the production of large numbers of fixed and organic acids, or the bicarbonate loss that accompanies chronic diarrhea. *(Figure 27-13)*

Metabolic Alkalosis, p. 1030

6. **Metabolic alkalosis** occurs when bicarbonate ion concentrations become elevated, as from extended periods of vomiting. *(Figure 27-14)*

The Detection of Acidosis and Alkalosis, p. 1031

7. Standard diagnostic blood tests such as blood pH, P_{CO_2} and bicarbonate levels are used to recognize and classify acidosis and alkalosis conditions as respiratory or metabolic in nature. *(Figure 27-15; Table 27-4)*

AGING AND FLUID, ELECTROLYTE, AND ACID–BASE BALANCE, p. 1032

1. Changes affecting fluid, electrolyte, and acid–base balance in the elderly include (1) reduced total body water content, (2) impaired ability to perform renal compensation, (3) increased water demands due to reduced ability to concentrate urine and reduced sensitivity to ADH and aldosterone, (4) a net loss of minerals, (5) reductions in respiratory efficiency that affect the ability to perform respiratory compensation, and (6) increased incidence of conditions that secondarily affect fluid, electrolyte, or acid–base balance.

REVIEW QUESTIONS

Level 1   Reviewing Facts and Terms

Match each numbered item with the most closely related lettered item. Use letters for answers in the spaces provided.

___ 1. ECF
___ 2. ADH
___ 3. aldosterone
___ 4. atrial natriuretic peptide
___ 5. oxidative phosphorylation
___ 6. overhydration
___ 7. dehydration
___ 8. sodium
___ 9. potassium
___ 10. carotid sinus
___ 11. calcitonin
___ 12. hydrochloric acid
___ 13. carbonic acid
___ 14. carbon dioxide
___ 15. fixed acid
___ 16. organic acid
___ 17. hypoventilation
___ 18. hyperventilation
___ 19. elevated bicarbonate ion concentration
___ 20. severe bicarbonate loss

a. reduces thirst
b. dominant cation in ECF
c. hyponatremia
d. weak acid
e. baroreceptors
f. promotes calcium loss
g. respiratory alkalosis
h. stimulates water conservation
i. sulfuric acid
j. volatile acid
k. ketone bodies
l. interstitial fluid
m. respiratory acidosis
n. metabolic acidosis
o. hypernatremia
p. metabolic alkalosis
q. dominant cation in ICF
r. conservation of sodium ions
s. strong acid
t. metabolic generation of water

21. A person is in fluid balance when
   (a) the ECF and ICF are isotonic
   (b) there is no fluid movement between compartments
   (c) the amount of water gained each day is equal to the amount lost to the environment
   (d) a, b, and c are correct

22. The primary components of the extracellular fluid are
   (a) lymph and cerebrospinal fluid
   (b) blood plasma and serous fluids
   (c) interstitial fluid and plasma
   (d) a, b, and c are correct

23. The principal ions in the ECF are
   (a) sodium and chloride
   (b) sodium and potassium
   (c) phosphate and chloride
   (d) phosphate and bicarbonate

24. All the homeostatic mechanisms that monitor and adjust the composition of body fluids respond to changes
   (a) in the ICF   (b) in the ECF
   (c) inside the cell   (d) a, b, and c are correct

25. Homeostatic adjustments to monitor fluid and electrolyte balance occur in response to changes in
   (a) pH of body fluids
   (b) autonomic nervous system activity
   (c) concentration of dissolved gases
   (d) plasma volume or osmolarity

26. A rise in body temperature accompanying a fever
   (a) decreases water losses
   (b) increases water losses
   (c) does not affect water gain or loss
   (d) causes the body to retain water

27. The most common problems with electrolyte balance are caused by an imbalance between gains and losses of
   (a) calcium ions   (b) chloride ions
   (c) potassium ions   (d) sodium ions

28. Angiotensin II produces a coordinated elevation in the ECF volume by
   (a) stimulating thirst
   (b) causing the release of ADH
   (c) triggering production and secretion of aldosterone
   (d) a, b, and c are correct

29. The rate of tubular secretion of potassium ions changes in response to
   (a) changes in pH
   (b) changes in aldosterone levels
   (c) alterations in the potassium ion concentration in the ECF
   (d) a, b, and c are correct

30. Respiratory acidosis develops when there is
   (a) elevated plasma pH due to a decreased plasma P_{CO_2}
   (b) decreased plasma pH due to an elevated plasma P_{CO_2}
   (c) elevated plasma pH due to an elevated plasma P_{CO_2}
   (d) decreased plasma pH due to a decreased plasma P_{CO_2}

31. Metabolic alkalosis occurs when
   (a) bicarbonate ion concentrations become elevated
   (b) there is a severe bicarbonate loss
   (c) the kidneys fail to excrete hydrogen ions
   (d) ketone bodies are generated in abnormally large quantities

32. What are fluid shifts? What is their function, and what factors can cause them?

33. Which three major hormones mediate major physiological adjustments that affect fluid and electrolyte balance? What are the primary effects of each hormone?

34. What influence does aldosterone have on the relationship between sodium and potassium?

35. Define and give an example of **(a)** a volatile acid, **(b)** a fixed acid, and **(c)** an organic acid. Which represent(s) the greatest threat to acid–base balance? Why?

Level 2 Reviewing Concepts

36. The higher the plasma concentration of aldosterone, the more efficiently the kidney will
   (a) conserve sodium ions
   (b) retain potassium ions
   (c) stimulate urinary water loss
   (d) secrete greater amounts of ADH

37. If the ECF is hypertonic with respect to the ICF
   (a) water will move from the ECF into the cells until osmotic equilibrium is reached
   (b) ions will move from the ECF into the cells until osmotic equilibrium is reached
   (c) ions will move from the cells into the ICF until osmotic equilibrium is reached
   (d) water will move from the cells into the ECF until osmotic equilibrium is reached

38. The osmolarity of the ECF decreases if the individual gains water without a corresponding
   (a) gain of electrolytes
   (b) loss of water
   (c) fluid shift from the ECF to the ICF
   (d) a, b, and c are correct

39. Chronic renal failure and congestive heart failure ultimately result in
   (a) hypernatremia   (b) hyponatremia
   (c) hyperkalemia   (d) hypokalemia

40. When pure water is consumed,
   (a) the ECF becomes hypertonic with respect to the ICF
   (b) the ECF becomes hypotonic with respect to the ICF
   (c) the ICF becomes hypotonic with respect to the plasma
   (d) water moves from the ICF into the ECF

41. When the plasma concentration of potassium decreases to a dangerous level, the result is
   (a) kidney failure
   (b) extensive muscular weakness, followed by paralysis
   (c) decreased blood-clotting capability
   (d) cessation of enzyme activity in cells

42. In a protein buffer system, if the pH rises,
   (a) the protein acquires a hydrogen ion from carbonic acid
   (b) hydrogen ions are buffered by hemoglobin molecules
   (c) a hydrogen ion is released, and a carboxyl ion is formed
   (d) a chloride shift occurs

43. Increasing or decreasing the rate of respiration alters pH by
   (a) lowering or raising the partial pressure of CO_2
   (b) lowering or raising the partial pressure of O_2
   (c) lowering or raising the partial pressure of N_2
   (d) a, b, and c are correct

44. Differentiate among fluid balance, electrolyte balance, and acid–base balance, and explain why each is important to homeostasis.

45. A sample of plasma contains chloride ions at a concentration of 250 mg/dl. How many mmol/l of chloride ions are in the sample of plasma?

46. Why should a person with a fever drink plenty of fluids?

47. Describe the effects of losses and gains of sodium on the ECF.

48. What are the three major buffer systems in body fluids? How does each system work?

49. How do pulmonary and renal mechanisms support the buffer systems?

50. Differentiate between respiratory compensation and renal compensation.

51. Distinguish between respiratory and metabolic disorders that disturb acid–base balance.

52. What is the difference between metabolic and respiratory alkalosis? What can cause these conditions?

53. The most recent advice from medical and nutritional experts is to decrease the intake of salt so that it does not exceed the amount needed to maintain a constant ECF composition. What effect does excessive salt and water ingestion have on **(a)** urine volume, **(b)** urine concentration, and **(c)** blood pressure?

54. Exercise physiologists recommend that adequate amounts of fluid be ingested before, during, and after exercise. Why is fluid replacement during extensive sweating important?

Level 3 Critical Thinking and Clinical Applications

55. After falling into an abandoned stone quarry filled with water and nearly drowning, a young boy is rescued. His rescuers assess his condition. They find that his body fluids have high P_{CO_2} and low P_{O_2} and that large amounts of lactic acid had been generated by the boy's muscles as he struggled in the water. As a clinician, diagnose the condition of the boy and recommend the necessary treatment to restore his body to homeostasis.

56. A patient who is hyperventilating and disoriented is brought into the emergency room. Blood analysis indicates that the patient has elevated levels of potassium (hyperkalemia). Is the patient most likely suffering from acidosis or alkalosis? Explain.

57. Willy suffers from chronic emphysema. A blood analysis indicates that his plasma pH is 7.4, although his plasma bicarbonate level is significantly elevated. How can this be?

58. While visiting a foreign country, Milly inadvertently drinks the water, even though she had been advised not to. She contracts an intestinal disease that causes severe diarrhea. How would you expect her condition to affect her blood pH, urine pH, and pattern of ventilation?

59. Yuka is suffering from dehydration, and her physician prescribes intravenous fluids. The nurse caring for Yuka becomes distracted and erroneously gives her a hypertonic glucose solution instead of normal saline. What effect will this mistake have on Yuka's plasma levels of ADH and urine volume?

60. Refer to the diagnostic flowchart in Figure 27-15●. Use information from the blood test results to categorize the acid–base disorders that affect the patients represented there.

	Patient 1	Patient 2	Patient 3	Patient 4
pH	7.5	7.2	7.0	7.7
P_{CO_2}	32	45	60	50
Na^{+2}	138	140	140	136
HCO_3^-	22	20	28	34
Cl^-	106	102	101	91
Anion gap*	10	18	12	11

*Anion gap = Na^+ concentration – (HCO_3^- concentration + Cl^- concentration). Refer to the *Applications Manual* for additional information.

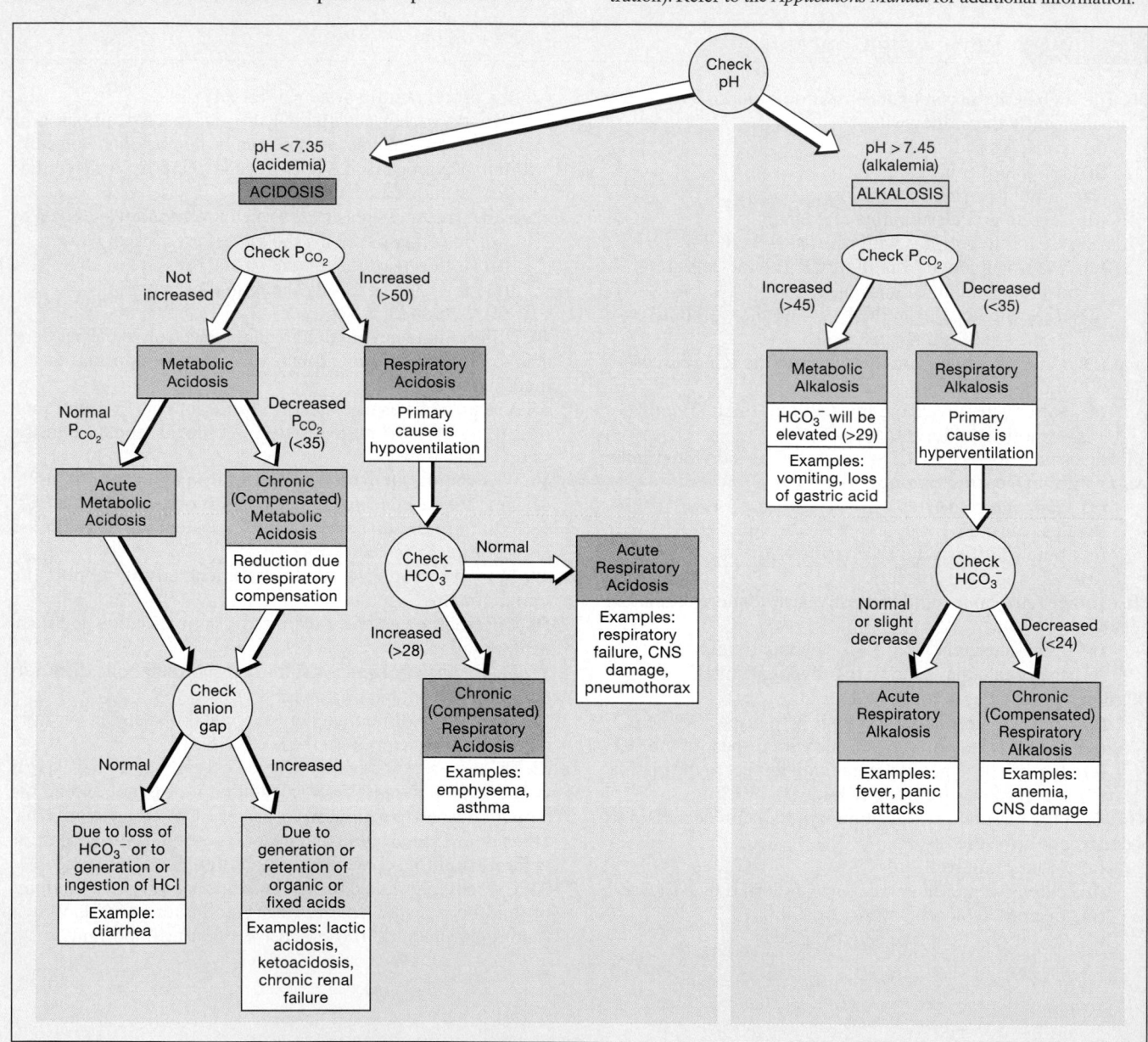

●*FIGURE 27-15* **A Diagnostic Chart for Acid–Base Disorders**

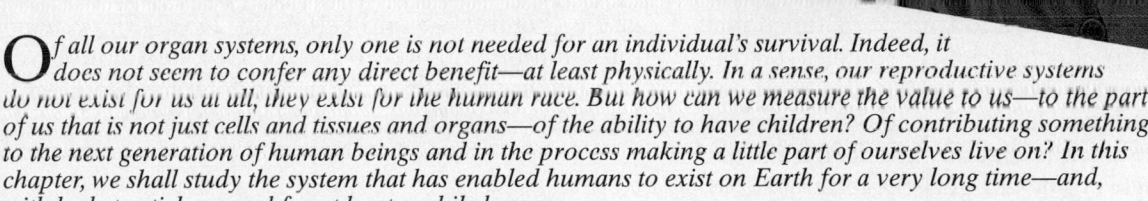

CHAPTER

28

The
Reproductive
System

*O*f all our organ systems, only one is not needed for an individual's survival. Indeed, it
does not seem to confer any direct benefit—at least physically. In a sense, our reproductive systems
do not exist for us at all, they exist for the human race. But how can we measure the value to us—to the part
of us that is not just cells and tissues and organs—of the ability to have children? Of contributing something
to the next generation of human beings and in the process making a little part of ourselves live on? In this
chapter, we shall study the system that has enabled humans to exist on Earth for a very long time—and,
with luck, to stick around for at least a while longer.

CHAPTER OUTLINE AND OBJECTIVES

The reproductive system is the only system that is not essential to the life of the individual. Instead, this system ensures the continued existence of the human species. The entire process of reproduction seems almost magical; many primitive societies failed to discover the basic link between sexual activity and childbirth and assumed that cosmic forces were responsible for the production of new individuals. Although our society has a much clearer view of the reproductive process, a sense of wonder remains. Sexually mature males and females produce individual reproductive cells that are brought together by the sexual act. The fusion of these cells starts a chain of events that lead to the appearance of an infant who will mature as part of the next generation.

In this and the next chapter, we shall consider the mechanics of this remarkable process. We shall begin by examining the anatomy and physiology of the reproductive system in this chapter. The human reproductive system produces, stores, nourishes, and transports functional male and female reproductive cells, or **gametes** (GAM-ēts). Chapter 29 begins with **fertilization**, also known as **conception**, in which a male gamete and female gamete unite. All the cells in a human body are the mitotic descendants of a single **zygote** (ZĪ-gōt), the cell created by the fusion of a *sperm* from the father and an *ovum* from the mother. The gradual transformation of that single cell into a functional adult, over a span of 15–20 years, is the process of *development*. Over this period, hormones produced by the reproductive system direct gender-specific patterns of development.

AN OVERVIEW OF THE REPRODUCTIVE SYSTEM

The reproductive system includes the following:

- **Gonads** (GŌ-nadz), or reproductive organs that produce gametes and hormones.
- Ducts that receive and transport the gametes.
- Accessory glands and organs that secrete fluids into these or other excretory ducts.
- Perineal structures associated with the reproductive system. These perineal structures are collectively known as the **external genitalia** (jen-i-TĀ-lē-uh).

The male and female reproductive systems are functionally quite different. In an adult male, the **testes** (TES-tēz; singular, *testis*), or male gonads, secrete sex hormones called *androgens*, (principally *testosterone*) and produce one-half billion sperm each day. (We introduced testosterone and other sex hormones in Chapter 18.) ∞ *[pp. 624–625]* During *emission*, mature sperm travel along a lengthy duct system, where they are mixed with the secretions of accessory glands. The mixture created is known as **semen** (SĒ-men). During *ejaculation*, the semen is expelled from the body.

In an adult female, the **ovaries**, or female gonads, typically release only one immature gamete, an **oocyte**, per month. This gamete travels along short **uterine tubes** *(oviducts)* that end in the muscular **uterus** (Ū-ter-us). A short passageway, the **vagina** (va-JĪ-nuh), connects the uterus with the exterior. During the sexual act, ejaculation introduces semen into the vagina, and the sperm ascend the female reproductive tract. If fertilization occurs, the uterus will enclose and support a developing embryo as the embryo grows into a *fetus* and prepares for eventual delivery.

We will now examine the anatomy of the male and female reproductive systems in detail and will consider the physiological and hormonal mechanisms responsible for the regulation of reproductive function. In earlier chapters, we introduced the anatomical reference points used in the following discussions; you may find it helpful to review the figures on the pelvic girdle (Figures 8-8 and 8-10•, pp. 243, 245), perineal musculature (Figure 11-13•, p. 338), pelvic innervation (Figure 13-11•, p. 429), and regional blood supply (Figures 21-28 and 21-34•, pp. 749, 755).

THE REPRODUCTIVE SYSTEM OF THE MALE

The principal structures of the male reproductive system are shown in Figure 28-1•. Proceeding from a testis, the sperm cells, or **spermatozoa** (sper-ma-tō-ZŌ-uh), travel within the **epididymis** (ep-i-DID-i-mus), the **ductus deferens** (DUK-tus DEF-e-renz), or *vas deferens*, the **ejaculatory** (ē-JAK-ū-la-to-rē) **duct,** and the **urethra** before leaving the body. Accessory organs—the **seminal** (SEM-i-nal) **vesicles**, the **prostate** (PROS-tāt) **gland**, and the **bulbourethral** (bul-bō-ū-RĒ-thral) **glands**—secrete into the ejaculatory ducts and urethra. The external genitalia consist of the **scrotum** (SKRŌ-tum), which encloses the testes, and the **penis** (PĒ-nis), an erectile organ through which the distal portion of the urethra passes.

The Testes

Each testis has the shape of a flattened egg that is roughly 5 cm (2 in.) long, 3 cm (1.2 in.) wide, and 2.5 cm (1 in.) thick. Each has a weight of 10–15 g (0.35–0.53 oz). The testes hang within the scrotum, a fleshy pouch suspended inferior to the perineum, anterior to the anus and posterior to the base of the penis (Figures 28-1 and 28-2•, p. 1040).

The Spermatic Cords

The **spermatic cords** consist of fascia layers, tough connective tissue, and muscle enclosing the blood vessels, nerves, and lymphatics that supply the testes. Each sper-

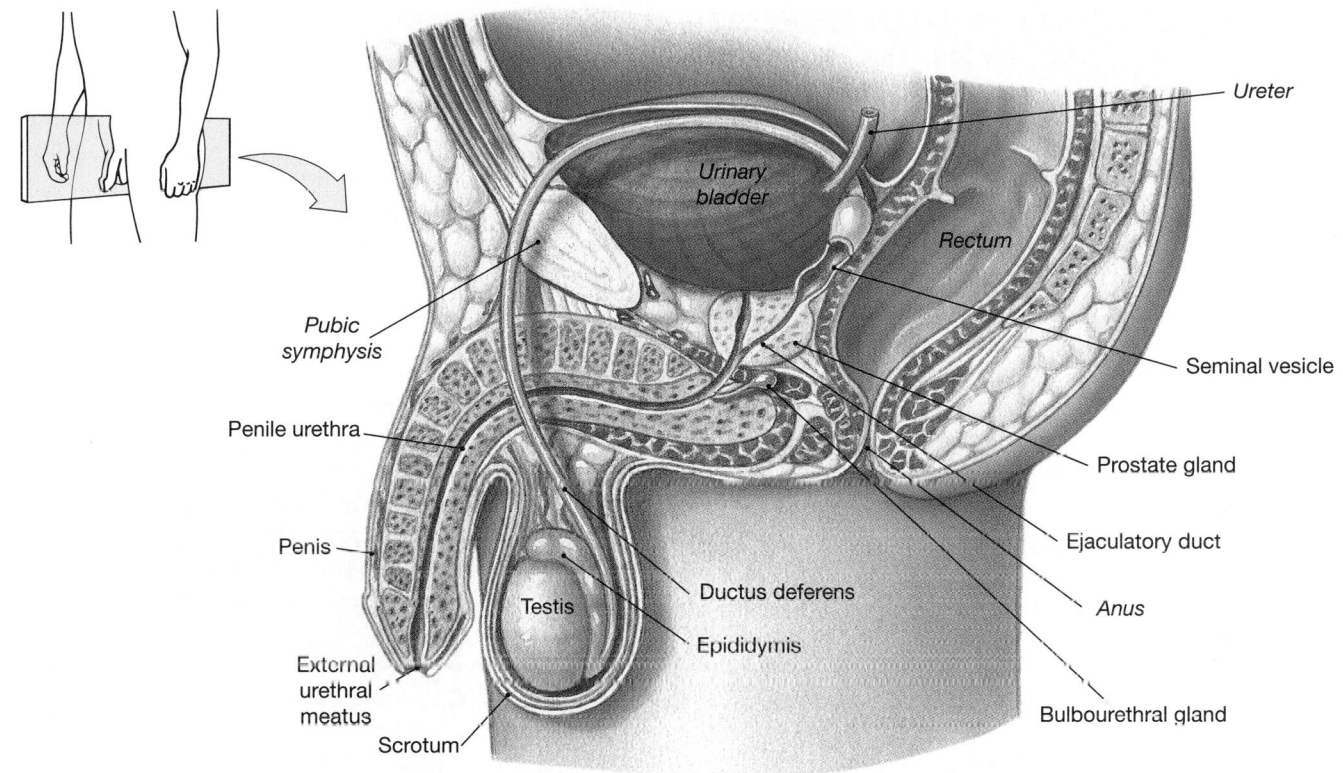

Ureter
Urinary bladder
Rectum
Pubic symphysis
Penile urethra
Seminal vesicle
Prostate gland
Ejaculatory duct
Penis
Anus
Testis
Ductus deferens
Epididymis
External urethral meatus
Bulbourethral gland
Scrotum

•**FIGURE 28-1** **The Male Reproductive System.** The male reproductive organs, diagrammatic sagittal section. ⬛ AM *Plate 6.7*

matic cord contains a *ductus deferens* and its *deferential artery*, a *testicular artery*, the **pampiniform** (pam-PIN-i-form; *pampinus*, tendril + *forma*, form) **plexus** of a testicular vein, and branches of the *genitofemoral nerve* from the lumbar plexus. Each spermatic cord begins at the *deep inguinal ring*, the entrance to the *inguinal canal* (a passageway through the abdominal musculature). ∞ *[p. 336]* After passing through the inguinal canal, the spermatic cord exits at the *superficial inguinal ring* and descends into the scrotum (Figure 28-2●).

The inguinal canals form during development as the testes descend into the scrotum; at that time, these canals link the scrotal cavities with the peritoneal cavity. In normal adult males, the inguinal canals are closed, but the presence of the spermatic cords creates weak points in the abdominal wall that remain throughout life. As a result, *inguinal hernias*, discussed in Chapter 11, are relatively common in males. ∞ *[p. 336]* The inguinal canals in females are very small, containing only the *ilioinguinal nerves* and the *round ligaments* of the uterus. The abdominal wall is nearly intact, and inguinal hernias in women are very rare.

Descent of the Testes

During development, the testes form inside the body cavity adjacent to the kidneys (Figure 28-3●, p. 1041). A bundle of connective tissue fibers extends from each testis to the posterior wall of a small, anterior and

inferior pocket of the peritoneum. These fibers constitute the **gubernaculum testis**. As growth proceeds, the gubernacula do not get any longer, and they lock the testes in position. As a result, the relative position of each testis changes as the rest of the body enlarges; the testis gradually moves inferiorly and anteriorly toward the anterior abdominal wall (Figure 28-3a●). During the seventh developmental month, fetal growth continues at a rapid pace, and circulating hormones stimulate a contraction of the gubernaculum testis. During this period, each testis moves through the abdominal musculature, accompanied by small pockets of the peritoneal cavity. This process is called the **descent of the testes**.

As it moves through the body wall, each testis is accompanied by the ductus deferens and the testicular blood vessels, nerves, and lymphatics (Figure 28-3b●). Together, these structures form the body of the spermatic cord. In **cryptorchidism** (kript-OR-ki-dizm; *crypto*, hidden), one or both of the testes have not descended into the scrotum by the time of birth. Typically, the cryptorchid testes are lodged in the abdominal cavity or within the inguinal canal. Cryptorchidism occurs in about 3 percent of full-term deliveries and in roughly 30 percent of premature births. In most instances, normal descent occurs a few weeks later, but the condition can be surgically corrected if it persists. Corrective measures should be taken before puberty, because cryptorchid (abdominal) testes will not produce sperm, and the individual will be sterile.

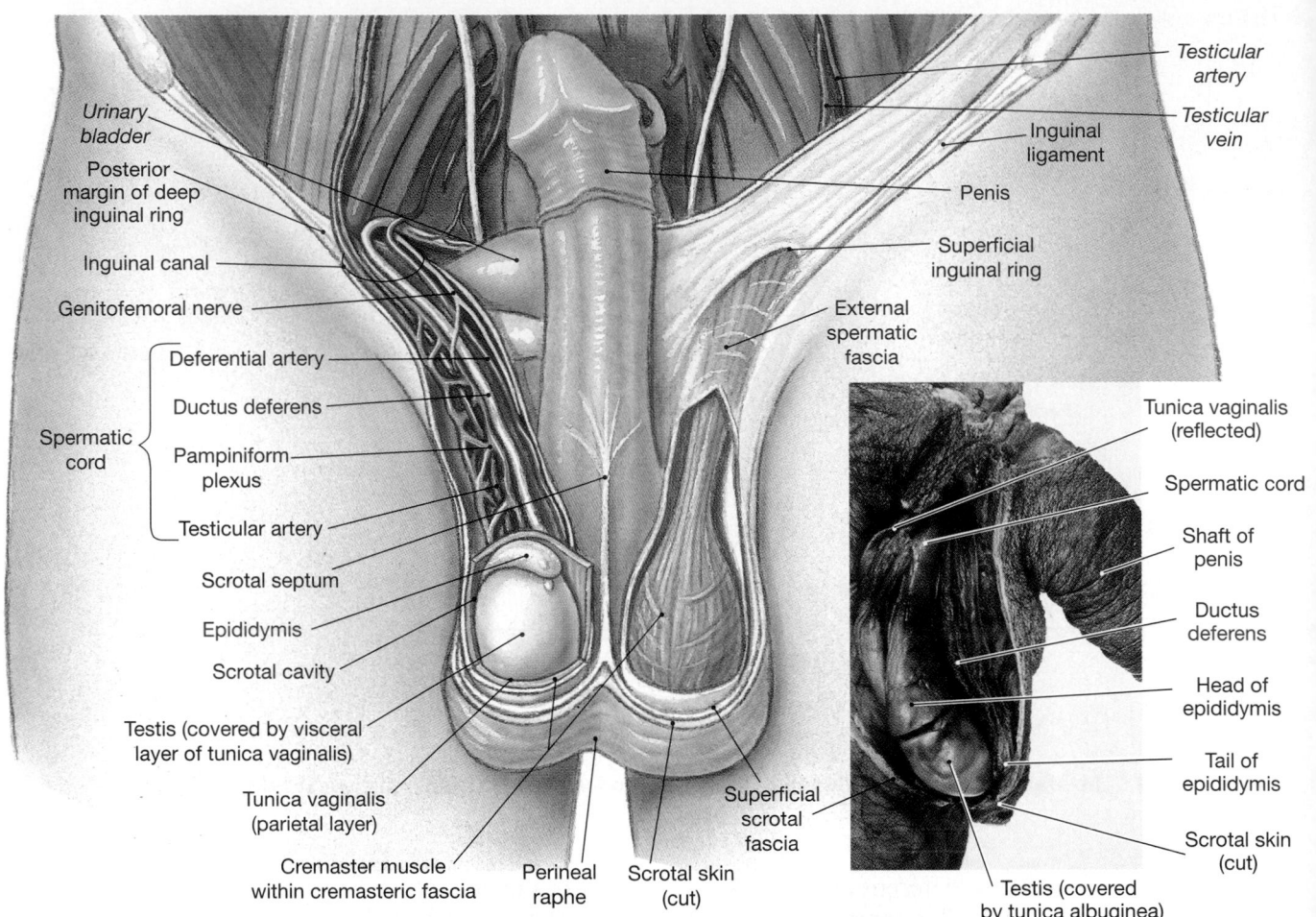

●**FIGURE 28-2** **The Male Reproductive System in Frontal View.** In the cadaver photograph, the testis has been moved posteriorly to expose the ductus deferens.

In most cases, if the testes cannot be moved into the scrotum, they will be removed, because about 10 percent of males with uncorrected cryptorchid testes eventually develop testicular cancer. This surgical procedure is called a *bilateral orchiectomy* (or-kē-EK-to-mē; *orchis*, testis).

The Scrotum and the Position of the Testes

The scrotum is divided internally into two chambers. The partition between the two is marked by a raised thickening in the scrotal surface known as the **perineal raphe** (RĀ-fē) (Figures 28-2 and 28-4a●, p. 1042). Each testis lies in a separate compartment, or **scrotal cavity**. Because the scrotal cavities are separated by a partition, infection or inflammation of one testis does not ordinarily spread to the other. A narrow space separates the inner surface of the scrotum from the outer surface of the testis. The **tunica vaginalis** (TOO-ni-ka vaj-i-NAL-is), a serous membrane, lines the scrotal cavity and reduces friction between the opposing parietal (scrotal) and visceral (testicular) surfaces. The tunica vaginalis is an isolated portion of the peritoneum that lost its connection with the peritoneal cavity after the testes descended, when the inguinal canal closed.

The scrotum consists of a thin layer of skin and the underlying superficial fascia. The dermis contains a layer of smooth muscle, the **dartos** (DAR-tōs). Resting muscle tone in the dartos causes the characteristic wrinkling of the scrotal surface. A layer of skeletal muscle, the **cremaster** (krē-MAS-ter) **muscle**, lies deep to the dermis. Contraction of the cremaster during sexual arousal, or in response to changes in temperature, tenses the scrotum and pulls the testes closer to the body. Normal sperm development in the testes requires temperatures about 1.1°C (2°F) lower than those elsewhere in the body. The cremaster relaxes or contracts to move the testes away from or toward the body as needed to maintain acceptable testicular temperatures. When air or body temperature rises, the cremaster relaxes and the testes move away from the body. Cooling of the scrotum, as occurs when a man enters a cold swimming pool, results in cremasteric contractions that pull the testes closer to the body and keep testicular temperatures from falling.

The scrotum is richly supplied with sensory and motor nerves from the **hypogastric plexus** and branches of the **ilioinguinal nerves,** the **genitofemoral nerves,**

•*FIGURE 28-3* **Descent of the Testes during Development.** (a) Diagrammatic sagittal sectional views of the positional changes involved in the descent of the testes. The size of the gubernaculum testis remains constant (see the scale bar at the left) while the rest of the fetus grows. This distinction is responsible for a shift in the relative position of the testis. (b) Frontal views during the descent of the testes and the formation of the spermatic cords.

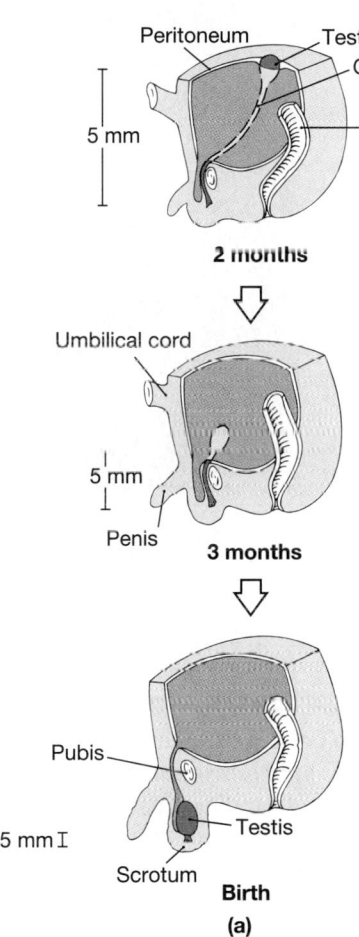

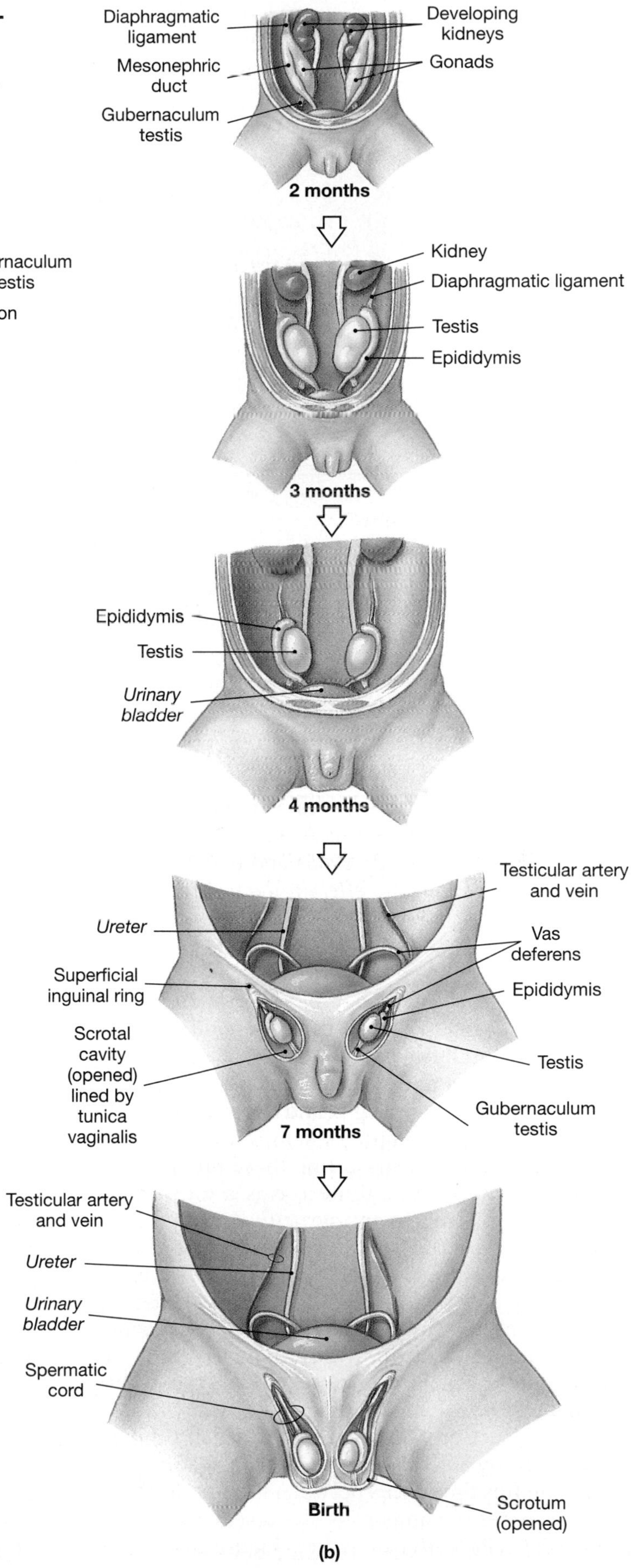

and the **pudendal nerves.** ∞ *[pp. 430, 528]* The vascular supply to the scrotum includes the **internal pudendal arteries** (from the internal iliac arteries), the **external pudendal arteries** (from the femoral arteries), and the **cremasteric branch** of the **inferior epigastric arteries** (from the external iliac arteries). ∞ *[pp. 747–749]* The names and distributions of the veins follow those of the arteries.

Structure of the Testes

Deep to the tunica vaginalis covering the testis lies the **tunica albuginea** (al-bū-JIN-ē-uh), a dense layer of connective tissue rich in collagen fibers. The fibers of this network are continuous with those surrounding the adjacent epididymis. The collagen fibers of the tunica albuginea also extend into the substance of the testis, forming fibrous partitions, or *septa* (Figure 28-4•). These

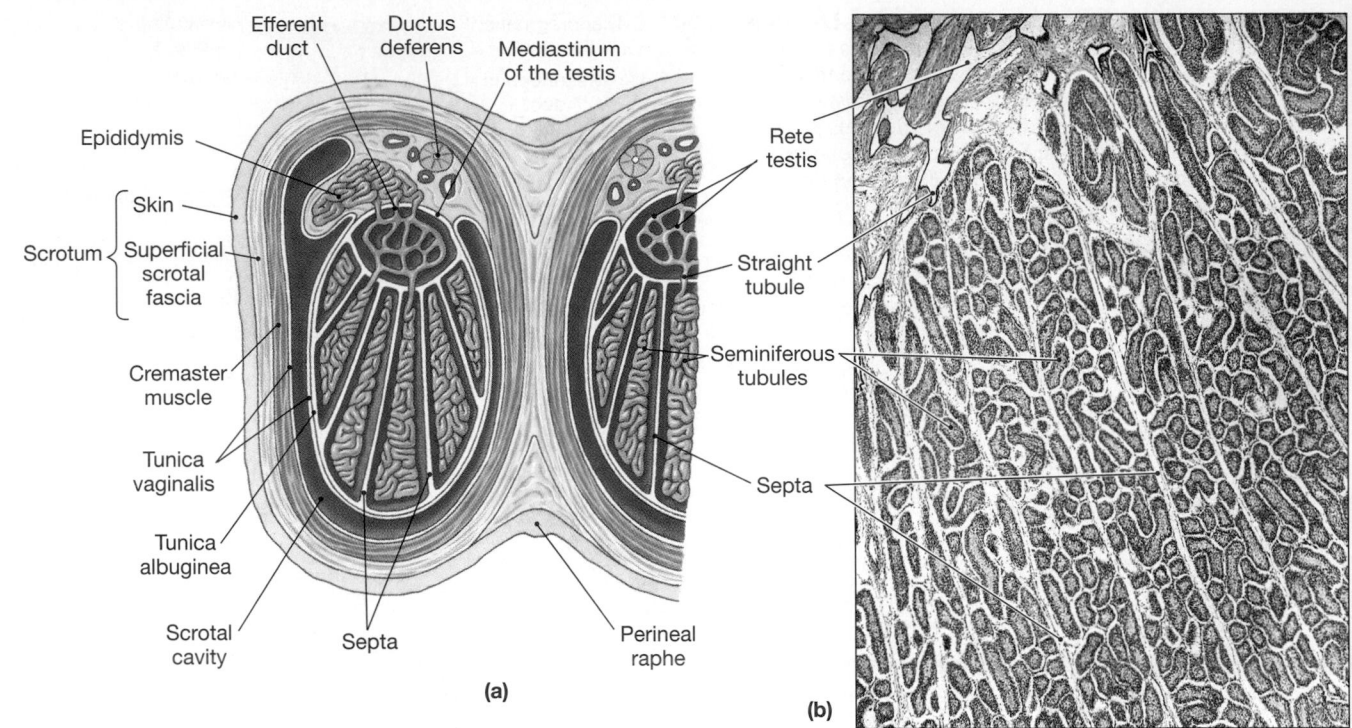

•FIGURE 28-4 **Structure of the Testes.** **(a)** Diagrammatic sketch (frontal section) and anatomical relationships of the testes. **(b)** Micrograph of a section through a testis. (LM × 26)

septa converge toward the area closest to the entrance of the epididymis. This region, located at the superior end of the testis, is called the **mediastinum** of the testis (or *mediastinum testis*). The connective tissues in this region support the blood vessels and lymphatics that supply the testis and the *efferent ducts*, which transport sperm to the epididymis.

Histology of the Testes

The septa subdivide the testis into a series of **lobules**. Roughly 800 slender, tightly coiled **seminiferous** (se-mi-NIF-e-rus) **tubules** are distributed among the lobules (Figures 28-4 and 28-5•). Each tubule averages about 80 cm (31 in.) in length, and a typical testis contains nearly one-half mile of seminiferous tubules. Sperm production occurs within these tubules.

Each seminiferous tubule forms a loop that is attached to a **straight tubule** *(tubuli recti)* at the mediastinum of the testis. The straight tubule is connected to a maze of passageways known as the **rete** (RĒ-tē; *rete*, a net) **testis** (Figure 28-4•). Fifteen to twenty large **efferent ducts** (or *efferent ductules*) connect the rete testis to the epididymis.

Because the seminiferous tubules are tightly coiled, most histological preparations show them in transverse section (Figure 28-5a•). Each tubule is surrounded by a delicate capsule, and loose connective tissue fills the spaces between the tubules. Within those spaces are numerous blood vessels and large **interstitial cells** *(cells of Leydig)* (Figure 28-5b•). In-

terstitial cells are responsible for the production of *androgens*, the dominant sex hormones in males. *Testosterone* is the most important androgen.

Sperm cells, or spermatozoa, are produced by the process of **spermatogenesis** (sper-ma-tō-JEN-e-sis). Spermatogenesis begins at the outermost layer of cells in the seminiferous tubules and proceeds toward the tubular lumen (Figure 28-5b,c•). Stem cells called **spermatogonia** (sper-ma-tō-GŌ-nē-uh) divide by mitosis to produce generations of daughter cells, some of which differentiate into **spermatocytes** (sper-MA-tō-sīts). Through *meiosis*, a specialized form of cell division involved only in the production of gametes (sperm in males, ova in females), spermatocytes give rise to **spermatids** (SPER-ma-tidz).

At each step in this process, the daughter cells move closer to the tubular lumen. The spermatids subsequently differentiate into spermatozoa. This differentiation process, called *spermiogenesis*, ends as the physically mature spermatozoa lose contact with the wall of the seminiferous tubule and enter the fluid in the lumen. Spermiogenesis is the last step in spermatogenesis.

Each seminiferous tubule contains spermatogonia, spermatocytes at various stages of meiosis, spermatids, spermatozoa, and large **sustentacular** (sus-ten-TAK-ū-lar) **cells** (or *Sertoli cells*). Sustentacular cells are attached to the tubular capsule and extend toward the lumen between the other cell types (Figure 28-5b,c•).

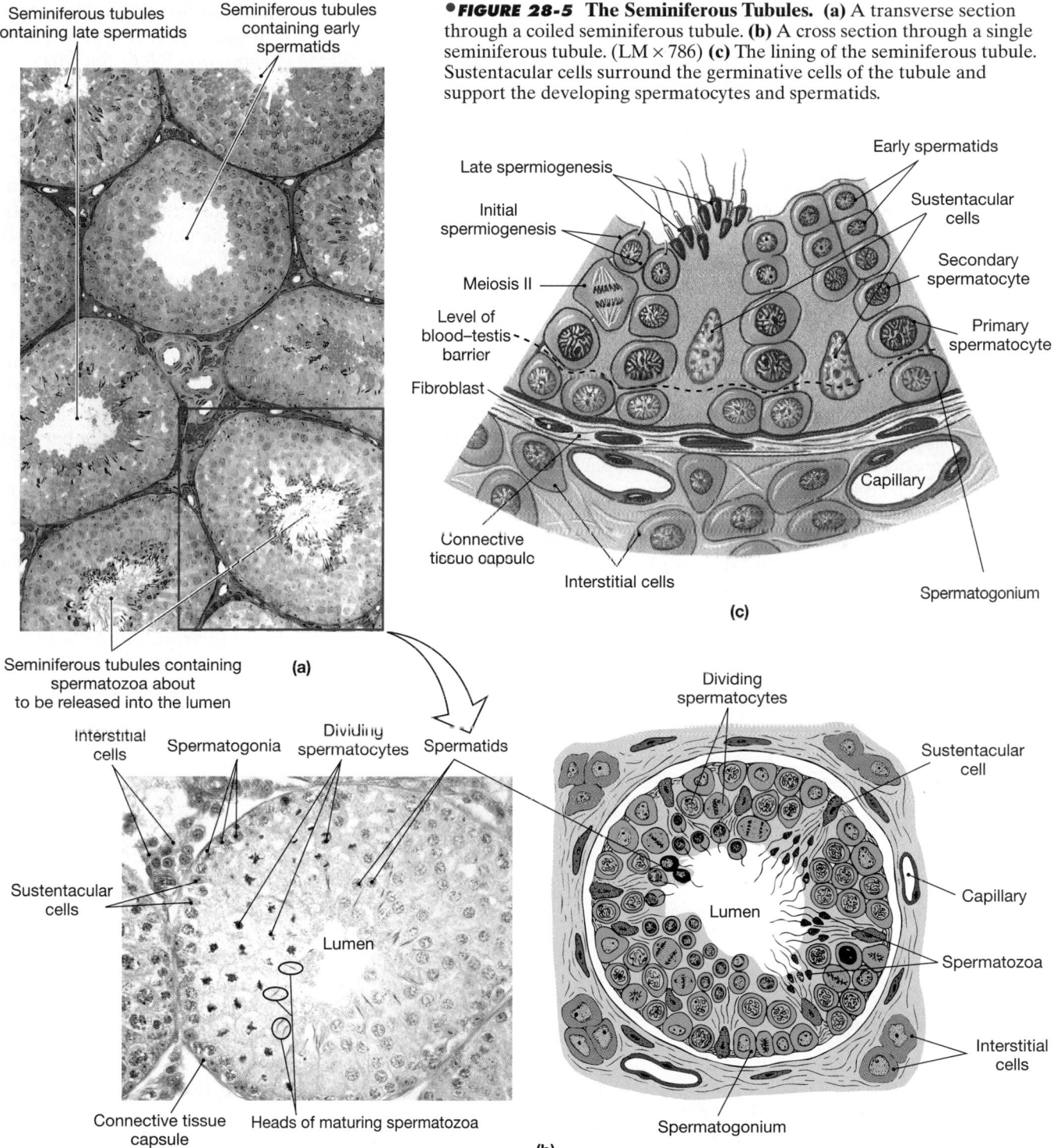

•FIGURE 28-5 **The Seminiferous Tubules.** **(a)** A transverse section through a coiled seminiferous tubule. **(b)** A cross section through a single seminiferous tubule. (LM × 786) **(c)** The lining of the seminiferous tubule. Sustentacular cells surround the germinative cells of the tubule and support the developing spermatocytes and spermatids.

Spermatogenesis

Spermatogenesis involves three integrated processes:

1. **Mitosis.** Spermatogonia undergo cell divisions throughout adult life. (You may wish to review the description of mitosis and cell division in Chapter 3. ∞ *[pp. 99–102]*) The cell divisions produce daughter cells that are pushed toward the lumen of the tubule. These cells differentiate into spermatocytes that prepare to begin meiosis.

2. **Meiosis.** Meiosis (mī-Ō-sis) is a special form of cell division involved in gamete production. Gametes contain half the normal chromosome complement. As a result, the fusion of the nuclei of a sperm and an ovum produces a cell that has the normal number of chromosomes (46) rather than twice that number. In the seminiferous tubules, the meiotic divisions of spermatocytes produce spermatids, or undifferentiated male gametes.

3. Spermiogenesis. Spermatids are small, relatively unspecialized cells. In the process of **spermiogenesis**, spermatids differentiate into physically mature spermatozoa. Spermatozoa are among the most highly specialized cells in the body. Spermiogenesis involves major changes in a spermatid's internal and external structure.

MITOSIS AND MEIOSIS Mitosis and meiosis differ significantly in terms of the events that take place in the nucleus. Mitosis is part of the process of somatic cell division, which produces two daughter cells, each containing 23 *pairs* of chromosomes (Figure 28-6a●). Each pair consists of one chromosome provided by the father and another by the mother at the time of fertilization. Because the daughter cells contain both members of each chromosome pair (for a total of 46 chromosomes), they are called **diploid** (DIP-loyd; *diplo*, double) cells. Meiosis involves two cycles of cell division (*meiosis I* and *meiosis II*) and produces four cells, each of which contains 23 *individual chromosomes* (Figure 28-6b●). Because these cells contain only one member of each pair of chromosomes, they are called **haploid** (HAP-loyd; *haplo*, single) cells. The events in the nucleus shown in Figure 28-6b● are the same whether you consider the formation of sperm or ova.

As a cell prepares to begin meiosis, DNA replication occurs within the nucleus as if the cell were about to undergo mitosis. As prophase of the first meiotic division, **meiosis I**, arrives, the chromosomes condense and become visible. As in mitosis, each chromosome consists of two duplicate *chromatids* (KRŌ-ma-tidz).

The corresponding maternal and paternal chromosomes now come together, an event known as **synapsis** (sin-AP-sis). Synapsis involves 23 pairs of chromosomes; each member of each pair consists of two chromatids. A matched set of four chromatids is called a **tetrad** (TET-rad; *tetras*, four). Some exchange of genetic material can occur between the chromatids of a chromosome pair at this stage of meiosis. This exchange, called *crossing-over*, increases genetic variation among offspring; we shall discuss it in Chapter 29.

The nuclear envelope disappears at the end of prophase I. As metaphase I begins, the tetrads line up along the metaphase plate. As anaphase I begins, the tetrads break up, and the maternal and paternal chromosomes separate. This is a major difference between mitosis and meiosis: In mitosis, each daughter cell receives one of the two copies of every chromosome, maternal and paternal, but in meiosis each daughter cell receives both copies of *either* the maternal chromosome *or* the paternal chromosome from each tetrad (compare Figure 28-6a and 28-6b●).

As anaphase proceeds, the maternal and paternal components are randomly distributed. As a result, telophase I ends with the formation of two daughter cells containing unique combinations of maternal and

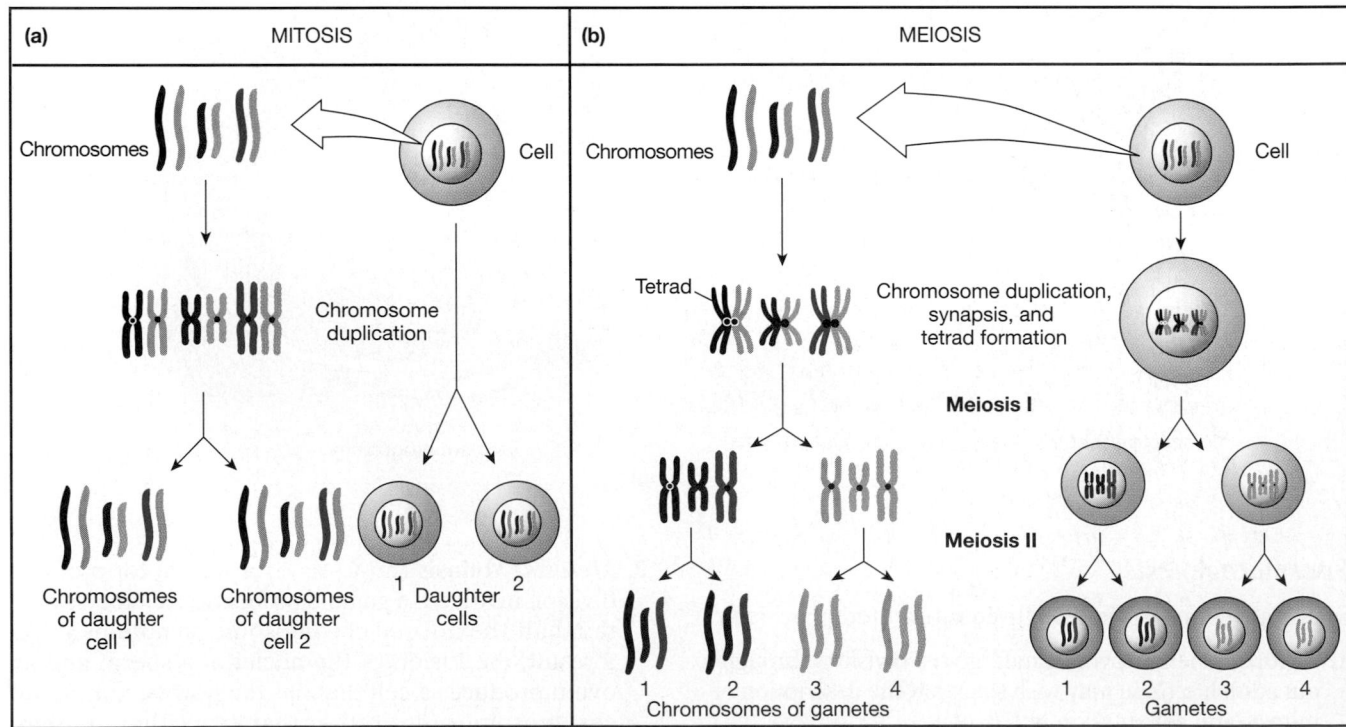

●**FIGURE 28-6** **Chromosomal Events during Mitosis and Meiosis.** **(a)** Steps in mitosis. (See Figure 3-27, pp. 100–101.) **(b)** Steps in meiosis.

paternal chromosomes. Both cells contain 23 chromosomes. Because the first meiotic division reduces the number of chromosomes from 46 to 23, it is called a **reductional division**. Each of these chromosomes still consists of two duplicate chromatids. These duplicates will separate during **meiosis II**.

The interphase separating meiosis I and meiosis II is very brief, and there is no DNA replication over this period. The cell then proceeds through prophase II, metaphase II, and anaphase II. During anaphase II, the duplicate chromatids separate. Telophase II thus yields *four cells*, each containing 23 chromosomes. Because the number of chromosomes has not changed, meiosis II represents an **equational division**.

We shall now consider meiosis and the production of spermatozoa. In a later section, we shall deal with *oogenesis* (the production of ova).

The mitotic divisions of spermatogonia produce **primary spermatocytes** (Figure 28-7 •). As meiosis begins, each primary spermatocyte contains 46 individual chromosomes. At the end of meiosis I, the daughter cells are called **secondary spermatocytes**. Every secondary spermatocyte contains 23 chromosomes, each of which consists of a pair of duplicate chromatids. The secondary spermatocytes soon enter prophase II. The completion of metaphase II, anaphase II, and telophase II yields *four spermatids*, each containing 23 chromosomes.

For each primary spermatocyte that enters meiosis, four spermatids are produced. Because cytokinesis (cytoplasmic division) is not completed in meiosis I or meiosis II, the four spermatids initially remain interconnected by cytoplasmic bridges. These connections assist in the transfer of nutrients and hormonal messages between the cells, thus helping ensure that the cells develop in synchrony. The interconnections are not broken until the last stages of physical maturation.

SPERMIOGENESIS Each spermatid matures into a single **spermatozoon** (sper-ma-tō-ZŌ-on), or sperm cell. This maturation process is called spermiogenesis (Figure 28-7 •). Developing spermatocytes undergoing meiosis and spermatids undergoing spermiogenesis are not free in the seminiferous tubules. Instead, they are surrounded by the cytoplasm of the sustentacular cells. As spermiogenesis proceeds, the spermatids gradually develop the appearance of mature spermatozoa. At *spermiation*, a spermatozoon loses its attachment to the sustentacular cell and enters the lumen of the seminiferous tubule. The entire process, from spermatogonial division to spermiation, takes approximately 9 weeks.

SPERMATOGENESIS AND SUSTENTACULAR CELLS
Sustentacular cells play a key role in the process of spermatogenesis. These cells have six important functions that directly or indirectly affect mitosis, meiosis, and spermiogenesis within the seminiferous tubules:

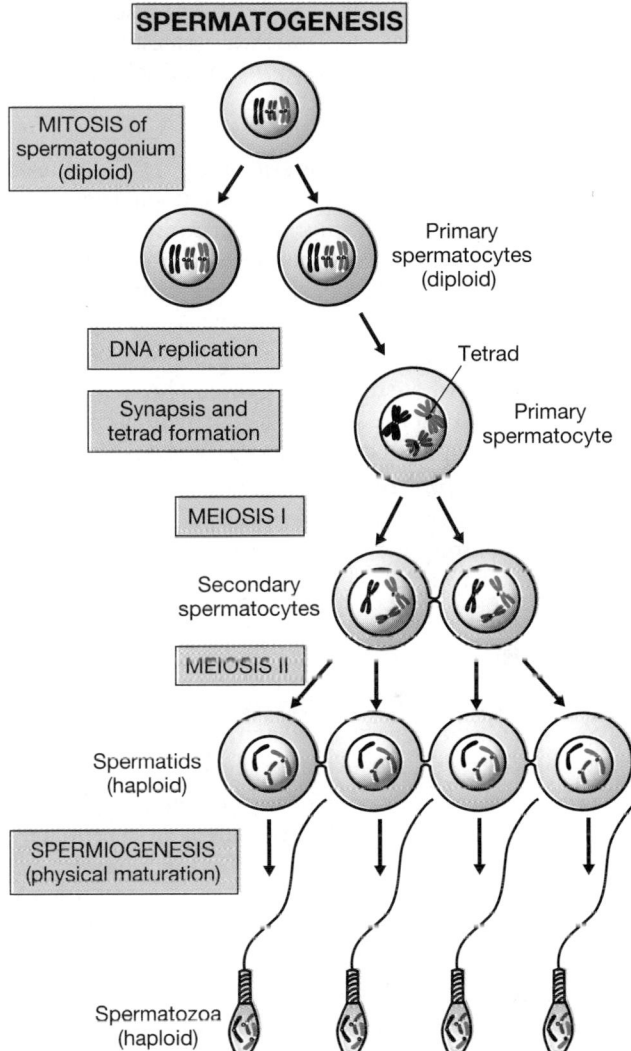

•FIGURE 28-7 **Spermatogenesis.** Schematic diagram of meiosis in the seminiferous tubules, showing the distribution of only a few chromosomes.

1. *Maintenance of the blood–testis barrier.* The seminiferous tubules are isolated from the general circulation by a **blood–testis barrier** comparable in function to the blood–brain barrier. Sustentacular cells are joined by tight junctions, forming a layer that divides the seminiferous tubule into an outer *basal compartment* that contains the spermatogonia and an inner *lumenal compartment* (or *adlumenal compartment*) where meiosis and spermiogenesis occur. Transport across the sustentacular cells is tightly regulated so that conditions in the lumenal compartment remain very stable. The fluid within the lumen of a seminiferous tubule is produced by sustentacular cells, which also regulate the fluid's composition. Tubular fluid is very different from the surrounding interstitial fluid. For example, tubular fluid is high in androgens, estrogens, potassium, and amino acids. The blood–testis barrier is essential to preserving the differences between tubular fluid and interstitial fluid. In addition, devel-

oping spermatozoa contain sperm-specific antigens in their cell membranes. These antigens, not found in somatic cell membranes, would be attacked by the immune system if the blood–testis barrier did not prevent their being detected.

2. *Support of mitosis and meiosis.* Spermatogenesis depends on the stimulation of sustentacular cells by circulating follicle-stimulating hormone (FSH) and testosterone. Stimulated sustentacular cells then in some way promote the division of spermatogonia and the meiotic divisions of spermatocytes.

3. *Support of spermiogenesis.* Spermiogenesis requires the presence of sustentacular cells. These cells surround and enfold the spermatids, providing nutrients and chemical stimuli that promote their development.

4. *Secretion of inhibin.* Sustentacular cells secrete *inhibin* (in-HIB-in), a peptide hormone, in response to factors released by developing sperm. Inhibin, introduced in Chapter 18, depresses the pituitary production of FSH and perhaps the hypothalamic secretion of gonadotropin-releasing hormone (GnRH). ∞ *[p. 604]* The faster the rate of sperm production, the greater the amount of inhibin secreted. By regulating FSH and GnRH secretion, sustentacular cells provide feedback control of spermatogenesis.

5. *Secretion of androgen-binding protein.* Androgen-binding protein (ABP) binds androgens (primarily testosterone) in the fluid contents of the seminiferous tubules. This protein is thought to be important in elevating the concentration of androgens within the tubules and stimulating spermiogenesis. The production of ABP is stimulated by FSH.

6. *Secretion of Müllerian-inhibiting factor.* Müllerian-inhibiting factor (MIF) is secreted by sustentacular cells in the developing testes. This hormone causes regression of the fetal *Müllerian ducts*, passageways that in females participate in the formation of the uterine tubes and the uterus. Inadequate MIF production during development leads to retention of these ducts and failure of the testes to descend into the scrotum.

TESTICULAR CANCER Testicular cancer occurs at a relatively low rate of about 3 cases per 100,000 males per year. Although there are only about 7200 new cases each year in the United States, testicular cancer is the most common cancer among males age 15–35. The incidence among Caucasian males has more than doubled since the 1930s, but the incidence among African Americans has remained unchanged. The reason for this difference is not known.

More than 95 percent of testicular cancers are the result of abnormal spermatogonia or spermatocytes rather than abnormal sustentacular cells, interstitial cells, or other cells of the testes. Treatment generally consists of a combination of orchiectomy and chemotherapy. The survival rate has increased from about 10 percent in 1970 to about 95 percent in 1997, primarily as a result of earlier diagnosis and improved treatment protocols.

Anatomy of a Spermatozoon

Each spermatozoon has three distinct regions: (1) the *head*, (2) the *middle piece*, and (3) the *tail* (Figure 28-8•). The **head** is a flattened ellipse containing a nucleus with densely packed chromosomes. At the tip of the head is the **acrosomal** (ak-rō-SŌ-mal) **cap**, a membranous compartment containing enzymes essential to the process of fertilization. During spermiogenesis, saccules of the Golgi apparatus fuse and flatten into an *acrosomal vesicle* that ultimately forms the acrosomal cap.

A short **neck** attaches the head to the **middle piece**. The neck contains both centrioles of the original spermatid. The microtubules of the distal centriole are continuous with those of the middle piece and tail. Mitochondria in the middle piece are arranged in a spiral around the microtubules. Mitochondrial activity provides the ATP that is needed to move the tail.

The **tail** is the only flagellum in the human body. A *flagellum*, an organelle introduced in Chapter 3, moves a cell from one place to another. ∞ *[p. 85]* Whereas cilia beat in a predictable, waving fashion, the flagellum of a spermatozoon has a complex, corkscrew motion.

Unlike other, less specialized cells, a mature spermatozoon lacks an endoplasmic reticulum, Golgi apparatus, lysosomes, peroxisomes, inclusions, and many other intracellular structures. Because the cell does not contain glycogen or other energy reserves, it must absorb nutrients (primarily fructose) from the surrounding fluid.

The Male Reproductive Tract

The testes produce physically mature spermatozoa that are, as yet, incapable of successful fertilization. The other portions of the male reproductive system are responsible for the functional maturation, nourishment, storage, and transport of spermatozoa.

The Epididymis

Late in their development, spermatozoa detach from the sustentacular cells and lie within the lumen of the seminiferous tubule. They have most of the physical characteristics of mature sperm cells but are functionally immature and incapable of coordinated locomotion or fertilization. Fluid currents then transport the cell along the straight tubule, through the rete testis (Figures 28-4a, p. 1042, and 28-9a•, p. 1048), and into the epididymis.

The epididymis lies along the posterior border of the testis (Figures 28-1, p. 1039, 28-4, p. 1042, and 28-9a•, p. 1048). It is firm and can be felt through the skin of the scrotum. The epididymis consists of a tubule almost 7 meters (23 ft) long, coiled and twisted so as to take up very little space. It has (1) a *head*, (2) a *body*, and (3) a *tail*. The superior **head** is the portion of the epididymis proximal to the mediastinum of the testis. The head receives spermatozoa from the efferent ducts that connect the rete testis to the epididymis.

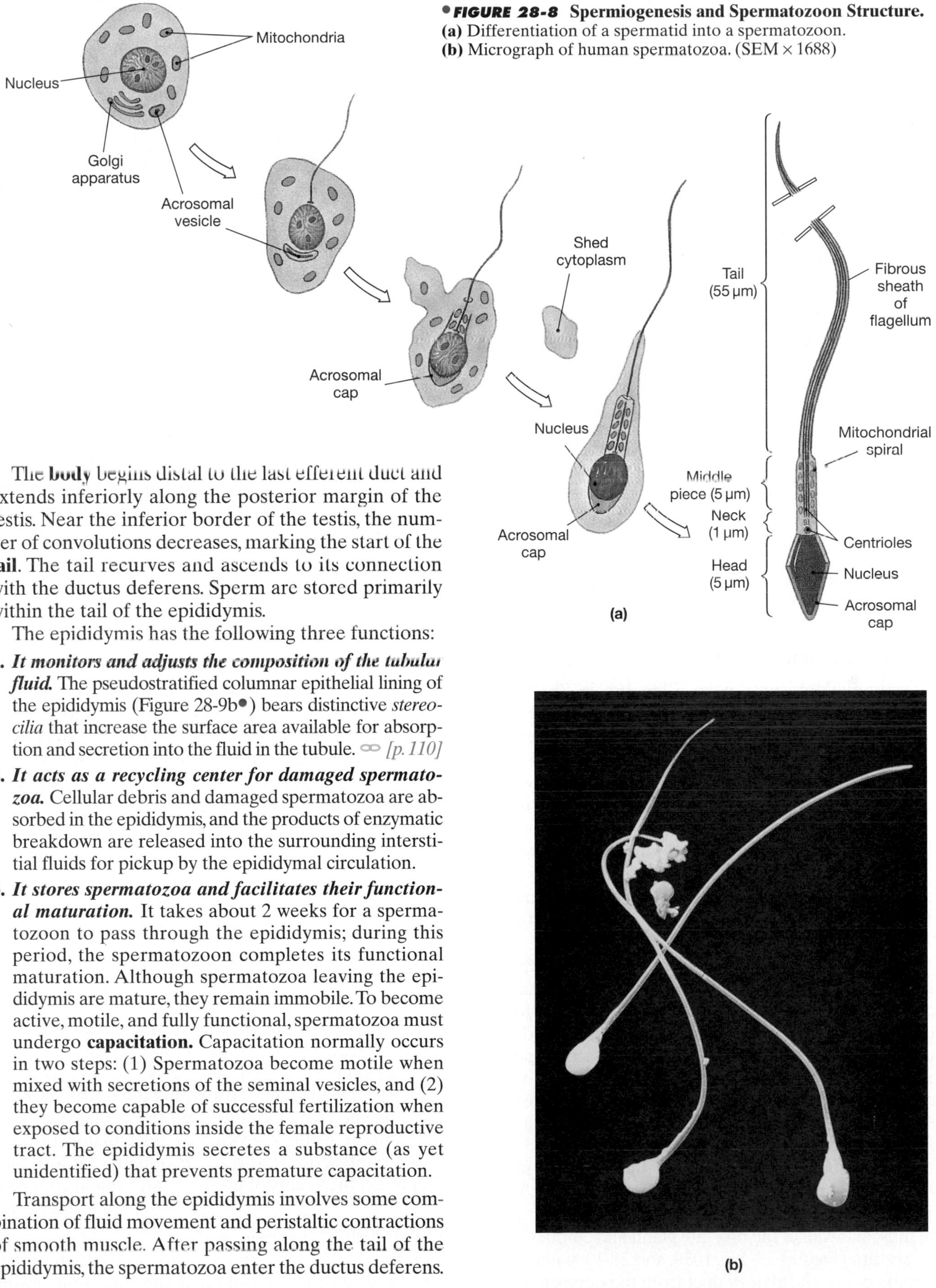

• FIGURE 28-8 Spermiogenesis and Spermatozoon Structure.
(a) Differentiation of a spermatid into a spermatozoon.
(b) Micrograph of human spermatozoa. (SEM × 1688)

Mitochondria

Nucleus

Golgi apparatus

Acrosomal vesicle

Acrosomal cap

Shed cytoplasm

Nucleus

Acrosomal cap

Tail (55 μm)

Middle piece (5 μm)

Neck (1 μm)

Head (5 μm)

Fibrous sheath of flagellum

Mitochondrial spiral

Centrioles

Nucleus

Acrosomal cap

(a)

The **body** begins distal to the last efferent duct and extends inferiorly along the posterior margin of the testis. Near the inferior border of the testis, the number of convolutions decreases, marking the start of the **tail**. The tail recurves and ascends to its connection with the ductus deferens. Sperm are stored primarily within the tail of the epididymis.

The epididymis has the following three functions:

1. *It monitors and adjusts the composition of the tubular fluid.* The pseudostratified columnar epithelial lining of the epididymis (Figure 28-9b•) bears distinctive *stereocilia* that increase the surface area available for absorption and secretion into the fluid in the tubule. ∞ *[p. 110]*

2. *It acts as a recycling center for damaged spermatozoa.* Cellular debris and damaged spermatozoa are absorbed in the epididymis, and the products of enzymatic breakdown are released into the surrounding interstitial fluids for pickup by the epididymal circulation.

3. *It stores spermatozoa and facilitates their functional maturation.* It takes about 2 weeks for a spermatozoon to pass through the epididymis; during this period, the spermatozoon completes its functional maturation. Although spermatozoa leaving the epididymis are mature, they remain immobile. To become active, motile, and fully functional, spermatozoa must undergo **capacitation.** Capacitation normally occurs in two steps: (1) Spermatozoa become motile when mixed with secretions of the seminal vesicles, and (2) they become capable of successful fertilization when exposed to conditions inside the female reproductive tract. The epididymis secretes a substance (as yet unidentified) that prevents premature capacitation.

Transport along the epididymis involves some combination of fluid movement and peristaltic contractions of smooth muscle. After passing along the tail of the epididymis, the spermatozoa enter the ductus deferens.

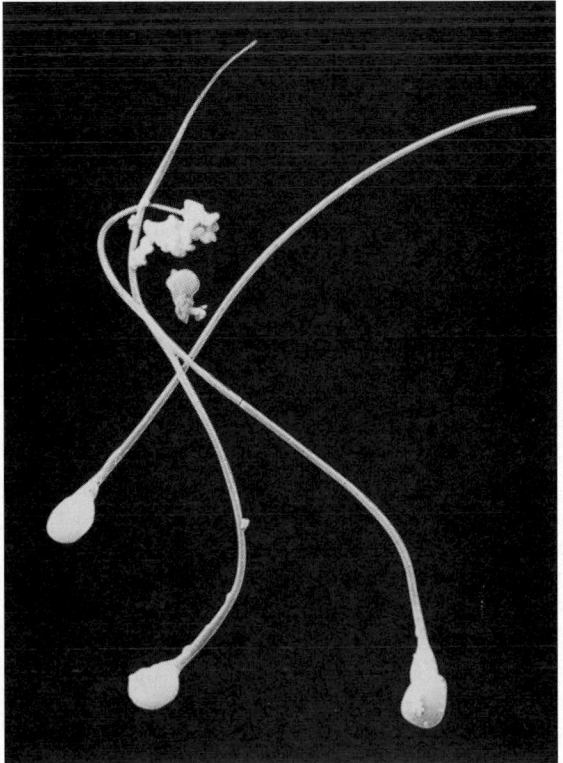

(b)

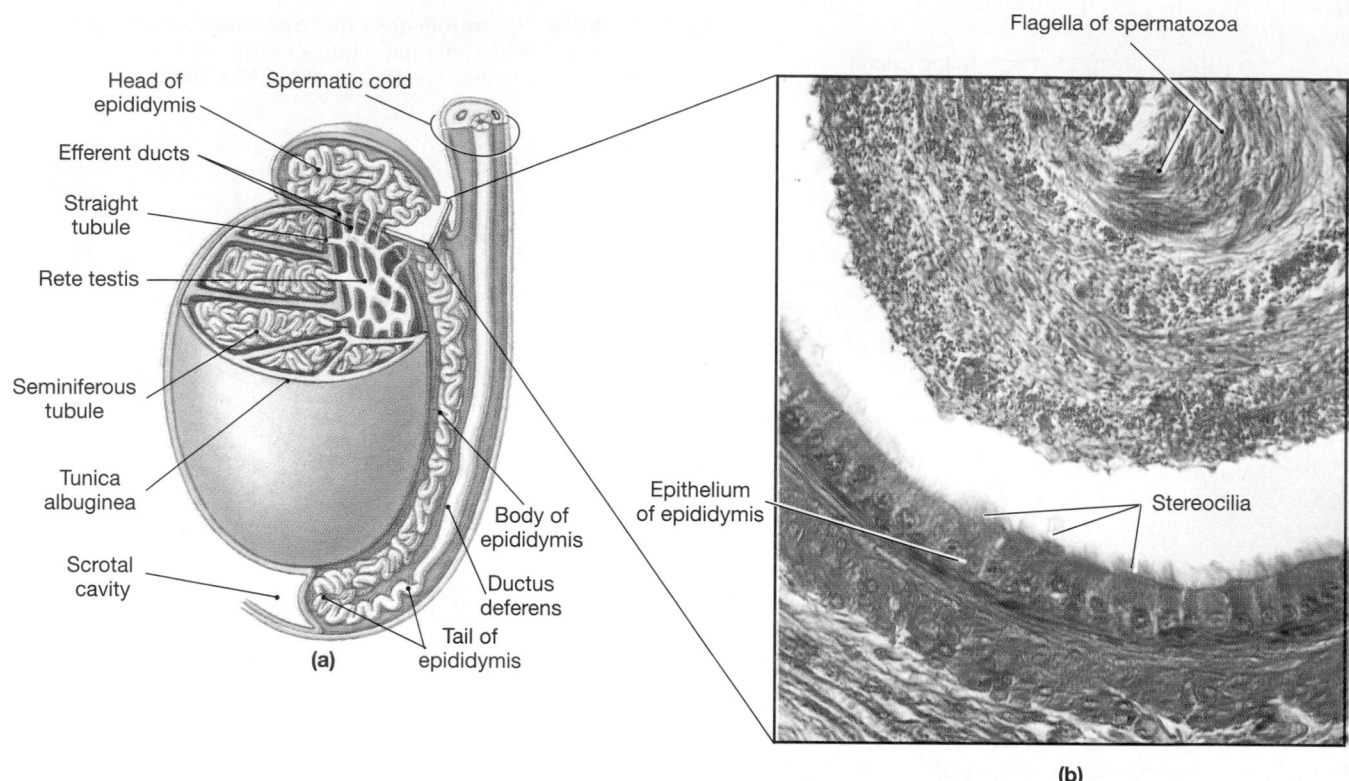

●**FIGURE 28-9** **The Epididymis.** (a) Diagrammatic view of the epididymis on gross dissection. (b) A micrograph showing epithelial features, especially the elongate stereocilia characteristic of the epididymis. (LM × 1304)

The Ductus Deferens

Each ductus deferens, or *vas deferens*, is 40–45 cm (16–18 in.) long. It begins at the tail of the epididymis (Figure 28-9a ●) and, as part of the spermatic cord, ascends through the inguinal canal (Figures 28-1 and 28-2 ●, pp. 1039, 1040). Inside the abdominal cavity, the ductus deferens passes posteriorly, curving inferiorly along the lateral surface of the urinary bladder toward the superior and posterior margin of the prostate gland. Just before the ductus deferens reaches the prostate gland and seminal vesicles, its lumen enlarges. This expanded portion is known as the **ampulla** (am-PŪL-uh) (Figure 28-10a ●).

The wall of the ductus deferens contains a thick layer of smooth muscle (Figure 28-10b ●). Peristaltic contractions in this layer propel spermatozoa and fluid along the duct, which is lined by a pseudostratified ciliated columnar epithelium. In addition to transporting sperm, the ductus deferens can store spermatozoa for several months. During this time, the spermatozoa remain in a state of suspended animation and have low metabolic rates.

The junction of the ampulla with the duct of the seminal vesicle marks the start of the ejaculatory duct. This short passageway (2 cm, or less than 1 in.) penetrates the muscular wall of the prostate gland and empties into the urethra (Figures 28-1, p. 1039, and 28-10 ●) near the opening of the ejaculatory duct from the opposite side.

The Urethra

The urethra of the male extends from the urinary bladder to the tip of the penis, a distance of 18–20 cm (7–8 in.). It is divided into *prostatic, membranous,* and *penile regions.* We considered those subdivisions in Chapter 26. ∞ *[p. 996]* In males, the urethra is a passageway used by both the urinary and the reproductive systems.

The Accessory Glands

The fluids contributed by the seminiferous tubules and the epididymis account for only about 5 percent of the volume of semen. The fluid component of semen is a mixture of the secretions of many different glands, each with distinctive biochemical characteristics. Important glands include the *seminal vesicles,* the *prostate gland,* and the *bulbourethral glands* (Figures 28-1, p. 1039, and 28-10a,c,d,e ●). Major functions of these glandular organs, which occur only in males, include (1) activating the spermatozoa; (2) providing the nutrients spermatozoa need for motility; (3) propelling spermatozoa and fluids along the reproductive tract, mainly by peristaltic contractions; and (4) producing buffers that counteract the acidity of urethral and vaginal contents.

The Seminal Vesicles

The ductus deferens on each side ends at the junction between the ampulla and the duct that drains

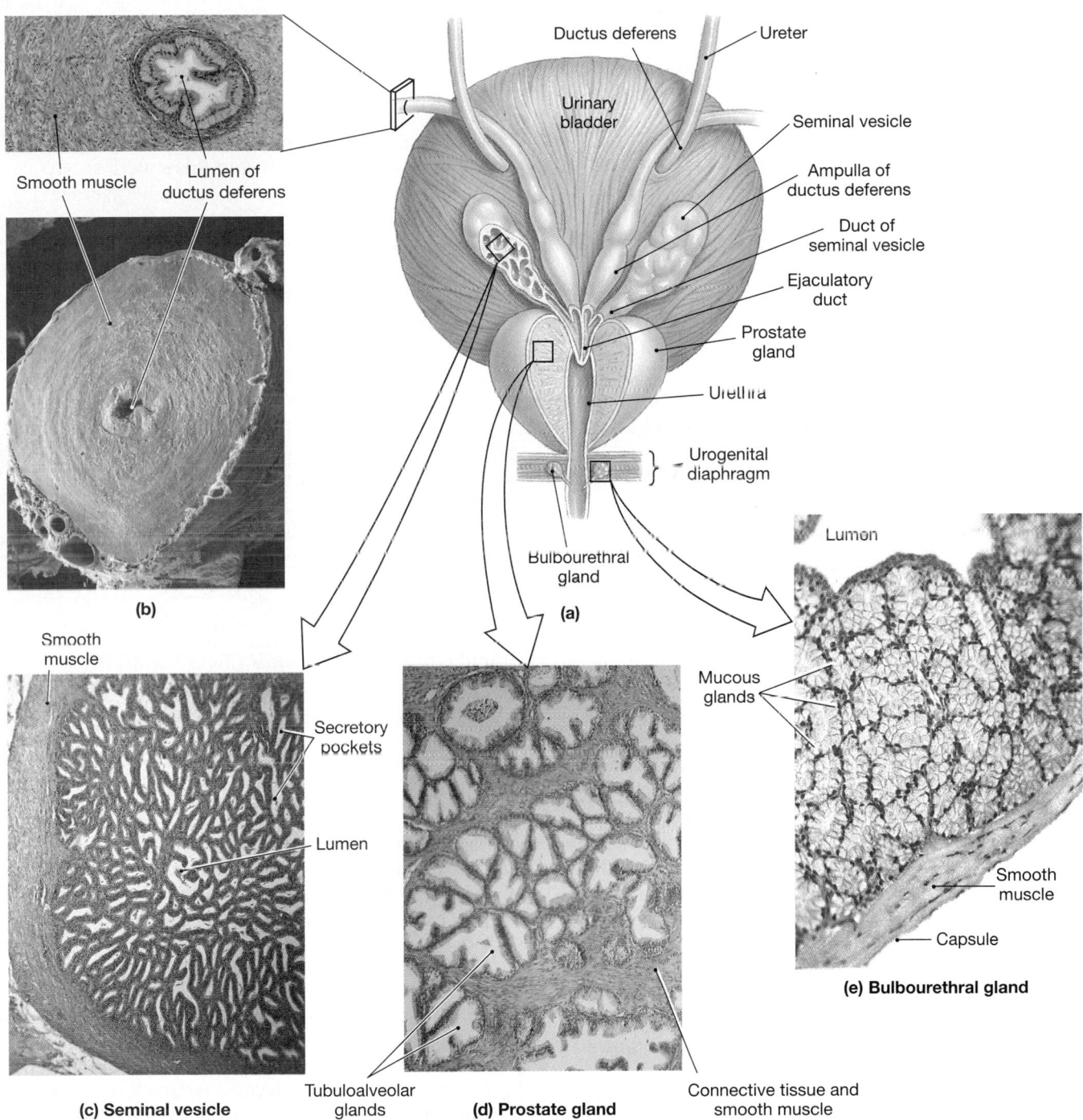

Smooth muscle

Lumen of ductus deferens

(b)

Smooth muscle

Ductus deferens

Ureter

Urinary bladder

Seminal vesicle

Ampulla of ductus deferens

Duct of seminal vesicle

Ejaculatory duct

Prostate gland

Urethra

Urogenital diaphragm

Bulbourethral gland

(a)

Secretory pockets

Lumen

Smooth muscle

(c) Seminal vesicle

Tubuloalveolar glands

(d) Prostate gland

Connective tissue and smooth muscle

Lumen

Mucous glands

Smooth muscle

Capsule

(e) Bulbourethral gland

•**FIGURE 28-10** **The Ductus Deferens and Accessory Glands.** (a) A posterior view of the prostate, showing subdivisions of the ductus deferens in relation to surrounding structures. (b) Micrographs of the ductus deferens, showing extensive layering with smooth muscle around the lumen. (LM [top] × 34, SEM × 42) (Reproduced from R. G. Kessel and R. H. Kardon, "Tissues and Organs: A Text-Atlas of Scanning Electron Microscopy," W. H. Freeman & Co., 1979.) (c) The appearance of the seminal vesicle. (LM × 44) (d) The glands of the prostate. (LM × 49) (e) A bulbourethral gland. (LM × 177)

the seminal vesicle (Figure 28-10a•). The seminal vesicles are embedded in connective tissue on either side of the midline, sandwiched between the posterior wall of the urinary bladder and the rectum. Each seminal vesicle is a tubular gland with a total length of about 15 cm (6 in.). The body of the gland has many short side branches. The entire assemblage is

coiled and folded into a compact, tapered mass roughly 5 cm × 2.5 cm (2 in. × 1 in.). You can see the location of the seminal vesicles in Figures 28-1, p. 1039 (lateral view); 28-10 (posterior and inferior views); and 28-11a•, p. 1052 (anterior view).

The seminal vesicles are extremely active secretory glands with an epithelial lining that contains ex-

Prostatic Hypertrophy and Prostate Cancer

In most cases, prostatic enlargement, or **benign prostatic hypertrophy**, occurs spontaneously in men over age 50. The increase in size occurs while androgen production by the interstitial cells decreases. At the same time, the interstitial cells begin releasing small quantities of estrogens into the circulation. The combination of lower testosterone levels and the presence of estrogen probably stimulates prostatic growth. In severe cases, prostatic swelling constricts and blocks the urethra and constricts the rectum. If not corrected, the urinary obstruction can cause permanent kidney damage. Partial surgical removal is the most effective treatment at present. In the procedure known as a **TURP** (*transurethral prostatectomy*), an instrument pushed along the urethra restores normal function by cutting away the swollen prostatic tissue. Most of the prostate remains in place, and there are no external scars.

Prostate cancer, a malignant, metastasizing cancer of the prostate, is the second most common cancer in men and the second most common cause of cancer deaths in males. In 1997, approximately 334,500 new cases of prostate cancer were diagnosed in the United States, and there were about 41,800 deaths. Most patients are elderly (average age 72 at diagnosis). There are racial differences in susceptibility that are poorly understood. The prostate cancer rates in the 50–54 age group are twice as high for African-American males as for Caucasian-American males. (The rates at all ages are about one-third higher for African Americans.) The prostate cancer rates for Asian males are relatively low compared with either Caucasian Americans or African Americans. For all age groups and all races, the rates of prostate cancer are rising sharply. The reason for the increase is not known.

Prostate cancer normally originates in one of the secretory glands. As the cancer progresses, it produces a nodular lump or swelling on the prostatic surface. Palpation of the prostate gland through the rectal wall,

---

tensive folds (Figure 28-10c●). The seminal vesicles contribute about 60 percent of the volume of semen. Although the vesicular fluid generally has the same osmotic concentration as blood plasma, the composition of the two fluids is quite different. In particular, the secretion of the seminal vesicles contains (1) relatively high concentrations of fructose, which is easily metabolized by spermatozoa; (2) prostaglandins, which may stimulate smooth muscle contractions along the male and female reproductive tracts; and (3) fibrinogen, which after ejaculation will form a temporary clot within the vagina. The secretions of the seminal vesicles are slightly alkaline. This alkalinity helps neutralize acids in the prostatic secretions and within the vagina. When mixed with the secretions of the seminal vesicles, previously inactive but functional spermatozoa begin beating their flagella, becoming highly mobile.

The secretions of the seminal vesicles are discharged into the ejaculatory duct at *emission*, when peristaltic contractions are under way in the ductus deferens, seminal vesicles, and prostate gland. These contractions are under control of the sympathetic nervous system.

The Prostate Gland

The prostate gland is a small, muscular, rounded organ with a diameter of about 4 cm (1.6 in.). The prostate gland encircles the proximal portion of the urethra as it leaves the urinary bladder (Figures 28-1, p. 1039, sagittal view; 28-10a, posterior view; and 26-19, p. 994, and 28-11a●, p. 1052, anterior views). The glandular tissue of the prostate consists of a cluster of 30–50 compound *tubuloalveolar glands* (Figure 28-10d●). ∞ *[p. 119]* These glands are surrounded by and wrapped in a thick blanket of smooth muscle fibers.

The prostatic glands produce **prostatic fluid**, a slightly acidic solution that contributes 20–30 percent of the volume of semen. In addition to several other compounds of uncertain significance, prostatic secretions contain **seminalplasmin** (sem-i-nal-PLAZ-min), an antibiotic that may help prevent urinary tract infections in males. These secretions are ejected into the prostatic urethra by peristaltic contractions of the muscular wall.

PROSTATITIS Prostatic inflammation, or **prostatitis** (pros-ta-TĪ-tis), can occur in males at any age, but it most commonly afflicts older men. Prostatitis may result from bacterial infections or in the apparent absence of pathogens. Individuals with prostatitis complain of pain in the lower back, perineum, or rectum, in some cases accompanied by painful urination and the discharge of mucous secretions from the external urethral meatus. Antibiotic therapy is effective in treating most cases that result from bacterial infection. Prostatitis is taken seriously, because the symptoms can resemble those of *prostate cancer*. Prostate cancer is the second most common cancer in men and the second leading cause of cancer deaths in males.

The Bulbourethral Glands

The paired bulbourethral glands, or *Cowper's glands*, are situated at the base of the penis, covered by the fascia of the urogenital diaphragm (Figures 28-1, p. 1039, sagittal view; 28-10a, p. 1049, posterior view; and 28-11a●, p. 1052, anterior view). The bulbourethral glands are

a procedure known as a *digital rectal exam* (DRE), is the easiest diagnostic screening procedure. *Transrectal prostatic ultrasound* (TRUS) can be used to obtain more detailed information about the status of the prostate but at significantly higher cost to the patient. Blood tests may also be used for screening purposes. The most sensitive is a blood test for *prostate-specific antigen (PSA)*. Elevated levels of this antigen, normally present in low concentrations, may indicate the presence of prostate cancer. The *serum enzyme assay,* which checks levels of the isozyme *prostatic acid phosphatase*, detects prostate cancer in comparatively late stages of development. Screening with periodic PSA tests is now being recommended for men over age 50.

If the condition is detected before the cancer cells have spread to other organs, the usual treatment is localized radiation or the surgical removal of the prostate gland. This operation, **prostatectomy** (pros-ta-TEK-to-mē), can be effective in controlling the condition, but it may have undesirable side effects including urinary incontinence and loss of sexual function. Modified surgical procedures can reduce the risks and maintain normal sexual function in almost three out of four patients.

The prognosis is much worse for prostate cancer diagnosed after metastasis has occurred, because metastasis rapidly involves the lymphatic system, lungs, bone marrow, liver, or adrenal glands. The survival rates at this stage become relatively low. Potential treatments for metastatic prostate cancer include widespread irradiation, hormonal manipulation, lymph node removal, and aggressive chemotherapy. Because the cancer cells are stimulated by testosterone, treatment may involve castration or hormones that depress GnRH or LH production. Until recently, the usual hormone selected was *diethylstilbestrol* (DES), an estrogen. Now there are two other options: (1) *Drugs that mimic GnRH:* These drugs are given in high doses, producing a surge in LH production followed by a sharp decline to very low levels, presumably as the endocrine cells adapt to the excessive stimulation. (2) *Drugs that block the action of androgens:* Several new drugs, including *flutamide* and *finasateride*, prevent stimulation of the cancer cells by testosterone. Despite these interesting advances in treatment, however, the average survival time for patients diagnosed with advanced prostatic cancer is only 2.5 years.

round, with diameters approaching 10 mm (less than 0.5 in.). The duct of each gland travels alongside the penile urethra for 3–4 cm (1.2–1.6 in.) before emptying into the urethral lumen. These are compound, tubuloalveolar mucous glands (Figure 28-10e●) that secrete a thick, alkaline mucus. The secretion helps neutralize any urinary acids that may remain in the urethra and provides lubrication for the *glans*, or tip of the penis.

Semen

A typical *ejaculation* releases 2–5 ml of semen. This volume of fluid, called an **ejaculate**, contains:

- *Spermatozoa.* A normal **sperm count** ranges from 20 million to 100 million spermatozoa per milliliter.
- *Seminal fluid.* **Seminal fluid**, the fluid component of semen, is a mixture of glandular secretions with a distinct ionic and nutrient composition. A typical sample of seminal fluid contains the combined secretions of the seminal vesicles (60 percent), the prostate (30 percent), the sustentacular cells and epididymis (5 percent), and the bulbourethral glands (less than 5 percent).
- *Enzymes.* Several important enzymes are present in the seminal fluid, including (1) a protease that may help dissolve mucous secretions in the vagina; (2) seminalplasmin, an antibiotic enzyme from the prostate gland that kills a variety of bacteria, including *Escherichia coli;* (3) a prostatic enzyme that converts fibrinogen to fibrin after ejaculation; and (4) *fibrinolysin*, which subsequently liquefies the clotted semen.

 SEMEN ANALYSIS Semen analysis is commonly performed to assess male fertility. A semen sample is donated after a 36-hour period of sexual abstinence. Standard tests include the following:

- *Volume.* Normal ejaculate volume is 2–5 ml. An abnormally low volume may indicate problems with the prostatic glands or seminal vesicles.
- *Sperm count.* There should be more than 60 million spermatozoa in an ejaculate. The concentration of spermatozoa should be over 20 million per milliliter of semen. Most individuals with lower sperm counts are infertile. A low sperm count may reflect inflammation of the epididymis, ductus deferens, or prostate.
- *Motility.* At least 60 percent of the spermatozoa in the sample should be beating their flagella and swimming actively.
- *Morphology.* At least 60 percent of the spermatozoa should have normal shapes. Common abnormalities are malformed heads and "twin" spermatozoa that did not separate at the time of spermiation.
- *Liquefaction.* Within a few minutes after ejaculation, semen coagulates, liquefying again after a variable period. The function of this clotting is unknown. Normal semen liquefies after 15–30 minutes. An extended liquefaction time may indicate problems with the secretions of accessory glands.

Determination of male fertility problems in the absence of abnormal semen analysis results may require additional tests. In what is often called the "hamster test," a semen sample is placed on a slide with the oocyte of a hamster. Normal human spermatozoa will fertilize the oocyte, although further development is impossible. If fertilization does not occur, there may be problems with the enzymes in the acrosomal cap.

A complete chemical analysis of semen appears in Appendix VI.

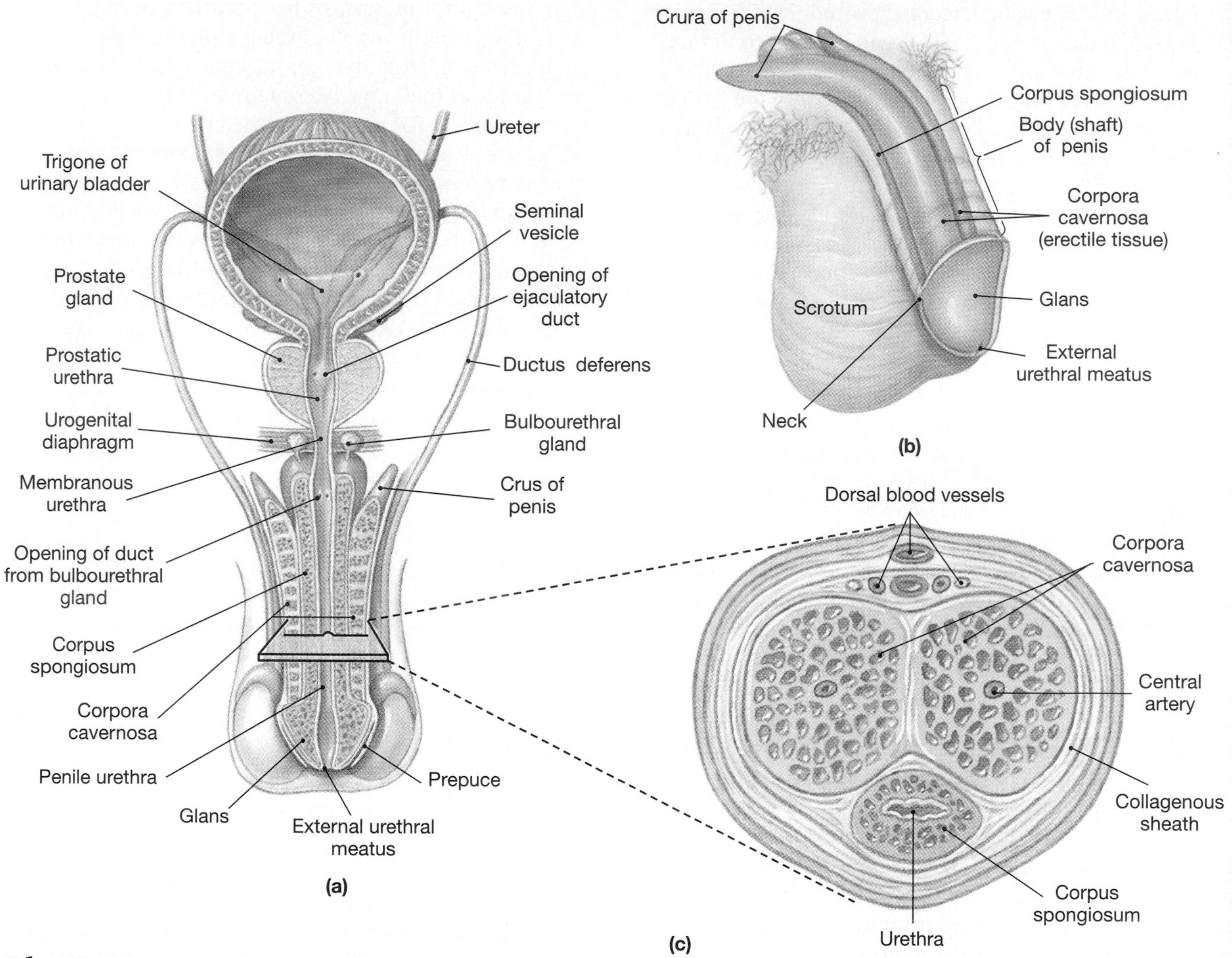

Crura of penis

Corpus spongiosum

Body (shaft) of penis

Corpora cavernosa (erectile tissue)

Glans

External urethral meatus

Scrotum

Neck

(b)

Ureter

Trigone of urinary bladder

Seminal vesicle

Prostate gland

Opening of ejaculatory duct

Prostatic urethra

Ductus deferens

Urogenital diaphragm

Bulbourethral gland

Membranous urethra

Crus of penis

Opening of duct from bulbourethral gland

Corpus spongiosum

Corpora cavernosa

Penile urethra

Prepuce

Glans

External urethral meatus

(a)

Dorsal blood vessels

Corpora cavernosa

Central artery

Collagenous sheath

Corpus spongiosum

Urethra

(c)

Corpora cavernosa

Dorsal blood vessels

Collagenous sheath

Urethra

Corpus spongiosum

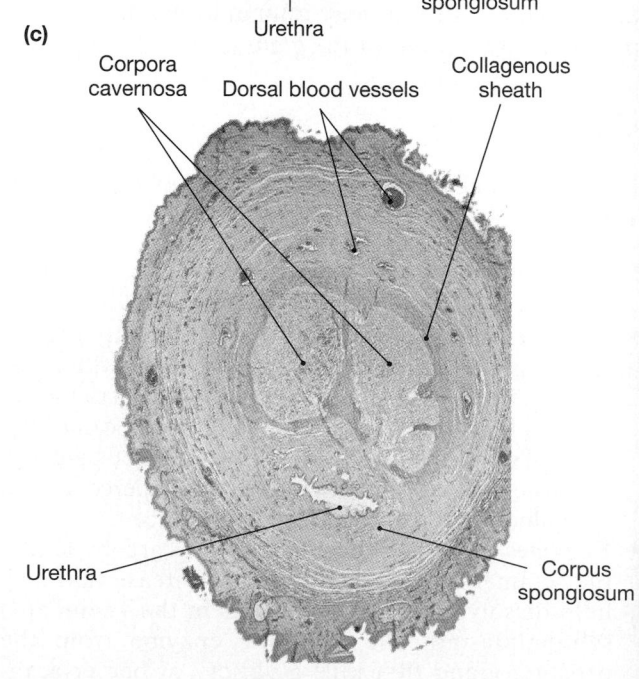

The Penis

The penis is a tubular organ through which the distal portion of the urethra passes (Figures 28-1, p. 1039, 28-2, p. 1040, and 28-11a ●). It conducts urine to the exterior and introduces semen into the female vagina during sexual intercourse. The penis is divided into three regions: (1) the *root*, (2) the *body*, and (3) the *glans* (Figure 28-11b ●). The **root** of the penis is the fixed portion that attaches the penis to the body wall. This connection occurs within the urogenital triangle immediately inferior to the pubic symphysis. The **body (shaft)** of the penis is the tubular, movable portion. Masses of *erectile tissue* are found within the body. The **glans** of the penis is the expanded distal end that surrounds the external urethral meatus. The *neck* is the narrow portion of the penis between the shaft and the glans.

The skin overlying the penis resembles that of the scrotum. The dermis contains a layer of smooth muscle, and the underlying loose connective tissue allows the thin skin to move without distorting underlying structures. The subcutaneous layer also contains superficial arteries, veins, and lymphatics.

●FIGURE 28-11 **The Penis. (a)** Frontal section through the penis and associated organs. **(b)** Anterior and lateral view of the penis, showing positions of the erectile tissues. **(c)** Sectional views through the penis.

A fold of skin called the **prepuce** (PRĒ-pūs), or *fore-skin*, surrounds the tip of the penis. The prepuce attaches to the relatively narrow **neck** of the penis and continues over the glans. There are no hair follicles on the opposing surfaces, but **preputial** (prē-PŪ-shal) **glands** in the skin of the neck and the inner surface of the prepuce secrete a waxy material known as **smegma** (SMEG-ma). Unfortunately, smegma can be an excellent nutrient source for bacteria. Mild inflammation and infections in this region are common, especially if the area is not washed thoroughly and frequently. One way of avoiding trouble is to perform a **circumcision** (ser-kum-SIZH-un), surgical removal of the prepuce. In Western societies (especially the United States), this procedure is generally performed shortly after birth. Although the practice of circumcision remains controversial, strong religious and cultural biases and epidemiological evidence[1] suggest that it will continue.

Beneath the loose connective tissue, a dense network of elastic fibers encircles the internal structures of the penis. Most of the body of the penis consists of three cylindrical columns of **erectile tissue** (Figure 28-11a,c●). Erectile tissue consists of a three-dimensional maze of vascular channels incompletely separated by partitions of elastic connective tissue and smooth muscle fibers. In the resting state, the arterial branches are constricted, and the muscular partitions are tense. This combination restricts blood flow into the erectile tissue. The parasympathetic innervation of the penile arteries involves neurons that release nitric oxide (NO) at their synaptic knobs. The smooth muscles in the arterial walls relax when NO is released, at which time (1) the vessels dilate, (2) blood flow increases, (3) the vascular channels become engorged with blood, and (4) **erection** of the penis occurs. The flaccid (nonerect) penis hangs inferior to the pubic symphysis and anterior to the scrotum, but during erection the penis stiffens and assumes a more upright position.

On the anterior surface of the flaccid penis, the two cylindrical **corpora cavernosa** (KOR-po-ruh ka-ver-NŌ-suh; singular, *cavernosum*) are separated by a thin septum and encircled by a dense collagenous sheath. The corpora cavernosa diverge at their bases, forming the **crura** (*crura*, legs; singular, *crus*) of the penis. Each crus is bound to the ramus of the ischium and pubis by tough connective tissue ligaments. The corpora cavernosa extend along the length of the penis as far as the neck. The erectile tissue within each corpus cavernosum surrounds a central artery (Figure 28-11c●).

The relatively slender **corpus spongiosum** (spon-jē-Ō-sum) surrounds the penile urethra (Figure 28-11a●). This erectile body extends from the superficial fascia of the urogenital diaphragm to the tip of the penis, where it expands to form the glans. The sheath surrounding the corpus spongiosum contains more elastic fibers than does that of the corpora cavernosa, and the erectile tissue contains a pair of small arteries.

Hormones and Male Reproductive Function

The hormonal interactions in males are diagrammed in Figure 28-12●, and we introduced major reproductive hormones in Chapter 18. ∞ *[p. 625]* The anterior pituitary releases *follicle-stimulating hormone* (**FSH**) and *luteinizing hormone* (**LH**). The pituitary release of these hormones occurs in the presence of *gonadotropin-releasing hormone* (**GnRH**), a peptide synthesized in the hypothalamus and carried to the anterior pituitary by the hypophyseal portal system.

GnRH secretion occurs in pulses rather than continuously. In adult males, small pulses occur at 60–90 minute intervals. As levels of GnRH change, so do the rates of secretion of FSH and LH (and testosterone, which is released in response to LH). Unlike the situation in women, which we will consider later in the chapter, the GnRH pulse frequency in adult males remains relatively steady from hour to hour, day to day, and year to year. As a result, plasma levels of FSH, LH, and testosterone remain within a relatively narrow range throughout adult life.

FSH and Spermatogenesis

In males, FSH targets primarily the sustentacular cells of the seminiferous tubules. Under FSH stimulation, and in the presence of testosterone from the interstitial cells, sustentacular cells (1) promote spermatogenesis and spermiogenesis and (2) secrete androgen-binding protein (ABP).

The rate of spermatogenesis is regulated by a negative-feedback mechanism involving GnRH, FSH, and inhibin. Under GnRH stimulation, FSH promotes spermatogenesis along the seminiferous tubules. As spermatogenesis accelerates, however, so does the rate of inhibin secretion by the sustentacular cells of the testes. Inhibin inhibits FSH production in the anterior pituitary and may also suppress secretion of GnRH at the hypothalamus.

The net effect is that when FSH levels become elevated, inhibin production increases until the FSH levels return to normal. If FSH levels decline, inhibin production falls, and the rate of FSH production accelerates.

LH and Androgen Production

In males, LH causes the secretion of testosterone and other androgens by the interstitial cells of the testes. Testosterone, the most important androgen, has numerous functions, such as (1) stimulating spermato-

---

[1]Uncircumcised males have a higher incidence of urinary tract infections and are at a greater risk of developing penile cancer than are circumcised males.

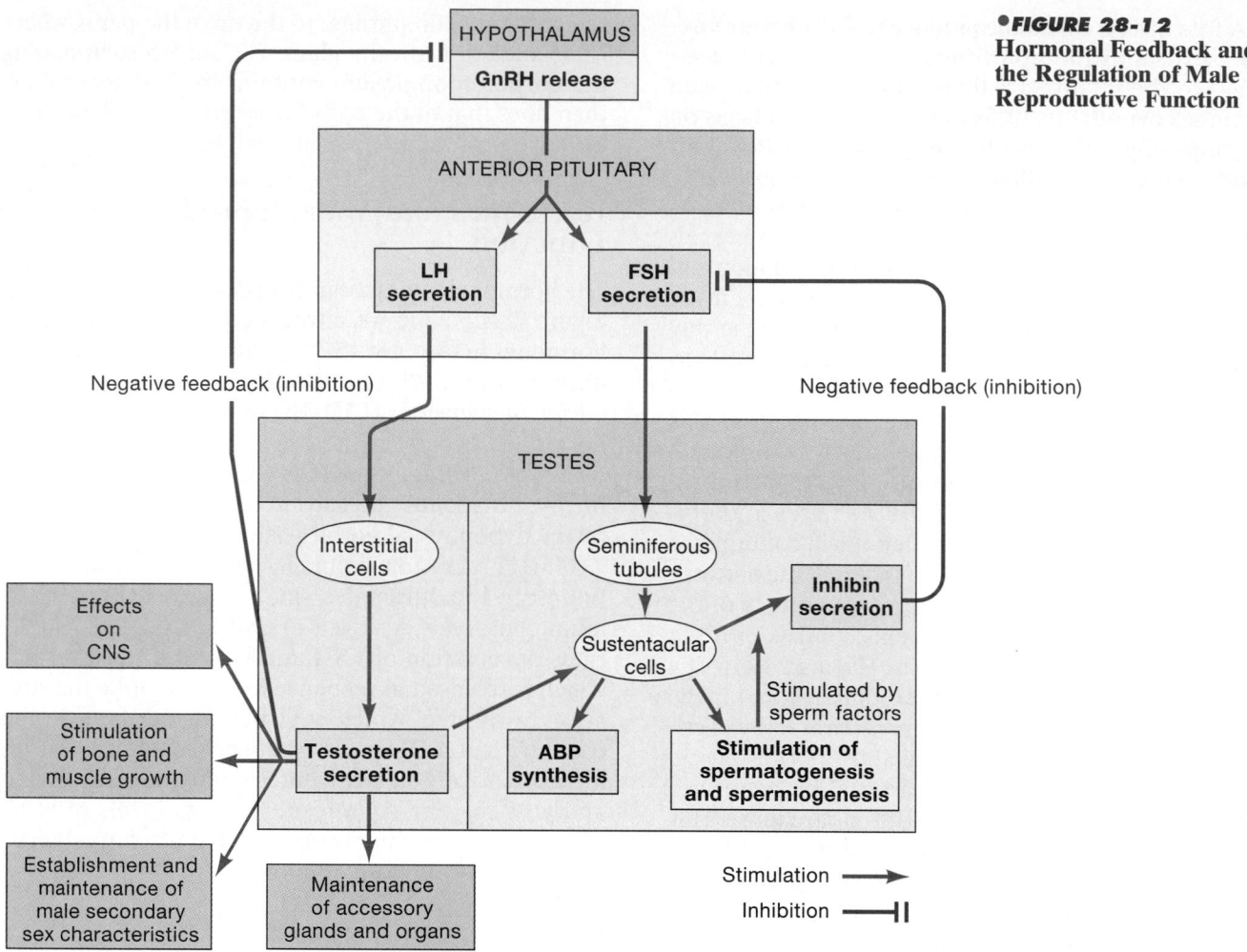

●*FIGURE 28-12*
**Hormonal Feedback and
the Regulation of Male
Reproductive Function**

genesis and promoting the functional maturation of spermatozoa, through its effects on sustentacular cells; (2) affecting CNS function, including the influence of libido (sexual drive) and related behaviors; (3) stimulating metabolism throughout the body, especially pathways concerned with protein synthesis and muscle growth; (4) establishing and maintaining the secondary sex characteristics, such as the distribution of facial hair, increased muscle mass and body size, and the quantity and location of characteristic adipose tissue deposits; and (5) maintaining the accessory glands and organs of the male reproductive tract.

Testosterone functions like other steroid hormones. ∞ *[p. 598]* It circulates while bound to transport proteins; *gonadal steroid-binding globulin* (GBG) carries roughly two-thirds of circulating testosterone, and the rest binds to albumins. Testosterone diffuses across the cell membrane and binds to an intracellular receptor. The hormone-receptor complex then binds to the DNA in the nucleus. In many target tissues, some of the arriving testosterone is converted to **dihydrotestosterone (DHT)**. A small amount of DHT diffuses back out of the cell and into the circulation, and DHT levels are usually about 10

percent of circulating testosterone levels. DHT can also enter peripheral cells and bind to the same hormone receptors targeted by testosterone. In addition, some tissues, notably those of the external genitalia, respond to DHT rather than to testosterone, and other tissues, including the prostate gland, are more sensitive to DHT than to testosterone.

Testosterone production begins around the seventh week of embryonic development and reaches a peak after roughly 6 months of development. Over this period, secretion of MIF by developing sustentacular cells leads to the regression of the Müllerian ducts. The early surge in testosterone levels stimulates the differentiation of the male duct system and accessory organs and affects CNS development. The best-known CNS effects occur in the developing hypothalamus, where testosterone apparently programs the hypothalamic centers involved with (1) GnRH production and the regulation of pituitary FSH and LH secretion, (2) sexual behaviors, and (3) sexual drive. As a result of this prenatal exposure to testosterone, the hypothalamic centers will respond appropriately when the individual becomes sexually mature. The factors responsible for regulating the fetal production of testosterone are not known.

Testosterone levels are low at birth. Up to puberty, background testosterone levels, although still relatively low, are higher in males than in females. Testosterone secretion accelerates markedly at puberty, initiating sexual maturation and the appearance of secondary sex characteristics. In adult males, negative feedback controls the level of testosterone production. Above-normal testosterone levels inhibit the release of GnRH by the hypothalamus. This inhibition causes a reduction in LH secretion and lowers testosterone levels.

The plasma of adult males also contains relatively small amounts of estradiol (2 ng/dl, versus 525 ng/dl of testosterone). Seventy percent of the estradiol is formed from circulating testosterone. The rest is secreted, primarily by the interstitial and sustentacular cells of the testes. The conversion of testosterone to estradiol is performed by an enzyme called *aromatase*. For unknown reasons, estradiol production increases in older men.

DHEA   DHEA, or *dehydroepiandrosterone*, is the primary androgen secreted by the zona reticularis of the adrenal cortex. ∞ *[p. 617]* As we noted in Chapter 18, these androgens, which are secreted in small amounts, are converted to testosterone (or estrogens) by other tissues. The significance of this small adrenal androgen contribution in both genders remains uncertain, but some people have suggested that DHEA could be used to treat a variety of conditions, including diabetes, heart disease, and depression. As yet no controlled clinical trials have been completed, and the effects of long-term, high-dose DHEA ad-

ministration remain unknown. However, you will recall from Chapter 18 that the long-term effects of androgen abuse can be quite serious. High levels of DHEA in women have been linked to masculinization, due to conversion of DHEA to testosterone, and to an increased risk of ovarian cancer.

Nevertheless, DHEA is being promoted as a wonder drug to increase vitality, strength, and muscle mass. Direct sales of the hormone were banned in 1984, because the side-effects were unknown, but food supplements prepared from wild Mexican yams are now being advertised as containing "DHEA precursors." These claims are false, as the compounds contained in these supplements have no effect on circulating levels of DHEA. The current recommendations are that (1) DHEA use be restricted to controlled, supervised clinical trials; and (2) that no one under age 30 use DHEA.

☑ On a warm day, would the cremaster muscle be contracted or relaxed? Why?

☑ What will occur if the arteries within the penis dilate?

☑ What effect would low levels of FSH have on sperm production?

THE REPRODUCTIVE SYSTEM OF THE FEMALE

A woman's reproductive system must produce sex hormones and functional gametes and also be able to protect and support a developing embryo and nourish the newborn infant. The principal organs of the female reproductive system (Figure 28-13•) are the *ovaries,*

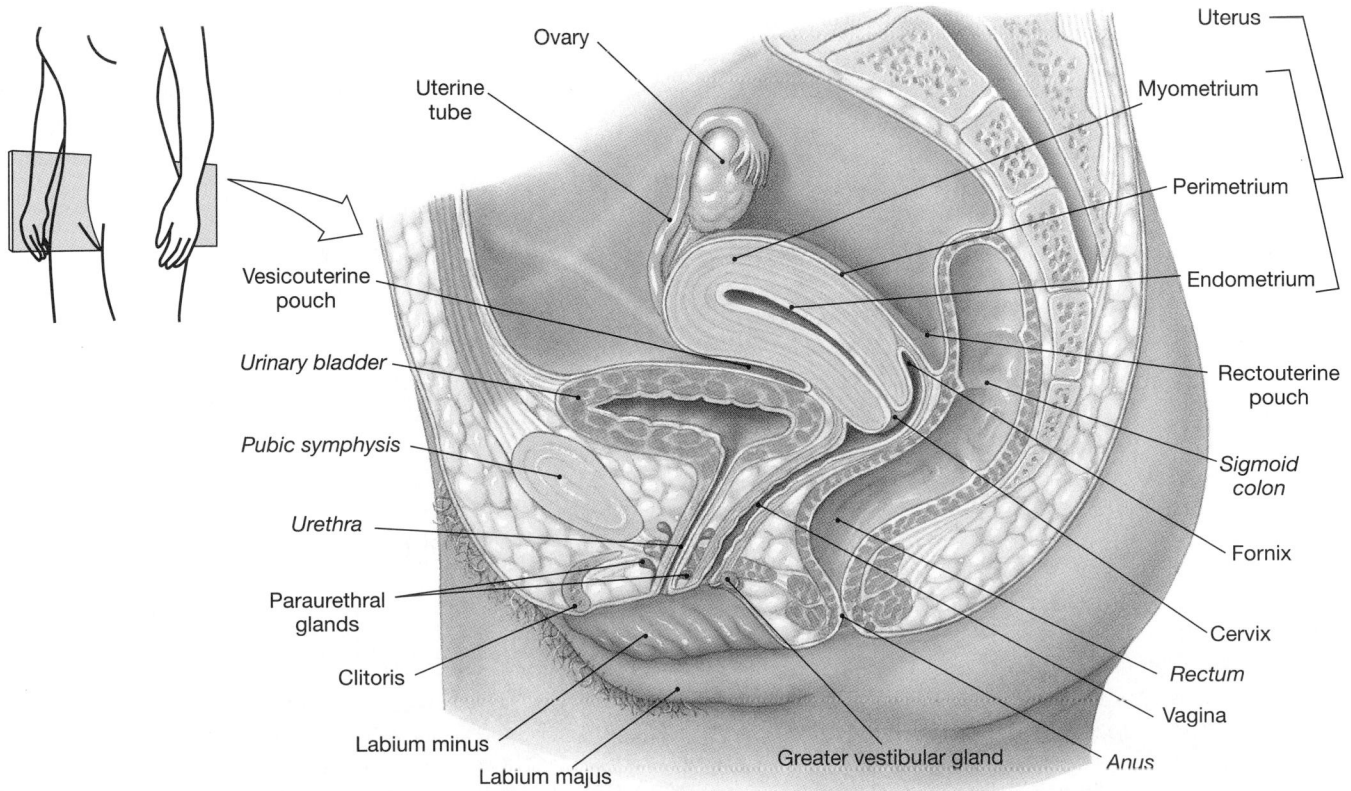

•**FIGURE 28-13** **The Female Reproductive System.** The female reproductive organs, diagrammatic sagittal section. [AM] *Plate 6.7*

the *uterine tubes (Fallopian tubes* or *oviducts)*, the *uterus*, the *vagina*, and the components of the external genitalia. As in males, a variety of accessory glands secrete into the female reproductive tract.

The ovaries, uterine tubes, and uterus are enclosed within an extensive mesentery known as the **broad ligament** (Figure 28-14●). The uterine tubes run along the superior border of the broad ligament and open into the pelvic cavity lateral to the ovaries. The **mesovarium** (mes-ō-VAR-ē-um), a thickened fold of mesentery, supports and stabilizes the position of each ovary. The broad ligament attaches to the sides and floor of the pelvic cavity, where it becomes continuous with the parietal peritoneum. The broad ligament thus subdivides this part of the peritoneal cavity. The pocket formed between the posterior wall of the uterus and the anterior surface of the colon is the **rectouterine** (rek-tō-ū-te-rin) **pouch**; that formed between the uterus and the posterior wall of the bladder is the **vesicouterine** (ves-i-kō-Ū-ter-in) **pouch**. These subdivisions are most apparent in sagittal section (Figure 28-13●).

Several other ligaments assist the broad ligament in supporting and stabilizing the position of the uterus and associated reproductive organs. These ligaments lie within the mesentery sheet of the broad ligament and are connected to the ovaries or uterus. The broad ligament limits side-to-side movement and rotation, and the other ligaments (described with the ovaries and uterus) prevent superior-inferior movement.

The Ovaries

The paired ovaries are small, almond-shaped organs located near the lateral walls of the pelvic cavity (Figures 28-13 and 28-14●). The ovaries (1) produce immature female gametes, or oocytes; (2) secrete female sex hormones, including *estrogens* and *progestins*; and (3) secrete inhibin, involved in the feedback control of pituitary FSH production.

The position of each ovary is stabilized by the mesovarium and by a pair of supporting ligaments: the *ovarian ligament* and the *suspensory ligament* (Figure 28-14●). The **ovarian ligament** extends from the uterus, near the attachment of the uterine tube, to the medial surface of the ovary. The **suspensory ligament** extends from the lateral surface of the ovary past the open end of the uterine tube to the pelvic wall. The suspensory ligament contains the major blood vessels of the ovary, the **ovarian artery** and **ovarian vein**. These vessels are connected to the ovary at the **ovarian hilum**, where the ovary attaches to the mesovarium.

A typical ovary is a flattened oval that is about 5 cm in length, 2.5 cm in width, and 8 mm in thickness (2 in. × 1 in. × 0.33 in.) and weighs 6–8 g (roughly 0.25 oz). An ovary is pink or yellowish and has a nodular consistency. The visceral peritoneum, or *germinal epithelium*, covering the surface of each ovary consists of a layer of columnar epthelial cells that overlies a dense connective tissue layer called the **tunica al-**

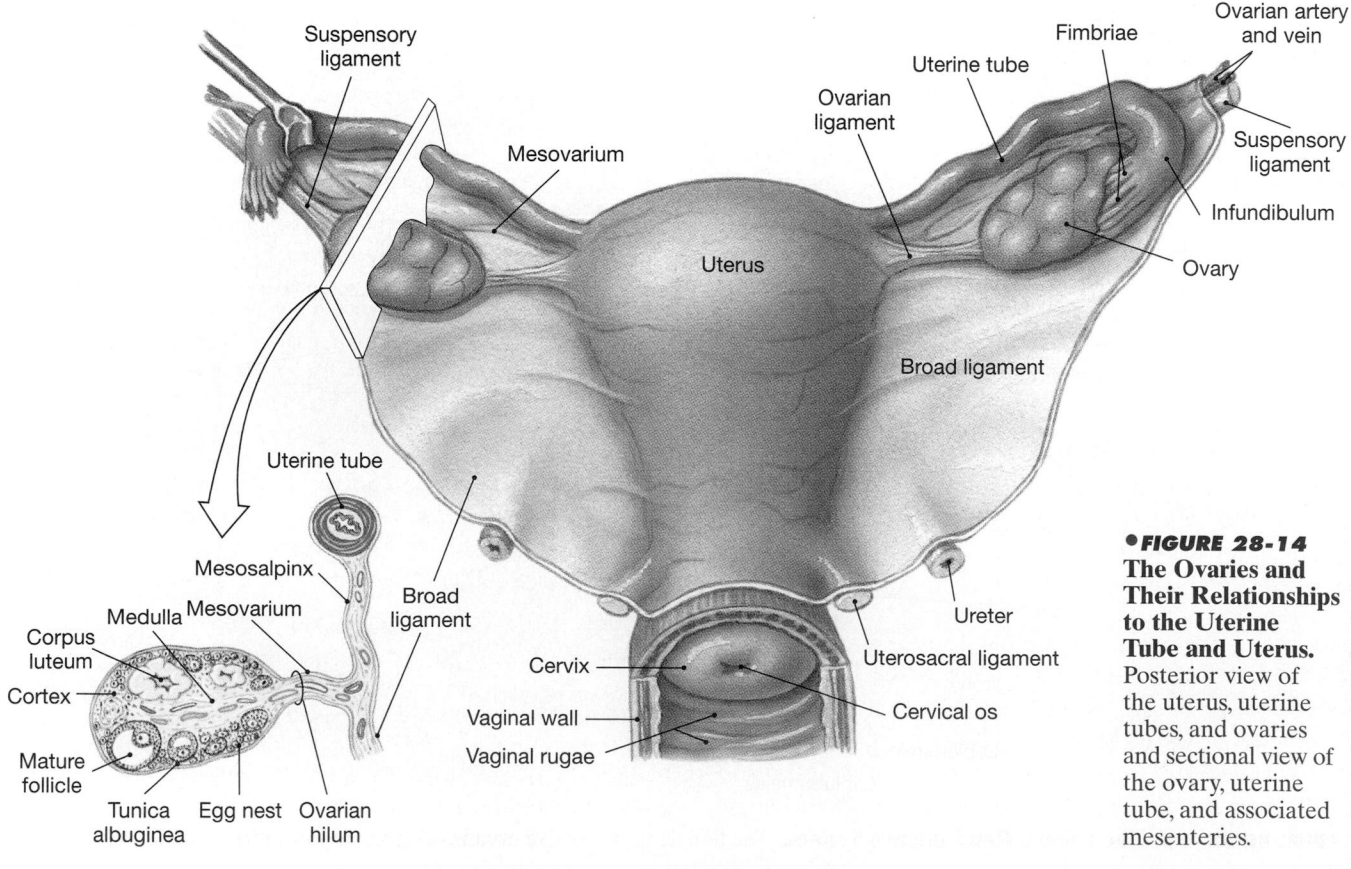

●**FIGURE 28-14**
The Ovaries and Their Relationships to the Uterine Tube and Uterus. Posterior view of the uterus, uterine tubes, and ovaries and sectional view of the ovary, uterine tube, and associated mesenteries.

buginea. We can divide the interior tissues, or **stroma**, of the ovary into a superficial *cortex* and a deeper *medulla*. Gametes are produced in the cortex.

Oogenesis

Ovum production, or **oogenesis** (ō-ō-JEN-e-sis), begins before a woman's birth, accelerates at puberty, and ends at *menopause*. Between puberty and menopause, oogenesis occurs on a monthly basis as part of the *ovarian cycle*.

Unlike the situation in the male gonads, the **oogonia** (ō-ō-GŌ-nē-uh), or stem cells of females, complete their mitotic divisions before birth. Between the third and seventh months of fetal development, the daughter cells, or **primary oocytes** (Ō-ō-sīts), prepare to undergo meiosis. They proceed as far as prophase of meiosis I, but at that time the process comes to a halt. The primary oocytes then remain in a state of suspended development until the individual reaches puberty. At that time, rising levels of FSH trigger the start of the ovarian cycle. Each month thereafter, some of the primary oocytes will be stimulated to undergo further development. Not all primary oocytes produced in development survive until puberty. There are roughly 2 million *primordial follicles* in the ovaries at birth, each containing a primary oocyte. By the time of puberty, that number has dropped to about 400,000. The rest of the primordial follicles degenerate in a process called **atresia** (a-TRĒ-zē-uh).

Although the nuclear events under way during meiosis in the ovaries are the same as those in the testes, the process differs in two important details:

1. The cytoplasm of the primary oocyte is unevenly distributed during the two meiotic divisions. Oogenesis produces one functional ovum, which contains most of the original cytoplasm, and two or three nonfunctional **polar bodies** that later disintegrate (Figure 28-15●).

2. The ovary releases a **secondary oocyte** rather than a mature ovum. The secondary oocyte is suspended in metaphase of meiosis II; meiosis will not be completed unless and until fertilization occurs.

The Ovarian Cycle

Ovarian follicles are specialized structures in which oocyte growth and meiosis I occur. The ovarian follicles are located in the cortex of the ovaries. Primary oocytes are located in the outer portion of the ovarian cortex near the tunica albuginea, in clusters called *egg nests*. Each primary oocyte within an egg nest is surrounded by a single squamous layer of *follicular cells*. The primary oocyte and its follicular cells form a **primordial follicle**. After sexual maturation, a different group of primordial follicles is activated each month. This monthly process is known as the **ovarian cycle**.

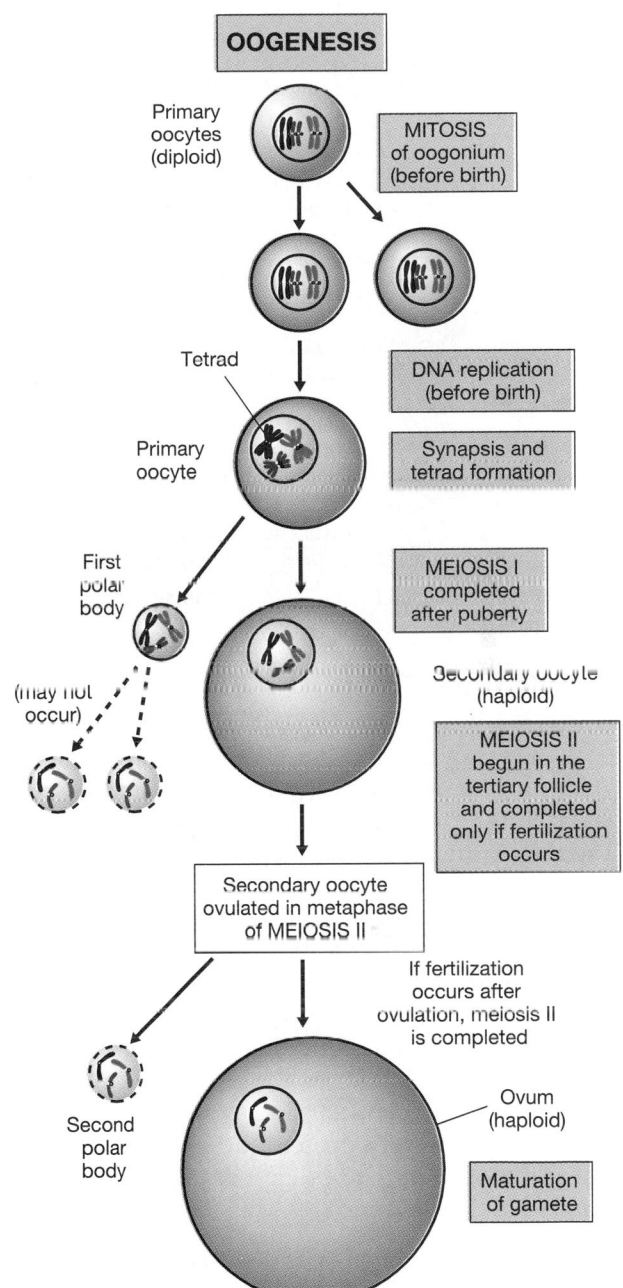

● FIGURE 28-15 **Oogenesis.** In oogenesis, a single primary oocyte produces an ovum and two or three nonfunctional polar bodies. Compare with Figure 28-7, p. 1045.

The ovarian cycle can be divided into a **follicular phase**, or *preovulatory phase*, and a **luteal phase**, or *postovulatory phase*. Important steps in the ovarian cycle (Figure 28-16●) can be summarized as follows:

STEP 1: Formation of primary follicles. Follicle formation is stimulated by FSH from the anterior pituitary. The ovarian cycle begins as activated primordial follicles develop into **primary follicles**. In a primary follicle, the follicular cells enlarge and undergo repeated cell divisions. The divisions create several layers of follicular cells around the oocyte. These follicle cells are now called **granulosa cells**.

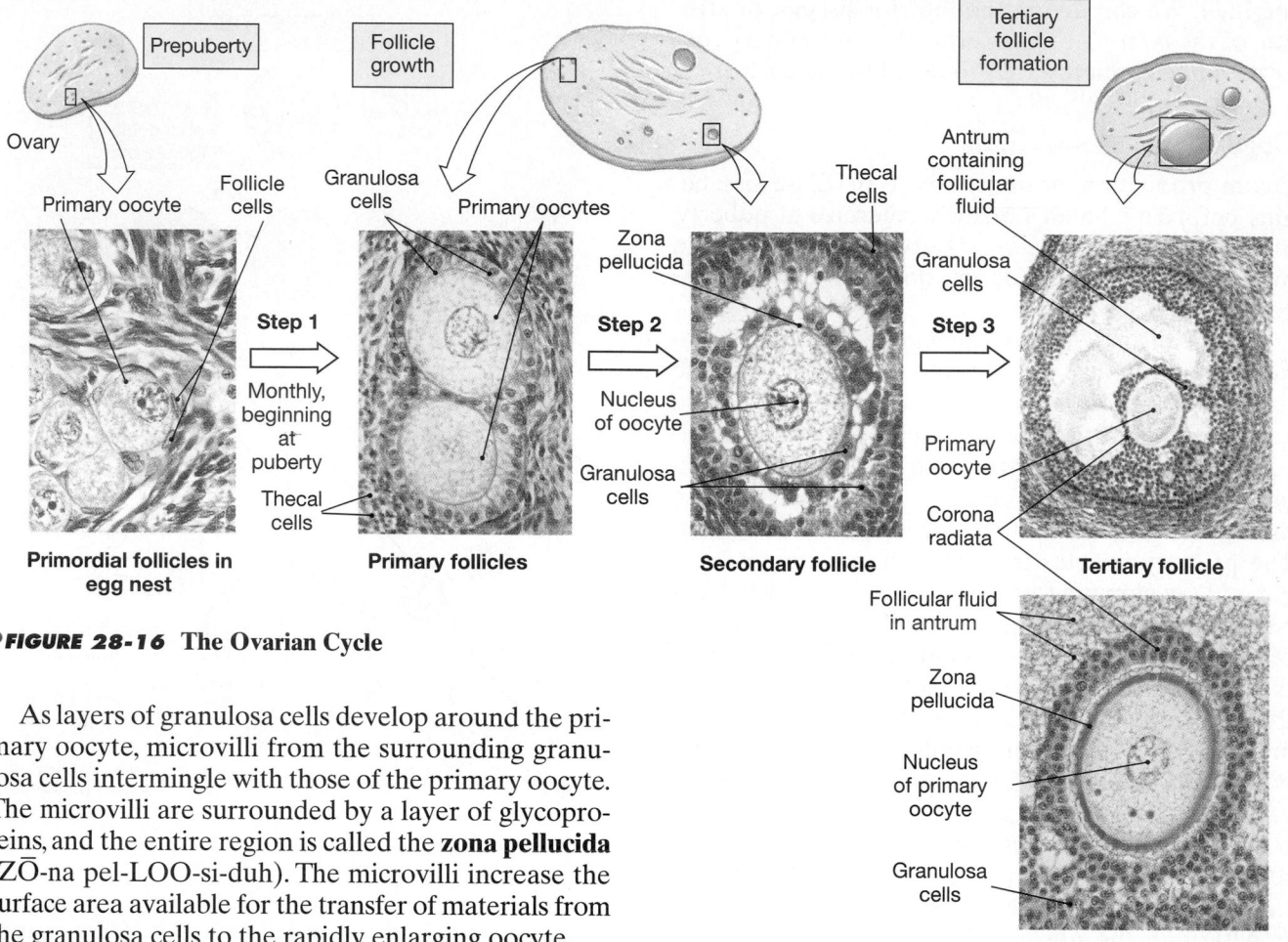

•*FIGURE 28-16* **The Ovarian Cycle**

As layers of granulosa cells develop around the primary oocyte, microvilli from the surrounding granulosa cells intermingle with those of the primary oocyte. The microvilli are surrounded by a layer of glycoproteins, and the entire region is called the **zona pellucida** (ZŌ-na pel-LOO-si-duh). The microvilli increase the surface area available for the transfer of materials from the granulosa cells to the rapidly enlarging oocyte.

The conversion from primordial to primary follicles and subsequent follicular development occurs under FSH stimulation. As the granulosa cells enlarge and multiply, adjacent cells in the ovarian stroma form a layer of **thecal cells** around the follicle. Thecal cells and granulosa cells work together to produce sex hormones called *estrogens.*

STEP 2: *Formation of secondary follicles.* Although many primordial follicles develop into primary follicles, only a few will proceed to the next step. The transformation begins as the wall of the follicle thickens and the granulosa cells begin secreting small amounts of fluid. This **follicular fluid**, or *liquor folliculi*, accumulates in small pockets that gradually expand and separate the inner and outer layers of the follicle. At this stage, the complex is known as a **secondary follicle.** Although the primary oocyte continues to grow at a steady pace, the follicle as a whole now enlarges rapidly because follicular fluid accumulates.

STEP 3: *Formation of a tertiary follicle.* Eight to ten days after the start of the ovarian cycle, the ovaries generally contain only a single secondary follicle destined for further development. By the tenth to fourteenth day of the cycle, that follicle has formed a **tertiary follicle**, or *mature Graafian* (GRAF-ē-an) *follicle*, roughly 15 mm in diameter. This complex spans the entire width of the ovarian cortex and distorts the

ovarian capsule, creating a prominent bulge in the surface of the ovary. The oocyte projects into the expanded central chamber, or **antrum** (AN-trum), of the follicle surrounded by a mass of granulosa cells.

Until this time, the primary oocyte has been suspended in prophase of the first meiotic division, a state it achieved before birth. As the tertiary follicle completes its development, LH levels begin rising, and the primary oocyte completes meiosis I. Instead of producing two secondary oocytes, the first meiotic division yields a secondary oocyte and a small, nonfunctional polar body. The secondary oocyte then enters meiosis II but stops once again on reaching metaphase. Meiosis II will not be completed unless fertilization occurs.

As the time of ovulation approaches, the secondary oocyte and the surrounding granulosa cells lose their connections with the follicular wall and drift free within the antrum. This event generally occurs on day 14 of a 28-day cycle. The granulosa cells immediately surrounding the secondary oocyte are now known as the **corona radiata** (ko-RŌ-nuh rā-dē-A-tuh).

STEP 4: *Ovulation.* At **ovulation**, the tertiary follicle releases the secondary oocyte. The distended follicular

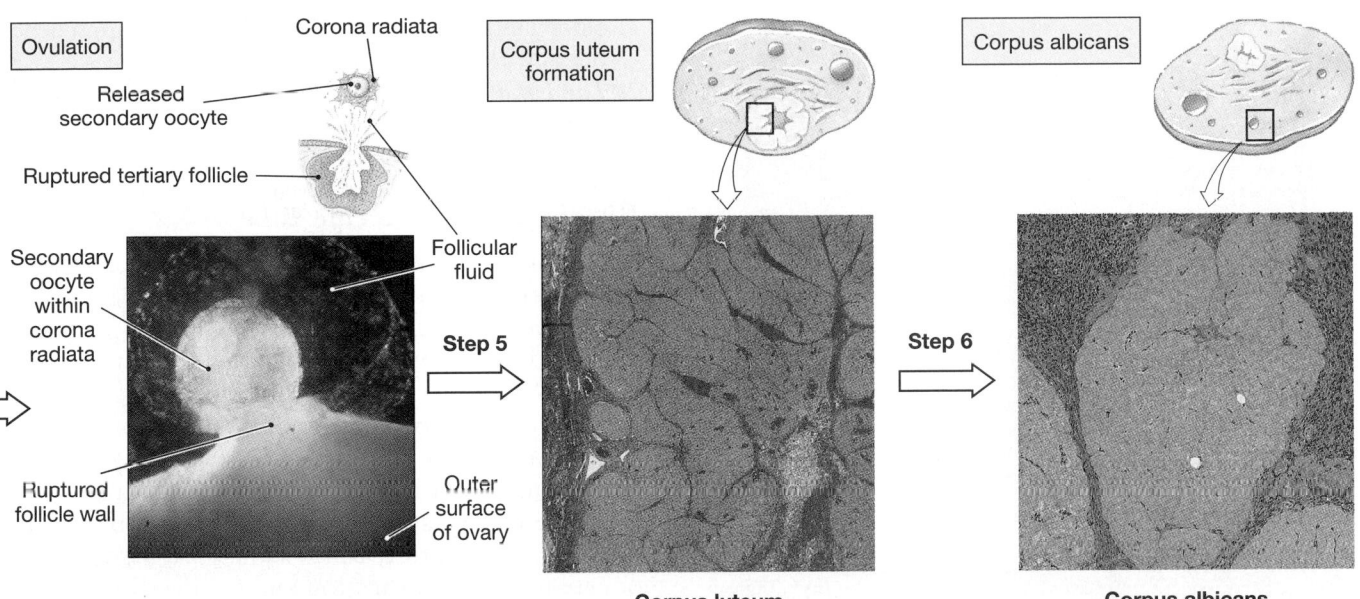

Corpus luteum

Corpus albicans

wall ruptures, releasing the follicular contents, including the secondary oocyte and corona radiata, into the pelvic cavity. The sticky follicular fluid keeps the corona radiata attached to the surface of the ovary, where direct contact with projections of the uterine tube or with fluid currents established by their ciliated epithelium can transfer the secondary oocyte to the uterine tube.

STEP 5: *Formation and degeneration of the corpus luteum.* The empty tertiary follicle initially collapses, and ruptured vessels bleed into the antrum. The remaining granulosa cells then invade the area, proliferating to create an endocrine structure known as the **corpus luteum** (LOO-tē-um), named for its yellow color (*lutea,* yellow). This process occurs under LH stimulation.

The lipids contained in the corpus luteum are used to manufacture steroid hormones known as **progestins** (prō-JES-tinz), principally the steroid **progesterone** (prō-JES-ter-ōn). Although moderate amounts of estrogens are also secreted by the corpus luteum, levels are not as high as they were at ovulation, and progesterone is the principal hormone in the interval after ovulation. Its primary function is to prepare the uterus for pregnancy by stimulating the maturation of the uterine lining and the secretions of uterine glands.

STEP 6: *Unless pregnancy occurs, the corpus luteum begins to degenerate roughly 12 days after ovulation.* Progesterone and estrogen levels then fall markedly. Fibroblasts invade the nonfunctional corpus luteum, producing a knot of pale scar tissue called a **corpus albicans** (AL-bi-kanz). The disintegration, or *involution,* of the corpus luteum marks the end of the ovarian cycle. A new cycle then begins with the activation of another group of primordial follicles.

Age and Oogenesis

Although many primordial follicles may have developed into primary follicles, and several primary follicles may have been converted to secondary follicles, generally only a single oocyte will be released into the pelvic cavity at ovulation. The rest undergo atresia. At puberty, there are about 200,000 primordial follicles in each ovary. Forty years later, few if any follicles remain, although only about 500 will have been ovulated during the interim.

OVARIAN CANCER A woman in the United States has a 1 in 70 chance of developing **ovarian cancer** in her lifetime. In 1997, an estimated 26,800 ovarian cancers were diagnosed, and 14,200 women died from this condition. Although ovarian cancer is the third most common reproductive cancer among women, it is the most dangerous, because ovarian cancer is seldom diagnosed in its early stages. The prognosis is relatively good for cancers that originate in the general ovarian tissues or from abnormal oocytes. These cancers respond well to some combination of chemotherapy, radiation, and surgery. However, 85 percent of ovarian cancers develop from epithelial cells, and sustained remission can be obtained in only about one-third of the cases of this type. For further discussion, see the *Applications Manual.* [AM] *The Diagnosis and Treatment of Ovarian Cancer*

The Uterine Tubes

Each uterine tube is a hollow, muscular tube measuring roughly 13 cm (5 in.) in length (Figures 28-13, sagittal view; 28-14 and 28-17•, posterior views). Each uterine tube is divided into three regions:

1. *The infundibulum.* The end closest to the ovary forms an expanded funnel, or **infundibulum**, with numerous fingerlike projections that extend into the pelvic cavity. The projections are called **fimbriae** (FIM-brē-ē). The inner surfaces of the infundibulum are lined with cilia that beat toward the middle segment of the uterine tube, the *ampulla.*

2. *The ampulla.* The thickness of the smooth muscle layers in the wall of the middle segment, or **ampulla,**

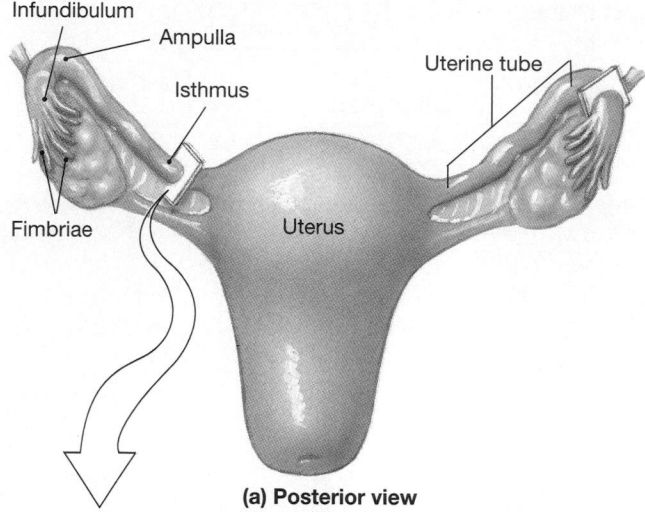

Infundibulum
Ampulla
Uterine tube
Isthmus

Fimbriae
Uterus

(a) Posterior view

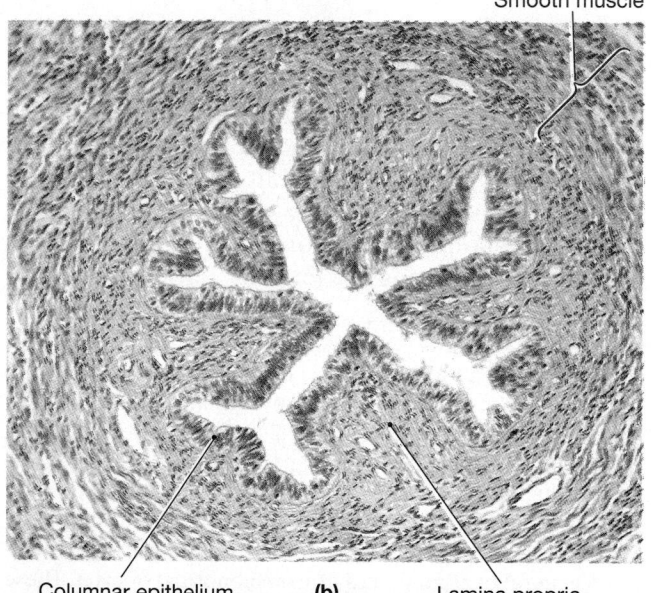

Smooth muscle

Columnar epithelium   **(b)**   Lamina propria

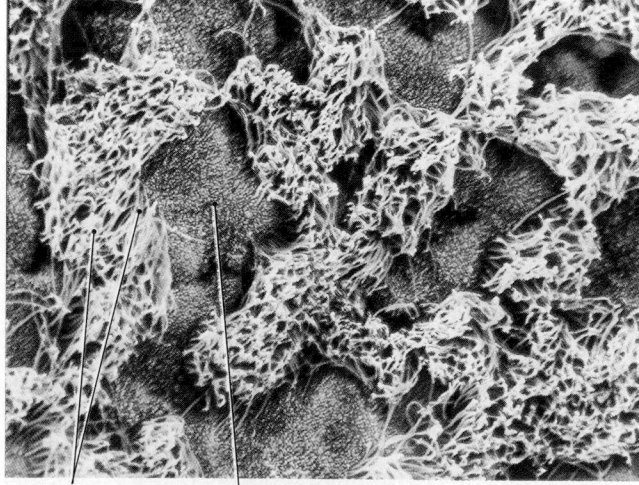

Cilia   Microvilli of mucous secreting cells

(c)

●*FIGURE 28-17* **The Uterine Tubes. (a)** Regions of the uterine tubes. **(b)** A sectional view of the isthmus. (LM × 122) **(c)** Colorized SEM of the ciliated lining of the uterine tube.

of the uterine tube gradually increases as it approaches the uterus.

3. ***The isthmus.*** The ampulla leads to the **isthmus** (IS-mus) (Figure 28-17b●), a short segment connected to the uterine wall.

Histology of the Uterine Tube

The epithelium lining the uterine tube is composed of ciliated columnar epithelial cells with scattered mucin-secreting cells (Figure 28-17c●). The mucosa is surrounded by concentric layers of smooth muscle (Figure 28-17b●). Oocyte transport involves a combination of ciliary movement and peristaltic contractions in the walls of the uterine tube. A few hours before ovulation, sympathetic and parasympathetic nerves from the hypogastric plexus "turn on" this beating pattern. The uterine tubes transport a secondary oocyte for final maturation and fertilization. It normally takes 3–4 days for an oocyte to travel from the infundibulum to the *uterine cavity*. *If fertilization is to occur, the secondary oocyte must encounter spermatozoa during the first 12–24 hours of its passage.* Fertilization typically occurs near the boundary between the ampulla and isthmus of the uterine tube.

Along with its transport function, the uterine tube also provides a nutrient-rich environment that contains lipids and glycogen. This mixture provides nutrients to both spermatozoa and a developing *pre-embryo* (the cell cluster produced by the initial divisions that follow fertilization). Unfertilized oocytes degenerate in the terminal portions of the uterine tubes or within the uterus.

PELVIC INFLAMMATORY DISEASE (PID) Pelvic inflammatory disease (PID) in women is a major cause of *sterility (infertility)*, the inability to have children. PID, an infection of the uterine tubes, affects an estimated 850,000 women each year in the United States alone. In many cases, sexually transmitted pathogens are involved. As much as 50–80 percent of all first cases may be due to infection by *Neisseria gonorrhoeae*, the organism responsible for symptoms of **gonorrhea** (gon-ō-RĒ-uh), a sexually transmitted disease. Invasion of the region by bacteria normally found within the vagina can also cause PID. Symptoms of PID include fever, lower abdominal pain, and elevated white blood cell counts. In severe cases, the infection may spread to other visceral organs or produce a generalized peritonitis.

Sexually active women in the 15–24 age group have the highest incidence of PID. Whereas the use of an oral contraceptive decreases the risk of infection, the presence of an intrauterine device (IUD) may increase the risk by 1.4–7.3 times. Treatment with antibiotics may control the condition, but chronic abdominal pain may persist. In addition, damage and scarring of the uterine tubes may cause infertility by preventing the passage of a zygote to the uterus. Recently, another sexually transmitted bacterium, belonging to the genus *Chlamydia*, has been identified as the probable cause of up to 50 percent of all cases of PID. Despite the fact that women with this infection may develop few if any symptoms, scarring of the uterine tubes may still produce infertility.

The Uterus

The uterus provides mechanical protection, nutritional support, and waste removal for the developing *embryo* (weeks 1–8) and *fetus* (from week 9 to delivery). In addition, contractions in the muscular wall of the uterus are important in ejecting the fetus at the time of birth.

The uterus is a small, pear-shaped organ about 7.5 cm (3 in.) long with a maximum diameter of 5 cm (2 in.). It weighs 30–40 g (1–1.4 oz). In its normal position, the uterus bends anteriorly near its base, a condition known as *anteflexion* (an-tē-FLEK-shun). In this position, the body of the uterus lies across the superior and posterior surfaces of the urinary bladder (Figure 28-13•, p.1055). If the uterus bends backward toward the sacrum, the condition is termed *retroflexion* (re-trō-FLEK-shun). Retroflexion, which occurs in about 20 percent of adult women, has no clinical significance. (A retroflexed uterus generally becomes anteflexed spontaneously during the third month of pregnancy.)

Suspensory Ligaments of the Uterus

In addition to the broad ligament, three pairs of suspensory ligaments stabilize the position of the uterus and limit its range of movement (Figure 28-18a,b•). The **uterosacral** (ū-te-rō-SĀ-kral) **ligaments** extend from the lateral surfaces of the uterus to the anterior face of the sacrum, keeping the body of the uterus from moving inferiorly and anteriorly. The **round ligaments** arise on the lateral margins of the uterus just posterior and inferior to the attachments of the uterine tubes. These ligaments extend through the inguinal canal and end in the connective tissues of the external genitalia. The round ligaments restrict primarily posterior movement of the uterus. The **lateral** *(cardinal)* **ligaments** extend from the base of the uterus and vagina to the lateral walls of the pelvis. These ligaments tend to prevent inferior movement of the uterus. Additional mechanical support is provided by the skeletal muscles and fascia of the pelvic floor. ∞ *[p. 338]*

Internal Anatomy of the Uterus

We can divide the uterus into two anatomical regions (Figure 28-18c•): the *body* and the *cervix*. The uterine **body**, or *corpus*, is the largest region of the uterus. The **fundus** is the rounded portion of the body superior to the attachment of the uterine tubes. The body ends at a constriction known as the uterine **isthmus**. The **cervix** (SER-viks) is the inferior portion of the uterus that extends from the isthmus to the vagina.

The tubular cervix projects about 1.25 cm (0.5 in.) into the vagina. Within the vagina, the distal end of the cervix forms a curving surface that surrounds the **cervical os** (*os*, an opening or mouth), or *external orifice* of the uterus. The cervical os leads into the **cervical canal**, a constricted passageway that opens into the **uterine cavity** of the body at the **internal os**, or *internal orifice* (Figure 28-18c•).

The uterus receives blood from branches of the **uterine arteries**, which arise from branches of the *internal iliac arteries*, and the *ovarian arteries*, which arise from the abdominal aorta inferior to the renal arteries. There are extensive interconnections among the arteries to the uterus. This arrangement helps ensure a reliable flow of blood to the organ despite changes in position and the changes in uterine shape that accompany pregnancy. Numerous veins and lymphatic vessels also supply each portion of the uterus. The uterus is innervated by autonomic fibers from the hypogastric plexus (sympathetic) and from sacral segments S_3 and S_4 (parasympathetic). Sensory information reaches the CNS within the dorsal roots of spinal nerves T_{11} and T_{12}. The most delicate anesthetic procedures used during labor and delivery, known as *segmental blocks*, target only spinal nerves T_{10}–L_1.

The Uterine Wall

The dimensions of the uterus are highly variable. In adult women of reproductive age who have not given birth, the uterine wall is about 1.5 cm (0.5 in.) thick. The uterine wall has a thick, outer, muscular **myometrium** (mī-ō-MĒ-trē-um; *myo-*, muscle + *metra*, uterus) and a thin, inner, glandular **endometrium** (en-dō-MĒ-trē-um), or *mucosa*. The fundus and the posterior surface of the uterine body and isthmus are covered by a serous membrane continuous with the peritoneal lining. This incomplete serosa is called the **perimetrium** (Figure 28-18c•).

The endometrium contributes about 10 percent to the mass of the uterus. The glandular and vascular tissues of the endometrium support the physiological demands of the growing fetus. Vast numbers of uterine glands open onto the endometrial surface. These glands extend deep into the lamina propria almost all the way to the myometrium. Under the influence of estrogen, the uterine glands, blood vessels, and epithelium change with the various phases of the monthly *uterine cycle*.

The myometrium, the thickest portion of the uterine wall, forms almost 90 percent of the mass of the uterus. Smooth muscle in the myometrium is arranged into longitudinal, circular, and oblique layers. The smooth muscle tissue of the myometrium provides much of the force needed to move a large fetus out of the uterus and into the vagina.

 CERVICAL CANCER Cervical cancer is the most common reproductive system cancer in women age 15–34. Roughly 14,500 new cases of invasive cervical cancer are diagnosed each year in the United States, and approximately 33 percent of the individuals will eventually die of this condition. Another 34,900 patients are diagnosed with a less aggressive form of cervical cancer. Cervical and other uterine tumors and cancers are discussed in the *Applications Manual*. **AM** *Uterine Tumors and Cancers*

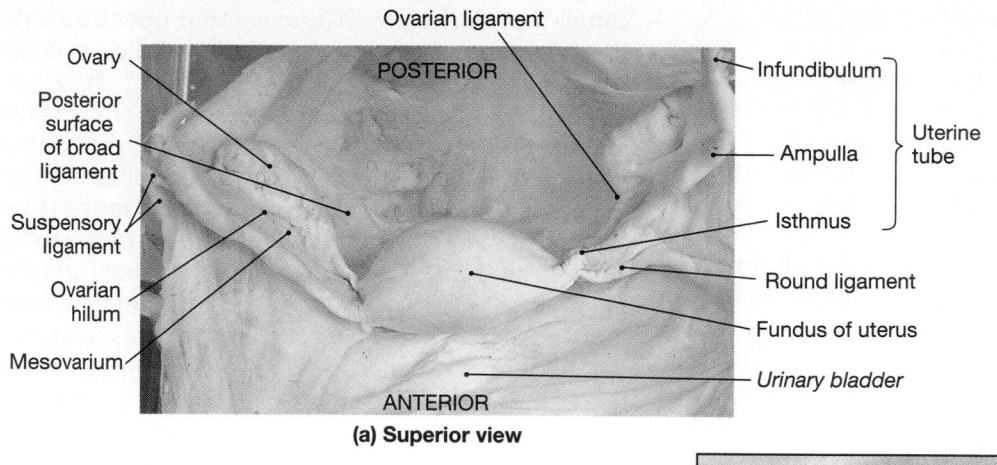

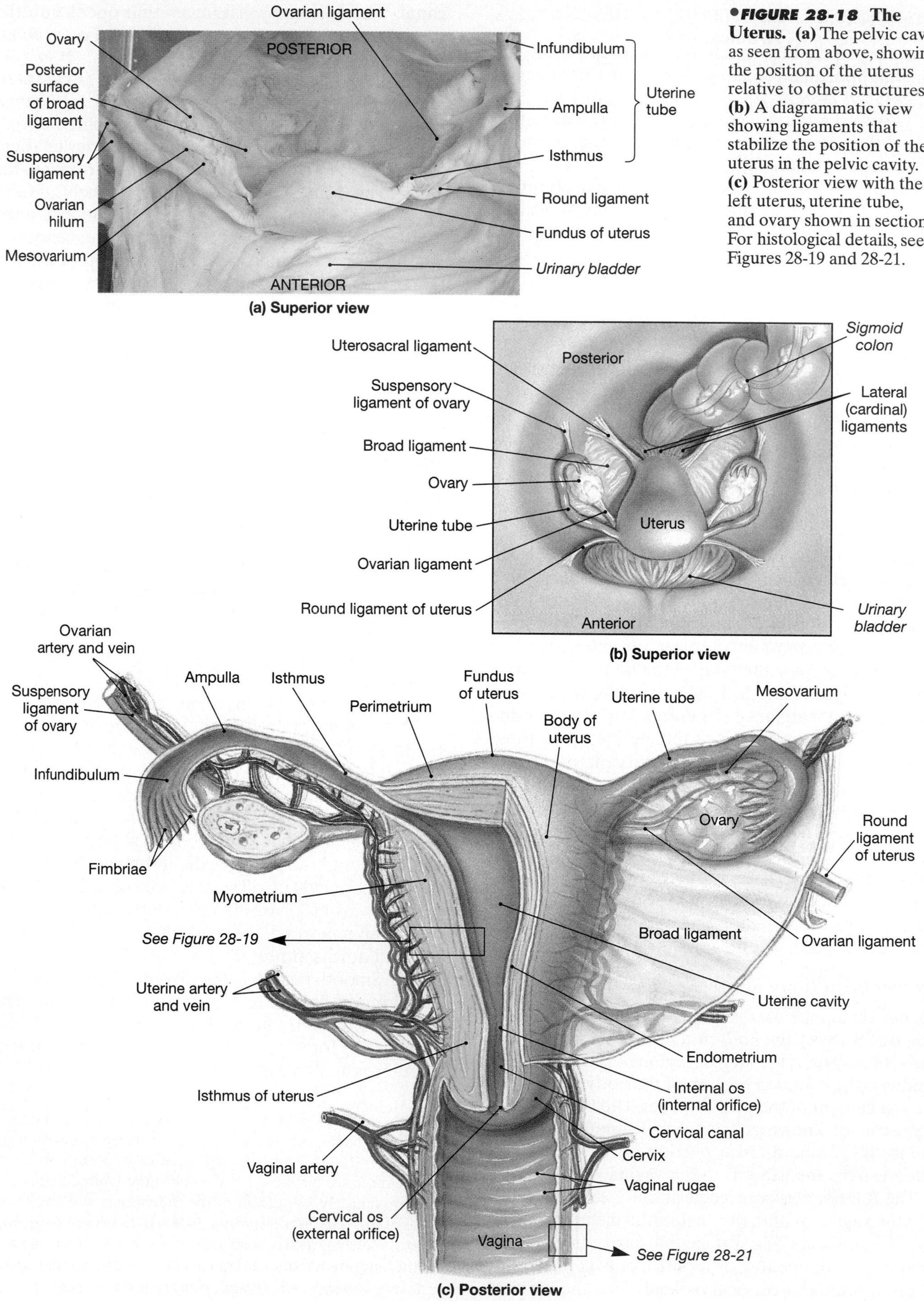

Ovarian ligament

POSTERIOR

Ovary

Posterior surface of broad ligament

Suspensory ligament

Ovarian hilum

Mesovarium

Infundibulum

Ampulla

Uterine tube

Isthmus

Round ligament

Fundus of uterus

Urinary bladder

ANTERIOR

(a) Superior view

•*FIGURE 28-18* **The Uterus. (a)** The pelvic cavity as seen from above, showing the position of the uterus relative to other structures. **(b)** A diagrammatic view showing ligaments that stabilize the position of the uterus in the pelvic cavity. **(c)** Posterior view with the left uterus, uterine tube, and ovary shown in section. For histological details, see Figures 28-19 and 28-21.

Uterosacral ligament

Suspensory ligament of ovary

Broad ligament

Ovary

Uterine tube

Ovarian ligament

Round ligament of uterus

Posterior

Sigmoid colon

Lateral (cardinal) ligaments

Uterus

Anterior

Urinary bladder

(b) Superior view

Ovarian artery and vein

Suspensory ligament of ovary

Infundibulum

Fimbriae

Myometrium

See Figure 28-19

Uterine artery and vein

Isthmus of uterus

Vaginal artery

Cervical os (external orifice)

Ampulla

Isthmus

Perimetrium

Fundus of uterus

Body of uterus

Uterine tube

Mesovarium

Ovary

Round ligament of uterus

Broad ligament

Ovarian ligament

Uterine cavity

Endometrium

Internal os (internal orifice)

Cervical canal

Cervix

Vaginal rugae

Vagina

See Figure 28-21

(c) Posterior view

HISTOLOGY OF THE UTERUS   We can divide the endometrium into (1) a **functional zone**, the layer closest to the uterine cavity, and (2) an outer **basilar zone** adjacent to the myometrium. The functional zone contains most of the uterine glands and contributes most of the endometrial thickness. The basilar zone attaches the endometrium to the myometrium and contains the terminal branches of the tubular glands (Figure 28-19a●).

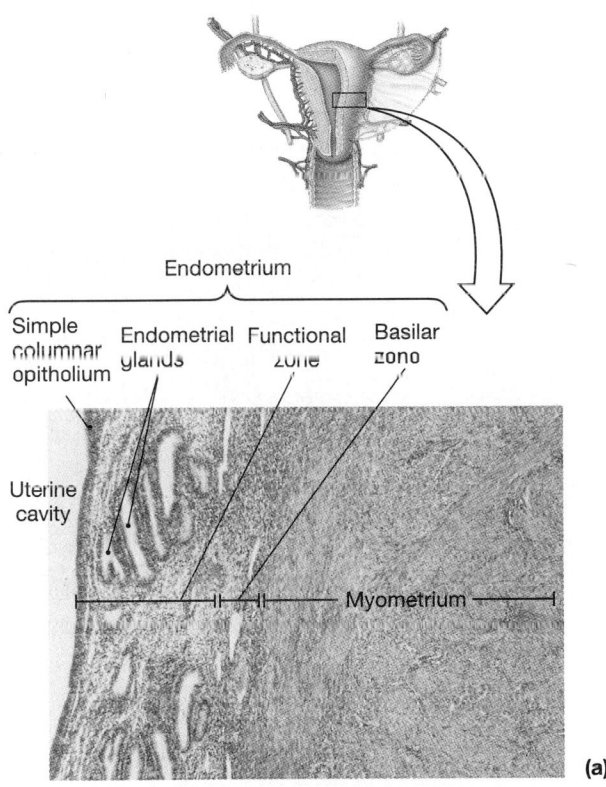

Endometrium

Simple columnar epithelium | Endometrial glands | Functional zone | Basilar zone

Uterine cavity

Myometrium

(a)

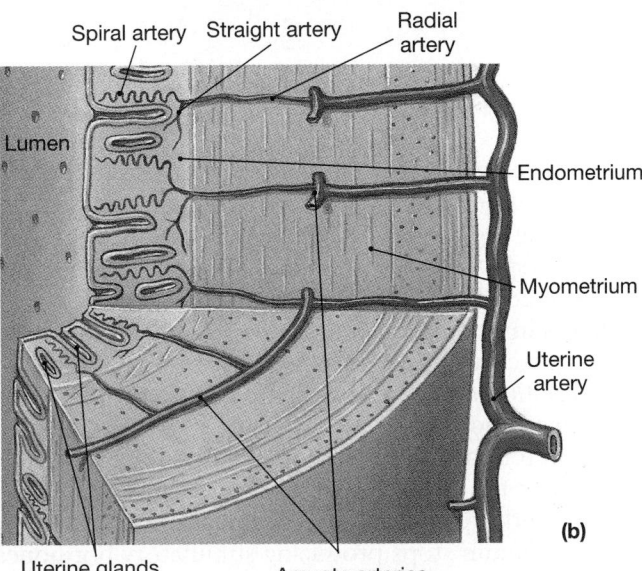

Spiral artery | Straight artery | Radial artery

Lumen

— Endometrium

— Myometrium

Uterine artery

(b)

Uterine glands

Arcuate arteries

●FIGURE 28-19 **The Uterine Wall.** **(a)** The basic structure of the endometrium. (LM × 32) **(b)** A sectional view of the uterine wall, showing the endometrial regions and the circulatory supply to the endometrium.

Within the myometrium, branches of the uterine arteries form **arcuate arteries** that encircle the endometrium. **Radial arteries** supply **straight arteries** that deliver blood to the basilar zone of the endometrium and **spiral arteries** that supply the functional zone (Figure 28-19b●).

The structure of the basilar zone remains relatively constant over time, but that of the functional zone undergoes cyclical changes in response to sex hormone levels. These cyclical changes produce the characteristic histological features of the uterine cycle.

The Uterine Cycle

The **uterine cycle**, or *menstrual* (MEN stroo al) *cycle*, is a repeating series of changes in the structure of the endometrium. The uterine cycle averages 28 days in length, but it can range from 21 to 35 days in healthy women of reproductive age. We can divide the cycle into three phases: (1) *menses*, (2) the *proliferative phase*, and (3) the *secretory phase*. The histological appearance of the endometrium during each phase is shown in Figure 28-20●. The phases occur in response to hormones associated with the regulation of the ovarian cycle. Menses and the proliferative phase occur during the follicular phase of the ovarian cycle. The secretory phase corresponds to the luteal phase of the ovarian cycle. We shall consider the regulatory mechanism in a later section.

MENSES   The uterine cycle begins with the onset of **menses** (MEN-sēz), an interval marked by the degeneration of the functional zone of the endometrium. The deterioration occurs in patches. It is caused by the constriction of the spiral arteries, which reduces blood flow to areas of endometrium. Deprived of oxygen and nutrients, the secretory glands and other tissues in the functional zone begin to deteriorate. Eventually, the weakened arterial walls rupture, and blood pours into the connective tissues of the functional zone. Blood cells and degenerating tissues then break away and enter the uterine lumen, to be lost by passage through the cervical os and into the vagina. Only the functional zone is affected, because the deeper, basilar zone is provided with blood from the straight arteries, which remain unconstricted.

The sloughing of tissue is gradual, and at each site, repairs begin almost at once. Nevertheless, before menses has ended, the entire functional zone has been lost (Figure 28-20a●). This process of endometrial sloughing is called **menstruation** (men-stroo-Ā-shun). Menstruation generally lasts from 1 to 7 days, and over this period roughly 35 to 50 ml of blood is lost. The process can be relatively painless. Painful menstruation, or **dysmenorrhea**, can result from

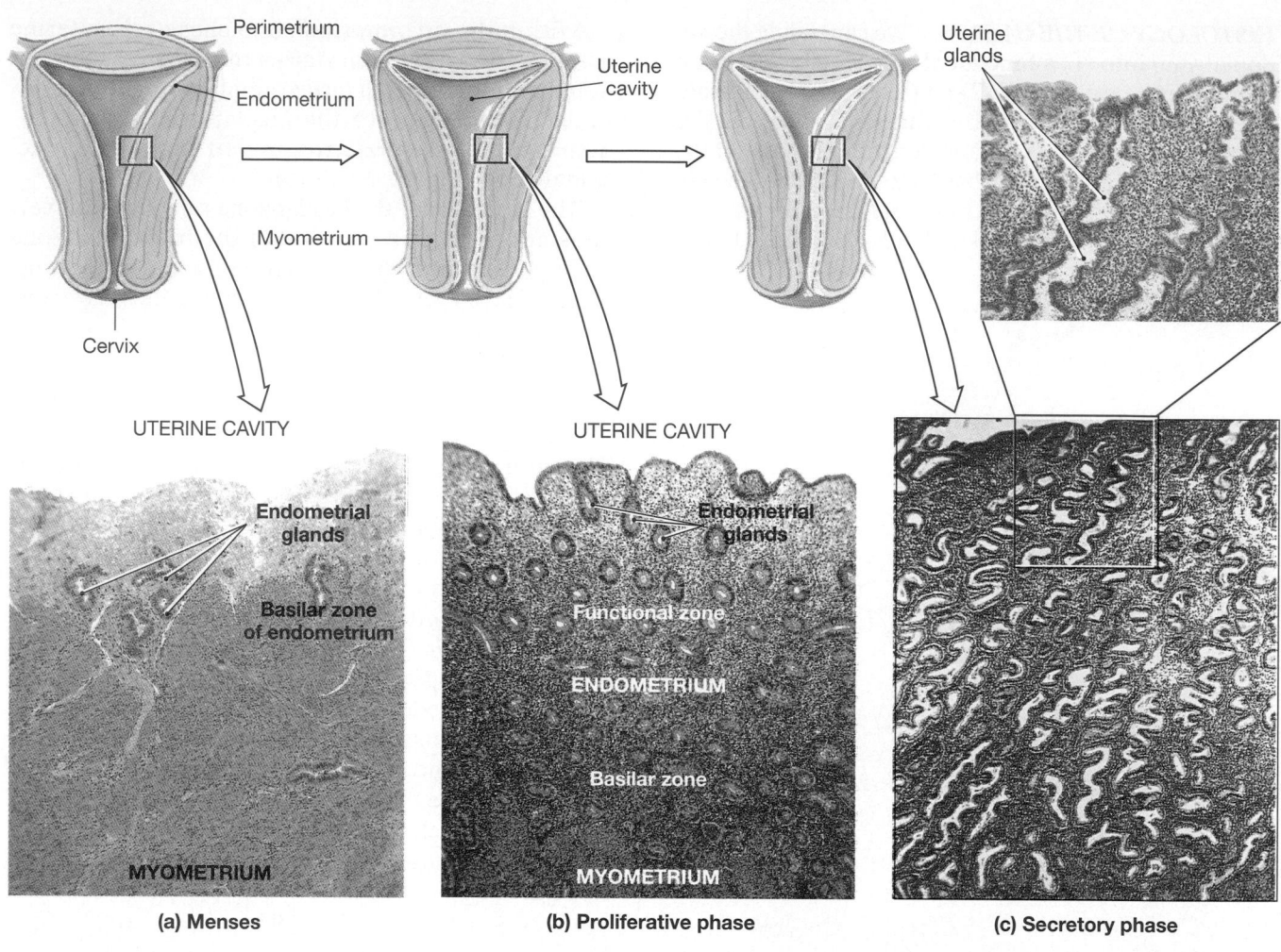

●*FIGURE 28-20* **The Uterine Cycle.** Micrographs showing the appearance of the endometrium **(a)** at menses (LM × 63) and during the **(b)** proliferative (LM × 66) and **(c)** secretory (LM × 52; blowup, LM × 150) phases of the uterine cycle.

uterine inflammation and contraction or from conditions involving adjacent pelvic structures.

THE PROLIFERATIVE PHASE The basilar zone, including the basal portions of the uterine glands, survives menses intact. In the days following the completion of menses, the epithelial cells of the glands multiply and spread across the endometrial surface, restoring the integrity of the uterine epithelium (Figure 28-20b●). Further growth and vascularization result in the complete restoration of the functional zone. As this reorganization proceeds, the endometrium is said to be in the **proliferative phase**. This restoration occurs at the same time as the enlargement of primary and secondary follicles in the ovary. The proliferative phase is stimulated and sustained by estrogens secreted by the developing ovarian follicles.

By the time ovulation occurs, the functional zone is several millimeters thick, and prominent mucous glands extend to the border with the basilar zone. At this time, the endometrial glands are manufacturing a mucus rich in glycogen. The entire functional zone is highly vascu-larized, with small arteries spiraling toward the inner surface from larger arteries in the myometrium.

THE SECRETORY PHASE During the **secretory phase** of the uterine cycle, the endometrial glands enlarge, accelerating their rates of secretion, and the arteries elongate and spiral through the tissues of the functional zone (Figure 28-20c●). This activity occurs under the combined stimulatory effects of progestins and estrogens from the corpus luteum. This phase begins at the time of ovulation and persists as long as the corpus luteum remains intact.

Secretory activities peak about 12 days after ovulation. Over the next day or two, the glandular activity declines, and the uterine cycle comes to a close as the corpus luteum stops producing stimulatory hormones. A new cycle then begins with the onset of menses and the disintegration of the functional zone. The secretory phase generally lasts 14 days. As a result, you can determine the date of ovulation after the fact, by counting backward 14 days from the first day of menses.

ENDOMETRIOSIS In *endometriosis* (en-dō-mē-trē-Ō-sis), an area of endometrial tissue begins to grow outside the uterus. The cause is unknown; because this condition is most common in the inferior portion of the peritoneum, one possibility is that pieces of endometrium sloughed during menstruation are in some way forced through the uterine tubes into the peritoneal cavity, where they have reattached. The severity of the condition depends on the size of the abnormal mass and its location. Abdominal pain, bleeding, pressure on adjacent structures, and infertility are common symptoms. As the island of endometrial tissue enlarges, the symptoms become more severe.

Diagnosis can generally be made by using a laparoscope inserted through a small opening in the abdominal wall. Using this device, a physician can inspect the outer surfaces of the uterus and uterine tubes, the ovaries, and the lining of the pelvic cavity. Treatment of endometriosis may involve hormonal therapy or surgical removal of the endometrial mass. If the condition is widespread, a *hysterectomy* (removal of the uterus) or *oophorectomy* (removal of the ovaries) may be required.

MENARCHE AND MENOPAUSE The uterine cycle begins with the **menarche** (me-NAR-kē), or first menstrual period at puberty, typically at age 11–12. The cycles continue until age 45–55, when **menopause** (MEN-ō-paws), the last uterine cycle, occurs. Over the intervening three and a half to four decades, the regular appearance of uterine cycles is interrupted only by unusual circumstances, such as illness, stress, starvation, or pregnancy.

If menarche does not appear by age 16, or if the normal uterine cycle of an adult woman becomes interrupted for 6 months or more, the condition of **amenorrhea** (ā-men-ō-RĒ-uh) exists. *Primary amenorrhea* is the failure to initiate menses. This condition may indicate developmental abnormalities, such as nonfunctional ovaries, the absence of a uterus, or an endocrine or genetic disorder. Transient *secondary amenorrhea* may be caused by severe physical or emotional stresses. In effect, the reproductive system gets "switched off" under these conditions. Examples of factors that can cause amenorrhea include drastic weight-reduction programs, anorexia nervosa, and severe depression or grief. Amenorrhea has also been observed in marathon runners and other women engaged in training programs that require sustained high levels of exertion and severely reduce body lipid reserves.

The Vagina

The vagina is an elastic, muscular tube. It extends between the cervix of the uterus and the *vestibule*, a space bounded by the female external genitalia (Figures 28-13, p. 1055, and 28-14•, p. 1056). The vagina has an average length of 7.5–9 cm (3–3.5 in.), but because the vagina is highly distensible, its length and width vary.

At the proximal end of the vagina, the cervix projects into the **vaginal canal**. The shallow recess surrounding the cervical protrusion is known as the **fornix** (FOR-niks). The vagina lies parallel to the rectum, and the two are in close contact posteriorly. Anteriorly, the

urethra extends along the superior wall of the vagina from the urinary bladder to the external urethral meatus, which opens into the vestibule. The primary blood supply of the vagina is via the **vaginal branches** of the internal iliac (or uterine) arteries and veins. Innervation is from the hypogastric plexus, sacral nerves S_2–S_4, and branches of the pudendal nerve.

The vagina has three major functions:

1. It serves as a passageway for the elimination of menstrual fluids.
2. It receives the penis during sexual intercourse and holds spermatozoa prior to their passage into the uterus.
3. It forms the lower portion of the *birth canal*, through which the fetus passes during delivery.

Histology of the Vagina

In sectional view, the lumen of the vagina appears constricted, forming a rough H. The vaginal walls contain a network of blood vessels and layers of smooth muscle (Figure 28-21•). The lining is moistened by the secretions of the cervical glands and by the movement of water across the permeable epithelium. The vagina and vestibule are separated by the **hymen** (HĪ-men), an elastic epithelial fold that partially or completely blocks the entrance to the vagina before the onset of sexual intercourse. The two *bulbospongiosus muscles* extend along either side of the vaginal entrance, and their contractions constrict the entrance. ∞ *[p. 338]* These muscles cover the *vestibular bulbs*, masses of erectile tissue on either side of the vaginal entrance. The vestibular

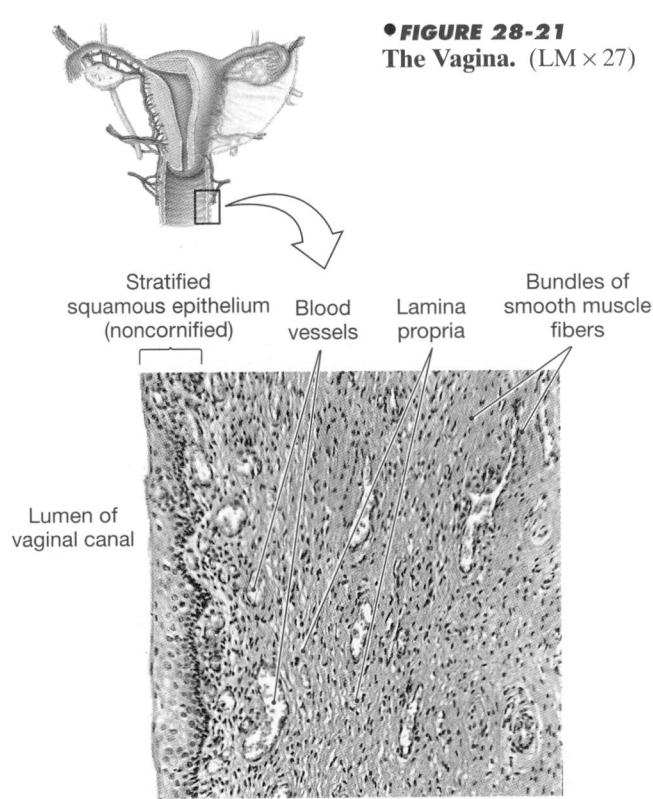

•**FIGURE 28-21**
The Vagina. (LM × 27)

Stratified squamous epithelium (noncornified)

Blood vessels

Lamina propria

Bundles of smooth muscle fibers

Lumen of vaginal canal

bulbs have the same embryological origins as the corpus spongiosum of the penis in the male.

The vaginal lumen (Figure 28-21●) is lined by a nonkeratinized stratified squamous epithelium that in the relaxed state is thrown into folds called *rugae*. The underlying lamina propria is thick and elastic, and it contains small blood vessels, nerves, and lymph nodes. The vaginal mucosa is surrounded by an elastic **muscularis** layer, with layers of smooth muscle fibers arranged in circular and longitudinal bundles continuous with the uterine myometrium. The portion of the vagina adjacent to the uterus has a serosal covering continuous with the pelvic peritoneum. Along the rest of the vagina, the muscularis layer is surrounded by an *adventitia* of fibrous connective tissue.

The vagina contains a population of resident bacteria, usually harmless, supported by nutrients in the cervical mucus. The metabolic activity of these bacteria creates an acidic environment, which restricts the growth of many pathogens. An inflammation of the vaginal canal, known as *vaginitis* (va-jin-Ī-tis), is caused by fungi, bacteria, or parasites. In addition to any discomfort that may result, the condition may affect the survival of sperm and thereby reduce fertility. [AM] *Vaginitis* An acidic environment also inhibits sperm motility; for this reason, the buffers in semen are important to successful fertilization.

The hormonal changes associated with the ovarian cycle also have an effect on the vaginal epitheli-

um. A *vaginal smear* is a sample of epithelial cells shed at the vaginal surface. By examining these cells, a clinician can estimate the corresponding stage in the ovarian and uterine cycles. This technique is an example of *exfoliative cytology*, a diagnostic procedure we introduced in Chapter 4. ∞ *[p. 117]*

The External Genitalia

The region containing the female external genitalia is the **vulva** (VUL-vuh), or *pudendum* (Figure 28-22●). The vagina opens into the **vestibule**, a central space bounded by small folds known as the **labia minora** (LĀ-bē-uh mi-NOR-uh; singular, *labium minus*). The labia minora are covered with a smooth, hairless skin. The urethra opens into the vestibule just anterior to the vaginal entrance. The **paraurethral glands,** or *Skene's glands,* discharge into the urethra near the external urethral opening. Anterior to the urethral opening, the **clitoris** (KLI-to-ris) projects into the vestibule. The clitoris, a small, rounded tissue projection, is the female equivalent of the penis; it is derived from the same embryonic structures (see the Embryology Summary on pp. 1073–1075). Internally, it contains erectile tissue comparable to the corpora cavernosa of the penis. The clitoris engorges with blood during sexual arousal. A small erectile *glans* sits atop the organ; extensions of the labia minora encircle the body of the clitoris, forming the **prepuce**, or *hood*, of the clitoris.

A variable number of small **lesser vestibular glands** discharge their secretions onto the exposed surface of

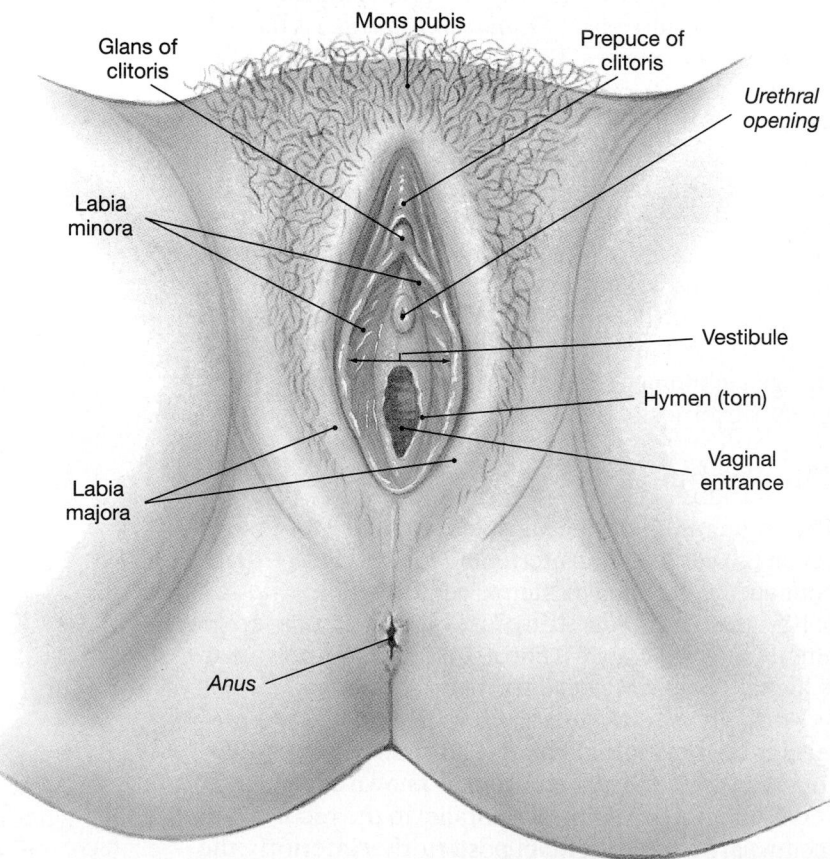

●**FIGURE 28-22** The Female External Genitalia

the vestibule, keeping it moistened. During arousal, a pair of ducts discharges the secretions of the **greater vestibular glands** *(Bartholin's glands)* into the vestibule near the posterolateral margins of the vaginal entrance. These mucous glands have the same embryological origins as the bulbourethral glands of males.

The outer limits of the vulva are established by the *mons pubis* and the *labia majora*. The bulge of the **mons pubis** is created by adipose tissue beneath the skin anterior to the pubic symphysis. Adipose tissue also accumulates within the **labia majora** (singular, *labium majus*), prominent folds of skin that encircle and partially conceal the labia minora and adjacent structures. The outer margins of the labia majora and the mons pubis are covered with coarse hair, but the inner surfaces of the labia majora are relatively hairless. Sebaceous glands and scattered apocrine sweat glands secrete onto the inner surface of the labia majora, moistening them and providing lubrication.

The Mammary Glands

A newborn infant cannot fend for itself, and several of its key systems have yet to complete their development. Over the initial period of adjustment to an independent existence, the infant gains nourishment from the milk secreted by the maternal **mammary glands**. Milk production, or **lactation** (lak-TĀ-shun), occurs in these glands, which in females are specialized organs of the integumentary system that are controlled primarily by hormones of the reproductive system and the *placenta*, a temporary structure that provides the embryo or fetus with nutrients.

On each side, a mammary gland lies in the subcutaneous tissue of the **pectoral fat pad** beneath the skin of the chest (Figure 28-23a•). Each breast bears a small conical projection, the **nipple**, where the ducts of underlying mammary gland opens onto the body surface. The reddish-brown region of skin around each nipple is the **areola** (a-RĒ-ō-luh). Large sebaceous glands beneath the areolar surface give it a grainy texture.

The glandular tissue of the mammary gland consists of separate lobes, each containing several secretory lobules. Ducts leaving the lobules converge, giving rise to a single **lactiferous** (lak-TIF-e-rus) **duct** in each lobe (Figure 28-23a,c•). Near the nipple, that lactiferous duct enlarges, forming an expanded chamber called a **lactiferous sinus**. Typically, 15–20 lactiferous sinuses open onto the surface of each nipple.

Dense connective tissue surrounds the duct system and forms partitions that extend between the lobes and the lobules. These bands of connective tissue, known as the **suspensory ligaments of the breast**, originate in the dermis of the overlying skin. A layer of loose connective tissue separates the mammary complex from the underlying pectoralis muscles. Branches of the *internal thoracic artery* supply blood to each mammary gland (Figure 21-23•, p. 742).

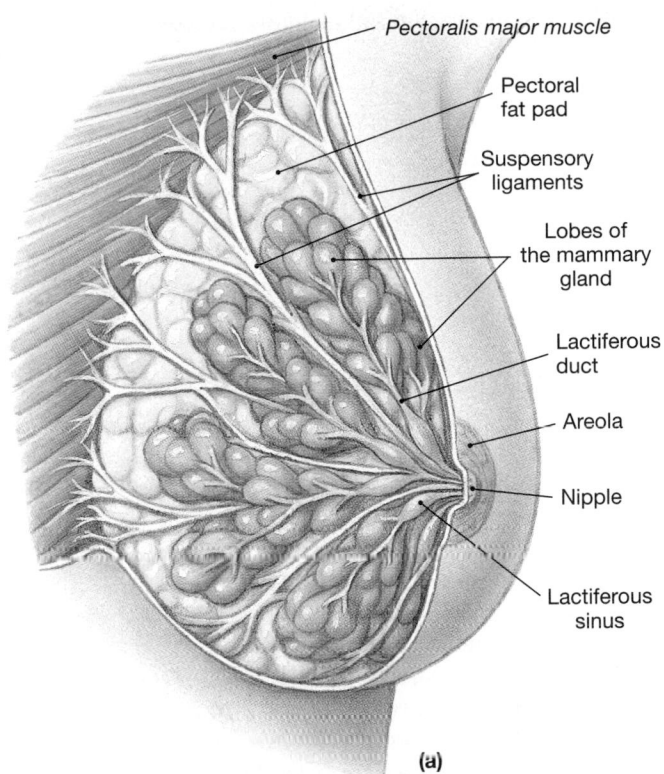

(a)

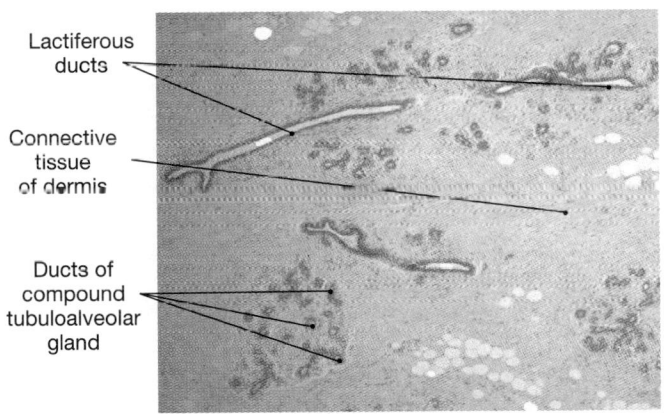

(b) Inactive mammary gland

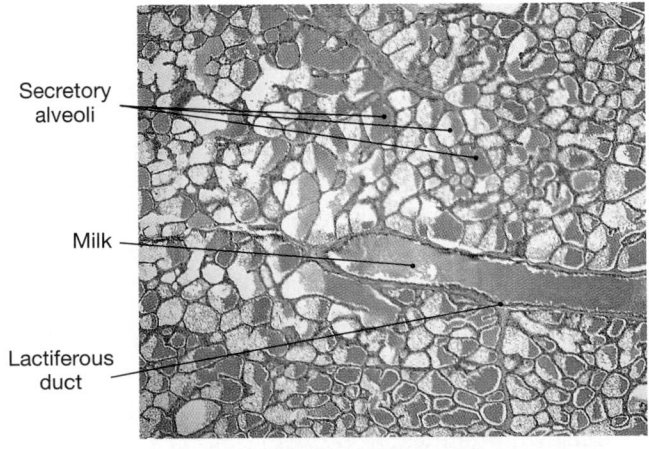

(c) Active mammary gland

•FIGURE 28-23 The Mammary Glands. (a) Structure of the mammary gland of the left breast. **(b)** An inactive mammary gland of a nonpregnant woman. (LM × 53) **(c)** An active mammary gland of a nursing woman. (LM × 119)

 Breast Cancer

The mammary glands are cyclically stimulated by the changing levels of circulating reproductive hormones that accompany the uterine cycle. The effects usually go unnoticed, but there can be occasional discomfort or even inflammation of mammary gland tissues late in the cycle. If inflamed lobules become walled off with scar tissue, **cysts** are created. Clusters of cysts can be felt in the breast as discrete masses, a condition known as **fibrocystic disease**. Because the symptoms are similar to those of breast cancer, biopsies may be needed to distinguish between this benign condition and breast cancer.

Breast cancer is a malignant, metastasizing cancer of the mammary gland. It is the leading cause of death in women between the ages of 35 and 45, but it is most common in women over age 50. There were approximately 44,200 deaths in the United States from breast cancer in 1997, and approximately 181,600 new cases were reported. An estimated 12 percent of women in the United States will develop breast cancer at some point in their lifetime, and the rate is steadily rising. The incidence is highest among Caucasians, somewhat lower in African Americans, and lowest in Asians and American Indians. Notable risk factors include (1) a family history of breast cancer, (2) a pregnancy after age 30, and (3) early menarche (first menstrual period) or late menopause (last menstrual period). Breast cancers in males are very rare, but about 300 men die from breast cancer each year in the United States.

Despite repeated studies, there are no proven links between oral contraceptive use, estrogen therapy, fat consumption, or alcohol use and breast cancer. It appears likely that multiple factors are involved. Most women never develop breast cancer—even women in families with a history of the disease. Adequate amounts of nutrients and vitamins and a diet rich in fruits and vegetables appear to offer some protection against the development of breast cancer. Mothers who breast-fed their babies have a 20 percent lower incidence of breast cancer after menopause than mothers who did not. The reason is not known. Adding to the mystery, nursing does not appear to affect the incidence of premenopausal breast cancer.

Early detection of breast cancer is the key to reducing mortalities. *Most breast cancers are found through self-examination*, but the use of clinical screening techniques has increased in recent years. **Mammography** involves the use of X-rays to examine breast tissues. The radiation dosage can be low,

because only soft tissues must be penetrated. This procedure gives the clearest picture of conditions within the breast tissues. Ultrasound can provide some information, but the images lack the detail of standard mammograms.

For treatment to be successful, the cancer must be identified while it is still relatively small and localized. Once it has grown larger than 2 cm (0.78 in.), the chances of long-term survival worsen. A poor prognosis also follows if the cancer cells have spread through the lymphatic system to the axillary lymph nodes. If the nodes are not yet involved, the chances of 5-year survival are about 82 percent, but if four or more nodes are involved, the survival rate drops to 21 percent.

Treatment of breast cancer begins with the removal of the tumor. Because in most cases the cancer cells begin to spread before the condition is diagnosed, part or all of the affected mammary gland is surgically removed:

- In a **segmental mastectomy**, or *lumpectomy*, only a portion of the mammary gland is removed.
- In a **total mastectomy**, the entire mammary gland is removed, but other tissues are left intact.
- In a **radical mastectomy**, the pectoralis muscles, the mammary gland, and the axillary lymph nodes are removed. In a *modified radical mastectomy*, the most common operation, the mammary gland and nodes are removed but the muscular tissue remains intact.

A combination of chemotherapy, radiation treatments, and hormone treatments may be used to supplement the surgical procedures. *Tamoxifen* is a controversial drug that is used to treat some cases of breast cancer. It is more effective than conventional chemotherapy for treating breast cancer in women over 50, and it has fewer unpleasant side effects. It can also be used in addition to regular chemotherapy in the treatment of advanced-stage disease. As an added bonus, tamoxifen prevents and even reverses the osteoporosis of aging. There are risks, however. When given to premenopausal women, tamoxifen can cause amenorrhea and hot flashes similar to those of menopause. Tamoxifen has also been linked to an increased risk of endometrial cancer and potentially to liver cancer as well. It has been proposed that this drug be used to prevent breast cancer rather than treat it. Large-scale trials are under way to determine whether the benefits of chronic low-dose tamoxifen therapy outweigh the risks of complications.

Development of the Mammary Glands during Pregnancy

Figure 28-23b,c• compares the histological organization of inactive and active mammary glands. The inactive, or *resting*, mammary gland is dominated by a duct system rather than by active glandular cells. The size of the mammary glands in a nonpregnant woman reflects primarily the amount of adipose tissue present, not the amount of glandular tissue. The secretory apparatus does not complete its development until pregnancy occurs. The active mammary gland is a *tubuloalveolar gland*, consisting of multiple glandular tubes that end in secretory alveoli. We shall discuss the hormonal mechanisms involved in the control of mammary gland function in Chapter 29.

☑ As the result of infections such as gonorrhea, scar tissue can block the lumen of each uterine tube. How would this blockage affect a woman's ability to conceive?

☑ What is the advantage of the normally acidic pH of the vagina?

☑ Which layer of the uterus is sloughed off during menstruation?

☑ Would blockage of a single lactiferous sinus interfere with delivery of milk to the nipple? Explain.

Hormones and the Female Reproductive Cycle

The activity of the female reproductive tract is under hormonal control that involves an interplay between pituitary and gonadal secretions. But the regulatory pattern in females is much more complicated than in males, because it must coordinate the ovarian and uterine cycles. Circulating hormones control the **female reproductive cycle**, coordinating the ovarian and uterine cycles to ensure proper reproductive function. If the two cycles cannot be coordinated normally, infertility results. For example, a woman who fails to ovulate will be unable to conceive, even if her uterus is perfectly normal. A woman who ovulates normally but whose uterus is not ready to support an embryo will be just as infertile. Because the processes are complex and difficult to study, many of the biochemical details of the female reproductive cycle still elude us, but the general patterns are reasonably clear.

As in males, GnRH from the hypothalamus regulates reproductive function in females. However, in females the GnRH pulse frequency and amplitude (amount secreted per pulse) change throughout the course of the ovarian cycle. If the hypothalamus were a radio, the pulse frequency would correspond to the radio frequency, and the amplitude would represent the volume control. We will consider changes in pulse frequency, as their effects are both dramatic and reasonably well-understood. Changes in GnRH pulse frequency are essential to normal FSH and LH production and thus to normal ovulation. If GnRH is ab-

sent or supplied at a constant rate (without pulses), FSH and LH secretion will stop in a matter of hours.

When you shift from one radio frequency to another, you change stations. You may then hear a very different message—from talk radio to jazz or from hard rock to classical. When the GnRH pulse frequency shifts, the pituitary gland hears a different message. The cells responsible for FSH and LH production are called *gonadotropes*. When the GnRH pulse frequency shifts, these cells change their pattern of FSH and LH production. For example, at one pulse frequency, the gonadotropes respond by preferentially secreting FSH, whereas at another, LH is the primary hormone released. FSH and LH production also occurs in pulses that follow the rhythm of GnRH pulses. The precise effects of these pulses on target cells are not known. Changes in GnRH pulse frequency are controlled primarily by circulating levels of estrogens and progestins. Estrogens increase the GnRH pulse frequency, and progestins decrease it.

Hormones and the Follicular Phase

Follicular development begins under FSH stimulation; each month some of the primordial follicles begin to develop into primary follicles. As the follicles enlarge, thecal cells start producing *androstenedione*, a steroid hormone that is a key intermediate in the synthesis of most sex hormones (Figure 28-24•). Androstenedione is absorbed by the granulosa cells and converted to estrogens. In addition, small quantities of estrogens are secreted by interstitial cells scattered throughout the ovarian stroma. Circulating estrogens are bound primarily to albumins, with lesser amounts carried by GBG (gonadal steroid-binding globulin).

There are three estrogens in circulation: *estradiol*, *estrone*, and *estriol*. All have similar effects on their target tissues. **Estradiol** (es-tra-DĪ-ol) is the most abundant estrogen and its effects on target tissues are most pronounced. It is the dominant hormone prior to ovulation. In estrogen synthesis, androstenedione is first converted to testosterone, which the enzyme *aromatase* converts to estradiol. The synthesis of estrone and estriol proceeds directly from androstenedione (Figure 28-24•).

Estrogens have multiple functions, affecting the activities of many different tissues and organs throughout the body. Important general functions include (1) stimulating bone and muscle growth, (2) maintaining female secondary sex characteristics such as body hair distribution and the location of adipose tissue deposits, (3) affecting CNS activity (especially in the hypothalamus, where it increases the sexual drive), (4) maintaining functional accessory reproductive glands and organs, and (5) initiating repair and growth of the endometrium. Figure 28-25•, which details the hormonal regulation of ovarian activity, includes an

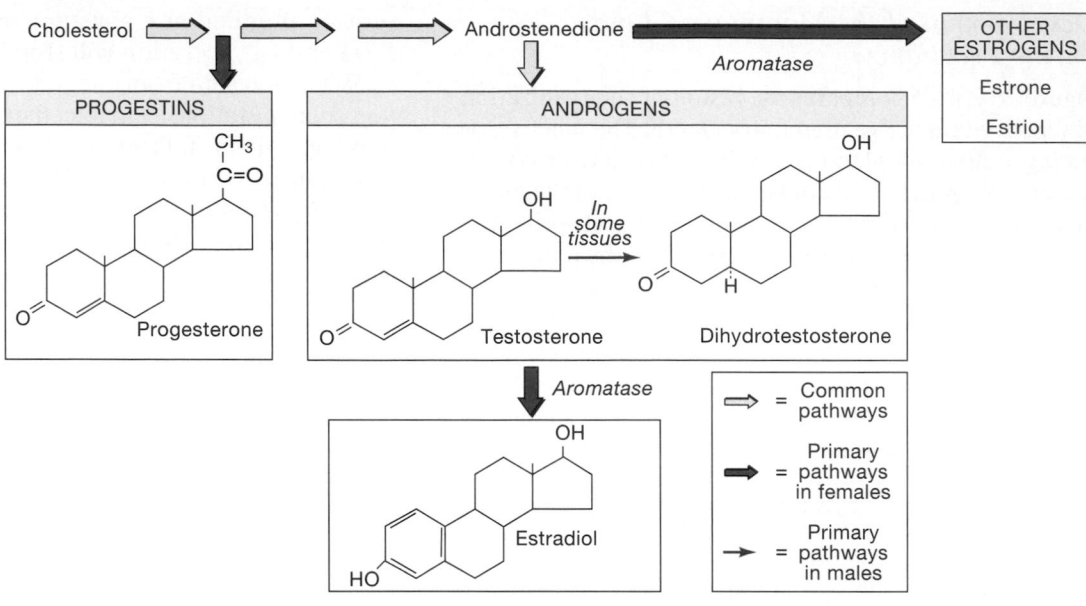

•FIGURE 28-24 Pathways of Steroid Hormone Synthesis. All the gonadal steroids are derived from cholesterol. In men, the pathway ends with the synthesis of testosterone, which may subsequently be converted to dihydrotestosterone. In women, an additional step past testosterone leads to estradiol synthesis. The synthesis of progesterone and the estrogens other than estradiol involve alternative pathways.

overview of the effects of estrogen on various aspects of reproductive function.

Figure 28-26a–d• shows the changes in circulating hormone levels that accompany the ovarian cycle. Early in the follicular phase, estrogen levels are low and the GnRH pulse frequency is 16–24 per day (one pulse every 60–90 minutes). At this pulse frequency, FSH is the dominant hormone released by the anterior pituitary gland; the estrogen released by developing follicles inhibits LH secretion. As secondary follicles develop, FSH levels decline due to the negative feedback effects of inhibin. Follicular development and maturation continues, however, supported by the combination of estrogens, FSH, and LH.

As one or more tertiary follicles begin forming, the concentration of circulating estrogens rises steeply. As a result, the GnRH pulse frequency increases to about 36 per day (one pulse every 30–60 minutes). The increased pulse frequency stimulates LH secretion. In addition,

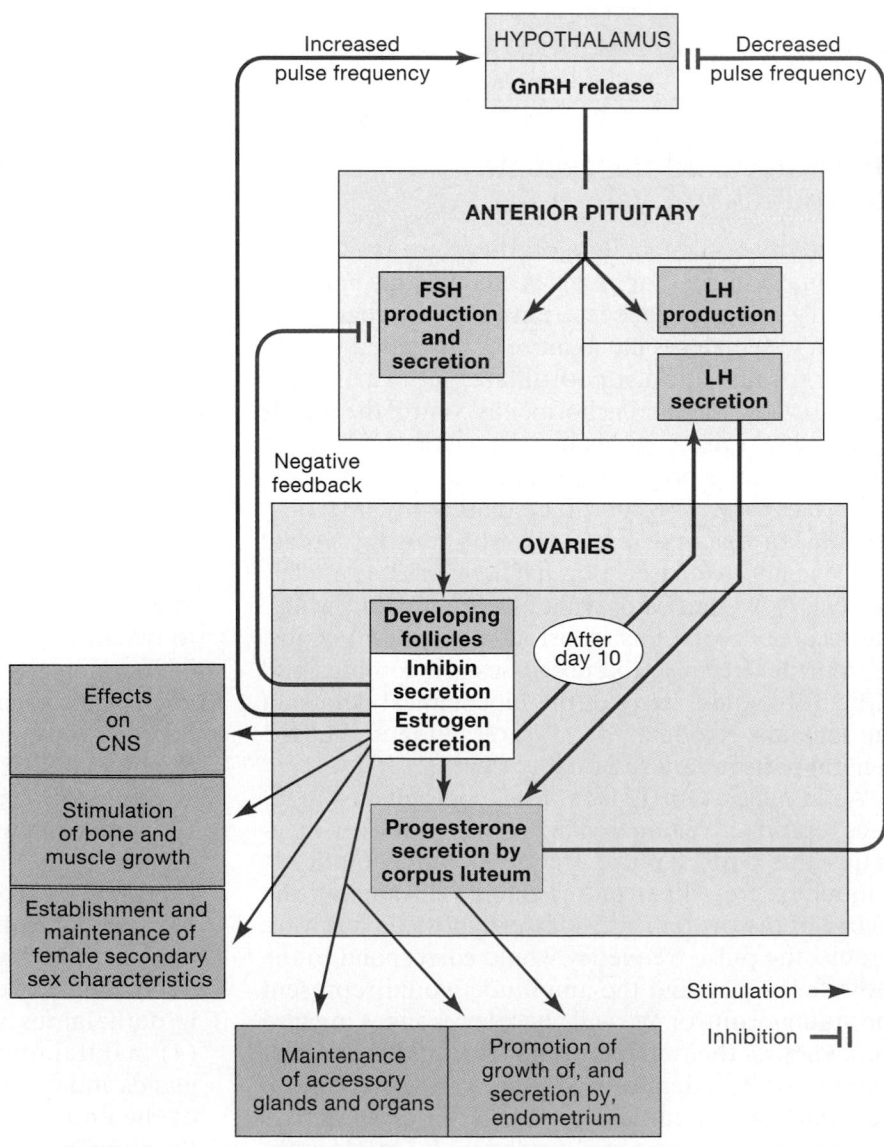

•FIGURE 28-25 Hormonal Regulation of Ovarian Activity

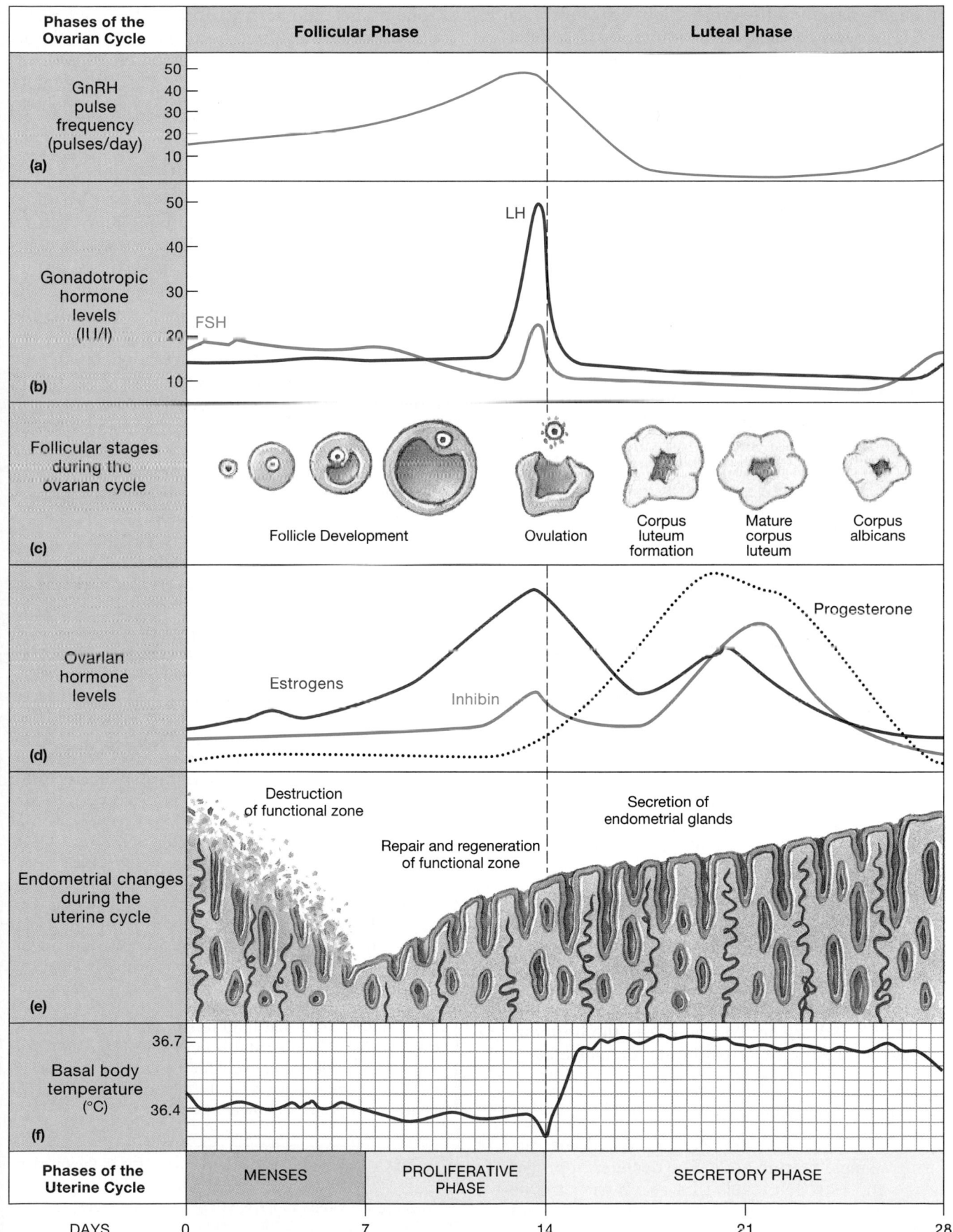

Phases of the Ovarian Cycle

Follicular Phase | **Luteal Phase**

(a) GnRH pulse frequency (pulses/day)

(b) Gonadotropic hormone levels (IU/l)

LH

FSH

(c) Follicular stages during the ovarian cycle

Follicle Development | Ovulation | Corpus luteum formation | Mature corpus luteum | Corpus albicans

(d) Ovarian hormone levels

Estrogens

Inhibin

Progesterone

(e) Endometrial changes during the uterine cycle

Destruction of functional zone

Repair and regeneration of functional zone

Secretion of endometrial glands

(f) Basal body temperature (°C)

36.7

36.4

Phases of the Uterine Cycle

MENSES | PROLIFERATIVE PHASE | SECRETORY PHASE

DAYS 0   7   14   21   28

• FIGURE 28-26 Hormonal Regulation of the Female Reproductive Cycle

at roughly day 10 of the cycle, the effect of estrogen on LH secretion changes from inhibition to stimulation. The switchover occurs only after rising estrogen levels have exceeded a threshold value for about 36 hours. (The threshold level and the time required vary among individuals.) High estrogen levels also increase gonadotrope sensitivity to GnRH. At about day 14, the estrogen level has peaked, the gonadotropes are at maximum sensitivity, and the GnRH pulses are arriving about every 30 minutes. The result is a massive release of LH from the anterior pituitary. This sudden surge in LH concentration triggers (1) the completion of meiosis I by the primary oocyte, (2) the rupture of the follicular wall, and (3) ovulation. Ovulation typically occurs 34–38 hours after the LH surge begins, roughly 9 hours after the LH peak.

Hormones and the Luteal Phase

The high LH levels that trigger ovulation also promote progesterone secretion and the formation of the corpus luteum. As progesterone levels rise and estrogen levels fall, the GnRH pulse frequency declines sharply, soon reaching 1–4 pulses per day. This frequency of GnRH pulses stimulates LH secretion more than it does FSH secretion, and the LH maintains the structure and secretory function of the corpus luteum.

Although moderate amounts of estrogens are secreted by the corpus luteum, progesterone is the main hormone of the luteal phase. Its primary function is to continue the preparation of the uterus for pregnancy by promoting the elaboration of the blood supply to the functional zone and stimulating the secretion of the endometrial glands. Progesterone levels remain relatively high for the next week, but unless pregnancy occurs, the corpus luteum then begins to degenerate. Roughly 12 days after ovulation, the corpus luteum becomes nonfunctional, and progesterone and estrogen levels fall markedly. The blood supply to the functional zone is restricted, and the endometrial tissues begin to deteriorate. As progesterone and estrogen levels decline, the GnRH pulse frequency increases. This increase stimulates FSH secretion by the anterior pituitary, and the cycle begins again.

The hormonal changes involved with the ovarian cycle in turn affect the activities of other reproductive tissues and organs. At the uterus, the hormonal changes are responsible for the maintenance of the uterine cycle.

Hormones and the Uterine Cycle

Figure 28-26e● follows changes in the endometrium during a single uterine cycle. The declines in progesterone and estrogen levels that accompany the degeneration of the corpus luteum (Figure 28-26c,d●) result in menses. The sloughing of endometrial tissue continues for several days, until rising estrogen levels stimulate the repair and regeneration of the function-al zone of the endometrium. The proliferative phase continues until rising progesterone levels mark the arrival of the secretory phase. The combination of estrogen and progesterone then causes the enlargement of the endometrial glands and an increase in their secretory activities.

Hormones and Body Temperature

The monthly hormonal fluctuations cause physiological changes that affect the core body temperature. During the follicular phase, when estrogen is the dominant hormone, *basal body temperature*, or the resting body temperature measured on awakening in the morning, is about 0.3°C lower than it is during the luteal phase, when progesterone dominates the endocrine picture. At the time of ovulation, basal body temperature declines noticeably, making the rise in temperature over the following day even more noticeable (Figure 28-26f●). As a result, by keeping records of body temperature over a few uterine cycles, a woman can often determine the precise day of ovulation. This information can be important for individuals who wish to avoid or promote a pregnancy, because fertilization typically occurs within a day of ovulation. Thereafter, oocyte viability and the likelihood of successful fertilization decrease markedly.

PREMENSTRUAL SYNDROME Several physical and physiological changes occur in women 7–10 days before the start of menses. Fluid retention, breast enlargement, headaches, pelvic pain, and an uncomfortable feeling of bloating are common symptoms. These sensations may be associated with psychological changes producing irritability, anxiety, and depression. This combination of symptoms has been called **premenstrual syndrome (PMS)**.

The mechanism responsible for PMS has yet to be determined. Changes in sex hormone levels may be involved directly, by action on peripheral organ systems, or indirectly, by the modification of neurotransmitter release in the CNS. There are no laboratory tests or procedures to diagnose PMS, but tracking the appearance of symptoms over a 2–3 month period can reveal characteristic patterns. Treatment at present is symptomatic and may involve exercise, dietary change, or medication, depending on the nature of the primary symptom. For example, if headache is the major problem, analgesics are prescribed; diuretics may be used to combat bloating and fluid retention. For severe PMS, drugs can be administered that block GnRH secretion and stop uterine cycles completely for 6 months or more. During the interim, estrogens can be administered to prevent symptoms of premature menopause.

✔ What changes would you expect to observe in the ovarian cycle if the LH surge did not occur?

✔ What effect would blockage of progesterone receptors in the uterus have on the endometrium?

✔ What event occurs in the uterine cycle when the levels of estrogens and progesterone decline?

 Development of the Reproductive System: Gender-Indifferent Stages

(WEEKS 3–6)

DEVELOPMENT OF THE GONADS

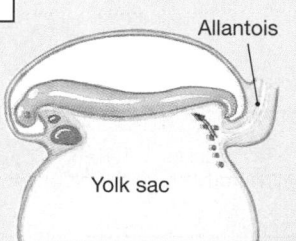

During the third week, endodermal cells migrate from the wall of the yolk sac near the allantois to the dorsal wall of the abdominal cavity. These primordial germ cells enter the **genital ridges** that parallel the mesonephros.

3 WEEKS

Each ridge has a thick epithelium continuous with columns of cells, the **primary sex cords**, that extend into the center (medulla) of the ridge. Anterior to each mesonephric duct, a duct forms that has no connection to the kidneys. This is the **paramesonephric** (Müllerian) **duct**; it extends along the genital ridge and continues toward the cloaca. At this gender-indifferent stage, male embryos cannot be distinguished from female embryos.

DEVELOPMENT OF DUCTS AND ACCESSORY ORGANS

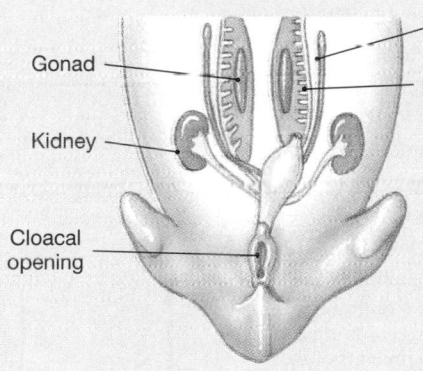

Both genders have mesonephric and paramesonephric ducts at this stage. Unless exposed to androgens, the embryo—regardless of its genetic gender—will develop into a female. In a normal male embryo, cells in the core (medulla) of the genital ridge begin producing testosterone sometime after week 6. Testosterone triggers the changes in the duct system and external genitalia that are detailed on the following page.

DEVELOPMENT OF EXTERNAL GENITALIA

4 WEEKS

6 WEEKS

After 4 weeks of development, there are mesenchymal swellings called **cloacal folds** around the **cloacal membrane** (the cloaca does not open to the exterior). The **genital tubercle** forms the glans of the penis in males and the clitoris in females.

Two weeks later, the cloaca has been subdivided, separating the cloacal membrane into a posterior *anal membrane*, bounded by the *anal folds*, and an anterior **urogenital membrane**, bounded by the **urethral folds**. A prominent **genital swelling** forms lateral to each urethral fold.

EMBRYOLOGY SUMMARY Development of the Male Reproductive System

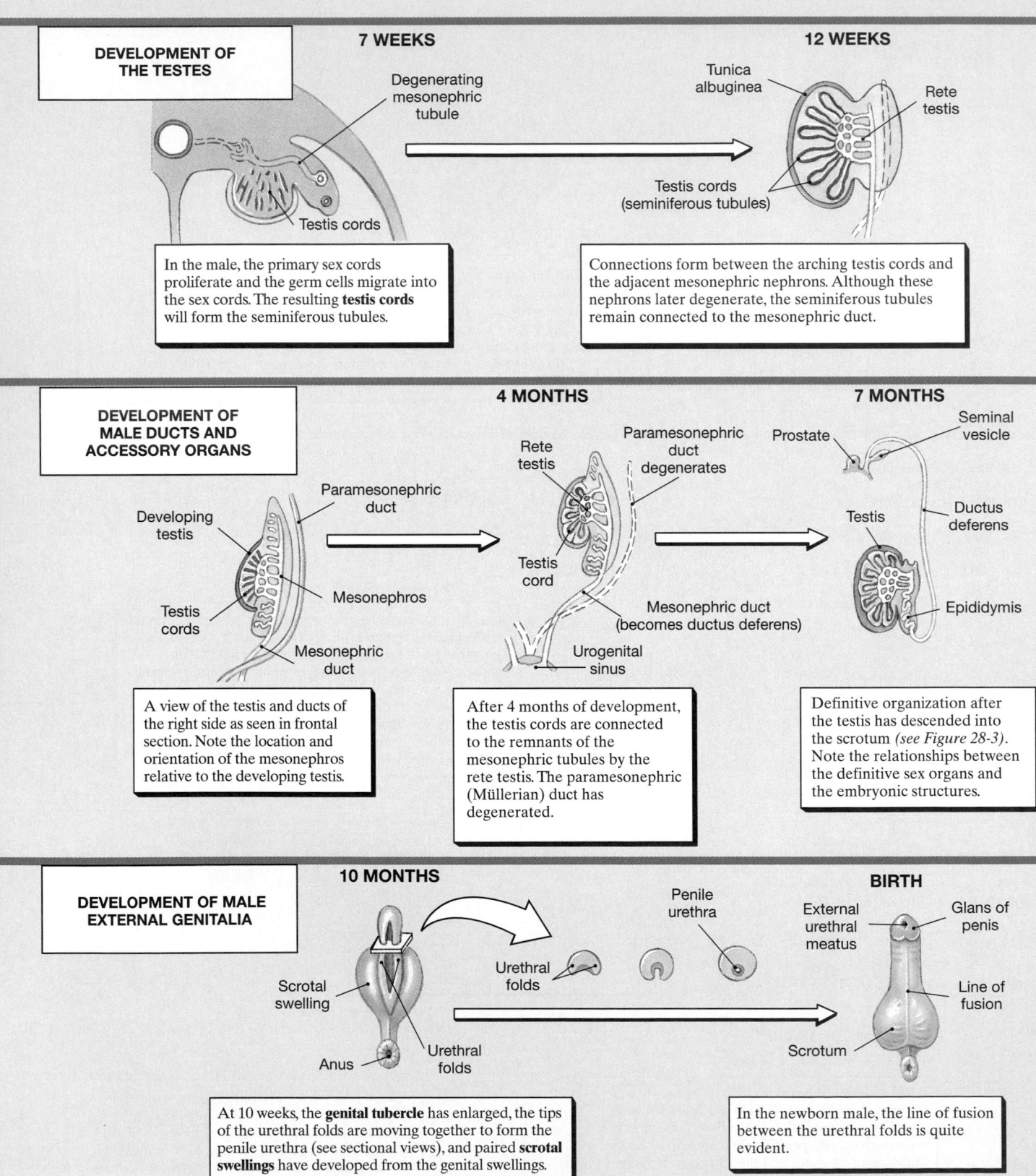

DEVELOPMENT OF THE TESTES

7 WEEKS

Degenerating mesonephric tubule

Testis cords

In the male, the primary sex cords proliferate and the germ cells migrate into the sex cords. The resulting **testis cords** will form the seminiferous tubules.

12 WEEKS

Tunica albuginea

Rete testis

Testis cords (seminiferous tubules)

Connections form between the arching testis cords and the adjacent mesonephric nephrons. Although these nephrons later degenerate, the seminiferous tubules remain connected to the mesonephric duct.

DEVELOPMENT OF MALE DUCTS AND ACCESSORY ORGANS

Developing testis

Paramesonephric duct

Testis cords

Mesonephros

Mesonephric duct

A view of the testis and ducts of the right side as seen in frontal section. Note the location and orientation of the mesonephros relative to the developing testis.

4 MONTHS

Rete testis

Paramesonephric duct degenerates

Testis cord

Urogenital sinus

Mesonephric duct (becomes ductus deferens)

After 4 months of development, the testis cords are connected to the remnants of the mesonephric tubules by the rete testis. The paramesonephric (Müllerian) duct has degenerated.

7 MONTHS

Prostate

Seminal vesicle

Testis

Ductus deferens

Epididymis

Definitive organization after the testis has descended into the scrotum (see Figure 28-3). Note the relationships between the definitive sex organs and the embryonic structures.

DEVELOPMENT OF MALE EXTERNAL GENITALIA

10 MONTHS

Scrotal swelling

Anus

Urethral folds

Urethral folds

Penile urethra

At 10 weeks, the **genital tubercle** has enlarged, the tips of the urethral folds are moving together to form the penile urethra (see sectional views), and paired **scrotal swellings** have developed from the genital swellings.

BIRTH

External urethral meatus

Glans of penis

Line of fusion

Scrotum

In the newborn male, the line of fusion between the urethral folds is quite evident.

Development of the Female Reproductive System

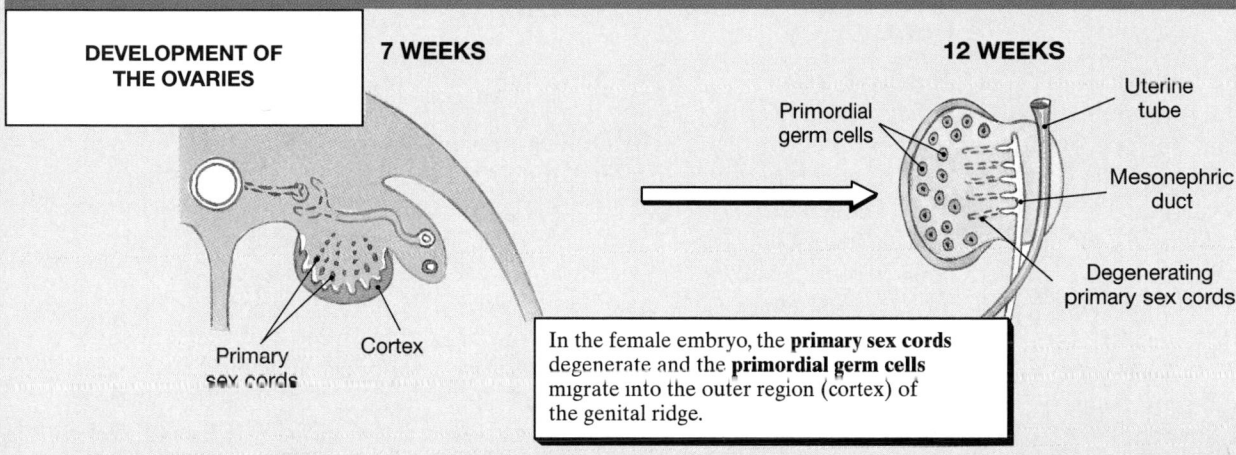

DEVELOPMENT OF THE OVARIES

7 WEEKS

12 WEEKS

Primordial germ cells

Uterine tube

Mesonephric duct

Degenerating primary sex cords

Primary sex cords

Cortex

In the female embryo, the **primary sex cords** degenerate and the **primordial germ cells** migrate into the outer region (cortex) of the genital ridge.

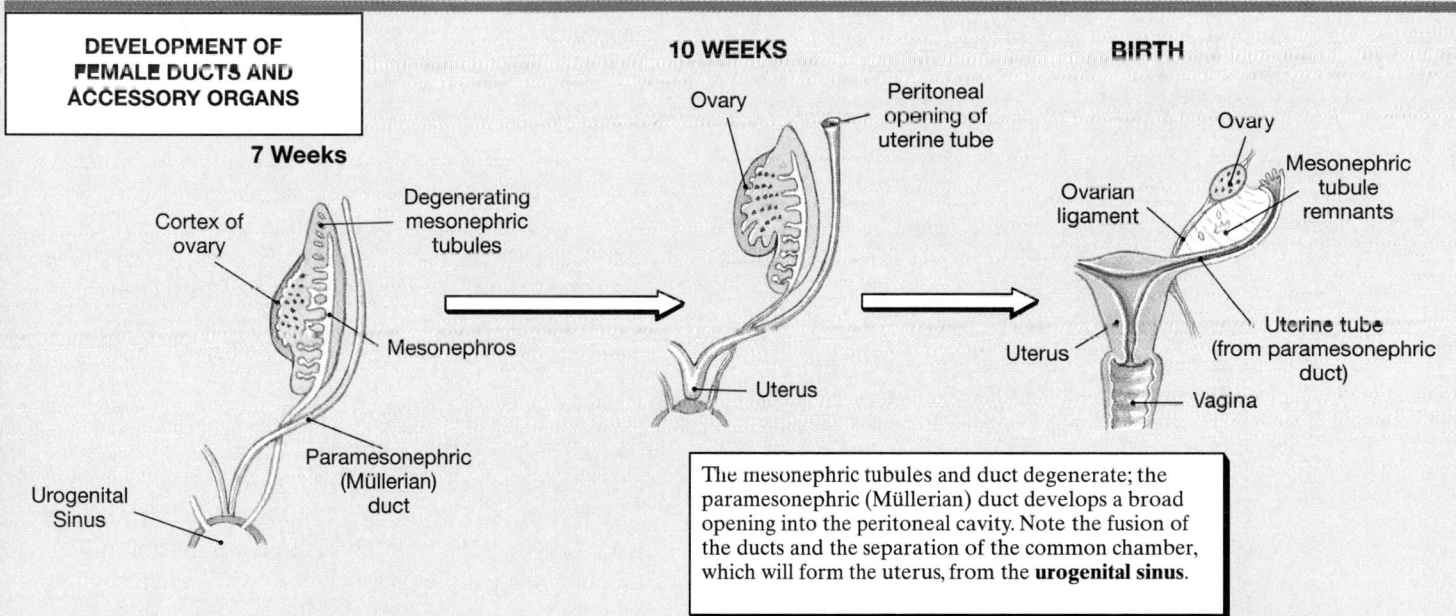

DEVELOPMENT OF FEMALE DUCTS AND ACCESSORY ORGANS

7 Weeks

10 WEEKS

BIRTH

Cortex of ovary

Degenerating mesonephric tubules

Mesonephros

Paramesonephric (Müllerian) duct

Urogenital Sinus

Ovary

Peritoneal opening of uterine tube

Uterus

Ovarian ligament

Ovary

Mesonephric tubule remnants

Uterine tube (from paramesonephric duct)

Uterus

Vagina

The mesonephric tubules and duct degenerate; the paramesonephric (Müllerian) duct develops a broad opening into the peritoneal cavity. Note the fusion of the ducts and the separation of the common chamber, which will form the uterus, from the **urogenital sinus**.

DEVELOPMENT OF FEMALE EXTERNAL GENITALIA

Comparison of Male and Female External Genitalia

Males	Females
Penis	Clitoris
Corpora cavernosa	Erectile tissue
Corpus spongiosum	Vestibular bulbs
Proximal shaft of penis	Labia minora
Penile urethra	Vestibule
Bulbourethral glands	Greater vestibular glands
Scrotum	Labia majora

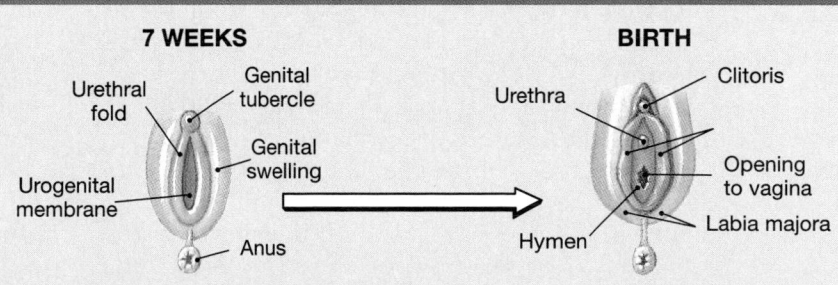

7 WEEKS

BIRTH

Urethral fold

Genital tubercle

Genital swelling

Urogenital membrane

Anus

Urethra

Clitoris

Opening to vagina

Labia majora

Hymen

In the female, the urethral folds do not fuse; they develop into the labia minora. The genital swellings will form the labia majora. The genital tubercle develops into the clitoris. The urethra opens to the exterior immediately posterior to the clitoris. The hymen remains as an elaboration of the urogenital membrane.

 Birth Control Strategies

For physiological, logistical, financial, or emotional reasons, most adults practice some form of conception control during their reproductive years. Well over 50 percent of U.S. women age 15–44 are practicing some method of contraception. In 1996, an estimated 50 million American women were taking oral birth control pills. When the simplest and most obvious method, sexual abstinence, is unsatisfactory for some reason, another method of contraception must be used to avoid unwanted pregnancies. The selection process can be quite involved, for many methods are available. Because each has specific strengths and weaknesses, the potential risks and benefits must be carefully analyzed on an individual basis. We shall consider only a few of the available contraception methods here.

Sterilization is a surgical procedure that makes an individual unable to provide functional gametes for fertilization. Either sexual partner may be sterilized. In a **vasectomy** (vaz-EK-to-mē), a segment of the ductus deferens is removed, making it impossible for spermatozoa to pass from the epididymis to the distal portions of the reproductive tract. The surgery can be performed in a physician's office in a matter of minutes. The spermatic cords are located as they ascend from the scrotum on either side; after each cord is opened, the ductus deferens is severed. After a 1-cm section is removed, the cut ends are tied shut. With the section removed, the cut ends do not reconnect. In time, scar tissue forms a permanent seal. Alternatively, the cut ends of the ductus deferens are blocked with silicone plugs that can later be removed. This more recent vasectomy procedure makes it possible to restore fertility at a later date. After a vasectomy, the man experiences normal sexual function, for the epididymal and testicular secretions normally account for only about 5 percent of the volume of the semen. Spermatozoa continue to develop, but they remain within the epididymis until they degenerate. The failure rate for this procedure is 0.08 percent. (*Failure* for a birth control method is defined as a resulting pregnancy.)

The uterine tubes can be blocked through a surgical procedure known as a **tubal ligation**. The failure rate for this procedure is estimated at 0.45 percent. Because the surgery requires that the abdominopelvic cavity be opened, complications are more likely than with vasectomy. As in a vasectomy, attempts may be made to restore fertility after a tubal ligation.

Oral contraceptives manipulate the female hormonal cycle so that ovulation does not occur. The contraceptive pills produced in the 1950s contained relatively large amounts of progestins. These concentrations were adequate to suppress pituitary production of GnRH, so FSH was not released and ovulation did not occur. In most of the oral contraceptive products developed subsequently, small amounts of estrogens have been added. Current *combination pills* differ significantly from the earlier products in that the hormonal doses are much lower, with only one-tenth the progestins and less than half the estrogens. The hormones are administered in a cyclic fashion, beginning 5 days after the start of menses and continuing for the next 3 weeks. Over the fourth week, the woman takes placebo pills or no pills at all. Low-dosage combination pills are sometimes prescribed for women who experience irregular uterine cycles, for they create a regular 28-day cycle.

At least 20 brands of combination oral contraceptives are now available, and more than 200 million women are using them worldwide. In the United States, 25 percent of women under age 45 use a combination pill to prevent conception. The failure rate for the combination oral contraceptives, when used as prescribed, is 0.24 percent over a 2-year period. Birth control pills are not without risks, however. For example, the combination pills may worsen problems associated with severe hypertension, diabetes mellitus, epilepsy, gallbladder disease, heart trouble, and acne. Women taking oral contraceptives are also at increased risk of venous thrombosis, strokes, pulmonary embolism, and (for women over 35) heart disease.

Two progesterone-only forms of birth control are now available. *Depo-provera*® is injected every 3 months. The Silastic® (silicone rubber) tubes of the *Norplant*® system are saturated with progesterone and inserted under the skin. This method provides birth control for approximately 5 years, but to date the relatively high cost has limited its use. Both Depo-provera and the Norplant system can cause irregular menstruation and temporary amenorrhea, but these methods are easy to use and extremely convenient.

The **condom**, also called a *prophylactic* or "rubber," covers the body of the penis during intercourse and keeps spermatozoa from reaching the female reproductive tract. Condoms are also used to prevent transmission of sexually transmitted diseases, such as syphilis, gonorrhea, and AIDS. The reported condom failure rate varies from 6 to 17 percent, depending on the study criteria.

Vaginal barriers, such as the *diaphragm* and *cervical cap*, rely on similar principles. A diaphragm, the most popular form of vaginal barrier in use at the moment, consists of a dome of latex rubber with a small metal hoop supporting the rim. Because vaginas vary in size, women choosing this method must be individually fitted. Before intercourse, the diaphragm is inserted so that it covers the cervical os. The diaphragm is usually coated with a small amount of spermicidal (sperm-killing) jelly or cream, adding to the effectiveness of the barrier. The failure rate of a properly fitted diaphragm is estimated at 5–6 percent. The cervical cap is smaller and lacks the metal rim. It, too, must be fitted carefully, but unlike the diaphragm, it may be left in place for several days. The failure rate (8 percent) is higher than that for diaphragm use.

An **intrauterine device (IUD)** consists of a small plastic loop or a T that can be inserted into the uterine cavity. The mechanism of action remains uncertain, but it is known that IUDs stimulate prostaglandin production in the uterus. The resulting alteration in the chemical composition of uterine secretions lowers the chances of fertilization and subsequent *implantation* of the zygote into the uterine lining. (We shall discuss implantation in Chapter 29.) IUDs are in limited use today in the United States, but they remain popular in many other countries. The failure rate is estimated at 5–6 percent.

The **rhythm method** involves abstaining from sexual activity on the days ovulation might be occurring. The timing is estimated on the basis of previous patterns of menstruation and sometimes by monitoring changes in basal body temperature and cervical mucus. The failure rate of the rhythm method is very high—almost 25 percent.

Sterilization, oral contraceptives, condoms, and vaginal barriers are the primary contraception methods for all age groups. But the proportion of the population using a particular method varies by age group. Sterilization is most popular among older women, who may already have had children. Relative availability may also play a role. For example, a sexually active female under age 18 can buy a con-

dom more easily than she can obtain a prescription for an oral contraceptive. But many of the differences are attributable to the relationship between risks and benefits for each age group.

When considering the use and selection of contraceptives, many people simply examine the list of potential complications and make the "safest" choice. For example, media coverage of the risks associated with oral contraceptives made many women reconsider their use. But complex decisions should not be made on such a simplistic basis, and the risks associated with use of contraceptives must be considered in light of their relative efficiencies. Although pregnancy is a natural phenomenon, it has its risks, and the mortality rate for pregnant women in the United States averages about 8 deaths per 100,000 pregnancies. That average incorporates a broad range; the rate is 5.4 per 100,000 women under 20, but 27 per 100,000 among women over 40. Although these risks are small, for pregnant women over age 35 the chances of dying from complications related to pregnancy are almost twice as great as the chances of being killed in an automobile accident and are many times greater than the risks associated with the use of oral contraceptives. For women in developing nations, the comparison is even more striking. The mortality rate for pregnant women in parts of Africa is approximately 1 per 150 pregnancies. In addition to preventing pregnancy, combination birth control pills have also been shown to reduce the risks of ovarian and endometrial cancers and fibrocystic breast disease.

Before age 35, *any contraceptive method is safer than pregnancy*. In general, over this period the risks are proportional to the failure rate of each method. The notable exception involves individuals who take the pill but also smoke cigarettes. Younger women are more fertile, so despite a lower mortality rate for each pregnancy, they are likely to have more pregnancies. As a result, birth control failures imply a higher risk in the younger age groups.

After age 35, the risks of complications associated with oral contraceptive use increase, but the risks of using other methods remain relatively stable. Women over age 35 (smokers) or 40 (nonsmokers) are therefore typically advised to seek other forms of contraception. Because each contraceptive method has its own advantages and disadvantages, research on contraception control continues. [AM]

Experimental Contraceptive Methods

THE PHYSIOLOGY OF SEXUAL INTERCOURSE

Sexual intercourse, also known as *coitus* (KŌ-i-tus) or *copulation*, introduces semen into the female reproductive tract. We shall now consider the process as it affects the reproductive systems of males and females.

Male Sexual Function

Sexual function in males is coordinated by complex neural reflexes that we do not yet understand completely. The reflex pathways utilize the sympathetic and parasympathetic divisions of the ANS. During **arousal**, erotic thoughts, the stimulation of sensory nerves in the genital region, or both lead to an increase in the parasympathetic outflow over the pelvic nerves. This outflow leads to erection of the penis (as detailed on p. 1053). The integument covering the glans of the penis contains numerous sensory receptors, and erection tenses the skin and increases their sensitivity. Subsequent stimulation may initiate the secretion of the bulbourethral glands, lubricating the penile urethra and the surface of the glans.

During intercourse, the sensory receptors of the penis are rhythmically stimulated. This stimulation eventually results in the coordinated processes of *emission* and *ejaculation*. **Emission** occurs under sympathetic stimulation. The process begins when the peristaltic contractions of the ampulla push fluid and spermatozoa into the prostatic urethra. The seminal vesicles then begin contracting, and the contractions increase in force and duration over the next few seconds. Peristaltic contractions also appear in the walls of the prostate gland. The combination moves the seminal mixture into the membranous and penile portions of the urethra. While the contractions are proceeding, sympathetic commands also cause the contraction of the urinary bladder and the internal urethral sphincter. The combination of elevated pressure inside the bladder and the contraction of this sphincter effectively prevents the passage of semen into the bladder.

Ejaculation occurs as powerful, rhythmic contractions appear in the *ischiocavernosus* and *bulbospongiosus muscles*, two superficial skeletal muscles of the pelvic floor. The ischiocavernosus muscles insert along the sides of the penis; their contractions serve primarily to stiffen that organ. The bulbospongiosus muscle wraps around the base of the penis, and its contraction pushes semen toward the external urethral opening. These contractions are controlled by somatic motor neurons in the lower lumbar and upper sacral segments of the spinal cord. (The positions of these muscles are shown in Figure 11-13•, p. 338.)

Ejaculation is associated with intensely pleasurable sensations, an experience known as male **orgasm** (OR-gazm). Several other noteworthy physiological changes occur at this time, including pronounced but temporary increases in heart rate and blood pressure. After ejaculation has been completed, blood begins to leave the erectile tissue, and the erection begins to subside. This subsidence, called **detumescence** (de-tū-MES-ens), is mediated by the sympathetic nervous system.

In summary, arousal, erection, emission, and ejaculation are controlled by a complex interplay between the sympathetic and parasympathetic divisions of the ANS. Higher centers, including the cerebral cortex, can facilitate or inhibit many of the important reflexes, thereby modifying the patterns of sexual function. Any physical or psychological factor that affects a single component of the system can result in male sexual dysfunction, also called **impotence**.

IMPOTENCE Impotence is an inability to achieve or maintain an erection. There are various potential physical causes, for erection involves vascular changes as well as neural commands. For example, low blood pressure in the arteries servicing the penis, due to a circulatory blockage such as a plaque, will affect the ability to obtain an erection. Drugs, alcohol, trauma, or illnesses that affect the ANS or CNS may have the same effect. But male sexual performance can also be strongly affected by the psychological state of the individual. The majority of clinical cases of impotence probably reflect psychological rather than anatomical problems. Temporary periods of impotence are relatively common in healthy individuals who are experiencing severe stresses or emotional problems. Depression, anxiety, and fear of impotence are examples of emotional factors that may result in sexual dysfunction.

Female Sexual Function

The phases of female sexual function are comparable to those of male sexual function. During sexual arousal, parasympathetic activation leads to an engorgement of the erectile tissues of the clitoris and increased secretion of cervical mucous glands and the greater vestibular glands. Clitoral erection increases the receptors' sensitivity to stimulation, and the cervical and vestibular glands provide lubrication for the vaginal walls. A network of blood vessels in the vaginal walls becomes filled with blood at this time, and the vaginal surfaces are also moistened by the fluid that moves across the epithelium from underlying connective tissues. (This process accelerates during intercourse as the result of mechanical stimulation.) Parasympathetic stimulation also causes engorgement of blood vessels at the nipples, making them more sensitive to touch and pressure.

During sexual intercourse, rhythmic contact of the penis with the clitoris and vaginal walls, reinforced by touch sensations from the breasts and other stimuli (visual, olfactory, and auditory), provides stimulation that eventually leads to orgasm. Female orgasm is accompanied by peristaltic contractions of the uterine and vaginal walls and, by means of impulses traveling over the pudendal nerves, rhythmic contractions of the bulbospongiosus and ischiocavernosus muscles. The latter contractions give rise to the sensations of orgasm.

SEXUALLY TRANSMITTED DISEASES **Sexually transmitted diseases (STDs)** are transferred from individual to individual, primarily or exclusively by sexual intercourse. A variety of bacterial, viral, and fungal infections are included in this category. At least two dozen different STDs are currently recognized. All are unpleasant. The bacterium *Chlamydia* can cause PID and infertility. Other types of STDs are also quite dangerous, and a few, including AIDS, are deadly. The incidence of STDs has been increasing in the United States since 1984, primarily in urban centers and in urban minority populations. Acute poverty, coupled with drug use, prostitution, and the appearance of drug-resistant pathogens all contribute to the problem. The *Applications Manual* contains a detailed discussion of the most common forms of STD, including *gonorrhea, syphilis, herpes, genital warts,* and *chancroid.* **AM** *Sexually Transmitted Diseases*

⧗AGING AND THE REPRODUCTIVE SYSTEM

Just as the aging process affects our other systems, it also affects the reproductive systems of men and women. The most striking age-related changes in the female reproductive system occur at menopause, whereas changes in the male reproductive system occur more gradually and over a longer period of time.

Menopause

Menopause is usually defined as the time that ovulation and menstruation cease. Menopause typically occurs at age 45–55, but in the years preceding it, the ovarian and uterine cycles become irregular. This interval is called *perimenopause*. A shortage of primordial follicles is the underlying cause of the cycle irregularities. It has been estimated that almost 7 million potential oocytes are in fetal ovaries after 5 months of development, but the number drops to about 2 million at birth and to a few hundred thousand at puberty. As perimenopause arrives, the number of follicles responding each month begins to drop markedly. As the number of available follicles decreases, levels of estrogen decline and may fail to rise enough to trigger ovulation. By age 50, there are often no primordial follicles left to respond to FSH. In **premature menopause**, this depletion occurs before age 40.

Menopause is accompanied by a decline in circulating concentrations of estrogen and progesterone and a sharp and sustained rise in the production of GnRH, FSH, and LH. The decline in estrogen levels leads to reductions in the size of the uterus and breasts, accompanied by a thinning of the urethral and vaginal epithelia. The reduced estrogen concentrations have also been linked to the development of osteoporosis, presumably because bone deposition proceeds at a slower rate. A variety of neural effects are also reported, including "hot flashes," anxiety, and depression. Hot flashes typically begin while estrogen levels are declining and cease when estrogen levels reach minimal values. These intervals of elevated body temperature are associated with surges in LH production. The hormonal mechanisms involved in other CNS effects of menopause are poorly understood. In addition, the risk of atherosclerosis and other forms of cardiovascular disease increases after menopause.

The majority of women experience only mild symptoms, but some individuals experience acutely unpleasant symptoms in perimenopause or during or after menopause. For most of those individuals, hormone replacement therapies involving a combination of estrogens and progestins can prevent osteoporosis and the neural and vascular changes associated with menopause. The hormones may be administered by pills, by injection, or by transdermal "estrogen patches." The synthetic hormone *etidronate* inhibits osteoporosis by suppressing osteoclast activity. When given at intervals over a 2-year period, it increases bone mass and reduces the rate of fracture incidence in postmenopausal women.

The Male Climacteric

Changes in the male reproductive system occur more gradually than do those in the female reproductive system. The period of change is known as the **male climacteric**. Levels of circulating testosterone begin to decline between ages 50 and 60, and levels of circulating FSH and LH increase. Although sperm production continues (men well into their eighties can father children), there is a gradual reduction in sexual activity in older men. This decrease may be linked to declining testosterone levels. Some clinicians are now tentatively suggesting the use of testosterone replacement therapy to enhance libido (sexual drive) in elderly men.

INTEGRATION WITH OTHER SYSTEMS

Figure 28-27● summarizes the relationships between the reproductive system and other physiological systems. Normal human reproduction is a complex process that requires the participation of multiple systems. The hormones discussed in this chapter play a major role in coordinating reproductive events (Table 28-1). The reproductive process depends on a variety of physical, physiological, and psychological factors, many of which require intersystem cooperation: The man's sperm count must be adequate, the semen must have the correct pH and nutrients, and erection and ejaculation must occur in the proper sequence; the woman's ovarian and uterine cycles must be properly coordinated, ovulation and oocyte transport must occur normally, and the woman's reproductive tract must provide a hospitable environment for sperm survival, movement, and subsequent fer-

tilization. For these steps to occur, the reproductive, digestive, endocrine, nervous, cardiovascular, and urinary systems must all be functioning normally.

Even when all else is normal and fertilization occurs at the proper time and place, a healthy infant will not be produced unless the zygote, a single cell the size of a pinhead, manages to develop into a full-term fetus that typically weighs about 3 kg. In Chapter 29, we shall consider the process of development, focusing on the mechanisms that determine both the structure of the body and the distinctive characteristics of each individual.

☑ Inability to contract the ischiocavernosus and bulbospongiosus muscles would interfere with which part(s) of the male sex act?

☑ What changes occur in the female during sexual arousal as the result of increased parasympathetic stimulation?

☑ Why does the level of FSH rise and remain high during menopause?

TABLE 28-1 Hormones of the Reproductive System

Hormone	Source	Regulation of Secretion	Primary Effects
Gonadotropin-releasing hormone (GnRH)	Hypothalamus	*Males:* inhibited by testosterone and possibly by inhibin	Stimulates FSH secretion, LH synthesis
		Females: GnRH pulse frequency increased by estrogens, decreased by progestins	As above
Follicle-stimulating hormone (FSH)	Anterior pituitary	*Males:* stimulated by GnRH, inhibited by inhibin	*Males:* stimulates spermatogenesis and spermiogenesis through effects on sustentacular cells
		Females: stimulated by GnRH, inhibited by inhibin	*Females:* stimulates follicle development, estrogen production, and oocyte maturation
Luteinizing hormone (LH)	Anterior pituitary	*Males:* stimulated by GnRH *Females:* production stimulated by GnRH, secretion by the combination of high GnRH pulse frequencies and high estrogen levels	*Males:* stimulates interstitial cells to secrete testosterone *Females:* stimulates ovulation, formation of corpus luteum, and progestin secretion
Androgens (primarily testosterone and dihydrotestosterone)	Interstitial cells of testes	Stimulated by LH	Establish and maintain secondary sex characteristics and sexual behavior; promote maturation of spermatozoa; inhibit GnRH secretion
Estrogens (primarily estradiol)	Granulosa and thecal cells of developing follicles; corpus luteum	Stimulated by FSH	Stimulate LH secretion (at high levels); establish and maintain secondary sex characteristics and sexual behavior; stimulate repair and growth of endometrium; increase frequency of GnRH pulses
Progestins (primarily progesterone)	Granulosa cells from mid-cycle through functional life of corpus luteum	Stimulated by LH	Stimulate endometrial growth and glandular secretion; reduce frequency of GnRH pulses
Inhibin	Sustentacular cells of testes and granulosa cells of ovaries	Stimulated by factors released by developing sperm (male) and developing follicles (female)	Inhibits secretion of FSH (and possibly of GnRH)

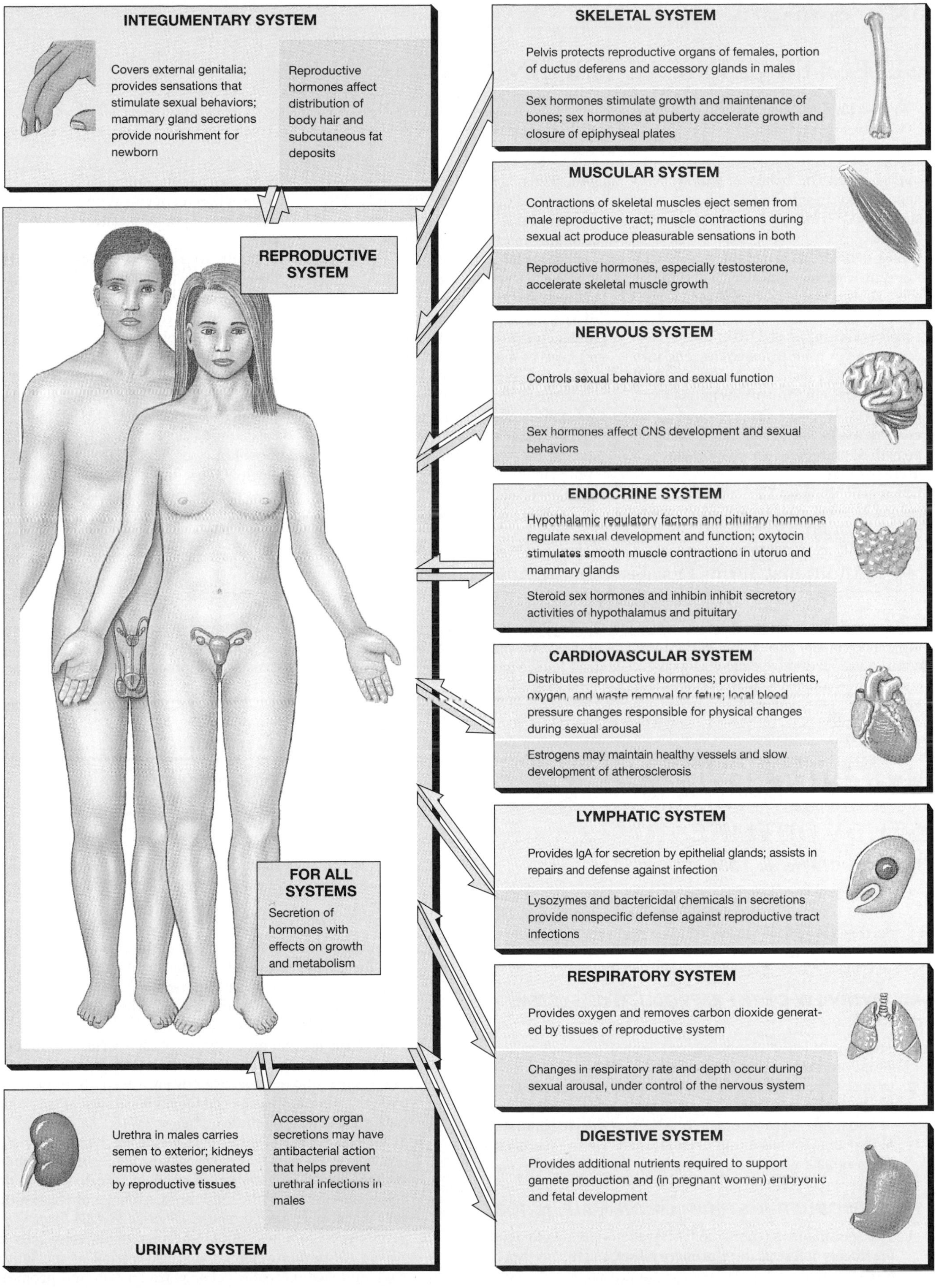

INTEGUMENTARY SYSTEM

Covers external genitalia; provides sensations that stimulate sexual behaviors; mammary gland secretions provide nourishment for newborn

Reproductive hormones affect distribution of body hair and subcutaneous fat deposits

SKELETAL SYSTEM

Pelvis protects reproductive organs of females, portion of ductus deferens and accessory glands in males

Sex hormones stimulate growth and maintenance of bones; sex hormones at puberty accelerate growth and closure of epiphyseal plates

MUSCULAR SYSTEM

Contractions of skeletal muscles eject semen from male reproductive tract; muscle contractions during sexual act produce pleasurable sensations in both

Reproductive hormones, especially testosterone, accelerate skeletal muscle growth

NERVOUS SYSTEM

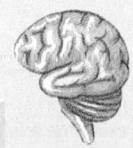

Controls sexual behaviors and sexual function

Sex hormones affect CNS development and sexual behaviors

REPRODUCTIVE SYSTEM

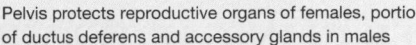

ENDOCRINE SYSTEM

Hypothalamic regulatory factors and pituitary hormones regulate sexual development and function; oxytocin stimulates smooth muscle contractions in uterus and mammary glands

Steroid sex hormones and inhibin inhibit secretory activities of hypothalamus and pituitary

CARDIOVASCULAR SYSTEM

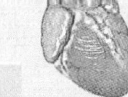

Distributes reproductive hormones; provides nutrients, oxygen, and waste removal for fetus; local blood pressure changes responsible for physical changes during sexual arousal

Estrogens may maintain healthy vessels and slow development of atherosclerosis

LYMPHATIC SYSTEM

Provides IgA for secretion by epithelial glands; assists in repairs and defense against infection

Lysozymes and bactericidal chemicals in secretions provide nonspecific defense against reproductive tract infections

FOR ALL SYSTEMS

Secretion of hormones with effects on growth and metabolism

RESPIRATORY SYSTEM

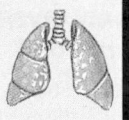

Provides oxygen and removes carbon dioxide generated by tissues of reproductive system

Changes in respiratory rate and depth occur during sexual arousal, under control of the nervous system

Urethra in males carries semen to exterior; kidneys remove wastes generated by reproductive tissues

Accessory organ secretions may have antibacterial action that helps prevent urethral infections in males

DIGESTIVE SYSTEM

Provides additional nutrients required to support gamete production and (in pregnant women) embryonic and fetal development

URINARY SYSTEM

• FIGURE 28-27 **Functional Relationships between the Reproductive System and Other Systems**

SELECTED CLINICAL TERMINOLOGY

Terms Discussed in This Chapter

amenorrhea: The failure of menarche to appear before age 16, or a cessation of menstruation for 6 months or more in an adult female of reproductive age. *(p. 1065)*

breast cancer: A malignant, metastasizing cancer of the mammary gland that is the primary cause of death for women age 35–45. *(p. 1068)*

cryptorchidism (kript-OR-ki-dizm): Failure of one or both testes to descend into the scrotum by the time of birth. *(p. 1039)*

dysmenorrhea: Painful menstruation. *(p. 1063)*

endometriosis (en-dō-mē-trē-Ō-sis): The growth of endometrial tissue outside the uterus. *(p. 1065)*

fibrocystic disease: Clusters of lobular cysts within the tissues of the mammary gland. *(p. 1068)*

gonorrhea (gon-ō-RĒ-uh): A sexually transmitted disease. *(p. 1060 and AM)*

impotence: An inability to achieve or maintain an erection. *(p. 1078)*

mammography: The use of X-rays to examine breast tissue. *(p. 1068)*

mastectomy: Surgical removal of part or all of a cancerous mammary gland. *(p. 1068)*

orchiectomy (or-kē-EK-to-mē): Surgical removal of a testis. *(p. 1040)*

pelvic inflammatory disease (PID): An infection of the uterine tubes. *(p. 1060)*

prostate cancer: A malignant, metastasizing cancer of the prostate that is the second most common cause of cancer deaths in males. *(p. 1050)*

prostatectomy (pros-ta-TEK-to-mē): Surgical removal of the prostate gland. *(p. 1051)*

prostate-specific antigen (PSA): An antigen whose concentration in the blood increases in prostate cancer patients. *(p. 1051)*

sexually transmitted diseases (STDs): Diseases transferred from one individual to another primarily or exclusively through sexual contact. Examples include gonorrhea, syphilis, herpes, and AIDS. *(p. 1079 and AM)*

vaginitis (va-jin-Ī-tis): Infection of the vaginal canal by fungal or bacterial pathogens. *(p. 1066 and AM)*

vasectomy (vaz-EK-to-mē): Surgical removal of a segment of the ductus deferens, making it impossible for spermatozoa to reach the distal portions of the reproductive tract. *(p. 1076)*

AM Additional Terms Discussed in the *Applications Manual*

---

cervical cancer: A malignant, metastasizing cancer of the cervix that is the most common reproductive cancer in women.
endometrial polyps: Benign epithelial tumors of the uterine lining.

leiomyomas (lī-ō-mī-Ō-maz), or **fibroids:** Benign smooth muscle tumors; myometrial leiomyomas are the most common tumors in women.

ovarian cancer: A malignant, metastasizing cancer of the ovaries that is the most dangerous reproductive cancer in women.

CHAPTER REVIEW

 On-line resources for this chapter are on our World Wide Web site at: http://www.prenhall.com/martini/fap

STUDY OUTLINE

INTRODUCTION, p. 1038

1. The human reproductive system produces, stores, nourishes, and transports functional **gametes** (reproductive cells). **Fertilization** is the fusion of male and female gametes to create a **zygote**.

AN OVERVIEW OF THE REPRODUCTIVE SYSTEM, p. 1038

1. The reproductive system includes **gonads**, ducts, accessory glands and organs, and the **external genitalia**.

2. In males, the **testes** produce sperm, which are expelled from the body in **semen** during *ejaculation*. The **ovaries** (gonads) of a sexually mature female produce an **oocyte** (immature *ovum*) that travels along **uterine tubes** toward the **uterus**. The **vagina** connects the uterus with the exterior.

THE REPRODUCTIVE SYSTEM OF THE MALE, p. 1038

1. The **spermatozoa** (sperm cells) travel along the **epididymis**, the **ductus deferens**, the **ejaculatory duct**, and the **urethra** before leaving the body. Accessory organs (notably the **seminal vesicles, prostate gland**, and **bulbourethral glands**) secrete into

the ejaculatory ducts and the urethra. The **scrotum** encloses the testes, and the **penis** is an erectile organ. *(Figure 28-1)*

The Testes, p. 1038

2. The **descent of the testes** through the *inguinal canals* occurs during development. The testes remain connected to internal structures via the **spermatic cords**. The **perineal raphe** marks the boundary between the two chambers in the scrotum. *(Figures 28-2, 28-3)*

3. The **dartos** muscle gives the scrotum a wrinkled appearance; the **cremaster muscle** pulls the testes close to the body. The **tunica albuginea** surrounds each testis. Septa extend from the tunica albuginea to the **mediastinum** of the testis, creating a series of **lobules**. *(Figure 28-4)*

4. **Seminiferous tubules** within each lobule are the sites of sperm production. From there, sperm pass through a **straight tubule** to the **rete testis**. **Efferent ducts** connect the rete testis to the epididymis. Between the seminiferous tubules are **interstitial cells** that secrete sex hormones. *(Figures 28-4, 28-5)*

5. Seminiferous tubules contain **spermatogonia**, stem cells involved in **spermatogenesis** (the production of spermatozoa), and **sustentacular cells**, which sustain and promote the development of spermatozoa. *(Figures 28-5–28-7)*

Anatomy of a Spermatozoon, p. 1046

6. Each **spermatozoon** has a **head, middle piece**, and **tail**. *(Figure 28-8)*

The Male Reproductive Tract, p. 1046

7. From the testis, the spermatozoa enter the epididymis, an elongate tubule with **head, body**, and **tail** regions. The epididymis monitors and adjusts the composition of the tubular fluid and serves as a recycling center for damaged spermatozoa. *(Figure 28-9)*

8. The ductus deferens, or *vas deferens*, begins at the epididymis and passes through the inguinal canal as one component of the spermatic cord. Near the prostate, it enlarges to form the **ampulla**. The junction of the base of the seminal vesicle and the ampulla creates the ejaculatory duct, which empties into the urethra. *(Figure 28-10)*

9. The urethra extends from the urinary bladder to the tip of the penis. It can be divided into *prostatic, membranous*, and *penile* regions.

The Accessory Glands, p. 1048

10. Each seminal vesicle is an active secretory gland that contributes about 60 percent of the volume of semen; its secretions contain fructose, which is easily metabolized by spermatozoa. The prostate gland secretes slightly acidic **prostatic fluid**. Alkaline mucus secreted by the bulbourethral glands has lubricating properties. *(Figure 28-10)*

Semen, p. 1051

11. A typical ejaculation releases 2–5 ml of semen (an **ejaculate**), which contains 20–100 million sperm per milliliter. The fluid component of semen is **seminal fluid**.

The Penis, p. 1052

12. The skin overlying the penis resembles that of the scrotum. Most of the **body** of the penis consists of three masses of **erectile tissue**. Beneath the superficial fascia are two **corpora cavernosa** and a single **corpus spongiosum** that surrounds the urethra. Dilation of the erectile tissue with blood produces an **erection**. *(Figure 28-11)*

Hormones and Male Reproductive Function, p. 1053

13. Important regulatory hormones include **FSH** *(follicle-stimulating hormone)*, **LH** *(luteinizing hormone)*, and **GnRH** *(gonadotropin-releasing hormone)*. *Testosterone* is the most important androgen. *(Figure 28-12)*

THE REPRODUCTIVE SYSTEM OF THE FEMALE, p. 1055

1. Principal organs of the female reproductive system include the ovaries, uterine tubes, uterus, vagina, and external genitalia. *(Figure 28-13)*

2. The ovaries, uterine tubes, and uterus are enclosed within the **broad ligament**. The **mesovarium** supports and stabilizes each ovary.

The Ovaries, p. 1056

3. The ovaries are held in position by the **ovarian ligament** and the **suspensory ligament**. Major blood vessels enter the ovary at the **ovarian hilum**. Each ovary is covered by a **tunica albuginea**. *(Figure 28-14)*

4. **Oogenesis** (ovum production) occurs monthly in **ovarian follicles** as part of the **ovarian cycle**. The ovarian cycle is divided into a **follicular** *(preovulatory)* **phase** and a **luteal** *(postovulatory)* **phase.**

5. As development proceeds, **primordial, primary, secondary**, and **tertiary follicles** are produced in turn. At **ovulation**, a **secondary oocyte** and the surrounding follicular cells of the **corona radiata** are released through the ruptured ovarian wall. The follicular cells remaining within the ovary form the **corpus luteum**, which later degenerates into a **corpus albicans** of scar tissue. *(Figures 28-15, 28-16)*

The Uterine Tubes, p. 1059

6. Each uterine tube has an **infundibulum** with **fimbriae** (fingerlike projections), an **ampulla**, and an **isthmus**. Each uterine tube opens into the uterine cavity. For fertilization to occur, the secondary oocyte must encounter spermatozoa during the first 12–24 hours of its passage from the infundibulum to the uterus. *(Figure 28-17)*

The Uterus, p. 1061

7. The uterus provides mechanical protection and nutritional support to the developing embryo. Normally, the uterus bends anteriorly near its base *(anteflexion)*. It is stabilized by the broad ligament, **uterosacral ligaments, round ligaments**, and **lateral ligaments**. *(Figure 28-18)*

8. Major anatomical landmarks of the uterus include the **body, isthmus, cervix, cervical os** *(external orifice)*, **uterine cavity, cervical canal**, and **internal os** *(internal orifice)*. The uterine wall consists of an inner **endometrium**, a muscular **myometrium**, and a superficial **perimetrium** (an incomplete serous layer). *(Figures 28-18, 28-19)*

9. A typical 28-day **uterine**, or *menstrual*, **cycle** begins with the onset of **menses** and the destruction of the **functional zone** of the endometrium. This process of **menstruation** continues from 1 to 7 days. *(Figure 28-20)*

10. After menses, the **proliferative phase** begins, and the functional zone undergoes repair and thickens. Menstrual activity begins at **menarche** and continues until **menopause**. *(Figure 28-20)*

The Vagina, p. 1065

11. The vagina is a muscular tube extending between the uterus and the external genitalia. Before the first sexual intercourse, a thin epithelial fold, the **hymen**, partially blocks the entrance to the vagina. *(Figures 28-21, 28-22)*

The External Genitalia, p. 1066

12. The components of the **vulva** are the **vestibule, labia minora, paraurethral glands, clitoris, labia majora**, and **lesser** and **greater vestibular glands**. *(Figures 28-13, 28-22)*

The Mammary Glands, p. 1067

13. A newborn infant gains nourishment from milk secreted by maternal **mammary glands**. *(Figure 28-23)*

Hormones and the Female Reproductive Cycle, p. 1069

14. Hormonal regulation of the **female reproductive cycle** involves the coordination of the ovarian and uterine cycles.

15. **Estradiol**, the most important *estrogen*, is the dominant hormone of the follicular phase. Ovulation occurs in response to a mid-cycle surge in LH. *(Figures 28-24, 28-25)*

16. The hypothalamic secretion of GnRH occurs in pulses that trigger the pituitary secretion of FSH and LH. FSH initiates follicular development, and activated follicles and ovarian interstitial cells produce estrogens. High levels of estrogen stimulate LH secretion, increase pituitary sensitivity to

GnRH, and increase the GnRH pulse frequency. **Progesterone**, one of the steroid hormones called **progestins**, is the principal hormone of the *luteal phase*. Changes in estrogen and progesterone levels are responsible for the maintenance of the uterine cycle. *(Figures 28-25, 28-26)*

THE PHYSIOLOGY OF SEXUAL INTERCOURSE, p. 1078

Male Sexual Function, p. 1078

1. During **arousal** in males, erotic thoughts, sensory stimulation, or both lead to parasympathetic activity that produces erection. Stimuli accompanying **sexual intercourse** (*coitus* or *copulation*) lead to **emission** and **ejaculation**. Contractions of the bulbospongiosus muscles are associated with **orgasm**.

Female Sexual Function, p. 1078

2. The phases of female sexual function resemble those of male sexual function, with parasympathetic arousal and skeletal muscle contractions associated with orgasm.

AGING AND THE REPRODUCTIVE SYSTEM, p. 1079

Menopause, p. 1079

1. Menopause (the time that ovulation and menstruation cease) typically occurs at age 45–55. Production of GnRH, FSH, and LH rise, whereas circulating concentrations of estrogen and progesterone decline.

The Male Climacteric, p. 1079

2. During the **male climacteric**, between ages 50 and 60, circulating testosterone levels decline, whereas levels of FSH and LH rise.

INTEGRATION WITH OTHER SYSTEMS, p. 1080

1. Hormones play a major role in coordinating reproduction. *(Table 28-1)*

2. In addition to the endocrine and reproductive systems, reproduction requires the normal functioning of the digestive, nervous, cardiovascular, and urinary systems. *(Figure 28-27)*

REVIEW QUESTIONS

Level 1   Reviewing Facts and Terms

1. Perineal structures associated with the reproductive system are collectively known as
- **(a)** gonads
- **(b)** sex gametes
- **(c)** external genitalia
- **(d)** accessory glands

2. Interstitial cells are responsible for
- **(a)** sperm production
- **(b)** maintenance of the blood–testis barrier
- **(c)** support of spermiogenesis
- **(d)** production of androgens

3. Sustentacular cells are responsible for the secretion of
- **(a)** inhibin
- **(b)** androgen-binding protein
- **(c)** Müllerian-inhibiting factor
- **(d)** a, b, and c are correct

4. During meiosis, when synapsis occurs, corresponding maternal and paternal chromosomes come together to produce
- **(a)** 46 pairs of chromosomes
- **(b)** 23 chromosomes
- **(c)** 23 pairs of chromosomes
- **(d)** the haploid number of chromosomes

5. The completion of meiosis in males produces four spermatids, each containing
- **(a)** 23 chromosomes
- **(b)** 23 pairs of chromosomes
- **(c)** the diploid number of chromosomes
- **(d)** 46 pairs of chromosomes

6. Erection of the penis occurs when
- **(a)** sympathetic activation of penile arteries occurs
- **(b)** arterial branches are constricted, and muscular partitions are tense
- **(c)** the vascular channels become engorged with blood
- **(d)** a, b, and c are correct

7. In males, the primary target of FSH is the
- **(a)** sustentacular cells of the seminiferous tubules
- **(b)** interstitial cells of the seminiferous tubules
- **(c)** cells of Leydig
- **(d)** epididymis

8. Testosterone and other androgens are secreted by the
- **(a)** hypothalamus
- **(b)** anterior pituitary gland
- **(c)** sustentacular cells
- **(d)** interstitial cells

9. The ovaries are responsible for
- **(a)** the production of female gametes
- **(b)** the secretion of female sex hormones
- **(c)** the secretion of inhibin
- **(d)** a, b, and c are correct

10. Gametes are produced in the _____ of the ovary:
- **(a)** germinal epithelium
- **(b)** medulla
- **(c)** cortex
- **(d)** tunica albuginea

11. Ovum production, or oogenesis, begins
- **(a)** before birth
- **(b)** after birth
- **(c)** at puberty
- **(d)** after puberty

12. In females, meiosis is not completed
- **(a)** until birth
- **(b)** until puberty
- **(c)** until fertilization occurs
- **(d)** until uterine implantation occurs

13. If fertilization is to occur, the ovum must encounter spermatozoa during the first _____ of its passage:
- **(a)** 1 to 5 hours
- **(b)** 6 to 11 hours
- **(c)** 12 to 24 hours
- **(d)** 25 to 36 hours

14. The part of the endometrium that undergoes cyclical changes in response to levels of sex hormones is the
- **(a)** serosa
- **(b)** basilar zone
- **(c)** muscular myometrium
- **(d)** functional zone

15. The secretory phase of the uterine cycle is influenced primarily by the stimulatory effects of progestins and estrogens from the
- **(a)** anterior pituitary
- **(b)** hypothalamus
- **(c)** endometrium
- **(d)** corpus luteum

16. A sudden surge in LH secretion causes
  (a) the onset of menses
  (b) the rupture of the follicular wall and ovulation
  (c) the beginning of the proliferative phase
  (d) the end of the uterine cycle

17. The principal hormone of the postovulatory phase is
  (a) progesterone   **(b)** estradiol
  (c) estrogen   **(d)** luteinizing hormone

18. At the time of ovulation, the basal body temperature
  (a) is not affected
  (b) increases noticeably
  (c) declines sharply
  (d) may increase or decrease a few degrees

19. Impotence is the inability to
  (a) produce sufficient amounts of sperm for fertilization
  (b) achieve or maintain an erection
  (c) achieve orgasm
  (d) ejaculate

20. Clitoral erection is caused by the influence of
  (a) parasympathetic activation
  (b) sympathetic activation
  (c) increased output of estrogen
  (d) increased output of progesterone

21. Menopause is accompanied by
  (a) sustained rises in GnRH, FSH, and LH
  (b) declines in circulating levels of estrogen and progesterone
  (c) thinning of the urethral and vaginal walls
  (d) a, b, and c are correct

22. Which reproductive structures are common to both males and females?

23. Trace the duct system that the sperm traverses from the site of production to the exterior.

24. Which accessory organs and glands contribute to the composition of semen? What are the functions of each?

25. What are the two primary cell populations in the testes that are responsible for functions related to reproductive activity? What are the functions of these cells?

26. What are the three primary functions of the epididymis?

27. Identify the three regions of the male urethra.

28. Describe the composition of a typical sample of ejaculate.

29. Enumerate the functions of testosterone in males.

30. What are the primary functions of the ovaries?

31. List and summarize the important steps in the ovarian cycle.

32. Describe the histological composition of the uterine wall.

33. What are the three major functions of the vagina?

34. What is the functional significance of the normally acidic pH of the vagina?

35. What is the role of the clitoris in the female reproductive system?

36. What is the function of the lesser and greater vestibular glands in the female?

37. Trace the route that milk takes from its site of production to the outside of the female.

Level 2 Reviewing Concepts

38. How does the human reproductive system differ functionally from all other systems in the body?

39. How are the male and female reproductive systems functionally different?

40. How is the process of meiosis involved in the development of the spermatozoon and the ovum?

41. Describe the erectile tissues of the penis. How does erection occur?

42. Using an average cycle of 28 days, describe each of the three phases of the uterine cycle.

43. Describe the hormonal events associated with the ovarian cycle.

44. Describe the hormonal events associated with the uterine cycle.

45. Summarize how arousal and orgasm are mediated. Do these processes differ in males and females?

46. How does the aging process affect the reproductive systems of men and women?

47. How do birth control pills prevent conception?

Level 3 Critical Thinking and Clinical Applications

48. Diane has an inflammation of the peritoneum (peritonitis), which her physician says resulted from a urinary tract infection. Why could this situation occur more readily in females than in males?

49. Rod suffers an injury to the sacral region of his spinal cord. Will he still be able to achieve an erection? Explain.

50. A 7-year-old girl develops an ovarian tumor that involves granulosa cells. What symptoms would you expect to observe?

51. How would blockage of the enzyme required for the production of testosterone affect a woman's ovarian cycle?

52. Women bodybuilders and women suffering from eating disorders such as anorexia nervosa commonly experience amenorrhea. What does this fact suggest about the relation between body fat and menstruation? What might be the benefit of amenorrhea under such circumstances?

29

Development and Inheritance

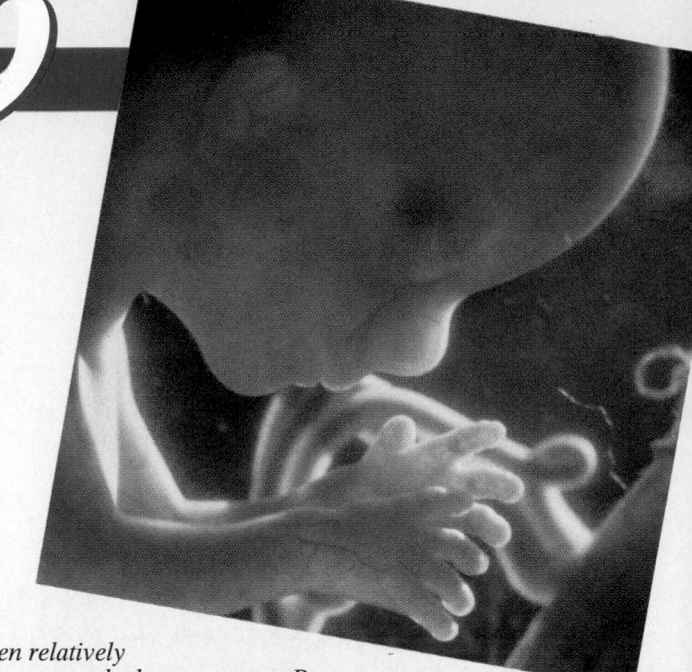

T he physiological processes we have studied thus far have been relatively
brief in duration. Many last only a fraction of a second; others may take hours at most. But
some important processes are measured in months, years, or decades. A human being develops in the womb
for 9 months, grows to maturity in 15 or 20 years, and may live the better part of a century. During that span
of time, he or she will always be changing. Birth, growth, maturation, aging, and death are all parts of a
single, continuous process. That process does not end with the individual, for human beings can pass at least
some of their characteristics on to their offspring. Thus, each generation gives rise to a new generation that
will repeat the cycle. In this chapter, we shall explore the continuity of life, from conception to death.

CHAPTER OUTLINE AND OBJECTIVES

Time refuses to stand still; today's infant will be tomorrow's adult. The gradual modification of anatomical structures during the period from fertilization to maturity is called **development**. The changes that occur during development are truly remarkable. What begins as a single cell slightly larger than the period at the end of this sentence becomes an individual whose body contains trillions of cells organized into highly specialized tissues, organs, and organ systems. The creation of different cell types required by this process is called **differentiation**. Differentiation occurs through selective changes in genetic activity. As development proceeds, some genes are turned off and others are turned on. The identities of these genes vary from one cell type to another.

A basic understanding of human development provides important insights into anatomical structures. In addition, many of the mechanisms of development and growth are similar to those responsible for the repair of injuries. In this chapter, we shall focus on major aspects of development. We shall consider highlights of the developmental process rather than describe the events in great detail. We shall also consider the regulatory mechanisms and how developmental patterns can be modified—for good or ill. Few topics in the biological sciences hold such fascination, and fewer still confront the investigator with so dazzling an array of technological, moral, and logistical challenges.

AN OVERVIEW OF TOPICS IN DEVELOPMENT

Development involves (1) the division and differentiation of cells and (2) the changes that produce and modify anatomical structures. Development begins at fertilization, or **conception**. We can divide development into periods characterized by specific anatomical changes. **Embryological development** comprises the events that occur during the first 2 months after fertilization. The study of these events is called **embryology** (em-brē-OL-o-jē). **Fetal development** begins at the start of the ninth week and continues until birth. Embryological and fetal development are sometimes referred to collectively as **prenatal development**, the primary focus of this chapter. **Postnatal development** commences at birth and continues to **maturity**, when the aging process begins.

Although all human beings go through the same developmental stages, differences in their genetic makeup produce distinctive individual characteristics. **Inheritance** refers to the transfer of genetically determined characteristics from generation to generation. **Genetics** is the study of the mechanisms responsible for inheritance. In this chapter, we shall consider basic genetics as it applies to inherited characteristics, such as sex, hair color, and various diseases.

FERTILIZATION

Fertilization involves the fusion of two haploid gametes, producing a zygote that contains 46 chromosomes, the normal number for a somatic cell. ∞ *[p. 1043]* The functional roles and contributions of the spermatozoon and the ovum are very different. The spermatozoon simply delivers the paternal chromosomes to the site of fertilization. It is the ovum that must provide all the nourishment and genetic programming to support embryonic development for nearly a week after conception. The volume of the ovum is therefore much greater than that of the spermatozoon (Figure 29-1a●). At fertilization, the diameter of the ovum is more than twice the entire length of the spermatozoon. The ratio of ovum volume to sperm volume is even more striking—roughly 2000:1.

The sperm arriving in the vagina are already motile, but they cannot perform fertilization until they have undergone *capacitation*, a process we discussed in Chapter 28. ∞ *[p. 1047]* It appears that a substance secreted by the epididymis prevents premature capacitation and that the mixing of sperm with the acidic secretions of the prostate gland is the first step in sperm activation. The precise mechanism of capacitation within the female reproductive tract remains uncertain.

Fertilization typically occurs near the junction between the ampulla and isthmus of the uterine tube, generally within a day after ovulation. Over this period of time, a secondary oocyte has traveled a few centimeters, but spermatozoa must cover the distance between the vagina and the ampulla of the uterine tube. An individual spermatozoon can propel itself at speeds of only about 34 μm per second, roughly equivalent to 12.5 cm (5 in.) per hour, so in theory it should take spermatozoa several hours to arrive in the upper portions of the uterine tubes. The actual passage time, however, ranges from 2 hours to as little as 30 minutes. Contractions of the uterine musculature and ciliary currents in the uterine tubes have been suggested as likely mechanisms for accelerating the movement of spermatozoa from the vagina to the fertilization site.

Even with transport assistance and available nutrients, this is not an easy passage. Of the roughly 200 million spermatozoa introduced into the vagina in a typical ejaculation, only about 10,000 enter the uterine tube, and fewer than 100 reach the isthmus. In general, a male with a sperm count below 20 million per milliliter is functionally sterile, because too few spermatozoa survive to reach and fertilize an oocyte. Large numbers of spermatozoa are required for successful fertilization, because a single sperm cannot penetrate the *corona radiata*, a layer of follicle cells that surrounds the oocyte. ∞ *[p. 1058]*.

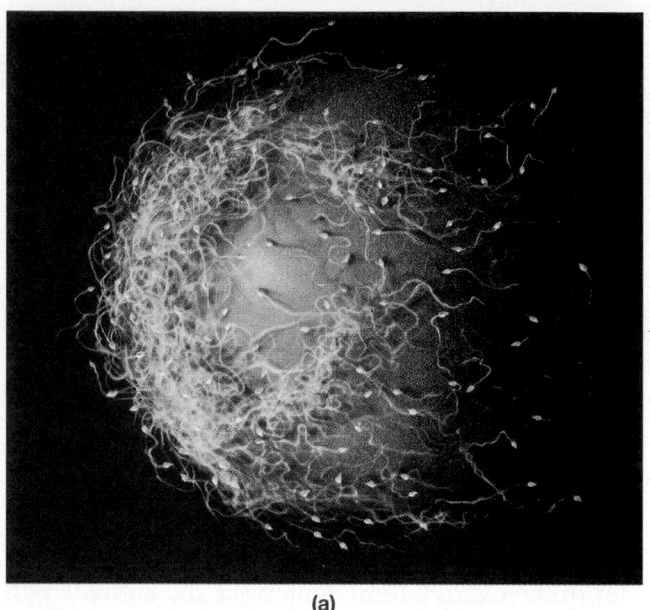

(a)

The Oocyte at Ovulation

Ovulation occurs before the oocyte is completely mature. The secondary oocyte leaving the follicle is in metaphase of the second meiotic division. The cell's metabolic operations have been discontinued, and the oocyte drifts in a sort of suspended animation, awaiting the stimulus for further development. If fertilization does not occur, the oocyte disintegrates without completing meiosis.

Fertilization is complicated by the fact that when it leaves the ovary, the secondary oocyte is surrounded by the corona radiata. Fertilization and the events that follow are diagrammed in Figure 29-1b•. The cells of the corona radiata protect the secondary oocyte as it passes through the ruptured follicular wall, across the surface of the ovary, and into the infundibulum of the uterine tube. Although the physical process of fertilization requires only a single spermatozoon in contact

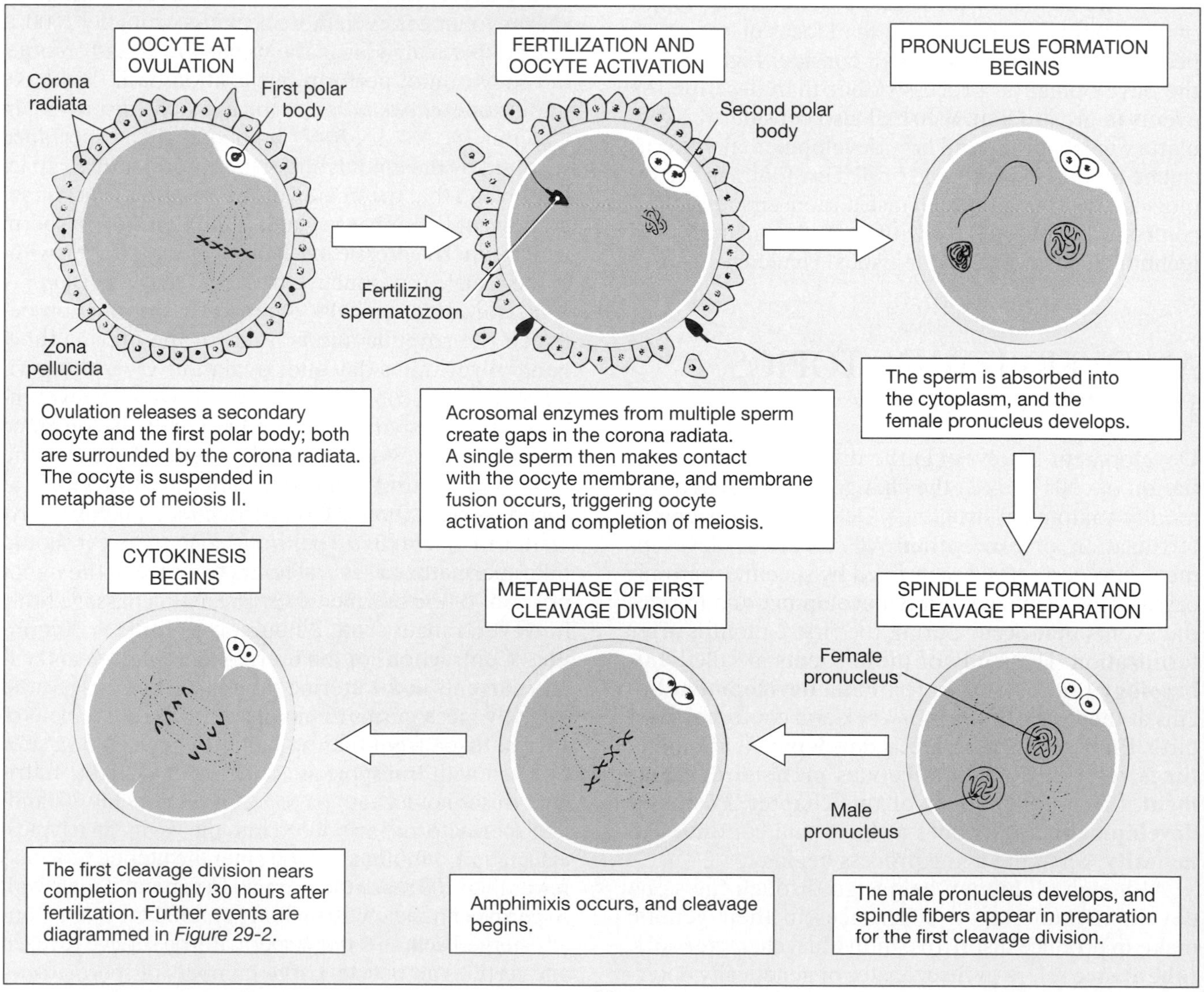

OOCYTE AT OVULATION

Corona radiata

First polar body

Zona pellucida

Ovulation releases a secondary oocyte and the first polar body; both are surrounded by the corona radiata. The oocyte is suspended in metaphase of meiosis II.

FERTILIZATION AND OOCYTE ACTIVATION

Second polar body

Fertilizing spermatozoon

Acrosomal enzymes from multiple sperm create gaps in the corona radiata. A single sperm then makes contact with the oocyte membrane, and membrane fusion occurs, triggering oocyte activation and completion of meiosis.

PRONUCLEUS FORMATION BEGINS

The sperm is absorbed into the cytoplasm, and the female pronucleus develops.

CYTOKINESIS BEGINS

The first cleavage division nears completion roughly 30 hours after fertilization. Further events are diagrammed in Figure 29-2.

METAPHASE OF FIRST CLEAVAGE DIVISION

Amphimixis occurs, and cleavage begins.

SPINDLE FORMATION AND CLEAVAGE PREPARATION

Female pronucleus

Male pronucleus

The male pronucleus develops, and spindle fibers appear in preparation for the first cleavage division.

(b)

•**FIGURE 29-1** **Fertilization.** **(a)** An oocyte and numerous sperm at the time of fertilization. Note the difference in size between the gametes. **(b)** Fertilization and the preparations for cleavage.

with the oocyte membrane, that spermatozoon must first penetrate the corona radiata. The acrosomal cap of the sperm contains several enzymes, including **hyaluronidase** (hī-al-ū-RON-a-dāz), which breaks down the intercellular cement between adjacent follicle cells. At least a hundred spermatozoa must release hyaluronidase before the connections between the follicular cells break down enough to permit fertilization.

No matter how many spermatozoa slip through the gap in the corona radiata, normally only a single spermatozoon will accomplish fertilization and activate the oocyte. The first step is the binding of the spermatozoon to *sperm receptors* in the zona pellucida. This step triggers the rupture of the acrosomal cap. The hyaluronidase and **acrosin,** another proteolytic enzyme, then digest a path through the zona pellucida toward the surface of the oocyte. When the sperm contacts that surface, the sperm and oocyte membranes begin to fuse. This step is the trigger for *oocyte activation,* a complex process we shall discuss in the next section. As the membranes fuse, the entire sperm enters the **ooplasm**, or cytoplasm of the oocyte.

Oocyte Activation and Amphimixis

Oocyte activation involves a series of changes in the metabolic activity of the oocyte. The trigger for activation is contact and fusion of the cell membranes of the sperm and oocyte. This process is accompanied by the depolarization of the oocyte membrane due to an increased permeability to sodium ions. The entry of sodium ions in turn causes the release of calcium ions from the SER. The sudden rise in calcium ion levels has several important effects, including the following:

- *The exocytosis of vesicles located just beneath the surface of the oocyte membrane.* This process, called the *cortical reaction,* releases enzymes that inactivate the sperm receptors and harden the zona pellucida. This combination prevents **polyspermy** (fertilization by more than one sperm). Polyspermy creates a zygote that is incapable of normal development. (Prior to completion of the cortical reaction, the depolarization of the oocyte membrane probably prevents fertilization by any sperm that penetrate the zona pellucida.)
- *The completion of meiosis II and formation of the second polar body.*
- *The activation of enzymes that cause a rapid increase in the cell's metabolic rate.* The ooplasm contains large numbers of mRNA strands, but until now they have been inactivated by special proteins. The mRNA strands are now released, and protein synthesis accelerates rapidly. Most of the proteins synthesized will be needed if development is to proceed.

After oocyte activation has occurred and meiosis has been completed, the nuclear material remaining within the ovum reorganizes as the **female pronucleus** (Figure 29-1b●). While these changes are under way,

the nucleus of the spermatozoon swells, becoming the **male pronucleus**. The male pronucleus then migrates toward the center of the cell. Spindle fibers form, and the two pronuclei fuse in a process called **amphimixis** (am-fi-MIK-sis). Fertilization is now complete, with the formation of a zygote that contains the normal complement of 46 chromosomes.

AN OVERVIEW OF PRENATAL DEVELOPMENT

During prenatal development, a single cell ultimately forms a 3–4 kg infant who in postnatal development will grow through adolescence and maturity toward old age and eventual death. One of the most fascinating aspects of development is its apparent order and simplicity. A continuity exists at all levels and at all times. Nothing leaps into existence, unheralded and without apparent precursors; differentiation and increasing structural complexity occur hand in hand.

Differentiation involves changes in the genetic activity of some cells but not others. In Chapter 3, we considered the exchange of information between the nucleus and the cytoplasm in a cell. ∞ *[p. 91]* Activity in the nucleus varies in response to chemical messages that arrive from the surrounding cytoplasm. In turn, ongoing nuclear activity alters conditions within the cytoplasm by directing the synthesis of specific proteins. In this way, the nucleus can affect enzyme activity, cell structure, and membrane properties.

In development, differences in the cytoplasmic composition of individual cells trigger alterations in genetic activity. These changes in turn lead to further changes in the cytoplasm, and the process continues in a sequential fashion. But if all the cells of the embryo are derived from cell divisions of a zygote, how do the cytoplasmic differences originate? What sets this process in motion? The important first step occurs before fertilization, while the oocyte is in the ovary.

Before ovulation, the growing oocyte accepts amino acids, nucleotides, and glucose, as well as more complex materials such as phospholipids, mRNAs, and proteins, from the surrounding granulosa cells. Because not all follicle cells are manufacturing and delivering the same nutrients and instructions to the oocyte, the contents of the ooplasm are not evenly distributed. After fertilization, subsequent divisions divide the cytoplasm of the zygote into ever smaller cells that differ from one another in their cytoplasmic composition. These differences alter genetic activity, creating cell lines with increasingly diverse fates.

As development proceeds, some of these cells will release chemical substances, such as RNAs, polypeptides, and small proteins, that affect the differentiation of other embryonic cells. This type of chemical interplay between developing cells is called **induction** (in-DUK-shun). Induction can work over very short distances, as when two different types of cells are in direct contact. It may also operate over longer distances, with the inducing chemicals functioning as hormones.

This type of regulation, which involves an integrated series of interacting steps, can control very complex processes. The mechanism is not without risk, however. The appearance of an abnormal or inappropriate inducer can throw the entire development plan off course.

AM *Teratogens and Abnormal Development*

The time spent in prenatal development is known as the period of **gestation** (jes-TĀ-shun). For convenience, we usually think of the gestation period as three integrated **trimesters**, each 3 months in duration:

1. The **first trimester** is the period of embryonic and early fetal development. During this period, the rudiments of all the major organ systems appear.

2. The **second trimester** is dominated by the development of organs and organ systems. During the second trimester, this process nears completion. The body proportions change, and by the end of this trimester, the fetus looks distinctively human.

3. The **third trimester** is characterized by rapid fetal growth. Early in the third trimester, most of the major organ systems become fully functional. An infant born 1 month or even 2 months prematurely has a reasonable chance of survival.

Many of the earlier chapters contain Embryology Summaries, in which we introduced key steps in embryonic and fetal development and traced the development of specific organ systems. We shall refer to those Summaries in the discussions that follow. As we proceed, you should return to the pages indicated and review the material. Doing so will help you understand the "big picture" as well as the specific details.

THE FIRST TRIMESTER

At the moment of conception, the fertilized ovum is a single cell with a diameter of about 0.135 mm (0.005 in.) and a weight of approximately 150 μg. By the end of the first trimester (twelfth developmental week), the fetus is almost 75 mm (3 in.) long and weighs perhaps 14 g (0.5 oz).

Many important and complex developmental events occur during the first trimester. We shall focus on four general processes: *cleavage, implantation, placentation,* and *embryogenesis:*

1. **Cleavage** (KLĒV-ij) is a sequence of cell divisions that begins immediately after fertilization and ends at the first contact with the uterine wall. During cleavage, the zygote becomes a **pre-embryo** that develops into a multicellular complex known as a *blastocyst.* (We introduced cleavage and blastocyst formation in an Embryology Summary in Chapter 4, p. 122.)

2. **Implantation** begins with the attachment of the blastocyst to the endometrium of the uterus and continues as the blastocyst invades the maternal tissues. While implantation is under way, other important events take place that set the stage for the formation of vital embryonic structures.

3. **Placentation** (plas-en-TĀ-shun) occurs as blood vessels form around the periphery of the blastocyst and the **placenta** develops. The placenta is a complex organ that permits exchange between the maternal and embryonic circulatory systems. It supports the fetus in the second and third trimesters, but it is lost at birth as the newborn becomes physically independent of the mother.

4. **Embryogenesis** (em-brē-ō-JEN-e-sis) is the formation of a viable embryo. This process establishes the foundations for all major organ systems.

These processes are both complex and vital to the survival of the embryo. Perhaps because the events in the first trimester are so complex, this is the most dangerous period in prenatal life. Only about 40 percent of conceptions produce embryos that survive the first trimester. For that reason, pregnant women are warned to take great care to avoid drugs and other disruptive stresses during the first trimester, in the hopes of preventing an error in the delicate processes under way.

Cleavage and Blastocyst Formation

Cleavage (Figure 29-2•) is a series of cell divisions that subdivides the cytoplasm of the zygote. The first cleavage division produces a pre-embryo consisting of two identical cells. Identical cells produced by cleavage divisions are called **blastomeres** (BLAS-tō-mērz). The first division is completed roughly 30 hours after fertilization, and subsequent cleavage divisions occur at intervals of 10–12 hours. During the initial cleavage divisions, all the blastomeres undergo mitosis simultaneously. As the number of blastomeres increases, the timing becomes less predictable.

After 3 days of cleavage, the pre-embryo is a solid ball of cells, resembling a mulberry. This stage is called the **morula** (MOR-ū-la; *morula*, mulberry). The morula typically reaches the uterus on day 4. Over the next 2 days, the blastomeres form a hollow ball, the **blastocyst**, with an inner cavity known as the **blastocoele** (BLAS-tō-sēl). The blastomeres are no longer identical in size and shape. The outer layer of cells, which separates the outside world from the blastocoele, is called the **trophoblast** (TRŌ-fō-blast). The function is implied by the name: *trophos*, food + *blast*, precursor. These cells are responsible for providing nutrients to the developing embryo. A second group of cells, the **inner cell mass**, lies clustered at one end of the blastocyst. These cells are exposed to the blastocoele but insulated from contact with the outside environment by the trophoblast. In time, the inner cell mass will form the embryo.

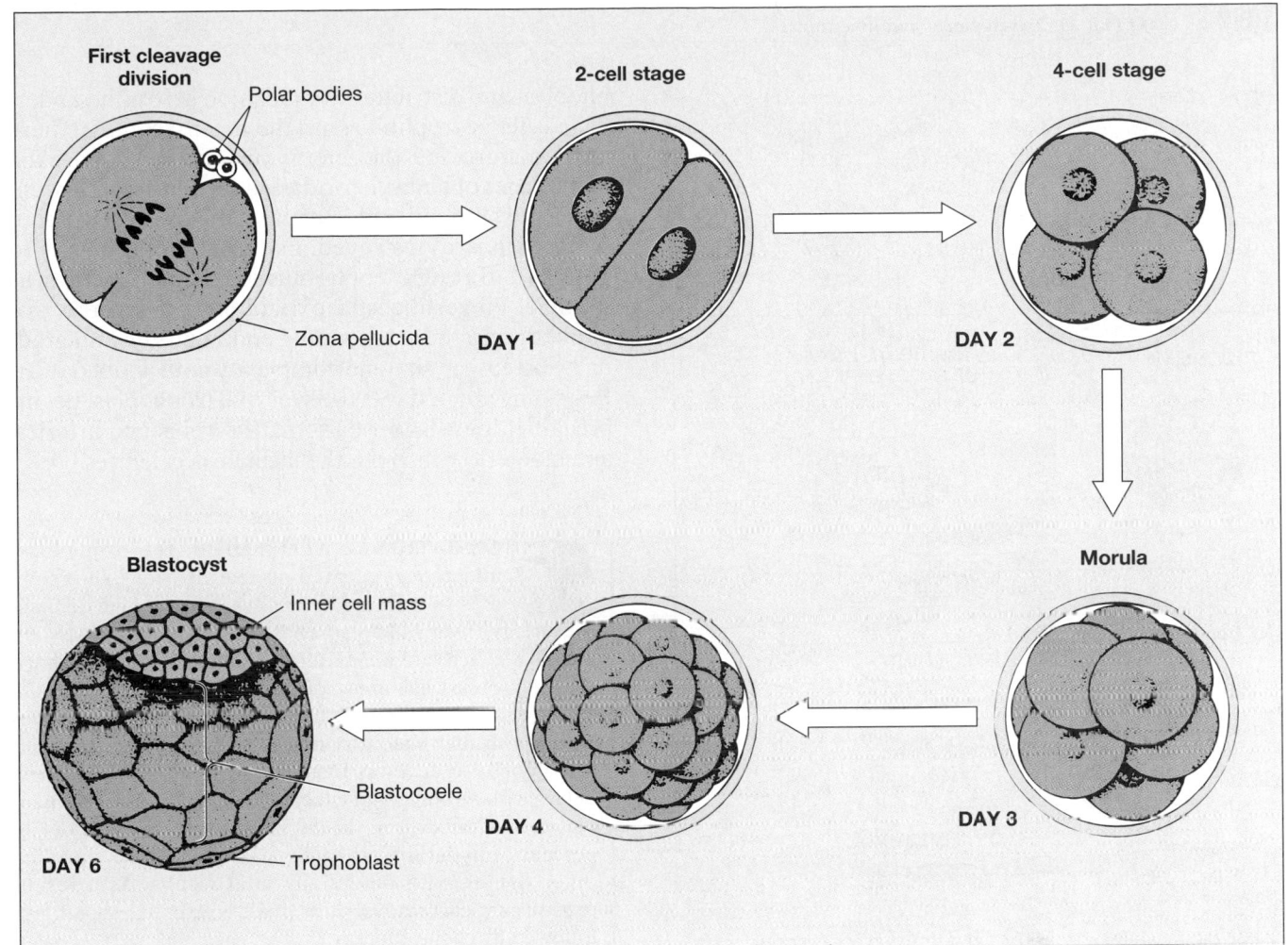

•FIGURE 29-2 Cleavage and Blastocyst Formation

Implantation

Roughly 4 days after fertilization, the morula arrives in the uterine cavity. Blastocyst formation occurs over the next 2–3 days. During this period, enzymes released by the trophoblast erode a hole through the zona pellucida, which is then shed in a process known as *hatching*. The blastocyst is now freely exposed to the fluid contents of the uterine cavity. This fluid, rich in glycogen, is secreted by the endometrial glands of the uterus. Over the past few days, the pre-embryo and then early blastocyst has been absorbing fluid and nutrients from its surroundings; that process now accelerates, and the blastocyst enlarges. When fully formed, the blastocyst contacts the endometrium, and implantation occurs. Stages in the implantation process are illustrated in Figure 29-3•.

Implantation begins as the surface of the blastocyst closest to the inner cell mass touches and adheres to the uterine lining (see *day 7*, Figure 29-3•). At the point of contact, the trophoblast cells divide rapidly, making the trophoblast several layers thick. The cells closest to the interior of the blastocyst remain intact, forming a layer of **cellular trophoblast**, or *cytotro-*

phoblast. Near the endometrial wall, the cell membranes separating the trophoblast cells disappear, creating a layer of cytoplasm containing multiple nuclei *(day 8)*. This outer layer is called the **syncytial** (sin-SISH-al) **trophoblast**, or *syncytiotrophoblast*.

The syncytial trophoblast erodes a path through the uterine epithelium by secreting hyaluronidase. This enzyme dissolves the intercellular cement between adjacent epithelial cells, just as the hyaluronidase released by spermatozoa dissolves the connections between cells of the corona radiata. At first, this erosion creates a gap in the uterine lining, but the migration and divisions of epithelial cells soon repair the surface. When the repairs are completed, the blastocyst loses contact with the uterine cavity, and development occurs entirely within the functional zone of the endometrium. ∞ *[p. 1063]*

In most cases, implantation occurs in the fundus or elsewhere in the body of the uterus. In an **ectopic pregnancy**, implantation occurs somewhere other than within the uterus, such as in one of the uterine tubes. Approximately 0.6 percent of pregnancies are ectopic pregnancies, which can be life-threatening. AM *Ectopic Pregnancies*

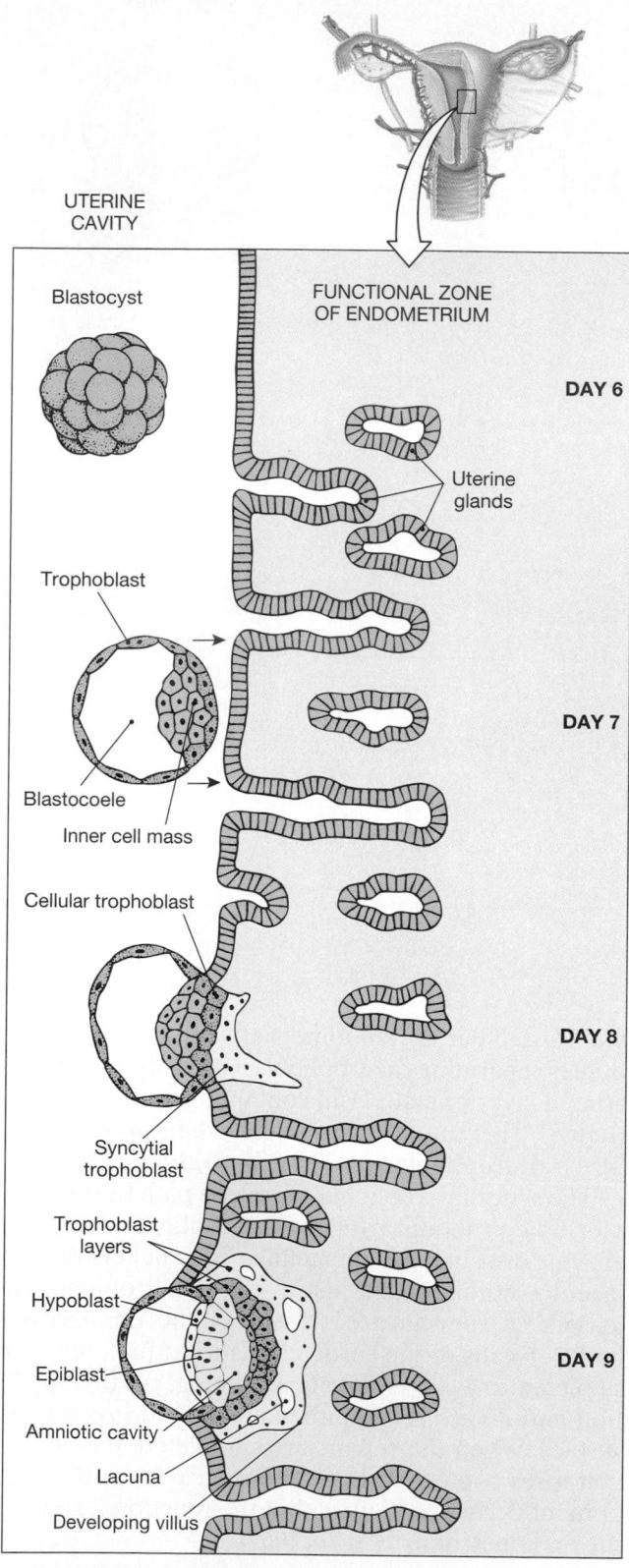

UTERINE
CAVITY

Blastocyst

FUNCTIONAL ZONE
OF ENDOMETRIUM

DAY 6

Uterine
glands

Trophoblast

DAY 7

Blastocoele

Inner cell mass

Cellular trophoblast

DAY 8

Syncytial
trophoblast

Trophoblast
layers

Hypoblast

DAY 9

Epiblast

Amniotic cavity

Lacuna

Developing villus

●*FIGURE 29-3* **Stages in the Implantation Process**

phoblast and distributed by diffusion across the underlying cellular trophoblast to the inner cell mass. These nutrients provide the energy needed to support the early stages of embryo formation. Trophoblastic extensions grow around endometrial capillaries. As the capillary walls are destroyed, maternal blood begins to percolate through trophoblastic channels known as **lacunae**. Fingerlike **villi** extend away from the trophoblast into the surrounding endometrium and gradually increase in size and complexity until about day 21. Beginning about day 9, the syncytial trophoblast begins breaking down larger endometrial veins and arteries, and blood flow through the lacunae accelerates.

GESTATIONAL NEOPLASMS The trophoblast undergoes repeated nuclear divisions, shows extensive and rapid growth, has a very high demand for energy, invades and spreads through adjacent tissues, and fails to activate the maternal immune system. In short, the trophoblast has many of the characteristics of cancer cells. In about 0.1 percent of pregnancies, something goes wrong with the regulatory mechanisms, and a normal placenta does not develop. Instead, the syncytial trophoblast behaves like a malignant cancer, forming a **gestational neoplasm**, or *hydatidiform* (hī-da-TID-i-form) *mole*. About 20 percent of hydatidiform moles will metastasize, invading other tissues, with potentially fatal results. As a result, prompt surgical removal of the mass is essential, sometimes followed by chemotherapy.

Formation of the Blastodisc

The inner cell mass has little apparent organization early in the blastocyst stage. Yet, by the time of implantation, the inner cell mass has separated from the trophoblast. The separation gradually increases, creating a fluid-filled chamber called the **amniotic** (am-nē-OT-ik) **cavity**. The trophoblast will later be separated from the amniotic cavity by layers of cells that originate at the inner cell mass and line the amniotic cavity. These layers form the *amnion*, a membrane we shall examine later in the chapter. The amniotic cavity can be seen in day 9 of Figure 29-3●; additional details from *day 10* to *day 12* are shown in Figure 29-4●. When the amniotic cavity first appears, the cells of the inner cell mass are organized into an oval sheet that is two cell layers thick. This oval, called a **blastodisc** (BLAS-tō-disk), initially consists of an epithelial layer, or **epiblast** (EP-i-blast), which faces the amniotic cavity, and an underlying **hypoblast** (HĪ-pō-blast), which is exposed to the fluid contents of the blastocoele.

Gastrulation and Germ Layer Formation

By day 12, a third layer begins to form through the process of **gastrulation** (gas-troo-LĀ-shun) (Figure 29-4●). During gastrulation, cells in specific areas of

As implantation proceeds, the syncytial trophoblast continues to enlarge and spread into the surrounding endometrium *(day 9)*. The erosion of uterine glands releases nutrients that are absorbed by the syncytial tro-

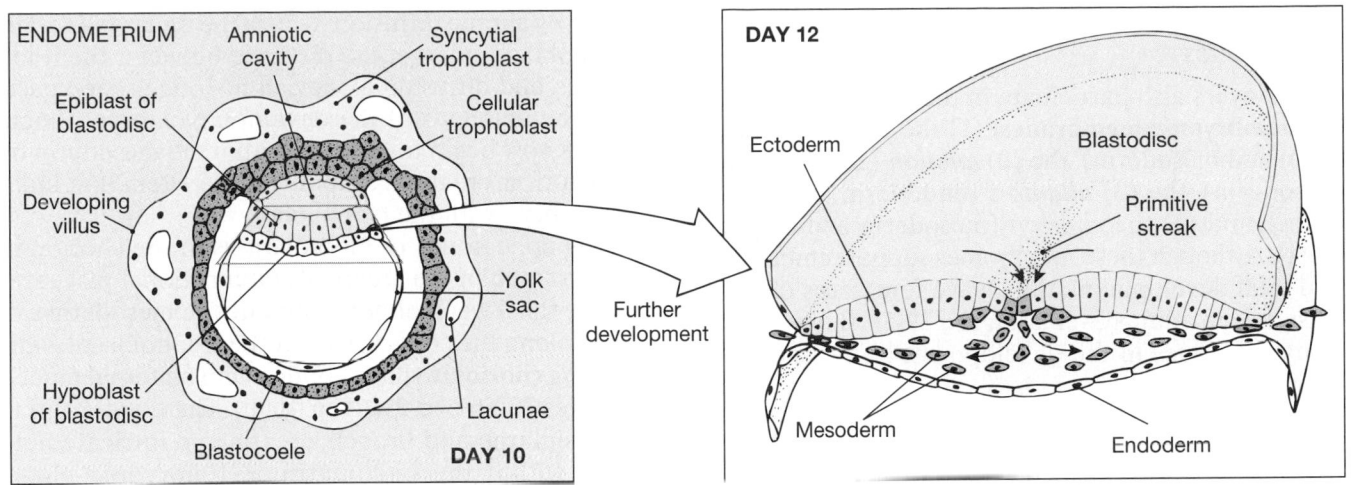

The blastodisc begins as two layers: the epiblast, facing the amniotic cavity, and the hypoblast, exposed to the blastocoele. Migration of epiblast cells around the amniotic cavity is the first step in the formation of the amnion. Migration of hypoblast cells creates a sac that hangs below the blastodisc. This is the first step in yolk sac formation.

Migration of epiblast cells into the region between epiblast and hypoblast gives the blastodisc a third layer. From the time this process (gastrulation) begins, the epiblast is called *ectoderm*, the hypoblast *endoderm*, and the migrating cells *mesoderm*.

•**FIGURE 29-4** Blastodisc Organization and Gastrulation

the epiblast move toward the center of the blastodisc, to a line known as the **primitive streak**. At the primitive streak, these migrating cells leave the surface and move between the epiblast and hypoblast. This movement creates three distinct embryonic layers: (1) the **ectoderm**, which consists of the epiblast cells that did not migrate into the interior of the inner cell mass; (2) the **endoderm**, which consists of the cells of the hypoblast; and (3) the **mesoderm**, which consists of the poorly organized layer of migrating cells between the ectoderm and the endoderm. We introduced the formation of mesoderm and the developmental fates of these **germ layers** in an Embryology Summary in Chapter 4, pp. 138–139. Table 29-1 contains a more comprehensive listing of the contributions each germ layer makes to the body systems described in earlier chapters.

After gastrulation is complete, the blastodisc is known as the **embryonic disc**. The embryonic disc will form the body of the embryo, while the rest of the blastocyst will form the *extraembryonic membranes*.

TABLE 29-1   The Fates of the Primary Germ Layers

Ectodermal Contributions

Integumentary system: epidermis, hair follicles and hairs, nails, and glands communicating with the skin (apocrine and merocrine sweat glands, mammary glands, and sebaceous glands)

Skeletal system: pharyngeal cartilages and their derivatives in the adult (portion of sphenoid bone, the auditory ossicles, the styloid processes of the temporal bones, the cornu and superior rim of the hyoid bone)*

Nervous system: all neural tissue, including brain and spinal cord

Endocrine system: pituitary gland and adrenal medullae

Respiratory system: mucous epithelium of nasal passageways

Digestive system: mucous epithelium of mouth and anus, salivary glands

Mesodermal Contributions

Skeletal system: all components except some pharyngeal derivatives

Muscular system: all components

Endocrine system: adrenal cortex, endocrine tissues of heart, kidneys, and gonads

Cardiovascular system: all components, including bone marrow

Lymphatic system: all components

Urinary system: the kidneys, including the nephrons and the initial portions of the collecting system

Reproductive system: the gonads and the adjacent portions of the duct systems

Miscellaneous: the lining of the body cavities (thoracic, pericardial, and peritoneal) and the connective tissues that support all organ systems

Endodermal Contributions

Endocrine system: thymus, thyroid, and pancreas

Respiratory system: respiratory epithelium (except nasal passageways) and associated mucous glands

Digestive system: mucous epithelium (except mouth and anus), exocrine glands (except salivary glands), liver, and pancreas

Urinary system: urinary bladder and distal portions of the duct system

Reproductive system: distal portions of the duct system, stem cells that produce gametes

*The neural crest is derived from ectoderm and contributes to the formation of the skull and the skeletal derivatives of the embryonic pharyngeal arches.

The Formation of Extraembryonic Membranes

Germ layers also participate in the formation of four **extraembryonic membranes:** (1) the *yolk sac* (endoderm and mesoderm), the (2) *amnion* (ectoderm and mesoderm), the (3) *allantois* (endoderm and mesoderm), and (4) the *chorion* (mesoderm and trophoblast). Although these membranes support embryonic and fetal development, they leave few traces of their existence in adult systems. Figure 29-5● shows representative stages in the development of the extraembryonic membranes.

THE YOLK SAC The first of the extraembryonic membranes to appear is the **yolk sac.** The yolk sac begins as the hypoblast cells spread out around the outer edges of the blastocoele to form a complete pouch suspended below the blastodisc. This pouch is already visible 10 days after fertilization (Figure 29-4●). As gastrulation proceeds, mesodermal cells migrate around this pouch and complete the formation of the yolk sac (Figure 29-5a●). Blood vessels soon appear within the mesoderm, and the yolk sac becomes an important site of blood cell formation.

THE AMNION The ectodermal layer enlarges, and ectodermal cells spread over the inner surface of the amniotic cavity. Mesodermal cells soon follow, creating a second, outer layer (Figure 29-5a●). This combination of mesoderm and ectoderm is the **amnion** (AM-nē-on). As development proceeds, the amnion and the amniotic cavity continue to enlarge. The amniotic cavity contains **amniotic fluid**, which surrounds and cushions the developing embryo or fetus (Figure 29-5b–e●).

THE ALLANTOIS The third extraembryonic membrane begins as an outpocketing of the endoderm near the base of the yolk sac (Figure 29-5b●). The free endodermal tip then grows toward the wall of the blastocyst, surrounded by a mass of mesodermal cells. This sac of endoderm and mesoderm is the **allantois** (a-LAN-tō-is). The base of the allantois later gives rise to the urinary bladder. (We detailed the formation of the allantois and its relationship to the urinary bladder in the Embryology Summary in Chapter 26, pp. 1000–1001.)

THE CHORION The mesoderm associated with the allantois spreads around the blastocyst, separating the cellular trophoblast from the blastocoele. This combination of mesoderm and trophoblast is the **chorion** (KOR-ē-on) (Figure 29-5a,b●).

When implantation first occurs, the nutrients absorbed by the trophoblast can easily reach the blastodisc by simple diffusion. But as the embryo and the trophoblast enlarge, the distance between them increases, and diffusion alone can no longer keep pace with the demands of the developing embryo. Blood vessels now begin to develop within the mesoderm of the chorion, creating a rapid-transit system that links the embryo with the trophoblast.

The appearance of blood vessels in the chorion is the first step in the creation of a functional placenta. By the third week of development, the mesoderm extends along the core of each of the trophoblastic villi, forming **chorionic villi** in contact with maternal tissues (Figures 29-5b and 29-6●, p. 1096). These villi continue to enlarge and branch, creating an intricate network within the endometrium. Embryonic blood vessels develop within each of the villi, and circulation through those chorionic vessels begins early in the third week of development, when the embryonic heart starts beating. The blood supply to the chorionic villi arises from the allantoic arteries and veins.

As the chorionic villi enlarge, more maternal blood vessels are eroded. Maternal blood now moves slowly through complex lacunae lined by the syncytial trophoblast. Chorionic blood vessels pass close by, and exchange between the embryonic and maternal circulations occurs by diffusion across the syncytial and cellular trophoblast layers.

Placentation

At first the entire blastocyst is surrounded by chorionic villi. The chorion continues to enlarge, expanding like a balloon within the endometrium. By the fourth week, the embryo, amnion, and yolk sac are suspended within an expansive, fluid-filled chamber (Figure 29-5c●). The **body stalk**, the connection between embryo and chorion, contains the distal portions of the allantois and blood vessels that carry blood to and from the placenta. The narrow connection between the endoderm of the embryo and the yolk sac is called the **yolk stalk**. (We detailed the formation of the yolk stalk and body stalk in the Embryology Summary in Chapter 24, pp. 910–911.)

The placenta does not continue to enlarge indefinitely. Regional differences in placental organization begin to develop as placental expansion creates a prominent bulge in the endometrial surface. This relatively thin portion of the endometrium, the **decidua capsularis** (dē-SID-ū-a kap-sū-LA-ris; *deciduus*, a falling off), no longer participates in nutrient exchange, and the chorionic villi in this region disappear (Figures 29-5d, 29-6a●). Placental functions are now concentrated in a disc-shaped area in the deepest portion of the endometrium, a region called the **decidua basalis** (ba-SA-lis). The rest of the uterine endometrium, which has no contact with the chorion, is called the **decidua parietalis**.

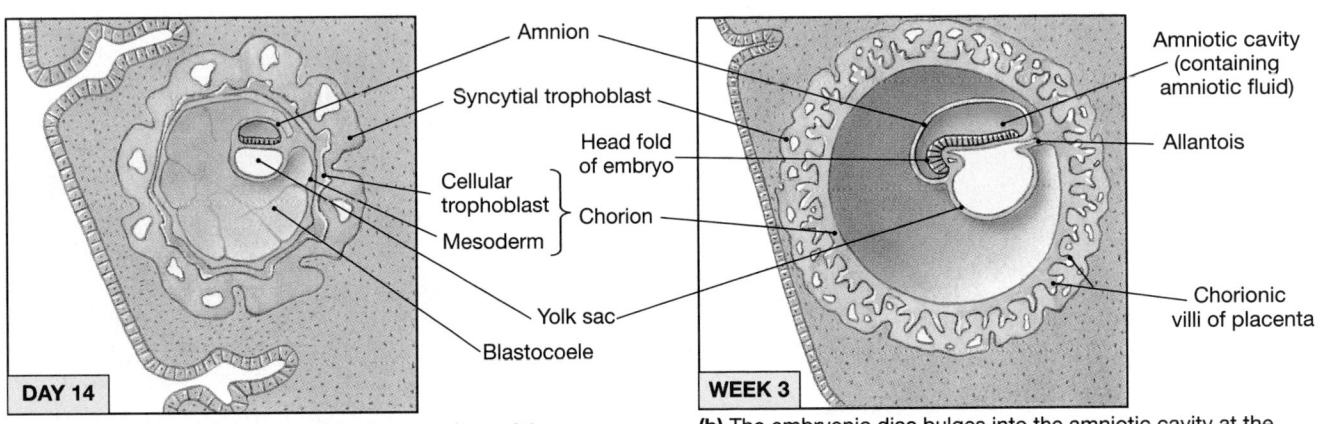

DAY 14

(a) Migration of mesoderm around the inner surface of the trophoblast creates the chorion. Mesodermal migration around the outside of the amniotic cavity, between the ectodermal cells and the trophoblast, creates the amnion. Mesodermal migration around the endodermal pouch below the blastodisc creates the yolk sac.

Amnion
Syncytial trophoblast
Cellular trophoblast
Mesoderm
Head fold of embryo
Chorion
Yolk sac
Blastocoele

WEEK 3

(b) The embryonic disc bulges into the amniotic cavity at the head fold. The allantois, an endodermal extension surrounded by mesoderm, extends toward the trophoblast.

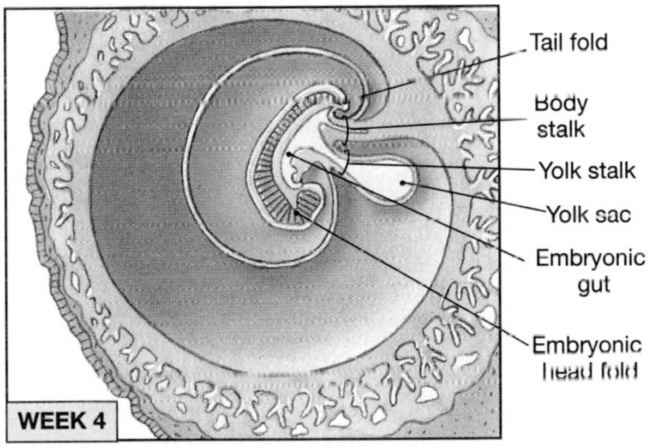

Amniotic cavity (containing amniotic fluid)
Allantois
Chorionic villi of placenta

WEEK 4

Tail fold
Body stalk
Yolk stalk
Yolk sac
Embryonic gut
Embryonic head fold

(c) The embryo now has a head fold and a tail fold. Constriction of the connection between the embryo and the surrounding trophoblast constricts the yolk stalk and body stalk.

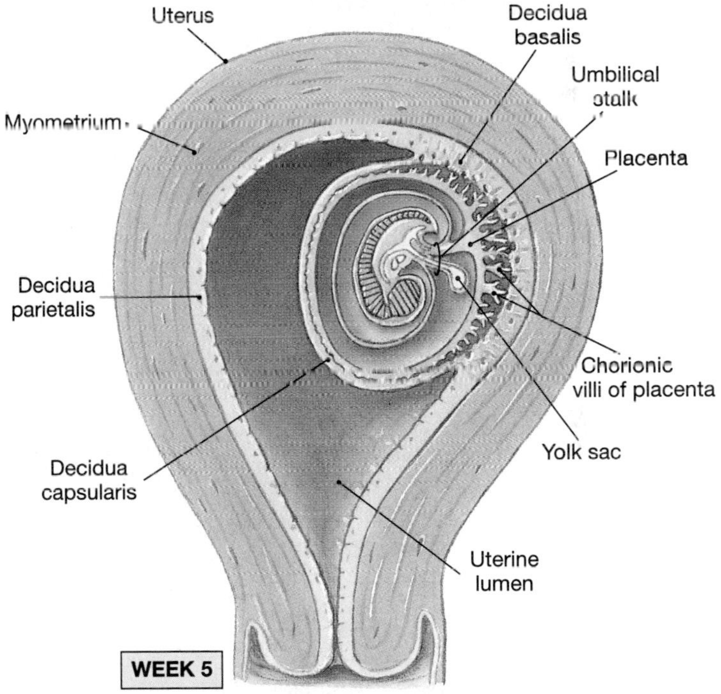

WEEK 5

Uterus
Myometrium
Decidua parietalis
Decidua capsularis
Decidua basalis
Umbilical stalk
Placenta
Chorionic villi of placenta
Yolk sac
Uterine lumen

(d) The developing embryo and extraembryonic membranes bulge into the uterine cavity. The trophoblast pushing out into the uterine lumen remains covered by endometrium but no longer participates in nutrient absorption and embryo support. The embryo moves away from the placenta, and the body stalk and yolk stalk fuse to form an umbilical stalk.

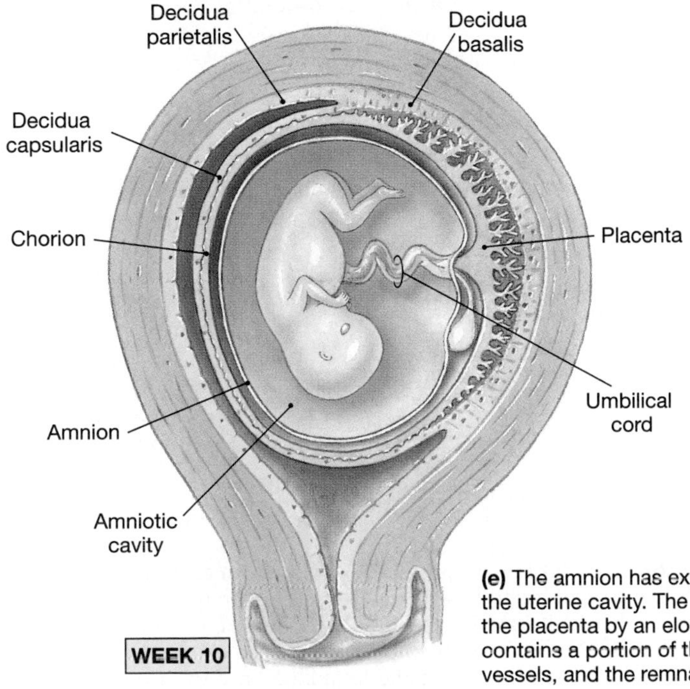

WEEK 10

Decidua parietalis
Decidua basalis
Decidua capsularis
Chorion
Placenta
Amnion
Umbilical cord
Amniotic cavity

(e) The amnion has expanded greatly, filling the uterine cavity. The fetus is connected to the placenta by an elongate umbilical cord that contains a portion of the allantois, blood vessels, and the remnants of the yolk stalk.

•FIGURE 29-5
Extraembryonic Membranes and Placenta Formation

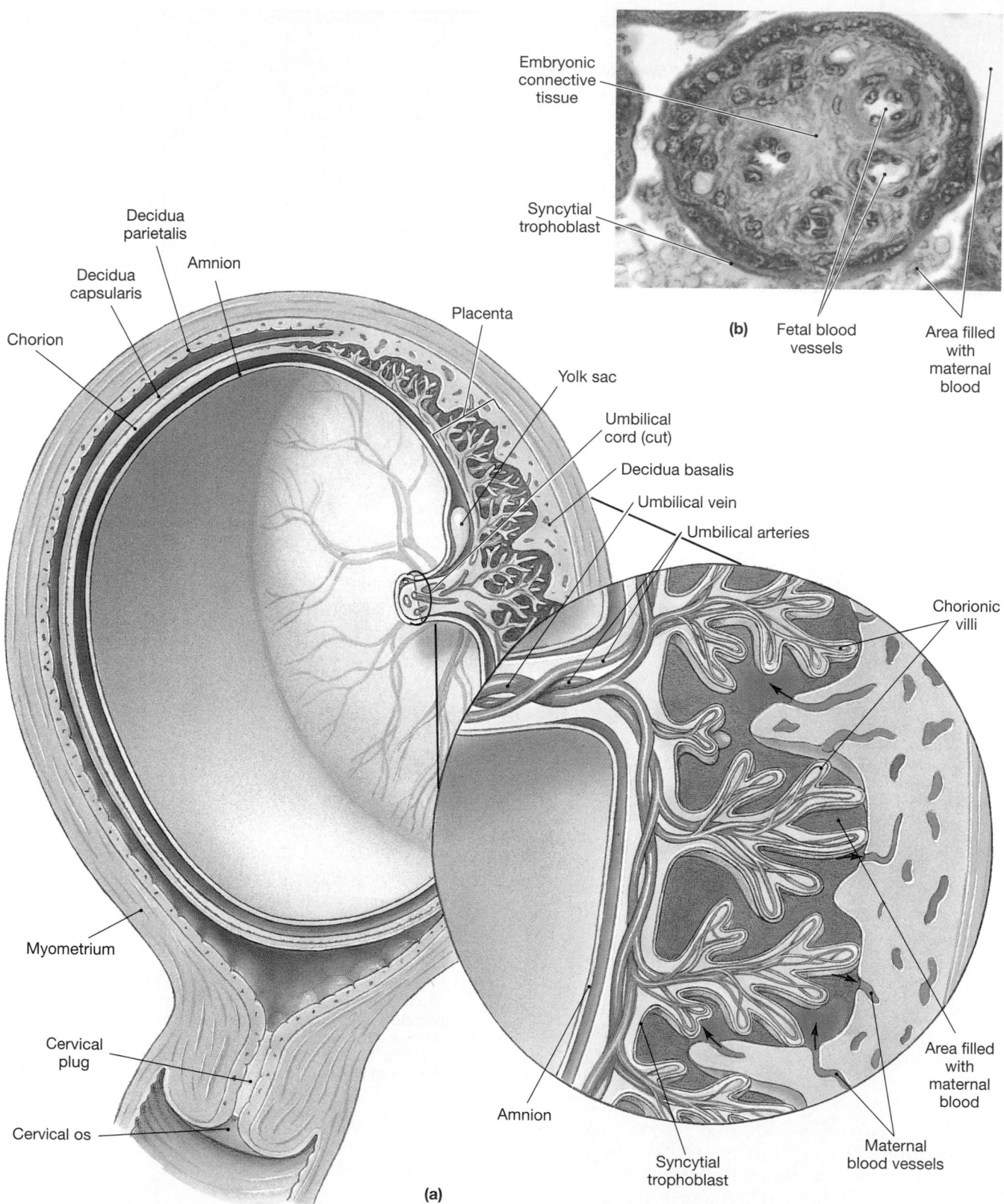

•FIGURE 29-6 A Three-Dimensional View of Placental Structure. (a) For clarity, the uterus is shown after the embryo has been removed and the umbilical cord cut. Arrows indicate the direction of blood flow. Blood flows into the placenta through ruptured maternal blood arteries. It then flows around chorionic villi, which contain fetal blood vessels. Fetal blood arrives over paired umbilical arteries and leaves through a single umbilical vein. Maternal blood reenters the venous system of the mother through the broken walls of small uterine veins. Note that no actual mixing of maternal and fetal blood occurs. **(b)** A cross section through a chorionic villus, showing the syncytial trophoblast exposed to the maternal blood space. (LM × 2045)

As the end of the first trimester approaches, the fetus moves farther from the placenta (Figure 29-5d,e•). The fetus and placenta remain connected by the **umbilical cord**, or **umbilical stalk**, which contains the allantois, the placental blood vessels, and the yolk stalk.

Placental Circulation

Figure 29-6a• diagrams circulation at the placenta near the end of the first trimester. Blood flows to the placenta through the paired **umbilical arteries** and returns in a single **umbilical vein**. ∞ *[p. 756]* The chorionic villi provide the surface area for active and passive exchange of gases, nutrients, and waste products between the fetal and maternal bloodstreams (Figure 29-6b•).

The placenta places a considerable demand on the maternal cardiovascular system, and blood flow to the uterus and placenta is extensive. If the placenta is torn or otherwise damaged, the consequences may prove fatal for both fetus and mother. [AM] *Problems with Placentation*

The Endocrine Placenta

In addition to its role in the nutrition of the fetus, the placenta acts as an endocrine organ. Hormones are synthesized by the syncytial trophoblast and released into the maternal circulation. The hormones produced include *human chorionic gonadotropin, human placental lactogen, placental prolactin, relaxin, progesterone,* and *estrogens.*

HUMAN CHORIONIC GONADOTROPIN The hormone **human chorionic** (ko-rē-ON-ik) **gonadotropin (hCG)** appears in the maternal bloodstream soon after implantation has occurred. The presence of hCG in blood or urine samples provides a reliable indication of pregnancy. Kits sold for the early detection of pregnancy are sensitive to the presence of this hormone.

In function, hCG resembles LH, for it maintains the integrity of the corpus luteum and promotes the continued secretion of progesterone. As a result, the endometrial lining remains perfectly functional, and menses does not occur. If it did, the pregnancy would end, for the functional zone of the endometrium would disintegrate.

In the presence of hCG, the corpus luteum persists for 3–4 months before gradually decreasing in size and secretory function. The decline in luteal function does not trigger the return of menstrual periods, because by the end of the first trimester the placenta is actively secreting both estrogens and progesterone.

HUMAN PLACENTAL LACTOGEN AND PLACENTAL PROLACTIN Human placental lactogen (hPL), or *human chorionic somatomammotropin* (hCS), helps prepare the mammary glands for milk production. It also has a stimulatory effect on other tissues comparable to that of growth hormone (GH). At the mammary

glands, the conversion from resting to active status requires the presence of placental hormones (hPL, **placental prolactin**, estrogen, and progesterone) as well as several maternal hormones (GH, prolactin [PRL], and thyroid hormones). We shall consider the hormonal control of mammary gland function in a later section.

RELAXIN Relaxin is a peptide hormone that is secreted by the placenta as well as by the corpus luteum during pregnancy. Relaxin (1) increases the flexibility of the pubic symphysis, permitting expansion of the pelvis during delivery; (2) causes dilation of the cervix, making it easier for the fetus to enter the vaginal canal; and (3) suppresses the release of oxytocin by the hypothalamus and delays the onset of labor contractions

PROGESTERONE AND ESTROGENS After the first trimester, the placenta produces sufficient amounts of progesterone to maintain the endometrial lining and continue the pregnancy. As the end of the third trimester approaches, estrogen production by the placenta accelerates. As we shall see in a later section, the rising estrogen levels play a role in stimulating labor and delivery.

Embryogenesis

Shortly after gastrulation begins, folding and differential growth of the embryonic disc produce a bulge that projects into the amniotic cavity (Figure 29-5b•). This projection is known as the **head fold**. Similar movements lead to the formation of a **tail fold** (Figure 29-5c•). The embryo is now physically as well as developmentally distinct from the blastodisc and the extraembryonic membranes. The definitive orientation of the embryo can now be seen, complete with dorsal and ventral surfaces and left and right sides. Table 9-2 provides an overview of the subsequent development of the major organs and body systems. The changes in proportions and appearance that occur between the second developmental week and the end of the first trimester are visually summarized in Figure 29-7• (p. 1100).

The first trimester is a critical period for development because events in the first 12 weeks establish the basis for organ formation, a process called **organogenesis**. In Embryology Summaries in earlier chapters, we described major features of organogenesis in each organ system. Important developmental milestones are indicated in Table 29-2.

☑ What is the fate of the inner cell mass of the blastocyst?

☑ Improper development of which of the extraembryonic membranes would affect the circulatory system?

☑ Sue's pregnancy test indicates elevated levels of the hormone hCG (human chorionic gonadotropin). Is she pregnant?

☑ What are two important functions of the placenta?

TABLE 29-2   **An Overview of Prenatal and Early Postnatal Development**

Gestational Age (Months)	Size and Weight	 Integumentary System	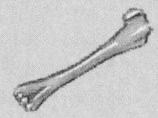 Skeletal System	 Muscular System	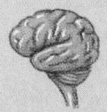 Nervous System	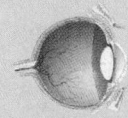 Special Sense Organs
	Background Material Ch. 4 Development of Tissues (p. 122) Development of Organ Systems (pp. 138–139)					
1	5 mm 0.02 g		(b) Formation of somites	(b) Formation of somites	(b) Neural tube formation	(b) Formation of eyes and ears
2	28 mm 2.7 g	(b) Formation of nail beds, hair follicles, sweat glands	(b) Formation of axial and appendicular cartilage	(c) Rudiments of axial musculature	(b) CNS, PNS organization, growth of cerebrum	(b) Formation of taste buds, olfactory epithelium
3	78 mm 26 g	(b) Epidermal layers appear	(b) Spreading of ossification centers	(c) Rudiments of appendicular musculature	(c) Basic spinal cord and brain structure	
4	133 mm 0.15 kg	(b) Formation of hair, sebaceous glands (c) Sweat glands	(b) Articulations (c) Facial and palatal organization	Fetus starts moving	(b) Rapid expansion of cerebrum	(c) Basic eye and ear structure (b) Formation of peripheral receptors
5	185 mm 0.46 kg	(b) Keratin production, nail production			(b) Myelination of spinal cord	
6	230 mm 0.64 kg			(c) Perineal muscles	(b) Formation of CNS tract (c) Layering of cortex	
7	270 mm 1.492 kg	(b) Keratinization, formation of nails, hair				(c) Eyelids open, retinae sensitive to light
8	310 mm 2.274 kg		(b) Formation of epiphyseal plate			(c) Taste receptor functional
9	346 mm 3.2 kg					
Postnatal development		Hair changes in consistency and distribution	Formation and growth of epiphyseal plates continue	Muscle mass and control increase	Myelination, layering, CNS tract formation continue	
Chapters containing relevant text and Embryology Summaries		5: pp. 166–167	6: pp. 179–182 7: pp. 216–217, pp. 228–229 8: pp. 240–241	11: pp. 362–363	12: p. 379 13: pp. 436–437 14: pp. 445–446, pp. 454–455	17: pp. 582–583

Note: (b) = beginning; (c) = completion.

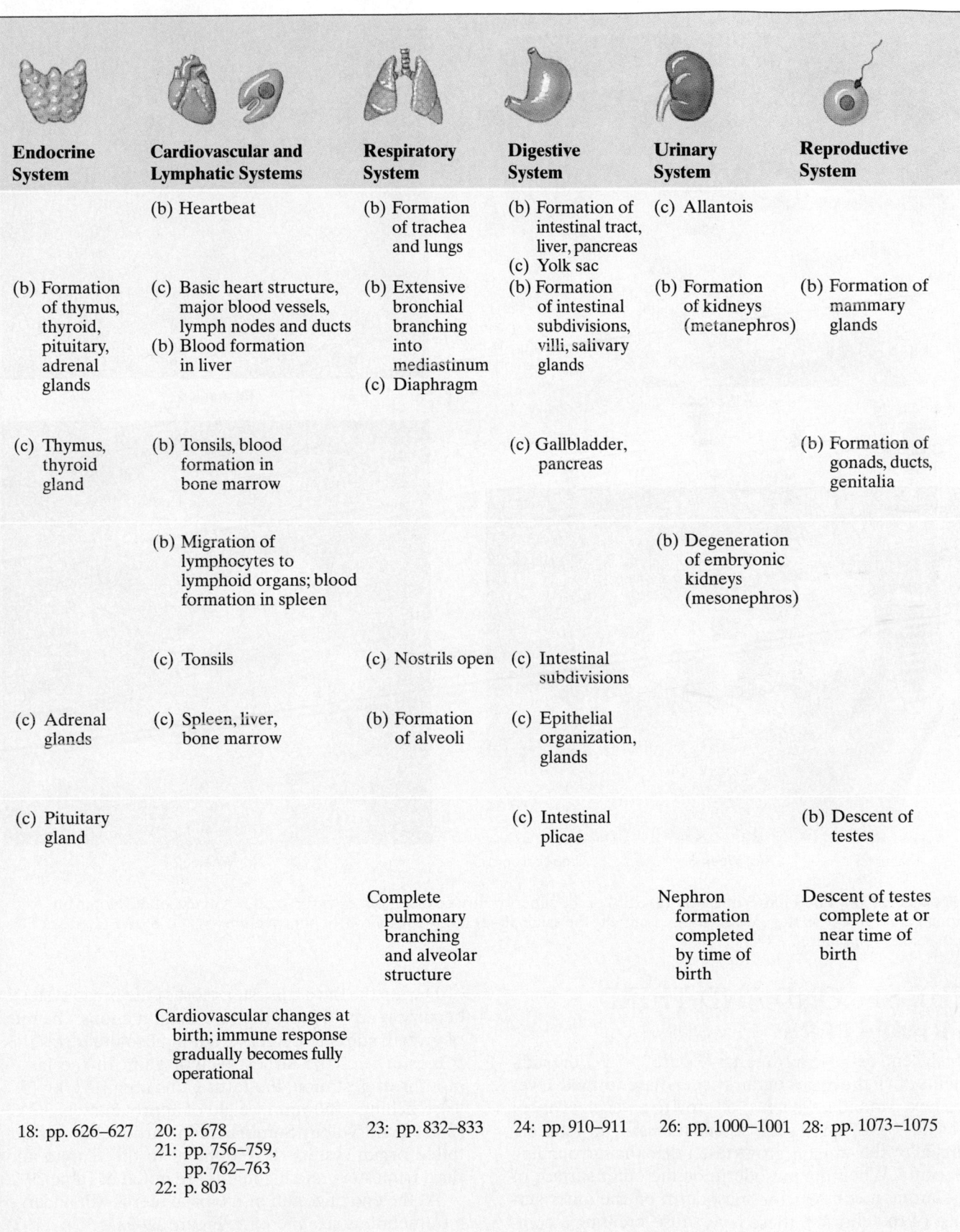

Endocrine System	Cardiovascular and Lymphatic Systems	Respiratory System	Digestive System	Urinary System	Reproductive System
	(b) Heartbeat	(b) Formation of trachea and lungs	(b) Formation of intestinal tract, liver, pancreas (c) Yolk sac	(c) Allantois	
(b) Formation of thymus, thyroid, pituitary, adrenal glands	(c) Basic heart structure, major blood vessels, lymph nodes and ducts (b) Blood formation in liver	(b) Extensive bronchial branching into mediastinum (c) Diaphragm	(b) Formation of intestinal subdivisions, villi, salivary glands	(b) Formation of kidneys (metanephros)	(b) Formation of mammary glands
(c) Thymus, thyroid gland	(b) Tonsils, blood formation in bone marrow		(c) Gallbladder, pancreas		(b) Formation of gonads, ducts, genitalia
	(b) Migration of lymphocytes to lymphoid organs; blood formation in spleen			(b) Degeneration of embryonic kidneys (mesonephros)	
	(c) Tonsils	(c) Nostrils open	(c) Intestinal subdivisions		
(c) Adrenal glands	(c) Spleen, liver, bone marrow	(b) Formation of alveoli	(c) Epithelial organization, glands		
(c) Pituitary gland			(c) Intestinal plicae		(b) Descent of testes
		Complete pulmonary branching and alveolar structure		Nephron formation completed by time of birth	Descent of testes complete at or near time of birth
	Cardiovascular changes at birth; immune response gradually becomes fully operational				
18: pp. 626–627	20: p. 678 21: pp. 756–759, pp. 762–763 22: p. 803	23: pp. 832–833	24: pp. 910–911	26: pp. 1000–1001	28: pp. 1073–1075

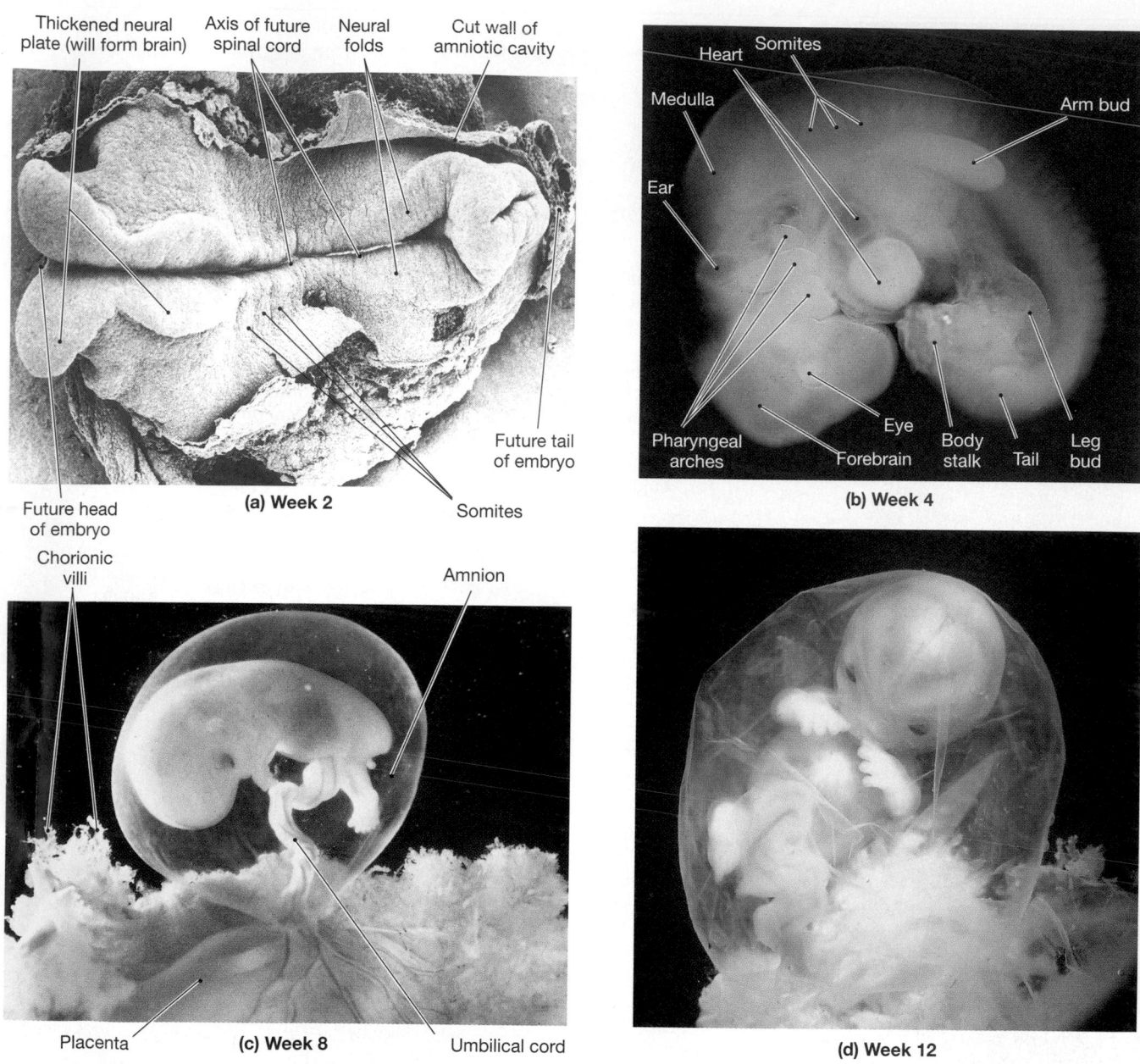

(a) Week 2

Thickened neural plate (will form brain) — Axis of future spinal cord — Neural folds — Cut wall of amniotic cavity — Future tail of embryo — Future head of embryo — Somites

(b) Week 4

Heart — Somites — Medulla — Arm bud — Ear — Pharyngeal arches — Eye — Forebrain — Body stalk — Tail — Leg bud

(c) Week 8

Chorionic villi — Amnion — Placenta — Umbilical cord

(d) Week 12

● **FIGURE 29-7 The First Trimester. (a)** SEM of the superior surface of a monkey embryo after 2 weeks of development. A human embryo at this stage would look essentially the same. **(b–d)** Fiber-optic views of human embryos at 4, 8, and 12 weeks. For actual sizes, see Figure 29-14, p. 1110.

THE SECOND AND THIRD TRIMESTERS

By the end of the first trimester (Figure 29-7d●), the rudiments of all the major organ systems have formed. Over the next 3 months, the fetus will grow to a weight of about 0.64 kg (1.4 lb). During this second trimester, the fetus, encircled by the amnion, grows faster than the surrounding placenta. When the mesoderm on the outer surface of the amnion contacts the mesoderm on the inner surface of the chorion, these layers fuse, creating a compound *amniochorionic membrane*. Figure 29-8a● shows a 4-month fetus; Figure 29-8b● shows a 6-month fetus.

During the third trimester, most of the organ systems become ready to fulfill their normal functions. The rate of growth starts to decrease, but in absolute terms this trimester sees the largest weight gain. In the last 3 months of gestation, the fetus gains about 2.6 kg (5.7 lb), reaching a full-term weight of approximately 3.2 kg (7 lb). Embryology Summaries in earlier chapters detailed organ system development in the second and third trimesters, and highlights are noted in Table 29-2.

At the end of gestation, a typical uterus will undergo a tremendous size increase. Figure 29-9a–c● (p. 1102) shows the positions of the uterus, fetus, and placenta from 16 weeks to full term. When the pregnancy is at term, the

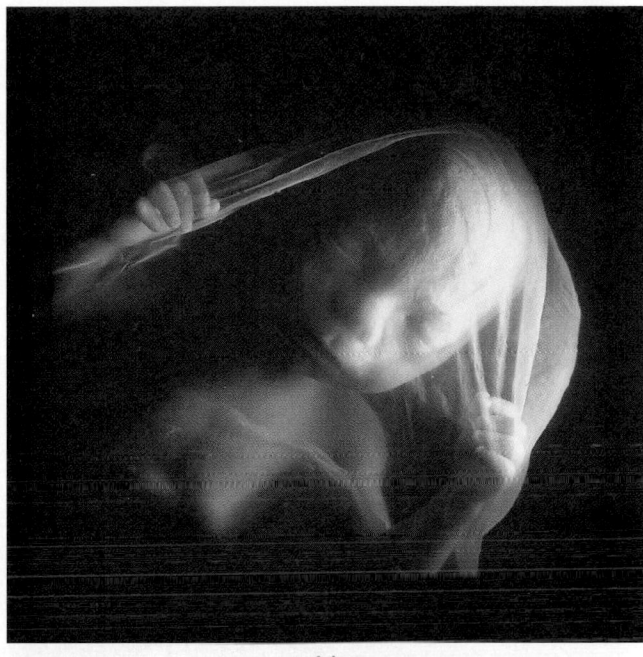

(a)

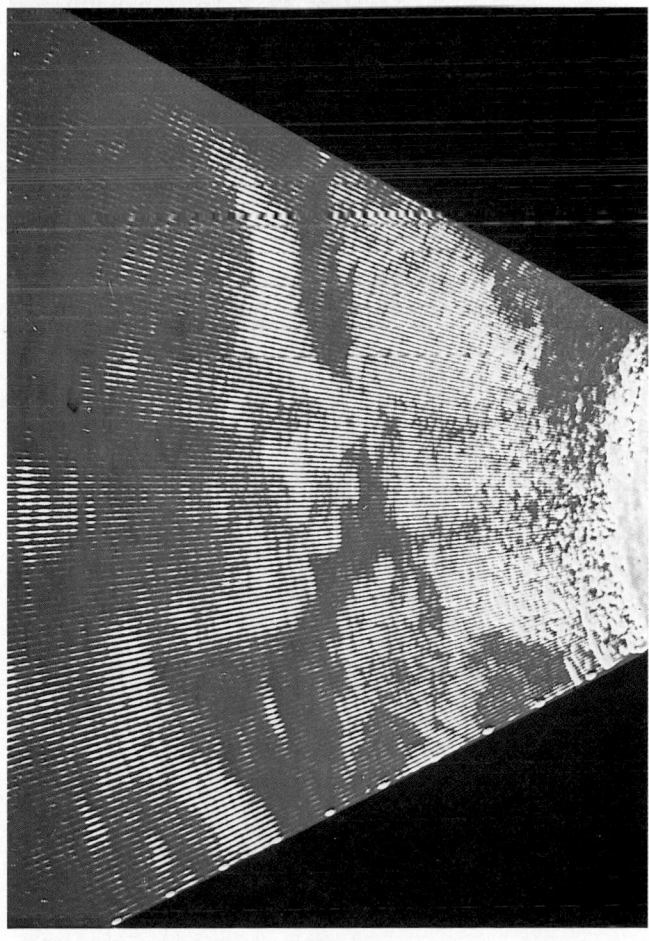

(b)

● **FIGURE 29-8** The Second and Third Trimesters.
(a) A 4 month fetus, as seen through a fiber-optic endoscope.
(b) Head of a 6-month fetus, as seen through ultrasound.

uterus and fetus push many of the maternal abdominal organs out of their normal positions (Figure 29-9c,d●).

Pregnancy and Maternal Systems

The developing fetus is totally dependent on maternal organ systems for nourishment, respiration, and waste removal. These functions must be performed by maternal systems in addition to their normal operations. For example, the mother must absorb enough oxygen, nutrients, and vitamins for herself *and* her fetus, and she must eliminate all the generated wastes. Although this is not a burden over the initial weeks of gestation, the demands placed on the mother become significant as the fetus grows. For the mother to survive under these conditions, the maternal systems must make major compensatory adjustments. In practical terms, the mother must breathe, eat, and excrete for two. Among the major changes that occur in maternal systems are the following:

- *The maternal respiratory rate goes up and the tidal volume increases.* As a result, the mother's lungs deliver the extra oxygen required and remove the excess carbon dioxide generated by the fetus.
- *The maternal blood volume increases.* This increase occurs because (1) blood flowing into the placenta reduces the volume in the rest of the systemic circuit, and (2) fetal activity lowers the blood P_{O_2} and elevates the P_{CO_2}. The combination stimulates the production of renin and erythropoietin, leading to an increase in maternal blood volume through mechanisms detailed in Chapter 18 (see Figure 18-17b●, p. 619). By the end of gestation, the maternal blood volume has increased by almost 50 percent.
- *The maternal requirements for nutrients and vitamins climb 10–30 percent.* Pregnant women, who must eat for two, are often hungry.
- *The maternal glomerular filtration rate increases by roughly 50 percent.* This increase, which corresponds to the increase in blood volume, accelerates the excretion of metabolic wastes generated by the fetus. Because the volume of urine produced increases and the weight of the uterus presses down on the urinary bladder, pregnant women need to urinate frequently.
- *The uterus undergoes a tremendous increase in size.* Structural and functional changes in the expanding uterus are so important that we shall discuss them in a separate section.
- *The mammary glands increase in size, and secretory activity begins.* Mammary gland development requires a combination of hormones, including human placental lactogen and placental prolactin from the placenta, and PRL, estrogens, progesterone, GH, and thyroxine from maternal endocrine organs. By the end of the sixth month of pregnancy, the mammary glands are fully developed, and the glands begin to produce secretions that are stored in the duct system.

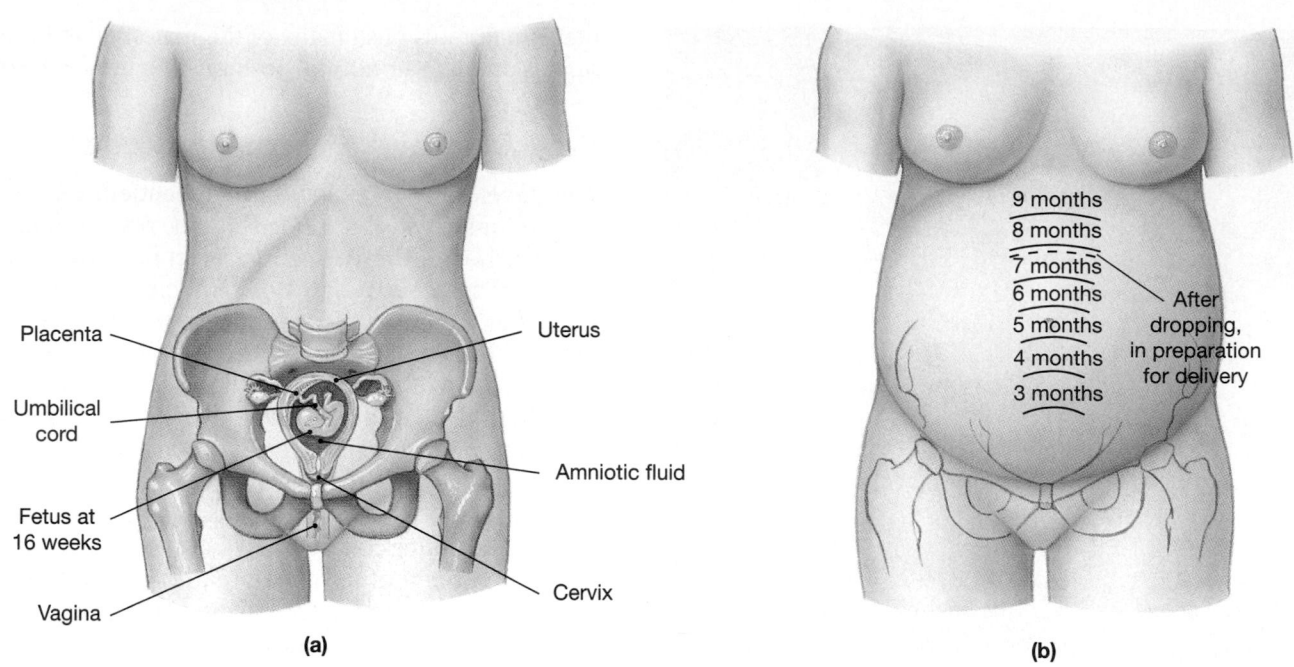

Placenta

Uterus

Umbilical cord

Fetus at 16 weeks

Amniotic fluid

Vagina

Cervix

(a)

9 months

8 months

7 months

6 months

5 months

4 months

3 months

After dropping, in preparation for delivery

(b)

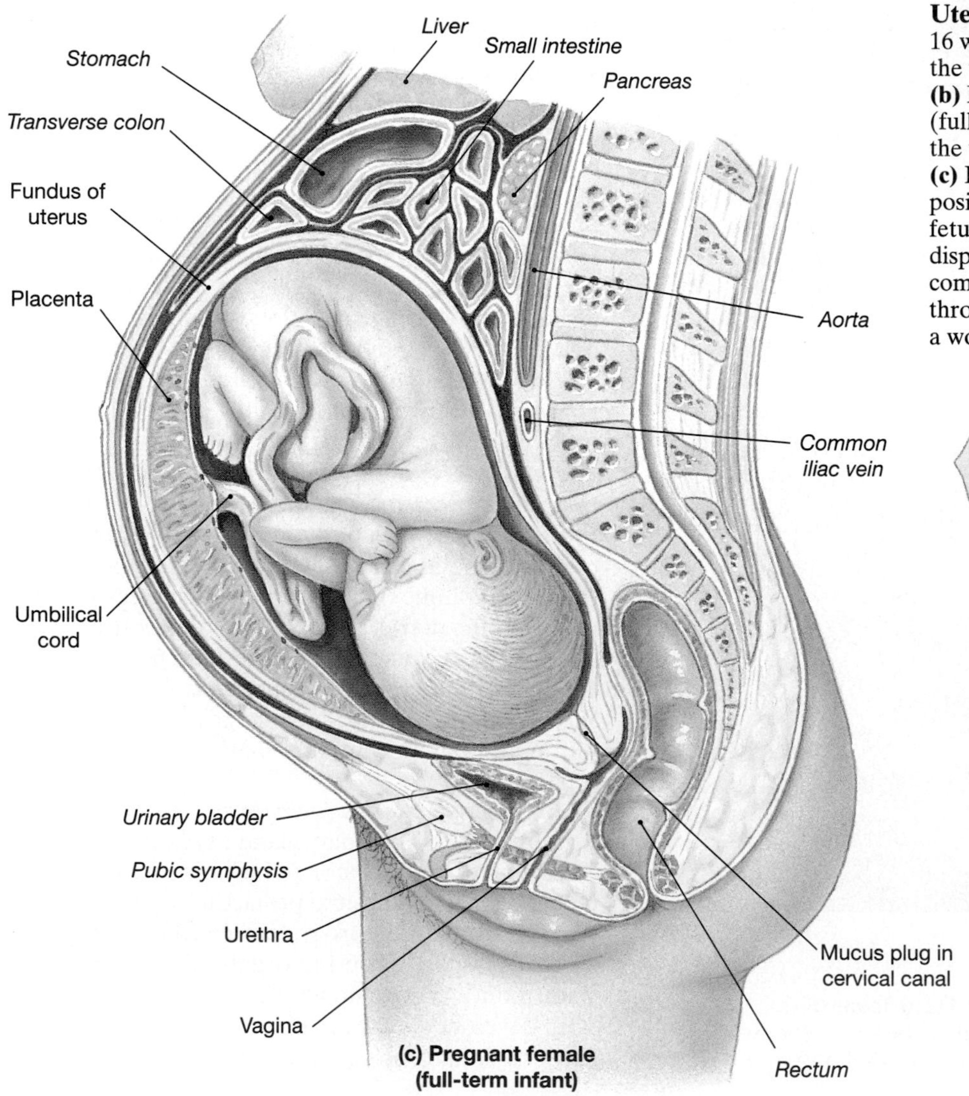

Stomach

Liver

Small intestine

Pancreas

Transverse colon

Fundus of uterus

Placenta

Aorta

Umbilical cord

Common iliac vein

Urinary bladder

Pubic symphysis

Urethra

Vagina

Mucus plug in cervical canal

Rectum

(c) Pregnant female (full-term infant)

(d) Nonpregnant female

●*FIGURE 29-9* **Growth of the Uterus and Fetus.** **(a)** Pregnancy at 16 weeks, showing the positions of the uterus, fetus, and placenta. **(b)** Pregnancy at 3 months to 9 months (full term), showing the position of the uterus within the abdomen. **(c)** Pregnancy at full term. Note the position of the uterus and full-term fetus within the abdomen and the displacement of abdominal organs, as compared with **(d)**, a sectional view through the abdominopelvic cavity of a woman who is not pregnant.

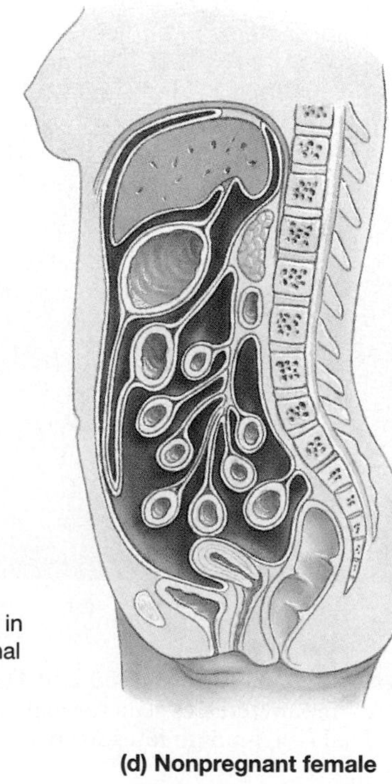

Technology and the Treatment of Infertility

Infertility is usually defined as an inability to achieve pregnancy after 1 year of appropriately timed intercourse. The problem has been the focus of media attention in recent years. The reason is simple: Problems with fertility are relatively common. An estimated 10–15 percent of U.S. married couples are infertile, and another 10 percent are unable to have as many children as they desire. It is thus not surprising that reproductive physiology has become a popular field, and the treatment of infertility has become a major medical industry.

An infertile, or *sterile*, woman is unable to produce functional oocytes or to support a developing embryo. An infertile man is incapable of providing a sufficient number of motile sperm capable of successful fertilization. Because sterility of either sexual partner will have the same result, diagnosis and treatment of infertility must involve evaluation of both partners. Approximately 40 percent of infertility cases are attributed to the female partner, 40 percent to the male partner, and 20 percent to both partners.

Recent advances in our understanding of reproductive physiology are providing new solutions to fertility problems. These approaches, called **assisted reproductive technologies (ART),** are diagrammed in Figure 29-10●:

- ▪ **Low sperm count.** In cases of male infertility due to low sperm counts, semen from several ejaculates can be pooled, concentrated, and introduced into the female reproductive tract. This technique, known as *artificial insemination*, may lead to normal fertilization and pregnancy. In special cases, in which a male's spermatozoa are unable to penetrate the oocyte, single-sperm fertilization has been accomplished with micromanipulation of the oocyte and corona radiata.

- ▪ **Abnormal spermatozoa.** If the male cannot produce functional sperm, sperm can be obtained from a *sperm bank,* a storage facility for donor sperm.

- ▪ **Hormonal problems.** If the problem involves the woman's inability to ovulate because her gonadotropin or estrogen levels are low or she is unable to maintain adequate progesterone levels after ovulation, these hormones can be provided.

- ▪ So-called *fertility drugs,* such as clomiphene (*Clomid*®), stimulate ovarian oocyte production. Clomiphene works by blocking the feedback inhibition of the hypothalamus and pituitary gland by estrogens. As a result, circulating FSH levels rise, and more follicles are stimulated to complete their development. Injected purified gonadotropins, such as *Pergonal*® (FSH and LH) and *Metrodine*® (FSH), are also used to accelerate ovum development. The chance that a single oocyte will be fertilized through sexual intercourse is about 1 in 3. Increasing the number of oocytes released increases the odds of fertilization and therefore the odds of a pregnancy. Unfortunately, it is not easy to determine just how much ovarian stimulation will be needed, so it is common for treatment with

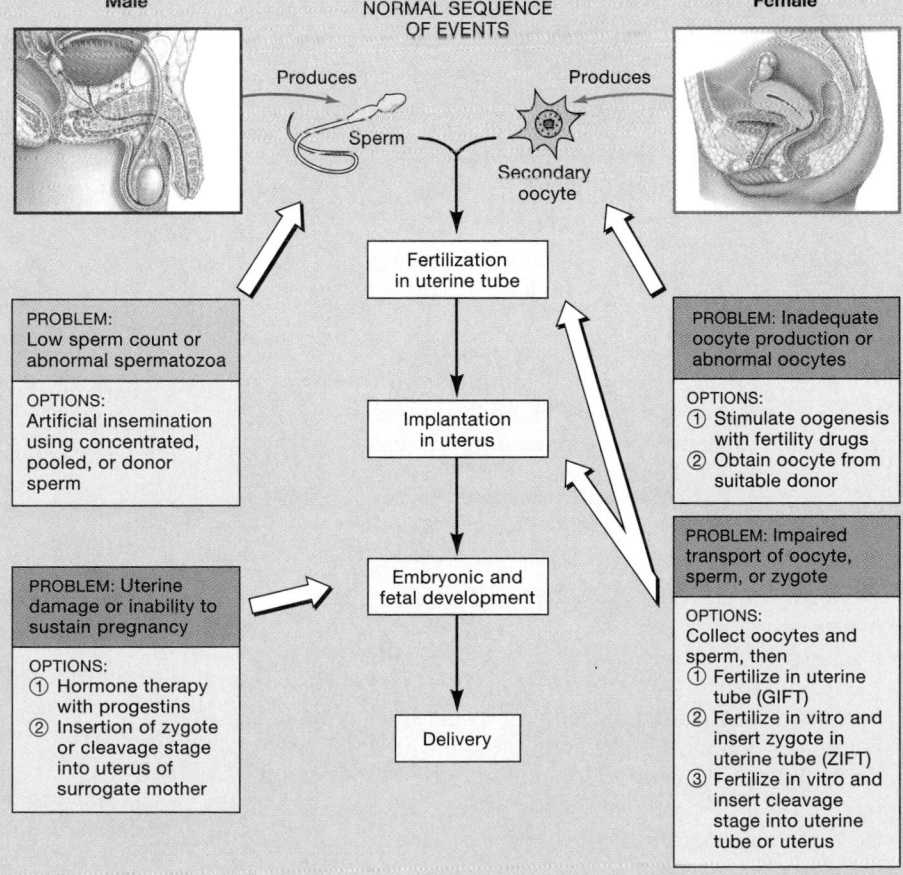

●**FIGURE 29-10** **The Treatment of Infertility**

fertility drugs to result in multiple births. Careful monitoring of follicle development somewhat reduces the chances of a multiple birth.

- *Problems with oocyte transport from the ovary to the uterine tube.* When there are problems with the transport of the oocyte from the ovary to the uterine tube, due to scarring of the fimbriae or other problems, a procedure called GIFT can be used. **GIFT** is short for *gamete intrafallopian tube transfer*. (*Fallopian tube* is another name for the uterine tube.) In this procedure, the ovaries are stimulated with injected hormones, and a large "crop" of mature oocytes is removed from tertiary follicles. The individual oocytes are examined for defects, inserted into the uterine tubes, and exposed to high concentrations of sperm from the husband or donor. The success rate of this procedure is less than that of natural fertilization (33 percent), and not every pregnancy produces an infant. The cost of a single procedure (successful or not) averages $5000.

- *Blocked uterine tubes.* Blockage of the uterine tubes or damage to the uterine lining can interfere with oocyte, sperm, and zygote transport. In the GIFT procedure, fertilization occurs in its normal location, within the uterine tube. This site is not essential, and fertilization can also take place in a test tube or petri dish. This process is called **in vitro fertilization**, or **IVF** (*vitro*, glass). If a carefully controlled fluid environment is provided, early development will proceed normally. One variation on the GIFT procedure, called **ZIFT** (*zygote intrafallopian tube transfer*), exposes selected oocytes to sperm outside the body and inserts zygotes or pre-embryos, rather than oocytes, into the uterine tubes. If multiple zygotes are available, some can be frozen and stored for later insertion in case the initial procedure fails to produce a successful pregnancy. The cost for a single ZIFT procedure ranges between $8000 and $10,000.

- Alternatively, the zygote can be maintained in an artificial environment through the first 2–3 days of development. This procedure is commonly selected if the uterine tubes are damaged or blocked. The cleavage-stage embryo is then placed directly into the uterus rather than into one of the uterine tubes. The cost of this procedure is comparable to that for ZIFT.

- *Abnormal oocytes.* If the oocytes released by the ovaries are abnormal in some way or if menopause has already occurred, viable oocytes can be obtained from a suitable donor. The donor may be anonymous or known; if anonymous, the donor usually receives a fee for the donation. Through treatment with fertility drugs, the donor's ovaries are stimulated to produce a large crop of oocytes. These are collected and fertilized in vitro, generally by the sperm of the recipient's husband. After cleavage has begun, the pre-embryo is placed in the recipient's uterus, which has been "primed" by progesterone therapy. The pregnancy rate of this procedure is roughly 33 percent for women over age 40, using oocytes donated by women in their early twenties. Oocyte donation has a much higher success rate for these women than ZIFT or GIFT, which have pregnancy rates of only about 4 percent. This difference suggests that age-related changes in the characteristics and quality of the oocytes, rather than changes in hormone levels or uterine responsiveness, are typically the primary cause of infertility in older women.

- *Abnormal uterine environment.* If fertilization and transport occur normally but the uterus cannot maintain a pregnancy, the problem may involve low levels of progesterone secretion by the corpus luteum. Hormone therapy may solve this problem. If the maternal uterus simply cannot support development, the zygote or cleavage-stage embryo can be introduced into the uterus of a *surrogate mother*. If the embryo survives and makes contact with the endometrium, development will proceed normally despite the fact that the mother has no genetic relationship with the embryo.

Surrogate motherhood, which sounds relatively simple and straightforward, has proven to be one of the most explosive solutions in terms of ethics and legality. Since 1990, several court cases have resulted from disputes over surrogate motherhood and who merits legal custody of the infant. Legal battles have also broken out over a variety of other complex questions, and some of them will take years to sort out. To understand the problem, consider the following questions:

- Do parents share property rights over frozen and stored zygotes? Can a husband have any of the stored zygotes implanted into the uterus of his second wife without the consent of his first wife, who provided the eggs? Can a wife use her husband's stored sperm to become pregnant after his death?

- If both donor egg and donor sperm are used, do adoption laws apply?

- If the husband provided the sperm that fertilized the oocyte of a donor who is not his wife, for implantation into a surrogate mother, can the wife, the surrogate mother, or the egg donor sue for custody of the child after a divorce?

- If a hospital stores frozen pre-embryos or sperm but the freezer breaks down, what is the hospital's liability? What is the monetary value of a frozen pre-embryo?

If you use your imagination, you can probably think of even more complex problems, many of which will probably be debated in courtrooms within the next decade.

ABORTION Abortion is the termination of a pregnancy. Most references distinguish among *spontaneous, therapeutic*, and *induced abortions*. **Spontaneous abortions**, or *miscarriages*, occur as a result of developmental or physiological problems. For example, spontaneous abortions may result from chromosomal defects in the embryo or from hormonal problems, such as inadequate LH production by the maternal pituitary gland (or reduced LH sensitivity at the corpus luteum), inadequate progesterone sensitivity in the endometrium, or placental failure to produce adequate levels of hCG. Spontaneous abortions occur in roughly 15 percent of all pregnancies. **Therapeutic abortions** are performed when continuing the pregnancy represents a threat to the life of the mother. In the *Applications Manual*, we consider several potential complicating factors during pregnancy. [AM] *Problems with the Maintenance of a Pregnancy*

Induced abortions, or *elective abortions*, are performed at the woman's request. Induced abortions remain the focus of considerable controversy. Each year approximately 1.5 million induced abortions are performed in the United States—roughly one abortion for every three births. Most induced abortions involve unmarried and/or adolescent women. The ratio between abortions and deliveries for married women is 1:10, whereas it is nearly 2:1 for unmarried women and adolescents. In most states, induced abortions are legal during the first three months after conception; with restrictions, induced abortions may be permitted until the fifth or sixth month.

Structural and Functional Changes in the Uterus

At the end of gestation, a typical uterus will have grown from 7.5 cm (3 in.) in length and 30–40 g (1–1.4 oz) in weight to 30 cm (12 in.) in length and 1100 g (2.4 lb) in weight. It may then contain almost 5 liters of fluid, giving the organ with contents a total weight of roughly 10 kg (22 lb). This remarkable expansion occurs through the enlargement (hypertrophy) and elongation of existing cells, especially smooth muscle fibers, rather than by an increase in the total number of cells.

The tremendous stretching of the uterus is associated with a gradual increase in the rates of spontaneous smooth muscle contractions in the myometrium. In the early stages of pregnancy, the contractions are weak, painless, and brief. There are indications that the progesterone released by the placenta has an inhibitory effect on the uterine smooth muscle, preventing more extensive and powerful contractions.

Three major factors oppose the calming action of progesterone:

1. *Rising estrogen levels.* Estrogens produced by the placenta increase the sensitivity of the uterine smooth muscles and make contractions more likely. Through-

out pregnancy, progesterone exerts the dominant effect, but as the time of delivery approaches, the production of estrogen accelerates, and the myometrium becomes more sensitive to stimulation. Estrogens also increase the sensitivity of smooth muscle fibers to oxytocin.

2. *Rising oxytocin levels.* Rising oxytocin levels stimulate an increase in the force and frequency of uterine contractions. Oxytocin release is stimulated by high estrogen levels and by the distortion of the uterine cervix. Uterine distortion occurs as the weight of the fetus increases.

3. *Prostaglandin production.* Estrogens and oxytocin stimulate the production of prostaglandins in the endometrium. These prostaglandins further stimulate smooth muscle contractions.

Late in pregnancy, some women experience occasional spasms in the uterine musculature, but the contractions are neither regular nor persistent. These contractions are called **false labor**. **True labor** begins when the biochemical and mechanical factors reach the point of no return. After 9 months of gestation, multiple factors interact to initiate true labor. Once the **labor contractions** have begun in the myometrium, positive feedback ensures that the contractions will continue until delivery has been completed.

Figure 29-11● diagrams important factors that stimulate and sustain labor. The actual trigger for the onset of labor may be events in the fetus rather than in the mother. When labor commences, the fetal pituitary secretes oxytocin that is released into the maternal bloodstream at the placenta. The resulting increase in myometrial contractions and prostaglandin production, on top of the priming effects of estrogens and maternal oxytocin, may be the "last straw."

LABOR AND DELIVERY

The goal of labor is the forcible expulsion of the fetus, a process known as **parturition** (par-tū-RISH-un), or birth. During true labor, each labor contraction begins near the top of the uterus and sweeps in a wave toward the cervix. These contractions are strong and occur at regular intervals. As parturition approaches, the contractions increase in force and frequency, changing the position of the fetus and moving it toward the cervical canal.

Stages of Labor

Labor has traditionally been divided into three stages (Figure 29-12●, p. 1107): (1) the *dilation stage*, (2) the *expulsion stage*, and (3) the *placental stage*.

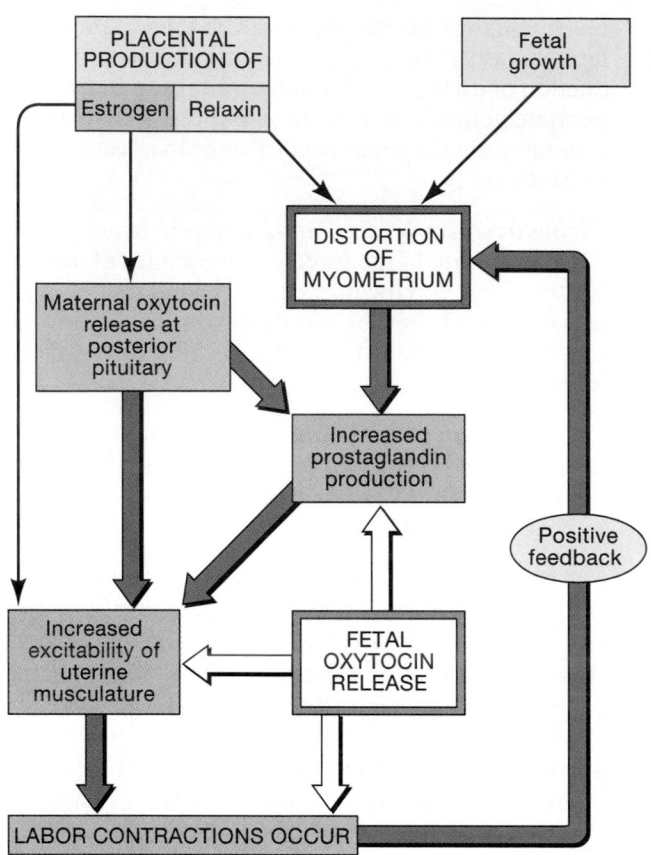

•FIGURE 29-11 **Factors Involved in the Initiation of Labor and Delivery**

The Dilation Stage

The **dilation stage** (Figure 29-12a•) begins with the onset of true labor, as the cervix dilates completely and the fetus begins to move into the cervical canal. This stage is highly variable in length but typically lasts 8 or more hours. At the start of this stage, the labor contractions last up to half a minute and occur at intervals of once every 10–30 minutes; their frequency increases steadily. Late in the process, the amniochorionic membrane ruptures, an event sometimes referred to as having the "water break."

The Expulsion Stage

The **expulsion stage** (Figure 29-12b•) begins as the cervix, pushed open by the approaching fetus, dilates completely. In this stage, contractions reach maximum intensity; they may occur at 2- or 3-minute intervals and last a full minute. Expulsion continues until the fetus has completed its emergence from the vagina; in most cases, it lasts less than 2 hours. The arrival of the newborn infant into the outside world is **delivery**, or birth.

If the vaginal canal is too small to permit the passage of the fetus and there is acute danger of perineal tearing, a clinician may temporarily enlarge the passageway by making an incision through the perineal

musculature. After delivery, this **episiotomy** (e-pēz-ē-OT-o-mē) can be repaired with sutures, a much simpler procedure than dealing with the bleeding and tissue damage associated with an extensive perineal tear. If unexpected complications arise during the dilation or expulsion stages, the infant may be removed by **cesarean section**, or "C-section." In such cases, an incision is made through the abdominal wall and the uterus is opened just enough to allow passage of the infant's head. This procedure is performed during 15–25 percent of the deliveries in the United States—more often than necessary, according to some studies. Over the last decade, efforts have been made to reduce the frequency of both episiotomies and cesarean sections.

The Placental Stage

During the **placental stage** of labor (Figure 29-12c•), muscle tension builds in the walls of the partially empty uterus, and the organ gradually decreases in size. This uterine contraction tears the connections between the endometrium and the placenta. In general, within an hour of delivery, the placental stage ends with the ejection of the placenta, or *afterbirth*. The disruption of the placenta is accompanied by a loss of blood (as much as 500–600 ml). Because the maternal blood volume has increased greatly during pregnancy, this loss can be tolerated without difficulty.

Premature Labor

Premature labor occurs when true labor begins before the fetus has completed normal development. The chances of newborn survival are directly related to the infant's body weight at delivery. Even with massive supportive efforts, infants born weighing less than 400 g (14 oz) will not survive, primarily because the respiratory, cardiovascular, and urinary systems are unable to support life without the aid of maternal systems. As a result, the dividing line between *spontaneous abortion* and **immature delivery** is usually set at 500 g (17.6 oz), the normal weight near the end of the second trimester.

Infants delivered before 7 months of gestation have been completed (weight under 1 kg) have less than a 50:50 chance of survival, and most survivors suffer from severe developmental abnormalities. A **premature delivery** produces a newborn weighing over 1 kg (35.2 oz). Its chances of survival range from fair to excellent, depending on the circumstances. AM *Complexity and Perfection*

Multiple Births

Multiple births (twins, triplets, quadruplets, and so forth) may occur for several reasons. The ratio of twin births to single births in the U.S. population is roughly 1:89. "Fraternal," or **dizygotic** (dī-zī-GOT-ik), twins

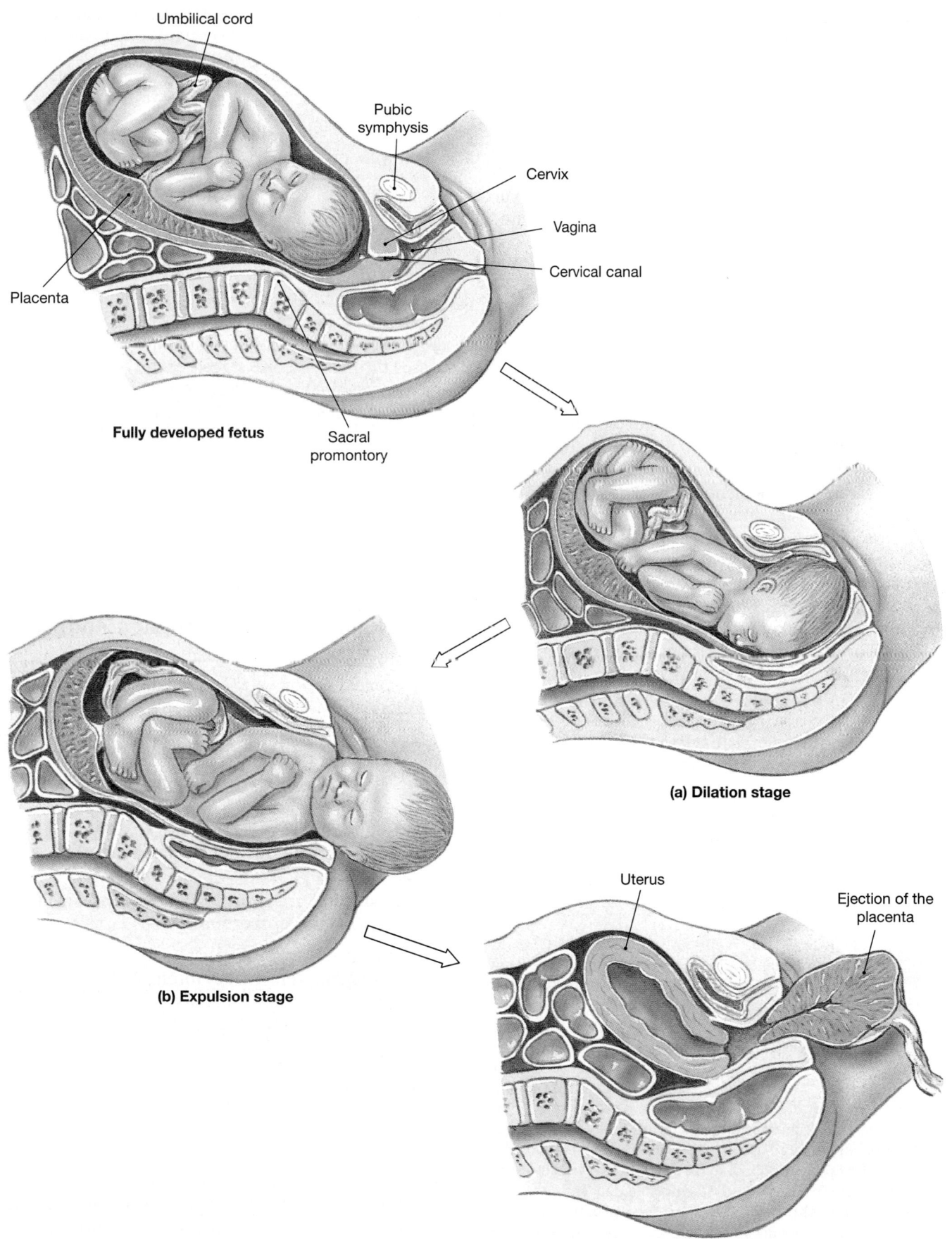

Umbilical cord

Pubic symphysis

Cervix

Vagina

Cervical canal

Placenta

Fully developed fetus

Sacral promontory

(a) Dilation stage

(b) Expulsion stage

Uterus

Ejection of the placenta

(c) Placental stage

●**FIGURE 29-12** The Stages of Labor

develop when two separate eggs were ovulated and subsequently fertilized. Because chromosomes shuffle during meiosis, the odds against any two zygotes from the same parents having identical genes exceeds 1:8.4 million. Seventy percent of twins are dizygotic.

"Identical," or **monozygotic**, twins may result either from the separation of blastomeres early in cleavage or by the splitting of the inner cell mass before gastrulation. In either event, the genetic makeup of the pair will be identical, for both formed from the same pair of gametes. Triplets, quadruplets, and larger multiples can result from multiple ovulations, blastomere splitting, or some combination of the two. For unknown reasons, the statistics for naturally occurring multiple births fall into a pattern: Twins occur at a rate of 1:89, triplets at a rate of $1:89^2$, quadruplets at $1:89^3$, and so forth. The incidence of multiple births can be increased by exposure to fertility drugs that stimulate the maturation of abnormally large numbers of oocytes (see the discussion of fertility and infertility on pp. 1103–1104).

Complete splitting of the embryonic portion of the blastodisc can produce identical twins. If the separation is not complete, **conjoined** *(Siamese)* **twins** may develop. These infants typically share some skin, a portion of the liver, and perhaps other internal organs as well. When the fusion is minor, the infants can be surgically separated with some success. Most conjoined twins with more extensive fusions fail to survive delivery.

Even normal multiple pregnancies pose special problems for maternal systems, for the strains are multiplied proportionately. The chances of premature labor are increased, and the risks to the mother are higher than for single births. The risks for the fetuses are also increased, both during gestation and after delivery, for even at full term the newborn infants have a lower average birthweight. They are also more likely to have problems during delivery. For example, in more than half of twin deliveries, one or both fetuses enter the vaginal canal in an abnormal position.

FORCEPS DELIVERIES AND BREECH BIRTHS In most pregnancies, by the end of gestation, the fetus has rotated within the uterus to transit the birth canal head first, with the face turned toward the mother's sacrum. In about 6 percent of deliveries, the fetus faces the mother's pubis instead. Although these infants can be delivered normally, given enough time, risks to infant and mother are reduced by a *forceps delivery*. The forceps resemble a large, curved set of salad tongs that can be separated for insertion into the vaginal canal one side at a time. Once in place, they are reunited and used to grasp the head of the fetus. An intermittent pull is applied, so that the forces on the head resemble those experienced during normal delivery.

In 3–4 percent of deliveries, the legs or buttocks of the fetus enter the vaginal canal first. Such deliveries are known as *breech births*. Risks to the infant are higher in breech births than in normal deliveries, because the umbilical cord may be-

come constricted and cut off placental circulation. Because the head is normally the widest part of the fetus, the mother's cervix may dilate enough to pass the baby's legs and body but not the head. Entrapment of the head compresses the umbilical cord, prolongs delivery, and subjects the fetus to severe distress and potential injury. If attempts to reposition the fetus or promote further dilation are unsuccessful over the short term, delivery by cesarean section will probably be required.

POSTNATAL DEVELOPMENT

Developmental processes do not cease at delivery, for the newborn infant has few of the anatomical, functional, or physiological characteristics of the mature adult. In the course of postnatal development, every individual passes through five **life stages**, each typified by a distinctive combination of characteristics and abilities.

These stages are a familiar part of human experience. You could probably identify the features and functions associated with the *neonatal period, infancy, childhood, adolescence*, and *maturity*. Although each stage has distinctive features, the transitions between them are gradual, and the boundaries are indistinct. Once maturity has arrived, development ends and the process of aging, or *senescence*, begins.

The Neonatal Period, Infancy, and Childhood

The **neonatal period** extends from the moment of birth to 1 month thereafter. **Infancy** then continues to 2 years of age, and **childhood** lasts until **adolescence**, the period of sexual and physical maturation. Two major events are under way during these developmental stages:

1. The major organ systems other than those associated with reproduction become fully operational and gradually acquire the functional characteristics of adult structures.
2. The individual grows rapidly, and there are significant changes in body proportions.

Pediatrics is a medical specialty focusing on postnatal development from infancy through adolescence. Because infants and young children cannot clearly describe the problems they are experiencing, pediatricians and parents must be skilled observers. Standardized testing procedures are used to assess an individual's developmental progress relative to average values. [AM] *Monitoring Postnatal Development*

The Neonatal Period

A variety of physiological and anatomical alterations occur as the fetus completes the transition to the status of a newborn infant, or **neonate**. Before delivery, dissolved gases, nutrients, waste products, hormones, and immunoglobulins were transferred across the placenta. At birth, the newborn infant must become relatively self-sufficient, with the processes of respiration, digestion, and excretion performed by its own specialized organs and organ systems. The transition from fetus to neonate may be summarized as follows:

1. The lungs at birth are collapsed and filled with fluid. Filling them with air involves a massive and powerful inspiratory movement. ∞ *[p. 534]*

2. When the lungs expand, the pattern of cardiovascular circulation changes due to alterations in blood pressure and flow rates. The ductus arteriosus closes, isolating the pulmonary and systemic trunks. Closure of the foramen ovale separates the atria of the heart, completing separation of the pulmonary and systemic circuits. We discussed these circulatory changes in Chapters 20 and 21. ∞ *[pp. 678, 756–759, 762–763]*

3. The typical heart rates (120–140 beats per minute) and respiratory rate (30 breaths per minute) of neonates are considerably higher than those of adults. In conjunction, a neonate's metabolic rate is roughly twice that of an adult for equivalent body weights.

4. Before birth, the digestive system remains relatively inactive, although it does accumulate a mixture of bile secretions, mucus, and epithelial cells. This collection of debris is excreted during the first few days of life. Over that period, the newborn infant begins to nurse.

5. As waste products build up in the arterial blood, they are excreted at the kidneys. Glomerular filtration is normal, but the urine cannot be concentrated to any significant degree. As a result, urinary water losses are high, and neonatal fluid requirements are much greater than those of adults, as we noted in Chapter 27. ∞ *[p. 1026]*

6. The neonate has little ability to control body temperature, particularly in the first few days after delivery. As the infant grows larger and its insulating subcutaneous adipose "blanket" gets thicker, its metabolic rate also rises. Daily and even hourly alterations in body temperature continue throughout childhood. ∞ *[p. 954]*

Throughout the neonatal period, the newborn is dependent on nutrients contained in the milk secreted by the maternal mammary glands.

LACTATION AND THE MAMMARY GLANDS By the end of the sixth month of pregnancy, the mammary glands are fully developed, and the gland cells begin to produce a secretion known as **colostrum** (ko-LOS-trum). Colostrum, which will be provided to the infant during the first 2 or 3 days of life, contains more proteins and far less fat than breast milk contains. Many of the proteins are immunoglobulins that may help the infant ward off infections until its own immune system becomes fully functional. In addition, the mucins present in both colostrum and milk can inhibit the replication of a family of viruses (*rotaviruses*) that can cause dangerous forms of gastroenteritis and infant diarrhea.

As colostrum production declines, the mammary glands convert to milk production. Breast milk consists of a mixture of water, proteins, amino acids, lipids, sugars, and salts. It also contains large quantities of *lysozymes*, enzymes with antibiotic properties. Human milk provides roughly 750 Calories per liter. The secretory rate varies with the demand, but a 5–6 kg (11–13 lb) infant usually requires about 850 ml of milk per day.

The secretion of the mammary glands is triggered when the infant begins to suck on the nipple. The stimulation of tactile receptors at that site leads to the stimulation of secretory neurons in the paraventricular nucleus of the maternal hypothalamus. These neurons release oxytocin at the posterior pituitary. When oxytocin reaches the mammary gland, this hormone causes the contraction of myoepithelial cells in the walls of the lactiferous ducts and sinuses. The result is the ejection of milk. This milk ejection, or **milk let-down reflex**, is diagrammed in Figure 29-13●.

The milk let-down reflex continues to function until weaning, typically 1–2 years after birth. Milk production ceases soon after, and the mammary glands gradually return to a resting state.

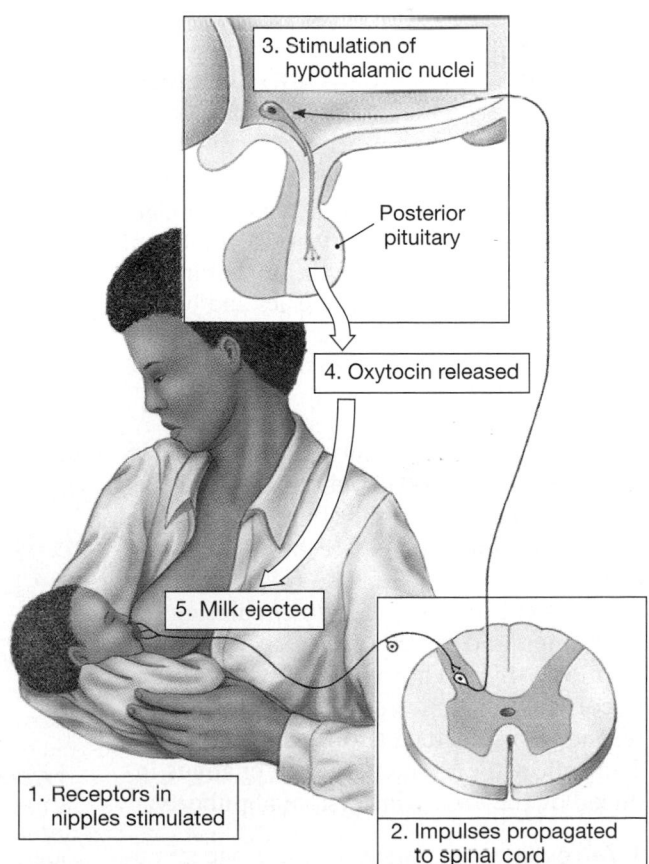

●*FIGURE 29-13* The Milk Let-Down Reflex

Infancy and Childhood

The most rapid growth occurs during prenatal development, and the rate of growth declines after delivery. Postnatal growth during infancy and childhood occurs under the direction of circulating hormones, notably pituitary growth hormone, adrenal steroids, and thyroid hormones. These hormones affect each tissue and organ in specific ways, depending on the sensitivities of the individual cells. As a result, growth does not occur uniformly, and the body proportions gradually change. The head, for example, is relatively large at birth but decreases in proportion with the rest of the body to adulthood (Figure 29-14●).

Adolescence and Maturity

Adolescence begins at **puberty**, the period of sexual maturation, and ends when growth is completed. Three major hormonal events interact at the onset of puberty:

1. The hypothalamus increases its production of gonadotropin-releasing hormone (GnRH). There is evidence that this increase is dependent on adequate levels of *leptin*, a hormone released by adipose tissues. ∞ *[p. 628]*

2. The anterior pituitary becomes more sensitive to the presence of GnRH, and there is a rapid elevation in the circulating levels of FSH and LH.

3. Ovarian or testicular cells become more sensitive to FSH and LH. These changes initiate (1) gametogenesis, (2) the production of sex hormones that stimulate the appearance of secondary sexual characteristics and behaviors, and (3) a sudden acceleration in the growth rate, culminating in the closure of the epiphyseal plates.

The age at which puberty begins varies. In the United States today, puberty generally occurs at about age 14 in boys and 13 in girls, but the normal ranges are broad (9–14 in boys, 8–13 in girls). Many body systems alter their activities in response to circulating sex hormones and to the presence of growth hormone, thyroid hormones, PRL, and adrenocortical hormones, so gender-specific differences in structure and function develop. The following are some of the changes induced by the endocrine system at puberty:

■ *Integumentary system.* Testosterone stimulates the development of terminal hairs on

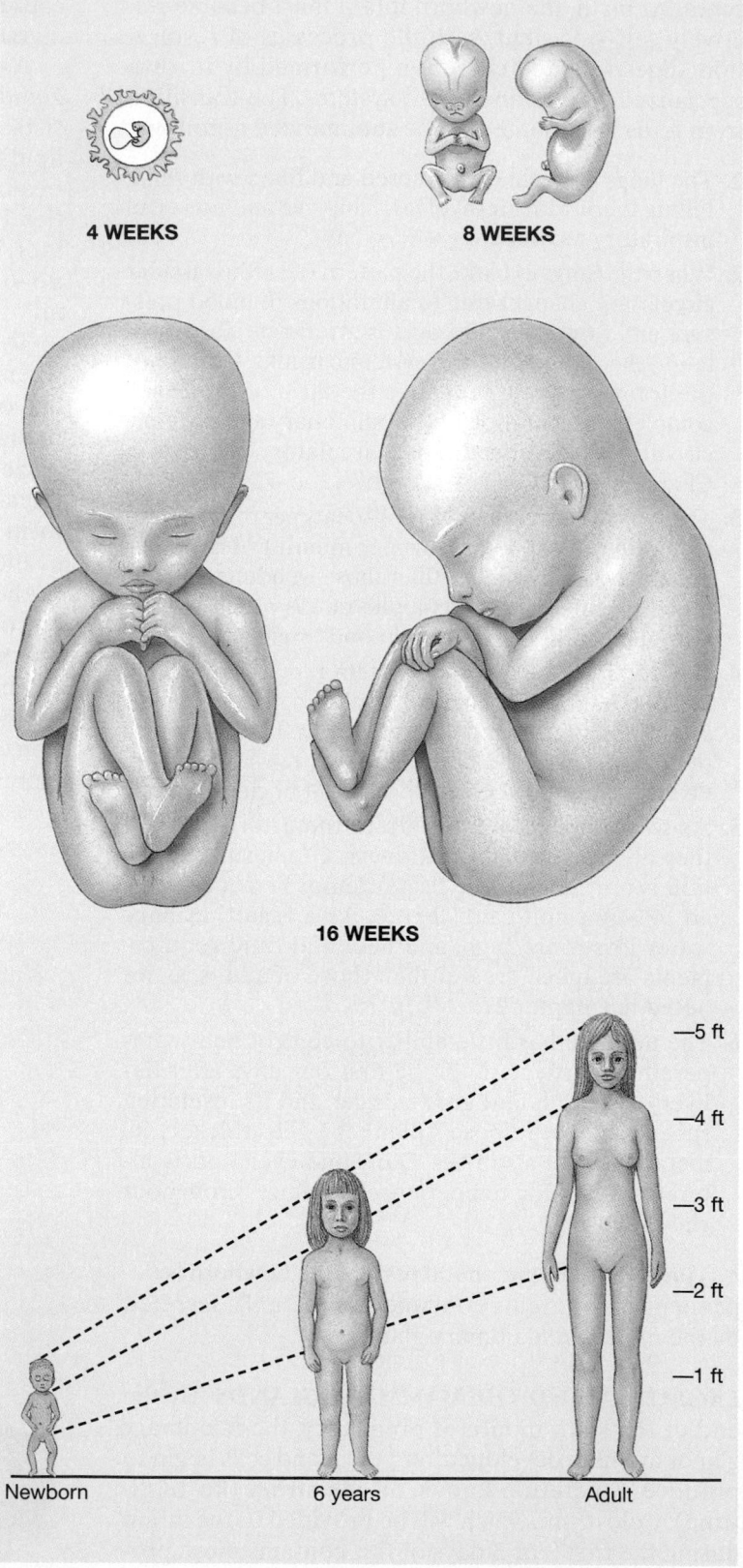

4 WEEKS

8 WEEKS

16 WEEKS

Newborn   6 years   Adult

—5 ft
—4 ft
—3 ft
—2 ft
—1 ft

●**FIGURE 29-14 Growth and Changes in Body Form.** The views at 4, 8, and 16 weeks are presented at actual size. Notice the changes in body form and proportions as development proceeds. These changes do not stop at birth. For example, the head, which contains the brain and sense organs, is relatively large at birth.

the face and chest, whereas those follicles continue to produce fine hairs under estrogen stimulation. The hairline recedes under testosterone stimulation. Both testosterone and estrogen stimulate terminal hair growth in the axillae and in the genital area. These hormones also stimulate sebaceous gland secretion, and they may cause acne. Adipose tissues respond differently to testosterone than to estrogens, and this difference produces changes in the subcutaneous distribution of body fat. In women, the combination of estrogens, PRL, growth hormone, and thyroid hormones promotes the initial development of the mammary glands. Although the duct system becomes more elaborate, true secretory alveoli do not develop, and much of the growth of the breasts during this period reflects increased deposition of fat rather than glandular tissue.

- *Skeletal system.* Both testosterone and estrogen accelerate bone deposition and skeletal growth. In the process, they promote the closure of the epiphyses and thus place a limit on growth in height. Estrogens cause more-rapid epiphyseal closure than does testosterone. In addition, the period of skeletal growth is shorter in girls than in boys, and girls generally do not grow as tall as boys. Girls grow most rapidly between ages 10 and 13, whereas boys grow most rapidly between ages 12 and 15.

- *Muscular system.* Sex hormones stimulate the growth of skeletal muscle fibers, increasing strength and endurance. The effects of testosterone greatly exceed those of the estrogens, and the increased muscle mass accounts for significant gender differences in body mass, even for males and females of the same height. The stimulatory effects of testosterone on muscle mass have produced an interest in anabolic steroids among competitive athletes of both genders.

- *Nervous system.* Sex hormones affect CNS centers concerned with sexual drive and sexual behaviors. These centers differentiated in gender-specific ways during the second and third trimesters, when the fetal gonads secreted either testosterone (in males) or estrogens (in females). The surge in sex hormone secretion at puberty activates these centers.

- *Cardiovascular system.* Testosterone stimulates erythropoiesis and increases the blood volume and the hematocrit. Estrogens and progesterone promote the movement of water from plasma into the interstitial fluid, leading to a late-cycle increase in tissue water content. Once uterine cycles have begun, the associated iron loss increases the risk of developing iron-deficiency anemia. Estrogens decrease plasma cholesterol levels and slow plaque formation. As a result, premenopausal women have a lower risk of atherosclerosis than do adult men.

- *Respiratory system.* Testosterone stimulates disproportionate growth of the larynx and a thickening and lengthening of the vocal cords. These changes cause a gradual deepening of the voice.

- *Reproductive system.* Testosterone stimulates the functional development of the accessory reproductive glands, such as the prostate and seminal vesicles, and helps promote spermatogenesis, as we detailed in Chapter 28. ∞ *[p. 1053]* At the uterus, estrogens promote a thickening of the myometrium, increase blood flow to the endometrium, and stimulate cervical mucus production. Estrogens also promote the functional development of the female accessory reproductive organs. Once uterine cycles have begun, the first few may not be accompanied by ovulation. After that initial period, the woman will be fertile, even though growth and physical maturation will continue for several years.

After puberty, the continued background secretion of estrogens or androgens maintains these gender-specific differences. In both genders, growth continues at a slower pace until age 18–21. By that time most of the epiphyseal plates have closed. The boundary between adolescence and maturity is very hazy, for it has physical, emotional, behavioral, and legal aspects. Adolescence is often said to be over when growth stops in the late teens or early twenties. The individual is then considered mature.

◪ Senescence

Although growth may cease at maturity, physiological changes continue. The gender-specific differences produced at puberty are retained, but further changes occur when sex hormone levels decline at menopause or the male climacteric. ∞ *[p. 1079]* All these changes are part of the process of **senescence**, or aging. Aging reduces the efficiency and capabilities of the individual. Even in the absence of other factors, such as disease or injury, senescence will ultimately lead to death.

Table 29-3 summarizes the age-related changes in physiological systems that we have discussed in earlier chapters. Taken together, these alterations reduce the functional abilities of the individual. They also affect homeostatic mechanisms. As a result, the elderly are less able to make homeostatic adjustments in response to internal or environmental stresses. The risks of contracting a variety of bacterial or viral diseases are proportionately increased as immune function deteriorates. This deterioration leads to drastic physiological alterations that affect all internal systems. Death ultimately occurs when some combination of stresses cannot be countered by existing homeostatic mechanisms.

Physicians attempt to forestall death by adjusting homeostatic mechanisms or removing the sources of stress. **Geriatrics** is a medical specialty concerned with the mechanics of the aging process; physicians trained in geriatrics are known as **geriatricians**. Problems commonly encountered by geriatricians include infections, cancers, heart disease, strokes, arthritis, and anemia. These conditions are directly related to the age-induced changes in vital systems. **AM** *Death and Dying*

TABLE 29-3 Effects of Aging on Organ Systems

The characteristic physical and functional alterations that are part of the aging process affect all organ systems. Examples discussed in previous chapters include:

- A loss of elasticity in the skin that produces sagging and wrinkling. ∞ *[p. 164]*
- A decline in the rate of bone deposition, leading to weak bones, and degenerative changes in joints that make them less mobile. ∞ *[p. 191]*
- Reductions in muscular strength and ability. ∞ *[p. 305]*
- Impairment of coordination, memory, and intellectual function. ∞ *[p. 509]*
- Reductions in the production of and sensitivity to circulating hormones. ∞ *[p. 631]*
- Appearance of cardiovascular problems and a reduction in peripheral blood flow that can affect a variety of vital organs. ∞ *[p. 760]*
- Reduced sensitivity and responsiveness of the immune system, leading to problems with infection and cancer. ∞ *[p. 807]*
- Reduced elasticity in the lungs, leading to decreased respiratory function. ∞ *[p. 856]*
- Decreased peristalsis and muscle tone along the digestive tract. ∞ *[p. 909]*
- Decreased peristalsis and muscle tone in the urinary system, coupled with a reduction of the glomerular filtration rate. ∞ *[p. 998]*
- Functional impairment of the reproductive system, which eventually becomes inactive when the menopause or male climacteric occurs. ∞ *[p. 1079]*

☑ Why does a mother's blood volume increase during pregnancy?

☑ What effect would a decrease in progesterone have on the uterus during late pregnancy?

☑ An increase in the levels of GnRH, FSH, LH, and sex hormones in the blood mark the onset of which stage of development?

GENETICS, DEVELOPMENT, AND INHERITANCE

We introduced chromosome structure and the functions of genes in Chapter 3. ∞ *[pp. 92–96]* Chromosomes contain DNA, and genes are functional segments of DNA. Each gene carries the information needed to direct the synthesis of a specific polypeptide. Every nucleated somatic cell in your body carries copies of the original 46 chromosomes present when you were a zygote. Those chromosomes and their component genes constitute your **genotype** (JĒN-ō-tīp).

Through development and differentiation, the instructions contained within the genotype are expressed in many different ways. No single living cell or tissue uses all the information and instructions contained within the genotype. For example, in muscle fibers the genes important for the formation of excitable membranes and contractile proteins are active, whereas a different set of genes is operating in cells of the pancreatic islets. Collectively, however, the instructions contained within your genotype determine the anatomical and physiological characteristics that make you a unique individual. Those anatomical and physiological characteristics constitute your **phenotype** (FĒN-ō-tīp). In architectural terms, the genotype is a set of plans and the phenotype is the finished building. Specific elements in your phenotype, such as your hair color and eye color, skin tone, and foot size, are called phenotypic *characters*, or *traits*.

Your genotype is derived from the genotypes of your parents. Yet you are not an exact copy of either parent, nor are you an easily identifiable mixture of their characteristics. Our discussion of genetics will begin with the basic patterns of inheritance and their implications. We will then examine the mechanisms responsible for regulating the activities of the genotype during prenatal development.

Genes and Chromosomes

The 46 chromosomes carried by each somatic cell occur in pairs; every somatic cell contains 23 pairs of chromosomes. At amphimixis, one member of each pair was contributed by the spermatozoon, and the other by the ovum. The two members of each pair are known as **homologous** (hō-MOL-o-gus) **chromosomes**. Twenty-two of those pairs are called **autosomal** (aw-tō-SŌ-mal) **chromosomes**. Most of the genes of the autosomal chromosomes affect somatic characteristics, such as hair color and skin pigmentation. The chromosomes of the twenty-third pair are called the **sex chromosomes**; one of their functions is to determine whether the individual will be genetically male or female. Figure 29-15● shows the chromosomes of a normal male.

Autosomal Chromosomes

The two chromosomes in a homologous autosomal pair have the same structure and carry genes that affect the same traits. Suppose that one member of the pair contains three genes in a row, with number 1 determining hair color, number 2 eye color, and number 3 skin pigmentation. The other chromosome will carry genes that affect the same traits, and the genes will be in the same sequence. The genes will also be located at equivalent positions on their respective chromosomes. A **locus** (LŌ-kus; plural, *loci*) is a gene's position on a chromosome.

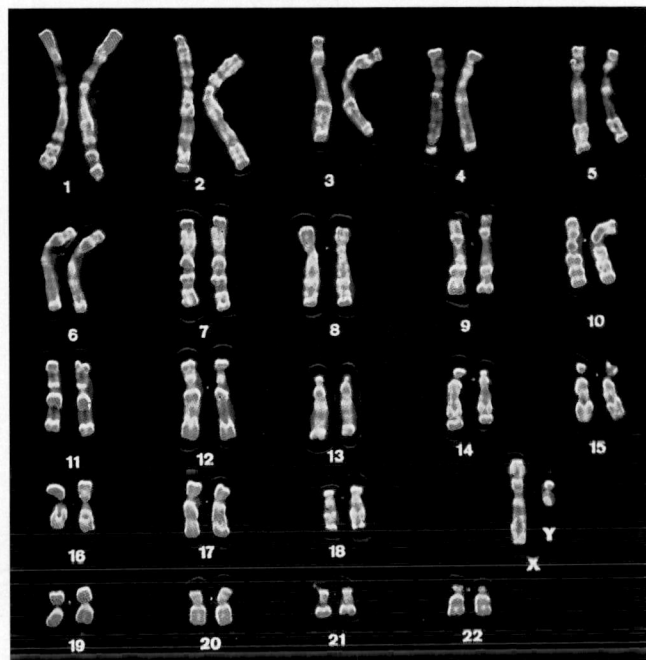

•FIGURE 29-15 Human Chromosomes. The 23 pairs of somatic-cell chromosomes of a normal male.

The two chromosomes in a pair may not carry the same *form* of each gene, however. The various forms of any one gene are called **alleles** (a-LĒLZ). It is these *alternate forms* that determine the precise effect of the gene on your phenotype. If the two chromosomes of a homologous pair carry the same allele of a particular gene, you are **homozygous** (hō-mō-ZĪ-gus) for the trait affected by that gene. For example, if you receive a gene for curly hair from your father and a gene for curly hair from your mother, you will be homozygous for curly hair. In general, about 80 percent of an individual's genetic complement consists of homozygous alleles.

INTERACTIONS BETWEEN ALLELES Because the chromosomes of a homologous pair have different origins, one paternal and the other maternal, they do not necessarily carry the same alleles. When you have two different alleles of the same gene, you are **heterozygous** (het-er-ō-ZĪ-gus) for the trait determined by that gene. The phenotype that results from a heterozygous genotype depends on the nature of the interaction between the corresponding alleles. For example, if you received a gene for curly hair from your father but a gene for straight hair from your mother, whether *you* will have curly hair, straight hair, or even wavy hair depends on the relationship between the alleles for those traits:

- An allele that is **dominant** will be expressed in the phenotype *regardless of any conflicting instructions carried by the other allele.* For instance, an individual with only one allele for freckles will have freckles, because that allele is dominant over the corresponding "non-freckle" allele.

- An allele that is **recessive** will be expressed in the phenotype only if it is present on *both* chromosomes of a homologous pair. For example, in Chapter 5 we described the albino condition, characterized by an inability to synthesize the yellow-brown pigment *melanin.* The presence of one normal allele will result in normal coloration. Two recessive alleles must be present to produce an albino individual. There can be many different alleles for a single gene, some dominant and others recessive.

- In **incomplete dominance**, heterozygous alleles produce a phenotype that is distinct from the phenotypes of individuals homozygous for one allele or the other. For instance, consider a gene that affects red blood cell shape. Individuals with homozygous alleles that carry instructions for normal adult hemoglobin A have red blood cells of normal shape. Individuals with homozygous alleles for hemoglobin S, an abnormal form, have red blood cells that will sickle in peripheral capillaries when the P_{O_2} decreases. These individuals develop *sickle cell anemia* ∞ *[p. 649]* Individuals who are heterozygous for this trait do not develop the anemia; they have red blood cells that will sickle only during extreme hypoxia.

- In **codominance**, an individual with a heterozygous allele for a given trait exhibits both the dominant and recessive phenotypes for that trait. Blood type in humans is determined by codominance. ∞ *[p. 653]* The alleles for types A and B blood are dominant over the allele for type O blood, but a person with one type A allele and one type B allele will have type AB blood, not A or B. Type AB blood has *both* type A and type B antigens. The distinction between incomplete dominance and codominance may be difficult. For example, a person heterozygous for hemoglobin A and hemoblogin S shows incomplete dominance for red blood cell shape but codominance for red blood cell hemoglobin; each RBC contains a mixture of hemoglobin A and hemoglobin S.

PENETRANCE AND EXPRESSIVITY Differences in genotype lead to distinctive variations in phenotype, but the relationships are not always easily predictable. The presence of a particular pair of alleles does not affect the phenotype in the same way in every individual. **Penetrance** is the percentage of individuals with a particular genotype that show the "expected" phenotype. The effects of the genotype in the other individuals may be overridden by the activity of other genes or by environmental factors. For example, *emphysema*, a respiratory disorder discussed in Chapter 23, has been linked to a specific abnormal genotype. ∞ *[p. 851]* Roughly 20 percent of the individuals with this genotype do not develop emphysema. The penetrance of this genotype is therefore approximately 80 percent. The effects of environmental factors are apparent; most people who develop emphysema are cigarette smokers.

If a given genotype *does* affect the phenotype, it can do so to varying degrees, again depending on the activity of other genes or environmental stimuli. For example, identical twins do not have exactly the same fingerprints, even though they have the same genotype. The extent to which a particular allele is expressed when it is present is termed its **expressivity**.

Environmental effects on genetic expression are particularly evident during embryonic and fetal development. Drugs, including certain antibiotics, alcohol, and the nicotine in cigarette smoke, can disrupt normal fetal development. Stimuli that result in abnormal development are called *teratogens*. Specific teratogens are considered in the *Applications Manual.*
AM *Teratogens and Abnormal Development*

PREDICTING INHERITANCE Not every allele can be neatly characterized as dominant or recessive. Several that can be are included in Table 29-4. If you consider the traits listed there, you can predict the characteristics of individuals on the basis of the parents' alleles.

In such calculations, dominant alleles are traditionally indicated by capitalized abbreviations, and recessives are abbreviated in lowercase letters. For a given trait, the possibilities are indicated by *AA* (homozygous dominant), *Aa* (heterozygous), or *aa* (homozygous recessive). Each gamete involved in fertilization contributes a single allele for a given trait. That allele must be one of the two contained by all cells in the parent's body. Consider, for example, the offspring of an albino mother and a father with normal skin pigmentation. Because albinism is a recessive trait, the maternal alleles are abbreviated *aa*. No matter which of her oocytes is fertilized, it will carry the recessive *a* allele. The father has normal pigmentation, a dominant trait. He may therefore be homozygous *or* heterozygous for this trait, because both *AA* and *Aa* will produce the same phenotype. Every sperm produced by a homozygous father will carry the *A* allele. A heterozygous father, however, will produce two different kinds of sperm: Some will carry the dominant allele *A*, and others will carry the recessive allele *a*.

A simple box diagram known as a **Punnett square** lets us predict the probabilities that children will have particular characteristics by showing the possible combinations of parental alleles they can inherit. In the Punnett squares shown in Figure 29-16●, the maternal alleles are listed along the horizontal axis and the paternal ones along the vertical axis. The possible combinations are indicated in the small boxes. Figure 29-16a● considers the possible offspring of an *aa* mother and an *AA* father. All the children must have the genotype *Aa*, so all will have normal skin pigmentation. Compare these results with those of Figure 29-16b●, for a heterozygous father *(Aa)*. The heterozygous male produces two types of gametes, *A* and *a*, and the secondary oocyte may be fertilized by

TABLE 29-4 The Inheritance of Selected Phenotypic Characters

DOMINANT TRAITS
One allele determines phenotype; the other is suppressed:
   normal skin pigmentation
   lack of freckles
   brachydactyly (short fingers)
   ability to taste phenylthiocarbamate (PTC)
   free earlobes
   curly hair
   presence of Rh factor on red blood cell membranes
   ability to roll the tongue into a U-shape
   color vision
Both dominant alleles may be expressed (codominance):
   presence of A or B antigens on red blood cell
     membranes
   particular structure of serum proteins (albumins,
     transferrins)
Two alleles produce intermediate traits (incomplete dominance):
   hemoglobin A and hemoglobin S production

RECESSIVE TRAITS
   albinism
   freckles
   normal digits
   inability to taste phenylthiocarbonate (PTC)
   attached earlobes
   straight hair
   blond hair
   red hair (expressed only if individual is also
     homozygous for blond hair)
   lack of A or B surface antigens (Type O blood)
   absence of Rh surface antigen
   inability to roll the tongue into a U-shape

SEX-LINKED TRAITS
   color blindness
   hemophilia

POLYGENIC TRAITS
   eye color
   hair colors other than pure blond or red

Note: For a listing of inherited clinical conditions, see Table 29-5.

either one. As a result, there is a 50 percent probability that a child of such a father will inherit the genotype *Aa* and so have normal skin pigmentation. The probability of inheriting the genotype *aa*, and thus having the albino phenotype, is also 50 percent. The Punnett square can also be used to draw conclusions about the identity and genotype of a parent. For example, a man with the genotype *AA* cannot be the father of an albino child *(aa)*.

SIMPLE INHERITANCE In **simple inheritance**, phenotypic characters are determined by interactions between a single pair of alleles. (The examples we have discussed

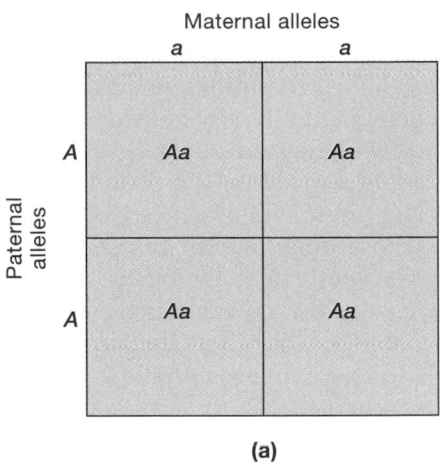

(a)

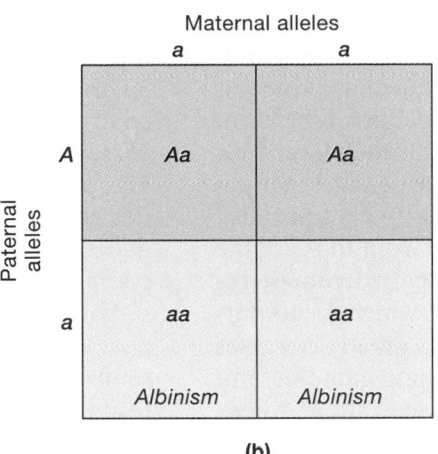

(b)

•*FIGURE 29-16* **Predicting Phenotypic Characteristics by Using Punnett Squares.** (a) All the offspring of a homozygous dominant father (*AA*) and a homozygous recessive mother (*aa*) will be heterozygous (*Aa*) for that trait. Their phenotype will be the same as that of the father. (b) The offspring of a heterozygous father and a homozygous recessive mother will be either heterozygous or homozygous for the recessive trait. In this example, half the offspring will have normal skin coloration, and the other half will be albinos.

thus far involve simple inheritance.) We can predict the frequency of appearance of an inherited disorder that results from simple inheritance by using a Punnett square. Although they are rare disorders in terms of overall numbers, more than 1200 different inherited conditions have been identified that reflect the presence of one or two abnormal alleles for a single gene. A partial listing is included in Table 29-5, along with the location where you can find additional information.

POLYGENIC INHERITANCE Many phenotypic characters are determined by interactions among several different genes. Such interactions are called **polygenic inheritance.** The resulting phenotype depends on how the genes interact. Because multiple alleles are involved, the frequency of occurrence cannot easily be predicted by using a simple Punnett square. In *suppression,* one gene suppresses the other, and the second gene has no ef-

fect on the phenotype. In *complementary gene action,* dominant alleles on two genes interact to produce a phenotype different from that seen when one gene contains recessive alleles. The risks of developing several important adult disorders, including hypertension and coronary artery disease, are linked to polygenic inheritance.

Many of the developmental disorders responsible for fetal mortalities and congenital malformations result from multiple genetic interactions. In these cases, the particular genetic composition of the individual does not by itself determine the onset of the disease. Instead, the conditions regulated by these genes establish a susceptibility to particular environmental influences. This means that not every individual with the genetic tendency for a particular condition will actually develop it. It is therefore difficult to track polygenic conditions through successive generations. However, because many inherited polygenic conditions are *likely* but not *guaranteed* to occur, steps can be taken to prevent a crisis. For example, you can prevent or reduce hypertension by controlling your diet and fluid volume, and you can prevent coronary artery disease by lowering your serum cholesterol concentrations.

SOURCES OF INDIVIDUAL VARIATION As we noted earlier, just as you are not a copy of either parent, you are not a 50:50 mixture of their characteristics. We noted one reason in Chapter 28: During meiosis, maternal and paternal chromosomes are randomly assorted, so each gamete contains a unique blend of maternal and paternal chromosomes. ∞ *[p. 1044]* Therefore, you may have an allele for curly hair from your father and an allele for straight hair from your mother, but your sister may have received an allele for straight hair from each of your parents.[1]

TABLE 29-5 **Fairly Common Inherited Disorders**

Disorder	Text Page and/or *Applications Manual*
Autosomal dominants	
Adult polycystic kidney disease	p. 970 and *AM*
Marfan's syndrome	pp. 123, 186 and *AM*
Huntington's disease	p. 508 and AM
Autosomal recessives	
Deafness	p. 585 and *AM*
Albinism	pp. 152, 1114
Sickle cell anemia	p. 649 and *AM*
Cystic fibrosis	p. 817 and *AM*
Phenylketonuria	p. 935 and *AM*
Tay-Sachs disease	p. 408 and *AM*
X-linked	
Duchenne's muscular dystrophy	p. 283 and *AM*
Myotonic muscular dystrophy	*AM*
Hemophilia (A and B)	p. 667
Testicular feminization syndrome	*AM*
Color blindness	p. 566

Note: The genetic bases have been identified for at least some forms of the diseases in blue.

[1]In very rare cases, an individual will receive both alleles from one parent. The few documented cases appear to have resulted when duplicate maternal chromatids failed to separate during meiosis II, and the corresponding chromosome provided by the sperm did not participate in amphimixis. This condition is called *uniparental disomy.* It is not known how often this occurs; in most cases, the event remains undetected because the individuals are phenotypically normal.

In addition, parts of chromosomes can become re-arranged during synapsis (Figure 29-17•). When tetrads are formed, adjacent chromatids may overlap, a process called **crossing-over**. Under these conditions, the chromatids may break, and the segments trade places. This process is called **translocation**. All the changes in chromosome structure that occur during meiosis produce gametes whose chromosomes differ from those of the parent. This reshuffling process is called **genetic recombination**. Genetic recombination, which occurs during meiosis in both males and females, greatly increases the range of possible variation among gametes, and thus among members of successive generations, whose genotypes are formed by the combination of gametes in fertilization. It can also complicate the tracing of inheritance of genetic disorders.

During recombination, portions of chromosomes may break away and be lost, or *deleted*. The effects of a deletion on a zygote depend on the nature of the lost genes. In some cases, the effects depend on whether the abnormal gamete is produced through oogenesis or spermatogenesis. This phenomenon is *genomic imprinting*. For example, the deletion of a specific segment of chromosome 15 is responsible for two very different disorders, *Angelman syndrome* and *Prader–Willi syndrome*. Symptoms of Angleman syndrome include hyperactivity, severe mental retardation, and seizures; this condition results when the abnormal chromosome is provided by the oocyte. Symptoms of Prader–Willi syndrome include short stature, reduced muscle tone and skin pigmentation, underdeveloped gonads, and less-marked retardation; this condition results when the abnormal chromosome is delivered by the sperm.

Recombination that produces abnormal chromosome shapes or numbers is lethal for the zygote in almost all cases. Roughly 10 percent of zygotes have **chromosomal abnormalities**—that is, damaged, broken, missing, or extra copies of chromosomes—but only about 0.5 percent of newborn infants have such abnormalities. Chromosomal abnormalities produce a variety of serious clinical conditions, in addition to contributing to prenatal mortality. *Few individuals with chromosomal abnormalities survive to full term.* The high mortality rate and severity of the problems reflect the fact that large numbers of genes have been added or deleted. The normal human chromosomal complement, or **karyotype**, is shown in Figure 29-15•.

Variations at the level of the individual gene may appear as a result of mutations that affect the nucleotide sequence of one allele. **Spontaneous mutations** are the result of random errors in the DNA replication process. Such errors are relatively common, but in most cases the error is detected and repaired by enzymes within the nucleus. Those that go undetected and unrepaired have the potential to change the phenotype in some way.

Mutations occurring during meiosis can produce gametes that contain abnormal alleles. These alleles may be dominant or recessive, and they may occur on autosomal chromosomes or sex chromosomes. The vast majority of mutations make the zygote incapable of completing normal development. Mutation, rather than chromosomal abnormalities, is probably the primary cause of the high mortality rate among pre-embryos and embryos. (Roughly 50 percent of all zygotes fail to complete cleavage, and another 10 percent fail to reach the fifth month of gestation.) [AM] *Complexity and Perfection*

If the abnormal allele is dominant but does not affect gestational survival, the individual's phenotype will show the effects of the mutation. If the abnormal allele is recessive and is on an autosomal chromosome, it will not affect the individual's phenotype, because the zygote will contain normal alleles contributed by the other parent at fertilization. Over generations, a recessive autosomal allele can spread through the population, remaining undetected until a fertilization occurs in which the two gametes contribute identical recessive alleles. This individual, who will be homozygous for the abnormal allele, will be the first to show the phenotypic effects of the original mutation. Individuals who are heterozygous for the abnormal allele but who do not show the effects of the mutation are called **carriers**. Genetic tests are available to determine if an individual is a carrier for one of several autosomal recessive disorders, including Tay-Sachs disease and Huntington's disease. The information can be useful in counseling prospective parents. For example, if both parents are carriers for the same disorder, there is a 25 percent probability that a child of theirs will have the disease. This information may affect their decision to conceive.

Sex Chromosomes

Unlike the other 22 chromosomal pairs, the sex chromosomes are not necessarily identical in appearance

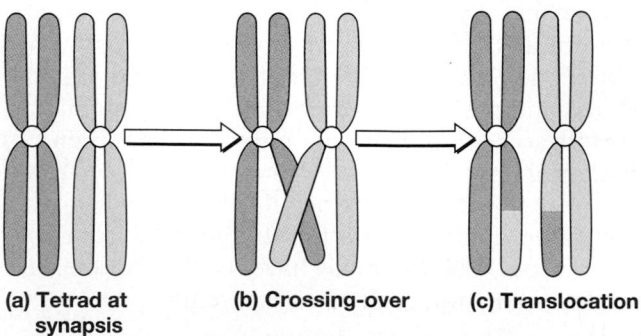

(a) Tetrad at synapsis   **(b) Crossing-over**   **(c) Translocation**

•FIGURE 29-17 **Translocation.** **(a)** Synapsis, with the formation of a tetrad during meiosis. **(b)** Crossing-over of homologous portions of two chromosomes. **(c)** The breakage and exchange of corresponding sections on the two chromosomes in **(b)**.

and gene content. There are two different sex chromosomes, an **X chromosome** and a **Y chromosome**. Y chromosomes are considerably smaller, and contain fewer genes, than X chromosomes. Among the genes on the Y chromosome are dominant alleles that specify that an individual with that chromosome will be a male. The normal sex chromosome complement of males is XY. Females do not have a Y chromosome, and their sex chromosome pair is XX.

All oocytes will carry an X chromosome, whereas sperm may carry an X or a Y chromosome. Thus, with a Punnett square you can show that the ratio of males to females in offspring should be 1:1. The birth statistics differ slightly from that prediction, with 106 males born for every 100 females. It has been suggested that more males are born because the sperm that carries the Y chromosome can reach the oocyte first, since they do not have to carry the extra weight of the larger X chromosome.

The X chromosome also carries genes that affect somatic structures. These characteristics are called **X-linked** (or *sex-linked*), because in most cases there are no corresponding alleles on the Y chromosome. The inheritance of characteristics regulated by these genes does not follow the pattern of alleles on autosomal chromosomes. The best known of these single-allele characters are those associated with noticeable diseases or functional deficits.

The inheritance of color blindness, a condition we discussed in Chapter 17, exemplifies the differences between sex-linked and autosomal inheritance. ∞ *[p. 566]* Red–green color blindness, a relatively common problem, is associated with the presence of a dominant or recessive gene on the X chromosome. Normal color vision is determined by the presence of a dominant allele, *C*, whereas red–green color blindness results from the absence of *C* and the presence of a recessive allele, *c*. A woman, with her two X chromosomes, can be either homozygous (*CC*) or heterozygous (*Cc*) and still have normal color vision. She will be unable to distinguish reds from greens only if she carries two recessive alleles, *cc*. But a male has only one X chromosome, so whichever allele that chromosome carries will determine whether he has normal color vision or is red–green color-blind. A Punnett square for an X-linked trait, as in Figure 29-18•, reveals that the sons produced by a father with normal vision and a heterozygous mother will have a 50 percent chance of being red–green color-blind, whereas any daughters will have normal color vision.

A number of other clinical disorders noted earlier in the text are X-linked traits, including certain forms of hemophilia, diabetes insipidus, and muscular dystrophy. In several instances, advances in molecular genetic techniques have enabled geneticists to local-

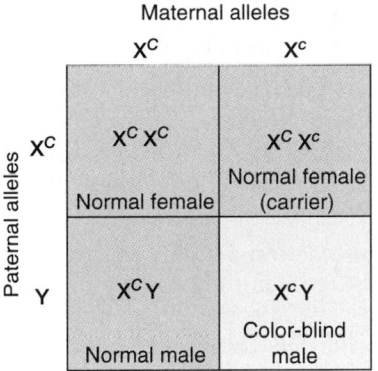

•*FIGURE 29-18* **X-Linked Inheritance**

ize the specific genes on the X chromosome. This technique provides a relatively direct method of screening for the presence of a particular condition before the symptoms appear and even before birth.

The Human Genome Project

Few of the genes responsible for inherited disorders have been identified or even associated with a specific chromosome. That situation is changing rapidly, however, due to the attention devoted to the **Human Genome Project**. This project, funded by the National Institutes of Health and the Department of Energy, is attempting to transcribe the entire human **genome**— that is, the full complement of genetic material— chromosome by chromosome and gene by gene. The project began in October 1990 and was expected to take 15 years. Progress has been more rapid than expected, and it may take less time.

The first step in understanding the human genome is to prepare a map of the individual chromosomes. **Karyotyping** (KAR-ē-ō-tīp-ing; *karyon*, nucleus + *typos*, mark) is the determination of an individual's chromosomal complement. Figure 29-15•, p. 1113, shows a set of normal human chromosomes. Each chromosome has characteristic banding patterns, and segments can be stained with special dyes. The banding patterns are useful as reference points for the preparation of more-detailed genetic maps. The banding patterns themselves can be useful, as abnormal banding patterns are characteristic of some genetic disorders and several cancers, including a form of leukemia.

As of 1997, a progress report includes the following:

- Eight chromosomes—chromosomes 3, 11, 12, 16, 19, 21, 22, and the Y chromosome—have been mapped completely, and preliminary maps have been made of all other chromosomes.
- More than 6800 genes have been identified. Although that is a significant number, it is only a small fraction of the estimated 100,000 genes in the human genome.

 Chromosal Abnormalities and Genetic Analysis

Embryos that have abnormal autosomal chromosomes rarely survive. There are, however, two types of abnormalities in autosomal chromosomes that do not invariably kill the individual before birth. These are *translocation defects* and *trisomy:*

1. In a **translocation defect**, crossing-over occurs between different chromosome pairs. For example, a piece of chromosome 8 may become attached to chromosome 14. The genes moved to their new position may function abnormally, becoming inactive or overactive. A translocation between chromosomes 8 and 14 is responsible for *Burkitt's lymphoma*, a type of lymphatic system cancer.

2. In **trisomy**, something goes wrong in meiosis II, so the gamete that contributes to the zygote carries an extra copy of one chromosome. Because there are three copies of this chromosome rather than two, the condition is termed *trisomy*. The nature of the trisomy is indicated by the number of the chromosome involved. Zygotes with extra copies of chromosomes seldom survive. Individuals with trisomy 13 and trisomy 18 may survive until delivery but rarely live longer than a year. The notable exception is trisomy 21.

Trisomy 21, or **Down syndrome**, is the most common viable chromosomal abnormality. Estimates of the frequency of appearance range from 1.5 to 1.9 per 1000 births for the U.S. population. The affected individual suffers from mental retardation and characteristic physical malformations, including a facial appearance that gave rise to the term *mongolism* once used to describe this condition. The degree of mental retardation ranges from moderate to severe. Few individuals with this condition lead independent lives. Anatomical problems affecting the cardiovascular system commonly prove fatal during childhood or early adulthood. Although some individuals survive to moderate old age, many develop Alzheimer's disease while still relatively young (before age 40).

For unknown reasons, there is a direct correlation between maternal age and the risk of having a child with trisomy 21. For a maternal age below 25, the incidence of Down syndrome approaches 1 in 2000 births, or 0.05 percent. For maternal ages 30–34, the odds increase to 1:900, and for ages 35–44 they go from 1:290 to 1:46, or more than 2 percent. These statistics are becoming increasingly significant, for many women have delayed childbearing until their mid-thirties or later.

Abnormal numbers of sex chromosomes do not produce effects as severe as those induced by extra or missing autosomal chromosomes. In **Klinefelter syndrome**, the individual carries the sex chromosome pattern XXY. The phenotype is male, but the extra X chromosome causes reduced androgen production. As a result, the breasts are slightly enlarged, the testes fail to mature, and affected individuals are sterile. The incidence of this condition among newborn males averages 1:750.

Individuals with **Turner syndrome** have only a single, female sex chromosome; their sex chromosome complement is abbreviated XO. This kind of chromosomal deletion is known as **monosomy**. The incidence of this condition at delivery has been estimated as 1:10,000. At birth, the condition may not be recognized, for the phenotype is normal female. But maturational changes do not appear at puberty. The ovaries are nonfunctional, and estrogen production occurs at negligible levels.

Fragile-X syndrome causes mental retardation, abnormal facial development, and increased testicular size in affected males. The cause is an abnormal X chromosome that contains a *genetic stutter*, an abnormal repetition of a single triplet. In this case, the problem is the repetition of the nucleotide sequence CCG at one site on the X chromosome. (A genetic stutter on another chromosome is responsible for Huntington's disease, a disorder we introduced in Chapter 15.) The presence of the stutter in some way disrupts the normal functioning of adjacent genes and so produces the symptoms of the disorder.

Chromosomal Analysis. In **amniocentesis**, a sample of amniotic fluid is removed, and the fetal cells that it contains are analyzed. This procedure permits the identification of over 20 different congenital conditions, including Down syndrome. The needle inserted to obtain a fluid sample is guided into position by using ultrasound. Unfortunately, amniocentesis has two major drawbacks:

1. Because the sampling procedure represents a potential threat to the health of the fetus and mother, amniocentesis is performed only when known risk factors are present. Examples of risk factors would include a family history of specific conditions or, in the case of Down syndrome, a maternal age over 35.

2. Sampling cannot safely be performed until the volume of amniotic fluid is large enough that the fetus will not be injured during the sampling process. The usual time for amniocentesis is at a gestational age of 14–15 weeks. It may take several weeks to obtain results once samples have been collected, and by the time the results are received, the option of therapeutic abortion may no longer be available.

In an alternative procedure known as **chorionic villus sampling**, cells collected from the chorionic villi during the first trimester are analyzed. Although it can be performed at an earlier gestational age than can amniocentesis, chorionic villus sampling has largely been abandoned, because the risks of misdiagnosis are high and placental damage can cause spontaneous abortion.

- The specific genes responsible for more than 60 inherited disorders have been identified, including 5 of the disorders listed in Table 29-5. Genetic screening can now be done for many of these conditions, as indicated in Figure 29-19●.

The Human Genome Project is attempting to determine the normal genetic composition of a "typical" human being. Yet we all are variations on a basic theme. As we improve our abilities to manipulate our own genetic foundations, we will face troubling ethical and legal decisions. For example, few people object to the insertion of a "correct" gene into somatic cells to cure a specific disease. (For a discussion of the process, see the discussion of genetic engineering in the *Applications Manual*.) But what if we could insert that modified gene into a gamete and change not only that individual but his or her descendants as well? And what if the gene did not correct or prevent a disorder but "improved" the individual by increasing intelligence, height, or vision or altering some other phenotypic characteristic? These and other difficult questions will not go away. In the years to come, we will have to find answers acceptable to us all.

☑ Curly hair is an autosomal dominant trait. What would be the phenotype of a person who is heterozygous for this trait?

☑ Why are children not identical copies of their parents?

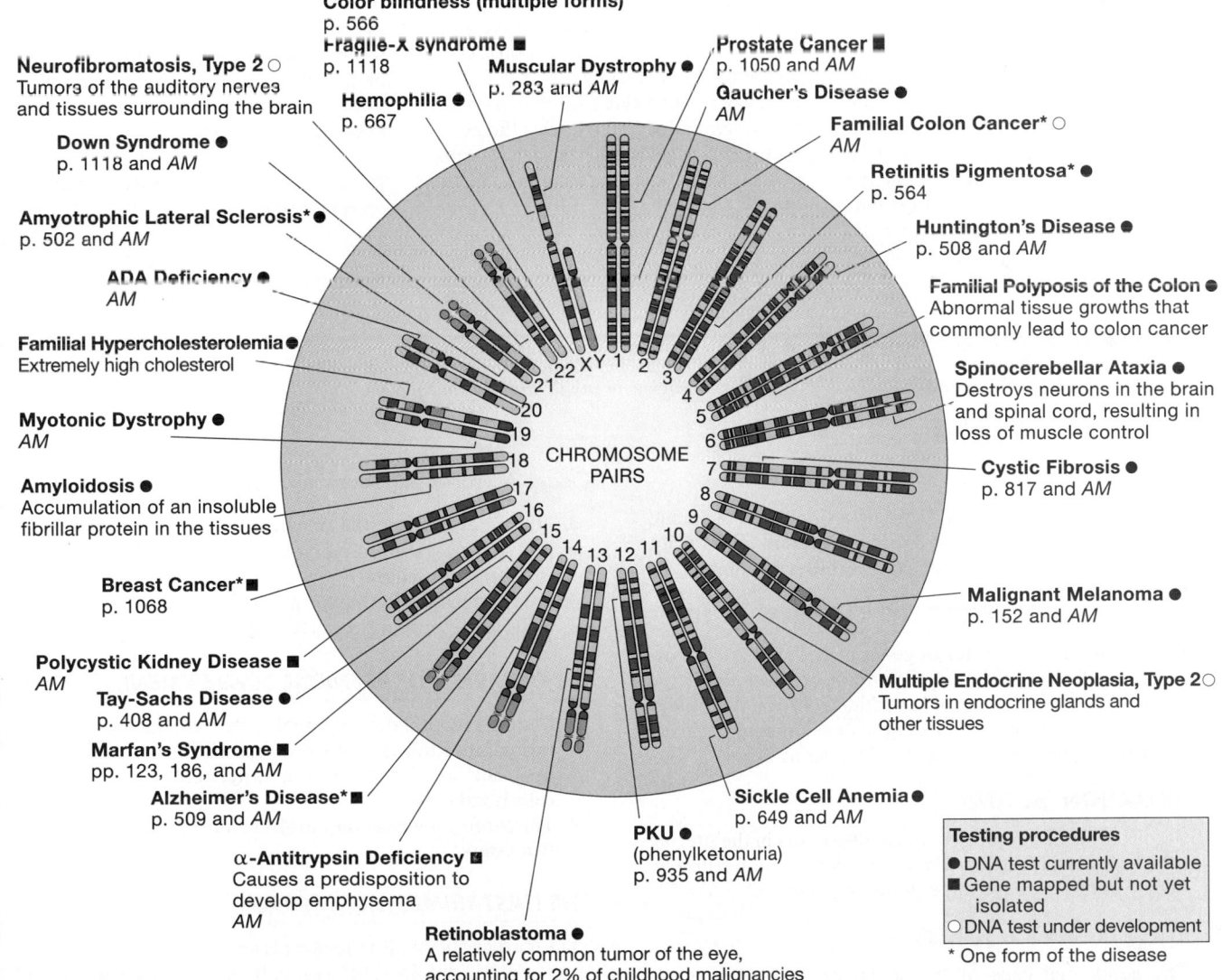

●FIGURE 29-19 A Map of the Human Chromosomes. The banding patterns of typical chromosomes in a male individual and the locations of the genes responsible for specific inherited disorders. The chromosomes are not drawn to scale.

SELECTED CLINICAL TERMINOLOGY

Terms Discussed in This Chapter

amniocentesis: An analysis of fetal cells taken from a sample of amniotic fluid. *(p. 1118)*

breech birth: A delivery during which the legs or buttocks of the fetus enter the vaginal canal first. *(p. 1108)*

chorionic villus sampling: An analysis of cells collected from the chorionic villi during the first trimester. *(p. 1118)*

ectopic pregnancy: A pregnancy in which implantation occurs somewhere other than the uterus. *(p. 1091 and AM)*

gestational neoplasm: A tumor formed by undifferentiated, rapid growth of the syncytial trophoblast; if untreated, the neoplasm may become malignant. *(p. 1092)*

infertility: The inability to achieve pregnancy after 1 year of appropriately timed intercourse. *(p. 1103)*

in vitro fertilization: Fertilization outside the body, generally in a petri dish. *(p. 1104)*

AM Additional Terms Discussed in the *Applications Manual*

---

abruptio placentae (ab-RUP-shē-ō pla-SEN-tē): Tearing of the placenta sometime after the fifth gestational month.

Apgar rating: A method of evaluating newborn infants; a test for developmental problems and neurological damage.

congenital malformation: A severe structural abnormality, present at birth, that affects major systems.

DDST: The Denver Developmental Screening Test, given to infants and children up to age 5 to monitor their development.

fetal alcohol syndrome (FAS): A neonatal condition resulting from maternal alcohol consumption; characterized by developmental defects typically involving the skeletal, nervous, and/or cardiovascular system.

placenta previa: A condition, resulting from implantation in or near the cervix, in which the placenta covers the cervix and prevents normal birth.

pseudohermaphrodite (soo-dō-her-MAF-rō-dīt): An individual who has the genotype of one gender but has some or most of the phenotypic characteristics of the opposite gender.

teratogens (TER-a-tō-jenz): Stimuli that disrupt normal development by damaging cells, by altering chromosome structure, or by altering the chemical environment of the embryo.

testicular feminization syndrome: A form of male pseudohermaphroditism resulting from an inability to respond to testosterone.

toxemia (tok-SĒ-mē-uh) **of pregnancy:** Disorders affecting the maternal cardiovascular system; includes *eclampsia* and *preeclampsia*.

CHAPTER REVIEW

On-line resources for this chapter are on our World Wide Web site at:
http://www.prenhall.com/martini/fap

STUDY OUTLINE

INTRODUCTION, p. 1087

1. **Development** is the gradual modification of physical and physiological characteristics from **conception** to maturity. The creation of different cell types is **differentiation**.

AN OVERVIEW OF TOPICS IN DEVELOPMENT, p. 1087

1. **Inheritance** is the transfer of genetically determined characteristics from generation to generation. **Genetics** is the study of the mechanisms of inheritance. **Prenatal development** occurs before birth; **postnatal development** begins at birth and continues to **maturity**, when aging begins.

FERTILIZATION, p. 1087

1. Fertilization, or conception, normally occurs in the uterine tube within a day after ovulation. Sperm cannot fertilize an oocyte until they have undergone *capacitation*.

The Oocyte at Ovulation, p. 1088

2. The acrosomal caps of the spermatozoa release **hyaluronidase** and **acrosin**, enzymes required to penetrate the corona radiata and zona pellucida. When a single spermatozoon contacts the oocyte membrane, fertilization begins and **oocyte activation** follows.

Oocyte Activation and Amphimixis, p. 1089

3. During activation, the oocyte completes meiosis II, and **polyspermy** is prevented by membrane depolarization and the *cortical reaction*.

4. After activation, the **female pronucleus** and the **male pronucleus** fuse in a process called **amphimixis**. *(Figure 29-1)*

AN OVERVIEW OF PRENATAL DEVELOPMENT, p. 1089

1. During prenatal development, differences in the cytoplasmic composition of individual cells trigger changes in genetic activity. The chemical interplay between developing cells is **induction**.

2. The 9-month **gestation** period can be divided into three **trimesters**.

THE FIRST TRIMESTER, p. 1090

1. **Cleavage** subdivides the cytoplasm of the zygote in a series of mitotic divisions; the zygote becomes a **pre-embryo** and then a **blastocyst**. During **implantation**, the blastocyst burrows into the uterine endometrium. **Placentation** occurs as blood vessels form around the blastocyst and the **placenta** develops. **Embryogenesis** is the formation of a viable embryo.

Cleavage and Blastocyst Formation, p. 1090

2. The blastocyst consists of an outer **trophoblast** and an **inner cell mass.** *(Figure 29-2)*

Implantation, p. 1091

3. Implantation occurs about 7 days after fertilization as the blastocyst adheres to the uterine lining. *(Figure 29-3)*

4. As the trophoblast enlarges and spreads, maternal blood flows through open **lacunae.** After **gastrulation,** the **blastodisc** becomes an embryo composed of **endoderm, ectoderm,** and an intervening **mesoderm.** It is from these **germ layers** that the body systems differentiate. *(Figure 29-4; Table 29-1)*

5. Germ layers help form four **extraembryonic membranes:** the yolk sac, amnion, allantois, and chorion. *(Figure 29-5)*

6. The **yolk sac** is an important site of blood cell formation. The **amnion** encloses fluid that surrounds and cushions the developing embryo. The base of the **allantois** later gives rise to the urinary bladder. Circulation within the vessels of the **chorion** provides a rapid-transit system that links the embryo with the trophoblast.

Placentation, p. 1094

7. **Chorionic villi** extend outward into the maternal tissues, forming an intricate, branching network through which maternal blood flows. As development proceeds, the **umbilical cord** connects the fetus to the placenta. The trophoblast synthesizes **human chorionic gonadotropin (hCG)**, estrogens, progesterone, **human placental lactogen (hPL)**, **placental prolactin**, and **relaxin.** *(Figure 29-6)*

Embryogenesis, p. 1097

8. The **first trimester** is critical because events in the first 12 weeks establish the basis for **organogenesis** (organ formation). *(Figure 29-7; Table 29-2)*

THE SECOND AND THIRD TRIMESTERS, p. 1100

1. In the **second trimester,** the organ systems increase in complexity. During the **third trimester,** many of the organ systems become fully functional. *(Figure 29-8, Table 29-2)*

2. The fetus undergoes its largest weight gain in the third trimester. At the end of gestation, the fetus and enlarged uterus push many abdominal organs out of their positions. *(Figure 29-9)*

3. **Infertility** is relatively common, but recent advances are providing new solutions. *(Figure 29-10)*

Pregnancy and Maternal Systems, p. 1101

4. The developing fetus is totally dependent on maternal organs for nourishment, respiration, and waste removal. Maternal adaptations include increased respiratory rate, tidal volume, blood volume, nutrient and vitamin intake, glomerular filtration rate, and uterine and mammary gland size.

Structural and Functional Changes in the Uterus, p. 1105

5. Progesterone produced by the placenta has an inhibitory effect on uterine muscles; its calming action is opposed by estrogens, oxytocin, and prostaglandins. At some point, multiple factors interact to produce **labor contractions** in the uterine wall. *(Figure 29-11)*

LABOR AND DELIVERY, p. 1105

1. The goal of labor is **parturition** (forcible expulsion of the fetus).

Stages of Labor, p. 1105

2. Labor can be divided into three stages: the **dilation stage,** the **expulsion stage,** and the **placental stage.** *(Figure 29-12)*

Premature Labor, p. 1106

3. **Premature labor** may result in **immature delivery** or **premature delivery**.

Multiple Births, p. 1106

4. Twin births may be **dizygotic** ("fraternal") or **monozygotic** ("identical").

POSTNATAL DEVELOPMENT, p. 1108

1. Postnatal development involves a series of **life stages:** the neonatal period, infancy, childhood, adolescence, and maturity. Senescence (aging) begins at maturity and ends in the death of the individual.

The Neonatal Period, Infancy, and Childhood, p. 1108

2. The **neonatal period** extends from birth to 1 month after. In the transition from fetus to **neonate,** the respiratory, circulatory, digestive, and urinary systems begin functioning independently. The newborn must also begin thermoregulation.

3. Mammary gland cells produce protein-rich **colostrum** during the neonate's first few days of life and then convert to milk production. These secretions are released as a result of the **milk let-down reflex.** *(Figure 29-13)*

4. Body proportions gradually change during **infancy** (from age 1 month to 2 years) and **childhood** (from 2 years to puberty). *(Figure 29-14)*

Adolescence and Maturity, p. 1110

5. **Adolescence** begins at **puberty:** (1) The hypothalamus increases its production of GnRH; (2) circulating levels of FSH and LH rise rapidly; and (3) ovarian or testicular cells become more sensitive to FSH and LH. These changes initiate gametogenesis, the production of sex hormones, and a sudden rise in growth rate. The hormonal changes at puberty, especially changes in sex hormone levels, produce gender-specific differences in the structure and function of many systems; these differences will be retained. Adolescence continues until growth is completed. Further changes occur when sex hormone levels decline at menopause or the male climacteric.

Senescence, p. 1111

6. **Senescence** then begins, producing gradual changes in the functional capabilities of all systems. *(Table 29-3)*

GENETICS, DEVELOPMENT, AND INHERITANCE, p. 1112

1. Every somatic cell carries copies of the original 46 chromosomes in the zygote; these chromosomes and their component genes constitute the individual's **genotype.** The physical expression of the genotype is the individual's **phenotype.**

Genes and Chromosomes, p. 1112

2. Every somatic human cell contains 23 pairs of chromosomes; each pair consists of **homologous chromosomes.** Twenty-two pairs are **autosomal chromosomes.** The chromosomes of the twenty-third pair are the **sex chromosomes;** they differ between the sexes. *(Figure 29-15)*

3. Chromosomes contain DNA, and genes are functional segments of DNA. The various forms of a gene are called **alleles.** If both homologous chromosomes carry the same allele of a particular gene, the individual is **homozygous;** if they carry different alleles, the individual is **heterozygous.**

4. Alleles are considered **dominant** or **recessive** depending on how their traits are expressed. *(Table 29-4)*

5. Combining maternal and paternal alleles in a **Punnett square** diagram allows us to predict the characteristics of offspring. *(Figure 29-16)*

6. In **simple inheritance**, phenotypic characteristics are determined by interactions between a single pair of alleles. **Polygenic inheritance** involves interactions among alleles on several genes.

7. **Genetic recombination**, the gene reshuffling, or **crossing-over** and **translocation**, that occurs during meiosis, increases the genetic variation of male and female gametes. *(Figure 29-17)*

8. There are two different sex chromosomes, an **X chromosome** and a **Y chromosome**. The normal sex chromosome complement of males is XY; that of females is XX. The X chromosome carries **X-linked** *(sex-linked)* **genes** that affect somatic structures but have no corresponding alleles on the Y chromosome. *(Figure 29-18)*

The Human Genome Project, p. 1117

9. The **Human Genome Project** has identified more than 6800 of our estimated 100,000 genes, including some of those responsible for inherited disorders. *(Figure 29-19; Table 29-5)*

REVIEW QUESTIONS

Level 1   Reviewing Facts and Terms

1. The gradual modification of anatomical structures during the period from conception to maturity is
   (a) development   (b) differentiation
   (c) embryogenesis   (d) capacitation

2. Human fertilization involves the fusion of two haploid gametes, producing a zygote that contains
   (a) 23 chromosomes
   (b) 46 chromosomes
   (c) the normal haploid number of chromosomes
   (d) 46 pairs of chromosomes

3. The secondary oocyte leaving the follicle is in
   (a) interphase
   (b) metaphase of the first meiotic division
   (c) telophase of the second meiotic division
   (d) metaphase of the second meiotic division

4. Chemical interplay between developing cells is
   (a) gestation   (b) induction
   (c) capacitation   (d) placentation

5. The stage of development that follows cleavage is the
   (a) blastocyst   (b) morula
   (c) trophoblast   (d) blastodisc

6. The process that establishes the foundation of all major organ systems is
   (a) cleavage   (b) implantation
   (c) placentation   (d) embryogenesis

7. What begins as a zygote arrives in the uterine cavity as a
   (a) morula   (b) trophoblast
   (c) lacuna   (d) blastomere

8. The first extraembryonic membrane to appear during development is the
   (a) allantois   (b) yolk sac
   (c) amnion   (d) chorion

9. The primitive streak appears during
   (a) cleavage   (b) implantation
   (c) gastrulation   (d) placentation

10. The part of the uterine endometrium that has no contact with the chorion is the
    (a) decidua capsularis   (b) decidua basalis
    (c) decidua parietalis   (d) allantois

11. The surface that provides for active and passive exchange between the fetal and maternal bloodstreams is the
    (a) yolk stalk   (b) chorionic villi
    (c) umbilical veins   (d) umbilical arteries

12. The placental hormone that appears in the maternal bloodstream soon after implantation is
    (a) human chorionic gonadotropin
    (b) human placental lactogen
    (c) placental prolactin
    (d) relaxin

13. The placental stage of labor ends with
    (a) the afterbirth
    (b) rupture of the amniotic sac
    (c) delivery of the newborn
    (d) parturition

14. Milk let-down is associated with
    (a) events occurring in the uterus
    (b) placental hormonal influences
    (c) circadian rhythms
    (d) reflex action triggered by suckling

15. An individual who has two different alleles for the same gene is _____ for that trait.
    (a) heterozygous   (b) homozygous
    (c) dominant   (d) recessive

16. If an allele must be present on both the maternal and paternal chromosomes in order to affect the phenotype, the allele is said to be
    (a) dominant   (b) recessive
    (c) complementary   (d) heterozygous

17. The percentage of individuals with a particular genotype who show the expected phenotype reflects
    (a) dominance
    (b) polygenic inheritance
    (c) spontaneous mutation
    (d) penetrance

18. For a trait *A*, the genotype of a homozygous recessive individual would be represented as
    (a) *AA*   (b) *Aa*
    (c) *aa*   (d) *Ab*

19. The normal sex chromosome composition of males is
    (a) XXXX   (b) XYXY
    (c) XXYXXY   (d) XY

20. Describe the changes that occur in the sperm and egg immediately after fertilization.

21. What is induction? Explain its significance.

22. Summarize the developmental changes that occur during the first, second, and third trimesters.

23. (a) What are the four extraembryonic membranes.
(b) How do they form, and what are their functions?
24. Indicate when each of the following appears during development, and describe each: blastodisc, blastocyst, morula, zygote.
25. What is the trophoblast, and what are its three major functions?
26. Identify the three stages of labor, and describe the events that characterize each stage.
27. What factors are involved in initiating labor contractions?

28. Identify the three life stages that occur between birth and approximately age 10. Describe the characteristics of each stage and when it occurs.
29. What hormonal events are responsible for puberty? What life stage does it initiate?
30. What occurs during the life stage of maturity?
31. What is senescence? Give some examples of how it affects organ systems throughout the human body.

Level 2   Reviewing Concepts

32. Relaxin is a peptide hormone that
  (a) increases the flexibility of the pubic symphysis
  (b) causes dilation of the cervix
  (c) suppresses the release of oxytocin by the hypothalamus
  (d) a, b, and c are correct
33. The production of prostaglandins in the endometrium
  (a) initiates release of oxytocin for parturition
  (b) stimulates smooth muscle contractions
  (c) initiates secretory activity in the mammary glands
  (d) a, b, and c are correct
34. During adolescence, the events that interact to promote increased hormone production and sexual maturation result from activity of the
  (a) hypothalamus
  (b) anterior pituitary
  (c) ovaries and testicular cells
  (d) a, b, and c are correct
35. What activity during oocyte activation prevents penetration by additional sperm?
36. In addition to its role in the nutrition of the fetus, what are the primary endocrine functions of the placenta?
37. Discuss the changes that occur in maternal systems during pregnancy. Why are these changes functionally significant?
38. Discuss the changes that occur in the uterus during pregnancy. How do these changes affect uterine tissues, and which hormones are involved?
39. Why are uterine contractions in the early stages of pregnancy weak, painless, and brief?

40. During true labor, what physiological mechanisms ensure that uterine contractions continue until delivery has been completed?
41. To what does the phrase "having the water break" refer during the process of labor?
42. What physiological adjustments must an infant make during the neonatal period in order to survive?
43. Distinguish between the following paired terms:
  (a) monozygotic and dizygotic
  (b) genotype and phenotype
  (c) heterozygous and homozygous
  (d) simple inheritance and polygenic inheritance
44. What would you conclude about a trait in each of the following situations?
  (a) Children who exhibit this trait have at least one parent who exhibits the same trait.
  (b) Children exhibit this trait even though neither parent exhibits it.
  (c) The trait is expressed more commonly in sons than in daughters.
  (d) The trait is expressed equally in daughters and sons.
45. Explain why more men than women are red–green colorblind. Which type of inheritance is involved?
46. Explain the impact of genetic recombination on the production of gametes and on the traits of individuals in future generations.
47. Explain the goals and possible benefits of the Human Genome Project.

Level 3   Critical Thinking and Clinical Applications

48. Hemophilia A, a condition in which the blood does not clot properly, is a recessive trait located on the X chromosome (Xh). A woman heterozygous for this trait marries a normal male. What is the probability that this couple will have hemophiliac daughters? What is the probability that this couple will have hemophiliac sons?
49. Joan is a 27-year-old nurse who is in labor with her first child. She remembers from her anatomy and physiology class that calcium ions can increase the force of smooth muscle contractions, and because the labor is prolonged, she asks her physician for a calcium injection. The surprised physician informs Joan that such an injection is definitely out of the question. Why?

50. A new mother tells you that when she nurses her baby, she feels as if she is having menstrual cramps. How would you explain this phenomenon?
51. Explain why the normal heart and respiratory rates of neonates are so much higher than those of adults, even though adults are much larger.
52. Women who are pregnant are advised against taking aspirin during their last trimester. Why?
53. Sally gives birth to a baby with a congenital deformity of the stomach. She swears that it is the result of a viral infection that she suffered during her third trimester of pregnancy. Do you think this is a possibility? Explain.

CHAPTER 1

Page 4

1. The ability of a cell or organism to respond to changes in its environment is called *responsiveness* or irritability.
2. A histologist investigates the structure and properties of tissues.
3. The study of the physiology of specific organs is called *special physiology*. In this particular case, the field of study is *cardiac physiology* (study of heart function). Because heart failure is typically caused by disease, this specialty would overlap or be closely related to *pathological physiology*.

Page 6

1. The ability to move around and produce heat are functions of the *muscular system*.
2. Defense against infection and disease is a function of the *lymphatic system*.

Page 16

1. Physiological systems can function normally only under carefully controlled conditions. Homeostatic regulation serves to prevent potentially disruptive changes in the body's internal environment.
2. When homeostasis fails, organ systems function less efficiently or begin to malfunction. The result is the state that we call *disease*. If the situation is not corrected, death may result.
3. Positive feedback is useful in processes that must move quickly to completion once they have begun, such as blood clotting. It is harmful in situations in which a stable condition must be maintained, because it will serve to increase any departure from the desired condition. For example, positive feedback in the regulation of body temperature would cause a slight fever to spiral out of control, with fatal results. For this reason physiological systems usually exhibit negative feedback, which tends to oppose any departure from the norm.

Page 24

1. The two eyes would be separated by a *midsagittal* section.
2. The body cavity inferior to the diaphragm is the *abdominopelvic* (or *peritoneal* or *abdominal*) *cavity*.

CHAPTER 2

Page 38

1. Atoms combine with each other so as to gain a complete set of eight electrons in their outer energy levels. Oxygen atoms do not have a full outer energy level, so they will readily react with many other elements to attain this stable arrangement. Neon already has a full outer energy level and thus has little tendency to combine with other elements.
2. There are three isotopes of hydrogen: hydrogen-1, with a mass of 1; deuterium, with a mass of 2; and tritium, with a mass of 3. The heavier sample must contain a higher proportion of one or both of the heavier isotopes.
3. A water molecule is formed by polar covalent bonds. Water molecules are attracted to one another by hydrogen bonds.

Page 41

1. Because this reaction involves a large molecule being broken down into two smaller ones, it is a *decomposition reaction*. Because energy is released in the process, the reaction can also be classified as *exergonic*.
2. Removing the product of a reversible reaction would keep its level low compared with the level of the reactants. Thus the formation of product molecules would continue, but the reverse reaction would slow down, resulting in a shift in the equilibrium toward the product.

Page 45

1. When it dissolves in water, salt dissociates into charged particles called ions that are capable of conducting an electrical current. Sugar molecules are held together by covalent bonds and do not dissociate in solution; thus there are no ions to carry a current.
2. Stomach discomfort is often the result of excess stomach acidity ("acid indigestion"). Antacids contain a weak base that neutralizes the excess acid.

Page 51

1. A C:H:O ratio of 1:2:1 would indicate that the molecule is a *carbohydrate*. The body uses carbohydrates chiefly as an energy source.
2. When two monosaccharides undergo a dehydration synthesis reaction, they form a disaccharide.
3. The most abundant lipid in a sample taken from beneath the skin would be a *triglyceride*.

4. Analysis of the lipid content of human cell membranes would indicate the presence of mostly phospholipids with small amounts of cholesterol and glycolipids.

Page 57

1. Proteins are composed of chains of small molecules called *amino acids*.
2. An agent that disrupts hydrogen bonding would affect the secondary level of protein structure.
3. The heat of boiling will break bonds that maintain the protein's tertiary and/or quaternary structure. The resulting change in shape will affect the ability of the protein molecule to perform its normal biological functions. These alterations are known as *denaturation*.
4. If the active site of an enzyme changes so that it better fits its substrate, the change will increase the level of enzyme activity. But if the change alters the active site to the extent that the enzyme's substrate can no longer bind or binds poorly, the enzyme's activity will decrease or be inhibited.

Page 61

1. The nucleic acid *RNA (ribonucleic acid)* contains the sugar ribose. DNA (deoxyribonucleic acid) contains the sugar deoxyribose.
2. Phosphorylation of an ADP molecule would yield a molecule of *ATP*.

CHAPTER 3

Page 72

1. The *phospholipid* bilayer forms a physical barrier between the internal environment of the cell and the external environment.
2. *Channel proteins* are integral proteins that allow water, some ions, and small molecules to pass through the cell membrane.

Page 76

1. The process of diffusion is driven by a *concentration gradient*. The greater the concentration gradient, the faster the rate of diffusion. The smaller the concentration gradient, the slower the rate of diffusion. If the concentration of oxygen in the lungs decreased, the concentration gradient between oxygen in the lungs and oxygen in the blood would decrease (assuming that the oxygen level of the blood remained constant). Thus oxygen would diffuse more slowly into the blood.
2. The 10 percent salt solution would be hypertonic with respect to the cells of the nasal lining, because it contains a higher concentration of salt than do the cells. The hypertonic solution would draw water out of the cells, causing the cells to shrink and loosening the mucus, thus relieving the congestion.

Page 82

1. In order to transport hydrogen ions against their concentration gradient—that is, from a region where they are less concentrated (the cells lining the stomach) to a region where they are more concentrated (the interior of the stomach)—energy must be expended. An *active transport* process must be involved.
2. If the cell membrane were freely permeable to sodium ions, more of these positively charged ions would move into the cell and the transmembrane potential would move closer to zero.
3. This process is an example of *phagocytosis*.

Page 86

1. The fingerlike projections on the surface of the intestinal cells are *microvilli*. They increase the cells' surface area, so the cells can absorb nutrients more efficiently.
2. Because the flagellum is an organelle of locomotion, sperm cells lacking a flagellum would be unable to move. (As a result, they would be unable to reach an oocyte and perform fertilization.)

Page 91

1. The function of mitochondria is to produce energy for the cell in the form of ATP molecules. A large number of mitochondria in a cell would indicate a high demand for energy.
2. SER functions in the synthesis of lipids such as steroids. Ovaries and testes would be expected to have a great deal of SER because they produce large amounts of steroid hormones.

Page 97

1. The nucleus of a cell contains DNA, which codes for the production of all the cell's polypeptides and proteins. Some of these proteins are structural proteins responsible for the shape and other physical characteristics

of the cell. Other proteins are enzymes that govern cellular metabolism, direct the production of cell proteins, and control all the cell's activities.

2. If a cell lacked the enzyme RNA polymerase, the cell would not be able to transcribe RNA from DNA.

Page 103

1. This cell would be in the G_1 phase of its life cycle.

2. The deletion of a base from a coding sequence of DNA would alter the entire base sequence after the deletion point. This change would result in different codons on the messenger RNA that was transcribed from the affected region. These codons, in turn, would result in the incorporation of a different series of amino acids into the polypeptide. Almost certainly the polypeptide product would be not functional.

3. If spindle fibers failed to form during mitosis, the cell would not be able to separate the chromosomes into two sets. If cytokinesis occurred, the result would be one cell with two sets of chromosomes and one cell with none.

CHAPTER 4

Page 112

1. The presence of many microvilli on the free surface of epithelial cells greatly increases the cell's surface area, allowing for increased absorption.

2. Gap junctions allow small molecules and ions to pass from cell to cell. Among epithelial cells, they help coordinate functions such as the beating of cilia. In cardiac and smooth muscle tissue, they are essential to the coordination of muscle cell contractions.

3. All these regions are subject to mechanical trauma and abrasion—by food (pharynx and esophagus), feces (anus), and intercourse or childbirth (vagina).

Page 119

1. No. A simple squamous epithelium does not provide enough protection against infection, abrasion, and dehydration and is not found on the skin surface.

2. The process described is *holocrine secretion*.

3. The gland is an *endocrine gland*.

Page 126

1. Collagen fibers add strength to connective tissue. We would therefore expect vitamin C deficiency to result in the production of connective tissue that is weaker and more prone to damage.

2. Antihistamines act against the molecule histamine, which is released from *mast cells*.

3. The tissue is *adipose* (fat) tissue.

Page 131

1. Cartilage lacks a direct blood supply, which is necessary for proper healing to occur. Materials that are needed to repair damaged cartilage must diffuse from the blood to the chondroblasts, a process that takes a long time and retards the healing process.

2. Intervertebral disks are composed of *fibrocartilage*.

3. The two connective tissues that contain a liquid matrix are *blood* and *lymph*.

Page 135

1. The pleural, peritoneal, and pericardial cavities are all lined by serous membranes.

2. This is an example of a *mucous membrane*.

3. The tissue is probably *fascia*, a type of dense regular connective tissue that attaches muscles to skin and bones.

Page 137

1. Because cardiac and skeletal muscle are both striated (banded), this must be *smooth muscle tissue*.

2. The cells are probably *neurons*.

3. New muscle cells can be produced by the division and fusion of *satellite cells*, mesenchymal cells that persist in adult skeletal muscle tissue.

CHAPTER 5

Page 151

1. Cells are constantly being shed from the outer layers of the *stratum corneum*.

2. The splinter is lodged in the *stratum granulosum*.

3. Because fresh water is hypotonic with respect to skin cells, water will move into the cells by osmosis, causing them to swell.

4. Sanding the tips of one's fingers will not permanently remove fingerprints. Because the ridges of the fingerprints are formed in layers of the

skin that are constantly regenerated, these ridges will eventually reappear. The actual pattern of the ridges is determined by arrangement of tissue in the dermis, which is not affected by sanding.

Page 154

1. When exposed to the ultraviolet radiation in sunlight or sunlamps, melanocytes in the epidermis and dermis synthesize the pigment *melanin*, darkening the color of the skin.

2. When skin is warm, more blood is diverted to the superficial dermis for the purpose of eliminating heat. Because the blood is red, it imparts a red cast to the skin.

3. The hormone *cholecalciferol* (vitamin D_3) is needed to form strong bones and teeth. The first step in this hormone's production involves the skin where cholesterol is exposed to specific wavelengths of ultraviolet light. When the body surface is covered, the UV light is not able to penetrate to the blood in the skin to begin vitamin D_3 production, resulting in fragile bones.

Page 156

1. The capillaries and sensory neurons of the skin are located in the *papillary layer* of the dermis.

2. The presence of elastic fibers and the resilience of skin turgor account for dermal elasticity.

Page 159

1. Contraction of the arrector pili muscles pulls the hair follicles erect, depressing the area at the base of the hair and making the surrounding skin appear higher. The result is known as "goose bumps."

2. Hair is a derivative of the epidermis, but the follicles are in the dermis. Where the epidermis and deep dermis are destroyed, new hair will not form.

Page 161

1. Sebaceous glands produce a secretion called *sebum*. Sebum lubricates and protects the keratin of the hair shaft, lubricates and conditions the surrounding skin, and inhibits the growth of bacteria.

2. *Apocrine sweat glands* produce a secretion containing several kinds of organic compounds. Some of these have an odor and others produce an odor when metabolized by skin bacteria. Deodorants are used to mask the odor of these secretions.

3. Apocrine sweat glands enlarge and increase secretion in response to the increase in sex hormones that occurs at puberty.

Page 162

1. The name given to the combination of fibrin clot, fibroblasts, and the extensive network of capillaries that is found in tissue that is healing is *granulation tissue*.

2. Skin can still regenerate effectively even after considerable damage has occurred because stem cells persist in both the epithelial and connective tissue components of skin. When injury occurs, cells of the stratum germinativum replace epithelial cells while mesenchymal cells replace cells lost from the dermis.

3. As a person ages, the blood supply to the dermis decreases and merocrine sweat glands become less active. These changes make it more difficult for the elderly to cool themselves in hot weather.

CHAPTER 6

Page 179

1. If the ratio of collagen to hydroxyapatite in a bone increased, the bone would be more flexible and less strong.

2. The presence of concentric layers of bone around a central canal is indicative of an osteon. Osteons make up compact bone. Because the ends (epiphyses) of long bones are primarily cancellous (spongy) bone, this sample most likely came from the shaft (diaphysis) of a long bone.

3. Because osteoclasts function in breaking down or demineralizing bone, the bone would have less mineral content and as a result it would be weaker.

Page 183

1. During the process of intramembranous ossification, fibrous connective tissue is replaced by bone or mesenchyme.

2. In endochondral ossification, cells of the inner layer of the perichondrium differentiate into osteoblasts.

3. Long bones of the body, such as the femur, have a plate of cartilage, called the *epiphyseal plate*, that separates the epiphysis from the diaphysis as long as the bone is still growing lengthwise. An X-ray would indicate whether the epiphyseal plate is still present. If it is, growth is still occurring; if it is not, the bone has reached its adult length.

Page 186

1. The larger arm muscles of the weight lifter will apply more mechanical stress to the bones of the upper limbs. In response to the stress, the bones will grow thicker. We would expect the jogger to have heavier thigh bones for similar reasons.
2. Growth continues throughout childhood. At puberty there is a growth spurt followed by epiphyseal closure. The later puberty begins, the taller the child will be when the growth spurt begins, and the taller the individual will be when growth is completed.
3. Increased levels of growth hormone prior to puberty will result in excessive bone growth, leading to taller stature.

Page 188

1. Children suffering from rickets have bones that are poorly mineralized and as a result are quite flexible. Under the weight of the body, the leg bones bend. The instability makes bending and walking difficult and can lead to other problems of the legs and feet.
2. Parathyroid hormone (PTH) stimulates osteoclasts to release calcium ions from bone. Increased PTH secretion would result in an increase in the level of calcium ions in the blood.
3. Calcitonin lowers blood calcium levels by (1) inhibiting osteoclast activity and (2) increasing the rate of calcium excretion at the kidneys.

Page 192

1. An external callus forms early in the healing process when cells from the endosteum and periosteum migrate to the area of the fracture and form an enlarged collar (external callus) that circles the bone in the area of the fracture.
2. The sex hormones known as estrogens play an important role in moving calcium into bones. After menopause, the level of these hormones decreases dramatically and as a result, it is difficult to replace the calcium in bones that is being lost due to normal aging. Males do not show a decrease in sex hormone levels (androgens).
3. Bones that form in the tendons of joints such as the knee and wrist are *sesamoid bones*.

CHAPTER 7

Page 213

1. The foramen magnum is located in the base of the *occipital bone*.
2. Tomás has fractured his right *parietal bone*.
3. The internal jugular veins pass through openings between the *occipital* and the *temporal bones*.
4. The sella turcica contains the pituitary gland and is located in the *sphenoid bone*.
5. Olfactory nerves, which deal with the sense of smell, pass through the olfactory foramina of the cribriform plate from the nasal cavity to the olfactory bulbs. If the cribriform plate failed to form, these sensory nerves could not reach the olfactory bulbs, and the sense of smell (olfaction) would be lost.

Page 220

1. The adult vertebral column has fewer vertebrae because the five sacral vertebrae fuse to form a single sacrum and the four coccygeal vertebrae fuse to form a single coccyx.
2. The secondary curves of the spine allow us to balance our body weight on our legs with minimal muscular effort. Without the secondary curves, we would not be able to stand upright for extended periods.
3. When you run your finger along a person's spine, you can feel the *spinous processes* of the vertebrae.

Page 225

1. The odontoid process is found on the second cervical vertebra, or *axis*, which is located in the neck.
2. The presence of transverse foramina indicate that this is a cervical vertebra.
3. The lumbar vertebrae must support a great deal more weight than vertebrae that are more superior in the spinal column. The large bodies allow the weight to be distributed over a larger area.

Page 227

1. True ribs are attached directly to the sternum by their own costal cartilage. False ribs either do not attach to the sternum at all (floating ribs) or attach by means of a common costal cartilage (the fusion of several).
2. Improper compression of the chest during CPR could and frequently does result in a fracture of the *sternum* or *ribs*.

CHAPTER 8

Page 234

1. The clavicle attaches the scapula to the sternum and thus restricts the scapula's range of movement. If the clavicle is broken, the scapula will have a greater range of movement and will be less stable.
2. The head of the *humerus* articulates with the glenoid cavity of the scapula.

Page 242

1. The two rounded projections on either side of the elbow are parts of the humerus (the *lateral* and *medial epicondyles*).
2. The radius is in a lateral position when the arm is pronated; it is in a medial position when the arm is supinated.
3. The first distal phalanx is located at the tip of the thumb, so Bill has broken his thumb.

Page 244

1. The three bones that make up the coxa are the *ilium*, *ischium*, and *pubis*.
2. The female pelvis is generally smoother and lighter and has less prominent markings. The pelvic outlet is enlarged, and there is less curvature on the sacrum and coccyx. The pelvic inlet is wider and more circular. The pelvis as a whole is relatively broad and low. The ilia project farther laterally, and the pubic arch is broader with an inferior angle between the pubic bones greater than 100°, as opposed to 90° or less. These differences adapt the pelvis for supporting the weight of the developing fetus and enabling the newborn to pass through the pelvic outlet at the time of delivery.
3. When you are seated, your body weight is borne by the *ischial tuberosities* of the pelvis.

Page 249

1. Although the fibula is not part of the knee joint and does not bear weight, it is an important point of attachment for many leg muscles. When the fibula is fractured, these muscles cannot function properly to move the leg, and walking is difficult and painful. The fibula also helps stabilize the ankle joint.
2. Joey has most likely fractured the *calcaneus* (heel bone).
3. The *talus* transmits the weight of the body from the tibia toward the toes.
4. Muscles that adduct the femur are attached at the linea aspera. As these muscles are strengthened, they will apply more stress to the femur at the linea aspera, and this portion of the femur will get larger and heavier.

CHAPTER 9

Page 258

1. All these joints (other than synostoses) consist of bony regions separated by fibrous or cartilaginous connective tissue.
2. Originally, the joint is a type of *syndesmosis*. When the bones interlock, they form *sutural joints*.
3. The articular cartilages lack a blood supply and rely on synovial fluid to supply nutrients and eliminate wastes. Impairing the circulation of synovial fluid would have the same effect as impairing a tissue's blood supply. Nutrients would not be delivered to meet the tissue's needs, and wastes would accumulate. Damage and ultimately death of the cells in the tissue would result.

Page 264

1. (a) *abduction*; (b) *supination*; (c) *flexion*
2. When you perform jumping jacks and move your lower limbs away from the midline of the body, the movement is *abduction*. When you bring the lower limbs back together, the movement is *adduction*.
3. *Flexion* and *extension* are the movements associated with hinge joints.

Page 267

1. Intervertebral discs are not found between the first and second cervical vertebrae, between sacral vertebrae in the sacrum, or between coccygeal vertebrae in the coccyx. An intervertebral disc between the first and second cervical vertebrae would prohibit rotation. The vertebrae in the sacrum and coccyx are fused.
2. (a) *flexion*; (b) *lateral flexion*; (c) *rotation*

Page 268

1. Ligaments and muscles provide most of the stability for the shoulder joint.
2. Because the subscapular bursa is located in the shoulder joint, the tennis player would be more likely to have inflammation of this structure (bursi-

tis). The condition is associated with repetitive motion that occurs at the shoulder, such as swinging a tennis racket. The jogger would be more at risk for injuries to the knee joint.

3. A shoulder separation is an injury involving partial or complete dislocation of the acromioclavicular joint.

4. Terry has most likely damaged his *annular ligament*.

Page 272

1. The iliofemoral, pubofemoral, and ischiofemoral ligaments are in the hip joint.

2. Damage to the menisci of the knee joint would result in a decrease in the joint's stability. The individual would have a harder time locking the knee in place while standing and would have to use muscle contractions to stabilize the joint. If the person had to stand for a long period, the muscles would fatigue, and the knee would "give out." We would also expect the individual to experience pain.

3. "Clergyman's knee" is a bursitis commonly found among members of the clergy, who spend a great deal of time kneeling. The work of carpet layers and roofers necessitates kneeling and sliding along on their knees, causing a similar inflammation of the bursae in the knees.

CHAPTER 10

Page 284

1. Because tendons attach muscles to bones, severing the tendon would disconnect the muscle from the bone, so when the muscle contracted, nothing would happen.

2. Skeletal muscle appears striated when viewed through the microscope because that muscle is composed of the myofilaments actin and myosin, which are arranged in such a way that they produce a banded appearance in the muscle.

3. You would expect to find the greatest concentration of calcium ions in the cisternae of the sarcoplasmic reticulum of the muscle.

Pages 291

1. Because the ability of a muscle to contract depends on the formation of cross-bridges between the myosin and actin myofilaments, a drug that would interfere with cross-bridge formation would prevent the muscle from contracting.

2. Because the amount of cross-bridge formation is proportional to the amount of available calcium ions, increased permeability of the sarcolemma to calcium ions would lead to an increased intracellular concentration of calcium and a greater degree of contraction. In addition, because relaxation depends on decreasing the amount of calcium in the sarcoplasm, an increase in the permeability of the sarcolemma to calcium could result in a situation in which the muscle would not be able to relax completely.

3. Without acetylcholinesterase, the motor end plate would be continuously stimulated by the acetylcholine, and the muscle would be locked into contraction.

Page 298

1. The ability of the muscle to contract depends on the ability to form cross-bridges between the actin and myosin. If the myofilaments overlap very little, then very few cross-bridges are formed, and the contraction is weak. If the myofilaments do not overlap at all, then no cross-bridges form, and the muscle cannot contract.

2. During treppe, there is not enough time between successive contractions to reabsorb all the calcium ions that were released during the prior contraction event. As a result, calcium ions accumulate in the sarcoplasm at higher than normal levels, allowing more cross-bridges to form and tension to increase.

3. A contraction occurs as thick and thin filaments interact. The entire muscle may shorten (isotonic, concentric), elongate (isotonic, eccentric), or remain the same length (isometric), depending on the relationship between the resistance and the tension produced by actin and myosin interactions.

Page 306

1. The sprinter requires large amounts of energy for a relatively short burst of activity. To supply this demand for energy, the muscles switch to anaerobic metabolism. Anaerobic metabolism is not as efficient in producing energy as aerobic metabolism is, and the process also produces acidic waste products. The combination of less energy and the waste products contributes to fatigue. Conversely, marathon runners derive most of their energy from aerobic metabolism, which is more efficient and does not produce the level of waste products that anaerobic respiration does.

2. We would expect activities that require short periods of strenuous activity to produce a greater oxygen debt, because this type of activity relies heavily on energy production by anaerobic respiration. Because lifting weights is more strenuous over the short term, we would expect it to produce a greater oxygen debt than swimming laps, which is an aerobic activity.

3. Individuals who are naturally better at endurance types of activities, such as cycling or marathon running, have a higher percentage of slow muscle fibers, which are physiologically better adapted to this type of activity than are fast fibers, which are less vascular and fatigue faster.

Page 311

1. Cardiac muscle cells are joined by gap junctions, which allow ions and small molecules to flow directly from one cell to another. This type of junction allows for action potentials generated in one cell to spread rapidly to adjacent cells. Thus all the cells will contract simultaneously as if they were one single unit (a syncytium).

2. Cardiac muscle and smooth muscle contractions are more affected by changes in the concentration of Ca^{2+} in the extracellular fluid than are skeletal muscle contractions because in cardiac and smooth muscles, the majority of the Ca^{2+} that trigger a contraction come from the extracellular fluid. In skeletal muscle, most of the Ca^{2+} come from the sarcoplasmic reticulum.

3. The actin and myosin filaments of smooth muscle are not as rigidly organized as they are in skeletal muscle. This loose organization allows smooth muscle to contract over a relatively large range of resting lengths.

CHAPTER 11

Page 319

1. A pennate muscle will contain more muscle fibers than a parallel muscle of the same size. A muscle that has more muscle fibers has more myofibrils and sarcomeres. As a result, the contraction of the pennate muscle generates more tension than a parallel muscle of the same size.

2. The opening between the stomach and the small intestine would be guarded by a circular muscle, or *sphincter*. The concentric circles of muscle fibers found in sphincters are ideally suited for opening and closing openings or for acting as valves in the body.

3. The joint between the occipital bone and the first cervical vertebra is a *first-class* lever system. The joint between the two bones, the fulcrum, lies between the skull, which is the resistance, and the neck muscles, which are the applied force.

Page 324

1. The origin of a muscle is the end that remains stationary during an action. Because the gracilis muscle moves the tibia, the origin must be on the pelvis (pubis and ischium).

2. Muscles A and B are *antagonists*, because they perform opposite actions.

3. The name *flexor carpi radialis longus* tells you that this is a long muscle that lies next to the radius and functions to flex the hand.

Page 330

1. Contraction of the masseter muscle raises the mandible, whereas the mandible depresses when the muscle is relaxed. If you were doing these movements, you would probably be in the process of chewing.

2. You would expect the *buccinator muscle*, which forms the mouth for blowing, to be well developed in a trumpet player.

3. Swallowing involves contractions of the palatal muscles, which raise the soft palate as well as portions of the superior pharyngeal wall. Elevation of the superior portion of the pharynx enlarges the opening to the auditory tube, which permits airflow to the middle ear and the inside of the eardrum. Making this opening larger facilitates airflow into or out of the middle ear cavity.

Page 337

1. Damage to the external intercostal muscles would interfere with the process of breathing.

2. A blow to the rectus abdominis would cause the muscle to contract forcefully, resulting in flexion of the torso. In other words, you would "double over."

3. The sore muscles are most likely the *sacrospinalis muscles*, composed of the longissimus and the iliocostalis of the lumbar region. These muscles would have to contract harder to counterbalance increased anterior weight, as when you carry a heavy box.

Page 351

1. When you shrug your shoulders, you are contracting your *levator scapulae muscles*.

2. The rotator cuff muscles include the *supraspinatus, infraspinatus, subscapularis*, and *teres minor*. The tendons of these muscles help enclose and stabilize the shoulder joint.

3. Injury to the flexor carpi ulnaris would impair the ability to flex and adduct the hand.

Page 357

1. Injury to the obturator muscle would interfere with your ability to rotate your thigh laterally.

2. The hamstring refers to a group of muscles that collectively function in flexing the leg. These muscles are the biceps femoris, semimembranosus, and semitendinosus.

3. The calcaneal tendon attaches the soleus and gastrocnemius muscles to the calcaneus (heel bone). When these muscles contract, they cause extension of the foot. A torn calcaneal tendon would make extension of the foot difficult, and the opposite action, flexion, would be more pronounced as a result of less antagonism from the soleus and gastrocnemius.

CHAPTER 12

Page 380

1. The afferent division of the nervous system is composed of nerves that carry sensory information to the brain and spinal cord. Damage to this division would interfere with a person's ability to experience a variety of sensory stimuli.

2. Most sensory neurons of the peripheral nervous system are unipolar. Thus these neurons are most likely sensory neurons.

3. *Microglial cells* are small phagocytic cells that occur in increased number in damaged and diseased areas of the CNS.

Page 396

1. Depolarization of the neuron membrane involves the opening of the sodium channels and the rapid influx of sodium ions into the cell. If the sodium channels were blocked, a neuron would not be able to depolarize and conduct an action potential.

2. If the extracellular concentration of potassium ions decreased, more potassium ions would leave the cell, and the electrical difference across the membrane (transmembrane potential) would be greater. This condition is called *hyperpolarization*.

3. Action potentials are propagated along myelinated axons by saltatory propagation at speeds much higher than those along unmyelinated axons. The axon with a propagation speed of 50 msec must be the myelinated axon.

Page 401

1. When an action potential reaches the presynaptic terminal of a cholinergic synapse, calcium ion channels are opened and the influx of Ca^{2+} triggers the release of ACh into the synapse to stimulate the next neuron. If the calcium channels were blocked, the ACh would not be released, and transmission across the synapse would cease.

2. Because of synaptic delay, the pathway with fewer neurons (three) will conduct impulses faster.

3. The effect of a neurotransmitter on postsynaptic membranes depends on the receptor molecule that binds it. In this case, although the neurotransmitter and target tissues are the same, the smooth muscle in blood vessels of skeletal muscle have different NE receptors than those in the blood vessels of intestines; hence the opposite effects.

Page 410

1. This pattern represents a *convergent* circuit.

2. For a severed axon to heal, it must come into contact with and grow into the new column of Schwann cells that forms distal to the injury site. If the axon fails to make this connection, the column of Schwann cells degenerates, the axon stops growing, and the connection will not be reestablished. Closely aligning the two ends of the axons after an injury increases the chance that the connection will be made and innervation reestablished.

3. Interneurons are found in the CNS and are responsible for analyzing sensory input and coordinating motor output. Without the interneurons, the nervous system would not be able to process sensory information or to make appropriate motor responses.

CHAPTER 13

Page 420

1. The *ventral root* of spinal nerves is composed of visceral and somatic motor fibers. Damage to this root would interfere with motor function.

2. The cerebrospinal fluid that surrounds the spinal cord is located in the *subarachnoid space*, which lies beneath the epithelium of the arachnoid layer and on top of the pia mater.

Page 423

1. Because the polio virus would be located in the somatic motor neurons, it would be in the *anterior gray horns* of the spinal cord, where the cell bodies of these neurons are located.

2. A disease that damages myelin sheaths would affect the columns of the spinal cord, because this part of the cord is composed of bundles of myelinated axons.

Page 430

1. The dorsal rami of spinal nerves innervate the skin and muscles of the back. The skin and muscles of the back of the neck and of the shoulders would be affected.

2. The phrenic nerves that innervate the diaphragm originate in the *brachial plexus*. Damage to this plexus, or more specifically to the phrenic nerves, would greatly interfere with the ability to breathe and might result in death.

3. Compression of the *sciatic nerve* produces the characteristic sensation that you perceive when your leg "falls asleep."

Page 435

1. The minimum number of neurons required for a reflex arc is two: One must be a sensory neuron, to bring impulses to the CNS, and the other a motor neuron, to bring about a response to the sensory input.

2. The suckling reflex is an innate reflex.

3. When the stretch receptors are stimulated by the gamma motor neurons, the spindles become narrower and less sensitive to stretch. As a result, it would take more force to get the muscles of the leg to contract for the knee jerk reflex. Thus the reflex would be slower.

Page 440

1. This response is the *tendon reflex*.

2. During the withdrawal reflex, the limb on the opposite side is extended. This response is called a *crossed extensor reflex*.

3. A positive Babinski reflex is abnormal for an adult and indicates possible damage of descending tracts in the spinal cord.

CHAPTER 14

Page 446

1. The three primary brain vesicles are the *prosencephalon, mesencephalon*, and *rhombencephalon*. The prosencephalon gives rise to the cerebrum and diencephalon; the mesencephalon does not subdivide further; the rhombencephalon develops into the cerebellum, pons, and medulla oblongata.

2. The response is controlled by the *mesencephalon*.

Page 456

1. If one of the interventricular foramina became blocked, cerebrospinal fluid would not be able to flow from the first or second ventricle into the third. Cerebrospinal fluid would continue to be formed, so the blocked ventricle would swell with fluid—a condition known as *hydrocephalus*.

2. Diffusion across the arachnoid granulations is the means by which cerebrospinal fluid reenters the bloodstream. If this process decreased, excess fluid would accumulate in the ventricles, and the volume of fluid in the ventricles would increase.

3. The blood–brain barrier, consisting of capillary endothelial cells stimulated by adjacent astrocytes, restricts and regulates the movement of water-soluble molecules from the blood to the extracellular fluid of the brain.

Page 462

1. The primary motor cortex is located in the *precentral gyrus* of the frontal lobe of the cerebrum.

2. Damage to the temporal lobe of the cerebrum would interfere with the processing of olfactory (smell) and auditory (sound) impulses.

3. The stroke has damaged Jake's *speech center* in the primary motor cortex of the frontal lobe.

Page 465

1. The extrapyramidal system is composed of pathways that control muscle tone and coordinate learned movement patterns. A person suffering an injury to these tracts would exhibit difficulty in walking and in making fluid, precise movements of the upper limbs.

2. Paul is probably having problems associated with the temporal lobe of the cerebrum, specifically the *hippocampus* and the *amygdaloid body*. His problems may also involve other parts of the limbic system that act as a gate for loading and retrieving long-term memories.

3. Terri's symptoms imply injury to the *mamillary bodies* of the hypothalamus. She may also have sustained injury to motor areas of the brain stem that control the muscles involved in these processes.

Page 469

1. The lateral geniculate nuclei are involved with processing visual information. Damage to these nuclei would interfere with the sense of sight.

2. Changes in body temperature would stimulate the *preoptic area* of the hypothalamus, a division of the diencephalon.

Page 474

1. The vermis and arbor vitae are part of the *cerebellum*.

2. Although the medulla oblongata is small, it contains many vital reflex centers, including those that control breathing and regulate the heart rate and blood pressure. Damage to the medulla oblongata can result in a cessation of breathing or in lethal changes of the heart rate and blood pressure.

3. Cells of the substantia nigra release the neurotransmitter dopamine at synapses with cerebral neurons. Dopamine produces IPSPs and decreases the activity of motor neurons. Decreased secretion would result in the motor neurons' being more active, producing symptoms including twitches, tremors, other types of involuntary contractions, and possibly tetany.

C H A P T E R   1 5

Page 496

1. The *fasciculus gracilis* in the posterior column of the spinal cord is what is responsible for carrying information about touch and pressure from the lower part of the body to the brain.

2. Nociceptors are stimulated by pain, so the action potentials generated by these receptors would be carried by the *lateral spinothalamic tracts*.

3. Impulses carried along the right fasciculus gracilis are destined for the primary sensory cortex of the left cerebral hemisphere.

Page 502

1. The anatomical basis for opposite-side motor control is that crossing-over (*decussation*) of axons occurs, so the pyramidal motor fibers innervate lower motor neurons on the opposite side of the body.

2. The superior portion of the motor cortex exercises control over the upper limb and upper portion of the lower limb. An injury to this area would affect the ability to control the muscles in those regions of the body.

3. Motor neurons of the red nucleus help control the muscle tone of skeletal muscles. Increased stimulation of these neurons would increase stimulation of the skeletal muscles, producing increased muscle tone.

Page 506

1. We would expect Tina's brain waves to be *beta waves*, which are characteristic of adults who are experiencing stress and/or tension.

2. An inability to comprehend the spoken or written word indicates a problem with the *general interpretive area* of the brain, which in most individuals is located in the left temporal lobe of the cerebrum.

3. Recalling information for an A & P test involves *fact memory*, specifically *secondary memory* (long-term memory).

Page 509

1. The RAS is responsible for rousing the cerebrum to a state of consciousness. If your RAS were stimulated, you would wake up.

2. A drug that increases serotonin levels would produce a heightened perception of certain sensory stimuli, such as auditory or visual stimuli, and hallucinations.

3. Some possible reasons for slower recall and loss of memory in the elderly include a loss of neurons (possibly those involved in specific memories), changes in synaptic organization of the brain, changes in the neurons themselves, and decreased blood flow, which would affect the metabolic rate of neurons and perhaps slow the retrieval of information from memory.

C H A P T E R   1 6

Page 516

1. Two neurons are required to carry an action potential from the spinal cord to the smooth muscle of the intestine. One neuron is required to carry

the action potential from the spinal cord to the autonomic ganglion, and a second to carry the action potential from the autonomic ganglion to the smooth muscle.

2. The *sympathetic division* of the autonomic nervous system is responsible for the physiological changes that occur in response to stress and increased activity.

3. The sympathetic division of the autonomic nervous system includes preganglionic fibers from the lumbar and thoracic portions of the spinal cord, whereas the parasympathetic division includes preganglionic fibers from the cranial and sacral portions.

Page 523

1. The neurons that synapse in the collateral ganglia originate in the lower thoracic and upper lumbar portion of the spinal cord and pass through the chain ganglia to the collateral ganglia.

2. Acetylcholine is the neurotransmitter released by all the preganglionic fibers of the sympathetic nervous system. A drug that stimulates ACh receptors would stimulate the postganglionic fibers of the sympathetic nerves, resulting in increased sympathetic activity.

3. Blocking the beta receptors on cells would decrease or prevent sympathetic stimulation of those tissues. Heart rate and the force of contraction and relaxation of the smooth muscle in the walls of blood vessels would decrease. These changes would contribute to lowering a person's blood pressure.

Page 526

1. The *vagus nerve (N X)* carries the parasympathetic fibers that innervate the lungs, heart, stomach, liver, pancreas, and parts of the small and large intestines (as well as several other visceral organs).

2. Muscarinic receptors are a type of acetylcholine receptor located in the postganglionic synapse of the parasympathetic nervous system. Stimulation of these receptors at the heart would cause potassium ion channels to open, resulting in hyperpolarization of the membrane and a decreased heart rate.

3. The parasympathetic division is sometimes referred to as the anabolic system because parasympathetic stimulation leads to a general increase in the nutrient content of the blood. Cells throughout the body respond to the increase by absorbing the nutrients and using them to support growth and other anabolic activities.

Page 532

1. Most blood vessels receive sympathetic stimulation, so a decrease in sympathetic tone would lead to a relaxation of the muscles in the walls of the vessels and hence to vasodilation. Blood flow to the tissue would, in turn, increase.

2. A patient who is anxious about an impending root canal would probably exhibit some or all of the following changes: a dry mouth, increased heart rate, increased blood pressure, increased rate of breathing, cold sweats, an urge to urinate or defecate, change in motility of the digestive tract (that is, "butterflies in the stomach"), and dilated pupils. These changes would be the result of anxiety or stress causing an increase in sympathetic stimulation.

3. A brain tumor that interferes with hypothalamic function would be expected to interfere with autonomic function as well. Centers in the posterior and lateral hypothalamus coordinate and regulate sympathetic function, whereas centers in the anterior and medial hypothalamus control parasympathetic function.

C H A P T E R   1 7

Page 546

1. The receptor with the smaller receptor field will provide the more precise sensory information—thus receptor A.

2. Nociceptors are pain receptors. If they were stimulated, you would perceive a painful sensation in the affected hand.

3. Proprioceptors relay information about limb position and movement to the CNS, especially to the cerebellum. Lack of this information would result in uncoordinated movements, and you would probably not be able to walk.

Page 550

1. By the end of the lab period, adaptation has occurred. In response to the constant level of stimulation, the receptor neurons have become less active, partially as the result of synaptic fatigue.

2. Your taste receptors (taste buds) are sensitive only to molecules and ions that are in solution. If you dry the surface of the tongue, there is no moisture for the sugar molecules or salt ions to dissolve in, and they will not stimulate the taste receptors.

3. You could tell your grandfather that the difference in the taste of his food is the result of several age-related factors. The number of taste buds declines dramatically after age 50, and those that remain are not as sensitive as they once were. In addition, the loss of olfactory receptors contributes to the perception of less flavor in foods.

Page 562

1. The first layer of the eye to be affected by inadequate tear production would be the *conjunctiva*. Drying of this layer would produce an irritated, scratchy feeling.
2. When the lens is round, you are looking at something close to you.
3. Renee will probably not be able to see at all. The fovea of the eye contains cones but no rods. Rods respond to light of low intensity, but cones need high light intensity to be stimulated. In a dimly lit room, the light would not be strong enough to stimulate these photoreceptors. As a result, Renee would not be able to see.
4. If the canal of Schlemm is blocked, the aqueous humor will not be able to drain, and glaucoma will develop. As the quantity of fluid increases, the pressure within the eye increases, distorting soft tissues and interfering with vision. If untreated, the condition will ultimately cause blindness.

Page 569

1. Even with a congenital lack of cone cells in the eye, you would still be able to see as long as you had functioning rod cells. But because cone cells function in color vision, you would see only black and white.
2. A deficiency or lack of vitamin A in the diet would affect the quantity of retinal the body could produce, and thus would interfere with night vision.
3. A decrease in phosphodiesterase activity would lead to a higher level of intracellular cGMP. This in turn would keep the gated sodium channels open and decrease the ability of receptor neurons to respond to photons.

Page 585

1. Without the movement of the round window, the perilymph would not be moved by the vibration of the stapes at the oval window, and there would be little or no perception of sound.
2. Loss of stereocilia (as a result of constant exposure to loud noises, for instance) would reduce hearing sensitivity and could lead to deafness.
3. The auditory tube allows pressure to equalize on both sides of the tympanic membrane (eardrum). If this tube is blocked, there will be greater pressure on the inside of the tympanic membrane, forcing it outward and producing pain.

CHAPTER 18

Page 601

1. Neural responses occur within fractions of a second and do not last long (short duration). Conversely, endocrine responses may be slow to appear but will last for minutes to days (long duration).
2. Adenylate cyclase is the enzyme that converts ATP to cAMP. A molecule that blocks this enzyme would block action of any hormone that required cAMP for a second messenger.
3. A cell's hormonal sensitivity is determined by the presence or absence of the necessary receptor complex.

Page 609

1. Dehydration increases the osmotic pressure of the blood. The increase in blood osmotic pressure would stimulate the posterior pituitary to release more ADH.
2. Somatomedins are the mediators of *growth hormone* action. If the level of somatomedins is elevated, we would expect to see the level of growth hormone elevated as well.
3. Increased levels of cortisol would inhibit the cells that control ACTH release from the pituitary; therefore, the level of ACTH would decrease. This is an example of a negative feedback mechanism.

Page 618

1. An individual whose diet lacks iodine would not be able to form the hormone thyroxine. As a result, we would expect to see the symptoms associated with thyroxine deficiency, such as decreased rate of metabolism, decreased body temperature, poor response to physiological stress, and an increase in the size of the thyroid gland (goiter).
2. Most of the thyroid hormone in the blood is bound to proteins called thyroid-binding globulins. These compounds represent a large reservoir of thyroxine that guards against rapid fluctuations in the level of this important hormone. Because there is such a large amount stored in this way, it

takes several days to deplete the supply of hormone, even after the thyroid gland has been removed.
3. Removal of the parathyroid glands would result in a decrease in the blood levels of *calcium ions*. This decrease could be counteracted by increasing the amount of vitamin D and calcium in the diet.
4. One of the functions of cortisol is to decrease the cellular use of glucose while increasing the available glucose by promoting the breakdown of glycogen and the conversion of amino acids to carbohydrates. The net result is an elevation in the level of glucose in the blood.

Page 621

1. One of the functions of the hormone atrial natriuretic peptide is to increase the rate of Na^+ excretion in the kidneys—an increase that would cause the amount of Na^+ excreted in the urine to increase.

Page 628

1. An individual with Type I or Type II diabetes has such elevated levels of glucose in the blood that the kidney cannot reabsorb all the glucose; some is lost in the urine. The water lost with the glucose elevates blood osmotic pressure and promotes thirst.
2. Glucagon stimulates the conversion of glycogen to glucose in the liver. Increased amounts of glucagon would lead to decreased amounts of liver glycogen.
3. Increased amounts of light would inhibit the production and release of *melatonin* from the pineal gland, which receives neural input from the optic tracts. The secretion of melatonin by the pineal gland is influenced by light/dark cycles.

Page 632

1. The type of hormonal interaction exemplified by the insulin and glucagon effects is *antagonism*. In this type of hormonal interaction, two hormones have opposite effects on their target tissues.
2. The hormones GH, thyroid hormone, PTH, and the gonadal hormones all play a role in formation and development of the skeletal system.
3. During the resistance phase of GAS, there is a high demand for glucose, especially by the nervous system. The GH-RH and CRH increase the levels of GH and ACTH, respectively. Growth hormone mobilizes fat reserves and promotes the catabolism of protein. ACTH increases cortisol, which stimulates the conversion of glycogen to glucose as well as the catabolism of fat and protein.

CHAPTER 19

Page 645

1. Venipuncture is a popular sampling technique because (1) superficial veins are easy to locate; (2) the walls of veins are thinner than those of arteries; and (3) blood pressure in veins is relatively low, so the puncture wound seals quickly.
2. A decrease in the amount of plasma proteins in the blood may cause (1) a decrease in plasma osmotic pressure, (2) a decreased ability to fight infection, and (3) a decrease in the transport and binding of some ions, hormones, and other molecules.
3. During a viral infection, you would expect to see an increase in the level of immunoglobulins (antibodies) in the blood.

Page 653

1. The hematocrit measures the amount of formed elements (mostly red blood cells) as a percentage of the total blood. In hemorrhage, the loss of blood, especially of red blood cells, would lower the hematocrit.
2. A decreased blood flow to the kidneys would trigger the release of erythropoietin. The elevated erythropoietin would lead to an increase in erythropoiesis (red blood cell formation). Thus, Dave's hematocrit should increase.
3. The liver conjugates bilirubin prior to excretion in the bile. Diseases that damage the liver, such as hepatitis or cirrhosis, would impair the liver's ability to perform this function. As a result, unconjugated bilirubin would accumulate in the blood, producing jaundice.

Page 656

1. A person with Type O blood can accept only Type O blood.
2. If a person with Type A blood receives a transfusion of Type B blood, the red cells will clump or agglutinate, potentially blocking blood flow to various organs and tissues.
3. Surface antigens on RBCs are glycolipids in the cell membrane.

Page 662

1. In an infected cut, we would expect to find a large number of *neutrophils*. Neutrophils are phagocytic white cells that are generally the first to arrive at the site of an injury and that specialize in dealing with infectious bacteria.

2. The type of white blood cell that produces circulating antibodies is the *B lymphocyte*; these would be found in increased numbers.

3. During an inflammatory response, basophils release a variety of chemicals such as histamine and heparin that exaggerate the inflammation and attract other types of white blood cells.

Page 668

1. Megakaryocytes are the precursors of platelets, which play an important role in hemostasis and the clotting process. A decreased number of megakaryocytes would result in fewer platelets, which in turn would interfere with the ability to clot properly.

2. The use of broad-spectrum antibiotics would lower the number of intestinal bacteria and thus the amount of vitamin K produced. This decrease in vitamin K would lead to a decrease in the production of several clotting factors, most notably prothrombin. As a result, clotting time would increase.

3. Activation of Factor XII initiates the series of events known as the intrinsic pathway.

C H A P T E R   2 0

Page 687

1. The semilunar valves on the right side of the heart guard the opening to the *pulmonary artery*. Damage to these valves would interfere with the blood flow through this vessel.

2. When the ventricles begin to contract, they force the AV valves to close, tensing the chordae tendineae. The chordae tendineae are attached to the papillary muscles, which begin contracting just before the rest of the ventricular myocardium does.

3. The wall of the left ventricle is more muscular than that of the right ventricle because the left ventricle must generate enough force to propel blood throughout all body systems except the lungs. The right ventricle must generate only enough force to propel the blood a few centimeters to the lungs.

Page 692

1. If these cells were not functioning, the heart would still continue to beat but at a slower rate.

2. If the impulses from the atria were not delayed at the AV node, they would be conducted through the ventricles so quickly by the bundle branches and Purkinje cells that the ventricles would begin contracting immediately, before the atria had finished their contraction. As a result, the ventricles would not be as full of blood as they could be, and the pumping of the heart would not be as efficient, especially during activity.

3. The cardioinhibitory center of the medulla oblongata is part of the parasympathetic branch of the autonomic nervous system. Damage to this center would result in fewer parasympathetic action potentials to the heart and an increase in heart rate due to sympathetic dominance.

4. An increase in extracellular calcium ions would result in an increased force of cardiac contraction. More calcium ions would enter heart cells when they were stimulated, due to the increased concentration gradient. The increased amount of intracellular calcium would lead to increased binding of troponin, more myosin heads binding to actin, and a stronger and longer-lasting contraction.

Page 699

1. When pressure in the left ventricle begins rising, the heart is contracting, but no blood is leaving the heart. During this initial phase of contraction, both the AV valves and semilunar valves are closed. The increase in pressure is the result of increased tension as the muscle contracts. When the pressure in the ventricle exceeds the pressure in the aorta, the aortic semilunar valves are forced open, and the blood is rapidly ejected from the ventricle.

2. An increase in the size of the QRS complex would indicate a larger-than-normal amount of electrical activity during ventricular depolarization. One possible cause would be an increase in the size of the heart. Because there is more muscle depolarizing, the magnitude of the electrical event would be greater.

Page 701

1. If the heart beats too quickly, there is not sufficient time for it to fill completely between the beats. The heart pumps in proportion to the amount of blood that enters: The less blood that enters, the less the heart can pump. If it beats too fast, very little blood will enter the circulation, and tissues will suffer damage from lack of blood supply.

2. Stimulating the acetylcholine receptors of the heart would cause the heart to slow down. The cardiac output is the product of stroke volume and the heart rate; if the heart rate decreases, so will the cardiac output (assuming no change in the stroke volume).

3. The venous return fills the heart with blood, stretching the heart muscle. According to the Frank–Starling principle, the more the heart muscle is stretched, the more forcefully it will contract (to a point). The more forceful the contraction, the more blood the heart will eject with each beat (stroke volume). Therefore, increased venous return will increase the stroke volume if all other factors are constant.

4. Increased sympathetic stimulation of the heart would result in an increased heart rate and increased force of contraction. The ESV represents the amount of blood that remains in a ventricle after a contraction (systole). The more forcefully the heart contracts, the more blood it will eject, and the lower the ESV will be. Therefore, increased sympathetic stimulation should result in a lower ESV.

5. SV = EDV − ESV, so SV = 125 ml − 40 ml = 85 ml

Page 704

1. Caffeine acts directly on the conducting system and contractile cells of the heart, increasing the rate at which they depolarize. The net result would be an increased heart rate.

2. Bradycardia (low heart rate) would likely result in a decreased cardiac output, because CO = HR × SV. It is possible that the stroke volume would increase enough to balance the decreased heart rate. This is not likely, however, because factors that decrease the heart rate also tend to decrease the force of myocardial contraction and thus stroke volume as well.

3. A drug that increases the time required for pacemaker cells to repolarize would decrease the heart rate, because the pacemaker cells would generate fewer action potentials per minute.

C H A P T E R   2 1

Page 718

1. The blood vessels are *veins*. Arteries and arterioles have a large amount of smooth muscle tissue in a thick, well-developed tunica media.

2. Blood pressure in the arterial system pushes blood into the capillaries. Blood pressure on the venous side is very low, and other forces help keep the blood moving. Valves prevent the blood from flowing backward whenever the venous pressure drops.

3. You would expect to find fenestrated capillaries in organs and tissues where small peptides move freely into and out of the blood. These would include endocrine glands, the choroid plexus of the brain, absorptive areas of the intestine, and filtration areas of the kidneys.

Page 727

1. In a healthy individual, the pressure should be greatest in the aorta and least in the venae cavae. Blood, like other fluids, moves along a pressure gradient from high pressure to low pressure. If the pressure were higher in the inferior vena cava, the blood would flow backward.

2. While a person stands for periods of time, blood pools in the lower extremities. This pooling decreases the venous return to the heart. In turn, the cardiac output decreases, sending less blood to the brain, causing light-headedness and fainting. A hot day adds to the effect, because body water is lost and blood volume is reduced, through sweating.

3. The nurse would have first heard the Korotkoff's sounds when the pressure in the cuff reached 125 mm Hg. At this point, the pressure in the vessel during systole is just enough to overcome the pressure in the cuff. Turbulent flow produced in the constricted vessel then produces the audible sounds.

Page 739

1. During exercise (1) blood flow to muscles increases, (2) cardiac output increases, and (3) resistance in visceral tissues increases.

2. Pressure at this site would decrease blood pressure at the carotid sinus, where the carotid baroreceptors are located. This decrease would cause a decreased frequency of action potentials along the glossopharyngeal nerve (IX) to the medulla, and more sympathetic impulses would be sent to the heart. The net result would be an increase in the heart rate.

3. Vasoconstriction of the renal artery would decrease both blood flow and blood pressure at the kidney. In response, the kidney would increase the amount of renin that it releases, which in turn would lead to an increase in the level of angiotensin II. The angiotensin II would bring about increased blood pressure and increased blood volume.

4. In circulatory shock, there is a decreased venous return to the heart. The decreased cardiac output that results from the decrease in venous return accounts for the weak pulse. Because the cardiac output is decreased, the baroreceptors are stimulated. In turn, there is increased sympathetic stimulation to the heart, causing the rapid heart rate. Although the heart is beating faster, there is less blood to pump, and pulse pressure remains low.

Page 749

1. The *left subclavian artery* is the branch of the aorta that sends blood to the left shoulder and arm.
2. The common carotid arteries carry blood to the head. A compression of the common carotid arteries would result in decreased blood pressure at the carotid sinus, triggering a reflexive increase in heart rate and blood pressure.
3. Organs served by the celiac artery include the stomach, spleen, liver, and pancreas.

Page 760

1. The vein that is bulging is the *external jugular vein*.
2. Blockage of the popliteal vein would interfere with blood flow in the *tibial* and *peroneal* veins (which form the popliteal vein) and the *small saphenous vein* (which joins the popliteal vein).
3. This blood sample must have come from the umbilical vein, which carries oxygenated, nutrient-rich blood from the placenta to the fetus.

CHAPTER 22

Page 780

1. The thoracic duct drains lymph from the area beneath the diaphragm and from the left side of the head and thorax. Most of the lymph enters the venous blood by way of this duct. A blockage of this duct would not only impair circulation of lymph through most of the body, but it would also promote accumulation of fluid in the extremities (lymphedema).
2. The thymosins from the thymus play a role in the differentiation of lymphoid stem cells into *T lymphocytes*. A lack of these hormones would result in an absence of T lymphocytes.
3. During an infection, the lymphocytes and phagocytes in the lymph nodes in the affected region undergo cell division to deal with the infectious agent more effectively. This increase in the number of cells in the node causes the node to become enlarged or swollen.

Page 787

1. A decrease in the number of monocyte-forming cells in the bone marrow would result in a decreased number of *macrophages* in the body, because all the different macrophages are derived from the monocytes. These include the Kuppfer's cells of the liver, Langerhans cells in the skin and digestive tract, and alveolar macrophages.
2. A rise in interferon would indicate a viral infection. Interferon is released from cells that are infected with viruses. It does not help an infected cell but "interferes" with the virus's ability to infect other cells.
3. Pyrogens stimulate the temperature control area of the preoptic nucleus of the hypothalamus. The result is an increase in body temperature, or fever.

Page 797

1. Abnormal peptides in the cytoplasm of a cell can become attached to MHC (major histocompatibility complex) proteins and displayed on the surface of the cell's membrane. Peptides presented in this manner are then recognized by T cells, which can initiate an immune response.
2. Cytotoxic T cells function in *cell-mediated immunity*. A decrease in the number of cytotoxic T cells would interfere with the ability to kill foreign cells and tissues as well as cells infected by viruses.
3. Helper T cells promote B cell division, the maturation of plasma cells, and the production of antibodies by the plasma cells. Without the helper T cells, the antibody-mediated immune response would probably not occur.
4. Because plasma cells produce and secrete antibodies, we would expect to see increased levels of circulating antibodies in the blood if the number of plasma cells were increased.

Page 808

1. The secondary response would be affected by the lack of memory B cells for a specific antigen. The ability to produce a secondary response depends on the presence of memory B cells and T cells that are formed during the primary response to an antigen. These cells are not involved in the primary response but are held in reserve against future contact with the same antigen.

2. The developing fetus is protected primarily by *natural passive immunity*, the product of IgG antibodies that cross the placenta from the mother's circulation.
3. Stress can interfere with the immune response by depressing the inflammatory response, reducing the number and activity of phagocytes, and inhibiting interleukin secretion.

CHAPTER 23

Page 822

1. The rich blood supply to the nose delivers body heat that warms the air as it passes through the nasal cavity. The heat also evaporates moisture from the epithelium to humidify the incoming air. The moisture is derived from the blood supply as well. The blood supply also brings nutrients and water to the secretory cells of the nasal mucosa.
2. The nasopharynx receives only air from the nasal cavity. The oropharynx and laryngopharynx receive air from the nasal cavity and food from the oral cavity. Ingested solids and liquids can be damaging to delicate cells; thus, in the areas in contact with food, there is a highly protective stratified squamous epithelium, like that of the exterior skin. The lining of the nasopharynx is the same as the nasal cavity, a pseudostratified ciliated columnar epithelium.
3. Increased tension in the vocal cord will cause a higher pitch in your voice.

Page 831

1. The tracheal cartilages are C-shaped to allow room for esophageal expansion when large portions of food or liquid are swallowed.
2. Without surfactant, surface tension in the thin layer of water that moistens the alveolar surfaces would cause the alveoli to collapse.
3. Air that passes through the glottis flows into the larynx and through the trachea. From there, the air flows into a primary bronchus, which supplies the lungs. In the lungs, the air passes to bronchi, bronchioles, a terminal bronchiole, a respiratory bronchiole, an alveolar duct, an alveolar sac, an alveolus, and ultimately to the respiratory membrane.

Page 834

1. The pulmonary arteries supply the exchange surfaces and the *external carotid arteries*, the *thyrocervical trunks*, and the *bronchial arteries* supply the conducting portions of the lungs.
2. The pleura is a serous membrane; pleural surfaces secrete *pleural fluid*, which lubricates the opposing parietal and visceral surfaces to prevent friction during breathing.

Page 842

1. Because the rib penetrates Mark's chest wall, atmospheric air will enter his thoracic cavity. This condition is a *pneumothorax*. Pressure within the pleural cavity is normally lower than atmospheric pressure. When air enters the pleural cavity, the natural elasticity of the lung may cause it to collapse. The resulting condition is *atelectasis*, or a collapsed lung.
2. Because the fluid produced in pneumonia takes up space that would normally be occupied by air, the vital capacity would be decreased.

Page 850

1. As skeletal muscles become more active, they generate more heat and more acid waste products, which lower the pH of the surrounding fluid. The combination of lower pH and higher temperature causes the hemoglobin to release more oxygen than it would under conditions of lower temperature and higher pH.
2. An obstruction of the airways would interfere with the body's ability to gain oxygen and eliminate carbon dioxide. Because most carbon dioxide is carried in the blood as bicarbonate ion that is formed from the dissociation of carbonic acid, an inability to eliminate carbon dioxide would result in an excess of hydrogen ions, thus lowering the body's pH.

Page 856

1. The pneumotaxic center inhibits the inspiratory center and the apneustic center. Exciting the pneumotaxic center in the brain stem would result in shorter breaths and a more rapid rate of breathing.
2. Chemoreceptors are more sensitive to carbon dioxide levels than to oxygen levels. When carbon dioxide dissolves, it produces hydrogen ions, thereby lowering pH and altering cell or tissue activity.
3. Johnny's mother shouldn't worry. When Johnny holds his breath, the level of carbon dioxide in his blood will increase. This increase will in turn lead to increased stimulation of the inspiratory center, forcing Johnny to breathe again.

Page 867

1. The mesenteries are double layers of serous membrane that support and stabilize the positions of organs in the abdominopelvic cavity and provide a route for the associated blood vessels, nerves, and lymphatics.
2. Peristalsis would be more efficient in propelling intestinal contents. Segmentation is essentially a churning action that mixes intestinal contents with digestive fluids. Peristalsis consists of waves of contractions that not only mix the contents but also propel them along the digestive tract.
3. Parasympathetic stimulation increases muscle tone and activity in the digestive tract. A drug that blocks this activity would decrease the rate of peristalsis.

Page 874

1. The oral cavity is lined by a stratified squamous epithelium. This type of lining is located in areas of the body that receive a great deal of friction or abrasion, and it is very protective.
2. Because the parotid salivary glands secrete salivary amylase, an enzyme that digests complex carbohydrates, damage to these glands would interfere with the digestion of carbohydrates.
3. The *incisors* are the type of tooth best suited for chopping, cutting, or shearing pieces of relatively rigid food, such as raw vegetables.

Page 877

1. The muscularis externa of the esophagus is unusual because (1) it contains skeletal muscle cells along most of the length of the esophagus and (2) it is surrounded by an adventitia rather than a serosa.
2. The fauces is the opening between the oral cavity and the pharynx.
3. The process that is occurring is *swallowing* (deglutition).

Page 884

1. The larger a meal (especially in terms of protein), the more stomach acid is secreted. The hydrogen ions for the acid come from the blood that enters the stomach; therefore, the blood leaving the stomach will have fewer than normal hydrogen ions and will be decidedly alkaline—that is, have a higher pH. This is referred to as the *alkaline tide*.
2. The vagus nerve contains parasympathetic motor fibers that can stimulate gastric secretions. This stimulation can occur even if food is not present in the stomach (cephalic phase of gastric digestion). Cutting the branches of the vagus that supply the stomach would prevent this type of secretion from occurring and decrease the chance of ulcer formation.

Page 898

1. The small intestine has several adaptations that increase surface area to increase its absorptive capacity. The walls of the small intestine are thrown into folds called the *plicae circulares*. The tissue that covers the plicae forms fingerlike projections, the villi. The cells that cover the villi have an exposed surface covered by small fingerlike projections called the microvilli. In addition, the small intestine has a very rich blood and lymphatic supply to transport the nutrients that are absorbed.
2. The CCK level in the blood would increase.
3. The hormone secretin, among other things, stimulates the pancreas to release fluid high in buffers to neutralize the chyme that enters the duodenum from the stomach. If the intestine did not secrete secretin, we would expect the pH of the intestinal contents to be lower than normal.
4. Damage to the exocrine pancreas would most affect the digestion of fats (lipids). Enzymes for carbohydrate digestion are produced by salivary glands and the small intestine as well as by the pancreas. Enzymes for protein digestion are produced by the stomach and the small intestine as well as by the pancreas. Even though digestion of carbohydrates and proteins would not be as complete as it is when the pancreas is functioning, some digestion would still take place. Because the pancreas is the primary source of lipases, lipid digestion would be most impaired.

Page 912

1. Chylomicrons are formed from the fats digested in a meal. A meal that is high in fat would increase the number of chylomicrons in the lacteals.
2. Removal of the stomach would interfere with the absorption of vitamin B_{12}. This vitamin requires intrinsic factor, which is produced by the parietal cells in the stomach.
3. Diarrhea is potentially life-threatening because a person could lose fluid and electrolytes faster than these substances can be replaced. This loss would result in dehydration and possibly death. Although it can be quite uncomfortable, constipation is not potentially life-threatening, because it does not interfere with any major body process that supports life. The few toxic waste products that are normally eliminated by way of the digestive system can move into the blood and be eliminated by the kidneys.

Page 928

1. The primary role of the TCA cycle in ATP production is to transfer electrons from substrates to coenzymes. These electrons carry energy that can then be used as an energy source for the production of ATP by the electron transport system.
2. The NADH produced by glycolysis cannot enter the mitochondria, where the enzymes of the electron transport chain are located. An intermediary in the mitochondrial membrane can, however, transfer the electrons from the NADH to a coenzyme within the mitochondria. In skeletal muscle cells, the intermediary transfers the electrons to FAD, whereas in cardiac muscle cells, a different intermediary is used that transfers the electrons to another NAD. In mitochondria, each NADH yields three molecules of ATP, whereas each $FADH_2$ yields just two molecules of ATP. The different intermediaries account for the difference in ATP yield.
3. A decrease in cytoplasmic NAD would lead to a decrease in the amount of ATP produced by the mitochondria. The mitochondria depend on a supply of pyruvic acid from glycolysis. Glycolysis in turn requires NAD. A decrease in NAD would decrease the available pyruvic acid for the TCA cycle and thus would decrease overall ATP production.

Page 937

1. Vitamin B_6 (pyridoxine) is an important coenzyme in the processes of deamination and transamination, the first steps in processing amino acids in the cell. A deficiency in this vitamin would interfere with the ability to metabolize proteins.
2. Uric acid is the product of purine degradation in the body. The macromolecules that contain purines are the nucleic acids. An increase in uric acid levels could indicate increased breakdown of nucleic acids.
3. HDLs are beneficial because they reduce the amount of fat (including cholesterol) in the bloodstream by transporting it back to the liver for storage or excretion in the bile.

Page 943

1. After a meal that is high in carbohydrates, you would expect increased *glycogenesis* (the formation of glycogen) to occur in the liver.
2. Urea is formed from byproducts of protein metabolism. During the postabsorptive state, many amino acids are being metabolized, and the ammonia produced by deamination is converted to urea in the liver. Thus, the amount of urea in the blood increases.
3. Excess acetyl-CoA is generally converted into compounds collectively known as *ketone bodies*.

Page 948

1. We would expect a person adding muscle mass to be in a positive nitrogen balance.
2. Bile salts are necessary for the digestion and absorption of fats and fat-soluble vitamins. Vitamin A is a fat-soluble vitamin. A decrease in the amount of bile salts in the bile would result in a decreased ability to absorb vitamin A from food and in a vitamin A deficiency.

Page 955

1. The BMR of a pregnant woman should be higher than her BMR when she is not pregnant due to the increased metabolism associated with support of the fetus as well as the added effect of fetal metabolism.
2. Vasoconstriction of peripheral vessels would decrease blood flow to the skin and decrease the amount of heat that the body can lose. As a result, the body temperature would increase.
3. Infants have higher surface-to-volume ratios than adults, and the temperature-regulating mechanisms of the body are not fully functional at birth. As a result, infants must expend more energy to maintain body temperature, and they get cold more easily than a healthy adult.

Page 972

1. The *renal corpuscle*, *proximal convoluted tubule*, *distal convoluted tubule*, and the proximal portions of the *loop of Henle* and *collecting duct* are all located within the renal cortex.
2. The pores of the glomerular capillaries will not allow substances the size of plasma proteins to pass into the capsular space, and the filtration slits of the podocyte will allow only the smallest plasma proteins to pass.
3. Damage to the juxtaglomerular apparatus of the nephrons would interfere with the hormonal control of blood pressure.

Page 981

1. The absorption of Ca^{2+} by the kidney is a countertransport mechanism in which sodium ions are traded for calcium. As more calcium is reabsorbed, more sodium enters the filtrate, increasing the sodium ion concentration of the filtrate.
2. When the plasma concentration of a substance exceeds its tubular maximum, the excess is not reabsorbed but is excreted in the urine.
3. Decreases in blood pressure would reduce the blood hydrostatic pressure within the glomerulus and decrease the GFR.

Page 990

1. Aldosterone promotes Na^+ retention and K^+ secretion at the kidneys. In response to increases in aldosterone levels, the K^+ concentration of the urine would increase.
2. The secretion of H^+ by the nephron involves a countertransport mechanism with Na^+. If the concentration of Na^+ in the filtrate decreased, fewer hydrogen ions could be secreted, and the result would be a tubular fluid with a higher pH.
3. If these nephrons lacked a loop of Henle, the kidneys would not be able to form a concentrated urine.
4. When the number of sodium ions in the filtrate passing through the DCT is low, the cells of the macula densa are stimulated to release renin. Renin activates angiotensin, and this activation brings about an increase in blood pressure.

Page 998

1. Urea, a nitrogenous waste, is formed during metabolism of amino acids that come from proteins. Thus, we would expect a high-protein diet to lead to an increased amount of urea. Fluid volume might also increase as a result of the need to flush the excess urea.
2. An obstruction of the ureters would interfere with the passage of urine between the *renal pelvis* to the *urinary bladder*.
3. In order to control the micturition reflex, you must be able to control the *external urinary sphincter*, a ring of skeletal muscle formed by the urogenital diaphragm, that acts as a valve.

C H A P T E R   2 7

Page 1018

1. Drinking a pitcher of distilled water would temporarily lower your blood osmolarity (osmotic pressure). Because ADH release is triggered by increases in osmolarity, a decrease in osmolarity would lead to a decrease in the level of ADH in your blood.
2. Consuming a meal high in salt would temporarily increase the osmolarity of your ECF. As a result, some of the water in the ICF would shift to the ECF.
3. Fluid loss through perspiration, urine formation, and respiration would increase the osmolarity of your body fluids.

Page 1032

1. A decrease in the pH of body fluids would have a stimulating effect on the respiratory center of the medulla. The result would be an increase in the rate of breathing. This would lead to an elimination of more carbon dioxide, which would tend to cause the pH to increase.
2. The kidney tubules modify the pH of the filtrate by secreting H^+ or reabsorbing HCO_3^-. The pH of the tubular fluid must be kept above about 4.5, because H^+ secretion cannot continue against a large concentration gradient. The buffers allow the filtrate to take more H^+ without decreasing the pH below the critical level.
3. In a prolonged fast, fatty acids are mobilized and large numbers of ketone bodies are formed. These molecules are acids that lower the body's pH. The lowered pH would eventually lead to ketoacidosis.
4. In vomiting, large amounts of stomach acid are lost from the body. This acid is formed by the parietal cells of the stomach by taking H^+ from the blood. Excessive vomiting would lead to excessive removal of H^+ from the blood to produce the acid, thus raising the body's pH and leading to metabolic alkalosis.

C H A P T E R   2 8

Page 1055

1. The cremaster muscle (as well as the dartos) would be relaxed on a warm day so that the scrotal sac could descend away from the warmth of the body and cool the testes.

2. Dilation of the arteries that serve the penis allows blood flow to increase and the vascular chambers to become engorged with blood, resulting in erection.
3. FSH is required for production of ABP, a protein that binds testosterone and keeps a high level of that hormone available to support spermatogenesis. Low levels of FSH would lead to low levels of testosterone in the seminiferous tubules and thus a lower rate of sperm production and a low sperm count.

Page 1069

1. A blockage of the uterine tube would cause sterility.
2. The acidic pH of the vagina helps prevent bacterial, fungal, and parasitic infections in this area.
3. The functional layer of the endometrium is sloughed off during menstruation.
4. Blockage of a single lactiferous sinus would not interfere with milk that flows to the nipple, because each breast generally has between 15 and 20 lactiferous sinuses.

Page 1072

1. If the LH surge did not occur during an ovarian cycle, ovulation and corpus luteum formation would not occur.
2. Progesterone is responsible for the functional maturation and secretion of the endometrium. Blockage of progesterone receptors would inhibit endometrial development and make the uterus unprepared for pregnancy.
3. A sudden decline in the levels of estrogen and progesterone at the end of the uterine cycle signals the beginning of the *menses*.

Page 1080

1. The inability to contract the ischiocavernosus and bulbospongiosus muscles would interefere with a male's ability to ejaculate and to experience orgasm.
2. As the result of parasympathetic stimulation in females during sexual arousal, there is engorgement of the erectile tissues of the clitoris, increased secretion of cervical and vaginal glands, increased blood flow to the walls of the vagina, and engorgement of the blood vessels in the nipples.
3. At menopause, circulating estrogen levels begin to drop. Estrogen has an inhibitory effect on FSH (and on GnRH). As the level of estrogen declines, the levels of FSH rise and remain elevated.

C H A P T E R   2 9

Page 1097

1. The inner cell mass of the blastocyst eventually develops into the embryo.
2. The cardiovascular system is formed from the mesoderm. Mesodermal cells migrate to form the *yolk sac*, in which blood vessels appear. The yolk sac becomes an important site of blood cell formation. Improper development of this extraembryonic membrane would thus affect the development and function of the circulatory system.
3. After fertilization, the developing trophoblasts and, later, the placenta produce and release the hormone hCG. Sue is pregnant.
4. Placental functions include (1) supplying the developing fetus with a route for gas exchange, nutrient transfer, and waste product elimination, and (2) producing hormones that affect maternal systems.

Page 1112

1. A mother's blood volume increases during pregnancy because blood flow through the placenta reduces the volume of blood in her systemic circuit, and this reduction stimulates an increase in maternal blood volume to compensate.
2. Progesterone reduces uterine contractions. A decrease in progesterone at any time during the pregnancy could lead to uterine contractions and, in late pregnancy, labor.
3. An increase in blood levels of GnRH, FSH, LH, and sex hormones would signal the onset of *puberty*.

Page 1119

1. A person who is heterozygous for curly hair would have one dominant allele and one recessive allele for that trait. The person's phenotype would be "curly hair."
2. One reason that children are not identical copies of their parents is that during meiosis, parental chromosomes randomly assort such that each gamete has a unique set of chromosomes. Also, mutations and the crossing-over that occurs during meiosis introduce new variations.

Accurate descriptions of physical objects would be impossible without a precise method of reporting the pertinent data. Dimensions such as length and width are reported in standardized units of measurement, such as inches or centimeters. These values can be used to calculate the **volume** of an object, a measurement of the amount of space it fills. **Mass** is another important physical property. The mass of an object is determined by the amount of matter it contains; on Earth the mass of an object determines its weight.

In the United States, length and width are typically described in terms of inches, feet, or yards; volumes in pints, quarts, or gallons; and weights in ounces, pounds, or tons. These are units of the **U.S. system** of measurement. Table 1 summarizes the familiar and unfamiliar terms used in the U.S. system. For reference purposes, this table also includes a definition of the "household units," popular in recipes and cookbooks. The U.S. system can be

very difficult to work with, because there is no logical relationship among the various units. For example, there are 12 inches in a foot, 3 feet in a yard, and 1760 yards in a mile. Without a clear pattern of organization, the conversion of feet to inches or miles to feet can be confusing and time-consuming. The relationships among ounces, pints, quarts, and gallons are no more logical than those among ounces, pounds, and tons.

In contrast, the **metric system** has a logical organization based on powers of 10, as indicated in Table 2. For example, a **meter (m)** is the basic unit for the measurement of size. For measurements of larger objects, data can be reported in units of **dekameters** (*deka*, ten), **hectometers** (*hekaton*, hundred), or **kilometers** (**km**; *chilioi*, thousand); for smaller objects, data can be reported in **decimeters** (0.1 m; *decem*, ten), **centimeters** (**cm** = 0.01 m; *centum*, hundred), **millimeters** (**mm** = 0.001 m; *mille*, thousand), and so forth. In the metric system, the same prefixes are

---

TABLE 1   The U.S. System of Measurement

Physical Property	Unit	Relationship to Other U.S. Units	Relationship to Household Units
Length	inch (in.)	1 in. = 0.083 ft	
	foot (ft)	1 ft = 12 in.	
		= 0.33 yd	
	yard (yd)	1 yd = 36 in.	
		= 3 ft	
	mile (mi)	1 mi = 5280 ft	
		= 1760 yd	
Volume	fluidram (fl dr)	1 fl dr = 0.125 fl oz	
	fluid ounce (fl oz)	1 fl oz = 8 fl dr	= 6 teaspoons (tsp)
		= 0.0625 pt	= 2 tablespoons (tbsp)
	pint (pt)	1 pt = 128 fl dr	= 32 tbsp
		= 16 fl oz	= 2 cups (c)
		= 0.5 qt	
	quart (qt)	1 qt = 256 fl dr	= 4 c
		= 32 fl oz	
		= 2 pt	
		= 0.25 gal	
	gallon (gal)	1 gal = 128 fl oz	
		= 8 pt	
		= 4 qt	
Mass	grain (gr)	1 gr = 0.002 oz	
	dram (dr)	1 dr = 27.3 gr	
		= 0.063 oz	
	ounce (oz)	1 oz = 437.5 gr	
		= 16 dr	
	pound (lb)	1 lb = 7000 gr	
		= 256 dr	
		= 16 oz	
	ton (t)	1 t = 2000 lb	

TABLE 2 The Metric System of Measurement

Physical Property	Unit	Relationship to Standard Metric Units	Conversion to U.S. Units	
Length	nanometer (nm)	1 nm = 0.000000001 m (10^{-9})	= 3.94×10^{-8} in.	25,400,000 nm = 1 in.
	micrometer (μm)	1 μm = 0.000001 m (10^{-6})	= 3.94×10^{-5} in.	25,400 μm = 1 in.
	millimeter (mm)	1 mm = 0.001 m (10^{-3})	= 0.0394 in.	25.4 mm = 1 in.
	centimeter (cm)	1 cm = 0.01 m (10^{-2})	= 0.394 in.	2.54 cm = 1 in.
	decimeter (dm)	1 dm = 0.1 m (10^{-1})	= 3.94 in.	0.254 dm = 1 in.
	meter (m)	standard unit of length	= 39.4 in.	0.0254 m = 1 in.
			= 3.28 ft	0.3048 m = 1 ft
			= 1.093 yd	0.914 m = 1 yd
	kilometer (km)	1 km = 1000 m	= 3280 ft	
			= 1093 yd	
			= 0.62 mi	1.609 km = 1 mi
Volume	microliter (μl)	1 μl = 0.000001 l (10^{-6}) = 1 cubic millimeter (mm^3)		
	milliliter (ml)	1 ml = 0.001 l (10^{-3}) = 1 cubic centimeter (cm^3 or cc)	= 0.0338 fl oz	5 ml = 1 tsp 15 ml = 1 tbsp 30 ml = 1 fl oz
	centiliter (cl)	1 cl = 0.01 l (10^{-2})	= 0.338 fl oz	2.95 cl = 1 fl oz
	deciliter (dl)	1 dl = 0.1 l (10^{-1})	= 3.38 fl oz	0.295 dl = 1 fl oz
	liter (l)	standard unit of volume	= 33.8 fl oz	0.0295 l = 1 fl oz
			= 2.11 pt	0.473 l = 1 pt
			= 1.06 qt	0.946 l = 1 qt
Mass	picogram (pg)	1 pg = 0.000000000001 g (10^{-12})		
	nanogram (ng)	1 ng = 0.000000001 g (10^{-9})	= 0.000000015 gr	66,666,666 mg = 1 gr
	microgram (μg)	1 μg = 0.000001 g (10^{-6})	= 0.000015 gr	66,666 mg = 1 gr
	milligram (mg)	1 mg = 0.001 g (10^{-3})	= 0.015 gr	66.7 mg = 1 gr
	centigram (cg)	1 cg = 0.01 g (10^{-2})	= 0.15 gr	6.67 cg = 1 gr
	decigram (dg)	1 dg = 0.1 g (10^{-1})	= 1.5 gr	0.667 dg = 1 gr
	gram (g)	standard unit of mass	= 0.035 oz	28.4 g = 1 oz
			= 0.0022 lb	454 g = 1 lb
	dekagram (dag)	1 dag = 10 g		
	hectogram (hg)	1 hg = 100 g		
	kilogram (kg)	1 kg = 1000 g	= 2.2 lb	0.454 kg = 1 lb
	metric ton (kt)	1 mt = 1000 kg	= 1.1 t	
			= 2205 lb	0.907 kt = 1 t

Temperature	Centigrade	Fahrenheit
Freezing point of pure water	0°	32°
Normal body temperature	36.8°	98.6°
Boiling point of pure water	100°	212°
Conversion	°C → °F: °F = (1.8 × °C) + 32	°F → °C: °C = (°F − 32) × 0.56

used to report weights, based on the **gram (g)**, and volumes, based on the **liter (l)**. This text reports data in metric units, in most cases with U.S. system equivalents. You should use this opportunity to become familiar with the metric system, because most technical sources report data only in metric units and most of the world outside the United States uses the metric system exclusively. Conversion factors are included in Table 2.

The U.S. and metric systems also differ in their methods of reporting temperatures. In the United States, temperatures are usually reported in degrees Fahrenheit (°F), whereas scientific literature and individuals in most other countries report temperatures in degrees centigrade or Celsius (°C). The relationship between temperatures in degrees Fahrenheit and those in degrees centigrade has been indicated in Table 2.

The following illustration spans the entire range of measurements that we will consider in this book. Gross anat-omy traditionally deals with structural organization as seen with the naked eye or with a simple hand lens. A microscope can provide higher levels of magnification and reveal finer details. Before the 1950s, most information was provided by *light microscopy*. A photograph taken through a **light microscope** is called a light micrograph **(LM)**. Light microscopy can magnify cellular structures up to about 1000 times and show details as fine as 0.25 µm. The symbol **µm** stands for **micrometer**; 1 µm = 0.001 mm, or 0.00004 inches. With a light microscope, we can identify cell types, such as muscle cells or neurons, and see large structures within a cell. Because individual cells are relatively transparent, thin sections taken

through a cell are treated with dyes that stain specific structures to make them easier to see.

Although special staining techniques can make the general distribution of proteins, lipids, carbohydrates, and nucleic acids in the cell visible, many fine details of intracellular structure remained a mystery until investigators began using *electron microscopy*. This technique uses a focused beam of electrons, rather than a beam of light, to examine cell structure. In *transmission electron microscopy*, electrons pass through an ultrathin section to strike a photographic plate. The result is a **transmission electron micrograph (TEM)**. Transmission electron microscopy shows the fine structure of cell membranes and intracellular structures. In *scanning electron microscopy*, electrons bouncing off exposed surfaces create a **scanning electron micrograph (SEM)**. Although it cannot achieve as much magnification as transmission microscopy, scanning microscopy provides a three-dimensional perspective of cell structure.

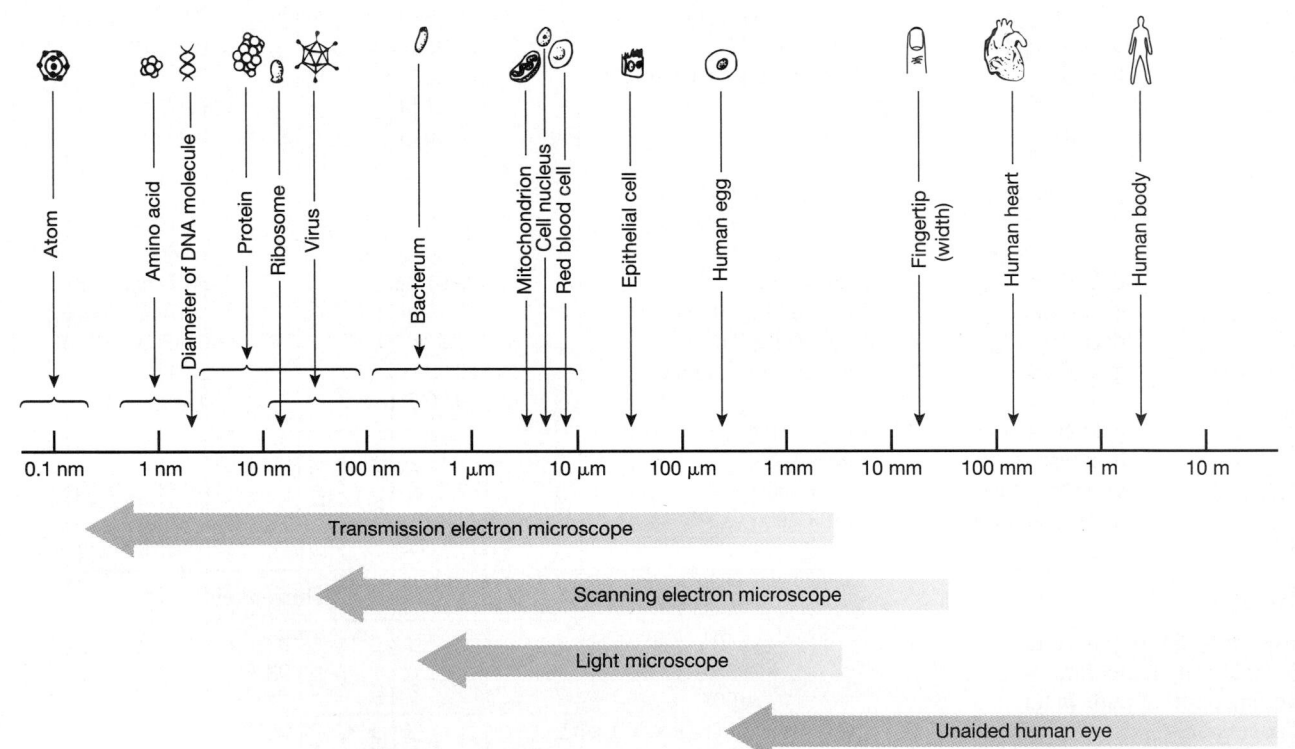

The **periodic table** presents the known elements in order of their atomic weights. Each horizontal row represents a single electron shell. The number of electrons in that row is determined by the maximum number of electrons that can be stored at that energy level. The element at the left end of each row contains a single electron in its outermost electron shell; the element at the right end of the row has a filled outer electron shell. Organizing the elements in this fashion highlights similarities that reflect the composition of the outer electron shell. These similarities are evident when you examine the vertical columns. All the gases of the rightmost column—helium, neon, argon, krypton, xenon, and radon—have full electron shells; each is a gas at normal atmospheric temperature and pressure, and none reacts readily with other elements. These elements, highlighted in blue, are known as the *noble*, or *inert*, *gases*. In contrast, the elements of the left-most column—lithium, sodium, potassium, and so forth—are silvery, soft metals that are so highly reactive that pure forms cannot be found in nature. The fourth and fifth electron levels can hold 18 electrons. Table inserts are used for the so-called *lanthanide* and *actinide series* to save space, as higher levels may store up to 32 electrons. Elements of particular importance to our discussion of human anatomy and physiology are highlighted in pink.

Legend:
- Atomic number — 1
- Chemical symbol — H
- Element name — Hydrogen
- Atomic weight — 1.01

Periodic Table

Number	Symbol	Name	Atomic weight
1	H	Hydrogen	1.01
2	He	Helium	4.00
3	Li	Lithium	6.94
4	Be	Beryllium	9.01
5	B	Boron	10.81
6	C	Carbon	12.01
7	N	Nitrogen	14.01
8	O	Oxygen	16.00
9	F	Fluorine	19.00
10	Ne	Neon	20.18
11	Na	Sodium	22.99
12	Mg	Magnesium	24.31
13	Al	Aluminum	26.98
14	Si	Silicon	28.09
15	P	Phosphorus	30.97
16	S	Sulfur	32.07
17	Cl	Chlorine	35.45
18	Ar	Argon	39.95
19	K	Potassium	39.10
20	Ca	Calcium	40.08
21	Sc	Scandium	44.96
22	Ti	Titanium	47.88
23	V	Vanadium	50.94
24	Cr	Chromium	52.00
25	Mn	Manganese	54.94
26	Fe	Iron	55.85
27	Co	Cobalt	58.93
28	Ni	Nickel	58.69
29	Cu	Copper	63.55
30	Zn	Zinc	65.39
31	Ga	Gallium	69.72
32	Ge	Germanium	72.61
33	As	Arsenic	74.92
34	Se	Selenium	78.96
35	Br	Bromine	79.90
36	Kr	Krypton	83.80
37	Rb	Rubidium	85.47
38	Sr	Strontium	87.62
39	Y	Yttrium	88.91
40	Zr	Zirconium	91.22
41	Nb	Niobium	92.91
42	Mo	Molybdenum	95.94
43	Tc	Technetium	(98)
44	Ru	Ruthenium	101.07
45	Rh	Rhodium	102.91
46	Pd	Palladium	106.42
47	Ag	Silver	107.87
48	Cd	Cadmium	112.41
49	In	Indium	114.82
50	Sn	Tin	118.71
51	Sb	Antimony	121.76
52	Te	Tellurium	127.60
53	I	Iodine	126.90
54	Xe	Xenon	131.29
55	Cs	Cesium	132.91
56	Ba	Barium	137.33
57	La	Lanthanum	138.91
72	Hf	Hafnium	178.49
73	Ta	Tantalum	180.95
74	W	Tungsten	183.85
75	Re	Rhenium	186.21
76	Os	Osmium	190.2
77	Ir	Iridium	192.22
78	Pt	Platinum	195.08
79	Au	Gold	196.97
80	Hg	Mercury	200.59
81	Tl	Thallium	204.38
82	Pb	Lead	207.2
83	Bi	Bismuth	208.98
84	Po	Polonium	(209)
85	At	Astatine	(210)
86	Rn	Radon	(222)
55	Cs	Cesium	132.91
87	Fr	Francium	(223)
88	Ra	Radium	226.03
89	Ac	Actinium	227.03
104	Rf	Rutherfordium	(261)
105	Db	Dubnium	(262)
106	Jl	Joliotium	(263)
107	Bh	Bohrium	(262)
108	Hn	Hahnium	(265)
109	Mt	Meitnerium	(266)
110	—	Unnamed	(269)
111	—	Unnamed	(272)
112	—	Unnamed	(272)

*Lanthanide series

Number	Symbol	Name	Atomic weight
58	Ce	Cerium	140.12
59	Pr	Praseodymium	140.91
60	Nd	Neodymium	144.24
61	Pm	Promethium	(145)
62	Sm	Samarium	150.36
63	Eu	Europium	151.96
64	Gd	Gadolinium	157.25
65	Tb	Terbium	153.93
66	Dy	Dysprosium	162.50
67	Ho	Holmium	164.93
68	Er	Erbium	167.26
69	Tm	Thulium	168.93
70	Yb	Ytterbium	173.04
71	Lu	Lutetium	174.97

†Actinide series

Number	Symbol	Name	Atomic weight
90	Th	Thorium	232.04
91	Pa	Protactinium	231.04
92	U	Uranium	238.03
93	Np	Neptunium	237.05
94	Pu	Plutonium	(244)
95	Am	Americium	(243)
96	Cm	Curium	(247)
97	Bk	Berkelium	(247)
98	Cf	Californium	(251)
99	Es	Einsteinium	(252)
100	Fm	Fermium	(257)
101	Md	Mendelevium	(258)
102	No	Nobelium	(259)
103	Lr	Lawrencium	(260)

Amino acids with acidic or basic side chain functional groups

Aspartic acid (asp)   Glutamic acid (glu)   Tyrosine (tyr)

Lysine (lys)   Arginine (arg)   Histidine (his)

Amino acids with unchanged but polar side chains

Serine (ser)   Threonine (thr)   Methionine (met)   Cysteine (cys)

Amino acids with hydrocarbon side chains

Glycine (gly)   Alanine (ala)   Valine (val)   Tryptophan (trp)   Asparagine (asn)

Glutamine (gln)   Leucine (leu)   Isoleucine (ile)   Phenylalanine (phe)   Proline (pro)

Eponym	Equivalent Term(s)	Individual Referenced
The Cellular Level of Organization (Chapter 3)		
Golgi apparatus		Camillo Golgi (1844–1926), Italian histologist; shared Nobel Prize in 1906
Krebs cycle	Tricarboxylic, or citric acid, cycle	Hans Adolph Krebs (1900–1981), British biochemist; shared Nobel Prize in 1953
The Skeletal System (Chapters 6–9)		
Colles' fracture		Abraham Colles (1773–1843), Irish surgeon
Haversian canals	Central canals	Clopton Havers (1650–1702), English anatomist and microscopist
Haversian systems	Osteons	Clopton Havers
Pott's fracture		Percivall Pott (1713–1788), English surgeon
Volkmann's canals	Perforating canals	Alfred Wilhelm Volkmann (1800–1877), German surgeon
Wormian bones	Sutural bones	Olas Worm (1588–1654), Danish anatomist
The Muscular System (Chapters 10, 11)		
Achilles' tendon	Calcaneal tendon	Achilles, hero of Greek mythology
Cori cycle		Carl Ferdinand Cori (1896–1984) and Gerty Theresa Cori (1896–1957), American biochemists; shared Nobel Prize in 1947
The Nervous System (Chapters 12–16)		
Broca's center	Speech center	Pierre Paul Broca (1824–1880), French surgeon
Foramen of Luschka	Lateral foramina	Hubert von Luschka (1820–1875), German anatomist
Foramen of Magendie	Median foramen	François Magendie (1783–1855), French physiologist
Foramen of Munro	Interventricular foramen	John Cummings Munro (1858–1910), American surgeon
Nissl bodies		Franz Nissl (1860–1919), German neurologist
Purkinje cells		Johannes E. Purkinje (1781–1869), Czechoslovakian physiologist
Nodes of Ranvier		Louis Antoine Ranvier (1835–1922), French physiologist
Island of Reil	Insula	Johann Christian Reil (1759–1813), German anatomist
Fissure of Rolando	Central sulcus	Luigi Rolando (1773–1831), Italian anatomist
Schwann cells		Theodor Schwann (1810–1882), German anatomist
Aqueduct of Sylvius	Mesencephalic aqueduct	Jacobus Sylvius (Jacques Dubois, 1478–1555), French anatomist
Sylvian fissure	Lateral sulcus	Franciscus Sylvius (Franz de le Boë, 1614–1672), Dutch anatomist
Pons varolii	Pons	Costanzo Varolio (1543–1575), Italian anatomist
Sensory Function (Chapter 17)		
Organ of Corti		Alfonso Corti (1822–1888), Italian anatomist
Eustachian tube	Auditory tube	Bartolomeo Eustachio (1520–1574), Italian anatomist
Golgi tendon organs	Tendon organs	Camillo Golgi (*see* Golgi apparatus *under* The Cellular Level of Organization)
Hertz (Hz)		Heinrich Hertz (1857–1894), German physicist
Meibomian glands		Heinrich Meibom (1638–1700), German anatomist
Meissner's corpuscles		Georg Meissner (1829–1905), German physiologist
Merkel's discs		Friedrich Siegismund Merkel (1845–1919), German anatomist
Pacinian corpuscles		Fillippo Pacini (1812–1883), Italian anatomist
Ruffini's corpuscles		Angelo Ruffini (1864–1929), Italian anatomist
Canal of Schlemm		Friedrich S. Schlemm (1795–1858), German anatomist
Glands of Zeis		Edward Zeis (1807–1868), German ophthamologist
The Endocrine System (Chapter 18)		
Islets of Langerhans	Pancreatic islets	Paul Langerhans (1847–1888), German pathologist
Interstitial cells of Leydig	Interstitial cells	Franz von Leydig (1821–1908), German anatomist

Eponym	Equivalent Term(s)	Individual Referenced
The Cardiovascular System (Chapters 19–21)		
Bundle of His		Wilhelm His (1863–1934), German physician
Purkinje cells		Johannes E. Purkinje (*see under* The Nervous System)
Starling's law		Ernest Henry Starling (1866–1927), English physiologist
Circle of Willis	Cerebral arterial circle	Thomas Willis (1621–1675), English physician
The Lymphatic System (Chapter 22)		
Hassall's corpuscles		Arthur Hill Hassall (1817–1894), English physician
Kupffer cells		Karl Wilhelm Kupffer (1829–1902), German anatomist
Langerhans cells		Paul Langerhans (*see* Islets of Langerhans *under* The Endocrine System)
Peyer's patches	Aggregate lymphoid nodules	Johann Conrad Peyer (1653–1712), Swiss anatomist
The Respiratory System (Chapter 23)		
Adam's apple	Thyroid cartilage	Biblical reference (laryngeal prominence)
Bohr effect		Niels Bohr (1885–1962), Danish physicist; won Nobel Prize in 1922
Boyle's law		Robert Boyle (1621–1691), English physicist
Charles' law		Jacques Alexandre César Charles (1746–1823), French physicist
Dalton's law		John Dalton (1766–1844), English physicist
Henry's law		William Henry (1775–1837), English chemist
The Digestive System (Chapter 24)		
Plexus of Auerbach	Myenteric plexus	Leopold Auerbach (1827–1897), German anatomist
Brunner's glands	Duodenal glands	Johann Conrad Brunner (1653–1727), Swiss anatomist
Kupffer cells	Stellate cells	Karl Wilhelm Kupffer (*see under* The Lymphatic System)
Crypts of Lieberkuhn	Intestinal crypts	Johann Nathaniel Lieberkuhn (1711–1756), German anatomist
Plexus of Meissner	Submucosal plexus	Georg Meissner (*see* Meissner's corpuscles *under* Sensory Function)
Sphincter of Oddi	Pancreaticohepatic sphincter	Ruggero Oddi (1864–1913), Italian physician
Peyer's patches		Johann Conrad Peyer (*see under* The Lymphatic System)
Duct of Santorini	Accessory pancreatic duct	Giovanni Domenico Santorini (1681–1737), Italian anatomist
Stensen's duct	Parotid duct	Niels Stensen (1638–1686), Danish physician/priest
Ampulla of Vater	Duodenal ampulla	Abraham Vater (1684–1751), German anatomist
Wharton's duct	Submandibular duct	Thomas Wharton (1614–1673), English physician
Foramen of Winslow	Epiploic foramen	Jacob Benignus Winslow (1669–1760), French anatomist
Duct of Wirsung	Pancreatic duct	Johann Georg Wirsung (1600–1643), German physician
The Urinary System (Chapter 26)		
Bowman's capsule	Glomerular capsule	Sir William Bowman (1816–1892), English physician
Loop of Henle		Friedrich Gustav Jakob Henle (1809–1885), German histologist
The Reproductive System (Chapters 28, 29)		
Bartholin's glands	Greater vestibular glands	Casper Bartholin, Jr. (1655–1738), Danish anatomist
Cowper's glands	Bulbourethral glands	William Cowper (1666–1709), English surgeon
Fallopian tube	Uterine tube/oviduct	Gabriele Fallopio (1523–1562), Italian anatomist
Graafian follicle	Tertiary follicle	Reijnier de Graaf (1641–1673), Dutch physician
Interstitial cells of Leydig	Interstitial cells	Franz von Leydig (*see under* The Endocrine System)
Glands of Littre	Lesser vestibular glands	Alexis Littre (1658–1726), French surgeon
Sertoli cells	Sustentacular cells	Enrico Sertoli (1842–1910), Italian histologist

Tables 3 and 4 present normal averages or ranges for the chemical composition of body fluids. These values are approximations rather than absolute values, as test results vary from laboratory to laboratory due to differences in procedures, equipment, normal solutions, and so forth. Blanks in the tabular data appear where data were not available; sources used in the preparation of these tables follow. Additional information concerning body fluid analysis can be found at the following locations in the text:

    Table 19-3 (p. 660) presents data on the cellular composition of whole blood.

    Table 26-2 (p. 973) compares the average compositions of plasma and urine.

    Tables 26-5 and 26-6 (p. 991) give the general characteristics of normal urine.

Sources

Braunwauld, Eugene, Kurt J. Isselbacher, Dennis L. Kasper, Jean D. Wilson, Joseph B. Martin, and Anthony S. Fauci, eds. 1994. *Harrison's Principles of Internal Medicine,* 13th ed. New York: McGraw-Hill.

Lentner, Cornelius, ed. 1981. *Geigy Scientific Tables,* 8th ed. Basel, Switzerland: Ciba–Geigy Limited.

Halsted, James A. 1976. *The Laboratory in Clinical Medicine: Interpretation and Application.* Philadelphia: W.B. Saunders Company.

Wintrobe, Maxwell, G. Richard Lee, Dane R. Boggs, Thomas C. Bitnell, John Foerster, John W. Athens, and John N. Lukens. 1981. *Clinical Hematology,* Philadelphia: Lea and Febiger.

TABLE 3   The Composition of Minor Body Fluids

Test	Normal Averages or Ranges					
	Perilymph	**Endolymph**	**Synovial Fluid**	**Sweat**	**Saliva**	**Semen**
pH			7.4	4–6.8	6.4*	7.19
Specific gravity			1.008–1.015	1.001–1.008	1.007	1.028
Electrolytes (mEq/l)						
Potassium	5.5–6.3	140–160	4.0	4.3–14.2	21	31.3
Sodium	143–150	12–16	136.1	0–104	14*	117
Calcium	1.3–1.6	0.05	2.3–4.7	0.2–6	3	12.4
Magnesium	1.7	0.02		0.03–4	0.6	11.5
Bicarbonate	17.8–18.6	20.4–21.4	19.3–30.6		6*	24
Chloride	121.5	107.1	107.1	34.3	17	42.8
Proteins (total)(mg/dl)	200	150	1.72 g/dl	7.7	386[†]	4.5 g/dl
Metabolites (mg/dl)						
Amino acids				47.6	40	1.26 g/dl
Glucose	104		70–110	3.0	11	224 (fructose)
Urea				26–122	20	72
Lipids (total)	12		20.9	‡	25–500[§]	188

* Increases under salivary stimulation.

[†] Primarily alpha-amylase, with some lysozomes.

‡ Not present in eccrine secretions.

[§] Cholesterol.

TABLE 4   The Chemistry of Blood, Cerebrospinal Fluid, and Urine

| Test | Normal Range | | |
	Blood*	CSF	Urine
pH	S: 7.38–7.44	7.31–7.34	4.6–8.0
Osmolarity (mOsm/l)	S: 280–295	292–297	500–800
Electrolytes		(mEq/l unless noted)	(urinary loss per 24-hour period[†])
Bicarbonate	P: 21–28	20–24	
Calcium	S: 4.5–5.3	2.1–3.0	6.5–16.5 mEq
Chloride	S: 100–108	116–122	120–240 mEq
Iron	S: 50–150 µg/l	23–52 µg/l	40–150 µg
Magnesium	S: 1.5–2.5	2–2.5	4.9–16.5 mEq
Phosphorus	S: 1.8–2.6	1.2–2.0	0.8–2 g
Potassium	P: 3.8–5.0	2.7–3.9	35–80 mEq
Sodium	P: 136–142	137–145	120–220 mEq
Sulfate	S: 0.2–1.3		1.07–1.3 g
Metabolites		(mg/dl unless noted)	(urinary loss per 24-hour period[‡])
Amino acids	P/S: 2.3–5.0	10.0–14.7	41–133 mg
Ammonia	P: 20–150 µg/dl	25–80 µg/dl	340–1200 mg
Bilirubin	S: 0.5–1.20.2	<0.2	0.02–1.9 mg
Creatinine	P/S: 0.6–1.2	0.5–1.9	1.01–2.5
Glucose	P/S: 70–110	40–70	16–132 mg
Ketone bodies	S: 0.3–2.0	1.3–1.6	10–100 mg
Lactic acid	WB: 5–20[§]	10–20	100–600 mg
Lipids (total)	S: 400–1000	0.8–1.7	0–31.8 mg
Cholesterol (total)	S: 150–300	0.2–0.8	1.2–3.8 mg
Triglycerides	S: 40–150	0–0.9	
Urea	P/S: 23–43	13.8–36.4	12.6–28.6
Uric acid	S: 2.0–7.0	0.2–0.3	80–976 mg
Proteins	(g/dl)	(mg/dl)	(urinary loss per 24-hour period[‡])
Total	S: 6.0–7.8	20–4.5	47–76.2 mg
Albumin	S: 3.2–4.5	10.6–32.4	10–100 mg
Globulins (total)	S: 2.3–3.5	2.8–15.5	7.3 mg (average)
Immunoglobulins	S: 1.0–2.2	1.1–1.7	3.1 mg (average)
Fibrinogen	P: 0.2–0.4	0.65 (average)	

* S = serum, P = plasma, WB = whole blood.

[†] Because urinary output averages just over 1 liter per day, these electrolyte values are comparable to mEq/l.

[‡] Because urinary metabolite and protein data approximate mg/l or g/l, these data must be divided by 10 for comparison with CSF or blood concentrations.

[§] Venous blood sample.

A

abdomen: The region of the trunk bounded by the diaphragm and pelvis.

abdominopelvic cavity: The portion of the ventral body cavity that contains abdominal and pelvic subdivisions; also contains the *peritoneal cavity*.

abducens (ab-DOO-senz): Cranial nerve VI, which innervates the lateral rectus muscle of the eye.

abduction: Movement away from the midline of the body, as viewed in the anatomical position.

abortion: Premature loss or expulsion of an embryo or fetus.

abruptio placentae (ab-RUP-shē-ō pla-SEN-tē): Premature loss of placental connection to the uterus, leading to maternal hemorrhaging and shock.

abscess: A localized collection of pus within a damaged tissue.

absorption: The active or passive uptake of gases, fluids, or solutes.

acclimatization: Physical adaption to a long-term environmental change, such as adaptions that accompany a change in season or in latitude.

accommodation: Alteration in the curvature of the lens of the eye to focus an image on the retina; decrease in receptor sensitivity or perception after chronic stimulation.

acetabulum (a-se-TAB-ū-lum): The fossa on the lateral aspect of the pelvis that accommodates the head of the femur.

acetylcholine (ACh) (as-ē-til-KŌ-lēn): A chemical neurotransmitter in the brain and peripheral nervous system; dominant neurotransmitter in the peripheral nervous system, released at neuromuscular junctions and synapses of the parasympathetic division.

acetylcholinesterase (AChE): An enzyme found in the synaptic cleft, bound to the postsynaptic membrane, and in tissue fluids; breaks down and inactivates ACh molecules.

acetyl-CoA: An acetyl group bound to coenzyme A, a participant in the anabolic and catabolic pathways for carbohydrates, lipids, and many amino acids.

acetyl group: —CH_3C=O.

achalasia (āk-a-LĀ-zē-uh): The condition that develops when the lower esophageal sphincter fails to dilate and ingested materials cannot enter the stomach.

Achilles tendon: *See* **calcaneal tendon**.

acid: A compound whose dissociation in solution releases a hydrogen ion and an anion; an acid solution has a pH below 7.0 and contains an excess of hydrogen ions.

acidosis (a-sid-Ō-sis): An abnormal physiological state characterized by a plasma pH below 7.35.

acinus/acini (AS-i-nī): A histological term referring to a blind pocket, pouch, or sac.

acne: A condition characterized by inflammation of sebaceous glands and follicles; commonly affects adolescents and in most cases involves the face.

acoustic: Pertaining to sound or the sense of hearing.

acquired immune deficiency syndrome (AIDS): A disease caused by the human immunodeficiency virus (HIV); characterized by the destruction of helper T cells and a resulting severe impairment of the immune response.

acromegaly: A condition caused by the overproduction of growth hormone in the adult, characterized by thickening of bones and enlargement of cartilages and other soft tissues.

acromion (a-KRŌ-mē-on): A continuation of the scapular spine that projects above the capsule of the scapulohumeral joint.

acrosomal cap (ak-rō-SŌ-mal): A membranous sac at the tip of a sperm cell that contains hyaluronic acid.

actin: The protein component of microfilaments that forms thin filaments in skeletal muscles and produces contractions of all muscles through interaction with thick (myosin) filaments; *see also* **sliding filament theory**.

action potential: A conducted change in the transmembrane potential of excitable cells, initiated by a change in the membrane permeability to sodium ions; *see also* **nerve impulse**.

activation energy: The energy required to initiate a specific chemical reaction.

active transport: The ATP-dependent absorption or excretion of solutes across a cell membrane.

acute: Sudden in onset, severe in intensity, and brief in duration.

adaptation: Alteration of pupillary size in response to changes in light intensity; in central nervous system, commonly used as a synonym for *accommodation*; physiological responses that produce acclimatization.

Addison's disease: A condition resulting from the hyposecretion of glucocorticoids; characterized by lethargy, weakness, hypotension, and increased skin pigmentation.

adduction: Movement toward the axis or midline of the body, as viewed in the anatomical position.

adenine: A purine; one of the nitrogenous bases in the nucleic acids RNA and DNA.

adenohypophysis (ad-e-nō-hī-POF-i-sis): The anterior portion of the pituitary gland, also called the **anterior pituitary**.

adenoids: The pharyngeal tonsil.

adenosine: A nucleoside consisting of adenine and a 5-carbon sugar.

adenosine diphosphate (ADP): A compound consisting of adenosine with two phosphate groups attached.

adenosine monophosphate (AMP): A nucleotide consisting of adenine plus a phosphate group (PO_4^{3-}); also known as adenosine phosphate.

adenosine triphosphate (ATP): A high-energy compound consisting of adenosine with three phosphate groups attached; the third is attached by a high-energy bond.

adenylate cyclase: An enzyme bound to the inner surfaces of cell membranes that can convert ATP to cyclic-AMP. Also called *adenyl cyclase* and *adenylyl cyclase*.

adhesion: Fusion of two mesenterial layers after damage or irritation of their opposing surfaces; this process restricts relative movement of the organs involved; the binding of a phagocyte to its target.

adipocyte (AD-i-pō-sīt): A fat cell.

adipose tissue: Loose connective tissue dominated by adipocytes.

adrenal cortex: The superficial portion of the adrenal gland that produces steroid hormones.

adrenal gland: A small endocrine gland that secretes steroids and catecholamines and is located superior to each kidney; also called the *suprarenal gland*.

adrenal medulla: The core of the adrenal gland; a modified sympathetic ganglion that secretes catecholamines into the blood after sympathetic activation.

adrenergic (ad-ren-ER-jik): A synaptic terminal that, when stimulated, releases norepinephrine.

adrenocortical hormone: Any of the steroids produced by the adrenal cortex.

adrenocorticotropic hormone (ACTH): The hormone that stimulates the production and secretion of glucocorticoids by the zona fasciculata of the adrenal cortex; released by the anterior pituitary in response to CRH.

adventitia (ad-ven-TISH-a): The superficial layer of connective tissue surrounding an internal organ; fibers are continuous with those of surrounding tissues, providing support and stabilization.

aerobic: Requiring the presence of oxygen.

aerobic metabolism: The complete breakdown of organic substrates into carbon dioxide and water, via pyruvic acid; a process that yields large amounts of ATP but requires mitochondria and oxygen.

afferent: Toward.

afferent arteriole: An arteriole that brings blood to a glomerulus of the kidney.

afferent fiber: An axon that carries sensory information to the central nervous system.

afterbirth: The distal portions of the umbilical cord and placenta that are ejected from the uterus during the placental stage of labor.

agglutination (a-gloo-ti-NĀ-shun): The aggregation of red blood cells due to interactions between surface antigens and plasma antibodies.

agglutinins (a-GLOO-ti-ninz): Immunoglobulins in plasma that react with antigens on the surfaces of foreign red blood cells when donor and recipient differ in blood type.

agglutinogens (a-gloo-TIN-ō-jenz): Antigens on the surfaces of red blood cells whose presence and structure are genetically determined.

aggregate lymphoid nodules: *See* **Peyer's patches**.

agonist: A muscle responsible for a specific movement.

agranular: Without granules; *agranular leukocytes* are monocytes and lymphocytes; the *agranular reticulum* is an intracellular organelle that synthesizes and stores carbohydrates and lipids.

AIDS: *See* **acquired immune deficiency syndrome**.

alba, albicans, albuginea (AL-bi-kanz) (al-bū-JIN-ē-uh): White.

albinism: The absence of pigment in hair and skin, caused by the inability of melanocytes to produce melanin.

aldosterone: A mineralocorticoid produced by the zona glomerulosa of the adrenal cortex; stimulates sodium and water conservation at the kidneys; secreted in response to the presence of angiotensin II.

aldosteronism: The condition caused by the oversecretion of aldosterone, characterized by fluid retention, edema, and hypertension.

alkalosis (al-ka-LŌ-sis): The condition characterized by a plasma pH of greater than 7.45; associated with relative deficiency of hydrogen ions or an excess of bicarbonate ions.

allantois (a-LAN-tō-is): One of the four extraembryonic membranes; provides vascularity to the chorion and is therefore essential to placenta formation; the proximal portion becomes the urinary bladder.

alleles (a-LĒLZ): Alternate forms of a particular gene.

allergen: An antigenic compound that produces a hypersensitivity response.

alpha-blockers: Drugs that prevent the stimulation of *alpha receptors*.

alpha cells: Cells in the pancreatic islets that secrete glucagon.

alpha receptors: Membrane receptors sensitive to norepinephrine or epinephrine; stimulation normally results in excitation of the target cell.

alveolar sac: An air-filled chamber that supplies air to several alveoli.

alveolus/alveoli (al-VĒ-o-lī): Blind pockets at the end of the respiratory tree, lined by a simple squamous epithelium and surrounded by a capillary network; sites of gas exchange with the blood; bony socket that holds the root of a tooth.

Alzheimer's disease: A disorder resulting from degenerative changes in populations of neurons in the cerebrum, causing dementia characterized by problems with attention, short-term memory, and emotions.

amacrine cells (AM-a-krīn): Modified neurons in the retina that facilitate or inhibit communication between bipolar cells and ganglion cells.

amenorrhea (ā-men-ō-RĒ-uh): The failure to commence menstruation at adolescence or the cessation of menstruation prior to menopause.

amination: The attachment of an amino group to a carbon chain; performed by a variety of cells and important in the synthesis of *amino acids*.

amino acids: Organic compounds whose chemical structure can be summarized as R—CHNH$_2$—COOH.

amino group: —NH$_2$.

amnesia: Temporary or permanent memory loss.

amniocentesis: Sampling of amniotic fluid for analytical purposes; used to detect certain genetic abnormalities.

amnion (AM-nē-on): One of the four extraembryonic membranes; surrounds the developing embryo/fetus.

amniotic fluid (am-nē-OT-ik): Fluid that fills the amniotic cavity; cushions and supports the embryo/fetus.

amphiarthrosis (am-fē-ar-THRŌ-sis): An articulation that permits a small degree of independent movement; *see* **interosseous membrane** and **pubic symphysis**.

amphicytes (AM-fi-sīts): Supporting cells that surround neurons in the peripheral nervous system; also called *satellite cells*.

amphimixis (am-fi-MIK-sis): The fusion of male and female pronuclei after fertilization.

ampulla/ampullae (am-PŪL-la): A localized dilation in the lumen of a canal or passageway.

amygdala/amygdaloid body (ah-MIG-da-loyd): A cerebral nucleus that is a component of the limbic system and acts as an interface between that system, the cerebrum, and sensory systems.

amylase: An enzyme that breaks down polysaccharides; produced by the salivary glands and pancreas.

anabolism (a-NAB-ō-lizm): The synthesis of complex organic compounds from simpler precursors.

anaerobic: Without oxygen.

analgesia: Relief from pain.

anal triangle: The posterior subdivision of the perineum.

anamnestic response (an-am-NES-tic): Sudden and exaggerated production of antibodies after a second exposure to a specific antigen; due to the activation of memory B cells.

anaphase (AN-a-fāz): The mitotic stage in which the paired chromatids separate and move toward opposite ends of the spindle apparatus.

anaphylaxis (a-na-fi-LAK-sis): The hypersensitivity reaction due to the binding of antigens to immunoglobulins (IgE) on the surfaces of mast cells; mast cell release of histamine, serotonin, and prostaglandins then causes widespread inflammation; a sudden decline in blood pressure may occur, producing *anaphylactic shock*.

anastomosis (a-nas-to-MŌ-sis): The joining of two tubes, usually referring to a connection between two peripheral vessels without an intervening capillary bed.

anatomical position: An anatomical reference position; the body viewed from the anterior surface with the palms facing forward; supine.

anatomy (a-NAT-o-mē): The study of the structure of the body.

anaxonic neuron (an-ak-SON-ik): A central nervous system neuron that has many processes but no apparent axon.

androgen (AN-drō-jen): A steroid sex hormone primarily produced by the interstitial cells of the testis and manufactured in small quantities by the adrenal cortex in either gender.

anemia (a-NĒ-mē-uh): The condition marked by a reduction in the hematocrit, the hemoglobin content of the blood, or both.

anencephaly (an-en-SEF-a-lē): The developmental defect characterized by incomplete development of the cerebral hemispheres and cranium.

anesthesia: The total or partial loss of sensation from a region of the body.

aneurysm (AN-ū-rizm): A weakening and localized dilation in the wall of a blood vessel.

angiogram (AN-jē-ō-gram): An X-ray image of circulatory pathways.

angiography: The X-ray examination of vessel distribution after the introduction of radiopaque substances into the bloodstream.

angiotensin I: The hormone produced by the activation of angiotensinogen by renin; converting enzyme converts angiotensin I into angiotensin II in the lung capillaries.

angiotensin II: A hormone that causes an elevation in systemic blood pressure, stimulates the secretion of aldosterone, promotes thirst, and causes the release of ADH; a converting enzyme in the pulmonary capillaries converts angiotensin I into angiotensin II.

angiotensinogen: The blood protein produced by the liver that is converted to angiotensin I by the enzyme renin.

anion (AN-ī-on): An ion bearing a negative charge.

ankyloglossia (ang-ki-lō-GLOS-ē-uh): A condition characterized by an overly robust and restrictive lingual frenulum.

annulus (AN-ū-lus): A cartilage or bone shaped like a ring.

anorectal canal (ā-nō-REC-tal): The distal portion of the rectum that contains the rectal columns and ends at the anus.

anorexia nervosa: An eating disorder marked by a loss of appetite and pronounced weight loss.

anoxia (a-NOKS-ē-uh): Tissue oxygen deprivation.

antagonist: A muscle that opposes the movement of an agonist.

antebrachium: The forearm.

anteflexion (an-tē-FLEK-shun): The normal position of the uterus, with the superior surface bent forward.

anterior: On or near the front, or ventral surface, of the body.

anterior pituitary: *See* **pituitary gland**.

anterograde amnesia: The inability to store memories of events that occur after a specific incident or time.

anthracosis (an-thra-KŌ-sis): "Black lung disease," a deterioration of respiratory exchange efficiency due to chronic inhalation of coal dust.

antibiotic: A chemical agent that selectively kills pathogenic microorganisms.

antibody (AN-ti-bod-ē): A globular protein produced by plasma cells that will bind to specific antigens and promote their destruction or removal from the body.

antibody-mediated immunity: Immunity resulting from the presence of circulating antibodies produced by plasma cells; also called *humoral immunity*.

anticholinesterase: The chemical compound that blocks the action of acetylcholinesterase and causes prolonged and intensive stimulation of postsynaptic membranes.

anticoagulant: A compound that slows or prevents clot formation by interfering with the clotting system.

anticodon: Three nitrogenous bases on a tRNA molecule that interact with an appropriate codon on a strand of mRNA.

antidiuretic hormone (ADH) (an-tī-dī-ū-RET-ik): A hormone synthesized in the hypothalamus and secreted at the posterior pituitary; causes water retention at the kidneys and an elevation of blood pressure.

antigen: A substance capable of inducing the production of antibodies.

antigen–antibody complex: The combination of an antigen and a specific antibody.

antigenic determinant site: A portion of an antigen that can interact with an antibody molecule.

antigen-presenting cell (APC): A cell that processes antigens and displays them, bound to MHC proteins; essential to the initiation of a normal immune response.

antihistamine (an-tī-HIS-ta-mēn): A chemical agent that blocks the action of histamine on peripheral tissues.

antipyretic agents: Chemicals that reduce fever.

antrum (AN-trum): A chamber or pocket.

anuria (a-NŪ-rē-uh): The cessation of urine production.

anus: The external opening of the anorectal canal.

aorta: The large, elastic artery that carries blood away from the left ventricle and into the systemic circuit.

aortic reflex: A baroreceptor reflex triggered by increased aortic pressures; leads to a reduction in cardiac output and a fall in systemic pressure.

Apgar test: A test used to assess the neurological status of a newborn infant.

aphasia: The inability to speak.

apnea (AP-nē-uh): The cessation of breathing.

apneustic center (ap-NOO-stik): A respiratory center whose chronic activation would lead to apnea at full inspiration.

apocrine secretion: A mode of secretion in which the glandular cell sheds portions of its cytoplasm.

aponeurosis/aponeuroses (ap-ō-nū-RŌ-sis): A broad tendinous sheet that may serve as the origin or insertion of a skeletal muscle.

apoplexy: A stroke (cerebrovascular accident).

appendicitis: Inflammation of the appendix.

appendicular: Pertaining to the upper or lower limbs.

appendix: A blind tube connected to the cecum of the large intestine.

appositional growth: Enlargement by the addition of cartilage or bony matrix to the outer surface.

aqueous humor: A fluid similar to perilymph or cerebrospinal fluid that fills the anterior chamber of the eye.

arachidonic acid: One of the essential fatty acids.

arachnoid (a-RAK-noyd): The middle meninges that encloses cerebrospinal fluid and protects the central nervous system.

arachnoid granulations: Processes of the arachnoid that project into the superior sagittal sinus; sites where cerebrospinal fluid enters the venous circulation.

arbor vitae: The central, branching mass of white matter inside the cerebellum.

arcuate (AR-kū-āt): Curving.

areflexia (ā-re-flek-sē-uh): The absence of normal reflex responses to stimulation.

areola (a-RĒ-ō-la): The pigmented area that surrounds the nipple of a breast.

areolar: Containing minute spaces, as in areolar connective tissue.

arrector pili (ar-REK-tor PI-lē): Smooth muscles whose contractions cause piloerection.

arrhythmias (ā-RITH-mē-az): Abnormal patterns of cardiac contractions.

arteriole (ar-TĒ-rē-ōl): A small arterial branch that delivers blood to a capillary network.

artery: A blood vessel that carries blood away from the heart and toward a peripheral capillary.

arthritis (ar-THRĪ-tis): Inflammation of a joint.

arthroscope: A fiber-optic device intended for visualizing the interior of joints; may also be used for certain forms of joint surgery.

articular: Pertaining to a joint.

articular capsule: The dense collagen fiber sleeve that surrounds a joint and provides protection and stabilization.

articular cartilage: The cartilage pad that covers the surface of a bone inside a joint cavity.

articulation (ar-tik-ū-LĀ-shun): A joint; the formation of words.

arytenoid cartilages (ar-i-TĒ-noyd): A pair of small cartilages in the larynx.

ascending tract: A tract carrying information from the spinal cord to the brain.

ascites (a-SĪ-tēz): The overproduction and accumulation of peritoneal fluid.

aseptic: Free from pathogenic contamination.

asphyxia: Unconsciousness due to oxygen deprivation at the central nervous system.

aspirate: To remove or obtain by suction; to inhale.

association areas: Cortical areas of the cerebrum responsible for the integration of sensory inputs and/or motor commands.

association neuron: *See* **interneuron.**

asthma (AZ-ma): The reversible constriction of smooth muscles around respiratory passageways, commonly caused by an allergic response.

astigmatism: A visual disturbance due to an irregularity in the shape of the cornea.

astrocyte (AS-trō-sīt): One of the glial cells in the central nervous system; responsible for maintaining the blood–brain barrier by stimulation of endothelial cells.

ataxia: The failure to coordinate muscular activities normally.

atelectasis (at-e-LEK-ta-sis): The collapse of a lung or a portion of a lung.

atherosclerosis (ath-er-ō-skle-RŌ-sis): The formation of fatty plaques in the walls of arteries, leading to circulatory impairment.

atom: The smallest stable unit of matter.

atomic number: The number of protons in the nucleus of an atom.

atomic weight: Roughly, the average total number of protons and neutrons in the atoms of a particular element.

atresia (a-TRĒ-zē-uh): The closing of a cavity, or its incomplete development; used to refer to the degeneration of developing ovarian follicles.

atria: Thin-walled chambers of the heart that receive venous blood from the pulmonary or systemic circuit.

atrial natriuretic peptide (ANP) (nā-trē-ū-RET-ik): The hormone released by specialized atrial cardiocytes when they are stretched by an abnormally large venous return; promotes fluid loss and reductions in blood pressure and in venous return.

atrial reflex: The reflexive increase in heart rate after an increase in venous return; due to mechanical and neural factors; also called the *Bainbridge reflex.*

atrioventricular (AV) node (ā-trē-ō-ven-TRIK-ū-lar): Specialized cardiocytes that relay the contractile stimulus to the bundle of His, the bundle branches, the Purkinje fibers, and the ventricular myocardium; located at the boundary between the atria and ventricles.

atrioventricular (AV) valve: One of the valves that prevent backflow into the atria during ventricular systole.

atrophy (AT-rō-fē): The wasting away of tissues from lack of use, ischemia, or nutritional abnormalities.

auditory: Pertaining to the sense of hearing.

auditory ossicles: The bones of the middle ear: *malleus, incus,* and *stapes.*

auditory tube: A passageway that connects the nasopharynx with the middle ear cavity; also called the *Eustachian* or *pharyngotympanic tube.*

autoantibodies: Antibodies that react with antigens on the surfaces of a person's own cells and tissues.

autodigestion: The digestion of tissues by digestive acids or enzymes from the stomach or pancreas.

autoimmunity: Immune system sensitivity to normal cells and tissues, resulting in the production of autoantibodies.

autolysis: The destruction of a cell due to the rupture of lysosomal membranes in its cytoplasm.

automatic bladder: Reflex micturition after stimulation of stretch receptors in the bladder wall; seen in patients who have lost motor control of the lower body.

automaticity: Spontaneous depolarization to threshold, a characteristic of cardiac pacemaker cells.

autonomic ganglion: A collection of visceral motor neurons outside the central nervous system.

autonomic nerve: A peripheral nerve consisting of preganglionic or postganglionic autonomic fibers.

autonomic nervous system (ANS): Centers, nuclei, tracts, ganglia, and nerves involved in the unconscious regulation of visceral functions; includes components of the central nervous system and the peripheral nervous system.

autopsy: The detailed examination of a body after death, normally performed by a pathologist.

autoregulation: Alterations in activity that maintain homeostasis in direct response to changes in the local environment; does not require neural or endocrine control.

autosomal (aw-to-SŌ-mal): Chromosomes other than the X or Y chromosome.

avascular (ā-VAS-kū-lar): Without blood vessels.

avitaminosis (ā-vī-ta-min-Ō-sis): A condition caused by inadequate intake of one or more essential vitamins.

avulsion: An injury involving the violent tearing away of body tissues.

axilla: The armpit.

axolemma: The cell membrane of an axon, continuous with the cell membrane of the soma and dendrites and distinct from any glial cell coverings.

axon: The elongate extension of a neuron that conducts an action potential.

axon hillock: In a multipolar neuron, the portion of the neural soma adjacent to the initial segment.

axoplasm (AK-sō-plazm): The cytoplasm within an axon.

azotemia (a-zō-TĒ-mē-uh): A condition resulting from impaired kidney function and the retention of nitrogenous wastes, especially urea.

B

Babinski sign: The reflexive dorsiflexion of the toes after the plantar surface of the foot is stroked; positive reflex (Babinski sign) is normal up to age 1.5 years; thereafter, a positive reflex indicates damage to descending tracts.

bacteria: Single-celled microorganisms, some pathogenic, that are common in the environment and in and on the body.

Bainbridge reflex: *See* **atrial reflex.**

baroreception: The ability to detect changes in pressure.

baroreceptor reflex: A reflexive change in cardiac activity in response to changes in blood pressure.

baroreceptors (bar-ō-rē-SEP-torz): The receptors responsible for baroreception.

basal metabolic rate: The resting metabolic rate of a normal fasting individual under homeostatic conditions.

base: A compound whose dissociation releases a hydroxide ion (OH^-) or removes a hydrogen ion (H^+) from the solution.

basement membrane: A layer of filaments and fibers that attach an epithelium to the underlying connective tissue.

basilar membrane: The membrane that supports the organ of Corti and separates the cochlear duct from the scala tympani in the inner ear.

basophils (BĀ-sō-filz): Circulating granulocytes (white blood cells) similar in size and function to tissue mast cells.

B cells: Lymphocytes capable of differentiating into the plasma cells that produce antibodies.

benign: Not malignant.

beta cells: Cells of the pancreatic islets that secrete insulin in response to elevated blood sugar concentrations.

beta oxidation: Fatty acid catabolism that produces molecules of acetyl-CoA.

beta receptors: Membrane receptors sensitive to epinephrine; stimulation may result in the excitation or inhibition of the target cell.

bicarbonate ions: HCO_3^-; anion components of the carbonic acid–bicarbonate buffer system.

bicuspid (bī-KUS-pid): A sharp, conical tooth, also called a *canine*.

bicuspid valve: The left atrioventricular valve, also known as the *mitral valve*.

bifurcate: To branch into two parts.

bile: The exocrine secretion of the liver that is stored in the gallbladder and ejected into the duodenum.

bile salts: Steroid derivatives in the bile; responsible for the emulsification of ingested lipids.

bilirubin (bil-ē-ROO-bin): A pigment, the byproduct of hemoglobin catabolism.

bioenergetics: The analysis of energy production and utilization by living cells.

biofeedback: The use of artificial signals to provide feedback about unconscious, visceral motor activities.

biopsy: The removal of a small sample of tissue for pathological analysis.

bipennate: A muscle whose fibers are arranged on either side of a common tendon.

bladder: A muscular sac that distends as fluid is stored and whose contraction ejects the fluid at an appropriate time; used alone, the term usually refers to the urinary bladder.

blastocoele (BLAS-to-sel): A fluid-filled cavity within a blastocyst.

blastocyst (BLAS-tō-sist): An early stage in the developing embryo, consisting of an outer trophoblast and an inner cell mass.

blastodisc (BLAS-tō-disk): A late stage in the development of the inner cell mass; includes the cells that will form the embryo.

blastomere (BLAS-tō-mēr): One of the cells in the morula; a collection of cells produced by the division of the zygote.

blockers/blocking agents: Drugs that block membrane pores or prevent binding to membrane receptors.

blood–brain barrier: The isolation of the central nervous system from the general circulation; primarily the result of astrocyte regulation of capillary permeabilities.

blood clot: A network of fibrin fibers and trapped blood cells.

blood–CSF barrier: The isolation of the cerebrospinal fluid from the capillaries of the choroid plexus; primarily the result of specialized ependymal cells.

blood pressure: A force exerted against the vascular walls by the blood, as the result of the push exerted by cardiac contraction and the elasticity of the vessel walls; usually measured along one of the muscular arteries, with systolic pressure measured during ventricular systole and diastolic pressure during ventricular diastole.

blood–testis barrier: The isolation of the seminiferous tubules from the general circulation, due to the activities of the sustentacular (Sertoli) cells.

Bohr effect: The increased oxygen release by hemoglobin in the presence of elevated carbon dioxide levels.

boil: An abscess of the skin, normally involving a sebaceous gland.

bolus: A compact mass; usually refers to compacted ingested material on its way to the stomach.

bone: *See* **osseous tissue**.

botulinus toxin (bot-ū-LĪ-nus): A toxin produced by the anaerobic bacterium *Clostridium botulinum* that can cause severe food poisoning.

bowel: The intestinal tract.

Bowman's capsule: The cup-shaped initial portion of the renal tubule; it surrounds the glomerulus and receives the glomerular filtrate.

Boyle's law: The principle that, in a gas, pressure and volume are inversely related.

brachial: Pertaining to the arm.

brachial plexus: A network formed by branches of spinal nerves C_5–T_1 en route to innervate the upper limb.

brachium: The arm.

bradycardia (brad-ē-KAR-dē-uh): A slow heart rate, below 50 bpm.

brain stem: The brain minus the cerebrum, diencephalon, and cerebellum.

brevis: Short.

Broca's center: The speech center of the brain, normally located on the neural cortex of the left cerebral hemisphere.

bronchial tree: The trachea, bronchi, and bronchioles.

bronchitis (brong-KĪ-tis): Inflammation of the bronchial passageways.

bronchodilation: The dilation of the bronchial passages; may be caused by sympathetic stimulation.

bronchodilators (brong-kō-dī-LĀ-torz): Drugs that produce bronchodilation; some are used clinically in treating asthma.

bronchoscope: A fiber-optic instrument used to examine the bronchial passageways.

bronchus/bronchi: One of the branches of the bronchial tree between the trachea and bronchioles.

buccal (BUK-al): Pertaining to the cheeks.

buffer: A compound that stabilizes the pH of a solution by removing or releasing hydrogen ions.

buffer system: Interacting compounds that prevent increases or decreases in the pH of body fluids; includes the carbonic acid–bicarbonate buffer system, the phosphate buffer system, and the protein buffer system.

bulbar: Pertaining to the brain stem.

bulbourethral glands (bul-bō-ū-Rē-thral): Mucous glands at the base of the penis that secrete into the penile urethra; the equivalent of the greater vestibular glands of the female; also called *Cowper's glands*.

bundle branches: Specialized conducting cells in the ventricles that carry the contractile stimulus from the bundle of His to the Purkinje fibers.

bundle of His (hiss): Specialized conducting cells in the interventricular septum that carry the contracting stimulus from the AV node to the bundle branches and then to the Purkinje fibers.

bursa: A small sac filled with synovial fluid that cushions adjacent structures and reduces friction.

bursectomy: The surgical removal of an inflamed bursa.

bursitis: The painful inflammation of one or more bursae.

C

calcaneal tendon: The large tendon that inserts on the calcaneus; tension on this tendon produces plantar flexion of the foot; also called the *Achilles tendon*.

calcaneus (kal-KĀ-nē-us): The heelbone, the largest of the tarsal bones.

calcification: The deposition of calcium salts within a tissue.

calcitonin (kal-si-TŌ-nin): The hormone secreted by C cells of the thyroid when calcium ion concentrations are abnormally high; restores homeostasis by increasing the rate of bone deposition and the renal rate of calcium loss.

calculus/calculi (KAL-kū-lī): Concretions of insoluble materials that form within body fluids, especially the gallbladder, kidneys, or urinary bladder.

callus: A localized thickening of the epidermis due to chronic mechanical stresses; a thickened area that forms at the site of a bone break as part of the repair process.

calorie (c) (KAL-o-rē): The amount of heat that is required to raise the temperature of one gram of water 1°C.

Calorie (C): The amount of heat that is required to raise the temperature of one kilogram of water 1°C; also called the *kilocalorie*.

calorigenic effect: The stimulation of energy production and heat loss by thyroid hormones.

calvaria (kal-VAR-ē-uh): The skullcap, formed of the frontal, parietal, and occipital bones.

calyx/calyces (KĀL-i-sēz): Cup-shaped divisions of the renal pelvis.

canaliculi (kan-a-LIK-ū-lī): Microscopic passageways between cells; bile canaliculi carry bile to bile ducts in the liver; in bone, canaliculi permit the diffusion of nutrients and wastes to and from osteocytes.

cancellous bone (KAN-sel-us): Spongy bone, composed of a network of bony struts.

cancer: A malignant tumor that tends to undergo metastasis.

cannula: A tube that can be inserted into the body; commonly placed in blood vessels prior to transfusion or dialysis.

canthus, medial and lateral (KAN-thus): The angles formed at either corner of the eye between the upper and lower eyelids.

capacitation (ka-pas-i-TĀ-shun): The activation process that must occur before a spermatozoon can successfully fertilize an oocyte; occurs in the vagina after ejaculation.

capillary: A small blood vessel, interposed between an arteriole and a venule, whose thin wall permits the diffusion of gases, nutrients, and wastes between the plasma and interstitial fluids.

capitulum (ka-PIT-ū-lum): A general term for a small, elevated articular process; used to refer to the rounded distal surface of the humerus that articulates with the radial head.

caput: The head.

carbaminohemoglobin (kar-bam-ē-nō-HĒ-mō-GLŌ-bin): Hemoglobin bound to carbon dioxide molecules.

carbohydrase (kar-bō-HĪ-drās): An enzyme that breaks down carbohydrate molecules.

carbohydrate (kar-bō-HĪ-drāt): An organic compound containing carbon, hydrogen, and oxygen in a ratio that approximates 1:2:1.

carbon dioxide: CO_2; a compound produced by the decarboxylation reactions of aerobic metabolism.

carbonic anhydrase: An enzyme that catalyzes the reaction $H_2O + CO_2 \rightleftharpoons H_2CO_3$; important in carbon dioxide transport, gastric acid secretion, and renal pH regulation.

carboxypeptidase (kar-bok-sē-PEP-ti-dās): A protease that breaks down proteins and releases amino acids.

carcinogenic (kar-sin-ō-JEN-ik): Stimulating cancer formation in affected tissues.

cardia (KAR-dē-uh): The area of the stomach surrounding its connection with the esophagus.

cardiac: Pertaining to the heart.

cardiac cycle: One complete heartbeat, including atrial and ventricular systole and diastole.

cardiac glands: Mucous glands characteristic of the cardia of the stomach.

cardiac output: The amount of blood ejected by the left ventricle each minute; normally about 5 liters.

cardiac reserve: The potential percentage increase in cardiac output above resting levels.

cardiac tamponade: Compression of the heart due to fluid accumulation in the pericardial cavity.

cardiocyte (KAR-dē-ō-sīt): A cardiac muscle cell.

cardiomyopathy (kar-dē-ō-mī-OP-a-thē): A progressive disease characterized by damage to the cardiac muscle tissue.

cardiopulmonary resuscitation (CPR): A method of artificially maintaining respiratory and circulatory function.

cardiovascular: Pertaining to the heart, blood, and blood vessels.

cardiovascular centers: Poorly localized centers in the reticular formation of the medulla of the brain; includes cardioacceleratory, cardioinhibitory, and vasomotor centers.

cardium: The heart.

carina (ka-RĪ-na): A ridge on the inner surface of the base of the trachea that runs anteroposteriorly, between two primary bronchi.

carotene (KAR-ō-tēn): A yellow-orange pigment found in carrots and in green and orange leafy vegetables; a compound that the body can convert to vitamin A.

carotid artery: The principal artery of the neck, servicing cervical and cranial structures; one branch, the internal carotid, represents a major blood supply for the brain.

carotid body: A group of receptors adjacent to the carotid sinus that are sensitive to changes in the carbon dioxide levels, pH, and oxygen concentrations of the arterial blood.

carotid sinus: A dilated segment at the base of the internal carotid artery whose walls contain baroreceptors sensitive to changes in blood pressure.

carotid sinus reflex: Reflexive changes in blood pressure that maintain homeostatic pressures at the carotid sinus, stabilizing blood flow to the brain.

carpus/carpal: The wrist.

cartilage: A connective tissue with a gelatinous matrix that contains an abundance of fibers.

castration: The removal of the testes; also called *bilateral orchiectomy*.

catabolism (ka-TAB-ō-lizm): The breakdown of complex organic molecules into simpler components, accompanied by the release of energy.

catalyst (KAT-uh-list): A substance that accelerates a specific chemical reaction but that is not altered by the reaction.

cataract: A reduction in lens transparency that causes visual impairment.

catecholamine (kat-e-KŌL-am-ēn): Epinephrine, norepinephrine, dopamine, and related compounds.

cathepsins (ka-THEP-sinz): Enzymes present in the sarcoplasm of skeletal muscle cells that can break down contractile proteins, providing amino acids that can act as a supplemental energy source.

catheter (KATH-e-ter): A tube surgically inserted into a body cavity or along a blood vessel or excretory passageway for the collection of body fluids, monitoring of blood pressure, or introduction of medications or radiographic dyes.

cation (KAT-ī-on): An ion that bears a positive charge.

cauda equina (KAW-da ek-WĪ-na): Spinal nerve roots distal to the tip of the adult spinal cord; they extend caudally inside the vertebral canal en route to lumbar and sacral segments.

caudal/caudally: Closest to or toward the tail (coccyx).

caudate nucleus (KAW-dāt): One of the cerebral nuclei of the extrapyramidal system; involved with the unconscious control of muscular activity.

cavernous tissue: Erectile tissue that can be engorged with blood; located in the penis and clitoris.

cecum (SĒ-kum): An expanded pouch at the start of the large intestine.

cell: The smallest living unit in the human body.

cell-mediated immunity: Resistance to disease through the activities of sensitized T cells that destroy antigen-bearing cells by direct contact or through the release of lymphotoxins; also called *cellular immunity*.

cellulitis (sel-ū-LĪ-tis): Diffuse inflammation, normally involving areas of loose connective tissue, such as the subcutaneous layer.

cementum (se-MEN-tum): Bony material covering the root of a tooth, not shielded by a layer of enamel.

center of ossification: The site in a connective tissue where bone formation begins.

central canal: Longitudinal canal in the center of an osteon that contains blood vessels and nerves, also called the *Haversian canal*; a passageway along the longitudinal axis of the spinal cord that contains cerebrospinal fluid.

central nervous system (CNS): The brain and spinal cord.

central sulcus: A groove in the surface of a cerebral hemisphere, between the primary sensory and primary motor areas of the cortex.

centriole: A cylindrical intracellular organelle composed of nine groups of microtubules, three in each group; functions in mitosis or meiosis by organizing the microtubules of the spindle apparatus.

centromere (SEN-trō-mēr): The localized region where two chromatids remain connected after chromosome replication; site of spindle fiber attachment.

centrosome: A region of cytoplasm that contains a pair of centrioles oriented at right angles to one another.

centrum: The body of a vertebra.

cephalic: Pertaining to the head.

cerebellum (ser-e-BEL-um): The posterior portion of the metencephalon, containing the cerebellar hemispheres; includes the arbor vitae, cerebellar nuclei, and cerebellar cortex.

cerebral cortex: An extensive area of neural cortex covering the surfaces of the cerebral hemispheres.

cerebral hemispheres: Expanded portions of the cerebrum covered in neural cortex.

cerebral nuclei: Nuclei of the cerebrum that are important components of the extrapyramidal system.

cerebral palsy: A chronic condition resulting from damage to motor areas of the brain during development or at delivery.

cerebral peduncle: A mass of nerve fibers on the ventrolateral surface of the mesencephalon; contains ascending tracts that terminate in the thalamus and descending tracts that originate in the cerebral hemispheres.

cerebrospinal fluid (CSF): Fluid bathing the internal and external surfaces of the central nervous system; secreted by the choroid plexus.

cerebrovascular accident (CVA): A stroke; occlusion of a blood vessel that supplies a portion of the brain, resulting in damage to the dependent neurons.

cerebrum (SER-ē-brum): The largest portion of the brain, composed of the cerebral hemispheres; includes the cerebral cortex, the cerebral nuclei, and the internal capsule.

cerumen: Waxy secretion of integumentary glands along the external auditory canal.

ceruminous glands (se-ROO-mi-nus): Integumentary glands that secrete cerumen.

cervical enlargement: Relative enlargement of the cervical portion of the spinal cord due to the abundance of central nervous system neurons involved with motor control of the upper limbs.

cervix: The lower part of the uterus.

cesarean section: The surgical delivery of an infant via an incision through the lower abdominal wall and uterus.

chalazion (kah-LĀ-zē-on): An inflammation and distension of a Meibomian gland on the eyelid; also called a *sty*.

chancre (SHANG-ker): A skin lesion that develops at the primary site of a syphilis infection.

charleyhorse: Soreness and stiffness in a strained muscle, normally involving the quadriceps group.

chemoreception: The detection of alterations in the concentrations of dissolved compounds or gases.

chemotaxis (kē-mō-TAK-sis): The attraction of phagocytic cells to the source of abnormal chemicals in tissue fluids.

chemotherapy: The treatment of illness through the administration of specific chemicals.

chloride shift: The movement of plasma chloride ions into red blood cells in exchange for bicarbonate ions generated by the intracellular dissociation of carbonic acid.

cholecystitis (kō-lē-sis-TĪ-tis): Inflammation of the gallbladder.

cholecystokinin (CCK) (kō-lē-sis-tō-KĪ-nin): A duodenal hormone that stimulates the contraction of the gallbladder and the secretion of enzymes by the exocrine pancreas; also called *pancreozymin*.

cholelithiasis (kō-lē-li-THĪ-a-sis): The formation or presence of gallstones.

cholesterol: A steroid component of cell membranes and a substrate for the synthesis of steroid hormones and bile salts.

choline: A chemical compound that is a breakdown product or precursor of acetylcholine.

cholinergic synapse (kō-lin-ER-jik): A synapse where the presynaptic membrane releases ACh on stimulation.

cholinesterase (kō-li-NES-ter-ās): The enzyme that breaks down and inactivates ACh.

chondrocyte (KON-drō-sīt): A cartilage cell.

chondroitin sulfate (kon-DROI-tin): The predominant proteoglycan in cartilage, responsible for the gelatinous consistency of the matrix.

chordae tendineae (KOR-dē TEN-di-nē-ē): Fibrous cords that stabilize the position of the AV valves in the heart, preventing backflow during ventricular systole.

chorion/chorionic (KOR-ē-on) (ko-rē-ON-ik): An extraembryonic membrane, consisting of the trophoblast and underlying mesoderm, that forms the placenta.

choroid: Middle, vascular layer in the wall of the eye.

choroid plexus: The vascular complex in the roof of the third and fourth ventricles of the brain, responsible for cerebrospinal fluid production.

chromatid (KRŌ-ma-tid): One complete copy of a DNA strand and its associated nucleoproteins.

chromatin (KRŌ-ma-tin): A histological term referring to the grainy material visible in cell nuclei during interphase; the appearance of the DNA content of the nucleus when the chromosomes are uncoiled.

chromosomes: Dense structures, composed of tightly coiled DNA strands and associated histones, that become visible in the nucleus when a cell prepares to undergo mitosis or meiosis; normal human somatic cells contain 46 chromosomes apiece.

chronic: Habitual or long term.

chylomicrons (kī-lō-MĪ-kronz): Relatively large droplets that may contain triglycerides, phospholipids, and cholesterol in association with proteins; synthesized and released by intestinal cells and transported to the venous blood by the lymphatic system.

chyme (kīm): A semifluid, acidic mixture of ingested food and digestive secretions that is found in the stomach in the early phases of digestion.

chymotrypsin (kī-mō-TRIP-sin): A protease found in the small intestine.

chymotrypsinogen: The inactive proenzyme secreted by the pancreas that is subsequently converted to chymotrypsin.

ciliary body: A thickened region of the choroid that encircles the lens of the eye; it includes the ciliary muscle and the ciliary processes that support the suspensory ligaments of the lens.

cilium/cilia: A slender organelle that extends above the free surface of an epithelial cell and generally undergoes cycles of movement; composed of a basal body and microtubules in a 9×2 array.

circulatory system: The network of blood vessels that are components of the cardiovascular system.

circumduction (sir-kum-DUK-shun): A movement at a synovial joint in which the distal end of the bone describes a circle but the shaft does not rotate.

circumvallate papilla (sir-kum-VAL-āt pa-PIL-la): One of the large, dome-shaped papillae on the dorsum of the tongue that form the V that separates the body of the tongue from the root.

cirrhosis (sir-RŌ-sis): A liver disorder characterized by the degeneration of hepatocytes and their replacement by fibrous connective tissue.

cisterna (sis-TUR-na): An expanded chamber.

citric acid cycle: *See* **TCA cycle**.

cleavage (KLĒ-vij): Mitotic divisions that follow fertilization of the ovum and lead to the formation of a blastocyst.

cleavage lines: Stress lines in the skin that follow the orientation of major bundles of collagen fibers in the dermis.

clitoris (KLI-to-ris): A small erectile organ of the female that is the developmental equivalent of the male penis.

clone: The production of genetically identical cells.

clonus (KLŌ-nus): Rapid cycles of muscular contraction and relaxation.

clot: A network of fibrin fibers and trapped blood cells; also called a *thrombus* if it occurs within the circulatory system.

clotting factors: Plasma proteins synthesized by the liver that are essential to the clotting response.

clotting response: The series of events that result in the formation of a clot.

coccygeal ligament: The fibrous extension of the dura mater and filum terminale; provides longitudinal stabilization to the spinal cord.

coccyx (KOK-siks): The terminal portion of the spinal column, consisting of relatively tiny, fused vertebrae.

cochlea (KOK-lē-uh): The spiral portion of the bony labyrinth of the inner ear that surrounds the organ of hearing.

cochlear duct (KOK-lē-ar): The central membranous tube within the cochlea that is filled with endolymph and contains the organ of Corti; also called the *scala media*.

codon (KŌ-don): A sequence of three nitrogenous bases along an mRNA strand that will specify the location of a single amino acid in a peptide chain.

coelom (SĒ-lōm): The ventral body cavity, lined by a serous membrane and subdivided during fetal development into the pleural, pericardial, and abdominopelvic (peritoneal) cavities.

coenzymes (kō-EN-zīmz): Complex organic cofactors; most are structurally related to vitamins.

cofactor: Ions or molecules that must be attached to the active site before an enzyme can function; examples include mineral ions and several vitamins.

colectomy (kō-LEK-to-mē): The surgical removal of part or all of the colon.

colitis: Inflammation of the colon.

collagen: A strong, insoluble protein fiber common in connective tissues.

collateral ganglion (kō-LAT-er-al): A sympathetic ganglion situated in front of the spinal column and separate from the sympathetic chain.

Colles' fracture (KOL-lēz): A fracture of the distal end of the radius and possibly the ulna, with posterior and dorsal displacement of the distal bone fragments.

colliculus/colliculi (kol-IK-ū-lus): A little mound; in the brain, used to refer to one of the thickenings in the roof of the mesencephalon; the superior colliculus is associated with the visual system, and the inferior colliculi with the auditory system.

colloid/colloidal suspension: A solution containing large organic molecules in suspension.

colon: The large intestine.

colonoscope (kō-LON-ō-skōp): A fiber-optic device for examining the interior of the colon.

colostomy (kō-LOS-to-mē): The surgical connection of a portion of the colon to the body wall, sometimes performed after a colectomy to permit the discharge of fecal materials.

colostrum (ko-LOS-trum): The secretion of the mammary glands at the time of childbirth and for a few days thereafter; contains more protein and less fat than the milk secreted later.

coma (kō-ma): An unconscious state from which the individual cannot be aroused, even by strong stimuli.

comedo (kō-MĒ-dō): An inflamed sebaceous gland.

comminuted: Broken or crushed into small pieces.

commissure: A crossing over from one side to another.

common bile duct: The duct formed by the union of the cystic duct from the gallbladder and the bile ducts from the liver; terminates at the duodenal ampulla, where it meets the pancreatic duct.

common pathway: In the clotting response, the events that begin with the appearance of thromboplastin and end with the formation of a clot.

compact bone: Dense bone that contains parallel osteons.

compensation curves: The cervical and lumbar curves that develop to center the body weight over the legs.

complement: A system of 11 plasma proteins that interact in a chain-reaction after exposure to activated antibodies or the surfaces of certain pathogens; complement proteins promote cell lysis, phagocytosis, and other defense mechanisms.

compliance: Distensibility; the ability of certain organs to tolerate changes in volume; a property that reflects the presence of elastic fibers and smooth muscles.

compound: A molecule containing two or more elements in combination.

concentration: Amount (in grams) or number of atoms, ions, or molecules (in moles) per unit volume.

concentration gradient: Regional differences in the concentration of a particular substance.

conception: Fertilization.

concha/conchae (KONG-kē): Three pairs of thin, scroll-like bones that project into the nasal cavities; the superior and medial conchae are part of the ethmoid, and the inferior conchae are separate bones.

concussion: A violent blow or shock; loss of consciousness due to a violent blow to the head.

conducted change: An action potential; a change in the transmembrane potential that spreads across an excitable membrane.

condyle: A rounded articular projection on the surface of a bone.

cone: A retinal photoreceptor responsible for color vision.

congenital (kon-JEN-i-tal): Already present at birth.

congestive heart failure (CHF): The failure to maintain adequate cardiac output due to circulatory problems or myocardial damage.

conjunctiva (kon-junk-TĪ-va): A layer of stratified squamous epithelium that covers the inner surfaces of the lids and the anterior surface of the eye to the edges of the cornea.

conjunctivitis: Inflammation of the conjunctiva.

connective tissue: One of the four primary tissue types; provides a structural framework for the body that stabilizes the relative positions of the other tissue types; includes connective tissue proper, cartilage, bone, and blood; always has cell products, cells, and ground substance.

contractility: The ability to contract; possessed by skeletal, smooth, and cardiac muscle cells.

contracture: A permanent contraction of an entire muscle after individual muscle cells have atrophied.

contralateral reflex: A reflex that affects the opposite side of the body from the stimulus.

conus medullaris: The conical tip of the spinal cord that gives rise to the filum terminale.

convergence: In the nervous system, the innervation of a single neuron by axons from several neurons; most common along motor pathways.

coracoid process (ko-RA-koyd): A hook-shaped process of the scapula that projects above the anterior surface of the capsule of the shoulder joint.

Cori cycle: The metabolic exchange of lactic acid from skeletal muscle for glucose from the liver; performed during the recovery period after muscular exertion.

cornea (KOR-nē-uh): The transparent portion of the fibrous tunic of the anterior surface of the eye.

corniculate cartilages (kor-NIK-ū-lāt): A pair of small laryngeal cartilages.

cornification: The production of keratin by a stratified squamous epithelium; also called *keratinization*.

cornu: Shaped like a horn.

corona radiata (ko-RŌ-na rā-dē-A-ta): A layer of follicle cells surrounding a secondary oocyte at ovulation.

coronoid (ko-RŌ-noyd): Hooked or curved.

corpora quadrigemina (KOR-po-ra quad-ri-JEM-i-na): The superior and inferior colliculi of the mesencephalic tectum (roof) in the brain.

corpus/corpora: Body.

corpus callosum: The bundle of axons that links centers in the left and right cerebral hemispheres.

corpora cavernosa (ka-ver-NŌ-suh): Two parallel masses of erectile tissue within the body of the penis (male) or clitoris (female).

corpus luteum (LOO-tē-um): The progestin-secreting mass of follicle cells that develops in the ovary after ovulation.

corpus spongiosum (spon-JĒ-ō-sum): The mass of erectile tissue that surrounds the urethra in the male penis and expands distally to form the glans.

cortex: The outer layer or portion of an organ.

Corti, organ of: The receptor complex in the cochlear duct that includes the inner and outer hair cells, supporting cells and structures, and the tectorial membrane; provides the sensation of hearing.

corticobulbar tracts (kor-ti-kō-BUL-bar): Descending tracts that carry information or commands from the cerebral cortex to nuclei and centers in the brain stem.

corticospinal tracts: Descending tracts that carry motor commands from the cerebral cortex to the anterior gray horns of the spinal cord.

corticosteroid: A steroid hormone produced by the adrenal cortex.

corticosterone (kor-ti-KOS-te-rōn): One of the corticosteroids secreted by the zona fasciculata of the adrenal cortex; a glucocorticoid.

corticotropin: *See* **adrenocorticotropic hormone (ACTH).**

corticotropin-releasing hormone (CRH): The releasing hormone secreted by the hypothalamus that stimulates secretion of ACTH by the anterior pituitary.

cortisol (KOR-ti-sol): One of the corticosteroids secreted by the zona fasciculata of the adrenal cortex; a glucocorticoid.

costa/costae: A rib.

cotransport: The membrane transport of a nutrient, such as glucose, in company with the movement of an ion, normally sodium; transport requires a carrier protein but does not involve direct ATP expenditure and can occur regardless of the concentration gradient for the nutrient.

countercurrent exchange: The diffusion between two solutions that travel in opposite directions.

countercurrent multiplication: Active transport between two limbs of a loop that contains a fluid moving in one direction; responsible for the concentration of the urine in the kidney tubules.

covalent bond (kō-VĀ-lent): A chemical bond between atoms that involves the sharing of electrons.

coxa/coxae: A bone of the hip.

cranial: Pertaining to the head.

cranial nerves: Peripheral nerves originating at the brain.

craniosacral division (krā-nē-ō-SAK-ral): *See* **parasympathetic division.**

craniostenosis (krā-nē-ō-sten-Ō-sis): A skull deformity caused by premature closure of the cranial sutures.

cranium: The braincase; the skull bones that surround and protect the brain.

creatine: A nitrogenous compound synthesized in the body that can bind a high-energy phosphate and serve as an energy reserve.

creatine phosphate: A high-energy compound present in muscle cells; during muscular activity, the phosphate group is donated to ADP, regenerating ATP. Also called *phosphocreatine* and *phosphorylcreatine*.

creatinine: A breakdown product of creatine metabolism.

crenation: Cellular shrinkage due to an osmotic movement of water out of the cytoplasm.

cribriform plate: A portion of the ethmoid bone of the skull that contains the foramina used by the axons of olfactory receptors en route to the olfactory bulbs of the cerebrum.

cricoid cartilage (KRĪ-koyd): A ring-shaped cartilage that forms the inferior margin of the larynx.

crista/cristae: A ridge-shaped collection of hair cells in the ampulla of a semicircular canal; the crista and cupula form a receptor complex sensitive to movement along the plane of the canal.

cross-bridge: A myosin head that projects from the surface of a thick filament and that can bind to an active site of a thin filament in the presence of calcium ions.

cruciate ligaments: A pair of intracapsular ligaments (anterior and posterior) in the knee.

cryosurgery: A surgical technique that involves freezing and killing cells in a localized area.

cryptorchid testis: An undescended testis that is in the abdominopelvic cavity rather than in the scrotum.

cuneiform cartilages (kū-NĒ-i-form): A pair of small cartilages in the larynx.

cupula (KŪ-pū-la): A gelatinous mass that sits in the ampulla of a semicircular canal in the inner ear and whose movement stimulates the hair cells of the crista.

curare: A toxin that prevents neural stimulation of neuromuscular junctions.

Cushing's disease: A condition caused by the oversecretion of adrenal steroids.

cutaneous membrane: The epidermis and papillary layer of the dermis.

cuticle: The layer of dead, cornified cells that surrounds the shaft of a hair; for nails, *see* **eponychium.**

cyanosis: A bluish coloration of the skin due to the presence of deoxygenated blood in vessels near the body surface.

cyst: A fibrous capsule containing fluid or other material.

cystic duct: A duct that carries bile between the gallbladder and the common bile duct.

cystitis: Inflammation of the urinary bladder.

cytochrome (SĪ-tō-krōm): A pigment component of the electron transport system; a structural relative of heme.

cytokinesis (sī-tō-ki-NĒ-sis): The cytoplasmic movement that separates two daughter cells at the completion of mitosis.

cytology (sī-TOL-o-jē): The study of cells.

cytoplasm: The material between the cell membrane and the nuclear membrane; cell contents.

cytosine: A pyrimidine; one of the nitrogenous base in the nucleic acids RNA and DNA.

cytoskeleton: A network of microtubules and microfilaments in the cytoplasm.

cytosol: The fluid portion of the cytoplasm.

cytotoxic: Poisonous to living cells.

cytotoxic T cells: Lymphocytes of the cellular immune response that kill target cells by direct contact or through the secretion of lymphotoxins; also called *killer T cells* and *T_C cells*.

D

daughter cells: Genetically identical cells produced by somatic cell division.

deamination (dē-am-i-NĀ-shun): The removal of an amino group from an amino acid.

decarboxylation (dē-kar-boks-i-LĀ-shun): The removal of a molecule of CO_2.

decerebrate: Lacking a cerebrum.

decomposition reaction: A chemical reaction that breaks a molecule into smaller fragments.

decubitis ulcers: Ulcers that form where chronic pressure interrupts circulation to a portion of the skin.

decussate: To cross over to the opposite side, usually referring to the crossover of the pyramidal tracts on the ventral surface of the medulla oblongata.

defecation (def-e-KĀ-shun): The elimination of fecal wastes.

deglutition: Swallowing.

degradation: Breakdown, catabolism.

dehydration: A reduction in the water content of the body that threatens homeostasis.

dehydration synthesis: The joining of two molecules associated with the removal of a water molecule.

delta cell: A pancreatic islet cell that secretes somatostatin.

dementia: A loss of mental abilities.

demyelination: The loss of the myelin sheath of an axon, normally due to chemical or physical damage to Schwann cells or oligodendrocytes.

denaturation: Irreversible alteration in the three-dimensional structure of a protein.

dendrite (DEN-drīt): A sensory process of a neuron.

denticulate ligaments: Supporting fibers that extend laterally from the surface of the spinal cord, tying the pia mater to the dura mater and providing lateral support for the spinal cord.

dentin (DEN-tin): The bonelike material that forms the body of a tooth; differs from bone in that it lacks osteocytes and osteons.

deoxyribonucleic acid (DNA) (dē-ok-sē-rī-bō-noo-KLĀ-ik): A nucleic acid consisting of a chain of nucleotides that contain the sugar deoxyribose and the nitrogenous bases adenine, guanine, cytosine, and thymine.

deoxyribose: A five-carbon sugar resembling ribose but lacking an oxygen atom.

depolarization: A change in the transmembrane potential from a negative value toward 0 mV.

depression: Inferior (downward) movement of a body part.

dermatitis: Inflammation of the skin.

dermatome: A sensory region monitored by the dorsal rami of a single spinal segment.

dermis: The connective tissue layer beneath the epidermis of the skin.

detrusor muscle (de-TROO-sor): A smooth muscle in the wall of the urinary bladder.

detumescence (dē-tū-MES-ens): The loss of a penile erection.

development: The growth and the acquisition of increasing structural and functional complexity; includes the period from conception to maturity.

diabetes insipidus: Polyuria due to inadequate production of ADH.

diabetes mellitus (mel-Ī-tus): Polyuria and glycosuria, most commonly due to the inadequate production of insulin with resulting elevation of blood glucose levels.

diabetogenic effect: An elevation in blood sugar concentrations after the secretion of growth hormone or glucagon.

dialysis: Diffusion between two solutions of differing solute concentrations across a semipermeable membrane that contains pores to permit the passage of some solutes but not others; to regulate the composition of blood.

diapedesis (dī-a-pe-DĒ-sis): The movement of white blood cells through the walls of blood vessels by migration between adjacent endothelial cells.

diaphragm (DĪ-a-fram): Any muscular partition; often used to refer to the respiratory muscle that separates the thoracic cavity from the abdominopelvic cavity.

diaphysis (dī-A-fi-sis): The shaft of a long bone.

diarrhea (dī-a-RĒ-uh): Abnormally frequent defecation, associated with the production of unusually fluid feces.

diarthrosis (dī-ar-THRŌ-sis): A synovial joint.

diastolic pressure: Pressure measured in the walls of a muscular artery when the left ventricle is in diastole.

diencephalon (dī-en-SEF-a-lon): A division of the brain that includes the epithalamus, thalamus, and hypothalamus.

differential count: The determination of the relative abundance of each type of white blood cell on the basis of a random sampling of 100 white blood cells.

differentiation: The gradual appearance of characteristic cellular specializations during development as the result of gene activation or repression.

diffusion: Passive molecular movement from an area of relatively high concentration to an area of relatively low concentration.

digestion: The chemical breakdown of ingested materials into simple molecules that can be absorbed by the cells of the digestive tract.

digestive system: The digestive tract and associated glands.

digestive tract: *See* **gastrointestinal tract**.

dilate: To increase in diameter; to enlarge or expand.

diploid (DIP-loyd): Having a complete somatic complement of chromosomes (23 pairs in human cells).

disaccharide (di-SAK-a-rīd): A compound formed by the joining of two simple sugars by dehydration synthesis.

dislocation: The forceful displacement of an articulating bone to an abnormal position, generally accompanied by damage to tendons, ligaments, the articular capsule, or other structures.

dissociation (di-sō-sē-Ā-shun): *See* **ionization**.

distal: Movement away from the point of attachment or origin; for a limb, away from its attachment to the trunk.

distal convoluted tubule (DCT): The portion of the nephron closest to the connecting tubules and collecting duct; an important site of active secretion.

diuresis (dī-ūr-Ē-sis): Fluid loss at the kidneys; the production of urine.

divergence: In neural tissue, the spread of excitation from one neuron to many neurons; an organizational pattern common along sensory pathways of the central nervous system.

diverticulitis (dī-ver-tik-ū-LĪ-tis): Inflammation of a diverticulum.

diverticulosis (dī-ver-tik-ū-LŌ-sis): The formation of diverticula in the wall of an organ.

diverticulum: A sac or pouch in the wall of the colon or other organ.

dizygotic twins (dī-zī-GOT-ik): Twins that result from the fertilization of two different oocytes.

DNA molecule: Two DNA strands wound in a double helix and held together by weak bonds between complementary nitrogenous base pairs.

dominant gene: A gene whose presence will determine the phenotype, regardless of the nature of its allelic companion.

dopamine (DŌ-pa-mēn): An important neurotransmitter in the central nervous system.

dorsal: Toward the back, posterior.

dorsal root ganglion: A peripheral nervous system ganglion containing the cell bodies of sensory neurons.

dorsiflexion: The elevation of the superior surface of the foot through movement at the ankle.

Down syndrome: A genetic abnormality resulting from the presence of three copies of chromosome 21; individuals with this condition have characteristic physical and intellectual deficits.

duct: A passageway that delivers exocrine secretions to an epithelial surface.

ductus arteriosus (ar-tē-rē-Ō-sus): A vascular connection between the pulmonary trunk and the aorta that functions throughout fetal life; normally closes at birth or shortly thereafter and persists as the ligamentum arteriosum.

ductus deferens (DUK-tus DEF-e-renz): A passageway that carries sperm from the epididymis to the ejaculatory duct.

duodenal ampulla: A chamber that receives bile from the common bile duct and pancreatic secretions from the pancreatic duct.

duodenal glands: *See* **submucosal glands**.

duodenal papilla: A conical projection from the inner surface of the duodenum that contains the opening of the duodenal ampulla.

duodenum (doo-AH-de-num): The proximal 25 cm of the small intestine that contains short villi and submucosal glands.

dura mater (DŪ-ra MĀ-ter): The outermost component of the meninges that surround the brain and spinal cord.

dynamic equilibrium: The maintenance of normal body orientation as sudden changes in position (rotation, acceleration, etc.) occur.

dynorphin (dī-NOR-fin): A powerful neuromodulator produced in the central nervous system that blocks pain perception by inhibiting pain pathways.

dyslexia: The impaired ability to comprehend written words.

dysmenorrhea: Painful menstruation.

dysmetria (dis-MĒT-rē-uh): Difficulty in performing movements due to problems with the interpretation and anticipation of the distance to be covered.

dysuria (dis-ū-rē-uh): Painful urination.

E

eccrine glands (EK-rin): Sweat glands of the skin that produce a watery secretion.

echocardiography (ek-ō-kar-dē-OG-ra-fē): Examination of the heart by using modified ultrasound techniques.

ectoderm: One of the three primary germ layers; covers the surface of the embryo and gives rise to the nervous system, the epidermis and associated glands, and a variety of other structures.

ectopic (ek-TOP-ik): Outside the normal location.

effector: A peripheral gland or muscle cell innervated by a motor neuron.

efferent: Away from.

efferent arteriole: An arteriole carrying blood away from a glomerulus of the kidney.

efferent fiber: An axon that carries impulses away from the central nervous system.

egestion: *See* **defecation**.

ejaculation (ē-jak-ū-LĀ-shun): The ejection of semen from the penis as the result of muscular contractions of the bulbospongiosus and ischiocavernosus muscles.

ejaculatory ducts (ē-JAK-ū-la-to-rē): Short ducts that pass within the walls of the prostate gland and connect the ductus deferens with the prostatic urethra.

elastase (ē-LAS-tās): A pancreatic enzyme that breaks down elastin fibers.

elastin: Connective tissue fibers that stretch and recoil, providing elasticity to connective tissues.

electrical coupling: A connection between adjacent cells that permits the movement of ions and the transfer of graded or conducted changes in the transmembrane potential from cell to cell.

electrocardiogram (ECG, EKG) (e-lek-trō-KAR-dē-ō-gram): A graphic record of the electrical activities of the heart, as monitored at specific locations on the body surface.

electroencephalogram (EEG): A graphic record of the electrical activities of the brain.

electrolytes (ē-LEK-trō-līts): Soluble inorganic compounds whose ions will conduct an electrical current in solution.

electron: One of the three fundamental particles; a subatomic particle that bears a negative charge and normally orbits the protons of the nucleus.

electron transport system (ETS): The cytochrome system responsible for most of the energy production in living cells; a complex bound to the inner mitochondrial membrane.

element: All the atoms with the same atomic number.

elephantiasis (el-e-fan-TĪ-a-sis): A lymphedema caused by infection and blockage of lymphatics by mosquito-borne parasites.

elevation: Movement in a superior, or upward, direction.

embolism (EM-bō-lizm): The obstruction or closure of a vessel by an embolus.

embolus (EM-bō-lus): An air bubble, fat globule, or blood clot drifting in the circulation.

embryo (EM-brē-ō): The developmental stage beginning at fertilization and ending at the start of the third developmental month.

embryology (em-brē-OL-ō-jē): The study of embryonic development, focusing on the first 2 months after fertilization.

emesis (EM-e-sis): Vomiting.

emmetropia: Normal vision.

emulsification (ē-mul-si-fi-KĀ-shun): The physical breakup of fats in the digestive tract, forming smaller droplets accessible to digestive enzymes; normally the result of mixing with bile salts.

enamel: Crystalline material similar in mineral composition to bone, but harder and without osteocytes, that covers the exposed surfaces of the teeth.

encephalitis: Inflammation of the brain.

endocarditis: Inflammation of the endocardium of the heart.

endocardium (en-dō-KAR-dē-um): The simple squamous epithelium that lines the heart and is continuous with the endothelium of the great vessels.

endochondral ossification (en-dō-KON-dral): The conversion of a cartilaginous model to bone; the characteristic mode of formation for skeletal elements other than the bones of the cranium, the clavicles, and sesamoid bones.

endocrine gland: A gland that secretes hormones into the blood.

endocrine system: The endocrine glands of the body.

endocytosis (EN-dō-sī-tō-sis): The movement of relatively large volumes of extracellular material into the cytoplasm via the formation of a membranous vesicle at the cell surface; includes pinocytosis and phagocytosis.

endoderm: One of the three primary germ layers; the layer on the undersurface of the embryonic disc; gives rise to the epithelia and glands of the digestive system, the respiratory system, and portions of the urinary system.

endogenous: Produced within the body.

endolymph (EN-dō-limf): The fluid contents of the membranous labyrinth (the saccule, utricle, semicircular canals, and cochlear duct) of the inner ear.

endometrial glands: The secretory glands of the endometrium.

endometrium (en-dō-MĒ-trē-um): The mucous membrane lining the uterus.

endomysium (en-dō-MĪS-ē-um): A delicate network of connective tissue fibers that surrounds individual muscle cells.

endoneurium: A delicate network of connective tissue fibers that surrounds individual nerve fibers.

endoplasmic reticulum (en-dō-PLAZ-mik re-TIK-ū-lum): A network of membranous channels in the cytoplasm of a cell that function in intracellular transport, synthesis, storage, packaging, and secretion.

endorphins (en-DOR-finz): Neuromodulators produced in the central nervous system that inhibit activity along pain pathways.

endosteum: An incomplete cellular lining on the inner (medullary) surfaces of bones.

endothelium (en-dō-THĒ-lē-um): The simple squamous epithelium that lines blood and lymphatic vessels.

enkephalins (en-KEF-a-linz): Neuromodulators produced in the central nervous system that inhibit activity along pain pathways.

enteritis (en-ter-Ī-tis): Inflammation of the intestinal tract.

enterocrinin: A hormone secreted by the duodenal lining after exposure to chyme; stimulates the secretion of the duodenal glands.

enteroendocrine cells (en-ter-ō-EN-dō-krin): Endocrine cells scattered among the epithelial cells that line the digestive tract.

enterogastric reflex: The reflexive inhibition of gastric secretion; initiated by the arrival of chyme in the small intestine.

enterohepatic circulation: The excretion of bile salts by the liver, followed by the absorption of bile salts by intestinal cells for return to the liver by the hepatic portal vein.

enterokinase: An enzyme in the lumen of the small intestine that activates the proenzymes secreted by the pancreas.

enzyme: A protein that catalyzes a specific biochemical reaction.

eosinophil (ē-ō-SIN-ō-fil): A microphage (white blood cell) with a lobed nucleus and red-staining granules; participates in the immune response and is especially important during allergic reactions.

ependyma (ep-EN-di-mah): The layer of cells lining the ventricles and central canal of the central nervous system.

epicardium: A serous membrane covering the outer surface of the heart; also called the *visceral pericardium*.

epidermis: The epithelium covering the surface of the skin.

epididymis (ep-i-DID-i-mus): A coiled duct that connects the rete testis to the ductus deferens; site of functional maturation of spermatozoa.

epidural block: Anesthesia caused by the elimination of sensory inputs from dorsal nerve roots after drugs are introduced into appropriate regions of the epidural space.

epidural space: The space between the spinal dura mater and the walls of the vertebral foramen; contains blood vessels and adipose tissue; a common site of injection for regional anesthesia.

epiglottis (ep-i-GLOT-is): A blade-shaped flap of tissue, reinforced by cartilage, that is attached to the dorsal and superior surface of the thyroid cartilage; folds over the entrance to the larynx during swallowing.

epimysium (ep-i-MĪS-ē-um): A dense layer of collagen fibers that surrounds a skeletal muscle and is continous with the tendons/aponeuroses of the muscle and with the perimysium.

epineurium: A dense layer of collagen fibers that surrounds a peripheral nerve.

epiphyseal plate (e-pi-FI-sē-al): The cartilaginous region between the epiphysis and diaphysis of a growing bone.

epiphysis (e-PIF-i-sis): The head of a long bone.

epistaxis (ep-i-STAK-sis): A nosebleed.

epithelium (e-pi-THĒ-lē-um): One of the four primary tissue types; a layer of cells that forms a superficial covering or an internal lining of a body cavity or vessel.

eponychium (ep-ō-NIK-ē-um): A narrow zone of stratum corneum that extends across the surface of a nail at its exposed base; also called *cuticle*.

equational division: The second meiotic division.

equilibrium (ē-kwi-LIB-rē-um): A dynamic state in which two opposing forces or processes are in balance.

erection: The stiffening of the penis due to the engorgement of the erectile tissues of the corpora cavernosa and the corpus spongiosum.

erythema (er-i-THĒ-ma): Redness and inflammation at the surface of the skin.

erythrocyte (e-RITH-rō-sīt): A red blood cell; a blood cell that has no nucleus and contains large quantities of hemoglobin.

erythrocytosis (e-rith-rō-sī-TŌ-sis): An abnormally large number of erythrocytes in the circulating blood.

erythropoietin (e-rith-rō-POY-ē-tin): A hormone released by tissues, especially the kidneys, exposed to low oxygen concentrations; stimulates erythropoiesis (red blood cell formation) in bone marrow.

Escherichia coli: A normal bacterial resident of the large intestine.

esophagus: A muscular tube that connects the pharynx to the stomach.

essential amino acids: Amino acids that cannot be synthesized in the body in adequate amounts and must be obtained from the diet.

essential fatty acids: Fatty acids that cannot be synthesized in the body and must be obtained from the diet.

estrogens (ES-trō-jenz): A class of steroid sex hormones that includes estradiol.

eupnea (ŪP-nē-uh): Normal quiet breathing.

evaporation: A movement of molecules from the liquid to the gaseous state.

eversion (ē-VER-shun): A turning outward.

excitable membranes: Membranes that conduct action potentials, a characteristic of muscle and nerve cells.

excitatory postsynaptic potential (EPSP): The depolarization of a postsynaptic membrane by a chemical neurotransmitter released by the presynaptic cell.

excretion: Elimination from the body.

exocrine gland: A gland that secretes onto the body surface or into a passageway connected to the exterior.

exocytosis (EK-sō-sī-tō-sis): The ejection of cytoplasmic materials by the fusion of a membranous vesicle with the cell membrane.

expiration: Exhalation; breathing out.

expiratory reserve: The amount of additional air that can be voluntarily moved out of the respiratory tract after a normal tidal expiration.

extension: An increase in the angle between two articulating bones; the opposite of flexion.

external auditory canal: A passageway in the temporal bone that leads to the tympanic membrane of the inner ear.

external ear: The pinna, external auditory meatus, external auditory canal, and tympanic membrane.

external nares: The nostrils; the external openings into the nasal cavity.

external respiration: The diffusion of gases between the alveolar air and the alveolar capillaries and between the systemic capillaries and peripheral tissues.

exteroceptors: General sensory receptors in the skin, mucous membranes, and special sense organs that provide information about the external environment and about our position within it.

extracellular fluid: All body fluids other than that contained within cells; includes plasma and interstitial fluid.

extraembryonic membranes: The yolk sac, amnion, chorion, and allantois.

extrafusal fibers: Contractile muscle fibers (as opposed to the sensory intrafusal fibers, or muscle spindles).

extrapyramidal system (EPS): Nuclei and tracts associated with the involuntary control of muscular activity.

extremities: The limbs.

extrinsic pathway: A clotting pathway that begins with damage to blood vessels or surrounding tissues and ends with the formation of tissue thromboplastin.

F

fabella: A sesamoid bone commonly found in the gastrocnemius muscle just behind the knee.

facilitated diffusion: The passive movement of a substance across a cell membrane by means of a protein carrier.

facilitation: The depolarization of a nerve cell membrane toward threshold, or making the cell more sensitive to depolarizing stimuli.

falciform ligament (FAL-si-form): A sheet of mesentery that contains the ligamentum teres, the fibrous remains of the umbilical vein of the fetus.

falx (falks): Sickle-shaped.

falx cerebri (falks ser-Ē-brē): The curving sheet of dura mater that extends between the two cerebral hemispheres; encloses the superior sagittal sinus.

fasciae (FASH-ē-ē): Connective tissue fibers, primarily collagenous, that form sheets or bands beneath the skin to attach, stabilize, enclose, and separate muscles and other internal organs.

fasciculus (fa-SIK-ū-lus): A small bundle; usually referring to a collection of nerve axons or muscle fibers.

fatty acids: Hydrocarbon chains that end in a carboxylic acid group.

fauces (FAW-sēz): The passage from the mouth to the pharynx, bounded by the palatal arches, the soft palate, and the uvula.

febrile: Characterized by or pertaining to a fever.

feces: Waste products eliminated by the digestive tract at the anus; contains indigestible residue, bacteria, mucus, and epithelial cells.

fenestra: An opening.

fertilization: The fusion of oocyte and sperm to form a zygote.

fetus: The developmental stage lasting from the start of the third developmental month to delivery.

fibrillation (fi-bri-LĀ-shun): Uncoordinated contractions of individual muscle cells that impair or prevent normal function.

fibrin (FĪ-brin): Insoluble protein fibers that form the basic framework of a blood clot.

fibrinogen (fī-BRIN-ō-jen): A plasma protein that is the soluble precursor of the fibrous protein fibrin.

fibrinolysis (fī-brin-OL-i-sis): The breakdown of the fibrin strands of a blood clot by a proteolytic enzyme.

fibroblasts (FĪ-brō-blasts): Cells of connective tissue proper that are responsible for the production of extracellular fibers and the secretion of the organic compounds of the extracellular matrix.

fibrocartilage: Cartilage containing an abundance of collagen fibers; found around the edges of joints, in the intervertebral discs, the menisci of the knee, etc.

fibrous tunic: The outermost layer of the eye, composed of the sclera and cornea.

fibula (fib-YOO-luh): The lateral, relatively small bone of the leg.

filariasis (fil-a-RĪ-a-sis): A condition resulting from infection by mosquito-borne parasites; may cause *elephantiasis.*

filiform papillae: Slender conical projections from the dorsal surface of the anterior two-thirds of the tongue.

filtrate: The fluid produced by filtration at a glomerulus in the kidney.

filtration: The movement of a fluid across a membrane whose pores restrict the passage of solutes on the basis of size.

filtration pressure: The hydrostatic pressure responsible for filtration.

filum terminale: A fibrous extension of the spinal cord that extends from the conus medullaris to the coccygeal ligament.

fimbriae (FIM-brē-ē): Fringes; the fingerlike processes that surround the entrance to the uterine tube.

fissure: An elongate groove or opening.

fistula: An abnormal passageway between two organs or from an internal organ or space to the body surface.

flaccid: Limp, soft, flabby; a muscle without muscle tone.

flagellum/flagella (fla-JEL-uh): An organelle structurally similar to a cilium but used to propel a cell through a fluid.

flatus: Intestinal gas.

flavin adenine dinucleotide: A coenzyme important in oxidative phosphorylation; it cycles between the oxidized ($FADH_2$) and reduced (FAD) states.

flavin adenine mononucleotide: A coenzyme important in oxidative phosphorylation; cycles between the oxidized ($FMNH_2$) and reduced (FMN) states.

flexion (FLEK-shun): A movement that reduces the angle between two articulating bones; the opposite of extension.

flexor: A muscle that produces flexion.

flexor reflex: A reflex contraction of the flexor muscles of a limb in response to an unpleasant stimulus.

flexure: A bending.

fluoroscope: An instrument that permits the body to be examined in real time with X-rays rather than with fixed images on photographic plates.

folia (FŌ-lē-uh): Leaflike folds; the slender folds in the surface of the cerebellar cortex.

follicle (FOL-i-kul): A small secretory sac or gland.

follicle-stimulating hormone (FSH): A hormone secreted by the anterior pituitary; stimulates oogenesis (female) and spermatogenesis (male).

folliculitis (fo-lik-ū-LĪ-tis): Inflammation of a follicle, such as a hair follicle of the skin.

fontanel (fon-tuh-NEL): A relatively soft, flexible, fibrous region between two flat bones in the developing skull; also spelled fontanelle.

foramen: An opening or passage through a bone.

forearm: The distal portion of the upper limb between the elbow and wrist.

forebrain: The cerebrum.

fornix (FOR-niks): An arch or the space bounded by an arch; in the brain, an arching tract that connects the hippocampus with the mamillary bodies; in the eye, a slender pocket situated where the epithelium of the ocular conjunctiva folds back on itself as the palpebral conjunctiva.

fossa: A shallow depression or furrow in the surface of a bone.

fourth ventricle: An elongate ventricle of the metencephalon (pons and cerebellum) and the myelencephalon (medulla) of the brain; the roof contains a region of choroid plexus.

fovea (FŌ-vē-uh): The portion of the retina providing the sharpest vision, with the highest concentration of cones; also called the *macula lutea.*

fracture: A break or crack in a bone.

frenulum (FREN-ū-lum): A bridle; *see* **lingual frenulum.**

frontal plane: A sectional plane that divides the body into an anterior portion and a posterior portion; also called the *coronal plane.*

fructose: A hexose (six-carbon simple sugar) found in foods and in semen.

fundus (FUN-dus): The base of an organ.

fungiform papillae: The mushroom-shaped papillae on the dorsal and dorsolateral surfaces of the tongue.

furuncle (FUR-ung-kl): A boil, resulting from the invasion and inflammation of a hair follicle or sebaceous gland.

G

gallbladder: The pear-shaped reservoir for the bile secreted by the liver.

gametes (GAM-ēts): Reproductive cells (sperm or oocytes) that contain half the normal chromosome complement.

gametogenesis (ga-mē-tō-JEN-e-sis): The formation of gametes.

gamma aminobutyric acid (GABA) (GAM-ma a-MĒ-nō-bū-TIR-ik): A neurotransmitter of the central nervous system whose effects are generally inhibitory.

gamma motor neurons: Motor neurons that adjust the sensitivities of muscle spindles (intrafusal fibers).

ganglion/ganglia: A collection of neuron cell bodies outside the central nervous system.

gangliosides: Glycolipids that are important components of cell membranes in the central nervous system.

gap junctions: Connections between cells that permit electrical coupling.

gaster (GAS-ter): The stomach; the body, or belly, of a skeletal muscle.

gastrectomy (gas-TREK-to-mē): The partial or total surgical removal of the stomach.

gastric: Pertaining to the stomach.

gastric glands: The tubular glands of the stomach whose cells produce acid, enzymes, intrinsic factor, and hormones.

gastric inhibitory peptide (GIP): A duodenal hormone released when the arriving chyme contains large quantities of carbohydrates; triggers the secretion of insulin and a slowdown in gastric activity.

gastrin (GAS-trin): A hormone produced by enteroendocrine cells of the stomach, after exposure to mechanical stimuli or vagal stimulation, and of the duodenum, after exposure to chyme that contains undigested proteins.

gastritis (gas-TRĪ-tis): Inflammation of the stomach.

gastroenteric reflex (gas-trō-en-TER-ik): An increase in peristalsis along the small intestine; triggered by the arrival of food in the stomach.

gastroileal reflex (gas-trō-IL-ē-al): Peristaltic movements that shift materials from the ileum to the colon; triggered by the arrival of food in the stomach.

gastrointestinal (GI) tract: An internal passageway that begins at the mouth, ends at the anus, and is lined by a mucous membrane; also known as the *digestive tract*.

gastroscope: A fiber-optic instrument that permits visual inspection of the stomach lining.

gastrulation (gas-troo-LĀ-shun): The movement of cells of the inner cell mass that creates the three primary germ layers of the embryo.

gene: A portion of a DNA strand that functions as a hereditary unit and is found at a particular site on a specific chromosome.

genetic engineering: Research and experiments involving the manipulation of the genetic makeup of an organism.

genetics: The study of mechanisms of heredity.

geniculate (je-NIK-ū-lāt): Like a little knee; the medial geniculates and the lateral geniculates are nuclei in the walls of the thalamus of the brain.

genitalia (jen-i-TĀ-lē-uh): The reproductive organs.

genotype (JĒN-ō-tīp): An individual's genetic complement, which determines the individual's phenotype.

germinal centers: Pale regions in the interior of lymphoid tissues or nodules, where cell divisions are under way.

gestation (jes-TĀ-shun): The period of intrauterine development.

gingivae (JIN-ji-vē): The gums.

gingivitis: Inflammation of the gums.

gland: Cells that produce exocrine or endocrine secretions.

glans: The expanded tip of the penis that surrounds the external urethral meatus; continuous with the corpus spongiosum.

glaucoma: An eye disorder characterized by rising intraocular pressures due to inadequate drainage of aqueous humor at the canal of Schlemm.

glenoid cavity: A rounded depression that forms the articular surface of the scapula at the shoulder joint.

glial cells (GLĒ-al): Supporting cells in the neural tissue of the central nervous system and peripheral nervous system.

globular proteins: Proteins whose tertiary structure makes them rounded and compact.

glomerular capsule: The expanded initial portion of the nephron that surrounds the glomerulus.

glomerular filtration rate: The rate of filtrate formation at the glomerulus.

glomerulonephritis (glo-mer-ū-lō-nef-RĪ-tis): Inflammation of the glomeruli of the kidneys.

glomerulus (glo-MER-ū-lus): A ball or knot; in the kidneys, a knot of capillaries that projects into the enlarged, proximal end of a nephron; the site of filtration, the first step in the production of urine.

glossopharyngeal nerve (glos-ō-fa-RIN-jē-al): Cranial nerve IX.

glottis (GLOT-is): Passageway from the pharynx to the larynx.

glucagon (GLOO-ka-gon): A hormone secreted by the alpha cells of the pancreatic islets; elevates blood glucose concentrations.

glucocorticoids: Hormones secreted by the zona fasciculata of the adrenal cortex to modify glucose metabolism; cortisol, cortisone, and corticosterone are important examples.

glucogenic amino acids: Amino acids that can be broken down, converted to pyruvic acid, and used in the synthesis of glucose (gluconeogenesis).

gluconeogenesis (gloo-kō-nē-ō-JEN-e-sis): The synthesis of glucose from protein or lipid precursors.

glucose (GLOO-kōs): A six-carbon sugar, $C_6H_{12}O_6$; the preferred energy source for most cells and the only energy source for neurons under normal conditions.

glucose-dependent insulinotropic hormone: *See* **gastric inhibitory peptide.**

glycerides: Lipids composed of glycerol bound to fatty acids.

glycogen (GLĪ-kō-jen): A polysaccharide that represents an important energy reserve; a polymer consisting of a long chain of glucose molecules.

glycogenesis: The synthesis of glycogen from glucose molecules.

glycogenolysis: Glycogen breakdown and the liberation of glucose molecules.

glycolipids (glī-cō-LIP-idz): Compounds created by the combination of carbohydrate and lipid components.

glycolysis (glī-KOL-i-sis): The anaerobic cytoplasmic breakdown of glucose into lactic acid by way of pyruvic acid, with a net gain of two ATP.

glycoprotein (glī-kō-PRŌ-tēn): A compound containing a relatively small carbohydrate group attached to a large protein.

glycosuria (glī-kō-SOO-rē-uh): The presence of glucose in the urine.

goblet cell: A goblet-shaped, mucus-producing, unicellular gland in certain epithelia of the digestive and respiratory tracts.

goiter: An enlargement of the thyroid gland.

Golgi apparatus (gol-jē): A cellular organelle consisting of a series of membranous plates that give rise to lysosomes and secretory vesicles.

Golgi tendon organ: *See* **tendon organ.**

gomphosis (gom-FŌ-sis): A fibrous synarthrosis that binds a tooth to the bone of the jaw; *see* **periodontal ligament**.

gonadotropic hormones: FSH and LH, hormones that stimulate gamete development and sex hormone secretion.

gonadotropin-releasing hormone (GnRH) (gō-nad-ō-TRŌ-pin): A hypothalamic releasing hormone that causes the secretion of FSH and LH by the anterior pituitary gland.

gonadotropins: Gonadotropic hormones.

gonads (GŌ-nads): Reproductive organs that produce gametes and hormones.

gout: A condition resulting from elevated uric acid concentrations in the blood and peripheral tissues.

granulocytes (GRAN-ū-lō-sīts): White blood cells containing granules visible with the light microscope; includes eosinophils, basophils, and neutrophils; also called *granular leukocytes.*

gray matter: Areas in the central nervous system dominated by neuron bodies, glial cells, and unmyelinated axons.

gray ramus: A bundle of postganglionic sympathetic nerve fibers that go to a spinal nerve for distribution to effectors in the body wall, skin, and extremities.

greater omentum: A large fold of the dorsal mesentery of the stomach that hangs anterior to the intestines.

greater vestibular glands: Mucous glands in the vaginal walls that secrete into the vestibule; the equivalent of the bulbourethral glands of the male.

greenstick fracture: A fracture in which a bone cracks and bends, most commonly involving the long bones of young children.

groin: The inguinal region.

gross anatomy: The study of the structural features of the human body without the aid of a microscope.

growth hormone (GH): An anterior pituitary hormone that stimulates tissue growth and anabolism when nutrients are abundant and restricts tissue glucose dependence when nutrients are in short supply.

growth hormone–inhibiting hormone (GH-IH): *See* **somatostatin.**

guanine: A purine; one of the nitrogenous bases in the nucleic acids RNA and DNA.

gustation (gus-TĀ-shun): Taste.

gynecologist (gī-ne-KOL-o-jist): A physician specializing in the female reproductive system.

gyrus (JĪ-rus): A prominent fold or ridge of neural cortex on the surfaces of the cerebral hemispheres.

H

hair: A keratinous strand produced by epithelial cells of the hair follicle.

hair cells: Sensory cells of the inner ear.

hair follicle: An accessory structure of the integument; a tube lined by a stratified squamous epithelium that begins at the surface of the skin and ends at the hair papilla.

hair root: A thickened, conical structure consisting of a connective tissue papilla and the overlying matrix, a layer of epithelial cells that produces the hair shaft.

hallux: The big toe.

haploid (HAP-loyd): Possessing half the normal number of chromosomes; a characteristic of gametes.

hapten: A partial antigen that can bind to an antibody but cannot stimulate antibody production; a foreign compound that has only one antigenic determinant site.

hard palate: The bony roof of the oral cavity, formed by the maxillary and palatine bones.

Hassall's corpuscles: Aggregations of epithelial cells in the thymus whose functions are unknown.

haustra (HAWS-truh): Saclike pouches along the length of the large intestine that result from tension in the taenia coli.

heat exhaustion: A condition characterized by excessive perspiration and resulting in dangerous fluid and salt losses.

heat stroke: A condition resulting from the failure of the normal temperature control mechanisms; characterized by a cessation of sweating and a potentially fatal elevation of body temperature.

Heimlich maneuver (HĪM-lik): A technique for removing an airway blockage by external compression of the abdomen and forceful elevation of the diaphragm.

helper T cells: Lymphocytes (T_H cells) whose secretions and other activities coordinate the cell-mediated and antibody-mediated immune responses.

hematocrit (he-MAT-ō-krit): The percentage of the volume of whole blood contributed by cells; also called the *packed cell volume* (PCV) or the *volume of packed red cells* (VPRC).

hematoma: A tumor or swelling filled with blood.

hematuria (hē-ma-TOO-rē-uh): The presence of abnormal numbers of red blood cells in the urine.

heme (hēm): A porphyrin ring containing a central iron atom that can reversibly bind oxygen molecules; a component of the hemoglobin molecule.

hemiplegia: Paralysis affecting one side of the body (arm, trunk, and leg).

hemocytoblasts: Stem cells whose divisions produce each of the various populations of blood cells.

hemodialysis (hē-mō-dī-AL-i-sis): Dialysis of the blood.

hemoglobin (hē-mō-GLŌ-bin): A protein composed of four globular subunits, each bound to a heme molecule, gives red blood cells the ability to transport oxygen in the blood.

hemolysis: The breakdown (lysis) of red blood cells.

hemophilia (hē-mō-FIL-ē-uh): A congenital condition due to the inadequate synthesis of one of the clotting factors.

hemopoiesis (hē-mō-poy-Ē-sis): Blood cell formation and differentiation.

hemorrhage: Blood loss.

hemorrhoids (HEM-o-roydz): Swollen, varicose veins that protrude from the walls of the rectum and/or anorectal canal.

hemostasis: The cessation of bleeding.

hemothorax: The entry of blood into one of the pleural cavities.

heparin (HEP-a-rin): An anticoagulant released by activated basophils and mast cells.

hepatic duct: The duct that carries bile away from the liver lobes and toward the union with the cystic duct.

hepatic portal vein: The vessel that carries blood between the intestinal capillaries and the sinusoids of the liver.

hepatitis (hep-a-TĪ-tis): Inflammation of the liver, resulting from exposure to toxic chemicals, drugs, or viruses.

hepatocyte (he-PAT-ō-sīt): A liver cell.

hernia: The protrusion of a loop or portion of a visceral organ through the abdominopelvic wall or into the thoracic cavity.

herniated disc: The rupture of the connective tissue sheath of the nucleus pulposus of an intervertebral disc.

heterotopic: Ectopic; outside the normal location.

heterozygous (het-er-ō-ZĪ-gus): Possessing two different alleles at corresponding sites on a chromosome pair; the individual's phenotype may be determined by one or both of the alleles.

hexose: A six-carbon simple sugar.

hiatus (hī-Ā-tus): A gap, cleft, or opening.

high-density lipoprotein (HDL): A lipoprotein with a relatively small lipid content; thought to be responsible for the movement of cholesterol from peripheral tissues to the liver.

hilum/hilus (HĪ-lum): A localized region where blood vessels, lymphatics, nerves, and/or other anatomical structures are attached to an organ.

hippocampus: A region beneath the floor of a lateral ventricle involved with emotional states and the conversion of short-term to long-term memories.

hirsutism (HER-soot-izm): Excessive hair growth in women that follows the distribution pattern typical of adult males; can be caused by the overproduction of androgens.

histamine (HIS-tuh-mēn): Chemical released by stimulated mast cells or basophils to initiate or enhance an inflammatory response.

histology (his-TOL-o-jē): The study of tissues.

histones: Proteins associated with the DNA of the nucleus; the DNA strands are wound around them.

holocrine (HO-lō-krin): A form of exocrine secretion in which the secretory cell becomes swollen with vesicles and then ruptures.

homeostasis (hō-mē-ō-STĀ-sis): The maintenance of a relatively constant internal environment.

homologous chromosomes (hō-MOL-o-gus): The members of a chromosome pair.

homozygous (hō-mō-ZĪ-gus): Having the same allele for a given phenotypic character on two homologous chromosomes.

hormone: A compound secreted by one cell that travels through the circulatory system to affect the activities of cells in another portion of the body.

human chorionic gonadotropin (hCG): The placental hormone that maintains the corpus luteum for the first 3 months of pregnancy.

human immunodeficiency virus (HIV): The infectious agent that causes acquired immune deficiency syndrome (AIDS).

human leukocyte antigen (HLA): *See* **MHC protein**.

human placental lactogen (hPL): The placental hormone that stimulates the functional development of the mammary glands.

humoral immunity: *See* **antibody-mediated immunity**.

hyaluronan: A carbohydrate component of proteoglycans in the matrix of many connective tissues.

hyaluronidase: An enzyme that breaks down hyaluronic acid; produced by some bacteria and found in the acrosomal cap of a sperm cell.

hydrocephalus: A condition resulting from excessive production or inadequate drainage of cerebrospinal fluid.

hydrogen bond: A weak interaction between the hydrogen atom on one molecule and a negatively charged portion of another molecule.

hydrolysis (hī-DROL-i-sis): The breakage of a chemical bond through the addition of a water molecule; the reverse of dehydration synthesis.

hydrophilic (hī-drō-PHIL-ik): Freely associating with water; readily entering into solution.

hydrophobic: Incapable of freely associating with water molecules; insoluble.

hydrostatic pressure: Fluid pressure.

hydroxyl group (hī-DROK-sil): OH^-.

hypercapnia (hī-per-KAP-nē-uh): High plasma carbon dioxide concentrations, commonly the result of hypoventilation or inadequate tissue perfusion.

hyperglycemia: Elevated plasma glucose concentrations.

hyperkalemia (hī-per-kā-LĒ-mē-uh): Abnormally high potassium concentrations in the extracellular fluid.

hypernatremia: Abnormally high sodium concentrations in the extracellular fluid.

hyperopia: Farsightedness, characterized by an inability to focus on nearby objects.

hyperplasia: An abnormal enlargement of an organ due to an increase in the number of cells.

hyperpnea (hī-perp-NĒ-uh): Abnormal increases in the rate and depth of respiration.

hyperpolarization: Movement of the transmembrane potential away from the normal resting potential and farther from 0 mV.

hyperreflexia: Abnormally exaggerated reflex responses to stimulation.

hypersecretion: The overactivity of glands that produce exocrine or endocrine secretions.

hypersensitivity: An overreaction to an allergen that results in tissue damage and inflammation.

hypertension: Abnormally high blood pressure.

hyperthermia: Excessively high body temperature.

hyperthyroidism: Excessive production of thyroid hormones.

hypertonic: In comparing two solutions, the solution with the higher osmolarity.

hypertrophy (hī-PER-trō-fē): An increase in the size of tissue without cell division.

hyperventilation (hī-per-ven-ti-LĀ-shun): A rate of respiration sufficient to reduce plasma P_{CO_2} to levels below normal.

hypervitaminosis (hī-per-vī-ta-min-Ō-sis): A clinical condition caused by the excessive ingestion and uptake of vitamins.

hypesthesia: An abnormally decreased sensitivity to stimuli.

hypoblast (HĪ-pō-blast): The undersurface of the inner cell mass that faces the blastocoel of the early embryo.

hypocapnia: An abnormally low plasma P_{CO_2}, often the result of hyperventilation.

hypodermic needle: A needle inserted through the skin to introduce drugs into the subcutaneous layer.

hypodermis: *See* **subcutaneous layer**.

hypokalemia (hī-pō-ka-LĒ-mē-uh): Abnormally low plasma potassium concentrations.

hyponatremia: Abnormally low plasma sodium concentrations.

hyponychium (hī-pō-NIK-ē-um): A thickening in the epidermis beneath the free edge of a nail.

hypophyseal portal system (hī-pō-FI-sē-al): The network of vessels that carry blood from capillaries in the hypothalamus to capillaries in the anterior pituitary gland (hypophysis).

hypophysis (hī-POF-i-sis): The anterior pituitary gland, which is subdivided into the pars distalis and the pars intermedia.

hyporeflexia: Abnormally depressed reflex responses to stimuli.

hyposecretion: Abnormally low rates of exocrine or endocrine secretion.

hypothalamus: The floor of the diencephalon; the region of the brain containing centers involved with the unconscious regulation of visceral functions, emotions, drives, and the coordination of neural and endocrine functions.

hypothermia (hī-pō-THER-mē-uh): An abnormally low body temperature.

hypothesis: A prediction that can be subjected to scientific analysis and review.

hypotonic: In comparing two solutions, the one with the lower osmolarity.

hypoventilation: A respiratory rate which is insufficient to keep plasma P_{CO_2} within normal levels.

hypovitaminosis: A clinical condition resulting from inadequate vitamin ingestion and uptake; vitamin deficiency.

hypovolemic (hī-pō-vō-LĒ-mik): An abnormally low blood volume.

hypoxia (hī-POKS-ē-uh): Low tissue oxygen concentrations.

I

ileocecal valve (il-ē-ō-SĒ-kal): A fold of mucous membrane that guards the connection between the ileum and the cecum.

ileostomy (il-ē-OS-to-mē): The surgical creation of an opening into the ileum; the opening created when the ilium is surgically attached to the abdominal wall.

ileum (IL-ē-um): The last 2.5 m of the small intestine.

ilium (IL-ē-um): The largest of the three bones whose fusion creates a coxa.

immunity: Resistance to injuries and diseases caused by foreign compounds, toxins, and pathogens.

immunization: The production of immunity by the deliberate exposure to antigens under conditions that prevent the development of illness but stimulate the production of memory B cells.

immunodeficiency: An inability to produce normal numbers and types of antibodies and sensitized lymphocytes.

immunoglobulin (i-mū-nō-GLOB-ū-lin): A circulating antibody.

immunosuppression (i-mū-nō-su-PRE-shun): The suppression of immune responses by the administration of drugs or exposure to toxic chemicals, radiation, or infection.

implantation (im-plan-TĀ-shun): The erosion of a blastocyst into the uterine wall.

impotence: The inability to obtain or maintain an erection.

inclusions: Aggregations of insoluble pigments, nutrients, or other materials in the cytoplasm.

incontinence (in-KON-ti-nens): The inability to control micturition (or defecation) voluntarily.

incus (IN-kus): The central auditory ossicle, situated between the malleus and the stapes in the middle ear cavity.

inducer: A stimulus that promotes the activity of a specific gene.

inexcitable: Incapable of conducting an action potential.

infarct: An area of dead cells that results from an interruption of circulation.

infection: The invasion and colonization of body tissues by pathogenic organisms.

inferior: Below, in reference to a particular structure, with the body in the anatomical position.

inferior vena cava: The vein that carries blood from the parts of the body below the heart to the right atrium.

infertility: The inability to conceive.

inflammation: A nonspecific defense mechanism that operates at the tissue level, characterized by swelling, redness, warmth, pain, and some loss of function.

inflation reflex: A reflex mediated by the vagus nerve (N X) that prevents overexpansion of the lungs.

infundibulum (in-fun-DIB-ū-lum): A tapering, funnel-shaped structure; in the nervous system, the connection between the pituitary gland and the hypothalamus; in the uterine tube, the entrance bounded by fimbriae that receives the oocytes at ovulation.

ingestion: The introduction of materials into the digestive tract by way of the mouth.

inguinal canal: A passage through the abdominal wall that marks the path of testicular descent and that contains the testicular arteries, veins, and ductus deferens.

inguinal region: The area near the junction of the trunk and the thighs that contains the external genitalia.

inhibin (in-HIB-in): A hormone produced by the sustentacular cells that inhibits the pituitary secretion of follicle-stimulating hormone.

inhibitory postsynaptic potential (IPSP): A hyperpolarization of the postsynaptic membrane after the arrival of a neurotransmitter.

initial segment: The proximal portion of the axon where an action potential first appears.

injection: The forcing of fluid into a body part or organ.

inner cell mass: Cells of the blastocyst that will form the body of the embryo.

inner ear: *See* **internal ear**.

innervation: The distribution of sensory and motor nerves to a specific region or organ.

insertion: A point of attachment of a muscle; the end that is easily movable.

inspiration: Inhalation; the movement of air into the respiratory system.

inspiratory reserve: The maximum amount of air that can be drawn into the lungs over and above the normal tidal volume.

insoluble: Incapable of dissolving in solution.

insomnia: A sleep disorder characterized by the inability to fall asleep.

insulin (IN-su-lin): A hormone secreted by the beta cells of the pancreatic islets; causes a reduction in plasma glucose concentrations.

integument (in-TEG-ū-ment): The skin.

intercalated discs (in-TER-ka-lā-ted): Regions where adjacent cardiocytes interlock and where gap junctions permit electrical coupling between the cells.

intercellular cement: Proteoglycans situated between adjacent epithelial cells.

intercellular fluid: *See* **interstitial fluid**.

interdigitate: To interlock.

interferons (in-ter-FĒR-onz): Peptides released by virus-infected cells, especially lymphocytes, that make other cells more resistant to viral infection and slow viral replication.

interleukins (in-ter-LOO-kinz): Peptides released by activated monocytes and lymphocytes that assist in the coordination of the cell-mediated and antibody-mediated immune responses.

internal capsule: The collection of afferent and efferent fibers of the white matter of the cerebral hemispheres, visible on gross dissection of the brain.

internal ear: The membranous labyrinth that contains the organs of hearing and equilibrium.

internal nares: The entrance to the nasopharynx from the nasal cavity.

internal respiration: The diffusion of gases between the interstitial fluid and the cytoplasm.

interneuron: An association neuron; neurons inside the central nervous system that are interposed between sensory and motor neurons.

interoceptors: Sensory receptors monitoring the functions and status of internal organs and systems.

interosseous membrane: The fibrous connective tissue membrane between the shafts of the tibia and fibula and between the radius and ulna; an example of a fibrous amphiarthrosis.

interphase: The stage in the life cycle of a cell during which the chromosomes are uncoiled and all normal cellular functions except mitosis are underway.

intersegmental reflex: A reflex that involves several segments of the spinal cord.

interstitial cell–stimulating hormone: *See* **luteinizing hormone**.

interstitial fluid (in-ter-STISH-al): The fluid in the tissues that fills the spaces between cells.

interstitial growth: A form of cartilage growth through the growth, mitosis, and secretion of chondrocytes inside the matrix.

interventricular foramen: The opening that permits fluid movement between the lateral and third ventricles of the brain.

intervertebral disc: A fibrocartilage pad between the bodies of successive vertebrae that acts as a shock absorber.

intestinal crypt: A tubular epithelial pocket lined by secretory cells and opening into the lumen of the digestive tract; also called an *intestinal gland*.

intestine: The tubular organ of the digestive tract.

intracellular fluid: The cytosol.

intrafusal fibers: Muscle spindle fibers.

intramembranous ossification (in-tra-MEM-bra-nus): The formation of bone within a connective tissue without the prior development of a cartilaginous model.

intramuscular injection: The injection of medication into the bulk of a skeletal muscle.

intraocular pressure: The hydrostatic pressure exerted by the aqueous humor of the eye.

intrapleural pressure: The pressure measured in a pleural cavity; also called the *intrathoracic pressure*.

intrapulmonary pressure (in-tra-PUL-mo-ner-ē): The pressure measured in an alveolus of the lungs; also called the *intraalveolar pressure*.

intrauterine: Within the uterus; during prenatal development.

intrinsic factor: A glycoprotein secreted by the parietal cells of the stomach that facilitates the intestinal absorption of vitamin B_{12}.

intrinsic pathway: A pathway of the clotting system that begins with the activation of platelets and ends with the formation of platelet thromboplastin.

inversion: A turning inward.

in vitro: Outside the body, in an artificial environment.

in vivo: In the living body.

involuntary: Not under conscious control.

ion: An atom or molecule bearing a positive or negative charge due to the donation or acceptance, respectively, of an electron.

ionic bond (ī-ON-ik): Molecular bond created by the attraction between ions with opposite charges.

ionization (ī-on-i-ZĀ-shun): Dissociation; the breakdown of a molecule in solution to form ions.

ipsilateral: A reflex response that affects the same side as the stimulus.

iris: A contractile structure made up of smooth muscle that forms the colored portion of the eye.

ischemia (is-KĒ-mē-uh): Inadequate blood supply to a region of the body.

ischium (IS-kē-um): One of the three bones whose fusion creates the coxa.

islets of Langerhans: *See* **pancreatic islets**.

isometric contraction: A muscular contraction characterized by rising tension production but no change in length.

isotonic: A solution with an osmolarity that does not result in water movement across cell membranes; of the same contractive strength.

isotonic contraction: A muscular contraction during which tension climbs and then remains stable as the muscle shortens.

isotopes: Forms of an element whose atoms contain the same number of protons but different numbers of neutrons (and thus differ in atomic weight).

isthmus (IS-mus): A narrow band of tissue connecting two larger masses.

J

jaundice (JAWN-dis): A condition characterized by yellowing of connective tissues due to elevated tissue bilirubin levels; normally associated with damage to the liver or biliary system.

jejunum (je-JOO-num): The middle part of the small intestine.

joint: An area where adjacent bones interact; an articulation.

juxtaglomerular apparatus: The macula densa and the juxtaglomerular cells; a complex responsible for the release of renin and erythropoietin.

juxtaglomerular cells: Modified smooth muscle cells in the walls of the afferent and efferent arterioles adjacent to the glomerulus and the macula densa.

K

karyotyping (KAR-ē-ō-tī-ping): The determination of the chromosomal characteristics of an individual or cell.

keratin (KER-a-tin): The tough, fibrous protein component of nails, hair, calluses, and the general integumentary surface.

keratinization (KER-a-tin-i-zā-shun): *See* **cornification**.

keto acid: A molecule that ends in —COCOOH; the carbon chain that remains after the deamination or transamination of an amino acid.

ketoacidosis (kē-tō-as-i-DŌ-sis): A reduction in the pH of body fluids due to the presence of large numbers of ketone bodies.

ketogenic amino acids: Amino acids whose catabolism yields ketone bodies rather than pyruvic acid.

ketone bodies: Keto acids produced during the catabolism of lipids and ketogenic amino acids; specifically acetone, acetoacetate, and beta-hydroxybutyrate.

ketonemia (kē-tō-NĒ-mē-uh): Abnormal concentrations of ketone bodies in the blood.

ketonuria (kē-tō-NŪ-rē-uh): Abnormal concentrations of ketone bodies in the urine.

ketosis (kē-TŌ-sis): A condition characterized by the abnormal production of ketone bodies.

kidney: A component of the urinary system; an organ functioning in the regulation of plasma composition, including the excretion of wastes and the maintenance of normal fluid and electrolyte balances.

killer T cells: *See* **cytotoxic T cells**.

kilocalorie (KIL-ō-kal-o-rē): *See* **Calorie**.

Krebs cycle: *See* **TCA cycle**.

Kupffer cells (KOOP-fer): Stellate reticular cells of the liver; phagocytic cells of the liver sinusoids.

kyphosis (kī-FŌ-sis): An exaggerated thoracic curvature.

L

labium/labia (LĀ-bē-uh): Lips; the labia majora and labia minora are components of the female external genitalia.

labrum: A lip or rim.

labyrinth: A maze of passageways; the structures of the inner ear.

lacrimal gland (LAK-ri-mal): A tear gland on the dorsolateral surface of the eye.

lactase: An enzyme that breaks down milk proteins.

lactation (lak-TĀ-shun): The production of milk by the mammary glands.

lacteal (LAK-tē-al): A terminal lymphatic within an intestinal villus.

lactic acid: A compound produced from pyruvic acid under anaerobic conditions.

lactiferous duct (lak-TIF-e-rus): A duct draining one lobe of the mammary gland.

lactiferous sinus: An expanded portion of a lactiferous duct adjacent to the nipple of a breast.

lacuna (la-KOO-nuh): A small pit or cavity.

lambdoidal suture (lam-DOYD-al): The synarthrotic articulation between the parietal and occipital bones of the cranium.

lamellae (la-MEL-lē): Concentric layers; the concentric layers of bone within an osteon.

lamina (LA-min-uh): A thin sheet or layer.

lamina propria (LA-min-uh PRŌ-prē-uh): The loose connective tissue that underlies a mucous epithelium and forms part of a mucous membrane.

laminectomy: The removal of the spinous processes of a vertebra to gain access and treat a herniated disc.

Langerhans cells (LAN-ger-hanz): Cells in the epithelium of the skin and digestive tract that participate in the immune response by presenting antigens to T cells.

laparoscope (LAP-a-ro-skōp): A fiber-optic instrument used to visualize the contents of the abdominopelvic cavity.

large intestine: The terminal portions of the intestinal tract, consisting of the colon, the rectum, and the anorectal canal.

laryngopharynx (la-rin-gō-FAR-inks): The division of the pharynx inferior to the epiglottis and superior to the esophagus.

larynx (LAR-inks): A complex cartilaginous structure that surrounds and protects the glottis and vocal cords; the superior margin is bound to the hyoid bone, and the inferior margin is bound to the trachea.

latent period: The time between the stimulation of a muscle and the start of the contraction phase.

lateral: Pertaining to the side.

lateral apertures: Openings in the roof of the fourth ventricle that permit the circulation of cerebrospinal fluid into the subarachnoid space.

lateral ventricle: A fluid-filled chamber within one of the cerebral hemispheres.

laxatives: Compounds that promote defecation through increased peristalsis or an increase in the water content and volume of the feces.

lens: The transparent body lying inferior to the iris and pupil and superior to the vitreous humor.

lesion: A localized abnormality in tissue organization.

lesser omentum: A small pocket in the mesentery that connects the lesser curvature of the stomach to the liver.

leukemia (loo-KĒ-mē-uh): A malignant disease of the blood-forming tissues.

leukocyte (LOO-kō-sīt): A white blood cell.

leukocytosis (loo-kō-sī-TŌ-sis): Abnormally high numbers of circulating white blood cells.

leukopenia (loo-kō-PĒ-nē-uh): Abnormally low numbers of circulating white blood cells.

ligament (LI-ga-ment): A dense band of connective tissue fibers that attach one bone to another.

ligamentum arteriosum: The fibrous strand found in the adult that represents the remains of the ductus arteriosus of the fetus.

ligamentum teres: The fibrous strand in the falciform ligament that represents the remains of the umbilical vein of the fetus.

ligate: To tie off.

limbic system (LIM-bik): The group of nuclei and centers in the cerebrum and diencephalon that are involved with emotional states, memories, and behavioral drives.

limbus (LIM-bus): The edge of the cornea, marked by the transition from the corneal epithelium to the ocular conjunctiva.

liminal stimulus: A stimulus sufficient to depolarize the transmembrane potential of an excitable membrane to threshold and thereby produce an action potential.

linea alba: The tendinous band that runs along the midline of the rectus abdominis muscle.

lingual: Pertaining to the tongue.

lingual frenulum: An epithelial fold that attaches the inferior surface of the tongue to the floor of the mouth.

lipase (LĪ-pas): A pancreatic enzyme that breaks down triglycerides.

lipemia (lip-Ē-mē-uh): Elevated concentration of lipids in the circulation.

lipid: An organic compound containing carbons, hydrogens, and oxygens in a ratio that does not approximate 1:2:1; includes fats, oils, and waxes.

lipofuscin (li-pō-FŪ-shun): A pigment inclusion of uncertain significance that is found in aging nerve cells.

lipogenesis (li-pō-JEN-e-sis): The synthesis of lipids from nonlipid precursors.

lipoids: Prostaglandins, steroids, phospholipids, glycolipids, and so on.

lipolysis: The catabolism of lipids as a source of energy.

lipoprotein (li-pō-PRŌ-tēn): A compound containing a relatively small lipid bound to a protein.

liver: An organ of the digestive system with varied and vital functions, including the production of plasma proteins, the excretion of bile, the storage of energy reserves, the detoxification of poisons, and the interconversion of nutrients.

lobule (LOB-ūl): The basic organizational unit of the liver at the histological level.

local hormone: *See* **prostaglandin.**

long-term memories: Memories that persist for an extended period.

loop of Henle: The portion of the nephron responsible for the creation of the concentration gradient in the renal medulla.

loose connective tissue: A loosely organized, easily distorted connective tissue that contains several fiber types, a varied population of cells, and a viscous ground substance.

lordosis (lor-DŌ-sis): An exaggeration of the lumbar curvature.

lumbar: Pertaining to the lower back.

lumen: The central space within a duct or other internal passageway.

lungs: Paired organs of respiration, situated in the left and right pleural cavities.

luteinizing hormone (LH) (LOO-tē-in-ī-zing): An anterior pituitary hormone that, in the female, assists FSH in follicle stimulation, triggers ovulation, and promotes the maintenance and secretion of the endometrial glands; in the male, stimulates spermatogenesis; formerly known as *interstitial cell–stimulating hormone* in males.

luxation (luks-Ā-shun): The dislocation of a joint.

lymph: The fluid contents of lymphatic vessels, similar in composition to interstitial fluid.

lymphadenopathy (lim-fad-e-NOP-a-thē): The pathological enlargement of the lymph nodes.

lymphatics: The vessels of the lymphatic system.

lymphedema (lim-fe-DĒ-ma): The swelling of peripheral tissues due to excessive lymph production or inadequate drainage.

lymph nodes: Lymphatic organs that monitor the composition of lymph.

lymphocyte (LIM-fō-sīt): A cell of the lymphatic system that participates in the immune response.

lymphokines: Chemicals secreted by activated lymphocytes.

lymphopoiesis (lim-fō-poy-Ē-sis): The production of lymphocytes.

lymphotoxin (lim-fō-TOK-sin): A secretion of lymphocytes that kills the target cells.

lysis (LĪ-sis): The destruction of a cell through the rupture of its cell membrane.

lysosome (LĪ-sō-sōm): An intracellular vesicle containing digestive enzymes.

lysozyme: An enzyme present in some exocrine secretions that has antibiotic properties.

M

macrophage: A phagocytic cell of the monocyte–macrophage system.

macula (MAK-ū-la): A receptor complex in the saccule or utricle that responds to linear acceleration or gravity.

macula densa (MAK-ū-la DEN-sa): A group of specialized secretory cells in a portion of the distal convoluted tubule adjacent to the glomerulus and the juxtaglomerular cells; a component of the juxtaglomerular apparatus.

macula lutea (LOO-tē-uh): *See* **fovea.**

major histocompatibility complex: *See* **MHC protein.**

male climacteric: The age-related cessation of gametogenesis in males due to reduced sex hormone production.

malignant cancer: A form of cancer characterized by rapid cellular growth and the spread of cancer cells throughout the body.

malleus (MAL-ē-us): The first auditory ossicle, bound to the tympanic membrane and the incus.

malnutrition: An unhealthy state produced by inadequate dietary intake of nutrients, calories, and/or vitamins.

mamillary bodies (MAM-i-lar-ē): Nuclei in the hypothalamus involved with eating reflexes and behaviors; a component of the limbic system.

mammary glands: Milk-producing glands of the female breast.

manus: The hand.

marrow: A tissue that fills the internal cavities in a bone; may be dominated by hemopoietic cells (red marrow) or adipose tissue (yellow marrow).

mass peristalsis: A powerful peristaltic contraction that moves fecal materials along the colon and into the rectum.

mass reflex: A hyperreflexia in an area innervated by spinal cord segments distal to an area of injury.

mast cell: A connective tissue cell that when stimulated releases histamine, serotonin, and heparin, initiating the inflammatory response.

mastectomy: The surgical removal of part or all of a mammary gland.

mastication (mas-ti-KĀ-shun): Chewing.

mastoid sinus: Air-filled spaces in the mastoid process of the temporal bone.

matrix: The ground substance of a connective tissue.

maxillary sinus (MAK-si-ler-ē): One of the paranasal sinuses; an air-filled chamber lined by a respiratory epithelium that is located in a maxillary bone and opens into the nasal cavity.

meatus (mē-Ā-tus): An opening or entrance into a passageway.

mechanoreception: The detection of mechanical stimuli, such as touch, pressure, or vibration.

medial: Toward the midline of the body.

mediastinum (mē-dē-as-TĪ-um or mē-dē-AS-ti-num): The central tissue mass that divides the thoracic cavity into two pleural cavities; includes the aorta and other great vessels, the esophagus, trachea, thymus, the pericardial cavity and heart, and a host of nerves, small vessels, and lymphatics; area of connective tissue attaching a testis to the epididymis, proximal portion of ductus deferens, and associated vessels.

medulla: The inner layer or core of an organ.

medulla oblongata: The most caudal of the five brain regions, also known as the *myelencephalon*.

medullary cavity: The space within a bone that contains the marrow.

medullary rhythmicity center: The center in the medulla oblongata that sets the background pace of respiration; includes inspiratory and expiratory centers.

megakaryocytes (meg-a-KAR-ē-ō-sīts): Bone marrow cells responsible for the formation of platelets.

meiosis (mī-Ō-sis): Cell division that produces gametes with half the normal somatic chromosome complement.

melanin (ME-la-nin): The yellow-brown pigment produced by the melanocytes of the skin.

melanocyte (me-LAN-ō-sīt): A specialized cell in the deeper layers of the stratified squamous epithelium of the skin; responsible for the production of melanin.

melanocyte-stimulating hormone (MSH): A hormone of the pars intermedia of the anterior pituitary that stimulates melanin production.

melanomas (mel-a-NŌ-mas): Dangerous malignant skin cancers that involve melanocytes.

melatonin (mel-a-TŌ-nin): A hormone secreted by the pineal gland; inhibits secretion of MSH and GnRH.

membrane: Any sheet or partition; a layer consisting of an epithelium and the underlying connective tissue.

membrane flow: The movement of sections of membrane surface to and from the cell surface and components of the endoplasmic reticulum, the Golgi apparatus, and vesicles.

membrane potential: *See* **transmembrane potential**.

membranous labyrinth: Endolymph-filled tubes that enclose the receptors of the inner ear.

memory: The ability to recall information or sensations; can be divided into short-term and long-term memories.

menarche (me-NAR-kē): The beginning of menstrual function; the first menstrual period, which normally occurs at puberty.

meninges (men-IN-jēz): Three membranes that surround the surfaces of the central nervous system; the dura mater, the pia mater, and the arachnoid.

meningitis: Inflammation of the spinal or cranial meninges.

meniscectomy: The removal of a meniscus.

meniscus (men-IS-kus): A fibrocartilage pad between opposing surfaces in a joint.

menses (MEN-sēz): The first portion of the uterine cycle, the portion in which the endometrial lining sloughs away.

merocrine (MER-ō-krin): A method of secretion in which the cell ejects materials through exocytosis.

mesencephalic aqueduct: The passageway that connects the third ventricle (diencephalon) with the fourth ventricle (metencephalon).

mesencephalon (mez-en-SEF-a-lon): The midbrain; the region between the diencephalon and pons.

mesenchyme: Embryonic/fetal connective tissue.

mesentery (MEZ-en-ter-ē): A double layer of serous membrane that supports and stabilizes the position of an organ in the abdominopelvic cavity and provides a route for the associated blood vessels, nerves, and lymphatics.

mesoderm: The middle germ layer that lies between the ectoderm and endoderm of the embryo.

mesothelium (mez-ō-THĒ-lē-um): A simple squamous epithelium lining one of the divisions of the ventral body cavity.

messenger RNA (mRNA): RNA formed at transcription to direct protein synthesis in the cytoplasm.

metabolic turnover: The continuous breakdown and replacement of organic materials within living cells.

metabolism (me-TAB-ō-lizm): The sum of all biochemical processes under way within the human body at a given moment; includes anabolism and catabolism.

metabolites (me-TAB-ō-līts): Compounds produced in the body as the result of metabolic reactions.

metacarpal bones: The five bones of the palm of the hand.

metalloproteins (met-al-ō-PRŌ-tēnz): Plasma proteins that transport metal ions.

metaphase (MET-a-fāz): The stage of mitosis in which the chromosomes line up along the equatorial plane of the cell.

metaphysis (me-TA-fi-sis): The region of a long bone between the epiphysis and diaphysis, corresponding to the location of the epiphyseal plate of the developing bone.

metarteriole (met-ar-Tē-rē-ōl): A vessel that connects an arteriole to a venule and that provides blood to a capillary plexus.

metastasis (me-TAS-ta-sis): The spread of cancer cells from one organ to another, leading to the establishment secondary tumors.

metatarsal bone: One of the five bones of the foot that articulate with the tarsals (proximally) and the phalanges (distally).

metencephalon (met-en-SEF-a-lon): The pons and cerebellum of the brain.

MHC protein: A surface antigen that is important to foreign antigen recognition and that plays a role in the coordination and activation of the immune response.

micelle (mī-SEL): A droplet with hydrophilic portions on the outside; spherical aggregation of bile salts, monoglycerides, and fatty acids in the lumen of the intestinal tract.

microcephaly (mī-krō-SEF-a-lē): An abnormally small cranium, due to the premature closure of one or more fontanels.

microfilaments: Fine protein filaments visible with the electron microscope; components of the cytoskeleton.

microglia (mī-KROG-lē-uh): Phagocytic glial cells in the central nervous system.

microphages: Neutrophils and eosinophils.

microtubules: Microscopic tubules that are part of the cytoskeleton and are found in cilia, flagella, the centrioles, and spindle fibers.

microvilli: Small, fingerlike extensions of the exposed cell membrane of an epithelial cell.

micturition (mik-tū-RI-shun): Urination.

midbrain: The mesencephalon.

middle ear: The space between the external and internal ear that contains auditory ossicles.

midsagittal plane: A plane passing through the midline of the body that divides it into left and right halves.

mineralocorticoid: Corticosteroids produced by the zona glomerulosa of the adrenal cortex; steroids such as aldosterone that affect mineral metabolism.

miscarriage: A spontaneous abortion.

mitochondrion (mī-tō-KON-drē-on): An intracellular organelle responsible for generating most of the ATP required for cellular operations.

mitosis (mī-TŌ-sis): The division of a single cell nucleus that produces two identical daughter cell nuclei; an essential step in cell division.

mitral valve (MĪ-tral): *See* **bicuspid valve**.

mixed gland: A gland that contains exocrine and endocrine cells, or an exocrine gland that produces serous and mucous secretions.

mixed nerve: A peripheral nerve that contains sensory and motor fibers.

mole: A quantity of an element or compound having a mass in grams equal to the element's atomic weight or to the compound's molecular weight.

molecular weight: The sum of the atomic weights of the atoms in a molecule.

molecule: A chemical structure containing two or more atoms that are held together by chemical bonds.

monoclonal antibodies (mo-nō-KLŌ-nal): Antibodies produced by genetically identical cells under laboratory conditions.

monocytes (MON-ō-sīts): Phagocytic agranulocytes (white blood cells) in the circulating blood.

monoglyceride (mo-nō-GLI-se-rīd): A lipid consisting of a single fatty acid bound to a molecule of glycerol.

monokines: Secretions released by activated cells of the monocyte–macrophage system to coordinate various aspects of the immune response.

monosaccharide (mon-ō-SAK-uh-rīd): A simple sugar, such as glucose or ribose.

monosynaptic reflex: A reflex in which the sensory afferent synapses directly on the motor efferent.

monozygotic twins: Twins produced through the splitting of a single fertilized egg (zygote).

morula (MOR-ū-la): A mulberry-shaped collection of cells produced through the mitotic divisions of a zygote.

motor unit: All of the muscle cells controlled by a single motor neuron.

mucins (MŪ-sins): Proteoglycans responsible for the lubricating properties of mucus.

mucosa (mū-KŌ-sa): A mucous membrane; the epithelium plus the lamina propria.

mucous: The presence or production of mucus.

mucous membrane: *See* **mucosa**.

mucus: A lubricating fluid composed of water and mucins produced by unicellular and multicellular glands along the digestive, respiratory, urinary, and reproductive tracts.

multipennate: A muscle whose internal fibers are organized around several different tendons.

multipolar neuron: A neuron with many dendrites and a single axon; the typical form of a motor neuron.

multiunit smooth muscle: A smooth muscle tissue whose muscle cells are innervated in motor units.

muriatic acid: Hydrochloric acid (HCl).

muscarinic receptors (mus-kar-IN-ik): Membrane receptors sensitive to acetylcholine (ACh) and to muscarine, a toxin produced by certain mushrooms; found at all parasympathetic neuroeffector junctions and at a few sympathetic neuroeffector junctions.

muscle: A contractile organ composed of muscle tissue, blood vessels, nerves, connective tissues, and lymphatics.

muscle tissue: A tissue characterized by the presence of cells capable of contraction; includes skeletal, cardiac, and smooth muscle tissues.

muscularis externa (mus-kū-LAR-is): Concentric layers of smooth muscle responsible for peristalsis.

muscularis mucosae: The layer of smooth muscle beneath the lamina propria; responsible for moving the mucosal surface.

mutagens (MŪ-ta-jenz): Chemical agents that induce mutations and may be carcinogenic.

mutation: A change in the nucleotide sequence of the DNA in a cell.

myalgia (mī-AL-jē-uh): Muscle pain.

myasthenia gravis (mī-as-THĒ-nē-a GRA-vis): A muscular weakness due to a reduction in the number of ACh receptor sites on the sarcolemmal surface; suspected to be an autoimmune disorder.

myelencephalon (mī-el-en-SEF-a-lon): *See* **medulla oblongata**.

myelin (MĪ-e-lin): An insulating sheath around an axon; consists of multiple layers of glial cell membrane; significantly increases conduction rate along the axon.

myelination: The formation of myelin.

myenteric plexus (mī-en-TER-ik): Parasympathetic motor neurons and sympathetic postganglionic fibers located between the circular and longitudinal layers of the muscularis externa.

myocardial infarction (mī-ō-KAR-dē-al): A heart attack; damage to the heart muscle due to an interruption of regional coronary circulation.

myocarditis: Inflammation of the myocardium.

myocardium: The cardiac muscle tissue of the heart.

myofibril: Organized collections of myofilaments in skeletal and cardiac muscle cells.

myofilaments: Fine protein filaments composed primarily of the proteins actin (thin filaments) and myosin (thick filaments).

myoglobin (MĪ-ō-glō-bin): An oxygen-binding pigment especially common in slow skeletal muscle fibers and cardiac muscle cells.

myogram: A recording of the tension produced by muscle fibers on stimulation.

myometrium (mī-ō-MĒ-trē-um): The thick layer of smooth muscle in the wall of the uterus.

myopia: Nearsightedness, an inability to accommodate for distant vision.

myosepta: Connective tissue partitions that separate adjacent skeletal muscles.

myosin: The protein component of the thick filaments.

myositis (mī-ō-SĪ-tis): Inflammation of muscle tissue.

N

nail: A keratinous structure produced by epithelial cells of the nail root.

narcolepsy: A sleep disorder characterized by falling asleep at inappropriate moments.

nares, external (NA-rēz): The entrance from the exterior to the nasal cavity.

nares, internal: The entrance from the nasal cavity to the nasopharynx.

nasal cavity: A chamber in the skull bounded by the internal and external nares.

nasolacrimal duct: The passageway that transports tears from the nasolacrimal sac to the nasal cavity.

nasolacrimal sac: A chamber that receives tears from the lacrimal ducts.

nasopharynx (nā-zō-FAR-inks): A region posterior to the internal nares, superior to the soft palate, and ending at the oropharynx.

N compound: An organic compound containing nitrogen atoms.

necrosis (nek-RŌ-sis): The death of cells or tissues from disease or injury.

negative feedback: A corrective mechanism that opposes or negates a variation from normal limits.

neonate: A newborn infant, or baby.

neoplasm: A tumor, or mass of abnormal tissue.

nephritis (nef-RĪ-tis): Inflammation of the kidney.

nephrolithiasis (nef-rō-li-THĪ-a-sis): A condition resulting from the formation of kidney stones.

nephron (NEF-ron): The basic functional unit of the kidney.

nerve impulse: An action potential in a neuron cell membrane.

neural cortex: An area where gray matter is found at the surface of the central nervous system.

neurilemma (noo-ri-LEM-muh): The outer surface of a glial cell that encircles an axon.

neuroeffector junction: A synapse between a motor neuron and a peripheral effector, such as a muscle, gland cell, or fat cell.

neurofibrils: Microfibrils in the cytoplasm of a neuron.

neurofilaments: Microfilaments in the cytoplasm of a neuron.

neuroglandular junction: A specific type of neuroeffector junction.

neuroglia (noo-RŌ-glē-ah): Cells of the central nervous system and peripheral nervous system that support and protect the neurons.

neurohypophysis (NOO-rō-hī-pof-i-sis): The posterior pituitary, or pars nervosa.

neuromodulator (noo-rō-MOD-ū-la-tor): A compound, released by a neuron, that adjusts the sensitivities of another neuron to specific neurotransmitters.

neuromuscular junction: A type of neuroeffector junction.

neuron (NOO-ron): A cell in neural tissue specialized for intercellular communication via (1) changes in membrane potential and (2) synaptic connections.

neurotransmitter: A chemical compound released by one neuron to affect the transmembrane potential of another.

neurotubules: Microtubules in the cytoplasm of a neuron.

neurulation: The embryological process responsible for the formation of the central nervous system.

neutron: A fundamental particle that does not carry a positive or a negative charge.

neutropenia: An abnormally low number of neutrophils in the circulating blood.

neutrophil (NOO-trō-fil): A microphage that is very numerous and normally the first of the mobile phagocytic cells to arrive at an area of injury or infection.

nicotinic receptors (nik-ō-TIN-ik): Acetylcholine receptors found on the surfaces of sympathetic and parasympathetic ganglion cells; will respond to the compound nicotine.

nipple: An elevated epithelial projection on the surface of the breast; contains the openings of the lactiferous sinuses.

Nissl bodies: The ribosomes, Golgi apparatus, rough endoplasmic reticulum, and mitochondria of the perikaryon of a typical nerve cell.

nitrogenous wastes: Organic waste products of metabolism that contain nitrogen, such as urea, uric acid, and creatinine.

nociception (nō-sē-SEP-shun): Pain perception.

node of Ranvier: The area between adjacent glial cells where the myelin covering of an axon is incomplete.

nodose ganglion (NŌ-dōs): A sensory ganglion of cranial nerve X.

noradrenaline: A catecholamine neurotransmitter secreted by the adrenal medulla, released at most sympathetic neuroeffector junctions and at certain synapses inside the central nervous system; also called *norepinephrine*.

norepinephrine (NE) (nor-ep-i-NEF-rin): A catecholamine neurotransmitter in the peripheral nervous system and central nervous system, and a hormone secreted by the adrenal medulla; also called *noradrenaline*.

nucleic acid (noo-KLĀ-ik): A polymer of nucleotides that contains a pentose sugar, a phosphate group, and one of four nitrogenous bases that regulate the synthesis of proteins and make up the genetic material in cells.

nucleolus (noo-KLĒ-ō-lus): The dense region in the nucleus that is the site of RNA synthesis.

nucleoplasm: The fluid content of the nucleus.

nucleoproteins: Proteins of the nucleus that are generally associated with the DNA.

nucleoside: A nitrogenous base plus a simple sugar.

nucleotide: A compound consisting of a nitrogenous base, a simple sugar, and a phosphate group.

nucleus: A cellular organelle that contains DNA, RNA, and proteins; a mass of gray matter in the central nervous system.

nucleus pulposus (pul-PŌ-sus): The gelatinous central region of an intervertebral disc.

nutrient: An organic compound that can be broken down in the body to produce energy.

nystagmus: An unconscious, continuous movement of the eyes as if to adjust to constant motion.

O

obesity: Body weight 10–20 percent above standard values as the result of body fat accumulation.

occlusal surface (o-KLOO-sal): The opposing surfaces of the teeth that come into contact when processing food.

ocular: Pertaining to the eye.

oculomotor nerve (ok-ū-lō-MŌ-ter): Cranial nerve III, which controls the extrinsic oculomotor muscles other than the superior oblique and the lateral rectus.

olecranon: The proximal end of the ulna that forms the prominent point of the elbow.

olfaction: The sense of smell.

olfactory bulb (ōl-FAK-tor-ē): The expanded ends of the olfactory tracts; the sites where the axons of the first cranial nerves (N I) synapse on central nervous system interneurons that lie beneath the frontal lobes of the cerebrum.

oligodendrocytes (o-li-gō-DEN-drō-sīts): Central nervous system glial cells responsible for maintaining cellular organization in the gray matter and providing a myelin sheath in areas of white matter.

oligopeptide (ol-i-gō-PEP-tīd): A short chain of amino acids.

oncogene (ON-kō-jēn): A gene that can turn a normal cell into a cancer cell.

oncologists (on-KOL-o-jists): Physicians specializing in the study and treatment of tumors.

oocyte (Ō-ō-sīt): A cell whose meiotic divisions will produce a single ovum and three polar bodies.

oogenesis (ō-ō-JEN-e-sis): Ovum production.

oogonia (ō-ō-GŌ-nē-uh): Stem cells in the ovaries whose divisions give rise to oocytes.

oophorectomy (ō-of-ō-REK-to-mē): The surgical removal of the ovaries.

oophoritis (ō-of-ō-RĪ-tis): Inflammation of the ovaries.

ooplasm: The cytoplasm of the ovum.

opsin: A protein; one structural component of the visual pigment rhodopsin.

opsonization: An effect of coating an object with antibodies; the attraction and enhancement of phagocytosis.

optic chiasm (OP-tik KĪ-asm): The crossing point of the optic nerves.

optic nerve: The second cranial nerve (N II), which carries signals from the eye to the optic chiasm.

optic tract: The tract over which nerve impulses from the retina are transmitted between the optic chiasm and the thalamus.

orbit: The bony cavity of the skull that contains the eyeball.

orchiectomy (or-kē-EK-to-mē): The surgical removal of one or both testes; also called *orchidectomy*.

orchitis: Inflammation of the testes.

organelle (or-gan-EL): An intracellular structure that performs a specific function or group of functions.

organic compound: A compound containing carbon, hydrogen, and in most cases oxygen.

organogenesis: The formation of organs during embryological and fetal development.

organs: Combinations of tissues that perform complex functions.

origin: A point of attachment of a muscle; the end that is not easily movable.

oropharynx: The middle portion of the pharynx, bounded superiorly by the nasopharynx, anteriorly by the oral cavity, and inferiorly by the laryngopharynx.

osmolarity (oz-mō-LAR-i-tē): The total concentration of dissolved materials in a solution, regardless of their specific identities, expressed in moles.

osmoreceptor: A receptor sensitive to changes in the osmolarity of the plasma.

osmosis (oz-MŌ-sis): The movement of water across a semipermeable membrane from one solution toward another solution that contains a higher solute concentration.

osmotic pressure: The force of osmotic water movement; the pressure that must be applied to prevent osmotic movement across a membrane.

osseous tissue: A strong connective tissue containing specialized cells and a mineralized matrix of crystalline calcium phosphate and calcium carbonate.

ossicles: Small bones.

ossification: The formation of bone.

osteoblast: (OS-tē-ō-blast): A cell that produces the fibers and matrix of bone.

osteoclast (OS-tē-ō-klast): A cell that dissolves the fibers and matrix of bone.

osteocyte (OS-tē-ō-sīt): A bone cell responsible for the maintenance and turnover of the mineral content of the surrounding bone.

osteogenic layer (os-tē-ō-JEN-ik): The inner, cellular layer of the periosteum that participates in bone growth and repair.

osteolysis (os-tē-OL-i-sis): The breakdown of the mineral matrix of bone.

osteon (OS-tē-on): The basic histological unit of compact bone, consisting of osteocytes organized around a central canal and separated by concentric lamellae.

otic: Pertaining to the ear.

otitis media: Inflammation of the middle ear cavity.

otoconia (otoliths) (ō-tō-KŌ-nē-uh): Aggregations of calcium carbonate crystals in a gelatinous membrane that sits above one of the maculae of the vestibular apparatus.

oval window: An opening in the bony labyrinth where the stapes attaches to the membranous wall of the vestibular duct.

ovarian cycle (ō-VAR-ē-an): The monthly chain of events that leads to ovulation.

ovary: The female reproductive organ that produces gametes.

ovulation (ov-ū-LĀ-shun): The release of a secondary oocyte, surrounded by cells of the corona radiata, after the rupture of the wall of a tertiary follicle.

ovum/ova (Ō-vum): The functional product of meiosis II, produced after the fertilization of a secondary oocyte.

oxidation: The loss of electrons or hydrogen atoms or the acceptance of an oxygen atom.

oxidative phosphorylation: The capture of energy as ATP during a series of oxidation–reduction reactions; a reaction sequence that occurs in the mitochondria and involves coenzymes and the electron transport system.

oxytocin (oks-i-TŌ-sin): A hormone produced by hypothalamic cells and secreted into capillaries at the posterior pituitary; stimulates smooth muscle contractions of the uterus or mammary glands in females but has no known function in males.

P

pacemaker cells: Cells of the sinoatrial node that set the pace of cardiac contraction.

Pacinian corpuscle (pa-SIN-ē-an): A receptor sensitive to vibration.

palate: The horizontal partition separating the oral cavity from the nasal cavity and nasopharynx; divided into an anterior bony (hard) palate and a posterior fleshy (soft) palate.

palatine: Pertaining to the palate.

palpate: To examine by touch.

palpebrae (pal-PĒ-brē): Eyelids.

pancreas: A digestive organ containing exocrine and endocrine tissues; the exocrine portion secretes pancreatic juice, and the endocrine portion secretes hormones, including insulin and glucagon.

pancreatic duct: A tubular duct that carries pancreatic juice from the pancreas to the duodenum.

pancreatic islets: Aggregations of endocrine cells in the pancreas; also called *islets of Langerhans*.

pancreatic juice: A mixture of buffers and digestive enzymes that is discharged into the duodenum under the stimulation of the enzymes secretin and cholecystokinin.

pancreatitis (pan-krē-a-TĪ-tis): Inflammation of the pancreas.

Papanicolaou (Pap) test: A test for the detection of malignancies based on cytological appearance of epithelial cells, especially those of the cervix and uterus.

papilla (pa-PIL-la): A small, conical projection.

paralysis: The loss of voluntary motor control over a portion of the body.

paranasal sinuses: Bony chambers lined by respiratory epithelium that open into the nasal cavity; the frontal, ethmoidal, sphenoidal, and maxillary sinuses.

parasagittal: A section or plane that parallels the midsagittal plane but that does not pass along the midline.

parasympathetic division: One of the two divisions of the autonomic nervous system; also known as the *craniosacral division*; generally responsible for activities that conserve energy and lower the metabolic rate.

parasympathomimetic drugs: Drugs that mimic the actions of parasympathetic stimulation.

parathyroid glands: Four small glands embedded in the posterior surface of the thyroid; responsible for parathyroid hormone secretion.

parathyroid hormone (PTH): A hormone secreted by the parathyroid gland when plasma calcium levels fall below the normal range; causes increased osteoclast activity, increased intestinal calcium uptake, and decreased calcium ion loss at the kidneys.

parenchyma (pa-RENG-ki-ma): The cells of a tissue or organ that are responsible for fulfilling its functional role; distinguished from the stroma of that tissue or organ.

paresthesia: A sensory abnormality that produces a tingling sensation.

parietal: Referring to the body wall or outer layer.

parietal cells: Cells of the gastric glands that secrete HCl and intrinsic factor.

Parkinson's disease: A progressive motor disorder due to degeneration of the cerebral nuclei.

parotid salivary glands (pa-ROT-id): Large salivary glands that secrete a saliva containing high concentrations of salivary (alpha) amylase.

pars distalis (dis-TAL-is): The large, anterior portion of the anterior pituitary gland.

pars intermedia (in-ter-MĒ-dē-uh): The portion of the anterior pituitary immediately adjacent to the posterior pituitary and the infundibulum.

pars nervosa: The posterior pituitary gland.

pars tuberalis: The portion of the anterior pituitary gland that wraps around the infundibulum superior to the posterior pituitary.

parturition (par-tū-RISH-un): Childbirth, delivery.

patella (pa-TEL-uh): The sesamoid bone of the kneecap.

pathogenic: Disease-causing.

pathologist (pa-THO-lo-jist): An M.D. specializing in the identification of diseases on the basis of characteristic structural and functional changes in tissues and organs.

pedicel (PED-i-sel): A slender process of a podocyte that forms part of the filtration apparatus of the kidney glomerulus.

pedicles (PE-di-kulz): Thick, bony struts that connect the vertebral body with the articular and spinous processes.

pelvic cavity: The inferior subdivision of the abdominopelvic (peritoneal) cavity; encloses the urinary bladder, the sigmoid colon and rectum, and male or female reproductive organs.

pelvis: A bony complex created by the articulations among the coxae, the sacrum, and the coccyx.

penis (PĒ-nis): A component of the male external genitalia; a copulatory organ that surrounds the urethra and serves to introduce semen into the female vagina; the developmental equivalent of the female clitoris.

pepsin: A proteolytic enzyme secreted by the chief cells of the gastric glands in the stomach.

peptidases: Enzymes that split peptide bonds and release amino acids.

peptide: A chain of amino acids linked by peptide bonds.

peptide bond: A covalent bond between the amino group of one amino acid and the carboxyl group of another.

perforating canal: A passageway in compact bone that runs at right angles to the axes of the osteons, between the periosteum and endosteum.

perfusion: The blood flow through a tissue.

pericardial cavity (per-i-KAR-dē-al): The space between the parietal pericardium and the epicardium (visceral pericardium) that covers the outer surface of the heart.

pericarditis: Inflammation of the pericardium.

pericardium (per-i-KAR-dē-um): The fibrous sac that surrounds the heart and whose inner, serous lining is continuous with the epicardium.

perichondrium (per-i-KON-drē-um): The layer that surrounds a cartilage, consisting of an outer fibrous and an inner cellular region.

perikaryon (per-i-KAR-ē-on): The cytoplasm that surrounds the nucleus in the soma of a nerve cell.

perilymph (PER-ē-limf): A fluid similar in composition to cerebrospinal fluid; found in the spaces between the bony labyrinth and the membranous labyrinth of the inner ear.

perimysium (pe-ri-MĪS-ē-um): A connective tissue partition that separates adjacent fasciculi in a skeletal muscle.

perineum (pe-ri-NĒ-um): The pelvic floor and its associated structures.

perineurium: A connective tissue partition that separates adjacent bundles of nerve fibers in a peripheral nerve.

periodontal ligament (per-ē-ō-DON-tal): Collagen fibers that bind the cementum of a tooth to the periosteum of the surrounding alveolus.

periosteum (per-ē-OS-tē-um): The layer that surrounds a bone, consisting of an outer fibrous and inner cellular region.

peripheral nervous system (PNS): All neural tissue outside the central nervous system.

peripheral resistance: The resistance to blood flow; primarily caused by friction with the vascular walls.

peristalsis (per-i-STAL-sis): A wave of smooth muscle contractions that propels materials along the axis of a tube such as the digestive tract, the ureters, or the ductus deferens.

peritoneal cavity: *See* **abdominopelvic cavity**.

peritoneum (per-i-tō-NĒ-um): The serous membrane that lines the peritoneal (abdominopelvic) cavity.

peritonitis (per-i-tō-NĪ-tis): Inflammation of the peritoneum.

peritubular capillaries: A network of capillaries that surrounds the proximal and distal convoluted tubules of the kidneys.

permeability: The ease with which dissolved materials can cross a membrane; if the membrane is freely permeable, any molecule can cross it; if impermeable, nothing can cross; most biological membranes are selectively permeable.

peroxisome: A membranous vesicle containing enzymes that break down hydrogen peroxide (H_2O_2).

perspiration, insensible: Evaporative water loss by diffusion across the epithelium of the skin or evaporation across the alveolar surfaces of the lungs.

perspiration, sensible: Water loss due to sweat gland secretion.

pes: The foot.

petrosal ganglion: A sensory ganglion of the glossopharyngeal nerve (N IX).

petrous: Stony; usually used to refer to the thickened portion of the temporal bone that encloses the inner ear.

Peyer's patches (PĪ-erz): Lymphoid nodules beneath the epithelium of the small intestine.

pH: The negative exponent of the hydrogen ion concentration, expressed in moles per liter.

phagocyte: A cell that performs phagocytosis.

phagocytosis (fa-gō-sī-TŌ-sis): The engulfing of extracellular materials or pathogens; movement of extracellular materials into the cytoplasm by enclosure in a membranous vesicle.

phalanx/phalanges (fa-LAN-gēz): Bones of the fingers or toes.

pharmacology: The study of drugs, their physiological effects, and their clinical uses.

pharynx: The throat; a muscular passageway shared by the digestive and respiratory tracts.

phasic response: A pattern of response to stimulation by sensory neurons that are normally inactive; stimulation causes a burst of neural activity that ends when the stimulus either stops or stops changing in intensity.

phenotype (FĒN-ō-tīp): Physical characteristics that are genetically determined.

phonation (fō-NĀ-shun): Sound production at the larynx.

phosphate group: PO_4^{3-}.

phospholipid (fos-fō-LIP-id): An important membrane lipid whose structure includes both hydrophilic and hydrophobic regions.

phosphorylation (fos-for-i-LĀ-shun): The addition of a high-energy phosphate group to a molecule.

photoreception: Sensitivity to light.

physiology (fiz-ē-OL-o-jē): Literally, the study of function; deals with the ways organisms perform vital activities.

pia mater: The tough, outer meningeal layer that surrounds the central nervous system.

pigment: A compound with a characteristic color.

piloerection: The "goosebumps" effect produced by the contraction of the arrector pili muscles of the skin.

pineal gland: Neural tissue in the posterior portion of the roof of the diencephalon; responsible for melatonin secretion.

pinna: The expanded, projecting portion of the external ear that surrounds the external auditory meatus.

pinocytosis (pi-nō-sī-TŌ-sis): The introduction of fluids into the cytoplasm by enclosing them in membranous vesicles at the cell surface.

pituitary gland: An endocrine organ situated in the sella turcica of the sphenoid bone and connected to the hypothalamus by the infundibulum; includes the posterior pituitary (pars nervosa) and the anterior pituitary (pars intermedia and pars distalis).

placenta (pla-SENT-uh): A temporary structure in the uterine wall that permits diffusion between the fetal and maternal circulatory systems; *see* **afterbirth**.

plantar: Referring to the sole of the foot.

plasma (PLAZ-muh): The fluid ground substance of whole blood; what remains after the cells have been removed from a sample of whole blood.

plasma cell: Activated B cells that secrete antibodies.

plasmalemma (plaz-ma-LEM-a): A cell membrane.

platelets (PLĀT-lets): Small packets of cytoplasm that contain enzymes important in the clotting response; manufactured in the bone marrow by megakaryocytes.

pleura (PLOO-ra): The serous membrane that lines the pleural cavities.

pleural cavities: Subdivisions of the thoracic cavity that contain the lungs.

pleuritis (ploor-Ī-tis): Inflammation of the pleura.

plexus (PLEK-sus): A network or braid.

plica (PLĪ-ka): A permanent transverse fold in the wall of the small intestine.

pneumotaxic center (noo-mō-TAKS-ik): A center in the reticular formation of the pons that regulates the activities of the apneustic and respiratory rhythmicity centers to adjust the pace of respiration.

pneumothorax (noo-mō-THŌ-raks): The introduction of air into the pleural cavity.

podocyte (PŌ-do-sīt): A cell whose processes surround the kidney glomerular capillaries and assist in the filtration process.

polar body: A nonfunctional packet of cytoplasm that contains chromosomes eliminated from an oocyte during meiosis.

polar bond: A covalent bond in which there is an unequal sharing of electrons.

polarized: Referring to cells that have regional differences in organelle distribution or cytoplasmic composition along a specific axis, such as between the basement membrane and free surface of an epithelial cell.

pollex: The thumb.

polymer: A large molecule consisting of a long chain of subunits.

polymorph: A polymorphonuclear leukocyte; a neutrophil.

polypeptide: A chain of amino acids strung together by peptide bonds; those containing more than 100 peptides are called *proteins*.

polyribosome: Several ribosomes linked by their translation of a single mRNA strand.

polysaccharide (pol-ē-SAK-uh-rīd): A complex sugar, such as glycogen or a starch.

polysynaptic reflex: A reflex with interneurons interposed between the sensory fiber and the motor neuron(s).

polyunsaturated fats: Fatty acids containing carbon atoms linked by double bonds.

polyuria (pol-ē-Ū-rē-uh): Excessive urine production.

pons: The portion of the metencephalon anterior to the cerebellum.

popliteal (pop-LIT-ē-al): Pertaining to the back of the knee.

porphyrins (POR-fi-rinz): Ring-shaped molecules that form the basis of important respiratory and metabolic pigments, including heme and the cytochromes.

porta hepatis: A region of mesentery between the duodenum and liver that contains the hepatic artery, the hepatic portal vein, and the common bile duct.

positive feedback: A mechanism that increases a deviation from normal limits after an initial stimulus.

postabsorptive state: A period that begins 4 hours after a meal; characterized by falling blood glucose concentrations and the mobilization of metabolic reserves.

postcentral gyrus: The primary sensory cortex, where touch, vibration, pain, temperature, and taste sensations arrive and are consciously perceived.

posterior: Toward the back; dorsal.

postganglionic neuron: An autonomic neuron in a peripheral ganglion, whose activities control peripheral effectors.

postovulatory phase: The secretory phase of the menstrual cycle.

postsynaptic membrane: The portion of the cell membrane of a postsynaptic cell that is part of a synapse.

potential difference: The separation of opposite charges; requires a barrier that prevents ion migration.

precentral gyrus: The primary motor cortex on a cerebral hemisphere, located rostral to the central sulcus.

prefrontal cortex: The rostral portion of each cerebral hemisphere; thought to be involved with higher intellectual functions, predictions, calculations, and so forth.

preganglionic neuron: A visceral motor neuron inside the central nervous system whose output controls one or more ganglionic motor neurons in the peripheral nervous system.

premolars: Bicuspids; teeth with flattened occlusal surfaces located anterior to the molar teeth.

premotor cortex: The motor association area between the precentral gyrus and the prefrontal area.

preoptic nucleus: The hypothalamic nucleus that coordinates thermoregulatory activities.

prepuce (PRE-pūs): The loose fold of skin that surrounds the glans penis (males) or the clitoris (females).

preputial glands (prē-PŪ-shal): Glands on the inner surface of the prepuce that produce a viscous, odorous secretion called *smegma*.

presbyopia: Farsightedness; an inability to accommodate for near vision.

presynaptic membrane: The synaptic surface where neurotransmitter release occurs.

prevertebral ganglion: *See* **collateral ganglion**.

prime mover: A muscle that performs a specific action.

proenzyme: An inactive enzyme secreted by an epithelial cell.

progesterone (prō-JES-ter-ōn): The most important progestin secreted by the corpus luteum after ovulation.

progestins (prō-JES-tinz): Steroid hormones structurally related to cholesterol; progesterone is an example.

prognosis: A prediction about the possibility or time course of recovery from a specific disease.

projection fibers: Axons carrying information from the thalamus to the cerebral cortex.

prolactin (prō-LAK-tin): The hormone that stimulates functional development of the mammary gland in females; a secretion of the anterior pituitary gland.

prolapse: The abnormal descent or protrusion of a portion of an organ, such as the vagina or anorectal canal.

proliferative phase: A portion of the uterine cycle; the interval of estrogen-induced repair of the functional zone of the endometrium through the growth and proliferation of epithelial cells in the uterine glands not lost during menses.

pronation (prō-NĀ-shun): Rotation of the forearm that makes the palm face posteriorly.

prone: Lying face down with the palms facing the floor.

pronucleus: An enlarged egg or sperm nucleus that forms after fertilization but before amphimixis.

properdin: The complement factor that prolongs and enhances non-antibody-dependent complement binding to bacterial cell walls.

prophase (PRŌ-fāz): The initial phase of mitosis; characterized by the appearance of chromosomes, breakdown of the nuclear membrane, and formation of the spindle apparatus.

proprioception (prō-prē-ō-SEP-shun): Awareness of the positions of bones, joints, and muscles.

prostaglandin (pros-ta-GLAN-din): A fatty acid secreted by one cell that alters the metabolic activities or sensitivities of adjacent cells; also called *local hormone*.

prostatectomy (pros-ta-TEK-to-mē): The surgical removal of the prostate gland.

prostate gland (PROS-tāt): An accessory gland of the male reproductive tract, contributing roughly one-third of the volume of semen.

prostatitis (pros-ta-TĪ-tis): Inflammation of the prostate gland.

prosthesis: An artificial substitute for a body part.

protease: *See* **proteinase**.

protein: A large polypeptide with a complex structure.

proteinase: An enzyme that breaks down proteins into peptides and amino acids.

proteinuria (prō-tēn-ŪR-ē-uh): Abnormal amounts of protein in the urine.

proteoglycan (prō-tē-ō-GLĪ-kan): A compound containing a large polysaccharide complex attached to a relatively small protein; examples include hyaluronic acid and chondroitin sulfate.

prothrombin: A circulating proenzyme of the common pathway of the clotting system; converted to thrombin by the enzyme thromboplastin.

proton: A fundamental particle bearing a positive charge.

protraction: To move anteriorly in the horizontal plane.

proximal: Toward the attached base of an organ or structure.

proximal convoluted tubule (PCT): The portion of the nephron between Bowman's capsule and the loop of Henle; the major site of active reabsorption from the filtrate.

pruritis (proo-RĪ-tus): Itching.

pseudopodia (soo-dō-PŌ-dē-ah): Temporary cytoplasmic extensions typical of mobile or phagocytic cells.

pseudostratified epithelium: An epithelium containing several layers of nuclei but whose cells are all in contact with the underlying basement membrane.

psoriasis (sō-RĪ-uh-sis): A skin condition characterized by excessive keratin production and the formation of dry, scaly patches on the body surface.

psychosomatic condition: An abnormal physiological state with a psychological origin.

puberty: A period of rapid growth, sexual maturation, and the appearance of secondary sexual characteristics; normally occurs at ages 10–15.

pubic symphysis: The fibrocartilaginous amphiarthrosis between the pubic bones of the coxae.

pubis (PŪ-bis): The anterior, inferior component of the coxa.

pudendum (pū-DEN-dum): The external genitalia.

pulmonary circuit: Blood vessels between the pulmonary semilunar valve of the right ventricle and the entrance to the left atrium; the blood circulation through the lungs.

pulmonary ventilation: The movement of air into and out of the lungs.

pulp cavity: The internal chamber in a tooth, containing blood vessels, lymphatics, nerves, and the cells that maintain the dentin.

pulpitis (pul-PĪ-tis): Inflammation of the tissues of the pulp cavity.

pupil: The opening in the center of the iris through which light enters the eye.

purine: A nitrogen compound with a double ring-shaped structure; examples include adenine and guanine, two nitrogenous bases common in nucleic acids.

Purkinje cell (pur-KIN-jē): A large, branching neuron of the cerebellar cortex.

Purkinje fibers: Specialized conducting cardiocytes in the ventricles of the heart.

pus: An accumulation of debris, fluid, dead and dying cells, and necrotic tissue.

putamen (pū-TĀ-men): The thalamic nucleus involved in the integration of sensory information prior to projection to the cerebral hemispheres.

P wave: A deflection of the ECG corresponding to atrial depolarization.

pyelogram (PĪ-el-ō-gram): A radiographic image of the kidneys and ureters.

pyelonephritis (pī-e-lō-nef-RĪ-tis): Inflammation of the kidneys.

pyloric sphincter (pī-LOR-ic): A sphincter of smooth muscle that regulates the passage of chyme from the stomach to the duodenum.

pylorus (pī-LOR-us): The gastric region between the body of the stomach and the duodenum; includes the pyloric sphincter.

pyrexia (pī-REK-sē-uh): A fever.

pyrimidine: A nitrogen compound with a single ring-shaped structure; examples include cytosine, thymine, and uracil, nitrogenous bases common in nucleic acids.

pyrogen (PĪ-rō-jen): A compound that promotes a fever.

pyruvic acid (pī-ROO-vik): A three-carbon compound produced by glycolysis.

Q

quadriplegia: Paralysis of the upper and lower limbs.

quaternary structure: The three-dimensional protein structure produced by interactions between individual protein subunits.

R

radiodensity: The relative resistance to the passage of X-rays.

radiographic techniques: Methods of visualizing internal structures by using various forms of radiational energy.

radiopaque: Having a relatively high radiodensity.

rami communicantes: Axon bundles that link the spinal nerves with the ganglia of the sympathetic chain.

ramus/rami: A branch.

raphe (RĀ-fē): A seam.

receptor field: The area monitored by a single sensory receptor.

recessive gene: An allele that will affect the phenotype only when the individual is homozygous for that trait.

recombinant DNA: DNA created by splicing together a specific gene from one organism into the DNA strand of another organism.

rectal columns: Longitudinal folds in the walls of the anorectal canal.

rectouterine pouch (rek-tō-Ū-te-rin): The peritoneal pocket between the anterior surface of the colon and the posterior surface of the uterus.

rectum (REK-tum): The inferior 15 cm (6 in.) of the digestive tract.

rectus: Straight.

red blood cell (RBC): *See* **erythrocyte**.

reduction: The gain of hydrogen atoms or electrons or the loss of an oxygen molecule.

reductional division: The first meiotic division, which reduces the chromosome number from 46 to 23.

reflex: A rapid, automatic response to a stimulus.

reflex arc: The receptor, sensory neuron, motor neuron, and effector involved in a particular reflex; interneurons may or may not be present, depending on the reflex considered.

refraction: The bending of light rays as they pass from one medium to another.

refractory period: The period between the initiation of an action potential and the restoration of the normal resting potential; during this period, the membrane will not respond normally to stimulation.

relaxation phase: The period after a contraction when the tension in the muscle fiber returns to resting levels.

relaxin: A hormone that loosens the pubic symphysis; a hormone secreted by the placenta.

renal: Pertaining to the kidneys.

renal corpuscle: The initial portion of the nephron, consisting of an expanded chamber that encloses the glomerulus.

renin: The enzyme released by cells of the juxtaglomerular apparatus when renal blood pressure or P_{O_2} declines; converts angiotensinogen to angiotensin I.

rennin: A gastric enzyme that breaks down milk proteins.

replication: Duplication.

repolarization: The movement of the transmembrane potential away from a positive value and toward the resting potential.

residual volume: The amount of air remaining in the lungs after maximum forced expiration.

respiration: The exchange of gases between living cells and the environment; includes pulmonary ventilation, external respiration, internal respiration, and cellular respiration.

respiratory minute volume (V_E): The amount of air moved into and out of the respiratory system each minute.

respiratory pump: A mechanism by which changes in the intrapleural pressures during the respiratory cycle assist the venous return to the heart; also called the *thoracoabdominal pump*

resting potential: The transmembrane potential of a normal cell under homeostatic conditions.

rete (RĒ-tē): An interwoven network of blood vessels or passageways.

reticular activating system (RAS): The mesencephalic portion of the reticular formation responsible for arousal and the maintenance of consciousness.

reticular formation: A diffuse network of gray matter that extends the entire length of the brain stem.

reticulospinal tracts: Descending tracts that carry involuntary motor commands issued by neurons of the reticular formation.

retina: The innermost layer of the eye, lining the vitreous chamber; also known as the *neural tunic*.

retinal (RET-i-nal): A visual pigment derived from vitamin A.

retraction: Movement posteriorly in the horizontal plane.

retroflexion (re-trō-FLEK-shun): A posterior tilting of the uterus that has no clinical significance.

retrograde flow (RET-rō-grād): The transport of materials from the telodendria to the soma of a neuron.

retroperitoneal (re-trō-per-i-tō-NĒ-al): Situated behind or outside the peritoneal cavity.

reverberation: A positive feedback along a chain of neurons such that they remain active once stimulated.

rheumatism (ROO-muh-tizm): A condition characterized by pain in muscles, tendons, bones, or joints.

Rh factor: An agglutinogen that may be present (Rh-positive) or absent (Rh-negative) from the surfaces of red blood cells.

rhizotomy: The surgical transection of a dorsal root, normally performed to relieve pain.

rhodopsin (rō-DOP-sin): The visual pigment found in the membrane disks of the distal segments of rods.

rhythmicity center: A medullary center responsible for the pace of respiration; includes inspiratory and expiratory centers.

ribonucleic acid (rī-bō-noo-KLĀ-ik): A nucleic acid consisting of a chain of nucleotides that contain the sugar ribose and the nitrogenous bases adenine, guanine, cytosine, and uracil.

ribose: A five-carbon sugar that is a structural component of RNA.

ribosome: An organelle containing rRNA and proteins that is essential to mRNA translation and protein synthesis.

right lymphatic duct: A lymphatic vessel delivering lymph from the right side of the head, neck, and chest to the venous system via the right subclavian vein.

rigor mortis: The extended muscular contraction and rigidity that occurs after death, as the result of calcium ion release from the sarcoplasmic reticulum and the exhaustion of cytoplasmic ATP reserves.

rod: A photoreceptor responsible for vision under dimly lit conditions.

rostral: Toward the nose; used in referring to relative position inside the skull.

rough endoplasmic reticulum (RER): A membranous organelle that is a site of protein synthesis and storage.

round window: An opening in the bony labyrinth of the inner ear that exposes the membranous wall of the tympanic duct to the air of the middle ear cavity.

rubrospinal tracts: Descending tracts that carry involuntary motor commands issued by the red nucleus of the mesencephalon.

Ruffini corpuscles (roo-FĒ-nē): Receptors sensitive to tension and stretch in the dermis of the skin.

rugae (ROO-gē): Mucosal folds in the lining of the empty stomach that disappear as gastric distension occurs.

S

saccule (SAK-ūl): A portion of the vestibular apparatus of the inner ear; contains a macula important for static equilibrium.

sagittal plane: A sectional plane that divides the body into left and right portions.

salivatory nucleus (SAL-i-va-to-rē): The medullary nucleus that controls the secretory activities of the salivary glands.

salt: An inorganic compound consisting of a cation other than H^+ and an anion other than OH^-.

saltatory conduction: The relatively rapid conduction of an action potential between successive nodes of a myelinated axon.

sarcolemma: The cell membrane of a muscle cell.

sarcoma (sar-KŌ-ma): A tumor of connective tissues.

sarcomere: The smallest contractile unit of a striated muscle cell.

sarcoplasm: The cytoplasm of a muscle cell.

satellite cdells: *See* **amphicytes**.

scala media: *See* **cochlear duct**.

scala tympani: *See* **tympanic duct**.

scala vestibuli: *See* **vestibular duct**.

scar tissue: The thick, collagenous tissue that forms at an injury site.

Schlemm, canal of: The passageway that delivers aqueous humor from the anterior chamber of the eye to the venous circulation.

Schwann cells: Glial cells responsible for the neurilemma that surrounds axons in the peripheral nervous system.

sciatica (sī-AT-i-ka): Pain felt along the peripheral distribution of the sciatic nerve.

sciatic nerve (sī-A-tik): A nerve innervating the posteromedial portions of the thigh and leg.

sclera (SKLER-uh): The fibrous, outer layer of the eye that forms the white area of the anterior surface; a portion of the fibrous tunic of the eye.

sclerosis: A hardening and thickening that commonly occurs secondary to tissue inflammation.

scoliosis (skō-lē-Ō-sis): An abnormal, exaggerated lateral curvature of the spine.

scrotum (SKRŌ-tum): The loose-fitting, fleshy pouch that encloses the testes of the male.

sebaceous glands (sē-BĀ-shus): Glands that secrete sebum; normally associated with hair follicles.

sebum (SĒ-bum): A waxy secretion that coats the surfaces of hairs.

secondary sex characteristics: Physical characteristics that appear at puberty in response to sex hormones but are not involved in the production of gametes.

secretin (sē-KRĒ-tin): A duodenal hormone that stimulates pancreatic buffer secretion and inhibits gastric activity.

semen (SĒ-men): The fluid ejaculate that contains spermatozoa and the secretions of accessory glands of the male reproductive tract.

semicircular ducts: The tubular components of the membranous labyrinth of the inner ear responsible for dynamic equilibrium.

semilunar valve: A three-cusped valve guarding the exit from one of the cardiac ventricles; the pulmonary and aortic valves.

seminal vesicles (SEM-i-nal): Glands of the male reproductive tract that produce roughly 60 percent of the volume of semen.

seminiferous tubules (se-mi-NIF-e-rus): Coiled tubules where sperm production occurs in the testis.

senescence: Aging.

septae (SEP-tē): Partitions that subdivide an organ.

serosa: *See* **serous membrane**.

serotonin (ser-ō-TŌ-nin): A neurotransmitter in the central nervous system; a compound that enhances inflammation and is released by activated mast cells and basophils.

serous cell: A cell that produces a serous secretion.

serous membrane: A squamous epithelium and the underlying loose connective tissue; the lining of the pericardial, pleural, and peritoneal cavities.

serous secretion: A watery secretion that contains high concentrations of enzymes; produced by serous cells.

serum: Blood plasma from which clotting agents have been removed.

sesamoid bone: A bone that forms within a tendon.

sigmoid colon (SIG-moyd): The S-shaped 18-cm-long portion of the colon between the descending colon and the rectum.

sign: The visible evidence of the presence of a disease.

simple epithelium: An epithelium containing a single layer of cells above the basement membrane.

sinoatrial (SA) node: The natural pacemaker of the heart; situated in the wall of the right atrium.

sinus: A chamber or hollow in a tissue; a large, dilated vein.

sinusitis: Inflammation of a nasal sinus.

sinusoid (SĪ-nus-oyd): An extensive network of vessels found in the liver, adrenal cortex, spleen, and pancreas; similar in histological structure to capillaries.

skeletal muscle: A contractile organ of the muscular system.

skeletal muscle tissue: A contractile tissue dominated by skeletal muscle fibers; characterized as striated, voluntary muscle.

sliding filament theory: The concept that a sarcomere shortens as the thick and thin filaments slide past one another.

small intestine: The duodenum, jejunum, and ileum; the digestive tract between the stomach and large intestine.

smegma (SMEG-ma): The secretion of the preputial glands of the penis or clitoris.

smooth endoplasmic reticulum (SER): A membranous organelle where lipid and carbohydrate synthesis and storage occur.

smooth muscle tissue: Muscle tissue found in the walls of many visceral organs; characterized as nonstriated, involuntary muscle.

soft palate: The fleshy posterior extension of the hard palate, separating the nasopharynx from the oral cavity.

sole: The inferior surface of the foot.

solute: Any materials dissolved in a solution.

solution: A fluid containing dissolved materials.

solvent: The fluid component of a solution.

soma (SŌ-ma): Body; the body of a neuron.

somatic (sō-MAT-ik): Pertaining to the body.

somatic nervous system (SNS): The efferent division of the nervous system that innervates skeletal muscles.

somatomedins: Compounds stimulating tissue growth; released by the liver after the secretion of growth hormone.

somatostatin: Growth hormone–inhibiting hormone (GH-IH); a hypothalamic regulatory hormone that inhibits growth hormone secretion by the anterior pituitary.

somatotropin: Growth hormone; produced by the anterior pituitary in response to growth hormone–releasing hormone (GH-RH).

sperm: *See* **spermatozoon**.

spermatic cord: Collectively, the spermatic vessels, nerves, lymphatics, and the ductus deferens, extending between the testes and the proximal end of the inguinal canal.

spermatids (SPER-ma-tidz): The product of meiosis in the male; cells that differentiate into spermatozoa.

spermatocyte (sper-MA-tō-sīt): A cell of the seminiferous tubules that is engaged in meiosis.

spermatogenesis (sper-ma-tō-JEN-e-sis): Sperm production.

spermatogonia (sper-ma-tō-GŌ-nē-uh): Stem cells whose mitotic divisions give rise to other stem cells and primary spermatocytes.

spermatozoon/spermatozoa (sper-ma-to-ZŌ-a): A sperm cell.

spermicide: A compound toxic to sperm cells; used as a contraceptive method.

spermiogenesis: The process of spermatid differentiation that leads to the formation of physically mature spermatozoa.

sphincter (SFINK-ter): A muscular ring that contracts to close the entrance or exit of an internal passageway.

spinal nerve: One of 31 pairs of nerves that originate on the spinal cord from anterior and posterior roots.

spindle apparatus: A muscle spindle (intrafusal fibers) and its sensory and motor innervation.

spinocerebellar tracts: Ascending tracts carrying sensory information to the cerebellum.

spinothalamic tracts: Ascending tracts carrying poorly localized touch, pressure, pain, vibration, and temperature sensations to the thalamus.

spinous process: The prominent posterior projection of a vertebra; formed by the fusion of two laminae.

spleen: A lymphoid organ important for red blood cell phagocytosis, the immune response, and lymphocyte production.

splenectomy (splē-NEK-to-mē): The surgical removal of the spleen.

sprain: A forceful distortion of an articulation that produces damage to the capsule, ligaments, or tendons but not dislocation.

sputum (SPŪ-tum): A viscous mucus that is transported to the pharynx by the mucus escalator of the respiratory tract and is ejected from the mouth.

squama: A broad, flat surface.

squamous (SKWĀ-mus): Flattened.

squamous epithelium: An epithelium whose superficial cells are flattened and platelike.

stapedius (sta-PĒ-dē-us): A muscle of the middle ear whose contraction tenses the auditory ossicles and reduces the forces transmitted to the oval window.

stapes (STĀ-pēz): The auditory ossicle attached to the tympanic membrane.

stenosis (ste-NŌ-sis): A constriction or narrowing of a passageway.

stereocilia: Elongate microvilli characteristic of the epithelium of the epididymis, portions of the ductus deferens, and the inner ear.

steroid: A ring-shaped lipid structurally related to cholesterol.

stimulus: An environmental alteration that produces a change in cellular activities; often used to refer to events that alter the transmembrane potentials of excitable cells.

stratified: Containing several layers.

stratum (STRA-tum): A layer.

stretch receptors: Sensory receptors that respond to stretching of the surrounding tissues.

stroma: The connective tissue framework of an organ; distinguished from the functional cells (parenchyma) of that organ.

subarachnoid space: A meningeal space containing cerebrospinal fluid; the area between the arachnoid membrane and the pia mater.

subclavian (sub-CLĀ-vē-an): Pertaining to the region under the clavicle.

subcutaneous layer: The layer of loose connective tissue below the dermis; also called the *hypodermis* or *superficial fascia*.

sublingual salivary glands (sub-LING-gwal): Mucus-secreting salivary glands situated under the tongue.

submandibular salivary glands: Salivary glands nestled in depressions on the medial surfaces of the mandible; salivary glands that produce a mixture of mucins and enzymes (salivary amylase).

submucosa (sub-mū-KŌ-sa): The region between the muscularis mucosae and the muscularis externa.

submucosal glands: Mucous glands in the submucosa of the duodenum.

subserous fascia: The loose connective tissue layer beneath the serous membrane that lines the ventral body cavity.

substrate: A participant (product or reactant) in an enzyme-catalyzed reaction.

sulcus (SUL-kus): A groove or furrow.

summation: The temporal or spatial addition of stimuli.

superficial fascia: *See* **subcutaneous layer**.

superior: Above, in reference to a portion of the body in the anatomical position.

superior vena cava (SVC): The vein that carries blood from the parts of the body above the heart to the right atrium.

supination (soo-pi-NĀ-shun): Rotation of the forearm so that the palm faces anteriorly.

supine (SOO-pīn): Lying face up, with palms facing anteriorly.

suppressor T cells: Lymphocytes that inhibit B cell activation and plasma cell secretion of antibodies.

suprarenal gland (soo-pra-RĒ-nal): *See* **adrenal gland**.

surfactant (sur-FAK-tant): A lipid secretion that coats the alveolar surfaces of the lungs and prevents their collapse.

sustentacular cells (sus-ten-TAK-ū-lar): Supporting cells of the seminiferous tubules of the testis; responsible for the differentiation of spermatids, the maintenance of the blood–testis barrier, and the secretion of inhibin, androgen-binding protein, and Müllerian-inhibiting factor.

sutural bones: Irregular bones that form in fibrous tissue between the flat bones of the developing cranium; also called *Wormian bones*.

suture: A fibrous joint between flat bones of the skull.

sympathectomy (sim-path-EK-to-mē): The transection of the sympathetic innervation to a region.

sympathetic division: The division of the autonomic nervous system responsible for "fight or flight" reactions; primarily concerned with the elevation of metabolic rate and increased alertness.

sympathomimetic drugs: Drugs that mimic the actions of sympathetic stimulation.

symphysis: A fibrous amphiarthrosis, such as that between adjacent vertebrae or between the pubic bones of the coxae.

symptom: An abnormality of function as a result of disease.

synapse (SIN-aps): The site of communication between a nerve cell and some other cell; if the other cell is not a neuron, the term *neuroeffector junction* is often used.

synaptic delay (sin-AP-tik): The period between the arrival of an impulse at the presynaptic membrane and the initiation of an action potential in the postsynaptic membrane.

synarthrosis (sin-ar-THRŌ-sis): A joint that does not permit relative movement between the articulating elements; *see* **lambdoidal suture**.

synchondrosis (sin-kon-DRŌ-sis): A cartilaginous synarthrosis, such as the articulation between the epiphysis and diaphysis of a growing bone.

syncope (SIN-kō-pē): A sudden, transient loss of consciousness; a faint.

syncytial trophoblast: The multinucleate cytoplasmic layer that covers the blastocyst; responsible for uterine erosion and implantation.

syncytium (sin-SI-shē-um): A multinucleate mass of cytoplasm, produced by the fusion of cells or repeated mitoses without cytokinesis.

syndesmosis (sin-dez-MŌ-sis): A fibrous amphiarthrosis.

syndrome: A discrete set of symptoms that occur together.

syneresis (sin-ER-ē-sis): Clot retraction.

synergist (SIN-er-jist): A muscle that assists a prime mover in performing its primary action.

synostosis (sin-os-TŌ-sis): A synarthrosis formed through the fusion of the articulating elements.

synovial cavity (si-NŌ-vē-ul): A fluid-filled chamber in a synovial joint.

synovial fluid: The substance secreted by synovial membranes that lubricates joints.

synovial joint: A freely movable joint where the opposing bone surfaces are separated by synovial fluid; a diarthrosis.

synovial membrane: An incomplete layer of fibroblasts confronting the synovial cavity, plus the underlying loose connective tissue.

synthesis (SIN-the-sis): Manufacture; anabolism.

system: An interacting group of organs that performs one or more specific functions.

systemic circuit: The vessels between the aortic semilunar valve and the entrance to the right atrium; the circulatory system other than vessels of the pulmonary circuit.

systole (SIS-tō-lē): The period of cardiac contraction.

systolic pressure: The peak arterial pressure measured during ventricular systole.

T

tachycardia (tak-ē-KAR-dē-uh): An abnormally rapid heart rate.

tactile: Pertaining to the sense of touch.

taenia coli (TĒ-nē-a KŌ-lī): Three longitudinal bands of smooth muscle in the muscularis externa of the colon.

tarsal bones: The bones of the ankle (the talus, calcaneus, navicular, and cuneiform bones).

tarsus: The ankle.

TCA (tricarboxylic acid) cycle: The aerobic reaction sequence that occurs in the mitochondrial matrix; in the process, organic molecules are broken down, carbon dioxide molecules are released, and hydrogen molecules are transferred to coenzymes that deliver them to the electron transport system; also called the *citric acid cycle* or the *Krebs cycle*.

T cells: Lymphocytes responsible for cell-mediated immunity and for the coordination and regulation of the immune response; includes regulatory T cells (helpers and suppressors) and cytotoxic (killer) T cells.

tears: The fluid secretion of the lacrimal glands that bathes the anterior surfaces of the eyes.

tectorial membrane (tek-TOR-ē-al): The gelatinous membrane suspended over the hair cells of the organ of Corti.

tectospinal tracts: Descending extrapyramidal tracts that carry involuntary motor commands issued by the colliculi.

tectum: The roof of the mesencephalon of the brain.

telencephalon (tel-en-SEF-a-lon): The forebrain or cerebrum, including the cerebral hemispheres, the internal capsule, and the cerebral nuclei.

telodendria (te-lō-DEN-drē-uh): Terminal axonal branches that end in synaptic knobs.

telophase (TĒL-ō-fāz): The final stage of mitosis, characterized by the disappearance of the spindle apparatus, the reappearance of the nuclear membrane, the disappearance of the chromosomes, and the completion of cytokinesis.

temporal: Pertaining to time (temporal summation) or pertaining to the temples (temporal bone).

tendinitis: Painful inflammation of a tendon.

tendon: A collagenous band that connects a skeletal muscle to an element of the skeleton.

tendon organ: A receptor sensitive to tension in a tendon.

tentorium cerebelli (ten-TOR-ē-um ser-e-BEL-ē): A dural partition that separates the cerebral hemispheres from the cerebellum.

teratogen (TER-a-tō-jen): A stimulus that causes developmental defects.

teres: Long and round.

terminal: Toward the end.

tertiary follicle: A mature ovarian follicle, containing a large, fluid-filled chamber.

tertiary structure: The protein structure that results from interactions between distant portions of the same molecule; complex coiling and folding.

testes (TES-tēz): The male gonads, sites of gamete production and hormone secretion.

testosterone (tes-TOS-te-rōn): The principal androgen produced by the interstitial cells of the testes.

tetanic contraction: A sustained skeletal muscle contraction due to repeated stimulation at a frequency that prevents muscle relaxation.

tetanus: A tetanic contraction; also refers to a disease state that results from the stimulation of muscle cells by certain bacterial toxins.

tetany: A tetanic contraction; also refers to abnormally prolonged contractions that result from disturbances in electrolyte balance.

tetrad (TET-rad): Paired, duplicated chromosomes visible at the start of meiosis I.

tetraiodothyronine (tet-ra-ī-ō-dō-THĪ-rō-nēn): T_4, or thyroxine, a thyroid hormone.

thalamus: The walls of the diencephalon.

thalassemia (thal-ah-SĒ-mē-uh): A hereditary disorder that affects hemoglobin synthesis and produces anemia.

theory: A hypothesis that makes valid predictions, as demonstrated by evidence that is testable, unbiased, and repeatable.

therapy: The treatment of disease.

thermogenesis (ther-mō-JEN-e-sis): Heat production.

thermography: A diagnostic procedure involving the production of an infrared image.

thermoreception: Sensitivity to temperature changes.

thermoregulation: Homeostatic maintenance of body temperature.

thick filament: A cytoskeletal filament in a skeletal or cardiac muscle cell; composed of myosin, with a core of titin.

thin filament: A cytoskeletal filament in a skeletal or cardiac muscle cell; consists of actin, troponin and tropomyosin.

thoracolumbar division (tho-ra-kō-LUM-bar): The sympathetic division of the autonomic nervous system.

thorax: The chest.

threshold: The transmembrane potential at which an action potential begins.

thrombin (THROM-bin): The enzyme that converts fibrinogen to fibrin.

thrombocyte (THROM-bō-sīt): In nonmammalian vertebrates, nucleated cells that are the equivalent of platelets in humans.

thrombocytopenia (throm-bō-sī-tō-PĒ-nē-uh): An abnormally low platelet count in the circulating blood.

thromboembolism (throm-bō-EM-bō-lizm): The occlusion of a blood vessel by a drifting blood clot.

thromboplastin: The enzyme that converts prothrombin to thrombin; formed by the intrinsic or extrinsic clotting pathway.

thrombus: A blood clot that develops at the lumenal wall of a blood vessel.

thymine: A pyrimidine; one of the nitrogenous bases in the nucleic acid DNA.

thymosins (THĪ-mō-sinz): Thymic hormones essential to the development and differentiation of T cells.

thymus: A lymphoid organ, the site of T cell formation.

thyroglobulin (thī-rō-GLOB-ū-lin): A circulating transport globulin that binds thyroid hormones.

thyroid gland: An endocrine gland whose lobes sit lateral to the thyroid cartilage of the larynx.

thyroid hormones: Thyroxine (T_4) and triiodothyronine (T_3), hormones of the thyroid gland; hormones that stimulate tissue metabolism, energy utilization, and growth.

thyroid-stimulating hormone (TSH): The anterior pituitary hormone that triggers the secretion of thyroid hormones by the thyroid gland.

thyroxine (TX) (thī-ROKS-ēn): A thyroid hormone; also known as T_4 or *tetraiodothyronine*.

tidal volume: The volume of air moved into and out of the lungs during a normal quiet respiratory cycle.

tissue: A collection of specialized cells and cell products that perform a specific function.

titer: The plasma antibody concentration.

tolerance: The failure to produce antibodies against antigenic compounds normally present in the body.

tonic response: An increase or decrease in the frequency of action potentials by sensory receptors that are chronically active.

tonsil: A lymphoid nodule beneath the epithelium of the pharynx; the palatine, pharyngeal, and lingual tonsils.

topical: Applied to the body surface.

toxic: Poisonous.

trabecula (tra-BEK-ū-la): A connective tissue partition that subdivides an organ.

trabeculae carneae (tra-BEK-ū-lē KAR-nē-ē): Muscular ridges projecting from the walls of the ventricles of the heart.

trachea (TRĀ-kē-a): The windpipe, an airway extending from the larynx to the primary bronchi.

tracheal ring: A C-shaped supporting cartilage of the trachea.

tracheostomy (trā-kē-OS-to-mē): The surgical opening of the anterior tracheal wall to permit airflow.

trachoma: An infectious disease of the conjunctiva and cornea.

tract: A bundle of axons inside the central nervous system.

tractotomy: The surgical transection of a tract, used in some cases to relieve pain.

transamination (trans-am-i-NĀ-shun): The enzymatic transfer of an amino group from an amino acid to another carbon chain.

transcription: The encoding of genetic instructions on a strand of mRNA.

transdermal medication: The administration of medication by absorption through the skin.

transection: The severing or cutting of an object in the transverse plane.

transfusion: The transfer of blood from a donor directly into the bloodstream of another person.

transient ischemic attack (TIA): A temporary loss of consciousness due to the occlusion of a small blood vessel in the brain.

translation: The process of peptide formation from the instructions carried by an mRNA strand.

transmembrane potential: The potential difference measured across the cell membrane, expressed in millivolts, that results from the uneven distribution of positive and negative ions across a cell membrane.

transudate (TRANS-ū-dāt): A fluid that diffuses across a serous membrane and lubricates opposing surfaces.

transverse tubules: *See* T tubules.

treppe (TREP-ē): A steplike increase in tension production after the repeated stimulation of a muscle, even though the muscle is allowed to complete each relaxation phase.

triad: The combination of a T tubule and two cisternae of the sarcoplasmic reticulum of a skeletal muscle fiber; a combination of branches of the hepatic portal vein, hepatic artery, and bile duct at a liver lobule.

tricarboxylic acid cycle (trī-kar-bok-SIL-ik): See **TCA cycle**.

tricuspid valve (trī-KUS-pid): The right atrioventricular valve, which prevents the backflow of blood into the right atrium during ventricular systole.

trigeminal nerve (trī-GEM-i-nal): Cranial nerve V, responsible for providing sensory information from the lower portions of the face (including the upper and lower jaws) and for delivering motor commands to the muscles of mastication.

triglyceride (trī-GLIS-e-rīd): A lipid composed of a molecule of glycerol attached to three fatty acids.

trigone (TRĪ-gōn): The triangular region of the urinary bladder bounded by the exits of the ureters and the entrance to the urethra.

triiodothyronine: T_3, one of the thyroid hormones.

trisomy: The abnormal possession of three copies of a chromosome; trisomy 21 is responsible for Down syndrome.

trochanter (trō-KAN-ter): Large processes near the head of the femur.

trochlea (TRŌK-lē-uh): A pulley; the spool-shaped medial portion of the condyle of the humerus.

trochlear nerve (TRŌK-lē-ar): Cranial nerve IV, controlling the superior oblique muscle of the eye.

trophoblast (TRŌ-fō-blast): The superficial layer of the blastocyst that will be involved with implantation, hormone production, and placenta formation.

tropomyosin (trō-pō-MĪ-ō-sin): A protein on the thin filaments that covers the active sites in the absence of free calcium ions.

troponin (TRŌ-pō-nin): A protein on the thin filaments that binds to tropomyosin and to G actin.

trunk: The thoracic and abdominopelvic regions.

trypsin (TRIP-sin): One of the pancreatic proteases.

trypsinogen: The inactive proenzyme secreted by the pancreas and converted to trypsin in the duodenum.

T tubules: The transverse, tubular extensions of the sarcolemma that extend deep into the sarcoplasm to contact cisternae of the sarcoplasmic reticulum; also called *transverse tubules*.

tuberculum (too-BER-kū-lum): A small, localized elevation on a bony surface.

tuberosity: A large, roughened elevation on a bony surface.

tubulin: A protein subunit of microtubules.

tumor: A tissue mass formed by the abnormal growth and replication of cells.

tunica (TOO-ni-ka): A layer or covering; in blood vessels, the tunica externa is the outermost layer of connective tissue fibers that stabilizes the position of the vessel; the tunica intima is the innermost layer, consisting of the endothelium plus an underlying elastic membrane; the tunica media is a middle layer containing collagen, elastin, and smooth muscle fibers in varying proportions.

turbinates: See **concha/conchae**.

T wave: A deflection of the ECG corresponding to ventricular repolarization.

twitch: A single stimulus-contraction-relaxation cycle in a skeletal muscle.

tympanic duct: The perilymph-filled chamber of the inner ear below the basilar membrane; pressure changes there distort the round window; also called the *scala tympani*.

tympanic membrane (tim-PAN-ik): The membrane that separates the external auditory canal from the middle ear; membrane whose vibrations are transferred to the auditory ossicles and ultimately to the oval window; also called the *eardrum*.

type A axons: Large myelinated axons.

type B axons: Small myelinated axons.

type C axons: Small unmyelinated axons.

U

ulcer: An area of epithelial sloughing associated with damage to the underlying connective tissues and vasculature.

ultrasound: A diagnostic visualization procedure that uses high-frequency sound waves.

umbilical cord (um-BIL-i-kal): The connecting stalk between the fetus and the placenta; contains the allantois, the umbilical arteries, and the umbilical vein.

umbilicus: The navel.

unicellular gland: Goblet cells.

unipennate muscle: A muscle whose fibers are on one side of the tendon.

unipolar neuron: A sensory neuron whose soma lies in a dorsal root ganglion or a sensory ganglion of a cranial nerve.

unmyelinated axon: An axon whose neurilemma does not contain myelin and across which continuous conduction occurs.

uracil: A pyrimidine; one of the nitrogenous bases in the nucleic acid RNA.

uremia (ū-RĒ-mē-uh): An abnormal condition caused by impaired kidney function; characterized by the retention of wastes and the disruption of many other organ systems.

ureters (ū-RĒ-terz): Muscular tubes, lined by transitional epithelium, that carry urine from the renal pelvis to the urinary bladder.

urethra (ū-RĒ-thra): A muscular tube that carries urine from the urinary bladder to the exterior.

urethritis: Inflammation of the urethra.

urinalysis: An analysis of the physical and chemical characteristics of the urine.

urinary bladder: The muscular, distensible sac that stores urine prior to micturition.

urination: The voiding of urine; micturition.

urobilinogen: A compound derived from the bilirubin excreted in the bile.

uterus (Ū-ter-us): The muscular organ of the female reproductive tract where implantation, placenta formation, and fetal development occur.

utricle (Ū-tri-kul): The largest chamber of the vestibular apparatus of the inner ear; contains a macula important for static equilibrium.

uvea: The vascular tunic of the eye.

uvula (Ū-vū-luh): A dangling, fleshy extension of the soft palate.

V

vagina (va-JĪ-na): A muscular tube extending between the uterus and the vestibule.

varicose veins (VAR-i-kōs): Distended superficial veins.

vasa vasorum: Blood vessels that supply the walls of large arteries and veins.

vascular: Pertaining to blood vessels.

vascularity: The blood vessels in a tissue.

vascular spasm: A contraction of the wall of a blood vessel at an injury site; may slow the rate of blood loss.

vasoconstriction: A reduction in the diameter of arterioles due to the contraction of smooth muscles in the tunica media; an event that elevates peripheral resistance and that may occur in response to local factors, through the action of hormones, or from the stimulation of the vasomotor center.

vasodilation: An increase in the diameter of arterioles due to the relaxation of smooth muscles in the tunica media; an event that reduces peripheral resistance and that may occur in response to local factors, through the action of hormones, or after decreased stimulation of the vasomotor center.

vasomotion: Alterations in the pattern of blood flow through a capillary bed in response to changes in the local environment.

vasomotor center: The medullary center whose stimulation produces vasoconstriction and an elevation in peripheral resistance.

vein: A blood vessel carrying blood from a capillary bed toward the heart.

vena cava (VĒ-na CĀ-va): One of the major veins delivering systemic blood to the right atrium; superior and inferior venae cavae.

ventilation: Air movement into and out of the lungs.

ventilatory rate: The respiratory rate.

ventral: Pertaining to the anterior surface.

ventricle (VEN-tri-kul): One of the large, muscular pumping chambers of the heart that discharges blood into the pulmonary or systemic circuits; one of four fluid-filled chambers within the brain.

ventricular escape: The initiation of ventricular contractions after a pause caused by impaired conduction of the contractile stimulus from the AV node.

ventricular folds: Mucosal folds in the laryngeal walls that do not play a role in sound production; the false vocal cords.

venule (VEN-ūl): Thin-walled veins that receive blood from capillaries.

vermis (VER-mis): A midsagittal band of neural cortex on the surface of the cerebellum.

vertebral canal: The passageway that encloses the spinal cord; a tunnel bounded by the neural arches of adjacent vertebrae.

vertebral column: The cervical, thoracic, and lumbar vertebrae, the sacrum, and the coccyx.

vertebrochondral ribs: Ribs 8–10; false ribs, connected to the sternum by shared cartilaginous bars.

vertebrosternal ribs: Ribs 1–7; true ribs, connected to the sternum by individual cartilaginous bars.

vertigo: Dizziness.

vesicle: A membranous sac in the cytoplasm of a cell.

vestibular duct: The perilymph-filled chamber of the inner ear above the vestibular membrane; pressure changes there result from distortions of the oval window; also called the *scala vestibuli.*

vestibular membrane: The membrane that separates the cochlear duct from the vestibular duct of the inner ear.

vestibular nucleus: The processing center for sensations that arrive from the vestibular apparatus of the inner ear; located near the border between the pons and the medulla oblongata.

vestibule (VES-ti-būl): A chamber; in the inner ear, the utricle, saccule, and semicircular canals.

vestibulospinal tracts: Descending tracts of the extrapyramidal system, carrying involuntary motor commands issued by the vestibular nucleus to stabilize the position of the head.

villus/villi: A slender projection of the mucous membrane of the small intestine.

virus: A pathogenic microorganism.

viscera (VIS-e-rah): Organs in the ventral body cavity.

visceral: Pertaining to viscera or their outer coverings.

visceral smooth muscle: A smooth muscle tissue that forms sheets or layers in the walls of visceral organs; the cells may not be innervated, and the layers often show automaticity (rhythmic contractions).

viscosity: The resistance to flow exhibited by a fluid due to molecular interactions within the fluid.

viscous: Thick, syrupy.

vital capacity: The maximum amount of air that can be moved into or out of the respiratory system; the sum of the inspiratory reserve, the expiratory reserve, and the tidal volume.

vitamin: An essential organic nutrient that functions as a coenzyme in vital enzymatic reactions.

vitreous humor: The gelatinous mass in the vitreous chamber of the eye.

vocal folds: Folds in the laryngeal wall that contain elastic ligaments whose tension can be voluntarily adjusted; the true vocal cords, responsible for phonation.

voluntary: Controlled by conscious thought processes.

vulva (VUL-vuh): The female pudendum (external genitalia).

W

Wallerian degeneration: The disintegration of an axon and its myelin sheath distal to an injury site.

white blood cells (WBCs): Leukocytes; the granulocytes and agranulocytes of the blood.

white matter: Regions inside the central nervous system that are dominated by myelinated axons.

white ramus: A nerve bundle containing the myelinated preganglionic axons of sympathetic motor neurons en route to the sympathetic chain or a collateral ganglion.

Wormian bones: *See* **sutural bones**.

X

X chromosome: The sex chromosome that is present in genetic males and genetic females.

xiphoid process (ZĪ-foyd): The slender, inferior extension of the sternum.

Y

Y chromosome: The sex chromosome whose presence indicates that the individual is a genetic male.

yolk sac: One of the four extraembryonic membranes, composed of an inner layer of endoderm and an outer layer of mesoderm.

Z

Zeis, glands of (ZĪS): Enlarged sebaceous glands on the free edges of the eyelids.

zona fasciculata (ZŌ-na fa-sik-ū-LA-ta): The region of the adrenal cortex responsible for glucocorticoid secretion.

zona glomerulosa (glo-mer-ū-LŌ-sa): The region of the adrenal cortex responsible for mineralocorticoid secretion.

zona pellucida (pel-LOO-si-duh): The region between a developing oocyte and the surrounding follicular cells of the ovary.

zona reticularis (re-ti-kū-LAR-is): The region of the adrenal cortex responsible for androgen secretion.

zygote (ZĪ-gōt): The fertilized ovum, prior to the start of cleavage.

I-10

glomerular filtration rate and, 976
glucocorticoids, 604, 616–617
gonadotropin-releasing hormone, 604
gonadotropins, 604
and G-proteins, 596–597
and growth, 629
growth hormone, 603, 606–607
growth hormone–inhibiting hormone, 607
growth hormone–releasing hormone, 607
and heart, 700–701, 702
heart, effects on, 702
hormone-responsive elements (HREs), 598
human chorionic gonadotropin, 625
of immune system, 798–800
inhibiting hormones, 602
insulin, 622–623
insulin-like growth factors (IGFs), 606
interaction with other systems, 632, 633
and intercellular receptors, 598
lipid derivatives, 592, 593–594
and liver, 894
local, 49
luteinizing hormone (LH), 604
of male reproductive system, 1053–1055
mechanisms of action, 594–596
melanocyte-stimulating hormone, 607
melatonin, 625
and metabolism, 940
mineralocorticoids, 616
norepinephrine, 618
oxytocin, 607–609
patterns of hormonal interaction, 628
peptide hormones, 592, 593
of placenta, 1097
progesterone, 625, 628
prolactin, 605
prolactin-inhibiting hormone, 605
regulatory hormones, 468, 599–600
relaxin, 625, 628
releasing hormones, 602
renin, 618, 619, 620
and skeletal muscles, 301
and stress, 629–631
testosterone, 625, 628
thyroid-stimulating hormone, 603, 604
Horner's syndrome, 520
Horns, spinal cord, 422
Howship's lacunae, 179
Human body, levels of organization, 4, 5–6
Human chorionic gonadotropin (hCG), 625, 1097
Human Genome Project, 1117–1119
Human immunodeficiency virus (HIV), 806, 808
Human leukocyte antigens, 790
Human physiology, 4
Human placental lactogen (hPL), 1097
Human tetanus immune globulin, 294
Humeroulnar joint, 237
Humerus, 182, 234, 236–237
Humoral immunity, 659, 774, 787
Humoral stimuli, 599
Huntington's disease, 508, 510
Hyaline cartilage, 128–130
Hyaluronan (hyaluronic acid), 111
Hyaluronidase, 1089, 1091
Hybridoma, 810
Hydatidiform mole, 1092
Hydration sphere, 42
Hydrocephalus, 452, 453, 485
Hydrochloric acid, 44, 879
Hydrocortisone, 617
Hydrogen, 31, 32
Hydrogen bonds, 36–37
Hydrogen ions, 43–44
in body fluids, 43–44
Hydrogen ion secretion, 983–984
Hydrogen peroxide, 658
Hydrolysis, 40, 41, 46, 47, 904

Hydrophilic molecules, 42
Hydrophobic molecules, 42–43
Hydrostatic pressure, 74, 721, 725
of the interstitial fluid, 725
in kidney, 974
Hydroxide ion, 43
Hydroxyapatite, 175
Hydroxyurea, 648
Hymen, 1065
Hyoglossus, 329
Hyoid arch, 217
Hyoid bone, 211
Hypaxial muscles, 363
Hypercalcemia, 701, 1017, 1018
Hypercapnia, 854–855, 857, 1027
Hyperchloremia, 1017
Hypercholesterolemia, 714
Hyperextension, joints, 259
Hyperkalemia, 406, 410, 1016, 1017, 1032
Hypermagnesemia, 1017
Hypernatremia, 1012, 1018, 1032
Hyperopia, 560, 561, 585
Hyperosmotic solution, 75
Hyperostosis, 194
Hyperparathyroidism, 613, 1018
Hyperphosphatemia, 1017
Hyperpnea, 840
Hyperpolarization, 386–387, 397
Hyperreflexia, 441
Hypersomnia, 510
Hypersplenism, 810
Hypertension, 723, 764
portal, 891
Hyperthyroidism, 613
Hypertonic solution, 75
Hypertrophy, skeletal muscles, 302–303
Hyperuricemia, 936, 955
Hyperventilation, 855, 1029
Hypervitaminosis, 947, 955
Hypervolemic, blood volume, 644, 668
Hypesthesia, 585
Hypoaldosteronism, 616
Hypoblast, 1092
Hypocalcemia, 701, 1017, 1018
Hypocalcemic tetany, 613, 632
Hypocapnia, 855, 857, 1029
Hypochloremia, 1017
Hypodermic needle, 156, 169
Hypodermis. See Subcutaneous layer
Hypogastric plexus, 527, 1040, 1061
Hypoglossal canals, 204
Hypoglossal nerves, 204, 329, 483, 484
Hypogonadism, 625
Hypokalemia, 701, 983, 1016, 1017, 1032
Hypomagnesemia, 1017
Hyponatremia, 1012, 1017, 1032
Hyponychium, 162
Hypoparathyroidism, 613
Hypophosphatemia, 1017
Hypophyseal fossa, 207
Hypophyseal portal system, 602
Hypophysis, 601
Hyporeflexia, 441
Hyposmotic solution, 75
Hyposplenism, 810
Hypotension, 723, 764
Hypothalamus, 13, 446, 468–469
and endocrine activity, 599, 601
functions of, 468
and thermoregulation, 952
Hypothermia, 952, 955
induced, 954
Hypothyroidism, 613
Hypotonic solution, 75
Hypoventilation, 855, 1027
Hypovolemic, blood volume, 644, 668
Hypoxia, 652, 668, 834, 857
Hysterectomy, 1065
H zone, 281

I

I band, 281
Ileocecal valve, 885, 898
Ileostomy, 912
Ileum, 885, 887
Iliac arteries, 745, 748
Iliac crest, 243
Iliac fossa, 243

Iliac tuberosity, 243
Iliacus, 352, 354
Iliac vein, 755–756
Ilio-, 321
Iliococcygeus, 339
Iliocostalis group, spine, 332, 334
Iliofemoral ligament, 269
Ilioinguinal nerve, 1039, 1040
Iliopsoas, 354
Iliotibial tract, 351
Ilium, 242
Image reversal, visual, 561
Immature delivery, 1106
Immediate hypersensitivity, allergies, 805
Immovable joints. See Synarthroses
Immune complex, 796, 797
Immune disorders, 808
allergies, 804, 805
autoimmune disorders, 804–805
immune complex disorders, 808, 969
immunodeficiency diseases, 804–805
Immune response, 653, 770
Immune surveillance, 659
Immune system, manipulation of, 800
Immunity, 770
acquired immunity, 788
and aging, 807–808
antibodies, 793, 794–797
B cells, 793–795
cell-mediated, 774, 787
and colony-stimulating factors, 800
complement system, 784–786
fever, 787
immunological surveillance, 783–784
inflammation, 785, 786–787
innate immunity, 788
integration with other systems, 808–809
interferons, 784
and interferons, 799
and interleukins, 798–799
macrophages, 781–783
and memory cells, 789
microphages, 781
natural killer (NK) cell activation, 783–784
nonspecific defenses, 780
passive immunity, 788
patterns of immune response, 800–801
phagocytes, 781, 785
phagocytosis, 783
physical barriers, 781
proteins of immune system, 52
resistance, development of, 801–803
response to antigens, 797–798
specific defenses, 780
specificity of, 789
and stress, 807
T cells, 789–793
tolerance, 789
and tumor necrosis factors, 799
versatility of, 789
Immunization
active and passive, 810
tetanus, 294
Immunodeficiency diseases, 804–805, 808
AIDS, 805, 806, 808
severe combined immunodeficiency disease, 804–805
Immunoglobulins, 645, 774. See also Antibodies
IgA, 795, 796, 802
IgD, 795, 796
IgE, 795, 796
IgG, 795, 796, 798, 802
IgM, 795, 796
Immunological competence, 801
Immunological escape, 783
Immunological surveillance, 774, 781
Immunosuppression, 793, 808
Immunosuppressive drugs, 793, 810
Immunotherapy, 143
Impacted teeth, 874
Implantation, prenatal development, 1090, 1091–1094

Impotence, 1078, 1082
Incision, 162
Incisive fossa, 209
Incisors, 873
Inclusions, 82
Incomplete dominance, 1113
Incomplete proteins, 944–945
Incomplete tetanus, 292
Incontinence, 998, 1002
Incus, 571
Indole, 902
Induced abortions, 1105
Induced active immunity, 788
Induced passive immunity, 788
Inducer T cells, 774
Induction, and development, 1089–1090
Inert gases, 33
Infancy, 1108, 1110
Infants, and thermoregulation, 954
Infarct, 667
Infection, 140
sepsis, 168
Inferior, 20
Inferior angle, scapula, 234
Inferior articular facets, 222, 225
Inferior articular processes, 219
Inferior cerebellar peduncles, 473
Inferior colliculus, 470, 500
Inferior demifacets, 222, 223
Inferior epigastric arteries, 1041
Inferior ganglion, 481
Inferioris, 321
Inferior meatuses, 818
Inferior mesenteric artery, 520
Inferior mesenteric ganglion, 520
Inferior mesenteric plexus, 527
Inferior nasal conchae, 210
Inferior nuchal lines, 204
Inferior oblique, 327
Inferior orbital fissure, 209
Inferior ramus, 243
Inferior rectus, 327
Inferior sagittal sinus, 451
Inferior section, 19
Inferior temporal lines, 204
Inferior vena cava, 675, 754–756
Infertility, 1060, 1103, 1120
and sperm count, 1087
Infertility treatment, 1103–1104
artificial insemination, 1103
fertility drugs, 1103
gamete intrafallopian transfer (GIFT), 1103
surrogate mother, 1104
in vitro fertilization, 1104
zygote intrafallopian tube transfer (ZIFT), 1104
Inflammation, 140–141, 781, 785, 786–787, 797
effects of, 786
glucocorticoids as anti-inflammatory, 617
process of, 786–787
Inflammatory bowel disease, 912
Inflammatory T cells, 774
Inflation reflex, 855
Information processing, nervous system, 402–408
reflex arc, 431
Infraorbital foramen, 209
Infraspinatus muscles, 267, 343, 344
Infraspinous fossa, 234
Infundibulum, 446, 468, 601, 1059
Ingestion, 862
Inguen, 18
Inguinal, 321
Inguinal canal, 336, 1039
Inguinal hernia, 336–337, 364, 1039
Inguinal region, 18
Inheritance, 1114–1116
individual variations in, 1115–1116
meaning of, 1087
polygenic inheritance, 1114, 1115
prediction of, 1114
simple inheritance, 1114–1115
Inhibin, 604, 625, 1046
and reproductive system, 1080
Inhibiting hormones, 603
growth hormone–inhibiting hormone, 607

SOME ABBREVIATIONS USED IN THIS TEXT

ACh	acetylcholine
AChE	acetylcholinesterase
ACTH	adrenocorticotropic hormone
ADH	antidiuretic hormone
ADP	adenosine diphosphate
AIDS	acquired immune deficiency syndrome
ALS	amyotrophic lateral sclerosis
AMP	adenosine monophosphate
ANP	atrial natriuretic peptide
ANS	autonomic nervous system
AP	arterial pressure
ARDS	adult respiratory distress syndrome
atm	atmospheric pressure
ATP	adenosine triphosphate
ATPase	adenosine triphosphatase
AV	atrioventricular
AVP	arginine vasopressin
BCDF	B cell differentiation factor
BCGF	B cell growth factor
BMR	basal metabolic rate
BCOP	blood colloid osmotic pressure
bpm	beats per minute
BUN	blood urea nitrogen
C	kilocalorie; centigrade
CABG	coronary artery bypass graft
CAD	coronary artery disease
cAMP	cyclic-AMP
CAPD	continuous ambulatory peritoneal dialysis
CCK	cholecystokinin
CCOP	capsular colloid osmotic pressure
CF	cystic fibrosis
CHF	congestive heart failure
C_{hp}	capillary hydrostatic pressure
CNS	central nervous system
CO	cardiac output; carbon monoxide
CoA	coenzyme A
COMT	cathecol-O-methyltransferase
COPD	chronic obstructive pulmonary disease
CP	creatine phosphate
CPK, CK	creatine phosphokinase
CPM	continual passive motion
CPR	cardiopulmonary resuscitation
CRF	chronic renal failure
CRH	corticotropin-releasing hormone
CSF	cerebrospinal fluid; colony-stimulating factors
CT	computerized tomography; calcitonin
CVA	cerebrovascular accident
CVS	cardiovascular system
DAG	diacylglycerol
D.C.	Doctor of Chiropractic
DCT	distal convoluted tubule
DDST	Denver Developmental Screening Test
DIC	disseminated intravascular coagulation
DJD	degenerative joint disease
DMD	Duchenne's muscular dystrophy
DNA	deoxyribonucleic acid
D.O.	Doctor of Osteopathy
DPG	diphosphoglycerate
D.P.M.	Doctor of Podiatric Medicine
E	epinephrine
ECF	extracellular fluid
ECG	electrocardiogram
EDV	end-diastolic volume
EEG	electroencephalogram
EKG	electrocardiogram
ELISA	enzyme-linked immunoabsorbent assay
EPSP	excitatory postsynaptic potential
ERV	expiratory reserve volume
ESV	end-systolic volume
ETS	electron transport system
FAD	flavin adenine dinucleotide

FAS	fetal alcohol syndrome
FES	functional electrical stimulation
FMN	flavin mononucleotide
FRC	functional residual capacity
FSH	follicle-stimulating hormone
GABA	gamma aminobutyric acid
GAS	general adaptation syndrome
GC	glucocorticoids
GFR	glomerular filtration rate
GH	growth hormone
GH-IH	growth hormone–inhibiting hormone
GHP	glomerular hydrostatic pressure
GH-RH	growth hormone–releasing hormone
GIP	gastric inhibitory peptide
GnRH	gonadotropin-releasing hormone
GTP	guanosine triphosphate
Hb	hemoglobin
hCG	human chorionic gonadotropin
HCl	hydrochloric acid
HDL	high-density lipoprotein
HDN	hemolytic disease of the newborn
hGH	human growth hormone
HIV	human immunodeficiency virus
HLA	human leukocyte antigen
HMD	hyaline membrane disease
HP	hydrostatic pressure
hPL	human placental lactogen
HR	heart rate
HTLV-III	human T-cell lymphotrophic virus type III (= HIV)
Hz	Hertz
ICF	intracellular fluid
IH	inhibiting hormone
IM	intramuscular
IP_3	inositol triphosphate
IPSP	inhibitory postsynaptic potential
IRV	inspiratory reserve volume
ISF	interstitial fluid
IUD	intrauterine device
IVC	interior vena cava
IVF	in vitro fertilization
kc	kilocalorie
LDH	lactate dehydrogenase
LDL	low-density lipoprotein
L-DOPA	levodopa
LH	luteinizing hormone
LLQ	left lower quadrant
LM	light micrograph
LSD	lysergic acid diethylamide
LUQ	left upper quadrant
MAO	monoamine-oxidase
MAP	mean arterial pressure
MC	mineralocorticoids
M.D.	Doctor of Medicine
mEq	milliequivalents
MHC	major histocompatibility complex
MI	myocardial infarction
mm Hg	millimeters of mercury
mmol	millimoles
mOsm	milliosmoles
MRI	magnetic resonance imaging
mRNA	messenger RNA
MS	multiple sclerosis
MSH	melanocyte-stimulating hormone
MSH-IH	melanocyte-stimulating hormone–inhibiting hormone
NAD	nicotinamide adenine dinucleotide
NCOP	net colloid osmotic pressure
NE	norepinephrine
NFP	net filtration pressure
NHP	net hydrostatic pressure
NRDS	neonatal respiratory distress syndrome
OP	osmotic pressure
Osm	osmoles
OT	oxytocin
PAC	premature atrial contractions
PAT	paroxysmal atrial tachycardia
PCT	proximal convoluted tubule

PCV	packed cell volume
PEEP	positive end-expiratory pressure
PET	positron emission tomography
PFC	perfluorochemical emulsions
PG	prostaglandin
PID	pelvic inflammatory disease
PIH	prolactin-inhibiting hormone
PIP	phosphatidylinositol
PKC	protein kinase C
PKU	phenylketonuria
PLC	phospholipase C
PMN	polymorphonuclear leukocyte
PNS	peripheral nervous system
PR	peripheral resistance
PRF	prolactin-releasing factor
PRL	prolactin
psi	pounds per square inch
PT	prothrombin time
PTA	post-traumatic amnesia; plasma thromboplastin antecedent
PTC	phenylthiocarbamide
PTH	parathyroid hormone
PVC	premature ventricular contraction
RAS	reticular activating system
RBC	red blood cell
RDA	recommended daily allowance
RDS	respiratory distress syndrome
REM	rapid eye movement
RER	rough endoplasmic reticulum
RH	releasing hormone
RHD	rheumatic heart disease
RLQ	right lower quadrant
RNA	ribonucleic acid
rRNA	ribosomal RNA
RUQ	right upper quadrant
SA	sinoatrial
SCA	sickle cell anemia
SCID	severe combined immunodeficiency disease
SEM	scanning electron micrograph
SER	smooth endoplasmic reticulum
SGOT	serum glutamic oxaloacetic transaminase
SIADH	syndrome of inappropriate ADH secretion
SIDS	sudden infant death syndrome
SLE	systemic lupus erythematosus
SNS	somatic nervous system
STD	sexually transmitted disease
SV	stroke volume
SVC	superior vena cava
T_3	triiodothyronine
T_4	tetraiodothyronine, also called thyroxine
TB	tuberculosis
TBG	thyroid-binding globulin
TEM	transmission electron micrograph
TIA	transient ischemic attack
T_m	transport (tubular) maximum
TMJ	temporomandibular joint
t-PA	tissue plasminogen activator
TRH	thyrotropin-releasing hormone
tRNA	transfer RNA
TSH	thyroid-stimulating hormone
TSS	toxic shock syndrome
TX	thyroxine
U.S.	United States
UTI	urinary tract infection
UTP	uridine triphosphate
UV	ultraviolet
$\dot{V}_A$	alveolar ventilation
V_D	anatomic dead space
$\dot{V}_E$	respiratory minute volume
V_T	tidal volume
VF	ventricular fibrillation
VLDL	very low-density lipoprotein
VPRC	volume of packed red cells
VT	ventricular tachycardia
WBC	white blood cell

Chapter 18   CO Focus on Sports, Inc.   **18-8b** Manfred Kage/Peter Arnold, Inc.   **18-10b, 14b** Copyright Lennart Nilsson, *Behold Man,* Little, Brown and Company.   **18-12b,d, 14c** Frederic H. Martini   **18-16c, 19b** Ward's Natural Science Establishment, Inc.   **18-16f** Prof. P. Motta, Dept. of Anatomy, University "La Sapienza," Rome/Science Photo Library/Photo Researchers, Inc.   **18-19c,d** Dr. Michael Ballo   **18-23a,b** Lester V. Bergman & Associates, Inc.   **18-23c** John Paul Kay/Peter Arnold, Inc.   **18-23d** Custom Medical Stock Photo   **18-23e** Biophoto Associates/Science Source/Photo Researchers, Inc.

Chapter 19   CO Lester Lefkowitz/Tony Stone Images   **19-1a** Martin M. Rotker   **19-2a** David Scharf/Peter Arnold, Inc.   **19-2b** Ed Reschke/Peter Arnold, Inc.   **19-2c** David Scharf/Peter Arnold, Inc.   **19-4a,b** Stanley Flegler/Visuals Unlimited   **19-9a-e** Ed Reschke/Peter Arnold, Inc.   **19-11** Frederic H. Martini   **19-13** Custom Medical Stock Photo

Chapter 20   CO Lunagrafix, Inc./Photo Researchers, Inc.   **20-3a** Ralph Hutchings   **20-4b, 14b** Copyright Lennart Nilsson, BEHOLD MAN, Little, Brown and Company   **20-4c** Ralph Hutchings   **20-5b** Ralph T. Hutchings   **20-6cL** Science Photo Library/Photo Researchers, Inc.   **20-6cR** Biophoto Associates/Science Source/Photo Researchers, Inc.   **20-7c** Peter Arnold, Inc.   **20-8b,d** Ralph T. Hutchings   **20-9a,b** DuPont Merck Pharmaceutical Company   **20-9c** Peter Arnold, Inc.   **20-14a** Larry Mulvehill/Photo Researchers, Inc.

Chapter 21   CO Professor P.M. Motta, A Caggiati, and G. Machiarelli/Science Photo Library   **21-2** Michael J. Timmons   **21-4a** B & B Photos/Custom Medical Stock Photo   **21-4b** William Ober/Visuals Unlimited   **21-6b** Biophoto Associates/Photo Researchers, Inc.

Chapter 22   CO Steven Rubin/The Image Works   **22-3b,c** *Bailey's Textbook of Microscopic Anatomy* by Kelly, Wood, & Enders. Copyright 1984, Williams & Wilkins   **22-4c** Ralph T. Hutchings   **22-6a** D. M. Phillips/Visuals Unlimited   **22-6b** Biophoto Associates/Photo Researchers, Inc.   **22-8c,d, 9c** Frederic H. Martini

Chapter 23   CO D. Kirkland/Sygma   **23-2a** Photo Researchers, Inc.   **23-2b** Frederic H. Martini   **23-4d** Phototake   **23-5c** John D. Cunningham/Visuals Unlimited   **23-7** Ralph Hutchings   **23-8a** University of Toronto   **23-8b** Ralph Hutchings   **23-10a & 23-11b,c** Adapted from Junqueira, Carneira, & Long, *Basic Histology,* 5/e. Norwalk, CT: Appleton-Century-Crofts, 1986.   **23-10b** Wards Natural Science Establishment, Inc.   **23-10c** Bloom & Fawcett, Textbook of Histology, 11/e. Copyright 1986, W. B. Saunders Publishers

Chapter 24   CO Erich Lessing/Art Resource NY   **24-7** Frederic H. Martini   **24-10a** Alfred Pasieka/Peter Arnold, Inc.   **24-10b** Astrid and Hanns-Frieder/Science Photo Library/Photo Researchers, Inc.   **24-10c** Prof. P. Motta, Dept. of Anatomy, University "La Sapienza," Rome/Science Photo Library/Photo Researchers, Inc   **24-12a** Ralph T. Hutchings   **24-13a** Prof. P. Motta, Dept. of Anatomy, University "La Sapienza," Rome/Science Photo Library/Photo Researchers, Inc   **24-13b** John D. Cunningham/Visuals Unlimited   **24-16a** Phototake NYC   **24-16e** M.I. Walker/Photo Researchers, Inc.   **24-17cL, C,R** Ralph T. Hutchings   **24-18b,c** Frederic H. Martini   **24-19a** Ralph T. Hutchings   **24-20c,d** Michael J. Timmons   **24-23a** Ralph T. Hutchings   **24-23c** Ward's Natural Science Establishment, Inc.

Chapter 25   CO L Don King/The Image Bank   **CO R** Robert Semeniuk/The Stock Market

Chapter 26   CO Kavaler/Art Resource NY   **26-2a, 3b** Ralph T. Hutchings   **26-5d** David M. Phillips/Visuals Unlimited   **26-8a** Ralph T. Hutchings   **26-8b** Reproduced from R. G. Kessel and R. H. Kardon, "Tissues and Organs: A Text-Atlas of Scanning Electron Microscopy," W. H. Freeman & Co., 1979   **26-17a** Peter Arnold, Inc.   **26-18** Photo Researchers, Inc.   **26-20a** Ward's Natural Science Establishment, Inc.   **26-20b** Frederic H. Martini   **26-20c** G. W. Willis, MD/Biological Photo Service

Chapter 27   CO Bell Helicopter Textron Inc.

Chapter 28   CO Barros & Barros/The Image Bank   **28-2** Ralph T. Hutchings   **28-4b** Frederic H. Martini   **28-5a** Dr. Don W. Fawcett   **28-5b** Wards Natural Science Establishment, Inc.   **28-8b** David M. Phillips/Visuals Unlimited   **28-9b** Frederic H. Martini   **28-10a** Ward's Natural Science Establishment, Inc.   **28-10b** Reproduced from R. G. Kessel and R. H. Kardon, "Tissues and Organs: A Text-Atlas of Scanning Electron Microscopy," W. H. Freeman & Co., 1979   **28-10c,d,e** Frederic H. Martini   **28-11c** Ward's Natural Science Establishment, Inc.   **28-16a,b,c,d** Frederic H. Martini   **28-16g** G. W. Willis, MD/Terraphotographics/Biological Photo Service   **28-17b** Frederic H. Martini   **28-17c** Custom Medical Stock Photo   **28-18a** Ralph T. Hutchings   **28-19a** Ward's Natural Science Establishment, Inc.   **28-20a,b** Frederic H. Martini   **28-20cB** Michael J. Timmons   **28-20cT** Frederic H. Martini   **28-21** Michael J. Timmons   **28-23b** Fred E. Hossler/Visuals Unlimited   **28-23c** Frederic H. Martini

Chapter 29   CO Petit Format/Nestle/Photo Researchers, Inc.   **29-1a** Francis Leroy, Biocosmos/Science Photo Library/Custom Medical Stock Photo   **29-6b** Frederic H. Martini   **29-7a** Dr. Arnold Tamarin   **29-7b,c,d, 8a** *A Child Is Born,* Dell Publishing Company. Copyright Boehringer Ingelheim International GmbH, photo Lennart Nilsson   **29-8b** Photo Researchers, Inc.   **29-15** SPL/Photo Researchers, Inc.

ART

All embryology summaries, system integrators, and anatomical paintings were rendered by William Ober and Claire Garrison, with the exception of the paintings noted below.

Illustrations by Ron Ervin

3-18b, 6-1, 7-1b, 7-2, 7-3, 7-4, 7-10b, 7-15, 7-16, 8-1, 8-5c, 8-8, 8-10, 8-14, 9-8, 9-10, 9-11, 11-6, 11-7, 11-13, 11-21a,b,c, 11-22a, 11-23, 13-18

Illustrations by Tina Sanders

7-14a, 7-13, 13-5a, 14-16a, 15-7, 16-3, 16-7, 17-9b, 25-20

Illustrations by Craig Luce

1-15, 23-3a

CREDITS

Figures 2-6a, 2-6b, 2-8, 2-21, 3-7, 3-11b, 22-10, 22-19a, 22-25, and **A-1** from Jacquelyn C. Black, *Microbiology: Principles and Applications,* 3/e. Upper Saddle River, NJ: Prentice-Hall, Inc., 1996 **(Figures 2.4, 2.6a, 2.7, 2.20 a&c, 4.27, 4.30, 17.10, 18.3a, 19.1, and 3.2)**

Figures 6-16a & b From Ganong, *Medical Physiology,* 16e, 1993, Appleton & Lange

Table 21-3 Data Sources: Asmussen and Kielson, *Acta Physiogogica Scandinavia,* 27:217 (1952) and Reindell et. al., *Schweiz, Zeitschr. f. Sportmed.,* 1:97 (1953)

Figure 29-16 Adapted from *Time,* January 17, 1994, pp. 50–51, graphic by Nigel Holmes, research by Leslie Dickstein, source: Dr. Victor A. McKusick, Johns Hopkins University